Observation of the Earth and Its Environment

Springer-Verlag Berlin Heidelberg GmbH

Herbert J. Kramer

Observation of the Earth and Its Environment

Survey of Missions and Sensors

With 522 figures and 857 tables

Springer

This book is again dedicated to my wife Mechtild
and to our daughters Monika, Ursula and Mechtild.

My family provided unequivocal support in my long-term book-writing efforts.
It is the reason why this edition could be completed.

Table of Contents

The sections P and Q are only part of the CD-ROM that comes along with the book. All information up to and including Appendix C are part of the 4th edition (a single volume) as well as part of the CD-ROM.

Foreword and Acknowledgements of the 4th Edition

An effort has been made to update the material presented in previous editions. In addition, a considerable amount of new information is provided. The 4th edition contains a number of changes in scope and layout.

- After the 3rd edition in 1996, I realized that some sort of "Earth observation history" was needed to put a lot of space-age facts into proper context for a better overview and understanding of the subject matter. I started by searching my documentation base for a number of themes and by reviewing the published literature with a more conscious view for the historic perspective. This resulted in a long-term and virtually daily update process of my history kernel. The initial emphasis was in collecting as many meaningful items of "technology introduction" as possible. Over the years, many computer printouts were given to colleagues from various space agencies and institutions with the intent of a review. Fortunately, I received some valuable comments and suggestions from experts in various fields. The compilation of the history represents indeed a major amount of time and effort on my part; I do hope it serves as a valuable introduction and reference to the readership and to all interested in the wide field of Earth observation.

I am quite aware of the fact that my "history" is incomplete and a bit lopsided in favor of the western space-faring nations. In particular, it does not reflect properly the accomplishments of the Soviet/Russian space program in the field of Earth observation. This is definitely not intended. The unavailability (and/or non-existence) of documentation on the Russian space program with regard to sensor developments is the prime reason for this deficit. My requests for history reviews, for critical comments and for background information on particular topics were only of very limited success. The language barrier, the secrecy of the past political system, the heritage of a military-dominated space program, and the working habits of the Russian research community are at the root of the problem. The nature of any reliable history depends, however, on access to multiple information sources for checks and cross checks for verification.

- Most emphases are given to the documentation of spaceborne missions and sensors, a field that grew considerably during the past five years. As a consequence, my effort to cover airborne instruments and campaigns had to be reduced considerably. With regard to the latter, updates were only made when readily available (I didn't search consciously for documentation of new airborne instruments because there was simply no time left to do so).

- A thematic re-ordering of the spaceborne missions was done (Part A to N). The different themes or categories were selected according to my own judgment. Obviously, most missions fit into several categories due to the functional/observational variety of the sensor complement. [A case in point is the Envisat mission of ESA with its sensor complement. It features ASAR as a SAR imaging instrument; the optical imaging instruments are: AATSAR and MERIS; the radar altimetry mission is covered with RA-2, MWR, DORIS and LRR; in addition there are atmospheric instruments with GOMOS, MIPAS and SCIAMACHY] Of course, many other arrangements are conceivable. I took the freedom of placing each mission into the category of my choice because of the lack of suggestions from anywhere else (I did ask for advice). - The new arrangement may lead to a bit more searching in the table of contents. Still, I feel the thematic ordering scheme is rather positive for the layout.

- My agreement with the publisher calls for a single volume limited to about 1500 pages plus a CD-ROM, the intent is to keep the documentation affordable. Hence, the book contains the spaceborne missions and sensors, the "history" and a few appendices. The CD-ROM, however, contains all information, including Part P: "Survey of Airborne Sensors" and Part Q: "Survey of Campaigns." The CD-ROM is intended as a search tool; it is not to be used as a copy tool.

- The spacecraft/instrument illustrations in the previous editions were mostly presented as line drawings. This standard could not be kept in the 4th edition because of lack of manpower. Hence, in the review cycle of my draft, I usually requested a spacecraft illustration from the information provider and converted this into my Interleaf (Quicksilver version) documentation system.

- The 4th edition contains also many internet references. I am aware of the fact that the behavior of many of these "addresses" is rather volatile and elusive. I inserted them to indicate the source of information.

- During the past years, spacecraft and instrument design and development has become part of the curriculum at many universities throughout the world, where a fresh look is being taken at many of the problems facing the future of spaceflight. SSTL (Surrey Satellite Technology Ltd) of Surrey, UK, since 1985 a commercial company whose principal shareholder is the University of Surrey, started out early with classroom teaching and getting "engaged" by solving practical problems. SSTL was the first to take a new look at affordable space flight by introducing its microsatellite program along with a wealth of new ideas for ways of doing things. Another example is SunSpace and Information Systems (Pty) Ltd. of the University of Stellenbosch, South Africa, which set up shop in 2000, following a successful launch of SUNSAT in Feb. 1999. - These university activities are certainly a new stimulant for technology introduction and demonstration in many fields. As a consequence, I integrated a number of these small missions into Part N of the book to reflect to some extent the activities in this field. Unfortunately, most of my partially documented missions (which didn't make it into the 4th edition due to a lack of communication or of interest by my university contacts) are part of this section. The US UNP (University Nanosatellite Program) of 10 nanosatellites, an initiative sponsored by the US Department of Defense, NASA, and industry and developed by a number of US universities, is such a case in point.

By its very nature, the book can only be an "approximation of Earth observation reality" - a fair try at short descriptions of past and current missions and some perspective of the future (a best-effort approach); it is also a product of many compromises, omissions and interpretations. There are still dozens of partially documented projects (mostly in an early project phase) that didn't make it into the book, a sign of continuous progress and development. Reality is much more bizarre, diverse and complex than can be portrayed in this technical volume. Still, I do hope that the general field of "Earth observation" has been given a boost with the completion of this edition.

Electronic mail (e-mail) was the dominant means of communication between myself and over 500 contacts worldwide to get most of the project information transferred (faxes and letters were the mainstay in the previous editions). In addition, I monitored the published literature and many internet websites for information on new space projects.

All editions of my book were written without a mandate or an endorsement from anyone. I could only count on the cooperative attitude of many individuals (hundreds) in numerous organizations, willing to share information and to provide reviews of my draft descriptions. On many occasions my requests were simply met with indifference or disregard, in spite of reminders. Of course, this has to be seen in relation to the global scope of the book, the working environment of the individual contacted, the pressure exerted by the project he or she was working on, the language barriers of the non-English-speaking world, and/or to local attitudes and experiences. Some institutions are simply not accustomed to open policies, to public outreach, to e-mail communications, or to any documentation standards. The best results were achieved with those contacts who were acquainted with previous editions of my book. They realized in particular the value of such a survey and they were eager to help and to get their project also into the new edition.

There is some coverage of US military service missions (navigation, weather) and technology missions as far as information could be found or was made available through contacts.

This time around, I experienced in particular good cooperation from NRL (Naval Research Laboratory) in Washington, DC. It was more difficult to get information on such missions as MightySat, ARGOS, TSX5 and PICOSAt. As a US citizen, I spent a day at SMC (Space and Missile Systems Center) and at JPO (Joint Program Office of GPS) in El Segundo, CA, trying to get a contact on several programs, but in the end there was no chance to meet anyone. A similar fate occurred to me at Kirtland AFB in Albuquerque, NM. In my experience, everything seems to be classified on European military missions in the field of Earth observation. I simply resigned after a few tries.

In my requests for information (not only to military establishments, but also to commercial entities and to some space agencies as well), I was often considered a potential security risk. However, my policy was always to respect the wishes of the information provider. I don't want to publish anything "considered secret," thereby compromising these sources. The security aspect of projects was one of the reasons why I was always asking for reviews. But I also insisted on certain standards of description and completion to present a text in context. I do admit that I didn't succeed in all cases.

The writing of the 4th edition turned out to be a major problem with regard to support. All previous editions were sponsored by DLR/DFD. But with the submission of the 3rd edition in early 1996, DFD management insisted on external support for continuation of the work. As a consequence, I applied for support at nearly all space agencies of the world, at German and European institutions, at the UN. All applications were met with negative replies or simply with silence. I was treated by all space agencies with benign neglect. My experience: All potential sponsors are very interested in a global view of Earth observation, but should it come to sponsoring, then they are only interested in the proper presentation of their own projects.

On the other hand, I received lots of encouragement and invigoration from the reader community. Many inquired about the 4th edition. In early 1999, I took the "early retirement option" from DLR (effective as of Feb. 2000), to be able to dedicate my work to the 4th edition. In this new arrangement, DFD generously provided an office, access to e-mail and internet services and a postal address for my mail, all other expenses (like a computer, licenses, data backup services, information trips to the US and Russia, needed documentation, etc.) were my own, including my time. DFD management gave notice in June 2001 requesting my office room back without any alternate accommodation possibility. As a consequence, I moved out on July 11, 2001.

In spite of all handicaps, I found my work exciting and fascinating being exposed to hundreds of projects and to an ever evolving field of technologies and applications in the wide field of Earth observation. It gave me a new interdisciplinary education in many aspects, since I had to understand the principles involved prior to explaining them to the reader. In particular, I wasn't afraid of asking the experts in the field. I am grateful for all comments and advice received to improve the quality of the book.

Looking back on a dozen years of Earth observation coverage, the book has become my way of life; it provided me with a sense of accomplishment and of self-assurance in spite of enormous work loads. But the 4th edition is definitely going to be my last. I can visualize an engagement on a rather reduced scale in the future, but I cannot afford to continue with book-writing at the required pace of personal time and monetary investment. My experience: growing a few vegetables in the garden is more profitable than book-writing. In particular, I want to be able to share more time with my family.

May the book be a rich source of information and a solid reference for all readers.

Acknowledgements:

I would like to thank the legion of individuals throughout the world who provided information or contributed their talents in other ways for the completion of this book. I have endeavored to list all of them below, in no particular order (but grouped to institutional affiliations), and want to thank every one of them, with apologies to anybody I may have accidentally forgotten. Naturally, there are considerable differences in the level of support provided. Again, I was very fortunate to enjoy the backing of some key contacts in several agencies, who opened doors that led to further contacts and/or contributed substantially with advice or reviews. Many of my contacts became my friends; only a few select treated me in a rather demeaning way (simply by disregarding all of my questions or pushing their particular interests of presentation), wielding their power and letting me feel my position as humble requester.

David L. Glackin, John S. Bohlson, Ernest Y. Robinson, David A. Hinkley, David J. Gorney, Robert J. DeLorenzo, Paul R. Straus, all of The Aerospace Corporation, El Segundo, CA; Miguel A. Garcia Primo, and Angel Martinez of INTA, Madrid, Spain; Michael Rast, Evert Attema, Andrew Wilson, Anthony Dickinson, Peter G. Edwards, Rudolf Halm, Manuel Martin-Neira, Hans-H. Fromm, Giorgio Saccoccia, Gotthard Oppenhäuser, Paul Ingmann, Massimo Cislaghi, Helge Rebhan, Guiseppe Racca, Hannah Tait, Mark Drinkwater, Leopold van Holtz, all of ESA/ESTEC, Noordwijk, The Netherlands; Mark Maier of University of Alabama in Huntsville, AL; Oliver Matthews of the University of Bremen, Germany; Seorim Lee, Dongseok Shin, Woo-Kyung Lee, Sungdong Park of KAIST/SaTReC, Taejon, Korea; Hongyul Paik and Young-Min Cho of KARI, Taejon, Korea; Nigel Wells, Paul Brooks and Clive Edwards of DERA, Farnborough, UK; Nigel Morris of RAL, Farnborough, UK; David J. Purll of Sira Electro-Optics Ltd, UK; Friedrich E. Jochim, Susan Giegerich, Roland Wattenbach, Christian Stelter, Bernd Kirchner, Ingo Walter, René Pischel, Klaus Brieß, Manfred Schröder, Gottfried Schwarz, Hannelore Krumbholz, Albert Huber, Dieter Sundermann, Jens Hammesfahr, Johann Furtner, Alberto Moreira, Ralf Horn, Thomas Werner, Ruth Titz, Dieter Oertel, Georg Landshammer, Hans-Jochen Lotz-Iwen, Franz Pätzold, Cornelia Varga, Mihai Datcu, Herwig Öttl, Bernd Aberle, Jörg Gredel, Stefan Dech, all of DLR, Germany;

Penina Axelrad, Elaine R. Hansen, Christopher K. Pankratz, Michael T. McGrath, of the University of Colorado at Boulder, CO; Christian Rocken of UCAR, Boulder, CO; James Janesick of Pixel Vision Inc., Huntington Beach, CA; Otto Hofmann, Brunntal, Germany (formerly MBB); Rainer Killinger, Reinhold Lutz and D. Meißner of Astrium GmbH (formerly DASA), Munich, Germany; Arnold Schoonwinkel of Stellenbosch University, South Africa; Otto Kessler of Naval Air Warfare Center, Patuxent River, MD; George Joseph, ISRO/SAC, Ahmedabad, India; David Chu, Paul Chen, K. P. Cheng, Jeng-Shing Chern, all of NSPO, Hsin-Chu City, Taiwan ROC; Dirk Offermann University of Wuppertal, Germany; Frank Ellmers, Amnon Ginati, Boris Penné of OHB-System, Bremen, Germany; J. Keith Tennant of Intermap Technologies Ltd., Calgary, Canada; Peter Bernath, University of Waterloo, Canada, Glen Rumbold of CSA, Canada; Lyne Rainville of MDA, Richmond, BC, Canada; Victor Larock, SAIT Systems, Brussels, Belgium; Robert Schulte, and Udo Renner, Technische Universität Berlin, Germany; Olle Norberg of IRF, Sweden; Peter Rathsman, Kaj Lundahl, Sytze M. Veldman of SSC, Solna, Sweden; Anders Gustavsson of FOA, Sweden; William L. Austin II and Natalie Clark of AFRL, Kirtland AFB, Albuquerque, NM; Janet C. Johnston of AFRL, Hanscom AFB, MA; Leonard John Otten III, Kestrel Corporation, Albuquerque, NM; Peter Schwintzer of GFZ, Potsdam, Germany; Marcos Machado, Daniel Caruso, Carlos Alsonso of CONAE, Argentina; Patrick Rosenbaum, of ImageSat Israel, Tel Aviv; Fred Ortenberg and Roni Waller of Asher Space Research Institute Technion, Haifa, Israel;

Michel Rouzé, Alain Baudoin, Patrick Vincent, Francois Parisot, Jean-Paul Aguttes, Bernard Tatry, Didier Massonnet, Robert Ecoffet, and Norbert Paluch, all of CNES, France; Pierre Touboul of ONERA, France; Fritz Primdahl of DTU/IAU, Lyngby, Denmark; Per Hoeg, DMI, Lyngby, Denmark; Per Lundahl Thomsen of DRSI, Copenhagen, Denmark; Finn Hass of Terma A/S, Birkeroed, Denmark; Michael D. Vanek, David M. Winker, William L. Smith, Allen M. Larar, William P. Chu, Jack Paden, Keisha R. Armistead, all of

NASA/LaRC; Brian R. Dennis, J. Bryan Blair, Jay Zwally, Dennis Chesters, Nicholas Speciale, Richard McPeters, Jay Herman, Abigail D. Harper, Michael D. King, Jim Closs, Raynor L. Taylor, Brian R. Dennis all of NASA/GSFC, Greenbelt, MD; Gordon I. Johnston and John L. LaBrecque of NASA/HQ; Rainer Bärs of Specim, Finland; John P. Burrows of the University of Bremen, Germany; Johannes Müller, Paolo Pili, of EUMETSAT, Darmstadt, Germany; Ijar M. Fonseca, Nalin B. Trivedi, Hisao Takahashi, Janio Kono, Thelma Krug, Luiz A. Bueno, Himilcon Carvalho, Eduardo Bergamini, all of INPE, Brazil; Takashi Hamazaki, Akio Yamamoto, Masanori Homma, Masaaki Mokuno, Hiroki Kohata, Takashi Moriyama, Makoto Suzuki, Naoto Matsuura, Araki.Kenichi, all of NASDA, Japan; Takeshi Manabe and Seiho Uratsuka of CRL, Tokyo, Japan; Takeo Kosugi of ISAS, Tokyo, Japan; Yasuhiro Sasano and Nobuo Sugimoto of NIES, Tsukuba, Japan; Paul G. Weber, Roger C. Wiens, Diane Roussel-Dupre, Stephen Knox, David J. McComas, all of LANL, Los Alamos, NM; Robert L. Lucke, Curtiss O. Davis, Rebecca Baugh, Lee J. Rickard, Justin Bobak, Herbert Gursky, Robert R. Conway, Christoph R. Englert, Russell A. Howard, Peter W. Gaiser, Michael F. Zedd, all of NRL, Washington, DC; Frantisek Fárnik, Ladislav Sehnal, Radek Peresty, Astronomical Institute, Ondrejov, Czech Republic; Steven Kilston, Tom Miers, Richard H. Munro, and Bary L. Mills of BATC, Boulder, CO; Charles P. Herring of EarthWatch Inc., Longmont, CO; Gary Heckman of NOAA/SEC, Darren Jackson of NOAA/CIRES, Albin J. Gasiewski, Madison J. Post, NOAA/ETL, Boulder, CO; R. Keith Raney, Robert Jensen, of JHU/APL, Laurel, MD; Barry Muhlfelder of Stanford University, Stanford, CA; Marc D. Rayman, Eastwood Im, Thomas P. Yunck, David J. Diner, Andrew McLean, Deborah G. Vane, Reinhard Beer, all of NASA/JPL, Pasadena, CA; Stephen A. Voels, NASA/JSC, Houston, TX; Steve Goodman, NASA/MSFC, Huntsville, AL; Philip W. Rosenkranz of MIT, Cambridge, MA; Donald Lencioni of MIT/LL, Lexington, MA; Rainer Sandau of LH-Systems, Heerbrugg, Switzerland; Jochen H. Hoffmann, Arlington, VA; Martin Kuhlmann, Broadcom Corp., Irvine, CA; Dominick M. DellaValle and Carl F. Schueler of Raytheon SBRS, Goleta, CA;

Gerald Zwirn and Ken Faller of Space Systems/Loral, Palo Alto, CA; Bizzarro Bizzarri of Rome, Italy; Filippo Graziani of the University of Rome, Italy; Andrea Bacchetta of Alenia Spazio, Torino, Italy; Megan Q. Trombley of SDL at USU (Utah State University), Logan, UT; Adam C. Harvey of VT (Virginia Polytechnic Institute), Blacksburg, VA; Jeff Kingwell of CSIRO, Canberra, Australia; Brian J. Fraser, University of Newcastle, Callaghan, NSW, Australia; E. Steve Seumahu, University of South Australia, Mawson Lakes, Australia; Craig Smith of Electro Optic Systems Pty Ltd., Queanbeyan NSW, Australia; Robert Horvath of ERIM International Inc. Ann Arbor, MI; Bill Falkenberg and Russell E. Kuba of SpectrumAstro, Gilbert, AZ;

Hitomi Miyamoto of JMA, Tokyo, Japan; Mark Hurwitz, Will Marchant, and Ellen Riddle Taylor, all of UCB/SSL, Berkeley, CA; Donald J. Blersch of IPO, Silver Spring, MD; Dennis Berry, Wilfred E. Mazur Jr., Michael Mignogno, Carol P. Welsch, all of NOAA; Jacinta M. Behne of McREL, Aurora, CO; Huabao Lin, and Lihua Zhang of CAST, Beijing, China; Lei Ding of SITP, Shanghai, China; Peter Alea and James Hendershot, Swales Aerospace, Beltsville, MD; Fred Tanner, Sebastian Riegger, Stefan Ochs, Thomas Mayer, Hans L. Trautenberg, Astrium GmbH, Friedrichshafen/ Munich, Germany; Rene Noordhoek of KNMI, De Bilt, The Netherlands; Andrew Hoskins of General Dynamics, Redmond, WA; Jay L. Smith of Weber State University, Ogden, UT; Gil Moore, Monument, CO; Gordon G. Shepherd, York University, Toronto, Canada; Jerry Sellers, US Air Force Academy, Colorado Springs, CO; Cliff Tsai, TRW, Redondo Beach, CA; Ignacio F. Tourné, Jesús Gonzalo, INSA, Madrid, Spain;

Boris G. Kutuza, Anatoli M. Shutko IRE-RAS, Moscow; Georgy N. Zastenker, Ian Ziman, IKI, Moscow; Vladimir Kharitonov, VNIIEM, Moscow; Alexei A. Koutcheiko of NPO Machinostroyenia, Reutov, Moscow Region; Vladimir Kuznetsov, Sergey A. Pulinets, Vladimir Dokukin, all of IZMIRAN, Troitsk, Moscow Region; Vladislav I. Pustovoit, Vitold E. Pozhar, and Vladimir N. Zhogun of STCUI-RAS, Moscow; Michael Ovchinnikov, Keldish

Institute of Applied Mathematics, Moscow; Manfred Krischke of RapidEye, Munich, Germany; Mark Pastrone of ORBIMAGE, Dulles, VA; Rick Fleeter, Aaron Jacobovits of AeroAstro, Herndon, VA; Manfred Leipold, Kayser Threde GmbH, Munich, Germany; William Lewis, SwRI, San Antonio, TX; Chris D. Hall, Virginia Polytechnic Institute, Blacksburg, VA; Mark Campbell, University of Washington, Seattle, WA; Ian Walkty, Bristol Aerospace Ltd., Winnipeg, Canada; Ahmad Sabirin Arshad, ATSB, Kuala Lumpur, Malaysia; Peter Sinander, Saab Ericsson Space AB, Göteborg, Sweden; Andrew O. Cameron, Intelligent Land Management LLC, Washington, DC; David M. Simpson of Astrium Ltd., Portsmouth, UK; Peter B. Anderson, Boston University, Boston, MA; Craig Underwood, Audrey Nice, John Cooksley, SSTL, Surrey, UK; David K. Schmidt, University of Colorado at Colorado Springs; Bob E. Schutz, University of Texas at Austin, TX; Michael Schaepman, University of Zürich, Switzerland.

History reviews or contributions: In particular, I am very grateful for the cooperation and contributions of the following individuals: William Bandeen of GSFC, Greenbelt, MD (Bill retired from GSFC in 1990 and provided actually two complete reviews of the history within a period of two years); Michael D. King and Leslie M. Cusick of GSFC; Friedrich E. Jochim and Boris S. Zhukov of DLR; Peter Schwintzer of GFZ, Potsdam; David L. Glackin of the Aerospace Corporation, El Segundo, CA; R. Keith Raney of JHU/APL, Laurel, MD; Lee J. Rickard and Michael Zedd of NRL, Washington, D.C.; Arnold Schoonwinkel of Stellenbosch University, Stellenbosch, South Africa; Dominique D. Crommelynck of IRMB, Brussels, Belgium; and Walter Flury of ESA/ESOC, Darmstadt, Germany. Ursula Weisgerber of Springer-Verlag, Berlin, was very helpful in the production of the CD-ROM.

At this point I would like to thank Ron Baalke of JPL. I consulted frequently his "Space Calendar Website" (http://www.jpl.nasa.gov/calendar/) for update information in particular on mission launches, meetings, etc. My thanks go also to Jonathan McDowell of Harvard University. I used his website "Jonathan's Space Report" for reference material.

I am dedicating this book also to the memory of Friedrich L. Porsch, a young colleague at DLR/DFD who died in an accident on Feb. 21, 1997 at the age of 39. Friedrich provided a lot of superb line drawings to all editions of this book, I owe him many thanks for his long-term cooperation. His last drawing was that of the AVHRR instrument.

Oberpfaffenhofen, August 2001

Herbert J. Kramer

1 Earth Observation Short-History

Prior to the space age (conventionally dated from 1957), humankind had never been able to take in the whole of a hemisphere in a single glance. In fact it had never had a global view of the world in which it lived. It was not until the first spacecraft went into orbit that our horizons expanded and we saw our planet as never before. During more than four decades of spaceflight, planet Earth has been rediscovered through the systematic collection and analysis of vast amounts of information. At the turn of the century/millennium, satellite-provided services in many fields of application (environmental monitoring, navigation, weather forecasting, communication, etc) are taken for granted. We've come to depend on the satellites in a way that would have been unimaginable a few decades ago.

Before the space age, remote sensing, although not named as such, was done exclusively with photographic cameras. The so-called aerial photo emerged in the 1850's, a mere dozen years after the invention of photography, with pictures taken from a tethered balloon - the French photographer, Gaspard Félix Tournachon (1820-1910), alias Nadar, obtained the first aerial photographs over Paris (Oct. 23, 1858) from an altitude of about 80 m. Thereafter, tethered balloons were used a few times during the US Civil War (1861-1865) by General George McClellan, to study enemy positions using aerial photographs. At the beginning of the twentieth century, the aeroplane proved its advantage as a civil and military observation platform. Aerial photography was extensively employed during both World Wars for military reconnaissance. In 1947, a few captured V-2 rockets were used by the US military to photograph the clouds from 110-165 km altitudes. [In the same year (Oct. 18, 1947), the Soviet Union launched its first LRBR (Long Range Ballistic Rocket) based on the German rocket A4 (V-2)]. The photographs demonstrated the immense potential of observing weather.[1)] After the wars and prior to 1960, the development of aerial color and color infrared film gave civilian remote sensing a distinct boost. The color infrared photography allowed some interpretation means for a rough classification of some vegetation types. High-speed cameras, combined with wide-angle lenses, provided greater opportunities to image Earth's surfaces.

Earth observation covers a wide field of remote sensing as well as of other sensing methods (in-situ), it encompasses the Earth itself (in particular its outer surface) and also the Earth's environment, including the study of interactions with the outside. Space-age Earth observation (although not named as such initially) started with the launch of Russia's first Sputnik satellite on Oct. 4, 1957 on the R-7 launch vehicle from Baikonur (satellite mass = 83.6 kg, diameter = 58 cm, perigee = 228 km, apogee = 947 km, inclination = 65.1°, period = 96 min, RF frequencies: 20.005 MHz and 40.002 MHz). The chief designer and general manager of the Sputnik project was Sergei Pavlovich Korolev. Sputnik-1 made measurements that permitted a first estimation of the density of the upper atmosphere. In the following period each new spacecraft launch produced new discoveries. The first successful US satellite was Explorer-1 (launch Jan. 31, 1958 with a S/C mass of 5 kg; orbit of 384 km x 1859 km, inclination of 32.2°) of the US Army, built by the Army Ballistic Missile Agency and by JPL. Explorer-1 (instruments included a cosmic ray and micrometeorite package, a micrometeorite impact microphone, micrometeorite erosion gages, and internal and external temperature gages) provided preliminary information on the environment and conditions in space outside Earth's atmosphere. It resulted in the discovery of the Van Allen radiation belts. Explorer-1 reentered the Earth's atmosphere on March 31, 1970.

The first spaceborne sensors flown were the film cameras, looking toward Earth and providing a "bird's eye view" from space. Initial efforts concentrated on the most obvious phenomenon to study, namely the weather. However, the major goals of the early US and Russian space programs were set to explore outer space, and not to look at Earth. Strangely enough,

1) "The conception, growth accomplishments and future of meteorological satellites," NASA Conference Publication 2257, 1980

Earth was somehow considered to be sufficiently known. It is interesting to note that planning for a deliberate and systematic approach to Earth observation, i.e. the survey and research of the Earth's surface (and many other items), did not start before the mid 1960s. The reason for the new interest in Earth was stimulated mainly by the study of some 1100 photographs (film imagery), taken of Earth during the Gemini missions and subsequently being used in preparation for the Apollo program (the objective was the study of possible lunar landing sites as seen from space). In any case, in trying to analyze and to interpret the Earth imagery at hand, it began to dawn upon some people (familiar with the basic physics of the electromagnetic spectrum) that these photographs might contain a wealth of information - worthy of systematic analysis. Quantitative interpretation schemes had to be developed to interpret the data! [2]

The planning for the first civil spaceborne Earth-surface imaging project was initiated at a press conference on Sept. 20, 1966 in Washington, DC. [3] [4] At this conference, Stewart Udall, the Secretary of the Department of the Interior (DOI), and William Pecora, Director of USGS (United States Geological Survey), announced plans for a program called Earth Resources Observation Satellites (EROS). Fortunately, President Johnson, the US Congress, and the US public supported this idea, there were strong objections voiced by DoD and the State Department. [5] This was indeed a new direction in the US space program at a time, when young NASA's foremost task was to get a man on the moon (a national goal), in the middle of the Cold War and a hot war (Vietnam). NASA was given the task to plan and build the newly designated ERTS (Earth Resources Technology Satellite) spacecraft (launch of ERTS-1 on July 23, 1972) that was later renamed to Landsat-1. [6]

New types of instruments were developed and flown, many of them capable of detecting and measuring radiation (such as radiometers) in the visible to microwave region of the spectrum. The creativity of the new space age generated an evolution in sensor technology resulting in a multitude of sensor types and many other innovations. The concepts of these new instruments were based on such divers fields as optics, solid-state electronics, pattern recognition, signal processing, computer technology, and communications. At the start of the 21st century it can safely be stated that the space age was instrumental of initiating and fostering new and vastly improved measurement technologies, accompanied by the greatest instrument-development spree in the history of science.

A most important space-age achievement (along with parallel developments in the wide field of electronics) is the ability, to observe the Earth and its environment as well as the universe in **the entire breadth of the electromagnetic spectrum**. This in turn opened a new era in experimentation and discovery in virtually all fields of the Earth and space sciences. For instance, synoptic observations over wide regions of the Earth and the capability to communicate, process and interpret vast amounts of information, practically in real-time, has revolutionized the way we do things and in which scientists study the atmosphere, oceans, land, vegetation, glaciers, sea ice, and other environmental aspects of the Earth's surface and their interactions. Earth observation has become the prime source of input for the considerable advances in the geosciences and many related disciplines. As a conse-

---

2) P. D. Lowman, "Landsat and Apollo: The Forgotten Legacy," PE&RS, Vol. 65, No 10, Oct. 1999, pp. 1143-1146

3) http://academic.emporia.edu/aberjame/remote/landsat/landsat.htm (Author: J. S. Aber)

4) Ironically, objections of the State Department and DoD against the distribution of civil high-resolution Earth imagery (in the optical and microwave regions) and the proliferation of space technology have been around ever since and continue to be a major issue in US space policy. Special rules (including shutter control) may be imposed in particular conflict situations to restrict US-based commercial remote-sensing firms from unauthorized distribution of their imagery. Special rules apply also to the export of space technology by US companies. However, with space-imaging technology readily available outside the USA, the US-internal control functions became more or less ineffective as of 2000.

5) The policies on US remote sensing technologies and their restrictions evolved from two primary sources: a) the secret capabilities first developed for the NRO (National Reconnaissance Office), and b) civil systems like the Landsat series instruments MSS and TM. Originally, commercial considerations were not a factor in either of these areas.

6) P. L. Hays, R. F. Houchin, "Commercial Spysats and Shutter Control: The Military Implications of the US Policy," Proceedings of AIAA Space 2000 Conference and Exposition, Long Beach, CA, Sept. 19-21, 2000

quence, Earth itself assumed a new clarity and gave us a better awareness of its dynamic nature. In the Sun-Earth system, space exploration provided for the first time a perception and understanding of the electromagnetic state of the interplanetary space between the Sun and the Earth.

Beside the research benefits, Earth observation has also evolved to become a technology driver and a mature service provider for a large spectrum of useful applications, ranging from weather forecasting over monitoring and managing of Earth resources (crop surveys, mineral surveys) to navigation-aid systems and monitoring of international treaty compliance. Also, more emphasis is being placed on such fundamental issues as the global environment and its changes (global interconnectivity of weather phenomena, preservation of the bio-environment, etc.). The goal is to develop predictive environmental, climate, natural disaster, and natural resource models to help ensure sustainable development and improve the quality of life on Earth. - At the turn of the century/millennium, the new concept of **"formation flying"** [a combination of Earth observation, navigation (on-board propulsion), inter-satellite communication, on-board autonomy, and on-board processing functions] has the potential of revolutionizing the way the space community conducts Earth observation missions. The space community is just beginning to understand the potential and perspectives of satellite formation flying. New distributed observation concepts of spaceborne **bistatic systems** are in the planning stage that may eventually permit more affordable spacecraft constellations for interferometric imagery. The utilization of solar energy from LEO or MEO space power stations and conversion of the electricity into microwave energy for transmission to Earth are other concepts being explored and investigated today (lasers are also under consideration for beaming the energy from space). Space solar power - a dream today - has a good chance of becoming a reality in the decades ahead. - In a long-term perspective, the past forty–five years of space flight at the start of the 21th century can be regarded as the "early or adolescence period" in the field of Earth observation. All indications on enabling technology developments lead to the conclusion, that the best is yet to come, leading to a better understanding of the total Earth system.

"Earth functions as a system - a large, complex, and dynamic one, but a system nonetheless. It is affected in measurable ways by external forces such as the sun and its variability, and by the internal forces that are shaped by variations in the atmosphere, oceans, continents, life, and the complex web of interactions among them.

We are the first generation with the ability to observe global-scale changes from the perspective of space and the scientific knowledge to link them with their causes and consequences. This ability to record and understand global change will be among the greatest gifts that we can offer our children and their children after them, for it will put in their hands the power to make informed decisions about the environmental challenges of the future.

The quest for a true predictive capability for Earth system changes requires a flexible and progressive space system architecture. That's why we need to design and establish a smart, autonomous, and flexible constellation of Earth-observing satellites that can be reconfigured based on the contemporary science and specific issues at hand." 7)

United Nations statistics in the 1990s reveal that less than half of the Earth's exposed land has been mapped at scales suitable for economic development measures. The need is particularly great for Africa. The answer lies in satellite image maps, which are accurate and can be quickly compiled and systematically updated. Satellite image maps have become the mapmaker's benchmark for small and medium scale mapping. Moreover, Earth observation satellites have become the precious allies of a new type of agriculture managed from the sky, especially in developing countries where agricultural management is still at an early stage. With a single glance they report on crop areas, identify soil types and inventory water

7) Daniel S. Goldin, "NASA in the 21st Century," Millennial Challenges Colloquium series, Oct. 10, 2000, the address was presented at JHU/APL, Laurel, MD

resources. All of which can be used to plan future agricultural development. Likewise, as the season changes, they monitor crop changes and enable early detection of diseases. Today, Earth observation satellites are becoming an extremely valuable monitoring and decision-making tool for disasters and environmental hazards (assessment of the extent of destruction, etc.).

The term "Digital Earth" was coined by US Vice President Al Gore, presented in a speech at the California Science Center in Los Angeles, on Jan. 31, 1998. His vision of the proposed concept model of "Digital Earth"refers to a multi-resolution, 3-D representation of Earth, into which geo-referenced data can be embedded. A "Digital Earth" could, for instance, provide a mechanism for users to navigate and search for geospatial information, etc. - Obviously, such an objective is so vast, that no one organization in government, industry or academia could undertake such a project. A vast standards infrastructure is needed to make it happen! The benefits of such a seamless system are apparent to the entire Earth Observation community. - Eventually, the data from "Digital Earth" may also contribute to a new genre of virtual reality applications.

During the 1990s, industry is gearing up as a commercial total system (space and ground segment owner and operator) service provider. The global research community is still a very large user of remotely sensed data, but an increasing amount of Earth observation data (information) permeates also into applications for everyday use. This development into a wider base is indeed a good perspective for a maturing service and utility environment.

At the turn of the 21st century, first attempts are being made by space agencies to design and demonstrate optical instruments to permit "remote sensing applications" from GEO (Geostationary Orbit). Earth observation missions (with spatial resolutions in the range of 1-4 km in the optical region of the spectrum) from GEO are rather challenging due to their enormous distance from the Earth's surface (about 36,000 km, or 45 times further away than from normal LEO altitudes of 800 km). Still, Earth observation missions from GEO are very attractive, offering the advantage of a continuous viewing capability, which so far has only been employed by GEO weather satellites (with much larger resolutions). NASA plans to fly GIFTS (Geosynchronous Imaging Fourier Transform Spectrometer) in 2004.

The intent of this chapter on "Earth Observation Short History" is to put some events, pertaining to the wide field of Earth observation, into proper context - the past has to be known and understood in order to plan for the future. The emphasis is on sensor technologies, system concepts, observation techniques, operational aspects, and navigation. Of interest is also the introduction/provision of general services and the start of international cooperation. The select nature of this EO-history overview precludes any claim for completeness. The scope of Earth observation is so immense, I simply do hope that some of the most important achievements are properly covered or even mentioned. It should also be pointed out that there is plenty of room left, as well as considerable needs, for creativity and innovation to continue to change things for the better. [8] - At the beginning of the 21st century, a better understanding is emerging of the relationship between technology development and the ability to do science.

In parallel to this chapter on "Earth Observation Short History," the interested reader should also refer to Part O of the document for more detailed descriptions and background on particular topics.

Of all satellite launches on a worldwide scale, 70-75% are commercial communication satellites, the rest are military and civil satellites for such services as surveillance, technology development, Earth observation and navigation. Hence, telecommunication is by far the

8) The fields of signal processing, data processing, interpretation and use (applications) are topics outside the scope of this text. The interested reader is given a good survey reference: Celebrating a half Century of Signal Processing, "The Past, Present and Future of Image and Multidimensional Signal Processing," IEEE Signal Processing Magazine, March 1998, pp. 21-58

most widespread application of space technology. [9] The field of Earth observation is in second place when compared by the number of spacecraft launches. Next to telecommunications, remote sensing may be the most significant commercial application in the space industry – one which, like satellite telecommunications, has the potential to fundamentally change the way certain industries operate.

1.1 Sensor/Technology Development

In the early years of Earth observation (in particular satellite meteorology) attention was focused on those phenomena which could be observed relatively directly in the visible and infrared bands of the spectrum. First images were obtained by photographic systems [10] such as automated still or movie cameras, followed eventually by vidicon electronic imaging systems [a framing system of TV heritage - the image from the photoconductive surface (detector) is raster scanned by an electron beam; example: RBV (Return Beam Vidicon), AVCS], and later by optomechanical scanner systems [examples: M-7 an airborne multispectral mapper of ERIM (first flown in 1971) and MSS on Landsat-1, launch July 23, 1972]. The TV cameras of the TIROS-1 satellite (launch April 1, 1960) provided daily low-resolution black and white pictures in the visible spectrum (panchromatic) of cloud cover and the Earth's surface where clear. A time sequence of these synoptic coarse-resolution images permitted a visual interpretation by the meteorological community of large weather patterns which moved slowly across a continent (inferring atmospheric motions) - a first application of the emerging field of Earth observation. The TV cameras of the very next TIROS satellites experimented already with the infrared spectrum. The resulting images permitted a first look at the Earth's heat distribution. In addition to the TV cameras, TIROS-2 and its successors experimented with the infrared spectrum by adding, variously, medium-angle, omni-directional and scanning radiometers. - The first years in Earth observation were dominated by such overall requirements for **repetitive coverage** (frequent observations) and the need of an **operational capability**, in particular for meteorological satellites. But it was also recognized that to create the new remote sensing technology, and to begin its utilization, would require an interdisciplinary effort by all parties involved. The challenges were monumental for research and technology development on all fronts, to create an infrastructure and to come up eventually with operational services.

The success of initial large-scale weather sensing started a planning and development period for better sensor systems, in particular with improved spatial and spectral resolutions, capable of land-surface imaging. Spectral resolution meant parallel sensing and detection in several bands of the visible and near-infrared spectrum. This technique was referred to as multispectral sensing, enhancing considerably the value (interpretability) of imagery. The MSS (Multispectral Scanner System) instrument of the Landsat series (for land surface imaging) is such an early multispectral instrument (1972), with the visible and near-infrared spectrum almost evenly divided into four bands (band specification for MSS was simply adopted from airborne photographic experience of films). [11]

Starting with Landsat-4 in 1982, a more sophisticated multispectral imaging sensor with the name of Thematic Mapper (TM) began its operation. It featured seven spectral bands and a ground resolution of 30 m in the VNIR bands. The TM band-selection process was on the basis of a comprehensive study and analysis of spectral reflection features for a variety of vegetation types. - The rather successful Landsat instruments eventually spawned other

9) Background Paper No 5: "Space Communications and Applications," UNISPACE-III Conference in Vienna, Austria, July 19-30, 1999, p. 112, A/CONF.184/BP/13 with the title: Space Benefits for Humanity in the Twenty-First Century, ISBN: 92-1-100818-2

10) Note: A photographic system is also referred to as a framing system. It means that all of the data in an image are acquired simultaneously.

11) In the Soviet Union, a parallel development to MSS (on Landsat) took place with the development of the MSS Fragment instrument at IKI (Space Research Institute), an 8-channel imager with a spectral range of 0.4-2.4 μm, flown on Meteor-Priroda-5 (launch June 18, 1980). Fragment operated successfully onboard the spacecraft for four years.

spaceborne Earth-surface imaging systems. Some of them are: The MOMS sensor missions on Shuttle (MOMS-01 in 1983), the SPOT satellite series sensors HRV (High Resolution Visible) of CNES (SPOT-1 launch in 1986), the Resurs-O1 series of Russia with MSU-E and MSU-SK (launch of Resurs-O1-1 in 1985), the IRS series of ISRO (launch of IRS-1A in 1988), JERS-1 of NASDA with OPS (launch of JERS-1 in 1992), ADEOS of NASDA with AVNIR (launch Aug. 17, 1996). See Table 1. At the beginning of the 21st century, the data from the Landsat S/C series constitute the longest record of the Earth's continental surfaces as seen from space. It is a record unmatched in quality, detail, coverage, and value.

Most of the instruments placed into orbit for the study of the Earth's atmosphere and surface have been of the type **passive sensors**, imagers and sounders operating in the visible, infrared and microwave spectral regions. Technology developments over the past four decades have allowed the capabilities of the current generation of passive sensors to advance far beyond those of the first instrument on Sputnik-1 and also the TV cameras on TIROS-1. - The first spaceborne **active sensors** were radar systems on Skylab (the instrument was named S-193, a combination of passive microwave radiometer with an active scatterometer, and radar altimeter) operated between May 1973 and Feb. 1974. The next radar altimeter was flown on GEOS-3 (launch April 9, 1975, E.7.3) of NASA (see Table 47).

Many items of technology introduction (the emphasis is on civil Earth observation programs, military programs are considered only when enough published information is available) are presented below. [12] [13] [14] The scope of description must be limited to abstract level detail due to the vast number of topics in the general field of Earth observation. The reader is invited to consult also the survey references of the published literature as well as the bracketed references in the text, referring to descriptions in other chapters of the book.

1.1.1 Concepts in Optical Observations

The principal detection methods used in optical observations are photographic (e.g., film), photoemissive (photomultipliers), and photoconductive (semiconductor).

- The first low-resolution space photograph of the Earth was taken and transmitted by the US satellite Explorer-6, a spin-stabilized S/C with a mass of 64 kg (launch Aug. 7, 1959 on a Thor vehicle from Cape Canaveral into a highly eccentric orbit, perigee = 237 km, apogee = 41,900 km, inclination = 47°, period = 765 minutes) in August of 1959. The camera system flown was a TV optical scanner, consisting of an optical unit containing a concave spherical mirror and phototransistor, a video amplifier, timing and logic circuits, and telemetry. The scanner's optical axis was directed 45° away from the S/C spin axis, which was parallel to the orbital plane. The vehicle's spin furnished the line scanning, and the spacecraft's forward motion along its trajectory provided the frame scanning. The first "television" photo, received in Hawaii, took nearly forty minutes to transmit.

- On June 7, 1967 the first color image of the Earth from space was taken by the Molniya-1 spacecraft of the USSR. In the US, the first color picture of the Earth from a near-synchronous orbit was taken on July 25, 1967 by the Dual Vidicon Camera system of the DODGE satellite (M.4).

- Spaceborne photographic film imagery for reconnaissance and mapping. The early US military program, endorsed by President Dwight D. Eisenhower in Feb. 1958, employed traditional camera technology for information gathering (to support SALT treaty verification). The US Discoverer-14 satellite [also known as the CORONA mission 9009 of NRO (DoD), launch Aug. 18, 1960 with a Thor vehicle from VAFB, CA] opened a new era in

12) "Perceiving Earth's Resources from Space," Special issue of Proceedings of the IEEE, Vol. 73, No. 6, June 1985, pp. 947-1128

13) Special Issue on Remote Sensing for Environmental Research, Proceedings of the IEEE, Vol. 82, No. 12, Dec. 1994, pp. 1771-1929

14) Special Issue on Remote Environmental Sensing, Proceedings of the IEEE, Vol. 57, No. 4, April 1969

spaceborne reconnaissance by successfully returning the film product in a reentry vehicle (capsule) and subsequent mid-air recovery on Aug. 19, 1960. A total of 95 successful CORONA missions took place over a period of 12 years, ending on May 25, 1972 with mission 1117 (D.8). The high-resolution Earth-surface imagery (5-12 m) of the programs CORONA, ARGON (12 missions between Feb. 17, 1961 and Aug. 21, 1964), and LANYARD (single mission in 1963) predated the imagery of the civil program Landsat. Declassification (i.e. availability to the civil community) of the imagery of the three programs was announced by the US government on Feb. 24, 1995 (Executive Order 12951, declassifying reconnaissance satellite images from the Corona program and two related subprograms - Argon and Lanyard). This, in turn, extends the baseline for systematic Earth surface coverage backwards by more than a decade. - Note: The US military reconnaissance program gave up the concept of high-resolution film imagery and film capsule recovery in the mid-seventies in favor of optoelectronic imaging (more practical for routine operations, but lower resolution than film) with the launch of the KH-11 (Keyhole designation) spacecraft series. The KH-11-1 satellite was launched Dec. 19, 1976. [15)] [16)] [17)]

• The former Soviet Union launched its first photoreconnaissance satellite, Zenit, on April 26, 1962 (built by NPO Energia). - A film camera of some sophistication on this spacecraft is MKF-6 (Multi-Kanal-Fotografie-6) or “multispectral film camera-6”, developed at VEB Carl Zeiss, Jena, German Democratic Republic. The camera employed a battery of six lenses (two rows of 3 lenses in parallel), looking into the same footprint, each for a different spectral band and furnished with a special filter, in the spectral range of 0.45 - 0.90 μm. Thus, color imagery with a high degree of geometric and radiometric accuracy and resolution was obtained in six spectral bands by using different films for each band. Un-perforated 70 mm films of 120 m length were used (shutter speeds of 7-56 ms, image format of 55 mm x 81 mm). MKF-6 was initially flown on aircraft (Antonov-30). The so-called “data processing” consisted of an opto-analog image analysis using MSP-4 (Multispectral Projector-4). This instrument assisted in the human visual inspection process by generating a color-coded synthesis, using multiple negative imagery of a multispectral set (analysis by superposition). A first spaceborne demonstration of MKF-6 took place on Soyuz-22 (launch Sept. 1976) within the framework of a Kosmos experiment called “Raduga” and within the Intercosmos program. During the 8-day flight of Soyus-22, Soviet Kosmonauts took over 2,500 Earth images with the hand-held MKF-6 camera. Thereafter, MKF-6 was also utilized on the Salyut-6 and -7 space stations. [18)]

• TV camera observations. Among the very early imaging sensors in space were shutter-style TV cameras, used to collect meteorological data. These so-called RBV (Return-Beam Vidicon) TV cameras employed a frame-sensor technique, where an electron beam scans the image on a photosensitive surface (raster scan of a complete image). In this concept, the image is exposed by a shutter device and stored on a photosensitive surface within each camera. This surface is then scanned in raster form by an internal electron beam to produce a video signal (the image is focused onto a photoconductor which causes the intensity of the electron beam, discharged from the electron gun, to vary with the intensity of the light). The values of the video signal are converted into digital information. Such RBV cameras were flown in several early satellite series such as: TIROS (TIROS-1 launch April 1, 1960), ESSA (launch of ESSA-1 on Feb. 3, 1966), Landsat-1 to -3 (LS-1 launch July 23, 1972), DMSP (Block IV series started in 1965), and Meteor (Meteor-1-1 launch March 23, 1969). - The pictures obtained in this fashion were relayed to the ground and required substantial signal processing, including oscilloscopes, to recover a satellite image.

15) R. A. McDonald, “CORONA: Success for Space Reconnaissance, A Look into the Cold War, and a Revolution for Intelligence,” PE&RS, Vol. 61, No. 6, 1995, pp. 689-719

16) http://www.nro.odci.gov/index5.html

17) http://www.fas.org/spp/eprint/mckinley.htm

18) “Atlas zur Interpretation Aerodynamischer Spektralaufnahmen - Methodik und Ergebnisse,” Akademie-Verlag, Berlin, 1982

• Solid-state (digital) imaging - a key technology in optical remote sensing capability which started a new era in quantitative information collection. The technique (introduced in the early 1960s) uses silicon-based photodetectors onto which radiation (photons) can be focused and an electronic readout scheme (scanner) for image capture. The new method, later referred to as an optomechanical system (later optoelectronics system with CCD technology), converts incident photons directly into electrical current, offering many advantages [true space-point (pixel based) information, high dynamic range, low power dissipation, low voltage operation, no geometric distortions, sampled signal output, rapid response, etc.] over the electron beam scanning vacuum tube technology, such as the vidicons, traditionally used in the RBV TV cameras. The new solid-state-oriented imaging technology provides considerable flexibility; in particular, it is suitable for digital processing, offering a variety of readout schemes [imaging arrays (1-D, 2-D) such as self-scanned photodiodes, CIDs, CCDs], and permitting the use of many detector element types. [19) 20)]

The new solid-state radiation detection concept employs scanner systems (see chapter O.3) permitting the acquisition of imagery within and outside the VNIR spectral region of photographic films, offering also a strategy of separating the received radiation into a number of spectral bands. This key technology provides the only practical means for obtaining high-accuracy radiometric information. The early scanners form an image successively on a cell-by-cell basis by the process of scanning. The first scanners employed were optomechanical whiskbroom systems where imaging occurs on a cell-by cell basis in the cross-track direction; these were followed by more efficient parallel line-scanned pushbroom (CCD) systems.

The newly available scanner technology (whiskbroom/pushbroom) [21)] offered a means of making imagery quantitatively available to computer processing methods; it stimulated in turn the development of new sensor systems with greatly enhanced performance and observation capabilities, it permitted also a better modular design concept of imaging sensors. An imager based on solid-state technology consists of a suitable lens system (the optics subsystem), a scanner subsystem with the detector positioned at the focal plane, and operating electronics. [22) 23)]

- Solid-state multispectral imaging in the 1960s. It signifies a great step forward in image interpretation capability, namely from visual or machine-sensed photointerpretation (analysis based on image characteristics) to quantitative interpretation of several bands of imagery (analysis based on spectral characteristics). The new quantitative approach relies on image processing methods of the evolving computer industry of the time.

  The pioneering airborne instrument of spatially-registered multispectral scanners is M-5, developed by the Willow Run Laboratories of the University of Michigan. The M-5 whiskbroom scanner is based on solid-state technology, and developed around an optomechanical scanning system called the S-5 (built by HRB Singer for the US Army). The S-5 had two optical channels (not registered) and recorded its data on film. From 1963 to 1965, two of these instruments were flown in tandem, each in a DeHavilland

19) P. K. Weimer, et al., "Multielement self-scanned mosaic sensors," IEEE, Spectrum, March 1969, pp. 52-65

20) Special Issue on Solid-State Imaging, IEEE Transactions on Electron Devices, Vol. ED-15, No. 4, April 1968, pp. 190-261

21) Conventional photographic films with high resolution imagery were not amenable for direct processing and transmission methods in remote sensing. Film material was not suitable for quantitative radiation measurements, only a small spectral range (VNIR, SWIR) could be covered with film.

22) Note: Realizing the value of multispectral data, there were a number of implementations in the early 1960s to obtain multispectral data (i.e. photographs) by special cameras. Spectral separation was accomplished by using photographic filters with multi-objective camera systems, each for a separate spectral range. This left the photo-interpreter with a dozen or so photos of the same scene, each at a different spectral band. The human eye finds it difficult to keep track of tonal variations over 16 or more levels of gray for even a modest number of different scene objects. - The discrimination problem was eventually solved with the availability of data from multispectral scanners and the development of proper algorithms for computer interpretation of imagery.

23) D. Landgrebe, "The Evolution of Landsat Data Analysis," PE&RS, July 1997, pp. 859-867

Beaver aircraft, to obtain four unregistered bands (selectable from UV through LWIR). From June 1963 through June 1964, around-the-clock (every 4 to 6 hours) imagery was collected once a month over a local fifty mile flight path, selected for natural and cultural diversity. These data were manually analyzed to form the initial basis for an understanding of day-night-seasonal spectral imaging phenomena. The M-5 instrument provided support of the developing ERTS (Landsat) program, in particular for the design of MSS (Multispectral Scanner). The M-7 multispectral scanner was a direct successor of M-5, it became operational in 1971, was upgraded several times, and provided a long-term source of remotely-sensed multispectral imagery (P.82.2). [24) 25)] A much later descendent of optomechanical whiskbroom technology was AVIRIS (Airborne Visible/Infrared Imaging Spectrometer) of NASA/JPL (P.42). This first airborne hyperspectral imager was introduced in 1987 and became operational in 1989, the first sensor to contiguously cover the 0.4-2.5 μm region with narrow spectral bands. At the start of the 21st century, its data is still very much in demand. - The short dwell-time problem of these cross-track whiskbroom scanners was solved by introducing a parallel coverage arrangement of detector arrays in the along-track direction. In this way, the wide along-track coverage permitted sufficient integration time for all parallel cells in each cross-track scan sweep.

Note: In some instances, preference is given to a whiskbroom imager design (the older imaging technology) because the optics for pushbroom operation must always cover FOV (the total field of view - meaning the full swath width) while the optics for whiskbroom operation deal with IFOV (instantaneous field of view) which is much smaller than FOV. Hence, there are less distortions at the swath edge for whiskbroom systems.

Spaceborne scanners. In 1968, SBRC (Santa Barbara Research Center) of Hughes Aircraft Company proposed to NASA to place a new type of imaging device on the planned ERTS spacecraft, namely a "scanner system." At the time, scanners were viewed with great skepticism by most scientists for two reasons. First, the scanner employed a moving part, an oscillating mirror, which was considered unreliable. Second, the scanner was not a full-frame imaging device; it created images from strips. Cartographers were suspicious of the scanner's geometric integrity.

The scanner did have one important advantage, namely its multispectral capability. Agricultural research had demonstrated the value of multispectral imagery. For example, the 0.63 to 0.68 μm band measures chlorophyll absorption; the 0.79 to 0.9 μm band indicates water content in leaves; and the 1.55 to 1.75 μm band displays soil moisture. SBRC initially designed a six-band scanner for the satellite. However, NASA's small satellite could not house the large scanner, so a more modest four-band MSS (Multispectral Scanner) was built for ERTS-1. The spectral bands were chosen to simulate false-color infrared photography. [26)] NASA put RBV cameras as well as the new scanner system (MSS) on-board the ERTS-1 satellite (in 1975 renamed to Landsat-1) to compare imagery from both types of systems. Within hours after launch of ERTS-1 (launch July 23, 1972), the first MSS images created a sensation with their amazing clarity and synoptic views of the landscape. Spatially, both systems (RBV and MSS) provided a resolution of about 80 m. Spectrally, each of the three-camera RBV system was designed to cover a band: the blue-green, yellow-red, and red/NIR. MSS provided a similar band set along with a fourth band to extend coverage into the infrared region. The imagery of MSS was in digital form. The scan mirror assembly of MSS was the key to providing wide-field and 'high-resolution' coverage. The new technique of scanning had the promise of more things to come. Thus, on the spaceborne side, MSS (with a

24) Information provided by B. Horvath of ERIM, who operated the M-5 imager and analyzed the data.

25) Note: In 1973 the Willow Run Laboratories team separated from the University of Michigan and became ERIM (Environmental Research Institute of Michigan). In 2000, ERIM became part of Veridian ERIM International

26) A. M. Mika, "Three decades of Landsat instruments," Photogrammetric Engineering & Remote Sensing, ASPRS, Vol. 63, No 7, July 1997, pp.839-852

cross-track whiskbroom scanner) ushered in an era of previously unimaginable synoptic knowledge of the Earth.

- The TM sensor on LS-4 (launch July 16, 1982) is a much improved successor in spatial and spectral resolutions to MSS. A good portion of land-surface observation history is intimately connected with this pioneering sensor (TM) and program (Landsat). All TM instruments (including ETM+of Landsat-7) employed the optomechanical whiskbroom technology. Other spaceborne optomechanical (whiskbroom or flying spot) instruments of note are: AVHRR (on the NOAA/POES series, all three generations of instruments, since 1978 starting with TIROS-N, see Table 34), OLS (flown on DMSP series, first introduced in 1976 - OLS uses a flying spot design), ASTER and MODIS (Terra, launch Dec. 18, 1999), and OSMI (Ocean Scanning Multispectral Imager, flown on KOMPSAT-1, launch Dec. 20, 1999). Note: Instrument designs like AVHRR and MODIS are of the optomechanical type because the extended focal planes, coupled with relatively short focal lengths, lead to unworkable field angles and geometric distortion problems (pushbroom designs are not well suited to moderate-resolution wide-field systems).

Launch Date	Satellite Platform (Agency or Company)	Earth Imaging Sensors of respective Payload	Comment (ops = operations)
Jul. 23, 1972	Landsat-1 (NASA)	MSS, RBV	MSS whiskbroom imaging End of service Jan. 6, 1978
Jan. 22, 1975	Landsat-2 (NASA)	MSS, RBV	S/C ops until 1983
Mar. 5, 1978	Landsat-3 (NASA)	MSS, RBV	S/C ops until Jan. 7, 1983
June 18,1980	Meteor-Priroda-5 (Russia) G.6	MSU-E, MSS Fragment	1st long-term use of CCD pushbroom technology
Jul. 16, 1982	Landsat-4 (NOAA)	MSS, TM	Standby since Dec. 1993
Mar. 1, 1984	Landsat-5 (NOAA)	MSS, TM	Operational as of 2001
Nov. 3, 1985	RESURS-O1-1 (Russia)	2 MSU-E, 2 MSU-SK	S/C ops until Nov. 11, 1986 Use of CCD technology
Feb. 22, 1986	SPOT-1 (CNES)	2 HRV	Use of CCD pushbroom technology
Mar. 17, 1988	IRS-1A (ISRO)	LISS-I, LISS-II A/B	Three CCD imagers S/C ops until 1995
Apr. 20, 1988	RESURS-O1-2 (Russia)	2 MSU-E, 2 MSU-SK	S/C ops until June 1, 1999
Jan. 22, 1990	SPOT-2 (CNES)	2 HRV	Operational as of 2000
Aug. 29, 1991	IRS-1B (ISRO)	LISS-I, LISS-II A/B	Operational as of 2001
Feb. 11, 1992	JERS-1 (NASDA)	OPS	S/C ops until Oct. 11, 1998
Oct. 5, 1993	Landsat-6 (NOAA)	ETM	Launch failure
Sept. 20, 1993	IRS-P1 (ISRO)	LISS-2, MEOSS (DLR)	Launch failure
Sept. 26, 1993	SPOT-3 (CNES)	2 HRV	S/C ops until Nov. 14, 1997 (loss of Earth lock)
Oct. 15, 1994	IRS-P2 (ISRO)	LISS-II A/B	
Nov. 4, 1994	RESURS-O-3 (Russia)	2 MSU-E, 2 MSU-SK	Operational as of 2000
Dec. 28, 1995	IRS-1C (ISRO)	PAN, LISS-III, WiFS	Operational as of 2001
Mar. 21, 1996	IRS-P3 (ISRO)	WiFS	Operational as of 2001
Aug. 17, 1996	ADEOS (NASDA)	AVNIR	S/C operations terminated June 30, 1997 (loss of power)
Apr. 26, 1996	MIR/Priroda (Russia)	MSU-E, MSU-SK, MOMS-2P (DLR)	The Priroda life ended with the de-orbit of MIR on March 23, 2001
Sept. 29, 1997	IRS-1D (ISRO)	PAN, LISS-III, WiFS	Operational as of 2000
Dec. 24, 1997	EarlyBird (EarthWatch)	EBP, EBM, 3 m resolution	The first commercial imaging S/C failed to operate
Mar. 24, 1998	SPOT-4 (CNES)	2 HRVIR, Vegetation	Operational as of 2000
July 10, 1998	RESURS-O-4 (Russia)	2 MSU-E, 2 MSU-SK	Operational as of 2000
April 15,1999	Landsat-7 (NASA)	ETM+	Optomechanical (whiskbroom) imaging, operational
April 27, 1999	Ikonos-1 (Space Imaging)	OSA (Optical Sensor Assembly)	Launch failure of commercial imaging mission

Launch Date	Satellite Platform (Agency or Company)	Earth Imaging Sensors of respective Payload	Comment (ops = operations)
Sept. 24,1999	Ikonos-2 (SI) Identical backup satellite and payload to Ikonos-1	OSA built by Kodak	A new era of 1 m spatial resolution imagery began for spaceborne instruments
Oct. 14, 1999	CBERS (China/Brazil)	HRCC, IRMSS, WFI	Ziyuan-1 (Chinese name)
Dec. 18, 1999	Terra of NASA)	ASTER (NASDA)	S/C operational as of 2001
Dec. 20, 1999	KOMPSAT-1 (KARI)	EOC	S/C operational as of 2001
Nov. 20, 2000	QuickBird-1 (EarthWatch)	BGIS-2000 (Ball Global Imaging System-2000)	Commercial imagery at 1 m GSD, swath = 22 km Launch failure
Nov. 21, 2000	EO-1 (NASA)	ALI (technology verification instrument)	Tandem orbits with Landsat-7 (1 minute apart)
Dec. 5, 2000	EROS-A1 of ImageSat International, Cayman Islands	PIC (Panchromatic Imaging Camera)	Commercial imagery at 1.8 m GSD, swath=12.5 km
2001 (planned)	OrbView-4 (Orbimage)	OHRIS with Pan & 4 MS bands, OHIS, 280 bands	Commercial imagery at 1 m Pan, 4 m MS, swath of 8 km 8 m resol., 5 km swath
2001 (planned)	QuickBird-2 (EarthWatch)	BGIS-2000 (Ball Global Imaging System-2000)	Commercial imagery at 0.61 m Pan, 2.5 m MS
2001 (planned)	IRS-P6 (ISRO)	LISS-IV, LISS-III', AWiFS	
2001 (planned)	CBERS-2 (China/Brazil)	HRCC, IRMSS, WFI	
2002 (planned)	OrbView-3 (OrbImage)	OHRIS (OrbView High Resol. Imaging System)	Commercial imagery at 1 m PAN, 4 m MS, 8 km swath
2002 (planned)	SPOT-5 (CNES)	2 HRG, HRS, Vegetation	5 m resolution for HRG PAN data, 3 m in supermode
2002 (planned)	IRS-P5 (ISRO) CartoSat-1	PAN-F, PAN-A	Along-track stereo imagery 2.5 m GSD, 30 km swath
2002 (planned)	EROS-B1 of ImageSat International	PIC-2 (Panchromatic Imaging Camera-2), TDI	Commercial imagery at 0.82 m GSD, swath=16 km
2003 (planned)	ALOS (NASDA)	PRISM, (stereo mapping) AVNIR-2	2.5 m GSD, 35 km swath 10 m MS, 70 km swath
2003 (planned)	ROCSat-2 (NSPO)	RSI (Remote Sensing Instrument)	2 m GSD, 24 km swath 8 m MS data (5 bands)
2003 (planned)	NEMO (NRL)	COIS, PIC	Coregistered data of sensors
2003 (planned)	Diamant-1 (OHB System, Bremen, Germany)	MSRS, 12 bands of MS data in VNIR	Commercial imagery at 5 m GSD, 26 km swath
2003 (planned)	TOPSAT (QinetiQ, RAL), UK	HIROC, 3 bands of MS, 1 Pan	Imagery at 2.5 m GSD Pan and 5 m MS, Swath 15 km
2003 (2 S/C) 2004 (2 S/C)	RapidEye (RapidEye AG, Munich, Germany)	REIS, 5 bands MS, 1 Pan band	Commercial MS and Pan imagery at 6.5 m, 150 km swath
2004 (planned)	KOMPSAT-2 (KARI)	MSC (Multi-Spectral Camera) Pan and MS	1 m GSD for Pan, 4 m GSD for MS, 15 km swath
2004 (planned)	IKONOS-3/4 (Space Imaging)		Commercial imagery at 0.5 m Pan, 2 m MS
2005 (planned) 2006 (planned)	Pléiades-1 (CNES) Pléiades-2 (CNES)	OHRI (Optical High-Resolution Imager)	Commercial imagery at 0.7 m Pan, 2.8 m for 4 MS bands

Table 1: Chronology of optical Earth-surface satellite imaging missions

• Charge-transfer devices (O.4.2). [27] [28] [29] The CCD (Charge-Coupled Device) detector technology was invented in 1969 by Willard Boyle and George Smith and first demonstrated (one-line eight pixel detector) by Gil Amelio, Mike Tompsett and George Smith at the Bell Labs (Bell Telephone Laboratories of AT&T, since 1996 of Lucent Technologies, Inc.) in Murray Hill, New Jersey, USA. **The silicon-based detector invention, along with the development of the planar process for integrated circuits (ICs), eventually led to a revolu-**

27) W. S. Boyle, G. E. Smith, "Charge-Coupled Semiconductor Devices," Bell System Technical Journal, Vol. 49, April 1970, pp. 587-593

28) G. F. Amelio, M. F. Tompsett, G. E. Smith, "Experimental Verification of the Charge Coupled Device Concept," Bell System Technical Journal, April 1970, pp. 593-600

29) G. F. Amelio, W. J. Bertram, M. F. Tompsett, "Charge-Coupled Imaging Devices: Design Considerations," IEEE Transactions on Electron Devices, Vol. ED-18, pp. 986-992, Nov. 1971

tion in optical imaging capabilities (resulting in the birth of digital photography), providing in addition a wide forum for multi-disciplinary applications. Early industry CCD manufacturers were: TI (Texas Instruments), RCA (Radio Corporation of America), Fairchild Semiconductor, GEC (General Electric Co., UK), and Thomson-CSF (France, since 1978). At the start of the 21st century, the CCD technology is still having profound effects in the fields of optical imaging, particle tracking, x-ray detection, as well as analog storage devices. The application spectrum ranges from security monitoring to high-definition television, from endoscopy to desktop videoconferencing, from Earth observation to astronomy and to cosmology.

A CCD is a photosensitive solid-state silicon-based detector implemented with large-scale integration technology based on MOS (Metal Oxide Semiconductor) capacitors that collect and transfer photon-generated charge. [30) 31)] A MOS capacitor is a three-layer sandwich formed by positioning a metal electrode, insulated by a layer of silicon dioxide, onto a silicon substrate (buried channel). Typically, the device is built on a p-type substrate. An n-type region (about 1 μm in thickness) is formed on the surface. The gate electrode consists of a thin silicon dioxide layer either on a metal surface or on a heavily doped polycrystalline silicon layer.

In remote sensing, the CCD pushbroom concept permits simultaneous measurements of an entire scan line (detector line array) for each spectral band (and serial shift-register readout), thereby improving the dwell time (or sampling time) over whiskbroom systems. (Note: very short dwell times, in the order of a few microseconds, limit the number of photons that can be detected without using very large optics). Advantages of the digital CCD technology lie in the real-time processing capabilities of the data, the high accuracy potential, the good radiometric characteristics, and the availability of relatively inexpensive system components. - The first astronomical CCD ground observation (of Jupiter, Saturn and Uranus) was done in early 1976 at the Mount Bigelow observatory of the University of Tucson, AZ (CCD detector array built by Texas Instruments). [32)] - The first experimental airborne opto-electronic instrument, featuring a pushbroom CCD detector line array, was EOS (Electro-Optical Scanner) of DLR (former DFVLR), designed and developed by Astrium GmbH (formerly MBB) of Munich in 1977, and flown on a DLR aircraft (DO-28) in early 1978 (chapter P.79). MEIS (P.127) of CCRS followed in 1978, and TIMS of Daedalus in 1981. ISRO built a single-band CCD camera in the late 1970s and had it flight-tested on an aircraft in 1980.

Spaceborne CCD detector technology was introduced with the **MSU-E** (Multispectral Scanning Unit-Electronic) flown on the Meteor-Priroda-5 (launch June 18, 1980) spacecraft of the former Soviet Union. [33)] MSU-E was built at ISDE (Russian Institute of Space Device Engineering) in Moscow; it featured a CCD line array of 1024 pixels, three parallel line arrays, each of 1024 elements, provided pushbroom imagery in three spectral bands (visible and near-infrared range); the spatial resolution was 28 m on a swath of 28 km. The CCD line arrays were designed and manufactured in the "Pulsar" plant, Moscow. The results obtained with MSU-E confirmed the potential of CCD technology for use in high-resolution multispectral monitoring. The MSU-E pushbroom instrument was also introduced on the Soviet Resurs series starting with Resurs-O1-1 (launch Oct. 3, 1985). [34)] Further early CCD instruments were: MOMS-01 of DLR (built by Astrium GmbH, and flown on Shuttle flights STS-7 in June 1983 and STS-41B in Feb. 1984), the HRV sensors on the SPOT series

30) C. J. S. Damerell, "Charge-coupled devices as particle tracking detectors," Review of Scientific Instruments, Vol. 69, No. 4, April 1998, pp. 1549-1573

31) D. F. Barbe, "Imaging Devices Using the Charge-Coupled Concept," Proceedings of the IEEE, Vol. 63, No. 1, Jan. 1975, pp. 38-67

32) I. S. McLean, "Electronic Imaging in Astronomy - Detectors and Instrumentation,"John Wiley & Sons, 1997, pp.128

33) A. S. Selivanov, Y. M. Tuchin, M. K. Naraeva, B. I. Nosov, "Experimental Satellite System for Earth Monitoring," Issledovanie Zemli iz Kosmosa, W 5, 1981

34) Information provided and document translation: courtesy of Boris Zhukov of DLR and Ian Ziman of IKI, Moscow

of CNES (launch of SPOT-1 on Feb. 22, 1986), and LISS on the IRS series of ISRO (launch of IRS-1A on March 17, 1988). - The first spaceborne CCD area array detectors were introduced in the **TVS** (TV System) instrument, a wide-angle TV camera (TVS, an IKI instrument, was developed in cooperation with France and Hungary) of the Soviet Vega interplanetary missions to Venus and later to Halley's Comet in 1984 (launched Dec. 15 and Dec. 21, 1984, respectively). On June 9 and 13 1985, each Vega S/C deployed a Venus capsule. In March 1986, each Vega S/C encountered Halley's Comet. TVS provided color images of Halley's coma and core [about 500 images from Vega-1 and about 700 images from Vega-2, each at closest approach (8000-9000 km)to Halley's Comet].

In the meantime CCDs have evolved to a sophisticated level of performance (high radiometric and spatial resolution). Starting with the 1990s CCDs have become the technology of choice for most imaging applications in remote sensing. The microelectronics industry provides CCD technology (area arrays) on a chip level. A high-resolution VNIR camera can be built around a CCD chip by combining it with a suitable lens system, a cooling method, and operating electronics. CCDs take the place of vacuum tube-based imagers and film in conventional cameras. They provide a wide field of applications in science as well as in the consumer market. An astronomical imaging system with the name of STIS (Space Telescope Imaging Spectrograph) on Shuttle flight STS-82 (launch Feb. 11, 1997) was installed aboard the orbiting Hubble Space Telescope; the key element of STIS is a very sensitive CCD array (developed by SITe) enabling faint radiation measurements of distant stars. Also in 1997 Philips Co. with several partners was testing a CCD area array of 9216 by 7168 detectors (pixels) on a chip. The pixel size is 12 μm x 12 μm, the chip size is 111 mm x 86 mm. - Some drawbacks of the CCD technology are: CCD's provide good image quality, but they are expensive, power hungry, and rather bulky with the required accessory chips.

- An alternate approach to the proven CCD-imaging concepts may be provided by the emerging **CMOS/APS** (Complementary Metal-Oxide Semiconductor/Active Pixel Sensor) technology, a second generation solid-state sensor technology (CSMT of JPL is the inventor of APS technology and a major developer in this field) which utilizes active transistors in each pixel to buffer the detected photon signal. The APS method basically provides a current as a function of the irradiance for each pixel. Normally, the output is proportional to the irradiance falling onto the APS detector (there are also APS versions providing a logarithmic output of the irradiance). The CMOS/APS technology employs an amplifier at each pixel site, thereby eliminating the bus capacitance and resistance problems of CIDs. It was not until sub-micron photolithography became available that APS imagers became useful. CMOS/APS imagers sense radiation (light) in the same way as CCDs. Both technologies convert incident photons into electronic charge (electrons) by the same photo-conversion process. The charge packets are not transferred, they are instead detected by the charge-sensing amplifiers, which are made from CMOS transistors. - The microprocessor (CMOS) compatibility of CMOS/APS concepts (widely available techniques) results in reduced fabrication costs over similar CCD devices. Other performance advantages of the CMOS/APS technology are: TTL-compatible operation (0-5V), only a single power supply is needed, electronic shuttering, readout windowing, variable integration time, on-pixel amplification, noise reduction, random access readout, a single CMOS/APS chip can integrate image capture, A/D conversion and digital processing. The CMOS detectors provide considerably higher radiation hardness over CCD detectors. However, current drawbacks of CMOS technology are seen in the small pixel sensitivity (low quantum efficiency and non-uniformity), the relatively high noise levels of relatively small pixel arrays, and in reduced calibration capabilities of the array. The CMOS/APS imager in silicon technology offers a spectral response in the UV/VNIR spectral regions. The CMOS/APS micro-technology implies low-power micro-instruments (imagers and other devices) for near-future microsatellites. Other fields of CMOS/APS technology applications may include: optical communications, multimedia machine vision, and biomedical imaging. Note: APS (Active Pixel Sensor) is also being referred to as APA (Active Pixel Array), both terms have the same meaning.

• The first spaceborne demonstration of a camera with CMOS/APS technology was probably VTS (Visual Telemetry System), built by DSS/OIP (Delft Sensor Systems/Optronic Instruments & Products), and IMEC, Belgium, and flown on TEAMSAT (M.26), an ESA low-cost satellite demonstrator mission, launched on Oct. 30, 1997 from Kourou (2nd qualification flight of an Ariane-5 vehicle). VTS, with a 512 x 512 pixel array for each of the three camera modules, performed visual post-launch monitoring of deployment structures. The TMSat (Thai-Phutt) microsatellite of Thailand, built by SSTL of Surrey, UK (launch July 10, 1998), carries a CMOS Video Camera with APS technology.- APS (Active Pixel Sensor) technology is also part of MICAS (Miniature Integrated Camera Spectrometer) flown on DS1 (Deep Space 1, launch Oct. 24, 1998, M.5) of NASA/JPL. The 3CSat (Three Corner Satellite) constellation of UNP (University Nanosatellite Program), consisting of three cooperating S/C (planned launch in 2001) – built by CU (University of Colorado) at Boulder, ASU (Arizona State University), and NMSU (New Mexico State University) and sponsored by DoD, NASA and US industry – utilizes CMOS/APA (Active Pixel Array) imaging technology in a distributed observation concept.

• APPS (Active Pixel Position Sensor). At the start of the 21st century, APPS represents an advanced type of APS detector with the added capability of very accurate position sensing/determination in conjunction with online processing (so-called "smart vision chips" provide the processing). The APPS detector concept provides a signal proportional to a moment equation of the irradiance falling onto the detector. The APPS detector itself computes the desired moment equations with appropriate circuitry. The APPS technology was developed at AFRL (Natalie Clark, with DARPA, AFSOR and NASA funding) in combination with adaptive optics methods to improve star tracking methods by higher accuracy centroiding. With CCD and APS methods, it is rather difficult to centroid on a star with a pixel accuracy of better than 1/10. The APPS method, on the other hand, provides centroiding in the order of 1/100 of a pixel accuracy. Further features of the APPS technology are:

- Detection of very low-light conditions
- APPS does not require the star light to be spread over an array of pixels; this enables much higher SNR, bandwidth and accuracy
- APPS enables a wide dynamic range permitting simultaneous imaging of faint and bright target objects (stars) in a single image frame
- APPS supports high-accuracy attitude sensing

The APPS scheme can of course also be used in support of conventional imaging applications. However, in this operations mode the detector signals are added rather than differenced.

• Detector technology extension. The measurement of electromagnetic radiation in the various spectral ranges, in particular in the infrared region, is at the heart of the problem. [35] Significant advances in Earth observation capability have come from the development of on-chip detector circuitry (see chapter O.4).

• The "digital camera," a commercial product of the photo industry (e.g., Kodak, Sony, Agfa, Canon, Fuji, Konica, Minolta, Nikon), is a natural application derivative of the CCD and CMOS/APS detector technologies, pioneered by the remote sensing and the electronics industry. The digital camera became a consumer product in the latter part of the 1990s when the detector densities (multi-megapixel detectors - approaching high-resolution quality) and storage capacities for image capture were sufficient and in addition economical for an evolving consumer market.
Since CCD detectors are inherently monochromatic, special filters are needed to separate the RGB radiation reflected by an object. A number of techniques are in use to capture color: a) RGB filter wheel (three exposures), b) tri-linear sensor (use of three linear sensors), c) multi-chip detectors, etc.

35) A good reference is: W. D. Rogatto, "The Infrared and Optical Systems Handbook, Vol. 3 Electro-Optical Components," Copublished by ERIM and SPIE, 1993, chapter 4.5.2

• Spatial resolution. The requirement of high spatial resolution (and with it the ability to discriminate between small objects), in particular to land surface imagery, has been around since the early days of remote sensing. In parallel, there are virtually always the needs for wide-swath coverage and high temporal resolution (i.e., frequent repeat coverage capability). High spatial resolution implies of course high data rates which conflicts in turn wide-area coverage. An overview of major land-surface imaging missions [36] (Figure 1, Table 2) demonstrates the trends and practical compromises taken with the best available technology of the time.

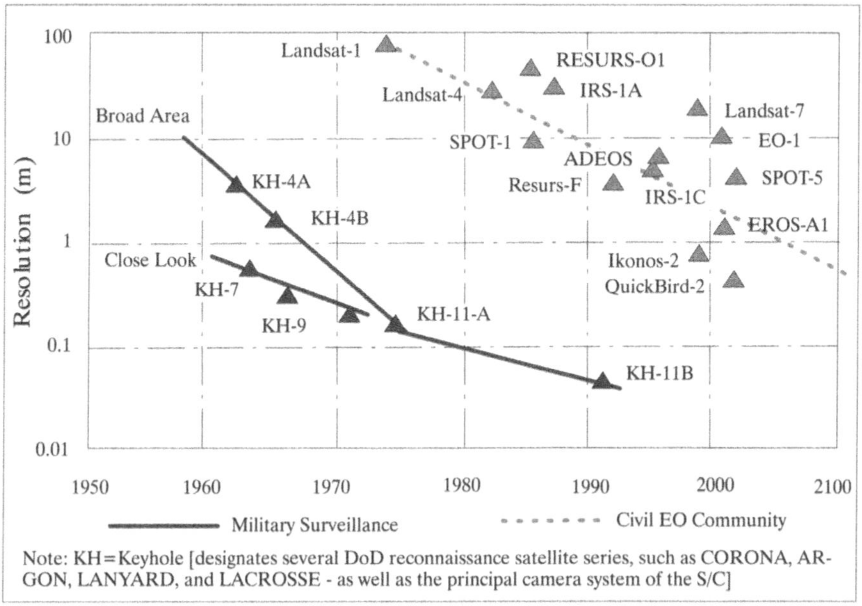

Figure 1: Spaceborne surface imaging resolution trends in the civil & military fields

Mission/Launch Date	Sensor	Spatial Resolution (best band)	Swath width	Data Rate
Landsat-1 / Jul. 23, 1972	MSS	80 m	185 km	15 Mbit/s
Landsat-4 / Jul. 16, 1982	TM	30 m	185 km	85 Mbit/s
Resurs-O1-1 / Nov. 3, 1985	2 MSU-E	45 m x 33 m	80 km (total)	
SPOT-1 / Feb. 22, 1986	2 HRV	10 m	117 km (total)	50 Mbit/s
IRS-1A / Mar. 17, 1988	2 LISS-II,	36 m	2 x 74 km	2 x 10.4 Mbit/s
Resurs-F / (series)	KFA-1000	6-8 m	70 km	Film
IRS-1C / Dec. 28, 1995	PAN	6 m	70 km	125 Mbit/s
ADEOS / Aug. 17, 1996	AVNIR	8 m	80 km	60 Mbit/s
Landsat-7 / Apr. 15, 1999 EO-1 / Nov. 21, 2000	ETM+ ALI	13 m x 15 m 10 m	185 km 185 km	150 Mbit/s 300 Mbit/s
Ikonos-2 / Sept. 24, 1999	OSA	1 m PAN 4 m MS	11-13 km	320 Mbit/s
QuickBird-1 / Nov, 20, 2000 (launch failure)	BGIS 2000	1 m PAN 4 m MS	27 km	320 Mbit/s
NEMO / 2003	COIS, PIC	30 m, 5 m	30 km	150 Mbit/s

Table 2: Spatial resolutions of major land-surface imagers

36) J. Pike, http://www.fas.org/eye/imint.htm

- A notable exception of the trend away from photographic film in spaceborne imaging technology was and is the RESURS-F satellite series of State Center Priroda, Moscow, Russia. The Resurs-F1 series (58 satellites in the time frame 1979-1993, Resurs-F2 series since 1995, see chapter D.31) on-board sensors are film cameras whose data (namely the films) are recaptured after the end of each mission. The film camera systems are returned to the ground in small spherical descent capsules which are reused an average of three times. The films are processed after each mission (average mission life is in the order of 14 to 30 days in orbits of 240 - 275 km). Note: At the turn of the millennium, film still holds the edge in resolution over CCD technology. However, digital technology can usually deliver images faster while allowing greater flexibility in storage, manipulation and distribution. The real-time aspect of digital imagery is of considerable importance for many applications.

- The directional pointing capability of a sensor (on command) in the cross-track direction was introduced in the SPOT series in 1986. The new pointing technique introduced the term "FOR (Field of Regard)", designating the angular coverage capability beyond (and including) the swath - it allowed the imaging of events of interest that happened to be outside the regular (nadir-centered) swath width.

- Hyperspectral imaging [an optical sensing technique breaking up the incoming radiation into numerous contiguous spectral bands (normally 20 - 200 narrow bands or more)], enabled primarily through improved filtering and CCD detector technologies, was introduced with such pioneering airborne instruments as AIS (1982, 128 spectral bands) and AVIRIS (1989, 224 spectral bands), both of JPL. - The first spaceborne hyperspectral imagers launched were: a) **UVISI** on MSX of DoD (launch Apr. 24, 1996; b) HSI and LEISA, both on the Lewis spacecraft (launch Aug. 23, 1997; however, Lewis never became operational and reentered the Earth's atmosphere on Sept. 28, 1997); c) FTHSI on MightySat II.1 of AFRL (Air Force Research Laboratory) with a launch July 19, 2000 (FTHSI demonstrates the advantage of Fourier systems over dispersive hyperspectral imagers, in that it can record the full spectra without any time delay and can decouple the spatial and spectral signatures). The Hyperion and LAC (LEISA Atmospheric Corrector) instruments on EO-1 (NASA/GSFC), of HSI and LEISA heritage, respectively, were launched Nov. 21, 2000. COIS (Coastal Ocean Imaging Spectrometer) of the NRL NEMO satellite has a planned launch date in 2003. Note: The hyperspectral imagers are also known by the term of "imaging spectrometers." [37] [38]

Typical examples of hyperspectral filter techniques include dispersive gratings, prisms, multi-order etalons, interference filters, Michelson interferometers, and acousto-optic tunable filters (AOTF). The other key feature is detector array technology which allows multiple spatial and/or spectral samples through 1-D or 2-D arrays.

The passive remote sensing technique of hyperspectral imaging offers an unparalleled spectral interpretation capability of the data (quantitative monitoring) for many applications, including land, water and atmospheric parameters (detection/discrimination of spectral fingerprints from objects that cannot be derived from coarser multispectral imagers). Land applications include vegetation studies (species identification, plant stress, leaf water content), soil science (erosion), geology (mineral identification and mapping, detection of underground/camouflaged structures), and hydrology (liquid/solid water differentiation, snow/grain size). Water applications include monitoring of water quality, bathymetry, etc. Atmospheric applications include the measurement of water vapor, ozone, aerosols, and cloud characteristics. - A disadvantage of the technique is the generation, communication, processing, interpretation and storage of immense data volumes.

The spectra retrieved with an imaging spectrometer show absorption features that are matter specific and depend on quantum mechanical interactions as well as molecular structure

37) G. Vane, A. F. H. Goetz, "Terrestrial Imaging Spectrometry: Current Status, Future Trends," Remote Sensing of Environment, Vol. 44, 1993, pp. 117-126

38) J. Kerekes, "Hyperspectral Sensors," http://daacdev1.stx.com/ELF/docs/PER/kerek18.html

and scattering properties of the observed material. Observing reflected solar illumination with such detailed spectral resolution enables direct identification of virtually all diagnostic absorption features in minerals and soils, the examination of physical and biological components in marine and inland waters, and the study of biochemical processes in vegetation.

Sensor (Agency/Company)	No. of Bands	Spectral Range (nm)	Bandwidth at FWHM (nm)	IFOV (mrad)	FOV (°)	Data Product	Period of Operation
AAS (ASTER) (DAIS-2815) (GER)	1 3 20	760-850 3000-5000 8000-12000	90 600-700 200	1.0, 2.5, or 5.0	28.8, 65, or 104	Image cube	since 1991
AHS,Daedalus	48	440-12700	20-1500	2.5	86	Image cube	1994
AIS-1 AIS-2 (NASA/JPL)	128 128	990-2100 1200-2400 800-1600 1200-2400	9.3 10.6	1.91 2.05	3.7 7.3	Image cube Image cube	1982-85 1986-87
AISA (Karelsilva Oy)	1-286	450-900	1.56-9.36	1	21	Image cube	since 1993
AMSS (GEOSCAN)	32 8 6	490-1090 2020-2370 8500-12000	170-240 430-440 550-590	2.1 x 3.0	92	Image cube	since 1985
ARES (AIP) (Lockheed)	75	2000-6300	25-70	1.17	3 x 3	Image cube	since 1985
ASAS upgraded ASAS (NASA/GSFC)	29 62	455-873 400-1060	15 11.5	0.80 0.80	25 25	Image cube Image cube	1987-91 since 1992
AVIRIS (JPL)	224	380-2500	9.7-12.0	1	30	Image cube	since 1989
CASI (Itres Research)	288 19	400-1000 (nominal)	650	1.3, 1.6	37.8 44.7	Profile image	since 1990
CIS (China)	64 24 1 2	400-1040 2000-2480 3530-3940 10500-12500	10 20 410 1000	1.2 x 3.6 1.2 x 1.8 1.2 x 1.2 1.2 x 1.2	80°	Image cube	since 1993
CHRISS, SAIC AAHIS,(SAIC)	40 288	430-860 440-880	11 3	0.05 1.0	10 11.5	Image cube Image cube	1992 1994
DAIS-7915 (GER/DLR)	32 8 32 1 6	400-1010 1500-1788 1970-2450 3000-5000 8700-12700	10-16 36 36 2000 600	3.3, 2.5, or 5.0	64-78	Image cube	since 1994
DAIS-16115 (GER)	76 32 32 6 12 2	400-1000 1000-1800 2000-2500 3000-5000 8000-12000 400-1000	8 25 16 333 333 stereo	3	78	Image cube	since spring 1994
DAIS-3715 (GER)	32 1 2 1 1	360-1000 1000-2000 2175-2350 3000-5000 8000-12000	20 1000 50 2000 4000	5	±45	Image cube	since 1994
FLI/PMI (Moniteq)	≥288	430-805	2.5	1.3	70	Profile image	1984-90
FTVHSI (Kestrel)	256	440-1150	67 cm^{-1}	0.8	15	Image cube	1996
GER-63 Channel Scanner (GER)	24 4 29 6	400-1000 1500-2000 2000-2500 8000-12500	25 125 17.2 750	2.5, 3.3, or 4.5	90	Image cube	since 1986

Sensor (Agency/Company)	No. of Bands	Spectral Range (nm)	Bandwidth at FWHM (nm)	IFOV (mrad)	FOV (°)	Data Product	Period of Operation
HYDICE (NRL/ERIM)	206	400-2500	7.6 - 14.9	0.5	8.94	Image cube	since 1994
ISM (DESPA/IAS/OPS)	64 64	800-1600 1600-3200	12.5 25.0	3.3 x 11.7	40 (selectable)	Image cube	since 1991
MAS (Daedalus)	50	547-14521	31-517	2.5	85.92	Image cube	since 1992
MISI (RIT)	60	400-1000	10	2	±45	Image cube	from 1996
MIVIS (Daedalus)	20 8 64 10	433-833 1150-1550 2000-2500 8200-12700	20 50 8 400-500	2.0	70	Image cube	1993
MUSIC (Lockheed)	90 90	2500-7000 6000-14500	25-70 60-1400	0.5	1.3	Image cube	1989
ROSIS-03 (DLR/GKSS)	115	430-830	4	0.56	16	Image cube	since 1993 ROSIS-03 since 1998
SFSI (CCRS)	115	1200-2400	10.4	0.4	9.4	Image cube	since 1994
SMIFTS (U. of Hawaii)	75 35	1000-5200 3200-5200	100 cm^{-1} 50 cm^{-1}	0.6	6.0	Image cube	since 1993
TRWIS-A TRWIS-B TRWIS-II TRWIS-III (TRW)	128 90 80 384	430-850 460-880 1500-2500 400-2450	3.3 4.8 12 5/6.25	1.0 1.0 0.5/1.0 0.9	15 15 7.5/15 15	Image cube Image cube Image cube Image cube	1990 1991 1992 1996
Hybrid VIFIS (U. of Dundee)	30 30	440-640 620-890	10-14 14-18	1 1	31.5 31.5	Image cube	Test phase 1994
WIS-FDU WIS-VNIR WIS-SWIR (Hughes SBRC)	64 129,265 81+90	400-1030 400-1000 1000-2500	10.3 12.5 & 6 30 & 23	1.36 0.66 0.66	10 & 15 19.1 12.0	Image cube Image cube Image cube	Test 1992 1995 1995

Table 3: Summary of some early hyperspectral airborne imaging spectrometers

- A spaceborne spectroradiometer (multi-channel scanning radiometer) by the name of CZCS (Coastal Zone Color Scanner), launched on Nimbus-7 (Oct. 24. 1978) provided first images of **ocean color** (chlorophyll and gelbstoff concentrations) distribution. The parameter measurements were optimized for use over water. The data of this first-generation ocean color instrument contributed greatly to our understanding of the marine environment and its biological, biochemical, and physical processes. Monitoring of marine phytoplankton is of great importance because it accounts for nearly half of the world's total primary productivity. Biomass turnover rates for plankton ecosystems are 100 times faster than those for terrestrial systems, leading to a close relationship between upper-ocean ecology and physical forcing.

The value of ocean-color data lies in the long-term monitoring of the marine environment, thereby providing a better understanding of the role of oceans in the global carbon cycle (climate research, etc.). The prime observables of ocean-color instruments are basically the chlorophyll and gelbstoff concentrations in the surface layer of the ocean. The concentration of chlorophyll is used to estimate the abundance of phytoplankton in ocean waters, and hence the abundance of ocean biota.

Sensor, Sensor Provider	Platform	Launch/Introduction
CZCS, NASA, USA	Nimbus-7	Oct. 24. 1978 - June 22, 1986
OCE, NASA/GSFC, USA	Shuttle (SIR-A mission), STS-2	Nov. 12-14, 1981
MOS, DLR, Germany	IRS-P3 (ISRO, India) Priroda (MIR module, Russia)	March 21, 1996 April 26, 1996

Sensor, Sensor Provider	Platform	Launch/Introduction
OCTS, NASDA, Japan POLDER, CNES, France	ADEOS	Aug. 17, 1996 - June 30, 1997 (some ocean color products)
SeaWiFS, OSC/Orbimage, USA	OrbView-2/SeaStar (previous name)	Aug. 1, 1997
OCI, NSPO, Taiwan	ROCSAT-1	Jan. 26, 1999
OCM, ISRO, India	IRS-P4 (OceanSat-1)	May 26, 1999
MODIS, NASA/GSFC, USA	Terra (previous name: EOS/AM-1)	Dec. 18, 1999
OSMI, KARI, Korea	KOMPSAT-1	Dec. 20, 1999
MERIS, ESA, Europe	Envisat	planned for Oct. 2001
MODIS, NASA/GSFC, USA	Aqua (formerly EOS/PM-1)	planned for Dec. 2001
OCS, SITP, China	HY-1 (Haiyang-1)	planned for 2002
GLI, NASDA, Japan Polder-2, CNES, France	ADEOS-II	planned for 2002
COIS, NRL, USA	NEMO (Navy EarthMap Observer)	planned for 2003
VIIRS, IPO, USA	NPP	planned for 2005

Table 4: Overview of spaceborne sensors for ocean color detection

Sensor	Ground Resolution (km)	Swath Width (km)	Spectral Bands (range nm)	Data Repetivity (days)	Sensor Tilt Capability	Absolute Radiomet. Accuracy	Polarization Sensitivity
CZCS	0.825	1600	5 bands 443 - 800 + 1 TIR	2			
OCE	0.90	180	8 bands 464 - 773				
MOS-A MOS-B MOS-C	1.57 x 1.40 0.52 x 0.52 0.52 x 0.64	195 200 192	4 (755-768) 13 (408-1011) 1 (1600)				
OCTS	0.70	1400	8 bands 402 - 885 + 4 IR	2	±20°	<10%	<2-5%
SeaWiFS	1.1 LAC 4.5 GAC	2800 LAC 1500 GAC	8 bands 402 - 885	2	±20°	<5%	<2%
MODIS	1	2330	9 bands 405 - 877	2	none	<3%	<2% (<5% at 412 nm)
OCM	0.36	1420	8 bands 402 - 885	2	±20°	<10%	<2%
OCI	0.80	691	6 bands 433 - 885				
OCS	1.1	1600	8 bands 402-885 +2 TIR	2		10-40%	
OSMI	1 0.85 nadir	800	8 bands 400 - 885		±45°		
MERIS	0.3 and 1.2	1150	15 bands 390 - 1040	3		<2%	<0.5%
GLI	1/0.25	1600	36 bands 375 - 2250	2			<2%
COIS	0.03/0.06	30	210 bands 400 - 2500	7 (2.5)	±30°	<5%	<5%

Table 5: Performance parameters of ocean color sensors (multispectral radiometers)

- Size/mass/power/cost reduction of imaging instruments. Miniaturization and the introduction of new technologies is the name of the game, in particular in the late 1990s and the early 21st century. The JHU/APL sensor MSI (Multispectral Imager) with a 537 x 244 CCD detector array, a spectral range of 0.4 to 1.1 μm, and a FOV of 2.9° x 2.25° (pixel resolution of 95 x 161 mrad), was flown on NEAR (Near Earth Asteroid Rendezvous/Shoemaker) of NASA (launch Feb. 17, 1996). A newer MSI design of JHU/APL with a lightweight reflective telescope, high-density solid-state memory and miniaturized on-chip-board electronics has a mass of 0.5 kg and a power of 1 W. 39)

Mission	Instrument	Instrument Mass, Power	Comment
Landsat-5 (launch Mar. 1, 1984), NASA	TM	285 kg, 385 W	Optomechanical instrument
IRS-1C (launch Dec. 28, 1995), ISRO	LISS-III	171 kg, 78 W	Optoelectronic instrument
NEAR/Shoemaker (launch Feb. 17, 1996), NASA	MSI	7.7 kg, 6.9 W	Optoelectronic instrument
DS1 (launch Oct. 24, 1998), NASA	MICAS	12 kg,	CMOS/APS technology
Landsat-7 (launch Apr. 15, '99), NASA	ETM+	425 kg, 590 W	Optomechanical instrument
IKONOS-2 (launch Sept. 24, '99), SI	OSA	171 kg, 350 W	Optoelectronic instrument
Terra (launch Dec. 18, 1999), NASA	ASTER	421 kg, 463 W	Optomechanical instrument
QuickBird-1 (launch Nov. 20, 2000), Earthwatch	BGIS 2000	380 kg, 430 W	Optoelectronic instrument (launch failure)
EO-1 (launch Nov. 21, 2000), NASA	ALI	106 kg, 100 W	Optoelectronic instrument
Diamant-1 (launch 2003), OHB-System	MSRS	70 kg, 120 W	Optoelectronic instrument
RapidEye (launch 2003 & 4) RapidEye	REIS	62 kg, 73 W	Optoelectronic instrument
PROBA (launch 2001), ESA	CHRIS	15 kg, 10 W	Optoelectronic instrument
SMART-1 (launch 2003) ESA	AMIE	0.45 kg, 0.5 W	Optoelectronic instrument
NPOESS (launch 2008)	VIIRS	160 kg, 134 W	Optomechanical instrument

Table 6: Mass/power reduction trends in typical spaceborne imaging instruments

• Phased array optical technology. 40) Optical phased arrays (or laser radars) are passive arrays, consisting of phase shifters (based on liquid-crystal technology). Optical phased arrays are the direct functional analogs of the microwave phased array. However, due to the orders-of-magnitude difference in wavelengths between the microwave and optical worlds, a different implementation approach is used for optical arrays. The current (late 1990s) generation of optical devices are 1-D, space-fed, passive, phase-only, apertures. An already formed beam is fed to the array of phase shifters, which then affects steering of the beam. The imposed phase profile steers, focuses, fans out, or corrects phase aberrations on the beam. High efficiency large-angle steering with phased arrays requires phase shifter spacing on the order of a wavelength or less; hence, addressing issues make 1-D optical arrays more practical than 2-D arrays. - Laser radars may be used for such applications as target detection and wind profiling.

• Spaceborne stereoscopic along-track imaging of the Earth's surface was introduced by MOMS-02, a three-line camera system on Shuttle flight STS-55 in April/May 1993, followed by a MOMS-02 reflight on MIR/Priroda (launch of Priroda April 23, 1996). 41) The OPS sensor of JERS-1 (launch Feb. 11, 1992) offers stereoscopic along-track imaging. The parallel HRV sensors of the SPOT series provide some stereo capability with their individual side-viewing feature (up to 27° cross-track pointing capability). The EarlyBird S/C (launch Dec. 24, 1997) of Earthwatch with the EBP sensor (along-track slewing capability) offered a stereo capability; however, the satellite could not be made operational by the ground system. - Stereo imagery is particularly suitable for topographic mapping applications (photogrammetry). Further instruments with stereoscopic imaging capability are operational (or planned) in the following missions:

- Ikonos-2 S/C of Space Imaging (launch Sept. 24, 1999) with the OSA sensor (along-track slewing capability of up to ±30°)

39) Binh Q. Le, P. D. Schwartz, Sharon X. Ling, et al., "A Low-Cost Miniaturized Scientific Imager Design with Chip-on-Board Technology for Space Applications," Johns Hopkins APL Technical Digest, Vol. 20, No 2, 1999, pp. 170-179

40) P. F. McManamon, et al., "Optical Phased Array Technology," Proceedings of the IEEE, Vol. 84, No.2, Feb. 1996, pp. 268-298

41) Note: The three-line observation concept results in forward, nadir and aft views of the CCD pushbroom array. The imagery from each scan line is assembled into strips. Relief displacement in the line perspective geometry of the strip approach differs from conventional nadir perspective geometry. Every object appears on all three strips. In contrast, on film imagery only about 60% of the area of any one photograph is in a triple overlap.

- Terra (formerly EOS/AM-1) of NASA with the NASDA instrument ASTER (along-track two-line system in VNIR), and with the NASA/JPL instrument MISR (along-track nine line camera system), launch Dec. 18, 1999
- BIRD (Bi-Spectral Infrared Detection) of DLR (planned launch in 2001) with the WAOSS-B (Wide-Angle Optoelectronic Stereo Scanner, BIRD version) instrument. A modified version of WAOSS flown on the Russian Mars-96 mission (launch on Nov, 17, 1996 from Baikonur; however, the S/C failed to obtain a Mars orbit). The along-track stereo imaging system of WAOSS-B is based on 3 CCD-lines (single optics and focal plate), operating in pushbroom mode, and taking images simultaneously in the forward-, nadir- and backward-pointing direction of the orbital ground track.
- IRS-P5 (CartoSat-1, launch in 2002) of ISRO flies a two-camera assembly. PAN-F (Panchromatic Forward-pointing Camera) is tilted at 26° forward from nadir while PAN-A (Panchromatic Aft-pointing Camera) is tilted 10° aft. A spatial resolution of 2.5 m is provided on a swath of 30 km.
- SPOT-5 of CNES with HRS (High Resolution Stereoscopic instrument, an along-track two-line camera system), projected launch in 2002. A pixel size of 5 m (along-track) by 10 m (cross-track is provided on a swath of 120 km.
- ALOS S/C of NASDA with PRISM (Panchromatic Remote-sensing Instrument for Stereo Mapping), an along-track three-line camera system, projected launch in 2003. A spatial resolution of 2.5 m is provided on a stereo swath width of 35 km. The nadir pointing camera has a swath of 70 km.
- On the airborne side, DLR developed HRSC (High-Resolution Stereo Camera) which features three-line stereo imaging. The instrument is flown since Feb. 1997. A completely automatic photogrammetric and cartographic processing procedure including digital image matching, digital terrain model (DTM) and ortho-image generation, mosaicking and merging of multispectral data has been developed at DLR (P.101)
- Another airborne example of a three-line stereo pushbroom camera is DPA (Digital Photogrammetric Assembly) of DASA/MBB Ottobrunn (P.71). DPA has been operational since the end of 1992 and is a parallel development to MOMS-02.

• **Acousto-Optic Tunable Filters (AOTF)** in acousto-optic imaging systems (O.4.5). An acousto-optic cell is a transparent birefringent crystal excited by a radio frequency transducer. Propagating acoustic waves inside the crystal create regular spatial variations of the refractive index. Under phase-matching conditions, light of a particular linear polarization and wavelength, incident on the crystal at a very specific angle, is diffracted by the moving grating produced by the acoustic wave. [42) 43) 44) 45)]

In the 1990s AOTF devices are based on solid-state technology, they are RF-tunable and random access devices, offering the important capability of spectral agility. When combined with a 2-D CCD array, the complete spectral, spatial, and polarimetric characterization of a scene can be measured with an imaging system that has no moving parts. An AOTF device, can switch from one spectral range to another in the time that it takes an acoustic wave to traverse the crystal, (typically in the μs range). By making the AOTF part of an imaging system [i.e. an AOS (Acousto-Optic Spectrometer)], and projecting the diffracted light onto a 2-D array, it is possible to form an image extracted from the particular spectral component of the incident radiation. CCD arrays have been utilized for this purpose. - Because the crystal is birefringent, the polarization of the light incident on the device affects the angle to which the light is diffracted. Diffraction changes the polarization of the incident

42) P. Ashcroft, "Acousto-Optic Tunable Filters for Imaging Spectrometry," http://daacdev1.stx.com/ELF/docs/PER/ashcr17.html

43) P. Katzka. "AOTF overview: Past, present, and future," Acousto-Optic, Electro-Optic, and Magneto-Optic Devices and Applications, Vol. 753, pp. 22-28. Society of Photo-optical Instrumentation Engineers, 1987

44) L. J. Cheng, T. H. Chao, M. Dowdy et al., "Multispectral imaging systems using acousto-optic tunable filters," Infrared and Millimeter-Wave Engineering, Vol. 1874, pp. 224-231, 1993. Society of Photo-optical Instrumentation Engineers

45) D. A. Glenar et al., "Acousto-optic imaging spectropolarimetry for remote sensing," Applied Optics, Vol. 33, No. 31, Nov. 1, 1994, pp. 7412-7424

radiation so that some of the initially ordinary polarization emerges from the device with extraordinary polarization, and some of the initially extraordinary polarization emerges from the device with ordinary polarization. **The result of this sensitivity to polarization is that an AOTF imager can be used to discern polarization information about a scene in addition to spectral information.** Because the two polarizations of incident light are diffracted differently, it is possible to use two CCD arrays in order to capture separately the image generated by each polarization. The birefringent material of TeO_2 (Tellurium dioxide) is frequently used due to its high acousto-optic figure of merit, and good transmission in the UV, visible and infrared (range of application 0.25-4.5 μm). Some airborne precursor instruments with AOS (Acousto-Optic Spectrometer) technology, leading toward AOTF capabilities are: POLAS (Polarization-sensitive Acousto-optic Spectrometer) of Russia introduced in 1989, AHSTRA (Airborne Heterodyne Spectrometer THz Astronomy) of the Max-Planck-Institute for Radio Astronomy, Bonn; AMSOS (Airborne Millimeter & Submillimeter-wave Observing System) since winter 1997/98, Univ. of Bern, Switzerland; SUMAS/ASUR (Submillimeter Atmospheric Sounder/Airborne Submillimeter SIS Radiometer), Institute of Environmental Physics at the University of Bremen, Germany; THOMAS (THz OH Measurement Airborne Sounder), DLR (Institute of Optoelectronics, Oberpfaffenhofen, Germany.

In the field of spaceborne Earth observation, the Russian instrument Trasser with AOTF technology was first flown on OKEAN-O1-N2 (or Cosmos 1869) with a launch on July 16, 1987. The instrument was operational until 1990. Another Trasser-O instrument was flown on OKEAN-O-1 (launch July 16, 1999, end of mission fall 2000). All Trasser instruments were designed and built by STCUI (Scientific Technological Center of Unique Instruments), Moscow. The Swedish SMR (Submillimeterwave Radiometer) instrument on ODIN (launch Feb. 20, 2001) features an AOS detector provided by France. The Japanese instrument SMILES (Superconducting Submillimeter-wave Limb-Emission Sounder) employs two acousto-optical spectrometers (AOS) with the objective to observe submillimeter-wave radiation (monitoring of trace gas distribution in the atmosphere). SMILES is an ISS payload planned to be launched in 2005 (L.2.15).

In spaceborne science, the NASA SMEX mission SWAS (Submillimeter Wave Astronomy Satellite) with a launch on Dec. 5, 1998 carries the instrument AOS (Acousto Optical Spectrometer), provided by the University of Cologne, Germany, to investigate the composition of dense interstellar clouds. [46] AOS on SWAS is used as a backend in combination with a heterodyne receiver to analyze the intermediate frequency signal. Inside AOS, the RF signals are converted to acoustic waves within a crystal causing pressure waves to travel through the crystal. When illuminated by laser light, the alternating patterns of compression and expansion within the crystal act like the finely spaced lines of a grating causing the laser light to be dispersed along one dimension with intensity variations along this direction that are proportional to the intensities within the input 1.4 to 2.8 GHz band.

1.1.2 Microwave Region, Active Observations (Radars)

• Active microwave instruments (providing their own illumination) of the type SLAR (Side-Looking Airborne Radar), RAR (Real Aperture Radar), and SAR (Synthetic Aperture Radar) were developed and flown. Originally discovered in the 1950s, SAR uses high-range resolution waveforms and synthetic aperture techniques to produce images of objects. Rather than depending on a large physical (real) aperture, a SAR forms a large virtual aperture in the processor when operating on an ensemble of received signals. In this way, an antenna only about 10 m in length on the S/C leads to image resolutions comparable to those of very large real antennas (several km in length). First radar systems for range measurements (tracking of aircraft, ships, etc.) were developed during World War II.
The first Earth-orbiting civilian spaceborne imaging radar instrument (SAR) was flown on

46) http://www.ph1.uni-koeln.de/spec.html

SEASAT (NASA/JPL, L-band SAR, launch June 27, 1978). A side-looking spaceborne RAR instrument with the name of RLSBO was first flown on the Cosmos-1500 satellite (launch Sept. 28, 1983), later also on the Okean series of the former Soviet Union (launch of Okean-O1-1, July 5, 1988).

The standard or conventional architecture of a spaceborne SAR instrument features a SAR antenna (analogous to a telescope of an optical instrument), mounted in a side-looking configuration (O.8.2 and Glossary) and tilted at a look angle, usually between 15-60°, into the swath. As the platform moves in its orbit, a continuous strip of swath width is mapped in the along-track direction of the satellite.

SAR instruments provide high-resolution imagery (typically 10-30 m spatial resolution) of selected polarization and frequency in the microwave region (typical wavelengths from a few cm to about 1m). SAR observation is transparent to cloud coverage and rain, and does not need any sun illumination.

Some background on early SAR developments: [47] [48] [49]
The earliest statement that Doppler frequency analysis could be used to obtain fine cross-range resolution is attributed to Carl Wiley of the Goodyear Aircraft Corporation in June 1951. Wiley observed that a one-to-one correspondence exists between the along-track coordinate of a reflecting object (being linearly traversed by a radar beam) and the instantaneous Doppler shift of the signal reflected to the radar by that object. In 1953, the Doppler-frequency technique was studied by the US Army in the Project Wolverine at the University of Michigan, resulting in an Army project to further develop the range-Doppler radar principle. Part of the program was to develop a practical data processor which could accept wideband signals and carry out the required Doppler-frequency analysis at each resolvable range interval. A group at the Willow Run Laboratories at the University of Michigan (ERIM since 1973) was assigned the problem of developing an optical computer for the purpose. - In the summer of 1957, the Willow Run researchers constructed an X-band radar and built an optical computer. The first fully focused SAR map was produced in August 1957. Soon afterwards, the Willow Run group, in cooperation with Texas Instruments, constructed the first demonstration system, AN/UPD-1, on Army request. Five radar systems were built and various demonstration flights on aircraft were conducted in early 1961.

On the civilian side, a three-wavelength SAR instrument was developed and flown by NASA on the Apollo-17 lunar mission (launch Dec. 7, 1972, moon landing on Dec. 11, return to Earth on Dec. 19, 1972). The objectives of ALSE (Apollo Lunar Sounder Experiment) were to detect subsurface geologic structures ("sounding"), to generate a continuous lunar profile, and to image the moon at radar wavelengths. [50] [51] The ALSE instrument, aboard the Apollo-17 CSM (Command and Service Module), operated at the wavelengths of 60 m, 20 m, and 2 m (corresponding to 5, 15, and 150 MHz). The instrument consisted of two distinct coherent radar subsystems, one operating at two frequencies in the HF-band and the second at a single VHF frequency. The combined HF/VHF radar had a mass of 49 kg and required 103 W of power. The radar data were recorded on 70 mm photographic film (optical recorder - CRT film type) in a conventional SAR format, and returned to Earth for processing. The HF-1 system (5 MHz) was capable of the deepest exploration. The HF-2 system (15 MHz) was operated simultaneously with the HF-1 system to provide partial overlap in the depth of exploration, trading off for improved resolution. The VHF system (150 MHz) was designed for shallow sounding and for surface imaging. All three frequen-

47) D. A. Ausherman, A. Kozma, J. L. Walker, H. M. Jones, E. C. Poggio, " Developments in Radar Imaging," IEEE Transactions on Aerospace and Electronic Systems, Vol. AES-20, No 4, July 1984, pp. 363-400

48) W. M. Brown, L. J. Porcello, "An introduction to synthetic-aperture radar," IEEE Spectrum, Vol. 6, No 9, Sept. 1969, pp. 52-60

49) C. A. Wiley, "Synthetic Aperture Radars - A Paradigm for Technology Evolution," IEEE Transactions on Aerospace and Electronic Systems, AES Vol. 21, 1885, pp. 440-443

50) http://nssdc.gsfc.nasa.gov/cgi–bin/database/www–nmc?72–096A–04

51) L. J. Porcello, et al., "The Apollo lunar sounder radar system," Proceedings of the IEEE, June 1974, pp. 768-783

cies were capable of surface profiling. Separate transmit/receive antenna systems were provided for the HF and VHF ranges. The returned data were optically processed and provided good profiles of the lunar surface/subsurface.

Mission	Launch Date	Spatial Resolution (m)	Swath Width (km)	Freq. GHz (Band)	Polarization	Look Angle	Data Rate Mbit/s	Altitude, Inclina. (km,º)	Instrument, Remarks
Seasat	27.6.78	25	100	1.28 (L)	HH	20º	110	800,108º	SAR
SIR-A	12.11.81	40	50	1.28 (L)	HH	47º	N/A	260, 38º	SAR, Shuttle
SIR-B	5.10.84	25	30	1.28 (L)	HH	15-60º	34	225,57º	SAR, Shuttle
Kosmos 1870	25.7.87	25-30	20-35	3.125(S)	HH	25-60º		270, 72º	SAR
Almaz-1	31.3.91	13-20	2 x 172	3.125(S)	HH	25-60º	100	350, 72º	SAR
ERS-1	17.7.91	30	100	5.3 (C)	VV	23º	105	785, 98.5º	AMI, 3 modes
JERS-1	11.2.92	18	75	1.27 (L)	HH	35.21º	60	568, 97º	SAR
ERS-2	21.4.94	30	100	5.3 (C)	VV	23º	105	785, 98.5º	AMI, 3 modes
SIR-C/X-SAR	9.4.94 30.9.94	15-25	30-100	1.28, (L) 5.3, (C) 9.6 (X)	VV,HH VV,HH HH	15-55º	46/Ch.	225, 57º	L-SAR, Scan X-SAR, Scan X-SAR, Scan
RADAR-SAT-1	4.11.95	25-100 (12x9)	100-170	5.3 (C)	HH	20-60º	110	800, 98.6º	SAR, ScanSAR
Priroda	23.4.96	50	50	1.28, (L) 3.28 (S)	HH, VV	35º	16	400, 52º	SAR Travers
SRTM	11. - 22. 2. 2000	30 30	225 50	5.3 (C) 9.6 (X)	VV,HH HH	45º 52º	180 90	225, 57º	C-RADAR X-RADAR Interferometry
Envisat	2001	<30	100	5.33 (C)	Polarimetric	15-45º	100	800, 98.5º	ASAR, ScanSAR
RADAR-SAT-2	2003	3-100	10 - 500	5.3 (C)	Polarimetric	10-50º	105	800, 98.6º	SAR, ScanSAR
ALOS	2003	10	70	1.27 (L)	HH or VV	9-60º	240	700, 98º	PALSAR, ScanSAR
Condor-E	2004	1-5 5-22	20-150 10-20	3.13 (S)	HH HH,VV	20-55º	61	600,	SAR-10
COSMO-SkyMed	2004	1, 3-15, 30, 100	10, 40 100-200	9.6 (X)	HH, VV,	20-60º	2 x 150	620	SAR-2000
TerraSAR -X1 & -L1	2006 2006	1 - 15	55--100	9.6 (X) 1.28 (L)	Polarimetric	20-55º	300	660, 98º	XSAR, LSAR

Table 7: Chronological survey of spaceborne imaging radars (SAR instruments)

The first high-quality images (from a ground-based instrument) of near-Earth space objects were obtained in the early 1970s with ALCOR (ARPA, Lincoln Laboratory, C-band, Observables Radar). The data of ALCOR were processed by Lincoln Laboratory of MIT (MIT/LL/) and Syracuse Research Corporation. Successful results were made possible by the range resolution of 50 cm, by the coherent recording, and by sufficient sensitivity to image low-altitude satellites. In the late 1970s, LRIR (Long-Range Imaging Radar), an X-band instrument, was an ALCOR successor system, built at MIT/LL. The objective: imaging of satellites at geosynchronous range. In particular, significant image processing developments were achieved with LRIR such as ECP (Extended Coherent Processing).

NASA/JPL[52)] started its coherent radar test program in 1968, resulting in the development of an airborne L-band SAR instrument, flown on the CV-990 aircraft of NASA/ARC since 1971. The objective was to observe ocean surface patterns and to study the relationships between the image and the surface effects by using synthetic aperture techniques. The L-band SAR (1.215 GHz, bandwidth of 10 MHz) featured a phased array antenna with HH and HV polarization. This airborne instrument was a predecessor to the L-band SAR on Seasat. -

52) W. E. Brown, C. Elachi, T. W. Thomson, "Radar Imaging of Ocean Surface Patterns," Journal of Geophysical Research, Vol. 81, No 15, May 20, 1976, pp. 2657-2667

Star-1 (of Intera, Canada, developed by ERIM) was another early civilian airborne SAR instrument in 1983. - See also Table 8 for airborne SAR instruments.

Instrument - Agency	Frequencies/Band (GHz)	Polarizations	Comment
AIRSAR - NASA/JPL	0.45/(P), 1.26/(L), 5.31/(C)	H, V, full polarimetric	since 1987
TOPSAR - NASA/JPL	5.288/(C)	VV	AIRSAR configuration
CARABAS - FOA Sweden	20-90 MHz/(VHF)	H	foliage/ground penetration
CASSAR - CAS/ISRA China	9.375/(X)	HH, VV, HV, VH,	
C/X-SAR - CCRS Canada	5.3/(C) 9.25/(X)	C-band polarimetric X-band V or H	C-band interferometry
DCS, ERIM	9.66/(X) 14.96/(Ku)	full polarimetric VV only	SAR single-pass interferometer
DO-SAR (Dornier, Germany)	3.2/(S) 5.3/(C) 9.6/(X) 35/(Ka)	VV, VH, HH, HV VV, VH, HH, HV VV, VH, HH, HV VV	polarimetric
DRA-SAR - DRA/UK	5.7/(C) 9.65/(X)	full HH	
EMISAR - EMI/ TUD Denmark	5.3/(C) 1.4/(L)	full, polarimetric full, polarimetric	since 1993 since 1994
E-SAR - DLR, Germany	1.3/(L) 5.3/(C) 9.6/(X) 0.450/(P)	full polarimetric VV or HH V or H full polarimetric	E-SAR since 1989 added in 1994
FOLPEN - SRI, CA	100-500 MHz/(VHF/UHF)	HH	foliage/ground penetration
IFSARE, ERIM	9.5 (X)	HH	interferometric mapping SAR
INGARA, Australia	9.375 (X)	HH	
MMW-SAR - MIT	33.56/(Ka)	V, H, full polarimetric	
NASAR (NASDA/ EORC)	1.291 (L-band)	full polarimetric	operation starting in fall 1996
NEC-SAR - NEC Jap.	9.53/(X)	HH	Interferometric mode
P-3/SAR - ERIM,Navy	0.350/(UWB) 1.25/(L) 5.30/(C) 9.35/(X)	VV, VH, HH, HV VV, VH, HH, HV VV, VH, HH, HV VV, VH, HH, HV	UBW added in 1994 existing, polarimetric
PHARUS - TNO-FEL Delft, The Netherlands	5.25/(C)	full, polarimetric	
PI-SAR, CRL, NASDA	9.55 (X), 1.27 (L)	full polarimetric in X,L Interferometric in X	since 1996, initially also called CLR-SAR
SAR Travers - Vega, Moscow	3.0/(S)	VV or HH	
RAMSES - ONERA France	1.6/(L), 3.0/(S), 6.2/(C) 9.5/(X), 14.5/(Ku), 35/(Ka) 95/(W)	VV, HH, VH, HV full polarimetric	
SASAR U. of Cape Town, South Africa	5.3/(C), 0.141 (VHF)	VV or HH	under construction, July 1996 first flight
STAR-1,2 - INTERA	9.375/(X)	HH	since 1983
IMARC - Vega Moscow	118 MHz/(VHF) 0.44/(L), 1.28/(L), 7.7/(X)	HH, VV, HV, VH full polarimetric	
Sandia SAR (SNL)	14.850 (Ku)	VV	
IFSAR (SNL)	15 (Ku)		interferometric

Table 8: Overview of some early airborne SAR systems

- Tilting capability of a SAR antenna. The SIR-B Shuttle mission (launch: Oct. 5, 1984) provided for the first time the ability to mechanically tilt the antenna over a range (15° - 55°) so that radar imagery from multiple angles of incidence could be obtained.

• First on-board recording of high-rate SAR data. The USSR missions Kosmos-1870 (launch July 25, 1987) and Almaz-1 (Launch March 31, 1991) were the first to record and dump SAR data during station passes, followed by JERS-1 (launch Feb. 11, 1992). JERS-1 is regarded the first long-term mission which provided an on-board SAR-recording capability for its user community on a regular basis. This enabled extended observations beyond the limits of available ground stations. JERS-1 is also the first S/C carrying two types of imaging instruments, namely a microwave imager (SAR) and an optical imager (OPS). [53)]

• Until 1978, SAR images were formed using analog techniques, incorporating optical lenses and photographic film. In 1978, the first reconstruction of a SAR image was formed on a digital computer. This SAR processor was developed by MDA (MacDonald Dettwiler) of Richmond, BC, Canada, for the purpose to manage SEASAT SAR data. Early digital SAR processors required all the processing power available of a system; they were installed on mainframe computers or on large dedicated hardware. - On the other hand, todays SAR images (since the late 1990s) can be formed on relatively inexpensive equipment like a workstation or a PC.

• ScanSAR. Refers to a radar imaging technique permitting extended observation coverage (e.g. wider swath). In ScanSAR mode, the scene size in range is extended by scanning the antenna in elevation and alternating illumination of several subswaths. Like in stripmap SAR, the scene extension in azimuth is principally only determined by the length of the observation period. The ScanSAR mode requires rapid electronic steering of the elevation beam pattern of the antenna (phased array antenna). The principle of ScanSAR is to share radar operational time between two or more separate subswaths in such a way, as to obtain full image coverage of each. However, scanning in elevation implies that the synthetic aperture length is reduced to the burst length (the burst length or burst duration is defined as the time interval for each target illumination in a ScanSAR cross-track sequence). A consequence of the shorter burst length (as compared to stripmap SAR) is that each target has a different Doppler history (depending on its azimuth position) resulting in a reduced ScanSAR azimuth resolution. The basic differences between ScanSAR and stripmap SAR processing are: [54)] [55)]

- The azimuth consists of several bursts, each of them corresponding to a fraction of the full synthetic aperture
- The azimuth resolution is limited by the burst duration and is no longer determined by the synthetic aperture length
- The Doppler history of each target within a burst has a different frequency offset depending on its azimuth location
- The azimuth multilook technique is constrained more by the scanning strategy than by the processing strategy
- The radiometric correction due to the azimuth antenna pattern is much more critical than in stripmap SAR since each target is illuminated by different parts of the azimuth antenna diagram
- For SAR data processing, an additional mosaic operation is needed in azimuth and range directions to join the azimuth bursts and the range subswaths, respectively.

The ScanSAR technique was initially developed by JPL and first flown on two Shuttle missions with the SIR-C/X-SAR payload in 1994, SRL-1 (STS-59, April 9 - 20, 1994) and SRL-2 (STS-68, Sept. 30 - Oct. 11, 1994). Further implementations of ScanSAR are on RADARSAT-1 (Canada, launch Nov. 4, 1995), SRTM/C-RADAR (NASA/JPL, launch Feb. 11-22,

53) Note: The two Shuttle imaging radars, SIR-A and SIR-B, had recorders, but data was collected only for select sites due to the short mission duration of one week each.

54) J. Mittermayer, A. Moreira, "A Generic Formulation of the Extended Chirp Scaling Algorithm (ECS) for Phase Preserving ScanSAR and Spot SAR Processing," Proceedings of the IEEE IGARSS 2000 Conference, Honolulu, HI, July 24-28, 2000

55) A. Moreira, J. Mittermayer, R. Scheiber, "Extended Chirp Scaling Algorithm for Air- and Spaceborne SAR Data Processing in Stripmap and ScanSAR Imaging Modes," IEEE Transactions on Geoscience and Remote Sensing, Vol. 34, No. 5, Sept. 1996, pp. 1123-1136

2000), Envisat/ASAR (ESA, launch 2001), and ALOS/PALSAR (Japan, launch in 2003). - Until RADARSAT-1, the maximum swath width of any previously orbiting SAR instrument had been 100 km, a parameter tightly constrained by the need to: a) "fill the aperture" and b) receive all the returned energy between transmitter pulses. RADARSAT-1 achieved with ScanSAR swaths up to 510 km with 100 m resolution, from an 800 km orbit, through the use of multiple electronically steered antenna beams and carefully synchronized transmitter pulse timing. [56) 57) 58) 59)]

The advantage of wider ScanSAR observation coverage is crucial in obtaining global repeat SAR imagery in several days, eventually on a daily basis (enhanced sampling capability) irrespective of cloudcover or sunlight conditions. [The narrow-swath high-resolution (<10 m) SAR imagery yields exciting periodic snapshot views of the 2-D ocean surface to study a number of applications (interactions with the atmosphere, long waves, and current)]. The wide-swath ScanSAR capability of RADARSAT (up to 510 km), Envisat (up to 400 km) and ALOS (up to 350 km) permit for the first time the study of larger-scale phenomena (mesoscale wind fields and circulation features, atmospheric boundary-layer processes, study of storm dynamics, etc.), thus providing a better monitoring capability that may eventually lead to operational SAR missions. The potential of SAR imagery wind-field monitoring alone is much higher than is possible with spaceborne scatterometers. A future operational SAR mission service can of course also provide several other services for land surface monitoring besides ocean surface monitoring, such as monitoring of natural hazards in a wide variety of applications: flooding, earthquakes, wildfires, severe weather events, etc. [60) 61)]

- In SpotSAR mode (see O.8.3 for definition of SAR imaging modes), the antenna is steered in the azimuth direction. During the formation of the synthetic aperture, the antenna is constantly looking into the scene center direction. The available synthetic aperture is in this case much longer than in stripmap mode and is denoted by spotlight aperture. The long spotlight aperture provides azimuth signals with a large Doppler bandwidth and therefore provides a high geometric resolution in azimuth.

- Phased array microwave technology (SAR antennas). [62)] The observation requirements of wide swaths for global coverage led eventually to the introduction of a new scanning concept referred to as 'phased array technology,' making use of electronic beam steering in the elevation domain (the use of electronic scanning allows flexibility in beam pointing and beam shape control for swath coverage). Naturally, the technology requires a lot of computer power to handle such complex radar functions as: beam switching, signal detection, frequency management, ranging, etc. The technique was first tested on active sensors (SARs) in the microwave region of the spectrum and later on laser radars. The SAR sensor of Seasat (launch June 27, 1978) pioneered this rapid-pointing method for spaceborne instruments. Other instruments with phased array technology followed or are being planned, such as: L-/C-band antenna of the SIR-C/X-SAR missions (Shuttle, SRL-1 April 1994, SRL-2 Sept. 1994), SAR on RADARSAT-1 (launch Nov. 4, 1995), PR on TRMM (launch Nov. 27, 1997), C-RADAR on SRTM (launch Feb. 11-22, 2000), ASAR on Envisat (launch 2001), and PALSAR on ALOS (launch 2003). The following instruments are operational on

56) R. C. Beal, "Toward an International Storm Watch Using Wide Swath SAR," JHU/APL Technical Digest, Vol. 21, No. 1, 2000, pp. 12-20

57) B. Holt, J. Hilland, "Rapid-Repeat SAR Imaging of the Ocean Surface: Are Daily Observations Possible?" JHU/APL Technical Digest, Vol. 21, No. 1, 2000, pp. 162-169

58) K. B. Katsaros, et al., "Wind Fields from SAR: Could They Improve Our Understanding of Storm Dynamics?" JHU/APL Technical Digest, Vol. 21, No. 1, 2000, pp. 86-93

59) S. Wu, A. Liu, et al., "Ocean Feature Monitoring with Wide Swath Synthetic Aperture Radar," JHU/APL Technical Digest, Vol. 21, No. 1, 2000, pp. 122-129

60) G. S. Young, "SAR Signatures of the Marine Atmospheric Boundary Layer: Implications for Numerical Forecasting," JHU/APL Technical Digest, Vol. 21, No. 1, 2000, pp. 27-32

61) J. Horstmann, S. Lehner, W. Koch, R. Tonboe., "Computation of Wind Vectors over the Ocean Using Spaceborne Synthetic Aperture Radar," JHU/APL Technical Digest, Vol. 21, No. 1, 2000, pp. 100-107

62) Note: JHU/APL was instrumental in developing the phased array radar technology for the US Navy whose construction began in 1959. A first prototype was installed on a Navy ship in 1961. See: A. Kossiakoff, "APL - Expanding the Limits," Johns Hopkins APL Technical Digest, Vol. 13, No. 1, 1992, pp. 8-27

the airborne side: FOLPEN (SRI International), PHARUS (TNO), CRL-SAR (CRL), and Sandia SAR (SNL). In the 1990s the phased array has become a key component in the design of advanced antenna systems, taking advantage of the considerable advances in microelectronics, transmitter/receiver architectures, miniaturized device (module) integration methods, and signal processing techniques.

- Digital chirp generator. The radar chirp concept of pulse frequency modulation was developed at the Bell Telephone Laboratories (at Whippany and Murray Hill, NJ) by J. R. Klauder et al. in the 1950s and first reported in the literature in 1960 (declassification of work supported by the US military). 63) 64) Early experimental work on radar signal design used the chirp, a linearly frequency-modulated (FM) pulse with a frequency range set by the required angular resolution, and with a pulse duration set for the required range resolution performance. By transmitting a modulated pulse, it was possible to obtain range resolutions that were smaller than the transmitted pulse length by very large factors. The reported technique employed the so-called "pulse compression technique" (covering a frequency range many times the inherent bandwidth of the envelope), permitting radars to transmit a pulse with about 100 times the energy of a short pulse with equivalent resolution and peak power.

Modern spaceborne SAR instruments are equipped with a digital chirp generator that allows to generate pulses with a programmable slope and bandwidth. 65) The resulting flexibility in terms of pulse duration and frequency-tuning capability allows the operation of the SAR system in different modes like a high-resolution mode and a wide swath, low resolution mode. Also, within one mode of operation it is useful to adjust the chirp bandwidth according to the incidence angle in order to maintain a more constant ground-range resolution and a more homogeneous signal-to-noise ratio over the access region. This method is used for example in the ASAR instrument of Envisat for the global monitoring and the wide swath mode with its 5 different swaths. The chirp waveform (linear FM) is also used for better on-board processing of the radar signals including compression schemes.

- SAR interferometry (O.10.1). Two interferometric observation concepts are in use: 'single-pass' and 'two-pass' (or multi-pass) interferometry. The interferometric phase, i.e. the phase difference between two images acquired with slightly different sensor positions, contains geometric information allowing the derivation of the 3-D position of the scattering element. SAR interferometry techniques promise to produce global topography maps in a similar fashion as stereo photogrammetry.

- Simultaneous observation of the same footprint by two antennas in parallel (single-pass interferometry) separated at a distance (baseline). Two separate images are received of the same ground track by two spatially-separated antenna systems (the interferometric baseline) to produce phase differences from slightly different viewing angles. DCS (Data Collection System of ERIM) is probably the first single-pass airborne INSAR instrument with first flights in 1987. TOPSAR (JPL) started its observations in 1991. On the spaceborne side, the SRTM (Shuttle Radar Topography Mission, J.25) of NASA represents the first fixed baseline single-pass spaceborne IFSAR (Interferometric SAR) system with simultaneous dual-polarization wide-swath scanning SAR and dual-frequency (C-band and X-band) coverage. The launch of the 11-day STS-99 Shuttle mission is planned for Jan. 2000. The SRTM digital topographic map production objective (referred to as C-RADAR) calls for a spatial pixel (30 m x 30 m) posting with a 16 m absolute vertical linear accuracy and 20 m absolute horizontal radial accuracy at 90%.

- Two-pass measurements of a single-antenna SAR platform. This refers to a superposition technique of imagery (in data processing) of fairly close repeat tracks. The ERS-1/-2

63) J. R. Klauder, A. C. Price, S. Darlington, W. J. Albersheim, "The Theory and Design of Chirp Radars," The Bell System Technical Journal, Vol. 39, No 4, July 1960, pp. 745-808

64) The term "chirp" was coined at the Bell Telephone Laboratories in the 1950s to designate a new and more effective radar signal generation method.

65) M. Suess, C. Schaefer, R. Zahn, "Discussion of the Introduction of On-board SAR Data Processing in Spaceborne Instruments," Proceedings of IEEE/IGARSS 2000, Honolulu, HI, July 24-28, 2000

tandem mission (start in Aug. 1995, end in May 1996) was flown with a one day delay of a coinciding ground track of the second S/C, providing good interferometric results. The SRL-1/2 missions in 1994 (April 9-20 for SRL-1, Sept. 30 - Oct. 11 for SRL-2) employed two-pass (and repeat pass) interferometry for topographic mapping (detection of topographic surface change in SRL-1 and SRL-2). The C-band SAR instruments of the ERS-1/2 and RADARSAT-1 (launch Nov 4, 1995) missions demonstrated the ability to detect cm-scale surface strain over large contiguous areas. The L-band SAR of JERS-1 (launch Feb. 11, 1992) demonstrated also two-pass SAR-interferometry for change detection.

- Scatterometers are normally active microwave instruments (radars that measure the target reflectivity or backscatter over a wide range of incidence angles; values are reported in terms of the normalized radar cross section) offering a capability of wind speed and direction determination. [66] Newer instruments under development, like Windsat on Coriolis and CMIS (Conical-scanning Microwave Imager/Sounder) on NPOESS, can also be passive microwave radiometers, capable and sensitive enough of measuring, among other parameters, also ocean surface wind speed and direction. The ocean surface wind vector is a key parameter in understanding the weather due to its dominant role in the energy exchange at the air-sea interface. Winds over the ocean modulate air-sea exchanges in heat, moisture, and gases regulating much of the global weather. - Scatterometry has its origin in early radar used in World War II. Early radar measurements over oceans were corrupted by sea clutter (noise). It was not known at the time that clutter was the radar response to the winds over the oceans. The radar response was first related to wind in the late 1960s - by realizing that the sea surface roughened by the wind modifies the surface backscatter (reflected signal or echo) properties.

Mission	Launch Date	Spatial Resolution (km)	Swath Width (km)	Frequency GHz (Band)	Polarization	Data Rate (kb/s)	Altitude, Inclination (km/°)	Sensor, Remarks
Skylab	14.5.73	16	180	13.9 (Ku)		5.33	435, 50°	S-193
Seasat	27.6.78	50	500	14.6 (Ku)	VV,HH		800/108°	SASS
Spacelab-1	28.11.83							MRSE, Shuttle
ERS-1	17.7.91	50	500	5.3 (C)	LV		785, 98.5°	AMI in Scat. mode
ERS-2	21.4.95	50	500	5.3 (C)	LV		785/98.5	AMI in Scat. mode
ADEOS	17.8.96	25,50	600/1200	13.99 (Ku)	VV,HH	2.9	797/98.6°	NSCAT
QuikSCAT	19.6.99	50	1800	13.4 (Ku)		40	803/98.6°	SeaWinds
ADEOS-II	2002	50	1800	13.4 (Ku)		20	803, 98.6°	SeaWinds
Coriolis	2003	25	1400	6.8, 10.7, 18.7, 23.8, 37.0 GHz	Circular	5000	830,98.7°	Windsat
MetOp-1	2005	50	2 x 550	5.255 (C)	VV	60	800, 98.5°	ASCAT
NPOESS	2008							CMIS

Table 9: Survey of spaceborne radar/microwave scatterometers (wind measurements)

Sensor	Sensor Provider	Frequency Band	Flown since
CRL Radar/Radiometer	CRL Japan NASA/GSFC/WFF	X-band, Ka-Band	1981 since 1985
Delta-K Spectrometer	IKI RAN, Moscow	X-band	1981
DUTSCAT	DUT/NLR, Holland	1.2-17.2 GHz (6 bands)	1983
ERASME	CNRS/CETP, France	C-band, X-band	1983
RACS	U. of Hamburg	C-band	1987
HELISCAT	U. of Hamburg	L, S, C, X, Ku	1989, 1990
HUTSCAT	HUT, Helsinki	C-band, X-band	1988
MINISCAT	HUT, Helsinki	C-band	1990

66) http://winds.jpl.nasa.gov/scatterometry/history.html

Sensor	Sensor Provider	Frequency Band	Flown since
C-SCAT	NASA/MIRSL	C-band	1990
RESSAC	CRPE, France	C-band	1990 (rebuilt ERASME)
NUSCAT	NASA/JPL	Ku-band	1991
RENE	CNRS/CETP, France	X-band	1992

Table 10: Overview of some airborne radar scatterometers

Wind vector retrievals by todays operational scatterometers (like AMI and SeaWinds) are based on indirect measurements, where the wind vector is inferred through a relationship between the backscattered power, the small-scale ocean surface roughness, and the local wind vector at the ocean surface. This relationship is empirical, based primarily on matches between sensor measurements, buoy wind measurements and analysis from numerical weather prediction models. [67)]

The first spaceborne scatterometers were S-193 on Skylab (launch 1973) and SASS on Seasat (launch June 28, 1978), followed by MRSE on Spacelab-1 and the AMI-SCAT (wind scatterometer mode) instrument on ERS-1 (launch July 17, 1991). See Tables 9 and 10. [68)] The newer (since 1999) spaceborne wind vector measurement instruments can be put into three classes: [69)]

- Broad-swath, dual-pencil beam active Ku-band scatterometers (e.g. SeaWinds)
- Dual swath, 3-look active C-band scatterometers (e.g., ASCAT)
- Dual- and single-look passive polarimetric radiometers (e.g., Windsat, CMIS)

Sensor Name	Sensor Provider	Introduction	Microwave Band
E-SLAR	DFVLR (DLR), Germany	1976	X-band
NIT	IKI RAN, Soviet Union	early 1980s	Ku-band
SLAR (part of MSS)	SSC, Sweden	1985	X-band
R-SLAR	RRL (CRL), Japan	1986	X-band
HUTSLAR	HUT, Finland	1988	X-band
SLAR	TNO/NLR, The Netherlands		X-band

Table 11: Some SLAR instruments

With the launch of ADEOS-II, there is the first opportunity since Seasat to obtain collocated active microwave radar with SeaWinds-II and passive microwave radiometer data of AMSR (Advanced Microwave Scanning Radiometer) , from the same platform.

Traditional **fan-beam** scatterometer designs such as SASS, AMI-SCAT, and NSCAT require significant power and mass and have antenna systems which are difficult to accommodate aboard any spacecraft. [70)] [71)] The fan-beam systems employ multiple fixed antennas to cast broad fan-shaped beams on the Earth's surface at multiple azimuth angles needed to measure wind. These designs have degraded wind performance limitations in particular in the near-nadir region of the measurement swath. The reason: the backscatter measurements in the region ±200 km to either side of the nadir track are at small incidence angles and are insensitive to wind direction, thus, creating a large "nadir gap" in their coverage. - Newer scatterometer designs employ the **conical-beam scanning technique** (with dual-

67) P. S. Chang, L. N. Connor, J. R. Carswell, R. S. Dunbar, "Operational Scatterometry: High Wind Speed Retrievals," Proceedings of IEEE/IGARSS Conference, Honolulu, HI, July 24-28, 2000

68) Note: Wind retrievals are also available from SSM/I, flown since 1987 on the DMSP series. However, the design of SSM/I, a microwave radiometer, only permits the retrieval of the speed component of the wind vector.

69) R. F. Millief, M. H. Freilich, et al., "Global Ocean Surface Vector Wind Observations from Space," Proceedings of OCEANOBS 99, Oct. 18-22, 1999, Saint-Raphael, France

70) M. R. Drinkwater, C. C. Lin, "Introduction to the Special Section on Emerging Scatterometer Applications," IEEE Transactions on Geoscience and Remote Sensing, Vol. 38, No 4, July 2000, pp. 1763-1764

71) W-Y. Tsai. S. V. Nghiem, J. N. Huddelston, et al., "Polarimetric Scatterometry: A Promising Technique for Improving Ocean Surface Wind Measurements," IEEE Transactions on Geoscience and Remote Sensing, Vol. 38, No 4, July 2000, pp. 1903-1921

beams). SeaWinds (of NASA/JPL on QuikSCAT, launch June, 19, 1999) introduced a scanning pencil-beam scatterometer with a dual-beam conical scan, provided by a 1 m diameter reflector (rotating dish) antenna. [72] Although the pencil-beam approach adopted by SeaWinds significantly lessens the nadir gap problem by making measurements at suitably high incidence angles, the problem is not completely eliminated. Measurements in the extreme inner and outer swath still suffer some degradation because the relative azimuth angles of backscatter measurements are too close together (approaching 0° for the outer swath), or are too far apart (approaching 180° for the inner swath) to determine the wind speed direction accurately. - A performance improvement is also introduced with the ASCAT instrument of ESA on MetOp-1, it employs a solid-state design and the use of a double-beam scanning technique. Scatterometer antenna and radar electronic concepts remain key technology issues in the design of efficient and low-cost instruments.

Beyond the original mission of spaceborne scatterometers to provide measurements of wind velocity over the ocean, a large number of applications of the radar data have emerged in such areas as: a) land applications, b) polar snow and ice, c) oceans and NWP (Numerical Weather Prediction), and d) polarimetric scatterometry.

The capabilities of future polarimetric scatterometers are of particular interest. They will be able to simultaneously measure the conventional co-polarized backscatter as well as the polarimetric correlation of the co- and cross-polarized radar returns from the ocean surface. The technique promises potential performance gains by improving in particular the ambiguity resolution.

- Microwave rain measurement instruments. First radiometer data evaluations with regard to rainfall started around 1975 using ESMR data of Nimbus-5 (launch Dec. 11, 1972) and later SMMR data of Seasat (launch June 27, 1978). The SSM/I radiometer on the DMSP series is flown since 1987, rainfall is determined for sea surface regions. The airborne radiometer AMR (NASDA, since 1995) is to confirm AMSR on ADEOS-II (planned launch in 2002). Active microwave instruments: CRL Radar/Radiometer (CRL Tokyo, since 1981), ARMAR (NASA/JPL, since 1992) is an airborne rain radar in support of the spaceborne PR (Precipitation Radar) instrument on TRMM (launch Nov. 27, 1997). The TRMM mission represents the first concerted effort (NASA/NASDA) to tackle the complex nature of rainfall with a complement of passive and active sensors. [73]

- The GPM (Global Precipitation Mission), a follow-up mission to TRMM with a reference constellation of eight microsatellites and advanced rain-measuring instruments, is in the proposal state as of 2000/1 (NASA/NASDA, ESA also considers to join the project) with a planned launch date in 2007. The GPM close-neighborhood constellation employs a completely new observation technology, namely the so-called interferometric cartwheel concept (first proposed by CNES), where a single spacecraft is equipped with an active radar and a passive radiometer, while the rest of the constellation is outfitted with just a radiometer (receiver). All observation instruments of the constellation look into the same footprint. Thus, all passive instruments in the constellation are also able to collect the echo information from the single active radar. The multi-location (or nodal) spacecraft data collected in this fashion provides automatically interferometric imagery after ground processing. Parasitic observation systems, like Cartwheel (Europe) and GPM, are under intensive study.

- Microwave medium-penetrating instruments, in particular with regard to soil and foliage penetration (see also: 'microwave signal penetration' in Glossary), were first introduced with airborne instruments (Table 12). Microwave observations of active and passive sensors in L-band, P-band, UHF, VHF and UWB are of interest. The microwave

72) V. Wismann, "Monitoring the Earth with Spaceborne Scatterometers," Proceedings of 32nd ESLAB Symposium on 'Remote Sensing Methodology for Earth Observation and Planetary Exploration,' ESA/ESTEC, Sept. 15-18, 1998 (SP-423 Dec. 1998), pp.189-199

73) L. J. Allison, et. al., "Tropical cyclone rainfall as measured by the Number 5 EMSR," BAMS, Vol. 55, pp. 1074 -1089, 1975

permittivity of water is an order of magnitude higher than that of any natural dry material. Radar penetration and resolution tend to be reduced by the EM attenuation, with a range that goes from a few meters in conductive media, to 50 m at most for low-conductivity (below 1 ms/m) materials such as sand, gravel, rock and fresh water. High-power pulses are essential in order to increase the ratio between signal and clutter (or noise). For this reason, instead of using a pulse radar, one may use swept-FM or step-frequency transmitters. The frequency synthesizer technique seems to overcome some of the limitations of pulse radars which makes them interesting for future developments. [74)]

System	Description
GPR (Ground Penetrating Radar) SRI, P.94.1	First airborne tests/studies (ice penetration) were conducted in 1974, GPR first flown in 1993
FOLPEN, SRI, since 1990, P.94	FOLPEN-I and-II are VHF impulse SAR systems with foliage penetration capability
CARABAS, FOA, Sweden, P.49	Penetration of vegetation/foliage and to some extent of the ground surface, since 1992
TOPSAR, JPL (upgrade 1994), P.13.1	Calculation of the differential penetration characteristics of the dual-frequency radar waves for different Earth terrain types
P3/SAR, ERIM/Navy (upgrade 1994), P.82.3	Support of foliage and ground penetration experiments/applications
LARSEN, CCRS, since 1985, P.111	A lidar for the measurement of shallow water depths
LFS, University of Oldenburg, P.116	Lidar for the analysis of the upper sea surface layers, since '93
SHOALS, USACE, 1994, P.184	A lidar for the measurement of shallow water depths
SASAR, U. of Cape Town, since 1996, P.181	The VHF-band of SASAR offers a surface/foliage penetrating capability
PBMR, GSFC/WFF, since 1983, P.160	L-band radiometer for the measurement of soil moisture, etc.
RADIUS, NPO Vega, since 1986, P.171	Microwave radiometer for the measurement of soil moisture, etc.
ESTAR, MIRSL/GSFC, 1988, P.87	L-band radiometer for the measurement of soil moisture and ocean salinity
AIMR, AES, Canada, since 1989, P.8	Microwave radiometer for the measurement of soil moisture and snow depth
PORTOS, CNES, since 1992, P.167	Microwave radiometer for the measurement of soil moisture
MIRAS, ESA, since 1996, P.132	L-band radiometer for the measurement of soil moisture and ocean salinity

Table 12: Overview of medium-penetrating airborne microwave instruments

Spaceborne L-band SAR instruments were flown on SIR-A (NASA/JPL, Shuttle, Nov. 12, 1981), SIR-B (Shuttle, Oct. 5, 1984), and SIR-C/X-SAR (Shuttle, SRL-1 April 1994, SRL-2 Sept. 1994), and on JERS (NASDA/MITI, launch Feb. 11, 1992).

In the late 1990s, the applications of medium-penetrating radars (a non-destructive monitoring technique), are becoming rather divers. The technique is employed in civil engineering for void detection, prediction of concrete deterioration from variations of its permittivity by the presence of moisture and chloride, pavement profiling for programmed road and bridge deck maintenance, reinforcing bar location, subgrade deterioration in railroad and airport runways. Object detection and classification by extent and permittivity contrast with the overburden, it is crucial in environmental engineering. Furthermore, the technique is useful for mapping hazardous wastes and buried contaminant containers, imaging and monitoring subsurface contaminants (e.g., gasoline and other hydrocarbon fuels). Stratigraphic and bedrock mapping are essential in geotectonic, archaeologic, and hydrogeologic applications, in site characterization, in mining planning (e.g., borehole profiling), in tunnel excavations, ice thickness profiling, permafrost mapping, etc. [75)]

74) L. Peters, J. J. Daniels, J. D. Young, "Ground Penetrating Radar as a Subsurface Environmental Sensing Tool," Proceedings of the IEEE, Vol. 82, No. 12, Dec. 1994, pp. 1802-1821

75) U. Spagnolini, "Ground Penetrating Radar," http://daacdev1.stx.com/ELF/docs/PER/spagnol6.html

1.1.3 Microwave Region, Passive Observations

• Radiometric imaging. The NOAA-2 S/C (launch Oct. 15, 1972) was the first operational weather satellite to rely solely upon radiometric imaging to obtain cloudcover data.

• Radiometers. The spectral range of remotely-sensed data was considerably enlarged by the use of radiometers in the microwave region [the overall microwave region is considered for wavelengths from about 1 mm (300 GHz) to 1 m (300 MHz); remote sensing is being conducted in the MMW (millimeter-wave) region (i.e., 1-10 mm or 300 GHz to 30 GHz); the frequency range from 1-20 GHz is mostly used for surface radiometry due to atmospheric attenuation]. Passive microwave (MW and MMW) Earth observation radiometry introduced such applications as temperature sounding for meteorological purposes, sea-ice mapping for navigation in polar regions, measurements of ocean surface temperature and ocean surface wind speeds. Eventually the applications of radiometers included also observations of land surface features and the measurement of atmospheric constituents (trace gases).- Two generic types of radiometers have evolved, profiling instruments (or sounders, also referred to as broadband radiometers), and surface imaging instruments (multichannel radiometers). In general, atmospheric window frequencies are used to study surface (ocean, ice, land) phenomena.

- Spaceborne microwave observations started September 23, 1968 with the launch of Cosmos 243 satellite (in sun-synchronous LEO) which carried a non-scanning, nadir-viewing 4-channel radiometer [wavelengths: 8.5 cm (3.5 GHz), 3.4 cm (8.8 GHz), 1.35 cm (22.2 GHz), and 0.8 cm (37.5 GHz)] with the objective to estimate atmospheric water vapor, liquid water, ice cover, and sea temperature. The same radiometer was also flown on Cosmos 384 with a launch in 1970. Nimbus-5 (launch Dec. 11, 1972) carried a radiometer with the name of NEMS (Nimbus-E Microwave Spectrometer). This represented the first step in in the application of the microwave spectrum to global sensing of the atmospheric temperature structure. Another radiometer on Nimbus-5 was ESMR (Electrically Scanning Microwave Radiometer). The Skylab (launch of Skylab-1 May 14, 1973) series flew two radiometers, S-193 and S-194. The NEMS instrument on Nimbus-5 was followed by SCAMS (Scanning Microwave Spectrometer), flown on Nimbus-6 (launch June 12, 1975), which led to the first MSU (Microwave Sounding Unit), flown on TIROS-N (launch Oct. 13, 1978). The SMMR sensor was flown on SEASAT (launch June 27, 1978) and on Nimbus-7 (launch Oct. 24, 1978). See also chapter O.7 and Table 546. [76] [77]

- Passive spaceborne radiometers have been primarily frequency-tuned to vertically sound and map specific constituents (gases and dust in atmospheres, and bulk properties on planetary surfaces). The first systems implemented were sounders (profiling systems). These were followed by the new type of surface imaging radiometers with the availability of focal plane antennas and better processing capabilities. The 1980s and 1990s mark a shift away from Dicke-switched superheterodyne radiometers to total-power systems, facilitating radiometer configuration simplification and sensitivity improvements. In the 1990s, superheterodyne receivers with mixers directly coupled to feedhorns and feeding low-noise IF amplifiers continue to dominate MMW radiometry. The mixers in superheterodyne radiometers remain the critical element for sensitivity. Mixers with whiskerless diodes yield significant improvements in reliability relative to mixers with whisker-contacted diodes.

- Future radiometer instruments are expected to include more receiver channels in a wider spectral range (MW, MMW, and sub-MMW). Improvements are likely to come from solid-state technologies based on MMIC (Microwave Monolithic Integrated Circuit) receiver chip development, providing direct amplification and detection in combination with

76) E. G. Njoku, "Passive Microwave Remote Sensing of the Earth from Space-A Review," Proceedings of the IEEE, Vol. 70, No. 7, July 1982, pp. 728-750

77) A. E. Basharinow, et al., "Some Results of Microwave Sounding of the Atmosphere and Ocean from the Satellite Cosmos 243," in COSPAR Space Research XI, Proceedings of Open Meetings of Working Groups of the Thirteenth Plenary Meeting of COSPAR, Leningrad May 20-29, 1970, Akademie-Verlag, Berlin, 1971, pp. 593-600

FPA (Focal Plane Array) imaging. In addition, the aperture synthesis method seems to be a promising development, providing a capability of 2-D radiometer imagery generation. Note: the MMIC technology is replacing the older MIC (Microwave Integrated Circuit) and TWTA (Traveling Wave Tube Amplifier) technology in RF electronics.

- Submillimeter-wave radiometers [300 GHz (1 mm)< frequency range >3000 GHz (0.1 mm)]. [78] At the end of the 1990s, new generations of microwave limb sounders (profilers) and imaging radiometers (or spectroradiometers) are emerging with advances in microwave detector technology, extending their measuring capability into the sub-mm wavelength region. The technology for these high-frequency ranges requires radiometers with superconducting heterodyne receivers or with FTS-based systems (the FTS technique provides a means to measure atmospheric radiance). [79] The cooling to superconducting temperatures of the detection system represents always a major investment in instrument design. The problem of past LTS (Low-Temperature Superconductivity) cooling seems to be gradually replaced by HTS (High-Temperature Superconductivity) cooling systems. - The interest in the short wavelengths is due to the fact that many molecules of importance in photochemistry exhibit emission lines in these wavelengths (e.g. HCl, OH). Submillimeter remote sensing has in addition advantages that complement visible and infrared techniques. Since the wavelength of submillimeter radiation is comparable to the size of ice particles in cirrus clouds, observed brightness temperature changes from cirrus are correlated to ice mass. [80] Some early airborne instruments in this class are: MIR (Millimeter-wave Imaging Radiometer) of NASA/GSFC flown on ER-2 which was upgraded in 1994 with three 325 GHz channels. THOMAS (THz OH Measurement Airborne Sounder, since 1994) of DLR, and ASUR (Airborne Submillimeter SIS Radiometer, since 1994) a cooperative sensor of SRON (Groningen, Netherlands) and the Institute of Environmental Physics at the University of Bremen. The NASA-sponsored FIRSC (Far Infrared Sensor for Cirrus) FTS instrument is flown on a T-39 (Sabreliner) aircraft since April 1998. [81] [82]

The first spaceborne sub-mmw instruments for the measurement of trace gases are: SMR (Submillimeterwave Radiometer) flown on the Swedish S/C ODIN (launch Feb. 20, 2001). There are sub-mm channels and one lower frequency channel on SMR. SMILES (Superconducting Submillimeter-wave Limb Emission Sounder) of NASDA/CRL is planned to fly on ISS/JEM in 2005. ESA/ESTEC has the following limb-sounding sensors (in the sub-mm region) under development: MASTER (Millimeter-wave Acquisition for Stratosphere Troposphere Exchange Research), SOPRANO (Sub-millimeter Observation of Processes in the Absorption Noteworthy for Ozone) and PIRAMHYD (Passive InfraRed Atmospheric Measurements of HYDroxyl). NASA is developing MLS (Microwave Limb Sounder), with three channels in the mm-range and two double sideband sub-mm channels, to fly on Aura (EOS-CHEM) in 2003. The new MLS instrument (frequencies of>600GHz) is a further development of MLS flown on UARS [Note: MLS on UARS has observed atmospheric thermal emissions from chlorine monoxide (ClO), ozone (O_3), water vapor (H_2O), sulfur dioxide (SO_2), and molecular oxygen (O_2), at frequencies of 63, 183 and 205 GHz].

- **Synthetic aperture radiometry.** Aperture synthesis is a relatively new technique for microwave remote sensing of the environment. The technique (pioneered in radio astronomy)

78) G. W. Schwaab, "Heterodyne spectrometers," Infrared Physics & Technology, Vol. 40, 1999, pp. 207-218

79) Note: The heterodyning technology is particularly suitable for the detection of very weak infrared and microwave radiation sources; heterodyning instruments (microwave limb sounders) play a crucial role for the monitoring performance of the Earth's atmosphere (chemical composition of the atmosphere, monitoring/deduction of the fine structure of the spectra) as well as in the study of astronomical objects. The heterodyning technique is applicable in the spectral range from about 10 µm (TIR) to the microwave region of about 10 mm wavelength.

80) K. F. Evans, S. J. Walter, et al., "Modeling of Submillimeter Passive Remote Sensing of Cirrus Clouds," Journal of Applied Meteorology, Vol. 37, 1998, pp. 184-205

81) K. F. Künzi, "Microwave Limb Sounders," Proceedings of 32nd ESLAB Symposium on 'Remote Sensing Methodology for Earth Observation and Planetary Exploration,' ESA/ESTEC, Sept. 15-18, 1998 (SP-423 Dec. 1998), pp.17-22

82) J. Mees, S. Crewell, et al., "ASUR - An Airborne SIS Receiver for Atmospheric Measurements of Trace Gases at 625 to 760 GHz," IEEE Transactions on Microwave Theory and Techniques, Vol. 43, No 11, Nov. 1995, pp. 2543-2548

generates high spatial resolution images by dividing the collection area of a telescope (or antenna) into smaller apertures spread out in a pattern covering several baselines. In microwave radiometry the concept employs an interferometric technique, in which the product from antenna pairs is sampled as a function of pair spacing. Substantial reductions in the antenna aperture needed for a given spatial resolution can be achieved with this technique. However, the advantages gained from aperture synthesis comes at the expense of reduced sensitivity resulting from the corresponding reduction in physical aperture. Detection sensitivity is an especially critical issue for measurements in LEO orbits (with orbital speeds of about 7 km/s) because of very limited instrument integration time availability per scene. - ESTAR [of NASA/GSFC and MIRSL (Univ. of MA, Amherst), P.87, first flown in 1988], a hybrid real-and-synthetic aperture combination, was the first radiometer built to test the concept of aperture synthesis. The experiences with the airborne ESTAR and MIRAS (ESA, P.132) instruments have demonstrated in particular the potential of aperture synthesis for remote sensing of **soil moisture** (Table 12). [83] While ESTAR is based on a 1-D (across-track) interferometer approach, MIRAS employed the 2-D synthetic aperture alternative. The new technology may eventually lead to a new generation of spaceborne passive microwave sensors by helping to overcome limitations set by antenna aperture size. The advantage of aperture synthesis is that it can achieve spatial resolutions equivalent to a total power radiometer with a large effective collecting area using relatively small antennas. The reduction in sensitivity that this entails can be restored because the synthetic aperture system does not need to scan and collect energy from many independent antenna pairs simultaneously.

Soil moisture. In this context, soil moisture is a fundamental parameter of the terrestrial environment. Its spatial distribution and temporal variation are crucial ingredients of hydrologic, ecologic, and climatic models, on regional and global scales. Spaceborne soil-moisture sensing may already get a start with two active instruments, namely ASAR (Advanced SAR) on Envisat (launch Oct. 2001), and with PALSAR (Phased Array L-band Synthetic Aperture Radar) on ALOS (launch in 2003). Each of these SAR instruments is dual-polarized, permitting the retrieval of soil moisture for bare soil surfaces and for surfaces with short vegetation cover.

A first spaceborne demonstration project, employing synthetic aperture radiometry (i.e. passive microwave sensing) with a sparsely filled 2-D antenna design, is **SMOS** (Soil Moisture and Ocean Salinity) of ESA (an approved science-driven demonstration mission in the ESA Explorer program). A launch of SMOS with a microwave imaging radiometer (the interferometric design employs aperture synthesis) is planned for 2005 (D.36). In this concept, the required spatial resolution of <50 km for global brightness temperature retrieval (3 day revisit) in the microwave range (L-band), demands in turn a large number of aperture synthesis radiometers elements - each having an independent receiver. The swath width of the radiometer is 934 km at an orbital altitude of 755 km, sufficient for a 3-day equatorial revisit time. The quest for still higher resolutions (in the order of 10 km x 10 km) of future missions will certainly increase the number of receivers in their designs. [84] - On the applications side such data is very much needed for the implementation of hydrologic models of large basins.

- Microwave sounding from GEO satellites. So far, infrared/microwave sounder instrument combinations, [like the TOVS instrument series with HIRS, SSU and MSU flown since TIROS-N (launch 1978), and the HIRS/AMSU combination first flown on NOAA-15 (launch May 13, 1998)], have only been flown on weather satellites in polar orbits. These combined observations have the advantage to provide more nearly all-weather soundings (recovery of vertical distribution of temperature and humidity soundings). Such S/C ob-

83) P. Racette, D. M. LeVine, "Synthetic Aperture Radiometry: Technology for Spaceborne Microwave Radiometers for the Future," http://daacdev1.stx.com/ELF/docs/PER/racett11.html

84) M. A. Fischman, A. W. England, "Sensitivity of a Direct Sampling Digital Correlation Receiver for Aperture Synthesis Radiometry," Proceedings of the IEEE IGARSS 2000 Conference, Honolulu, HI, July 24-28, 2000

servations have had a significant impact on weather forecasting accuracy, especially in region where in-situ observations are sparse. Infrared sounders alone do not have adequate cloud-penetration ability to satisfy most forecast requirements, while microwave sounders alone do not satisfy observations in the infrared region. - On the geostationary (GEO) side, NASA was planning the GEO/SAMS (Geostationary Synthetic Aperture Microwave Sounder) instrument to be flown on the EO-3 mission (Earth Observing-3, launch in 2003) in geostationary orbit. The GEO/SAMS concept employes the synthetic aperture radiometer technique (a sparsely populated antenna array is substituted for a filled array). However, in 2000 the GEO/SAMS proposal was not accepted. [85] [86]

1.1.4 Optical Region, Active Observations (Lidars)

• Laser radars (active instruments), or lidars, combine the principles of submillimeter-wave radar and optics. The introduction of lidar (Light Detection and Ranging) technology, generally used in the spectral range of 0.3 μm to about 10 μm or the equivalent frequency range of 1000 - 30 THz, provided a new capability to study fine-scale phenomena of the atmosphere (with respect to structure, dynamics, models and climatology) as well as of the Earth's surface (texture, terrain profiling, shallow water depth sounding, etc.). Some lidar measuring techniques employed are: incoherent backscatter, DIAL (Differential Absorption Lidar), coherent Doppler (wind), laser altimetry, etc. (see O.8.5). Airborne lidars are flown since about 1977 [AOL (1977), CALS 1979), ALEX (1979), ALPHA-1 (1979), etc.)], most instruments were developed during the latter 1980s and the 1990s. MACAWS of NASA/MSFC (since 1995) is the first airborne 'coherent atmospheric wind lidar' measuring 2-D, 3-D or vertical wind fields. - On the spaceborne side, the first lidars were introduced in 1994 for atmospheric applications (detection of stratospheric and tropospheric aerosols; measurement of the planetary boundary layer, cloud top heights, atmospheric temperature and density); these were followed by lidar altimeters, mainly used for ocean surface or terrestrial surface monitoring.

- Atmospheric lidars: LITE (Lidar In-Space Technology Experiment) of NASA/LaRC, a triple wavelength lidar, was test-flown on Shuttle (STS-64) in Sept. 1994. Balkan-1 of Atmospheric Optics, Tomsk (Russia), flown on the MIR/Spektr module (launch May 20, 1995). ALISSA, a CNES-developed lidar flown on MIR/Priroda (launch April 23, 1996). In Europe, the instruments ATLID (Atmospheric Lidar) of ERM (Earth Radiation Mission) and ALADIN (Atmospheric Laser and Doppler Instrument) of ADM (Atmospheric Dynamics Mission) are part of ESA's spaceborne lidar program. ALADIN will probably be the first spaceborne instrument in a sustained mission using the incoherent DWL (Doppler Wind Lidar) technique of direct detection of wind profiles on an optical detector (with no LO). In this measurement scheme the backscattered optical signal field is analyzed and dispersed in an interferometric filter (or in diffraction grating) prior to detection. A combined Mie and Rayleigh backscattering fringe-imaging receiver is employed to analyze aerosol and cloud backscatter.

Instrument name	LITE	Balkan-1	ALISSA	ALADIN	ATLID
Platform/Mission	Shuttle STS-64	MIR/Spektr	MIR/Priroda	ADM (ESA)	ERM (ESA)
Orbit altitude (km)	260	350-400	350-400	400	400
Laser type	Nd:YAG	Nd:YAG	Nd:YAG	Nd:YAG	Nd:YAG
Wavelength (nm)	1064,532,355	532	532	355	1064
PRF (Hz)	10	0.18	50	100	100
Telescope diameter	100 cm	27 cm	40 cm	110 cm	60 cm

85) J. Gurka, J. Heil, "The US National Weather Service Operational Requirements for the Evolution of future NOAA Operational Geostationary Satellites," Proceedings of the EUMETSAT Meteorological Satellite Data User's Conference, Copenhagen, Denmark, Sept. 6-10, 19999, pp. 69-76

86) W. C. Boncyk, W. J. Wilson, B. H. Lambrigtsen, "The Enabling Technologies of the Geostationary Synthetic Aperture Microwave Sounder (GEO/SAMS)," Proceedings of IEEE/IGARSS Conference, Honolulu, HI, July 24-28, 2000

Instrument name	LITE	Balkan-1	ALISSA	ALADIN	ATLID
Range resolution	35 m	3 m	150 m	100-2000	15 m
Scanning					±20°
Instrument mass	990 kg	150 kg		263 kg	240 kg
Instrument power	2000 W	250 W	3000 W	304 W	520 W

Table 13: System performance parameters of spaceborne atmospheric lidars

- Altimeter lidars: [87] [88] [89] SLA-1 (Shuttle Laser Altimeter), was flown on STS-72 (Jan. 11-20, 1996), SLA-2 was flown on STS-85 (Aug. 7-19, 1997) to acquire altimetric samples of land topography (see J.17). GLAS (Geoscience Laser Altimeter System) is being built by NASA/GSFC (to be flown on the ICESat mission in 2001, see D.16) to determine the mass balance of the polar ice sheets and their contributions to global sea level change. The MBLA (Multi-Beam Laser Altimeter) lidar instrument is flown on the VCL (Vegetation Canopy Lidar Mission), a NASA/GSFC mission with a launch in mid 2002. The objective of the SLA and VCL instruments (D.39) is to provide lidar altimetry for Earth surface monitoring (in particular land surface processes and interactions). - Advantages of the laser technology include small footprint size, individual pulse processing of surface elevations, and dense ground sample spacing.

The early EOS (Earth Observing System) program of NASA considered the development of LAWS (Lidar Atmospheric Wind Sounder), a coherent Doppler wind lidar with a carbon dioxide laser. However, the project was cancelled in 1994 due to severe budget cuts in the EOS program.

1.1.5 Sounding of the Atmosphere

- Sounding (see also: Limb/occultation sounding in Glossary). Airborne and spaceborne Earth observation adapted fairly early the ancient technique of sounding ('to find bottom' - originally used for the measurement of shallow water depths) into the development of suitable sensors (sounders) to measure conditions of the medium 'atmosphere.' Sounders provide profiles of state parameters (a series of measurements of temperature, pressure, moisture, etc.) in a particular plane of observation (nadir or limb configuration) at various heights. The early sounder measurements were obtained from filter radiometers with a spectral resolving power ($\lambda/\Delta\lambda$) typically in the order of 100. At the start of the 21st century the spectral resolving power of typical radiometers is more in the range of 1000.

Later applications included also the sounding of aerosols, trace gas constituents ("absorptive" or "extinctive" occultation monitoring), and of wind components with a variety of instruments, employing such techniques as FTS (Fourier Transform Spectrometry) and Doppler. The technique of sounding has also been extended far beyond the Earth's atmosphere to measure plasma densities of the solar wind. Active airborne sounders (namely lidars) are also used to fathom the depths of shallow waters (FLASH since 1989, SHOALS since 1994), to detect sea surface pollution, and to measure surface emissions. Naturally, the entire spectrum may be used for sounding measurements, but the microwave region is the traditional and dominant arena for sounders.

The presence of clouds in the field of view of sounders has a detrimental effect on the quality of a retrieval. The absorption properties of cloud droplets and ice particles at infrared sounding wavelengths are so strong that even thin clouds contaminate the measurement of radiances. As a result, a number of post-processing techniques have been developed to minimize the effects of clouds on soundings. These usually require some way of identifying cloudy scenes to arrive at an equivalent clear-sky radiance quantity. Methods accounting for the effects of clouds on the data are generally based on higher-resolution visible and infrared imaging data that are required to supplement the sounding channels.

87) Special Issue on Laser Radar, Proceedings of the IEEE, Vol. 84, No. 2, February 1996, pp. 99-298

88) J. A. McKay, D. Rees, "Space-based Doppler Wind Lidar: Modeling of Edge Detection and Fringe Imaging Doppler Analyzers," Advances in Space research, Vol. 26, No. 6, 2000, pp. 883-891

89) Courtesy of A. Ginati of OHB-System, Bremen

• The first airborne devices used were self-registering instruments (referred to as sondes) on balloons to record meteorological data. The MTS (Microwave Temperature Sounder of GSFC) instrument was first flown on aircraft in 1976. - A large number of daily atmospheric profiles from spaceborne sounders [SSH on the DMSP series, launch of F-1 Sept. 11, 1976; TOVS (TIROS Operational Vertical Sounder), on TIROS-N consisting of HIRS/2, SSU and MSU, launch Oct. 13, 1978; SBUV/TOMS on Nimbus-7, launch Oct. 24, 1978], ILAS and IMG on ADEOS (launch Aug. 17, 1996) and ILAS-II on ADEOS-II (launch in 2002) began to find their way into global forecasting models.

Today's sounders like AMSU (Advanced Microwave Sounding Unit), first flown on NOAA-15 (launch May 13, 1998), provide atmospheric temperature profiles with a sensitivity of about 1 K and a vertical resolution of about 2 km in the troposphere. SSMIS (Special Sensor Microwave Imager Sounder) of the DMSP series, first flown on F16 (launch 2001), features a similar performance as AMSU. Current generation sounder performance (AMSU, SSMIS) still falls short of the requirements for NWP (Numerical Weather Prediction). Some examples of next-generation sounders are: AIRS (Atmospheric Infrared Sounder) on Aqua (launch 2001), ATMS (Advanced Technology Microwave Sounder) and CrIS (Cross-track Infrared Sounder) on NPP (launch in 2005). Their data is planned to be used in NWP models.

• **Aerosol monitoring** plays an important role in global climate change. Spaceborne monitoring of aerosols started with SAM of NASA/LaRC [SAM (Stratospheric Aerosol Measurement) was flown on ASTP (Apollo-Soyuz Test Project), July 15-24, 1975, to perform the first successful solar occultation measurement of stratospheric aerosol], SAM II (launch Oct. 24, 1978 on Nimbus-7, see M.17.7), SAGE I, SAGE II). Both SAM and SAM-II were single spectral instruments measuring the aerosol extinction near the 1000 nm wavelength region. Multiple spectral measurements began with SAGE-I, with a launch on the AEM-2 (Application Explorer Mission-2) satellite, Feb. 18, 1979 (see A.5). SAGE-II is an advanced version of SAGE-I with 7 channels at 385, 448, 453, 525, 600, 940, and at 1020 nm. SAGE-II was flown on ERBS with a launch on Oct. 5, 1984 (see A.13). The measurements of SAM-II, SAGE-I and SAGE-II have provided long-term observations of aerosol and ozone for over 20 years.

Instrument	Mission	Comment
ISAMS, HALOE	UARS (launch Sept. 13, 1991), NASA	ISAMS failure in July 1992
ORA	EURECA-1 (1992 - 1993), ESA	Retrievable carrier platform
LITE	STS-64 (Sept. 9-20, 1994), NASA	Technology experiment
GOME	ERS-2 (launch Apr. 21, 1995), ESA	Spectral range of 240-790 nm for ozone, aerosols, etc.
SPIRIT-III	MSX of DoD (launch Apr. 24, 1996)	S/C built and operated by JHU/APL
POLDER, CNES and ILAS of JEA	ADEOS (launch Aug. 17, 1996), NASDA	S/C failure June 30, 1997
MOS of DLR	IRS-P3 (launch Mar. 21, 1996), ISRO	also MOS-P on Priroda
SLA of NASA	SLA-1 (Jan. 11-20, 1996 on STS-72) SLA-2 (Aug. 7-19, 1997 on STS-85)	Shuttle missions
POAM-II of NRL	SPOT-3 (launch Sep. 26, 1993), CNES	S/C operations until Nov. 16, 1996
POAM-III of NRL	SPOT-4 (launch Mar. 24, 1998), CNES	Aerosols in altitudes of 10-30 km
AVHRR/3 of NOAA	NOAA-15 (launch May, 13, 1998)	First flight of AVHRR/3 instrument
MODIS, MISR MODIS	Terra (launch Dec. 18, 1999), NASA Aqua (launch 2001), NASA	EOS (Earth Observing System) program
OSIRIS of CSA	ODIN (launch Feb. 20, 2001), Sweden	UV/VIS/IR limb sounder
CHRIS	PROBA (launch 2001), ESA	Aerosol studies
MERIS, MIPAS, GOMOS, SCIAMACHY	Envisat (launch Oct. 2001), ESA	
GLAS	ICESat (launch 2001), NASA	Nadir-viewing lidar
ILAS-II of JEA	ADEOS-II (launch 2002), NASDA	
ACE-VNIRI	SCISAT/ACE (launch 2002), CSA	
HIRDLS	Aura (launch 2003), NASA	EOS program

Instrument	Mission	Comment
EPIC	Triana (launch 2004), NASA	Measurement of aerosol amounts from Lagrangian point L1
GOME-2	MetOP-1 (launch 2005), EUMETSAT	
ALADIN (ESA)	ADM (launch 2006/7)	Incoherent direct detection lidar
ABI (Advanced Base-line Imager)	GOES-series, starting with GOES-Q in 2008, NASA/NOAA	Aerosol measurement from GEO

Table 14: Some aerosol-monitoring instruments of spaceborne missions

SAGE-III is an Earth limb-scanning grating spectrometer (part of NASA's EOS mission) scheduled to fly on Meteor-3M-1 (Russia) with a launch in 2001. An important objective of SAGE-III is to characterize tropospheric and stratospheric aerosols and upper tropospheric and stratospheric clouds, and investigate their effects on the Earth's environment, including radiative, microphysical, and chemical interactions. SAGE-III is also scheduled to fly on ISS in 2004. - Note: There is an inherent coverage weakness in the solar occultation technique by offering only two measurement opportunities per orbit. It has been estimated that six spaceborne SAGE-III instruments on various orbits would be needed to achieve a weekly coverage of the Earth.

- Spatial Heterodyne Spectroscopy (SHS), see O.11 and J.16. SHS provides the first practical approach to extend interference spectroscopy into the FUV (Far Ultraviolet, 1200-2000 Å) spectral range. The technique appears to be offering high-resolution spaceborne spectroscopy applications in astronomy (detection of faint interstellar emission lines, study of the dynamics of hot interstellar gases) as well as in Earth observation. Examples in this field include the measurement of vertical density profiles of the hydroxyl (OH) radical in the middle atmosphere (30-100 km), study of the distribution of aerosols in the mesosphere, and to investigate the role of OH in the photochemistry of water vapor and ozone in the presence of aerosols in the mesosphere. NRL (Naval Research Laboratory) in Washington, DC has been cooperating with UWM (University of Wisconsin-Madison) for the last years to develop an instrument. A first implementation of the SHS concept is realized in SHIMMER (Spatial Heterodyne Imager for Mesospheric Radicals). The instrument is planned of be flown on Shuttle in 2002.

- **Refractive occultation monitoring,** also referred to as **radio occultation monitoring** (Earth-observation applications of navigation systems and use of GPS receivers as science instruments).[90) 91) 92) 93)] In 1995 retrieval of atmospheric limb soundings (moisture and temperature distributions) was demonstrated by GPS receiver sounding techniques referred to as "refractive occultation monitoring." The technique makes use of occulting navigation signals of the GPS constellation they pass through the atmosphere and are intercepted by a GPS receiver on a LEO S/C. The highly accurate measurements obtained with these special-function GPS receivers permit the derivation of vertical profiles of the temperature, pressure and humidity in the atmosphere, as well as profiles of electron content in the ionosphere.

90) Note: The radio occultation technique was initially invented in the early 1960s by NASA/JPL in its planetary exploration programs to Venus, Mars (Mariner-IV flyby in 1964) and later to the outer planets (the first relevant proposal came from Stanford University in 1962). Applying this technique to the Earth's atmosphere and using the GPS constellation as a source and a GPS receiver in LEO was first suggested in the late 1980s in two papers: a) A. S. Gurvich, T. G. Krasilnikova, "Navigation Satellites for Radio Sounding of the Earth's Atmosphere," Issled. Zemli Kosmosa, Vol. 6, 1986 (in English: Soviet Journal of Remote Sensing, 6, 1990), and b) T. P. Yunck et al. "The role of GPS in precise Earth observation, Proceedings of the IEEE Position, Location, and Navigation Symposium, Orlando, FLA, 1988.

91) T. P. Yunck, C. H. Liu, R. Ware, "A History of GPS Sounding," Special issue of TAO (Terrestrial, Atmospheric and Oceanic Science, Vol. 11, No 1, March 2000, pp. 1-20

92) W. G. Melbourne, et al., "The Application of Spaceborne GPS to Atmospheric Limb Sounding and Global Change Monitoring, JPL Pub. 94-18, Pasadena, CA, 147 pp., 1994

93) W. Bertiger, Y. Bar-Sever, S. Desai, et al., "Precise Orbit Determination for the Shuttle Radar Topography Mission using a New Generation of GPS Receiver," ION GPS-2000, Sept. 19-22, 2000, Salt Lake City, UT, pp.1646-1654

The GPS/MET instrument of UCAR/JPL on OrbView-1/Microlab-1 (launch April 3, 1995, B.8.1) introduced this alternate sounding technique to Earth observation, starting a new era of atmospheric profiling technology. GPS/MET temperature retrieval accuracies of better than 1K have been demonstrated in the altitude range of 10-30 km during the proof-of-concept phase of the mission with vertical resolutions of better than 1 km. This all-weather sounding technique promises to greatly contribute to global atmospheric monitoring, offering an abundance of data for operational meteorology and climatology. [94] The TRSR (TurboRogue Space Receiver) instruments of GFO (Geosat Follow-On, launch Feb. 10, 1998), SUNSAT and Ørsted (both were launched Feb. 23, 1999) missions are a further development of the GPS/MET instrument on OrbView-1/Microlab-1. The TRSR instruments of NASA/JPL were manufactured by Allen Osborne Associates Inc. of Westlake Village, CA.

Mission	Instrument	Comment
OrbView-1/Microlab (launch Apr. 3, 1995)	GPS/MET	NASA
GFO (launch Feb. 10, 1998)	TRSR (JPL)	US Navy
SUNSAT (launch Feb. 23, 1999) Ørsted (launch Feb. 23, 1999)	TRSR TRSR	University of Stellenbosch, South Africa Danish geomagnetic mission
SRTM (Feb. 11-22, 2000) Shuttle mission	BlackJack (JPL)	NASA, DLR, ASI
CHAMP (launch Jul. 15, 2000)	BlackJack	GFZ, DLR
TiungSat (launch Sept. 26, 2000), D.40.19	Experimental GPS	ATSB (Malaysia) S/C built by SSTL. Refractive sounding of the ionosphere
SAC-C (launch Nov. 21, 2000)	BlackJack/ GOLPE	CONAE, NASA
GRACE (launch Nov. 2001)	BlackJack	US-German mission
ICESat (launch 2001)	BlackJack	NASA
PICOSat (launch Aug. 2001) of USAF/SSP	IOX (SMC)	Ionospheric Sounding Experiment measures electron profiles (TEC) with GPS IOX is of TRSR heritage
FedSat (launch 2002), TEC (Total Electron Content)	BlackJack (AstroNav)	Australian mission (CRCSS), single antenna configuration, TEC monitoring
ROCSat-3/COSMIC (launch 2005) Constellation of six microsatellites	BlackJack	Taiwanese-US mission
ACE (Atmosphere Climate Experiment) Cluster of six microsatellites, (launch 2005)	GRAS (Saab Ericsson Space)	ESA mission
MetOp-1 (launch 2005)	GRAS	EUMETSAT/ESA mission
GCOM-A1 (launch 2006)	GRAS	NASDA/Japanese mission
NPOESS (launch 2008)	GPSOS (Saab Ericsson Space)	NOAA/NASA/DoD mission
WATS (Water vapor in Atmospheric Troposphere and Stratosphere), launch 2008	GRAS	ESA mission (constellation of 8 microsatellites)

Table 15: Overview of missions with refractive occultation measurements

The following missions use the BlackJack configuration (built by SpectrumAstro of Gilbert, AZ, for JPL - AstroNav is the SpectrumAstro product name for BlackJack) a new generation of GPS flight receiver and a TRSR successor: SRTM (Feb. 11-22, 2000), CHAMP (launch July 15, 2000), SAC-C (launch Nov. 21, 2000), GRACE (launch 2001), Jason-1 (launch 2001), ICESat (launch 2001), , and VCL (2003). As an integrated service system (offering the functions of tracking mode, occultation mode, and altimetry mode support), the full BlackJack configuration is using a multiple antenna design. - Other missions using refractive radio occultation techniques are: a) MetOp-1 (launch 2005) with GRAS (GNSS Receiver for Atmospheric Sounding), and b) ROCSat-3/COSMIC (Constellation Observing System for Meteorology, Ionosphere and Climate), a collaborative US/Taiwanese six-microsatellite constellation, planned for launch in 2005. ROCSat-3/COSMIC (A.24) will demonstrate - for the first time - the usefulness of microsatellite constellations in obtaining global atmospheric "snapshots" in near-real time. The system has the potential to furnish

94) G. Hajj, E. R. Kursinski, et. al., "Sensing the Atmosphere From a Low-Earth Orbiter Tracking GPS: Early Results and Lessons From the GPS/MET Experiment," Proceedings of ION GPS-95, Sept. 12-15, 1995, pp. 1167-1174

valuable data for weather prediction, global climate-change analysis and research, and ionospheric research and prediction. [95) 96)]

Background: The deployment of GNSS (Global Navigation Satellite Systems) constellations such as GPS and GLONASS with their network of navigation signals made radio occultation techniques possible in the Earth's atmosphere for applications of weather and climate studies. Prior to GPS, there was a general lack of radio sources suited to meet the performance requirements of refractive radio occultation.
The basic functions of the radio occultation technique include: a) reception of signals that have crossed the atmosphere at varying altitudes by means of two antenna arrays; b) acquisition of such signals, also during the rise (ascending) of occultation events, when a signal first appears after crossing dense tropospheric layers causing large dynamics in amplitude and phase; c) signal tracking to provide precise amplitude and phase measurements; d) onboard processing to support occultation-event predictions, also to aid the tracking. [97)]

Instrument (Agency)	Platform	Launch Date	Comment
BUV (NASA)	Nimbus-4 (NASA) AE-E (NASA)	Apr. 8, 1970 Nov. 20, 1975	2 Ebert-Fastie-type monochromators AE-E reentered on June 10, 1981
SSH, SSH-2 (DoD)	DMSP (DoD) series	Sep. 11, 1976	Starting with F1 satellite
SBUV/TOMS (NASA)	Nimbus-7	Oct. 24,1978	Nadir-viewing Ebert-Fastie spectrometer of TOMS. Swath width of 2700 km (scanning). TOMS failed in May 1993. SBUV failed in 1990
SAGE-I (NASA)	AEM-2 (NASA)	Feb. 18, 1979	
UVSP (NASA/MSFC)	SMM (NASA)	Feb. 14, 1980	
UV Ozone Experiment Airglow Instrument Solar UV Monitor	SME (NASA)	Oct. 6, 1981	
UV Spectrometer	EXOS-C (ISAS)	Feb. 14, 1984	
SBUV/2 (NASA)	NOAA-9 (NOAA) NOAA-11 (NOAA)	Dec. 12, 1984 Sep. 24, 1988	S/C service ended Aug. 3, 1995 S/C service ended Sep. 9, 1994
SSBUV (NASA)	STS-34 (NASA)	Oct. 19, 1989	Coincident observations with SBUV/2 on NOAA-9 and NOAA-11
TOMS (NASA)	Meteor-3-6 (Russia)	Aug. 15, 1991	TOMS operation until Dec. 1994
HALOE, MLS, CLAES, ISAMS	UARS (NASA)	Sep. 13, 1991	Sun occultation method (HALOE) Heterodyne limb sounder (MLS)
POAM-II (NRL)	SPOT-3 (CNES)	Sep. 26, 1993	SPOT-3 entered safehold Nov. 14, '97; Solar occultation through the Earth's atmospheric limb
GOME (ESA) a scanning optical double spectrometer	ERS-2 (ESA)	Apr. 21, 1995	DOAS (Differential Optical Absorption Spectroscopy) measurement concept
Ozon-M	Priroda (Russia)	Apr. 23. 1996	Priroda, a module of the MIR station
TOMS (NASA)	TOMS-EP	July 2, 1996	Operational as of 2000
TOMS (NASA), RIS	ADEOS (NASDA)	Aug. 17, 1996	ADEOS failed on June 30, 1997
MAHRSI (NRL)	CRISTA-SPAS-2	Aug. 7-19, 97	STS-85 Shuttle, OH interaction with ozone and other trace gases
POAM-III (NRL)	SPOT-4 (CNES)	Mar. 24, '98	Operational
OLME (FACH)	FASat-Bravo (Chile)	July 10, 1998	Total column ozone measurements
OM-2	TechSat/Gurwin-II	July 10, 1998	Technion (Israel Institute of Techn.)
OSIRIS (CSA)	ODIN (Sweden)	Feb. 20, 2001	Detection of aerosols and trace gases
TOMS-5 (NASA)	QuikTOMS	2001	
GOMOS , SCIAMACHY (ESA)	Envisat (ESA)	2001	Star occultation measurement method, DOAS and BUV in parallel

95) P. Silvestrin, P. J. Baptista, P. Hoeg, " Radio Occultation Data Analysis: From Planetary Atmosphere Sounding to Operational Meteorology," Proceedings of 32nd ESLAB Symposium on 'Remote Sensing Methodology for Earth Observation and Planetary Exploration,' ESA/ESTEC, Sept. 15-18, 1998 (SP-423 Dec. 1998), pp.179-187

96) P. Silvestrin, "Earth-Observation Applications of Navigation Satellites," ESA Bulletin, No 102, May 2000, pp. 101-106

97) P. Silvestrin, R. Bagge, M. Bonnedal, et al., "Spaceborne GNSS Radio Occultation Instrumentation for Operational Applications," ION GPS 2000, Sept. 19-22, 2000, Salt Lake City, UT, pp. 872-880

Instrument (Agency)	Platform	Launch Date	Comment
SAGE-III (NASA)	Meteor-3M-1 (Russia)	2001	Self-calibrating solar and lunar occultations (9 spectral channels)
GLI , ILAS-II	ADEOS-II (NASDA)	2002	Swath width of 1600 km
ACE-FTS (CSA)	SCISAT-1/ACE (CSA)	2002	FTS instrument for trace gases
HIRDLS, MLS, TES (NASA), OMI (NIVR)	Aura (EOS/CHEM) (NASA)	2003	Limb sounder (HiRDLS), FTS (TES), Hyperspectral capabilities (OMI)
EPIC	Triana (NASA)	2004	Lagrangian point 1 measurements
GOME-2, IASI (ESA/Eumetsat)	MetOp-1 (Eumetsat)	2005	DOAS measurements
ODUS , SOFIS	GCOM-A1 (NASDA)	planned 2006	Nadir-viewing grating spectrometer (ODUS), Solar occultation (SOFIS)
OMPS IPO/NASA	NPOESS (IPO)	planned 2008	Limb-viewing sensor suite

Table 16: Major spaceborne instruments for the global measurement of ozone

• **Monitoring of ozone** (global coverage). Detection of short-term and long-term (seasonal) changes in ozone with measurements of total column amounts and stratospheric and tropospheric profiles of ozone. See also Table 430 for a survey of Shuttle flights with the SSBUV payload.

Atmospheric ozone has several environmental implications, it can be classified according to atmospheric layers (or altitude). Long-term trend measurements are of great importance.

- Stratosphere. About 90% of the atmospheric ozone is contained in the stratosphere. Ozone plays a critical role in absorbing UV radiation and preventing it from reaching Earth's surface. The so-called "ozone hole" is a consequence of stratospheric ozone depletion over the poles of the Earth.
- Troposphere. In the upper and middle troposphere, ozone is a major greenhouse gas, causing inhomogeneous radiative forcing.
- Troposphere. Ozone is an oxidizing power in the lower and middle troposphere.
- Surface air. Ozone is a pollutant, toxic to humans and to vegetation.

• Airborne cloud investigations (properties, interactions, phases, droplets, microphysics, etc.) with radiometers, lidars, radars, hyperspectral imagers, etc. have been performed since the mid 1970s (with most observations in the 1990s) on various scales with a number of instruments, such as: ARES (NASA/JSC), AWSR (NOAA), AMMS, CALS and CAR (NASA/GSFC), CVI (MISU), Deimos and MARSS (UKMO), ALEX, OLEX, H_2O-DIAL and Microlidar (DLR), MPIR (SNL), CDL (LLNL), ELDORA/ASTRAIA (NCAR/CRPE), LASAL (GSFC), LEANDRE (CNRS/CNES), MCR and MIR (GSFC), MTP (JPL), NAILS (NCAR), OVID (MPIfM Hamburg), RAMS (NASA/NOAA), MAKREL-2 and M2M (Tomsk), APDOR-95 (MIRSL).

- The first project to use in-service aircraft for trace-constituent data collection took place in 1968 with carbon monoxide (CO) measurements made from Lufthansa B-707 aircraft. A vertical CO gradient was observed in the tropopause, this resulted in an estimation of the size of the stratospheric CO sink. [98] (The same group of researchers employed a manned laboratory inside a container on 10 German cargo service flights between 1981 and 1987 to measure CO).

- The meridional distribution [99] [100] of tropospheric ozone concentrations was studied by MPAe (Max-Planck-Institut für Aeronomie, Katlenburg-Lindau, Germany) on 37 commercial flights between Northern Europe (Frankfurt) and South Africa (Cape Town) during the period 1970-1974.

98) W. Seiler, C. Junge, "Carbon Monoxide in the Atmosphere," Journal of Geophysical Research, Vol. 75, No. 20, April 20, 1970, pp. 2217-2226

99) P. Fabian, P. G. Pruchniewicz, "Meridional Distribution of Ozone in the Troposphere and its Seasonal Variations," Journal of Geophysical Research, Vol. 82, No 15, May 20, 1977, pp. 2063-2073

100) H. K. Tiefenau, P. G. Pruchniewicz, P. Fabian, "Meridional Distribution of Tropospheric Ozone from Measurements Aboard Commercial Airliners," Pure and Applied Geophysics, Vol. 106-108, 1973, pp. 1036-1040

- GASP (Global Atmospheric Sampling Program) was introduced by NASA in 1975 to measure trace gases on commercial airliners. [101] Four B-747 (two Pan American, one United, and one Qantas) aircraft were equipped with instruments to routinely measure ozone, carbon monoxide, water vapor, aerosols, temperature, and horizontal winds. GASP was over the period 1975-1979 with over 6900 flights.

- The Japanese ASE (Automatic Air-Sampling Equipment) [102] program with instrumentation on a Boeing 747 aircraft between Japan and Australia was introduced in 1993. It lasted until 1996.

- Also in 1993, the EU started its MOZAIC (Measurement of Ozone by Airbus In-Service Aircraft) program. During MOZAIC-I (1993- Sept. 1996), fully automated devices, developed by CNRS (France) and Forschungszentrum Jülich (Germany), were flown on five Airbus aircraft in normal airline serve (Air France, Qantas, Lufthansa, Sabena). MOZAIC-II started in Oct. 1996 with the aim to continue the ozone and water vapor measurements, several new instruments were added to measure CO and NO_y. Between Sept. 1994 and Dec. 1997, 7500 flights (54,00 flight hours) were made in the MOZAIC program over the continents (Europe, North America, Asia, South America, and Africa) and the Atlantic Ocean. A MOZAIC-III program is currently being carried out in the Fifth Framework Program of the EU. [103] [104]

- NOXAR (Nitrogen Oxides and ozone measurements along Air Routes) [105] was conducted aboard 540 flights of Swissair B-747 aircraft in the period from May 1995 to May 1996. The instruments recorded data at a temporal resolution of 3s.

- Another German program, CARIBIC (Civil Aircraft for Remote-Sensing and In-Situ-Measurements in Troposphere and Lower Stratosphere Based on the Instrumentation Container Concept), started in 1996.

1.1.6 Sounding of the Ionosphere

Ground-based echo sounding of the ionosphere has been conducted since about the 1920s to obtain information about the upper atmosphere and the effects of solar emissions. Spaceborne soundings of the ionosphere started very probably with the launch of Alouette-1 in 1962.

• **Topside sounding.** The term refers to RF (Radio Frequency) sounders looking from a LEO spacecraft into the nadir direction for ionospheric electron density observations. Other ionospheric parameters such as TEC (Total Electron Content), critical frequency (f_0F2) and F2-layer peak height, plasma temperature, and ion composition may be derived from the topside ionograms (Note: the ionospheric F-layer is extending from about 160 km to about 1000 km with a peak electron concentration around 250-300 km altitude). The ionosphere below the peak electron concentration is referred to as bottomside ionosphere, while the outer part of the ionosphere, extending to about 2000 km (Note: at altitudes below 2000 km in altitude the plasma is simply part of the "ionosphere"), is referred to as the topside ionosphere. A sounder, installed onboard a LEO satellite (800-1000 km) and practically immersed in the medium it is measuring, is able to see only the uppermost portion of the

101) G. D. Nastrom, "Ozone in the Upper Troposphere From GASP Measurements," Journal of Geophysical Research, Vol. 84, No C7, July 20, 1979, pp. 3683-3688

102) H. Matsueda, et al., "Carbon Monoxide in the upper troposphere over the western Pacific between 1993 and 1996," Journal of Geophysical Research, Vol. 103, 1998, pp. 19,093-19,110

103) A. Marenco, et al., "Measurement of ozone and water vapor by Airbus in-service aircraft: The MOZAIC airborne program, An overview," Journal of Geophysical Research, Vol, 103, No D19, Oct. 20, 1998, pp. 25,631-25,642. The same volume, D19, contains a special section, pp. 25,631-25,737

104) J. Y. N. Cho, et al., "Trace Gas Study Accumulates Forty Million Frequent-Flyer Miles for Science," EOS Transactions of AGU, Vol. 80, No. 34, Aug. 24, 1999, pp. 377-384

105) D. Brunner, et al., "Large-scale nitrogen oxide plumes in the tropopause region and implications for ozone," Science, Vol. 282, 1998, pp. 1305-1309

ionosphere down to about the peak electron height, i.e., the topside ionosphere; hence, such a sounder is referred to as a "**topside sounder.**" [106] [107] [108]

The Canadian spacecraft Alouette-1 (launch Sept. 29, 1962 from VAFB) can be regarded as the first satellite (with a circular orbit of 1000 km altitude and an inclination of 80.5°) carrying a topside sounder to investigate the ionosphere (Canada/USA cooperative venture). The primary purpose of the Alouette-1 mission was to investigate the geographic, seasonal, and diurnal properties of the topside ionosphere up to 1000 km in altitude. The Alouette-1 swept-frequency sounder revealed that an equatorial anomaly extends high into the topside ionosphere; it also demonstrated its usefulness in observing HF electron resonances. The Alouette-1 ionospheric sounder had the following parameters: Frequency range of 1-12 MHz, transmitter power of 100 W, pulse width of 100 μs, pulse repetition frequency of 62 Hz, and a frequency sweep rate of 1 MHz/s. In a supporting experiment, an untuned VLF receiver operated in the range of 400 Hz to 10 kHz. The VLF receiver permitted the measurement of relative ion abundances in conjunction with the sounder. A worldwide network of 22 ground receiving stations collected the sounding data of Alouette-1 and -2. [109] [110]

Alouette-2 followed in 1965, also with a topside sounder. [111] Other spacecraft in the early era of spaceflight of countries or agencies with various techniques of topside sounding implementations were: Explorer-20 of NASA (launch Aug. 25, 1964), the basic sounder instrumentation comprised a two-frequency transceiver coupled to an electrically short dipole antenna (three transceivers were used for sounding at six fixed frequencies between 1.5-7.22 MHz); Explorer-22 of NASA (launch Oct. 10, 1964); ISIS-I (International Satellite for Ionospheric Studies-I) of NASA with a launch in 1969 (the ISIS program was actually an extension of the Alouette program); Cosmos-381 (launch in 1970) of the Soviet Union; ISIS-2 of NASA (launch April 1, 1971); ISS-A (Ionospheric Sounding Satellite-A) of NASA with a launch in 1976; ISS-B of NASA (launch Feb. 16, 1978); EXOS-B (Exospheric Satellite-B or Jikiken) of NASDA (launch Sept. 16, 1978); ISEE-1/-2 of NASA/ESA (launch Oct. 22, 1977); Intercosmos-19 of the Soviet Union (launch Feb. 27, 1979); EXOS-C (Ohzora, launch Fe. 14, 1984) of NASDA; Cosmos-1809 (launch Dec. 1986) of the Soviet Union; Coronas-I of Russia (launch March 2, 1994 from Plesetsk), and the orbital MIR Space Station (the topside sounder was installed on the Priroda module in 1998).

Topside sounding from a satellite reveals the response of the ionosphere to disturbances such as a geomagnetic storm. The pulses emitted by the sounder on consecutive frequencies are reflected on different heights depending on electron density height distribution. The time delay dependence of emitted pulses on the frequency is called the height-frequency characteristic of the ionosphere, referred to as the ionogram. With the help of a special algorithm, which takes into account the propagation characteristics of electromagnetic waves in magnetized plasma, the measurements are transformed into density height distributions from the satellite altitude down to the maximum F2-layer density distribution, which is usu-

106) S. A. Pulinets, "Prospects of Topside Sounding," Chap. 3 of 'World Ionosphere/Thermosphere Study,' WITS Handbook, Vol. 2, edited by C. H. Liu, Dec. 1989

107) Proceedings of the IEEE, Special Issue on Topside Sounding and the Ionosphere, Vol. 57, June 1969, pp. 859-1240

108) S. A. Pulinets, R. F. Benson, "Radio-Frequency Sounders in Space," Review of Radio Science, ed. by W. Ross. Stone, Oxford University Press, Chapter 28, 1999, p. 711-733.

109) Note: Swept-frequency sounding is a technique in which a measurement is made of the frequency shift, phase shift, or time delay between the transmitted signal and its echo). Because of the Earth's magnetic field, the ionosphere is birefringent with the result that a transmitted electromagnetic wave normally splits into two characteristic waves which travel independently at different velocities and different polarizations. These are called the ordinary (O) and the extraordinary (X) waves and are typically elliptically polarized. As the sounding frequency is increased, the electron number density required to reflect the transmitted signal increases until reflection occurs at a region or height of maximum ionization. Above the critical frequency corresponding to the electron number density at the peak of the F2-layer, reflection can no longer take place and the ionosphere becomes transparent to the sounding signal.

110) C. A. Franklin, M. A. Maclean, "The Design of Swept-Frequency Topside Sounders," Proceedings of the IEEE, Vol. 57, No 6, June 1969, pp. 897-929

111) C. D. Florida, "The Development of a Series of Ionospheric Satellites," Proceedings of the IEEE, Vol. 57, No 6, June 1969, pp. 867-875

ally located at altitudes of 250 to 350 km. [112] In equatorial regions, especially in geomagnetically disturbed conditions, the F2-layer peak height may extend to altitudes of 500 km and even more.

Direct observations of ionospheric features are crucial for various applications in the fields of communications, navigation, early warning, radar, etc. - In spite of the numerous early missions, topside global sounding of the ionosphere has not become an operational service at the turn of the 21 century, rather it can still be put into the experimental category of missions.

- Possible future applications of topside ionospheric monitoring may involve the capability to diagnose the effect of anomalous atmospheric electric fields that penetrate into the ionosphere. The data analysis from the topside sounder on the Intercosmos-19 (1979) spacecraft showed strong variations in the vertical structure of the ionosphere over regions of impending earthquakes. [113] The ionosphere rises over a seismic active region forming a dome of density depletion. One of the main sources of atmosphere-ionosphere modification over the regions of preparing earthquakes is the emanation of various gases (radon, hydrogen and helium) from the ground as well as sub-micron aerosols. These may lead to the changes in the electrodynamic properties of the atmosphere-ionosphere that can be observed by the topside sounder.

- The COMPASS microsatellite (launch in fall of 2001) of IZMIRAN is a new topside sounding mission dedicated to the study of the structure of the geodynamic processes (in particular above deep tectonic faults and earthquake regions). The sensor complement of the mission employs ionospheric tomography (monitoring of electromagnetic fields of various intensities and frequencies as well as heat distributions). The combined measurements are processed and analyzed in search of possible earthquake indicators.

- Hyperspectral imaging of the global ionosphere is provided by the HIRAAS (High-Resolution Airglow/Aurora Spectroscopy) instrument package of NRL flown on the ARGOS satellite of DoD (launch Feb. 23, 1999, M.1). The imagery of the ionosphere obtained is "hyperspectral" at wavelengths from 50 to 310 nm. The HIRAAS, GIMI (Global Imaging Monitor of the Ionosphere) and EUVIP (Extreme Ultraviolet Imaging Photometer) instruments on the ARGOS S/C provide a powerful remote sensing exploration of the global ionosphere. This unusual combination of hyperspectral imagers and imagers provides a unique database of global structure and variability of ionospheric structure.

1.1.7 Some Instrument/Observation Techniques

- TDI (Time Delay Integration) is a concept of increasing the effective dwell time of CCD detector arrays in the pushbroom imaging mode. In scanned infrared imaging systems, TDI is one of the most important functions performed by CCDs. It was developed along with the early CCD technology at several places in the US industry and at NRL (DoD) in the later 1970s as well as at RSRE (Royal Signal and Radar Establishment, Malvern, UK) in 1982. RSRE introduced TDI into an infrared detector system by the name of SPRITE (Signal Processing In The Element). The TDI technique refers to a cumulative exposure concept of each ground image line by a CCD detector array (or any equivalent array). In TDI operation, signals generated by a moving scene-segment in successive detectors are sequentially delayed and added to the CCD. The main objective is to improve the SNR (signal-to-noise ratio) value which is one of the most important issues for high resolution imaging. [Since the total detector noise is given by the root mean summation while the total sig-

112) S. A. Pulinets, "Seismic activity as a source of the ionospheric variability," Advances in Space Research, Vol. 22, 1998, No 6, pp. 903-906

113) S. A. Pulinets, " Strong Earthquake Prediction Possibility with the help of Topside Sounding from Satellites," Advances in Space Research, Vol. 21, No. 3, 1998, pp. 455-458

nal is given by a simple sum, an increase in SNR equal to the square root of the number of detector rows is obtained]. [114] [115] [116]

Spaceborne instruments with TDI implementations are: The SeaWiFS instrument of SeaStar (the renamed OrbView-2 mission, launched Aug. 1, 1997), the ISIR sensor of NASA/GSFC flown on Shuttle flight STS-85 (Aug. 7-19, 1997), the OSA instrument of the Ikonos-2 satellite (launch Sept. 24, 1999), BGIS 2000 of QuickBird-1 (launch Nov. 20, 2000 - launch failure), PIC-2 of the EROS-B satellite program of ImageSat International, Cayman Islands - with a planned launch EROS-B1 in 2002/3), and MSRS (Multi-Spectral high Resolution System) of the Diamant satellite of OHB-System, Bremen (launch 2003) - are examples of more recent TDI implementations in 2-D CCD line-array technology in the visible range of the spectrum. Note: For extrinsic detector arrays (such as silicon detectors in the visible range) there are two possible observation approaches: TDI or "staring array." (see also O.3.6)

- Introduction of polarimetric measurements in the optical and microwave (radar) regions of the spectrum (retrieval of scattering matrix).

- Polarimetric SAR (microwave): In addition to measuring the amplitude of the radar return from a target, a polarimetric radar measures the relative phase difference between the four linear polarizations. This allows calculation of the target scattering matrix that can be used to optimize the polarization combination of radar for various applications. [117]

First airborne radar (SAR) polarimeters started to be flown in the time frame 1988 to 1990 with such instruments as C/X-SAR (CCRS), ARMAR (NASA/JPL), DO-SAR (Dornier), HUTSCAT (HUT), MMW-SAR (MIT/LL), NUSCAT (NASA/JPL), P-3/SAR (ERIM), IMARC (NPO Vega), and RAMSES (ONERA). First spaceborne polarimetric data were obtained from the L/C-band SAR (NASA/JPL) of the SIR-C/X-SAR Shuttle missions (SRL-1 in April 1994, SRL-2 in Sept/Oct. 1994). See also Table 8.

- Polarimetric radiometry: Refers to microwave and optical observations and interpretation of polarized microwave/optical emission from natural surfaces. The technique represents an extension of the measurement of vertically and horizontally polarized microwave/optical brightness temperatures by considering, in addition, the third and fourth Stokes parameters. Collectively, the four Stokes parameters provide a complete characterization of Gaussian-random electromagnetic field. - On the airborne side, there is PSR (Polarimetric Scanning Radiometer), a multi-channel conical-scanning instrument of NOAA/ETL and Georgia Tech, flown since 1997, providing fully polarimetric imagery in the microwave and optical regions. POLDER on ADEOS (launch Aug. 17, 1996) is an example of a passive spaceborne polarimetric imaging radiometer in the optical region.

- Introduction of FTS (Fourier Transform Spectrometer) technology in sensor development, usually in combination with an interferometer (Michelson, Sagnac, Fabry-Perot, etc.) for the measurement of the composition of the atmosphere (trace gases, air pollution monitoring,monitoring of accidental releases of toxic gases at industrial plants, etc.). See Tables 17 and 18 as well as chapter O.9. The first spaceborne FTS instrument [sometimes also referred to as FTIR (Fourier Transform Infrared) spectrometer, due to the fact that most observation occurs in the IR region] flown was IRIS on Nimbus-3 (launch April 14, 1969) and on Nimbus-4 (launch April 8, 1970), followed by SI-1 (also referred to as SI-GDR) and SI-2 aboard the USSR satellites Meteor 25 (also referred to as Meteor-Priroda-2, launch May

114) D. F. Barbe, "Charge-Coupled Devices," Topics in Applied Physics, Vol. 38, Springer-Verlag, Berlin, 1980

115) J. A. Cox, "Signal-to-noise ratio dependence on frame time, time delay and integration (TDI), and pulse shaping," Optical Engineering, May/June 1982, Vol. 21, No. 3, pp. 528-536

116) W. L. Wolfe, G. L. Zissis, "Time Delay and Integration in CCD Signal Processing," The Infrared Handbook, prepared by the Environmental Research Institute of Michigan for ONR (US Navy), 1978, pp. 12-66 and 12-28

117) Y. Kim, J. van Zyl, "On the Relationship between Polarimetric Parameters," Proceedings of the IEEE/IGARSS 2000 Conference, Honolulu, HI, July 24-28, 2000

15, 1976) and Meteor 28 (also referred to as Meteor-Priroda-3, launch June 29, 1976). [118] ATMOS of NASA/JPL was an FTS sensor flown on the Shuttle (STS-17) in 1985. Mark-IV of JPL collected first ground-based spectra in April 1985, in 1987 the sensor was flown on a balloon and in aircraft campaigns. The interested reader should also consult the following reference. [119]

The first FTS instrument to be flown in GEO is GIFTS (Geosynchronous Imaging Fourier Transform Spectrometer) on EO-3 (Earth Observing-3), a NASA technology demonstration mission with a planned launch in 2004. The GIFTS measurement concept combines a number of advanced technologies, including LFPA (Large-area format Focal Plane detector Array) and a compact, light-weight Fourier Transform Spectrometer (FTS). The GIFTS design offers high sounding resolutions, about 4 km (IR) vertical and 1-2 km (VIS) in the horizontal direction, for temperature and water vapor profiles. Dynamic observations of temperature, water vapor, and wind enable a better understanding of climate physics, hydrology and the water cycle, the transport of chemical species, and weather processes.

Sensor	Aircraft, Platform	Spectral Range	Observation Direction	Start of Operation Comments
HIS (U. of Wisconsin)	ER-2	3.6-4.63 μm 5.1-9.6 μm 9.3-16.4 μm	Nadir viewing	1986
FIRS-2 (SAO)	Balloon	14-25 μm 48-125 μm	Limb viewing	1987
MkIV (JPL)	Balloon, DC-8	1.8 - 16 μm	Limb and zenith	1987
MIPAS-FT (KfK)	Transall	7.4 - 8.7 μm 10.0 - 13.2 μm	Limb viewing	1991
MIPAS-B (KfK)	Balloon	7.25-8.45 μm 10.3-13 μm	Limb viewing	1992
SMIFTS (U. of Hawaii)	Helicopter	1.0-5.2 μm	Nadir	1993
AES (JPL)	DC-8, P-3	2.4-15.4 μm	Nadir viewing	1994
FTS (NPL)	BAe Jetstream	3-13 μm		1994
FTVHSI (Kestrel)	Cessna TU-206	0.44-1.15 μm	Nadir viewing	1995
MIPAS-B2 (FZK)	Balloon	5.25-5.48 μm 6.0-6.36 μm 7.4-8.8 μm 10.2-13.2 μm	Limb viewing	1995
MIPAS-STR (FZK)	Strato-2C	5.15 - 5.42 μm 6.08 - 6.31 μm 7.30 - 8.33 μm 10.0 - 13.0 μm	Limb viewing	1996
MIROR (DLR)	VFW 614	2-18 μm	Aircraft jet engine	1995 (trace gas emissions)
SAFIRE-A (CNR/IROE)	Balloon, M-55 Geophysika	62.5-125 μm 20-350 μm future	Limb viewing	1996, polarizing FTS
UAV-AERI (U. of Wisconsin)	Perseus-A	3-25 μm	Nadir viewing	1996

Table 17: Survey of airborne Fourier Transform Spectrometer (FTS) instruments

Sensor	S/C or Platform	Spectral Range	Observation Direction	Start of Operation Comments
IRIS, (NASA/ GSFC)	Nimbus-3 Nimbus-4	5 - 20 μm 6.25 - 25 μm	Nadir viewing	launch, Apr. 14, 1969 launch, Apr. 8, 1970
SI-GDR (IKF)	Meteor-P-2 Meteor-P-3 Meteor-P-4	6.25 - 25 μm	Nadir viewing	launch, May 15, 1976 launch, Jun. 29, 1977 launch, Jan. 25, 1979

118) V. Kempe, D. Oertel, R. Schuster, H. Becker-Ross, H. Jahn, "Absolute IR-spectra from the measurement of Fourier spectrometers aboard Meteor 25 and 28," Acta Astronautica, Vol. 7, 1980, pp. 1403-1416

119) M. J. Persky, "A review of spaceborne infrared Fourier transform spectrometers for remote sensing," Review of Scientific Instruments, Vol. 66, No. 10, October 1995, pp. 4763-4797

Sensor	S/C or Platform	Spectral Range	Observation Direction	Start of Operation Comments
ATMOS (NASA/ JPL)	Spacelab-2,-3, ATLAS-1,-2,-3	2.2-16.0 µm	Limb viewing	1985, (2) 92, 93, 94
CIRRIS (DoD/ USAF)	STS-39 Shuttle Discovery	2.5-25 µm	Limb viewing	Apr. 28 - May 6, 1991
IMG (NASDA, JA-ROS)	ADEOS	3.3-16.7 µm	Nadir viewing	launch, Aug. 17, 1996
MIRIAM (Free U. of Berlin)	MIR/Priroda	2.5-20 µm	Limb viewing	1996
SPIRIT-III (BMDO, USU)	MSX	2.5-28 µm	Limb viewing	launch, Apr. 24, 1996 FTS+scanning radiometer
FTHSI (AFRL, Kestrel Corp.)	Mightysat II.1	0.475-1.05 µm	Nadir viewing	launch July 19, 2000
HESSI (NASA)	HESSI		Sun viewing	2001
MIPAS (ESA,FZK)	Envisat	4.15-14.6 µm	Limb viewing rear-wards & sideways	2001
ACE-FTS (CSA) ACE-VNIR (CSA)	SCISAT/ACE	2-14 µm 0.5-1 µm	Limb viewing	2002
TES (JPL)	Aura (EOS/ Chem-1)	2.3-15.4 µm	Limb or nadir viewing	2003
GIFTS (NASA/LaRC)	EO-3	VIS, IR (4-16 µm)	Nadir viewing (directional)	2004
IASI (CNES/ASI)	MetOp-1	3.62-15.5 µm	Nadir viewing	2005
SOFIS (JEA)	GCOM-A1	3-13 µm	Limb viewing	2006

Table 18: Survey of spaceborne Fourier Transform Spectrometer (FTS) instruments

• TDLAS (Tunable Diode Laser Absorption Spectrometer). TDLAS is an alternate technique of measuring atmospheric trace gases. The in-situ measurement technique makes use of the existence of rotational-vibrational molecule transitions in the mid IR spectrum and monitors a single absorption line. The airborne instrument FAST (Frequency-modulated Absorption Spectroscopy by TDLAS) of MPICh (Max-Planck-Institut für Chemie) of Mainz, Germany, flown since 1992, is an example of the TDLAS technique.

• Wedge Imaging Spectrometer (WIS) technology. WIS is a spectral separation technique in which the spectral separation filters are mated to the detector array to achieve two-dimensional sampling of the combined spatial/spectral information passed by the filter. The technique obviates the need for a complex aft-optics assembly; it was developed by Hughes SBRC and sponsored by ARPA (introduced in 1992, see chapter P.207). The technology (hyperspectral imaging in the VNIR and SWIR regions) was initially flown on airborne sensors (WIS). It is also employed for the spaceborne sensor AC (Atmospheric Corrector) of the EO-1 mission (launch Nov. 21, 2000).

• On-board sensor calibration. The goal of any calibration procedure is to determine a functional relationship between the target source flux and the sensor output signal. Instrument calibration provides a means for consistent long-term data set interpretation by reducing the assumptions to be made. - The very early spaceborne instruments (cameras) didn't provide a calibration capability. But the need for on-board calibration became evident from the very start of regular observations. Today's sensor calibration involves usually comparing the detector of a particular spectral range against a known reference. For the VNIR region, a calibration source may be internal lamps or the sun (external source). In the infrared region, a suitable calibration reference may be an internal black body or an external source like deep (cold) space that may be viewed periodically. A calibration 'period' may be a scan sequence, each orbital eclipse period, the pass over a dedicated ground calibration site, or it may be a month or more, depending on need or opportunity. Some sensors employ several calibration techniques (for instance: GOME on ERS-2 retrieves ozone distributions by exploiting the traditional backscatter calibration approach, as well as by differential optical absorption spectroscopy). The ETM+ sensor on Landsat-7 uses three independent on-board calibration systems [the two new systems are: FAC (Full Aperture Calibrator), and PAC (Partial Aperture Calibrator)], representing a significant step forward in absolute radiometric calibration accuracy.

- Another (rather costly) method of data calibration occasionally employed is that of airborne underflights of spaceborne missions. Many spaceborne sensors (in particular since the 1980s) have their airborne predecessors, which makes underflight calibrations a natural extension for independent parameter checks. Very important observations of a mission (proving a concept, etc.) are usually verified with parallel campaigns to compare spaceborne, airborne and ground-based observations. An example of such a scenario are the SRL-1/2 (Space Radar Laboratory) missions in 1994 (see SRL Campaigns). ESA's ERS-1 mission is another example of extensive underflight campaigns. A very special instrument among the airborne underflight sensors is NASIC (P.148), operated by NASA/GSFC since 1988 (built in 1980), capable of providing independent calibration data for several long-term satellite sensors, such as: AVHRR (on several NOAA satellites), TM (Landsat), VAS (GOES satellites), and CZCS (Nimbus-7).

- Adaptive optics. [120] [121] [122] A technology referring to the physical modification of components of an optical system for the purpose of compensating for the distortion of electromagnetic radiation (refractive index variations) as it passes through the turbulent atmosphere and the optical system. Although the problem of aberration was long recognized in the field of optics (in particular in astronomy), two initial solutions were proposed by H. Babcock (1953, USA) and by V. Linnik (1957, USSR). They independently published papers with the conjecture that the aberrations caused by atmospheric turbulence could be corrected in real-time by (what became later known as) an "adaptive optical system." Essential elements of an adaptive optical system are a high-speed wavefront sensor (sensing the turbulence-induced aberrations), a thin flexible mirror whose surface can be electronically controlled to correct for aberrations, and a computer controller that converts the wavefront measurements into deformable mirror commands. The mirror is physically deformed in real-time by a large number of piezoelectric actuators. The device that measures the distortions in the incoming wavefront of light is called a "wavefront sensor" (analysis of wavefront, estimation of the shape of the original wavefront, derivation of the correction signals for the deformable mirror). High update rates are an essential ingredient for distortion removal.
Examples of early installations with adaptive optics technology in the field of astronomy: In 1997, a 3.5 m telescope by the name of ALFA (Adaptive optics with a Laser For Astronomy system), a cooperative project of MPIA Heidelberg and MPE Garching, was installed at the Calar Alto Observatory in Almería (Southern Spain). In 1997/8, some demonstrations of the adaptive optics technique were conducted with MMT (Multiple Mirror Telescope) at the University of Arizona in the field of astronomy. [123] A 10 m telescope of the Keck Observatory in Arizona started operations in fall of 1999, using an adaptive optics system.

The technology of adaptive optics is also being considered for Earth observation instruments (looking from space toward Earth). The RedEye (Regional Environmental Dynamics Active-aperture Infrared Imager) instrument proposed for NASA's EO-3 mission (Earth Observing-3, launch in 2003) in geostationary orbit demonstrates a deformable mirror design to correct for residual wavefront errors in a large segmented aperture optical system. However, as of spring 2000 the RedEye proposal was not accepted by NASA management.

Not all adaptive optics implementations are used to correct for atmospheric turbulence. The technique is also being used to compensate in general for optical aberrations in the de-

120) M. C. Roggemann, V. P. Lukin, V. E. Zuev, "Adaptive optics: introduction to the feature issue," Applied Optics, July 20, 1998, Vol. 37, No. 21

121) P. M. Hinz, et al., "Imaging circumstellar environments with a nulling interferometer," Nature Vol. 395, Sep. 17, 1998, pp. 251-253

122) G. Stix, "Shading the Twinkle," Scientific American, Dec. 1998, p. 20

123) G. Schilling, "Technique for Unblurring The Stars Comes of Age," Science, Vol. 286, Nov. 19, 1999, pp. 1504-1506

sign of very small and light-weight optical systems to improve the optical characteristics (including focal length, aberrations, removal of misalignments, improvement of wider FOV, etc.) of a system subjected to space environment variations. Not all adaptive optics implementations use piezoelectric actuators in combination with computers for mirror control. Newer adaptive optics instrument designs employ so-called MOEM (Micro Optoelectronic-Mechanical) devices (tiny chips), able to physically react to a non-normal situation, for micro-mirror control (i.e., actuation) at high update rates (about 1 kHz). A star tracker with these capabilities is being developed at AFRL and planned to be flown on the TechSat-21 mission of DoD (the star tracker has a pointing accuracy of better than 0.20 arcsec). [124]

• MEMS (Micro-Electro-Mechanical System) - also known as "microsystems" in Europe or "micromachines"in Japan. MEMS technology first emerged in the late 1960s and early 1970s. MEMS devices, tiny machines virtually invisible to the naked eye (they range in size from a few μm to mm), fashioned largely from silicon with techniques adapted from the microchip industry, are able to **"touch" the physical world and act upon it**. There is a trend toward complete microsystems - merging sensing, actuating, computing and communications. Microtechnology represents essentially a synergetic combination between thin film technology, semiconductor technology, silicon micromechanics, ultra precision engineering, LIGA technique, laser lithography, and several technologies for packaging and assembly. The new microsystems offer enhanced levels of perception, control, and performance (nanoscale design, low power quantum electronics, high bandwidth photonics, etc.). The evolving list of MEMS features includes: provision of inputs and outputs for information systems, permitting multiple and mixed technology integration (CMOS/MEMS integration, integrated trench technology, etc.); in particular, there is the ability **to sense and measure changes of physical parameters on the micro-scale (atomic) level,** orders of magnitude better, than traditional techniques. At the end of the 1990s, the technology of miniaturization experiences a tremendous growth. It is destined to become a natural pick for a divers number of applications in all fields, in particular for the design of new instruments in the space industry. The potential of MEMS as an enabling technology has a definite impact on new designs, similar to the impact of integrated circuits for the electronics industry during the 1960s and 1970s. Some examples of space applications are: [125] [126] [127] [128]

- MEMS technology is being applied to a variety of optical applications, ranging from adaptive mirrors to integrated optical benches, grating spectrometers, optical scanners, fluid pumps, etc.
- MEMS technology is making possible the development of many different types of thermal sensors for spaceborne instruments and operational hardware. At the start of the 21st century, these are becoming available with high performances and in the case of arrays (linear or matrix) make possible a new class of imaging and spectrographic instruments.
- Integrated microaccelerometers and gyros
- Uncooled infrared detectors
- MEMS technology is used to form miniature antennas for millimeter-wave applications
- MEMS technology is applied for distributed sensors and actuators (closed-loop control is a viable option)
- MEMS is being used for electromechanical signal processing

124) N. Clark, P. Furth, S. Horan, "Intelligent Star Tracker," Proceedings of the 14th Annual AIAA/USU Conference on Small Satellites, Logan, UT, Aug. 21-24, 2000, SSC00-III-1

125) I. Amato, "Fomenting a Revolution in Miniature," Science, Vol. 282, Oct. 16, 1998, pp. 402-405

126) Special Issue on Integrated Sensors, Microactuators, and Microsystems (MEMS), Proceeding of the IEEE, Vol. 86, No. 8, August, 1998

127) M. C. Wu, "Micromachining for Optical and Optoelectronic Systems," Proceedings of the IEEE, Vol. 85, No 11, Nov. 1997, pp. 1833-1856

128) M. Lacher, W. Ehrfeld, "Microproducts for Space Applications," Proceedings of the 2nd International Conference on MicroNanotechnology for Space Applications, Apr. 11-15, 1999, Pasadena, CA, Vol. 1, pp. 4-25

- In the early 1980s, the Institute of Microstructure Technology of KfK (Kernforschungszentrum Karlsruhe (Nuclear Research Center, Karlsruhe, Germany) initiated/developed a micromachining technology by the name of LIGA [Lithographie, Galvanoformung und Abformung (lithography, electroplating and moulding)]. LIGA is widely used throughout industry and the research community for advanced MEMS applications. LIGA is a three-stage process which can be used for the manufacture of high aspect ratio, 3-D microstructures in a wide variety of materials (e.g. metals, polymers, ceramics and glasses). By using the penetrating power of x-rays from a synchrotron, LIGA allows the fabrication of structures which have vertical dimensions from hundreds of microns to millimeters and horizontal dimensions which can be as small as microns. SUMMiT (Sandia Ultra-planar Multi-level MEMS Technology) is another fabrication process developed by SNL (Sandia National Laboratories).
- MEMS technology is applied to micropropulsion systems. [129)]
- In 1993 the term **nanosatellite** was first coined and defined by Siegfried W. Janson, Henry Helvajian, and Ernest Y. Robinson of the Center for Microtechnology at The Aerospace Corporation, El Segundo, CA. [130)] In this paper, the proposed nanosatellite architecture employs MEMS and ASIMs (Application Specific Integrated Microinstruments). The integration of MEMS with microelectronics for data processing, memory, signal conditioning, power conditioning and communications results in standalone ASIMs (chip-to-chip wireless integration). [131)] ASIMs with silicon or other semiconductor substrate can be applied to S/C systems, i.e. tiny instruments, for such functions as guidance, navigation and control, attitude sensing, attitude control, thermal control primary propulsion, power and communication. In this concept, the nanosatellite denotes a S/C built almost entirely of ASIMs. The silicon satellites are spacecraft that utilize single-crystal silicon wafers for electronic substrates, mechanical structure, thermal control system, and radiation shield.

• Optical (laser) free-space communications. Laser links have the potential to offer much higher transmission rates (up to 10 Gbit/s) for LEO-to-ground communications than conventional TT&C radio frequency (RF-based) links which are limited to rates of about 250 Mbit/s. [132)] The optical information bandwidth of a signal is only a fraction of the RF carrier frequency, typically 0.1 that of RF systems. Extremely high 'antenna' gains with relatively modest apertures are obtained in optical communication systems, resulting in very low carrier power (for instance 60 mW in SILEX data transmissions of 50 Mbit/s over a distance of 42,000 km). The optical spectrum for TT&C applications may range in bandwidth from about 532 nm (563 THz) to about 1550 nm (194 THz), providing for many GHz of information transfer on optical carriers. Another advantage of laser links is that by their very nature, there can be no interference between optical and radio transmissions. The technical challenge of the technology demonstration involves in particular alignment and stabilizing issues, it requires pointing errors of < 10 μrad (or < 0.0005°). This pointing accuracy is several orders of magnitude lower (better) than open-loop pointing of a typical platform. Also, platform jitter caused by the operation of other payload instruments, must be mitigated. Other constraints to be considered deal with eye-safety issues and with weather and atmospheric effects. The optical signals can be greatly attenuated by the weather and the atmosphere. - Numerous applications of optical communication links are envisioned, such as:

129) J. Mueller, I. Chakraborty, et al., "MEMS-Micropropulsion Activities at JPL," Proceedings of the 2nd International Conference on MicroNanotechnology for Space Applications, Apr. 11-15, 1999, Pasadena, CA, Vol. 1, pp. 175-200

130) Siegfried W. Janson, Henry Helvajian,Ernest. Y. Robinson, "The Concept of "Nanosatellite" for Revolutionary Low Cost Space Systems," Proceedings of the 44th Congress of the International Astronautics Federation, Oct. 16-22, 1993, Graz, Austria, IAF-93-U.5.573

131) S. W. Janson, "Silicon Satellites for 21 st Century Missions," Proceedings of the 2nd International Conference on MicroNanotechnology for Space Applications, Apr. 11-15, 1999, Pasadena, CA, Vol. 1, pp. 535-544

132) Note: In Feb. 1996, researchers at Fujitsu Laboratories Ltd (Japan) achieved for the first time data rates of 1.1 Tbit/s using WDM (Wavelength Division Multiplexing) technology in an optical transmission experiment over a distance of 150 km (with 50 km amplifier spacing). In the experiment, 55 optical channels in the 1.55 μm range (0.6 nm spacing) were used, each at 20 Gbit/s, for a total data rate of 1.1 Tbit/s.

Space-to-ground, interorbit (LEO-GEO), intersatellite (LEO-LEO or GEO-GEO), or deep space to ground. Some demonstrations are:

- **LCE** (Laser Communication Equipment) [133) 134)], a CRL (Japan) payload flown on ETS-6 (Experimental Test Satellite-6) of NASDA and NTT. Launch of ETS-6 on Aug. 28, 1994. Note: ETS-6 attained a highly elliptical orbit (instead of a planned GEO) due to a failure in the launch vehicle propulsion system. In spite of this misfortune, some LCE experiments were conducted with ground stations in Japan and at JPL.
- **SILEX** (Semiconductor Intersatellite Link Experiment), an ESA experiment built by MMS. SILEX consists of two optical terminals, a LEO and a GEO terminal, namely **PASTEL** (PAssager SPOT de Télécommunication Laser), a LEO terminal on-board SPOT-4 (launch March 22, 1998) on the anti-Earth side of the SPOT platform, and **OPALE** (Optical Payload for Intersatellite Link Experiment) a GEO terminal mounted on ESA's geostationary satellite ARTEMIS. ARTEMIS was launched on July 12, 2001. The ARTEMIS downlink to the ground segment at CNES Toulouse uses a conventional radio transmission (Ka-band). See also SILEX under D.37.2.
- **LUCE** (Laser Utilizing Communications Equipment), a Japanese payload (built by NEC) flown on OICETS (Optical Interorbit Communications and Engineering Test Satellite) of NASDA (planned launch in early 2002). OICETS in LEO conducts interorbit communications experiments with ARTEMIS of ESA in GEO.
- **Lasercom** (Laser Communication Experiment), an experiment on the AFRL mission TSX-5 and its payload STRV-2 (launch on June 6, 2000), see M.29.

With constantly increasing communication rates, optical systems are also being considered for small satellites. - Apart from governmental/institutional research in the area, a significant driver for the development of advanced TT&C systems is the prospect of commercial LEO communication systems, such as the Teledesic/Celestri constellations which include laser communication crosslinks as part of their baseline design. [135)]

- QWIP (Quantum-Well Infrared Photodetector). QWIP represents a new infrared photoconductive detection technology (using 2-D focal plane arrays), holding great promise for a variety of infrared imaging applications in the spectral region of 6-25 μm which includes TIR (Thermal Infrared). The technology employs large-bandgap compound semiconductor materials such as GaAs (Gallium Arsenide) and AlGaAs (Aluminum Gallium Arsenide) with different aluminum compositions. The potential advantages of QWIPs (GaAs and AlGaAs) include the use of standard manufacturing methods based on GaAS growth and processing techniques. However, the QWIP technology requires low-temperature cooling for good spectral performance. Note: Infrared detection may also be implemented on QWIP structures operated in photovoltaic mode (as opposed to photoconductive mode). - QWIP structures are intrinsically very fast photodetectors since the re-capture time of optically excited carriers is very short, in the order of a few picoseconds (ps). Besides thermal imaging applications, QWIP structures are also interesting for heterodyne applications, where the high electrical bandwidth results in an improved spectral range of a heterodyne spectrometer. - The first experimental demonstration of 2-D QWIP imaging arrays took place in 1985. First imaging instruments of the QWIP technology are: [136) 137)]

- The first demonstration model of a QWIP FPA camera with a 128 x 128 pixel array was developed by B. F. Levine and his colleagues at AT&T Bell Laboratories, Murray Hill, NJ, in 1991.

133) M. Shikatani, et al., "Ground system development for the ETS-6/LCE laser communication experiment," Proceedings of SPIE, 20-21 Jan. 1993, Los Angeles, Vol. 1866, pp. 21-29

134) Y. Arimoto, et al., "Preliminary Results on Laser Communication Experiment using ETS-6," Proceedings of SPIE, Vol. 2381, 1995

135) M. Enoch, S. Herrin, et al., "Optical Tracking Telemetry and Commanding (TT&C) for Small Satellites," Proceedings of the 13th AIAA/USU Conference on Small Satellites, Aug. 23-26, 1999, Logan. Utah, SSC99-IIb-4

136) B. F. Levine, C. G. Bethea, K. G. Glogovsky, et al., "Long-wavelength 128 x 128 GaAs quantum well infrared photodetector arrays," Semiconductor Science Technology, Vol. 6, pp. C114-C119, 1991, IOP Publishing Ltd., UK

137) B. F. Levine, "Quantum-well infrared photodetectors," Journal of Applied Physics, Vol. 74, No. 8, Oct. 15, 1993, pp. R1 - R81, part of the series: Applied Physics Reviews; it includes an extensive review of the QWIP literature.

- In 1996, JPL/CSMT of NASA and Amber (of Raytheon) demonstrated the QWIP technology with the development of a hand-held infrared camera (256 x 256 pixel FPA of QWIPs to detect radiation at 8.5 μm). [138]
- In 1997, FhG/IAF (Fraunhofer Gesellschaft / Institut für Angewandte Festkörperphysik) of Freiburg in partnership with AIM (AEG Infrarot Module) of Heilbronn, Germany, demonstrated also a QWIP-based camera system (256 x 256 pixel FPA with a peak wavelength of 8.1 μm, the FPA is cooled to 65 K. [139] [140] [141] - Note: Objects at room temperature glow brightest in the wavelength range of 8-10 μm). The camera exhibits extremely low noise levels corresponding to NEDT (Noise Equivalent Differential Temperature) of 9.5 mK over the entire array. The research was supported by the German Ministry of Defence. The same camera, with the name of AIM-640Q, is also commercially available.
- A spaceborne QWIP camera, funded by BMDO and built by JPL, is flown on the STRV-1d (launch Nov. 16, 2000) satellite of DERA (UK), to demonstrate the radiation detection performance within the 11-17 μm spectral region (M.25.2.2).

• DOAS (Differential Optical Absorption Spectroscopy). [142] [143] [144] [145] [146] [147] A measurement technique of atmospheric spectra, first developed by Dieter Perner at KfA Jülich (now Forschungszentrum Jülich, Germany) in the early 1970s for the analysis of spectral windows. The distinguishing feature of DOAS (as compared to other spectrometers measuring atmospheric constituents) is that it aims to measure differential rather than absolute spectra. This means that broadband absorption or scattering effects arising from aerosols or gases can be separated (in ground analysis) from differential features which are characteristic of the trace gases absorbing in a given window. The distinguishing feature of a DOAS instrument is that it records simultaneously the entire spectrum of a selected spectral window. The DOAS technique was first applied to long-path tropospheric investigations by D. Perner, U. Platt and co-workers. This resulted in an experimental ground-based DOAS-UV and VIS spectrograph installation at KfA Jülich in 1978, Germany (measurement of atmospheric CH_2O, O_3, and NO_2). Further ground-based DOAS instruments were built. Two airborne instruments, DOAS-VIS, developed by the Institute of Environmental Physics at the University of Heidelberg, and DOAS-UV of MPI for Chemistry (MPICh), Mainz, are flown since Jan. 1991 on a C160 Transall aircraft. A DOAS-type experiment by the name of "Mapping Atmospheric Pollution" was first proposed to ESA in March 1985 by J. P. Burrows, D. Perner, and P. J. Crutzen to fly on the EURECA free-flyer platform (EURECA was launched on STS-46 on July 31, 1992 and retrieved with STS-57 on July 1, 1993). - First spaceborne DOAS instruments are GOME (Global Ozone Monitoring Experiment)

138) S. Gunapala, M. Sundaram, S. Bandara, ""Quantum Wells stare at long-wave IR scenes," Laser Focus World, June 1996 p. 233

139) M. Walther, F. Fuchs, et al., "III-V Semiconductor Quantum Well and Superlattice Detectors," Infrared Technology and Applications XXIV, B. F. Andresen, M. Strojnik, Editors, Proceedings of SPIE, Vol. 3436, 1998, pp. 348-358

140) M. Walther, F. Fuchs, et al., "Electrical and optical properties of 8-12μm GaAs/AlGaAs quantum well infrared photodiodes in 256x256 focal plane arrays," in 'Intersubband transitions in quantum wells: physics and devices,' ed. by S. Li and Y. K. Su, pp. 207 - 212, Kluver Academic Publishers, Boston, 1998

141) http://www.iaf.fhg.de/ir/qwip/camera.html

142) J. P. Burrows, 1994 "GOME and SCIAMACHY," in Ozone Layer Observation by Satellite Sensors - *Proceedings of the International Workshop on Global Environment and Earth Observing Satellite Sensors Tokyo, Dec. 8-9, 1993* pp 67-74

143) J. P. Burrows, M. Buchwitz, M. Eisinger, V. Rozanov, M. Weber, A. Richter, A. Ladstätter-Weißenmayer 1998, "The Global Ozone Monitoring Experiment (GOME), Mission, Instrument Concept, and first scientific results", *Proceedings of the third ERS Symposium Florence, Italy, March 18-23, 1997. Space at the service of our Environment, 1997.* ISBN 92-9092-656-2, ESA SP 414, pp. 585-590

144) U. Platt, D. Perner, H. W. Pätz, "Simultaneous measurements of atmospheric CH_2O, O_3, and NO_2 by differential optical absorption," Journal of Geophysical Research, Vol. 84, 1979, pp. 6329-6335

145) J. F. Noxon, "Nitrogen dioxide in the stratosphere and Troposphere. Measurement by ground-based absorption spectroscopy," Science, 189, 1975, p. 547

146) H. Edner, P. Ragnarson, S. Spännere, S. Svanberg, "Differential optical absorption spectroscopy (DOAS) system for urban atmospheric pollution monitoring," Applied Optics, Jan. 20, 1993, Vol. 32, No. 3, pp. 327-333

147) J. Frerick, H. Bovensmann, S. Noel, J. P. Burrows, M. R. Dobber, "SCIAMACHY on-ground/in-flight calibration, performance verification, and monitoring concepts," Proceedings of SPIE, Vol. 3117, pp. 176-187, Earth Observing Systems II, William L. Barnes; Ed. Publication Date: 09/1997

flown on ERS-2 (launch April 21, 1995) and SCIAMACHY (Scanning Imaging Absorption Spectrometer for Atmospheric Cartography) to be flown on Envisat of ESA (planned launch in Oct. 2001). Since the launch of ERS-2, there are also several ground-based DOAS stations in Europe providing ground truth measurements for GOME and SCIAMACHY. [148)]

- ALOS of NASDA (launch 2003, D.3) is the first satellite in Earth observation capable to obtain optical and SAR imagery of the same area (70 km swath) at the same time.

- Supermode acquisition of imagery. The term "supermode" refers to an acquisition process, introduced by CNES for the HRG (High Resolution Geometric) instrument of SPOT-5 (planned launch in 2002), through which an image, sampled at 2.5 m, may be obtained from two 5 m resolution panchromatic images acquired simultaneously, keeping within the same borders as the two 5 m resolution images. See chapter D.37.3.

The new sampling concept of imagery, technically referred to as **quincunx sampling pattern** (an arrangement of five things with one at each corner and one in the middle of an area), employs a linear dual-array CCD detector in the focal plane, offset by one half pixel in one direction and 3.5 pixels in the other to avoid overlapping. [149)] This configuration is sufficient to improve the sampling grid without doubling each array's acquisition rate. The new sampling concept is based on Shannon's theory of information which states that "the sampling frequency must be equal to or greater than twice the maximum signal frequency" to obtain clean images. A specific image processing software, developed by CNES, is used to reconstruct the final image after three processing steps: interleaving, interpolation and restoration.

- Lightning detection instruments (event-based monitoring). In the early 1980s, the NOSL (Night-time and daytime Optical Survey of Lightning) instrument of NASA/MSFC was flown in the Space Shuttle program on flights STS-2 (Nov. 12-14, 1981), STS-4 (Jun. 27-July 4, 1982) and STS-6 (Apr. 4-9, 1983). This was followed by MLE (Mesoscale Lightning Experiment) on Shuttle flights in the late 1980s (STS-26, Sept. 29- Oct. 3, 1988; STS-30, May 4-8, 1989; STS-34, Oct. 18-23, 1989; STS-32, Jan. 9-20, 1990). The OLS (Operational Linescan System) of the DMSP weather satellite series of DoD is also able to detect lightning activity (data sets exist starting from 1973). [150)] An airborne sensor with a capability for lightning detection is LIP (NASA/MSFC, an upgraded instrument as of 1990). A further spaceborne sensor with a capability for lightning detection/distribution is Blackbeard (an RF experiment), flown on ALEXIS of LANL and SNL (launch of ALEXIS April 25, 1993). The OTD (Optical Transient Detector) of NASA/MSFC is being flown on OrbView-1/Microlab-1 (launch April 3, 1995, B.8.1). This was followed by simultaneous observations of optical and RF emissions of lightning activity with the instruments: OLS (Optical Lightning Subsystem) and the RF-System, both flown on the FORTE spacecraft (LANL/SNL, launch Aug. 29, 1997, A.16). The LIS (Lightning Imaging Sensor, NASA/MSFC) is flown on TRMM (launch Nov. 27, 1997, A.31). A further instrument, LMS (Lightning Mapper Sensor) of NASA/MSFC is planned to be flown on the GOES spacecraft series starting with an experimental flight on GOES-O in 2004 (F.4.6). [151)] The LMS on GOES represents the first lightning instrument to monitor lightning from GEO, permitting the study of the electrosphere over dimensions ranging from the Earth's radius down to individual thunderstorms. The student-built Emerald microsatellite of Stanford and Santa Clara University (launch in late 2001) as part of UNP (University Nanosatellite Program) flies LDE (Lightning Detection Experiment), a VLF instrument. The objective is to detect electromagnetic radiation and lightning. The ROCSat-2 mission of NSPO (Taiwan, planned launch in 2003) flies the

148) J. P. Burrows, K. V. Chance, A. P. H. Goede, R. Guzzi, B. J. Kerridge, C. Muller, D. Perner, U. Platt, J.-P. Pommereau, W. Schneider, R. J. Spurr, H. van der Woerd, "Global Ozone Monitoring Experiment: Interim Science Report", *ESA SP-1151 Edited by T. D. Guyenne and C. J. Readings ISBN 92-9092-041-6 European Space Agency 1993*

149) "The Secrets of SPOT-5 Supermode," SPOT Magazine, No 31, 2000, pp. 21-23

150) http://thunder.msfc.nasa.gov/ols/

151) http://thunder.msfc.nasa.gov/

instrument ISUAL (Imager of Sprite Upper Atmospheric Lightning) with the objective to observe the natural upward lightning discharge phenomenon toward the ionosphere on top of the troposphere.

1.1.8 Cooling Techniques and HTS (High-Temperature Superconductivity)

- Cryocoolers (active cryogenic devices). Depending on the temperatures required, two different cooling technologies are commonly used to actively cool focal planes. These are:

1) TEC (Thermoelectric Coolers). The TEC scheme is capable of cooling focal planes to 180 K starting from an ambient temperature of 300 K. TEC instruments are small, lightweight, and, being solid-state, have no vibration. TECs have the main disadvantage of having an efficiency that drops off rapidly with decreasing temperature and tending towards zero around 180 K. - Examples of thermoelectric cooler implementations are: RBV (Return Beam Vidicon Camera) flown on early Landsat spacecraft; GERB (Geostationary Earth Radiation Budget) flown on MSG (Meteosat Second Generation) satellites of EUMETSAT (launch of MSG-1 in 2002);

2) Mechanical coolers, such as Stirling cycle engines and pulse tube coolers. They can produce temperatures below 20 K and are more efficient than TECs. However, mechanical coolers are much larger than TECs and can produce vibrations that must be canceled or isolated.

A Stirling cycle engine can operate at very low temperatures [about 5 K to 80 K range, they are usually closed cycle cooler (CCC) based cryostats]. Mechanical cryocoolers compress and expand a gas, usually helium, through heat exchangers, which lowers the temperature. A typical application is the provision of a cold environment for infrared sensor detector systems. - Early cryocoolers in space were rather heavy devices (due to the dewar construction) with relative limited cooling capacities (about 0.5 W) and life times of less than one year. The basic types of cryocoolers in use are: a) the Stirling cycle cooler, b) the Joule-Thomson cryocooler, and c) hybrid designs (of Stirling and Joule-Thomson technologies). The Stirling designs of the 1990s seem to be all of "Oxford cryocooler" heritage, flown on several US and European programs (O.5).

- The Oxford cryocooler [152)], designed and built by Oxford Instruments Plc., UK, was first flown on the NASA UARS satellite (launch Sept. 12, 1991) to provide cooling for the ISAMS (Improved Stratospheric and Mesospheric Sounder) instrument, a limb-sounding radiometer of Oxford University. Its design cooling power was 0.8 W at 80 K. Its mass about 4.3kg. Two coolers were installed for vibrational balance. Another Oxford cryocooler was part of the NRL HTSSE-II instrument flown on ARGOS (launch Feb. 23, 1999).
- SHOOT (Superfluid Helium On-Orbit Transfer), a NASA/GSFC cooling demonstration experiment, was part of the space shuttle STS-57 mission in June, 1993. [153)] The experiment objectives included: transfer of superfluid between two dewars in a low gravity environment at different flow rates; operation of two different liquid acquisition devices within the dewars; liquid/vapor phase separation for normal liquid helium as well as superfluid at varying venting rates; accurate mass gauging and flow metering; and autonomous control of the transfer process by an expert system aboard a computer on the Shuttle.
- The CSE (Cryo System Experiment) cooler of Hughes, first flown on STS-63 (Feb. 3-11, 1995), provides 1.2 W of cooling at 65 K, it is the first US-built long life (1.5 years of continuous operation) Stirling cooler to operate in space.

- Pulse tube cooler. A mechanical refrigerator capable of achieving cryogenic temperatures of <20 K in the 0.5-3.5 W cooling capacity range. The principal benefits of a pulse tube

152) Oxford cryocooler information provided by Manny Tward of TRW, Redondo Beach, CA

153) J. G. Tuttle, M. J. DiPirro, P. J. Shirron, "Liquid/gas phase separators for the Superfluid Helium On-Orbit Transfer (SHOOT) project," Advances in Cryogenic Engineering, 39,1994, p. 121

cooler are greater reliability and lower cost compared to the Stirling cooler and an order of magnitude lower mass, lower cost, and longer life than the current state-of-the-art coolers. A pulse tube cooler is also more efficient than a TEC-type instrument, however, as a mechanical cooler, it is much larger than a TEC. In addition, mechanical coolers can produce vibrations that must be canceled or isolated.
The NASA Lewis satellite (launch Aug. 23, 1997 - however, the S/C was lost and reentered the atmosphere Sept. 28, 1997) was the first S/C in civil Earth observation featuring pulse tube cryocooler technology (built by TRW) in its HSI (Hyper-Spectral Imager) for low-noise SWIR cooling. Already in 1994, TRW delivered its first pulse tube instrument, built under contract to the Air Force's Phillips Laboratory. Other instruments planned with pulse tube cooling technology are AIRS (Atmospheric Infrared Sounder) of NASA's Aqua satellite (launch in 2001) and TES (Tropospheric Emission Spectrometer) on the Aura mission (launch in 2003).

- Optical cooling by fluorescence is an alternate and new concept in refrigeration that represents possible applications in small cryocooler technology, in particular for infrared devices and spectrometers (O.5.4). The method uses anti-Stokes fluorescence to remove heat from a glass or crystal that is pumped with laser light. A first demonstration instrument by the name of LASSOR (Los Alamos Solid-State Optical Refrigerator) was developed at LANL (Los Alamos National Laboratory). [154) 155)] In Oct. 2000, NASA/LaRC awarded to BATC (Ball Aerospace & Technologies Corporation) a contract to develop an optical cryocooler (infrared focal plane demonstrator). [156)] Optical cryocooling is a feasible method for cooling focal planes and has a distinct niche in extended solid-state cooling to those temperatures that cannot be achieved efficiently (or at all) by TEC methods. Optical cryocoolers have a clear advantage over mechanical coolers in vibration, ruggedness, EMI (Electromagnetic Interference) and magnetic field, and cooler mass. Nevertheless, they are at a disadvantage to mechanical coolers based on efficiency. Optical cryocoolers have an advantage over TECs in minimum operating temperature, magnetic field, and ruggedness. They are at a disadvantage to TECs in cooler mass, however.

Active cryogenic refrigerators offer an alternative to passive radiators. Active coolers offer greater capacity and provide additional freedom in packaging and locating the sensor on the spacecraft, since there are no preferred orientations or constraints on view factors. These benefits are provided at the expense of fairly high power consumption, added mass, and diminished reliability. Indeed, there is a crossover point in size/mass efficiency: Passive coolers tend to be the better choice for modest heat loads (<1 W) at temperatures above approximately 80 K; active coolers are more attractive for higher heat loads (>2 W) at temperatures of 65 K or colder.

- On spacecraft, focal planes are sometimes cooled with passive radiators, the performance of which is highly dependent on mission parameters and spacecraft configuration. Passive radiative coolers are well developed and provide exceptional reliability for detector arrays. The size of these coolers is, however, governed by the first-order physics of Planck's radiation law. The required minimum area for the radiative surface is governed by the relationship $W = AesT^4$, where W is the radiated power, A is the area of the radiator, e is the effective emissivity of the radiator's surface, s is the Stefan-Boltzmann constant, and T is absolute temperature. This relationship can be expressed as $A = W/(esT^4)$. Thus, the area of the radiator needed to achieve a given cryogenic temperature depends directly upon the amount of cooling power required and upon the inverse fourth power of temperature. This is a highly nonlinear relationship; for example, it takes four times as much radiator area to reach 60 K as to reach 85 K. Consequently, application of this comparatively straightfor-

154) T. R. Gosnell, "Laser Cooling of a Solid by 65 K Starting from Room Temperature," Optical Letters, Vol. 24, No. 15, 1999, pp. 1041-1043

155) B. C. Edwards, J. E. Anderson, R. I. Epstein, "Solid State Optical Cooler Developments, Proceedings of the International Cryocooler Conference, Keystone, CO, June 2000, Paper No 76

156) G. L. Mills, A. J. Mord, P. A. Slaymaker, " Design and Predicted Performance of an Optical Cryocooler for a Focal Plane Application," Proceedings of the International Cryocooler Conference, Keystone, CO, June 2000

ward technology is limited to relatively modest levels of cooling power at temperatures of about 65 K or warmer. This is more than adequate for sensors such as MODIS, but passive cooling would be a challenge for a high-resolution pushbroom longwave infrared imager, for example. [157]

• Use of uncooled detector arrays in the spectral region of TIR (Thermal Infrared) for surface imaging. The technology employs silicon micromachined microbolometer focal plane arrays. The NASA/GSFC instrument ISIR (Infrared Spectral Imaging Radiometer) demonstrated the potential of the concept on Shuttle flight STS-85 (Aug. 7-19, 1997, see J.7). Note: [158] The first micromachined MBAs (Microbolometer Array) were probably developed in the USA in the late 1980's and early 1990's under DARPA's HIDAD (High Density Array Development) program and built by the Honeywell Sensor and System Development Center. This early MBA had about 80,000 pixels (336 x 240) with an average room-temperature sensitivity (NEΔT) of less than 50 mK, measured at a 30 Hz frame rate, with f/1.0 optics and a broadband (8-14 μm) filter.

• HTS (High-Temperature Superconductivity), see also "superconductivity" in the Glossary. The characteristics/performance of superconductive electronic devices have exceptionally low loss, high speed, high dynamic range, and low noise. Superconducting materials (in hybrid components, hybrid digital circuits, semiconductors, sensors) are promising to reduce loss by several orders of magnitude at RF frequencies, increasing the sensitivity and precision of signal discrimination in advanced electronic systems. A new detector type, based on SQUID (Superconducting Quantum Interference Device) technology, offers promising results in such applications as absolute magnetic field measurements.

- The first program to demonstrate the operation and survivability of simple HTS electronic components in space was initiated by NRL (Naval Research Laboratory) in 1988 within the framework of HTSSE-I (High Temperature Superconducting Space Experiment I). [159]
- The DoD ARGOS satellite (launch Feb. 23, 1999), operated by SMC, carries HTSSE-II (High Temperature Superconducting Space Experiment II) of NRL to demonstrate the performance of superconducting (semiconductors and RF) components at cryogenic temperatures of 70-80 K (see M.1). Reductions in power by several orders of magnitude as well as signal speed increases of an order of magnitude are expected with this technology. HTSSE-I of NRL, with relatively simple HTS devices, the majority being passive components made from a single HTS film, were to be placed in orbit to investigate the durability of HTS in space. However, the DoD satellite with HTSSE-I aboard experienced a launch failure in 1993. Despite the unfortunate loss of on-orbit data from HTSSE-I, the program did conclusively demonstrate that viable and robust HTS microwave devices could be fabricated, packaged and space-qualified. - Note: The HTSSE project started at NRL in December 1988 as a long-term joint venture with other labs/industry [which in turn started CSE (Consortium on Superconducting Electronics) in 1989].
- Another HTS demonstration, namely SUPEX (Superconductivity Experiment), based on thin-film technology is flown on TechSat/Gurwin-II (launch July 10, 1998), a satellite built by the Haifa-based Technion Israel Institute of Technology. N.20
- A magnetometer based on SQUID (Superconducting Quantum Interference Device) technology is flown on Gravity Probe-B (planned launch for 2002).
- SMILES (Superconducting Submillimeter-wave Limb Emission Sounder) of NASDA/CRL is planned to fly on ISS/JEM in 2005. The superconductive technique enables 3-D global observation of trace gases in the stratosphere in the frequency band of 640 GHz. A heterodyne receiver detects very low level signals in the submillimeter-wave band ra-

157) http://www.nationalacademies.org/ssb/smallsatappendb.htm

158) R. A. Wood, N. A. Foss, "Micromachined Bolometer Arrays," Laser Focus World, 30, pp.101-106, 1994

159) J. C. Ritter, M. Nisenoff, G. Price, S. A. Wolf, IEEE Transactions on Magnetics, Vol. 27, 1991, p. 2533

diated from trace gases using a sensitive SIS (Superconductor-Insulator-Superconductor) mixer method to perform high-resolution spectral analysis.

- A European consortium by the name of HIFI (Heterodyne Instrument for FIRST), led by SRON (Groningen, The Netherlands), is designing HET (Heterodyne Instrument) to be flown on ESA's FIRST (Far InfraRed and Submillimeter Telescope) satellite (launch 2005-7). HET takes very high-resolution spectra of astronomical objects in thousands of frequencies simultaneously. HET features a large telescope (> 3 m diameter), a large dewar for liquid Helium, and SIS mixers covering the frequency range of 500-1200 GHz continuously, and possibly a Hot Electron Bolometer (HEB) mixer operating at about 1.8 THz.
- SRON is also developing the ESA instrument PIRAMHYD (Passive InfraRed Atmospheric Measurements of HYDroxyl) which features an HTS transition edge bolometer to measure the signal from a spectrally isolated emission line of the OH molecule in the far infrared spectrum of wavelength 84.420 μm.

1.2 Fundamental Science Limits in Space Flight and Earth Observation

All system design is subject to the fundamental limits imposed by the laws of physics, especially those formulated by Kepler, Newton, Maxwell, Airy, Nyquist and Planck. [160)]

1) **Kepler's laws of planetary motion** (his first two laws were announced in 1609, the third law in 1619), after Johannes Kepler (1571-1630) a German astronomer, who discovered that the Earth and planets travel about the sun in elliptical orbits. [161)] Kepler's laws, based on extraordinarily accurate observations of the 16th century Danish astronomer Tycho Brahe, describe the motion of a body subject to gravitational forces - and thus describe a satellite in orbit about Earth, the sun or the moon. Kepler's deduction of the three fundamental planetary laws (which he solved geometrically) were key elements that later enabled Isaac Newton to formulate his theory of gravitational force.

- Kepler's 1st law: A planet orbits the sun in an ellipse with the sun at one focus
- Kepler's 2nd law: A ray directed from the sun to a planet sweeps out equal areas in equal times
- Kepler's 3rd law: The square of the period of a planet's orbit is proportional to the cube of that planet's semimajor axis; the constant of proportionality is the same for all planets.

For a satellite in Earth orbit, the time required to complete one revolution is determined primarily by its mean radial distance. LEO spacecraft with typical altitudes of 700-800 km require about 100 minutes for one orbit. Many Earth observation missions require global coverage (see also chapters 1.7.6 and O.12.1). The laws of orbital mechanics, in combination with the daily Earth rotation, demand a space-time trade-off: a shorter revisit interval implies less dense spatial coverage. Under such constraints, it may not be possible with only one spacecraft to achieve the simultaneity required for fine spatial coverage and short revisit intervals. In this case, a constellation may be the only solution.

2) **Newton's laws of motion** (first published in the Principia in 1687), named after Isaac Newton (1642-1727), English physicist and mathematician. Newton's laws of motion (two of them were already discovered experimentally by Galileo) describe the response of a body to its own inertia and to forces applied from internal or external sources - relative to an inertial reference frame. Both sorts of forces impact all spacecraft. Examples of internally applied forces are:

160) Excerpts are reprinted with permission from "Assessment of Mission Size Trade-offs for NASA's Earth and Space Science Missions," pp. 6-12, National Academy Press, 2000, Copyright (2000) by the National Academy of Sciences. Courtesy of the National Academy Press, Washington, D. C.

161) Kepler is considered a founder of modern astronomy, he formulated the famous three laws of planetary motion. They comprise a quantitative formulation of Copernicus's theory that the planets revolve around the sun.

- S/C pointing by the attitude control subsystem. Observation requirements might call for a particular orientation of the S/C along its orbital path.
- In addition to S/C pointing, there may be a requirement for subsystem pointing. This is for instance the case when the solar panels are rotated to keep them facing into the sun for maximum power generation.

  In general, instruments that must provide very fine angular resolution require in turn that their host S/C satisfy very stringent angular stability requirements. The unwanted reactive movements generated by these subsystems are in turn offset by reaction wheels or torque rods. - Examples of externally generated forces are:
- All S/C must be launched from Earth. Newton's laws and the characteristics of the available propulsion system impose strict limits on the payload mass that can be lofted into Earth orbit or beyond. These limits are expressed in Newton's second law of motion.
- The solar radiation pressure (a surface force) changes the S/C orbit over long periods of time, depending on S/C mass. The drag in LEO orbits is also a surface force. Both, radiation pressure and drag, represent non-gravitational orbit perturbations.
- Orbital maneuvers. A S/C propulsion subsystem might change the orbital altitude.

3) Maxwell equations (publication of Treatise on Electricity and Magnetism in 1873), named after James Clerk Maxwell (1831-1879) a Scottish physicist. Maxwell's equations describe the behavior of the electromagnetic waves as they propagate in space. Portions of the electromagnetic spectrum are used by all space missions. Observation instruments as well as the communication subsystems for the transfer of data through space must be designed within the constraints of the Maxwell equations. The first and most obvious constraint is that radio waves travel at the speed of light (c, the speed of light in vacuum, is 299,792,458 m/s). Even at this speed, light travel time imposes substantial delays on all communication between Earth and satellites. This is already quite noticeable for two-way GEO communications, it is in particular apparent for deep-space probes. A further consequence of the Maxwell equations is that all radiation gets weaker in proportion to the square of the distance between the radiation source and the observer. [162)] Hence, very distant radiation sources become very faint and therefore require much larger viewing apertures. The system design takes the signal weakening into account through the link budget of the communication system (with corresponding antenna designs and sizes). - The same physical principles as discussed above apply of course also to a satellite's electrical power generating capability.

4) Airy diffraction (ca. 1835), named after George Biddell Airy (1801-1892), an English astronomer. The problem: Observed imagery can never be perfect in the geometrical sense. It is either degraded by geometrical aberrations, which are a function of the lens parameters, and/or by diffraction, which is a physical optics effect. The diffraction degradation is due to the wave nature of light.

The Airy diffraction limit enforces a lower limit on the resolution (or beam width) of any device that radiates or receives electromagnetic energy. The principle applies of course also to all S/C imaging instruments (the resolution of an imaging system depends on the size of the collecting aperture). The diffraction limit requires that the optical aperture diameter must be directly proportional to the satellite's distance from the observed surface. - As an illustration, the optics system on a S/C at GEO altitude (of about 36,000 km) must have an aperture 45 times larger than a similar instrument on a spacecraft at typical LEO altitudes (of about 800 km) to obtain the same surface resolution. In addition, the GEO sensor's internal optical path length also has to grow in proportion to the aperture diameter if similar performance to the LEO instrument is to be obtained. As a consequence, instruments at GEO tend to be larger than their LEO counterparts.

162) Note: The "signal weakening" effect is due to the widening cross-sectional area of the light ray as it propagates through space - resulting of course in ever fewer photons per unit area of cross-section, hence, of measurable energy.

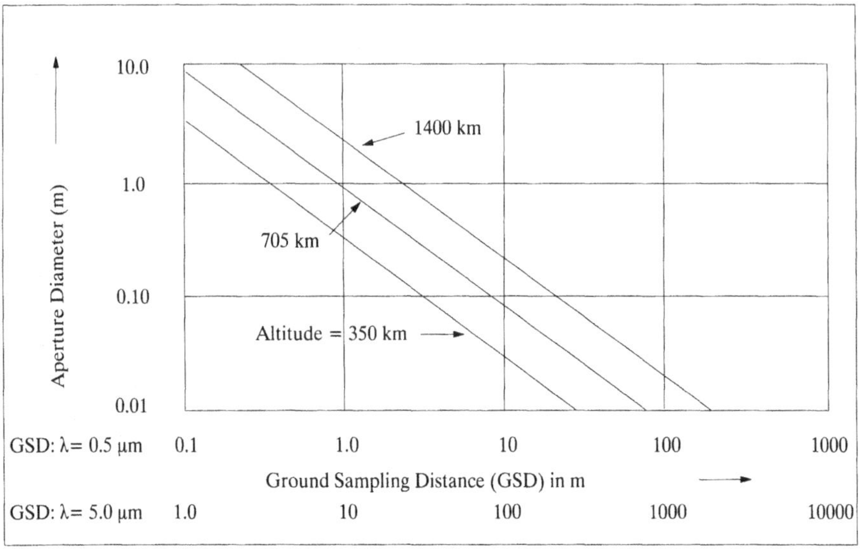

Figure 2: Diffraction limit as a function of aperture diameter, GSD and altitude

The diffraction limit on aperture size has deeper implications as well. For any device that sends or receives energy, the size of the aperture must be proportional to the wavelength it uses. The so-called optical spectrum extends from 0.01 μm to 1000 μm (or 1 mm), i.e., from the UV to the FIR region inclusively. The microwave region (radar) of the electromagnetic spectrum is generally considered from 1 mm to 1m wavelengths. In general, there are at least three orders of magnitude in wavelength between the optical and microwave instruments. Hence, for radar systems the diffraction limit implies apertures in the order of a thousand times larger than their optical counterparts. To meet a given level of performance, an instrument's minimum aperture size is dictated by wavelength and observable distance, it cannot be reduced by technology. Thus, even when a major SAR design goal is smallness, the SAR antenna size cannot be reduced to fit that objective. Any mission that relies on SAR to conduct science observations will continue to require a medium-sized or large S/C platform. - Future distributed (complex and expensive) apertures in special circumstances may to some extent circumvent this limit.

Enhanced performance can be obtained under favorable conditions. For example, a thinned array (also referred to as a sparse array or a sparse aperture) may substitute for a filled array if the viewing objective is relatively sparse. The angular resolution of such a thinned array, however, is governed by the Airy diffraction limit, and by the radiometric sensitivity of the array (proportional to the areas summed of all contributing sub-apertures). Some examples are:

- ESTAR (Electronically Steered Thinned Array Radiometer) a passive airborne instrument developed by NASA/GSFC and the Microwave Remote Sensing Laboratory of the University of Massachusetts at Amherst (P.87). The key feature of a thinned array is that all the required baselines are obtained with fewer antennas than there are spacings.
- The high angular resolution of radio telescopes (in astronomy) is achieved by using the principles of interferometry to synthesize a very large effective aperture from a number of small elements. In a simple two-element radio interferometer, the signals from an unresolved, or "point" source alternately arrive in phase and out of phase as the Earth rotates and causes a change in the difference in path from the radio source to the two elements of the interferometer.

- Each SAR instrument is by definition an aperture synthesis device. The resolution is limited by the synthetic aperture length, and so is the effective received power. Of course, a SAR instrument's antenna is "real."

5) **Nyquist frequency** (1929), named after Harry Nyquist, a US physicist of Swedish descent and a pioneer in the field of communication theory (1889 -1976). - The Nyquist frequency dictates the minimum number of samples per second, required to transport a given amount of data over a communication channel of certain bandwidth (analog-to-digital conversion begins with sampling, or measuring the amplitude of the analog waveform at equally spaced discrete instants of time). The sample frequency is also determined by how often a point on the surface must be sampled to resolve the variations in the physical processes over time. In general, more information implies more detail but also more data. If those data have to be transferred rapidly, then the data rate must increase in proportion. The Nyquist sampling theorem in its simplest form states that a signal can be uniquely represented by periodic discrete samples whose period is no larger than 1/2 of the signal's inverse bandwidth.

6) **Planck's radiation law**, named after Max Planck (1858-1947), a German physicist who originated quantum theory. In 1900, Planck formulated an equation to explain the spectral energy distribution of thermal radiation from a perfect absorber (blackbody) and showed that the formulation required a discontinuous process of emission or absorption involving discrete quantities of energy. Planck's radiation law relates the energy of radiation to its frequency ($E=h \times \nu$). The energy E of each quantum, or of each photon, equals Planck's constant h multiplied by its radiation frequency ν.
Within the electromagnetic spectrum the spectral radiance (i.e. energy per unit time, area, solid angle, and wavelength) varies considerably and with it the ability to detect (measure) radiation. In general, photons are considerably more energetic at shorter wavelengths than at longer wavelengths.

All instrument observation techniques based on radiation measurements (there are indeed many instruments in this category) are constrained to Planck's radiation law. Chapter O.4.4 provides some information as to the radiation detection limits in the various spectral ranges. To illustrate the spectral energy distribution, let us consider three imaging instruments on the same spacecraft, all looking into the same footprint, with each instrument observing incoming radiation in a different spectral range, namely:

- The VIS range (0.4 - 0.7 μm) with photon energy levels of 3.1 - 1.8 eV
- The FIR range (10 - 1000 μm) with photon energy levels of 124 - 1 meV
- The MW range (1 mm - 1 m) with photon energy levels of 1 meV - 1 μeV

The photon energy levels for each spectral range decrease practically by three orders of magnitude from VIS to FIR and again from FIR to MW. This vast photonic energy range of 10^6 puts in particular great demands on the detector technology in the microwave range to measure such minute energy levels and to discriminate between signal differences.

1.3 Spacecraft Systems

- Spacecraft design. Early satellites featured custom designs (in shape, size, etc.) to suit the requirements of a particular mission. It meant continued change for the S/C builders to incorporate general support functions over and over again along with specific mission instruments. The availability of a standard S/C bus (for a particular observation capability) was practically unknown until the end of the 1970s. The Magsat mission (launch Oct. 1979), a dual-spin satellite, demonstrated to NASA in particular the usefulness of a general-purpose bus for science applications. The bus could support a variety of bolt-on experiments. The concept has been used extensively ever since. - Integration of flight hardware and subsystems has become an important aspect of modular design. Introduction of such items/ concepts as: GaAs solar cells, light-weight composite materials, multi-purpose platforms.

Satellite design lives have improved considerably they are now approaching seven to ten years (depending on applications), mini- and microsatellites. In the 1990s, onboard processing is becoming a service.

Spacecraft design (for Earth observation) has experienced interesting trends. Early space-age satellites were small due to the limitations of launch capabilities. The current classes of microsatellites (10 - 100 kg) and minisatellites (100 - 500 kg) are a fitting description for these early space-age satellites (only with regard to mass). As the space age progressed into the 1970s and 1980s, the capabilities of launch vehicles increased constantly. Hence, satellites became much larger and much more complex. In parallel, electronics became much more dense, many subsystems could be combined into a single instrument on a satellite. The new complexity, along with its demands for reliability and quality assurance, had to be managed, generating bureaucracy and large organizations. As a consequence, overall creativity suffered considerably while costs increased rapidly. It meant a reduction in launches with longer in-between periods. Large projects like ISS, EOS, and Envisat encountered planning (and re-planning) phases alone that came close to a decade. For the experimenter, it meant a reduction in flight opportunities for observations as well as a reduction in technology advances.

Organization	S/C Bus	Mission
NASA	MMS (Multimission Modular Spacecraft) SMEX (Small Explorer Program) S/C bus SMEX-Lite	SMM, launch Feb. 14, 1980 SAMPEX, launch Jul. 3, 1992 Triana, planned launch in 2004
ESA	PPF (Polar Platform) bus, based on an enlarged version of the SPOT Mk2 bus AMM (Advanced Microsatellite Mission) platform	Envisat, launch 2001 Metop-1, launch 2005
CNES	Spot Mk1; Mk2; and Mk3 bus; Proteus (Alcatel Space), mass of 100-500 kg	SPOT-1 to 3, ERS-1/2; SPOT-4; SPOT-5; Jason-1
SSTL (UK)	MicroBus since UoSAT-3 and -4 (launch Jan. 22, 1990); MiniBus since UoSAT-12	UoSAT-1, launch Oct. 6, 1981 UoSAT-12 launch, Apr. 21, 1999
Spectrum Astro	SA-series platforms	MSTI-1 to -3, (1992 to 1996) DS1 (launch Oct. 24, 1998)
OSC	MicroStar platform (payloads up to 68 kg), mission life between 3-5 years	OrbView-1 (Microlab), Apr. 3, 1995 QuikTOMS, planned launch in 2001
Astrium GmbH Germany	1) FLEXBUS (minisatellite bus); S/C mass of 100-1000 kg, power of 150-1000 W average 2) MiniFlex bus, S/C mass of 80-250 kg	CHAMP, launch July 15, 2000 GRACE, planned launch in 2001
ASI	1) PRIMA (Reconfigurable Italian Platform for Multiple Applications) of Alenia Aerospazio. Support of payload masses up to 1000 kg 2) MITA (Minisatellite for Advanced Technology) of Carlo Gavazzi Space.	RADARSAT-2, planned launch in 2003 MITA S/C launch Jul. 15, 2000
Stanford University	SQUIRT (Satellite Quick Research Testbed)	ORION, planned launch in 2001
Astrium Ltd., UK	Snapdragon (a deployable bus)	TerraSAR-L1 (launch in 2006)
CNES	Myriade microsatellite bus for missions of <120 kg	DEMETER (launch in 2003), PARASOL and Microscope (launch in 2004), Picard (launch 2006)

Table 19: Introduction of some S/C bus designs

A rethinking of the situation started with the re-invention of microsatellites in the early 1980s. SSTL [Surrey Satellite Technology Ltd (University of Surrey, UK)] was such a pioneer of microsatellite design, including a modular and flexible platform (MicroBus), instruments, looking for new concepts in functionality, services, and cost reductions. Miniaturization techniques of solid-state electronics, sensors, optics, actuators (i.e. miniature mechanisms), etc., are important enabling factors in microsatellite design. Short development times from project approval to S/C launch were the result. This new approach seems to gain momentum, everywhere. There is also a realization that small-project financing can easier find support in tight budgets, this applies to government sponsorship of research projects as well as to the commercial sector. Calculated risks are being taken again. Many organizations are 're-organizing' to improve the conditions for innovation and creativity. The future seems to have room for microsatellites as well as for larger-class satellites, depending on applications. 163)

The early microsatellites of the 1980s were simple S/C, they were for instance built without a propulsion system, due to the cost and complexity of such a system. Attitude control was typically performed using magnetic torquers and gravity stabilization, at a later time reaction or momentum wheels were introduced. Also, propulsion for attitude control using tiny thrusters could be implemented using cold pressurized gas. However, for orbit changes, such a cold gas system remained simply too inefficient. - Small satellite projects have also small budgets. Thus, by their very nature, they depend on launch opportunities as secondary payloads which are offered by larger satellites. As a consequence, a microsatellite project has only the choice of taking the same orbit of the main satellite payload. - In the late 1990s the miniaturization technology is considered viable for integration of electric propulsion systems onto small satellites - to obtain orbit maneuverability. Examples: STRV-1a (launch June 17, 1994) a microsatellite of DERA (S/C mass of 52 kg), made flight tests of the xenon gas flow control system, developed for the UK-10 IPS (Ion Propulsion System), with associated solenoid valves, orifices, and valve actuating electronics (M.25.1.1). Deep Space 1 (launch Oct. 24, 1998) a minisatellite of NASA/JPL with a mass of 490 kg, carried IPS (Ion Propulsion System) to demonstrate deep space propulsion. UoSAT-12, a minisatellite of SSTL (launch April 12, 1999, S/C mass of 325 kg) carries an electric propulsion system, a 100 W resistojet, which uses nitrous oxide as its working fluid.

Launch Date	**Nanosatellite**	Comment
Nov. 3, 1997	Sputnik-II, built by French students from the l'Aeroclub of France and staff from the Russian Aeronautical Federation	A S/C of only 3 kg deployed from the MIR space station (with VHF-FM beacon transmitter)
July 7, 1998	TUBSAT-N and -N1 of TU Berlin, Germany	A dual underwater/space launch of data collection satellites
Jan. 27, 2000	ASUSat-1 (Arizona State University)	S/C mass of 5.9 kg (launch on JAWSAT)
June 28, 2000	SNAP-1 of SSTL (UK) and an academic team of Surrey University	S/C mass of 6.5 kg. SNAP-1 was deployed from the Nadezhda mother ship
Nov. 21, 2000	Munin of IRF Sweden (S/C mass of 6 kg)	Space science and space weather activity,
Launch Date	**Picosatellite**	Comment
Jan. 27, 2000	All picosats were ejected from OPAL of Stanford University (as mother ship) StenSat, of amateur enthusiasts Artemis, of Santa Clara University PICOSAT1.0, a tethered 2 S/C system from the Aerospace Corporation	 No ground communication with StenSat No ground communication with Artemis Communication established with ground and in cross-link
July 19, 2000	PICOSAT1.1, a tethered system from the Aerospace Corporation was launched and flown on MightySat II.1 (as mother ship)	Each PICOSAT has a mass of 0.275 kg. The system will be ejected from MightySat II.1 one year after launch.

Table 20: Chronology of early nanosatellite and picosatellite launches

At the start of the 21st century, the spectrum of microsatellite services is by all means as impressive as that of their bigger brother satellites, but at considerably reduced costs. 164) Overall, microsatellites experienced an impressive evolution from flying gadgets to real and advanced service providers. In fact, microsatellites make it possible to open up new fields of services previously considered too expensive (in particular technological missions are in this category). As a consequence, the space agencies as well as the military establishments

163) C. E. Willey, B. Huettl, D. Dowen, S. W. Hill, "Miniature Mechanisms Tool Kit for Micro Spacecraft," JHU/APL Technical Digest, Vol. 22, No 2, 2001, pp. 115-119

164) A. L. Lew, B. Q, Le, P. D. Schwartz, et al., "Microsatellites: An Enabling Technology for Government and Commercial Aerospace Applications," JHU/APL Technical Digest, Vol. 22, No 2, 2001, pp. 124-134

of the world are re-evaluating their programs, in favor of smaller systems, to offer a solution for ever tighter budgets.

• S/C instrument mounting. A cartwheel mounting configuration was realized in TIROS-9 (launch Jan. 22, 1965). For the first time, the two TV-WA cameras were mounted into the sidewalls of the S/C cylinder, 180° opposite to each other. This cartwheel concept allowed each camera to observe the same ground target during each cycle of rotation. Prior to this time, instruments were generally mounted axially into spin-stabilized S/C.

• Spacecraft stabilization. Early S/C employed spin-stabilization techniques for S/C attitude control (i.e. S/C orientation in space), the simplest method of S/C stabilization (enough S/C spin provided gyroscopic stiffness, hence the S/C axis remained fixed in inertial space). However, such a spinning spacecraft introduced some limitations to Earth observations by the sensor complement. Note: The world's first weather satellite, TIROS-1 (launch April 1, 1960) and the follow-up TIROS and ESSA series S/C (up to ESSA-9, launch Feb. 26, 1969) were all spin-stabilized.

- A dual-spin configuration [Advanced Vela [165] satellite series of USAF (launch of Vela-7 on April 28, 1967; six S/C, Vela-7 to -12 were launched from 1967 to 1970) and SAS-1 (launch Dec. 12, 1970)] followed the single-spin concept, where a portion of the platform rotated, while the payload portion was despun to allow instrument Earth observation (or to ensure proper antenna communications toward Earth). The first EO missions with dual-spin Earth-pointing were METEOSAT-1 of ESA (launch Nov. 23, 1977) and MAGSAT of APL/NASA/USGS with a launch on Oct. 30, 1979.

- The gravity-gradient boom technique, particularly suitable for microsatellites, was first introduced by the LEO Transit navigation satellite series of the US Navy (starting with the launch of Transit-5A-3 on June 16, 1963), followed by other missions: the GGSE (Gravity Gradient Stabilization Experiment) satellite series of DoD started on Jan. 11, 1964 with GGSE-1, GGSE-4 and -5 were launched May 31, 1967 along with Timation-1 (NRL, launch May 31, 1967); ATS-2 (NASA, launch April 6, 1967). GGTS-1 (Gravity Gradient Test Satellite-1 with a mass of 47 kg) of the USAF was launched June 16, 1966 from Cape Canaveral. The UoSAT microsatellite series of SSTL, UK (launch of UoSAT-1 on Oct. 6, 1981) is using the gravity-gradient concept (in LEO) extensively. The DODGE S/C (launch July 1, 1967, see M.4) was the first to study a number of advanced biaxial and triaxial gravity-gradient stabilization techniques at near-synchronous altitudes.

- This was followed by three-axis stabilization (offering better size and space options, in particular with the mounting of deployable solar panels). The Nimbus-1 satellite (launch Aug. 28, 1964) was the first three-axis stabilized spacecraft (followed by the Nimbus series). The Improved TIROS Operational Satellite, ITOS-1 (launch Dec. 11, 1970, the NOAA-1 satellite that followed it (launch Dec. 11, 1970), and all follow-up satellites of the series) featured also three-axis stabilization. The three-axis control concept is still the most applied stabilization method in space flight.

- The first passive magnetic field stabilization system of a satellite was introduced at JHU/APL on the US Navy's Transit-1A (launch Sept. 17, 1959), Transit-1B (Apr. 13, 1960, see H.6) and DME-A (Direct Measurement Explorer - A) with a launch on Nov. 29, 1965. The principle of operation: If three mutually orthogonal electromagnets are placed in a S/C, a magnetic dipole **M** can be created in any direction. The satellite can determine the Earth's magnetic field **H** by means of a vector magnetometer. By activating the electromagnets appropriately in response to the magnetometer measurements, a torque $\mathbf{T} = \mathbf{M} \times \mathbf{H}$ in any direction can be obtained. This torque can be used for S/C attitude control (including spin/despin). Essentially, the satellite acts like the rotor of an electric motor, with the

165) Note: The Vela (meaning "watchman" in Spanish) S/C series of DoD was designed to monitor worldwide compliance with the 1963 nuclear test ban treaty. Vela-1 was launched Oct. 17, 1963, Vela-6 was launched July 20, 1965.

Earth's magnetic field being the stator. The method has been used many times since (see also Table 42). Examples:

a) The ESRO satellites ESRO-1/Aurora (launch Oct. 3, 1968) and ESRO-1/Boreas (launch Oct. 1, 1969) as well as the German Azur-1 (launch Nov. 11, 1969).
b) The Magion-1, -2, and -3 subsatellites of the Czech Republic (launch of Magion-1 on Oct. 24, 1978) all employed the technique of magnetic field stabilization (the Magion-4 and -5 S/C were spin-stabilized). The Magion S/C were part of the USSR mission ACTIVE (K.2);
c) The AMSAT OSCAR microsatellite series [OSCAR 5 (launch Jan. 23, 1970), 16, 17, 18, 19, 26, and 30, for example] employ magnetic field stabilization with its SAPPHIRE ACS (Attitude Control Subsystem);
d) the Munin nanosatellite of IRF (Sweden, launch Nov. 21, 2000).

In the late 1990s, an evolving micro-technology is permitting the design of single-board spacecraft, referred to as nanosatellites (1 kg < mass <10 kg) and picosatellites (mass <1 kg). In such single-board designs, there is no physical separation between platform and payload. Naturally, the capabilities and performance of these tiny pioneering monsters are still very limited and "inferior" to those of their bigger brothers, the micro- and minisatellites (because they have less pointing accuracy, less power, less communication capability, etc. - than the larger micro- and minisatellites). However, the main advantages of nano- and picosatellites are their very low cost and the speed of designing/building a satellite practically from off-the-shelf components; these are indeed strong arguments, even for a limited set of objectives that can be achieved. In particular, such applications as technology demonstrations are favored within the class of nano- and picosatellites (an example is the introduction of such concepts as spacecraft constellations (networks or clusters) for distributed Earth observations or for communication purposes in LEO orbits). [166] The satellite classification of the Glossary (table 562), is simply repeated here for better reference to the reader.

Satellite Class		Mass
Large satellite (observatory, etc.)		> 1000 kg
Medium size satellite		500 - 1000 kg
Minisatellite	Small Satellite Class (or LightSats)	100 - 500 kg
Microsatellite		10 - 100 kg
Nanosatellite		1 - 10 kg
Picosatellite		< 1 kg

Table 21: Satellite classification by mass criterion

• Most existing lightsats, in particular nanosatellites, feature one of two types of primary bus structures: 1) a load-bearing shell structure, or 2) a stack of component trays with stiffeners. [167] The designs use either a sandwich construction, stiffened plates, or thin-walled trays (also referred to as cast design). The materials chosen are either fiber-reinforced composites or aluminum as primary structures. Sandwich structures are usually composed of two face sheets bonded to a core. The most common core type material is aluminum honeycomb, constructed of bonded strips of aluminum foil which are expanded to create hexagonal cells. Both metals and fiber-reinforced composites can be used in honeycomb panels. At the beginning of the 21st century, the sandwich design approach seems to be the most efficient S/C structure design technique, offering compact and modular choices with minimal mass. On the other hand, the cast structure design approach seems to be particularly suitable for mass production of nanosatellites that are needed for larger constellations. The cast design is based on different modules attached to the main structure. Some advantages of the cast or tray structure are: modular design, flexibility, each assembly can be handled independently, minimum assembly time, less number of parts.

166) R. Fleeter, "Being Disruptive," Launchspace Magazine, Volume 3.01, Feb/Mar 1998

167) B. Shirgur, D. Shannon, "The Design and Feasibility Study of Nanosatellite Structures for Current and Future FSI Micro-missions," 14th AIAA/USU Conference on Small Satellites, Logan, UT, Aug. 21-24, 2000, SSC00-VII-5

• Spacecraft busses (as listed in Table 19) come in all shapes and sizes. A very special bus in this listing is **Snapdragon**, a deployable bus of Astrium Ltd., Portsmouth, UK, developed under ESA contract. The design of Snapdragon resulted from the requirement to fly rather large structures, like a SAR antenna, and to accommodate this payload structure elegantly into the available space of a launch vehicle fairing (the smaller the launch vehicle, the lower the launch cost). In Snapdragon, the accommodation problem was approached from the payload's point of view. It resulted in a series of "diametric" SAR accommodations, meaning that the SAR antenna is stowed (two hinged halves) across the diameter of the launcher fairing. The basic snapdragon configuration features two similar structural assemblies to either side of the diameter of the launcher payload fairing. A single central deployment mechanism is placed at the foldline between the two halves of the spacecraft, with the rotation axis in the separation plane of the launcher-satellite interface. A series of HRMs (Hold-down and Release Mechanism)up each side of the S/C ensures structural integrity during launch. Thus, the Snapdragon concept represents essentially a re-packaging of existing hardware into a new structure. A typical Snapdragon deployment sequence is illustrated in chapter B.7.2, it demonstrates also vividly the apt name giving of Snapdragon. Several design variations make the Snapdragon concept very attractive for a number of potential applications with a range of options (not only for large structures) in Earth observation as well as in communication payloads. For instance, the Snapdragon concept is capable to support other missions than SAR. It can carry a large optical telescope simultaneously with radar, multi-frequency radars; and when fitted with fixed arrays and a control-moment gyro it can be very agile having a deployed S/C stiffness >10 Hz. Snapdragon configurations may also be used for large communication antennas as well as antenna farms. A hinge removal provides the option of flying two similar spacecraft. [168)]

• Spacecraft on-board recorders. There was a need for on-board data recording from the very beginning of the space age. TIROS-1 with a launch on April 1, 1960, had already a tape recorder. The first satellite electronic memory in space (384 bits of magnetic core register - required to store its own orbit ephemeris) was flown on Transit-3B (Feb. 21, 1961). Science data storage into solid-state memory followed much later this trend. Some S/C with the provision of solid-state memory are: SAMPEX (NASA, launch July 3, 1992), ALEXIS (LANL, launch Apr. 25, 1993), TEMISAT (Telespazio, launch Au. 31, 1993), PoSAT-1 (SSTL, launch Sept. 26, 1993), FASat (SSTL, launch, Aug. 31, 1995).

An on-board storage capacity of 1 Gbit was first realized with FAST (NASA, launch Aug. 21, 1996), followed by ACE (NASA, launch Aug. 25, 1997) with 2 Gbit, and Equator-S (MPE, launch Dec. 2, 1997) with 1.5 Gbit. EarlyBird (Earthwatch, launch Dec. 24, 1997) had a storage capacity of 16 Gbit. STEX of DoD/NRO (launch Oct. 3, 1998) has a 51 Gbit recorder and uses 64 Mbit DRAM technology. ARGOS of DoD (launch Feb. 23. 1999) provides a storage volume of 64 Gbit as well as Ikonos-2 (Space Imaging, launch Sept. 24, 1999). The Landsat-7 storage capacity (NASA, launch April 15, 1999) is 378 Gbit. The EO-1 (Earth Observing-1 of NASA, launch Nov. 21, 2000) demonstrates with WARP (Wideband Advanced Recorder Processor) a number of high density electronic board advanced packaging techniques. WARP utilizes advanced integrated circuit packaging (3-D stacked memory devices) and "chip-on-board" bonding techniques to obtain very high density memory storage per board (24 Gbit/memory card). It also includes a Mongoose V processor which can perform on-orbit data collection, compression and processing of land image scenes. WARP consists of a high-rate (up to 840Mbit/s capability), high-density (48 Gbit storage), low weight (< 25 kg) solid-state recorder/processor with X-band modulation capability. ALOS of NASDA (launch 2003) provides a storage capacity of 768 Gbit for its high-resolution optical and SAR imagery.

• Solar arrays (S/C power generation). Over the years, the main effort has been to increase the conversion efficiency of silicon-based semiconductor technology; another im-

168) D. M. Simpson, "The Snapdragon Family," Proceedings of the European Conference on Spacecraft Structure, Materials & Mechanical Testing, ESA/ESTEC, Noordwijk, The Netherlands, Nov. 29 - Dec. 1, 2000, pp.337-344

portant aspect is to extend their life times, the end-of-life (EOL) power generation capability per unit mass (W/kg). Conventional solar arrays of the late 1990s provide power/mass ratios of <40 W/kg. In the 1980s, silicon cells offered EOL efficiencies ranging from 13-16%, while the costlier GaAs cells delivered efficiencies of 16-19%. In the 1990s the silicon cell efficiency improved to 17-18%, the GaAs cells (and GaAs/Ge) to that of 19-21%. 169) - In 1997, AEC-Able Engineering Co., of Goleta, CA (in cooperation with Entech Inc. and Spectrolab, the program was sponsored by BMDO and NASA/LeRC), developed and introduced a new variant of solar cells, referred to as SCARLET (Solar Concentrator Array with Refractive Linear Element Technology). The method employs curved, glass-composite optics to concentrate the solar energy on a strip of photovoltaic cells (use of cylindrical silicone Fresnel lenses to concentrate sunlight onto $GaInP_2/GaAs/Ge$ cells). Efficiencies of 22% are obtained. Some inherent benefits of concentrator arrays are: a) high array efficiency, b) protection from space radiation effects, and c) minimized plasma interactions. A disadvantage of this technique is that the solar panels have to be pointed and stabilized (e. g., articulated toward the sun) for maximum output. 170) SCARLET-II is flown on Deep Space 1 (launch Oct. 24, 1998), it uses dual-junction solar cells. DS1 is the first S/C to rely exclusively on refractive concentrator arrays, it is among the first to use only multi-bandgap cells. The first dual-junction cells were introduced in 1997 (Spectrolab of Hughes Electronics Corporation and HSC, who built PAS-5) with an efficiency of about 21%. Spectrolab produced also triple-junction solar cells in 1998 to be flown in 1999 (efficiency of 27%). - At the turn of the 20th century the state of the art in solar array technology provides specific power levels of 50-80 W/kg and packing efficiencies of about 8 kgW/m^3. with the highest demonstrated power of 15 kW. These systems represent a power generation cost of approximately $1000-2000/W.

System/Technology	Comment
Thin GaAS/Ge solar cells	Mass saving, flown on many S/C such as UoSat series (since UoSat-5, launch: July 17, 1991)
Multijunction solar cells	MightySat-I (launch of STS-88, Dec. 4, 98) uses InP (Indium Phosphide) dual-junction solar cells; STEX (launch Oct. 3, 1998) uses dual-junction solar cells PAS-5 (launch July 30,1997) is the first commercial S/C to use dual-junction GaAs solar cells
Concentrators for solar cells	DS1 (launch Oct. 24, 1998) employs refractive concentrator arrays (SCARLET-II) in combination with cylindrical lenses and dual-junction $GaInP_2/GaAs/Ge$ solar cells
Lightweight solar arrays	Cells on flexible Kapton substrate and inflatable Torus solar array. LFSA (Lightweight Flexible Solar Array) is a technology demonstration on EO-1 (launch Nov. 21, 2000)
Micromachined blue-red reflective cover glasses	Increase of solar cell efficiency

Table 22: Emerging solar cell technologies at the start of the 21st century

An alternative approach is seen in thin-film solar cell technology which offers large reductions in power cost and in weight over traditional cells. However, due to their relatively low efficiency the needed thin-film arrays turn out to be quite large. At the start of the 21st century the US military (AFRL) is developing a new PowerSail (PS) program to meet the power requirements of future missions (Space Based Radar, Space Based Lidar, etc.). The goal of the PS initiative is to develop a scalable solar power generation array technology with performance levels of up to 100 kW; the specific power increases with the new technology are estimated to be 3-5 times higher than conventional systems, there is also a reduction in packing volume and cost. The PS concept replaces the thin polyimide blanket with an

169) M. Williamson, "Advancing Satellite Technology," Space & Communications, March-April 1998, pp. 3-7

170) Note: Solar concentrators reflect radiation so as to expose the cells to more radiation, together with multi-junction devices that capture a larger slice of the spectrum (UV, VNIR, and IR). Efficiencies of about 30% and more are expected to be achieved in the early years of the next decade.

FTFPV (Flexible Thin-Film Photovoltaic) solar cell blanket. An FTFPV blanket consists of a thin-layer deposition of amorphous or polycrystalline semiconductors onto a lightweight polymer or steel substrate. This configuration permits the FTFPV structure to be pointed (directed into the sun) and to be used in a similar fashion as the Kapton sail. Four key technologies are targeted to support this program: large deployable/inflatable structures, FTFPV blankets, advanced GNC (Guidance Navigation and Control), and electric propulsion. The PS flight demonstration is based on a 50 kg system package and is expected to be flown in the time frame of 2007. [171)]

- Electrical power subsystem (EPS). [172)] [173)] An EPS of a spacecraft normally consists of the following elements: a solar array for power generation, a power conditioning electronics unit providing a regulated or unregulated bus voltage, and a battery for energy provision to all required satellite functions during eclipse phases of the orbit. The EPS mass is generally dominated by the solar array and battery mass portions (the electronics unit represents the smallest portion of the mass). Nickel Cadmium (NiCd) batteries have been the battery standard for a long time. Higher efficiency Nickel Hydrogen (NiH_2) are being used since the 1980s (NiH_2 batteries have the advantage of a high number of charge-discharge cycles). The Lithium-Ion (LI) battery still holds greater promises in efficiency at the turn of the 21st century.

Battery type	Time frame	Energy/unit mass
NiCd (Nickel Cadmium)	Up to the early 1980s and later	20-30 Wh/kg
NiH_2 (Nickel Hydrogen)	In the early 1980s	36-40 Wh/kg
NiH_2 (Nickel Hydrogen)	At the end of the 1990s	45-50 Wh/kg
NiMH (Nickel Metal Hydride)	First use probably on Falcon Gold of USAF Academy and UCCS (launch Oct. 24, 1997), SEDSAT-1 (launch Oct. 24, 1998), UniSat (launch Sept 26, 2000)	50-60 Wh/kg
LI (Lithium Ion)	Introduction at turn of the 21st century	60-120 Wh/kg
Polymer battery	Introduction on PICOSat (planned launch Aug. 2001)	20-90 Wh/kg

Table 23: Overview of S/C battery evolution

JHU/APL offers an innovative architecture in EPS design for small spacecraft systems (micro and nanosatellites). The patented system is called IPS (Integrated Power Source). IPS includes highly integrated electronics technologies as well as the integration of functions (normally implemented in separate elements) which combines energy storage (a matrix of LI battery cells), solar array electronics (dual-junction solar cells), and processor-based charge control electronics into a single structural element. The final IPS product is a multipurpose panel that may even be used as a S/C side panel. [174)]

- Satellite structure vibration/jitter damping. The operation of cryocoolers, needed for the cooling of detectors in the infrared region (to lower the thermal noise threshold), as well as attitude actuators (momentum wheels, power thrusters, etc.) introduce unwanted microvibrations into the satellite structure which in turn degrade the quality of instrument measurements. - Examples of experiments involving vibration damping are: The STRV-1b satellite of DERA (launch June 17, 1994, see M.25.1.2) is flying CVSE (Cryocooler Vibration Suppression Experiment), provided by BMDO and JPL, to demonstrate a new vibration suppression design by employing piezo materials. The NASA/JPL CSE (Cryo System Experiment) demonstrated vibration suppression on Shuttle flight STS-63 (Feb. 3-11, 1995) in the launch phase and on-orbit by using the Hughes-built ISSC (Improved Standard Space-

171) T. Meink, K. Reinhardt, K. Luu, et al., "PowerSail - A High Power Solution," Proceedings of the AIAA Space 2000 Conference and Exposition, Long Beach, CA, Sept. 19-21, 2000

172) Ch. F. Hoeber, D. J. Kim, "The Continued Evolution of Communication Satellites," Acta Astronautica, Vol. 47, No 2-9, July-November 2000, pp. 65-89, Special issue: Space an Integral Part of the Information Age

173) V. Venugopalan, "Lithium-Ion Cells for Space Applications," Journal of Spacecraft Technology, Vol. 11, No 1, Jan. 2001, pp. 1-73

174) P. D. Schwartz, A. F. Hepp, et al., "Spacecraft Miniaturization: Integrated Power Source," JHU/APL Technical Digest, Vol. 22, No. 2, 2001, pp. 106-109

craft Cryocooler) Stirling cooler and an experimental diode oxygen heat pipe. [175] The TSX-5 mission of AFRL (launch June 6, 2000) flies VISS (Vibration, Isolation, Suppression and Steering System), a self-contained precision vibration control device designed to provide an ultra-quiet environment for sensitive optical sensors, laser transmitters, and other detection and measurement devices (M.29). The NASDA spacecraft OICETS (Optical Inter-orbit Communications Engineering Test Satellite) with a launch in early 2002 is carrying MVE (Micro-Vibration Measurement Equipment) to sense the vibration environment of the spacecraft.

• The first university/student developed microsatellite was NUSAT-1 (Northern Utah Satellite), built as a senior class project at Weber State University, Ogden, Utah. NUSAT was deployed from a modified GAS (Get Away Specials) canister on Shuttle flight STS-51B (Spacelab-3 mission, April 29 - May 6, 1985). The 54 kg microsatellite orbited for 20 months (reentry in Dec. 1986). NUSAT-1 was designed to study high-altitude radar field patterns for FAA (Federal Aviation Agency). - During the 1990s, satellite and payload development projects have become the program of choice for challenging (multi-year) training courses in quite a few engineering departments at universities throughout the world (see Part N).

• Reusable space platforms. Flown on Shuttle flights as free-flyer structures for experiments/payloads with special requirements, and offering a number of service functions for operational autonomy. SPAS-1 (Shuttle Pallet Satellite) on STS-7 initiated the free-flyer scenario.

Platform	Mission	Comments
SPAS-1 (Shuttle Pallet Satellite)	STS-7, June 18-24, 1983 STS-41B, Feb. 3-11, 1984	MOMS-01 imaging instrument of DLR (MOMS-01 and SPAS-1 built by MBB, Munich)
LDEF (Long Duration Exposure Facility)	STS-41-C, Apr. 7, 1984	NASA/LaRC free-flying S/C. Retrieval of LDEF on STS-32, Jan. 12, 1990
SPARTAN-1 (Shuttle Pointed Autonomous Research Tool for Astronomy)	STS-51G, Jun. 17-24, 1985	NASA free-flyer platform
SPARTAN-Halley	STS-51-L, Jan. 28, 1986	Challenger accident
SPAS-II	STS-39, Ap. 28 - May 6, '91	DoD payload
EURECA-1 (European Retrievable Carrier)	STS-46, Jul. 31, 1992	ESA platform. Retrieval during STS-57 mission, June 21-July 1, 1993
SPARTAN-201-1	STS-56, Apr. 8-17, 1993	Solar physics mission
ORFEUS-SPAS-1	STS-51,Sept. 12-22, 1993	ORFEUS-1 is a DARA/NASA science mission
SPARTAN-201-2	STS-64, Sept. 9-20, 1994	Solar physics mission coordinated with Ulysses south polar pass
CRISTA-SPAS-1	STS-66, Nov. 3-14, 1994	Joint flight of CRISTA-1 and ATLAS-3 mission
SPARTAN-204	STS-63, Feb. 3-11, 1995	
SPARTAN-201-3	STS-69, Sept. 7-18, 1995	Solar physics mission coordinated with Ulysses north polar pass
SPARTAN-206	STS-72, Jan. 11-20, 1996	Also retrieval of SFU (Space Flyer Unit)
SPARTAN-207	STS-77, May 19-29, 1996	Inflatable Antenna Experiment (IAE)
ORFEUS-SPAS-2	STS-80, Nov. 19, Dec. 7, 96	ORFEUS-2 is a DARA/NASA science mission
CRISTA-SPAS-2	STS-85, Aug. 7-19, 1997	Reflight of CRISTA-1 mission
SPARTAN-201-4	STS-87, Nov. 19-Dec. 7, 97	Solar physics mission coordinated with SOHO
SPARTAN-201--5	STS-95, Oct. 29 Nov. , 98	Solar physics mission coordinated with SOHO

Table 24: Chronology of retrievable free-flyer structures on Shuttle missions

• Inflatable antenna structures. [176] Inflatable structures have the advantages of low mass, low stored volume, and of low cost. They also have the potential to deploy much more reliably than the conventional mechanical systems used for deploying rigid structures. In-

175) http://www–techtrans.jpl.nasa.gov/success/stories/cse.html

176) C. E. Willey, R. S. Bokulic, W. E. Skullney, R. C. Schulze, "Ka-band Hybrid Inflatable Dish Antenna," JHU/APL Technical Digest, Vol. 22, No 2, 2001, pp. 110-111

flatable antennas can be used in various frequency ranges (in particular VHF, UHF, MW) for such diverse applications as measuring soil moisture and ocean salinity or communications and power generation for satellites. First examples:

- The airborne SAR instrument CARABAS (Coherent All Radio Band Sensing) of FOA (National Defense Research Establishment, Linköping, Sweden), first flown in 1992, used two inflatable and flexible canvas sleeves which were trailed behind the aircraft in flight and employed as its VHF antenna system (P.49).
- On the spaceborne side, IAE (Inflatable Antenna Experiment) was flown on SPARTAN-207 (STS-77, May 19 - 29, 1996) to demonstrate the performance of a 14 m diameter inflatable deployable antenna reflector structure in orbit.
- The US AFRL at Kirtland AFB, NM, launched an inflatable sphere, named OCSE (Optical Calibration Sphere Experiment), on the JAWSAT mission (launch Jan. 27, 2000 on a Minotaur launcher into LEO) of Weber State University, Ogden, UT. The objective is to refine and support spacecraft SLR tracking from the ground.

• STR (Spring-Tape-Reinforced) booms. [177] Besides antennas, inflatable structures may also be employed for other components of space structures such as: solar arrays, sunshades, solar concentrators, and telescope reflectors. The STR technology, invented at JPL, represents another enabling element of inflatable structures, including self-rigidizable booms made of stretched aluminum laminates. A typical boom consists of a tube, formed of aluminum laminate sheet, and seamed by a Kapton tape and two end caps. A number of distinct advantages of the STR booms have been identified and demonstrated during the development process. These include: simplicity of the design, self-rigidizable in space, high-load carrying capability, high packing efficiency, and low inflation deployment pressure.

1.3.1 Spacecraft Operations

• **Satellite on-board autonomy** (in the context of unattended operations). The provision of enhanced functional autonomy demonstrations has been a goal of a number of technology missions by various agencies/institutions and satellite integrators. The intent is always to reduce operating costs, in particular to free resources of routine operations for long-term follow-up missions, with minimum ground involvement. The nature of "autonomy provision" turns out to be rather complex (in the end it requires nothing less but fault-tolerant systems on all levels). One aspect of autonomy is the automatic performance of routine operations over a limited period of time, requiring on-board monitoring, analysis and self-correcting measures for each functional aspect on the instrument level, prior to updates from ground control. The other aspect of autonomy is the provision of realistic backup scenarios to handle failure situations. - It has long been realized by all parties involved that so-called "cheapsats" are not really cheapsats, if overall costs (all operations + S/C costs + launch) cannot be reduced. And operations happen to be an important and persistent long-term item in the overall bill. The requirement for an autonomous on-board navigation function seems to be on the agenda of many missions (see also autonomy under Navigation in chapter 1.7.8). The reason: on-board knowledge of satellite position in ground coordinates is needed for many operational functions such as: activation (start/stop) of Earth observation instruments, data transmission, satellite attitude control. Examples of partial autonomy demonstrations are:

- Redundancy design. Satellite reliability is an important consideration for continued provision of operational services. In Transit-3-A (launch in 1960), a US Navy navigation satellite, JHU/APL integrated for the first time a passively redundant battery system that could automatically switch to a backup battery. - Further redundant designs followed for such subsystems as RF and telecommand.
- Enhancements on the various generations of GPS satellites. The block I S/C series (with launches from Feb. 1978 to Oct. 1985) was able to sustain unattended operations

177) M. Lou, H. Fang, L. M. Hsia, "Development of Space Inflatable/Rigidizable STR Aluminum Laminate Booms," Proceedings of the AIAA Space Conference and Exhibition, Long Beach, CA, Sept. 19-21, 2000

for up to 3.5 days between navigation message uploads from the ground. The latest block IIR S/C series features autonomous satellite operations for a period of up to six months without ground control corrections (first launch of a block IIR S/C on July 23, 1997).

- The PoSAT-1 microsatellite mission of SSTL (launch, Sept. 26, 1993) reached a certain degree of operational autonomy through its on-board orbit determination capability (first orbital elements).
- For NASA's Clark satellite (launch in 1998), a GPS receiver with the name of GADFLY was used to demonstrate autonomous orbit determination and to use GADFLY also for coarse attitude knowledge. Note: At the end of Feb. 1998, NASA cancelled the Clark mission due to severe cost overruns and launch delays.
- The DS1 (Deep Space 1) technology mission of NASA/JPL (launch Oct. 24, 1998) introduced the concept of an autonomous **"Remote Agent,"** a technology demonstration package which includes an onboard mission manager with a mission plan expressed in terms of high-level goals. A planning and scheduling engine uses the goals, along with a comprehensive knowledge of the S/C status, and constraints on S/C operations - to generate a set of time-based or event-based activities. The autonomous Remote Agent concept was successfully tested on May 17, 1999 when the primary S/C command was given over to Remote Agent for three days of S/C operations. In this period, Remote Agent successfully planned DS1 activities on-board and then carried out the plan without ground intervention. The software detected, diagnosed and fixed simulated problems, showing that it can make decisions to keep the mission on track. (M.5).
- The ARGOS technology mission (M.1, launch Feb. 23, 1999) of DoD employs an on-board automated mission planning system to optimize on-board data handling and power.
- The UoSAT-12 minisatellite of SSTL (launch April 21, 1999) in Surrey, UK, carries a closed-loop autonomous control system that enables orbit operations to be performed without the need of any ground segment. The autonomous control system consists of a GPS receiver, an on-board propulsion system EPS (Electric Propulsion System), and an orbit maintenance software package of Microcosm Inc., El Segundo, CA, referred to as OCK (Orbit Control Kit).[178) The objective of the control part of the software is to ensure that the orbital altitude of the satellite never falls outside of a prescribed window due to drag. The orbit is described using a set of epicycle parameters which provide an analytic model of LEO orbits. The parameters in this model are estimated onboard the satellite using a Kalman filter. In addition, the software provides control, and an estimation of the orbit parameters by including drag in the model. The satellite was maneuvered into an exact 7-day repeat (ground track) pass orbit for the technology demonstration. The new orbit maintenance software then maintained the satellite mean radius in its resonant orbit for a test period of 27 days within an error band of ±5 m (3 sigma), and slowly maneuvered the S/C into a frozen orbit.[179)]
- The EO-1 mission of NASA/GSFC (launch Nov. 21, 2000) requires demonstrations of autonomous navigation/instrument operation with an on-board software package referred to as EFF (Enhanced Formation Flying). EFF is capable of autonomously planning, executing, and calibrating routine spacecraft maneuvers to maintain satellites in their respective constellations and formations. The EFF features include an innovative use of fuzzy logic decision making capabilities and natural language to resolve multiple conflicting constraints.

178) G. Gurevich, R. Bell, J. R. Wertz, "Autonomous On-Board Orbit Control: Flight Results and Applications," Proceedings of AIAA Space 2000 Conference and Exposition, Long Beach, CA, Sept. 19-21, 2000

179) M. Aorpimai, Y. Hashida, P. Palmer, "Autonomous Control System for Precise Orbit Maintenance," Proceedings of the 14th AIAA/USU Conference on Small Satellites, Logan, UT, Aug. 21-24, 2000, SSC00-IX-6

- NASA's TIMED mission (launch 2001), designed and operated by JHU/APL, introduces operational autonomy by separating payload/instrument operations from all S/C system operations activities. In this way, the instrument teams are able to control all of the instrument modes, operations and science data return at their own choice without explicit interactions or approvals by the S/C project team. A combination of on-board GPS processing and the use of the Internet move the data from the APL ground station to each investigator's home site. In addition, the Internet is used by each investigator to control his instrument directly (the packetized messages are integrated into the uplink command structure in an automated fashion). The ultimate goal for TIMED is to develop a "lights-out" concept of operations as the missions progresses.
- ESA's PROBA minisatellite mission (launch in 2001, M.20). The attitude control and avionics subsystems accommodate the core technologies for S/C autonomy. Demonstration of autonomous operations for orbit and attitude determination with a GPS receiver (SRG-20). The other autonomous subsystem is ASC (Advanced Stellar Compass) of DTU, Denmark. ASC is capable to reconstruct autonomously the S/C inertial attitude starting from the condition "lost in space." The mission operations concept provides considerable flexibility in the allocation of on-board resources and in scheduling of operations.
- A NASA/GSFC technology initiative is developing (1998) a concept referred to as IA/GNC (Image-Aided/Guidance, Navigation and Control) to support S/C autonomy and to reduce operating costs. [180] The IA/GNC scheme uses 2-D images from the S/C science instrument or a secondary low-cost camera and generates attitude control and image stabilization information that enhances pointing performance and ultimately permits autonomous onboard geo-referencing and geometric rendering of data. IA/GNC relies heavily on image correlation tracking (ICT) techniques which compute the translational offset between an instantaneous image and a stored reference image derived from a previous image or a model.
- The STEREO mission of NASA (launch 2004, K.26) will demonstrate the feasibility of autonomous solar navigation (the sun is used as a reference). The technique employs a dual-mode imaging system (designed by JHU/APL) for measuring the direction of the sun using a CCD camera which captures the image of the sun against a star background (the stars serve as a direction reference). The two-mode design, based on a DSAD (Digital Solar Attitude Detector) imaging system, permits the control of vast the brightness contrast differences. The S/C state vector is determined by on-board processing the solar observation data. [181]
- Autonomous station keeping of S/C. Up to the late 1990s, the station keeping of satellites (in GEO and LEO) has primarily been ground based, involving the Control Center personnel in all phases of operations, including orbit maintenance and station keeping. Current GEO satellite operations have evolved so as to take advantage of the stationary nature of the satellite position relative to the ground stations. The favorable geometry provides a continuous window for ranging, tracking, and commanding, thereby minimizing the computational burden on the on-board processors. The LEO satellites (with intermittent ground contacts), on the other hand, always required an on-board processing capability to provide some limited autonomy in navigation. Functional capabilities in the direction of autonomy improved considerably with the availability of on-board GPS systems and advanced feedback control techniques. New orbit control designs are based on the concept of continually minimizing the position and velocity error between a satellite and its target orbital position.
- The Emerald two-spacecraft mission of Stanford and Santa Clara University within UNP (launch 2001) provides a two-level on-board autonomy. At the lowest level,

180) http://gnctech.gsfc.nasa.gov/gto/library/staif/staif171.html

181) Y. Guo, "Autonomous Solar Navigation System," JHU/APL Technical Digest, Vol. 22, No 2, 2001, pp. 119121

formation-flying algorithms provide relative position determination and control. At the highest level, the on-board autonomy functions and fleet-level commanding are demonstrated. An expert system is used to execute model-derived analysis rules for robust anomaly management. The execution system provides intersatellite command synchronization and planning. The objective is to enable fleet-level commanding (i.e., a single high-level command to the fleet will cause coordinated fleet activity) and to support opportunistic observations (i.e., the ability of all S/C to detect "events of interest" and to react by starting coordinated observation activities). In addition, the distributed beacon system validates these methods by providing the anomaly notification function of the cluster to the ground.

Autonomous on-board control offers several operational capabilities not available to missions without this feature (according to reference 178):

- The position of the spacecraft at all future times is known as far in advance as desirable
- The ground track (or inertial track) of the spacecraft can be made to follow a predefined pattern which may be changed at user request
- The process for computing future positions is simple and can be included in any ground-based equipment that uses a general-purpose microprocessor
- There is a longer planning horizon dealing with the potential problems of RF or physical interference with other spacecraft or debris
- Disturbance torques are much lower than with more traditional orbit control processes, such that the size and responsiveness of control actuators can be lessened and restrictions on the timing of stationkeeping maneuvers can be reduced or eliminated.

In summary, autonomous on-board orbit control can fundamentally change the way space missions operate. It is a key component in extending the philosophy of "faster, better, cheaper" to 21st century ground operations.

Today's requirements of future space-system functionality call for nothing less but **system intelligence** (error-tolerant systems) with the following characteristics: (see reference 7)

- Autonomy to think for themselves
- Self-reliance to identify, diagnose, and correct internal problems and failures
- Self-repair to overcome damage
- Self-assembly and evolvability to adapt to and explore new and unknown environments
- Extreme efficiency to operate with very limited resources and without input from mission control.

• **Internet access**. A demonstration of Internet access to flying spacecraft was first performed by SSTL (Surrey Satellite Technology Ltd) using its UoSAT-12 (launch April 21, 1999) minisatellite.[182)] [183)] The demonstrations consisted of uploading an IP (Internet Protocol) software stack to the UoSAT-12, including simple modifications to the SSTL ground station, and a series of tests to measure the performance of various Internet applications (UoSAT-12 was assigned an IP address). The UoSAT-12 S/C was reconfigured on-orbit. The initial tests included basic network connectivity (PING), automated clock synchronization with NTP (Network Time Protocol), and FTP (File Transfer Protocol) transfers.

The UoSAT-12 Internet demonstration was carried out in the time frame April-June 2000 within the OMNI (Operating Missions as Nodes on the Internet) project of NASA/GSFC. The UoSAT-12 spacecraft was selected because of its ability to support the HDLC (High-Level Data Link Control) framing in hardware for link-level protocol on space-to-ground links. This allowed simple and straightforward interfacing with existing commercial routers. UoSAT-12 was an ideal test platform because it already used HDLC framing for its AX.25 protocol. - Follow-on work is to demonstrate the following functions and additional proto-

182) J. Rash, R. Parise, K. Hogie, E. Criscuolo, J. Langston, C. Jackson, H. Price, "Internet Access to Spacecraft," Proceedings of the 14th AIAA/USU Conference on Small Satellites, Logan UT, Aug. 21-24, 2000, SSC00-IX-4

183) http://ipinspace.gsfc.nasa.gov/general/

cols: http file delivery, mobile IP, security, store-and-forward commanding, data delivery using SMTP (Simple Mail Transfer Protocol), and VPN (Virtual Private Network) to enable automated, operational S/C communication. [184)]

- Autonomous ground stations and systems.

- The first ground stations that could be operated autonomously were probably the APT (Automatic Picture Transmission) stations for NOAA satellite data reception. The APT broadcast service was first introduced with TIROS-8 (launch Dec. 21, 1963). Simple APT ground stations could soon be purchased to support autonomous data acquisition.

- The broadcast of HRPT data started with TIROS-N (launch Oct. 13, 1978). Along with this service came simple HRPT stations supporting autonomous data acquisition.

- True and continuous autonomous ground station operations has probably first been realized with the introduction of systematic spaceborne data collection services (DCS) from the ground segment. Each remote ground station (on land, on water, or in the air) had to be operationally autonomous to participate in such a venture. The first such system in LEO (polar orbit) was IRLS (Interrogation, Recording, and Location System), employing a range-only platform location technique and flown on Nimbus-3 (launch April 14, 1969, see M.17.3). The French-US Eole experiment measured ambient air temperature and pressure and determined 200 mb winds over the Southern Hemisphere by tracking constant-density balloons, employing a range and range-rate location technique (launch Aug. 16, 1971). A simplified approach, using only random-access range-rate data, was employed in the TWERLE (Tropical Wind Energy conversion and Reference Level Experiment) flown on Nimbus-6 (launch June 12, 1975, see M.17.6). These systems led eventually to the French-developed **Argos** Data Collection and Location System, first flown on TIROS-N (launch Oct. 13, 1978, see C.1). A similar system, COSPAS, was developed by the Soviet Union and first launched on June 29, 1982. The US, Canada, France, and the Soviet Union banded together to form COSPAS-S&RSAT (Search and Rescue Satellite Aided Tracking System), an international humanitarian search and rescue system that became fully operational in 1984 (see I.6). As of April 2001, over 11,000 persons have been rescued by COSPAS-S&RSAT services since its inception.
  The first DCS in GEO orbit was flown on SMS-1 (Synchronous Meteorological Satellite-1, launch May 17, 1974). See chapter 1.8.

- Autonomy is also invading the mission control room. Traditionally, satellite operations have been preformed from a fixed set-up of computers in a dedicated control room at the space agency's or satellite owner's premises. The control room was staffed around the clock in order to continuously monitor and operate the spacecraft. This 24 hour staffing approach, however, was very expensive. For small scientific satellites on a constrained budget, this type of support may only be justified during the initial commissioning phase (typically a few weeks) and in case of contingencies or special campaigns.
  - Future missions of satellites will increasingly rely on autonomy for routine operations and automatic downlinking of housekeeping and instrument data. Both the on-board computer and computers at the ground station or a dedicated control facility will continuously monitor the health of the spacecraft and alert the operator-on-call in case of anomalies.

- Increased use of Internet services in S/C (or mission) operations, in particular for small S/C to save costs over the conventional leased-line communications approach. Furthermore, with the first LEO commercial telephone networks in operation, S/C operators (in particular student developed satellites) are designing their prime communications

184) J. Rash, R. Parise, K. Hogie, E. Criscuolo, J. Langston, "Internet Technology on Spacecraft," Proceedings of the AIAA Space Conference and Exhibition, Long Beach, CA, Sept. 19-21, 2000

needs for uplink and downlink data transmissions entirely on the service provision of these distributed cellular telephone networks. This eliminates the conventional dedicated ground segment to space segment communication links for a space mission.

- Munin (N.9), a Swedish nanosatellite of IRF (Swedish Institute of Space Physics), students at Umeå University (RYP), and students at Luth (Luleå University of Technology) with a launch as a secondary payload on NASA's EO-1 mission (launch Nov. 21, 2000), feeds its science data (space weather parameters) directly into the Internet at reception, available to anyone.
- The ION-F (Ionospheric Observation Nanosatellite-Formation, three satellite constellation of USU (Utah State University), UW (University of Washington, and VT (Virginia Polytechnic Institute) with a planned launch in 2001, employs also commercial cellular phone communications in all links.
- etc.

• Spacecraft data transmission (telemetry and telecommand) using CCSDS protocols. The first recommendations for a CCSDS (Consultative Committee for Space Data Systems) standard were released in 1987, allowing information from different parts of a satellite to be placed into packets and to be sent to a ground station (or to be uplinked as telecommand data). By using a standard method of processing and distributing satellite data, one ground station can handle data from different satellites. [185] - The ERS-1 mission of ESA was probably the first one to introduce CCSDS formats and protocols. The SAMPEX mission was the first in USA to implement the CCSDS communication standards. TEAMSAT (launch Oct. 30, 1997) is the first ESA spacecraft flown with telemetry and telecommand systems both fully compatible with CCSDS standards; it is also the first spacecraft to exploit the self-adaptive asynchronous telemetry capabilities successfully that the CCSDS standards support. [The communication system excepts user data as randomly occurring squirts of various sizes or as fully asynchronous individual bytes. The bandwidth on the link is shared among the users according to allocation ratios. Any bandwidth not taken up by one user is offered to the other users in the proportionate ratios].

Satellite (Agency)	Launch Date
ERS-1 (ESA)	July 17, 1991
SAMPEX (NASA in SMEX program)	July 3, 1992
EURECA (ESA)	July 31, 1992, retrieval June 21-July 1, 1993
MSTI-1 (DoD), MSTI-2, MSTI-3	November 21, 1992, May 9, 1994, May 16, 1996
Spacelab-D2 (DLR/ESA)	April 26 - May 6, 1993
STRV-1a and -1b (DERA, UK)	July 17, 1994
ERS-2 (ESA)	April 21, 1995
RADARSAT-1 (CSA)	November 4, 1995
SOHO (ESA/NASA)	December, 2 1995
Cluster-I (ESA)	June 4, 1996
TEAMSAT/YES (ESA)	October 30, 1997
TRMM (NASDA/NASA)	November 27, 1997
ETS-VII (NASDA)	November 28, 1997
SNOE (University of Colorado, Boulder)	February 26, 1998
DS1 (NASA/JPL)	October 24, 1998
ROCSAT-1 (NSPO, Taiwan)	January 26, 1999
Ørsted (DRSI, Denmark)	February 23, 1999
Landsat-7 (NASA)	April 15, 1999
QuikSCAT (NASA)	June 19, 1999
MTSAT (NASDA/JMA, Japan)	November 15, 1999
Terra (EOS/AM-1 of NASA)	December 18, 1999
KOMPSAT-1 (KARI, Korea)	December 20, 1999
MTI (DOE)	March 12, 2000
IMAGE (NASA)	March 25, 2000

185) P. Shaki, "Industry, Agencies Adopt Data Collection Standard," Space News, Oct. 26 - Nov. 1, 1998, p. 6

Satellite (Agency)	Launch Date
MITA (ASI)	July 15, 2000
CHAMP (GFZ/DLR)	July 15, 2000
Cluster-II (ESA)	July 16, (1st launch), Aug. 9, 2000 (2nd launch)
QuickBird-1 (EarthWatch)	November, 20, 2000
STRV-1c and -1d (DERA, UK)	November 16, 2000
EO-1 (NASA)	November 21, 2000
TIMED (NASA), Jason-1 (CNES)	planned for 2001
PROBA (ESA), BIRD (DLR)	planned for 2001
ICESat (NASA), CATSAT (UNH)	planned for 2001
GRACE (NASA/DLR/GFZ)	planned for 2001
Aqua - formerly EOS/PM1 (NASA)	planned for 2001
ADEOS-II (NASDA)	planned for 2002
GP-B (Gravity Probe-B), NASA/Stanford	planned for 2002
SCISAT/ACE (CSA)	planned for 2002
MTSAT-1R (NASDA/JMA, Japan)	planned for 2003
Diamant-1 OHB-System, Bremen	planned for 2003
COSMO-SkyMed (ASI, ESA)	planned for 2004
MetOp-1 (EUMETSAT)	planned for 2005
Pléiades (CNES, ESA)	planned for 2005
TerraSAR (Astrium, ESA)	planned for 2006

Table 25: Some CCSDS protocol implementations on satellite missions

The CCSDS standard offers a number of significant advantages over the more traditional 'mission-specific', fixed-format data frames, such as:

- Very reliable uplink, with automatic retransmission capability
- Cost reducing, commercial off-the-shelf onboard components
- Potential for inter-operability with other ground stations/segments
- A CCSDS protocol design in combination with a MIL-STD-1553B communications bus can provide substantial advantages in terms of spacecraft/payload decoupling. Such a design is for instance implemented in the IMAGE satellite of NASA.

As commercial satellite manufacturers adopt the CCSDS standard, the scope for international inter-operability will offer significant opportunities for collaborative programs resulting in a reduction in ground control facility costs.

- The communications bus MIL-STD-1553B has become an established spacecraft bus standard with roots in military aircraft avionics. [186)]The bus was first introduced in 1973 and experienced a number of revisions. The structure of the bus consists of a single bus controller connected to remote terminals (up to 31 max can be used). The use of the MIL-STD-1553B communications bus in combination with CCSDS protocols is increasingly being used in the design of small satellite command and data handling systems. In this way, they function as standard spacecraft-to-payload interfaces and protocols. In addition, the IEEE 1394 high-speed and low-power serial bus standard (subsystem communications), in combination with MIL-STD-1553B, is gaining acceptance in spacecraft design. IEEE 1394 defines techniques for serial digital data communications among multiple nodes. [187)] Example of IEEE 1394 I/F: The VIIRS (Visible/Infrared Imager and Radiometer Suite) instrument of NPOESS (R.147.3) utilizes an IEEE 1394 cable.

- Store & Forward (S&F) satellite communication. The UoSAT-2 S/C (SSTL, launch March 1, 1984) carried the first digital S&F payload in orbit using a protocol suite by the name of PACSAT.

186) R. Killough M. McLelland, "Designing Command and Telemetry Systems Using MIL-STD-1553 and CCSDS," Proceedings of the 14th AIAA/USU Conference on Small Satellites, SSC00-XI-4, Aug. 21-24, 2000, Logan, UT

187) M. E. Fraeman, "Advanced Spacecraft Architectures: 1394 Serial Bus," JHU/APL Technical Digest, Vol. 22, No 2, 2001, pp. 114-115

• Automatic docking maneuvers in space. ETS-VII, a Japanese technology demonstration satellite (launch Nov. 28, 1997) of NASDA, conducted the first successful release, tracking, and capture of a subsatellite without help from the ground on Sept. 1. 1999. The ETS-VII system consists of two spacecraft: a chaser satellite (Hikoboshi), and a target satellite (Orihime). Engineers commanded the Orihime to float about 200 mm away from the Hikoboshi main satellite body. Then, using a robotic arm and grapple attached to the larger craft, Orihime was retrieved and held firmly in place. The entire capture procedure, including the activation of the robotic arm and extension, was conducted automatically without help from ground control computers. [188] - Previous automatic docking activities, such as those conducted with the Russian Soyuz and Progress S/C, have used either operator assistance or ground control computers. On Oct. 30, 1967, the USSR accomplished the first docking of two unmanned satellites in LEO (Cosmos-186 and Cosmos-188).

• Communication downlink experiment. XPAA (X-band Phased Array Antenna) is a communication experiment on EO-1 (launch Nov. 21, 2000) of NASA/GSFC with the objective to demonstrate link-pointing capability with the use of a body-fixed low-mass and low-cost phased array antenna (data rate of 105 Mbit/s). XPAA is composed of a flat grid of 64 radiating elements whose transmitted signals are combined spatially to produce the desired antenna directivity. An inherent advantage of the body-fixed design (mass of 5.5 kg) is to permit simultaneous capture and transmission of data, avoiding perturbations to instrument measurements.

• **On-board data compression systems**. Vast amounts of source data, generated by satellite high-resolution data instruments, such as hyperspectral imagery or SAR imagery, require data reduction algorithms to alleviate data handling and processing problems. These problems are severe (a limiting operational factor) in particular for the functions of on-board storage and high-volume data transmissions to the ground segment. They also permeate throughout the ground segment in communications, data processing (memory) and archiving.

The development of compression algorithms requires a complete understanding of the characteristics and the use of the data. In particular, data compression techniques require internal redundancy in the data sequence if compression is to work. [189] [190] Streamlined approaches of compression are in use by various reversible (lossless) as well as irreversible (lossy) methods for reduction of data. Statistical codes (e.g. Huffman, Shannon-Fano, etc.), statistical methods (e.g. runlength and arithmetic coding), vector quantization (e.g. block truncation coding, JPEG) encoding by transformations (e.g. Fourier and wavelet transformations) and fractal approaches (e.g. contracting mappings) depict some examples of the first, second and third generation approaches for data compression. As of 1999, JPEG-LS (JPEG lossless) is the new lossless/near-lossless compression standard for continuous-tone images, ISO - 14495-1/ITU-T.87. [191] [192] The standard is based on the LOCO-I algorithm (LOw COmplexity LOssless COmpression for Images) developed at Hewlett-Packard Laboratories. - Speed is the overriding issue for all real-time compression applications. Some examples of data compression implementations in past, current, and planned spaceborne systems are:

188) http://www.space.com/news/international/japan_satcapture.html

189) P. Hou, M. Petrou, C. Underwood, "Advanced On-board Image Compression in Conjunction with Cloud Detection for Microsatellite Optical Imaging," Proceedings of the 13th AIAA/USU Conference for Small Satellites, Aug. 23-26, 1999, Logan UT, SSC99-IV-5

190) P. Trinadh, R. Seshaiah, U. N. Das, V. Nalanda, "Effect of Transmission Channel Errors on ADPCM and JPEG Compression," Journal of Spacecraft Technology, Vol. 9, No 1, 1999, pp. 23-36

191) Note: In "lossy compression" information is thrown away during compression, so that the original data cannot be recovered by decompression. The decompression produces an approximation to the original data, with the level of approximation dependent on the compression ratio. In "lossless compression" the original data is reproduced exactly by decompressing the compressed stream.

192) B. V. Brower, A. Lan, J. M. McCabe, "Hyperspectral lossless compression," Proceedings of SPIE, Imaging Spectroscopy V, Vol 3753, Denver , CO, July 19-21, 1999, pp. 247-257

- The SPOT series satellites of CNES employ DPCM (Differential Pulse Code Modulation) on the PAN data stream of the HRV instruments on SPOT-1, -2 and -3 (launch SPOT-1 on Feb. 22, 1986) and on the PAN and MS data streams of HRVIR on SPOT-4 (launch March 24, 1998). SPOT-5 (launch 2002) uses DCT (Discreet Cosine Transform).
- The MOMS-02 (Modular Optoelectronic Multispectral Scanner), a stereoscopic along-track imaging system of DLR - flown on Shuttle flight STS-55 in April/May 1993, followed by a MOMS-02 reflight on MIR/Priroda (launch of Priroda April 23, 1996) - employed DPCM to compress its data from 8 bit to 6 bits.
- PoSAT-1 (launch Sept. 26, 1993), UoSAT-12 (launch April 21, 1999), Tsinghua-1 (launch June 28, 2000) of Tsinghua University in Beijing, China - all S/C built by SSTL, use a compression algorithm by the name of AMPBTC (Adaptive Moment-Preserving Block Truncation Coding), achieving compression ratios of 2.5:1 and 4:1.
- TRACE of NASA/GSFC (launch April 2, 1998) uses a data compression scheme.
- KITSAT-3 of KAIST/SaTReC, Korea (launch May 26, 1999) uses on-board image data compression in JPEG, GIF and DPCM.
- Kodak developed an algorithm by the name of ADPCM (Adaptive Differential Pulse Code Modulation) for imagery compression. It is applied to Ikonos-2 (launch Sept. 24, 1999) and QuickBird-1 (launch Nov. 20, 2000 - launch failure) imagery. The real-time technique compresses 11-bit digital imagery to average values of 2.6 bits per pixel (reduction of 4.25 : 1) with little detectable loss in image quality. The compression rate enables more efficient on-board storage and downlink transmission of the data.
- The MTI (Multispectral Thermal Imager) mission of DOE (launch March 12, 2000) employs image data compression with the USES (Universal Source Encoder for Science data) chip for lossless compression (Rice coding algorithm, developed at JPL) [193) 194)]; the compression ratio is 2.5:1.
- The EO-1 satellite of NASA (launch Nov. 21, 2000) carries WARP (Wideband Advanced Recorder Processor) to demonstrate a number of high density electronic board advanced packaging techniques. Besides storage functions, it provides also a data compression of imagery.
- The MSG-1 (Meteosat Second Generation -1) satellite of EUMETSAT (launch in early 2002) uses encryption and JPEG compression on the HRIT (High Rate Information Transmission) data stream.
- IRS-P5 (CartoSat-1) of ISRO (launch 2002) uses an on-board ADPCM/JPEG compression algorithm of 3.2 : 1 to reduce the source data rate of 338 Mbit/s to 105 Mbit/s in the downlink.
- ESA sponsored the development of a quasi-lossless data compressor ASIC (Application Specific Integrated Circuit) which exploits the advantages of a wavelet-based data compression. The ASIC, developed by Saab Ericsson of Sweden, employs the Rice algorithm defined by CCSDS (with some additional features improving the functionality but maintaining the compatibility with the standard). The ASIC is baselined for use on the PROBA mission (launch in 2001) and considered for other projects as well.
- NASA/GSFC is developing a high-performance lossy data compression technique to support high-speed pushbroom frame-based imaging applications. [195)] The algorithm is based on MLT (Modulated Lapped Transform) and DCT techniques combined with bit-plane encoding. Flight qualified hardware implementations are in development. A functional chip set is expected by the end of 2001. The chip set is being designed to compress data in excess of 20 Msamples/s and support quantizations from 2-16 bits.

---

193) Note: "Rice" is an adaptive variable-length compression scheme on images, an algorithm developed by Robert F. Rice of JPL and implemented by Frank Rabe of the Technical University in Braunschweig, Germany for the Mars Pathfinder Lander IMP imaging system.

194) Robert F. Rice, "Some Practical Universal Noiseless Coding Techniques, Part III, Module PSI14,K+," JPL Publication 91-3, November 15, 1991.

195) P. Yeh, J. Venbrux, P. Bhatia, W. H. Miller, "A Real-Time High Performance Data Compression Technique for Space Applications," Proceedings of IEEE/IGARSS Conference, Honolulu, HI, July 24-28, 2000

- Adaptive Spectral Signature Recognition. The NEMO S/C of NRL (see D.25, planned launch in 2003) employs **ORASIS** (Optical Real-time Adaptive Signature Identification System), an automated and patented end-to-end HSI (Hyperspectral Imaging) algorithm and advanced processing system that performs on-board processing, thereby significantly reducing (compression factor >10) the amount of source data for on-board storage and ground transmission. The approach is to analyze each spectra in the scene sequentially, discarding duplicate spectra, and working only with the unique spectra and a map of their location in the scene. The ORASIS algorithm preserves 97-98% of the data fidelity. [196)]
- The ALOS satellite of NASDA (launch in 2003) uses a lossy data compression technique DCT (Discrete Cosine Transformation) and Huffman coding on PRISM (Panchromatic Remote-sensing Instrument for Stereo Mapping) to reduce the source data rate (1 Gbit/s) to about 240 Mbit/s in the downlink. DPCM is used on AVNIR-2 data (reduction from 160 Mbit/s to 120 Mbit/s in the downlink).

• Relay satellites for data communication from Earth observation and space exploration S/C. The introduction of space agency relay satellite constellations (in GEO) opened up entirely new coverage capabilities in data communication. Operationally, it represents a definite advantage (although more expensive) over the conventional distributed ground-station concept in combination with S/C recorders. Relay stations in space permit extended or continuous (three satellites) coverage between a S/C in LEO and the ground. This can be a decisive advantage to some missions (for instance to Shuttle missions, space station operations, etc.). Extended contact periods permit also the transfer of vast amounts of data.

- The TDRS (Tracking and Data Relay Satellite) series of NASA (built by TRW) pioneered this relay technology. The communication services provided by TDRS include such items as: a) simultaneous tracking of multiple satellites, b) multiple access schemes of S-band and Ku-band, etc. The TDRS antenna module consists of seven antennas: two single-access antennas that support both Ku-band and S-band user S/C communications (each 4.9 m diameter); a multiple access S-band phased array consisting of 30 helix-antenna elements; a 2 m diameter space-ground link antenna in K-band; an S-band omnidirectional antenna for TT&C support; a C-band antenna for commercial communications, and a K-band commercial communications antenna. Starting with TDRS-H (TDRS-8 in orbit, launch of TDRS-H on June 30, 2000), an additional link in Ka-band is provided to the existing S-band and Ku-band communication service capabilities. The receive data rates are 300 Mbit/s for Ku- and Ka-band and 6 Mbit/s for S-band. Transmit data rates are 25 Mbit/s for Ku- and Ka-band, and 300 kbit/s for S-band. The TDRS-8 S/C (dry mass = 2910 kg, 2.04 kW power) carries two steerable 5 m disk antennas for C-, Ku- and Ka-band support. In addition, a phased array antenna in C-band can receive signals from five different S/C simultaneously, while transmitting to one of them.

- Russia established also a relay satellite network, similar to NASA's TDRS system. The first satellite in the series, Luch-1 (Kosmos-1700), was launched Oct. 25, 1985. It was used to relay communication to the new MIR station (launched in 1986). A controlled reentry of MIR occurred on March 23, 2001 over the Pacific Ocean.

- Japan (NASDA, CRL, JST, etc.) started its geostationary data relay and tracking satellite (DRTS) system with the launch of a prototype satellite, COMETS (Communications and Broadcasting Engineering Test Satellite), on Feb. 21, 1998. COMETS provides S-band and Ka-band communication services. In addition, the satellite is used to test the following technologies: a) Inter-orbit communications for relay of communications between observation satellites or space stations in low-altitude circular orbits (LEO) and Earth stations; b) Advanced satellite broadcast technology for broadband region-specific broadcasts and high definition television (HDTV) broadcasts using Ka-band frequency bands. Note: a sec-

196) J. Bowles et al., "New results from the ORASIS/NEMO compression algorithm," Proceedings of SPIE, Imaging Spectroscopy V, Vol 3753, Denver , CO, July 19-21, 1999, pp. 226-234

ond-stage problem failed to put COMETS into GEO. - Two DRTS are being built, with planned launches in 2002 (DRTS-W) and 2004 (DRTS-E), to complement the system.

- ARTEMIS (Advanced Relay and Technology Mission Satellite) of ESA has been developed for testing and operating new telecommunications techniques in GEO environment, a prototype data relay system for the planned DRS (Data Relay Satellite) system. ARTEMIS features S-band and Ka-band communication services. In addition, ARTEMIS is used to test the following technologies: a) an L-band land mobile payload allows two-way communications, via satellite, between fixed Earth stations and land mobiles - trucks, trains or cars - anywhere in Europe and North Africa; b) a laser-optical payload with a relay terminal called SILEX (Semiconductor Intersatellite Link Experiment), M.2. - Some users of the ARTEMIS' communication services are: SPOT-4 (CNES), OICETS (NASDA), Envisat (ESA), and ISS.

Satellite	Comment
TDRS-1 (NASA), launch Apr. 4, 1983 (STS-6)	Position at 41° W, also support of VLBI experiment demonstrations
Luch-1 (NPO PM) Russia, launch Oct. 25, 1985	Position at 95° E
TDRS-B (NASA), launch Jan. 28, 1986 (STS-51L)	Challenger accident !!
TDRS-3 (NASA), launch Sept. 29, 1988 (STS-26)	Position at 171° W
TDRS-4 (NASA), launch March 13, 1989 (STS-29)	Position at 41° W, replaced TDRS-1. Introduction of commercial C-band services
TDRS-5 (NASA), launch Aug. 2, 1991 (STS-43)	Position at 174° W, replaced TDRS-3
TDRS-6 (NASA), launch Jan. 13, 1993 (STS-54)	Position at 46° W
TDRS-7 (NASA), launch Jul. 13, 1995 (STS-70)	Position at 150° W (deletion of K- and C-band packages)
COMETS (NASDA), launch Feb 21, 1998, H-II launch vehicle	Position at 121° E, a prototype data relay and tracking satellite (DRTS) system with S-band and Ka-band services
TDRS-8 (NASA) launch June 30, 2000 (TDRS-H)	Position at 47° W,S-band, Ku-band, and Ka-band
ARTEMIS-1 (ESA), launch July 12, 2001	Position at 9-16° E, a prototype DRS with S-band, Ka-band, and optical communication tests
DRTS-W (NASDA), planned launch in 2002	Position at 90° E (DRTS-W), S-band and Ka-band
DRTS-E (NASDA), planned launch in 2004	Position at 170° W (DRTS-E),S-band and Ka-band

Table 26: Overview of relay satellites operated by space agencies

- ETS-VIII (Engineering Test Satellite - VIII), a GEO technology spacecraft of NASDA (planned launch in 2003). The objective is to demonstrate mobile satellite communications and broadcasting system technologies (in S-band) with small-scale ground terminals such as hand-held terminals. ETS-VIII features LDR (Large-scale Deployable Reflector) to support the experiments.

- Disaster monitoring. Although a spaceborne (LEO) service of disaster monitoring is of great interest to a global community of civil protection authorities, there has been no satellite so far dedicated to the task of service provision during periods of natural disasters (volcano eruptions, floods, etc.). However, there have been imagery services, in particular for flood events, from a number of instruments on various satellites of existing space missions. Examples are: ERS-1/2 (AMI), JERS-1 (SAR, OPS), RADARSAT-1 (SAR), IRS-1C and -1D (LISS-3, PAN), Landsat-5 (TM), NOAA (AVHRR) series, and the SPOT (HRV) series. - Future missions are: BIRD a demonstration mission of DLR (planned launch in 2001); FOC/FUEGO (Fire Observation Constellation), in the definition phase by ESA and the EU, is planned to deal in particular with forest fire observations.

1.3.2 Special S/C Maneuvers and Rescue Operations: Examples

- S/C rotation maneuver of RADARSAT-1. During AMM (Antarctic Mapping Mission), in the time period of September to November 1997, the RADARSAT-1 spacecraft underwent a successful maneuver to rotate its normally right-looking radar array into a left-looking attitude. The shift involved rotating the satellite by 180° in yaw. Although other missions do regular yaw maneuvers to reorient S/C, RADARSAT-1 was the first satellite in remote sensing history to perform this maneuver twice (back to nominal right-looking configuration at the end of AMM). The left-looking configuration allowed to map the entire Antarctic continent for the first time.

- S/C rescue operations - GOES-10. In May 1997, the solar array mechanism of NOAA's newly launched GOES-10 (launch April 25, 1997 as an orbit backup) satellite, designed to rotate once each day to keep itself pointed normally at the sun, slowed down and shortly thereafter stopped working entirely. The subsequent rescue work by NOAA and NASA engineers led to an unusual solution by turning the spacecraft into an upside down orientation (a yaw-flip maneuver). This maneuver turned out to be highly successful in saving the spacecraft for regular operations, because it permitted the solar array to work in a backwards-rotating orientation. As a consequence, the on-board software was changed so that data from the instruments are provided to users as if the spacecraft were in the normal upright position. GOES-10 was activated on July 21, 1998 to replace the GOES-9 spacecraft. [197)]

- MIR accident. In the summer 1997, an emergency occurred that had considerable impact on MIR's fate. While docking with the MIR space station, a Progress cargo ship failed to respond to commands and rammed the Spektr module. As a result of this first collision between operating spacecraft, the Spektr module was punctured and leaked oxygen. The international crew aboard MIR retreated to a docked Soyuz spacecraft to prepare for emergency evacuation. However, the situation was brought under control. The punctured module's hatch was closed and the oxygen leak was stopped. - For the next two years, MIR crews did extensive maintenance and brought the space station back into full working order.

- SOHO rescue operations (launch of SOHO on Dec. 2, 1995). [198)] Operational control of the SOHO S/C was lost on June 25, 1998 (when SOHO spun out of control and communication was lost). Subsequent investigations by a joint NASA/ESA/MMS team showed that the loss of contact with SOHO had been preceded by a routine calibration of the spacecraft's three roll-control gyros. - On July 23, first S/C recovery operations were performed with the Arecibo radio telescope in Puerto Rico. The 305 m diameter dish antenna was used to transmit S-band signals to SOHO, while NASA's DSN at Goldstone functioned as receiver, thereby locating the spacecraft's echo. On Aug. 3, contact with SOHO was re-established and reception of the carrier signal by DSN. Attitude recovery was established on Sept. 16, 1998, resulting in a SOHO lock to the sun. SOHO was finally brought back to normal operating mode on Sept. 25. The only equipment failures at S/C level were in two of the three gyros. Instrument re-commissioning started on Oct. 5 with SUMER and ended with CELIAS on Oct. 24, 1998. - On Dec. 21, 1998, the last on-board gyro failed during the preparation of a routine orbit-correction and wheel management maneuver.

In January 1999, a **gyroless mode of operation** was devised and installed with a new software patch for a modification of the AOCS (Attitude and Orbit Control Subsystem), making SOHO the first three-axis-stabilized S/C of ESA to be operated without a gyro. The software patch was developed by MMS (Matra Marconi Space), now Astrium SAS of France (an ESA patent is pending). The software allows to determine SOHO's drift by measuring the changes in the speed of the spacecraft's momentum wheels (the software triggers the momentum wheels instead of the faulty gyroscopes). [199)] Final recovery from ESR (Emergency Sun Re-acquisition mode) and full S/C operation of SOHO was regained Feb. 1, 1999. A joint ESA/NASA investigation board came to the conclusion that the loss of the SOHO

---

197) G. J. Dittberner, "NOAA's Geostationary Operational Environmental Satellite (GOES) Systems and Plans," Proceedings of the EUMETSAT Meteorological Satellite Data User's Conference, Copenhagen, Denmark, Sept. 6-10, 1999, pp. 33-38

198) F. C. Vandenbussche, "SOHO's Recovery - An Unprecedented Success Story," ESA Bulletin, Nr. 97, March 1999, pp. 39-47

199) J. Bates, "Software Extends Satellite Missions When Gyroscopes Fail," Space News, Feb. 5, 2001, p. 8

S/C was a direct result of operational errors, a failure to adequately monitor S/C status, and an erroneous decision which disabled part of the on-board autonomous failure detection.

- PseudoGyro - is a software package of The Aerospace Corporation (El Segundo, CA) to save satellites from failure (US patent as of Feb. 1, 2000). The PseudoGyro package emulates a hardware gyro through software processes. The technique was successfully demonstrated in 1999 on a classified satellite of NRO (National Reconnaissance Office) which had experienced failures of its primary and secondary hardware gyros. PseudoGyro works in conjunction with orbital-attitude-estimation filtering techniques and takes advantage of all available sensors, including the vehicle itself, in determining attitude. It uses the principle of conservation of momentum to accurately determine the angular velocity of a spacecraft. It can be applied to virtually all types of space vehicles that are controlled by any type of momentum storage device. The technology can be integrated before or after a satellite is in orbit. In general, gyroscope-mimicking software might not replace replace the real thing, but it can reduce the workload on gyros, thereby extending their lives. PseudoGyro addresses also the need to reduce the risks and costs associated with new technology gyros. [200]

- S/C rescue operations - DS1 extended mission. In November 1999, two months after the end of its extremely successful primary mission and early in its extended bonus mission, the star tracker of DS1, responsible for the spacecraft's orientation, ceased operating. Rather than abandon the project, NASA engineers managed a deep-space rescue. They sent new software to DS1 (at a distance of 321 million km from Earth), turning an onboard camera (MICAS) into a navigation instrument. The challenging task was completed in June 2000 to resume thrusting in time to give DS1 a chance to encounter comet Borrelly in September 2001.

1.3.3 Cooperative Distributed Systems, Satellite Formations

Distributed space systems are multi-satellite systems that work together to perform a unified mission. Such systems are an alternative to monolithic satellite missions in which all on-orbit activities are performed on a single platform. Distributed space systems can range from global constellations offering extended service coverage to clusters of highly coordinated vehicles that perform distributed sensing. Global constellations of communication and navigation satellites are considered "loosely coupled" systems, they are in existence for many years. On the other hand, the use of spacecraft clusters for remote-sensing applications represents a new and rather unexplored type of constellation, namely "tightly coupled" systems with a new quality of service and support capabilities. [201]

- **Virtual satellite** - a new paradigm for a distributed remote sensing space segment (cooperative constellation of satellites). The idea of the EO mission concept is to use LEO clustered microsatellites (10 kg< S/C mass < 100 kg) and nanosatellites (1 kg < S/C mass < 10 kg), capable to cooperate and to share resources and functions (processing, communications, payload, and/or observation/mission functions) with each other, in order to perform the functions of a large single satellite. Sensors on satellites, flying in formation, may function as a single large sensor, offering entirely new observation capabilities [more frequent coverage through wider swaths, stereoscopic observations of imaging instruments (multifold coverage of common target regions), parallel provision of similar measurements at higher time-and-space sample rates (altimetry applications), virtual large aperture radar, etc.]. Distributed S/C control architectures are characterized by interactions between S/C, cooperation between S/C, and collective behavior among S/C within a constellation or formation. The introduction of concepts such as decision making, hierarchical control, decentralized control, etc., within the management of the formation enables spacecraft to cooperate with one another. Autonomy is a critical technology in this scenario that impacts

200) http://www.aero.org/news/current/pseudogyro.html

201) J. Bistrow, D. Folta, K. Hartman, "A Formation Flying Technology Vision," Proceedings of AIAA 2000 Space Conference and Exposition, Long Beach, CA, Sept. 19-21, 2000

every level of the design, from the subsystem, to the instrument, the platform, as well as the entire constellation. If a cooperative constellation includes a variety of basic, versatile instruments, for example UV, VIS and IR spectrometers, then 'virtual platforms' for different applications can be formed in space, on the fly, and "disassembled" (or reconfigured) later for other uses - flexibility to adapt to the evolving science needs and to implement the newest technologies during the course of an on-going mission are the basic requirements. In this new observation concept of **formation flying**, GPS plays a key role as a navigation instrument in the maintenance and control of the constellation (measurement of formation states, i.e., the relative orientations, positions and velocities of the vehicles). The ultimate goal of "virtual satellite" is a **satellite on a chip** providing all functions of conventional S/C (telemetry, tracking, and control). [202] [203] [204]

Examples of planned S/C cluster missions are:

- A program initiative, TechSat-21 (Technology Satellite of the 21st Century), was started in 1998 by various US DoD departments/directorates, namely AFRL (Air Force Research Laboratory), AFOSR (Air Force Office of Scientific Research), and DARPA (Defense Advanced Research Projects), with the objective to exploit the new paradigm and enabling technologies, such as MEMS (Micro-Electro-Mechanical Systems) and overall component miniaturization and function integration - that "involves satellites flying in tight formation that operate cooperatively to perform a surveillance mission." Under this effort, a variety of application missions are being considered including surveillance, passive radiometry, terrain mapping, and communications. A University Nanosatellite Program (UNP) was set up for this purpose by AFSOR/DARPA/NASA to fund ten university 'research projects centered on the design and demonstration of nanosatellites.' [205] [206] The funded projects within TechSat-21/UNP are:
- Emerald (Electromagnetic Radiation and Lightning Detection), a two-satellite project of Stanford University and Santa Clara University. A launch on a Shuttle mission is planned for 2001.
- 3CSat (3 Corner Sat), a constellation of three nanosatellites in formation. 3CSat is a joint project of the University of Colorado at Boulder, ASU (Arizona State University), and NMSU (New Mexico State University). The cluster is to be launched in 2002 with the objective to cooperate on science experiments (stereo imaging). While most satellites use a dedicated radio link, these will be the first to utilize cellular phone technology to communicate.
- ION-F (Ionospheric Observation Nanosatellite-Formation), a three-satellite project of USU (Utah State University), UW (University of Washington), and VT (Virginia Polytechnic Institute). Formation flying is a primary objective. Introduction of new technologies including micro-thrusters, magnetic gimbaled attitude control, advanced tether system, and an Internet-based operations center.
- Constellation Pathfinder, a one to three satellite mission of Boston University.
- Solar Blade Heliogyro Nanosatellite, a mission of CMU (Carnegie Mellon University), Pittsburgh, PA to demonstrate the solar sail concept.

Further projects of planned cluster missions and formation flying are:

- GRACE (Gravity Recovery and Climate Recovery), a cooperative US-German dual-minisatellite mission with a launch in 2001. The mission concept makes use of measurements of the inter-satellite range and its derivatives between two co-planar satellites (in

202) G. H. Fountain, et al., "A Technology Path to Distributed Remote Sensing," IAA 2nd International Symposium on Small Satellites for Earth Observation, Berlin, April 12-16, 1999, pp. 189-193

203) J. Esper, P. V. Panetta, et al., "NASA/GSFC Nano-Satellite Technology for Earth Science Missions," IAA 2nd International Symposium on Small Satellites for Earth Observation, Berlin, April 12-16, 1999, pp. 219-225

204) F. H. Bauer, K. Hartman, J P. How, et al., "Enabling Spacecraft Formation Flying through Spaceborne GPS and Enhanced Automation Technologies," Proceedings of the ION-GPS Conference, Nashville TN, Sept. 15, 1999

205) W. Ferster, "Tiny Satellite Fleet May Function as One Craft," Space News, Aug. 17-23, 1998, p. 7

206) http://www.vs.afrl.af.mil/vsd/techsat21/

low-altitude and polar orbits), using a microwave tracking system and a GPS receiver system for precision orbit determination.

- DMC (Disaster Monitoring Constellation), a five microsatellite constellation of SSTL (Surrey Satellite Technology Ltd), Surrey, UK. The overall objective is to provide global daily imagery at 36 m resolution for the monitoring and mitigation of natural and man-made disasters and dynamic Earth observation. DMC is planned for launch in late 2002.
- The cooperative missions of CloudSat and ESSP-3, flying active and passive instruments, will be launched in 2004.
- Demonstration of close-proximity formation flying and sparse aperture sensing. TechSat-21 of DoD/AFRL is a constellation of three microsatellites (each of about 120 kg) flying in formation, each equipped with an X-band radar instrument (the TechSat-21 system is being built by ITN Energy Systems Inc. of Denver, CO). The objective is to demonstrate the MTI (Ground Moving Target) capability by operating as a virtual satellite with X-band transmit and receive payloads on each platform to form a large, sparse aperture system. A crosslink system (with carrier phase differential GPS positioning, intersatellite ranging and communication, phased array antennas, etc.) as well as a micropropulsion PPT (Pulsed Plasma Thruster) system, is provided for each spacecraft. A launch of TechSat-21 is planned for 2003/4.
- Trailblazer/ST5 (Space Technology 5) is a NASA mission in the New Millennium Program of three microsatellites, flying in formation, with a planned launch in 2003. The objective is to test new technologies in hardware and software.
- ROCSat-3/COSMIC (Constellation Observing System for Meteorology, Ionosphere and Climate), is a collaborative US/Taiwanese (led by UCAR and NSPO) microsatellite constellation, planned for launch in 2005. In the scenario, a LEO constellation of six spacecraft collects atmospheric remote sensing data for operational weather prediction, climate, ionospheric (space weather monitoring), and gravity research (refractive GPS radio occultation measurements).
- ACE (Atmospheric Climate Experiment), an ESA Explorer mission to obtain profiles of atmospheric parameters with occultation measurements using a cluster of six LEO microsatellites. The baseline configuration of the cluster considers two orbital planes. A launch is planned for 2005.
- On the military side, [207] the US NRO (National Reconnaissance Office) awarded a contract in Sept. 1999 to Boeing Space & Communications Group, Seal Beach, CA, to develop a constellation of imaging (optical and radar) microsatellites, based on the "Future Imagery Architecture." Planned first launches are in 2005.
- NASA/GSFC is defining MagCon (Magnetospheric Constellation), a mission of about 100 nanosatellites (deployed from a single mother ship) to acquire vector images of the magnetic and plasma flow fields in highly elliptical orbits (from 3 to 50 R_E).Simultaneous, multi-point observations of the Earth's magnetospheric environment. The overall science objectives are to a) determine the equilibria of the magnetotail, b) understand the responses of the magnetotail to the solar wind, c) reveal the instabilities of the magnetotail. A launch of MagCon is planned for 2006.
- NASA/JPL is planning to fly its ST-3 (Space Technology 3, formerly known as Deep Space-3) mission in 2006, using two S/C in heliocentric orbit. In early 2001, ST-3 was renamed to StarLight. The objective is to demonstrate separated spacecraft interferometry (first long baseline optical interferometer in space). StarLight tests and validates technologies and concepts requiring precise formation flying (separation distances of up to 1 km - requiring accuracies in relative range and attitude of 1 cm and 1 arcmin, respectively) in order to act as one giant telescope system.
-

207) W. Ferster, "NRO Awards Giant Satellite Contract to Boeing," Space News, Sept. 13, 1999, p. 1

• Intersatellite communication. [208) 209)] A prerequisite to formation flying is intersatellite communication (also referred to as cross-communication) for the control of the spacecraft cluster. Such a service may be provided by the GPS constellation offering a range of sensing options for a formation including: absolute positioning, relative vehicle positioning, attitude estimation, and precise timing. This approach is based on a combination of CDGPS (Carrier-phase Differential GPS) and GPS-like transceivers (the challenge of using CDGPS is in estimating the cycle ambiguity to sufficient accuracy). The transceivers are RF-based cross-communication devices that can also be used as local ranging systems. Examples of such transceivers include:

- CLT (Cross Link Transceiver) of NASA designed and developed by JHU/APL. The CLT provides both an absolute and relative navigation solution (position and velocity) as well as precision time recovery and a steered one pulse-per-second output. Orbit determination is provided by the reception of GPS signals and processing by an extended Kalman filter. In addition, crosslink signals support relative navigation , with the potential for both, direct solutions as well as relative GPS solutions which rely upon the computation of double differences of GPS data in a paired manner among spacecraft. [210)]
- SPTC (Stanford Pseudolite Transceiver), a relative navigation and crosslink system for formation-flying spacecraft
- Star Ranger of AeroAstro Inc. of Herndon, VA.
- LPT (Low Power Transceiver) of ITT, Fort Wayne, IN (and NASA). The LPT integrates TDRS S-band two-way communications and GPS navigation in a compact package. The first demonstration flight is expected aboard the Shuttle in 2002.
- AFF (Autonomous Formation Flying) sensor of JPL. AFF is planned to be flown on the ST-3 (Space Technology-3) with a launch in 2003.
- NCLT (Nanosatellite Cross Link Transceiver) of CLT heritage is under development at APL. The UNP University Nanosatellite Program) satellites (of DoD/NASA) employ this device as well as AFRL's TechSat-21 program and NASA's ST-5 (Space Technology-5) constellation. NCLT is a multicard, stand-alone system that implements three fundamental functions: on-board processing, GPS receiver, and crosslink communications. NCLT supports multiple-access communications architectures (FDMA/CDMA).

The first onboard experiment with a relative GPS navigation receiver was probably conducted on the NASDA spacecraft ETS-VII (Engineering Test Satellite VII) with a launch on Nov. 28, 1997. The instrument RGR (Relative GPS Receiver) was used successfully in the docking demonstrations (see M.8.1).

• Other distributed observation strategies of formation flying. This category considers so-called "trains of satellites performing similar observation objectives with their payloads in closely spaced sequences." Their orbits are optimized to coincide (within a few minutes) with those of other satellites. Examples are:

- The Landsat-7, EO-1, SAC-C and Terra "morning constellation" train (in formation as of early 2001). There is 1 minute separation between Landsat-7 and EO-1, a 15 minute separation between EO-1 and SAC-C, and a 1 minute separation between SAC-C and Terra. The objective is to compare coincident imagery from the ETM+ and ALI instruments. The "paired scene" images are used to evaluate the performance of ALI.
- The TOPEX/Poseidon and Jason-1 spacecraft are planned to fly in tandem (as soon as Jason-1 is launched in 2001) with a one minute separation time enabling calibration to

208) C.-W. Park, J. P. How, L. Capots, "Sensing Technologies for Formation Flying Spacecraft in LEO using CDGPS and Inter-Spacecraft Communication System," ION GPS 2000, Sept. 19-22, 2000, Salt Lake City, UT, pp. 1595-1607

209) E. A. Olsen, P. A. Stadler, M. S. Asher, "Long-Baseline Differential GPS based Relative Navigation for Spacecraft with Crosslink Ranging Measurements," ION GPS 2000, Sept. 19-22, 2000, Salt Lake City, UT, pp. 1612-1621

210) A. A. Chacos, P. A. Stadter, W. S. Devereux, "Autonomous Navigation and Crosslink Communication Systems for Space Applications," JHU/APL Technical Digest, Vol. 22, No 2, 2001, pp. 135-143

be performed. [211] After the checkout/verification phase, the tandem mission will fly in an interleaved orbit (i.e., the TOPEX/Poseidon orbit will be moved to an orbit that produces interleaved groundtracks with a 1.4° longitude spacing from the Jason-1 tracks). The intent is to address the following topics: a) double the spatial sampling to enhance the resolution of the Rossby waves and eddies; b) quadruple the crossover points for the estimation of the current velocity vector; c) enhanced coastal tide models; and d) obtain improved global change detection.

- The Aqua, ESSP-3, PARASOL, CloudSat and Aura spacecraft will form the so-called "A-train" (Aqua in the lead and Aura at the tail) or afternoon constellation (formation flight starting in the 2004 time period). The objective is to provide a coincident set of data on aerosol and cloud properties, radiative fluxes and atmospheric state essential for accurate quantification of aerosol and cloud radiative effects.

1.3.4 Bistatic Systems in Remote Sensing

The bistatic remote-sensing concept refers to a measurement geometry in which the transmitter and receiver locations of an active system (radar, etc.) are separated by a distance comparable to that of the target distance. In addition, bistatic observations require accurate time synchronization; antenna pointing between transmitter and receiver may also be involved. In contrast, a **monostatic system** refers to a measurement arrangement, in which both transmitter and receiver are collocated, often the same antenna is used for both functions. The great majority of all radars (SAR instruments, Doppler radars, lidars, etc.) in use today (and flown on spaceborne missions) are of the monostatic type, providing their own illumination and measuring the echo at the same location. The dual monostatic or repeat-pass approach (for interferometric operation) has been demonstrated by the ERS-1/2 tandem mission (Aug. 1995 to May 1996). In a tandem mission in monostatic mode, the two spacecraft have equal status, there is no direct link between them. In the bistatic mode concept, one satellite (the master) generates the radar pulses, and the reception of the echoes by both master and slave must be synchronized. [212] [213]

It is interesting to note here that all early radar experiments, conducted by experimenters in various countries like in France, Germany, Italy, Japan, Russia, UK, and USA (from about 1905 on until after World War II), were of the bistatic type. In the USA, for instance, an observation of the radar effect was made in 1922 at the U.S. Naval Research Laboratory (NRL) in Washington, D.C. NRL researchers conducted an experiment by positioning a radio transmitter on one shore of the Potomac River and a receiver on the other. A ship sailing on the river caused fluctuations in the intensity of the received signals when it passed between the transmitter and receiver. Today, such a configuration is called a **bistatic** radar system. In spite of the promising results of this experiment, U.S. Navy officials were unwilling to support further work at the time. Practically all early military radar systems until the end of World War II) were of the bistatic type. - In the space age era, bistatic radar systems employing an orbiting transmitter and an Earth-based receiver have been used for lunar and planetary exploration.

Since the 1990s, bistatic systems are also finding ever increasing applications in the wide field of Earth observation. A typical example is atmospheric research where ground-based bistatic multiple-Doppler networks have significant scientific and economic advantages accruing from the use of only single sources of illumination. Individual spatial volumes are viewed simultaneously from multiple look angles, minimizing storm evolution induced errors. The passive receivers in a bistatic network do not require expensive transmitters, moving antenna hardware, or operators. Thus they require only a small percentage of the invest-

211) L. L. Fu, Y. Menard, "Summary of the Third Joint TOPEX/Poseidon and Jason-1 Science Working Team Meeting," The Earth Observer, Jan/Feb. 2001, Vol. 13, No 1, pp. 17-18

212) A. Moccia, N. Chiacchio, A. Capone, "Spaceborne bistatic Synthetic Aperture Radar for remote Sensing applications," International Journal of Remote Sensing, Vol. 21, No 18, 2000, pp. 3395-3414

213) N. J. Willis, "Bistatic Radar," Artech House, Boston, 1991, ISBN: 0-89006-427-X

ment needed to field traditional transmitting radars. In 1993, NCAR (National Center of Atmospheric Research) developed such a dual-Doppler weather radar network at its facilities in Boulder, CO. The network provides three-dimensional fields of full vector winds, including directly measured vertical precipitation particle velocities for numerous applications in meteorological research, aviation, forecasting, media, and education. [214) 215) 216)]

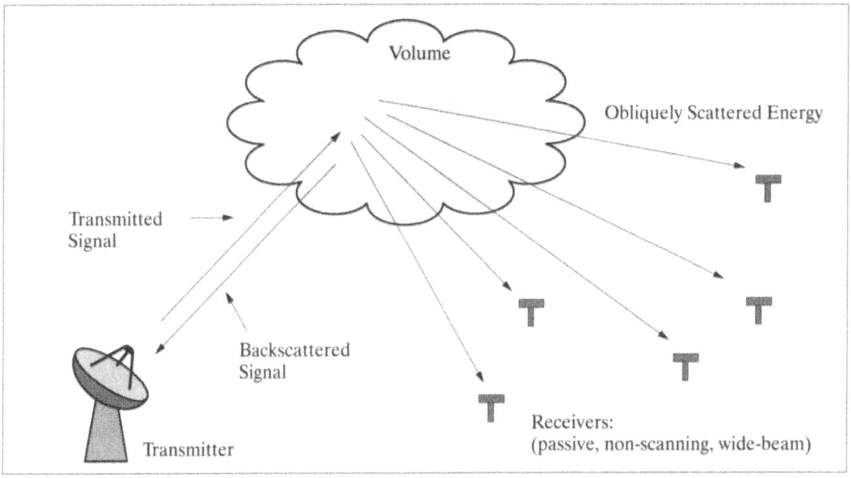

Figure 3: Schematic illustration of a ground-based bistatic multiple Doppler network

Bistatic or multistatic radar systems offer some advantages: [217)]

- A significant cost reduction can be achieved in the design and operation of a bistatic system (passive radiometers are significantly less expensive than active SAR instruments). Microsatellites may be used to accommodate the passive receiver antennas.
- The physical separation of the transmitter and receiver function permits for instance designs, in which the number of participating receivers can be any number in an observation system; thus, the receiver function is of a parasitic nature, simply passively participating in an observation environment/arrangement, it is in a sense "independent" of the transmitter function (not in the time domain for SAR applications). Systems with two or more receiving sites are also referred to as multistatic systems. They employ overlapping coverage of the illumination footprint and combine target data coherently and non-coherently at a central location.
- Bistatic systems offer the potential of upgrading existing monostatic missions. For instance, an existing SAR mission may be complemented with a constellation of passive receiver spacecraft in its neighborhood. This improvement results in added value to the master mission offering bistatic SAR data. Interferometric processing of the bistatic data may be unrealistic, due to rather large antenna separations; however, cross-track and along-track interferometry may be achieved. See also the Cartwheel mission below.

There may be also some disadvantages in upgrading an existing monostatic system: a) the transmitting antenna and satellite are non-cooperative; i.e., they are unaware of the new

214) J. Wurman, "Vector Winds from a Single Transmitter Bistatic Dual-Doppler Radar Network," Bulletin of the American Meteorological Society, July 1994

215) J. Wurman, M. Randall, C. Frush, E. Loew, C. Holloway, "Design of a Bistatic Dual-Doppler Radar for Retrieving Vector Winds using One Transmitter and a Remote Low-Gain Passive Receiver," Proceedings of IEEE, Dec. 1994

216) J. Wurman, S. Heckman, D. Boccippio, "A Bistatic Multiple-Doppler Radar Network: Part 1, Theory," Journal of Applied Meteorology, Dec. 1993

217) http://aaron.ou.edu/bistatic/

satellite with its receiver; in this case, time-synchronization of the receiving-only antenna may pose added complexities on the data processing side. Hence, long-term plans are needed for upgrading options.

- The GPS and GLONASS navigation constellations are also examples of bistatic systems. By its very nature, each navigation constellation provides the transmitter function while the "signal user community" at its numerous instant locations plays the role of the receiver function. The direct reception of GNSS signals can be used a) as navigation function, b) as timing reference, and/or c) as remote sensing function (for refractive sounding, etc.).
- Lately, indirect or reflected GNSS signals are also being used as information. [218] In radar remote-sensing nomenclature, GPS reflection data (of the ocean surface or of any ground surface) can be characterized as a "bistatic radar scatterometer." The forward scatter from the transmitter in the direction of the receiver is recorded. An entire new spectrum of remote sensing applications is beginning to unfold with the measurement of this new source of reflection data. Initial experiments with airborne GPS receivers were conducted in the early 1990s (see also "ocean reflection measurements of GNSS signals" in 1.7.5).
- Reflected ocean wave spectra, caused by Bragg scattering, were already performed with LORAN-A (Long-Range Navigation) radionavigation systems in the 1960s. The LORAN-A bistatic experiments demonstrated that the Doppler shift of the radar echo is exactly equal to the wave frequency; it is also related to the bistatic angle. [219] See also LORAN in Glossary.
- **Bistatic SAR systems.** SRTM (Shuttle Radar Topography Mission), a cooperative project of NASA, DLR and ASI on STS-99 (Feb. 11-22, 2000, see J.25), was the first spaceborne bistatic SAR mission where both SAR systems operated with the main antenna of each instrument located in the open cargo bay of the Shuttle, and a second receive antenna mounted on a deployable outboard mast, respectively (a single-pass interferometer on a 60 m baseline).

Interferometric capabilities are becoming a key requirement at the turn of the 21st century to obtain topographic performance. A number of projects under development are planning to test the bistatic concepts in the spaceborne arena of close formation-flying satellite missions (offering larger baselines than SRTM). A viable option in such a distributed concept is for example a master-slave implementation. A single so-called master spacecraft is equipped with an active radar (i. e., a transmitter and receiver function), while the rest of the constellation (slave spacecraft) is outfitted with just a radiometer (receiver). All observation instruments of the constellation are directed to look into the same footprint. Thus, all passive instruments in the constellation are also able to detect coherently the bistatic responses from the single active radar of the master spacecraft as well. [220] The multi-location (or nodal) spacecraft data collected in this fashion provides automatically interferometric imagery after ground processing. The distributed apertures improve also the performance of the system with regard to detection and resolution capabilities. A wealth of independently sampled angle-of-arrival data can be collected, the constellation forms a large but sparse coherent array. By virtue of its sparseness, the independent apertures look through different parts of the ionosphere, thus temporal and spatial variations on the scale of their separation could adversely affect their operation. This innovative bistatic observation concept has the potential to reduce the mass, power and cost of the constellation as a whole.

Examples of the first collaborating constellations demonstrating the new bistatic SAR observation technologies are under intensive study at the start of the 21st century.

218) A. Komjathy, J. L. Garrison, V. Zavorotny, "GPS: A New Tool for Ocean Science," GPS World, April 1999, pp. 50-56

219) A. M. Peterson, C. C. Teague, G. C. Tyler, "Bistatic-radar observation of long-period, directional ocean-wave spectra with LORAN-A," Science, Vol. 170, 1970, pp. 158-161

220) Note: Coherent detection requires synchronization between master and slaves. The synchronization may either be achieved using a crosslink from master to slaves, or by accurate synchronization of all involved to the same source.

- In Europe, CNES and DLR (in cooperation with ESA) are planning a so-called **Interferometric Cartwheel** (ICW) mission constellation of three passive microsatellites, each equipped with a receiver dish antenna (a slave receiver), to co-orbit with a soon-to-come conventional SAR mission (as host or "illuminator" or master satellite). A Cartwheel launch is planned for late 2004 followed by a four month commissioning phase prior to the operational phase in 2005. The candidate conventional SAR missions considered for Cartwheel are: Envisat, RadarSat-2, TerraSAR, and ALOS (in fact, any radar satellite mission provides a suitable context for a cartwheel mission). An "interferometric cartwheel orbit" of the passive constellation is employed. In such a cartwheel concept, the locations of the passive satellites are planned to be ahead or behind of the transmitter satellite at a fairly constant angle. Their orbits are the same as the transmitter spacecraft orbit, but with an eccentricity slightly different from that of the transmitter spacecraft orbit (the three passive satellites are moving relatively to a common reference point, in motion and coplanar with the transmitter S/C; the common reference is either lagging or preceding the transmitter S/C orbit at a constant angle). The new orbital concept is considered to offer a good geometric stability of the baselines, both vertically and horizontally, essential for interferometric processing (see O.12.8 for a definition of the Cartwheel orbit). [221) 222) 223) 224)]
- The Canadian Space Agency (CSA) is implementing an interferometric SAR monitoring concept for its RadarSat-2 (launch 2003) and RadarSat-3 tandem missions (launch 2005). The interferometric capability (beyond 2005) is a key requirement on both missions. As a consequence, the RadarSat-3 design is based on that of RadarSat-2. As of 2001, either a bistatic or a "pursuit dual monostatic" implementation are still under consideration.
- GPM (Global Precipitation Mission), a follow-up mission to TRMM with a constellation of eight microsatellites and advanced rain-measuring radar instruments, is in the proposal state as of 2000/1 (NASA/NASDA). A possible launch date is 2007.

1.3.5 On-orbit Propulsion

• On-orbit chemical propulsion systems. Traditionally, low-thrust propulsion tasks are accomplished with cold gas or hot gas thrusters (bi-propellant technology). These thrusters, which are quite simple in principle, are perfectly suited to most of the applications (attitude control, etc.). There are some drawbacks to the system. Of course, all gases need to be stored in pressurized tanks. This means that extra weight is added to the spacecraft due to the heavy structure, hermetic seals, piping etc. The specific impulse (Isp) of cold gas thrusters is very low (in the order of two orders of magnitude lower than in electric propulsion systems). Low specific impulse means a relatively large amount of propellant to deliver a given total impulse. The total amount of propellant that can be carried poses a limit to the mission lifetime.

• **On-orbit electric propulsion systems** use electrical power provided by the spacecraft's solar panels to accelerate a propellant to high velocity creating a reaction which causes the movement of the spacecraft. Although the trust generated in this fashion is rather small, electric propulsion systems nonetheless provide the capability of raising a satellite's orbit and/or to maintain platform attitude - thereby offering the potential for considerably extended mission life times. A further advantage is spacecraft mass reduction (the fuel load on

221) H. Runge, R. Bamler, J. Mittermayer, F. Jochim, D. Massonnet, E. Thouvenot, "The Interferometric Cartwheel for Envisat," 3rd International Symposium of IAA, Berlin, April 2-6, 2001, pp.187-190

222) The term "cartwheel orbit" was initially coined by D. Massonnet of CNES in 1997

223) D. Massonnet, "Capabilities and Limitations of the Interferometric Cartwheel," CNES paper presented at the CEOS Workshop in Toulouse, October 1999

224) J. Mittermayer, G. Krieger, M. Wendler, A. Moreira, E. Thouvenot, T. Amoit, R. Bamler, "Preliminary Interferometric Performance Estimation for the Interferometric Cartwheel in Combination with ENVISAT ASAR," CEOS Workshop, Tokyo, Japan, April 2-5, 2001

conventional propulsion systems is considerable for long-term missions). The electric propulsion technology can also be used for missions requiring drag compensation. Several thruster technologies have been demonstrated or are being demonstrated. A new era in space propulsion is evolving, gaining momentum in the late 1990s. The basic thruster technologies are: (see also chapter O.14)

1) Electrothermal propulsion systems: arcjets, resistojets
2) Electrostatic propulsion systems: Gridded ion engine, Hall-effect thrusters (free electron drift propulsion), FEEP (Field Emission Electric Propulsion)
3) Electromagnetic propulsion systems: Pulsed Plasma Thruster (PPT), Stationary Plasma Thruster (SPT), and Magneto Plasma Dynamic (MPD) thrusters.

Background/Applications:

- Research on Hall-effect thrusters started in the early 1960s notably in the Soviet Union. The first spaceborne plasma thruster was flown on the USSR satellite Zond-2 (launch Nov. 30, 1964). The Zond-2 S/C, on its way to Mars, was equipped with secondary electric thrusters. The Yantar-1 satellite (launch Oct. 1, 1966 from Kapustin Yar into a 400 km orbit), a USSR military surveillance S/C built by TsKB, flew the first argon ion engine. In Feb. 1972, the Meteor-1-10 S/C (launch Dec. 29, 1971) was the first satellite to correct its orbit with an SPT engine. Since 1971, Soviet/Russian spacecraft (more than 15 missions) have used low-power xenon plasma thrusters, referred to as SPT (Stationary Plasma Thrusters), for station-keeping and on-orbit maneuvering functions on the Express communication satellite series. The early SPT family engines (SPT-60 and SPT-70) were manufactured by Fakel Company of Kaliningrad, Russia.

- In 1964, a pair of NASA/LeRC ion engines were launched on a Scout rocket from Wallops Island, VA, under the name SERT-1 (Space Electric Rocket Test); one of the two thrusters onboard did not work, but the other operated for 31 minutes. A follow-up mission, SERT-2, which carried two ion thrusters, one operating for more than five months and the other for nearly three months.

- The navigation satellites of the US Navy (designed and built by JHU/APL), namely Triad-2 (also referred to as TIP-2, launched Oct. 11, 1975) and Triad-3 (TIP-3, launched Sept. 1, 1976), were each equipped with a redundant pulsed-plasma thruster (PPT) electric propulsion system, used for drag compensation (drag-free satellite).

- SCATHA (Spacecraft Charging at High Altitude) satellite of the USAF (launch Jan. 30, 1979) flew the first xenon ion engine, built by Hughes Research Laboratories in Malibu, CA.

- The Hall thruster technology has been under intensive [governmental and industrial (TsNIIMASH)] investigation in Russia since the early 1990s. The first US spacecraft with a Hall thruster is the STEX (Space Technology Experiment) mission of DoD (launch Oct. 3. 1998). STEX flies EPDM (Electric Propulsion Demonstrator Module), built by NASA/GRC, to provide orbit-raising and station-keeping functions (see O.14.2). The TAL-D55 (Thruster with Anode Layer) Hall effect thruster used in EPDM was developed by the Russian Central Research Institute of Machine Building (TsNIIMASH).

- DS1 (Deep Space 1) of NASA/JPL (launch Oct. 24, 1998) employs IPS (Ion Propulsion System), designed and built by the NSTAR team. It is the first time that IPS technology was demonstrated as the spacecraft's primary engine (smooth and continuous IPS operation from Nov. 24, 1998 for over 200 days) throughout the prime mission which ended in Sept. 1999. - The EO-1 spacecraft of NASA/GSFC (launch Nov. 21, 2000) uses PPT technology.

- ARGOS of DoD (SMC/TE at Kirtland AFB, NM), launch Feb. 23, 1999. The spacecraft uses arcjet propulsion technology for orbit transfer, maneuvering capability, and attitude adjustment. The electric propulsion system is ESEX (Electric Propulsion Space Experiment), built by TRW. Demonstration of the largest electric propulsion system (26 kW of power input) so far.

- UoSAT-12, the first minisatellite of Surrey (launch April 21, 1999), employs a resistojet electric propulsion system. The thruster can impart velocity changes of up to 10 m/s to the S/C.

- SNAP-1 (Surrey Nanosatellite Applications Program-1), launch June 28, 2000. SNAP-1 of SSTL carries a micro-propulsion system, the size of a pencil, using butane as a cold gas system. The propulsion system is used to maneuver SNAP-1 to rendezvous with Tsinghua-1 of the University of Beijing (see D.40.17).

- The AMSAT-P-3D [Amateur Satellite-Phase-3D, now called AMSAT-Oscar 40, launch Nov. 15, 2000 from Kourou into HEO (Molniya-type orbit)] spacecraft employs an arcjet thruster by the name of ATOS (Arcjet Thruster on OSCAR Satellite) for on-orbit fine control of the S/C. ATOS was developed at IRS (Institut für Raumfahrtsysteme) of the University of Stuttgart, Germany and funded by DLR. ATOS is a 750 W ammonia arcjet with a specific impulse Isp of 4600 m/s. Orbit maneuvers are achieved by a separate 400 N chemical thruster.

- EO-1 (Earth Observing-1) of NASA/GSFC (launch Nov. 21, 2000). A PPT (Pulsed Plasma Thruster) of General Dynamics, Ordinance and Tactical Systems, Aerospace Operations [formerly Primex Aerospace (formerly Olin (formerly Rocket))] is used to demonstrate on-orbit electromagnetic propulsion technology and to provide a spacecraft precision-pointing capability.

- The ARTEMIS data relay satellite of ESA (launch in 2001), built by Alenia Spazio of Rome, Italy, includes also a xenon electric propulsion thruster system (redundant systems: RIT-10 of DASA, Germany, and EIT of MMS, UK). The RIT-10 system was test-flown on the EURECA-1 mission (launch July 31, 1992 - retrieval July 1, 1993).

- The ESA mission SMART-1 (Small Mission for Advanced Research in Technology) to the moon with a planned launch in 2002, employs the STS (Stationary Plasma Thruster) concept, referred to as PPS-1350, to demonstrate electric propulsion as the primary engine of the S/C. The PPS-1350 electric propulsion engine of SNECMA Moteurs (Paris, France) has also been selected for France's Stentor technology demonstration satellite.

- On the commercial side (communications satellites), on-orbit electric propulsion makes great inroads. The Lockheed Martin S7000 platform series communication satellites, such as Intelsat-8, TelStar-IV, and Echostar, use hydrazine arcjets for station-keeping functions. Hughes is supplying xenon ion engines as an option on high-powered versions of the HS-601 platform. PanAmSat-5 (PAS-5) of PanAmSat Corp., a telecommunications satellite launched July 30, 1997 (and launch of PanAmSat-1R launch Nov. 15, 2000), use XIPS (Xenon Ion Propulsion System, built by HSC of Hughes Electronics Corp., Los Angeles, CA) on the HS-702 platform for station keeping and attitude maintenance. In the latter case, XIPS is the spacecraft's sole means of station keeping.[225] By the end of 2000, 12 communication satellites were using XIPS and variations thereof (Note: as of Oct. 2000 the Boeing Co. acquired three units within Hughes Electronics Corporation). SSL (Space Systems/Loral) and ASI (Alcatel Space Industries) are manufacturing communication satellites (Telstar-8 and Astra-1K, respectively) which use SPT (Stationary Plasma Thrusters) type electric propulsion systems to support such functions as stationkeeping. The Russian company Fakel Design Bureau in Kaliningrad produces the SPT-100 electric propulsion system for many applications in the commercial satellite market. The function of electric orbit raising is also being considered for large communication satellites. An arcjet electric propulsion system of General Dynamics (Primex) had its first flight in 1994. As of January 2001, there were over 100 Primex arcjets in orbit.

- **Solar sails** [refer to S/C concepts which utilize the momentum transfer of photons (solar radiation pressure) on large, highly reflecting sails in space for passive propulsion and

225) http://www.hughespace.com/factsheets/xips/xips.html

orbit transfer functions]. In 1990 RSC Energia of Korolev (Moscow region), Russia, as the lead sponsor, formed SRC (Space Regatta Consortium - an association of 15 enterprises) with the objective to promote, develop and demonstrate satellite-based solar sails, so-called solarcraft (large deployable structures based on thin film sails). The first practical step toward this end was Znamya-2 (Banner), a demonstration experiment conducted on Feb. 4, 1993, when a segmented disk sail (20 m in diameter) was deployed by centrifugal forces from the Progress M-15 spacecraft, docked at the MIR space station (the experiment included verification of the system concept, stability and control of the structure, etc.). The Znamya-2 experiment represents the first solar sail deployment in space.

In parallel to the Znamya-2 deployment, a further experiment, "Novey Svet" (New Light) was conducted on the same solar sail structure with the objective to illuminate Earth from space. The first illumination from space took place in the early morning hours before sunrise over Western Europe. A reflected spot of light of about 5 km in diameter travelled from southern France through Switzerland, Germany, Czech Republic, and Poland, it then disappeared in the early sunlight of Belarus.

A second Znamya experiment, called Znamya-2.5, from MIR (actually the departing service vehicle Progress M-40) took place on Feb. 4, 1999 with a 25 m diameter sail (solar reflector) structure. The objective of Znamya-2.5 was to demonstrate beaming of solar power (reflected sun light) from space to Earth for artificial target illumination (spot light of 5-7 km in diameter) under controlled conditions; also test of film structure and test of prolonged illumination. However, a problem in the deployment mechanism resulted in an abort of the demonstration.

The Cosmos-1 Solar Sail Project, a privately funded venture of the Planetary Society, Pasadena, CA. The objective is to conduct a solar sail demonstration flight. In a first step, a sub-orbital test flight of an inflatable tube sail deployment system (600 m^2 of mylar sail) on a Volna launch vehicle (a converted SS-N 18 ICBM) took place on July 23, 2001 (submarine launch from Barents Sea). The Russian Babakin Space Center is the prime contractor and spacecraft integrator of the Planetary Society. The sub-orbital flight of the 40 kg S/C is from the Barents Sea to the Kamchatka peninsula (solar sail deployment at 400 km altitude). Initial flight analysis revealed a capsule separation failure (neither the solar sail deployment test nor the re-entry capsule inflation sequence that were planned for this sub-orbital test flight were carried out). The Cosmos-1 test craft was to separate from its capsule and unfurl its two 15 m long, wing-like blades before making a soft landing on Kamchatka. - In a second step, an orbital flight of Cosmos-1, also on a Volna launch vehicle, is planned for the end of 2001. [226)]

A mission of CMU (Carnegie Mellon University), Pittsburgh, PA is called "Solar Blade Heliogyro Nanosatellite" (planned launch in 2002). The objective is to demonstrate the solar sail concept for attitude precession, spin rate management, and orbital adjustments. The mission is supported within UNP (University Nanosatellite Program) of DoD and NASA.

A NASA/JPL and DLR [227) 228) 229)] proposal of ODISSEE (Orbital Demonstration on an Innovative Solar Sail driven Expandable structure Experiment) plans a technology demonstration mission (ASAP microsatellite on Ariane-5). A solar sail ground deployment demonstration, conducted by a DLR, ESA and INVENT project team in Dec. 1999, was highly successful.

226) http://www.planetary.org/solarsail/missions/planetary_solar_sai.html

227) M. Leipold, C. E. Garner, et al., "ODISSEE -A Proposal for Demonstration of a Solar Sail in Earth Orbit," Proceedings of Third IAA Conference on Low-Cost Planetary Missions, Pasadena, CA, April 27 - May 1, 1998

228) M. Leipold, C. E. Garner, "Solar Sails - Exploiting the Space Resource of the Solar Radiation Pressure," ESA Conference on 'Engineering and Economic Aspects into the 21st Century,' 20-22, Oct. 1998, Cagliari, Italy

229) M. Leipold, M. Eiden, C. E. Garner, et al., "Solar Sail Technology Development and Demonstration," Proceedings of the 4th IAA International Conference on Low-Cost Planetary Missions, JHU/APL, Laurel, MD, May 2-5, 2000

1.3.6 Space Environment Experiments

All spacecraft inevitably interact with their environments. [230] Besides the interactions one immediately thinks of in space (zero-g, solar heating, atmospheric drag, expansion into vacuum conditions, etc.) other interactions are also important. Those of interest to spacecraft designers so far may be grouped under several headings:

- Plasma interactions and spacecraft charging
- Debris and micrometeoroids impact
- Chemical reaction with neutral species
- Radiation degradation

• Satellite surface charging experiments (see also Glossary). The ionizing radiation environment in space represents one of the most severe environment loads to space hardware and can cause a large number of problems such as spurious errors or permanent damage to electronics, erroneous signals in detectors, electrostatic charging of spacecraft or health damage to astronauts. In particular, the electrostatic charge on satellite surfaces can pose a hazard, when differential charging is leading to potential gradients. In some cases this potential build-up causes discharge arcing. [231] [232] Electrostatic charging by the natural space radiation environment is regarded as the most serious source of anomalies of S/C electronics. Hence, in-situ measurement of the space radiation environment is an important prerequisite for the enhancement of the static radiation environment models towards dynamic models which are required for the optimized design of future advanced systems. Examples of missions and experiments are:

- GEOS-1/-2 (Geostationary Satellite) missions of ESA, two reference S/C for IMS (International Magnetosphere Study). Launch of GEOS-1 on April 20, 1977; launch of GEOS-2 on July 14, 1978. The GEOS-1/-2 were the first S/C anywhere to carry a totally conductive coating - even over their solar panels. An electron beam experiment and a pair of probes, 40 m apart, provided independent measurements of the electric field surrounding the S/C. The S/C surface-treatment technology was confirmed. E.6
- SCATHA (Spacecraft Charging at High Altitude), the S/C is also referred to as P78-2. A NASA/JPL mission (launch Jan. 30, 1979; orbit: 28000 km x 42000 km, 8.3° inclination, S/C mass = 360 kg). [233] Experiment SC1 (Engineering+VLF and HF Receivers) measured surface potentials (RF waves between 0-300 kHz and 2-30 MHz) of various S/C materials. SC2 (Spacecraft Sheath Fields + Energetic Ions) measured low energy electrons and ions, energetic protons, and electrons. SC3 (High Energy Particle Spectrometer) measured high energy electrons and protons. SC4 (Satellite Electron and Positive Ion Beam System) used ion and electron beam guns to control spacecraft surface potential. SC5 (Rapid Scan Particle Detector) measured electrons and ions. .. SC11 (Magnetic Field Monitor) measured DC and ELF magnetic fields. ML12 (Spacecraft Contamination + Thermal Control Materials Monitoring) measured contamination rates and property changes of several thermal control material samples.
- GEOTAIL
- STRV-1a and -1b (Space Technology Research Vehicle, of DERA, UK, see M.25), launch June 17, 1994. Both S/C flew CAE (Charge Alleviation Experiment) to demonstrate an active S/C surface charge alleviation system. Other space environment instruments flown on STRV-1a are: SCDE (Surface Charge Detector Experiment), CID (Cold Ion Detector), LPI (Langmuir Probe Experiment), CREDO-II (Cosmic Radiation Environment and Dosimetry Experiment), and RDRS (Radiation Dose Rate Sensor). Other space environment instruments flown on STRV-1b are: SEE (Space En-

230) http://powerweb.lerc.nasa.gov/pvsee/publications/TheBasics.html

231) D. C. Wilkinson, M. A. Shea, D. F. Smart, " A Case History of Solar and Galactic Space Weather Effects on the Geosynchronous Communication Satellite TDRS-1," Advances in Space Research, Vol. 26, No. 1, 2000, pp. 27-30

232) V. I. Degtjarev, G. V. Popov, A. D. Johnstone, "Solar Wind Control of Spacecraft Charging Conditions in Geostationary Orbit during Magnetic Storms," Advances in Space Research, Vol. 26, No 1, 2000, pp. 37-40

233) http://www.newspace.com/ref/msl/QuickLooks/scathaQL.html

vironmental Effects), and REM (Radiation Environment Monitor). Note: The REM instrument was also flown on the MIR space station starting in Nov. '94 until Nov. '96.
- STRV-1c and -1d (launch Nov. 16, 2000) are follow-up missions of DERA, UK with the environment objectives focused on emerging technology hardware.
- Equator-S (launch Dec. 2, 1997, see K.9) carried PCD (Potential Control Device) to reduce the potential caused by the ambient plasma charging of the outer surfaces.
- SPOT-4 (launch March 24, 1998, see D.37.2) carried SILLAGE to measure electrostatic potentials on the outside of the SPOT satellite which are due to the "wake" effect.
- Cluster-II (launch July 16, 2000 of first pair of S/C; on Aug. 9, 2000 the second pair was launched, K.7). The four identical Cluster S/C feature conductive outer surfaces. An extremely low S/C-generated electromagnetic background noise is mandatory for accurate electric field and cold plasma measurements. The solar arrays consist of BSR (Back-Surface-Reflection) cells, arranged in self-compensating formations to minimize the generation of DC magnetic fields. The conductive coating on the cell cover glass minimizes the build-up of differential charge potentials.
- TSX-5 (Tri-Service Experiments Mission 5) of AFRL (launch June 6, 2000) flies two payloads: a) STRV-2 (Space Technology Research Vehicle) of DERA and AFRL with SAMMES (Space Active Modular Materials Experiment System), and b) CEASE (Compact Environmental Anomaly Sensor) of AFRL.

• Radiation-hardened components. The adverse effect of the 'space weather' (solar wind acting against the Earth's magnetic field, energetic particle flux, etc.) on space hardware, (in particular electronic components, sensors, power and communication systems) increases with radial distance of the S/C orbit. It implies that the long-term performance and reliability of hardware in GEO is much more subjected to a severe environment than a spacecraft in MEO or LEO orbits. - As a consequence, radiation-hardened components (CMOS chips, solar cells, etc.) were developed to account for the space weather and to improve hardware life times. The introduction of radiation-hardened components was first realized in GEO satellites.

• Space environment experiments - collection of cosmic dust, particles and debris from platforms. Dust is ubiquitous in the solar system, with dust particles abundant in regions as diverse as interplanetary space, the terrestrial and martian atmospheres, comets, asteroids, satellite surfaces, and planetary rings. The dynamics of these short-lived particles are particularly sensitive to their environment, so they serve as probes of a variety of processes and are strongly effected by electromagnetic forces. - Since the early 1980s several programs were initiated with the aim of systematic collection and analysis/curation of cosmic dust. Also study of radiation degradation effects on some types of materials. Initial airborne dust collections (since 1981) were made with instruments mounted underneath a wing using NASA WB-57F aircraft from NASA/JSC. Several space agencies (NASA, ESA, NASDA, CNES, etc.) maintain special laboratories to analyze impact damage, instruments are being developed and flown to measure orbital micro-debris sources. Long-term data on the space environment and its effects on space systems and operations is a constant need of spacecraft and instrument designers. In addition, the study of recovered long-duration space structures like LDEF and EURECA, and other recovered hardware (solar array from Hubble, etc.), is of major interest. In particular, large surface structures such as solar arrays, can serve as passive detectors for the study of debris impacts. [234] [235]

In 1995, NASA started its SEE (Space Environments Effects) program with the objective to collect, develop and to disseminate the technologies required to design, manufacture and operate more reliable, cost-effective spacecraft for the government and commercial sectors. In partnership with industry, academia, and other government agencies, the SEE program defines the space environments and advocates technology development to accommo-

234) http://setas-www.larc.nasa.gov/
235) N. L. Johnson, "Monitoring and Controlling Debris in Space," Scientific American, Aug. 1998, pp. 42-47

date or mitigate these harmful environments on spacecraft. The ESEM (Evaluation of Space Environment and Effects on Materials) payload of STS-85, in cooperation with NASDA, is one of many experiments in SEE. 236)

Mission or Instrument	Description
LAC (Large Area Collector), NASA/JSC	LAC is flown on NASA ER-2 aircraft since 1989, P.110
EURECA-1 launched on STS-46 July 31, 1992 and retrieved on STS-57 July 1, 1993	ESA mission. The experiment TICCE on EURECA-1 collected debris, meteoroids, and cosmic dust. See J.5.1
LDEF (Long Duration Exposure Facility) see J.8	NASA/LaRC S/C deployed from Shuttle in April 1984 (STS-41-C) and recovered on STS-32 Jan. 12, 1990 Exposure of many systems to cosmic radiation and collection of cosmic dust by Exp. A0138-2 of CERT/ONERA, and by Exp. A0201, S001, etc. of NASA/LaRC
MIR Space Station	On three occasions (1988, 1995, 1997) dust collection detectors were used on the outside of MIR.
Ulysses mission (ESA/NASA), launch Oct. 6. 1990	DUST experiment of MPIK (Heidelberg, Germany), see K.28
APEX (Advanced Photovoltaic & Electronics Experiment), S/C of DoD, launch Aug. 3, 1994 (no particle collection on APEX)	Study of long-term radiation effects in an orbit of 361 km x 2528 km, inclination = 70°. Experiments included: PASP (Photovoltaic Array Space Power Diagnostics), CRUX (Cosmic Ray Upset Experiment), and FERRO (Thin-film Ferro-electric Experiment)
MEEP (MIR Environmental Effects Payload) deployed on STS-76 (Mar. 22, 96) and retrieved on STS-86 (Oct. 1, 1997)	NASA/LaRC/MSFC conducted a series of experiments such as POSA-I & -II (Passive Optical Sample Assembly), PEC (Passive Experiment Carrier), PPMD (Polished Plate Meteoroid Detector), ODC (Orbital Debris Collector)
CDCE (Cosmic Dust Collection Experiment) and AOE (Atomic Oxygen Experiment), both on STS-85 (Aug. 7-19, 1997)	NASA/LaRC experiments in ESEM program. CDCE: Use of aerogel to capture cosmic dust. AOE: candidate materials for ISS (lubricants, paints, thermal coatings, etc.)
MightySat-1 of AFRL, launch Dec. 4, 1998, see M.11.1	MPID (Micro-Particle Impact Detector)
Stardust (NASA minisatellite in Discovery program), launch Feb. 7, 1999	Rendezvous with a comet and return of comet tail samples to Earth (encounter with comet Wild-2 in Jan. 2004), in 2006 return of interstellar dust grains in a reentry capsule to Earth
ARGOS (DoD, launch Feb. 23, 1999)	SPADUS (Space Dust Experiment)
SUNSAT-1 of Stellenbosch University, South Africa, launch Feb. 23, 1999	MIS (Meteoroid Impact Sensor) of NASA/LaRC
STRV-1c and -1d, DERA, launch Nov. 16, 2000	CDMS (Compact Debris and Micrometeoroid Sensor) characterizes of the debris and micrometeoroid environment in GTO
STRV-2 of AFRL and DERA (launch June 6, 2000) on TSX-5 mission of AFRL	MDIM (Meteoroid and Debris Impact Monitor) of NASA/LaRC.

Table 27: Overview of some cosmic dust/particle collection platforms

• **Orbital debris.** 237) 238) 239) 240) 241) 242) 243) The use of outer space presents a number of hazards to S/C owners and operators. Temperature extremes, radiation, solar flares, and micrometeoroids have long been essential elements to consider in spacecraft and mission design. Increasing use of space has brought a new source of risk collisions with manmade objects. The proliferation of debris in space generated by spacecraft and launchers is likely to become a serious issue in the near future. Such debris, originating from S/C, upper stage

236) http://see.msfc.nasa.gov/

237) "Dossier: Orbital debris," CNES Magazine No 4, Jan. 1999, pp. 11-28

238) http://www.aero.org/cords/index.html

239) M. J. Meshishnek, "Overview of the Space Debris Environment," The Aerospace Corporation Report No. TR-95(5231)-3, SMC Report No. SMC-TR-95-9, March 15, 1995. Also at: http://www.aero.org/publications/papers/pdfs/TR–95–5231–3.pdf

240) W. Flury, "Space Debris: An Overview," Earth Space Review, Vol. 9, No 4, 2000, pp. 40-47

241) M. L. Fudge, "The Effect of Orbital Debris on Commercial Satellites," Earth Space Review, Vol. 9, No 4, 2000, pp. 48-56

242) N. L. Johnson, "Man-Made Debris In And From Lunar Orbit," Earth Space Review, Vol. 9, No 4, 2000, pp. 57-65

243) B. Reijnen, "Space Debris: A Responsibility of States," Earth Space Review, Vol. 9, No 4, 2000, pp. 66-70

explosions and from the rising number of non-operational and abandoned satellites, represent a hazard for all operational missions (manned and unmanned). The GEO and LEO orbits are in particular littered with debris. Some examples are:

- Micro-debris originating from objects launched from Earth were first encountered in 1976 on the US Skylab Space Station, where aluminum oxide exhaust from apogee kick motors was identified.

- In 1986, the spent third stage of the Ariane launcher, used to place SPOT-1 into orbit (launch Feb. 22, 1986), exploded after nine months in orbit (LEO). The explosion generated over 700 fist-sized fragments. In addition, thousands of micro-particles of all sizes were released.

- Meteoroid impact on a satellite. In 1993 a Perseid meteoroid (tiny grains of dust) hit a solar panel on the European Olympus communications satellite. The project management is convinced the hit caused a pulse that sent false signals to the control system of the S/C. As a consequence, the satellite went into the wrong automated sequencing causing the loss of the spacecraft (it took all the remaining propellant to regain control, resulting in the end of the mission).

Particle diameter class	Effects on S/C structure	S/C protection	Estimated Nr. of particles on orbit
<0.01 cm	Cumulative effect over long periods of time	Unnecessary	Very large number of micro particles
0.01 - 1 cm	Serious structural damage, perforations. Consequences depend on element affected	Shielding needed	35,000,000 objects
1 - 10 cm	Very severe damage	No solution so far	110,000 objects
>10 cm	Objects are cataloged, catastrophic consequences	Avoidance maneuvers	>8700 objects

Table 28: Overview of space debris classes that may impact on operational S/C

- On July 24, 1996, the French microsatellite CERISE collided with a piece of debris (a suitcase-sized piece of the old Ariane rocket of 1986 - the first official collision between two catalogued objects), cutting in half the 6 m long stabilizing boom of CERISE. Fortunately, the French ground controllers regained control over the satellite following the collision.

- On Oct. 26, 1999 ISS (International Space Station) made a maneuver to avoid a potentially dangerous piece of orbital debris (the US Space Command and NORAD had reported that a spent Pegasus upper stage would pass ISS to within 1.4 km). Hence, thrusters of ISS were fired to increase the stations altitude by about 1.5 km.

- A major contributor to the orbital debris background have been breakups of rocket upper stages and satellites. More than 150 breakups have been verified, and more are believed to have occurred. Breakups generally are caused by explosions and collisions. Explosions can occur when propellant and oxidizer inadvertently mix, residual propellant becomes overpressurized due to heating, or batteries become overpressurized.

These incidents show that all aspects of debris-related issues, including some regulation standards, have to be dealt with in the future. The SSN (Space Surveillance Network) of the US Space Command and NORAD (North American Aerospace Defense Command) in Colorado Springs, CO, maintains orbital element sets of some 8764 objects (of 10 cm diameter or larger) - still many more are too small or in too eccentric orbits to be cataloged. Of the 8764 objects being tracked (as of Oct. 20, 1999), there are 2634 satellites (any payload operational or non-operational currently in Earth orbit - meaning LEO, GEO, GTO, MEO, HEO, or any other Earth orbit), 6040 objects are termed "debris" (meaning all man-made objects in Earth orbit), and 90 objects are termed "space probes" (referring to payloads that are not in Earth orbit - such as deep space missions). The historical catalog status of the US Space Command lists on Oct. 20, 1999: payloads decayed = 2,501, debris decayed = 14,680, total = 17,181 objects. [244] - In this context, it is also of interest to look at the estimated distribution of orbital debris in the following scheme: Operating satellites = 6%, spent satellites = 22%, upper stages = 17%, operational debris = 13%, and fragments (of rocket upper stages, etc.) = 42%.

The debris in orbit totals an estimated 4000 tons in 1999, with 175 tons added every year. [245] [246] Complications will arise from the ever-increasing use of space. The operational networks of major commercial satellite constellations (initial launches started in 1998) increases the crowding of space in LEO very rapidly. The Iridium constellation has 77 in orbit (nominal of 66 operational), Orbcomm has 35 (constellation completed in Dec. 1999), the Teledesic constellation has 288, Globalstar 48, and Skybridge 64, when fully deployed. These are only a few of the commercial entries. In Earth observation the picture is pretty much the same. Constellations or clusters of small satellites are planned by a number of agencies to enter a new dimension of formation flight observations.

Debris in the geostationary ring. [247] Since the launch of Syncom-1 in Feb. 1963, more than 800 satellites and rocket upper stages have been inserted into GEO or its vicinity. As of 2000, only about 250-270 of these satellites are used operationally. GEO satellites are therefore increasingly at risk of colliding with uncontrolled objects. Contrary to the situation with LEO satellites, there are no effective natural removal mechanisms for GEO objects. Routine surveillance with ground-based radars of SSN (Colorado Springs, CO) is able to catalog objects of 1 m in size or larger in GEO. Observations with ESA's Zeiss telescope at the Teide Observatory in Tenerife, with an optimized debris-detection system, have shown that there is a significant population of about 1600 small-debris objects (10-100 cm size) in the geostationary ring.

Collision risk monitoring. Starting in about 1995, space agencies (CNES, ESA, NASA, etc.) are beginning to monitor close encounters of catalogued objects with their own S/C. CNES experienced in this process, for instance, that on average, an object passes within less than 1500 m of a S/C every two weeks. Even closer approaches (under 800 m) were observed on 19 occasions.

Debris, reentering the atmosphere, presents a further problem, in particular to population safety. In 1999, it is estimated that about 17,000 objects of debris have reentered the atmosphere since Sputnik - virtually all of them disintegrated in the reentry process, very few actually hit Earth, although some particularly large or well-shielded objects have been known to get through. Controlled de-orbiting and reentry procedures over oceans or sparsely populated regions are the rule for larger returning S/C that can still be steered. At the Skylab uncontrolled reentry in 1979, with a mass of 82 metric tons, fragments fell into the Indian Ocean and the western region of Australia. This uncontrolled reentry definitely caused many alerts throughout the world. The MIR space station of Russia (with 140,000 kg of mass) was deorbited in a controlled manner in March 2001 with a final reentry into the Pacific Ocean on March 23, 2001. Many governments had their national control centers working for several weeks. An atmosphere of "we are doing something, and everything is under control" was portrayed publicly to keep the people calm. Fortunately, the Russian Control Center did an excellent job to bring MIR to the intended splashdown region. [Russia had asked NASA (incl. radars of the US Space Surveillance Network) and ESA (incl. the radar facilities of FGAN near Bonn, Germany) for support, which was granted. ESOC was the

244) http://www.peterson.af.mil/usspace/boxscore.htm

245) N. L. Johnson, Joseph, P. Loftus, "Reducing Orbital Debris: Standards and Practices," Launchspace, March/April 1999, p. 24

246) M. Bille, D. Dickey, "A Microsatellite 'Space Guard' Force," Proceedings of the 13th AIAA/USU Conference on Small Satellites, Aug. 23-26, 1999, Logan UT, SSC99-II-6

247) W. Flury, A. Massart, T. Schildknecht, U. Hugentobler, J. Kuusela, Z. Sodnik, "Searching for Small Debris in the Geostationary Ring," ESA Bulletin, No 104, Nov. 2000, pp. 92-100

center of a data exchange network among technical centers (TSUP, NASA/JSC, FGAN, and other space centers in Europe)].

A principal forum for debating orbital debris issues and new practices (including debris mitigation measures) is the "Inter-Agency Space Debris Coordination Committee" (IADC), created in 1993 by government representatives from USA (NASA), Europe (ESA), Russia (RKA), Japan (ISAS, NASDA and NAL), India (ISRO), China (CNSA), and the Ukraine (NSAU).

- **Spacecraft removal from orbit at end-of-life.** In the late 1990s, NASA adopted a new policy, a "NASA Management Instruction" with regard to spacecraft development (concerning debris prevention and end-of-life disposition). A so-called "debris assessment" is needed to determine if requirements are met for waiving the controlled reentry at the end of mission life. Because of lack of global regulations, waivers are likely to be issued for expensive measures that effect the projects of the own agency. But this is not special for the US: most spacefaring nations behave in the same way. Sometimes the US government insists on the strict application of its standard (NASA Management Instruction) as in the case of the Iridium constellation (condition for obtaining the license). - The planned NPP (NPOESS Preparatory Project) spacecraft and the follow-up NPOESS series, for instance, are expected to have sufficient debris that survives reentry; hence, they feature the capability of a controlled reentry (propulsive maneuvers) to place the debris in a pre-determined location in the ocean.

ESA came out with an "ESA Space Debris Mitigation Handbook," Release 1, April 7, 1999, and a draft of the "European Space Debris Safety and Mitigation Standard" in 2000 (Sept. 27, 2000, Issue 1, Rev. 0), which discusses the issues of in-orbit debris and gives various scenarios for end-of-life (EOL) disposal. - The two TerraSAR spacecraft of Astrium (XSAR and LSAR mission with a planned launch in 2006), within the framework of ESA Earth Watch missions, are planning thruster firings at EOL (using budgeted fuel) to obtain an elliptical orbit with a perigee of about 300 km or lower. This generates sufficient atmospheric drag to bring the S/C eventually into a reentry trajectory.

Further references on space debris are: "Position Paper on Orbital Debris," of IAA (International Academy of Astronautics), Paris, 2001, and "Technical Report on Space Debris," UN, New York, 1999. The problem with debris policies is that the standard in its entirety will only be applied when everybody in space will abide by the rules. Since no international regulation concerning space debris exists, this will take some time. Such a regulation must necessarily be issued by the UN, where the deliberations in the UN Committee on the Peaceful Uses of Outer Space (UNCOPUOS) on debris are ongoing since 1994. All agreements must be reached by unanimity. It took several years to reach agreement to have space debris on the agenda.

Among the space agencies there is practically a consensus that S/C and rocket upper stages in Low Earth Orbit (LEO - orbits <2000 km) should be removed from orbit within 25-50 years after mission completion. This is simply to avoid further proliferation and generation of debris (through collision). A regulation is however needed to force everybody, including commercial users/operators to apply the same rules. [248)]

The situation is different with objects which will reenter at some time, but which will not completely burn up. There is already now international space law, the Liability Agreement, which clearly stipulates the liability of the owner in case of damage on ground or in air–space. Therefore, more and more satellites which could be a risk on ground, will be deorbited in a controlled way. Examples: MIR, Compton Gamma Ray Observatory.

-

248) Information provided by Walter Flury of ESA/ESOC, Darmstadt, Germany

1.3.7 Tether Experiments

- Tether experiments in space (M.27). Although the idea of using space tethers goes back to the 1960s, it was the launch of TSS (Tethered Satellite System) in 1993 on Shuttle mission STS-46, which gave a decisive impetus to tether technology development and projects, by realizing the potential of tethers to meet a broad range of applications in various fields. [249) 250)] Tether deployment techniques and stable operations were the dominant themes of the early years including the 1990s. Potential tether applications include: propulsion (orbit changes), payload capture and release, power generation, atmospheric studies [a) reaching otherwise unaccessible flight regions with downward deployed tethers; b) active experimentation with the surrounding plasma], and gravity experiments.

Mission (Agency)	Launch (Timeframe)	Orbit	Tether Length	Comment
Gemini-11(NASA)	Sep. 11-15,'66	LEO	30 m	Spin-stabilized, 0.15 rpm
Gemini-12(NASA)	Nov. 11-15,'66	LEO	30 m	Local vertical, stable swing
H-9M-69	1980	Suborbital	500 m	Partial deployment
S-520-2	1981	Suborbital	500 m	Partial deployment
Charge-1	1983	Suborbital	500 m	Full deployment
Charge-2	1984	Suborbital	500 m	Full deployment
ECHO-7	1988	Suborbital		Magnetic field aligned
Oedipus-A (NRC/ NASA,CRC, CSA)	Jan. 30, 1989	Suborbital	958 m	Spin-stabilized, 0.7 rpm (Canada/USA)
Charge-2B	1992	Suborbital	500 m	Full deployment
TSS-1 (STS-46) (ASI/NASA)	Jul. 31-Aug. 8, 1992,	LEO	< 270 m	Electrodynamic tether, partial upward deployment and retrieval of tether
SEDS-1 (NASA)	Mar. 29,1993	LEO	20 km	Nonconducting tether, downward deployment, swing and cut
PMG (DoD) Plasma Motor Generator	June 26, 1993	LEO	500 m	Electrodynamic tether, upward deployment, tether current in both directions. Demonstration of hollow cathode plasma contactors for current collection and emission
SEDS-2 (NASA)	Mar. 9, 1994	LEO	20 km	Local vertical stable, downward deployment
Oedipus-C (NRC/ NASA,CRC, CSA)	Nov. 6, 1995	Suborbital	1 km	Spin-stabilized, 0.7 rpm (Canada/USA)
TSS-1R (STS-75) (ASI/NASA)	Feb. 22-Mar. 9, 1996	LEO	19.6 km	Electrodynamic tether, severed prior to full deployment
TiPS (NRO/NRL)	May 12, 1996 Jun. 20, 1996 deployment	LEO	4 km	Long life tether (tracking of one year)
YES (Technical Univ. of Delft)	Oct. 30, 1997	GTO	35 km	YES (of TEAMSAT) was not deployed to avoid a collision with other S/C (M.26)
STEX/ATEx (NRO/NRL)	Oct. 3, 1998 Jan. 16 1999 deployment	LEO	6 km	Nonconductive tether, planned demonstration of tether system stability and control. However, a deployment malfunction caused an experiment failure
PICOSAT1.0 of the Aerospace Corporation	Jan. 27, 2000 on OPAL of Stanford	LEO	30 m	PICOSAT1.0 was ejected OPAL on Feb 6, 2000. The system of two S/C operated for 3 days when battery power decayed.
PICOSAT1.1 of the Aerospace Corporation	July 19, 2000 on MightySat II.1 of AFRL	LEO	30 m	PICOSAT1.1 will be ejected from Mightysat II.1 one year after launch (in Sept. 2001) to test storage effects on the system.
METS (MIR Electrodynamic Tether System)	2001	LEO	5 km	First practical application of non-rocket propulsion in space

249) Note: The objective of the Gemini tether missions was to see if tethers could be used for rendezvous and docking in preparation for the future moon missions. A parachute cable was used as the tether, which was attached to the Agena by a spacewalking astronaut. The other mission (Gemini-11) experimented with rotation about the CM (Center of Mass) to see if it was stable - it was. The Gemini missions were not electrodynamic in nature, nor were they performed for scientific purposes - as were the TSS missions.

250) http://infinity.msfc.nasa.gov/Public/ps01/ps02/table1.html

Mission (Agency)	Launch (Timeframe)	Orbit	Tether Length	Comment
ProSEDS (NASA)	2001	LEO	25 km	Upward deployment (producing drag and electrical power)
TSE (ESA)	2002	LEO	35 km	Demonstration of ISS sample return

Table 29: Chronology of tether missions

As of the late 1990s spaceborne tethers are indeed a reality (there are numerous projects) and their potential is far from being fully appreciated.

1.4 ISS (International Space Station) Built-up

• The ISS (International Space Station) built-up era in LEO began in 1998. - The program of an international space station was first announced/initiated by President Reagan in his 'State of the Union' address on Jan. 25, 1984. A total of 45 US/Russian missions are planned up to 2005 to assemble more than 100 elements that will complete the International Space Station (total mass of 460 tons of structures and equipment). The ISS target orbit altitude is 410 - 450 km, orbit height when docking is lowered to 350 km, orbit inclination is 51.6°. - Other nations, participating in the ISS project (besides the USA and Russia), include: Brazil, Belgium, Canada, Denmark, France, Germany, Italy, Japan, the Netherlands, Norway, Spain, Sweden, Switzerland and the United Kingdom. ISS is the largest cooperative space research program in the history of space flight. When assembled ISS will have dimension of: 108 m in length and 74 m in width. The total mass will be more than 415 tons. - Three of the four operational Shuttles - Discovery, Atlantis and Endeavour - have been modified for increased lift capability (16,000 kg max payload into orbits of 51.6° inclination) to meet the projected needs of ISS.

- The first component of ISS, namely FGB (Functional Cargo Block), also referred to as Zarya (meaning 'sunrise'), was launched Nov. 20, 1998 on a Proton vehicle from the Baikonur Cosmodrome in Kazakhstan. Zarya serves as the cornerstone of ISS, it is a 20,040 kg pressurized module. Zarya was built for NASA by KhSC ((Khrunichev State Research and Production Space Center) of Moscow in cooperation with Russian space industry, under contract with the US Boeing Company.

- The second component of ISS, Unity (11,800 kg, 11 m in total length), referred to as node 1, with six berthing ports, was launched Dec. 4, 1998 on Shuttle (STS-88, Endeavour) and mated to Zarya three days later.

- The STS-96 Shuttle flight on Discovery (May 27 - June 6, 1999) carried a) internal logistics and resupply cargo for station outfitting, and b) an external Russian cargo crane that was mounted to the exterior of the Russian station segment and used to perform spacewalking maintenance activities.

- The STS-101 flight on Atlantis (May 19 - 29, 2000) services were: a) prepared the station for the arrival of the Zvezda Service Module, b) installed four new batteries, 10 new smoke detectors and four new cooling fans on the Zarya module, c) installed the final parts of the Strela crane on Pressurized Mating Adapter 1, d) removed and replaced the early communications system antenna, and e) installed handrails on the Unity module.

- The Russian-built Zvezda (meaning "star") Service Module was launched on July 12, 2000 by a Proton-K vehicle from Baikonur (ISS 1R flight). Zvezda was docked to ISS on July 26, 2000. The Zvezda Service Module provided a) the first primary living quarters for three-person crews, b) primary docking port for Progress-type cargo resupply vehicles, c) power and steering (propulsive attitude control and reboost) capability to keep ISS in a safe orbit.

- Aug. 6, 2000 launch of Russian Soyuz-2 from Baikonur on ISS flight 1P with the Progress M1-3 resupply ship for ISS. Progress docked with ISS on Aug. 9, 2000.

- The STS-106 flight of Atlantis (Sept 8-20, 2000) provided a) internal logistics and resupply cargo for station outfitting, b) a space walk was conducted to perform maintenance on the station.

- STS-92 flight of Discovery (Oct. 11-24, 2000, ISS-05-3A flight) provided: ITS (Integrated Truss Structure) Z1, PMA-3 (Pressurized Mating Adapter-3), a Ku-band communications system, and CMGs (Control Moment Gyroscopes) for non-propulsive attitude control. ITS-Z1 is an early exterior framework to allow first U.S. solar arrays on flight 4A to be temporarily installed on Unity for early power. The Z1 truss is the first permanent lattice-work structure for ISS.

- A Russian Soyuz rocket (launch Oct. 31, 2000) lifted the first expedition crew to ISS from Baikonur, referred to as ISS-2R. [251] Objectives: a) to establish the first station manning with a three-person crew (Commander Bill Shepherd; Soyuz Commander Yuri Gidzenko, Flight Engineer Sergei Krikalev), b) provides Russian assured crew return capability without the Shuttle present, c) **the "permanent" habitation of ISS has begun.**

- STS-97 flight of Endeavour (Nov. 30 - Dec. 11, 2000). The ISS assembly flight 4A provided the P6 Integrated Truss Segment, a 7700 kg package containing the PV (Photovoltaic) Module, plus S4 and S5 radiators. The US-provided power system increases the available electrical power of ISS by a factor of five. The P6 consists of two identical PVAAs (Photovoltaic Array Assembly), each of which is made up of an SAA (Solar Array Assembly) and a SAW (Solar Array Wing). The two SAWs have a total power generation capability of about 64 kW. Each SAW has a structure size of 35 m x 11.58 m and a mass of 1087 kg. Each SAW contains 33,000 solar arrays, each of size 8 cm x 8 cm with 4,100 diodes. The entire assembly represents the largest structure deployed in space so far (73 m from tip to tip) as well as the largest space-installed power supply.

- STS-98 flight of Atlantis (Feb. 7 - 20, 2001). The ISS assembly flight 5A carried the Destiny Laboratory Module, an aluminum structure of 8.5 m in length and 4.3 m in diameter. The module provides an initial US user capability in the field of near-zero gravity research.

- STS-102 flight of Discovery (March 8-21, 2001). The ISS logistics and resupply flight 5A.1 provided the Leonardo MPLM (Multi-Purpose Logistics Module). The module is about 6.4 m long and 4.6 m in diameter containing equipment racks. Exchange of first crew by a new crew.

- STS-100 flight of Endeavour (April 19 - May 1, 2001). The flight 6A provided the Raffaelo MPLM (lab outfitting) for the US Lab, the UHF antenna (for EVA support), and delivery/installation of the Canadian SSRMS (Space Station Remote Manipulator System), the cornerstone of ISS's Mobile Servicing System (MMS), also referred to as Canadarm2 (Canadarm1 is on Shuttle). SSRMS, is currently attached to the Destiny module and used as "construction crane" of ISS, lifting payloads and performing maintenance work.. With the installation of Canada's Mobile Base System (MBS) in 2002, SSRMS will be able to slide along the main truss of the station (lateral mobility), providing an additional functional capability with four Power Data Grapple Fixtures. A further upgrade of Canadarm2 is planned for 2003/4 with SPDM (Special Purpose Dextrous Manipulator). This smaller two-armed robot has the capability to perform even more delicate tasks at ISS. SSRMS is 17.6 m long when fully extended and has seven motorized joints. This arm is capable of handling large payloads and assisting with docking the Shuttle.

- A Russian Soyuz spacecraft arrived at ISS on April 30, 2001 with the first space tourist ever on-board, namely Dennis Tito, a well-to-do US citizen who paid $20 Million to Rosaviakosmos, the Russian Space Agency (one week stay at ISS). There was considerable criticism and protest from NASA officials prior to the launch, but Rosaviakosmos insisted on Tito's mission. The agency wanted and needed the cash badly due to the abysmal lack of

251) http://spaceflight.nasa.gov/station/assembly/flights/2000/2r.html

governmental funding for its space programs. Further space tourists are waiting in the wings for a flight to ISS.

- STS-104 flight of Atlantis (July 12-24, 2001). Installation of an ISS Joint Airlock. The airlock is the primary path for ISS space walk activity (entry and departure of astronauts or cosmonauts). The airlock is a pressurized flight element consisting of two cylindrical chambers attached end-to-end by a connecting bulkhead and hatch. The airlock structure, built at MSFC, is 6 m in length and 4 m in diameter with a mass of about 6500 kg. There are two main components to the airlock: a crew airlock and an equipment airlock for storing EVA gear and EVA preflight preparations. STS-104 also carries a spacelab pallet with four High Pressure Gas Assembly containers that were attached to the exterior of the airlock.

1.5 Overview of Operational Meteorological Missions

Spaceborne operational meteorological services, namely the measurement of short-term weather variations in the lower regions of the atmosphere (in particular the troposphere as well as the Earth's surface), are provided by two types of satellite constellations: LEO and GEO.

"Today we can make 3-5-day weather forecasts with nearly 80% accuracy. That's fantastic. In 10 year's time, with all we can learn from our upcoming missions, we intend to push that out to 7 to 10 days with the same accuracy. With the right investments in observations and computational modeling, we might even push that out to 14 days perhaps 20 years from now." (Daniel Goldin, NASA Administrator, at JHU/APL address, Oct. 10, 2000, Ref. 7)

1.5.1 LEO (Low Earth Orbit) Satellite Missions

LEO S/C in polar sun-synchronous orbits provide wide-swath repeat observations (same sun illumination angles) of a given Earth surface region at the same time of the day (due to same latitudinal crossings).

- The first meteorological LEO services started with NASA's launch of TIROS-1 (Television Infrared Observation Satellite) on April 1, 1960, the first true weather satellite. Since then, five generations of polar-orbiting meteorological S/C were developed by NASA and flown by NOAA as outlined in Table 30. - The NOAA 4th-generation polar program (ATN), which started in 1978 with the launch of TIROS-N, is based on the services of two operational satellites flying in complementary sun-synchronous orbits, one in a "morning or AM" orbit, and the second in an "afternoon or PM" orbit.

1st generation 1960 - 1965	TIROS series [TIROS-1 (1960) to TIROS-10 (1965)]
2nd generation 1966 - 1969	TOS (TIROS Operational System) series as pre-launch designation. The in-orbit satellite designation was ESSA [ESSA-1 (1966) to ESSA-9 (1969)], after the S/C operating agency.
3rd generation 1970 - 1976	ITOS (Improved TIROS Operational System) series as pre-launch designation. The in-orbit satellite designation was NOAA [NOAA-1 (1970) to NOAA-5 (1976)]
4th generation 1978 - 1994	ATN [Advanced TIROS-N) series. After TIROS-N (1978) the pre-launch designation changed to NOAA-A (the corresponding inflight name was NOAA-6). The pre-launch letter designation was kept throughout. NOAA-8 through NOAA-14 were designated ATN (Advanced TIROS-N) spacecraft, equipped for S&R (Search and Rescue) and growth instruments.
5th generation 1998 - 2007	NOAA-15 (pre-launch NOAA-K), NOAA-L, -M, -N, and -N' incorporate advanced instrumentation and are designated "NOAA next."

Table 30: Overview of the US civilian polar meteorological programs

- The US military services of DoD built their own polar-orbiting meteorological satellite series in parallel to the civil system of NASA/NOAA, referred to as DMSP (Defense Meteorological Satellite Program). The first generation of DMSP S/C started with the launch of a Block 4 series satellite on Jan. 19, 1965 and ended with the launch of a Block 5C S/C on Feb 19, 1976. The various generations of DMSP satellites are listed in Table 330. The OLS (Operational Linescan System) instrument on DMSP provides in addition to its normal daytime (visible) imagery also nighttime imagery with its IR channels.

Satellite Series (Agency)	Launch	Major Instruments	Comment
NOAA-2 to -5 (NOAA)	21.10. 1971, 29.7. 1976	VHRR	2580 km swath
TIROS-N (NOAA-POES) NOAA-15 to -L, M, N, N'	13.10.1978 13.5.1998 - 2007	AVHRR AVHRR/3	>2600 km swath >2600 km swath
DMSP Block 5D-1 (DoD) DMSP Block 5D-2 (DoD) DMSP Block 5D-3 (DoD)	11.9.1976 - 14.7.1980 18.6.1987 12.12.1999	OLS OLS, SSM/I OLS, SSM/I,	3000 km swath SSMIS replaces SSM/I starting with F-16 (2001)
Meteor-3 series of Russia	24.10.1985	MR-2000M, MR-900B	3100 km, 2600 km swath
Meteor-3M series of Russia	2001 (Meteor-3M-1)		
FY-1A, 1B, 1C (CMA, China)	7.9.1988, 3.9.1990, 10.5, 2000	MVISR	2800 km swath
MetOp-1 (EUMETSAT)	2005	AVHRR/3, MHS, IASI	PM complement to NOAA-POES series
NPP , (NASA/IPO)	2005	VIIRS, CrIS, ATMS	(NPOESS Preparatory Project)
NPOESS (IPO)	2008	VIIRS, CMIS, CrIS,	Successor of NOAA-POES and DMSP series

Table 31: Overview of polar-orbiting meteorological satellite series

• The Meteor-1 series of the former USSR started with the launch of Meteor-1-1 on March 23, 1969; the 1st generation series lasted until 1978 with the launch of Meteor-1-28. This was succeeded by the Meteor 2 series from 1975 to 1990 (Meteor-2-1 to 2-21), and further by the Meteor-3 series from 1985 to the present time (see Table 361).

• The Chinese polar-orbiting meteorological program started in 1977 followed by the launch of the FY-1A (Feng-Yun-1) satellite of CMA (China Meteorological Administration) on Sept. 7, 1988 and FY-1B on Sept. 3, 1990.

• EUMETSAT, the European meteorological service provider, was planning on a polar-orbiting satellite series since the mid 1980s. Since the early 1990s, NOAA and EUMETSAT have been discussing/planning future polar cooperation with increased European responsibility for the "morning orbit" to ensure continuity of the POES (Polar-orbiting Operational Environmental Satellites) services. The basic intent is to join the space segment of the emerging MetOp program of EUMETSAT with the existing POES program of NOAA into a fully coordinated service, thus sharing the costs of a program for synergetic reasons. The plans came to a common baseline and agreement, referred to as IJPS (Initial Joint Polar System), in 1998. IJPS comprises two series of independent, but fully coordinated polar satellite systems, namely POES and MetOp, to provide for the continuous and timely collection and exchange of environmental data from space. EUMETSAT plans to include its satellites MetOp-1, -2 and -3 for the morning orbit, while NOAA is starting with its NOAA-N and N' spacecraft for the afternoon orbit of the coordinated system. Plans call for MetOp-1 to be launched in 2005. - In this context, Europe (i.e. ESA, EUMETSAT) refers to its application-oriented missions (this includes operational meteorological and climate monitoring missions) in LEO and GEO as "Earth Watch Missions" (MetOp, MSG). The Earth Watch concept was introduced in 1994/5, dealing in particular with prototype operational missions and serving applications-oriented needs in partnership of industry and/or public agencies. Long-term operational service provision is the ultimate aim of the ESA Earth Watch program.

• NPP (NPOESS Preparatory Project). NASA and IPO (Integrated Program Office) plan the NPP mission (launch in 2005) with the objective to demonstrate/validate the operation of three advanced instruments (VIIRS, CrIS, ATMS) scheduled to fly on NPOESS. ATMS (Advanced Technology Microwave Sounder) is a NASA-provided instrument with the objective to combine the passive-microwave observation capabilities of three heritage instruments, namely AMSU-A1/A2 and MHS, into a single instrument with a correspondingly reduced mass and power consumption and with advanced microwave-receiver electronics technologies.

• NPOESS (National Polar-orbiting Operational Environmental Satellite System). NPOESS (G.10) represents the future US meteorological polar-orbiting system (merger of NOAA-POES and DMSP programs) with the objective to provide a single national remote-sensing capability for meteorological, oceanographic, climatic and space environmental data. The NPOESS data sets will contain a number of variables that are currently not included in operational measurements (such as: radiation budget, total ozone, wind speed and direction, ocean topography, and ocean color) and will offer improved quality for some variables now being measured (such as: atmospheric moisture and temperature profiles, all-weather SST, and vegetation indices). The IPO plan for NPOESS is to launch its first satellite in the time frame 2008. Initial plans for the merger of the two services started in 1993. The following NPOESS sensor payloads are under development: CMIS (Conical-scanning Microwave Imager/Sounder) of SSM/I (Special Sensor Microwave Imager), SSMIS (Special Sensor Microwave Imager Sounder) and TMI (TRMM Microwave Imager) heritage, CrIS (Cross-Track Infrared Sounder) of HIRS/4, AIRS and IASI heritage, OMPS (Ozone Mapping and Profiler Suite) of SBUV/2, TOMS and GOME heritage, VIIRS (Visible/Infrared Imager and Radiometer Suite) of OLS, AVHRR/3, MODIS, and SeaWiFS heritage, GPSOS (GPS Occultation Sensor) of GRAS and GPS/MET heritage, and SESS (Space Environment Sensor Suite) of SEM-2 heritage.

LEO meteorological instruments: An overview cannot be complete without giving credit to at least one workhorse instrument, AVHRR (Advanced Very High Resolution Radiometer).

• AVHRR (Advanced Very High Resolution Radiometer) of the NOAA-POES satellite series is the spaceborne instrument with the longest service period, the widest data distribution and data analysis in the history of operational meteorology, oceanography, climatology, vegetation monitoring, and land and sea ice observation. The instrument provides wide-swath (>2600 km, FOV $= \pm 56^{\circ}$) multispectral imagery of about 1.1 km spatial resolution from LEO polar orbits (nominal altitude of 833 km). The resolution of 1.1 km is still quite high for the wide-swath measurement of large-scale meteorological phenomena. The imagery is used in a great variety of applications, such as: investigation of clouds, land-water boundaries, snow and ice extent, sea surface temperature, day and night cloud distribution, vegetation index, etc. The benefit of AVHRR data lies in its high temporal frequency of global coverage. The AVHRR instrument was initially designed for meteorological applications. - It is of interest to note that initial objectives of AVHRR were to develop a system that would provide a more efficient way to track clouds, estimate snow cover extent, and estimate sea surface temperature. It wasn't until a few years after the launch of the first AVHRR instrument that its usefulness in monitoring global vegetation became obvious.

Three generations of AVHRR whiskbroom scanning instruments (built by ITT Aerospace of Fort Wayne, IN) provided daily global coverage starting from 1978 (TIROS-N) to the turn of the millennium and beyond. The first 3rd generation instrument (AVHRR/3) with six channels was launched on May 13, 1998 as part of the NOAA-15 sensor complement. Although described in G.14.1, G.14.11, and G.15.1, it is worthwhile to repeat some tables of this remarkable instrument.

AVHRR is of VHRR (Very High Resolution Radiometer) heritage, a two-channel instrument flown on the ITOS (Improved TIROS Operational System) series, starting with NOAA-2 (ITOS-D), launch of ITOS-D on Oct. 15, 1972. The last one of the series flying the VHRR instrument was NOAA-5 (ITOS-H) with a launch July 29, 1976.

Channel No.	TIROS-N	NOAA-6, -8, -10	NOAA-7, -9, -11, -12, -14	IFOV (mrad)	Principal use of channel
1	0.550 - 0.90	0.550 - 0.68	0.550 - 0.68	1.39	Day cloud and surface mapping
2	0.725 - 1.10	0.725 - 1.10	0.725 - 1.10	1.41	Surface water delineation, vegetation mapping
3	3.550 - 3.93	3.550 - 3.93	3.550 - 3.93	1.51	SST and fire detection
4	10.50 - 11.50	10.50 - 11.50	10.30 - 11.30	1.41	SST and night time cloud mapping
5	repeat of Channel 4	repeat of Channel 4	11.50 - 12.50	1.30	Surface temperature and day/night cloud mapping
	AVHRR/1		AVHRR/2		Instrument configuration/designation

Table 32: **Spectral channels of AVHRR**

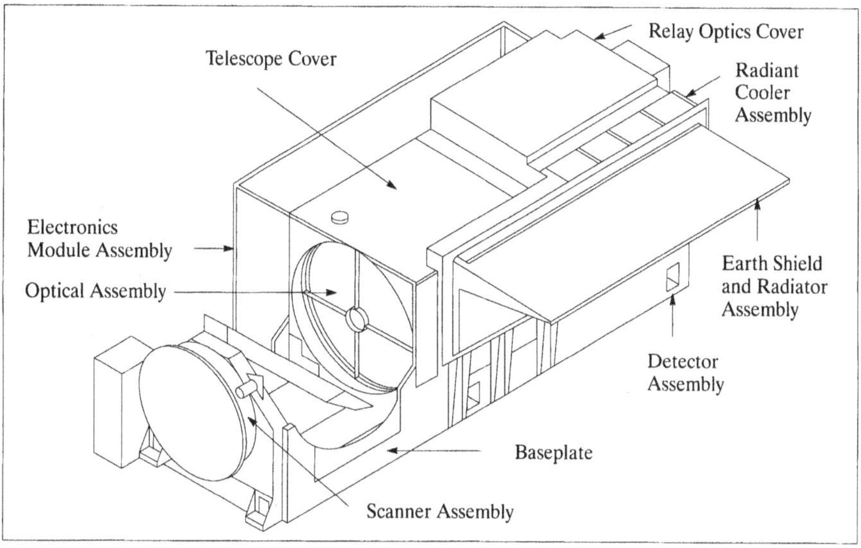

Figure 4: **Illustration of the AVHRR instrument**

Channel	Center wavelength (μm)	Spectral Range FWHM (μm)	Channel Noise	Detector Type
1	0.630	0.58 - 0.68	SNR ≥ 9:1 @ 0.5% albedo	Silicon
2	0.862	0.725 - 1.00	SNR ≥ 9:1 @ 0.5% albedo	Silicon
3a	1.61	1.58 - 1.64	SNR ≥ 20:1 @ 0.5% albedo	InGaAs
3b	3.74	3.55 - 3.93	NEΔT ≤ 0.12 K @ 300 K	InSb
4	10.80	10.30 - 11.30	NEΔT ≤ 0.12 K @ 300 K	HgCdTe
5	12.00	11.50 - 12.50	NEΔT ≤ 0.12 K @ 300 K	HgCdTe

Table 33: **Spectral parameters of AVHRR/3 (starting with NOAA-15)**

S/C	Launch Date	Ascending Node	Descending Node	S/C Service Period	
TIROS-N	Oct. 13, 1978	15:00	03:00	Oct. 19, 1978	Jan. 30, 1980
NOAA-6	Jun. 27, 1979	19:30	07:30	Jun. 27, 1979	Nov. 16, 1986
NOAA-7	Jun. 23, 1981	14:30	02:30	Aug. 24, 1981	Jun. 7, 1986
NOAA-8	Mar. 28, 1983	19:30	07:30	May 3, 1983	Oct. 31, 1985
NOAA-9	Dec. 12, 1984	14:20	02:20	Feb. 25, 1985	Aug. 3, 1995
NOAA-10	Sep. 17, 1986	19:30	07:30	Nov. 17, 1986	March 1995
NOAA-11	Sep. 24, 1988	13:30	01:40	Nov. 8, 1988	Sep. 9, 1994
NOAA-12	May 14, 1991	19:30	07:30	May 14, 1991	Present

S/C	Launch Date	Ascending Node	Descending Node	S/C Service Period	
NOAA-13	Aug. 9, 1993	Failure 12 days after launch, NOAA lost contact with the S/C			
NOAA-14	Dec. 30, 1994	13:40	01:40	Dec. 30, 1994	Present
NOAA-15	May 13, 1998				

Table 34: Temporal AVHRR coverage of NOAA POES series satellites

• Sea Surface Temperature (SST) measurements from space (a physical parameter derived from microwave radiometer data). The analysis of SST observations is an important indicator of the coupling between the ocean surface and the atmosphere - used in climate modeling and in many other fields (meteorology/oceanography). The ability to monitor global and regional surface temperature has improved so that it is now possible to use SST observations as indicators of regional- to basin-scale change, as well as for forecasting stress on the natural flora and faunal assemblages. - The longest data set of SST observations is based on observations initially made from ships (capturing buckets of seawater from over the sides of ships and measuring the temperature with a thermometer). From about 1870 onwards the ship observations were sufficiently frequent to permit a global SST analysis. [252] - First spaceborne SST data was provided by the VHRR (Very High Resolution Radiometer) instrument flown on NOAA-2 (launch Oct. 15, 1972). Better spaceborne SST retrievals became available from channels 3 and 4 of the AVHRR/1 instrument flown on the TIROS-N S/C (launch Oct. 13, 1978). More accurate SST retrievals became available with the introduction of AVHRR/2, first flown on NOAA-7 (launch June 23, 1981) and subsequent NOAA/POES missions. The AVHRR/3 instrument generation (with 6 bands) was first flown on NOAA-15 (launch May 13, 1998). AATSR of ESA (of ATSR heritage on ERS-1) provides a radiometric resolution of 0.1 K, an SST accuracy <0.5 K, and a spatial resolution of 1 km.

The accuracy of SST data is of great importance for climate modeling and climate change detection/prediction. The energy (gas) transfer between the ocean surface and the atmosphere is highly dependent on the "skin temperature" of the surface waters (surface winds are another important factor). For illustration: An error of 0.1 K in SST knowledge corresponds to an error in the global flux of CO_2 of 15%, while an error of 0.5 K leads to an error of 75% in CO_2 flux.

Passive microwave radiometers offer an all-weather observation capability for measuring SST. The challenge with microwave instruments is to increase the spatial resolution of the data (which are in the ranges of 20-100 km depending on channel frequency). The first such instruments are: MSMR (Multifrequency Scanning Microwave Radiometer) on IRS-P4 (ISRO), AMSR-E on Aqua (NASA) and AMSR on ADEOS-II (NASDA). The CMIS (Conical-scanning Microwave Imager/Sounder) of the planned NPOESS series, an advanced instrument of SSM/I and SSMIS heritage of the DMSP (US/DoD) satellite series, is a dual- and single look passive polarimetric microwave radiometer. SST is going to be one of the CMIS data products with a medium-scale spatial resolution of about 25 km.

Mission	Instrument	Comment
NOAA/POES series starting with NOAA-2 (launch Oct. 15, 1972)	VHRR, AVHRR/1, AVHRR/2, AVHRR/3	Operational series
Meteor-2 series starting with Meteor-2-4 (launch Oct. 25, 89)	Klimat	Pre-operational instrument of PLANETA, Russia
ERS-1 (launch July 17, 1991) ERS-2 (launch Apr. 21, 1995)	ATSR ATSR-2	Pre-operational instrument Pre-operational instrument
IRS-P4 (launch May 26, 1999)	MSMR of ISRO	Pre-operational instrument
FY-1B (launch Sept. 3, 1990) FY-1C (launch May 10, 1999)	MVISR (Multichannel Visible and IR Scanning Radiometer)	Pre-operational instrument of CMA, China

252) R. W. Reynolds, "Specific Contributions to the Observing System: Sea Surface Temperature," Proceedings of OCEANOBS 99, Oct. 18-22, 1999, Saint Raphael, France

Mission	Instrument	Comment
OKEAN-O-1 (launch Jul. 16, 1999)	Delta-2D (Conical Scanning Microwave Radiometer)	OKEAN-O-1 operations lasted until fall 2000
Terra (launch Dec. 18, 1999) Aqua (launch 2001)	MODIS MODIS, AMSR-E	Pre-operational instrument Pre-operational instruments
Envisat (launch 2001)	AATSR	Pre-operational instrument
ADEOS-II (launch 2002)	AMSR (Advanced Microwave Scanning Radiometer), GLI	Pre-operational instrument
EPS-MetOp series (starting with MetOp-1, launch 2005)	AVHRR/3	Operational series
NPP (NPOESS Preparatory Project), launch in 2005	VIIRS (Visible/Infrared Imager and Radiometer Suite)	Pre-operational instrument
NPOESS (launch in 2008)	VIIRS, CMIS	

Table 35: Overview of spaceborne instruments suitable for SST retrieval

1.5.2 GEO (Geosynchronous Orbit) Weather Satellites

The geostationary orbit provides new observation concepts. Satellites in geostationary orbits exhibit a fixed-position, constant-signal and continuous-coverage relationship with large area coverage between the satellite and its ground segment. The observation data, provided by the S/C instruments from GEO locations (about 45 times further away than LEO systems at altitude of 800 km), are generally of coarse resolution (in the order of kilometers). However, the data are ideal to monitor large-scale weather phenomena. - Since the mid 1960s a global network of geostationary meteorological satellites has been built up continuously around the equator by the various national agencies of the world to provide an ever-increasing service to its user community. The data of the various satellite series also contribute to GARP (Global Atmospheric Research Program) of WMO.

S/C Series (Agency)	Launch	Major Instruments	Comment
ATS-1 to ATS-6 (NASA) GOES-1 to -7 (NOAA) GOES-8 to -12 (NOAA)	6.12.1966, 12.8.1969 16.10.1975, 26.2.1987 13.4.1994, 23.7. 2001	SSCC (MSSCC ATS-3) VISSR GOES-Imager, Sounder	Tech. Demonstration 1st generation 2nd generation
GMS-1 to -5 (JMA) MTSAT-1 (JMA, et al.) MTSAT-1R (JMA)	14.7.1977, 18.3.1995 Nov. 15, 1999 (launch failure of H-2 vehicle) re-planned for 2003	VISSR Imager (GOES heritage) JAMI	1st generation 2nd generation
Meteosat-1 to -7 (Eumetsat) MSG-1 (Eumetsat)	23.11.1977, 3.9.1997 early 2002	VISSR SEVIRI, GERB	1st generation 2nd generation
INSAT-1B to -1D (ISRO) INSAT-2A to -2E (ISRO) INSAT-3B (ISRO) INSAT-3A (ISRO) MetSat-1 (ISRO)	30.8.1983 - 12.6.1990 9.7.1992 - 2.4.1999 21.3.2000 planned for 2001 planned for Oct. 2001	VHRR VHRR/2 VHRR/2	 Starting with -2E Communications only Weather satellite only
GOMS-1 (Russia/Planeta) Electro-M (Russia)	31.10.1994 2005/6	STR	1st generation 2nd generation
FY-2A to -2B (CMA), China	10.6.1997, 26.6.2000	S-VISSR	
AVStar (AstroVision Inc.), Pearl River, MS, USA	2003	Suite of 5 cameras	First commercial GEO weather satellite

Table 36: Overview of geostationary meteorological satellites

- The NASA ATS (Application Technology Satellite) series set the stage for demonstrations (in particular communication experiments, first meteorological observations, etc.) in geostationary orbits. ATS-1 (launch Dec. 6, 1966) flew SSCC (Spin-Scan Cloudcover Camera) to provide continuous cloudcover patterns of the full-disk Earth view. The telescope photomultiplier assembly could be tilted in discrete steps to $\pm 7.5^o$ to a north-south scan (equivalent to $\pm 52^o$ latitude). The east-west scan was provided by the S/C spin of 100 rpm. A ground resolution of about 4 km was obtained. SSCC operated on ATS-1 until Oct. 16, 1972.

ATS-1 (Applications Technology Satellite), a spin-stabilized S/C (mass of 352 kg), built by Hughes Aircraft Company, was the first experimental near-geostationary weather satellite (position at 150° W longitude) with the ability to "**see weather systems**" with SSCC, an instrument built by Hughes SBRC. The optical system consisted of a two-element Cassegrain-type telescopes. SSCC provided imagery of the Earth in the visible spectrum. Within the first month of the availability of ATS-1 imagery, a time-sequence movie of mesoscale cloud patterns in motion was shown on TV (the animated imagery revealed atmospheric motion and their potential use for research and operations). By the early 1970s ATS imagery was being used in US operational forecast centers. [253) 254)] ATS-1 also served as a platform for several communication experiments:

- C-band communications experiment (also used for international TV broadcasts)
- VHF communications package. The VHF experiment tested the ability to act as a link between ground stations and aircraft, demonstrated collection of meteorological data from remote terminals, and evaluated the feasibility of using VHF signals for navigation.
- A WEFAX (Weather Facsimile) system experiment was flown for the first time with the intention to test satellite retransmissions of meteorological data products to participating ground stations.

The ATS-3 S/C (launched Nov. 5, 1967, positioned at 70° W, spin stabilized, mass = 365 kg, equipped with a mechanically despun antenna) flew for the first time the **MSSCC** (Multicolor Spin-Scan Cloud Camera) instrument (built by SBRC) providing full-disk Earth-cloud images in color. [255)] A cartwheel mounting configuration was realized with MSSCC (the concept was first introduced on TIROS-9), i. e, the instrument was mounted with its optical axis perpendicular to the S/C spin axis, permitting to view the Earth through a special aperture in the S/C cylindrical wall.

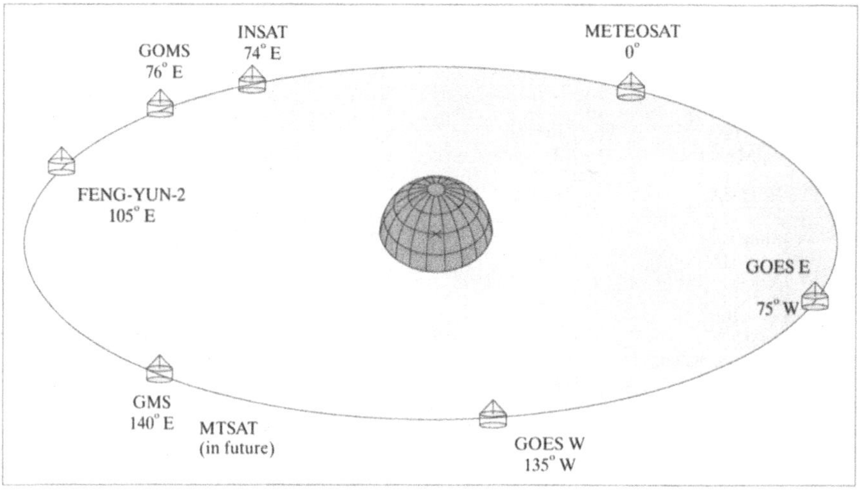

Figure 5: The geostationary meteorological satellite families

253) W. P. Menzel, "Cloud Tracking with Satellite Imagery: From the Pioneering Work of Ted Fujita to the Present," Bulletin of the American Meteorological Society, Vol. 82, No 1, Jan. 2001, pp. 33-47

254) T. Fujita, W. A. Bohan, "Detailed Views of Mesoscale Cloud Patterns Filmed from ATS-1 Pictures," a 16 mm film of 9 minute length is available from Walter A. Bohan Co,, P. O. Box 736, Park Ridge, IL 60068-0736, USA

255) J. R. Greaves, W. E. Shenk, "The Development of the Geosynchronous Weather Satellite System," Monitoring Earth's Ocean, Land, and Atmosphere from Space - Sensors, Systems, and Applications, Progress in Astronautics and Aeronautics, AiAA, Volume 97, 1985, pp. 150-181

• NASA's demonstration of two Synchronous Meteorological Satellites (SMS) began with the launch of SMS-1 on May 17, 1974. NOAA's operation of the GOES series followed with the launch of GOES-1 in October 1975. The SMS satellites carried VISSR (Visible Infrared Spin-Scan Radiometer) as prime instruments. VISSR provided high-quality day/night cloudcover data and made radiance temperature measurements of the Earth-atmosphere system. - The evolution of ATS into SMS and eventually into the GOES series permitted new services like the routine tracking of clouds with infrared imagery available for cloud height determination and nighttime tracking capabilities.

• Japan started its geostationary meteorological program with the launch of GMS-1 (Geostationary Meteorological Satellite-1, referred to as Himawari-1 in Japan) of JMA (Japan Meteorological Agency) and NASDA on July 7, 1977 (see F.3). The newest entry into the ring, MTSAT-1 (Multifunctional Transport Satellite-1) with a launch Nov. 15, 1999 (however, a launch failure of the H-2 vehicle occurred), provides the double service of an "aeronautical mission" (providing navigation data to air-traffic control services in the Asia-Pacific region) and a "meteorological mission." In the latter function, MTSAT is a successor program to the GMS series. A replacement satellite, MTSAT-1R, is planned for launch in 2003. The prime instrument of the meteorology mission on MTSAT-1R is JAMI (Japanese Advanced Meteorological Imager), see F.9.1.2.

• The METEOSAT program of Europe was initiated by ESA in 1972 followed by a launch of METEOSAT-1 (a demonstration satellite) on Nov. 23, 1977. The EUMETSAT convention was signed on May 24, 1983 by 16 countries. On January 1, 1987, responsibility for the operation of the METEOSAT spacecraft was transferred from ESA to EUMETSAT (F.7). The MSG (METEOSAT Second Generation) series with a first launch in early 2002 is expected to provide considerable improvements.

• India started planning of the INSAT series in the 1970s with a launch of INSAT-1B on April 30, 1983 (see F.6). INSAT is a multipurpose operational satellite system series (geostationary) employed for meteorological observation over India and the Indian Ocean, as well as for domestic telecommunications (nationwide direct TV broadcasting, TV program distribution, meteorological data distribution, etc.). The ISRO-developed prime instrument VHRR (Very High-Resolution Radiometer) was enhanced several times providing high-quality data of 2 km spatial resolution in the visible band and 8 km resolution in the IR and TIR bands.
As of 2000, the Government of India approved a dedicated GEO weather satellite named MetSat (Meteorological Satellite). The MetSat-1 launch is planned for October 2001. MetSat-1 is expected to fill the void caused by two important meteorological payloads, namely VHRR/2 on INSAT-2E which failed already in 1999. On Nov. 4, 2000, ISRO was forced to retire INSAT-2B, after it ran out of station-keeping fuel (in July 2000, INSAT-2B completed its design life of seven years of operation). This action stopped of course also the VHRR instrument on INSAT-2B, used for operational meteorology.

• Russia launched GOMS-1 (Geostationary Operational Meteorological Satellite-1) on October 31, 1994. STR (Scanning TV Radiometer) is the prime instrument to observe clouds and underlying surface in VIS and IR bands. GOMS-1, also referred to as Electro-1, ended operations in Nov. 2000. Russia plans to launch Electro-M (Modified) in the time frame 2005/6. Until that time, the Russian weather service is dependent on the services provided by Meteosat of EUMETSAT with regard to GEO data.

• China joined the geostationary meteorological club in 1997 with the launch of FY-2A (Feng-Yun-2A) on June 10, 1997. The prime sensor, S-VISSR (Stretched - Visible and Infrared Spin-Scan Radiometer) is an optomechanical system, providing observations in three bands (at resolutions of: 1.25 km in VIS, 5 km in IR and water vapor).

- As of 2000/1, a US commercial GEO weather satellite, AVStar, is being built by AstroVision Inc. (located at NASA's Stennis Space Center in Pearl River, MS).[256] The overall objective is to monitor the weather over the Americas (North and South) and to provide meteorological data products to a customer base. One goal is to produce quasi-live regional imagery with a narrow-field instrument to permit researchers to monitor quickly the formation of major weather patterns. A launch of AVStar-1 is planned for 2003. The sensor suite consists of five cameras: 1) a video camera, 2) wide-field camera with RGB imagery of the full disk, 3) narrow-field camera, a pointing capability exists to track weather phenomena of interest 4) a panchromatic low-light camera, and 5) a 10-channel multispectral camera.

1.6 Solar-Terrestrial Connection

- X-ray sensor technology and applications. Devices measuring in the X-ray spectrum are being used in a variety of applications. Examples are: to observe X-ray sources in the universe (astronomy, study of particle transport mechanisms in the interplanetary medium), to measure the sun's X-ray radiation (study of solar flares, warning services of solar flares for possible S/C damage, etc.), to diagnose the chemistry of a planet's surface, to monitor the Earth's atmosphere for nuclear explosions in support of the Nuclear Non-proliferation Treaty and the Atomic Test Ban, and to monitor the morphology and spectra of energetic electron precipitation and its effect on the Earth's atmosphere. The X-ray spectrum is generally considered from 60 nm to about 10^{-8} nm (Soft-band X-rays: 60 - 0.1 nm; medium-band X-rays: 10^{-2} - 10^{-3} nm; hard-band X-rays: 10^{-3} - 10^{-8} nm).

S/C Mission	X-Ray Instruments	Comment
ALEXIS (launch Apr. 25, 1993)	ALEXIS	Soft X-ray monitor (6 telescopes)
ARGOS (launch Feb. 23, 1999)	USA	X-ray timing, time-resolved spectroscopy, Feasibility tests of X-ray S/C navigation
Bhaskara (launch Nov. 20, 1981)	X-ray monitor experiment	Study of transient and long-term X-ray sources
CORONAS-I (Mar. 2, 1994) CORONAS-F (Jul. 31, 2001)	TEREK-C, RES-C, DIOGENESS, HELICON, IRIS,	Study of the solar atmosphere structure
DMSP series	SSB, SSB/X-2,	Array-based systems for the detection of X-rays emitted by the atmosphere
GOMS (launch Oct. 31, 1994)	RMS	Study of solar X-ray radiation
GPS/Navstar (Block-II satellites)	IONDS, CXD is follow-up sensor	Monitoring of nuclear explosions, CXD (Combined X-ray detector and Dosimeter)
HESSI (launch Jul. 4, 2000)	HESSI	Study of the sun
IMP-8 (launch Oct. 26, 1973)	CPME, EPE	Solar X-rays
Interball (launch Aug, 3, 1995)	SKA-2, DOK-2X, RF-15, SOSNA-2	Tail Probe instruments
IRS-P3 (launch May 21, 1996)	X-Ray Payload	Study of X-ray sources
ISEE-3 (launch Aug. 12, 1978)	ANH,	Study of solar flares
MTI (launch March 12, 2000)	HXRS	Measures solar X-ray emissions in 8 bands
NOAA-GOES series	SEM package, XRS, SXI (GOES-M)	Monitoring of the solar-terrestrial environment
POLAR (launch Feb 24, 1996)	PIXIE (pinhole camera concept)	Study of energetic electron precipitation on and its effect on the atmosphere
SMART-1 (planned for 2003)	D-CIXS, XSM	Moon surface chemistry, swept charge detector design, instrument mass of 3 kg
SMM (launch Feb 14, 1980)	GRS, HXBRS (study of electron acceleration of solar flares), HXIS, XRP	GRS made important contributions to the international study of Supernova 87A, which in Feb. 1987 provided astronomers with their first opportunity to study such an explosion
SNOE (Launch Feb. 26, 1998)	SXP	Monitoring of soft X-ray flux from the sun
SOHO (launch Dec. 2, 1995)	CDS, EIT, UVCS, LASCO	Study of the composition of the solar corona, etc.

256) Information provided by Malcolm A. LeCompte of AstroVision Inc.

S/C Mission	X-Ray Instruments	Comment
Solar-A (launch Aug. 30, 1991)	HXT, SXT,	Study of high-energy solar phenomena
Solar-B (launch 2004)	XRT	Study of high-energy solar phenomena
SNOE (launch Feb. 26, 1998)	SXP	Solar soft X-ray flux measurement
SORCE (planned for 2002)	XPS (SXP heritage)	Solar EUV measurement
TRACE (launch Apr. 2, 1998)	TRACE	Study of coronal mass ejections, etc.
UARS (launch Sept 12, 1991)	AXIS,	Measurement of Bremsstrahlung X-rays from Earth
Ulysses (launch Oct. 6, 1990)	GRB	X-ray and gamma-ray measurement of solar and cosmic origin

Table 37: Overview of some missions flying X-ray instruments (alphabetical order)

- Solar observations are performed by a number of dedicated missions, normally with strong international cooperation (Section K). There are also some Earth-observation missions with instruments dedicated for sun observations. Coriolis of NRL with SMEI (Solar Mass Ejection Imager) is an example of such a mission.

1.6.1 Earth Radiation Budget and Solar Constant

The solar constant, also referred to as TSI (Total Solar Irradiance), is the radiative power received normally onto a unit area at the mean Earth/Sun distance. Along with Earth's global average albedo and the emitted long wave, it determines Earth's global average equilibrium temperature. It is the primary power source driving our climate system. Hence, knowledge of the long-term solar constant variability is essential in understanding the past and future climate changes. Solar irradiance varies slightly over an 11-year cycle. This cycle of the sun's magnetic activity alters its energy output, as well as the occurrence of sunspots, flares, and CMEs (Coronal Mass Ejections). - Since the 1880s, routine solar irradiance measurements were made from the ground to detect an eventual variability. The irradiance was referred to as the "solar constant" because atmospheric transmittance variability, of the order of several percent, prevented the detection of systematic irradiance changes from the Earth's surface.

The continuous observation of the solar irradiance at the highest possible precision and accuracy is an important objective of the Earth climate change program. It requires high quality metrology in the space environment linked to ground solar radiometric comparisons supported by laboratory characterization activities. [257]

The Earth radiation budget (ERB) is the difference between the Earth's absorbed solar radiation and the longwave radiation emitted by Earth and its atmosphere. The concept applies globally to the whole Earth and locally to any place , with important consequences due to its gradients. Similarly, the net radiation at TOA (Top of the Atmosphere) is derived as a function of the planetary albedo and the longwave radiation emitted by Earth into space. ERB measurements seek to contribute to two key scientific goals: 1) the determination of how long- and shortwave fluxes are distributed in the atmosphere and how they vary in time, and 2) the development of a quantitative understanding of the links between the radiation budget and the properties of the atmosphere and the surface related to the energy budget and its processes. [258] [259]

First Earth radiation budget measurements of the space age go back to the late 1950s, with instruments flown on Soviet and US satellites, such as Explorer-6 (launch Aug. 7, 1959 - last meaningful data on Aug. 25, 1959). The 'TV Optical Scanner' flown on Explorer-6, an im-

257) D. Crommelynck, S. Dewitte, "Metrology of Total Solar Irradiance Monitoring," Advances in Space Research, Vol. 24, No 2, 1999, pp. 195-204

258) R. B. Lee III, M. A. Gibson, R. S. Wilson, S. Thomas, "Long-term total solar irradiance variability during sunspot cycle 22," Journal of Geophysical Research, Vol. 100, No A2, pp. 1667-1675, Feb. 1, 1995

259) E. N. Parker, "The Physics of the Sun and the Gateway to the Stars," Physics Today, June 2000, pp. 26-31

proved version of the TV system first employed on Pioneer 2, can be regarded as one of the first instruments, providing at least some very crude data (low-resolution daylight cloudcover photographs) for initial Earth radiation budget estimation (because of the poor spatial and temporal sampling without Earth emitted radiation observations).

In the early 1970s, the importance of improving knowledge on the Earth's radiation budget was recognized [i.e., the balance between incoming energy from the sun and the outgoing thermal (longwave) and reflected (shortwave) energy from the Earth] and its relationship with the Earth's climate. NASA built a multichannel absolute radiometer instrument (blackbody calibrated) by the name of ERB (Earth Radiation Budget) which was flown for the first time on Nimbus-6 (launch June 12, 1975, M.17.6). However, the analysis of the ERB data failed to detect any irradiance variability due to degraded responses of the ERB radiometer. An improved version of ERB was subsequently flown on Nimbus-7 (launch Oct. 24, 1978, M.17.7). This radiometer was stable enough to detect short-term and long-term solar irradiance variability. ERB was the first long term solar monitor utilizing the ESCC (Electrically Self Calibrating Cavity) technique. The design of this absolute cavity radiometer technique is based on the principle of substitution of electrical power for radiative power (the idea of the cavity design is to come as close as possible to a "perfect absorber" of solar radiation). TSI data from ERB on Nimbus-7 were available until the end of 1993. [260)]

The success of ERB on Nimbus-7 resulted in the development of a number of TSI instruments for various missions. Since the solar constant is one of the prime determining factors of the Earth's climate (together with the cloud cover and the Earth's thermal emission), the data obtained is used in such applications as climate change, solar physics studies, and Earth radiation budget estimates. Solar-constant instruments ought to be very sensitive and accurate radiometers for the measurement of solar radiation. Most sensor designs employ the method of "active cavity detector geometry" pyrheliometers (detector in equilibrium with incident solar radiation) and variations thereof. Following is a survey of "Earth radiation budget missions," most of them with absolute or active cavity radiometers for the measurement of TSI; there are also instruments (such as HCMM, ScaRaB and GERB) that are not of the absolute radiometer type. [261)]

- A first comprehensive thermal survey of the Earth's surface was provided by HCMM (Heat Capacity Mapping Mission, launch April 26, 1978, end of mission Sept. 30, 1980) of NASA. The satellite featured a Heat Capacity Mapping Radiometer (HCMR) for the measurement of the heat budget, in particular the thermal inertia of the land surface for geological studies.

- SMM (Solar Maximum Mission, launch Feb. 14, 1980, [262)] ACRIM data until 1989) flew ACRIM-I (Active Cavity Radiometer Irradiance Monitor) of NASA/JPL and provided data until 1989. The ACRIM-I and ERB/Nimbus-7 instruments detected a short-term (less than 27-day solar rotational period) irradiance variability of as much as 0.2% and a long-term variability of about 0.1%.

- A new generation instrument was realized with ERBE (Earth Radiation Budget Experiment) of NASA/LaRC, first flown on ERBS (Earth Radiation Budget Satellite, launch Oct. 5, 1984), then on NOAA-9 (launch Dec. 12, 1984), and NOAA-10 (launch Sept. 17, 1986). ERBE was able to provide more accurate and systematic parameters for estimating the Earth's radiation budget. ERBE flew both wide-field-of-view, flat-plate radiometers and the narrow-field-of-view scanning radiometric telescopes measuring with three chan-

260) J. R. Hickey, et al., " Total solar irradiance measurements by ERB/Nimbus-7, a review of nine years," Space Science Review, Vol. 48, 1988, pp. 321-342

261) C. Fröhlich, "Observations of Irradiance Variations," pp. 15-24 in Solar Variability and Climate, Editors: E. Friis-Christensen, C. Fröhlich, J. D. Haigh, M. Schüssler and R. von Steiger, Kluwer Academic Publishers, ISBN 0-7923-6741-3, 2000

262) D. V. Hoyt, H. L. Kyle, J. R. Hickey, R. H. Maschhoff, " The Nimbus-7 solar total irradiance: A new algorithm for its derivation," Journal of Geophysical Research, Vol. 97, 1992, pp. 51-63

nels the "short wave", the "long wave" and the "total" radiances. The improved spatial resolution of flux data achieved with this scanner, by taking into account the bidirectional reflection functions at the TOA, is perhaps the most important advance offered by ERBE, because it led to better estimates of the difference between cloudy and clear-sky fluxes and thus a better estimate of the effect of clouds on Earth's radiation budget. [263)]

- The instruments ACRIM-II, SOLSTICE (Solar/Stellar Irradiance Comparison Experiment) and SUSIM (Solar Ultraviolet Spectral Irradiance Monitor), all with the objective to measure solar radiation, are flown on NASA's UARS (Upper Atmosphere Research Satellite) mission. Launch Sept. 12, 1991, the S/C is operational as of 2001.

- EURECA-1 (European Retrievable Carrier). An ESA long-term free-flyer platform (J.5.1) launched on Shuttle flight STS-46 on July 31, 1992 and retrieved with STS-57 on July 1, 1993. Two instruments aboard EURECA were dedicated to solar observations, namely SOVA (Solar Constant and Variability Instrument) with D. Crommelynck as PI, and SOSP (Solar Spectrum Instrument) with G.Thuillier as PI.

- The ScaRaB (Scanner for Radiation Budget) program of CNES (France, Russia and Germany are program partners), started in 1987 with the objective to measure the terms of the Earth's radiation budget. Two of three instruments, built by CNRS/LMD and operated by CNES, were flown already on Meteor-3-7 (launch Jan. 25, 1994, ScaRaB stopped operating March 5, 1995) and Resurs-O1-4 (launch July 10, 1998 - failure of the backup downlink on Apr. 8, 1999 causing a termination of ScaRaB operations), respectively. The third instrument is planned to be flown on a future mission (currently a CNES/ISRO climate mission in the tropics is considered the most likely candidate, namely Megha-Tropiques). [264)] Note: ScaRaB performs no measurements of TSI. However, the overall radiation budget analysis of ScaRaB uses solar constant values from other sources (such as ISP-2).

- VIRGO (Variability of Solar Irradiance and Gravity Oscillations), an ESA AO instrument package flown on SOHO (launch Dec. 2, 1995, operational as of 2001). The subunits DIARAD (Differential Absolute Radiometer) and PMO6-V are the absolute radiometers in this assembly. Regularly updated observations of the solar constant can be found at the following reference: [265)]

- ISP-2 (Izmeritel Solnechnoy Postoyannoy-2 - a Solar Constant Instrument) flown on Russia's RESURS-O1-4 (launch July 10, 1998). The intention is to complement the ScaRaB measurements.

- CERES (Clouds and the Earth's Radiant Energy System), an instrument of ERBE heritage of NASA/LaRC, is flown on TRMM (launch Nov. 27, 1997) as a single cross-track radiance sensor of short, long- and total wave. CERES is also being flown on NASA's Terra mission (launch Dec. 18, 1999) as a dual-track scanner (two radiometers). Another CERES instrument is scheduled to fly on further missions such as Aqua (formerly EOS/PM1).

- ACRIM-III on ACRIMSAT (NASA), launch Dec. 20, 1999 (A.2)

- GERB (Geostationary Earth Radiation Budget) is an ESA AO instrument of the EUMETSAT MSG-1 (METEOSAT Second Generation) mission, a two-channel broadband radiometer, provided by a consortium led by the UK (NERC), Belgium (SSTC) and Italy (ASI). A launch of MSG-1 is planned for early 2002. The instrument measures the radiances leaving the Earth at TOA (Top of the Atmosphere); it will provide an image of the reflected sunlight and infrared radiation at a five-minute sampling rate, in particular the shortwave (0.32 - 4.0 μm) and longwave (4.0 - 30 μm) regions of the spectrum. The overall radiation budget analysis of GERB uses solar-constant values from other sources. Because

263) E. F. Harrison, P. Minnis, B. Barstrom, et al., "Seasonal variation of cloud radiative forcing derived from tire ERBE," Journal of Geophysical Research, Vol. 95: 1990, pp. 18667-18703

264) Information provided by M. Rouzé of CNES

265) http://remotesensing.oma.be/solarconstant/solar.html

the evaluation of the TOA Earth radiation balance terms (conversion of radiances to TOA fluxes) will be made available about three hours after observation by the IRMB, the GERB observations can make a unique contribution to the understanding of the Earth's climate balance (diurnal sampling), since such measurements have never been carried before from geostationary orbit (F.8). GERB instruments are planned to be flown on the MSG-2 and MSG-3 missions.

- NASA is sponsoring TIM (Total Irradiance Monitor), developed at LASP (Laboratory for Atmospheric and Space Physics) of the University of Colorado, to be flown on SORCE (Solar Radiation and Climate Experiment). SORCE is part of NASA's ESE (Earth Science Enterprise) program with a planned launch in 2002. Besides very precise and accurate solar irradiance measurements with TIM, the other instruments of SORCE, namely SOLSTICE (Solar/Stellar Irradiance Comparison Experiment), SIM (Spectral Irradiance Monitor), and XPS (XUV Photometer System), complement the solar spectral irradiance measurements at wavelengths extending from the far ultraviolet to the near infrared. - A TIM instrument, referred to by the name of TSIS (Total Solar Irradiance Sensor), is also planned to fly on NASA's NPP (NPOESS Preparatory Project) with a launch in 2005 and on NPOESS (National Polar-orbiting Operational Environmental Satellite System) of IPO (Integrated Program Office) with a first launch projected for 2008.

- NISTAR (National Institute of Standards and Technology Advanced Radiometer) is a three-channel cavity radiometer of the NASA mission Triana (planned launch in 2004) which measures the Earth's "irradiance" in the small angle as seen from a halo orbit at L1 (Lagrangian point 1), about 1.5 million km from Earth in the direction of the sun. The NISTAR TSI measurements offer a great opportunity to compare its radiances with instruments in LEO (Low Earth Orbit).

- Solar-A (Solar Monitoring Observatory) is an ESA experiment package consisting of three instruments, namely SOVIM (Solar Variability and Irradiance Monitor), SOLSPEC (Solar Spectral Irradiance Measurements), and SOL-ACES (Solar Auto-Calibrating EUV/UV Spectrophotometers). The overall objective is to measure the solar spectral irradiance with unprecedented accuracy (L.2.16). The three instruments cover the combined wavelength range from 17-3000 nm. The Solar-A package will be mounted on the ESA-developed CPD (Course Pointing Device), located on EPA of ISS. A launch of Solar-A is planned for the timeframe of 2004.

- Picard is a CNES solar-terrestrial microsatellite mission (launch in 2006). Two instruments measure the solar constant: SOVAP (Solar Constant Variability, Picard) of IRMB (Royal Meteorological Institute of Belgium) in Brussels, and PREMOS (Precision Monitoring of Solar variability) photometric spectral band measurements of WRC (World Radiation Center) of Davos, Switzerland. The measurement of the solar diameter is done by SODISM (Solar Diameter Imager and Surface Mapper), a CNRS/SA (Verrières-le-Buisson, France) instrument in collaboration with ESA/ESTEC.

- Periodic (short-term) measurements were conducted in particular with SOLCON (Solar Constant Sensor) on a number of Shuttle flights. The instrument is a cooperative effort of IRMB (Belgium), Space Science Dept. of ESA, and NASA/LaRC. [266) 267)]

Launch Vehicle	Date	Mission	Instrument
STS-9 (Shuttle)	Nov. 28 - Dec. 8, 1983	Spacelab-1 (ESA)	SOLCON-1
STS-45	March 24 - April 2, 1992	ATLAS-1 (NASA)	SOLCON-2, ACRIM-II
STS-56	April 8-17, 1993	ATLAS-2 (NASA)	SOLCON-2, ACRIM-II
STS-66	Nov. 3-14, 1994	ATLAS-3 (NASA)	SOLCON-2, ACRIM-II
STS-85	Aug. 7-19, 1997	Hitchhiker (NASA)	SOLCON-2

266) http://sspp.gsfc.nasa.gov/
267) http://estirm2.oma.be/solarconstant/articles/article1.html#references

Launch Vehicle	Date	Mission	Instrument
STS-95	Oct. 29 - Nov. 7, 1998	Hitchhiker (NASA)	SOLCON-2
STS-107	Oct. 2001 (planned)	Hitchhiker (NASA)	SOLCON-2

Table 38: Chronology of Shuttle-based solar-constant measurements with absolute radiometers

1.6.2 Solar Wind Observation

The phenomenon of the solar wind is a discovery of the space age by data analysis first conducted from the Soviet Luna spacecraft series [Luna-1 (launch Jan. 2, 1959), Luna-2 (launch Sept. 12, 1959), and Luna-3 (Oct. 14, 1959)] on their flight to the moon. The first tentative evidence of the solar wind was observed with an experiment of Konstantin I. Gringauz (1918-1993, USSR) and his team onboard the Luna-2 and Luna-3 spacecraft. The total electric charge of arriving ions was measured in the "ion trap experiment." Gringauz noted that the signal fluctuated as the spacecraft spun around its axis, suggesting an ion flow was entering the instrument whenever it faced the sun. [268) 269)] In 1961, James Dungey (UK) proposed a mechanism for transmitting solar wind energy to the magnetosphere by direct magnetic linkage between the two. The magnetopause, the boundary between magnetosphere and the solar wind, was observed for the first time by Explorer-12 in 1961. The first variations in the speed of the solar wind (27 days intervals) were observed in 1962 from the US Mariner-II on its way to Venus. The data analysis of the IMP-1 (Interplanetary Monitoring Platform 1, launch Nov. 27, 1963, the mission is also known as Explorer-18) revealed a large bow shock formed in the solar wind ahead of the magnetosphere, and a long magnetic tail on the night side of the Earth. In 1983, the ISEE-3 (International Sun-Earth Explorer 3) mission explored the distant magnetotail, before heading for comet Giacobini-Zinner.

Mission	Instruments	Comment
ACE, launch Aug. 25, 1997	SWIMS, SWICS, SWEPAM	Focus on solar wind composition and acceleration
AMPTE, launch Aug. 16, 84		Release of trace gases into the solar wind
Cluster-II, launch Jul. 16, 00	CIS, HIA, STAFF	Study of solar wind bow shock
DS1, launch Oct 24, 1998	PEPE	Solar wind energy spectrum
Equator-S, launch Dec. 2, 1997		Solar wind measurements with WIND and SOHO
Genesis, launch Aug. 8, 2001	Collector Array, Solar Wind Concentrator	Solar wind measurement, collection and return of solar wind samples. Halo orbit at L1
GEOTAIL, launch Jul. 24, 1992	LEP, CPI	Measurement of fluctuations in the solar wind
IMAGE, launch Mar. 25, 2000	LENA, MENA, HENA	Response of magnetosphere with solar wind
IMP-8, launch Oct. 26, 1973	GAF, MAP	Plasma field environment for magnetospheric studies
INTERBALL, launch Aug. 3, 1995 and Aug. 29, 1996	Monitor-3, etc.	Solar wind energy and interaction with the Earth's magnetosphere
ISEE-1, -2, launch Oct. 2, 77	SWE, EGD, OGM, SHM,	Study of Earth's bow shock
ISEE-3, launch Aug. 12, 78	BAH, OGH, SBH	Measurement of solar wind/plasma fields First ever cometary encounter 7800 km tailward of comet Giacobini-Zinner.
NOAA-15, launch May 13, 1998	SEM-2 package	An operational space weather warning system
POLAR, launch Feb. 24, 1996	MFE, TIMAS,	Coupling of the solar wind and magnetosphere, study of particle populations
SOHO, launch Dec. 2, 1995	UVCS, SWAN, CELIAS,	Study of solar wind and energetic particles, interaction with the Earth. First tracing of the slow-speed solar wind

268) T. I. Gombosi, "Modeling Gringauz's legacy from the solar wind to weakly magnetized solar system bodies," International Symposium on Space Plasma Studies by In-Situ and Remote Measurements (Gringauz Symposium), Moscow, Russia, June 1-5, 1998.

269) http://www–spof.gsfc.nasa.gov/Education/whsolwi.html

Mission	Instruments	Comment
STEREO, launch in 2004	IMPACT, SWEA, PLASTIC, SWAVES,	Measurements of solar wind in a heliocentric elliptical orbit in the ecliptic plane at 1 AU
TIMED, launch in 2001	TIDI	Solar wind structure in MLTI region
TIROS-N, launch 1978 and on all S/C of NOAA POES	SEM (NOAA/SEC package)	Measurement of solar wind particle flux
Triana (2004)	PlasMag	Study of solar wind in halo orbit at L1
Ulysses, launch Oct. 6, 1991	SWICS, SWOOPS, UARP,	Study of the solar wind
Viking, launch Feb. 22, 1986	V1, V2, V3, V4L	Solar wind interaction with the magnetosphere
WIND, launch Nov. 1, 1994	MFI, WAVES, SWE, SMS, EPACT, PLASMA,	Study of solar wind mass momentum and energy. Halo orbit at L1

Table 39: Some solar wind experiments/studies (alphabetic order of missions)

- SEM (Space Environment Monitor). SEM is an operational NOAA instrument package with the objective to provide "**space weather**" on a regular basis to the user community by measuring the solar wind particle flux and its variations. The package is provided by NOAA/SEC of Boulder, CO. SEM-1 instruments were initially introduced on satellites in geostationary orbit starting in 1974. SMS-1, a predecessor of the NOAA-GOES series, was the first satellite to carry SEM. NOAA introduced SEM on the POES series (G.14.1) as well starting with TIROS-N in 1978. An upgraded instrument package, SEM-2, was introduced into the POES series with the launch of NOAA-K (NOAA-15) on May, 13, 1998 (G.15.2).
- The ISEE-3 (International Sun-Earth Explorer-3) mission, with a launch in 1978, orbited the Earth for nearly four years. It was then directed to study the geomagnetic tail for a first-ever exploration of that region. This resulted in the discovery of gigantic plasmoids that were ejected from the near-Earth magnetosphere. After exploring the nature of the geomagnetic tail, ISEE-3 was sent off for the first spacecraft encounter with a comet in September 1985. When ISEE-3 was near Comet Giacobini-Zinner, first measurements of the solar wind's interaction with a comet were obtained.
- The Ulysses S/C (launch Oct. 6, 1991, K.28) of ESA/NASA is the first S/C having left the ecliptic plane to observe the poles of the sun (sling shot past Jupiter). The three major objectives are to study the sun, the solar wind, and interstellar space. Although Ulysses is the first S/C to probe the sun's polar regions, it does not travel near the sun. In 1994, it was reported that near the south pole the solar wind is flowing away from the sun at nearly twice the speed that is typically observed near the sun's equator.
- The IMAGE S/C (launch Mar. 25, 2000, K.15) of NASA carries three ENA (Energetic Neutral Atom) imagers (LENA, MENA, and HENA) whose combined energy coverage permits the detection of ENAs with energies ranging from 1 eV to 500 keV per atomic mass unit (amu). [270] Each neutral atom instrument generates images showing the intensity and spatial distribution of ENA emissions produced in the inner magnetosphere through charge-exchange reactions between geocoronal neutral hydrogen and various magnetospheric ion populations. Neutral atom imaging of the ionosphere and magnetosphere is possible because the Earth's geocorona acts like an imaging screen for magnetospheric and ionospheric ions. [The neutral atoms in the space environment are measured against the large and ubiquitous UV background which can produce high noise count rates in MCP detectors]. To date, four different techniques have been developed to allow neutral atom detection and imaging against the UV background:

- A thick foil which blocks the UV
- An ultra thin charge conversion foil to ionize ENAs (which are then passed through an electrostatic analyzer)
- A charge exchange surface from which ENAs can reflect as ions which are then analyzed

270) D. J. McComas, "Two Wide-Angle Imaging Neutral-Atom Spectrometers,"

- Transmission gratings which block UV but allow ENAs to pass.

Mission	ENA Instruments
ATLAS-1 of NASA on STS-45 (launch Mar. 24, 1992)	ENAP (Energetic Neutral Atom Precipitation), of the University of Texas, Dallas
POLAR of NASA, (launch Feb. 24, 1996)	CEPPAD/SEPS (Comprehensive Energetic-Particle Pitch Angle Distribution / Source Loss Cone Energetic Particle Spectrometer)
SAC-B of CONAE (launch Nov. 4, 1996), the 3rd stage failed to separate from the S/C	ISENA (Imaging Spectrometer for Energetic Neutral Atoms) of CNR/IFSI, Italy
IMAGE of NASA (launch March 25, 2000)	LENA, MENA, HENA
TWINS of NASA, LANL, etc. (launch of 1st S/C in 2002, launch of 2nd S/C in 2004)	TEI (TWINS ENA Imager)

Table 40: Overview of ENA instruments flown on various missions

The IMAGE spacecraft and its sensor complement represent a new era of magnetospheric observation capability. While traditional instruments provide in-situ observations of the magnetospheric plasma environment, the new ENA devices are able to image the plasma motions by detecting the particles and their traveling directions (the imaging is done using "electromagnetic wave sounding").

Using neutral atoms (hydrogen, helium, oxygen), which speed through the magnetosphere unimpeded by magnetic forces, to image the magnetosphere was an idea first realized by researchers at JHU/APL in the late 1980s.

- The Genesis mission of NASA (launch Aug. 8 2001) has the objective to collect solar wind particles at L1 (Lagrangian point 1) and return these particles to Earth (by a sample return capsule) after mission completion.
- The TWINS two-spacecraft mission of NASA (launch 2003) has the objective to demonstrate the new capability of stereoscopic imaging of the magnetosphere. By imaging the ENAs (produced by charge-exchange reactions between the geocoronal neutral hydrogen and the plasma) over a broad energy range (about 1-100 keV) using two identical instruments [TEI (TWINS ENA Imager)] on two widely spaced high-altitude and high-inclination spacecraft, TWINS will enable the three-dimensional visualization and the resolution of large scale structures and dynamics within the magnetosphere for the first time.

Mission	Instrument	Objective or Measurement
AE-C (NASA), Launch Dec. 16, 1973	VAE (Visible Airglow Experiment)	Measurement of airglow and aurora features
APEX (Active Plasma Experiment), launch Dec. 18, 1991	Suite of sensors	Study of auroral-ionospheric relationships
ARGOS (DoD), launch Feb. 23, 1999	EUVIP, HIRAAS, LORAAS, GIMI	Characterization of the aurora
Astrid-1 (IRF-K), Sweden, launch Jan. 24, 1995 Astrid-2, launch Dec. 10, 1998	MIO (Miniature Imaging Optics) PIA of MPAe	Auroral emissions Auroral imaging
ATLAS-1 (NASA) STS-45, launch March 24, 1992	AEPI (Atmospheric Emissions Photometric Imaging)	Images of natural and induced aurorae and airglow
DE-1 (Dynamic Explorer) of NASA, launch Aug. 3, 1981	EICS (Energetic Ion Mass Spectrometer), SAI (Spin-Scan Auroral Imager)	Coupling of magnetosphere/ionosphere, auroral images in UV and VIS
DMSP series of DoD	OLS (Operational Linescan System), SSULI (Special Sensor Ultraviolet Limb Imager), SSUSI (Special Sensor Ultraviolet Spectrographic Imager)	OLS auroral images (nighttime) since the mid-1970s, SSULI and SSUSI starting with F-16 in 2001
DODGE (DoD) launch July 1, 1967	Dual Vidicon Cameras	Some measurement of airglow and aurorae
EXOS-A (ISAS, Japan) launch Feb. 4, 1978 EXOS-D, launch Feb. 22, 1989	UV-TV camera, UV Glow Spectrophotometer ATV (Auroral TV)	Measurement of auroral activity, with ATV auroral imagery in UV and VIS

Mission	Instrument	Objective or Measurement
FREJA (Sweden), Oct. 6, 1992	F5 (Auroral UV Imager)	Study of aurorae
GEOTAIL (ISAS/NASA), launch July 24, 1992	CPI (Comprehensive Plasma Investigation)	Auroral imaging for magnetotail plasma dynamics
HILAT (DoD) or P83-1, launch Jun. 27, 1983	AIM (Auroral Ionospheric Mapper) built by APL	Imaging of the sunlit aurora
IMAGE (NASA), launch May, 25, 2000	WIC (Wideband Imaging Camera), SI (Spectrographic Imager), MENA (Medium-Energy Neutral Atom Imager), HENA, EUV	Broadband auroral imaging, different types of aurorae, etc. Imaging of ENAs, substorms, and ion populations of the cusp
INTERBALL (Russia, etc.), launch of Auroral Probe Aug. 29, 1996	UFSIPS, UVAI	Study of aurorae
Lewis (NASA) launch Aug. 23, 1997, contact to Lewis was lost on Aug. 26, 1997	LEISA	Study of nightglow aurorae
MSX (DoD), launch Apr. 24, 1996	UVISI with 4 cameras and 5 imaging spectrometers	Study of aurorae in FUV-VIS, tomographic imagery
POLAR (NASA), launch Feb. 24, 1996	UVI (UV Imager), VIS (Visible Imaging System)	Study of dayside and nightside aurorae
ROCSat-2 (NSPO), launch in 2003	ISUAL (Imager of Sprite Upper Atmospheric Lightning)	Study of aurorae and airglow
SNOE (U. of Colorado) launch Feb. 26, 1998	AP (Auroral Photometer)	Study of auroral emissions
STS-39, launch Apr. 28, 1991	CIRRIS (DoD)	Study of aurorae and airglow
TIMED (NASA), launch in 2001	GUVI (Global UV Imager)	Horizon-to-horizon imagery in five bands with coverage of one limb (away from the sun)
TWINS (NASA, LANL, APL, SwRI, USC, etc.), planned launches in 2002 and 2004	TEI (TWINS ENA Imager)	3-D visualization of large-scale structures in the magnetosphere, study of geomagnetic storms, etc.
Viking (Sweden), launch Feb. 22, 1986	V5 (Auroral Imaging Experiment)	Study of dynamic behavior of aurorae

Table 41: Some space science missions/instruments relating to auroral imaging

1.7 Navigation

Navigation is the art of establishing position and velocity of an object in space (distances, angles and time to known references). The object (platform) attitude is also of fundamental importance, since it serves in general as a reference for most sensor observations. Of particular interest are the ground-based and on-board instruments which provide inputs for orbit and/or attitude determination. In general, orbit determination is an iterative process, building upon the results of previous solutions, and on tracking data inputs from various sources over significant time periods. In spite of these obstacles, orbit determination is becoming an increasingly automated process due to better computing capabilities and ever-improving algorithms (and filters) in the latter 1990s.

Spacecraft	Launch Date	Innovation
Sputnik-1	Oct. 4, 1957	Satellite Doppler tracking
Transit-1A	Sept. 17, 1959	Yo-Yo spin/despin mechanism
Transit-1B	Apr. 13, 1960	**Dual-frequency Doppler tracking for correcting ionospheric error** First attitude-controlled S/C using permanent magnets First solar attitude detectors
Transit-2A	June 22, 1960	First dual-payload launch (Transit-2A and Solrad-1 of NRL) and first piggyback separation, demonstrated before on Transit-1B
Transit-3B	Feb. 21, 1961	First satellite electronic memory in space (384 bits of magnetic core register - required to store its own orbit ephemeris)
Transit-4A	June 29, 1961	First nuclear power generator tested in a spacecraft (RTG)
TRAAC	Nov. 15, 1961	Damping of satellite vibration by lossy spring-and-mass technique. (TRAAC= Transit Research And Attitude Control) satellite
ANNA-1B	Oct. 31, 1962	First geodetic satellite (ANNA = Army Navy, NASA, Air Force) which also flew the first gallium arsenide cell in space.
Transit-5A-1	Dec. 19, 1962	First uplink authentication system
Transit-5A-3	June 16, 1963	First successful gravity-gradient stabilization (plus spring-and-mass damping) to maintain Earth pointing for one side of the S/C. - Automatic temperature S/C control
Transit-5C-1	June 4,1964	Demonstration that hysteresis rods, used previously for damping magnetic stabilization, were also effective for gravity-gradient stabilization
DME-A	Nov. 29, 1965	Magnetic spin/despin system (DME=Direct Measurement Explorer)
SAS-1	Dec. 12, 1970	Dual-spin control of satellite pointing (Small Astronomy Explorer-1)
Triad (Transit-improved DISCOS)	Sept. 2, 1972	First satellite compensated for drag and radiation pressure. The drag-free concept was realized with DISCOS (Disturbance Compensation Device). First demonstration of single-frequency refraction-free satellite navigation using pseudonoise modulation.
DODGE	July 1, 1967	First yaw stabilization of a satellite using a 'pitch axis wheel' [constant-speed 'momentum wheel'] DODGE = DoD Gravity Experiment, M.4
GEOS-3	April 9, 1975	Demonstration of first satellite-to-satellite tracking. By closed-loop tracking with a S-band transponder, the position of GEOS-3 was measured relative to that of ATS-6 of known position.
TIP-II	Oct. 12, 1975	Crystal oscillator as on-board timing system with all drift removed by a programmable synthesizer. TIP = Transit Improvement Program
SeaSat SAR downlink	Jun. 27, 1978	Quadrifilar helix antenna with beam shaping to compensate for slant range
Magsat	Oct 30, 1979	First attitude and command systems using microprocessors
Landsat-4	Jul. 16, 1982	GPSPAC (see 1.7.8) the first spaceborne GPS receiver in history
Geosat-A	Mar. 12, 1985	Bifilar helix antenna
MSX	April 24, 1996	First spaceborne hyperspectral imager, UVISI (Ultraviolet/Visible Imaging and Spectrographic Imaging)

Table 42: Some technology innovations introduced by JHU/APL

1.7.1 Tracking Techniques

The two-way support of the communication function requires the satellite to carry a simple transceiver. In order to also accomplish the tracking function, the satellite must carry a transponder which, in addition to providing two-way communications, also returns the tracking antenna's transmitted ranging signal, thus, permitting the quick determination of the distance between the ground antenna and the spacecraft. - Ground-based navigation is with us from the very beginning of the space age. The conventional tracking methods are "ranging," "Doppler velocity," and "angle-only" (when no transponder is on-board) determination along the line of sight. [271] Ranging (radial distance) is derived form transit time, namely the round-trip light time of ranging signals from the spacecraft transponder, while the ground antenna pointing direction provides angular information. More precise angle measuring methods are those of "differenced Doppler" and VLBI. The Doppler shift is a measure of object velocity. A two-way coherent transmission mode permits the measurement of the induced Doppler shift to within 1 Hz, since the uplink frequency is known with great precision. The Doppler shift is directly proportional to the radial component of the spacecraft's velocity. - The measurement accuracy is dependent on the frequency band selected. Conventional ranging and Doppler measurement capabilities are in the microwave region, such as S-band with typical accuracies in the order of about 0.5 m. Satellite laser ranging (SLR) techniques between a ground-based laser station and a satellite use the much shorter wavelengths of visible light, resulting in a single-shot precision of <2 cm.

- Doppler tracking. Soon after the launch of Sputnik-1 (Oct. 4, 1957), JHU/APL (Johns Hopkins University/Applied Physics Laboratory) researchers W. H. Guier and G. C. Weiffenbach discovered that they could **determine a satellite's orbit solely from RF Doppler measurements** made on a single pass over their laboratory. In early 1958, F. T. McClure of APL inverted the problem: **if the position of the satellite were accurately known, then Doppler data could tell an observer on the ground his unknown position** (Note: the inverse problem became later known as the "navigation problem"). This led to the conceptual design of the first satellite Doppler navigation system, namely "Transit" for the US Navy (H.6). [272] [273] [274] [275]

- S/C tracking techniques, tracking system behavior and geodetic studies were the objective of three dedicated NASA missions, namely GEOS-1 (launch Nov. 6, 1965, E.7.1), GEOS-2 (launch Jan. 11, 1968) and GEOS-3 (launch April 9, 1975). Doppler instruments on-board and on-ground were employed for systematic range and range-rate measurements. Doppler shift measurements of a spaceborne Doppler instrument were also used to establish the structure of the Earth's gravity field to a fairly good accuracy.

- Satellite-to-satellite tracking technique (SST). The concept is based on tracking the "relative motion" between two satellites. SST is employed in particular in geodetic applications (gravity missions) to obtain highly accurate orbits for LEO satellites, and from these, by applying orbit perturbation analysis, the structure of the Earth's gravity field. The lack of sufficient coverage in ground station tracking capability requires such measures. SST was demonstrated for the first time ever in 1968, mapping the near side gravity field of the moon (with Earth being considered a satellite of the moon). With regard to the Earth's gravity field, SST was first demonstrated between GEOS-3 (in LEO) and ATS-6 (Applied Technology Satellite), a geostationary S/C of NASA in April 1975 (closed-loop tracking with a S-band transponder, the position of GEOS-3 was measured relative to that of ATS-6 of known position). However, few results were obtained due to the relatively high altitude of the lower satellite (E.7.3, GEOS-3 perigee of 818 km, apogee of 858 km). Later in 1975, there were SST measurements between the Apollo-18 S/C (launch of Apollo-18 ASTP on July 15,

271) Note: In the very early period of space flight, the technique of optical tracking was employed by the use of the Baker-Nunn camera.

272) E. J. Hoffman, "Spacecraft Design Innovations in the APL Space Department," Johns Hopkins APL Technical Digest, Vol. 13, No. 1, 1992, pp. 167-181

273) R. B. Kershner, "Technical Innovations in the APL Space Department," Johns Hopkins APL Technical Digest, Vol. 1, No. 4, 1980, pp. 264-278

274) Note: The very concept of being able to compute a location on Earth by observing the change in frequency of a spaceborne transmitter during a single pass was initially ridiculed by a number of reputable scientists.

275) The Legacy of Transit, Special issue of Johns Hopkins APL Technical Digest, Jan.-March 1998, Volume 19, No. 1

1975) in LEO and ATS-6. A further SST demonstration was conducted between the GPS constellation (in MEO) and the TOPEX/Poseidon altimeter satellite in LEO [the GPS receiver (6 channels, two frequencies, code and phase) aboard TOPEX/Poseidon is providing GPS-SST data since Dec. 1992; there exits an almost continuous data set for 1993. In addition to the GPS receiver, TOPEX/Poseidon was tracked with TDRS (Tracking Data Relay Satellite), DORIS (on-board the S/C) and SLR (Satellite Laser Ranging). Comparison of the tracking information from the four sources has demonstrated a satellite position determination capability in the <5 cm range. This level of accuracy is achieved through post-pass processing of the GPS data obtained from the satellite's GPS receiver. [276) 277)] Unfortunately, the artificial anti-spoofing GPS signal degradation switched on again in 1994/5, permitted only occasional GPS observations, not advantageous for routine precise orbit determination]. The contribution of Topex/Poseidon GPS-SST data to existing gravity-field models were not very significant, due to the relative high altitude of the TOPEX/Poseidon (1334 km) orbit. But whenever anti-spoofing-off periods were available, then GPS observations provided an orbit restitution with centimeter-level accuracy, an essential requirement for the quality of an altimeter mission. [278) 279)]

Note: SST methods for applications in gravity field recovery favor low-altitude orbits (referred to as low-low SST), in which two orbiters flying close-proximity circular trajectories, perform relative velocity or range measurements of high accuracy, at the lowest possible altitude (in the co-orbiting satellite approach, the basic observed quantity can be a distance or a Doppler frequency shift, or both). - The other viable approach (at the end of the 1990s) is continuous on-board GPS position measurements with high-quality receivers, providing a second and independent satellite-to-satellite tracking method, in combination with low-low SST. The global long-wave gravity field recovery by GPS is based primarily on the combined carrier phase measurement on-board the LEO S/C, and a network of ground receivers which allow the recovery of the GPS trajectories. High-low GPS SST is also applied for on-board navigation purposes. [280)]

1.7.2 Gradiometry , Accelerometry and Magnetometry

- Gradiometry. Measurement of the Earth's gravity field from space. Two basic and complementary approaches are in use: 1) SST (Satellite-to-Satellite Tracking Technique) and 2) SGG (Satellite Gravity Gradiometry). SGG measures some derivatives of the gravity vector, called gravity gradients, on-board a S/C in various directions with the use of a gradiometer, consisting of several accelerometers operated in differential mode. Gravity gradients are highly sensitive to the local features of the gravity field in the proximity of the measurement location. SST is best at providing the long and medium wavelength of the geopotential, while SGG performs best at the shorter wavelengths as a result of the measurement bandwidth characteristics of the accelerometers. Both approaches favor low-altitude orbits in which air drag modeling is critical, because the non-gravitational orbit perturbations have to be separated from the purely gravitational ones.

- Drag-free satellite. A drag-free satellite, free falling along a space-time geodetic, is the ideal test bench for some of the most advanced gravitation theories, from the general relativity validation to the search for experimental evidence of the existence of gravitational waves. - Triad-1 of the US Navy and built by JHU/APL (launched Sept. 2, 1972) is considered the first satellite to fly a completely gravitational orbit, free from all surface forces such as drag and radiation pressure. The orbit could in fact be predicted for up to 60 days (see

276) J. Rush, "Current Issues in the Use of the Global Positioning System Aboard Satellites," Acta Astronautica, Vol. 47, No 2-9, 2000, pp. 377-387

277) W. Bertiger, P. Abusali, et al., "The First Low Earth Orbiter with Precise GPS Positioning: TOPEX/Poseidon," ION Proceedings, Sept. 1993

278) P. Argentiero, et al., "Results of GEOS 3/ATS-6 Satellite-to-Satellite Tracking Orbit Determination Experiment," Journal of Geophysical Research, Vol. 84, No. B8, pp. 3921–3925, 1979.

279) Information provided by P. Schwintzer of GFZ Potsdam

280) http://www.estec.esa.nl/vrwww/explorer/GRAVITY.html#introduction

also Table 42 and H.6). Triad-2 (launch Oct. 11, 1975) and Triad-3 (launch Sept. 1, 1976) were follow-up drag-free missions to Triad-1. The Gravity Probe-B mission (launch 2002), and the ESA GOCE (Gravity Field and Steady-State Ocean Circulation Explorer) mission with a launch in 2005, are further satellites employing the drag-free concept. In general, a drag-free system consists of accelerometers (a free-floating proof mass), micro-thrusters, and a corresponding control system, often referred to as DFC (Drag-Free Control) system, to measure the non-gravitational influences.

Mission	Comment
Triad series of US Navy (Triad-1 launch Sept. 2, 1972); Triad-2 launch Oct. 11, 1975	DISCOS obtained a residual acceleration of $5x10^{-11}$ ms^{-2}. Triad-2 was equipped with a redundant pulsed-plasma thruster (PPT)
CHAMP of Germany, launch July 15, 2000	STAR of ONERA with a proof mass
GRACE, a dual minisatellite mission, launch 2001	SuperSTAR, a proof-mass instrument of ONERA to measure non-gravitational accelerations
GP-B (Gravity Probe-B) of of NASA/Stanford, launch 2002	Drag-free control with proportional helium thrusters and a drag-free proof mass ($<10^{-12}$ g).
Microscope of CNES (launch in 2004)	Drag-free S/C for a test of EP (Equivalence Principle) with an accuracy of one in 10^{15}
GOCE of ESA, launch 2005	DFACS compensates drag with thrusters
LISA (Laser Interferometer Space Antenna) of NASA, ESA, etc. with a launch in 2005/6	The LISA constellation of 3 S/C represents a giant interferometer to be used to detect gravitational waves. Each S/C contains a proof mass.

Table 43: Overview of some drag-free and non-gravitational measurement missions

• Accelerometry. [281] The measurement of spacecraft surface forces requires ingenious designs due to various forces (drag, solar and Earth radiation pressure) and orientation knowledge needed for proper results. An accelerometer at the S/C center of mass can measure these forces that cause non-gravitational orbit perturbations. [A large accelerometer class employs the concept of force balance: a frame contains a moving proof-mass, the servo control detects its position and exerts a force to maintain the proof-mass motionless with respect to this frame. The combination of position sensing and force actuating results in accelerometer designs based on the following principles: piezoelectric, piezoresistive, acoustic, capacitive (electrostatic), and magnetic.] At the beginning of the 21st century, sensitivity requirements of accelerometers for space applications generally call for acceleration measurements below the pico-g (10^{-12} g) level. The electrostatic accelerometer type is seen as a most likely candidate to provide these performance levels. [282]

- CACTUS. A first version of an electrostatic accelerometer (measuring differential accelerations between the external surface and a small ball placed at the center of mass), referred to as CACTUS (Capteur Accélérométrique Capacitif Triaxial Ultra Sensible), developed and built by ONERA of Chatillon, France, was flown on the CASTOR D-5B S/C (of CNES, launch May 17, 1975, with nominal S/C operations until 1979) for atmospheric density studies. The measured precision was 10^{-10} m/s^2 (or 10^{-11} g). [283]

- Another ONERA accelerometer by the name of ASTRE (Accéléromètre Spatial Triaxial Electrostatique), was part of the ESA Microgravity Measurement Assembly (MMA), and flown on STS-55 (Apr. 26 - May 6, 1993), STS-83 (Apr. 4-8, 1997) and on STS-94 (Jul. 1-17, 1997). The resolution achieved by ASTRE was 10^{-9} g in the measurement bandwidth DC to 1 Hz (monitoring of the low-frequency acceleration environment). The ASTRE working principle is based on keeping a proof mass motionless in its nominal posi-

281) P. Touboul, B. Foulon, E. Willemenot, "Electrostatic Space Accelerometers for Present and Future Missions," Acta Astronautica, Vol. 45, No. 10, 1999, pp. 605-617

282) V. Josselin, P. Touboul, R. Kielbasa, "Capacitive detection scheme for space accelerometer applications," Sensors and Actuators, Vol. 78, 1999, pp. 92-98

283) C. J. Koblinsky, P. Gaspar, and G. Lagerloef, editors, "The Future of Spaceborne Altimetry: Oceans and Climate Change," Joint Oceanographic Institutions Inc., Washington, DC, 1992, pp. 72-73

tion and attitude by means of electrostatic suspension, such that the required electrostatic forces are a direct measure of the three acceleration components. 284)

- In 1996, a Czech micro-accelerometer by the name of MACEK (Mikroakcelerometr), designed and developed at the Institute of Astronomy of the Academy of Sciences of the Czech Republic at Ondrejov, was flown as a technology experiment on Shuttle flight STS-79 (Sept. 16 - 26, 1996). MACEK was placed inside of Spacehab, about 2 m away from the center of gravity of the Shuttle (measurement of Shuttle vibrations during the orbital flight phase, performance tests of MACEK). MACEK employs the concept of an electrostatically compensated proof-mass with a measurement accuracy of 10^{-9} m/s^2. The first version of MACEK has been flown on the Russian satellite Resurs-F1 in 1992 (launch June 23, 1992, proof of concept flight for MACEK). 285) - An upgraded version of the MACEK (10^{-10} m/s^2) instrument is planned to fly on the Czech MIMOSA (Microaccelerometric Measurements of Satellite Accelerations) satellite mission in 2002 (see E.17).

- The GOCE (Gravity field and steady-state Ocean Circulation Explorer) mission of ESA with a planned launch date in 2005, flies EGG (Electrostatic Gravity Gradiometer), developed by ONERA, in an SGG (Satellite Gravity Gradiometer) configuration, combined with an SST configuration. See E.11.

Gravity/Mag.-field missions	Description
Cosmos-26 (Soviet Union)	Launch March 18, 1964. The USSR Cosmos-26 satellite provided the first global magnetic mapping of the Earth's surface.
Prognoz-6 (Soviet Union) Magnetic field measurements	Prognoz-6 (launch Sept. 26, 1977), IZMIRAN (data provider) Prognoz-7 (launch Nov. 11, 1978), IZMIRAN Prognoz-9 (launch July. 1, 1983), IZMIRAN
ISEE-1/2 (NASA/ESA)	ISEE-1/2 (launch Oct. 2, 1977). The RUM/RUD magnetometer experiment of UCLA studied the dynamic plasma field of the Earth
GEOS-2 (ESA)	GEOS-2 (launch July 24, 1978, E.6.2) into a near GEO. Objective: Measurement of fluctuations in the Earth's magnetic field and waves and particles. Two years of data.
MAGSAT (APL,NASA, USGS)	MAGSAT (launch Oct. 30, 1979, operated to June 1980) was the first mission to systematically measure the Earth's magnetic field. It provided the first IGRF (International Geomagnetic Reference Field) model based on global scalar and vector data of high accuracy, determining the core radius and mapping fluid motions at the core mantle boundary.
Interball (IKI, Russia)	Interball (launch Aug. 3, 1995) constellation of 4 S/C carries the FM-3I fluxgate magnetometers (solar wind interaction with the magnetosphere).
POLAR (NASA)	POLAR (launch Feb. 24, 1996) flies MFE (Magnetic Fields Experiment of UCLA to study the coupling of the solar wind and the magnetosphere
FAST (NASA)	FAST (launch Aug. 21, 1996) flies MFI (Magnetic Fields Instrument) of UCLA to measure the vector DC and AC magnetic fields.
ASTRID-2 (SSC, IRF-K, Sweden)	ASTRID-2 (launch Dec. 10, 1998) with the objective to perform high-resolution E-field and B-field measurements in the auroral region
Ørsted (TUD, Denmark)	Ørsted (launch Feb. 23, 1999) with the objective to perform highly accurate and sensitive measurements of the geomagnetic field
SACI-1 (INPE, Brazil)	SACI-1 (launch Oct. 14, 1999). The MAGNEX instrument investigates the phenomena related to current alignment with the trans-equatorial field and the plasma electrodynamics involving the Earth, specifically in the region of the South Atlantic Anomaly.
SAC-C (CONAE, NASA, DSRI, etc.)	SAC-C launch Nov. 21, 2000. The objective of Ørsted-2 (Magnetic Mapping Payload) instrument suite is to map the Earth's magnetic field.
IMAGE (NASA, SwRI)	IMAGE launch on Mar. 25, 2000. The objective is to study the global response of the Earth's magnetosphere to changes in the solar wind. NASA's first S/C for making ENA measurements of the terrestrial magnetosphere.
CHAMP (GFZ, DLR, Germany)	CHAMP launch July 15, 2000. Gravity field and magnetic field mapping (MIAS and DIDM). Use of SST/GPS + laser for orbit restitution and accelerometry for non-gravitational force measurements

284) http://esapub.esrin.esa.it/microgra/micrv8n2/natv8n2.htm

285) L. Sehnal, R. Peresty, L. Pospisilova, A. Kohlhase, "Dynamical Microaccelerometric Measurements on board Space Shuttle," Acta Astronautica, Vol. 47, No 1, 2000, pp. 27-34

GRACE (NASA/DLR)	A dual minisatellite mission (launch 2001) with accelerometers and GPS receivers for SST (+laser), like CHAMP plus ultra-precise low-low SST
FEDSAT (CSIRO) Australia	FEDSAT is due for launch in early 2002. The objectives of the NewMag instrument are to measure electrical currents and perturbations in the Earth's magnetic field.
Microscope (CNES, ONE-RA)	A microsatellite mission (2004) with the objective to measure EP (Equivalence Principle) to an accuracy of one part in 10^{15}
GOCE (ESA)	A gravity field mission (2005) with three-axis satellite gravity gradiometry (SGG) provided by EGG (Electrostatic Gravity Gradiometer) and SSTI (Satellite to Satellite Tracking Instrument) for SST
Planned but not flown missions	
Gravity/Mag.-field missions	Description (the listing of the following missions is kept for historical reasons). All the gravity objectives, concepts and goals which were elaborated for them are still valid.
GRAVSAT (NASA)	A mission defined in the 1980s and cancelled in 1987 due to technical difficulties and budget constraints
GGM (Gravity Gradiometer Mission), NASA also known as SGGM	Objective: to map the Earth's gravity and magnetic fields using two drag-free S/C orbiting in polar orbit at a very low altitude. The project was cancelled in 1987.
GRADIO (CNES) (a gradiometer instrument - not a mission)	Under study by CNES and terminated in the 1980s. GRADIO was a satellite gravity gradiometer experiment aimed at measuring the full set of gravity gradients (and separating it from the spacecraft attitude disturbances) by means of eight three-axis micro-accelerometers of a new generation.
ARISTOTELES (ESA/NASA)	Mission studies started in 1989 with the objective to fly a high-low SST mission using GPS and satellite gravity gradient measurements (GRADIO). ARISTOTELES was terminated in 1994/5 due to budget constraints
GAMES (NASA/CNES)	A two (co-orbiting) S/C mission was designed with laser measurements between S/C. A GPS receiver on one S/C was planned for LEO orbit determination. Also measurement of the magnetic field. GAMES was cancelled in the 1990s due to budget constraints.

Table 44: Overview of gravity/magnetic field missions flown (and not flown)

- Magnetometers are flown to measure the attitude of the S/C relative to the Earth's magnetic field. As such, they are normally part of the on-board attitude system. Magnetometers, in particular triaxial fluxgate instruments, are also flown as a science instrument to measure the geomagnetic field. Table 45 represents a chronology of fluxgate magnetometers flown as science instruments on various missions.

Mission	Launch	Instrument	Comment
DODGE (DoD)	July 1, 1967	Triaxial magnetometer	S/C was operational for over 3 years
GEOS-2 (NASA))	Jan. 11, 1968	Uniaxial fluxgate	
TRIAD-1 (US Navy) TRIAD-2 TRIAD-3	Sept. 2, 1972 Oct. 11, 1975 Sept. 1, 1976	Triaxial magnetometer	
IMP-8 (NASA)	Oct. 26, 1973	GNF (triaxial fluxgate)	S/C was operational in 1997
GEOS-1 (ESA) GEOS-2 (ESA)	Apr. 20, 1977 July 14, 1978	S-331 (triaxial fluxgate) S-331 (triaxial fluxgate)	S/C was operated until Apr. 1980 S/C operated until Oct. 1985
MAGSAT (APL, NASA,USGS)	Oct. 30, 1979	Scalar and vector magnetometers	
DE-1/DE-2 (NASA)	Aug. 3, 1981	MAG-1 (triaxial)	Combined launch of S/C
IRM (Germany) UKS (UK) CCE (USA)	Aug. 16, 1984 Aug. 16, 1984 Aug. 16, 1984	Triaxial magnetometer Triaxial magnetometer Triaxial magnetometer	AMPTE mission with 3 subsatellites
Viking (Sweden)	Feb. 22, 1986	V2 (triaxial)	End of mission May 12, 1987
EXOS-D (ISAS)	Feb. 22, 1989	MFG (triaxial)	Akebono mission (Japan)
Magion-2 (Czech)	Sept. 28, 1989	SGR-7 (triaxial)	Magion is subsatellite of ACTIVE
Ulysses (ESA/NASA)	Oct. 6, 1990	FGM/VHM	Boom-mounted magnetometers
UARS (NASA)	Sept. 13, 1991	VMAG part of PEM	
APEX (USSR) Magion-3 (Czech)	Dec. 18, 1991	SGR-5 (triaxial) SGR-7 (triaxial)	Magion-3 is subsatellite of APEX
FREJA (Sweden)	Oct. 6, 1992	F2 (triaxial)	Boom-mounted

Mission	Launch	Instrument	Comment
POLAR (NASA)	Feb. 24, 1996	MFE (triaxial)	Two instruments
FAST (NASA)	Aug. 21, 1996	MFE (triaxial)	
ACE (NASA)	Aug. 25, 1997	Magnetic Field Monitor	Triaxial fluxgate
DMSP/F7 (DoD) DMSP/F12 (DoD) DMSP/F13 (DoD) DMSP/F15 (DoD)	Dec. 18, 1983 Aug. 29, 1994 Mar. 24, 1995 Dec. 12, 1999	SSM (triaxial) SSM (triaxial) SSM (triaxial) SSM (triaxial)	NASA instrument
Equator-S (MPI)	Dec. 2, 1997	MAM (triaxial)	Boom-mounted
ASTRID-2 (Sweden)	Dec. 10, 1998	EMMA	
Ørsted (Denmark)	Feb. 23, 1999	Vector+scalar magnetometer	Both instruments are boom-mounted
SACI-1 (INPE)	Oct. 14, 1999	MAGNEX (triaxial)	Boom-mounted
OPAL (Stanford)	Jan. 27, 2000		
CHAMP (GFZ)	July 15, 2000	MIAS (scalar+triaxial)	Boom-mounted package of OVM and FGM
Cluster-II (ESA)	July 16, 2000 Aug. 9, 2000	FGM	Boom-mounted (2) on all S/C
SAC-C (CONAE)	Nov. 21, 2000	Ørsted-2	Boom-mounted
FedSat (CSIRO)	2002	NewMag	Boom-mounted

Table 45: Chronology of some fluxgate magnetometers flown on geomagnetic field missions

1.7.3 Satellite Laser Ranging

- The SLR (Satellite Laser Ranging) technique was first successfully demonstrated with a 'laser tracking reflector experiment' flown on the following NASA S/C (E.7.1): GEOS-1, (launch Nov. 6, 1965), GEOS-2 (launch Jan. 11, 1968) also referred to as Explorer-36, GEOS-3 (launch Apr. 9, 1975) and SEASAT (launch June 27, 1978). International cooperation and participation of many agencies has successively led to a global network of SLR ground stations (see H.4.3.6). - SLR measurements are applied to such fields as: global tectonic plate motion, regional crustal deformation near plate boundaries, Earth's gravity field and the orientation of its polar axis and its rate of spin. - A passive satellite laser-ranging measurement capability (entire S/C consists of laser corner reflectors) from ground to space was first provided with the CNES Starlette satellite (launch, Feb. 6, 1975).

- A future extension of the SLR (ground to satellite) technique inverts the traditional SLR system with the ranging hardware being placed onboard a satellite to range to the ground. This will be attempted with the GLAS (Geoscience Laser Altimeter System) instrument of the EOS program on ICESat (planned launch in 2001) for high-resolution ice and land topography mapping (realized already in the Mars Explorer Mission).

S/C or Mission (instrument)	Launch	Comment
GEOS-1 (Laser Tracking Reflector), NASA	Nov. 6, 1965	322 cubes were mounted on fiberglass panels on the bottom rim of the S/C
GEOS-2 (Laser Tracking Reflector), NASA	Jan. 11, 1968	Identical system as flown on GEOS-1
GEOS-3 (Laser Tracking Reflector), NASA and JHU/APL	Apr. 9. 1975	264 quartz cube corner reflectors mounted on a 45° conic frustum
Starlette, CNES	Feb. 6, 1976	First mission where the entire S/C consists of laser corner reflectors
LAGEOS-I, NASA, MEO orbit	May 4, 1976	
SEASAT (Laser Tracking Reflector), NASA/JPL	June 27, 1978	96 fused silica 3.75 cm hexagonal corner cube retroreflectors
EGS (Ajisai), NASDA, Japan	Aug. 12, 1986	318 mirrors and 120 laser reflector assemblies (1436 corner cube reflectors)
GEO-IK, NPO PM, Krasnojarsk, USSR The series of GEO-IK S/C is incomplete due to lacking information	May 30, 1988 Aug. 28, 1989 1990 Nov. 24, 1994	Orbit:1500 km altitude, inclin. =73.6° Orbit:1500 km altitude, inclin.=73.6°
ETALON-1, USSR, MEO orbit ETALON-2, USSR, MEO orbit	Jan. 10, 1989 May 31, 1989	2140 laser reflectors

S/C or Mission (instrument)	Launch	Comment
ERS-1 (LRR), ESA	July 17, 1991	
TOPEX/Poseidon (LRA), NASA/CNES	Aug. 10, 1992	
LAGEOS-II of ASI and NASA, MEO orbit	Oct. 22, 1992	
Stella, CNES	Sept. 26, 1993	Stella is an exact twin of Starlette
MSTI-2 (RRA), BMDO/SMC	May 8, 1994	
GFZ-1 of GFZ, Potsdam, Germany	April 9. 1995	Low-altitude and slowly decaying orbit
ERS-2 (LRR), ESA	April 21, 1995	
Meteor-2-22, Russia	Aug. 31, 1993	Fizeau retroreflector array
RESURS-01-3 (RRA), Russia	Nov. 4, 1994	Fizeau retroreflector array
TiPS (Tether Physics and Survivability), DoD/ NRL, see M.27	June 20, 1996	Retroreflectors are mounted on the exterior surfaces Ralph and Norton
ADEOS (RIS), NASDA, JEA	Aug. 17, 1996	Design of a hollow and flat cube-corner retroreflector
WESTPAC, EOS Australia (RESURS-O1-4)	July 10, 1998	Fizeau retroreflector array
Techsat/Gurwin-II (SLRRE), Technion Israel Institute of Technology, launched with RESURS-O1-4 satellite, see N.20	July 10, 1998	Array of laser retroreflectors, corner-cube mounted on the Earth-viewing panel
STEX/ATEx, DoD/NRL	Oct. 3, 1998	
SUNSAT, Stellenbosch University, SA	Feb. 23, 1999	Eight retroreflectors
CHAMP (LRR), GFZ, Germany	July 15, 2000	
Jason-1 (LRA), CNES/NASA	planned 2001	
REFLECTOR, US/Russian (AFRL),	planned 2001	Nanosatellite on Meteor-3M-1, retroreflector array of 32 corner cubes
Envisat (LRR), ESA	planned 2001	
ICESat (GLAS), NASA/GSFC	planned 2001	
GRACE (LRA), NASA/DLR/GFZ	planned 2001	
ADEOS-II, NASDA	planned 2002	
Gravity Probe-B, NASA, Stanford University	planned 2002	
VCL (RRA), NASA	planned 2002	
ALOS, NASDA	planned 2003	

Table 46: Chronology of some SLR systems

• Modulation of retroreflectors. The state of the art of modulating retroreflectors has progressed considerably in the 1990s. The advent of ferro-electric liquid crystals (FLCs) and the technology to produce MEMS (Micro-Electro-Mechanical Systems) make new electro-optical systems available. The most significant development has been the increase in the switching speeds that now allow data to be sent at tens of kilohertz up to multi-megahertz with very low power consumption. The technologies used for these devices are:

- Ferro-electric liquid crystal (FLC) - Switches the polarization by reversing the polarity of the bias voltage
- Multiple Quantum Well Device (MQW) - Switches resonance to pass or reject optical beam by electro-absorption
- Atomic Absorption Cell (AAC) - Quantum electronic filter
- MEMS Spoiled Corner Cube - Diffuse or specular surface reflection on one surface of the retroreflector
- Switchable Gratings (PLZT) - Uses piezo-electric transducers to momentarily create diffracting surface on one surface of the retroreflector.

1.7.4 Attitude and Control Instruments

• From a historical point of view, the first on-board navigation instruments were mainly involved with the measurement of attitude and the provision of guidance. Early satellites were spin-stabilized to maintain orientation. Conventional attitude sensor types for the measurement of orientation are: gyroscope, horizon sensor, sun sensor, star sensor, magnetometer, etc; attitude control instruments (actuators) are: flywheels (reaction/momen-

tum wheels), magnetic torque rods, thrusters, etc. Satellites in the 1990s usually employ a multisensor navigation payload, this was not available technology in the early days of spaceflight. At the turn of the century, there is a tendency to replace conventional Earth horizon sensors and also sun sensors of the attitude control system by star trackers. Stars provide a much finer reference (point source) thus keeping the satellite pointing more accurately. Examples of advanced star sensor systems are: a) ASC (Advanced Stellar Compass) flown on the Danish satellite Ørsted (launch Feb. 23, 1999) with an attitude precision of a few arcseconds; b) ATS (Autonomous Star Tracker) of the DoD/NRO STEX (Space Technology Experiment) satellite (launch Oct. 3, 1998). ATS provides attitude information to an accuracy of 0.005°; c) ASC and PASS (Payload Autonomous Star Sensor) are both flown on ESA's PROBA (planned launch 2001) mission. PASS has a wide FOV (19.3° x 14.4°) and about 3 arcseconds of pointing accuracy in pitch and yaw.
Present day (1997-2000) star sensors are designed to image only a portion of the sky and report the location of any stars that are tracked. Among other factors, the accuracy of these devices is historically limited by three design parameters: 1) A small FOV in the order of 8° x 8°, 2) The capability to track 5 stars simultaneously, and 3) A CCD detector array with a 512 pixel square imaging area. These choices have led to a limiting measurement accuracy of about 5 arcseconds per star and per frame.

As of 2000, AFRL at Kirtland AFB (Albuquerque, NM) is developing an "intelligent star tracker" with a pointing accuracy of better than 0.20 arcsec and NEA (Noise Equivalent Angle) better than 0.10 arcsec. The MEMS instrument mass is about 0.2 kg with a power consumption of <2 W. An adaptive optics catadioptric telescope design is used in combination with APPS (Active Pixel Position Sensor) detector technology. [286)]

- **Inertial sensors.** Gyroscopes and accelerometers are inertial sensors to obtain information on orientation (attitude) and accelerations relative to an inertial reference frame. Attitude measurements of a gyroscope provide angular information, while acceleration measurements (which are corrected for gravity and, integrated twice for distance) provide linear information.

The classical gyroscope is a mechanical device with a rapidly spinning mass (solid body or rotor of substantial moment of inertia), supported on a mount (e.g. a gimbal structure) that allows freedom of tilt of the spin axis relative to the base on which it is mounted. The momentum of such a rotor causes the gyro to retain its attitude when the mount is tilted. The ability of the gyro to maintain its orientation (due to the stored energy, namely the angular momentum) is employed in a number of applications (compass, direction finder, steering mechanisms, inertial guidance systems, etc.). - A practical gyroscope suffers from a phenomenon known as drift due to undesired torques resulting from imperfections in machining, bearing and lubrication frictional torques, material impurities, etc.

Since the early 1960s, new generations of gyroscopes were developed (in particular through research funded by the military), working on different principles than the conventional mechanical gyroscope. The Ring Laser Gyro (RLG) and the Fiber-Optic Gyro (FOG) are such new developments based on optical (interferometric) principles. [287)] [288)] [289)]

- **RLG** (Ring Laser Gyro). RLG is a Sagnac effect optical gyro which uses an active HeNe lasing medium in a mirror-defined ring cavity. Two contra-rotating beams of coherent (laser) light circulate within the quartz ring cavity (the RLG technique was first demonstrated in 1962, it took another 20 years of testing for operational use of the system in commercial aircraft). The objective is the measurement of angular rate. The concept is

286) N. Clark, P. Furth, S. Horan, "Intelligent Reconfigurable Integrated Satellite Processor," Proceedings of the 14th AIAA/USU Conference on Small Satellites, Logan, UT, Aug. 21-24, 2000, SSC00-VI-1

287) W. K. Burns, "Fiber Optic Gyroscopes - Light is Better," Optics & Photonics News, May 1998, pp. 28-32

288) K. Hotate, "Fiber Optic Gyros Put in New Spin on Navigation," Photonics Spectra, April 1997, pp. 108-112

289) P. J. Klass, "Fiber-Optic Gyros Now Challenging Laser Gyros," Aviation Week & Space Technology, July 1, 1996, pp. 62-64

based on the principles of general relativity, which predict that the distance around a closed optical path in a rotating frame of reference depends on the direction the path is traversed. A beam of light traveling around the path in the direction of rotation has to travel farther than one traveling in the opposite direction of rotation. The pathlength difference (and frequency change) is used as a measure of angular rotation.
Some characteristics of an RLG are: 1) it offers long-term stability and it is insensitive to the environment; 2) it can operate continuously and has no start-up time; 3) it can withstand large angular rates (800°/s); 4) it has a digital output; 5) it has no moving parts as the conventional gyro (however, in practice it is kept in motion to minimize some error sources); 6) it requires very little power for operation.

- **FOG** (Fiber-Optic Gyro). A solid-state device with no moving parts for the measurement of angular rate (first demonstrated in 1976). In a FOG, the two light beams rotate in a spool of optical fiber. The measurement concept is based on the Sagnac effect, which defines an optical phase shift between counter-propagating beams in a rotating coil of fiber. This phase shift is proportional to the rotation rate about the axis of the coil, with a scale factor related to the coil geometry and the optical wavelength. NASDA's TR1-A rocket (a microgravity mission, Japan), launched in 1991, marked the first spaceborne experimental application of a FOG instrument. The Japanese MUSES-B (Mu Space Engineering Satellite - VLBI measurements in astronomy) mission of ISAS, launch Feb. 12, 1997, uses a hybrid FOG/radio-wave guidance system as well as an open-loop FOG for rate control (providing a drift rate of 0.05°/h). - In the late 1990s, performance range sensitivities (drift rates) of FOG devices vary from 0.001°/hr to 100°/hr with dynamic ranges (usable measuring range) of 10^4 to 10^5.

- Vibratory microgyroscope. Starting in 1997, a new generation of microgyroscopes, based on MEMS principles, was developed out of a technology cooperation agreement between NASA/JPL, UCLA, and HSC (Hughes Space and Communications Company). The instrument (1999), an all-silicon chip, has a size of 4 mm x 4 mm and a mass of less than one gram (the mass of the entire 3-axis microgyroscope is < 30 gram, power of about 1 W, size of 1.5 cm x 1.5 cm x 3 cm, rate of > 360°/s, etc.). The new microgyroscope is lighter, cheaper, higher-performing and less complex than its conventional counterparts, while uniquely designed for continuous and long-term space operation. The instrument concept is based on a cloverleaf design that is tied down and vibrates at a very high speed. The gyroscope senses the Coriolis acceleration, which results from the vibratory motion in a rotating reference frame, to detect rotation. The design represents a qualitative step forward in gyroscope development.

The applications of angular rate gyros include navigation of manned and unmanned platforms, location of fleet vehicles, and attitude heading measurements. Fiber-optic gyros can also provide antenna pointing and tracking. Following are some typical classes of gyro applications:

- General low-precision navigation (such as for automobiles) is done with gyros providing drift rates of 10°/hr - 100°/hr
- Attitude and heading references for airplanes use gyros with drift rates of 1°/hr
- Precision inertial navigation as used in commercial airplanes and on some space platforms use gyros with drift rates of 0.01°/hr - 0.001°/hr.
- High-precision inertial navigation with a drift rate of<0.001°/hr is used for satellite pointing and tracking applications.

Three parameters define the performance of optical gyroscopes:

- Bias drift (a shift in the instrument zero point in the absence of rotation)
- Optical scale factor (a proportionality constant between applied rate and instrument output)
- Output noise (due to signal processing electronics, noise in the optical detection process, and thermal fluctuations in the optical fiber).

-

- The first spaceborne horizon sensor was flown on TIROS-2 (launch Nov. 23, 1960). A further new feature on this S/C was the introduction of magnetic attitude control, i.e., adjustment of spin axis orientation along the orbital path. The attitude device consisted of 250 cores of wire wound around the outer surface of the spacecraft. The interaction between the induced magnetic field in the spacecraft and the Earth's magnetic field provided the necessary torque for attitude control.

- Magnetic damping. In a satellite, magnetic damping can be realized with hysteresis rods (nickel-iron rods which are easily magnetized, even by the Earth's weak magnetic field. Such rods, rotated in a magnetic field, dissipate a fixed amount of energy for every revolution). The concept was invented by R. E. Fischell of JHU/APL and first flown on Transit-5A-3 in 1963. Such a rod produces a constant retarding torque which opposes any rotation. Since this opposing torque remains constant, even for a small rotation, hysteresis damping is effective even at low angular velocities.

- Improvements in spaceborne attitude control with instruments like star sensors and flywheels were introduced in the 1970s.

- Magnetometers traditionally have been used on spacecraft for sensing of coarse attitudes (in the order of 1°) and for input to the momentum control logic (magnetic torquers). However, in combination with accurate calibration and the availability of accurate gyroscope data, attitudes accuracies of better than 0.1° can be obtained using only magnetometer data and rate data. This was done at GSFC with UARS data (launch of UARS, Sept. 13, 1991).

- Programmed yaw steering of the S/C. The ERS-1 S/C (launch July 17, 1991) of ESA introduced the concept of yaw steering (a form of dynamic attitude control) to compensate for Earth rotation (about 4° per orbit). The vertical axis of the S/C is pointed towards the geodetic nadir to optimize the observation performance of its payload instruments. This involves a carefully controlled rotation about the yaw axis over the course of each orbit. As a result, the natural coordinate system of the radar instrument (AMI) remains orthogonal to the ground track of the S/C. In addition, ERS-1 introduced a roll-tilt mode in which the entire S/C can be rotated about its horizontal axis. This permits radar imaging at different incidence angles over limited periods of time.

- Spacecraft/platform and instrument pointing is an important aspect of navigation (see also same heading in Appendix A). The ability to locate a sensor measurement with the required accuracy on the Earth's surface (or to point at a faint celestial body from an orbiting platform) has made great strides in four decades of spaceflight. Precision pointing is a function of platform stability through suitable attitude sensing and control mechanisms. Early spacecraft had pointing accuracies of about $\pm 1^{\circ}$ which is in the same order of magnitude as spacecraft with gravity-gradient boom stabilization in the 1990s. Three-axis stabilized spacecraft, such as the NOAA/POES series (starting with TIROS-N) and the early SPOT series (SPOT-1,-2,-3) offer pointing accuracies of 0.1°. IPS (Instrument Pointing System) of ESA (built by Dornier, Friedrichshafen), first flown on Shuttle flight STS-51-F as Spacelab-2 in July/Aug. 1985, then on STS-35 as Astro-1 (Dec. 2-10, 1990), and again on STS-67 as Astro-2 (March 2-18, 1995), is providing a pointing stability within ± 1.2 arcseconds for all sensors on the platform (J.6). [290] [291] NASA/LaRC conducted precision pointing experiments with MACE (Middeck Active Control Experiment) flown on Shuttle (STS-67, March 2-18, 1995). MACE consisted of three components: a multibody platform, experiment support module, and the handheld terminal. MACE-II (of AFRL) is a follow-up experiment to the original MACE flight. The AFRL objective is use adaptive control algorithms for preci-

290) http://www.esrin.esa.it/htdocs/esa/progs/mg.html
291) http://liftoff.msfc.nasa.gov/Shuttle/Astro2/description/ips/ips.html

sion structural control. MACE-II is manifested on STS-106 (launch in 2003).[292] The imaging sensor OSA (Optical Sensor Assembly) on Ikonos-2 (launch Sept. 24, 1999) is providing an absolute ground location accuracy of 12 m (without the use of ground control points), the relative accuracy is 2 m. The imagery of such a sensor may certainly be used for cartographic applications (see Table 563). A separate class of pointing accuracy, namely <20 milliarcseconds, is required of Gravity Probe-B (Relativity Mission, planned launch in 2001), a spin-stabilized spacecraft. This is three orders of magnitude higher than IPS or close to five orders of magnitude higher than normal three-axis stabilized spacecraft.

- The ARGOS mission of DoD (launch Feb. 23, 1999) is flying an instrument by the name of USA (Unconventional Stellar Aspect) also referred to as NRL-801 experiment. USA conducts feasibility tests of X-ray satellite navigation and new computational approaches to autonomous parameter estimation that includes GPS inputs and a variety of redundant truth measures. There are reasons why X-ray navigation might prove to be attractive in the future. The advantages are associated mainly with drawbacks of optical methods or with potential advantages of X-ray characteristics that have no exact analogs in the optical wavelengths (M.1).

1.7.5 Altimetry

The third dimension in navigation, the measurement of changes in orbital height, is provided by such instruments as altimeters and accelerometers.

- Altimeters are active microwave instruments for the accurate measurement of vertical distances. The technology determines the two-way delay of the radar echo from the Earth's surface to a very high precision (in the 1990s to less than a nanosecond). It has also the capability to measure the power and the shape of the reflected radar pulses. The data provide in particular information on ocean topography [the gravity grids reveal also all of the major structures of the ocean floor. The "equipotential surface" of the ocean bulges outward and inward mimicking the topography of the ocean floor. The actual ocean surface deviates by up to 100 m from the ideal ellipsoid. The bumps and dips (troughs) in geoid height are caused by minute variations in the Earth's gravitational field]. - First spaceborne measurements of ocean surface heights started in 1973 with S-193 on Skylab at an altitude of 435 km. Using a pulse width of 0.1 microseconds this system was able to get a resolution of 15 m.

In altimetry the variation of the return signal strength (as the cube power of height) and the footprint area with the altitude requires the use of the pulse compression for higher altitudes.[293] Since GEOS-3 (launch Apr. 9, 1975), flying at an altitude of 840 km, all altimeters have used pulse compression - the resolution due to this compression has been improved. GEOS-3 offered significant improvements over the Skylab altimeter (S-193), including improved performance as well as greater global coverage. The SeaSat (1978) ALT instrument, designed and built by JHU/APL, is regarded the first high performance altimeter (first use of the full-deramp technique; introduction of the capability of waveform sampling). In contrast with previous designs, the samples are now an integral part of the altitude tracking process and are used in such a way that the system adapts as a function of wave height to tracker performances. Despite the premature demise of the SeaSat S/C after only 100 days in orbit, the altimeter performance clearly demonstrated the significance of the signal processing innovation, it was adopted as the standard approach for new radar altimeters. GEOSAT (1985 - 1990) altimetry introduced also measurements of sea-state and ocean surface winds. The "exact repeat mission" of GEOSAT, in which it retraced the SEASAT ground tracks, provided the first long-term high-quality altimeter measurements to the scientific community. More missions were launched in the 1990s: The RA-1 altimeter on ERS-1 (1991 - 1996)

292) R. R. Ninneman, "Middeck Active Control Experiment Reflight (MACE-II) Program: Lessons Learned," Proceedings of AIAA Space 2000 Conference and Exposition, Long Beach, CA, Sept. 19-21, 2000

293) R. F. Gasparovic, R. K. Raney, R. C. Beal, "Ocean Remote Sensing Research and Applications at ALP," JHU/APL Technical Digest, Vol. 20, No 4, pp. 600-610, 1999

and on ERS-2 (since 1995) was designed to optimize its performance over all types of surfaces. The ALT instrument on Topex/Poseidon (launch Aug. 10, 1992), designed to optimize performances over ocean surfaces, is regarded the first dual-frequency altimeter, permitting the estimation of ionospheric range delay. The Topex/Poseidon mission has demonstrated that the time variation of ocean topography (dynamic and thermodynamic) can be determined with an accuracy and precision of a few centimeters. For the first time, the seasonal cycle and other temporal variabilities of the ocean have been determined globally with high accuracy, yielding fundamentally important information for testing ocean circulation models. Topex/Poseidon was the first mission specifically designed for observing ocean circulation. The missions ERS-1/2 and TOPEX/Poseidon are part of international oceanographic and meteorological programs, such as WOCE (World Ocean Circulation Experiment) and TOGA (Tropical Ocean and Global Atmosphere), both linked to WCRP (World Climate Research Programme).

Mission/ Sensor	Launch Date	Range Precision (cm)	Frequency (GHz)	Uncompressed Pulse Width (μs)	Compressed Pulse Width (μs)	Peak Power (W)	Beam Limited Footprint (km)	Pulse Limited Footprint (km)	Orbit altitude, inclination (km/°)
Skylab, S-193	14.5.73	90	13.9			2000		8	435, 50
GEOS-3, ALT	9.4.75	50	13.9	1	12.5	2000	38	3.5	843,115
SEASAT, ALT	27.6.78	10	13.5	3.2	3.1	2000	22	1.7	800,108
GEOSAT, Radar Alt	12.3.85	5	13.5	102.4	3.1	20	29	1.7	800,108
ERS-1, RA-1	17.7.91	5-7	13.8	20	3	55	18	1.7	785,98.5
Topex/Pos ALT SSALT	10.8.92	2.4 2.5	13.6/5.3 13.65	102.4	3.1	20 5	26/65	2.2	1334, 63.5
ERS-2, RA-1	21.4.95	5-7	13.8	20	3	55	18	1.7	785,98.5
Priroda (Greben)	23.4.96	10	13.76	1.7	12.5	40	13	2.3	400, 52
GFO-1, RA	10.2. 98	3.5	13.5					2	800, 108
Jason-1 Poseidon-2	2001		13.575/ 5.3	105		58			1336, 66
Envisat RA-2	2001	<4.5	13.575 3.2	20		60 60		1.7	800, 98.55
ICESat-1 GLAS	2001	<10						0.7	705, 94
CryoSat SIRAL	2003/4		13.6	51		25			720, 92
Jason-2	2004								

Table 47: Chronological survey of spaceborne radar altimeters

- Delay/Doppler radar altimeter. While the conventional altimeter technique is to measure the distance between the satellite and the mean ocean surface, the delay/Doppler method differs from those instruments in two ways:
  - Pulse-to-pulse coherence and full Doppler processing to allow for measurement of the along-track position of the range measurement
  - Use of two antennas and two receiver channels that allow for measurement of the across-track angle of the range measurement.

The central innovation of the delay/Doppler technology is that the returns from a group of transmissions along-track are coherently processed together, rather than incoherently as is customary. The coherent along-track processing allows much more of the instrument's radiated power to be converted into height measurement data. - The phase between two otherwise identical coherent signals is a direct measure of the time shift between them (referred to as phase-monopulse technique). If the geometry of the observation space is known, then the observed phase shift can be inverted to an angular offset. The phase-monopulse technique uses this principle to estimate the angle of arrival between two signals collected through separated antennas. [294)]

An airborne instrument named D2P (Delay/Doppler Phase-monopulse Radar) was developed and built by JHU/APL which demonstrated the technology in varies test flights in the spring and summer of 2000 (P.62). In addition, a S/C mission (WITTEX) with the same objective was proposed. [295)] [296)] SIRAL (SAR Interferometer Radar Altimeter), the CryoSat mission altimeter of ESA (planned launch in 2003/4), will employ the concepts that have been demonstrated with the D2P radar. The SIRAL dual-beam concept offers for the first time some swath information, a significant advance over conventional pencil-beam altimeters (thus, offering a better capability of ocean current pattern recognition). [297)]

- **Ocean reflection measurements of GNSS signals as sources of opportunity** (a passive altimetry technique). [298)] [299)] Background: The detection of GPS signals scattered of the ocean surface from the ground and from aircraft were obtained experimentally in 1991 and reported by a French team (Dassault Electronique) in 1994. [300)] Earlier work by Boeing had revealed the existence of such signals in connection with multipath effects (Boeing Co. report FAA-RD-73-57-V of 1973). In 1993, ESA/ESTEC presented a multistatic altimeter (see Glossary for "bistatic surface scattering") [301)] concept, referred to as PARIS (Passive Reflectometry and Interference System), based on an interferometric approach which combines the direct and ocean-reflected GNSS signals. That work included accuracy analysis and discussions of alternative altimetry constellations. [302)] [303)] Stephen J. Katzberg and James L. Garrison of NASA/LaRC first analytically predicted the change in GPS signal structure following an ocean reflection. They then demonstrated the intentional tracking of reflected GPS signals with an airborne GPS receiver.

On the spaceborne side, GNSS reflections were first observed by a JPL team after the SRL-02 (Shuttle Radar Laboratory-02) mission (STS-68, Sept 30 - Oct. 11, 1994).

The concept of GPS (or better: GNSS) signal reflection measurements (as a source for bistatic surface scattering) is based on ocean surface roughness which results in multipath signal reflections emanating from the GPS and GLONASS satellite constellations. [304)] [305)] These signal reflections can in turn be picked up by a GPS receiver (a delay-Doppler-mapping receiver system) and may be used to determine such factors as wave height, wind speed, and wind direction. The strength of the reflected signals (normally regarded as surface clut-

294) http://sd-www.jhuapl.edu/Intro/2.6ocean.html#3

295) R. K. Raney, "The Delay/Doppler Radar Altimeter," IEEE Transactions on Geoscience and Remote Sensing, Vol. 36, No 5, Sept. 1998, pp. 1578-1588

296) R. K. Raney, D. L. Porter, "WITTEX: An Innovative Three-Satellite Radar Altimeter Concept," Proceedings of IEEE/IGARSS Conference, Honolulu, HI, July 24-28, 2000

297) http://www.estec.esa.nl/explorer/cryosat/index.html

298) S. T. Lowe, J. L. LaBrecque, C. Zuffada, L. J. Romans, L. E. Young, G. A. Hajj, "First spaceborne observations of an earth-reflected GPS signal," submitted to Radio Science, July 2000.

299) J. S. LaBrecque, L. Loewe, L. Young, E. Caro, S. Wu, L. Romans, "Recent Advances in the study of GPS Earth surface reflections from orbiting receivers," UNAVACO Community Meeting, 1998

300) J.-C. Auber, A. Bibaut, J.-M. Rigal, "Characterizations of Multipath on Land and Sea at GPS Frequencies," Proceedings of ION GPS-94, Vol. 2, pp. 1155-1171

301) Note: A multistatic altimeter can be regarded as a multistatic radar for which the transmitters and the receivers belong to different systems.

302) M. Martin-Neira, "A Passive Reflectometry and Interferometry System (PARIS): Application to Ocean Altimetry," ESA Journal, 17 (4), 1993, pp. 331-355

303) M. Martin-Neira, et al., "ESA's activities on GPS reflected signals for Earth observation," Proceedings of Ionospheric Determination and Specification for Ocean Altimetry and GPS Surface Reflections Workshop, JPL, Pasadena, Dec. 1997

304) J. L. Garrison, S. J. Katzberg, M. I, Hill, "Effect of Sea Roughness on Bistatically Scattered Range Coded Signals from the Global Positioning System," Geophysical Research Letters, Vol. 25, No 13, 1998, pp. 2257-2260

305) Information provided by T. P. Yunck of NASA/JPL

ter or simply as noise) is also a discriminator between wet and dry surface areas and, therefore, can be applied to coastal and wetland mapping. In this concept, high spatial resolutions are obtained with wide-beam and low-gain receiving antennas by using the GPS carrier phase and its PRN code modulation. This is similar to the SAR (Synthetic Aperture Radar) technique in which the Doppler frequency shifts and pulsed-signal time delays are used to create a small footprint on the mapped surface. - By combining code-range and Doppler measurements, a receiver can distinguish particular patches of the ocean surface illuminated by GPS signals.

In some ways, the above GPS applications discovery is similar to that of using GPS to map the ionosphere's TEC (Total Electron Content) by measuring propagation delay differences between the L1 and L2 frequencies. [306)]

The GNSS-based concept of altimetry with reflective pseudorange noise (PRN) operates in a bistatic geometry (i.e., the transmitted and received signals travel two different paths) with the following features:

- Up to 20 ocean height measurements are possible simultaneously with a single receiver for the measurement of reflected GNSS signals. In comparison, traditional altimetry is limited to looking in the nadir direction and obtaining one height observation at a time below the altimeter.
- GNSS reflection coverage is very dense but random, while in traditional altimetry, the coverage is regular according to a certain repeat pattern. However, the reflected signal retrieval produces only modest SNRs with low-gain antennas. Hence, exploitation of GNSS reflections observed from space is still rather speculative at the start of the 21st century. It will require high gain antennas (preferably >26 dB gain) with multiple independently steerable beams. A single platform might typically see 10-12 reflections at once over the oceans. Measurement precision will be modest (about 1 m in a few seconds for altimetry), hence, massive averaging is required to get to the 5-cm level that will be needed for ocean circulation, and that may require a dozen or so platforms at once.
- When the GNSS signal impinges on the ocean surface, it is scattered randomly with maximum power in the direction of the specular reflection. Since the ocean reflectance depends on the surface temperature and salinity, in principle, measuring the return power with sufficient accuracy could eventually by used to infer these parameters.
- The GNSS reflection relies on detecting the amount of energy return of the GNSS signal by correlating the pseudo-random code of the reflected signal and a delayed model of the same code generated in the receiver. By examining the power return at different delays and Doppler shifts, different parts of the ocean are sensed. The footprint of GNSS reflection is determined by the intersection of iso-Doppler and iso-range contours.
- The BlackJack GPS receiver configurations/implementations of the missions CHAMP (launch July 15, 2000) and SAC-C (launch Nov. 21, 2000) have the added capability of bistatic ocean reflection measurements, originating from the GPS and GLONASS constellations (besides refractive atmospheric occultation monitoring). This relatively new technique of "GNSS reflected pseudorange altimetry" recovery with on-board GPS (or GNSS) receivers will eventually permit applications such as oceanic circulation sensing (particularly eddy-scale currents that are missed by today's altimetry missions), and scatterometry (measuring the shape of the reflected pulse) for determining wind speeds. The CHAMP and SAC-C GNSS reflection experiments are considered simple engineering tests of opportunity with antenna gains of only 6-9 dB. Thus, long averaging times are needed just to detect the signals. - Essential to the success of GNSS-recovered altimetry and scatterometry is the ability to see multiple reflected signals si-

306) A. Komjathy, J. L. Garrison, V. Zavorotny, "GPS: A New Tool for Ocean Science," GPS World, April 1999, pp. 50-56

multaneously. Note: The Blackjack instrument on SAC-C is also referred to as GOLPE (GPS Occultation and Passive reflection Experiment).

-

1.7.6 Orbits

A number of different Earth orbit types were introduced to cover particular observational and/or operational requirements (see also O.12).

1) **LEO** (Low Earth Orbit - altitude range between 300 - 2000 km). The most common and natural orbit for Earth observation. The very first satellites, like Sputnik-1 and most successors, had LEO orbits and were launched in an eastward direction (thereby gaining a boost with the Earth's rotation), and obtaining an inclination of the launch site's latitude. [307)]

Global LEO observation coverage was an early requirement in spaceflight. This led to new strategies in mission planning - in particular with regard to orbital plane inclinations and to orbital altitudes.

- The US satellite Discoverer-1 (launch on a Thor vehicle from VAFB on Feb. 28, 1959, apogee = 968 km, perigee = 163 km, inclination = 89.7°, period = 96 min) was the first spacecraft in **polar orbit**. It flew a military payload. This was followed by the "Transit" navigation satellite series of the US Navy. Starting with the launch of Transit-5A-1 (and follow-up satellites of the series) on a Scout vehicle from VAFB, CA, on Dec. 18, 1962.

- In the field of Earth observation, the orbital strategy of **polar orbit** was introduced with Nimbus-1 (launch Aug. 28, 1964), then with TIROS-9 (launch Jan. 22, 1965). The polar orbit enabled the provision of full repetitive global coverage. Depending on sufficient swath width, some sensors (such as AVHRR) were able to provide daily global coverage, while other sensors needed several days for a repeat cycle.

- The **sun-synchronous polar orbit** was first realized on TIROS-10 (launch July 2, 1965). This new orbit introduced the concept of repeat observations (same sun illumination angles) of a given Earth surface region at the same time of the day (due to same latitudinal crossings), improving considerably the conditions for data analysis. An additional advantage of the sun-synchronous orbit resulted in a better power provision for the spacecraft.

2) **GEO** (Geostationary Earth Orbit - approximately 35,786 km above the Earth's surface at the equator). [308)] The GEO orbit provides a fixed-position relation between the satellite and the ground, a continuous viewing capability of the same footprint is given (however, the polar regions of the Earth are not accessible from GEO). Although the GEO concept is quite extensively utilized by meteorological and communication satellites at the end of the 20th century, it still remains challenging for remote sensing applications in terms of proper spatial resolutions for Earth surface imagery due to its radial distance of at least 36,000 km. The great distance from Earth implies tougher constraints on imaging instruments and on communication systems.

The concept of a geostationary orbit seems to have several independent authors - men of vision and imagination:

a) Hermann Potocnic [Dec. 22, 1892 - Aug. 27, 1929, born in Pola (Pulj), Moravia, Austrian-Hungarian Empire, an engineer (Vienna) and retired captain of the army] published his

307) Note: Satellites (Ofeq series) launched from Israel (Palamchim Air Force Base south of Tel Aviv) orbit from east to west, as opposed to the traditional west to east direction, as Israel can only safely launch rockets to the west, over the Mediterranean Sea.

308) Note: With regard to the GEO concept, credit is almost universally given to only one author in the literature, namely to A. C. Clarke, who published his first article almost 20 years after that of Hermann Potocnic. In fact, some space age historians talk about the "Clarke Orbit" when referring to a geostationary Earth orbit. This is an injustice to the accomplishments of Hermann Potocnic.

concepts of future space travel in a book with the title: "Das Problem der Befahrung des Weltraums - Der Raketen-Motor," Richard Carl Schmidt & Co., Berlin W62, 2. Auflage, 1929. [309] [310] He used the pseudonym of Hermann Noordung for the publication, since he was afraid to be ridiculed by his peers. In the book (page 98), Noordung proposed a stationary orbit of a space station (referred to as "Raumwarte" for astronomical observations) at radius 42,300 km from Earth center (or about 35,900 km above the surface), in the equator plane, with a chosen orbital speed of about 3080 m/s - resulting in a free orbit with an angular velocity equal to that of Earth rotation. An orbiting body would seem to be stationary over a certain point at the equator.

b) Arthur C. Clarke (born Dec. 16, 1917 in Somerset, England) published an article in 1945 with the title: "Extra-terrestrial Relays: Can Rocket Stations Give World-wide Radio Coverage?" (Wireless World, Oct. 1945) where he described the principles of satellite communication in geostationary orbit. Clarke suggested that if a satellite were placed above Earth's equator at just the right height - 35,888 km - it would orbit Earth exactly once in 24 hours - and seem to stay put in the sky above some point on the ground.
In 1947, Arthur C. Clarke proposed how three satellites, suspended in orbits 35,888 km above the equator, could serve as relay platforms for the entire Earth. He cautioned that the concept was based on the theory that microwave radio signals would not bounce off the atmosphere like shortwave signals, something that wasn't known for certain in 1947.

- Syncom-2 (launch July 26, 1963 on a Thor Delta vehicle, built by Hughes Corp.) was the first successful experimental communication satellite to achieve geostationary orbit. Positioned over the Atlantic, it demonstrated the feasibility of geostationary satellite communications. Note: Syncom-1 (launched Feb. 14, 1963) achieved GEO orbit but contact was lost after is was placed in orbit.

- The objective of ATS (Applications Technology Satellite) program, a series of six NASA S/C, was to explore and test new technologies and concepts for communications, meteorological and navigation satellites (investigation of the geostationary orbit environment). Note: ATS-2 and ATS-4 were MEO satellites, the rest of the ATS program were GEO satellites.

- In the 1990s the geostationary orbit (GEO) is indeed the most densely populated orbit of satellites. This is due in particular to the needs of the communication industry.

3) **IGSO** (Inclined Geosynchronous Orbit), see O.12.2. A geosynchronous satellite with an inclination different from zero is not "stationary" when observed from the Earth. This particular orbit has the property of remaining stationary over a particular meridian (there is no longitudinal precession for successive orbits but latitudinal motion along the meridian). An inclination of 30° causes a latitudinal displacement of ±30° during one orbital period (one day). First applications of IGSO were considered in the planning phase with GalileoSat (Europe's space segment of a navigation system).

4) **GTO** (Geosynchronous Transfer Orbit - an elliptical orbit with perigees usually between 300 - 600 km and an apogee of 36,000 km). In the normal case, GTO is the short-term initial orbit of a S/C after launch to be placed into GEO by a sequence of orbit maneuvers that are executed by the satellite's own propulsion system to circularize the orbit into GEO - all GEO satellites (weather or communication) go through this procedure. Secondly, GTO may be used as a long-term orbit, usually selected by S/C with payloads that measure the harsh solar radiation environment. GTO implies that a S/C passes through the Van Allen radiation belts (these are regions located at about 1.4-1.5 R_E and 4.5-6 R_E where many energetically charged particles from the solar wind are trapped in the Earth's magnetic field) four times per day, causing them to be exposed,

309) Note: A 1st edition of the book could not be located anywhere. The reprint of "Das Problem der Befahrung des Weltraums, " Ausgabe von 1929, is available at Turia + Kant Verlag: ISBN: 3-85132-060-3

310) http://www.ijs.si/slo/country/culture/potocnik.html

in a twelve month period, to levels of radiation equivalent to about 8-10 years in GEO or LEO. GTO gives also exposure to other environmental effects such as atomic oxygen erosion at perigee and electrostatic charging at apogee. - The accelerated life testing of radiation effects provided by GTO is a unique opportunity to evaluate new technology solar panels quickly and cost-effectively. Examples of S/C in long-term GTO are: CRRES of NASA/DoD (launch July 25, 1990, A.11); STRV-1a and -1b of DERA, UK (launch June 17, 1994, M.25); STRV-1c/1d of DERA (launch Nov. 16, 2000); MDS-1 of NASDA (planned launch 2000/1), and IMEX (Inner Magnetospheric Explorer) of the University of Minnesota at Minneapolis, (planned launch in 2001).

5) **MEO** (Medium Earth Orbit - altitude range of about 5000 km to about 25,000 km above surface) provide considerably longer contact times with the user. Examples of typical MEO S/C are:
   - Some laser ranging satellites (passive systems) such as: LAGEOS-1 of NASA (launch May 4, 1976, altitude 5900 km). ETALON-1 of USSR (launch Jan. 10, 1989, altitude about 19,000 km).
   - Navigation satellite constellations of GPS (GPS-1 launch Feb. 22, 1978, altitudes of about 20,000 km, period of about 12 hours) and GLONASS (GLONASS-1, -2, -3 launch, Oct. 12, 1982; altitudes of about 19,000 km, period of about 11.25 hours). Prior to the GPS constellation, the launch of NTS-2 (July 14, 1974) had a MEO orbit of 13900 km.
   - Some communication satellites: Inmarsat-P series, ATS-2 and ATS-4.

6) **HEO** (Highly-elliptical Earth Orbit - a typical HEO has a perigee of about 500 km and an apogee of 5-10 R_E and more). The observation objectives in HEO orbits involve mainly investigations of the energy transport between the sun and the Earth in the interplanetary plasma (energetic particle fluxes, etc.), solar-terrestrial interactions (in particular in the auroral regions), as well as magnetic field studies. Examples of satellites in HEO are:
   The IMP-4, -5, -6, -7 and -8 (Interplanetary Monitoring Platform series of NASA - also referred to as Explorers 34, 41, 43, 47, and 50, respectively). IMP-4 was launched on May 24, 1967 (apogee of 34 R_E, perigee of 250 km). IMP-8 was launched Oct. 26, 1973. A number of the Prognoz spacecraft series of the USSR (magnetospheric research, etc.) had apogees ranging from 200,000 - 800,000 km (or about 30-125 R_E) such as Prognoz-6 (launch Sept. 26, 1977, Prognoz-7 (launch Nov. 11, 1978), and Prognoz-9 (launch July 1, 1983). Onboard the Prognoz-9 spacecraft was also the Relict-1 experiment to investigate the large-scale anisotropy of relict radiation.. The ISEE-1 and -2 satellites (K.18) had a joint launch on Oct. 2, 1977 (apogee at 23 R_E). The AMPTE satellites (IRM, UKS, CCE, see K.4) were launched on Aug. 16, 1984. GEOTAIL, launch July 24, 1992. WIND, launch Nov. 1, 1994. INTERBALL (launch of Tail probe Aug. 3, 1995, Auroral Probe launch Aug. 29, 1996). POLAR, launch Feb. 24, 1996. Equator-S, launch Dec. 2, 1997. Chandra X-ray Observatory, launch on Shuttle flight STS-93 on July 23, 1999. The XMM (X-Ray Multi-Mirror Mission - Note: XMM was officially renamed to "Newton" in Feb. 2000) of ESA, launch Dec. 10, 1999 with an Ariane-5 vehicle from Kourou, has an operational orbit of: perigee = 7000 km, apogee = 114,000 km, an inclination of 40° and a period of 47.86 hours. The IMAGE (Imager for Magnetopause-to-Aurora Global Exploration) mission of NASA/GSFC (launch Mar. 25, 2000), has a polar orbit of: perigee = 1000 km, apogee = 7 R_E (44,647 km). The location of the apogee changes during the course of the two-year mission, both in latitude and, because of the Earth's revolution about the sun, in local time.

7) **Halo orbits** (an orbit around the sun/Earth Lagrangian Point (L1), about 1.5 million km from Earth (in the direction Earth - sun), in which the sun/Earth gravitational forces are balanced. L1 is an ideal outlook point for Earth observation as well as for monitoring of "space weather" parameters. Examples of satellites in L1 halo orbits are (see also

O.12.11): ISEE-3 (NASA/GSFC, launch Aug. 12, 1978); SOHO (ESA/NASA, launch Dec. 2, 1995); WIND (NASA, launch Nov. 1, 1994 - WIND was placed into a halo orbit in Nov. 1996); ACE (NASA, launch Aug. 25, 1997). Further planned missions into a halo orbit at L1 are: Genesis of NASA/JPL with a launch in 2001, Triana of NASA with a launch in 2004, Geostorm of NASA/NOAA with a launch in 2003.
Astronomy missions to L2 are (see Figure 394): a) MAP (Microwave Anisotropy Probe) of NASA/GSFC with a launch June 30, 2001; the objective is to probe conditions in the early universe by measuring the properties of the cosmic microwave background radiation over the full sky; b) FIRST (Far Infrared Submillimeter Telescope) of ESA with a planned launch in 2006.
Note: the L2 point is the same distance away as L1 but at the opposite side of the Earth. This implies that a spacecraft positioned at L2 is always located in the Earth's shadow. An advantage of L2 is astronomical viewing (looking toward the universe) due to constant lighting conditions. - The first three Lagrangian points of the sun/Earth system, namely L1, L2 and L3, lie along an axis with Earth and the sun, with L3 on the opposite side of the sun from Earth. The other Lagrangian points, L4 and L5, are in the same orbit plane around the sun as Earth; however, L4 is 60° ahead of Earth while L5 is trailing Earth by 60°.

8) **EEO** (Elliptical Earth Orbits). Refer to orbits chosen in such a fashion that maximum service provision (or viewing time) can be provided to a particular region of the Earth. A prominent example are the numerous Molniya-type orbits (apogee = 39950 km, perigee about 100-300 km, inclination = 63.4°) with periods of 12 hours, whose apogee (by far the longest duration period of the orbit) occurs always in the intended region, namely the northern hemisphere. The first launch of the Molniya-1 satellite series occurred April 23, 1965 (over 20 S/C of the series until 1974), followed by the Molniya-2 series with first launches in 1972 and countless S/C of the Molniya-3 series. The Molniya series represents also the first communication satellites of the USSR. The TWINS dual-spacecraft mission of NASA (planned launch in 2003) employs Molniya-type orbits to achieve long periods in apogee.

• Close encounters/rendezvous of satellites with celestial bodies. The art of satellite navigation has evolved steadily over the years, resulting occasionally in close flyby maneuvers. Examples are:

- On July 29, 1999, the DS1 spacecraft of NASA/JPL successfully performed a close flyby of asteroid 9969 Braille using the AutoNav system. At about 27 km separation, it was by far the closest flyby of an asteroid so far attempted.

- The NEAR (Near Earth Asteroid Rendezvous S/C)/Shoemaker satellite of NASA, in orbit around asteroid 433 Eros since Feb. 2000, experienced his closest encounter with Eros on Oct. 26, 2000. [311] The flyby, with corresponding surface observation, occurred at an elevation of about 5 km, planned and supervised by the mission control center at JHU/APL. The NEAR/Shoemaker S/C then performed year-long observations of Eros (322 million km from Earth). **On Feb 12, 2001, it was commanded by JHU/APL to land on the asteroid** (the S/C was not built for landing). The final touchdown speed of the craft was 1.9 m/s after some descent maneuvers. The S/C instruments kept operating after touchdown. The GRS (Gamma Ray Spectrometer) is detecting key chemical signatures of a "planetesimal" - one of the original building blocks of planets. NEAR/Shoemaker thus became the first spacecraft ever to land on an asteroid.

1.7.7 On-Orbit Servicing

Space logistics, the ability of on-orbit service provision, is a natural consequence of high-value assets in space (like HST and ISS) as well as of ever increasing satellite fleets in all

311) http://spacescience.com/headlines/y2000/ast26oct_2.htm?list65492

fields of space endeavors. Without any logistics, satellites have severe limitations affecting their operational life and functional capabilities. The conventional practice in spacecraft and instrument building puts very high demands on reliability resulting in redundant and expensive satellites. With this approach, a spacecraft may still be lost even if the majority of its components are operational. Redundancy in turn produces heavy spacecraft which require extra lift capacity for launch services. The emerging technologies of the 21st century may change the conventional practices of the past (at least to some degree) with the introduction of on-orbit serving capabilities. The solutions offered by the space industry must be affordable in order to be accepted. [312)]

• Shuttle repair/service missions. The first Shuttle repair mission took place in April 1984 (STS-41C). The SMM (Solar Maximum Mission) S/C became the first satellite to be retrieved, repaired, and redeployed in orbit by a Shuttle crew. The first Hubble repair mission was launched Dec. 2, 1993 on STS-61 [the HST (Hubble Space Telescope) was launched April 24, 1990 on STS-31]. The second Hubble service flight was on STS-82 (Feb. 11-21, 1997). The third Hubble service flight was on STS-95 (Oct. 29 - Nov. 7, 1998). The forth Hubble service flight on STS-103 is took place Dec. 19-27, 1999. Objective: replacement of gyroscopes, a fine guidance sensor and a S/C computer. Installation of a voltage/temperature kit for the S/C batteries. Installation of a new transmitter, solid state recorder (12 Gbit), and thermal insulation blankets. - A further routine servicing mission to Hubble is planned for November 2001. Without periodic on-board servicing, Hubble would have been a disaster and would not have produced all of the great science it has.

• The capability of on-orbit servicing is of course not limited to Shuttle flights (based on manned servicing techniques developed in the 1980s) resulting in very high service bills for a single mission. Small robotic spacecraft in combination with new technologies may in fact be the answer to potential future servicing mission capabilities in such fields as: a) mission repair, b) maintenance and scheduled repair, c) replenishment of consumables, and d) preplanned product improvements and upgrades. First experiments with on-orbit servicing have already been conducted.

- SVS (Space Vision System). SVS is a Canadian-built camera system, sponsored by CSA and NASA with its maiden flight (prototype) on Shuttle STS-52 (Oct. 22 - Nov. 1, 1992). Since then, SVS has completed four Shuttle flights. The Shuttle mission STS-88 (Dec. 4 - 15, 1998), an assembly flight with the module Unity to ISS, marked the first operational use of SVS. The primary objective of SVS is to help astronauts perform so-called blind mates of the space station modules. SVS is a computer-aided vision system that works in conjunction with four stationary cameras located inside the Shuttle's cargo bay and two cameras attached to the robotic arm itself. [313)]

- AERCam (Autonomous EVA Robotic Camera), a free-flying space camera (of NASA/MSFC). The first demonstration occurred on STS-87 (November 19 to December 5, 1997). The free-flyer has a self contained cold gas propulsion system giving it the capability to be propelled with a 6 degrees of freedom control system. Onboard the free-flyer are rate sensors to provide data for an automatic attitude hold capability.

- Inspector platform family of Astrium GmbH (formerly DASA). [314)] [315)] The program started in 1993 as a joint project of DASA/Space Infrastructure, RSC Energia, and The Boeing Company. The objective was to monitor events in space (docking maneuvers, etc.) and to extend the inspection capabilities for MIR as well as for ISS. At the start of the 21st century the Inspector family consists of X-MIR and three more service struc-

312) G. Leisman, "Analysis of on-board Servicing Architectures using Microsatellites, Advanced Propulsion, Secondary Opportunities and the Military Spaceplane Concept," Proceedings of the AIAA 2000 Space Conference and Exhibition, Long Beach, CA, Sept. 19-21, 2000

313) B. Berger, "Astronauts Use Space Vision to Assemble Station," Space News, Feb. 1, 1999, p. 10

314) "The Inspector Product Family," brochure provided by F. Steinsiek of DASA/Space Infrastructure, Bremen

315) http://www.dasa.com/dasa/index_e.htm

tures (ISS Inspector, Visitor, and Operator) in varying phases of development:
X-MIR Inspector was launched on its first demonstration flight to the Mir station on Oct. 5 , 1997. The prime objectives were to verify the video navigation system and the visual inspection capabilities of X-MIR in the neighborhood of MIR. On Dec. 17, 1997, X-MIR started its demonstration flight as a free-flying observer service module, by orbiting MIR. The successful X-MIR demonstration lasted from 1997 to 1998 providing high-quality video imagery of MIR as well as of the docking sequences of the service vehicle Progress. X-MIR consists of: a) space vehicle, b) monitoring and control station, and c) transport launch container. [316)] X-MIR was remotely controlled from MCS (Monitoring and Control Station) on-board the MIR station. MCS consisted of a laptop PC, a video display, a video recorder and an electronic module for radio communications, power distribution and navigation.
The prime objective of the ISS Inspector is the inspection of the surfaces of the ISS or visiting transportation vehicles (US Space Shuttle, Proton vehicle, etc.). In 1997 NASA selected the Inspector for the AETD (Advanced Engineering and Technology) program. The prime objective of the Inspector for ISS services is the inspection of space vehicle surfaces (video and infrared observations), this includes the ISS and visiting spacecraft. The Inspector payload performs also leakage, contamination and radiation environment monitoring of the ISS.
The Visitor concept is a co-orbiting multi-purpose platform of ISS to be used for various payload operations in the fields of science observations, technology or commercial applications. Visitor is able to carry various kinds of payloads under very low gravity conditions. The payloads are mounted onto standard payload modules, referred to as ORU (Orbital Replaceable Unit) or onto ExPA (Express Pallet Adapter).

- MMS (Mobile Servicing System) of ISS provided by Canada. MMS is a robotics system consisting of the elements SSRMS (Space Station Remote Manipulator System), SPDM (Special Purpose Dextrous Manipulator), and MBS (Mobile Base System). In April 2001, SSRMS was flown and installed on ISS. The system will be used to assemble and maintain the ISS (International Space Station).

- Inspection services during launch and early-orbit phases are already practiced with on-board miniature cameras providing visible proof to the mission operators of such functions as stage separation, antenna deployment, secondary vehicle deployment, and many other on-board activities and events. An example of inspection services is TEAMSAT (M.26), an ESA/ESTEC low-cost satellite demonstrator mission (launch Oct 30, 1997) with VTS (Visual Telemetry System).

- The SNAP-1 nanosatellite mission of SSTL (Surrey Satellite Technology Ltd), UK, with a launch on June 28, 2000 (along with Tsinghua-1 of China) is quipped with MVS (Machine Vision System) permitting SNAP-1 to function as a remote inspector.

- Most work on autonomous on-orbit serving has been confined to very high-value assets such as ISS. Servicing spacecraft developments considered for ISS are: a) CTV (Cargo Transfer Vehicle) of NASA; b) ATV (Automated Transfer Vehicle) of ESA, a cargo resupply vehicle for ISS to be launched by Ariane-5; and c) HOPE of NASDA.

- Designs for on-orbit servicing of average Earth-orbiting satellites are in the proposal state as of 2001. [317)]

1.7.8 Satellite Radionavigation Systems

Radionavigation systems employ the concept of triangulation with line-of-sight radio signals from different satellites to find position - in the same fashion that angular measure-

316) H. Günther, "From MIR to ISS - The Inspector Mission," Proceedings of the 4th International Symposium on Small Satellites Systems and Services, Sept. 14-18, 1998, Antibes Juan les Pins, France

317) E. Lamassoure, D. E. Hastings, "Generalized Metrics for Optimization of Space Systems Cost-Effectiveness," Proceedings of the AIAA Space Conference and Exhibition, Long Beach, CA, Sept. 19-21, 2000

ments of distant stars were made (and are still being made) throughout navigation history to find position.[318)] [319)] Instead of angular measurements to natural stars, the new systems use a radio-ranging measurement technique, a triangulation to a constellation of satellites. The first generation spaceborne navigation systems, Transit (USA) and Tsikada (USSR), were mostly reserved for military use; however, they were of pivotal importance due to the insights gained in the nature of navigation and the considerable advances in early space-age technology. These were followed by second generation navigation systems, namely GPS and GLONASS, which were again developed, built and operated by the military services; but their use was so revolutionary and universal, that the civil community was permitted to use this new utility as well.

• Transit navigation program (US Navy, see H.6). The first operational US navigation satellite was Transit-1B (designed and built at JHU/APL), launched April 13, 1960 aboard a Thor-Able rocket from Cape Canaveral. Starting with Transit-5A-3 (launch June 16, 1963), each satellite in the series featured a gravity-gradient boom for stabilization and had a total mass of about 55 kg (polar circular orbits at altitudes of about 1100 km). Transit is generally credited with demonstrating the feasibility of using artificial satellites as navigational aids. By the end of 1962, a first position fix could be performed by a Polaris vessel.[320)] The Transit system was used by the US Navy in 1964, it became fully operational in 1966 (12 S/C constellation with seven operational and five stored S/C in orbit). The satellites broadcast their signals on two frequencies: 150 and 400 MHz. The dual-frequency method was introduced for the first time (on Transit-1B) to correct for ionospheric refraction effects (the 2nd frequency enabled the distortion to be cancelled out). - The Navy's Transit (2-D) navigation system allowed the user to determine position by measuring the Doppler shift of the received signal (constant tone broadcast - the frequency of 150 MHz was transmitted to correct for ionospheric delay). The 2-D system did not permit velocity determination. - In 1967 the system was released for non-military purposes, a benefit to broad ocean navigation. The Transit series reached a peak utilization of about 100,000 commercial and military users in the late 1980s, and was decommissioned on Dec. 31, 1996.[272)] [321)]

One of the technologies required to make Transit possible was a considerable improvement on time and frequency standards. To realize the fixed site survey accuracy of 1 m, the satellite Doppler signal at 400 MHz must be measured to 0.0005 Hz (a resolution of 5×10^{-12}). In 1960, the best standards of frequency were only good to resolve 5×10^{-10} and could only measure the 400 MHz carrier to 0.2 Hz. APL established a precision time and frequency facility to support the necessary development of various equipment needed for Transit.

• Timation (Time Navigation) program. A US Navy satellite navigation system, initiated in 1964, designed by NRL (Naval Research Laboratory), with the objective to explore the idea of continuous navigation - of providing both accurate position and precise time to passive terrestrial observers (passive ranging). Timation-1, a small (39 kg) gravity-gradient stabilized satellite, was launched May 31, 1967 (VAFB, Thor Agena-D) into a 810 km polar orbit (as a secondary payload, along with GGSE-4, -5, Calsphere-3, -4, NRL-PL-153, -154, and -159). Timation-2 was launched Sept. 30, 1969; both S/C flew with stable quartz oscillators. The STR (Side Tone Ranging) signals were transmitted with a continuous Doppler tracking beacon at about 400 MHz. The experience with the new time system of Timation was a technology demonstrator for future missions.

• In 1964, SAMSO (Space and Missile System Organization) of the USAF in El Segundo, CA initiated a parallel navigation system program to that of the Navy, referred to as

318) B. W. Parkinson, T. Stansell, R. Beard, K. Gromov, "A History of Satellite Navigation," Navigation, ION, Vol. 42, No. 1, Special Issue, Spring 1995, pp. 109-164

319) B. W. Parkinson, "GPS Eyewitness: The Early Years," GPS World, September 1994, pp. 32-45

320) Note: Some benefits of satellite navigation are: precise, all-weather, worldwide availability, timekeeping capability, and unified reference coordinates.

321) Note: To achieve the required accuracy of the Transit system position measurement, APL had to develop a time standard several orders of magnitude more precise than existing devices (time frame of 1958/59).

'Project 621 B', with the objective to use a constellation of satellites for navigation signal transmission from highly eccentric orbits of 24-hours periods. First signal propagation tests were conducted in 1969 using ATS-5 of NASA. Then in 1972, a four-channel airborne receiver was tested with signals transmitted from the ground (satellite signal simulation) at Holloman AFB, NM. The demonstration used a new type of signal, modulated with a PRN (Pseudo Random Noise) code, to provide ranging and timing data. The signal modulation technique used a repeated digital sequence of random bits - that permitted a navigation user device to detect a start ("phase") of the repeated sequence. Recognition of the repeat sequence allowed to determine the range to the signal emitter (satellite). The PRN (spread spectrum) technology turned out to be a key ingredient for GPS signal ranging.

- Start of the NAVSTAR/GPS program in 1973 (a joint-services program on DoD direction) by merging of the Timation and 621 B programs. The proposed new system concept integrated the basic concept of Timation, employing passive ranging precision clocks (time-of-propagation measurements by the user), and the PRN signal technology of Project 621 B - for an eventual constellation of multiple satellites in circular orbits for global coverage. The new GPS measurement technique employs the "time difference-of-arrival" concept (requiring the simultaneous view of four satellites) rather than the Doppler shift of its predecessor system, Transit, to determine position.

- The first spaceborne atomic clocks were flown in the GPS pre-series concept validation program of NRL, referred to as NTS (Navigation Technology Program - previous name was 'Timation'). NTS-1 (a renamed and modified Timation-2A satellite), a three-axis stabilized S/C, was launched on July 14, 1974 into a 13900 km orbit, the payload included two modified commercial rubidium oscillators (Efratom, Munich). The navigation signals were transmitted in L-band. NTS-2 was launched June 23, 1977 from VAFB and placed into a 12 hour (semisynchronous) orbit - later used by the GPS satellites. The NTS-2 payload included two cesium clocks (Frequency and Time Inc.) and the PRN code generator. The signals were in L-band and modulated with a PRN code. NTS-2 was actually the first test satellite that contained the basic system features of the soon-to-follow GPS satellites.

- Launch of the first GPS-series navigation satellite, called NAVSTAR-1 (built by Rockwell International), on February 22, 1978 and declared operational March 29, 1978 (see Table 396 of GPS launches). The first full GPS constellation with 21 operational satellites was reached in 1994. On February 17, 1995, the FAA announced that GPS is now operational and is an integral part of US air traffic control system. The US Air Force Space Command declared the GPS system operational as of July 17, 1995 for the international user community. At the end of the 1990s, GPS is rapidly becoming an integral component of an emerging global information infrastructure. [322] See also chapter H.4 for a description of GPS.

- Launch of the first three GLONASS-series navigation satellites, called GLONASS-1, 2 and -3 (Cosmos 1413, 1414 and 1415 deployed), on October 12, 1982 (see chapter H.3 and Table 394).[323] Note: The first USSR navigation satellite, Tsyklon, was launched into a circular orbit of 750 km, on Cosmos flight 192, November 23, 1967. Tsyklon navigation was based on Doppler shift techniques. Tsyklon was followed by a six-satellite constellation with the name of Parus (also Tsikada-M). A virtually identical civilian navigation network with the name of Tsikada began deployments in 1976 with Kosmos 883, the constellation employs four orbital planes separated by 45°.

- Position and time measurements with ground-based GPS and GLONASS receivers have been made with partial GPS and GLONASS constellations since the first launch in each constellation (first institutionally, then by the civil community as commercial receivers became available). The evolution of the GPS receiver: In 1978, Texas Instruments decided to make available one of the first GPS receivers for civil use. Its price tag was $153,000. In

322) M. Shaw, P. Levin, J. Martel, "The DoD: Stewards of a Global Information Resource, The NAVSTAR Global Positioning System (GPS)," Proceedings of ION GPS-97, Sept. 16-19, 1997, Kansas City, MO, pp. 1237-1243

323) N. L. Johnson, "GLONASS Spacecraft," GPS World, November 1994, pp. 51-58

1988, Magellan offered the first hand-held GPS receiver. Its cost was $3,500. In 1997, Magellan offered the first GPS receiver to break the $100 price barrier. Called the GPS Pioneer, it is a 12-channel, 7-ounce mobile unit powered by two AA batteries. The global GPS receiver market produced over 1 million units in 1997.

- Demonstration of GPS receiver position and velocity measurements from LEO satellites. [324] The Landsat-4 (launch July 16, 1982) and Landsat-5 (launch March 1, 1984) spacecraft are flying the first experimental GPS receivers referred to as GPSPAC [GPS receiver and processor Package (the first on-board GPS receivers in history had to cope with a partial constellation of GPS satellites)]. GPSPAC (designed at JHU/APL, NASA as co-sponsor, built by Magnavox) is an integrated GPS receiver/processor assembly (R/PA), whose tracking data is used for spacecraft time system synchronization and for post-event (on-ground) orbit determination. Despite the sparse GPS constellation in orbit, GPSPAC demonstrated navigational accuracy of better than 50 m over 10- to 30-minute arcs on 88% of the revolutions. Since those early days, many GPS receivers have been installed on all types of satellites.

- Introduction of the GPS carrier-frequency phase measurement technique. [325] In the pre-operational period of GPS service provision [Initial Operating Capability (IOC) of the GPS system was declared in December 1993], the civil user community had to live with measurement accuracies in the 50-100 m range provided by the SPS (Standard Positioning Service) signals. The pseudorange measurement technique (the intentional downgrading of GPS signal quality) for positioning services, introduced the errors caused by atmospheric effects and SA (Selective Availability). In the early 1980s, two MIT radio astronomers (Counselman and Shapiro) suggested to the GPS user community, to use the **carrier frequency to make phase measurements**, instead of using the prescribed pseudorange modulation technique. This change in measurement technique opened up a window of greatly increased accuracies - namely centimeter-range GPS - provided one could resolve the so-called integer ambiguity (integer number of whole wavelengths). In response to this new opportunity, a number of commercial receiver manufacturers developed high-precision geodetic GPS receivers capable of relative positioning accuracies of a few centimeters. As a consequence, new applications became possible in such fields as geodesy, oceanography, land surveying, astronomy and many others. By 1987 it was known (within a small community) that unclassified "quasi-codeless" receivers with broad-beam antennas could recover dual-frequency GPS phase with the requisite millimeter precision, and that a strategy of concurrent observing from multiple sites would permit "double differencing" and related techniques to eliminate selective availability and other clock errors. [326] Thus, in the 1990s, the technique of carrier phase tracking of GPS signals has resulted in a revolution in land surveying as well as in other fields such as refractive occultation monitoring.

With regard to occultation monitoring, it was then a small step to see that analog techniques could be applied directly to occultation processing to remove clock errors. It thus emerged that the one-way GPS observing constraint, which at first seemed to demand stable clocks everywhere in the system, could be artfully adapted through concurrent observations to eliminate stable clocks altogether. This enabled both accurate retrievals and a reduction in instrument cost. Moreover, it was becoming clear that the basic techniques of GPS geodesy could be extended to provide orbit determination on the level of a few centimeters for LEO spacecraft, adequate for occultation analysis. [327]

- Introduction of DGPS (Differential GPS) services. DGPS is a further approach by the civilian GPS user community to overcome the large errors provided by pseudorange signal

324) http://gauss.gge.unb.ca/gradssunil/sgps.htm

325) V. Ashkenazi, "Galileo - Challenge and Opportunity," Galileo's World, Vol. 1, No 1, Winter 2000, pp. 42-44

326) J. T. Wu, "Elimination of clock errors in a GPS based tracking system," AIAA-84-2052, AIAA/AAS Astrodynamics Conference, Seattle, WA, August. 1984

327) T. P. Yunck, C. H. Liu, R. Ware, "A History of GPS Sounding," Special issue of TAO (Terrestrial, Atmospheric and Oceanic Science, Vol. 11, No 1, March 2000, pp. 1-20

measurements. The DGPS concept is based on the principle that a user is affected by satellite ephemeris, atmospheric propagation, SA, and clock synchronization errors to the same extent as a relatively nearby reference station. By predicting the reference receiver's position to a high degree of accuracy, one can set up a system that can be used to calculate corrections to the pseudoranges measured to the various GPS satellites and to transmit these corrections to a multitude of users in the vicinity (100 km radius and more) of the reference station. DGPS accuracies are generally in the range of 1-2 m. - Further advances by the civilian user community led to the development of RTK (Real-Time Kinematic) tracking, enabling the receiver to make carrier-frequency phase measurements while in motion. See also H.4.3, H.4.4.

- Demonstration of GPS high-precision orbit determination. [328)] The GPSDR (GPS Demonstration Receiver) of Motorola was flown on two missions: EUVE (Extreme Ultraviolet Explorer, launch June 7, 1992, of NASA/GSFC) and TOPEX/Poseidon (launch Aug. 10, 1992) to demonstrate/validate accurate orbit determination with GPS data. Two GPSDR instruments (dual-frequency receivers producing long cycle-slip free carrier phase passes as well as pseudorange measurements) on TOPEX/Poseidon were the first spaceborne receivers obtaining high-accuracy range measurements and demonstrated DGPS techniques with a set of IGS (International GPS Service) ground reference stations. A comparison of orbit parameters from three independent techniques (retroreflectors for SLR, DORIS and GPS), all available on TOPEX/Poseidon, obtained GPS orbit differences <25 cm (3D rms). These results of GPS orbit reconstitution are indeed impressive in terms of accuracy. [329)]

- Spaceborne orbit determination with a GPS receiver (initial validation flight tests). The goal is the provision of a real-time service of autonomous on-board functions (orbit, etc.). The provision of mean orbital elements is a first step toward operational autonomy. The orbital elements are being used on the ground for ground station operations and other support functions; they are also being used on-board for S/C operations. In general, however, there are further functional services needed for autonomous on-board S/C operations, such as: position, velocity and time; GPS receiver data logging; S/C clock synchronization; data logging; instrument triggering by position; status monitoring, and more. - The introduction of on-board autonomy has certainly great potential of reducing operating costs of future missions.

Background: Orbit determination is generally based on two models, namely the dynamic model (describing the forces acting on the satellite), and the observation model (providing the relationship between the measurements of the tracking system). This requires considerable on-board processing and storage capability. [330)] A recursive Kalman filter algorithm provides a sequential approach to combine the inputs of both models, namely the satellite's motion (in the dynamic Earth model) with the GPS measurements (observation model) as they become available, resulting in best estimates of position and velocity (continuous comparison and update of actual with estimated values).

- PoSAT-1 (launch Sept. 26, 1993, built at SSTL, UK) [331)] is the first microsatellite to make use of a GPS receiver in orbit, and to autonomously determine its orbit through the processing of GPS data into orbital elements. The GNU (GPS Navigation Unit) consists of a TANS Vector-II 6-channel C/A code receiver of Trimble and is operated with a software package run on a Transputer Data Processing Unit (a T800 32-bit RISC

328) R. Muellerschoen, S. Lichten, U. Lindqwister, W. Bertiger, "Results of an Automated GPS Tracking System in Support of Topex/Poseidon and GPS/MET," Proceedings of ION GPS-95, Sept. 12-15, 1995, pp. 183-193

329) W. Bertiger, et al., "The First Low Earth Orbiter with precise GPS Positioning: Topex/Poseidon," Proceedings of ION GPS-93, Salt Lake City, Utah, Sept. 22-24, 1993

330) V. Ashkenazi, W. Chen, et al., "Real-Time Autonomous Orbit Determination of LEO Satellites using GPS," Proceedings of ION GPS-97, Sept. 16-19, 1997, Kansas City, MO, pp. 755-761

331) M. Unwin, M. Sweeting, "A Practical Demonstration of Low Cost Autonomous Orbit Determination Using GPS," Proceedings of ION GPS-95, Sept. 12-15, 1995, Palm Springs, CA, pp. 579-587

microprocessor). The GNU is operated intermittently to conserve power. Some services demonstrated on PoSAT-1 are: a) orbital elements: they provide a prediction accuracy of 1-10 km for two weeks, and b) the variables of position, velocity and time can be requested at any time.

- Since 1993 NASA/JSC is conducting different GPS experiments to validate concepts required to fly GPS as an in-line avionics component on Shuttle missions. [332] A series of seven flights were conducted on Endeavour which started with STS-56 (April 1993) using a Rockwell Collins 3M receiver operating in SPS mode. The next phase of Shuttle/GPS operations used a PPS (Precise Positioning Service) receiver as a single string navigation device (Rockwell Collins MAGR receiver). The first flight was on STS-79 (Sept. 16-26, 1996). During flight the navigation data was downlinked in real-time for display. During landing, the GPS system performance was compared to TACAN, the primary Shuttle navigation device.
- The REX-II satellite of USAF (see below, launch March 9, 1996) provides GPS-generated orbital elements in the downlink.
- The ORFEUS-SPAS-II free-flyer payload on Shuttle flight STS 80 (November 19-December 7, 1996) flew the first GPS Tensor receiver system (SS/L and LABEN) capable of providing on-board orbit parameters with an adapted Kalman filter algorithm along with Earth model software.
- Autonomous navigation on DS1 (Deep Space 1). This on-board system enables a S/C to determine its location in the Solar System as well as its flight path without help from controllers on Earth. With the knowledge of on-board time, AutoNav computes the position of the asteroids. By measuring where the asteroids appear relative to the stars, it computes where the S/C must be. It then can project its path to its destination and use its propulsion system to make any course changes that are required.

• The first GPS attitude determination systems - proof-of-concept demonstrations - were conducted on research aircraft, and on a commercial airliner.

- Three GPS attitude measurement flight tests were conducted in April/May 1991 on a DC-3 aircraft of Ohio University by using an Ashtech-3DF (Three-dimensional Direction Finding) 24-channel receiver configured in four 6-channel sections, and four microstrip patch antennas (fuselage and wing-tip mounted). Attitude data were computed, stored and displayed in realtime by two PCs. [333]
- A Stanford University attitude demonstration was conducted with a Trimble TANS Vector GPS receiver on a Piper Dakota aircraft in 1991. [334] Four strip-mounted antennas were used for attitude sensing. A 486 laptop computer provided both real-time display and data recording for post-flight analysis.
- On January 12, 1993, a flight experiment was conducted on a NASA/Ames King Air 200 aircraft with the objective to evaluate the dynamic response of a GPS attitude system. The attitude equipment consisted of two independent systems: INU (inertial Navigation Unit) and a six-channel C/A code GPS receiver (TANS Vector) and four microstrip patch antennas (fuselage and wing-tip mounted). [335] Attitude data from both systems were recorded for post-flight analysis.
- On Dec. 10, 1994, a test flight of a TANS Vector receiver on board a DC-10 commercial airliner (United Airlines) demonstrated the potential for SATCOM antenna pointing,

332) F. H. Bauer, J. R. O'Donnell, "Space-Based GPS 1996 Mission Overview," Proceedings of ION GPS-96, Sept. 17-20, 1996, Kansas City MO, pp. 1293-1302

333) F. van Graas, M. Braasch, "GPS Interferometric Attitude and Heading Determination: Initial Flight Test Results," Navigation ION, Vol. 38, No. 4, Winter 1991-92, pp. 297-316

334) C. E. Cohen, B. W. Parkinson, "Aircraft Applications of GPS-based Attitude Determination," Proceedings of ION GPS-92, Albuquerque, NM, Sept. 1992, pp. 775-782

335) C. E. Cohen, B. W. Parkinson, D. McNally, "Flight Tests of Attitude Determination Using GPS Compared Against an Inertial Navigation Unit," Navigation ION, Vol. 41, No. 1, Spring 1994, pp. 83-97

by providing aircraft attitude measurements (Note: Proper antenna pointing requires knowledge of aircraft position, satellite position and aircraft attitude). The aircraft used a Honeywell SATCOM system beam-steering antenna, two INS (Delco Carousel IV - the Carousel is a free azimuth INU which contains four-gimbal IRUs), a vertical gyro and a compass. In addition to the TANS Vector receiver, the aircraft was equipped with two modified TNL-8100 GPS receivers. The TANS Vector receiver exhibited expected performance during all phases of the test flight when compared to the reference measurements. [336)]

• Demonstration of spaceborne GPS orbit and attitude measurements. A number of tests for attitude determination with "prototype GPS receivers" have been carried out (or are planned) on Shuttle and on various commercial and military satellites. Table 48 lists a chronology of these instruments (see also chapter H.4.3.4).

- The CRISTA-SPAS-1 mission of DARA/NASA on Shuttle fight STS-56 (Nov. 3-14, 1994) provided the first demonstration of real-time attitude determination on the SPAS free-flyer platform of DASA (one day receiver operation). [1311)] See also chapter J.1.2.

- The REX-II satellite of USAF (launch March 9, 1996), with a GPS experiment by the name of ADACS, can be regarded the first successful demonstration of closed-loop attitude control application using real-time GPS carrier phase-based attitude measurement as sensor input (first time on March 29, 1996). It is also the first mission to provide orbital elements in the downlink.

Launch Date	Mission	GPS Receiver	Comment
June 25, 1993	RADCAL (USAF)	TANS Quadrex (Trimble)	2 cross-strapped receivers, attitude solutions in post-processing
Sept. 12-22, 1993	ORFEUS-SPAS-1 (DARA)	Alcatel/SEL receiver,	Shuttle STS-51, free flyer C/A code +L1 carrier phase
Aug. 3, 1994	APEX (USAF)	TANS Vector	Trimble Navigation instrument
Nov. 3-14, 1994	CRISTA-SPAS-1 (DARA/NASA)	Alcatel/SEL Rx+ TANS Vector + IRU	Shuttle STS-66, SPAS free-flyer
Jan. 11-20, 1996	GADACS (NASA)	TANS Vector (2)	Shuttle STS-72, on SPARTAN
March 9, 1996	REX-II (USAF)	TANS Vector (2) named ADACS	First long-term mission with GPS orbit/attitude determination
May 19-29, 1996	GANE (NASA)	TANS Vector +IRU	Shuttle STS-77
Nov.19-Dec 7, 96	ORFEUS-SPAS-2 (DARA/NASA)	Alcatel/SEL Rx+ GPS Tensor (Laben) TANS Quadrex	Shuttle STS-80, SPAS free-flyer; also relative navigation experiment ARP of ESA/ESTEC
May15-24, 1997	ARP (ESA) Shuttle/MIR	GPS Tensor (Laben)	STS-84 Shuttle - MIR rendezvous
Sep. 25-Oct. 6, 97	ARP (ESA) Shuttle/MIR	GPS Tensor (Laben)	STS-86 Shuttle - MIR rendezvous
Aug. 23, 1997	SSTI-Lewis (NASA)	GPS Tensor (2) named GADFLY	Lewis could not be operated and reentered Sept. 28, 1997
Aug.7-19, 1997	CRISTA-SPAS-2	Alcatel/SEL Rx+ GPS Tensor	Shuttle STS-85, SPAS free-flyer
Dec. 24, 1997	EarlyBird-1 (Earthwatch)	Vector and Viceroy	S/C lost contact with ground
Feb. 14, 98 (1st four satellites)	Globalstar (constellation of 48 satellites)	GPS Tensor on SS/L LS-400 platform	Big LEO communication system of Globalstar L. P., San Jose, CA
1998	SSTI-Clark (NASA)	TANS Vector (2) named GADFLY	NASA cancelled the Clark mission in Feb. 1998
Feb. 23, 1999	ARGOS (USAF)	Embedded receiver	
Nov. 15, 2000	AMSAT-3D (AMSAT)	TANS Vector (2)	
2001	Gravity Probe B (NASA)	GPS Tensor	
1998-2003	ISS (NASA, ESA, CSA, NASDA, and RKA)	SIGI (Space Integrated GPS/INS)	Start of SIGI operation on ISS is planned for 2001

Table 48: Overview of early spaceborne GPS attitude receivers flown on varies missions

336) L. Kruczynski, J. Delucchi, T. Iacobacci, "Results of DC-10 Tests using GPS Attitude Determination," Proceedings of ION GPS-95,

- The first spaceborne flight demonstration of **SIGI** (Space Integrated GPS/INS) took place on Shuttle flight STS-101 (May 19-29, 2000) under the name of SOAR (SIGI Operational Attitude Readiness). A reflight of SOAR occurred on STS-106 (Sept. 8-20, 2000). Two units of SIGI are planned to be installed on ISS., one on ISS itself, the other on CRV (Crew Return Vehicle). CRV is scheduled to become an operational part of the station in 2004/5 (CRV is designed to be used as a life-boat in case of an emergency evacuation from ISS). A short description of SIGI is provided in H.4.3.4. [337)]

At the beginning of the 21st century it is expected that lightweight spaceborne GPS receiver systems, providing the functions of: position, velocity, attitude, attitude rate and time, are going to replace a number of conventional attitude measurement devices such as horizon sensors and sun trackers. The GPS orbit/attitude receiver in combination with a control scheme (actuator) offers sound technical and economical attitude-control solutions. Previous experiments with GPS attitude determination have demonstrated a potential for coarse attitude determination in the 0.1^o range. Also spaceborne GPS receivers have demonstrated a time transfer capability of <100 ns, thus making very precise, coordinated time available to spacecraft systems.

• Flight demonstrations of relative navigation using GPS observables. ESA, in cooperation with NASA, DARA, and RKA (now Rosaviakosmos), conducted three flight demonstrations in its ARP (ATV Rendezvous Pre-development) program with proximity navigation approaches [ATV is the future Automated Transfer Vehicle, the ESA resupply vehicle for ISS (International Space Station)].

- Deployment and retrieval maneuvers of SPAS release/approach with Shuttle. [338)] [339)] The ORFEUS-SPAS-2 free-flyer on Shuttle flight STS-80 (Nov. 19 - Dec. 7, 1996) carried, in addition to its prime payload, an ARP secondary payload of ESA/ESTEC, consisting of a GPS Tensor receiver and R-GPS (Relative-GPS) navigation algorithms (developed by MMS). The Shuttle orbiter carried also a GPS receiver (TANS Quadrex), implemented on the fixed part of WSF (Wake Shield Facility) within the cargo bay, and an optical sensor TCS (Trajectory Control Sensor). TCS (of NASA, a laser based sensor) verified the R-GPS measurements by tracking a retro-reflector mounted on the x-face of ORFEUS-SPAS.
- Shuttle - MIR flight approach and departure phases. ARP proximity navigation approaches were demonstrated with an instrument package of GPS Tensor (Laben) in combination with the R-GPS algorithm, and a close-range laser sensor, referred to as RVS (Rendezvous Sensor), built by DASA Jena Optronik - on Shuttle flights STS-84 (May15-24, 1997) and STS-86 (Sep. 25-Oct. 6, 97). The GPS receiver of the MOMS-NAV payload (DARA) on MIR/Priroda provided GPS measurements from the other direction. The Shuttle RVS used three retro-reflectors installed on MIR.

• Integrated GPS+GLONASS receiver technology. [340)] The combination of GPS and GLONASS signals enhances the overall navigation solution in three ways: availability, integrity and accuracy. The GLONASS constellation has no deliberate signal degradation; hence, the GPS+GLONASS horizontal accuracy is 15-20m compared to 100m for GPS-only (with SA on).

- In 1996/7 first GPS+GLONASS receivers came onto the market. Some models: GG24 (Ashtech, Sunnyvale, CA), GePOS RG24 (Carl Zeiss Geodetic Systems, Jena, Germa-

337) J. Um, E. G. Lightsey, "Space Flight Test Results for the SOAR Experiment," ION-2000, Salt Lake City, UT, Sept. 19-22, 2000, pp. 2243-2251

338) M. Cislaghi, U. Thomas, M. Lellouch, J. M. Pairot, "Development and Verification of Automated Rendezvous for ATV," Proceedings IAF-96-T.2.08, Oct. 7-11, 1996, Beijing

339) M. Cislaghi, U. Thomas, M. Lellouch, G. Limouzin, "ATV - Pre-development Program - Flight Demonstrations," IAF-97.T.2.03

340) Note: Prior to measurement integration of GPS and GLONASS into one receiver, proper transformations must be established between the two time scales [UTC(USNO) and UTC(SU)] ,and the two coordinate systems [WGS84 and PE-90 respectively]

ny), ASN-22 (DASA, Ulm Germany), GSR2400 (Sokkia Corp., Overland Park, KS), GNSS-300 (3S Navigation, Laguna Hills, CA).[341)]

- En-route navigation in civil aviation. [342)] Airborne testing against a proven flight truth source (Z-12 dual-frequency GPS + ground-based reference station). In September 1996, Ashtech Inc. and Universal Avionics made the world's first transoceanic flight with GPS+GLONASS technology, using an Ashtech GG24 receiver for an overseas flight (from Shannon, Ireland to Teterboro, NJ, USA). During the flight the 24-channel receiver computed positions using up to 17 satellites. Horizontal positions within an accuracy of 16 m were achieved 95% of the time. The potential benefits to civil aviation (navigation and positioning operations) in general have become increasingly obvious.
- Spaceborne testing. a) A GPS+GLONASS receiver [LAGRANGE (Laben GNSS Receiver for Advanced Navigation, Geodesy and Experiments) of Laben S.p.A., Vimodrone, Italy] is part of the SAC-C satellite payload with a launch Nov. 21, 2000. b) an ESA GPS+GLONASS receiver by the name of GRAS (GNSS Receiver for Atmospheric Sounding) is part of the MetOp-1 satellite.

• Starting in about 1994/5, spaceborne/airborne GPS/GLONASS receiver applications developed into two major functional directions:

- Use of the receiver as a navigation sensor - to determine relative and absolute spacecraft attitude, orbit and time. The receiver is expected to provide an engineering environment of real-time, autonomous onboard support to the spacecraft (mission-critical functions).
- Use of the receiver as a science instrument in Earth observation - to perform atmospheric sounding measurements, gravity measurements, and ionospheric sounding measurements, etc. These science applications require extensive postprocessing of the data to achieve the desired results.
  Receivers are included on virtually all new Earth-observation missions. Parallel to orbit determination, the technology opened new applications in atmospheric profiling (GPS/MET on OrbView-1/Microlab-1, launch April 3, 1995) and possibly in ocean altimetry (SAC-C). NASA/JPL-provided dual-frequency GPS receivers are part of four geomagnetic missions: Ørsted (launch Feb. 23, 1999), SUNSAT (launch Feb. 23, 1999), CHAMP (launch July 15, 2000), and SAC-C, launch Nov. 21, 2000) - and also of such missions as: SRTM (Shuttle Radar Topography Mission, launch Feb. 11-22, 2000), Jason, VCL (Vegetation Canopy Lidar), and GRACE (Gravity Recovery and Climate Experiment).

• Demonstration of GPS measurements in satellite orbits outside of the MEO GPS constellation of 20,000 km altitudes.

- The Motorola GPS Viceroy receiver of the Equator-S (launch Dec. 2, 1997) satellite - with a near-equatorial HEO orbit of 10 R_E distance apogee and low perigee at 500 km - was able to measure carrier phase and/or C/A code of up to three GPS satellites at altitudes of 34,000 km (Dec. 3. 1997 while in transfer orbit). The GPS receiver measurements were considered for post-event (on-ground) orbit determination at GSOC. In addition, GPS side-lobe measurements were verified. [343)]
- AMSAT-3D (launch Nov. 15, 2000) in a HEO orbit of 4000 x 47,700 km, inclination = 63.4°. The AMSAT-GPS experiment is to demonstrate real-time GPS attitude and orbit determination above the GPS constellation (two Trimble TANS Vector receivers

341) GPS World Receiver Survey, GPS World, January 1998, pp. 46-59

342) J. G. Murphy, W. V. Cottrell, "Airborne Testing of GPS+GLONASS Positioning Sensor Against A Proven Test Truth Source," Proceedings of ION GPS-97, Sept. 16-19, 1997, Kansas City, MO, pp. 1047-1054

343) O. Balbach, et al., "Tracking of GPS Above GPS Satellite Altitude: Results of the GPS Experiment on the HEO Mission Equator-S," Proceedings of ION GPS 1998 Conference, Nashville, TN, 1998

along with two sets of four GPS antennas, GEC Plessey front-end chipset and AMD 29200 embedded RISC processor board). An additional experiment is to map the GPS signal patterns outside the GPS constellation.

- GPS measurements in GEO. In general, GPS receivers in GEO cope with unfavorable conditions which are: 1) poor visibility and geometrical distribution of GPS signals, and 2) weak GPS signal power (low SNR). In spite of these handicaps, GPS pseudorange measurements were already demonstrated in GEO for orbit determination in US military applications. [344)] Since geostationary orbit determination is a very attractive option for commercial GEO satellite operations, solutions are being introduced to a new generation of GPS receivers for any satellites in GEO. Two examples: Astrium GmbH offers an instrument, MosaicGNSS, using a software and hardware correlation. [345)] The TOPSTAR 3000 GPS receiver of ASI (Alcatel Space Industries) employs a filter and integration techniques to analyze weak GPS signals. TOPSTAR (a 4-antenna 24-channel C/A code GPS receiver) is part of the STENTOR spacecraft (launch in 2001) to be used as a navigation tool for GEO spacecraft. [346)]

• Second frequency for civil GPS users (Block IIF series). Uninterrupted access to "carrier phase" of the L2 frequency will be provided as a second civil frequency. This agreement was announced Feb. 27, 1997, by DoT and DoD. The goal of the second civil frequency (designated L5) is to enhance GPS civil capabilities by: 1.) providing a redundant signal for civil use in the event that the GPS L1 were to become unavailable (due to jamming or interference), and 2.) enhancing civil GPS performance by providing a second civil frequency for ionospheric delay calculations, a critical design requirement for the FAA's Wide Area Augmentation System (WAAS).

• The GPS-on-a-chip project (NASA) miniaturizes instrumentation.

• CDGPS (Carrier-phase Differential GPS), see H.4.3. CDGPS sensing techniques provide very precise measures of relative position (1-2 cm level) and attitude between vehicles in formation. Given GPS measurements at two nearby antennas, relative position between these antennas can be estimated to a high degree of accuracy based on tracking the relative phase. Sample implementations: 1) CDGPS is flown on the Orion microsatellite of Stanford University (planned launch at the end of 2001). The Orion CDGPS receiver (total instrument mass of 1.07 kg, including processors) consists of a single 6-antenna attitude and relative navigation receiver using carrier-differential GPS. 2) As of 2000/1, AeroAstro Inc. of Herndon, VA is developing "Star Ranger" for the TechSat-21 mission of AFRL. The Star Ranger design concept utilizes Ku-band for its operation (intersatellite communication), DSSS (Direct Sequence Spread Spectrum) for precise ranging, a two-PN code technique for multiple access within the three-satellite formation, and CDGPS for the determination of relative position and attitude between the formation flying satellites. The mass of Star Ranger is expected to be <2 kg. [347)]

• PRARE and DORIS are tracking systems requiring a host satellite (for the space segment instrument) and global ground-based tracking networks for precise orbit determination. Both systems (PRARE and DORIS) require an orbit determination process using conventional dynamic techniques (with physical models) whose accuracies depend on the quality of the models.

344) J. D. Kronman, "Experience Using GPS for Orbit Determination of a Geosynchronous Satellite," Proceedings of ION GPS-2000, Sept. 19-22, 2000, Salt Lake City, UT, pp. 1622-1626

345) M. Mittnacht, W. Fichter, "Real-Time On-board Orbit Determination of GEO Satellites using Software and Hardware Correlation," Proceedings of ION GPS-2000, Sept. 19-22, 2000, Salt Lake City, UT, pp. 1976-1984

346) Ch. Mehlen, D. Laurichesse, "Real-time GEO orbit determination using TOPSTAR 3000 GPS Receiver," Proceedings of ION GPS-2000, Sept. 19-22, 2000, Salt Lake City, UT, pp. 1985-1994

347) R. Zaenick, K. Kohlhepp, "GPS Micro Navigation and Communication System for Clusters of Micro and Nanosatellites," Proceedings of the 14th AIAA/USU Conference on Small Satellites, Logan, UT, Aug. 21-24, 2000, SSC00-VI-8

PRARE (H.7.2) was initially installed on ERS-1 (launch July 17, 1991), but could not be operated. Further PRARE uses: Meteor-3-7 (launch Jan. 25, 1994), ERS-2 (April 21, 1995). PRARE is a two-way tracking system, broadcasting dual-frequency signals (2200 and 8500 MHz) from a transmitter on-board the host satellite to receiver-transponders on the ground. The signals are modulated by pseudonoise ranging codes to permit both range and range-rate measurements. The 8500 MHz signal is coherently transponded back to the host satellite (at 7200 MHz) for on-board range and range rate extraction. The 2200 MHz signal is received and tracked on the ground to provide an ionospheric correction.

DORIS was a prototype payload on SPOT-2 (launch Jan. 22, 1990) and SPOT-3 (launch Sept. 26, 1993); then flown on: Topex/Poseidon (launch Aug. 10, 1992). DORIS was initially a one-way Doppler system (E.21.1) which broadcasts continuously at two frequencies: 401 and 2036 MHz. A two-way upgrade is planned for use on Envisat (launch in 2001). The DORIS instrument on the host satellite observes individual ground beacons in sequence and measures the Doppler frequency of the received signals.

- Spaceborne GPS wide-area augmentation service. Japan is the first country to use a spaceborne GPS augmentation system to improve the management of its civil air traffic. The "aeronautical mission" of MTSAT (Multifunctional Transport Satellite) in GEO (launch of MTSAT on Nov 15, 1999, but a launch failure of the H-2 vehicle occurred) provides an operational system that satisfies every aspect of the ICAO (International Civil Aviation Organization) requirements in terms of reliability and accuracy to aviation enroute through precision approach navigation.

- GNSS-1 (Global Navigation Satellite System-1). The first generation GNSS-1 comprises the following elements: GPS, GLONASS and their augmentation systems [WAAS (Wide Area Augmentation System) of the US, EGNOS (European Geostationary Navigation Overlay System) of Europe, and MSAS (Multi-Transport Satellite Augmentation System) of Japan]. The three segments of GNSS-1 are expected to be operational by 2003.

- Galileo. In early 1999 the EU proposed a strategy with the goal to design, implement and operate its own (civil) constellation of navigation satellites with the appropriate terrestrial infrastructure within a program by the name of Galileo. The space segment of the overall Galileo program is referred to as GalileoSat. Galileo is considered an element in a future GNSS-2 (Global Navigation Satellite System-2) - currently comprising GPS, GLONASS and their future augmentation systems. The rationale for Galileo is the provision of a service with a certifiable service performance level, which neither GPS nor GLONASS can presently do. The goal is the support of safety-critical civilian applications, especially in civil aviation, marine navigation, and road transport. The goal is to complete implementation of the Galileo system by 2008.

- GNSS-2 (Global Navigation Satellite System-2). The second generation GNSS comprises all elements of GNSS-1 plus Galileo. GNSS-2 is planned to be operational by 2008.

- **Removal of the SA (Selective Availability)** feature for the GPS constellation signal on May 2, 2000 (the SA levels were set to zero at 0400 UT). [348) 349)] The Presidential Directive (of US President Bill Clinton) permits civil users worldwide general access to the highest-possible accuracy of GPS signals. [350)] - Background: The SA service has been used to degrade the GPS signal for the civil community while retaining the higher-accuracy signals for US and allied military forces (as well as for approved receivers of NASA satellite projects). A 1996 Presidential Directive promised to review the SA issue every year starting in 2000, and to remove SA in 2007 at the latest.

348) R. E. Neilan, A. Moore, T. Springer, J. Kouba, J. Ray, Ch. Reigber, "International GPS Service 2000: Life without SA," ION GPS 2000, Slat Lake City, UT, Sept. 19-22, 2000, pp. 438-446

349) http://www.igeb.gov

350) P. B. de Selding, "Europe Cheers While Questioning End of GPS Selective Availability," Space News, May 15, 2000, pp. 4 and 26

• Joint GPS/LEO navigation receiver. [351] [352] As of 2000/1 new dual-use tracking concepts of GPS and LEO constellation signals are being considered by the communications industry. Commercial LEO satellite systems, such as Orbcomm, Iridium, and Globalstar, provide a global coverage of their data and/or voice services. In addition to these prime services, the LEO constellations may also serve as "guide posts in space" to complement and enhance the GPS navigation performance. The combined use of positioning and two-way communication in LEOs may eventually lead to such applications ranging from emergency roadside assistance to location-based merchandising. The overall objective is to achieve precision (cm-level) navigation performance using pre-existent LEO transceiver hardware. In this concept, the LEOs provide additional ranging signals, which improves GDOP, availability of navigation solutions, and availability of RAIM (Receiver Autonomous Integrity Monitoring) geometry. The geometric diversity, achieved mainly by the motion of the LEOs, enables the GPS/LEO receiver to resolve the integer cycle ambiguities on the GPS constellation signals as well as parameters related to the cycle ambiguities on the LEO signals. First experimental tests have been conducted with the Orbcomm constellation in conjunction with GPS.

1.8 Services

The rendering of any sustained or long-term service implies the provision of an operational capability. The early introduction of the NASA/NOAA broadcast service policy of free access to polar-orbit weather satellite data reception generated a totally new participative/cooperative research and application environment for a global user community (laboratories and research institutes in particular). Real-time data reception of AVHRR data became affordable to many with the installation of a simple receiving station. Eventually a network of thousands of small ground stations was realized, and the AVHRR sensor became the best known sensor in the world. The most important service aspect was probably the provision of a timely, reliable and repetitive data stream to the user community which in turn accelerated the pace of exploration and of technological development in the various fields of applications.

- TIROS-1 (launch April 1, 1960) is regarded the first true weather satellite.
- Starting with TIROS-8 (launch Dec. 21, 1963), real-time observational data were transmitted (broadcast) continuously in APT (Automatic Picture Transmission) mode to ground stations. Eventually, APT pictures could be received on fairly simple ground stations anywhere in the world.

• Introduction of APT and HRPT (High Resolution Picture Transmission) broadcast modes (VHF link for ATP and L-band link for HRPT) in parallel with TIROS-N (launch Oct. 13, 1978). Further broadcast services of 'weather data' followed:

- The Terra satellite (launch Dec. 18, 1999) of NASA's EOS (Earth Observing System) program introduced a direct broadcast in X-band for instrument data (MODIS) to the user community.
- The Aqua (formerly EOS/PM-1) S/C of NASA (planned launch in late 2001) provides a direct broadcast in X-band for all its instrument data, including MODIS
- MetOp-1 of EUMETSAT (planned launch in 2005) supports a real-time broadcast of instrument data to local users by means of LRPT (72 kbit/s in VHF for selected instruments) links and HRPT (3.5 Mbit/s in L-band) links.
- NPP (NPOESS Preparatory Project) of NASA (planned launch in Dec. 2005) provides broadcasts of instrument data.

• WEFAX (Weather Facsimile) services. The first installation of WEFAX on ATS-1 (launch Dec. 6 1996) worked well until 1972. It demonstrated that WEFAX transmissions

351) M. Rabinowitz, B. W. Parkinson, K. Gromov, C. H. Cohen, "Architectures for Joint GPS/LEO Satellite Carrier Phase Receivers Designed for Rapid Robust Resolution of Carrier Cycle Ambiguities on Mobile Platforms," ION GPS 2000, Slat Lake City, UT, Sept. 19-22, 2000, pp. 881-890

352) M. Rabinowitz, B. W. Parkinson, J. J. Spilker, "Some Capabilities of a Joint GPS-LEO Navigation System," ION GPS 2000, Slat Lake City, UT, Sept. 19-22, 2000, pp. 225-265

were practical. The original transponder received on 149.22 MHz and transmitted on 135.6 MHz. Due to technological design improvements and interference with the aeronautical band, these frequencies were re-allocated to the S-band frequencies. The most pertinent one of these to date is the 1691 MHz WEFAX frequency.

- Presumably, NOAA-6 (launch June 27, 1979) is regarded as the first **operational** satellite of the TIROS-N series.

- The on-board store & forward data concept (with a data recorder and a data dump during a station pass) was introduced fairly early in the game (TIROS-1 with a launch in 1960 had a tape recorder). The lack of world-wide coverage (ground stations) for data downlinks dictated this strategy.

- Global coverage once per day was first provided by AVHRR on TIROS-N (launch Oct. 13, 1978)

- Introduction of search and rescue services on polar-orbiting satellites with the launch of COSPAS-1 on June 29, 1982 and with S&RSAT flown on NOAA-8, March 28, 1983. The first geostationary S&RSAT (demonstration) system, referred to as GEOS&R (Geostationary Search and Rescue), was flown on GOES-7 (launch Feb. 26, 1985).

- Introduction of spaceborne data collection systems (DCS). The first DCS anywhere was IRLS (Interrogation, Recording, and Location System) flown on Nimbus-3 (launch April 14, 1969, see M.17.3). DCS were also flown on the first three Landsat S/C, LS-1, LS-2, and LS-3 (LS-1 launch July 23, 1972). The first Argos DCS in LEO (polar orbit) was flown on TIROS-N (launch Oct. 13, 1978). The first DCS in GEO orbit was flown on SMS-1 (launch May 17, 1974). In the meantime all geostationary meteorological satellites series (METEOSAT, NOAA-GOES, GMS (JMA), etc. use DCS or variations thereof. The TEMISAT microsatellite (launch Aug. 31, 1993) of Telespazio represents further advancements in data collection services for the user community with regard to access scheme and functionality (see Table 133 for DCS).- The systematic collection of ground truth data from a large number of systems in the ground segment provides a reference (and verification) for the analysis of primary sensor data.+

- Satellite navigation data, such as position, velocity, and time, of GPS (GPS-1 launch in 1978) and GLONASS (first GLONASS launch in 1982) systems. The GPS constellation became operationally available in 1994 (officially in 1995). The real-time global availability of navigation data has spawned numerous positioning and navigation applications that have surpassed initial expectations. They are fast becoming an indispensable part of people's everyday lives. The navigation services on all levels (civil and military aviation, coastal and ocean ship navigation, automobile navigation, surveying, etc.) are becoming increasingly essential to the world's infrastructure. As a consequence, a large commercial market of equipment and service providers is unfolding, responding to the new needs of society.

- GPS time is being utilized as a cost-effective standard time source by many operators. Networks like Internet and many TTPs are being synchronized by GPS time. The GPS time service is provided by USNO (US Naval Observatory). GPS provides two types of time:[353)]

  - GPS time. Defined by its 'composite clock.' It consists of an ensemble of more than 20 GPS space and ground-based atomic frequency standards. GPS time is referenced to UTC (Universal Time Coordinated) as maintained by USNO. Time updates are maintained to within 1 µs of UTC by the operators of the GPS Master Control Station at Falcon Air Force Base, Colorado Springs. Note: GPS time differs from UTC by the number of leap seconds accumulated since January 1980.
  - UTC. To obtain UTC from GPS, users must apply the GPS-UTC (USNO) correction, available in the navigation message, to transition from GPS time to an estimate of UTC (USNO). The estimate of UTC (USNO) is called UTC (GPS).

353) F. Vannicola, "The Time Is Now," GPS World Showcase, Dec. 1997, p. 40

Starting in the mid 1980s a number of commercial services have come into the Earth observation arena to complement (or replace) the institutional services.

- Spacecraft operators and data distributors like Eosat (Landsat-4, -5)

- In 1990, SSTL of Surrey (UK) started a unique microsatellite technology transfer program, providing on-the-job training of engineers and scientists of foreign national organizations in cooperative programs [KAIST (Korea), LNETI (Portugal), FACH (Chile), TMSC (Thailand), etc.]. Affordable access to space is the overall theme of the service. - The design, building and operating experience of their own microsatellites gave these organizations a means to start/continue their own involvement in national space programs. The 'Surrey Space Centre' of SSTL (since 1992) is a European center of excellence, a facility which accommodates the activities for the technology transfer service of academic and post-graduate research.

- Distributors of imaging data products from a particular spacecraft and sensor (SPOT Image, Eurimage, etc.)

- The RADARSAT-1 satellite of CSA, Canada, is an Earth observation mission for SAR data, operated on a commercial basis by RSI.

- Real-time ERS altimetry data are distributed via Internet. The NOAA/NODC Laboratory for Satellite Altimetry (LSA) receives ERS altimetry data and generates RGDRs (Real-time Geophysical Data Records). These RGDRs are distributed via Internet and may be used in oceanographic analysis and model assimilation studies. [354)]

Mission	Data Availability	Sensor	Band	Spatial Resolution
Landsat-4,-5	1982	TM	0.45 - 2.35 μm (MS)	30 m
Landsat-7	1999	ETM+	0.50 - 0.90 μm (Pan)	15 m
SPOT-1, -2, -3	1986	HRV	0.51 - 0.73 μm (Pan)	10 m
Resurs-O1-1, 2, 3	1985	MSU-E	0.50 -0.90 μm (MS)	45 m x 33 m
JERS-1	1992	OPS	0.52 - 2.40 μm (MS)	18 m x 24 m
IRS-1C, IRS-1D	1996, 1997	PAN	0.50 - 0.75 μm (Pan)	6 m

Table 49: Availability of long-term high-resolution imaging data of major Earth surface missions

- A multitude of commercial enterprises provide their services in the airborne arena, most with a full service (aircraft, sensor(s), data recording and processing, etc.).

- **Shuttle Small Payload Program (SSPP).** A flight service package offered by NASA for payload masses in the range between 23 kg to about 2270 kg. SSPP is for "small, self-contained payloads" with the objective to ensure divers user groups (educational, commercial, government, foreign experimenters, etc.) to have access to space at reasonable costs. Payloads are accommodated by providing various carrier systems [GAS, Hitchhiker, Hitchhiker Jr., and SEM (Space Experiment Module)] in the Shuttle's unpressurized payload bay. SSPP started in 1984. [355)]

Note: In general a S/C launch from Shuttle has advantages and disadvantages. Typically the low orbit implies a relatively short lifetime for the mission. This may be of interest in itself, as the orbital decay can be studied. But the Shuttle is the only launcher that really allows a payload to be viewed being released into space. In addition the g-loads are modest compared with other launchers.

- **GAS** (Get Away Special). GAS is a carrier system concept with standards and conditions relating to GAS payloads (they must fit in a standard container of 0.14 m^3 in volume with a payload mass not exceeding 90 kg, two or more experiments may be included in a

354) J. Lillibridge, "Real-time ERS altimetry at NOAA," Earth System Monitor, Vol. 9, No. 3, March 1999, pp. 6-9

355) http://sspp.gsfc.nasa.gov/gas.html

single container). In addition, GAS payloads must be self-powered and be easy to handle for the payload crew. The GAS container is made of aluminum with circular end plates. It can be pressurized (or evacuated) to suit experiment requirements.

GAS payloads are mounted during flight in the Shuttle payload bay, on the sidewall, or on a cross-bay truss structure (referred to as "getaway bridge"). The aluminum bridge fits across the payload bay of the orbiter and offers a convenient and economical way of flying several GAS canisters. The getaway bridge, capable of holding up to 12 canisters, made its maiden flight on STS-61-C (Jan. 12-18, 1986).

The service was initially announced by NASA in 1976, five years prior to the first Shuttle flight in 1981. The first GAS demonstration payload, FVP (Flight Verification Payload), was flown in 1982 on STS-3 (Mar. 22-30, 1982). The first customer GAS payload, G-001 of Utah State University, flew on STS-4 (June 27 - July 4, 1982). STS-95 (Oct. 29 - Nov. 7, 1998) is the 33rd Shuttle mission with the GAS payload service. Almost 200 individual GAS canisters have been flown in these 33 missions for a very divers user community.

- **Hitchhiker carrier system.** The Hitchhiker concept is based on a modular and expandable carrier with the provision of extended functional features (standard power, data, and command services for customer equipment). The structure can carry equipment mounted in canisters but also has mounting plates of various sizes for user equipment (provision of options). The carrier provides electrical power (28 VDC), command signals (1200 baud), and downlink data interfaces (various data rates: 1200 baud asynchronous low-rate, 1-1400 kbit/s medium rate). Hitchhiker customers are able to operate their payloads from the ground segment (GSFC) using their own ground support equipment (usually a PC) to send commands and display data. The ACE (Advanced Carrier Equipment) package provides such standard services devices as: power distribution unit, remote interfaces unit, Hitchhiker central unit, digital storage unit, Hitchhiker video interface unit, and lightweight avionics plate. NASA initiated the service in 1984. The first flight was on STS-61-C (Jan. 12-18, 1986) with the Hitchhiker-G1 payload [consisting of IEH-1 (International Extreme Ultraviolet Hitchhiker, CAPL-2 (Combined Pumped Loop-2), TES (Thermal Energy Storage), etc.]. The SLA-1 (Shuttle Laser Altimeter) experiment on STS-72 (Jan. 11-20, 1996) was also a Hitchhiker payload.

- **Hitchhiker Jr. carrier system.** A limited Hitchhiker version, designed for payloads which do not need the functions of command and telemetry interfaces from the experiment to the ground (GSFC). The CONCAP-IV-3 payload on STS-69 (Sept. 7-18, 1995) was the first to use the Hitchhiker Jr. service package. The experiment was activated and de-activated by the Shuttle crew via a laptop computer. This requires autonomous payload operations for the duration of the experiment. The SOLSE/LORE (Shuttle Ozone Limb Sounding Experiment/Limb Ozone Retrieval Experiment), a NASA instrument package flown on STS-87 (Nov. 19 - Dec. 5, 1997), used also the Hitchhiker Jr. service version.

- **SEM** (Space Experiment Module) carrier system. The SEM program is an upgrade version of the GAS program. It uses GAS canisters with the added feature of installed power provision and more. On behalf of frequent student requests, NASA funded in 1995 the SEM program and designed a standard power supply. As a result, the standard SEM consists of subsystems which function together to provide containment, structural support, power, experiment command and data storage capabilities for experiment support. This new functional capability/availability of the carrier system permitted the students to focus their energies on creating their own experiments. The very first flight of the SEM system, SEM-1, took place on STS-80 (Nov. 19 - Dec. 7, 1996) with many experiments, built by students in cooperation with their mentors of various High Schools and Universities across the USA.

- Commercial spaceborne imaging missions with high-resolution data (1 m) are being introduced in the latter nineties (1998) by several companies and/or consortia (Space Imag-

ing EOSAT, Earthwatch, Resource 21, OrbView, etc.). The consortia provide also a full ground segment with corresponding archives and services. Their imaging products are being offered to anyone who pays for the service.
The provision of commercial imaging is considered a major shift in Earth observation policies - from government-sponsored research institutes toward private enterprise.

- Underwater/space launch. July 7, 1998, a Russian nuclear submarine of the Northern Fleet launched two environmental nanosatellites, TUBSAT-N and -N1 of the Technical University of Berlin, into Earth orbit (launch site of western Barents Sea). The launch represents the world's first underwater/space launch of a satellite into Earth orbit on the basis of a commercial service.

- First commercial satellite constellations in LEO (Low Earth Orbit). In 1998, the first two major satellite telephone systems, so-called "Big LEO Systems" of global handheld telephone service as well as mobile fax and data services, were launched. Iridium, a network developed by Motorola, completed its planned 66-spacecraft constellation in LEO, and started service. Globalstar, created by Loral, successfully launched its first 8 satellites. However, 12 other spacecraft were lost in a failure of Ukraine's Zenit rocket. In all, Iridium featured 10 launches and Globalstar had three.
On the "Little LEO Systems" front of non-voice messaging and data relay (i.e. store & forward) services, Orbcomm of Dulles VA, had the complete constellation of 35 satellites in orbit (with altitudes of 825 km) on Dec. 4, 1999. The initial launch of this constellation started on April 3, 1995.

1.9 Start of International Cooperation

The space age in general and Earth observation in particular turned out to be a natural field for all types of national and international cooperations/participations - unmatched in history. Initial cooperation (with NASA) started with permissions to operate ground stations in various countries. Later on, foreign ground station operators contributed to NASA missions through tracking and data receiving services. International cooperation in the early 1960s manifested itself also in such policies of flying sensors (experiments) of non-US scientists on NASA missions or in providing launches for foreign satellites. This evolved eventually in the common design and construction of spacecraft and instruments. In January 1970, NASA had official cooperations with agencies/institutes of 35 foreign countries (agreements of ground stations, exchange of personnel, etc. brought the total cooperations to 74 countries). Some of the first cooperations involving research satellites are listed below: [356)]

- Ariel-1, a product of USA/UK cooperation, was launched April 26, 1962 on a Delta vehicle from Cape Canaveral (60 kg S/C, 389 km x 1214km inclined orbit at 54°) with the objective to measure parameters of the ionosphere and of the sun.

- Alouette-1, a Canada/USA cooperative venture, was launched September 29, 1962 from VAFB, CA aboard a Thor-Agena B launch vehicle (Alouette-1 mass = 145 kg) with the objective to investigate the ionosphere.

- San Marco-1 (mass of 24 kg), an Italian satellite, was launched December 15, 1964 with a US launcher (from a floating platform off Kenya) to investigate atmospheric densities.

- Asterix-1, the first French satellite launched on November 26, 1965 on a French launch vehicle, Diamant, from Hammaguir, Algeria (Elliptical orbit, apogee of 1736 km, perigee of 530 km, inclination of 34.3°). The 41.7kg satellite transmitted for two days. FR-1, the first operational French satellite, was launched into a 780 km orbit on Dec. 6, 1965 on a Scout launcher from Vandenberg AFB. The satellite was used to study the ionosphere.

- WRESAT (Weapons Research Establishment Satellite), an Australian/US S/C involved the development and launch from Woomera, Australia, of a small scientific satellite

356) W. Buedeler, "Geschichte der Raumfahrt," Sigloch Edition, Künzelsau, 1979,

(mass of 50 kg) on Nov. 29, 1967 (reentry Jan. 10, 1968). The satellite was placed in a near-polar orbit (perigee = 198 km, apogee = 1252 km, inclination = 83.3°, period = 99.3 min) by a US Redstone rocket. The objective was to monitor solar radiation in the upper atmosphere and to demonstrate an Australian capability for developing a satellite.

• ESRO-1/Aurora, an ESRO (Europe)/NASA cooperative venture (86 kg S/C mass), was launched Oct. 3, 1968 from VAFB to investigate auroras and the ionosphere. Prior to this, ESRO-2 (a 75 kg S/C) was launched on May 17, 1968 on a Scout launcher from VAFB with a payload measuring cosmic rays and solar X-rays. - ESRO-1/Boreas, almost identical to ESRO-1/Aurora was launched Oct. 1, 1969 from VAFB and performed similar measurements simultaneously with its sister spacecraft Aurora.

• Azur-1, a German satellite (71 kg), was launched Nov. 11, 1969 on a Scout vehicle from VAFB (387 km x 3150 km sun-synchronous orbit inclined at 103°), to investigate radiation belts, solar particles and polar lights.

• ANS-1, Netherlands satellite (129 kg), launched August 30, 1974 on a Scout vehicle from VAFB, CA (258 km x 1173km inclined orbit at 98°). ANS-1 studied UV spectra of young stars, and hard and soft X-rays from cosmic sources.

• Intasat, Spain/USA satellite, was launched November 15, 1974.

• COS-B, the first ESA/NASA satellite, was launched August 9, 1975 to investigate stellar x-ray and gamma-ray radiation.

• Meteosat-1, the first geostationary ESA weather satellite, was NASA-launched on November 23, 1977.

• In the USSR, the program **Intercosmos** was created in 1967 with the objective to invite cooperation/participation of Soviet-affiliated countries in the Soviet space program with their own national contributions. An important area of participation was in remote sensing, building sensors for specific missions, dissemination and scientific interpretation of data, etc. The new policy fostered a number of collaborative science projects among its nine members as well as with other nations. The Intercosmos satellite series began with the launch of Intercosmos-1 on October 14, 1969. The payload featured, beside Soviet, also Czech and East-German instruments for the measurement of UV and x-ray radiation in the upper atmosphere. Up to 1991, there were a total of 25 Intercosmos satellites.

• The first foreign-built satellites launched from a Russian launch site, were from France.

- Aureole-1 was launched from Plesetsk by an SL-8 vehicle on Dec. 27, 1971 into an orbit of 410 km x 2500 km, inclination of 74°. S/C mass of 300 kg. Objective: study of the aurora borealis and ionosphere.
- SRET-1 was launched from Plesetsk on Apr. 4, 1972 along with the Molniya-1 spacecraft. The French satellite was placed in a 460 km x 39248km orbit and was used to study radiation effects on solar cells.
- Aureole-2 was launched from Plesetsk by an SL-8 vehicle on Dec. 26, 1973. S/C mass of 400 kg. Orbit: 400 km x 1975 km, inclination of 74°.
- April 19, 1975. Launch of the first Indian satellite, Aryabhata of ISRO, by a Cosmos launch vehicle.
- SRET–2, a technological research and study satellite, was launched piggyback on a Molniya satellite from Plesetsk on June 5, 1975. The 29.6kg satellite was used to test the passive cryogenic radiation system for Meteosat cooling, and to study of the aging of thermal casings and plastic films.

• The Soviet Union (A. Kosygin) and the USA (R. Nixon) signed formal agreements of cooperation on May 24, 1972 (of talks that started in 1969) in Moscow, concerning the Apollo-Soyus Test Project (ASTP), leading to a common spaceflight (docking of both spacecraft, Apollo 18 and Soyus-19) on July 17, 1975 (the flight was from July 15 - 24). This represented the first international meeting of men in orbit. The main objects were to test the compatibility of rendezvous and docking systems for American and Soviet spacecraft, to open the way for international space rescue as well as future joint manned flights. The Apollo spacecraft was nearly identical to the one that orbited the Moon and later carried astronauts to Skylab. The Soyuz craft was the primary Soviet spacecraft used for manned flight since its introduction in 1967. A docking module was designed and constructed by NASA to serve as an airlock and transfer corridor between the two craft.

- First Shuttle docking at the MIR space station. A new era of international cooperation in space began on June 30, 1995 with the docking of both S/C. The event occurred on Shuttle flight STS-71 (Atlantis, June 27 - July 7, 1995).

- International partnerships have also been very successful in solar and space physics. NASA is the lead agency for the GGS (Global Geospace Science) program which includes the WIND and POLAR satellites, both of which have important international components. Conversely, ESA is the lead agency for Ulysses, SOHO (Solar and Heliospheric Observatory) and Cluster, while ISAS of Japan is the lead for Yohkoh (Solar-A), and Geotail. Within STP (Solar Terrestrial Probes) program, NASA has the lead with TIMED and STEREO while ISAS has the lead of Solar-B.

- In the latter part of the 1990s, the cooperations between partners in the Earth observation community have reached new dimensions. They are truly global in nature - a network of interrelations - they are so numerous and on so many levels (permeating many facets of society and affecting our every-day lives) - too complex for the scope of this writing.

- Public/private partnerships are vital for the continued growth and commercialization of the space sector. Historically, spaceborne systems have become a business only after the high-risk technologies and markets have been developed, most often through government initiatives and with public funding. First government/industrial project partnerships in the area of Earth observation were being introduced as early as 1985. The overall objective is cost-sharing of ever tighter government space budgets with commercial companies in projects that require new technology introduction. The investing companies are given some incentives (commercial data rights and/or ownership of the S/C, etc.) to recover their investments and to make a profit. An effective framework in such partnerships includes benefits to all partners. Some examples in satellite development are: [357) 358)]

  - In Sept. 1985 the US company Eosat was selected by NOAA (government) to operate the Landsat system (LS-4 and LS-5), to market LS data, and to build and launch LS-6.
  - SPIN-2 venture of Russia (since 1992). SPIN-2 is a joint venture (company), located in Washington DC, of Interbranch Association SOVINFORMSPUTNIK (Moscow, Russia), Aerial Images, Inc. (Raleigh, NC), and Central Trading Systems, Inc., (Huntington Bay, NY). The objective is to market high-resolution panchromatic imagery data (2 m) of past Russian missions, in particular data from the Resurs-F series.
  - RADARSAT-1 (launch Apr. 11, 1995) and RADARSAT-2 (launch in 2003) are jointly-funded SAR missions of CSA (Canadian Space Agency) and MDA (MacDonald Dettwiler Associates Ltd. of Richmond, BC). CSA is providing approximately 75% of the funding for the development of the satellites and MDA is investing the difference. MDA owns and operates the satellites. CSA's investment will be recovered through the supply of imagery to a number of Canadian government agencies during the mission lifetime.
  - OrbView-1/Microlab-1 and OrbView-2/SeaStar are commercially built and operated small satellites of OSC (Dulles, VA) flying government-sponsored instruments, OTD

357) D. L. Glackin, "International Earth remote sensing: overview 1980-2010," Proceedings of the SPIE International Symposium on Optical Science and Technology, San Diego, July 2000

358) "International Space Cooperation: Solving Global Problems," Report of an AIAA, UN/OOSA, CEAS, CASI Workshop, April 1999 (printed and distributed by AIAA)

of NASA/MSFC and GPS/MET of UCAR in the case of OrbView-1, and SeaWiFS of GSFC on OrbView-2. Data of these instruments is provided to government agencies as well as commercially sold (SeaWiFS).

- In Europe, CNES is introducing cost-sharing programs with commercial S/C builders. The SPOT-5 satellite development is such a joint venture of CNES with French industry, namely MMS (Matra Marconi Space), now Astrium SAS. Earlier S/C in the SPOT series of CNES (government) started data distribution arrangements with private companies like Spot Image.
- The US NEMO (Navy EarthMap Observer) satellite is sponsored by ONR (Office of Naval Research) and DARPA as well as with commercial investments. A consortium led by STDC (Space Technology Development Corporation) of Alexandria, VA is the S/C operator and commercial distributor of the imagery.
- The LightSAR project of NASA was conceived as a public/industry partnership. However, in July 1999, NASA cancelled the project due to lack of interest from industry. Industry officials could not see a sufficiently large commercial market for L-band imagery (a NASA requirement) to justify the required investments. [359)]
- ISRO of Bangalore, India, [360)] announced in Nov. 2000 a new policy by inviting private sector investments with corresponding customer equipment and service-sharing arrangements on its GEO program of INSAT satellites. User demand is mostly expected in the field of communications.
- TerraSAR alliance. As of 2001, Infoterra/TerraSAR is a proposed cooperative Public-Private Partnership (PPP) program between the Infoterra Company (Astrium) and ESA. The fundamental objective of the Infoterra/TerraSAR initiative is to establish a self-sustaining geo-information business built on European strengths in SAR satellite technology, in SAR applications expertise and in the provision of services based on Earth observation data sources. TerraSAR is planned to be an element of the ESA Earth Watch program. The initial program calls for two satellites, featuring high-resolution X-band and L-band SAR imagery, respectively. A launch of TerraSAR-X1 and TerraSAR-L1 are planned for 2006.
- COSMO-SkyMed (Italy) and Pléiades (France) program alliance. France and Italy signed an agreement on Jan. 29, 2001 to jointly develop four radar satellites and two optical satellites. The first two COSMO-SkyMed high-resolution radar satellites will be launched in 2004. Pléiades is the SPOT successor program of CNES. The first two high-resolution optical satellites are planned for launch in 2005 and 2006. COSMO-SkyMed as well as Pléiades are also proposed PPP programs within ESA's Earth Watch and GMES initiatives.

359) W. Ferster, "Weak Industry Response Brings End to LightSAR," Space News, Aug. 9, 1999, p. 3

360) K. S. Jayaraman, "India Plans New Insat Design Around Private Sector Needs," Space News, Dec. 18, 2000, pp. 4, 44

Part A Atmosphere/Radiation/Aeronomy Missions

A.1 ACE (Atmosphere Climate Experiment)

An ESA Explorer mission, proposed in 1998, with the overall objective to obtain profiles of atmospheric parameters with occultation measurements using **a constellation of six LEO microsatellites.** The concept involves the systematic gathering of data over a five-year period. The profiles are used in such applications as climate modeling and climate prediction techniques to improve the understanding of the driving forces behind climate change and variability. The ACE science team is composed of researchers from the Meteorological Institutes in Denmark, Sweden, France and the United Kingdom. The ACE project is in phase A as of 2001. [361) 362) 363) 364) 365)]

The baseline configuration of the cluster considers two orbital planes. The six microsatellites are equipped with the GRAS receiver, flown on all S/C; they employ the technique of refractive radio occultation measurements of the GPS constellation, providing profiles of temperature, pressure, and humidity.

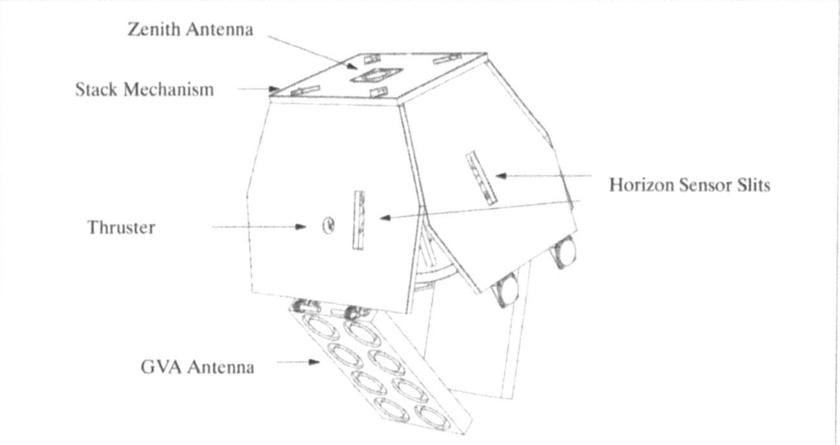

Figure 6: Illustration of the ACE spacecraft

The S/C bus is based on the SMART-1 bus of SSC (Swedish Space Corporation). The basic S/C body is trapezium-shaped with a size of 1 m x 0.62 m and a height of 0.6 m. Each S/C is nadir pointing and three-axis stabilized with a momentum bias system. Attitude is sensed by an Earth horizon sensor and a magnetometer; actuation is provided by a momentum wheel and three magnetorquers. The pointing accuracy is 1°, the rate error<0.025°/s. The S/C

361) S. M. Veldman, K. Lundahl, "Atmospheric Climate Experiment ACE - a Constellation of Microsats for Atmospheric Sounding," ISU International Symposium on Smaller Satellites: Bigger Business? May 21-23, 2001, Strasbourg, France

362) P. Hoeg, J. Guldberg, K. Lundahl, "Atmosphere Climate Experiment," IAA 2nd International Symposium on Small Satellites for Earth Observation, Berlin, April 12-16, 1999, pp. 343-346

363) K. Lundahl, S. Veldman, "Atmospheric Climate Experiment ACE," Proceedings of the 51st International Astronautical Congress, Rio de Janeiro, Brazil, Oct. 2-6, 2000, IAF-00-B.2.08

364) http://www.ssc.se/ssd/msat/ace/ace.html

365) S. Veldman, K. Lundahl, P. Hoeg, F. Hass, P. Sinander, "Atmospheric Climate Experiment ACE," Proceedings of the 3rd International Symposium of IAA, Berlin, April 2-6, 2001, pp. 95-98

mass is 87 kg, the solar power is 65 W (EOL) as orbital average provided by regulated solar arrays (unregulated bus of 28 V ±4V). An NiCd battery of 6 Ah is used for solar eclipse phases. Orbit control is provided by an on-board propulsion system for constellation maintenance. The ACE avionics system is also based on the SMART-1concept. This includes CCSDS compatible TM/TC, CAN bus for internal data handling, failure-tolerant design, FPGA (Field Programmable Gate Arrays) and an ERC-32 processor. The ERC-32 computer is able to support a number of on-board processing functions including science data processing.

Communication is provided by an S-band system with data rates of 4.8 kbit/s in uplink and 1 Mbit/s in downlink. An on-board storage capacity of 2 Gbit is available. The mission control is based on SCOS (Satellite/Mission Control System), a generic software package developed through ESA (by Terma A/S of Birkeroed, Denmark) and configured for several on-going ESA missions. SCOS 2000 has the functionality to support automated execution of satellite, the ground segment commanding, evaluation of the S/C housekeeping telemetry, and the ground segment monitor function.

A launch of the ACE constellation is planned for 2005. Two launches inject the satellites (stacked launch of three S/C) into two perpendicular near-polar orbits.

Orbit: The six microsatellites fly in two orthogonal orbital planes (three satellites per plane) with an altitude of 700-800 km; inclination =80-100°. An even time coverage is obtained with two orbits with each S/C sequentially spaced at 90°. In this arrangement each longitude can be revisited after six hours.

Sensor complement:

GRAS (GNSS Receiver for Atmospheric Sounding) of MetOp-1 heritage but with considerably reduced mass and power consumption. The instrument will be built by Saab-Ericsson Space and Austrian Aerospace. The instrument uses the occultation principle where the active instrument is mounted on a LEO satellite, and the monitored signal comes from the GPS and GLONASS constellations. The receiver will lock on transmitted GPS signals and monitor the phases of these signals. As the GPS satellite sets below (or rises above) the horizon, the influence of various layers in the atmosphere can be determined in the form of vertical profiles of density, pressure, humidity and temperature. GRAS is a 16-dual-frequency-channel GPS occultation instrument. It comprises two subsystems: the antenna subsystem with three antennas, GVA (GRAS Velocity Antenna), GAVA (GRAS Anti-Velocity Antenna), and GEU (GRAS Electronic Unit) with RF (Radio Frequency) front-end, RF/IF, A/D converter, frequency generation, channel processing, digital signal processor and power supply (see GRAS description under G.2). The GVA and GAVA are directed in the velocity and anti-velocity directions of the spacecraft, mechanically tilted by an angle of 30° below the horizontal plane to point to the Earth horizon. The zenith antenna is used for precise orbit determination. To enable the high degree of integration, the third generation AGGA-3 (Advanced GPS/GLONASS ASIC) chip will be developed. Each GRAS instrument has a mass of 8.2 kg with a power consumption of 20 W.

Ground segment: The ground segment is composed of two main facilities, the SDAC (Satellite Data Acquisition and Control) network and DPMCC (Data Processing and Mission Control Center). The SDAC consists of the main tracking and control station on Svalbard and the NMC (Network Management Center) at Kiruna, Sweden. The NMC supports alternative tracking and control stations especially during LEOP (Launch and Early Orbit Phase).

The DPMCC implements three main functional groups. The first group is mission control, which is based on SCOS 2000, a generic product developed through ESA. The second group is "application data acquisition and processing." Data acquisition includes satellite data ingestion, validation and reformatting and also retrieval and similar processing of data from ground based sources such as fiducial stations, such data being essential for the correction of occultation data before derivation of refraction profiles. Processing involves the use of a precise orbit determination tool based on the Bernese tool from the University of Bern, Switzerland, for the calculation of the orbits of the constellation satellites. This tool has been developed in the context of the ACE mission and the MetOp mission. In addition a clock extraction tool is developed for the precise estimation of the GPS and constellation satellite oscillator clocks. The third group is "data distribution and archiving," which makes different levels of data products available to meteorologists, e.g., for assimilation in weather prediction and climate models.

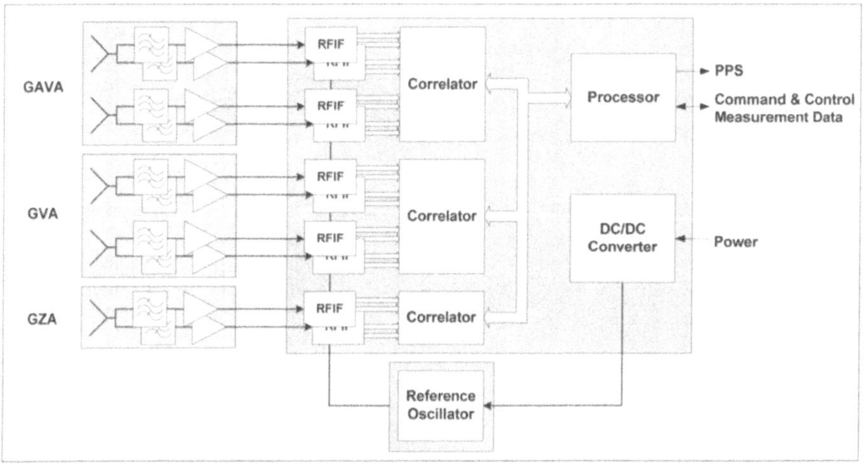

Figure 7: **Electronic unit of GRAS**

A.2 ACRIMSAT (Active Cavity Radiometer Irradiance Monitor)

ACRIMSAT is a NASA funded minisatellite to monitor the amount of total solar energy received (part of EOS program). The spacecraft builder/integrator is OSC of McLean, VA. The structure is based on the Ministar bus. In stowed configuration the S/C dimensions are: 79 cm in diameter x 69 cm in height. The S/C is sun pointing. It carries a flight processor used on previous missions. S/C mass = 213.3 kg, power =47 W. The mission lifetime is 5 years. The communications links are in S-band, uplink frequency at 2065.5 MHz, uplink data rate at 2 kbit/s; downlink frequency at 2250.0 MHz with data rates of 3.6, 28.8, 57.6 or 115.2 kbit/s. [366) 367) 368)]

A launch of ACRIMSAT took place on Dec. 20, 1999 on a Taurus launch vehicle (together with KOMPSAT-1).

Orbit: Sun-synchronous circular orbit, altitude = 685 km, inclination = 98.13°, equator crossing at 10:50 AM on descending node.

Sensor complement:

ACRIM-III (Active Cavity Radiometer Irradiance Monitor), built at JPL. The ACRIM-III instrument has a heritage of ACRIM-I, flown on the SMM satellite (launch Feb. 14, 1980)

366) Information provided by A. McLean of NASA/JPL
367) http://acrim.jpl.nasa.gov/mission/missionindex.html
368) http://acrim.jpl.nasa.gov/

and ACRIM-II, flown on UARS (launch Sept. 1991). The objective is to monitor the variability of total solar irradiance (TSI) with state-of-the-art accuracy and precision, thereby extending the high-precision database compiled by NASA since 1980 by other ACRIM experiments as part of the Earth Radiation Budget Program in the National Climate Program. Note: The first ACRIM sensor was flown on the Solar Maximum Mission (Launch: Feb. 14, 1980), follow-up missions: UARS, Spacelab-1, and ATLAS.

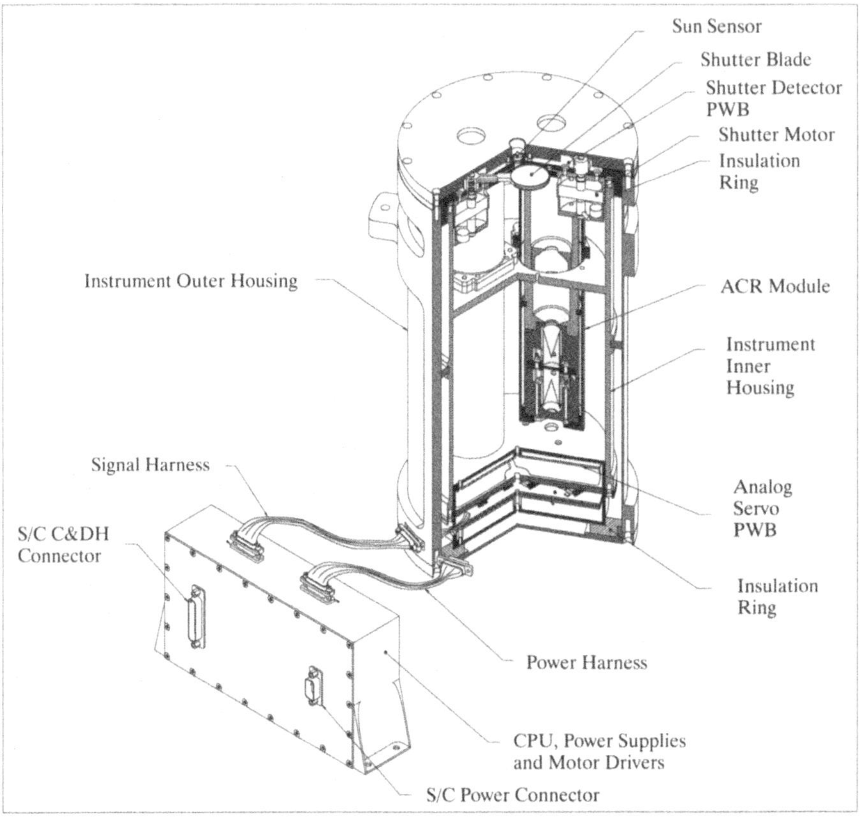

Figure 8: Illustration of the ACRIM-III instrument

Spectral coverage	0.2 - 2000 nm
Observation coverage	Minimum of 55 minutes per orbit of full solar disk data
View angle (FOV)	5°
Pointing accuracy	±0.25° on S/C to sun line
Detectors (ACR)	Three, each with 30° right-circular cone painted specular black on the sun-viewing face
Accuracy	0.1% of full scale
Instrument mass, power	9.2 kg, 10 W
Instrument dimensions	Sensor housing: 18.5 cm diameter x 36 cm in length Central processing unit: 20.5 cm x 11.5 cm
Sampling interval	2 minutes (approx.)
Data rate	600 bit/s (about 600 kBytes of data are collected per day)

Table 50: Specification of the ACRIM instrument

The ACRIM-III sensor points toward the sun; the assembly is mounted on a two-axis tracker to observe the solar disk during each orbit. The instrument contains three independent active cavity radiometer (ACR) solar-monitoring sensors and a sun-position sensor. One ACR monitors solar irradiance, the other two are used to calibrate the optical degradation of the first instrument.

A.3 ADM (Atmospheric Dynamics Mission)

ADM is an ESA Earth Explorer Core Mission (a science-oriented mission) with the primary objective to provide wind profile measurements for an improved analysis of the global three-dimensional wind field. Such knowledge is crucial to the understanding of the atmospheric dynamics, including the global transport of energy, water, aerosols, chemicals and other airborne materials - to be able to deal with many aspects of climate research and climate and weather prediction.

The ADM mission makes use of a single observation instrument, referred to generically as DWL (Doppler Wind Lidar). The retrieval of wind speed relies on direct measurement along the LOS (Line-of-Sight) by lidar using Doppler shift information from atmospheric molecules and particles advected by wind. [369)]

The following description reflects the status of definition at the end of the phase A study. [370)] The detailed design is under revision. The S/C structure, consisting of aluminum honeycomb elements, uses a conventional box-shaped spacecraft design with a central cone, upon which the observation instrument is mounted via three isostatic bipods. The electronic boxes of the bus and the associated satellite equipment are mounted on the side panels. The S/C is three-axis stabilized with AOCS (Attitude and Orbit Control Subsystem), using reaction wheels and magnetorquers as actuators, and magnetometers, star trackers, and a GPS receiver as sensors. The orbit is maintained by two 1 N thrusters. Electric power is provided by two fixed solar arrays of 8.4 m^2 of surface area (960 W power at EOL). A NiCd battery of 18 Ah is used for eclipse phases and the commissioning phase. The S/C mass at launch is 785 kg, including 100 kg of fuel. The design life is 3 years. [371)]

Type	Equipment	Nr.	Main characteristics	Technology	Redundancy
Sensors	Magnetometer	2	Three-axis	Fluxgate	Cold redundancy
	GPS antennas & pre-amps	2	Half-space coverage	Patch antenna	Cold redundancy
	GPS receiver	2	C/A GPS receiver	RF ASIC and micro processor	Cold redundancy
	Star tracker	2	Large FOV	CCD	Cold redundancy
	Inertial reference unit (IRU)	1	3 measurement axis	HRG	Internal
Actuators	Reactions wheels	4	10 Nms/0.1 Nm	Ball bearings	3 among 4
	Magnetic torquers	3	100 Am^2	Windings	Cold redundancy

Table 51: AOCS elements of ADM

On-board data handling can be performed by an ESA-developed ERC-32 radiation tolerant processor; the subsystems are linked via serial links to the central processor. A solid-state memory provides a capacity of 2 Gbit on-board data storage. TT&C communications are based on standard S-band links, the uplink data rate is 2 kbit/s the downlink data rate is up to 64 kbit/s. The science data are dumped via a HRPT (High-Resolution Picture Transmission) link, as defined for METOP, in L-band at a data rate of 3.5 Mbit/s (an HRPT sta-

369) P. Ingmann, J. Fuchs, J. Pailleux, A. Stoffelen, "The Atmospheric Dynamics Mission," ESA Earth Observation Quarterly, No 66, July 2000, pp. 12-17

370) "Atmospheric Dynamics Mission," ESA publication SP-1233 (4), July 1999

371) http://www.estec.esa.nl/explorer/

tion is a small remotely controlled facility with an antenna disk of 2.4 m and a G/T better than 6 dB/K). S/C operations are performed at ESOC using the Kiruna TT&C station. Two L-band stations stations at Kiruna (67.88° N, 20.25° E) and Barrow, Alaska (71.3° N, 156.7° W), are considered for science data dumps.

A Rockot launch vehicle has been baselined for a launch in 2006/7.

Orbit: Sun-synchronous orbit, altitude = 408 km (mean), inclination = 96.99°, local equator crossing time at 6:00 and 18:00 (dawn-dusk orbit).

The DWL (Doppler Wind Lidar) operation principle of ALADIN: The DWL sensor is an active instrument which fires laser pulses towards the atmosphere and measures the resulting Doppler shift of the return signal, backscattered at different levels in the atmosphere. The frequency shift results from the relative motion of the scatter elements along the sensor line of sight. This motion relates to the mean wind in the observed volume (cell). The measurement volume is determined by the maximum ground integration length of 50 km, the required height resolution and the width of the laser footprint. The measurements are repeated at intervals of 200 km.

Light is scattered either by interaction with aerosol or cloud particles (Mie scattering) or by interaction with air molecules (Rayleigh scattering). The two scattering mechanisms exhibit different spectral properties and different wavelength dependencies such that instruments evaluating only one signal type or both in separate processing chains can be constructed.

There are two detection techniques to measure the Mie and Rayleigh backscattering effects, namely a coherent-detection heterodyne system, and an incoherent or direct-detection interferometric system. The operation of each system exhibits profound differences in principle (see Chapter O.4.1 for heterodyne definitions).

- Coherent-detection heterodyne systems combine the weak (observed) optical signal at nominal frequency f_1 with a strong optical reference beam, the frequency-stable LO at frequency f_2, on a wideband square-law detector, thereby producing radio frequency beats at the frequency difference $f_1 - f_2$. The resultant beat-frequency signal is analyzed in a post-detection step to provide the Doppler frequency.
- Incoherent-detection heterodyning refers to direct detection of an optical signal on an optical detector, with no LO present. The backscattered optical signal field is analyzed and dispersed in an interferometric filter (or in diffraction grating) prior to detection. The measurement accuracy of a direct-detection interferometric system depends only on the total scattered signal, it is not dependent on the energy of individual pulses, but on the total laser energy.

Observation configuration: The satellite is flown with the instrument pointing toward Earth in a plane quasi-perpendicular to the flight path and 35° offset from nadir in the anti-sun direction. The measurement geometry is depicted in Figure 606. The LOS is oriented such that the relative velocity at the intersection with the Earth is zero (yaw steering). All measurements are taken along the LOS. The Doppler shift of the backscatter signal reflects the relative wind speed along the LOS and has to be processed to a horizontal wind speed component (HLOS) referenced to the ground.

The measurement volume of the return signal from a single shot is defined by the lateral extension of the transmitted beam (a few meters in diameter) and the time gating of the receiver, which is adopted to the desired vertical resolution (several 100 m to 2 km). Due to the fact that the signal from a single shot is too weak for the evaluation, several shots along a ground track of up to 50 km have to be accumulated and integrated.

Measurement profile: The on-board instrument is operated at a duty cycle of 25% to obtain wind profile separation. An active operation cycle lasts 6.93 seconds (equivalent to a 50 km ground track), followed by a gap in observations of 20.79 seconds (equivalent of 150 km ground track). Winds can be measured in clear air (i.e., above or in the absence of thick clouds), and within and through thin clouds (e.g., cirrus).

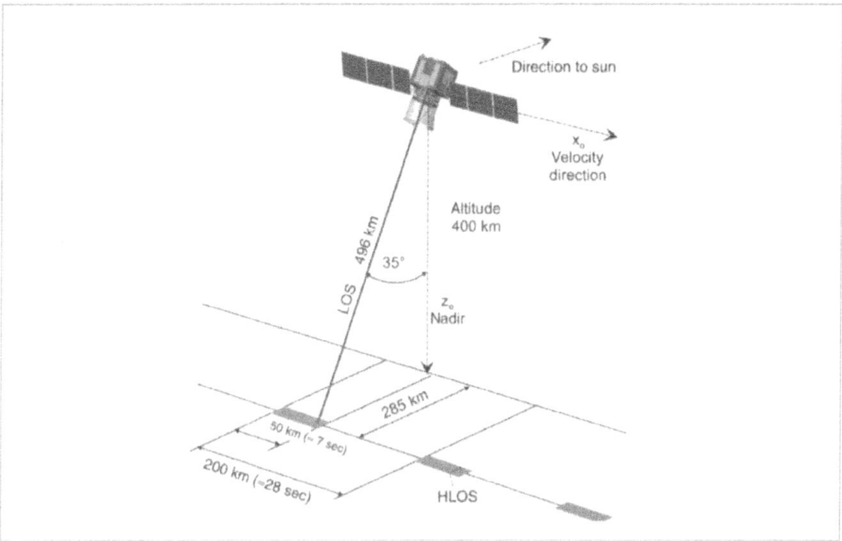

Figure 9: **Nominal measurement geometry and coverage of ADM**

Sensor complement:

ALADIN (Atmospheric Laser Doppler Instrument). [372) 373)] ALADIN is an incoherent direct detection lidar incorporating a fringe-imaging receiver (analyzing aerosol and cloud backscatter) and a double-edge receiver (analyzing molecular backscatter). The lidar emits laser pulses towards the atmosphere, then acquires, samples, and retrieves the frequency of the backscattered signal. The overall ALADIN instrument architecture is based on a 130 mJ diode-pumped frequency-tripled Nd:YAG laser operating in the ultraviolet (solid-state laser technology); the instrument consists of three major elements: a transmitter, a combined Mie and Rayleigh backscattering receiver assembly, and the opto-mechanical subsystem (a 1.1 diameter telescope).

Transmitter configuration: The power laser is composed of a medium power oscillator, a single amplifier, and a frequency tripler. The oscillator is actively Q-switched by a Pockels cell. A seed laser is used as frequency reference. The injection seeding technique is used to achieve a single frequency mode with a low-power continuous wave (CW) single frequency laser. The power laser is conductively cooled via heat pipes.

Receiver configuration: A combined Mie and Rayleigh backscattering receiver is implemented.

- The Mie receiver consists of a Fizeau spectrometer. The received backscatter signal produces a linear fringe whose position is directly linked to the wind velocity. The resolution of the Fizeau interferometer is 100 MHz (equivalent to 18 m/s). The wind value is determined by the fringe centroid position to better than a tenth of the resolution. The backscattered signals are detected by a thinned back-illuminated silicon CCD detector working in an accumulation mode which allows photon counting.

372) F. Fabre, A. Heliere, et al., "Direct Detection Doppler Wind Lidar Prototype: Design and Preliminary Results," Proceedings of the 2000 EUMETSAT Meteorological Satellite Data Users' Conference, Bologna, Italy, May 29-June 2, 2000, pp. 239-242

373) D. Morancais, F. Fabre, "Incoherent Doppler Wind Lidar ADM Concept and Related Prototype," Proceedings of the EUMETSAT Meteorological Satellite Data User's Conference, Copenhagen, Denmark, Sept. 6-10, 1999, pp. 85-92

- The Raleigh receiver employs a double Fabry-Perot etalon with a 2 GHz resolution and 5 GHz spacing. It analyzes the wings of the Rayleigh spectrum with a CCD. The etalon is split into two zones, which are imaged separately on the detector. The wind velocity is proportional to the relative difference between the intensities of the two etalons.

The optomechanical subsystem of ALADIN uses a Cassegrain afocal telescope for both functions of laser emission and backscatter reception. The telescope design employs isothermal and lightweight techniques based on e.g. SiC (Silicon Carbide) type ceramic mirrors and structures. This concept provides the needed optical quality and stability without a focusing or alignment mechanism. Star trackers for attitude sensing are mounted on the telescope structure to minimize the misalignment between the optical axis and the telescope's line-of-sight.

	ADM requirements			ADM predicted performances		
Parameter	Planetary Boundary Layer (PBL)	Tropo-sphere	Strato-sphere	PBL	Tropo-sphere	Strato-sphere
Vertical domain	0-2 km	2-16 km	16-20 km	0 - 2 km	2 - 16.5 km	16.5-26.5 km
Vertical resolution	0.5 km	1.0 km	2.0 km	0.5 km	1.0 km	2.0 km
Horizontal domain	global coverage			80° S to 85° N		
Number of profiles/hour	100			100		
Profile separation	> 200 km			> 200 km		
Temporal sampling	12 hours			12 hours		
Accuracy of wind velocity	2 m/s	2-3 m/s	3 m/s	2 m/s	2-3 m/s	3 m/s
Horizontal integration	50 km			50 km		
Error correlation	0.01			0.01		
Reliability	95%			>95%		
Timeliness of data availability	3 hours			3 hours		
Length of observational data set	3 years			3 years		

Table 52: Observational requirements and performance of ADM

Instrument type	Diode-pumped Nd:YAG lidar with active Q-switch
Transmitter	
Emission wavelength	355 nm
Pulse energy	130 (150 mJ goal)
Pulse Repetition Frequency (PRF), pulse width	100 Hz, 15 ns
Line width	30 MHz
Duty cycle	25%
Receiver	
Line width of Fizeau spectrometer	30 MHz (Mie scattering)
Line width of double Fabry-Perot receiver	2 GHz (Rayleigh scattering)
Spacing of double Fabry-Perot receiver	5 GHz (Rayleigh scattering)
Optical efficiency (Mie/Rayleigh receivers)	3.1% / 4.6%
Detector quantum efficiency (Mie/Rayleigh)	75%
Signal detection (Mie receiver)	Silicon CCD detector in accumulation mode
Signal detection (Rayleigh receiver)	Silicon CCD detector and two read-outs
Signal processing capabilities	
Altitude range (Mie + Rayleigh)	-1 to 26.5 km (extendable)
Vertical resolution	1 km (adjustable)
On-chip horizontal accumulation length	3.5 km (adjustable) in along-track direction
Processing integration length	50 km
Opto-mechanical subsystem	
Telescope diameter, f number	1.1 m aperture Cassegrain type telescope, f/0.9
Optical efficiency	0.8
Instrument mass, power	263 kg, 304 W average power (25% duty cycle)
Instrument data rate	11 kbit/s (max)

Table 53: Major instrument parameters of ALADIN

The instrument transmits raw source data consisting of the accumulated spectra from the Mie receiver and the flux intensities from the Rayleigh receiver. These data are provided for strips of 50 km length and a horizontal resolution down to 3.5 km. In the vertical direction, many layers or volume cells of the various altitude bins (-1 km to 16.5 km height for the Mie channel, and 0.5 km to 26.5 km for the Rayleigh channel) are measured; the instrument looks into a fixed direction (quasi perpendicular to the flight path and 35° away from nadir) and provides a vertical wind profile along the line of sight. In addition to these source data, laser internal calibration and attitude data are transmitted, as well as the receiver response calibration data.

The instrument performance considers the SNR error for each channel at the indicated altitude range. In addition, systematic bias errors are taken into account. When no ground echo is retrieved, the measurement bias is not cancelled; the total measurement error is slightly deteriorated. - For the Mie channel, the LOS (Line-of-Sight) wind error is below the requirement of 0.6 m/s for altitudes from o to 2 km in height. For the Rayleigh channel, the LOS wind error is below the requirement (except a marginal performance around 16 km).

	Mie channel @ 2 km	Rayleigh channel @ 10 km	Comment
Measurement accuracy (noise+bias)	0.67 m/s 0.85 m/s	1.05 m/s 1.23 m/s	LOS with ground echo LOS without ground echo

Table 54: ALADIN instrument overall LOS performance

A programmable sequencer is implemented for the detector permitting configuration changes with regard to vertical altitude resolution and range coverage. The vertical resolution can be varied from 250 m to 2 km or more. However, the measurement accuracy is only obtained for the nominal vertical resolution of 1 km. The altitude range is limited to 35 km. The horizontal (along-track) accumulation length can also be changed (nominal of 3.5 km) - from 100 m (Mie or Raleigh) to 50 km without significant performance degradation for the required along-track integration length of 50 km.

In addition to the line-of-sight (LOS) velocity measurements, ALADIN is able to provide information on cloud characteristics over the depth of the atmosphere, as well as aerosol measurements in the troposphere. These include:

- Cloud top height (notably cirrus top and base)
- Cloud cover
- Cloud and aerosol extinction and optical thickness
- Identification of multi-layer clouds
- A measure of clear air turbulence or wind variability
- Lower troposphere aerosol stratification
- The height of the tropopause
- The height of the PBL (Planetary Boundary Layer).

A.4 AE (Atmosphere Explorer)

A NASA satellite program (with two aeronomy and three atmosphere S/C) within the early Explorer Program. The main objective was the study of the upper atmosphere (ionospheric and thermospheric densities, temperatures, winds/drifts, and radiative emissions). [374]

The satellites AE-C, -D, and -E were flying nearly identical in-situ experiments in complementary orbits, thereby producing a global database of many ionospheric and thermospheric parameters. Mass spectrometers were flown in each mission to determine the response of the lower thermosphere and ionosphere to radiation. In addition to determining the gas

374) http://nssdc.gsfc.nasa.gov/nmc/sc–query.html

composition, consisting of helium, atomic oxygen, atomic nitrogen, molecular nitrogen, molecular oxygen, and argon, a modification of the mass spectrometer instrument allowed wind and gas kinetic temperature to be measured over an altitude range of 130 to 500 km.

A.4.1 AE-A (Aeronomy-A, Explorer 17)

A NASA/GSFC aeronomy satellite, a spin-stabilized sphere of 0.95 m in diameter. The S/C was vacuum sealed to prevent contamination of the local atmosphere. The battery power failed on July 10, 1963. Launch of AE-A (Explorer 17) from Cape Canaveral on a Thor-Delta vehicle on April 3, 1963. The AE-A S/C mass was 183.7 kg.

Orbit: Elliptical orbit, perigee = 255 km, apogee = 916 km, inclination = 57.6°, eccentricity = 0.04743, period = 96.39 minutes.

Sensor complement:

Neutral Mass Spectrometer (PI: C. A. Reber). Two identical double-focusing magnetic mass spectrometers were used. Objective: measurement of the concentrations of the major neutral particle constituents of the upper atmosphere - atomic and molecular oxygen, atomic and molecular nitrogen, helium, and water vapor. The neutral particles were ionized by electron bombardment. Measurements of the six different ion currents and the total current were made sequentially for four seconds in high sensitivity and for four seconds in low sensitivity (64 s for the entire measurement cycle). Included in the cycle was an operation to correct for any dc drift of the zero voltage level in the output signal. One spectrometer produced useless data due to malfunction. The other detector system experienced intermittent degeneration of the amplifier output. Consequently, the data were good only during certain periods.

Langmuir Probe (L. H. Brace). Two Langmuir probe systems were flown. One was used to provide measurements of the positive ion density, the other measured the electron temperature. Each system used a two-element sensor consisting of an outer cylindrical guard electrode of 10 cm length which was concentric with an inner collector electrode of 0.056 cm in diameter and 23 cm in length. The potentials of the electrodes were varied with respect to the satellite shell. The electron temperature probe was swept at 10 sweeps per second over two voltage intervals, 0 to 0.75 V and 0 to 1.5 V. The ion density probe was swept from -3 to +2 V in 2 seconds. The instrument operated nominally until July 10, 1963, when the S/C batteries failed.

Pressure Gauge (G. P. Newton). The instrument consisted of two Redhead (cold cathode) and two Bayard-Alpert (hot filament) ionization vacuum gauges to measure the neutral particle density and ambient pressure of the upper atmosphere between 260 km and 960 km (pressure range from 10^{-4} torr to 10^{-11} torr). The instruments were operated for 4-minute periods during a pass over a ground station. The neutral particles were ionized by electron bombardment, the resulting ion currents were detected and converted to voltages suitable for telemetry.

A.4.2 AE-B (Aeronomy-B, Explorer 32)

A NASA/GSFC aeronomy satellite with the objective to measure temperatures, composition, densities, and pressures in the global upper atmosphere. The S/C was a spin-stabilized stainless steel (spin axis normal to the orbit plane, spin rate = 30 rpm), vacuum-sealed sphere of 0.9 m in diameter. The on-board equipment included optical and magnetic aspect sensors, magnetic attitude and spin-rate control systems, and a tape recorder. Power was provided by silver-zinc batteries and a solar array. Two identical pulse-modulated telemetry systems and a canted turnstile antenna were used for communication. The S/C was launched from Cape Canaveral on a Delta vehicle on May 25, 1966. S/C mass = 224.5 kg.

Orbit: Elliptical orbit, perigee = 276 km, apogee = 2725 km, inclination = 64.67°, eccentricity = 0.15532, period = 116 minutes.

Sensor complement:

RF Ion Mass Spectrometer (PI: H. C. Brinton). Objective: Measurement of the ion species concentrations (atomic hydrogen, helium, nitrogen, and oxygen) in the topside ionosphere as a function of time, location, and solar and geomagnetic activity. The spectrometer consisted of a 5-3 cycle ceramic tube with 5 mm grid spacing and an external guard ring assembly. Two RF frequencies, 3.7 MHz and 9.0 MHz, were used to cover the ion mass range 12 to 19, and 1 and 4 atomic mass units (amu) assuring detection of the prime ionic constituents of the topside ionosphere. A mass scan was done in 208 seconds followed by a recycling of the sweep voltage and a second measurement of the high mass range. The stopping potential and the guard ring potential controlled the sensitivity of the spectrometer. The ion current of the spectrometer was measured by a series of five-decade amplifiers with a particle sensitivity range from about 10 to 10^6 ions/cm^3. An automatic calibrator functioned once during each turn-on to supply two known signals to the amplifier system and to the sweep monitor. - The useful S/C lifetime of 10 months permitted a global study of the diurnal variation of the atmosphere during nearly two complete diurnal cycles, since the orbit plane precessed one revolution each 5.5 months.

Neutral Particle Magnetic Mass Spectrometer (PI: C. A. Reber). Two double-focusing mass spectrometers were used to measure the composition of the neutral atmosphere between 285 km and 1000 km. One instrument was mounted on the equator the spherical S/C normal to the spin axis, the other instrument was mounted on the top of the S/C, parallel to the spin axis. The neutral particles were ionized by electron bombardment and separated according to the mass-to-charge ratio (m/q) in the analyzer of the instrument (one collector cup for each of the seven ion species). An electrometer amplifier with two sensitivity ranges differing by a factor 100, sampled the seven collectors sequentially. The dwell time on a specific mass and sensitivity range was 2.4 s. The first four of the fifteen 2.4 s steps of a cycle were used to correct any zero drift of the electrometer and to record the low- and high-sensitivity zero levels. The ion currents were then measured in the high sensitivity for m/q equal to 2 (molecular hydrogen), 4 (helium), and 14 (atomic nitrogen), and in high and low sensitivity for m/q equal to 28 (molecular nitrogen), 32 (molecular oxygen), 16 (atomic oxygen), and 18 (water vapor). The time for a complete cycle was 36 seconds. - Electronic malfunctions caused one instrument to fail after four days in orbit, the other failed after 7 days.

Atmospheric Drag (PI: Wulf-Mathies). The spherical shape of the S/C permitted to determine the upper atmospheric density as a function of altitude, latitude, season, and solar activity. Density values near perigee were deduced from sequential observations of the S/C position, using optical and radio and/or radar tracking methods.

Electron Temperature and Density (L. H. Brace). The objective of this experiment was to measure the distribution of electron temperature and densities from 10^3 to 10^6 electrons/cm^3 using a swept voltage electron tube.

Pressure Gauge (G. P. Newton). Three cold-cathode magnetron type density gauges (Redhead ionization gauges), each which its own high voltage supply and output electrometer, were flown to measure the density of the neutral atmosphere as a function of altitude, time, latitude, and solar and geomagnetic activity. One gauge was designated as NRC-528, the other two as GCA-R5 to reflect different origins. Mounted on the satellite equator was one gauge of each designation, with the third gauge mounted 55° above the equator. The metal–ceramic GCA-R5 gauges had an internal magnetic field of about 0.1 T and contained radioactive material deposited on the anode to permit operation at low atmospheric densities ($<10^{-17}$ g/cm^3) with the anode potential fixed at 3500 V. The GCA-R5 equatorially mounted gauge had a linear range switchable electrometer output, and high-resolution current measurements were obtained. The remaining two gauge outputs were through logarithmic electrometers. All electrometers were calibrated once each turn-on. The time resolution of the measurements was 2 s, which was equal to the satellite spin period and corresponded to a spatial resolution of 6 km along the orbit path.

A.4.3 AE-C (Atmosphere Explorer-C, Explorer 51)

A NASA/GSFC atmosphere mission with the objective to investigate the thermosphere with emphasis on energy transfer and processes. The S/C was a multi-sided polyhedron with a diameter of 1.4 m. The S/C could be operated in either of two modes: spinning at a nominal 4 rpm or despun to 1 revolution per orbit. The spin axis was perpendicular to the orbit plane. Power was supplied by a solar cell array. The S/C used a PCM telemetry data system that operated in real time or in a tape recorder mode. The S/C mass was 658 kg including 85 kg of instrumentation. The S/C was launched from Cape Canaveral on a Delta vehicle on Dec. 16, 1973.

Orbit: Elliptical orbit, perigee = 149 km, apogee = 4,294 km, inclination =68.1°, eccentricity = 0.24086, period = 132.30 minutes. - The initial elliptical orbit was altered many times in the first year of life by means of an onboard propulsion system employing a 3.5-lb thruster. The purpose of these changes was to alter the perigee height to 129 km. After this period, the orbit was circularized and was raised periodically to about 390 km when it would decay to 250 km altitude. During the first year, the latitude of perigee moved from about 10° up to 68° north and then down to about 60° south. During this period about two cycles through all local times were completed.

Sensor complement:

CEP (Cylindrical Electrostatic Probe - PI: L. H. Brace). Two identical instruments designed to o measure electron temperatures, electron and ion concentrations, ion mass, and spacecraft potential. One probe was oriented along the spin axis of the S/C (normally perpendicular to the orbit plane), and the other radially so that it could observe in the direction of the velocity vector once each 15-s spin period. Each instrument was a retarding potential Langmuir probe device that produced a current-voltage (I-V) curve for a known voltage pattern placed on the collector. Electrometers were used to measure the current. There were two systems of operation (one with two modes and another with three modes) using collector voltage patterns between plus and minus 5 V. Most modes involved an automatic or fixed adjustment of collector voltage limits (and/or electrometer output) such that the region of interest on the I-V profile provided high resolution. Each system was designed for use with only one of the probes, but they could be inter-switched to provide backup redundancy. The best measurements in the most favorable modes provided 1-s time resolution; electron temperature between 300 and 10^3 K (10% accuracy); ion density between 10^3 and 10^6 ions/cm^3 (10-20% accuracy); electron density between 50 and 10^5 electrons/cm^3; and ion mass at ion densities above 10^3 ions/cm^3. Each probe had a collector electrode extending from the central axis of a cylindrical guard ring. The 2.5 cm long guard ring was at the end of a 25 cm boom, and the collector extended another 7.5 cm beyond the guard ring. The boom, guard, and collector were 0.2 cm in diameter.

MESA (Miniature Electrostatic Analyzer - PI: K. S. W. Champion). An atmospheric density accelerometer with the objective to measure the neutral density of the atmosphere in the altitude range of 120 to 400 km from the measurements of satellite deceleration due to aerodynamic drag. The instrument consisted of three single-axis accelerometers, mounted mutually at right angles, two in the S/C x-y plane and the other along the z-axis. The instrument determined the applied acceleration from the electrostatic force required to re-center a proof mass. The output of the device was a digital pulse rate proportional to the applied acceleration. The measurements allowed determination of the density of the neutral atmosphere, monitored the thrust of the orbit-adjust propulsion system (OAPS), determined the satellite minimum altitude, measured spacecraft roll, and provided some attitude-sensing information. S/C nutations of less than 0.01° were monitored. The instrument had three sensitivity ranges: 8 x 10^{-4} Earth's gravity (g) in OAPS monitor mode; 4 x 10^{-5} g between 120 (± 2%) and 280 km (± 10%); and 2 x 10^{-6} g between 180 km (± 2%) and 400 km (± 10%). Numbers in parentheses represent errors; in addition, there may be a systematic error of up to plus or minus 5% due to drag coefficient uncertainty. The highest measurement altitude was determined assuming the instrument could sense to 0.2% of full scale.

PES (Photoelectron Spectrometer - PI: J. P. Doering). The experiment was designed to provide information on the intensity, angular distribution, energy spectrum, and net flow along field lines, of electrons in the thermosphere with energies between 1 and 500 eV. The instrument consisted of two identical oppositely directed hemispherical electrostatic analyzers, and 30 operating modes. Each spectrometer had a relative energy resolution of plus or minus 2.5% and a geometric factor on the order of 0.001 cm^2 sr, independent of electron energy. Three separate energy ranges could be sensed: 0 to 25, 0 to 100, or 0 to 500 eV. Measurements from these intervals could be sequenced in five different ways. Data could be taken from either sensor separately, or alternately with time resolution varying from 0.25 to 8 s. There were two deflection voltage scan rates determined by spacecraft clock. This voltage was changed in 64 steps, and was done at 4 or 16 steps per telemetry frame. With 16 frames/s, this allowed a choice of either one 64-point spectrum, or four 16-point spectra in 1 s. The longest (8 s) cycle of data involved observations using increasing voltage steps for the lowest, middle, lowest, then highest energy ranges (in that order) for 1 s each. A repeat for decreasing voltage step completed the cycle.

RPA (Retarding Potential Analyzer/Drift Meter - PI: W. B. Hanson). The experiment was designed to determine vector ion drift velocities, ion concentration and temperature, and S/C potential. An ionospheric irregularity index was also obtained from the ion concentration sensor. The experiment consisted of a retarding potential analyzer with four planar sensor heads. The sensor head used for ion drift measurements was collocated with another head, and all were spaced nearly equally, looking outward from the satellite equator. Since the satellite spin axis was perpendicular to the orbit plane, these heads could observe along the spacecraft velocity vector in either the spin or despun mode of the S/C. The primary objective of this experiment was to provide accurate ion temperatures with other measurements being of secondary importance. Three of the sensor heads were similar. They had two grounded entrance grids, two retarding grids, a suppressor grid, a shield grid, and a collector. A linear sweep voltage (32 or 22 to 0 V, up or down) was normally applied to the retarding grids in 0.75 s.

ESUM (Extreme Solar UV Monitor - PI: D. F. Heath). ESUM made absolute broadband spectro-radiometric measurements of the solar EUV flux from 200 Å to Lyman-alpha at 1216 Å and made precise measurements of the temporal variability - approximately 1% per solar rotation. The instrument consisted of two identical windowless EUV photodiodes with aluminum oxide cathodes and a filter wheel containing two sets of unbacked metallic filters (aluminum, tin, indium) and an open position. A visible light diode measured the pinhole transmittance of the filters to determine the white light background. The tilt angle of the instrument relative to the +z S/C axis was optimized for the maximum viewing time of the sun in both spinning and despun spacecraft modes. The instrument field of view was 60°. The nominal bandwidths (for 50% of signal) were 270 - 550 Å, 570 - 584 Å, 800 - 935 Å, and 1216 Å.

EUVS (Solar EUV Spectrometer - PI: H. E. Hinteregger). EUVS was used to observe the variations in the solar EUV flux in the wavelength range from 140 to 1850 Å and the atmospheric attenuation at various fixed wavelengths. This provided quantitative atmospheric structure and composition data. The instrument consisted of 24 grazing-incidence grating monochromators, using parallel-slit systems for entrance collimation and photoelectric detectors at the exit slits. Twelve of these monochromators had wavelength scan capability, each with 128 selectable wavelength positions, which could also automatically step scan through these positions. The other 12 monochromators operated at fixed wavelengths with fields of view smaller than the full solar disk to aid in the atmospheric absorption analysis. The spectral resolution varied from 2 to 54 Å depending upon the particular instrument. The FOV varied from 60 x 60 arcmin down to 3 x 6 arcmin. All 24 monochromator-entrance axes were co-aligned parallel. A solar pointing system could point to 256 different positions, execute a 16-step one-dimensional scan or a full 256-step raster. The time resolution varied from 0.5 s for observing 12 fixed wavelengths up to 256 s for programming the EUVS through all possible modes.

OSS (Open-Source Neutral Mass Spectrometer - PI: A. O. C. Nier). OSS mass = 7 kg, power = 10 W. The objective of this experiment was to contribute to a study of the chemical, dynamic, and energetic processes that control the structure of the thermosphere by providing direct, in situ measurements of both major and minor neutral atmospheric constituents having masses in the range from 1 to 48 atomic mass units (amu). A double-focusing Mattauch-Herzog magnetic deflection mass spectrometer with an impact ion source was flown. Two ion collectors were included to measure ions differing in mass by a factor of 8, i.e., the two mass ranges covered were 1 to 6 amu and 6 to 48 amu. In the ion source the neutral species were ionized by means of electron impact. At altitudes > 380 km, ion currents were measured with an electron multiplier counting individual ions. Counts were accumulated for 1/20 s before automatically switching to a different mass number. While complete mass spectra could be swept, in the common mode of operation peak stepping was employed, with readings on the principal peaks in the mass spectrum being repeated approximately every 0.5 s and on other species less frequently. Data below 380 km were measured using an electrometer. In addition to the peak stepping mode, there were several other operating modes which were selected by ground command. In the fly-through mode, the ion source voltages were adjusted so that there was no electric field to draw ions out of the electron beam when they were formed. Ambient particles striking the ion source retained energies less than 0.1 eV, which was not high enough to overcome the negative space charge potential holding the ions in the beam. Those ambient particles that did not strike the ion source retained their incoming energy of several eV after ionization and escaped into the accelerating region of the analyzer. The electron accelerating potential was 75 eV in normal mode operation and was 25 eV in the fly-through mode.

CSS (Closed-Source Neutral Mass Spectrometer -PI: D. T. Petz). CSS mass = 8 kg, power = 18 W. CSS measured in situ the spatial distribution and temporal changes of the concentrations of the neutral atmospheric species. In addition, new insight into in situ measurement techniques was obtained from comparisons of these measurements with those obtained from other on-board experiments; namely, OSS, EUVS, and MESA. CSS had a gold-plated stainless steel thermalizing chamber and ion source, a hyperbolic-rod quadrupole analyzer, and an off-axis electron multiplier. Five different sequences of mass selections were available and, expressed in atomic mass units (amu); they were (a) geophysical -1, 2, 4, total, 16, 28, 32, selected, 40, (b) analytical -12, 14, 18, 20, 22, 30, 44, calibrate, zero, (c) individual -selected, selected, selected, (any mass 1 to 44), (d) sweep digital -1, 2, 3, 4, 5, .. 45 (in 3/16 amu steps), and (e) sweep analog -2, 3, 4, 5, ... 45 (continuous). Spatial resolution was determined primarily by the mode of S/C operation. In orbit, the pre-sealed spectrometer was opened, and the atmospheric constituents passed through a knife-edged orifice into the thermalization chamber and ion source. Selected ions left the quadrupole analyzer through a weak focusing lens and were accelerated into a 14-stage electron multiplier, where they were turned 90° to strike the first dynode. For each impacting ion, the multiplier output was a pulse of 2×10^6 electrons.

NATE (Neutral Atmosphere Temperature Experiment - PI: N. W. Spencer). NATE mass = 7 kg, power = 10 W. NATE measured the kinetic temperature of the neutral atmosphere by determining the instantaneous density of molecular nitrogen in a spherical chamber coupled to the atmosphere through a knife-edged orifice. Analysis of the measured molecular nitrogen density variation over a spin cycle with a knowledge of the satellite's motion and orientation led to a determination of the ambient temperature, independent of scale height. A measurement of the ambient nitrogen density was also obtained. An alternate measurement of neutral temperature was also undertaken, using a baffle inserted in front of the orifice to intercept a portion of the gas particle stream entering the chamber. When the satellite was in the despun mode, the baffle was made to oscillate in the stepwise fashion to interrupt the particle stream seen by the orificed chamber. These chamber density variations were interpreted to yield the neutral gas kinetic temperature. A dual-filament ion source sampled the thermalized molecular nitrogen in the chamber and produced an ion beam density proportional to the nitrogen chamber density. From the source, this ionized nitrogen beam was directed from a quadrupole analyzer, tuned to pass those particles whose mass-to-charge ratio (m/q) is 28, on to an electron multiplier. The output pulses were amplified and counted in a 16-bit accumulator. The experiment also provided measurements of neutral atmospheric composition, when commanded into the appropriate mode and, for the first time measured the local wind (vertical motions). The wind values were determined by measurement of the "stream" position relative to the satellite velocity. When the S/C was in the despun mode, the nitrogen density was measured except when the particle stream was interrupted by the baffle.

MIMS (Magnetic Ion-Mass Spectrometer - PI J. H. Hoffman). MIMS mass = 7.5 kg, power = 6 W. The MIMS objective was to measure in-situ the concentrations of the ambient ion species in the mass range from 1 to 90 atomic mass units (amu). MIMS was mounted on the S/C equator normal to the spin axis, and the entrance aperture faced forward when the S/C was in the despun mode. The electric and magnetic fields were arranged to produce a mass spectrum along the focal plane following the magnetic analyzer. Three slits were placed along the focal plane in appropriate places to simultaneously collect ions in the mass ratios 1 to 4 to 16. Ionospheric ions were accelerated into the analyzer system by a negative voltage that varied from -1060 to -225 V. The three mass ranges measured simultaneously were 1-4, 4-16, and 16-90 amu. Following each slit was an electron multiplier and a logarithmic electrometer-amplifier detector. The detector output could be measured directly for an analog output, or it could be fed to a "peak" circuit that determined the amplitude of each peak in the spectrum. Only the amplitude of each peak was telemetered in the primary peaks mode, and in this mode the time required to simultaneously sweep all three mass ranges was 1 s.

BIMS (Bennett Ion-Mass Spectrometer - PI: H. C. Brinton). BIMS mass = 3 kg, power = 2 W. The BIMS objective was to measure, throughout the AE-C orbit, the individual concentrations of all thermal ion species in the mass range of 1 to 72 atomic mass units (amu), and in the ambient density range from 8 to 5×10^6 ions/cm^3. Any combination of the following three mass ranges, expressed in amu, were selected by ground command: range A, -1 to 4, range B, -2 to 18, range C, -8 to 72. Each range was normally scanned in 1.7 s (approximately 12 km along orbit). Normal operation consisted in sequence ABCABC (1 to 72 amu in 5.1 s). Laboratory and inflight determination of spectrometer efficiency and mass discrimination permitted direct conversion of measured ion currents to ambient concentrations. The experiment's four primary mechanical components were guard ring and ion-analyzer tube, collector and preamplifier assembly, vent, and main electronics housing. The guard ring was normally at ground potential, but it could be placed at -6 V by command if desirable, e.g., if the S/C acquired a positive charge. A three-stage Bennett tube with 7 to 5 cycle drift spaces was flown and was modified to permit ion concentration measurements to be obtained at low altitudes. The frequency of the 30 V peak-to-peak RF voltage varied with the mass range measured: range A, -10 MHz, range B, -5 MHz, and range C, -2.5 MHz. Primary analog instrument output was a compressed ion current spectrum which displayed the full dynamic range of the amplifier system on a single telemetry channel.

LEE (Low-Energy Electron Experiment - PI: J. H. Hoffman). This experiment provided direct measurements of the energy input into the upper atmosphere due to electrons and protons in the energy range of 0.2 to 25 keV. The experiment acquired differential measurements of the energy influx and angular distribution. There were two detectors measuring electrons and protons from 0.2 to 25 keV in 16 logarithmically spaced steps, and one detector measuring 5 keV electrons continuously. Each detector consisted of a cylindrical electrostatic analyzer for species and energy selection, and a Spiraltron electron multiplier for particle detection. Energy distributions were obtained by applying different fixed or stepped voltages to the deflection plates. Distributions in angle were measured using the spacecraft spin and the analyzers' positions on the S/C. In the despun modes, measurements were obtained at 45° to the S/C equator, and radially away from the Earth. Detector look angles were chosen to give optimum magnetic pitch-angle coverage when the S/C was moving either poleward or equatorward. All detectors were identical in construction and used 1 by 6 mm entrance apertures. Counts were accumulated over 55.7 ms and read out each main telemetry frame (62.5 ms). The two stepped detectors moved one energy step once each main frame with the same accumulation time, requiring about 1 s for a complete cycle of steps.

UVNO (Ultraviolet Nitric-Oxide Experiment - PI: C. A. Barth). UVNO consisted of a two-channel fixed-grating Ebert-Fastie spectrometer which measured the airglow in the (1, 0) Gamma band in a 15- Å region centered at 2149 Å. The observed intensity was produced by resonance fluorescence of sunlight by the nitric – oxide molecules in the instrument's FOV. The intensity profiles obtained yielded altitude profiles of nitric-oxide density as a function of time and location. Profiles were measured along-track of the S/C at times when it was on the sunlit side of the Earth. The remote sensing character of the UVNO experiment permitted measurements of nitric-oxide to be made at altitudes both above and below S/C perigee. As the S/C spun, the spectrometer, which looked outward through the rim of the satellite, repeatedly had its FOV carried down through the atmosphere onto the Earth's limb, and altitude profiles of the emitted airglow intensity were obtained. Below some altitude the measured signal at 2149 Å was contaminated by Rayleigh-scattered sunlight. To correct for this contamination, a second channel measured only scattered light intensity in a 12 Å region centered at 2190 Å. The two channels were optically and electrically independent. Nitric-oxide airglow intensity was determined by taking the difference between these two measurements. The sensor's spherical fused-quartz telescope mirror had a 125 mm focal length, and focused incident light on the entrance slit of the spectrometer. From this slit the light struck one half of the Ebert mirror and was collimated onto the grating. The 3600-lines-per-mm grating returned it collimated to the other half of the mirror, and the light was focused on two exit slits. The spectrometer FOV was 4° by 0.25°, with the long axis parallel to the S/C spin axis, and therefore parallel to the viewed limb. In normal operation each channel was integrated for 20.8 ms and was read out alternately at 10.4 ms intervals.

VAE (Visible Airglow Experiment - PI: P. B. Hays). The experiment contained a filter photometer designed to measure various airglow and auroral features in the spectral range between 3000 and 7500 Å. The primary information obtained from this experiment was the rates of excitation of the atomic and molecular constituents of the thermosphere. For the AE-C mission, the following six specific lines and bands were chosen for study since they play an important role in the photochemical energy balance of the atmosphere: 3371, 4278, 5200, 5577, 6300, and 7319 Å. The emissions were measured in pairs: 5577 and 6300, 7319 and calibration, 3371 and 5577, 5200 and 7319, 4278 and 3371, calibration and 5200, and 6300 and 4278. Two optical systems viewed at right angles to each other. Each one employed a combination of a simple objective lens and field stop to define the FOV, and each contained a multistage light baffle. The wide-angle high-sensitivity system (designated channel 2) had a FOV of 3° half-angle, and was used to measure the nightglow, dayglow above the S/C, and other weak emission features. The less sensitive system (designated channel 1) had a FOV of approximately 0.75° half-angle and was used for dayglow and nightglow horizon measurements, as well as discrete auroral features which showed strong spatial gradients. Both optical channels had a diameter of 2.2 cm. They shared a filter wheel that contained six interference filters at the wavelengths identified above, and two other positions. One was a dark position for noise measurements, and the other was a calibrate position. The dynamic range of the instrument was 10^{16} photons/m^2 (10^6 Rayleigh).

PSA (Pressure Sensor A - PI: C. J. Rice). PSA, the cold cathode ion gauge, was primarily an engineering experiment to provide data on S/C operation. However, data from this experiment was correlated with accelerometer and capacitance manometer data to evaluate satellite drag performance. PSA measured atmospheric pressure in the region between 120 and 370 km above the Earth's surface for values of atmospheric pressure between 1.3 x 10^{-3} to 1.3 x 10^{-7} mb. The estimated accuracy of the PSA was ±20%. The cylindrically shaped sensor package consisted of a wedge-shaped orifice, a cathode near ground potential, an anode operating at about 1300 vdc, and a permanent magnetic field of about 1600 gauss. The gage contained no primary source of ionizing electrons. The discharge was initiated by field emission and was self-sustaining at a pressure above 1.3 x 10^{-7} mb. The ion current was collected at the cathode. The sensor was mounted on the S/C, with the orifice perpendicular to the S/C spin axis which was normal to the orbital plane. The instrument could be operated in two modes, spinning or despun. When the S/C was in a spinning mode, the PSA alternately sampled the ram and wake pressure. When the S/C was in the despun mode, the PSA faced 30° from the direction of motion.

PSB (Pressure Sensor B - PI: C. J. Rice). PSB mass = 1.14 kg, power = 3 W. The capacitance manometer was primarily an engineering experiment to provide data on S/C operations. However, data from this experiment were also correlated with accelerometer and ion gauge data in evaluating satellite drag. PSB measured atmospheric pressure in the region below 200 km. The accuracy of the PSB gauge varied from about 10% at 120 km to about 40% at 180 km. The PSB consisted of two spherical, thermally controlled chambers, separated by a thin membrane stretched flat and under radial tension. Any deflection of the diaphragm caused by a pressure differential between the two sides caused a change in capacitance between the diaphragm and an adjacent electrode which was measured by a ac bridge circuit. Air was permitted into one of the chambers through two ports 180 deg apart and perpendicular to the spacecraft spin axis. Thus, the wave-ram pressure differential was sampled twice each spacecraft revolution.

Temperature Alarm - PI: P. S. Caruso. The engineering experiment measured the impact temperature during low perigees. [375) 376)]

A.4.4 AE-D (Atmosphere Explorer-D, Explorer 54)

A NASA/GSFC mission with the objective to continue the investigation begun by AE-C of the chemical processes and energy transfer mechanisms that control the structure and behavior of the Earth's atmosphere and ionosphere in the region of high absorption of solar energy. This mission was planned to sample the high latitude regions at the same time that the AE-E mission was sampling the equatorial and low latitude regions. The same type of spacecraft as AE-C was used, the payload consisted of the same instruments except for deletion of EUVS and BIMS. They were part of the AE-E payload. The AE-D S/C was launched Oct. 6, 1975 on a Delta vehicle from Point Arguello, USA. The AE-D S/C mass was 681 kg.

Orbit: Polar elliptical orbit, perigee = 154 km, apogee = 3,816 km, inclination =90.1°, eccentricity = 0.21881, period = 126.9 minutes. The polar orbit provided the sampling of all latitudes and the perigee moved through all latitudes in 3 months and all local times in 4 months. Unfortunately, a failure in the solar power panels resulted in the termination of operations on January 29, 1976, after slightly less than 4 months of useful life. However, all the regions at the perigee altitudes were sampled during this time. The S/C re-entered the

375) E. W. Young, Jr., P. S. Caruso Jr., "Satellite temperature-flux monitor for low perigee applications," ISA ASI 75225, 1975, pp. 133–143

376) P. S. Caruso Jr., C. R. Naegeli, "Low perigee aerodynamic heating during orbital flight of an Atmosphere Explorer," NASA TM D–8308, Sept. 1976

atmosphere about 1 month after cessation of telemetry. To continue the correlated observations with the AE-E mission, AE-C was reactivated on February 28, 1976, to replace AE-D.

Sensor complement: all sensors from AE-C (deletion of EUVS and BIMS). Note, the **CSS** instrument of mission AE-C was renamed to **NACE** (Neutral Atmosphere Composition Experiment) - PI: A. E. Hedin).

A.4.5 AE-E (Atmosphere Explorer-E, Explorer 55)

A NASA/GSFC mission with the objective to investigate the chemical processes and energy transfer mechanisms that control the structure and behavior of the Earth's atmosphere and ionosphere in the region of high absorption of solar energy at low and equatorial latitudes. The simultaneous sampling at higher latitudes was carried out by the AE-D S/C until its failure on January 29, 1976, and then by AE-C, until it reentered on December 12, 1978. The same type of spacecraft as AE-C was used, and the payload consisted of the same types of instruments except that the LEE and UVNO experiments were deleted and a backscatter UV spectrometer was added to monitor the ozone content of the atmosphere. The two experiments that were deleted were more appropriate for the high-latitude regions. The perigee swept through more than six full latitude cycles and two local time cycles during the first year after launch when the orbit was elliptical and the perigee height was varied between 130 and 400 km. The circularization of the orbit around 390 km was made on November 20, 1976 and the S/C was raised to this height whenever it would decay to about 250 km. AE-E reentered on June 10, 1981

The AE-E spacecraft was launched on Nov. 20, 1975 on a Delta vehicle from Cape Canaveral (operational for 5 1/2 years). The AE-E spacecraft mass was 735 kg.

Orbit: Elliptical orbit, perigee = 156 km, apogee = 2,983 km, inclination = 19.7°, eccentricity = 0.17774, period = 117.9 minutes.

Sensor complement: All sensors from AE-C (except LEE and UVNO) plus BUV as described below.

BUV (Backscatter UV Spectrometer - PI: D. F. Heath). BUV monitored the spatial distribution of atmospheric ozone by measuring the intensity of the UV radiation backscattered from the Earth's atmosphere. To obtain this ozone distribution, the BUV subsystem measured direct solar radiation and backscattered UV radiation from the daytime sun-illuminated atmosphere. The instrument consisted of a spectrometer (monochromator) and a photometer. The monochromator measured the intensity of UV radiation backscatter and reflected radiation from the Earth's atmosphere in 12 wavelengths (2555 to 3398 Å) in which ozone attenuation occurs. The photometer measured the reflected UV radiation in a single wavelength span in which attenuation by ozone does not occur. The BUV had four operating modes.

A.5 AEM-2 (Applications Explorer Mission-2)

NASA/LaRC mission (also referred to as 'SAGE' or 'Explorer 60') with the objective to monitor stratospheric aerosol and its influence on the climate (study of aerosol sources and sinks, aerosol transport, aerosol radiative and climatological implications, etc.). Launch: Feb 18, 1979 by Scout vehicle from Wallops Island, VA. The S/C experienced power problems after May 15, 1979. However, operations continued until November 19, 1981. [377] [378]

Application: stratospheric chemistry related to aerosol, ozone, and nitrogen dioxide. Provision of almost global profiles by solar occultation techniques.

377) M. P. McCormick, P. Hamill, T. J. Pepin, W. P. Chu, T. J. Swissler, L. R. McMaster, "Satellite Studies of the Stratospheric Aerosol," Bulletin of the American Meteorological Society, Vol. 60, No. 9, September 1979, pp. 1038-1046

378) L. R. McMaster, M. W. Rowland, "SAGE-I Data User's Guide," NASA Reference Publication 1275, Aug. 1992

Orbit: Near-circular orbit, perigee = 548 km, apogee = 660 km, inclination = 55°, period = 96.8 min.

Sensor:

SAGE-I = Stratospheric Aerosol and Gas Experiment. The instrument is a sun photometer[379] [of SAM (on Apollo-Soyuz) and SAM-II (on Nimbus-7) heritage].

Objective: measurement of solar intensity profiles during each sunrise and sunset event (about 30 sampling opportunities per day, sunrise + sunset). The SAGE instrument is a radiometer (mass = 18 kg) consisting of a Gregorian telescope and a detector subassembly measuring the attenuation of solar radiation at four wavelengths (0.385, 0.45, 0.6, and 1.0 μm) during solar occultation. The sensor scanned the atmosphere from the horizon up, and measured the attenuation of solar radiation at different atmospheric layers as the S/C emerged from the Earth's shadow. This procedure was repeated during S/C sunset. Two vertical scannings were obtained during each orbit, with each scan requiring approximately 1 minute to cover the atmosphere above the troposphere. The instrument FOV was about 0.15 mrad, resulting in a vertical resolution of about 1 km.

The aerosol profiles can be interpreted to give concentrations of ozone, nitrogen dioxide, and total molecular density. A number of ground truth measurements in the US, Japan, Europe, and Nimbus-7 data are an integral part of the SAGE mission.

Data:
The instrument data include vertical profiles of stratospheric aerosol, ozone, and nitrogen dioxide. The SAGE-I aerosol data were validated by comparison with correlative lidar and dustsonde in situ measurements, the ozone data were validated by comparison with balloon electrochemical cell ozonesonde and sounding rocket measurements, the nitrogen dioxide measurements were compared with climatology data. All data are archived at NSSDC (National Space Science Data Center) at GSFC.

A.6 Aura Mission (EOS/Chem-1)

See description under D.11.3

A.7 CLOUDS (Cloud and Radiation Monitoring Satellite)

CLOUDS is a mission study (as of 2000), conducted by 12 European partners (7 institutes and 5 companies) and with NOAA/ETL participation, sponsored by the EC (European Commission) in its 4th Framework Program under the topic: "Environment and Climate." The study project lasted two years, from 1998 to 2000. [380] [381] [382]

CLOUDS is proposed as a monitoring mission to perform measurements (accurate, comprehensive, consistent and frequent information on cloud structures and associated radiative parameters) necessary to describe cloud-radiation interaction in operational models for climate and long-term weather prediction. Its strategic objective is to extend the overall European service of climate monitoring from Space, beyond what is achievable by the in-

379) Note: A photometer is usually a broadband instrument capable of measuring thermal continuum radiation (i.e. flux) thereby permitting the study of energy balance and surface composition (also detection of infrared roughness of surface features)

380) B. Bizzarri, P. Spera, E. Maggi, et al., "Instruments and system for CLOUDS - a Cloud and Radiation monitoring satellite," the EOS/SPIE Symposium on Remote Sensing, Sept. 25-29, 2000, Barcelona, Spain, SPIE Vol. 4163-43

381) B. Bizzarri, M. Desbois, C. Stanfuss, J. Murray, J. Russell, C. Naud, A. Gasiewski, K. Künzi, et al., "Scientific background for CLOUDS - a Cloud and Radiation Monitoring Satellite, the EOS/SPIE Symposium on Remote Sensing, Sept. 25-29, 2000, Barcelona, Spain, SPIE Vol. 4168-08

382) ftp://romatm9.phys.uniroma1.it/pub/clouds/report/

strumentation presently foreseen for MSG and MetOp/EPS, whose mission definition has been driven by nowcasting and short-medium term weather prediction.

The CLOUDS bus is derived from the PRIMA bus of ASI for small-medium class satellites. Commonality of the ground segment with MetOp/EPS is envisaged.

Note: In 2000, the CLOUDS proposal was submitted to ESA in response to its "Call for Ideas" on future Earth Explorer Core Missions. But unfortunately, CLOUDS was not selected because of very tight budgets. Instead, the EarthCARE (Earth Clouds Aerosol and Radiation Explorer) mission proposal was selected (however, not yet approved) which is considered more promising because of its passive and active payload mix of instruments, expected to be "more advanced" and to yield a better observation data return. EarthCARE seems to be on its way to become a cooperative ESA/NASDA venture. - ESA stated that the CLOUDS proposal should again be submitted, but in the context of its Earth Watch program (the outcome will be known in mid 2002). In the meantime (2001), an attempt is being made to fly the various well-defined CLOUDS instruments on different smaller missions, all of which are in their proposal states. - Against my practice of including only "approved missions" into my documentation, I am including the CLOUDS project because of its clarity in concept and its balanced instrument complement on the topic of "clouds." A lot of effort has been invested into the definition phase. The CLOUDS instruments will eventually show up in other missions. Hence, their description in this context may still be of benefit to all concerned.

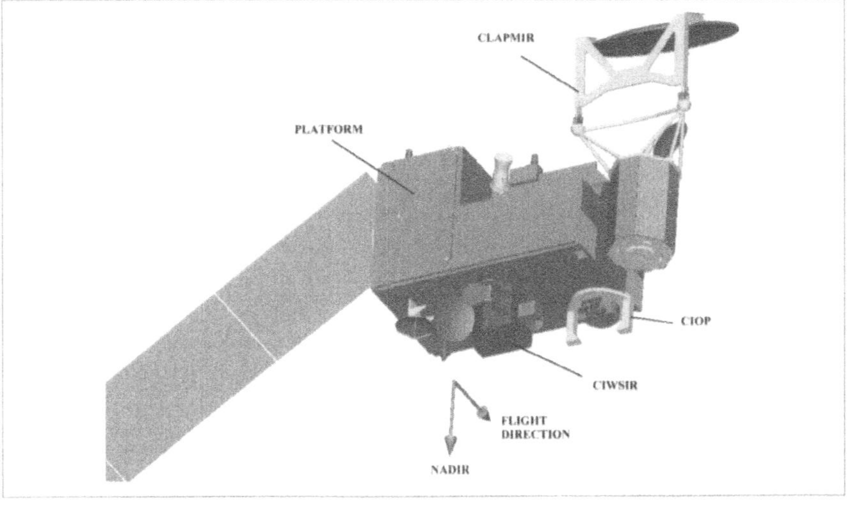

Figure 10: Illustration of the proposed CLOUDS spacecraft

Payload resource requirements	
Payload total mass	350 kg (including payload data handling and transmission)
Electrical power requirement	400 W
Data rate at instrument output	747.8 kbit/s
Spacecraft parameters	
Spacecraft dimensions, design life	3.2 m x 2.0 m x 1.5 m, 5 years
Satellite mass	900 kg
S/C power	1600 W (peak), 1400 W (average), 10 m^2 solar array size
S/C stabilization Attitude sensors Attitude actuators Pointing performance	Three-axis stabilized 2 magnetometers, 2 star sensors, 4 gyros 3 magnetic torquers, 4 reaction wheels Accuracy: 350 µrad, stability = 20 µrad/s

Communication capacity	
Mass memory of temporary storage	670 MByte (including RS-coding)
Data rate for global downlink	30 Mbit/s (X-band)
Uplink data rate	2 kbit/s (S-band)
Data rate of local read-out	1.1 Mbit/s in S-band (including Viterbi-coding)

Table 55: Overview of spacecraft parameters

Orbit: Sun-synchronous circular orbit, altitude = 840 km, inclination = 98°, period = 101.7 minutes, local equator crossing time at 10:00. The mission scenario calls for CLOUDS to co-orbit with MetOp, de-phased by 30 minutes.

Sensor complement:

The CLOUDS payload includes six radiometers, composed of four instruments integrated into one assembly, named CIOP (CLOUDS Integrated Optical Payload), and two MW/Sub-mm instruments. Use is made of: a) passive radiometry over a wide region of the electromagnetic spectrum (UV to MW), b) different polarizations to determine 2-3 Stokes parameters in a number of channels, c) fore- and aft-viewing capability for all channels and multi-angle viewing for one channel. For reasons of synergy, all instruments are of the conical-scanning type. In addition, all instruments feature in-flight calibration, and co-registration with accuracy requirement, ranging from 1/16 to 1/2 of a pixel.

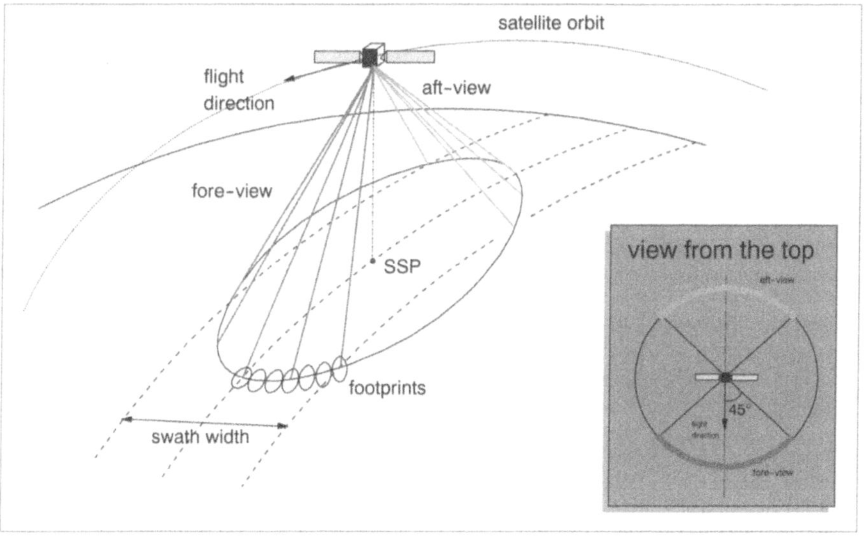

Figure 11: Geometry of conical scanning

CIOP instruments: The four CIOP instruments, covering the range UV/VIS/NIR/SWIR/TIR/FIR, are integrated into a single rotating drum performing conical scanning with a scan speed of 1 scan per 8 seconds. The optical viewing geometry corresponds to: 45° off-nadir (MAVIR actually covers the range 21-57°) observing the target under a zenith angle ζ of 53.2° (MAVIR in the 23.9 - 71.7° range), with two arcs over ±45° fore- and aft-views for a swath width of about 1400 km (the MAVIR swath changes with viewing angle). When viewing the side arcs, calibration devices are observed.

CASIR (Clouds and Aerosol Shortwave Imaging Radiometer). The 10-channel instrument collects all narrow-band short-wave channels. The main objectives are:

- To perform basic clouds observations with emphasis on liquid/ice discrimination and drop size evaluation at the cloud top, and cloud optical thickness

- To observe aerosol over sea and land, with possible inference of gross vertical distribution
- To complement the broadband coarse-resolution Earth radiation budget observation in the short-waves, by observing components at finer spatial and spectral resolution.

Additionally, some Earth surface parameters are observed such as: albedo, PAR (Photosynthetically Active Radiation), FPAR (Fractional PAR), NDVI (Normalized Difference Vegetation Index), LAI (Leaf Area Index), and short-wave radiative fluxes (in association with BERIR and MAVIR).

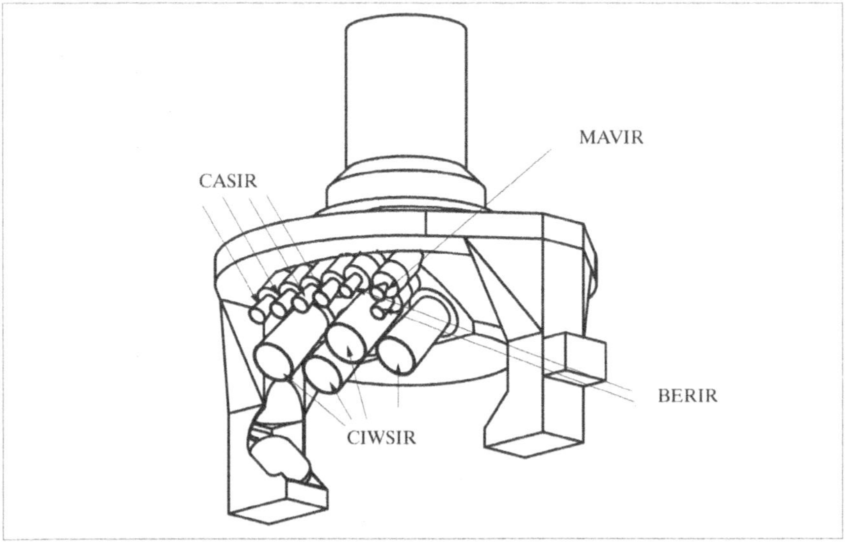

Figure 12: Schematic illustration of the CIOP instrument package

IFOV	0.175° providing a footprint of 6.5 km x 3.9 km (ellipse), equivalent to a 5 km circle				
Scanning	Conical: (α=45°, ζ=53.2°), fore- and aft-views by >±45° in azimuth, swath width = 1400 km				
Sampling	1 scan per 8 seconds, 8 detectors/channel, readings at 5 ms intervals				
Optics	Aperture diameter = 20 mm, f/2.5 (four objectives)				
Detectors	Hybrid Si-CMOS (UV, VIS, NIR), Peltier-cooled CMT-CMOS (SWIR)				
Date rate	307.2 kbit/s (+HK)				
Spectral parameters					
λ (center wave)	$\Delta\lambda$	SNR @ 10% albedo (required)	SNR @ 10% albedo (estimated)	Polarizations	No of columns for TDI
340 nm, UV	6 nm	300	310	not required	2
388 nm, UV	6 nm	300	530	not required	2
443 nm, VIS	20 nm	200	1060	not required	-
555 nm, VIS	20 nm	200	620	three	-
670 nm, VIS	20 nm	200	430	three	-
865 nm, NIR	20 nm	200	680	not required	-
940 nm, NIR	50 nm	200	460	not required	-
1,240 nm, SWIR	50 nm	200	320	three	-
1,375 nm, SWIR	30 nm	200	1020	not required	-
1,610 nm, SWIR	30 nm	200	630	not required	-

Table 56: CASIR instrument parameters and expected performances

CASIR is composed of four different objectives, each with its detectors in the focal plane. The UV, VIS and NIR channels operate with 2-D array Si-CMOS detectors, the SWIR channels are cooled to -25°C, using 2-D CMT arrays. Band separation and polarization are obtained by bandpass filters and polarizers located closely to the FPA of each detector. Two UV bands feature the TDI (Time Delay Integration) scheme to meet the SNR requirements.

CIRIR (Clouds InfraRed Imaging Radiometer). The 10-channel instrument collects the narrow-band channel radiation in TIR and FIR. The objectives are:

- To observe clouds with emphasis on cloud top height, cloud emissivity, liquid/ice discrimination at cloud top, and thin cirrus cloud detection
- To observe layer-integrated water vapor, including low-level
- To complement the broadband coarse-resolution Earth radiation budget observation in the long waves, by observing components at finer spatial and spectral resolution.

Additionally, some Earth surface parameters are observed such as: SST (Sea Surface Temperature), ATI (Apparent Thermal Inertia), indicative of soil moisture, derived in association with CASIR and, in association with BERIR, long-wave radiative fluxes.

The CIRIR instrument features uncooled microbolometer detectors. However, long integration times are needed (>10 ms) with this design. For the WV (Water Vapor), TIR (Thermal Infrared) and FIR (Far Infrared) channels, an in-field bands separation is realized by means of different bandpass filters disposed close to different regions of each array detector. One blackbody and the view of space through a mirror provide calibration during the lateral arcs of the scanning path not used for measurements. In addition, CIRIR is able to view the calibration unit of BERIR for intercalibration.

The MIR channel of CIRIR is sampled at 5 km resolution to improve the co-registration accuracy between CIRIR and CASIR, all other channels have a 10 km spatial resolution. Full synergy (co-registration and identical observing conditions) with cloud observation in short-wave channels (CASIR) is preserved by the use of the same scanning and viewing scheme.

IFOV	Cha. 3.7 μm: 0.175°, providing a footprint of 6.5 km x 3.9 km, equ. to a 5 km circle All other cha: 0.35°, providing a footprint of 13 km x 7.8 km, equ. to a 10 km circle
Scanning	Conical: (α=45°, ζ=53.2°), fore- and aft-views by >±45° in azimuth, swath width = 1400 km
Sampling	Channel 3.7 μm: 1 scan per 8 s, 8 detectors/channel, readings at 5 ms intervals All other channels: 1 scan per 8 s, 4 detectors/channel, readings at 10 ms intervals
Optics	Four objectives, all f/1.0; MIR: L=66 mm, WV & TIR: L=60 mm, FIR: =75 mm
Detectors	Channel 3.7 μm: linear PbSe detector Peltier-cooled to -25°C All other channels: array microbolometers (uncooled)
Data rate	62.4 kbit/s + HK

Spectral parameters

λ (center wave)	Δλ	NEΔT (required)	NEΔT (estimated before TDI)	NEΔT (estimated after TDI)	No of TDI columns
3.74 μm, MIR	0.40 μm	0.10 K @ 300 K	0.10 K @ 300 K	0.10 K @ 300 K	-
6.25 μm, WV	1.00 μm	0.30 K @ 250 K	0.26 K @ 250 K	0.26 K @ 250 K	-
7.35 μm, WV	0.50 μm	0.30 K @ 250 K	0.35 K @ 250 K	0.25 K @ 250 K	2
8.70 μm, WV	0.50 μm	0.10 K @ 280 K	0.19 K @ 280 K	0.10 K @ 280 K	4
9.66 μm, TIR	0.50 μm	0.30 K @ 220 K	0.51 K @ 220 K	0.29 K @ 220 K	3
10.8 μm, TIR	1.00 μm	0.10 K @ 300 K	0.09 K @ 300 K	0.09 K @ 300 K	-
12.0 μm, TIR	1.00 μm	0.10 K @ 300 K	0.11 K @ 300 K	0.11 K @ 300 K	-
13.4 μm, TIR	0.50 μm	0.30 K @ 280 K	0.32 K @ 280 K	0.32 K @ 280 K	-
18.2 μm, FIR	2.00 μm	0.50 K @ 220 K	0.47 K @ 220 K	0.47 K @ 220 K	-
24.4 μm, FIR	2.00 μm	0.50 K @ 220 K	0.72 K @ 220 K	0.51 K @ 220 K	2

Table 57: CIRIR instrument parameters and expected performances

BERIR (Broadband Earth Radiation Imaging Radiometer). The instrument features two broadband channels, one for total reflected radiation, the other for total energy measurement from UV to FIR. The overall objective is to observe the outgoing short-wave and long-wave radiative flux at the top of the atmosphere. Additionally, Earth surface radiative fluxes are observed in association with CASIR, CIRIR and MAVIR.

The BERIR design is derived from CERES of NASA. Since the detectors employed (bolometers) exhibit relatively slow response times and low detectivity, an integration time of 40 ms is needed along with an equivalent IFOV of 40 km. No polarizations are needed. The BERIR total-wave and short-wave (SW) channels are realized using two optical heads of CERES with minor modifications. Also, the CERES ICM (Internal Calibration Module) is used at each scan stroke. ICM contains a grooved blackbody source for the total-wave channel and a tungsten lamp for the short-wave channel. The total-wave channel also looks at the CIRIR blackbody and at a space reference during each scanning stroke for better offset evaluation and to permit an intercalibration between all infrared channels.

IFOV	1.4° providing a footprint of 52 km x 31 km, equivalent to a 40 km circle			
Scanning	Conical: (α=45°, ζ=53.2°), fore- and aft-views by >±45° in azimuth, swath width = 1400 km			
Sampling	1 scan per 8 s, 1 detector/channel, readings at 40 ms intervals			
Optics	L=18 mm, f/1.8			
Detectors	One uncooled thermistor/bolometer per channel			
Data rate	0.6 kbit/s + HK			
Spectral parameters				
Channel	NEΔT (required)	NEΔT (estimated)	Calibration (required)	Calibration (estimated)
0.3 - 4.0 μm, SW	0.3 $Wm^{-2} sr^{-1}$	0.15 $Wm^{-2} sr^{-1}$	1.0 $Wm^{-2} sr^{-1}$	0.8 $Wm^{-2} sr^{-1}$
0.3 - 100 μm, total	0.3 $Wm^{-2} sr^{-1}$	0.15 $Wm^{-2} sr^{-1}$	0.5 $Wm^{-2} sr^{-1}$	0.6 $Wm^{-2} sr^{-1}$

Table 58: BERIR instrument parameters and expected performance

Full synergy (accurate co-registration and identical observing conditions) with the narrow-band channels is maintained with the same scanning mode and viewing geometry.

MAVIR (Multi-Angle VIS Imaging Radiometer). MAVIR is complementary to BERIR. The instrument operates in a single, relatively broad channel, mostly covering the VIS range. Its objective is to complement the broadband and short-wave solar reflected radiation information by the measurement of bi-directional reflectance properties at a spatial resolution consistent with cloud structures. In addition, Earth surface radiative fluxes are observed in association with CASIR, specifically the short-wave radiative parameters: PAR, FPAR, NDVI, and LAI.

IFOV	0.175° providing a footprint of 6.5 km x 3.9 km, equivalent to a 5 km circle (α = 45°)		
Scanning	Conical, four viewing angles, fore- and aft-views by >±45°, swath changing with viewing angle (1400 km for α=45°)		
Sampling	1 scan per 8 s, 8 detector/channel, readings at 5 ms intervals		
Optics	L=3.2 mm, f/2.7		
Detectors	Linear CCD		
Data rate	86.4 kbit/s		
Spectral parameters			
Channel	Viewing angles	Radiometric accuracy (required)	Radiometric accuracy (estimated)
0.4 - 1.0 μm (VNIR)	α = 21°, ζ = 23.9° α = 33°. ζ = 38.1° α = 45°, ζ = 53.2° α = 57°, ζ = 71.7°	SNR=200 @ 10% albedo	SNR=273 @ 10% albedo

Table 59: MAVIR instrument parameters and expected performance

MAVIR is implemented using a linear silicon CCD array in the FPA of a wide-angle telescope covering nadir angles in the range of 20° - 60°. Four series of eight detectors are active, centered in correspondence of the selected nadir angles. MAVIR synergy with BERIR and all other instruments is provided with one viewing angle being the same (α = 45°) for the conical scanning geometry.

MW/Sub-mm instruments: The two MW/Sub-mm instruments perform conical scanning at a uniform rate of 1 scan per 2 seconds. The viewing geometry is identical to that of the optical payload (45° off-nadir - observing the target under a zenith angle ζ of 53.2°, with two arcs over ±45° fore- and aft-views for a swath width of about 1400 km). The scan-rate of one scan per 2 s provides contiguous along-track coverage by a single feed for most channels. The 89 GHz channel, required at 5 km IFOV, implies the use of two parallel feeds. Along-track sampling assures a 10% overlap between adjacent IFOVs by using 1.25 ms intervals (5 km) or 2.5 ms (10 km). For low frequencies, the IFOV degrades with the diffraction law; hence, the integration time is progressively increased.

CIWSIR (Cloud Ice and Water-vapor Sub-mm Imaging Radiometer). The instrument objectives are:

- To observe cloud ice with a higher penetration capability than in the optical range
- To be able to discriminate the water phase in the cloud interior
- To infer convective penetration in the troposphere through differential water vapor optical depth.

IFOV	0.35° providing a footprint of 13 km x 7.8 km, equivalent to a 10 km circle			
Scanning	Conical (α=45°, ζ=53.2°), fore- and aft-views by ±45° in azimuth, swath=1400 km			
Sampling	1 scan per 2 s, 1 feed/channel, readings at 1.25 ms intervals			
Antennas	L=40 cm for channels 150-220 GHz, L=16 cm for channels 463-874 GHz			
Detection	Subharmonic Schottky mixers for mm-channels, pumped mixers for sub-mm channels			
Resources	Mass = 79 kg; cylindrical volume: 110 cm diameter, height=43 cm; electrical power = 110 W; data rate = 83.2 kbit/s + HK			
Channel center frequency (GHz)	Bandwidth Δv	Polarizations	NEΔT (required)	NEΔT (estimated)
874.38±6.0	3.0 GHz	two	1.0 K @ 240 K	2.0 K @ 240 K
682.95±6.0	3.0 GHz	two	1.0 K @ 240 K	1.2 K @ 240 K
462.64±3.0	2.0 GHz	two	1.0 K @ 240 K	0.9 K @ 240 K
220.50±3.0	2.0 GHz	two	1.0 K @ 240 K	0.9 K @ 240 K
183.31±1.0	1.0 GHz	one	1.0 K @ 240 K	1.2 K @ 240 K
183.31±3.0	2.0 GHz	one	1.0 K @ 260 K	0.9 K @ 260 K
183.31±7.0	4.0 GHz	one	1.0 K @ 280 K	0.6 K @ 280 K
150	4.0 GHz	two	1.0 K @ 300 K	0.6 K @ 300 K

Table 60: CIWSIR instrument parameters and expected performance

The CIWSIR instrument is a six-frequency, 13-channel microwave radiometer with seven channels in the mm-range (water vapor) and six channels in the sub-mm-range (ice in clouds). Two quasi-optical multiplexers are implemented for the beam-splitting function (one for mm channels, one for sub-mm channels) using dichroic plates. The separation of the two polarizations (H and V) employs wire grids which offer low losses and almost frequency-independent performance.

The instrument features also two built-in calibration systems utilizing ambient temperature loads and cold space (2.7 K) as references. The sub-mm channels employ pumped Schottky mixers for maximum sensitivity (quasi-optical diplexers are used for LO coupling). Subharmonic mixers are employed for the mm-wave channels. Synergy with the optical payload is provided with the same viewing geometry.

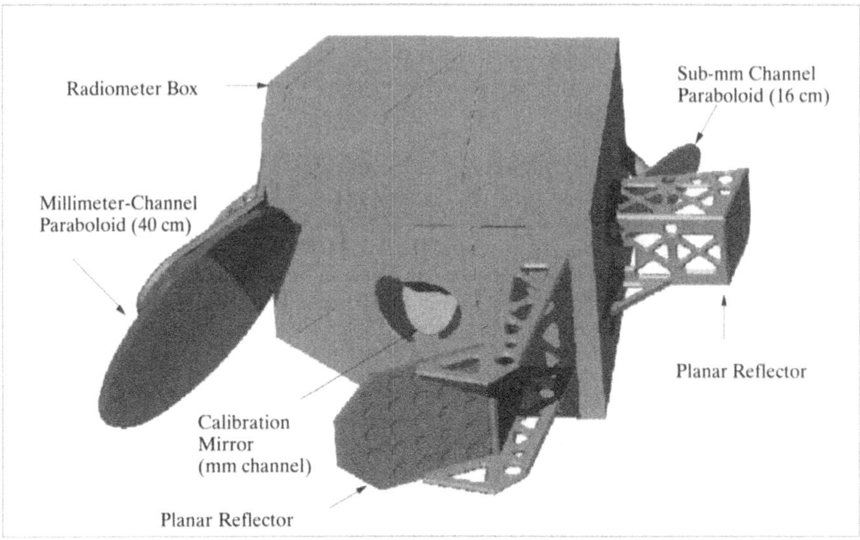

Figure 13: Schematic view of the CIWSIR instrument

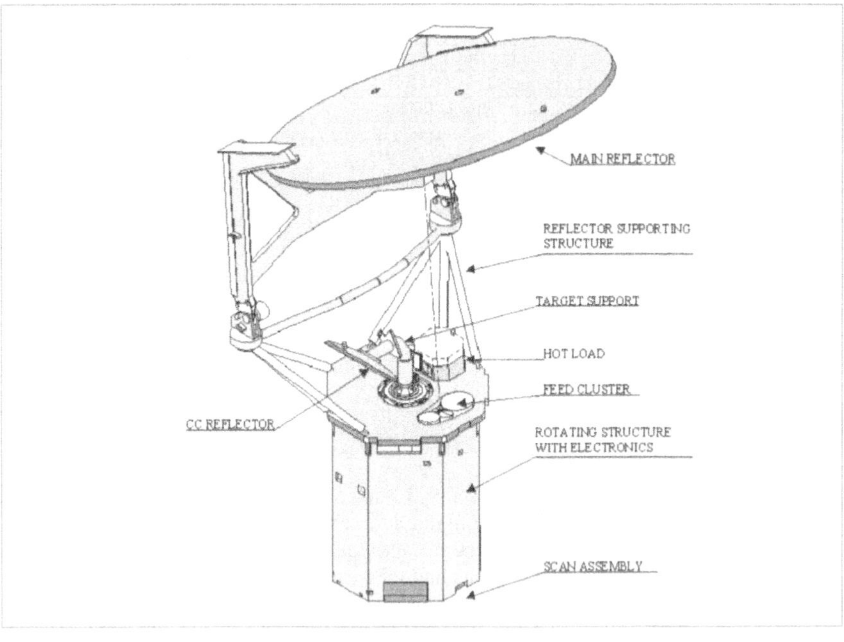

Figure 14: Schematic view of the CLAPMIR instrument

CLAPMIR (Cloud Liquid-water And Precipitation Microwave Imaging Radiometer). The instrument includes MW channels in window and absorption bands. The main objectives are:

- To observe cloud liquid water, precipitating and non-precipitating, with inference of drop size
- To be able to discriminate the water phase in the cloud interior and to measure total-column vapor
- To infer a gross vertical profile of liquid/precipitating water cores as linked to air temperature.

The instrument features an offset parabolic reflector of 1.6 m x 1.4 m in size, illuminated by a cluster of feeds. The reflector and feed horn antenna are mounted on a drum containing the receivers, a digital data unit, mechanical balancing subsystem, and power supply. The drum assembly is rotated about the instrument axis by a coaxially mounted bearing and brushless DC motor. All information (commands, timing and telemetry signals) and power pass through the rotating part of the stator via a dedicated assembly. A calibration system with two calibration targets (cold space view and an internal hot blackbody load) permits a period end-to-end calibration of the receiving chains for all channels. The receiving channels are configured as total-power radiometers with direct detection. The polarimetric channels are designed to provide the third Stokes parameter by means of adding analog correlators. Synergy with the optical payload and with CIWSIR is provided with the same viewing geometry.

IFOV	At 89 GHz: 0.175° providing a footprint of 6.5 km x 3.9 km, equiv. to a 5 km circle Bands 118 and 55 GHz: 0.35° - a footprint of 13 km x 7.8 km, equiv. to a 10 km circle All other channels: changing with frequency according to diffraction limits								
Scanning	Conical (α=45°, ζ=53.2°), fore- and aft-views by ±45° in azimuth, swath=1400 km								
Sampling	1 scan per 2 s, 1 feed/channel (two feeds @ 89 GHz), readings at 1.25-10 ms intervals								
Antenna	L=160 cm								
Detection	Total power radiometers								
Resources	Mass = 160 kg; stowed volume: cylindrical with 90 cm diameter, height = 180 cm; power = 170 W; data rate = 208 kbit/s + HK								
Channel center frequency (GHz)	Bandwidth (GHz)	IFOV Along-track (km)	IFOV Along-scan (km)	IFOV Average (km)	Samples/ scan	Integration time (ms)	Polarization	NEΔT (required) (K)	NEΔT (estimated) (K)
118.75±1.0	1.0	13.0	7.8	10	800	2.5	one	0.5 @ 230	1.0 @ 230
118.75±1.5	1.0	13.0	7.8	10	800	2.5	one	0.5 @ 250	1.0 @ 250
118.75±2.0	1.0	13.0	7.8	10	800	2.5	one	0.5 @ 270	1.0 @ 270
118.75±4.0	1.0	13.0	7.8	10	800	2.5	one	0.5 @ 290	1.2 @ 290
89.0	3.0	6.5	3.9	5	1600	1.25	four	1.0 @ 300	1.0 @ 300
55	0.5	13.0	7.8	10	800	2.5	one	0.5 @ 230	1.0 @ 230
54	0.5	13.0	7.8	10	800	2.5	one	0.5 @ 250	1.0 @ 250
53	0.5	13.0	7.8	10	800	2.5	one	0.5 @ 270	1.0 @ 270
50	0.5	13.0	7.8	10	800	2.5	one	0.5 @ 290	1.0 @ 290
36.5	1.0	15.8	9.5	12	800	2.5	four	0.7 @ 300	0.6 @ 300
23.8	0.4	24.3	14.6	19	400	5	two	0.6 @ 250	0.6 @ 250
18.7	0.2	30.1	18.6	24	400	5	four	0.5 @ 300	0.6 @ 300
10.6	0.1	54.6	32.7	42	200	10	four	0.4 @ 300	0.4 @ 300
6.9	0.3	83.8	50.3	65	200	10	two	0.3 @ 300	0.3 @ 300

Table 61: CLAPMIR instrument parameters and expected performance

A.8 CloudSat

A NASA/CSA (USA/Canada) cooperative mission in the (ESSP) Pathfinder program. CloudSat's primary goal is to furnish data needed to evaluate and improve the way clouds are represented in global models, thereby contributing to better predictions of clouds and thus to their poorly understood role in climate change and the cloud-climate feedback [focused on understanding the role of optically thick clouds on the Earth's radiation budget (a balance of solar energy reaching the Earth and lost to space that ultimately controls the temperature of the Earth)]. In the cooperative arrangement, NASA provides the S/C (a variant of the BCP 2000 bus of BATC), major portions of the cloud-profiling radar including integration and test of the radar system, and the S/C launch, while CSA (Canadian Space Agency) provides additional components of the advanced cloud-profiling radar. Other partners in the program are: CIRA (Cooperative Institute for Research in the Atmosphere) of CSU (Colorado State University) handles data processing and distribution during the mission operations phase (archiving after the mission will be the responsibility of a NASA DAAC); ground operations and communications are provided the USAF/STP (Space Test Program), and the U.S. Department of Energy provides ground and aircraft-based observations for validation. In addition, members of the Science Team, including non-USA members, are contributing additional ground and aircraft-based measurements for validation. The CloudSat mission and payload development is managed by NASA/JPL. 383) 384) 385) 386)

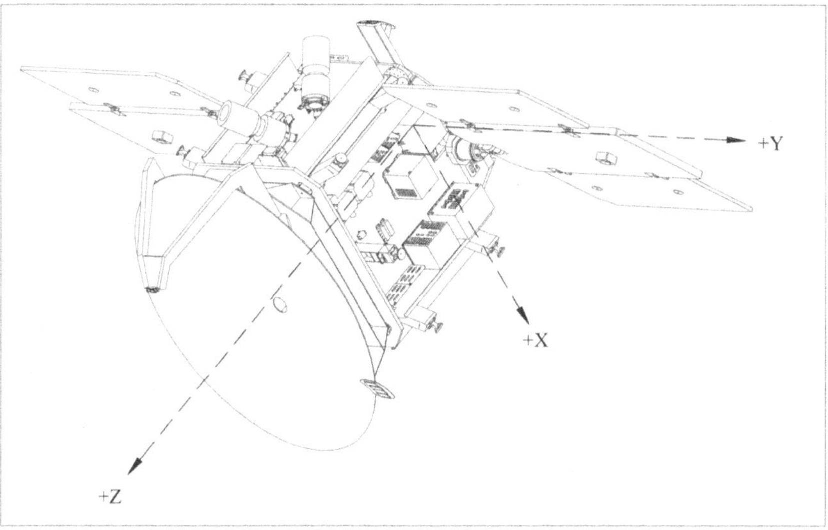

Figure 15: Illustration of the CloudSat spacecraft

CloudSat fills a significant gap in the existing and planned Earth observation missions by measuring the vertical profile of clouds using active remote sensing (94 GHz radar). The CloudSat S/C flies on-orbit in tight formation with a NASA lidar satellite, which carries a dual-wavelength backscattering (polarization-sensitive) lidar. The CloudSat orbit will be adjusted to hold a fixed distance of about 460 km (about 60 s delay) with respect to the lidar satellite, with an option to decrease this distance to a 15 second separation. The sensors of both S/C point into the same ground track. The footprint of the radar is expected to overlay the footprint of the lidar by about 50%, or more, of the time and will always be within 2 km. It is planned that the two satellites fly in formation with the Aqua S/C (EOS-PM) such that

383) F. K. Li, E. Im, S. L. Durden, R. Girard, G. Sadowy, C. Wu, "Cloud Profiling Radar (CPR) for the CloudSat Mission," Proceedings of IEEE/IGARSS 2000, Honolulu, HI, July 24-28, 2000

384) http://cloudsat.atmos.colostate.edu/cs.4.html

385) G. L. Stephens, D. G. Vane, S. J. Walter, "The CloudSat Mission: A new Dimension to space-based Observations of Cloud in the coming Millennium," paper presented at the GCSS-WGNE Workshop, Fort Collins, CO, Nov. 9-13, 1998

386) http://essp.gsfc.nasa.gov/cloudsat.html

the radar and lidar footprints will always fall in the central 36 km of the Aqua-MODIS swath.

The CloudSat S/C (built by Ball Aerospace) uses a RS200 series bus of QuikSCAT and ICESat heritage. The S/C is three-axis stabilized, the ADCS (Attitude Determination and Control Subsystem) uses a star tracker, reaction wheels, and a GPS receiver with C/A code; the pointing accuracy is < 0.07°, two-axes, 99.7% probable; the geolocation knowledge of the radar footprint position is better than 1 km on the geoid, two-axes, 99.7% probable. The S/C mass is <999 kg. The planned mission operational life of CloudSat is 2 years to enable more than one seasonal cycle to be observed. The design life of both the spacecraft and radar exceeds two years, and consumables are carried for a three-year mission, enabling an extended mission if funding is provided. A launch of CloudSat on a Delta 7420-10 launch vehicle, co-manifested with the NASA lidar mission, is projected for 2004.

Orbit: Sun-synchronous near-circular "frozen" orbit [same orbit as Aqua (EOS/PM)], altitude = 705 km, inclination = 98.2°.

Formation flying: Formation flying is a navigation strategy that enables CloudSat to closely track the groundtracks of the lidar spacecraft and Aqua in a very precise way. After launch, maneuvers within the first 45 days of the mission will bring the CloudSat spacecraft into formation with the other two spacecraft. The CloudSat orbit will be adjusted and monitored to hold the CloudSat spacecraft at a fixed distance/time from the lidar S/C. CloudSat will be controlled so that the mean separation between the satellites is approximately 460 kilometers which corresponds to approximately 60 seconds delay between lidar and radar measurements. The chosen delay is a compromise between the desire to minimize the time delay between the radar and lidar measurements and the need to hold the frequency of formation flying maintenance maneuvers to an interval of once per week or longer. In this way, the radar footprint will spatially overlay the lidar footprint much of the time, creating coordinated and essentially simultaneous measurements. It is believed that this time separation can, with further study, be reduced to as little as 15 seconds.

RF communications: CloudSat is controlled from a USAF facility using the Air Force Satellite Control Network (AFSCN). The payload data is downlinked at 1 Mbit/s and relayed to the control center, where it is sent to CIRA for processing into data products.

Sensor complement:

CPR (Cloud Profiling Radar), joint development by NASA/JPL and CSA. The objective is to provide information on the vertical structure of all cloud systems. CPR is a 94 GHz nadir-looking radar that measures the power backscattered by clouds as a function of distance from the radar. The overall design of the CPR consists of the following subsystems: RFES (Radio Frequency Electronics Subsystem), HPA (High Power Amplifier), Antenna Subsystem (Quasi-Optical Transmission Line), and DSS (Digital Subsystem). The RFES consists of an up-converter which generates a pulsed signal and up-converts it to 94 GHz. The signal is amplified to about 200 mW by a state-of-the-art MMIC power amplifier. The receiver portion of the RFES down-converts the signal to an IF (Intermediate Frequency). The IF signal is detected using a logarithmic amplifier (high dynamic range). The receiver noise level is critical in achieving the required sensitivity. The HPA consists of EIK (Extended Interaction Klystron) and a high-power voltage supply; it amplifies the transmitted pulse to a power level of 1.5 kW. This family of EIKs has been used extensively in existing ground-based and airborne 94 GHz cloud radars. The antenna is a fixed 1.85 m diameter reflector of composite graphite material to reduce mass. The antenna provides >62 dBi gain, has a beamwidth of <12°, and has sidelobes < about -50 dB for angles >7° from boresight. The quasi-optical transmission line (QOTL) replaces the conventional waveguide and circulator for sending power from the HPA to the reflector and sending received power to the receiver subsystem. Waveguides are replaced by mirrors and free-space propagation, while the circulator duplexing function is handled by a polarizer and Faraday rotator. The advantage of the QOTL over conventional waveguide and circulator is reduced loss, important for meeting sensitivity requirements. The DSS provides the command, control, and telemetry interface to the S/C. It includes a Control and Timing Unit and a data handling unit that accepts the analog signal from the RFES logarithmic detector. It digitizes it and performs the required sample averaging of 0.16 s.

To detect the low reflectivity of clouds, the CPR averages many samples of the measured power and subtracts the estimated system noise level. The number of independent samples can be increased by increasing the PRF (Pulse Repetition Frequency). However, the maximum PRF is given by the range ambiguity considerations. For CPR, the nominal range window size is set to 30 km, permitting the capture of the surface return and cloud return up to an altitude of 25 km. The system noise level is estimated using the clear air radar return from 25 to 30 km altitude. The radar footprint is 1.4 km, and is averaged over 0.16 seconds to produce an effective footprint of 2.5 km (along-track) by 1.4 km (cross-track).

Parameter	Short Pulse (SP) Operation
Nominal frequency	94 GHz
Pulse width	3.3 μs
PRF	about 4300 Hz
Data window	0-25 km
Vertical range resolution Cross-track resolution Along-track resolution	500 m 1.4 km 2.5 km
Antenna size	1.85 m
Dynamic range	70 dB
Peak power	1.5 kW
Bandwidth	0.3 MHz
Integration time	0.16 s
Along-track sampling	1.1 km
Data rate	20 kbit/s
Instrument mass, power	230 kg, 270 W

Table 62: CPR instrument parameters

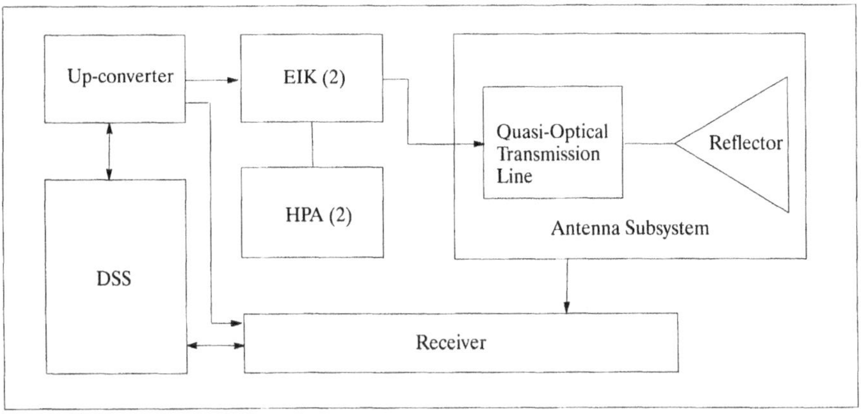

Figure 16: Simplified block diagram of CPR

Validation of the data products will use remote sensing measurements from surface and airborne platforms together with in-situ aircraft measurements of relevant cloud parameters as well as matching aircraft data with satellite data after launch. The validation strategy also involves the exploitation of existing cloud data bases. The CloudSat validation plan benefits from the systematic measurement programs of ARM as well as selected sites within Europe, regular aircraft radar measurement activities within the USA, Japan and Europe, measurement capabilities at a number of universities, and field-experiment activities representing targets of opportunity are planned in the coming years.

A.9 COMPASS

COMPASS [387] is a Russian microsatellite mission, funded by Rosaviakosmos, the Russian Academy of Sciences, and the Science Ministry; it is designed and developed at IZMIRAN (Institute of Terrestrial Magnetism, Ionosphere and Radio Wave Propagation of the Russian Academy of Sciences), Moscow, and at Makeev's Federal Rocket Center. The overall objective is to demonstrate possible topside sounding techniques that may eventually lead to the forecasting of natural disasters such as earthquakes. Specific objectives are:

- To develop coordinated monitoring methods (spaceborne and ground-based) to observe pre-earthquake phenomena
- To develop methods to monitor man-made catastrophes in general
- To study the dynamic coupling of the atmosphere, ionosphere and magnetosphere systems
- To acquire technical experience and know-how in the design and development of low-cost microsatellite building. [388] [389]

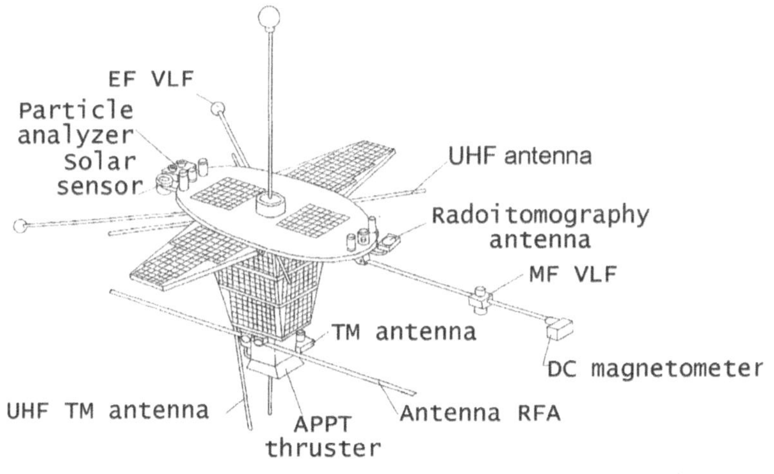

Figure 17: Illustration of the COMPASS microsatellite

The COMPASS spacecraft bus has the shape of an inverted (nadir looking) pyramid containing the instrumentation. Two deployable solar panel wings (total area of 1 m^2) are zenith facing. The S/C is three-axis stabilized. Attitude is sensed by sun sensors and by a DC

387) V. S. Dokukin, V. N. Oraevsky, et al., "The General Conception of the Microsatellite Compass to be launched from Submarine to the Study of Earthquake Forerunners," Proceedings of the 2nd IAA Symposium on Small Satellites for Earth Observation, Berlin, April 12-16, 1999, pp. 327-329

388) Information provided by Vladimir S. Dokukin of IZMIRAN, Troitsk, Russia

389) V. N. Oraevsky, V. S. Dokukin, V. A. Alekseev, Yu. Ra. Ruzhin, V. G. Degtiar, V. A. Danilkin, "The General Conception of the Microsatellite COMPASS to the Study of Earthquake Forerunners," presented at the 11th Meeting of the US/Russian Earth Sciences Joint Working Group, April 23-26, 2001, Washington, D.C., USA

magnetometer. Actuation is provided by three magnetic torquers and by a gravity-gradient boom. A pointing knowledge of about 1° is provided. Real-time orbit determination and on-board timing services are provided by a GLONASS/GPS receiver system. The solar power provided is 50 W (EOL) and 27 V. The NiCd battery subsystem has a capacity of 4 Ah. The OBC (On-board Computer) provides all control and processing functions required. On-board data storage of 512 Mbit is available. The S/C mass is 64 kg. The design life is five years.

A launch of COMPASS is scheduled for 2001 on a Zenit-2 launch vehicle from the Baikonur Cosmodrome, Kazakhstan. The prime payload on this flight is Meteor-3M-1. COMPASS is a secondary payload along with Badr-2 satellite of Pakistan, Maroc-TUBSAT of Morocco, and REFLECTOR (USA/Russia).

Orbit: Sun-synchronous circular polar orbit, mean altitude = 830 km, inclination = 98.85°, period = 102 minutes; local time of ascending node is 9:15 AM

RF communication: A VHF (8 W, 137 MHz transmitter) amateur-standard data link is provided with a data rate of 64 kbit/s. The functions of data acquisition and mission control are performed at IZMIRAN, Troitsk (Moscow region).

Sensor complement:

NVK-ONCH (Low Frequency Wave Analyzer). The instrument is a cooperative development of IZMIRAN, Eötvös University, Budapest, and the Lvov Center of Space Research, Ukraine. The objective is the measurement of the electric and magnetic vector components of the low-frequency radiation in the range of 8 Hz to 20 kHz.

RFA (High Frequency Wave Analyzer), a cooperative experiment of IKI (Space Research Center), Moscow, the Polish Academy of Sciences, Warsaw, and IZMIRAN. The objective is the high-frequency spectrum analysis of the electromagnetic radiation in the range of 0.1 - 15.1 MHz.

FM-3K (Fluxgate Magnetometer), developed at IZMIRAN. The objective is the measurement of the three vector components of the magnetic field.

GIT-12T (GLONASS/GPS Receiver System) of IZMIRAN. The objective is to provide radio-tomography (measurement of the TEC distribution) and orbit determination (position to within 30 m, velocity to within 10 cm/s). GIT-12T has 12 channels to track 12 GPS or GLONASS satellites in parallel.

PR41/53 (UHF Receivers), developed at IZMIRAN and the Technical Institute, Athens, Greece. The objective is the measurement of radio waves at 41 MHz and at 53 MHz.

A.10 Coriolis

Coriolis is a US DoD satellite demonstration mission within STP (Space Test Program), sponsored by ONR (Office of Naval Research) and NPOESS/IPO, and managed and implemented by NRL (Naval Research Laboratory). The overall objective is to conduct an operational verification of new instrument technology, aimed at validating space-based multi-channel polarimetric radiometry as a reliable and cost effective means for achieving wind vector measurements that meet NPOESS requirements. A further objective of the Air Force is to monitor solar activity for better predictive models of solar storms. [390)]

Coriolis is a fully redundant three-axis stabilized spacecraft (nadir pointing) built by Spectrum Astro Inc. of Gilbert, AZ. The S/C bus is based on the SA-200HP catalog bus, an Al honeycomb/Al facesheet deck structure, as flown on DS1, with specific tailoring to suit the

390) http://www.spectrumastro.com/PDFs/Coriolis.PDF

Coriolis requirements. The overall dimensions are: 4.69 m height (without solar array), 1.34 m diameter. Power is provided by a solar array assembly (1.174 kW) and by a single 50 Ah NiH_2 battery for ecliptic period operations. Attitude sensing is provided by autonomous sun acquisition using coarse sun sensors and magnetometers, actuation is provided by reaction wheels. A hydrazine propulsion system is used for orbit transfer. The S/C has a total mass of 827.4 kg (S/C bus of 377 kg), the design life is three years.

A launch of Coriolis is planned for spring 2002, aboard a Titan 2 rocket.

Communications: An on-board solid-state recorder of 30 Gbit capacity is provided. TT&C communications are provided in S-band. Instrument data are downlinked in X-band at 51.2 Mbit/s.

Orbit: Sun-synchronous circular orbit, altitude = 830 km, inclination = 98.7°, revisit time of 8 days, the longitude of the ascending node shifts westward by 138 km daily.

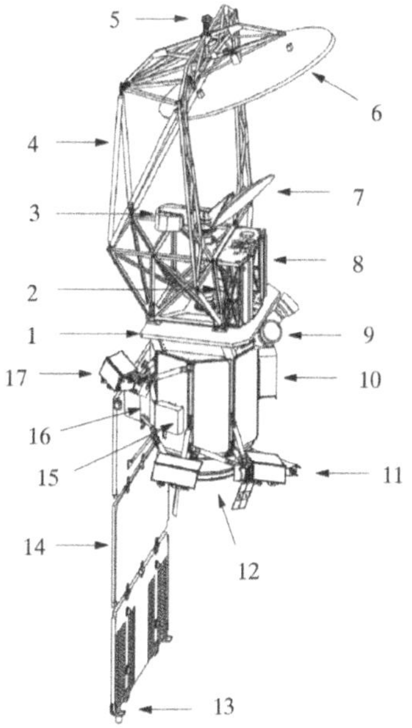

1) WindSat Stationary Deck
2) Bearing and Power Transfer Assembly
3) WindSat Warm Load
4) WindSat Spinning Composite Truss
5) WindSat GPS Antenna
6) WindSat Reflector
7) WindSat Cold Sky Reflector
8) WindSat Feed Through Horns
9) Attitude Control Bench
10) 19 Slot VME Chassis
11) Propulsion Boom
12) Propulsion Module
13) X-band Antenna
14) Solar Array Assembly
15) Data Handler Unit of SMEI
16) GPS Antenna (2)
17) SMEI Camera (3)

Figure 18: The Coriolis S/C model

Sensor complement:

WindSat (Wind Microwave Radiometer), a polarimetric microwave radiometer developed and built by NRL. [391] WindSat is also considered a precursor instrument to CMIS (Conical-scanning Microwave Imager/Sounder), a radiometric polarimeter to be flown on NPOESS (National Polar-orbiting Operational Environmental Satellite System). The objective of WindSat is to conduct an operational demonstration of the WindSat system and to measure the ocean surface wind vector (speed and direction) for a number of practical applications (secondary measurements are sea surface temperature, rain rate and water vapor). WindSat also provides insight into such problems as the upwind/downwind asymmetry and boundary layer conditions.

391) http://windsat.pxi.com/

The measurement principle makes use of the natural microwave emissions of the sea surface which vary with the degree of sea surface roughness. The rougher the seas, the more intense the emissions. The wind direction is obtained from the relationship between the horizontal and vertical polarization characteristics of the received signal and the anisotropic distribution of wind-driven waves.

WindSat is a 22-channel polarimetric radiometer operating at the frequencies of 6.8, 10.7, 18.7, 23.8, and 37 GHz. The receivers at 10.7, 18.7 and 37.0 GHz are fully polarimetric; that is, they establish all four Stokes parameters by measuring the six principal polarizations. The 6.8 GHz channel is dual-polarimetric (vertical and horizontal) and provides sea surface temperature (SST) as a secondary product. The 23.8 GHz channel is also dual-polarimetric, its purpose is to correct for atmospheric water vapor which is unpolarized.. The WindSat frequencies are useful beyond the measurement of ocean surface wind vectors and SST. Other parameters can be extracted/estimated from the data such as: atmospheric water vapor, cloud liquid water, rain rate, sea ice (age, concentration, and boundary), snow cover extent, and water content of snow. [392)]

Capability	NPOESS requirements	WindSat
Horizontal resolution	20 km	25 km
Mapping accuracy	5 km	5 km
Measurement range	3 - 25 m/s, 0 - 360° (direction)	3 - 25 m/s, 0 - 360° (direction)
Measurement precision Speed Direction	 1 m/s 10°	 1 m/s 10°
Measurement accuracy Speed Direction	 greater of ±2 m/s or ±20% ±20°	 greater of ±2 m/s or ±20% 3-5 m/s: ±20°

Table 63: Performance comparison of NPOESS requirements and WindSat parameters

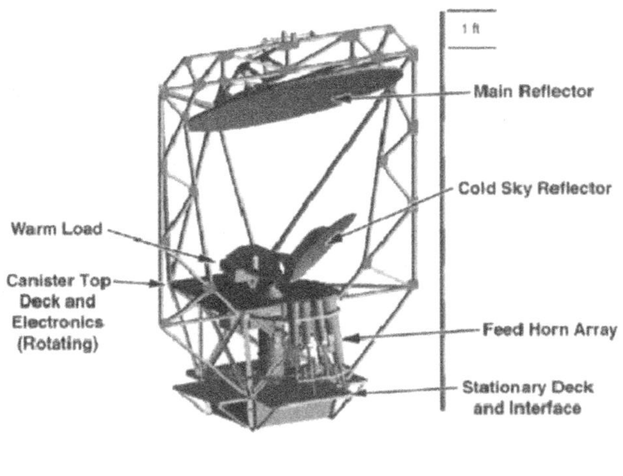

Figure 19: Illustration of the WindSat instrument

392) K. M. St. Germain and P. W. Gaiser, "Space borne polarimetric radiometer and the WindSat Coriolis mission,",Proceedings of the IEEE Aerospace Conference, IEEE Catalog No. 00TH8484C, March 2000.

WindSat uses a conically scanned 1.83 m parabolic offset reflector with multiple feeds. A forward swath width of 1025 km is obtained from an orbital altitude of 830 km. The placement of external calibration targets allows for an aft swath of 400 km. The availability of fore and aft swaths enables evaluation of one-look versus two-look wind vector retrievals. The WindSat instrument has not-to-exceed limits of 340 kg for mass and 350 watts of power.

SMEI (Solar Mass Ejection Imager) is being designed and built by an international consortium comprising the University of Birmingham, UK, the US AFRL (Air Force Research Laboratory), and the University of California, San Diego (UCSD). Funding is provided by NASA, AFRL and the University of Birmingham. SMEI) is an all-sky camera experiment capable of imaging CMEs (Coronal Mass Ejections) as they propagate from the sun through the solar wind. The objective is to monitor solar activity (detection of transient clouds of hot ionized gases - referred to as plasma) with the goal of more accurately predicting geomagnetic disturbances as they can have a serious effect on Earth-orbiting satellites, electrical power distribution networks, and long-distance radio communication. [393] [394] [395] [396]

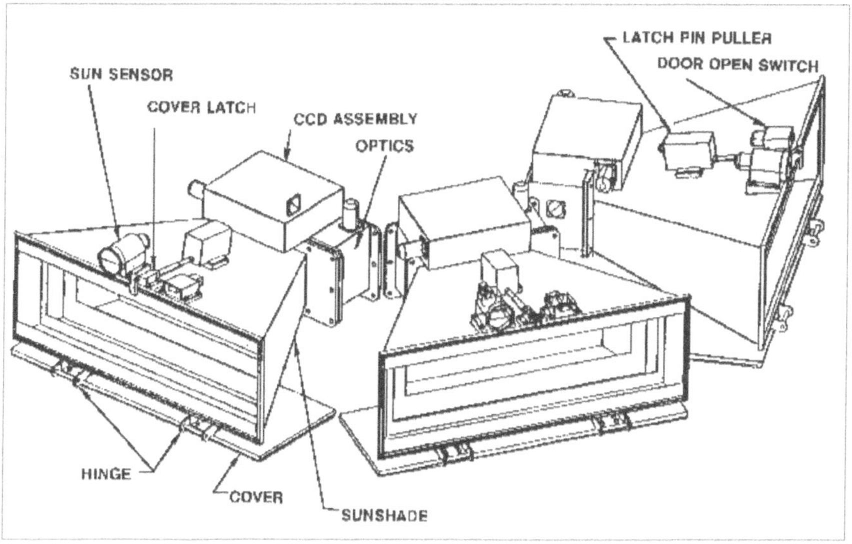

Figure 20: Illustration of the SMEI camera configuration

SMEI is designed to monitor solar optical radiation which is scattered off the plasma cloud as it moves out from the sun (measurement of the size, shape, speed, frequency, energy flux, mass, and momentum of CMEs). The SMEI instrument consists of three wide-angle CCD cameras each of which has a fan-beam field of view of size 3° x 60°. The cameras with their baffles are mounted onto the S/C in such a way that the combined 180° long fan-beam of the three cameras sweeps around the entire sky once per orbit. The baffles reduce the amount of scattered sunlight reaching a camera by about 10 orders of magnitude when it is viewing a region near the sun. The CCDs are cooled to about -35°C by passive radiators. A common data handling unit (DHU), interfacing to the S/C telemetry and telecommand systems, col-

393) http://www.sr.bham.ac.uk/instrument/smei.html

394) http://www.vsbs.plh.af.mil/projects/smei/smei.html

395) http://www.vs.afrl.af.mil/factsheets/SMEI.html

396) B. V. Jackson, A. Buffington, P. Hick, S.W. Kahler, S. L. Keil, R. C. Altrock, G. M. Simnett, D. F. Webb, "The Solar Mass Ejection Imager," Phys. Chem. Earth, 22, 441, 1997

lects the data from all three cameras and controls their operation (a MIL-STD-1553B bus is used for communications). The DHU includes a powerful processor system which is based around the digital signal processor SMJ320C50 of TI, operating at 19 MIPS making extensive use of field programmable gate arrays. The flight software controls the data transfer between the cameras, data compression, data transfer from SMEI to the S/C telemetry system, and operational commands to SMEI from the ground. - The SMEI data is used to estimate the outbreak of CMEs (success means that 1-3 day warnings can be given with regard to Earth-intersecting solar shocks, etc.).

After the contribution of stars is removed, SMEI's basic product will be a 100 minute (one orbit) cadence of all-sky maps of heliospheric brightness flux with an angular resolution of about 1° and a photometric precision equivalent to the brightness of a 12th magnitude star.

A.11 CRRES (Combined Release and Radiation Effects Satellite)

The CRRES program is a joint NASA/MSFC DoD/USAF mission in the field of solar-terrestrial energy transport with the objective to study ionospheric and magnetospheric transport phenomena and to study the effects of the natural radiation environment on microelectronic components (DoD). [397),398)] The satellite mass is 1753 kg (payload = 678 kg including the chemical canisters), nominal spin rate = 2 rpm. The spacecraft was launched on an Atlas/Centaur vehicle from Cape Canaveral on July 25, 1990. The nominal mission duration is 3 years, the extended mission is planned for 6 years [in support of the GGS (Global Geospace Science) mission]. The joint NASA/DoD mission is scheduled to last one year after launch, followed by a two year DoD science mission. After the third year of operation the CRRES mission will become part of the GGS program. The satellite failed on October 9, 1991.

Orbit: Initial elliptical orbit of 350 km x 35584 km, referred to as the Geosynchronous Transfer Orbit (GTO), inclination = 18.15°, orbit period = 10.56 hrs. The CRRES orbit plane is designed to precess in local time coordinates [taking advantage of perturbations (J2 perturbation due to the Earth's oblateness and the apparent 1°/day motion of the sun) resulting in a new rotation of the orbit perigee and apogee toward earlier local time, as the mission proceeds. Apsidal rotation also produces a periodic variation (36° peak to peak) in the latitude of perigee with a period of about 525 days. These two motions, given the initial local time of apogee, determined when and where, in local time and latitude, significant mission events such as the chemical releases occurred.

Investigation	Experiment Code	Mass (kg)	Data Rate (bit/s)	Principal Investigator
Chemical Releases	NASA	890		16 PIs from 13 institutions
Low Altitude Satellite Studies of Ionosphere Irregularities (LASSII)	NRL-701	28.5	13616	P. Rodrriguez, NLR
Space Radiation Experiment	AFGL-701	168.6	11027	E. G. Mullen, AFGL
Energetic Particles and Ion Composition (EPIC)	ONR-307	34.4	2674	R. Vondrak, Lockheed
Solar Flares II	ONR-604	15	750	J. Simpson, U. of Chicago
High Efficiency Solar Panel (HESP)	AFAPL-801	5.2	63	T. Trumble, Wright Patterson AFB

Table 64: The CRRES S/C science payload

CRRES carries a complement of chemical release canisters to be released at certain times over ground observation sites, and diagnostic facilities. These releases form large clouds of

397) R. A. Cooper, D. H. Burks, "Space Physics Missions Handbook," NASA, Office of Space Science and Applications, Feb. 1991

398) Special Section on CRRES, Journal of Spacecraft and Rockets, Vol. 29, No. 4 July-Aug. 1992, pp. 555-617

metal vapor that interact with the ionospheric and magnetospheric plasma and the Earth's magnetic field. These interactions will be studied with optical, radar, plasma wave and particle instruments from the ground, from aircraft, and from the CRRES spacecraft.

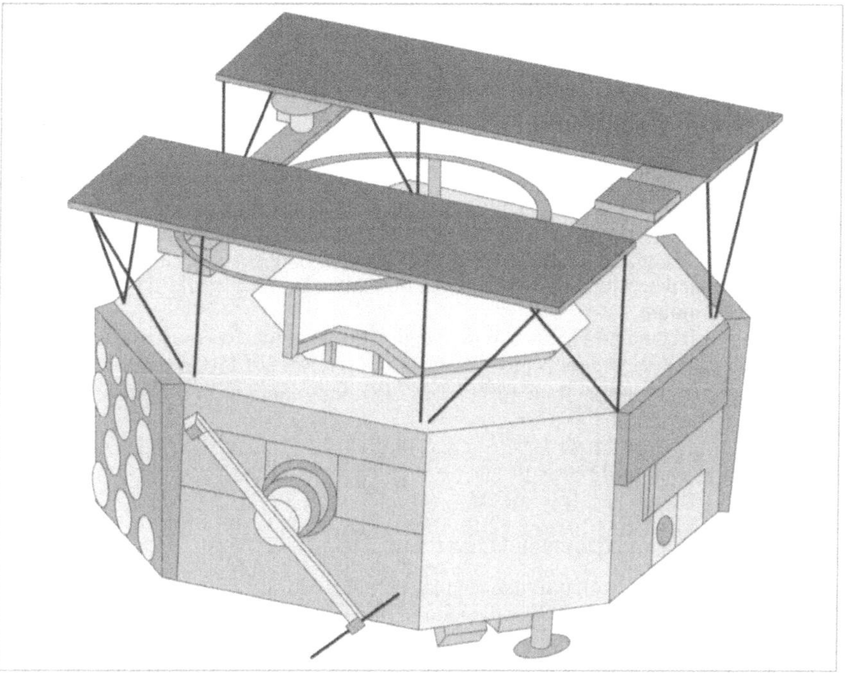

Figure 21: The CRRES S/C model

The experiments performed near perigee are intended to further understanding of the interaction of plasmas with magnetic fields, the coupling of the upper atmosphere with the ionosphere, the structure and chemistry of the ionosphere, and the structure of low-altitude electric fields. Those experiments which are performed near apogee in the Earth's magnetosphere are intended to study the formation of diamagnetic cavities, coupling between the magnetosphere and ionosphere, and the effects of artificial plasma injections on the stability of the trapped particles in the radiation belts.

The on-board scientific experiments consist of 46 electronic boxes, 10 booms, and 24 chemical canisters (Ba, Li, Ca, and Sr). There are 16 PI's responsible for the 16 separate chemical releases (24 canisters) at various times during the CRRES mission. The three chemical release campaigns were:

1) low-altitude releases (near perigee) over the South Pacific in September 1990.
2) high-altitude releases (from 6000 to about 33,500 km) over North America in January and February 1991.
3) low-altitude releases over the Caribbean in July and August of 1991.

Sensor complement:

LASSII = Low-Altitude Satellite Studies of Ionosphere Irregularities (PI: P. Rodriguez, NRL). Objective: monitoring naturally occurring and artificially produced ionospheric perturbations and the effects of ionospheric perturbations on communication paths. Measurements are conducted near perigee of selected orbits. LASSII is composed of three plasma experiments:

- P^3 = Pulsed Plasma Probe. A Langmuir probe experiment that operates as two pulsed plasma probes, measuring ionospheric electron densities and temperatures.
- QIMS = Quadrupole Ion Mass Spectrometer. Measurement of the densities of positive ions in both the natural ionosphere and in the chemical release experiments. Energy range from 500 keV to 100 MeV.
- ELFWA = Extremely Low Frequency Wave Analyzer. Measurement of the spectrum of the electric and magnetic field fluctuations (electric field spectra from 10 to 250 Hz, and single-axis magnetic field spectra from 10 Hz to 125 Hz.

EPIC = Energetic Particles and Ion Composition Experiment. EPIC consists of four instruments:

- IMS-HI = Ion Mass Spectrometer - High/Medium Energy. Measurement of energy spectra and pitch angle distributions of magnetospheric ions and neutral particles.
- IMS-LO = Ion Mass Spectrometer - Low Energy. Two instruments measure the composition of plasmas that are the sources of radiation belt particles, and provide data on the origin and acceleration processes of these plasmas. E/q = 0.11-35 keV/e, and M/q from 1-32 amu/e.
- EPAS = Electron and Proton Wide-Angle Spectrometer. It covers the energy range from 21 - 285 keV, and for protons from 37 keV - 3.2 MeV. EPAS consists of two units to measure electrons simultaneously in ten directions and ions in four directions over a total angle of about 100°.
- SEP = Spectrometer for Electron and Protons. SEP measures the energy and pitch angle distribution of energetic electrons and protons throughout the S/C orbit.

A.12 Dynamics Explorers (DE-1 and DE-2)

NASA/GSFC program with international cooperation on the payload.[399] Combined DE-1 and DE-2 launch on Aug. 3, 1981 with Delta vehicle from Vandenberg. Objectives: Study of the flow of solar energy and matter from space through Earth's magnetic field into the upper atmosphere (coupling of energy, electric currents, electric fields, and plasmas between the magnetosphere, ionosphere, and the atmosphere). The DE S/C have a spin rate of 10 rpm, mass = 424 kg (at launch), payload mass = 105 kg

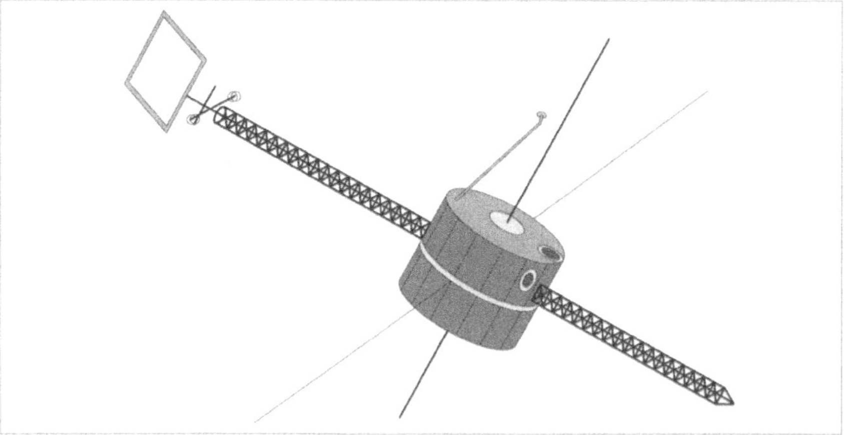

Figure 22: The Dynamics Explorer S/C model

399) R. A. Hoffman, G. D. Hogan, R. C. Maehl, "Dynamics Explorer Spacecraft and Ground Operating Systems," Space Science Instrumentation, 5, 1981, pp. 349-367

Orbit: Coplanar polar orbits;

- DE-1 (high altitude mission) 464 km x 23370 km highly elliptical orbit ; inclination = 90°
- DE-2 (low altitude mission) of 309 x 1012 km; inclination = 90°; period = 101 min; DE-2 re-entered the atmosphere on Feb. 19, 1983.

The orbits of the two spacecraft rotate in local time, precess and decay throughout the mission. This permits investigations in a multitude of constellations.

A.12.1 DE-1 Instruments (High Altitude Mission)

MAG-1 = Magnetometer (PI: Seguria, MSFC).[400)]
Measurement of the vector magnetic field. Objectives include: observation of field-aligned currents, magnetospheric equatorial currents, and ULF waves.
The instrument is a triaxial fluxgate magnetometer. Resolution = ±1.5 nT in basic mode. In addition there are two modes of higher resolution: ±0.25 nT and ±20 pT. The sampling rate is 16 vector samples per second in all modes.

PWI = Plasma Wave Instrument (PI: S. D. Shawhan, NASA).[401)]
Measurement of electromagnetic waves and electrostatic waves (PWI can distinguish between electromagnetic and electrostatic phenomena and determines wave polarization and wave propagation direction). The quasi-static electric fields are measured parallel to the spacecraft spin axis in a range of 2 mVm^{-1} to 2 mVm^{-1}, and perpendicular to the spin axis 0.5 mVm^{-1} to 2 mVm^{-1} at 16 samples per second. The AC electric field sensors include a 200 m tip-to-tip long wire antenna. Frequency range of AC fields: 1 Hz - 2 MHz. Integrated into this instrument is a Linear Wave Receiver (LWR, from Stanford) measuring wave amplitude in the frequency range 1.5 - 16 kHz.

HAPI = High Altitude Plasma Instrument (PI: J. L. Burch, SWRI).[402)]
Objectives: measurement of the velocity-space distributions of electron and positive ions. Information on the hot-plasma effects of the coupling of energy, mass, and momentum within the Earth's atmosphere, ionosphere, and magnetosphere. HAPI consists of five electrostatic analyzers mounted in a fan-shaped angular array at angles of 45°, 78°, 90°, 102°, and 135° with respect to the S/C spin axis, each making simultaneous measurements of electrons and positive ions at up to 64 energy steps from 5 eV to 32 keV.

RIMS = Retarding Ion Mass Spectrometer (PI: C. R. Chappel, MSFC).[403)]
Measurement of thermal ion density, temperature bulk velocity, and ion composition. RIMS consists of three nearly identical sensor heads which contain a retarding potential analyzer, followed by a magnetic ion mass spectrometer with two channel electron multiplier detectors.

EICS = Energetic Ion Mass Spectrometer (heritage of ISEE and GEOS-1,-2 spectrometers, PI: E. G. Shelley, Lockheed).[404)] Objective: measurement of the ion composition [energy and pitch angle distributions of the principal mass constituents (O^+, H^+) of the upward flowing ions from the auroral acceleration region]. EICS is a high sensitivity (≈ 1 cm^2 -sr-eV) high resolution ($M/\Delta M \gtrsim 10$ FWHM at focus) instrument. It covers the entire mass range from <1 amu/e to >150 amu/e in 64 mass channels at each of 32 energy per charge steps covering the range from 0 to ~ 17 keV/e. Data are used to investigate the coupling between the magnetosphere and the ionosphere via the auroral acceleration region.

400) W. H. Farthing, L. J. Cahill, et al., "Magnetic Field Observations on DE-A and -B," Space Science Instrumentation, 5, 1981, pp. 551-560
401) S. D. Shawhan, R. A. Helliwell, et al., "The Plasma Wave and Quasi-Static Electric Field Instrument (PWI) for Dynamics Explorer-A," Space Science Instrumentation, 5, 1981, pp. 535-550
402) J. L. Burch, R. A. Hoffman, et al., "High-Altitude Plasma Instrument for Dynamics Explorer-A," Space Science Instrumentation, 5, 1981, pp. 455-463
403) C. R. Chappell, J. H. Hoffman, et al., "The Retarding Ion Mass Spectrometer on Dynamics Explorer-A," Space Science Instrumentation, 5, 1981, pp. 477-491
404) E. G. Shelley, et al., "The Energetic Ion Composition Spectrometer (EICS) for the Dynamics Explorer-A," Space Science Instrumentation, 5, 1981, pp. 443-454

SAI = Spin-scan Auroral Imager (PI: L. A. Frank, U. of Iowa).[405] Measurement of visual and UV global auroral images. The instrument consists of three spanning photometers to view the dim emissions from Earth in the presence of strong stray light sources. The rotation of the S/C and the instrument scanning mirror provide a 2-D array of pixels comprising an image frame.

A.12.2 DE-2 Instruments (Low Altitude Mission)

MAG-2 = Magnetometer. Measurement of the vector magnetic field (see MAG-1).

VEFI = Vector Electric Field Instrument (PI: N. C. Maynard, GSFC)[406]
Measurement of the vector electric field. Instrument: Six 11 m cylindrical antennas are deployed along three orthogonal axes to form the sensor array. Measurements of the DC electric field are made 16 times per second. A 20 channel comb filter spectrometer monitors the AC electric fields, concentrating on frequencies below 1 kHz.

NACS = Neutral Atmosphere Composition Spectrometer (of Pioneer heritage, PI: G. R. Carignan, U. of Michigan). Measurement of the abundances of neutral species (atomic oxygen, helium, argon, molecular nitrogen) over an altitude range of at least 100 km to 300 km upward in the Earth's atmosphere. The NACS instrument is a quadrupole mass spectrometer with electron bombardment ionization (an electron multiplier in the counting mode is used for detection). [407]

WATS = Wind and Temperature Spectrometer (PI: N. W. Spencer, GSFC).[408]
Measurement of vertical, zonal and meridional components of neutral wind. In addition, measurements of the concentration and velocity of the ambient thermal ions are possible. Instrument accuracies: concentration of neutral particles ~20%; their temperature ~5 K.

FPI = Fabry-Perot Interferometer (PI: P. B. Hays, U. of MI).[409] Objective: Study of cross-sectional views of the dynamic and thermodynamic state of the thermosphere. Measurement of drift and temperature of neutral and ionic atomic oxygen in the thermosphere. FPI has a FOV of 0.53° (semi-cone angle) defined by a doublet objective lens and a field stop.

IDM = Ion Drift Meter (PI: R. A. Heelis, U. of Texas).[410]
Measurement of the components of the ion drift normal to the spacecraft velocity. Data are used in studies of the convection pattern and energy deposition in the ionosphere.

RPA = Retarding Potential Analyzer (PI: W. B. Hansen, Univ. of Texas).[411]
Measurement of bulk ion velocity in the direction of spacecraft motion, the constituent ion concentrations and the ion temperature along the satellite path. In addition, the spectral characteristics of irregularities in the total ion concentration are determined.

LANG = Langmuir Probe (PI: L. H. Brace, GSFC).[412]
Measurement of electron temperature and density and ion density in the Earth's ionosphere (heritage of Pioneer Venus mission).

LAPI = Low Altitude Plasma Instrument (PI: R. Hoffman, GSFC) .[413]
Objectives: measurement of high-resolution velocity-space distributions of positive ions

405) L. A. Frank et al., "Global Auroral Instrumentation for the Dynamics Explorer Mission," Space Science Instrumentation 5, 1981, pp. 369-393
406) N. C. Maynard et al., "Instrumentation for Vector Electric Field Measurements from DE-B," Space Science Instrumentation, 5, 1981, pp. 523-534
407) G. R. Carignan, et al., "The Neutral Mass Spectrometer on Dynamics Explorer B," Space Science Instrumentation, 5, 1981, pp. 429-441.
408) N. W. Spencer, et al., "The Dynamics Explorer Wind and Temperature Spectrometer," Space Science Instrumentation, 5, 1981, pp. 417-428
409) P. B. Hays, et al., "The Fabry-Perot Interferometer on Dynamics Explorer," Space Science Instrumentation, 5, 1981, pp. 395-416
410) R. A. Heelis, W. B. Hanson, et al., "The Ion Drift Meter for Dynamics Explorer-B," Space Science Instrumentation, 5, 1981, pp. 511-521
411) W. B. Hanson et al., "The Retarding Potential Analyzer for Dynamics Explorer-B," Space Science Instrumentation, 5, 1981, pp. 503-510

and electrons from 5 eV to 30 keV, and from 0° to 180° in pitch angle. The instrument contains an array of 15 parabolic electrostatic analyzers and two Geiger-Mueller counters mounted on a one-degree of freedom platform.

Data: On-board tape recorder (2 for each S/C). Science data transmission rate of 16.384 kbit/s.

A.13 ERBS (Earth Radiation Budget Satellite)

ERBS = Earth Radiation Budget Satellite. NASA Earth Radiation Budget Experiment (ERBE) Research Program (at GSFC). Launch of the free-flyer ERBS satellite from Space Shuttle Challenger: October 5, 1984 (STS-41G). ERBS was a three-axis momentum-biased S/C (1° pointing using magnetic torquers and a hydrazine backup system) built by Ball Aerospace Systems of Boulder, CO. The ERBS S/C structure was composed of three basic modules: the keel module, the base module, and the instrument module. The keel module was a torque-box structure providing structural support for the propulsion system, the solar array panels, and the antennas. The base module provided a direct interface to the Shuttle. S/C mass = 249 kg, power = 470 W. ERBS subsystems included TCS (Thermal Control Subsystem), EPS (Electrical Power Subsystem) which consisted of two 50 Ah, 22 cell NiCd batteries; PCU (Power Unit) for regulating electrical power; C&DH (Command and Data Handling Subsystem) for collection of instrument and S/C data for real-time transmission; CS (Communications Subsystem), which included NASA TDRSS transponders and antennas; AC&DS (Attitude Control and Determination Subsystem), a three-axis, momentum system for attitude pointing, maneuvers, and thruster control; and OAPS (Orbit Adjust Propulsion System), a non-propellent hydrazine propulsion system used for raising ERBS to its operating orbit after launch from the Shuttle. ERBS was held primarily in the Earth-pointing mode for most of the mission. Downlink data rate at 128 kbit/s, uplink through TDRSS using electrically steerable spherical array antenna. [414)]

Objective: Measurement of reflected and emitted energy at various spatial levels. Observations provide useful data for studies of geographical-seasonal variations of the Earth's radiation budget. Data are correlated with NOAA polar-orbiting TIROS S/C.

Orbit: Non-sun-synchronous circular orbit, altitude = 600 km, inclination = 57°, period = 96.4 min.

Status: As of June 2001 the ERBS mission is still operational after more than 16 years in orbit. However, the S/C is operational on only one battery. SAGE-II is operational, ERBE is only operational in periods of enough sunlight. Of the ERBE instrument only the non-scanner portion is still functioning (the scanner portion failed Feb. 28, 1990). ERBE is no longer capable of movement to the internal calibration position. The ERBE nonscanner was a real value, since it outlived it's design lifetime of 3 years by a factor of 5! It continues to acquire data. The non-scanning data will be processed until Sept. 2001. Although the ERBS spacecraft would probably function beyond 2010, de-orbit plans are being developed for 2003. [415)]

Sensor complement:

ERBE (Earth Radiation Budget Experiment). ERBE is a multimission system of NASA/LaRC to measure the Earth's radiation budget [i.e., the balance between incoming energy

412) J. P. Krehbiel, L. H. Brace, W. H. Pinkus, R. B. Kaplan, et. al., "The Dynamics Explorer Langmuir Probe Instrument," Space Science Instrumentation, 5, 1981, pp. 493-502

413) J. D. Winningham, R. A. Hoffman, et al., "The Low Altitude Plasma Instrument (LAPI)," Space Science Instrumentation, 5, 1981, pp. 465-475

414) J.A.Dezio, C.A. Jensen, "Earth Radiation Budget Satellite," in Monitoring Earth's Ocean, Land, and Atmosphere, Vol. 97 by AIAA, 1985, pp. 261-292

415) Information provided by Jack Paden and Bob Lee of NASA/LaRC

from the sun and outgoing thermal (longwave and reflected shortwave) energy from the Earth]. The instrument package (designed and built by TRW, Redondo Beach, CA) consists of a 'scanner' and a 'nonscanner' unit, providing measurements on several spatial and temporal scales.

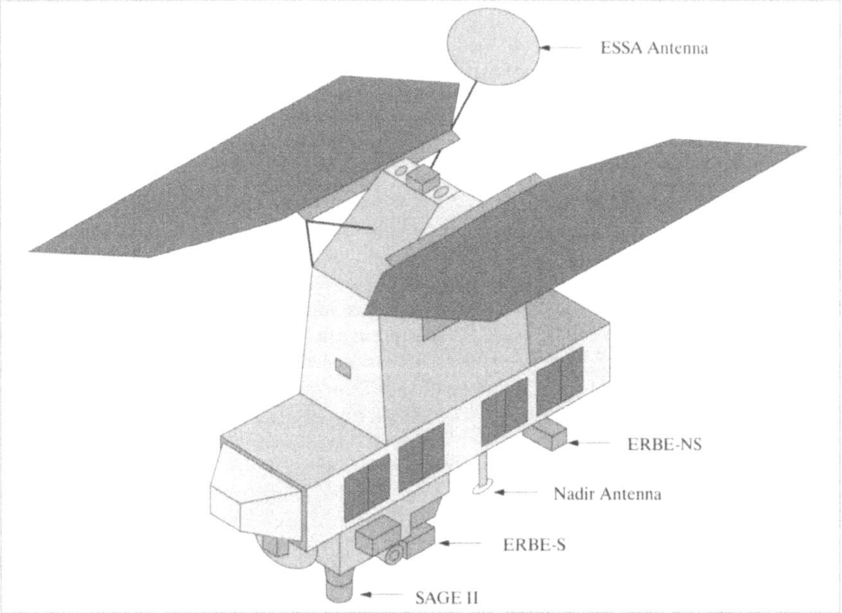

Figure 23: The ERBS S/C model

The nonscanner unit features four Earth-viewing channels and a solar monitor used for solar calibration measurements. The Earth-viewing channels have two spatial resolutions: a horizon-to-horizon view (wide FOV or WFOV), and a FOV limited to about 1000 km in diameter (also referred to as medium FOV, or MFOV). For MFOF and WFOV there is a total spectral channel sensitive to all wavelengths, and a shortwave channel which uses a high purity, fused silica filter dome to transmit only the shortwave radiation from 0.2 to 5 μm. All five channels of the nonscanner are active cavity radiometers. Data rate: 160 bit/s.

Nonscanner			
Channel Nr.	Spectral bands (μm)	Filter	Field of View (FOV)
1 (Wide FOV)	0.2 - 50+	None	Horizon - Horizon
2 (Wide FOV)	0.2 - 5	Suprasil-W Dome	Horizon - Horizon
3 (Medium FOV)	0.2 - 50+	None	10°
4 (Medium FOV)	0.2 - 5	Suprasil-W Dome	10°
5 (Solar monitor)	0.2 - 50 +	None	18° conical
Scanner			
6	0.2 - 5 (shortwave)	Suprasil-W Dome	3° x 4.5° (IFOV)
7	5 - 200 (longwave)	Diamond plus shortwave cutoff	3° x 4.5° (IFOV)
8	0.2 - 200		3° x 4.5° (IFOV)

Table 65: ERBE instrument parameters

The scanner instrument package contains three co-planar detectors (longwave, shortwave and total energy). Each detector scans the Earth perpendicular to the groundtrack from horizon to horizon. The detectors are thermistors (bolometers) which use space on every scan as a reference point to guard against drift. They are located at the focal point of a f/1.84 Cassegrain telescope whose aluminum-coated filters have been coated to enhance UV reflectivity. The IFOV for each channel is hexagonal, with an angular size of 3° (across track) x 4.5° (along track). Data rate = 960 bit/s.[416) 417)]

SAGE-II (Stratospheric Aerosol and Gas Experiment II). Earth limb-scanning grating spectrometer (with a Dall-Kirkham telescope, two-axis gimbaled system capable of rotating in azimuth). Objective: monitoring of concentrations and distributions of stratospheric aerosols, nitrogen dioxide, and water vapor. SAGE-II is a 7 channel radiometer. Spectral range: 0.385 - 1.020 μm. Data rate = 6.3 kbit/s. The instrument provides self-calibrating near global measurements. - The measurements are inverted using the "onion-peeling" approach to yield 1 km vertical resolution profiles of aerosol extinction (at 0.385, 0.453, 0.525, and 1.02 μm). The focus of the measurements is on the lower and middle stratosphere.

A.14 ESSP-3 (Earth System Science Pathfinder-3)

ESSP-3 is the temporary name of the former PICASSO-CENA (Pathfinder Instruments for Cloud and Aerosol Spaceborne Observations / Climatologie Etendue des Nuages et des Aerosols) minisatellite mission, a collaborative NASA/CNES project in the ESSP (Earth System Science Pathfinder) program of NASA. A new name for the mission couldn't be found until manuscript submission (objections were voiced by the Picasso family). A launch of ESSP-3 is planned in 2003 (co-manifested with CloudSat on a Delta-II vehicle). Other project partners (algorithm development) are Hampton University in Hampton, VA, and IPSL (Institut Pierre Simon Laplace), Jussieu, France. The overall science objective of the mission is to profile the vertical distribution of clouds and aerosols and their role in the heating/cooling of the Earth (improve of estimates for direct and indirect radiative forcing, improve accuracy of long-wave radiative fluxes at the Earth's surface and within the atmosphere, assessment of cloud feed back in the climate system). [418) 419) 420) 421) 422) 423)]

The mission is managed by NASA/LaRC. CNES supplies the S/C and an IR imaging sensor (of IASI heritage). NASA/LaRC provides the main instrument suite (a Ball-built lidar sensor) and S/C launch services aboard a Taurus launch vehicle. CNES provides S/C operations, the payload operations center is located at LaRC.

The S/C employes a Proteus bus, provided by CNES. It is three-axis stabilized. The S/C mass is 478 kg at launch, power = 232 W (average). An on-board data storage capability of 24 Gbit is provided.

Orbit: Near sun-synchronous orbit, altitude = 705 km, inclination = 98.05°. Observation strategy: The PICASSO-Cena orbit is optimized to form the so-called "A-train" (Aqua in the lead and Aura at the tail) or afternoon constellation (formation flight starting in the

416) URL: http://eosweb.larc.nasa.gov/HBDOCS/sensor_info.html

417) B. R. Barkstrom and J. B. Hall, Jr., "Earth Radiation Budget Experiment (ERBE): An Overview", J. Energy, Vol. 6, 1982, pp. 141–146

418) D. M. Winker, B. A Wielicki, "The PICASSO-CENA Mission," Part of the EUROPTO Conference on Sensors, Systems and Next Generation Satellites, Proceedings of SPIE, Vol. 3870, Florence, Italy, Sept. 20-24, 1999, pp. 26-36

419) J. Blouvac, B. Lazaed, J. M. Martinuzzi, " CNES Small Satellites Earth Observation Scientific Future Missions, IAA 2nd International Symposium on Small Satellites for Earth Observation, Berlin, April 12-16, 1999, pp. 11-14

420) J. Reagan, D. Winker, "PICASSO-CENA: Combined Active-Passive Sensing from Space," Proceedings of IGARSS'99, Vol. 1, pp. 240-242

421) D. M. Winker, "Global Observations of Aerosols and Clouds from combined Lidar and passive Instruments to improve Radiation Budget and Climate Studies," The Earth Observer, Vol. 11, No 3, May/June 1999, pp. 22-25

422) http://www–picasso–cena.larc.nasa.gov/picasso.html

423) D. Q. Robinson, "PICASSO-CENA Satellite-Based Research Mission: K-12 Education and Public Outreach (Student use of remote sensing for research validation)," Proceedings of the IEEE/IGARSS 2000 Conference, Honolulu, HI, July 24-28, 2000

2004 time period). The train will consists of the following S/C: Aqua, ESSP-3, PARASOL, CloudSat and Aura. The objective is to provide a coincident set of data on aerosol and cloud properties, radiative fluxes and atmospheric state essential for accurate quantification of aerosol and cloud radiative effects. The ESSP-3 orbit is maintained so that a target area can be observed within 6 minutes of each other (ESSP-3 and Aqua). The relevant Aqua instruments for ESSP-3 are: CERES, MODIS, AIRS, and AMSR-E.

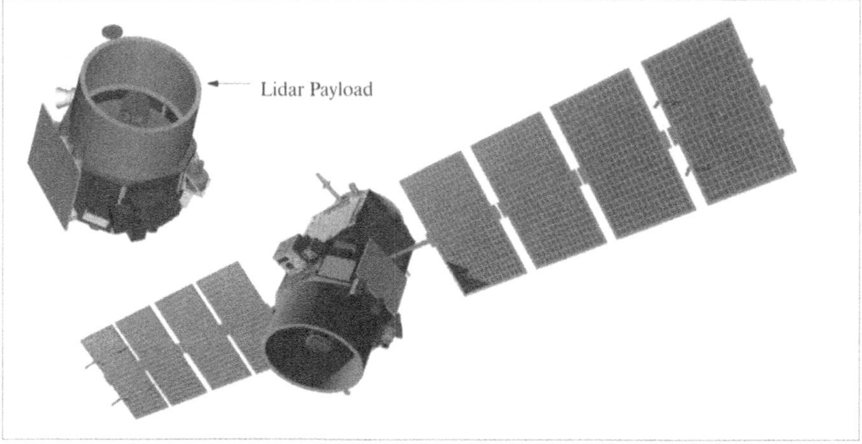

Figure 24: Illustration of the ESSP-3 spacecraft

Sensor complement: The payload consists of four co-aligned nadir viewing instruments.

Lidar, provided by LaRC (built by Ball Aerospace). The objective is to acquire vertical profiles of elastic backscatter (distributions of aerosols and clouds, cloud particle phase) at dual-wavelength frequencies from a nadir-viewing geometry during the day and night phases of the orbit. Two orthogonal polarization components (parallel and perpendicular to the polarization plane of the transmitted beam) are measured at 532 nm. Vertical profiles are measured from -2 to +40 km with the 532 nm channel, with the upper region used for normalization/calibration. The other two channels span -2 to +26 km, covering the cloud and aerosol measurement region with polar stratospheric clouds occurring toward the top of the range. The maximum vertical resolution is 30 m, the footprint diameter on the ground is 70.5 m. The horizontal spacing between footprint centers is 250 m. At altitudes above 5 km, resolutions are reduced by on-board averaging.

Lidar type	Nd:YAG, diode-pumped, Q-switched, frequency-doubled
Wavelength (2)	532 nm and 1064 nm
Energy/pulse	105 - 115 mJ
PRF (Pulse Repetition Frequency)	27 Hz
Pulse width	24 ns nominal
Telescope aperture diameter	1.0 m
Horizontal/vertical resolution	250 m/30 m
Data rate	279 kbit/s
Instrument mass, power	135 kg, 190 W

Table 66: Parameters of the Lidar instrument

IIR (Imaging Infrared Radiometer), provided by CNES. IIR features a microbolometer array technology. The objective is to measure calibrated radiances at 10.5 and at 12 μm over a 40 km swath (the two wavelengths are chosen to optimize the joint Lidar/IIR retrievals of cirrus emissivity and particle size).

Spectral bands (2)	10.5 µm and at 12.0 µm
Spectral resolution	0.8 µm
Spatial resolution (IFOV)	624 m
FOV or swath width	42 km
Data rate	20.5 kbit/s

Table 67: IIR specifications

WFC (Wide-Field Camera), built by Ball Aerospace. The objective is to provide meteorological context and highly accurate spatial registration between the ESSP-3 and the Aqua mission. WFC has a spectral coverage of 620-670 nm providing images of a 25 km swath with a spatial resolution of 125 m. The source data rate is 28.3 kbit/s.

A.15 FBM (French-Brazilian Microsatellite)

FBM is a cooperative mission between CNES and INPE. The overall objectives are to study the Earth's plasma environment and to test several technologies.[424) 425)] The FBM satellite design is part of the CNES Myriade series which encompasses the CNES microsatellite program with DEMETER, Microscope, PARASOL, and Picard. In this cooperative concept, INPE provides the launch of the S/C, the satellite structure, the TCS (Thermal Control Subsystem) and the PSS (Power Supply Subsystem); CNES provides OBDH (On-Board Data Handling), TMTC (Telemetry & Telecommand) and ACS (Attitude Control Subsystem).

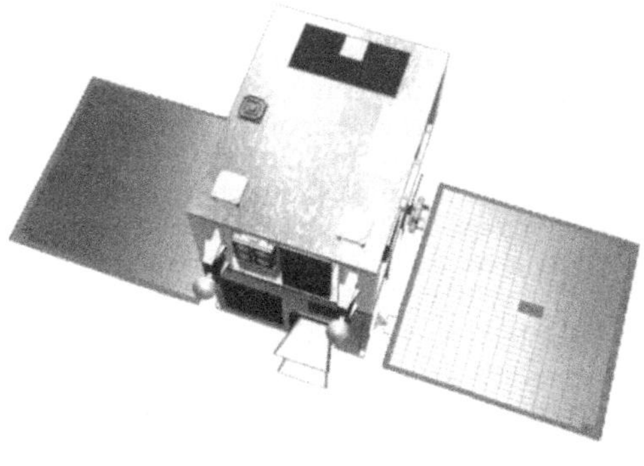

Figure 25: Illustration of the FBM satellite

The FBM satellite is composed of the payload module and the platform module, the later comprising the structure, TCS, ACS, OBDH, TMTC and PSS. The spacecraft is a three-axis stabilized (size of structure: 60 cm x 70 cm x 80 cm). ACS employs reaction wheels and torque rods for actuation, and a star sensor, a coarse sun sensor and a magnetometer for attitude sensing. A navigation software package supports various modes of operation. PSS comprises a solar generator (180 W at EOL) in two identical panels, a NiCd battery of 10 Ah capacity, a non-regulated bus power conditioning unit and a power distribution unit.

424) H. Carvalho, "The French Brazilian Microsatellite," Proceedings of the 51st IAF Congress, Rio de Janeiro, Brazil, Oct. 2-6, 2000, IAA-00-IAA.11.1.04

425) M. N. Barbosa, S. Plattard, "New opportunities for international cooperation - FBM, a French-Brazilian microsatellite to study the sun," CNES Magazine No 9, June 2000, pp. 33-34

OBDH consists of a computer (used also for ACS) and a 1 Gbit solid-state memory. The S/C mass is 93 kg, the design life is 13 months. A launch of FBM is scheduled for 2003 from the Alcantara launch site in Brazil on VLS (Veiculo Lancador de Satellites), Brazil's launch vehicle.

RF communications are provided in S-band (redundant transmitter/receiver and antennas). The downlink data rates are 200 kbit/s (nominal) and 25 kbit/s in safe mode. The modulation is in QPSK, the downlink frequency is 2255.2 MHz. The uplink data rate is 20 kbit/s. The CCSDS protocol is used for all data transmission and formatting.

Orbit: Circular near-equatorial orbit, altitude = 750 km, inclination = 6°.

The FBM ground segment consists of a ground station in Natal (Brazil) and a SCC (Satellite Control Center) at INPE. CNES acquires the spacecraft science data to its users.

Sensor complement:

APEX (Alpha, Proton and Electronics Experiment) an INPE instrument with the objective to monitor alpha, proton and electron particle fluxes and their temporal variations in the inner magnetosphere.

- Flux measurements of He ion events in the energy range of 3.3 to 29 MeV/nuc, in 4 discrete channels
- Flux measurements of proton events in the energy range of 2.2 to 29 MeV, in 5 discrete energy channels
- Flux measurements of electron events in the energy range of 0.1 to 3.6 MeV, in 4 discrete energy channels

The APEX instrument consists of telescopes. Instrument dimensions: 20 cm x 10 cm x 14 cm, mass = 3 kg, power= 3 W.

PDP (Plasma Diagnostic Package), an INPE instrument in collaboration with Japan. The objective is to measure the temperature, density and structure of the ionospheric plasma at low altitude near the equator [study of the electrodynamics and nonlinear processes of the equatorial ionosphere-thermosphere system (EITS)]. The spaceborne observations are supported by ground-based ionospheric and thermospheric measurements in Brazil and in equatorial West Africa using a network of diagnostic instruments that include radars, digisondes/ionosondes, optical imagers and interferometers and the available GPS satellites. High-resolution measurements of the plasma density, plasma temperature, spectral distribution of the irregularities using the following experiments:

- HFCP (High Frequency Capacitance Probe) for measuring the plasma density
- A Langmuir Probe (LP) for measuring the electron density and temperature variations and the spectral distribution of plasma irregularities
- Electron Temperature Probe for measuring the kinetic temperature of the ionospheric electrons.

The PDP instrument mass is 3.2 kg, power = 5.5 W.

CPL (Capillary Pumped Loop) an INPE instrument. The aim is to space-qualify a capillary-pumped loop in microgravity conditions. CPL is designed to transport up to 40 W from the evaporator plate to the radiator, using ammonia as the working fluid. The absorber plate is designed to absorb up to 30 W of solar energy. The CPL is tested in the following modes:

- On/off power input of 3 W in the reservoir and solar power input up to 30 W during the entire mission of the satellite.
- On/off power input of 3 W in the reservoir, power input of 15 W in the electrical resistance of the capillary pump and a solar power input up to 30 W, during special modes reserved to the CPL experiment
- Testing of the CPL without any kind of external control (solar power input = 30 W).

The instrument operating temperature is 280-300 K, the pressures is 550-1061 kPa, the capillary pumping pressure is 3-10 kPa. The measurement accuracy is ±0.5 K. The instrument mass is 2.9 kg.

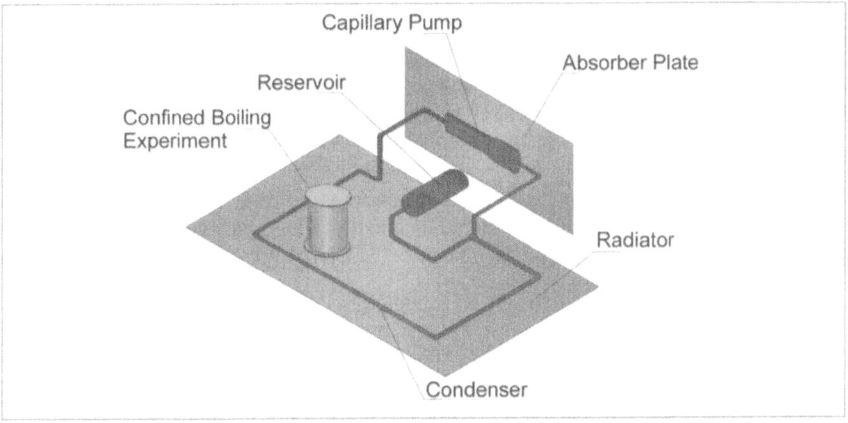

Figure 26: Schematic diagram of the CPL experiment

CBEMG (Confined Boiling Experiment under Microgravity), an INPE package. The objective is to study nucleate boiling and critical heat flux under microgravity conditions. Transient and steady-state tests will be conducted. The measured data will provide very useful information for designing and selection of devices for thermal control of satellites which employ two-phase heat transfer processes like heat exchangers, heat pipes and capillary pumps. [426)]

FluxRad (Fluxmeter Radiometer), an INPE instrument package. The objective is instrument qualification and to measure the net flux and its distribution on the outer surfaces of the spacecraft. Several HFM (Heat Flux Meter) groups with distinct surface emissivities are mounted on the spacecraft surfaces (sunlit and deep-space directions); in addition there are slug radiometers. The mean HFM temperatures and the slug radiometer temperatures are measured by a thermistors. The mass of FluxRad is 0.270 kg, power = 2 W.

CHADOCC (Circuit Hydraulique a Ammoniaque Diphasique en Orbite a Pompage Capillaire et Centrifuge). A CNES technology experiment to test and qualify an active thermal control concept (able to transfer up to 10 W) that makes use of a capillary fluid loop. The experiment will test two capillary devices: a Loop Heat Pipe (LHP) and micro Heat Pipe (μHP). [427)]

Debris, a CNES experiment to collect data about the distribution of orbital debris. The objective is to aid in the mapping of the space environment, modeling and to simulate it, besides the study of the effects on the specific mission.

Therme, a CNES experiment. The objective is to study the degradation of thermal coatings subjected to the space environment (charged particles, atomic oxygen, UV rays etc.).

MAGI, a CNES technology experiment to study and qualify a gyroscopic sensor that uses interferometric principles for angular measurements.

426) Information provided by Himilcon Carvalho of INPE
427) Information provided by Christophe Bastien-Thiry of CNES

A.16 FORTE (Fast On-Orbit Recording of Transient Events)

FORTE is a DOE-sponsored spacecraft designed and built by the Los Alamos National Laboratory (LANL) and by Sandia National Laboratories (SNL). The S/C features an advanced radio frequency (RF) impulse detection and characterization experiment. The prime objective is the measurement of electromagnetic pulses (EMP), primarily due to lightning, within a noise environment dominated by continuous wave (CW) carriers, such as TV and FM stations. The goal is to develop an understanding of the correlation between the optical flash and the VHF (30-300 MHz) emissions from lightning. Lightning is the main source of electromagnetic transients in this frequency range (nuclear weapon detonations produce also RF transients). [428) 429) 430)]

FORTE also conducts ionospheric physics experiments. The effects of large scale structures within the ionosphere are being studied [such as traveling ionospheric disturbances and horizontal gradients in the total electron content (TEC) on the propagation of broad bandwidth signals.

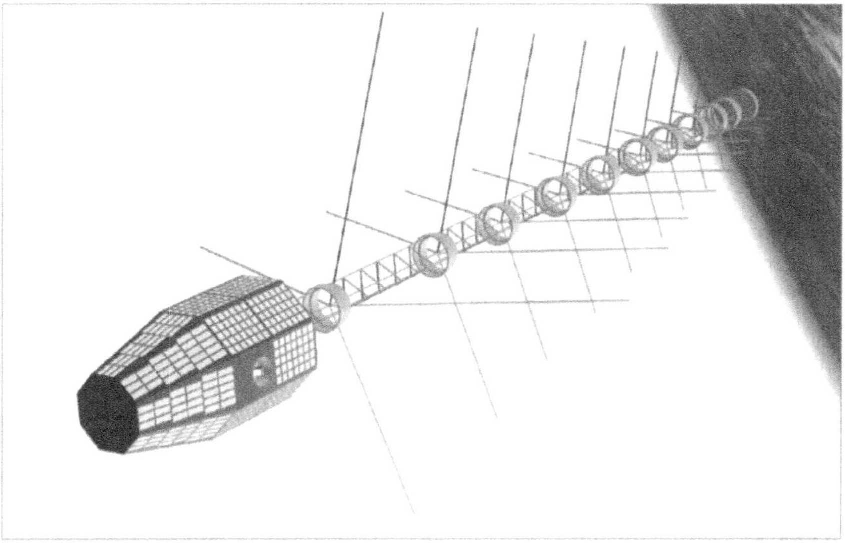

Figure 27: Illustration of the FORTE satellite

The S/C, developed in partnership by LANL and COI (Composite Optics Incorporated), features an all-composite structure (of 42 kg). The primary structure consists of six major structural components: three structural trusses and three structural instrument decks, along with 24 solar array substrate (SAS) panels.

The S/C has a total mass of 236 kg and an average power of 50 Watt (body-mounted solar cells). The S/C is nadir-pointing with biased momentum wheel and gravity-gradient stabilization. The attitude is sensed by magnetometers and an Earth horizon sensor. Satellite op-

428) C. I. Grastataro, T. A. Butler, et al., "Development of a Composite Satellite Structure for FORTE,". Proceedings of The Tenth International Conference on Composite Materials, Whistler, British Columbia, Canada, 1995,. LA-UR-95-1016

429) T. C. Thompson, C. I. Grastataro, et al., "Development of an All-Composite Spacecraft Bus for Small Satellite Programs,". Proceedings of The Eighth Annual AIAA/USU Conference on Small Satellites, Logan, UT, 1994

430) K. K. Ruud et al., "FORTE Hardware-in-Loop Simulation," Proceedings of AIAA/USU Conference on Small Satellites, 1997, pp. 1-9

erations are conducted from SNL in Albuquerque, New Mexico. A Pegasus XL launch took place on Aug. 29, 1997 from VAFB, CA.

Communications: The S/C is operated in store&forward fashion with an on-board data recorder of 1.4 Gbit. The uplink and downlink communications are in S-band. The downlink data rate is 2 Mbit/s.

Orbit: near-circular orbit of 830 km x 800 km altitude, inclination = 70°, orbital period of about 100 minutes.

Sensor complement: The FORTE payload consists of a suite of instruments, each individually useful, and together very useful, for the study of lightning (simultaneous observation of optical and RF emissions of lightning activity in the electromagnetically noisy environment of near-Earth space). [431]

The observation of "lightning events" by an RF system and by an OLS (Optical Lightning Subsystem) requires an EC (Event Classifier) as part of the FORTE payload. In addition, the on-board data acquisition system must have the capability to be triggered by an RF trigger, an optical trigger, or by the FPC (Flight Payload Controller). The EC is connected to the FPC (Flight Payload Controller). The FPC triggers the EC only when it is needed by enabling power to the EC. [432]

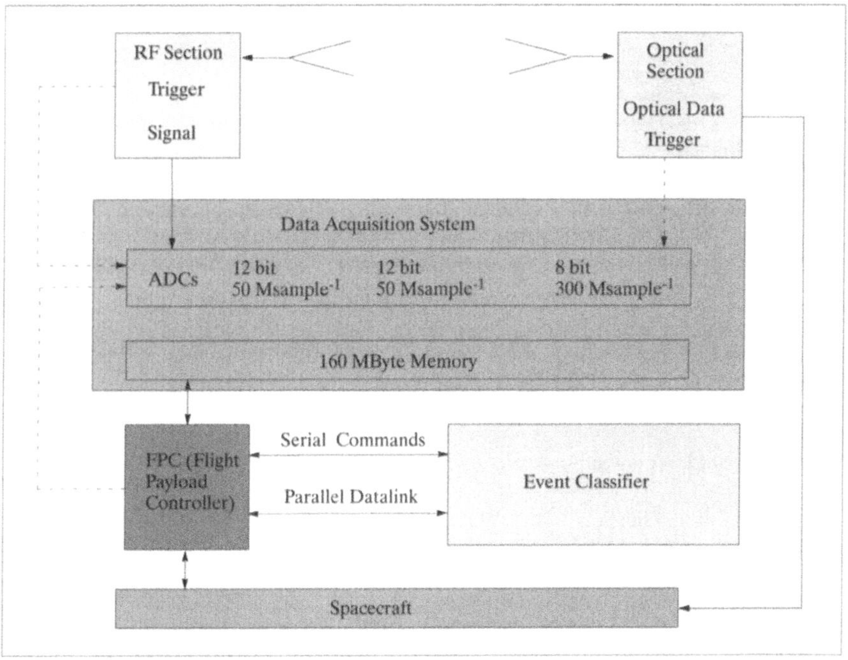

Figure 28: Schematic diagram of the FORTE experiment

EC (Event Classifier). [433] The EC is an experiment to determine the feasibility of telemetry reduction by onboard classification. The EC provides on-orbit characterization of im-

431) http://nis–www.lanl.gov/nis–projects/forte_science/

432) K. R. Moore, P.C. Blain, et al., "Classification of rf transients in space using digital signal processing and neural network techniques,". Applications and Science of Neural Networks, Proceedings SPIE, Vol. 2492, pp. 995-1006, 1995

pulsive RF events which satisfy the trigger criteria (extensive digital signal processing is involved). FORTE features a 10 m pseudo-log-periodic boom antenna coupled to a DSP (Digital Signal Processor) to capture the VHF (30-300 MHz) electromagnetic transients.

The EC, a DSP of TI (SMJ3320C30), executes neural network-based classification schemes. Communicating only with the FPC, the EC receives raw data and returns analysis reports via a 16-bit parallel port, it receives commands uplinks and state-of-health reports via a serial port.

There are three steps to EC demonstration testing: 1) Processing and classification algorithms are applied to digitized waveforms downlinked by the FORTE payload through the ground station to desktop computers. 2) Testing of promising algorithms in the FORTE engineering unit on the ground. 3) After successful ground testing, the EC is then commanded to perform the same analysis in space.
Successful demonstration of onboard event classification by the EC establishes confidence limits of classifications and enables follow-on missions to eliminate the costly data downlink and ground analysis steps.

RF (Radio Frequency) System. The RF system was developed and built by LANL. The RF system provides the following features: [434) 435)]

- Three broad bandwidth RF receivers (10 to 100 MHz bandwidth) covering the frequency range from 30 - 300 MHz. A multi-band coincidence trigger with perpetual recording of power background in all 16 trigger subbands (each 1 MHz wide).
- High-speed, low power digitizers (300 M samples/s)
- Broad bandwidth, dual polarization-selective VHF antenna
- Extensive on-board signal processing (adaptive discrimination, signal categorization by attributes)

The RF system derives its signals from either of two active monopoles, or alternatively from two mutually orthogonal, multi-element, passive, moderate-gain log-periodic antennas on nadir-directed deployed monopoles, or alternatively from two mutually orthogonal, multi-element, passive, moderate-gain log-periodic antennas on a nadir-directed deployed boom.

Both the RF System and the OLS (Optical Lightning Subsystem) are configurable for signal-triggered or for time-triggered digitization. There are also options for cross-triggering of PDD and LLS and for cross-triggering of PDD and the RF payload. With 160 MByte, there is ample memory on board for thousands of events to be stored and then downloaded. The downloading is done up to several times per day, at both Sandia and the University of Alaska.

OLS (Optical Lightning Subsystem), developed by SNL. The OLS instrument consists of:

- LLS (Lightning Location System) which employs a CCD imager with a ground resolution of 10 x 10 km (FOV = 80°, 500 frames/s)
- A fast time-response PDD (Photodiode Detector) for the recording of individual light curves. [436) 437)]

Photodiode/photometer systems are characterized by excellent temporal resolution (e.g. 10-100 μs) and generally poor spatial resolution (100-1000 km). Conversely, CCD imagers are primarily utilized to provide accurate geolocation and 2-D imagery of lightning events.

433) S. Briles, K. Moore, et al., "Innovative Use of DSP Technology in Space: FORTE Event Classifier,".1998

434) K. R. Moore, J. F. Wilkerson, et al., "A Space-based Classification System for RF Transients". Proceedings of the International Workshop on Artificial Intelligence in Solar-Terrestrial Physics, Lund, Sweden, p. 205, 1993

435) A. R. Jacobson, S. O. Knox, et al., "FORTE observations of lightning radio-frequency signatures: Capabilities and basic results," Radio Science, Vol. 34, 1999, pp. 337-354

436) D. M. Suszcynsky, T. E. Light S. Davis, J. L. Green, et al., "Coordinated Observations of Optical Lightning from Space Using the FORTE Photodiode Detector and Imager," Reg. LA-UR-00-341,

437) M. W. Kirkland, et al., "Observations of terrestrial lightning at optical wavelengths by the photodiode detector on the FORTE satellite," Rep. LA-UR-98-4098, LANL, 1998

They are characterized by excellent spatial resolution (1-10 km), low temporal resolution (1-10 ms), and a lack of waveform information.

The PDD employs an unfiltered, single element silicon photodiode with a sensing area of 1 cm^2. It is responsive to wavelengths of 400-1100 nm, with a peak response near 850 nm. A sunshade of 15 cm length provides a circular FOV of 80° (the total footprint is about 1200 km in diameter)to match that of LLS. The PDD measures the intensity of the incoming light and triggers on impulse events, thus providing a means to record the light-intensity time history. The PDD employs 12-bit sampling covering a dynamic range of more than four orders of magnitude. Two 128 x 128 pixel CCD (segmented array) detectors image the lightning; the input is the bright narrowband-filtered line emission centered at 777.4 nm wavelength.. In the "autonomous mode" operation, the instrument produces 1.92 ms long waveforms with 15 μs time resolution. The high temporal resolution of the PDD (about 15 μs) and the high spatial resolution of the LLS imager (about 10 km) combine to give a detailed satellite-based picture of both the spatial and temporal evolution of terrestrial lightning on both stroke (pulse) and flash timescales. OLS augments the RF system to develop high performance lightning discriminator/locator capability.

A.17 GCOM (Global Change Observation Mission)

GCOM is a NASDA program and a Japanese initiative with the overall objective to extend the observations (Earth climate monitoring), introduced by the ADEOS missions, on a long-term basis for the study and prediction of global-change phenomena to support the preservation of the environment. Three main goals are pursued: [438) 439)]

- Understanding of the energy cycle
- Study of the atmosphere/ocean interaction and radiative forcing, leading to the prediction of medium-to-long-term climate changes
- Understanding of the ozone and greenhouse gas circulation mechanism leading to the prediction of ozone layer and atmospheric composition variabilities.

The GCOM program, as defined in 2000, was previously referred to as ADEOS-III. The overall GCOM program consists of the GCOM-A and GCOM-B satellite series with their corresponding ground segments.

A.17.1 GCOM-A1

The emphasis is on atmospheric monitoring. [440) 441)] The mission objective is to support the Kyoto protocol that was adsopted at COP3/UNFCCC (3rd session of the conference in the framework of climate change) in 1997. The protocol calls for a reduction of greenhouse gases, in particular CO_2; it requires all parties to reduce their emissions by 5% below the level of the year 1990, for the period of 2008-2012. - The GCOM-A1 mission calls for the study of the transport mechanisms of greenhouse gases (CO_2, methane, CFC and halon), transport process of ozone and ozone layer depletion.

The S/C bus is three-axis stabilized with a structure size of 2.6 m x 2.1 m x 3.0 m. The AOCS (Attitude & Orbit Control Subsystem) is based on a zero-momentum design, attitude is sensed by Earth sensors, a star tracker and a GPS receiver. The EPS (Electrical Power Subsystem) uses a 28 V unregulated bus, the solar panels are of rigid padpole design with 1300

438) S. Sobue, N. Tomii, T. Moriyama, et al., NASDA's Future Earth Observation Satellite Plan," Proceedings of the IEEE/IGARSS 2000 Conference, Honolulu, HI, July 24-28, 2000

439) http://www.eorc.nasda.go.jp/GCOM/GCOM.PDF

440) M. Suzuki, K. Shibasaki, H. Shimoda, T. Ogawa, " Overview of GCOM-A1 Satellite Program," Proceedings of the IEEE/IGARSS 2000 Conference, Honolulu, HI, July 24-28, 2000

441) K. Shibasaki, M. Suzuki, Y. Yamamoto, "Ozone Dynamics Ultraviolet Spectrometer (ODUS) on Board GCOM-A1," Proceedings of the IEEE/IGARSS 2000 Conference, Honolulu, HI, July 24-28, 2000

W of power (EOL), and two NiCd batteries with energy of 35 Ah for solar eclipse operations. The overall S/C mass is about 1200 kg with an instrument mass of 256 kg. The overall design life is three years with propellant sized for five years. The planned launch date for GCOM-A1 is 2006.

Orbit: Non-sun-synchronous inclined circular orbit, altitude = 650 km, inclination = 69°, period = 98 min.

Communication: Science data is received and level-0 processed at NASDA/EOC (Earth Observation Center) in Hatoyama, Japan. The downlink is provided in X-band (8 GHz) with a data rate of 60 Mbit/s. The TT&C data link is DRTS compatible.

Sensor complement:

ODUS (Ozone Dynamics Ultraviolet Spectrometer). ODUS is a NASDA nadir-looking core instrument with the primary objective to measure the global total column ozone on a daily basis (target accuracy is 5% nominal and 2% after calibration/validation with a precision of 2%). Secondary objectives call for the measurement of volcanic sulfur dioxide (SO_2), of aerosols, NO_2, HCHO, BrO, and stratospheric OClO. In addition, cloud top heights are determined. ODUS employs the BUV (Backscattered Ultraviolet) technique to measure ozone and atmospheric trace gases. The design uses a Fastie-Ebert type spectrograph, measuring in the spectral range of 306 - 420 nm with 0.5 nm spacing. The spectral resolution is 0.5-0.7 nm. The instrument consists of a scanning subsystem, a spectrometer subsystem with detector assembly and calibration components, and a signal processing unit.

Instrument optics	Fastie-Ebert type grating (blazed holographic) monochromator
Spectral range, spectral resolution	306 - 420 nm, 0.5 nm
Detector, number of elements	CMOS type silicon photodiode array, 232 (with quantum efficiency >50%)
SNR	40 at 306 nm, >100 at 310 nm
FOV, IFOV (spatial resolution)	1.8° x 120° (swath of 2500 km), 1.8° x 1.8° (or 20 km x 20 km at nadir)
Instrument mass, power, size	<50 kg, <100 W, 650 mm x 650 mm x 750 mm
Data rate	<100 kbit/s

Table 68: ODUS instrument characteristics

The whiskbroom scanning technique is used for wide-field global coverage (80° N to 80° S) on a daily basis.

SOFIS (Solar-Occultation FTS for Inclined-orbit Satellite), sponsored by the Ministry of Environment and developed by NEC as the prime contractor. [442)] SOFIS is of ILAS (Improved Limb Atmospheric Spectrometer) and ILAS-II heritage, flown on ADEOS and ADEOS-II, respectively. SOFIS employs the solar occultation technique by measuring the vertical profiles of atmospheric constituents with a spectral resolution of 0.2 cm^{-1} in the spectral range of 3-13 μm with 1 km vertical resolutions. The main science objective of SOFIS is to monitor ozone and its minor constituents, thereby obtaining the global distribution of the following constituents/parameters: O_3, HNO_3, NO_2, N_2O, CH_4, H_2O, CO_2, CFC-11, CFC-12, $ClONO_2$, aerosol extinction, atmospheric pressure and temperature. The combination of SOFIS and ODUS instrument measurements onboard GCOM-A1 provide in effect 3-D global ozone distributions.

The SOFIS instrument employs a dual-pass flexible blade Michelson FTS (Fourier Transform Spectrometer) design as well as a diode laser sampling system to reduce the instrument size and mass. Two photovoltaic (PV) HgCdTe detectors and a pulse-tube cooler pro-

442) A. Kuze, H. Nakajima, J. Tanii, Y. Sasano, "Conceptual Design of Solar Occultation FTS for Inclined-orbit Satellite (SOFIS) on GCOM-A1," Proceedings of SPIE 45th Conference, San Diego, Jul. 30 to Aug. 4, 2000, Vol. 4131-30 (Remote Sensing and Infrared Systems), 2000

vide high linearity and low-noise level performance. FTS acquires a two-sided interferogram for phase information and for accurate inverse-transform performance (there is in addition the capability of separate interferogram acquisitions in two directions). The FTS controller uses a laser fringe counter that is updated by the initial scan with a white-light lamp prior to any measurements to enable the FTS to sample at the same fringe position during the same occultation. The instrument features also a visible (O_2 A band) grating spectrometer for pressure and temperature retrieval. A sun-edge sensor is used for the detection of the tangent height position. The SOFIS gimbal mechanism provides an azimuth scan of ±90°. A sun-tracking flat mirror and sun sensors are mounted on the two-axis gimbal. The sun sensors detect the radiometric center of the sun within ±5° and within ±0.5° at 875 nm and 1050 nm, respectively.

SOFIS configuration/accommodation parameters	
Measurement technique	Solar occultation of sunrise and sunset configurations
FOV	Altitude range of 5 km to 150 km (outer space)
IFOV	Tangent heights of 1.1 km (FWHM) at 6.5 μm, 0.8 km (interval)
Instrument calibration	Full level: pointing at the sun through outer space Zero level: pointing at deep space
Sun tracking	Radiometric center tracking with 2-axis gimbal
Instrument design life	3 years
Instrument data rate	45 Mbit/s
Instrument mass, power, size	190 kg, 250 W, 120 cm x 100 cm x 80 cm
FTS (Fourier Transform Spectrometer) parameters	
Spectral region -2 bands in PV-MCT design	3.25-6.5 μm or 3080-1540 cm^{-1} (no undersampling), 6.5 - 13 μm or 1540-770 cm^{-1} (interval 2 under sampling)
Spectral resolution (moderate)	0.2 cm^{-1} unapodized (interval), 0.24 cm^{-1} (FWHM)
Interferogram scan	0.23 s per interferogram, both sides, both ways
Zero path difference detection	White light
OPD (Optical Path Difference)	±2.5 cm;
Scan direction detection	Two polarized phase-shifted diode lasers and InGa detectors
Scan speed stability	±1% rms with servo control
SNR	>300 with 16 bit ADC
Sampling reference	Diode laser of 1550 nm wavelength
Beam splitter and compensator	ZnSe
FTS type	Double corner cubes, double path
Gimbal and telescope parameters	
Tracking range	Azimuth = 180°, elevation = 25°
Speed, acceleration	1°/s, 20°/s^2
Motor	DC brushless
Sun sensor	Coarse: Si position sensitive device Fine: InGaAs quadrant device
Resolver (data quantization)	18 bit
Telescope	Aperture diameter of 60 cm, F=5 Au coated
Sun edge sensor summary	
Image type	Reflected slit image
Detector	5000 pixel array using a silicon CCD (2500 pixels are used in elevation and 2500 pixels are in azimuth)
Data quantization	12 bit
Sun edge detection	Two images per interferogram
Solar image	Solar disk scan with gimbal elevation motion
Visible-range grating spectrometer parameters	
Spectral range, spectral resolution	753 - 784 nm, 0.06 nm
Optics, grating	Reflective (telecentric), 1800 grooves/mm (holographic)
Detector, data quantization	1024 pixels Si CMOS, 16 bit

Table 69: Overview of SOFIS instrument parameters

The FTS technique for solar occultation measurements requires a rapid interferogram acquisition method to achieve a high vertical resolution. The new technologies of high-speed A/D converters and precision linear motors offer such performance capabilities. By replacing the grating spectrometer with an FTS device and the uncooled thermal detector with a the cooled quantum detector, SOFIS is able to provide a 50 times higher spectral resolution than ILAS-II on ADEOS-II.

SWIFT (Stratospheric Wind Interferometer For Transport studies), sponsored by CSA (Canadian Space Agency) and designed/developed by a team from York University, Meteorological Service of Canada, and EMS Technologies Canada Ltd.. In 1998, the SWIFT concept was proposed to an ESA competition and was selected as one of five instruments for the Earth Explorer Mission series. ESA is supporting the accommodation of SWIFT on GCOM-A1, NASDA is providing the S/C and data acquisition. The scientific rationale for flying SWIFT is driven by the urgent need for global observations of vector-resolved vertical wind profiles between 18 and 40 km in the Earth's stratosphere. The intent is to address a) the factors/mechanisms influencing and controlling the stratospheric wind patterns, and b) to study stratospheric wind patterns and their role in controlling the global distributions of ozone and secular variations in ozone densities. The specific scientific objectives are: [443)]

- To study the ozone transport, including the Brewer Dobson circulation and horizontal ozone fluxes, from co-located wind and ozone measurements
- To determine the transport across the subtropical mixing barrier which is relevant to green house gases and halogen lifetimes
- To study tropical stratospheric winds, including QBO (Quasi Biennial Oscillation), SAO (Semi Annual Oscillation) and equatorial waves
- To assimilate the data with sophisticated GCMs (General Circulation Models), including model statistics, and a quality-controlled four-dimensional dataset.

The SWIFT wind and ozone measurements will be used to develop and refine general circulation models to provide more reliable predictions on global change phenomena. The SWIFT data are also used in data assimilation analysis to assess how stratospheric wind measurements would improve medium-term and long-term weather forecasts. SWIFT, combined with ODUS and SOFIS will provide major synergistic enhancements in the science return from all three instruments.

SWIFT is a spaceborne limb-viewing instrument (heritage of WINDII on UARS) designed to image the Doppler shift from thermal ozone emissions in the altitude range from 18 to 40 km (simultaneous observation of wind and ozone concentration profiles in the stratosphere). The operational concept employs optical Doppler imaging phase-stepping interferometry using a field-widened Michelson interferometer. Separation of the two components of the horizontal wind velocity of a given region on the limb is accomplished by having two fields of view at 45° and 135° to the spacecraft velocity vector. The measurements allow a reconstruction of the horizontal wind vectors from the two orthogonal components. A single thermal emission line from stratospheric ozone near 8.9 μm is isolated using two thermally stabilized and passively cooled 0.8 nm germanium Fabry-Perot etalons in series.

In the SWIFT concept, two off-axis telescopes view the limb in the fore and aft directions. Each has a FOV at the limb of 50 km x 100 km as seen from a S/C altitude of 600 km. Thermal radiation from the Earth's surface and warm baffles is prevented from entering the instrument via intermediate field stops and internal re-imaging of the telescope mirrors. The fore and aft FOVs are combined into one with a reflecting prism. This combined image is re-collimated and passed on to the Michelson interferometer with an optical path difference of 15 cm. For each wind measurement the mirror in one arm of the interferometer is piezoelectrically scanned over a single fringe in quarter-wave steps. The phase is then calculated for each pixel of the cryogenically cooled HgCdTe detector (128 x 128 pixel array,

443) G. G. Shepherd, I. C. McDade, W. A. Gault, Y. I. Rochon, A. Scott, et al., "The Stratospheric Wind Interferometer for Transport Studies (SWIFT)," 33rd COSPAR Scientific Assembly, Warsaw, Poland, July 16-23, 2000

cooled to 65 K using a mechanical Stirling cycle cooler), optimized for the observed wavelength, yielding a wind accuracy of 5 m/s. The instrument concept depends on a set of solid germanium etalon filters with clear apertures of 5 cm, bandwidths of 8 Å and 20 Å, finesses of at least 20, and which are cooled to about 150 K.

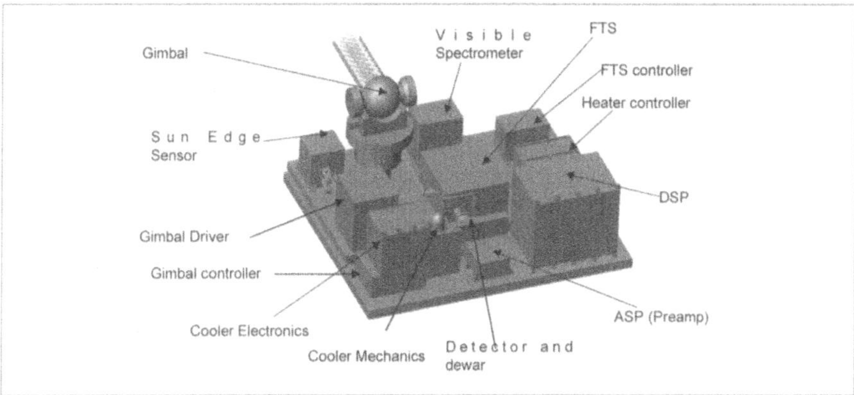

Figure 29: SOFIS instrument configuration

Limb observations are performed by a pointing mirror in SWIFT which directs the relatively narrow IFOV (1° vertical by 2° horizontal) towards the desired tangent height range (15 - 65 km) on the limb. Active pointing of the mirror ensures that the region of interest stays within the field of view. The mirror also periodically (as often as once every 12 s) turns the instrument's IFOV toward deep space to accurately monitor the thermal background flux. In addition, a periodic view to an internal phase calibration gas cell and to a radiometric (blackbody) calibration source is done to verify the instrument performance.

The volume of the SWIFT instrument is about 0.5 m^3, its mass is 75 kg, the power consumption is 75 W. The data rate is about 100 kbit/s.

A.18 HCMM (Heat Capacity Mapping Mission)

NASA Mission, also known as **AEM-1** (Application Explorers Mission-1). Launch: April 26, 1978, end of operation: Aug. 1980. The S/C was launched on a Scout-F vehicle from VAFB, CA. 444) 445)

The HCMM spacecraft was made of two distinct modules: 1) an instrument module, containing the heat capacity mapping radiometer and its supporting gear, and 2) a base module, containing the data handling, power, communications, command, and attitude control subsystems required to support the instrument module. The spacecraft was spin stabilized at a rate of 14 rpm. S/C mass = 134 kg. With no on-board data storage capability, only real-time data were transmitted when the satellite came within reception range of seven ground stations.

Orbit: Sun-synchronous near-polar orbit, apogee = 646 km, perigee= 558 km, inclination = 97.6°, period = 96.7 minutes, repeat cycle = 16 days, local equator crossing at 2 pm (ascending node). On Feb. 21-23, 1980 the HCMM orbit was lowered from 620 km to 540 km altitude to stop the drift of the orbit plane to unfavorable sun angles which in turn reduced the power collection capability of the solar panels.

444) P. Slater, 'Remote Sensing' Optics and Optical Systems, Addison-Wesley, 1980, pp. 462 - 465
445) HCMM System in 'Manual of Remote Sensing,' Second Edition, American Society of Photogrammetry, 1983, pp. 663-670

Application: Measurement of heat budget, surface temperature, soil moisture, urban heat islands, etc. HCMM was an experimental remote sensing mission to test thermal inertia to help discriminate between different surface materials, and to identify different states, such as degree of soil wetness.

Sensor:

HCMR (Heat Capacity Mapping Radiometer). Instrument mass = 38.9 kg. The radiometer was similar to the surface composition mapping radiometer (SCMR) of Nimbus-5. The HCMR had a small instantaneous geometric field of view (IFOV) of 0.83 mrad, high radiometric accuracy, and a wide 716 km swath coverage on the ground so that selected areas were covered within the 12 hour period corresponding to the maximum and minimum of temperature observed. The instrument operated in two channels and utilized a radiation cooler to cool the two Hg:Cd:Te detectors to 115 K. The experiment included an analog multiplexer that accepted the analog outputs of the detectors and multiplexed them in a form suitable for transmission by the spacecraft S-band transmitter. The instrument performed satisfactorily until the spacecraft operations terminated on September 30, 1980.

Spectral range (channel 1)	0.55 - 1.11 μm (solar reflection)	pixel size = 500 x 500 m
Spectral range (channel 2)	10.5 - 12.5μm (thermal emission)	pixel size = 600 x 600 m
Image size	2016 lines (1680 pixels/line)	8 bit resolution

Table 70: Parameters of HCMM

HCMM data were stored and processed at GSFC. Data products: geometrically corrected and calibrated images of radiance or equivalent black-body temperature.

A.19 Megha-Tropiques

Megha-Tropiques [446)] is a French/Indian cooperative mission of CNES and ISRO with the objective to study the convective systems (water cycle and energetic exchanges) that affect the intertropical zone. [447)] The satellite uses the French Proteus bus, the S/C integration is performed at CNES while the payload integration is done at ISRO. On the operational side, CNES is providing the S/C operations via its own ground station; ISRO's ground station at Bangalore is used as the Mission Science Center. The science mission is also supported by CNRS/LMD (Laboratoire de Météorologie Dynamique) of Palaiseau, France. A statement of intent was signed on Nov. 21, 1999 in Bangalore between ISRO and CNES. [448)] [449)] [450)]

Data gathered by the mission will be used to predict tropical cyclones. The goals are to:

- Provide simultaneous measurements of several elements of the atmospheric water cycle: water vapor, clouds, condensed water in clouds, precipitation and evaporation
- Measure the corresponding radiative budget at the top of the atmosphere
- Ensure high temporal sampling in order to characterize the life cycle of the convective systems.

The satellite uses the Proteus bus of CNES/Alcatel. It is three-axis stabilized providing a pointing accuracy of 0.1°. The S/C has a mass of 570 kg (260 kg of payload mass), power =

446) Megha means "cloud" in Sanskrit; Tropiques is the French word for "tropics."

447) J. Blouvac, B. Lazaed, J. M. Martinuzzi, " CNES Small Satellites Earth Observation Scientific Future Missions, IAA 2nd International Symposium on Small Satellites for Earth Observation, Berlin, April 12-16, 1999, pp. 11-14

448) J. P. Aguttes. J. Schrive, C. Goldstein, G. Raju, M. S. Narayanan, M. Desbois, "Megha-Tropiques, A Satellite for Studying the Water Cycle and Energy Exchanges in the Tropiques," IAF Congress, Rio de Janeiro, Brazil, Oct. 2-6,-2000

449) J. P. Aguttes, J. Schrive, Ch. Goldstein, M. Rouzé, G. Raju, "MEGHA-TROPIQUES, a satellite for studying the water cycle and energy exchanges in the tropics," IEEE/IGARSS Conference 2000, Honolulu, HI, July 24-28, 2000

450) Illustration courtesy of Nadia Karouche of CNES

591 W (291 W for payload). The design life is 3 years. A launch on an ISRO PSLV launcher from SHAR is planned for 2006.

Orbit: Circular orbit, altitude = 867 km, inclination = 20°, period = 100 min.

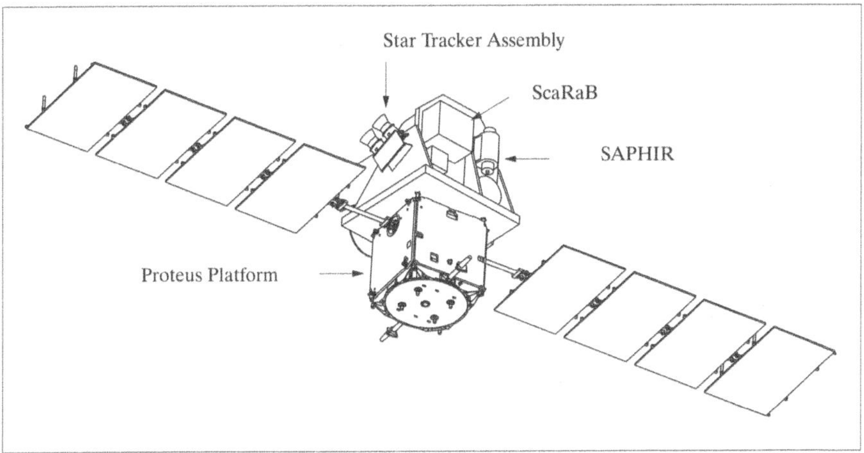

Figure 30: Illustration of the Megha-Tropiques satellite

Sensor complement:

ScaRaB (Scanner for Radiation Budget), sponsored by CNES and developed at LMD. SCARAB is of Meteor-3-7 heritage. Its objective is the collection of data on shortwave and longwave radiation (reflected solar and emitted thermal radiation) to estimate the Earth's radiation budget at the top of the atmosphere on global and regional scales. The instrument is a cross-track scanning radiometer featuring four channels. Channels 2 and 3 are considered the main channels, while channels 1 and 4 are auxiliary channels. The optical subsystem features four parallel telescopes, one telescope per channel, they are identical except for their filters. 451) 452) 453)

Nr.	Spectral band (μm)	Band description	Filter	Signal W m^2 sr^{-1}	Noise W m^2 sr^{-1}
1	0.5 - 0.7	Visible channel: scene identification	Interference	120	<1
2	0.2 - 4	Solar channel: derivation of Earth radiation budget parameters	Fused silica	425	<0.5
3	0.2 - 200	Total radiation channel	None	500	<0.5
4	10.5 - 12.5	Atmospheric channel scene identification	Interference	30	<0.5

Table 71: Spectral bands of ScaRaB

ScaRaB uses BARNES pyroelectric detectors for all bands (placed at the focus of a spherical aluminium mirror), which are sensitive only to the AC component of the signal (i.e., the modulated energy). Hence, chopping is needed for each pixel. This reduces the influence of the self radiation of the telescope and filters. Two mechanical choppers are used (one for two channels), providing a 10 Hz chopping frequency. The four channels, the two choppers, and a filter wheel dedicated to channel 2 and 3, are mounted on a scanning optical bench

451) http://scarab.cnes.fr:8020/

452) J. L. Monge, R. Kandel, L. A. Pakhomov, B. Bauche, "ScaRaB Earth radiation budget scanning radiometer," SPIE, Vol. 1490 , 'Future European and Japanese Remote Sensing Programs,' 1991

453) J. Mueller, et al., "Ground Characterization of the Scanner for Radiation Budget (ScaRaB) Flight Model 1," Journal of Atmospheric and Oceanic Technology, Vol. 14, No 4, pp.802-813, 1997.

(rotor). The telescopes are swiveled by the optical bench so that no extra mirror for the scanning is needed. This reduces the likelihood of offsets dependent on the scanning angle.

The spatial resolution of ScaRaB data is 48 x 48 mrad, scan angle = 100°, swath width = 3200 km. ScaRaB points to nadir and scans the full field of view (FOV) within six seconds. In this cross-track mode data are generated continuously.

Parameter	Value	Parameter	Value
IFOV (spatial resolution)	48 mrad x 48 mrad (60 km x 60 km at nadir)	Sampling interval Sampling period	34 mrad 62.5 ms
FOV (swath)	100° (3200 km)	Scan period	6 s
Pixels per scan	51	Useful scan time	3.18 s
Dynamic range (solar)	up to 425 W m^{-2} sr^{-1}	Instrument mass, power	40 kg, 42 W (average)
Dynamic range (total)	up to 500 W m^{-2} sr^{-1}	Instrument size (mm)	614 x 512 x 320

Table 72: ScaRaB instrument parameters

Calibration subsystem: Gray lamps and blackbodies are used for on-board gain calibration; deep space is used for offset calibration. That subsystem comprises a set of two reference blackbodies for channels 3 and 4, and a set of gray calibration lamps for channels 1, 2 and 3. There is continuous thermal control of the blackbodies. The gray lamps are turned on during the calibration session (typically once per day). In addition, there are short wave references, consisting of two lamps for the calibration of channels 2 and 3 (typical use is once per month). On the ScaRaB/Meteor-3-7 mission, however, the lamp system was damaged so that actual calibration was performed by using the instrument temperature and a prelaunch established gain-temperature law. The remaining lamps were then used to verify this calibration. During one year of operation, no significant sensor degradation was observed.

ScaRaB has a duty cycle of 100%, data rate=3 kbit/s, data volume=18 Mbit/orbit. An instrument mass memory provides data storage for up to 12 hours. The mass of the instrument is 20 kg, the maximum power use is 33 W.

The data processing system is based on algorithms for transforming the instantaneous measurements of radiances, filtered by the optics and detectors, into estimates of the monthly mean values of the radiant excitations in the solar and thermal domains, at the top of the atmosphere. This requires corrections for non-flat spectral response, anisotropic, and diurnal variations. The estimates are provided on a spatial grid of 250 km.

SAPHIR (Sondeur Atmospherique du Profil d'Humidite Intertropicale par Radiometrie), developed by CETP (Centre d'etude des Environnements Terretre et Planetaires), by DEMIRM (Department de Radioastronomie Millimetrique de l'Observatoire de Paris), and by LMD. SAPHIR is a cross-track sounding instrument with the objective to measure water vapor profiles in the troposphere (water absorption band at 183.3 GHz) in six layers from 2-12 km altitudes and spatial resolutions of 10 km. The measurement sensitivity is 1.5 K with an inter-channel calibration of 0.5 K. The measurement polarization is H.

Cha. Frequency	Bandwidth	Spatial resolution	Measurement objective
18.7 GHz	±100 MHz	<40 km x 66 km	Rain above oceans
23.8 GHz	±200 MHz	<40 km x 63 km	Integrated water vapor
36.5 GHz	±500 MHz	<40 km x 63 km	Liquid water in clouds, rain above sea
89 GHz	±1 GHz	<10 km x 16 km	Convective rain regions over land & sea
157 GHz	±1 GHz	< 6 km x 9 km	Ice and cloud tops, window/SAPHIR

Table 73: Specification of the MADRAS channels

MADRAS (Microwave Analysis & Detection of Rain & Atmospheric Structures) developed by ISRO with CNES providing the RF assembly (dish, horns, receivers). The main objective of the microwave imager is to measure precipitation and cloud properties. MADRAS is a five-frequency (9 channel) conical-scanning radiometer. The frequencies of 89 and 157 GHz are responding to ice particles in cloud tops, thus permitting the detection of convective rain regions over land and sea. The lower frequency channels are employed over oceanic regions for the measurement of cloud liquid water and precipitation (absorption at 18.7 and 36.5 GHz, integrated water vapor at 23.8 GHz, and estimation of the sea surface wind speed with the 18.7 GHz channel. The polarization is in H+V, except for the 23.8 GHz channel which is H or V polarized. The sensitivity requirement is 1 K for the 89 and 157 GHz channels and 0.5 K for the other channels. Inter-channel calibration is 0.5 K. The dish effective diameter is 65 cm. The beam efficiency is 0.95. A single horn is used for the three channels: 18.6, 23.8 and 36.5 GHz. The incidence angle is between 50°-55° (upper limit for each channel). The mass of MADRAS is 150 kg, power consumption is 180 W.

A.20 ODIN

Odin is a Swedish minisatellite mission for astronomical and atmospheric research (previously named MOSES). The mission design life is 2 years minimum. The project is carried out jointly by the space agencies of Sweden (SNSB), Canada (CSA), Finland and France (CNES). Swedish Space Corporation (SSC) is responsible for system design and project management. A launch took place on Feb. 20, 2001 on a Russian Start-1 launch vehicle from the Svobodny Cosmodrome, Russia.

The S/C structure is of aluminum honeycomb with carbon reinforced plastic for the reflector support structure. ODIN is three-axis-stabilized with reaction wheels, star trackers and gyros; mass = 250 kg (80 kg payload); size: 2.0 m height, 1.1 m width stored and 3.8 m in flight configuration; power =260 W from deployable fixed solar arrays, during eclipse ODIN is powered by 6-Ah NiCd batteries; cooling: closed Stirling cycle coolers; pointing = ±15 arcsec in staring mode, ±1.2 arcmin in scanning mode; datalink: S-band at 720 kbit/s to Esrange tracking station; on-board storage >100 MByte in solid state memory.

Applications: astronomy and aeronomy (atmospheric research: stratospheric ozone chemistry, mesospheric ozone science, summer mesospheric science, coupling of atmospheric regions). [454] [455] [456]

Objectives: Measurements in the wavelengths of 0.5 - 0.6 mm and 2.5 mm. These contain emission lines from molecules such as water vapor, molecular oxygen, ozone and carbon monoxide which are important for the study of atmospheric processes as well as for the study of astronomical objects. Complementary information on the Earth's atmosphere comes from spectral lines at UV and VIS wavelengths. Major scientific issues relate to star formation processes, interstellar chemistry and atmospheric ozone balance.

Orbit: Sun-synchronous polar orbit; altitude = 625 km; inclination = 97.8°; with an ascending node at 18:00 hrs; period = 97.6 minutes.

Operating modes: For aeronomy the spacecraft follows the Earth limb, scanning the atmosphere up and down from 15 km to 120 km at a rate of up to 40 scans per orbit. When observing astronomical sources Odin continuously points towards the object for up to 60 minutes. Note: Both sensors point into the same direction.

Sensor complement:

SMR = Submillimeterwave Radiometer [Swedish instrument in cooperation with Finland (119 GHz channel), and France, provision of the AOS (Acousto-Optic Spectrometer) de-

454) F. v. Scheele, "Star Formation and Ozone Depletion: The Swedish ODIN Satellite to Eye Heaven and Earth," Nordic Space Activities, No. 5, 1994, pp. 44-46

455) "ODIN - A Small Satellite for Astronomy and Atmospheric Research," SSC/SNSB brochure

456) http://www.ssc.se/ssd/

tector]. This is a passive microwave limb sounder with one receiver at a wavelength $\lambda = 3$ mm and additional four bands within the submillimeter range (0.5 - 1.0 mm). Antenna reflector type: offset Gregorian telescope [off-axis system, 1.1 m diameter, surface: 10 μm rms, material: carbon fiber reinforced plastic (CFRP)]. SMR is used in the astronomy as well as in the atmospheric research mission to detect molecular transitions. The signal coming from the telescope is routed to the receivers, split and filtered by optics consisting of combinations of mirrors, grids and meshes. Diplexers and sideband filters are based on tunable polarizing Michelson interferometers.

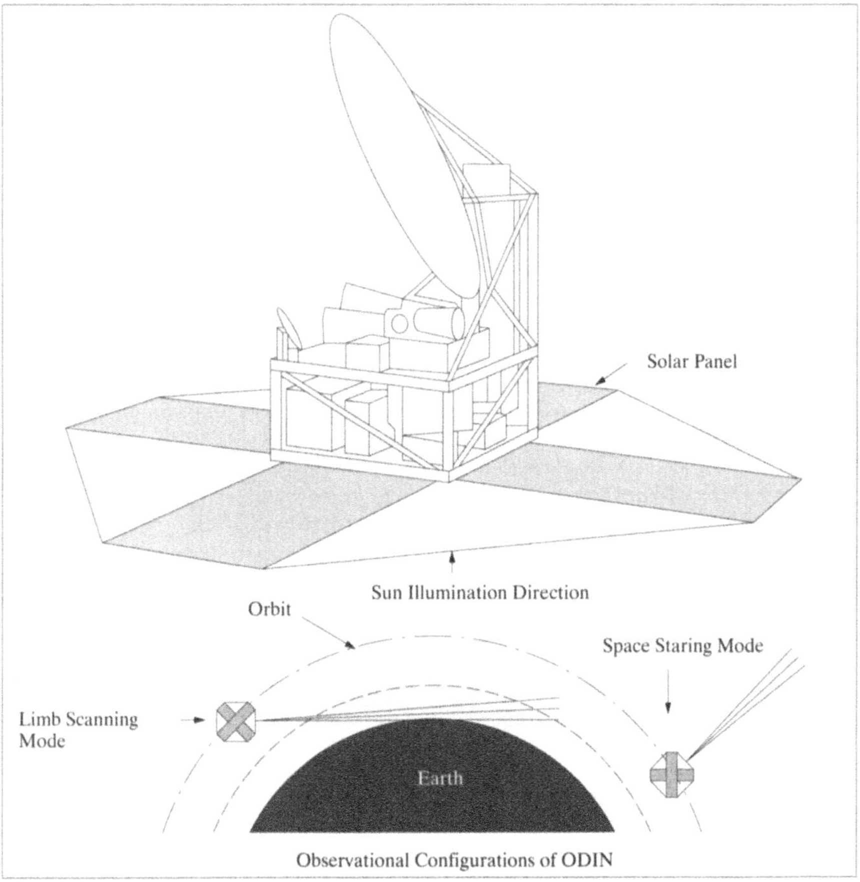

Figure 31: The ODIN S/C model and observational configurations

The five single sideband heterodyne receivers in SMR are continuously switched between a reference source of known signal strength and the signal from the telescope. The telescope is periodically targeted towards well-known celestial objects. These procedures ensure both stability and good calibration.

Spectral lines: The SMR instrument covers transitions of aeronomical interest from the following molecules: ClO, CO, NO_2, N_2O, H_2O_2, HO_2, H_2O, $H_2{}^{18}O$, NO, HNO_3, O_3, and O_2; and atomic and molecular transitions of astrophysical interest from: Cl, $H_2{}^{18}O$, H_2O, H_2S, NH_3, H_2CO, O_2, CS, ^{13}CO, H_2CS, SO, SO_2

Type	Single sideband heterodyne receivers,
Frequencies	118.25 - 119.25 GHz, 486.1 - 503.9 GHz, 541 - 580.4 GHz
Bandwidth	100 MHz to 1 GHz
Resolution	0.1 MHz to 1 MHz
Sensitivity	1 K in 1 MHz with S/N = 5 after 15 minutes
Mixers	Cooled Schottky mixers
Local Oscillators	LO based on Gunn diodes and frequency multipliers
LNA	Cooled HEMT low noise amplifiers
Spectrometers	two hybrid autocorrelators and one AOS (Acousto-Optic Spectrometer)

Table 74: Specification parameters of SMR

OSIRIS = Optical Spectrograph and Infrared Imaging System (Canadian sensor built by Routes of Kanata, Ontario, with funding by CSA).457) OSIRIS has a dual-purpose objective of detecting aerosol layers and to detect abundances of species such as O_3, NO_2, OClO, and NO. The instrument is a UV/VIS/IR limb sounder, in effect a double instrument, mounted in a common optical housing and supported by common electronics. The UV/VIS (optical) imaging spectrograph uses compact reflective optics (off-axis system, folded design, aperture = 36 mm x 36 mm) and an aspherical ruled grating along with UV-enhanced CCD arrays; the IR imager consists of three infrared telescopes. The Optical spectrograph measures species in the altitude range from 15 to 80 km (measurement of atmospheric airglow as well as utilization of the DOAS technique on scattered and subsequently absorbed moonlight). The FOV of the UV/VIS and the IR channels are aligned (and co-aligned with SMR) to produce simultaneous measurements. Instrument mass = 12 kg, power=20W. On-orbit spectral calibration for the UV/VIS instrument is done by observing artificial sources emanating from the Earth, such as discrete lines produced by mercury and sodium street lights. Radiometric calibration is not provided internally. For the IR imager, the IR shutter assembly contains a built-in incandescent lamp providing the radiometric calibration signal.

UV/VIS Spectrograph	
Wavelength	280 - 800 nm
Wavelength resolution	1 nm from 300-450 nm, <2nm from 450-800 nm
Slit orientation	perpendicular to the orbit plane (horizontal)
FOV	0.02° (vertical) x 0.8° (horizontal) or 1 km x 40 km
Spatial resolution	1 km at the limb
Pointing direction	aligned with SMR boresight
AD converter	14 bit
Detector type	Si CCD array (1353 x 143 pixels)
Species to be detected	O_3, NO_2, BrO, OCLO, O_2, aerosols
IR Imager	
Bandpass filter center wavelengths	1.263 µm, 1.273 µm, 1.520 µm
Bandwidth (FWHM)	10 nm (1.263 and 1.273), 40 nm (1.520)
FOV	118 km (vertical) x 2 km (horizontal)
Spatial resolution	1 km in the vertical direction
Telescope aperture	23 mm diameter
Detector type (linear arrays with hybrid multiplexers)	3 InGaAs, each detector has 128 pixels
Detector pointing direction	in the orbit plane (vertical)
Detector temperature	-40°C (or less)
Temperature stability during exposure	±0.1°C

Table 75: OSIRIS specification parameters

457) G. D. Warshaw, D. Desaulniers, D. Degenstein, "Optical Design and Performance of the ODIN UV/Visible Spectrograph and Infrared Imager Instrument," Proceedings of the 10th Annual AIAA/Utah State University Conference on Small Satellites, Sept. 16-19, 1996

A.21 OrbView-1/Microlab-1

See description under B.8.1

A.22 QuikSCAT (Quick Scatterometer Mission)

A NASA ESE (Earth Science Enterprise) program satellite built by BATC (Ball Aerospace & Technologies Corporation) of Boulder, CO. The QuikSCAT mission was initiated in the wake of the lost NSCAT instrument measurements aboard NASDA's ADEOS-1 satellite. The ADEOS-1 S/C ceased functioning June 30, 1997. The objective is to restart NASA's ocean-wind measurement program, needed for improved weather forecasts and climate research. The NASA/GSFC contract to Ball was awarded under its new Rapid Spacecraft Acquisition (RSA) procurement program. QuikSCAT is a mission designed to complete turnaround from conception to launch in a very short period of time (one year). JPL's NSCAT/SeaWinds Program Office has been assigned responsibility and provides overall project management, as well as science, ground processing systems, and the SeaWinds instrument. NASA/GSFC manages the satellite development and operation [data acquisition is facilitated at Wallops Flight Facility (WFF), Poker Flats (AGS), Svalbard Ground Station (SGS) Norway, and McMurdo (MGS) Ground Station, Antarctica, Hatoyama, Japan (contingency station)].

Figure 32: The QuikSCAT S/C model

The S/C uses a BCP 2000 (Ball Commercial Platform) series bus with dimensions: 2.2 m x 1.7 m x 1.4 m; it is three-axis stabilized, the ADCS (Attitude Determination and Control Subsystem) uses a star tracker, IRU, reaction wheels, and a GPS receiver with C/A code; the pointing accuracy is < 0.1° absolute per axis; the pointing knowledge is <0.05° per axis. The propulsion system uses anhydrous hydrazine (N_2H_2) blowdown. S/C mass (wet) = 970 kg, power 874 W (average), payload mass = 205 kg, payload power = 250 W (average). Design life > 2 years with 3 years for expendables. The on-board data recorder has a capacity of 8 Gbit. Downlink/uplink communications are provided in S-band at a data rate of 2 Mbit/s for payload data. Housekeeping data are at data rates of 4, 16, and 256 kbit/s, the uplink data rate is 2 kbit/s. The launch of the S/C took place on June 19, 1999 atop a Titan II vehicle (LM) from VAFB, CA.

Orbit: Circular sun-synchronous polar orbit with a local equator crossing time at the ascending node of 6:00 AM ± 30 minutes, altitude = 803 km, inclination = 98.6°.

Sensor complement:

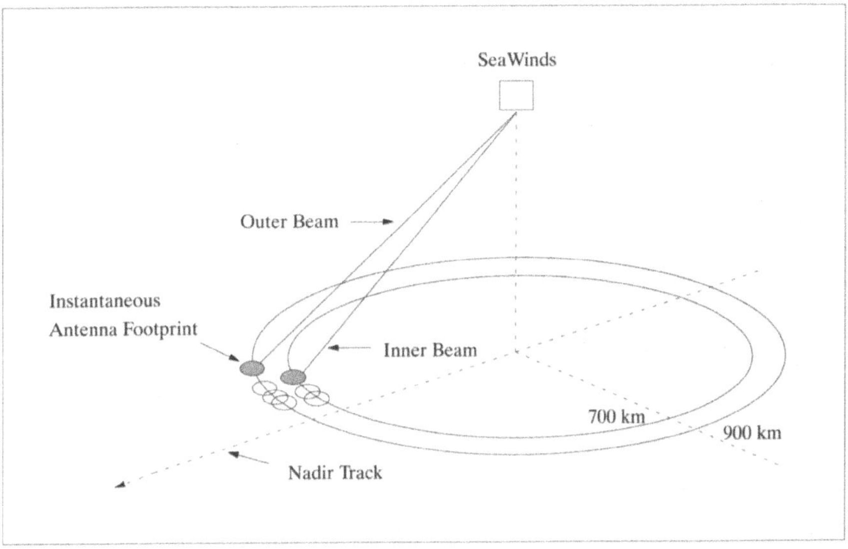

Figure 33: Illustration of the dual-beam scanning geometries

SeaWinds = NASA Scatterometer II (PI: M. Freilich, NASA/JPL and Oregon State University, Corvallis, OR). [458) 459)] The old designation of the instrument was: NSCAT II. Objective: to acquire accurate, high-resolution, global measurements of sea-surface wind vectors in 1 to 2 day repeat cycles. Applications: studies of tropospheric dynamics and air-sea interaction processes, including air-sea momentum transfer. The instrument is an active microwave radar (a conically scanning pencil-beam scatterometer) with dual-beam, 40° (inner beam) and 46° look angle from nadir (outer beam), conical scan 1 m diameter reflector (rotating dish) antenna, operating in Ku-band at 13.402 GHz (110 W pulse at 189 Hz PRF). The antenna is conically scanned such that each point on the Earth within the inner 700 km of the swath is viewed from four different azimuth directions (twice by the inner beam looking forward then aft and twice by the outer beam in a similar fashion). Measurement of wind speeds between 3-20 m/s to an accuracy of 2 m/s, wind vector directions to an accuracy of 20°. The dish antenna is rotated about the satellite nadir axis at 18 rpm. Data is collected in a continuous 1800 km swath, centered about nadir (about 400,000 measurements daily covering 90% of Earth's surface). Spatial resolution = 50 km; FOV = ±52° from nadir; instrument mass = 200 kg; power = 220 W; duty cycle = 100%; average data rate = 40 kbit/s; thermal operating range is 5-40°C; pointing knowledge to 500 arcseconds. - SeaWinds data

458) B. D. Boller, et al., "The Development of the SeaWinds Scatterometer Electronics Subsystem (SES)," Proceedings of IGARSS'96, Vol. 1, pp. 269-272

459) http://winds.jpl.nasa.gov/missions/quikscat/quikindex.html

products consist of global multiazimuth normalized radar cross section measurements and 50-km-resolution ocean vector wind maps.

The antenna subsystem consists of a one-meter parabolic reflector antenna mounted to a spin activator assembly, which causes the reflector to rotate at 18 rpm. The activator assembly provides very accurate spin control and precise position or pointing information to the CDS (Command and Data Subsystem). The antenna spins at a very precise rate, and emits two beams about 6° apart, each consisting of a continuous stream of pulses. The two beams are necessary to achieve accurate wind direction measurements. The pointing of these beams was calibrated before launch for accurate echo location determination.

A.23 REX-II (Radiation Experiment Satellite II)

The REX satellites in STP (Space Test Program) of the USAF are designed to study scintillation effects of the Earth's atmosphere on RF transmissions. REX-I was launched June 29, 1991 on a Scout G1 vehicle from VAFB, CA (orbit: 770 km x 870 km, 89.6° inclination; S/C mass = 85 kg). Built by Defense Systems Inc. (DSI), the spacecraft was gravity gradient stabilized (approx. 5 degrees control) using a 6 m boom and 2.5 kg tip mass. Minor control was provided by three axis torque coils.

REX-II is the name of a minisatellite and a payload experiment, managed by SMC (Space & Missile Systems Center) at Los Angeles AFB. The objective of the mission is to study the physics of electron density irregularities that cause disruptive scintillation effects on radio signals. The S/C (built by CTA of McLean, VA) is stabilized using a passive gravity-gradient boom (6.4 m long) and two magnetic hysteresis rods for damping. In addition the S/C is actively controlled by electromagnetic coils and a pitch-axis reaction wheel which provides momentum bias. The S/C structure consists of an eight-sided box, 76 cm diameter, 56 cm height, total mass = 113 kg. Power (30 W) is provided by three NiCd batteries and by four solar arrays. Data transmission is via UHF-band during ground-station passes at rates of 2.4 kbit/s (uplink and downlink). The S/C was launched on March 9, 1996 on a Pegasus-XL vehicle from VAFB, CA.

Orbit: circular polar orbit, altitude=830 km, inclination=90°.

Experimental payload:

REX (Radiation Experiment) is the primary payload sponsored by the Rome Laboratory of of the USAF/RL, Griffis AFB, NY. The objective is to study (and overcome) the physics of the electron density irregularities that cause disruptive scintillation effects on radio signals.

Three-axis magnetometer.

ADACS (Attitude Determination and Control Subsystem).[460) 461)] The objective is to employ attitude knowledge in order to perform on-orbit attitude determination (maintaining a nadir pointing S/C orientation) using a GPS receiver/antenna system in closed-loop control with the ground control center. Three-axis attitude knowledge (0.3° on 1 m baseline) is provided by redundant TANS Vector receivers (Trimble Advanced Navigation Sensor) with software modifications/additions by GSFC and Stanford University. The GPS measurement concept uses four patch antennas, mounted coplanar with the S/C top plate. A multiplexed frontend allows the receiver to measure the differential carrier phase between the master and each of the slave antennas. - Several operating modes are defined within ADACS to coordinate attitude sensing and corresponding attitude control. Results

460) D. Freesland,et al., "GPS Based Attitude Determination, The REX II Flight Experience," Proceedings of the 10th Annual AIAA/Utah State University Conference on Small Satellites, Sept. 16-19, 1996

461) E. G. Lightsey, E. Ketchum, T. W. Flatley, J. L. Crassidis, et al., "Flight Results of GPS Based Attitude Control on the REX-II Spacecraft," Proceedings of ION GPS-96, Sept. 17-20, 1996, Kansas City, MO, pp. 1037-1046

of the technique (comparison with magnetometer measurements) indicate that the mean three-axis error of ADACS performance is 2.66° with a standard deviation of 1.16°, which is well within specifications (5° per axis).

A.24 ROCSat-3/COSMIC (Constellation of Six Microsatellites)

The ROCSat-3/COSMIC (Republic of China Satellite-3/Constellation Observing System for Meteorology, Ionosphere and Climate) is an international collaborative project between NSPO (National Space Program Office) of Taiwan and UCAR (University Corporation for Atmospheric Research) of the United States of America. Initiated in December 1997, the project will launch a LEO constellation of six microsatellites to collect atmospheric remote sensing data for operational weather prediction, climate, ionospheric (space weather monitoring), and geodesy research. NSPO is the prime sponsor and owner of the satellites. UCAR, located at NCAR in Boulder, CO, is primarily sponsored by NSF (National Science Foundation). Other partners in the project include JPL, NRL, USAF, NOAA, CWB (Central Weather Bureau of Taiwan), industry from both countries, universities, and other research organizations from the US, Taiwan, and other countries. [462) 463) 464)]

Figure 34: The ROCSat-3/COSMIC spacecraft

The overall objective of ROCSat-3/COSMIC is to extend the low-cost research approach of refractive GPS radio occultation measurements (to derive important weather and climate research parameters, including atmospheric temperature, moisture, and pressure), that began with the GPS/MET instrument on Microlab-1 (launch April 3, 1995), to the next step by testing the ability of a constellation of six "ROCSat-3/COSMIC microsatellites with GPS/

462) L.C. Lee, C. Rocken, "Applications of Constellation Observing System for Meteorology, Ionosphere & Climate", R. Kursinski (Ed.), Springer, 2000, ISBN 962-430-135-2

463) C. Rocken, Y. H. Kuo, W. S. Schreiner, D. Hunt, S. Sokolovskiy, C. McCormick, "COSMIC System Description," Special issue of TAO (Terrestrial, Atmospheric and Oceanic Science), Vol. 11, No. 1, March 2000, pp.21-52

464) G. A. Hajj, L. C. Lee, X. Pi, L. J. Romans, et al., COSMIC GPS Ionospheric Sensing and Space Weather," Special issue of TAO (Terrestrial, Atmospheric and Oceanic Science), Vol. 11, No. 1, March 2000, pp.235-272

MET heritage" to provide the data needed to fully evaluate the impact of this promising new observational tool. In addition to carrying an advanced version of the JPL-developed GPS receiver for occultation measurement, each satellite will carry two tiny, simple secondary instruments (tri-band-beacon and photometer) which synergistically enhance the accuracy and utility of the ionospheric observations. A global data collection network and operations center will process space and ground observations and deliver products to users in real-time for operational impact studies. The ROCSat-3/COSMIC constellation will be launched, using one Minotaur launch vehicle, in 2005, and be operated for at least two years.

Orbital Sciences Corporation (OSC) of Dulles, VA, is selected as the prime contractor for the satellites by NSPO. A joint team of the OSC and NSPO will design and manufacture the satellites, with the early phase of the work performed at OSC and integration and test performed at NSPO. Several local industry companies of Taiwan are selected to provide satellite components. The S/C structure is of cylindrical shape of 1 m diameter. All spacecraft are identical, with a mass of about 60 kg. Each S/C features on-board propulsion to reach its final destination orbit. The ADCS (Attitude Determination and Control Subsystem) provides attitude knowledge with an Earth limb sensor and a magnetometer. Power is provided by a solar array and 10 Ah batteries. The propulsion system consists of two tanks to store propellant (hydrazine) and 4 small monopropellant thrusters. The communication subsystem consists of an S-band receiver, an L-band transmitter, and a set of S-band and L-band antennas. [465] [466] [467] [468]

Downlink Parameters		Uplink Parameters	
Receiver frequency	1690-1700 MHz (L-band)	Transmitter frequency	2025-2120 MHz
Polarization	Left Hand Circular	Polarization	Left Hand Circular
Data rates	Up to 2 Mbit/s	Data rate	Up to 32 kbit/s
Protocol	CCSDS compatible	Protocol	CCSDS compatible

Table 76: ROCSat-3/COSMIC communication characteristics

The ground segment of ROCSat-3/COSMIC consists of three ground TT&C stations and a MOC (Mission Operations Center). The three stations are located in Fairbanks (Alaska), Kiruna (Sweden), and Taiwan. The MOC is embedded into NSPO's MMC (Multi-Mission Center). The MOC performs all S/C operations. All science and some telemetry data is sent to CDAAC (COSMIC Data Analysis and Archive Center) at Boulder, CO, and to TACC (Taiwan Analysis Center for COSMIC), a mirror site of CDAAC in Taiwan. The centers also receives data from a global network of ground GPS and TBB (Tri-Band Beacon Transmitter) receiving sites (the so-called fiducial network). The centers analyze the received data and distribute it to the principal investigators and to the science community for operational evaluation and research.

During the initial two years of the mission, the science data will be made available through the ROCSat-3/COSMIC project websites without charge.

Orbit: Circular orbit, altitude = 700 km, inclination = 72°. The six satellites are being positioned into six orbital planes .
The initial orbital altitude is 500 (-25/+100) km. Then different satellites will be boosted by

465) Y. K. Kuo, L. C. Lee, "A Constellation of Microsatellites Promises to Help in a Range of Geoscience Research," EOS Transcriptions, AGU, Vol. 80, No. 40, Oct. 5, 1999, pp. 467-471

466) http://www.cosmic.ucar.edu/

467) Information provided by Paul Chen of NSPO

468) The GPS radio occultation technique is based on the following principles: As a signal travels through the atmosphere it is retarded and bent. This results in a phase and Doppler shift, which can be measured very accurately by the GPS receiver aboard the LEO ROCSat-3/COSMIC satellites. Since the transmitter and receiver positions and velocities are accurately know from precise orbit determination, the signal bending angle alpha as a function of impact parameter, can be computed from the Doppler shift observed at LEO. From the basic bending angle versus impact parameter data, vertical profiles of refractivity as a function of tangent point radius can be derived. Further analysis converts refractivity to electron density in the ionosphere.

on-board thrusters to different altitudes ranging up to 700 km. The corresponding different rates of orbit nodal precession will then gradually drift the orbit planes apart, until a more-or-less even distribution of six orbit planes is achieved. The orbit-adjustment process is estimated to take about 13 months, during which time the satellites will already start collecting atmospheric soundings.

Sensor complement:

The ROCSat-3/COSMIC constellation produces about 3000 soundings of bending angle and refractivity globally in all weather each day for at least one year after the spacecraft are placed in their final orbits. From these soundings, estimates of electron density in the ionosphere and temperature, water vapor and pressure in the stratosphere and troposphere will be derived. Desirable characteristics of these data include such items as: high accuracy, high vertical resolution, all weather (clouds and aerosols do not affect measurements), no calibration of instrument required, no instrument drift, require no first guess, modest cost.

Parameter	Science requirement	Comment
Vertical GPS occ. resolution	0.3-1.5 km	0.3 km near surface, 1.5 km at 45 km
Horizontal GPS occ. resolution	300-600 km	Smaller for fronts with large slope
Bending angle profile	1.5 x 10^{-6} rad	Limited mainly by residual ionospheric error and data noise
Refractivity profile	<1%	As good as 0.2% between 10-30 km
Temperature profile	1° C	In 0-40 km height range, assuming dry air
Water vapor profile	<1-10%	Between 0-5 km, assuming model temp. error of 1-2°, errors between 5-10 km are larger
Geopotential height vs pressure	10-20 m	Between 0-30 km (worse at solar maximum)
Electron density profile	<1-20%	Profile quality near the F2 peak, can be worse elsewhere; limited by horizontal gradients
GPS TEC	0.001 TECU	Relative TEC (absolute about 3-5 TECU)
TBB TEC	0.003 TECU	Relative TEC (absolute about 1-3 TECU)
TIP peak for F2 density, NmF2	<1-10%	Performs better at high electron densities Works only on the night side of the globe
Magnetometer measurement	10 nT	Precision at 1 Hz rate, 500 NT accuracy
Scintillation GPS	N/A	100 Hz SNR data used to determine S4
Scintillation TBB	N/A	50-1000 Hz SNR data
LEO position knowledge	about 10 cm	3-D rms position error (not critical)
LEO velocity knowledge	<0.1 mm/s	Relative velocity error for paired satellites

Table 77: Science requirements of the ROCSat-3/COSMIC constellation

Number of occultation measurements	About 2500 soundings/day on a global scale
L1/L2 phase measurements	About 2 mm (10 s interval for precise orbit determination)
GPS phase sampling rate	0.1-50 Hz
GPS vertical range neutral atmosphere	Surface to 60 km (50 Hz sampling)
GPS vertical range ionosphere	90-800 km (10 Hz sampling)
TBB phase measurement	<32 mm at 150 MHz (ground receiver)
TBB sampling rate	>50 Hz (ground receiver)
TIP measurement	<10% (uncertainty in photon count)
TIP footprint	125 km x 25 km (at 400 km height of F2 layer)
TIP resolution	0.1-10 s averaging
Magnetometer	10 nT precision, 500 nT accuracy

Table 78: Observational requirements of ROCSat-3/COSMIC

A **BlackJack GPS occultation receiver** is the primary science instrument of the ROCSat-3/COSMIC constellation. The particular receiver chosen for ROCSat-3/COSMIC is a refined version of the BlackJack occultation receiver developed by the Jet Propulsion Laboratory for the CHAMP, SAC-C, and GRACE missions. The BlackJack receivers on ROC-Sat-3/COSMIC will be able to track all GPS satellites in view simultaneously, including two or more occulting satellites. It will operate fully autonomously, scheduling when to track which satellites and at what sampling rate based on its own known position and those of the GPS satellites. The instrument will report high-rate (50 Hz) dual frequency carrier phase measurements on the occulting links with sub-millimeter precision for accurate, high resolution profiling. Lower rate (0.1 Hz) phase measurements of all satellites in view will be collected for precise orbit determination at the 5-10 cm level. The receiver must track both GPS carrier frequencies to separate the frequency-dependent (dispersive) ionospheric delay from the non-dispersive refractive delay of the neutral atmosphere. A patented "semi-codeless" technique is used to obtain precise measurements of the L2 signal, both carrier phase and pseudorange, with anti-spoofing turned on. In addition to these measurements, the GPS instrument can record GPS signal amplitudes for on-orbit ionospheric scintillation monitoring and correction of signal diffraction effects in post-processing. The instrument mass is 3 kg; size of about 20 cm x 20 cm x 10 cm; power of 13 W, antenna inputs: 4.

TIP (Tiny Ionosphere Photometer), designed and built at NRL (Naval Research Laboratory), Washington, DC. TIP and TBB provide measurements of electron density, an important parameter of the upper atmosphere. The readings of TIP and TBB complement the primary BlackJack instrument so that 3-D fields of electron density between 90 and 750 km can be inferred.

In particular TIP provides horizontal gradients in electron density at the peak of the F2 layer, along the satellite orbit track. TIP measure the naturally occurring nighttime emission of neutral oxygen at 135.6 nm. This emission is produced by the recombination of O + ions and electrons and is proportional to the square of the electron density in the ionospheric F region. Since horizontal gradients of electron density are a limiting error source for occultation inversions in the ionosphere, the combined analysis of TIP and GPS data promises improved retrievals of nighttime ionospheric profiles.

TBB (Triband Beacon Transmitter), designed and built at NRL. TBB transmits phase data measurements at 150, 400 and 1067 MHz which can be received at ground stations worldwide. These data are converted to line-of-sight total electron content (TEC) observations that can be processed with 2-dimensional ionospheric tomography techniques. TBB data can also be combined with the other ionospheric observations in tomographic and physical data assimilation models to compute global four-dimensional electron density fields.

The ROCSat-3/COSMIC instrument suite permits three-dimensional tomography of the ionosphere with unprecedented resolution and accuracy. ROCSat-3/COSMIC data will be highly complementary to other satellite sounding systems, including radiometric sounders on the POES and GOES series satellites of NOAA. The independence and the high-vertical resolution of the radio occultation soundings complement the high horizontal resolution of the radiometric soundings and together the two systems can likely be combined to yield composite soundings of temperature and water vapor with unprecedented accuracy, horizontal and vertical resolution, and global coverage.

A.25 SAN MARCO D/L

San Marco D/L[469),470)] is a cooperative mission of CRA (University of Rome/Italy) and NASA/GSFC with sensor contributions from several agencies. NASA Scout launch vehicle, spin-stabilized satellite (axis parallel to Earth's axis), payload mass = 234 kg, satellite diameter = 0.940 m. Launch site: San Marco Range, Kenya; Mission duration: March 25, 1988 until Dec. 6, 1988.

469) Payload Definition Document for SAN MARCO D/L Satellite, CRA, Oct. 1987
470) G. Schmidtke, H. Doll, C. Wita, and S. Chakrabarti, "Solar EUV/UV and equatorial airglow measurements from San Marco-5," Journal of Atmospheric and Terrestrial Physics, Vol. 53, No. 8, pp. 781-785, 1991

Objective: to explore the possible relationship between solar activity and meteorological phenomena by investigating the structure, dynamics, and aeronomy of the equatorial thermosphere, in particular:

- The relationship between density, composition, temperature, and airglow as a function of solar flux
- The interhemispheric transport of ions and neutral particles
- The dependence of neutral wind, ion drag, temperature, and particle densities on local time
- The relationship between neutral wind and the electric field
- Ionosphere irregularities and special events, and other parameters keyed to longitude
- The equatorial distribution of ozone
- Magnetic storm induced variations in the measured parameters
- The dependence of meridional wind and ion drift on the magnetic field
- The variation of the midnight temperature maximum

Orbit: Near-equatorial orbit, apogee = 619 km, perigee = 262 km, inclination = 2.9°.

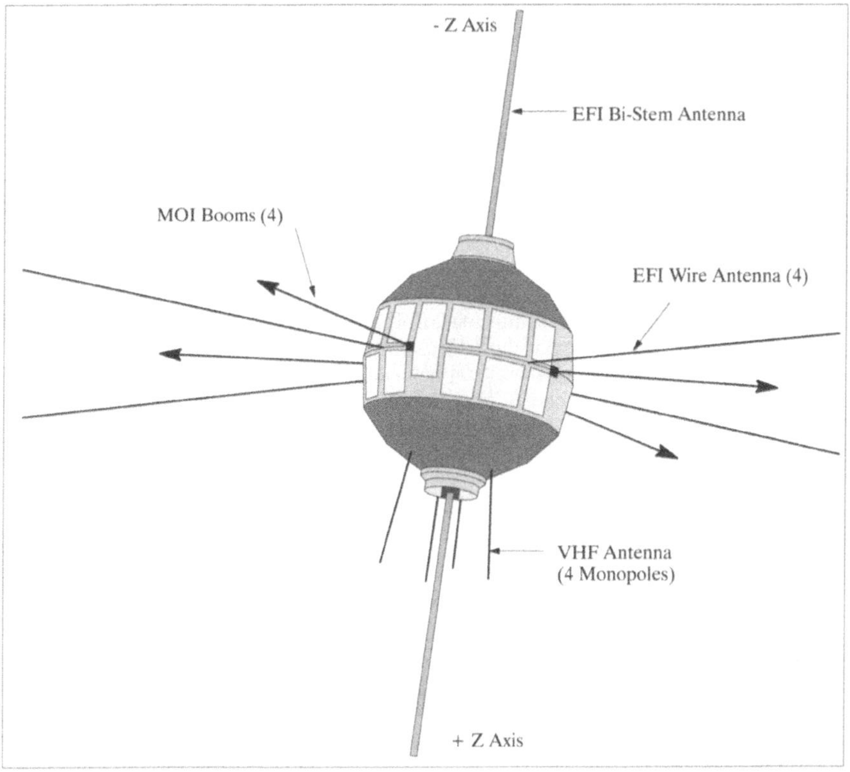

Figure 35: San Marco D/L spacecraft model

Sensor complement:

DBI = Drag Balance Instrument (PI: L. Broglio, CRA). Measurement of the total aerodynamic drag force on the S/C.

ASSI = Airglow Solar Spectrometer Instrument (PI: G. Schmidtke, FhG/IPM, Fraunhofer Gesellschaft/ Institut für physikalische Meßtechnik, Freiburg, Germany).

Measurement of airglow, solar radiation, reflected solar radiation, and radiation from interplanetary and intergalactic sources. Four scanning spectrometers with solar pointing control cover a broad spectral range [from 20 - 700 nm (EUV - VIS)] using channels with 18 overlapping wavelength ranges with a spectral resolution of 0.7 - 4 nm (each spectrometer contains four or five detectors for a total of 18 detectors, there are two units of ASSI, namely ASSI-A and ASSI-B). Large dynamic ranges up to 10^{11} permit the measurement of very faint airglow or interplanetary radiation as well as intense solar emissions.
The ASSI-A and ASSI-B units are mounted on separate solar pointing devices, viewing 180° away from each other from the equatorial plane of the spacecraft.

WATI = Wind and Temperature Spectrometer (PI: N. Spencer, GSFC)
Measurement of two wind components (horizontal and normal to orbit plane), along with the kinetic temperature.

EFI = Electric Field Instrument (N. Maynard, PHG; T. Aggson, GSFC)
Plasma transport measurements. EFI consists of two orthogonal 40 m wire antenna pairs that are extended in the spin plane of the satellite, and a shorter antenna pair in the S/C spin axis. DC and electric field measurements.

IVI = Ion Velocity Instrument (PI: W. Hanson, University of Texas)
IVI measures the 3-D bulk velocity of the ambient ions in the S/C velocity frame. In addition IVI measures ambient plasma concentration and ion temperature.

Data: On-board recorder, VHF-band data link, data rate = 6 kbit/s. Sensor data at CRA and at World Data Center (GSFC).

The San Marco D/L mission is the fifth mission in Italy's San Marco program series.[471)]

- San Marco 1 - Launch: Dec. 15, 1964 from Wallops. Orbit 194 x 697 km, incl. = 38°. Measured atmospheric density until its decay on Sept. 13, 1965
- San Marco 2 - Launch: April 26, 1967, Scout vehicle from San Marco (floating platform base in the Indian Ocean off the coast of Kenya). Orbit 185 x 211 km, incl. = 3°. Observation of atmospheric density. S/C re-entry: Oct. 14, 1967.
- San Marco 3 - Launch: April 24, 1971 from San Marco. Orbit: 222 x 718 km, incl. = 3°.
- San Marco 4 - Launch: Feb. 18, 1974 from San Marco. Orbit: 231 x 910 km, incl. = 3°. Observation of auroral zone over the equator. S/C re-entry on May 4, 1976.

A.26 SCISAT-1/ACE (Science Satellite/Atmospheric Chemistry Experiment)

In the time frame 1996/97 CSA (Canadian Space Agency) initiated the SCISAT program with the objective to provide opportunities for Canadian scientists to define and conduct space experiments in the following fields: Earth sciences, space astronomy, and solar-terrestrial relations. Mission selection procedures in the program are conducted via an AO (Announcement of Opportunity) process and peer reviews. [472)] [473)] [474)] [475)]
In addition, the SCISAT program is also part of a CSA/NASA collaboration program, consisting of two missions. Under the terms of the cooperative agreement, each agency provides a spacecraft and instrumentation, to be co-launched on an expendable vehicle. The AO for the Canadian elements of the first SCISAT (SCISAT-1) was released in 1997. The ACE mission was selected for flight in November of 1998.

471) Jane's Spaceflight Directory 1988-89, pp. 35-36

472) H. Dahl, W. Eliuk, G. Rumbold, R. Shelly, "ACE - A Canadian Small Satellite Mission," Proceedings of the 13th AIAA/USU Conference on Small Satellites, Aug. 23-26, 1999, Logan UT, SSC99-V-7

473) http://www.ace.uwaterloo.ca/

474) I. Walkty, J. Petersen. T. Doherty, B. Whitehead, "SCISAT-1 ACE Mission C&DH Unit Development," Proceedings of the 14th Annual AIAA/USU Conference on Small Satellites, Logan, UT, Aug. 21-24, 2000, SSC00-I-5

475) P. Bernath, "Atmospheric Chemistry Experiment (ACE): An Overview," Spectroscopy from Space, J.Demaison, editor, Kluver, 2001

The SCISAT/ACE mission is based at the University of Waterloo, Waterloo, Ontario (Mission Scientist: P. Bernath). The overall objective is to monitor and analyze the chemical processes that control the distribution of ozone in the upper troposphere and stratosphere. A comprehensive set of simultaneous measurements of trace gases, thin clouds, aerosols and temperature are being collected by solar occultation from low earth orbit. Detection of more than 30 molecules such as: O_3, N_2O, CH_4, HNO_3, H_2O, HCl, Hf, NO, NO_2, $ClNO_3$, CO, CO_2, CCl_3F, CCl_2F_2, and $N2O_5$.

The ACE minisatellite structure, built by Canadian industry (Bristol Aerospace Ltd of Winnipeg, Manitoba), uses a circular instrument/component aluminum mounting plate (1.1 m in diameter) as the main structure of the three-axis stabilized platform. Attitude control is based on a bias momentum stabilization approach. The subsystem consists of a momentum wheel, torque rods along all three body-fixed axes, one fine sun sensor, a magnetometer and a set of six coarse sun sensors. All sensors and actuators are off-the-shelf components with flight heritage. In addition, the GyroWheel of Bristol is flown for technology validation (the GyroWheel has the ability to provide the S/C simultaneously with stored angular momentum, function as a 3-axis torque actuator, and measure the S/C angular rates in two axes). The S/C is always in a sun-pointing configuration. S/C power (71 W average) is generated by a single body-mounted solar panel. In addition, there are Lithium-Ion batteries for the orbital eclipse phase operations. On-board source data recording of up to 1.5 GByte is provided. The C&DH (Command & Dat Handling) unit was developed for small satellites. It is responsible for all onboard data handling, monitoring and recording. The unit features low-power (<10 W), low-mass (<4 kg) and a radiation-tolerant core in a single string architecture (a UTMC 80C196 16-bit processor is used to perform all S/C operations). A storage capacity of 1.5 Gbit is provided. The S/C mass is about 150 kg, the design life is two years with a goal of five years. A NASA-sponsored launch of SCISAT/ACE is scheduled for early summer 2002 on a Pegasus-XL vehicle.

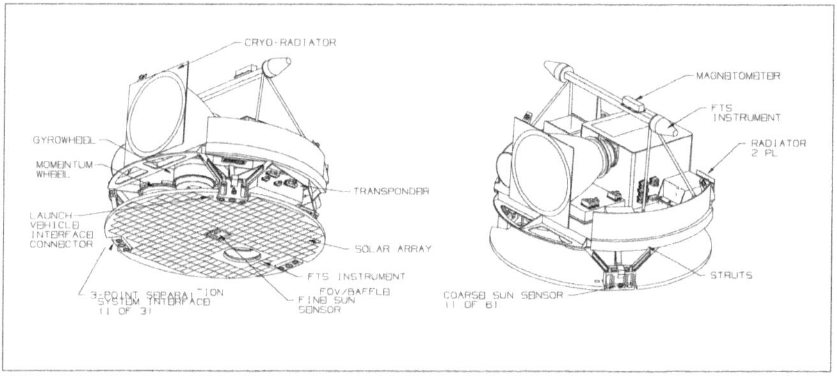

Figure 36: Schematic illustration of the SCISAT-1/ACE spacecraft (two views)

S/C communications are provided in S-band using the CCSDS protocol (variable-rate telemetry up to 5 Mbit/s can be supported). The downlink uses OQPSK modulation, and the uplink is compatible with NASA's STDN (Satellite Tracking and Data Network) standards. The maximum data rate (occurring during an occultation) is about 9.6 Mbit/s, and almost all of it come from the ACE-FTS instrument. The uplink data rate is 4 kbit/s. Mission operations of SCISAT/ACE are conducted at CSA in St. Hubert near Montreal.

Orbit: Circular orbit, altitude = 650 km, inclination = 73.9°.

Sensor complement:

ACE-FTS (ACE-Fourier Transform Spectrometer), is being built by ABB Bomem Inc. of Quebec City, Quebec. The objective is to measure the vertical distribution of trace gases, in particular of the regional polar O_3 budget, as well as pressure and temperature (derived from CO_2 lines). The instrument is an adapted version of the classical sweeping Michelson interferometer, using an optimized optical layout. It consists of the following components: the FTS, a VNIR (Visible Near Infrared) imager, a sun tracker, the instrument electronics, and a power supply. An SNR >100 is achieved; IFOV (FTS) = 1.25 mrad; an aperture diameter of 100 mm and a measurement period of 2 s. Instrument mass = 40 kg, power =40 W operating, and 15 W in standby.

- The FTS spectrometer looks at the sun through the atmosphere (occultation or limb-viewing geometry) at different tangent heights, providing a series of spectra that are used to deduce the vertical distribution of trace gases and temperature. The spectral range of the instrument is from 2-13 µm (750 - 4100 cm^{-1}) in two bands, and the maximum resolution is 0.025 cm^{-1}. InSb (1800-4100 cm^{-1}) and HgCdTe (750-1800 cm^{-1}) detectors are used. Both detectors are cooled below 110 K. The spectrometer transforms the spectra into a modulated signal, the interferogram, in which all of the IR bands are present simultaneously. The spectrometer output consists of such interferograms for each observed scene. The interferograms are Fourier-transformed into spectra on the ground to provide vertical profiles of atmospheric constituents at resolutions of 3-4 km.

Parameter	Value	Parameter	Value
Spectral range	750-4100 nm	Noise equiv. radiance	<0.5% of the radiance of a blackbody at 5800 K
Spectral resolution cm^{-1}	<0.028, 0.056, 0.11, 0.55	Detectors	InSb, HgCdTe
Sweep duration	2, 1, 0.5, 0.1 s	Detector cooling	Passive cooling <110 K
Spectral stability (relat.)	3 x 10^{-7} rms for 180 s	FOV (Field of View)	1.25 mrad

Table 79: Characteristics of the ACE-FTS instrument

- VNIRI (Visible Near Infrared Imager). Objective: monitoring of aerosols using the method of extinction of solar radiation. VNIRI provides sun images in two spectral bands at 0.525 µm and at 1.02 µm. Refractive index distortion of the solar image for low altitude measurements is monitored with a large photodetector array of 128 x 128 effective elements covering 30 mrad with a pixel separation of 0.25 mrad (the IFOV is more than four times smaller than the IFOV of the FTS). These measurements have an SNR>1000 for all sun-illuminated pixels in a two-second observation time. The sun tracker keeps the instruments (FTS and VNIRI) automatically pointed at the sun's radiometric center.

MAESTRO (Measurement of Aerosol Extinction in the Stratosphere and Troposphere Retrieved by Occultation), designed and built in a partnership between MSC (Meteorological Service of Canada) and EMS Technologies of Ottawa. MAESTRO is a dual optical spectrograph in the spectral region of 285-1030 nm. The objective is to measure ozone, nitrogen dioxide and aerosol/cloud extinction. The use of two overlapping spectrographs (280 - 550 nm, 500 - 1030 nm) improves the stray-light performance. The spectral resolution is about 1-2 nm. The detectors are linear EG&G Reticon photodiode arrays with 1024 elements. The instrument design is based on a simple concave grating with no moving parts. The entrance slit is held horizontal to the horizon during sunrise and sunset by controlling the spacecraft roll with a startracker and a momentum wheel on the satellite bus. The vertical resolution of the MAESTRO data is about 1 km; the SNR is >1000. The high vertical resolution may help to distinguish between various atmospheric layers. The oxygen A-band at 762 nm (as well as the B-band and gamma-band) are used to make an independent determination of atmospheric temperature and pressure. MAESTRO is also able to make some near-nadir solar backscatter measurements with a separate backscatter port. The mass of the instrument is about 7 kg, power = 15 W (operating), 7 W (standby), the data rate = 3 Mbit/s.

A.27 SORCE (Solar Radiation and Climate Experiment)

SORCE is a NASA-sponsored microsatellite mission. The overall objectives to continue the precise, long-term measurements of TSI (Total Solar Irradiance) at UV and VNIR wavebands. The science objectives include:

- Daily measurements of the solar UV (120-300 nm) with a spectral resolution of 1 nm. To achieve an absolute accuracy of better than ±5%, and a precision and long-term relative accuracy of ±0.5%.
- To make the first precise measurements of the visible solar irradiance suitable for future climate studies - to obtain a daily measurement of the solar irradiance, between 0.3 and 2 µm, with a spectral resolution ($\lambda/\Delta\lambda$) of at least 1/30, an absolute accuracy of ±0.3% (3 sigma), and a precision and relative accuracy of better than ±0.01%/year.
- Improve our understanding of how and why the variability occurs at the Sun and how the variable irradiance affects our atmosphere and climate. To use this knowledge to estimate past and future solar behavior and climate response.
- To make precise and accurate measurements of TSI, connect them to previous TSI measurements to form the long-term climate record. To provide absolute accuracy of 0.03 % (300 parts per million) based on SI units and a relative accuracy of 0.001%/yr.

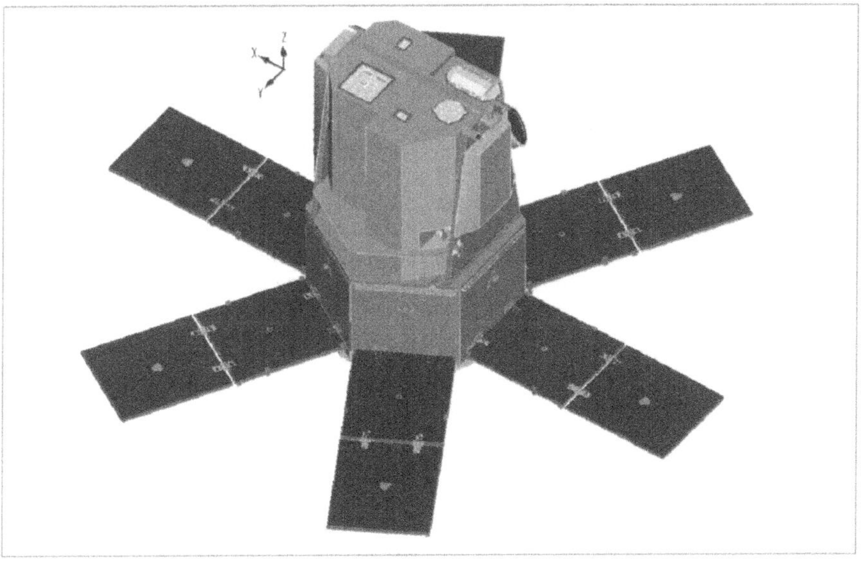

Figure 37: The SORCE S/C model

SORCE extends the broad data set gathered by the ACRIM instrument series while exploring a new capability to measure solar irradiance in two discreet spectral bands (ten-fold measurement accuracies are required over existing spaceborne radiometer and spectrometer performances). The SORCE project is designed, managed and operated by LASP (Laboratory for Atmospheric & Space Physics) of the University of Colorado in Boulder, CO (PI: G. J. Rottman). The payload instruments are designed and built by LASP. [476)]

The SORCE spacecraft is a three-axis stabilized, momentum-bias system using redundant reaction wheel assemblies, active magnetic control, and redundant sun sensors (provided by the instrument payload). The S/C is sun-pointed during the dayside orbit and star-

476) http://lasp.colorado.edu/sorce/

pointed during the nightside portion of the orbit. The S/C pointing values are: accuracy of 0.24 arc-minutes, knowledge of 0.20 arc-minutes, stability of 33 arc-minutes/min, and jitter of 3.0 arc-seconds/s. The slew rate is 2.2°/s. The S/C dimensions are: 1.48 m in height and 1.12 m in diameter; power = 730 W (EOL), provided by six fixed GaAs solar arrays; a 20 Ah NiH battery provides power during eclipse phases. The S/C mass is 268 kg, the payload mass is 86 kg, the S/C design life is six years. An on-board data storage of 1024 Mbit is provided. S/C communications are in S-band with an uplink of 2 kbit/s and a downlink of 1.25 Mbit/s (maximum data volume of about 570 Mbit/day). Mission operations are performed from LASP, using two ground stations located at Wallops and at Hawaii. Nominal SORCE operations include data downloads twice a day. A launch is planned for mid 2002 on a Pegasus XL vehicle from Cape Canaveral, FL.

Orbit: Circular orbit, altitude = 645 km, inclination = 40°.

Sensor complement:

TIM (Total Irradiance Monitor). TIM is an active cavity radiometer of ACRIM heritage, but with significant improvements in sensor and electrical design, particularly in thermal control. Objective: Measurement of TSI directly traceable to SI units (Système International d'Units) with an absolute accuracy of 0.03% and relative accuracy of 0.001% per year. The complete wavelength range is covered.

The TIM instrument consists of a cylindrical housing containing four identical, right circular, 15° apex angle, conical cavity detectors cantilevered from a single support. The mount provides thermal impedance between the cavities and the heat sink which surrounds them. One cone, the measurement cavity, views the total solar disk through a precision aperture, and any of the other cones may be used as the reference cavity. The cones are made from 1 mm thick electro-deposited 99.99% pure silver. The cone interiors and the heat sink are coated with a Ni-P black and etched to produce a diffuse black surface which has a nominal visible reflectance of 0.3%. - Six miniature (1 mm^2) diamond-substrate chip heaters are soldered to the back of each cone. Gold electrode spinel chip thermistors mounted between the apex of each cone and its support structure are used as sensors for the electrical bridge circuit which controls the measurement cavity temperature. Baffles, located behind the primary aperture, shield the measurement cavity from earth albedo and off axis infrared radiation. Small Si and InAs photodiodes, located just behind the heat sink, view the measurement cavity at an oblique angle to measure the change in its reflectance. The heat sink is thermally isolated from the TIM channel case and its temperature is controlled to 0.001°C.

The ESR (Electrical Substitution Radiometer) of TIM has dual bolometers with autobalancing through an AC, digital, feedback loop.

TIM incorporates four cavities (cones) and adheres to the basic concepts of Electrical Substitution Radiometers (ESRs), but employs modern, state-of-the-art electronics and materials. The four cavities provide multiple redundancy and added duty-cycling capability. TIM looks at the sun every spacecraft orbit. Each measurement consists of multiple samples taken over the course of a single orbit, providing 15 measurements per day. The TIM shutters, one for each radiometer, are driven by brushless torque motors. In its normal operational mode, the TIM shutter is cycled 50% open and 50% closed every 100 seconds throughout the orbit. Periodic (approximately once per week) field of view maps are obtained by offset pointing the S/C by ±15° about the sun vector. - Instrument mass = 5.9 kg, power = 13.1 W (average), size = 17.7 cm x 27.9 cm x 27.2 cm (H x W x D), data rate = 1.661 kbit/s.

SOLSTICE (Solar/Stellar Irradiance Comparison Experiment). The 2nd generation instrument is of SOLSTICE heritage flown on UARS. SOLSTICE is a two-channel grating spectrometer with the objective to measure UV radiation in the spectral coverage of 115 to 320 nm with a spectral resolution between 0.1 nm and 0.2 nm. The device compares the ultraviolet output of the sun with similar radiation produced by about 20 stable, bright blue stars. The stars constitute the standards (calibration) against which the solar irradiance is measured. The standard may therefore serve to directly compare the solar observations done by SORCE as well as by UARS, or by any other future SOLSTICE-type observation.

The SOLSTICE design features two identical spectrometers, mounted at right angles to each other to use perpendicular dispersion directions in order to facilitate verification of stellar pointing . A plane mirror reflects the beam onto one of two diffraction gratings mounted in a precision single-axis gimbal. Rotating the gimbal to a specified angle causes the illuminated grating to diffract a small wavelength band of the original beam toward a second plane mirror. The diffraction grating disperses the radiation and directs a specific wavelength onto one of the two photomultiplier detectors. - About 15 solar spectra and multiple stellar observations are provided each calendar day, resulting from about 15 orbits/day with measurements taken in the daytime (sun) and nighttime (stars) portions of the orbit. [477]

Instrument type	Modified Monk-Gilleison spectrometers
Primary detector type	Photomultiplier tubes
FOV (Field of View)	1.5° x 1.5°
Wavelength range	115 - 180 nm and 170 - 320 nm
Spectral resolution	0.1 - 2.2 nm
Accuracy, precision	5%, <1% over mission life
Instrument dimensions (HxWxD)	18.3 cm x 38.7 cm x 84.6 (x2) cm
Instrument mass ,power, nominal data rate	36 kg, 33.2 W, 738 bit/s

Table 80: SOLSTICE instrument parameters

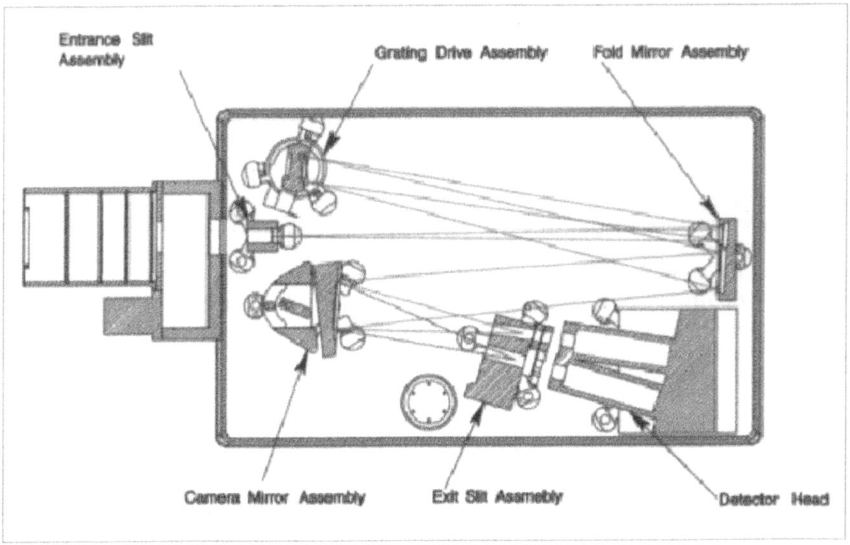

Figure 38: Optical layout of the SOLSTICE instrument

In-flight instrument calibration includes 1) long-term multi-wavelength observations of bright, blue stars, 2) redundant channel capability, and 3) anticipated underflight calibrations by almost identical instruments on sounding rockets. The wavelength calibration for

477) William McClintock, "2nd Science Team Meeting for the Solar Radiation and Climate Experiment - SORCE," The Earth Observer, Nov./Dec. 2000, Vol. 12, No 6, pp.22-27

the in-flight data uses the many well-known solar emission features to determine the wavelength scale.

SIM (Spectral Irradiance Monitor). Objective: Measurement of the solar spectral irradiance in the 200 to 2000 nm range. SIM is a dual-redundant spectrometer with a single optical element; it provides spectral measurements by rotating a small prism and recording the spectrum at the exit slit using a new high-sensitivity ESR (Electrical Substitution Radiometer). The ESR measurement technique measures heat flux. Two identical sensors, one active and the other used as reference, are connected so that they are in the same environment and at the same temperature. Joule heat is supplied to each sensor by an "actively controlled" heater circuit. The sensors have very high absorptivity in order to efficiently collect radiation, so that all photon energy incident on the detector is converted into heat. As radiation is allowed to fall on the active sensor, a corresponding amount of Joule heat must be removed from that sensor in order to restore the heat flux balance. This change in Joule heat to the active sensor is equivalent to the amount of radiation incident upon it. [478)]

The SIM instrument consists of two identical mirror-image spectrometers (A and B). This dual-channel design provides redundancy, self-calibration capability, and duty cycling necessary to meet the science objectives. In normal operation, spectrometer A is used for daily solar irradiance measurements while spectrometer B is used to make solar measurements on a much lower duty cycle (perhaps once per month). Additionally, a periscope system couples the two spectrometers and is used in conjunction with two photodiodes (Si and extended InGaAs) on a rotation turntable to provide an in-flight measurement of prism.

Parameter	Value	Parameter	Value
Spectral range	200 - 2000 nm	Spectral resolution	0.25 - 34 nm
Spectrometer f number	f/16	Solar f number	f/115
Prism aperture	25 mm x 12.5 mm	Effective focal length	400 mm
Prism vertex angle	34.5^{o}	Front surface radius	435.3 mm
Back surface radius	445.4 mm (aluminized)	Scan range (focal plane)	40 mm
Optical aberration at exit slit	5 μm	Diffraction correction Slit sizes	0.3 - 2.2% 0.3 mm x 7 mm

Table 81: SIM instrument S-channel specification

In normal operation with the shutter open, sunlight enters the 0.3 mm x 7 mm entrance slit and follows a f/115 beam to the prism, 400 mm from the slit. The focusing prism forms a nearly flat spectrum at the focal plane. For each spectrometer, four separate detectors measure the solar irradiance; the primary detector is an Electrical Substitution Radiometer (ESR), and additional silicon and extended InGaAs photodiodes are used for the UV, Visible, and the IR portion of the spectrum respectively. The ESR and the photodiodes, behind separate exit slits, simultaneously measure radiation at four neighboring wavelengths.

Prism drive: The solar spectral intensity at S-channel resolution typically varies 3% per slit width (average value over the entire spectral range), requiring that the prism drive reproducibility be better than 1 μm (one slit width/300). The prism drive achieves 0.3 μm precision by rotating the prism on a flex pivot with a flex-suspended voice coil motor. Closed-loop control of prism rotation uses a small concave mirror on the prism table to focus a second, reference solar beam back to the focal plane and onto a 12,000 element (78 mm active length), 6.5 m^2 pixel size, 6.5 m^2 pitch linear CCD (Thomson TH7834C). The CCD detector at the focal plane compensates for motion and any bending of the optical bench. For ESR operation, spectra are measured in a step-stop-integrate mode with 3 steps per slit width (100 m^2 steps) to eliminate spectral aliasing of the integrated spectrum. The steep portions of the solar spectrum are benchmarks for wavelength adjustment, and the local maxima and

478) G. Rottman, G. Mount, G. Lawrence, T. Woods, J. Harder, S. Tournois, "Solar spectral Irradiance measurements: visible to near-infrared regions," Metrologica, Vol 35, 1998, pp. 707-712

minima in the spectrum provide the most reproducible measurements of spectral irradiance.

The ESR provides a long-term, absolute detector for the S-channel and maintains the sensitivity standard for the working photodiodes. The ESR optical head consists of dual bolometers each at the center of an optical quality isolation hemisphere, which reflects light back onto and thermally shields the detectors.

XPS (XUV Photometer System), of SXP (Solar X-ray Photometer) heritage flown on SNOE. The objective is to measure the XUV (Extreme Ultraviolet) solar irradiance from 1 to 35 nm. The instrument package consists of twelve silicon XUV photodiodes. Each photodiode has a thin-film filter to provide an approximately 5 nm spectral bandpass. These thin film filters are deposited directly on the photodiode to avoid using delicate metal foil filters which are difficult to handle, prone to develop pin holes, and degrade with time. The set of twelve XUV photometers is packaged together with a common filter wheel mechanism, which can rotate a closed aperture, a fused silica window, or an open aperture in front of any given photometer. The fused silica windows on this filter wheel permit accurate subtraction of the background signal, if any, from visible and near UV light.

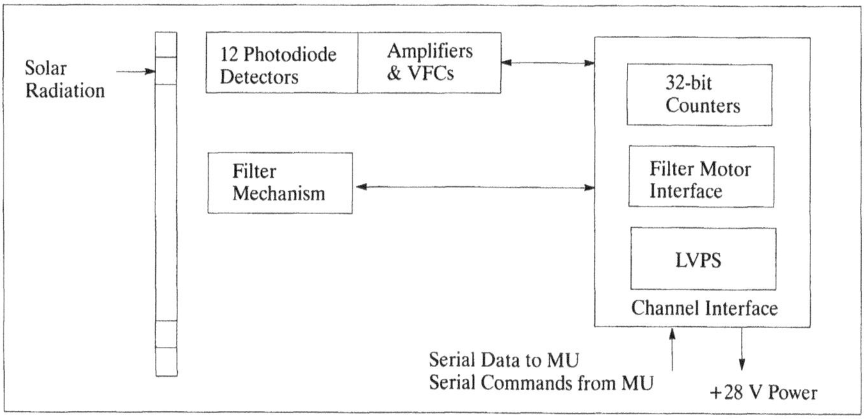

Figure 39: Block diagram of the XPS instrument

Instrument type	XUV photometer
Primary detector type	EUV photodiodes
Spectral range, resolution	0.1 - 31 nm, 5-10 nm
Accuracy, precision	20%, 4% over mission life
Instrument dimensions (HxWxD)	15.6 cm x 18.7 cm x 17.2 cm
Instrument mass, power, nominal data rate	2.6 kg, 8.6 W, 267 bit/s
FOV (Field of View)	8° (full cone angle)

Table 82: XPS instrument parameters

The 12 XUV photometers (XPs) are grouped into three sets. Each set of four XPs is arranged in a circle for use with the filter wheel mechanism. The filter wheel, which has 3 different rings of filters for the three sets of XPs, has 8 positions : four blocked for dark measurements, two clear for solar XUV measurements, and two with fused silica windows for solar visible background measurements. An observation run is a sequence of measurements from five consecutive filter wheel positions, normally starting and ending with dark measurements. The electronics for each photodiode include only a current amplifier and a voltage-to-frequency (VTF) converter.

Calibration: The XPs are silicon photodiodes with thin film filters to measure the integrated solar UV irradiance from 0.1 to 35 nm with a typical bandpass of 5 nm. The XPS instrument has redundant XPs in order to have a working photometer for daily measurements and a lower duty cycle (reference) photometer for degradation analysis of the working photometer. The reference photometer is used for a solar measurement every week; therefore, the cadence of the reference photometer is once per 105 measurements by the daily channel assuming 15 orbits per day.

A.28 TIMED (Thermosphere, Ionosphere, Mesosphere Energetics and Dynamics)

TIMED is a NASA exploratory mission, managed by JHU/APL (Johns Hopkins University/Applied Physics Laboratory) for NASA. [479] [480] [481] The primary objective is to investigate and understand the energetics of the Mesosphere and Lower-Thermosphere/Ionosphere (MLTI), the region in the Earth's atmosphere from about 60 to 180 km in altitude. The measurements of TIMED will provide data defining the basic states of the MLTI region and its thermal balance. Specific missions goals are:

- to determine the temperature, density and solar wind structure in the MLTI region, including the seasonal and latitudinal variations
- to determine the relative importance of the various radiative, chemical, electrodynamical, and dynamical sources or sinks of energy for an understanding of the thermal structure of the MLTI. [482]

The TIMED S/C structure is an aluminum and framework/honeycomb design with a MIL-STD-1553B data bus; three-axis stabilization, attitude knowledge is 0.03° ($3\ \sigma$, all stellar), attitude control is 0.5° ($3\ \sigma$, three-axis momentum-bias); mission design life = 2 years; S/C mass = 660 kg, power = 426 W (solar panels, 50 Ah Ni-H_2 battery). Integrated Electronics Module (IEM) consisting of: 2 DC/DC converter, CMD&TLM processor, 2.5 Gbit solid-state recorder, GNS (GPS Navigation System), CMD&TLM interface, S-band uplink and S-band downlink (each consisting of a plug-in card). The transmission protocol is CCSDS compliant. Of particular interest is GNS, developed at APL. GNS is the spacecraft's navigation and timekeeping system. It is designed to autonomously provide position, velocity, time, sun vector, and defined orbital event notifications (e.g., terminator crossings and SAA (South Atlantic Anomaly) region encounters in real time. GNS consists of a zenith-oriented antenna, a preamplifier, RF downconverter, GTA (GPS Tracking ASIC - with a 12 channel GPS signal tracking capability), and a dual-processor-based computer system. [483] [484]

A launch of TIMED is planned for 2001 aboard a Delta-2 launcher from VAFB, CA (co-manifest with Jason-1).

Orbit: Circular orbit, altitude = 625 km; inclination = 74.1° with a 720° per year nodal regression (this means the local time of the equator crossing varies from dawn to dusk in three months).

Areas of applications: The MLTI is a very poorly understood region of the atmosphere. The results of the TIMED mission will enable the scientific community to establish the first quantitative MLTI baseline and will serve as a basis for future investigations.

479) Information provided by G. E. Cameron and by K. J. Heffernan of JHU/APL
480) http://www.timed.jhuapl.edu/
481) D. Y. Kusnierkiewicz, "A description of the TIMED spacecraft," American Institute of Physics (AIP) Conference Proceedings, 387, Part One, pp. 115-121, 1997
482) R. S. Bokulic, et al., "A Highly Integrated S-Band Transceiver System with Two-Way Doppler Tracking Capability," Proceedings of AIAA/USU Conference on Small Satellites, 1997, pp. 1-8
483) A. A. Chacos, P. A. Stadter, W. S. Devereux, "Autonomous Navigation and Crosslink Communication Systems for Space Applications," JHU/APL Technical Digest, Vol. 22, No 2, 2001, pp. 135-143
484) Ch. C. DeBoy, M. J. Reinhart, "A Flexible, Transceiver-based RF Communications System for Small Satellites," Proceedings of the 3rd International Symposium of IAA, Berlin, April 2-6, 2002, pp. 363-366

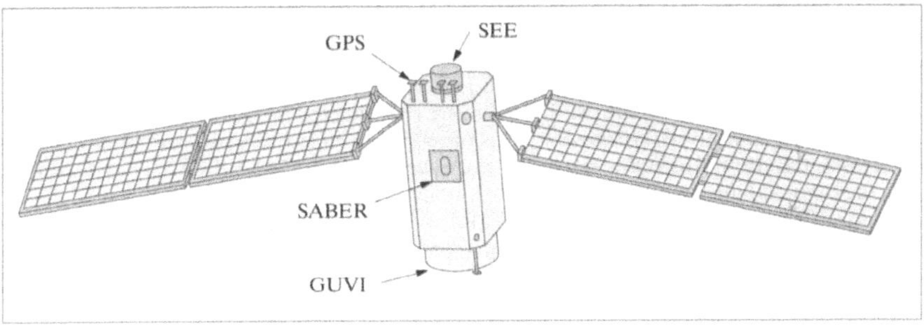

Figure 40: The TIMED S/C model

RF communication: On-board storage capability of 2.5 Gbit in solid-state memory; communication in S-band (CCSDS standard). The scalable architecture of the S-band RF transceiver system is part of the IEM. The transceiver consists of two RF cards, each of size 15 cm x 22 cm. The IEM approach provides the integration of functionalities that are normally part of the system (inclusion of a real-time critical command decoder). The downlink card outputs a CCSDS-compatible downlink signal and support data rates up to 4 Mbit/s. The downlink card also includes an on-board data framer and Reed-Solomon encoder.

Non-coherent navigation: In addition, the transceiver system cards provide the capability for highly accurate Doppler tracking, using a two-way non-coherent technique. The APL-developed technique obviates the need for coherency between the uplink's carrier tracking oscillator and the downlink carrier. In this technique, the uplink carrier signal is received and compared with the receiver's on-board reference oscillator. This operation results in a set of phase comparison counts, placed in the telemetry and transmitted to the ground. There, the ground station continues tracking the S/C as if it were a coherent transponder. The uplink and downlink error counts are compared with the telemetered phase comparison counts. Tracking accuracies to 0.1 mm/s (velocity error) are achieved.

RF output frequency, output power	2214.97 MHz (S-Band), 3 W
High-rate modulation	4 Mbit/s with DQPSK modulation
Low-rate modulation	9 kbit/s residual carrier PM
Coding	Differential (select or bypass) Convolutional (select or bypass) Reed-Solomon (select or bypass)
Downlink framer	CCSDS compatible protocol
Bus interface	PCI
Card size, mass	15.3 cm x 23 cm, <0.6 kg
DC power	12.7 W conditioned power (transmit) 1.0 W conditioned power (standby)

Table 83: Downlink card characteristics

Mission operations are performed from JHU/APL in Laurel, MD. The payload operations concept requires each PI to perform his own instrument operations from his home site. A combination of on-board GPS processing and the use of Internet move the data from the APL ground station to each investigator's home site. By separating instrument operations from all S/C system activities, the instrument teams are able to control all of the instrument modes, operations and science data return at their own choice without explicit interactions or approvals by the S/C project team. In addition, the Internet is used by each investigator to control his instrument directly (the packetized messages are integrated into the uplink command structure in an automated fashion).

Sensor complement:

SEE = Solar EUV Experiment (PI: T. N. Woods; built by LASP of the University of Colorado). [485] The solar sensors in SEE are **EGS** (EUV Grating Spectrometer) and **XPS** (XuV Photometer System). The objective is to measure the absolute fluxes of solar UV, EUV, and XUV radiation and to determine the rates of energy deposition, dissociation and ionization. EGS features a Rowland-circle grating design; detector: 64 x 1024 MCP/CODACON (microchannel plates with coded anode position array); spectral range of 25 to 200 nm; resolution of 0.4 nm (0.17 nm per anode); FOV = $6^\circ \times 12^\circ$. - XPS has 12 XUV silicon photodiodes (thin film coatings on diodes) as detectors. The spectral range is 0.1 to 40 nm, the resolution is 0.4 nm above 25 nm and about 3 nm below 25 nm, FOV = 20° in diameter. Instrument mass = 26 kg, power=16 W (average). Look direction: solar pointing; data rate = 210 bit/s.

SABER = Sounding of the Atmosphere using Broadband Emission Radiometry (PI: J. M. Russell, Hampton University; instrument builder: SDL of Utah State University, Logan, UT). SABER (heritage of LIMS, SAMS, CIRRIS, ATMOS, HALOE, CLAES, ISAMS, and SME) is a 10-channel radiometer with the objective to measure the IR emissions emitted at altitudes generally below 100 km by limb scanning to drive vertical distributions of temperature and concentrations of energetically important species (O_3, H_2O, NO, NO_2 CO, CO_2) as well as radiative energy loss. The telescope is an on-axis Cassegrain design (rejection of stray light outside IFOV). Instrument mass=61.7 kg, power = 64 W; detector types: five HgCdTe (photoconductive mode of operation), two InSB (photovoltaic mode of operation), and three InGaAs (photovoltaic mode of operation). Spectral coverage: discrete ranges from 1.27 - 17 μm (7865 cm^{-1} to 650 cm^{-1}). A scan depressing angle of -20.0° to -13.9° from the S/C horizontal is used; vertical resolution of 2 km; the telescope is cooled to 240 K; the focal plane assembly is cooled to 75 K; data rate: 4 kbit/s. SABER scans up and down the Earth's horizon once every 58 seconds, collecting data over an altitude range from about 180 km down to the Earth's surface (vertical profiles of elemental constituents) and temperature.

Instrument drifts due to changes in telescope and focal plane base temperatures are corrected by an in-flight calibration system.

Altitude range (km)	Parameter measured	Spectral range (μm)	Application
10-130	CO_2	14.9 & 15.2	Temperature, density, IR cooling rates, P(z), non-LTE
15-100	O_3	9.6	O_3 concentration, cooling rates, solar heating and dynamics studies
50-105	O_2	1.27	O_3 concentration (day), inferred O at night, energy loss for solar heating efficiency
85-150	CO_2	4.3	CO_2 concentration, mesosphere solar heating, tracer
80-100	OH (v)	2.0 & 1.6	HO_y chemistry, chemical heat source, dynamics, inference of O and H, PMC studies
90-180	NO	5.3	Thermosphere cooling, NO_x chemistry
15-80	H_2O	6.9	HO_y source gas, dynamical tracer

Table 84: Overview of SABER measurements and applications

TIDI = TIMED Doppler Interferometer (PI: T. L. Killeen, U. of Michigan). Objective: investigation of the dynamics and energetics of MLTI . TIDI measures the VIS/NIR emissions emitted at altitudes between 60 and 300 km by limb scanning techniques to determine the temperature and horizontal winds with the use of the Doppler effect. The instrument makes also density measurements, mostly on the day side of the orbit. TIDI comprises three subsystems: four identical telescopes (off-axis Gregorian type, aperture 7.5 cm, f/2.2 FOV = 2.5° horizontal x 0.05° vertical), a Fabry-Perot interferometer (fixed-gap single etalon)

485) http://lasp.colorado.edu/see/see_instrument.html

with a CCD detector (passively cooled), and an electronics box. TIDI views emissions from OI 557.7 nm, OI 630.0 nm, OI 732.0 nm, O_2 (0-0), O_2 (0-1), Na D, OI 844.6 nm, and OH to determine Doppler wind and temperature throughout the altitude range. Instrument mass=42 kg; power=19.3 W; spectral coverage: selected lines between 550 - 900 nm; look direction: limb scan; a scan depression angle of 23.2° to - 16.8° from the S/C horizontal is used; data rate = 2.336 kbit/s.

GUVI = Global Ultraviolet Imager (PI: A. B. Christensen, The Aerospace Corporation). The instrument is of SSUSI heritage and a joint collaboration between JHU/APL and The Aerospace Corporation, El Segundo, CA. The objective is to monitor three general regions: the daytime low- to mid-latitude thermosphere, the nighttime low- to mid-latitude ionosphere, and the high-latitude auroral zone. The goal is to obtain a detailed quantitative and predictive understanding of auroral phenomena. The instrument consists of the following elements: a scan mirror feeding a parabolic telescope and Rowland circle spectrograph, with a wedge-and-strip detector at the focal plane. GUVI is a FUV scanning imaging spectrograph providing horizon-to-horizon images in five selectable bands or "colors." The colors chosen are: H 121.6 nm, O 130.4 nm, O 135.6 nm, and N_2 Lyman-Birge-Hopefield (LBH) bands at 140 to 150 nm and 165 to 180 nm. GUVI uses a scan mirror to sweep its FOV of 11.78° through an arc of up to 140° (scan from 80° to -60°) in the cross-track direction. This FOV is mapped via an f/3 Rowland circle spectrograph (with a toroidal grating) into 14 spatial and 160 spectral "pixels." The scan sweep time is 22 s. The detector is a microchannel plate (MCP) intensified wedge-and-strip anode which provides a 2-D readout.
The SIS (Scanning Imaging Spectrograph) of GUVI consists of a cross-track scanning mirror at the input to the telescope (a 75 mm focal-length off-axis parabola system with a 25 x 50 mm clear aperture) and a Rowland circle spectrograph. Two 2-D photon-counting detectors are located at the focal plane of the spectrograph. The operating detector is selected by a "pop-up" mirror that is moved into or out of the optical path to direct radiation from the grating onto one of two detectors. The detectors employ a position-sensitive anode to determine the photon event location.

GUVI has both cross-track disk and limb scan modes; it measures the UV radiation emitted at altitudes generally above 150 km up to 520 km by limb-scanning and imaging to determine during daytime conditions the concentrations of N_2, O_2, O, and the temperature; infers the fluxes of precipitating auroral particles. Instrument mass=19.2 kg, power=24 W; swath width of about 300 km; spectral coverage: 110 - 180 nm; look direction: cross-track scan (disk + limb); data rate = 8.0 kbit/s.

The GUVI observations are compared with data collected by the ground stations for "ground truthing." GUVI products include maps of the auroral oval, the characteristic energy and flux of the electrons which excite it, F-region ionospheric electron density profiles, and dayside neutral composition information.

Parameter	H (121.6 nm)	O (130.4 nm)	O (135.6 nm)	N_2 (LBH-1)	N_2 (LBH-2)
Dayside limb	H profile and escape rate	O profile	O altitude profile	Solar EUV inputs	N_2, O_2, temperature
Dayside disk	H column	Composition	O/N_2	O/N_2	
Nightside limb	H column	F-region electron density profile			
Nightside disk	H column	F-region total electron content			
Auroral zone	Proton auroral boundaries	Auroral boundaries	Effective energy flux Q	Effective average energy of precipitating particles	

Table 85: GUVI environmental parameters measured by different colors

A.29 TOMS Missions

Within NASA's Earth Probes Program several TOMS (Total Ozone Mapping Spectrometer) or TOMS/NSCAT missions are planned. The objectives are:

- to continue the global ozone data set that began in 1978 with the flight of TOMS on NIMBUS-7
- to continue observation of environmentally important areas such as the Antarctic ozone hole
- to observe sulfur dioxide clouds resulting from volcanic eruptions

A.29.1 TOMS-EP

TOMS-EP (Earth Probe) is a free-flying spacecraft (built by TRW, Redondo Beach, CA) with a payload of the TOMS instrument. It is nadir pointing and provides a contiguous survey of the Earth every day. The S/C bus is of modular design (TRW lightsat bus) with a stowed size of 1.16 m in diameter and 1.77 m in height, three-axis stabilized with a 3 sigma pointing accuracy of $<0.25^{\circ}$ (knowledge) and $<0.5^{\circ}$ (control). Design life = 2 years, S/C mass = 295 kg, power = 130 W. The S/C carries an orbit adjust module which fully integrates the propulsion subsystem (54 kg of propellant). In addition, the S/C features an autonomous operation/redundancy management providing 24 hours of autonomous operation and autonomous safing modes with ground-support recovery. Launch vehicle: Pegasus XL. Launch: July 2, 1996.

Communication: S-band (Omni-spherical coverage) with an uplink rate of 2 kbit/s and three downlink rates of 1.1 kbit/s, 50 kbit/s, and 200 kbit/s.

Orbit: circular, polar, sun-synchronous, altitude = 500 km, inclination = 99.3°, equator crossing time of 12 noon. - Starting Dec. 4, 1997, the TOMS-EP orbit was re-boosted during a 12-day period to an altitude of 750 km. The object is to: 1) increase the daily coverage, and 2) reach a stable orbit for a period of 4 years.

Sensor: **TOMS** (Total Ozone Mapping Spectrometer)

- 105° scan of $3x3^{\circ}$ FOV, TOMS scanner is not pointed at sun
- Ebert-Fastie monochromator
- Wavelengths: 308.6 nm, 312.5 nm, 317.5 nm, 322.3 nm, 331.2 nm, and 360.0 nm
- Ozone trend measurement capacity: 0.1% yearly (goal)
- Swath width: 2722 km
- Spatial resolution: Nadir - 50 km x 50 km; average: 62 km x 62 km
- Dimensions: 15 x 30 x 27 cm
- Mass: 34 kg; power: 21 W
- Data rate: 600 - 700 bit/s (average). Data is available to the worldwide community of global change researchers through the National Space Science Data Center (NSSDC).

There is a new on-board calibration system involving illumination of the diffuser plate from a mercury lamp behind a fluorescing screen. - The electronics of TOMS have been modified/updated to include a reprogrammable microprocessor. Side-scanning can now be under the old 35 position mode or in a 37 position mode (used for ADEOS).

A.29.2 TOMS/NSCAT on ADEOS

TOMS and NSCAT instruments will be flown on the Japanese ADEOS mission in 1996 (see ADEOS mission).

Orbit: Sun-synchronous polar and circular orbit, altitude = 797 km, inclination = 98.6°

TOMS and NSCAT data will be available to the worldwide community of global change researchers through the Earth Science Data and Information System (ESDIS). Note: TOMS on ADEOS failed on June 30, 1997 when the S/C suffered several malfunctions and stopped functioning after an apparent loss of power (solar panel failed).

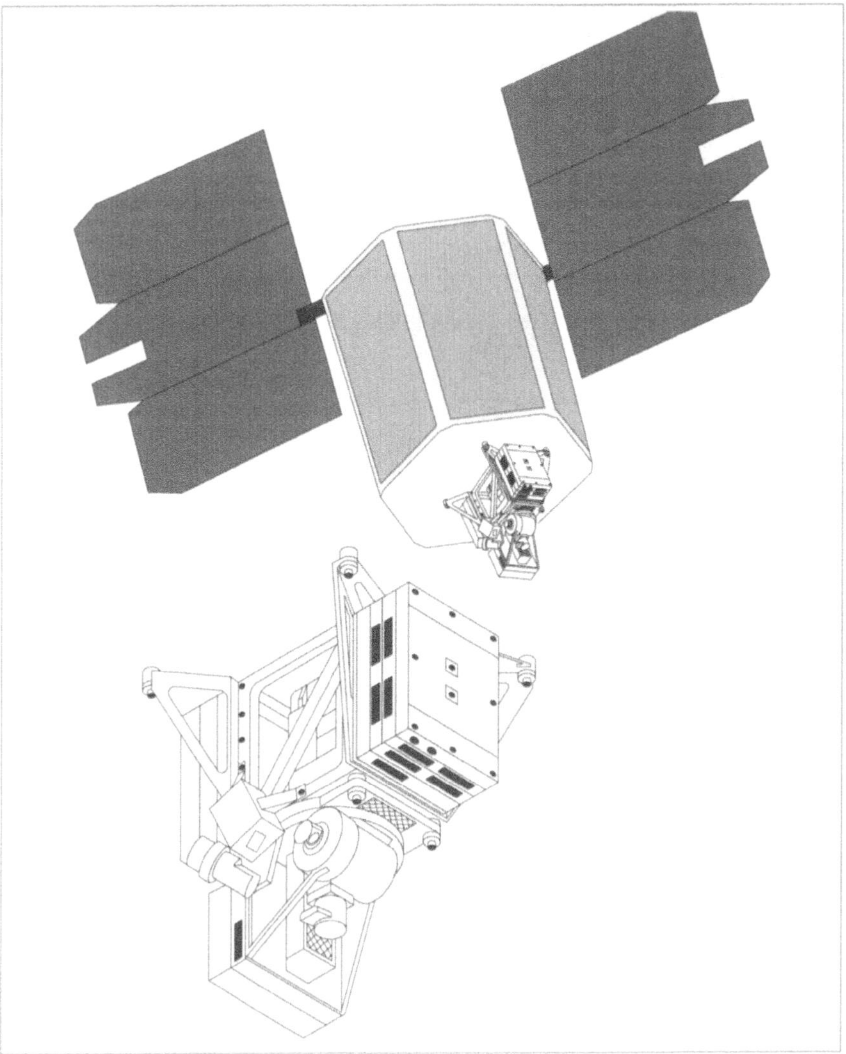

Figure 41: The TOMS-EP S/C and sensor

A.29.3 QuikTOMS (Quik Total Ozone Mapping Spectrometer)

A NASA mission, part of the Earth Explorer program, with the objective to continue long-term ozone measurements with the TOMS instrument beyond the TOMS-EP mission. [486] The TOMS-5 instrument on QuikTOMS was originally scheduled to be flown on the Russian Meteor-3M mission; however, constant delays of the Meteor-3M launch caused NASA to set up an alternate mission to avoid potential coverage gaps. NASA selected OSC of Dulles, VA, as prime contractor for the QuikTOMS mission. That includes S/C provision, sensor integration, launch and two years of mission operations.

486) http://quiktoms.gsfc.nasa.gov/

The QuikTOMS S/C is a modified MicroStar platform of OSC. The bus is comprised of two rings stacked together vertically a core ring (a dual-faced cylinder consisting of a honeycomb structure) housing all primary spacecraft systems and a payload support module and the second ring carries four propulsion tanks and two solar arrays. The average payload power is 107 W (S/C power of 228 W EOL). Nickel hydrogen (NiH_2) batteries provide 10 Ah of energy. The S/C is three-axis stabilized with pointing control of 0.5°, a pointing knowledge of < 0.25°, and a rate stability of < 0.01°/s. The S/C design life is three years (with consumables of five years). The S/C mass is 166 kg. A launch of QuikTOMS as a secondary payload to OrbView-4 is planned for March 2001 on a Taurus launch vehicle form VAFB, CA.

Communications: The on-board data recorder has a capacity of 64 MByte SSR (Solid State Recorder). S-band communications is used for uplink (2 kbit/s) and downlink (500 kbit/s).

Orbit: Sun-synchronous circular polar orbit, altitude = 800 km, inclination = 98°, descending node equator crossing at 10:30 AM.

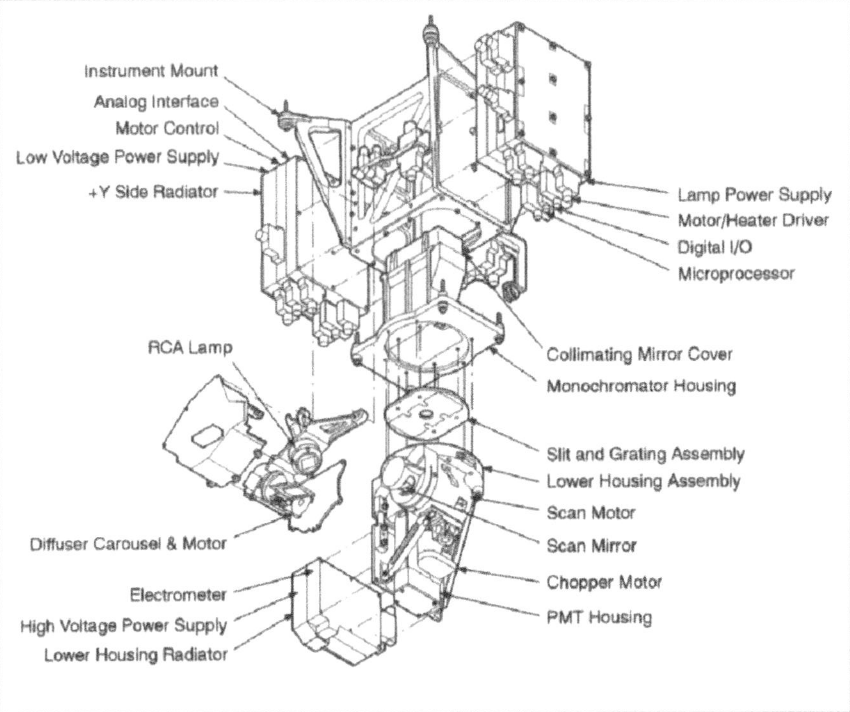

Figure 42: Illustration of the TOMS-5 instrument

Sensor complement:

TOMS-5 (Total Ozone Mapping Spectrometer), built by OSC, Pomona, CA (former Perkin Elmer subsidiary bought by OSC). TOMS-5 is a f/5 single grating Ebert-Fastie type monochromator with multiple exit slits for six different wavelengths centered at: 308.6, 312.5, 317.5, 322.3, 331.2, and 360 nm. The nominal bandwidth is 1.2 nm for each band. The wavelength is selected with a chopper wheel near the slit plate which gates the pair of the entrance and exit slits for each wavelength band. IFOV = 3° x 3° or a footprint of about 42 km x 42 km. The instrument scans 18 scenes from the nadir to each cross-track direction in a 3°

interval to cover a swath of about 2300 km (max scan angle of ±54°, scan cycle time = 6.3 s). The scanning axis of the scanner is 45° from the scanner normal and coincides with the optical axis of the instrument system so that the incidence angle on the scanner mirror is constant regardless of the scan angle. A depolarizer is placed immediately after the scanner mirror to minimize sensitivity to the polarization of the incoming radiation. A baffle with a square opening is placed before the objective lens as an aperture stop and the objective lens images the baffle to the grating. A PMT (Photomultiplier) detector is used for radiation detection.

The instrument employs a diffuser assembly to measure the solar flux and uses three onboard calibration features in orbit. The diffuser assembly has three ground aluminum surfaces, one of which is selected for sun illumination at the northern terminator. The diffuser surface generates a radiance source from the solar irradiance and the radiant source fills the field of view of TOMS-5. Two other surfaces are less frequently used for the solar viewing to minimize the degradation. The wavelength monitor is used to monitor a wavelength shift of the TOMS-5 bands using the mercury 296.7 nm line. Two narrow entrance slits, slightly offset, are imaged over the 312.5 nm exit slit when the 296.7 nm radiation illuminates them. The chopper wheel selects one of the two entrance slit images alternately. Any center wavelength change due to the image shift in the exit plane of the monochromator can be detected by comparing the relative strength of the two alternating signals. The electronic calibration (ECAL) is used to monitor the gain of the electronic amplifiers. A precision current source injects a current at several different levels to the front end of the electronics. The reflectance calibration (RCAL) is for monitoring the reflectivity of the diffuser surfaces. The Reflectance Calibration Assembly (RCA), a light source for RCA, is made of a mercury pen lamp with a reflector which has a window coated with a mixture of phosphors. This phosphor mixture produces a broad spectral emissions over the TOMS wavelength range when excited by the mercury ultraviolet radiation. The TOMS scanner mirror views the RCA lamp window and the diffuser surface illuminated by the RCA lamp alternately. The relative reflectance of the diffuser surfaces is monitored from the comparison of these two signals.

IFOV (scanner)	3° x 3° (42 km x 42 km)
Maximum scan angle, scan cycle time	±54°, 6.3 s
Scan dwell time, scan steps	150 ms, 37
Spectral bands: Band 1, Band 2, Band 3 Band 4, Band 5, Band 6	360 ±0.1 nm, 331.2±0.1, 322.3±0.1 nm, 317.5±0.1, 312.5±0.1, 308.6±0.1
Instrument mass, power	32 kg, average power is 25 W, peak power is 57 W

Table 86: Some parameters of the TOMS-5 instrument

A.30 Triana

A NASA mission named after Rodrigo de Triana, the sailor who first spotted the New World on Columbus's first expedition in 1492. Triana is a mission to L1, the Lagrangian liberation point (or balanced gravity point between the Earth and the sun), with the objective to continuously observe the sunlit Earth (full disk). The overall objective of the mission is to demonstrate the innovative use of a satellite to serve as a pathfinder to climate studies and global Earth scenes. The data products are planned to be made available on the Internet for global distribution, as one of Triana's goals is to provide high quality educational products that will aid in increasing public awareness about planet Earth. A Shuttle launch was planned for early 2002 from KSC (Kennedy Space Center).

However, with the new Bush Administration in office, launch plans of Triana were changed in the spring of 2001 (after all, the project was initially proposed by Al Gore who came in second in the 2000 US elections). Following is the official NASA version: "The constrained

Shuttle flight rate of six per year, established in the 2001 NASA budget planning process, requires that NASA give priority to the primary International Space Station (ISS) payloads, Hubble Space Telescope (HST) reboost, and Microgravity experiments. Consequently, it is not possible at this time to identify a definitive launch date for Triana. - As a result, NASA is working to complete the integration and qualification testing of the Triana Mission flight hardware and to place those assets into storage to await the determination of a future launch opportunity. Efforts associated with preparing for both the launch and operation of the mission are rapidly being suspended. System hardware qualification is anticipated to be completed by the end of September, 2001. No Fiscal year 2002 (FY 02) funding has been set aside for the project. Launch opportunities are being investigated within the Space Transportation System (STS) planning manifest process and additionally as a possible Expendable Launch Vehicle (ELV) payload. Restart plans will be dependent upon the definition of the mission launch date and approach." [487) 488) 489) 490) 491)]

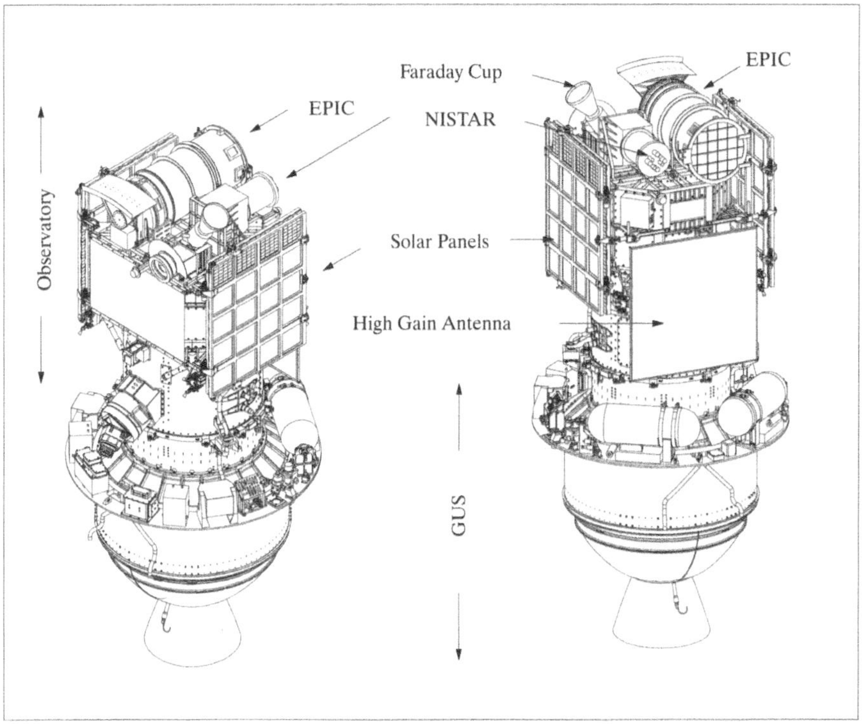

Figure 43: Illustration of the Triana observatory (two views of the undeployed S/C)

A likely launch date of Triana could be in 2004.

The Triana spacecraft will have two major components within the Shuttle: a) the Observatory, and b) GUS (Gyroscopic Upper Stage). The Observatory design is based on the SMEX-

487) F. P. J. Valero, J. Herman, P. Minnis, W. D. Collins, R. Sadourny, W. Wiscombe, D. Lubin, K. Ogilie, "Triana - a Deep Space Earth and Solar Observatory," Report prepared for the National Academy of Sciences by the Triana Science Team (SIO, NASA/GSFC, NIST, LaRC, ARC, NCAR, LMD, LM, LANL, VT).

488) J. G. Watzin, "The Triana Mission - A Pathfinder Mission to Explore the Utility of using Deep Space in Conducting Earth Observation," Proceedings of the 51st IAF Congress, Rio de Janeiro, Brazil, Oct. 2-6, 2000

489) S. A. W. Gerstl, F. P. J. Valero, "The Triana Satellite Mission from L1 for Global Vegetation Monitoring," Proceedings of IEEE/IGARSS'99, Vol. I, Hamburg, June 28-July 2, 1999

490) http://triana.gsfc.nasa.gov/

491) http://www.earth.nasa.gov/ebn/triana/index.html

Lite architecture bus which ties all the S/C functions together. The spacecraft is three-axis stabilized, equipped with solar arrays, a reaction-wheel-based GN&C (Guidance, Navigation and Control) subsystem [consisting of four reaction wheels, a gyro, a Startracker and sun sensors], and a hydrazine thruster system. The propulsion module is used to a) de-spin the Observatory (following its separation from GUS), b) to provide thrust for mid-course corrections and for L1 orbit maintenance. The OBC consists of a radiation-hardened 32-bit computer. Communications between the payload and the bus is provided by a redundant serial bus (MIL-STD-1553). The Startracker of Ball Aerospace contains an internal catalog of 2000 stars. The Startracker acquires, tracks and identifies stars, computes Triana's orientation, then transfers that information to the GN&C. The total mass of the Observatory is 564.5 kg, including the bus (461.3 kg) and the payload mass (103.2 kg).

GUS, also known as the kick stage, provides the power and thrust to boost the Observatory from LEO to the LOI (Lissajous Orbit Insertion) point. Once GUS has completed its solid rocket burn, it will separate from the Observatory. Prior to being deployed from the Space Shuttle, Triana will be spun up to 60 rpm by the IRIS (Italian Research Interim Stage) spin table, developed by Alenia Aerospazio of Italy (IRIS also provides the physical supporting structure for Triana while it's in the Space Shuttle Payload Bay). Approximately 45 minutes after Shuttle deployment, GUS fires its kick motor to begin the journey to L1. The total GUS mass at Shuttle deployment is 2424.4 kg.

Orbit: A halo orbit into the Lagrangian point L1, balanced gravitationally between the sun and the Earth. From this viewpoint, instruments have a continuous view of the entire sunlit face of the Earth, from sunrise to sunset and pole to pole. Every point on the Earth rotates under the view of the instruments providing a full day's data. L1 is located about 1.5 million km from Earth, amounting to 1/100 the distance from Earth to the sun. The Observatory will arrive at L1 approximately three months after launch.

Communications: The Observatory has three antennas, one high-gain and two omni antennas. The downlink data rates are 100-200 kbit/s, the uplink data rate is 2 kbit/s, both in S-band. The CCSDS protocol is used for all data transmissions.

Sensor complement:

EPIC (Earth Polychromatic Imaging Camera). This primary instrument on-board Triana is managed by SIO (Scripps Institution of Oceanography) at UCSD (University of California at San Diego) and designed and built by Lockheed Martin Advanced Technology Center (ATC). The objective is to measure ozone amounts, aerosol amounts, cloud height and phase, hotspot land properties (a view of the land from angles where shadows are a minimum), and UV radiation estimates at the Earth's surface. EPIC is able to view the entire sunlit Earth from sunrise to sunset at an almost constant scattering angle between 165-178°. The EPIC channels were selected to match closely with TOMS in the UV region and with MODIS in the visible range; hence, the data products will be very similar and can be directly compared. These comparisons will validate both the calibration and data reduction algorithms.

The EPIC instrument is of Cassegrain design (30.5 cm aperture diameter of telescope) featuring a ten-filter spectroradiometer with a spectral range of 0.317 to 0.905 μm (UV to NIR). The 10 filters are contained in two filter wheels. The sub-nadir spatial resolution of the observed data is 8 km. The measurement frequency is 15 minutes for RGB images (consisting of band 6, 7, and 8) and one hourly frame for the 10-channel radiometer data, each frame consisting of 4 million calibrated Earth-located radiances. Views of the moon are used to calibrate EPIC.

Band No	Center Wavelength (nm)	Bandwidth (nm) FWHM	Measurement Objective
1	317.5	1	Ozone, SO_2
2	325	1	Ozone, SO_2
3	340	3	Aerosols
4	388	3	Aerosols, clouds
5	393.5	1	Cloud height
6	443	10	Aerosols (blue)
7	551	10	Aerosols (green)
8	645	10	Aerosols (red)
9	869.5	15	Clouds, vegetation
10	905	15	Water vapor

Table 87: Spectral bands of the EPIC instrument

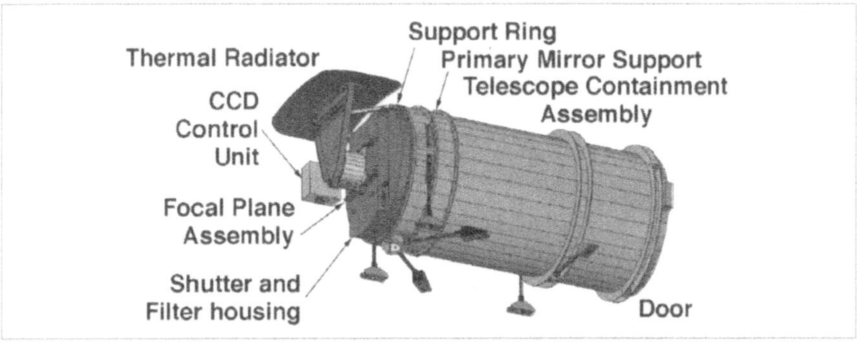

Figure 44: Illustration of the EPIC instrument

Telescope: Aperture, effective focal length FOV, wavefront error	Cassegrain type with adjustable secondary for on-orbit focus 30.5 cm diameter, 282 cm 0.61°, 0.054 waves rms at 633 nm on-axis
Shutter	Individual exposure times of 2 ms, 10 ms and 40 ms to >1 min Multiple exposures for timings between 2 ms and 40 ms at 2 ms resol.
Focal plane assembly: CCD format; pixel size CCD type; spectral range Pixel full well depth Digital intensity conversion Readout; pixel readout rate CCD operating temperature Dark current; readout noise	 2048 x 2048 pixels; 15 µm x 15 µm, 100% fill factor Thinned backside illuminated; 200-950 nm (QE>25%) >80,000 electrons 0-4095, 12 bits at 20 electrons per bit Single or dual (opposite corners); 500 kHz -40° C using passive cooling <5 electrons per second per pixel; <20 electrons rms
Minimum image cadence	< 20 s
Image output format	Raw (bit map) and 12 bit JPEG/JFIF
Instrument power; total mass	32 W (electronics), 30 W (operational heaters); 63.2 kg

Table 88: EPIC performance parameters

NISTAR (National Institute of Standards and Technology Advanced Radiometer), developed by NIST and Ball Aerospace. The objective is to measure the radiance output from the sunlit Earth over a broad portion of the spectrum (UV and VIS as reflected radiation, IR as emitted radiation) in order to detect changes in the Earth's energy balance (climate studies). The instrument package consists of three active-cavity electrical-substitution radiometers and one silicon photodiode channel to measure the "total Earth reflected and emitted radiant power" in the direction of the Triana spacecraft with a goal of 0.1% accuracy and a precision (relative changes) of 0.03% (at L1 the measurement of the Earth's reflected and emitted radiance is a true challenge due to its minuteness; it is in fact $<10^{-5}$ than that of the sun). These levels of accuracy and precision are necessary for detecting the small changes in Earth's radiances corresponding to small changes occurring now that can be of major importance for humankind far into the future. The accuracy goal is based on the high sensitivity of the NISTAR instrument and on the ability to use long integration times (minutes) at the L1 location. - Besides broadband measurements of the total Earth irradiance, filtered channels on NISTAR allow separation of the reflected solar from the Earth-emitted infrared radiation. Techniques developed at NIST for SNR improvements in active-cavity radiometers were essential in making feasible the NISTAR measurement performance. [492)]

NISTAR has a FOV of 1°, sufficient to see and image the full Earth disk whose FOV is approximately 0.5° as seen from L1. The photodiode channel has been included in order to obtain a faster time series (<1 s) than what can be obtained by the cavity radiometers. The channel is used to provide for the tracking stability of the filters, to verify co-alignment of NISTAR and EPIC, and to continuously observe the solar reflected broadband radiation from Earth with high temporal resolution. - A PTC (Positive Temperature Coefficient) thermistor and a wire-wound low-temperature coefficient heater are bonded to the outside of the 30° conical receiver cavity. The absolute cavity radiometers are designed for optimum power measurements (in the tens of μW range). The optical signal incident on the receiver is only 1 μWcm^{-2}, however, the emission from the receiver cavity to space is estimated to be 30 μWcm^{-2} when the shutter is open. There are four digital control loops, three receiver cavity control loops and one for the heat sink. The PTC temperature sensor resistance measurements are performed with AC-bridge circuits operating between 35 and 155 Hz. The NISTAR absolute cavity radiometers are designed for a noise floor of less than 10 nW (defined as the level at which the SNR is equal to one for a single one second measurement). The NISTAR electronics have a measurement resolution of 10 mΩ and internal equivalent noise of less than that. - The NISTAR total instrument mass is 23.5 kg.

The NISTAR halo orbit data will complement the large database of LEO Earth radiation budget measurements of such instruments as: ERB, ERBE, ScaRab, CERES, and ISP-2, as well as those of the GERB instruments flown on EUMETSAT's MSG series in GEO. The NISTAR radiance measurements will be used in several ways:

- To make integrative measures of global change (longwave channel)
- To estimate the Earth albedo
- To attempt to interpret the near-infrared to visible albedo ratio
- To test CERES algorithms in an integral sense
- To provide an extensive integral test of how well radiative transfer in the Earth-atmosphere system is understood.

Channel	Range	Comment
UV+VIS+ FIR Band-A	0.2 - 100 μm	Total radiant power in the UV, VIS and FIR wavelengths (longwave channel). Band-A is unfiltered
UV+VIS+SWIR Band-B	0.2 - 4 μm	Solar channel to measure reflected solar radiance
NIR+SWIR Band-C	0.7 - 4 μm	IR channel to measure reflected IR solar radiance.
VIS+NIR	0.3 - 1 μm	A non-absolutely calibrated silicon photodiode channel to calibrate the spectroradiometer

Table 89: Overview of NISTAR spectral bands

PlasMag (Plasma Magnetometer). The objective of this sun-viewing instrument is to measure the solar wind and magnetic field parameters (composition and energy) at high time resolution. The instrument serves also as an early-warning device for solar-event storms that could damage satellites and electrical equipment on Earth. The PlasMag instrument

492) J. P. Rice, S. R. Lorentz, T. M. Jung, "The next generation of active cavity radiometers for space-based remote sensing," 10th Conference on Atmospheric Radiation, Madison, WI, 1999, American Meteorological Society

suite is an advanced but smaller version of the ACE instrumentation (NASA S/C with a launch on Aug. 25, 1997). The PlasMag instrument package has a mass of 16 kg. The instrument suite was designed and developed (optimized for small size, low power, and very large dynamic range) by GSFC and MIT, consisting of a Faraday cup, a magnetometer and an electron analyzer.

- The Faraday cup measures the 3-D distributions of proton and alpha particles of the solar wind with a time resolution of 90 ms. The Faraday cup has a 60° FOV and uses multiple detectors, permitting the measurement of the full range of solar wind deflections (about 15° in all directions), allowing a full 3-D velocity distribution in parallel. The entire distribution function remains in the FOV, so that more accurate measurements can be made, especially at high densities.
- The triaxial fluxgate magnetometer provides magnetic vector field measurements in 30-40 ms intervals. The objective is to investigate solar-wind magnetic fields with a sensitivity of >0.1 nT. The instrument achieves a dynamic range of almost 8 orders of magnitude in field measurement capability.
- The tophat electron analyzer provides 3-D electron velocity distributions in intervals of 800 ms in the energy range of 3 eV to 2 keV. The 3-D measurements are accomplished by electronically simulating the data sampling of an electron spectrometer on a spinning S/C. The instrument has a set of anodes distributed uniformly in azimuth, each with a FOV of 50° x 70° in azimuth and elevation.

In addition to studies of the sun and solar wind, the PlasMag data are also being used to provide early warning (less than 1 hour) of solar events that might cause damage to power generation, communication, and other satellites.

A.31 TRMM (Tropical Rainfall Measuring Mission)

TRMM [493] [494] is a joint NASA/NASDA mission and program with a low-inclination (equatorial) orbit. NASA/GSFC provides the satellite, four passive sensors, and mission operations, NASDA the launch vehicle (H-II rocket) and the precipitation radar instrument. Each agency processes the data from its own instruments. Design life = 3 years. The launch of TRMM [along with ETS-7, composed of two spacecraft: Chaser (Orihime) and Target (Ikoboshi)] took place on Nov. 27, 1997 from Tanegashima Space Center (TNSC), Japan.

TRMM S/C mass = 3620 kg (including 725 kg of fuel); three-axis stabilization, power = 1100 W; data rate = 170 kbit/s average and 2 Mbit/s on playback; TDRSS S-band communications (8.5 minutes/orbit playback time).

Orbit: Non-sun-synchronous circular orbit, altitude = 350 km (approx.), inclination = 35°, period = 96 min. An observation coverage of the tropics was chosen since the volume of rainfall in the tropics accounts for about two-thirds of the total rainfall on the earth.

Operational staus as of June 2001: TRMM is fully operational after 3.5 years in orbit. In fact, the project expects to be operational for at least another five years.

Objectives: Global change studies, especially in developing an interdisciplinary understanding of atmospheric circulation, ocean-atmospheric coupling, and tropical biology. General circulation models require detailed data on the latent heating of equatorial air masses, and the forcing and propagation speed of waves involved in the 30- to 60-day tropical oscillations.

- Measure diurnal variation of precipitation and evaporation in the tropics to provide increased understanding of how substantial rainfall affects global climate patterns.

493) "The Early Observing System Reference Handbook, ESAD Missions 1990-1997," NASA/GSFC, pp. 62-64
494) T. Keating,T. Ryan, "Tropical Rainfall Measuring Mission (TRMM): US/Japan Science Operations," AIAA-92-0594

- Obtain a minimum of three years of climatologically significant observations of rainfall in the tropics.
- In tandem with cloud models, provide accurate estimates of the vertical distributions of latent heating in the atmosphere.

Sensor complement:

PR = Precipitation Radar (NASDA instrument in cooperation with CRL, Japan). [495), 496)] Active phased array microwave radar operating with horizontal polarization (orbit permits monthly sampling over the complete diurnal cycle), minimum measurable rain rate of 0.5 mm/h. Objective: 3-D rainfall distribution over land and oceans (combined with TMI sensor). Periodically the PR performs an external calibration with ARC (Active Radar Calibrator) and an internal loop calibration to measure the transfer function of the PR receiver.

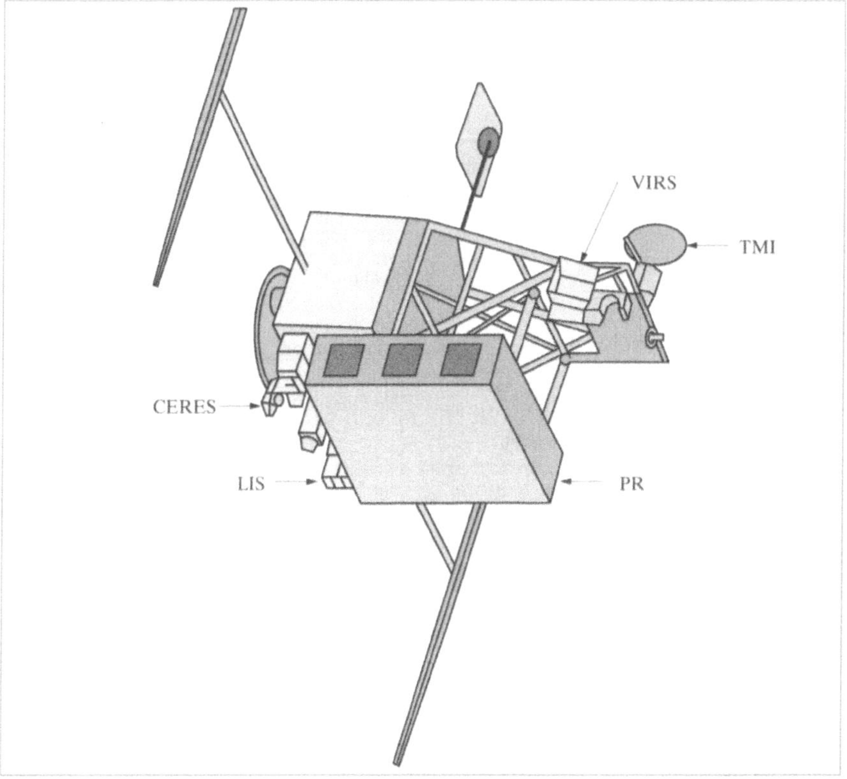

Figure 45: The TRMM S/C model

The radar echo of the instrument consists of the following three components: rain echo, surface echo, and mirror image echo. The surface echo is measured for estimating the total path attenuation and for providing the range of the surface along the radar beam. The mirror image, which is the rain echo received through the double reflection at the surface, may be useful to estimate rain rate retrieval; it is measured at nadir incidence.

495) T. Kozu, M. Kojima, K. Oikawa, K. Okamoto, T. Ihara, T. Manabe, "Development Status of Rain Radar for Tropical Rainfall Measuring Mission," IEEE IGARSS '92, Volume II, pp. 1722-1724

496) NASA paper provided by ESAD and OSSA.

Type of radar echo	Description
Nominal rain and surface echoes	Surface to 15 km altitude, 250 m interval, all scan angles
Mirror image	0 - 5 km altitude, 250 m interval, at nadir only
Oversampled surface echo	Surface echo peak ±0.5 km,125 m interval, scan angles within ±10°
Oversampled rain echo	Surface echo peak to 7.5 km alt., 125 m interval, scan within ±3.5°

Table 90: PR radar echo collection parameters

The PR is a 128-element active phased array system. The transmitter/receiver consists of 128 Solid-State Power Amplifiers (SSPA) LNAs and PIN-diode phase shifters. Each transmitter/receiver element is connected to a 2 m slotted waveguide (WG) antenna (2 m x 2 m planar array). The PR uses a frequency-agility technique to obtain 64 independent samples (N_S) with a single PRF of 2776 Hz, in which a pair of 1.6 μs pulses - having two different frequencies at 6 MHz apart from each other - are transmitted. There are 49 angle bins with the angle-bin interval of 0.71° over the swath width; 32 pairs of pulses are transmitted for each angle bin. [497)]

Item	Value	Item	Value
Frequency	13.796 and 13.802 GHz	Antenna type Beam width Aperture Scan angle Samples cross-track Samples vertically	128 element WG planar array 0.71° x 0.71° 2.0 m x 2.0 m ±17° (cross track scan) 49 80
Sensitivity	≤0.7 mm/h	Transmitter/receiver Peak power Pulse width PRF	SSPA &LNA (128 channels) ≥500 W (antenna input) 1.6 μs x 2 channel 2776 Hz
Swath width	215 km	Dynamic range	≥70 dB
Observable range	Surface to 15 km	N_S (No. indep. sample)	64
Horizontal resolution	4.3 km (nadir)	Data rate	93.2 kbit/s
Range resolution	250 m (nadir)	Instrument mass, power	465 kg (max), 250 W (max)

Table 91: Instrument parameters of PR

Instrument Mode	Description
Observation	Nominal science observation ±17° cross-track scan
External calibration	Special oversample scan (center scan angle ±1.1°) or fixed beam position
Internal calibration	Internal-loop calibration for receiver I/O transfer function measurement
Standby	Temporal RF radiation stop, phase shifter data load and dump
Analysis	LNA functional check using surface return
Health check	On-board computer ROM/RAM function check
Safety	Instrument power off

Table 92: Summary of PR operational modes

VIRS = Visible Infrared Scanner (NASA/GSFC instrument, built by Hughes SBRC) a passive cross-track scanning radiometer which measures scene radiance in five spectral bands: 0.63 μm (±0.05), 1.6 μm (±0.03), 3.75 μm (±0.05), 10.8 μm (±0.05), and 12 μm (±0.05). Horizontal resolution = 2 km at nadir. Telescope=2-mirror Cassegrain-type focusing the image onto five collocated detectors, each with its own bandpass interference filter. Spectral separation: discrete filters mounted on a cooled focal plane. Swath width = 720 km (FOV=±45°). Applications: Data will be used in conjunction with data from CERES to determine cloud radiation. VIRS will enable "calibration" of precipitation indexes derived from data of other sources (rain estimation from brightness temperature). Data rate = 50 kbit/s (day) = 28.8 kbit/s (night), instrument mass = 34.5 kg, power = 40 W. In-flight radio-

497) T. Kozu, et al., "TRMM Precipitation Radar: Calibration and Data Collection Strategies," Proceedings of IGARSS '94, Volume IV, pp. 2215-2217

metric calibration is provided by an on-board blackbody, a solar diffuser, and a space view for a zero radiance calibration reference. Radiometric accuracy of at least 5% in the thermal bands and 10% in the visible region.

Channel	1 & 2	3 & 4	5	6 & 7	8 & 9
Frequency (GHz)	10.65	19.35	21.3	37.0	85.5
Polarization	V, H	V, H	V	V, H	V, H
Bandwidth (MHz)	100	500	200	2000	3000
IFOV (km x km)	63 x 37	30 x 18	23 x 18	16 x 9	7 x 5
Samples/scan	104	104	104	104	208

Table 93: Performance parameters of TMI

TMI = TRMM Microwave Imager (NASA instrument). TMI is a passive multichannel/ dual-polarized microwave radiometer (heritage of SSM/I on DMSP series) with frequencies in five discrete channels at 10.65, 19.35, 21.3, 37.0, and 85-5 GHz. Resolution (IFOV): 7 km x 5 km to 63 km x 37 km depending on the frequency used. Swath width = 760 km. Applications: Data is related to rainfall rates over oceans (vertically integrated rainfall distribution). Data rate = 8.8 kbit/s, instrument mass =65 kg, power = 50W.

CERES (Clouds and the Earth's Radiant Energy System), an EOS program funded NASA instrument. See CERES description in chapter D.11.1.

Sensor	Observation Objectives	Frequency	Horizontal Resolution	Swath Width
PR	3-D rainfall distribution	13.8 GHz	4.3 km (nadir)	215 km
TMI	Vertically integrated rainfall distribution	10.7, 19.4, 21.3, 37, and 85.5 GHz	5-45 km	760 km
VIRS	Cloud distribution and height, rain estimates from brightness temp.	0.63, 1.6, 3.75, 10.7, and 12 μm	2 km	720 km
CERES	Radiation from top of clouds and Earth, energy budget	0.3 - 5 μm 8.0 - 12.0 μm 0.3 - 50 μm	10 km (nadir)	Scan angle: ±78° global
LIS	Lightning distribution	0.7774 μm	4 km (nadir)	600 x 600 km

Table 94: Overview of TRMM sensor complement and objectives[498)]

LIS = Lightning Imaging Sensor (NASA/MSFC instrument, PI: H. J. Christian).[499)] LIS is an EOS-funded instrument. Objective: Measurement of lightning distribution and variability over the Earth, its correlation with rainfall, and its relationship with the global electric circuit. Measurement approach: LIS is an optical staring telescope/filter imaging system that detects the rate, position, and radiant energy of lightning flashes. LIS detects intracloud and cloud-to-ground lighting with storm-scale resolution. Applications: study of mesoscale phenomena such as storm convection, dynamics, and microphysics. These will be related to global rates, amounts, and distributions of convective precipitation, as well as to the release and transport of latent heat, which are all influenced by global-scale processes. Further applications are: cloud characterization, hydrologic cycle studies, storm convection, microphysics and dynamics, seasonal and interannual variability of thunderstorms.

The LIS instrument consists of two main elements: a) a telescope with a CCD detector matrix, b) the real-time data processing unit. LIS uses an expanded optics wide-FOV lens, combined with a narrow-band interference filter that focuses the image on a small, high-speed CCD focal plane.

- special filter to image at 777.4 nm (OI line) onto a 128 x 128 high-speed CCD array detector

498) Courtesy of K. Maeda, NASDA
499) EOS Reference Handbook, NASA/GSFC, 1993

- event processor to subtract out the bright background during daylight (sensor records data during day and night)
- location coverage of lightning flashes within 5 km over a FOV of 600 km x 600 km
- spatial resolution: 5 - 10 km
- temporal resolution: 2 ms

Accommodation parameters: view direction: nadir, instrument mass: 21 kg, power: 33 W, data rate: 6 kbit/s, FOV: 80° x 80°, IFOV: 0.7° [500)] [501)]

Primary products of TRMM mission data:

- Average monthly rainfall over the tropics and subtropics for at least 3 years

Secondary Products:

- Cloud cover (VIRS, CERES)
- Rain rates (TMI)
- Rain rate vertical profile (PR)
- Path-averaged rain rate and liquid water content (PR)
- Lightning distribution and variability (LIS)

Data Validation Program:

- Rain rate spatial distribution (surface radars)
- Rain rate point measurements (in situ measurements)

Application	Total Precipitation Rate		
	Spatial Aver.	Time Average	Accuracy
Climate models	500 x 500 km	monthly mean	1 mm/day (10% in heavy rain)
Diurnal cycle over ocean	20° longitude	bimonthly	10% first harmonic amplitude 20% second harmonic amplitude
General circulation model vertical distribution	500 m	N/A	N/A
Tropical rain systems structure and evolution	20 km	N/A	30 - 50%

Table 95: TRMM scientific accuracy requirements

A.32 UARS (Upper Atmosphere Research Satellite)

UARS = Upper Atmosphere Research Satellite.[502)] The first NASA mission in the series: 'Mission to Planet Earth.' Launch date: Sept. 12, 1991 with Space Shuttle (Discovery, STS-48; deployment date: Sept. 15, 1991). UARS is a free-flying laboratory; three years nominal lifetime; GSFC = POCC. The UARS S/C has a mass of 6795 kg (size = 4.6 m diameter and 9.8 m length), it is three-axis stabilized via reaction wheels and torque rods (accuracy of 36 arcsec). Attitude knowledge to 20 arcsec using star trackers, Earth sensors, IRU, and sun sensors. The single solar array has a size of 1.5 m x 3.3 m, it generates 1.6 kW of power. In addition there are three 50 Ah batteries. A hydrazine propulsion system is used for orbit insertion and maintenance.
Status: as of 2001 the UARS spacecraft is operational.

Application: Measurement of energy flux (input and loss) in the upper atmosphere; global photochemistry in the upper atmosphere, in particular in the stratosphere and mesosphere (trace gases and temperature profiles); dynamics of the upper atmosphere; transport phenomena of the different processes; correlations between the upper and lower atmosphere and their changes.

500) Z. Kawasaki, S. Yoshihashi, "TRMM/LIS observations of Lightning Activity," Proceedings of the 11th International Conference on Atmospheric Electricity (ICAE), June 7-11, 1999, NASA/CP-1999-209261, pp. 176-179

501) http://thunder.nsstc.nasa.gov/lis/

502) "UARS Seen as Earth Observing System's Dress Rehearsal," Space News September 9-15, 1991, p. 24

- Atmospheric chemistry and temperature
- Atmospheric winds
- Solar energy
- Energetic particles

Orbit: Circular orbit, 57° inclination; altitude = 585 km, period = 97 minutes.

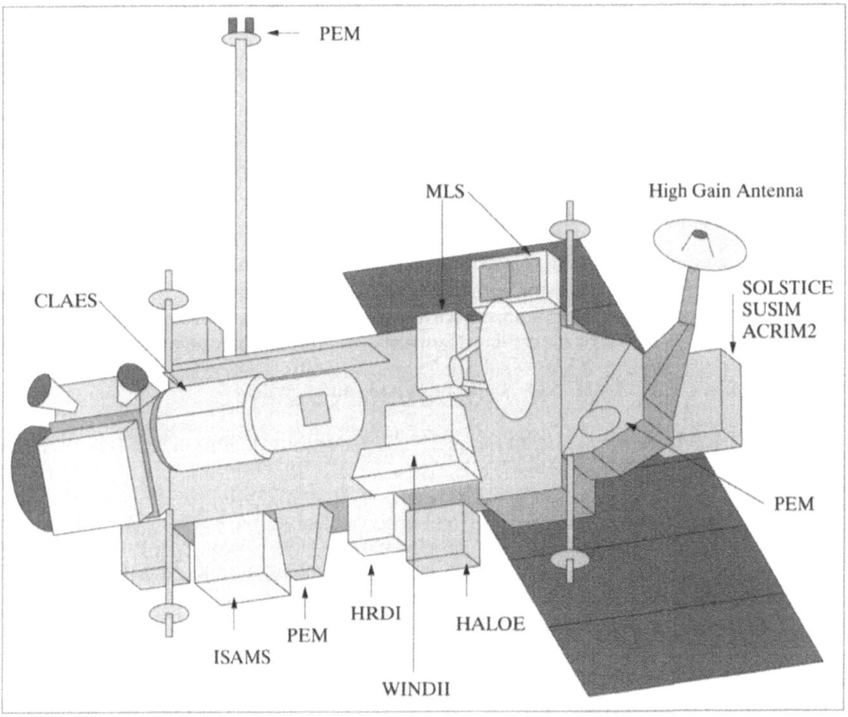

Figure 46: The UARS S/C model

Sensor complement: 503) 504)

CLAES = Cryogenic Limb Array Etalon Spectrometer (Solid-hydrogen cooled spectrometer sensing atmospheric infrared emissions); NASA sensor, (A. E Roche, Lockheed Palo Alto Research Lab)
Measures with four Etalons and eight filters the following wave spectrum: 3.5 μm, 6 μm, 8 μm and 12.7 μm. Because the detectors and optics generate their own thermal emissions, they must be cooled to temperatures which suppress this emission. The coolant inside CLAES is expected to be depleted after about 18 months of service.
CLAES measures concentrations of members of the nitrogen and chlorine families, as well as ozone, water vapor, methane, and carbon dioxide. To obtain a vertical profile of species concentration, CLAES utilizes a telescope, a spectrometer, and a linear array of 20 detectors to make simultaneous measurements at 20 altitudes ranging from 10 to 60 km.
Observables: N_2O, NO, NO_2, HNO_3, CF_4, CF_2Cl_2, $CFCl_3$, HCl, O_3, $ClONO_2$, CO_2, H_2O,

503) Portion of a UARS publication put out by NASA (provided by B. Needham of NOAA)
504) "Upper Atmosphere Research Satellite," Summaries of papers presented at the Optical Remote Sensing of the Atmosphere Topical Meeting, Feb. 12-15, 1990, Optical Society of America, Volume 4, pp. 1-22

ClO, CH_4 and temperature.
Ops status: CLAES ran out of oxygen in May 1993, as planned.

ISAMS = Improved Stratospheric and Mesospheric Sounder (mechanically cooled spectrometer sensing atmospheric emissions); (PI: F. W. Taylor, Oxford University, UK).
This is a SAMS instrument successor using filter radiometry and pressure modulation techniques in the following ranges: 4.6 - 16.6 μm (medium infrared band). The ISAMS instrument is provided by UK (it is an improved version of SAMS, which operated aboard Nimbus-7 from 1978 to 1983).
ISAMS is a filter radiometer employing eight detectors. It observes infrared molecular emissions by means of a movable off-axis reflecting telescope. In addition to scanning the atmosphere vertically, the telescope can also be commanded to view regions to either side of the UARS observatory, thus providing increased geographic coverage. One feature of ISAMS is that it carries samples of some of the gases to be measured in cells within the instrument. Atmospheric radiation collected by the telescope passes through these cells on its way to the detectors (spectra matching).

The limb-sounding instrument, ISAMS, uses a combination of pressure-modulated and wide-band infrared channels to measure the concentrations of nitrogen chemical species, as well as ozone, water vapor, methane, and carbon monoxide in the middle atmosphere. Typically, ISAMS produces vertical profiles of constituents and temperature every 200 km along the tangent track, with an IFOV of about 2.4 km vertically. Data rate: 1.25 kbit/s. Observables: CO, H_2O, CH_4, N_2O_5, NO, N_2O, O_3, HNO_3 and aerosols.

Status: the ISAMS instrument experienced a chopper motor failure in late July 1992, preventing its collection of atmospheric chemistry data.

HALOE = Halogen Occultation Experiment (J. M. Russell, LaRC)
Gas filter radiometer correlation in sun occultations. Infrared range = 2.43 - 10.25 μm. Spatial resolution: vertical = 1.6 km at limb; horizontal = 6.2 km at limb. Measurement of the vertical distribution of hydrofluoric and hydrochloric acids as well as of methane, carbon dioxide, ozone, water vapor, and members of the nitrogen family. The HALOE experiment uses samples of the gases to be observed as absorbing filters in front of the detectors to obtain a high degree of spectral resolution.
During every UARS orbit, at times of S/C sunrise and sunset, HALOE is pointed toward the sun to measure the absorption of energy along this line of sight. There are 28 solar occultation opportunities per day, providing data for 14 different longitudes in each of the northern and southern hemispheres. Data rate = 4 kbit/s. Observables: HF, HCl, CH_4, NO, H_2O, O_3, NO_2 and pressure.

MLS = Microwave Limb Sounder (J. W. Waters, JPL). MLS is a microwave radiometer which measures atmospheric thermal emission from selected molecular lines at mm wavelengths. Three-channel heterodyne limb sounder. Measurement frequencies: 63 GHz (1 band), 183 GHz (2 bands) and 205 GHz (3 bands, corresponding to wavelengths of 4.8, 1.64, and 1.46 mm, respectively). Each band is 500 MHz wide.

MLS is a microwave radiometer providing global measurements of chlorine monoxide (key reactant that destroys ozone), hydrogen peroxide, water vapor, and ozone. The MLS observations will provide, for the first time, a global data set on chlorine monoxide in the upper atmosphere. MLS will also determine the altitudes of atmospheric pressure levels. Spatial resolution in each band: ~400 km horizontal and 4 km vertical; measurement are made along the tangent track of the limb view, with no cross-track scanning, swath width: 5 - 85 km (vertical limb coverage). Data rate: 1.25 kbit/s.
Observables: O_3, ClO, H_2O_2, H_2O and pressure.

Calibration: radiometric views of cold space, and an on-board ambient-temperature blackbody target. Calibration stability: better than 1% over mission lifetime.

SOLSTICE = Solar/Stellar Irradiance Comparison Experiment (G. J. Rottman, University of Colorado at Boulder). Three-channel grating spectrometer for measuring solar and stellar irradiation (UV radiation in the wavelength range from 115 to 430 nm with a resolution of 0.12 - 0.25 nm). The device compares the ultraviolet output of the sun with similar radiation produced by 30 stable, bright blue stars. The stars constitute the standards against which the solar irradiance is measured.
The experiment consists of a spectrometer with three spectral channels, each with a separate grating and photomultiplier tube. SOLSTICE will be pointed toward the sun during the daylight portion of the orbit, and toward one of the calibration stars during most of the nighttime orbit. Data rate: 250 bit/s.

SUSIM = Solar Ultraviolet Spectral Irradiance Monitor (G. E. Brueckner, Naval Research Lab, Washington, DC). SUSIM is a double-dispersion scanning spectrometer which measures solar irradiance (UV radiation in the spectral range from 120 - 400 nm with a resolution of 0.1 nm). SUSIM incorporates two spectrometers, seven detectors, and a set of four deuterium UV calibration lamps. One spectrometer observes the sun and measures variation in solar UV flux as a function of time. The second spectrometer monitors independently the calibration lamps. Data rate: 2 kbit/s.

PEM = Particle Environment Monitor (J.D. Winningham, Southwest Research Institute). PEM represents a collection of four instruments called AXIS, HEPS, MEPS and VMAG. PEM is mounted at three separate locations on UARS, including the boom.

- AXIS (X-ray, proton and electron spectrometer). Measurement of Bremsstrahlung X-rays from Earth, solid-state detection. Energy range from 3 keV to 100 keV. Spectral band: 0.012 to 0.41 nm
- HEPS (charged particle spectrometer). Measurement of in situ electron energies from 0.04 MeV to 5 MeV, proton energies from 0.07 MeV to 139 MeV. Measurement technique: solid state identification.
- MAPS (charged particle spectrometer). Measurement of electrons and protons. Energy range of electrons: 1 eV - 32 keV, energy range of protons: 1 eV to 32 keV; measurement technique; electric deflection with mechanical separation.
- VMAG (three-axis fluxgate magnetometer). VMAG DC: center at 0 nT, width: ±65500 nT. VMAG DC: center at 0 nT with width dependent on AXIS. Measurement of the Earth's magnetic field and the type, number and distribution of charged particles flowing into the upper atmosphere from space.

HRDI = High Resolution Doppler Imager (P. B. Hays, U, of Michigan).
Mapping of atmospheric winds at altitudes below 45 km. HRDI observes the Doppler shifts of spectral lines within the atmospheric band system of molecular oxygen to determine the wind field. There are no sharp emissions lines in the radiance of the Earth's limb at such altitudes, but the oxygen bands contain many lines that appear as deep absorption features in the brilliant spectrum of scattered sunlight.
A triple-etalon Fabry-Perot interferometer, serving as a high-resolution spectral filter, will ensure efficient rejection of the intense emission continuum outside the absorption lines. HRDI will exploit these daytime absorption features to provide wind data for the stratosphere and upper troposphere to an accuracy of 5 m/s or better.
At altitudes above about 60 km, HRDI observes emission lines of neutral and ionized atomic oxygen in the visible and near-infrared regions. Unlike the molecular absorption lines, however, the emission lines are observable both day and night. Mesosphere and thermosphere wind profiles to an accuracy of 15 m/s or better are observed.

Measurement in 13 spectral bands with center wavelength and bandwidth in brackets, all in nm: 557.5 (0.84), 630.0 (.74), 630.5 (1.48), 686.8 (0.88), 687.6 (0.88), 692.3 (1.06), 692.9 (1.36), 723.5 (0.85), 760.7 (0.92), 763.5 (1.45), 764.9 (0.86), 766.6 (1.01), 775.6 (0.93).

The interferometer is tunable within the spectral width and can simultaneously measure 31 samples in a 0.5 cm^{-1} spectral region at a resolution of 0.05 cm^{-1}. Data rate = 4.75 kbit/s.

WINDII = Wind Doppler Imaging Interferometer (G. G. Shepherd, York University, Canada; Canadian/French Sensor, 145 kg mass). WINDII is a field-widened Michelson interferometer measuring the Doppler shift and line broadening of atmospheric emission in the visible and near-infrared. The airglow emissions observed are: O^1S at 557.7 nm (green line), O^1D at 630 nm (red line), O^+ at 732 nm, OH $(8,3)P_1(3)$ at 734.1 nm, and O_2 Atm (0,0) at 763.2 nm. O^+ is observed only on the dayside orbit while OH only during the nightside of the orbit. Measurement of airglow emissions rates and wind conditions in the thermosphere and upper mesosphere. Temperatures in the upper mesosphere are determined from the Rayleigh scattering of sunlight. [505) 506) 507)]
WINDII utilizes emission lines for the basic Doppler-shift measurements. In addition to lines of neutral and ionized oxygen, these include two lines of the OH molecule and a molecular oxygen line. WINDII obtains measurements both day and night at altitudes above 80 km.

The instrument consists of a telescope, the interferometer, and a CCD detector array. The telescope has two fields of view, each 4° x 6° looking at the airglow above the Earth's limb at 45° and 135° from the S/C velocity vector simultaneously. In normal operation, the detector provides a vertical resolution of 20 km. Wind velocity accuracy is within 10 m/s in the altitude range between 80 and 300 km. Data rate: 2 kbit/s. The WINDII spectral filter is a high-resolution Michelson interferometer.

ACRIM-2 = Active Cavity Radiometer Irradiance Monitor (R.C. Willson, JPL; see ACRIM in EOS)

Data:

There are two on-board recorders available (each capable of recording two orbits of data at 32 kbit/s). Tape recorder playback = 512 kbit/s.
Data transmission via TDRSS with 32 kbit/s (R/T) and 512 kbit/s (P/B).

Correlative data sets are provided in parallel, in particular NOAA SUBV data sets, as well as ground truth measurements (balloons, sounding rockets, Shuttle flights) and meteorological data.

505) "Wind Imaging Interferometer (WINDII) for the UARS Mission," Optical Remote Sensing of the Atmosphere, 1990 Technical Digest Series of the Optical Society of America, Volume 4, pp. PD3-1 to 4

506) W. A. Gault, W. E. Ward, et al., "Optical Doppler Imaging of Atmospheric Winds," Proceedings of IGARSS'99, Vol. III, Hamburg, Germany, June 28 - July 2, 1999, pp. 1612-1615

507) "Windii To Read Upper Atmosphere In Depth," Space News September 16-22, 1991, p. 8

Part B Commercial Imaging Satellites

The provision of high-resolution imagery in the optical and/or microwave regions – on a commercial basis by several companies with their own space- and ground segments (spacecraft, sensor complement, control center, ground receiving stations, archives, distribution networks, and extensive software), and the operation of these entities along with the provision of general and/or customized service arrangements – represents a new milestone in the field of space-flight exploitation. The investment in each venture is considerable; every major player in this game forms a consortium (or international alliances) to pool expertise, technology, and resources–and to share the risks. The overall objective in every satellite-project constellation is the provision of low-cost and high-quality data products along with expedient services for a large customer base. The strategic approach taken by industry with regard to concept design is that virtually all major system components are based on proven technologies; experience from past defense contract engagements is of great value. The introduction of a sensor-pointing capability beyond its rather limited swath-width indeed represents a new concept outside of the defense community; it extends the potential imaging coverage to the so-called field of regard (FOR), in order to make data rates manageable. The consequence of this pointing capability is "scheduled instrument operation," permitting the imaging of scattered targets to suit customer requirements. Industry is betting on a competitive market, whose demand for high-resolution imagery goes far beyond the utilization of the rather 'coarse-resolution' interpretations of current satellite data. The aim is to address the wide field of public/private applications (in particular cartographic use), as well as the needs of the research community. There is also a considerable demand for high-resolution imagery by the defense and intelligence agencies of the world.

An observation prerequisite for all optical imaging acquisition from space is a cloudless sky. This limitation is somewhat counterbalanced by the periodic revisit capabilities of polar-orbiting satellites; still, optical observation is rather constrained in high-latitude regions of the world due to almost permanent cloud cover. - Providing optical high-resolution imagery is obviously only the first step into the commercial data arena; an extension of services to other spectral ranges (microwave region for SAR imagery) is already on its way.

Reliable and consistent service provision in any business depends very much on back-up solutions in case of need. This proven concept also applies to the providers of high-resolution imagery. Every consortium in this game has a complete back-up satellite/payload in store for the contingency of a launch failure or for a terminal instrument malfunction in orbit; it is also a neat solution in the event of unexpected demand in imagery, to be able to launch a second satellite at short notice.

The Resource21 mission, a long-planned (and sometimes dormant) commercial project for spaceborne crop monitoring of Resource21 of Englewood, CO, is missing in this chapter (Resource21 was already part of the 3rd edition in 1996). As of the summer 2001, the company is planning a smaller-scale project (than previously planned), the new plans call for a one spacecraft mission for a launch not before 2003. [508]

B.1 Condor Series of NPO Machinostroyenia

The Russian enterprise NPO Machinostroyenia (or NPO Machine-builder), headquartered in the Moscow suburban town of Reutov, has been engaged in the field of remote sensing and space exploration since the start of the space age. The Almaz-1 spacecraft (launch March 31, 1991) with its provision of SAR instrumentation and data was a major engagement and experience in this direction. With the drastically changed economic situa-

508) J. Bates, "Resource21 Moves Ahead With Smaller-Scale Project," Space News Aug. 13, 2001, p. 3 and p. 28

tion in Russia during the past decade, the company began to re-orient its market strategies and is working for some time now on modern service solutions and projects by utilizing a small/medium sized spacecraft family (up to 1000 kg), combined with new-technology instruments, operational concepts, and a low-cost launch option. The overall objective is to offer high-resolution imagery (SAR imagery and eventually also optical imagery) on a commercial basis to worldwide customers as well as to institutional users in Russia for a variety of applications. From a service provider commitment and view point, the Condor project offers a long-term and pragmatic perspective to the customer base.

The idea is to overcome the large-spacecraft philosophy/practice of the past, exemplified by the Salyut spacecraft of Almaz-1 and Almaz-1B, each with launch mass of over 18,500 kg. If Almaz-1B was going to be launched by a Proton heavy-lift launch vehicle today, such a payload would amount to launch costs of $60-80 million, making any low-cost project unprofitable. Hence, the project Almaz-1B, a planned follow-up mission to Almaz-1 for a long period of time, was finally cancelled in favor of the new concept of Condor-E; it represents a first step into the light-weight and small-size direction. [509) 510)]

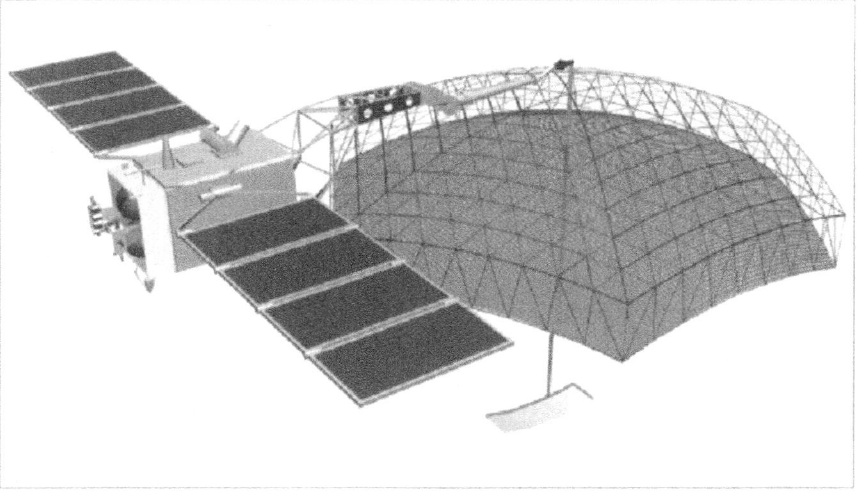

Figure 47: Illustration of the Condor-E-1 spacecraft

B.1.1 Condor-E (Condor Experimental)

The Condor-E spacecraft employs a standardized bus, the S/C is three-axis stabilized. Attitude sensing is provided by a gyroscope and star senors, actuation is provided by reaction wheels and torquers. The pointing accuracy of the S/C is equal to or less than 6 arcmin with an angular drift of <0.001°/s. On-board orbit determination is provided by a GLONASS/GPS receiver. Electrical power is provided by two solar panels with a total surface area of 9.2 m^2. The payload power consumption is in the range of 240-1500 W, depending on service provided.. In addition there are batteries for power provision during orbital eclipse phases. Condor-E-1 has a spacecraft total mass of 800 kg (250 kg of payload mass); the design life is 5 years with a goal of 7 years.

509) V. Viter, V. Petrovsky, A. Koutcheiko, "Space-based radars designed by NPO Machinostroyenia, Novosti Kosmonnavtiki (Cosmic News), Vol 218, No 3 (218), Jan. 31, 2001, Vol. 11

510) V. Viter, "NPO Machinostroyenia - Advanced technology, reasonable economic policy and addressing practical problems of developing countries," Russian Air Force, Aircraft & Space Review, No 17, June 2000, pp. 58-59

A launch of Condor-E-1 is planned in the 2003/4 time frame on a Strela ("arrow") launch vehicle from the Svobodny Cosmodrome in the Amur region (Russian Far East). The Strela launcher is a converted RS-18 ICBM (NATO code name SS-19 Stiletto) utilizing a silo launch.

Orbit: Sun-synchronous polar or inclined circular orbit (final orbit selection at a later date), altitude = 500-900 km, inclination = 52-98°. The revisit time capability is 2-3 days.

RF data handling and communications: The instrument source data rate is up to 960 Mbit/s. On-board data compression is provided prior to data recording onto a 192 Gbit solid-state recorder. Downlink communications are in X-band. The data rate is 61 Mbit/s to regional data centers, and 245 Mbit/s to the central data center, located at NPO Machinostroyenia. The functions of spacecraft operations and data distribution are also performed at NPO Machinostroyenia.

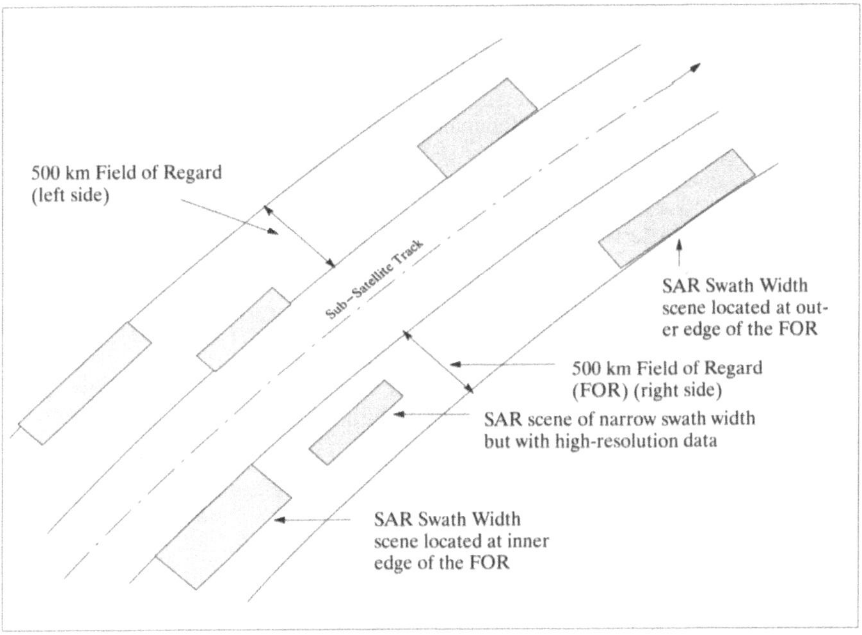

Figure 48: Observational coverage geometries of the SAR-10 instrument

Sensor complement:

SAR-10 (Synthetic Aperture Radar-10), designed and developed at NPO Vega. SAR-10 is an S-band instrument with a wavelength of 9.6 cm (or a frequency of 3.13 GHz). The overall SAR design concept employs a space-deployed parabolic dish antenna to save weight and to permit a cross-track pointing capability (a similar antenna has already been successfully tested on-board the MIR station). Unlike a phased-array antenna with a one-sided viewing capability from the spacecraft, the parabolic dish antenna design permits the observing equipment to be dynamically redirected in the cross-track direction. Thus, a Field of Regard (FOR) within the incidence-angle range of 20-55° on either side of the spacecraft may be observed. This concept is also employed on the US DoD Lacrosse satellite series.

Wavelength, (frequency)	9.6 cm, (3.13 GHz), S-band
Incidence angles	20-55° to either side of the spacecraft
Field of Regard (FOR)	500 km on either side (left or right) of the spacecraft
Swath width of imagery	20-150 km (medium swath), or 10-20 km (narrow swath, high resolution data)
Spatial resolution	5-22 m (medium swath) 1-5 m (narrow swath)
Length of a scene	Up to 4000 km
Parabolic dish antenna	6 m diameter
Antenna polarization (transmit/receive)	HH for medium swath, HH or VV for narrow swath
SNR (Signal-to-Noise Ratio)	>2
Instrument calibration	Performed in HF part of the equipment

Table 96: Main characteristics of the SAR-10 instrument

B.2 Diamant, OHB Bremen

Diamant is a cooperative low-cost minisatellite system with an end-user oriented infrastructure, designed and built a consortium of German (OHB-System, GAF) and Israeli (ELOP) companies. The overall objective of the commercial venture is to provide high-resolution imagery (spatial and spectral in combination with good radiometric sensitivity) with short revisit and delivery times to a user community. Applications of the image data are seen in particular in the fields of environmental and vegetation monitoring, hazard warning and damage assessment. Diamant has a direct relevance to the objectives of the ESA's "Living Planet Program" and is especially in line with the "Earth Explorer" research objectives. The final Diamant constellation will consist of three satellites. [511) 512) 513)]

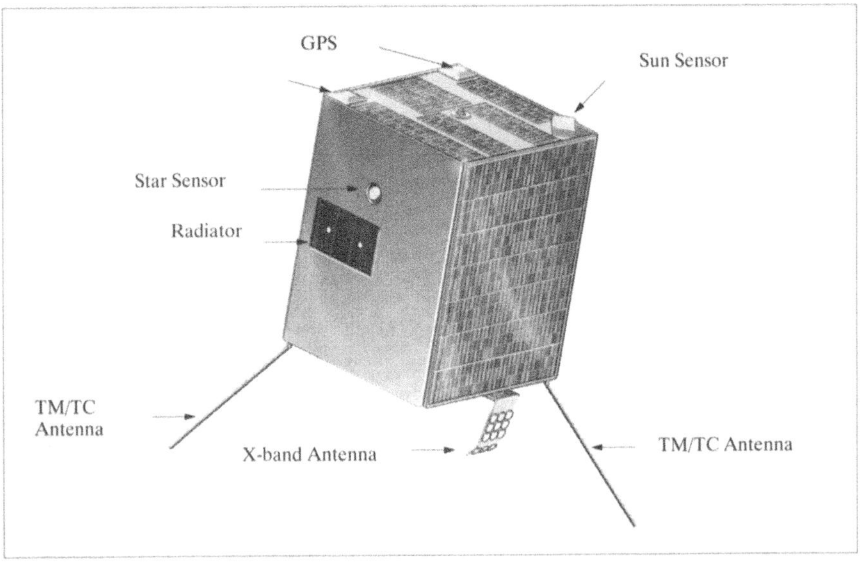

Figure 49: Illustration of the Diamant spacecraft

511) A. Ginati, M. Fuchs, M. Kassebom, "Commercial Earth Observation with Small Satellites at OHB-System," Proceedings of the 13th AIAA/USU Conference on Small Satellites, Aug. 23-26, 1999, Logan UT, SSC99-IV-7

512) A. Ginati, M. Fuchs, M. Kassebom, "Earth Observation Program with Small Satellites at OHB-System," Acta Astronautica Vol. 46, No 2-6, 2000, pp. 297-306

513) http://www.fuchs–gruppe.com/ohb–system/

The Diamant satellite uses the UniSat (Universal Satellite Platform) bus of OHB-System, Bremen and NPO VNIIEM of Moscow, Russia. The S/C geometry is a boxlike structure of size: 71 cm x 71 cm x 91 cm. The overall S/C mass is <250 kg, power = 141 W (peak) 55 W orbital average. The S/C is three-axis stabilized providing a stability of <2 pixels/scene (or 3.2 arcsec) after 6 s; the target error is <10% of a 30 km swath this corresponds to 3 km or 0.26° (reaction wheels as actuators). A GPS receiver provides the S/C time reference and position determination. A body-pointing technique (slewing capability of the S/C) is employed for instrument pointing with a field of regard of ±30° in the along-track as well as in the cross-track direction. The angular slew rate is sufficient to perform both wide-area monoscopic and same-pass stereo imagery collections. The S/C design life is 3 years with a goal of five years.

A launch of Diamant-1 is planned for 2003/4 as a secondary payload (with a Cosmos launch vehicle).

Orbit: Sun-synchronous circular orbit, altitude = 670 km, inclination = 98.1°, equatorial nodal crossing at 11:30 AM on descending node, revisit time of 3 days above a latitude of ±27° (revisit time of 4 days above latitudes of ±20°). The orbit decay is <35 km in five years.

Sensor complement:

MSRS (Multi-Spectral high Resolution System), built by ELOP (El-Op Electro-Optics Industries of Rehovot, Israel) in cooperation with OHB-System with some funding provided by the EC (European Commission). [514)] The overall objective is to demonstrate an advanced Earth observation system in a user-integrated approach (along with an end-user oriented infrastructure for commercial Earth observation). The MSRS imagery concept combines the qualities of high spatial resolution with high radiometric resolution using twelve finely defined multispectral bands in the VNIR (Visible Near-Infrared) spectral region.

Band No	Band center (nm)	Commercial applications	Science applications
1	415	Water quality mapping road detection, land cover, urban areas	Desert soils, biogenetic crusts, dissolved organic material
2	445	Real natural blue color for mass media, water quality, urban areas, landcover,	Low chlorophyll concentration, maximum chlorophyll-A absorption at 450 nm
3	490	Water monitoring, road detection, lithology and soils, crops and farming	Non-linear chlorophyll, medium chlorophyll concentration
4	510	Soils, landcover, surveying, water quality, coastal zones	Low suspended sediment, high pigment concentration
5	555	Real natural green color for mass media, landcover, crops and farming	Green peak of chlorophyll, null band of chlorophyll variation
6	645	Real natural red color for mass media, landcover, lithology and soils, forestry	Chlorophyll absorption, vegetation indices, high suspended sediment
7	670	Crops and farming, pest control, security	Max chlorophyll absorption, vegetation indices, 2nd peak of chlorophyll-A absorption
8	700	Vegetation general, nature conservation	Red edge of vegetation
9	740	Vegetation general, nature conservation	Red edge of vegetation
10	790	Vegetation general, nature conservation	Red edge of vegetation
11	865	Lithology and soils, crops and farming, forestry, landcover, nature conservation	Max. chlorophyll reflectance, vegetation indices, water vapor, atmospheric correction
12	905	Lithology and soils, nature conservation, wetlands, coastal waters	Max. chlorophyll reflectance, atmospheric correction

Table 97: Spectral bands of the MSRS instrument and data application scenarios

The MSRS instrument consists of the following elements: optical subsystem, detector subsystem, a data compression and storage unit, and an instrument control computer. MSRS

514) A. Ginati, B. Penné, A. Blasberger, P. Volk., "MSRS - Multi Spectral High Resolution System," IAF-1998-B3.10, 49th IAF Congress, Melbourne, Australia, Sept. 28 - Oct. 2, 1998

features the pushbroom imaging concept with twelve CCD detector channels (400 - 1000 nm) in a focal plane configuration; each channel is furnished with a separate interference filter for wavelength band definition and a TDI (Time Delay Integration) type detector array. The detector array for each channel has a size of 5200 pixels (cross-track direction) x 32 rows (used for TDI), thus offering a cumulative exposure concept to improve the SNR value for multispectral imaging. Three of such (5200 x 32 pixel) detector units are integrated on a single chip. Two of the detectors with two different spectral regions (6 spectral bands) are cemented to the exit surface of one BSGB (Beam-Splitter Glass Block). Two separate BSGBs are used to cover the spectral region of 12 bands. The spectral coverage is contiguous in the visible region (435 - 750 nm) with bandwidths of 10-50 nm, permitting good discrimination of image features. [515)]

A spatial resolution of 5 m at nadir (IFOV = 0.075 mrad) is achieved for all channels of the imagery. The focal length of the catadioptric telescope (a modified Ritchey-Chretien instrument with an entrance pupil diameter of 250mm) is 1750 mm providing a FOV of 2.3° (26 km swath at an orbital altitude of 670 km). A body-pointing capability of ±30° in the cross-track and along-track directions is provided by the S/C, offering a short revisit period of 3 days for requested cross-track monitoring applications; an alternate service of this S/C agility is the provision of along-track stereo imagery.

Instrument mass, power, size	70 kg, 120 W peak, 60 cm x 58 cm x 58 cm
Spectral bands, range, spectral resolution	12, 400 - 1000 nm (VIS and VNIR), 10-50 nm
Spatial resolution	5 m (IFOV = 0.075 mrad)
TDI capability	4 TDI detectors with 3 channels each
Data quantization	10 bit/pixel
Source data rate	4 x 300 Mbit/s peak
Scene size	26 km x 40 km (nominal), up to 26 km x 400 km
Scene rate	Up to 140 scenes/day
On-board recorder capacity	210 Gbit

Table 98: Some instrument performance parameters of MSRS

On-board calibration uses the sun for direct calibration purposes. A small mirror, whose size controls the required attenuation level, reflects a defocused image of the sun onto the detectors via the MSRS telescope.

Data communication: The source imagery may be stored onto a 210 Gbit solid-state on-board recorder. Downlink communication in X-band (frequency = 8.2 GHz, modulation scheme = OQPSK) is provided at data rates of 100 Mbit/s (optionally at 320 Mbit/s). An online JPEG data compression scheme is employed to reduce the number of bits per pixel without compromising radiometric accuracy. The CCSDS protocols are used for all data communication.

The ground segment for the operational support of the Diamant mission consists of the following elements: Control Station, Data Archive & Distribution Center, and Data Receiving Stations (fixed and mobile). The services to the customer base consist of a number of data products (level 1 through level 5).

B.3 COSMO-SkyMed and Pléiades Programs

In 1996 the Italian Government provided initial funding for the realization of a national Earth observation program. In 1997 the general guidelines for the 1998-2002 Italian space plan were approved including the activities on Earth observation. The strategic element in this plan is the **COSMO-SkyMed** (Constellation of Small Satellites for Mediterranean ba-

515) Information provided by B. Penné of OHB System, Bremen

sin Observation) dual-use program of ASI (Agenzia Spaziale Italiana). [516] The overall objective of this program is global Earth observation and the relevant data exploitation for the needs of the military community as well as for the civil (institutional, commercial) community.

Sample applications of COSMO-SkyMed data are seen the following fields:

- Defense and security applications: Surveillance, intelligence, mapping, damage assessment, vulnerability assessment, target detection/localization
- Risk management applications: Floods, droughts, landslides, volcanic/seismic, forest fire, industrial hazards, water pollution
- Other applications: Marine and coastal environments, agriculture, forestry, cartography, environment, geology and exploration, telecommunication, utilities and planning.

Since 1997 CNES (France) is studying the use of smaller satellites (a medium spacecraft size of about 500 kg instead of 3 tons as for the SPOT-5 S/C), resulting in the "3S" platform concept (Small Satellite System) standing for "Suite de Systeme du SPOT" or for "SPOT Successor System." The focus is on cost reduction, technological innovation, user services, and performance upgrades for a new generation of satellites, referred to as **Pléiades.** [517] [518] [519]

On Jan. 29, 2001, an intergovernmental agreement (memorandum of understanding) was signed during the Turin meeting between Italy (Guiliano Amato) and France (Lionel Jospin). The objective of this agreement is the cooperation of France and Italy on a "dual high-resolution Earth observation system," comprising a **two-satellite constellation in the optical region under the leadership of France**, and a **four-satellite constellation under Italian leadership in the microwave region of the spectrum (initially X-band SAR)**. The intent of this agreement is to provide a long-term perspective on a number of high-quality data products and services on the commercial market for a wide range of applications in the fields of cartography, agriculture, forestry, hydrology, and geological prospecting. The dual service concept is seen in the data requirements of the defense and civilian communities with an option of a daily revisit capability. The agreement calls for funding and development of the space segment by each country and a common sharing of the ground segment. [520] [521]

In this framework the Italian and French Governments started a cooperation with the goal of Earth observation for dual-use applications (military and civil) with SAR and optical instruments based on the on-going COSMO-SkyMed and Pléiades small satellite programs, respectively. This dual-use scenario calls for missions that offer advanced observation capabilities in several modes of operation, permitting to meet the objectives of the military and civil communities at the same time. The development of innovative and complementary instrumentation in the radar field (e.g., multi-mode and flexible-support SAR's offering high-resolution data) and in the optical field (e. g., hyperspectral sensor with capabilities of variable spatial resolutions as well as high detection sensitivities in the visible and infrared spectral regions) are major objectives of the programs.

As of mid-2001, **COSMO-SkyMed** as well as **Pléiades** are proposed cooperative Public Private Partnership (PPP) initiatives between ASI and ESA and between CNES and ESA, respectively. Approval of the programs is expected by the European Council of Ministers in

516) C. Galeazzi, C. Portelli, "The SkyMed/COSMO Platform Preliminary Definition," Proceedings of the 4th International Symposium on Small Satellites Systems and Services, Sept. 14-18, 1998, Antibes Juan les Pins, France

517) "Pléiades to succeed SPOT," CNES Magazine No 9, June 2000, p. 8

518) A. Baudoin, "The Current and Future SPOT Program," Proceedings of the ISPRS Joint Workshop 'Sensors and Mapping from Space 1999,' Sept. 27-30, 1999, Hannover, Germany

519) In Greek mythology, Pléiades refers to the seven daughters of the Titan Atlas and the Oceanid Pleione: Maia, Electra, Taygete, Celaeno, Alcyone, Sterope, and Merope. They all fell in love with gods (except Merope, who loved a mortal) and were the mothers of gods. The Pléiades eventually formed a constellation.

520) "Earth Watch Program: COSMO-SkyMed Element," ESA/PB-EO (2001) 55, May 18, 2001, with Annex: Draft Italian Earth Watch Program Proposal - ESA participation to COSMO-SkyMed Mission (Radar Component)

521) http://www.alespazio.it/cosmo/skymed.htm

Nov. 2001. Both programs are set within ESA's Earth Watch program and the European GMES (Global Monitoring for Environment and Security) initiative.

The COSMO-SkyMed / Pléiades space segment is based on a constellation of small satellites combined with a fast data reception capability. The provision of data on an operational basis (of continuity and quality) is essential for the system.

The constellation will be deployed in two orbital planes, one orbit plane for the SAR satellites and one orbit plane for the optical satellites. This permits a) the SAR satellites to collect the maximum solar radiation for S/C power demands in sun-synchronous dawn-dust orbits, and b) the optical satellites to operate under optimal illumination conditions in a sun-synchronous near noon orbit.

The COSMO-SkyMed / Pléiades ground segment provides all the infrastructures needed to support the mission in a dual-use scenario (functions/operations in terms of constellation control and global data management). The various elements are:

- CPCM (Centro Planificazione e Controllo Missione - Mission Planning and Control Center). Coordination of on-board and ground activities, mission planning, and resource allocation.
- CCS (Centro Controllo Satelliti - Satellite Control Center). Provides the monitor and control function of the constellation including flight dynamics.
- TT&C stations. Primary service link between the space and ground segments (dedicated communications network).
- CREDO (Centro Ricezione ed Elaborazione Dati Operativi). CREDO operates X-band stations for high-rate data acquisition and provides data archiving and processing. To the user side there is a two-fold function:
  - As a civil support facility which handles civil user requests
  - As a military support facility to handle the military service needs.

B.3.1 COSMO-SkyMed

The COSMO-SkyMed space segment is composed of a constellation of four SAR satellites. The PRIMA [Piattaforma Riconfigurabile Italiana Multi-Applicativa (Reconfigurable Italian Platform for Multiple Applications)] bus of Alenia Spazio is employed. Alenia Spazio is also the prime contractor of the space segment (ASI funding). The S/C is three-axis stabilized, it consists of the main body (bus), two deployable solar arrays, and a SAR antenna. The bus provides all support functions like: AOCS, electrical power (power generation, storage and distribution), data handling, thermal control, RF communications, and on-orbit propulsion for orbit injection and maintenance. The platform mechanical configuration consists of two elements (boxes), namely a) SVM (Service Module) at the bottom of the bus which contains all bus subsystems including the propulsion module, and b) PLM (Payload Module) at the top, dedicated to the payload complement, the PDHT (Payload Data Handing and Transmission) subsystem, and the AOCS (Attitude and Orbit Control Subsystem) with star trackers gyros actuators. The bus structure material is CFRP (Carbon Fiber Reinforced Plastic) while SVM and PLM consist of aluminum alloys. The interfaces of the SAR antenna, star trackers and gyros are mounted on the CFRP structure for pointing precision and stability. The SAR antenna bore sight pointing with an incidence angle about 38° to the right side of the S/C ground track. AOCS provides an antenna steering capability of ±2° in yaw as well as for a re-pointing capability to the left side of the ground track.

Orbit: Circular sun-synchronous dawn-dusk orbit, nominal altitude = 619.6 km, inclination = 97.86°, with LTAN (Local Time of Ascending Node) at 6:00 AM, 14.8125 rev./day (or 14 13/16). All spacecraft of the SAR constellation will be positioned in the same orbital plane with a phasing outlined in Table 99. The nominal repeat cycle is 16 days; however, each single satellite will have a near revisit time of 5 days.

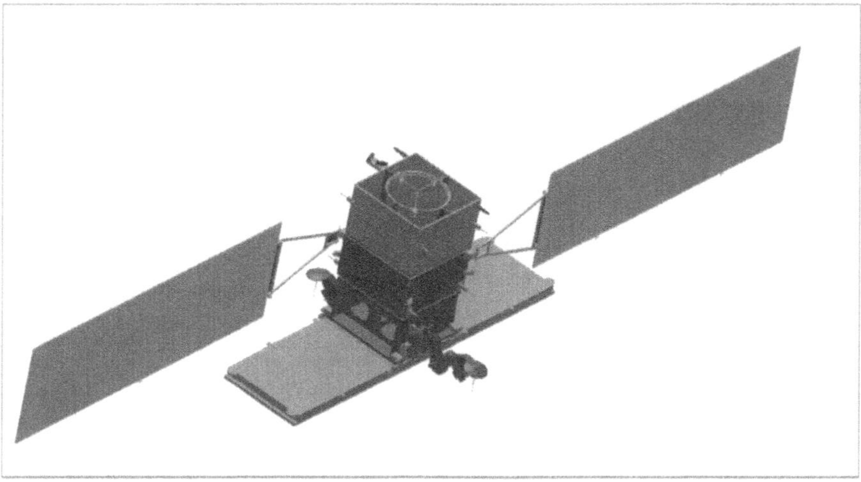

Figure 50: Illustration of the COSMO-SkyMed spacecraft

The phasing of the S/C in the orbit plane has been selected to achieve optimum performance in terms of accessibility and revisit time with respect to the number of satellites. Table 100 summarizes the achievable observation performance capabilities of the SAR satellites within the latitudinal coverage of ±20° - 60°.

Satellite number	True Anomaly
1	0°
2	0°, 180°
3	0°, 120°, 240°
4	0°, 90°, 180°, 270°

Table 99: SAR satellite phasing scheme

Constellation	Right Looking		Left and Right Looking		Comment
	Nominal	Extended	Nominal	Extended	
1 satellite	37 to 64 h	25 to 44 h	18 to 35 h	12 to 23 h	Mean revisit time
	<252 h	<120 h	<156 h	<60 h	Max. revisit time
	38% (24 h)	55% (24 h)	67% (24 h)	85% (24 h)	Access % vs time
2 satellites	19 to 35 h	13 to 24 h	9 to 18 h	6 to 12 h	Mean revisit time
	<108 h	<60 h	<60 H	<36 h	Max. revisit time
	41% (12 h) 64% (24 h)	62% (12 h) 84% (24 h)	60% (12 h) 81% (24 h)	77% (12 h) 92% (24 h)	Access % vs time
3 satellites	13 to 24 h	9 to 16 h	6 to 12 h	4 to 8 h	Mean revisit time
	<60 h	<36 h	<36 h	<36 h	Max. revisit time
	62% (12 h) 84% (24 h)	88% (12 h) 98% (24 h)	84% (12 h) 96% (24 h)	98% (12 h) 99.97% (24 h)	Access % vs time
4 satellites	10 to 18 h	6 to 12 h	5 to 9 h	3 to 6 h	Mean revisit time
	<60 h	<24 h	<24 h	<12 h	Max. revisit time
	80% (12 h) 95% (24 h)	99% (12 h) 100% (24 h)	97% (12 h) 100% (24 h)	100% (12)	Access % vs time

Table 100: Accessibility parameters for the various constellation configurations

The phase B2 of COSMO-SkyMed project was completed in Dec. 2000. Launch/deployment of the constellation is planned to be in the time frame of late 2003 or early 2004.

Support Mode	Configuration	Comments
SAR observation side with respect to the nadir ground track	Right-side looking (nominal) Left-side looking for limited periods	Providing better performance of northern hemisphere coverage Use intended to satisfy ASAP priority requests
Duty cycle (estimated on a daily basis)	75 min of operation in Stripmap or ScanSAR 150 images in Spotlight mode	A duty cycle is needed to overcome the S/C power limitations
Max. imaging duty cycle (on an orbit basis)	10 min of continuous operations in Stripmap or ScanSAR 20 consecutive images in Spotlight mode	
Image acquisition management	On a 24 h basis	Capability of time-tagged commands for SAR-2000
PDHT max duty cycle	20% per orbit	
PDHT storage capacity PDHT storage strategy	256 Gbit a) Delete after output (nominal) b) Retain after output	
PDHT support modes	- Store only - Downlink only - Store and downlink - Pass through (near real time)	Both in right and left looking Only in right looking Only in right looking Only in right looking
PDHT downlink rates	- Downlink only - Store and downlink - Pass through (near real time)	2 separate links at 150 Mbit/s 1 link at 150 Mbit/s 1 link at 150 Mbit/s

Table 101: Overview of operational support modes

Note: It was not possible to receive any information from the COSMO project at Alenia Spazio although this was definitely promised in April 2001 with a due date of arrival on June 29, 2001, at the latest. On this date, however, I was simply informed that COSMO had just been classified concerning all information on technical parameters as well as on applications. Hence, a number of parameters could not be obtained such as: S/C mass and power, S/C dimensions, design life, instrument power, instrument frequency. Although the first constellation satellite SAR instruments (SAR-2000) are expected to observe in X-band (probably at 9.6 GHz with a wavelength of 3.1 cm), multi-mode scenarios (X-, C- L- and P-band) are planned for the future.

Sensor complement:

SAR-2000 (Synthetic Aperture Radar-2000) instrument, developed at Alenia Spazio. The objectives call for the following design features:

- Very large instantaneous bandwidth
- Electronic beam steering in range and azimuth
- Multi-polarization support
- Programmable PRF, pulse width and bandwidth
- Multiple imaging mode support
- On-board hardware calibration techniques
- On-board data compression processing techniques (both analog and digital)

The instrument is composed of two major elements: a phased array antenna subsystem (external equipment) and of the central electronics module (internal equipment). The central electronics module in turn consists of three parts:

- RFA (Radio Frequency Assembly). RFA has four units: XDU (X-band Driver Unit), XSU (X-band path Switch Unit), DCU (X-band Receiver), and FGU (Frequency Generation Unit).
- DESS (Digital Electronics Subsystem)
- CPSU (Central Power Supply Subsystem

Operational modes with one polarization selectable among HH, VV, HV, or VH	
Spotlight (also referred to as "Frame")	Spatial resolution: <= 1 m Spot observation area: 10 km x 10 km
HIMAGE (Stripmap)	Spatial resolution: 3-15 m Swath width: 40 km
WideRegion (ScanSAR)	Spatial resolution: 30 m Swath width: 100 km
HugeRegion (ScanSAR)	Spatial resolution: 100 m Swath width: 200 km
Operational modes with two polarizations selectable among HH, VV, HV, or VH	
Ping Pong (Stripmap)	Spatial resolution: 15 m Swath width: 30 km km

Table 102: Some performance characteristics of the X-band SAR payload

The antenna subsystem is a large deployable planar phased array. It consists of 40 identical tiles arranged in five identical electrical configurations (5 electrical panels). Each tile consists of T/R (Transmit/Receive) modules. Within a tile there are all functions needed for beam forming and reception. Each tile is a self-consistent active antenna element including all functional support (thermal, RF, digital and power). The following operational acquisition modes are supported: Stripmap, ScanSAR, and Spotlight. The SAR data localization accuracy is 25 m without GCP (Ground Control Point).

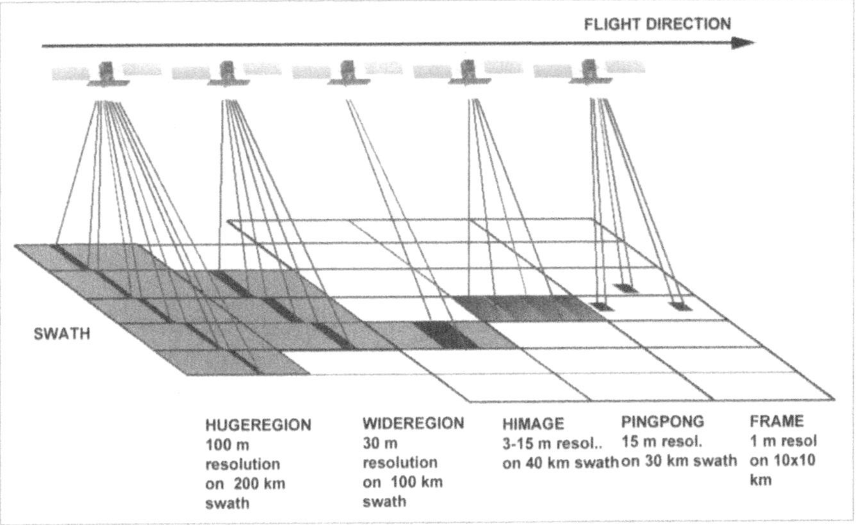

Figure 51: Illustration of the SAR operational modes

The single orbital plane constellation offers a number of interferometric applications/features such as:

- A full interferometric accessibility (i.e. a double acquisition of all sites with the same incidence angle within the 16 day orbit repeat cycle
- No degradation of the mean revisit time with respect to the performances achievable with the same number of satellites
- A fixed time period of one day between two interferometric acquisitions.

SAR product examples: Multidate basic products, stereo pairs for radargrammetry, interferograms (for DEM and DTM) or differential interferogram, coherence products, etc.

B.3.2 Pléiades

The Pléiades spacecraft structure is shown in Figure 52. The S/C is three-axis stabilized; attitude is sensed by star trackers (capable of finding attitude starting from a "lost-in-space" condition) and by fiber-optic gyros. The S/C permits a body-pointing capability with roll and pitch maneuvers, each up to 60°, within a period of 25 s. S/C power is provided by rigid AsGa solar arrays and by LI (Lithium-Ion) batteries. The total mass of the S/C is about 900 kg.[522)] [523)]

A launch of Pléiades-1 is planned for the end of 2005, Pléiades-2 will follow a year later.

Orbit: Sun-synchronous phased orbits (180° phasing, 14 and 15/26 rev./day), altitude = 695 km, inclination = 98.2°, local equator crossing time on ascending node around 10:30. The two-satellite constellation in combination with agile cross-track pointing provides the capability of daily revisits.

RF communication: An X-band downlink provides a payload transmission rate of 600 Mbit/s in 4 channels, each of 150 Mbit/s capacity. All source data are compressed prior to on-board storage (mass memory of 600 Gbit).

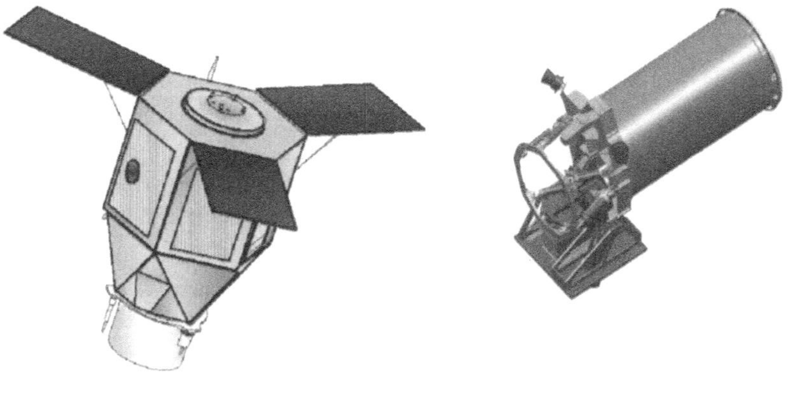

Spacecraft — Schematic view of the Korsch telescope

Figure 52: Illustration of the Pléiades spacecraft

Sensor complement:

OHRI (Optical High-Resolution Imager). The objective is to provide high-resolution multispectral imagery with geo-location accuracy. The instrument employs a Korsch telescope with TMA (Three Mirror Anastigmatic) optics. The focal plane for the Pan band makes use of TDI (Time Delay Integration), the size of the detector array assembly is 5 x 6000 (30,000 pixels in cross-track). Five detector line arrays, each of 1,500 elements (7,500 pixels in cross-track), are utilized for the four MS channels.

522) "Proposal for an Earth Watch Wide Field / Superspectral Element," ESA/PB-EO (2001) 59, Annex-1, May 16, 2001

523) The SPOT project management provided a set of viewgraphs in early June 2001. After my draft submission to the project, however, I did not receive any response (in spite of several reminders). In my experience, this disregard of small requested services seems to be a trademark of SPOT management.

Spectral bands	Pan, MS (RGB, NIR)
Spatial resolution	0.7 m for Pan, 2.8 m for MS bands
Swath width	21 km
FOR (Field of Regard)	60°
Coding	10 bit
Image location accuracy	1 m (with ground control points), 20 without GCP
Data compression	Wavelet compression algorithm with an average compression factor of 4
Source data rate	600 Mbit/s

Table 103: Characteristics of the OHRI instrument

The basic data products of Pléiades are level 0 and level 1 imagery (up to orthorectified imagery). Possible intermediate products are: Mosaicked imagery, DTM (Digital Terrain Model) extracted from stereo pairs.

The main application fields identified for the Pléiades program data fall into the following categories: Cartography, agriculture, forestry, hydrology, geological prospecting, dynamic geology and risk management.

Pléiades is a multi applications, multi sensor and multi partnership program. The first generation will only be furnished with an optical high-resolution imager. Wide-field, superspectral, hyperspectral and thermal (TIR) system observation capabilities are options for future Pléiades implementations (with launches beyond 2006).

Sensor	Resolution (m)	Swath Width (km)	Nr. of Bands	Revisit Time (days)	Main applications
Wide Field	2-5	40-100	3-4	3-7	Cartography, geology, agriculture, forest, hydrology
Optical HR	<= 1	10-30	3-4	1-2	Cartography, risk, forest, geology
Superspectral	3-10	100-300	6-20	1-2	Agriculture, forest, geology
Hyperspectral	5-20	50-300	30-200	2-7	Geology
Thermal	1-40	100	TBD	<1	Forest fires, geology, ocean

Table 104: Optical system identification for possible user needs

B.4 EarthWatch (DigitalGlobe) Imaging Satellites

EarthWatch Inc. is changing its name to "DigitalGlobe Inc." by Oct. 2001.

B.4.1 EarlyBird

EarlyBird is an imaging spacecraft designed, built and operated by EarthWatch Inc. of Longmont, CO, along with its major partners: Ball Aerospace Corporation, Boulder, CO; CTA Inc. of McLean, VA; Hitachi Ltd. of Tokyo, Japan; and Telespazio of Rome, Italy. The EarlyBird S/C is three-axis stabilized; the attitude is sensed by a star tracker, position knowledge by GPS; design life = 3 years, 5 years of on-board fuel; S/C mass = 310 kg; payload mass = 150 kg; power = 90 W; on-board storage capability of a 16 Gbit solid-state recorder. 524)

A launch of EarlyBird-1 took place on December 24, 1997 with a Start-1 launch vehicle from the Svobodny Cosmodrome in Eastern Russia (note: the Start-1 rocket is based on the SS20 and SS25 intercontinental ballistic missiles with proven launch performance). EarthWatch controllers lost contact with the S/C on Dec. 28, 1997, stopping the commencement of operations.

Orbit: Sun-synchronous polar orbit, altitude = 470 km, inclination = 97.3°, period = 94 min, 10:30 AM equator crossing, descending node (1:30 PM equator crossing for the second

524) Information provided by D. B. Gerull, R. N. Herring, and B. Wientzen of EarthWatch, Longmont, CO.

satellite), repeat cycle = 20 days (max), revisit time = 1.5 - 2.5 days (with a two satellite configuration).

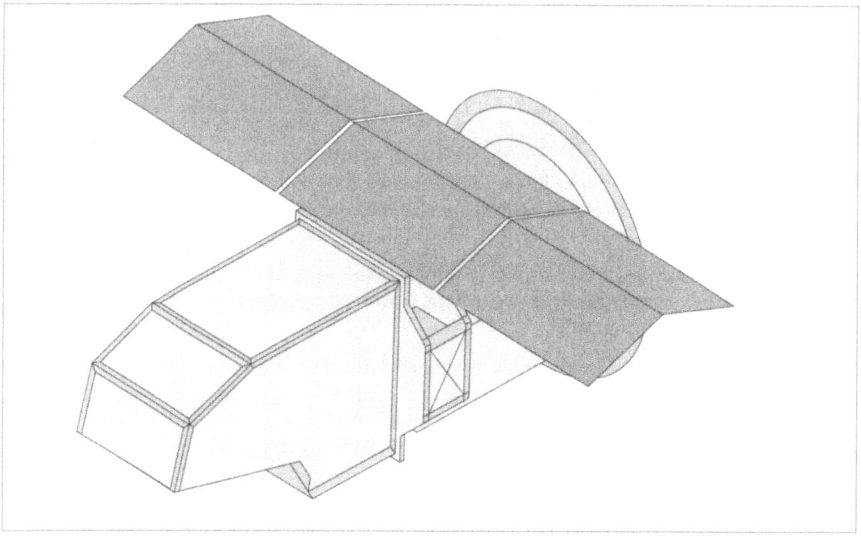

Figure 53: The EarlyBird S/C Model

Data: The downlink of imaging data is encrypted and provided in X-band at data rates of 25 Mbit/s to EarthWatch-owned ground receiving stations in the USA, Europe and Asia. The TT&C up- and downlinks are in UHF-band. The objective of EarthWatch is to become the first global supplier of commercial high-resolution imagery and related geographic information products by creating and maintaining a *Digital Globe*™ product database (the master archive, a gateway, and the satellite control center are located in Longmont, CO). A number of customized service options are provided with product delivery times ranging from 30 minutes to 48 hours after acquisition.

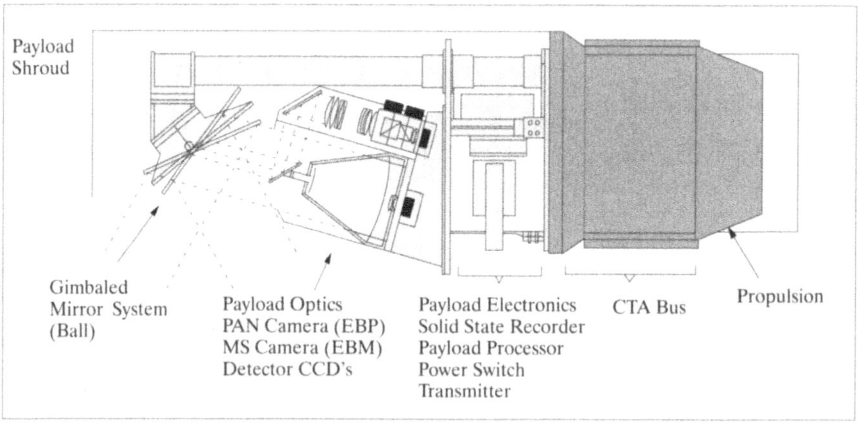

Figure 54: The EBP and EBM optical design concept

Sensor complement:

EBP = EarlyBird Panchromatic, and **EBM** = EarlyBird Multispectral (built by Earth-Watch, same concept design as on Clark S/C, but with enhanced capabilities). The instrument provides high-resolution panchromatic and multispectral imagery simultaneously. A staring focal plane array detector technology (Kodak designed and built) is employed featuring a gimbaled mirror design (see Figure 54) with a 30° pointing capability from nadir into any direction. The fast steering mirror permits exposures of a single frame or of a matrix of images (point and shoot). The panchromatic image scene is always contained within the larger matrix of the multispectral image scene, in a fixed location. However, by taking a multispectral image and then moving the gimbal around to take additional multiple exposures, one can cover the area of the multispectral image with panchromatic images, or there is the ability to collect a panchromatic image in any desired location within a multispectral image. The cameras may also be used for stereo imaging by slewing the gimbal mirror fore and aft in the S/C flight direction. The on-board processor provides real-time radiometric/geometric calibration and image compression for all imaging data.

Parameter	Panchromatic camera (EBP))	Multispectral camera (EBM)
Spectral range(s)	0.45 - 0.80 μm	0.45-0.59, 0.61-0.68, 0.79-0.89 μm
Spatial resolution (at nadir)	3 m	15 m
Footprint (or scene)	3 km x 3 km per patch (4 patches exposed simultaneously) over a swath ~11.1 km	15 km x 15 km per patch (4 patches exposed simultaneously) over a swath ~55.5 km
Instrument (mirror) pointing	±30° (along-track/across-track) providing a 560 km wide field of regard	
Pointing accuracy	6 m horizontal, 4 m vertical (relative)	
Stereo imaging	In along-track direction, maximum base/height ratio = 1	
Data quantization	8 bit	
Array size	4 10^6 pixel	4 10^6 pixel (x 3)
Data size/scene	4 MByte (uncompressed)	12 MByte (uncompressed)
Data volume	~2000 patches (500 scenes) per orbit	

Table 105: Performance characteristics of the EarlyBird imagers

B.4.2 QuickBird-1

QuickBird-1 is a next-generation EarthWatch satellite, offering commercial imagery at 1 m (PAN) and at 4 m (MS) resolution. This high-resolution imaging capability requires a change in instrument technology from a staring array design to a pushbroom/large-telescope technique, resulting in a new spacecraft design. QuickBird-1 uses the BCP 2000 (Ball Commercial Platform 2000) satellite bus design.[525)] It is 3-axis stabilized. The ADCS (Attitude Determination and Control subsystem) uses two star trackers, redundant IRUs, sun sensors and magnetometers for attitude sensing. Attitude control is provided by low-vibration reaction wheels (0.68 Nm, 20 Nms), three torque rods, and four hydrazine thrusters. Position knowledge is provided by redundant GPS receivers. The pointing accuracy is ±0.016° (3 sigma steady state in all three axes; the attitude pointing knowledge is ±0.0008° (3 sigma steady state in all three axes), geolocation knowledge <15 m (3 sigma) after ground processing. S/C design life of 5 years. S/C mass = 931 kg at launch. The BCP 2000 configuration uses a simple panel-post aluminum honeycomb structure. The total bus mass (wet) is 641 kg. Two solar panels (GaAs/Ge cells), each of 3.2 m^2 area and single axis drive, provide a S/C power of 1500 W. The NiH_2 battery provides energy of 40 Ah for ecliptic operations.

A launch of QuickBird-1 took place on Nov. 20, 2000 on a Cosmos-3M vehicle from Plesetsk, Russia. Unfortunately the launch ended in a failure. No contact could be established with the spacecraft.

525) "Ball Commercial Platform 2000 (BCP 2000)," Technical Description, Jan. 2000, provided by Tom Miers of BATC

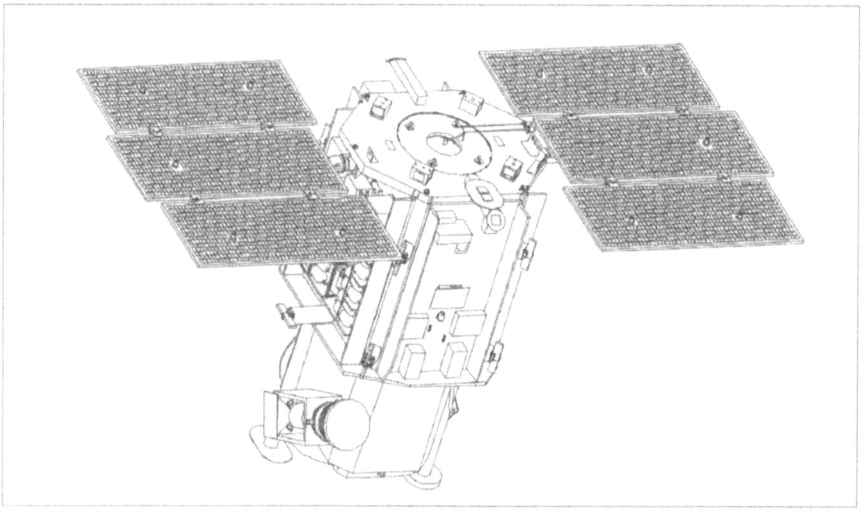

Figure 55: Illustration of QuickBird configured with BGIS-2000

Orbit: Circular orbit (non-sun-synchronous), altitude = 600 km, inclination = 66°, the equator crossing time is variable, average revisit time = 1 to 5 days depending on latitude.

Sensor complement:

BGIS 2000 (Ball Global Imaging System 2000).[526] BGIS is a BATC-developed imager - the instrument-bus combination is called BGIS 2000. The bus itself is called BCP 2000. The camera instrument of BGIS-2000 is called **BHRC 60** (Ball High Resolution Camera 60). BHRC 60 consists of the following elements: Optical subsystem, FPU (Focal Plane Unit) and the DPU (Digital Processing Unit), with FPU and DPU designed and custom-built by Kodak (same, except for size, as that used on IKONOS). BHRC 60 has a design life of >5 years, achieved with a redundant architecture. Instrument mass = 380 kg, instrument power = 250 W silicon and 430 W for GaAs (orbital average).

- The optical subsystem, mounted on an optical bench (with sunshield and internal baffling to suppress stray light), is of Ball design (telescope aperture of 60 cm diameter, lightweight structure, focal length of 8.8 m, f/14.7, the telescope mass is 138 kg, telescope size: 115 cm x 141 cm x 195 cm), providing a FOV (Field of View) of 2.12°, obtained with an unobscured off-axis three-mirror-anastigmatic (TMA) optical form. A fourth mirror is used to fold the light bundle for compact telescope packaging. The enlarged FOV of the BHRC 60 instrument offers a ground swath of 15 km at 400 km orbital altitude or 34 km at 900 km altitude with a GSD varying between 0.5-1.5 m, respectively.
- The detector subsystem employs the pushbroom imaging technique. The CCD detector array features 27,568 pixels in the cross-track direction for the panchromatic band and 6892 pixels for each of the four multispectral bands. The spectral ranges of the multispectral BGIS bands correspond to the first four bands of the ETM+ instrument on Landsat-7, the PAN band is also identical to PAN on ETM+.
- The TDI (Time Delay Integration) concept is employed for PAN imagery. TDI levels of 10, 13, 18, 24, and 32 are available and selectable. The TDI arrays prevent exposure saturation while maximizing the SNR over a wide range of angles and Earth albedos.

526) "Ball High Resolution Camera 60 (BHRC 60), Technical Description, Jan. 2000, provided by Tom Miers of BATC

Parameter	Panchromatic imagery	Multispectral imagery (4 bands)
Spectral range(s)	0.45 - 0.90 μm, grayscale	0.45-0.52, 0.52-0.60, 0.63-0.69, 0.76-0.90 μm
Spatial resolution, IFOV	1 m (GSD), 1.37 μrad	4 m (GSD), 5.47 μrad
Swath width, FOV	22 km (600 km altitude), 2.12°	
Camera	Pushbroom array (11 bits pixels)	Pushbroom arrays (11 bits x 4)
Detector array	27,000 pixels	6,700 pixels x 4
S/C body pointing capability	±30° (along-track/across-track) providing a 704 km wide field of regard	
Pointing accuracy	<0.5 mrad absolute per axis (200 to 450 m)	
Geolocation of data	<15 m (3 sigma) after ground processing	
Pointing agility	Maneuver of 10° in 20 s, maneuver of 50° in 45 s	
Imaging modes	Snapshot: 22 km x 22 km Strip mode: 22 km x 200 km Mosaic patterns: Stereo: 20 km x 20 km typically; in along-track direction (single pass)	
Data quantization	11 bits	
Data size/ PAN scene	8 Gbit (uncompressed), 1.5 Gbit (compressed)	
Data volume	64 - 100 scenes per orbit	

Table 106: Performance parameters of the BGIS 2000 instrument on QuickBird-1

The instrument provides high-resolution panchromatic and multispectral imagery simultaneously. The pushbroom imager is rigidly aligned with the S/C axis, providing a nominal body-pointing capability of ±30° into the along-track and cross-track directions (45° max). The panchromatic and multispectral image scenes are coincident. BGIS 2000 may also be used for stereo imaging by slewing the S/C fore and aft. The on-board processor provides real-time radiometric/geometric calibration and image compression for all imaging data. The compression technique employed is ADPCM (Adaptive Differential Pulse Code Modulation) of Kodak.

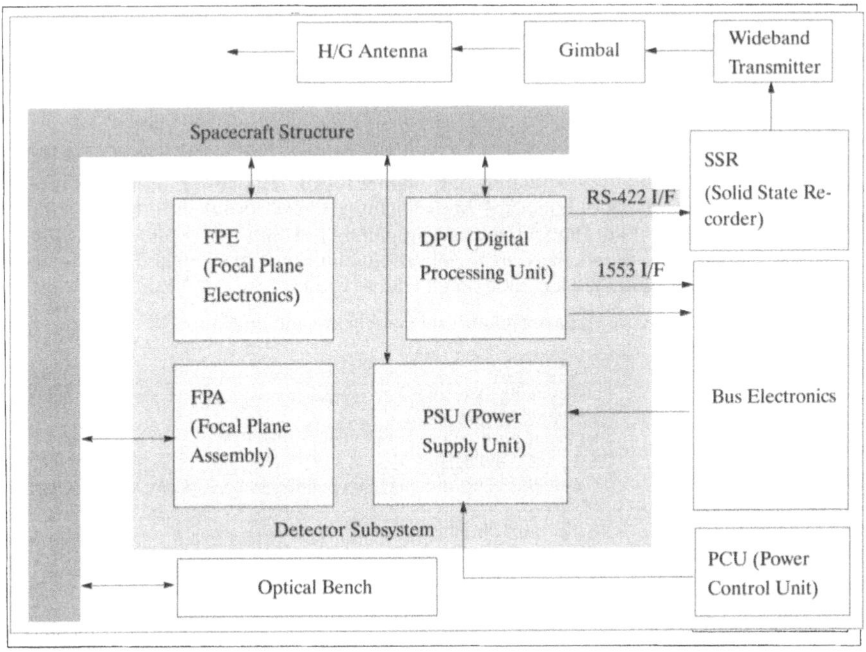

Figure 56: Functional block diagram of the BHRC 60 detector subsystem

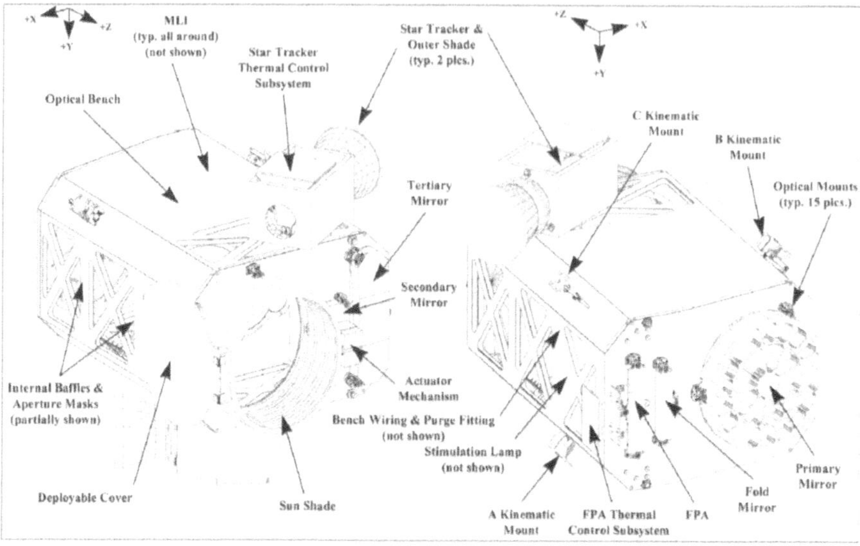

Figure 57: Schematic illustration of the BHRC 60 instrument

S/C communications: An on-board image data storage capability of 137 Gbit is provided in solid-state memory. The downlink of all imaging data is provided in X-band at data rates of up to 320 Mbit/s to EarthWatch-owned ground receiving stations in the USA, Europe and Asia. The real-time X-band data channel is PCM/PSK/PM modulated, while the playback data channel is PCM/PM modulated. The TT&C functions are provided in S-band at data rates of 4-16 kbit/s in downlink and 2 kbit/s in uplink. The realtime narrow-band data is downlinked at these rates on the subcarrier, while the stored engineering data is downlinked on the carrier at a rate of 256 kbit/s.

B.4.3 QuickBird-2

QuickBird-2 is an almost identical follow-up optical imaging mission of QuickBird-1, owned and operated by EarthWatch Inc. of Longmont, CO. A higher ground resolution of the imagery is obtained with QuickBird-2 by selecting a lower orbital altitude, naturally at the expense of swath width. The S/C carries enough fuel to adjust the lower orbit more frequently (due to drag influences) over the operational lifetime of the mission. A launch of QuickBird-2 is scheduled for Oct. 2001 on a Delta-II vehicle from VAFB, CA. [527] [528]

Orbit: Sun-synchronous circular orbit, altitude = 450 km, inclination = 98°, average revisit time of 1-3.5 days.

S/C pointing capability	Body-pointing S/C with ±30° nominal along-track and cross-track pointing capability, 45° maximum
S/C pointing accuracy	Standalone geolocation: 23 m Circular Error (CE), 17 m Linear Error (LE), 90% confidence
S/C autonomy	Autonomous Line-of-Site (LOS) pointing using GPS receivers and attitude sensors
S/C mass, size, design life	951 kg, 3 m in height, 5 years
On-board data storage	128 Gbit

Table 107: Performance features of QuickBird-2

527) http://www.digitalglobe.com/products/quickbird1.shtml
528) Information provided by Charles P. Herring of EarthWatch Inc., Longmont, CO

Parameter	Panchromatic imagery	Multispectral imagery
Spectral bands	Pan: 450-900 nm	450 to 520 nm (blue) 520 to 600 nm (green) 630 to 690 nm (red) 760 to 890 nm (VNIR)
Spatial resolution	0.61 m (GSD at nadir)	2.5 m (GSD at nadir)
Swath width (FOV)	17 km (450 km altitude), 2.12°	
Field of Regard (FOR)	544 km	
Imaging modes (at nadir)	Snapshot: 17 km x 17 km Area: 32 km x 32 km (typically) Strip: 17 km x 225 km (typically) Stereo: 15 km x 15 km (typically)	
Data quantization	11 bit	

Table 108: Performance parameters of the BGIS 2000 instrument on QuickBird-2

B.5 EROS (Earth Remote Observation System), ImageSat International

EROS is a program of ISI (ImageSat International, N.V.), formerly WIS (West Indian Space) Ltd., Cayman Islands (a Netherlands Antilles-registered joint venture). ImageSat's ownership includes former WIS partners: Israel Aircraft Industries (IAI) of Tel-Aviv (owned by the Israeli government), Elbit Systems Ltd. of Haifa, CORE Software Technology (CST) Inc. of Pasadena, CA, and El-Op Electro-Optics Industries of Rehovot, Israel. The objective is to launch and operate a constellation of satellites under a license from the government of Israel, and to offer high-resolution imagery, collected by EROS satellites, on a commercial basis to a large customer base. Commercial spaceborne remote sensing technology, developed by IAI, was approved by the government of Israel in Oct. 1996. [529)] [530)]

The first satellite, EROS-A, launched on Jan. 22, 1998 from the Yavne launch site in Israel, failed to reach orbit due to a Shavit launcher failure.

An enhanced EROS-A1 was launched on a Russian Start-1 launcher on Dec. 5, 2000 (Svobodny Cosmodrome). This is to be followed by EROS-A2 in 2001. The first EROS-B1 S/C is scheduled for launch in 2002. A constellation of six Eros-B satellites (B1 to B6) are planned to be operational by the year 2004. The operational constellation provides a global daily revisit for any point on Earth. [531)]

Note: In Feb. 2001, ISI changed plans in its launch sequence with the consequences that the launch of EROS-A2 was scratched in favor of a launch of EROS-B1. The B-series offers higher resolution imagery (<1 m versus 1.8 m for the A-series instrument). The decision to move directly to the B-series (with platform and instrument) results in a new launch date for EROS-B1 in 2003. In July 2001, ImageSat awarded a contract to IAI to build a new spacecraft. [532)]

The EROS program offers a combination of services and products tailored to meet specific customer requirements and budgets: [533)]

- Communication, computing and networking technology are designed to augment existing information sources including regional cartographic, aerial imagery, as well as commercially available information sources.
- Provision of low-cost imagery products to support applications in many fields.

529) ISI is incorporated in the Netherlands Antilles and headquartered in Cyprus
530) "EROS Program Presentation," Jan. 12, 1999, an IAI paper
531) Information provided by P. Rosenbaum of IAI
532) B. Opall, "ImageSat Initiates Production of New Craft," Space News, Aug. 13, p. 3 and p. 28
533) http://www.imagesatintl.com/

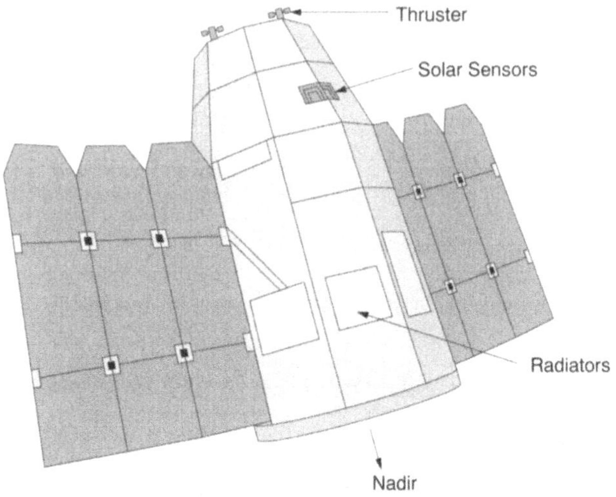

Figure 58: The EROS-A S/C Model

B.5.1 EROS-A Spacecraft Series

EROS-A1 (of OFEQ-3 heritage)[534] is an Israeli high-resolution commercial imaging satellite designed and built by IAI (Israeli Aircraft Industries). The S/C structure consists of low-mass composite material with passive thermal control, it is three-axis stabilized, attitude control and navigation is performed using horizon sensors, sun sensors, gyros and magnetometer, four reaction wheels and thrusters. The pointing accuracy is $<0.1°$ in all three axes, attitude stabilization is <40 mrad/s, the jitter is < 0.2 mrad. S/C mass = 250 kg (174 kg S/C bus, 36 kg of instrument mass and 30 kg of hydrazine), solar panel power (silicon array, fixed panels) = 450 W (EOL), plus 14 Ah NiCd batteries for eclipse operation, the bus power consumption is 100 W. S/C design life = 4 years.

Communications: Imagery is transmitted in X-band at a rate of 70 Mbit/s (RF downlink) to the ground receiving stations, using a 1.5 W transmitter and one of the two existing two-axis gimbaled directional antennas. The EROS satellites are monitored/operated in S-Band (TT&C) via a single ground control station (GCS), located at IAI/MBT in Israel (4 to 5 passes per day and per satellite are in station visibility). The S-band date rate is variable between 2.5 - 15 Kbit/s.

In addition there is a network of existing ground receiving stations (ImageSat has a ground segment infrastructure of at least ten stations), located around the world, for real-time image data acquisition.

Image data reception, processing (geocoding, value-added services), archiving and distribution is provided by CORE Software Technology (CST) of Pasadena, CA through CORE's online global network with the name of 'ImageNet'. Operational capabilities/services include such items as:

- SOP (Satellite Operating Partner). This service provides a dedicated regional satellite with local customer tasking. SOP receiving ground stations are able to plan, to generate and to transmit imaging commands to the satellite.
- PAS (Priority Access Satellite). The service provides priority tasking of EROS S/C on a global scale.

534) Note: OFEQ-1 was launched in September 1988, OFEQ-2 in April 1990; both had a lifetime of 6 months.

- AAD (Acquisition, Archiving and Dissemination). The service is intended for existing, experienced satellite imagery reception, archiving and distribution operations to support worldwide commercial imagery and derivative product sales.

Parameter	Value	Parameter	Value
Data transmission	Real-time downlink	Antenna gain	19 dBi
RF frequency	X-Band, 8200 GHz	Antenna polarization	RHCP
Modulation	QPSK	Ant. pointing accuracy	1.5°
Bit rate	70 Mbit/s	Nr. of channels	1 (+1 redundant)
Error correction code	Convolutional, R=4/5	Transmitted power	1.5 W (31.8 dBm)

Table 109: Performance characteristics of the EROS-A downlink

Orbit for EROS-A1: Circular sun-synchronous orbit, altitude = 480 km, inclination = 97.3°, period = 94.2 minutes, local time of descending node = 10:30 AM. The revisit capability of any point on Earth within a 15° cone is within seven days. A revisit period/satellite of any point at latitudes $> \pm 40°$ is 2-7 days within a 30° cone, and only two days within a 45° cone.

Operational status of EROS-A1 as of June 2001: All subsystems are operating nominally.

Sensor:

PIC (Panchromatic Imaging Camera), built by El-Op Electro-Optics Industries Ltd. of Rehovot, Israel. PIC is a body-mounted instrument, featuring CCD pushbroom technology. PIC uses a Cassegrain telescope with an aperture of 30 cm in diameter and a focal length of 3.5 m. FOV = 1.5°. The CCD detector array provides 7000 pixels per line. The scan speed is 30 - 750 line/s, this corresponds to 18 to 1.3 ms of integration time, respectively. Asynchronous pushbroom scanning is provided for panchromatic imagery in the spectral range of 0.5 - 0.9 µm. The ground sampling distance (GSD) is 1.8 m, the swath width is 12.5 km (from a 480 km orbital altitude). The data is quantized at 12 bits (samples) and transmitted at 11 bits. SNR <2/2,048 gray levels after quantization. Proper observations require sunlight conditions with a sun-over-the-horizon angle of >20°. Instrument mass = 36 kg (includes telemetry unit), power = 140 W (average).

Type of Scene	Scene Size	Max. Nr. of Scenes
Monostrips (30° inclination to ground track)	120 km x 12.5 km	5
Monostrips (43° inclination to ground track)	217 km x 12.5 km	3
Discrete scenes (nadir view along the ground track)	12.5 km x 12.5 km	28
Mosaics	25 km x 25 km	7
Stereo monostrips (nadir)	40 km x 12.5 km	5
Stereo scenes (24° fore and aft pointing)	12.5 km x 12.5 km	10
Stereo scenes (40° fore and aft pointing)	12.5 km x 12.5 km	7

Table 110: Observation options for image scenes of a satellite pass

A body-pointing technique is employed for instrument pointing (the entire S/C is pointed into the desired direction), permitting a field of regard of ±45° from nadir into any direction. The physical pointing of the S/C is provided by the four reaction wheels (the pointing acceleration is $0.4° s^{-2}$ with a maximum angular rate of 2.2°/s).

The non-availability of on-board source data storage permits only observations in real-time in view of a GRS (Ground Receiving Station). The following observation options (scenes) are possible within a single 10-minute pass:

B.5.2 EROS-B Spacecraft Series

The EROS-B series S/C structure is identical to that of EROS-A. The additional on-board fuel (up to 60 kg) is sufficient for S/C operations of up to 10 years. This increases the S/C mass to about 350 kg. The S/C attitude is sensed and controlled as done with EROS-A1. In addition, there is a star sensor for the B satellites. The S/C design life is six years.

Orbit: Sun-synchronous circular orbit, altitude = 600 km, inclination = 97.8°, local time of descending node 10:45 AM.

Sensor:

PIC-2 (Panchromatic Imaging Camera-2). The EROS-B series imager instrument features CCD technology in combination with a TDI (Time Delay and Integration) scheme in its focal plane, a cumulative expose concept of each ground image line by a CCD detector array, to improve the SNR value (an important issue for high-resolution imaging). The instrument uses also a Cassegrain telescope with an aperture of 50 cm in diameter and a focal length of 5 m. FOV = 1.5°. The CCD detector array provides 9,666 pixels per line and a total of 96 lines. The scan speed is 3,050 lines/s (max) using VSU (Video Storage Unit) and 1,525 lines/s without VSU. Asynchronous pushbroom scanning is provided for panchromatic imagery in the spectral range of 0.5 - 0.9 μm. The ground sampling distance (GSD) is 0.82 m, the swath width is 16 km. The data is quantized at 8 bits (samples) and transmitted at 8 bits. SNR <2/256 gray levels after quantization.

Type of Scene	Scene Size	Max. Scenes at 3,050 lines/s (scan rate)	Max. Scenes at 1,525 lines/s (scan rate)
Monostrips		4 of 305 km x 16 km or 5 of 246 km x 16 km	5 of 130 km x 16 km
Discrete scenes (computed along the ground track)	16 km x 16 km	40	30
Mosaics	32 km x 32 km	12	8
Stereo monostrips	160 km x 16 km	2	N/A
	68 km x 16 km	8	4
Stereo scenes	16 km x 16 km	20	10

Table 111: Observation options for image scenes of a B-satellite pass

Parameter	EROS-A1 and -A2	EROS-B1	EROS-B2 to B6
Orbit	Sun-synchronous, altitude = 480 km circular	Sun-synchronous, altitude = 600 km circular	Sun-synchronous, altitude = 600 km circular
Orbit repeat cycle	1 week	1 week	1 week
Equator crossing times	10:30, 10:30, 13:00, 13:30, and 14:00		
GSD	1.8 m	0.82 m	0.82 m
Swath width	12.5 km	16 km	16 km
Scanning	Asynchronous up to 750 lines/s	Asynchronous up to 3,050 lines/s	Synchronous and asynchronous up to 9000 lines/s
Sensor type	CCD	CCD/TDI, selectable: 1, 4, 8, 16, 32, 48, 96	CCD/TDI, selectable: 1, 4, 8, 16, 32, 48, 96
Spectral band	0.5 - 0.9 μm	0.5 - 0.9 μm	0.5 - 0.9 μm
Data quantization	12 bits (11 bits transmission)	8 bits	11 bits
Pointing accuracy	<100 m	<100 m	<100 m
Pixel to pixel relative accuracy	<100 m within 12.5 km x 12.5 km scene	<50 m within a 16 km x 16 km scene	<50 m within a 16 km x 16 km scene
Multispectral capability	No	No	Yes
Signal processing unit	No	No	Yes, loss-less compression of source data prior to transmission
Downlink data rate	70 Mbit/s	280 Mbit/s	280 Mbit/s
S/C mass	280 kg	<=350 kg	

Table 112: Performance characteristics of EROS satellites

The imager instrument of the EROS-B series (B2 to B6) satellites can be operated in either asynchronous or in synchronous imaging mode. In synchronous mode, the S/C platform keeps a constant pointing angle toward the Earth's surface. In asynchronous mode, imaging by the detector array is performed in a "step-and-stare" fashion, i.e., by slewing the S/C platform in the along-track direction (this permits also stereo imaging).

The non-availability of on-board source data storage permits only observations in real-time in view of a GRS (Ground Receiving Station). The following observation options (scenes) are possible within a single 12-minute pass:

B.6 IKONOS (Space Imaging Inc., Thornton, CO)

Background: CRSS (Commercial Remote Sensing System) was a remote sensing imaging satellite project of Lockheed Martin that started in 1991. [535] In 1994 a new company was formed for this venture, namely Space Imaging Inc. (of Thornton, CO), with the following partners: Lockheed Martin (Sunnyvale, CA: space segment, satellite operations, and tasking of ground segment), Raytheon/E-Systems (Garland TX: communications, image processing and customer service center). Eastman Kodak Co. of Rochester, NY designed and built the digital camera/sensor. The overall objective is to offer commercial high-resolution (1 m GSD panchromatic and 4 m GSD multispectral) imagery with excellent location knowledge in near real-time and offline. In 1997 the CRSS satellite was renamed by Space Imaging to "IKONOS-1," supposedly a variant of the Greek word 'eikon' (icon), meaning "image." [536]

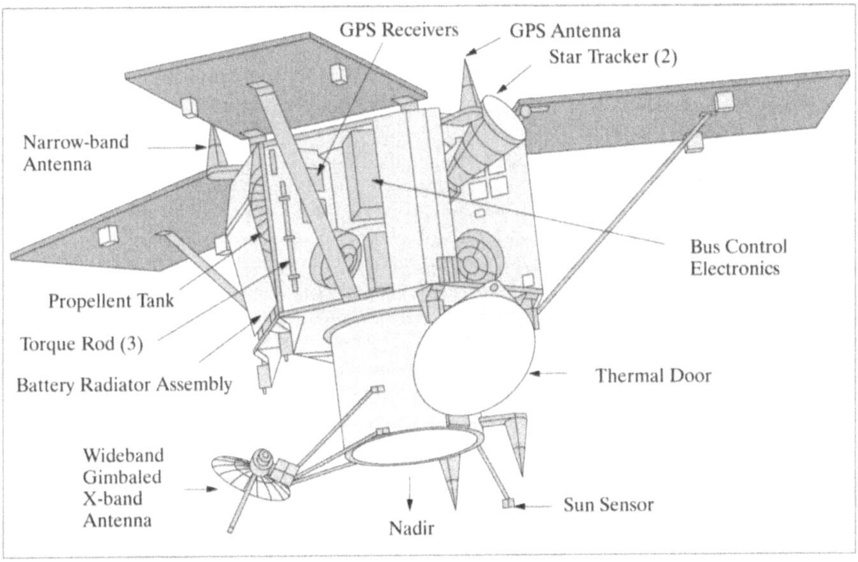

Figure 59: The IKONOS S/C Model

B.6.1 IKONOS-1

IKONOS-1 is a 3-axis stabilized spacecraft, the attitude is measured by two star trackers and a sun sensor and controlled by four reaction wheels; location knowledge is provided by

535) Information provided by S. Kilston, formerly of Lockheed Martin, Palo Alto, CA

536) Note: Space Imaging acquired EOSAT (a joint venture of Lockheed Martin and Hughes Aircraft) in 1995. The new company was subsequently renamed into: Space Imaging EOSAT. Eventually, the became simply: Space Imaging

GPS; LM900 satellite bus system; design life = 7 years; S/C body size = 1.83 m x 1.57 m (hexagonal configuration); S/C mass = 728 kg; power = 1.5 kW.

Note: A launch of IKONOS-1 on LLV2 from VAFB was initially planned for December 1997. However, there were several launch delays, the last one to replace the laser ring gyro. Finally, IKONOS-1 was launched April 27, 1999 from VAFB aboard an Athena 2 launcher. Unfortunately, the rocket's nose cone failed to separate as planned at 4 minutes, 27 seconds into flight - resulting in a complete loss of the satellite. With the protective shroud still attached, the rocket's upper stage and satellite did not have enough speed to reach a stable orbit around Earth. The vehicle then reentered the atmosphere over the South Pacific Ocean. [537)]

B.6.2 IKONOS-2

IKONOS-2, identical to IKONOS-1 and built in parallel to IKONOS-1, was launched successfully on Sept. 24, 1999 from VAFB aboard an Athena 2 launcher of Lockheed Martin.

Orbit: Sun-synchronous near-polar circular orbit, altitude = 681 - 709 km, inclination = 98.1°, period = 98 min, repeat cycle = 14 days (max), revisit cycle = 1-3 days. Local equator crossing time is at 10:30 AM on descending node.

Sensor complement:

OSA (Optical Sensor Assembly), designed and custom-built by Kodak Co. of Rochester, NY (Space Imaging owns the design of OSA). The instrument features a Cassegrain-type telescope with a 70 cm diameter primary mirror, a 10 m focal length (folded optics design). OTA (Optical Telescope Assembly) captures imagery across a swath of 11-13 km, it uses five mirrors to reflect the imagery to the imaging sensor arrays at the back end of the telescope. Three of the mirrors are powered (curved), and are of TMA (Three Mirror Anastigmatic) design. Note: TMA refers to lenses that are able to form approximately point images of target (object) points. The other two mirrors are flat, and serve to 'fold' or bounce the imagery across the width of the telescope. Pushbroom detector technology (a large focal plane detector array, generation of 6500 lines/s of panchromatic image data) is employed. Simultaneous imaging in panchromatic and multispectral modes is provided. The pixel size on the detector array is 12 μm for the panchromatic (PAN), and 48 μm for the multispectral (MS) detectors. The MS bands correspond to those of TM on Landsat in the visible range of the spectrum. The instrument light level is governed by a 70 cm aperture and a choice of 10, 13, 18, 24, or 32 TDI (Time Delay and Integration) stages for panchromatic (gray-scale) imaging. The detector array offers a cumulative exposure concept for panchromatic imaging. On-board electronics provide low-loss data compression of the original 11-bit data using ADPCM (Adaptive Differential Pulse Code Modulation). - The OSA instrument design features lightweight materials and advanced manufacturing techniques. The mass of the primary mirror was reduced by cutting a honeycomb pattern into its core using abrasive waterjet technology, and fusing thin mirror plates to each face.

Optical telescope assembly	Assembly size: 1.524 m x 0.787 m (1 m^3 volume) Assembly mass without the focal plane unit: 109 kg Focal length / focal ratio: 10 m / f14.3 Primary mirror aperture diameter: 0.7 m
Imaging detectors & electronics	Focal plane unit, unit size: 25 cm x 23 cm x 23 cm Detector array: 13,500 pixels cross-track (PAN) Detector array: 3375 pixels cross-track (MS), pixel size: 48 x 48 μm
Digital processing unit	Unit size: 46 cm x 19 cm x 31 cm ADPCM data compression, compression rate of 4:25 : 1
Power supply unit	Unit size: 18 cm x 20 cm x 41 cm
Total instrument mass, power	171 kg, 350 W

Table 113: OSA instrument layout

537) http://www.spaceimage.com/

A body-pointing technique with antenna gimbals and reaction wheels is employed for instrument pointing (the entire S/C is pointed into the desired direction), permitting a field of regard of ±30° into any direction. The angular slew rate is sufficient to perform both wide-area monoscopic and same-pass stereo collections. The location knowledge accuracy of the imagery is 2 m horizontal (relative) i.e. with ground control points, and 12 m (absolute), i.e. without the use of ground control points. Smooth scanning is provided with accurate gyros, low disturbance torques (smooth antenna gimbals and reaction wheels), and a rigid high-frequency structure of the satellite. Radiometric calibration provides relative and absolute corrections for detector channel responsivity differences.

The S/C may also be rotated about its imaging axis for proper (broadside) detector array orientation. This technique permits, for instance, the full-swath imaging of a particular feature of interest on the Earth's surface, such as a coastline, which traverses under some angle through the in-track direction.

Parameter	Value	Parameter	Value
Spectral range PAN	0.45 - 0.90 μm	Off-nadir pointing angle	±30° in any direction
Spectral range MS (μm)	0.45-0.52, 0.52-0.60, 0.63-0.69, 0.76-0.90	Stereo capability	along-track
Spatial resolution	≤1 m PAN, ≤4 m MS	Swath width, scene	13 km, 13 km x 100 km
Pixel quantization	11 bit	Field of regard (FOR)	±350 km at 1 m GSD

Table 114: Some performance parameters of the OSA instrument

Data: The downlink of all imaging data is in X-band at a rate of 320 Mbit/s to dedicated ground stations located around the world (on-board data recording capacity is 64 Gbit in solid-state memory). The TT&C function is provided in S-band (uplink of tasking and command data at 2 kbit/s, downlink of housekeeping data and metadata at 32 kbit/s).

Ground image data processing provides geocoding along with image compensation algorithms [misregistration, image motion, radiometric correction, MTF (modulation transfer function) compensation, etc.]. Space Imaging Inc. introduced a global archive (of digital imagery and services) under the trade name CARTERRA, which in turn is made up of regional archives, operated by regional partners. A great variety of image products and services are provided. Standard products are:

- Radiometrically corrected images
- Geometrically corrected images
- Orthorectified images and mosaics
- Digital terrain model (DTM) data
- Multispectral images

B.6.3 Kodak Model 1000 Camera System

As of July 1999, Kodak/C&GS (Commercial & Government Systems) is offering a "Model 1000™ camera system" of OSA camera heritage as an off-the-shelf product - at a 30% discount, deliverable within 24 months of order placement. This Model 1000 system design is owned by Kodak, containing some design changes with respect to OSA (reduced telescope aperture and instrument mass to fit onto minisatellites). [538) 539) 540)]

The Model 1000 camera system consists of the following elements: OTU (Optical Telescope Unit), FPU (Focal Plane Unit), DPU (Digital Processing Unit), PSU (Power Supply Unit), and CU (Cabling Unit). The total mass of the system is <100 kg.

538) Information provided by Michael J. Richardson of Eastman Kodak Company, Rochester, NY

539) "Kodak Introduces 1-Meter-Resolution Remote Sensing Camera In An Off-The-Shelf, Fixed Price Configuration," Kodak press release of July 19, 1999

540) T. Delaney, "Satellite Imagery in Land Development Applications," EOM, Oct. 1999, pp. 47-48

- The OTU is an all-reflective three mirror anastigmatic design with two flat fold mirrors to decrease package volume. The optical components are made from high quality, low thermal expansion glass substrates. The metering and mounting structures are made from low thermal expansion materials to match the expansion properties of the glass components. OTU has a mass of 45 kg, the power consumption is 15 W.
- The FPU includes the PAN and MS detectors and A/D converters. Timing and command signals are received from the DPU, power is received from the PSU. The mass of FPU is 16 kg, power = 85 W.
- The DPU generates the timing for the sensor electronics via a master clock. DPU accepts S/C commands over a standard 1553 bus and routes the information to the FPU and PSU. The DPU compresses the 11 bit digitized image data to about 2.5 bits/pixel using the Kodak proprietary algorithms of ADPCM (Advanced Differential Pulse Code Modulation). The DPU can format data for interface with an on-board storage unit and data downlink. DPU mass = 14 kg, power = 130 W.
- The PSU filters, regulates and generates the unregulated S/C power to the DPU and PSU. Mass = 8 kg, power 75 W. There is full redundancy.
- The CU provides the cabling between the various electronic boxes. Mass = 5kg, power = 10W.

Spectral range PAN (panchromatic)	0.45 - 0.90 μm
Spectral range MS (multispectral)	0.45-0.52 μm, 0.52-0.60 μm, 0.63-0.69 μm, 0.69-0.90 μm
Spatial resolution (GSD)	0.88 m PAN, 3.52 m MS, orbital altitude of 600 km
Swath width	12.2 km
Design life	5 years
Optical system parameters	
Clear aperture of primary mirror	44.8 cm diameter
Effective focal length; Focal ratio (f/number)	8.0 m / f/17.9
FOV along-track, FOV cross-track	0.75°, 1.19°
Panchromatic focal plane detector array Detector material, array type Pixel size Number of cross-track pixels Line rate TDI (Time Delay and Integration)	 Silicon, CCD 12 μm x 12 μm 13,816 6500 lines/s 10, 13, 18, 24, 32
Multispectral focal plane detector array Detector material, array type Pixel size Number of cross-track pixels Line rate Spectral filters	 Silicon, photodiode 48 μm x 48 μm 3,454 1625 lines/s Multi-layer on glass
Imaging performance parameters	
MTF @ Nyquist PAN (41.6 lp/mm)	0.09 (camera geometric mean of in- and cross-track)
SNR (80% scene reflectance, 20% background reflectance, 2.66 mW/cm^2 sr μm, 30° sun angle), PAN (24 TDI stages)	>45
Data quantization	11 bits
Data compression technique	ADPCM, 2.5 bits/pixel

Table 115: Specification of the Model 1000 camera system

Figure 60: Illustration of Kodak's Model 1000 camera system

B.7 Infoterra/TerraSAR (Astrium, ESA)

As of June 2001, Infoterra/TerraSAR [541] is a proposed cooperative Public-Private Partnership (PPP) program between TEC (TerraSAR Exploitation Company) and ESA (approval of the program is expected by the Council of Ministers in Nov. 2001). The Infoterra/TerraSAR initiative seeks to exploit the growing market for geo-information products and services through the establishment of a new global geo-information company, the TEC part of which is Infoterra (an Astrium company), founded on European partnerships. The TEC business concept is to service this market on a sustainable commercial basis primarily through the exploitation of Earth Observation (EO) information and to establish a viable share of the market through the introduction of new information products and services, optimized to satisfy the commercial and strategic needs of both private - and public - user communities. Many of these new products and services will be based on the exploitation of a spaceborne radar system, TerraSAR; a source of high-quality, reliable and secure data, superior to currently available data and specifically aimed to meet market needs.

TerraSAR is a space and ground system, which will consist of an initial deployment of two satellites (one X-band and one L-band). In the Infoterra business concept, TerraSAR is the key element to capturing a significant proportion of the existing market and, more importantly, to open new market opportunities from 2006 onwards when it becomes operational. To achieve operational advantage, Infoterra will employ TerraSAR whenever possible to provide data of the required quality. Infoterra will initially use current data sources to generate pilot products to provide early access to the market; these will be superseded by operational products as data from TerraSAR become available.

The Infoterra business concept and the technical concept of TerraSAR have evolved through feasibility and definition studies (initiated in 1997 and conducted in 1999/2000) led

541) Infoterra is the generic name for different legal entities. As of early 2001 two companies have been founded: Infoterra GmbH, a subsidiary of Astrium GmbH, and Infoterra Ltd., a subsidiary of Astrium Ltd. Additional companies in other countries may be founded in the future. It is planned to combine all national Infoterra companies under an umbrella of an Infoterra Holding Company. Infoterra will be a shareholder within TEC.

by German and UK industry, namely Astrium, [the former industrial partners DSS (Dornier Satellitensysteme) and MMS (Matra Marconi Space, UK) together with NRSC (National Remote Sensing Center), are now all part of Astrium], and supported by government and expertise on the institutional side (DLR/BNSC).

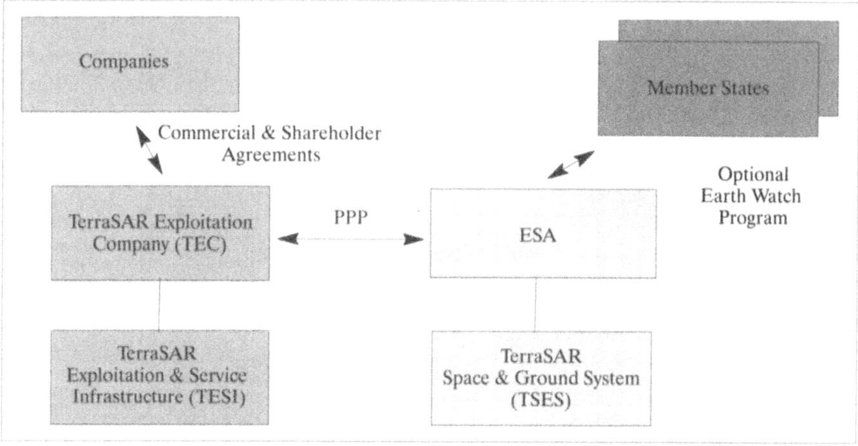

Figure 61: Overview of ESA - TEC partnership

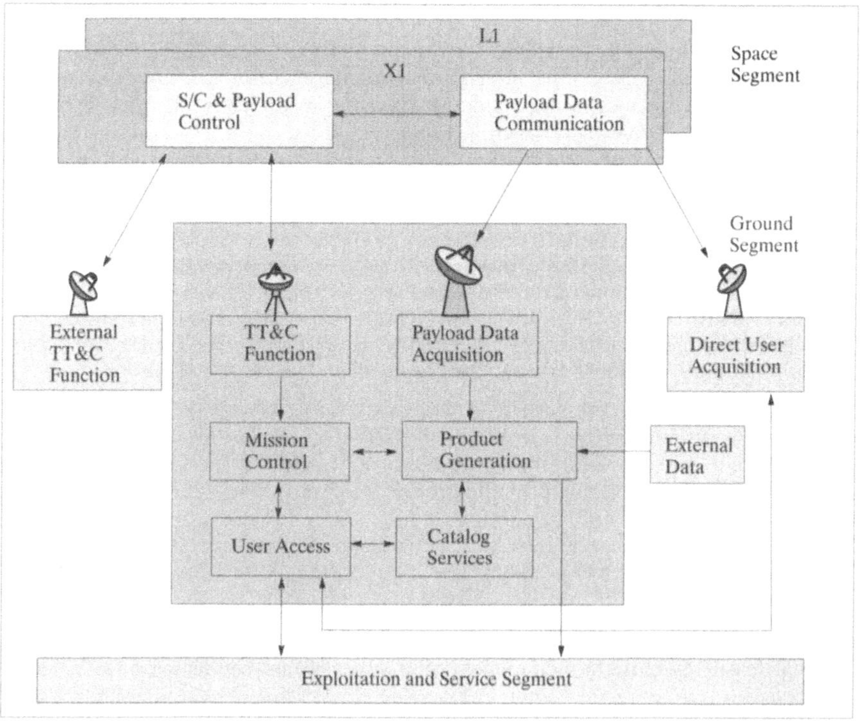

Figure 62: The system functional architecture of TerraSAR

The Infoterra/TerraSAR initiative is planned to become an integral part of the ESA Earth Watch Program. The TEC company started trading in January 2001. In the overall setup, Infoterra is the global service provider/distributor to the private and public customer base of geo-information services and SAR data products. These products may be applied to such fields as: agriculture, forestry, geology, cartography, and risk management.

The TerraSAR system has specifically been designed for land monitoring based on a dual frequency (L-band and X-band) SAR data source with multiple polarization capabilities. The X-band element will feature resolutions in the 1 m class for land mapping while the L-band, with its unique strengths for land monitoring, will enable discrimination of land cover types due to its penetrative ability, and with polarimetry enhancing both land cover classification and soil moisture retrieval.

The sensor system capabilities include stereo in X-band and repeat pass interferometry in L-band. Simultaneous dual frequency operation in combination with image sharpening will provide high resolution high quality thematic images. [542) 543) 544)]

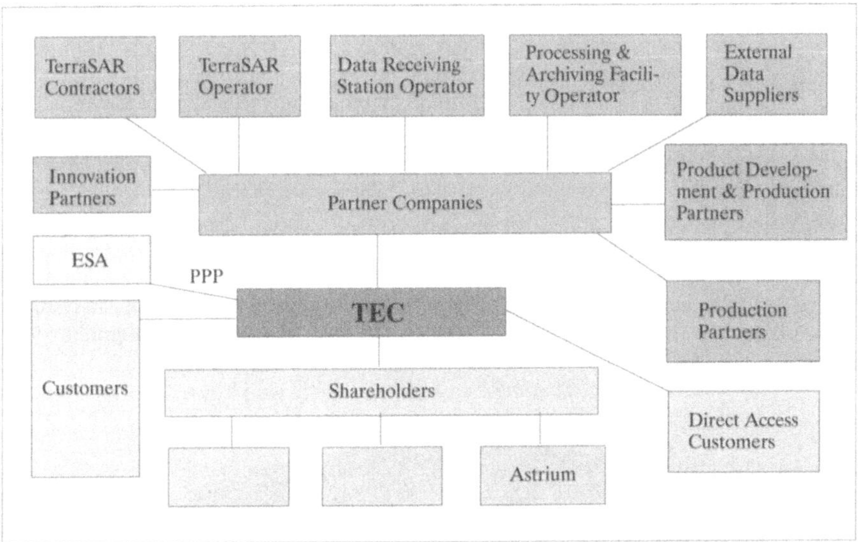

Figure 63: Infoterra & TerraSAR partnerships and opportunities

Based on the Infoterra business concept and the TerraSAR system definition, the fundamental objective of the Infoterra/TerraSAR initiative is to establish a self-sustaining geo-information business built on European strengths in SAR satellite technology, in SAR applications expertise and in the provision of services based on Earth observation data sources. The overall TerraSAR program represents a logical long-term continuation of European commitment comprising the development, launch and operations of SAR missions, as well as the needed infrastructure for the provision of a wide range of services. The first pair of satellites, TerraSAR-X1 and TerraSAR-L1, are planned to be launched in 2006, each with an operational life of five years (follow-up missions, i.e. TerraSAR-X2 and TerraSAR-L2, are planned beyond 2010). Both satellites will be launched into sun-synchronous dusk-dawn orbits with the same ground track and a spacing of 12 minutes, establishing a

542) "Formal Program Proposal for Infoterra/TerraSAR," submitted by DLR and BNSC, May 11, 2001, ESA/PB-EO (2001) 23, Rev. 2, Annex 1

543) "ESA Strategy for Earth Observation," ESA/PB-EO (98) 13, rev. 2

544) "Resolution on the Agency's framework for PPPs," ESA/C/CXLVII/Res. 1 (final), Oct. 19, 2000

so-called tandem mission. The nodal crossing time of X1 and L1 are shifted by ± 6.25 minutes to achieve the desired separation.

B.7.1 TerraSAR-X1 (X-band Satellite)

The TerraSAR X-band satellite (X1 is led by Astrium GmbH, Friedrichshafen, Germany) employs a mission-tailored Flexbus design with a total wet mass of approximately 1300 kg. The S/C structure features a central cylinder as the main load carrying element; it is of CFRP material with a diameter of 0.94 cm. The cross-sectional view of Figure 65 illustrates the mounting concept of radiator, solar array, and SAR antenna elements. [545] The outer shape of the spacecraft is mainly driven by the accommodation of the X-band SAR instrument and the geometrical limitations of the launcher fairing (PSLV of ISRO). The body-mounted solar array delivers an orbit average power of 875 W (EOL). The solar array is of size 7 m^2, the cells are of GaAs type with a size of 41 mm x 42 mm. The attitude control system is based on reaction wheels for fine-pointing, with magnetorquers for desaturation, and a propulsion system also capable of attitude control in order to achieve rapid rate damping during initial acquisition. Attitude measurement is performed with a GPS/Star Tracker system during nominal operation and a CESS (Coarse Earth and Sun Sensor) in safe mode situations, initial acquisition respectively (CESS is of CHAMP and GRACE heritage). A laser gyro and a magnetometer serve to support rate measurements in all mission phases. In fine pointing mode, a pointing accuracy of 30 arcsec is achieved. The S/C design life is 5 years of operation with a goal of 6.5 years (de-orbiting is planned at the end of the useful life time.). A launch of TerraSAR-1 is planned for 2006 - very probably on a PSLV launch vehicle from SHAR, India.

Orbit: Sun-synchronous circular dawn-dusk orbit (local time ascending node at 18:00 hours), altitude = 629 km, inclination = 98°, revisit period of 18 days with 3-4 days of sub-repeats. Global coverage within the latitudes ± 80°. - Spacecraft reentry (ESA requirement): At the end of its mission life the spacecraft orbit will be lowered to about 300 km (perigee) resulting eventually in enough air drag for a reentry (and a complete disintegration of the S/C in the atmosphere).

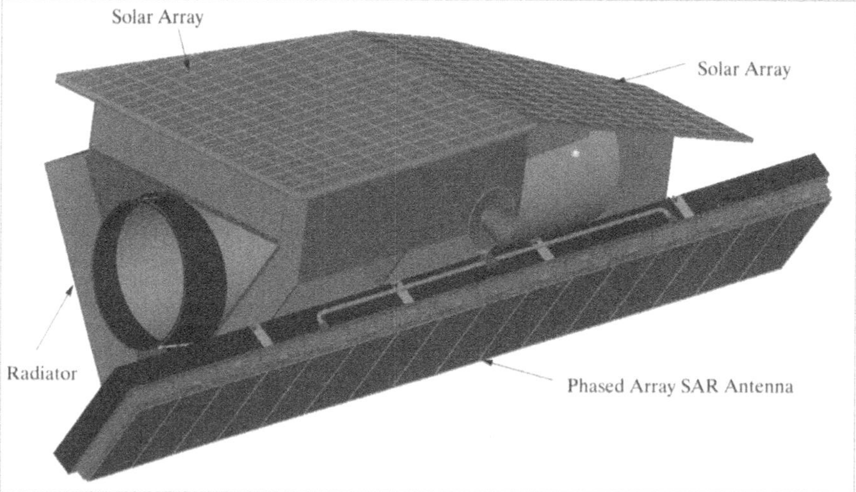

Figure 64: Illustration of the TerraSAR-X1 spacecraft

545) Information provided by Sebastian Riegger of Astrium GmbH, Friedrichshafen

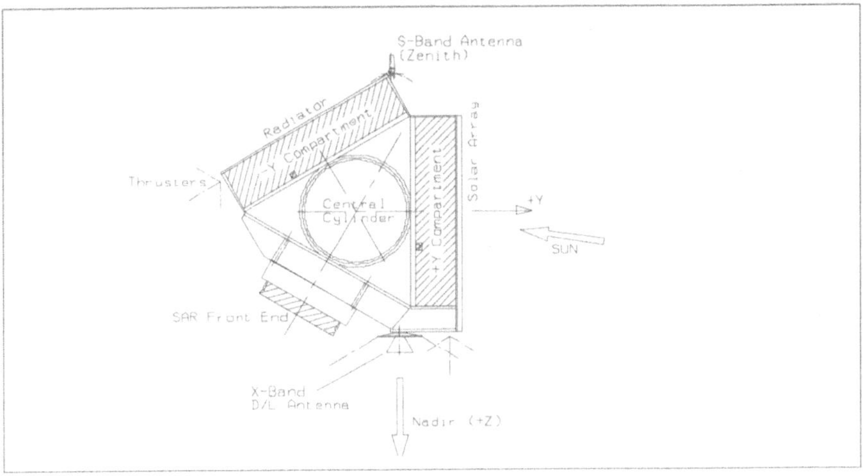

Figure 65: Cross-sectional view of the TerraSAR-X1 spacecraft

RF communications: A standard S-band TT&C System with 360° coverage in uplink and downlink (assuming switching) is used for satellite command reception and telemetry transmission. Generated payload (SAR) data are stored on-board in a Mass Memory Unit of 576 Gbit capacity prior to transmission via an X-band downlink at a data rate of 300 Mbit/s. The X-band downlink is encrypted. The on-board SAR raw data are compressed using the BAQ (Block Adaptive Quantization) algorithm, a standard SAR procedure. The compression factor is 2 (more efficient techniques can only be applied to processed SAR imagery). Both communication links are designed according to the ESA CCSDS Packet Telemetry Standard.

The Svalbard X/S-band station is considered the prime TT&C and payload acquisition station in the ground segment.

Sensor complement:

XSAR (X-band SAR instrument). XSAR is an active phased array X-band system providing high-resolution and multipolarization SAR imagery (H and V). The body-fixed SAR antenna has dimensions of 6.5 m (length) x 0.85 m (width) and 0.30 m in depth. The mass of XSAR is about 490 kg (370 kg of antenna mass and about 120 kg for instrument electronics accommodated within the service module). The antenna features 512 T/R (Transmit and Receive) modules each capable of operation in two polarizations. Beam steering is provided in azimuth (± 0.6°) and elevation (± 22°). The inherent flexibility due to amplitude and phase settings in the transmit/receive modules allows individual beam pattern pointing and shaping. This flexibility is mainly used in the elevation direction (cross-track) over the total range of incidence angles. A variable FOR (Field of Regard) is provided depending on support mode.

Legend to Table 116: NESZ= Noise Equivalent Sigma Nought. NESZ is that value of sigma nought of a uniform scene that produces a processor output signal to noise of unity. The value XXX is a swath length selectable up to 1600 km. The "access range" is the cross-track accessible (viewable) range on the ground provided by electronic beam steering. The instantaneous range of XSAR is e.g. only 40 km in stripmap mode and 100 km for ScanSAR mode.

Parameter/ Support Mode	Spotlight Mode 1	Spotlight Mode 2	StripMap Mode	ScanSAR Mode
Spatial resolution - Cross-track - Along-track	 1.2 m (slant range) 1 m	 1.2 m (slant range) 2 m	 <3 m 3.2 m	 15 m 15 m
Product coverage Along x cross track	 5 km x 10 km	 10 km x 10 km	 XXX km x 40 km	 XXX km x 100 km
Access range (FOR) (full performance)	20° - 55° 570 km	20° - 55° 570 km	20° - 45° 350 km	20° - 45° 350 km
Access range (FOR) (Data collection)	15° - 60° 760 km	15° - 60° 760 km	15° - 60° 760 km	20° - 60° 700 km
Sensitivity (NESZ) - Typical - Worst case	 -23 dB -19 dB	 -23 dB -19 dB	 -22 dB -20 dB	 -21 dB -19 dB
Source data rate, mean (8/4 BAQ)	415 Mbit/s	415 Mbit/s	816 MBit/s	816 Mbit/s

Table 116: Performance overview of the XSAR instrument

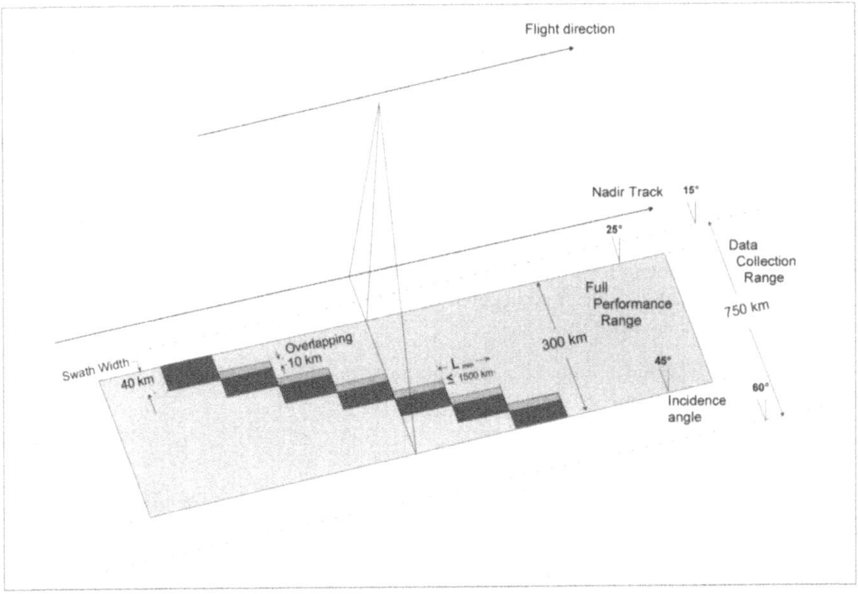

Figure 66: Coverage concept of the stripmap mode operation of XSAR

XSAR (X-band SAR instrument)	Antenna: active phased array
	X-band center frequency of 9.6 GHz (3.1 cm), chirp bandwidth: 150 MHz
	Polarization: HH/VV/HV/VH (selectable or twin)
	Beam steering in azimuth (± 0.6^o) and elevation (± 22°)
	512 T/R modules; PRF= 2.4 - 5.0 kHz
	Antenna dimensions: 650 cm x 85 cm x 30 cm (L x W x H)
	Body-fixed antenna: 370 kg antenna mass, 120 kg electronics
	Nominal radar duty cycle: 13% (20% max)
	Data quantization: 8 bit I, 8 bit Q
	Instrument power: 3.650 kW (for imaging)

TerraSAR-X1 spacecraft (Flexbus)	S/C wet mass: 1285 kg (bus=713 kg, payload=490 kg, propellant of 85 kg)
	S/C dimensions: 2.5 m (H) x 6.5 m (L) x 2.6 m (W)
	S/C 875 W of orbit average power (EOL); peak power=1.6 kW; energy storage of 40 Ah capacity (26 NiH cells)
	Power distribution: 55-83 V unregulated bus; converter to 28 V and converter to 115 V 30 kHz AC for XSAR front end
	S/C pointing control: 30 arcsec
	Communication: X-band of 300 Mbit/s payload data downlink with QPSK; S-band uplink of 4 kbit/s (2025-2110 MHz), BPSK modulation; S-band downlink of 32 kbit/s to 1 Mbit/s (2200-2400 MHz), BPSK modulation

Table 117: Overview of X1 spacecraft and instrument characteristics

B.7.2 TerraSAR-L1 (L-band Satellite)

The TerraSAR L-band satellite (L1 is led by Astrium Ltd. of Portsmouth, UK) uses a deployable Snapdragon platform design with a total wet mass of about 2600 kg. The L-band SAR antenna is large, at 11m x 2.5 m and presents formidable launcher accommodation problems if configured as a conventional deployable array. The Snapdragon platform of Astrium Ltd. elegantly avoids this problem by arranging the SAR array in two parts across the launcher fairing diameter, providing the deployment and deployed structure functions within the platform itself. In this way, a medium (Delta-2) class launcher can be selected, largely on a mass basis, rather than a large (Ariane V) class launcher on a stowed volume basis. Figure 68 illustrates the key deployment feature of the Snapdragon spacecraft, which is covered by European and US patents. 546)

S/C power is provided by solar arrays (GaAs) on each Snapdragon half totaling 1720 W (EOL). The arrays are fixed but could be made deployable (steerable array not required due to the dawn/dusk orbit) for an optimum solar radiation aspect. The S/C voltage is 33 V, the instrument operates at 50 V. Lithium Ion batteries are used: 42 Ah/33 V for the subsystems of the platform; 29 Ah/50 V for the leading side of the hingeline; and 21 Ah/50V on the trailing half. The S/C is three-axis stabilized. The attitude control system is based on a four-axis set of momentum wheels for fine-pointing, with magneto-torquers for desaturation. A propulsion system is only used for orbit control and end-of-life de-orbiting. Attitude and position measurement is performed with a star tracker system coupled with a solid-state inertial reference unit, a three axis magnetometer and GPS. A pointing accuracy of <1 arcmin is predicted, which is sufficient to achieve the L-SAR objectives. The S/C design life is 5 years of operation with a goal of 6.5 years (de-orbiting is planned at end of the useful life).

A launch of TerraSAR-L1 is planned for 2006. A Soyuz-Fregat ST, launched from the Baikonur Cosmodrome, is the current baseline launch vehicle.

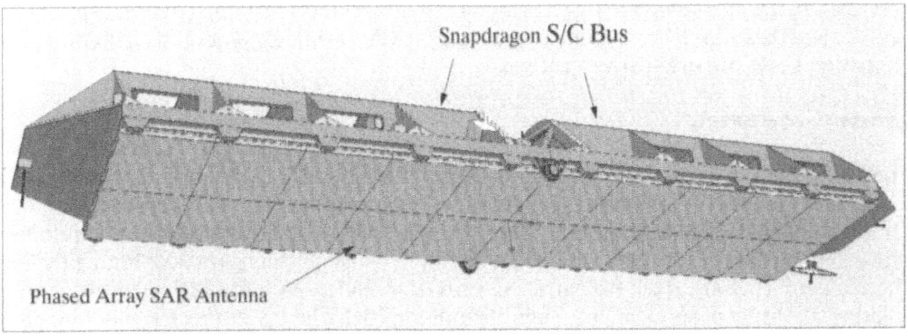

Figure 67: Illustration of the TerraSAR-L1 S/C (shown without thermal blankets)

546) Information provided by David M. Simpson of Astrium Ltd.

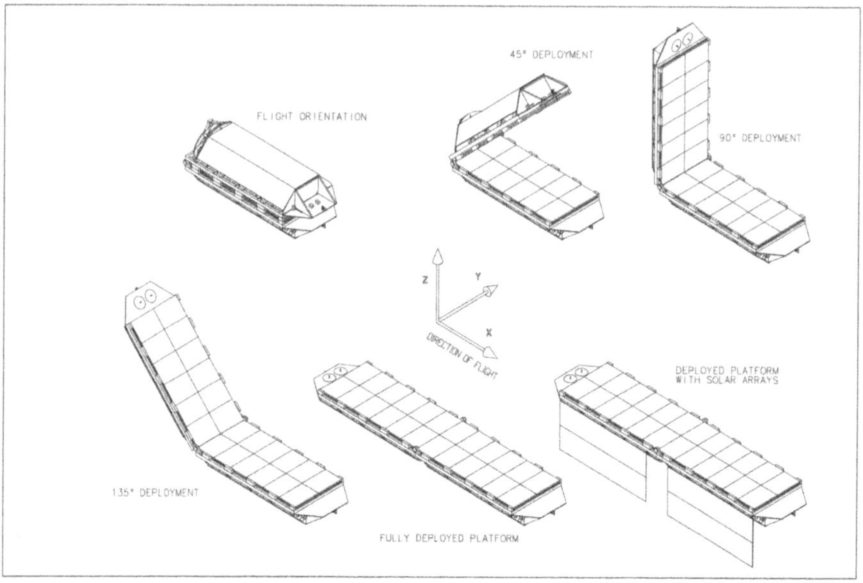

Figure 68: Separation deployment sequence of a Snapdragon spacecraft

Orbit: Sun-synchronous circular dawn-dusk orbit (local time ascending node at 18:00 hours), altitude = 629 km, inclination = 98°; orbital period of about 97 minutes; revisit period of 18 days with 3-4 days of sub-repeats. Global coverage within the latitudes ± 80°. The TerraSAR-L1 spacecraft follows the same orbital track as the X1 spacecraft, which means that the two are displaced in crossing time by about 6.25 minutes. Otherwise the orbital parameters are identical.

At EOL, S/C reentry is initiated with thruster firings (using budgeted fuel) to obtain an elliptical orbit with a perigee of about 300 km or lower. This generates sufficient atmospheric drag to bring the S/C eventually into a reentry trajectory.

RF communication: A standard TT&C S-band system is used for spacecraft and payload services. Instrument data are stored on-board in a 384 Gbit (beginning of life) capacity recorder and downlinked in X-band at a rate of 300 Mbit/s with encryption via a uniform illumination hemispherical pattern antenna.

Sensor complement:

LSAR (L-band SAR instrument), designed and developed by Astrium Ltd., Portsmouth, UK. LSAR is an active phased array L-band system providing fine resolution and fully polarimetric imagery. The antenna is of size: 11 m x 2.5 m or 27.5 m^2. It is segmented into two rows and ten columns of tiles, where tiles are interchangeable units (of ASAR architecture heritage on Envisat). Each tile carries seven T/R modules, each of which feeds its own radiating sub-array of six elements. Discrete units on each tile service the T/R modules, providing power, control, and temperature compensation processing. Antenna beam steering in elevation (cross-track) is achieved with the use of 14 rows of T/R modules (each rows has 10 T/R modules) with independent phase control. The instrument calibration scheme is implemented entirely within the T/R modules. Single, dual, and quad polarization is achieved.

Center frequency	1.2575 GHz (L-band) corresponding to a wavelength of 24 cm
Nr of sub-arrays	140 arranged as 20 panels with 7 sub-arrays per panel
Polarization	Horizontal (H) and Vertical (V) independently selectable for transmit, simultaneous on receive
Chirp bandwidth	55 MHz
Receive channel bandwidth (Rx)	27.5 MHz, 55 MHz selectable for adaptation to incidence angle
T/R module output power	46 dBm
Quantization of radar signal	8 bit for I and Q
SAR data compression	Online BAQ (Block Adaptive Quantization), 8:4, 8:3, 8:2 and 8:1 selectable
Data storage capacity	384 Gbit
Output data rate to X-band	300 Mbit/s encrypted and formatted (CCSDS) data, simultaneous to data input at max. rate

Table 118: Summary of LSAR instrument parameters

Parameter/ Support Mode	Quad StripMap	Dual StripMap	Low resolution L (Dual ScanSAR 200 km)	Low resolution XL (Dual ScanSAR 100 km)
Polarization	Alternating H&V on Tx; Co&cross pol on Rx	Single pol. on Tx; Co&cross pol on Rx	Single pol. on Tx; Co&cross pol on Rx	Single pol. on Rx Co&cross pol on Rx
Range resolution	9 m	9 m	30 m	30 m
Swath width	40 km	60 km	200 km	100 km
Sensitivity	-30 dB	-30 dB	-28 dB	-28 dB
DTAR	-17 dB	-17 dB	-17 dB	-17 dB
Field of Regard (FOR)	20° - 32°	20° - 45°	23° - 45°	20° - 45°

Table 119: Summary of instrument performance by support mode

Legend to Table 119:

- DTAR: Distributed Target Ambiguity Ratio (a combination of azimuth & range ambiguities)
- Dual StripMap: In conventional SAR operation the "transmit" pulse occurs on one polarization (e.g. H) while the "receive" echo signal is measured in both polarizations, H and V.
- Quad StripMap: The transmit pulses are alternatively H and V polarized. The receive function is in both polarizations, H and V simultaneously, thus providing polarimetric measurements (HH, HV, VV, VH), permitting the removal of the atmospheric Faraday rotation effect as well as enhanced land-use classification capabilities.
- Low resolution L: This mode provides a wide swath (L-band data only)
- Low resolution XL: This support mode offers almost co-temporal measurements (about 6 minutes apart) of the same location (target) by the L-band and X-band SAR satellite tandem. Individual swath widths are constrained by the narrowest of the two instruments, which is the XSAR; hence, a narrower total swath width is resulting (when both instruments are operating using same swaths/bandwidths etc..). Almost co-temporal data at two frequencies far apart in spectrum enables unique applications.

The LSAR subsystem ICE (Instrument Central Electronics) uses the existing "core radar" architecture (Note: core radar is the name of an Astrium product. The name comes from the idea that the same "core architecture" can be applied to various radar missions, not necessarily to SAR or to space for that matter.) as developed for Radarsat-2 and modified to L-band with the extended front-end subsystem control. A single transmit channel (without chirp switching) is employed with two receive channels. The core architecture uses a CAN (Controller Area Network) bus supplemented with timing strobes and SSB (Setting Selector Bus), allowing a relatively low-speed instrument control and monitoring bus when combined with ASICs to the T/R modules. A MIL-STD-1553B bus interface is used from LSAR to the platform.

The LSAR instrument has a mass of 925 kg and a max. operating power consumption of 3.517 kW. This high power demand results in LSAR operation limitations to about 9.5 minutes (max)/orbit (in an period of about 97 minutes), or a maximum duty cycle of about 10%.

B.8 OrbView Satellite Series of ORBIMAGE

OrbView® is the name of an imaging satellite series of Orbital Image Corporation (ORBIMAGE) of Dulles, VA, an affiliate of Orbital Sciences Corporation (OSC). The objective of this commercial satellite series is to acquire affordable high-quality imagery of the Earth for a variety of customers that include local governments, telecommunication companies, architects, civil engineers, real estate managers, farmers and environmental monitoring agencies.

ORBIMAGE renamed its satellites in 1997, affecting S/C in development and in operational service. A renaming list is provided to avoid confusion depending on what references are cited of the available literature.

New S/C Name of ORBIMAGE as of 1997	Launch Date of S/C	Remarks/Sensors	Old S/C Name (given in literature)
OrbView-1	April 1, 1995	GPS/MET, OTD	Microlab-1
OrbView-2	August 1, 1997	SeaWiFS	SeaStar
OrbView-3	2002	commercial imaging PAN +MS	Not applicable
OrbView-4	Sept. 2001	commercial imaging PAN+MS+HS	OrbView-3

Figure 69: ORBIMAGE change of S/C naming

B.8.1 OrbView-1

OrbView-1 is an imaging satellite managed and operated by Orbital Imaging Corporation (ORBIMAGE) of Dulles, VA. ORBIMAGE and is a commercial provider of Earth imagery acquired from a family of imaging satellites that it owns and operates. The OrbView-1 satellite includes two sensor systems, OTD (Optical Transient Detector) a lightning imaging system, and GPS/MET (GPS/Meteorology), an atmospheric measurement system.

OrbView-1 was built and launched on April 3, 1995 by OSC. The mission design life is two years with a goal of four years. The primary payload of the OrbView-1 spacecraft is the OTD (Optical Transient Detector), a lightning instrument package provided by NASA/MSFC. The secondary payload on OrbView-1 is GPS/MET (GPS/Meteorology), an instrument provided by UCAR (University Consortium for Atmospheric Research, with collaboration by JPL, Stanford University, and U. of Arizona) which uses GPS signals to measure Earth atmospheric properties. Science data processing of OTD and GPS/MET is done at MSFC and UCAR, respectively. OrbView-1 mass = 69 kg, power = 42 W average, data rate = 2 Mbit/s (one OSC tracking station support). The S/C shape in orbit resembles a butterfly (a cylinder of 1 m diameter and 37 cm in height, whose covers open like the wings of a bird; the gravity gradient boom points toward the Earth's center). [547] [548]

Orbit: circular orbit, altitude = 785 km, inclination = 70°. The attitude control system is an active magnetic gravity-gradient system for nadir pointing within ±5° (1 σ). The nominal pointing accuracy of OrbView-1 pointing knowledge is ±2° (1σ), depending on the model-

547) Information provided by Mark Pastrone of ORBIMAGE
548) Information provided by G. Moody and D. Finn of OSC, and by W. J. Koshak of NASA/MSFC

ing accuracy of the Earth's magnetic field. The short-term pointing stability is 0.10°/s. Yaw steering is used to maximize sun exposure of solar arrays.

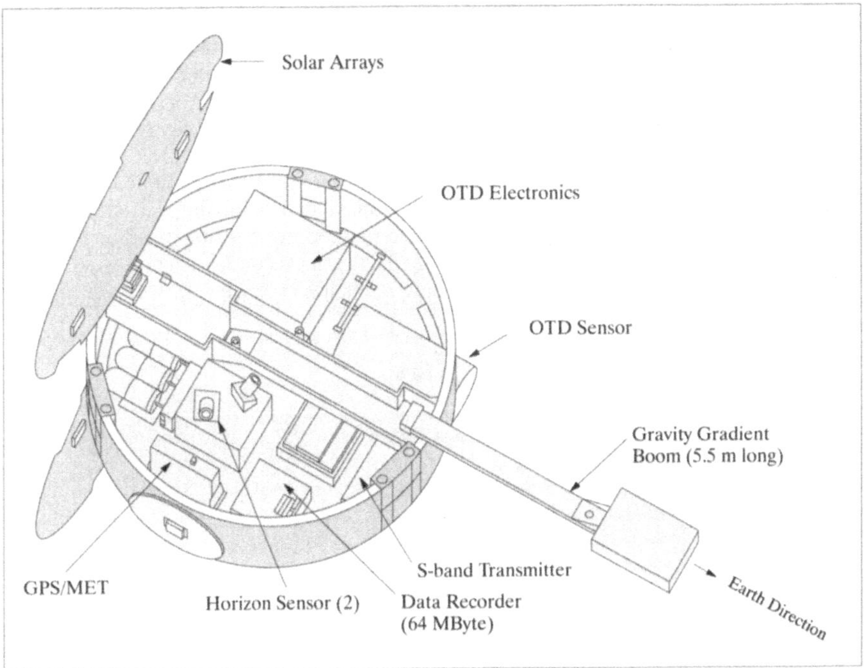

Figure 70: The OrbView-1 S/C model

Operational status as of June 2001: OrbView-1 continues to operate nominally after six years in orbit. The satellite has far surpassed its original design life of three years.

Sensor complement:

OTD = Optical Transient Detector (NASA/MSFC instrument). [549] Objectives: observation of the global distribution of lightning, leading to the formation of a climatic database of the spatial and temporal distribution of thunderstorms and lightning. The goal is to better understand thunderstorm activity on a global scale. OTD is a staring imager for the detection of lightning over a large region of the Earth's surface with storm scale resolution; it marks the time of occurrence and location of the lightning and measures its radiant energy. Both intracloud and cloud-to-ground discharges can be detected during daytime and nighttime conditions. OTD images a scene much like a television camera; however, daytime detection of highly transient lightning sources against a bright cloud-top background makes actual data handling and processing much more involved than that required by a simple imager. The OTD instrument is actually the LIS engineering model of TRMM.

OTD is composed of six major subsystems: an imaging system, a focal plane assembly (including a CCD array detector, preamplifiers, and multiplexers), a Real-Time Event Processor (RTEP) and background remover, an event processor and formatter, power supply, and interface electronics. The imaging system is a simple telescope consisting of a beam expander, an interference filter, and reimaging optics. The filter is narrow band (8.4Å) and is cen-

549) _http://thunder.nsstc.nasa.gov/otd/

tered about a prominent neutral oxygen emission triplet in the lightning spectrum to optimize SNR in the presence of a bright solar-lit cloud top.

OTD parameters: view direction: nadir; FOV = 80° x 80°; spatial resolution = 10 km; temporal resolution = 2 ms; wavelength = 777.4 nm; sensor mass = 2 kg; power = 3 W.

GPS/MET = GPS Meteorology (proof-of-concept system). Objective: global remote sensing of the atmosphere using the radio occultation measurement method. GPS/MET is a modified commercial high-precision GPS receiver (TurboRogue of Allen Osborne Associates, Westlake Village, CA) with the capability of simultaneously receiving GPS signals on L1 and L2 frequencies by tracking up to eight GPS satellites. The receiver can track both C/A and P codes and also implements a codeless carrier recovery technique. The accurate measurement in the change of carrier phase permits to determine the atmospheric refractive index as a function of altitude. Pressure and temperature profiles can then be derived using the gas law. All observables are sampled 50 times per second. A single low-gain antenna (a standard 'patch' on a ground plane) points roughly in the anti-velocity direction. [550) 551) 552) 553)]

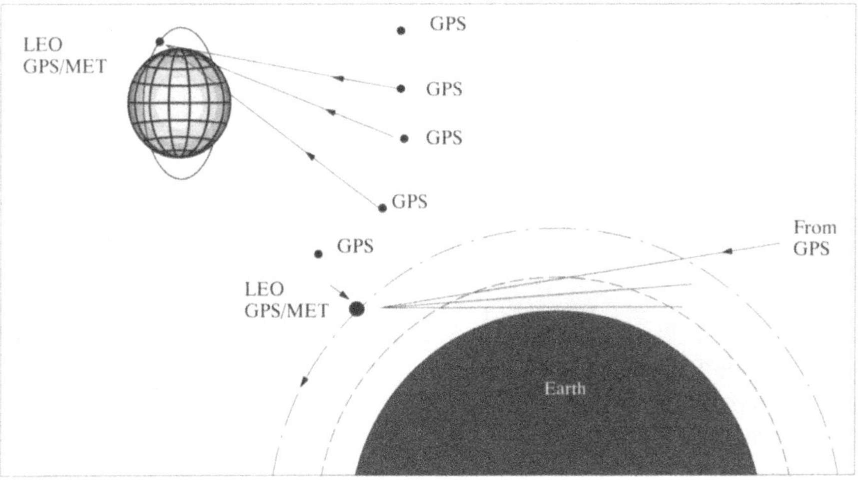

Figure 71: Observational geometries of the GPS/MET occultation measurements

High resolution atmospheric soundings can be retrieved when the radio path between the LEO GPS receiver and one GPS satellite traverses the Earth's atmosphere. When the path of the GPS signal begins to transect the mesopause at about 85 km altitude, it is sufficiently retarded so that a detectable delay in the order of 1 mm in the dual frequency carrier phase observations is obtained by the LEO GPS receiver. As the signal path descends through successively denser layers of the atmosphere, the delay increases to approximately 1 km at the Earth's surface. Thus the atmosphere creates a unique signal with over six orders of magnitude in dynamic range. A single LEO GPS receiver can observe about 200 such occultations per day. The key observables are atmospheric temperature and moisture distributions. GPS/MET measurement resolutions:

550) W. G. Melbourne, et al., "The application of spaceborne GPS to atmospheric limb sounding and global change monitoring." JPL Publication 94-18, 147 pp.

551) E. R. Kursinski, et al., "Observing Earth's atmosphere with radio occultation measurements using the Global Positioning System," Journal of Geophysical Research, Vol. 102, No. D19, Oct. 20, 1997,pp. 23,429-23,465

552) S. S. Leroy, "The Measurement of Geopotential Heights by GPS Radio Occultation. Journal of Geophysical Research, Vol. 102, No. D6, March 27, 1997, pp. 6971-6986

553) R. Ware, M. Exner, et al., "GPS Sounding of the Atmosphere from Low Earth Orbit: Preliminary Results. Bulletin of the American Meteorological Society, Vol. 77, 1996, pp. 19-40

- Precision of phase measurements: order of 1 mm
- Velocity accuracy: 10^{-4} m/s
- Position accuracy: 1 m (vertical: < 1 km; horizontal: < 200 km)

GPS/MET temperature retrieval accuracies of better than 1K have been demonstrated in the altitude range of 10-30 km during the proof-of-concept phase of the mission with vertical resolution of better than 1 km.

B.8.2 OrbView-2 (renamed from SeaStar in 1997)

OrbView-2 is an imaging satellite managed and operated by Orbital Imaging Corporation (ORBIMAGE) of Dulles, VA. The OrbView-2 satellite includes the SeaWiFS imaging system, an 8-channel multispectral imaging instrument with a one kilometer spatial resolution.

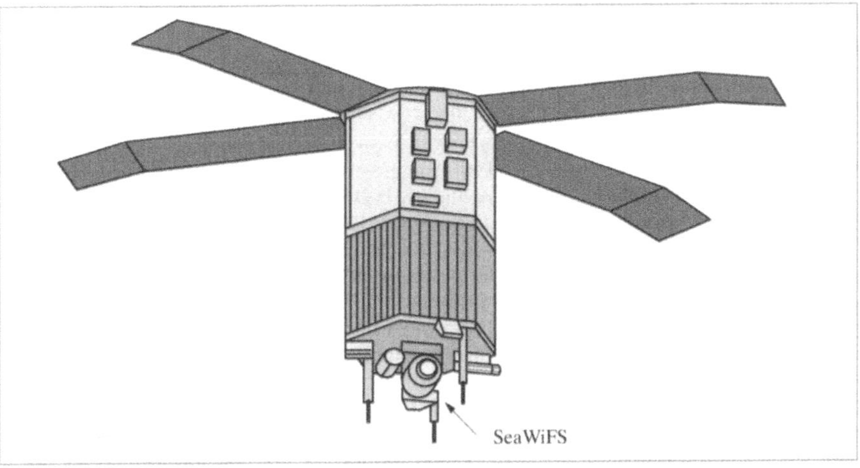

Figure 72: The OrbView-2 S/C model

The S/C design is based on PegaStar, a multi-purpose bus of OSC. The platform combines functions common to both satellites and launch vehicles, including guidance and control, power, communications, and data systems. The S/C is three-axis stabilized (0.5° with 0.08° knowledge) using three orthogonal magnetic torque rods for roll and yaw control and two momentum wheels for pitch stabilization. Attitude is sensed with three 2-axis sun sensors, two scanning horizon sensors, and two 2-axis magnetometers. Redundant GPS receivers are used for orbit determination. The propulsion system consists of two subsystems, a reaction control system and a hydrazine propulsion system. Nominal design life = 5 years. S/C mass = 390 kg. A hydrazine propulsion system using four thrusters is used for orbit raising and orbit maintenance. Four deployed solar panels with zenith-facing cells and two body-mounted side-facing solar panels produce 165 W orbit-average power after 5 years. 554) 555) 556) 557)

The OrbView-2 S/C was air-launched on August 1, 1997 by a Pegasus XL rocket (also of OSC) from Vandenberg AFB, CA. The OrbView-2 satellite operations as well as the commercial data processing and distribution functions are performed at ORBIMAGE. The

554) "Orbital Sciences Captures $120 Million in Business, Pegasus Launches Ocean Satellite Ordered," Space News, March 11-17, 1991, p. 7

555) "OSC Reviews Seastar Design," Space News, Oct. 28 - Nov. 3, 1991, p. 22

556) P. R. Leygraaf, "OrbView-2 (SeaStar) Flight Operations and Data Delivery," Proceedings of the 11th AIAA/USU Conference on Small Satellites, Sept. 15-18, 1997, Logan, UT

557) http://seawifs.gsfc.nasa.gov/SEAWIFS/SEASTAR/SPACECRAFT.html

processing and distribution of data for research applications is handled by the SeaWiFS Project at NASA/GSFC.

Application: Ocean-color data, ocean biology and ecology, phytoplankton concentrations and growth, pollution, algae blooms, etc. The data may help scientists to understand the role of ocean plant life in the Earth's carbon cycle.

Orbit: Sun-synchronous polar orbit, altitude = 705 km, inclination =98.2°, equator crossing time at local noon (descending node), successive orbit equatorial crossing longitude = -24.721°, period = 98.2 minutes, orbital repeat time = 16 days (233 orbits)

Operational status as of June 2001: OrbView-2 continues to perform exceedingly well on-orbit. For 2000, it achieved a 98.7% data availability rate and has achieved a 100% data availability rate for the first quarter of 2001. Because of its superb on-orbit performance, ORBIMAGE is developing plans for an extended mission beyond the original five-year mission which ends August 2002 .

Sensor:

SeaWiFS (Sea-Viewing Wide Field-of-View Sensor) built by Raytheon SBRC of Goleta, CA. [558) 559)] Objective: measurement of reflected visible sunlight from the ocean surface for the derivation of ocean-color data. The SeaWiFS sensor may be regarded as a next generation instrument of CZCS (Nimbus-7 sensor which ceased operations in 1986, after an eight-year mission). The SeaWiFS instrument employs scanning mechanisms to drive an off-axis (afocal) folded-optics scanning telescope and a rotating half-angle mirror that is phase-synchronized with, and rotating at half the speed of, the folded telescope. The scanning telescope design minimizes the polarization sensitivity and, correspondingly, maximizes the radiometric fidelity of its ocean color measurements. The SeaWiFS scanning concept employs a 360° uni-directional cross-track rotating telescope scanner. A small counter-rotating flat scan mirror between the telescope and the focal planes eliminates image rotation which would otherwise occur as the telescope rotates. The absence of FOV rotation permits the use of a multichannel, time delay integration (TDI) processing in each of the eight spectral bands to achieve the required SNR. Incoming scene radiation is collected by the folded telescope and reflected onto the rotating half-angle mirror. The collected radiation is then relayed through dichroic beam splitters to separate the radiation into four wavelength intervals (each wavelength encompassing two of the eight SeaWiFS spectral bands). The radiation in the four separate wavelength intervals is directed by four corresponding aft-optics assemblies through two separate spectral bandpass filters that further separate the radiation into the eight required SeaWiFS spectral bands. The aft-optics also image each of the resultant bands of radiation onto four silicon detectors that are aligned in the scan direction. The detected signals are then amplified for TDI processing in the electronics module.

Radiometric accuracy	< 5% absolute each band
Calibration/stability monitor	solar diffuser, lunar view
Relative precision	< 1% linearity of signal output to radiance
Between-band precision	< 5% relative band/band over 50 - 90% of saturation
Polarization sensitivity	< 2% worst case, all scan and tilt angles
Dynamic range	15,000:1 using bilinear gain
Bright target recovery	< 10 samples
Location knowledge	0.5 km, at 1-sigma level for instrument

Table 120: SeaWiFS parameters

The off-axis telescope (with an aperture of 7.6 cm; f/2) rotates at six revolutions per second (or 360 rpm) in the cross-track direction (for HRPT format compatibility) to provide con-

558) "System Concept for Wide-Field-Of View Observations of Ocean Phenomena from Space," NASA-NOAA-Eosat publication, 1987

559) H. v.d. Piepen, V. Amman, R. Doerffer, "Remote Sensing of Substances in Water," GeoJournal 24.1, pp. 24-27, 1991 (May) by Kluwer Academic Publishers

tiguous scan coverage. Data quantization = 10 bit; spatial resolution = 1.6 mrad (1.13 km at nadir); an active scan angle of ±58.3° is used. A scanner tilt mechanism enables the entire sensor to be oriented in the along-track direction to +20°, 0°, or -20° to avoid specular sun reflection (sun glint from the sea surface). Instrument mass = 45 kg; design life = 5 years.

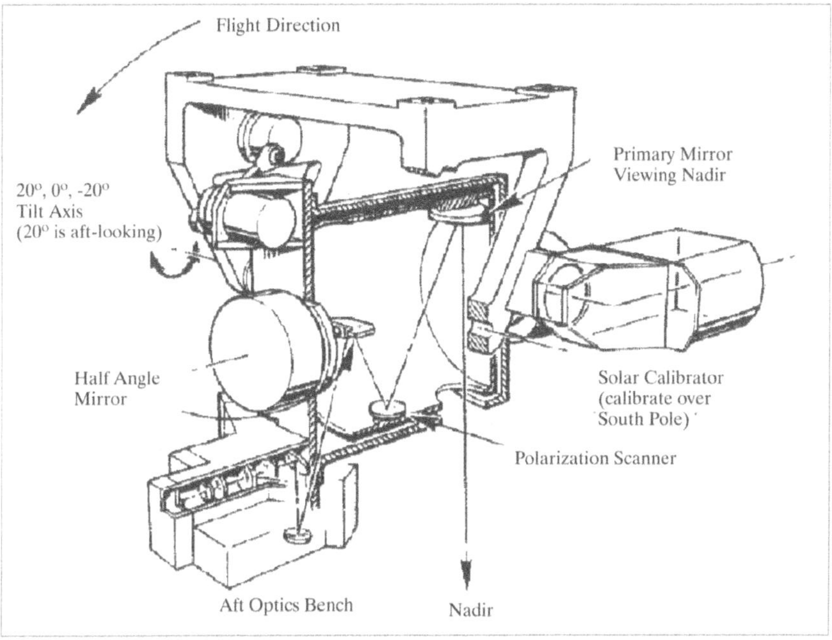

Figure 73: Schematic illustration of the SeaWiFS sensor

Calibration: SeaWiFS uses short-period solar calibration (for a few orbits) and long-term lunar calibration (for a few months and longer). Solar calibration employs a solar radiation diffuser and an input port located in a fixed position outside of the 58.3° SeaWIFS scene-scan interval. Lunar calibration is accomplished by a S/C maneuver to view the moon during the nighttime portion of an orbit. Of particular importance to the postlaunch validation period are field measurements from MOBY (Marine Optical Buoy, see P.139) located off the coast of Lanai, Hawaii. [560) 561) 562)]

Data: [563)]
On-board storage of sensor data is provided. With two dumps/day, this allows global observations at reduced spatial resolution to be recorded, and a limited amount (about 20 min/day) of high-resolution Local Area Coverage (LAC) data. The real-time LAC data stream is merged with S/C health and instrument telemetry at a rate of 665.4 kbit/s; this is transmitted at L-band with a frequency of 1702.56 MHz. To increase the coverage of LAC data, NASA encourages the operation of HRPT (High Resolution Picture Transmission) stations by the user community throughout the world. These HRPT stations can collect real

560) G. Valenti, "Sea-viewing Wide Field-of-view Sensor," The Earth Observer, March/April 1998, Vol 10, pp. 20-22
561) Note: the reflected radiance is related to the concentration of chlorophyll and other plant pigments present, since chlorophyll is a green pigment and the color of the water changes from blue to green as the concentration of chlorophyll increases. If the concentration of chlorophyll is known, the amount of phytoplankton, or 'ocean color,' may be calculated.
562) SeaWiFS Project Home Page at http://seawifs.gsfc.nasa.gov/SEAWIFS.html
563) "Roles and Responsibilities of HRPT Stations for SeaWiFS," SeaWiFS Project Office, GSFC, Dec. 19, 1991

time LAC data via direct broadcast whenever the spacecraft is in view. The other telemetry stream consists of stored GAC and selected LAC data, along with S/C health and instrument telemetry, at 2.0 MBit/s; this is transmitted at S-band with a frequency of 2275.5 MHz.

Sensor Parameter	GAC Mode (Global Area Coverage)	LAC Mode (Local Area Coverage)
Scan width (degree)	45°	58.3°
IFOV at sensor (mrad)	1.5835	1.5835
Ground IFOV at nadir (km) - spatial resolution	4.5	1.13
Ground swath width, km	1500	2800
Pixels along-scan	248	1285
Pixel numbers	(147 - 1135, by 4)	(1-1285)
Scans/second	1.5	6
Scan plane tilt	20°	20°

Table 121: SeaWiFS sensor/transmission characteristics

Band	Wavelength (nm)	Measurement Parameter	Saturation Radiance (mW/cm^2)	Input Radiance (mW/cm^2)	SNR (Measured at Input Radiance)
1	402 - 422	Gelbstoffe	13.63	9.10	499
2	433 - 453	Chlorophyll absorption	13.25	8.41	674
3	480 - 500	Pigment concentration	10.50	6.56	667
4	500 - 520	Chlorophyll absorption	9.08	5.64	640
5	545 - 565	Sediments/hinge point	7.44	4.57	596
6	660 - 680	Atmospheric aerosols	4.20	2.46	442
7	745 - 785	Atmospheric aerosols	3.00	1.61	455
8	845 - 885	Atmospheric aerosols	2.13	1.09	467

Table 122: SeaWiFS spectral performance summary

To protect the commercial interests of the builder and operator of the mission, ORBIMAGE, SeaWiFS real time LAC data are encrypted. All data users require a license from OSC (a decryptor is furnished with the license). Commercial data users pay a fee for their license, research licences are issued at no cost, but are subject to certain restrictions. Commercial stations requesting HRPT stations for research data use must obtain prior approval from NASA (SeaWiFS Project Office at GSFC).

Satellite transmitting power	5 Watt
HRPT L-band frequency	1702.56 MHz
Frequency stability	± 20 ppm
Modulation	Split-Phase
Modulation rate	0.6654 Mbit/s
Satellite transmitter antenna loss	2.0 dB
Satellite antenna gain	1.8 - 2.1 dBiC
EIRP	31 - 37 dBm
Slant range	2900 km
Free space loss	166.1 dB
Fading and rain margin	5 dB

Table 123: Requirements for a SeaWiFS LAC downlink ground station HRPT

B.8.3 OrbView-3

OrbView-3 is an imaging satellite owned and operated by Orbital Imaging Corporation (ORBIMAGE) of Dulles, VA, a commercial provider of Earth imagery acquired from a family of imaging satellites that it owns and operates. The S/C includes one imaging camera, the OrbView High Resolution Imaging System capable of collecting one meter resolution panchromatic and four meter resolution multispectral imagery.

The spacecraft of the OrbView-3 mission uses the flight-proven OSC bus design. The S/C body consists of a cylinder of about 1 m in diameter and 2.3 m in length. A solar panel is mounted to the "top side" of the cylinder. OrbView-3 is three-axis stabilized providing a real-time pointing accuracy of better than 100 arcseconds (knowledge). The post-processed pointing accuracy allows pixels to be located to within 12 m for certain product types. The ACS (Attitude Control Subsystem) employs Orbital-built gyroscope, star sensors, sun sensors, and a magnetometer for attitude sensing, and four reaction wheels (3 always active, one is a cold spare) and thrusters for all actuation functions. The S/C has the capability to perform body-pointing of ±50° into the in-track and cross-track directions thereby increasing the FOR (Field of Regard) for all observations. The S/C mass is about 300 kg, the design life is five years. [564)]

Figure 74: OrbView-3 spacecraft

OrbView-3 is planned to be launched in 2002 on an OSC launch vehicle from VAFB.

Orbit: Circular sun-synchronous orbit, altitude = 470 km, inclination = 97.3°, equator crossing at 10:30 AM on a descending node. The revisit capability is 3 days (depending on latitude).

Data: On-board storage capability of 32 Gbit (the onboard storage is available for customers requiring imagery of regions where receiving stations are unavailable). Real-time downlink in X-band at 150 Mbit/s.

Sensor complement:

OHRIS (OrbView High Resolution Imaging System) of ORBIMAGE designed and built by Northrop Grumman, Baltimore, MD. The objective is to provide global high-resolution imagery (1 m panchromatic and 4 m multispectral) on a commercial basis. OHRIS is identical to the instrument flown on OrbView-4. OHRIS is an optoelectronic instrument based on a three-mirror anastigmatic telescope design. The nominal scene size of the imagery is 8 km x 8 km with a spatial resolution of 1 m (Pan) and 4 m (MS) at nadir. The source data is generated with 11-bit quantization and compressed to 2 bits/pixel.

564) Information provided by Mark Pastrone of ORBIMAGE

Imaging Mode	Panchromatic	Multispectral (MS)
Spatial resolution	1 m	4 m
Imaging bands	1	4 MS
Spectral range	Pan: 450 - 900 nm	MS1: 450-520 nm MS2: 520-600 nm MS3: 625-695 nm MS4: 760-900 nm
Silicon detector size (linear array)	8000 pixels	2000 x 4

Table 124: OHRIS instrument parameters

Ground segment: The ORBIMAGE Operations Center (OOC) is located in Dulles, VA providing full-time service. The OOC generates the satellite tasking commands, which are uplinked to the OrbView satellites from two ground stations; one located in Point Barrow, Alaska and the other in Dulles, Virginia. These same ground stations receive and route the telemetry and imagery data downlinked from the satellite back to the OOC for processing and distribution. The imagery data routed to the OOC is processed, archived, and distributed to customers. In parallel, ORBIMAGE franchises these service tasks (data reception, archiving, product generation, distribution, requests for regional image acquisition scheduling, etc.) out to official regional distributors or to large-volume customers. ORBIMAGE offers several license options for its distributors and large-volume users with regard to their satellite tasking capabilities and priorities in their respective regions of coverage.

B.8.4 OrbView-4

OrbView-4 is a imaging minisatellite owned and operated by ORBIMAGE of Dulles, VA. The S/C includes two imaging instruments, OHRIS and OHIS (OrbView Hyperspectral Imaging System). The satellite is designed and built by Orbital Sciences Corporation. [565)] The spacecraft structure (body) consists of a six-sided cylinder of 102 cm in diameter and 152 cm in length. Orbview-4 is three-axis stabilized; attitude is sensed by star trackers, sun sensors, a magnetometer, and a gyro; actuation is provided by four reaction wheels (3 are always active, one is a cold spare) and thrusters. The attitude system provides a real-time pointing accuracy of better than 100 arcseconds. The post-processed pointing accuracy allows pixels to be located to within 12 m for certain product types. An agile body-pointing technique, using reaction wheels, is employed for instrument pointing, i.e., the entire S/C is pointed into the desired direction, permitting a field of regard of $\pm 50^o$ into any direction. A GPS receiver provides orbit determination and all on-board timing services. Electrical power is provided by six solar panels, mounted in a star configuration on the cylinder base. In addition, NiH batteries serve as a power source for eclipse operations. The satellite mass is about 360 kg, the design life is five years. The satellite is planned to be launched on a Taurus vehicle in September 2001 along with QuikTOMS of NASA as a secondary payload from VAFB. [566)]

An on-board recorder of 32 Gbit capacity provides storage for non-contact periods. RF communications of the imagery is provided in X-band. The downlink data rates is 150 Mbit/s. Mission operations are provided by ORBIMAGE via its Dulles-based operations center and two uplink/downlink ground stations located in Point Barrow Alaska and Dulles, Virginia. Data distribution is provided by ORBIMAGE and its worldwide Regional Distributors.

Orbit: Sun-synchronous circular orbit, altitude = 470 km, inclination = 97.3^o, equator crossing on descending node 10:30 AM. The large off-nadir slewing capability of Orbview-4 offers a worst-case revisit time of three days.

565) Information provided by Mark Pastrone of ORBIMAGE

566) W. Ferster, "Orbimage Restructuring Delays OrbView-4 Launch," Space News, Aug. 27, p. 3 and p. 27

Figure 75: Illustration of the OrbView-2 and -4 satellites

Sensor complement: Both instruments can be operated in parallel.

OHIS (OrbView Hyperspectral Imaging System). The objective is to demonstrate the utility of hyperspectral technology for the support of various applications. The instrument is a body-mounted optomechanical instrument (whiskbroom scanning imager), built by Northrop Grumman. The spectral range of the instrument is 0.4 - 2.5 µm. Up to 200 contiguous spectral bands can be defined within the spectral region. A typical scene size is 5 km (swath width) x 20 km (length) with a spatial resolution of 8 m (GSD) at nadir. The data quantization is at 12 bit. No data compression is employed.

Parameter	Detector Array Size	Type of Array	Band Spacing	Operating Temperature	Wavelength Range (µm)	Nr. of Bands
VNIR	40 x 640	Si	11.4 nm	257 K	0.45 - 0.905	40
NIR	80 x 640	HgCdTe	11.4 nm	257 K	0.83 - 1.74	80
SWIR	80 x 640	HgCdTe	11.4 nm	257 K	1.58 - 2.49	80

Table 125: Characteristics of the OHIS instrument

Note: OHIS had originally an additional MWIR band in the wavelength range 3.0 - 5.0 µm, with HgCdTe detectors (80 x 640), a band spacing of 25 nm, an operating temperature of 90 K, and 80 bands. This option was cancelled in 2000.

OHRIS (OrbView High Resolution Imaging System) of ORBIMAGE designed and built by Northrop Grumman. The instrument is identical to OHRIS flown on OrbView-3, the description is provided under OrbView-3.

B.9 RapidEye Satellite Constellation

RapidEye [567) 568)] is a planned constellation of four minisatellites of RapidEye AG of Munich, Germany with the objective to provide high-resolution (6.5 m) multispectral imagery along with an operational GIS (Geographic Information System) service on a commercial basis. The overall objectives are to provide a range of Earth-observation products and services to a global user community. Typical fields of application and services are:

567) M. Krischke, W. Niemeyer, et al., "RapidEye - Satellite Based Geo-Information System," IAA 2nd International Symposium on Small Satellites for Earth Observation, Berlin, April 12-16, 1999, pp. 249-252

568) Information provided by Manfred Krischke of RapidEye

- Agriculture - crop monitoring and mapping, yield prediction
- Cartography - satellite based maps, ortho photos, DEM
- Other markets - disaster assessment, 3D-visualization
- Service spectrum at completion of constellation: Guaranteed daily revisit, global coverage, product delivery to the customer within 24 hours, possibility of dedicated programming, capability of direct transmission and imagery transfer within hours, global digital database of "ortho maps" of 1:25,000 scale and DEMs of 20 m x 20 m resolution. The service permits also the merging of multi-temporal imagery with information from other sources.

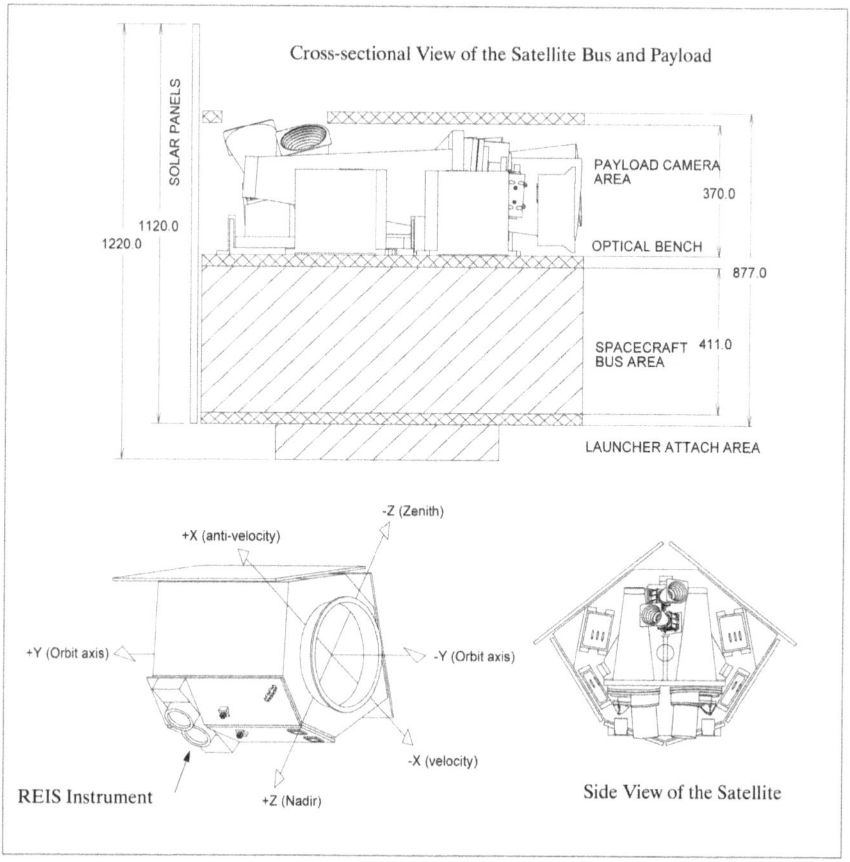

Figure 76: Schematic illustration of the RapidEye satellite

The satellites are being built by SSTL (Surrey, UK) as the prime contractor on its Minibus platform. The RapidEye spacecraft has a modified cylindrical barrel shape, with rectangular side facets, and two end panels. The launch vehicle attach fitting is mounted onto one of the end panels; two of the Earth-facing side facets contain the camera apertures and communications antennas. The solar panels are wrapped around the three zenith-facing facets of the structure to maximize the power generation of the S/C. The solar arrays are integrated on aluminium honeycomb substrates. They provide a power of up to 740 W when optimally illuminated. Three packs of NiCd batteries are used with a combined storage capacity of 21 Ah at 28 V. - The S/C structure is 877 mm in height, with additional height for the attach fitting, and the solar panels, which are 1120 mm tall. The mounting frames are circular with a diameter of 500 mm. With the solar panels included, the cylinder (barrel) has a dimension of 560 mm in diameter.

Each S/C is three-axis stabilized. Attitude sensing is provided by star sensors and magnetometers, actuation is provided by reaction wheels and magnetorquers (for wheel momentum dumping). The spacecraft uses a redundant GPS receiver (model SRG-20 of SSTL) for orbit determination and on-board time provision. The attitude system operates nominally in three-axis reaction-wheel control mode and flies the S/C in a so-called barrel-roll configuration, with the cameras pointing towards nadir. A body-pointing capability in the roll axis of the spacecraft exists which permits a ±30° FOR (Field of Regard) for camera observations into any direction (however, the feature is planned to be used only in the cross-track direction). On-orbit geo-referencing of the imagery is provided using inputs of the autonomous navigation and attitude determination system. In post processing the images will be referenced to at least one pixel accuracy (±6.5m). The wet mass of each S/C is 315 kg, the design life is seven years.

The RapidEye constellation is planned to be installed in two separate launches (two S/C each) on a Russian launch vehicle in 2003 and in 2004.

Orbit: Sun-synchronous orbit (all four satellites are evenly spaced in one orbital plane), altitude = 600 km (±10 km), inclination = 98°, local equator crossing time at 12:00 hours (noon) on the ascending node. A regular full repeat coverage is provided in 3-4 days depending on latitude, a revisit time of one day can be obtained with body-pointing techniques.

The on-board storage capacity of data is 126 Gbit. RF communication of imagery is provided in X-band (8.25-8.40 GHz) while TT&C communications are in S-band. The X-band downlink has a data rate of 150 Mbit/s.

Sensor complement:

REIS (RapidEye Earth Imaging System), designed and developed by Sira Electro-Optics Ltd of Chislehurst, Kent, UK. REIS is a multispectral pushbroom instrument, providing a swath width of 158 km with continuous observation coverage of up to 1500 km in length. The spatial resolution (GSD) of the imagery is 6.5 m. REIS is a camera with a dual optical-system design resulting in a dual-swath observation, each swath is of width 79 km with a slight overlapping configuration.

The telescope is of catadioptric design with a focal length of 74 cm. The instrument employs a linear silicon CCD detector array of size 6 x 12288 for each focal plane. The instrument mass is 62 kg, power = 73 W (orbit average).

The source data of REIS are compressed in real-time (lossless or lossy compression) prior to on-board storage and/or downlink transmission. In addition, a capability is provided to perform on-board pixel binning (2 x 2 or 3 x 3 pixels) prior to data compression.

Band Number	Band Name	Spectral Coverage (nm)
1	Blue	440-510
2	Green	520-590
3	Red	630-685
4	Red edge	690-730
5	NIR 1	760-900
6	Pan	550-850

Table 126: Spectral parameters of REIS

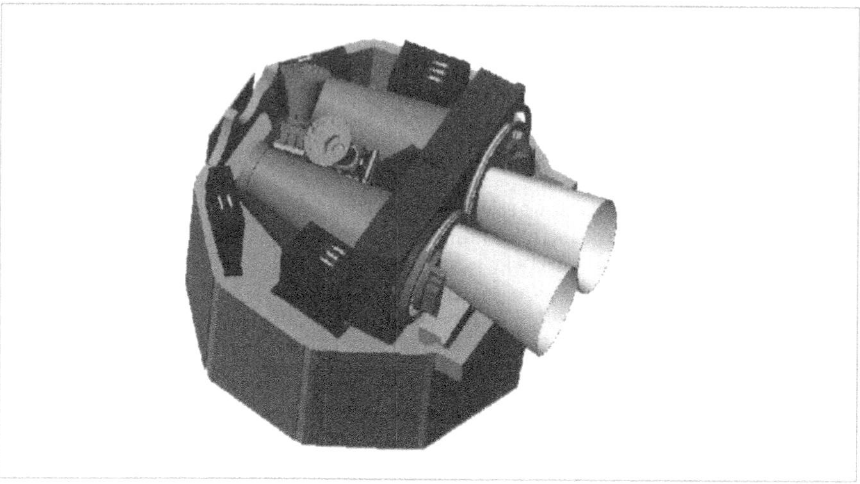

Figure 77: Schematic illustration of the REIS instrument

Part C Data Collection (Messaging) Systems

C.1 ARGOS (Data Collection System)

ARGOS is a joint program of CNES, NASA and NOAA, started in 1974 for the purpose of long-term continued global satellite data collection services (in particular environmental data) from fixed and mobile platforms located anywhere in the world. The ARGOS system package has been flown on all TIROS-N family satellite since 1978 (see chapter G.15.4). The space segment comprises two NOAA satellites in simultaneous orbit.

The ARGOS system exploitation, i.e. the data collection and distribution function and user interface, is a service provided commercially by CLS (Collecte Localisation Satellites), a CNES daughter in Toulouse, France, and by Service Argos of Landover MD, USA (a CLS subsidiary). Service to the user community has been continuously provided since fall 1978.

Orbit: the ARGOS payload on NOAA satellites is in sun-synchronous polar orbit, altitude = 830 - 870 km, inclination = 98 - 99°, period = 102 minutes (approximately 14 orbits/day). The circle of visibility (or the footprint) is 5000 km in diameter at 5° elevation, which is identical with the swath width.

ARGOS Space Segment:

Each ARGOS payload is equipped with a DCLS (Data Collection and Location System), also referred to simply as DCS, which receives all transmissions from the platforms in view during a pass. Functionally a DCLS is comprised of the following subsystems:

- housekeeping equipment, power supply and DCLS command interface
- receive assembly (receiver and search unit, both with full redundancy)
- signal processing assembly (four identical Data Recovery Units (DRUs, eight DRUs are planned for the next ARGOS series), and command unit, telemetry encoder, and buffer memory). All data are tape recorded on board the spacecraft.

ARGOS Ground Segment:

A set of user platforms, fixed or mobile, deployed at sea, on land, or in the air. All platforms reporting to the ARGOS system must carry a certified PTT (Platform Transmitter Terminal) package for satellite uplink communication. Each PTT outputs a short message (of 0.36 to 0.92 seconds duration, or of 32 bits to 256 bits maximum length) modulating a carrier frequency. Message transmission intervals range from 90 - 300 s, depending on the application.

The ground segment of the service provider consists of two NOAA/NESDIS CDA (Command and Data Acquisition) stations, one at Wallops Island VA, the other at Fairbanks, AK. In addition there is a downlink station at CMS (Centre de Météorologie Spatiale) Lannion, France. All these stations also provide real-time data during the pass. ARGOS provides two GPCs (Global Processing Centers), one in Landover, MD, the other in Toulouse, France. Each GPC receives data from all platforms but processes only the data that belong to "its" users. Both centers will, however, immediately process all data in case of necessity, thereby ensuring full redundancy.

Communication Concept:

Collection Uplink: ARGOS provides a total of four (eight in next series) parallel receiving channels for data collection, each at a rate of 400 bit/s. Each PTT in the ground segment transmits encoded messages at regular intervals (fixed platforms at 200 - 300 seconds, drifting or mobile platforms in the order of 90 - 150 seconds).

Note: the search unit is a spectrum analyzer that scans a 24 kHz band centered at 401.650 MHz. The next series of DCLS will have 80 kHz of bandwidth (100 kHz allocated, two safeguard bands of 10 kHz at each end). Time tagging and frequency measurements are made by the DRUs and processed on the ground for location determination.

Downlink: The data received by the ARGOS DCLS is multiplexed on-board by the TIP processor and transmitted to the ground via three paths:

- Real-time: the TIP output (8.32 kbit/s, see Figure 216) directly modulates a VHF beacon which transmits continuously.
- Real-time: the TIP output is multiplexed on-board the satellite with HRPT data and transmitted in S-band
- Delayed Transfer: the TIP output is also recorded by a tape recorder, and each time the satellite passes over one of the ground stations, the recorded data is dumped via S-band telemetry.

The ARGOS communication capability is limited to the function of data collection from the PTTs. The concept does not offer a remote configuration control capability of the data collection platforms in the ground segment.

Access Method: The on-board DCLS receiver picks up messages from the transmitting platforms in its area of visibility. The receiving system can discriminate between message arrival times and between frequency shift due to the Doppler effect. Up to four (eight in next series) messages may be processed simultaneously.

The ARGOS access scheme employs 'pure (i.e. unslotted) ALOHA.' Messages from the PTTs are received on-board on a random access basis. The ARGOS Doppler system provides a position fix for drifter (or mobile) platforms. This setup requires between three and five successful transmissions, which must occur within one pass (footprint).

Within an average footprint of 10 minute duration, each platform in the ground segment usually has a number of attempts to make contact with the DCLS in the space segment.

- Fixed platforms: the number of transmission attempts of fixed platforms is three at a repetition rate of 200 seconds (average = 3).
- Drifting (mobile) platforms: the repetition rate is 90 - 150 seconds, hence the maximum number of transmission attempts possible within a footprint is 5 - 6 (average = 5). [About 80% of the possible position fixes are actually achieved by the system; 20% are rejected during ground processing for various reasons, mainly geometrical configuration: number of messages, pass duration, distance to the track, etc., according to CLS ARGOS].

The nature of random access very much degrades data collection performance by the space segment. The scheme of pure ALOHA permits under normalized offered channel traffic a maximum channel throughput rate of 18%. Any two signals overlapping in time and frequency may interfere, with the loss of both. The principal parameter that affects the performance of the ARGOS data relay system is "interference": it occurs when the demand for service exceeds the system's capability. The result is loss of data from system 'blockage'. The maximum number of platforms that a single ARGOS DCLS can actually service within a footprint is in the order of 650. In this number, there is a certain mix of fixed (collection service only) platforms and drifting (collection and location services) platforms, a further assumption is a certain message length.[569),570)]

The probability of good message reception is 67% with a traffic density of 2.6 Erlang, and 8.3 Erlang for the next improved DCLS series which is scheduled to be launched starting in 1998 with NOAA-K.

The total number of platforms registered as active in the ARGOS system globally is around 4000, out of which around 2300 are transmitting every day. The remaining platforms trans-

569) Note: the figure of 650 serviceable platforms in a footprint was provided by 'CLS ARGOS' of Toulouse

570) "A Definition Study of an Advanced Data Collection and Location System (ADCLS)," prepared for GSFC by ECOSYSTEMS International Inc., January 1986

mit once every two or three days, or less. This information was provided by CLS ARGOS (6/1993), the service provider of the system.

C.2 FAISAT (Final Analysis Inc. Satellite)

FAISAT-1[571)] is a store-and-forward/data-collection/communication minisatellite of Final Analysis Inc. (FAI) of Greenbelt, MD, which was launched as a secondary payload from Russia (Plesetsk) on a Cosmos launch vehicle on January 24, 1995. FAISAT-1 is an experimental version of the satellites in a constellation of 26 LEO satellites planned by FAI within the next years. The objective of the FAISAT program is to provide a short digital messaging service (< 512 bytes per transmission) to customers around the globe. The capability of this two-way messaging service includes a range of support options and applications, such as: tracking, electronic mail, paging, environmental monitoring of autonomous remote ground stations (collection of ground truth), location determination, and distress signalling. Most other applications are anticipated to evolve as the system potential becomes obvious to the user community.

The FAISAT-1 S/C has the shape of an octagonal cylinder of 45 cm diameter and 90 cm in length. Spacecraft stabilization is provided by a gravity boom and magnetic torquers providing an attitude accuracy of 5° in yaw and 1° in the pitch direction. The outside of the S/C cylinder is covered with solar cells, S/C mass=114.5 kg. In addition to its own subsystems, FAISAT-1 carries the SSTT (Small Satellite Thermal Technology) secondary payload, an Air Force Flight Experiment of the Phillips Laboratory. The SSTT is a two-phase capillary-pumped loop thermal device which is being developed to provide thermal control for future space missions. Communication parameters are:

- Transmitter power: 10 W, 20 W selectable
- UHF downlink frequency: synthesized 387 - 407 MHz in 50 kHz steps; FCC experimental licence range: 400.595 - 400.645 MHz; Air Force allocation: 399.8375 MHz with a 38 kHz bandwidth
- VHF uplink frequency: synthesized 131-151 MHz in 10 kHz steps; FCC experimental license range: 148.65-148.75 MHz and 149.25-149.35 MHz; Air Force allocation: 141.775 MHz with a 25 kHz bandwidth
- Modulation methods:
  - CPFSK (Continuous Phase Frequency Shift Keying) @ 4800 bit/s
  - GMSK (Gaussian Minimum Shift Keying) @ 9600 bit/s
- Data storage capacity: 7.5 MByte

Orbit: Polar near-circular orbit, apogee = 1041 km, perigee = 988 km, inclination = 82.9°, period = 105 minutes.

FAISAT-1 contains a synthesized receiver capable of tuning from 131-151 MHz in 10 kHz steps. The experimental license granted to FAI allows for uplink operation in two 100 kHz wideband segments. With the on-board receiver scanning these segments in 10 kHz steps, each segment has nine potential channels (148.66, 148.67, 148.68, 148.69, 149.70, 140.71, 148.72, 148.73, and 148.74 MHz for segment 1).

The FAISAT ground segment consists of the Master Ground Station (MGS) and the Remote Terminals (RTs). The MGS provides the command and control functions of the satellite and the downlink data collection center. The RTs collect data from the monitoring instruments and transmit it to the satellite for store-and-forwarding. The access scheme for the monitoring task in the ground segment uses a modified TDMA technique.

- The satellite emits a beacon to signal the RTs in its footprint (polling scheme). The message includes the RTs ID code, transmission time and transmission frequency

571) Information provided by J. T. Skladany of Final Analysis Inc., Greenbelt, MD

- The RTs in the ground segment power up their transmitters upon receiving the ID code and set the transmitter to the assigned frequency
- The satellite RT receivers collect data from 50 RTs polled in series
- The satellite continues taking data in 10-second cycles until it is out of range. Data retransmission requests occur if the RT's transmission data is corrupted. In such cases the RT will be requested to retransmit 10 seconds later
- The RT powers down to stand-by mode after a transmission sequence is successfully completed.

An RT must have a synthesized transceiver and some intelligence provided by a simple microprocessor. A future extension of satellite communication capability is the provision of a transponder mode used for those cases when sender and receiver of a message are in the same footprint, or for the purpose of sending messages from one RT to another. The maximum footprint of FAISAT communication is 5650 km in diameter, assuming an orbit altitude of 1000 km and a 5° elevation angle. With a typical orbital pass of 10 minutes across such a footprint, and a polling scheme chosen to interrogate 50 RTs in 10 second intervals, this amounts to 3000 RTs per footprint for each satellite in the FAISAT constellation of 26 satellites.

FAISAT-1 was not able to establish satisfactory remote terminal demonstrations and communication links. It was eventually switched off in May 1996 when it was decided that different frequency bands were to be used for the constellation.

The FAISAT experimental messaging S/C, **FAISAT-2V**, was launched on Sept. 24, 1997 on a Cosmos-3M launch vehicle from Plesetsk. An on-board GPS receiver is part of the FAISAT-2V payload (and planned for all successive launches). VITA of Arlington, VA is leasing communications capacity on FAISAT-2V.

C.3 LLMS (Little LEO Messaging System)

LLMS is a demonstration store-and-forward messaging payload of ESA, flown on the Russian satellite RESURS-O-N4 (launch July 10, 1998), with the overall objective to provide a simple and low-cost communication access scheme (e-mail, etc.) on a commercial basis to organizations with personnel in remote locations; an additional service objective is data collection from environmental platforms in the ground segment. The global e-mail service function of LLMS is referred to as **IRIS** (International Retrieval of Information via Satellite). An IRIS subscriber needs a small, relatively inexpensive satellite terminal (about half the size of a portable PC) to communicate with his platform via IRIS. [572) 573) 574) 575)]

Background: The LLMS/IRIS program, an initiative of Belgium's space industry, is mainly financed by the Federal Office for Scientific, Technical and Cultural Affairs (OSTC) of Belgium, with a contribution from its German counterpart (DLR). The development of this advanced telecommunication platform (a turnkey demonstrator system) has been carried out within the context of an ESA development initiative, under SAIT-Systems S. A. of Brussels as the prime contractor. Other industrial participants in the program are: OHB-System of Bremen, Germany, Alcatel Bell of Belgium, the SEMA Group SAE of Spain, and Warberry Communications, UK. LLMS/IRIS is the first Little-LEO communications system using spread-spectrum modulation in the service links to and from the S/C as well as on-

572) V. Larock, A. Jongejans, "IRIS, First Operations," presented at the Conference on "Small Satellites Systems and Services," Sept. 14-18, 1998, Antibe-Juan les Pins, France

573) C. van Himbeek, I. Deman, B. Clarenne, "The LLMS DS/SS (Direct Sequence/Spread Spectrum) Payload," accepted for ISSSA (International Symposium on Spread-Spectrum Techniques & Applications), Darmstadt 1996

574) V. Larock, A. Ginati, "IRIS: Going Commercial with High-Tech European LEO Microsatellites," Proceedings of the IAF '94

575) "First European Payload for Worldwide E-Mail Service Launched," ESA Bulletin 95, Aug. 1998, p. 179

board demodulation techniques. SAIT Systems has also a contract from ESA to operate LLMS/IRIS for a period of three years. 576)

Orbit: Sun-synchronous polar orbit, altitude = 830 km, inclination = 98°, period = 101.3 min.

LLMS/IRIS payload and ground segment elements:

The payload consists of four redundant and interconnected units: OBC (On-Board Computer), TT&C, the RF unit and the SS (Spread Spectrum) unit with CDMA modem and UHF transceiver. The payload is planned to be available eventually in two polar orbiting S/C for reasons of redundancy. In the overall configuration, messaging is uploaded during contact periods from the HUB (or gateway) station in the ground segment via satellite to its destination and vice versa (remote subscriber terminal at the one end and fixed user at the other end). Each HUB station pass enables the exchange of all data collected by the satellite (and by the HUB station) since the last contact by store-and-forward techniques. The HUB station, located at the high-latitude site of Svalbard (78° N), Spitzbergen, Norway, for reasons of frequent satellite contact times (visibility of each S/C orbit), is connected via public network to the user community. A dedicated TT&C station provides in addition operations control of the payload(s). The message storage and retrieval functions of OBC are handled by an adaption of a transputer-based architecture already flown on BREMSAT-1 and on SAFIR-1/2 satellites. The total mass of the attached LLMS/IRIS payload is 67 kg, its power consumption is 72 W, of which 44 W is for the communication payload (RF+SS).

LLMS/IRIS Communication Concept:

LLMS/IRIS is employing a data transmission scheme referred to as SSMA (Spread Spectrum Multiple Access) in the UHF frequency band (uplink at 388 MHz, downlink at 400.6 MHz), permitting shared operations for customer services from remote user terminals. SSMA is used in both directions (uplink and downlink) offering such advantages as: a) interference-tolerant transmission, b) use of anti-jamming features, c) the random access scheme provides very good channel discrimination for multiple access, and d) it permits the measurement of propagation time between payload and user terminal (in combination with the Doppler estimation this measurement enables the system to localize the user terminal). The chip-based timing resolution of the spread-spectrum downlink is 0.2 μs. Localization accuracies projected: a)<3 km for coarse and b) <1 km for fine localization.

The communication payload can be put into one of two different communication modes:

- Service link: The service link is used for communication between the user terminal and the payload (or vice versa).
- Feeder link. The feeder link serves as high-rate communication link between the payload and the HUB station.

Parameter	Service downlink	Service uplink	Feeder downlink	Feeder Uplink
F-chip (MHz)	0.538	1.077	0.538	1.077
L (chips)	128	512	32	64
Efficiency (bit/symbol)	2	1	6	6
Byte/slot	134	134	345	345
Symbol rate (s/s)	4207	2103	16,827	16,827
Brute bit rate (bit/s)	8,414	2103	100,962	100,962
Net bit rate (bit/s)	4,709	1,177	71,112	71,112
Slot duration (ms)	127.4	509.7	27.3	27.3
Modulation scheme	O-QPN	B-PN	O-QPN	B-PN

Table 127: Modulation parameters employed on the different link modes

576) P. B. de Selding, "Belgium Firm Seeks Investors for IRIS E-Mail Venture," Space News, Sept. 28 - Oct. 4, '98, p. 22

A user terminal (fixed or mobile) design resorts to custom (ASIC) integration to process the spread-spectrum signals used by LLMS/IRIS. Typical terminal characteristics are:

- Standard RS-232 interfaces (one RS-232 opto isolated)
- Antenna: 1/4 wave groundplane or drooped-dipole
- Batteries: 6 x UM-3; standby: autonomy for 100 hrs, and continuous receive for 5 hrs
- Transmit power of <3 W
- Terminal size: 200 mm x 150 mm x 50 mm; mass < 1 kg.

C.4 Orbcomm Satellite System

Orbcomm is a family of store-and-forward communication microsatellites designed and built by OSC and operated by Orbcomm (Orbital Communications Corporation) of Dulles, VA, USA (project financing by OSC, Telglobe Inc. of Canada, and Technology Resources Industries Berhad of Malaysia). The objective is the provision of global two-way (store-and-forward) messaging services between any two users (subscribers) in the ground segment. The capability of this two-way messaging service includes a range of applications, such as: tracking, paging services, e-mail interconnection, remote data collection, location determination, and emergency alerts. The initial deployment configuration consists of two satellites, Orbcomm FM1 and FM2, which were launched April 3, 1995 on a Pegasus launch vehicle. [577) 578)]

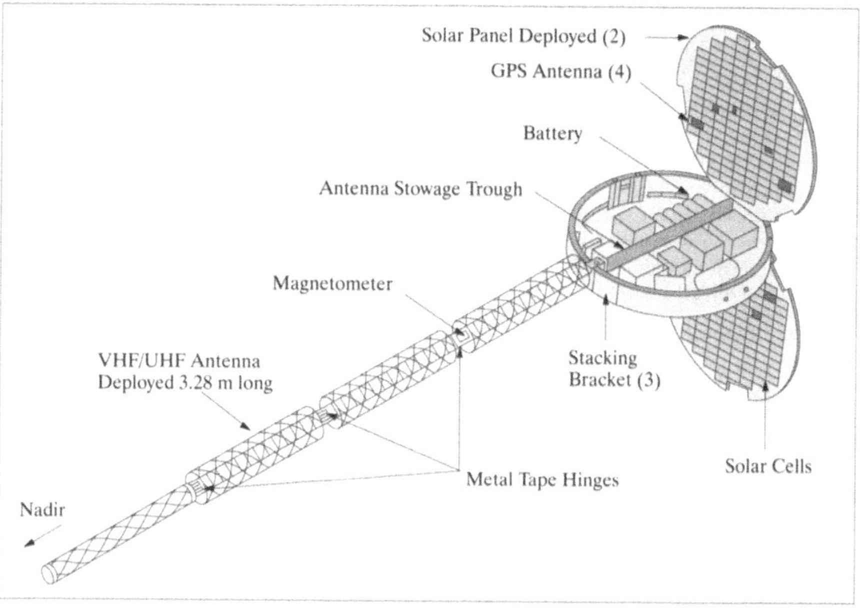

Figure 78: The Orbcomm S/C model (deployed configuration)

Orbit of FM1 and FM2: Polar circular orbit, altitude = 785 km, inclination = 70°, period =100 minutes.
Orbcomm is licensed to operate up to 36 satellites. The space segment has a total of up to 35 satellites in four inclined orbital planes (eight satellites per plane, and three spares) provid-

577) Information provided by M. Deckett of Orbcomm, Dulles, VA
578) http://www.orbcomm.com/our_technology/satellites.htm

ing near-overlapping footprint coverage of the world - resulting in near-continuous communications capability.

Spacecraft parameters: The Orbcomm family uses the OSC bus design (a cylinder of 1 m diameter and 17 cm height with deployed solar panels in orbit, resembling a butterfly); the S/C is stabilized with a nadir-pointing gravity boom and magnetic torquers (1 σ nadir pointing is within $\pm 5^{\circ}$); S/C mass = 42 kg; power = 70 W; S/C expected life = 4-5 years; on-board data storage is into solid-state memory, up to 2 MByte. The antenna FOV is 120° providing a footprint of 5000 km in diameter (5° elevation limit). FM1 and FM2 are flying GPS receivers (Trimble TANS II) with the objective to demonstrate real-time on-board orbit determination. 579)

Communication parameters of the S/C:

- Data rates: 2400 bps inbound (from ST to NCC), 4800 bps outbound (from NCC to ST)
- Frequencies: 148.00-150.05 MHz (VHF) data uplink; 137.00-138.00 MHz (VHF) TT&C up/downlink and 400.05-400.15 MHz (UHF) data downlink
- The access scheme between a subscriber terminal (ST) and the satellite is by FDMA (Frequency Division Multiple Access). Other links are either continuous or by TDMA.
- Number of parallel communication channels: 18
- Addressing: X.400 (CCITT 1988) standard
- Message size: 6-250 bytes typical length (no maximum length limit)
- Each packet contains overhead information and error detection bits

Spacecraft	Launch	Comments
FM1, FM2	April 3, 1995	Experimental S/C, altitude 785 km, inclination= 70°
FM5 to FM12 (8 S/C)	Dec. 23, 1997, Wallops	Orbit altitude of 810 km, inclination = 45°
FM3, FM4	Feb. 10, 1998, VAFB	Orbit altitude of 790 km, inclination = 108°
FM13 to FM20 (8 S/C)	Aug 2, 1998,	
FM21 to FM28 (8 S/C)	Sept. 23, 1998,	
FM30 to FM36 (7 S/C)	Dec. 4, 1999,	

Table 128: Orbcomm satellite constellation

Meanwhile, operational experience with FM1 and FM2 attitude control has led to an augmented ACS (Attitude Control System) design which includes substantially improved attitude determination and control relative to the experimental S/C, FM1 and FM1. A vertically oriented reaction wheel (0.63 kg) has been added providing improved torque inputs. In addition, residual magnetic dipole states in the attitude estimator algorithms were added. The resulting pointing accuracy of the S/C is more than doubled. 580)

Communication/messaging concept:

The Orbcomm ground system has a prevailing star architecture where a Network Control Center (NCC) functions as a central base station monitoring and controlling the network of a geographical region (like the USA, or Europe). Each NCC uses several unattended Gateway Earth Stations (GES) which relay messages between the satellites and the NCC in both directions. [A GES consists of two independent tracking antennas (HPA/LNA units), each enclosed in a radome, and auxiliary equipment]. The number and location of the GESs in any network are chosen to achieve the needed availability for a given region in the network. The USA network covers initially (1995) the contiguous 48 states, it consists of four GESs located in the states of New York, Georgia, Arizona and Washington, controlled and monitored by the NCC at Orbcomm, VA. The NCCs control only the communication of their networks. In addition, there is one Satellite Control Center, collocated with the USA NCC,

579) D. A. Steffy, "Orbcomm Satellites Launch and Initial Flight Operations," Proceedings of the AIAA/USU Conference on Small Satellites, Sept. 16-19, 1996, Logan, UT

580) M. R. Krebs, "A New Attitude Control Mechanism for LEO Satellites," Proceedings of the AIAA/USU Conference on Small Satellites, Sept. 16-19, 1996, Logan, UT

which monitors and controls the bus subsystems (power, attitude, thermal household, etc.) of all Orbcomm satellites.

Orbcomm satellites do not communicate with each other (i.e., there are no intersatellite links). Each GES tracks one of the satellites whenever it is in view; during this period (a pass may be up to 15 minutes long) there is continuous two-way communication between them. If more then one satellite is visible to a GES, then the controlling NCC decides which satellite should be tracked. There may also be an occasion when more than one GES is tracking the same satellite. This happens when the satellite can be seen by GESs located in different networks (and controlled by different NCCs). In this case, the uplinks from the GESs to the satellite are operated in TDMA mode; the downlink is continuous and received by all NCCs.

Remote stations in the ground segment are referred to as STs (Subscriber Terminals); they are always in listening mode. STs scan through a list of frequencies used by the satellites for downlink transmissions. Each satellite has a specific frequency for this purpose, and transmits continually. When an ST receives one of these signals at some threshold level, and determines that the satellite is in communication with the NCC controlling the territory in which the ST is located, it stops scanning and stays on that channel until the signal level drops below the threshold, whereupon it starts scanning again. There are 18 such channels, allocated to the satellites such that co-frequency satellites are never close to each other.

If an outbound message is addressed to a fixed ST, then the NCC knows which satellite to use. If the ST is mobile, then the system must know (or be told) which network it can be in, and a search routine is followed. The extend of the search, including the number of retries, depends on the contractual agreement with the customer. The Orbcomm facilities are used only for communications between an ST and the NCC of the region in which the ST is located. Communications between the NCC and the base station use the public switched network or other non-Orbcomm link.

The probability that at any given moment an ST can see a satellite that is also in contact with a GES is termed the availability. Availability varies with latitude and with the position of the ST relative to the GESs in the network. The orbits have been selected to maximize availability in the temperate latitudes. With a 26-satellite constellation, availability varies from about 70% in the equatorial zone to about 95% in the temperate zone. It is about 12% at the polar caps. - The number of STs that can be served by a network depends on the usage statistics and the acceptable queuing delays. About 250,000 STs can be served by each satellite (depending on the types of the STs).

The following types of data collection and messaging services are provided:

Data communication service: This service refers to the periodic or discrete collection of digital data from remote stations (STs) in the ground segment that are known to the system. In environmental monitoring this may mean the collection of ground truth data from a buoy network or from other sensors (meteorological stations, balloons, etc.) distributed throughout the world. For a utility company, it may mean the reading of meters of their customers. - All of these 'passive' sensors or meters (STs) must have an intelligent interface (communicator) as well as a small memory for storage capability. The space segment communicates with a remote station via a standard RS 232 dataport (using X.400 addressing and other standard protocols); data may be communicated in both ways (depending on need), inbound to configure the remote station, outbound to collect data from the remote station.

One application is position determination of a remote station or any other remote subscriber unit. The objective is to track objects. In the field of science this may apply to Lagrangian experiments (drifting buoys in the ocean or balloons in the atmosphere). In the commercial world, tracking may involve in locating stolen or other vehicles (ships, barges, trucks, planes, etc.).

Further applications are in the field of emergency alerts and confirmation (distress signalling by people in remote areas, alarms due to an intrusion into protected property, simple limit sensing of a particular device, etc.). In all cases an alarm message is initiated by a remote communicator and transmitted periodically through the system until confirmation of receipt is provided.

Messaging service: A personal two-way messaging service combines the functions of the data communications service into one package. A message can be sent by a user to a remote subscriber communicator through any personal computer using standard communication protocols (X.400, X.25, dial-up). The user (A) simply connects to the NCC and sends a message using a computer keyboard. The NCC routes and relays the message via the appropriate satellite to the remote subscriber (B). This messaging service supports any type of paging throughout the world.

C.5 SAFIR (Satellite For Information Relay)

SAFIR (**SA**tellite **F**or **I**nformation **R**elay) is a commercially operated data collection system (DCS) developed and built by OHB-System GmbH of Bremen, Germany, and sponsored in part by DARA. The system is intended to offer data collection and positioning services on a global scale for an international user community. SAFIR systems operations are intended to be provided by OHB-Teledata (an associated company).[581]

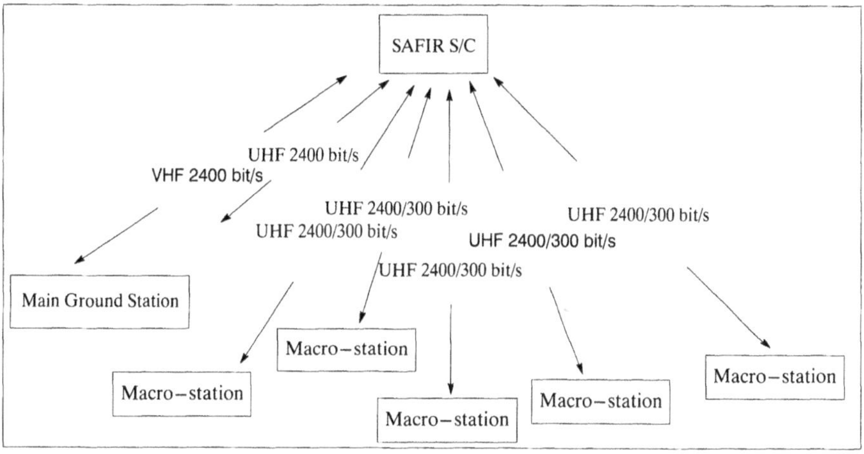

Figure 79: Concept overview of the SAFIR system

C.5.1 SAFIR-1

The space segment of SAFIR consists in phase 1 (1994-95) of an independent and autonomous SAFIR payload on the Russian RESURS-O1-3 spacecraft (referred to as SAFIR-R), launched November 4, 1994, for technological verification of the system concept. The nominal life of SAFIR-1 is 5 years. Phase 2 considers an autonomous polar-orbiting microsatellite, called SAFIR-2.
Operational status: As of 2001 SAFIR is fully operational with over 100 customers.

The SAFIR-1 satellite consists of a baseplate with several electronic units which is attached to RESURS-O1-3. SAFIR is equipped with an electronic box containing two on-board

581) Brochures and documentation provided by OHB-System

computers, two data storage units (5 MByte), and two TM/TC transmitter/receiver units for redundancy plus antennas, a GPS receiver, and a clock. The satellite is capable of measuring the frequency difference of any ground station based on the Doppler effect which serves in turn as input into the on-board computer in the form of integrated Doppler counts.

The ground segment consists of a main ground station at OHB-Teledata (for central scheduling of all customer traffic, distribution of all customer data, and for satellite control), and small portable ground stations that collect the data at customer-selected sites.

Orbit: (SAFIR-1) polar orbit, perigee = 663 km, apogee = 691 km, inclination = 98°, period = 98 minutes.

User Ground Stations:

The data collection concept considers macro-stations (in the order of 0.5 kg), which provide a storage capacity of 190 kByte for user data; they can collect and store data from their own monitoring instruments or from the satellite. Macro-stations may be able to transmit their data at 300 bit/s or 2400 bit/s to the satellite, depending on communication requirements. A standard macro-station consists of a CPU board, a HF board, and an interface board. The I/F board provides the following external interfaces: RS 232, digital input channels, and digital output channels. The macro-station may optionally be equipped with a GPS receiver board (for position monitoring), a battery pack, and a telephone modem.

The ground concept consists of a number remote macro-stations as well as a 'home station' for each user. The home station consists of a macro-station with a PC interface, a battery pack, and a loading unit.

Each ground station knows the next nominally scheduled contact time. In case of failure (no nominal contact between the ground station and SAFIR) the ground station uses the next possible contact time that is known in advance and stored in its memory. A minimum power level of 5 W EIRP is required at the ground station to communicate with the satellite. A ground station may be configured go into a 'sleep mode' during noncontact periods (in order to save power; in this case only real-time measurements at contact time are possible); it goes back into listening mode according to a calculated schedule. It is also possible to configure a station to remain in operational (listening) mode and to record measurements during noncontact periods if called for by the requirements.

Communication Concept:

The principal task of the satellite is scheduled data collection when in view of user ground stations, intermediate on-board storage of the data, and relay of the data to the main ground station (store-and-forward mode with the use of a mailbox). The bidirectional communication principle also permits data transmission from the main ground station or from the user home station via SAFIR to individual user ground stations (for configuration control, to reprogram the remote station, etc.). Users are normally equipped with a home station which permit communication with their remote stations in store-and-forward or relay mode, depending on visibility conditions within the satellite link.

SAFIR operates with two communication frequencies:

- VHF-band (137.175 MHz) link between the satellite and the main ground station at a transmission rate of 2.4 kbit/s.
- UHF-band (401.00 MHz uplink, and 400.75 MHz downlink) between the ground stations and the satellite, and also between the main ground station and the satellite, at transmission rates of 300 or 2400 bit/s, respectively.

The digital transmission of messages between the satellite and the ground stations is based on an HDLC protocol for error-free communication (error-free in the sense that the SAFIR communication concept provides either correct data or no data). The protocol permits any system user to select an individual message length for a ground station. The transfer frame length depends on the total amount of user data. Shorter messages are transmitted with short frames. The maximum frame length is 1 kByte.

Access Method:

SAFIR-1 is equipped with one receive and one transmit channel for servicing the user ground segment. Each ground station has a scheduled time slot within the time of visibility in which there is communication with SAFIR. A communication link is established by addressing the user station by its individual identification number. The addressed user station must respond to this addressing prior to message transfer. The link is closed by SAFIR after the data transfer is completed. The overall efficiency of this access method is affected by synchronization delays and the error rate of the channel. The access method of SAFIR-1 is TDMA.

With an average contact time of 6 seconds per remote link (5 seconds of data transfer on the average plus 1 second of synchronization time) SAFIR is capable of servicing up to 100 remote stations within a 10 minute circle of view (of 5000 km diameter). A Doppler location service requires three to six consecutive contacts for each remote station (depending on accuracy requirements), which amounts to a SAFIR serviceability of 34 or 17 moving stations, respectively, within a circle of view (instead of the 100 normal stations).

The swath of SAFIR provides the following coverage conditions (changing with latitude):

- 3 times of daily coverage at equatorial latitudes (min)
- 4 times of daily coverage at 40-45° latitudes (min)
- 5 times of daily coverage at 50° latitude (min)
- > 10 times of daily coverage in polar regions

Communication between Main Ground Station and Satellite		Communication between Macro-station and Satellite	
		Frequency uplink	401.00 - 401.10 MHz
Frequency	137.175 - 137.275 MHz	Frequency downlink	400.75 - 400.85 MHz
Modulation	BPSK, $\Phi = 0, 180^\circ$	Modulation	FSK, Δf = 9 kHz
Transmission rate	2400 bit/s	Transmission rate	300/2400 selectable
Coding	Manchester code	Coding	Manchester code
On-board computer		Transputer-based multiprocessor	
SAFIR-1 mass		38.5 kg	
SAFIR-1 stabilization		provided by RESURS-01-3	
SAFIR-1 power consumption		20 W average, 80 W maximum	

Table 129: Some SAFIR specification parameters

C.5.2 SAFIR-2

SAFIR-2 is an advanced and follow-up satellite (commercially operated little LEO store-and-forward communication system of the Fuchs Group, built by OHB-System, and operated by OHB-Teledata, Bremen, Germany) to continue and enhance the services provided by SAFIR-1.[582] [583] The overall objective is to provide communication/data collection services for logistical (tracking), environmental (monitoring) and industrial (meter reading) applications. The SAFIR-2 microsatellite is of cubical structure (0.5 m x 0.5 m x 0.5 m) with four surface-mounted solar panels, providing power (average/peak) of 20/50 W. In addition, NiH_2 batteries of 4 Ah capacity provide power during eclipse the phase of the orbit. Total S/C mass of 65 kg. The S/C is passively stabilized with a 6 m gravity-gradient boom and by an active control loop consisting of magnetic coils in three dimensions, two magnetometers, and a horizon sensor. A pointing accuracy of $\pm 10^\circ$ is provided. The S./C undergoes a very

582) B. Brünjes, F. Ellmers, M. Fuchs, et al., "SAFIR-2 Small Commercial Telecommunication Satellite," Proceedings of the 49th IAF Congress in Melbourne, Australia, Sept. 28 - Oct. 2, 1998

583) OHB company brochure of SAFIR-2 provided by F. Ellmers

slow spin rate of about seven revolutions in one orbit. Two OBC (T800 CPU transputer) operate in master/slave configuration. A total of four MByte on-board user data storage is provided.

The S/C was launched as a secondary payload on a Zenit-2 vehicle (along with the primary payload of RESURS-O1-4 of Russia, Thai-Phutt of Thailand, the Israeli TechSat/Gurwin-II, WESTPAC of Australia, and Chilean FASat-Bravo) from the Baikonur Cosmodrome on July 10, 1998. Status as of June 2001: SAFIR-2 is fully operational.

Orbit: Sun-synchronous polar orbit, altitude = 830 km, inclination = 98°, period = 101.3 min.

Communication Concept:

SAFIR-2 uses a combined UHF/VHF Earth-pointed antenna system (5 or 10 W transmit power). The UHF/A (analog) assembly is dedicated to user communication and redundant satellite operation, while the UHF/D (digital) assembly is dedicated to mass user communication. The modulation of all VHF and UHF/A telemetry is in FSK with Manchester encoding and RH circular polarization. Selectable data rates of 300, 600, 1200, 2400, or 4800 bit/s are provided for the various services. The modulation of UHF/D data is in BPSK for uplink and FSK for downlink transmission with data rates of 3200 bit/s (200 bit/s per station, up to 16 stations simultaneously). The newly developed FHT (Frequency Hopping Telemetry) access technology is implemented for UHF/D mobile user stations in parallel to the TDMA access scheme in order order to permit 16 user stations to transmit their data within the same time slot in the same frequency channel. The SAFIR-2 system has the capacity of collecting 10.000 telegrams/day and per 50 kHz bandwidth from a single defined coverage area (communication footprint). Forward error correction (FEC) is implemented on all transmissions. Bidirectional semi-duplex operation is possible within each of the telemetry assemblies.

The ground segment includes one HUB station in Bremen to operate the satellite and three types of user ground terminals: 1) user home stations, 2) 1-channel user mobile stations, and 3) 16-channel user mobile stations based on advanced DSP (Digital Signal Processing) technology.

- Service A provides a direct global bidirectional point-to-point user communication (arbitrary data files) in store-and-forward fashion between any user home stations. Connection of all user home stations in the network by a single S/C.
- Service B provides global unidirectional transmission in store-and-forward fashion of position telegrams and short messages to the S/C from 16-channel user mobile stations (moving assets like trucks, containers, etc.). The position telegrams include asset ID, time and GPS position. The telegrams are received by the HUB station and distributed to the customer.

Service type	Frequency allocation	Communication service
Bidirectional between user stations (Service A)	400 MHz downlink, 401 MHz uplink, UHF	Between S/C and HUB, user home station and 1-channel user mobile stations
Unidirectional from moving assets to user stations (Service B)	400 MHz downlink, 400.00 MHz uplink, UHF	Between 16-channel user mobile stations and the S/C. Transmission of position telegrams and short messages to the S/C.
Bidirectional S/C operations	137 MHz, VHF band	Between S/C and HUB

Table 130: Frequency allocations for service types

C.6 SCD-1 (Satélite de Coleta de Dados)

SCD-1 is Brazil's first experimental Data Collection Satellite, a portion of MECB (Missao Espacial Completa Brasileira). INPE's responsibility in MECB was the space segment and associated ground segment of the program [six satellites were planned within this program: four for environmental data collection (SCD-1, SCD-2, SCD-2A and SCD-3) and two for remote sensing of the Earth (SSR1 and SSR2)]. The Brazilian Space Activities Program has been reorganized in 2000, now, the satellites are part of the Application Satellites Program.

A DCP network in Brazil is equipped with sensors suitable for the fields of meteorology, hydrology, oceanography and atmospheric chemistry. Spin-stabilized S/C; mass = 115 kg; power = 90 W; design life = 1 year (projected life is now 2.5 years). Launch of SCD-1: Feb. 9, 1993 from NASA's Kennedy Space Center by OSC of Fairfax VA (Pegasus launch vehicle). SCD-1 is performing nominally. [584) 585)]

The shape of SCD-1 is an octagonal prism, 80 cm tall, whose base fits into a 1 m diameter circle. The outer surface of the S/C is covered with solar cells. S-band TT&C communication is provided by a TM/TC subsystem for housekeeping telemetry transmission, telecommand and ranging (S/C mission control functions). On-board storage capability for TT&C data.

Orbit: Near-circular orbit; altitude = 750 km; inclination = 25°; period = 98.8 minutes.

The SCD-1 satellite is a real-time repeater of environmental data gathered on the ground by automatic data collection platforms (DCPs) providing random access service for DCPs. The real-time repeater concept implies that there is no on-board processing and storage capability but simply a relay function of data by the satellite. Data reception on the ground is at a single station in Cuiabá, Brazil every time SCD-1 comes into view. This concept permits a data collection capability from all sites in Brazil and well beyond (coverage region of 3000 km radius). It is estimated that the system setup is capable of servicing up to 1000 DCPs randomly located in the coverage area with a probability of 85% of acquiring the data of any DCP at least once per day. The DCP data received by the Cuiabá station are processed by the MCC in Cachoeira Paulista and subsequently stored in a database for user access.

The random access method employed by SCD-1 for data collection is similar to and compatible with the ARGOS system. This means:

- Access scheme is pure (i.e. unslotted) ALOHA, without a return channel
- Message format, data rates, and transmission frequencies are ARGOS compatible (message length of 32-256 bits, 400 bit/s uplink rate, 401.650 MHz and 401.620 MHz uplink frequencies).
- Brazilian DCPs in the ground segment can also be serviced by other satellites carrying ARGOS.

As of 2001 there are 400 DCPs operational in Brazil; others are in the phase of installation or are being planned for the future (including those from other South American countries).

Note with regard to the SCD-1 access scheme: [586)]
SCD-1 uses a "random time-division multiple access" scheme which is similar to "unslotted ALOHA." However, SCD-1 and also ARGOS access are simpler than ALOHA due to the lack of a return channel from the satellite to the DCP. This means in fact loss of data due to interference (interference occurs when the demand for service exceeds the system's capability; the result is loss of data from system 'blockage'). However, the system works satisfactorily in spite of this disadvantage for the following reasons:

- interference is mitigated by Doppler shifts, which tends to spread the DCP transmissions
- the system operates at a low input rate (400 bit/s) so that a data packet has at least a probability of 80-90% of getting through without interference

584) INPE brochure 'SCD-1 Data Collection Satellite,' and fax information from Prof. P. M. Fagundes, Rio de Janeiro
585) "SCD-1 Satellite Description," and "The Brazilian Data Collecting System," papers provided by C. E. Santana of INPE, May/June 1992
586) Information provided by C. E. Santana and by J. Kono of INPE

- there is a lot of redundancy in the data of consecutive transmissions from the same DCP, so that successful reception of one out of two or even three transmissions is still satisfactory
- there is no catastrophic effect for subsequent communications if some data are lost.

C.6.1 SCD-2

The launch of SCD-2 took place Oct. 23, 1998 on a Pegasus launcher (with OSC's L-1011 Stargazer aircraft) from Cape Canaveral, FL. The launch site of SCD-2 was at 29.0° N and 78.3° W. SCD-2 is a follow-on satellite with the same objectives as SCD-1, but with improved data collection performances, due to modifications in the antenna- and attitude control subsystems. SCD-2 has a mass of 117 kg and power of 110 W. SCD-2 is performing nominally as of 2001.

Orbit of SCD-2: perigee = 743 km, apogee = 768 km, inclination = 25.0°.

The SCD-3 satellite will carry, along with the same payload as SCD-1, another data collection payload, capable of supporting advanced DCPs (the intent is the transmission of longer messages from the DCPs). SCD-3 will be in an equatorial orbit, collecting data from the Amazon region with a revisit period of two hours. The launch of SCD-3 is planned for 2006.

C.7 TEMISAT (Telespazio MicroSatellite)

TEMISAT[587),588),589)] is a data collection and distribution satellite for environmental monitoring of autonomous ground stations. The satellite and a portion of the corresponding ground segment are owned and operated by Telespazio of Rome, Italy; it was built by Kayser Threde of Munich, Germany. TEMISAT-1 was launched (piggyback on Meteor-2-24) on August 31, 1993 on a Russian Cyclone launcher from Plesetsk. The spacecraft is essentially a cube of 0.35 m side length with a mass of about 40 kg, and passive magnetic field stabilization, power = 62 W, expected lifetime = 5 years.

A second microsatellite (TEMISAT-2) was built together with the TEMISAT-1. As of the end of 1995 TEMISAT-2 is still in storage; however, it is going to be launched to increase in-orbit service capacity.

Application: worldwide commercial data collection, in particular for the science community in remote areas.

Orbit: Polar circular orbit, altitude = 950 km, inclination = 82.5°, period = 112 minutes.

The ground segment consists of a Mission Control Center (MCC, a major task being central scheduling and configuration control of all customer traffic), a number of User Control Stations (UCS, for those users who operate under their own control a sizable number of remote terminals), Remote Terminals (RTs, which control the monitoring instrument(s), receive and store data from the monitoring instrument(s), and function as communication nodes with the satellite), and finally the monitoring instruments themselves.

The TEMISAT polar orbit provides about a 10-12 minute contact period for all the monitoring platforms in the instantaneous area of visibility, or the footprint, of its orbit. TEMISAT is constantly polling the ground segment for possible customers (RTs, UCSs, or MCC) to initiate communication opportunities. It is foreseen that all the platforms with their RTs in this footprint will have a chance to communicate their data to the satellite in due time.

All instrument location knowledge is provided with a built-in GPS receiver in the monitoring instrument configuration (this applies to moving instrumentation in the ground seg-

587) "Telespazio Readies Temisat Satellite for Summer Launch," Space News, April 19-25, p. 24
588) "Temisat," Kayser Threde paper
589) "Blackbird: A Family of Microsatellites for Communications and Remote Sensing," Kayser Threde brochure

ment such as drifting buoys, or other moving objects). The GPS concept offers position accuracy of about 30-100 m.

Communication Concept: TEMISAT covers the VHF and UHF frequency spectrum allocated by WARC 92 for micro satellite LEO communication applications. A total of eight parallel receiving channels and two simultaneous transmitting channels are provided for the monitoring task of the ground segment. With this capability TEMISAT operates as an intelligent communications node in orbit with two different operational support modes:

- Transparent regenerative transponder mode (data relay function when data sender and receiver are in the same coverage area of the satellite)
  In this mode, data from the receiving channels (eg., from the ground segment platforms) are decoded, forward corrected, then again recoded and formatted for retransmission to the ground in real time (either to the MCC or to the UCS, or to both).

- Store-and-forward mode (operational function when data sender and receiver are in different coverage areas of the satellite)
  The received data are decoded, forward corrected, and stored on board (total storage capability of 30 MByte RAM) until the intended receiving station is in satellite visibility. Then the specific data file is recoded, formatted, and transmitted to the ground.

Data can generally be collected by the MCC and provided/distributed to the user community via electronic interface.

No. of 'Receive' channels, data collection & TC	10 (8 + 2)
Operational	8 data/1 TC (all uplink ground - S/C)
Receive rate	2400 bit/s
Data rate	1200 bit/s
Coding	Viterbi 1/2
Modulation	MSK
Frequency	9 channels in VHF, 1 experimental (UHF)
No. of 'Transmit' channels	3 + 3 cold redundant (3 channels downlink)
Operation	1 TM (S/C housekeeping) / polling, 2 data
Transmission rate (TM/polling)	2400 bit/s
Data rate (TM/polling)	1200 bit/s
Coding (TM/polling)	Viterbi 1/2
Modulation (TM/polling)	MSK
Transmission rate	9600 bit/s (downlink)
Data rate	4800 bit/s
Coding	Viterbi 1/2
Modulation	GMSK
RF-output power (per channel)	>5W / 2.5 W
Frequency	VHF
Number of antennas	5
Antenna type	quarter-wave monopole
Polarization	linear
On-board computer	transputer-based multiprocessor
Mass memory	2 x 12.5/8.5 MB SECDED protected
S/C dimensions	350 x 350 x 350 (mm)
S/C total mass	40 kg
S/C stabilization	passive, magnetic field
Power consumption operational (max)	60 W
Power consumption standby	2 W
Average available power	18-20 W

Table 131: Summary of TEMISAT specification

Access Methods: TEMISAT uses a TDMA/SCPC (Time Division Multiplexing Access/ Single Channel per Carrier) access scheme for the monitoring task of the ground segment. TDMA/SCPC is centrally controlled by MCC (computer scheduling of customer traffic up to four times per day), which avoids data collision in the uplink and maximizes the channel throughput rate.

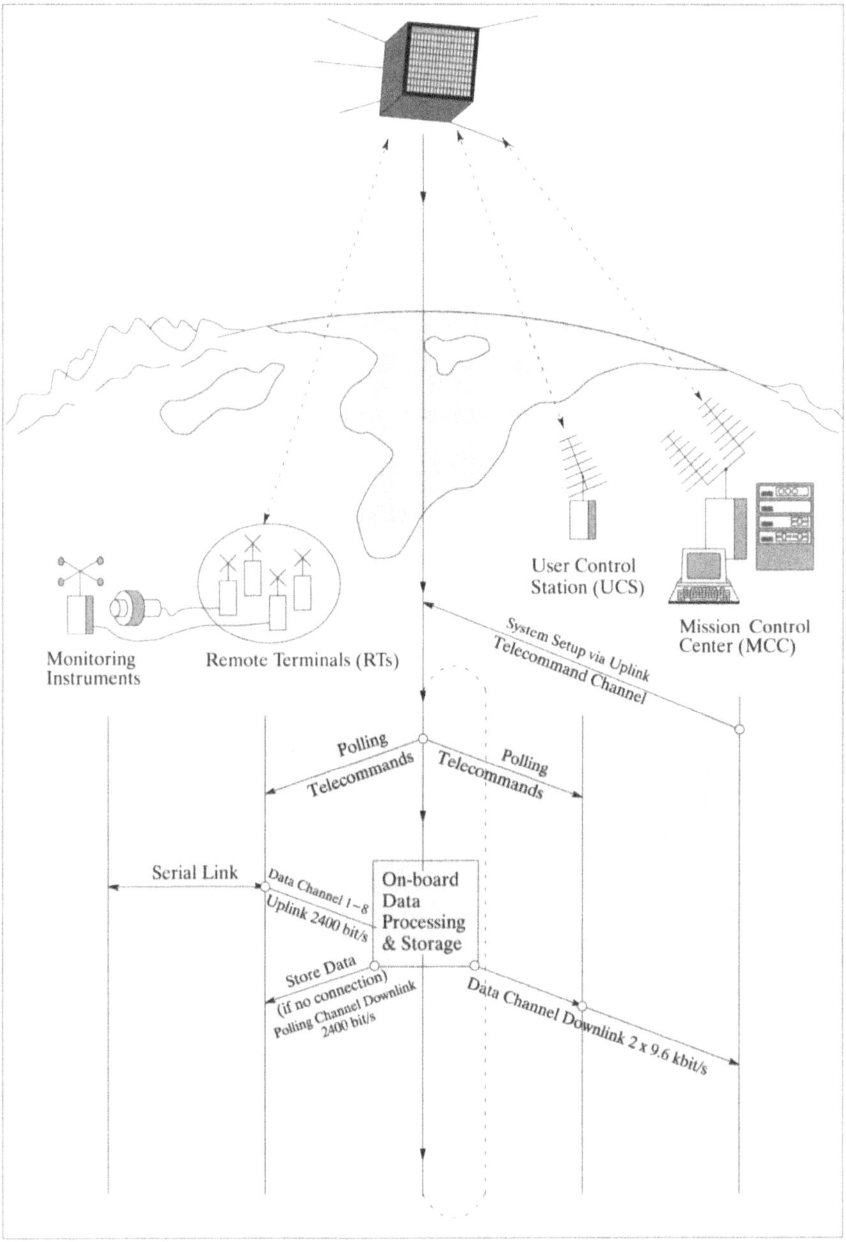

Figure 80: Scenario/Communication logic of the TEMISAT Data Collection System

The TDMA/SCPC access scheme permits a data rate of 1200 bit/s (2400 bit/s viterbi-encoded) for every one of the eight channels. The MCC has configuration control over the channel assignment and time slot allocation for each customer.

VHF Uplink Frequencies (MHz)	UHF Uplink Frequencies (MHz)	VHF Downlink Frequencies (MHz)
149.6625	402.3375	137.645
149.6875	402.3625	137.670
149.7125	402.3875	137.695
149.7375	402.4125	137.720
149.7625	402.5875	137.745
149.7875	402.6125	
149.8125	402.6375	
149.8375	402.6625	
149.8625		
149.8875		

Table 132: Operational frequencies of TEMISAT (allocated by WARC)

Note: The TDMA/SCPC access scheme allows a system throughput capacity of over 80% which is practically independent of the number of simultaneous users (30 or 300); on the other hand, the access scheme of ARGOS is 'unslotted ALOHA,' which permits under 0.5 normalized offered channel traffic a channel throughput rate of about 18% (max).

All RTs in the ground segment are always in listening mode, ready to establish contact with the TEMISAT polling channels at the scheduled time slot. If an average contact time of 5 seconds per RT is considered, in which the communication link (about 200 ms of sync time) is established and the data transmitted (uplinked), then TEMISAT is capable of servicing up to 960 RTs within a 600 second period with eight parallel channels [10 minute pass period over any ground point in the TEMISAT swath width of 5000 km (10° elevation above the horizon)].

The wide swath of TEMISAT visibility provides the following coverage conditions:

- 3 times of daily coverage at equatorial latitudes (min)
- 4 times of daily coverage at 40-45° latitudes (min)
- 5 times of daily coverage at 50° latitude (min)
- >10 times of daily coverage in polar regions

TEMISAT recognizes every remote terminal (RT) in the ground segment by its unique address in the message header. The message length for an uplink transmission can be variable, depending on the individual RT requirements. With an available net data rate of 1200 bit/s and an actual transmission time of four seconds in a contact, this amounts to 600 bytes of data in a single transmission. All user data is packetized with 600 bits per packet (2 packets per second in the transmission link). Users have the freedom to insert data into the frame according to their specific requirements (this may be sensor data along with a time stamp plus GPS data, there may be super- or subcommutations, etc.). A transmission can consist of some real-time data and some stored data. In general every RT stores its collected sensor data and provides it for the next available uplink opportunity. Data can be sent twice (i.e. on the next possible pass) in case of an unsuccessful transmission connection.

The concept and technology of TEMISAT operation represent a new dimension - in fact a new generation - in data collection services for the user community. This applies in particular to the introduction of the following features:

- Selection of the most efficient access scheme technology (TDMA/SCPC)
- The availability of sufficient data rates for the data collection function along with eight communication channels for instantaneous parallel service assignments
- The bidirectional store-and-forward communication concept, which offers a remote control capability for every remote sensing instrument to the user (or to the MCC on user request).
- Provision of real-time/stored data transmission which offers full-time (round-the clock) coverage (as compared to spot coverage) of the measured parameters.

- The inclusion of position data into every packet, if needed. There are no constraints of several contacts per RT within the same footprint for Doppler position fixing (as is the case for ARGOS PTTs).

C.8 Store-and-Forward (S&F) Systems

Automatic S&F data collection of ground truth from satellites has grown over the years from the first operation of the ARGOS system (on polar orbiting satellites) in the seventies to an Armada of systems and services (institutional and commercial) in the mid-nineties. The following table serves as an overview; descriptions of the different systems are provided in the text. [590)]

System Designation (Service Provider)	Satellite/Platform	Orbit	Service offered	Start of Operation, (Description in text)
ADEOS II DCS	ADEOS II	LEO	environmental monitoring	2002, (D.2)
ARGOS (CLS Argos)	NOAA-POES series		environmental monitoring, emergency signalling	1978, (C.1, G.15.4)
CBERS DCS	CBERS	LEO	environmental monitoring	1999, (D.7)
FAISAT (FAISAT)	FAISAT series	LEO	commercial DCS	1995, (C.2)
FedSat (Australia)	FedSat (ADAM)	LEO	environmental monitoring	2002, (M.9)
GMS (JMA)	GMS series	GEO	environmental monitoring	1985, (F.3.1)
GOES DCS(NOAA)	NOAA-GOES series	GEO	environmental monitoring	1974, (F.4.1)
GOMS DCS (Planeta)	GOMS series	GEO	environmental monitoring	1995, (F.5) no info available
INSAT DCS (ISRO)	INSAT series	GEO	environmental monitoring	1981 (F.6)
KAISTSAT DCS	KAISTSAT-4	LEO	environmental monitoring	2002 (D.19.3)
LLMS/IRIS (ESA)	LLMS	LEO	two-way messaging	1998 (C.3)
Meteosat DCS (Eumetsat)	Meteosat series	GEO	environmental monitoring	1977, (F.7.1, F.7.2)
MetOp DCS-2 (Eumetsat)	MetOp series	LEO	environmental monitoring	2005, (G.2.1)
MTSAT (JMA)	MTSAT-1R (JCAB)	GEO	environmental monitoring	2003, (F.9.1.3)
OKEAN Kondor (Planeta, NSAU)	OKEAN, SICH series	LEO	environmental monitoring	(D.26.2) no info available
Orbcomm (Orbcomm)	FM series	LEO	commercial DCS, two-way messaging	1995, (C.4)
SAC-C (CONAE)	SAC-C	LEO	environmental monitoring	2000, (D.34)
SAFIR (OHB)	SAFIR series	LEO	commercial DCS	1994, (C.5)
SCD1 (INPE)	SCD series	LEO	environmental monitoring	1993, (C.6)
UoSAT/SSTL	UoSAT	LEO	S&F, messaging	1984, (D.40)
TEMISAT (Telespazio)	TEMISAT	LEO	commercial DCS	1993, (C.7)

Table 133: Survey of data collection/S&F data communication systems and services

The capabilities of S&F systems have grown with the general developments in the satellite communications industry. Many commercial entries into this nonvoice communication market view the data collection part of environmental monitoring only as an additional source of income; their interest is mainly in global two-way data messaging services for a variety of applications to a large customer base. The provision of operational commercial (private) networks with their own space segments (constellations of microsatellites, also referred to as 'little LEO systems') and ground segments may be regarded as 'emerging and experimental' in early 1996; however, the real potential of this service market is considered promising. There are several projects on the drawing boards; others are in varies phases of completion. However, the future may mean licensing problems for the little LEO system services from ITU (WARC-95 did not allocate sufficient bandwidth for these services).

590) "Firms Battle for Spectrum," Space News, November 27 - December 3, 1995, p. 3 and p. 20

Part D Earth Observation/Monitoring Missions

D.1 ADEOS (Advanced Earth Observing Satellite)

ADEOS = Advanced Earth Observing Satellite. Japanese (NASDA) satellite mission.[591] Objective: Global observation of land, ocean and atmospheric processes (ocean color and sea surface temperature). In addition, communication experiments are planned for the study (feasibility) of interorbit links, called IOCS (Inter-Orbital Communication Subsystem). Launch with H-II launch vehicle from Tanegashima Space Center, Japan, on August 17, 1996 [592]

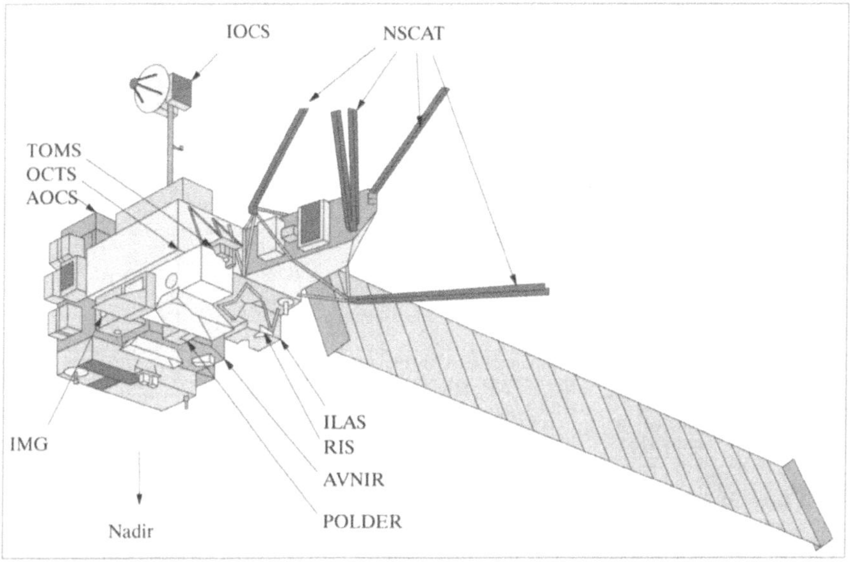

Figure 81: The ADEOS S/C model

Satellite mass = 3560 kg at lift-off, payload mass = 1300 kg, power = 4.5 kW, size: 4 m x 4.5 m. The AOCS employs a three-axis strap-down attitude detection system and a zero-momentum attitude control system, attitude error < 0.3°, attitude stability < 0.003°/s. Mission design life=3 years. The IOCS transmits observation data via ETS-VI (Engineering Test Satellite-VI) and COMETS (Communications and Broadcast Engineering Test Satellite).[593] [594] [595]

Orbit: Sun-synchronous sub-recurrent polar orbit; apogee = 804.6 km; perigee = 789.0 km; orbital period = 100.8 min, 10:30 AM local sun orbit (descending node). Ground repeat cycle=41 days (subcycle = 3 days), inclination = 98.625°.

ADEOS S/C operations are terminated: ADEOS operated nominally for some period, but then suffered several malfunctions and stopped functioning June 30, 1997 after an apparent loss of power (solar panel failure).

591) NASDA handout at the CEOS WGD-10 Meeting in Annapolis MD, April 16-19, 1991

592) In Japan the ADEOS satellite is also referred to as 'MIDORI', meaning 'green'.

593) "ADEOS," NASDA brochure, 1993

594) "Special issue on ADEOS," IEEE Transactions on Geoscience and Remote Sensing, Vol. 37, No 3, May 1999, Part II of two parts

595) ADEOS Reference Handbook, 1996, online available at http://www.eorc.nasda.go.jp/ADEOS/Products/Handbook.html

Sensor complement:

OCTS = Ocean Color and Temperature Scanner (mechanical scanning radiometer, NASDA core sensor). Objectives: Ocean color and sea surface temperature measurements (ocean primary productivity, interaction between the ocean and the atmosphere and environmental studies). OCTS offers 12 measurement bands from 0.402 - 12.5 μm. Swath width = 1400 km; ±20° along-track tilting. Spatial resolution: approx. 700 m. Operation requirements: global observation of the Earth during daytime (TIR channel during night if required).

Band Number	Spectral Band (μm)	Bandwidth (μm)	Radiance (W/m/sr/μm)	SNR
1	0.402 - 0.422	0.020	145	450
2	0.433 - 0.453	0.020	150	500
3	0.480 - 0.500	0.020	130	500
4	0.511 - 0.529	0.018	120	500
5	0.555 - 0.575	0.020	90	500
6	0.660 - 0.680	0.020	60	500
7	0.745 - 0.785	0.040	40	500
8	0.845 - 0.885	0.040	20	450
			NEDT	
9	3.55 - 3.88	0.33	0.15K	
10	8.25 - 8.80	0.55	0.15K	
11	10.3 - 11.4	1.1	0.15K	
12	11.4 - 12.7	1.3	0.20K	
	IFOV	0.85 mrad (~700 m)	Quantization	10 bit/pixel
	Scanning angle	~±40° (1400 km)	Tilting angle	-20°, 0°, +20°
	Polarization sensitivity	Band 1 ≤5% Band 2-8 ≤2%	Calibration VIS/NIR IR	Solar, internal lights, deep space,blackbody

Table 134: Definition of OCTS parameters

The OCTS sensor consists of a scanning radiometer with an optical system, a detector module and an electrical unit. OCTS employs a catoptric optical system and a mechanical rotating scanning method with a mirror (due to wide spectral coverage). OCTS can be tilted about the along-track axis to prevent sea surface sun glitter. The IR detectors are cooled to 100 K by a large radiant cooler facing deep space. Calibration: solar and internal lamp for VNIR, deep space and black body for IR.

AVNIR = Advanced Visible and Near-Infrared Radiometer (NASDA core sensor). Electronic scanning radiometer. Objective: Land and coastal zone observations, measurement of reflected sunlight from the Earth's surface. Scanning method: electronic (CCD). Spectral range: 5 bands from 0.42 - 0.89 μm (multispectral bands: 0.42-0.50, 0.52-0.60, 0.61-0.69 and 0.76-0.89 μm, panchromatic band (visible): 1 band 0.52-0.69 μm). Spatial resolution: multispectral bands: ~16 m (IFOV=20μrad), panchromatic band: about 8 m (IFOV =10 μrad). Swath width=80 km (FOV=5.7°). Observation requirements: regional observation according to user requests; simultaneous multispectral and panchromatic operation.

The AVNIR instrument has the capability to tilt the observation field by ±40° about the across-track axis. The 0.42-0.50 μm band is useful for coastal zones and lakes. The scanning radiometer unit uses a catadioptric Schmidt optical system to reduce aberration in a wide field of view. Calibration of sensor using solar light and internal lamps. The radiometric absolute accuracy is ±10%, the on-board calibration accuracy =±5%. The large linear array CCDs offer 5,000 and 10,000 detector elements for high spatial resolution. Instrument mass=250 kg, power=300W.

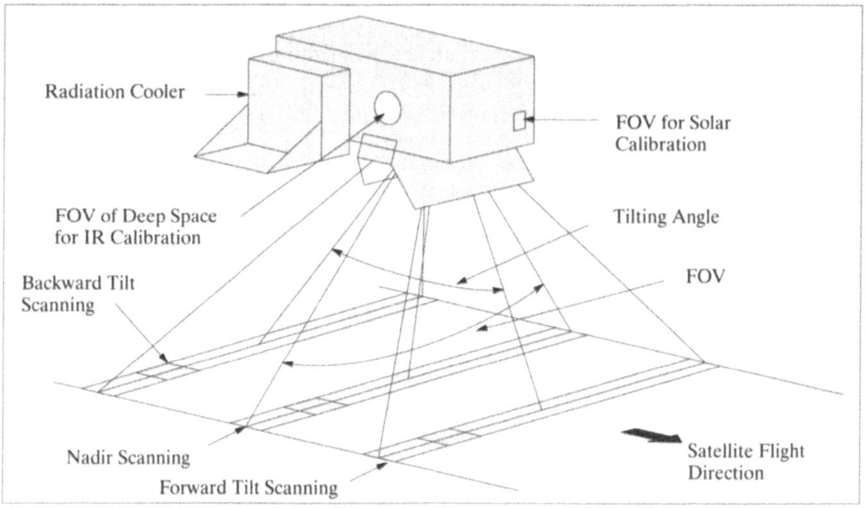

Figure 82: The observation concept of the OCTS instrument

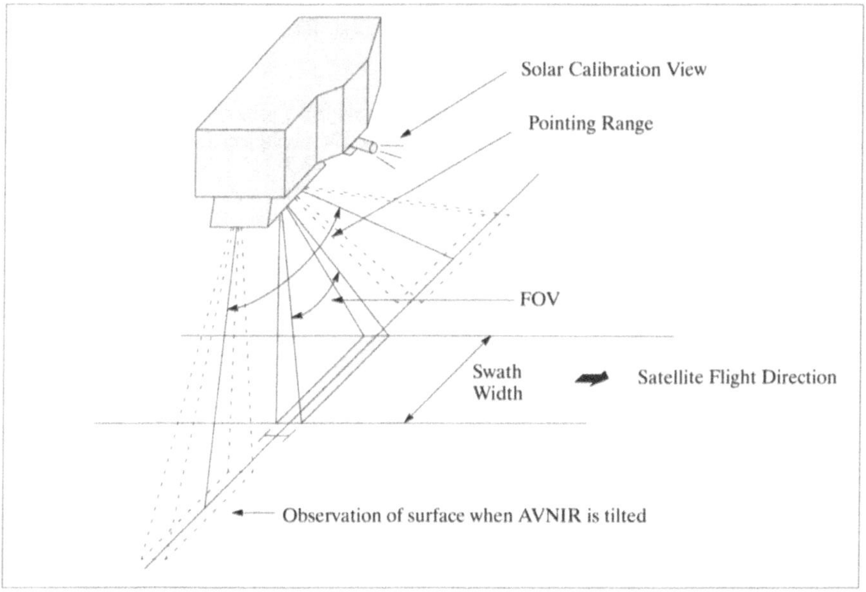

Figure 83: The observation geometries of the AVNIR instrument

NSCAT = NASA Scatterometer (NASA/JPL sensor).[596] Objective: Measurement of surface wind speeds and directions over the global oceans, coverage every two days under all weather and cloud conditions. NSCAT is a microwave radar instrument (a fan-beam Doppler scatterometer), using an array of six antennas that radiate microwave pulses at a frequency of 13.995 GHz across broad regions of the Earth's surface. An array of six, 3 m

596) F. M. Naderi, M. H. Freilich, D. G. Long, "Spaceborne Radar Measurement of Wind Velocity Over the Ocean - An Overview of the NSCAT Scatterometer System," Proceedings of IEEE, Vol. 79, No. 6, June 1991, pp. 850-866

long antennas scan two swaths of 600 km width each, one band to each side of the flight path, separated by a gap of 330 km at nadir. Wind speed accuracy = 2 m/s (rms), direction accuracy = 20° (rms), resolution = 50 km; antenna polarization: 6 V, 2 H; antenna beam width=25° (3 dB broad beam) =0.42° (3 dB narrow beam); instrument mass=300 kg; power=275 W; data rate=2.9 kbit/s; peak transmit power=110 W; pulse width=5 ms; PRF=62 Hz. Operation requirements: continuous operation for global observation of the oceans (about 190,000 wind measurements/day). Note: NSCAT is an upgraded version of the Radar Scatterometer (SASS) on SEASAT (see D.35).

NSCAT transmits microwave pulses and receives a backscattered echo from the ocean surface. Changes in wind velocity cause changes in ocean surface roughness, modifying the radar cross section of the ocean and the magnitude of the backscattered power. Multiple collocated measurements acquired from several directions can thus be used to solve for wind speed and direction simultaneously.

Wind velocity	2 m/s 10%	3 - 20 m/s 20 - 30 m/s
Wind direction	20° (rms)	3 - 30 m/s
Spatial resolution	25 km 50 km	σ_0 cells wind cells
Location accuracy	25 km (rms) 10 km (rms)	absolute relative
Coverage	90% of ice-free ocean surface every two days	
Instrument mass	300 kg	
Instrument power	275 W	

Table 135: NSCAT instrument parameters

TOMS = Total Ozone Mapping Spectrometer (NASA/GSFC sensor). Objective: Observation of total ozone changes, evaluation of changes in UV radiation and the observation of sulfur dioxide. Measurement wavelengths: 308.6, 312.5, 317.5, 322.3, 331.2 and 360 nm with 1 nm bandpass. Swath width: 2795 km. IFOV= 50 km at nadir; cross-track scan = 111° (37 3° steps). Operation requirements: global observation of illuminated part. TOMS mass=34 kg, power=24 W, data rate=700 bit/s.

TOMS measures the albedo of the Earth's atmosphere at six narrow spectral bands in the near-ultraviolet region. The albedo is measured by comparing the radiance of the Earth with the radiance of a calibrated on-band diffuser plate. Total ozone is derived from the differential albedo in three pairs of spectral bands, which are selected to function at all latitudes and solar illumination conditions.

POLDER = Polarization and Directionality of the Earth's Reflectances (passive optical imaging radiometer, CNES sensor as passenger instrument on ADEOS). [597] [598] Objectives: Observation of bidirectionality and polarization of the solar radiation reflected by the atmosphere: tropospheric aerosols (inversion of the physical properties); ocean color (accurate determination of sea surface reflectances); land surfaces (determination of surface BRDF and improvement in the correction of the surface bidirectional and atmospheric effects on vegetation indices); Earth radiation budget (determination of cloud BRDF and classification of clouds according to their bidirectional properties).

a) measurement of polarized reflectance in VIS/NIR
b) observation of the Earth's target reflectance from 12 directions during a single S/C pass
c) operation in two dynamic modes for high SNR and wide dynamic range

597) P. Y. Deschamps, F. M. Bréon, et al., "The POLDER mission: Instrument characteristics and scientific objectives," IEEE Transactions on Geoscience and Remote Sensing, Vol. 32, 1994, pp. 598-615

598) P. Y. Deschamps, M. Herman, A. Podaire, M. Leroy, M Laporte, P. Vermande, "A Spatial Instrument for the Observation of Polarization and Directionality of Earth Reflectances: POLDER," IGARSS '90 Conference Proceedings, Washington, D. C.

Six of POLDER's eight frequencies are optimized for observing atmospheric aerosols, clouds, ocean color, and land surfaces. The other two frequencies are centered on the H_2O and O_2 absorption bands for retrieving atmospheric water vapor amount and cloud top height, respectively.

Measurement channels: 15 channels (three channels for each polarized band) Swath width = 1440 km x 2200 (across-track) km, ground spatial resolution of 7 km x 6 km at nadir. Data rate: 0.882 Mbit/s, 12 bit quantization. Operation requirements: global observation of the Earth with more than 15° of solar elevation, simultaneous operation with OCTS.

POLDER is a pushbroom, wide-field-of-view, multi-band imaging radiometer/polarimeter. Multi-angle viewing is achieved by the along-track migration of the S/C of a quasi-square footprint. The optical detection unit of the POLDER consists of a telecentric lens, a rotating wheel supporting filters and polarizers, and a matrix CCD sensor (242 x 548 photo-elements, the pixels are binned one by two, resulting in 242 x 274 sensitive areas). The instrument has a focal length of 3.57 mm, f/4.6, FOV=±43° along-track, FOV=±51° across-track, FOV=±57° diagonal. Instrument mass= 33 kg, power=42 W.

Wavelength nm (FWHM)	Bandwidth (nm)	Polarization	Dynamic Range (Normalized Radiance) High	Low	Main Measurement Objective
443	20	no	NA	0.05-0.22	Ocean color
443	20	yes	0.05-1.1	NA	Aerosols, ERB
490	20	no	NA	0.034-0.17	Ocean color
565	20	no	NA	0.019-0.11	Ocean color
670	20	yes	0.013-1.1	0.013-0.27	Vegetation, aerosols, ERB
763	10	no	0.007-1.1	0.007-0.25	Cloud top temperature
765	40	no	0.007-1.1	0.007-0.25	Aerosols, CTP
865	40	yes	0.007-1.1	0.007-0.25	Vegetation, aerosols, ERB
910	20	no	0.007-1.1	0.007-0.25	Water vapor content

Table 136: Spectral characteristics of POLDER

IMG = Interferometric Monitor for Greenhouse Gases, [MITI sensor developed by JAROS, this is a nadir-looking Fourier transform infrared spectrometer (FTIR)]; Objectives: Observation of detailed spectra of IR radiation of the Earth's surface and the atmosphere, mapping greenhouse gases on a global scale (CO_2, CH_4, N_2O). Continuous measurements of the infrared spectrum in the range from 3.3 - 14.0 μm with a very fine spectral resolution of 0.1 cm^{-1} (apodized) absolute accuracy of measurement= <1 K, stability of measurement <0.1K, IFOV=10 mrad (~64 km^2 footprint); vertical resolution about 2-6 km depending on species; interferogram scan time ≤10 s. Operation requirements: full time operation for 3 days out of 13 days. A mechanical cryogenic coolant system is used to regulate the temperature of the quantum detectors. An image motion compensation mirror is used to compensate for the satellite orbital motion. Measurements are made in 20 km swaths at 8 km resolution.

Spectral range (in 3 bands)	600 - 3030 cm^{-1} (3.3 - 16.7 μm)
Band 1 Band 2 Band 3	2335 - 3030 cm^{-1} (3.3-4.3 μm) 2000 - 2326 cm^{-1} (4.3 - 5.0 μm) 600 - 2000 cm^{-1} (5.0 - 16.7 μm)
Nr. of samples	Band 1: 1.6 x 10^5; band 2 & 3: 1.0 x 10^5
Spectral resolution	<0.15 cm^{-1} (band 3); <0.25 cm^{-1} (band 2)
Measurement stability	<0.43 K (band 3); <1.59 K (band 2), < 1.58 K (band 1)
Data rate	882 kbit/s
Instrument mass, power	130 kg , 150 W
Instrument size	1150 mm x 930 mm x 650 mm

Table 137: Some characteristics of the IMG instrument

IMG is a Michelson-type Fourier Transform Spectrometer (FTS) with two mirrors and a beam splitter. The incident radiation is divided by the beam splitter into two paths. One mirror is moved so that the two paths produce an interference pattern when recombined. The signal measured by the detector, the interferogram, can be Fourier-transformed to obtain the incident spectrum. The diameter of the entrance aperture is 10 cm. The moving mirror is suspended on magnetic bearings and scans a 10 cm long path in 10 seconds.[599)]

ILAS = Improved Limb Atmospheric Spectrometer (sensor of the Environment Agency of Japan, EA), heritage of LAS on Ohzora (EXOS-C).[600), 601), 602)] Objectives: Measurement of atmospheric trace gases and the pressure and temperature profiles by the solar occultation technique (vertical profiles of O_3, NO_2, aerosols, H_2O, CFC11, CH_4, N_2O). ILAS has two grating spectrometers, one is used for the IR-band trace gases, the other measures aerosols, air temperature and pressure in the VIS-band. The ILAS occultation measurements are performed only at high latitudes in both hemispheres (due to orbit characteristics). ILAS keeps looking at the sun during sunrise and sunset, it measures the sequence of spectral intensity changes in the solar light which pass through the various tangent heights of the atmosphere. Since atmospheric trace gases have their own characteristic absorption spectra, the concentration of gases can be derived from these absorption measurements.

Spectrometer	2 grating spectrometers (VIS and IR) with linear array detectors
Spectral coverage	44 infrared channels in the range from 6.21-11.77 μm (850-1610 cm^{-1}) and 1024 visible channels in the range of 0.753 - 0.784 μm
Telescope	Cassegrain, 12 cm diameter
Sun tracker	Pointing mirror: 18 x 18 cm; sun edge sensor: 512 element array
IFOV (at tangent height)	IR channels: 2 km (vertical) x 13 km (horizontal) VIS channels: 2 km (vertical) x 2 km (horizontal)
Observation parameters	O_3, HNO_3, NO_2, N_2O, H_2O, CFC-11, CH_4, aerosols, pressure, air temp.
Observation region	Latitudinal zones 56° - 70° N and 63° - 88° S
Data rate	517 kbit/s, sampling rate at 12 Hz
Instrument operation	10 minutes per occultation (limb observations)
ILAS mass, power, size	130 kg, 78 W, 800 x 1630 x 550 mm

Table 138: ILAS instrument parameters

RIS = Retroreflector in Space (JEA sensor).[603)] RIS is a single element corner-cube reflector for Earth-satellite-Earth laser long-path absorption measurements of atmospheric trace gases. RIS has a hollow reflector structure with an effective diameter of 0.5 m. A spherical mirror with a very small curvature is used for one of the three mirrors forming the corner cube, in order to optimize the ground pattern of the beam reflected by RIS.[604)] Instrument mass=43 kg.

Objective: Used for laser long-path absorption measurements of atmospheric trace species and to support two color laser ranging experiments. Measurements of ozone (O_3), CFC12, HNO_3, methane (CH_4), CO, and other gases by the laser beam absorption technique. Wavelength range: 0.4 - 14 μm. - In the RIS experiments a laser beam is transmitted from a SLR ground station, reflected by RIS, and received back at the ground station. The absorption of the intervening atmosphere is measured in the round-trip optical path (measure-

599) H. Kobayashi, T. Ogawa, et al., "IMG, precursor of the high-resolution FTIR on the satellite," SPIE Proceedings, Vol. 3501, Optical Remote Sensing of the Atmosphere and Clouds, Beijing, Sept. 15-17, 1999, pp. 23-33

600) "Upper Atmosphere Monitoring with ADEOS - ILAS and RIS," EA/NIES brochure provided by Y. Sasano of NIES

601) "Ozone Layer Observation by Satellite Sensors," Proceedings of the International Workshop on Global Environment and Earth Observing Satellite Sensors, December 8-9, 1993, Tokyo, Japan

602) Y. Sasano, et al., "ILAS and RIS for ADEOS," SPIE, Vol. 1490, 1991, pp. 233-242

603) "Retroreflector-In-Space for ADEOS: Earth-Space-Earth Laser Long-Path Absorption Measurements of Atmospheric Trace Species," Optical Remote Sensing of the Atmosphere, 1990 Technical Digest Series of the Optical Society of America, Volume 4, pp. 488-490

604) A. Minato, N. Sugimoto,, S. Sasano, "Optical Design of Cube-Corner Retroreflectors Having Curved Mirror Surfaces," Applied Optics, Vol. 31, 1992, pp. 6015-6020

ment of absorption spectra by the Doppler shift of the reflected beam). The column contents and vertical profiles of atmospheric trace gases are obtained from the measured spectra. The RIS experiments are performed when ADEOS passes over the ground stations in Japan. The ground stations use two single-longitudinal-mode TEA-CO_2 lasers for the spectroscopic measurements.

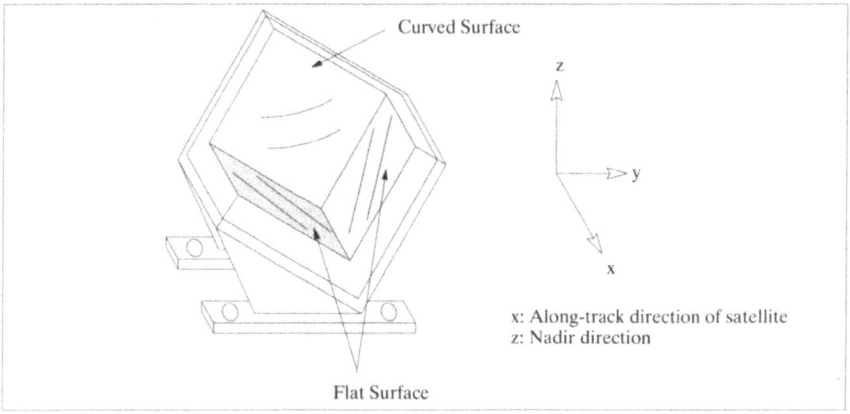

Figure 84: Schematic configuration of the RIS corner retroreflector

ADEOS carries in addition TEDA (Technical Engineering Data Acquisition Equipment) to measure the space environment surrounding the S/C. TEDA monitors radiation absorption, memory malfunction, static charge voltage, and contamination, and analyzes heavy ions.

Data rates of ADEOS:

ADEOS provides on-board recording (MDR=Mission Data Recorder (three instruments) and LMDR = Low Speed Mission Data Recorder).
Observation data rates: AVNIR (M): 60 Mbit/s, AVNIR (P): 60 Mbit/s, OCTS: 3.0 Mbit/s. Polder: 0.882 Mbit/s. IMG: 0.9 Mbit/s. ILAS: 0.517 Mbit/s. NSCAT: 2.9 kbit/s. TOMS: 0.7 kbit/s.

Data links: Uplink frequency (S-band) = 2.0 GHz (CMD and ranging), downlink frequency = 2.2 GHz (TLM and ranging), command bit rate = 500 bit/s.
Science data transmission: 3 X-band links (8.15, 8.25, 8.35 GHz) with QPSK modulation. IOCS frequencies = S-band (low rate mission data), Ka-band (120 Mbit/s max).

D.2 ADEOS-II (Advanced Earth Observing Satellite-II)

A Japanese (NASDA) mission planned for launch in Feb. 2002 with a mission life of three years minimum [(five years goal), FedSat of Australia is a secondary payload on the mission]. [605] As a successor mission to ADEOS, the objectives of ADEOS-II are to acquire data contributing for international global change research (carbon cycle and the water and energy cycle), as well as for applications in such fields as meteorology and fishery. ADEOS-II is the Japanese contribution in the framework of the International Earth Observation System (IEOS). Other parts of IEOS are EOS (USA), and the ENVISAT and MetOp programs of ESA and EUMETSAT respectively. The ADEOS-II mission is dedicated to the following programs: WCRP/GEWEX & CLIVER, IGBP and GCOS.

605) "Monitoring the Earth Environment from Space," NASDA Bulletin

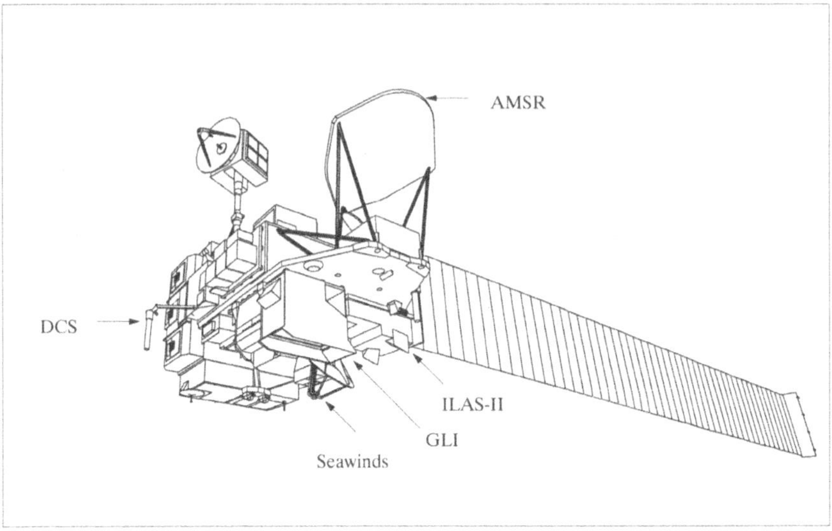

Figure 85: Illustration of the ADEOS-II spacecraft model

The ADEOS-II S/C employs the general design of ADEOS to reduce costs. Dimension of main S/C body: approximately 6 m x 4 m x 4 m. S/C mass = 3700 kg, payload mass = 1300 kg, power = 5 kW (EOL), launch vehicle = H-IIA rocket, launch site = TNSC (Tanegashima Space Center). Attitude and Orbit Control: The Attitude and Orbit Control Subsystem (AOCS) employs a three-axis strap down attitude detection system and zero momentum attitude control system achieving an attitude pointing error of less than 0.3°.

Communication: Mission data are downlinked in X-band to ground receiving stations. S-band is used for TT&C support. In addition there is communication via DRTS (Data Relay and Test Satellite) in Ka-band for mission data and S-band for TT&C data.

Orbit: Sun-synchronous subrecurrent orbit, altitude = 802.9 km, inclination = 98.62°, period = 101 minutes, recurrent period = 4 days, local time = 10:30 AM ±15 minutes.

Sensor complement:

AMSR = Advanced Microwave Scanning Radiometer (passive NASDA core sensor) of MSR heritage on MOS-1 and -1B satellites. Objectives: measurement of sea surface temperature (SST), soil water content, sea wind speed, water equivalent of snow cover, precipitation intensity, sea ice distribution, precipitable water, etc. Microwave emission from the atmosphere, ocean, sea ice, and land are measured at multiple frequencies. From this information a number of geophysical data related to the Earth environment, such as water vapor content, water content of clouds, water equivalent of the snow cover, etc. are measured.[606) AMSR has a high-temperature calibration source (about 340 K) and a small reflector to acquire the radiant temperature of deep space (about 2.7 K). Each feed horn, from 6.9-89 GHz sees the calibration sources once per scan period. The total power receiver technique is employed for radiometric sensitivity. The antenna scans conically the Earth's surface at an incidence angle of 55° (a scan drive motor rotates the antenna, rotating mass is nearly 200 kg, momentum and torque compensation is achieved with momentum wheels).

606) M. Nakajima, Y. Ito, H. Maejima, Y. Kojima, "The Development of AMSR and GLI for ADEOS-II," presented at the 45th Congress of the International Astronautical Federation, October 9-14, 1994, Jerusalem, Israel

Center frequencies, (GHz)	6.925	10.65	18.7	23.8	36.5	89.0	50.2	53.8
Bandwidth, (MHz)	350	100	200	400	1000	3000	200	400
Radiometric resolution	0.34 K	0.7 K	0.7 K	0.6 K	0.7 K	1.2 K	1.8 K	1.6 K
Spatial resolution (km)	50		25		15	5	10	
Antenna	offset parabolic, aperture diameter = 2 m, conical scan at 55° incid. angle							
Polarization	H/V (horizontal and vertical)						H (vertical)	
Antenna scan rate	40 rpm							
Cross-polarization	less than -20 dB							
Swath width	1600 km							
Dynamic range	2.7 K to 340 K							
Absolute accuracy	1 K (1 sigma)							
A/D quantization	12 bit	10 bit						
Data rate	130 kbit/s (CCSDS format, 14 channels of data acquisition)							
Instrument mass, power	320 kg, 400 W							

Table 139: AMSR parameter requirements

GLI = Global Imager (NASDA core sensor) of OCTS heritage on ADEOS. Objectives: Biological and physical processes, stratospheric ozone. GLI is for studying and monitoring the carbon cycle in the ocean, principally as to biological processes. Multispectral observations from the near UV to the near IR reflected solar radiation from the Earth's surface including land, ocean and clouds. Determination of chlorophyll pigment, phycobilin and dissolved organic matter (DOM) in the ocean; classification of phytoplankton according to their pigment. Measurement of sea surface temperature (SST), cloud distribution, land coverage, vegetation index, etc.

Spectral bands Visible and near infrared (VNIR) Short wavelength infrared (SWIR) Middle and thermal IR (MWIR, TIR)	 23 bands (380 - 830 nm), 18 channels with 10 nm bandwidth 6 bands (1050 - 2215 nm) 7 bands (3.715 - 11.95 µm) channels 0.33-1.0 µm bandwidth
IFOV	1.25 mrad (or 1 km at nadir) for 28 bands 0.3125 mrad (or 250 m at nadir) for 6 bands
FOV	±43° (swath = 1600 km)
A/D quantization	12 bit
Polarization sensitivity	<2%
Registration among total bands	<0.2 pixels
Data rate	4 Mbit/s (for 1 km resolution), 16 Mbit/s (250 m resolution)
Tilt angle	+20°, 0°, -20°
GLI scan mirror shape, size mass reflectance	Material: beryllium, surface coated with silver and SiO_2 Elliptic flat mirror with both faces (600 mm x 260 mm) 2 kg (mirror), 0.8 kg (axis) Over 90% (370 - 400 nm) Over 95% (400 - 550 nm) Over 90% (550 nm - 14 µm)
Instrument mass, power	450 kg, 400 W

Table 140: GLI parameter specification

GLI is an imaging spectrometer featuring a cross-track mirror and an off-axis parabolic mirror as the collecting optics and focal planes in which the detectors are arrayed in the along-track direction with spectral interference (dichroic) filters. The scan mirror rotates at 16.7 Hz. GLI can tilt the scan mirror ±20° from nadir in order to avoid sun glitter. GLI has five focal planes, two for VNIR, two for SWIR, and one for MWIR/TIR. Two VNIR focal planes have detector arrays for 13 and 10 bands respectively. Two SWIR focal planes have detector arrays for 4 and 2 bands, while the MWIR/TIR regions have one focal plane with a detector array for 7 bands. One SWIR and the MWIR/TIR focal planes are cooled to 220 K and 80 K by a multistage Peltier element and Stirling cycle mechanical cooler, respectively. The VNIR detector material is Si, the SWIR is InGaAs, the MWIR/TIR material is CMT. -

GLI employs piecewise linear method with cascade amplification for signal processing on four bands in order to meet requirements for automatic observation of objects with large radiance differences (ocean color and land vegetation) exhibiting a wide dynamic range. [607]

SeaWinds = NASA Scatterometer II (PI: M. Freilich, NASA/JPL - old designation: NSCAT II). Objective: to acquire accurate, high-resolution, global measurements of sea-surface wind vectors in 1 to 2 day repeat cycles. Applications: studies of tropospheric dynamics and air-sea interaction processes, including air-sea momentum transfer. Improvement of weather forecasts near coastlines by using wind data in numerical weather- and wave-prediction models. SeaWinds consists of three major parts: SAS (SeaWinds Antenna Subsystem), SES (SeaWinds Electronics Subsystem), and CDS (Command and Data Subsystem). The instrument is an active microwave radar (a scanning pencil-beam scatterometer) with dual-beam, 40° and 46° look angle from nadir, conical scan 1 m diameter reflector (dish) antenna, operating in Ku-band at 13.402 GHz (110 W pulse at 189 Hz PRF). Measurement of wind speeds between 3-20 m/s to an accuracy of 2 m/s, wind vector directions to an accuracy of 20°. The dish antenna is rotated about the satellite nadir axis at 18 rpm. Data is collected in a continuous 1800 km swath, centered about nadir. Spatial resolution = 50 km; IFOV = ±52° from nadir; mass = 200 kg; power = 220 W; duty cycle = 100%; average data rate = 40 kbit/s; thermal operating range is 5-40°C; pointing knowledge to 500 arcseconds. - SeaWinds data products consist of global multiazimuth normalized radar cross section measurements and 50-km-resolution ocean vector wind maps. [608] [609]

ILAS-II = Improved Limb Atmospheric Spectrometer-II, the sensor is [funded by MOE (Ministry of the Environment)] of ILAS heritage on ADEOS. The spectrometer uses gratings for solar occultation measurements of polar stratospheric ozone, atmospheric trace gases (O_3, HNO_3, NO_2, N_2O, CH_4, H_2O, CFC-11, CFC-12, $ClONO_2$, etc.), aerosols, temperature and pressure. ILAS-II is used to monitor and study changes in the stratosphere which are triggered by emissions of CFC gases. [610] [611]

Spectrometer		4 grating spectrometers (VIS and IR)
Spectral coverage	Band 1	44 IR channels from 6.21-11.76 μm (850-1610 cm^{-1}) with 0.1296 μm interval
	Band 2	22 IR channels from 3.0 - 5.7 μm (1754-3330 cm^{-1})
	Band 3	22 IR channels from 12.78-12.85 μm (778.2-782.4 cm^{-1}) with 0.2 cm^{-1} resol.
	VIS	1024 channels from 0.753 - 0.784 μm with 0.1 nm FWHM resolution
Telescope		Cassegrain, 13 cm diameter
Sun tracker		pointing mirror: 20 (az.) x 17 (el.) cm; sun edge sensor, 8 arcsec resolution per pixel
IFOV (at tangent height)		IR channels: 1 km (vertical) x 13 km (horizontal) (Band 1 and Band 2) 1 km (vertical) x 21.7 km (horizontal) (Band 3) VIS channels: 1 km (vertical) x 2 km (horizontal)
Observation parameters		O_3, HNO_3, NO_2, N_2O, CH_4, H_2O, CFC-11, CFC-12, $ClONO_2$, aerosols, pressure and temperature
Observation region		Latitudinal zones 56° - 70° N and 63° - 88° S
Data rate		453.7 kbit/s, sampling rate at 10 Hz, CCSDS packeting
Instrument operation		12 minutes per occultation (limb observations)
ILAS mass, power, size		< 138 kg, < 120 W, 950 mm x 1670 mm x 600 mm

Table 141: ILAS-II instrument parameters

607) T. Y. Nakajima, et al., "Optimization of the Advanced Earth Observing Satellite II Global Imager channels by use of radiative transfer calculations," Applied Optics, Vol. 37, No. 15, May 20, 1998, 3149-3163

608) M. W. Spencer, C. Wu, D. G. Long, "Tradeoffs in the Design of a Spaceborne Scanning Pencil Beam Scatterometer: Application to SeaWinds," IEEE Transactions on Geoscience and Remote Sensing, Vol. 35, No 1, Jan. 1997, pp. 115-120

609) B. D. Boller, et al., "The Development of the SeaWinds Scatterometer Electronics Subsystem (SES)," Proceedings of IGARSS'96, Vol. 1, pp. 269-272

610) Information provided by Y. Sasano of NIES (National Institute for Environmental Studies)

611) Y. Sasano, et al., "ILAS-II Instrument and Data Processing System for Stratospheric Ozone Layer Monitoring", Proceedings of SPIE, Vol.4150, pp.106-114, 2001

The instrument consists of the following elements: a two-axis gimbal mirror which is controlled to track the radiometer center of the sun, a 13 cm diameter Cassegrain telescope, beam splitters, and reflective transfer optics, three IR spectrometers, a VIS spectrometer, a sun-edge sensor, and signal processing units. The band 1 and 2 spectrometers employ a Czerny-Turner type spectrograph design with a plane grating in 30 gr./mm for band 1; the detector material for both bands is $PbTiO_3$. The band 3 spectrometer employs an echelle grating with 23.2 gr./mm. The VIS spectrometer uses a holographic concave grating (f/8.0, f=400 mm, 1800 lines/mm) with a 1024 pixel MOS photo diode array detector. The VIS spectrometer is self-calibrating using the information on the solar Fraunhofer lines. Instrument spectral coverage: 3-12.85 µm and 753-784 nm, spatial coverage = 10 - 60 km, vertical resolution = 1 km, observation accuracy = 5% (1% for ozone). The prime contractor for ILAS-II is MEI (Matsushita Electric Industrial Co. Ltd.).[612)]

POLDER (Polarization and Directionality of the Earth's Reflectances), passive optical imaging radiometer of CNES.

Receiving Frequency	401.65 MHz±0.0405 MHz
Receiving Signal Bit Rate	400bit/s
Receiving Signal Modulation Mode	PCM(Bi phi -L)/PM
Receiving Signal Bit Error	below 1×10^{-5}
Transmitting Frequency	465.9875 MHz
Transmitting Power	over 5 W
Transmitting Signal Bit Rate	200 bit/s
Transmitting Signal Modulation Mode	PCM(Bi-L)/PM
UHF Antenna	Formed broad beam pattern
G/T	over −36.6 dBk
EIRP	over 27.1 dBm

Table 142: Some DCS characteristics

DCS = Data Collection System (NASDA/CNES joint development). The DCS offers worldwide capabilities for location and environmental data collection for fixed and moving platforms. The downlink frequency = 460-470 MHz with a data rate of 200 bit/s is added to the existing ARGOS. The received frequency of the DCP (Data Collection Platform)= 401.65 MHz, the data rate of the DCP=400 bit/s. Total DCS instrument mass = 76 kg, power consumption = 60 W.

D.3 ALOS (Advanced Land Observing Satellite)

ALOS is a Japanese Earth-observation satellite, developed by NASDA, manufactured by NEC, Toshiba, and Mitsubishi Electric Corp. The objectives call for an optical and microwave sensor payload who's high-resolution data may be used for such applications as cartographic mapping, environmental and hazard monitoring (within 48 hours). The intent is to provide the user community with data of sufficient resolution to be able to generate 1:25,000 scale maps. This in turn requires observational data of 2.5 m horizontal resolution for the determination of land conditions, and a 3-5 m vertical accuracy for contour mapping. Multispectral data with 10 m horizontal resolution is needed for the classification of land cover (vegetation, forests, etc.). Short-term hazard monitoring (within 24 hours on average) will be accommodated by the use of pointing mechanisms. Particular application features of the ALOS mission are:

- The mapping of land areas (without the need for ground control points)
- The monitoring of disasters (as a complement to the capabilities of other spacecraft)

612) http://www.ilas2.nies.go.jp/

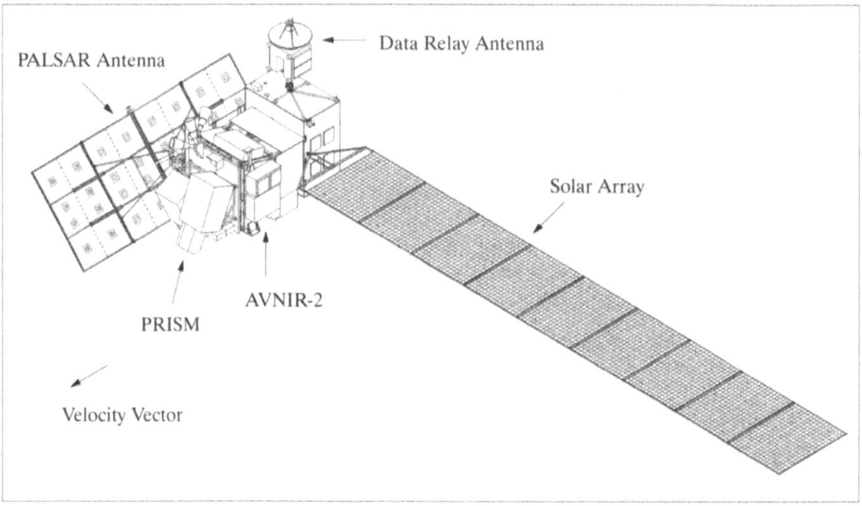

Figure 86: Illustration of the ALOS S/C model

The S/C structure consists of CFRP (Carbon Fiber Reinforced Plastic) trusses and aluminum fittings. The approximate S/C dimensions in the stowed configuration are: 6.4 m x 3.4 m x 4.3 m (x, y, z); the in-orbit configuration dimensions are: 8.9 m x 27.4 m x 6.2 m (x, y, z, where x is in the velocity direction and z is toward nadir). The ALOS S/C requires in particular precise attitude and position determination to minimize image quality degradation. This is provided by star trackers, carrier phase tracking of a GPS receiver, and by an IRU (Inertial Reference Unit). The actuators used are: reaction wheels, magnetic torque rods and hydrazine thrusters. A short-term attitude stability of ±0.00002°/0.37 ms (3 sigma) is provided; the long-term stability is ±0.0002°/5 s (3 sigma).). The pointing accuracy knowledge is ±0.0002°, and the spacecraft position accuracy knowledge is ±1.0 m. The dual-frequency carrier-phase tracking GPS receiver of Toshiba Corp. is used for orbit determination. The S/C mass is 4000 kg (at lift-off), the largest satellite ever for Japan. Its 22 m long solar array generates power of 7 kW (EOL). The design life is 3 years with a goal of 5 years. [613) 614)]

A launch of ALOS is planned for mid-2003 by a Japanese H-IIA rocket from Tanegashima Space Center, Japan. [615) 616)]

Orbit: Sun-synchronous near-recursive circular orbit, altitude = 692 km, inclination = 98.16°, repeat cycle = 46 days (with a sub-cycle of 2 days), local time at descending node 10:30 AM (±15 min).

Communication: The primary data transmission link is via DRTS (Data Relay and Test Satellite of Japan) in Ka-band for mission data at 240 Mbit/s, and S-band for TT&C data. In addition there is an X-band downlink for maximum data rates of 120 Mbit/s. This is considered a backup only for AVNIR-2 data. A further Ka-band downlink at 120 Mbit/s is considered via the Artemis relay satellite of ESA. - ALOS is equipped with a solid-state recorder with a capacity of 768 Gbit using 64 Mbit DRAM technology. The storage capacity is suffi-

613) http://alos.nasda.go.jp/index–e.html

614) http://www.eorc.nasda.go.jp/ALOS/set_about.html

615) T. Hamazaki, "Overview of the Advanced Land Observing Satellite (ALOS): Its Mission Requirements, Sensors, and a Satellite System," presented to ISPRS Joint Workshop "Sensors and Mapping From Space 1999," International Society for Photogrammetry and Remote Sensing (ISPRS), Sept. 27-30, 1999

616) Y. Osawa, H. Wakabayashi, K. Toda, T. Hamazaki, "Advanced Land Observing Satellite (ALOS): Mission Requirements, Payloads and Satellite System," paper of NASDA provided by K. Misawa

cient for a 50 minute recording of a 240 Mbit/s data stream. The data rate capabilities of the recorder are: 360 Mbit/s for recording and 240 Mbit/s for readout of playback data.

Sensor complement:

PRISM (Panchromatic Remote-sensing Instrument for Stereo Mapping). The objective is to obtain high-resolution stereo data (pixel size of 2.5 m) for cartographic applications (extraction of highly accurate digital elevation models, etc.). The instrument is a "three-line sensor" with three independent catoptric systems for nadir, forward and backward-looking to achieve along-track stereoscopy. Each of the three telescope employs a three mirror type optics design (30 cm aperture diameter and 2 m focal length) and several CCD detectors for pushbroom scanning. Eight silicon CCDs (5000 pixels each) are physically aligned at each telescope's focal plane. Of the 40,000 pixels per telescope, 14,000 pixels are electronically selected and transferred to the ground station. Thus, a triplet image consists of three (fore, nadir aft-looking) 14,000 pixels/line. - The nadir-looking telescope provides a swath of 70 km width (28,000 pixels per band), each of the fore and aft-looking telescopes provides a swath of 35 km (14,000 pixels per band). The fore and aft telescopes are inclined by ±24° from nadir to realize a base to height/ratio of one at an orbital altitude of 692 km.

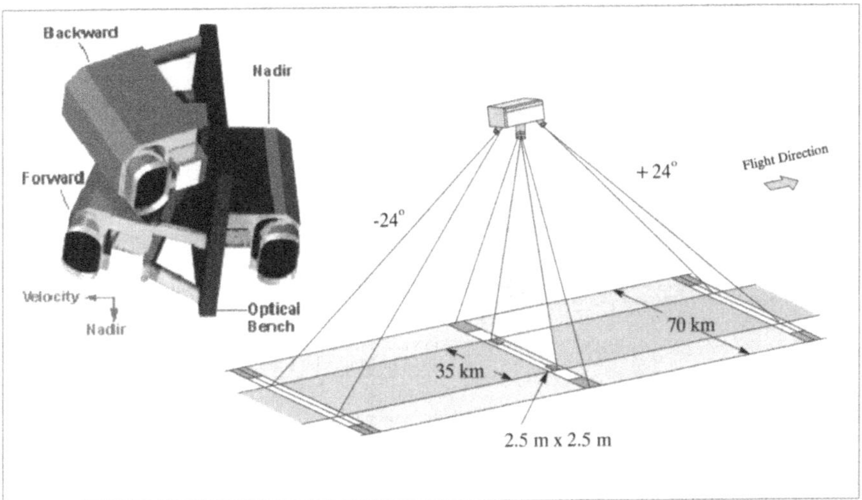

Figure 87: Illustration of the PRISM instrument and imaging configuration

Parameter	Panchromatic Sensor
Spectral band (panchromatic)	0.52-0.77 μm (nadir, forward, backward)
SNR, MTF	>70, >0.2
IFOV	2.5 m (3.57 μrad)
Swath width	35 km (triplet stereo observations), 70 km for nadir observations)
Gimbal angle	±1.5° (cross-track)
Data quantization	7 bits
Data rate	1 Gbit/s of raw data, a lossy (DCT and Huffman coding) data compression technique reduces the actual downlink data rate of PRISM to either 240 Mbit/s (1/4 reduction) or to 120 Mbit/s (1/8 reduction)

Table 143: PRISM parameters

The PRISM optics are mounted on a rigid optical bench which is thermally controlled within ±3° C to minimize distortions in the optics system.

Calibration: PRISM has electrical calibration sources on-board. They are used to estimate gain and offset of the amplifiers. For absolute and relative radiometric calibration, PRISM uses AVNIR-2 data. AVNIR-2 is calibrated using internal calibration lamps.

Note: PRISM is now a separate sensor with stereo capability what used to be the Panchromatic Sensor of AVNIR-2.

AVNIR-2 (Advanced Visible and Near-Infrared Radiometer - 2), a NASDA instrument of AVNIR heritage, built by Mitsubishi Electric Corporation. Objective: provision of high-resolution (10 m) multispectral data. The instrument optics are of "folding Schmidt" type. The telescope aperture is 24 cm in diameter, it has a focal length of about 800 mm. AVNIR-2 features a pointing capability of ±44° in the across-track direction, thereby providing a wide field of regard for disaster monitoring. The silicon CCD detector arrays have 7000 pixels per line. Applications: monitoring of regional environment (land coverage and land-use maps, etc.). A quasi-lossless data compression technique of DPCM (Differential Pulse Code Modulation) with Huffman coding is employed for a source data reduction from 160 Mbit/s to 120 Mbit/s.

Calibration: AVNIR-2 uses two on-board calibration lamps which are used for absolute and relative calibration sequences. In addition, internal electrical calibration is provided.

Parameter	Multispectral Sensor
Spectral bands (4) Band 1 Band 2 Band 3 Band 4	 0.42-0.50 μm 0.52-0.60 μm 0.61-0.69 μm 0.76-0.89 μm
SNR	>200
IFOV (spatial resolution)	10 m (at nadir, 14.3 μrad)
Swath width	70 km
Gimbal angle	±44° (cross-track)
MTF (Modulation Transfer Function)	Band 1-3 =>0.25; band 4 =>0.20
Data rate	about 160 Mbit/s of raw data, a quasi-lossless (DPCM) data compression technique reduces the actual downlink data rate of AVNIR-2 to 120 Mbit/s (3/4 reduction)

Table 144: Some characteristics of the AVNIR-2 instrument

PALSAR (Phased Array L-band Synthetic Aperture Radar), developed jointly by NASDA and JAROS (Japan Resources Observation System Organization), heritage of SAR on JERS-1. PALSAR is a side-looking phased array L-band instrument with a pointing capability from 8 to 60° of incidence angle. The antenna is of size: 8.9 m (length, along-track direction) and 3.1 m in width. An array of 80 transmitting/receiving modules is mounted behind the antenna panels. PALSAR features a ScanSAR mode (employing the active phased array technique) permitting extended observation coverage of at least 250 km swath width. The instrument performance parameters for high-resolution mode at a standard incidence angle of 39° are: 10 m spatial resolution, 70 km swath width, and -23 dB NE sigma-zero.

Parameter	High-resolution Mode		ScanSAR Mode	Polarimetry
Center frequency	1270 MHz (L-band)			
Bandwidth	30 MHz			
Polarization	HH or VV	HH/HV or VV/VH	HH or VV	HH/HV+VV/VH
Spatial resolution (range)	7.0-44.3 m	14.0-88.6 m	≤ 100 m	24.1-88.6 m
Number of looks		2	8	
Swath width	40 - 70 km		250-350 km	30 km
Incidence angle	8° - 60°		18° - 43°	8° - 30°

Parameter	High-resolution Mode	ScanSAR Mode	Polarimetry
Data quantization	3/5	5	3/5
Data rate	120 Mbit/s / 240 Mbit/s		240 Mbit/s

Table 145: Parameters of the PALSAR instrument

The ALOS instruments are capable to observe the surface of the entire world within the following limits:

- Any place within two days
- Around the equator: about 60% of the area within one day
- At latitudes of 35°: about 70% of the area within one day
- At latitudes larger than 55°: any place every day (provided there is no cloud cover for the optical instruments).

Daytime observation modes: PRISM (fore, nadir & aft) and AVNIR-2 simultaneously

Nighttime observation modes: PALSAR (Note: AVNIR-2 and PALSAR are able to operate simultaneously).

Data policy:

The significance of ALOS is its ability to provide global coverage of land observation and to transmit this huge amount of data acquired. NASDA seeks to implement the ALOS Data Node concept for processing and utilizing ALOS data through international cooperation. The Data Node consists of one representative organization in each region and cooperative organizations including the data distributors. All the ALOS data is downlinked to NASDA's Earth Observation Center (EOC) via a data relay satellite. NASDA will transfer the regional level-0 data it has obtained to the ALOS Data Node. The ALOS Data Node is responsible for providing regional level-1 data upon request from data users, including NASDA, throughout the ALOS mission life. The current plan is to divide the entire world into four regions: Asia, Europe and Africa, North and South America, and Australia and Oceania. The Data Nodes would serve both as a regional ALOS data center and a scientific research center. [617)]

The main concept of the ALOS data policy should be that each Data Node would be allowed to handle their regional data according to its own data policy. International talks on this concept are about to produce an agreed inter-Node distribution policy defining an interface where the respective regional policies will meet. The whole mechanism will be a combination of de-centralized research and non-research data distribution. A private consortium may be established for promoting commercial data use.

At a multilateral level, nations have not been able to reach a coherent, effectively functioning data policy despite decades of lengthy discussions. If this Data Node concept is successfully implemented, it may demonstrate an effective model of a program-oriented approach to an integrated, disseminated data policy.

D.4 ALMAZ Program

Background: The ALMAZ (="diamond") Earth Resources Satellite program of the former Soviet Union did not start with the launch of ALMAZ-1 in March 1991. At NPO Machinostroyenia in Reutov near Moscow, S/C of the ALMAZ series have been developed for more than 25 years. The program is based on the technical experience gained at NPO Machinostroyenia when developing manned and automatic space stations - the Salyut Space Stations of the 60's and 70's were a very essential part of the ALMAZ program. The former

617) Information provided by the NASDA ALOS team (Naoto Matsuura) during the review process of my ALOS draft

USSR needed an all-weather remote sensing capability based on SAR technology, due to the cloud coverage conditions predominant in the northern latitudes of its land mass. Even today (1993), ALMAZ automatic (i.e. unmanned) spacecraft are actually of the type 'Salyut space station' - a proven technology - such a S/C can even be visited by a crew in case of serious problems.

Some of the ALMAZ program objectives:

- Development of a series of automatic space stations on the basis of a universal ALMAZ platform
- Development of a sensor complement with upgrading capability in sensor performance characteristics with regard to spatial resolution, coverage, spectral characteristics, etc. Increase in the number of instruments and the number of solvable tasks with every mission.
- Development of a data system on-board and on-ground and a corresponding infrastructure to cope with the vast volumes of data.
- Provision of an all-weather remote sensing capability to support the following tasks:
  - Prospecting for natural resources for many sectors of the national economy
  - Collecting of remote sensing data for ecological analysis and applications
  - Observing natural catastrophes/disasters of immediate interest.

D.4.1 COSMOS-1870 (also Kosmos-1870)

This mission is regarded as the first USSR radar mission (ALMAZ prototype mission). The S/C was launched on July 25, 1987 with a Proton vehicle from the Baikonur Cosmodrome launch facility; the COSMOS-1870 mission ended July 30, 1989.

Orbit: altitude = 275 km, inclination = 73°, orbital period = 92 minutes

Spacecraft mass = 18550 kg, payload mass = 1950 kg, design life = 2 years, attitude precision = 15-20', stabilization precision = 4-6'.

Sensor:

- **SAR** = Synthetic Aperture Radar (S-band; freq. = 3.125 GHz, instrument built by NPO Vega, Moscow). Wavelength = 9.6 cm, spatial resolution = 20-25 m, swath width = 20 km.

The S/C carried two SAR instruments, one on each side for surveying. The S/C had the capability to roll about its axis, thereby extending the pointing range of the SAR antennas in the cross-track direction to 250 km field of regard (swath width is nearly constant at 20 km). The S/C had three tons of fuel on board permitting many roll maneuvers for this type of operational support.

Data: On-board recording capability and subsequent stored data dump during passes over ground stations, or real-time data transmission. Data transmission rate: 90 Mbit/s.

D.4.2 ALMAZ-1

ALMAZ-1 [618), 619), 620), 621), 622), 623)] was launched on March 31, 1991 (Proton booster from the Baikonur Cosmodrome launch facility, ALMAZ S/C designer and builder: NPO Machinostroyenia, Reutov, Moscow Region), the operational phase started in May 1991. End of mission: Oct. 17, 1992 (controlled descent into the Pacific ocean due to lack of fuel).

618) "Soviets Launch Largest Earth Resources Satellite on Modified Salyut Platform," Aviation Week & Space Technology/April 8, 1991, pp. 21-22
619) "Almaz to add Dimension to Earth Study," Space News, March 18-24, 1991, p. 1
620) "ALMAS - Sowjetischer Erdsatellit mit Synthetic Aperture Radar zur Erderkundung," IKF Berlin, 1990 , aus der Reihe: Informationen aus der internationalen Zusammenarbeit.
621) "Almaz to add Dimension to Earth Study," Space News, March 18-24, 1991, p. 1
622) "Sowjetisches Weltraumauge sammelt Ströme digitaler Daten," VDI Nachrichten, 21. Dez., 1990, Seite 20

The ALMAZ-1 S/C has a total mass of 18550 kg, and a payload mass of 3420 kg. Attitude precision = 15-20', stabilization precision = 4-6'. The stabilization precision during the SAR operation is 1 arcmin.

ALMAZ-1 is considered an important operational satellite (after SEASAT in 1978, SIR-A in 1981, SIR-B in 1984 and Cosmos-1870 in 1987) with a radar sensor for Earth observation. The Soviet space agency Glavkosmos has contractual agreements (Western marketing rights for Almaz data) with Almaz Corp., a subsidiary of the Houston-based Space Commerce Corp. (SCC). The Radar data of ALMAZ are complementary to the SPOT and Landsat data. ALMAZ-1 data are also considered to be in direct competition with ERS-1 and JERS-1 data.

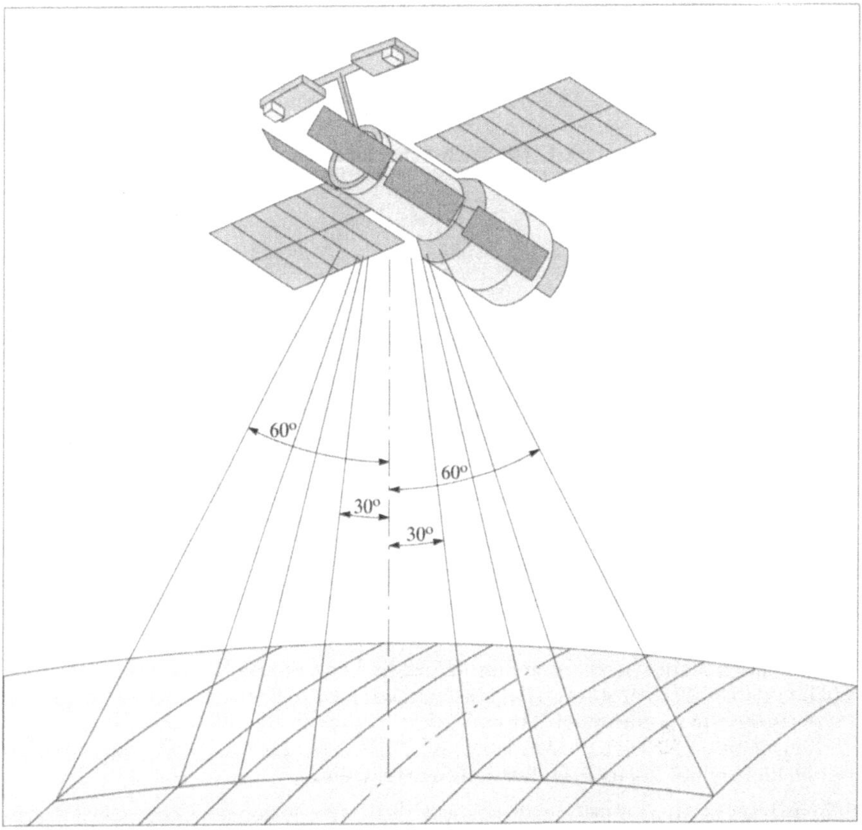

Figure 88: The Almaz-1 S/C model and its SAR sensor's observation geometry

Application: oceanology (study of the distribution and dynamics of currents and hydrospheric fronts, the spatial structure of wave formations and turbulence, the evaluation of surface winds and hurricanes, the topography of the ocean floor and its effects on wave patterns at the surface, the identification of oil spills and other forms of pollution, the state of the ice cover and its seasonal variations, boundaries of water exchange of rivers with the ocean, etc.); geology (structure of geological formations (such as folds, valleys and fractures) and the nature of volcanic activity, survey of mineral deposits, etc.); cartography and

623) "Almaz Falls from Orbit," Space News, Oct 26-Nov. 1, 1992, p. 1

geophysics (topographical maps, climatic changes (such as ice thawing and desertification)); agriculture and forestry (large scale evaluation of agricultural lands and crops, overall volume of the biomass in a region, soil moisture conditions, etc.); statistics, ecology, etc.

Orbit: Inclination = 72.7°; altitude = 270-380 km, Orbital period = 92 Minutes;

Sensor complement:

- **SAR** = Synthetic Aperture Radar (S-band; freq. = 3.125 GHz, wavelength = 9.6 cm, built by NPO Vega).
  - Resolution = 10-15 m (depending on range and azimuth).
  - SAR images can be taken from each side of the satellite; the swath width of each SAR is 40 km within a field of regard (FOR) of 350 km (obtained by rolling the S/C)
  - Observation (incidence) angles: 30-60°
  - Radiometric resolution: 2-3 dB
- **UHF Radiometer**. Wavelengths of 0.8 cm, 5 cm, 11 μm, 12 μm, and 13.7 μm (or 37.5 GHz, 6 GHz, 2.72 THz, 2.5 THz, and 2.19 THz, respectively); swath width = 10-30 km, FOR = 500 km; temperature resolution = 0.1 - 0.3 K, spatial resolution = 5km. Objective: compilation of an Earth surface temperature map.

Data: Ground station in Moscow Region. The ALMAZ-1 schedule calls for a transmission of 60 images (=scenes per day). A scene = 40 km x 300 km. Data rate: 10 Mbit/s. Operational modes: on-board data recording and subsequent transmission via relay satellite to a DRP (Data Reception Point).

D.5 BADR-B

BADR-B, a follow-up of the Pakistani BADR-A microsatellite, in collaboration with European industry and science institutes. SIL (Space Innovations Limited) of Newburry, UK provided the satellite bus, the spacecraft integration was performed by SUPARCO (Pakistan Space and Upper Atmosphere Research Commission), demonstrating the use of relatively inexpensive microsatellite missions in the field of space technology transfer. Prime objectives of the Badr-B mission are: [624) 625)]

- Commencement of an indigenous development capability of low-cost satellites and creation of a needed infrastructure for future development in this field
- Acquisition of know-how in all fields of satellite and instrument design/development.

Background: BADR-A, a microsatellite of 52 kg (an experimental digital communications satellite built by SSTL of Surrey, UK), was launched July 16, 1990 as a secondary payload on a Chinese booster (Long March 2) from Xichang, China, into an elliptical orbit (perigee of 208 km, apogee of 988 km, inclination of 28.5°). The S/C operated until Aug. 21, 1990; it ceased functioning due to an on-board subsystem failure.

BADR-B is a gravity-gradient stabilized small Earth Observation satellite (the S/C is spin-stabilized prior to boom deployment). The S/C bus structure is polyhedral in shape fabricated from aluminum alloy (T6061). The bus has a size of 51 cm x 51 cm x 46.5 cm. Most subsystems are mounted onto a central shelf, which is attached to a thrust tube; its base contains a strong-ring, forming part of the separation system. The boom has a length of 6 m, it is mounted within the trust tube and deploys through the strong-ring. The Earth-pointing face of the S/C accommodates the communication antennas (VHF, UHF, and S-band), the magnetometer, and the CCD camera. S/C power is provided with GaAs solar panels mounted on the external surface of the S/C bus. Attitude sensing is provided by a pair of two-axis digi-

624) I. Iqbal, A. V. Qureshi, A. S. Ahmed, "SUPARCO BADR Satellite," International Workshop on Low-Cost Space Missions, Islamabad, Pakistan, Nov. 24 to Dec. 4, 1999

625) http://www.sil.com/PROJECTS.htm

tal sun sensors and by a three-axis fluxgate magnetometer. Actuation is achieved by a pair of magnetorquer rods and by the gravity-gradient boom (4 kg of tip mass). The satellite mass is 70 kg, and produces 25W of average power.

RF communications: S-band telemetry and telecommand, VHF/UHF store-and-forward operations.

S-band uplink freq.	2061.976 MHz	**UHF uplink frequency**	449.850 MHz
Subcarrier modulation	PSK	Subcarrier modulation	FSK
Carrier modulation	PM	Carrier modulation	FM
Data rate	4 kbit/s max.	Data rate	1.2 kbit/s
Line coding	SP-L	Line coding	NRZ-L
S-band downlink	2239.250 MHz	**VHF downlink**	143.625 MHz
Subcarrier modulation	None	Subcarrier modulation	FSK
Carrier modulation	BPSK	Carrier modulation	FM
Data rate	150 kbit/s max	Data rate	1.2 kbit/s
Line coding	SP-L	Line coding	NRZ-L
S-band beacon	2250 MHz on +17 dBm carrier		

Table 146: Overview of communication parameters

BADR-B will be launched on a Zenit-2 launch vehicle from Baikonur as a secondary payload in 2001. The prime payload on this flight is METEOR-3M-1 of Russia; other secondary payloads are: Compass of IZMIRAN, Moscow, Maroc-Tubsat of Morocco, and REFLECTOR (Russian/US microsatellite).

Orbit: Sun-synchronous polar orbit, altitude = 830 km, inclination = 98.85°, period = 102 minutes.

Sensor complement:

EIC (Earth Imaging Camera), developed at RAL. EIC utilizes a wide-angle CCD array camera with an instrument mass of 2.5 kg. The camera operates at visible wavelengths and uses a large format CCD to achieve 250 m x 250 m resolution on the ground with a field of view of ±8.5° (equivalent to 194 x 144 km on the ground). The challenge in designing and building this instrument has been the need to maintain performance with restrictions on the available space, mass, power and schedule. Snapshot imagery can be provided from any part of the globe, stored on-board and later transmitted to the ground. [626)]

CDE (Compact Dosimeter Experiment). The objective is to measure ionizing radiation levels encountered in polar orbits. The total accumulated dose is measured with miniature RadFET (Radiation-sensitive Field Effect Transistor) sensors based on pMOS technology, installed at various spacecraft locations (total of 8 RadFETs).

S&FE (Store & Forward Experiment). The objective is to provide point-to-point communications for 5-7 users. It consists of a UHF/VHF communications system linked to the on-board computer. The system permits uplinking of encoded information along with recipient address. The information is downlinked when the S/C is in reach of the recipient user location.

Battery End-of-Charge Detector. The objective is to monitor the battery charge status. The technique employed is to monitor the temperature of the battery. It detects a small rise in temperature at the end of the battery charge sequence and acts for overcharge protection.

626) http://www.ssd.rl.ac.uk/ssd/ccdtg/BADR–B.htm

Figure 89: Illustration of the camera head and the compact electronics control unit

D.6 Bhaskara

In 1975, ISRO (Indian Space Research Organization) started the design and development of two experimental Earth observation satellites, Bhaskara-I (launch June 7, 1979), and Bhaskara-II (launch Nov. 20, 1981). The objective of the program was to develop an initial capability of satellite/instrument building, to gain hands-on experience in satellite operations, and to develop a capability/infrastructure for data reception, processing, evaluation and interpretation on practical remote sensing applications (oceanography, atmosphere, etc.). [627) 628)]

The Bhaskara satellites were spin-stabilized [heritage and design of the first Indian satellite, **Aryabhata**, [629)] launch April 19, 1975, orbital altitude of 400 km, inclination of 50.7°, mass = 360 kg, Aryabhata was instrumented to explore the Earth's ionosphere, measure neutrons and gamma rays from the sun, and perform investigations in x-ray astronomy], generated 47 W of power from silicon solar panels, and used NiCd batteries for orbital eclipse periods. The S/C spin axis was maintained at right angles to the orbital plane. The S/C structure was a quasi-spherical polyhedron, 1.59 m in diameter and a height of 1.19 m. The mass of each S/C was 444 kg, the design life was one year. The attitude control system employed cold gas jets in combination with a nutation damper and magnetic bias coils. The Bhaskara satellites were launched on a Vostok vehicle from the Kapustin Yar Cosmodrome (provided by the USSR) within the Intercosmos program.

S/C Communication: VHF antenna; low bit rate telemetry at 224 bit/s in PCM/FM/PM modulation; high bit rate at 91.4 kbit/s in PCM/PSK modulation. Telecommand in PCM/FSK/AM modulation. Data reception at SAC and at SHAR (Sriharikota, East cost launch facility of ISRO), mission control and operations from SAC (Space Applications Center, Ahmedabad). The Bhaskara-I satellite was formally turned off in March 1981.

Orbit: Near circular orbit, altitude = 557 km (572 km for Bhaskara-II), inclination = 50.7°, period = 95.2 minutes.

627) Information provided by George Joseph of ISRO

628) "Bhaskara - Satellite for Earth Observations," ISRO publication, June 1979

629) Aryabhata (476-550) and Bhaskara (1114-1185) were two ancient mathematicians and astronomers of India. Aryabhata is the earliest Hindu mathematician whose work and history are available to modern scholars. He was one of the first known to use algebra. Bhaskara ("The Learned") was the leading mathematician of the 12th century, who wrote the first work with full and systematic use of the decimal number system.

Sensor complement:

Vidicon TV Camera, PI: George Joseph. Objective: Collection of Earth surface imagery for use in geology, forestry and land use. The two camera assembly was of RBV type, providing framing imagery with a slow scan vidicon coupled to an image intensifier (exposure is achieved by electronically gating the intensifier). The optical axis of the camera was normal to the spin axis of the S/C, providing a 'look' along the local vertical once every spin. Focal length of the imaging lens = 18.46 mm, FOV = 49.37°, f/1.9. Spectral channels: 0.54 - 0.66 μm (camera-1) and 0.75 - 0.85 μm (camera-2). An image frame had the size of 341 km x 341 km for an orbital altitude of 525 km (typical overlap of imagery about 10%), the spatial resolution was about 1 km. The instrument provided commandable exposure control (1, 1.5 or 2 ms). Power = 225 W, mass = 44 kg.

SAMIR (Satellite Microwave Radiometer), developed at ISRO/SAC, PI: P. N. Calla. [630] A nadir-pointing three-channel Dicke-type passive radiometer with the objective to measure atmospheric water vapor (liquid content), surface rain and wind (ocean surface studies). Frequencies at 19, 22, and 37 GHz. Ground resolutions (footprint) were 125 km for the 19-GHz channel, 150 km for the 22-GHz channel, and 230 km for the 37-GHz channel. Swath widths were 300 km and 230 km.

In addition, the following experiments (secondary payloads) were conducted on the Bhaskara platform: Heat Pipe Experiment, Solar Cell Experiment, X-Ray Sky Monitor Experiment (objectives: investigations on transient X-ray sources and on the long term variability of steady X-ray sources in the energy range of 2 to 10 keV), Thermal Control Coating Experiment, and a DCS (Data Collection System) experiment (downlink frequency at 136.43 MHz, uplink frequency at 402.5 MHz) for data collection from remote platforms in India.

D.7 CBERS (China/Brazil - Earth Resources Satellite)

CBERS is a cooperative program between the People's Republic of China, PRC (CAST) and Brazil (INPE), it was started in 1988 to establish a complete remote sensing system (space and ground segment) to supply both countries with multispectral remotely sensed imagery. Prior to the international cooperative program of CBERS, there existed already a Chinese program, referred to as **Zi Yuan** or simply **ZY**, meaning **resource** in Chinese. The satellite (ZY-1) of this program came to be known as CBERS-1. [631] [632] [633] [634] [635] [636]

D.7.1 CBERS-1

The first satellite, CBERS-1, was launched Oct. 14, 1999 (along with SACI-1 as a piggyback payload) by a Chinese launch vehicle (CZ-4 series) from the Taiyuan launch site in China's Shangxi province.

The CBERS-1&2 S/C is composed of two modules. The payload module houses the optical and electronic systems used for Earth observation and for data collecting. The service module incorporates the equipment that ensures the power supply, the control, the telecommunications and all other functions needed for the satellite operation. The main S/C character-

630) In the Sanskrit language, SAMIR means "breeze"

631) Zhu Yilin, "Ziyuan-1, China's First Earth Resources Satellite (CBERS)," Earth Space Science Review, July-September 1994, Vol. 3, No. 3, pp. 16-19

632) "The China-Brazil Earth Resources Satellite Program," paper provided by G. Santana of INPE

633) "CBERS Spacecraft: Conception and Design," paper presented by E. A. Parada Tude of INPE and by C. Quinnan of CAST at the 1st Brazilian Symposium of Aerospace Technology, Sao Jose dos Campos, Aug. 27-31, 1990

634) G. K. Rayalu, et al., "Multispectral and Multitemporal Optical Sensors of CBERS," INPE internal paper

635) http://www.inpe.br/programas/cbers/english/index.html

636) C. de Oliveira Lino, M. G. Rodrigues Lima, G. L. Hubscher, "CBERS - An International Space Cooperation Program," Acta Astronautica, Vol. 47, No 2-9, 2000, pp. 559-564

istics are: three-axis stabilized, satellite mass = 1450 kg, power = 1100 W (solar), in addition there are two NiCd batteries with 30 Ah of energy. S/C body dimensions: 1.8 m x 2.0 m x 2.2, the solar array has a size of 6.3 m x 2.6 m. The hydrazine propulsion system has 16 thrusters of 1 N and 2 thrusters of 20 N each. The nominal design life is 2 years.

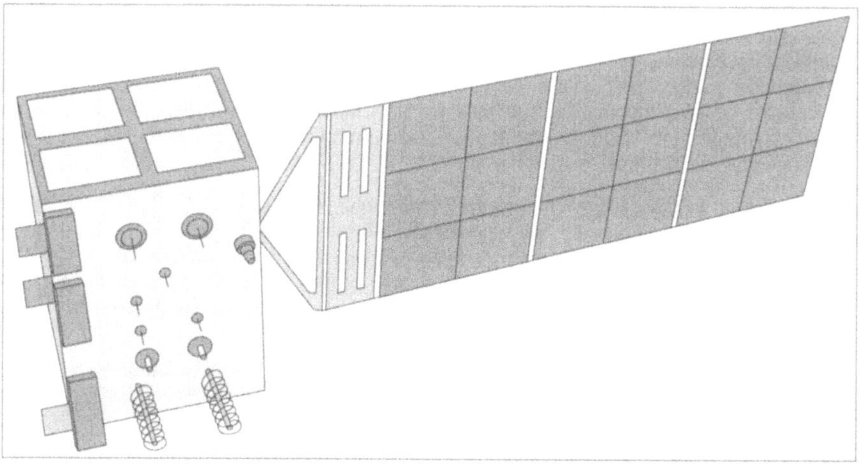

Figure 90: The CBERS/Ziyuan-1 spacecraft model

Orbit: Sun-synchronous orbit; altitude = 778 km; inclination = 98.5°; nodal period = 100.26 min; mean local solar time at descending node=10:30 AM; revisit period=26 days.

Objectives: Observation and monitoring of the Earth's resources and environment with a multi-sensor imaging payload providing different spatial resolutions.

Sensor complement: CBERS features a multisensor payload with imaging sensors of different spatial resolutions and collection frequencies.

Parameter	Band	Wavelength	Radiance (W/m^2 sr)		SNR (dB)	
			Maximum	Minimum	Maximum	Minimum
MS band	B1	0.42-0.52 µm	28.7	4.6	48	32
	B2	0.52-0.59 µm	30.1	3.7	50	31
	B3	0.63-0.69 µm	25.9	2.4	48	26
	B4	0.77-0.89 µm	35.6	2.7	52	29
PAN band	B5	0.51-0.73 µm	55.6	9.0	53	37

Table 147: Spectral parameters of HRCC

HRCC (High Resolution CCD Camera), prime sensor on CBERS, developed and built by the Beijing Institute of Space Machines and Electricity and sponsored by CAST (Chinese Academy of Space Technology). [637] The multispectral instrument consists of the following main subsystems: optical, beam-splitting, detection, and an electronics unit; the auxiliary subsystems are: the side-looking reflector, relative calibration mechanism, focusing device, and a thermal control subsystem. Telescope focal length=520 mm (all refracting lens system); aperture=f/4; spatial resolution = 19.5 m (at nadir); IFOV=25 µrad; FOV= 8.4°; temporal resolution=26 days (nadir view), off-nadir view (pointable side-looking capability of ±32°, providing an increased observation frequency for a given region - any phenomenon or event detected by WFI may be zoomed by HRCC) minimum time lag of 3 days;

637) D. Lin, S. Cui, "CCD Camera for CBERS," Proceedings of the Asian Conference on Remote Sensing, Hongkong, Nov. 1999, pp. 285-288

swath width = 113 km; objective: high resolution imaging. The pushbroom detector subsystem uses a Fairchild detector array (CCD-143A) with a pixel size of 13 μm x 13 μm. There are a total of five bands with 5812 pixels per band. Calibration: a tungsten filament lamp of known spatial profile provides illumination of the CCD arrays. Absolute calibration is made using a ground test site.

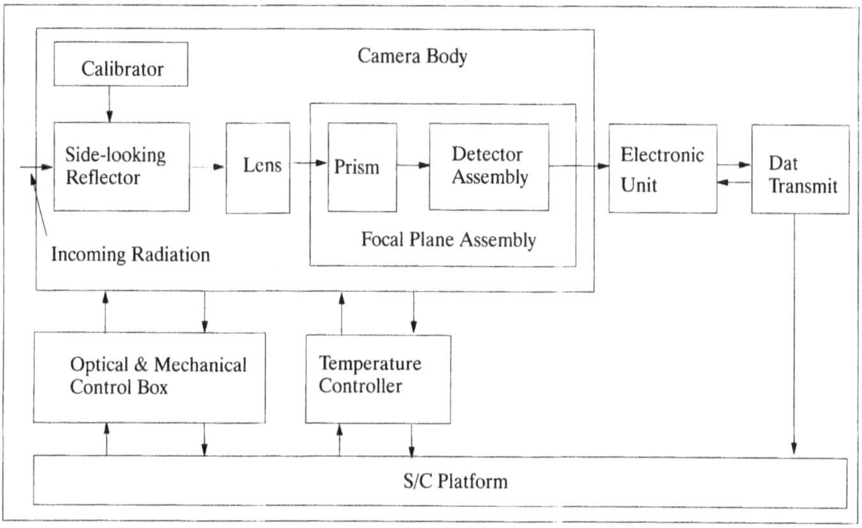

Figure 91: Block diagram of HRCC

IRMSS (Infrared Multispectral Scanner), developed and built by the Beijing Institute of Space Machines and Electricity and sponsored by CAST. [638] The instruments an optomechanical system, consisting of a main module [scan mirror, primary optics (Ritchey Chretien telescope), prime focal plane assembly, relay optics, detectors, etc.], an electronic unit, an on-board calibration assembly, a radiative cooler, and a thermal control unit. The telescope has an aperture diameter of 250 mm and a focal length of 1 m for the Pan and SWIR bands, and 0.5 m for the TIR (Thermal Infrared) band. The telescope mirrors are made of fused silica and held by an invar structure. There are four spectral bands: 0.50-1.1 μm (panchromatic), 1.55-1.75 μm (SWIR), 2.08-2.35 μm (SWIR), 10.40-12.50 μm (TIR); spatial resolution: = 78 m for (0.5 - 2.35 μm ranges), 156 m for the TIR band; temporal resolution = 26 days; swath width = 120 km; IFOV=0.1 mrad (0.2 mrad for TIR band); FOV = 8.78°; objective: medium resolution imaging (complementary information to HRCC). The data rate of IRMSS is 6 Mbit/s.

The linear scan mirror performs cross-track scanning at a scan rate of 5.39 Hz and a scan angle of ±2.5°. Each of the four bands employs an eight-element staggered detector array in the along-track direction. For the TIR band, four of the eight detectors are spares. The Pan band detectors use silicon photodiodes, photovoltaic HgCdTe detectors are used for the SWIR bands, while the TIR band utilizes photoconductive HgCdTe detectors. The detector size is 0.1 mm x 0.1 mm for all bands.

The radiation from the prime focal mirror is divided into three parts by relay optics. IRMSS features three focal plane assemblies (Pan, SWIR and TIR). The Pan band is in the warm FPA while the SWIR and TIR bands are located in the cooled FPAs with temperatures of 148 K and 101 K, respectively. Active and passive thermal control methods are employed.

638) W. Huaiyi, M. Wenpo, "The IRMSS for CBERS," Proceedings of the Asian Conference on Remote Sensing, Hongkong, Nov. 1999, pp. 902-905

The on-board calibration system is used to correct the output changes of the scanner in flight. It includes an internal calibrator (lamp and blackbody) and a solar calibrator. The calibration precision is 3% for both, the band-to-band and the channel-to-channel calibration. The absolute calibration accuracy is <10%. Internal calibration occurs at every return stroke of the scan mirror, sun calibration is performed occasionally on ground command.

While IRMSS employs the same principles as the MSS and TM instruments on the Landsat series, it differs in many ways in the implementation approach such as in the three-level FPAs, the SWIR detectors, the relay optics, and the calibration system.

Parameter	HRCC	IRMSS	WFI
Spectral bands (μm)	0.51 - 0.73 (PAN) 0.45 - 0.52 0.52 - 0.59 0.63 - 0.69 0.77 - 0.89	0.50 - 1.10 (PAN) 1.55 - 1.75 (SWIR) 2.08 - 2.35 (SWIR) 10.4 - 12.5 (TIR)	0.63 - 0.69 0.76 - 0.90
Spatial resolution	20 m	80 m (PAN & SWIR) 160 m (TIR)	260 m
Swath width (FOV)	113 km (8.32°)	120 km (8.78°)	885 km (60°)
Temporal resolution	26 days	26 days	3-5 days
Cross-track pointing	±32°		
Data rate	2 x 53 Mbit/s	6.13 Mbit/s	1.1 Mbit/s
Carrier frequency (X-band)	8.103 and 8.321 GHz	8.216 GHz	8.203 GHz
EIRP	43 dBm	39.2 dBm	31.8 dBm
Modulation	QPSK	BPSK	QPSK
Tracking beam frequency	8.196 GHz	8.196 GHz	8.196 GHz

Table 148: Overview of CBERS imaging instruments

WFI = Wide-Field Imager (INPE). WFI is also a complementary sensor to HRCC. It consists of an electro-optics module (optics, spectral filters, CCD detectors, and radiometric calibration device), a signal processor and a modulator. Spectral bands: 0.63-0.69 μm, 0.77-0.89 μm; spatial resolution = 260 m (3456 pixels per line); temporal resolution = 3-5 days; swath width = 885 km (FOV = 60°); data quantization = 8 bit; data rate = 1.1 Mbit/s; objective: acquisition of low-resolution wide-swath imaging (WFI provides a synoptic view). Note: The WFI instrument experienced a malfunction in May 2000, after seven months of service.

In addition to the imaging payload, the satellite carries a Data Collection System (DCS) for environmental monitoring; a Space Environment Monitor (SEM), and an experimental High Density Tape Recorder (HDTR) to record imagery on-board. Each DCP (Data Collection Platform) in the ground segment transmits randomly short digital data messages at a fixed rate (in UHF band). The DCS retransmits these messages in UHF and S bands to Earth. The DCS concept is similar to that of SCD-1 (see C.6).

DCS system access type	Random
Uplink carrier frequency, EIRP	401.635 MHz ±30 kHz (UHF), 33 dBm
Downlink carrier frequencies	2267.52 MHz (S-band); EIRP = 20 dBm 462.5 MHz (UHF-band); EIRP = 35 dBm

Table 149: The DCS parameters

SEM = Space Environment Monitor. Objective: detection of high-energy radiation. The CAST instrument consists of the following subsystems: PRD (Particle Radiation Detector), APD (Aurora Particle Detector), CRED (CMOS-IC Radiation Effect Detector), SDM (SEM Data Memory) and a 4.8 kHz subcarrier modulator.

Data: X-band downlink transmission of all science data in four channels, two for HRCC (8103 and 8321 MHz) at data rates of 2 x 53 Mbit/s, one for IRMSS (8216.84 MHz) at a data rate of 6.13 Mbit/s, and one for WFI data (8203.35 MHz) at a data rate of 1.1 Mbit/s. The modulation is QPSK for HRCC and WFI data streams and BPSK for IRMSS. The TT&C subsystem uses S-band and UHF-band data transmission.

Both countries, China and Brazil, plan widespread use of CBERS image data in their countries and offer the data also to other nations.

D.7.2 CBERS-2

CBERS-2 is planned to be launched in Dec. 2001 from the Taiyuan launch site in China. The spacecraft carries the identical payload as CBERS-1. The intend is to provide continuity of observations.

Orbit: Sun-synchronous orbit; altitude = 778 km; inclination = 98.5°; nodal period = 100.26 minutes; mean local solar time at descending node = 10:30 AM

D.7.3 CBERS-3 & 4

The planned cooperative CBERS 3&4 program of CAST and INPE employs enhanced versions of spacecraft and instruments. The specification of the project is still under discussion, but the new satellites CBERS 3&4 will be compatible with CBERS 1&2. The planned launch date for CBERS-3 is 2004. [639)]

Orbit: Sun-synchronous orbit; altitude = 778 km; inclination = 98.5°; nodal period = 100.26 minutes; mean local solar time at descending node = 10:30 AM

Sensor complement:

PANCAM (Panchromatic Camera). The objective is to provide imagery for cartographic applications. The instrument features a spectral range of 0.51-0.73 μm with a ground resolution of 5 m and a swath width of 60 km. In addition, an instrument cross-pointing capability is provided to extend the FOR (Field of Regard) of the instrument.

MUXCAM (Multispectral Camera) of CAST and of HRCC heritage. The spectral coverage is identical to HRCC with the Pan band eliminated. The ground resolution is 20 m with a swath width of 120 km. The additional instrument feature is a cross-pointing capability.

Parameter	MUXCAM	PANCAM	IRCAM	WFI
Spectral bands (μm)	0.45-0.52 Blue 0.52-0.59 Green 0.63-0.69 Red 0.77-0.89 VNIR	0.51-0.73	0.80-1.10 NIR 1.55-1.75 SWIR 2.08-2.35 SWIR 10.40-12.50 TIR	0.52-0.59 Green 0.63-0.69 Red 0.77-0.89 VNIR
Spatial resolution (m)	20	5	20/40/40/80 (TIR)	160
Swath width (km)	120	60	120	890
Revisit capability (days)	4	5		
Normal revisit time (days)	26	1-26	26	5
Data quantization (bits)	8	8	8	8
Data rate (Mbit/s)	64	140	26	6

Table 150: Basic characteristics of the CBERS-3 & 4 instruments

IRCAM (Infrared Camera) of CAST of IRMSS heritage. Three of the four bands are identical to those of IRMSS (0.80-1.10; 1.55-1.75; 2.08-2.35; 10.4-12.5) μm. The spatial resolution is halved with regard to IRMSS.

WFI = Wide-Field Imager (INPE). WFI features three spectral bands (0.52-0.59; 0.63-0.69; 0.77-0.89) μm, with ground resolution of 160 m and ground swath of 890 km.

639) Information provided by Luiz A. Bueno of INPE

D.8 CORONA

Declassification of all imagery, acquired by three early US military/CIA reconnaissance programs (CORONA, ARGON, and LANYARD), was announced by the US government (Vice President Al Gore) on February 24, 1995. CORONA is the program name of the first and most extensive operational DoD satellite photo-reconnaissance series, managed jointly by the USAF and by the CIA, during the observation period 1960 - 1972. 640) 641)

Camera system	KH-1	KH-2	KH-3	KH-4	KH-4A	KH-4B
Camera manufacturer	Fairchild Camera Co.		Itek Corporation			
Period of operation	1959 - 1960	1960 - 1961	1961 - 1962	1962 - 1963	1963 - 1969	1967 - 1972
Total missions Successful missions (usable photography)	10 1	10 4	6 4	26 21	52 49	17 16
Mission designations	9000 series				1000 series	1100 series
Mission series life	1 day	2-3 days	1-4 days	6-7 days	4-15 days	19 days
Orbit (average values) Perigee Apogee Inclination	 190 km 820 km 80°	 250 km 700 km 81.5° - 83°	 215 km 230 km 81° - 82°	 210 km 410 km 81° - 83°		

Table 151: Overview of the operational CORONA reconnaissance program

The industry team of the CORONA program included: Lockheed Missiles & Space Co. (S/C integrator), Itek with Fairchild Camera & Instrument Corporation (camera system), General Electric (recovery capsule), and Kodak (film material).

CORONA's first successful mission on Aug. 18, 1960 [Discoverer-14 (mission 9009) on a Thor launch vehicle from VAFB, on-orbit mass = 850 kg] opened a new era in spaceborne reconnaissance (successful air recovery of the first photo reconnaissance product, stowed away in a reentry capsule, on Aug. 19, 1960). This film product became the first object in history to be recovered in mid-air by an aircraft (USAF C-119). The cassette contained observations stored on film material of about 1000 m in length.

The total coverage of Earth surface imagery acquired by the CORONA program amounts to nearly 2 billion km^2 (95% abroad and 5% of domestic US coverage) of high-resolution data, collected over a 12 year period. After declassification in 1995, it represents a significant amount of imagery to the civil community. 642) 643)

Sensor complement:

The users of the imagery referred to the reconnaissance satellites and their imagery by the KH (KeyHole) designators that were assigned to the camera systems.

The quality of CORONA's imagery improved significantly over the life of the program from KH-1 to KH-4B. The resolution of the KH-4 systems could be enlarged routinely by a factor of 20, for the KH-4A and -B systems, enlargement up to a factor of 40 was possible. Starting with KH-3 missions, a secondary sensor system in the form of horizon, stellar and/or index cameras, was installed in parallel. The imagery acquired of the horizon camera was used to determine the attitude of the major imaging camera system. The imagery of the star camera was used for accurate determination of pitch, roll, and yaw angles during operational cycles. The index (or terrain) camera acquired correlative imagery used for proper position of the prime imagery.

640) R. A. McDonald, "CORONA: Success for Space Reconnaissance, A Look into the Cold War, and a Revolution for Intelligence," PE&RS, Vol. 61, No. 6, 1995, pp. 689-719

641) R. A. McDonald, "Opening the Cold War Sky to the Public: Declassifying Satellite Reconnaissance Imagery," PE&RS, Vol. 61, No. 4, 1995, pp. 385-390

642) J. Leachtenauer, K. Daniel, T. Vogl, "Digitizing Satellite Imagery: Quality, and Cost Considerations," PE&RS January 1998, pp. 29-34

643) J. T. Richelson, "Scientists in Black," Scientific American, Feb. 1998, pp. 38-45

KH-1 and KH-2 camera systems. These were panchromatic nadir-looking reciprocating film cameras, providing a 70° scan (FOV) in the cross-track direction. The KH-1 camera provided constant-velocity image motion compensation, while for the KH-2 camera, image motion compensation was variable throughput the orbit.

KH-3 cameras. There were five design changes. 1) the structural design was modified to avoid the negative impact of thermal differentials, 2) the camera controls were made more reliable, 3) the method of metering film and achieving/maintaining camera focus was improved, 4) the scan arm design was improved, and 5) a faster lens system was used, which in turn permitted the use of slower, finer grain film.

KH-4 stereo camera system. The KH-4 system consisted of two KH-3 cameras on a common mount, one aft-looking at 15°, the other forward-looking at 15° from nadir, providing a 30° view angle for stereo photography. The cameras were mounted back-to-back and scanned the target region in opposite directions.

KH-4A and KH-4B camera systems. The KH-4A system featured increased film load with a second film recovery bucket - thereby offering improved coverage of imaging targets. The KH-4B camera, a further improvement over KH-4A, could accommodate a variety of film types and operate more effectively under varying exposure conditions.

Camera system (data user designation)	KH-1	KH-2	KH-3	KH-4	KH-4A	KH-4B
Camera model (engineering designation)	C	C'	C'''	Mural	J or J-1	J-3
Imaging capability	mono	mono	mono	stereo	stereo	stereo
FOV (mono) FOV (stereo)	70°	70°	70°	70° 30°	70° 30°	70° 30°
Shutter					focal plane	focal plane
Lens	f/5 Tessar	f/5 Tessar	f/3.5 Petzval	f/3.5 Petzval	f/3.5 Petzval	f/3.5 Petzval
Focal length	62 cm	62 cm	62 cm	62 cm	62 cm	62 cm
Ground resolution (m) Film resolution (l/mm)	12.5 50-100	8 50-100	4-8 50-100	3-8 50-100	2.7-8 120	2 160
Image size (km x km)					20 x 265	16 x 215
Film base	acetate	polyester	polyester	polyester	polyester	polyester
Film width	5.3 cm	5.3 cm	5.7 cm	5.7 cm	5.7 cm	5.7 cm
Image format (cm x cm)	5.3	5.5	5.7 x 76	5.5 x 76	5.5 x 76	5.5 x 76

Table 152: Some KH (Keyhole) camera parameters

D.9 ENVISAT (Environmental Satellite)

The ENVISAT program of ESA is a renamed successor program to the original POEM (Polar Orbit Earth-Observation Missions) utilizing the Polar Platform (PPF). 644) 645) 646) 647) 648) 649)

Background: Originally, two parallel and staggered mission series were proposed in the POEM program. The first series was planned with a Morning Orbit (M), the second series

644) "The ESA Earth Observation Programme and its Role in Global Remote Sensing," P. Goldsmith, Proceedings of the Twenty-Third International Symposium of Remote Sensing of the Environment," Vol. I, ERIM, Ann Arbor, MI, pp. 125-137.

645) Programme Proposal for the first Polar Orbit Earth-Observation Mission using the Polar Platform, Part 1 , ESA paper, 31-08-89

646) Objectives and Strategy for the Earth-Observation Programme of the European Space Agency,ESA, Oct. 88

647) Polar Platform Concept Evaluation, ESA paper, Sept. 25, 1989

648) Programme Proposal for the first ESA Polar Platform, ESA/PB-EO (89) 32, Sept. 1, 1989

649) Programme Proposal for the Development and Exploitation of the First Polar Orbit Earth-Observation Mission (POEM-1) using the Polar Platform, ESA/POEM 1, Issue 1, Oct. 28, 1991, Part 1, Issue 1, Oct. 30, 1991, Part 2

with an (After) Noon Orbit (N). The objectives of the M-series were dedicated studies in meteorology/atmosphere/ocean/ice/environment, while the N-Series was considered for the study of land resources/atmosphere/environment. This scenario was followed in 1991 by the POEM-1 mission (a single satellite mission) which fulfilled both environmental and meteorological mission objectives.

The updated POEM program of 1992/93 considers a split scenario of two dedicated mission series called **ENVISAT** (Environmental Satellite) and **MetOp** (Meteorology Operational Programme). The key element of the split scenario is the transfer of the meteorological instrument package onto a separate platform with a morning orbit devoted primarily to operational meteorology and climate monitoring. This allows the ENVISAT mission to focus on environmental issues that are more research-oriented with a package of essentially pre-operational instruments.

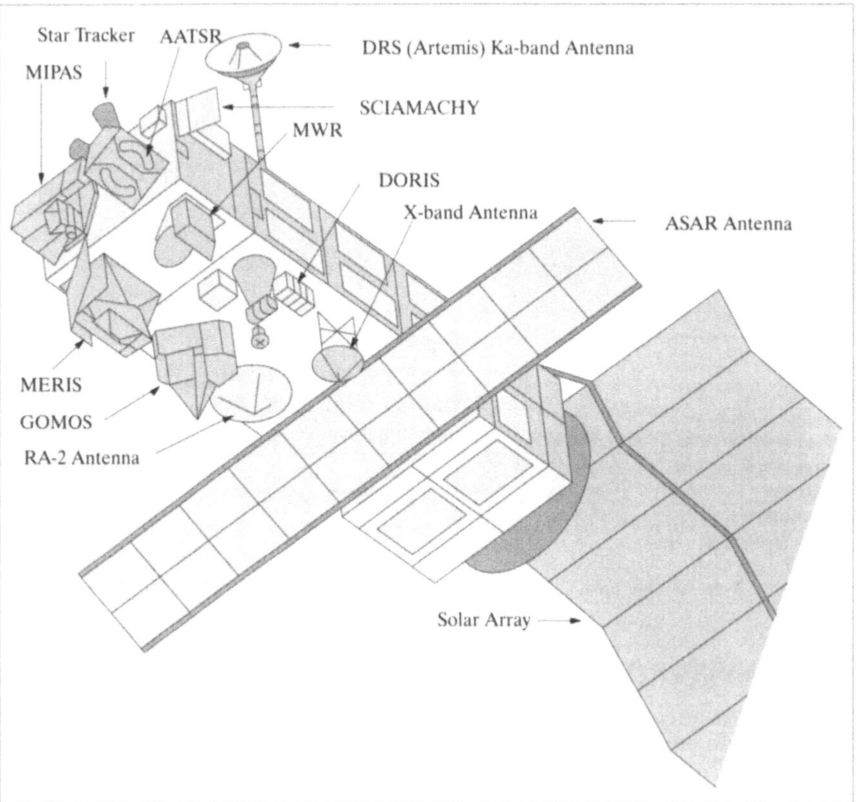

Figure 92: Illustration of the ENVISAT spacecraft

ENVISAT overall objectives: Studying and monitoring the Earth's environment on various scales, from local through regional to global. [650] [651] [652] Monitoring and management of the Earth's resources, both renewable and nonrenewable. Continuation and improvement

650) ENVISAT Special Issue, ESA Bulletin No 106, June 2001

651) "ENVISAT - Europe's Earth Observation Mission for the new Millennium," ESA Earth Observation Quarterly, No. 60, 1998

652) http://envisat.esa.int/

of the services provided to the worldwide operational meteorological community. Contribution to the understanding of the structure and dynamics of the Earth's crust and interior. Major disciplines covered: meteorology, climatology, environment, atmospheric chemistry, vegetation, hydrology, land use, ocean and ice processes.

Spacecraft: 653) 654) 655)

Configuration: The ENVISAT satellite comprises the PPF bus and the payload complement. The PPF itself comprises: a) the service module (SM) providing the main satellite support functions, and b) the payload module (PLM), providing accommodation and supporting subsystems (data handling, power, communications) to the payload complement.

The SM is derived from the SPOT Mk-II bus design. The PPF is being developed and integrated by MMS, Bristol, UK as the prime contractor.

The payload module features externally mounted antennas (of ASAR and RA-2) and the following instruments (MERIS, MIPAS, GOMOS optical assembly, SCIAMACHY, MWR and DORIS) on a payload carrier. In addition there is a payload equipment bay which houses internally-mounted instruments (ASAR, RA-2, GOMOS) and subsystems like instrument control and data handling as well as X-band and Ka-band communications. The industrial team for the ENVISAT payload is being led by DASA/DSS.

Orbit: Near-circular sun-synchronous orbit; altitude = 800 km; descending node at 10 AM local time, inclination = 98.55°, orbit period = 100.6 minutes, repeat cycle = 35 days, same ground track as ERS-2.

The launch of ENVISAT is projected for Nov. 2001 on an Ariane-5 launch vehicle, nominal lifetime of 5 years (including 6 months' commissioning phase). Spacecraft operations are being conducted at ESA/ESOC, the payload center is at ESA/ESRIN.

Parameter	Description
S/C dimensions	Launch configuration: 10.5 m length; 4.57 m envelope diameter In-orbit configuration: 26 m x 10 m x 5 m
S/C stabilization	Three-axis stabilized, attitude pointing: $<0.1^{\circ}$ (3 sigma); attitude measurement: $<0.03^{\circ}$ (3 sigma)
S/C mass	8140 kg including 319 kg of hydrazine
Payload mass	2150 kg (instruments and interfacing hardware)
S/C power	6.5 kW (EOL) with single sided solar arrays, size of solar panel: 5 m x 14 m Eclipse power provided by eight 40 Ah NiCd batteries
Payload power	1.9 kW sunlight and eclipse average (peak = 4.1 kW)
On-board data storage	Three tape recorders of 30 Gbit each, a LBR (Low Bit Rate) recording rate capability of 4.6 Mbit/s and a replay rate of 50 Mbit/s. One solid-state recorder of 60 Gbit, for LBR as well as ASAR high rate and MERIS full resolution
Payload data transmission	2 Ka-band channels each at 50/100 Mbit/s via DRS (Artemis) for payload data 2 X-band channels each at 50/100 Mbit/s (direct transmission to ground)
RF links	S-band direct to/from ground via DRS for TT&C Ka-band to ground via DRS for payload data X-band direct to ground for payload data
RF link operation	Half link operation uses BPSK modulation, full link operation with QPSK
Command & Control	2 kbit/s uplink and 4.096 kbit/s downlink in S-band

Table 153: Overview of ENVISAT spacecraft parameters

Data transmission:

The satellite is capable of providing simultaneous operation of X-band and Ka-band channels.

653) J. Louet, "The Envisat Mission and System," ESA Bulletin No 106, June 2001, pp. 11-25

654) P. A. Dubock, F. Spoto, J. Simpson, D. Spencer, E. Schutte, H. Sonntag, "The Envisat Satellite and its Integration," ESA Bulletin, No 106, June 2001, pp. 26-45

655) "ENVISAT: Mission and System Summary," ESA brochure, March 1998

- Global mission data (low rate) = 4.6 Mbit/s - 'global mission' involves the operation of all low data instrument modes, whenever observation conditions permit, with on-board data recording. Nominal operation involves one tape dump per orbit at 50 Mbit/s.
- Regional mission data (the term 'regional mission' implies real-time data acquisition via DRS or via X-band ground stations.). This includes:
  - ASAR data in its imaging modes with a real-time dedicated channel at 100 Mbit/s
  - MERIS full resolution data multiplexed with low rate instrument data. This includes half of a 50 Mbit/s channel. The other half of the channel is used for tape recorder dump. In the absence of a dump, this channel is operated at half rate (BPSK modulation at 50 Mbit/s versus QPSK for 100 Mbit/s).

Sensor operations strategy:

Continuous operations around the orbit of GOMOS, MIPAS, SCIAMACHY, AATSR, RA-2, MWR, and DORIS. On-board recording of all the above mentioned low-rate instrument data outside the visibility range of ESA X-band stations. [656) 657)]

ASAR can be operated in high rate mode up to 30 minutes per orbit, including 10 minutes in the eclipse phase. MERIS can be operated only out of eclipse and when the sun zenith angle at the subsatellite point is less than 80° (this corresponds to about 43% of the orbit).

Sensor complement:

1) Instruments provided by ESA

- MERIS (Medium Resolution Imaging Spectrometer)
- MIPAS (Michelson Interferometer for Passive Atmospheric Sounding)
- RA-2 (Radar Altimeter-2, a Laser Retro-Reflector (LRR) is included)
- MWR (Microwave Radiometer)
- ASAR (Advanced SAR)
- GOMOS (Global Ozone Monitoring by Occultation of Stars), chemistry and dynamics of upper atmosphere

2) Announcement of Opportunity (AO) Instruments

- SCIAMACHY (Scanning Imaging Absorption Spectrometer for Atmospheric Cartography, provided by Germany and The Netherlands)
- AATSR (Advanced Along Track Scanning Radiometer, provided by the UK and Australia), see also ATSR (ERS-1, -2)
- DORIS (see E.21.1) provided by France

MERIS = Medium Resolution Imaging Spectrometer, built by Alcatel Space Industries, France (formerly Aerospatiale). [658) 659) 660)] MERIS is a passive optical pushbroom wide-field instrument (CCD technology) intended to measure reflected radiation the Earth's surface, the ocean, and from clouds with high spectral resolution (high radiometric resolution in the VIS range). MERIS features individual gain settings to optimize the dynamic range of each of its fifteen bands (ability to reprogram the position and width of the spectral bands in flight).

MERIS objectives/applications:

656) P. Soerensen, A. Rudolph, L. O'Rourke, T. Beck, X. Marx, et al., The Flight Operations Segment, ESA Bulletin, No 106, June 2001, pp. 88-95

657) F. Martin Crespo, J.-P. Guignard, C. Garrido, M. Irle, "The Payload Data Segment," ESA Bulletin, No 106, June 2001, pp. 96-102

658) J.-L. Bézy, S. Delwart, M. Rast, "MERIS - A New Generation of Ocean-Color Sensor onboard ENVISAT," ESA Bulletin, No 103, August 2000, pp. 48-56

659) "MERIS Medium Resolution Imaging Spectrometer," ESA brochure

660) J.-L. Bézy, G. Gourmelon, R. Bessudo, G. Baudin, H. Sonntag, S. Weiss, "The ENVISAT Medium Resolution Imaging Spectrometer (MERIS), Proceedings of IGARSS'99, Vol. 2, pp. 1432-1434, Hamburg, June 28 - July 2, 1999

- Ocean: monitoring marine biophysical and biochemical parameters (chlorophyll, gelbstoff concentrations, suspended particles). Mapping of sea pollution, coastal erosions, sea ice.
- Atmosphere: monitoring cloud distribution, cloud altitude, water vapor column content, aerosols
- Land: monitoring vegetation/biomass, inland water, agriculture/forestry. Mapping snow and ice, vegetation stress and calamities (fires) [661]

Spectral characteristics:	
Spectral range	0.39 - 1.04 μm (VNIR)
Spectral sampling interval	1.25 nm
Spectral bandwidth	1.25 - 25 nm (programmable)
Registration between bands	< 0.1 pixel (FR)
Band transmission capability	15 bands programmable in position and width
Band center knowledge	< 1 nm
Radiometric characteristics:	
Polarization sensitivity	< 0.5% over the full spectral range
Radiometric accuracy	< 2% in reflectance (relative to the sun irradiance)
Spectral band/band accuracy	< 0.05%
Pixel to pixel accuracy	< NEDL = Noise Equivalent Detected Radiance at top of atmosphere with input signal Locean (upwelling radiance from ocean surface)
Dynamic range	up to albedo 1.0 (in appropriate spectral bands)
Spatial resolutions:	
Open ocean observation	1040 m x 1200 m at nadir (reduced resolution: RR)
Land and coastal zones	260 m x 300 m at nadir (full resolution: FR)
Integration time/frame	about 44 ms
TFOV, swath width	±34.2° or about 1150 km, providing global coverage every 3 days.
Operational modes:	
Full resolution: FR (300 m)	Data rate = 24.0 Mbit/s (no on-board recording is considered), duty cycle of about 43%
Reduced resolution (RR)	Data rate = 1.6 Mbit/s, full coverage duty cycle (1.2 km resolution)
Data localization accuracy	< 2 km without use of landmarks
Instrument mass, power, size	200 kg, 175 W (average), 1.8 m x 0.9 m x 1.0 m

Table 154: MERIS instrument performance parameters

The MERIS instrument is a programmable, high-spectral resolution imaging spectrometer measuring in the spectral region of 390 - 1040 nm. The total swath is covered by five identical modules (cameras) each having a 14° FOV, with 0.4° overlap between adjacent cameras (1150 km total swath, with a FOV of 68.5°). The cameras are arranged in a fan-shaped configuration and view the Earth through five depolarizing windows with a common depointing prism. An image is formed at the entrance slit of a grating imaging spectrometer, which disperses the image spectrum on an area array CCD detector matrix (576 lines x 780 columns, the element size is 22.5 μm x 22.5 μm). The CCDs (thinned backside illuminated) are operated in frame-transfer mode. The CCDs are coupled to a Peltier cooler to keep the temperature at -22°C. A five-channel VEU (Video Electronic Unit) performs 12 bit A/D conversion of the video signals provided by the five CCDs (each CCD has 740 useful pixels per line and 576 lines). After onboard digital conversion, the data is corrected in real-time with the calibration coefficients to produce raw image data in 15 spectral bands. [662] [663]

The MERIS optics subsystem consists of an external window, a folding mirror, an off-axis catadioptric-ground imager, and a spectrometer of concentric design. The windows are used to scramble the incident polarized radiation thereby making the instrument less sensitive to polarization changes. The dispersive element of the spectrometer is a low-groove-density holographic grating.

661) M. Morel, J. L. Bézy, F. Montagner, A, Morel, J. Fischer, "Envisat's Medium-Resolution Imaging Spectrometer," ESA Bulletin, No. 76, November 1993, pp. 40-46

662) G. Levrini, E. Attema, "The Commissioning Phase and the Calibration/Validation Activities," ESA Bulletin, No 106, June 2001, pp. 109-117

663) G. Levrini, E. Attema, "The Envisat Calibration and Validation Approach," ESA Earth Observation Quarterly, No 67, Oct. 2000, pp. 9-16

Instrument calibration is performed at the orbital south pole, were the calibration flat-plate diffuser is illuminated by the sun by rotating a calibration mechanism (the reference for absolute calibration is based on an assumed solar irradiance at the time of calibration). Four in-flight calibration sequences are defined:

- Dark calibration, During calibration the signal is recorded with the Earth and sun aperture closed to provide, in the case of radiometric calibration, a uniform (spectrally and spatially) radiance source, and in the case of spectrometric calibration, a radiance source with a spectral signature.
- Radiometric gain calibration (a white diffuser plate, sun illuminated, is inserted at the cross-over point of the five cameras FOVs). The diffuser provides a reflectance standard across the entire spectral range and FOV.
- Diffuser ageing characterization (exposure to the sun for 1 hr during the mission life)
- Wavelength referencing (another diffuser is used with known absorption peaks)

MERIS operation: The MERIS instrument can only observe out of eclipse and when the sun zenith angle at the subsatellite point is below 80°. The instrument can be operated either in "averaging" or in "direct and averaging" observation modes. In averaging mode, the data are spatially averaged on-board to produce an additional but separate data stream at 1200 m resolution, referred to as RSR (Reduced Spatial Resolution). When in direct and averaging mode, the instrument delivers, in addition to the RSR data, FSR (Full Spatial Resolution) data at 300 m resolution in the same 15 spectral bands. FSR data is available for up to 20 minutes per orbit.

Band No.	Band center (nm)	Bandwidth (nm)	Application
1	412.5	10	Yellow substance and turbidity
2	442.5	10	Chlorophyll absorption maximum
3	490	10	Chlorophyll and other pigments
4	510	10	Turbidity, suspended sediment and red tides
5	560	10	Chlorophyll, suspended sediment
6	620	10	Suspended sediment
7	665	10	Chlorophyll absorption
8	681.25	7.5	Chlorophyll fluorescence, red edge
9	705	10	Aerosol, red edge transition
10	753.75	7.5	Oxygen absorption reference band, vegetation
11	760	2.5	O_2 absorption R-branch
12	775	15	Aerosol, vegetation
13	865	20	Aerosols correction over ocean
14	890	10	Water vapor absorption reference
15	900	10	Water vapor absorption, vegetation

Table 155: Nominal spectral band specification of MERIS

MIPAS = Michelson Interferometer for Passive Atmospheric Sounding. [664] [665] [666] [667] [668] MIPAS is a high-resolution Fourier-transform spectrometer designed to measure concentration profiles of atmospheric constituents on a global scale. MIPAS is a limb emission sounding instrument operating in the spectral region between 4.15 and 14.6 μm (685-2410 cm^{-1}), observing during day and night phases of the orbit, measuring simultaneously a large

664) M. Endemann, P. Garé, D. J. Smith, R. Geßner, "The ENVISAT Michelson Interferometer for Passive Atmospheric Sounding (MIPAS)," Proceedings of IGARSS'99, Vol. 2, pp. 1435-1437, Hamburg, Germany, June 28 - July 2, 1999

665) M. Endemann, H. Fischer, "Envisat's High-Resolution Limb Sounder: MIPAS," ESA Bulletin 76, November 1993, pp. 47-52

666) W. Posselt, "Michelson Interferometer for Passive Atmospheric Sounding," Proceedings of the Twenty-fourth International Symposium on Remote Sensing of the Environment, May 27-31, 1991, Rio de Janeiro, Volume II, pp. 737-748, ERIM, Ann Arbor MI.

667) http://envisat.estec.esa.nl/

668) M. Endermann, Ph. Gare, D. J. Smith, K. Hoerning, B. Fladt, R. Geissler, "MIPAS: Design Overview and Current Development Status," Proceedings of SPIE, Vol. 2956, pp. 124-135, Sept. 24-26, 1996,

variety of relevant trace gases (over 20), including the complete family of nitrogen-oxygen compounds and several CFCs, especially in the stratosphere and in cloud-free regions in the upper troposphere. In 1989 ESA selected MIPAS as an 'ESA developed instrument' for the POEM/ENVISAT program. The MIPAS satellite instrument development (by DASA/DSS) results from a long heritage of airborne MIPAS instruments (see P.129).

Major MIPAS objectives:

- Chemistry of the stratosphere: O_3 (global and polar), NO, NO_2, HNO_3, HNO_4, N_2O_5, $ClONO_2$, COF_2, HOCl.
- Climate research: (global distribution of relevant parameters and clouds) eg., O_3, CH_4, H_2O, N_2O, CFC_s, (F11, F12, F22, CCl_4, CF_4), CO, OCS, aerosols and clouds.
- Transport processes (tropospheric-stratospheric exchange): O_3, CH_4, N_2O, C_2H_2, C_2H_6, SF_6.
- Upper tropospheric chemistry: NO, CO, CH_4, O_3, HNO_3.

MIPAS makes measurements in either of two pointing regimes: rearwards within a 35°-wide range (anti-flight direction), and sideways within a 30°-wide range on the anti-sun side. The rearward viewing geometry is the prime measurement mode of MIPAS. The sideways observation mode is intended to cover special events (volcanic eruptions, etc.). As a result of the limb viewing geometry, the distance between the instrument and the tangent point of the target area is between 3000 and 3300 km, depending on the tangent height.

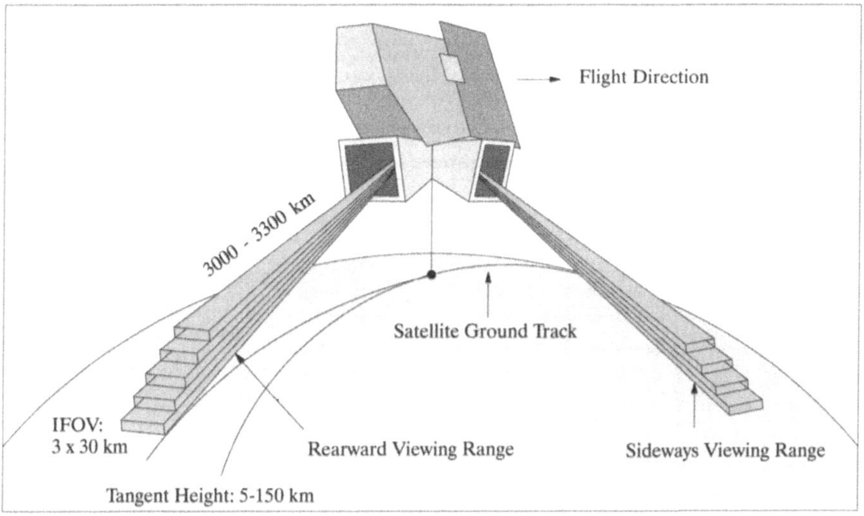

Figure 93: The MIPAS limb-observation geometry from a satellite

MIPAS instrument:

MIPAS measures a series of spectra from different tangent heights. The spectrometer transforms the spectra into a modulated signal, the interferogram, where many IR bands are present simultaneously. The spectrometer output consists of such interferograms for each observed scene. One basic elevation scan sequence comprises 16 high-resolution spectra (or up to 75 spectra with reduced spectral resolution) and takes 75 seconds. A typical elevation scan starts at about 50 km tangent height and descends in 3 km steps to 8 km (any limb heights from 150 km downwards in 5 km steps can be performed). Instrument pointing accuracy is of utmost importance to measure at a predetermined limb height with a standard deviation below 600 m. This in turn requires a S/C pointing knowledge of >0.01° (1 sigma) to nadir.

The MIPAS instrument has a total mass of 327 kg and a power consumption of 210 W. It consists of the following basic modules:

- Optics module: with front-end optics (comprising the azimuth and elevation scan units and a receiving telescope), Michelson interferometer, focal plane subsystem, calibration blackbody, and reference laser.
- Electronics module: with signal processing subsystem, instrument control unit, and Stirling cooler, on a common carrier plate.

Observation geometry	
IFOV	3 x 30 km (height x width) - 3 km height correspond to a 0.05° viewing angle
Elevation scan range	5 - 150 km above the Earth horizon (tangential height)
Azimuth scan range (with respect to flight direction)	between 160°-195° (rearwards), and 80°-110° (sideways)
Spectral coverage	
Spectral range	685 - 2410 cm^{-1} (14.6 - 4.15 μm)
Spectral resolution (unapodized)	0.035 cm^{-1}
Spectral resolution submodes	selectable in range from 0.035 - 0.35 cm^{-1}
Radiometric parameters	
Radiometric sensitivity (NESR)	50 (at 685 cm^{-1}) - 4.2 (at 2410 cm^{-1}) $nW/(cm^2 \, sr \, cm^{-1})$
Absolute radiometric accuracy	2% (at 14.6 μm) - 5% (at 4.15 μm) of input radiance with an offset of 2 NESR
Measurement duration	
Time per spectrum	4.5 s (full spectral resolution); 1.0 s (1/10 spectral resolution)
Time per elevation scan	75 s (500 km ground trace)
Spectra per elevation scan	16 (full spectral resolution); 75 (1/10 spectral resolution)
In-orbit lifetime	4 years
Instrument operation	continuously over full orbit
Data rate	533 kbit/s; raw data mode = 8 Mbit/s
Mass, power	320 kg, 210 W
Detector operating temperature	65-75 K

Table 156: Summary of MIPAS performance parameters

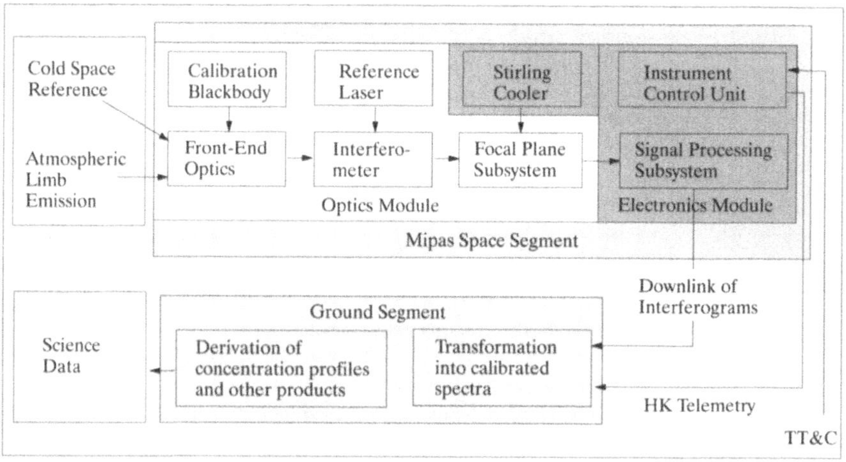

Figure 94: Schematic of MIPAS instrument elements and data flow

From the telescope, the radiation is directed to the interferometer. The input signals are divided at the beamsplitter inside the interferometer and directed to cube corners moving at a constant velocity of 25 mm/s along a path of 100 mm. This permits a spectrum to be recorded in 4s. Depending on the optical path difference in the two interferometer arms, the recombined signal is an intensity-modulated interferogram. [The Michelson interferometer provides an unapodized spectral resolution of better than 0.035 cm^{-1} throughout the spectral range. This is necessary to resolve the lines in the spectrum and to reduce the interference of overlapping spectral features. With this high spectral resolution, MIPAS provides a total of about 60,000 independent spectral samples in each spectrum.] - The output signal (interferogram) enters the focal-plane subsystem where beam size matching, beam splitting and optical filtering is performed. After optical filtering the input spectrum is separated into eight narrow spectral bands for detection by eight HgCdTe detectors operating at about 70 K for maximum sensitivity. The detectors are of photoconductive type in the spectral region from about 7-14.6 μm, and of photovoltaic type in the region 4-7 μm. The detectors are cooled by a pair of synchronized Stirling cycle coolers. The detector output is filtered and compressed in the signal processing subsystem, which outputs the interferograms and ancillary data for ground processing.

The calibration blackbody is mounted in the azimuth scan unit, it is used for the in-flight calibration of the instrument responsivity. MIPAS performs regular in-orbit calibration sequences.

- Radiometric offset calibration: This is done every two to four scans and prior to every elevation scan sequence to correct for the instrument self-emission
- Gain calibration: This occurs about once a week, applying a two-point calibration method, where radiances from deep space and an internal blackbody are observed in sequence.
- Line of sight calibration (used for inflight determination of the line-of-sight pointing direction): Performed every week. This is based on star observations crossing the instrument FOV and subsequent correlation of actual and predicted time of star crossing.
- Spectral calibration: Performed every two scans. The calibration parameters are retrieved from subsets of the atmospheric limb measurements, so that the routine scene data acquisition is not interrupted.
- Instrument line shape calibration is performed once a week. This involves retrieval from subsets of the routine scene data.

ASAR (Advanced SAR). [669] [670] [671] [672] [673] ASAR, of AMI heritage heritage flown on ERS-1/2 (built by MMS, UK, now Astrium Ltd.), measures the radar backscatter of the Earth's surface at C-band with a choice of five polarization modes: VV, HH, VV/HH, HV/HH, or VH/VV. The ASAR instrument comprises two major functional groups, the Antenna Subassembly (ASA), and the Central Electronics Subassembly (CESA). The antenna array contains 20 tiles (each 1 m x 0.65 m in size) with 16 transmit/receive (Tx/Rx) modules each. The transmit pulse characteristics are: the output is a linear FM up-chirp pulse centered at 124 MHz of the IF carrier. In the RF subsystem the pulse is up-converted to the RF frequency (5.331 GHz) and amplified. The signal is then passed to the Tx/Rx modules in the tile subsystem. - In general, the use of digital technologies for signal generation and processing is implemented. These permit chirp versatility in terms of pulse duration and bandwidth required for the operational modes and various swaths of the instrument. On reflection, AMI of ERS-1/2 used SAW devices for analog chirp generation.

The ASAR active antenna is a phased array of size 1.3 m x 10 m with distributed Tx/Rx modules arranged across the antenna, such that by adjusting individual module phase and gain, the transmit and receive beams may be steered and configured (independent control of the phase and amplitude of the transmitted radiators from different regions of the antenna sur-

669) M. Zink, C. Buck, J. L. Suchail, R. Torres, et al., "The Radar Imaging Instrument and Its Applications: ASAR," ESA Bulletin No 106, June 2001, pp. 46-55

670) Y. L. Desnos, C. Buck, et al., "ASAR - Envisat's Advanced Synthetic Aperture Radar," ESA Bulletin, No. 102, May 2000, pp. 91-100

671) J. L. Suchail, C. Buck, J. Guijarro, R. Torres, "The ENVISAT Advanced Synthetic Aperture Radar Instrument," Proceedings of IGARSS'99, Vol. 2, pp. 1441-1443, Hamburg, Germany, June 28 - July 2, 1999

672) "ASAR Advanced Synthetic Aperture Radar," ESA brochure

673) S. Karnevi, E. Dean, D. J. Q. Carter, S. S. Hartley, "Envisat's Advanced Synthetic Aperture Radar: ASAR," ESA Bulletin, No. 76, November 1993, pp. 30-35

face). Use of GaAs MMIC (Monolithic Microwave Integrated Circuit) technology, based on field effect transistors, for amplifiers, switches, attenuators, and phase shifters. ASAR also provides independent weighting of the received signal to each of these regions. This offers great flexibility in the generation and control of the radar beam, giving the ASAR instrument the capability to operate in a number of different modes (37 different and mutually exclusive high-rate operating modes). These modes use two principal methods of taking measurements; the ASAR instrument may operate as a conventional stripmap SAR or as a ScanSAR.

ASAR stripmap modes (Image and Wave):
When operating as a stripmap SAR, the phased array antenna gives ASAR the flexibility to select an imaging swath by changing the beam incidence angle and the elevation beamwidth. In addition, the appropriate PRF required to ensure acceptable ambiguity performance and to suppress unwanted nadir returns is selected.

- **Image mode:** ASAR operates as an imaging radar collecting data in one of seven predetermined but relatively narrow swaths (up to 100 km within a viewing area of about 485 km) with high spatial resolution (30 m). The image mode provides continuous coverage over a single swath that can be pointed anywhere with 15-45° incidence angles. The transmit and receive polarization may be either HH or VV.
- **Wave mode:** imaging of small areas of 5 km x 5 km (referred to as vignettes) in frequent intervals over along-track distances of 100 km on ocean surfaces. Measurement of the change in radar backscatter from the sea surface due to ocean surface waves. The wave mode uses the same swaths and polarizations as image mode. The intermittent operation provides a low data rate, such that the data can be stored on board the satellite, rather than being downlinked immediately to a ground station.

Operating Mode ----> Parameter	Image Mode	Wide Swath Mode	Alternating/ Cross Polariz.	Wave Mode	Global Monitoring
Polarization	VV or HH	VV or HH	VV/HH, HH/HV or VV/VH	VV or HH	VV or HH
Spatial resolution (along-track and across-track)	28 m x 28 m	150 m x 150 m	29 m x 30 m	28 m x 30 m	950 m x 980 m
Radiometric resolution	1.5 dB	1.5-1.7 dB	2.5 dB	1.5 dB	1.4 dB
Swath width	up to 100 km 7 subswaths	400 km 5 subswaths	up to 100 km 7 subswaths	5 km vignette 7 subswaths	>=400 km 5 subswaths
Ambiguity ratio (point) along-track across-track	 26-30 dB 32-46 dB	 22-29 dB 26-34 dB	 19-28 dB 26-41 dB	 27-30 dB 31-46 dB	 27-29 dB 25-32 dB
Ambiguity ratio (distrib.) along-track across-track	 23-25 dB 17-39 dB	 20-25 dB 17-31 dB	 18-25 dB 17-39 dB	 23-25 dB 21-48 dB	 25-28 dB 17-31 dB
Radiometric stability	0.32-0.40 dB	0.32-0.42 dB	0.50-0.55 dB	0.55-0.60 dB	0.46-0.53 dB
Noise equivalent sigma zero	-22 to -22 dB	-21 to -26 dB	-19 to -22 dB	-20 to -22 dB	-32 to -35 dB
Incidence angle range	15-45°		15-45°	15-45°	
Center frequency	5.331 GHz (C-band)				
PRF	1650 to 2100 Hz				
Chirp bandwidth	up to 16 MHz				
Antenna size	10 m x 1.3 m (consisting of five 1.3 m x 2 m panels)				
Operation (duty cycle)	up to 30 min/orbit			rest of orbit	
Data rate	up to 100 Mbit/s			0.9 Mbit/s	
Power	1365 W	1200 W	1395 W	647 W	713 W
Instrument mass	832 kg				

Table 157: ASAR instrument parameters

ASAR ScanSAR modes (Alternating polarization, Wide swath, and Global monitoring)
The ScanSAR principle achieves swath widening by the use of an antenna beam which is

electronically steerable in elevation. Radar images can then be synthesized by scanning the incidence angle and sequentially synthesizing images for the different beam positions. The area imaged from each particular beam forms a sub-swath. The principle of the ScanSAR is to share the radar operation time between two or more separate sub-swaths in such a way as to obtain full image coverage of each. The system transmits pulses to and receives echoes from a sub-swath for a period long enough to synthesize a radar image of the area within the beam footprint at the required resolution. It then switches beams to illuminate a different sub-swath and continues in this manner until the full wide swath is covered. At this point the system returns to the original sub-swath and the scanning cycle is repeated.

- **Alternating polarization mode:** provides a choice of HH or VV or cross-polarized imaging of the same scene by interleaving looks with each polarization along track within the synthetic aperture. The mode employes a modified ScanSAR technique. Instead of scanning between different elevation sub-swaths, the alternating polarization mode (co-polar) scans between two polarizations, HH and VV, within a single swath (which is preselected, as for image and wave modes).
- **Wide swath mode:** continuous image coverage over a swath width of 405 km (or more), divided into five subswaths ranging from 60 to 100 km in width; the spatial resolution is 150 m.
- **Global monitoring mode:** provides continuous along-track sampling across a 405 km swath. Allows ASAR to be operated in a reduced spatial resolution mode of 1000 m (with a corresponding reduced data rate for on-board recording). Global monitoring of features such as ice or snow coverage, deforestation, desertification or humidity.

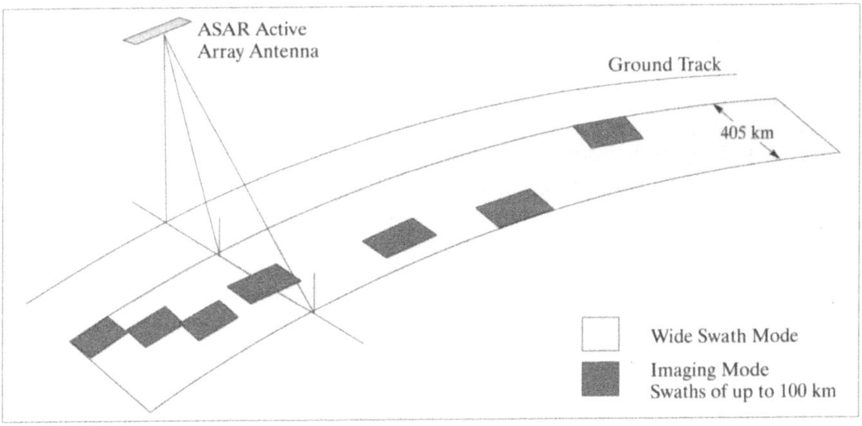

Figure 95: ASAR observation geometries (simplified)

Application: The main objective of ASAR is to provide information on: ocean waves, sea ice extent and motion, snow and ice extent, surface topography, land surface properties, Earth's biomass (especially deforestation in equatorial zones), surface soil moisture and wetland extent. Important applications by ASAR to the global mission include: [674)]

- Measuring sea-state conditions at various scales
- Mapping ice-sheet characteristics and dynamics
- Mapping sea-ice distribution and dynamics
- Detecting large-scale vegetation changes
- Monitoring natural and man-made pollution over the oceans.

Some ASAR applications for the regional mission are in the following areas:

674) E. Attema, Y-L. Desnos, G. Duchossois, "Synthetic Aperture Radar in Europe: ERS, Envisat, and Beyond," JHU/APL Technical Digest, Vol. 21, No. 1, 2000, pp. 155- 161

- Offshore operations in sea ice
- Snow and ice mapping
- Coastal protection and pollution monitoring
- Ship traffic monitoring
- Agriculture and forest monitoring
- Soil moisture monitoring
- Geological exploration, topographic mapping
- Predicting, tracking and responding to natural hazards

Compared with AMI of ERS-1/2, a single-channel fixed-geometry instrument, ASAR represents a step forward in system flexibility and the science value of the data sets. The new features of ASAR are:

- A digital chirp generator and an improved linear dynamic range
- Flexible swath positioning offers the choice among several image swath positions at various distances from the sub-satellite track, i.e. with different incidence angles
- The dual polarization capability offers simultaneous operation at H, V, and cross-polarization combinations
- Wide-swath coverage offers a 405 km swath with 150 m or 1 km resolution imagery
- The enhanced wave-mode acquires imagettes at 100 km intervals in along-track
- An extended duty cycle (operating time) is available at high resolution (30 min of operation, 10 min eclipse)
- Global SAR coverage is possible by using a solid-state on-board recorder and a data relay satellite (ARTEMIS).

RA-2 (Radar Altimeter-2, extended capability of ERS-1/2). RA-2 is a nadir-pointing pulse-limited radar altimeter which transmits frequency-modulated (chirp) pulses. Alenia Spazio SpA of Rome, Italy is the prime contractor for the ESA instrument. Frequencies: 13.575 GHz (Ku-band) and a second frequency at 3.2 GHz (S-band) for the measurement and correction of ionospheric delays. Adaptable width/resolution windows (measurements over ice surfaces), autonomous resolution control. [675) 676) 677) 678) 679)]

The main objectives of RA-2 are high-precision measurements of distance (time delay) from the satellite to the Earth surface (topography) and the measurement of the power and the shape of the radar echoes from ocean, ice and land. Ground processing of the radar echo power and shape data enables the determination of wind speed and significant wave height in the observed target area (supporting studies of ocean circulation, bathymetry, gravity anomalies, and marine geoid characteristics). Sea ice monitoring is possible through frequent temporal coverages.

RA-2 transmits radio frequency pulses and measures the run time of the echoes. A model-free tracker in the on-board signal processor keeps the radar echoes within the sampling window. Adaptive height resolution operation is implemented by selecting the bandwidth of the transmitted pulses. As a result, measurements over ocean surfaces are carried out with improved accuracy. Internal calibration measurements are performed periodically in parallel to the tracking functions. For this the transmit pulse is coupled into the receiver by means of a calibration coupler. Echo samples are processed on ground to account for correction and calibration data.

In parallel to the nominal collection of echo waveforms, which are collected on-board at 19 kHz rate, RA-2 has also the capability (on ground command) to record streams of non-aver-

675) J. Benveniste, M. Roca, G. Levrini, P. Vincent, S. Baker, O. Zanife, C. Zelli, O. Bombaci, "The Radar Altimetry Mission: RA-2, MWR, DORIS, and LRR," ESA Bulletin, No 106, June 2001, pp. 67-76

676) C. Zelli, et al., "RA-2 Radar Altimeter: Instrument EM Model Performance Results," IGARSS'97, Vol. 1, pp. 18-20

677) G. Angino, et al., "High Spatial Resolution Radar Altimetry for Global Earth Topography Mapping," IGARSS'97, Vol. 1, pp. 15-17

678) A. Resti, "Envisat's Radar Altimeter: RA-2," ESA Bulletin, No. 76, November 1993, pp. 58-60

679) A. Resti, et al., "The Envisat Radar Altimeter System (RA-2)," ESA Bulletin No. 98, June 1999, pp. 94-101

aged "individual" waveforms. The feature is used to study the physics of the backscattering process.

Altitude	764 - 824 km range with an accuracy of <4.5 cm
Backscatter coefficient	-10 dB to +50 dB with an accuracy of <0.4 dB (bias)
Wave height	0.5 m to 20 m with an accuracy of <5% or 0.25 m
Measurement datation	±100 μs with respect to UTC
Operating center frequencies	13.575 GHz (Ku-band), and 3.2 GHz (S-band)
Chirp bandwidths	320, 80, 20 MHz CW (Ku-band), 160 MHz (S-band)
Pulse repetition frequency (PRF)	1795.33 Hz (Ku-band), 448.83 Hz (S-band) interleaved operation
Pulse length	20 μs
Peak transmission power	60 W (Ku-band), 60 W (S-band)
Antenna diameter	1.5 m
Antenna beamwidth	1.33° (Ku-band), 5.25° (S-band)
IF center frequency	6.4 MHz
A/D conversion	8 bit
Instrument operation	continuously over full orbit
FFT resolution	128 complex points (samples) at 16 bit
Antenna diameter, beamwidth	1.2 m, 1.84° (Ku-band), 6.2° (S-band)
Total mass, power, data rate	110 kg, 161 W, 100 kbit/s

Table 158: RA-2 instrument parameters

LRR (Laser Retro Reflector), built by Alcatel Space Industries, France. LRR is mounted on the nadir panel close to the RA-2 antenna to support satellite laser ranging and RA-2 altitude measurement calibration. The LRR is a passive device which is used as a reflector by groundbased satellite laser ranging (SLR) stations. The operating principle is to measure the round-trip run time of laser pulses reflected from an array of corner cubes at the satellite. The corner cubes are designed to reflect the incident laser beam back directly, keeping the reflected beam parallel to the incident beam within a few arcseconds. The corner cubes of LRR work at two specified wavelengths: $\lambda = 694$ nm and $\lambda = 532$ nm. Laser ranging to LRR is possible for view angles of 360° in azimuth and 60° elevation around the perpendicular to the satellite's Zs Earth panel. The total mass of LRR is 2 kg.

GOMOS = Global Ozone Monitoring by Occultation of Stars (built by MMS, France). GOMOS is a UV/Visible/Near-infrared limb viewing grating spectrometer, operating in a star occultation mode. [680) 681) 682) 683) 684) 685) 686)] GOMOS relies on the self-calibrating occultation method (which minimizes the impact of instrument degradation). GOMOS introduces a new measurement principle: using stars rather than the sun or the moon as light sources (occultation measurement method), it achieves permanent, homogeneous global coverage through the measurement of 25 to 40 stars per orbit. The optical configuration, based on two grating spectrometers (range: 250 - 950 nm), permits the measurement of stratospheric profiles of O_3, H_2O, NO_2, NO_3, aerosols and temperatures with a 1.7 km vertical resolution.

Objectives: High accuracy (altitude-resolved) and global long-term monitoring of stratospheric and mesospheric ozone. Global survey of related stratospheric trace gases (NO_2,

680) H. Nett, J. Frerick, T. Paulsen, G. Levrini, "The Atmospheric Instruments and their Applications: GOMOS, MIPAS and SCIAMACHY," ESA Bulletin, No 106, June 2001, pp. 77-87

681) T. Paulsen, A. F. Popescu, G. Ratier, G. Uguen, Ch. Lemercier, "The Global Ozone Monitoring by Occultation of Stars (GOMOS), Proceedings of IGARSS'99, Vol. 2, pp. 1438-1440, Hamburg, Germany, June 28 - July 2, 1999

682) G. Ratier, G. Levrini, et. al., "GOMOS: Envisat's Contribution to Measuring Long-Term Trends in Ozone and Other Trace Gases, ESA Bulletin, Nr. 97, March 1999, pp. 20-27

683) A. Popescu, P. Ingmann, "Envisat's Global Ozone Monitoring by Occultations of Stars Instrument: GOMOS," ESA Bulletin, No. 76, November 1993, pp. 36-39

684) GOMOS handout from 'Atmospheres Panel Meeting' in Washington DC, Feb. 26-27, 1991

685) "GOMOS - Global Ozone Monitoring by Occultation of Stars," ESA brochure

686) J. L. Bertaux, E. Kyrölä, T. Wehr, "Stellar Occultation Technique for Atmospheric Ozone Monitoring: GOMOS on Envisat," ESA Earth Observation Quarterly, No 67, Oct. 2000, pp. 17-20

NO_3, H_2O), aerosols and temperature. Other species (OClO, BrO, ClO) under perturbed chemistry conditions.

Parameter	UV/VIS Spectrometer	IR1 Spectrometer	IR2 Spectrometer	Photometers PHOT1, PHOT2
Spectral bands	248-371/387-693 nm	750-776 nm	915-956 nm	644-705 & 466-528
Spectral resolution	0.89 nm	0.12 nm	0.12 nm	broadband
SNR	>12	>6	>3	>15
Linearity	0.5%	<0.33%	<0.34%	N/A
Spectral stability	0.04 nm	0.007 nm	0.008 nm	N/A
Altitude range	20 - 100 km			
Vertical resolution	1.7 km			
Operation	continuously over full orbit			

Table 159: GOMOS characteristics

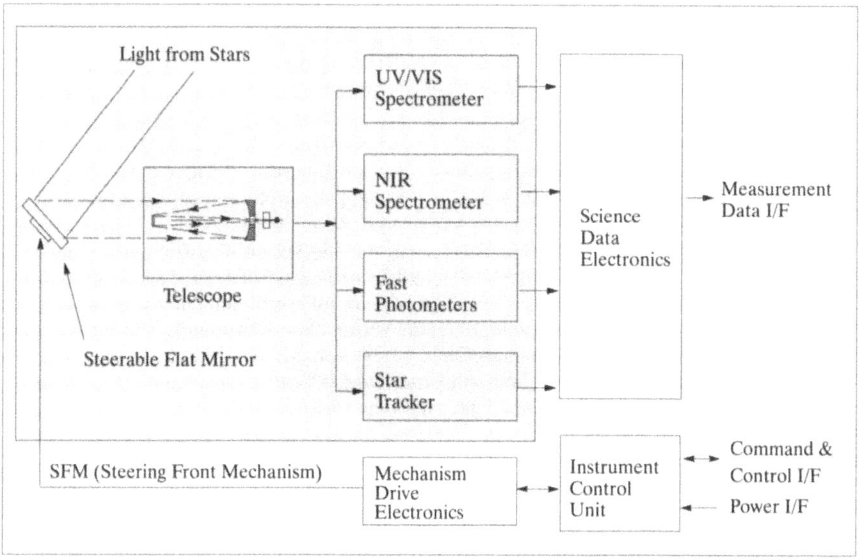

Figure 96: Functional block diagram of GOMOS

GOMOS instrument: The instrument's optical design (all-reflective system) is based on a single telescope concept: a Cassegrain telescope simultaneously feeding a UV/VIS medium-resolution spectrometer, NIR high resolution spectrometers, two photometers (monitoring the input signal scintillation) and two (redundant) star trackers. The photometer measurements allow to correct the spectral data from the high-frequency component introduced by the atmospheric scintillations. The telescope, the optics, all sensors and their associated front-end electronics are mounted on a thermally controlled CFRP optical bench. By using a steerable flat mirror (30 cm x 40 cm) in front of the telescope, GOMOS is able to acquire and track stars down to a magnitude of 5 over a very large angular range (Figure 96). The two-stage Steering Front Mechanism (SFM) is used to point the line of sight towards the selected star and to track it with very high accuracy as it sets through the atmosphere. SFM has an angular steering range of 100° in azimuth and 8° in elevation, and a pointing accuracy of some 20 µrad at a bandwidth of 5 Hz (a typical scan is 50 s in duration). During the occultation phase, SFM is operated over a reduced FOV (8° in azimuth and 6° in elevation) to provide fine centering of the star image within the spectrometer entrance slit.

The SFM relies on a combination of linear motors and flexible joints for its operation. The accurate tracking function is performed via a digital closed control loop using star tracker information read at 1000 Hz sampling rate. [687) 688)]

A special frame-transfer CCD has been developed for the spectrometers to meet the needs of the GOMOS mission. A thinned, backside-illuminated, anti-reflection coated CCD (2 x 143 lines and 1353 columns with 20 x 27 µm pixels), operating in MPP (Multi Pin Phase) mode, serves as primary detector array. A total of eight CCDs are used (two for the UV/VIS, two for the NIR, two for the fast photometers, and two for the star trackers). - The entire spacecraft and external GOMOS instrumentation (optomechanical assembly: OMA) is covered by an optomechanical cover responsible for protecting the instrument from light coming from a different direction other than the defined angular range, and for ensuring a stable, defined thermal environment.

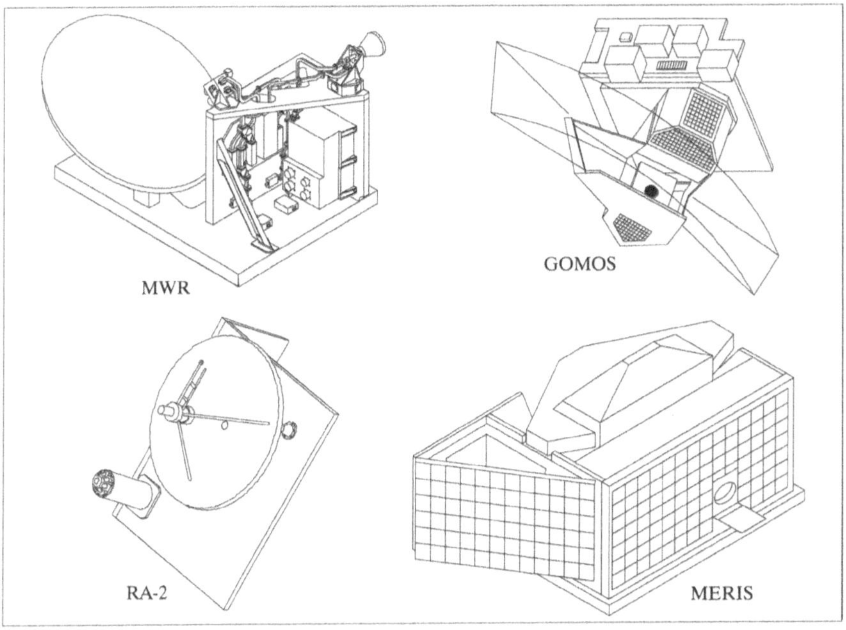

Figure 97: Illustration of some ENVISAT instruments

Requirements	Typical GOMOS Performance	Design Driver for
Nr. of occultations/orbit (45 on average)	about 820.000 occultations during 4 year mission	Lifetime of pointing mechanism
Wide angular coverage	-11° to +91° with respect to flight direction	Large FOV of pointing mechanism
Altitude resolution/accuracy	1.7 km/30 m, pointing stability better than 40 µrad	High angular pointing accuracy (pointing servo electronics)
Spectral resolution of spectrometer (FWHM)	0.72 / 0.89 nm UV/VIS, min/max 0.12/0.14 nm in NIR, min/max	High-transmission optics, high detector sensitivity in the UV
Photometer frame rate	1 kHz in the spectral range: 466-528 nm and 644-705 nm	Fast, high-sensitivity detectors

Table 160: GOMOS instrument performances and major technical challenges

687) J. Langen, "Envisat's Contribution to Atmospheric Chemistry Studies," Proceedings of IGARSS'99, Hamburg, Vol. III, June 28-July 2, 1999, pp. 1497-1499

688) J. Bertaux, "Vertical Profiles of Ozone from Envisat Space Platform with GOMOS Instrument," Proceedings of IGARSS'99, Hamburg, Vol. III, June 28-July 2, 1999, pp. 1616-1618

The occultation-mode operation of GOMOS involves pointing the instrument (telescope) toward preselected stars, acquiring and then tracking them with high accuracy. GOMOS observes stars whose lines-of-sight are tangential to the Earth's limb. For each individual star, the spectrum measured outside the atmosphere is compared to the spectrum seen through the atmosphere as the star sinks below the horizon. The difference reveals the presence of ozone and other trace gases. - The instrument has three additional calibration modes to meet the stringent pointing requirements; these are:

- A linearity monitoring mode, allowing calibration for possible nonlinearity of the sensor chains within the large dynamic range.
- A spatial-spread monitoring mode, allowing optimization of position of the CCD read-out regions with respect to that of the stellar spectrum.
- A uniformity monitoring mode, allowing measurement of the sensitivity difference between the various CCD pixels of a spectral sampling interval.

Some GOMOS instrument parameters: Telescope: 30 cm x 20 cm rectangular aperture, FOV: 0.6°, Cassegrain design based on aspherical primary and secondary mirrors; cooled 2-D CCD array; 2-axis stabilized platform; star tracker: ±0.3° FOV; total mass: 163 kg; power: 146 W; data rate: 222 kbit/s; data quantization: 12 bits.

MWR = Microwave Radiometer (of ERS-1/2 heritage, built by Alenia Spazio, Italy). A nadir-viewing, two-channel Dicke-type radiometer (passive sensor) operating at 23.8 (K-band) and 36.5 (Ka-band) GHz. The main objectives of MWR is the measurement of atmospheric humidity as supplementary information for tropospheric path correction of the RA-2 signal, which is influenced by the integrated atmospheric water content and by liquid water. In addition, MWR measurements are useful for the determination of surface emissivity and soil moisture of land, for surface energy budget estimations, investigations to support atmospheric studies, and for ice characterization. [689)]

Some instrument parameters: FOV = 20 km diameter at nadir; bandwidth = 650 MHz; dynamic range = 3 K - 330 K; absolute radiometric accuracy <3 K, radiometric sensitivity <0.4 K; operation of instrument: continuously over full orbit, data rate = 16.7 kbit/s, instrument mass = 24 kg, power = 18 W. The microwave radiation is received by an offset feed parabolic reflector antenna (fully integrated into the instrument structure), routed through a Dicke switch assembly to a down converter which translates the K- and Ka-band signals to a suitable IF range. The antenna subsystem includes a 60 cm reflector with a focal length of 350 mm and an offset-angle of 47°. Two feeds are used such that the 23.8 GHz channel is pointing in the forward direction and the 36.5 GHz channel in the aft direction, both beams with a footprint of about 20 km (1.5° beam). These frequencies are separately routed into the RF front-end where to two-point calibration scheme is adopted, namely with hot and cold references. MWR is controlled by a common MWR-DORIS instrument control unit. MWR has independent thermal-control elements (heaters and thermostats) to give its electronic circuits optimum performance. - Retrieval of antenna and brightness temperature values from measurement data is accomplished by ground processing. This includes ground calibration data, antenna characteristics and in orbit characterization data.

AATSR = Advanced Along Track Scanning Radiometer, built by the prime contractor Astrium Ltd., UK, and funded by DETR (UK Department of the Environment, Transport and the Regions), DISR (Australian Department of Industry, Science and Resources), and by NERC (UK Natural Environment Research Council). Under DETR's agreements with the Commonwealth of Australia, DISR have provided certain instrument components via contracts with Australian industry, and CSIRO (Commonwealth Science and Industrial Research Organization) of Canberra have been the interface for science issues. Technical support on data processing is provided by RAL (Rutherford Appleton Laboratory) of Chilton, UK.

689) J. Guijarro, A. Auriol, et al., "MWR and DORIS - Supporting Envisat's Radar Altimetry Mission," ESA Bulletin, No 104, Nov. 2000, pp. 41-46

AASTR is an imaging radiometer operating in four infrared channels (1.6, 3.7, 10.8, and 12.0 μm) and with three visible/reflected channels (0.55, 0.66, 0.87 μm). AATSR retains the full capability of ATSR (of ERS-1,2). [690]

Objectives of AATSR: a) Continuation of precise measurement of Sea Surface Temperature (SST) started with ATSR-1 and ATSR-2; b) To provide high-quality imagery of TOA (Top-of-Atmosphere) brightness temperature at 1 km resolution covering all parts of the globe except the polar regions, c) To enable scientific studies of ocean dynamics, land surface properties and the properties of clouds.

Channel	Center wavelength	Bandwidth	Primary application
0.55 μm	0.555 μm	20 nm	Chlorophyll
0.66 μm	0.659 μm	20 nm	Vegetation index
0.87 μm	0.865 μm	20 nm	Vegetation index
1.6 μm	1.61 μm	300 nm	Cloud clearing
3.7 μm	3.70 μm	300 nm	SST
11 μm	10.85 μm	1000 nm	SST
12 μm	12.00 μm	1000 nm	SST
Spatial resolution		1 km x 1 km	
Radiometric resolution		0.1 K	
Data quantization		12 bit	
SST accuracy		better than 0.3 K	
Swath width		500 km	
Instrument operation		Continuously over full orbit	
Data rate, instrument mass, power		625 kbit/s, 101 kg, 100 W	

Table 161: AASTR instrument parameters

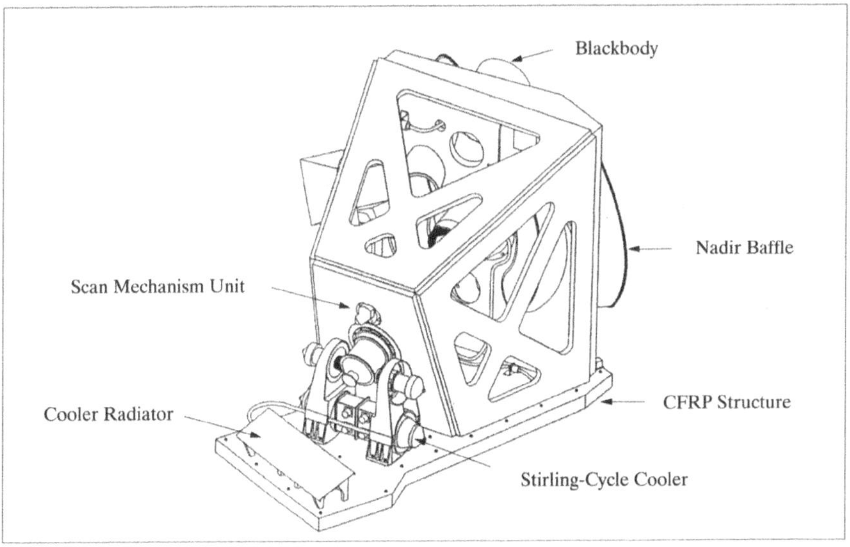

Figure 98: Illustration of the AATSR instrument

AATSR consists of the following elements: the IR/VIS radiometer, the electronics unit (providing the signal processing function, the scan mirror drive control and temperature sensor conditioning),the blackbody electronics unit (provides control of the blackbody

690) J.-P. Huot, H. Tait, M. Rast, S. Delwart, J.-L. Bézy, G. Levrini, "The Optical Imaging Instruments and their Applications: AATSR and MERIS," ESA Bulletin, No 106, June 2001, pp. 56-66

heaters and collects temperature sensor data), the cooler control unit (provide the control function for the Stirling-cycle coolers), the DEU (Digital Electronics Unit) for instrument control and data formatting functions, and the instrument harness. [691)]

The VIS/IR radiometer includes an inclined plane scan mirror which is rotated continuously in front of a reflecting telescope to provide a conical scan. The scan axis projects downwards (nadir) and ahead in the along-track direction (about 47°), achieving scanning across the satellite track in two regions: 1) close to the nadir, and 2) at an angle to the sea surface normal looking along the track (47°). The two-angle viewing technique provides improved atmospheric corrections. The focal plane assembly (FPA) is equipped with the detectors, one per channel, converting the radiation to electrical signals. A pair of Stirling cycle coolers keep the detector temperature at 80 K. Two blackbody infrared calibration targets, controlled by the blackbody electronics unit and a diffuse reflector visible calibration target, are viewed between the nadir and along-track scans.

SSTs are derived using the 11 and 12 μm channels for day time data, and the 11, 12 and 3.7 μm channels for night time data. In addition, the two VIS channels centered at 0.67 and 0.87 μm provide measurements of the vegetation index in the same way as AVHRR. An additional channel at 0.55 μm is provided, to indicate, from chlorophyll content, the growth stage and health of vegetation.

The Level 1B product (calibrated and geolocated BT/reflectance) is provided at 1km x 1km resolution at nadir. The Level 2 product containing SST and land parameters is available at 1 km x 1km resolution for regional studies, or spatially averaged over 10 arc min and 30 arc min or 17 km and 50 km cells for global measurements.

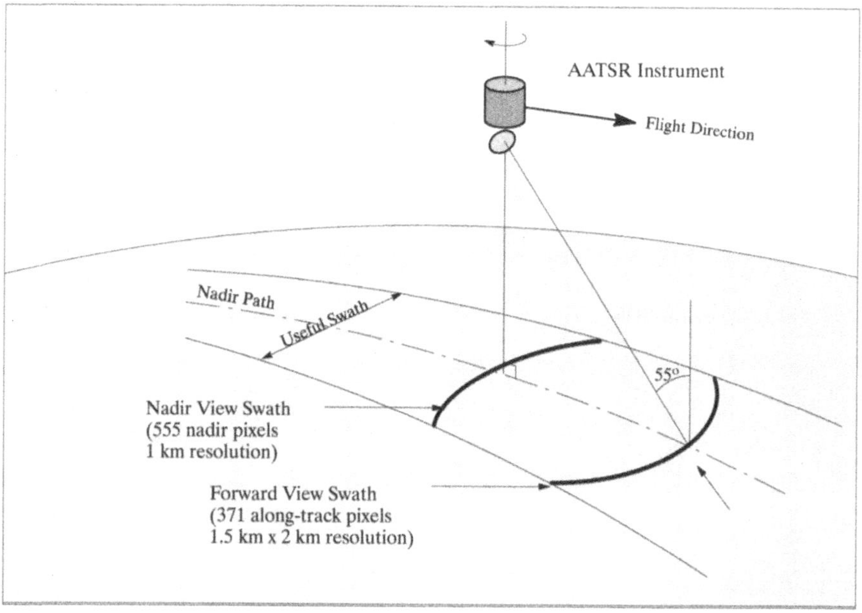

Figure 99: Typical AATSR viewing geometries

SCIAMACHY = Scanning Imaging Absorption Spectrometer for Atmospheric Cartography [a cooperative project funded by DLR/NIVR; PIs: J. P. Burrows, University of Bremen;

691) D. Lllewellyn-Jones, M. C. Edwards, C. T. Mutlow, A. R. Birks, I. J. Barton, H. Tait, "AATSR: Global-Change and Surface Temperature Measurements from Envisat," ESA Bulletin, 105, Feb. 2001, pp. 11-21

H. van Dop, University of Utrecht; built by DASA/DSS (electrical assembly) and Fokker Space, TNO and SRON (optical assembly and radiant cooler)]. SCIAMACHY is a passive spectrometer (of GOME heritage) for the measurement of trace gases in the troposphere and stratosphere, operating by differential optical absorption spectroscopy (DOAS) of solar and lunar radiation in the spectral region between 240 nm and 2400 nm. Parallel application of DOAS and BUV (Backscatter UV) measuring techniques. Viewing in three different modes: occultation (sun/moon), nadir, and limb scattering. Data rate: 400 kbit/s for nadir and limb observations; 1.867 Mbit/s for sun occultations.

SCIAMACHY observation objectives:

- Global measurement (abundance and distribution) of trace gases in the troposphere and stratosphere.
- Measurement of aerosol, cloud altitudes and spectral reflection of the surface.
- Troposphere: O_3, O_2, O_4, NO_2, N_2O, CO, CO_2, CH_4 and H_2O. Under special conditions also: HCHO, SO_2 and NO_3
- Stratosphere: O_3, O_2, NO, NO_2, NO_3, N_2O, CO, CO_2, CH_4, H_2O, and BrO. Observation of the NO column above 40 km. Under ozone hole conditions measurement of OClO and ClO in the lower stratosphere are possible.

Spectral Band	Spectral Range (nm)	Spectral Resolution	Detector Operating Temperature	Measured Constituents
1	240-314	0.24 nm	200 K	O_3, NO, (ClO)
2	309-405	0.26 nm	200 K	O_3, O_4, BrO, NO_2, HCHO, SO_2
3	394-620	0.44 nm	200 K	O_2, O_3, O_4, OClO, NO_2, NO_3
4	604-805	0.48 nm	200 K	O_2, O_4, NO_3, H_2O
5	785-1050	0.54 nm	200 K	
6	1000-1750	1.48 nm	200 K	O_4, H_2O
7	1940-2040	0.22 nm	135 K	CO_2, H_2O
8	2265-2380	0.26 nm	135 K	CO, CH_4, N_2O, H_2O
PMD 1-7	310 - 2380	67 to 137 (channel dependent)		

Table 162: Spectral parameters of SCIAMACHY

The instrument consists of the following systems: optical assembly [scanner module (elevation and azimuth scanners), telescope (off-axis parabolic mirror), spectrometer (grating for each band), detector modules (photodiode array detectors), thermal hardware (cooling of telescope)]; SCIAMACHY Radiant Cooler [radiator, thermal busses]; electronic assembly [power, mechanism and thermal control unit; science data processing unit; instrument control unit; and the digital bus unit]. Instrument mass = 198 kg; power = 122 W. Instrument operation: continuously over the full orbit. IFOV=0.045° x 1.8°. [692)]

a) Limb-scanning mode (into the flight direction). Objective: vertical distribution of trace gases and aerosols. In this mode the instrument is scanning across-track with a swath width of ±500 km with respect to the subsatellite track.. Scan layers of 3 km vertical thickness (and 100 km pixel width) are measured up to an altitude of 100 km.

b) Nadir-viewing mode. Objective: global distribution and abundance of trace gases, aerosols, cloud cover (observation at least once within three days); swath width = 1000 km; scan period=4 s in forward and 1 s in backward motion; pixel size at nadir (resolution) = 32 km along-track x 16 km across-track.

c) Solar and lunar occultation modes (measurements during each orbit). Objective: high-resolution absorption spectroscopy (DOAS). IFOV = 0.07° (azimuth) x 0.014° (elevation).

692) A. P. H. Goede, H, Schrijver, "SCIAMACHY: An Atmospheric Chemistry Instrument on ENVISAT," Proceedings of IGARSS'99, Hamburg, Vol. III, June28-July 2, 1999, pp. 1609-1611

The sun or moon are either tracked or a vertical scan over the complete sun/moon surface is performed. The obtained spectra can then be compared with suitable calibration spectra to yield the differential absorption of the atmosphere.

In the optical assembly the radiation from the atmosphere is fed by the scanner (consisting of an azimuth and an elevation scanner) into the telescope which directs it onto the entrance slit of the spectrometer. The spectrometer contains a pre-disperser which separates the radiation into three bands, followed by a series of dichroic mirrors which further divide the radiation into eight bands. Each band features a grating to diffract the radiation into a high-resolution spectrum which is then focused onto the detectors (each band has its own 1024 pixel detector array). The pre-disperser also serves as a Brewster window to separate polarized light, a portion of which is sensed by the PMD (Polarization Measurement Device). The PMD output is later used to correct for the polarization effects.

Calibration: In order to guarantee long-term stability of the scientific measurements, SCIAMACHY can perform many different calibration measurements. The calibration system includes internal calibration lamps; SCIAMACHY is radiometrically calibrated for backscattered UV measurements by direct observation of the sun and the full moon (occasionally). Direct measurements of the sun are also used to obtain the background solar source function for the DOAS measurements. Calibrations can in principle be made every orbit.

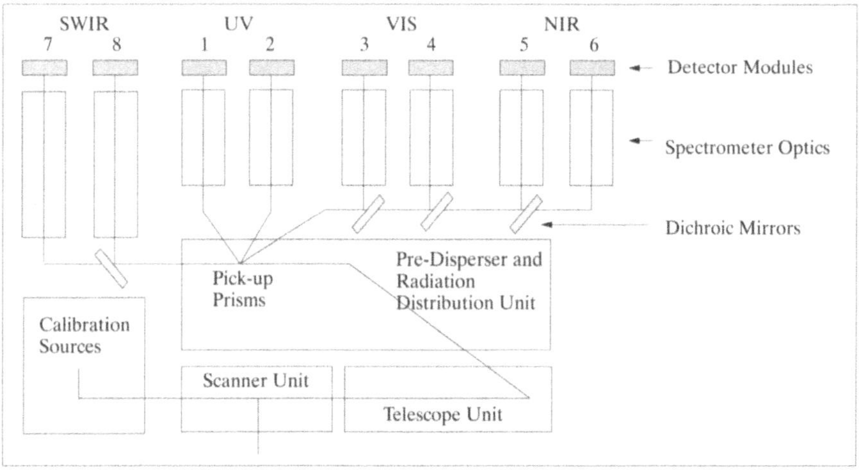

Figure 100: Optical concept of the SCIAMACHY double spectrometer

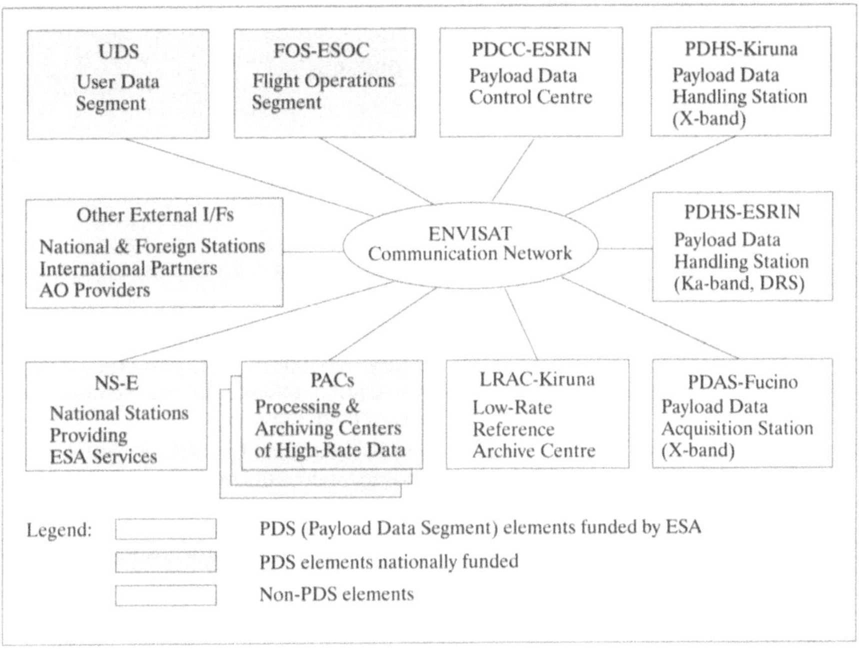

Figure 101: The ENVISAT ground segment concept

D.10 EO-1 (Earth Orbserving-1)

EO-1 is a technology mission. See description under M.6.

D.11 EOS (Earth Observing System)

EOS is the centerpiece of NASA's Earth Science Enterprise (ESE). It consists of a science component and a data system supporting a coordinated series of polar-orbiting and low inclination satellites for long-term global observations of the land surface, biosphere, solid Earth, atmosphere, and oceans. [693] [694] [695] [696]

Background: The EOS program is a NASA initiative of the US Global Change Research Program (USGCRP). Planning for EOS began in the early 1980s, and an AO (Announcement of Opportunity) for the selection of instruments and science teams was issued in 1988. Early in 1990 NASA announced the selection of 30 instruments to be developed for EOS. Major budget constraints imposed by the US Congress forced the EOS program into a restructuring process in the time frame of 1991-92. In addition a rescoping of the EOS program occurred in 1992 leading to just half of the 1990 budget allocation (the HIRIS sensor was eliminated). The instruments adopted as part of the restructured/rescoped EOS program were chosen to address the key scientific issues associated with global climate change. This action reduced the required instruments to 17 that needed to fly by the year 2002 (six

693) G. Asrar, R. Greenstone (editors), "MTPE/EOS Reference Handbook 1995," NASA/GSFC
694) "Earth Observing System," Reference Handbook 1990, and 1991, NASA/GSFC
695) "Optical Remote Sensing of the Atmosphere," 1990 Technical Digest Series of the Optical Society of America, Volume 4, pp. 23-58
696) G. Asrar, D. J. Dokken, "EOS Reference Handbook," March 1993, NASA

were deferred and seven instruments were deselected from the original 30). Furthermore, a shift occurred in the conceptual design of the EOS satellite platforms from "large observatories" to intermediate and smaller spacecraft that may be launched by smaller and existing launch vehicles. The EOS program experienced a further rebaselining process in 1994, due to a budget reduction of about 9%. This resulted in the cancellation of the combined EOS Radar and Laser Altimeter Mission (but rephasing the latter as two separate missions), deferring the development of some sensors and spreading the launch of missions by increasing the basic re-flight periods of missions from 5 to 6 years, and flying some EOS instruments on missions of partner space agencies (NASDA, RKA, CNES, ESA) in a framework of international cooperation. The EOS program includes instruments provided by international partners (ASTER, MOPITT, HSB, OMI) as well as an instrument developed by a joint US/UK partnership (HIRDLS).

The overall goal of the EOS program is to determine the extent, causes, and regional consequences of global climate change. The following science and policy priorities are defined for EOS observations (established by the EOS investigators working group and in coordination with the national and international Earth science community):

- Water and Energy Cycles: Cloud formation, dissipation, and radiative properties which influence the response of the atmosphere to greenhouse forcing, large-scale hydrology, evaporation
- Oceans: Exchange of energy, water, and chemicals between the ocean and atmosphere, and between the upper layers of the ocean and the deep ocean (including sea ice and formation of bottom water)
- Chemistry of the Troposphere and Lower Stratosphere: Links to the hydrologic cycle and ecosystems, transformations of greenhouse gases in the atmosphere, and interactions including climate change
- Land-Surface Hydrology and Ecosystem Processes: Improved estimates of runoff over the land surface and into the oceans. Sources and sinks of greenhouse gases. Exchange of moisture and energy between the land surface and the atmosphere. Changes in land cover
- Glaciers and Polar Ice Sheets: Predictions of sea level and global water balance
- Chemistry of the Middle and Upper Stratosphere: Chemical reactions, solar-atmosphere relations, and sources and sinks of radiatively important gases
- Solid Earth: Volcanoes and their role in climate change.

EOS Satellites

The original EOS mission elements (AM S/C series, PM S/C series, Chemistry S/C series) was redefined again in 1999. The EOS program space segment elements are now: Landsat-7, QuikSCAT, Terra, ACRIMSat, Aqua, Aura and ICESat (see also Table 179 in D.12).

Parameter	Terra (EOS/AM-1) S/C	Aqua (EOS/PM-1 S/C)
Downlink center frequency	8212.5 MHz	8160 MHz
EIRP	14 W	27.2 W
Bandwidth	26 MHz	15 MHz
Data modulation	OQPSK	SQPSK
Data format	NRZ-L	NRZ-L
I/Q power ratio (nominal)	1:1	1:1
Operational duty cycle	100%	100%
Antenna coverage from nadir	±64°	±64°
Antenna polarization	RHCP	RHCP
Data rate	13 Mbit/s	15 Mbit/s
Data protocol standard	CCSDS	CCSDS
Instrument data provided	MODIS	MODIS, AIRS, AMSU-A, CERES, HSB, AMSR-E

Table 163: Specification of Direct Broadcast (DB) service of Terra and Aqua satellites

EOS policy includes providing Direct Broadcast (DB) service to the user community; this applies to real-time MODIS data from the Terra spacecraft, as well as to the entire real-time data stream of the Aqua satellite. These data may be received by anyone with the appropriate receiving station, without charge. The broadcast data are transmitted in X-band. A 3 m antenna dish (minimum) should be sufficient for X-band data reception.

D.11.1 Terra Mission (EOS/AM-1)

Terra [697] is a joint Earth observing mission within NASA's ESE (Earth Science Enterprise) program between the United States, Japan, and Canada. The US provided the spacecraft, the launch, and three instruments developed by NASA (CERES, MISR, MODIS). Japan provided ASTER and Canada MOPITT.

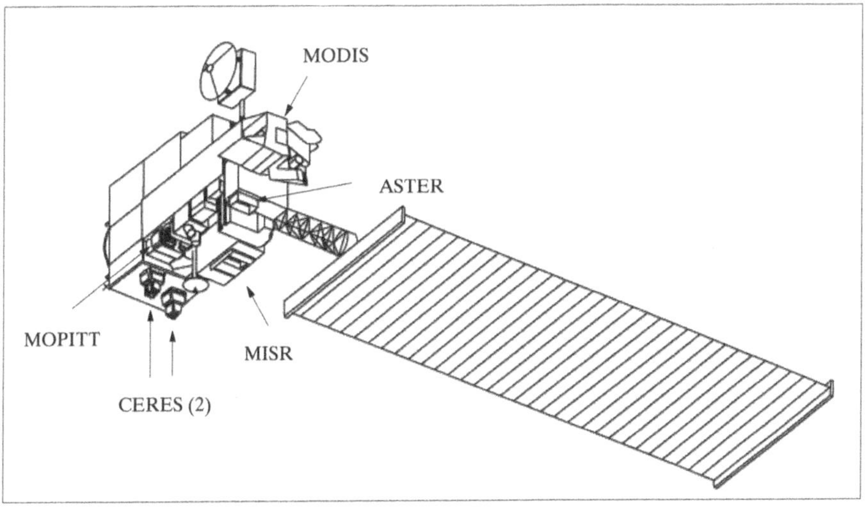

Figure 102: Illustration of the Terra spacecraft

The objective of the mission is to obtain information about the physical and radiative properties of clouds (ASTER, CERES, MISR, MODIS); air-land and air-sea exchanges of energy, carbon, and water (ASTER, MISR, MODIS); measurements of trace gases (MOPITT); and volcanology (ASTER, MISR, MODIS). The science objectives are:

- To provide the first global and seasonal measurements of the Earth system, including such critical functions as biological productivity of the land and oceans, snow and ice, surface temperature, clouds, water vapor, and land cover;
- To improve the ability to detect human impacts on the Earth system and climate, identify the "fingerprint" of human activity on climate, and predict climate change by using the new global observations in climate models;
- To help develop technologies for disaster prediction, characterization, and risk reduction from wildfires, volcanoes, floods, and droughts
- To start long-term monitoring of global climate change and environmental change.

Terra consists of a spacecraft bus built by Lockheed Martin Missiles and Space, Valley Forge, PA. It is three-axis stabilized with a bus size of 6.8 m (length) x 3.5 m (diameter) and a total launch mass of 5,190 kg. The total payload mass is 1155 kg. The average power of the satellite is 2.53 kW provided by a GaAs solar array. The design life is six years. The average

697) The EOS/AM-1 satellite was renamed by NASA to "Terra" in Feb. 1999

data rate of the payload is 18.545 Mbit/s (109 Mbit/s peak); on-board recorders for data collection of one orbit. Mission operations are performed at GSFC.

RF communications: [698] The primary Terra telemetry data transmissions are via TDRSS (Tracking & Data Relay Satellite System). Command and engineering telemetry data are transmitted in S-band. The science data recorded onboard are transmitted via Ku-band at 150 Mbit/s. The nominal mode of operation is to acquire two 12 minute TDRSS contacts per orbit. During each TDRSS contact, both S-band and Ku-band transmission is being used.

Besides Ku-band and S-band communication, Terra is also capable of downlinking science data via X-band. The X-band communication can be operated in three different modes, Direct Broadcast (DB), Direct Downlink (DDL) and Direct Playback (DP). DB and DDL is used to directly transmit real-time MODIS and ASTER science data respectively to users. The Direct Access System (DAS) provides a backup option for direct transmission in X-band. DAS supports transmission of data to ground stations of qualified EOS users around the world. These users fall into three categories:

- EOS team participants and interdisciplinary scientists
- International meteorological and environmental agencies
- International partners who require data from their EOS instruments

The launch of Terra took place on Dec. 18, 1999 from the VAFB, CA, on an Atlas-Centaur IIAS rocket.

Orbit: Sun-synchronous circular orbit, altitude = 705 km, inclination = 98.5°, period = 99 minutes (16 orbits per day, 233 orbit repeat cycles). The descending nodal crossing is at 10:30 AM.

Sensor complement: (ASTER, CERES (2), MISR, MODIS, MOPITT)

ASTER (Advanced Spaceborne Thermal Emission and Reflection Radiometer), a Japanese instrument sponsored by MITI (cooperative project with NASA). The ASTER team leaders are Hiroji Tsu of ERSDAC (Japan) and Anne B. Kahle of JPL. ASTER management is provided by JAROS (Japan Resources Observation System Organization). A Joint US/Japan Science Team is responsible for instrument design, calibration, and validation. Previous instrument name: ITIR (Intermediate Thermal Infrared Radiometer). Objective: Provision of high-resolution and multispectral imagery of the Earth's surface and clouds for a better understanding of the physical processes that affect climate change. Applications: studies of the surface energy balance (surface brightness temperature), plant evaporation, vegetation and soil characteristics, hydrologic cycle, volcanic processes, etc. [699] [700]

The ASTER instrument consists of three separate instrument subsystems; each subsystem operates in a different spectral region, has its own telescope(s), and is built by a different Japanese company. The subsystems are in the VNIR (Visible Near Infrared), SWIR (Shortwave Infrared) and TIR (Thermal Infrared) spectral regions. ASTER is pointable in the cross-track direction such that any point on the globe may be observed at least once within 16 days in all 14 bands and once every 5 days in the VNIR bands. The absolute temperature accuracy is 3K in the 200-240 K range, 2K in the 240-270 K range, and 2 k in the 340-370 K range for TIR bands. Instrument mass=421 kg; power=463 W average, 646 W peak; data rate = 8.3 Mbit/s average and 89.2 Mbit/s peak; thermal control by 80 K Stirling cycle coolers, heaters, cold plate/capillary pumped loop, and radiators; pointing accuracy: for control = 1 km on ground (all axes), knowledge= 342 m on ground (per axis), stability=2 pixels for 60 seconds.

698) Special issue on EOS/AM-1 Platform, Instruments and Scientific Data, IEEE Transactions on Geoscience and Remote Sensing, Vol. 36, No 4, July 1998

699) http://asterweb.jpl.nasa.gov/

700) ASTER, EOS Reference Handbook, 1999, pp. 102-105

Measurement Region	Measurement	Instruments used
Atmosphere	Cloud properties Radiative energy flux Tropospheric chemistry Aerosol properties Atmospheric temperature Atmospheric humidity	MODIS, MISR, ASTER CERES, MODIS, MISR MOPITT MISR, MODIS MODIS MODIS
Land surface	Land cover and land use change Vegetation dynamics Surface temperature Fire occurrence Volcanic effects	MODIS, MISR, ASTER MODIS, MISR, ASTER MODIS, ASTER MODIS, ASTER MODIS, MISR, ASTER
Ocean	Surface temperature Phytoplankton and dissolved organic matter	MODIS MODIS, MISR
Cryosphere	Land ice change Sea ice Snow cover	ASTER MODIS, ASTER MODIS, ASTER

Table 164: Overview of major physical process measurements of the Terra instruments

Parameter	Band No. VNIR	Band No. SWIR	Band No. TIR
Spectral bands in μm	1 0.52 - 0.60 2 0.63 - 0.69 3N 0.76 - 0.86 3B 0.76 - 0.86 Stereoscopic viewing capability along-track	4 1.600 - 1.700 5 2.145 - 2.185 6 2.185 - 2.225 7 2.235 - 2.285 8 2.295 - 2.365 9 2.360 - 2.430	10 8.125 - 8.475 11 8.475 - 8.825 12 8.925 - 9.275 13 10.25 - 10.95 14 10.95 - 11.65
Ground resolution	15 m	30 m	90 m
IFOV (nadir)	21.5 μrad	42.6 μrad	128 μrad
Date rate	62 Mbit/s	23 Mbit/s	4.2 Mbit/s
Cross-track pointing	±24° (±318 km)	±8.55° (116 km)	±8.55° (116 km)
Swath width	60 km	60 km	60 km
Detector type	Si (CCD of 5000 elements, 4000 are used)	PtSi-Si Schottky barrier linear array, cooled to 80 K (Stirling cooler)	HgCdTe cooled to 80 K (Stirling cooler)
Quantization	8 bit	8 bit	12 bit
Radiometric accuracy	4%	4%	

Table 165: ASTER instrument parameters

	ASTER		TM on Landsat 4/5	
Wavelength Region	Band No.	Spectral Range (μm)	Band No.	Spectral Range (μm)
VNIR			1	0.45 - 0.52
	1 2 3	0.52 - 0.60 0.63 - 0.69 0.76 - 0.86	2 3 4	0.52 - 0.60 0.63 - 0.69 0.76 - 0.90
SWIR	4 5 6	1.60 - 1.70 2.145 - 2.185 2.185 - 2.225	5	1.55 - 1.75
	7 8 9	2.235 - 2.285 2.295 - 2.365 2.360 - 2.430	7	2.08 - 2.35
TIR	10 11 12 13 14	8.125 - 8.475 8.475 - 8.825 8.925 - 9.275 10.25 - 10.95 10.95 - 11.65	6	10.4 - 12.5

Table 166: Spectral range comparison of ASTER and TM (on Landsat)

The VNIR subsystem, built by NEC Corporation, is a reflecting-refracting improved Schmidt design. VNIR features two telescopes, one nadir-looking with a three-spectral-band detector, and the other backward-looking with a single-band detector. The backward-looking telescope provides a second view of the target area in band 3B for stereo observations. Cross-track pointing is accomplished by rotating the entire telescope assembly. Band separation is through a combination of dichroic elements and interference filters that allow all three bands to view the same ground area simultaneously. Calibration of the nadir-pointing detectors is performed with two halogen lamps.

The SWIR subsystem, built by MELCO (Mitsubishi Electric Company), uses a nadir-pointing aspheric refracting telescope. Cross-track pointing is accomplished by a pointing mirror. The size of the detector/filter combination requires a wide spacing of the detectors, causing in turn a parallax error of about 0.5 pixels per 900 m of elevation. This error is correctable if elevation data (DEM) are available. Two halogen lamps are used for calibration. The maximum data rate is 23 Mbit/s.

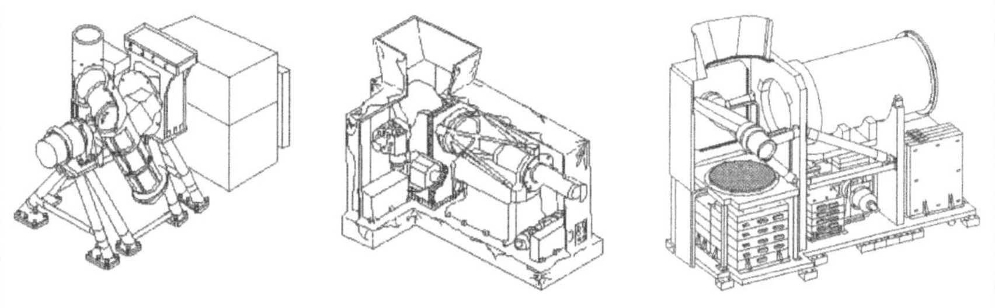

Figure 103: Illustration of the ASTER subsystems

CERES (Clouds and the Earth's Radiant Energy System), PI: B. Barkstrom, NASA/LaRC, built by TRW. Objective: Long-term measurement of the Earth's radiation budget and atmospheric radiation from the top of the atmosphere to the surface; provision of an accurate and self-consistent cloud and radiation database (input to WCRP international programs like TOGA, WOCE, and GEWEX). Retrieval of cloud parameters in terms of measured areal coverage, altitude, liquid water content, and shortwave and longwave optical depths. Specific science objectives are: [701)]

- For climate change analysis, provide a continuation of the ERBE record of radiative fluxes at the top of the atmosphere (TOA), analyzed using the same algorithms that produced the ERBE data.
- Double the accuracy of estimates of radiative fluxes at TOA and the Earth's surface.
- Provide the first long–term global estimates of the radiative fluxes within the Earth's atmosphere.
- Provide cloud property estimates that are consistent with the radiative fluxes from surface to TOA.

The CERES instrument assembly (of ERBE heritage) consists of a pair of broadband scanning radiometers (two identical instruments), one instrument operates in the cross-track mode for complete spatial coverage from limb to limb; the other one operates in a rotating scan plane (biaxial scanning) mode to provide angular sampling. The cross-track radiometer measurements are a continuation of the ERBS mission. The biaxially scanning radiometer provides angular flux information to improve model accuracy. A single cross-track CERES instrument is flown on TRMM, while the dual scanner instrument flies on Terra (EOS/AM-1) and Aqua (EOS/PM-1).

Instrument parameters (2 scanners): mass = 90 kg , power = 103 W (average), data rate = 20 kbit/s, duty cycle = 100%, thermal control by heaters and radiators, pointing knowledge

701) NASA/LaRC CERES brochure, NP-1999-04-069-GSFC

= 180 arcsec. CERES measures longwave and shortwave infrared radiation using thermistor bolometers to determine the Earth's radiation budget. There are three channels in each radiometer:

- Total radiance in the spectral range of 0.35 - 505 μm. Measurement accuracy of 0.5%.
- Shortwave: 0.3 - 55 μm. Measurement of reflected sunlight to an accuracy of 1%.
- Longwave: 85 - 125 μm. Measurement of Earth-emitted radiation to an accuracy of 0.3%.

Limb-to-limb scanning with a nadir IFOV (Instantaneous Field of View) of 14 mrad, FOV = ±78° cross-track, 360° azimuth. Spatial resolution = 20 km at nadir.

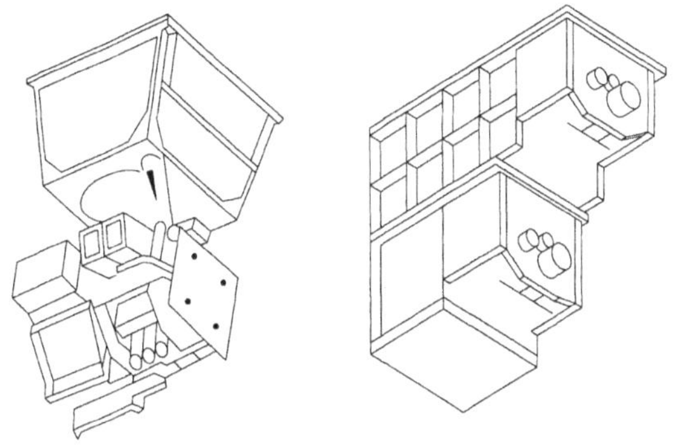

Figure 104: Schematic illustration of some EOS instruments

MISR (Multi-angle Imaging SpectroRadiometer), PI: D. J. Diner, NASA/JPL. Objective: provision of multiple-angle continuous sunlight coverage of the Earth with high spatial resolution (multidirectional observations of each scene within a time scale of minutes). MISR uses nine CCD cameras to observe the Earth at nine discrete viewing angles: one at nadir, plus eight other symmetrical views at 26.1°, 45.6°, 60.0°, and 70.5° forward and aft of nadir. Images at each angle are obtained in four spectral bands centered at 0.446, 0.558, 0.672, and 0.866 μm. Each of the 36 instrument data channels (i.e. four spectral bands for each of the nine cameras) is individually commandable to provide ground sampling of 275 m, 550 m, or 1100 m. The swath is 360 km; multi-angle coverage (repeat cycle) of the entire Earth in nine days at the equator, and in two days at higher latitudes.

Camera	View angle	Boresight angle	Swath offset angle	Effective focal length
Df	70.3° forward	57.88°	-2.62°	123.67 mm
Cf	60.2° forward	51.30°	-2.22°	95.34 mm
Bf	45.7° forward	40.10°	-1.71°	73.03 mm
Af	26.2° forward	23.34°	-1.06°	58.90 mm
An	0.1° nadir	-0.04°	0.04°	58.94 mm
Aa	26.2° aftward	-23.35°	1.09°	59.03 mm
Ba	45.7° aftward	-40.06°	1.76°	73.00 mm
Ca	60.2° aftward	-51.31°	2.24°	95.33 mm
Da	70.6° aftward	-58.03°	2.69°	123.66 mm

Table 167: MISR as-built camera pointing specifications

Application: MISR provides global maps of planetary and surface albedo (brightness temperature), and aerosols and vegetation properties. Monitoring of global and regional trends in radiatively important optical properties (eg., opacity, single scattering albedo, and scattering phase function) of natural and anthropogenic aerosols. [702]

Parameter	Description
Mission life	6 years
Global coverage time	Every 9 days, with repeat coverage between 2-9 days depending on latitude
Cross-track swath width	360 km common overlap of all 9 cameras, FOV = ±60° along-track and ±15° cross-track.
Nine CCD cameras	Named An, Af, Aa, Bf, Ba, Cf, Ca, Df, and Da where fore, nadir, and aft viewing cameras have names ending with letters f, n, a respectively and four camera designs are named A, B, C, D with increasing viewing angle respectively
View angles at Earth surface	0°, 26.1°, 45.6°, 60.0°, and 70.5°
Spectral coverage	Four bands centered at 0.446, 0.558, 0.672, and 0.866 μm (blue, green, red, and NIR)
Spatial resolution	275 m, 550 m, or 1.1 km, selectable in-flight
Detectors	CCDs, each camera with 4 independent line arrays (one per filter),1504 active pixels per line
Radiometric accuracy	3% at maximum signal
Detector temperature	-5 ±0.1° C (cooled by thermo−electric cooler) of focal plane
Structure temperature	5° C
Instrument mass, power	149 kg, 131 W peak and 83 W average
Instrument size	0.9 m x 0.9 m x 1.3 m
Data rate	3.3 Mbit/s average, 9.0 Mbit/s peak

Table 168: MISR instrument specification

Figure 105: Illustration of the MISR observing concept from Terra

MISR images are acquired in two observing modes: global and local. The global mode provides continuous planet-wide observations, with most channels operating at moderate reso-

702) http://www−misr.jpl.nasa.gov/

lution; some selected channels operate at the highest resolution for cloud screening and classification, image navigation, and stereo-photogrammetry. The local mode provides data at the highest resolution in all spectral bands and all cameras for selected 300 km x 300 km regions. In addition to data products providing radiometrically calibrated and geo-rectified images, global mode data will be used to generate two standard (level 2) science products: TOA (Top-of-Atmosphere)/Cloud Product and the Aerosol/Surface Product.

MISR on-orbit radiometric calibration is performed bi-monthly, using deployable white Spectralon panels to reflect diffuse sunlight into the cameras, and a set of photodiodes to measure the reflected radiance. Additionally, vicarious calibrations using field and Air-MISR data are done on six-month intervals. Geometric calibration of the cameras is done using ground control points.

MODIS (Moderate-Resolution Imaging Spectroradiometer), Science team leader: V. Salomonson, NASA/GSFC; prime contractor is Raytheon SBRS; MODIS algorithm development by an international team of scientists from USA, UK, Australia, and France; there are four discipline groups: Atmosphere, Land, Oceans, and Calibration. [703] [704] [705]

The instrument is flown on the Terra and Aqua satellites (prime instrument). Objective: to measure biological and physical processes on a global basis on time scales of 1 to 2 days. MODIS is an optomechanical imaging spectroradiometer, consisting of a cross-track scan mirror (continuously rotating double-sided scan mirror assembly) and collecting optics, and a set of linear detector arrays with spectral interference filters located in four focal planes. The optical arrangement provides imagery in 36 discrete bands between 0.4 and 14.5 μm (21 bands within 0.4-3.0 μm range, 15 bands within 3-14.5 μm range). The spectral bands provide a spatial resolution of 250 m, 500 m, or 1 km at nadir. MODIS heritage: AVHRR (POES), HIRS (POES), TM (Landsat), CZCS (Nimbus-7). In fact, the MODIS instrument is considered to be a next-generation AVHRR instrument, having 36 bands (AVHRR/3 has 6) and a spatial resolution of 250 m (AVHRR has 1 km).

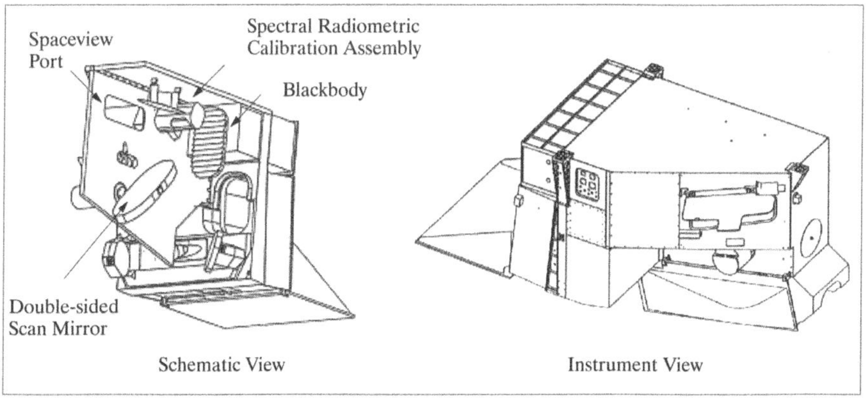

Figure 106: MODIS instrument illustration

A high-performance passive radiative cooler provides cooling to 83 K for the infrared bands on two HgCdTe FPAs (Focal Plane Assemblies). A new photodiode-silicon readout technology for the VNIR range provides unsurpassed quantum efficiency and low-noise readout with a very good dynamic range.

703) MODIS brochure of NASA/GSFC provided by M. D. King
704) http://terra.nasa.gov/About/MODIS/about_modis.html
705) http://modis.gsfc.nasa.gov/

Primary Use	Band No.	Bandwidth (μm)	Spectral Radiance NESR (W m^{-2} μm^{-1} sr^{-1})	Required SNR (Required NEΔT in K)	Spatial Resolution at nadir
Land/Cloud Boundaries	1	0.620 - 0.670	21.8	128	250 m
	2	0.841 - 0.876	24.7	201	
Land/Cloud Properties	3	0.459 - 0.479	35.3	243	500 m
	4	0.545 - 0.565	29.0	228	
	5	1.230 - 1.250	5.4	74	
	6	1.628 - 1.652	7.3	275	
	7	2.105 - 2.155	1.0	110	
Ocean Color/ Phytoplankton/ Biogeochemistry	8	0.405 - 0.420	44.9	880	1000 m
	9	0.438 - 0.448	41.9	838	
	10	0.483 - 0.493	32.1	802	
	11	0.526 - 0.536	27.9	754	
	12	0.546 - 0.556	21.0	750	
	13	0.662 - 0.672	9.5	910	
	14	0.673 - 0.683	8.7	1087	
	15	0.743 - 0.753	10.2	586	
	16	0.862 - 0.877	6.2	516	
Atmospheric Water Vapor	17	0.890 - 0.920	10.0	167	
	18	0.931 - 0.941	3.6	57	
	19	0.915 - 0.965	15.0	250	
Surface/Cloud Temperature	20	3.660 - 3.840	0.45	(0.05)	
	21	3.929 - 3.989	2.38	(2.00)	
	22	3.929 - 3.989	0.67	(0.07)	
	23	4.020 - 4.080	0.79	(0.07)	
Atmospheric Temperature	24	4.433 - 4.598	0.17	(0.25)	
	25	4.482 - 4.549	0.59	(0.25)	
Cirrus Clouds	26	1.360 - 1.390	6.00	150	
Water Vapor	27	6.535 - 6.895	1.16	(0.25)	
	28	7.175 - 7.475	2.18	(0.25)	
	29	8.400 - 8.700	9.58	(0.25)	
Ozone	30	9.580 - 9.880	3.69	(0.25)	
Surface/Cloud Temperature	31	10.780 - 11.280	9.55	(0.05)	
	32	11.770 - 12.270	8.94	(0.05)	
Cloud Top Altitude	33	13.185 - 13.485	4.52	(0.25)	
	34	13.485 - 13.785	3.76	(0.25)	
	35	13.785 - 14.085	3.11	(0.25)	
	36	14.085 - 14.385	2.08	(0.35)	

Table 169: MODIS spectral performance parameters

MODIS polarization sensitivity < 2% for the visible range out to 2.2 μm; the performance goal for SNR (Signal-to-Noise Ratio) and NEΔT (Noise-Equivalent Temperature Difference) values is 30-40% better than the required values in Table 169.; absolute irradiance accuracy of 5% for < 3 μm and 1% for > 3 μm; absolute temperature accuracy of 0.2 K for oceans and 1 K for land; daylight reflection and day/night emission spectral imaging; swath width of 2330 km at 110° FOV; scan rate = 20.3 rpm across track; instrument mass = 250 kg; duty cycle = 100%; power = 225 W (average); data rate = 6.2 Mbit/s (average), 10.6 Mbit/s (peak daytime), 3.2 Mbit/s (night); quantization = 12 bit. Instrument IFOV (spatial resolution) = 250 m (bands 1-2), =500 m (bands 3-7), = 1000 m (bands 8-36).

MODIS provides global coverage every 1 to 2 days. It provides specific global survey data, which includes the following (some standard data products):

- Cloud mask: at 250 m and 1 km resolution by day and at night
- Aerosol concentration and optical properties: at 5 km resolution over oceans and 10 km over land during the day
- Cloud properties: characterized by optical thickness, effective particle radius, cloud droplet phase, cloud-top altitude, cloud-top temperature
- Vegetation and land-surface cover, conditions, and productivity, defined as:
  - Vegetation indices corrected for atmospheric effects, soil, polarization, and directional effects

- Surface reflectance
- Land-cover type with identification and detection of change
- Net primary productivity, leaf-area index, and intercepted photosynthetically active radiation

- Snow and sea-ice cover and reflectance
- Surface temperature with 1 km resolution, day and night, with absolute accuracy goals of 0.3-0.5°C for oceans and 1°C for land surfaces.
- Ocean color: defined as ocean-leaving spectral radiance within 5% from 415-653 nm, based on adequate atmospheric correction from NIR sensor channels
- Concentration of chlorophyll-a within 35% from 0.05 to 50 mg/m^3 for case 1 waters
- Chlorophyll fluorescence within 50% at surface water concentrations of 0.5 mg/m^3 of chlorophyll-a.

Parameter	Value	Parameter	Value
Instrument type	Opto-mechanical design (whiskbroom scanner)	Data rate	10.6 Mbit/s (peak daytime), 6.1 Mbit/s (orbital average)
Scan rate	20.3 rpm	Data quantization	12 bit
Telescope	17.8 cm diameter off-axis, afocal (collimated) with intermediate field stop	Spatial resolution	250 m (bands 1-2) 500 m (bands 3-7) 1000 m (bands 8-36)
Size	1.0 m x 1.6 m x 1.0 m	Swath width, FOV	2330 km, 110°
Mass	229 kg	Swath length/scan	10 km (along track)
Power	162.5 W	Design life	6 years

Table 170: Some specification parameters of the MODIS instrument

MODIS on-board calibration employs various techniques for comprehensive verification of spectral, radiometric and spatial measurements. They include:[706)]

- Spectroradiometric Calibration Assembly (SRCA)
  - Spectral calibration of reflective channel channel bandpasses
  - Verification of spectral band registration
  - DC restoration on every scan using a direct view of space
  - Lunar calibration via the space-view port as well as periodic rotations of the S/C to enable full scans across the moon through the active scan aperture
- Blackbody calibration of thermal bands on every scan (a v-groove blackbody)
- Solar Diffuser (SD) reference
- Solar Diffuser Stability Monitor (SDSM)

MOPITT (Measurement of Pollution in the Troposphere), PI: J. Drummond, University of Toronto, Canadian sensor supported by CSA, built by COMDEC. [707)] [708)] [709)] MOPITT is the first satellite sensor to use gas correlation spectroscopy. The instrument measures emitted and reflected infrared radiance in the atmospheric column. Analysis of these data permit retrieval of tropospheric CO profiles and total column CH_4. Objective: study of how these gases interact with the surface, ocean, and biomass systems (distribution, transport, sources and sinks). Measurements are performed on the principle of correlation spectroscopy utilizing both pressure-modulated and length-modulated gas cells, with detectors at 2.3, 2.4, and 4.7 μm. Vertical profile of CO (carbon monoxide) and total column of CH_4 (methane) are to be measured; CO concentration in 4 km layers with an accuracy of 10%; CH_4 column abundance accuracy is 1%. Swath width = 616 km, spatial resolution = 22 x 22 km; instrument mass = 182 kg; power = 243 W; duty cycle = 100%; data rate = 25 kbit/s; thermal control by an 80 K Stirling cycle cooler, capillary-pumped cold plate and passive radiation; thermal operating range = 25° C (instrument) and 100 K (detectors).

706) Information provided by C. Schueler and J. Thunen of Hughes SBRC (now Raytheon SBRS)
707) MOPITT brochure of CSA/NASA
708) http://www.science.sp–agency.ca/J1–MOPITT(Eng).htm
709) http://www.atmosp.physics.utoronto.ca/MOPITT/home.html

MOPITT is designed as a scanning instrument. IFOV = 1.8° x 1.8° (22 km x 22 km at nadir). The instrument scan line consists of 29 pixels, each at 1.8° increments. The maximum scan angle is 26.1° off-axis which is equivalent to a swath width of 640 km. - MOPITT data products include gridded retrievals of CH_4 with a horizontal resolution of 22 km and a precision of 1%. Gridded CO soundings are retrieved with 10% accuracy in three vertical layers between 0 and 15 km. Three-dimensional maps to model global tropospheric chemistry.

D.11.2 Aqua Mission (EOS/PM-1)

The focus of the Aqua mission objective is the multi-disciplinary study of the Earth's interrelated processes (atmosphere, oceans, and land surface) and their relationship to Earth system changes. The data sets of Aqua provide information on cloud formation, precipitation, and radiative properties, air-sea fluxes of energy, carbon, and moisture (AIRS, AMSU, AMSR-E, HSB, CERES, MODIS); and sea ice extent (MODIS). A launch of Aqua is planned for Dec. 2001 with a Delta-2 7920-10L vehicle from VAFB.

The Aqua [710] (EOS/PM-1) S/C is based on TRW's modular, standardized AB1200 bus design with common subsystems. The satellite dimensions are: 2.68 m x 2.47 m x 6.49 m (stowed) and 4.81 m x 16.70 m x 8.04 m (deployed). Aqua is three-axis stabilized, with a total mass of 2,934 kg at launch, S/C mass of 1,750 kg, payload mass = 1,082 kg, propellant mass = 102 kg; power = 4.86 kW (EOL). Propulsion: hydrazine blow-down system; 4 pairs of thrusters. The spacecraft mass is 6.468 kg, the design life is six years.

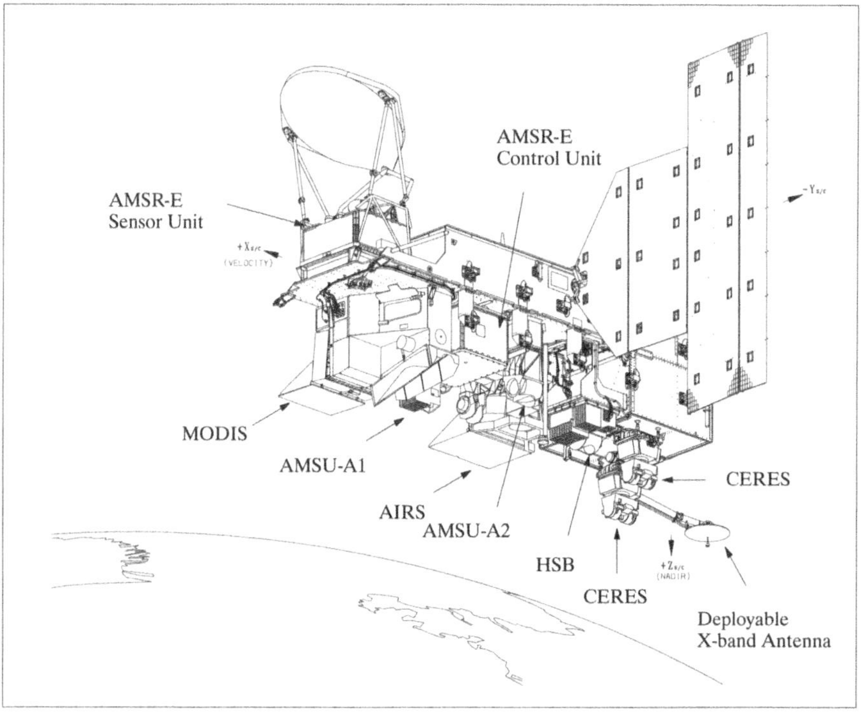

Figure 107: Illustration of the Aqua satellite

Communication: X-band. S-band (TDRSS and Deep Space Network/Ground Network compatible).

710) Note: NASA renamed the EOS/PM-1 satellite to Aqua on Oct. 18, 1999

Orbit: Sun-synchronous circular orbit, altitude = 705 km (nominal), inclination = 98.2°, local equator crossing at 13:30 (1:30 PM) on ascending node, period = 98.8 minutes.

Sensor complement: (AIRS, AMSU, AMSU-E, CERES, HSB, MODIS). The descriptions of CERES and MODIS can be found under Terra.

AIRS (Atmospheric Infrared Sounder), NASA/JPL instrument, PI: M. T. Chahine; prime contractor is LMIRIS (Lockheed Martin Infrared Instrument Systems), Lexington, MA. AIRS, along with AMSU and HSB, is of HIRS and MSU heritage flown on the NOAA POES series. Objective: High-spectral-resolution measurement of global temperature/humidity profiles in the atmosphere. Measurement of the Earth's upwelling infrared radiances in the spectral range of 3.74 - 15.4 μm, simultaneously at 2378 frequencies (bands). A limited number of visible wavelength channels are also present. [711)]

The AIRS instrument consists of an array grating spectrometer that provides coverage in the infrared region (spectral range of 3.74 - 15.4 μm) with a spectral resolution of 1200 ($\lambda/\Delta\lambda$). The high spectral resolution permits the separation of unwanted spectral emissions and, in particular, provides spectrally clean "super windows," ideal for surface observations. - This is supplemented by a VNIR photometer of four bands in the range between 0.4 and 1.0 μm. The VNIR channels are used to discriminate between low-level clouds and different terrain and surface covers, including snow and ice. The AIRS infrared bands have an IFOV of 1.1° and FOV = ± 49° scanning capability perpendicular to the spacecraft ground track (swath width = 1650 km, 13.5 km horizontal resolution in nadir, 1 km vertical). It takes 22.41 ms for each footprint of 1.1° in diameter (or 13.5 km). Each IR scan produces 90 footprints across the flight track and takes 2.67 s. The VNIR channels have a footprint of 0.185° or about 2.3 km on the ground, nine VNIR footprints are within a 40 km swath. The VNIR photometer is boresighted to the spectrometer to allow simultaneous VNIR observations.

The IR detectors array (HgCdTe) is positioned at the focal plane, it is cooled to 60K by redundant Stirling cycle cryocoolers. The IR spectrometer is also cooled to 150 K by a two-stage radiative cooler. The VNIR photometer uses optical filters to define the four spectral bands. It operates at ambient temperatures (293-300 K). Inflight calibration is performed during each scan period. In addition, AIRS uses four independent cold-space views.

Spectral coverage	3.74 - 15.4 μm for the array grating spectrometer (IR bands) 0.4 - 1.0 μm for photometer (4 VNIR bands)
Spectral resolution	1200 ($\lambda/\Delta\lambda$) array grating spectrometer, 2378 bands
Spatial resolution	13.5 km horizontal at nadir for IR bands (IFOV = 1.1°), 1 km vertical resol. 2.3 km x 2.3 km for VNIR bands (IFOV = 0.185°)
Swath width	1650 km (FOV=±49.5°) for IR bands 40 km for VNIR bands
Instrument mass, power	140 kg, 224 W
Date rate, duty cycle	1.42 Mbit/s, 100%

Table 171: Overview of some AIRS parameters

Thermal control by redundant 60 K Stirling cycle coolers, heater, mini thermal bus, two-stage radiator; thermal operating range = 20-25 °C.

The major data products derived from AIRS are atmospheric temperature profiles, humidity profiles (from channels in the 6.3 μm water vapor band and the 11 μm windows, sensitive to the water vapor continuum), and land skin surface temperature.

AIRS has been selected to fly on the Aqua (EOS/PM-1) satellite with two operational microwave sounders: NOAA's AMSU and Brazil's HSB (Humidity Sounder Brazil). Together, the three sensors constitute the advanced operational sounding system on future NOAA missions.

711) http://www-airs.jpl.nasa.gov/

AMSU/HSB (Advanced Microwave Sounding Unit (NOAA Instrument)/ (Humidity Sounder for Brazil), provided by INPE. Both instruments operate in conjunction.
AMSU was designed and developed by Aerojet of Azusa, CA (a GenCorp company). AMSU primarily provides temperature soundings, whereas HSB provides humidity soundings. AMSU is a 15-channel microwave radiometer. AMSU and HSB have a total of 19 channels, 15 are assigned to AMSU, each having a 3.3° beamwidth, and four are assigned to HSB, each having a beamwidth of 1.1°. AMSU comprises two separate units: AMSU-A-1 (channels 3-15), and AMSU-A2 (channels 1 and 2). Channels 3 - 14 use the 50 to 60 GHz oxygen band to provide data for vertical temperature profiles up to 50 km. The "window" channels (1, 2, and 15) provide data to enhance the temperature sounding by correcting for surface emissivity, atmospheric liquid water, and total precipitable water. HSB channels 17 - 20 use the 183.3 GHz water vapor absorption line to provide data for the humidity profile.

AMSU-A1 measures temperature profiles from the surface up to 50 km in 15 channels. Temperature resolution: 0.25 - 1.2 K. The AMSU-A1 instrument has two 15 cm diameter antennas (reflectors with momentum compensation), each with a 3.3° nominal IFOV at the half power points or FWHM (Full width Half maximum). Each antenna provides a cross-track scan of ±49.5° from nadir with a total of 30 Earth views (scan positions) per scan line. The total scan period is eight seconds. The footprint (resolution) at nadir is 40 km. The swath width is approximately 1650 km. Internal calibration is performed with internal warm loads and cold space.

AMSU-A-2 has a single 28 cm diameter antenna (reflector without momentum compensation) with a 3.3° nominal IFOV. All other instrument/observation parameters are the same as those of AMSU-A1.
AMSU parameters: mass = 100 kg; power = 125 W; data rate = 3.2 kbit/s; thermal control by heater, central thermal bus, radiator; thermal operating range= 0-20° C.

Sensor	Channel	Center Frequency (GHz)	Bandwidth (MHz)	Sensitivity NEΔT (K)
AMSU-A2	1	23.8	280	0.3
(2 channels)	2	31.4	180	0.3
AMSU-A1	3	50.300	180	0.4
(13 channels)	4	52.800	400	0.25
	5	53.596± 0.115	170	0.25
	6	54.400	400	0.25
	7	54.940	400	0.25
	8	55.500	330	0.25
	9	57.290344 = Flo	330	0.25
	10	Flo ± 0.217	78	0.4
	11	Flo ± 0.3222, (±0.048)	36	0.4
	12	Flo ± 0.3222, (±0.022)	16	0.6
	13	Flo ±0.3222, (± 0.010)	8	0.8
	14	Flo ±0.3222, (± 0.0045)	3	1.2
	15	89.000	6000	0.5
HSB	17	150.0	2000	1.0
(4 channels)	18	183.31±1.00	1000	1.1
	19	183.31±3.00	2000	1.0
	20	183.31±7.00	4000	1.2

Table 172: Spectral parameters of the AMSU-A and HSB instruments

HSB (Humidity Sounder for Brazil). An INPE-provided instrument of AMSU-B heritage [built by MMS of Bristol, UK (now Astrium Ltd) with participation of Equatorial Sistemas of Brazil], and sponsored by AEB (Brazilian Space Agency). [712] HSB is a microwave radiometer which the objective to measure atmospheric radiation, to obtain atmospheric water vapor profile measurements and to detect precipitation under clouds with 13.5 km horizontal nadir resolution (humidity profiles for weather foresting). HSB is a four-channel self-calibrating instrument (passive sounder) providing a humidity profiling capability in the

712) http://www.dss.inpe.br/programas/hsb/ingl/index.html

frequency range of 150 - 190 GHz, spanning the height from surface to about 42 km. The measured signals are also sensitive to a) liquid water in clouds (cloud liquid water content) and b) graupel and large water droplets in precipitating clouds (qualitative estimate of precipitation rate). HSB scans in the cross-track direction at a rate of 2.67 seconds in continuous mode. The instrument features a momentum-compensated scan mirror system. HSB is operated in combination with AMSU-A, they have a total of 19 channels: 15 are assigned to AMSU-A, each having a 3.3° beamwidth, and four assigned to HSB, each having a 1.1° beamwidth. The HSB receiver channels are configured to operate in DSB (Double Sideband). 713)

Nr. of channels	4 (total), Ch 17 at 150 GHz, Ch 18: 183.31 ±1 GHz, Ch. 19: 183.31 ±3 GHz, and Ch 20: 183.31 ±7 GHz
Swath width, scan period	1650 km, 2.67 s
FOV	±49.5° cross track from nadir (+90° to -49.5° for calibration)
IFOV (spatial resolution)	1.1° (13.5 km at nadir)
Instrument pointing	Control = 3600 arcseconds, knowledge = 360 arcseconds, stability = 74 arcseconds/s
Thermal control, operating range	Radiator, 13 - 35°C
Instrument power	80 W average, 154 W peak
Instrument mass, size, data rate	50.5 kg, 73 cm x 69 cm x 47 cm, 4.2 kbit/s
Temperature accuracy (data profile)	1.0 - 1.2 K, coverage (twice daily) of land and ocean surfaces, resolution of 50 km (horiz.) and 1 km (vertical), up to 100 mb
Humidity accuracy (data profile)	20%, global coverage (twice daily), res. = 50 km, 1 km (vertical)
Radiance accuracy (data profile)	1-1.2 K, global coverage (twice daily), res. = 15 km (average)

Table 173: Specification of the HSB instrument

AMSR-E (Advanced Microwave Scanning Radiometer-EOS), a NASDA/NASA cooperative instrument, of AMSR heritage, built by Mitsubishi Electronics Corporation (PIs: A. Shibata, R. W. Spencer). The objective is the measurement of geophysical parameters such as: cloud properties, radiative energy flux, precipitation, land surface wetness, sea ice, snow cover, sea surface temperature (SST), and sea surface wind fields. AMSR-E is a modified design of AMSR on ADEOS-II (Japan). AMSR-E is a twelve channel, six frequency total power passive microwave radiometer system. It measures brightness temperatures at 6.925, 10.65, 18.7, 23.8, 36.5, and 89.0 GHz. Vertically and horizontally polarized measurements are taken at all channels. 714)

Center frequency (GHz)	6.925	10.65	18.7	23.8	36.5	89.0
Bandwidth (MHz)	350	100	200	400	1000	3000
Sensitivity (K)	0.3	0.6	0.6	0.6	0.6	1.1
IFOV (km x km) footprint	75 x 43	51 x 29	27 x 16	31 x 18	14 x 8	6 x 4
Sampling rate (km x km)	10 x 10	10 x 10	10 x 10	10 x 10	10 x 10	5 x 5
Integration time (ms)	2.6	2.6	2.6	2.6	2.6	1.3
Main beam efficiency (%)	95.3	95.0	96.3	96.4	95.3	96.0
Beamwidth (°)	2.2	1.4	0.8	0.9	0.4	0.18

Table 174: Performance parameters of AMSR-E

AMSR-E consists of an offset parabolic reflector 1.6 m in diameter, fed by an array of six feedhorns. The reflector and feedhorn arrays are mounted on a drum which contains the radiometers, digital data subsystem, mechanical scanning subsystem, and power subsystem. The reflector/feed/drum assembly is rotated about the axis of the drum by a coaxially mounted bearing and power transfer assembly. All data, commands, timing and telemetry signals, and power pass through the assembly on slip ring connectors to the rotating assembly. The AMSR-E instrument has a mass of 324 kg, power = 350 W, a duty cycle of 100%, and an average data rate of 87.4 kbit/s.

713) Information provided by Janio Kono of INPE, Sao José dos Campos, Brazil
714) http://wwwghcc.msfc.nasa.gov/AMSR/

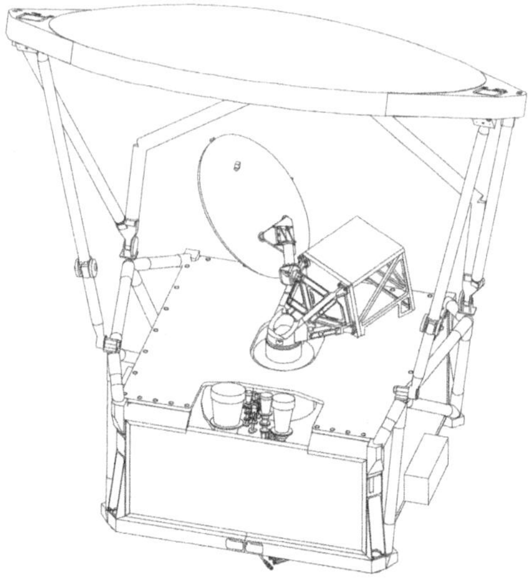

Figure 108: Illustration of the AMSR-E instrument

The AMSR-E instrument rotates continuously about an axis parallel to the local spacecraft vertical at 40 rpm. At an altitude of 705 km, it measures the upwelling scene brightness temperatures over an angular sector of ± 61° about the subsatellite track, resulting in a swath width of 1445 km. During a period of 1.5 seconds the S/C subsatellite point travels 10 km. Even though the IFOV for each channel is different, active scene measurements are recorded at equal intervals of 10 km (5 km for the 89 GHz channels) along the scan. The half cone angle at which the reflector is fixed is 47.4° which results in an Earth incidence angle of 55.0°.

The AMSR-E instrument rotates continuously about an axis parallel to the local spacecraft vertical at 40 rpm. At an altitude of 705 km, it measures the upwelling scene brightness temperatures over an angular sector of ± 61° about the subsatellite track, resulting in a swath width of 1445 km. During a period of 1.5 seconds the S/C subsatellite point travels 10 km. Even though the IFOV for each channel is different, active scene measurements are recorded at equal intervals of 10 km (5 km for the 89 GHz channels) along the scan. The half cone angle at which the reflector is fixed is 47.4° which results in an Earth incidence angle of 55.0°.

Instrument calibration. The radiometer calibration accuracy budget, exclusive of antenna pattern correction effects, is composed of three major contributors: warm load reference error, cold load reference error, radiometer electronics nonlinearities and errors.

D.11.3 Aura Mission (EOS/Chem-1)

Aura [715] is the chemistry mission of NASA with the overall objective to study the chemistry and dynamics of Earth's atmosphere from the ground through the mesosphere. The goal is to monitor the complex interactions of atmospheric constituents from both natural sources, such as biological activity and volcanoes, and man-made sources, such as biomass burning, are contributing to global change and effect the creation and depletion of ozone. The Aura mission will provide global surveys of several atmospheric constituents which can be classi-

715) Note: The EOS/Chem-1 mission was renamed in 2000 to Aura.

fied into anthropogenic sources (CFC types), radicals (e.g., ClO, NO, OH), reservoirs (e.g., HNO, HCl), and tracers (e.g., N_2O, CO_2, H_2O). Temperature, geopotential heights, and aerosol fields will also be mapped. A launch of Aura on a Delta-2 7920 vehicle from VAFB is planned for 2003.

The Aura spacecraft, like Aqua, is based on TRW's modular, standardized AB1200 bus design with common subsystems. The S/C dimensions are: 2.68 m x 2.34 m x 6.85 m (stowed) and 4.71 m x 17.03 m x 6.85 m (deployed). Aqua is three-axis stabilized, with a total mass of 2,967 kg at launch, S/C mass of 1,767 kg, payload mass =1,200 kg. The S/C design life is six years. RF communications are in S-band.

Orbit: Sun-synchronous circular orbit, altitude = 705 km, inclination = 98.7°, with a local equator crossing time of 13.45 (1:45 PM) on the ascending node.

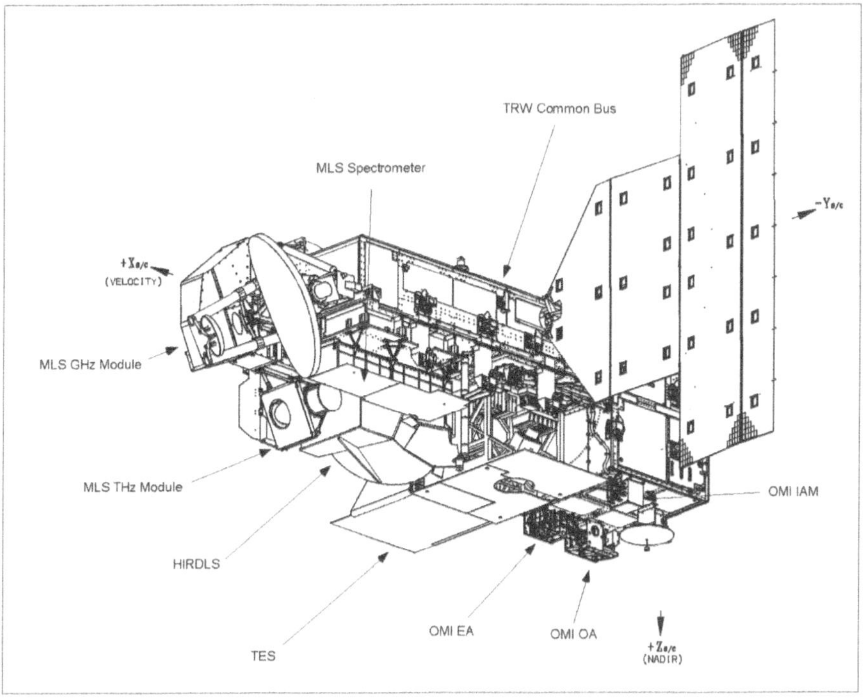

Figure 109: Illustration of the Aura spacecraft

Sensor complement: (HIRDLS, MLS, OMI, and TES)

HIRDLS (High-Resolution Dynamics Limb Sounder), a joint instrument between the University of Colorado at Boulder and Oxford University, Oxford, UK. PIs: J. Gille of the University of Colorado and J. Barnett, Oxford University; prime contractors are Lockheed Martin and Astrium Ltd., UK. - Spectral range: 6 - 18 µm. The instrument is selected to fly on Aura. HIRDLS observes global distributions of temperature and trace gas concentrations of O_3, H_2O, CH_4, N_2O, HNO_3, NO_2, N_2O_5, CFC_{11}, CFC_{12} $ClONO_2$, and aerosols in the upper troposphere, stratosphere, and mesosphere plus water vapor, aerosol, and cloud tops. The swath width is 2000-3000 km (typically six profiles across swath). Complete Earth coverage (including polar night) can be obtained in 12 hours. High horizontal resolution is obtained with a commandable azimuth scan which, in conjunction with a rapid elevation scan, provides profiles up to 3,000 km apart in an across-track swath. Spatial resolution: standard profile spacing is 500 x 500 km horizontally (equivalent to 5° longitude x 5° latitude) x 1 km vertically; averaging volume for each data sample 1 km vertical x 10 km across x 300 km along line-of-sight. [716)]

HIRDLS is an infrared limb-scanning radiometer designed to sound the upper troposphere, stratosphere, and mesosphere. HIRDLS performs limb scans in the vertical at multiple azimuth angles, measuring infrared emissions in 21 channels (temperature distribution) ranging from 6.12 - 17.76 μm. Four channels measure the emission of CO_2. Taking advantage of the known mixing ratio of CO_2, the transmittance is calculated, and the equation of radiative transfer is inverted to determine the vertical distribution of the Planck black body function, from which the temperature is derived as a function of pressure. Winds and potential vorticity are determined from spatial variations of the height of geopotential surfaces.

Data rate = 65 kbit/s; mass = 220 kg; duty cycle = 100%; power = 220 W(average), 239 W (peak); instrument size: 154.5 cm x 113.5 cm x 130 cm; FOV(scan range): elevation from 22.1° to 27.3° below horizontal, azimuth: -21° (sun side) to +43° (anti-sun side). Detector IFOV: 1 km vertical x 10 km (2.5°) horizontal. The instrument has 21 photoconductive HgCdTe detectors cooled to 65 K; each detector has a separate bandpass interference filter. Thermal control by paired Stirling cycle coolers, heaters, sun baffle, radiator panel; the thermal operating range is 20-30° C.

MLS (Microwave Limb Sounder), of UARS MLS heritage, PI: J. W. Waters, NASA/JPL. MLS is selected to fly on the Aura mission. The instrument measures thermal emissions from the atmospheric limb in submillimeter and millimeter wavelength spectral bands and is intended for studies of the following processes and parameters: [717)]

- Chemistry of the lower stratosphere and upper troposphere. Measurement of lower stratospheric temperature and concentrations of: H_2O, O_3, ClO, BrO, HCl, OH, HO_2, HNO_3, and HCN, and N_2O. Measurement of upper tropospheric H_2O and O_3 (radiative forcing on climate change).
- Chemistry of the middle and upper stratosphere. Monitoring of ozone chemistry by measuring radicals, reservoirs, and source gases in chemical cycles which destroy ozone
- The effects of volcanoes on global change. MLS measures SO_2 and other gases in volcanic plumes.

Measurements are performed continuously, at all times of day and night, altitude range from the upper troposphere to the lower thermosphere. The vertical scan is chosen to emphasize the lower stratosphere and upper troposphere. Complete latitude coverage is obtained each orbit. Pressure (from O_2 lines) and height (from a gyroscope measuring small changes in the FOV direction) are measured to provide accurate vertical information for the composition measurements.

Measurement approach: passive limb sounder; thermal emission spectra collected by offset Cassegrain scanning antenna system; Limb scan = 0 - 120 km; spatial resolution = 3 x 300 km horizontal x 1.2 km vertical. MLS contains heterodyne radiometers in five spectral bands.

FOV: boresight 60-70° relative to nadir; IFOV = ±2.5° (half-cone, along-track); spatial resolution: measurements are performed along the suborbital track; the resolution varies for different bands, at 640 GHz the spatial resolution is: 1.5 km vertical x 3 km cross-track x 300 km along-track at the limb tangent point; a typical resolution is: 3 km vertical x 5 km cross-track x 500 km along-track. Spectral bands at millimeter and submillimeter wavelengths. Instrument mass = 430 kg, power = 530 W; duty cycle = 100%; data rate = 100 kbit/s; thermal control via radiators and louvres to space as well as heaters; thermal operating range is 10-35°C.

716) http://aura.gsfc.nasa.gov/hirdls/

717) http://aura.gsfc.nasa.gov/mls/

Spectral band center	Measurement objective
118 GHz	Primarily for temperature and pressure
190 GHz	Primarily for H_2O, HNO_3, and continuity with UARS MLS measurements
240 GHz	Primarily for O_3 and CO
640 GHz	Primarily for N_2O, HCl, ClO, HOCl, BrO, HO_2, and SO_2
2.5 THz	Primarily for OH

Table 175: MLS instrument frequency bands

OMI (Ozone Monitoring Instrument), [718) 719) 720)] the instrument is a contribution of NIVR (Netherlands Institute for Air and Space Development) of Delft in collaboration with FMI (Finnish Meteorological Institute), Helsinki, Finland, to the EOS Aura mission. The PI is Pieternel Levelt of KNMI; co-PIs are: Gilbert W. Leppelmeier of FMI and Ernest Hilsenrath of NASA/GSFC. OMI is of GOME heritage, flown on ERS-2, as well as of SCIAMACHY and GOMOS heritage, flown on Envisat. The overall objective is to monitor ozone and other trace gases (continuation of the TOMS measurement series) and to monitor tropospheric pollutants worldwide. The OMI measurements are highly synergistic with the HIRDLS and MLS instruments on the Aura platform. The OMI observations provide the following capabilities and features:

- Mapping of ozone columns at 13 km x 24 km and profiles at 36 km x 48 km (continuation of TOMS and GOME ozone column data records and the ozone profile records of SBUV and GOME)
- Measurement of key air quality components: NO_2, SO_2, BrO, OClO, and aerosol (continuation of GOME measurements)
- Distinguish between aerosol types, such as smoke, dust, and sulfates
- Measurement of cloud pressure and coverage
- Mapping of the global distribution and trends in UV-B radiation
- A combination of processing algorithms is employed including TOMS version 7, DOAS (Differential Optical Absorption Spectroscopy), Hyperspectral BUV retrievals and forward modeling to extract the various OMI data products
- Near real-time production of ozone and other trace gases.

OMI is a wide-angle, non-scanning and nadir-viewing instrument measuring the solar backscattered irradiance with a swath width of 2600 km. The telescope has a FOV of 114°. The instrument is designed as a compact UV/VIS imaging spectrograph, using two array CCDs for simultaneous spatial and spectral registration (hyperspectral imaging in push-broom mode). The instrument has two channels measuring in the spectral range of 270-500 nm. The Earth is viewed in 1500 bands in the along-track direction providing daily global coverage. OMI employs a polarization scrambler to depolarize the incoming radiance. The radiation is then focussed by the secondary telescope mirror. A dichroic element separates the radiation into a UV and a VIS channel. The UV channel is split again into two subchannels UV1 (270-314 nm) and UV2 (306-380 nm). In the UV1 subchannel, the spatial sampling distance per pixel is a factor two larger than in the UV2 subchannel. The idea is to increase the ratio between the useful signal and the dark current signal, hence, to increase SNR in UV1. The resulting IFOV values of a pixel in the cross-track direction are 6 km for UV1 and 3 km for UV2 and VIS. The corresponding spatial resolution is twice as good as the sampling distances. Groups of 4 or 8 CCD detector pixels are binned in the cross-track direction. The basic detector exposure time is 0.4 s, corresponding to an along-track distance of 2.7 km. In OMI, five subsequent CCD images are electronically co-added, resulting in a

718) P. F. Levelt, B. van den Oord, E. Hilsenrath, G. W. Leppelmeier, et al., "Science Objectives of EOS-Aura's Ozone Monitoring Instrument (OMI)," Proceedings of the Quadrennial Ozone Symposium, Sapporo, Japan, 2000, pp. 127-128

719) E. Laan, J. de Vries, B. Kruizinga, H. Visser, et al., "Ozone Monitoring with the OMI Instrument," Proceedings of 45th Annual Meeting of SPIE, San Diego, CA, July 30 to Aug. 4, 2000, Paper No 4132-41, pp. 334-343

720) E. Laan, J. de Vries, P. Levelt, P. Stammes, H. Saari, J. Lundell, A. Maelkki, et al., "Ozone Monitoring in the Next Millennium with the OMI Instrument," Proc. IAF Congress, Oct. 2-6, 2000, Rio de Janeiro, IAF-99-B.2.09

FOV of 13 km in the along-track direction. In addition, one column (wavelength) of each CCD data is downlinked without co-adding (monitoring of clouds, ground albedo). The pixel binning and image co-adding techniques are used to increase SNR and to reduce the data rate.

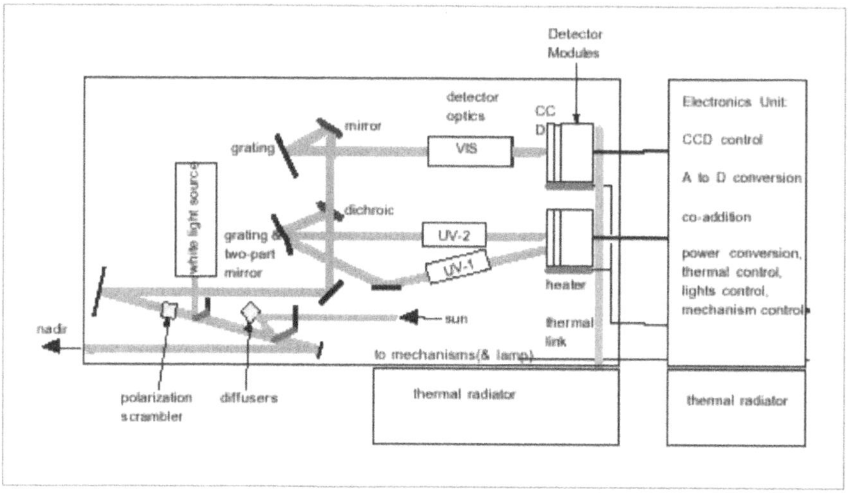

Figure 110: Conceptual design of the OMI instrument

Visible band; two UV bands	UV1: 270-314 nm, UV-2: 306-380 nm, VIS: 350-500 nm
Spectral resolution	0.42 - 0.65 nm FWHM (Full width Half Maximum)
Spectral sampling	1.3 - 3 for FWHM
Telescope FOV	114° (providing a surface swath width of 2600 km)
IFOV (spatial resolution)	13 km x 24 km; or 36 km x 48 km (depending on the product) Two zoom modes: 13 km x 13 km (detection of urban pollution)
Detector	CCD pushbroom with 780 x 576 (spectral x spatial) pixels
Instrument mass, power, data rate	65 kg, 66 W, 0.8 Mbit/s (average)
Instrument size	50 cm x 40 cm x 35 cm
Duty cycle	60 minutes on the daylight side of the orbit

Table 176: OMI instrument parameters

The CCD detector arrays are of the back-illuminated and frame-transfer type, each with 576 x 780 pixels in the image section and the same amount in the storage or readout section. The frame transfer layout allows simultaneous exposure and readout of the previous exposure. This in turn permits fair pixel readout rates (130 kHz) and good data integrity. There are two zoom modes, besides the global observation mode, for regional studies with a spatial resolution of 13 km x 13 km. In one zoom mode, the swath width is reduced to 725 km; in the other zoom mode, the spectrum is reduced to 306 - 432 nm. Cloud coverage information is retrieved with a high spatial resolution, independent of the operational mode.

Instrument calibration: Daily measurements of the sun are taken with a set of reflective diffusers for absolute radiometric calibration. Relative radiometric calibration is performed using a WLS (White Light Source) and two LEDs per spectral (sub-) channel. The two LEDs fairly uniformly illuminate the CCD's. Spectral calibration is being performed using Fraunhofer features in the sun and nadir spectra. This is supported by a dedicated spectral correction algorithm in the level 0-1 b software. Dark signal calibration is performed at the dark side of the orbit using long-exposure time dark measurements. Straylight is always monitored at dedicated rows at the side of the images; there are also covered CCD pixels measuring dark current and, at the top and bottom of the image, exposure smear.

TES (Tropospheric Emission Spectrometer), PI: R. Beer, NASA/JPL. [721)]
The instrument is selected to fly on the Aura mission. Heritage: ATMOS (ATLAS), and AES. TES is a high-resolution infrared imaging Fourier transform spectrometer with spectral coverage from 3.2 - 15.4 μm. TES has the capability to make both limb and nadir observations. Limb mode: height resolution = 2.3 km, height coverage = 0 - 34 km. In the nadir modes, TES has a spatial resolution of 0.53 km x 5.3 km with a swath of 5.3 km x 8.5 km. TES is a pointable instrument; it can access any target within 45° of the local vertical, or produce regional transects up to 885 km in length without any gaps in coverage. TES employs both, the natural thermal emission of the surface and atmosphere, and reflected sunlight, thereby providing day and night coverage anywhere on the globe.

Observations from TES will further understanding of long-term variations in the quantity, distribution, and mixing of minor gases in the troposphere, including sources, sinks, troposphere-stratosphere exchange, and the resulting effects on climate and the biosphere. TES will provide 3-D global maps of tropospheric ozone (primary objective) and its photochemical precursors (chemical species involved in ozone formation and destruction). Other objectives:

- Simultaneous measurements of NO_y, CO, O_3, and H_2O, determination of the global distribution of OH.
- Measurements of SO_2 and NO_y as precursors to the strong acids H_2SO_4 and HNO_3
- Measurements of gradients of many tropospheric species
- Determination of long-term trends in radiatively active minor constituents in the lower atmosphere.

The following key features are part of the TES instrument design:

- Back-to-back cube corner reflectors to provide the change in optical path difference
- Use of KBr (potassium bromide) material for the beam splitter-recombiner and the compensator
- Only one of two input ports for actual atmospheric measurements. The other input views an internal, grooved, clod reference target
- A diode-pumped solid-state Nd:YAG laser for interferogram sampling control
- Cassegrain telescopes for condensing and collimating wherever possible to minimize the number of transmissive elements in the system.
- A passive space-viewing radiator to maintain the interferometer and optics at 180 K.
- A two-axis gimbaled pointing mirror operating at ambient temperature to permit observation of the full field of regard (a 45° cone about nadir plus the trailing limb).
- Two independent focal plane assemblies maintained at 65 K with active pulse-tube coolers.

The TES instrument operates in a step-and stare configuration when in downlooking mode. At the limb the instrument points to a constant tangent height. Thus, the footprint is smeared along the line-of-sight by about 110 km during the 16 s limb scan (this is comparable to the effective size of the footprint itself). Hence, atmospheric inhomogeneity in the atmosphere becomes an issue; it must be dealt with in data processing (usually through a simplified form of tomography).

The routine operating procedure for TES is to make continual sets of nadir and limb observations (plus calibrations) on a one-day on, one-day off cycle. During the off-days, extensive calibrations and spectral product observations are made.

721) R. Beer, T. A. Glavich, D. M. Rider, "Tropospheric emission spectrometer for the Earth Observing System's Aura satellite," Applied Optics, Vol. 40, No 15, May 20, 2001, pp. 2356-2367

Parameter	Specification	Comments
Spectrometer type	Connes-type four-port FTS (Fourier Transform Spectrometer)	Both limb and nadir viewing capability essential
Spectral sampling distance	Interchangeably 0.0592 cm^{-1} down-looking and 0.0148 cm^{-1} at the limb	Unapodized
Optical path difference	Interchangeably±8.45 cm^{-1} down-looking and ± 33.8 cm at the limb	Double-sided interferograms
Overall spectral coverage	650-3050 cm^{-1} (3.2-15.4 μm)	Continuous, but with multiple sub-ranges typically 200-300 cm^{-1} wide
Individual detector array coverage	1A, 1900-3050 cm^{-1} 1B, 820-150 cm^{-1} 2A, 1100-1950 cm^{-1} 2B, 650 -900 cm^{-1}	All MCT photo voltaic (PV) detectors at 65 K.
Array configuration	1 x 16	All four arrays optically conjugated
Aperture diameter	5 cm	Unit magnification system
System étendue (per pixel)	9.45 x 10^{-5} cm^2 sr	Not allowing for small central obscuration from the Cassegrain secondaries
Modulation index	>0.7; 650-3050 cm^{-1}	>0.5 at 1.06 μm (control laser)
Spectral accuracy	±0.00025 cm^{-1}	After correction for finite FOV, off-axis effects, Doppler shifts, etc.
Channeling	<10% peak to peak; <1% after calibration	All planar transmissive elements wedged
Spatial resolution	0.5 km x 0.5 km at nadir 2.3 km x 2.3 km at limb	IFOV IFOV
Spatial coverage	5.3 km x 8.5 km at nadir 37 km x 23 km at limb	
Pointing accuracy	75 μrad pitch, 750 μrad yaw, 1100 μrad roll	Peak-to-peak values
Field of regard	45° cone about nadir plus trailing limb	Also views internal calibration sources
Scan (integration) time	4 s nadir and calibration, 16 s limb	Constant-speed scan, 4.2 cm/s (optical path difference rate)
Max stare time at nadir	208 s	40 downlooking scans
Transect coverage	885 km maximum	
Interferogram dynamic range	<=16 bit	Plus four switchable gain steps
Radiometric accuracy	<= 1 K, 650-2500 cm^{-1}	Internal, adjustable, hot blackbody plus cold space
Pixel-to-pixel cross talk	<10%	Includes diffraction, aberrations, carrier diffusion, etc.
Spectral SNR	As much as 600:1, 30:1 min requirement	Depends on spectral region and target. General goal is to be source photon shot-noise limited
Instrument lifetime	5 year on orbit	Plus 2 years before launch
Size	1.0 m x 1.3 m x 1.4 m	Earth shade stowed
Power	334 W (average, 361 W (peak)	
Instrument mass	385 kg	
Instrument data rate	4.5 Mbit/s (average), 6.2 Mbit/s (peak)	Science only

Table 177: TES performance characteristics

D.12 ESE (Earth Science Enterprise)

ESE, formerly called MTPE (Mission to Planet Earth), is a NASA-initiated but US-wide framework, concept and strategy of spaceborne, airborne, and ground-based measurement/observation systems, which began in the 1980s. It represents in a sense the US 'Earth Observation Program', and is as such a contribution to the U.S. Global Change Research Program (USGCRP) and to related international efforts to better understand Earth and how humans may be affecting it. Aside from NASA, other US agencies are participating in and contributing to ESE and to USGCRP such as NOAA (DOC), USGS (DOI), NSF/NCAR, DOE, USDA, EPA, DoD, a number of US universities, and others (by designing and/or operating satellites, providing facilities, participating in numerous campaigns, building instruments, maintaining archives, analyzing data, etc.). The ESE framework is in fact so vast and complex that it also has a considerable international component. [722) 723) 724) 725)]

The goal of ESE is to advance our understanding of the entire Earth system and its interactions. The ESE science objectives address the physical, chemical, and biological phenomena that govern the dynamic Earth system. The emphasis of ESE process research is on the following Earth system topics:

- Hydrologic processes that govern the interactions of land and ocean surfaces with the atmosphere through the transport of heat, moisture, and momentum
- Biogeochemical processes which contribute to the formation, dissipation, and transport of trace gases and aerosols, and their global distributions.
- Atmospheric processes which control the formation, dissipation, and distribution of clouds and aerosols and their interaction with solar radiation
- Ecological processes which are affected by and/or will affect global change, and their response to such changes through adaptation
- Geophysical processes which have shaped or continue to modify the surface of the Earth through tectonics, volcanism, and the melting of glaciers and sea ice.

ESE is regarded as a long-term program consisting of spaceborne, airborne, and ground-based measurement systems, of the corresponding infrastructure and operational services (of data processing and distribution centers, archives, communication, etc.), and of the elements of data analysis and interpretation. Phase 1 of ESE is considered the time frame from 1990 to 1998, consisting of ongoing Earth observation missions. Phase 2 of ESE starts with the launch of the first EOS (Earth Observing System) satellite in 1999 (Terra). The EOS program is regarded as the principal element of ESE, providing systematic and continuous observations from low Earth orbit for a minimum time period of 15 years.

NASA Missions/Sensors	Mission Objectives
ERBS (operational)	Earth radiation budget, aerosol, and ozone data from 57° inclination orbit
UARS (operational)	Stratospheric and mesospheric chemistry and dynamic processes
Spacelab series (92-94)	Shuttle-based experiments such as ATLAS, LITE, SRL-1 and SRL-2
TOPEX/Poseidon (op.)	Ocean circulation (joint mission with CNES)
Lageos-2 (with Italy)	Satellite laser ranging for monitoring the crustal motions of the Earth
SeaStar/SeaWiFS (1997)	Ocean color data to monitor ocean productivity
TOMS/EP (1996)	Ozone mapping and monitoring
NSCAT/ADEOS (1996)	Ocean surface wind vectors (with NASDA)
TOMS/ADEOS (1996)	Ozone mapping and monitoring (with NASDA)
TRMM (1997)	Precipitation, clouds, and radiation processes over tropics (with NASDA)
Non-NASA Missions	Mission Objectives
POES series (NOAA)	Earth's VNIR radiation, IR atmospheric sounding, ozone measurements
Landsat-4, -5 (Eosat)	High spatial resolution VNIR radiation and terrestrial surfaces
DMSP series, (DoD)	VNIR and passive microwave atmospheric and surface measurements
ERS-1, op. (ESA)	SAR data, microwave altimeter, sea surface winds and temperature
JERS-1 (NASDA)	SAR data and high spatial resolution surface VNIR radiation
ERS-2, op. (ESA)	Same as ERS-1, plus ozone mapping and monitoring
Radarsat, 1995 (CSA)	SAR data of Earth's surface (joint CSA/NASA mission)
ADEOS (NASDA)	Surface VNIR and microwave radiation, winds, and atmospheric chemistry

Table 178: ESE Phase 1 survey of missions

722) Note: As of January 1998 MTPE was renamed by NASA to "Earth Science Enterprise" (ESE)

723) M. D. King, R. Greenstone (editors), "1999 EOS Reference Handbook, NASA publication

724) MTPE/EOS Reference Handbook, 1995, NASA, G. Asrar and R. Greenstone (editors)

725) Earth Observation from Space, Report of 'Committee on Earth Studies,' 'Space Studies Board,' 'Commission on Physical Sciences, Mathematics and Applications,' 'National Research Council,' National Academy Press, Washington, D. C., 1995

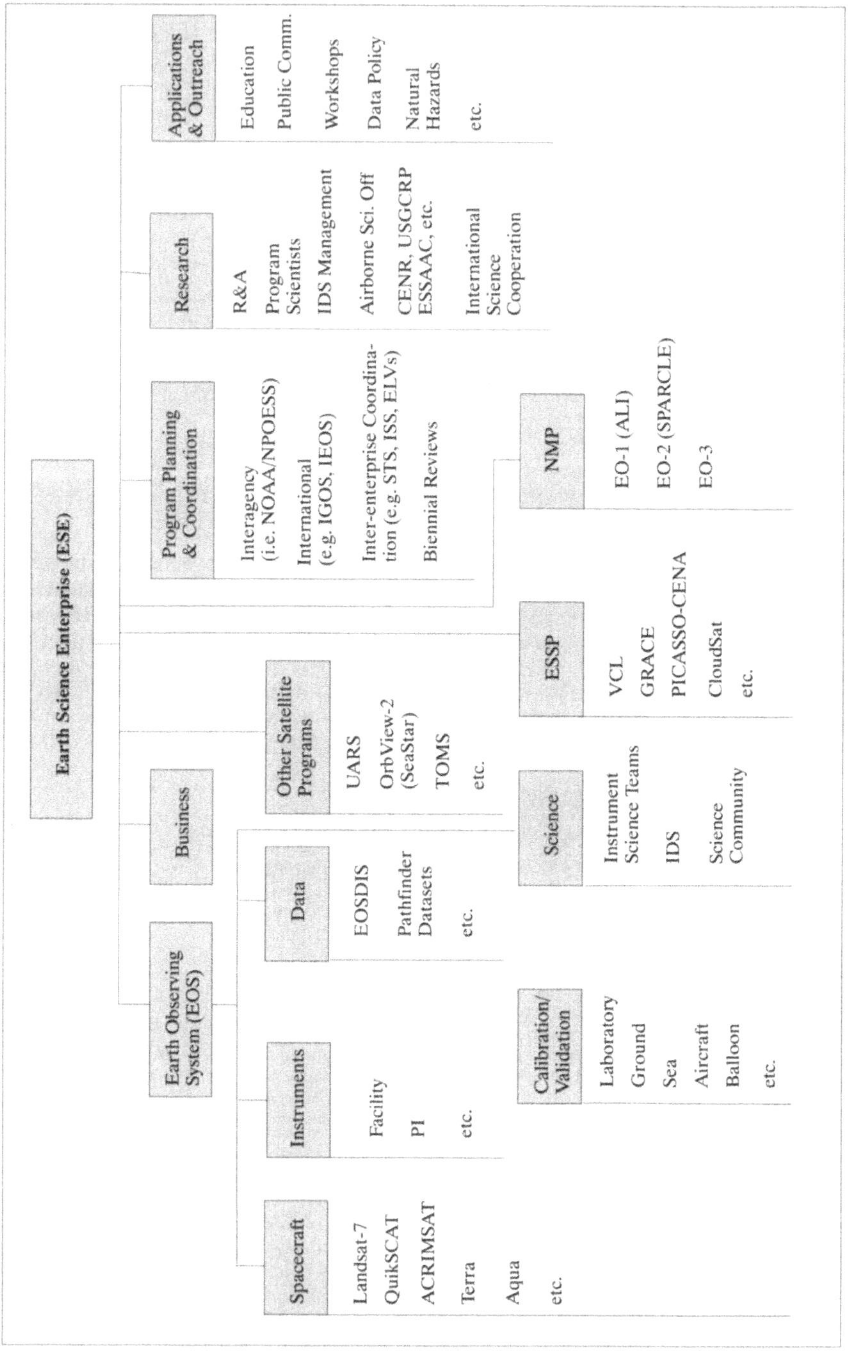

Figure 111: Overview of NASA's Earth Science Enterprise

A survey of ESE Phase 1 missions is given in Table 178, showing the pre-EOS missions up to 1998. The EOS missions for ESE Phase 2 are illustrated in Figure 111 and Table 179 on page 399. There are a number of international elements in the EOS program. In addition there are further international missions fitting into the category of Phase 2 (EOS era) Earth observing missions; they are referred to as IEOS (International EOS). These are: ESA's ENVISAT; NASDA's ADEOS-II, GCOM series (GCOM-A1, GCOM-B1), and ALOS (Advanced Land Observation Satellite); and EUMETSAT's MetOp series.

Landsat-7 (15.4.1999)	ETM+, land surface monitoring in the VIS and infrared regions
QuikSCAT (19.6.1999)	SeaWinds, ocean surface wind vectors
Terra (18.12.1999)	Clouds, aerosols, and radiation balance, characterization of the geo-ecosystem, land use soils, terrestrial energy/moisture, tropospheric chemical composition, volcanoes, ocean primary productivity, etc.
AcrimSat (20.12.1999)	ACRIM-III, monitoring of TSI (Total Solar Irradiance)
EO-1 (21.11.2000)	New ALI instrument. Scene comparison between ALI and ETM+
SAGE-III on Meteor-3M (2001)	Mid-inclination flight of SAGE-III. Global total ozone as well as measurement of SO_2 in volcanic plumes to investigate the effects of volcanic injections into the atmosphere
Jason-1 (2001)	NASA/CNES altimetry mission, ocean circulation
Envisat (2001)	ESA, atmospheric chemistry and marine biology, ERS continuation
Aqua (EOS/PM-1) (2001)	Atmospheric temperature and humidity; clouds, precipitation, and radiative balance; characterization of terrestrial and oceanic processes including productivity; air-sea fluxes of energy and moisture; and sea-ice and snow-cover extent (includes Brazilian and Japanese instruments)
GRACE (2001)	US-German dual-minisatellite gravity mission with the overall objective to obtain long-term data with unprecedented accuracy for global (high-resolution) models of the mean and the time-variable components of the Earth's gravity field.
ICESat-1 (2001)	Ice sheet mass balance and cloud-top and land-surface topography
SORCE (2002)	Total and spectral solar irradiance
VCL (2002, TBD)	To characterize the 3-D structure of land cover for ecosystem/climate modeling, monitoring, and prediction (mapping of biomass, height of vegetation layer, density profile of the vegetation layer).
Aura (EOS/CHEM-1) (2003)	Atmospheric chemical composition; tropospheric-stratospheric exchange of energy and chemicals; chemistry-climate interactions; air quality (includes joint Netherlands/Finland and joint UK/US instruments)
ISS/SAGE-III (2003) International Space Station	Mid-inclination, low-altitude flight of SAGE-III
PICASSO-CENA (2003) NASA/CNES mission	To profile the vertical distribution of clouds and aerosols and their role in the heating/cooling of the Earth
ALOS (2003) Japanese mission	High-resolution optical and microwave payload for land surface, cartography, and disaster monitoring
CloudSat (2003)	Increase the understanding of the role of optically thick clouds on the Earth's radiation budget
MetOop-1 (2005) EUMETSAT with NOAA	Operational meteorology and climate monitoring, with the future objective of operational climatology
GCOM of NASDA (2006)	Study of global change phenomena (energy cycle, atmosphere/ocean interaction, understanding of ozone and greenhouse mechanisms

Table 179: Overview of EOS era (first series) remote sensing satellites within ESE

D.13 ERS-1 (European Remote-Sensing Satellite)

ERS = European Remote Sensing Satellite (ESA Program). ERS-1 launch: July 17, 1991 from Kourou (Ariane IV vehicle, Mission Control at ESOC). The ERS program was initiated in 1981; it also includes the build-up of a corresponding ground segment for the exploitation of all data.

The ERS-1 S/C is a three-axis-stabilized Earth-pointing satellite [zero momentum bias with control to 0.11° (pitch/roll) and 0.21° (yaw)]. The platform is a SPOT program platform

(SPOT MK1 bus) modified to meet the needs of the ERS missions. The payload support module has dimensions of 2 m x 2 m x 3 m (high). Attitude is measured by a number of sensors (horizon sensor, narrow-field sun sensors, gyroscopes, wide-field sun sensors). Attitude control by a set of momentum wheels. The payload module consists of PEM (Payload Electronics Module) and ASS (Antenna Support Structure). Thermal control is a passive system complemented by an active heater. Satellite mass = 2380 kg (payload mass about 1000 kg). The single solar panel is 11.7 m x 2.4 m and supports peak payload power of 2600 W. Battery storage capacity is 2650 Wh.

There is worldwide cooperation in the area of data services (reception and exchange). DLR/DFD provides, among other European nation's agencies, a contribution in the area of science data reception (O'Higgins station in Antarctica), and in the area of science data processing at a 'Processing and Archiving Facility' (D-PAF). The other PAFs in the ground segment are provided by France (F-PAF), Italy (I-PAF) and the United Kingdom (UK-PAF).[726) 727) 728) 729)]

Application: A broad range of disciplines is covered. Observation of oceans, polar ice, land ecology, geology, forestry, wave phenomena, bathymetry (water depth), atmospheric physics, meteorology, etc. Scientific research: PIPOR (Program for International Polar Ocean Research); PISP (Polar Ice Sheet Proposal). Demonstration of concept and technology for space and ground segments (performance and operational capability).

Orbit: Sun-synchronous polar orbit, 98.52° inclination, 785 km altitude. Nominal mission duration of 2 years, 3rd year reduced mission; there are the following orbit coverage cycles:
1. Reference orbit - 3 day repeat cycle (high repetition change monitoring with dedicated calibration sites; this orbit was used during the commissioning phase)
2. Ice-Orbit - similar 3 day repeat cycle with a slightly different longitudinal phase. The main limitations of a 3 day cycle are the restricted global coverage for the imaging SAR and the wide separation of the RA-1 tracks.
3. Mapping-Orbit - 35 days repeat cycle (guaranteeing full Earth coverage). This enables SAR imaging of every part of the Earth's surface, with at least twice the frequency of the coverage at middle and high latitudes.

ERS-1 status: the satellite was in hibernation since June 1996. Operations were limited to the monitoring of the platform vital elements, the battery maintenance and the periodic checkout of the entire system every 70 days. - To maintain the ERS-1 batteries performance, the SAR image mode was activated once or twice a day. This was used to perform limited ERS-1/2 SAR interferometry acquisitions.

The ERS-1 mission ended on March 10, 2000 by a failure of the on-board attitude control system (ERS-1 was lost when a failed gyro prevented the S/C from maneuvering into an emergency acquisition mode with its solar panels pointed toward the sun to keep the batteries charged). This gives a service life of about 8 1/2 years, more than three times its planned lifetime. The most exiting results of the ERS-1 mission have been in the field of SAR interferometry, where for the first time precise topographic information could be produced in a tandem mission with ERS-2.

Sensor complement:

AMI (Active Microwave Instrument)[730)] Synthetic Aperture Radar (SAR), built by MMS, France. Two separate radars are incorporated within the AMI, a SAR for 'Image and Wave mode' operation, and a scatterometer (SCAT) for 'Wind mode' operation. This instrument can operate in either one of the following modes:

726) ESA Bulletin No. 65 Feb. 1991
727) W. Markwitz, "Das ERS-1 Bodensegment, Empfang, Verarbeitung und Archivierung von SAR Daten," Die Geowissenschaften, 9. Jahrgang, Heft 4-5, April-Mai 1991, pp. 111-115
728) D. Gottschalk, "ERS-1 Mission and System Overview," Die Geowissenschaften, 9. Jahrgang, Heft 4-5, April-Mai 1991, pp. 100-101
729) M.F. Buchroithner, J. Raggan, D. Strobl "Geokodierung und geometrische Qualitätskontrolle," Die Geowissenschaften, 9. Jahrgang, Heft 4-5, April-Mai 1991, pp. 116-112

a. AMI in **Imaging mode**. Measurement in C-band (frequency = 5.3 GHz (equivalent to 5.66 cm wavelength), bandwidth = 15.55 MHz; polarization = Linear Vertical (LV); PRF range = 1640-1720 Hz in 2 Hz steps; long pulse = 37.12 μs; compressed pulse = 64 ns; peak power = 4.8 kW; antenna size = 10 m x 1 m; look angle = 23°; radiometric resolution = 5 Bit on raw data (SAR mode), which corresponds to about 30 m spatial resolution; swath width = 100 km. Data rate = 105 Mbit/s.
Imaging mode operating time per orbit = 12 minutes ((12% duty cycle) including four minutes in eclipse).
b. AMI in **Wave Mode**. Measurement of the changes in radar reflectivity of the sea surface due to surface waves. Provision of images (5 km x 5 km), also referred to as "imagettes," at regular intervals of 200 km along track. These imagettes are transformed into spectra providing information about the lengths and directions of the ocean wave systems. Characteristics: frequency = 5.3 GHz, polarization = Linear Vertical (LV); incidence (look) angle = 23°; wave direction: 0 - 180°.; wavelength = 100-1000 m; direction accuracy = ±20°; length accuracy = ±25%; spatial sampling: 5 km x 5 km every 200-300 km; resolution = 30 m; data rate = 370 kbit/s; duty cycle of 70 %.
c. AMI in **Wind Scatterometer Mode (AMI-SCAT).**[731] Use of three separate sideways-looking antennas (fore, mid and aft beams, see Figure 112) to measure sea surface wind speed and direction. Characteristics: wind direction range = 0 - 360°; accuracy = ±20°; wind speed range = 4-24 m/s; accuracy = 2m/s or 10%, spatial resolution = 50 km; grid spacing = 25 km; swath width = 500 km (same side as SAR imaging); swath stand-off = 200 km to side of orbital track; frequency = 5.3 GHz ±200 kHz; polarization = LV; peak power = 4.8 kW; incidence angle range = 16-42° (mid), 22-50° (fore), 22-50° (aft); antenna length = 2.3 m (mid), 3.6 m(fore), 3.6 m (aft); data rate = 500 kbit/s. Operation over all oceans. Note: AMI-SCAT cannot be operated in parallel with the AMI SAR imaging mode; however, parallel operation of the wind and waves modes is possible.

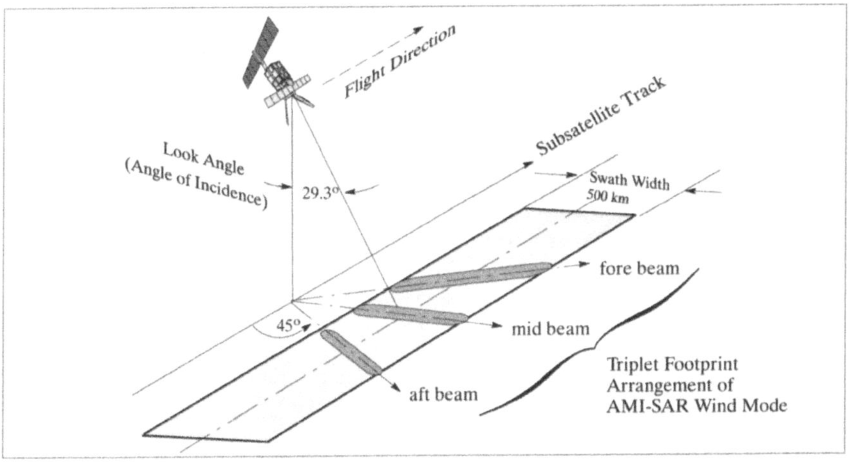

Figure 112: ERS-1 Wind Scatterometer observation geometries

The three antenna beams continuously illuminate a swath of 500 km each measuring the radar backscatter from the sea surface for overlapping 50 km resolution cells using 25 km grid spacing. The result is three independent backscatter measurements relating to cell center nodes on a 25 km grid (three different viewing directions, separated by a very small time

730) E. P. W. Attema, "The Active Microwave Instrument On-Board the ERS-1 Satellite," Proc. IEEE, Vol. 79, No.6, June 1991, pp. 791- 799
731) ERS-1 User Handbook, ESA SP-1148, May 1992, pp. 6-7

delay). This permits surface wind vector determination using 'triplets' within the mathematical model. AMI is an instrument providing data for a wide range of research disciplines such as: climatology, oceanography, glaciology, land processes, operational meteorology.

Radar Altimeter (RA-1), operating in Ku-band, consisting of a reflector, waveguide feed, tripod plus supporting structure, horn feed and the waveguide (built by Alenia Spazio, Italy). RA-1 is a nadir-pointing pulse radar taking precise measurements of the echos from the ocean and ice surfaces. Frequency = 13.8 GHz; pulse length = 20 μs; pulse repetition frequency = 1020 Hz; chirp bandwidth = 330 MHz (for ocean mode) and 82.5 MHz (for ice mode); RF transmit power = 55 W peak; antenna diameter = 1.2 m; max. data rate = 15 kbit/s; instrument mass = 96 kg; power = 130 W. RA-1 operates in 2 modes: ocean mode and ice mode. Beam width = 1.3°; foot print = 16 - 20 km (depending on sea state). RA-1 operates by timing the two-way delay for a short duration radio frequency pulse, transmitted vertically downwards. The required level of range measurement accuracy (better than 10 cm) calls for a pulse compression technique (chirp). The instrument employs frequency modulation and spectrum analysis of the pulse shape. RA-1 provides measurements leading to the determination of: [732),733),734)]

- precise altitude (ocean surface elevation for the study of ocean currents, the tides and the global geoid)
- significant wave height
- ocean surface wind speed
- various ice parameters (surface topography, ice types, sea/ice boundaries)

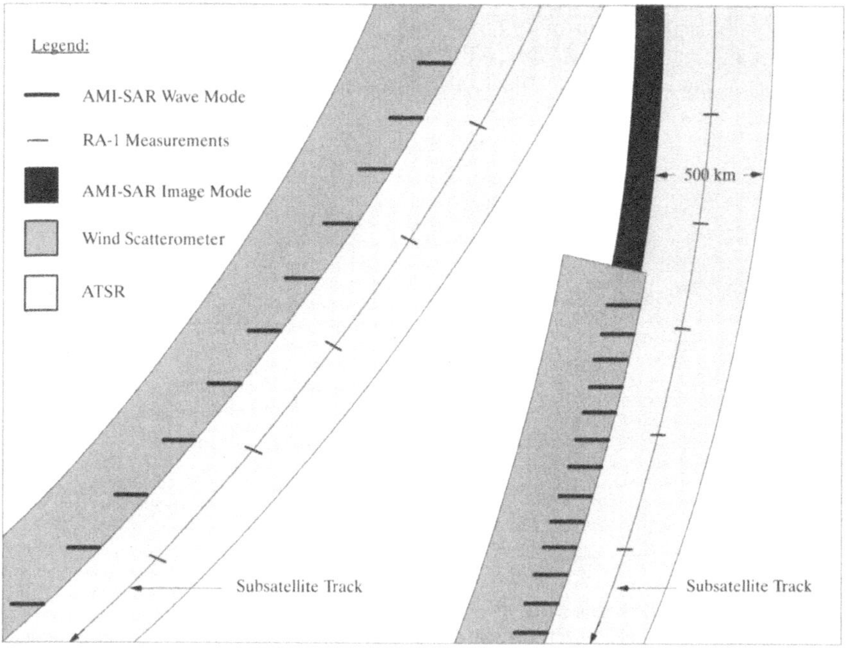

Figure 113: Schematic swath coverages for ERS-1 sensors

732) G. Schreier, K. Maeda, B. Guindon, "Three Spaceborne SAR Sensors: ERS-1, JERS-1, and RADARSAT- Competition or Synergism?," Geo Informationssysteme, Heft 2/1991, Wichmann Verlag, Karlsruhe, pp. 20 - 27

733) R. Winter, D. Kosmann "Anwendungen von SAR-Daten des ERS-1 zur Landnutzung," Die Geowissenschaften, 9. Jahrgang, Heft 4-5, April-Mai 1991, pp. 128-132

734) W. Kühbauch, "Anwendung der Radarfernerkundung in der Landwirtschaft," Die Geowissenschaften, 9. Jahrgang, Heft 4-5, April-Mai 1991, pp. 122-127

ATSR (Along-Track Scanning Radiometer and Microwave Sounder). ATSR [built by RAL, UK (British Aerospace as prime contractor); CRPE, France, and CSIRO, Australia] consists of two instruments: the MWR (Microwave Radiometer) and the IRR (Infrared Radiometer). A major objective of ATSR is to measure the global SST (Sea Surface Temperature) with the high accuracy required by the climate change research community.

MWR characteristics:
The MWR instrument uses a 60 cm Cassegrain offset-fed antenna to view the Earth in nadir direction in the frequencies of 23.8 and 36.5 GHz. The signals received are compared with those from a reference source at a known temperature to minimize the effects of short-term variations. Additional features are used to calibrate MWR: the sky-horn antenna is pointed toward cold space; the hot reference is obtained internally. IFOV = 20 km (= resolution); each channel has a bandwidth of 400 MHz. Prime objective of MWR is measurements of atmospheric water-vapor and liquid content in order to improve the accuracy of the sea surface temperature measurements and also to provide accurate tropospheric range correction for the RA-1.

IRR characteristics:
The IRR imager has 4 spectral channels: 1.6 (SWIR), 3.7, 10.8 and 12 µm. Spatial resolution = 1 km x 1 km (IFOV at nadir). Radiometric resolution < 0.1 K. Absolute accuracy < 0.5 K by averaging over a 50 km x 50 km area for SST with 80 % cloud cover; radiometric resolution < 0.1 K; swath width = 500 km. The scanning technique enables the Earth's surface to be viewed at two different angles (0° and 47°) in two curved swaths 500 km wide and separated, along track, by about 800 km. Successive scans in the cross-track direction are displaced by about 1 km (along-track) due to the satellite's motion. A rotating mirror scans the two tracks once every 150 ms (total of 2000 pixels per scan, 555 for nadir-view data and 371 for forward-view data) Measurements of: [735]

- cloud-top temperature and cloud cover
- sea-surface temperature (prime objective of IRR)

The thermal channels of IRR use an advanced detector cooling system and can be calibrated. Of particular interest is the inclined scanning configuration of IRR. The sensor records a line of off-nadir pixels at a view zenith angle of about 55° and some 900 km along-track. About 2 minutes later, a nadir view is obtained when the S/C is directly over the target. The resultant image data set (after resampling both the nadir and forward view data) consists of two co-registered images with a 1 km spatial resolution on a 500 km swath. The data permit accurate atmospheric correction over oceans by exploiting the information on atmospheric properties derived from successive images of the same area within a short period.

LRR (Laser Retro-Reflector). A passive optical device for accurate satellite tracking from ground (laser ranging stations of the SLR network) to support instrument data evaluation. LRR characteristics: wavelength = 350 - 800 nm (optimized for 532 nm), efficiency: greater than 0.15 at end-of-life, reflection coefficient: > 0.8 end-of-life, FOV: elevation half-cone angle 60°, azimuth of 360°, diameter: ≤ 20 cm.

PRARE [736] (Precise Range And Range-Rate Equipment) see also chapter H.7.2. Precise satellite range determination will lead to higher-accuracy altitude measurements that will extend the mission to ocean circulation studies and geodetic applications such as sea-surface topography and crustal dynamics.

Data: The payload data are transmitted by the IDHT (Instrument Data Handling and Transmission) subsystem. The instruments generate data in the form of source packets

735) F. M. Danson, N. A. Higgins, N. M. Trodd, "Measuring Land-Surface Directional Reflectance with the Along-Track Scanning Radiometer," PE&RS, Vol 65, No 12, Dec. 1999, pp. 1411-1417

736) **Note: The on-board PRARE instrument of the ERS-1 payload could not achieve operational status after launch. The instrument worked nominally for five days after launch (five contacts with the command station showed nominal telemetry). A thorough failure analysis came to the conclusion that the most likely cause of the PRARE failure is RAM damage due to radiation (destructive RAM latch-up).**

which in turn are put into transport frames for transmission. Three data streams are transmitted from the IDHT in X-band: Link 1 contains high-rate real-time SAR data at a rate of 105 Mbit/s (8140 MHz); Link 2 contains low-rate real-time data (AMI wave and wind data, RA-1 and ATSR data) at a rate of 1.093 Mbit/s (8040 MHz); Link 3 contains recorder data (all of Link 2) at a rate of 15 Mbit/s. Link 1 is dedicated onto one X-band link, while Link 2 and 3 share the second X-band link. The modulation scheme for Links 1 is QPSK (Quadrature Phase-Shift Keying). The low-rate link uses UQPSK (Unbalanced Quadrature Phase-Shift Keying) to modulate Link 2 and Link 3 data onto a single link. With no recorder dump data BPSK (Bi-Phase-Shift Keying) is used for the real-time data.

The on-board recorder has a capacity of 6.5 Gbit. AMI SAR data cannot be recorded but are downlinked in real-time. Image size for SAR data: 100 km x 100 km (4 Looks Full Digital Image Processing). TT&C operations are in S-band at a data rate of 2 kbit/s.

Ground stations: ESA stations at Kiruna, Fucino, Maspalomas (Canary Islands), and Gatineau (Canada); national facilities, like the Canadian "Prince Albert" station, the DLR/DFD "O'Higgins" station (in Antarctica) as well as a portable station which can be set up anywhere (DTXS = DFD Transportable X-band Station), the CNES Aussaguel station, the Japanese stations "Hatoyama," "Kumamoto" and "Syowa" (Antarctica), the Indian (ISRO) station Hyderabad, the Alaska SAR Facility, Fairbanks (NASA), Alice Springs and Hobart (Australia), Tromsö (Norway), Cuiaba (Brazil, INPE), Cotopaxi (Ecuador), Miyun (China, CAS), Ryadh (Saudi Arabia), Bangkok (Thailand), Pretoria (South Africa, CSIR), Chung-li, Taoyuan (Taiwan), West Freugh (Scotland, BNSC), Tel Aviv (Israel, ISA), Parepare (Indonesia, LAPAN), Islamabad (Pakistan), Norman (Oklahoma, Eosat), Singapore (University of Singapore), etc.

Ground stations are equipped with "Fast Delivery" SAR processors, capable of generating quicklook images after reception of the pass. These "Fast Delivery Products" (FDP) are directly mailed to the national PAF's (Processing and Archiving Facility).

- D-PAF = DLR/DFD in Oberpfaffenhofen, Germany
- F-PAF = CERSAT, Brest, France
- I-PAF = ASI, Matera, Italy
- UK-PAF = RAE, Farnborough, UK

The Earthnet ERS Central Facility (EECF) in Frascati, Italy coordinates all product requests from customers.

SAR Image Products	SAR Wave Mode Products	Altimeter Products	Wind Scatterometer Products
Annotated Raw Data	Fast Delivery Product	Fast-Delivery Product	Fast-Delivery Product
Fast-Delivery Image	Fast Delivery Product Copy	Fast-Delivery Copy	Fast-Delivery Product Copy
Fast Delivery Image Copy		Ocean Product	
Single-Look Complex	Intermediate Product Copy	Quicklook Ocean Product	
Precision Image	Complex Imagette	Waveform Product	
Ellipsoid Geocoded Image	Detected Imagette and its Spectrum	Sea Surface Height	
Terrain Geocoded Image	Imagette Precise Spectrum	Sea Surface Topography	
		Ocean Geoid	
Orbit Products	Earth Gravity Products	ATSR Products	
Preliminary Orbit	ERS-1 Gravity Model 1st Generation	IR Brightness Temp.	
Precise Orbit	ERS-1 Gravity Model 2nd Generation	Sea Surface Temp.	
		Precision SST	

Table 180: ERS-1 data products

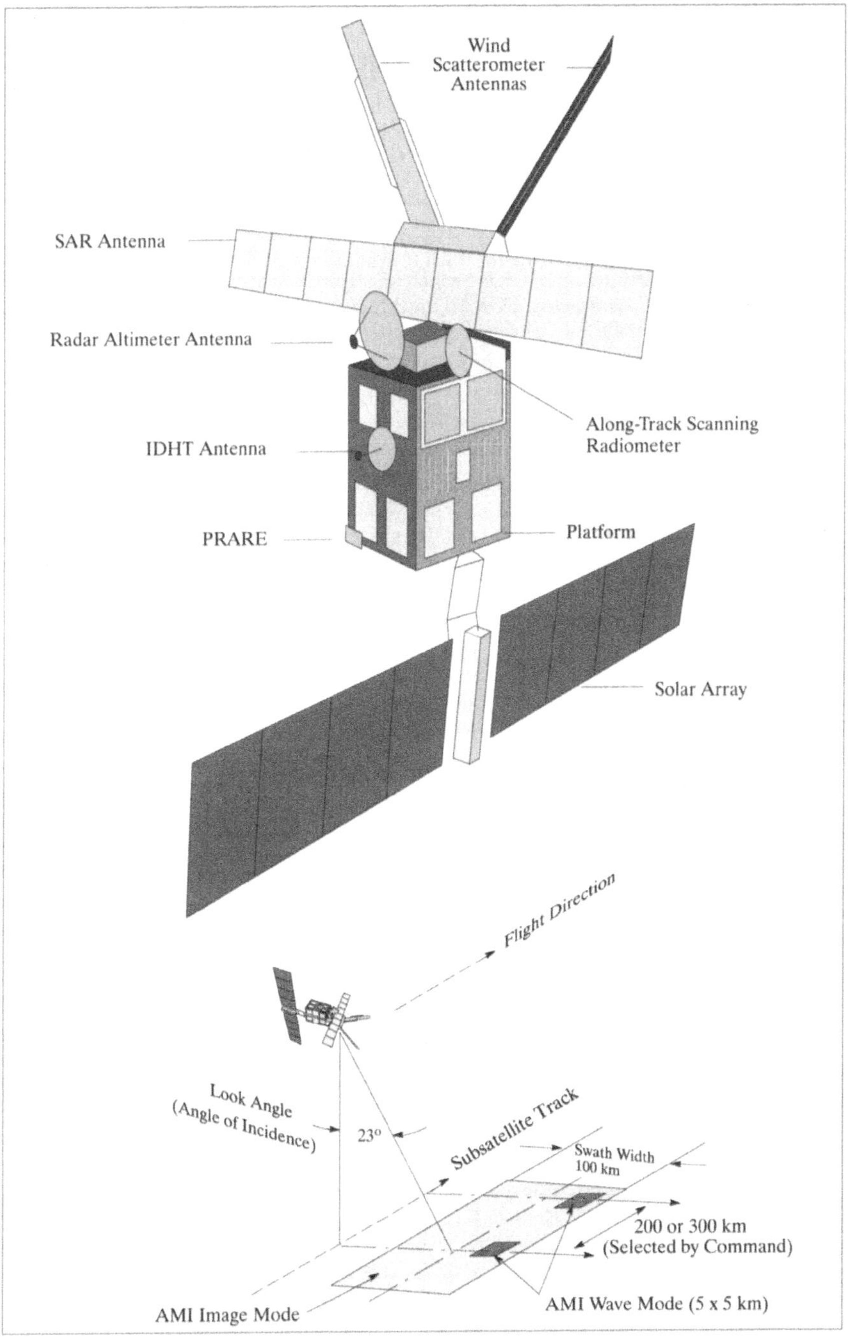

Figure 114: The ERS-1 satellite model and SAR (AMI) observation geometry

Commercial ERS-1 Data Distributors

In 1992 ESA selected three distributors for ERS-1 Data. These are [737]:

- Radarsat International of Ottawa, Canada (responsible for commercial sales in Canada and USA)
- EURIMAGE of Rome, Italy (markets in Europe, North Africa, and the Middle East)
- SPOT Image of Toulouse, France (rest of the world)

D.14 ERS-2

ERS-2 is the follow-up mission of ERS-1 which was launched April 21, 1995 (3 year mission design life, prime S/C contractor: DASA/Dornier, Friedrichshafen). ERS-2 has the same mission objectives as ERS-1, plus an atmospheric chemistry mission objective (with GOME). [738] [739]

Orbit: Sun-synchronous polar orbit (retrograde orbit). Altitude = 780 km (mean); inclination = 98.5°, local crossing time of the equator = 10:30 AM, orbital period is about 100 minutes. Repeat cycle = 3 days.

In the orbital tandem configuration ERS-2 follows ERS-1 with an approximate delay (referred to as orbit phasing) of 35 minutes. Due to the phasing and the Earth's rotation, the ground-track patterns of ERS-2 are slightly shifted westwards with respect to those of ERS-1. The orbit phasing is adjusted in such a way that the ERS-2 track coincides exactly with that of ERS-1 24 hours earlier. The benefits of tandem operations are the following:

- When the same area of the Earth is observed by the same instrument on both satellites with a ground-site revisiting interval of 1 or 8 days, the improvements compared to a single satellite are:
  - the number of acquisitions in a repeat cycle (35 days) is doubled
  - the time interval between two successive acquisitions is shorter, being 1 to 8 days rather than 35 days
- For global applications, such as wind scatterometer data, coverage is doubled within a given time period
- When the ground-track pattern of nadir-looking instruments on one satellite overlap ground tracks of side-looking instruments on the other satellite, then their information can be combined, providing synergistic benefits.

The greatest benefit of the tandem mission is seen in SAR interferometry providing excellent temporal (1 day) and spatial coherence. This technique requires two separate images taken of the same ground track by two spatially-separated antennas (the interferometric baseline) to produce phase differences from slightly different viewing angles. For each pixel corresponding to the same area of the ground in both images, the phase values (depending on the satellite-to-ground pixel length) are subtracted to produce a phase difference image known as the 'interferogram.' With the orbit of both satellites known, the phase interferogram can be used to generate a DEM (Digital Elevation Model) of the surface with an accuracy of about 10 m. An additional product of interferometry is the so-called 'coherence image,' which shows bright areas where the coherence between to SAR images is high, indicating no variation in backscatter between the acquisition times of the two images. Dark regions indicate areas where changes have occurred. ERS-1 and ERS-2 have been in tandem operations since mid-August 1995. The tandem mission was completed after nine months of successful operation in May 1996. About 110,000 ERS SAR pairs of data were acquired during the tandem mission, covering nearly the whole global land surface. Over South America and part of Southeast Asia, just one data pair was acquired. As many as five

737) 'ESA Signs Long-awaited Imagery Sales Deal,' Space News, Feb. 10-16, 1992, p. 4

738) C. R. Francis, G. Graf, et al., "The ERS-2 Spacecraft and its Payload," ESA Bulletin, No. 83, Aug. 1995, pp. 13-31

739) G. Duchossois, P. Martin, "ERS-1 and ERS-2 Tandem Operations," ESA Bulletin, No. 83, August 1995, pp. 54-60

or six interferometric pairs are available for Europe and North America, allowing greater accuracies to be achieved. [740)]

Sensor complement:

Same as ERS-1 (AMI, ATSR-2, RA, PRARE (2))+ GOME (atmospheric chemistry)

ATSR-2 (Along-Track Scanning Radiometer and Microwave Sounder). The IRR instrument of ATSR was upgraded for ERS-2 by adding three more bands in the visible range to provide data for vegetation studies. The channels are accommodated by the addition of a second focal plane assembly (FPA). The center wavelengths of the VIS channels are at 0.555, 0.659, and 0.865 μm (the IR channels are at 1.6, 3.7, 10.85 and 12 μm). The VIS channels are calibrated through a complicated procedure using sunlight.[741)]

GOME = (ESA instrument, PI: J. P. Burrows, prime contractor: Officine Galileo, Florence, Italy). [742) 743) 744) 745) 746) 747)]
GOME is a cross-track scanning optical double spectrometer. The double spectrometer operates in the spectral range of 240 nm to 790 nm (1st stage: prism, 2nd stage: grating). The spectral range is split into four channels, each equipped with a 1024-pixel linear array detector. The resulting spectral resolution is 0.2 nm in UV and 0.4 nm in the VNIR spectrum. The objective of GOME is to observe upwelling solar radiation reflected or scattered in the Earth's atmosphere and from its surface. The measured spectrum contains absorption features which can be used to derive quantitative information on the presence of ozone, and of a number of other atmospheric species. In addition to the improved backscattering technique, GOME exploits the full capabilities of the enhanced ATSR-2. The GOME measurement concept is based on 'Differential Optical Absorption Spectroscopy' (DOAS), a technology proven in balloon flights.

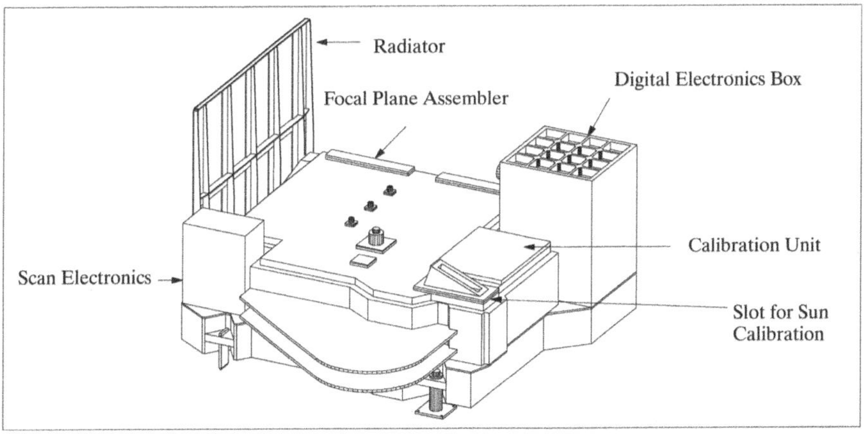

Figure 115: The GOME instrument model

740) "Case Study 16: SAR Interferometry," pp. 107-115, in 'Further Achievements of the ERS Missions,' ESA SP-1228, Dec. 1998, ISBN: 92-9092-508-6

741) N. Stricker, A. Hahne, et al., "ATSR-2: The Evolution in its Design from ERS-1 to ERS-2," No. 83, August 1995, pp. 32-37

742) ESA 1998: GOME Special, Earth Observation Quarterly No. 58, March 1998

743) C. Zehner, G. Pittella, "Preparing atmospheric applications for future ESA Earth-observation missions in the frame of the data user program, ESA Earth Observation Quarterly No. 61, Feb. 1999, pp. 1-6

744) C.J. Readings, 'The Interim GOME Science Report,' Feb. 1990,

745) 'The Global Ozone Monitoring Experiment (GOME) and ERS-2,' Earth Observation Quarterly, ESA periodical No. 32 Dec. 1990

746) A. Hahne, et al., "GOME: A New Instrument for ERS-2," ESA Bulletin, No. 73, February 1993, pp. 22-29

747) GOME Global Ozone Monitoring Experiment, Interim Science Report, ESA SP-1151, September 1993

GOME objectives: measurement of total column amounts and stratospheric and tropospheric profiles of ozone. - In addition: measurement of column amounts of H_2O and other gases involved in ozone photochemistry (like NO_2, OClO, BrO, and possibly ClO in anticyclonic conditions, and pollutants like SO_2 and HClO). GOME can also be used to investigate the distribution of atmospheric aerosols and clouds-plus-surface spectral reflectance.

The GOME instrument consists of the following modules: spectrometer, four FPA (Focal-Plane Assembly), calibration unit, SEU (Scan Electronics Unit) assembly, PMD (Polarization Measurement Device), DDHU (Digital Data Handling Unit), optical bench structure, and thermal control unit.

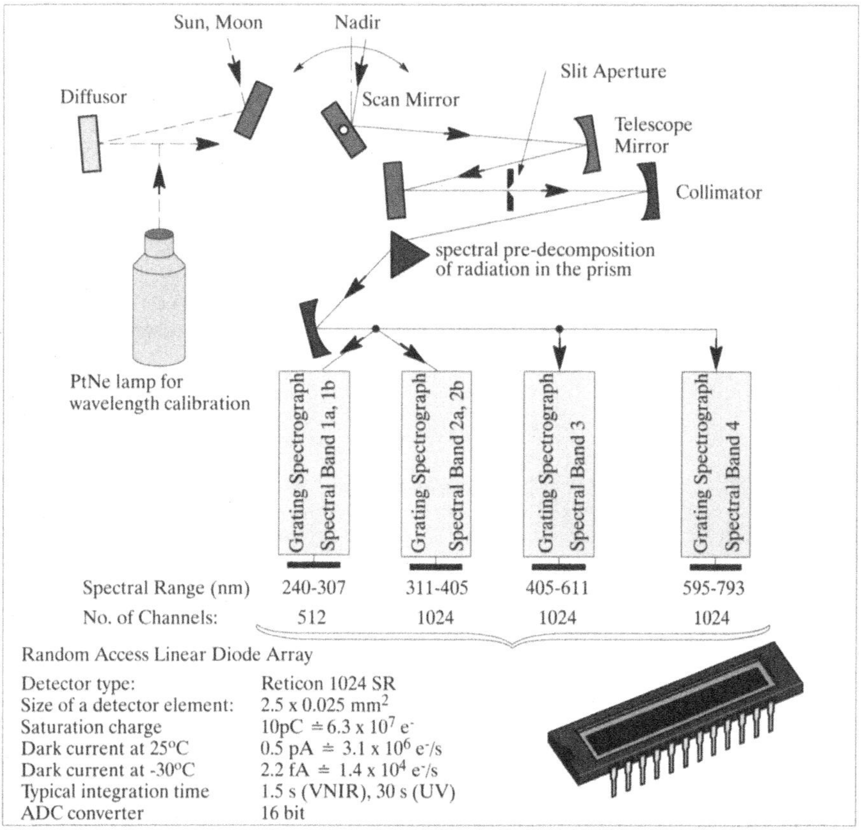

Figure 116: Optical principle of the GOME double-spectrograph

Its main mode of operation is nadir looking, but it is also able to look at the sun and the moon for calibration purposes. For on-board calibration, GOME does not depend on one technique, but seeks to exploit several. Both absolute radiometric and wavelength calibrations are performed. This means that GOME retrieves ozone distributions by exploiting the traditional backscatter approach, as well as by the more novel differential optical absorption spectroscopy.

The IFOV, corresponding to the projection of the spectrometer slit on the Earth, is a narrow rectangle 40 km (along-track) x 1.7 km (across-track). A scan mirror sweeps the IFOV in the cross-track direction in three steps (normal mode). Maximum swath width = 960 km with corresponding pixel sizes of 40 km x 320 km. Smaller pixel sizes of 40 km x 40 km can be commanded (smaller swath). With the 960 km swath global coverage can be achieved within three days. Calibration is performed with regular views of the sun; in addition, a wavelength calibration lamp is used for wavelength stability. As the instrument is sensitive to polarization of incoming radiation, a polarization detector monitors one polarization direction in the broadband channels corresponding essentially to detectors 2-4. GOME data rate = 40 kbit/s. GOME data are processed, archived and distributed at DLR/DFD.

GOME spectral bands:

- Channel 1 (240 -295 nm), 512 channels; [the Hartley bands of O_3 and NO emission are the dominant features]
- Channel 2 (290 - 405 nm), 1024 channels; [a variety of target species absorbs in regions such as O_3 (Huggin's band), O_4, NO_2, HCHO, SO_2, BrO and OClO]
- Channel 3 (400 - 605 nm), 1024 channels; [the target molecules for this region are: NO_2, OClO, O_2, O_3, (Chappius band), O_4 and H_2O]
- Channel 4 (590 - 790 nm), 1024 channels; [the target molecules are: O_3 (Chappius band), NO_3, H_2O, and O_2]

A combination of channels 2 to 4 can be used to make measurements of O_2 and O_4 column amounts (radiation penetration depths, cloud top heights).

D.15 HY-1 (Haiyang-1/Ocean-1)

Haiyang-1 is a Chinese minisatellite under development by CAST (Chinese Academy of Space Technology) sponsored by the China Ocean Administration (COA). This satellite was developed on a common minisatellite platform, namely CAST-968, also employed by the SJ-5 satellite. HY-1 is planned to be launched along with FY-1D on a LM-4B launch vehicle in the second half of 2001 (in Aug. 2001 delayed to 2002) from the Taiyuan launch site in China.

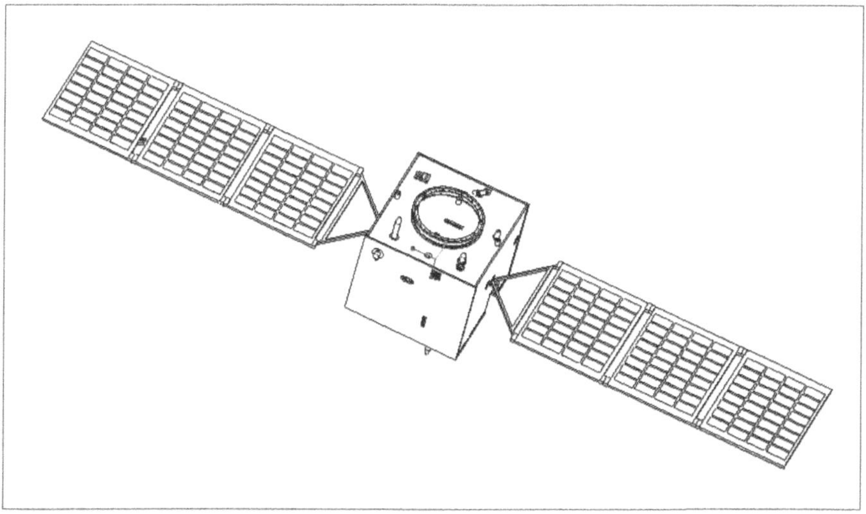

Figure 117: Illustration of the HY-1 satellite

The structure of the spacecraft is of hexahedron shape with overall dimensions of 1.2 m x 1.1 m x 0.94 m. The total deployed length of HY-1 is about 7.5 m. The S/C is three-axis stabilized

(Earth pointing) using three magnetorquers and hydrazine propulsion for attitude control and orbit change maneuvers; the attitude accuracy is 0.4° in roll and pitch and 0.5° in yaw. Attitude sensing is provided by a sun sensor and an infrared Earth sensor. S/C power of 450 W (BOL) and 320 W (EOL) is provided by two solar arrays of size 5.67 m^2. In addition, there are two packs of NiCd batteries of 23 Ah for night phase power supply. Electrical heaters are used for active thermal control. The S/C mass is 365 kg (including 13kg of propellant); payload mass = 87 kg; the design life is two years. [748)]

RF communication. The TT&C communication is in S-band, the downlink date rate is 4 kbit/s, the uplink rate is 2 kbit/s. An on-board storage capacity of 80 Gbit is provided. All payload data are downlinked in X-band using QPSK modulation. Data rate = 5.32 Mbit/s. EIRP = 39.4 dBm.

Orbit: Sun-synchronous orbit, altitude = 798 km, inclination =98.8°; nodal period = 100.8 minutes; the local time at the descending node is between 10:16 and 10:41 AM (at the end of two year's life). The revisit time for the Ocean Color Scanner is three days; the revisit time for the CCD Camera is seven days.

Sensor complement: The data user of both instruments is COA (China Ocean Administration).

OCS (Ocean Color Scanner), developed by SITP (Shanghai Institute of Technical Physics) of CAS (China Academy of Sciences). The objective is global ocean color monitoring for the study of biological oceanography. OCS consists of the following subsystems: optics, scanner, FPA (Focal Plane Array) and an electronics box. Imagery is provided in 8 VNIR (Visible Near Infrared) and in two TIR (Thermal Infrared) bands with a spatial resolution of 1.1 km. The swath width is 1600 km. The instrument data rate is 0.67 Mbit/s.

Pixel resolution	1.1 km (nadir)	
Pixels per scan line	1024	
Focal length of optical system	650 mm for VNIR, 190 mm for TIR	
Detectors	Si for VNIR, HgCdTe for TIR	
Deviation of central wave length	2 nm	
Data quantization	10 bit	
Cooling temperature	80 K for TIR band	
Instrument mass, power	50 kg, average = 29.3 W, peak = 41.7 - 71.7 W	
Spectral range (μm)	SNR	Dynamic range (%)
0.402 -0.422	440	40
0.433 - 0.453	600	35
0.480 - 0.500	590	30
0.510 - 0.530	560	28
0.555 - 0.575	525	25
0.660 - 0.680	390	20
0.745 - 0.785	400	15
0.845 - 0.885	415	15
10.30 - 11.40 (TIR)	NEΔT = 0.2 K (at 300 K)	
11.30 - 11.40 (TIR)	Measurement of SST (Sea Surface Temperature)	

Table 181: Specification of the OCS instrument

CCS (CCD Camera System), a multispectral instrument developed by the Beijing Institute of Space Machines and Electricity, a facility of CAST. The objective is to provide in particular imagery of coastal regions (estuaries, tidal waters, etc.)

748) Information provided by Lihua Zhang of CAST, Beijing, China

Spectral ranges (μm), 4 bands	0.42 - 0.50, 0.52 - 0.60, 0.61 - 0.69, 0.76 - 0.89
Pixel resolution, number of pixels per line	250 m (nadir), 2048
Pixel size (μm)	10 x 10
Focal length of optical system	32 mm
FOV (Field of View), swath width	36°, or 500 km swath width
Deviation of center wavelength	<5 nm
SNR	641
Data quantization	12 bit
Instrument mass, power	15 kg, average = 6 W, peak = 30 W
Instrument data rate	2.67 Mbit/s

Table 182: Major characteristics of CCS

D.16 ICESat (Ice, Cloud and land Elevation Satellite)

A NASA/GSFC mission within the ESE (Earth Science Enterprise) program with the prime objective to monitor the mass balance of the polar ice sheets and their contributions to global sea level change. Secondary goals are to measure cloud heights and the vertical structure of clouds and aerosols in the atmosphere, further to measure roughness, reflectivity, vegetation heights, snow-cover, and sea-ice surface characteristics, and to map topography of land surfaces. Note: ICESat is the renamed former "Laser Altimetry-1" mission. A launch is planned on a Delta-2 vehicle (along with CATSAT) for Dec. 2001 from VAFB, CA. 749) 750) 751)

The mission has been developed by a partnership of NASA, industry, and university teams. The S/C integrator is BATC (Ball Aerospace and Technologies Corp.) of Boulder, CO. The ICESat structure is three-axis stabilized and is based on the BCP 2000 (Ball Commercial Platform 2000) series. The S/C configuration uses a simple panel-post aluminum honeycomb structure. The command and data handling subsystem consists of a S/C Control Computer (SCC), a Command and Telemetry Unit (CTU) and two Solid State Recorders (SSR) with a capacity of 32 Gbit. The attitude control and determination subsystem (ACDS) uses two star trackers, redundant inertial reference units (IRU), sun sensors and magnetometers for attitude determination. The pointing accuracy is 100 mrad per axis (1 sigma), the pointing knowledge is 50 mrad per axis (1 sigma), and the pointing stability is 10 μrad/s. The geolocation knowledge is <15 m (3 sigma) after ground processing (BlackJack GPS receivers). Four, low-vibration reaction wheels, and three torque rods with redundant windings and drivers are used for control. The electrical power and distribution subsystem employs a fully redundant power control unit, a 40 Ah NiH2 battery with a spare cell, two 3.2 m single-axis drive solar array panels, and distribution electronics. The S/C design life is 5 years. S/C mass = 970 kg (wet), payload mass = 298 kg. S/C power = 730 W (orbital average) with 350 W for payload operations. The S/C bus permits off-nadir pointing of up to 5°.

Communication: Two separate, redundant X-band downlink transmitters and gimbaled antennas are provided for payload data. The SSR has direct links to the X-band downlink data transmitter. The X-band downlink data rate is 40 Mbit/s at a frequency of 8.100 GHz, modulation type PCM NRZ-M, SQPSK. The CCSDS protocol used is: version 2, grade 3, VCDU header and CRC. TT&C data can be transmitted in S-band or X-band at data rates of: 1, 4, 16 or 256 kbit/s. The uplink is in S-band at 2 kbit/s.

Orbit: Near polar LEO orbit, altitude = 600 km, inclination = 94°. Two primary orbits are used:

749) B. E. Schutz, "Spaceborne Laser Altimetry: 2001 and Beyond," published in: H. P. Plag (ed.), 1998, Book of Extended Abstracts, Wegener-98, Norwegian Mapping Authority, Honefas, Norway

750) http://icesat.gsfc.nasa.gov/

751) http://www.csr.utexas.edu/glas/

- Verification/validation orbit (90-120 days): this orbit has a ground track repeat cycle of 8 days to enable overflights of specific locations on the Earth which will be instrumented to support measurement verification or data product validation
- Mission orbit: this orbit has a ground track repeat cycle of 183 days to enable uniform sampling of the surface with high resolution. At the equator, the separation between ascending tracks will be about 15 km after 183 days.

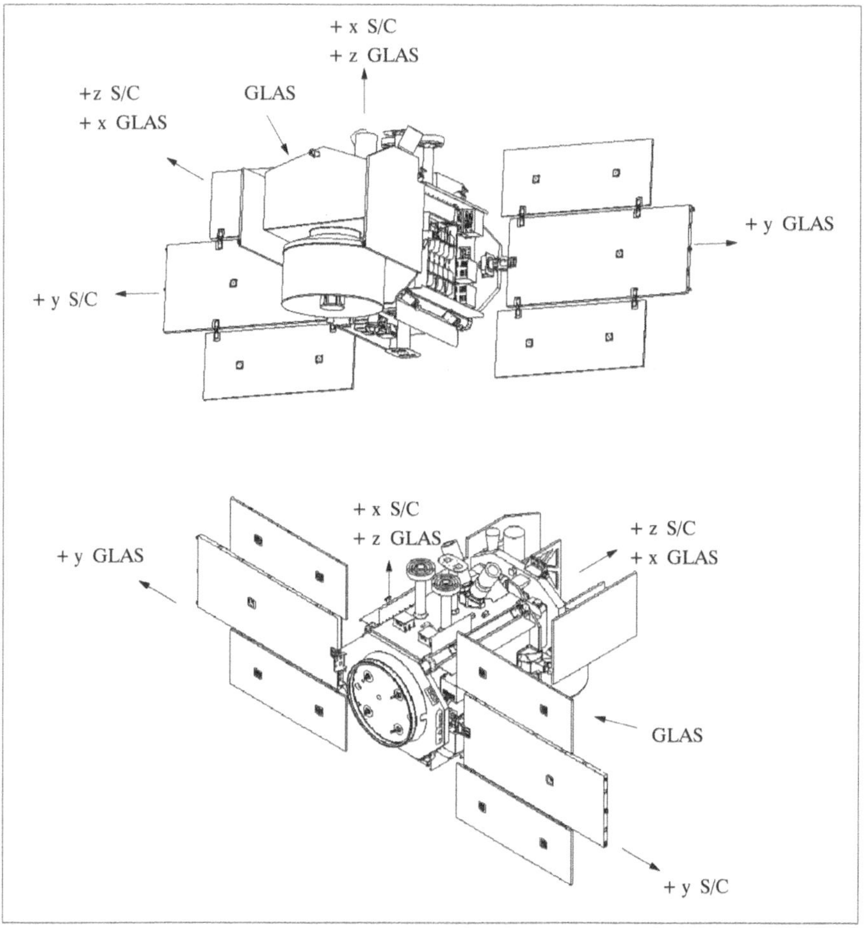

Figure 118: Illustration of the ICESat spacecraft (+z view top, -z view bottom)

Sensor complement:

GLAS (Geoscience Laser Altimeter System), NASA/GSFC instrument (Science team leader: B. E. Schutz, University of Texas at Austin). GLAS is a descoped version of the former GLRS (Geoscience Laser Ranging System). Objective: GLAS measures ice sheet topography, cloud heights, planetary boundary layer heights, and aerosol vertical structure. In addition, operation of GLAS over land and water provides along-track topography.

GLAS consists of a laser system to measure distance, a GPS receiver, and a star-tracker attitude-determination system. The laser transmits short pulses (4 ns) of infrared light (at 1064 nm) and visible green light (at 532 nm). The instrument is a nadir-viewing frequency-doubled, Q-switched, solid-state Nd:YAG laser with energy levels of 100 mJ (1064 nm) and 50 mJ (532 nm). The pulse repetition rate is 40 pulses/s, and the beam divergence is approximately 0.1 mrad. The infrared pulse is used for surface altimetry, the green pulse is used for atmospheric measurements. The height measurements are determined from the round-trip pulse time. [752) 753)]

Parameters or constituents measured	Cloud and aerosol data (extracted from 532 nm laser pulse)
Laser type and transmitter wavelength	Nd:YAG laser with 1064 and 532 nm wavelength, a nadir-viewing frequency-doubled, Q-switched, solid-state Nd:YAG laser
PRF (Pulse Repetition Frequency)	40 Hz
Laser energy/pulse	100 mJ for 1064 nm and 50 mJ for 532 nm

Table 183: Some parameters of the GLAS instrument

Instrument mass = 298 kg, power = 330 W average, duty cycle = 100%, data rate = 450 bit/s, thermal control by radiators supplemented by heaters, heat pipes, thermal operating range: 20° ± 5° C, FOV of telescope = 375 µrad (nadir view only), FOV of instrument = 66 m laser footprint at nadir at 1064 nm, telescope size = 100 cm in diameter (height = 1.75 m).

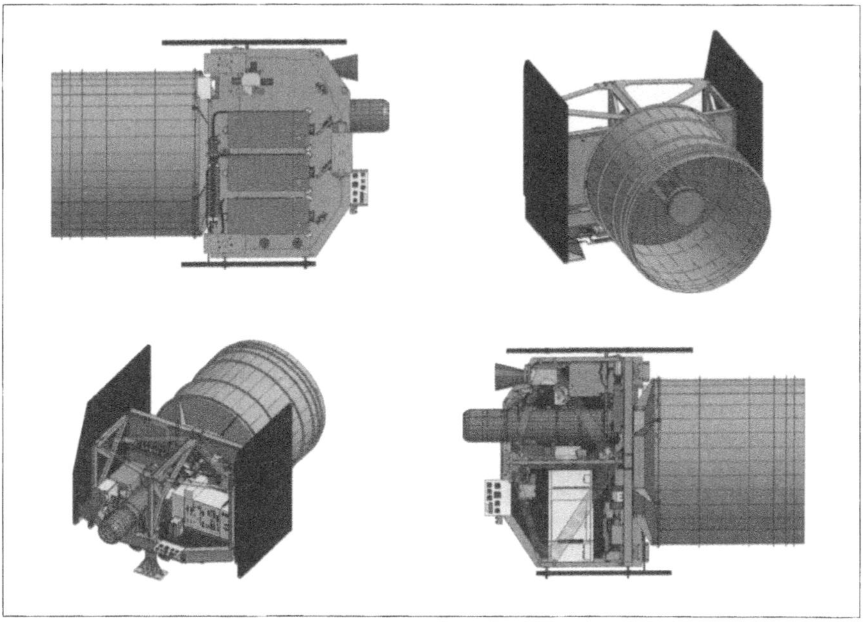

Figure 119: Various views of the GLAS instrument [754)]

Pointing requirements (platform+instrument):

- Control (3 sigma): 30 arcsec roll, 60 arcsec pitch, 1° yaw
- Post-processed pointing knowledge (1-sigma): 1.5 arcsec (roll and pitch axes, provided by instrument-mounted star trackers laser reference sensor gyroscope)

At 40 pulses per second, the centers of 66 m diameter footprints are separated in the along track direction by 170 m for a 600 km altitude orbit; cross track resolution is determined by

752) "GLAS Geoscience Laser Altimeter System," ESE Reference Handbook, 1999, NASA/GSFC, pp. 113-114
753) B. E. Schutz, "Laser Altimetry and Lidar From ICESat/GLAS," IGARSS 2001, Sydney, Australia, July 9-13, 2001
754) Illustration provided by Michael D. King of NASA/GSFC

the 183 day ground track repeat cycle which yields 15 km track spacing at the equator and 2.5 km at 80° latitude.

A small portion of the outgoing laser energy is used for instrument pointing. The extracted energy is input into two CCD cameras: LRS (Laser Reference Sensor) and LPA (Laser Profiling Array). The laser image in the LRS is formed at 10 Hz while the LPA uses the 40 Hz PRF of the instrument. A FOV of 0.5° x 0.5° of stars in the zenith direction is used in LRS. Thus, the LRS provides a combined image of the transmitted laser pulse superimposed on the star background. A third CCD camera, mounted on the GLAS optical bench, is IST (Instrument Star Tracker) at 10 Hz and a FOV of 8° x 8°. IST is used for the spatial orientation of the optical bench with respect to the stars referred to as CRF (Celestial Reference Frame). The LPA provides a shot-by-shot record of the laser far field pattern and the beam direction with respect to to the optical bench; the LRS provides a calibration link between the CRF and the optical bench. The system provides a laser pointing accuracy of 1.5 arcsec (knowledge).

The ICESat GPS receiver system is a spinoff of the BJ (BlackJack) dual-frequency receiver developed by NASA/JPL for satellite-based science applications as flown on Jason. Post-processing of the pseudorange and carrier phase recorded by BJ support the determination of the radial component of the of the GLAS position to an accuracy of 5 cm rms. The ground-based SLR (Satellite Laser Ranging) network provides in addition tracking data to support the validation of the GPS-determined orbit.

D.17 IRS (Indian Remote Sensing Satellites)

IRS is an integrated element of India's NNRMS (National Natural Resources Management System) with the objective to provide a long-term spaceborne operational capability to India for the observation and management of the country's natural resources (applications in agriculture, hydrology, geology, drought and flood monitoring, marine studies, snow studies, and land use). The intend of the program is to create an environment of new perspectives for the Indian research community as a whole, to stimulate the development of new technologies and applications, and to utilize the Earth resources in more meaningful ways.

The program started in the mid 1980s, the first launch of IRS-1A occurred in March 1988. [755] Eventually, a continuous supply of synoptic, repetitive, multispectral data of the Earth's land surfaces was obtained (similar to the US Landsat program). In 1995, IRS imagery was made available to a larger international community on a commercial basis. The initial program of Earth-surface imaging was extended by the addition of sensors for complementary environmental applications. This started with the IRS-P3 satellite which is flying MOS (Multispectral Optoelectronic Scanner) for the measurement of ocean color. The IRS-P4 mission is dedicated to ocean monitoring. [756] [757] [758]

755) Note: The availability of Landsat imagery created a lot of interest in the science community. The Hyderabad ground station started receiving Landsat data on a regular basis in 1978. The Landsat program with its design and potentials was certainly a great model and yardstick for the IRS program.

756) G. Joseph, B. L. Deekshatulu, "Evolution of Remote Sensing in India," Space in Pursuit of New Horizon, National Academy of Sciences publication, (editor: R. K. Verma and others), Allahabad, 1992, pp. 331-354

757) K. Kasturirangan, G. Joseph, et al., "IRS Mission," Current Science, Vol. 61, No. 3 and 4, Aug. 25, 1991, pp. 136-151

758) P. S. Goel, "Spacecraft Technology Development in India," Space Forum, Vol. 5, No 1-3, 2000, pp. 5-38

Satellite	Launch Date	Earth Imaging Sensors	Spectral Bands (μm)	Spatial Resolution (m)	Swath Width (km)	Repeat Cycle (days)
IRS-1A	Mar. 17, '88	LISS-I, and LISS-II A/B (3 sensors)	0.45-0.52 0.52-0.59 0.62-0.68 0.77-0.86	72.5 m LISS-I 36 m LISS-II	148 74x 2 (swath of 148 km)	22
IRS-1B	Aug. 29, '91	LISS-I and LISS-II A/B	same as for IRS-1A		148 74 x 2	22
IRS-P2	Oct. 15, '94	LISS-II M	0.45-0.52 0.52-0.59 0.62-0.68 0.77-0.86	32 m x 37 m	66 x 2 (131 km for combined swaths)	24
IRS-1C	Dec. 28, '95	LISS-III	0.52-0.59 0.62-0.68 0.77-0.86 1.55-1.70	23.5 23.5 23.5 70	142 142 142 148	24
		PAN	0.50-0.75	5.8	70	24 (5)
		WiFS	0.62-0.68 0.77-0.86	188	804	5
IRS-P3	Mar. 21, '96	WiFS	0.62-0.68 0.77-0.86 1.55-1.70	188	804	5
IRS-1D	Sept. 29, '97	Satellite and instruments are identical to those of IRS-1C				
IRS-P6 ResourceSat-1	2001	LISS-IV	0.52-0.59 0.62-0.68 0.77-0.86	5.8 5.8 5.8	70	24 (5)
		LISS-III*	0.52-0.59 0.62-0.68 0.77-0.86 1.55-1.70	23.5 23.5 23.5 23.5	140	24
		AWiFS	0.62-0.68 0.77-0.86 1.55-1.70	70 70 70	740	5
IRS-P5 CartoSat-1	2002	PAN-F PAN-A	0.50-0.75 0.50-0.75	2.5 2.5	30 30	stereo imagery

Table 184: Chronology of IRS-series Earth surface imaging missions and instruments

D.17.1 IRS-1A

IRS-1A (Indian Remote Sensing Satellite-1A). Launch: March 17, 1988 by ISRO (Soviet launch vehicle Vostok from the Baikonur Cosmodrome. The IRSO S/C control center is in Bangalore. TT&C function is provided by ISTRAC (ISRO Tracking Network), supported by DLR (GSOC, Weilheim), NOAA (Fairbanks), ESA (Malindi) and the USSR (Bearslake) ground stations. Status: nominal S/C operations until the end of 1995. [759) 760)]

The S/C is box-shaped (1.6 m x 1.6 m x 1.5 m) with two solar panels (8.6 m^2) as shown in Figure 120. The S/C structure is made of aluminum/aluminum honeycomb. The satellite is three-axis stabilized utilizing a zero momentum system. Hydrazine thrusters (80 kg fuel) are also used for control and momentum dumping. The IRS series satellites are built around a zero-momentum reaction wheel based system. Gyro-based attitude reference using quaternion propagation with attitude updates from Earth sensors and sun sensors (CCD-based is used for yaw angle measurements) provide the high-pointing accuracy and stability required for the imaging payload. Attitude is sensed by Earth sensor, sun sensor, star sensor and dynamically tuned gyros. The actuators are reaction wheels (4), magnetic torquers, and hydrazine thrusters (sixteen 1-newton thrusters). A pointing accuracy of 0.3° is achieved in pitch/roll and 0.5° in yaw. Attitude determination accuracy of ±0.1°. Total S/C mass = 975 kg (at launch), power = 700 W, two NiCd batteries (40 Ah) provide power for the eclipse phase of the orbit. The design life is three years.

759) "Indian Remote Sensing Satellite and Associated Data Products," A.K.S. Gopalan, Proceedings of the Twenty-Third International Symposium of Remote Sensing of the Environment, Vol. I, p. 71, ERIM, Ann Arbor, MI, 1990

760) IRS NewsLetter, ISRO, Vol. 2 No. 1, March 1991

Orbit: sun-synchronous orbit, nominal altitude = 904 km, inclination = 99.49°, period = 103.2 minutes; the repeat cycle = 22 days; equator crossing at 10:26 AM.

Application: Land use, agriculture, forestry, hydrology, soil classification, coastal wetland mapping, natural resources (in particular pinpointing likely groundwater locations), disaster monitoring, cartography, etc.

Sensor complement:

Parameter	LISS-I	LISS-II A/B
Focal length	162.2 mm	324.4 mm
FOV, IFOV	9.4°, 80 μrad	4.7°+ 4.7°, 40 μrad
Spectral bands (μm)	0.46 - 0.52 (blue) 0.52 - 0.59 (green) 0.62 - 0.68 (red) 0.77 - 0.86 (NIR)	0.46 - 0.52 (blue) 0.52 - 0.59 (green) 0.62 - 0.68 (red) 0.77 - 0.86 (NIR)
Integration time	11.2 ms	5.6 ms
Ground resolution	72.5 m (each band)	36.25 m (each band)
Swath width	148 km	2 x 74 km
Radiometric resolution	7 bit	7 bit
Data rate	5.2 Mbit/s	2 x 10.4 Mbit/s
Instrument mass, power	38.5 kg, 34 W	2 x 80.8 kg, 2 x 34 W

Table 185: Specifications of the LISS-I and -II instruments

LISS = Linear Imaging Self-Scanning Sensor (total of three cameras). [761] LISS-I and LISS-II are two multispectral camera assemblies, each with a different resolution providing a swath of about 150 km. Each LISS camera consists of the collecting optics, imaging detectors, inflight calibration system, the processing electronics, and data formatting electronics. LISS-I employs four 2048-element linear CCD detector arrays with spectral filters (Fairchild CCD 143A). All cameras use refractive type collecting optics with spectral selection by appropriate filters. The refractive optics were chosen to obtain a large FOV (Field of View). A lens assembly for each spectral band is used for better performance and effective utilization of the full dynamic range of the CCDs. Two LEDs (Light Emitting Diodes) per band are provided for inflight calibration. A LISS-I scene is 148 km x 174 km. The LISS-II A/B assembly features eight 2048-element linear CCD detector arrays with spectral filters (2 parallel swaths of 74 km each for the LISS-II A/B assembly with 3 km overlap, the total swath is 145 km). Four LISS-II scenes cover the area of one LISS-I scene. [762]

Data of LISS-I are transmitted in S-band (2217.6 MHz) at 5.2 Mbit/s, PCM/BPSK modulation. Data of the LISS-II assembly are transmitted in X-band, 2 x 10.4 Mbit/s, PCM/QPSK modulation.

IRS-1A data products are being acquired, processed and disseminated by NRSA (National Remote Sensing Agency) Data Center, Hyderabad. These data products compete directly with Landsat TM and MSS data as well as with SPOT Image data on the market.

D.17.2 IRS-1B

IRS-1B is a follow-up satellite of IRS-1A. Launch of the S/C was by a Russian Vostok vehicle from Baikonur on August 29, 1991 (same tracking support configuration as for IRS-1A). The IRS-1B S/C and instruments are practically identical with those of IRS-1A

761) G. Joseph, IRS-1A Camera - Its Evolution and Realization," brochure of NNRMS (National Natural Resources Management System), Bangalore, India

762) Note: At the time of project initiation, CCD arrays with maturity of production were limited to 2048 elements. Hence the swath of LISS-I was limited to about 150 km. Since LISS-II has a better resolution by a factor of two compared to the LISS-I camera, two LISS-II cameras were needed to produce a swath similar to that of LISS-I.

(satellite mass = 975 kg). IRS-1B is fully operational as of 2001. [763) 764)]

Orbit: Polar sun-synchronous orbit; altitude = 904 km, inclination = 99.49°, period = 103.2 minutes. Repeat cycle: 22 days.

ISRO's policy allows ground stations from other countries to have direct access to the Indian satellite imagery. Any existing ground station equipped to receive data from SPOT or from Landsat will be able to receive IRS-1B satellite data with very minor changes (ISRO can supply the upgrade).

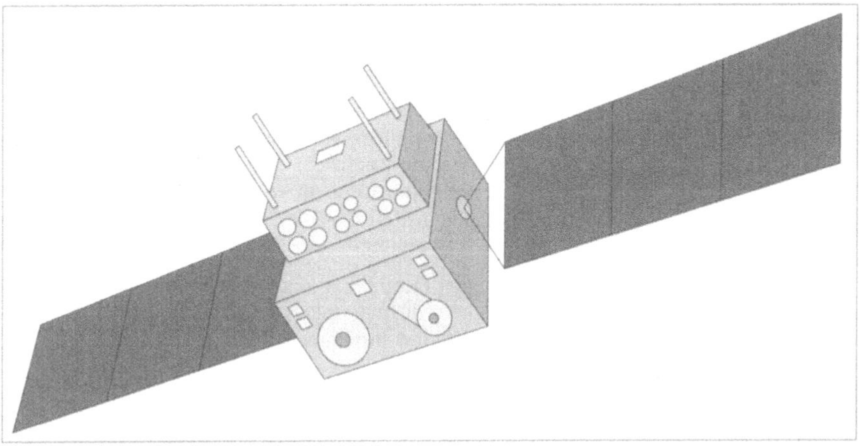

Figure 120: The IRS-1B S/C model

D.17.3 IRS-1E (P1)

A technological mission with the primary objective to test an ISRO-developed launch vehicle by the name of PSLV (Polar Space Launch Vehicle). The launch occurred on Sept. 20 1993, but the satellite failed to achieve orbit (PSVL failure of 2nd stage).Orbit: Sun-synchronous polar orbit with equatorial crossing at 10:30 AM descending node, altitude = 904 km, period = 103 min, repeat cycle = 22 days. [765) 766)]

Sensor complement:

LISS-I = Linear Imaging Self-Scanning System. Same definition as under IRS-1A.

MEOSS = Monocular Electro-Optical Stereo Scanner (DLR sensor) with pushbroom CCD technology. MEOSS is a stereo camera system capable of recording three images simultaneously with a single lens by means of linear scanning. The scanner operates in the spectral range of 0.57-0.7 µm. Resolution = 50 m along track, 158 m across track, 45 m in vertical direction; swath width = 510 km, 8 bit quantization. Application: stereo view capability to study topography, geology, terrain analysis and modelling, snow/ice mapping, meteorology (cloud height and movement), etc.

D.17.4 IRS-P2

An ISRO mission with the objective to acquire remote sensing data for oceanographic, land, and atmospheric applications (demonstration of the IRS-1C/1D bus). Launch date:

763) "India Expands Access to Imagery," Space News Aug. 26 - Sept. 8, 1991, p. 22
764) "India Calls IRS-1B Launch a Success," Space News, September 9-15, 1991, p. 12
765) IRS-1E MEOSS Utilization Plan, ISRO, July 1991
766) Note: The satellite designations P1, P2, P3, etc. stand for the launches carried out by PSLV (Polar Satellite Launch Vehicle), the launch vehicle developed by ISRO

October 15, 1994 from SHAR (East coast launch facility of India) with a PSLV launcher. Three-axis stabilized S/C is using conical scanning Earth sensors and dynamically tuned gyros for attitude sensing, and reaction wheels, magnetic torquers and monopropellent hydrazine thrusters as actuators. Inertial attitude referencing with the star sensor updates is one of the major ISRO developments for astronomical X-ray observations. The attitude control system of IRS-P2 provides multipurpose functions by pointing for remote sensing, and inertial pointing for the X-ray observations. S/C mass = 870 kg, solar power = 510 W, two 21 Ah NiCd batteries. [767)]

Orbit: Sun-synchronous circular orbit with an equatorial crossing at 10:30 AM descending node, altitude = 817 km, inclination = 98.7°, repeat cycle = 24 days, period = 101 min.

Sensor:

- **LISS-IIM** = Linear Imaging Self-Scanning System (Modified). Same definition as under IRS-1A. The LISS-IIM instrument is realized with a single optical feed utilizing the full ±5° FOV (there are 4 lenses in one optical head). The instrument employs push-broom scanning and a linear CCD line detector array with 4 spectral bands in VNIR (0.45-0.52 μm, 0.52-0.59 μm, 0.62-0.68 μm, 0.77-0.86 μm. Focal length = 324.4 mm. The ground resolution is 32 m (across) x 37 m (along-track). The combined swath is 131 km.

Data: No on-board data storage capability. Downlink broadcast of R/T science data in X-band (8.316 GHz, PCM/QPSK/PM modulation, data rate = 2 x 10.4 Mbit/s) to a dedicated ground station network and to a general user community. TT&C operations in S-band.

D.17.5 IRS-1C/1D

IRS-1C [768) 769) 770)] is an ISRO-built (Indian Space Research Organization) second generation remote sensing satellite with enhanced capabilities in terms of spatial resolution and spectral bands. IRS-1C is a three-axis body-stabilized S/C. The attitude is sensed by star sensors, Earth sensors and gyros. Actuators: four reaction wheels, two magnetic torquers, 16 one Newton hydrazine trusters; one 11 Newton thruster. Pointing accuracy: roll and pitch = ±0.15°, yaw = ±0.2°. The S/C structure consists of the main platform and the payload platform (aluminum honeycomb structure). The S/C bus consists of a central cylinder attached to the bottom and the top decks on four sides of the S/C. In addition, a thermally isolated payload deck is introduced (better alignment stability for payload instruments). The S/C mass is 1250 kg; power = 813 W (two sun-tracking type solar panels of 9.6 m^2) plus two NiCd batteries of 21 Ah each; S/C design life = 3 years; the on-board recorder capacity is 62 Gbit with an input/output data rate of 42.45 Mbit/s. [771) 772) 773)]

Orbit: Polar sun-synchronous orbit; altitude = 817 km, inclination = 98.71°, period = 101.23 min; equatorial local crossing time = 10:30 AM, descending node; 22 day revisit period. Note: the IRS-1D went into an initial elliptical orbit of 817 km x 320 km (slight underperformance of the 4th stage of PSLV). In spite of this handicap, ISRO controllers were able to place the satellite into a functional sun-synchronous (near-circular) orbit of 737 km perigee and 821 km apogee.

The launch of IRS-1C took place on December 28, 1995 with a Russian launcher (Molniya) from Baikonur. The **IRS-1D** launch with an ISRO PSLV-C1 launcher from Sriharikota (SHAR, India) took place on September 29, 1997.

767) Document on Configuration of IRS-P2 and MOS and their Interfaces, ISAC, Bangalore, Nov. 1992
768) IRS-1C Executive Summary, IRS-1C/1D Project, May 1990, ISRO
769) "India's IRS-1C Satellite to offer sharper Images," Space News, May 25-31, 1992 p. 11
770) "India Readies Sharper IRS-1C for Molniya Launch," Space News, January 9-15, 1995, p. 3
771) S. Kalyanaraman, "Technologies Developed for IRS Program," Journal of Spacecraft Technology, Vol. 9, No 1, 1999, pp. 1-13
772) K. Kasturirangan, et al., "Indian remote sensing satellite (IRS) – 1C – The beginning of a new era," Current Science, Vol. No. 7, April 10, 1996, pp. 495–500
773) "IRS-1C Data Users Handbook," NRSA (India) document provided by Euromap (of GAF), September 1995

Downlink transmission data rate: 40 W of output power (X-band) is used to transmit 42.5 Mbit/s of LISS-III and WiFS data and 85 Mbit/s of PAN data, both links are in X-band with QPSK modulation, TT&C in S-band with PCM/FSK/FM/PM modulation (ISTRAC mission control center at Bangalore, the NRSA data center at Hyderabad is the focal agency for user interfaces).

Availability of IRS data. Space Imaging of Thornton, CO (the Landsat-4 and -5 data distributor) bought the worldwide commercial distribution rights for IRS-1C/D imagery and follow-up IRS series data from Antrix Corp. of Bangalore (the commercial arm of India's Department of Space) in 1995. GAF of Munich, Germany, bought from Space Imaging the IRS distribution rights for Europe. Data reception of IRS-1C data for GAF started in early summer 1996 at the DLR/DFD station Neustrelitz.

Sensor complement:

The three sensors are cameras operating in the pushbroom scanning mode using solid state charge-coupled device (CCD) detectors. 774)

Spectral range	0.5 - 0.75 μm
Spatial resolution	≤ 10 m (6 m at nadir)
Swath width	70 km nadir view (91 km swath for the most inclined cross-track view)
Off-nadir viewing capability	± 26° swath steering range (with this a revisit cycle of 5 days is achieved)
Repetition cycle	24 days
SNR at saturation radiance	> 64
Data quantization	6 bit
Integration time	0.8836 ms
Image data rate	84.9 Mbit/s
Instrument mass, power	230 kg, 55-65 W

Table 186: Specifications of the PAN camera

PAN = Panchromatic Camera. PAN uses an all-reflective (off-axis f/4.5) folded mirror telescope (focal length = 982 mm) along with three separately mounted 4096-element CCD arrays, adding up to 12,288 pixels in the cross-track direction (there is some overlapping of the three subscenes of the image requiring special processing). Each detector array has separate interference filters and four LEDs along with a cylindrical lens. Two LEDs are for optical biasing and two are for inflight calibration of the sensor (calibration of CCDs excluding optics). A calibration cycle comprises 2048 lines (1.8 s for calibration cycle). The PAN instrument has a cross-track pointing capability of ±26° (body-pointing mechanism), providing a FOR (Field of Regard) coverage of ± 398 km. 775)

LISS-III = Linear Imaging Self-Scanning Sensor. Continuous service multispectral imagery. Application: Land and water resources management. 776) The pushbroom camera uses refractive optics in four spectral bands (separate optics and detector array for each band). The collecting optics consists of eight refractive lens elements with interference filter in front. A linear CCD array of 6000 silicon-based elements is used for each VNIR band. The SWIR-band device has a 2100 element InGaAs linear array (temperature controlled at -10°C with a passive radiative cooler and an on-off heater control). The SWIR device itself consists of a lattice mismatched heterojunction photodiode array for the detection of SWIR radiation and silicon-based CCD multiplexers for signal readout [seven identical modules are butted together to form a linear array of 2100 elements; each module consists of a two-

774) G. Joseph, et al., "Cameras for Indian remote sensing satellite IRS–1C," Current Science, Vol. 70, No. 7, April 10, 1996, pp. 510–515

775) K. Jacobsen, "Geometric Potential of IRS-1C PAN-Camera," Proceedings of ISPRS Symposium on Earth Observation Systems for Sustainable Development, Feb. 25-27, 1998, pp. 131-136, ISRO, Bangalore

776) A. S. Kirankumar, P. N. Babu, R. Bisht, "A Study of On-Orbit Behavior of InGaAs SWIR Channel Device of IRS-1C/1D LISS-III Camera," Proceedings of International Symposium on Earth Observation System for Sustainable Development," Feb. 25-27, 1998, Bangalore, India, pp. 303-307

sided InGaAs die of 300 photodiodes and two 150-element silicon-based CCD arrays on either side of the photodiode die to multiplex the signal from the photodiodes]. Instrument mass = 171 kg, power = 74-78 W.

Inflight calibration of the LISS-III camera is realized with LEDs (1.55 μm) as illuminating source and operated in pulsed mode to generate six intensity levels. A LED is mounted on either side of the photodiode array. Each LED is followed by diverging optics. A calibration cycle comprises 2048 scan lines (7.3 s for one cycle) and includes six intensity levels. Calibration is normally performed during the eclipse period of the satellite pass.

Spectral bands (4)	0.52 - 0.59 μm (VNIR), 0.62 - 0.68 μm (VNIR), 0.77 - 0.86 μm (VNIR), 1.55 - 1.75 μm (SWIR)
Spatial resolution	23.5 m (VNIR) and 70.5 m (SWIR)
Swath width	142 km (VNIR) and 148 km (SWIR), FOV=±5°
Repetition cycle	24 days
SNR at saturation radiance	>128
Data quantization	7 bit (resolution of 128 grey levels)
Integration time	3.55 ms for VNIR, 10.65 ms for SWIR
Data rate	35.79 Mbit/s for VNIR, 1.39 Mbit/s for SWIR data

Table 187: Specifications of the LISS-III camera

Spectral bands	0.62 - 0.68 μm, 0.77 - 0.86 μm
Spatial resolution	188 m
Swath width	810 km
Repetition cycle	5 days
SNR at saturation radiance	>128
Data quantization	7 bit (radiometric resolution of 128 grey levels)
Integration time	28.42 ms
Data rate	2.06 Mbit/s

Table 188: Specifications of the WiFS camera

WiFS = Wide Field Sensor (camera), similar to LISS-1. Application: Vegetation index mapping. The camera provides two spectral bands in the VNIR range. The total swath is formed by using two optical heads (i.e., two lenses and two CCDs) per band. The WiFS camera employs refractive collecting optics consisting of eight refractive lens elements with interference filter and neutral density (ND) filter in the frontend assembly. A 2048 element linear CCD array is used. FOV=±26°. The two lenses are mounted with their optical axes canted at 13° either side of nadir for each imager. Instrument mass = 41 kg, power = 22 W.

D.17.6 IRS-P3

A follow-up mission to IRS-P2 with the PSLV launcher which was launched March 21, 1996 from Sriharikota, India. [777] The IRS-P3 S/C structure is of IRS-P2 heritage. The bus structure consists of four vertical panels and two horizontal decks supported on a central load-bearing cylinder of 930 mm diameter and 1188 mm height. The payload is accommodated on the outer side of the upper deck, it is oriented in flight direction. The solar arrays (9.6 m^2) provide 810 W of power at end-of-life. The S/C is three-axis stabilized using conical scanning Earth sensors, sun sensors, star sensors, and dynamically tuned gyros, reaction wheels, magnetic torquers and reaction control system. An Earth pointing accuracy of better than 0.20° in all axes and better than 0.05° in all axes for stellar pointing (x-ray observation mode) is provided. Total S/C mass = 922 kg, a hydrazine propulsion system (84 kg of fuel sufficient for three years) with 16 thrusters is used for orbit maneuvers.

777) K. Thyagarajan, A. Neumann, G. Zimmermann, "The IPS-P3 Remote Sensing Mission," Small Satellites for Earth Observation, International Symposium of IAA, Berlin, Nov. 4-8, 1996, Walter de Gruyter

The TT&C-system is operating in S-band with PCM/FSK/FM/PM modulation. The telemetry system uses PCM/PSK modulation in S-band (2203 MHz). The payload data is transmitted in S-band (2280 MHz) with BPSK modulation at a data rate of 5.2 Mbit/s.

Orbit: Sun-synchronous circular orbit, altitude = 817 km, inclination = 98.8°, repeat cycle = 24 days, period = 101.3 min, equatorial crossing at 10:30 AM (descending node).

Sensor complement:

WiFS = Wide Field Sensor (camera) of IRS-1C heritage. WiFS is an extended version of 3 channels: 0.62 - 0.68 µm, 0.77 - 0.86 µm, with an additional channel at 1.55-1.75 µm (SWIR). Each band has two detectors centered at a FOV of ±13.6° to achieve a swath of 770 km (repeat cycle of 5 days). The optics system consists of eight lenses with spectral bandpass and neutral density filters for each spectral band. The dynamic range in each gain is 7 bits. The absolute radiometric accuracy is better than 10% with relative in-band accuracy of 2%. The data rate for the VNIR data (2 channels) is 2.6 Mbit/s, for the SWIR data it is 1.73 Mbit/s. WiFS has a mass of 25 kg and uses 50 W.

MOS = Multispectral Optoelectronic Scanner (see also MOS-P under PRIRODA, D.28). An advanced imaging spectrometer for the VNIR and SWIR spectrum. MOS is provided by DLR/ISST, Berlin. Objective: Image generation of the Earth surface (surface-atmosphere interaction, ocean color, phytoplankton, regional and global distributions of man-made aerosols and their links to gaseous admixtures, spectral and spatial cloudiness characteristics, etc.) in the VNIR/SWIR region of 0.4 - 1.6 µm. [778)]

The sensor apparatus consists of three complementary instruments: (see also Figure 143). MOS operation requires at least one calibration per month (with respect to the sun).

- MOS-A is an atmospheric spectrometer with four narrow channels in the O_2A-absorption band at about 760 nm for the measurement of atmospheric turbidity. The data from MOS-A are used for correction of the atmospheric influence (scattering) on the multispectral data of MOS-B. In addition the O_2A-method permits the measurement of aerosol content and profile.
- MOS-B is a 13-channel spectrometer in the range of 408 to 1010 nm. The center wavelengths of the channels are chosen with the objective for the quantitative retrieval of ocean and coastal zone parameters. They also provide a capability for vegetation signature determination (red edge) and estimation of H_2O (water vapor) content in the atmosphere from the NIR-measurements.
- MOS-C is a line camera at 1.6 µm with a bandwidth of ± 50 nm. The SWIR channel data is used for improved surface term and roughness estimation. In addition the data of the SWIR channel may be used for the following applications: cloud/snow/ice discrimination, cloud type discrimination, estimation of sea surface roughness, and for the improvement of atmospheric correction algorithms.

MOS calibration: In-orbit calibration measurements are performed using internal reference lamps (prior to each data take). In addition sun calibration measurements are performed once a month. This is achieved with a diffuser in front of the entrance optics of the sensor. The following calibration functions are performed:

- DSNU (Dark Signal Non-Uniformity) and PRNU (Photo Response Non-Uniformity)
- Absolute sensitivity calibration
- Linearity control
- Spectral alignment control

MOS in-orbit intercalibrations with sensors from other missions are attempted when orbital opportunities arise for a common target area or test sites. Examples are: MOS on IRS/P3 with MOS on Priroda, or with SeaWiFS on Seastar, or with OCTS on ADEOS.

778) G. Zimmermann, A. Neumann, "The Imaging Spectrometer Experiment MOS on IPR-P3 - Three Years of Experience," Journal of Spacecraft Technology, Vol. 10, No 1, Jan. 2000, pp. 1-9

MOS data are received at the following ground stations:: Hyderabad (ISRO, India), Neustrelitz (DLR, Germany), Maspalomas (ESA, Spain - since 1998), Wallops Island (NASA, USA - since April 1999).

Parameter	MOS-A	MOS-B	MOS-C
Spectral range (nm)	755 - 768 nm	408 - 1010	SWIR
No. of channels	4	13	1
Wavelengths (nm)	756.7; 760.6; 763.5; 766.4 (O_2A-band)	408; 443; 485; 520; 570; 615; 685; 750; 870; 1010; 815; 945 (H_2O-vapor)	1600
Spectral resolution	1.4 nm (FWHM)	10 nm (FWHM)	100 nm (FWHM)
FOV along-track FOV across-track	0.344° 13.6°	0.094° 14.0°	0.14° 13.4°
Swath width	195 km	200 km	192 km
No. of pixels per row	140	384	299
Spatial resolution	1.57 km x 1.4 km	0.52 km x 0.52 km	0.52 km x 0.64 km
Measuring range L_{min}-L_{max} [$\mu Wcm^{-2}nm^{-1}sr^{-1}$]	0.1 - 40	0.2 - 65	0.5 - 18
Data quantization	16 bit		
Data rate	1.3 Mbit/s		

Table 189: Specifications of the MOS instruments

X-Ray Payload. An ISRO astronomy instrument package with the objective to study periodic and aperiodic intensity and spectral variations in x-ray sources. Source observation is achieved by 'pointed mode observations,' employing an array of three co-aligned collimated **PPC** (Pointed Proportional Counter). The system operates in mutual anti-coincidence fashion for significant reduction of background noise (cosmic rays and Compton interaction of gamma rays). - Another objective is the study of light curves and the spectral evolution of transient and flaring x-ray sources as well as long-term intensity monitoring of known binary x-ray stars and other bright x-ray sources. This is achieved by means of **XSM** (X-ray Sky Monitor), based on the principle of a pin hole placed above a position sensitive to PPC in anti-coincidence mode.

PPC		XSM	
Energy range	2 - 20 keV	Energy range	2 - 8 keV
FOV	2° x 2°	FOV	90° x 90°
No of PPC	3	Pin hole size	1 cm^2
No of layers per PPC	3	Distance to detector	16 cm
No. of anode cells/layer	18	Detector	32 proportional counters
Size of cell	1.1 cm x 1.1 cm	Detector cell size	1 cm x 1 cm x 32 cm
Entrance window	25µm , 500 Å, Al coated	Window	25 µm Mylar, Al coated
Filling gas	Ar+CH_4, at 800 torr	Filling gas	Ar+CH_4

Table 190: PPC and XSM instrument specification

D.17.7 IRS-P4 (OceanSat-1)

The objectives of the ISRO IRS-P4 mission are directed at ocean- and atmosphere-related applications, hence it is also referred to as OceanSat-1. The S/C structure (aluminum/aluminum honeycomb with CFRP elements) is of IRS-1C/1D, -P3 heritage. The platform is three-axis stabilized, the pointing accuracies are ±0.15° in pitch/roll and ±0.20° in yaw; the drift rate is <0.0003°/s. Attitude is measured by Earth sensors, sun sensors, a magnetometer and gyros. Actuators: reaction wheels, magnetic torquers and hydrazine thrusters. An on-board storage capability of 320 Mbit is provided. S/C mass = 1036 kg; power = 800 W at EOL; two NiCd batteries provide 2 x 21 Ah. The S/C size is: 2.8 m x 1.98 m x 2.57 m. The mission design life is 5 years. A launch of IRS-P4 on a PSLV-C2 launcher from SHAR (along with the secondary payloads of KITSAT-3 and DLR/TUBSAT) took place on May, 26, 1999. This launch service of secondary payloads is regarded as India's "first commercial launch." It is also the first time that an ISRO launcher carried multiple satellites. [779] [780] [781] [782] [783]

The TT&C functions are carried out by the ISRO/ISTRAC control center with ground stations at Bangalore, Lucknow and Hyderabad (payload data and TT&C) in India and from its station in Mauritius. External ground stations at Bearslake (OKB/MEI), Biak (Indonesia) and the DLR station in Weilheim (Germany). The payload data reception stations are operated by NRSA (National Remote Sensing Agency) at Hyderabad and Shadnagar. The real-time data from OCM and MSMR playback data are downlinked in X-band, while the real-time MSMR data as well as housekeeping data are transmitted in S-band. Status: IRS-P4 is operational as of 2001.

Mechanical system	
Structure	Al/Al honeycomb with CFRP elements
Thermal system	Passive/semi-active thermal control with paints, blankets, OSRs and closed loop temperature controllers
Thermal control	Payloads: 15±2°C for OCM; 20±10°c for MSMR; battery: 5±5°C; electronics: 0 to 40°C
Mechanisms	Solar panel deployment, OCM hold and release & OCM tilt; MSMR antenna scanning
Power system	
Solar panels	9.6 m² rigid, sun-tracking type-800 W power at EOL
Chemical battery	2 x 21 Ah NiCd batteries
Power electronics	Two raw buses (28-42V) providing power to all subsystems, modular DC/DC converters
TT&C system (S-band)	
Telecommand	PCM/FSK/FM/PM modulation, time tag command facility
Telemetry	PCM/PSK/PM modulation; ASIC-based system
Transponder	Uplink frequency: 2028.70 MHz; downlink frequency: 2203.20 MHz
AOCS	
Sensors	Earth sensors, sun sensors, magnetometer and gyros
Actuators	Reaction wheels, magnetic torquers, hydrazine thrusters (1N & 11N)
AOCE	Microprocessor-based system
Pointing accuracy	0.15° (pitch and roll); 0.20° (yaw)
Drift rate	$<3.6 \times 10^{-4}$ °/s
Data handling system	
Baseband data rate	20.8 Mbit/s
RF system	QPSK modulated transmitter
Frequency	X-band (8350 MHz)
On-board recorder	320 Mbit for OCM (coarse resolution) and MSMR data
S/C positioning system	
Positional accuracy	100-150 m (autonomous mode)

Table 191: Overview of IRS-P4 S/C characteristics

Orbit: Sun-synchronous circular orbit, altitude = 720 km, inclination = 98.28°, orbital period = 99.31 min, repeat cycle = 2 days, equator crossing time at 12 noon.

IRS-P4 is the first S/C of the IRS series providing successfully GPS data with an 8-channel C/A code receiver (L1 frequency), referred to as SPS (Satellite Positioning System). The instrument is mounted in the anti Earth view face of the payload to track GPS signals continuously. SPS computes on-board the instantaneous orbit (position, Velocity, time). In addition, pseudo range and delta pseudo range are provided. The on-board GPS measure-

779) R. N. Tyagi, "IRS-P4 mission," Current Science, Vol. 77, No 8, Oct. 25, 1999, pp. 1033-1037
780) M. Rao, V. Jayaraman, G. Joseph, "Earth Observation Programme of India - Catering to National Needs of Sustainable Development," Proceedings of the International Symposium on Earth Observation System for Sustainable Development, Feb. 25-27, 1998, Bangalore, India, pp. 277-292
781) R. N. Tyagi, "Indian Remote Sensing Satellite (IRS)-P4 (OCEANSAT-1), NNRMS Bulletin 22, May 1998, pp. 5-12
782) ISRO brochure of IRS-P4 (OceanSat-1), provided by George Joseph
783) SAC Courier, Vol. 24, No 2, July 1999, the issue focuses on IRS-P4 (OceanSat-1), instruments and applications

ments along with the GPS positions from the ground reference receiver are used to estimate the IRS orbit. [784)]

Sensor complement: The payload includes the OCM (Ocean Color Monitor) and MSMR (Multi-frequency Scanning Microwave Radiometer) instruments. Both instruments are very different from the ones flown in earlier missions and needed new developments in many areas like tilt mechanism (to avoid sun glint) and hold-down and release mechanism for OCM, a CFRP (Carbon Fibre Reinforced Plastic) support structure for MSMR and the mission specific data handling system.

OCM (Ocean Color Monitor). OCM is a solid-state radiometer providing observations in eight spectral bands in the VNIR region (nm): B1: 402-422, B2: 433-453, B3: 480-500, B4: 500-520, B5: 545-565, B6: 660-680, B7: 745-785, and B8: 845-885. The instrument employs pushbroom scanning technology with linear CCD detector arrays (191 6K CCD) of 6000 elements (3700 active detectors, the rest are used to correct for dark current). A swath of 1440 km is provided. An along-track instrument tilt capability of ±20° is provided to avoid sun glint. OCM optics is based on one lens per band (wide angle telecentric lens design, refractive system). A scheme of using unused CCD pixels is being utilized as a reference for calibration. The ground resolution is 360 m in the along-track and 236 m in the cross-track direction. Applications are: [785)] [786)]

- Measurement of concentrations of phytoplankton pigments, suspended sediments, Gelbstoff concentration in the euphotic layer. Assessment of their distribution, both spatially and temporally; detection of algal blooms and their dynamic behavior.
- Identification of potential fishery zones
- Delineation of coastal currents and eddies
- Estimation of optical properties and phytoplankton abundance for marine resource and habitat assessment
- Observation of pollution and sediment inputs to the coastal zone and their impact on marine food
- Sediment dynamics, dynamics of estuarine/tidal inlets, prediction of shoreline changes, coastal circulation and dispersion patterns.

Parameter	Value	Parameter	Value
Spectral range (8 bands)	0.4 - 0.9 μm (VNIR)	SNR	> 512 (saturation)
Scan plane tilt	±20°,to avoid sun glitter	Effective focal length	20 mm
Camera MTF	>20% at Nyquist frequency	Absolute radiometric accuracy	<10%
IFOV at nadir	360 m x 236 m (VNIR)	No of CCD elements	6000
FOV (swath)	1420 km (±43°)	CCD element size	10 μm x 7 μm
Tilt capability	±20° along track	Integration time	52.4 ms
Data quantization	12 bit	Exposure levels (gain)	16
Data rate (real-time)	20.8 Mbit/s	On-board calibration	2 LEDs per band

Table 192: Specification of the OCM instrument

The two modes of OCM data transmission are real-time and playback. The playback mode provides a coarser spatial resolution of 2.16 km. The data are averaged on-board (8 scan lines x 4 pixels) and transmitted in the eclipse phase of the orbit at a rate of 0.3 Mbit/s. The instrument mass is 75 kg.

The imaging concept of OCM is an extension of the imaging concept adopted for the IRS LISS instruments. Unlike SeaWiFS or OCTS, OCM does not have common collecting op-

784) A. S. Ganeshan, S. A. Rathnakara, et al., "Precise Position Determination of IRS-P4 Using GPS Measurements," Journal of Spacecraft Technology, Vol. 10, No 1, Jan. 2000, pp. 16-24

785) M. S. Kumar, A. S. Kumar, "Ocean Color Monitor (OCM) of IRS-P4," NNRMS Bulletin-22, May 1998, pp.13-19

786) P. S. Desai, H. Honne Gowda, K. Kasturirangan, "Ocean research in India: Perspective from space," Current Science, Vol. 78 No. 3, Feb. 2000, pp, 268-278

tics coupled to a scan mechanism for realizing the wide swath. OCM has separate wide-angle optics and a linear array CCD detector for each band. The issue of spectral response variation with large incidence angles is overcome by the choice of a telecentric design and the use of a spectral selection filter close to the linear array detector. - The pushbroom approach has enabled the use of a 12 bit digitizer to cover the instruments dynamic range. [787]

Band	Band center and bandwidth (nm)	Saturation radiance (mW cm^{-2} sr^{-1} μm^{-1})	Applications	NESR (mW cm^{-2} sr^{-1} μm^{-1})
B1	412, 20	35.5	Yellow substance and turbidity	0.026
B2	442, 20	28.5	Chlorophyll absorption	0.022
B3	489, 20	22.8	Chlorophyll and other pigments	0.017
B4	512, 20	25.7	Turbidity, suspended sediment	0.017
B5	557, 20	22.4	Chlorophyll, suspended sediment	0.015
B6	670, 20	18.1	Chlorophyll absorption	0.01
B7	768, 40	9.0	O_2 absorption	0.005
B8	867, 40	17.2	Aerosol optical thickness, vegetation, water vapor reference over the ocean	0.008

Table 193: Spectral band definition in the VNIR range

MSMR (Multifrequency Scanning Microwave Radiometer). [788] Objective: to complement the biological ocean parameters of OCM with some geophysical parameters such as: sea surface temperature (SST), sea surface wind speed (SSW), atmospheric water vapor (WV), and cloud liquid water content (CLW). MSMR is a dual-polarized four-frequency radiometer. It features a conical scanning mechanism with a constant look angle of 43.13° and a scan rate of 11.16 rpm. The various receivers are Dicke-switched with a two-point on-board calibration using hot termination and a cold-space horn antenna. The eight-channel MSMR employs six independent receivers 6.6, 10.6, 18, 21 GHz. MSMR observations are continuous over the mission life. Instrument mass = 65 kg, power = 76 W, the data rate is 6.4 kbit/s.

Parameter	Value	Parameter	Value
Frequencies (GHz)	6.6, 10.65, 18, 21	Antenna diameter	0.862 m x 0.800 m
Polarization	V&H, V&H, VH, VH	3 dB beam width	4.2°, 2.6°, 1.6°, 1.4°
Swath width	1360 km	Spatial resolution (km)	105x68, 66x43, 40x26, 34x22 (for frequency sequence)
Sampling interval	9 ms	Temperature resolution	<1.0 K
Data quantization	12 bit	Dynamic temp. range	3.7 - 330 K

Table 194: MSMR instrument parameters

MSMR has an antenna reflector of size 860 mm x 800 mm, an off-axis parabola, and a corrugated feed to receive the emitted radiation from the Earth and its atmosphere. The antenna reflector is rotated at 11.16 rpm to get a circular scan of 1360 km on the Earth's surface with a constant incidence angle of 49.7°. The receiver signal is detected by a square-law detector and synchronously demodulated. The amplitude of the incoming signals are quite low and hence very sensitive to the temperature stability of the components (often a fraction of a degree). To overcome this difficulty, an onboard calibration scheme has been implemented wherein for each antenna scan, temperature values of critical components are recorded along with payload data. These data are used along with look-up tables on the ground to compensate for the temperature variations on-board, thus improving the determination of the brightness temperature. The MSMR instrument provides a two-day repetivity thus providing applications in fisheries, meteorology, sea-state, etc. Research applications, like rainfall, soil moisture, and snow cover are also being tested.

787) R. R. Navalgund and A. S. Kiram Kumar, "Ocean Color Monitor (OCM) on Indian Remote Sensing Satellite IRS-P4," ISRO/SAC internal paper provided by George Joseph

788) S. S. Rana, "Multifrequency Scanning Microwave Radiometer of IRS-P4," NNRMS Bulletin-22, May 1998, pp.20-23

The following data products of MSMR are generated on a routine basis:

- Level 0 antenna temperature data
- Level 1 brightness temperature data
- Level 2 geophysical parameter data

The MSMR realization required the following indigenous technical developments: a) a multifrequency polarization feed with a frequency ratio of 1:3.2; b) microwave blackbody targets with an emissivity>99.9% complete with cryogenic control; c) a precision thermistor-based temperature sensor with an accuracy of 0.1 K and a precision of 0.01 K.

Parameter	Channel	Accuracy	Resolution	Value Range
WV	21 with 18 & 10 (H&V)	0.3 g/cm^2	50 km x 36 km	0.2 - 7.5 g/cm^2
CLW	21 with 18 & 10 (H&V)		50 km x 36 km	0 - 80 mg/cm^2
SSW	10 with 6.6, 18 & 21 (H&V)	1.5 m/s	75 km x 75 km	2 - 24 m/s
SST	6.6 with 10, 18, & 21 (H&V)	1.5 K	150 km x 146 km	273 - 303 K

Table 195: Retrievable geophysical parameters from MSMR

D.17.8 IRS-P5 (CartoSat-1)

The objectives of the IRS-P5 mission are directed at geo-engineering (mapping) applications, calling for high-resolution panchromatic imagery with high pointing accuracies. Hence, IRS-P5 is also referred to as CartoSat-1. [789)]

The S/C structure is of IRS-1C/1D and -P3 heritage, having a diameter of 2.7 m and a height of 2.9 m. The S/C mass is 1500 kg (including fuel for five years of operation). The platform is three-axis stabilized (star sensors in loop, magnetic bearing reaction wheels, 16 nozzles with 1 N thrusters, 4 nozzles with 11 N thrusters), the pointing accuracies are ±0.05°, the stability is 5×10^{-5} °/s. A launch of the IRS-P5 mission is planned on a PSLV launcher from SHAR for 2002.

Payload data transmission is in X-band (2 carriers, QPSK modulated, single polarized) at 105 Mbit/s. A new electronically steerable beam phased-array antenna system is being implemented.

Orbit: Sun-synchronous circular orbit, altitude = 617 km, inclination =97.5°.

Sensor complement:

The payload instrumentation consists of two panchromatic cameras of PAN heritage as flown on the IRS-1C/D satellites. The objective is to obtain fore-aft stereo imagery with two fixed (body-mounted) instruments. The discrimination of elevation differences of better than 5 m make the data particularly suitable for map-making and terrain modeling. **PAN-F** (Panchromatic Forward-pointing Camera), featuring a fixed forward tilt of 26°. **PAN-A** (Panchromatic Aft-pointing Camera), it is fixed at an aft tilt of 10°. Each camera provides a spectral range of 0.5 - 0.75 µm, a spatial resolution of 2.5 m, a swath width of 30 km, and data quantization of 10 bits. The on-board source data rate of the two cameras is 338 Mbit/s. An on-board ADPCM/JPEG compression algorithm of 3.2 : 1 is applied reducing the data rate to 105 Mbit/s (i 52.5 + q 52.5).

The optical system of each PAN camera is designed with a three-mirror off-axis reflective telescope with an off-axis concave hyperboloidal primary mirror and an off-axis concave ellipsoidal tertiary mirror - to meet the required resolution and swath width. The overall size of each PAN camera is 150 cm x 850 cm x 100 cm with a mass of 200 kg.

789) Note: ISRO is the only Space Agency anywhere that did not provide any imagery electronically of its spacecraft or of its instruments (in spite of many requests).

D.17.9 IRS-P6 (ResourceSat-1)

The overall objectives of the IRS-P6 mission (**ResourceSat-1**) are for the support of agricultural applications. The LISS-IV camera is the prime instrument of this payload.

The S/C mainframe is of IRS-1C/1D -P3 heritage. The overall S/C size is about 2 m in diameter and 2.1 m in height. The total S/C mass is 1200 kg, including fuel for five years of operation. The platform is three-axis stabilized (star sensors in loop, magnetic bearing reaction wheels, 16 nozzles with 1 N thrusters, 4 nozzles with 11 N thrusters), the pointing accuracies are ±0.15°, the stability is 3×10^{-4} °/s. A solid-state on-board recorder provides 60 Gbit capacity to store about 10 minutes of LISS-IV data. A launch of the IRS-P6 satellite is planned on a PSLV launcher from SHAR in late 2001.

Orbit: Sun-synchronous circular orbit, altitude = 817 km, inclination = 98.8°.

Sensor Complement:

LISS-IV (Linear Imaging Self-Scanning Sensor-IV). LISS-IV is a three-band pushbroom camera of LISS-III heritage (same spectral VNIR bands as LISS-III) with a spatial resolution of 6 m and a swath of 70 km. LISS-IV features a ±20° steering capability in the cross-track direction. The optoelectronic module for LISS-IV is identical to that of the PAN camera of IRS-1C/1D.

LISS-III* (Linear Imaging Self-Scanning Sensor-III*). The instrument is identical to LISS-III on IRS-1C/1D (with regard to lens modules, detectors, and electronics) in the three VNIR bands, each with a spatial resolution of 23.5 m. The resolution of the SWIR band is now also of 23.5 m on a swath of 140 km. The optics design and the detector of the SWIR band are modified to suit the required resolution.

AWiFS (Advanced Wide Field Sensor). AWiFS is a wide-angle medium resolution (70 m) camera with a swath of 740 km (FOV = ±25°) of WiFS heritage. The pushbroom instrument operates in three spectral bands which are identical to two VNIR bands (0.62 - 0.68 μm, 0.77 - 0.86 μm) and the SWIR band (1.55-1.70 μm) of the LISS-III camera. The wide swath coverage enables AWiFS to provide a five-day repeat capability.

Sensor	Spectral Bands (μm)	Spatial Resolution	Swath Width	Bit Quantization
LISS-IV	0.52-0.59, 0.62-0.68, 0.77-0.86	6 m	70 km	7 bit
LISS-III*	0.52-0.59, 0.62-0.68, 0.77-0.86 1.55-1.70 (SWIR)	23.5 m 23.5 m	140 km 140 km	7 bit 7 bit
AWiFS	0.62-0.68, 0.77-0.86, 1.55-1.70	70 m	740 km	10 bit

Table 196: Summary of the IRS-P6 instrument parameters

D.18 JERS-1 (Japan Earth Resources Satellite)

JERS-1 [790) 791)] is a joint NASDA/MITI/STA project (NASDA/STA developed the satellite (built by Mitsubishi Electric Co.); MITI sponsored the instruments). The JERS-1 satellite was launched on Feb. 11, 1992 from Tanegashima Space Center by the H-I launch vehicle. JERS-1 carries a two-instrument payload, SAR and OPS (Optical Sensor). JERS-1 (built by Mitsubishi Electric) consists of a net rectangular bus (0.9 m x 1.8 m x 3.2 m) with a single 2 kW solar array (3.5 m x 7.0 m) and an eight-segmented SAR antenna. The S/C is three-axis stabilized with zero momentum bias system using reaction wheels and magnetotorquers. The attitude is sensed by an Earth sensor, an inertial reference unit and two sun sensors

790) The Japanese nickname for JERS-1 is Fuyo-1, the name of a Japanese flower.

791) Y. Nemoto, et al., "Japanese Earth Resources Satellite-1 Synthetic Aperture Radar," Proceedings of the IEEE, Vol. 79, No. 6, June 1991, pp. 800-809

(0.3° attitude knowledge). Hydrazine propulsion system. Nominal mission duration = 2 years. Satellite mass = 1340 kg. Status: On Oct. 11, 1998 a malfunction occurred on JERS-1 causing a termination of satellite operations. Hence, a total of about 6 1/2 years of observation time could be achieved with JERS-1. [792)]

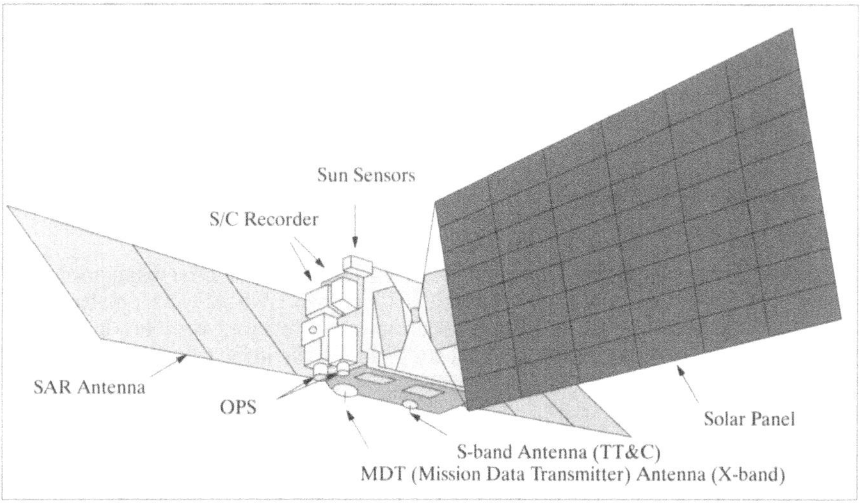

Figure 121: JERS-1 spacecraft model

Application: Survey of geological phenomena, land usage (agriculture, forestry), observation of coastal regions, geologic maps, environment, disaster monitoring, etc.

Objectives: Generation of global data sets with SAR and OPS sensors in order to survey resources; establishing an integrated Earth observation system, verifying instrument/system performances.

Orbit: Sun-synchronous subrecurrent polar orbit; inclination = 97.7°; repeat cycle = 44 days (westward); altitude = 568 km; period = 96 minutes, local mean time = 10:30 - 11:00 AM (descending node).

Sensor complement:

SAR (Synthetic Aperture Radar). Measurement in L-band (1.275 GHz) in HH polarization. The instrument consists of the following subsystems: antenna, transmitter/receiver, and signal processor. The SAR antenna is composed of eight 2.2 m x 1.5 m panels, measuring 2.2 m x 11.9 m in total. A honeycomb sandwich structure is used. The S/C recorder has a capacity of 7.2 x 10^4 Mbit (recording time of up to 20 minutes of SAR and OPS data). The signal processor performs several functions. First, it performs the A/D conversion of I and Q video signals from the transmitter/receiver. Secondly, after formatting the A/D converted I and Q data plus other data necessary for processing on the ground, it transmits the I and Q channel data. Finally, it controls itself and the transmitter/receiver based on commands received from the ground stations.

Calibration: SAR on-board calibration is for receiver gain. The parameter derived from the SAR image mode is the normalized radar backscattering coefficient, sigma-zero. The absolute accuracy of the image product is dependant upon the calibration efforts undertaken at each processing facility.

792) JERS-1 Data User's Handbook, provided by NASDA/EOC

Center frequency	1.275 GHz (L-band, 23 cm wavelength)
Bandwidth	15 MHz
Spatial resolution	18 m (range) x 18 m (azimuth, 3 looks)
Swath width	75 km
Transmitting power	1100-1500 W
Pulse width	35 μs
PRF (Pulse Repetition Frequency)	1505.8 - 1606.0 Hz
Antenna Polarization Look angle Antenna gain Signal to ambiguity ratio	Array of 1024 microstrip radiation elements HH 35.21° >33.5 dB >14 dB
Data quantization	3 bits
Data rate	60 Mbit/s

Table 197: Performance parameters of the JERS-1/SAR instrument

OPS = Optical Sensor. OPS is a high-resolution radiometer with the objective to measure sunlight reflected by the Earth's atmosphere in eight bands (color imagery). OPS consists actually of two independent radiometers: the VNIR band instrument and the SWIR radiometer. Swath width = 75 km, spatial resolution = 18.3 m (cross-track direction) x 24.2 m (along-track direction). Band 4 is for off-nadir viewing (15.33° forward in flight direction); bands 3 and 4 make a stereo pair. The stereo images are useful for geomorphology and digital elevation model (DEM) construction. Data from the 4 SWIR bands allow hydrothermal, vegetation, and soil moisture mapping as well as discrimination between snow and cloud coverage. IFOV=32.2 μrad, FOV=7.55°, stereoscopic view angle=15.33° (B/H=0.3), Nr. of pixels=4096 per band, data rate=30 Mbit/s in two channels (each), data quantization=6 bits. The following spectral bands are defined:

- Band 1 = 0.52 - 0.60 μm
- Band 2 = 0.63 - 0.69 μm
- Band 3 = 0.76 - 0.86 μm
- Band 4 = 0.76 - 0.86 μm (forward viewing)
- Band 5 = 1.60 - 1.71 μm (SWIR)
- Band 6 = 2.01 - 2.12 μm (SWIR)
- Band 7 = 2.13 - 2.25 μm (SWIR)
- Band 8 = 2.27 - 2.40 μm (SWIR)

The VNIR radiometer has a telescope with aspherical lenses (refractive optics) and a dichroic prism. The telescope aperture is 43 mm in diameter with an f/Nr.5 and a focal length of 215 mm. The Si-CCD detectors have 4096 elements. There are four detectors, one for each band. A reference light source unit emits a calibration light during the eclipse period of the orbit to check the radiometric performance of the radiometer. The VNIR instrument mass is 32 kg.

The SWIR radiometer telescope uses reflective and refractive optical systems. The aperture is 200 mm in diameter with an f/Nr.1.55 and a focal length of 310 mm. Detectors: PtSi Schottky CCDs with 2048 pixels per row and 4096 pixels per band. The CCDs are cooled to 77-82 K. A reference light source provides sensor calibration. The SWIR mass is 60 kg. The total instrument mass (common electronics = 82 kg) is 174 kg. The total instrument power is 250 W.

Data:

Mission (sensor) data is downlinked in X-band (f_1 = 8.15 GHz, f_2 = 8.35 GHz) in QPSK modulation at a data rate of 60 Mbit/s by a right-hand circular polarized antenna onboard the S/C. Provision of on-board data recording. The JERS-1 ground system consists of TACC (Tracking and Control Center) at the Tsukuba Space Center, NASDA/EOC (Earth Observation Center) for data acquisition, processing, distribution and archiving, and a number of licensed foreign receiving stations. Some ground stations which receive JERS-1 data are:

NASDA/EOC (Hatoyama, Japan)
Tokai University (Kumamoto, Japan)
National Institute of Polar Research, Japan (Syowa, Antarctica)
Alaska SAR Facility (ASF, Fairbanks, Alaska)
Canada, CCRS (Gatineau, Prince Albert)
ESA (Kiruna, Tromsoe)
National Research Council of Thailand (Bangkok, Thailand)
ACRES (Australia)
DLR/DFD station (O'Higgins, Antarctica)
etc.

JERS-1 SAR Data Products	
Level 0	Unprocessed signal data product
Level 1	Partially processed signal data product
Level 1.1	Basic image product
Level 2.0	Bulk image product
Level 2.1	Standard georeference image product
Level 3.0	Precise corrected image product
Level 4	Geocoded with terrain correction
JERS-1 OPS Data Products	
Level 0	Unprocessed data product
Level 1	Radiometric corrected image product
Level 2	System corrected image product
Level 3	Precise corrected image product
Level 4	Registration corrected image product
Level 5	Stereo image product

Table 198: Data Products of JERS-1 SAR and OPS instruments [793)]

D.19 KITSAT Program

In the late 1980s, the Satellite Technology Research Center (SaTReC) of the Korea Advanced Institute of Science and Technology (KAIST) was given the task of defining a national satellite program. The major objectives of the program were to develop a capability and to create an environment for state-of-the-art space age technology and research. Economic considerations implied an affordable program in microsatellite technology. A need in project experience and associated activities for all phases of successful space missions resulted in an on-the-job traineeship/development of a Korean microsatellite for a team of SaTReC engineers at SSTL (University of Surrey), UK. The first satellite in this cooperative program was named KITSAT-1 (Korea Institute of Technology Satellite-1), it was launched August 10, 1992 on Ariane-4 ASAP (V.52, piggyback), along with the Topex/Poseidon S/C as the prime payload (KITSAT-1 is defined under D.40.6).

D.19.1 KITSAT-2 (Korea Institute of Technology Satellite-2)

KITSAT-2 was designed and built at KAIST/SaTReC in Korea (based on the UoSAT bus). It was launched on Ariane-4 (along with PoSAT-1 and HealthSAT-2), September 26, 1993 as an auxiliary payload on the SPOT-3 launch from Kourou.

The microsatellite has dimensions of 35.2 x 35.6 x 67 cm and a mass of 47.5 kg. GaAs solar panels on four sides of the structure provide 30 W of power. The S/C is spin-stabilized with

793) K. Maeda, M. Nakai, O. Ryuguji, "JERS-1/ERS-1 Verification Program and Future Verification Program," Advanced Space Research, Vol. 12, No. 7, pp. 327-331, 1992

the help of a gravity-gradient boom and three magnetorquers (attitude knowledge from one Earth sensor, two sun sensors, and two horizon sensors). Data transmission via three VHF uplinks and two UHF downlinks (amateur radio services are provided). S/C operations are performed at SaTReC. Nominal operations are conducted as of 6/1998.

Orbit: Sun-synchronous, perigee = 791.1 km, apogee = 805.4 km, inclination = 98.64°, period = 100 minutes.

Sensor complement:

CEIS (CCD Earth Imaging System). CEIS consists of two planar CCD array cameras: a narrow angle camera (NAC) with 200 m ground resolution and a wide angle camera (WAC) with 2 km ground resolution. WAC is a color camera developed in cooperation with Samsung Electronics.
The NAC system uses a filter with 80 nm bandwidth at 650 nm with 60% peak transmission (spectral range from 610 - 690). FOV = 6°, providing a target area of 120 x 100 km. The WAC system uses a single-chip color chip (KC73125) and detector array of 511 x 492 with one of four selectable optical filters. The main bandpass filter which passes the visible band (with optical low pass filter) is located between the sensor and the lens system. The output is sampled as 8 bit per pixel. Transform algorithms are used to convert the 8-bit gray level color matrix information into RGB color information. The WAC system has a mean ground pixel resolution of 2 km before color transformation and 4 km resolution after transformation into the color image. FOV = 68°. [794)]

DSPE (Digital Signal Processing Experiment). The sensor is very similar to DSPE on PoSAT. DSPE uses software digital signal processing techniques to implement a high speed modem. The experiment aims to improve the quality of communications between the satellite and the ground, and also provides a software-controlled method for selecting the data rate and the modulation technique to be used.

IREX (Infrared Experiment). IREX monitors the effects of the space environment on a KAIST-developed infrared sensor. The experiment is designed to operate in a minimum temperature environment with the assistance of a passive cooling system.

LEED (Low Energy Electron Detector). LEED measures the distribution of low energy electrons (less than 6 keV) at the orbit altitude of 800km. This experiment uses a 3000 V spiral electron multiplier with 16 measurement channels. The data from this experiment is used for studying the effects of solar activities on the space environment around the Earth.

DSFCE (Digital Store-and-Forward Communication Experiment).

KASCOM (KAIST Satellite Computer). KASCOM is an experimental on-board computer system based on an 8 MHz 32 bit Intel 80960 microprocessor with 10 MByte CMOS SRAM. The experiment tests the operations of this new computer in the space environment and provides data for future systems to be used in the next generation satellites.

Downlink frequency	Transponder/Beacon	Mode
435.175 MHz 436.500 MHz	T T	9.6 kbit/s FSK FM digital 9.6 kbit/s FSK FM digital
Uplink frequency	Transponder/Beacon	Mode
145.870 MHz 145.980 MHz	T T	9.6 kbit/s FSK FM digital 9.6 kbit/s FSK FM digital

Table 199: Kitsat-2 frequencies

D.19.2 KITSAT-3 (Korea Institute of Technology Satellite-3)

KITSAT-3, or Wooribyul-3 ('our star'), is a KAIST (Korean Advanced Institute of Science and Technology)/SaTReC (Satellite Technology Research Center) microsatellite develo-

794) SK Yoo, S. Lee, et al., "The KITSAT-2 CCD Earth Imaging Experiment," Proceedings of SPIE Conference on Small Satellite Technology and Applications IV, Vol. 2317, Rome, September 1994

ped on a new modular bus structure (no heritage to the KITSAT-1 and -2 bus of SSTL). The main objective of KITSAT-3 is to test the new satellite bus and its payloads. KITSAT-3 is three-axis stabilized providing a pointing accuracy of <0.5°. Attitude of the box-like satellite (495 mm x 604 mm x 852 mm) is sensed by a sun sensor (two-axis), Earth horizon sensor, star sensor (CCD type), fiber optic gyro, and a three-axis magnetometer; control is provided by a three-axis magnetorquer and a reaction wheel as actuators. S/C mass = 110 kg, power = 150 W (deployable solar panels). KITSAT-3 was launched on an ISRO PSLV-C2 launch vehicle as a piggyback payload from SHAR, India, on May 26, 1999 (the prime payload was IRS-P4, also referred to as OceanSat-1 of ISRO). [795] [796]

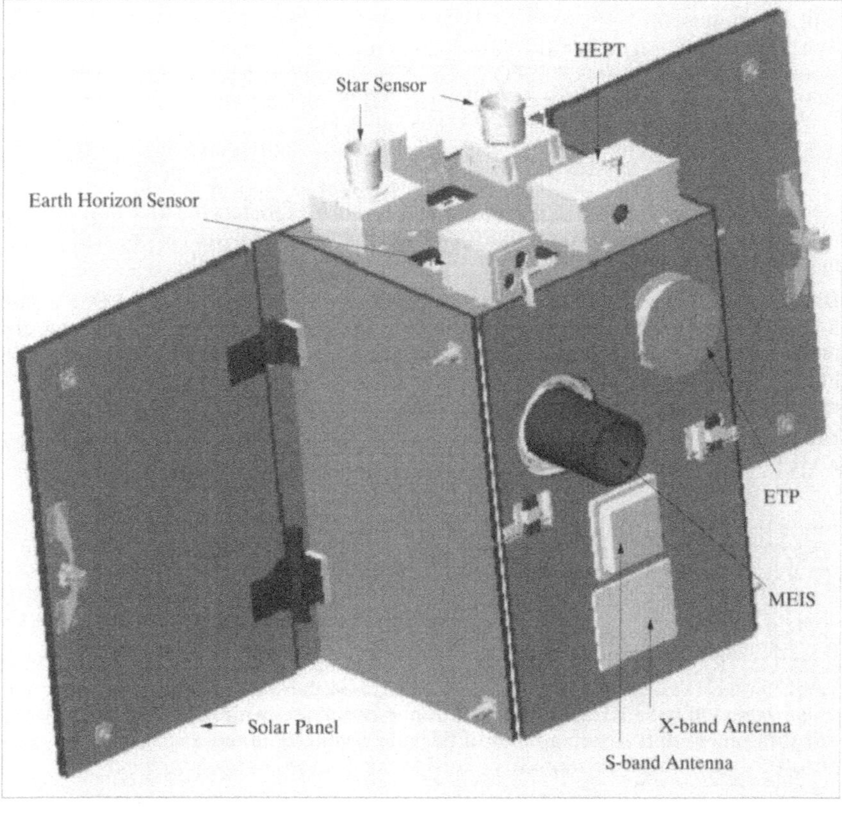

Figure 122: Illustration of KITSAT-3

Communication: VHF receiver at 1200/9600 bit/s, AFSK/FSK modulation, frequency of 148 MHz in uplink. UHF transmitter at 1200/9600 bit/s, AFSK/FSK modulation, frequency of 401 MHz downlink. S-band transmitter at 9.6/38.4 kbit/s, FSK modulation, frequency of 2.2 GHz. X-band transmitter (for imagery transmission) at 3.2 Mbit/s, QPSK modulation, frequency of 8.2 GHz. On-board image data compression in JPEG, GIF and DPCM. An on-board data storage capability of 10 Gbit is provided (2 Gbit SRAM, 8 Gbit flash memory).

795) KITSAT-3 brochure provided by Dongseok Shin of SaTReC, Taejon, Republic of Korea

796) B. J. Kim, H. Lee, S. D. Choi, "Three-Axis Reaction Wheel Attitude Control System for KITSAT-3 Microsatellite," Pergamon, Space Technology, Vol. 16, No 5/6, pp. 291-296, 1996

Orbit: Sun-synchronous circular orbit, altitude = 720 km, inclination = 98.37°.

Sensor complement: (one remote sensing instrument and an experiment package referred to as SENSE (Space Environment Scientific Experiments), consisting of REME, HEPT, SMAG, and ETP.

MEIS (Multispectral Earth Imaging System), developed by SaTReC in cooperation with Stellenbosch University, South Africa (see the HRI instrument of SUNSAT, N.18). The objective is to acquire multispectral high-resolution Earth surface imagery. MEIS consists of a single 10 cm diameter optical tube assembly (a catadioptric telescope), a lens system, pentaprism with a dichroic color splitter, three vertically mounted linear CCD detectors (TC104 of Texas Instruments), and their clock drivers and output buffers. The camera optics parameters are: 570 mm focal length, 100 mm aperture diameter and f/5.7 relative aperture. The camera employs a beam splitter and three linear CCD detector arrays with 3456 pixels in the cross-track direction (one CCD per band), offering a swath width of 50 km (FOV=3.8°) with a spatial resolution of 15 m (IFOV = 19.1 μrad). Spectral band coverage: 510-620, 620-690 and 730-900 nm. For operational purposes the imagery is subdivided into manageable volumes of data, a scene is defined as 50 km x 30 km (or 3456 pixels x 2048 lines along-track), a maximum of 12 consecutive scenes may be scanned during any one pass. The source data rate is 40 Mbit/s. Further instrument parameters: instrument mass = 6.5 kg, power = 15 W.

REME (Radiation Effect on Micro-Electronics). REME is designed to test non-space qualified electronic components, such as memory ICs and discrete components. This experiment consists of two sub-experiments, namely: **TDE** (Total Dose Effect) and **SEU** (Single Event Upset) monitoring.

- **TDE** measures the long-term accumulated ionizing radiation dose (in SiO_2) for three locations on-board the S/C. The sensing elements are RADFET dosimeters. Each RADFET sensor consists of a matched pair of p-channel MOSFETs. One of them is biased during exposure, the other remains unbiased. Exposure to radiation causes the formation of trapped holes (positive charge) in the gate oxide, which in turn causes a gradual shift in the threshold voltage with accumulated dose. The radiation shielding effect can be calculated and modeled according to the sensor locations.
- **SEU.** The objective is to monitor trapped particles in the radiation belts, solar flare protons, and galactic cosmic rays (all of which can cause single-event upsets within microelectronic devices). A series of tests are performed with memory chips to measure the SEU characteristics. Discrete components are connected to appropriate measurement circuits, and variations of their characteristics are monitored together with the total dose measurement.

HEPT (High Energy Particle Telescope). The objectives are to study the physics of energetic particles in the Earth's radiation belts (sources, sinks, transport, and lifetimes). The HEPT module provides seven measurement channels. The pointable instrument features a cone-shaped structure (funnel) which limits the incident angles of particles to be measured. The signals from four parallel-positioned sensors are combined and processed to identify the energy and type of a particle. Measured pitch angle distributions (pitch angles are calculated using the sensor's look directions and the magnetic field direction, determined by the on-board magnetometer) can be used for developing models of the populations of both the static and dynamic radiation belts.

SMAG (Scientific Magnetometer). The objectives are to provide look angle information (1° pitch angle resolution) for the particle experiments, to utilize it as a diagnostic tool for the detection of geomagnetic disturbances, to measure low-frequency waves in the ambient environment, and to provide a secondary source for S/C attitude information. SMAG measures the projections of the ambient field in three orthogonal direction with a resolution of about 5 nT.

ETP (Electron Temperature Probe). The objectives are to measure the electron (plasma) temperature in high latitude regions, in particular in the polar cusp, to study anomalous heating in SAA (Southern Atlantic Anomaly), and to find a relation with other plasma parameters. The electron temperature is determined from the voltage difference between the modulated probe potential and the floating potential.

D.19.3 K-4 (KAISTSAT-4)

KAISTSAT-4 (Korean Advanced Institute of Science and Technology Satellite-4) is a low-cost KAIST/SaTReC minisatellite technology demonstration mission, funded by the Ministry of Science and Technology of Korea, a follow-up mission in the KITSAT program. [797) 798)] The objective is to develop and test a minisatellite bus, along with a new star sensor in a three-axis attitude control subsystem, to demonstrate the performance of new science instruments, and to deploy a newly developed DCS (Data Collection System). The spacecraft mass is about 120 kg, power = 200 W, the mission design life is three years. A launch of KAISTSAT-4 (as a secondary payload) is planned for 2002.

The S/C structure resembles a box of approximate size: 60 cm x 60 cm x 80 cm. It is three-axis stabilized. The S/C pointing requirements call for a pointing accuracy of 0.5°, an attitude knowledge of 5 arcmin, and a stability of about 5 arcmin/s. In addition, the S/C requires a complex set of attitude maneuverability in support of its mission objectives. Hence, the ADCS (Attitude Determination and Control Subsystem) consists of four fiber optic gyros (FOG), two precision star trackers, a coarse sun sensor and two three-axis fluxgate magnetometers for attitude sensing. [799)] Magnetic torquer coils are used for momentum unloading of the four reaction wheels as well as for spin-rate control for the initial phase after spacecraft separation. A GPS receiver is used to provide S/C position, velocity and timing. Attitude determination is based on an extended Kalman filter algorithm considering the bias drift of the gyroscope.

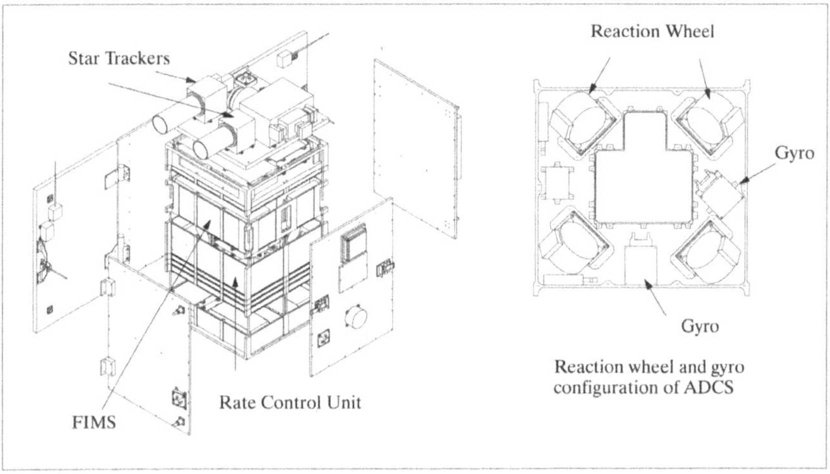

Figure 123: Exploded view of KAISTSAT-4 and ADCS

797) J. Seon, K. I. Deon, S. H. Kim, et al., "Brief Reports on KAISTSAT-4 Mission Analysis," Journal on Astronomy and Space Sciences, Vol. 17, No. 2, 2000, pp. 1-9

798) J. Seon, H. S. Kim, B. J. Kim, Y. S. Chang, K.-M. Park, et al., " Preliminary results from mission analysis on KAISTSAT-4," SaTReC paper provided by Woo-Kyung Lee

799) H.-W. Lee, B. J. Kim, M.-J. Tahk, D.-J. Park, "Attitude Determination and Control of KAISTSAT-4 Satellite," internal paper of SaTReC provided by Woo-Kyung Lee

There are four operational modes of KAISTSAT-4 to achieve the scientific objectives; these are: a) pointed observation mode (observation of selected and extended galactic sources with FIMS during eclipse phases of the orbit), b) sky-survey mode (observation of the entire sky, the S/C spins around the axis parallel to the slit of FIMS), c) aurora observation mode (FIMS is being pointed in the nadir direction over the north and south poles), and d) air-glow mode (FIMS is being pointed to an inertial or to a nadir direction).

Orbit: Sun-synchronous circular orbit, altitude = 800 km, inclination = 92°.

RF communication: All TT&C communications are provided in UHF (prime). In addition, an S-band link is installed (first attempt in KITSAT series). The S-band downlink data rate is 38.4 kbit/s (uplink 9.6 kbit/s). An X-band downlink is used for payload downloads with a data rate of 3.2 Mbit/s. The on-board recorder has a capacity of 2 Gbit. All mission operations are performed at SaTReC, including data acquisition, distribution and archiving.

Sensor complement:

FIMS (FUV Imaging Spectrograph). The objective of FIMS observations is to study the diffuse hot interstellar matter. The overall goals of FIMS are 1) to map the spatial distribution of hot Galactic plasmas through a one-year sky survey, 2) to determine the physical states of hot interstellar matter such as superbubbles and supernova remnants with pointed observations, and 3) to test the models presently available for the Galactic evolution.

The instrument allows the detailed mapping of the spatial distribution of the hot galactic plasmas and the determination of the physical states of hot interstellar matters as well as the detection of the various emission lines from the Earth's upper atmosphere. FIMS employs a dual bandpass (900-1175 Ä and 1335-1750 Ä), high resolution (1.5 Ä and 2.5 Ä, respectively) imaging spectrograph with a 8° x 5' FOV (Field of View) and a 5' (five minutes) spatial (angular) resolution. FIMS is sensitive to emission line fluxes that are fainter than any previous detection by an order of magnitude. The observation data permits the determination of the thermal and ionization equilibrium state in hot Galactic plasmas.

The FIMS instrument consists of two elements, the spectrograph and the electronics unit. The spectrograph has the following subassemblies: Contamination Door Mechanism, Entrance Baffle Assembly, Filter Wheel Assembly, and Detector Assembly. The electronics unit houses circuits for experiment power, control and communication. FIMS generates about 500 Mbit of data per day. The instrument mass is 20 kg, the power is 20 W.

PIP (Plasma In-situ Package) of three instruments: SST (Solid State Telescope), LP (Langmuir Probe), and SM (Scientific Magnetometer). The science objectives are to provide a fast-sampling capability of high-energy magnetospheric plasma components, of cold ionospheric plasmas, and to measure the Earth's magnetic fields. Particular objectives are:

- Detection of directly penetrating solar wind plasmas and up-flowing, cold ionospheric electrons
- Investigation of sub-km scale structures of the Earth's polar regions
- Comparison of the in-situ measurements with the FIMS spectrographic images ($\Delta\lambda/\lambda$ of about 500) of the Earth's aurora in the far-ultraviolet ranges.

SST measures the high-energy components of auroral particles. It permits research on particle acceleration mechanisms in the magnetosphere. In addition, plasma fluxes are studied flowing in and out of the Earth's magnetosphere.

DCS (Data Collection System). A DCS is being developed in a cooperative effort between SaTReC and ITR (Institute for Telecommunications Research) of the University of South Australia in Adelaide and CRCSS in conjunction with FedSat-1. In FedSat-1 terminology, the system is called ADAM (Advanced Data Acquisition and Messaging System). The objectives of the KAISTSAT-4 DCS are:

- Deployment of a satellite-based data collection system for environment monitoring, wildlife tracking and transportation monitoring
- Demonstration of LEO satellite communication technology by adopting new protocols for bi-directional messaging, on-board modem algorithms and architectures
- Establishment of bi-directional communication between the on-board DCS and the ground segment, consisting of many DCPs (Data Collection Platforms). The DCPs are also referred to as MTs (Mobile Terminals). A TDMA (Time Division Multiple Access) protocol is being used in the uplink to collect the data from various DCPs simultaneously.
- Acquisition of the combined operations between KAISTSAT-4 DCS, MTs, data receiving stations and those of FedSat-1 (Austraila).

D.20 KOMPSAT (Korea Multi-Purpose Satellite)

The KOMPSAT program was initiated in 1995 as a major space investment in Korea. Its objective is the development of a national space segment in Earth observation along with an efficient infrastructure and ground segment to provide valuable services to remote sensing users in various fields of applications. According to the national long-term space development plan by the Korean government, eight satellites of the KOMPSAT series are to be developed and launched until 2015.

D.20.1 KOMPSAT-1

KOMPSAT-1 is a minisatellite, funded by the government of Korea and developed by KARI (Korea Aerospace Research Institute) in Taejon, along with seven Korean companies/institutes as well as by TRW of Redondo Beach, CA. [800)] [801)]

The KOMPSAT-1 spacecraft is based on a TRW-developed standardized bus (including a MIL-STD-1555B communications bus) that can be easily modified to support a variety of mission requirements. A core module houses the housekeeping functions - power, telemetry and data handling, attitude control - needed on all space missions. A payload module hosts mission-unique equipment, KOMPSAT's imaging instruments and space physics experiments. A third module, housing a monopropellant hydrazine propulsion system, provides on-orbit thrust and station-keeping.

The S/C size of KOMPSAT-1 is 1.33 m in diameter and 2.33 m in length. The platform uses a three-axis stabilized bus with a reaction control subsystem module, providing a 2-sigma pointing accuracy of 0.08° (knowledge) and 0.10° (control). The S/C mass is 510 kg (including 73 kg hydrazine), power = 630 W (500 W EOL), the S/C design life is three years. The KOMPSAT-1 satellite was launched on a Taurus booster (along with ACRIMSAT) from VAFB, CA on Dec. 20,. 1999.

Communication: S-band, downlink data rates are at 1.5 Mbit/s (playback) or 2.048 kbit/s (real-time), the uplink data rate is at 2 kbit/s. A second downlink in X-band provides data rates of 45 Mbit/s. The CCSDS protocol standards of CEOS are implemented for all communications. KOMPSAT-1 features a 1 Gbit on-board mass memory storage and a 2.5 Gbit solid-state recorder.

Orbit: Sun-synchronous circular polar orbit, altitude = 685 km, inclination = 98.13°, period = 98.46 min, ascending node at 10:50 AM, 14 17/28 orbits per day, revisit time = 28 days.

Status as of May 2001. KOMPSAT-1 is fully operational.

800) http://www.kari.re.kr/
801) Information provided by H. Paik and G. H. Choi of KARI

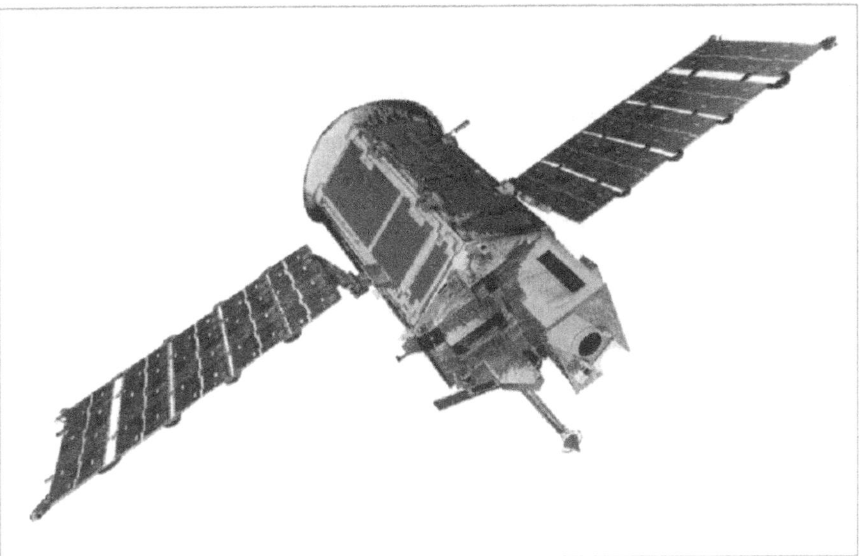

Figure 124: Illustration of the KOMPSAT-1 S/C

Sensor complement.

EOC (Electro-Optical Camera). The objective is to obtain cartographic imagery of Korea (may be extended to other regions of the globe) at 1/25.000 scale. EOC collects panchromatic imagery (spectral region of 510 - 730 nm) with a ground sample distance (GSD) at nadir of 6.6 m and a swath width of 17 km by pushbroom scanning. The S/C features a cross-track pointing capability (body pointing) of up to $\pm 45^{o}$, thereby extending the field of regard for imagery collection. Some instrument parameters: MTF >10% at Nyquist frequency, SNR >50 over entire FOV, FPA has a CCD line array of 2592 pixels, data quantization of 8 bits, total instrument mass is 35 kg, power = 46 W, the source data rate is equal to or less than 25 Mbit/s. - Calibration: use of a built-in dark calibration function. In addition, ground reference targets of known brightness can be used for white calibration.

OSMI (Ocean Scanning Multispectral Imager). [802) 803)] The objective is global ocean color monitoring for the study of biological oceanography. OSMI is a whiskbroom-type [804)] imager generating imagery in up to six spectral bands (in the region of 400 - 900 nm) with a capability for bandwidth and band center selection on command. The center of each band can be varied within steps of 2.6 nm, and bandwidth ranges from 5.2 nm (min) to 166.4 nm (max) can be assigned. The ground resolution is 1 km in a swath of 800 km (0.85 km at nadir and about 1 km at the edge of the swath). The scanner assembly features a line array of 96 detectors (Si), positioned in the along-track direction, thus providing an instantaneous parallel ground coverage of 96 km in one cross-track scan with the whiskbroom configuration. This wide along-track coverage permits sufficient integration time for all cells in each scan sweep of about 15 seconds. Naturally there is some overlap for successive scans. The ensuing collecting optics perform the spectral separation of the radiation received. The bands B0

802) Y. M. Cho, S. S. Yong, et. al., "Ocean Scanning Multispectral Imager (OSMI)," Proceedings Fifth International Conference on Remote Sensing for Marine and Coastal Environments, San Diego, CA, Oct. 5-7, 1998

803) Y. M. Cho, "Ocean Scanning Multispectral Imager (OSMI), Post-launch Radiometric Responsivity Analysis," Proceedings of IEEE/IGARSS 2000, Honolulu, HI, July 24-28, 2000

804) Note: Preference is given to a whiskbroom imager (the older imaging technology) because the optics for pushbroom operation must always cover FOV (the total field of view) while the optics for whiskbroom operation deal with IFOV (instantaneous field of view) which is much smaller than FOV. Hence, there are less distortions at the swath edge.

through B4 provide ocean color data, the other bands provide information for atmospheric (aerosol) corrections. Some instrument parameters: MTF is about 20% at Nyquist frequency, SNR is between 350 and 450 over entire FOV, data quantization = 10 bits, instrument mass = 15 kg, power = 30 W, the source data rate is 600 kbit/s. The duty cycle for image collection is 20%. - Calibration: A solar calibration is performed with an on-board spectralon once per orbit (note: a spectralon is a kind of optical diffuser which can be used as a Lambertian source). Dark calibration is performed before and after each imaging period.

Spectral bands	B0	B1	B2	B3	B4	B5	BX	B6
Center wave length (nm)	412	443	490	510	555	670	765	865
Nominal bandwidth (nm)	20	20	20	20	20	20	40	40

Table 200: Spectral parameters of the OSMI instrument

SPS (Space Physics Sensor). The SPS package consists of two instruments: **HEPD** (High Energy Particle Detector) and **IMS** (Ionosphere Measurement Sensor). The objective for HEPD is to characterize the low-altitude high-energy particle environment and to study the effects of the radiation environment on microelectronics. The objective of IMS is to measure in-situ densities and temperature of electrons in the ionosphere.

- HEAD in turn consists of the following instruments: PES (Proton and Electron Spectrometer), LET (Linear Energy Transfer Spectrometer), TDM (Total Dose Monitor), and SEM (Single Event Monitor).
  - PES identifies the particle types detected and measures their energy in seven channels. Objective: distribution of energetic electrons and proton in the radiation belts.
  - LET measures particle populations, providing information of radiation effects on electronic components (comparison with TDM and SEM data).
  - TDM measures with RADFET dosimeters the long-term ionizing dose of radiation (in SiO_2) accumulated at SPS. The threshold voltage provides information with respect to dose accumulation.
  - The objective of SEM is to test exposed non-space qualified static RAM components in the radiation environment of KOMPSAT-1 for future space missions. The experiment uses four static RAM chips (4 Mbit), connected to appropriate circuitry, to measure SEU (Single Event Upset) characteristics.

Channel	Particle	Energy Range (MeV)
pE1	Proton	30-38
pE2	Proton	15-30
pE3	Proton	6.4-15
eE1	Electron	2.0 <
eE2	Electron	0.72-2.0
eE3	Electron	0.25-0.7
AA	Alpha particle	15-60

Table 201: Specification of PES channels

- The IMS instrument contains an electron temperature sensor (ETS) and an electron density sensor (EDS), measuring temperature and density distributions of electrons (ionospheric plasma). EDS is a modification of a Langmuir Probe. The measurement range for ETS is 0 - 1 eV, for EDS the range is 10 - 10^6 e/cm^3.

D.20.2 KOMPSAT-2

KOMPSAT-2 is being developed by KARI to continue the observation program of KOMPSAT-1. The main mission objectives of the KOMPSAT-2 are to provide a surveillance capability for large-scale disasters by acquiring high-resolution imagery for GIS (Geographic Information Systems) applications. A launch of the KOMPSAT-2 spacecraft is planned for 2004. [805]

The spacecraft design is based on the KOMPSAT-1 heritage making extensive use of existing hardware, software, tools, and facilities; it allows parallel integration of the payload, equipment, and propulsion modules. The attitude and orbit control subsystem provides three-axis stabilization with high accuracy for roll, pitch and yaw pointing. Star trackers, gyro reference assemblies, and reaction wheels are used for attitude control. The spacecraft bus offers a cross-track body-pointing capability through roll maneuvers. The propulsion subsystem makes use of re-used components. A MIL-STD-1553B data bus interfaces most of the spacecraft bus and payload to the on-board computer. GPS receivers provide the time, satellite position and velocity. The spacecraft mass is <800 kg (including propellant), the power is 955 W (EOL).

Orbit: Sun-synchronous circular orbit, altitude = 685 km, inclination = = 98.13°, period = 98.46 min.

RF communications: S-band and X-band communications are provided for all data transmission to the ground.

Sensor complement:

MSC (Multi-Spectral Camera). The objective is to collect high-resolution panchromatic and multispectral imagery of the Earth's surface. MSC is an optoelectronic pushbroom instrument with a single nadir-pointing telescope. The Pan band has a spectral range of 500-900 nm, the four MS bands are in the 450-900 nm range. The spatial resolution of the Pan data is 1 m (GSD), the MS data has a GSD of 4 m. The swath width is 15 km. A FOR (Field of Regard) in the cross-track direction is provided through body pointing.

Note: The project management of KOMPSAT-2 denied any further information on the specification of the spacecraft and its sensor complement.

805) Information provided by Young-Min Cho of KARI

D.21 LANDSAT

A pioneering US land remote sensing satellite program which has provided a continuous supply of synoptic, repetitive, multispectral data of the Earth's land surfaces since 1972. Over the years a large international user community evolved along with the Landsat series. The program opened entire new fields of research, providing insights into geologic, agricultural, and land-use surveys, and led eventually to new paths of resource exploration - in all, for a better understanding of the Earth system. The success of the Landsat program stimulated new approaches to data analysis and gave impetus to new sensor designs. In addition international participation was fostered on many levels which spawned other Landsat-like programs such as the SPOT series of France, the Resurs series of Russia, and the IRS series of India. [806)]

Background: NASA launched ERTS-1 (Earth Resources Technology Satellite) on July 23, 1972. This satellite was subsequently renamed Landsat-1. The Landsat system remained a research/experimental NASA program until 1983. The system was then declared operational and its management was turned over to NOAA (LS-4 in 1983 and LS-5 in 1984). The Land Remote-Sensing Commercialization Act of 1984 almost immediately authorized a phased commercialization of remote sensing data. Following to a contentious commercialization effort, the Land Remote-Sensing Policy Act of October 1992 reversed the 1984 decision to commercialize the Landsat system. The 1992 Policy Act created a new "Landsat Program Management" under NASA and DoD leadership. Due to difficulties with the Landsat-7 development program, DoD withdrew from the Landsat Program Management in early 1994. In spring 1994 the Landsat-7 program was restructured and in May 1994 it was put under joint NASA/NOAA/USGS management by Presidential Decision Directive, with NASA responsible for the development of the LS-7 S/C, instrument, and ground system; NOAA is responsible for LS-7 S/C and ground system operations; and USGS is responsible for maintaining the national archive of LS data and data distribution to the user community (EDC is prime US receiving station of LS-7 data).[807)]

Date	Landsat Program Event/Action
Mar. 1, 1984	NASA launches LS-5, NOAA provides LS-4 and LS-5 operation
Sept. 1985	Eosat selected by NOAA to operate the LS system (LS-4 and LS-5), to market LS data, and to build and launch LS-6
Sept. 1992	Eosat loses capability of processing MSS data (no further acquisition of MSS in USA)
Oct. 1992	Land Remote Sensing Policy Act (Congress) LS program management under NASA and DoD (develop and launch LS-7)
Oct. 1992	DoD signs contract with Martin Marietta Astro Space to build HRMSI for LS-7
Oct. 5, 1993	LS-6 launch failure
Spring 1994	DoD withdraws from the LS program (HRMSI is withdrawn from LS-7)
April 1994	LS program management and Eosat agree for continued operation of LS-4 and LS-5
Aug. 10, 1994	Management Plan for the Landsat Program (LS-7 objectives and beyond). LS program responsibilities are with NASA (S/C), NOAA (operations), and USGS (archive)
Dec. 1994	NOAA and Eosat sign a contract to extend LS-5 operation
Nov. 1995	Eosat completed maneuvers to correct the equatorial crossing time of LS-5 (to 9:30 AM)

Table 202: Chronology of major Landsat program events since 1984

The Landsat satellite family is managed and procured by NASA [built by GE Astro Space of East Windsor, NJ (S/C), and Raytheon Santa Barbara Remote Sensing, formerly a Hughes Santa Barbara Research Center Division. (sensors)]. In 1985 the functions of Landsat S/C operation and data handling were handed over from NOAA to a commercial contractor company: Eosat (Earth Observation Satellite Company, of Lanham, MD; Eosat is a Division of Lockheed Martin Space Imaging. The science data (of LS-4 and LS-5) are processed

806) Special Issue: 25th Anniversary of Landsat, PE&RS Vol. LXIII, No. 7, July 1997, pp. 829-905

807) E. J. Sheffner, "The Landsat Program: Recent History and Prospects," PE&RS, Vol. 60,, 1994, pp. 735-744

and archived by Eosat and commercially distributed. The LS-4 and LS-5 S/C and sensors are regarded as second generation technology in the LS program, following the early MSS sensor series. MSS was carried for the last time on LS-5, and LS-7 carries an improved Thematic Mapper sensor known as the Enhanced Thematic Mapper Plus (ETM+). [808] [809] [810]

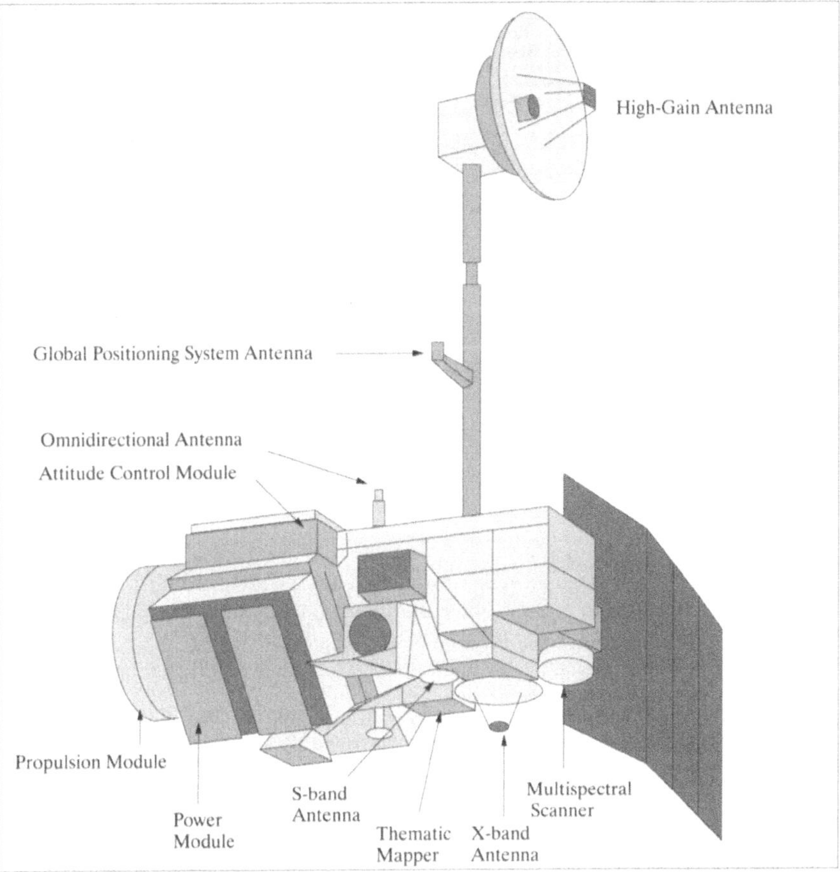

Figure 125: Model of the Landsat-4 and 5 spacecraft

Application: Land use, agriculture, forestry, geology, water resources, mapping, etc. Landsat data are particularly suited to long-term estimation and monitoring of standing vegetation biomass, biological productivity, and the movement of fragile ecosystem boundaries.

The purchase of Landsat imagery became rather expensive to the user community with S/C operations and data distribution in the hands of Eosat. NOAA and Eosat agreed in November 1990 that all Landsat data older than 10 years return to the "Public Domain" category, hence affordable for research. The "Public Domain" data is provided and distributed by the EROS Data Center (EDC), Sioux Falls, S D. All MSS data is in the public domain at EDC, effective as of February 1993.

808) "Taschenbuch zur Fernerkundung," F. Strathmann, Wichmann Verlag, 1990
809) Monitoring Earth's Ocean, Land, and Atmosphere from Space, Volume 97, AIAA, 1985, Chapter 3
810) A. F. Goetz, J. B. Wellman, W. L. Barnes, "Optical Remote Sensing of the Earth," Proceedings of the IEEE, Vol. 73, No. 6, June 1985, pp. 950-969

Orbit for LS-1 to LS-3: Sun-synchronous polar orbit (AM Orbit), altitude = 907- 915 km, inclination = 99.2°, period = 103 minutes, repeat coverage = 18 days.

Orbit for LS-4 and LS-5: Sun-synchronous polar orbit (AM Orbit), altitude = 705 km, inclination = 98.2°, Period = 99 minutes, repeat coverage = 16 days.

S/C	S/C Launch	Sensor Complement	Data Resolution (m)	Data Communications	Orbital Altitude	S/C Operator(s)	End of Service/ Status
LS-1 (ERTS)	Jul. 23, 1972	RBV MSS, DCS	80 80	DD (Direct Downlink) 2 WBVTR	907 km	NASA	Jan. 6, 1978
LS-2	Jan. 22, 1975	RBV MSS,DCS	80 80	DD with 2 WBVTR	908 km	NASA	1983
LS-3	Mar. 5, 1978	RBV MSS,DCS	30 80	DD with 2 WBVTR	915 km	NASA	Jan. 7, 1983
LS-4	Jul. 16, 1982	MSS TM, GPS	80 30	DD TDRSS	705 km	NOAA ('83)/ Eosat ('85)	standby Dec. 1993
LS-5	Mar. 1, 1984	MSS TM, GPS	80 30	DD TDRSS	705 km	NOAA ('84)/ Eosat ('85)	operational as of 2001
LS-6	Oct. 5, 1993	ETM	15 (PAN) 30 (MS)	DD with recorders	launch failure (contact lost during launch)		
LS-7	Apr. 15, 1999	ETM+	15 (PAN) 30 (MS)	DD with recorders	705 km	NOAA	

Table 203: Overview of Landsat S/C series

D.21.1 Landsat-1 to -5

The LS-1 satellite was a modified version of Nimbus-4, a three-axis stabilized Earth-oriented platform. An advanced attitude control system, consisting of horizon scanners, sun sensors, and a command antenna combined with a Freon gas propulsion system permitted the spacecraft's orientation to be maintained within plus or minus 0.7° in all three axes. Spacecraft communications included a command subsystem operating at 154.2 and 2106.4 MHz and a PCM narrow-band telemetry subsystem, operating at 2229.5 and 137.86 MHz, for spacecraft housekeeping, attitude, and sensor performance data. Video data from the three-camera RBV system was transmitted in both real-time and tape recorder modes at 2265.5 MHz, while information from the MSS was constrained to a 20 MHz RF bandwidth at 2229.5 MHz. Two WBVTR (Wideband Video Tape Recorder) were installed on LS-1 through LS-3. Each WBVTR was capable of recording 30 minutes of either 3.2 MHz video (analog) data from RBV, or 15 Mbit/s digital data from the MSS multiplexer. On LS-1 the WBVTR-1 lasted until July 1974, while WBVTR-2 operated for only 10 days.

S-band image data transmission characteristics: frequency = 2229.5 MHz (LS 1-3), 2265.5 MHz (LS 4-5); transmitter power = 10 W; MSS data rate = 15.06 Mbit/s.

Data: image size: 2583 lines x 5500 pixel (EDIPS format); image size: 2286 lines x 3600 pixel (Telespazio format).

The LS-1 to -3 satellites were modified Nimbus S/C. The different S/C had the following on-orbit dry masses: LS-1 = 1800 kg; LS-2 = 816 kg; LS-3 = 960 kg; LS-4 = 1407 kg; LS-5 = 1407 kg; LS-6 = 2750 kg. A new S/C design (2nd generation) was introduced with LS-4. The flight segment consisted of two major systems:

- The instrument module, containing the instruments together with the mission unique subsystems, such as the solar array and drive, the TDRS antenna, the wide-band module, and the global positioning system (GPS)
- The multimission modular spacecraft that contained the modularized and standardized power, propulsion, attitude control, and communications and data handling subsystems. The flight segment was designed with 3 years nominal lifetime.

The DCS systems of LS-1 to -3 were replaced by experimental GPS systems (DoD program) on LS-4 and LS-5. The GPS assembly on LS-4 operated in two phases. The first phase (approximately 90 days) was an experimental one to validate and calibrate the position and timing information provided by the GPS assembly. The second phase called for operational use of the GPS data by LS-4.

Sensor complement: [811] [812]

MSS (Multispectral Scanner built by SBRC of Hughes) on LS 1-5. MSS is an opto-mechanical scanning instrument. The sensor operates by repeatedly scanning a 24-element fiber-optic array (arranged in 6 x 4 elements) from west to east across the Earth's surface, the orbital motion provides the natural north-south scanning motion. A separate binary-number array is generated for each spectral band. Mirror scan rate: 13.6 Hz (74 ms period); telescope: 22.9 cm diameter, f/3.6, Ritchey-Chretien type telescope; size 53 x 58 x 127 cm; six detectors were employed in each of the four spectral bands; [813] bands 4 to 6 used photomultiplier tubes (PMT) as detectors, and band 7 used silicon photodiodes; instrument mass = 64 kg; power = 50 W; quantization = 6 bit; spatial resolution = 80 m across-track and 56 m along-track; swath width = 185 km; four spectral bands: 0.5 - 0.6 μm, 0.6 - 0.7 μm, 0.7 - 0.8 μm, and 0.8 - 1.1 μm; a gray-lamp sensor calibration is performed during every second retrace period of the scan mirror. Note: LS-1-3 orbits were at an average altitude of 908 km while LS-4 and LS-5 orbits were positioned at an altitude of 705 km. For reasons of compatibility the optics of LS-4 and LS-5 were adjusted to keep the spatial (cross-track) resolution at 80 m.

Channel designation on Landsat S/C		Spectral Range (nm)
LS-1 through LS-3	LS-4 and LS-5	
Channel 4	Channel 1	500 - 600
Channel 5	Channel 2	600 - 700
Channel 6	Channel 3	700 - 800
Channel 7	Channel 4	800 - 1100
Channel 8 (channel 8 only on LS-3 for test purposes until July 11, 1978)		10400 - 12600

Table 204: MSS spectral channels on Landsat series

RBV (Return Beam Vidicon Camera). This instrument was only flown on LS-1 to LS-3. It consisted of three coaligned television cameras, one for each spectral band. RBV measurements of reflected solar radiation were only conducted in daylight. The three earth-oriented cameras were mounted to a common base, which was structurally isolated from the spacecraft to maintain accurate alignment. Each camera contained an optical lens, a 5.08 cm RBV, a thermoelectric cooler, deflection and focus coils, a mechanical shutter, erase lamps, and sensor electronics. The cameras were similar except for the spectral filters contained in the lens assemblies that provided separate spectral viewing regions. The viewed ground scene, 185 by 185 km in area, was stored on the photosensitive surface of the camera tube, and, after shuttering, the image was scanned by an electron beam to produce a video signal output. Each camera was read out sequentially, requiring about 3.5 s for each of the spectral images. The cameras were operated every 25 s to produce overlapping images along the direction of spacecraft motion. Video data from the RBV were transmitted (2265.5 MHz) in both real-time and tape recorder modes.

A later variant of the RBV system for LS-3 utilized two cameras with panchromatic spectral response and higher spatial resolution (40 m) to complement the multispectral coverage provided by MSS. The vidicon camera was equipped with an electron tube in which the opti-

811) S. C. Freden, F. Gordon, "Landsat Satellites," Chapter 12 of 'Manual of Remote Sensing,' 2nd edition, Vol I, published by the American Society of Photogrammetry, 1983, pp. 517-570

812) A. M. Mika, "Three Decades of Landsat Instruments," PE&RS, July 1997, pp. 839-852

813) Note: the line array of six detectors was positioned in the along-track direction, thus providing an instantaneous parallel ground coverage of 336 m in one cross-track scan with the whiskbroom configuration. This wide along-track coverage permits sufficient integration time for all cells in each scan sweep.

cal image is projected onto a photoconductive detector scanned by the electron beam. Note: Starting with LS-4 the MSS bands were renumbered to bands 1 through 4 (from 4 through 7 on LS-1 to 3).

RBV Instrument	
Spectral bands (μm)	1) 0.48 - 0.58, 2) 0.58 - 0.68, 3) 0.70 - 0.83 (bands 1 through 3)
Spatial resolution	80 m
Three coaligned cameras	image of 185 km x 185 km (framing cameras)
Radiometric signal	Analog video signal transmission, 33 dB SNR in bands 1 and 2, 30 dB in band 3
MSS Instrument	
Spectral bands (μm)	4) 0.5 - 0.6, 5) 0.6 - 0.7, 6) 0.7 - 0.8, 7) 0.8 - 1.1 (bands 4 through 7)
Spatial resolution	80 m
Swath	185 km (continuous strip image)
Radiometric signal	Digital video signal transmission; 6 bit per pixel, linear coding;; logarithmic coding also available on bands 4), 5), and 6)

Table 205: Specification of RBV and MSS instruments

DCS (Data Collection System) on LS-1 through LS-3. Heritage of IRLS (Interrogation, Recording, and Location System- of Nimbus-3). An experimental system with the objective to collect and retransmit data from remote data collection platforms (DCPs). The on-board DCS initially serviced only a pilot group of only six DCPs, with user agencies procuring, instrumenting, and developing additional platforms according to their needs. The DCP uplink frequency to the Landsat S/C was at 401.55 MHz. The on-board DCS equipment, essentially a receiver, received and retransmitted data (at 2287.5 MHz, DCS data on 1.024 MHz subcarrier) to selected ground receiving stations. There was no signal multiplexing or data processing on the satellite. From a nominal orbital altitude of approximately 900 km, the S/C was capable of acquiring data from DCPs within a radius of approximately 3000 km from the subsatellite point, thus allowing data to be obtained from any remote platform at least once every 12 hours.

TM (Thematic Mapper, built by SBRC of Hughes).[814) 815)] TM is a multispectral mechanically scanning optical imager operating in the visible and infrared ranges. The instrument consists of the following subsystems: Scan Mirror Assembly, Telescope, Scan Line Corrector, Primary Focal Plane Array, Relay Optics, Cooled Focal Plane Array, and the Internal Calibrator. The telescope is of the type Ritchey-Chretien with 40.6 cm in diameter [primary mirror clear aperture diameter = 41.15 cm, secondary mirror baffle diameter = 15.7 cm, telescope clear aperture = 1056 cm^2, effective focal length = 243.8 cm (f/6)]. The scan line corrector compensates for the forward motion of the S/C, allowing the scan mirror to produce usable data (parallel scans) in both scan directions. The bidirectional scan and the use of detector arrays for each spectral band provides scan efficiency. Scan period = 7 Hz cross-track; FOV = $\pm 7.2^{\circ}$ (swath width = 185 km); IFOV = 42.5 mrad; overall instrument size of 2.0 m x 1.1 m x 0.7 m with a mass = 258 kg; power = 385 W (peak); quantization = 8 bit. Design life = 2 years (with a goal 3 years). Instrument inflight calibration is done at the start and end of each scan by using solar and lamp-based approaches (three lamps) for the solar reflective bands. On-board blackbodies are used to calibrate the thermal bands. TM instruments are operational since 1982 (first flight on LS-4).

Background: The Thematic Mapper got its name from the intended "thematic" applications of its data. Its images were used to produce maps tailored to different Earth-observation themes, such as agriculture, hydrology, geology, etc. In addition, the TM bands were chosen on the basis of a comprehensive analysis of spectral reflection features for a variety of vegetation types and surface minerals. Spectral classification accuracy was a key determi-

814) "Landsat-4 Data Users Handbook," USGS/NOAA, 1984

815) P. N. Slater, "Remote Sensing Optics and Optical Systems," Addison-Wesley, Reading, MA, 1980

nant for selecting the specific band edges and bandwidths. In this context, band number 7 (2.08 - 2.35 μm) was added to the TM specifications much later in the design process then the other bands. It was an extra for the geology community, therefore it required contract modifications.

Applications: LS TM data may be used in fields such as: global change research, agriculture, forestry, geology, resources management, geography, water quality, and oceanography.

- Band 1: Coastal water mapping, soil/vegetation differentiation, deciduous/coniferous differentiation, chlorophyll absorption
- Band 2: Green reflectance, peak of healthy vegetation, plant vigor
- Band 3: Chlorophyll absorption, plant type discrimination
- Band 4: Biomass surveys, water body delineation
- Band 5: Vegetation moisture measurement, snow/cloud differentiation
- Band 6: Plant heat stress, thermal mapping, soil mapping
- Band 7: Hydrothermal mapping, geology

Band No.	Bandwidth (μm)	Detectors	Resolution (m)	SNR (average)
1	0.45-0.52 (VIS, blue)	SiPD (16)	30	60
2	0.52-0.60 (VIS, green)	SiPD (16)	30	60
3	0.63-0.69 (VIS, red)	SiPD (16)	30	46
4	0.76-0.90 (NIR)	SiPD (16)	30	46
5	1.55-1.75 (SWIR)	InSb (16)	30	36
7	2.08-2.35 (SWIR)	InSb (16)	30	28
6	10.4-12.5 (TIR)	HgCdTe (4)	120	

Table 206: TM parameter definition (LS-4/5)

TM data image size: 185 x 172 km; 5760 lines x 6928 pixels. Transmission: frequency = 8215.5 MHz (X-band); data rate = 84.9 Mbit/s (246 MByte per scene). There is no capability for on-board recording of TM data. Data transmission from the S/C to the ground is in realtime via a network of licensed X-band stations. An alternate data transmission link is via TDRSS (Ku-band) to GSFC (Note: LS missions with TDRSS support do not provide an on-board recorder).

As of 2000 Landsat-5 is still able to acquire TM data. Landsat-5 has no operational TDRSS support. Data transmission to the ground is by X-band direct downlink only. On the other hand, Landsat-7 has been providing data since April 1999.

D.21.2 Landsat-6

The LS-6 payload[816)] was launched on October 5, 1993 with a Titan 2 booster from Vandenberg Air Force Base, CA. S/C builder and integrator: Martin Marietta Astro Space (formerly General Electric Astro Space), S/C procurement and management by Eosat. The S/C structure consists of aluminum with graphite struts. Hydrazine propulsion system. Single solar array with 1-axis articulation produces 1430 W, two NiCd batteries provide 100 Ah total. Data stored on-board using tape recorders for direct downlink to ground stations at 85 Mbps. LS-6 is three-axis stabilized, zero momentum with control to 0.01° using reaction wheels. - The satellite failed to achieve its orbit; communication with the satellite was never established. A formal review was conducted by NOAA to investigate the failure.

Orbit: Altitude = 705 km, polar sun-synchronous orbit, inclination = 98.2°, period = 99 min, repetition cycle (repeat coverage) = 16 days, equatorial crossing time: 9:45 AM. Nominal life = 5 years.

816) "Satellite Loss Raises Questions for Eosat's Future," Space News, October 11-17, 1993, p. 3

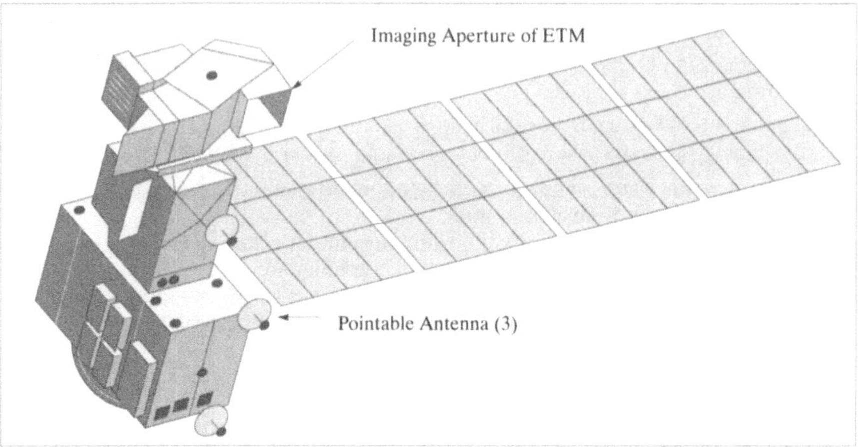

Figure 126: The Landsat-6 S/C model

Sensor:

LS-6 was designed to carry a single sensor, the Enhanced Thematic Mapper (ETM), which includes several new features to significantly improve data quality. Instrument mass (scanner assembly) = 288 kg, auxiliary electronic module = 81 kg, power = 490 W (max).

ETM (Enhanced Thematic Mapper)

LS-6 functional capabilities: Note: an attempted pushbroom version upgrade for the instrument failed due to severe budget constraints.

- ETM (seven TM bands, plus an additional panchromatic band). An instrument upgrade had the following features: Use of a single, monolithic silicon detector array for all the VNIR spectral bands. All detectors were on a common silicon substrate, this approach provided a better band-to-band geometric registration and stability (TM used four separate detector arrays).
- Two on-board recorders (playback to ground in X-band); each recorder is capable of recording/reproducing at a rate of 85 Mbit/s, and each can store $\approx$ 15 minutes worth of image data, or 29 scenes.
- Three pointable antennas
- Simultaneous acquisition of TM and PAN data
- ETM generates three different data streams:
  - Mode 1 Seven spectral bands
  - Mode 2 Panchromatic band (PAN) + channels 4, 5, and 6
  - Mode 3 Panchromatic band (PAN) + channels 4, 6, and 7

Spectral ranges:

- Channel PAN (8) : 500 - 900 nm, 15 m resolution; applications: cartography
- Channel 1: 450 - 520 nm (VIS, blue); applications: water penetration, bathymetry (water depth), chlorophyll absorption, distinguishes deciduous/coniferous forests
- Channel 2: 520 - 600 nm (VIS, green); applications: matches green reflectance, peak of healthy vegetation, assessment of plant vigor
- Channel 3: 600 - 690 nm (VIS, red); applications: chlorophyll absorption, plant type discrimination
- Channel 4: 760 - 900 nm (VNIR); applications: plant cell structure, plant vigor, complete absorption by water, shoreline mapping
- Channel 5: 1550 - 1750 nm (SWIR); applications: moisture content, soil mapping, thin cloud penetration

- Channel 7: 2080 - 2350 nm (SWIR); applications: hydroxyl ion absorption, geology
- Channel 6: 10.4 - 12.5 μ (TIR); applications: brightness temperature, soil moisture, plant heat stress.

Ground pixel resolution: 15 m (panchromatic), 30 m (channels 1-5, channel7) 120 m (channel 6). Ground image size: 185 x 185 km.

Band No.	Wavelength (μm)	Detectors	IFOV (μrad)	Ground Res. (m)
PAN (8)	0.50 - 0.90	SiPD (32)	18.5 x 21.3	13 x 15
1	0.45 - 0.52	SiPD (16)	42.5	30
2	0.52 - 0.60	SiPD (16)	42.5	30
3	0.63 - 0.69	SiPD (16)	42.5	30
4	0.76 - 0.90	SiPD (16)	42.5	30
5	1.55 - 1.75	InSb (16)	42.5	30
7	2.08 - 2.35	InSb (16)	42.5	30
6	10.4 - 12.5	HgCdTe (4)	170	120

Table 207: Summary of Landsat-6 ETM bandwidth specifications

Data: Landsat-6 has three X-band frequencies for downlink (8082.5, 8212.5, and 8342.5 MHz) to allow combinations of up to two real time data links (TM and/or a panchromatic mode) and two tape recorder dumps. The ETM outputs two 85 Mbit/s serial composite streams, each consisting of the digital image area, internal ETM calibration data, image-related timing data, S/C time code, ephemeris, and attitude data. The ancillary timing, calibration, and attitude data are used by the Landsat-6 Image Data Processing System S/W to perform geometric image corrections. The X-band downlink communications subsystem permits simultaneous downlinks to three stations using pointable antennas.[817)]

As of 1995 Landsat-5 is the only remaining Landsat system acquiring TM data (S/C operated by Eosat). Landsat-5 has no operational TDRSS support and no on-board storage. Data transmission to the ground is by X-band direct downlink only.

D.21.3 Landsat-7

The Landsat-7 satellite is part of NASA's ESE (Earth Science Enterprise) program. The mission is to extend and improve upon the long-term record of images of the Earth's continental surfaces provided by the earlier Landsat satellites. The S/C and payload were developed under NASA/GSFC management/procurement responsibility (Landsat Project Scientist: D. Williams). The LS-7 satellite was built by LMMS (Lockheed Martin Missiles and Space) at the facility in Valley Forge, PA. The S/C features the Landsat-6 bus; an on-board recorder in solid state memory (378 Gbit capacity to capture data beyond the range of ground receiving stations, recording rates of 150 Mbit/s, playback with 300 Mbit/s), and a single observation instrument: ETM+. The HRMSI instrument was removed from the S/C due to funding constraints when DoD withdrew from the "Landsat Program Management" team in 1994. The HRMSI instrument is kept in this text for reference only. A launch of Landsat-7 took place on a Delta 2 vehicle from VAFB on April 15, 1999.

The Landsat-7 S/C is very similar in design to the Landsat-6 S/C. It features three-axis stabilization with a pointing capability of 180 arcsec (3 sigma) and a pointing knowledge of 45 arcsec (1 sigma). Attitude control with four reaction wheels and two torque rods, attitude is sensed with a static Earth sensor, 2 magnetometers and gyros. S/C mass = 2200 kg, dimensions: 4.3 m in length and 2.8 m in diameter, power = 1550 W (about 1000 W average, provided by a silicon cell solar array and a nickel hydrogen battery power subsystem). S/C design life = 7 years. Landsat-7 communication:

- S-band (2 omni-directional antennas), 5 W, with real-time telemetry data rates of 1.2 kbit/s and 4.8 kbit/s, and 256 kbit/s of playback data, 2 kbit/s of command data. S-band frequencies of 2106.4 MHz (uplink) and 2287.5 MHz (downlink).

817) EOSAT Landsat Technical Notes, September 1992

- X-band (3 steerable antennas), 3.5 W; each antenna transmits data on two channels, with each channel carrying 75 Mbit/s (total of 150 Mbit/s per antenna); up to three separate links are supported. X-band frequencies: 8082.5 MHz, 8212.5 MHz, 8342.5 MHz.

Orbit: Sun-synchronous polar orbit (AM orbit), altitude = 705 km, inclination = 98.2°, period = 99 minutes, repeat coverage = 16 days (same as Landsat-5), a nominal 10 AM descending equator crossing time. The ground track is referenced to WRS (worldwide Reference System) with a repeat accuracy of ±5 km. [818]

A coordinated tandem orbit of Landsat-7 with EO-1 (a few minutes apart) is considered by NASA for reasons of data calibration and synergetic use of data.

Sensor:

ETM+ = Enhanced Thematic Mapper Plus (built by Raytheon Santa Barbara Remote Sensing, Goleta, CA. [819] [820] ETM+ is a whiskbroom scanning radiometer.[821] The principal functional differences between the ETM and the former TM series are the addition of a 15 m resolution panchromatic band and two 8-bit "gain" ranges. The ETM+ adds a 60 m resolution thermal band, replacing the 120 m band on ETM/TM (Band No. 6, see Table 207). Design life = 7 years. The ETM+ also includes a number of radiometric enhancements to achieve an absolute radiometric uncertainty of <5% (bands 1-4). Two new calibration devices were added: FAC (Full Aperture Calibrator), and PAC (Partial Aperture Calibrator). ETM+ uses three independent on-board calibration systems (plus preflight calibration), representing a significant step forward in absolute radiometric calibration accuracy.

- A full-aperture solar diffuser (FAC) on the inner surface of the aperture door that illuminates the focal planes with diffusely reflected solar energy when commanded into position
- A partial-aperture solar reflector (PAC) that illuminates the focal planes with attenuated solar energy, once per orbit
- Calibration lamps that project calibrated energy onto the focal planes via the main calibration shutter, once per scan, during the scan mirror turnaround.

Band No.	Wavelength (μm)	Detectors	IFOV (μrad)	GSD (m)	SNR (at min signal radiance)
8 PAN	0.50 - 0.90	SiPD (32)	18.5 x 21.3	13 x 15	15
1 VIS	0.45 - 0.52	SiPD (16)	42.5	30 x 30	32
2 VIS	0.53 - 0.61	SiPD (16)	42.5	30	35
3 VNIR	0.63 - 0.69	SiPD (16)	42.5	30	26
4 VNIR	0.78 - 0.90	SiPD (16)	42.5	30	32
5 SWIR	1.55 - 1.75	InSb (16)	42.5	30	25
7 SWIR	2.09 - 2.35	InSb (16)	42.5	30	17
6 TIR	10.4 - 12.5	HgCdTe (8)	85.2	60	0.5 K

Table 208: Landsat-7 ETM+ bandwidth specifications

818) K. Dolan, P. Sabelhaus, D. Williams, "Landsat-7 Extending 25 Years of Global Coverage," Proceedings of Information for Sustainability, 27th International Symposium on Remote Sensing of Environment, Tromsoe, Norway, June 8–12, 1998, pp. 622-625

819) B. L. Markham, et al., "Radiometric Calibration of the Landsat-7 Enhanced Thematic Mapper Plus," Proceedings of IGARSS '94, Volume IV, pp. 2004-2006

820) K. Thome, B. Markham, J. Barker, P. Slater, S. Biggar, "Radiometric Calibration of Landsat," PE&RS, July 1997, pp. 853-858

821) Note: The detector line arrays (16 for VNIR bands, 32 for PAN, and 8 detectors for TIR) of the whiskbroom scanner are oriented in the along-track direction. This arrangement provides a parallel coverage of 480 m along-track in one scan sweep (cross-track direction). The wide along-track coverage permits sufficient integration time for all cells in each scan sweep.

Scanning method	Bidirectional cross-track, scan frequency = 7 Hz
Swath width	185 km (15° FOV from 705 km orbit)
Telescope	40.6 cm aperture, Ritchey-Chretien
Instrument size	Scanner assembly: 1.5 m x 0.7 m x 2.5 m Auxiliary electronics module: 0.4 m x 0.7 m x 0.9 m
Instrument mass	Scanner assembly: 298 kg, AEM = 103 kg, cable harness = 20 kg
Power	590 W
Data quantization	9 bit A/D conversion, 8 bit/pixel transmitted (2 gain states)
Data rate	150 Mbit/s (2 x 75) by each of three directional X-band antennas, CCSDS format

Table 209: Some parameters of the ETM+ instrument

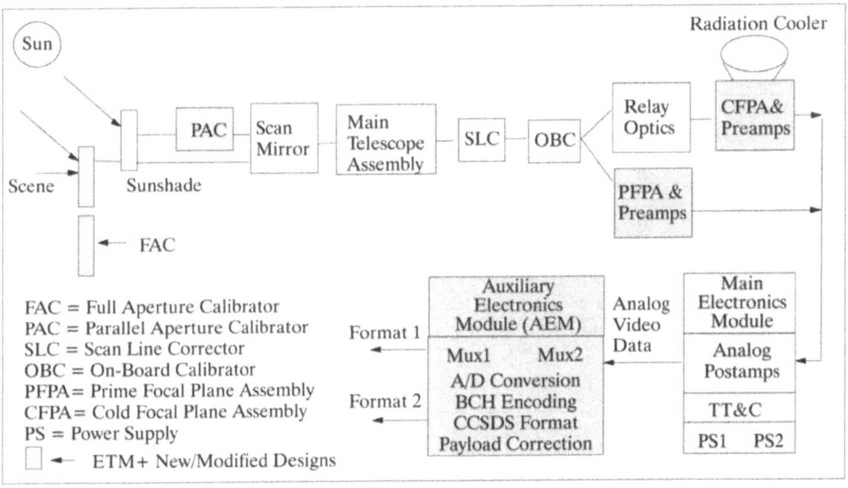

Figure 127: ETM+ block diagram

HRMSI = High-Resolution Multispectral Stereo Imager (built by SBRC - this sensor is no longer being built, it is simply here for reasons of reference). A pushbroom instrument providing stereo imaging at 5 m spatial resolution in the in-track direction (±30° from nadir); cross-track instrument pointing capability up to ±38°; swath width = 60 km; telescope diameter = 18 cm (unobscured reflecting triplet), size = 82.5 x 81.3 x 64.8 cm; instrument mass = 45.4 kg; power = 125 W (145 W when slewing); data rate (compressed) = 75 Mbit/s for panchromatic data and 75 Mbit/s for VNIR bands.

Band	Bandwidth (μm)	Detectors	Resolution (m)	SNR (average)
1	0.45-0.52	3050 SiPD	10	32
2	0.52-0.60	3050 SiPD	10	32
3	0.63-0.69	3050 SiPD	10	26
4	0.76-0.90	3050 SiPD	10	32
5 (PAN)	0.50-0.90	6100 SiPD	5	2

Table 210: HRMSI parameter specifications

LS-7 data handling policy:

Landsat-7 data is received and distributed to the user community by USGS (capturing and processing 250 Landsat scenes per day and delivering at least 100 of the scenes to users each day). NASA/GSFC performs on-orbit mission operations until until Oct. 1, 2000; after that responsibility for flight operations transfers to the USGS. The operating philosophy changed to the effect that ETM+ data covering the global continental surfaces are being archived in the USA. The ETM+ archive is continually being updated as data become available. This data policy differs from the past (Landsat-4 and -5), where data was only acquired from the S/C on the basis of customer requests. The new archiving policy will substantially increase the amount of data available to the user community. [822] [823] [824]

Landsat sensor	MSS (LS-1-5)	TM (LS-4/5)	ETM (on LS-6)	ETM+ (on LS-7)
Spectral bands (all bands in µm)	1) 0.5 - 0.6 2) 0.6 - 0.7 3) 0.7 - 0.8 4) 0.8 - 1.1	1) 0.45 - 0.52 VNIR 2) 0.52 - 0.60 VNIR 3) 0.63 - 0.69 VNIR 4) 0.76 - 0.90 VNIR 5) 1.55 - 1.75 SWIR 7) 2.08 - 2.35 SWIR 6) 10.4 - 12.5 TIR	P) 0.52 - 0.90 VNIR 1) 0.45 - 0.52 VNIR 2) 0.52 - 0.60 VNIR 3) 0.63 - 0.69 VNIR 4) 0.76 - 0.90 VNIR 5) 1.55 - 1.75 SWIR 7) 2.08 - 2.35 SWIR 6) 10.4 - 12.5 TIR	P) 0.52 - 0.90 VNIR 1) 0.45 - 0.52 VNIR 2) 0.53 - 0.61 VNIR 3) 0.63 - 0.69 VNIR 4) 0.78 - 0.90 VNIR 5) 1.55 - 1.75 SWIR 7) 2.09 - 2.35 SWIR 6) 10.4 - 12.5 TIR
Swath width	185 km	185 km	185 km	185 km
Spatial resolution	80 m	30 m VNIR/SWIR 120 m TIR	15 m PAN, 30 m VNIR/SWIR, 120 TIR	15 m PAN 30 m VNIR/SWIR 60 m TIR
Radiometric resolution	6 bit	8 bit	9 bit (8 bit transmitted)	9 bit (8 bit transmitted)
Band-to-band registration		0.2 pixel (90%)	0.2 pixel (90%)	0.2 pixel (90%)
Geodetic accuracy without ground control		500 m (90%)	1000 m (90%)	400 m (90%)
Data rate	15 Mbit/s	85 Mbit/s	2 x 85 Mbit/s	2 x 75 Mbit/s
Instrument mass	64 kg	258 kg	288 kg scanner, plus 81 kg AEM	318 kg scanner, plus 103 kg AEM, plus 20 kg cable harness
Average power	50 W	332 W	490 W	590 W
Telescope aperture	23 cm	40.6 cm	40.6 cm	40.6 cm

Table 211: Overview of Landsat series imaging instrument parameters

The Landsat-7 ground system design includes an Image Assessment System (IAS) to provide users with ancillary information needed to generate useful, radiometrically calibrated and geometrically corrected ETM+ digital imagery. Another aspect of the new data handling policy is that ETM+ data will be distributed from the archive in an essentially raw form. Users are responsible for the task of preprocessing their imagery (i.e. radiometric and geometric corrections). The price tag for Landsat-7 imagery is substantially lower than for the commercial products of Landsat-4 and -5.

The direct downlink service to a global network of existing Landsat ground stations (in X-band via each of three directional antennas) is maintained. All real-time image data (within view of a licensed ground station) are directly downlinked via the three X-band links. TDRSS may be used for TT&C data relays in S-band (backup). The prime (US) ground receiving station for the Landsat-7 archive is located at the EROS Data Center (EDC) in Sioux Falls, South Dakota. A second (US) reception facility near Fairbanks, AK, acquires coverage of Alaska and international coverage using the on-board recorder. An additional receiving station in Spitzbergen (Norway) provides backup reception. All data received at either Fairbanks or Spitzbergen will be shipped to EDC for archiving and distribution.

822) J. R. Irons, D. L. Williams, B. L. Markham, "Landsat-7 ETM+ On-Orbit Calibration and Data Quality Assessment," Proceedings IGARSS '95, Vol. II, pp. 1573-1575

823) W. C. Draeger, T. M. Holm, D. T. Lauer, R. J. Thompson, "The Availability of Landsat Data: Past, Present and Future," PE&RS, July 1997, pp. 869-875

824) R. A. Williamson, "The Landsat Legacy: Remote Sensing Policy and the Development of Commercial Remote Sensing," PE&RS, July 1997, pp. 877-885

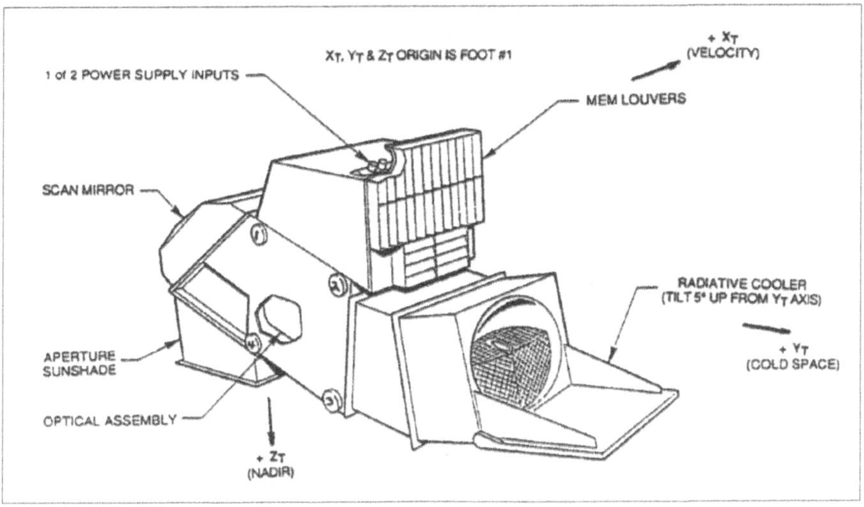

Figure 128: Schematic illustration of the ETM+ instrument

D.22 Lewis and Clark Missions

The Lewis and Clark[825)] satellite missions are part of the NASA SSTI (Small Spacecraft Technology Initiative) program which in turn has the following objectives:

- Demonstrate a new approach to technology integration that includes performance-based design specifications with the ability to incorporate commercial standards into the design and qualification process
- Proof the concept of a small spacecraft designed to envelope a range of mission requirements and develop standard hardware and software interfaced for various applications
- Enable low-cost common bus manufacturing concepts with large savings to future mission applications.

Within this framework NASA awarded in June 1994 two contracts to two industry teams, one led by TRW of Redondo Beach, CA and the other by CTA of McLean, VA, with the objective to design, build, launch and operate one spacecraft each, with a minimum set of requirements (no phase A or B design with the customer) and a contract-to-launch period of ≤ two years (each S/C < $ 60 million). A large, broad-based team of industry, small business, NASA centers and universities will use the data to develop commercial products and to stimulate further research. The SSTI concept required NASA to streamline its contract management and procurement procedures in many ways.

D.22.1 Lewis S/C

The Lewis S/C was designed and built by the team led by TRW, Redondo Beach, CA. The S/C features a small-satellite modular design, pulse tube cryocooler technology, solid-state recorder (2 Gbit); fiber-optic data bus; WFOV star tracker; magnetically suspended reac-

825) The satellite missions are named in honor of Meriwether Lewis (1774-1809) and William Clark (1770-1838), who headed the first overland expedition of about 40 persons (1804-06) to the Pacific coast and back, starting in St. Louis, Missouri. The expedition was initiated by President Thomas Jefferson, who wanted a first survey (information in the form of maps and diaries) of the territory west of the Mississippi acquired by the Louisiana Purchase in 1803 from France.

tion wheel; a lightweight structure with integrated thermal control; GPS attitude determination; three-axis stabilized (zero momentum bias); design life of 3 years, mass = 288 kg; power = 550 W. The S/C was launched successfully on August 23, 1997 with a Lockheed Martin Launch Vehicle (LMLV-1) from VAFB, CA. [826)]

The GNCS (Guidance Navigation and Control Subsystem) consists of the following components: 1) the coarse sun sensor (TOMS heritage) provides full view to locate the sun; 2) NSTA (Narrow-FOV Star Tracker) tracks up to six stars with an accuracy >8 arcsec; 3) TAM (Three Axis Magnetometer); 4) GRA (Gyro Reference Assembly) three tuned-rotor two-axis systems; 5) Earth sensor assemblies, used for roll/pitch determination; 6) GADFLY (GPS Attitude Determination Flyer), a NASA experiment consisting of the following hardware: four GPS antennas and pre-amplifiers, cross-strapped to two GPS Tensor (SS/L) receivers; 7) reaction wheels and magnetic torquers; 8) single axis solar drive array assemblies, used to continuously point the solar arrays at the sun; 9) valve drive electronics, the propulsion system is used for orbit raising. All functions are controlled by an on-board computer (R3000 CPU). [827)]

The initial nominal orbit was 300 km. On-orbit check-out commenced in preparation for its final orbit transfer maneuver. However, on Aug. 26 (three days after launch), an inflight anomaly led to the loss of attitude control and a discharged battery (contact was lost when Lewis entered a spin that disrupted the satellite's power-generating capability. The spin was thought to be caused by a stray firing of one of the attitude control thrusters). Controllers were unable to rescue to mission. The S/C was lost and reentered the atmosphere Sept. 28, 1997. The review board found that the S/C was lost as a direct result of the implementation of a technically flawed safe mode in the attitude control system, compounded by the limited control and monitoring of the S/C after launch.

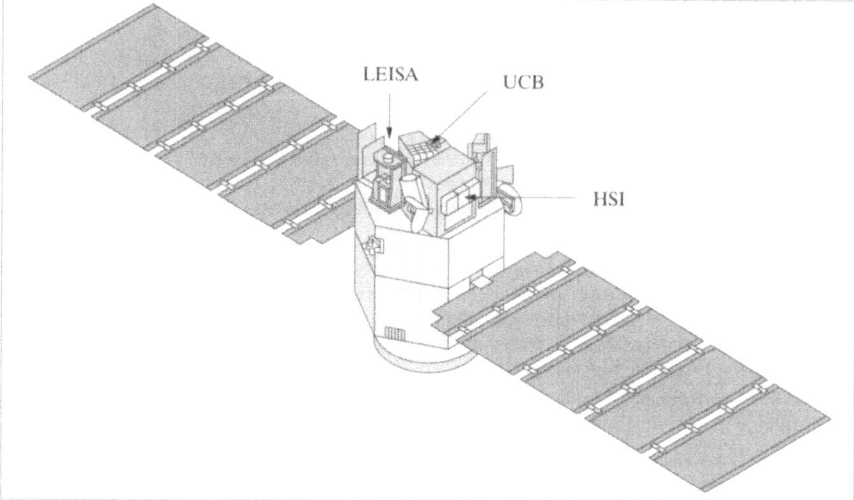

Figure 129: The Lewis S/C model

Applications: the S/C data may be used in a number of areas such as agriculture (cropstress management, crop maturity, etc.), forestry, land management, environmental assessment, resource exploration, etc.

826) Information provided by J. S. Pearlman and S. K. Manlief of TRW, Redondo Beach, CA

827) P. Parry, "The SSTI Lewis Better, Faster, Cheaper Guidance<Navigation, and Control Subsystem," Proceedings of the 10th AIAA/USU Conference on Small Satellites, Sept. 16-19, 1996, Logan, UT

Orbit: Sun-synchronous circular orbit, altitude = 523 km, inclination = 97.45°

Sensor Complement:

HSI (Hyper-Spectral Imager, TRW instrument)
The spaceborne HSI instrument has evolved from the TRWIS airborne sensor family. HSI is a nadir-pointing imaging spectrometer generating 384 spectral bands in the VNIR and SWIR regions (0.4 - 2.5 μm) and in addition a visible panchromatic image (the 384 spectral bands can be selected in any combination). The instrument has a minimum design life of 3 years (goal 5 years), it consists of three independent imaging systems providing parallel near-instantaneous image capture. HSI features:

- Foreoptics: 12.5 cm aperture, 1 m effective focal length), three mirror design. The optical design utilizes shared foreoptics to generate three line images separated slightly in field
- The spectrometers provide constant bandwidth
- Pushbroom dual-grating spectrometers, common design, 2 slits
- PAN (panchromatic visible channel): 2592 x 1 CCD, 10 μm pixels (the panchromatic channel allows for additional spatial sharpening of the 30 m hyperspectral images during ground processing)
- VNIR: 128 bands of 5 nm, 256 x 128 CCD array, 60 μm aggregated pixels. The CCD focal planes are cooled slightly, to 273 K, to reduce dark current noise.
- SWIR: 256 bands of 6.38 nm, 256 x 256 HgCdTe detector array, 60 μm pitch, 250 Hz frame rate, cooled at 115 K
- Mechanisms: cover/diffuser, pulse tube cryocooler for low noise in SWIR
- Electronics: FPA readout and A/D conversion, spectral band selection
- Instrument pointing: HSI can be pointed across-track up to ±20° from nadir
- Calibration: Cover in calibration position (Lambertian diffuser inside cover). Three current-controlled lamps illuminate the diffuser over the full dynamic range (0.8-2.5 μm); sun reflects off diffuser for solar calibration (0.4-2.5 μm); calibration of entire optical train; spectral calibration uses atmospheric absorption in SWIR, solar spectra in VNIR. The absolute radiometric accuracy of HSI is better than 6%.

Parameter	PAN	VNIR	SWIR
Spectral range	0.48 - 0.75 μm	0.4 - 1.0 μm	0.9 - 2.5 μm
No. of spectral bands	1	128	256
Spectral resolution	-	5 nm	6.38 nm
Spatial resolution (IFOV in μrad)	5 m (10)	30 m (60)	30 m (60)
Swath width (at 523 km altitude, nadir)	12.9 km	7.7 km	7.7 km
Saturation radiance (% albedo)	100	100	100
Quantization	8 bit	12 bit	12 bit
Radiometric accuracy			
Pre-launch (%absolute)	15	5	5
Lifetime drift on orbit (%)	5	2	2
Drift between on-orbit calibrations	1	0.5	0.5
Data selectivity	any combination of bands		
Maximum data rate	327 Mbit/s		
Instrument mass, power	23 kg (including cryocooler), 50 W		

Table 212: HSI instrument parameters

LEISA (Linear Etalon Imaging Spectrometer Array, NASA/GSFC sensor, PIs: D. C. Reuter and D. Jennings). The major science objectives are to map reflectance spectra of surface and atmospheric features (eg., vegetation types and health, snow/ice/cloud discrimination, possibly snow depth, grain size, age), industrial pollution (eg., oil spills on water), cloud studies (phase, particle seize), cloud editing for HSI, aerosols, and nightglow/aurora studies. LEISA is also a technology demonstration for a similar spectrometer planned for a future Pluto mission.

LEISA is a very low-mass, low-power infrared hyperspectral spectral imager; spectral region: 1 - 2.5 μm; spectral resolving power $\lambda/\Delta\lambda \geq 250$; nadir pixel size =300 m (IFOV). The instrument is based on a newly-developed spectral filter known as LVE (Linear Variable Etalon) which is placed directly over a 256 x 256 IR [NICMOS 3 technology (NICMOS= Near Infrared Camera and Multi-Object Spectrometer)] detector array (HgCdTe).[828] The active area of this array has a size of 1.024 cm by 1.024 cm. The LEISA instrument is being pointed with an optical pointing assembly and cooled with a pulse-tube cooler (both are provided by TRW).

An image is formed having 2-D spatial and 1-D spectral information content (an image cube with a 256 x 256 spatial dimension and 256 spectral images). A complete spectral image is formed by sweeping the array in a cross-track direction and by the forward motion of the S/C. It takes about 9 seconds to complete a full image cube of a given scene. Swath width = 77 km x yyy km in length, where yyy is determined by the on-board memory available (4 Gbit max) and by the amount of data compression used. The instrument has a steerable mirror allowing off-track viewing. Note: a complete spectral image consists of 256 frames - for an area of 77 km x 77 km this amounts to about 270 Mbit of data. Within the infrared region the LEISA data is regarded as complementary to HSI.

UCB (Extreme Ultraviolet Cosmic Background Explorer; built at the Space Sciences Laboratory of the University of California at Berkeley, PI: J. Edelstein). Objectives: Measurements of diffuse emission line spectra of EUV radiation with a sensitivity exceeding that of previous work by several orders of magnitude. Studies of physical conditions of matter in the local interstellar medium, cosmological theories of dark matter, observation of the solar-excited interstellar medium, measurement of energetic geocoronal plasma phenomena.

The instrument has a full bandpass coverage from 55 - 105 nm, with prime bandpass in the range from 58.5 - 95.0 nm. Spectral resolution in the prime bandpass of 0.5 nm, 3σ sensitivity to diffuse line emission ≤ 2000 photons/s/cm^2/sr. Observations are integrated over the orbital eclipse, viewing into the anti-sun direction. FOV = 8.4° x 25.6°.
The UCB design features the following technology: unique diffuse spectrograph optical design, metal substrate diffraction grating, lightweight ceramic-metal microchannel plate photon counting detector, energetic particle anticoincidence scintillator/PMT system, low noise detector charge amplifiers with particle overload peak restoration and charge-pulse anode particle cleaning.

Photons enter the optics chamber by way of a 0.15 mm x 60 mm slit. The slit is gated by a rotating filter-wheel mechanism. The filter wheel has four operational positions: 1) closed, 2) open, 3) MgF crystal, 4) Al thin film. Measurements of: a) detector background noise, b) low-energy EUV radiation signal, c) 1216 Å airglow radiation noise, d) high-energy EUV radiation signal. Light in the optics chamber encounters baffles, a diffraction grating, more baffles, and then the sensor. The slit, baffles and grids are polarized at low voltage to further reject charged particle noise.

Data: Downlink telemetry in S-band at a transmission rate of 2 Mbit/s to a number of ground stations (GSTDN compatible). This rather low data rate limits the operation of the HSI and LEISA instruments to selected areas only.

Title of Experiment	Short Description	PI / Agency
Data Compression	Demonstrate on-orbit compression of image data with and without loss	W. Miller/ GSFC
S/C Loads & Acoustic Measurements (SLAM)	Measure S/C environment during launch sequence	GSFC
Radiation & SEU Monitoring	Measure LET (Linear Energy Transfer) spectrum with gas proportional counter; Correl. With processor memory SEUs	J. Shinn/ LaRC

828) Note: The NICMOS3 array is being developed for the next-generation IR instruments for the Hubble Space Telescope.

Title of Experiment	Short Description	PI / Agency
GPS Attitude Determination (GADFLY)	Demonstrate a 0.15° attitude determination by interferometric measurement of the GPS signal	F. Bauer/ GSFC
Wide FOV Star Tracker	Demonstrate a 20° FOV star tracker with a 6-star capability	Hughes
Micromachined Accelerometers	Demonstrate the use of micromachined technology for S/C loads measurements	U. of Cincinnati
R-3000 Processor and 4-Gbit Solid State Recorder	Demonstrate on-orbit operation of R-3000 processor/SSR combination	TRW
Fiber Optic Data Bus	Demonstrate on-orbit operation of a high speed fiber optic data bus	HI/TRW
Pulse-Tube Cryocooler	Demonstrate on-orbit operation of a compact, low-vibration cryocooler	TRW
Optical Pointing Assembly	Demonstrate on-orbit operation of a precision steerable mirror	TRW
Advanced Solar Cell Experiment (ASCE)	Measure on-orbit performance of multi-junction GaAs and amorphous silicon solar cells	E. Gaddy/ GSFC/TRW
Metal Matrix Composition Heat Strap	Measure on-orbit performance of graphite/aluminum matrix for thermal control	G. Castro/ GSFC
Magnetically Suspended Reaction Wheel	Demonstrate on-orbit performance of magnetically suspended reaction wheel	TRW
Autonomous Orbit Control	Demonstrate on-orbit autonomous maintenance (ΔV) capability (calculation only)	Microcosm
Cloud/Feature Identification	Use HSI and LEISA data to develop algorithms for cloud, snow, etc. identification	R. Davis/ LaRC
Enhanced ACS Experiment	Demonstrate on-orbit MIMO (Multi-Input/Multi-Output) attitude control S/W	P. Maghami/ LaRC
On-Orbit Identification Experiment	Assess S/C dynamic performance	K. Elliot/ LaRC
EPS Regulator	Measure on-orbit performance of a regulator designed for small satellites	LeRC
Advanced Packaging Experiment	Measure on-orbit operation of RH-32 processor using MCM (Multichip Module) technology	JPL/TRW

Table 213: Technology demonstrations on Lewis S/C

D.22.2 Clark S/C

The Clark S/C is designed and built by a consortium headed by CTA Space Systems of McLean, VA. The S/C has an integrated graphite structure and two solar array wings driven by a solar array drive. Attitude knowledge is on the order of 5 arc-seconds. GPS is used for autonomous orbit determination and for coarse attitude knowledge (GADFLY). A new coarse sun sensor is used for sun pointing. 829) A star tracker provides accurate attitude knowledge; a gyro is used to interpolate between star fixes. The S/C is three-axis stabilized (0.5° to 2° of control and 0.03° to 0.15° of attitude knowledge), mass = 305 kg (including fuel), power = 235 W (orbit average), propulsion = hydrazine blowdown, design life = 1 year for mission with a 3-year life of the S/C. A launch was projected for June 1996 aboard an LLV1 launch vehicle from VAFB, CA.830)

Note: At the end of Feb. 1998, NASA cancelled the Clark mission due to severe cost overruns and launch delays. NASA will keep the S/C and its instruments and use them for other projects. 831)

The Clark S/C represents a demonstration of 36 advanced technologies, among them: minimizing of ground processing for instrument data, on-board data compression with lossy algorithms, on-board imaging processing, use of new detectors in space (silicon and CdZnTe), 3-D tomographic mapping of the atmosphere, etc.

829) J. Benton, "Pyramyd Coarse Sun Sensing for NASA SSTI Clark Safe-Hold Mode," Proceedings of the 10th AIAA/ USU Conference on Small Satellites, Sept. 16-19, 1996, Logan, UT

830) Information provided by J. Jacobi of CTA, McLean, VA and by R. J. Hayduk of NASA/HQ, Washington, DC

831) A. Lawler, "Faster, Cheaper, Better is Also Harder," Science, Vol. 29, March 6, 1998

Orbit: Sun-synchronous circular orbit, altitude = 496 km, inclination = 97.3°, period = 92.25 minutes, nonrepeating ground track, nodal crossing time = 11:25 AM descending node, revisit time = 4.7 days for off-nadir and 20 days for nadir.

Data: An on-board digital recorder is employed with a storage capacity of 172 MByte. Downlink imagery communications via X-band at 8.03-8.04 GHz and a data rate of 25 Mbit/s; imagery downlink to mobile users in S-band at 2300 MHz and a data rate of 500 kbit/s; the TT&C data are in UHF-band at 402 MHz and a data rate of 19.2 kbit/s.

CTA and Earthwatch provide the ground stations for telemetry reception and S/C control. CTA's master control station is located in McLean, VA; Earthwatch's stations are in Fairbanks, AK, Tromsö, Norway, and Longmont, CO. Image data collection and processing is provided by Earthwatch.

Clark is designed to support proof-of-concept technology demonstrations rather than production operations for the acquisition and distribution of large quantities of image data. Archived imagery may be obtained from the EROS Data Center, Sioux Falls, SD.

Sensor complement:

Earthwatch Imager (built by Earthwatch of Longmont, CO). The instrument is the prime S/C sensor with the objective to provide high-resolution panchromatic and multispectral imaging data (simultaneously) of the Earth's surface. The imager employs a staring focal plane array technology and features a gimballed mirror with a 30° pointing capability from nadir into any direction. The registration of the panchromatic and multispectral scenes is fixed, but it is possible to take first a multispectral scene and then a panchromatic scene, so in effect the panchromatic scene can be placed anywhere within the multispectral scene. The cameras may also be used for stereo imaging by slewing the gimbal mirror fore and aft in the S/C flight direction. Images are stored on-board in a solid state memory and dumped at 25 Mbit/s to one of several ground stations upon overflight (store-and-forward operation). The downlink is encrypted. The on-board processor provides real-time radiometric/ geometric calibration and image compression for all imaging data.

Parameter	Panchromatic Camera	Multispectral Camera
Spectral range	0.45 - 0.80 μm	0.50-0.59, 0.61-0.68, 0.79-0.89 μm
Spatial resolution (GSD)	3 m at nadir, 3.5 m off-nadir	15 m at nadir, 17.3 m off-nadir
Footprint (or scene)	6 km x 6 km per exposure	30 km x 30 km per exposure
Swath width	12 km in strip mode (note: with step stare, swath width is variable so a wider swath may be collected over short intervals)	
Instrument (mirror) pointing	±30° (along-track/across-track) providing a 560 km wide field of regard	
Stereo imaging	maximum base/height ratio = 1	
Data quantization	8 bit	
Array size	4 10^6 pixel	4 10^6 pixel (x 3)
Data size/scene	4 MByte (uncompressed)	12 MByte (uncompressed)
Data volume	~50 scenes per day	

Table 214: Performance characteristics of the Earthwatch Imager

MicroMAPS (Micro-Measurement of Air Pollution from Satellites), built by Resonance Ltd of Alliston, Ontario, Canada, PI: G. Walberg. The miniaturized sensor got its name from the much larger MAPS instrument which was flown on Shuttle missions STS-2 (1981), STS-59 (April 1994), and STS-68 (Oct. 1994). The objective of MicroMAPS is to measure the CO content of three altitude bands in the troposphere and to provide N_2O cloud mapping. The instrument provides the following capabilities:

- Nadir-viewing gas filter radiometer
- Operation in 4.67 μm CO band
- Simultaneously views of several spectral lines
- High spectral resolution with high throughput

- Internal comparison of incoming radiation to that seen through gas cells containing gases of interest
- Obtaining mixing ratios of CO and N_2O (for cloud editing)

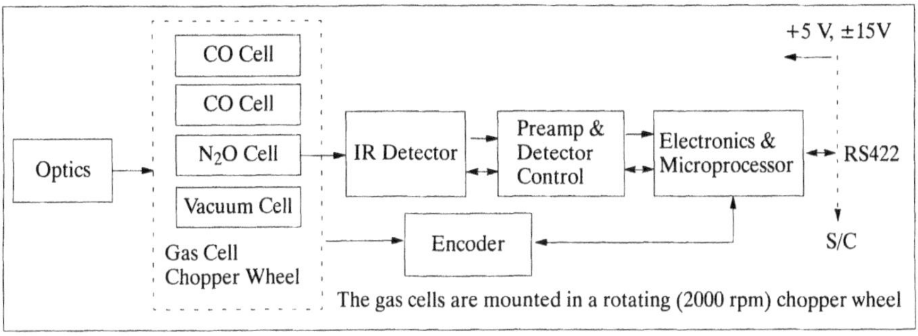

Figure 130: Functional diagram of MicroMAPS

Parameter	Value	Parameter	Value
Telescope aperture	5.2 cm^2	Element/material coating	Ge/AR coated >99% T at 4.67 μm
Telescope FOV	2.4°	Etendue (AΩ)	7.2 10^{-3} cm^2 sr
Optical magnification	27	Gas cell clear aperture	2 cm circular
Ray collection efficiency	>90%	Beam size at gas cell chopper	0.5 cm (approx. square)
No. of elements	5	Detector size/type/temp	0.75 mm^2/ PbSe/ -70°C

Table 215: Parameters of MicroMAPS

Instrument	MAPS (on Shuttle)	MicroMAPS (on Clark)
Spectral band	2080 to 2220 cm^{-1} (4.5 - 4.8 μm)	4.67 μm band
Resolution (NER)	1.7 10^{-7} W cm^{-2} sr^{-1} in the rad. channel 5.7 10^{-9} W cm^{-2} sr^{-1} in the difference channel	
FOV	5°	1°
No. of altitude bands	1	3
Swath width	20 km (corresponds to spatial resolution)	
Data rate	43 bit/s	9600 bit/s
Data storage	19 kByte	512 kByte
Earth coverage	sparse, ±57° latitude	continuous global
Mission duration	7 to 14 days	≥1 year
Sensor mass	40.4 kg	6.13 kg
Power	65 W	12 W

Table 216: Performance parameters of MAPS versus MicroMAPS

XRS (X-Ray Spectrometer), built by NASA/GSFC, PI: R. Starr. The objective is to demonstrate state-of-the-art detector technology using two new solid-state sensors that operate near room temperature: APD (Avalanche Photodiode) and CdZnTe (Cadmium, Zinc,Telluride) detectors. The instrument is mounted for the zenith-pointing direction in order to view the sun each orbit for about 10-15 minutes. Energy range = 2-25 keV (APD), 10-80 keV (CdZnTe), resolution = 1 keV (ADP), ~6 keV (CdZnTe), FOV = ±20° in transit direction, instrument mass = 5.2 kg, power = 5 W, detector array size = 8 APD and 12 CdZnTe, each 8 mm x 8 mm.

ATOMS (Atmospheric Tomography Experiment), built by IsComp Systems Inc., PI: J. Butts. Three-dimensional atmospheric tomography is an absorption technique used in identifying, monitoring, tracking, and modeling air pollutants generated by industry or in large population areas. In an operational system, a laser mounted on a satellite transmits a beam through the atmosphere to the ground where a network of retroreflectors reflect the beam back to the satellite receiver. Multiple-pass-monitoring over dedicated sites enables the generation of 3-D images of selected atmospheric constituents based on long-path absorption measurements. The resolution of these 3-D images depends on the number of retroreflectors on the ground and on other system parameters (see also RIS in the ADEOS sensor complement, D.1).

ATOMS is a low-cost proof-of-concept demonstration program on Clark employing the atmospheric tomography technique in an 'upside-down' fashion. This means the corner reflector is on the S/C rather than on the ground; the experiment uses a ground-based laser which is reflected off the S/C corner reflector. Two-dimensional maps (planes) of trace gas measurements are generated during each pass of the S/C over the ground station. The ATOMS experiment uses the AMOS (Air-Force Maui Optical Station) site, located at the summit of Mount Haleakala on the island of Maui, Hawaii, to take advantage of existing laser tracking mounts and receivers. Measurements will be made with the DIAL (Differential Absorption Lidar) technique and by exploiting the Doppler shift of the reflected beam due to the varying range rate of the satellite pass. The abundance of CO is measured using a frequency-doubled CO_2 laser at 4.65 μm, matching the strong CO absorption line. Absorption measurements are obtained along many different lines-of-sight, receiving density profiles of CO versus altitude.

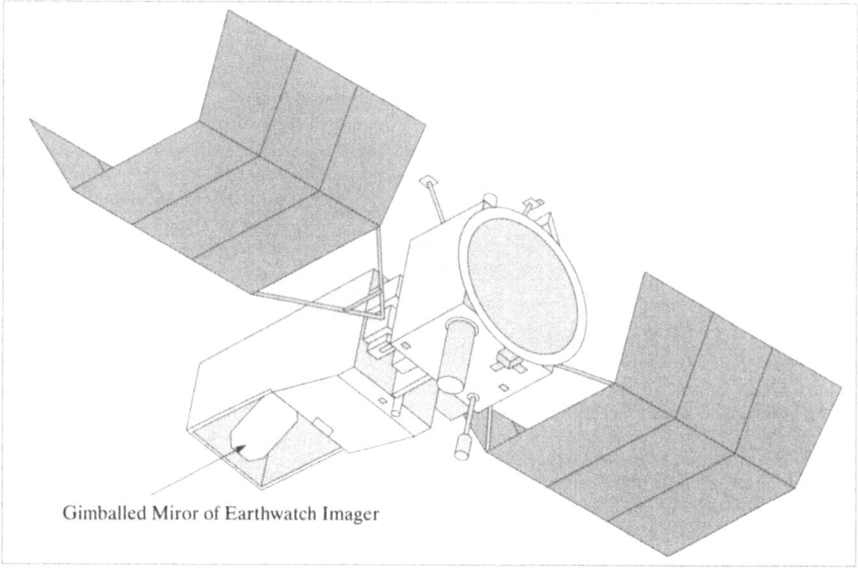

Figure 131: The Clark S/C model

D.23 MOS (Marine Observation Satellite)

MOS = Marine Observation Satellite. MOS-1 is Japan's (NASDA) first experimental Earth observation mission, also referred to as Momo-1 (peach tree). MOS-1 has been operational since Feb. 1987 (launch Feb. 19, 1987). The MOS-1 S/C ceased mission operations in April 1995. ESA/Earthnet is a data distributor of MOS-1.

The follow-up mission was **MOS-1B**, launch: Feb. 7, 1990. The operational life of MOS-1B mission was up to April 17, 1996.

Application: Observation of ocean surfaces (color), vegetation, land ecology, measurement of water vapor in the stratosphere, measurement of surface temperatures, etc.

The MOS S/C (NEC Corp. prime contractor) is three-axis stabilized with momentum wheels and hydrazine thrusters, consisting of a box-shape bus (1.3 m x 1.5 m x 2.4 m) with a single solar array (2.0 m x 4.5 m), its mass is 745 kg. Design life = 2 years.

Orbit: Sun-synchronous with a repeat cycle of 17 days; inclination = 99.1°; altitude = 908 km;. period = 103 min; local sun time = 10:15 for MOS-1 and 10:33 for MOS-1B.

Sensor complement: (MOS-1 and MOS-1B)

- **MESSR** (Multispectral Electronic Self-Scanning Radiometer) two instruments are installed; 50 m resolution on ground, measurement in four channels, visible and near-infrared ranges. Bands: 0.51-0.59 µm, 0.61-0.69 µm, 0.73-0.80 µm, 0.80-1.10 µm. Swath width = 100 km (185 km when both camera systems are operating). MESSR is a CCD array instrument. The swaths of both instruments are slightly overlapped (15 km) providing stereo viewing in this region. Instrument mass = 70 kg (one unit).
- **VTIR** (Visible and Thermal Infrared Radiometer, mechanically scanning mirror instrument); resolution = 900 m on the ground in the visible range and 2700 m within the three infrared ranges. Bands: 0.5-0.7 µm, 6.0-7.0 µm, 10.5-11.5 µm, 11.5-12.5 µm. Swath width = 1500 km for all bands. Instrument mass = 25 kg.
- **MSR** (Microwave Scanning Radiometer, mechanical); MSR is a microwave sensor measuring (scanning) the Earth's surface in flight direction. Frequency bands: 23.8 GHz, 31.4 GHz. Resolution: 23 km (at 31 GHz) and 32 km (at 23 GHz). Swath width = 320 km. Instrument mass = 54 kg.

Data:

- MESSR = 9 Mbit/s
- VTIR = 0.8 Mbit/s
- MSR = 2 kbit/s

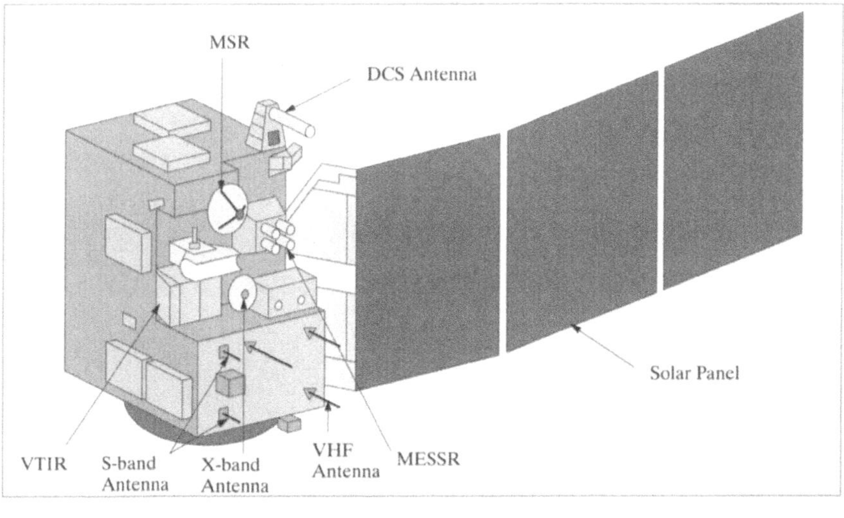

Figure 132: The MOS-1B S/C model

D.24 MTI (Multispectral Thermal Imager)

MTI is a US DOE-funded (Department of Energy) satellite mission, an R&D program and a demonstration mission of DOE/NN (Nonproliferation and National Security) with the objective to develop a broad range of new technologies, and to demonstrate the efficacy of highly accurate multispectral imaging for passive characterization of industrial facilities and related environmental impacts from space. The MTI program is managed by SNL (Sandia National Laboratories), LANL (Los Alamos National Laboratory) and SRTC (Savannah River Technology Center), representing DOE National Laboratories. Other program participants include: USAF Space Test Program (launch provision), AFRL (Air Force Research Laboratory), and universities. Major industrial partners are: Raytheon Optical Systems, Ball Aerospace, SBRC (Santa Barbara Research Center), and TRW. [832) 833)]

The MTI satellite, with the bus built by Ball Aerospace, is three-axis stabilized. S/C dimensions (stowed for launch) are 1.4 m diameter and 2.4 m in height. The S/C attitude is measured by multiple sun sensors, a horizon sensor, and gyros. Actuators are reaction wheels and torque rods. A pointing accuracy of 0.25° is provided. The S/C is agile and can be body-pointed up to ±50° in all axis. This feature provides a FOR (Field of Regard) of ±200 km for cross-track coverage (the nominal swath width of the imager is 12 km). The body-pointing feature may also be used for along-track stereo imaging support. The S/C mass is 650 kg, average total power of 575 W, with 11 battery cells (plus one spare) for ecliptic operations support. The S/C design life is three years. A launch of MTI on a Taurus vehicle from VAFB occurred on March 12, 2000.

Orbit: Sun synchronous polar orbit, altitude = 575 km, inclination = 97.52°, with a local crossing time of 13:00 on ascending node. The S/C has no orbit maintenance capability, resulting in an orbit decay over time.

Figure 133: Illustration of the MTI satellite

832) P. G. Weber, B. C. Brock, A. J. Garrett, et al., "Multispectral Thermal Imager Mission Overview," Proceedings of SPIE, Imaging Spectroscopy V, Vol 3753, Denver , CO, July 19-21, 1999, pp. 340-346

833) http://nis-www.lanl.gov/nis-projects/mti/

Communication: Downlink data transmission in X-band. The CCSDS protocol is used for uplink and downlink communication. S/C operations are performed at SNL (ground station). The DPAC (Data Processing and Analysis Center) function is provided by LANL.

Sensor complement:

MTI (Multispectral Thermal Imager). [834] [835] The overall instrument design objective is to support and evaluate advanced multispectral and thermal imaging, image processing and associated technologies. The instrument consists of the following elements: a telescope assembly, a cryogenically cooled focal plane with 15 linear spectral-sensitive pushbroom detector arrays, a complex built-in calibration hardware designed to collect radiometrically calibrated images with accuracies currently achievable only in the laboratory, and associated readout and control electronics. MTI features 15 spectral bands, ranging from the visible to the thermal infrared (TIR). The nominal FOV = 12 km x 12 km.

Imaging of a nominal nadir ground sample time in the along-track direction takes 715 μs for bands A-D, and 2.86 ms for bands E-O. The integration times for all bands are variable, up to 12 μs less than the pixel readout rate. A single detector array scans a 12 km image in about 1.7 seconds. The detectors are staggered in the along-track direction, so the system images a given ground point at slightly different times for each spectral band. A single 12 km x 12 km image requires about 4.5 s. - An average revisit time of one week for all target areas considered is provided with the body-pointing feature of the S/C.

Instrument calibration: The strategy is based on accurate pre-launch calibration in the LOs Alamos Radiometric Calibration Facility, plus on-orbit maintenance using sources built into the payload plus vicarious calibrations. The built-in calibration system uses instrumented ground sites, the sun, the moon and cold space looks to maintain long-term calibration. A more limited limited calibration sequence is performed before and after each imaging sequence. [836]

The spectral coverage of MTI permits daytime and nighttime observations. The MWIR and TIR bands are in particular suited for nighttime measurements. During its three-year demonstration mission, MTI records periodically images of participating government, industrial, and natural sites, mostly within the continental US. These sites are instrumented to record ground-truth data by a team of scientists and technicians from SRTC (Savannah River Technology Center) and by site owners and operators.

Spectral band	Spectral range (μm)	GSD (Ground Sample Distance)	Radiometric accuracy (%)	NER W/(cm^2 sr μm)
A	0.45 - 0.52 VIS	5	3	4.4 10^{-5}
B	0.52 - 0.60 VIS	5	3	2.8 10^{-5}
C	0.62 - 0.68 VIS	5	3	3.2 10^{-5}
D	0.76 - 0.86 VIS	5	3	2.3 10^{-5}
E	0.86 - 0.90 NIR	20	3	3.0 10^{-5}
F	0.91 - 0.97 NIR	20	3	6.0 10^{-6}
G	0.99 - 1.04 NIR	20	3	1.7 10^{-5}
H	1.36 - 1.39 SWIR	20	3	5.2 10^{-7}
I	1.55 - 1.75 SWIR	20	3	4.8 10^{-8}
J	3.50 - 4.10 MWIR	20	1	1.3 10^{-7}
K	4.87 - 5.07 MWIR	20	1	6.6 10^{-7}
L	8.00 - 8.40 TIR	20	1	4.4 10^{-6}
M	8.40 - 8.85 TIR	20	1	3.2 10^{-6}
N	10.2 - 10.7 TIR	20	1	3.9 10^{-6}
O	2.08 - 2.35 SWIR	20	3	1.2 10^{-6}

Table 217: Spectral coverage of MTI bands

834) R. Rex Kay, S. C. Bender, T. D. Henson, D. A. Byrd, et al., "Multispectral Thermal Imager (MTI) Payload Overview," Proceedings of SPIE, Imaging Spectroscopy V, Vol 3753, Denver , CO, July 19-21, 1999, pp. 347-358

835) T. Henson, S. Bender, W. Rappoport, et al., "Multispectral Thermal Imager Optical Optical Performance and Integration of the Flight Focal Plane Assembly," SPIE Vol. 3753, Denver , CO, July 19-21, 1999, pp. 359-368

836) W. B. Clodius, et al., "MTI On-Orbit Calibration," SPIE Vol. 3753, Denver , CO, July 19-21, 1999, pp. 380-391

Optical telescope assembly	36 cm aperture, off-axis TMA (Three Mirror Anastigmatic), f/3.5, FOV = 1.38° in the cross-track direction, and 1.82° in along-track
FPA (Focal Plane Assembly)	Consists of SCAs (Sensor Chip Assemblies), optical filters, and a vacuum enclosure; six SCAs are mounted on 3 motherboards; 3 SCA pairs are needed to provide cross-track coverage in all bands
Detectors at 75 K	Monolithic silicon PIN diodes for VNIR bands Backside illuminated photovoltaic InSb for NIR/SWIR/MWIR bands Backside illuminated photovoltaic HgCdTe for TIR bands
Cryogenic cooler	Pulse-tube cryogenic cooler (>3 W of cooling at 65 K)
Calibration sources	Blackbody, QLCS (Quick Look Calibration Source), QTH (Quartz Tungsten Halogen)
Readout/control electronics	12 bit data quantization, 266 Mbit/s of source data rate (all bands), Image data compression by USES (Universal Source Encoder for Science data) chip for lossless compression (Rice coding algorithm). A compressed 2-look image is about 500 Mbit with a 2.5:1 compression ratio
Mass storage unit	4.1 Gbit capacity
Instrument mass, power	240 kg, 350 W

Table 218: Performance parameters of the MTI instrument

HXRS (Hard X-Ray Spectrometer). The instrument is a joint endeavour by the Astronomical Institute of the Academy of Sciences of the Czech Republic and NOAA/SEC (Space Environment Center) in Boulder, CO. The instrument was built in Prague by Space Devices, Ltd. HXRS has the following objectives: a) to record with a high time-resolution a rare species of solar flare associated with high-energy proton storms known to damage satellites and potentially endanger astronauts, b) to evaluate the efficiency of this type of instrument to predict interplanetary proton events, and c) to test the effectiveness of new shielding methods. NOAA hopes to obtain information needed to design a system capable of forecasting such storms. [837] [838]

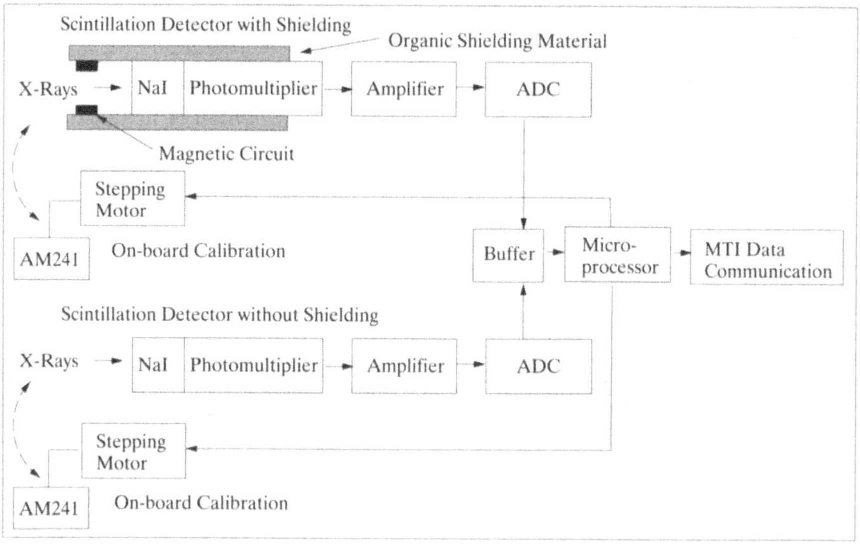

Figure 134: Schematic illustration of HXRS instrument components

837) F. Fárnik, H. Garcia, A. Kiplinger, "Solar Broad-band Hard X-Ray Spectrometer Onboard the MTI Satellite," Proceedings of 'A Crossroads for European Solar & Heliospheric Physics Conference,' Tenerife, March 23-27, 1998, pp. 305-308, ESA SP-417

838) H. A. Garcia, F. Fárnik, A. L. Kiplinger, "Hard x-ray spectroscopy for proton flare prediction," Proceedings of the SPIE Conference on Missions to the Sun II, San Diego, CA, July 1998, Vol. 3442, pp. 210-216

HXRS measures hard X-ray emissions of the sun in eight energy bands with time resolutions of up to 200 ms. The instrument consists of two identical scintillation detectors, one with shielding and one without, mounted side by side. Shielding is provided by a strong permanent magnet and a thick layer of plastic material. Each detector consists of a NaI crystal (25 mm diameter, 4.5 mm thick) coupled with a photomultiplier (Hamamatsu). Note: NaI (Tl) is a crystal of sodium-iodide activated by thallium. Such crystals are commonly used as X-ray detectors (coupled with photomultipliers).

Each detector is automatically calibrated in-flight by means of a radioactive source (AM241) on a movable arm. A 16-bit microprocessor controls all instrument functions and operations. The instrument data rate depends very much on the solar activity, i.e. on the number of hard X-ray events. During an event when all energy bands register with 200 ms time resolution, there are 10 channels that are read at 5 times/s (or 50 words/s). During quiet periods the data rate is about 1/10. The HXRS instrument mass is 7.695 kg, the power consumption is 5 - 6.5 W.

Incident X-ray photons create optical pulses in the crystals, they are converted into electrical signals and amplified by the photomultipliers. Pulse height (amplitude) discriminators measure the signal which is proportional to the incident photon energy. The number of pulses in each energy band counted per observation time interval is registered and send to the telemetry processor.

The MTI S/C spends a major portion of its time in standby mode (i.e., when the MTI instrument is not observing); during standby the S/C is always sun-pointing within 1° to collect solar energy. HXRS observations coincide with this standby mode. [839)]

Band Number	Energy Range (keV)	Max. Counts	Band Number	Energy Range (keV)	Max. Counts
S0	12.6 - 19.0	65535	S5	100.2 - 147.2	2047
S1	19.0 - 29.0	32767	S6	147.2 - 219.5	1023
S2	29.0 - 44.0	16383	S7	219.5	511
S3	44.0 - 67.2	8191	S8	249.5	255
S4	67.2 - 100.2	4095	SA	12.6	65535

Table 219: Definition of the HXRS energy bands

The "S" in "Band Number" of Table 219 signifies a "shielded detector." In parallel, the same energy bands are also part of the other (non-shielded) detector of the instrument.

D.25 NEMO (Navy EarthMap Observer)

The NEMO program (started in 1997) is a cooperative US government/industry satellite program, sponsored by ONR (Office of Naval Research) and DARPA as well as commercial investments, between the Navy's NRL (Naval Research Laboratory) on the government side as the program manager in partnership with commercial developers, a consortium led by STDC (Space Technology Development Corporation) of Alexandria, VA. In 1999, STDC was acquired by Earth Search Sciences Inc. (ESSI) of Kalispell, MT. [840)]

The overall objective is the development/demonstration of hyperspectral technology and the collection/processing of moderate-resolution hyperspectral imagery for military and commercial use. The emphasis is on broad-area, synoptic, and unclassified hyperspectral imagery. The prime military interest is to obtain imagery of the coastal regions on a global scale (littoral modeling, shallow water bathymetry, topography, bottom type composition,

839) http://www.asu.cas.cz/english/new/HXRS_descr.htm

840) T. Wilson, C. Davis, "Naval EarthMap Observer (NEMO) Satellite," Proceedings of SPIE, Vol. 3753, Denver, CO, July 19-21, 1999, pp. 2-11

detection of underwater hazards, water clarity and visibility, etc.), while the civil needs call for imagery supporting such applications as land use management, agriculture, environmental studies, and mineral exploration. Contract agreements call for STDC and its industry partners to provide the commercial satellite bus, selected flight avionics components, launch services, and long-term flight operations. NRL provides the design and integration of the NEMO sensor imaging payload with the commercial satellite bus, bus modifications, the on-board processor, ORASIS (Optical Real-time Adaptive Signature Identification System), and systems engineering. NRL has responsibility for calibration software and calibrated radiances; development of at-launch algorithms including those for atmospheric correction; on-board data compression software and related software for the ground data system; some data processing; and data archiving. The planned products of at-launch algorithms are water clarity (probably diffuse attenuation coefficient at 490 nm), concentration of chlorophyll, absorption coefficient of colored dissolved organic matter, concentration of suspended sediments, bathymetry, and bottom characteristics. [841] [842] [843] [844]

Subsystem	Short Description
ADCS (Attitude Determination and Control Subsystem)	3-axis stabilized for all modes; 0.07° control Geolocation: 30 m with a Circular Error of Probability (CEP) of 0.9 2 star trackers and 1 inertial reference unit for attitude determination 4 reaction wheels and 2 electromagnetic torquers for attitude control
EPS (Electrical Power Subsystem)	2 gimbaled (single axis) solar array panels 4 for 3 redundant power control box 64 Ah nickel hydrogen battery Approximately 1500 W power
RCS (Reaction Control Subsystem)	Monopropellant hydrazine blowdown system 1 propellant tank (76.5 kg capability) 5 x 1N thrusters for orbit correction/orbit maintenance
TCS (Thermal Control Subsystem)	Passive thermal control with heater augmentation Battery and payload panels thermally isolated from bus
Structural Subsystem	Primary structure is a combination of a rigid aluminum tubular frame and aluminum honeycomb shear panels Kinematically decoupled optical bench S/C mass: 496 kg dry; 547 kg wet; payload mass: 292 kg
Mechanisms Subsystem	Optical bench launch restraints Contamination covers for sensors Solar array drive mechanisms for gimballing arrays Launch vehicle separation Two redundant main electrical umbilical connectors
CT&DH (Command, Telemetry, and Data Handling Subsystem)	On-Board Processing Electronics (OBPE) primary controller: MIL-STD-1750A radiation-hardened processor Payload controller: R3000 processor (12 MIPs) Imagery On-Board Processor (IOBP): multiple Super Harvard Architecture RISC Computing (SHARC) processors in a parallel array Solid state data recorder with a capacity of 56 Gbit
Communications Subsystem	Data downlink in X-band, data rate of 131 Mbit/s; 20 watt; 8025-8400 MHz, with embedded housekeeping telemetry Command uplink in S-band, data rate of 2 kbit/s, 2074.1 MHz Low-rate downlink in S-band, data rate at 1.024 Mbit/s, 2252.5MHz, supports tactical imagery demonstrations
Software Subsystem	On-board task scheduling capability, FDIR (Embedded fault Detection, Isolation, and Recovery)

Table 220: Spacecraft subsystem characteristics

A launch of NEMO is planned for 2003. Launch vehicle is TBD.

The satellite consists of eight hardware subsystems and a software subsystem that collectively accommodate the payload and meet the mission and science requirements (Table

841) http://nemo.nrl.navy.mil/public/index.html

842) C. O. Davis, K. Carder, "Requirements Driven Design of an Imaging Spectrometer System for Characterization of the Coastal Environment," Proceedings of SPIE, Vol. 3118, San Diego, CA, 1997

843) C. O. Davis, "The Hyperspectral Remote Sensing Technology (HRST) Program," NRL White Paper, 1997

844) C. O. Davis, K. Carder, "Requirements Driven Design of an Imaging Spectrometer System for Characterization of the Coastal Environment," Proceedings of SPIE, Vol. 3118, San Diego, CA, 1997

220). The satellite platform selected is a commercial bus (LS-400) from SS/L. The satellite is three-axis stabilized and consists of a trapezoidal main body and two deployable solar arrays. STDC has responsibility for ground data processing system architecture, some data processing and merging, data archiving, and a web-based database interface. The S/C design life is five years (minimum of three year on-orbit mission life).

Orbit: Sun-synchronous circular orbit, altitude = 605 km, inclination = 97.81°, nodal equator crossing at 10:30 AM (ascending node), repeat cycle = 7 days with a 2.5 day global average re-access capability when S/C pointing (±30°) in the cross-track direction is used.

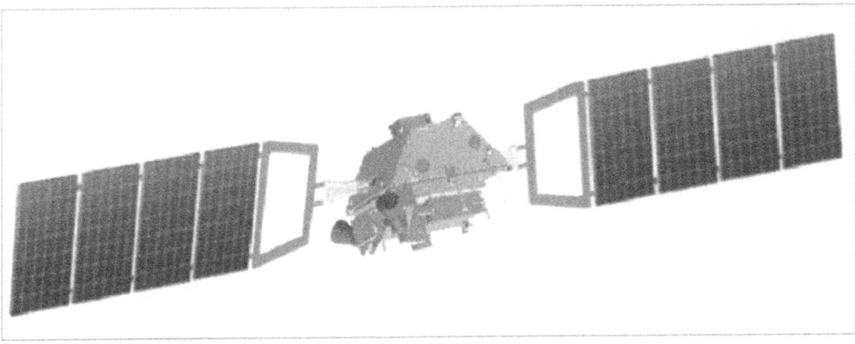

Figure 135: Illustration of the NEMO S/C

Sensor complement:

COIS (Coastal Ocean Imaging Spectrometer), a hyperspectral (ocean color - land surface) imager provided by NRL (built by SAIC). The instrument employs a TMA (Three-Mirror-off-Axis) anastigmatic design, a 15 cm diameter aperture telescope, a dichroic beam splitter, and two spectrometers (VNIR and SWIR). The overall spectral range of COIS is from 400 to 2500 nm with 10 nm spectral resolution (210 bands). A FOV of 2.86° provides a swath width of 30 km. COIS uses Offner convex grating spectrometers. The two spectrometers are:

- A Visible Near Infrared (VNIR) spectrometer that disperses the 400 to 1000 nm radiation into 60 spectral bands (10 nm wide) and onto a Focal Plane Array (FPA). This provides a resolution of 1.0 - 1.5% of the band and 60 spectral bins.
- A Short-Wave Infrared (SWIR) spectrometer that disperses the 1000 to 2500 nm radiation into 150 spectral bands (10 nm wide) and onto an FPA. This provides a resolution of about 0.6% of the band and 150 spectral bins. A cryocooler provides active cooling at 110 K.

The COIS instrument provides very high SNR environmental products for imaging the low-reflectivity ocean surface. The NEMO platform can implement Ground Motion Compensation (GMC) sufficient to reduce the apparent ground speed by a factor of 5. This provides the required dwell time to give a high SNR at a GSD of 30 m. COIS is also capable of producing 60 meter GSD, high SNR data products without GMC by using spatial binning of the hyperspectral FPAs. - COIS calibration is maintained by on orbit by monthly imaging of the full moon and weekly imaging of uniform ground reference targets. [845)]

A typical COIS scene is 30 km wide by 200 km in length, with 30 m by 30 m pixel resolution, yielding nine image strips per orbit (number of strips limited by on-board data storage capacity). An alternate pixel size is 60 m x 60 m, yielding a 30 km wide swath as long as 8000 km per orbit.

845) M. Corson, "Calibration of the NEMO sensor imaging payload," SPIE Proceedings, Vol. 3437, 1998

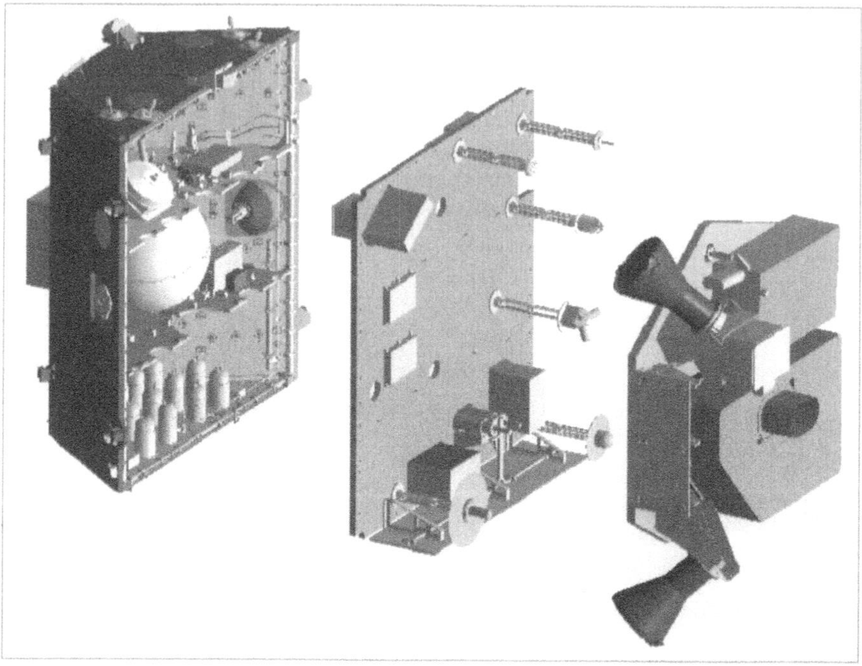

Figure 136: Exploded bus view with COIS instrument

Spectral coverage	400 - 2500 nm with a 10 nm spectral resolution (total of 210 spectral bands)
Spatial resolution	30 m or 60 m (for high SNR data products)
SNR (Signal-to-Noise)	>200 for VNIR data (5% reflectance target which is typical of the ocean) >100 for SWIR data at 30% reflectance (typical)
Pointing capability	±30° in the cross-track and along-track directions (body pointing) The pointing is used to increase integration time when sampling in the 30 m mode
Detector arrays	FPA of 1024 x 1024 pixels for VNIR (silicon CCD) and SWIR (MCT detector)
Pixel summing	Pixels are summed by 6 in VNIR and SWIR to achieve the desired spectral resolution
Data volume	Estimated average daily imagery is about 227 Gbit

Table 221: Some performance parameters of COIS

Parameter	COIS - VNIR	COIS - SWIR	PIC
GSD (Ground Sample Distance)	30 m with 5:1 GMC 60 m with no GMC	30 m with 5:1 GMC 60 m with no GMC	5 m (with or without GMC)
Spectral range	0.4 - 1.0 μm	1.0 - 2.5 μm	0.45 - 0.69 μm
Spectral bands	60	150	1
FOV, swath width	2.86°, 30 km	2.86°, 30 km	2.86°, 30 km
Swath length	200 km at 30 m GSD Continuous at 60 m GSD	200 km at 30 m GSD Continuous at 60 m GSD	Continuous
Aperture diameter	15 cm	15 cm	16.4 cm
Focal length, f/Nr.	36 cm, f/2.4	36 cm, f/2.4	120 cm, f/7.32
Pixel size	18 μm	18 μm	10 μm
Nr. of pixels/band	6	6	N/A
Data quantization	12 bits	12 bits	12 bits
Detector material	Silicon	MCT	Silicon

Table 222: Characteristic parameters of COIS and PIC

PIC (Panchromatic Imaging Camera). The instrument uses an off-axis telescope to simultaneously image the same 30 km swath as the COIS instrument onto a 6000 pixel CCD linear array (PIC data is co-registered with COIS data). The spectral range is from 0.45 - 0.69 μm. The panchromatic imagery has a resolution of 5 m GSD. The moderate COIS GSD matches the spatial scale of natural objects in the littoral zone, while the high resolution PIC provides simultaneous context and sharpening and supports civil land imaging requirements. The 30 km ground swath of COIS and PIC spans the near-shore littoral region of land, beach, and ocean. COIS and PIC use a single optical bench to minimize optical boresight drift. [846)]

Mission operations:

NEMO uses a combination of commercial/industry operations and an existing DoD infrastructure ground segment to provide the user communities (military and civil) with data services. The ground segment consists of the following elements:

- ASOC (Advanced S/C Operations Center), located in Washington, DC area. ASOC is the central hub for all mission planning, S/C operations, and data processing
- TAGS (Transportable Autonomous Ground Stations), located at Fairbanks, AK and elsewhere. Downlinked data is maintained on mass storage devices at the TAGS until final receipt of the IPC is verified.
- IPC (Image Processing Center). The IPCs perform quick-look data processing to level 1B and data archive functions.
- NCC (NRL Control Center). NCC is the science hub for the NEMO program. Also data distribution function for the military side.

Adaptive Spectral Signature Recognition.

The NEMO S/C employs **ORASIS** (Optical Real-time Adaptive Signature Identification System) developed by NRL, an automated and patented end-to-end HSI (Hyperspectral Imaging) algorithm and advanced processing system that performs on-board data compression, thereby significantly reducing the capacities needed for on-board storage and for downlink transmission.

ORASIS identifies the spectral signatures corresponding to physical objects in the scene without supervision or a priori knowledge. The approach is to analyze each spectrum in the scene sequentially, discarding duplicate spectra, and working only with the unique spectra. Using convex set methods and orthogonal projection techniques, each observed spectrum is then analyzed in terms of the set of spectra that represent the physically meaningful patterns or end-members that have combined to make the observed spectrum. Matched filters (Filter Vectors) are created, they can be used to de-mix the image. [847)]

The ORASIS processor, hosted on IOBP (Imagery On-Board Processor), is based on the SHARC (Super Harvard Architecture Computer), it has 32 processors for data handling and a processing capability of up to 2.5 GFLOPS. The on-board processing minimizes subsequent ground processing and data exploitation, and enables the on-board production of data products, such as maps of identified features. An important benefit of ORASIS processing is a **greater than tenfold data compression**, relieving the bottlenecks of on-board data storage, transmission to the ground, and beyond.

D.26 OKEAN-O

A USSR/Ukrainian (prior/after 1992) satellite series (OKEAN=Ocean) for the operational monitoring of ocean surfaces (sea surface temperatures, wind speed, sea color, status of

846) A. Myers, "NEMO satellite sensor imaging payload," SPIE Proceedings, Vol. 3437, 1998

847) J. Bowles, et al., "Hyperspectral Data Compression and Science Algorithms for the NEMO Satellite," Proceedings of 1st EARSeL Workshop on Imaging Spectroscopy, University of Zürich, Switzerland, Oct. 6-8. 1998, pp. 183-190

ice coverage, cloud coverage and precipitation) day and night operation. The Okean-O series with their polar orbits provides valuable complementary data on the arctic and antarctic regions, which are not visible from geostationary meteorological satellites. Launches of the series are conducted from the Plesetsk Cosmodrome by the Tsyklon-3 launch vehicle.

Orbit: Sun-asynchronous polar orbit, altitude = 660 km, inclination 82.5°, period = 98 minutes.

The OKEAN program started in 1979 with the pre-operational (i.e. experimental) Cosmos program.[848] Mission objective: a major interest was in the operation of active and passive microwave sensors. The active MW sensor provides all-weather observation capability.

D.26.1 Experimental Cosmos Program

The predecessor series of the Okean series was the experimental Cosmos series.

- Cosmos 1076 (also referred to OKEAN-E); launch from Plesetsk on a Tsyklon vehicle on Feb 12, 1979; orbit: altitude = 660 km; inclination = 82.6°; S/C mass = 1950 kg. A test mission of the Okean satellite without the radar. Objective: Development of methods for obtaining operational information on the Pacific Ocean. Operations were terminated on March 13, 1980.
- Cosmos 1151 (OKEAN-E); launch from Plesetsk on a Tsyklon vehicle on Jan. 23, 1980; orbit: altitude = 660 km; inclination = 82.6°; S/C mass = 1950 kg. The second test mission of the OKEAN satellite without the radar. Operations were completed on Oct. 13, 1981.
- Cosmos 1500 (also referred to as OKEAN-OE); launch from Plesetsk on a Tsyklon vehicle on Sept. 28, 1983; orbit: 649 x 679 km, inclination = 82.6°; S/C mass = 1950 kg. The S/C was operational until July 16, 1986.
- Cosmos 1602 (also referred to as OKEAN-OE); launch from Plesetsk on a Tsyklon vehicle on Sept. 28, 1984; orbit: 629 x 664 km, inclination = 82.5°; S/C mass = 1950 kg. The S/C was operational until Dec. 5,1986.
- Cosmos 1766 (also referred to as OKEAN-N1); launch from Plesetsk on a Tsyklon vehicle on July 29, 1986; orbit: 601 km x 631 km, inclination = 82.5°; S/C mass = 1950 kg. The S/C was operational until Oct. 24, 1988.
- COSMOS 1869 (also referred to as OKEAN-N2); launch from Plesetsk on a Tsyklon launcher on July 16, 1987; orbit: 602 km x 632 km, inclination = 82.5°; S/C mass = 1950 kg. The S/C was operational until May 30, 1989.

Cosmos satellites 1076 and 1151 carried the following sensors:

The simultaneous spectral measurements in several wavelengths provide a means of evaluating the hydro-physical parameters of the sea surface, the meteorological parameters of the atmosphere as well as of the sea surface. The parameters are: sea surface temperature, wind speed at the sea surface boundary, liquid vapor content in clouds, integrated water vapor, etc.

- **Device** ν = Passive MW Radiometer (NPO Vega) measuring the emission radiation of the atmosphere/ocean system at the following (micro) wavelengths/frequencies: 0.8 cm (37.5 GHz), 1.35 cm (22.22 GHz), 3.2 cm (9.37 GHz), and 8.5 cm (3.53 GHz). Nadir pointing.
- **Device** π = a non-scanning polarimeter measuring radiation at wavelength 3.2 cm (with two orthogonal polarizations at a look angle of 53° relative to nadir, and with one polar-

848) Verbal information provided by B. Kutuza of IRE (Russian Academy of Sciences), Moscow

ization at nadir). Along-track footprints of 12 m distance and 6 m normal to the orbital track are generated.

- **Device 174 K** = IR spectrometer.
- **MSU-M**. A four-band (VIS) multispectral scanner for monitoring ocean color. Swath width = 1900 km, resolution = 1.8 km.

S/C Name	Launch Date	Remarks	Instrument complement
Cosmos-1500 (OKEAN-OE-N1)	Oct. 28, 1983	June 12, 1986	RLSBO(3.2 cm), RM-0.8, 3-cha. MWR (0.8, 1.35 and 8.5 cm), RTVK
Cosmos-1602 (OKEAN-OE-N2)	Oct. 28, 1984	Nov. 10, 1985	RLSBO(3.2 cm), RM-0.8, 3-cha. MWR (0.8, 1.35 and 8.5 cm), RTVK
Cosmos 1766 OKEAN-O1-N1	July 29, 1986	operational until 1988	RLSBO(3.2 cm), RM-0.8, MWP (3.2 cm), RTVK, MSU-SK, Kondor
Cosmos 1869 OKEAN-O1-N2	July 16,, 1987	operational until 1989	RLSBO(3.2 cm), RM-0.8, MWP (3.2 cm), RTVK, MSU-SK, Kondor (SLAR did not work)
OKEAN-O1-N3	July 5, 1988	operational until 1990	RLSBO(3.2 cm), RM-0.8, MWP (3.2 cm), RTVK, MSU-SK, Trasser, Kondor
OKEAN-O1-N4	June 9, 1989	launch failure	RLSBO(3.2 cm), RM-0.8, MWP (3.2 cm), RTVK, MSU-SK, Kondor
OKEAN-O1-N5	Feb. 28, 1990	operational until 1991	RLSBO(3.2 cm), RM-0.8, MWP (3.2 cm), RTVK, MSU-SK, Kondor
OKEAN-O1-N6	June 4, 1991	operational until 1993	RLSBO(3.2 cm), RM-0.8, MWP (3.2 cm), RTVK, MSU-SK, Kondor
OKEAN O1-N7	Oct. 11, 1994	operational 1/1996	RLSBO(3.2 cm), RM-0.8, MWP (3.2 cm), RTVK, MSU-SK, Kondor
SICH-1 (OKEAN-O1-N8)	Aug. 31, 1995	operational 1/1996 Russian/Ukrainian joint venture	RLSBO(3.2 cm), RM-0.8, MWP (3.2 cm), RTVK, MSU-SK, Kondor
OKEAN-O-1	July 17, 1999	OKEAN-O-1 stopped operations in the fall of 2000 due to S/C attitude control problems	RLSBO (2 units), MSU-M, MSU-SK, MSU-V, Delta-2D, R225, R-600, Trasser-O

Table 223: OKEAN-O1 series designations

The Cosmos satellite series 1500 - 1869 was equipped with a subset of instruments of the first series, i.e. one or two passive MW radiometers per satellite measuring the emission radiation of the atmosphere/ocean system. Cosmos 1500 was the first Soviet satellite equipped with a side-looking, all-weather radar, namely **RLSBO** (or SLAR).

D.26.2 OKEAN-O1 Operational Series

A S/C series, built by Yuzhnoye (Research and Production Association) of Dnepropetrosvk, Ukraine. The S/C bus is a three-segmented, vertically oriented cylinder, three meters tall with a base diameter of 1.4 m and an upper diameter of 0.8 m. OKEAN's primary structure is pressurized and maintained at normal temperatures to protect the support system and payload electronics housed within. Two small, rotatable solar arrays (1.6 m x 2.0 m) provide a modest 110–270 W average daily power to the payload. The S/C is three-axis stabilized (nadir pointing), stabilization is also partially provided by a gravity-gradient boom extended from the top of the satellite. At the bottom, four large panels (1.0 m x 2.9 m), attached at 90° intervals, support a number of payload receivers and transmitters. A narrow, 11 m long radar antenna is fixed to the base of one panel. The overall S/C mass is about 1950 kg with a payload capacity of 550 kg.

Operational series (design life is 2 years):

- Okean-O1-1; launch: July 5, 1988; orbit: 635 x 666 km; inclination = 82.5°
- Okean-O1-2; launch: Feb. 28, 1990; orbit: 639 x 666 km; inclination = 82.5°
- Okean-O1-3; launch: June 4, 1991; orbit: 664 x 684 km; inclination = 82.5° (conditional operation since Jan. 1992, RLSBO is not working!)
- Okean-O1-4; launch: October 4, 1994
- Sich-1: launch: August 31, 1995

Sich is the Ukrainian name for 'Owl'. Sich-1 has the same orbit and sensor complement as the Okean-O1-1 to O1-4 S/C series. As of January 1996 all instruments of Sich-1 are operational.

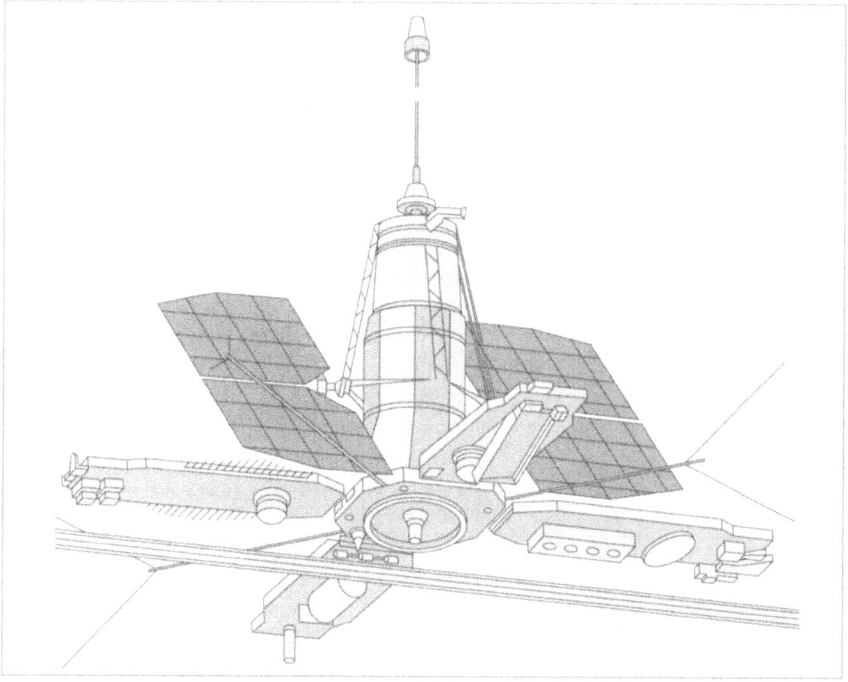

Figure 137: The OKEAN-O1 series S/C model

Sensor complement: (Okean-O1 Series)

RLSBO = Side Looking Real Aperture Radar (Kharkov IRE, Ukraine, PI: Kalmykov), prime sensor of the Okean series (11.1 m antenna length). Wavelength/frequency: 3.2 cm/9.7 GHz, X-band; resolution = 2.1 km - 2.8 km in flight direction, = 1.2 km - 0.7 km in cross track direction; swath = 450 km.

- impulse duration = 3 μs
- impulse peak power = 100 kW
- pulse repeat frequency = 100 Hz
- polarization = V (vertical)

RM-08 = Passive Microwave Scanning Radiometer (Kharkov IRE, Ukraine). Wavelength/ frequency: 0.8 cm / 36.6 GHz; resolution = 15 x 20 km, swath = 550 km. Objective: monitoring atmospheric vapor, sea ice, and ocean surface temperature with an accuracy of 1-2 K.

MSU-M = Multispectral Scanner of low resolution (ISDE, Moscow). Resolution = 1.0 x 1.7 km, swath = 1900 km. Objective: cloud monitoring and sea surface temperature. Spectral ranges: 0.5 - 0.6 μm; 0.6 - 0.7 μm; 0.7 - 0.8 μm; 0.8 - 1.1 μm

MSU-S = Multispectral Scanner of moderate resolution (ISDE, Moscow). Two spectral ranges: 0.6 - 0.7 μm, 0.7 - 0.9 μm; resolution = 250 m; swath = 1100 km.

Kondor = DCPs data collection system

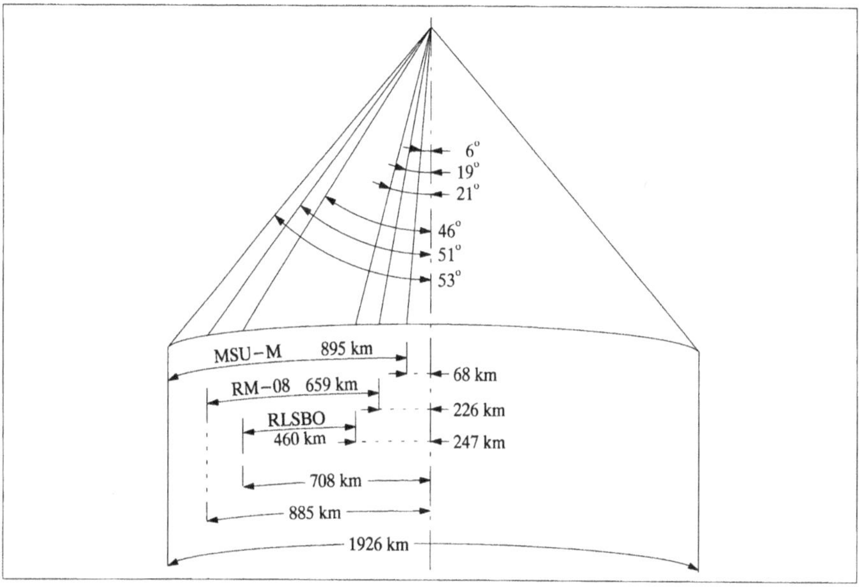

Table 224: Observation geometries for OKEAN instruments (shown to one side only)

One of the features of Okean type satellites is the possibility of direct data reception by various users having only simple hardware and software for data processing. The downlink frequency of 137.4 MHz (carrier frequency with 2.4 kHz subcarrier frequency) is used for direct broadcasts of RLSBO and MSU-M imagery with reduced resolution to any of the CIS APT stations as well as to foreign users. The second science data downlink has a carrier frequency of 466.5 MHz.

Data collected are transmitted to receiving stations at Moscow, Novosibirsk, and Khabarovsk (Pacific region). Maps can be readied in 10 days and ice forecasts sent directly to ships at sea.

D.26.3 OKEAN-O Series

The OKEAN-O series, a joint Ukrainian/Russian program, is an upgrade over the OKEAN-O1 series with regard to the sensor complement (the S/C is also considerably larger). The OKEAN-O-1 S/C was sponsored by RKA and NKAU (National Space Agency of Ukraine), designed by the Yangel Yuzhnoye State Design Office, and built by NPO Yuzhny Machine-Building Plant Production Association in Dnepropetrovsk, Ukraine. The satellite is three-axis stabilized with an attitude pointing accuracy of ±10 arcmin and an angular stabilization velocity of ±0.0015°/s. Solar power of up to 3.5 kW. S/C mass of 6150 kg, the mass of the on-board special complex (BSC) is 1520 kg. The design life is three years. [849) 850)]

OKEAN-O-1 was launched on a Zenit-2 vehicle from Plesetsk on July 17, 1999.

Orbit: Near-circular orbit, 620 km (perigee) x 652 km (apogee), inclination = 82.5°, the revisit time is between 4 to 16 days.

Operational status as of April 2001: OKEAN-O-1 stopped all operations in the fall of 2000 due to attitude control problems of the spacecraft. Both space agencies, NKAU and RKA (or Rosaviakosmos), do not provide any funding for the development of a further OKEAN spacecraft due to the current economic situation in the Ukraine and in Russia.

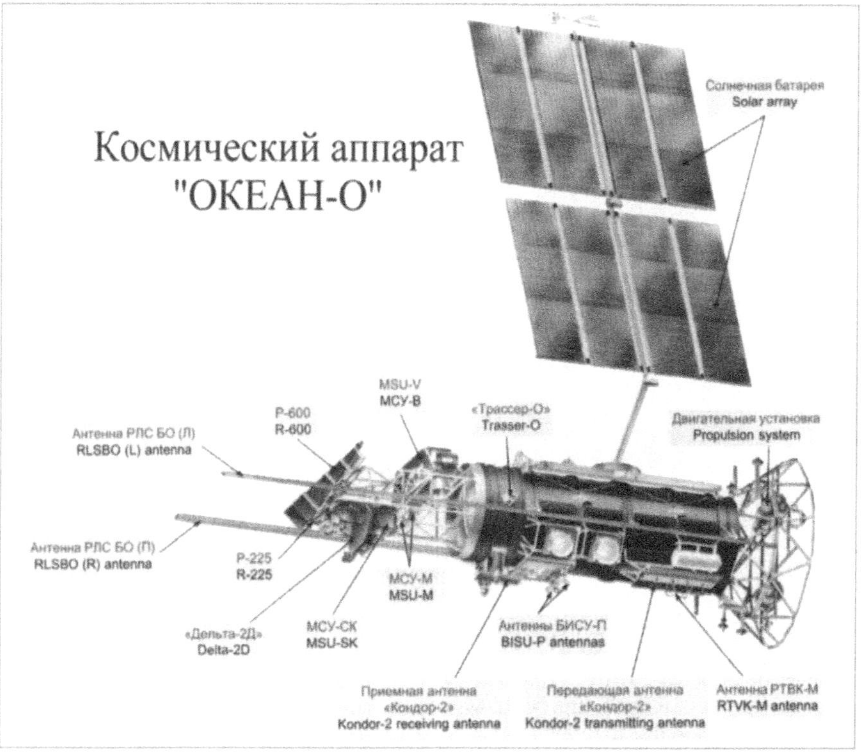

Figure 138: The OKEAN-O-1 spacecraft illustration

S/C Communication.

1) BISU-P (On-board Information System Unified). This is a digital system with an X-band downlink, the frequency is 8.2 GHz. The transmission rate is 61.44 Mbit/s (4R) or 15.36 Mbit/s. The on-board date recorder has recording rates of 15.36 Mbit/s or 0.96 Mbit/s.
2) RTVK-M (Radio and Television Complex). This is an analog system in VHF-band with a downlink frequency of 137.4 MHz and a bandwidth of 2.4 kHz.
3) BITS-2-7 (On-board Information and Telemetry System). This is a digital system in the UHF-band with a downlink frequency of 600 MHz.

849) OKEAN-O Earth Observation Spacecraft, a brochure of RKA and NKAU provided by B. Kutuza of IRE, Moscow
850) http://www.okean–o.dp.ua/en_satellite.html

Sensor complement:

The payload instruments are part of BSC (On-board Special Complex). 851) 852)

MSU-V = High-Resolution Multispectral Scanner (Kazan) with photodiode detectors. The instrument has a swath width of 195 km.

Band Nr.	Spectral Range	Spatial resolution
Band 1	0.45 - 0.52 μm (VIS)	50 m
Band 2	0.54 - 0.61 μm (VIS)	50 m
Band 3	0.63 - 0.73 μm (VIS)	50 m
Band 4	0.78 - 0.92 μm (VIS)	50 m
Band 5	0.91 - 0.997 μm (NIR)	50 m
Band 6	1.47 - 1.62 μm (SWIR)	100 m
Band 7	2.06 - 2.38 μm (SWIR)	300 m
Band 8	10.6 - 12.0 μm (TIR)	250 m

Table 225: Spectral parameters of MSU-V

MSU-SK = Multispectral Scanner of Moderate Resolution with conical scanning (ISDE, Moscow). Swath width = 620 km for each instrument; there are 5 spectral bands: 0.53 - 0.59 μm, 0.59 - 0.72 μm, 0.72 - 0.81 μm, 0.81 - 1.1 μm, 10.4 - 12.6 μm (TIR). Spatial resolution = 157 m x 245 m (across-track and along-track) in VNIR and 590 m x 820 m in TIR. There are two units on-board: MSU-SK1 is a forward-looking instrument, while MSU-SK2 is an aft-looking instrument - the off-nadir angle for each instrument is 39°.

MSU-M = Multispectral mechanical scanner of low resolution (ISDE, Moscow). Resolution = 1.5 km x 1.8 km, swath = 2000 km. Objective: cloud monitoring and sea surface temperature. Spectral ranges: 0.5 - 0.6 μm; 0.6 - 0.7 μm; 0.7 - 0.8 μm; 0.8 - 1.1 μm. There are two units - prime and spare.

RTVK-M (Radio and TV Complex), part of MSU-M (ISDE, Moscow). The MSU-M and RLSBO instruments are integrated into one complex. Data from RLSBO (R or L) and MSU-M (1 among 4 channels) may be selected. Their information may be transmitted via a VHF (137.4 MHz) downlink to users in real-time. There are over 1000 VHF receiving stations in the ground segment.

RLSBO = Side Looking Real Aperture Radar. Wavelength = 3 cm (X-band). Resolution: 1.3 km cross-track and 2.5 km along-track. The swath width is 455 km for each instrument. There are two instruments, one looking to either side of the ground track. The near angle of incidence is ±20°. Both instruments operate in parallel.

Trasser-O (Polarization Spectroradiometer-OKEAN), designed and built by STCUI-RAS (Scientific Technological Center of Unique Instruments - Russian Academy of Sciences), Moscow. 853) A non-scanning demonstration device with an off-nadir angle of 20°. Trasser introduces the technology of AOTF (Acousto-Optic Tunable Filters) in acousto-optic spectrometers (AOS). The measurement concept is based on collinear diffraction of light by ultrasonic waves propagating in an anisotropic crystal (optical radiation diffraction on bulk phase grating being generated in the quartz crystal). The AOTF technique offers a means of high-accuracy spectral monitoring in the spectral region from about 250 nm (UV) to about 4.4 μm (MWIR). The relatively new AOTF technique of optical spectra detection (by absorption, emission or scattering of light) may be employed for a number of applications such as: monitoring of atmospheric trace gases, of sea surface characteristics (biopro-

851) Information provided by B. Kutuza of IRE, Moscow, and translated by B. Zhukov of DLR, Oberpfaffenhofen

852) I. V. Bragin, V. P. Sgibnew, K. A. Pobedonostsev, A. V. Evtushenko, et. al., "Space-Based Remote Sensing Complexes," Proceedings of the 29th European Microwave Conference, Munich, Sept. 1999, pp. 388-390

853) Information provided by V. I. Pustovoit, V. E. Pozhar, and V. N. Zhogun of STCUI-RAS

ductivity of ocean surfaces, detection of polluted ocean surfaces), and of vegetable canopies on land. [854)]

Background: The first spaceborne AOS instrument by the name of Trasser was flown on OKEAN-O1-N3 with a launch on July 5, 1988. The instrument was operational until 1990. In parallel to Trasser, there was an airborne AOS instrument by the name of **POLAS-128** (Polarization-sensitive Acousto-optic Spectrometer) which was also used for underflights of Trasser. Both instruments operated in the visible band. They were developed and built by VNIIFTRI (All-Russian Scientific Research Institute of Physical and Radio-Engineering Measurements). Later on (1996) the Trasser work was continued at STCUI-RAS.

Spectral range	430 - 800 nm
Polarization channels	2 (H and V)
Spectral channels	62 x 2 (for each polarization channel)
Bandwidth	From 3 nm at 430 nm to 12 nm at 800 nm
Detection range on radiance spectral density	0.13-26 μW/ (cm^2 x nm x sr)
Sampling time	3.3 s per spectrum
Input angle	2° (corresponding to a spatial resolution of 45 km at spacecraft altitude of 668 km)
Off-nadir angle	20°
Detector type	PMT (Photomultiplier Tube)
Data quantization	12 bit
Sample rate	2 Hz
Size of optical head	300 mm diameter x 702 mm
Data rate	14 kbit/channel
Mass of instrument unit	15 kg optical head, 16.5 kg signal processing unit, 5 kg control unit, 9 kg power supply commutator
Instrument mass, power	30 kg, 285 W

Table 226: Technical characteristics of Trasser-O

Trasser-O consists of the following elements (see Figure 140):

- A double-channel (H,V) spectrometer
- A calibration system consisting of an internal gray lamb and a sun view through a milky glass
- A three-position movable mirror
- AOTF (Acousto-Optic Tunable Filter - virtually a tunable monochromator) consisting of two polarizers (H, V) and an acousto-optic cell. An external RF reference signal is introduced via a piezotransducer for elasto-optic crystal excitation.

AOTF principle: AOTF is a spectral element containing an elasto-optic crystal with piezo-transducer attached and two polarizers. It allows selecting radiation in the narrow spectral band with the wavelength controlled by the frequency of ultrasound supplied to the crystal. In AOTF, an anisotropic diffraction of the light beam takes place at phase volume grating caused by a running acoustic wave. Two features of this physical effect are critical for the selection. The first one is the resonance nature of acousto-optic interaction: the synchronism wavelength of light is tuned by acoustic frequency. The second feature is changing the polarization and direction of diffracted wave that permits to separate it by polarizers or diaphragms.

The current non-scanning design of Trasser instruments offers a rather limited observation coverage, a footprint of about 25 km in diameter which progresses in the along-track direction as the spacecraft moves in its ground track. This renders a coverage scheme very similar in scope to altimetric measurements. However, the overall goal of Trasser-type instruments

854) V. I. Pustovoit, V. E. Pozhar, "Acousto-optical spectrometers for Earth remote sensing," Proceedings of SPIE 44th Annual Meeting, International Symposium on Optical Science, Engineering, and Instrumentation, Denver, CO, July 18-23, 1999

is to provide eventually an alternate measurement approach to conventional hyperspectral imaging systems, namely AOS imagers. For Trasser, this requires the addition of an optical front-end to the current instrument configuration, permitting a spatial cross-track scan to obtain spatial/spectral image cubes.

Figure 139: The Trasser-O instrument along with power and electronic units

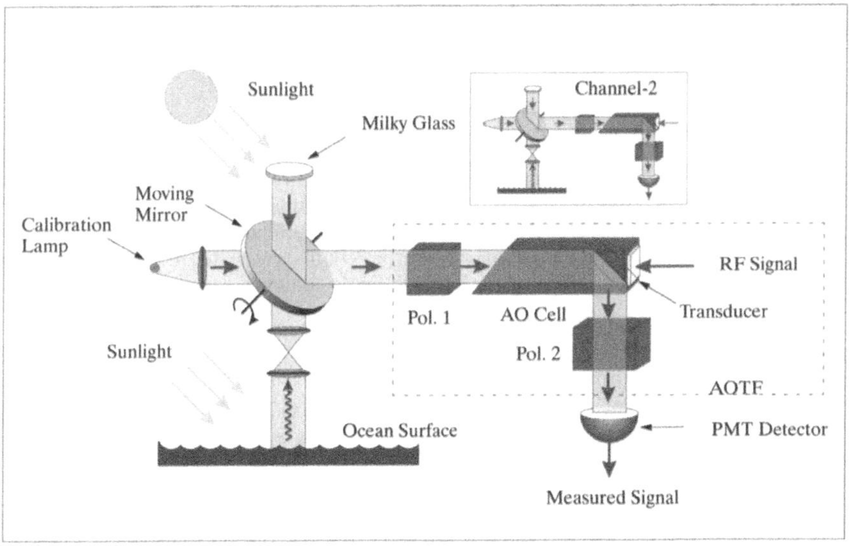

Figure 140: Optical scheme of a collinear AOS (Acousto-Optical Spectrometer)

Delta-2D = Scanning Microwave Radiometer with conical scanning (SRB/MPEI, Moscow). Objective: Measurement of sea surface temperatures (SST) and monitoring of Earth surface temperatures, characteristics of arctic ice, etc. The instrument is an improved version of IKAR-P heritage flown on Priroda, providing four frequency bands. The swath width of Delta-2D is 1130 km, there are two polarizations, the off-nadir angle is 40°.

Frequency (GHz)	Wavelength (cm)	Polarization	Resolution	Noise temp. measurement error
37.5	0.8	V, H	17 km x 21 km	0.15 K
22.3	1.35	V, H	28 km x 37 km	0.15 K
13.0	2.2	V, H	49 km x 65 km	0.1 K
6.9	4.5	V, H	91 km x 120 km	0.1 K

Table 227: Parameters of the Delta-2D instrument

R-600 = Passive Microwave Radiometer, built by MPEI, Moscow. Objective: measurement of emissive microwave radiation at the atmosphere/sea surface interface. Wavelength (frequency) = 6 cm (5 GHz); The footprint of the instrument is about 165 km in diameter. There are two polarizations, the off-nadir angle is 42°.

R-225 = Passive Microwave Radiometer (built by MPEI, Moscow). Wavelength (frequency) = 2.25 cm (13.3 GHz); The footprint of the instrument is about 130 km in diameter. There are two polarizations, the off-nadir angle is 42°.

Kondor-2 is an on-board data collection system. The objective is to provide collection services for the Kondor-1 DCPs (Data Collection Platform) at 1533 MHz as well as to Kondor-3 receiving stations at 460 MHz. The data receiving rate of Kondor-2 is 500 bit/s, the transmission rate is 2 kbit/s. Kondor-2 has an on-board storage capacity of 64 kbit. - The positioning accuracy of Kondor-1 DCPs is about 3.5 km x 2.5 km in polar regions, and about 3.5 km x 20 km in the other regions.

D.27 OrbView-2/SeaStar

See OrbView-2 description under B.8.2.

D.28 PRIRODA

PRIRODA [855) 856) 857)] (=Nature) is a multisensor research module for Russia's orbital MIR station dedicated to the observation of the environment. Major institutes involved: RKA, RKK Energia, IRE-RAN, NPO Planeta, DLR/DFD. The launch of the Priroda module took place on April 23, 1996 aboard a Proton-K launch vehicle from Baikonur (MIR docking on April 26, 1996). Nominal mission life = 2-3 years.

Application: Geoecology (radiophysical methods for the study of the environment), infrared-spectroradiometric methods for the study of spatial IR-radiation of the atmosphere, spectroradiometric methods for the study of reflected solar irradiation (VNIR) in particular in the region 'atmosphere - Earth surface,' oceans, hydrology, meteorology, geology, etc.

The Priroda mission has an extensive science program. The basic mission objectives are of a research nature in the following fields:[858)]

- Land surface exploration
  - Snow cover
  - Soil investigations
  - Vegetation

855) "PRIRODA," Ein Forschungsmodul der sowjetischen Orbitalstation MIR zur Fernerkundung der Erde, Wissenschaftliche Nutzlast Technische Beschreibung, Institut für Kosmosforschung (IKF), Berlin, 1990
856) "PRIRODA-Experimente," Programm zur Beschaffung, Verarbeitung, Bewertung und Anwendung von Daten des Multisensorsystems PRIRODA der sowjetischen Orbitalstation MIR, 1992 -94, DARA, Berlin, Mai 1991
857) "Complex for Remote Sensing of the Earth," Science Program, DLR paper 1991
858) Orbital Station MIR, Complex of Remote Sensing of the Earth "PRIRODA," Scientific Program, IRE brochure, Moscow, 1991

- Earth surface mapping at different wavelength ranges for solid Earth research

- Ocean exploration
  - Global survey by using passive sensing techniques
  - Sea surface temperature determination
  - Wind speeds and direction
  - Surface roughness due to wind
  - Ocean color characteristics
  - Ocean bioproductivity investigations
  - Ice cover determination

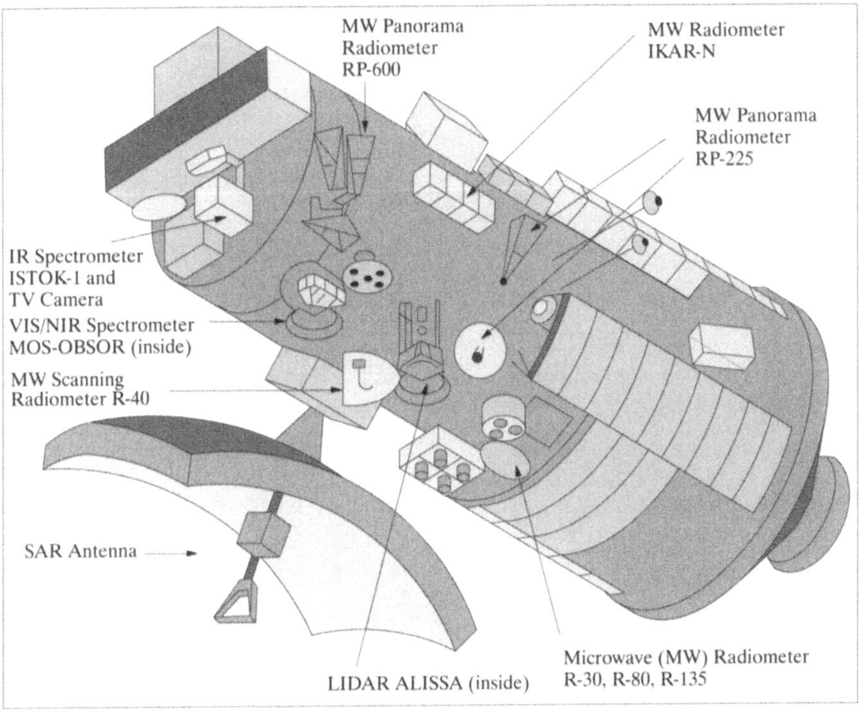

Figure 141: Model of the PRIRODA module on the MIR station

- Atmosphere investigations
  - Large-scale atmospheric processes
  - Atmosphere-ocean interactions
  - Lower stratosphere and troposphere physical parameters
  - Ozone experiment

- Ecological investigations

The great variety of observation techniques employed by the various instruments provides synergies on many levels in the evaluation and interpretation of the data (for instance: the combination of SAR and radar altimeter data with MW radiometer data in ocean research).

Orbit (same as that of MIR): Altitude = 300 to 400 km, 51.6° inclination.

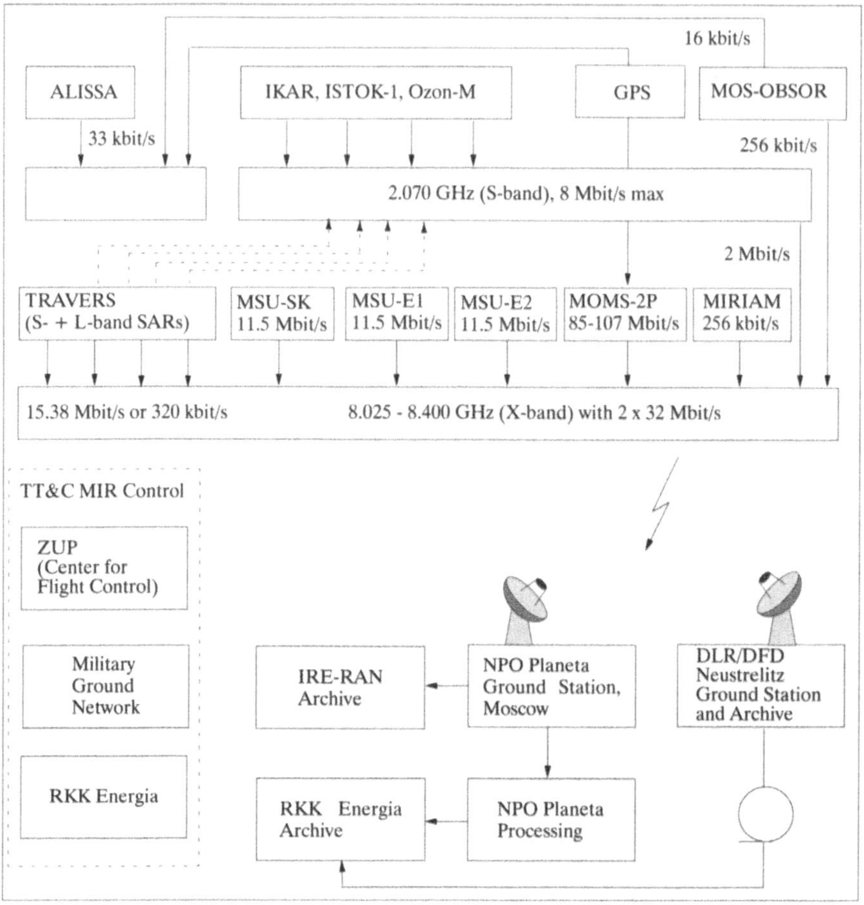

Figure 142: PRIRODA data distribution

Sensor complement:

1) **PRIRODA passive microwave instruments** (developed and built by SRB/MPEI (Special Research Bureau/Moscow Power Engineering Institute), consisting of the following instrument blocks: IKAR-N, IKAR-D, IKAR-P, and a scanning radiometer.[859]

IKAR-D (DELTA): - a three-channel scanning radiometer at wavelengths of 0.3, 0.8, and 1.35 cm and a look angle of 40° from nadir.

IKAR-P: - (Panorama) a three-channel scanning microwave radiometer (RP-225) at a wavelength of 2.25 cm, and a five-channel panorama radiometer RP-600 at 6.0 cm. Both with a look angle of 40° against nadir. Objective: Measurement of sea surface temperatures (SST). Scanning is performed by a rotating reflector, a non-symmetric truncated paraboloid with a diameter of 700 mm. Conical horns serve as exciters, they separate each waveband. The reflector (mass <0.3 kg) consists of two layers of glass fiber and a radio-reflecting layer in the form of nickel coating. The look angle is 40° against nadir. The swath width is about 680 km. A full scan cycle is 2.56 s. Calibrations are performed during each scan cycle

859) G. Zimmermann, "Mission PRIRODA," German Proposals to Scientific Program, DARA Bulletin, Dec. 1991

by internal noise generators and antenna views to cold space. A flat mirror, not interfering with the measurement process, is used for cold space calibrations. [860)]

Frequency (GHz)	Polarization	Beamwidth angle (3 dB)	Spatial resolution (km)	Noise temperature measurement error
36.5	H, V	1.1°	8 x 12	0.5 K
22.3	V	1.7°	15 x 20	0.5 K
13.0	H, V	3°	25 x 33	0.5 K

Table 228: Parameters of the IKAR-P instrument

The receivers are Dicke-type radiometers. The control of modulators, the synchronization of drive engine operations and the control of standard noise generators is supervised by a digital data processing unit.

IKAR-N (Nadir): - five microwave radiometers (R-30, R-80, P-135, R-225P, and RP-600) at wavelengths of 0.3 cm, 0.8 cm, 1.35 cm, 2.25 cm, and 6.0 cm in nadir direction. R-225P can also measure different polarizations. Objective: Measurement of microwave radiation (emission) at the atmosphere/sea surface interface. The IKAR-N instrument is similar to the radiometers of the IKAR-P complex. Some differences are in the construction of the receiving equipment (rectangular horns, are used for each frequency).

Frequency (GHz)	Polarization	Beamwidth angle (3 dB)	Noise temperature measurement error
89.0	V	9°	0.5 K
36.5	V, H	8°	0.5 K
22.3	V, H	8°	0.5 K
13.0	V, H	7.5°	0.5 K

Table 229: Parameters of the IKAR-N instrument

R-400: - a one-channel polarization scanner microwave radiometer. Wavelength of 4.0 cm (7.5 GHz). Look angle also 40° against nadir. Objective: Measurement of the thermal radiation of the Earth surface (radio wave range).

2) PRIRODA active microwave equipment

Travers (RSA) = Sideview radar system with synthetic aperture (SAR, measurement in two frequencies). Russian sensor, built by MEI (Moscow Power Engineering Institute). Objective: generation of radar images of the Earth's surface (land and ocean). Wavelengths of scanning beam: 9.2 cm (3.28 GHz, S-band), and 23 cm (1.28 GHz, L-band).

Parameter	S-band SAR	L-band SAR
Frequency	3.28 GHz	1.28 GHz
Wavelength	9.2 cm	23 cm
Bandwidth	5 MHz	
Peak power	300 W	1000 W
Impulse power	250 W	600 W
Compressed pulse width	0.4 ms (effective pulse), 25 ms (total duration)	
Pulse repetition frequency	3 kHz (depends on the field of regard (FOR))	
Antenna gain	37 dB	33 dB
Antenna size	2.8 m x 6 m	
Polarization	VV or HH	
Quantization	complex 4 bit	
Swath width (H=400 km)	50 km	
Resolution in azimuth: Resolution in azimuth: Range resolution:	150 m (for satellite processing) 20 m (for ground processing) 100 m	

860) I. V. Bragin, V. P. Sgibnew, et al., "Space-Based Remote Sensing Complexes," Proceedings of the 29th European Microwave Conference, Munich, Germany, Oct. 5-7, 1999, Vol. 2, pp. 388-390

Parameter	S-band SAR	L-band SAR
Look angle	35°	
Data rate	16 Mbit/s (ground processing)	16 Mbit/s (ground processing)
Data rate	2.56 Mbit/s (on-board processing)	2.56 Mbit/s (on-board processing)
Orbit repetition cycle	6 days (coverage)	
Operation mode	L- and S-band operate in parallel	

Table 230: Specification of SAR Travers parameters

3) PRIRODA optical equipment, consisting of ISTOK-1, MOS-A, MOS-B, MSU-SK, MSU-E, TV camera, Ozon-M, and a lidar ALISSA.

ISTOK-1 = IR Spectroradiometric System (sounder). The ISTOK-1 apparatus consists of the following modules:

- Opto-Mechanical Block (OMB),
- Electronic Block of the IR-Spectrometer system (EBS),
- The data processing block (digital processing of video data),
- The computer block (built in Rumania)

Objective: Measurement of atmospheric irradiation for different observation angles.

- Vertical temperature profiles of the atmosphere (up to 30 km with an accuracy of 4 - 5 K) and its CO_2 band radiation.
- Altitude profile of pressure and temperature (with 0.2 km resolution).
- Humidity profiles from the measurement of the water vapor band (6.3 μm)
- Measurement of ozone content in the 9.6 μm band
- Measurement of ocean surface temperature
- Atmospheric chemistry
- Study of convective processes in the atmosphere

The IR-spectrometer consists of two identical diffraction polychromatographs for the spectral ranges of 4 - 8 μm, and 8 - 16 μm.

- Number of spectral channels: 64
- Spectral resolution (bandwidth): 0.125 μm (4-8 μm region), 0.025 μm (8 - 16 μm region)
- Sensitivity: = 1 x 10 (W/cm/sr/m) for range 1; = 5 x 10 (W/cm/sr/m) for range 2
- Radiometric resolution: 14 bit
- Spatial resolution: 0.75 - 3 km (in flight direction)
- Look angles: -15° to 132° in orbit plane; 0° to 90° in azimuth plane
- Maximum rotation rate of instrument: 2.4 degrees/s
- Data rate: 10 kbit/s

MSU-SK = Optical Multispectral Scanner - moderate resolution. (heritage: from Meteor and Cosmos 1639, the sensor is being built by the Institute of Space Device Engineering, ISDE). MSU-SK measures in four channels of the visible spectrum: (0.5 to 0.6 μm, 0.6 to 0.7 μm, 0.7 to 0.8 μm, and 0.8 to 1.1 μm) with a resolution of about 120 m; MSU-SK also measures in the infrared spectrum (10.4 to 12.6 μm, TIR) with a resolution of about 400 m. The swath width is 320 km (at altitude of 400 km).

- Look angle = 39°
- SNR for channel 1 = 70
- SNR for channels 2-5 = 100
- Noise level for channel 5 = 0.5 K
- Range of radio temperatures for channel 5 = 210 - 320 K (TIR)
- Data rate for channels 1-4 = 2.56 Mbit/s
- Data rate for channel 5 = 1.28 Mbit/s

MSU-E = Optical Multispectral Scanner (electronic scanning, CCD technology, built by ISDE). Measures in three visible channels (0.5 to 0.6 μm, 0.6 to 0.7 μm, and 0.8 to 0.9 μm);

the resolution is 20 m; the average swath width is 2 x 27 km (two devices measure in parallel, one to each side of the ground track).

- Look angles: ±32° (the instrument may be pointed in the cross-range direction, thereby extending FOV to the Field of Regard (FOR)
- Radiometric resolution: 8 bit
- Data rate for each channel: 3.84 Mbit/s

Both scanners (MSU-SK and MSU-E) are operationally coupled (for instance: two channels may be switched on MSU-SK plus three channels on MSU-E).

MOS-P (Modular Optoelectronic Scanner - Priroda).MOS is a passive pushbroom imaging spectrometer of medium spatial resolution for the visible and near-infrared spectrum (VNIR) of 0.4 - 1.01 μm. It is a nadir-viewing instrument, built by DLR/ISST, Berlin, Germany. MOS is of MKS, and MKS-M heritage which were flown on Intercosmos-21, and the space stations Salyut-7 and MIR. Another version of MOS is flown on the ISRO mission IRS-P3. [861)]

Parameter	MOS-A	MOS-B
Spectral range	755 - 768 nm	400 - 1010 nm
No. of channels	4	13
Spectral channels	756.7, 760.6, 763.5, 766.4 nm	408, 443, 485, 520, 570, 615, 650, 685, 750, 815, 870, 945, 1010 nm
Spectral resolution	1.4 nm	10 nm
Swath width (H = 400 km)	80 km	82 km
Spatial resolution	2.8 km	0.7 km
Quantization	12 bit	12 bit
Measurement range	0.1 - 40 W/cm²/ nm/sr	0.2 - 60 W/cm²/ nm/sr
Data rate	4 kbit/s	210 kbit/s

Table 231: Specification of some MOS-A and MOS-B parameters

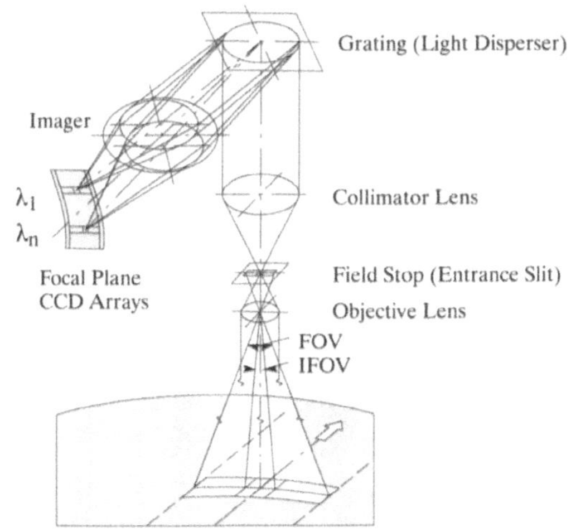

Figure 143: Principle of the MOS-P electronic imaging spectrometer

Objective: Image generation of the Earth's surface (surface-atmosphere interaction, ocean color, phytoplankton, regional and global distributions of aerosols and its links to gaseous

861) A. Neumann, "Spaceborne Imaging Spectrometers for Ocean Color Remote Sensing, MOS-Priroda and MOS-IRS," DLR/ISST paper presented at the IOC Ocean Color Workshop, Victoria, BC, September 21-22, 1995

admixtures, spectral and spatial cloudiness characteristics, etc.) in the VNIR region. The sensor apparatus consists of two complementary instruments: MOS-A and MOS-B (see also Figure 143). MOS-A is a four-channel diffraction-scanner spectrometer, specialized for observing the atmosphere-ocean system in the O_2A absorption band around 760 nm. The objective is to estimate the aerosol optical thickness and stratospheric aerosols. - MOS-B is a 13-channel spectrometer in the range of 408 to 1010 nm. The center wavelengths of the channels are chosen with the objective for the quantitative retrieval of ocean an coastal zone parameters. They also provide a capability for vegetation signature determination (red edge) and estimation of H_2O (water vapor) content in the atmosphere from the NIR-measurements. An additional calibration mechanism permits the direct measurement of solar radiation.

TV Camera
Objective: surveying the region of measurement with 256 x 256 pixels of about 900 x 900 m edge. Spectral ranges: 0.4 - 0.75 μm, look angle = 15°, spatial resolution = 300 m.

Ozon-M = Sounder apparatus for the measurement of ozone concentration and other trace gas elements in the atmosphere. (Russian sensor, Organization: INTEGRAL).

ALISSA[862] = l'Atmosphere par LIdar Sur SAliout (French sensor. Note: ALISSA did not fly on Salyut). ALISSA is conceived as a simplified fixed frequency lidar, using Mie scattering to detect clouds and aerosols. The primary objective is to study the impact of altitude determination on the description of the cloud field provided by geostationary satellites. Operation is during orbital night to minimize the difficulty of accurate system alignment. The main characteristics of the sensor are as follows:

Emitter:

- Second harmonic of Nd-Yag lasers: λ = 532 mm
- Energy pulse: 40 mJ
- Repetition rate: 50 Hz
- Natural divergence: 10^{-3} rad
- Divergence after collimator: 10^{-4} rad

Receiver:

- Cassegrain telescope of area A = 0.12 m^2
- Field of view: 10^{-3} rad
- Filter bandwidth: 0.5 nm

Energy requirements: 3 kW. The signal is analyzed by a pulse counting system with a gate of 1 μs, which provides a height resolution of: Δ_z = 150 m. The integration of six consecutive pulses (to increase the SNR) gives a horizontal resolution along the satellite track of: $\Delta_{x,y}$ = 1 km. The pulse counting system has 512 channels of 1 μs, which can be adjusted at 2, 4, 8 μs.

Energy E_0 T^2 (window)	24 mJ 83%	40 mJ 83%	24 mJ 40%	40 mJ 40%
Cumulus	200	340	50	80
Cirrus (τ = 0.2)	20	30	4	7
Boundary layer	5	8	1	2
Stratospheric aerosol (R = 2)	1	2	0.2	0.4

Table 232: Number of photoelectrons expected for six pulses

ALISSA science objectives:

- The cloud radiation problem: improving retrieval of large-scale cloud cover and cloud property data.

862) M. L. Chanin, M. Desbois, A. Hauchecorne, "ALISSA a French Russian cooperation in the PRIRODA mission." Paper of CNRS - Service d'Aeronomie

- The cloud radiation problem: interpreting small-scale structures of cloud tops
- Measurement of tropospheric aerosols and the boundary layer.
- Stratospheric studies.

MIRIAM (MIR Infrared Atmospheric Measurements). MIRIAM is a sun occultation experiment for atmospheric trace gas spectroscopy. The instrument (German/Russian) was developed by the "Freie Universität, Berlin" (PI: R. Furrer) and by the St. Petersburg State University, St. Petersburg (PI: A. V. Poberovsky), Russia, using the FTIR (Fourier Transform Infrared) spectrometer technique (capable of sun radiation slant path IR absorption measurements during orbital sunsets) from satellites.[863), 864)]

The objective of MIRIAM is to gather long-term datasets (vertical profiles) of various trace gases, like: O_3, N_2O, CO, CH_4, NO, SO_2, NO_2, HNO_3, HF, HCl, NH_3, OCS, HCN, CH_3Cl, C_2H_2, H_2O, HBr, ClO, HOCl, H_2O_2, OH, H_2CO, C_2H_6, etc.

The FTIR-spectrometer is mounted outside the PRIRODA module in a hermetic (N_2-pressurized) container. Solar radiation enters the spectrometer via an active suntracker and a telescope. The light passing the beamsplitter (potassium bromide) separates into two beams prior to entering the retro-reflector arrays. The amplified detector signal is noise-filtered before A/D conversion (14 bit resolution). Prior to Fourier transformation the signal biases are removed; the interferograms are apodized with a Hamming window.

Type of instrument	Fourier Michelson Interferometer
Design	rapid scan double pendulum with multi-pass capability due to retro-reflectors
Spectral range	2.5 - 20 μm (or 4000 - 500 cm^{-1} wavenumber range)
Spectral resolution	0.01 cm^{-1}, double sided, apodized
Vertical resolution	1 - 2 km
Bandwidth	ca. 4000 cm^{-1}
Scan time	1 second
Detector type	CMT
Beamsplitter	KBr with Germanium coating
Mirrors	Zerodur with gold coating
Telescope	Cassegrain, 70 mm diameter, copper with gold coating
Spectrometer size	85 cm x 50 cm x 20 cm
Repeat coverage	2 months
Instrument mass	100 kg
Data points/spectra	max. 1,000,000

Table 233: Technical specification of MIRIAM

4) MOMS-2P (Modular Optoelectronic Multispectral/Stereo Scanner - 2 Priroda)

Mission context: MOMS-2P is a German imaging instrument on Priroda, a module of the Russian MIR space station. Part of the MOMS-2P instrument (the optical module, which is mounted on the outside of PRIRODA) was launched by a Progress M-31 launch vehicle on May 5, 1996 (docking of Progress at MIR on May 7, 1996). The Progress launch was a regular service flight to MIR, it was preceded by the PRIRODA launch on a Proton vehicle on April 23, 1996. MOMS-2P on Priroda is a modified MOMS-02 (of D2 Shuttle mission heritage, see J.13) instrument featuring an additional integrated navigation package (GPS receiver + Gyro), referred to as MOMSNAV (MOMS Navigation) for more accurate attitude and orbit determination, hence ground positioning of the data. MOMS-2P is a DLR instrument, sponsored by DARA, built by DASA/DSS and by Kayser Threde GmbH (MOMSNAV). Portions of the MOMS-2P (camera system) and the MOMSNAV package (two GPS

863) R. Furrer, H. Rubin, M. Schaale, A. V. Poberovsky, A. V. Mironenkov, Y. M. Timofeyev, "MIRIAM - A Spaceborne Sun Occultation Experiment for Atmospheric Trace Gas Spectroscopy," GeoJournal 32.1, January 1994, pp. 17-27

864) "MIRIAM 1995-1998 MIR-Infrared Atmospheric Measurements - Untersuchung der Atmosphäre aus der Raumstation MIR," Institut für Weltraumwissenschaften an der Freien Universität Berlin, 1994

antennas, two redundant gyro sensor blocks, one gyro electronics unit) are mounted onto the outside of the Priroda module. MOMS-2P is being operated from DLR/GSOC via ZUP, Moscow. 865) 866) 867) 868)

MOMS-2P is a modular instrument with a total of five optical systems: three are dedicated for stereoscopic imagery, two are employed for multispectral imagery, one (central lens) is used for high-resolution data. Two linear CCD (Charge-Coupled Device) arrays are optically combined in the focal plane of the center lens in order to obtain a sufficiently wide swath. Along-track stereo imagery is obtained with the center lens (ch. 5) and two tilted (forward and backward) lenses (ch. 6 and ch.7). The focal length of the tilted lenses was chosen in such a way as to obtain an integral relationship between the ground pixel sizes seen by the center lens and the two tilted lenses; this ratio is 1:3. Each of two MS lenses (ch. 1,2,3,4) also optically combine two linear CCD arrays in the focal plane for a wider swath. Data quantization: 8 bit with seven gain steps [uncompressed for MS and stereo, 6 bit compressed for stereo and HR (High Resolution) data]. Each CCD array features 6000 elements (10 μm per element).

Channel	Mode	Orientation	Bandwidth	Ground Pixel Size	IFOV	Swath Width	Focal Length
1	MS	nadir	449-511 nm	18 m x 18 m	45.45 μrad	105/50 km	220 mm
2	MS	nadir	532-576 nm	18 m x 18 m	45.45 μrad	105/50 km	220 mm
3	MS	nadir	645-677 nm	18 m x 18 m	45.45 μrad	105/50 km	220 mm
4	MS	nadir	772-815 nm	18 m x 18 m	45.45 μrad	105/50 km	220 mm
5 (PAN)	HR	nadir	512-765 nm	6 m x 6 m	15.15 μrad	60/50 km	660 mm
6 (PAN)	Stereo	+21.4°	524-763 nm	18 m x 18 m	42.16 μrad	105/50 km	237.2 mm
7 (PAN)	Stereo	-21.4°	524-763 nm	18 m x 18 m	42.16 μrad	105/50 km	237.2 mm

Table 234: Performance parameters of MOMS-2P (400 km orbit)

MOMS-2P operations are limited by the data rate of the on-board tape recorder of 100 Mbit/s. It implies that all channels cannot be operated simultaneously. A set of four operational modes are defined combining different channels for various applications. Table 235 summarizes the four modes with the corresponding numbers of pixels per imaging line. The tape recorder allows a maximum tape capacity of 48 GByte corresponding to a recording time of 80 minutes for an average data rate of 10 MByte/s.

The MOMS-2P objectives cover applications in the fields of photogrammetry, land cover, geomorphology, ecology, basic research in the spectral signatures of rocks, soil, vegetation, etc. The following capabilities are provided:

- Three-line linear stereo imagery of high spatial resolution and pointing accuracy
- Along-track stereo capability in panchromatic or in defined multispectral (MS) modes
- Detection of surface materials by optimized position bands and detection of small-scale textures of the Earth's surface.

The MOMS-2P instrument is being operated by means of timelines stored onboard the Priroda module. Mission planning selects target areas for imaging along with all information required for execution. A set of four operation modes have been defined, combing different channels to be selected for various applications. The following modes are being operated during the mission:

865) German User Requirements to PRIRODA Mission, Annex 1 of Protocol to MOMS-2 for the PRIRODA Mission, DLR paper of PRIRODA Workshop, May 1991

866) Protocol of the Meeting of Specialists of USSR and Germany on MOMS-2 for the PRIRODA Mission. DLR paper, May 1991

867) D. Meißner, et. al., "The MOMS-2P Instrument and its Mission on Priroda/MIR Station," IAF-96-B.4.03, 47th International Astronautical Congress, Oct. 7-11, 1996, Beijing, China

868) DASA Endbericht, "MOMS-02P auf Priroda/MIR," Doc. No. M2P-DAS-100-RP-001.0, Dec. 12, 1996

Mode/Channel	1	2	3	4	5a	5b	6	7	Swath (km)
Mode A/No. of pixels					4152	4152	2976	2976	50
Mode B/No. of pixels	5800	5800	5800	5800					105
Mode C/No. of pixels		3220	3220	3220	6000				36 (58)
Mode D/No. of pixels	5800			5800			5800	5800	105

Table 235: Operational modes of MOMS-2P

Mode A: Bands 5a, 5b, 6, and 7 allow for the calculation of three-band stereo models, i.e. high precision DTM (full stereo mode).

Mode B: Bands 1, 2, 3, and 4 in various combinations as color composites as well as digital data serve as base for thematic applications like classification relative to lithology, pedology, vegetation etc. (full spectral mode).

Mode C: Bands 2, 3, 4, and 5a fulfil the requirements for the generation of various standard image processing products by use of the three spectral bands, the additional application of the high resolution panchromatic band, and consideration of suitable algorithms (multispectral channel plus HR nadir channel).

Mode D: Makes use of 2 multispectral channels plus two stereo channels (channels 1,4,6,7). It permits DTM generation and thematic applications.

The MOMS-2P source data (along with housekeeping and navigation data) are recorded onto an on-board tape recorder and are transmitted to a ground receiving station (DFD Neustrelitz) during overpasses.

Other MOMS-2P observation restrictions are due to energy budget limitations of the MIR station and/or particular MIR orientations unsuitable for remote sensing.

MOMS-2P calibration:
Geometric calibration. Refers to the generation of exact object side angles for a precise photogrammetric three-dimensional reconstruction of the Earth's surface (considering all geometric scanner influences such as irregularities of the CCD arrays, bending of the focal planes, optical distortions, etc.).
Geometric inflight calibrations are performed on a set of precisely known ground targets in conjunction with navigation data (MOMSNAV). The Catalonia test site in Spain is being used (several times per month) during the commissioning phase.

Radiometric calibration. This method employs a set of ground targets with known spectral signatures.

Sun calibration. The cover of the MOMS-2P camera is coated on the inside with a diffuse reflecting coating. When opening the cover it can be locked at an angle of 20°. A sun calibration is performed by pointing this cover toward the sun. The measurements obtained are compared with data from inflight calibration for reasons of verification.

MOMSNAV (MOMS Navigation). [869)] An integrated navigation package built by Kayser-Threde (Munich) with the objective to provide accurate location knowledge to MOMS-2P imagery. MOMSNAV is a DGPS-based navigation package consisting of the following elements: two GPS antennas, two redundant gyro sensor blocks, one gyro electronic unit, and an electronic box. The instrument provides navigation data which is used in post-processing in combination with the image data. MOMSNAV location knowledge accuracy of the imagery is ≤ 5 m horizontal (1 σ), the relative attitude accuracy is ≤ 10 arcseconds (1 σ). Instrument mass = 41 kg, power = 70 W (average).

869) S. Föckersperger, et al., "MOMSNAV: Location of the Russian Space Station MIR with Differential GPS," Proceedings of the 2nd ESA International Conference on GNC, ESTEC, 12-15 April 1994, pp. 159-165

Sensor Type	IR - Spectrometer ISTOK-1	Imaging Spectrometer MOS-A	MOS-B	VIS Scanners MUS-SK	MSU-E	TV Camera	Lidar ALISSA
Wavelength	4-16 μm	755-768 nm	408-1010 nm	0.5-1.1 μm 8-12.5 μm	0.5-0.9 μm	0.4 - 0.7 μm	527 nm
No. of channels	64	4	13	5	3	1	1
Geo. Resolution (km)	1 x 6	2.8x2.8	0.7x0.65	0.12x0.12 IR:0.3x0.3	23x25 m	0.3x0.3	1.0 vertical 0.15 horiz.
Swath (km)	6	83	83	350	2x27	90	3'
Spectral Resolution	.125 (4-8μ) .25 (8-16μ	1.4 nm	10nm	0.1μm	0.1μm		
Look Angle	0-90°	Nadir	Nadir	39°	Nadir		

Table 236: Overview of PRIRODA optical instruments (without MOMS)

Sensor Type	Passive MW-Radiometers				Active MW Sensors
	IKAR-N	IKAR-D	IKAR-P	R-400	SAR Travers
Wavelength (cm) Frequency (GHz)	cm - GHz 0.3 - 100 0.8 - 37.5 1.35 - 22.22 2.25 - 13.76 6.0 - 5.0	cm - GHz 0.8 - 37.5 1.35 - 22.22 2.25 - 13.76	cm - GHz 2.25 - 13.76 6.0 - 5.0	cm - GHz 4.0 - 7.5	cm - GHz 9.2 - 3.28 23.0 - 1.30
Sensitivity (K)	0.15	0.4-1.5	0.15		
FOV (km)	60	400	750	400	70-100
Geo. Resolution (km)	60	5-15	50-75	50	0.1-0.15
Look Angle	0°	40°	40°	40°	30-40°

Table 237: Overview of PRIRODA microwave instruments

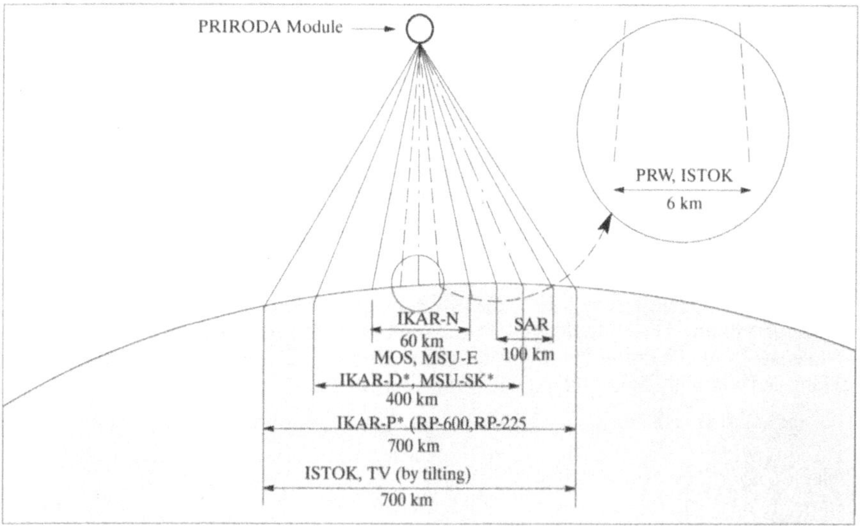

Figure 144: Scheme of overlapping FOV's of PRIRODA sensors [870)]

Data Distribution: Priroda has six on-board data recorders with a capacity 6 Gbit each. One recorder is used for the recording of SAR data, four recorders are used for all other data,

870) IKAR-D, -P and MSU-SK with forward look angle (in flight direction) of 40° against nadir

one recorder is in reserve. PRIRODA sensor data are transmitted to two ground stations, one at RKK Energia in Obninsk, the other at DLR/DFD Neustrelitz, Germany. RKK Energia preprocesses all sensor data and provides all required ancillary data. The time delay for preprocessing is currently projected for about 1 week after data reception.

Agreements relating to data sharing have been concluded between IRE and its partners (the sensor providers and countries participating in the PRIRODA mission) who have primary access to all science data and ancillary information. MOS-OBSOR and MOMS-2P sensor data is being processed and archived by DLR/DFD Neustrelitz as the responsible data center for all German users. Thematic processing may be done at the user sites (SAR Travers data calibration is being planned at DLR in Oberpfaffenhofen). Data transfer from Moscow to Germany may be provided via suitable transportable storage media (Exabyte) or via Eutelsat satellite communication.

D.29 RADARSAT

The origin of Canada's RADARSAT program resides in the necessity to monitor those parts of Canada that are covered with clouds and ice for large portions of the year.

D.29.1 RADARSAT-1

RADARSAT-1 [871)] is a collaborative Canadian developed and operated (CSA = Canadian Space Agency) Earth observation satellite with a microwave (SAR) instrument. Launch: November 4, 1995. NASA is a major partner in the mission, providing launch services with a Delta II vehicle from Vandenberg AFB, in exchange NASA receives RADARSAT data at ASF, Fairbanks Alaska. The S/C prime contractor is Spar Aerospace Ltd. of Toronto, Ontario (Ball Aerospace & Technologies Corp. of Boulder, CO provided the spacecraft bus). [872)]

Field of application	Description
Agriculture and land cover monitoring	Measure of soil moisture, assess crop conditions, update land cover maps
Forestry (boreal and tropical)	Detect clear cuts, update forest inventories, map depletions, map forest fires, assess regeneration, monitor land use changes
Geology	Detect structural and lithologic features, assess geo-hazards (landslides), extract geomorphological information
Hydrology	Measure flood extent, improve hydrological modeling, assess flood damage
Coastal zones and oceans	Detect and track vessels, detect wind and wave spectra, detect mesoscale ocean features, monitor coastline changes
Cartography and land use	Update topologic maps, create digital elevation models, create land use and land cover maps

Table 238: Overview of some key RADARSAT applications

S/C characteristics: three-axis-stabilized with attitude sensors (horizon scanners, sun sensors and magnetometers), momentum wheels and yaw maneuver capability (the S/C uses a set of orthogonally mounted reaction wheels operating around a nominal momentum set point to achieve positive three axis control at all times); momentum desaturation is accomplished with magnetic torque rods. S/C pointing accuracy of 0.1°, pointing knowledge of 0.05°. S/C mass = 3200 kg, payload mass = 1500 kg; solar array power = 3.4 kW BOL (2.9 kW EOL); battery: 3 NiCd (48 Ah); TT&C link in S-band; fuel: 67 kg hydrazine (orbit adjustment and restitution); design life = 5 years.

871) R. K. Raney, A.P. Luscombe, E.J. Langham, S. Ahmed "RADARSAT," reprint from Proceedings of the IEEE, Vol. 79, No. 6, June 1991

872) RADARSAT Annual Review 1997/98, CSA brochure, p. 19

Orbit: Near circular sun-synchronous dawn-dusk orbit at an altitude of 798 km; inclination = 98.6°, ascending node at 18:00 (6 PM equatorial crossing); period = 100.7 min; repeat cycle = 24 days (subcycles of 7 and 17 days); the dawn-dusk orbit has the following advantages:

- The platform is in near continuous solar illumination, providing equivalent opportunities for image acquisition during ascending and descending passes (energy consumption).
- A late equatorial crossing time reduces reception conflicts with other remote sensing satellites

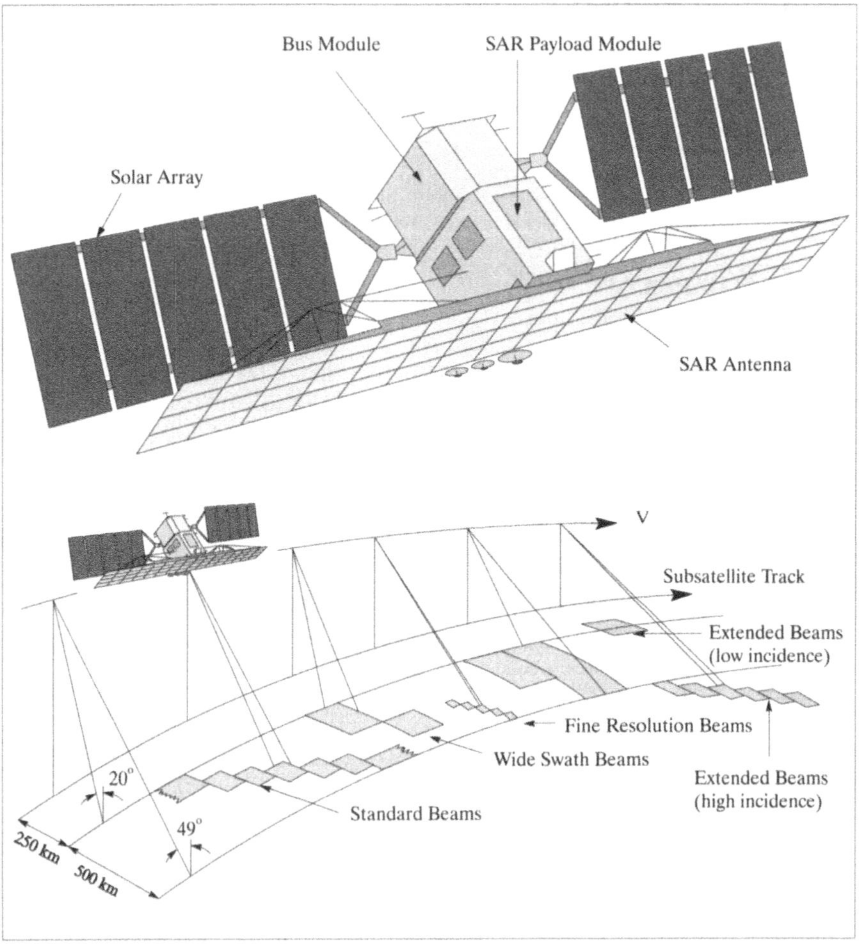

Figure 145: The RADARSAT-1 S/C model and illustration of observation geometries

Sensor:

SAR (Multi-Mode instrument; observation in C-band (5.3 GHz, 5.6 cm wavelength, RF bandwidth: 11.6 MHz, 17.3 MHz, or 30.0 MHz; transmitter power: up to 5 kW), transmitting and receiving horizontally polarized radiation. Choice of three transmit pulses and selection of numerous beams permit a wide range of swath widths, incidence angles and image resolutions. The SAR instrument can operate at a ‘high duty cycle’ capable of imaging for up to 28 minutes per orbit (orbital period of 101 minutes).

Special SAR design features: calibration, rapid data processing, a phased array antenna to provide controlled beam steering, and the first satellite implementation of a radar technique known as ScanSAR. The SAR antenna dimensions are: 1.5 m x 15 m (built by CAL Corp.). The antenna comprises one fixed and four deployable panels, each with 32 slotted waveguides fed by beam-forming networks which produce the varied coverage swaths of the radar.

ScanSAR is a technique permitting extended observation coverage (wider swath) on command. In this mode, rapid steering of the elevation beam pattern of the antenna is essential. Extended range coverage can be obtained by using a set of contiguous beams, enabling images to swath widths of up to 500 km. This is accomplished at no increase in mean data rate from the sensor, but at the cost of degraded resolution of the resulting image. The principle of ScanSAR is to share radar operational time between two or more separate subswaths in such a way as to obtain full image coverage of each.

Mode	Resolution m)* (Range x Azimuth)	Looks**	Swath Width (km)	Incidence Angle (°)
Standard	25 x 28	4	100	20 - 49
Wide (1)	48-30 x 28	4	165	20 - 31
Wide (2)	32-45 x 28	4	150	31 - 39
Fine Resolution	11-9 x 9	1	45	37 - 48
ScanSAR (N)	50 x 50	2-4	305	20 - 40
ScanSAR (W)	100 x 100	4-8	510	20 - 49
Extended (H)	22-19 x 28	4	75	50 - 60
Extended (L)	63-28 x 28	4	170	10 - 23

Table 239: RADARSAT imaging modes [873)]

Data:

On-board recording capability of the total stream for up to 10 minutes (satellite outside of receiving station range). Downlink: two parallel X-band channels at 105 Mbit/s for real-time data and at 85 Mbit/s of recorded data. Ground stations: Prince Albert (Saskatchewan), Gatineau (Quebec), Fairbanks (Alaska). The RADARSAT-1 reception stations are compatible with ESA's ERS series. Data distribution by RADARSAT International Ltd. (RSI) of Ottawa. RSI also distributes SPOT, Landsat, and JERS-1 data in Canada.

AMM (Antarctic Mapping Mission). During the period September 9-11, 1997, RADARSAT-1 underwent a successful maneuver to rotate its normally right-looking radar array into a left-looking attitude. This shift involved rotating the satellite by 180° in yaw. Although other missions do regular yaw maneuvers to reorient S/C, RADARSAT-1 was the first satellite to perform this maneuver to map the entire Antarctic continent, including the South Pole. Note: Complete coverage of Antarctica was not possible with existing or previous spaceborne sensors because of their orbit inclination and/or field of view capability. Almost 70% of the Earth's fresh water is contained in the Antarctic region - changes in this enormous reservoir directly influence world sea levels and climate. The mapping of Antarctica began September 26 and lasted 18 days, followed by another yaw maneuver. Routine operations with the right-looking mode resumed November 4, 1997.

D.29.2 RADARSAT-2

RADARSAT-2 is a jointly-funded satellite mission of CSA (Canadian Space Agency) and MDA (MacDonald Dettwiler Associates Ltd. of Richmond, BC), representing a Canadian government/industry partnership in a commercial venture. In Feb. 1998, CSA awarded a

873) * Nominal: range dependent and processor dependent; ** Nominal: ground range resolution varies with range

contract to MDA to built RADARSAT-2. The contract calls for MDA to develop, own and operate RADARSAT-2 and related infrastructure. CSA provides a fixed financial contribution to MDA (about 75%), in exchange for imagery allocation from the S/C to government agencies. Also, ORBIMAGE, an affiliate of Orbital Sciences Corporation, is a significant participant in this program as the exclusive distributor of RADARSAT-2 imagery to U.S. customers. [874] [875] [876]

US space policy, regulations related to the export of satellite and rocket technology as well as to the distribution of high-resolution imagery, caused CSA in January 2000 to cancel an existing satellite bus contract with OSC and to award a new contract to Alenia Aerospazio of Rome, Italy.

RADARSAT-2 is an advanced follow-on satellite mission of RADARSAT-1 with the objective to: a) continue Canada's RADARSAT program and to develop an Earth Observation satellite business through a private sector-led arrangement with the federal government, b) provide data continuity to RADARSAT-1 users and to offer data for new applications tailored to market needs..

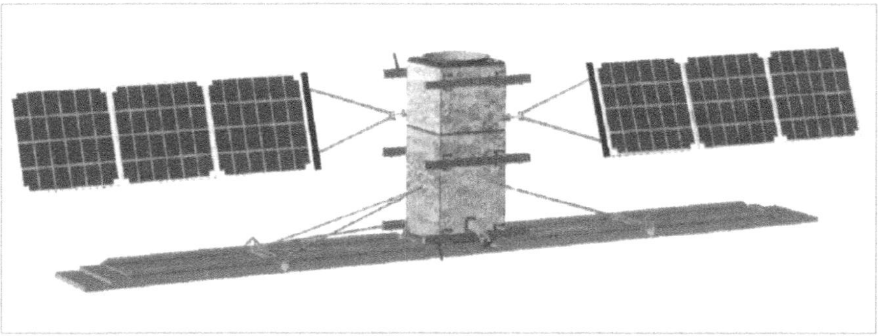

Figure 146: Illustration of the RADARSAT-2 satellite

The S/C, consisting of a bus module and a payload module, is built by Alenia Aerospazio (based on PRIMA, a reconfigurable bus developed for the Italian Space Agency) with bus dimensions of 3.7 m (height) by 1.36 m (diameter). The phased array SAR antenna, built by EMS of Montreal, has 8192 radiating elements fed by 512 T/R modules The overall antenna size is 15 m by 1.5 m. The S/C is three-axis stabilized. ACS (Attitude Control Subsystem) is using star trackers for precision pointing. Attitude knowledge is ±0.03°, attitude control is ±0.05° (3 sigma in each axis). The GPS receiver provides real-time position knowledge of ±60 m. Image location knowledge is <300 m at downlink, and <100 m for processed imagery. S/C mass at launch is 2600 kg., power of 915 W (two solar panels with dimensions: 3.73 m x 1.8 m, each), one Nickel-hydrogen battery with 89 Ah. The design life is seven years. RADARSAT-2 provides several new imaging modes, such as polarimetric imagery (retrieval of full vectorial polarization information), ultra-fine (3 m resolution) beams, in addition to preserving all RADARSAT-1 modes. A launch of RADARSAT-2 is planned for 2003.

Orbit: Sun-synchronous polar orbit, altitude = 798 km, inclination = 98.6°, period = 100.7 min (14 orbits/day), ascending node = 18:00 hrs, repeat cycle = 24 days. RADARSAT-2 is in the same orbit as RADARSAT-1, separated by 30 minutes (and having the same ground track and repeat cycle as RADARSAT-1). This facilitates proper ground station scheduling. In addition, the tandem flight configuration of RADARSAT-1 and -2 provides a wealth of interferometric applications.

874) http://radarsat.mda.ca

875) http://www.space.gc.ca/csa_sectors/earth_environment/radarsat/default.asp

876) P. Fox, "The RADARSAT-II Mission," Proceedings of IGARSS'99, Hamburg, Vol. III, June28 - July 2, 1999, pp. 1500-1502

Sensor complement:

SAR (Synthetic Aperture Radar). The SAR instrument specification is outlined in Tables 240 and 241 with the following functional capabilities:

- Operationally, the SAR instrument is identical to that of RADARSAT-1 (with regard to mission control, mission planning, and data collection and processing)
- Fully integrated into the RADARSAT-1 ground segment
- Supports tandem mission operations with RADARSAT-1
- All RADARSAT-1 image quality specifications are met or exceeded
- A 3 m ultra-fine resolution is provided
- A selectable incidence angle and the choice of **left- and right-looking imaging capability** is provided with the ability to maneuver the S/C on command. This feature offers much quicker revisit times for handling user requests.
- Provision of fully polarimetric imaging modes
- An increased downlink power permits the use of 3 m antenna dishes for ground receiving stations
- The encrypted downlink science data stream maintains service confidentiality
- The utilization of on-board GPS receivers offers real-time position knowledge within 60 m
- The functional exploitation of solid-state recording technology permits the provision of new customer services (selection of image scenes and receiving station)

Frequency	5.3 GHz, C-band (5.6 cm wavelength)
RF bandwidth	11.6, 17.3, 30 or 100 MHz
Transmitter power (peak)	1650 W (normal mode), 2280 W (ultra fine mode)
Antenna	Size: 15 m x 1.5 m; type: active phased array antenna
Antenna polarization	Simultaneous reception of two polarizations: H and V or LHC and RHC. Selectable transmission of one H, V, LHC or RHC polarization. Quad polarization beams allow dual transmit polarization and dual receive polarization.
Imaging mode delay time	10 ms
X-band downlink margin at 5° elevation	9 dB (allows 3 m receiving ground station antenna)
Data rate (max)	105 Mbit/s [recorded (encrypted) and realtime]

Table 240: SAR instrument parameters

	Beam modes	Nominal swath width	Incidence angles to left or right side	Nr. of looks	Spatial resolution (approx.)
RADARSAT-1 modes with selective polarization Transmit H or V Receive H or V or (H and V)	Standard	100 km	20°-50°	1x4	25 m x 28 m
	Wide	150 km	20°-45°	1x4	25 m x 28 m
	Low incidence	170 km	10°-20°	1x4	40 m x 28 m
	High incidence	70 km	50°-60°	1x4	20 m x 28 m
	Fine	50 km	37°-48°	1x1	10 m x 9 m
	ScanSAR wide	500 km	20°-50°	4x2	100 m x100m
	ScanSAR narrow	300 km	20°-46°	2x2	50 m x 50 m
New RADARSAT-2 modes					
Polarimetry	Standard Quad polariz.	25 km	20°-41°	1x4	25 m x 28 m
	Fine Quad polarization	25 km	30°-41°	1	11 m x 9 m
Selective single polarization Transmit H or V Receive H or V	Triple fine	50 km	30°-50°	3x1	11 m x 9 m
	Ultra-fine wide	20 km	30°-40°	1	3 m x 3 m
	Ultra-fine narrow	10 km	30°-40°	1	3 m x 3 m

Table 241: SAR imaging modes of RADARSAT-2

Agreements between CSA and MDA make MDA the S/C owner and operator. The RADARSAT-2 ground segment is also owned and operated by MDA. This includes the re-use of the existing RADARSAT-1 infrastructure where possible. CSA's investment will be recovered through the supply of imagery to a number of Canadian government agencies during the mission lifetime. RSI (as well as others) are a commercial distributors of imagery.

Data communication: An on-board recorder (capacity = 2 x 128 Gbit BOL and 100 Gbit EOL) provides recording of the source data outside of the receiving station range. Downlink of two parallel X-band channels at 105 Mbit/s for real-time data reception.

D.30 RADCAL (Radar Calibration Satellite)

A microsatellite built by CTA Space Systems of McLean, VA for the USAF under the Small Test and Small Launch Vehicle (ST&SLV) program. The satellite is hexagonal in shape with 762 mm diameter, it has a total mass of 87kg. It is passively stabilized to a nadir-pointing attitude by a gravity gradient boom and magnetic nutation dampers (magnetorquers and magnetometer). A prime objective is to provide space-based radar cross-sectional area calibration (via two C-band transponders) for more than 70 military and civilian radars operating in C-band (the S/C position is known to a fair degree of accuracy, thus allowing ground-based users to calibrate their systems by acquiring the RADCAL satellite and comparing their position with the "known" position). A further objective is to demonstrate spaceborne GPS-based attitude determination. RADCAL also carries a peak power tracker and a UHF Store and Forward payload. The design mission life is three years.

The S/C was launched on June 25, 1993 on a Scout launch vehicle from VAFB, CA. Mission design life is 1 year with a 3-5 year goal. The satellite transmits/receives at 306.775 MHz. [877) 878)]

Initial S/C operations were performed from SMC (USAF/Space & Missile Systems Center at Los Angeles AFB). Since November 1996 RADCAL is controlled from Kirtland AFB, NM (SMC/TEO).

Orbit: Near-circular polar orbit, altitude = 815 km x 765 km, inclination = 89.5°.

Sensor complement:

TANS Quadrex GPS receiver (Trimble Navigation). Objective: to demonstrate post processing attitude determination Two receivers (channels: 6+6, L1 C/A) were modified by Stanford University to measure the differential phase of GPS signals arriving at four microstrip patch antennas mounted on the zenith face of the vehicle and canted outward 17.5°. The phase differences between each of three slaves and a master antenna are downloaded for post-processing.

SSPSR (Small Satellite Power System Regulator), built by PL. The objective is to test improved methods for battery charging.

D.31 RESURS-F

The Russian/CIS Resurs-F[879)] (Resource-F, F1 and F2) satellite series is a photoreconnaissance S/C with short mission lifetimes, in the order of 14 to 30 days [built by TsSKB-Progress (Central Specialized Design Bureau Progress), Samara]. The on-board sensors are film cameras whose data (namely, the films) are recaptured after the end of each mission (Vos-

877) L. M. Ward, P. Axelrad, "A Combined Filter for GPS-Based Attitude and Baseline Estimation," Navigation: Journal of The Institute of Navigation, Vol. 44, No. 2, Summer 1997, pp. 195-213

878) L. M. Ward, P. Axelrad, "Spacecraft attitude estimation using GPS: Methodology and results for RADCAL,." Navigating the 90s: Technology, Applications, and Policy, Proceedings of The Institute of Navigation, National Technical Meeting, Anaheim, Calif., 18-20 January, The Institute of Navigation, Alexandria, Va., pp. 813–825.

tok-based film-return craft, the film camera systems are returned in small spherical descent capsules, which are reused an average of three times). The altitude of the orbit is changed during the mission from 400 km to 170 km.[880)]

Objective: Provision of high-resolution photographic images of the Earth's surface. Monitoring of natural resources, ecology studies, etc.

Basic Orbit:
F1: Near-circular polar orbit (non-sun-synchronous). Altitude = 275 km, inclination = 82.3°.
F2: Near-circular polar orbit (non-sun-synchronous). Altitude = 240 km, inclination = 82.3°

The Resurs-F1 S/C was space-tested in 1975-80, the Resurs F2 S/C in the period of 1989-90. The F1 typically flies missions of two weeks' duration (up to 11 days in standby mode) with three KFA-200 and 2 KFA-1000 cameras on-board. F2 undertakes 30 day missions in lower orbits using the MK-4 camera apparatus.

The S/C angular orientation is in an orbital coordinate system with an accuracy of ±1.5° for roll, pitch and yaw angles and ±0.01°/s for angular velocities. Both types of S/C are equipped with an orbit correction engine. The launch site for F1 and F2 is Plesetsk with a Soyuz launch vehicle. Launches of the Resurs-F series are carried out mainly in the spring and summer seasons. Annually between two or three F1 and one or two F2 S/C are launched, depending on user requirements.[881)]

Since October 1990 the data products of the Resource-F series have been commercially distributed by State Center 'Priroda' (4b Str. Pervomajskaja, Moscow; in Germany by GAF = Gesellschaft für Angewandte Fernerkundung, in the US through Central Trading Systems, Fort Worth, Texas).

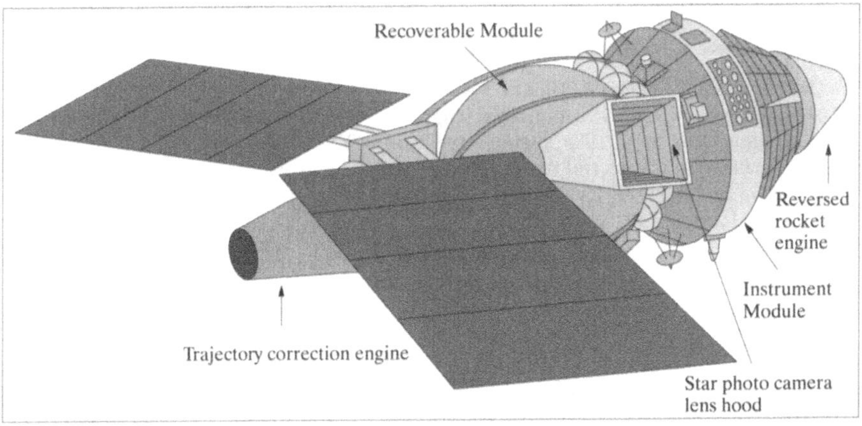

Figure 147: The Resurs-F2 S/C model

Sensor complement:

- **KFA-200** Camera System. Focal length = 200 mm, swath width = 0.9 altitude (245 km), duty cycle = 12%. KFA-200 provides images in three ranges: 0.5 - 0.6 µm, 0.6 - 0.7 µm,

879) 'Sowjetisches kosmisches System zum Studium der Naturschätze der Erde und zur Umweltkontrolle - der heutige Stand und die Perspektiven für den Zeitraum 1991 -1995,' the paper is a translation of a presentation given by L. Dessinow of the USSR Academy of Sciences in 1989.

880) Interavia Space Directory 1990-91, p. 436

881) E. L. Lukashevich, "The Space System Resurs-F for the Photographic Survey of the Earth," Space Bulletin, Vol. 1, No. 4, 1994, pp. 2-4

and 0.7 - 0.9 µm. The cameras provide synchronous images of the Earth surface in an image format size of 180 x 180 mm to a scale of 1 : 10^6. Spatial resolution is between 25 and 30 m. These images provide very suitable correlation data for color synthesis in thematic data products. The geometric properties of the images permit photogrammetric analysis. The precision of altitude determination (method of image triangulation) is in the order of 50 m.

- **KFA-1000** Camera System. Focal length = 1000 mm, swath width = 0.3 altitude (82 km for one camera), duty cycle = 12%. KFA-1000 provides spectro-zonal and panchromatic images. The concept employs a fan-like mounting configuration of two cameras (with a large focal length and wide format) in order to achieve an extra wide swath wide of 120 to 165 km. The film width is 300 mm. The scale of the images (focal length of 1000 mm) is 1 : 200 000 - 1 : 270 000, the resolution on ground (terrain) is 5 - 10 m. The total number of images of one film is about 1800. Spatial resolution = 8 - 10 m.

Parameter	No. of channels	Spectral Bands	Filter	Film	Film Length
KFA-200 (3)	Camera 1 Camera 2 Camera 3	510 - 600 nm 700 - 840 nm 600 - 700 nm	ZHS-18+SZS-23 KS-19 KS-10	T-42 I-840K T-38	230-250 m 230-240 m 230-250 m
KFA-1000 (2)	1 (right side) 2 (left side)	570 - 800 nm 570 - 800 nm	OS-14 OS-14	SN-10 SN-10	580 m 580 m
MK-4 (4 of 6 possible channels)	Channel 1 Channel 2 Channel 3 Channel 4 Channel 5 Channel 6	640 - 690 nm 810 - 860 nm 515 - 565 nm 460 - 510 nm 610 - 750 nm 435 - 680 nm	KS-13 KS-17 interference filter interference filter OS-14 ZHS-11	T-30M I-840K T-30M T-30M CN-10 TsN-4	500 m 500 m 500 m 500 m 500 m 250 m
MK-4M	Channel 1 Channel 2 Channel 3 Channel 4	640 - 690 nm 520 - 560 nm 800 - 870 nm 610 - 760 nm	KS-13 interference filter KS-17 OS-14	T-92 T-92 I-840K SN-18	500 m 500 m 500 m 500 m

Table 242: Photocamera characteristics of the Resurs-F S/C series[882)]

- **MK-4** = Four-Channel Camera System. Apparatus with four photographic lenses. Image frame = 180 x 180 mm, focal length = 300 mm, swath width = 0.6 altitude (144 km), duty cycle up to 19%, spatial resolution = 12 - 14 m.
  Provision of multispectral data in three channels and spectro-zonal image data (colored) in one channel. The operational imaging program allows selection of four zones (channels) out of the six available.

The apparatus operation is based on photometric concepts. This assures images with high-quality geometric references. The sum average distortion of coordinate points in an image is max. ± 9 µm.

Parameter	Resurs-F1M		Resurs F2M
Lifetime	19 days of operation, up to 6 days in standby		30 days of operation
Orbit	altitude = 235 km, inclination =82.3°		altitude = 240 km
Coverage	once/orbit (lat. 0 - 83°, 1/2 overlapping)		once/orbit (lat. 0-83°, 2/3 overlapping)
Photocameras	KFA-200 (1)	KFA-1000 (3)	MK-4M (4-channel camera)
Image size	180 mm x 180 mm	300 mm x 300 mm	180 mm x 180 mm
Focal length	200 mm	1000 mm	300 mm
Swath width	0.9 altitude	0.3 x altitude (1 camera)	0.6 x altitude
Duty cycle	up to 12%	up to 10%	up to 19%
Type of data	B&W	Spectro-zonal B&W	Multispectral in 3 channels, spectro-zonal in 1 channel
Resolution	23-25 m	6-8 m	7 -10 m 8 - 11 m

Table 243: Main characteristics of the modernized Resurs-F S/C

882) Information provided by the State Center "PRIRODA," Moscow

No.	Satellite Type	TASS Launch No.	Launch Date	Inclination	Altitude (km)
1	Resurs-F1	Kosmos-1127	05.09.1979	82.3°	275
2		Kosmos-1185	06.06.1980		275
3		Kosmos-1203	31.07.1980		275
4		Kosmos-1209	03.09.1980		240
5		Kosmos-1280	02.07.1981		275
6		Kosmos-1301	27.08.1981		275
7		Kosmos-1369	25.05.1982		280
8		Kosmos-1376	08.06.1982		280
9		Kosmos-1401	20.08.1982		280
10	Resurs-F1	Kosmos-1440	10.02.1983	82.3°	280
11		Kosmos-1462	17.05.1983		280
12		Kosmos-1468	07.06.1983		280
13		Kosmos-1483	20.07.1983		280
14		Kosmos-1487	05.08.1983		275
15		Kosmos-1498	14.09.1983		275
16		Kosmos1537	16.02.1984		280
17		Kosmos1572	15.06.1984		280
18		Kosmos-1575	22.06.1984		280
19		Kosmos-1582	19.07.1984		280
20	Resurs-F1	Kosmos-1590	16.08.1984	82.3°	280
21		Kosmos-1591	30.08.1984		280
22		Kosmos-1653	22.05.1985		280
23		Kosmos-1657	06.06.1985		280
24		Kosmos-1663	21.06.1985		280
25		Kosmos-1672	07.08.1985		280
26		Kosmos-1678	29.08.1985		280
27		Kosmos-1708	13.12.1985		275
28		Kosmos-1746	28.05.1986		275
29		Kosmos-1762	10.07.1986		275
30	Resurs-F1	Kosmos-1768	02.08.1986	82.3°	275
31	Resurs-F1	Kosmos-1789	31.10.1986		340
32	Resurs-F1	Kosmos-1848	21.05.1987		340
33	Resurs-F1	Kosmos-1882	15.09.1987		275
34	Resurs-F1	Kosmos-1920	18.02.1988		340
35	Resurs-F1	Kosmos1951	31.05.1988		275
36	Resurs-F1	Kosmos-1957	07.07.1988		275
37	Resurs-F2	Kosmos-1965	23.08.1988		275, 355
38	Resurs-F1	Kosmos-1968	09.09.1988		275
39	Resurs-F2	Kosmos-1990	12.01.1989		275, 180
40	Resurs-F1		25.05.1989	82.3°	275
41	Resurs-F1		27.06.1989		275
42	Resurs-F1		18.07.1989		275
43	Resurs-F2		15.08.1989		275
44	Resurs-F1		06.09.1989		275
45	Resurs-F1		29.05.1990		275
46	Resurs-F2		17.07.1990		275
47	Resurs-F1		16.08.1990		275
48	Resurs-F1		07.09.1990		275
49	Resurs-F2		21.05.1991		240
50	Resurs-F1		28.06.1991	82.3°	275
51	Resurs-F1		23.07.1991		275
52	Resurs-F2		21.08.1991		245
53	Resurs-F2		29.04.1992		240
54	Resurs-F1		23.06.1992		240
55	Resurs-F1		19.08.1992		240
56	Resurs-F2		21.05.1993		240
57	Resurs-F1		25.06.1993		240
58	Resurs-F1		24.08.1993		240
59	Resurs-F2		26.09.1995		240
	Resurs F-1M		18.11.1997		
	Resurs-F1		10.7.1998		
	Resurs-F-1M		28.9.1999		

Table 244: Survey of Resurs-F series satellite launches in chronological order [883) 884)]

883) Courtesy of E. L. Lukashevich of State Center Priroda, Moscow

884) Note: For S/C No. 37 and (39), the orbit was changed from an altitude of 275 km (275 km) to an altitude of 355 km (180 km), respectively

D.32 RESURS-O

Resurs-O1[885] is an operational Russian/CIS satellite series (operator: NPO Planeta; S/C integrator: VNIIEM, Moscow). The objective is observing and monitoring natural resources (similar in function and objectives to the Landsat series). Operation of the Resource-O1 series was started in 1985 and is planned to continue at least until 2000 (and beyond). Applications: observation of the state of agricultural crops, assessment of hydrological conditions, forest and tundra fires, pollution monitoring.[886]

Satellite	Launch Date	Orbit	Sensor Complement or payload	Remarks
Resurs-O1-1	Oct. 3,1985 to Nov. 11,1986	Perigee= 574 km Apogee= 663 km Inclination= 98°	MSU-E, MSU-SK, SAR-Travers	Analog sensor output and digital transmission
Resurs-O1-2	April 20, 1988 (operational until June 1, 1999)	Perigee= 620 km Apogee= 678 km Inclination= 98°	MSU-E, MSU-SK	Analog sensor output and digital transmission
Resurs-O1-3	November 4, 1994 Operational	Perigee=660 km Apogee=690 km Inclination=98°	MSU-E, MSU-SK, RRA with SAFIR/R satellite	Analog sensor output and digital transmission
Resurs-O1-4	July 10, 1998 Operational	Sun-synchronous Altitude=835 km Inclination=98.7°	MSU-E1, MSU-SK1, MP-900B, RMK-2, ScaRaB, ISP-2, NINA, IRIS	Digital sensor output and digital transmission
Resurs-O1-5	2003	Sun-synchronous Altitude=630 km Inclination=98.7°	RLSBO, optical and microwave sensors	

Table 245: Overview of Resurs-O mission program

Orbit: sun-synchronous orbit, altitude = about 650 km, inclination = 98°, period = 97 minutes. Orientation accuracy = 6 arc/min along the velocity vector; service life =2 (3) years, payload mass = 420 kg (Resurs-O1) series. Note: the Resurs and Meteor series S/C have the same platforms, they are being designed and built by VNIIEM, Moscow. Note 2: the Resurs-O-1 S/C and up have a higher orbit than the Resurs-O1-1 to -O1-3 series. Resurs-O1-4: Sun-synchronous orbit, altitude = 835 km, inclination = 98.7°. Communication: X-band downlink at 8192 MHz, data rate of 7.68 Mbit/s.

Sensor complement: [887]

MSU-E = High-Resolution Multispectral Scanner with Electronic Scanning (for Resurs-O1-1 to O1-3). Three spectral ranges of 0.5 - 0.6 μm, 0.6 - 0.7 μm, 0.8 - 0.9 μm are provided. Spatial resolution = 45 m x 33 m; swath = 45 km (for one device), and 80 km for both devices (some overlapping), repeat cycle = 18 days. MSU-E has a look angle of ±32° (the instrument may be pointed in the cross-range direction thereby extending potential coverage to a 600 km wide strip (±300 km to each side of the S/C). The observation direction of the instrument can be set by ground commands in steps of 2°. The scan rate of MSU-E is 200 lines/s.

MSU-E1 (Note: Resurs-O-1 has a higher orbit than Resurs-O1-1 to -3). For Resurs-O-1 and up, there are changes with regard to: a) look angle of ±30°; and b) swath width = 60 km (for one instrument), the two MSU-E1 devices provide a total swath of 105 km. Otherwise same specification as for MSU-E.

MSU-SK = Multispectral Scanner of moderate resolution with a conical scanning device and a look angle of 39°. This gives the sensor the advantage that although it is a wide field

885) T.M. Wasjuchina, A.M. Wolkow, "Zustand und Perspektiven der Entwicklung Kosmischer Systeme zur Erforschung natürlicher Ressourcen der Erde und der Hydrometeorologie," Moscow 1988, translated into German by R. Müller, 1989 (IKF)

886) COSPAR-90-Paper by A. Karpov, USSR State Committee for Hydrometeorology, Moscow. Title of paper: "Hydrometeorological, Oceanographic and Earth-Resources Satellite Systems operated by the USSR."

887) Information provided by B. Kutuza of IRE, Moscow, and translated by B. Zhukov of DLR, Oberpfaffenhofen

sensor, it produces constant resolution and viewing angle for all pixels, resulting in increased radiometric accuracy. Spectral bands (5): 0.5-0.6, 0.6-0.7, 0.7-0.8, 0.8-1.1, and 10.4 - 12.6 μm. Spatial resolution = 170m (in VIS) and 600 m (TIR). Swath = 600 km, potential repeat cycle = 3-5 days, orbit repeat cycle = 21 days. The scan rate is 50 lines/s in channels 1-4 and 12.5 lines/s in channel 5. The unusual scanner allows the MSU-SK instrument to have an absolute radiometric error of just 1%, a Signal-to-Noise Ratio (SNR) of 80 dB and a noise equivalent temperature of 0.5 K. Data quantization is at 8 bits. Instrument mass = 21kg. Downlink limitations: four out of five bands may be downlinked simultaneously.

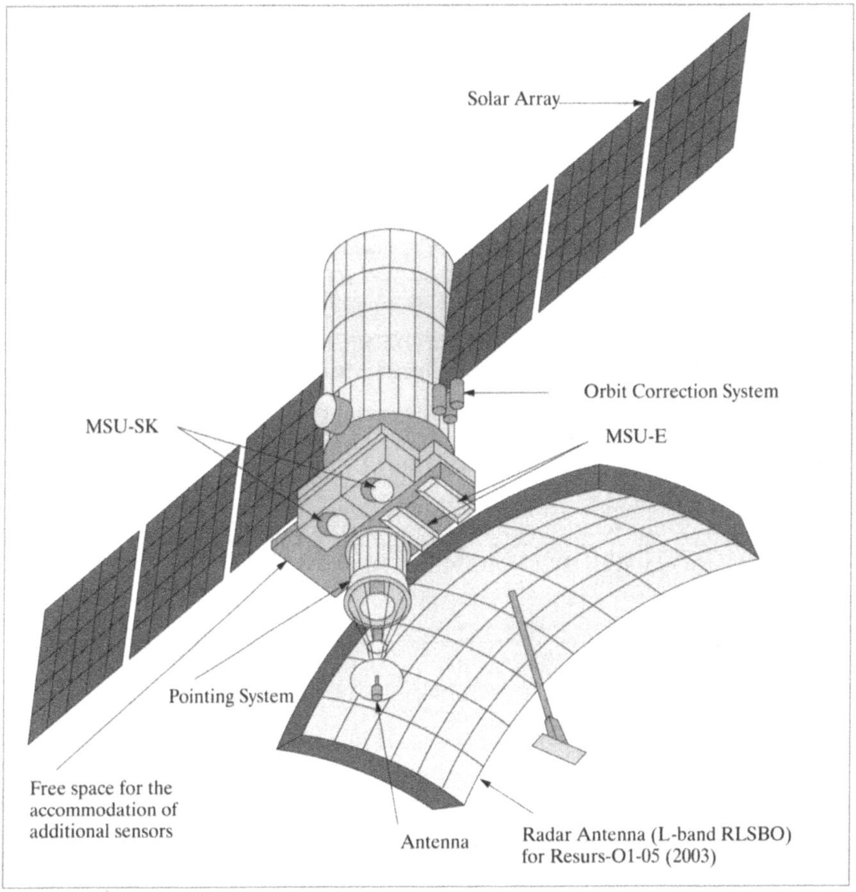

Figure 148: The Resurs-O S/C model

MSU-SK1. Instrument for Resurs-O-1 and up. Spatial resolution = 210 m (VIS) and 700 km (TIR). Otherwise same specification as for MSU-SK. An additional spectral band (3.5 - 4.1 μm) was used for Resurs-O1-4 only.

ISP-2 (Izmeritel Solnechnoy Postoyannoy-2 - a Solar Constant Instrument). The objective is the measurement of the integral solar irradiation (solar constant) and of the shortwave reflected radiation of the Earth system. It is being used to complement the ScaRaB measurements. The ISP-2 radiometer consists of two channels with bolometer detectors:

- Solar constant channel: spectral band of 0.2 - 10.5 μm, FOV = 5°, accuracy of one measurement = 0.05%, accuracy of day-averaged measurements = 0.01%.
- Reflected radiation channel: spectral band of 0.3 - 3.0 μm, FOV = 60°, accuracy = 1%.

RMK-M (Radiation Measurement Control), of RMK-2 heritage flown on the Meteor-3 series satellites. Objectives: Registration of flux densities of protons in the 1-90 MeV and electrons in the 0.17-3.0 MeV energy regions. Measurement of galactic radiation with energies above 600 MeV.

NINA (New Instrument for Nuclear Analysis). NINA is a compact telescope, developed by INFN (Italian National Institute of Nuclear Physics) of Rome, Italy and MEPhI (Moscow Engineering and Physics Institute). [888] The objective is to measure fluxes of charged particles, in particular to detect cosmic ray nuclei of galactic, solar, or other origin from hydrogen to iron, between 10 and 200 MeV/n. The silicon detector telescope is composed of 16 X-Y planes, giving information on the energy of the crossing particle and its incident angle. Each of the 32 sensitive elements consists of two n-type silicon detectors, 60 mm x 60 mm, divided in 16 strips and connected to a supporting ceramic frame under lateral strips (1 and 16). Each couple of detector is glued orthogonal in order to provide X and Y independent view information. The thickness of the detector is 150±15 μm for the first plane, and 380±15 μm for the remaining 15 planes. The geometric factor of the instrument ranges from 8.6 cm 2 sr for low energy particles to 1 cm 2 sr for particles crossing the detector. The instrument mass is 40 kg, power = 40 W. - NINA was flown on RESURS-O1-4 only.

ScaRaB (Scanner for Radiation Budget), of CNES [France (CNES, LMD), Russia (Planeta, RKA), Germany (GKSS) are program partners]. See description under Meteor-3 series.

MP-900M TV Camera. The objective is the monitoring (imaging) of reflected radiation of the Earth's surface and atmosphere (clouds) in the VIS and NIR spectral regions. The spatial resolution of the imagery is 1.6 km x 1.8 km.

RESURS-O1-4 was launched from the Baikonur Cosmodrome on July 10, 1998. Secondary payloads on the Zenit-2 vehicle were: TMSat (also known as Thai-Phutt) of Thailand, the Israeli TechSat/Gurwin-II, WESTPAC of Australia, SAFIR-2 of OHB-System, Bremen, and the Chilean FASat-Bravo.

Parameter		Resurs-O1 Series
S/C mass	including payload mass payload mass alone	1900 kg 500 kg
Earth pointing accuracy	 along the velocity vector	10 arc min 30 arc min
Power supply	average per day peak value	500 W 1200 W
Power supply	voltage range	23-34 V
Lifetime		2 years

Table 246: Resurs-O1 series satellite characteristics

Note: Operations of RESURS-O1-4 continued until April 1999 when a communication failure disrupted all service.

D.33 ROCSat (Republic of China Satellite)

ROCSat is a satellite program of NSPO (National Space Program Office) of Taiwan. A major program objective was to provide a national (NSPO, industry, university, etc.) capability

888) R. Sparvoli, et al., "Launch in orbit of the telescope NINA for cosmic ray observations: preliminary results," Proceedings of The Sixth Topical Seminar on 'Neutrino and Astro-Particle Physics,' Centro Studi 'I Cappuccini' in San Miniato al Todesco, Italy, May 17-21, 1999

and to create an infrastructure and environment for state-of-the-art space age technology and research.

D.33.1 ROCSat-1

The S/C, ROCSat-1, was developed/built by TRW (Space & Electronics Group) of Redondo Beach, CA, as a cooperative development project between TRW and NSPO, offering training capabilities and participation for NSPO engineers in S/C design, testing, and operation/control. The joint development effort started in June 1994. In May 1997, the S/C was returned to NSPO for integration and testing. Some satellite components (OBC, remote interface unit, filter/diplexer and S-band antenna, and solar panels) and instruments/experiments were developed at industry and institutes/facilities in the Republic of China (ROC). [889) 890) 891) 892)]

The ROCSat-1 S/C structure (bus) is a hexahedron in shape and of modular design, accommodating payload growth. The hexahedron has a size of 2.1 m (height) x 1.1 m, the deployed vehicle has an overall length of 7.2 m. The S/C is three-axis stabilized with a reaction control subsystem module, providing pointing control of 0.5°, pointing knowledge of 0.1° and a stability of 0.01°/s. S/C dimensions: 1.10 m diameter and 2.1 m in height (hexahedron shape). Solar arrays of size 1.16 m x 2.46 m provide 450 W of power (645 W max). A reaction control module provides orbit trim capability. Total S/C mass is 402 kg (including propellant), the S/C design life is two years with a goal of 4 years. ROCSat-1 was launched Jan. 26. 1999 (UTC) from Cape Canaveral on an Athena-1 vehicle (Lockheed Martin). S/C operations and data processing are performed from NSPO facilities at Hsin-Chu City, there are two tracking stations at Chung-Li and Tainan, Taiwan.

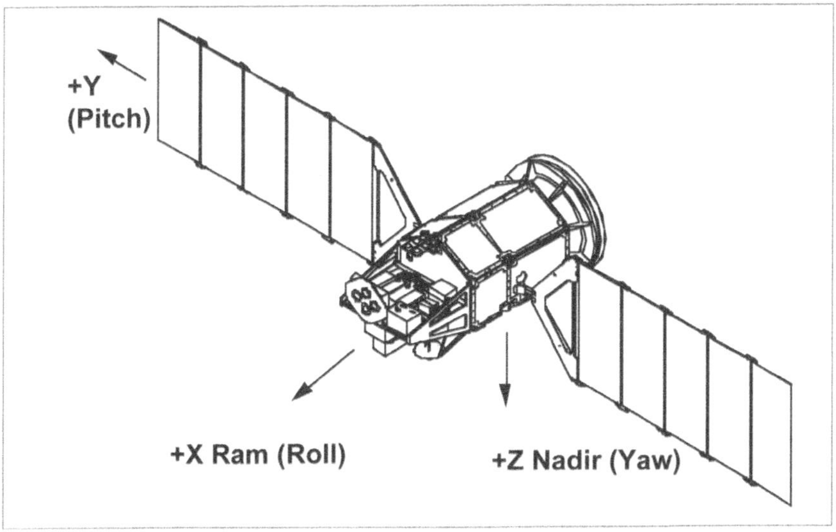

Figure 149: Illustration of the ROCSat-1 spacecraft

RF communications are provided in S-band (uplink frequency = 2.039 GHz, downlink frequency = 2.215 GHz, bandwidth = ±15 MHz, RHCP polarization, power = 6 W). The

889) http://www.nspo.gov.tw/e40/welcome.htm
890) http://www.nspo.gov.tw/e–html.v30/welcome.html
891) W. Ferster, "ROCSat Set to Launch Taiwan's Space Program," Space News, Feb. 1, 1999, p. 7
892) http://tours.oce.ntou.edu.tw/crocsat.htm

uplink data rate is 2 kbit/s, the downlink data rate is 1.4 Mbit/s. On-board storage is provided for 2 Gbit of data.

Orbit: Circular orbit, altitude = 600 km, inclination = 35°, period = 96.7 min.

Sensor complement:

OCI (Ocean Color Imager), built by NEC Corp. of Japan. Objective: Collection of visible and near-infrared radiances over low latitude oceans. The science goals are: mapping of pigment distribution, study of marine productivity and the dynamics of meso-scale eddies, and estimation of influence of the atmospheric aerosols in remote sensing. The instrument is an all-refractive spectral radiometer designed to sense in six spectral bands (443 - 865 nm) with 800 m spatial resolution. Swath width = 700 km (FOV=60.4°).

Spectral bands	B1	B2	B3	B4	B5	B6
Center wave length (nm)	443	490	510	555	670	865
Tolerance (nm)	±3	±4	±4	±4	±4	±5
$\Delta\lambda$ (FWHM in nm) bandwidth	20	20	20	20	20	40
Minimum SNR at EOL	450	450	450	450	350	350
Instrument operating modes	FB, NI-A, NI-B, RGB, CA, SBY, OFF, SUV					
Instrument mass, power	15.2 kg, 33 W (peak), 18 W (standby)					
Date rate	655.5 kbit/s in mode FB via RS-422 (inclusive CCSDS overhead)					
Data quantization	12 bit					
Imaging method, detectors	Pushbroom scanning, linear CCD arrays with 1728 cells per line					
Pixels per band	896 pixels (832 double+64 single cells)					
Absolute radiance accuracy	5% or better at BOL					
Integration time	115.8 ms					
Co-registration error (band 4)	Along-track: <= 0.65 IFOV; cross-track: <=0.97 IFOV					
Instrument optics	Four telecentric dioptric systems (lens system)					
Polarization sensitivity	<=2%					

Table 247: Parameters of the OCI instrument

IPEI (Ionospheric Plasma and Electrodynamics Instrument). Objective: Measurement of the Earth's upper atmosphere, in particular the ionospheric plasma and electrodynamic effects at 600 km altitudes to gain a better understanding of the ionospheric structure and effects on radio communications in the low and middle latitudes (up to 35°). IPEI is a cooperative project between the National Central University of ROC and the University of Texas at Dallas (UTD).

Instrument/Subsystem	RPA	IDM	IT
Measurement	Ion temperature and RAM velocity	Ion transverse velocity component	Total ion concentration
Sampling rate	Normal mode: 332/s Fast mode: 64/s	Normal mode: 32/s Fast mode: 1024/s	Normal mode: 32/s Fast mode: 1024/s
Envelope dimension	SEP: 19 cm x 47 cm x 41.9 cm; MEP: 12.8 cm x 27.9 cm x 13.9 cm		
Instrument mass, power	9.4 kg, 10 W		
FOV	45° x 45°		
Operating temperature	-10°C to about 40°C		
Date rate	Normal mode: 2.2 kbit/s; Fast mode: 53.38 kbit/s		
Duty cycle	Normal mode: 85%, Fast mode: 15%		

Table 248: Parameters of the IPEI instrument

ECP (Experimental Communication Payload), built by MTI (Microelectronic Technology Inc.) of ROC and NEC Corp. of Japan. ECP is a microwave instrument with the objective to conduct various low-altitude satellite communication experiments, in particular the collection of system characteristics and their relationship to different communication technologies (Ka-band telecommunications).

Antenna coverage	±65° conic
Duty cycle (averaged)	8% of an orbital period
Bandwidth	27 MHz
Power consumption	Operating: 85 W peak, on-orbit power: 25 W
Frequencies	Uplink: 28.25 GHz, downlink: 18.45 GHz, beacon: 19.5 GHz
Instrument mass	18.53 kg
Volume	Transponder: 20 microwave boxes distributed in two sides of the payload adapter. Antennas: 25.4 cm diameter (transmit), 20.3 cm (receive)
Transmitter power	10 W (communication channel), 1.77 W (beacon signal)
Polarization	Transmit: LHCP; receive: RHCP

Table 249: Parameters of the ECP instrument

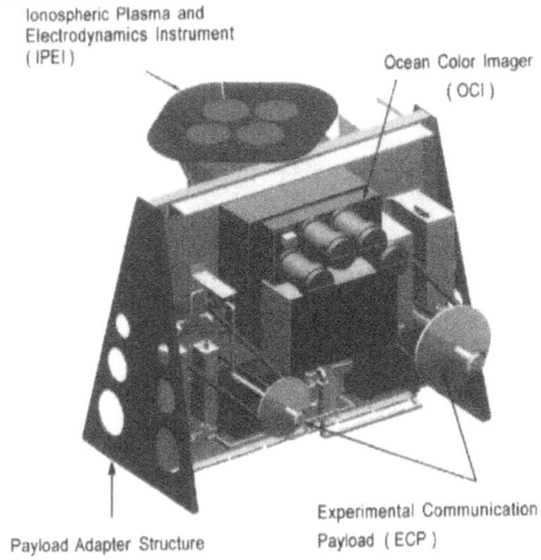

Figure 150: Illustration of the ROCSat-1 payload

D.33.2 ROCSat-2

ROCSat-2 is an NSPO (Taiwan) Earth imaging satellite with the objective to collect high-resolution panchromatic (2 m) and multispectral (8 m) imagery for a great variety of applications such as in land use, agriculture and forestry, environmental monitoring, natural disaster evaluation, and in support of research interests, in particular with the ISUAL instrument. Daily image coverage of Taiwan and the surrounding region is required.

Background: A contract was signed in May 1999 between NSPO and DASA/DSS of Germany to build a high-resolution optical imaging satellite. However, the German government refused to give DASA an export licence for the S/C (the People's Republic of China was protesting the deal). The stalemate was resolved in Dec. 1999 when NSPO signed a new contract with MMS (now Astrium SAS of France). Quick approval of the export of ROCSat-2 was provided by the French government.

The S/C bus is built by Astrium SAS of Vélizy, France, based on the Leostar 500 XO family. The S/C structure consists basically of a hexagonal body of 1.5 m side length. The satellite is three-axis stabilized. The upper deck of the platform carries the payload (RSI and ISUAL) and also part of AOCS (Attitude and Orbit Control Subsystem), namely the star sensors and gyroscopes. The lower deck carries the four reaction wheels and the autonomous propulsion module. The fixed solar array uses GaAs cells and consists of two deployable flaps. The entire S/C architecture is designed in such a way as to provide a low roll inertia, a key factor for satellite agility and instrument line-of-sight stability. The design life is five years. A launch of ROCSat-2 is planned for 2003 (Taurus vehicle from OSC). [893) 894)]

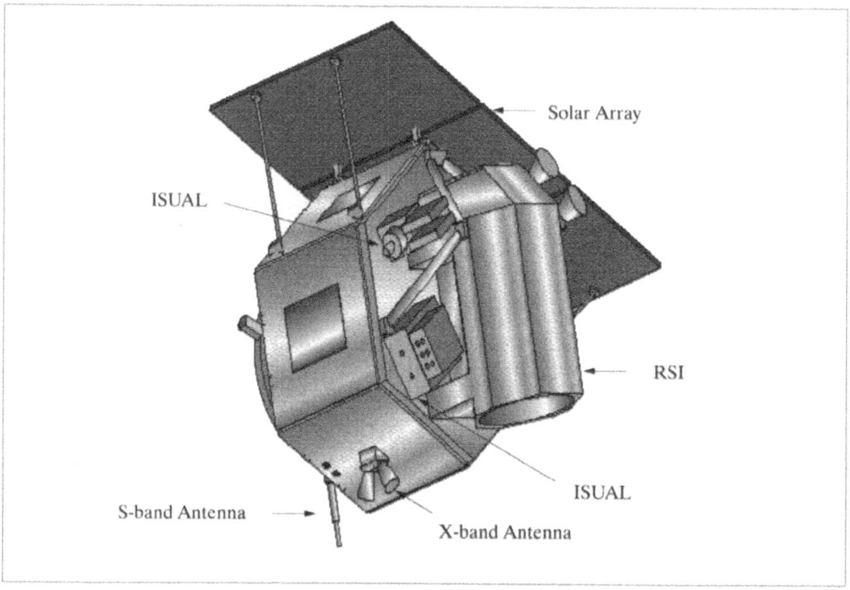

Figure 151: Illustration of ROCSat-2

S/C mass, power	712 kg, 693 W (EOL)
S/C pointing accuracy, knowledge	0.038°, 0.20°
RF communication S-band (TT&C function) RF communication in X-band X-band modulation, polarization X-band link margin at 20° elevation angle X-band station G/T at 20° elevation angle	Data rate: 4 kbit/s uplink, 1.6 Mbit/s downlink Data rate: 120 Mbit/s in downlink NRZ-L, RHCP/LHCP (selectable) >6 dB for clear sky >31.3 dB for clear sky
On-board data storage capability	42 Gbit

Table 250: Some performance parameters of ROCSat-2 spacecraft

Orbit: Sun-synchronous circular orbit, altitude = 891 km, inclination = 98.99°, the equatorial descending local time is between 9:45 to 10:00 AM (14 orbits/day).

The basic elements of the ROCSat-2 ground segment are the MMC (Multi Mission Center) and the XAS (X-band Acquisition System). MMC in turn consists of MOC (Mission Operations Center), MCC (Mission Control Center), SCC (Science Control Center), FDF (Flight Dynamics Facility), and GCN (Ground Communications Network). ROCSat-2 X-band imagery reception is also made available to third parties (international partners) with their own ground stations through cooperative agreements.

893) H. C. Wang, L. C. Lee, J. Ling, A. M. Wu, "ROCSat-2 Remote Sensing Mission," Proceedings of the 51st IAF Congress, Rio de Janeiro, Brazil, Oct. 2-6, 2000, IAF-00-B.1.09

894) J. S. Chern, A. M. Wu, J. Ling, "Some Aspects of ROCSat-2 System Engineering," Proceedings of the 3rd International Symposium of IAA, Berlin, April 2-6, 2001, pp. 57-60

Sensor complement:

RSI (Remote Sensing Instrument), built by Astrium SAS. RSI is made up of the camera and IPU (Instrument Processing Unit). The camera itself consists of the optical sub-assembly, the primary structure, and the FPA (Focal Plane Assembly). The video electronics, sequencer, DC/DC converter, and compression cards comprise the IPU.

Spectral bands: Panchromatic band Four MS bands (nm)	460-890 nm (670 nm center) B1= 460-520, B2 = 540-580, B3 = 630-680, B4 = 770-900 nm
Spatial resolution (GSD)	2 m for PAN, 8 m for MS imagery
Swath width	24 km
S/C body pointing capability, FOR (Field of Regard)	±45° in the roll and pitch axis (providing a cross-track observation capability for disaster monitoring)
Optics, focal length, Aperture diameter	Cassegrain type optics with refractive corrector, 2896 mm, 600 mm
Detector types	CCD: TH 7834 for PAN, THX 31547 quad-linear CCD for MS
Integration time	0.308 ms for PAN, 1.232 ms for MS
Processing rate	10 Mpixel/s for PAN, 5 Mpixel/s for MS
Pixel quantization	12 bit
Data compression ratio	2.8 and 3.8 for PAN, 1.7 and 3.8 for MS
Instrument mass, power consumption	130 kg, 163 W for imaging, 78 W for standby

Table 251: Performance characteristics of RSI

ISUAL (Imager of Sprite Upper Atmospheric Lightning). ISUAL is a joint research program by NSPO, UCB (University of California at Berkeley), National Cheng Kung University, and Tohoku University. The objective is to observe the natural upward lightning discharge phenomenon toward the ionosphere on top of the troposphere.

The instrument consists of four elements: sprite imager, spectrophotometer, array photometer, and electronics package. The imager is a staring-type frame CCD camera, taking 180 frames/s with a resolution of 512 x 80 pixels in a FOV of 20° x 3.15°. ISUAL is operated in three modes: 1) the sprite continuous mode to take images with a sample rate of 100 Hz, 2) the sprite burst mode with a sample rate of 1000 Hz, and 3) the auroral mode at a constant rate of 1 sample/s. ISUAL monitoring occurs during each nightside pass. The source data are stored and compressed in a separate 128 MByte memory. The data are downlinked in S-band. The total data volume is about 1 Gbit/day.

An ISUAL observation sequence of sprites has the following performance characteristics:

- Pointing: Dip down to reduce range and limb view for large coverage
- Location: Storms
- Imager wavelength: N2 1st positive, 427.8 nm
- Photometer wavelength: UV bands, N2 2nd positive, 427.8 nm, N2 1st positive
- Spatial resolution: about 2 km
- Time resolution: 1-30 ms (programmable)
- Mode: Burst mode and sprite mode
- Sensitivity: SNR>10 at 100 MR at 30 ms

An ISUAL observation sequence for a survey of aurora and airglow has the following performance characteristics:

- Pointing: Limb view; Location: All latitudes
- Imager wavelength: All wavelengths
- Photometer wavelength: All wavelengths
- Spatial resolution: about 2 km; Time resolution: 1 s (programmable)
- Mode: Continuous aurora mode
- Sensitivity: SNR>10 at 10 kR

D.34 SAC-C (Satélite de Aplicaciones Científicas-C)

SAC-C is a cooperative Earth observation mission of Argentina, USA, Denmark, Brazil, Italy and France with CONAE and NASA as the main partners. CONAE provides the S/C bus and payload instrumentation, NASA provides launch services and a GPS flight receiver (BlackJack of JPL). INPE of Brazil is providing the system-level environmental testing facilities and its test support capabilities. The overall objective is to study the structure and dynamics of the Earth's surface, atmosphere, ionosphere and geomagnetic field. Specific science objectives are to:

- Provide multispectral images of the Earth in order to monitor the condition and dynamics of the terrestrial and marine biosphere and environment
- Develop and utilize new GPS-based techniques to globally measure atmospheric phenomena for the study of weather, seasonal, inter-annual and long-term climate change
- Enhance the understanding of the Earth's magnetic field and related sun-Earth interactions
- Measurement of the high-energy radiation environment, trapped particles intensities and energy distribution and correlation with degradation of electronic components.

A SAC-C launch on a Delta-2 launcher (secondary payload to EO-1) from VAFB took place on Nov. 21, 2000. [895] [896] [897]

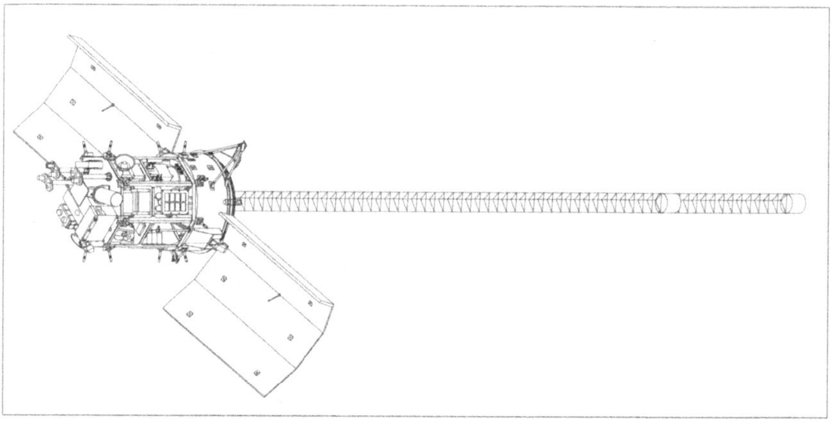

Figure 152: Illustration of the SAC-C spacecraft

The minisatellite structure has dimensions of 1.85 m x 1.68 m x 2.4 m with a total mass of 485 kg (payload = 104 kg, bus = 340.3 kg, fuel = 12.5 kg). The main body of the S/C features an aluminum structure. The antenna booms employ a CFRP (Carbon Fiber Reinforced Plastic) structure. Two fixed solar panels provide 360 W of EOL power (average power consumption of 221 W) and two redundant 12 Ah NiH batteries providing 100 W during eclipse phases of the orbit. The S/C is three-axis stabilized and nadir-pointing. The pointing accuracy required is 1.5° with a post-facto knowledge of 0.2° on three axis, the pointing stability is 0.1°/s. Attitude is maintained by two reactions wheels, three redundant torque rods, eight thrusters for orbit injection/maintenance (12.5 kg hydrazine), and a boom of 8 m in length; while attitude is measured by two partially redundant star trackers [referred to as ASC (Ad-

895) C. Alonso, "SAC-C Mission," presented at the Euro-Latin-American Space Days in Mexico DC in November 1997

896) R. Colomb, C. Alonso, I. Nollmann, "SAC-C Mission and the International AM Constellation for Earth Observation," Proceedings of the 3rd International Symposium of IAA, Berlin, April 2-6, 2001, pp. 433-437

897) CONAE-NASA Workshop, Volume I and II, Dec. 1-2, 1993 - paper provided by J. L. LaBreque of NASA-HQ

vanced Stellar Compass)], two redundant DGPS receivers, two horizon sensors, and a coarse sun sensor. All on-board communications are conducted via a Mil-STD-1553B bus. The S/C design life is 4 years.

Orbit: Sun-synchronous circular polar orbit, altitude = 705 km, inclination =98.2°, equatorial crossing on descending node at 10:15 AM, repeat cycle = 16 days (with sub-cycles of 7 and 9 days).

RF communications: X-band and S-band data downlinks provide real-time and playback data transmissions. There are two S-band transmitters for MMRS low-resolution imagery transmission in real-time with a data rate of 0.9435 Mbit/s, recorder data are transmitted at 1.887 Mbit/s. In addition there are two X-band transmitters for high-resolution MMRS imagery at 3.774 Mbit/s. The on-board storage capacity is 768 Mbit. The spacecraft mission is operated by the CONAE Control Center at Cordoba, Argentina.

As of early 2001, the SAC-C spacecraft is part of a "morning constellation" train, a formation consisting of the NASA spacecraft Landsat-7, EO-1, Terra, and SAC-C (separation times are between 1-15 minutes). The objective is to explore new synoptic observational capabilities and to compare coincident imagery.

Sensor complement:

MMRS (Multispectral Medium Resolution Scanner). MMRS is provided by CONAE. Objective: Monitoring of agricultural production and the influence of climatic anomalies over agriculture and forestry. The instrument features a 5-band CCD linear array pushbroom imager. The swath width of the imagery is 360 km. There are two modes of operation: a high-resolution mode with a spatial resolution of 175 m, and a low-resolution mode of 350 m resolution (real-time transmission with low-resolution data).

Spectral bands (5) in nm	480-500, 540-560, 630-690, 795-835, 1550-1700
FOV (swath), IFOV (spatial resolution)	29.42° (or 360 km), 0.014° (or 175 m)
Optics system: effective focal length, f/number	40 mm, f/3.1
Data quantization	8 bits
Detectors	2048 x 4 silicon array (VNIR) , 2100 x 1 (SWIR)
Data compression	Onboard JPEG data compression at rates of 4:1 to 10:1
Co-registration of bands	±2 pixels (life time), ±0.25 pixel (one orbit)
Data rate (downlink transmission)	0.943 Mbit/s (low-rate mode) 3.3.774 Mbit/s (high-rate mode)
Instrument mass, power	22 kg, 25 W (peak)

Table 252: Technical characteristics of MMRS

Instrument pre-flight and on-orbit calibration of MMRS is provided. The on-board calibration system uses sunlight (in the south pole region of the orbit) for calibration. A cover mechanism in front of the instrument optics is partially opened for this purpose to reflect the sunlight into the optics and to perform a calibration.

HRTC (High-Resolution Technological Camera). A CONAE panchromatic instrument with the objective to acquire high-resolution imagery over portions of MMRS scenes to aid in the data analysis. HRTC has a spectral range from 400 - 900 nm, a spatial resolution of 35 m, and a swath width of 90 km. A scene size of HRTC is 90 km x 1150 km. Imagery of HRTC is regularly stored onto the on-board recorder prior to downlink transmission. Some instrument characteristics: the CCD detector array has 2592 pixels; data quantization = 8 bit; IFOF = 0.0027°; FOV= 7°; instrument mass = 8.5 kg; power = 10.5 W. HRTC has a mirror capable of cross-track pointing. The objective is to locate the HRTC swath of 90 km anywhere within the bounds of the MMRS observation of 360 km width.

HSC (High Sensitivity Camera). A small CCD imaging camera of CONAE with the objective to study a number of phenomena on the night pass of the orbit such as light intensities in urban regions, lightning in thunder storms, forest fires, and polar auroras. HSC has a spectral range from 450 - 850 nm, a spatial resolution of 300 m, and a swath width of 700 km.

Ørsted-2 (Magnetic Mapping Payload). Primary objective: mapping of the Earth's magnetic field. DSRI (Danish Space Research Center) is providing the Ørsted-2 suite (of Ørsted heritage) sponsored by Denmark. Ørsted-2 consists of a deployable boom (8 m in length) mounted with a vector magnetometer and a star imager (0.5 arcseconds orientation), further included are the associated support electronics to control the sensors and the boom deployment, a power control unit and a command and data handling unit. A scalar helium magnetometer placed on the tip of the boom and associated support electronics from JPL completes the Ørsted-2 instrument package. The magnetic field measurements have a resolution of 1 nT at scalar and 2 nT at vector. - The overall objectives of Ørsted-2 are to perform highly accurate measurements of the Earth's magnetic field:

- To determine models of the main magnetic fields and its secular variation
  - To study the physical properties of the fluid core and the electrical conductivity of the mantle
  - To investigate the correlation between the geomagnetic fields and variations in the length of the day
  - To study the lithospheric structure and evolution
- To study the interaction between the Earth's magnetic field and the solar wind
  - Structure and variability of high-altitude fields and currents
  - Relationships between field-aligned and ionospheric currents in the cleft and cusp
  - External magnetic fields as functions of local time, season, and solar wind conditions
  - Ionospheric signatures of localized processes in the outer magnetosphere
  - Substorm processes.

IST (Italian Star Tracker). An ASI-funded technology demonstration instrument developed by Alenia Spazio. [898]) The objective is to test a fully autonomous system for attitude determination using a star tracker. IST is able to provide in real-time the estimated attitude of the instrument boresight and its rotation rate, without any support from ground or other attitude sensors. IST is a prototype for a future multi-head system where a single EU (Electronic Unit) drives up to three Optical Heads (OH) to output absolute attitude and angular rate with high reliability and accuracy. - The prototype demonstrated accuracies in the order of a few arcsecs, continuity of operation, capability to recover from "lost in space" within a few seconds. The system is based upon a library of algorithms for autonomous attitude determination including reduced star catalogues, pattern recognition, estimation and self-calibration algorithms. IST is a single-head prototype composed of:

- One EU, developed by Laben, is based on an ERC32 chip-set processor. The CPU is based on the TEMIC ERC32 chip set (SPARC V, 32-bit RISC, 10 Mips) 12 MHz, 2 Mbyte RAM, and 512 kByte EEPROM. An STD-MIL-1553B external bus interface is provided.
- One OH (including baffle) of Alenia Difesa Officine Galileo. The optical system employs radiation hardened lenses (double Gauss modified), FOV= 8.23^{o} x 10.96^{o}, focal length = 46 mm, f/1.5 number, spectral range of 550-950 nm, CCD array of 288 x 384 pixels, a Peltier cooling system (3 stages), IFOV of 103 arcsec, integration time of 10 ms to 1 s (in 10 ms steps), data quantization of 16 bit (100 kHz), CCD readout time of 35 ms.
- The software package with attitude determination algorithms is developed by Alenia Spazio and EICAS Automazione.

IST provides as output the attitude quaternion and angular rate estimates with a frequency of 10 Hz. The average accuracy of the system, estimated after the first months of operation,

898) Information provided by Andrea Bacchetta of Alenia Spazio, Torino, Italy

is: 12 arcsec (x axis), 4 arcsec (y axis), and 2 arcsec (z axis). The IST instrument mass is: 3 kg (EU)+2.5 kg (OH)+ 0.6 kg baffle, power of 19-26 W.

INES (Italian Navigation Experiment). INES is composed of two separate systems: the GPS Tensor receiver and LAGRANGE (Laben GNSS Receiver for Advanced Navigation, Geodesy, and Experiments) of Laben S. p. A. Vimodrone, Italy. Both instruments are funded by ASI.

- LAGRANGE is a 12-channel, dual GPS/GLONASS receiver that operates at L1 and L2 frequencies. It can provide raw data, including differential L1/L2 measurements that can be used for such applications as geodesy and atmospheric profiling. Navigation accuracies are 80 m for GPS and 10 m for GLONASS. Integrated Doppler precision, sampled at 1 Hz, is listed as 1 mm for GPS and GLONASS.
- The GPS Tensor unit is utilized by the SAC-C S/C as the primary AOCS sensor providing navigation and attitude solutions. The prime objective is to validate the attitude determination capabilities of the GPS Tensor instrument and to demonstrate new filtering techniques in combination with instrument solutions provided.

GOLPE (GPS Occultation and Passive reflection Experiment), a NASA/JPL instrument called BlackJack (see E.1) with the objective to observe ionospheric electron content, to provide atmospheric soundings permitting the derivation of atmospheric profiles of density, pressure, and temperature (refractive occultation monitoring), and to study GPS signals reflected from the Earth's surface.

The GPS receiver assembly of GOLPE consists of a multi-antenna system (four independent high gain antennas pointed in the zenith direction, in the fore and aft direction, and in the nadir direction, respectively). The zenith-pointing antenna is used for simultaneous tracking of GPS satellites. The receiver performs phase measurements of both GPS frequencies L1 and L2, and derives pseudoranges from these measurements for orbit determination.

- In limb radio-occultation monitoring the high gain and aft antennas receive the signals from the setting and rising satellites of the GPS constellation. This permits to determine atmospheric temperature and water vapor at about 500 occultations per day which are fairly uniformly distributed over the globe.
- The nadir pointing antenna is used to determine the utility of GPS signals reflected from Earth's surface to characterize the elevation and roughness of the Earth's surface for applications such as the determination of oceanic circulation and surface winds.
- The dual frequency capability of the instrument receiver permits the determination of ionospheric total electron content (TEC) along the direct, refracted and reflected ray paths, thereby providing data valuable for the tomographic determination of ionospheric structure.
- The ionospheric TEC measurements capability is an important complementary measurement to the Ørsted-2 (MMP) instrument data, thus permitting the derivation of parameters as: ionospheric and magnetospheric currents and conductivity.
- The data of GOLPE also help to improve the estimates of ionospheric delay which degrade the accuracy of single-frequency radar altimeters.

ICARE (Influence of Space Radiation on Advanced Components), a radiation monitoring instrument of CNES, France. The objectives are to support: [899)]

- The improvement of risk estimation models for radiation effects on the latest generations of IC technologies
- Updates of environment models for radiation responsible for electronic component degradations and breakdowns
- Real-time monitoring of environmental conditions for preventive measures.

899) Information provided by Robert Ecoffet of CNES

ICARE is composed of a set of radiation detectors (electrons, protons, ions) associated with a component test board. The radiation detectors are made of silicon fully depleted solid state detectors used in single and coincident mode. The on-board measurements consist in accumulating energy loss spectra in the junctions over a programmable accumulation period. The spectra are generated through signal amplitude classification using 8 bit A/D converters and resulting in 256 channels histograms. The reference levels of the discriminators and gain of the amplifiers are programmable to provide for possible on-board tuning optimization. The accumulation time for the spectra can be programmed from 4 to 256 s. - The component test board functions independently of the detector modes. It can be switched on or off in each mode. The purpose of this board is to measure single event effects rates and total dose parametric drifts on a set of advanced components. In its present version, the test processor is an AD2101 signal processor, single event test devices, commercial sub-micron 1 to 4 Mbit SRAMs, 16 to 64 Mbit DRAMs, and total dose test devices bipolar and CMOS operational amplifiers. The component test board is a separate module that is assembled on the main instrument box by 6 screws. The module can be changed from one mission to another. The instrument is also able to function if no module is connected. - An ICARE communication interface is available for dialog through a single RS-422 or a MIL STD 1553B bus interface. The component experiment module downloads its results in an asynchronous way when its result buffer is filled with a certain amount of results. ICARE has a mass of 2.4 kg, a size of 28 cm x 15.5 cm x 7.1 cm, and a power consumption of 3 W. The data rate is 2.6 kbit/s (max.).

WTE (Whale Tracker Experiment). A CONAE experiment (in cooperation with the Argentinean Secretariat of Natural Resources and Environment). The objective is to study the migration behavior of the Franca Australis Whale, to learn about its migration paths and to protect the species. WTE is in effect a regular data collection system consisting of two elements: a data collection module aboard the satellite and a number of data modules, the DCP (Data Collection Platforms) function, in the ground segment. A data module has been "installed" on a number of whales, each one containing a miniature GPS receiver to determine the whale location at any given time. In addition, the module contains two transducers to measure the water temperature and pressure. Communication with the space segment is provided by a transmitter and a data management unit. - The satellite module consists of a receiver to acquire the DCP data and the needed logic to handle the data.

DCS (Data Collection System). A "regular" CONAE-provided system (in addition to WTE) with the objective to collect data from user DCPs in the ground segment.

D.35 SEASAT

NASA/JPL Earth observation mission (built by Lockheed and Ball Aerospace Systems). Launch: June 27, 1978 (Atlas-Agena from Vandenberg AFB, CA); end of mission (abrupt power system failure in its 1,502 orbit) October 10, 1978. Mission duration: 70 days (data generation) of 106 operational days. The satellite utilized the Agena upper stage to provide satellite bus functions, including power, telemetry (S-band), attitude control, and command and control functions. A sensor package containing the mission's five experiments was attached to the Agena, as were the experiments' antenna systems. Seasat was three-axis stabilized using momentum wheels and horizon sensors. The vehicle was oriented with the SAR and other antennas remaining nadir pointing and the Agena rocket nozzle and solar panels zenith pointing. S/C size: 21 m length, 1.5 m diameter, total S/C mass=2290 kg.

Application: Ice and ocean monitoring (sea-surface winds, sea-surface temperatures, wave heights, internal waves, sea-ice features, ocean features, ocean topography, and the marine geoid), land use, geology, forestry, and mapping.

Orbit: Polar near circular orbit, 108° inclination, apogee = 799 km, perigee =769 km, period = 101 minutes.

Sensor complement:

SAR (Synthetic Aperture Radar) in HH polarization, look angle = 20°; pixel size = 25 x 25 m (spatial resolution on the surface at 4 looks); radiometric resolution = 5 bit raw data. Sensor transmission frequency: 1.275 GHz (L-band); wavelength= 23.5 cm; swath width=100 km. Antenna: 1024-element phased array antenna of size 10.74 m x 2.16 m; PRF= 1464 to 1640 Hz; pulse duration = 33.9 s; bandwidth (linear FM) = 19.077 MHz; transmitted peak power = 1 kW (nominal). [900] [901] [902]

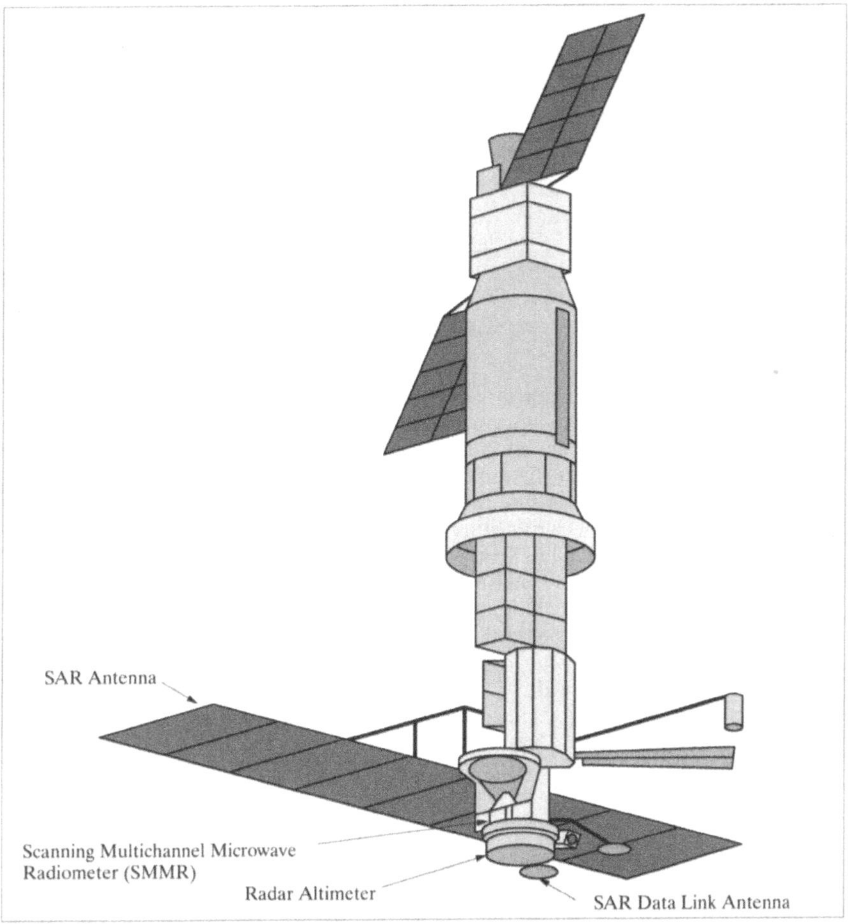

Figure 153: The Seasat S/C model

The Seasat SAR sensor is regarded as the first imaging SAR system used in Earth orbit. A SAR antenna is mounted on the S/C with its boresight oriented at 20° from the vertical

900) Lee-Lueng Fu, B. Holt, "Seasat Views Oceans and Sea Ice With Synthetic Aperture Radar," JPL publication 81-120, February 15, 1982

901) Ch. Elachi, "Spaceborne Imaging Radar: Geologic and Oceanographic Applications," Science, Vol. 209, No. 4461,, September 5, 1980, pp. 1073-1082

902) R. L. Jordan, "The Seasat-A synthetic-aperture radar systems," IEEE Journal of Oceanic Eng., Vol. OE-5, pp. 154-164,1980.

direction (look angle), pointing to the right of the flight path. The antenna beamwidth measures 6.2° in elevation and 1° in azimuth. A footprint of 100 km x 15 km (3 dB contour) is provided (the swath extends from 240 km to 340 km to the right of the S/C ground track). The received radar echoes are downlinked in S-band (analog data link at 2.265 GHz) to a total of five ground receiving stations in real-time (no on-board recording capability) located at: Goldstone, CA, Fairbanks, AK, Merrit Island, FL, Shoe Cove, Newfoundland, and Oakhanger, UK.

SMMR (Scanning Multichannel Microwave Radiometer); Objectives: Monitoring sea surface temperatures, wind speeds, rain rate, atmospheric water content (mapping of columnar water vapor distribution over the global oceans) and ice conditions. SMMR is a multispectral, dual-polarization microwave radiometer observing at the following frequencies: 6.6 GHZ (45.4 mm), 10.7 GHz (28 mm), 18.0 GHz (16.6 mm), 21.0 GHz (14.2 mm), and 37.0 GHz (8.1 mm). Six Dicke-type radiometers were utilized. Those operating at the four longest wavelengths measured alternate polarizations during successive scans of the antenna; the others operated continuously for each polarization. The antenna was a parabolic reflector offset from the nadir by 42°. Motion of the antenna reflector provided observations from within a conical volume along the ground track of the spacecraft. SMMR had a swath width of about 600 km and the spatial resolution ranged from about 22 km at 37 GHz to about 100 km at 6.6 GHz. The absolute accuracy of sea surface temperature obtained was 2 K with a relative accuracy of 0.5 K. The accuracy of the wind speed measurements was 2 m/s for winds ranging from 7 to about 50 m/s. An identical instrument was flown on Nimbus-7 (launch Oct. 24, 1978). [903) 904)]

ALT (Radar Altimeter). [905)] Heritage of S-193 flown on Skylab and ALT flown GEOS-3. Objective: Determination of sea surface profiles, currents, wind speeds and wave heights (first attempt to achieve 10 cm altitude precision from orbit). ALT was a Ku-band compressed pulse radar altimeter (first use of the full-deramp technique). With this new full-deramp technique no compression filter is required in the receiver. From SEASAT onwards, all altimeters have been using this technique, achieving a significant improvement in the resolution. Two of its unique features were a linear FM transmitter with a 320 MHz bandwidth, which yielded a 3.125 ns time-delay resolution, and microprocessor-implemented closed-loop range tracking, automatic gain control, and real-time estimation of significant wave height. This instrument flew the first microprocessor (8080-based controller/tracker) in space. The altimeter operated at 13.5 GHz using a 1-m parabolic antenna pointed at nadir and had a swath width which varied from 2.4 to 12 km, depending on sea state. The precision of the height measurement was 10 cm (rms). The estimate of significant wave height was accurate to 0.5 m or 10%, whichever was greater, the ocean backscatter coefficient had an accuracy of 1 dB.

In the SEASAT design the number of echo samples is increased (compared to GEOS-3). The samples are spaced 3.125 ns apart to encompass the anticipated spread in ocean return for wave heights up to 20 m. In this case waveforms sampling is implemented by a bank of filters with 312.5 kHz bandwidth and spacing. In contrast with previous designs, the samples are an integral part of the altitude tracking process and are used in such a way that the system adapts as a function of wave height to optimize tracker performances. The altitude tracking loop is closed in two parts: a coarse adjustment of the local oscillator pulse timing in 12.5 ns step, and a fine adjustment. - The ALT instrument was designed and built at JHU/APL.

903) E. Njoku, et al., "The Seasat Scanning Multichannel Microwave Radiometer (SMMR): instrument description and performance," IEEE Journal of Oceanic Eng., Vol. OE-5, pp. 100-115, 1980

904) P. N. Swanson, A. L. Riley, "The SeaSAT Scanning Multichannel Microwave Radiometer (SMMR): Radiometric calibration algorithm development and performance," IEEE Journal of Ocean Engineering, Vol 5 No.2, 1980, pp. 116-124

905) W. Townsend, "An initial assessment of the performance achieved by the Seasat-1 radar altimeter," IEEE Journal of Oceanic. Eng., Vol. OE-5, pp. 80-92, 1980

SASS = Seasat-A Scatterometer System
SASS (of S-193 heritage on Skylab) is a fan-beam dual-polarized Doppler scatterometer with the objective of radar backscatter measurements (sigma naught) over ocean surfaces for estimation of the wind field. Pulse transmit frequency of 14.599 GHz (Ku-band). SASS illuminated the sea surface with four fan-shaped beams (two orthogonal beams, each 500 km wide, on each side of the ground track). Doppler filters were used to discriminate resolution cells in the long dimension of the fan beam, resulting in 500 km swaths on either side of the satellite. The high wind swaths added an additional 250 km to each side. The spatial resolution was 50 km over a region of 200 to 700 km on either side of the spacecraft. [906)]

VIRR = Visible and Infrared Radiometer
VIRR is a supporting instrument on Seasat (of SR heritage on NOAA-1, G.13.1) with the objective to provide images of visual reflection and thermal infrared emission from oceanic, coastal, and atmospheric features that might aid in interpreting the data from the other Seasat sensors (also some quantitative measurements of SST and cloud top height). Scanning is accomplished by a rotating mirror mounted at 45° to the optical axis of the collecting telescope (scan angle=±51.2°). VIRR uses a 12.7 cm diameter Cassegrain-type telescope, focusing the radiation onto a field stop. A relay optical system transmits the radiation to a dichroic beamsplitter, which separates it into the visible and infrared wavelengths.[907)]

Parameter	Visible Channel	Infrared Channel
Spectral region	0.49 - 0.94 μm	10.5 - 12.5 μm
Resolution		
angular	2.8 mrad	5.3 mrad
ground (nadir)	2.3 km	4.4 km
Sensitivity (NEΔT)	Not applicable	4 K with a scene at 185 K 1 K with a scene at 300 K
Dynamic range	65 - 10,000 fL (scene brightness)	180-330 K (scene temperature)
Detector	silicon photovoltaic	thermistor bolometer
Scan rate	48 rpm	48 rpm

Table 253: VIRR instrument parameters

Laser Tracking Reflector. A device to support precision orbit determination for Seasat 1. Laser corner reflectors, composed of 96 fused silica 3.75 cm hexagonal corner cube retroreflectors, and ground-based laser systems were used to obtain precise satellite tracking information. The retroreflector array was configured as a single ring of cube corners 1.27 m in diameter. Sixteen of the cube corners were tilted away from the axis of the ring by an angle of 25° and the remaining 80 cubes by an angle of 50°. Because of the great distance of the array from the center of mass of the satellite, the range correction varied from 5.28 m at zenith to 3.08 m near the horizon. When illuminated by laser light pulses from the ground, each retroreflector cube in the array reflected the light pulses back to a telescope/receiver on the ground. A digital counter recorded the time of flight of the laser light pulses from the ground to the satellite and back to the ground. Range was determined from this time. NASA, USAF, SAO (Smithsonian Astrophysical Observatory) and foreign laser tracking stations tracked this satellite.

SAR Data:

Image size: 100 x 100 km (JPL); 100 x 75 km (DLR); four looks full digital image processing with 12.5 m pixel spacing; 8000 lines x 8000 pixels (JPL). Transmission frequency= 11.25 GHz (X-band); data rate = 110 Mbit/s.

Seasat provided global statistics on ocean mesoscale variability, detected strong currents, and showed the potential to detect the longer wavelengths of the circulation.

906) J. W. Johnson, et al., "Seasat-A satellite scatterometer instrument evaluation," IEEE Journal of Oceanic Eng.,Vol. OE-5, pp. 138-144, 1980

907) P. McClain, R. Marks, G. Cunningham, A, McCulloch, "Visible and Infrared Radiometer on Seasat-1," IEEE Journal on Oceanic Engineering, Vol. OE-5, No. 2, April 1980, pp 164-168

D.36 SMOS (Soil Moisture and Ocean Salinity)

SMOS is an approved ESA Explorer Opportunity mission, a technology demonstration minisatellite mission, in cooperation with CNES and CDTI (Center for Technological and Industrial Development) Madrid, Spain. 908) 909) 910) 911) 912) 913) 914)

As of June 2001 SMOS is undergoing phase A activities. The concept review took place in March 2001, and phase A will end in December 2001. The launch date is scheduled for 2005.

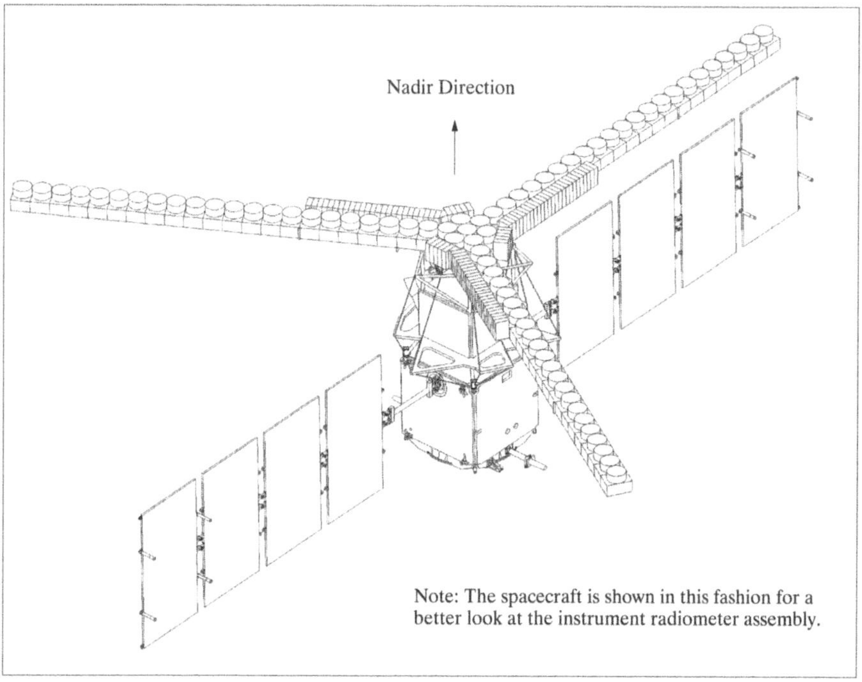

Figure 154: Illustration of the SMOS satellite

The main science objective of the SMOS mission is to demonstrate observations of SSS (Sea Surface Salinity) over oceans and SM (Soil Moisture) over land to advance climatologic, meteorologic, hydrologic, and oceanographic applications. Soil moisture is a key variable in the hydrologic cycle. Over land, water and energy fluxes at the surface/atmosphere interface are strongly dependent upon soil moisture. SM is an important variable for numerical weather and climate models as well as in surface hydrology and in vegetation monitoring.

908) P. Silvestrin, M. Berger, Y. H. Kerr, J. Font, "ESA's Second Earth Explorer Opportunity Mission: The soil Moisture and Ocean salinity Mission – SMOS." IEEE Geoscience and Remote Sensing Newsletter (118), 2001, pp.11-14

909) J. Blouvac, B. Lazaed, J. M. Martinuzzi, " CNES Small Satellites Earth Observation Scientific Future Missions, IAA 2nd International Symposium on Small Satellites for Earth Observation, Berlin, April 12-16, 1999, pp. 11-14

910) M. Martin-Neira, J. Font, M. Srokosz, I. Corbella, A. Camps, "Ocean Salinity Observations with SMOS Mission," Proceedings of the IEEE IGARSS 2000 Conference, Honolulu, HI, July 24-28, 2000

911) Y. H. Kerr, J. Font, P. Waldteufel, M. Berger, "The Soil Moisture and Ocean Salinity Mission -SMOS," ESA Earth Observation Quarterly, No 66, July 2000, pp. 18-26

912) Y. H. Kerr, P. Waldteufel, J. P. Wigneron, J. Font, "Description of the Soil Moisture and Ocean Salinity Mission," COST 712 -WG 3 report, 2001, European Union, Brussels

913) http://www.cesbio.ups–tlse.fr/indexsmos.html

914) J. Font, Y. Kerr, M. Berger, "Measuring Ocean Salinity from Space: the European Space Agency's SMOS Mission," Backscatter (Alliance for Marine Remote Sensing Association), Vol. 11, No 3, 2000, pp. 17-19

[915] [916] Knowledge of the global distribution of salt in the oceans and of its annual and interannual variability, is crucial for understanding the role of the ocean and the climate system. Ocean circulation is mainly driven by the momentum and heat fluxes through the atmosphere/ocean interface, it is dependent on water density gradients, which in turn can be traced by the observation of SSS and SST (Sea Surface Temperature).

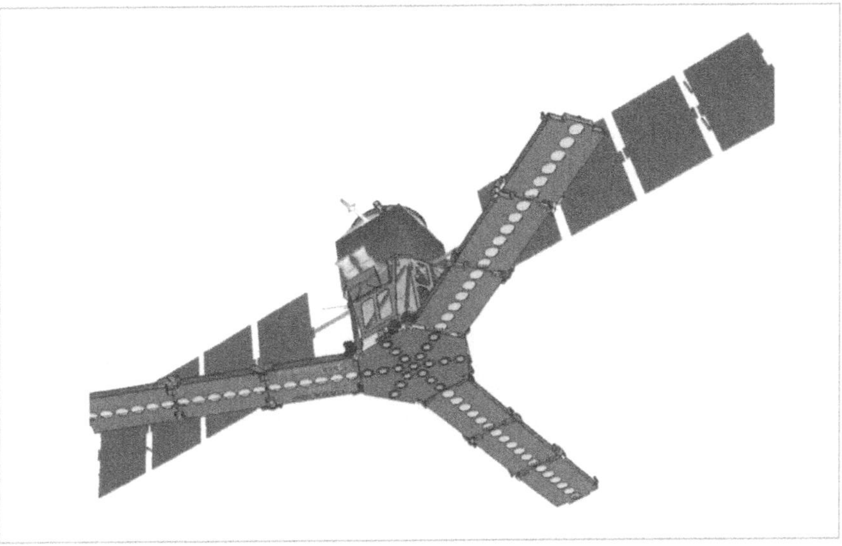

Figure 155: SMOS baseline S/C after PCR (Preliminary Concept Review, April 2001)

Soil moisture can be retrieved from brightness temperature observations. Due to the large dielectric contrast between dry soil and water, the soil emissivity ε at a particular microwave frequency depends upon the moisture content. At L-band in particular, the sensitivity to soil moisture is very high, whereas sensitivity to atmospheric disturbances and surface roughness is minimal. - Mission requirements call for 1) a soil moisture accuracy of 0.04 m^3/m^3 (i.e. 4%) or better; 2) a spatial resolution of <50 km, relevant for large-scale studies; and 3) a revisit time of 3-5 days. [917] [918]

For sea water, the dielectric constant is determined by the electrical conductivity and the microwave frequency. The ocean surface emissivity is a function of the dielectric constant and the state of the surface roughness. In principle it is possible to retrieve SSS from brightness temperature observations. - Mission requirements call for typical values to resolve specific phenomena:

- Barrier layer effects on the tropical Pacific heat flux: accuracy of 0.2 PSU (Practical Salinity Unit) [919], with a spatial resolution of 100 km x 100 km, and a revisit time of 30 days
- Halosteristic adjustment of heat storage from the sea level: 0.2 PSU, a spatial resolution of 200 km x 200 km, and a repeat cycle of 7 days

915) Y. H. Kerr, P. Waldteufel, J.-P. Wigneron, J. Font, "The Soil Moisture and Ocean Salinity Mission: The Science Objectives of an L-band 2-D Interferometer," Proceedings of the IEEE IGARSS 2000 Conference, Honolulu, HI, July 24-28, 2000

916) Y. Kerr, J. Font, et al., "Next Generation Radiometers: SMOS - A Dual Pol L-band 2-D Apertures Synthesis Radiometers," 2000 IEEE Aerospace Conference, March 2000, Montana, USA

917) J. P. Wigneron, A. Chanzy, P. Waldteufel, J. C. Calvet, O. Marloie, J. P. Hanocq, Y. H. Kerr, "Retrieval capabilities of L-Band 2-D interferometric radiometry over land surfaces (SMOS Mission), VSP, Netherlands, 2000

918) J. P. Wigneron, P. Waldteufel, A. Chanzy, J. C. Calvet, Y. H. Kerr, "Two-D microwave interferometer retrieval capabilities of over land surfaces (SMOS Mission)," Remote Sensing Environment, Vol. 73, No 3, 2000, pp. 270-282

919) Note: SSS is defined in practical salinity units (1 PSU=0.1%) and ranges from 32 to 37 PSU

- North Atlantic thermohaline circulation: 0.1 PSU, a spatial resolution of 100 km x 100 km, and a repeat cycle of 30 days
- Surface freshwater flux balance: 0.1 PSU, a spatial resolution of 300 km x 300 km, and a revisit time of 30 days.

The SMOS satellite uses the Proteus bus developed by CNES and ASI (Alcatel Space Industries). This standard platform has been designed to accommodate a wide field of missions, orbits, attitudes, instruments, and launch vehicles. Proteus has simple, well defined interfaces. The platform architecture is generic. Adaptations are limited to minor changes in software modules and launch vehicle interface. The S/C bus is a box, nearly 1 m per side, with all the equipment units accommodated on four lateral panels and the lower plate. Electrical power is generated by two symmetric wing arrays covered with silicon cells which provide 619 W orbital average after 3 year mission (EOL). The power is distributed through a single non-regulated primary electrical bus (23/36 V) using a recurrent SPOT-4 NiCd battery. The S/C is three-axis stabilized. Typical pointing performance of better than 0.05° (3 sigma) is provided by control system with four reaction wheels and gyro-stellar attitude determination. Coarse sun sensor and magnetometers provide attitude measurement and magnetic torquers generate torque. In addition, two of the four reaction wheels are used to provide gyroscopic stiffness. The on-board command and data handling relies on a fully centralized architecture. The DHU (Data Handling Unit) performs most of the tasks through the central processor running the satellite software. It also supports the management of the communication links with all the satellite units either via discrete point-to-point lines or via a MIL-STD-1553B bus. The S/C bus is designed to operate in five distinct satellite modes: 1) normal operations mode, 2) safe hold mode, 3) star acquisition mode, 4) orbit correction mode with 2 thrusters, and 5) orbit correction mode with 4 thrusters. SMOS has a total mass of about 500 kg. The design life is three years with a goal of five years.

Orbit: Sun-synchronous polar orbit, altitude = 755 km, inclination =98.43°, local equator crossing time at 6 AM on ascending node.

RF communications: An on-board solid-state recorder has a capacity of 2 Gbit for payload and TT&C data. Standard S-band communications are used.

Sensor complement:

The **SMOS** sensor is a dual polarized 2-D interferometer operating at L-band (1.41 GHz) and based on the MIRAS (Microwave Imaging Radiometer with Aperture Synthesis) concept. The design of the passive spaceborne instrument uses three coplanar arms consisting of an elementary antenna regularly spaced (0.875 λ, maximum redundancy) in a Y-shaped configuration (also referred to as a sparsely populated antenna, a thinned array is substituted for a filled array). In this concept, an interferometric Fourier synthesis is applied to derive images from the correlations between each pair of antenna elements (small independent receivers) operating in the microwave region. The 2-D SMOS interferometer permits the brightness temperature to be measured simultaneously at different incidences, and at two polarizations. Moreover, the instrument records an entire scene instantaneously. As the S/C moves, a given point within the 2-D FOV is observed from different view angles. A series of independent measurements is obtained permitting the derivation of surface parameters with improved accuracy. The brightness temperature field of such a design is reconstructed with a resolution corresponding to the spacing between the outmost receivers. 920) 921)

The SMOS instrument, operating at L-band (1.4-1.427 GHz, or 21 cm wavelength), employs a Y-shaped antenna arrangement, consisting of three arms, separated at 120°. Every

920) P. Waldteufel, E. Anterrieu,J. M. Goutoule, Y. H. Kerr, "Field of view characteristics of a 2-D interferometric antenna, as illustrated by the MIRAS/SMOS L-band concept, VSP, 2000

921) Y. H. Kerr, J. Font, P. Waldteufel, A. Camps, J. Bará, et al., "Next Generation Radiometers: SMOS A dual pol L-band 2-D Aperture Synthesis Radiometer," IEEE Aerospace Conference, Big Sky, Montana, March 18-25, 2000

antenna arm supports a row of evenly spaced receivers along its span. Each arm consists of three deployable segments. The total deployed arm length is 3.2 m, each providing 25 receivers, 18 are mounted on the deployable arm with six positioned on the hub (non-deployable part). The hub interfaces with the platform and houses some common equipment, such as a total power radiometer for absolute calibration. The instrument structure (modular design) uses CFRP (Carbon Fiber Reinforced Plastic) material for reasons of stiffness.

Background on L-band radiometry: Operating the radiometer in the L-band provides the maximum sensitivity of the emissivity to both SM (Soil Moisture) and OS (Ocean Salinity). However it implies antenna diameters of several meters in order to meet the spatial resolution requirement for SM mapping which, for the design of SMOS, has been set at 50 km. A large real-aperture antenna on a LEO satellite poses a number of practical difficulties. A problem also arises from the need to achieve global coverage of the Earth within three days, which implies a large instrument swath. Microwave imaging by aperture synthesis provides an effective alternative to more classical solutions like the use of mechanically or electrically steered antennas and of pushbroom instruments. This interferometric approach to microwave radiometry, inspired by the techniques developed in radio-astronomy over several decades, is based on the use of many small antenna/receiver units, geometrically arranged so as to sample the signal that would have been received by a real-aperture antenna. This sampling in the spatial domain, enabled by the use of multiple antenna elements, provides a scanning capability through a wide swath by means of (ground) signal processing only.

Background on main instrument characteristics: Each antenna/receiver unit is based on a patch antenna without dielectric substrate with about 70° half-power beamwidth, directivity of about 8 dB and provides both H and V polarizations with excellent cross-polarization characteristics (co-polarization/cross-polarization ratio > 25 dB). A single receiver chain per antenna element is available, so each unit can operate on either H or V polarization upon command from a control unit. In each receiver the antenna signal is filtered to the selected bandwidth (1404 to 1423 MHz, in a region of the spectrum reserved for passive measurements), amplified and finally sampled and converted to a 1-bit digital signal. The output data stream, combining both I and Q components and at a rate of about 130 Mbit/s, is transmitted to the correlator unit by means of an optical fibre link. Each element also receives (via a second optical fibre link) a centrally-generated reference clock signal in order to perform the frequency down-conversion and the sampling with phase coherence among all elements. An oversampling by a factor of about 2 with respect to the Nyquist criterion is achieved in each receiver, which improves the radiometric sensitivity. In the correlator unit, after conversion from optical to electrical signals, a massive bank of 1-bit/2-level correlators implemented in dedicated integrated circuits performs the cross-correlations between all signals. Horizontal and vertical polarization images are interlaced and the cross-correlation for each polarization is performed over a 0.3 s period. Up to 5 images are then averaged, so two images (one per polarization) are available every 3 s.

Each antenna arm provides 3 x 6 receivers with a polarization capability in H+V. The antenna features four balanced feeds to obtain a co/crosspolarized ratio >25 dB over an angular extension of ±30°.; there is also an intermediate stage for E- and H-radiation pattern alignment. Each receiver outputs the I- and Q-digitized signals. Each receiver is furnished with a suitable filter to reject interference from adjacent bands and to shape the frequency response. A multiplexer converts the eight parallel I- and Q-level bit streams from each group of four receivers into a ten-bit serial data stream (quantization) which are fed into a central demultiplexer for serial/parallel conversion. The digital correlator, at the end of the processing chain, performs a complex correlation (at zero delay) between each pair of antenna elements of the interferometric array, resulting in the so-called "visibility function" at the spatial frequency defined by that particular antenna element baseline. The visibility function is ideally the Fourier transform of the brightness temperature of the scene, weighed by the element gain pattern, and recovered by the inverse Fourier transform. The output is the visibility map.

A calibration subsystem, consisting of a set of noise sources that are being fed into the receivers, corrects the phase and modulus errors of the receivers. The calibration frequency depends on the thermal gradient and the receiver behavior with the temperature. The operating temperature of MIRAS is kept within a range of 0 to 50°C.

A typical SMOS receiver includes the following elements: [922] 1) RF device (with an input switch, a filter and an amplifier), 2) a phase/quadrature frequency down converter, 3) two two-level quantizers and samplers. The RF bandwidth is limited to 1400-1427 MHz, the input RF filter band is fixed at 1404-1423 MHz (interference avoidance). The I (In-phase) and Q (Quadrature) outputs of all receivers are processed by a matrix of 1-bit digital correlators, located in the hub.

Parameter	Desired		Acceptable
	Soil Moisture (SM)	SSS (Sea Surface Salinity)	
Coverage	Global (separate)	Global	Global (separate)
Mission duration	3-5 years	3-5 years	3 years
Frequency	L-band	L+C-Band	L-band
Polarization	H+V	-	H+V
Spatial resolution	10 km	20 km	50 km
Revisit time	1-3 days	1-10 days	3 days
Radiometric accuracy	1 K	0.5 K	2 K
Radiometric sensitivity	1 K	0.25 K	3 K

Table 254: Science requirements for the SMOS instrument

The SMOS configuration provides a swath width of 934 km at an orbital altitude of 755 km, sufficient for a 3-day equatorial revisit time. The antenna boresight is tilted at an incidence angle of 33° in the forward along-track direction for nominal radiative flux observations. The instrument tilt maximizes the footprint area for a given resolution goal and antenna spacing ratio. The tilted antenna footprint is represented by the sum of the overlapping individual antenna receivers, each representing a pixel of a distorted hexagon shape with curved sides; all pixel projections are of various sizes and incidence angles. A snapshot brightness temperature map of the FOV is taken every 0.3 seconds with an average resolution of about 5 K over 200 K. Due to the platform motion in orbit, each pixel is measured several times with different spatial and radiometric resolutions and incidence angles. The minimum FOV along-track dimension is about 800 km, this corresponds to 380 snapshots at 0.3 s each. Ground data processing must account for all observation variations in size, shape, angles, overlapping conditions, weighing and averaging schemes, etc., resulting eventually in a brightness temperature map for each snapshot as well as for the accumulated and averaged along-track observation incidences within FOV. The repetitive measurement scheme of radiation, from varying footprints and incidences, has a similar summation effect on the retrieval of the overall signal, as a TDI (Time Delay Integration) scheme for an optical imager.

The overall instrument mass is 305 kg, power =313 W. SMOS data processing requires the introduction of corrections due to atmospheric, ionospheric, and galactic effects over ocean surfaces. In addition, SSS retrieval requires a knowledge of sea surface temperature and sea roughness. For observations over land surfaces, knowledge of the surface temperature is needed with an accuracy of 2 K. These reference parameters have to be obtained from instruments of other missions.

922) I. Corbella, F. Torres, et al., L-band Aperture Synthesis Radiometry: Hardware Requirements and System Performance," Proceeding of the IEEE IGARSS 2000 Conference, Honolulu, HI, July 24-28, 2000

D.37 SPOT (Système Pour l'Observation de la Terre)

The SPOT Earth observation system was designed by CNES (Centre National d'Etudes Spatiales) of France, and developed with the participation of Sweden (SNSB) and Belgium (OSTC). The manufacturers of spacecraft series are: MMS (Matra Marconi Space), Aerospatiale, and Sodern of France. All SPOT launches were on Ariane from Kourou. The SPOT program was initiated by France in 1978, the SPOT S/C series services represent a major provider of long-term and high-resolution imagery for the world. The SPOT program evolved into a worldwide network for receiving and disseminating data. [923] [924] [925] [926]

S/C	Launch	Sensor Complement	Comment
SPOT-1	Feb. 22, 1986	2 HRV	Tape recorder failed in Sept. 1986. Operational activities ceased at the end of 1990, SPOT-1 was reactivated on March 20, 1992). Stopped tracking on August 2, 1993. SPOT-1 was reactivated in March 1994.
SPOT-2	Jan. 22, 1990	2 HRV, DORIS	Tape recorder failed in 1991 and early 1993. Operational as of 2000
SPOT-3	Sep. 26, 1993	2HRV, POAM-II, DORIS	An ACS failure occurred (loss of Earth lock) on November 14, 1997, terminating the operational service life of SPOT-3.
SPOT-4	Mar. 24, 1998	2HRVIR, Vegetation, SILEX, PASTEC, POAM-3, DORIS	Second generation S/C featuring an additional service module for passenger payloads (5 year design life instead of 3)
SPOT-5	planned 2002	2 HRG, HRS, Vegetation,	

Table 255: Overview of SPOT series missions

The spacecraft is 3-axis stabilized with a pointing accuracy of 0.1° using three momentum wheels (unloaded by two magnetic coils). Attitude is sensed by digital Earth sensors (FOV of 2° x 2°). A single solar panel (8.14 m in length) generates power of 1 kW, three NiCd batteries provide 24 Ah for eclipse operations. A hydrazine propulsion system provides orbit maintenance (eight 15.6 N thrusters). Total S/C mass at launch = 1907 kg; payload mass = 790 kg; solar power = 1 kW; nominal design life time of SPOT-1,2,3 = 3 years; S/C dimensions: 2m x 2 m x 4.5 m (solar panel span of 8.14 m); on-board recorders with a capacity of 22 minutes. The S/C structure features the SPOT Mk1 bus for SPOT-1 to SPOT-3 satellites. The SPOT S/C are operated at CNES in Toulouse.

Application: Land use, agriculture, forestry, geology, cartography, regional planning, etc.

Orbit: Sun-synchronous polar orbit, altitude= 832 km, inclination = 98.7°, period = 101 minutes, repeat coverage = 26 days, 10:30 AM local time (descending) equator crossing. Duty cycle: daylight coverage only.

Data transmission: carrier frequency = 8.253 GHz (X-band downlink); data rate = 2 x 25 Mbit/s. A second downlink capability is by tracking beacon at 8.307 GHz.

Sensor complement:

HRV = Haute Resolution Visible (High-Resolution Visible Sensor), using pushbroom CCD detector technology (array of 4 x 1728 pixels). The HRV sensor operates in either of two modes, namely the Multispectral Mode (MS) and the Panchromatic Mode (P). Each SPOT payload consists of two identical HRV imaging instruments that are pointable in the

923) CNES viewgraphs of 1991

924) Jane's Spaceflight Directory 1988-89, Fourth Edition, pp. 22-23

925) Note: SPOT-1 was retired from normal operations in Sept. 1990. Both of its recorders are defect. SPOT Image wants to reactivated SPOT-1 to meet increased demand for satellite imagery. See Space News Dec. 4, 1991, p. 4

926) Note: The board of inquiry investigating the failure of SPOT-3 reported that the successive failure of three of the spacecraft's six gyroscopes caused the satellite to lose attitude control, ran out of power and then shut down within a period of hours.

cross-track direction up to 27° from nadir (each imager instrument is featuring its own tilting mirror for independent off-nadir pointing). This new capability dramatically improves the availability of data for a given area, and also enables (repeat pass) stereo imaging, a key feature for the establishment of digital elevation models.

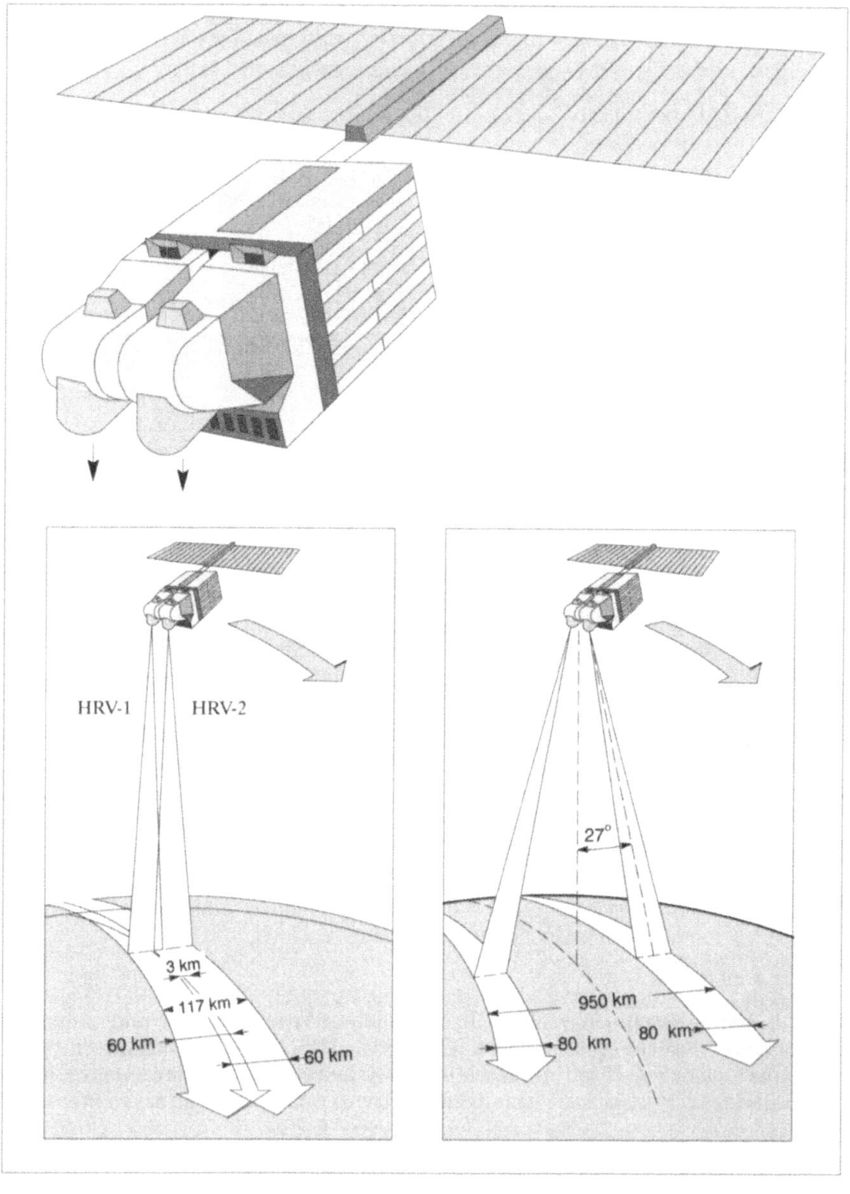

Figure 156: The SPOT S/C model and its observation geometry bounds

Each HRV sensor consists of a telescope with a pointing mirror and in the focal plane a series of CCD detector arrays. Two IFOVs are available, 20 m in the MS bands and 10 m in panchromatic band. Each instrument covers a swath width of 60 km when vertically positioned (both instruments have a TFOV of 117 km, 3 km overlap, nadir pointing). The swath width is 80 km (each) when the instrument is angled 27° to the vertical. Measurement angle: ± 27° from the vertical (in 45 steps). Radiation from the scene being viewed enters the HRV instrument via a plane mirror that is steerable by ground control. The viewing axis can thus be oriented as required in the plane perpendicular to the orbit. This off-nadir viewing capability covers a range of ±27° relative to the vertical (in 45 steps of 0.6° each). The pointing feature increases the revisit capability at intervals ranging from one to several days.

- **HRV Multispectral mode** (MS) three spectral bands (coregistered) : 500 - 590 nm, 610 - 680 nm, and 790 - 890 nm. Spatial resolution in MS = 20 m.
- **HRV panchromatic mode** (black and white images), spectral band: 510 - 730 nm, spatial resolution = 10 m.

The FOV = 4.13° in both modes with 3000 pixels per multispectral line and 6000 pixels per panchromatic line. With its side-viewing feature, the HRV instruments are capable of plotting contours by taking stereoscopic images, as well as imaging a particular location more frequently, without waiting for an overhead pass. - An on-board data compression scheme DPCM (Differential Pulse Code Modulation) is employed for the PAN data stream on the SPOT-3 and -4 missions.

Data: (commercial data distribution by SPOT-Image Corp. of Toulouse and of Reston VA)

- Image size: 60 x (60 - 85) km
- Image size: MS mode: (3000 - 4900 Lines) x (3000 - 5200 pixels)
- Image size: P mode: (6000 - 9800 Lines) x (6000 - 10400 pixels)

DORIS (on SPOT-2 as an experiment and prototype, later on SPOT-3, ff.). DORIS is a CNES-developed tracking system for the determination of precise orbits (see DORIS description under TOPEX/Poseidon, chapter E.21.1).

Station Location	Country	Commissioning Date	Organization
Toulouse/Aussaguel	France	February 1986	CNES
Kiruna	Sweden	February 1986	ESA
Prince Albert	Canada	June 1986	CCRS
Gatineau	Canada	June 1986	CCRS
Hyderabad	India	May 1987	NRSA
Maspalomas	Spain, Canary Islands	November 1988	ESA
Cuiaba	Brazil	April 1988	INPE
Lad Krabang	Thailand	May 1988	NRCT
Hatoyama	Japan	October 1988	NASDA
Islamabad	Pakistan	June 1989	SUPARCO
Hartebeetshoek	South Africa	August 1989	CSIR
Riyadh	Saudi Arabia	October 1989	KACST
Alice Springs	Australia	May 1990	NATMAT
Tel Aviv	Israel	February 1991	IAI

Table 256: Some of the SPOT series direct receiving stations around the world

D.37.1 SPOT-3

SPOT-3 was launched September 26, 1993. The S/C is equipped with 2 HRV imaging sensors + DORIS + POAM-II. The satellite entered safehold on November 14, 1996.

POAM-II = Polar Ozone and Aerosol Measurement (atmospheric chemistry experiment). The instrument was provided by the Naval Research Laboratory (NRL) of Washington, DC (built by ThermoTrex Corp., San Diego, CA); it is sponsored by the Innovative Science and Technology Office of the Ballistic Missile Defense Organization (BMDO) and the Space Test Program of DoD. The objective is to measure vertical profiles of polar ozone, aerosols, water vapor, nitrogen dioxide, atmospheric density and temperature in the stratosphere and upper troposphere. The measurement technique is by solar occultation through the Earth's atmospheric limb at nine wavelengths from the UV to NIR. There are separate filtered optics for each of the nine channels, which are co-aligned with each other and the sun tracker. FOV = 0.01° by 0.75° (slit). Observation of the full width of the solar disk with a vertical resolution of 0.6 km (at the tangent point in the Earth's atmosphere). Instrument mass = 25 kg. 927)

Channel	Primary Measurement	Center Wavelength (nm)	Bandwidth at FWHM (nm)
1	Aerosols	353.0	5
2	NO_2 off	442.0	2
3	NO_2 on	448.3	2
4	O_3	600.0	15
5	O_2 on	760.8	2
6	O_2 off	780.0	15
7	H_2O off	920.0	2
8	H_2O on	935.5	2
9	Aerosols	1059.0	10

Table 257: Spectral coverage of the POAM-II instrument

Parameter	Height Range (km)	Accuracy (%)
Aerosols	10 - 40	5 - 15
O_3	10 - 60	5
H_2O	15 - 40	5 - 7
N_2O	20 - 40	5 - 10
Temperature	10 - 60	1 - 2

Table 258: POAM-II measurement capabilities

POAM-II observes 14 sunrise and 14 sunset events per day. The primary measurements are through the Earth's limb for each of the nine wavelengths as a function of tangent altitude. These measurements are inverted to yield vertical profiles of ozone, aerosols, nitrogen dioxide, water vapor, and oxygen.

The POAM data rate is 32.768 kbit/s. Downlinks are via a dedicated POAM antenna. The downlink is received by a network of ground stations operated by the US Air Force. Uplinks are part of the general SPOT uplinks. Once validated, the reduced data (transmittances) are archived in the BMDO's Background Data Center at NRL, and will be publicly available.

When NASA's TOMS instrument failed to provide operational service aboard Russia's Meteor-3-6 satellite after Dec. 27, 1994, POAM became the only operating US ozone sensor.

D.37.2 SPOT-4

The SPOT-4 S/C was taken out of storage and launched on March 24, 1998 on an Ariane-4 vehicle from Kourou. SPOT-4 is considered a 2nd generation satellite. The most important advance is the addition of the "Vegetation" instrument, with four spectral bands to allow continuous, worldwide crop monitoring. The data may be used for crop forecasts and environmental studies.

927) R. M. Bevilacqua, et al., "Polar Stratospheric Studies with the Polar Ozone and Aerosol Measurement Experiment (POAM-II)," Proceedings of the American Meteorological Society, Eighth Conference on Atmospheric Radiation, January 23-28, 1994, Nashville, TN

The SPOT-4 S/C design differs from the earlier SPOT series (SPOT-1,2,3) in the following aspects: five years design lifetime instead of three; a new extended platform design (the SPOT-4 satellite bus (SPOT MK2, provided by MMS) and service module accommodates twice the payload of the SPOT 3 bus). The propulsion module consists of a frame made of aluminum bars and two capillary tanks holding 158 kg of hydrazine. Attitude is sensed by an inertial platform consisting of four rate gyros, two digital Earth sensors, and two digital sun sensors. The actuators employed are three magnetic-bearing reaction wheels, two magnetic torquers (to control the speed of the reaction wheels), and two types of hydrazine thrusters each producing a force of 3.5 N or 15 N. SPOT-4 offers increased on-board storage capacity (from 22 min to 40 min). In addition, a 10 Gbit solid-state memory was added to increase the overall onboard recording capability. S/C mass = 2755 kg; power = 2.2 kW; pointing accuracy $<0.15^\circ$ (3σ). [928) 929)]

Orbit: Sun-synchronous circular orbit, altitude = 820 km, inclination = 98.8°, repeat cycle of 26 days, period = 101.5 min, descending node at 10:30 AM (equatorial crossing time).

Sensor complement: Improvements: introduction of electronic sensor gains matching according to landscape type and season, thus ensuring greater dynamic range. In addition, the imaging instruments are no longer susceptible to glare or affected by the polarization of the incident light.

HRVIR (two units)= High-Resolution Visible and Infrared sensor (improved version of HRV). Spectral ranges: B_1=0.50 - 0.59 µm, B_2=0.61 - 068 µm, B_3=0.79 - 0.89 µm, SWIR= 1.58 - 1.75 µm (a supplementary band in the 20 m multispectral mode at 1.58-1.75 µm is provided). The panchromatic band (0.51 - 0.73 µm of HRV on SPOT-1,-2,-3) has been replaced by band B_2, enabling the 10 and 20 meter resolution data to be co-registered on-board the satellite instead of on the ground. Same pointing geometries and resolution capabilities as HRV. In addition, the two HRVIR sensors can be programmed for independent image acquisition (viewing directions of both sensors are independent). Both HRVIR instruments are protected against polarization or blinding by direct sunlight.

The main elements of HRVIR are: the telescope, the detection unit, the image processing electronics, and the viewing direction control mechanism. The telescope (catadioptric Schmidt telescope type) has a focal length of 1.08 m and an aperture of 3.5. The incoming beam is split into four spectral channels by a beam-splitter consisting of prisms and filters, then focused onto four rows of linear CCD detectors [one line of 6000 (PAN) and three lines of 3000 (MS) pixels, the four rows of detectors simultaneously generate four lines of "registered" pixels]. The SWIR band uses InGaAs/InP linear arrays developed specifically for the purpose. FOV = 4.13° corresponding to 60 km of ground coverage.
The entrance to each HRVIR features a strip-selection mirror enabling the viewing direction to be adjusted through $\pm 27^\circ$ about nadir. Adjustment is controlled by a stepper motor moving through increments of 0.3°. For each position, the viewing direction is accurate to within 200 m on the ground.

HRVIR calibration: The calibration system allows image signals to be corrected in two ways: (used at regular intervals to check and, if necessary, adjust the instrument response).

- In-band calibration (also referred to as CCD detector response normalization), the aim is to balance the response of the 3000 detectors in each band while the instrument views a perfectly uniform landscape
- Absolute calibration to measure the instrument's dynamic responsivity by establishing a precise relationship between a perfectly stable external source (the sun) and the instrument's output signal.

928) F. Achard, J. P. Malingreau, T. Phulpin, G. Saint, B. Saugier, B. Segun, D. Vidal-Madjar, "The Vegetation Instrument on Board SPOT-4 - A Mission for Global Monitoring of the Continental Biosphere," LERTS brochure, Toulouse, 1990

929) http://spot4.cnes.fr/

VEGETATION. An additional sensor for SPOT-4, called VEGETATION or **VMI** (Vegetation Monitoring Instrument), with a ground swath width of 2200 km, and a resolution close to 1 km [European project with the cooperation of the EC, France (CNES), Sweden (SNSB), Belgium (OSTC) and Italy (ASI), designed by Aerospatiale as the prime contractor and built by Sodern of Limeil-Brivannes], providing the capability of wide-area monitoring of the Earth's vegetation (VMI features advanced optics that allow perfect geometrical rectification of the pictures despite the wide swath width. The VMI optics virtually cancel the curvature of the Earth to provide directly usable geographical information). Calibration accuracy: interband and multitemporal =3%, absolute = 5%; TFOV = 101° (swath width of about 2200 km); pixel size = 1.15 km at nadir; location accuracy < 0.5 km, local distortion < 0.3 pixel, collocation with HRVIR sensor data = 0.3 km for simultaneous acquisitions. Spatial coverage: about 90% of the equatorial regions are imaged each day (total daily imaging for latitudes > 35°).

The instrument collects radiation reflected by the earth's surface. Silicon linear detector arrays are used for spectral band B0 (blue), B2 (red) and B3 (near infrared), while InGaAs photodiodes (Thomson CSF) are used for the SWIR (Short Wave Infrared) band. Each array features 1728 individual CCD detectors. The output voltage signals are fed to a multiplexer (prior to this, SWIR voltages are adjusted for dark current values), then to a single analog to digital converter(ADC). All sensors are processed by a single high precision 11 bit ADC, thus completely by-passing the problem of electronic interband calibration.

The VEGETATION package is an independent add-on payload to the SPOT satellite [the package includes the sensor, a solid-state recorder (up to 97 minutes of imagery), an X- and L-band telemetry subsystem, and a computer]. The instrument design uses four cameras, one for each spectral band, each camera is covering the TFOV. VEGETATION payload parameters: mass = 152 kg, power = 200 W, instrument size = 0.7 m x 1 m x 1 m, data transmit frequencies in X-band (8153 MHz) at 3.4 Mbit/s, and L-band (1704 MHz), data rate of sensor (L-band) = 510 kbit/s. [930) 931) 932)]

Three spectral bands of VEGETATION (B2, B3 and SWIR) and HRVIR (Red, NIR, SWIR) are identical, they register perfectly allowing data to be interpreted at several scales. The combination of high and low spatial resolutions (of HRVIR and VEGETATION) provides a major contribution to the measurement of temporal changes over a few points or at certain time periods for a proper determination of the influence of the various kinds of ground cover. [933)]

Channel	Spectral Range	Surface Reflection Range (albedo)	Radiometric Resolution NEΔR	Optimized to detect
B0 (Blue)	0.43 - 0.47 μm	0.0 - 0.5	0.003	chlorophyll
B2 (Red)	0.61 - 0.68 μm	0.0 - 0.5	0.001 albedo ≤0.1 0.003 albedo of 0.5	vegetation
B3 (NIR)	0.78 - 0.89 μm	0.0 - 0.7	0.003	vegetation, atmospheric correction
SWIR	1.58 - 1.75 μm	0.0 - 0.7	0.003	vegetation, atmospheric correction

Table 259: Spectral parameters of the VEGETATION instrument

The VEGETATION processing center (CTIV) is at VITO (Vlaamse instelling voor technologisch onderzoek – Flemish institute for technological research), located in Mol, Belgium. The commercial distribution of VEGETATION products is provided on a global scale by SPOT Image (France), with CLEO (Belgium), SSC Satellitbild (Sweden) and Telespazio (Italy).

930) Information provided by T. Genet of CNES, Toulouse

931) http://sirius–ci.cst.cnes.fr:8080/

932) http://www.cnes.fr/WEB_UK/activites/programmes/Vegetation/VEGETATION.html

933) R. H. Frazer, Z. Li, R. Landry, "SPOT VEGETATION for characterizing boreal forest fires," International Journal of Remote Sensing, Vol. 21, No 18, 2000, pp. 3525-3532

DORIS (Doppler Orbitography and Radiopositioning Integrated by Satellite), a satellite-based orbit determination and radiopositioning system, see E.21.1. The SPOT-4 DORIS package includes experimental software (DORIS/DIODE experiment) to determine the S/C position in real-time to within < 0.3 m on the radial component.

SILEX (Semiconductor Intersatellite Link Experiment), an ESA experiment built by MMS. SILEX consists of two optical terminals, namely PASTEL on board SPOT-4 (a LEO terminal) on the anti-Earth side of the SPOT platform, and OPALE mounted on ESA's geostationary satellite ARTEMIS (a GEO terminal). ARTEMIS is to be launched in early 2000. The ARTEMIS downlink to the ground segment link uses conventional radio transmission (Ka-band). The SILEX payload on SPOT-4 (namely PASTEL) has a mass of 160 kg, power = 150 W. [934) 935)]

- **PASTEL** (PAssager SPOT de Técommunication Laser). A joint ESA/CNES passenger demonstration experiment. PASTEL is a prototype high data-rate intersatellite transmission system based on laser technology. The objective is to transmit imaging data from SPOT-4 to ARTEMIS. The aim of the experiment is to validate the PASTEL concept design in an operational environment. PASTEL is a gimbal-mounted assembly consisting of a telescope, an optical bench with a fine pointing system, communication detectors with avalanche photodiodes, a thermal control system for precision temperature control, a two-axis gimbal mechanism, and the launch locking mechanisms needed during the launch phase. The telescope mirrors and main structural elements are made of Zerodur. The acquisition and tracking sensors use CCD detectors. The laser diodes are of the GaAlAs type. The SPOT-4 - ARTEMIS optical links operate at wavelengths of 830 nm. Data to be transmitted include: HRVIR image data, pseudo-noise (PN) code, PASTEL telemetry.
- **OPALE** (Optical Payload for Intersatellite Link Experiment) terminal, mounted on the geostationary satellite ARTEMIS (a GEO terminal).

Laser links have the potential to offer much higher transmission rates (up to 10 Gbit/s) than conventional radio links which are limited to rates of about 250 Mbit/s. Another advantage of laser links is that by their very nature, there can be no interference between optical and radio transmissions. The technical challenge of the technology demonstration involves alignment and stabilizing issues, it requires pointing errors of < 10 μrad. This pointing accuracy is several orders of magnitude lower than open-loop pointing of a typical platform.

The divergence tolerance of the communication beam for the SILEX configuration is 8 μrad (or about 0.00046°). PASTEL and OPALE use a dedicated acquisition sequence. Initially, both terminals (PASTEL and OPALE) coarsely point to each other. This is done when OPALE scans a wide-angle (750 μrad) beacon beam in the direction of PASTEL. On illumination of PASTEL by the beacon beam, it rapidly corrects its line of sight and directs in turn a narrow communication beam towards OPALE. Similarly, OPALE detects the incoming PASTEL signal, aligns its line of sight, and transmits its narrow communication beam towards PASTEL. The two terminals then remain locked on each other in closed-loop tracking, permitting subsequent communication.

PASTEC. A technology demonstration passenger payload to study the orbital environment. PASTEC contains the following experiment packages (PASTEC mass = 50 kg, power = 50W):

- CEDRE: Contamination and degradation of thermal control coatings in space. The objective is to study the variation with time of the thermo-optical properties of thermal control coatings using a calorimetric method combined with contamination measurement using quartz-crystal microbalances.

934) T. Tolker-Nielsen, J. C. Guillen, "SILEX: The First European Optical Communication Terminal in Orbit," ESA Bulletin 96, Nov. 1998, pp. 42-44

935) A. F. Popescu, B. Furch, "Status of the European developments for laser intersatellite communications," SPIE, Vol. 1866, 1993, pp. 10-20

- ERCOS: Cosmic radiation experiment. The objective is to study the effects of heavy ion fluxes on VLSI electronic components.
- ERDOS: Radiation dosimetry experiment. The aim is to measure cumulative radiation doses with MOS dosimeters.
- MEDY: Measurement of satellite dynamics. The objective is to characterize the vibration environment during the launch phase
- MicroMEDY: The objective is the characterization of the in-flight microvibration environment.
- SILLAGE: Measurement of electrostatic potentials on the outside of the SPOT satellite which are due to the "wake" effect. The objective is to evaluate the behavior of the materials used and to gain insights into the electrostatic discharge hazard comparable with our current understanding of the corresponding phenomena affecting geostationary spacecraft.
- THERME: same aims as CEDRE, but highly simplified. The aim is to study the ageing of thermal control coatings depending on their exposure to solar and terrestrial radiation and on their orientation with respect to the satellite's velocity vector (exposure to and effects of atomic oxygen) using a calorimetric method based on simplified sensors.

POAM-III (Polar Ozone and Aerosol Measurement).[936] POAM-III is an NRL instrument (now sponsored by ONR instead of BMDO as done before), an improved version of POAM-II flown on SPOT-3. The objective is the measurement of the vertical distribution of atmospheric ozone, water vapor, nitrogen dioxide, aerosol extinction, and temperature. Improvements were made in the electronic system that provides lower noise and more accurate measurements. POAM-III is a nine-channel photometer device with a light-weight and low-power design. The instrument mounts on the exterior of SPOT-4, has a size of 22 cm x 23 cm x 34 cm, and a mass of 13.9 kg. A PCEM (Primary Control and Electronics Module) with a mass of 11 kg is mounted inside the S/C. The optical head assembly of the outside instrument includes an azimuth-elevation gimbal mount to track the sun. The azimuth and elevation axes move the line-of-sight parallel and perpendicular to the Earth's horizon. The range of elevation is from -10° to -35°, the range of azimuth motion is ±173° from a nominal zero position. - The measurement technique is by solar occultation through the Earth's atmospheric limb at nine wavelengths from the UV to NIR. There are separate filtered optics for each of the nine channels, which are co-aligned with each other and the sun tracker. The new optical filters have a greater ability to withstand the space environment. The wavelengths and bandwidths of the science channels differ slightly from those in POAM-II and are given in Table 260.

Channel	Primary Measurement	Center Wavelength (nm)	Bandwidth at FWHM (nm)
1	Rayleigh scattering	353.4	9.7
2	NO_2 present	439.6	2.1
3	NO_2 absent	442.2	2.1
4	O_3	603	17.7
5	O_2 present	761.3	2.3
6	O_2 absent, aerosol	779	10.2
7	H_2O absent, aerosol	922.4	2.6
8	H_2O present	935.9	2.6
9	Aerosol	1018	11.6

Table 260: Spectral coverage of the POAM-III instrument

Each of the nine wave bands is measured by a separate optical channel. An image of the sun is formed with a fused silica plano-convex lens having a clear aperture of 11 mm, a nominal focal length of 63 mm, and an f-number of f/5.7. The detectors are silicon photodiodes operated in photovoltaic mode (data quantization of 15 bits). The instrument tracks the sun as it

936) R. L. Lucke, D. R. Korwan, et. al., "The Polar Ozone and Aerosol Measurement (POAM-III) instrument and early validation results," Journal of Geophysical Research, Vol. 104, D15, Aug. 20, 1999, pp. 18,785 - 18,799

rises and sets through the atmosphere, using a four-cell sun tracker with a 1° FOV. For POAM-III, the 1° sensor is supplemented with a 10° sensor to accommodate initial acquisition.

POAM-III measurement extent in altitude ranges: Aerosols = 10-30 km; O_2, O_3 = 10-60 km; H_2O = 10-40 km; NO_2 = 20-40 km.

D.37.3 SPOT-5

Background: In 1994 the SPOT-5 program was approved by the French government, consisting initially of two identical S/C in orbit and two new optical imagers for these S/C, called HRG (High Resolution Geometric) and HRS (High Resolution Stereoscopic) to provide a ground resolution of 5m. [937)] In 1996, the SPOT-5 program was downsized (to one orbiting S/C) and redefined to improve the spatial resolution of the imagery below 5 m. As a result, an innovative image acquisition and processing scheme has been developed by CNES to obtain spatial resolutions of about 3 m from two 5 m images. In January 1999, a further functional improvement was introduced giving the HRS (High Resolution Stereoscopic) instrument a full stereoscopic capability. It was also decided to continue to fly the VEGETATION instrument. The SPOT-5 satellite program continues the partnerships of France (CNES), Belgium (OSTC) and Sweden (SNSB) as established at the beginning of the SPOT program.

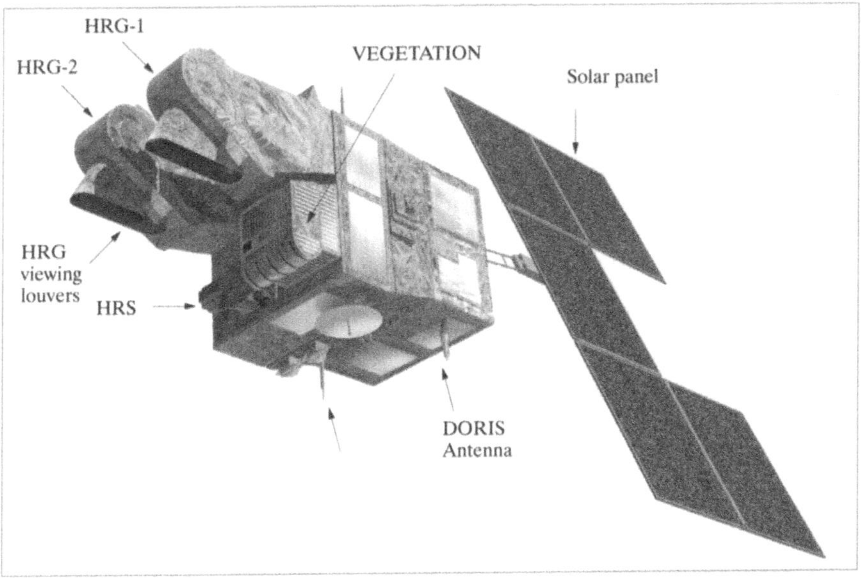

Figure 157: The SPOT-5 S/C model

The overall mission objectives are: [938)]

- To provide image acquisition and service continuity consistent with previous SPOT satellites to satisfy the user investments. Hence, the same sun-synchronous orbit is used providing the existing functional instrument capabilities with a 26 day repeat cycle, the

937) A. Ammar, A. Baudoin, D. Assemat, M. Arnaud, "The SPOT Programme, An Operational Earth Observation System," Proceedings 45th Congress of the International Astronautical Federation, October 9-14, 1994, Israel

938) A. Baudoin, "The Current and Future SPOT Program," Proceedings of the ISPRS Joint Workshop 'Sensors and Mapping from Space 1999,' Sept. 27-30, 1999, Hannover, Germany

same off-track viewing capability of ± 27° about nadir, the same spectral band selection, and the same 60 km double swath.

- To improve the spatial resolution of the imagery to < 3 m in the panchromatic band and to 10 m in the multispectral mode. The SWIR band imagery remains at 20 m.
- To offer in parallel a stereoscopic along-track observation capability (instead of the previously provided cross-track capability). The intend is to offer high-resolution imagery to be used for DEM (Digital Elevation Model) generation with an accuracy of 10 m.

The SPOT-5 satellite configuration takes advantage of the SPOT-4 bus design, using the extended platform design (SPOT MK2, provided by MMS) and service module accommodates twice the payload of the SPOT 3 bus. The ACS (Attitude Control Subsystem) provides a pointing accuracy of 0.05° and an attitude restitution of 6×10^{-5} radians. The S/C structure has dimensions of: 3.4 m x 3.1 m x 6 m (excluding the solar array). The S/C mass is 3000 kg, power = 2400 W (EOL), batteries = 4 x 40 Ah, design life = 5-7 years. A launch of SPOT-5 is planned for early 2002.

Orbit: Sun-synchronous circular orbit (identical to those of SPOT-1 to -4), altitude = 818-833 km, inclination = 98.7°, period = 101 min, equator crossing on descending node at 10:30 AM, repeat cycle = 26 days, revolutions/day = 14 5/26.

Sensor complement:

HRG (High Resolution Geometric), built by MMS of Vélizy, France to continue to improve the HRVIR service of SPOT-4. Two HRG instruments are provided in the conventional SPOT-series double-observation configuration, each with a FOV of 4.13° and the same cross-track pointing capabilities of ±27° as the HRVIR imager on SPOT-4. The observation coverage of each HRG is 60 km in the nadir direction and >80 km in the oblique configuration (same two-swath coverage as before).
HRG features a new linear detector array configuration geometry for the panchromatic band using two parallel rows (i.e., a dual array) of 12000 silicon CCD detectors (6.5 μm in size) for each instrument. The two PAN detector lines are offset in the focal plane in such as way as to provide coincident imagery of the same instantaneous cross-track area, each at a spatial resolution of 5 m. [Note: The dual array in the focal plane (offset by one half pixel in one direction and 3.5 pixels in the other to avoid overlapping) is sufficient to improve the sampling grid without doubling each array's acquisition rate. The new sampling concept is based on Shannon's theory of information which states that "the sampling frequency must be equal to or greater than twice the maximum signal frequency" to obtain clean images using interpolation.] The CCD integration time is within 0.75 ms for a dual-array observation in cross-track of 60 km in which the S/C is moving 5 m in the along-track direction.

Bands	Spectral range	SPOT-1,-2,-3	SPOT-4	SPOT-5	
		Spatial resolution		On-board	Supermode
PA-1 (PAN)	0.49-0.69 μm	10 m		5 m	3 m (on-ground)
PA-2 (PAN)	0.49-0.69 μm			5 m	
B0 (Blue)	0.43-0.47 μm		1 km		
B1 (Green)	0.49-0.61 μm	20 m	20 m	10 m	
B2 (Red)	0.61-0.68 μm	20 m	20 m	10 m	
B3 (NIR)	0.78-0.89 μm	20 m	20 m	10 m	
SWIR	1.58-1.75 μm		20 m	20 m	

Table 261: Overview of SPOT series spectral continuity and resolution improvement

Supermode: The term refers to an acquisition process, specific to the HRG instrument of SPOT-5, through which an image sampled at 2.5 m may be obtained from two 5 m resolution panchromatic images acquired simultaneously, keeping within the same borders as the two 5 m resolution images. It is possible to combine the two 5 m pixel image samples into four new slightly overlapping image samples of about 3 m pixel size. A specific image processing software, developed by CNES is used to reconstruct the final image after three processing steps: interleaving, interpolation and restoration. This new type of image, simulated with all its geometric and radiometric characteristics, has been compared by several users to other types of digital images at different ground resolutions. These users (cartographers, urban planners, environmental experts, foresters, agronomists,) have concluded that the supermode resolution is about 3 m (between 2.5 m and 3.5 m depending on applications and analyzed features). [939)]

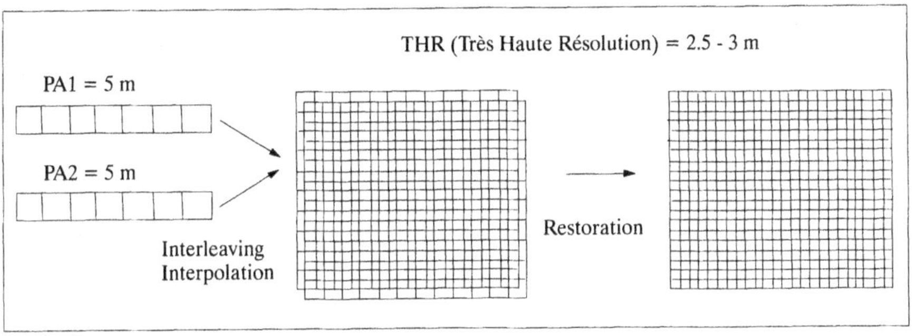

Figure 158: Supermode processing scheme

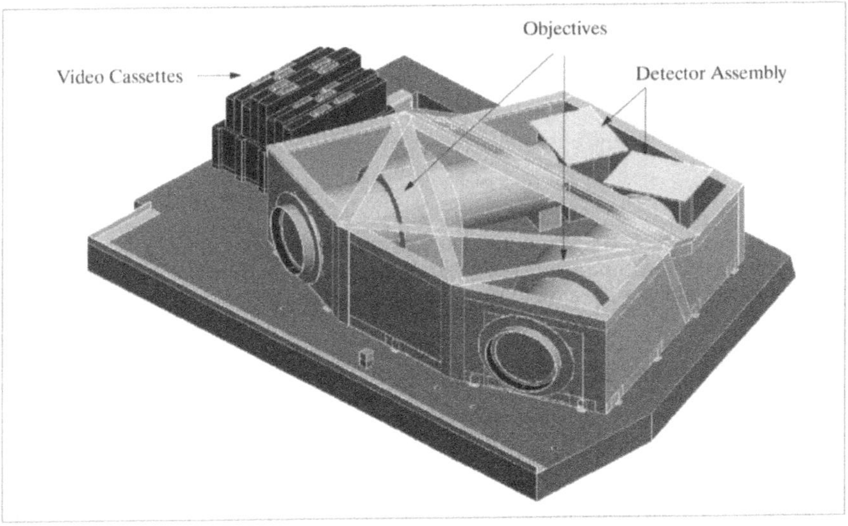

Figure 159: Illustration of the HRS instrument

HRS (High Resolution Stereoscopic), developed and built by MMS, sponsored by CNES and SPOT IMAGE. The objective is to provide large-area along-track stereoscopic imagery with good altimetric accuracy (5-10 m relative and 10-15 m absolute). Applications of the stereo imagery are seen in various fields such as map making and in the generation of DTMs (Digital Terrain Model) The panchromatic band (0.51-0.73 μm) of SPOT-1,-2,-3 is being reintroduced. Parallel observation of forward and aft swaths (120 km width) in the along-track direction at pointing angles of ±20°. A spatial resolution of 10 m is provided in cross-track and 5 m (parallax direction) in along-track. The stereo acquisition mode can be sus-

939) SPOT 5 brochure, "Supermode," of CNES and SPOT Image, May 1999

tained for scene lengths of up to 600 km. HRS uses the same CCD line detector design as for the HRG instrument.

Parameter	SPOT-1,-2,-3	SPOT-4	SPOT-5
Prime sensor	2 x HRV	2 x HRVIR	2 x HRG
Spectral bands PAN	PAN (0.51-0.73 μm) at 10 m resolution	PAN (0.61-0.68 μm) at 10 m resolution	PA-1 (0.49-0.69 μm),5 m PA-2 (0.49-0.69 μm),5 m
Spectral bands MS and resolutions	B_1 (50-0.59 μm) B_2 (0.61-0.68 μm) B_3 (0.79-0.89 μm) all at 20 m resolution	B_1 (0.50-0.59 μm), 20 m B_2 (0.61-0.68 μm), 10 m B_3 (0.79-0.89 μm), 20 m SWIR (1.58-1.7μm), 20 m	B_1 (0.49-0.61 μm), 10 m B_2 (0.61-0.68 μm), 10 m B_3 (0.78-0.89 μm), 10 m SWIR(1.58-1.7μm), 20 m
FOV (swath) per sensor	4.13° (60 km)	4.13° (60 km)	4.13° (60 km)
S/C mass (at launch)	1907 kg	2755 kg	3000 kg
S/C size (main structure)	2 m x 2 m x 4.5 m	2 m x 2 m x 5.6 m	3.4 m x 3.1 m x 6 m
Solar panel span, power	8.14 m, 1 kW	8.032 m, 2.1 kW	2.40 kW
Detector line array (Si) (Si) (InGaAs/InP)	6000 PAN, 3000 MS	6000 PAN, 3000 MS (2 lines) 3000 SWIR (1 line)	12000 PAN (2 lines) 6000 MS (3 lines) 3000 SWIR (1 line)
Recording capability	2 x 22 min	2 x 40 min	2 x 40 min (+solid state)
On-board data compression technique	DPCM (3/4) for PAN data only	DPCM (3/4) MS and PAN	DCT
Data rate (X-band)	2 x 25 Mbit/s	50 Mbit/s	150 Mbit/s
X-band frequency	8.253 GHz	8.253 GHz	
Design life	3 years	5 years	5-7 years

Table 262: Overview of performance parameters of the SPOT family

VEGETATION-2. With some minor improvements regarding instrument operations, the VEGETATION-2 instrument is identical in its technical specification to VEGETATION flown on SPOT-4.

DORIS

Communication: SPOT-5 provides an on-board storage capacity of 90 Gbit in solid state memory. The data compression algorithm used is DCT (Discrete Cosine Transform). 940) Compression ratios of 3 are used. Downlink data transmission is in X-band.

D.38 SSR1 (Satelite de Sensoriamento Remoto)

SSR1 is Brazil's Remote Sensing Satellite (INPE). Objective: repetitive monitoring of Brazil through spaceborne imagery (in particular the Amazon rainforest). A launch is planned for the end of 2004. 941) 942) 943) 944)

A multi-mission platform (MMP) is being developed for the SSR program supporting a payload mass of up to 190 kg and power up to 180 W (80 W during eclipse phase of orbit). The S/C is three-axis stabilized using a star and a sun sensor for attitude measurement; momentum wheels and magnetic rods are used as actuators The pointing accuracy <0.1° in several pointing modes (Earth, anti-Earth, inertial and sun). A GPS receiver provides on-board position data. SSR1 has a total mass of 290 kg. Nominal design life of 4 years.

940) P. Lier, G. Moury, C. Latry, F. Cabot, "Selection of the SPOT-5 Image Compression Algorithm," Proceedings of SPIE, Vol. 3439,70, 1998

941) H. Carvalho, J. Kono, M. M. Quintino, C. E. Santana, "The Amazon Rainforest Monitoring Satellite - SSR," Proceedings of the 3rd International Symposium of IAA, Berlin, April 2-6, 2001, pp. 19-21

942) C. H. Santana, C. E. Kono, M. M. Quintino, "SSR Amazon Rainforest Observation System," IAA 2nd International Symposium on Small Satellites for Earth Observation, Berlin, April 12-16, 1999, pp. 49-52

943) "The first Brazilian Earth Observation Satellite (SSR)," paper by C. E. Santana and J. Kono of INPE

944) "Satellite Launch to Advance Brazilian Space Program," Space News Aug. 31-Sept. 6, 1992, p. 43

Orbit: Circular equatorial LEO orbit; altitude = 905 km; inclination = 0°, the orbit provides a revisit rate < 2 hours with up to 5 useful passes during daylight.

Sensor complement:

WFI = Wide-field Imager (CCD Linear Array Camera)
Objective: medium resolution VIS/NIR/SWIR remote sensing (in particular over areas of extensive vegetation in the Amazon region). Operation of the camera is only considered over Brazil (direct sensor data transmission to the ground).

Spectral bands	B1	B2	B3	B4	B5	B6
Spectral region (μm)	0.440-0.505	0.530-0.575	0.650-0.680	0.845-0.885	0.895-0.990	3.4 - 4.2
Spatial resolution at nadir	100 m				300 m	600 m
Swath width	2200 km (from latitudes of 5° to -15°), or FOV = ±44°					
Data quantization	8 bits					

Table 263: Parameters of the WFI instrument

The MCC function will be performed at INPE over a single ground station located in Cuiabá (Brazil). Data will be pre-processed and retransmitted at Cuiabá. Sensor image data from SSR1 is in X-band and TT&C communications via S-band. SSR1 on-board storage capability of 5 Gbit. Data distribution to the user community is provided through a GEO communications satellite.

D.39 VCL (Vegetation Canopy Lidar Mission)

VCL is a NASA/GSFC minisatellite mission in the ESSP (Earth System Science Pathfinder) program (selected in the spring of 1997) with the overall objective to characterize the 3-D structure of land cover for ecosystem/climate modeling, monitoring, and prediction (mapping of biomass, height of vegetation layer, density profile of the vegetation layer). The principal long-term data source for landcover studies (in the horizontal plane) thus far has been that of AVHRR on the NOAA POES S/C series. The goals of the VCL mission, namely a quantitative description of landcover and global productivity, try to extend the quality of current monitoring capabilities with a new altimetric multi-beam measurement concept, yielding survey data on a finer scale than before (also in the vertical plane), with sufficient coverage and repetitive observation cycles. The new approach promises improved interpretation methods for parameter estimations and modeling.

- Significant improvement (5-10 fold over past system capabilities) in canopy height monitoring for better biomass estimation
- Significant improvement in surface coverage (2% of the tropical land surface)
- Direct estimation of degraded areas, areas of regrowth, and intact forests for quantifying biomass degradation
- Comprehensive assessment of structure and organization at landscape level for biodiversity studies
- Generation of an (almost) global reference grid of land topography. The dataset represents a network of closely-spaced topographic control points at 1 m accuracy.

The S/C structure (built and integrated by OSC, Dulles, VA) is comprised of two octagonal cylinders of 1.1 m max. diameter. The bus structure is based on the LeoStar design of OSC with three elements: core module, propulsion module, and separation system. The S/C structure consists of modular Al honeycomb plates with stringers; thermal control is provided by a passive radiation system with backup heaters; three-axis stabilization provides 23 arcsecond (3 sigma) pointing knowledge and 0.1° (3 sigma) attitude control; on-board hydrazine propulsion is feeding four 4.5 N thrusters with 45 kg propellant; power is provided by GaAs solar arrays of 3.8 m^2 and 30 Ah NiH_2 batteries; S/C design life of two years; the S/C mass is 260 kg plus 133 kg of payload mass. On-board control and data handling (C&DH) uses a 80C186 based computer an S-band transponder and an X-band transmitter. Use of AstroNav GPS receivers. [945] [946]

A launch of VCL, initially planned for Sept. 2001 on the Athena-1 vehicle from Kodiak Island, AK, is being re-planned as of mid-2001. There is also a reassessment of the science and the associated risk with the laser development. A new launch approval from NAS/HQ is required. [947]

Data and S/C communications: On-board data storage of 32 Gbit is provided; science data downlink in X-band at 28 Mbit/s; TT&C in S-band; data volume of 2.2 GByte/day compressed (5.5 GByte/day depacketized). Mission operations and data processing is performed at the University of Maryland at College Park, MD. Data distribution is provided through the EROS Data Center of USGS, Sioux Falls, SD.

Data products are gridded at two different spatial and temporal resolutions:

- Monthly products, gridded to 2 km x 2 km (level 2)
- Six-month products: Updated and accumulated level 2 data for a 2 km grid cell are gridded and released as level 3 products.

Orbit: Circular orbit, altitude = 390 - 410 km (the low orbit requires monthly reboosts to 410 km, using monopropellant hydrazine, to maintain the nominal orbital altitude of 400 km), inclination = 67°.

Sensor complement:

MBLA (Multi-Beam Laser Altimeter). MBLA [GSFC instrument of SLA (Shuttle Laser Altimeter, flown on STS-72, Jan. 11-20, 1996) and MOLA (Mars Orbiter Laser Altimeter) heritage; PI: R. Dubayah, University of Maryland] utilizes a pulsed laser radar (i.e. a lidar) system for continuous global remote sensing of tree canopy height ($< \pm 1$m), vertical distribution of intercepted surfaces in the canopy, and ground topography elevations ($< \pm 1$m). MBLA features five laser transmitters (arranged in a circular configuration) in a single altimeter instrument, a large receiver telescope (made of beryllium), a set of five detectors and laser-pulse analysis electronics, computer/data command electronics, and a pointing angle measuring system. Instrument power = 220 W; mass = 133 kg.

Lasers	3-5 Nd:YAG diode-pumped pulsed lasers, operating at 1064 nm wavelength
Laser pulse rate and energy	242 pps (land), 10 mJ per pulse
Laser pulse duration	5 ns
Telescope	0.9 m diameter f/1 parabolic mirror, 20 mrad TFOV, 0.3 mrad IFOV
Detectors	Silicon avalanche photodiodes
Waveform digitization	250 Msamples/s
Samples per waveform	10-200, average of 50
Swath width	8 km
Resolution	25 m (60 μrad) footprint diameter (at 400 km altitude)
Track spacing	2 km
Elevation accuracy	$< \pm 1$ m in low slope terrain
Vegetation height accuracy	$< \pm 1$ m limited by 100:1 pulse detection, dynamic range and cal/val
Data quantization	10 bits

Table 264: MBLA parameter specification

945) R. Dubayah, B. Blair, J. Bufton, D. Clarke, et al., "The Vegetation Canopy Lidar Mission," presented at ASPRS, Washington, D.C., 1997

946) http://essp.gsfc.nasa.gov/vcl.html

947) Information provided by Nick Chrissotimos of NASA/GSFC

The active MBLA instrument employs the pulse time-of-flight measurement technique and a laser shape analysis. Spatial coverage in the cross-track direction is provided with multiple beams (5) arranged in a pentagon pattern, each beam traces a separate (almost contiguous) ground track, sufficient to characterize horizontal spatial variability and vertical height distributions. The footprint of each beam describes a circle of 25 m in diameter(60 µrad FOV), adequate to resolve vegetation height and structure. Five parallel tracks (spaced at 2 km intervals across-track) of Earth surface coverage are simultaneously monitored, providing a (perforated) swath of 8 km as illustrated in Figure 160.

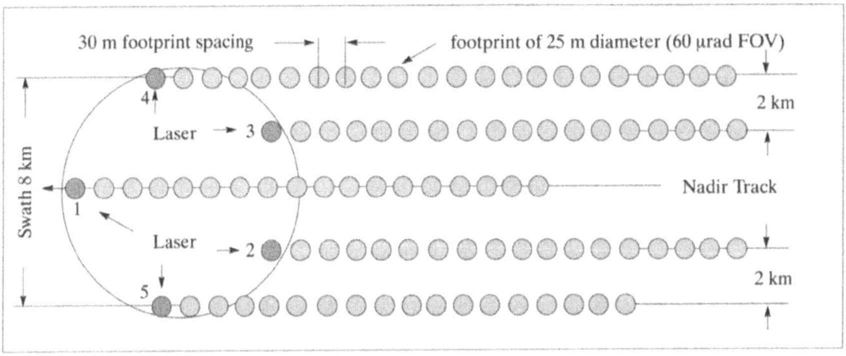

Figure 160: Typical MBLA lidar configuration/ground measurement pattern of beams

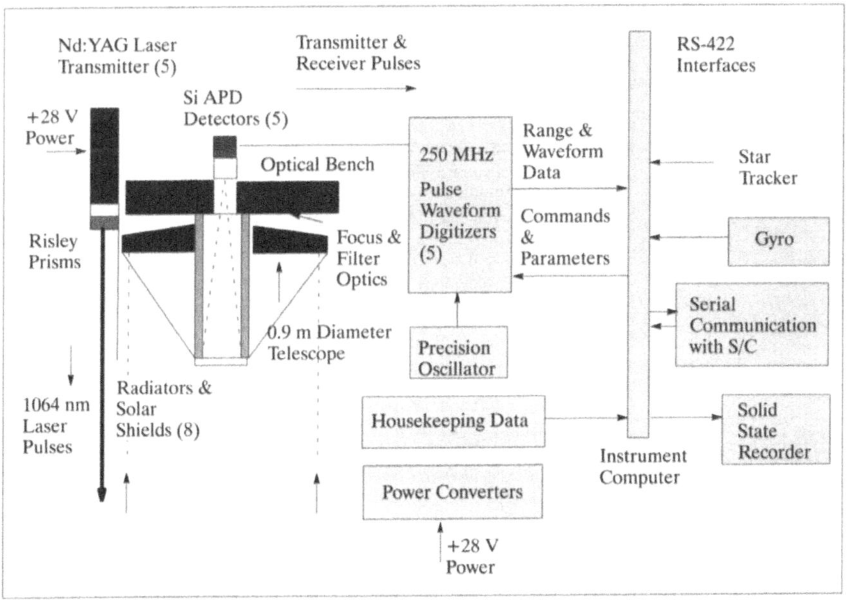

Figure 161: Schematic illustration of the MBLA instrument

MBLA makes simultaneous measurements of range by synchronous triggering of five laser pulse transmitters and detection with a single telescope that is staring at nadir and is equipped with multiple silicon avalanche detectors in its focal plane. Surface echoes from the five beams are digitized in the MBLA electronics at 250 Msamples/s.

Calibration. Calibration of ground height measurements requires on-orbit bias and drift retrievals. These are obtained through comparisons with results from the TOPEX/POSEIDON altimeter at ocean orbital crossover points between VCL and TOPEX/POSEIDON, as well as comparisons of VCL ocean altimetry with high resolution mean sea surface elevation maps.

RRA (Retro-Reflector Array). The corner cubes are made of the fused silica. Their performance is optimized at the green wavelength (532 nanometers). The corner cubes are symmetrically mounted on a hemispherical surface with one nadir-looking corner cube in the center, surrounded by an angled ring of eight corner cubes. The design is identical to the array used on the MSTI-2 Satellite.

D.40 UoSAT/SSTL Microsatellite Missions

The Center of Satellite Engineering (CSER) at the University of Surrey in Guildford, United Kingdom, has been pioneering the use of microsatellites in low Earth orbits (LEOs) since the late seventies. The numerous microsatellite missions so far are simple and low-cost designs featuring digital store-and-forward communications systems and services on a global scale. Communications payloads form the basis of communication links for both experimental and operational microsatellite missions and perform a range of missions from remote sensing data transfer and experimental data downlinks to remote site data collection and messaging. UoSAT stands for University of Surrey Satellites.[948), 949)]

SSTL (Surrey Satellite Technology Limited) was formed in 1985 by the University of Surrey as a 'technology transfer company' in order to make its extensive research capability available to external organizations in an efficient and professional manner. Successful technology transfer and training programs have been conducted with numerous foreign government organizations.

S/C	Launch	Communications Payload and Objective	Uplink; Downlink
UoSAT-2	1984	First Store-& Forward (S&F) mission, 1200 baud AFSK, 128 kByte memory. - Amateur service	VHF; VHF/ UHF
UoSAT-3 (OSCAR-14)	1990	S&F, 4 MByte file store, FSK digitally filtered, HealthNet/ VITA. - Experimental	VHF; UHF
UoSAT-5 (OSCAR-17)	1991	S&F, 13 MByte file store, 9600 baud, FSK digitally filtered. - Amateur service	VHF; UHF
S-80/T	1992	S&F, 16 MByte file store, 9600 baud, FSK digitally filtered, transparent measurement transponder. - Experimental	VHF; UHF (WARC-92)
KITSAT-1 (OSCAR-23)	1992	S&F, 16 MByte file store, 9600 baud, FSK digitally filtered, DSP experiment using TMS320C25/C30, capability for adaptive modulation. - Amateur service	VHF; UHF
PoSAT-1	1993	S&F, 16 MByte file store, 9600/38400 baud, FSK digitally filtered, DSP experiment includes MSK downlink option and adaptive modulation. - Experimental	VHF; UHF
KITSAT-2	1993	S&F, 16 MByte file store, 9600 baud, FSK digitally filtered,	VHF; UHF
HealthSat-2	1993	S&F, 3x16 MByte file store, 9600/38400 baud, FSK digitally filtered, adaptive transmitter power under on-board computer control	VHF; UHF (WARC-92)
CERISE	1995	S&F, 9600/38400 baud, FSK digitally filtered,	VHF; UHF
FASat-Alfa	1995	S&F, 9.6/76.8 kbaud, FSK digitally filtered, DSP communications payload. Solid-state recorder at 256 MByte	VHF; UHF
FASat-Bravo	1998	S&F,	
TMSat	1998	S&F "email" communications,	

Table 265: UoSAT/SSTL store-and-forward payloads

948) Information provided by E. Milton and M. Fouquet of SSTL

949) URL address - http://www.ee.surrey.ac..uk/CSER/UOSAT

Store & Forward (S&F) Communication. [950], [951], [952] The UoSAT-2 S/C carried the first digital S&F payload in orbit. Since then all further UoSAT/SSTL microsatellites have operated with S&F communication technique. A protocol suite called PACSAT was developed for this purpose. PACSAT uses packet radio techniques in the microsatellite system to transmit its data over the satellite RF link. Several layers of protocol are implemented in the PACSAT suite, at the lower level HDLC (High-Level Data Link Control) and X.25 provide the functions of packet multiplexing, error detection and ARQ (Automatic-Repeat Request) error correction. PACSAT is a point-to-multipoint protocol (broadcast); small ground terminals in the satellite footprint receive/send the data. The SSTL research aim in this area is to provide reliable service from distributed data collection terminals.

Aside from its many research and S/C-building activities, SSTL operates a command and control center in support of all its missions. The autonomous downlink data acquisition during a S/C pass includes housekeeping, experiment and point-to-point communications data.

The first UoSAT satellites were spin-stabilized and nadir-pointing S/C, employing attitude control by a combination of actuators: passive gravity-gradient boom, and 3-axis active magnetorquing system (switched currents under algorithmic computer control). Attitude is generally sensed by a combination of magnetometers (3-axis), horizon sensors, and/or sun sensors. The pointing accuracy is in the order of 2-3° rms.

SSTL microsatellite bus. The platform or bus is a modular tray design to exploit the technology upgrade opportunities presented by frequent missions (each launch represents in effect a new design configuration). The MicroBus was originally developed in 1988 and first flown on UoSAT-3 and -4 in 1990. The MicroBus platform is a boxlike structure (spin or three-axis stabilized) comprising service modules [such as: battery, solar power, communication (uplink/downlink, telemetry/telecommand), on-board computer (OBC), and attitude determination and control (ADCS)] and payload modules. All SSTL ADCS have been based on 3-axis magnetometers, computer-controlled 3-axis magneto-torquing, and by a gravity-gradient boom. Pointing accuracy was improved with the addition of sun sensors, horizon sensors, and star cameras. [953]

Satellite	Launch	Horizon Sensor	Sun Sensors	Star Imager	Reaction Wheel
UoSAT-3	1990				
UoSAT-5	1991	x			
KITSAT-1	1992	x			
S-80/T	1992	x	x		
PoSAT-1	1993	x	x	x	
HealthSat-2	1993				
CERISE	1995	x	x		
FASat-Alfa	1995		x		x

Table 266: Introduction of advanced ADCS techniques into the UoSAT series

D.40.1 UoSAT-1 (University of Surrey Satellite-1)

The first experimental microsatellite from the University of Surrey. Piggyback launch of the S/C on a NASA Thor Delta vehicle on October 6, 1981 from VAFB, CA (prime payload was

950) J. W. Ward, "Microsatellites for global electronic mail networks," Electronics and Communications Engineering Journal, December 1991, Vol. 3, No. 6, pp. 267-272
951) J. W. Ward, H. E. Price, "The UoSAT-2 Digital Communications Experiment," Journal of the Institute of Electronic and Radio Engineers, 1986
952) UoSAT internet home page
953) J. W. Ward, A. da S. C., "An Evolutionary Approach to Small Satellite Technology Development." Proceedings of the 9th AIAA/USU Conference on Small Satellites, Sept. 18-21, 1995, Logan, UT

SME). Objectives: Scientific and educational - investigation into low-cost microsatellite engineering techniques. S/C mass = 50 kg, power = 30 W peak.[954)]

Orbit: Sun-synchronous polar orbit, altitude = 534 km, inclination =97.7°. The S/C re-entered the Earth's atmosphere on October 13, 1989 after eight years in orbit.

Sensors/Payload: CCD camera, Geiger-Müller radiation detectors, amateur radio communications, synthesized speech.

D.40.2 UoSAT-2

UoSAT-2[955), 956)] was built by the University of Surrey within a 6-month period. Objective: proof-of-concept for LEO store-and-forward communications. UoSAT-2 was launched on March 1, 1984 (piggyback on the NASA payload Landsat-D' which became Landsat-5 in orbit), it is operating nominally as of 1/1996. S/C mass =68 kg, power = 30 W peak (20 W orbit average).

Orbit: Sun-synchronous polar orbit, perigee = 655.7 km, apogee = 671.1 km, inclination = 97.98°

Spacecraft data is downlinked in VHF (145.825 MHz) at a rate of 1200 baud. In addition a secondary S-band beacon (2401.5 MHz) is operated with an output power of 180 mW. The S/C is operated by SSTL and is mainly providing amateur radio communication service.

Sensors/Payload: S&F (Store & Forward) communications, CCD camera, space dust detectors, amateur radio communications, synthesized speech.

The prototype CCD camera was based on a P8602 area-array CCD from EEV Ltd., array size: 384 x 288 pixels, 7 bit quantization; optics 4.8 mm, f/1.8.

D.40.3 UoSAT-3 (HealthSat-1)

UoSAT-3 was launched January 22, 1990 (piggyback on an Ariane flight V35 ASAP with SPOT-2 as primary payload), S/C mass = 50 kg, size: 650 x 300 x 300 mm, power = 35 W peak (25 W orbit average). The primary 80C186 on-board computer failed in May 1997. UoSAT-3 supports S&F communications for use by various humanitarian and medical groups around the world (VITA), in addition to its sensor operations. Sponsors: University of Surrey, Arianespace, VITA, AMSAT, RACAL.

Orbit: Sun-synchronous polar orbit, perigee = 784.4 km, apogee = 800.0 km, inclination = 98.58°

Sensor:

CREDO (Cosmic Radiation Environment and Dosimetry Experiment), of DRA. The objective is the monitoring of the near-Earth radiation environment. CREDO consists of two subexperiments: **CPE** (Cosmic Particle Experiment) and **TDE** = Total Dose Experiment (description under PoSAT). The **CPE** sensor is based around an array of large-area pin diodes (10 cm^2) which are connected to a charge pulse height analyzer monitoring the linear energy transfer of particles as they pass through the diodes. CREDO instrument mass=2kg, power=0.5W, data=37.5 kByte/day.

D.40.4 UoSAT-4

UoSAT-4 was launched alongside UoSAT-3 (same mass, size, and power as UoSAT-3). UoSAT-4 carried technology demonstration payloads for ESA and DRA (Defense Research

954) UoSAT-1: Special issue of The IERE Journal, Vol. 52, No. 8/9, August 1982

955) J. M. Radbone, "The UoSAT-2 Spacecraft CCD Imaging and Digital Store/Read-out Experiments," The IERE Journal, Vol. 57, No. 5, September 1987, ISSN 0267-1689

956) M. N. Sweeting, "UoSAT microsatellite missions," Electronics & Communication Engineering Journal, IEE, June 1992

Agency), Malvern, UK. After operating perfectly immediately following launch, communications were lost with UoSAT-4 and the mission had to be abandoned.

D.40.5 UoSAT-5

Objective:[957), 958)] operational support for amateur satellite service, providing store-and-forward communications. Several space science and imaging experiments are supported in addition. S/C mass = 50 kg, size: 650 x 300 x 300 mm, power = 35 W peak (25 W orbit average). The S/C was topped by a 6 m antenna. UoSAT-5 was launched on July 17, 1991 (ASAP auxiliary payload to ERS-1 on Ariane flight V44). Nominal operations are conducted as of 1/1996.

Orbit: Sun-synchronous polar orbit, perigee = 763.7 km, apogee = 773.5 km, inclination = 98.42°

The S/C suffered a crash of its on-board computer system on January 29, 1995 (due to power problems). The S/C was diagnosed by ground control and returned to full operational state by February 1, 1995.

Sensors:

TDE = Total Dose Experiment (description under PoSAT)

EIS = Earth Imaging System (description under PoSAT). The images from UoSAT-5, KITSAT-1 and KITSAT-2 are freely available on amateur radio downlinks to suitably equipped and licensed radio amateurs.

SCTE = Solar Cell Technology Experiment. The experiment is designed to evaluate the effects of the space environment on 27 test samples of GaAs (Gallium Arsenide), Si (Silicon) and InP (Indium Phosphate) solar cells from a variety of manufacturers.

D.40.6 KITSAT-1 (Korea Institute of Technology Satellite)

A cooperative project between SSTL and KAIST (Korean Advanced Institute of Science and Technology) involving a S/C design based on the UoSAT platform (MicroBus) and concepts and a mission. On-the-job training of KAIST engineers at SSTL. A ground station was installed in South Korea (KAIST/SaTReC) for the purpose of tracking Kitsat S/C as well as to monitor other amateur satellites. S/C stabilization was realized with 3-axis magnetorquers and a gravity gradient boom. The KITSAT-1 S/C, the Korean name is Uribyul-1, (mass=46.6 kg) was launched August 10, 1992 on Ariane-4 ASAP (V.52, piggyback), along with Topex/Poseidon S/C as the prime payload. Nominal KITSAT-1 operations are conducted as of 6/1998. [959)]

Orbit: Circular orbit, perigee = 1302.5 km, apogee = 1327.8 km, inclination =66.08°, period=111.9 minutes.

Sensors:

CRE = Cosmic Ray Experiment. CRE monitors the space radiation environment and its effects on S/C semiconductor electronics. CRE contains a PIN diode of 900 mm^2 and a multichannel analyzer capable of detecting energetic particles (5-600 MeV) with a wide range of linear energy transfer. The large number of channels and the fast response time of

957) M. N. Allery, J. J. Sellers, M. N. Sweeting, "Results of University of Surrey on-orbit microsatellite experiments," Proceedings of the International Symposium on Small Satellite Systems and Services, Biarritz, France, June 27-30, 1994

958) M. Fouquet, "The UoSAT-5 Earth Imaging System - in-orbit results," 2nd Conference on Small Satellite Technologies and Applications, SPIE Symposium on Aerospace Sensing, Orlando, FL, April 20-22, 1992

959) I. Lee, D. K. Sung, S. D. Choi, "Experimental Multimission Microsatellites - KITSAT Series," Proceedings of the 7th AIAA/USU Conference on Small Satellites, Set. 13-16, 1993

the sensor allow CRE to build up the spectrum of observed energies of particles. CRE is the first UoSAT instrument to have its own built-in on-board compression software, enabling the system to carry out nearly 295,000 observations per day.

CEIS = CCD Earth Imaging System (of EIS heritage). Two video CCD imaging cameras with two optical assemblies. The cameras are aimed in the same direction but offer a different FOV. The wide angle camera (WAC) image has a ground resolution of 2 km and a FOV (68°) of 1500 km x 1050 km (NIR 810-900 nm filter; 4.8 mm focal length). The objective is to provide images with good contrast between land, sea, clouds, and ice or snow. - The narrow angle camera (NAC) has a ground resolution of 200 m and a FOV of about 150 km x 100 km (red 610 - 690 nm filter, 50 mm focal length). The cameras feature a monochromatic design with optical filters chosen to contrast ground properties, the NAC highlights variations in soil moisture content and vegetation density. An image is taken in a snapshot or staring mode (as opposed to a mechanically scanned pushbroom array) and is read from each CCD by a Transputer Data Processing Experiment (TDPE) and, after processing, provided to the On-Board Computer (OBC) and recorded. Camera sensor (both cameras): EEV CCD04-06 image sensor + chip set; 578 x 576 array (interleaved fields); anti-blooming and electronic integration control. Operational Earth imaging is performed on a scheduled snapshot basis.

DSPE = Digital Signal Processing Experiment. The DSPE consists of two Texas Instruments processors (C25 and C30 from TM320 series). The DSPE can be used as a programmable communications modem to modulate downlink data from, or uplink data to the OBC, thus enabling experiments with new modulation techniques to be optimized for LEO satellite mobile communications.

Communication: uplink = 9.6 kbit/s (FM) modulated FSK AX.25, 145.850 MHz, 145.900 MHz; downlink = 9.6 kbit/s FSK (FM) modulated AX.25, 435.175 MHz. In addition, **DSFCE** (Digital Store and Forward Communications Experiment).

D.40.7 S-80/T

A SSTL-built satellite under contract to Matra Marconi (France) for CNES. The S/C uses a standard ARIANE-4 compatible bus. The S-80/T satellite is a technology demonstrator for the S-80 LEO communications system. The objective is to evaluate the use of VHF frequencies for a full-scale LEO communications service (S-80) planned for the late 1990s. S-80/T measures in conjunction with a mobile ground station the effect of VHF spectrum noise and any signals interfering with the ability to communicate with a satellite in LEO. Frequencies: uplink=148 MHz, downlink=137 MHz. S/C mass= 50 kg, power = 35 W peak (25 orbit-average). The S/C was launched August 10, 1992 (along with KITSAT-1). SSTL provides S/C operations for CNES. Nominal operations as of 1/1996.

Orbit: Circular orbit, perigee = 1302.5 km, apogee = 1327.8 km, inclination =66.08°, period=111.9 minutes.

D.40.8 HealthSat-2

A satellite built by SSTL for the SatelLife organization (VITA, Arlington, VA, USA) with the objective to provide inexpensive or free S&F communication links to doctors and health-workers beyond the reach of the global information infrastructure. HealthSAT-2 was launched on Ariane [along with six other microsatellites, among them: PoSAT-1, KITSAT-2, EyeSat, and ItamSat] September 26, 1993 as an auxiliary payload on the SPOT-3 launch from Kourou.

HealthSAT-2 is considered the operational successor to UoSAT-3, providing a second orbital element for the HealthNet and VITA networks.

Orbit: Sun-synchronous polar orbit, perigee= 792.5 km, apogee = 805.3 km, inclination =98.64°

D.40.9 PoSAT-1 (Portuguese Satellite)

PoSAT[960),961)] is Portugal's first satellite, built by a joint team of Portuguese and SSTL engineers at SSTL within a technology transfer program between UK and Portugal. The primary objective of the program is to stimulate a Portuguese space industry; the purpose of the PoSAT-1 mission is to generate a nucleus of engineers with first-hand space technology experience for Portugal's possible future satellite programs.

PoSAT-1 was launched from Kourou on Ariane, September 26, 1993 as an auxiliary payload on the SPOT-3 satellite. PoSAT-1 uses the modular UoSAT microsatellite platform. The S/C is stabilized by an Earth-pointing gravity boom and by a 3-axis magnetic torquing system. S/C mass = 50 kg, power = 20 W, downlink data rate = 9.6 or 38.4 kbit/s, VHF and UHF bands, ground receiving stations are in Portugal (Sintra) and the UK. S/C operation via Sintra station. The Portuguese S/C operator is 'Consorcio PoSAT of LNETI (Laboratorio Nacional de Engenhario e Technologia Industrial).' S/C design life = 5 years. On-board data recording capability of 32 MByte into solid-state memory. As of 1/1996 nominal operations are conducted.

Orbit: Sun-synchronous, perigee = 791.1 km, apogee = 805.4 km, inclination = 98.64°, period = 100.8 minutes.

Downlink frequencies: 145.975 or 145.925 MHz	Mode: JD 9.6 kbit/s FSK (primary or secondary)
Uplink frequencies: 435.075 or 535.050 MHz	Mode: JD 9.6 kbit/s FSK (primary or secondary)

Table 267: PoSAT-1 transponder frequencies

Sensor complement:

EIS = Earth Imaging System. Two video CCD imaging cameras with two optical assemblies. The cameras are aimed in the same direction but offer a different FOV. The wide angle camera (WAC) image has a ground resolution of 2 km and a FOV (68°) of 1500 km x 1050 km (NIR 810-900 nm filter; 4.8 mm focal length). The objective is to provide images with good contrast between land, sea, clouds, and ice or snow. - The narrow angle camera (NAC) has a ground resolution of 200 m and a FOV of about 150 km x 100 km (red 610 - 690 nm filter, 50 mm focal length). The cameras feature a monochromatic design with optical filters chosen to contrast ground properties, the NAC highlights variations in soil moisture content and vegetation density. An image is taken in a snapshot or staring mode (as opposed to a mechanically scanned pushbroom array) and is read from each CCD by a Transputer Data Processing Experiment (TDPE) and, after processing, provided to the On-Board Computer (OBC) and recorded. Camera sensor (both cameras): EEV CCD04-06 image sensor + chip set; 578 x 576 array (interleaved fields); anti-blooming and electronic integration control. Operational Earth imaging is performed on a scheduled snapshot basis.

SIS = Star Imaging System (attitude control). The sensor is based on the EIS camera but with suitable optics for imaging the faint light from stars. Spectral range: 400 - 1000 nm, the FOV is 10° x 7°, IFOV = 0.1° (in the Y-direction of the S/C). The star sensor image is analyzed by the TDPE and measurement data is returned to the OBC.

GPS Receiver (based on the Trimble TRANS-II 6-channel receiver). GPS data provides onboard position, velocity and time reference. The data is used by OBC to generate an orbital element set and to provide scheduling and synchronization to other on-board computers,

960) Information provided by J. Radbone of SSTL, University of Surrey, UK
961) "First PoSAT images," Space, Vol. 9, No. 9, December 1993, p. 6

and to allow ground stations equipped with a GPS receiver to experiment with applications for real-time DGPS. Posat is the first microsatellite to make use of a GPS receiver in orbit. It is the only satellite which has yet been able to autonomously determine its orbit through the processing of GPS data into orbital elements.

Satellite	EIS Camera configuration	Spectral Band	Orbit Altitude	Mean spatial resolution	Area of Image
UoSAT-5	single	600-615 nm	800 km	2 km	1500 x 1080 km
KITSAT-1	WAC NAC	810-890 nm 810-890 nm	1300 km	3.5 km 350 m	2550 x 1800 km 220 x 160 km
PoSAT-1	WAC NAC	810-890 nm 610-690 nm	800 km	2 km 200 m	1500 x 1050 km 150 x 100 km
KITSAT-2	WAC NAC	810-890 nm 610-690 nm	800 km	2 km 200 m	1500 x 1050 km 120 x 100 km
FASat-Alfa	WAC NAC	810-890 nm 810-890 nm	816 km	2 km 120 m	1500 x 1050 km 93 x 62 km
Thai-Phutt	WAC NAC	810-890 nm 810-890 nm 610-690 nm 510-590 nm	800 km	2 km 100 m	1500 x 1050 km 100 x 100 km

Table 268: Versions of EIS video camera configurations on different spacecraft [962)]

CRE = Cosmic Ray Experiment (description under KITSAT-1)

TDE = Total Dose Experiment. TDE is part of CRE, monitoring the total accumulated ionizing dose in a solid-state RADFET dosimeter. The larger memory devices in the computers are regularly 'washed' to detect and log Single Event Upset (SEU) information.

DSPE = Digital Signal Processing Experiment. The DSPE consists of two Texas Instruments processors (C25 and C30 from TM320 series). The DSPE can be used as a programmable communications modem to modulate downlink data from, or uplink data to the OBC, thus enabling experiments with new modulation techniques to be optimized for LEO satellite mobile communications.

D.40.10 KITSAT-2

KITSAT-2 was designed and built at KAIST/STRC in Korea. It was launched on Ariane-4 (along with PoSAT-1 and HealthSAT-2), September 26, 1993 as an auxiliary payload on the SPOT-3 launch from Kourou. For a description, see chapter D.19.1.

D.40.11 CERISE

CERISE (Caracterisation de l'Environnement Radioelectrique par un Instrument Spatial Embarque) was procured by the French Ministry of Defense (DME) through its prime contractor Alcatel Espace of France, and built by SSTL. Objective: characterization of the Earth's radio environment. CERISE was launched on July 7, 1995 [Ariane ASAP, secondary payload to Helios-1A (a French military photo-surveillance S/C), flight number 75]. S/C mass = 50 kg, size: 650 x 300 x 300 mm, power = 35 W peak (25 W orbit average). The CERISE payload is classified (designed to eavesdrop on weak high-frequency communications).

On July 24, 1996, the CERISE S/C collided with a piece of debris (a suitcase-sized piece of an old Ariane rocket), cutting in half the 6 m long stabilizing boom of CERISE.[963)] French ground controllers regained control over the satellite following the collision.

962) M. Fouquet, A. Brewer, "The Role of Microsatellites for Earth Observation, Eight years of orbital experience at the University of Surrey," in Small Satellites for Remote Sensing, Proceedings of Space Congress, Bremen, Germany, May 24-25, 1995, pp. 133-144

Orbit: Sun-synchronous polar orbit, altitude = 657.5 km x 680.3 km, inclination = 98.07°.

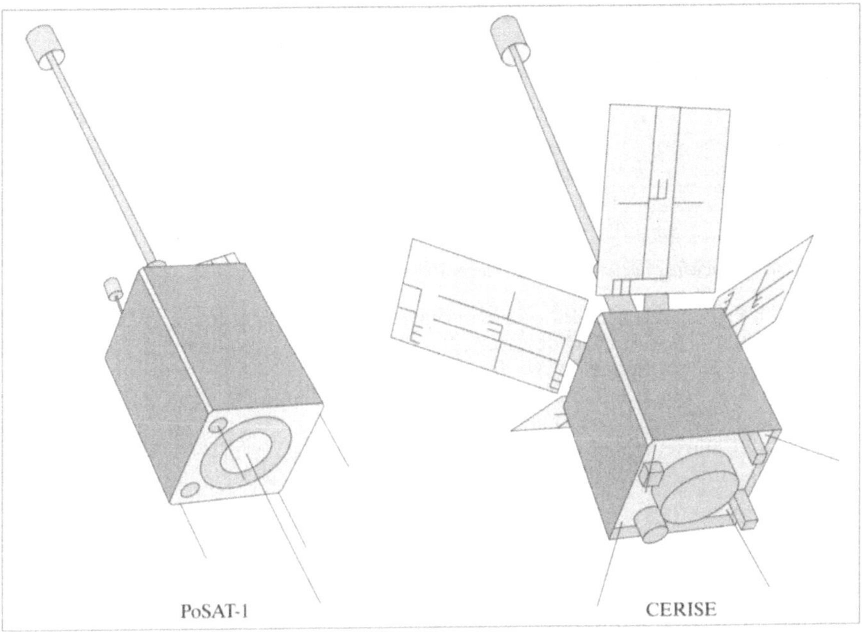

Figure 162: Typical UoSAT S/C models

D.40.12 FASat-Alfa (Fuerza Aerea Satellite - Alfa)

A microsatellite of Chile designed and built within one year under a Technology Transfer Program between the Chilean Air Fore (FACH = Fuerza Aerea de Chile) and SSTL. The primary goal of the program is to obtain for Chile the basic scientific and technological experience required to continue with more advanced steps. Specific objectives of the FASat-Alfa mission are to create a group of engineers with aerospace experience and to operate the Mission Control Station (ECM-Santiago) in Chile.

FASat uses the proven modular UoSAT bus design; the S/C is stabilized by an Earth-pointing gravity boom and by a three-axis magneto-torquing system. First introduction of a yaw-axis reaction wheel into the micro bus to provide a nadir-pointing non-spinning platform. The FASat structure consists of 11 module trays including the battery box. Eight of the module trays are used for the platform systems, three module trays for payloads. S/C mass = 50 kg, power = 35W peak (25 W orbit-average).

A launch (secondary payload to SICH-1) was performed August 31, 1995 on a Russian Cyclone (Tsyklon) vehicle. **Unfortunately, the separation mechanism to release the microsatellite from SICH-1 failed to operate.** This implied non-operation (or hibernation) for FASat-Alfa. Some time after the launch, FACH and SSTL have declared the spacecraft to be lost. The SICH-1 S/C is able to perform its observation functions unharmed with the FASat S/C permanently attached to it.

A replica, FASat-Bravo, is under construction at SSTL as of 1996 for a launch in 1998.

963) "Space Debris Damages French Defense Satellite," Space News, August 26 - September 1, 1996, p. 4 and p. 19

Orbit: Sun-synchronous near-circular orbit, perigee=651 km, apogee=682 km, inclination = 82.5°, period = 98.7 minutes.

Sensor complement:

EIS = Earth Imaging System (description under D.40.9).

WAC		NAC	
Number of active pixels	568 x 560	Number of active pixels	568 x 560
Focal lens length	4.8 mm	Focal lens length	75 mm
FOV	102 x 73°, 1150 x 850 km	FOV	6.5 x 4.8°, 74 x 55 km
IFOV (nadir)	0.18 x 0.13°, 2.0 x 1.5 km	IFOV (nadir)	0.011 x 0.085°, 130x97 m

Table 269: Some parameters of the EIS camera system for FASat orbit

OLME = Ozone Layer Monitoring Experiment. The instrument consists of two UV cameras, one operating with CCD detectors, the other with UV photodiodes. The objective is ozone layer monitoring by measuring the backscattered UV solar radiation in frequency bands of 308 - 318 nm and 385 - 395 nm (total column ozone measurements). The measurements of OLME are in particular dedicated to overflights of the antarctic and sub-antarctic regions of Chile (correlations with ground-based observations of the UV radiometric network). The UV cameras use the EEV CCD02 (384 x 576 pixels) in the same circuit as the EIS cameras.

Parameter	Value	Parameter	Value
Number of active pixels	378 x 560	Focal lens length	12.5 mm
FOV	39° x 29° (450 x 330 km)	Resolution at nadir	0.1 x 0.05° (1.2 x 0.6 km)
Mean resolution (nadir)	0.9 km		

Table 270: Parameters of the OLME UV cameras

DTE = Data Transfer Experiment (S&F mode operation). The instrument consists of two redundant receivers and a redundant DSP (Digital Signal Processing) unit. The objective of DTE is to provide the space segment infrastructure (hardware, interfaces, firmware and software) to allow a variety of communication experiments of the following nature: a) uplink experiments, b) duplex/two-way experiments, c) data collection, d) data transfer (collection and distribution), e) downlink experiments, f) data distribution, g) technology demonstration, h) enhanced modulation/data rates/small terminals.

GPS Receiver. GPS data provides on-board position, velocity and time reference (Trimble TANS II with LEO firmware). Generation of Keplerian elements using a 32-bit transputer system (see description under PoSAT).

SSDRE = Solid-State Data Recorder Experiment. Provision of 256 MByte of RAM memory to all systems of the S/C (SSDRE was built by Lockheed Martin of Nashua, NH). The recorder is a highly integrated processing unit able to communicate with the host system, to manage storage operations and to perform internal diagnostic functions.

EdEx = Educational Experiment. EdEx is intended to promote direct participation in space by Chilean schools to communicate with the FASat S/C. EdEx in turn uses DTE to generate telemetry signals that can be received by low-cost receivers and a PC. DSP chips with the DTE produce digitized voice data that may be received by simple receivers. EdEx takes place on a scheduled basis. The curricular benefit for the students arises from the following activities:

- S/C tracking (learning of some basic orbital parameters)
- Handling of satellite communications (telemetry reception)
- FASat satellite telemetry analysis (in particular those parameters related to electric and thermodynamic parameters).

Data: The RF system consists of two redundant transmitters, three receivers and the associated modulators/demodulators. Downlink data rates are between 9.6 to 76.8 kbit/s to a single ground station in Chile. MSC-Santiago (Mission Control Station - Santiago at the Los Cerrillos Air Force Base) is in charge of controlling all aspects of satellite operations.

D.40.13 FASat-Bravo (Fuerza Aerea Satellite - Bravo)

FASat-Bravo is the second Chilean microsatellite in orbit, built under a technology transfer program between the Chilean Air Fore (FACH = Fuerza Aerea de Chile) and SSTL. The program's primary objective is that of acquiring the basic scientific and technological experience required to take more advanced steps. The mission includes the same sensors/experiments as those of FASat-Alfa.

FASat-Bravo was launched successfully as a secondary payload (along with Thai-Phutt, TechSat/Gurwin-II, WESTPAC, and SAFIR-2 - and RESURS-O1-4 as primary payload) from the Baikonur Cosmodrome on a Zenit-2 launcher on July 10, 1998.

Orbit: Sun-synchronous circular orbit, altitude = 830 km, inclination = 98°.

Sensor complement: see FASat-Alfa description.

D.40.14 UoSAT-12

UoSAT-12 is an SSTL-funded proof-of-concept (technology) mission in low-cost minisatellite engineering techniques. Invited collaborative payloads are from NTU (Nanyang Technological University) in Singapore and from ESA. The design of a new standard minisatellite platform, MiniBus, started in 1996. The objective is to demonstrate and qualify the new MiniBus platform to fly an Earth observation package and to qualify new technology for flight on subsequent missions.

General MiniBus characteristics:

The MiniBus structure is essentially cylindrical with a height of about 1 m and a diameter of 0.6 m, supporting a payload mass of 50 to 150 kg. Payloads can either be carried in one or more standard SSTL MiniBus modules in the Module Stacks, or in the large payload frame with aperture access to the Earth-facing facet. Payloads for Earth observation can be carried directly on the nadir pointing Earth frame, and space-facing payloads are accommodated on the zenith-facing facet. The thrust bearing column of the minibus consists of three stacks of Surrey's standard microsatellite module trays. The stacks (340 mm x 340 mm x 500 mm) are joined around a triangular central void and placed on a honeycomb panel of about 1.1 m diameter.[964)] [965)]

The OCS (Orbit Control and Determination System) comprises a GPS receiver, and a cold gas thruster system. Both are also an integral part of ADCS (Attitude Control and Determination System). The ADCS actuators include: two Surrey momentum wheels, one Ithaco momentum wheel, cold-gas jets, magnetorquers, and a gravity gradient boom. The sensors include: three three-axis magnetometers, four two-axis analog sun sensors, a GPS receiver (SGR-20) and two star cameras, a Servo Earth sensor and an Ultra magnetometer and a solid-state gyro. The three-axis control system maintains Earth pointing to an accuracy of 0.5°, with an experimental target of 0.1°, for Earth observation payloads and communication antennas. The ADCS can be operated in momentum-biased mode, zero momentum mode, or in gravity-gradient mode (for EOL mission extension). ADCS and OCS are supported by an attitude control processor, which can be run on any of the four primary on-

964) J. Ward, M. Sweeting, "First In-Orbit Results from the UoSAT-12 Minisatellite," Proceedings of 13th Annual AIAA/USU Conferences on Small Satellites, Logan, Utah, Aug. 23-26, 1999, SSC-99-I-2

965) W. Sun, M. N. Sweeting, "In-Orbit Results from UoSAT-12 Earth Observation Minisatellite Mission," Proceedings of the 3rd International Symposium of IAA, Berlin, April, 2-6, 2001, pp. 79-82

board processors. Additionally, UoSAT-12 carries an electric propulsion system, a 90 W resistojet producing 93 mN of thrust, which uses nitrous oxide as its working fluid. 966) 967)

The power system comprises a set of nine deployed solar panels and battery charge regulators, three 6Ah battery packs, a centralized power switching system, and a distributed power conditioning system. Power is distributed and switched via an unregulated 28V bus. The solar panels employ highly efficient GaAs cells, supplying an orbit average power of 150 W to the platform. Panels are deployed at an angle depending on the orbit and power profile required for the mission. In order to provide prolonged battery cycle life, NiCd cells are used in the 18Ah battery.

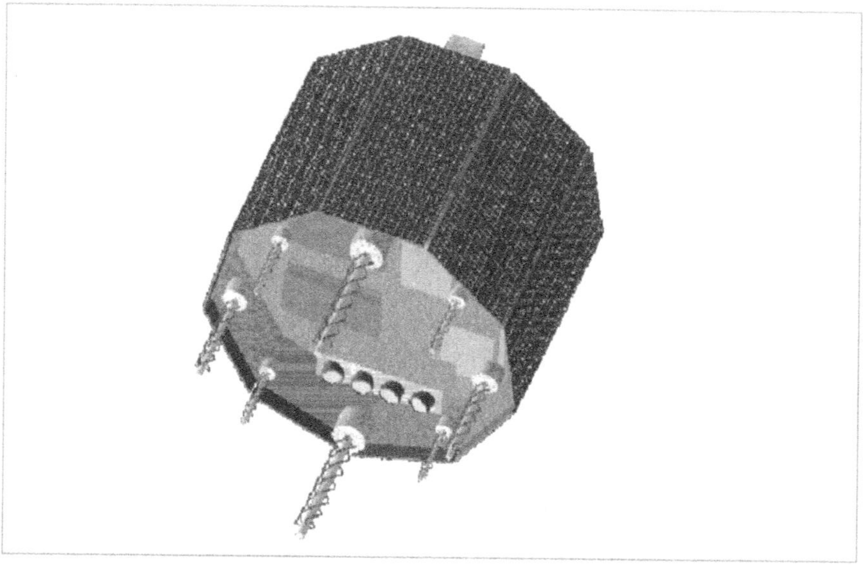

Figure 163: The UoSAT-12 minisatellite

Communication: UoSAT-12 carries five uplink receivers. Three are 10 kbit/s FSK channels in the VHF-band. Two are L-band receivers integrated in the Merlion 1 Mbit/s transponder. The communication system employs VHF and UHF uplinks and downlinks for standard telemetry and telecommand. These also provide the standard S&F (Store and Forward) digital transponder service. An S-band 1Mbit/s downlink supports payloads that generate large amounts of data, such as Earth Observation payloads. The three UHF downlinks support 9.6 and 38.4 kbit/s with CPFSK modulation. The dual redundant S-band downlink provides 1Mbit/s BPSK modulation, with optional Viterbi encoding.

Telemetry and telecommand is implemented as a distributed system based on a microcontroller with ISO-11898 CAN (Controlled Area Network) support. Each payload and bus system contains one or more nodes, each of which is implemented using a CAN microcontroller. The network is dual redundant and DC isolated. The TT&C master is implemented in software, and its function is to collect, format and downlink telemetry and issue telecommands, either direct or programmed. The TT&C master would nominally be the primary on-board-computer, however this architecture allows any other computer, and indeed any of the nodes to perform this function.

966) M. Fouquet, M. Sweeting, "UoSAT-12 Minisatellite for High Performance Earth Observation at Low Cost," Acta Astronautica, Vol. 41, No. 3, pp. 173-182, 1997

967) A. Wicks, A. da Silva-Curiel, J. Ward, M. Fouquet, " Advancing Small Satellite Earth Observation: Operational Spacecraft, Planned Missions and Future Concepts," Proceedings of the 14th Annual AIAA/USU Conference on Small Satellites, Logan, UT, Aug. 21-24, 2000, SSC00-I-8

UoSAT-12 was launched April 21, 1999 by a Russian launcher, Dnepr, of Kosmotras International Space Co., Moscow, from the Baikonur Cosmodrome, Kazakhstan. S/C mass = 325 kg. The launch was the first commercial mission for the former Soviet Union's arsenal of SS-18 Satan missiles (all of which have to be destroyed or used for peaceful purposes under the START arms reduction treaty). The Dnepr vehicle was created in 1997 under an agreement between the Russian and Ukrainian space agencies.

Orbit: Circular orbit, altitude = 650 km, inclination = 64.57°.

Sensor/payload complement of UoSAT-12

The payload imaging systems include high-performance processor modules to provide autonomous image assessment and compression. The processing units are based on the Inmos T805 transputer and are an integral part to the operation of the imaging instruments (they manage all aspects of of the imagers including scheduling, image capture, analysis preprocessing and compression). The compression algorithm used is AMPBTC (Adaptive Moment-Preserving Block Truncation Coding), achieving compression ratios up to 4:1. Although the compression is relatively modest, the AMPBTC routine preserves fine detail more faithfully than for instance JPEG.

WAC (Wide Angle Camera). The instrument is of UoSAT-5 and PoSAT-1 heritage under the name of EIS (Earth Imaging System). An extremely wide angle lens is used (4.8 mm focal length) to offer wide coverage. The camera is fitted with an optical filter giving NIR sensitivity and providing strong contrast between land, sea and cloud. The WAC instrument often serves as a spotter camera to assist in locating scenery from the other cameras.

SHC (Surrey High-resolution Camera). SHC uses COTS (Commercial Off-The-Shelf) components. The panchromatic instrument features a Kodak-built silicon CCD detector of size: 1024 x 1024 pixels and a Leica lens with an aperture of 560 mm (f/8). The footprint is 10 km x 10 km per image (step-and stare operation of a 2-D detector array). The camera can also be programmed to capture a sequence of images, thus allowing long swaths of imagery to be collected. The mass of SHC is 10 kg.

Camera	Spectral Band	Orbit	Mean Resolution	Coverage (Image Size)
WAC	810–890 nm	650 km	975 m	1000 x 1000 km
SHC	Panchromatic	650 km	10 m	10 x 10 km
MSI-0	Bands 1,2,3,4 & 8	650 km	32.5 m	33 x 33 km
MSI-1	Bands 1,2,3,4 & 8	650 km	32.5 m	33 x 33 km
Star0	400-1000 nm	650 km	0.02°	20° x 15° (FOV)
Star1	400-1000 nm	650 km	0.02°	20° x 15° (FOV)

Table 271: Parameters of UoSAT-12 camera systems

MSI (Multispectral Imager). There are two MSI instruments (MSI-0 and MSI-1) on-board working in tandem and angled in the cross-track direction (3°) to provide a combined coverage of 60 km (cross-track) x 30 km (along-track). Each imager has a 2-D CCD silicon detector of size 1024 x 1024 pixels and a lens of 180 mm aperture (f/3.4). The spatial resolution is 32.5 m. The MSI instruments use rotating filter wheels to image in four spectral bands (0.45 - 0.52; 0.52 - 0.60; 0.63 - 0.69; and 0.76 - 0.90 μm), identical to bands 1-4 of Landsat's TM (Thematic Mapper). The transputer support of MSI makes it possible to image contiguous swaths or to support step-and-stare observations. The mass of each MSI instrument is 3 kg.

Star cameras (Star0 and Star1). There are two star cameras aboard acting as precision attitude sensors. The FOV of one camera is 20° x 15° to ensure coverage of enough stars, the other camera has a FOV of 10° x 7° providing a fine resolution. The optics for the star cameras are: f/0.85 25 mm and f/1.0 50 mm, respectively. Both cameras use CCD detectors. In order to operate under very low light conditions, such techniques as long integration times (up to a second), reduced output noise bandwidth, a non-linear transfer function, and low-noise amplification are used to obtain usable signals from faint stars.

The spacecraft imagers have been performing excellently, their success has been the return of spectacular 10m resolution panchromatic and 32m multispectral images from orbit demonstrating "a breakthrough in the order of magnitude reduction in cost for high resolution space imagery."

SGR (Space GPS Receiver). SGR-20 is an ESA-funded instrument in partnership with SSTL. The objective is to provide an experimental test-bed for GPS orbit and attitude determination. The SSTL design uses a MITEL chip set and an ARM60 RISC processor (32 bit). Radiation tolerance is implemented through an EDAC (Error Detection And Correction) memory and a latch-up protected power switch. SGR is a 24-channel instrument, it features five GPS antennas, four of which can be used simultaneously to track GPS signals. The carrier phase of L1 signals are observed at each of the four antennas. One antenna is sufficient for orbit determination. With the signals from the multiple antennas it is possible to determine the S/C attitude by measuring phase differences and applying interferometric techniques. [968)]

EPS (Electric Propulsion System). EPS is an experimental resistojet electric propulsion system. Water on nitrous oxide is super-heated over a resistive heater element, the resulting hot gas is expelled through a nozzle to produce low-level thrust at moderate specific impulse. The thruster provides a thrust of 93 mN, using 90 W of input power, the Isp is 1250 m/s. A total Δv of 10.4 m/s is provided by a 2-liter tank of self-pressurized nitrous oxide.

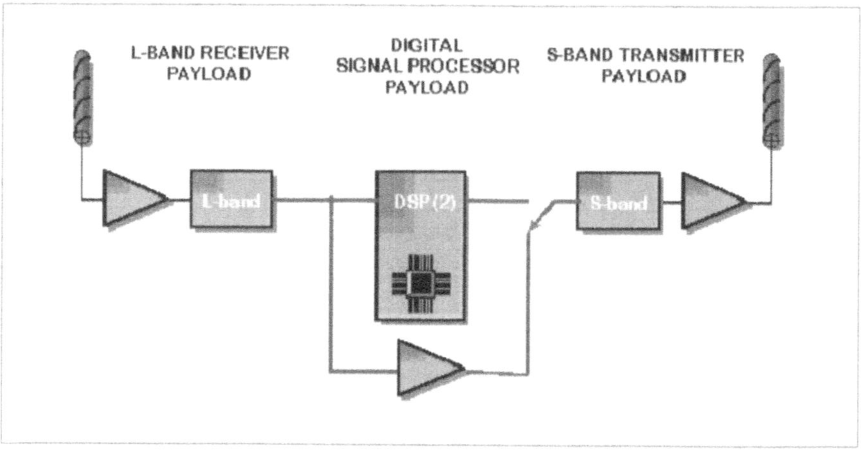

Figure 164: The Merlion L-band to S-band transponder

Merlion Communications Package of NTU (Nanyang Technological University), Singapore in cooperation with SSTL. The objective is to demonstrate a new communications technology (radio amateur L/S-band transponder) for LEO systems to improve coverage in particular for equatorial regions. A later constellation of several suitably phased satellites in such an orbit can provide continuous coverage for the lower latitude regions. The use of L-band and S-band amateur frequencies is to stimulate more extensive use of these bands (by experimenting with various access schemes and operating modes). A ground station at

968) S. Purivigraipong, M. J. Unwin, Y. Hashida, "Demonstrating GPS Attitude Determination from UoSat-12 Flight Data," ION-2000, Salt Lake City, UT, Sept. 19-22, 2000, pp. 2625-2633

NTU is testing these schemes. The payload combines an analog and digital regenerative transponder with L-band uplink and S-band downlink. [969)]

Downlink: The frequency-agile digital S-band downlink is used for high-speed data transfer and performs a variety of communication experiments. It is capable of low rate spread spectrum communications, and has Viterbi coding options to investigate its performance in the highly dynamic LEO environment. The downlink is also used to perform link characterization at these frequencies. The modulation scheme is NRZ-L/BPSK at 256 kbit/s, 512 kbit/s and 1 Mbit/s with square root raised cosine shaping. The data is processed via various scramblers and encoders. Data is scrambled using a standard V.35 scrambler, or via a programmable 32-bit universal scrambler. A 1/2 rate Viterbi encoder (k=7, G1=171, G2=133) is also included, and a spread spectrum mode permits experimentation with lower data rates for smaller ground stations while keeping the chip rate at 1Mbit/s. The downlink power is 7W.

Uplink: The digital uplink consists of a frequency-agile L-band 9600 bit/s FSK receiver. Both IF outputs and baseband outputs are presented to a TI C31 Digital Signal Processor, for modulation and coding experiments. The frequency scanning functions and variable bandwidth IF filtering permit the receiver to be used in conjunction with the DSP for narrow band channel characterization and transponder measurements. In conjunction with the on-board-computers, the digital uplink and downlink system can be configured into a high rate S&F transponder.

The satellite antennas are shaped in such a way as to provide equal power flux density across the satellite footprint. Quadrifilar helices are employed with downlink LHCP and uplink RHCP.

D.40.15 TMSat (Thai-Microsatellite)

TMSAt, also known as Thai-Phutt, is the result of a two-year in-depth collaborative technology transfer program (focusing on the construction and launch of TMSat) between the UK and Thailand. TMsat is also the name of a company, namely of TSC (Thai Satellite Communication) and of MUT (Mahanakorn University of Technology), both of Bangkok, Thailand. MUT and TSC are investing in TMSat in order to develop satellite technology in Thailand. The TMSat application objectives are in the fields of Earth observation and store-forward communications. [970)]

The S/C platform is based on the latest version of the 'Micro-Bus' of SSTL, a boxlike three-axis stabilized (Earth-pointing) structure with gravity boom. The bus consists of a series of identical module trays stacked one on top of the other to form a body, onto which solar panels and instruments are mounted. The ADCS (Attitude Determination and Control Subsystem) module controls the alignment of S/C in the Earth-pointing direction. Magnetometers measure the Earth's magnetic field vector, they are used together with sun sensors for achieving the attitude information. A S/W package (using estimation theory and robust control theory) provides for proper attitude determination and control (attitude is maintained within 1° of nadir and determined within 0.001°). Magnetorquers and a momentum wheel are applied as actuators. Power is provided by four body-mounted GaAs panels (35 V, 1 A). Batteries are used to provide power during eclipse phases. There are 10 NiCd commercial cylindrical cells with F size cells with a capacity of 7Ah. The S/C mass is 50 kg.

TMSat was launched successfully as a secondary payload (along with FASat-Bravo, Tech-Sat/Gurwin-II, and SAFIR-2 - and RESURS-O1-4 as primary payload) from the Baikonur Cosmodrome on a Zenit-2 launcher on July 10, 1998.

969) Tai Wei Chua, et al., "Merlion L&S-band System," Proceedings of 13th Annual AIAA/USU Conferences on Small Satellites, Logan, Utah, Aug. 23-26, 1999, SSC-99-I-1

970) http://www.ee.surrey.ac.uk/CSER/UOSAT/missions/tmsat/info/index.html

Communication: 1) uplink: three redundant receivers and four monopole antennas, three demodulators (CPFSK). VHF uplink frequency, UHF downlink frequency, downlink data rate at 9.6 kbit/s, FSK modulation. 2) downlink: two redundant transmitters, four monopole antennas, two modulators (CPFSK). UHF downlink frequency, downlink data rates of 9.6 kbit/s and 38.4 kbit/s (data).

Orbit: Sun-synchronous circular orbit, altitude = 835 km, inclination = 98°.

Sensor complement:

EIS = Earth Imaging System. EIS consists of two basic elements:

- **WAC** (Wide Angle Camera), a CCD sensor made by EEV Ltd. with an active imaging pixel array of 568 x 560 (same as for FASAT-Bravo). WAC is is fitted with an ultra wide angle lens (focal length 4.8 mm) to provide a coverage area of 1500 km x 1050 km at a mean resolution of 2 km. The camera is fitted with a near-IR optical filter (810-890 nm) to provide strong contrast between land, sea and clouds. The camera is recording light from the sun, reflected off the Earth's surface and atmosphere. A WAC image is 330 kByte in volume.
- **NAC** (Narrow Angle Camera). The camera design is a new development for the TMSat mission and uses a 1020 x 1020 pixel CCD sensor manufactured by Eastman-Kodak. The three NACs are electrically identical, as are the lenses (75 mm focal length) used to provide 100 m resolution for an 835 km orbit. The cameras employ 2-D CCD arrays with digitized data of 8 bits, offering 100 km swath widths and a capability to collect images contiguously along the flight path. The images are in three spectral bands (red, green and VNIR) providing information on Earth resources, land use, and effects of natural disasters. Each NAC image is 1 MByte in volume.

WAC		NAC	
Spectral range	810 - 890 nm (VNIR)	Spectral range, NAC0 NAC1 NAC2	810 - 890 nm (VNIR) 610 - 690 nm (red) 510 - 590 nm (green)
Number of active pixels	568 x 560	Number of active pixels	1020 x 1020
Focal lens length	4.8 mm	Focal lens length	75 mm
FOV	102° x 73°, or 1500 km x 1050 km	FOV	6.5° x 6.5°, or 100 km x 100 km
IFOV (nadir)	0.18 x 0.13°, 2.0 x 1.5 km	IFOV (nadir)	0.085° x 0.085°, 100 m

Table 272: Some parameters of the EIS camera system for the TMSat orbit

CMOS Video Camera with APS (Active Pixel Sensor) technology.

CCD detector array	382 x 287 pixels
Spatial resolution, FOV	400 m, 10.5° x 7.8°
Ground coverage	150 km x 110 km
Spectral range	350 - 750 nm (VIS range)
Lens	25 mm, f/8
Pixel size, array size	12 nm x 12 nm, 4.6 mm x 3.4 mm
Frame rate	50 frames/s, CCIR TV format
Exposure	Automatic (On-Chip H/W Controlled - 25,000:1 Dynamic Range)
Instrument power	200 mA at 5V = 1W continuous operation
Instrument data storage	4 MByte (= 30 images)

Table 273: Parameters of the CMOS Video Camera

DSPE = Digital Signal Processing Experiment. The objective is to provide a programmable DSP capability to perform any communication and signal processing experiments. The

DSPE consists of two Texas Instruments processors (TM320 series). The system is equipped with 128kbyte program memory with Error Detection and Correction (EDAC). Also 1MB of unprotected data memory is provided. The design permits a number of experiments to be conducted via DSPE which can be summarized as follows:

- Real-time voice communication
- Experimental digital modem design
- In orbit signal analysis
- A scanning receiver
- Digital voice broadcaster
- etc.

GPS Receiver. The objective is to provide: a) a real-time clock to synchronize the satellite and ground station, b) information of the satellite position which is used by the ground station to determine the microsatellite orbit.

D.40.16 SNAP (Surrey Nanosatellite Applications Program)

A program of SSTL and CSER (started in 1998) with the objective to develop a modular, multi-mission nanosatellite bus (mass range of 1-10 kg), to demonstrate the use of miniature electrical and mechanical COTS (Commercial-Off-The-Shelf) product technologies in space and their use as autonomous robots for observing orbiting space vehicles. SSTL intends to market SNAP-type satellites commercially. Future applications of nanosatellites are seen in remote inspection of satellites and monitoring of deployment systems in orbit, and carrying small space science instruments requiring measurements with spatial diversity. 971) 972) 973)

D.40.17 SNAP-1

SNAP-1 was designed and built as a low-cost research mission by a joint academic-commercial team at the Surrey Space Center and at SSTL, funded entirely by SSTL. The objective of SNAP-1 is to demonstrate the feasibility of a standardized modular nanosatellite bus.

- Provide a test-bed for novel microelectronic technologies - in particular a new GPS navigation system, APS camera technologies and RISC processors. The intent is to use SNAP-1 as a "remote inspector demonstrator."
- Provide experimental data and imagery to the radio-amateur/amateur-scientific communities
- Provide a vehicle for the education and training of students in spacecraft engineering at undergraduate and post-graduate level
- The SNAP program is also intended to demonstrate the feasibility of using clusters of low-cost satellites that can fly in formation and conduct multipoint remote sensing.

The SNAP-1 nanosatellite is three-axis stabilized by a momentum wheel and magnetorquers as actuators; the S/C attitude is sensed by a three-axis magnetometer and sun sensors. A GPS receiver (model: SGR-05 of SSTL) is used for autonomous orbit determination (onboard navigation parameters and timing). The SGR-05 performs differential orbit determinations in conjunction with Tsinghua-1. The S/C bus features a modular design based on a tray structure (up to three trays). A micro-propulsion system (liquified gas) is used for orbit maneuvers (rendezvous with other S/C). Four body-mounted solar cells provide an average power of 6.5 W each. SNAP-1 has 4 W of average orbit power and 9.1 W of peak power. A six cell NiCd battery provides eclipse-phase energy of 1.4 Ah. The S/C mass is 6.5 kg (the all up mass is 8.3 kg with about 1.8 kg for the attach fitting for mating onto the carrying vehicle).

971) Data sheets provided by Craig Underwood of SSTL

972) J. Singer, "US Eyes British Demonstration Satellite," Space News, Oct. 23, 2000, pp. 3 and 19

973) A. Cropp, "The SNAP-1 NanoSat Project at Surrey - A New Generation of Satellites," Proceedings of the 49th IAF Congress, Melbourne, Australia, Sept. 1998

S/C mass	6.5 kg (1.2 kg of payload mass)
S/C size	175 mm x 200 mm (diameter) hexagonal prism
Solar panels	Each panel 7 x 8 cells (40 mm x 40 mm GaAs)
S/C power	2.5 W at 5.0 V regulated, 6 cell NiCd (an advanced form of NiCd)
RF communications	VHF uplink, 750 mW, S-band downlink
On-board data handling	Asynchronous uplink (9.6 kbit/s), downlink (76.8 or 38.4 kbit/s selectable)
Data rate	Uplink: 9600 bit/s, FSK modulated Downlink: 38.4 kbit/s nominal, 76.8 kbit/s max; BPSK & QPSK modulation, convolutional encoding on QPSK
On-Board Computer (OBC)	A 32-bit StrongARM OBC includes 16 MB FLASH and 4 MB of EDAC protected code memory. CAN-bus on-board data handling network
ADCS	Navigation, magnetometer, sun sensors, magnetorquers. GPS receiver accuracy <15 m. Attitude is estimated using a Kalman filter
Payloads	- SRG-05 based on the Orion GPS receiver of MITEL Semiconductors, - APS camera (referred to as MVS)

Table 274: Specification of the SNAP-1 satellite

The launch of SNAP-1, along with Tsinghua-1 of China (a microsatellite of 50 kg mass) as secondary payloads, took place on June 28, 2000 on a Russian Cosmos-3M launcher from the Plesetsk Cosmodrome. The primary payload on this launch was Nadezhda, a Russian S&RSAT (Search & Rescue Satellite) of COSPAS.

Communication: SNAP-1 carries a VHF receiver and an S-band downlink transmitter, as well as an intersatellite link, a UHF receiver, tuned to the downlink frequency of Tsinghua-1 (when within range, data can be transmitted from Tsinghua-1 to SNAP-1). The SNAP-1 S-band data rate is 38.4 kbit/s nominal or 76.8 kbit/s maximum. The S-band modulation is BPSK or QPSK (selectable), the convolutional encoding on QPSK is selectable. The uplink data rate is 9.6 kbit/s with FSK modulation. The mass of the S-band transmitter is 0.5 kg, the size is: 120 mm x 160 mm x 20 mm; DC power of 5.2 V @ 670 mA (330 mA idle). [974] [975]

Orbit: Sun-synchronous orbit, 700 km altitude, inclination=98°, period about 98.7 minutes.

Instrument complement:

MVS (Machine Vision System). The objective of MVS is to enable SNAP-1 to act as a remote inspector. The intent is to obtain imagery of TV-quality of other S/C in the local vicinity of SNAP-1 and to relay this information to the ground. MVS, a CMOS video system with APS (Active Pixel Sensor) technology, consists of the following components: [976] [977] [978]

- Three wide-angle CMOS cameras, each with a 350 x 288 pixel detector, and each with a 90° FOV to cover an arc of 270°
- A single narrow-angle camera (350 x 288 pixels) co-aligned with the center wide-angle camera providing the capability of finer feature inspection
- Video digitizer circuitry to convert the composite camera video output into digital data
- A recording capability of 8 Mbit of 70 ns SRAM for code and image storage
- A processor of the type StrongARM 1100 (220 MHz) for video/image compression. The processor may also being used for optical navigation functions in support of target tracking or automated optical docking.
- A CAN (Control Area Network) bus interface for communication with the spacecraft.

MVS features a software digitizer design due to the lack of implementation time (six months from project start to launch). The video signals of the cameras are fed into an ADC

974) Z. A. Wahl, K. L. Walker, J. Ward, "Modular and Reusable Miniature Subsystems for Small Satellites: An Example Describing Surrey's Nanosatellite S-Band Downlink," Proceedings of the 14th Annual AIAA/USU Conference on Small Satellites, Logan, UT, Aug. 21-24, 2000, SSC00-IX-4

975) H. Steyn, et al., "An Attitude Control System and Commissioning Results of the SNAP-1 Nanosatellite," Proceedings of the 14th AIAA/USU Conference on Small Satellites, Logan, UT, Aug. 21-24, 2000, SSC00-VIII-8

976) R. Lancaster, C. Underwood, "The SNAP-1 Machine Vision System," Proceedings of the 14th Annual AIAA/USU Conference on Small Satellites, Logan, UT, Aug. 21-24, 2000, SSC00-II-6

977) http://www.ee.surrey.ac.uk/EE/CSER/UOSAT/missions/SNAP/nanosat/index.htm

978) http://www.sstl.co.uk/services/subpage_services.html

(Analog Digital Converter) and recorded. The MVS processor scans the digital representation and extracts the video data. - MVS has performed to specifications. The images captured in flight show in detail the complete deployment sequence of both the SNAP-1 and the Tsinghua-1 satellites.

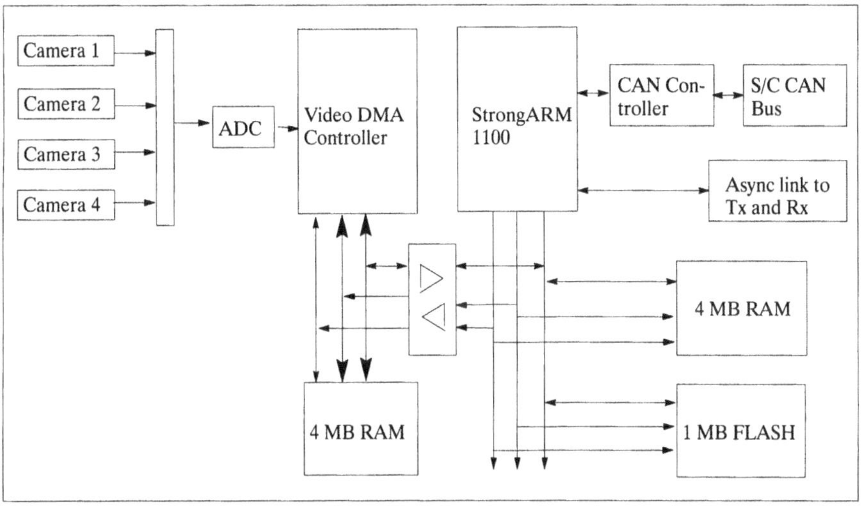

Figure 165: Schematic layout of the MVS

MPS (Micro-Propulsion System). [979] MPS is an SSTL-funded technology development with the objective to demonstrate the capability of SNAP-1 as an inspection vehicle (rendezvous with Tsinghua-1). The MPS is a stand-alone instrument consisting of three boxes and interconnected in such a way as to form a triangular structure (140 mm side length). The truster selected is a cold gas system from Polyflex Aerospace Ltd. with a maximum thrust of 100 mN at a chamber pressure of 4 bar (minimum impulse of < 1 mNs). It utilizes butane as propellant, operating in a cold gas mode. The propellant (32.6 grams) is stored in a formed titanium tube (coiled tube of 1.1 m in length with a volume of 65 cm^3). The thruster valve is a solenoid-operated valve. The drive electronics are operated from a microcontroller, deriving its commands and feedback telemetry via a CAN (Control Area Network) interface. The microcontroller also controls the manifold heater, which ensures that the propellant is vaporized. MPS is rated as a 30 mN thruster with a Δv capacity of 3 m/s.

Initially, the propulsion system was also intended to be used to maneuver SNAP-1 to rendezvous with Tsinghua-1 of the University of Beijing (sometime in Nov. 2000). However, the rendezvous did not occur in November. Still, MPS demonstrated the new nanosatellite's unique capability for controlled orbital maneuvers using its miniature propulsion system.

SNAP-1 and Tsinghua-1 were both released from the Cosmos launcher in different directions specifically to avoid the possibility of accidental re-contact. SNAP-1 ended up in an orbit about 2 km below that of Tsinghua-1 and, being relatively light (6.5kg), suffered more from the effects of atmospheric drag than the much heavier (50 kg) Tsinghua-1 microsatellite. [980] This meant that, relative to Tsinghua-1, SNAP-1 dropped in altitude more quickly. This was exacerbated by a very active sun (at solar maximum), causing the atmospheric den-

979) D. Gibbon, J. Ward, N. Kay, "The Design, Development and Testing of a Propulsion System for the SNAP-1 Nanosatellite," Proceedings of the 14th Annual AIAA/USU Conference on Small Satellites, Logan, UT, Aug. 21-24, 2000

980) Information provided by Craig Underwood of SSTL, Surrey, UK

sity at 700 km to be higher than normal. SSTL measurements showed that on average, SNAP-1 was falling about 10 m per day with respect to Tsinghua-1.

Thus MPS was used firstly to demonstrate orbit control (the primary objective) by maintaining its altitude by overcoming the relative atmospheric drag effects, and then also to climb back up to an altitude about 1 km higher than that of Tsinghua-1. Most of the propellent was used for this climb in December 2000. In this period Tsinghua-1 separated from SNAP-1 by more than 10,000 km along their orbital tracks. Once SNAP-1 was higher than Tsinghua-1, the along-track gap began to close. In January 2001, SNAP-1 was about 300 m higher than Tsinghua-1 with the gap closing.

SGR-05 (Space GPS Receiver-05). [981)] The instrument is essentially based on the Orion GPS GP2000 applications receiver of MITEL Semiconductors. SGR-05 comprises solely the core GPS engine (no features of the SGR-10/20 series). It uses a single antenna with 12 channels. The features of EDAC protection, CAN-bus support, and telemetry and command capability were disengaged in the trimmed-down version.

D.40.18 Tsinghua-1

Tsinghua-1 is a microsatellite of Tsinghua University, developed and built in a joint venture between SSTL of Guildford, Surrey, UK, and Tsinghua University in Beijing, China. The TSRC (Tsinghua Space Research Center) was set up in Oct. 1998 with the goal to integrate all space research activities at Tsinghua University and to provide a means and facilities for S/C building. The joint-venture company in Beijing is referred to as T-SSSC (Tsinghua-Surrey Small Satellite Company). The cooperative program is to develop and build microsatellites (Tsinghua-1) and nanosatellites (THNS-1) and to provide integrated training in small satellite design. [982) 983) 984)]

Tsinghua-1 is a demonstrator microsatellite in the 50 kg class of size: 35 cm x 35 cm x 64 cm. The overall objective is to provide daily high-resolution imaging for disaster monitoring and mitigation on a worldwide scale. A further goal of Tsinghua-1 is to conduct communications research in LEO. The design uses standard platform modules or trays to carry the subsystems and payload. The three payload modules include the GPS receiver, Transputer, and DSP/DTE (Digital Signal Processing/Data Transfer Experiment) unit. The cameras and wheels are accommodated in the Earth Observation Compartment. Two GPS antennas are accommodated on the space facing side of the S/C..

Attitude control	Magnetorquer-assisted gravity gradient, 3-axis reaction wheels and 3 magnetorquers	±0.3° roll/pitch, nadir pointing ±0.3° in yaw axis pointing accuracy: ±15° roll nadir
Solar panels	GaAs, about 35 W/panel (BOL)	Four fixed body-mounted panels
Batteries	7 Ah NiCd	
OBC	80C186+16 MByte, 80C386+64 MByte	Primary and secondary OBC
On-board data	9.6 kbit/s serial bus; High-speed CAN bus	Single network Dual network
Communication links	9.6 kbit/s VHF uplink 9.6 - 38.4 kbit/s UHF downlink	3 single cha. RCV, 2 synthesized RCV 2 synthesized transmitters

Table 275: Some parameters of the Tsinghua-1 S/C

The S/C is three-axis stabilized using a combination of passive (gravity-gradient boom) and active (magnetorquers, reaction wheels) actuator elements. The platform is nadir pointing.

981) M. J. Unwin, P. L. Palmer, Y. Hashida, C. I. Underwood, "The SNAP-1 and Tsinghua-1 GPS Formation Flying Experiment," ION GPS 2000, Sept. 19-22, 2000, Salt Lake City, UT, pp. 1608-1611

982) You Zheng, Gong Ke, M. Sweeting, "Tsinghua Micro/Nanosatellite research and its application," Proceedings of the 13th AIAA/USU Conference on Small Satellites, Aug. 23-26, 1999, Logan UT, SSC99-IX-3

983) http://www.ee.surrey.ac.uk/CSER/UOSAT/press/cjv.htm

984) Y. Zheng. M. Sweeting, "Initial Mission Status Analysis of 3-axis stable Tsinghua-1 Microsatellite," Proceedings of the 14th Annual AIAA/USU Conference on Small Satellites, Logan, UT, Aug. 21-24, 2000

Attitude is sensed by sun sensors and a magnetometer. The body-pointing platform has the capability to perform fast slew maneuvers within ±15° about the roll axis (or ±180° about the yaw axis). An off-nadir pointing configuration can be sustained for up to half an orbit.

A launch of Tsinghua-1 as a secondary payload to Nadezhda, a Russian S&RSAT (Search & Rescue Satellite) of COSPAS, took place on June 28, 2000 on a Russian Cosmos-3M launcher from the Plesetsk Cosmodrome.

Orbit: Sun-synchronous circular orbit, altitude = 700 km, inclination = 98°.

Sensor complement:

MEIS (Multispectral Earth Imaging System). The objective is to acquire multispectral Earth surface imagery with a spatial resolution of 50 m. The MEIS camera assembly consists of two cameras mounted at an angle of 15° from nadir so that the yaw angle can be selected to offer the required off-pointing angle of ±15° from nadir. The image swath is 80 km, each camera can collect four images contiguously along the flight path.

Imaging detector array	1024 x 1024 pixels, Kodak KAI-1001 non-interlaced sensor
Optics	Color-corrected Nikon lenses with 150 mm aperture
GSD (Ground Sampling Distance)	50 m x 50 m (resolution)
Swath width	400 km
Spectral bands (selectable in μm)	0.5 - 0.59 (green), 0.61 - 0.69 (red)) 0.81 - 0.89 (NIR)
Exposure control	Electronic integration time & gain (1000:1)
Radiometric resolution	8 bit - video digitization is synchronized with pixel stream producing 8-bit quantization (9 bit linearity)
SNR	>35 dB at 100% (about 2000:1)
Image raw data size	1 MByte per spectral band and per frame
Image compression	Scene-dependent compression ratios of 3:1 to 5:1 using AMPBTC (Adaptive Moment-Preserving Block Truncation Coding)
On-board processing	2 x T805 20 MHz Transputers + 32 MByte SRAM
On-board data storage	Up to 150 compressed multispectral images

Table 276: Specification of the MEIS instrument

D.40.19 TiungSat

TiungSat-1 (Tiung is the name of a small bird in Malaysia) is an SSTL-built (Surrey, UK) microsatellite for the Malaysian government (developed through a collaborative program of technology transfer and in-depth training). A team of eight Malaysian engineers from ATSB (Astronautic Technology SB) of Kuala Lumpur completed the construction during 1998. ATSB was established for this purpose and is wholly owned by the government of Malaysia. The overall objective is to generate a cadre of trained personnel in Malaysia to establish a national capability for sophisticated microsatellite and minisatellite design and fabrication for a variety of commercial and scientific research applications. The TiungSat-1 specific applications are in the following fields: [985] [986]

- Collection of imagery for environmental and meteorological use
- Digital S&F (Store & Forward) communications
- Technology demonstration
- Space science
- Amateur radio access

The TiungSat-1 S/C structure comprises eleven module trays used to house the electronics for the bus and payload systems. The box-like spacecraft has a size of 690 mm x 366 mm x 366mm; it is three-axis stabilized using a gravity-gradient boom, two 3-axis magnetorquers,

985) http://www.atmsb.com.my/
986) TiungSat-1 data sheet of SSTL provided by Craig Underwood

and a momentum wheel. Attitude is sensed by two 3-axis magnetometers and by two-axis analog sun sensors. In addition, there are UED (Underneath Earth Detector) and SOD (Sun Overhead Detector). The pointing knowledge is $\pm 1^{o}$ in roll and pitch and $\pm 3^{o}$ in yaw (3 sigma values). The S/C power is 35 W per panel, provided by four surface-mounted GaAs solar panels and by a 10-cell NiCd battery (7 Ah). The power subsystem provides regulated voltage supplies at +5V and ±10V along with an unregulated supply which fluctuates between 12-14 V. Autonomous functions, safe modes and data are handled by two OBCs (On-Board Computer), OBC-186 and OBC-386. The on-board data storage capacity is 1 Gbit. The S/C mass is 50 kg (platform = 35 kg, payload = 15 kg). The design life is three years.

TiungSat-1 was launched Sept. 26, 2000 on a Dnepr-1 vehicle (the other payloads were: SaudiSat-1A/-1B of SISR (Saudi Institute for Space Research), UniSat of the University of Rome, and MegSat-1 of the MegSat Space Division of "Gruppo Meggiorin," Bresia, Italy) from the Baikonur Cosmodrome.

RF communications: TiungSat uses conventional AMSAT frequencies, thereby giving amateur radio operators access to its data (imagery and communication capabilities). The uplink is in VHF-band with a data rate of 9.6 kbit/s; three receivers are used: Rx1 operates at 144.46 MHz, Rx2 and Rx3 operate at 145.86 145.925 MHz, selectable. The downlink is dual-redundant in UHF-band (435 to 438 MHz range with 437.300, 437.325, 437.350, 437.375 MHz, selectable) with data rates of 9.6 and 38.4 kbit/s. An error-protected digital packet communications protocol is used. All spacecraft operations are performed at ATSB in Kuala Lumpur.

Orbit: Circular orbit, altitude = 642 km, inclination = 6454°, period = 97.43 minutes.

Sensor/experiment complement:

MSEIS (Multi-Spectral Earth Imaging System). MSEIS is a NAC (Narrow Angle Camera) system of three cameras (green, red and near-infrared) in parallel, each with a 75 mm focal length optic and 100 mm aperture diameter. The NAC system provides a 70 m ground spatial resolution in three spectral bands: 510-590 nm, 610-690 nm and 810-890 nm (green, red, and near infrared). A 2-D CCD staring array detector with 1024 x 1024 pixels is used providing snapshot imagery of scene size 70 km x 70 km. Up to four contiguous images can be collected along the flight path. The data are digitized to 8 bits radiometric resolution (256 levels).

MEIS (Metrological Earth Imaging System). MEIS is a single-band WAC (Wide Angle Camera) system with 4.8 mm focal length optics. It provides NIR imagery (810-890 nm) with a 1.2 km spatial resolution. The CCD area array detector has a size of 1024 x 1024 elements (pixels) for snapshot observations. Data are quantized to 8 bits radiometric resolution. An image has the size of 1200 km x 1200 km.

S&F (Digital Store & Forward Communications). The subsystem provides global, frequency-agile, communications for any form of digitized data: e-mail, voice-mail, scientific data exchange, fax, imagery, or even Internet mail for remote regions. The low cost and direct access offered by the TiungSat-1 microsatellite in orbit also makes it ideal for use by scientists, engineers and students based in institutes, universities and even schools throughout the world. - **DSPE** (Digital Signal Processing Experiment). The DSPE consists of a TM320C31 low power DSP suitable for special or general purpose signal processing tasks on LEO satellites. The VHF scanner operates in the 140-150 MHz range. A built-in FSK decoder is used. The system is capable to detect signals from a pre-set signal strength threshold within selected bar. DSP can be used for processing audio transmission for re-broadcast.

CEDEX (Cosmic Ray Energy Deposition Experiment). The objective of CEDEX is to characterize the TiungSat-1 orbit radiation environment in terms of the observed particle LET (Linear Energy Transfer) spectrum at the spacecraft. The primary sensor consists of a

30mm x 30mm PIN diode detector 300 microns in depth, housed in a separate screened aluminium unit mounted on the CEDEX module box (three area PIN-diode detectors are mounted in a "telescopic" arrangement; hence, information pertaining to directions of the energy particles detected can be derived.). This is connected to a charge amplifier and a pulse-shaping circuit which, in turn, are connected to an event-driven, hardware-logic controlled pulse-height multi-channel analyzer. CEDEX is controlled autonomously by a CAN-microcontroller with its own data-storage RAM and built-in data-compression software. This sends data to an internal CAN-controller which formats and sends them on to the primary OBC via the spacecraft's CAN (Controlled Area Network) bus. CEDEX is a multichannel analyzer with 512 channels and a 0.5 pC (picocoulomb) charge resolution. The instrument charge range is between 0.2 -24 pC, equivalent to a normal incidence particle LET range of about 60 - 7500 MeV cm^2 g^{-1} (200,000 particles/s).

The CEDEX data obtained are comparable directly with such instruments as CREDO-II flown on STRV-1c (launch Nov. 16, 2000) and CRE (Cosmic Ray Experiment) flown on KitSat-1 (launch Aug. 10, 1992) and PoSat-1 (launch Sept. 26, 1993).

Experimental Microsatellite GPS, an SSTL/ESA collaboration. An advanced 12-channel GPS receiver with two GPS patch antennas is installed for several objectives: 1) on-board generation of Keplerian orbital elements (NORAD experiment, this was first performed on PoSat), 2) on-board navigation, attitude, and timing services, and 3) refractive sounding of the ionosphere. The instrument is primarily used for orbit and position determination and for precise on-board timing services. In parallel, the instrument is also employed for refractive ionospheric monitoring. The TEC (Total Electron Content) occultation observations of the instrument provide slant range measurements which can be converted into vertical profiles.

D.40.20 DMC (Disaster Monitoring Constellation)

See description of DMC under I.3.

D.40.21 TOPSAT

See description of TOPSAT under M.28.

Part E Geodynamic/Earth-System Missions

E.1 CHAMP (Challenging Minisatellite Payload)

CHAMP is a German BMBF-funded geophysical minisatellite mission of GFZ (GeoForschungsZentrum, Potsdam, Germany) in cooperation with DLR. The satellite was built by the German space industry with the intent to foster high-tech capabilities especially in the East-German space industry. The S/C prime contractor is DJO (Jena Optronic GmbH) in Jena, a daughter of DASA (now Astrium). The overall science objectives are in the following fields of investigation:

- Global long- to medium-wavelength recovery of the static and time variable earth gravity field from orbit perturbation analyses for use in geophysics (solid Earth), geodesy (reference surface), and oceanography (ocean currents and climate), supported by a feasibility test of GPS altimetry for ocean and ice surface monitoring
- Global Earth magnetic field recovery (solid Earth and solar-terrestrial physics)
- Atmosphere/ionosphere sounding by GPS radio occultation with applications in weather forecasting, navigation, space weather, and global climate change.

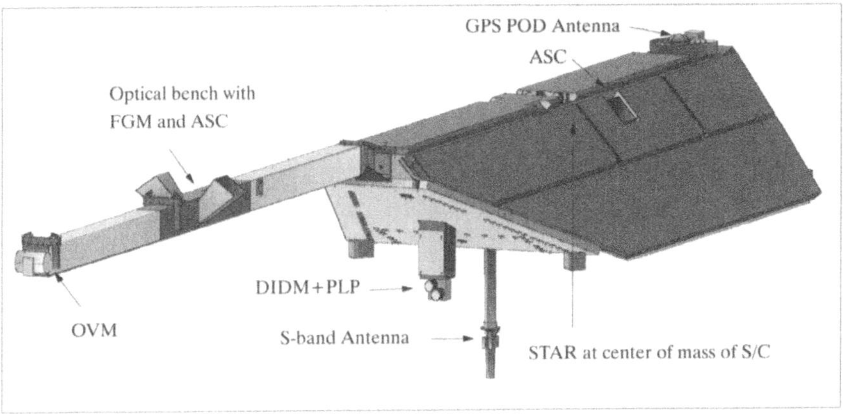

Figure 166: Front view of the CHAMP spacecraft

The structure of the satellite (FLEXBUS of Astrium GmbH) features a trapezoidal body with overall dimensions of 4.3 m length x 0.75 m height x 1.6 m/0.4 m width at base/top. A boom of 4 m length accommodates the magnetometry assembly (the boom is nominally pointed into the flight direction). The S/C is three-axis stabilized (Earth pointing) with three magnetorquers and cold gas propulsion for attitude and orbit change maneuvers; attitude control is $< 2°$, attitude knowledge is provided by two star sensors on the boom and on the body, respectively, each one with two sensor heads (0.01° orientation accuracy), the star sensor instrument is referred to as ASC (Advanced Stellar Compass) and is manufactured by DTU, Denmark [dual-head CCD (hot), Quad A/D (2 cold) and single DPU]; S/C design life = five years; total S/C mass = 522 kg (including 30 kg of propellant); payload mass = 32 kg; S/C power = 150 W (50 W for payload) provided by a solar array of 6.9 m^2 in size; in addition there are 10 cells of NiH_2 batteries (16 Ah) for night phase operations. Passive thermal control is provided by paints and multi-layer insulation. A S/C launch took place on July 15, 2000 from Plesetsk on a Russian Cosmos-3M launcher (along with the MITA satellite of ASI, Italy).

CESS (Coarse Earth-Sun Sensor). An Astrium-developed patented attitude package consisting of six single sensor heads arranged on the S/C in such a way that an omnidirectional and unobstructed view to space is given. Each head features two equally sized optical surfaces of different properties in the spectral ranges (the IR absorptance is the same but the VIS absorptance is different). Thermistors serve as detectors. The hardware configuration permits to take temperature measurements driven by the Earth and sun illumination. The measurements are weighted, averaged, and extrapolated to steady-state equilibrium temperatures. An algorithm derives overall heat flux based on the input temperatures and corrects for albedo. The output is a computed Earth vector and a computed sun vector in the S/C coordinate system. CESS pointing performance is within 6° for sun vector knowledge and within 15° for Earth vector knowledge. On CHAMP, CESS serves as a low-cost and low-mass demonstration package. The intend of the CESS design is to permit future S/C to fall back on a safe attitude control in case of a failure or degraded service of components of the regular attitude control system. CESS offers also its services for initial acquisition and coarse pointing in general. [987)]

Data: On-board storage capacity of 1 Gbit. All data links are in S-band using BPSK modulation. Spacecraft science data are received by the DLR/DFD station in Neustrelitz (downlink frequency of 2.28 GHz at a data rate of 1 Mbit/s). S/C operations are provided at DLR/GSOC in Oberpfaffenhofen via the Weilheim station (TT&C data in S-band; uplink at 2.093 GHz and 4 kbit/s; downlink at 2.28 GHz and 32 kbit/s). [988), 989) 990) 991)]

Initial orbit: perigee = 418 km, apogee = 474 km (the orbit is decaying to about 300 km at end of the design life of five years), inclination = 87.275°.

Sensor complement:

BlackJack (GPS Flight Receiver), a new generation instrument of TRSR (TurboRogue Space Receiver) and of GPS/MET heritage, first flown on SRTM. BlackJack is being provided and built by NASA/JPL. Objectives: a) BlackJack is a dual-frequency GPS receiver system for precise (cm accuracy range) orbit determination and continuous coverage; b) The instrument observes in parallel ionospheric electron content, and provides atmospheric soundings permitting the derivation of atmospheric vertical profiles of density, pressure, temperature (of water vapor) and ionospheric electron density profiles (refractive occultation monitoring); c) Experimental use of the GPS signals reflected from the ocean's surface for altimetric measurements.

The LEO CHAMP S/C and the MEO GPS satellite-constellation establish a so called high-low satellite-to-satellite (SST) link for the BlackJack instrument.

BlackJack features a new DSP (Data Signal Processor) ASIC, capable of tracking 12 GPS signals at once; there are four ASICs in each receiver so the instrument can track up to 48 signals in parallel, in any mode (C/A, P, or codeless). Two new ASIC capabilities (among several) are: 1) a patented new enhanced "quasi-codeless" tracking of L2, providing a measurement error comparable to that achieved by the TRSR instrument with P-code tracking, and 2) special software for atmospheric occultation and ocean reflection applications. This particular BlackJack implementation is the first 4-antenna version. The BlackJack receiver system consists of the RPA (Receiver/Processor Assembly) containing the RF down-converter sections, an internal bus and the cold-redundant baseband processor cards, the RF

987) A. Zaglauer, W. Pitz, "CHAMP-The First FLEXBUS in Orbit," Proceedings of the 3rd International Symposium of IAA, Berlin, April 2-6, 2001, pp. 105-109

988) Ch. Reigber, P. Schwintzer, "A Challenging Microsatellite Payload for Geophysical Research and Application," in: Small Satellites for Remote Sensing, Space Congress '95, Bremen, May 24-25, 1995, pp. 83-89, European Space Report, Munich, 1995

989) Ch. Reigber, R. Casper, W. Päffgen, "The CHAMP Geopotential Mission," IAA 2nd International Symposium on Small Satellites for Earth Observation, Berlin, April 12-16, 1999, pp. 25-28

990) Ch. Schmitt, H. Bauer, "CHAMP Attitude and Orbit Control System," IAA 2nd International Symposium on Small Satellites for Earth Observation, Berlin, April 12-16, 1999, pp. 269-272

991) http://op.gfz-potsdam.de/champ/systems/index_SYSTEMS.html

coaxial cables to the 4 antennas and the antennas itself. The RPA antennas are pointed: in the zenith direction, two in the aft direction, and in the nadir direction. The zenith-pointing antenna is used for simultaneous tracking of up to twelve GPS satellites to derive CHAMP's trajectory. The receiver performs phase measurements of both GPS frequencies L1 and L2, and code pseudoranges. The GPS output is the complete navigation solution, i.e., position, velocity and time for use in on-board orbit prediction and attitude control, in addition to the direct phase and pseudorange measurements from GPS-LEO satellite-to-satellite tracking which are used for precise offline orbit restitution. The receiver has the following measurement modes: [992) 993)]

- Tracking mode (default), 0.1 Hz
- Occultation mode: in this mode the receiver software schedules every 50 Hz tracking of setting occultations of up to four GPS satellites
- Altimetry mode: in this mode the nadir antenna collects specular reflections of GPS signals from the surface of the oceans.

The BlackJack aft antennas are used for limb sounding. High resolution atmospheric soundings can be retrieved when the radio path between the LEO GPS receiver and one GPS satellite traverses the Earth's atmosphere. This occultation technique employs dual frequency carrier-phase observations of retarded signals (atmospheric path delays) which permit the derivation of atmospheric profiles of density, pressure, and temperature (or water vapor) in an altitude range from 85 km to the ground. All observables are sampled 50 times per second. In parallel to limb sounding a non-occulted GPS satellite is tracked, which allows correction of system inherent biases. Ionospheric occultation is performed with 1 Hz samples.

A nadir-pointing antenna is used experimentally to receive specular reflections of the GPS signals from the ocean's surface. In this case, the receiver does not track the direct GPS signals, but instead performs an "open-loop" sampling of the GPS spectrum for later analysis on the ground. Therefore, any and all reflected GPS signals that happen to be present in sampled spectrum are being captured, though that may not generally be more than one or two. The sample rate for this operations mode is about 100 samples/s, or about the same as in occultation mode. Knowing both the position of CHAMP and the transmitting GPS satellite, it is possible to derive the precise position of the S/C with respect to the ocean's surface and thus to gain information about its topology.

Computed position in telemetry (navigation solution)	< 60 m
Time calibration accuracy	< 1 µs from GPS time (resolution 0.1 ns)
Dual-frequency range and integrated carrier for POD (Precise Orbit Determination) at a 1s interval:	
Phase (ionosphere–free) Range (ionosphere–free)	< 0.2 cm < 30 cm
Dual-frequency integrated carrier phase and amplitude for atmospheric occultation:	
Phase (ionosphere–free) Range (ionosphere–free)	< 0.05 cm < 50 cm
Limb-sounding observables (prior to atmospheric de-focusing):	
L1 carrier phase L2 carrier phase	< 0.05 cm (1 s) < 0.15 cm (1 s)

Table 277: Performance characteristics of BlackJack

The BlackJack data rate is between 0.4 - 17 kbit/s, depending on occultation events (0, 1, or 2 occultations); the mean data rate is 2.5 kbit/s, which amounts to about 27 MByte/day.

STAR (Space Three-axis Accelerometer for Research mission), an accelerometer system provided by CNES and developed by ONERA, France. The objective is to measure all non-

992) Information provided by T. P. Yunck of NASA/JPL

993) T. Meehan, C. Duncan, et al., "GPS On A Chip - An Advanced GPS Receiver for Spacecraft," Proceedings of the ION GPS-97 Conference, Sept. 16-19, 1997, Kansas City, MO, pp. 1509-1517

gravitational accelerations of the satellite (drag, solar and Earth radiation pressure) in order to determine the Earth's gravity field from purely gravitational orbit perturbations (orbit from BlackJack). The accelerometer measurement principle is based on electrostatic suspension of a proof-mass in a cage. Instantaneous position of the proof-mass is measured by three capacitive sensors which permit a determination of the acceleration vector. The instrument has a dynamic range of $\pm 10^{-4}$ ms^{-2}, a resolution of better than $\pm 3 \times 10^{-9}$ ms^{-2}, and a frequency range of 10^{-1} to 10^{-4} Hz. The STAR instrument is positioned at the center of gravity of CHAMP to minimize the influence of measurement disturbances due to rotational accelerations and gravity gradients. STAR is also connected to a star sensor (ASC of DTU, Denmark) with two heads to provide the accelerometer's axes orientation. The accelerometer proof-mass is positioned at the center of gravity of CHAMP. [994)]

LRR (Laser Retro Reflector), developed at GFZ. A passive optical device for accurate satellite tracking from ground laser ranging stations of the SLR network. LRR consists of four cube corner prisms to reflect short laser pulses back to the transmitting SLR ground station. Laser pulses from the ground typically have durations of 35 - 100 picoseconds, providing a single-shot range accuracy of 1 to 2 cm without any ambiguities. LRR data are used for a) precise orbit determination in connection with BlackJack for gravity field recovery, b) calibration of the on-board BlackJack system, and c) technological experiments such as two-color ranging to achieve millimeter ranging accuracies.

MIAS (Magnetometer Instrument Assembly System). A boom-mounted package consisting of an Overhauser scalar magnetometer (OVM, built by LETI, France), two fluxgate vector magnetometers (FGM), and two star imagers to provide attitude information for FGM (FGM and the two star sensors are built by DTU, Denmark).

The measurement method of OVM is based on proton Larmor precession in a weak magnetic field (the Larmor frequency's being directly proportional to the magnetic field permits the derivation of the absolute total Earth magnetic field intensity). The absolute intensity is used for calibrating the FGM observations. OVM measurement range = 16,000 - 64,000 nT, resolution = 0.1 nT, absolute accuracy = 0.5 nT, sampling rate = 1 Hz.

The measurement principle of FGM is based on 'vector feedback geometry' (three sets of coaxial coils are mounted equidistant on the surface of a 9 cm diameter sphere, with the three sets orthogonal to each other - the coil currents are a measure of the three magnetic field components). Two FGMs are located on the boom, 60 cm apart, permitting simultaneous measurement of both FGMs in a 'gradiometer' arrangement. The FGMs have a measurement range of ±64,000 nT and a resolution of 1 to 2 nT. The sampling rates are at 32 Hz (both sensors active) or 64 Hz (one sensor active), corresponding to spatial resolutions of about 200 or 100 m, respectively. - The FGMs are rigidly mounted on an optical bench also housing two star sensors.

ASC (Advanced Stellar Compass), designed and built by DTU (Technical University of Denmark in Lyngby). The objective is to provide an attitude reference instrument with a precision of a few arcseconds aboard the satellite (to measure the magnetic field with a vector precision of a fraction of a nT). The design of this star imager is based on a new development presently flown on the Ørsted satellite. On CHAMP, there are two ASC systems, each consisting of two Camera Head Units (CHU) and a common Data Processing Unit (DPU). One ASC is part of the magnetometry optical bench unit on the boom, the other provides high precision attitude information for the instruments fixed to the S/C body. Additionally the ASCs serve as sensors for the satellite attitude control system. The ASC is a fully autonomous system that directly outputs the attitude quaternions in a preselectable coordinate system.

ASC measurement principle: An image of the stars within the FOV is acquired by integrating the light focused onto a photosensitive CCD array. The pattern is serially read out, digi-

994) P. Touboul, B. Foulon, G. M. LeClerc, "STAR, The Accelerometer of the Geodesic Mission CHAMP," Proceedings of the 49th IAF Congress, Melbourne, Australia, Sept. 28 - Oct. 2, 1998, IAF-98-B.3.07

tized and fed to a microprocessor. The digital image is than sifted for stars brighter than mV = 6 and corrected for lens distortions resulting in a list of calculated star centroids with sub-pixel precision. The determined star centroids are subsequently matched against real star positions derived from an on-board HIPPARCOS star catalog. The star attitude which fits best the observed pattern is the output result. For further improvement of the attitude solution a GPS updated orbit model is maintained to correct for astronomical aberration and also the epoch of star constellation is taken into account.

The ASC on the boom provides the high attitude precision needed for the magnetic field vector measurements. The ASC on the spacecraft body provides attitude data primarily for the STAR and DIDM instruments. This information is also required for the proper reduction of the GPS data, laser ranging data and for attitude control.

Attitude determination precision	4 arcsec (3 sigma, BOL)
FOV (Field of View)	18.4° x 13.4°
Sampling rates	1 Hz (nominal), 0.5 Hz, 2 Hz
Magnetometer moment CHU	10^{-5} A/m^2
Power	8 W
Mass: CHU, DPU	0.20 kg (excluding baffles), 0.80 kg
Dimensions: CHU, DPU	50 mm x 50 mm x 45 mm, 100 mm x 100 mm x 100 mm

Table 278: Performance characteristics of ASC

DIDM (Digital Ion Drift Meter, the instrument is provided by AFRL, Hanscom Air Force Base, Bedford, MA). DIDM is an enhanced version of the first DIDM instrument, flown on STP-4 (Space Test Program-4) with a launch in Oct. 1997. DIDM on CHAMP includes a charge collecting plate to account for spacecraft charging. The objective of DIDM is to measure the Earth's electric field parallel to the magnetic field (in-situ measurements of the ion distribution and its moments within the ionosphere). DIDM measures magnitude and direction of the incoming ion flux. The electric field is derived from the relationship between electric field, measured ion drift velocity and measured magnetic field strength.

Note: SSI/ES (Special Sensor Ionospheric Plasma Drift/Scintillation) instruments have been flown on DMSP series satellites for a long period of time (20 years) to monitor the space plasma. These instruments rely on analog technology and older electronics. As a result they have limited dynamic range, are physically large and heavy, and have significant power requirements. The follow-on US meteorological satellite program (NPOESS), which will replace DMSP in the first decade of the 21st century, requires higher resolution, sensitivity, accuracy, and dynamic range for its plasma measurement devices. The goal is for a next generation DIDM instrument to be the standard drift meter aboard the NPOESS spacecraft series. [995)]

DIDM utilizes miniaturized state-of-the-art detector components and on-board digital signal processing. The instrument consists of two side-by-side ion detectors (MCP type) arranged within one unit of 2.2 kg mass. Both apertures are facing the S/C ram direction. Both detectors can measure the normal and perpendicular velocity components of incident ions. Ions entering through a pinhole aperture fall into a Retarding Potential Analyzer (RPA) cup. Those ions with energies higher than a certain threshold potential are electrostatically focused onto charge multiplying microchannel plates (MCPs) and high-resolution wedge and strip charge detecting anodes. The images created by these impacting ions on the anode are digitally processed on-board to reconstruct the full 3-D ion velocity vector.

A **PLP** (Planar Langmuir Probe) is operated in combination with DIDM. This device provides auxiliary data needed to interpret the ion drift measurements. Quantities that can be derived from the PLP sweeps are S/C potential, electron temperature, and density. The

995) http://www.vsbs.plh.af.mil/projects/didm/didm.html

PLP voltage is swept for 1 s every 15 seconds typically between ±2.5 V in 32 steps. A selectable bias voltage can be added to account for the S/C potential. By interpreting the measured current/voltage characteristic the plasma parameters are being determined. The S/C floating potential is measured during the remaining 14 seconds of the measuring cycle.

Direct measurements	Ion density: 10^2 - 10^6 ions/cm^3 Ion resolution: <1° direction, <130 m/s in speed Ion energy: 0 - 32 eV
Indirect measurements	Ion drift velocity: 0 - 6 km/s, Electric field resolution: <4 mV/m Ion temperature: 200 - 55 000 K
Sample rates DM (Digital Meter) mode RPA (Retarding Potential Analyzer) mode PLP (Planar Langmuir Probe) mode	 0, 1, 2, 4, 8, 16 Hz 0, 8, 16 Hz 0, 1/15 Hz
Instrument data rates	5 kbit/s (peak), 1 kbit/s (nominal)
Power, mass, dimensions	5 W, 2.2 kg, 153 mm x 150 mm x 109 mm

Table 279: DIDM instrument parameters

E.2 CryoSat

CryoSat is the first Earth Explorer Opportunity Mission in ESA's Living Planet Program to be launched in 2003/4. CryoSat is a radar altimetry mission dedicated to observations of the polar regions. The science objectives are to determine variations in the thickness of Arctic sea ice and elevation changes of ice sheets, ice caps and glaciers that ring the Arctic Ocean. Sea ice plays a central role in Arctic climate. The CryoSat mission is to further knowledge of the wider role of the Earth's cryosphere and its effect on the sea level. The CryoSat mission was proposed by Duncan Wingham of the University College London and an international science team. Duncan Wingham is also the mission PI. A mission duration of three years is planned. [996] [997]

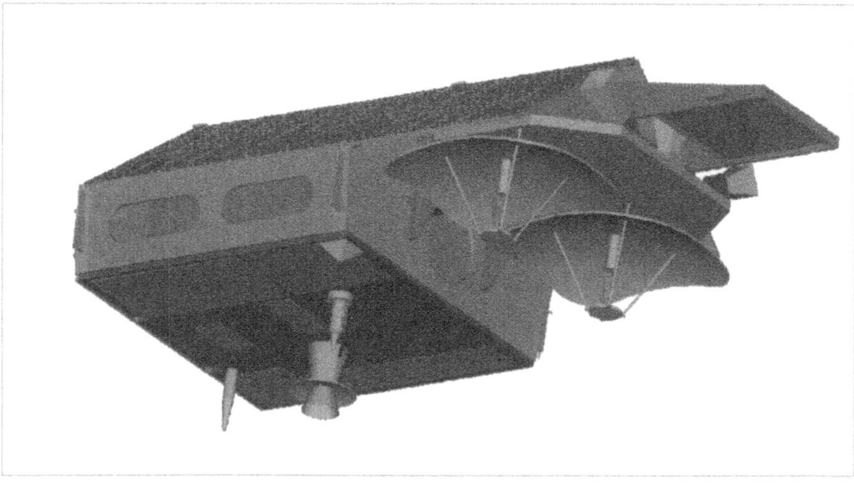

Figure 167: Illustration of the CryoSat spacecraft

996) D. J. Wingham, "The first of ESA's Opportunity Missions: CryoSat," ESA Earth Observer Quarterly, No 63, Sept. 1999, pp. 21-24

997) http://www.estec.esa.nl/explorer/cryosat/

The CryoSat spacecraft is being built by Astrium GmbH of Friedrichshafen, Germany, as prime contractor. The spacecraft structure consists of a long rectangular main platform body, surmounted by fixed solar arrays in the form of a tent. The two SIRAL instrument antenna dishes are mounted on a separate rigid bench in the forward section of the S/C. In addition, a dedicated SIRAL radiator is mounted at the nose tip. A slight nose-down attitude of the S/C is chosen to ensure minimum attitude correction due to gravity-gradient disturbances. The S/C has a total length of 4.6 m and a width of 2.34 m. The S/C mass is 628 kg, the design life is 3 years.

The AOCS (Attitude Orbit and Control Subsystem) comprises the following elements:

- A cold gas system for attitude control and orbit transfer and maintenance maneuvers, 16 attitude control thrusters (10 mN) and 4 orbit control thrusters (40 mN). Nitrogen is used as propellant (132 l tank).
- A set of 3 magnetometers for compensation of environmental torques
- A set of three star tracker heads (also a part of the payload) providing autonomous inertial attitude determination for the spacecraft. The multiple configuration makes the sensor system one-failure tolerant, except for the rare occurrence of simultaneous sun and moon blinding of two heads, to which the system software is tolerant. Consequently, two camera heads are operated in parallel at all times. The star tracker attitude serves also as reference for determining the orientation of the SIRAL interferometric baseline.
- Real-time DORIS measurements of satellite position, velocity and time
- CESS (Coarse Earth-Sun Sensor). Provides attitude measurements with respect to the sun and Earth for initial acquisition and coarse pointing
- A set of three-axis magnetometers are used for magnetorquer control and as rate sensors.

The CDMU (Control and Data Management Unit), consisting of a processor and a hardware-based fault detection system, handles all on-board command and control functions including telecommand decoding and the AOCS processing functions. A MIL-STD-1553B communications bus is used as payload interface (for SIRAL and DORIS). The on-board solid-state memory has a capacity of 126 Gbit. S/C power of 414 W is provided by fixed solar panels (GaAs cells). Lithium-ion batteries are used for operations support during eclipse phases.

Orbit: Non sun-synchronous circular orbit, altitude = 720 km, inclination =92°. Ground track repeat cycles are of 1 year duration with subcycles of one month. This configuration allows a sufficient coverage for the polar regions. The CryoSat mission requirements include:

- An orbit of 92° inclination. Since this orbit is not sun-synchronous, it results in a nodal rotation period, through all mean local solar times, of about 8 months
- An orbit change is required during the mission with the objective to visit twice a validation orbit, approximately 6 km lower in altitude than the science phase orbit
- The payload must be operated in various modes, as a function of geographical region, such that the orbital operations, and data sets collected, on successive orbits are dissimilar
- The payload utilization demands very precise orbit and attitude restitution. Minimum operations of three years are required.

RF communication: The S-band link is used for all TT&C communications. The X-band downlink provides a payload transfer rate of 100 Mbit/s.

The CryoSat mission will be operated at ESA/ESOC. The Kiruna ground station in Sweden functions as the prime command and data acquisition facility. The payload data segment (data processing, archiving and distribution) function is also located at the Kiruna station.

Sensor complement:

SIRAL (SAR Interferometer Radar Altimeter) designed and developed by ASI (Alcatel Space Industries), France. The objective is to observe ice sheet interiors, the ice sheet margins, for sea ice and other topography. The SIRAL design features two receiving antennas forming an interferometer in the cross-track direction. In addition, the return signal in along-track direction is processed to construct a synthetic aperture for enhanced ground resolution. The instrument is a Ku-band radar altimeter which uses the full-deramp range compression technique of conventional altimeters. However, it introduces two features that make it different from previous spaceborne altimeter implementations: [998] [999]

1) The instrument has two antennas (including pulse-to-pulse phase coherence) and two receive chains, permitting an interferometric mode of operation

2) SIRAL operates at high PRF (Pulse Repetition Frequency), ensuring coherent along-track sampling for aperture synthesis (PRF>Doppler bandwidth)

The instrument consists of three major subsystems, two of these are in discrete electronic boxes:

- DPU (Digital Processing Unit), it serves all digital altimeter functions, including the digital chirp generation, the full sequencing functions of the altimeter, and the receive and processing functions of the echo
- RFU (Radio Frequency Unit). It contains all analog IF and RF electronics and a solid-state power amplifier with an RF peak power of 25 W.

The antenna subsystem consists of two Cassegrain antennas, mounted side-by-side and forming the interferometric cross-track. Both antennas are identical; one is used to transmit and receive, whereas the other antenna is used to receive echoes. The primary super-elliptic reflectors are about 1.1 m x 1.2 m in size. They are supported by a composite sandwich plate. A high thermoelastic stability is needed to meet the interferometric instrument performance. The new features of SIRAL have been demonstrated with the airborne D2P (Delay/Doppler Phase-monopulse Radar) of JHU/APL.

Parameter	Value
RF frequency	13.575 GHz (Ku-band)
Pulse bandwidth	320 MHz, (40 MHz for tracking only in SARIn)
PRF (Pulse Repetition Frequency)	1.97 kHz in LRM, 17.8 kHz in SAR and in SARIn
Pulse duration	51 µs
Timing	Regular PRF in LRM, burst mode in SAR/SARIn
Samples/pulse	128 in LRM and SAR, 512 in SARIn
RF peak power	25 W
Antenna size	2 reflectors 1.2 m x 1.1 m, side-by-side
Antenna beamwidth	1° x 1.2° (cross-track)
Antenna footprint	15 km
Range resolution	About 45 cm
Along-track resolution	250 m (SAR/SARIn)
Data rate	60 kbit/s for LRM, 12 Mbit/s in SAR, 2 x 12 Mbit/s in SARIn
Mass (with antennas)	60 kg non-redundant, 90 kg redundant
Instrument power	100 W LRM, 135 W for SAR/SARIn

Table 280: SIRAL key instrument parameters

The science requirements demand of CryoSat to measure variations in ice thickness of perennial sea and land ice fields to the limit allowed by natural variability, on spatial scales

998) L. Rey, P. de Chateau-Thierry, L. Phalippou, C. Mavrocordatos, R. Francis, "SIRAL, a High Spatial Resolution Radar Altimeter for the CryoSat Mission," Proceedings of IGARSS 2001, Sydney, Australia, July 9-13, 2001

999) L. Phalippou, L. Rey, P. de Chateau-Thierry, "Overview of the Performances and Tracking Design of the SIRAL Altimeter for the CryoSat Mission," Proceedings of IGARSS 2001, Sydney, Australia, July 9-13, 2001

varying over three orders of magnitude. The natural variability of sea and land ice depends on fluctuations in the supply of mass by the atmosphere and ocean, and snow and ice density. The precisions of the measurements are expressed in terms of cm of yearly ice equivalent thickness variations. These are:

- Arctic sea-ice: 1.6 cm/year vertical measurement accuracy at 10^5 km^2 scale (equivalent to 300 km x 300 km cells). Temporal sampling: 1 month
- Land ice (small scale): 3.3 cm/year at 10^4 km^2 (equivalent to 100 km x 100 km cells). Temporal sampling : 1 year
- Land-ice (large scale): 0.17 cm/year over 13.8 x 10^6 km^2 (about the area of Antarctica). Temporal sampling: 1 year.

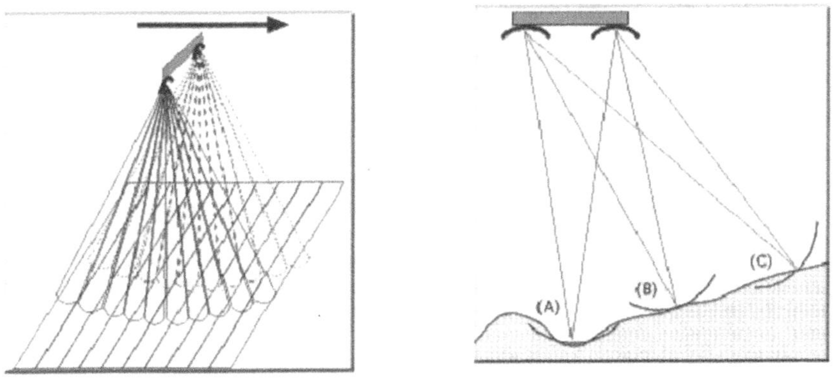

Left image: Synthetic aperture processing is performed on an otherwise conventional Ku-band pulse-limited altimeter to form a set of high-resolution, along-track beams. As the altimeter progresses along its orbit, these beams sequentially scan each sub-satellite area. The data from all beams can be combined successively, leading to one height measurement for any given ground location.

Right image: Over ice sheets a second synthetic aperture system is added to form an interferometer across the satellite track. The angle of the echo at each range may be determined, and this, together with the range, determines the elevation and across-track location of the surface. When the across-track echo direction is unambiguous, as in (B) or (C), its coherence is high. Echoes at ranges in ambiguous directions, as in (A), have low coherence, and may be masked.

Figure 168: SAR observation principle of SIRAL

SIRAL operations: SIRAL provides the following operational modes for different observational support types. The complex waveform data stream from the CryoSat altimeter requires a sophisticated processing scheme in particular for exploiting the synthetic aperture and interferometry techniques over ocean and ice surfaces.

- LRM (Low Resolution Mode) operation support: LRM uses a single receive channel and low PRF for conventional pulse-limited operation for ice sheet interiors/open oceans. The transmitted pulse length and the transmitted bandwidth are set to the same value as that for Envisat in a similar mode (51 s, 320 MHz). The PRF is kept constant over the orbit at a value around 2 kHz to ensure the decorrelation of received echoes. The averaging for tracking and ground processing is performed after the FFT (Fast Fourier Transform).

  The LRM mode is useful over surfaces where the topography is homogeneous, at least as large as the antenna footprint of about 15 km. The altimeter echoes have a predictable shape and the mean surface level of this area can be derived by an appropriate model.

- SAR (Synthetic Aperture Radar) support mode: SAR uses a single channel and a high PRF. Closed burst timing is employed to ensure a high along-track resolution. The PRF is chosen higher than the Doppler bandwidth over the half-power beamwidth to avoid aliasing in the ground processing of the data (the PRF is about 10 times higher than that of LRM to ensure coherence between the echoes of successive pulses). Bursts of 64 pulses at a PRF of 18.5 kHz with a burst repetition frequency of 85 Hz are transmitted.

In the SAR mode, the resolution of the radar is improved in the along-track direction. This is achieved by exploiting the Doppler properties of the echoes as they cross the antenna beamwidth. The result is equivalent to decomposing the main antenna beam into a set of 64 narrower synthetic beams in the along-track direction. The footprints of the different sub-beams over a flat surface are adjacent rectangular areas, about 250 m wide in along-track and as large as the antenna's cross-track footprint (up to 15 km). Hence, a larger number of independent measurements are available over a given area; this property is used to enhance the accuracy of the measurements over sea ice. The echoes are transmitted to the ground segment in the time domain, prior to any averaging. Hence, the data rate in SAR mode is significantly higher than that for LRM.

- SARIn (SAR Interferometric) support mode. This mode is used mainly over ice sheet margins with high surface slopes. Both receive channels are operating simultaneously at high PRF to ensure the availability of a high cross-track resolution used for ice sheet margins and coastal areas (accurate determination of the arrival direction of the echoes in along-track and in cross-track). This is needed to derive the height of the surface from the range measurement of the radar. Narrow-band tracking pulses, transmitted in-between successive wide-band measurement bursts are used in this range-tracking concept to cope with abrupt height variations.

  In the SARIn mode, the addition of the interferometric feature to the SAR further improves the echo localization capabilities, as the cross-track direction angle of the echoes can be determined. This is achieved by comparing the phase of one receive channel with respect to the other.

Parameter / Mode of Operation	LRM	SAR	SARIn
Receive chain	1	1	2
Samples per echo	128	128	512
Range window	60 m	60 m	240 m
Bandwidth	350 MHz	350 MHz	350 MHz
PRF	1970 Hz	17.8 kHz	17.8 kHz
Tx pulse length	51 µs	51 µs	51 µs
Useful echo length	44.8 µs	44.8 µs	44.8 µs
Burst length	N/A	51 µs	51 µs
Pulses/burst	N/A	64	64
Burst repetition interval	N/A	11.7 ms	46.7 ms
Azimuth looks (46.7 ms)	92	240	60
Tracking pulse bandwidth	350 MHz	350 MHz	40 MHz
Samples per tracking echo	128	128	128
Size of tracking window	60 m	60 m	480 m
Averaged tracking pulses (46.7 ms)	92	32	24
Data rate	51 kbit/s	12 Mbit/s	2 x 12 Mbit/s
Power	100 W	135 W	130 W
Mass (non redundant)	62 kg		

Table 281: Summary of instrument parameters for operational mode support

The innovative technical features of SIRAL are:

- The capability to operate in all measurement modes
- Digital chirp generation with pulse to pulse coherence for Doppler processing
- Solid State Power Amplifier (SSPA) in Ku band with high performance (25 W),

- Dual antennas forming an interferometer, mounted on an optical bench together with star-tracker heads, ensuring the accurate knowledge and stability of the interferometric baseline orientation
- Two receive chains matched together with very low distortions.

The novel feature of SIRAL, as compared with conventional altimeters, is the capability to locate a resolution cell in the 3 dimensional space. The SIRAL concept is based on a Ku-band nadir-looking radar which can be operated in the conventional mode over oceans. Over terrain (ice or land) the "advanced mode" uses Doppler filtering for the enhancement of the along-track resolution. A second antenna and receiving channel provides a second take of the scene which is used for surface height retrieval as it is usually done with SAR interferometry.

DORIS (Doppler Orbitography and Radiopositioning Integration by Satellite). DORIS is an uplink radio frequency tracking system based on the Doppler principle. The CNES instrument provides accurate measurements for a precise orbit determination. Knowledge of the orbit is essential for exploitation of the altimeter data and the overall performance. The onboard receiver measures the Doppler shift of uplink beacons in two frequencies (2.03625 GHz for Doppler measurement and 401.25 MHz for the ionospheric correction) which are transmitted continuously by the ground stations. One measurement is used to determine the radial velocity between spacecraft and beacon, the other to eliminate errors due to ionospheric propagation delays. The 401.25 MHz frequency is also used for measurements of time-tagging and auxiliary data transmission. The DORIS instrument comprises:

- A fixed omni-directional dual-frequency antenna
- A receiver performing the Doppler measurements every ten seconds. The nominal mode of operation is an autonomously programmed mode in which the receiver tracks the beacon signals according to information provided by the navigation software (DIODE) based on an on-board table of beacon data.
- An USO (Ultra Stable Oscillator) delivering the reference frequency with a stability of 5×10^{-13} over a period of 10 to 100 s.

The mass of DORIS is 15 kg (including the antenna of 160 mm diameter). The instrument requires 20 W of power, the data rate is 4 kbit/s.

The following DORIS services are used for CryoSat operations:

- Real-time orbit determination for spacecraft attitude and orbit control (on-board)
- Provision of a precise time reference based on TAI (International Atomic Time); in addition a precise 10 MHz reference signal is used (on-board)
- Provision of on-ground POD (Precise Orbit Determination) and ionospheric modelling.

LRR (Laser Retroreflector). The objective is to use LRR as an additional tool and backup for precise orbit determination with the aid of the international laser tracking network. LRR is accommodated in the nadir plate of the spacecraft, suitable for range measurements above 20° elevation angles at all azimuths.

E.3 EGS (Experimental Geodetic Satellite, Ajisai)

Japanese (NASDA) geodetic mission. Launch: Aug. 12, 1986 from Tanegashima Space Center with the H-1 launch vehicle. The primary objective was performance confirmation of the H-1 launch vehicle. Objectives of EGS: provision of long-range geodetic applications aimed at rectifying Japan's domestic geodetic triangular network, determining the exact position of many Japanese islands and establishing Japan's geodetic point of origin.

EGS is a passive spherical satellite of 2.15 m diameter and a mass of 685.2 kg, carrying 318 mirrors and 120 laser reflector assemblies (1436 corner cube reflectors) for precise satellite laser ranging (SLR) measurements from ground-based laser ranging stations. The reflectors and mirrors on the satellite sphere reflect laser beams back to the source, regardless of the angle of incidence (solid Earth studies, crustal movements, plate tectonics). The reflection of solar light is used to determine the direction to the satellite from an observation site.

The body of the satellite is a hollow sphere made of glass-fiber-reinforced plastics. The surface is covered with corner cube reflectors (CCR's) and solar light reflectors. Twelve CCR's form a set of laser reflectors (LR's), and 120 LR's are distributed almost uniformly over the surface. The remaining surface is covered with 318 solar reflectors. The base of the mirrors is made of an aluminum alloy and the surface has a protective coating of silicon oxide. [1000)]

Orbit: circular orbit with an altitude of 1500 km, inclination = 50°, period = 116 minutes.

Observation method of EGS: The basic concept for geodetic use of the EGS is to simultaneously determine directions through photography and measure distances by laser ranging. The principle is to determine the relative geodetic location of two observation sites by simultaneously measuring the distance and direction to the EGS at a known station (named the base station) and an unknown station (primary station) of which the precise geodetic location is unknown. The distance from a site to the EGS is measured by SLR. To determine the location of a number of unknown stations around the base station, at least one Transportable SLR (TSLR) is used.

The laser ranging surveys are conducted by the 'Geophysical Survey Institute' of the Ministry of Construction, and the 'Hydrography Department of the Maritime Safety Agency', Ministry of Transport.

E.4 ETALON

ETALON ("standard") = Soviet Geodynamical Satellites. ETALON-1 was launched on January 10, 1989 from Baikonur together with two GLONASS satellites. (ETALON-1 is also known as Cosmos-1989). ETALON-2 (identical S/C) was launched on May 31, 1989.

Orbits (ETALON-1 and -2): Near-circular orbit, altitude = 19130 km, eccentricity = 0.00068, inclination = 64.8°, nodal period = 10874.7 days (29.8 years), perigee period = 94402.9 days (258.6 years), one revolution period = 675 min (11.25 h) [1001)]

Features of ETALON S/C.
Shape = sphere with a diameter of 1.294 m, mass = 1415 kg, laser reflectors = 2140, distance from the geometrical center to the plane of probable reflection = 558 mm.

Objectives: ETALON is a passive satellite system dedicated entirely to laser ranging [see also LAGEOS (E.15)]. Solid Earth studies: geodynamic processes, development of high-accuracy global references, long-period disturbances, geopotential modelling, etc.

Three laser stations in Russia (at Ternopol, Yevpatoria, and Maydanak) provide laser ranging services. In addition, a network of 10 sites outside Russia are performing ranging measurements. GFZ of Potsdam, Germany, is the data collection and distribution center for ETALON laser measurements in Europe.

E.5 GEO-IK

GEO-IK[1002)] = Space Geodetic Complex. A Russian S/C for Solid Earth Research, launched May 30, 1988. Builder and operator of GEO-IK S/C is NPO PM, Krasnojarsk.

1000) M. Sasaki, H. Hashimoto, " Launch and Observation of the Experimental Geodetic Satellite of Japan', IEEE Transactions on Geoscience and Remote Sensing, Volume 25, No. 5, Sept. 1987

1001) S.K. Tatevian, A.N. Zakharov, "The Geodynamical Satellite ETALON," CSTG Bulletin No. 11, Title: New Satellite Missions for Solid Earth Studies, 1989, pp. 3-9

Each mission is planned for an operational life of 5 years. More GEO-IK satellites were launched in 1989 (Aug. 28), 1990 and 1994 (Nov. 29).

Orbit: Near-circular orbit, altitude = 1500 km, inclination = 74°, period = 116 minutes.

The satellite is equipped with a Doppler transmitter, a flashlight system and laser corner reflectors. The S/C is oriented with the long axis toward the Earth's mass center, stabilization is realized towards the orbital axis.
The frequency range of the Doppler transmitter is 150 and 400 MHz

Objectives: geodetic measurements (geodetic connection of islands, development and control of local geodetic networks, improvement of general ellipsoid parameters, connection of initial geodetic dates, etc.). The scientific program is carried out by the Astronomical Council of the Russian Academy of Sciences.

The geodetic complex GEO-IK consists of the satellite itself, the ground-based tracking sites, and the S/C control center.

E.6 GEOS (GEOstationary Satellite)

The GEOS [1003] missions (1 and 2) of ESA were reference spacecraft for IMS (International Magnetosphere Study, 1976-79). GEOS was designed for GEO operations to study the particles, fields and plasmas of the Earth's magnetosphere.

E.6.1 GEOS-1

GEOS-1 mission (ESA). Launch April 20, 1977 by Delta vehicle from Cape Canaveral, FL (a booster/separation problem meant that geostationary orbit could not be achieved). The S/C performed measurements in this orbit for 14 months.

The S/C structure consisted of a cylindrical bus, 132 cm high and 164.5 cm in diameter, equipped with several booms. The S/C was spin-stabilized. An apogee kick motor, with 269 kg of propellant, was used for injection into GEO. Six 15 N thrusters provided reaction control. Attitude was measured by sun and Earth sensors and an accelerometer. Surface-mounted solar cells provided energy of >110 W (BOL). S/C mass = 574 kg. S/C integration by British Aircraft Corporation as prime contractor. [1004]

Orbit: GEOS-1 was left in a GTO orbit due to a stage 2/3 separation problem; apogee of 38,475 km, perigee of 2683 km, inclination of 26.6°.

Application: Detailed study of the magnetosphere's radial distribution of plasma, particles and waves (in conjunction with ground-based observations in Scandinavia, Iceland, Antarctica, and Alaska). Magnetospheric transport phenomena.

Sensors: [1005] (simultaneous waves and particle experiments along with electric and magnetic fields and total plasma density).

- **S 300** = Search Coil Magnetometer (CRPE, France; ESTEC; Danish Space Research Institute). Objectives: study of magnetospheric wave phenomena in both electric and magnetic domains; measurement of AC magnetic fields up to 30 kHz; DC/AC electric fields and plasma resonances up to 80 kHz; mutual and self-impedance

1002) S.K. Tatevian, "The Space Geodetic Complex GEO-IK," CSTG Bulletin No. 11, Title: New Satellite Missions for Solid Earth Studies, 1989, pp. 9-11

1003) JANE's Spaceflight Directory, 1988-89, pp. 332-333

1004) "More than Thirty Years of Pioneering Space Activities," ESA publication BR-142, 1999, compiled by Andrew Wilson, pp. 46-49

1005) GEOS - Projects under Development, ESA Report to COSPAR, Jan. 1977, pp. 112-123

- **S 302** = 2 Electrostatic Analyzers (Mullard Space Science Lab, UK). Objectives: measurement of thermal plasma (electron, protons) up to 500 eV.
- **S 303** = Combined Electrostatic and Magnetic Analyzer (University of Bern; MPI, Garching). Objectives: measurement of the composition (1-140 amu) and energy spectra of ions up to 16 keV.
- **S 310** = 10 Electrostatic Analyzers (Kiruna Geophysical Observatory, Sweden). Objectives: measurement of the pitch-angle distribution of electrons and protons in the 0.2-20 keV energy range.
- **S 321** = Magnetic Deflection System followed by Solid-state Detectors (MPI, Lindau). Objectives: measurement of the pitch-angle distribution for electrons (20-300 keV) and protons (20 keV - 3 MeV).
- **S 329** = Tracing of Electron Beam over one or more gyrations (MPI Garching). Objectives: measurement of DC electric field grad $|\mathbf{B}|$.
- **S 331** = Fluxgate Magnetometer (CNR, Frascati, Italy and NASA/GSFC). Objectives: measurement of the three components of the DC and ULF magnetic field. The instrument was boom-mounted.

The GEOS-1/-2 were the first S/C anywhere to carry a totally conductive coating - even over their solar panels. An electron beam experiment and a pair of probes, 40 m apart, provided independent measurements of the electric field surrounding the S/C. The S/C surface-treatment technology was confirmed.

Data: Transmission rate of 100 kbit/s (continuously at a frequency of 2299.5 MHz, no on-board data storage) containing high-speed data (wide-band analog and correlator data from the wave experiment), and low-speed data, consisting of low-frequency data from the wave experiment and the data from all other experiments. Uplink at 149.48 MHz.

E.6.2 GEOS-2

The GEOS-2 satellite (ESA) carried the same payload (sensor complement) as GEOS-1 and is regarded a replacement for GEOS-1 (GEOS-2 was originally considered as backup). GEOS-2 was launched July 14, 1978 from Cape Canaveral. Orbit: 25,640 km perigee, 35,592 km apogee at 0.77° inclination, positioned at 37° East (initially).
Objective: Measurement of fluctuations in the magnetic field which are picked up by small antennas (about 25 cm). GEOS-2 had six antennas to measure the three components of the field in two frequency bands.[1006)]

GEOS-2 provided two years of data, was placed in hibernation for eight months, then revived for eight months in 1981 to support the EISCAT program of upper atmosphere motion measurements. GEOS-2 remained in use until the end of 1983. Periodic monitoring support (1984) of the chemical releases of the AMPTE mission.

E.7 GEOS

The NASA mission initiative of three S/C was of a study nature to test tracking system behavior experimentally and to unify the world's tracking datums to the 5-10 m level of uncertainty with respect to the geocenter. In this context the GEOS acronym has two definitions, 1) Geodetic Earth Orbiting Satellite, and 2) Geodynamics Experimental Ocean Satellite.

E.7.1 GEOS-1 (Geodetic Earth Orbiting Satellite)

S/C launch Nov. 6, 1965 on a Delta vehicle from Cape Canaveral, FL. The GEOS 1 S/C (also known as Explorer 29) was a gravity-gradient-stabilized, solar-cell powered satellite (387 kg

1006) "GEOS," Interavia Space Directory 1992-93, pp. 155-156 of on-orbit mass), designed exclusively for geodetic studies. It was the first successful active spacecraft of the National Geodetic Satellite Program. Instrumentation included 1) four optical beacons, 2) laser reflectors, 3) a radio range transponder, 4) Doppler beacons, and 5) a range and range rate transponder. These were designed to operate simultaneously to fulfill the objectives of locating observation points (geodetic control stations) in a three dimensional Earth center-of-mass coordinate system within 10 m of accuracy, of defining the structure of the Earth's irregular gravitational field and refining the locations and magnitudes of the large gravity anomalies, and of comparing results of the various systems onboard the spacecraft to determine the most accurate and reliable system. Acquisition and recording of data were the responsibility of the GSFC Spaceflight Tracking and Data Network (STDN). Ten major observing networks were used.

Orbit: Elliptical orbit, perigee = 1113 km, apogee = 2275 km, inclination = 59.4°, eccentricity = 0.07193, period = 120.3 minutes.

Sensor complement:

Optical Beacon System. The optical beacon system, used for geometric geodesy, consisted of four xenon 670 W (1580 candle-second/flash) flash tubes housed in reflectors. These tubes were programmed to flash sequentially in a series of five or seven flashes at times when they could be optically observed from Earth. Observations were made by STDN and SPEOPT MOTS 1 m and 0.6 m cameras, SAO (Smithsonian Astrophysical Observatory) Baker-Nunn and geodetic 0.9 m cameras, USAF PC 1000 cameras, and U.S. C&GS (Coastal and Geodetic Survey) BC-4 cameras. Telescopic cameras at three or more stations simultaneously photographed the flashes against the star background. Satellite position and angle of elevation from each station could then be determined by using star charts as guides. If two of the three stations had known positions, the coordinates of the third could be calculated by triangulation. The instrument operated satisfactorily from November 18, 1965, to December 1, 1966, when command capability was lost.

Laser Tracking Reflector. Laser corner reflectors composed of fused quartz cubes with silvered reflecting surfaces, were used for determining the spacecraft's range and angle. The 322 cubes were mounted on fiberglass panels on the bottom rim of the spacecraft and provided a total reflecting area of 0.18 m^2. The reflectors conserved the narrow beamwidth of incoming light and reflected a maximum signal to the ground, almost exactly to where it originated. About 50% of the light which struck the prism area at a 90° angle was reflected within a beam of 20 arcseconds. Reflected light received by ground telescopes was amplified by a photomultiplier tube that converted the optical impulse to an electrical signal. The time for the beam to return to the Earth was recorded by a digital counter. The reflected laser pulse was also photographed against the stellar background. Total time traveled by the light pulses was also considered in the optical laser tracking system.

Radio Doppler System. The Doppler technique of timing and measuring the frequency shift of radio transmissions from a moving S/C was used to help establish the structure of the Earth's gravitational field to an accuracy of about five parts in 100 million. Three transmitters were operated on frequencies of 162, 324, 972 MHz. Timing markers (bursts of 60° phase modulation) of 0.3 s duration once each minute were carried by the 162 and 324 MHz transmitters. Synchronization of the markers was to an accuracy of 0.4 ms. The system began operating during November 1965. The 972 MHz Doppler operated until January 14, 1967, when an intermittent command system response caused a permanent turn-off. The two lower frequency beacons operated continuously until December 1967, when operations became intermittent and progressively weaker. Operations were terminated on Jan. 15, 1968.

SECOR (Sequential Collation of Range). The system operated by the Army Map Service (AMS) was used for the S/C radio range system. A 3.6 kg transponder received and retransmitted ground radio signals (421 MHz receiver and 224.5 and 449.0 MHz transmitter).

Ground-based equipment included phase-modulated transmitters, range-data receivers, and electronic phase-meters. The transponder provided valuable ranging data for four US SECOR stations to make possible inter-comparison tests from December 29, 1965, to May 1, 1966. Inter-datum and inter-island ties were completed from Tokyo to Hawaii using data from SECOR tracking stations and other geodetic observations made between May 24, 1966, and February 8, 1967. The transponder failed on February 8, 1967.

Radio Range/Rate System. Objective: Measurement of range and the radial velocity of the S/C by making phase shift and Doppler measurements. The onboard transponder, which operated on 2271 (receiver) and 1705 MHz (transmitter), and a conical antenna were the spacecraft components of the NASA/GSFC Range and Range/Rate System. The antenna, mounted on the Earth-facing portion of the S/C, could receive and transmit data. Its beam width was 150°. Data received from this instrument by three S-band stations were used to augment other geodetic data and to provide a comparison of this system with others used in tracking the S/C. The command system failure occurred on December 1, 1966.

Minitrack System. The minitrack interferometer tracking system (136 MHz) was used in combination with the NASA Range and Range/Rate System to establish the GEOS 1 orbit. Raw data from the two systems were used for early orbit determinations and routine orbit updating. Minitrack acquired 16,271 data points (station axis crossings). The Minitrack stations also participated in mutual visibility events that involved tracking with other stations for tracking system intercomparison experiments. The system operated nominally from Nov. 6, 1965, until January 14, 1967.

E.7.2 GEOS-2 (Geodetic Earth Orbiting Satellite)

The S/C launch was on Jan. 11, 1968 on a Delta vehicle from VAFB, CA (also known as Explorer 36). Objectives: optimization of optical station visibility periods and provision of complementary data for inclination-dependent terms established by GEOS 1 gravimetric studies. A a gravity-gradient-stabilized, solar-cell-powered S/C (469 kg of on-orbit mass) that carried electronic and geodetic instrumentation. The geodetic instrumentation systems included 1) four optical beacons, 2) two C-band radar transponders, 3) a passive radar reflector, 4) a sequential collation of range radio range transponder, 5) a range and range rate transponder, 6) laser reflectors, and 7) Doppler beacons. Non-geodetic systems included a laser detector and a Minitrack interferometer beacon. The S/C was placed into a retrograde orbit to accomplish these objectives. Operational problems occurred in the main power system, optical beacon flash system, and the spacecraft clock, and adjustments in scheduling resulted in nominal operations.

Orbit: Elliptical polar retrograde orbit, perigee = 1082 km, apogee = 1570 km, inclination = 105.8°, eccentricity = 0.03165, period = 112 minutes.

Sensor complement:

Optical Beacon System. Identical system as flown on GEOS-1. Data were obtained until Jan. 31, 1970.

Laser Tracking Reflector. Identical system as flown on GEOS-1.

Radio Doppler System. Identical system as flown on GEOS-1. The US Navy Doppler Tracking Network (TRANET) monitored the spacecraft for Doppler data. Observations made from three or more known stations allowed deduction of orbital parameters.

SECOR (Sequential Collation of Range). Identical system as flown on GEOS-1. The range measurements were made by measuring the phase shift of the ranging sidetones that modulated the CW carrier. By using trilateration techniques, the unknown position of one of the four stations could be accurately determined.

Radio Range/Rate System. Identical system as flown on GEOS-1. The system is also known under the name of GRARR (Goddard Range and Range Rate).

C-band Radar Transponder. The system was used for experimental range radar calibration and data recording to determine the accuracy of the system for geometric and gravimetric investigations. For redundancy, two transponders, each operating on 5690 MHz (receiver) and 765 MHz (transmitter) were carried on the S/C. One transponder had a 5 ms interval time delay, and the other had a near-zero internal delay that allowed for real-time identification by the C–band participants. The transponders were operated on a select-call basis to conserve S/C power. A C-band passive reflector was used in conjunction with the transponders for precise calibration of the internal time delay and to provide passive C-band tracking capabilities.

Minitrack System. Identical system as flown on GEOS-1.

Precipitating Electron Detector. The instrument consisted of an electrostatic deflection device and channeltron detector intended to measure electrons in the energy range 2 - 10 keV.

Magnetometer. The instrument consisted of a uniaxial fluxgate magnetometer oriented perpendicular to the orbital plane. Although the principal function of the magnetometer was to serve as an attitude sensor, a very limited amount of scientifically useful data on fluctuations in the range 0.03 to 3.0 cps were obtained through use of a filter.

E.7.3 GEOS-3 (Geodynamics Experimental Ocean Satellite)

GEOS-3 is a NASA mission.[1007] Launch: April 9, 1975 with a Thor-Delta launch vehicle from VAFB, CA. Operation of satellite for over 3 1/2 years. The S/C structure was an octahedron, topped by a truncated pyramid, with a parabolic reflector for a radar altimeter on the flat bottom side. A metal ribbon boom with end mass extended upward approximately 6.1 m from the top of the pyramid. Passive laser retroreflector cubes were mounted in a ring around the parabolic reflector with the normal vector from each cube facing 45 deg outward from the direction of the Earth. A turnstile antenna for VHF and UHF frequencies and separate antennae for Earth-viewing 324 MHz Doppler, C-band, and S-band transponders were mounted separately on flat surfaces next to the parabolic reflector. The dimension across the flats of the octahedron was 1.22 m, the spacecraft was 1.11 m high. S/C on-orbit mass = 340 kg.

Mission objectives were to perform a satellite altimetry experiment in orbit, to support further the calibration and position determination of NASA and other agency C-band radar systems, and to perform a satellite-to-satellite tracking experiment with the ATS-6 spacecraft using an S-band transponder system. This system was also used for periodic GEOS-3 telemetry data relay through ATS-6, to support further the intercomparison of tracking systems, to investigate the solid Earth dynamic phenomena through precision laser tracking, to refine further orbit determination techniques and determine inter-datum ties and gravity models, and to support the calibration and position determination of NASA Spaceflight Tracking and Data Network (STDN) S-band tracking stations.

Orbit: Near-circular polar retrograde orbit, apogee = 858 km, perigee = 818 km, inclination = 115°, period =101.82 minutes.

Application: Determination of oceanographic and geophysical parameters, satellite altimetry. Estimation of significant wave height. Geometric, gravimetric and other geodetic investigations. First estimations of surface wind speeds with altimeter data. Provision of the first comprehensive data set for most areas of the world's oceans.

Sensor complement:

ALT (Radar Altimeter). Multimode radar system with two distinct data-gathering modes (global and intensive). The global mode used a long pulse with two altimetric readings per

1007) H.R. Stanley, "The GEOS 3 Project," Journal of Geophysical Research, July 30, 1979, pp. 3779-3783 second, while the intensive mode used a short pulse with six readings per second. Both modes operated at 13.9 GHz frequency, used a parabolic antenna, had a maximum range acquisition time of 6 s, and had an altitude granularity of plus or minus 0.2 m. First radar altimeter to provide sea surface height measurements (50 cm range precision in the global mode and 20 cm in the intensive mode). This radar altimeter was of S-193 heritage on Skylab; however, with improved accuracy and the ability to operate over extended areas for greater periods of time, thereby providing the capability of examining the Earth over longer arcs and observing extensive ocean areas.

Laser Tracking Reflector (JHU/APL) consisting of a spaceborne laser retroreflector subsystem and ground-based laser-ranging systems (provision of precision satellite-ranging data). The laser reflector system consists of 264 quartz cube corner reflectors mounted on a 45° conic frustum. When illuminated by a laser light pulse from the ground, each retroreflector cube in the array reflected the light ray back to a special telescope receiver on the ground. The reflected light was picked up by the telescope, and the optical impulses converted to an electrical signal. A digital counter recorded the time when the light beam was returned to the ground. The total travel time of the light pulses, from ground to satellite and back to the ground, measured the distance to the satellite, thus forming the basis of the satellite optical laser system. The following observational systems acquired the necessary data: NASA/Wallops Laser Ranging Systems, SAO (Smithsonian Astrophysical Observatory) Laser Ranging Systems, GSFC Laser Ranging Systems, and other national and international laser stations.

Doppler System GEOS-3 (US Navy) of GEOS-1 heritage, consisting of two spaceborne transmitters and ground Doppler receiving stations. The dual frequencies (162 and 324 MHz) were coherently related and utilized in conjunction with ground Doppler receiving stations to obtain precision satellite range-rate data. The dual frequencies were generated by a highly stable oscillator driving two frequency multipliers. Both frequencies were used simultaneously to provide comparison data of the effect of the ionosphere on the signals. Observations made from three or more known stations allowed deduction of orbital parameters. Data rate = 15.6 kbit/s or 1.56 kbit/s.

S-Band Tracking System. The transponder subsystem consisted of a single-channel transponder, a power amplifier, a diplexer, and an Earth-viewing and ATS-viewing antenna system. The transponder operated in the following three modes: 1) satellite-to-satellite tracking, 2) direct unified S-band (ground-station tracking of GEOS-3), and 3) direct GRARR (Goddard Range and Range Rate) ground-station tracking of GEOS-3.

C-band Radar Transponder. Identical to the system on GEOS-2.

Satellite-to-Satellite Tracking (SST). The experiment consisted of 1) the ground-based ATS ranging system, 2) the wideband communication transponder on the ATS-6 geosynchronous spacecraft, and 3) the ranging transponder on the LEO satellite. [1008)]

E.8 GEOSAT (Geodetic/Geophysical Satellite)

US Navy Altimeter Mission.[1009),1010)] Mission control, data handling and processing, archiving and distribution at JHU/APL, Navy and NOAA. The S/C was launched on March 12, 1985 on an Atlas-E vehicle from VAFB, CA. GEOSAT consisted of two mission phases. The first was the classified Geodetic Mission (GM); mission duration = 18 months (until Sept. 1986). The second mission phase is known as the 'Exact Repeat Mission' (ERM),

1008) P. Argentiero, et al., "Results of GEOS 3/ATS-6 Satellite-to-Satellite Tracking Orbit Determination Experiment," Journal of Geophysical Research, Vol. 84, No. B8, pp. 3921–3925, 1979.
1009) "The Navy GEOSAT Mission: An Overview," Johns Hopkins APL Technical Digest, Volume 8, No. 2, 1987
1010) "The Navy GEOSAT Mission Radar Altimeter Satellite Program," in Monitoring Earth's Ocean, Land, and Atmosphere from Space, Volume 97, 1985 AIAA, pp. 440-463

which was unclassified; it started Oct. 1, 1986 and ended in January 1990. APL was the prime contractor for the S/C and radar altimeter and provided S/C operations.

Objectives: Provision of a dense global grid of altimeter data for Navy use in the areas of geodesy (Earth's gravitational models), the study of fronts and eddies, winds, waves and ice topography, physical oceanography in the 'Exact Repeat Mission.'

Orbit:
1. GM mission: Sun-synchronous polar orbit, inclination=108.1°, apogee = 814 km, perigee = 757 km, period = 100.6 minutes. Orbit of nonrepeating ground tracks in order to obtain a densely sampled map of the marine geoid.
2. ERM mission: Starting Nov. 8, 1986 the orbit was changed to an exact repeat cycle of 17.05 days for the observation of geodetic parameters of the oceans. ERM continued until January 1990, when the mission was terminated due to the degradation of the altimeter's output power.

The basic structure of the GEOSAT is similar to the GEOS-3 satellite: The design consists of a conical structure below the core for the structural attachment of the velocity control system. The GEOSAT attitude control subsystem (gravity gradient stabilization) was designed to point the radar altimeter to within 1° of nadir 98% of the time. The system components were a 6 m scissors boom with 45 kg end mass, redundant momentum wheels for roll and yaw stiffness, and pitch and roll attitude control thrusters. Attitude sensing was provided through the use of three digital sun-attitude detectors and a three-axis vector magnetometer. Spacecraft command was accomplished via a VHF uplink from the APL ground station. The telemetry subsystem consisted of two S-band transmitters, two tape recorders, and two encryption units. The GEOSAT was equipped with two dual-track high-density tape recorders (Odetics) that independently recorded the 10.205 kbps telemetry stream and played it back at 833 kbps for transmission to the ground. S/C mass = 635 kg. [1011)]

Sensor:

Radar Altimeter. (mass = 86.5 kg, power = 125 W). Frequency = 13.5 GHz; range precision = 5 cm. Range measurement between satellite and subsatellite point (at nadir) of the orbit with high measurement precision. Orbit determination with a two-frequency Doppler Tracking System (TRANET). The altimeter's two main subsystems are the RF section and a signal processor. The RF section consists of a 5 cm thick honeycomb panel with various subsystems attached to one surface and a parabolic dish attached to the opposite surface. The signal system generates a linear FM (chirp) pulse waveform for transmission by a traveling wave tube 20 W amplifier. The on-orbit dry mass of the S/C is 635 kg. [1012)]

Data:[1013)]

On-board recording up to 12 hours. S-band downlink to APL tracking facility.
Data products: Raw data were processed into Geophysical Data Records (GDRs) by APL and NOAA. The GDRs include height data derived from average echoes at a rate of 10 Hz, and mean height values at 1 Hz (every 6.7 km on the surface). Although the primary objective of the Geosat mission was to operate over oceans, echoes were also collected over ice and land surfaces.

NOAA/NODC in Washington, DC provides Geosat data of the ERM period on CD-ROM (Geosat altimeter crossover differences) as of Dec. 1992.

1011) J. J. Jensen, F. R. Wooldridge, "The Navy GEOSAT Mission: An Introduction"; McConathy, D. R. and C. C. Kilgus, "The Navy GEOSAT Mission: An Overview", and W. E. Frain, M. H. Barbagallo, R. J. Harvey, "The Design and Operation of GEOSAT", all in Johns Hopkins APL Technical Digest, Volume 8, No. 2, 1987

1012) J. L MacArthur, P. C. Marth, Jr., J. G. Wall, "The GEOSAT Radar Altimeter," Johns Hopkins APL Technical Digest, Volume 8, No. 2, 1987

1013) D. R. Mantripp, J. K. Ridley, C. G. Rapley, "Antarctic map from the Geosat Radar Altimeter Geodetic Mission," ESA Earth Observation Quarterly, No. 37-38, May-June 1992, pp. 6-10

E.9 GFO-1 (Geosat Follow-On Program)

A US Navy spacecraft, designed and built by Ball Aerospace Corp. of Boulder, CO. The objective of the GFO program is to provide operational altimetry data for the US Navy. The S/C structure consists of a conductive graphite composite; three-axis stabilized; attitude knowledge is derived from horizon scanners; sun sensors for yaw knowledge; attitude control accuracy is 0.25° (3σ); location knowledge by redundant GPS receivers with an rms accuracy of 10 cm (radial component); a Doppler beacon transmitter is also carried to support orbit determination. A hydrazine propulsion system for orbit maintenance. S/C mass = 340 kg; power = 319 W; design life = 8 years; on-board recorder of 384 Mbit capacity; the S/C is 2.56 m in length and 0.78 m in diameter. The launch of GFO-1 took place on Feb. 10, 1998 on a Taurus launcher from Vandenberg Air Force Base, CA.[1014)]

Areas of application: The altimetry data from the GFO program will be used to obtain ocean topography measurements which can be used to derive the location of fronts, eddies, and current data. This information is pivotal for the development of global structure models.

Orbit: Non-sun-synchronous polar orbit providing an exact repeat ground track; inclination=108°; perigee = 775 km, apogee = 878 km repeat period = 101.4 minutes.

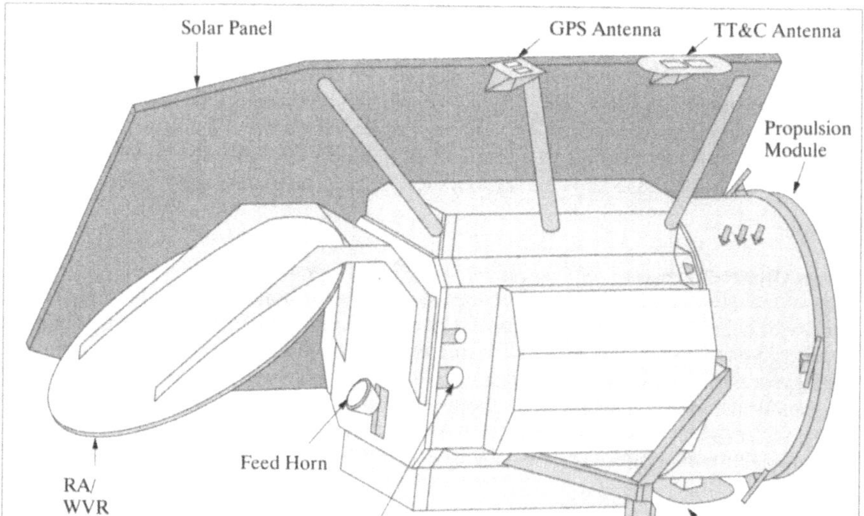

Figure 169: The GFO-1 S/C Model

Sensor complement:

- **RA** = Radar Altimeter. Frequency = 13.5 GHz ; range precision ≤ 3.5 cm. Range measurement between satellite and subsatellite point (at nadir) of the orbit with high measurement precision (performance equivalent to GEOSAT altimeter). RA consists of the following modules: RF unit, digital unit, solid-state power amplifier, T/R switch, combined RA/WVR feed horn, and integrated antenna. RA measures continuously along the subtrack with an approximately 2 km pulse-limited footprint. On-board calibration of radar cross section and height. Instrument mass = 24.7 kg; power = 75.5 W; data rate = 8 kbit/s.

1014) Information provided by B. Barry of Ball Aerospace, Boulder, CO, and by Ch. Kilgus of JHU/APL

- **WVR** = Water Vapor Radiometer. Measurement of the water vapor content along the altimeter pulse path. A two-channel instrument (22 and 37 GHz) that provides water vapor time corrections for the altimeter. The instrument consists of the electronics module, power supply, cold horns, combined RA/WVR feed horn and integrated antenna. WVR data rate = 160 bit/s.
- **GPS** = Global Positioning System. Redundant GPS receiver systems referred to as **TRSR** (TurboRogue Space Receiver) are used for precision orbit determination.

Data:
S-band downlink at 16 kbit/s (direct) and 512 kbit/s in dump mode to ground receiving stations. All data links to/from GFO-1 are encrypted. The science data will be made available to the general user community through NOAA. Mission operations by NAVSOC (Naval Satellite Operations Center) with ground stations at Prospect Harbor, ME, and Laguna Peak, CA.

E.10 GFZ-1 (GeoForschungsZentrum-1 Geodesy Satellite)

GFZ-1 is a German (GFZ Potsdam) geodesy mission (passive satellite) with the objective to improve knowledge of the Earth's gravity field. Kayser-Threde GmbH of Munich is the GFZ general contractor for the design and construction of the satellite as well as for the arrangement of a MIR launch by RKK Energia. GFZ-1 has a mass of 20 kg and a diameter of 215 mm. The launch of GFZ-1 was April 9, 1995; insertion of GFZ-1 into its own orbit: April 19, 1995. - GFZ-1 reentered the atmosphere in June 1999. The last track by a German ground station occurred on June 23, 1999 at an orbital altitude of 121 km.

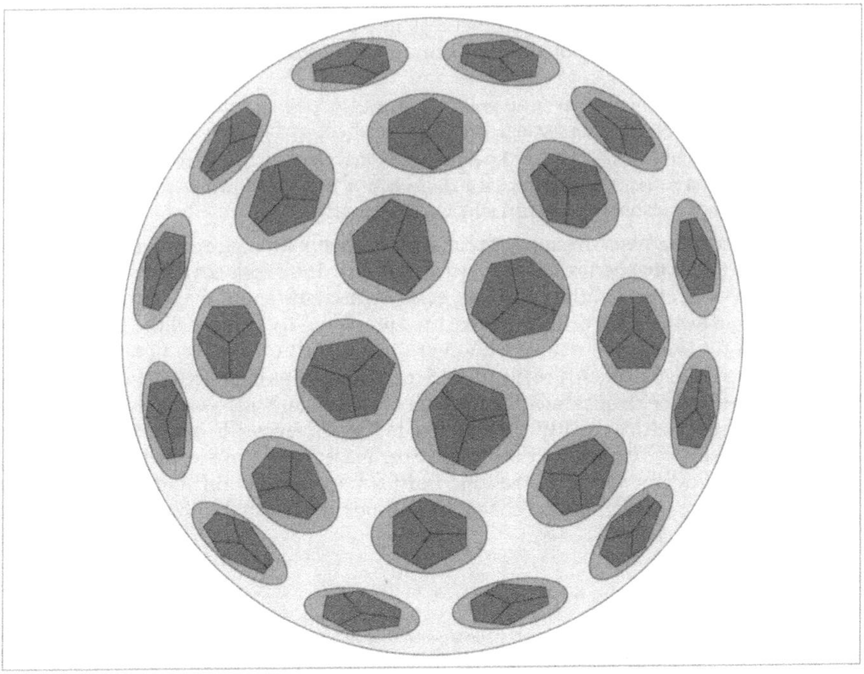

Figure 170: Illustration of the GFZ-1 S/C model

The small GFZ-1 satellite is of spherical shape whose outer surface is covered with 60 laser (corner cube) reflectors. The low Earth orbits of GFZ-1 was selected for high-accuracy determination of the Earth's gravity field. [1015)]

Orbit: Near-circular orbit, eccentricity □ 0.01, altitude 400 km (initial orbit of MIR) to about 150 km (end of mission), inclination = 51.6°, mission life □ 3 years. The low-altitude and slowly decaying orbit is chosen for high-sensitivity response in the Earth's gravity field. As such, GFZ-1 is the lowest orbiting geodynamics satellite in its class.

The precise orbit of GFZ-1 is determined by laser ranging (cm range accuracy). GFZ operates its own SLR (Satellite Laser Ranging) station in Potsdam and supports a further station at Santiago de Cuba (in cooperation with Centro Nacional de Investigaciones Sismologicas). The international SLR community, comprising a worldwide network of about 30 SLR stations, is invited to track GFZ-1.

E.11 GOCE (Gravity field and steady-state Ocean Circulation Explorer)

GOCE is a combined SGG (Satellite Gravity Gradiometry) and SST (Satellite-to-Satellite Tracking) mission. It was selected as a core mission in the ESA Earth Explorer Program (selected at the Granada meeting Oct. 12-14, 1999). The mission objectives are to determine the stationary gravity field - geoid and gravity anomalies with high accuracy (1 cm of geoid heights and 1 mgal) at spatial grid resolutions of 100 km over the Earth's surface. The GOCE mission serves to support the following multi-disciplinary science objectives: [1016)] [1017)] [1018)]

- To provide a new understanding of the physics of the Earth's interior including geodynamics associated with the lithosphere, mantle composition and rheology, uplifting and subduction processes
- To permit, for the first time, a precise estimate of the marine geoid, needed for the quantitative determination, in combination with satellite altimetry, of absolute ocean circulation and transport of mass
- To estimate the thickness of the polar ice sheets through a combination of bedrock topography, derived from space gravity, and ice sheet surface elevation (from altimetry)
- To provide a high-accuracy global height reference system for datum connection. This may serve as a reference surface for the study of topographic processes, including the evolution of ice sheets and land surface topography. [1019)] [1020)]

The overall mission objective is to obtain measurements with high spatial resolution and high accuracy (homogeneous accuracy) such that global and regional models of the (static) Earth's gravity field and of the geoid (the equipotential surface of the Earth's gravity field potential) can be deduced with unprecedented precision. Knowledge if the Earth's gravity field allows for exact orbit determination of satellites with regard to a unique reference plane, the geoid. This is then directly related to topics such as high-accurate point positioning using satellite techniques and mapping of ocean and land surfaces. A second argument to determine the Earth's gravity field is related to Earth sciences: To better understand processes that take place within the Earth's interior, and on and above its surface. Knowledge of the geoid allows for studies of the solid Earth's mass distribution, interpretation of sea-level changes, ocean water flows/ocean heat transport and related with these, climate studies- and predictions.

1015) Information provided by Ch. Reigber and R. König of GFZ Potsdam

1016) http://www.estec.esa.nl/explorer/goce

1017) H. Rebhan, M. Aguirre, J. Johannessen, "The Gravity Field and Steady-State Ocean Circulation Explorer Mission - GOCE," ESA Earth Observation Quarterly, No. 66, July 2000, pp. 6-11

1018) Information provided by Mark Drinkwater of ESA/ESTEC

1019) "Gravity Field and Steady-State Ocean Circulation Mission," ESA publication SP-1233 (1), July 1999

1020) http://www.estec.esa.nl/vrwww/explorer/GRAVITY.html#introduction

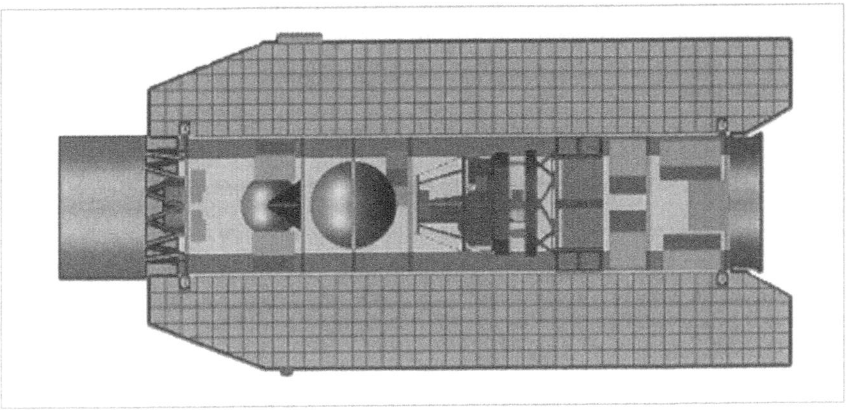

Figure 171: Schematic illustration of the GOCE S/C (structural and payload elements)

Application	Accuracy		Spatial Resolution half wavelength
	Geoid (cm)	Gravity (mgal)	
Solid Earth:			
- Lithosphere and upper mantle density structure		1-2	100 km
- Continental lithosphere			
• Sedimentary basins		1-2	50-100 km
• Rifts		1-2	20-100 km
• Tectonic motions		1-2	100-500 km
• Seismic hazards		1	100 km
Ocean lithosphere & interaction with asthenosphere		0.5-1	100-200 km
Oceanography:			
- Short scale	1-2		100 km
	0.2		200 km
- Basin scale	0.1 (approx.)		1000 km
Ice sheets:			
- Rock basement		1-5	50-100 km
- Ice vertical movements	2		100-1000 km
Geodesy:			
- Levelling by GPS	1		100-1000 km
- Unification of worldwide height systems	1		100-20000 km
- Inertial navigation system		1-5 (approx.)	100-1000 km
- Orbits (1 cm radial orbit error for altimetric satellites	1-3 (approx.)		100-1000 km
Sea-level change	Many of the above applications, with their specific requirements, are relevant to sea-level studies		

Table 282: Measurement requirements in terms of geoid height and gravity anomaly accuracies

The GOCE satellite will be built by a consortium led by Alenia Spazio of Italy as the prime contractor, Astrium GmbH is responsible for the platform. The S/C structure consists of a long slender (octagonal) prism, with a cross sectional area of 0.9 m^2 (featuring total symmetry to minimize disturbances, there are no deployable appendages) and a length of about 4.5 m. Within the structure there are several platforms upon which the payload modules are mounted, and which subdivide the platform into 3 modules for ease of integration. The lower module contains AOCS/DFACS (Attitude and Orbit Control System/Drag-Free and Attitude Control System), and Ion Thruster including the Xenon tank. The central module houses the the EGG assembly and its electronics. In fact, the EGG assembly is located close to the center of mass of the S/C (and will stay within 10 cm of the center of mass throughout the S/C lifetime). The upper module contains the electrical equipment, data-handling and radio-frequency equipment, and a nitrogen gas tank. Electric power of 1 kW EOL is generated by fixed body-mounted solar arrays (about 5.0 m^2) with GaAs cells. NiCd batteries of 25 Ah provide energy storage. The S/C thermal design and control is based on passive insulation and radiation techniques. The S/C has a launch mass of about 1000 kg, including up to 125 kg of propellant.

The key element of the on-board AOCS/DFACS system is the drag-free attitude and orbit control. The DFACS is designed to compensate for the effects which atmospheric drag forces and torques have upon the gradiometer measurements using ion thrusters and micro-thrusters. The total error budget for the gradiometer is of the order of 4 $mEHz^{-1/2}$ (Note: 1 E = 1 Eötvös = 10^{-9} s^{-2}, a unit of gravity gradient).). S/C attitude control is provided with an absolute pointing accuracy of 0.38 mrad. Attitude is sensed by a sun sensor, a three-axis magnetometer, a star sensor plus SSTI (Satellite to Satellite Tracking Instrument) and EGG. Two 20 mN RF ion thrusters and four pods of four 1 mN cold gas proportional thrusters are used as actuators.

A conventional data handling system is used on-board, based on the MIL-1553B bus, an ERC 32 processor, and 2.5 Gbit data storage. S/C communications are in S-band (two coherent S-band transponders, two antennas and a radio frequency distribution unit, 1 W RF power) with data rates of 2 kbit/s in the uplink and up to 850 kbit/s in the downlink. The ground receiving station (TT&C) is Kiruna. Mission operations and control of GOCE are at ESOC.

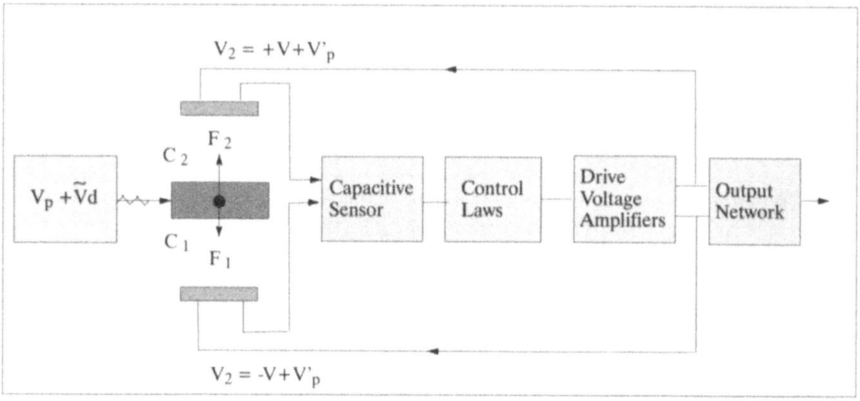

Figure 172: Single accelerometer control loop for proof mass position and feedback to AOCS/DFACS

A nominal mission duration of 20 months is planned. A Rockot-class launch vehicle from Plesetsk, Russia is planned as a reference launcher. The expected launch date is 2005.

Orbit: Sun-synchronous low Earth orbit, altitude = 250 km, inclination = 96.5°, equatorial crossing at 6:00 hours (dawn-dusk orbit) or 18:00 hours (dusk-dawn orbit) at the ascending node. Global coverage outside the polar caps is reached after about 30-40 days.

Sensor complement:

Technical concept: Satellite gradiometry is the measurement of acceleration differences between the test masses of an ensemble of accelerometers inside a satellite. The measured signal is the difference in gravitational acceleration inside the spacecraft, where the gravitational signal reflects the pull of the Earth's varying gravity field (caused by varying masses of mountains and valleys, ocean ridges and trenches, subduction zones and mantle inhomogeneities, etc.). The measured signals correspond to the second derivatives of the gravitational potential. - The gradiometer measurements are supplemented by SST (Satellite-to Satellite Tracking) measurements.

The two core instruments are SSTI and EGG. SSTI incorporates a geodetic GPS/GLONASS receiver for high-low tracking between the satellites of the two navigation constellations, and the low-flying GOCE spacecraft, referred to as SST-hl. The EGG is a three axis satellite gravity gradiometer (SGG). The gradiometer principle is based upon differential accelerometry. Drag and attitude control together with some fundamental properties of gradiometry - allow the separation of the gravitational signal from non-gravitational satellite skin forces and angular motion. Time variable effects of eigen-gravitation will be kept below the instrument noise level. The SSTI allows the retrieval of the long wavelength terms of the gravity field, while the EGG function is devoted to the medium and short wavelength terms. The instruments overlap in the low frequency range, around 0.005 Hz.

From the measurement principle point of view, the GOCE mission concept is unique by meeting four fundamental criteria for gravity field missions, namely:

- Uninterrupted tracking in three spatial dimensions
- Continuous compensation of the effect of non-gravitational forces
- Selection of a low orbital altitude for a strong gravity signal
- Counteracting of the gravity field attenuation at altitude

EGG (Electrostatic Gravity Gradiometer). The main objective of EGG is to measure the three components of the gravity-gradient tensor (i.e. gradiometer data). The EGG instrument, designed and developed at ONERA and to be built at Alcatel, is based on an ambient temperature, closed loop, capacitive accelerometer concept. EGG is a three-axis gradiometer consisting of 3 pairs of three-axis servo-controlled capacitive accelerometers on an ultra-stable carbon-carbon structure. The thermal control (passive with heaters) provides 10 mK stability during 200 s. The performance is better than 4 $mEHz^{-1/2}$ (or 4×10^{-13} g $Hz^{-1/2}$). The EGG assembly has a mass of 150 kg and requires up to 75 W of electric power.

The principle of operation of the EGG is based on the measurement of the forces needed to maintain a proof mass at the center of a cage. A six degree of freedom servo-controlled electrostatic suspension provides control of the proof mass in terms of translation and rotation. A pair of identical accelerometers, mounted on the ultra-stable structure, 50 cm apart, form a "gradiometer arm." The difference measured between accelerations measured by each of the two accelerometers, in the direction joining them, is the basic gradiometric datum (differential measurement), while half the sum is proportional to the externally induced perturbing drag acceleration (common mode measurement). Three identical arms are mounted orthogonal to one another and, the axes so defined are nominally aligned to the along-track, cross-track and vertical directions. The three differential accelerations provide direct, independent measurements: not only of the diagonal gravity components, but also of the perturbing linear and angular accelerations.

Design bandwidth (MBW)	5×10^{-3} to 10^{-1} Hz
Baseline length	0.5 m
Sensitivity (detection noise) Measurement bandwidth (MBW) Extended bandwidth (10^{-5} to 1 Hz)	 $<10^{-12}$ m s^{-2} $Hz^{-1/2}$ $<10^{-10}$ m s^{-2} $Hz^{-1/2}$
Proof-mass positioning error	6×10^{-8} m $Hz^{-1/2}$
Absolute / relative scale factors	10^{-3} / 10^{-5}
Absolute / relative misalignment	10^{-3} rad / 10^{-5} rad

Table 283: EGG performance parameters

In-orbit calibration of EGG: Calibration involves carefully planned coordination with S/C maneuvers and feedback from the gradiometer to the AOCS.. Such calibrations will be repeated, to check parameter stability with respect to thermal drifts and fluctuations. The objective of in-orbit calibration is to enhance the level of balancing to 10^{-5} in both scale-factor matching and alignment.

SSTI (Satellite to Satellite Tracking Instrument). The objective is to provide the SST-hl (Satellite-to-Satellite Tracking - high/low) contribution to the gravity field recovery, by the simultaneous tracking of up to 12 GPS/GLONASS satellite signals. In addition, SSTI provides data for precise orbit determination; it is also used for real-time on-board navigation and attitude-reference-frame determination. The instrument has a 12-channel dual-frequency receiver with a codeless tracking capability. It processes, demodulates and decodes the signals from GPS and GLONASS satellites, received through a hemispherical antenna pointing in the zenith direction. The frequency bands L1 and L2 signals are used to allow the compensation of ionospheric delays by ground post-processing. Each channel of SSTI receives either GPS or GLONASS signals and provides the following measurements: C/A (coarse acquisition) pseudo range (L1), L1 and L2 carrier phase, P1 and P2 code pseudo range (L1 and L2), L1-L2 differential carrier phase and P1-P2 differential pseudo range. In addition, SSTI provides the following capabilities:

- Position and velocity measurements from GPS (GLONASS optional) and corresponding UTC time
- One pulse per second output synchronized with GPS time
- Measurement time-tagging with respect to instrument internal time
- Redundant communication interface
- The ability to turn off unused measurement channels for power saving.

The carrier phase noise is better than 1 mm. The receiver mass is about 15 kg with a peak power demand of <40 W.

SREM (Standard Radiation Environment Monitor). The objective is to provide radiation environment measurements. Since the EGG instrument of the primary payload is sensitive to electrical charging, SREM data can be used to correlate its measurements with encountered electron and proton fluxes. The SREM instrument of ESA/ESTEC has already flown on the DERA mission STRV-1c (Space Technology Research Vehicle-1c). The SREM detector unit features two heads, each with a 20° half-cone FOV. The electronics unit comprises three particle detectors for electron and proton spectroscopy (measurement error <1%), cosmic-ray events counting and radiation-dose measurements. SREM has a mass of 2.6 kg and a power demand of 2.6 W.

LRR (Laser Retro Reflector). LRR provides a supplementary data set of range observations (satellite laser ranging by the SLR ground network) as backup for precise orbit determination post-processing. The LRR is a corner-cube array capable of reflecting laser pulses back along the incident light path. LRR has a total mass of 2.5 kg.

RITA (Radio-Frequency Ion Thruster Assembly). RITA consists of a quartz discharge chamber around which an RF field coil is wrapped, which induces the internal ionizing electric field. Separate Xenon propellant flows feed the discharge chamber and a hollow-cathode neutralizer. A positive voltage on the screen grid attracts electrons into the discharge chamber from the neutralizer plasma, to initiate the discharge. A flat triple-grid system is used to extract the ion beam, with the thruster grid at +1200 V, the acceleration grid at -500 V, and a grounded deceleration grid. To minimize erosion, the acceleration grid is made from graphite. The RITA system on GOCE is operated in the drag control range; it goes from 100 W for 1 mN to 500 W for 12 mN. The 20 mN required for orbit reboost require 625 W of power input. The total mass of RITA is about 60 kg. See also Figure 400 in O.14.2.

The RITA system was already test-flown on the EURECA-1 mission (launch July 31, 1992 - retrieval July 1, 1993). The ARTEMIS data relay satellite of ESA (launch in 2001), also employs a RITA propulsion system.

GOCE/GRACE (Gravity Recovery And Climate Experiment) mission comparison:

The GRACE mission (launch in 2001) complements GOCE by providing extremely high precision gravity measurements (an order of magnitude better than GOCE) at half-wavelengths exceeding 250 km. The advantage of GRACE data analysis is to recover temporal variations of the gravity field at these relatively longer spatial scales. The high resolution and accurate gravity field derived from GOCE in the 80 - 250 km half-wavelength range may also help to de-alias the shorter wavelengths of the gravity field of the GRACE analysis. A combination of the GRACE and GOCE results will permit construction of a gravity field model of the required precision on all relevant spatial scales.

- GRACE: Designed to measure the time variability of the gravity field at a low spatial resolution at the Earth's surface (typical values for half lambda are 1000 - 200 km).
- GOCE: Designed to measure the static gravity field at a high spatial resolution at the Earth's surface (typical values for half lambda are 200 - 80 km).

	1/2 wavelength =1000 km	1/2 wavelength =260 km	1/2 wavelength =133 km	1/2 wavelength =80 km
GRACE	<0.001 cm	= 0.15 cm	= 15 cm	unmeasurable
GOCE	= 0.04 cm	=0.15 cm	= 0.8 cm	= 10 cm

Table 284: GRACE/GOCE performance in terms of cumulative geoid error at various spatial scales

E.12 GP-B (Gravity Probe B - Relativity Mission)

A NASA gravitational physics mission, under development at Stanford University (PI: C. W. F. Everitt; Stanford is building the experiment apparatus, Lockheed Martin the S/C and payload). The objective is to provide two extremely precise new tests of Einstein's general theory of relativity by means of observations on gyroscopes in Earth orbit. The experiment will check, very precisely, tiny changes in the directions of spin of four gyroscopes contained in an Earth-orbiting polar satellite. The gyroscopes will be so free of disturbances that they will provide an almost perfect space-time reference system. They will measure how space and time are warped by the presence of the Earth, and, more profoundly, how the Earth's rotation drags space-time around with it. These effects, though small for the Earth, have far-reaching implications for the nature of matter and the structure of the Universe.[1021], [1022] Since GP-B is a "controlled physics experiment" it required many years of challenging work and sophistication in technology. [1023] [1024] [1025] [1026]

The S/C is spin stabilized (0.1 to 1 rpm) using proportional helium thrusters (drag-free control). Attitude measurements are provided with a cryogenic science telescope and control gyros for angular rates. The attitude reference platforms are mounted on the stable graphite rings of the dewar. Each platform has a star tracker and a rate gyro providing the information to orient the S/C initially to acquire the guide star with the science telescope, and to maintain accurate pointing during 45 minutes of occultation of the guide star by the Earth during each orbit. One star sensor provides roll information (<1 arcsec). The pointing control □ 20 milliarcseconds rms. The S/C structure has a length of about 6 m, total mass = 3,000 kg, power = 500 W, design life = 3 years. A launch of GP-B is scheduled for May 2002 on a Delta-2 vehicle from VAFB.

The S/C has an on-board solid-state recorder of 1.9 Gbit capacity. There are two communication links, a direct downlink via S-band at 2.5 Mbit/s, and a TDRS MA (Multiple Ac-

1021) "Testing Einstein with Orbiting Gyroscopes, Gravity Probe B," Stanford University brochure

1022) Information provided by C. W. F. Everitt of Stanford University, Stanford, CA

1023) S. Buchman, C. W. F. Everitt, B. Parkinson, et al., "The Gravity Probe B Relativity Mission," Advances in Space Research, Vol. 25, No. 6, 2000, pp. 1177-1180

1024) J. A. Lipa, D. H. Gwo, R. K. Kirschman, "Status of the cryogenic inertial reference system for the Gravity Probe B mission," SPIE, Vol. 1765 Cryogenic Optical Systems and Instruments V, 23-24 July 1992, San Diego, pp. 85-93

1025) C. W. F. Everitt, D. Bardas, Y. M. Xiao, et al., "Three Papers on Gravity Probe B," presented at The Sixth Marcel Grossmann Meeting on Relativity, Kyoto, Japan, June 23-29, 1991

1026) M. Tapley, et al., "Gradiometry Coexperiments to the Gravity Probe B and Step Missions," Advanced Space Research, Vol. 11, No. 6, 1991, pp. 179-182 cess) link at 1.032 kbit/s (return) and 250 bit/s (forward). GP-B will be operated from Stanford University.

Orbit: Polar circular orbit, altitude = 640 km, inclination = 90°, period about 90 minutes, eccentricity e = 0.00134, the orbit plane is aligned with the guide star direction.

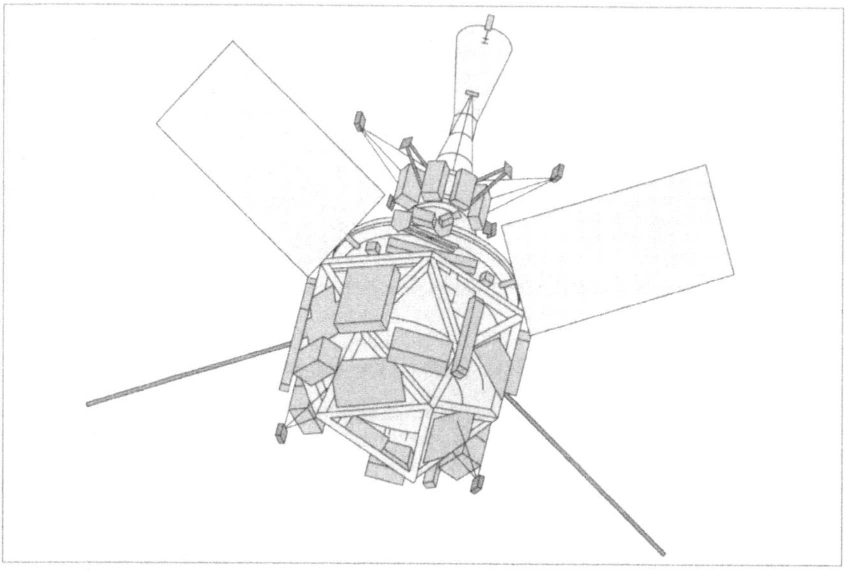

Figure 173: The GP-B S/C model

Background: [1027] Gravity Probe A (GP-A) was launched on June 18, 1976 on a Scout-D launch vehicle from Wallops Island, VA. The probe attained a ballistic (sub-orbital) flight path with an apogee of 10,000 km. The principal payload was an atomic hydrogen-maser oscillator system with the objective to measure directly the effect of the gravitational potential on the frequency of a proper clock, a hydrogen maser. The objective of the experiment was to determine whether time progressed at a different rate in conditions where gravity is weaker. Hence, the sub-orbital clock tests included the Shapiro time delay experiment and redshift experiments. During this short trip, GP-A transmitted accurate measurements of slight changes in the clock's rate in lower gravity. The analysis of the GP-A experimental data verified a portion of Einstein's gravitation and relativity theories which are referred to as the "Principle of Equivalence."

GP-B background: A gyroscope orbiting the Earth experiences two distinct space-time processes – namely the Frame-Dragging effect and the Geodetic Effect (orthogonal effects) – gradually changing its direction of spin.

- Frame Dragging: A rotating massive body drags space and time around with it. A gyroscope orbiting Earth tends to tilt away from the plane of its orbit because the Earth is dragging it.
- Geodetic Effect: According to Einstein's General Theory of Relativity, space and time in the vicinity of a massive body is distorted. For a gyroscope orbiting near the Earth, this distortion leads to a tilting of the gyroscope's spin axis in the plane of the orbit. This effect is predicted by general relativity theory to be 150 times larger than the frame dragging.

1027) R. F. C. Vessot, M. W. Levine, "A Test of the Equivalence Principle Using a Space-Borne Clock," General Relativity and Gravitation, Vol. 10, No. 3, 1979, pp. 181-204

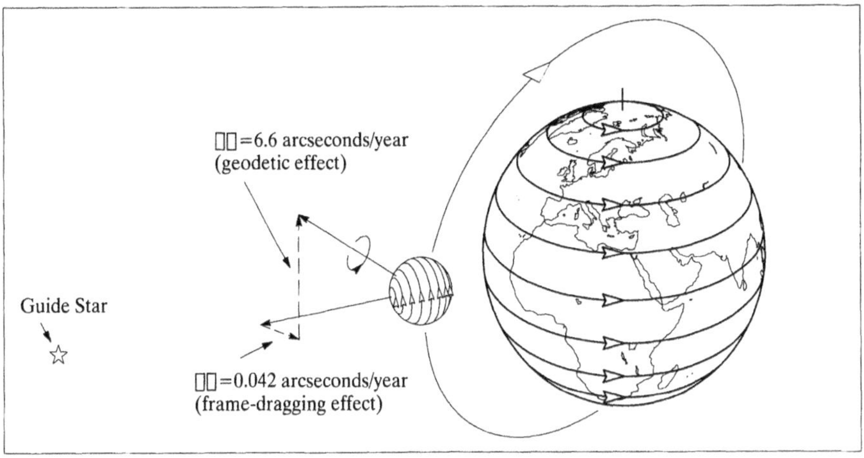

Figure 174: Relativistic precessions of an orbiting gyroscope

In late 1959, these concepts were independently conceived and formulated by the physicists, Leonard Schiff of Stanford University and George Pugh of DoD. According to calculations by L. Schiff, the frame-dragging effect (rotation of space-time) should turn the gyroscope with the Earth through an angle of 42 milliarcseconds in a time period of one year. G. Pugh provided a complete error analysis of a possible experiment. In particular, he suggested for the first time the far-reaching concept of a drag-free (or drag-compensated) satellite. [1028] The GP-B project evolved eventually from these early thoughts. - GP-B will measure this frame-dragging effect to a precision of □ 1%.

According to Einstein, a second, much larger change in spin direction, the geodetic effect, follows from the gyroscope's motion through this space-time curvature. The predicted effect for a gyroscope is a rotation in the orbit-plane of 6,600 milliarcseconds per year – quite a large angle by relativistic standards (see Figure 174). GP-B will measure this change to 1 part in 10,000 or better.

The measurement of Schiff's two effects demands much more than placing a gyroscope in a satellite. Six distinct technical requirements have to be simultaneously satisfied:

1) A drift-free gyroscope: a gyroscope having an absolute drift-rate (i.e. change in spin direction from non-relativistic disturbances) of □ 10^{-11} degrees/hour. This is six orders of magnitude (10^6) better than the best inertial navigation gyros available.
2) A gyro readout: a method for determining the changes in spin angle to 0.1 milliarcsecond without disturbing the gyroscope.
3) A stable reference: a means (telescope and mechanical structure) of referring the gyro readout to the guide star.
4) A trustworthy guide star: a bright, properly located star whose motion with respect to inertial space is known.
5) A technique for separating relativity effects: an orbit and a data processing method that together allow the frame-dragging and geodetic effects to be separated.
6) A credible calibration scheme: a scheme of in-flight calibration tests to ensure that the gyroscopes – and the entire instrument – are free of errors that might masquerade as relativity signals.

All six prerequisites are met by GP-B.

1028) Note: The first drag-free satellite to fly a completely gravitational orbit was Triad-1 of the US Navy, built by JHU/APL and launched Sept. 2, 1972. The development of the drag-free system on Triad-1 is a direct flight experiment of George Pugh's proposal.

E.12.1 Mechanical Systems

Two mechanical systems, essential to the success of the relativity measurement, are decidedly unusual: 1) a set of seven mass-trim mechanisms is used to balance and align the principal axis of the spacecraft, and 2) the use of helium proportional thrusters. [1029] [1030]

Mass trim mechanisms. Each mass trim mechanism consists of a 20 kg mass mounted in a closed rectangular box; they are adjusted in position by a lead screw driven by a stepper motor. These mechanisms have been subject to a very extensive qualification program. Four of the masses are mounted transversely to move the axis of rotation laterally until it coincides with the line through the center of the gyroscopes to within 0.8 mm. The other three masses are mounted parallel to the S/C axis, they adjust the direction of the principal axis of inertia. Trimming is performed intermittently as the helium is depleted (this occurs probably a few times during the course of a year).

Helium proportional thrusters for attitude and translational control: In most S/C with gas-jet attitude control systems, constraints on gas consumption set by mass dictate the use of on-off valves that are fired only on demand. To apply this method to GP-B with its very fine pointing requirement of ±20 milli-arcseconds would take a space-qualified valve capable of reseating perfectly hundreds of millions of times; this is obviously a severe reliability problem. Fortunately there is a solution. Already on board is a supply of gas that must be vented, namely the helium boil-off from the dewar. By directing this gas continuously through pairs of opposed nozzles operated as "proportional thrusters," one obtains a control system that is at once smoother and mechanically more reliable than the conventional kind. This design was proposed by Stanford graduate students (J. S. Bull, J.-H. Chen, P. Wiktor and Y. Jafry) and eventually further developed and built by Lockheed Martin.

S/C control is maintained in all three support modes, namely pointing, drag-free, and roll, using signals derived from:

- The science telescope
- Either of two science gyroscopes operated drag-free
- Conventional rate-gyroscopes mounted on the spacecraft

The signals are updated from a 'star-blipper' picking up a band of stars spread over the heavens at an angle to the roll axis. The pointing accuracy is better than 10 milli-arcsec, the residual cross-track average acceleration is measured to better than 10^{-12} g, and the roll rate to about 1 part in 10^5. The translational controller is designed to force the center of rotation of the S/C into coincidence with the line through the gyro centers. The mass trim mechanism brings it within 0.8 mm; the translational controller brings it to 50 µm, that is, to within the limit to which the gyros are aligned. - The spacecraft has 16 helium thrusters mounted on fixed struts, and extending from the S/C, with geometrical and internal redundancies such that any four systems can be allowed to fail with no loss of control performance.

Note, acceleration levels on satellites at the GP-B altitude (of 640 km) with normal mass/area ratios are typically of the order of 10^{-8} g. In GP-B, as a drag-free spacecraft rolling about the common axis of the gyroscopes, the mean cross-track acceleration will be below 10^{-12} g.

E.12.2 Experiment Payload

The GP-B payload is a 'thermos bottle' in space. The science payload consists of a superfluid (2,400 l) helium dewar, within which a high-vacuum probe is installed, containing the **SIA** (Science Instrument Assembly), made up of a Quartz Block Assembly (QBA), which is op-

1029) C. W. F. Everitt, S. Buchman, D. B. DeBra, G. M. Keiser, J. M. Lockhart, B. Muhlfelder, B. W. Parkinson, J. P. Turneaure, "Gravity Probe B: Countdown to Launch," NASA contract NAS8-39225 paper, 2000

1030) G. M. Reynolds, R. H. Vassar, R. T. Parmley, et al., "Payload and Spacecraft Technology for GP-B," Advances in Space Research, Vol. 25, No 6, 2000, pp. 1193-1197 tically contacted to a fused quartz Cassegrainian telescope. The QBA consists of four gyroscopes and a drag-free proof mass sensor mounted in a precision-machined fused-quartz block. The spin axes of the four gyroscopes are aligned parallel to the line of sight to the guide star. The SIA is maintained at at a stable temperature of 1.8 K which provides necessary stability of the precision alignment between the gyroscopes and the telescope. [1031]

Each gyroscope (with the world's roundest sphere) is a 38 mm diameter sphere of homogeneous fused quartz with a uniform superconducting layer of niobium (Nb) coated on its surface. The coating thickness is 1.25 μm. The gyroscope is assembled inside a housing composed of two halves also made of fused quartz. Each housing half has three pairs of electrodes for electrostatic suspension of the gyroscope and raised support lands around each electrode to prevent electrical contact (when not suspended) between the gyroscope and the electrodes. In addition, the readout half has a thin-film Nb pickup loop located on its parting plane; the spin-up half has a spin-up channel with an inlet and an outlet for He gas spin-up of the gyroscope.

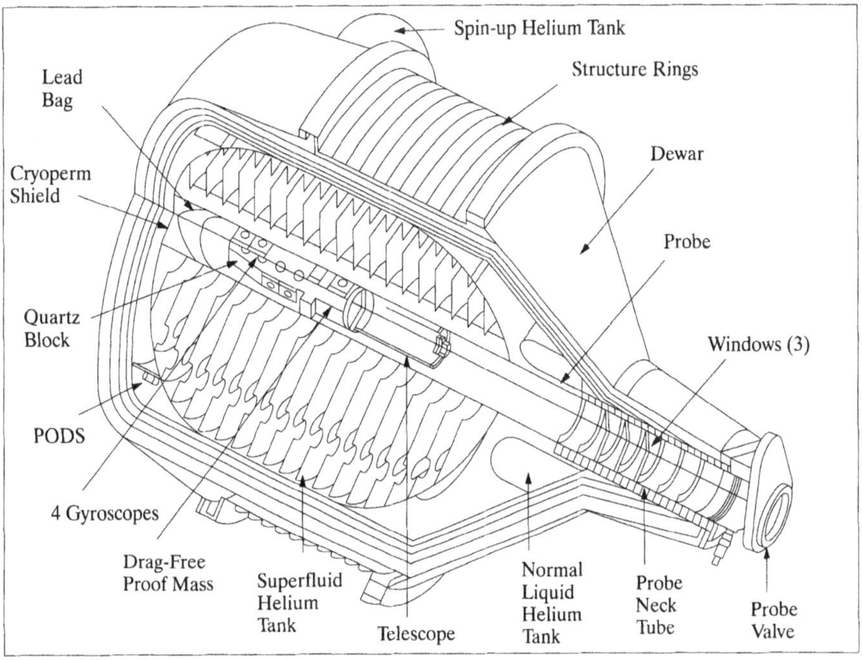

Figure 175: Gravity Probe B experiment payload [1032]

The four gyroscopes are aligned parallel to the telescope axis, two spinning clockwise and two counterclockwise, each gyroscope measuring both relativity effects. The difference in spin direction does not change the predicted effects; it does provide one of the many validation checks on the experiment. The gyroscopes are electrostatically suspended and then spun up to about 100 Hz by means of a helium gas at a temperature of 6.5 K. They are then evacuated to a pressure of □ 10^{-11} torr which, in conjunction with the averaging effect of rolling the S/C about the line-of-sight to the guide star, sufficiently minimizes torques due to differential gas damping. The proof mass sensor controls thrusters which compensate for the effects of residual drag on the S/C, and achieves an average acceleration at the gyroscopes of □ 10^{-11} g.

1031) S. Buchman, C. W. F. Everitt, B. Parkinson, et al., Gyroscopes and Charge Control for the Relativity Mission Gravity Probe B," Advances in Space research, Vol. 25, No 6, 2000, pp. 1181-1184

1032) Note: PODS = Passive Orbital Disconnect Struts

Two characteristics of the superconducting Nb play important roles in the performance of the gyroscope. First, the Nb coating makes an ideal equipotential surface for the electrostatic suspension; secondly, the London [1033] magnetic dipole moment provides a means to readout the direction of the gyro spin axis. Note: The spinning superconducting rotor develops a magnetic 'London moment' whose north-south axis exactly coincides with the rotor's spin axis. This direction is measured by a magnetometer of extreme sensitivity, known as the SQUID (Superconducting Quantum Interference Device), coupled to a superconducting loop on the gyro housing. The SQUID has the capability to detect a field change of 5×10^{-14} Gauss within a few days, which corresponds to a gyro tilt of 0.1 milliarcseconds.

The gyroscope spin directions are related via a quartz block to the telescope viewing the guide star. Like the rest of the instrument, the telescope is made of fused quartz. Its aperture (i.e. primary mirror diameter) is 14.2 cm, length = 35.5 cm, folded optics design, Cassegrainian-type, focal length = 3.70 m.

Within the dewar, surrounding the probe, is a shield formed of superconducting lead bags (foil) which almost completely excludes the Earth's magnetic field. Thus, the gyroscopes operate: a) at low temperature, b) at low pressure, c) in a low magnetic field, and d) in the low gravity of space.

The spin of the spacecraft is needed to average out potential error sources in GP-B (drifts in the telescope readout, Newtonian drift torques on the gyroscopes, etc.). Thus, the spin offers an important symmetry effect in the overall measurement concept.

The SIA is enclosed in a cylindrical cryogenic vacuum tube, 0.3 m in diameter and 2.4 m in length. The inside of the dewar vessel contains 2400 l of superfluid helium. The dewar, which maintains cryogenic temperatures on orbit for 18 months, serves as the main structural element of the S/C. The boil-off gas from the dewar, vented through proportional thrusters, provides thrust for attitude, translational and roll control of the S/C.

Magnetic shielding (DC and AC) is an essential feature of the experiment. For the London moment readout to work, the field trapped in the gyro rotor during cooling through its superconducting transition temperature must be below the value of 3×10^{-10} T (this is well below the limit achievable by conventional ferromagnetic shields). Equally important is to eliminate the AC disturbances from external magnetic fields such as the Earth's. The required AC shielding factor is 10^{12}. These requirements are met by surrounding the instrument with a nested series of conventional and superconducting magnetic shields.

E.13 GRACE (Gravity Recovery And Climate Experiment)

An international cooperative US-German dual-minisatellite SST (Satellite-to-Satellite Tracking) mission with the overall objective to obtain long-term data with unprecedented accuracy for global (high-resolution) models of the mean and the time-variable components of the Earth's gravity field (a new model of the Earth's gravity field every 30 days for five years). GRACE is also part of NASA's ESSP (Earth System Science Pathfinder) program. Some science objectives are: [1034] [1035] [1036]

- To enable a better understanding of ocean surface currents and ocean heat transport
- To measure changes in the sea-floor pressure
- To study ocean mass changes
- To measure the mass balance of ice sheets and glaciers

1033) When a superconductor like niobium spins, it generates a magnetic field effect known as the 'London moment,' after physicist Fritz London (1900-1954).

1034) C. W. Hughes, C. Wunsch, V. Zlotnicki, "Satellite Peers through the Oceans from Space," EOS Transmissions of AGU, Vol. 81, No. 7, Feb. 15, 2000, p. 68

1035) http://www.csr.utexas.edu/grace/

1036) http://essp.gsfc.nasa.gov/grace/

• To monitor changes in the storage of water and snow on the continents

The mission concept makes use of measurements of the inter-satellite range and its derivatives between two co-planar satellites (in low-altitude and polar orbits), using a microwave tracking system. The orbits of the two separately flying S/C are perturbed differently in the Earth's gravity field, leading to inter-satellite range variations. In addition, each S/C carries a GPS receiver of geodetic quality and high-accuracy accelerometers to enable accurate orbit determination, spatial registration of gravity data and the estimation of gravity field models. The fluctuations in the strength of the Earth's gravity field reflect in turn changes in the distribution of mass in the ocean, atmosphere, and solid Earth, and in the storage of water, snow, and ice on land. Since ocean bottom pressure represents a column integral of the mass of the atmosphere plus ocean, this measurement technique permits the deduction of ocean bottom pressure changes from space.

The GRACE mission is led by B. Tapley (PI) of the University of Texas at Austin and by Ch. Reigber (Co-PI) of GFZ, Potsdam. NASA/JPL leads the S/C development in partnership with DASA/DSS (Dornier Satellitensysteme) and SS/L (Space Systems/Loral). DSS provides major elements of two flight satellites based on the existing CHAMP S/C. SS/L provides the attitude control system, microwave instrument electronics and system and environmental testing. DLR/GSOC performs mission operations with tracking stations at Weilheim and Neustrelitz. Science data distribution/processing is managed in a cooperative approach by JPL and UTA/CSR (University of Texas at Austin/Center for Space Research) in the US and GFZ in Germany. Germany provides also the Eurockot launch vehicle.

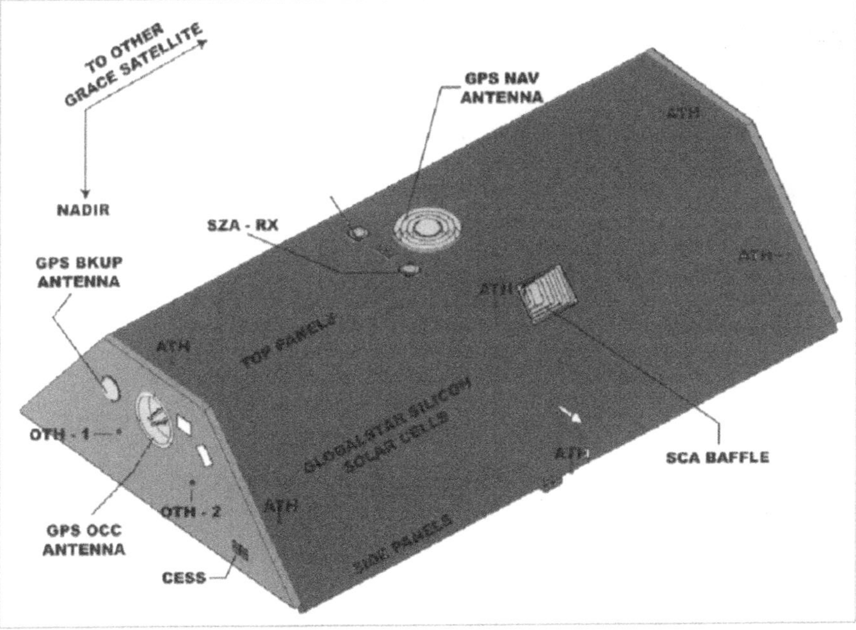

Figure 176: The GRACE S/C model

Both S/C structures are of identical design. The shape of each satellite is trapezoidal in cross section, based on the FLEXBUS design of DSS (length = 3122mm, height = 720 mm, bottom width = 1942 mm, top width = 693 mm). Each Earth-pointing S/C is three-axis stabilized by AOCS (Attitude and Orbit Control System) consisting of sensors, actuators and software. The sensors include CESS (Coarse Earth Sun Sensor) for omni-directional, coarse attitude measurement in the initial acquisition, survival and stand-by modes of the satellite. The high precision sensors are SCA (Star Camera Assembly) of ASC heritage (flown on Ørsted), and the BlackJack (GPS Flight Receiver), see description under E.1. An IMU (Inertial Measurement Unit) provides 3-axis rate information in survival modes. A boom-mounted Förster magnetometer provides additional rate information. The actuators include a cold gas system (with 12 attitude control thrusters and two orbit control thrusters, each rated at 40 mN) and three magnetorquers. Each S/C has a mass of of 432 kg (science payload = 40 kg, fuel = 34 kg); the S/C power is 150-210 W (science payload = 75 W). The top and side panels of each S/C are covered with strings of silicon solar cells; NiH batteries with 16 Ah provide power storage. The S/C design life is five years.

A dual-launch on an Eurockot vehicle is planned from Plesetsk, Russia, for Nov. 2001.The re-ignitable third stage, BREEZE, is used to place both satellites in the same nominal orbit.

Orbit: Circular polar co-planar orbit (non-repeat ground track); the initial altitude is 485 km at launch (near a solar maximum), decaying to about 300 km (near a solar minimum) after five years; inclination = 89°. The two satellites in tandem formation are loosely controlled, they are separated at distances between 170 to 270 km apart.

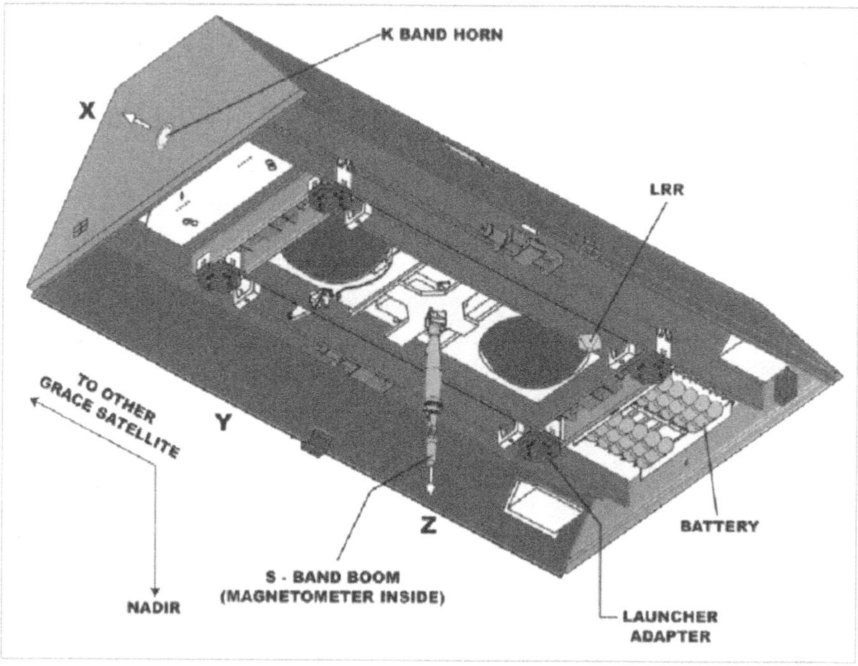

Figure 177: Bottom view of GRACE

Communications: The TT&C activities are carried out using a pyro-deployed S-band receive and transmit antenna, mounted on a nadir-facing deployable boom. A backup zenith receive antennae and a backup nadir transmit antenna (SZA-Tx), along with the appropriate RF electronics assembly, complete the telemetry and telecommand subsystem. The daily science data volume is about 50 MByte, including gravity data and GPS occultation data. CCSDS protocols are used for all data communication. The S-band frequencies for the two satellite system are:

- Downlink: 2211.0 MHz for satellite 1 and 2260.8 MHz for satellite 2. Modulation: BPSK/NRZ is modulated onto the subcarrier which is PM modulated onto the uplink carrier. The data rate is 32 kbit/s for real-time data and 1 Mbit/s for dump data.
- Uplink: 2051.0 MHz for satellite 1 and 2073.5 MHz for satellite 2. Modulation: BPSK/NRZ.

Sensor/payload complement:

The science instruments are mounted on a CFRP (Carbon Fiber Reinforced Plastic) bench in the S/C interior, as are the fuel tanks and the batteries and other satellite subsystems.

The **SIS** (Science Instrument System) includes all elements of the inter-satellite ranging system, the GPS receivers required for precision orbit determination and occultation experiments, and associated sensors such as SCA. SIS also coordinates the integration activities of all sensors, assuring their compatibility with each other and the satellite.

KBR (K-Band Ranging) instrument assembly of JPL. KBR is the key science instrument of the GRACE mission. The objective is ultra-precise satellite-to-satellite tracking (SST) in low-low orbit. Variations in the gravity field cause the range between the two satellites to vary. The relative range is measured by KBR (a microwave link which is integrated with a GPS receiver). The measured range variations are corrected for non-gravitational effects by SuperSTAR. KBR consists of the following elements: USO (Ultra Stable Oscillator), the microwave assembly, the horn, and IPU (Instrument Processing Unit).

USO (of JHU/APL) serves as the frequency reference. The microwave assembly, or sampler, is used for up-converting the reference frequency to 24 and 32 GHz; down-converting the received phase from the other satellite; and for amplifying and mixing the received and the reference carrier phase. The horn is used to transmit and receive the carrier phase between the satellites. The IPU is used for sampling and digital signal processing of not only the K-Band carrier phase signal, but also the signals received by the GPS antenna and the star cameras. Each satellite transmits carrier phase to the other at two frequencies, allowing for ionospheric corrections. The transmit and receive frequencies are offset from each other by 0.5 MHz in the 24 GHz channel, and by 0.67 MHz in the 32 GHz channel. This shifts the down-converted signal away from DC, enabling more accurate measurements of the phase. The 10 Hz samples of phase change at the two frequencies are downlinked from each satellite, where the appropriately decimated linear combination of the sum of the phase measurements at each frequency gives an ionosphere-corrected measurement of the range change between the satellites.

SuperSTAR (Super Space Three-axis Accelerometer for Research mission) of ONERA, France (of STAR heritage on CHAMP, with a resolution a factor 10 higher than that on CHAMP). The objective is the measurement of non-gravitational accelerations (drag, solar and Earth radiation pressure) acting on the S/C. SuperSTAR is mounted at the CG (Center of Gravity) of the satellite. SuperSTAR consists of the following elements: SU (Sensor Unit, EEU (Electromagnetic Exciting Unit), ICU (Interface Control Unit), and a harness. SU consists of a metallic proof mass, suspended inside an electrode cage of gold-coated silica. The proof mass motion is servo-controlled using capacitive sensors, and is a measure of the non-gravitational accelerations acting on the satellite. The mass and electrode cage core is enclosed by a sole plate and a housing in which vacuum is maintained using a getter. The SU vacuum unit is surrounded by analog electronics. The EEU is used to deliver a 10 mg acceleration, and is used only in case of an SU start-up problem. The ICU supplies power to the SU and EEU, and operates the accelerometer through a micro-controller board.

SCA (Star Camera Assembly). The objective is the precise measurement of satellite attitude. SCA consists actually of two DTU (Technical University of Denmark) star camera assemblies, each with a FOV of 18° x 16°. Both assemblies are rigidly attached to the accelerometer, and view the sky at a 45° angle with respect to the zenith, on the port and starboard sides. The SCA is used for both: science as well as AOCS; the two assemblies provide the primary precise attitude determination for each satellite. The baffles are used to avoid the degradation due to solar heating.

LRA (Laser Corner-cube Reflector Assembly), provided by GFZ. LRA is mounted on the underside of the spacecraft to permit orbit verification from terrestrial laser tracking networks.

BlackJack (GPS Flight Receiver), a new generation instrument of TRSR (TurboRogue Space Receiver) heritage, provided by JPL (see description under E.1). The objective is to use the GPS instrument for navigation (precise orbit determination) and radio-occultation (refractive occultation monitoring) applications. BlackJack features three antennas, the main zenith crossed dipole antenna is used to collect the navigation data. In addition, a bakkup crossed dipole antenna and one helix antenna on the aft panel are used for back-up navigation and atmospheric occultation data collection, respectively. This system is capable of simultaneously tracking up to 24 dual frequency signals. In addition, this system provides digital signal processing functions for the KBR and SCA instruments as well.

E.14 Jason-1 (Joint Altimetry Satellite Oceanography Network)

Jason [1037] [1038] is the name of a joint CNES/NASA oceanography mission series with the objective to monitor global ocean circulation, discover the tie between the oceans and atmosphere, improve global climate predictions, and to monitor events such as El Niño conditions and ocean eddies. The oceanography mission series is considered a cornerstone of GCOS (Global Ocean Observing System). The Jason-1 satellite carries a radar altimeter, a follow-on mission to the TOPEX/Poseidon mission. Jason-1 is planned to be launched on a Delta-2 7920 vehicle from VAFB in Dec. 2001 (along with TIMED). [1039] [1040] [1041] [1042]

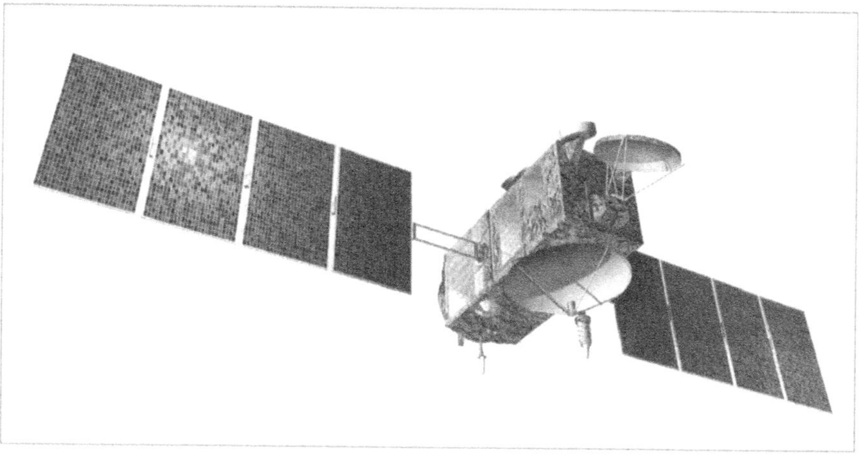

Figure 178: Illustration of the Jason S/C

1037) Jason-1 is named after the mythological hero who led the Argonauts on the adventurous and hazardous search for the Golden Fleece which they found and returned. "Jason" symbolizes both the hard-fought quest for a worthy goal and civilization's fascination with the ocean and its mysteries.

1038) M. Lefebvre, "A New Voyage for Jason," CNES/AVISO Newsletter Nr. 5, April 1997

1039) F. Parisot, T. Lafon, "The JASON-1 Satellite Design and Development Status," Proceedings of the 4th International Symposium on Small Satellites Systems and Services, Sept. 14-18, 1998, Antibes Juan les Pins, France

1040) P. Escudier, G. Kunstmann, F. Parisot, et al., "Jason System Overview and Status," CNES/AVISO Newsletter No. 7, Jan. 2000, pp. 9-15

1041) http://topex-www.jpl.nasa.gov/jason1/

1042) T. Lafon, F. Parisot, "The Jason-1 satellite design and development status," Proceedings of the AIAA/USU Conference on Small Satellites, Logan, UT, Aug./Sept. 1998, SSC98-V-6

The Jason S/C uses the Proteus minisatellite bus, a multimission platform of CNES/Alcatel Space Industries (partnership). The overall architecture uses redundancy by sharing two half satellites, each under the control of one data handling unit. The satellite consists of the payload module (which accommodates the various instruments) and the Proteus platform, a boxlike structure of 1 m side lengths, which provides on-board services. The hydrazine propellant system uses a 28 l tank with 4 x 1 N thrusters (propulsion is only used for orbit acquisition and maintenance). A WFOV star tracker for attitude sensing is located on the payload module to minimize attitude errors. The normal in-orbit attitude control is based on a gyro-stellar concept. Four reaction wheels are used as actuators (they are desaturated using magneto-torquers). Power (450 W) is provided by two solar panels (silicon cells). Power is distributed by a non-regulated primary electrical bus, using a single battery (40 Ah NiCd). S/C mass = 490 kg (245 kg bus dry mass), the S/C bus size is: 954 mm x 954 mm x 1000 mm, the payload module size is: 954 mm x 954 mm x 1218 mm, design life = five years, data storage capacity of 2 Gbit (EOL), pointing accuracy = 0.035°. Communication: downlink rate at 650 kbit/s (S-band, QPSK modulation), uplink at 4 kbit/s (S-band). The CCSDS communication protocol standard is used.

The Jason-1 S/C is monitored and controlled from CNES in Toulouse up to the end of the assessment phase (30/50 days). Then the POCC (Project Operation Control Center) of JPL takes over. The baseline ground network includes stations at Aussaguel (France), Fairbanks (AK, Poker Flats location), and at Wallops Island, VA. In addition, the Altimetric and Orbitography Mission Center (SSALTO), located in Toulouse and the Jason Science Data System at JPL in Pasadena, form the mission ground segment.

Orbit: Circular non-sun-synchronous orbit; 1336 km altitude (2 hour period), inclination = 66°, 9.9-day repeat orbits (127 revolutions), ground track repeatability = ±1 km cross-track at the equator. The plan is to fly Jason-1 and Topex/Poseidon along identical orbital tracks (about a minute apart) to perform cross calibration. Topex/Poseidon was launched August 10, 1992 and is now well beyond its design life of five years.

Sensor Complement: [1043)]

Poseidon-2 (Solid-State Radar Altimeter). Poseidon-2 is a dual-frequency nadir-looking radar altimeter with the objective to map the topography of the sea surface for calculating ocean surface current velocity and to measure ocean wave height and wind speed. Poseidon-2 is a CNES-sponsored instrument developed by ASI (Alcatel Space Industries) as prime contractor. [1044)] Poseidon-2 is of SSALT heritage on TOPEX/Poseidon (chapter E.21). SSALT is being upgraded for the Jason radar altimeter mission by adding a second C-band frequency of 5.3 GHz (Ku-band at 13.575 GHz), as well as changing to digital technology and using a new radiation-hardened microprocessor (the altimeter electronics are split into two boxes, the processing unit and the RF unit). The processing unit includes a chirp generator, baseband demodulator, spectrum analyzer, instrument control unit and interfaces. The RF unit performs the up-conversion to Ku- and C-bands, the high power solid-state amplification, the low-noise amplification of the received echoes, and its mixing with a reference chirp. The C-band channel provides direct ionospheric correction for the primary Ku-band measurement and uses a 1.2 m antenna on the nadir side of the S/C.
Instrument parameters: transmitted pulse width of 105 μs; PRF of 2100 Hz (1800 for Ku-band and 300 for C-band); maximum RF output power to antenna of 38.4 dBm (Ku-band) and 42 dBm (C-band); mass = 70 kg for dual-frequency configuration; duty cycle = 100%; power = 78 W, thermal control by conduction to mounting surface and by radiation within the instrument; thermal operating range = -5-35°C; IFOV = 20° cone centered at nadir.

The major evolutions of the Poseidon-2 altimeter with respect to SSALT are:

1043) "Jason-1," ESE Reference Handbook of NASA, 1999, pp. 120-123

1044) L. Rey, G. Carayon, et al., "Poseidon-2, the new generation altimeter for JASON mission," Proceedings of IGARSS'99, Hamburg, Vol. III, June 28-July 2, 1999, pp. 1503-1505

- Addition of the second frequency (C Band) to obtain direct ionospheric correction.
- The digital chirp generator to allow a selectable bandwidth necessary for the C band channel and a better phase knowledge
- Increase of the emitting power to keep an acceptable link budget (the size of the antenna has been decreased with respect to Topex/Poseidon)
- A 128 points FFT transform to obtain a better mispointing evaluation and then reduce the pointing requirements
- Use of a rad hard microprocessor to provide a better radiation tolerance and thus increase the operational availability of the sensor. This microprocessor is also more powerful to increase the on-board processing capability of the altimeter.

Transmission frequencies	5.3 GHz (C-band), 13.575 GHz (Ku-band)
Transmitted pulse width	105.6 μs
Bandwidth	320 MHz (Ku-band), 320/100 MHz (C-band)
PRF (Puls Repetition Frequency)	1800 Hz (Ku-band), 300 Hz (C-band)
Peak output power	8 W for Ku-band, 25 W for C-band
Max. RF power output to antenna	38.4 dBm Ku-band, 42 dBm for C-band
Noise figure	3.2 dB (Ku-band), 0.9 dB (C-band)
Data rate	22.5 kbit/s including waveform data and onboard parameters

Table 285: Poseidon-2 parameters

DORIS (Doppler Orbitography and Radiopositioning Integrated by Satellite), a CNES/Thomson development. DORIS is a precision orbit determination system providing position and ionospheric correction for Poseidon-2. The DORIS flight segment consists of a two-channel, two-frequency (401.25 MHz and 2036.25 MHz) Doppler receiver capable of tracking signals from a worldwide network of about 50 ground beacons. The Jason DORIS receiver is the same second generation device as the one developed for the ENVISAT mission. Its main functional improvements over first-generation receivers are its capability to receive two beacons simultaneously and to produce on-board the orbit ephemeris in real time with a precision of 1 m. The receiver is controlled by an ultra-stable oscillator delivering the reference frequency with a stability of 5×10^{-13} over a 10-100 second interval and delivering an on-board time output within 0.1 ms accuracy. Two DORIS instrument mass is 31 kg, power = 30 W. For the DORIS concept description, see also chapter E.21.1

JMR (Jason Microwave Radiometer), a JPL instrument of TMR heritage, see chapter E.21). JMR is a passive microwave radiometer measuring the brightness temperatures in the nadir column at at 18.7, 23.8, and 34 GHz, providing path delay correction for the altimeter (the brightness temperatures are converted to path-delay information). The 23.8 GHz channel is the primary water vapor sensor, the 34 GHz channel provides a correction for non-raining clouds, and the 18.7 GHz channel provides the correction for effects of wind-induced enhancements in the sea surface background emission. The JMR receivers employ MMIC (Microwave Monolithic Integrated Circuit) technology (built by TRW) for high reliability, low power and low mass. Three temperature controlled noise diodes are used for gain calibration in all the radiometers. JMR consists of a collecting aperture shared with Poseidon-2, a multifrequency feed assembly that illuminates the collecting aperture, multi-channel microwave receivers, a data unit, power supplies, and ground support equipment. Instrument parameters: beamwidth = 1.2° at 18.7 GHz, 1.0° at 23 GHz, and 0.7° at 34 GHz; temperature resolution □ 1 K; system temperature = 800 K; mass=27 kg; power = 31 W; data rate = 100 bit/s; thermal control by conduction through the mounting feet of the satellite structure; thermal operating range = 5-35°C.

BlackJack (GPS Flight Receiver). The instrument is of GPS/MET (Microlab) heritage (of a design as flown on CHAMP) and is being provided by NASA/JPL and built by Spectrum Astro Inc. of Gilbert, AZ. BlackJack is a 16-channel GPS receiver with the objective to provide supplementary positioning data to DORIS in support of the POD (Precision Orbit De-

termination) function and to enhance and/or improve gravity field models. Radial accuracies of 1-2 cm are obtained in post-processing. BlackJack is a fully redundant unit. Each unit is comprised of an omnidirectional antenna, low-noise amplifier, crystal oscillator, sampling down-converter, and a baseband digital processor assembly, communicating through a 1553 bus interface. Instrument mass = 10 kg (2), power = 17.5 W.
BlackJack can track up to 16 GPS satellites simultaneously in dual-frequency mode. From these signals, BlackJack acquires measurements of the GPS carrier phase providing range measurements with an accuracy of about 1 mm; the absolute pseudo range (defined as the absolute range plus receiver time offset from GPS time) has an accuracy of about 10 cm. BlackJack provides also on-board solutions for S/C position and time, accurate to about 50 m and 150 ns, respectively.

LRA (Laser Retroreflector Array), a JPL instrument of TOPEX/Poseidon heritage, built by ITE Inc. under GSFC contract. LRA provides a reference target for laser tracking measurements, which are necessary to calibrate the POD system and the altimeter throughout the mission. The LRA is placed on the nadir face of the satellite. It is a totally passive unit that consists of nine quartz corner cubes arrayed as a truncated cone with one in the center and the other eight distributed azimuthally around the cone. LRA provides a FOV of about 100°. The LRA instrument mass is 2.2 kg.

Parameter	Topex/Poseidon	Jason-1
S/C mass	2500 kg	490 kg
S/C power	1000 W	450 W
Platform mass	980 kg	245 kg
Platform power	500 W	300 W
Payload mass	385 kg	120 kg
Payload power	380 W	147 W
Altimeter mass	230 kg	70 kg
Altimeter power	260 W	78 W

Table 286: Comparison of mass/power budgets for the Topex/Poseidon and Jason-1 S/C [1045)]

E.15 LAGEOS-I (Laser Geodynamics Satellite)

LAGEOS-I [1046)] = Laser Geodynamics Satellite (NASA). Launch: May 4, 1976 with a Delta launch vehicle from Vandenberg.
Orbit: Near-circular orbit, altitude = 5950 km, inclination = 110°.

Objective: First NASA satellite dedicated wholly to laser ranging. LAGEOS was designed to act as a permanent reference point so that the Earth's progress could be tracked relative to the satellite (in contrast to the traditional system of tracking satellites relative to the Earth). The USGS uses LAGEOS to measure continental drift (plate tectonics, crustal deformations).

The first four years until 1980 were devoted to determining Lageos' precise orbit and to building up a global network of 14 Earth stations. By accurately measuring the time for a laser pulse to travel to the satellite and return, the position of the laser system could be determined to about 10 cm. Under NASA's Crustal Dynamics Project (started in 1979), 56 investigators from 12 countries were making repeated measurements between their locations and Lageos.

The Lageos satellite is an aluminum sphere with a brass core. Its 426 prisms, called cube-corner reflectors, give it an appearance of a golf ball (60 cm diameter and 411 kg mass). The

1045) Y. Menard, Ph. Escudier, "Cruising the Ocean from Space with Jason-1," EOS/AGU Transactions, Vol. 81, No 34, Aug. 22, 2000, p. 1 and pp. 390-391
1046) Jane's Spaceflight Directory 1988-89, Fourth Edition, pp. 83-84

three-dimensional prisms reflect laser beams back to the source, regardless of the angle from which they come.

Since the LAGEOS-I orbit is known to extremely high accuracy, the location of a laser ranging station on the surface of the Earth can be determined to a precision of less than 1 cm (by measuring the time for a laser pulse to travel from the laser ranging station to the satellite and return).

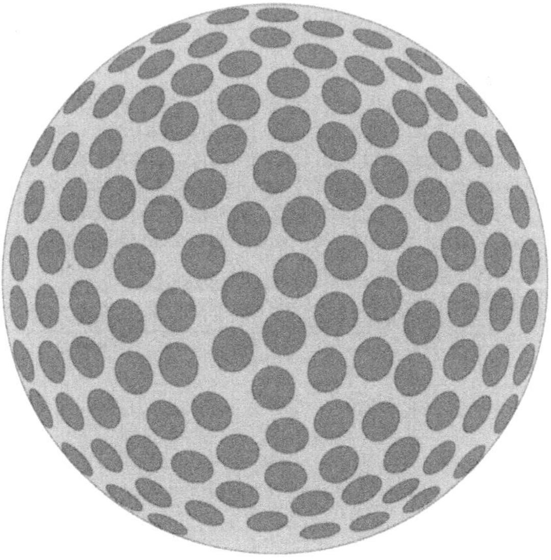

Figure 179: Model of the LAGEOS sphere with its reflectors

E.15.1 LAGEOS-II

LAGEOS-II[1047),1048),1049)] is a collaborative NASA-ASI mission (ASI built LAGEOS-II based on the same design as the NASA-produced LAGEOS-I), a follow-up of LAGEOS-I. LAGEOS-II has a mass of 405 kg, a diameter of 60 cm, and a total of 426 laser reflectors. Shuttle (Columbia, STS-52) launch from Cape Canaveral: Oct. 22, 1992 (NASA).

Italy developed and provided IRIS (Italian Research Interim Stage), a solid-fueled booster, which carried the satellite from the Shuttle's parking orbit into the required Lageos II orbit.

Orbit: LAGEOS-II and LAGEOS-I are deployed in prograde (LAGEOS-II: 52.6° inclination) and retrograde (LAGEOS-I: 110° inclination) orbital planes. Near-circular orbit, altitude = 5950 km.

Objectives: LAGEOS-II is an integral part of the Crustal Dynamics Project (CDP). Study of the Earth's crust in the Mediterranean region. Research in solid Earth geophysics [study of global and local tectonic processes, polar motion and Earth rotation, determination of Universal Time (UT-1), the recovery of Earth and ocean tidal parameters, and geopotential modelling].

1047) R. Kolenkiewicz, S. Zerbini, "LAGEOS-II: A collaborative NASA-ASI Mission," CSTG Bulletin No.11, Title: New Satellite Missions for Solid Earth Studies,, June 1989, pp. 13-18

1048) "Columbia Successfully Lofts Italian Lageos Satellite," Space News, Oct. 26-Nov. 1, 1992, p. 13

1049) NASA/ASI Lageos II brochure

LAGEOS-II is an identical S/C to LAGEOS-I. LAGEOS-I and -II are passive satellites dedicated exclusively to laser ranging (no on-board attitude control).

Both LAGEOS satellites are tracked by a global network of fixed and transportable lasers from some 65 sites. The current precision for laser systems varies from 15 cm to less than 1 cm for single-shot range measurements.

The data available to the investigators consist of both preprocessed and analyzed data (i.e. station positions, baselines, and Earth rotation parameters as a function of time). Data will be archived in the Crustal Dynamics Data Information System (CDDIS) at GSFC. The facility is accessible for interactive log-ins through SPAN, INTERnet, etc.

E.16 MAGSAT

MAGSAT is a Johns Hopkins APL/NASA/USGS satellite mission for the survey of the Earth's Magnetic Field. Launch: Oct. 30, 1979 on a Scout vehicle from VAFB, CA. Objectives: collection of data for improved modeling of the time-varying magnetic field generated within the core of the Earth, and to map variations in the strength and vector characteristics of crustal magnetization. [1050] [1051] [1052]

The S/C consists of two distinct parts: the payload module and the base module (containing the subsystems for data handling, power, communications, commanding, and attitude control). The base module is comprised of spare parts from the SAS-C (Small Astronomy Satellite) S/C. The magnetometers were deployed after launch to a position 6 m behind the spacecraft.

Orbit: Sun-synchronous polar orbit; inclination = 96.76°; perigee = 352 km, apogee = 578 km, period = 93.9 minutes. The satellite remained in orbit for seven and a half months until June 11, 1980.

MAGSAT collected scalar (total-field) and three orthogonal vector components of the magnetic field. During its seven and one-half months in orbit (October 1979 to June 1980), this satellite provided the most accurate measurements of the global field ever obtained, and the first measurements of the vector field in low-Earth orbit.

Sensors: the sensors were mounted on an instrument platform at the end of the 6 m magnetometer boom to eliminate the effect of spacecraft fields. The basic MAGSAT mission required knowledge of the magnetic field orientation to a total system accuracy of < 20 arcseconds.

- **Scalar Magnetometer** (PI: R. A. Langel), cesium vapor type, with an accuracy of about 0.5 nT. The sensor developed internal oscillations in its lamp circuitry shortly after launch, which prevented full data recovery and which, at times, slightly degraded its accuracy. Its data were, however, sufficient to provide in-flight calibration for the vector magnetometer (8 magnetic field strength measurements/s), except during the last few weeks of operation.
- **Vector Magnetometer** (built at GSFC), fluxgate type, with an accuracy of < 3 nT for each component <3 nT (after in-flight calibration). The instrument consisted of three fluxgate ring-core sensors mounted on a MACOR (machinable ceramic) structure.

The on-board attitude determination system required two star cameras, two Attitude Transfer System (ATS) optical heads (for pitch/yaw and roll). Two ATS (pitch/yaw and roll)

1050) F. F. Mobley, L. D. Eckard, G. H. Fountain, G. W. Ousley, "Magsat - A New Satellite to Survey the Earth's Magnetic Field," IEEE Transactions on Magnetics, Vol. Mag. 16, No. 5, September 1980, pp. 758-760

1051) R. Langel, G. Ousley, J. Berbert, "The MAGSAT Mission," Geophysical Research Letters, Vol. 9, No. 4, April 1982, pp. 243-245

1052) R. Langel, "The Magnetic Earth as Seen from Magsat, Initial Results," Geophysical Research Letters, Vol. 9, No.4, April 1982, pp. 239-242 mirrors were mounted on the back of the vector magnetometer at the end of the boom, providing a reflected beam of light for accurate magnetometer axis determination. A precision sun sensor was mounted in addition on the vector magnetometer for redundant measurements.

Data: MAGSAT carried two on-board tape recorders to allow 10 hours of recorded data to be dumped to a NASA ground station during a pass. The data are available in several formats from the National Space Science Data Center (NSSDC) at GSFC and at NOAA/NGDC (subset of NASA data).

E.17 MIMOSA (Microaccelerometric Measurements of Satellite Accelerations)

MIMOSA is a microsatellite of the Czech Republic, designed and developed at the Astronomical Institute of Ondrejov. The project is funded by the Grant Agency of the Czech Republic. The overall mission objective is to obtain total density distributions in space and time of the upper ionosphere by sensitive measurements of the non-gravitational orbital perturbations (atmospheric drag, solar radiation pressure). The data obtained will serve for analytical model improvements of the upper atmosphere. The project builds on non-gravitational research conducted with the instrument MACEK in previous short-duration missions on the Russian satellite Resurs-F1 (launch June 23, 1992) and on Shuttle flight STS-79 (Sept. 16 - 26, 1996). An improved version of MACEK is the only science instrument onboard MIMOSA. [1053) 1054) 1055) 1056) 1057) 1058)]

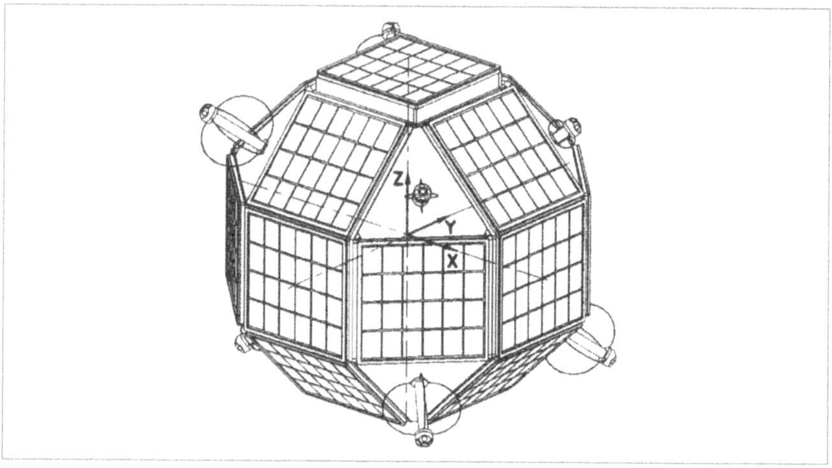

Figure 180: Illustration of the MIMOSA spacecraft

1053) R. Peresty, L. Sehnal, M. Chvojka, P. Dostal, "MIMOSA Satellite," Acta Astronautica, Vol. 46, No 2-6, 2000, pp. 345-349

1054) L. Sehnal, R. Peresty, L. Pospisilova, P. Dostal, "Software Features for the Orbital Dynamics of the MIMOSA Satellite," Proceedings of the 51st IAF Congress, Oct. 2-6, 2000, Rio de Janeiro, Brazil, IAF-00-A.5.03

1055) L. Sehnal, R. Peresty, L. Pospisilova, A. Kohlhase, "Mission Analysis of the MIMOSA Satellite," Proceedings of the 49th IAF Congress, Sept. 28 - Oct. 2, 1998, Melbourne, Australia

1056) L. Sehnal, L. Pospisilova, R. Peresty, P. Dostal, A. Kohlhase, " MIMOSA - A Satellite Measuring Orbital and Attitudinal Accelerations caused by Non-Gravitational Forces," Advances in Space Research, Vol. 23, No 4, 1999, pp. 705-714

1057) http://www.asu.cas.cz/~ macek/nep.html

1058) I. Sehnal, R. Peresty, L. Pospisilova, "Project MIMOSA - Final Stage of the Satellite Fabrication," 3rd International Symposium of IAA, Berlin, April 2-6, 2001, pp. 241-244

The design of the MIMOSA spacecraft structure employs a 26-sided symmetrical polyhedron (maximum diameter of 617 mm, the shape is nearly spherical to eliminate the effects of a possible lift) with a high area-to-mass ratio for overall measurement sensitivity (smooth surface for drag measurements). The outer surface is covered with 17 solar panels of 200 mm x 200 mm in size. The inside of the S/C features six subsystems: Balancing masses, ACS (Attitude Control Subsystem), PDU (Power Distribution Unit), Data Management and Store Unit, Telemetry and Telecommand System, and Service Module.

- The balancing mass system consists of three movable masses (1.5 kg each) in each of the main S/C axes with a moving range of ±20 mm. The objective is to aline the center of S/C gravity with that of the proof-mass cavity to better than 0.5 μm.
- For all gravity measurement operation periods, the S/C is not stabilized, it is orbiting and rotating freely. However, the tumbling or spin of the spacecraft can be readjusted/retarded by magnetic coils (a telecommand action by ground control) whenever the spin of 0.01 radians/s is exceeded. ACS employes eight solar sensors to determine the sun direction to an accuracy of 0.5°. In addition, there is a fluxgate magnetometer and for attitude measurement. Actuation is provided by three magnetic coils to eliminate the initial spin rate. MACEK measurement periods require a residual spin below 0.01 radians/s. In addition, a GPS receiver [MPE-1 (Miniature PLGR Engine) model, where PLGR (Precision Lightweight GPS Receiver)] in PPS mode is being used for orbit determination.

A S/C power of 11 W is provided. Lithium-ion batteries with a capacity of 14 Ah are used during orbital eclipse phases. The S/C mass is 55 kg.

RF communication. The uplink is in UHF-band at 450 MHz the data rate is 1.2 kbit/s. The dual downlink employs S-band (2.2 GHz, data rate at 20/80 kbit/s, bi-phase FM modulation) and VHF-band (137 MHz, data rate at 5 kbit/s). Data of up to 2 MByte can be stored on-board in solid-state memory. The spacecraft control center and ground station is located at the Astronomical Institute. The ground station is in Panska Ves, Czech Republic (the Panska Ves station served already as the ground station for the MAGION satellites).

A launch of MIMOSA is planned for 2002 on an Eurockot launch vehicle from Plesetsk.

Orbit: elliptical orbit, initial perigee = 350 km, initial apogee = 1400 km, inclination = 63.5°. The MIMOSA spacecraft mission is subjected to a decaying orbit, measuring the orbital and attitudinal accelerations (non-gravitational forces) with an expected life time of about nine years. The elliptical orbit selection emphasizes the predominance of drag forces in the lower altitudes, as well as the overriding influence of the solar radiation pressure in regions near the apogee.

Sensor complement:

MACEK (Mikroakcelerometr) is the Czech name for the electrostatically compensated microaccelerometer instrument. MACEK is an accelerometer with six degrees of freedom and a cubic proof-mass. The objective is to detect accelerations to a magnitude of 10^{-10} m/s^2.

Parameter	SI value of acceleration	Value expressed in "g - notation"
Input range	$\pm 4 \times 10^{-4}$ ms^{-2}	±40 μg (micro g)
Input frequency	1 mHz to 100 mHz (milliHz)	1 mHz to 100 mHz
Scale factor	25 kV/ms^{-2}	0.25 V/μg
Bias	$< 1 \times 10^{-5}$ ms^{-2}	<1 μg
Temperature coefficient	2.6×10^{-8} ms^{-2}/° C	2.6 x ng/° C
Measurement resolution	10^{-10} ms^{-2}	0.01 ng
Axis misalignment	5 μrad	5 μrad

Table 287: Performance parameters of the MACEK instrument

The major element of the MACEK accelerometer is the cubic proof-mass (29.96 mm) moving in a cubic cavity of 30 mm size. The cavity is created by six prism blocks. The cube and prism blocks are made of a special quartz glass and slightly chrome-plated such that appropriate patterns on the prism blocks and the cube make up the electrodes. The cube is kept within the geometric center of the cavity. During periods of measurement, the cube position is detected by capacitive sensors and stabilized in all six degrees of freedom. The signal of the stabilizing forces provides the magnitude and direction of the acceleration measured.

The total instrument mass is 5.61 kg (2.67 kg sensor head, 2.94 kg electronics box). The sensor head power is 0.6 W, the power for the electronic box is 5.9 W. The sensor head dimensions are: 15 cm x 15 cm x 18 cm. The electronic box dimensions are: 14 cm x 20 cm x 23 cm.

E.18 Ørsted

A Danish geomagnetic research microsatellite mission [named in honor of the Danish scientist Hans Christian Ørsted (1777-1851) who discovered electromagnetism in 1820] developed by a consortium of organizations including the University of Copenhagen, the Technical University of Denmark (DTU), the Danish Meteorological Institute, the Danish Space Research Institute (DSRI), and Computer Resources International. The mission objective is to perform highly accurate and sensitive measurements of the geomagnetic field and global monitoring of the high energy charged particles in the Earth's environment. The data is used to improve geomagnetic models, to study auroral phenomena, and for correlation with Earth-based measurements in order to study the relationship between the external field and the energy coupling of the magnetosphere-ionosphere system. [1059] [1060] [1061] [1062] [1063]

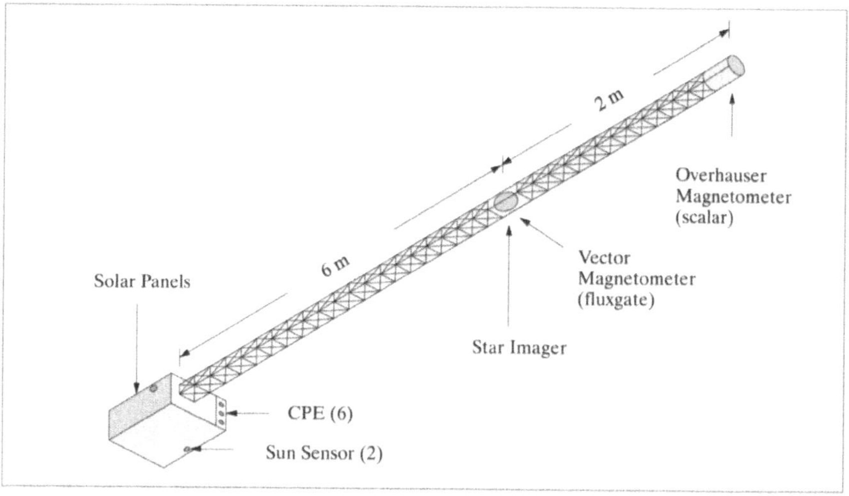

Figure 181: The Ørsted S/C model

Ørsted is in the low-cost microsatellite class, it was launched Feb. 23, 1999 (after almost two years in storage) as a secondary payload to ARGOS of the US Air Force on a Delta II rocket

1059) P. Lundahl Thomson, F. Hansen, "Danish Ørsted Mission In-Orbit Experiences and Status of the Danish Small Satellite Program," Proceedings of the 13th Annual AIAA/USU Conference on Small Satellites, Aug. 23-26, 1999, Logan UT, SSC99-I-8

1060) Information provided by F. Primdahl of DTU, Lyngby, Denmark

1061) P. Donaldson, "Mapping Magnetism," Space, April 1993

1062) T. Neubert, M. Mandea, G. Hulot, R. von Frese, F. Primdahl, et al., "Ørsted Satellite Captures High-Precision Geomagnetic Field Data," AGU/EOS, Vol. 82, No 7, Feb. 13, 2001

1063) http://www.dmi.dk/projects/oersted/

from VAFB. S/C mass = 61 kg, power = 54 W (22 W average load), S/C stabilization by a gravity gradient boom (8 m boom containing the star imager and magnetometers) and active magnetic torquing (three-axis magnetorquer coils maintain yaw to within 10°). Attitude determination via ASC (Advanced Stellar Compass) a star imager camera, with sun sensors and magnetometers as backup. The satellite has a box-like shape (34 x 45 x 68 cm) and is covered by solar panels (GaAs), 54 W of average power. NiCd batteries provide power in eclipse. Design life of at least one year (three year goal). Ørsted is supposed to fill the gap in magnetic measurements after Magsat (launch of Magsat in 1979, see E.16).

Data: Science data sampling rates: 1 to 100 samples/s. On-board data storage and downlinking every 12 hours or less. Uplink and downlink in S-band (2.114 GHz and 2.296 GHz, respectively). Maximum downlink rate is 256 kbit/s. There are three downlink stations in Denmark.

Orbit: Slowly drifting elliptical polar orbit, perigee = 646 km km, apogee=864 km, inclination = 96.5°. Local equator crossing times at 2:30 AM/PM, drifting at -23 minutes/month.

S/C body dimensions	340 mm x 450 mm x 680 mm
Power	GaAS body-mounted solar panels, 54 W EOL. Two power control units, two 6-cell NiCd battery packs, 6 Ah
Primary structure	H-beam with honeycomb platforms
Attitude control	Gravity gradient and magnetorquers. ACS
Communication	S-band with ESA standard packet TM/TC. Cold redundant TX/RX units
On-board computers	Two, each with 16 MByte storage, 80C186, 16 MHz CPU
Orbit position	Redundant GPS
Boom	8 m deployable 3 longeron (the boom is pointing in the zenith direction)
Thermal control	Passive
S/C mass	61 kg

Table 288: Overview of the Ørsted satellite parameters

Sensor complement:

Scalar Magnetometer (of LETI and CNES, the sensor is provided free of charge by France). Overhauser proton-precession magnetometer [coils for proton resonance excitation and detection, and a resonator for Electronic Spin Resonance (ESR) pumping of a nitro-oxide solution] for measuring magnetic field scalar values with an absolute measurement error of less than 1 nT. Measurement range: 16,000 - 64,000 nT; sampling rate = 1 Hz. Provides in-flight calibration for the vector magnetometer. Sensor mass = 1 kg. The instrument is boom-mounted at a distance of 8 m. The instrument mass is 2.5 kg, power = 3 W.

Vector Magnetometer (DTU). The instrument is a compact spherical coil (CSC) triaxial fluxgate magnetometer which measures magnetic field vectors at an angular resolution of 1 arcsec and an absolute measurement error of less than 1 nT. Dynamic range: ±65,536 nT; resolution: 0.25 nT, 100 vector samples/s burst mode, 10 vector samples/s normal mode. Sensor mass = 300 g. The sensor is boom-mounted at a distance of 6 m from the satellite. The instrument mass is 2.1 kg, power consumption = 1.1 W.

CPE = Charged Particle Experiment (Danish Meteorological Institute). The instrument consists of six solid-state particle detectors for high energy measurements of electrons (30 keV - 1 MeV), protons (200 keV - 30 MeV), and □-particles (1-100 MeV). Detectors look in different directions with a FOV of $15-45^{\circ}$.

ASC (Advanced Stellar Compass) of DTU, Lyngby, Denmark. The objective is to provide an attitude reference instrument with a precision of a few arcseconds aboard the satellite (to measure the magnetic field with a vector precision of a fraction of a nT). The instrument is a CCD star imager for determining the pointing vector for the CSC fluxgate magnetometer with a resolution of 6 arcsec (3 □) or less. ASC is boom-mounted close to the position of the vector magnetometer. The camera of the star imager is a 752 x 588 pixel CCD device. ASC consists of a camera head unit (CHU) connected to a DPU (Data Processing Unit), i.e. a microcomputer fitted to a frame-grabber. The CHU acquires star images within its FOV, while the DPU provides the processing power to perform image analysis, pattern recognition, data reduction, and communication. The total mass of ASC is 1.647 kg, the power consumption is 4.5 W. [1064] [1065] ASC is also flown (or going to be flown) on the following missions: Astrid-2, TEAMSAT, CHAMP, SAC-C, ADEOS-II, GRACE, and PROBA.

TRSR (TurboRogue Space Receiver), a special GPS receiver provided by NASA/JPL (Trimble Tans II + modified JPL Turbo Rogue). The objective is to a) accurately determine the position of the satellite, and b) the instrument observes in parallel ionospheric electron content, and provides atmospheric soundings permitting the derivation of atmospheric profiles of density, pressure, and temperature (refractive occultation monitoring). TRSR supports both C/A and P code and P-codeless (cross-correlation) operation. It provides parallel dual-frequency code and cross-correlation tracking, and data output of up to eight GPS satellites simultaneously. The TRSR sampling rate is 10 Hz. The instrument has a mass of 4 kg, power = 7-15 W.

E.19 Starlette

Starlette [1066] is a CNES 'Solid Earth' mission, a passive satellite dedicated to geodetic and geophysical studies with SLR (Satellite Laser Ranging) observation support. Starlette was launched Feb. 6, 1975 from Kourou on a Diamant B vehicle. Starlette is the world's first passive laser satellite for solid Earth research.

Orbit: Altitude varies between 810 km (perigee) and 1105 (apogee) km, inclination = 50°, projected orbital life = 2000 years.

Objectives: Studies of the gravity field of the Earth (both long term average values and temporal variations due to Earth's tides).

Satellite: sphere with a radius of 12 cm; mass = 47.29 kg; coefficient of reflectivity = 1.1 (approx.). The core parameters are:

- shape: isocaedron
- metal: alloy of uranium 238 and 0.2% of vanadium
- density: 18.7
- satellite mass: 35.5 kg

The skin features are:

- shape: twenty spherical caps with triangular bases, fixed on the faces of the isocaedron
- metal: alloy of aluminum and 5% magnesium

Measurement campaigns require the support of the worldwide SLR network.

E.20 Stella

Stella is a CNES satellite mission launched on Ariane as a passenger experiment (piggyback) along with the SPOT-3 launch on September 26, 1993. Stella was put on top of Ariane's the third stage with a special device, including a spring to give the required increment of velocity and a spin of 5 to 8 rev/minutes (spin axis perpendicular to the ecliptic).

1064) J. L. Joergensen, C. C. Liebe, "The Advanced Stellar Compass, Development and Operations," Acta Astronautica, Vol. 39, No. 9-12, 1996, pp. 775-783

1065) A. Eisenman, C. C. Liebe, "Operation and Performance of a Second Generation, Solid-State, Star Tracker, The ASC, Acta Astronautica, Vol. 39, No 9-12, 1996, pp. 697-705

1066) M. Lefebvre, "Stella," CSTG Bulletin No. 11, Title: New Satellite Missions for Solid Earth Studies, 1989, pp. 25-32

Orbit: Sun-synchronous quasi-circular polar orbit, altitude = 780-800 km, inclination = 98.2°, orbital life: several centuries.

Objectives: Studies of the gravity field of the Earth. Contribution to the modelling of non-gravitational forces. Contribution to the modelling of Earth and ocean tides in conjunction with other passive laser satellites (LAGEOS-I and -II, ETALON-1, and -2, Starlette, AJI-SAI). Contribution to modeling temporal variations of the gravity field due to geophysical causes like post glacial rebound, allowing determinations of mantle viscosity and possible geographical variations, again in conjunction with tracking data from other laser satellites.

Satellite: Stella is an exact twin of Starlette. Sphere with a diameter of 24 cm, mass of 47.29 kg, etc. Tracking requirement of the worldwide SLR network (international cooperation and active participation is desired).

E.21 TOPEX/Poseidon

TOPEX/Poseidon (Topography Experiment for Ocean Circulation). A NASA(JPL)/CNES joint Earth observation mission. Launch: Aug. 10, 1992 (Ariane 4 launch vehicle from Kourou). Poseidon was originally a separate CNES Mission, but was later combined with TOPEX (1985). TOPEX/Poseidon is the heart of the WOCE (World Ocean Circulation Experiment) program; it was also used for TOGA/COARE (see Part Q). Nominal Life: 3-5 years. TOPEX is regarded as the SEASAT successor mission. Satellite mass = 2402 kg, solar power = 3.4 kW (plus three 50 Ah batteries). Hydrazine propellant system for orbit maintenance. [1067) 1068) 1069) 1070) 1071) 1072)]

The TOPEX/Poseidon S/C design is based on the Fairchild Multimission Modular Satellite (MMS). The S/C is three-axis stabilized (nadir pointing) via reaction wheels and torque rods. Attitude determination via Earth sensors, sun sensors, star cameras, IRU, magnetometers. The S/C consists of the MMS platform and the instrument module housing the sensors. The MMS consists of the following elements: a command and data handling subsystem which includes the main on-board computer; the attitude determination and control subsystem, for maintaining the spacecraft attitude; and the electrical power subsystem, which contains the solar array and three batteries. The command and data handling subsystem houses three tape recorders for collecting engineering telemetry and instrument data. The S/C has a variety of communication antennas (steerable high-gain antenna dish and two omni antennas) to link the mission with TDRSS, with the DORIS tracking system and with GPS.

Objectives: Dedicated altimetry mission. Combination of high altimetric precision and high orbital accuracy for the purpose of ocean topographic mapping. Measurements of sea surfaces for modeling of global changes in ocean circulation and sea level (global panoramic maps of sea-surface topography). Development of climate models for long-term forecasts (in the order of a season or longer). Requirements: Precision orbit.

Orbit: Circular non-sun-synchronous orbit; 1334 km altitude (2 hour period), inclination = 66°, 10-day repeat orbits.
Note: The satellite orbit tracking coverage provided by the laser and DORIS is not continuous in time, hence, orbit computation based on dynamical equations is required to produce precise orbit for the mission. Observed orbit accuracy: 3 cm (rms for the radial distance).

1067) 'Topex-Poseidon Partners Discuss Sequel', Space News, Aug. 17-23, 1992, p. 3

1068) "Predicted Topex Positioning Accuracy with Differential GPS Techniques," presented at, and published in the 'Proceedings of the first International Symposium on Precise Orbit Positioning with GPS' April 15, 1985

1069) Lee-Lueng Fu, M. Lefebvre, "TOPEX/Poseidon: Precise Measurement of Sea Level From Space," CSTG Bulletin No. 11, Title: New Satellite Missions for Solid Earth Missions, June 1989, pp. 51-54

1070) 'Currents' - the JPL Topex/Poseidon Newsletter, March 1990, Issue 1

1071) Topex/Poseidon Science Investigation Plan, NASA (Document Resource Facility), Sept. 1, 1991

1072) Ch. A. Yamarone, et al., "TOPEX/Poseidon Mission Global Measurements of Sea Level at Unprecedented Accuracy," 45th Congress of the International Astronautical Federation, Oct. 9-14, 1994, Jerusalem

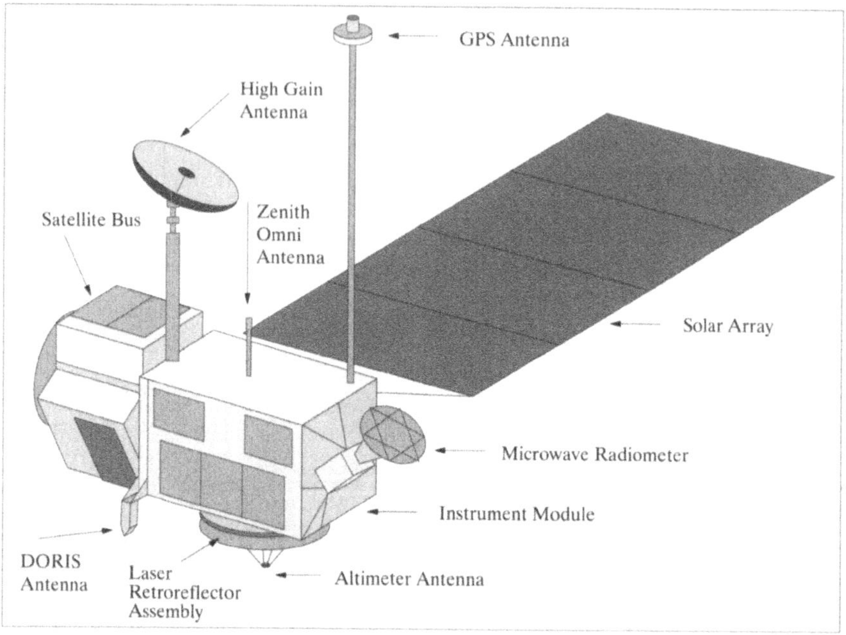

Figure 182: The Topex/Poseidon S/C model (zenith view)

As of August 1995 the primary TOPEX/Poseidon mission has been completed. The data set from this prime mission has provided oceanographers with the first global data set on the Earth's oceans. The satellite subsystems and sensors are healthy, offering the opportunity of four to seven years of extended operational service.[1073)]

Sensor complement:

- **ALT** = Radar Altimeter (of GEOS-3, Seasat, and Geosat heritage, ALT is also referred to as NRA = NASA Radar Altimeter). ALT uses a linear FM chirp pulse centered at 13.6 GHz (Ku-band) and at 5.3 GHz (C-band), the dual-frequency design is used to correct for ionospheric path delays (first spaceborne dual-frequency altimeter). The range difference measured at these two frequencies provided a first-order correction for the influence of the ionosphere. The ALT instrument consisted of a signal processor, an RF section, and an antenna assembly. ALT was designed and built by JHU/APL, managed by GSFC. It is the prime sensor for the measurement of sea surface heights, wave heights, and surface wind speed. The Ku-band and C-band bandwidth is 320 MHz, with C-band selectable at 100 MHz; the antenna control pointing accuracy is 0.14° (1-sigma). ALT mass = 206 kg, power = 237 W, altitude measurement accuracy of 2.4 cm. Simultaneous measurements at both frequencies so that ionospheric range delay can be directly estimated from the two measurements. [1074)]
  Note: The altimeter antenna (1.5 m parabolic antenna, beam width = 1.1° for Ku-band and 2.7° for C-band) is shared between ALT and SSALT - with ALT using 88% of the time. The data rate of ALT is 9.8 kbit/s.
- **TMR** = Topex Microwave Radiometer (JPL). TMR is of Nimbus-7 heritage. Operation at 18, 21, and 37 GHz to measure the total water vapor content along the altimeter

1073) TOPEX/Poseidon Internet homepage

1074) A. R. Zieger, et al., "NASA Radar Altimeter for the TOPEX/Poseidon Project," Proceedings IEEE, Vol. 79, No. 6, June 1991, pp. 810-826 pulse path to correct for water-vapor-induced range delay. The uncertainty in the altimeter range measurement made by such a system under normal ocean conditions is expected to be less than 5 cm at 7 km spatial resolution (3 cm at 100 km resolution). Mass = 50 kg, power = 25 W. TMR used a 79 cm diameter nadir-pointed offset paraboloid reflecting antenna to measure atmospheric emissivity. It also used a cold blackbody calibration with a cold sky horns looking at right angles to the sun line. The instrument consisted of RF/data modules, power supply modules, cold horns, and multifrequency feed horn. The RF/data modules and power supply modules were located directly beneath the feed horn. The instrument was thermally controlled by louvers and replacement heaters. The footprint of the 21 GHz channel was about 35 km. Measurements of the columnar water vapor along the satellite ground track had an accuracy of 0.2 gm/ cm^2 over a range of 0.2 to 6.0 gm/cm^2. TMR data provided corrections to altimeter height data for the effects of atmospheric water vapor to an uncertainty of 1.2 cm.

- **GPSDR** = GPS Demonstration Receiver system for direct position measurement, JPL (manufacturer: Motorola); Demonstration of GPS differential ranging as an experiment. The GPS receiver measures incoming signals from the GPS satellites and uses in addition ITRF (International Terrestrial Reference Frame, i.e. a set of reference ground stations) measurements for DGPS results. GPSDR operates at 1227.6 MHz and at 1575.4 MHz. Mass = 28 kg, power = 29 W, altitude accuracy < 10 cm.
  TRANET orbit determination (with the ground network) was dropped in favor of DORIS and laser approaches. (Note: GPSDR is not able to receive P-code measurements; full accuracy is achieved when GPS antispoofing is turned off).
- **SSALT** =Single-Frequency Solid-State Altimeter. (1-frequency of 13.65 GHz). Experimental sensor (of CNES, built by Alcatel/Espace) to demonstrate the concept of low-power, low-mass, low data rate (1/7 the rate of ALT due to extensive on-board processing) and low-cost altimeter for future Earth observing missions. Ionospheric range correction is provided by a model that makes use of simultaneous DORIS measurements. Measurement accuracy = 2.5 cm. SSALT mass = 24 kg, power = 49W. SSALT consists of two packages: the processing and control unit (PCU), and the radio frequency unit (RFU). The Poseidon altimeter sends a pulse generated by a surface acoustic wave generator with a 300 MHz bandwidth and 900 MHz frequency to the sea surface. The pulse is frequency converted to 13.65 GHz after generation. The pulses are amplified before transmission. The received pulses are amplified and an FFT analyzer computed the power spectrum of the signal. The signal is then sent through a microprocessor which performs the waveform computations. The output data is related to the sea state and wind speed.
- **LRA** = Laser Reflector Array, JPL. This passive sensor is used by a ground laser network (of 10-15 SLR stations) to track the position of the satellite for precision orbit determination (verification of altitude measurements). Mass = 29 kg, LRA accuracy depends on laser station characteristics.
- **DORIS** = Doppler Orbitography and Radiopositioning Integrated by Satellite (CNES-GRGS-IGN development). [1075] [1076]

E.21.1 DORIS

DORIS is a one-way microwave tracking system for the determination of precise orbits (3-10 cm radial distance and 0.3 mm/s in range rate accuracy). The concept is based on a ground segment (of globally positioned tracking stations) and a space segment (i.e. DORIS as a passenger payload in a satellite consisting of a receiver, an ultra-stable oscillator, and an antenna). There is also a control center as part of the ground segment, located at CNES.

1075) "Other Satellite-Based Microwave Systems," Lecture Notes in Earth Sciences - The Interdisciplinary Role of Space Geodesy, Springer Verlag I. Mueller, S. Zerbini, chap. 5, p. 161
1076) DORIS - Precision Satellite-Based Orbit Determination, CNES brochure

The on-board receiver measures the Doppler shift of 'uplink beacons' in two frequencies (f_1 = 2036.25 MHz, f_2 = 401.25 MHz), which are transmitted continuously by the DORIS ground network of stations (50). One measurement is used to determine the radial velocity between spacecraft and beacon, the other to eliminate errors due to ionospheric propagation delays. Only one beacon can be received by the space segment at any time (Note: a second generation receiver starting with DORIS/ENVISAT will be able to treat simultaneously two beacons). Orbit determinations with a precision of □ 5 cm on the radial distance component are achieved on the TOPEX/Poseidon mission.

The DORIS on-board package comprises a receiver, (or radial velocity measurement unit, consisting essentially of two receiving chains; total mass = 17 kg, power = 20 W; size = 385 x 280 x 210 mm), an ultrastable crystal oscillator, and an omnidirectional antenna. IFOV = 125° centered on nadir; data rate = 200 bit/s; duty cycle = 100%; thermal control by conduction to mounting surface and by radiation within the instrument module; thermal operating range = -10-50°C.

The DORIS ground segment comprises:

- the DORIS Control Center (DCC) at CNES
- a beacon installation and management center, managed by IGN. A network of Orbit-Determination Beacons (ODBs) is positioned throughout the world.
- Precision orbit determination computations performed by CNES (Earth's gravitational field computation on the basis of DORIS data by GRGS).

An ODB comprises two transmitters (one operating at 401.25 MHz, the other at 2036.25 MHz), an ultrastable oscillator, and a microprocessor performing the necessary control and management functions, transmission of timing, housekeeping, and failure diagnosis. An ODB also includes an antenna and three meteorological sensors (atmospheric pressure, air temperature, and relative humidity); these parameters are needed for atmospheric propagation delays. An ODB message carries meteorological data, the beacon ID, and information concerning the beacon operating status. The complete message lasts 0.8 seconds and is repeated once every 10 seconds.

A second class of beacons is termed Ground Location Beacons (GLBs). These are at positions that are either unknown or not known to sufficient accuracy. GLBs use the results of high-precision orbit determination as input for the precise determination of ground positions. GLBs are functionally identical to ODBs. Each GLB transmits independently of all others for 10 seconds, once, twice, or three times every minute, but only while the satellite is in range. These beacons are used for precise positioning applications. Since October 1994 DORIS measurements are included in IERS (International Earth Rotation Service).

The master beacon (MB) is the link between DCC and the on-board package. On each pass the DCC transmits instructions for on-board programming and time-tagging information. DORIS time is based on the on-board receiver clock which is related to UTC using master beacon measurements.

DORIS application: All-weather global tracking, ground-beacon positioning, estimate of the total content of ionospheric free electrons. DORIS is flying on-board SPOT-2 and -3, TOPEX/Poseidon; it is scheduled to be on SPOT-4 and on ENVISAT.

Data: The geophysical data produced by the Topex/Poseidon mission (as well as by SPOT-2, -3, etc.) are accessible to the international scientific community through US and French national data centers (IERS/DORIS database operated by IGN). The French data center for oceanography is called AVISO (CNES). The data products include:

- Sea surface topography
- Significant wave height
- Surface wind speed
- Ocean tides
- Vertically integrated atmospheric water vapor

- Vertically integrated ionospheric electron content

E.22 WESTPAC (Western Pacific Satellite)

WESTPAC, formerly known as WPLTN (Western Pacific Laser Tracking Network) satellite, is a commercial Australian communications and research satellite owned by Electro Optic Systems Pty Limited (EOS), Queanbeyan, NSW, Australia. The S/C was designed and built by ISDE (Institute of Space Device Engineering) of Moscow. A wide range of research capabilities have been included into the S/C design, in addition to its communications function. The objective of the passive satellite is to enhance the contribution of satellite laser ranging (SLR), in particular to space geodesy in the Western Pacific, to serve in the development of free space optical communications, and also to study the Fizeau effect that occurs when laser light is reflected from a satellite travelling at orbital velocities. Its primary mission is to act as a target with intrinsically very stable range biases for calibrating ground station biases and satellite signatures. WESTPAC was launched as a secondary payload to RESURS-O-4 on July 10, 1998 on a Zenit launcher from Baikonur, along with other secondary payloads [FASat-Bravo (Chile), TMSat (Thai Microsatellite), SAFIR-2 (OHB Bremen), TechSat/Gurwin-II (Israel)]. [1077] [1078]

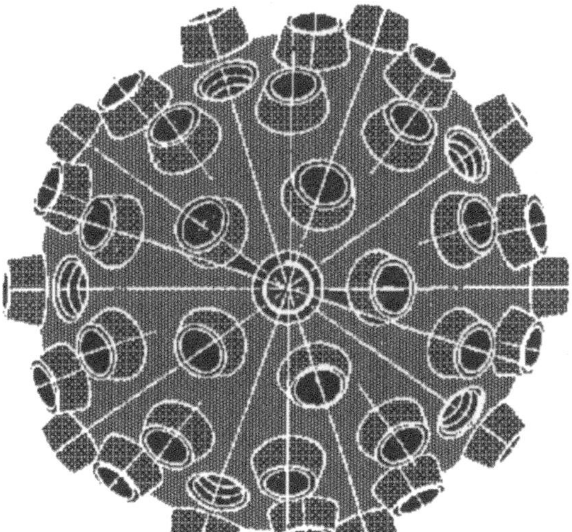

Figure 183: The WESTPAC-1 S/C model

The WESTPAC satellite is spherical in shape (24.5 cm in diameter, mass of 23.8 kg) with 60 laser reflectors mounted and evenly spaced to its outer surface (each separated by 26°). The retroreflector design utilizes the Fizeau effect to correct for velocity aberration effects instead of normal dihedral angle spoiling techniques. The Fizeau corner-cubes have a re-

1077) Information provided by C. Smith of EOS Pty Limited, Canberra, Australia

1078) Armand Fizeau (1819 - 1896) is a French physicist noted for the first experimental determination of the speed of light in 1849. He used a beam of light reflected from a mirror 8 km away. The beam passed through the gaps between the teeth of a rapidly rotating wheel. The speed of the wheel was increased until the returning light passed through the next gap and could be seen. Then "c" was calculated to be 315,000 km/s. Leon Foucault improved on this a year later by using rotating mirrors and got the much more accurate answer of 298,000 km/s. Fizeau's technique was good enough to confirm that light travels slower in water than in air.

fractive index of 1.62, they were built by ISDE. Each cube is recessed so that the average 0.7 cubes retroreflect on any given shot. The prism mirrors reflect any incident laser light back in the incoming direction, irrespective of the incident angle. The distinguishing features of the WESTPAC design include:

- Only a single corner-cube reflects on any shot; in fact on average only 0.7 cubes are active
- Its response is optimized for 1.54 μm wavelength, to provide for fully eyesafe ranging at any power
- A new process was developed to obtain the center-of-mass correction within 0.5 mm accuracy (the retroreflectors are positioned 91.0 mm from the center of mass)
- The corner-cube design and material assumes the Fizeau Effect which, if real, will decrease return signal levels dramatically at 532 mm wavelength.

To minimize size variations due to temperature changes as the satellite travels from sun to shade during its orbit, WESTPAC is covered in a special thermal stabilizing white coating.

Orbit: Sun-synchronous circular orbit, altitude = 835 km, inclination = 98°.

Part F Meteorology - GEO (Geosynchronous Earth Orbit) Missions

F.1 Elektro-M-1(Elektro-Modified-1)

Elektro-M-1 is the planned geostationary meteorological mission of Rosaviakosmos, a successor spacecraft to GOMS (also known as Elektro-1). The overall mission objectives are:

- To provide multispectral imagery (hydro-meteorological data) of the atmosphere (including the cloud-covered sky) and of the Earth's surface within the coverage region (visible disk) of the spacecraft
- To collect heliospheric, ionospheric, and magnetospheric data
- To provide the needed communication services for the transmission/exchange of all data with the ground segment
- To provide the services of data collection for the DCPs (Data Collection Platforms) in the ground segment.

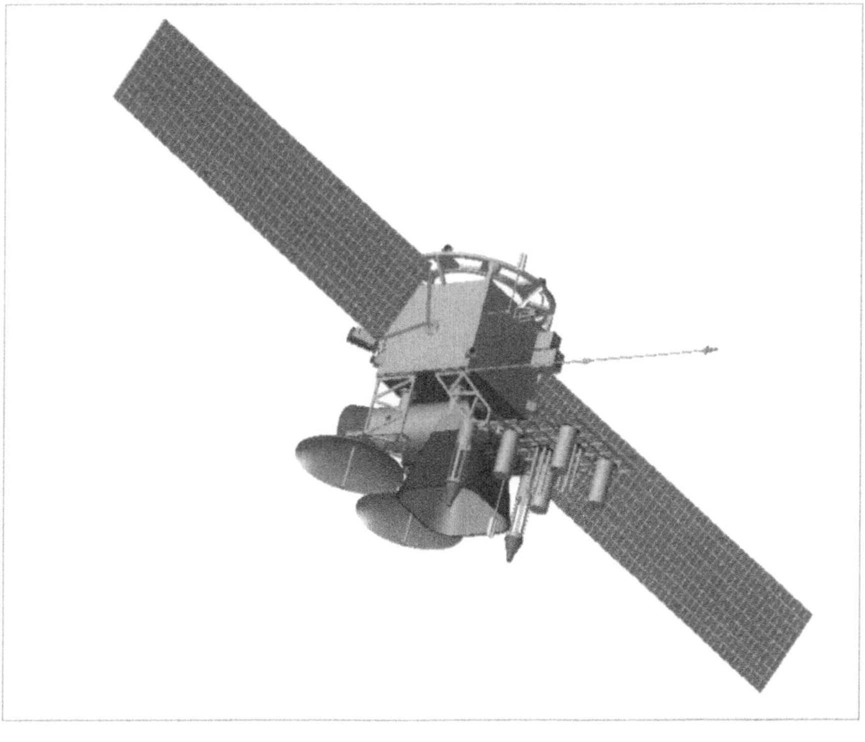

Figure 184: Illustration of the Electro-M-1 spacecraft

The Electro-M-1 spacecraft is being built by NPO Lavochkin (Chimki) of Moscow. The S/C is three-axis stabilized. A pointing accuracy of $<0.05^{\circ}$ is provided; the angular drift is $<5 \times 10^{-4}$ °/s. Two deployable solar arrays provide a power of 1 kW (the mean power consumption of the S/C is about 700 W). The total total mass of the spacecraft is 870 kg. The S/C design life is =>10 years. A launch of Electro-M-1 is planned for 2006.

Orbit: geostationary orbit at the location of 76° East (over the Indian Ocean).

BRTK (On-board RF Communications System). The objectives are to provide all data transmission, relay, and retransmission services with the ground segment. These consist of the following functions:

- The sensor data downlink to CPPI (Ground Acquisition and Distribution Center) is in X-band (7.5 GHz) at a data rate of 2.56-15.36 Mbit/s
- Data reception from ground segment DCPs at 400 MHz (UHF), or DCP data relayed via LEO satellites at a frequency of 470 MHz. This data is transmitted from Elektro-M-1 to the CPPI in S-band at 1.7 GHz.
- On-board reception of processed hydro-meteorological data products in X-band (8.2 GHz) and relay of this data in HRIT LRIT and WEFAX formats (in S-band at 1.7 GHz) to all customers
- Exchange of hydro-meteorological data and remote sensing data between regional centers in X-band (at 8.2 and 7.5 GHz) with data rates of up to 15.36 Mbit/s
- Data reception of COSPAS-S&RSAT messages at 406 MHz and retransmission of these messages at 1.54 GHz.

Sensor complement: [1079)]

Multispectral radiometer. The objectives are to obtain solar reflected imagery and brightness temperature measurements from TOA (Top of Atmosphere) and of the Earth's surface (ocean and land). In addition, the tropospheric moisture content is determined. An image of a sub-region of the disk of size yy km x zz km can be obtained in parallel. The instrument data rate is 2.56 - 15.36 Mbit/s.

Band name	Spectral range	Spatial resolution	Comment
VNIR	0.5-0.8 μm (2 channels)	1 km	Cloud cover (solar reflective band)
MWIR1	3 - 4 μm	1 km	Nighttime and cloud cover
MWIR2	5.7 - 7 μm	4 km	Water vapor
TIR1	10.2 - 11.2 μm	4 km	SST (Sea Surface Temperature) and water vapor
TIR2	11.2 - 12.5 μm	4 km	SST and water vapor

Table 289: Some radiometer instrument specifications

RMS (Radiation Measurement System). The objective of this instrument suite is to obtain particles (particle count of protons, electrons and alpha particles), to measure the sun's X-ray radiation, the solar constant, and to measure the magnetic field components.

The translation of Boris Zhukov gives simply a few measurement ranges. I do want a description of the various instruments, including their names.

Registration of:
Electrons: 0.05–50 keV, 0.15–3.2 MeV
Protons: 0.05–50 keV, 5–40 MeV, >25 MeV, >90 MeV, >600 MeV
X–rays: 3–10 keV
Solar UV radiation at 121.6 nm,
Magnetic field – <300 nT

Solar constant instrument.

- Instrument1:
- Instrument2:
- Instrument3:

Data Collection System:

1079) The scant information provided by the Russian contact doesn't permit a real description of the project

F.2 Feng-Yun-2 (Geostationary Satellite Series)

Feng-Yun-2, or FY-2, is a geostationary meteorological satellite series of the Peoples Republic of China (PRC), organized and operated by NSMC (National Satellite Meteorological Center) of CMA (China Meteorological Administration) and built by the Shanghai Institute of Satellite Engineering. The S/C is spin-stabilized at 100 rpm with a despun antenna platform (similar in design and operation to the GMS series of Japan). The FY-2 spacecraft bus is 2.1 m in diameter with a height of about 1.6 m, S/C on-orbit mass = 593 kg. Power (280 W) is generated by surface-mounted solar cells. The observation objectives are:

- Acquiring daylight visible cloud maps, day-and-night infrared cloud maps and water vapor distribution maps
- Data collection from meteorological, oceanic and hydrological observational platforms
- Broadcasting of stretched digital cloud maps, low resolution cloud maps and weather map information
- Obtaining cloud top and sea surface temperatures as well as the wind field distribution by data processing.

Note: [1080] The first FY-2 satellite series, FY-2A, was undergoing final check-out on 2 April 1994, before being mated to its launch vehicle, when a fire and explosion erupted, destroying the vehicle.

Figure 185: The FY-2 S/C model

F.2.1 FY-2A

A second identical satellite, also referred to as FY-2A, was built by the Shanghai Institute of Satellite Engineering and launched by a Long March 3 (CZ-3) booster from Xichang on June 10, 1997. The first imagery of S-VISSR was received on July 13, 1997. FY-2A data can be received by all international users.

1080) A. Lawler, "Blast at Xichang Derails China's Weather Effort", Space News, May 2-8, 1994, p. 3

On 8 April 1998, FY-2A ceased transmission of S-VISSR images, to resume for a few hours on 10 April, due to a problem with the S-band antenna on the spacecraft. CMA managed to regain Earth lock and ranging with the S-band antenna for 12 hours on 14 April 1998, but there were no regular S-VISSR re-transmissions during the summer of 1998. However, CMA could still get raw image downlinks a few times per day, and they would make one full-earth picture on the ground each day, but not broadcast it. As of October 1998, CMA resumed a partial but regular communication service of six transmissions per day. [1081)]

Orbit: Geosynchronous orbit at position: 105° eastern longitude.

Sensor:

S-VISSR (Stretched - Visible and Infrared Spin-Scan Radiometer), an optomechanical system. Scanning modes: a) normal scanning; b) optional scanning; c) single line scanning. A picture frame (normal scanning) = 30 minutes, a scan (North-South) = 2500 steps. The telescope collects radiation during each scan and focuses it on the detectors in the focal plane using primary and secondary mirrors. The Si photodiode detector array has four elements mounted normal to the scan direction, providing a 5 km wide instantaneous observation coverage on each scan line, matching the scan line width of the IR detectors with only one detector element. The HgCdTe detectors are cooled by radiation coolers to a temperature of 100 K. A complete scan of 20° x 20°, covering the full Earth disk, is obtained every 30 minutes by means of the S/C spin motion (100 rpm from E-W) and the step action (2500 steps from north to south) of the scan mirror.

Parameter	Visible	Infrared	Water vapor
Spectral band (µm)	0.55 - 1.05	10.5 - 12.5	6.2 - 7.6
No. of detectors/channels	4 (+4 spare)	1 (+1 spare)	1 (+1 spare)
Scan lines	2500 x 4	2500	2500
Spatial resolution	1.25 km x 1.25 km IFOV=35 µrad	5 km x 5 km IFOV=140 µrad	5 km x 5 km IFOV = 140 µrad
Detectors	Si photodiode	HgCdTe	HgCdTe
Data quantization	6 bits	8 bits	8 bits
Noise performance	SNR=6.5 (albedo=2.5%) SNR= 43 (albedo=95%)	NEDT=0.5-0.65 K (300 K)	NEDT= 1 K (300 K)
Scan step increment	140 µrad (N-S scanning)		
Temporal resolution	30 minutes (frame time)		

Table 290: Some characteristics of S-VISSR

The FY-2 ground segment (geostationary satellites) consists of the following facilities: 1) CDAS (Command and Data Acquisition Station); 2) DPC (Data Processing Center); 3) SOCC (Satellite Operations and Control Center); 4) ranging stations (one primary and three secondary stations, including one in Australia); 5) DCPs (Data Collection Platforms); 6) MDUS (Medium-scale Data Utilization Station) and SDUS (Small-scale Data Utilization Station) stations.

Channel	Frequency	Modulation	Power	Bandwidth
S-VISSR	1687.5 MHz	PCM/BPSK, NRZ-M	57 dBm	2 MHz
LR-Wefax	1691.0 MHz	AM/FM	57 dBm	260 kHz
S-Fax	1699.5 MHz	AM/FM	46 dBm	26 kHz

Table 291: Some downlink characteristics of FY-2A

FY-2 series communication: The FY-2 satellites broadcast their data to the ground, this includes S-VISSR, WEFAX and S-FAX data. S-VISSR data are received at MDUS, while

1081) http://202.106.103.181/s-vissr.htm

WEFAX and S-Fax (domestic use only) data are retransmitted to SDUS stations only. The S-FAX link transmits processed satellite image data, other weather data and administrative information via S-band to domestic users in China.

Digital and cloud map transponder: Frequency: 1.7/2.0 GHz; EIRP (dB_W) original cloud map and stretched cloud map: 57.5; EIRP (dB_W) weather map broadcast: 46. The data rate of S-VISSR is 660 kbit/s. Since the signal characteristics of the S-VISSR instrument data are the same as those of the VISSR instrument of the Japanese GMS series (except the frequency), the user stations receiving GMS VISSR data are also capable of receiving S-VISSR data, simply by changing the antenna pointing and frequency of the receiver local oscillator.

Data Collection System (DCS) of the FY-2 Series

The DCS provides operational collection services for data collection platforms (DCPs) within its field of view. The overall system is composed of the following subsystems:

- The space segment: a DCS platform on the FY-2 satellite consisting of: a UHF/S-band transponder, a mechanically despun S-band antenna, and a UHF receive antenna. Data collection transponder:Frequency = 401/468 MHz; EIRP = 47.3 dB_W; No. of channels: 133, 100 of these channels are for domestic use and 33 international channels. DCS offers a capability and a new digital S-band fax service (CCITT G3) for domestic distribution of charts and imagery.
- The deployed DCPs (Data Collection Platforms) in the ground segment. The regional DCPs are stationary DCPs installed on buoys, islands, rivers, mountains or ships.
- The ground-based NSMC (National Satellite Meteorological Center) which collects and processes the DCP data and distributes it via GTS to the user community.

F.2.2 FY-2B

The second Chinese GEO satellite FY-2B was launched on June 26, 2000 from the Xichang launch center with the Long March 3 vehicle. [1082)]

The FY-2B S/C is identical to the previous FY-2A satellite and carries the same payload (S-VISSR, and DCS). The S/C is spin-stabilized with a spin rate of 100 rpm. The on-orbit dry mass is 600 kg, the launch mass is 1200 kg. The S/C design life is 3 years.

The daytime visible image can be used to derive the reflectance of cloud top and the earth surface, the nighttime infrared image can observe the thermal radiance, while the water vapor image can estimate the water vapor amount in the atmosphere. The satellite also broadcasts low resolution WEFAX image and collects data from automatic data collection platforms. Free transmission of S-VISSR data and WEFAX imagery are open to all users within the FY-2B transmission coverage.

Orbit: Geosynchronous orbit at position: 105° eastern longitude.

Channel	FY-2, A, B	FY-2, C, D, E
VIS (Visible)	0.50 - 1.05 µm	0.50 - 0.75 µm
IR1 (Infrared 1)	10.5 - 12.5 µm	10.5 - 11.5 µm
IR2		11.5 - 12.5 µm
IR3		3.5 - 4.0 µm
WV (Water Vapor)	6.3 - 7.6 µm	6.3 - 7.6 µm

Table 292: Spectral layout of the VISSR instrument beyond FY-2B

Current plans call for VISSR instrument improvements beyond FY-2B, featuring five channels instead of three. In addition, the power supply of each S/C will be increased to accommodate the demand.

1082) http://202.106.103.181/

F.3 GMS (Geostationary Meteorological Satellite)

GMS = Geostationary Meteorological Satellite of JMA (Japan Meteorological Agency) and NASDA. GMS is also known by the name of Himawari. The GMS series satellites represent a link in World Weather Watch (WWW) sponsored by the World Meteorological Organization (WMO). The operational meteorological program has the following satellites:

GMS-1 F1	Launch: July 14, 1977	NASA Delta 2914 vehicle	Position: de-orbit
GMS-2 F2	Launch: Aug. 11, 1981	N-II vehicle of NASDA	Position: de-orbit
GMS-3	Launch: Aug. 3, 1984	N-II vehicle of NASDA	Position: 120° East
GMS-4	Launch: Sept. 6, 1989 Operations until Feb. 2000	H-I vehicle of NASDA	Position: 140° East
GMS-5	Launch: Mar. 18, 1995	H-II vehicle, NASDA	Position: 140° East

Table 293: GMS program overview

The mission objectives of the GMS program are:

1) Weather watch by VISSR: a) imaging the Earth's surface, cloud distribution and atmospheric water vapor distribution, and observation of meteorological phenomena such as typhoons, cyclones, fronts, etc. b) meteorological parameter extraction such as temperature on both Earth surface and cloud tops, cloud top height, cloud amount, cloud motion, etc.

2) Collection of meteorological data from DCPs (Data Collection Platforms) installed on ships, buoys, aircraft, and other land weather stations.

3) Direct broadcast of cloud images: a) dissemination of digital image data (stretched VISSR data) via satellite repeater to MDUS (Medium Scale Data Utilization Station) users. b) dissemination of analog processed image data (WEFAX data) via satellite to SDUS (Small Scale Data Utilization Station) users.

4) Monitoring of solar particles

GMS-4 is a spin-stabilized satellite (100 rpm) weighing 725 kg (satellite diameter = 214.6 cm, height = 345 cm). The S/C consists of a despun Earth-oriented antenna assembly and a spinning section rotating at 100 rpm. Despun S-band and UHF antennas provide high gain for on-orbit communications with ground stations.[1083)]
Attitude information is provided by Earth horizon and sun sensors. The spacecraft transmits its data in L-band (1.7 GHz) to JMA (Japan Meteorological Agency).

Sensor complement:

- **VISSR** (Visible and Infrared Spin-Scan Radiometer), built by Hughes SBRC. Spectral bands: 0.50 - 0.75 μm (visible), and 10.5 - 12.5 μm (infrared). Resolution: 1.25 km (visible) and 5 km (infrared). The VISSR is used to obtain visible and infrared spectrum mappings of the Earth and its cloud cover. Four photomultiplier tube detectors (PMTs) and redundant sets convert the visible spectra into four-channel analog signals. A cooled (95 K) HgCdTe detector (also redundant) converts infrared spectra into an infrared signal (10.5 – 12.5 microns) at 5 km resolution. The resulting data stream is then fed into the VISSR Digital Multiplexer (VDM) unit. A complete 20° x 20° scan area to cover the full Earth disk can be accomplished in 30 minutes utilizing the GMS's 100 rpm spin motion for west-east scans and 2500 steps (140 μrad each) of motor-actuated VISSR mirror scan for north-south scans. Imaging takes 25 minutes, mirror retrace another 2.5 minutes, and a spacecraft stabilization 2.5 minutes.

1083) M. Homma, M. Minowa, M. Kobayashi, M. Harada, "Geostationary Meteorological Satellite System in Japan" in 'Monitoring Earth's Ocean, Land, and Atmosphere from Space,' Volume 97 AIAA, 1985, pp. 570-583

VDM (VISSR Digital Multiplexer). VDM is a high-speed pulse code modulation (PCM) encoder. It converts VISSR signals into digital form and multiplexes them with image and message synchronization data.

- **SEM** (Space Environment Monitor, NOAA sensor). Spectral bands: 1 - 500 MeV (protons); 8 - 390 MeV (alpha particles); >2 MeV (electrons). SEM measures three kinds of solar particles: protons, alpha particles, and electrons.
- **DCS** (Data Collection System) from ground platforms (see METEOSAT and NOAA-GOES)

F.3.1 GMS Data Collection System (DCS)

The GMS DCS provides operational collection services for data collection platforms within its field of view. The overall system is composed of the following subsystems:[1084]

1) The space segment: a DCS platform on GMS satellites consisting of: a UHF/S-band transponder, a mechanically despun S-band antenna, and a UHF receive antenna.
2) The deployed DCPs (Data Collection Platforms) in the ground segment
3) The ground-based MSC (Meteorological Satellite Center)

The DCS on-board GMS provides a transponder with a passband that is divided into 133 communication channels separated at intervals of 3 kHz. The system has the ability of simultaneous communication through all channels. The 133 channels are divided into two blocks: 100 are for regional use, and 33 are for international use. Each DCP is allocated one of the 133 parallel channels. Since the DCPs transmit either according to a schedule, or under 'alert' conditions, many platforms can use a single 3 kHz channel sequentially. A message from a DCP in the ground segment is relayed via the GMS on-board DCS to a ground-based MSC which is responsible for the coverage area. - The platforms (DCPs) transmit (uplink) their data in the 402 MHz UHF band to the GMS on-board DCS, where it is converted to S-band (1694 MHz) and retransmitted (downlinked) to a ground facility CDAS (Command and Data Acquisition Station) and relayed via GTS to MSC. There, the DCP messages are collected, processed, disseminated, formatted into bulletins and distributed via GTS to the platform owners.

As of the end of 1995, 448 DCPs in the ground segment are regularly serviced by GMS. The DCPs which can be operated by the system are of two types:

- self-timed DCPs which transmit their data automatically at preset time slots driven by an internal stable clock
- interrogational DCPs which transmit their data only when an interrogation command is received at the DCP.

The GMS DCS is part of an international network in the framework of WWW (together with Meteosat and GOES series satellites); this allows a further classification:

- International DCPs (IDCPs) which are mobile (Lagrangian type free-floating in the medium) and whose data may be collected by any one of the satellite operators. The allocated frequency band for the IDCP operation is 402.0-402.1 MHz for the uplink and 1694.3-1694.4 MHz for the downlink.
- Regional DCPs (RDCPs) which are under the responsibility of and operated by a specific satellite operator. The allocated frequency band for the RDCP operation is 402.1 - 402.4 MHz for the uplink and 1694.4 - 1694.7 MHz for the downlink.

The data transmission rate for both types of platforms is 100 bit/s.

1084) Information provided by T. Hiraki of JMA, Tokyo

Communication Link	UHF (Uplink to GMS DCS)	S-band (DCS downlink)
Frequency	402.0-402.4 MHz	1694.3-1694.7 MHz
Type of modulation	PCM/PSK (±60°)	PCM/PSK (±60°)
EIRP	43.0 dBm	11.3 dBm
Free space loss	176.7 dBm	188.3 dBm
RCVR off-beam loss	0.4 dB	0.6 dB
RCVR input power level	-134.1 dBm	-177.6 dBm
RCVR system noise temperature G/T	-19.0 dB/K	27.2 dB/K
RCVR C/T	-153.1 dB/kHz	-150.4 dB/kHz

Table 294: DCP link budget for GMS operation

F.4 GOES (Geostationary Operational Environmental Satellite)

A NASA/NOAA weather satellite series. NASA, under contract from NOAA, manages the program acquisition (design, development and launch of the satellite); NOAA owns and operates the satellites and provides the services to the user community. All GOES first and second-generation satellites have an operational design life of nominally 5 years.

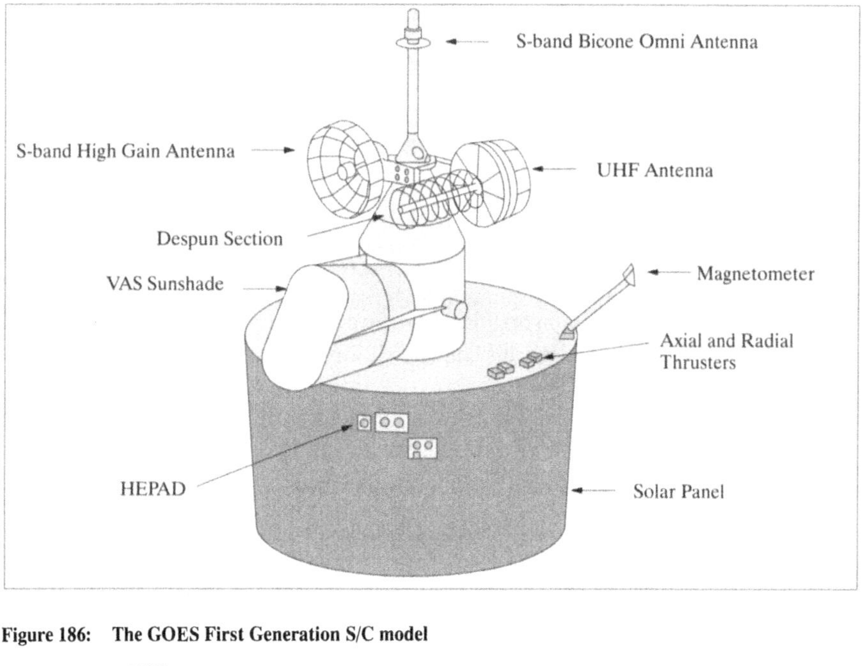

Figure 186: The GOES First Generation S/C model

Background:[1085] ATS-1 (Applications Technology Satellite) was the first US geostationary satellite (launched December 6, 1966) with the ability to "see weather systems" by imaging the full-disk Earth every half hour with a spin scan camera. Operational use of ATS-3 (launched November 6, 1967) imagery at NSSFC (National Severe Storm Forecast Center) and at NHC (National Hurricane Center) followed in 1972. A total of six spacecraft were part of the ATS series. The prime objective of the series was to test new technologies (spin stabilization, gravity gradient stabilization, demonstration of data collection from remote terminals, etc.) for communications and to investigate the geostationary orbit environment.

1085) W. P. Menzel, J. F. W. Purdom, "Introducing GOES-I: The first of a Generation of new Geostationary Operational Environmental Satellites," Bulletin of the American Meteorological Society, Vol. 75 No. 5, May 1994, pp. 757-781

NASA developed two Synchronous Meteorological Satellites (SMS-1, -2), which were launched on May 17, 1974 and February 6, 1975. A total of eight additional "GOES" satellites in the same series were funded by NOAA, the first being GOES-1 launched on October 16, 1975. The last satellite in this initial series (1st generation) was GOES-7.

Spacecraft	Launch	Comment
SMS-1 (SMS-A)	May 17, 1974	First operational geostationary weather satellite, S/C mass = 627 kg, full-disk images of western hemisphere every 30 min.
SMS-2 (SAMS-B)	Feb. 6, 1975	
GOES-1 (GOES-A)	Oct. 16, 1975	S/C mass of 293 kg.
GOES-2 (GOES-B)	Jun. 16, 1977	
GOES-3 (GOES-C)	Jun. 16, 1978	
GOES-4 (GOES-D)	Sept. 9, 1980	GOES-4 VAS failed in Nov. 1982. The VAS instrument of GOES-4 was able to provide the first vertical temp. and moisture measurements from GEO
GOES-5 (GOES-E)	May 22, 1981	
GOES-6 (GOES-F)	Apr. 28, 1983	
GOES-G	May 3, 1986	Launch failure
GOES-7 (GOES-H)	Feb. 26,1987	First GOES with an S&RSAT system flown
GOES-8 (GOES-I)	Apr. 13, 1994	First second generation S/C (3-axis stabilized); S&RSAT became operational on GOES-8
GOES-9 (GOES-J)	May 23, 1995	
GOES-10 (GOES-K)	April 25, 1997	In the initial mission, the S/C suffered a solar array malfunction. A recovery procedure was performed to invert the S/C and operate the solar array in the reverse direction. The S/C became operational in July 1998. It replaces GOES-9 which had problems with its attitude control system. GOES-10 is positioned at 135° W.
GOES-11 (GOES-L)	May 3, 2000	GOES-11 was launched into a storage orbit
GOES-12 (GOES-M)	July 23, 2001	GOES-M is the last of the 2nd generation S/C built by SS/L
GOES-N	2003	
GOES-O	2005	
GOES-P	2007	
GOES-Q	2010	
GOES-R	2012	

Table 295: GOES S/C launches [1086)]

The GOES system serves a region covering the central and eastern Pacific Ocean, North, Central, and South America, and the central and western Atlantic Ocean to the coast of Africa. Pacific coverage includes Hawaii and the Gulf of Alaska. This is accomplished by a two operational satellite configuration, GOES-East (75° West) and GOES-West (135° West). A common ground station at Wallops, Virginia, supports the interface to both satellites. The NOAA/NESDIS SOCC (Satellite Operations Control Center) in Suitland Maryland, provides spacecraft operations and all services to the user community. The GOES ground segment also uses seven satellite field service stations, located at: Anchorage, Honolulu, Kansas City, Miami, New Orleans, San Francisco, and Washington, D.C.

Delivery of products involves ground processing of the raw instrument data for radiometric calibration and Earth location information, and retransmission to the satellite (GOES) for relay to the data user community. The processed data are received by the National Weather Service (NWS, at Camp Springs, MD) and the NWS forecast offices, including the National

1086) Note: For a number of years the designation GOES Next (N) was used to identify the first of the satellites that would follow the GOES I-M series. It was also thought this GOES N would be the start of a totally new generation satellite series. However, during the time frame of about 1992-95, NOAA has come to realize that a new satellite series would take at least a decade to develop, manufacture, and launch. This new situation made NOAA realize it would need a few more clones of the current GOES I-M series to maintain continuity of GOES service prior to the GOES Next being available. What has evolved is a program that will likely buy 3-4 additional GOES I-M satellites beyond the GOES I-M series. These would then be labeled GOES N through Q. GOES R would be the first of a new generation of three-axis stabilized satellites. (Information: R. Heymann of NOAA).

Hurricane Center at Miami, FL, and the National Severe Storms Forecast Center at Kansas City, Missouri. Processed SOCC data are also received by DoD installations, universities, and numerous private commercial users.

Note: The GOES-I-M series[1087)] is a change from the spinning S/C design of GOES-7 and before to a three-axis stabilized S/C to increase the observational dwell times and to achieve other improvements.

Application: Weather forecast, meteorology, derivation of temperature and moisture profiles, monitoring severe storms and tropical cyclones (hurricanes), etc.

According to NOAA convention all satellites are designated with a letter prior to achieving orbit. The letter name is converted into a number on achieving orbit. For instance, GOES-J, launched on May 23, 1995, was redesignated as GOES-9 on May 31, 1995 on achieving geosynchronous orbit.

Sensor complement of first-generation S/C series:

VISSR = Visible Infrared Spin Scan Radiometer (built by Hughes SBRC). Measurement (scanning) in eight channels in the visible range and two channels in the infrared spectrum. VISSR acquires imagery as it spins west-to-east at 100 rpm aboard a spin-stabilized S/C, step scanning north to south on each spin to provide cloud imagery and data for determining cloud and surface temperatures and wind fields. The instrument uses a Ritchey-Chretien telescope of 40.6 cm diameter, f/7.2; spectral bands: 0.55 - 0.72 μm (VIS) and 10.5 - 12.6 μm (TIR); detectors: S-20 (PMT (8), HgCdTe (2 at 90 K); resolution = 0.9 km in VIS and 7 km in TIR. Instrument mass = 63.2 kg for the scanner module and 7.0 kg for the electronics module. The VISSR instrument has been flown on five US satellites (SMS-1, -2, GOES-1 through -3) and five GMS (NASDA/JMA) satellites (GMS-1 through -5).

VAS= VISSR Atmospheric Sounder (built by SBRC). [1088)] [1089)] [1090)] This VISSR successor instrument has been flown on GOES-4 through -7). VAS is a radiometer with eight visible channel detectors and six thermal detectors that detect IR radiation in 12 spectral bands (two of the detectors are used primarily for imaging and four for atmospheric sounding). The instrument uses a Ritchey-Chretien telescope of 40.6 cm diameter, f/7.2 (VIS), f/1.0 (IR); detectors: S-20 (PMT (8), InSb; HgCdTe; spectral bands: 0.55 to 0.70 μm (VIS) and 12 channels in IR in μm [1) 14.734, 2) 14.480, 3) 13.984, 4) 13.984, 5) 1.333, 6) 4.525, 7) 12.658, 8) 11.173, 9) 7.261, 10) 6.725, 11) 4.484, 12) 3.945];the spatial resolution is 0.9 km in the VIS range and 7-14 km in the IR range. Instrument mass = 68 kg. Full Earth-disk coverage is accomplished by spinning in the west to east direction at 100 rpm. The VAS sensor has three operational modes:

1) The VISSR operating mode (same as that of the original VISSR)
2) The multispectral imaging (MSI) operating mode. Measurement of relatively frequent (eg., half hourly) full Earth-disk imagery of atmospheric water vapor, temperature, and cloud distribution, as well as variations in the surface skin temperature of the Earth.
3) In the dwell sounding mode (DS) up to 12 spectral filters can be positioned in sequence into the optical train while the scanner dwells on a single N-S scan line. This allows data to be obtained on the Earth's atmospheric water vapor and CO_2 bands. Note: the DS mode cannot be used simultaneously with either the VISSR or MSI mode.

DCS = Data Collection System. This on-board system gathers and relays environmental data transmissions that originate at remote automatic Data Collection Platforms (DCPs)

1087) E. P. Mercanti, "Need for Expanded Environmental Measurement Capabilities in Geosynchronous Earth Orbit," Proceedings of the Twenty-Fourth International Symposium on Remote Sensing of the Environment,' ERIM, Volume I, pp. 45-55

1088) R. Koffler, L. Spayd, "30 Years of Operational Environmental Satellites: A Retrospective and Future View of the United States Program," presented at the Twenty-Third International Symposium on Remote Sensing of the Environment, Bangkok, Thailand, April 18-25, 1990, pp. 95-97

1089) J.R. Greaves, W.E. Shenk, 'The Development of the Geosynchronous Weather Satellite System,' in Monitoring Earth's Ocean, Land, and Atmosphere from Space, Volume 97, 1985, pp. 150-181

1090) Space Sensors, brochure of Hughes Santa Barbara Research Center (SBRC), January 1994 for obtaining in situ data. These platforms can be set to transmit data in four modes: at fixed times; in response to interrogation; in an emergency mode during which the platform signals for a GOES interrogation within 1 minute; and in an adaptive random reporting mode. The DCS collected data are retransmitted from satellite to small ground-based regional data utilization centers. The system also allows for the retransmission of narrow-band **WEFAX** (Weather Facsimile) data to existing small, ground-based APT (Automatic Picture Transmission) receiving stations from a larger facility.

Background: The WEFAX service has been a feature of the GOES system since 1975. The service uses the S/C as a transponder to transmit low-resolution imagery sectors as well as conventional weather maps to users with low-cost reception equipment. There are literally hundreds of WEFAX reception stations that use the GOES WEFAX service. With the GOES VAS system, the WEFAX service was provided at UHF frequencies and was not continuous since it interfered with other S/C functions. The WEFAX service with GOES I-M is a continuation of the previous service, however, transmissions are continuous, and the transmission frequency has changed from UHF to S-band at 1691.0 MHz, with a signal bandwidth of about 30 kbit/s. Up to 13 maps or images can be transmitted through the GOES I-M system every half hour.

F.4.1 NOAA-GOES Data Collection System (DCS)

The Data Collection System (DCS)[1091),1092),1093)] on NOAA-GOES satellites provides an operational data collection service for a large number of land-, air- and sea-based data collection platforms (DCPs). Geostationary satellites (like GOES) offer the advantage of continuous coverage; however, their coverage is not global, but limited to the GOES service region. The overall system is composed of the following subsystems:

1) The space segment: a DCS platform on two operational GOES satellites (GOES-East, and GOES-West). Each GOES S/C is equipped with two DCS transponders, one active and one as backup.
2) The deployed Data Collection Platforms (DCPs) in the ground segment
3) The NESDIS DCS ground receive system CDA (Command and Data Acquisition Station) at Wallops Island, Virginia.
4) The DAPS (Data Collection System Automatics Processing System) at CDA.

The Space Segment: DCS

The DCS on-board GOES uses the GOES S/C for the relay of data from remotely located in-situ sites at or near the Earth's surface to properly equipped receiving stations in radio view of the GOES satellites. Each GOES provides an RF link between the DCP and CDA. They up-convert DCP data from UHF (401.9 MHz) to S-band (1694.5 MHz) for transmission to CDA, and down-convert the CDA -DCP interrogate signal from S-band (2034.9 MHz) to UHF (468.8 MHz).

The DCS uses 400 kHz of satellite transponder bandwidth. This bandwidth is subdivided (by frequency division multiplexing) into 200 domestic channels with a 1.5 kHz channel separation (channels 1-200), and 33 international channels (channels 202 - 266, even numbered only) with a 3 kHz channel separation. The 33 international channels are common to those of METEOSAT, GMS and GOMS spacecraft.

The DCS has the capacity for handling at least 25,320 messages from DCP sites via the spacecraft transponder in each one-hour period. This figure is based on the present capabil-

1091) "The Geostationary Operational Satellite Data Collection System," NOAA Technical Memorandum NESDIS 2, June 1983

1092) "Users Guide for Random Reporting - An Introduction to GOES Random Reporting Services," NOAA, April 1985

1093) User Interface Manual, Version 1.1, for the 'Data Collection System Automatic Processing System (DAPS),' Integral Systems Inc., Sept. 1990 ity to assign 30-second transmission windows for each of the 200 domestic channels, i.e., 120 windows per hour over 200 channels (120 x 200 = 24,000), plus 33 international channels (33 x 40 = 1320), 1320+24,000=25320. - With advances being made in the stability and accuracy of timing oscillators, transmission windows of 15 seconds may become standard in the future. The data transmission rate for all operational modes is currently (1995) 100 bit/s. Prototype 300/1200 bit/s DCPs are being developed and should become operational by 1996.

Data Collection Platforms in the Ground Segment

All DCPs serviced by GOES must be type-certified by NOAA-NESDIS. The types available to the user community have the following functional capabilities:

- Self-timed DCPs (this is the most common mode of use)
- Self-timed and random reporting DCPs
- Random reporting DCPs
- Interrogated DCPs
- Self-timed and interrogated DCPs
- Random and interrogated DCPs

1) Self-timed DCPs. - These are simple platforms containing a transmitter and a pre-programmed self-timer to report at a specific time and intervals to the GOES. The sensor measuring cycle is independent of the DCP reporting cycle. Sensor data may be recorded at the DCP prior to a message transmission to GOES. The reporting frequency from a DCP may vary from a few minutes to a few hours.

2) Self-timed and random reporting DCPs. - Same functionality of 1) with the addition of capability of transmitting over a secondary channel when environmental conditions require more frequent reporting than offered under self-timed operation.

3) Random reporting DCPs. - The platform contains a transmitter that broadcasts at random time (threshold reporting, set by the user). During normal periods a random reporting DCP is expected to report up to three times per day that it is properly functioning.

4) Interrogated DCPs. - Scheduled transmissions initiated from the CDA at NESDIS (the schedule is stored at CDA). This type of DCP contains a receiver and transmitter. The receiver detects its own DCP ID upon message reception from CDA. Upon detection of its ID the DCP transmits all data accumulated since the last reporting sequence. Interrogated DCPs may be interrogated as often as every 5 minutes or as infrequently as once per day. In addition some interrogated DCPs have a second or alert reply channel that may be used for threshold reporting. Interrogated DCPs may be reconfigured by CDA command when requested by the user.

5) Self-timed and interrogated DCPs. - This type of DCP operates in a self-timed mode under normal conditions. When a message is not received at the DCP's normal reporting time, an interrogation address is sent via the DAPS to the DCP and a reply is expected over the interrogate reply channel.

6) Random and interrogated DCPs. - This mode of operation is used to monitor special events (such as seismic events). When the event condition reaches a pre-set level, a random transmission is sent through the GOES S/C. Only the DCP ID (address) is transmitted. When the ID is received by the DCS system at the CDA, an interrogate ID is transmitted through the GOES S/C. The interrogate ID can be used as a notification of the event, to change a DCP's mode of operation, or to cause certain functions to be performed.

NOAA-NESDIS Ground Segment

The NOAA ground system consists of the RF front-end equipment and the DCS computer equipment or DAPS (GOES DCS Automatic Processing System). The messages received from the DCPs are routed to the DAPS, which provides all required processing and dissemination functions including data storage and distribution to the user community.

DCS Access Method

Data transmission from a DCP in the ground segment to the DCS on-board GOES is generally initiated (i.e. self-triggered) by the ground segment DCPs (except for the few interrogated DCPs). All of the 200 channels on-board GOES are in listening mode, being available all the time. The assigned transmission times of the DCPs are actually time intervals (slots) of 1 minute duration.

As of the end of 1993, NOAA service records list 7728 activated DCPs being serviced by GOES S/C, out of a total of 11,927 DCPs that are assigned to the GOES user community at large.[1094] - There are six DCPs of the type 'interrogated' serviced by NOAA/NESDIS and used for seismic and tsunamic warning. In addition there are another 15 DCPs which use the interrogate mode as backup to the self-timed mode and to command these DPCs to perform specific functions under specific conditions.

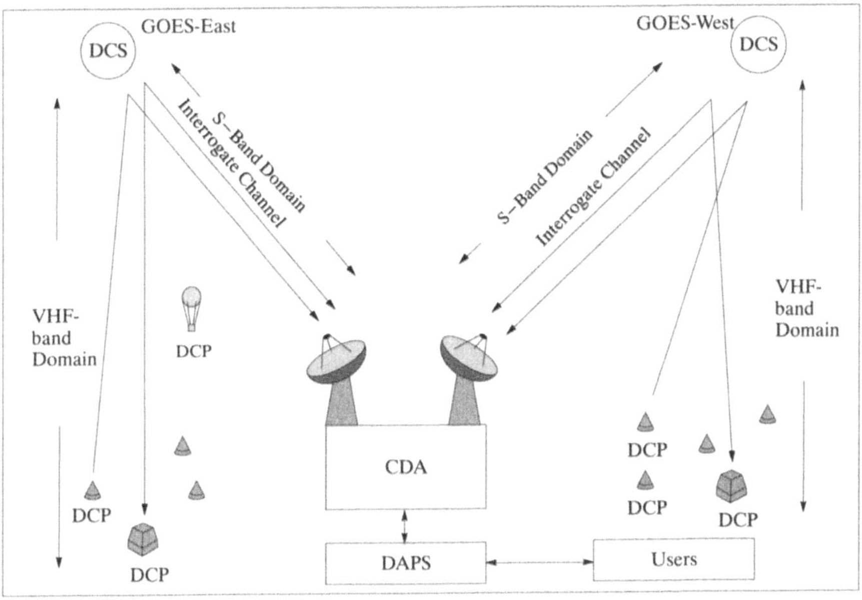

Figure 187: Schematic overview of the GOES Data Collection System

F.4.2 NOAA-GOES SEM Instruments

The **SEM** = Space Environment Monitor instrument package on each of the GOES satellites provides data which are processed by SEL (Space Environment Laboratory, Boulder CO) of NOAA in near real-time for space environment forecasts. The SEM instruments survey the sun and monitor the solar-terrestrial electromagnetic environment that affects a number of Earth activities [operational reliability of ionospheric radio (satellite communications, navigation systems), over-the-horizon radar, electric power transmission, crews on-board of high-altitude aircraft and space stations].

Each SEM package is composed of four separate sensor systems: two energetic particle sensors (EPS, HEPAD), a magnetometer, and a solar soft X-ray sensor. Starting with GOES-L

1094) Information provided by M. J. Nestlebush of NOAA/NESDIS the SEM package will be extended by a fifth instrument, the SXI. Note: the GOES series SEM package differs considerably from the POES series SEM package.

EPS = Energetic Particle Sensor. Objective: monitoring solar protons and alpha particles produced during flares, and continuous monitoring of electrons at geostationary orbit. Measurement in seven channels for high proton fluxes, six channels for alpha particle fluxes, and in three channels for electron flux (note: electron fluxes were not measured from GOES-6 and GOES-7).

HEPAD = High Energy Proton and Alpha Particle Detector. Objective: monitoring the very high energy protons and alpha particles in large solar flares, and continuous monitoring of galactic cosmic rays. HEPAD uses the phenomenon of Cerenkov radiation which occurs when an incident particle travels at speeds greater than the local speed of light in a high refractive index material, in this case fused silica.

Channel Designation	Particle Type	Nominal Energy Range (MeV)	GOES-7 Nominal Energy Range (MeV)
P1	Proton	≤ 0.8 - 4	0.6 - 4.2
P2		4 - 9	4.2 - 8.7
P3		9 - 15	8.7 - 14.5
P4		15 - 40	15 - 44
P5		40 - 80	39 - 82
P6		80 - 165	84 - 200
P7		165 - 500	110 - 500
A1	Alpha	4 - 10	3.8- 9.9
A2		10 - 21	9.9 - 21.3
A3		21 - 60	21.3 - 61.0
A4		60 - 150	60 - 180
A5		150 - 250	160 - 260
A6		300 - 500	330 - 500
E1	Electron	≥ 0.6	
E2		≥ 2.0	≥ 2.0
E3		≥ 4.0	

Table 296: Summary of energetic particle sensor outputs (GOES-I - M)

Channel Designation	Particle Type	Nominal Energy Range (MeV)	GOES-7 Nominal Energy Range (MeV)
P8	Proton	350 - 420	365 - 430
P9		420 - 510	430 - 505
P10		510 - 700	505 - 685
P11		> 700	>685
A7	Alpha	2560 - 3400	
A8		>3400	

Table 297: Summary of HEPAD channel outputs (GOES-I - M)

Magnetometer of SEM package. Objective: measurement of the magnitude and direction of the ambient magnetic field. Three boom-mounted orthogonal sensors for the three components. Sensitivity of 0.2 nT, measurement range of ±1000 nT.

XRS = Solar X-Ray Sensor. Objective: measurement of the solar X-ray spectrum in two broad energy bands: 1-8 and 0.5-4 Å. The X-ray sensor continuously looks into the sun direction. The collimator and ion chamber assembly are mounted on a single axis positioner, which in turn is mounted on the S/C solar yoke.

SXI = Solar X-Ray Imager (starting with GOES-M), developed at MSFC. Objective: provision of full disc solar imagery in the soft X-ray spectrum allowing identification of multiple solar features including solar flares, active regions, coronal holes, etc. SXI uses a grazing incidence mirror to image the sun in the spectral range from 6 - 60 Å. Image resolution: 512 x 512 pixels. Each pixel has a FOV of 5 arc seconds. Data rate: 100 kbit/s. SXI normally produces an image every 10 minutes with an option of producing one image per minute. A filter wheel permits the pass band of the system to be modified to provide additional spectral information.

F.4.3 NOAA-GOES Second Generation

GOES-I-M series sensors: Starting with GOES-I (GOES-8 on-orbit) each satellite carries two separate so-called second generation instruments, an imager and a sounder, providing simultaneous imaging and sounding capabilities. All GOES S/C (starting from GOES-I) are of three-axis, body-stabilized design capable of continuously pointing the optical line of sight of the imaging and sounding radiometers to the Earth. GOES S/C are built and integrated by Space Systems/Loral (SS/L) for NASA/GSFC; the sensors are provided by ITT Aerospace of Fort Wayne, IND. NOAA is the owner and operator of all GOES S/C. [1095) 1096) 1097)]

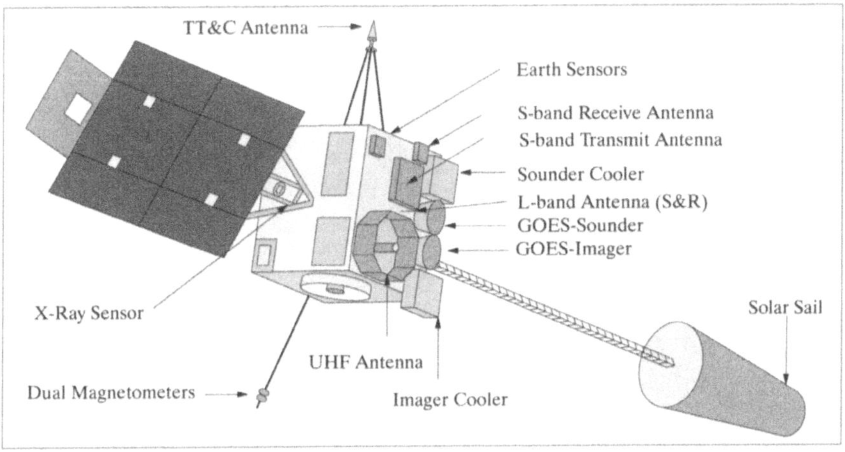

Figure 188: The GOES Second Generation S/C model starting with GOES-I

A single-wing solar array (silicon cells) on the S/C rotates about the satellite pitch axis to track the sun during orbital motion, generating a minimum of 1057 W at summer solstice. A conical solar sail on top of the 17 m boom of the S/C is used to balance the torque caused by the solar radiation pressure (a trim tab on the solar wing provides fine control). [1098)]

Orbit: Geostationary at an altitude of about 35,786 km above the equator. Two satellites in two different locations are operated in parallel. GOES-E is positioned at 75° western longitude, GOES-W at 135° western longitude.

The GOES I-M system performs the following basic functions:

- Acquisition, processing, and dissemination of imaging and sounding data
- Acquisition and dissemination of SEM (Space Environment Monitor) data
- Reception and relay of data from ground-based DCPs (Data Collection Platforms)
- Continuous relay of WEFAX and other data to users
- Relay of distress signals (alerts via S&RSAT) from people, aircraft, or marine vessels to the search & rescue ground stations.

1095) J. Savides, "Geostationary Operational Environmental Satellite GOES I-M," System Description, Space Systems/Loral, Palo Alto, CA, Dec. 1992

1096) "The GOES I-M Series Satellites - A brief description and Status Report," NOAA draft paper, March 1993

1097) http://rsd.gsfc.nasa.gov/goes/text/goes.databook.html

1098) "GOES I-M Data Book" by Space Systems/Loral

S/C stabilization	Three-axis stabilized	
S/C design life	7 years (5 year mission)	
S/C body size	2 m x 2 m x 2 m	
S/C on-orbit dimensions Array to body Solar sail to body Magnetometer to body Overall length Magnetometer boom	 6.1 m 17.7 m 3.0 m 26.9 m 5.9 m	
S/C mass Deployment mass Dry mass Propellant and pressurant	GOES-I/J/K/L 2105 kg 977 kg 1128 kg	GOES-M 2270 kg 1042 kg 1128 kg
S/C power (solar array) Batteries	1057 W EOL at summer solstice, bus voltage of 42 V 2 NiCd batteries, each with 28 cells, 12 Ah, 400 W in eclipse	
Pointing accuracy (3 sigma) of antenna Payload pointing (roll, pitch, yaw)	±0.25° in roll, pitch and yaw ±9.1 μrad, ±9.4 μrad, ± 73.3 μrad in 90 minutes	

Table 298: General S/C parameters

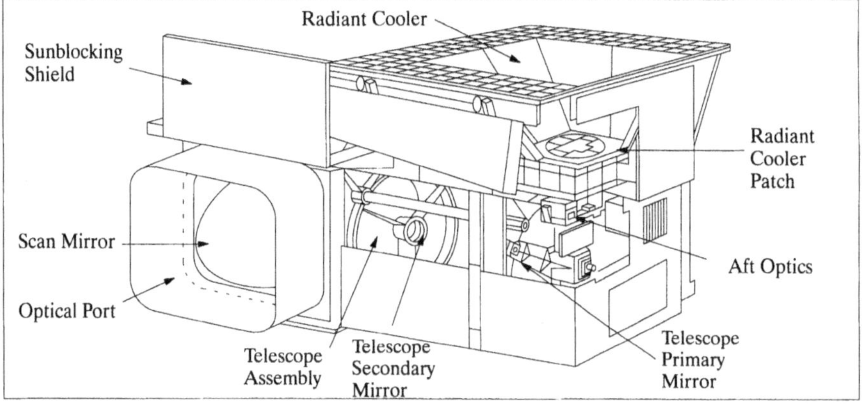

Figure 189: The GOES Imager instrument

Sensor complement:

GOES Imager = Multispectral Imaging Radiometer, built by ITT A/CD of Fort Wayne, IN. Objective: Operational meteorology and climatology (one channel in the VIS and four channels in the IR range). The GOES Imager measures radiant and solar-reflected energy from sampled areas of the Earth. The instrument consists of electronics, power supply, and sensor modules. The sensor module containing the telescope, scan assembly, and detectors, is mounted on a baseplate external to the spacecraft, together with the shields and louvers for thermal control. By means of a servo-driven, two-axis gimballed mirror scan system in conjunction with a Cassegrain telescope, the imager's channels can simultaneously sweep an 8 km north-to-south swath along an east-to-west/west-to-east path, at a rate of 20° (optical) east-west per second. Resolution: = 1km in the VIS range and 4 km in the IR region (except for the 6.5-7.0 μm MWIR channel at 8 km resolution). A normal full Earth disk scan (18° x 18°) is done in 25 minutes (a 3000 km x 3000 km area can be scanned in 3.1 minutes, a 1000 km x 1000 km area in 40 seconds). The instrument can produce full-Earth disk images, sector images that contain the edges of the Earth, and various sizes of area scans completely enclosed within the Earth scene using a new flexible scan system. Scan selection permits rapid continuous viewing of local areas for monitoring of regional phenomena. The key features of the imager are summarized in Tables 299 and 301. The pointing accuracy of the GOES Imager is within a 4 km radius at nadir. [1099]

Spectral Channels	1 (VIS)	2 (MWIR)	3 (MWIR)	4 (TIR)	5 (TIR)
Prime Measurement Purpose	Cloud Cover	Nighttime Clouds, Fires, Volcanoes	Water Vapor	Surface Temp. Clouds	Low-level moisture, clouds
Wavelength (μm)	0.55 - 0.75	3.80 - 4.00	6.50 - 7.00	10.20 - 11.20	11.5 - 12.50
Measurement Range	1.6 to100% albedo	4 to 320 K	4 to 320 K	4 to 320 K	4 to 420 K
S/N or NEΔT	150:1	1.4 K at 300 K	1.0 K at 230 K	.35 K at 300 K	.35 K at 300 K
Detector Type	Silicon	InSb	HgCdTe	HgCdTe	HgCdTe
Spatial Resolution (μrad)	28 (1 km x 1 km at nadir) (+0-10%)	112 (4 km x 4 km at nadir) (+0-10%)	224 (8 km x 8 km at nadir) (+0-25%)	112 (4 km x 4 km at nadir) (+0-25%)	112 (4 km x 4 km at nadir) (+0-25%)

Table 299: GOES Imager performance

Channel No	Spectral Range (μm)	Measurement Range	Measurement Objective
1	0.55 - 0.75	1.6 to100% albedo	Cloud cover
2 (GOES-I/J/K) 2 (GOES-L/M)	3.80 - 4.00 3.80 - 4.00	4 - 320 K 4 - 335 K	Nighttime clouds (space-340 K) Nighttime clouds (space-340 K)
3 (GOES-I/J/K/L) 3 (GOES-M)	6.50 - 7.00 13.0 - 13.7	4 - 320 K 4 - 320 K	Water vapor (space-290 K) Cloud cover and height
4	10.20 - 11.20	4 - 320 K	SST and water vapor (space-335 K)
5 (GOES (I/J/K/L) 5 (GOES-M)	11.50 - 12.50 5.8 - 7.3	4 - 320 K 4 - 320 K	SST and water vapor (space-335 K) Water vapor

Table 300: GOES Imager channel allocations for the various S/C in the series

GOES Sounder = Infrared sounder for operational meteorology and climatology (19 channel discrete-filter radiometer). Objective: atmospheric soundings. Data products are: vertical temperature and moisture profiles, layer mean temperature and moisture, cloud height and amount, total precipitable water, surface temperatures, lifted index, gradient and wind derived from horizontal temperature and moisture fields. Resolution: 8 km. The GOES Sounder is of HIRS/2 heritage providing similar instrument operation and performance. A filter wheel rotating ten times per second provides for data sampling at ten steps per second (the filter wheel brings the spectral filters into the optical path of the detector array, thereby providing channel definition). The detector filter arrangement makes use of four spectral bands (see Table 302). The Sounder's multi-element detector array assemblies simultaneously sample four separate fields or atmospheric columns, i.e., four detectors are simultaneously irradiated in each band, providing an output from four Instantaneous Geometric Fields of View (IGFOV). In this manner the system can sample four 8 km IGFOV each 0.1 seconds. - The data products of the sounder are planned for use in numerical weather forecast models. They provide data coverage in oceanic regions lacking conventional observations.

Both Imager and Sounder employ a servo-driven, two-axis gimballed mirror systems in conjunction with a 31 cm Cassegrain telescope (separate sensors with independent operation). Each instrument has a flexible scan control, enabling coverage of small areas as well as global scenes (Earth's full disk), and close-up, continuous observations of severe storms.

1099) K. A. Hursen, et al., "The GOES Imager: overview and evolutionary development," SPIE, Vol. 2812, pp.160--173, 1996

Feature	GOES Imager	GOES Sounder
Optical aperture	31.1 cm	31.1 cm
Telescope type	Cassegrain	Cassegrain
Total step & sample time	N/A	0.1 s (0.2 s and 0.4 s optional)
Methods of scan	2-axis, continuous, linear E-W, 64 μrad (2.3 km) line step N-S, 224 μrad (8 km)	2-axis, step & dwell; E/W 280 μrad steps; N/S 1120 μrad steps (optional 2240 μrad steps, 0.2 s dwell)
Scan rate	20°/s optical	40 soundings/s
Slew rate	10°/s mechanical	10°/s mechanical
Spatial resolution (μrad)	VIS = 28, Ch. 2,4, & 5 = 112, Ch. 3 = 224	all channels = 242 μrad (diameter)
Sampling	1.75/IGFOV VIS, Ch. 2, Ch. 4 & Ch. 5 3.5/IGFOV Ch. 3	4 IGFOVs sampled simultaneously
Sampling rate	183.3 μs/pixel (IR), 45.8 μs/pixel (VIS)	0.1 s
Chan. co-registration	± 28 μrad	Within 22 μrad of channel 8
Star sensing	Uses visible array SNR 6 for 4th mag stars (400 samples)	Separate visible array SNR 6 for 4th mag stars (each sample)
Data output	10 bit quantization	13 bit quantization
Data rate	2.6208 Mbit/s	40 kbit/s
Data format	NRZ-S, PN code	NRZ-S, PN code
Patch temperature	Regulated at 94 K, 101 K or 104 K	Regulated at 94 K, 101 K or 104 K
Time between space looks	2.2 s large frame 9.2 or 36.6 s smaller frame	2 min
Time between BB calibrations	10-30 min (can override or inhibit)	20 min (can override or inhibit)
Priority frame select	1 level normal; 2 levels priority 1 level star sense	1 level normal; 2 levels priority 1 level star sense

Table 301: GOES second generation instrument parameters

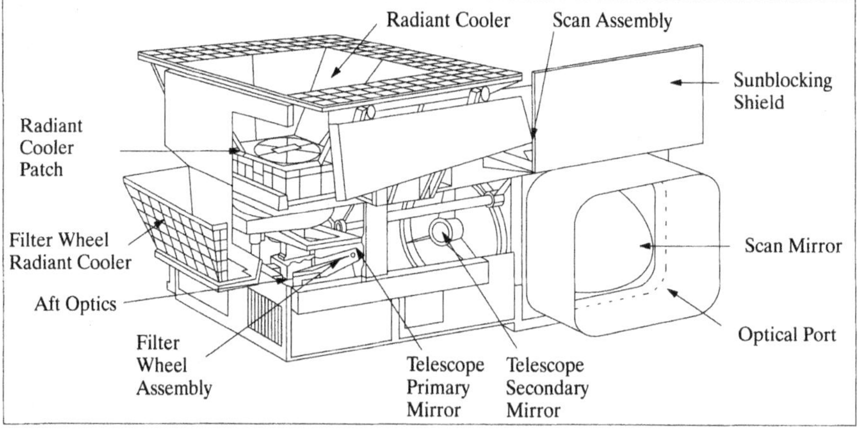

Figure 190: The GOES Sounder instrument

An on-board 'Image Navigation and Registration' (INR) system provides good pointing accuracy, i.e. geographical location capability of the imager and sounder pixels in near real-time.[1100] The INR system removes the apparent image-to-image motion resulting from orbital motion, satellite attitude deviations, and satellite thermal distortions. This is achieved by continuously steering the instrument scan mirrors to compensate for this motion. Compensation signals are generated in the attitude and orbit control system based on predicted

1100) Note: Image navigation refers to the determination of the location of a pixel within an image in terms of Earth's longitude and latitude; registration refers to stability of maintaining pointing of each pixel to a specific Earth location within an image and between repeated images.

orbits and attitude parameters that are transmitted to the satellite at least once each day from a ground station. The parameters are generated using image landmarks, star views, and satellite range data collected throughout the day.

Detector Channel (Band)		Center Wavelength (μm)	NEΔN ($mW/m^2/sr/cm^{-1}$)	Measurement Objective
HgCdTe1		14.71	0.66	Temperature Sounding
(Longwave)	2	14.37	0.58	Temperature Sounding
	3	14.06	0.54	Temperature Sounding
	4	13.96	0.45	Temperature Sounding
	5	13.37	0.44	Temperature Sounding
	6	12.66	0.25	Temperature Sounding
	7	12.02	0.16	Surface Temperature
HgCdTe8		11.03	0.16	Surface Temperature
(Midwave)	9	9.71	0.33	Total Ozone
	10	7.43	0.16	Water Vapor Sounding
	11	7.02	0.12	Water Vapor Sounding
	12	6.51	0.15	Water Vapor Sounding
InSb	13	4.57	0.013	Temperature Sounding
(Shortwave)	14	4.52	0.013	Temperature Sounding
	15	4.45	0.013	Temperature Sounding
	16	4.13	0.008	Temperature Sounding
	17	3.98	0.0082	Surface Temperature
	18	3.74	0.0036	Surface Temperature
Silicon (Visible)	19	0.70	0.10% Albedo	Cloud
Silicon (Star Sense)		0.65	6:1 SNR	4th magnitude stars

Table 302: GOES Sounder performance requirements

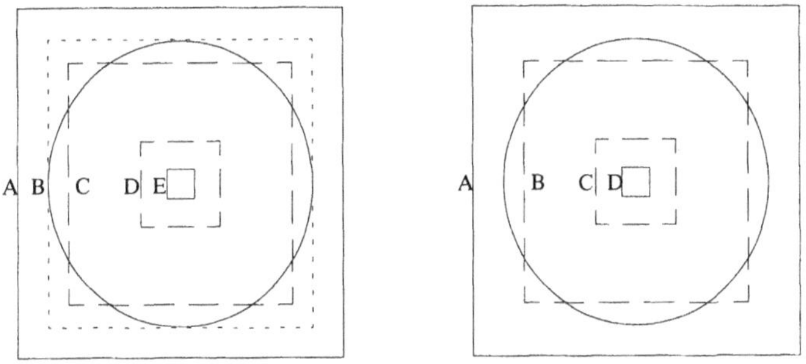

GOES Imager Scan Field

A. Maximum Frame Field 19.2° E-W by 19° N-S
Maximum Star Field 23° E-W by 21 N-S
B. Full Earth Limb Scan in 26 minutes
C. Earth Scan 60° N-S by 60° E-W in 22 min
D. Area Scan 3000 km x 3000 km in 3.1 min
E. Small Area Scan 1000 km x 1000 km in 40 s

GOES Sounder Scan Field

A. Maximum Frame Field 23° E-W by 21° N-S
B. Full Earth Sounding 60° N-S x 60° E-W
464 minutes
C. Area Sounding 3000 km x 3000 km in 42 min
D. Small Area Sounding 1000 x 1000 km in 5 min

Figure 191: GOES second generation scan operations

In addition, the GOES I-M missions carry the **SEM** subsystem (see description under F.4.2), **S&RSAT, DCS** and **WEFAX**. The SEM consists of the following instruments:

- Magnetometer, similar to the versions on previous GOES satellites (except for two redundant magnetometers), each with three separate orthogonal probes.
- Solar XRS (X-Ray Sensor), similar to the previously flown designs; however, it is continuously pointed at the sun by virtue of being mounted on the solar array yoke, and on top of a positioner that tracks the sun in the north-south direction.
- The EPS and HEPAD are functionally identical to those flown on previous GOES satellites, except for the additional two electron channels and a reduced entrance aperture of the D4 dome detector.

- The SXI is part of the SEM subsystem starting with GOES-M.
  SXI = Solar X-Ray Imager. Objective: Solar storm warning. SXI measures solar X-ray emissions.

Product Type	Vert. Resolution	Horiz. Resolution	Absolute Accuracy	Relative Accuracy
Temperature				
Profile	3-5 km	50 km	2-3 K	1 K
Land	---	10 km	2 K	1 K
Sea	---	10 km	1 K	0.5 K
Moisture				
Profile	2-4 km	50 km	30%	20%
Total	---	10 km	20%	10%
Motion	3 layers	50 km	6 m/s	3 m/s
Cloud				
Height	2 layers	10 km	50 mb	25 mb
Amount	total	10 km	15%	5%
Ozone				
Total	---	50 km	30%	15%
Motion	1 layer	50 km	10 m/s	5 m/s
IR Flux	total	50 km	10 W m^{-2}	3 W m^{-2}

Table 303: Sounder products, resolutions and accuracy

S&RSAT on GOES, referred to as **GEOS&R** (Geostationary Search and Rescue):[1101) S&R capability (the distress signal detection function) was first introduced on GOES-7 as a research/demonstration program to introduce the technology into the COSPAS-S&RSAT system (see I.6). S&R is now an operational GOES service of the GOES-I - M series.

The S&RSAT design differs somewhat between GOES-7 and the new GOES-8 series. On GOES-7, -8, and -9, S&RSAT receives signals with the GOES UHF antenna. On GOES-7, both the DCP data and the S&RSAT data are transmitted back through the same downlink transmitter, the S-band antenna. The GOES-7 downlink frequency is 1698 MHz. On GOES-8 (and follow-on satellites), the S&RSAT downlink transmission is through a separate S&RSAT (L-band) helix antenna. On GOES-8, the uplink frequency for S&RSAT is 406 MHz and the downlink frequency is 1544.5 MHz; and for the DCP reporting mode (DCPR), the uplink frequency is 401 MHz with the DCPR downlink at 1694 MHz. - The GEOS&R system is also being considered for future Meteosat as well as for the GMS satellite series.

Data Link Type	Source	Uplink (MHz)	Downlink (MHz)	Destination
Command	CDA/DSN	2034		S/C
Telemetry + SEM	S/C		1694	CDA, SEL-Boulder CO, DSN
Ranging	DSN, STDN	2034	2209	S/C, DSN
WEFAX	CDA	2033	1691	Users/APT
DCPI	CDA	2034	468	DCP
DCPR	DCP	401	1694	CDA/Users
S&RSAT	ELT/EPIRB	406	1544	S&RSAT Ground Station
MDL (diagnostic data)	S/C		1681.5	SOCC/SEL
PDR (Processed Data Relay at 2.111 Mbit/s)	CDA	2027.7	1685.7	Users (via S/C)
Image Raw Data (2.62 Mbit/s)	S/C		1676	CDA

Table 304: Overview of GOES communication links

GOES Data Transmission: Direct broadcast and relay in S-band. - UHF-band at 401, 406 and 468 MHz. The UHF-band is used to receive environmental data from the DCPs. It is also used for S&R service provision.

1101) M. J. Nestlebush, "The Geostationary Operational Environmental Satellite Data Collection System," NOAA Technical Memorandum NESDIS 40, June 1994

Operating Mode	Imaging & Sounding	SEM Package	Communications	Attitude Control
Sunlight (Normal)	on	on	on	Earth pointing with INR
Eclipse	off	on, except XRS off	on, except WEFAX and MDL off	Earth pointing without INR
Launch Operations (Transfer Orbit)	off	off	off	Earth pointing during maneuvers, otherwise sun pointing
On-Orbit Storage	off	off	off	Earth pointing without INR

Table 305: GOES second generation satellite on-orbit operating modes

F.4.4 Next-Generation Imager for the NOAA GOES Series

NOAA and NASA plan to replace the current GOES Imager in the long-term perspective with a new instrument, called **ABI** (Advanced Baseline Imager). As of 2000 the project is in the definition/requirements phase. NOAA anticipates that the first ABI instrument will be available for flight on GOES-Q, with a projected launch readiness date of 2008. [1102)]

The following four requirements of the NWS (National Weather Service) are considered with highest priority:

- Continuous instrument operation capability during eclipse phases of the GEO orbit
- Simultaneous observation capability for the modes "full-disk" and "CONUS" (Contiguous USA).
- Improvement of the temporal instrument imagery resolutions.
  - Full-disk Earth observation within 15 minutes
  - CONUS, or the equivalent of a nadir-viewed rectangle (3000 km x 500 km) every 5 minutes (goal of 1 minute)
  - Imagery of minimum size 1000 km x 1000 km (nadir) every 30 seconds
  - A capability must exist to observe concurrently the CONUS and full-disk imagery along with all other imaging activities, such as space locks, blackbody calibrations, and star observations
- Improvement of the spatial resolution of the imagery. The current GOES Imager spatial resolution (1 km in VIS and 4 km in IR) must be doubled for ABI. The intent is to allow for better identification and tracking of cloud and moisture signatures.

The band selection has been optimized to meet all cloud, moisture, and surface observation requirements. The phenomena observed and the various applications are:

- VIS band (0.64 μm): Daytime cloud imaging, snow and ice cover, severe weather onset detection, low-level cloud drift winds, fog, smoke, volcanic ash, flash flood analysis, hurricane analysis, winter storm analysis
- SWIR band (1.6 μm): Daytime cloud/snow/ice discrimination, total cloud cover, aviation weather analysis for icing, smoke from low-burn-rate fires
- MWIR band (3.9 μm): Fog and low-cloud discrimination at night, fire identification, volcanic eruption and ash, daytime reflectivity for snow/ice
- MWIR band (7.0 μm): Middle-tropospheric water vapor tracking, jet stream identification, hurricane track forecasting, mid-latitude storm forecasting, severe weather analysis
- TIR band (11.2 μm): Continuous day/night cloud analysis for many general forecasting applications, precipitation estimates, severe weather analysis and prediction, cloud drift winds, hurricane strength and track analysis, cloud top heights, volcanic ash, winter storms, cloud phase/particle size (in mid-band products)
- TIR band (12.3 μm): Continuous cloud monitoring for numerous applications, low-level moisture, volcanic ash trajectories, cloud particle size (in mid-band products)
- TIR band (13.3 μm): Cloud top height assignments for cloud-drift winds, cloud products for ASOS supplement, tropopause delineation, cloud opacity.

1102) "GOES Advanced Baseline Imager, Technical Requirements Document," GSFC, S-415-200, March 2000

Requirement		Threshold
Spatial resolution	VIS (0.64 µm) All other bands	0.5 km (14 µrad IFOV) 2 km (56 µrad IFOV)
Spatial coverage	Full disk CONUS (3000 km x 5000 km) Mesoscale (1000 km x 1000 km)	4 per hour 12 per hour Every 30 s, may impact CONUS or full disk
Operation during eclipse		Yes
Data timelines (when scanning is complete)		CONUS: <1 minute Full disk: <6 minutes
Simultaneity		Within 5 s for all bands at any FOV Within 30 s for any adjacent (n/S) pixels Within 15 s for any adjacent (E/W) pixels
Number of spectral bands		8 (threshold), 12 (goal) Star sensing capability required
Spectral bands (µm), 8 bands at minimum		0.64±0.05, 1.61±0.03; 3.9±0.1; 6.15±0.45; 7.0±0.2; 10.7±0.5; 12.0±0.5; 13.3±0.3
Spectral bands (µm), goal of four additional bands		0.86±0.05, 1.375±0.015, 8.5±0.2, 10.35±0.25
Navigation		<1.0 km (28 µrad)
Registration within frame		<1 km (28 µrad)
Line-to-line registration		<25 km (7 µrad)
Registration image to image, VIS, IR		<0.75 km (<21 µrad), <1.0 km (<28 µrad)
Band-to-band co-registration, VIS-IR, IR-IR		<1.0 km (<28 µrad), <15% of FOV
On-orbit calibration	VIS, (goal) IR	Pre-launch to ±5%, On-board to ±3% 0.2 K repeatability, 1.0 K absolute accuracy
IR band linearity		±1%
Total instrument mass, power, volume		<125 kg, <256 W, <0.8 m^3
Instrument data rate		<15 Mbit/s
Lifetime	Ground storage On-orbit storage Mean mission life Design life	2 years 2 years 8.4 Years 10 years

Table 306: Requirements overview for the ABI instrument

Application spectrum of the four additional bands.

- VIS band (0.86 µm): This band provides synergy with AVHRR/3 band 2. The band is used for determining vegetation amount, aerosols and ocean/land studies.
- SWIR band (1.38 µm): This band is similar to a MODIS band. It does not see into the lower troposphere due to water vapor sensitivity, thus it provides excellent daytime sensitivity to very thin cirrus.
- TIR band (8.5 µm): This band permits the detection of volcanic cloud with sulfuric acid aerosols, thin cirrus in conjunction with 11 µm band and determination of cloud microphysical properties with the 11.2 µm and 12.3 µm bands. This includes a more accurate delineation of ice from water clouds during day or night
- TIR band (10.3 µm): The band permits the determination of microphysical properties of clouds with the 11.2 and 12.3 µm bands. This includes a more accurate determination of cloud particle size during the day or night.

F.4.5 Next-Generation Sounder for the NOAA GOES Series

No information could be obtained from the NOAA/GOES project. The GIFTS instrument on the EO-3 mission is a technology demonstration for the next-generation GOES sounder.

F.4.6 LMS (Lightning Mapper Sensor)

The LMS mission consists of an optical imaging instrument of GHCC (Global Hydrology and Climate Center) at NASA/MSFC (Huntsville, AL) to be flown on a GOES spacecraft in 2004. The prime objective is to measure from a geosynchronous orbit the total lightning activity on a continuous basis (under both day and nighttime conditions) over the Americas (North and South) and portions of the adjoining oceans. LMS in GEO permits the study of the electrosphere over dimensions ranging from the Earth's radius down to individual thunderstorms. The instrument is capable of detecting all types of lightning phenomena at a nearly uniform coverage. Near real-time data transmission to MSFC is required for processing and quality assurance and redistribution of the data within 20 seconds of reception.[1103)]

The LMS instrument consists of a staring imager optimized to detect and locate lightning. The major subsystems of the instrument are: an imaging system, a focal plane assembly, real-time event processors, a formatter, power supply, and interface electronics. The imaging subsystem is a fast f/1.2 telescope with a 12 cm aperture diameter and a 1 nm bandwidth interference filter. A broadband blocking filter is placed on the front surface of the filter substrate to maximize the effectiveness of the narrowband filter.

Spectral wavelength	777.4 nm
FOV, IFOV	8° x 5°, 8 km at nadir
Threshold	<4.0 J m^{-2} sr^{-1} A lightning event is identified whenever the selected signal difference threshold is exceeded.
SNR, dynamic range	6, >100
Detector array size	700 x 560 pixels
Detection efficiency, false alarm rate	>90% of total events, <5% of total events
Measurement accuracy	Location: 1 pixel; intensity: 10%; time tag at frame rate
Instrument mass, power	35 kg, 110 W
Data downlink communication	Data rate: 80 kbit/s; modulation: PCM; quantization = 12 bit

Table 307: Specification of the LMS instrument

Event filtering approaches: The daytime lightning signals tend to be buried in the background noise; hence, special techniques are implemented to maximize the lightning signal relative to this background noise.

- Spatial filtering is used which matches the IFOV of each detector element in the LMS focal plane array to the typical cloud-top area illuminated by a lightning stroke (i.e., in the order of about 10 km). This results in an optimal sampling of the lightning scene relative to the background illumination.
- Spectral filtering is obtained by using a narrowband interference filter centered on a strong optical emission line (e.g., OI at 777.4 nm) in the lightning spectrum. This method further maximizes the lightning signal relative to the reflected daylight background.
- LMS employs temporal filtering which takes advantage of the difference in lightning pulse duration which is on the order of 400 μs versus the background illumination which tends to be constant on the time scale of seconds. In an integrating sensor, such as LMS, the integration time specifies how long a particular pixel accumulates charge between readouts. The lightning SNR improves as the integration period approaches the pulse duration. An integration time of 2 ms (technological limit) is used to minimize pulse splitting and maximize lightning detectability.
- Since the ratio of the background illumination to the lightning signal often exceeds 100 to 1 at the focal plane, a fourth technique, a modified frame-to-frame background subtraction is implemented to remove the slowly varying background signal from the raw data coming off the LMS focal plane. Each real-time event processor generates an estimate of the background scene imaged at each pixel of its section of the focal plane array. This background scene is updated during each frame readout sequence and, at the same time, the background signal is compared with the off-the-focal-plane signal on a pixel-by-pixel basis. When the difference between these signals exceeds a selected threshold, the signal is identified as a lightning event and an event processing sequence is initiated.

1103) http://thunder.msfc.nasa.gov/lms/

RTEP (Real-Time Event Processor). The large data rate of about 5 Gbit/s is read out from the focal plane of LMS into several RTEPs for event detection and data compression. Each RTEP detects weak lightning flashes from the intense but slowly evolving background. The RTEP continuously averages the output from the focal plane over a number frames on a pixel-by-pixel basis to generate a background estimate. It then subtracts the average background estimate of each pixel from the current signal of the corresponding pixel. The subtracted signal consists of shot noise fluctuating about zero with occasional peaks due to lightning events. When a peak exceeds the level of a variable threshold, it triggers comparator circuits and is processed by the rest of the electronics as a lightning event.

F.5 GOMS (Geostationary Operational Meteorological Satellite)

GOMS-1 = Geostationary Operational Meteorological Satellite (referred to as '**Elektro-1**' in Russia).[1104), 1105), 1106), 1107), 1108)] Russian RKA program series (ROSGIDROMET/ Planeta operator), built by VNIIEM, Moscow. The first GOMS satellite (GOMS-1) was launched October 31, 1994 (by a Proton carrier vehicle) from Baikonur.

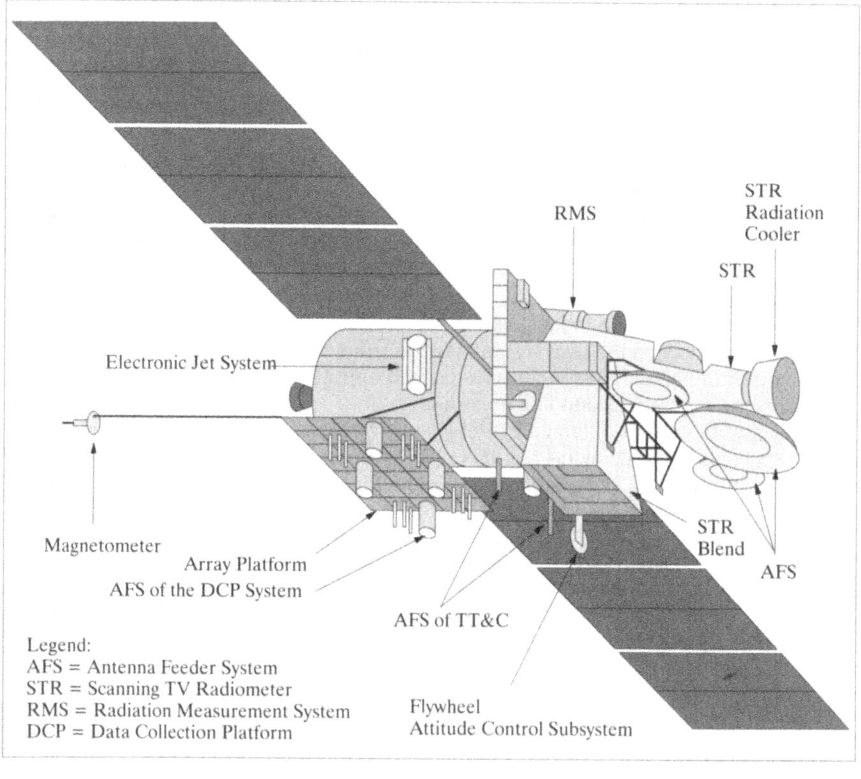

Figure 192: The GOMS S/C model

1104) "Space System with Geostationary Meteorological Satellite (GOMS)," Paper of Planeta, Moscow, Nov. 1990

1105) S. A. Stoma, Yu. V. Trifonov, "Geostationary Space System 'ELECTRO' (GOMS): Preconditions for Creation and Structure," Space Bulletin, Vol. 2, No. 3, 1995, pp. 2-6

1106) O. M. Miroshnik, et. al., "A Drama in Orbit with a Happy Ending," Space Bulletin, Vol. 2, No. 3, 1995, pp. 7-10

1107) Yu. V. Trifonov, "S/C ELECTRO On-board Control Complex," Space Bulletin, Vol. 2, No. 3, 1995, pp. 11-14

1108) Yu. V. Trifonov, A. V. Gorbunov, "Prospects for the ELECTRO Space System Development," Space Bulletin, Vol. 2, No. 3, 1995, pp. 14-15

Satellite: mass = 2650 kg (including payload of 950 kg), power = 1.7 kW (average); GOMS is designed as a three-axis stabilized (fly-wheel) spacecraft with orientation to the Earth and along the velocity vector; orientation knowledge is through the polar star sensor and the sun sensors. Design life = 3 years (replacement on operational need). Orientation accuracy = ±2-5 arcmin.

The GOMS-1 S/C experienced orientation/stabilization problems during its initial phase of operation. These problems were eventually fixed by February 1, 1995. Since then GOMS has been in its operational testing phase. Operational status: GOMS-1, also referred to as Elektro-1, ended all spacecraft operations in Nov. 2000.

GOMS-1 Orbit: Geostationary, position = 76° East (over the Indian Ocean)

Objectives:

- to acquire television images in real time of the Earth's surface and cloud cover within a radius of 60° centered at the subsatellite point in the visible and IR regions of the spectrum
- to measure temperature profiles of the Earth surface (land and ocean) as well as top-of-cloud-cover temperatures
- to measure the radiation state and magnetic field of the space environment at the geostationary position
- to transmit via digital radio channels television images, temperature and radiation as well as magnetometric information to the main and regional data receiving centers
- to acquire information from Russia and international data collection platforms (DCPs), located in the GOMS visibility range, and to transmit the information obtained to all receiving and processing centers (RPCs)
- to call for the DCPs to transmit information to the satellite
- to retransmit the processed meteorological data in the form of facsimile or alphanumerical information from the receiving processing centers to the receiving stations of the independent data users via GOMS
- to provide an exchange of high-speed digital data (retransmission via GOMS) among all regional centers within Russia (Federal Service for Hydrometeorology and Environmental Monitoring).

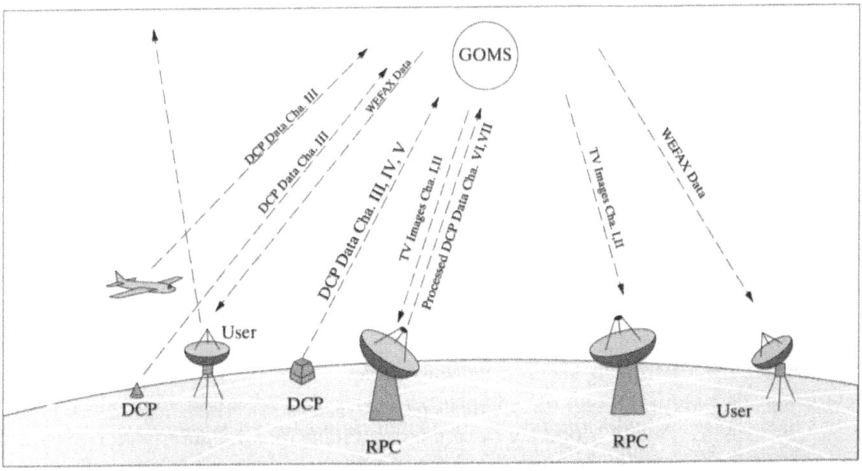

Figure 193: Data collection and distribution scenario for GOMS

Sensor complement:

STR = Scanning TV Radiometer providing imagery in VIS and IR bands. Objectives: observation of clouds and underlying surface in VIS and IR spectra and temperature data of underlying surface, determination of top of clouds.
Spectral bands: 0.46 - 0.7 μm (VIS), 10.5 - 12.5 μm (IR-1), 6-7 μm (IR-2, starting with GOMS-2); Number of scan lines per frame = 8000 (VIS) and 2500 (IR); IFOV (VIS) = 31.5 μrad, IFOV (IR)=160 μrad; FOV = 13500 x 13500 km; spatial resolution = 1.25 km (VIS) and 6.5-8.0 km (IR). The imaging session frequency is 30 min; the length of one take (frame time) is 15 minutes. Direct data transmission rate = 2.56 Mbps.

RMS (Radiation Measurement System), RMS consists of instruments for heliogeophysical measurements, radiation and magnetic parameters). FOV = 13,500 x 13,500 km. Objective: registration of particles (protons, electrons, α-particles), measurement of X-ray radiation from the sun, measurement of magnetic field vector components.

- Density of electron fluxes with energies in four bands from 0.04 - 1.7 MeV
- Density of proton fluxes with energies in four bands from 0.5 - 90 MeV
- Density of alpha particles with energies from 5 - 12 MeV
- Intensity of the galactic cosmic radiation with energies > 600 MeV
- Solar X-ray radiation intensity with energies from 3 - 8 keV
- Intensity of solar UV radiation in four wave bands up to 1300 Å
- Magnetic induction vector component quantities along 3 axis with ± 180 nT interval

F.5.1 Radio Complex for Data Collection, Transmission and Relay

- GOMS Communication Characteristics

- transmission of TV images and heliogeophysical information from the satellite to the receiving and processing centers - RPCs (Radio channels I and II).
- carrier frequencies: 1685 MHz (S-band) and 7465 MHz (X-band)
- data transmission rate = 2.56 Mbps
- Radio channel III: transmission of data from DCPs; frequencies: 401 - 403 MHz; transmission of data is possible through 33 international and 100 Russian channels at rates of 100 b/s.
- Radio channels IV and V: retransmission of information obtained from DPCs to RPCs; frequencies: 1697±1 MHz (S-band) and 7482 ±1 MHz (X-band)
- Radio channels VI and VII: transmission of facsimile information in standard WEFAX format and alphanumerical data from RPCs to GOMS. Frequencies: 2115±1.5 MHz and 8195±1.5 MHz. Data transmission rate = 1200 b/s
- Radio channel VIII: retransmission of facsimile and alphanumerical data from GOMS to independent receiving stations. Frequency: 1691±1.5 MHz
- Radio channel IX: transmission of high-speed digital information from RPC to GOMS. Frequency: 8190±5 MHz; data rate up to 0.96 Mbit/s
- Radio channel X: transmission of high-speed digital information from GOMS to the RPCs. Frequency: 7.465±2.5 MHz; data rate up to 0.96 Mbit/s
- Radio channel XI: calling for DCP from GOMS. Frequency: 469±1 MHz
- Radio channel XII: transmission of DCP request from RPC to GOMS. Frequency: 2119± 1 MHz

- GOMS Data Receiving and Transmitting Modes

- Channels I and II operate 24 - 48 times per day; each session lasts 15 minutes
- Channels III, IV and V operate by commands calling for DCP information
- Channels VI, VII and VIII operate continuously
- Channels IX and X function in the interval when channels I and II are not active
- Channels XI and XII operate on request

F.6 INSAT

INSAT[1109]= Indian National Satellite system is a joint venture, conceived in the 1970s, of the following departments/agencies of the government of India: DOS (Department of Space), DOT (Department of Telecommunications), IMD (Indian Meteorological Department), AIR (All India Radio). DOS is responsible for the operation of the INSAT space segment. IMD operates a Meteorological Data Utilization Center (MDUC, at Delhi) for the dissemination and distribution of INSAT meteorological images and ancillary data.

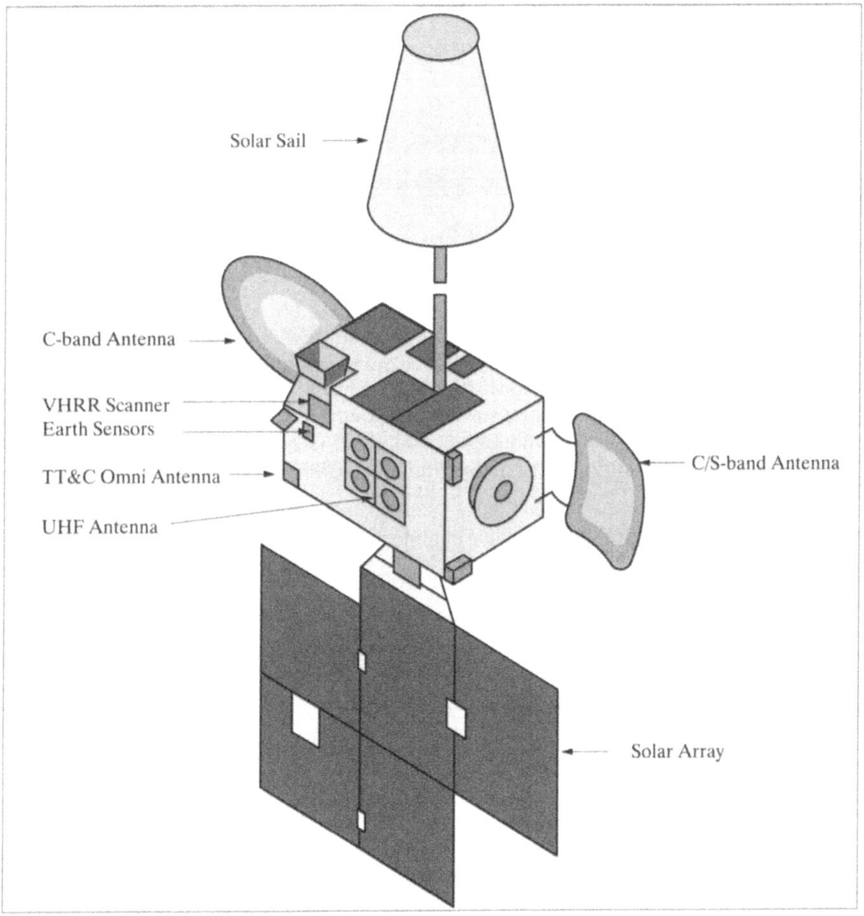

Figure 194: The INSAT S/C model

F.6.1 INSAT-1 Satellite Series

INSAT is a multipurpose operational satellite system series (geostationary) employed for meteorological observation over India and the Indian Ocean, as well as for domestic telecommunications (nationwide direct TV broadcasting, TV program distribution, meteorological data distribution, etc.). The INSAT-1 series satellites were built by Ford Aerospace Corporation (USA) to Indian specifications.

1109) 'Space Applications,' DOS Annual Report 1990-91. pp. 13-23

Sensor: (Meteorological Package)
VHRR = Very High-Resolution Radiometer. Spectral ranges: 0.55 - 0.75 μm (VIS) and 10.5 - 12.5 μm (IR). Resolutions: 2.75 km for VIS and 11 km for IR channels, respectively. Scanning time/image = 30 minutes. Scanned field: full frame mode (20° E-W and 20° N-S); sector scan mode (20° E-W and 4.5° N-S).

Data Collection System (**DCS**): This system gathers and relays environmental data (meteorological, hydrological and oceanographic) from unattended land- and ocean-based automatic data collection platforms (DCPs). The global data receive frequency (platforms to satellite) is 402.75 MHz (UHF-band). As of 1991 over 100 DCPs were installed.

Mission	Comment
INSAT-1A	Launch: April 1981;
INSAT-1B	Launch: Aug. 30, 1983; position: 74° East; the satellite is used as a standby for INSAT-1D.
INSAT-1C	Launch: July 22, 1988 (Ariane vehicle). The satellite lost Earth lock on Nov. 22, 1989 and is inoperable since then.
INSAT-1D	Launch: June 12, 1990 (US Delta vehicle); position: 74° E

Table 308: Overview of INSAT-1 satellite series

F.6.2 INSAT-2 Satellite Series

This satellite family is considered the second generation INSAT series for the nineties (built by ISRO). Like the INSAT-1 series, the INSAT-2 series is a multi-purpose satellite family for telecommunication, television broadcasting and meteorological services. The S/C is three-axis stabilized and uses a box-shaped structure based around a central cylinder of 0.89 m diameter with asymmetrical five-panel solar wings (23 m span). Power = 1.18 kW (EOL). Orbit maneuvers are supported by an integral apogee boost motor, there are also 16 attitude control thrusters. On-orbit mass is 911 kg.

The INSAT-2 operational satellites have enhanced capabilities. The INSAT-2 meteorological sensor VHRR has a 2 km resolution in VIS and 8 km resolution in the IR spectral ranges. The INSAT-2C launch with Ariane occurred on Dec. 6, 1995, the INSAT-2D launch with Ariane 44L occurred on June 3, 1997 (along with Inmarsat 3F4), the INSAT-2E launch took place on April 2, 1999. The meteorological instrument on the spacecraft features an additional band in the water vapor absorption spectrum of 6.8 μm.

VHRR = Very High-Resolution Radiometer. VHRR is an improved version over that of the INSAT-1 series. The instrument can be operated in any one of the following modes: [1110)]

- Full scan mode (20° x 20°): This covers the entire Earth disk and some space around in 33 minutes (can be positioned in steps of 0.35° in east-west direction).
- Normal scan mode: Coverage of at least 50° N to 40° S latitude (in 23 minutes)
- Sector scan mode: The sector can be positioned anywhere in the Earth disk along the N-S direction (in 7.2 minutes). [1111)]

The VHRR instrument consists of three packages: the EOM (Electrooptics Module), the electronics package-1, and the electronics package-2. The electrooptics module houses the scan assembly, the optics assembly, and the radiant cooler assembly. The VIS and IR detectors with their amplifiers are also part of the electrooptics module. The electronics package-1 contains the signal processors for all channels, data formatter, power distribution, and timing logic circuits. The scanner electronics, the DC/DC converter regulators, patch temperature controller and the analog telemetry processor are part of the electronics package-2.

1110) G. Joseph, et al., "Very high resolution radiometers for INSAT-2," Current Science, Vol. 66, No. 1, Jan. 10, 1994, pp. 42-56

1111) G. Joseph, et al., "INSAT-2 Very High Resolution Radiometer for Meteorological Observations," Journal of Spacecraft technology, Vol. IV, No. 1, Jan. 1994, pp. 183-207

The optical subsystem uses an all-reflective Ritchey-Chretien telescope (folded optics, the primary mirror is a concave hyperboloid with an aperture of 202 mm and f/2.0, the secondary mirror is a convex hyperboloid with an aperture of 62 mm; focal length of the system is 1594 mm). The convergent light beam, reflected from the secondary mirror, is incident on the dichroic beam splitter which is mounted at 45° to the telescope axis. The VIS radition uses two arrays of four silicon photodiodes each providing a 2 km resolution at nadir (scan mirror is oscillating is 0.5 Hz). The IR detector assembly consists of two HgCdTe photoconductive detectors mounted in the passive radiant cooler. The IR detector is operated at 105 K. The scan mirror is mounted on a two-axis gimballed scan structure which generates 2-D images by sweeping the detector IFOV in the E-W and N-S directions. Each E-W/W-E scan generates one IR line of 8 km width and four VIS lines each of 2 km width.

Spectral band: VIS Spectral band: TIR	0.55 - 0.75 μm; Integrated out-of-band response <3% Inter detector mismatch <5% 10.5 - 12.5 μm; Integrated out-of-band response <3% Out-of-band response peak <0.1%
Spatial resolution VIS Spatial resolution TIR	56 μrad (or 2 km x 2 km) 224 μrad (or 8 km x 8 km)
Radiometric performance: SNR Radiometric performance: NEDT	>6 for VIS at 2.5% albedo <0.25 K at 300 K for TIR channel
Dynamic range of IR channel Dynamic range of VIS channel	4-340 K 0-100%
Misregistration between VIS and TIR	<56 μrad
Modulation Transfer Function (MTF)	>21% for TIR; >23% for VIS channel

Table 309: Specification of the VHRR instrument

Mission	Comment	Payload (Meteorological)	Power, Battery
INSAT-2A	Launch: July 9, 1992, Ariane-IV, at 74° East, Mass = 905 kg	VHRR, DCS, SAS&R	1.16 kW summer solstice NiCd, 18 Ah, 28 cells
INSAT-2B	Launch: July 22, 1993, Ariane-IV, at 93.5° East, Mass=905 kg	VHRR, DCS, SAS&R	
INSAT-2C	Launch: Dec. 6, 1995, Mass = 980 kg, co-located with -2B		1.735 kW, summer solstice NiCd, 24 Ah, 28 cells
INSAT-2D	Launch: June 3, 1997, Ariane 44L, Mass = 980 kg, the S/C **lost Earth lock in October 1997**	VHRR, DCS, SAS&R	
INSAT-2DT	Transponder leased from ArabSat to continue the INSAT-2D services to the users		
INSAT-2E	Launch: Apr. 2, 1999, Ariane 4, Mass = 1150 kg, 12 year design life with a goal of 15 years	VHRR, CCD Camera, DCS, SAS&R	2.5 kW at equinox NiH_2, 60 Ah, 27 cells

Table 310: Overview of INSAT-2 satellite series

INSAT VHRR Data Distribution: The INSAT-1/2 VHRR data is distributed in near real-time to/from 22 SDUCs (Secondary Data Utilization Centers) throughout the country.

INSAT Data Availability: As of Dec. 16, 1997 a major shift in ISRO data policy occurred when representatives of NASA, NOAA and ISRO signed an agreement in a Washington DC meeting, for ISRO to provide full-sized INSAT weather data to the world meteorological community in digital format and without delay. The agreement calls also for the USA to make Earth and atmospheric data available to India. [1112)]

The INSAT-2 series flies a data collection system (DCS) with a data relay transponder (DRT) for environmental data. It is also furnished with **SAS&R** (Satellite Aided Search and Rescue) system. SAS&R provides an emergency alert capability for the Indian subcontinent and beyond as part of the international satellite aided search and rescue program (see also chapter I.6). India has signed an agreement with the international COSPAS-S&RSAT council for the use and operation of LUTs (Local User Terminals) and an MCC.

1112) W. Ferster, "Policy Shift Paves Way for India-U.S. Imagery Deal," Space News, Jan. 5-11, 1998, p. 18

F.6.2.1 INSAT-2E

General description of the INSAT-2E satellite bus system. The S/C mass at launch is 2550 kg (dry mass of 1146 kg). The S/C power is 2.5 kW at equinox provided by solar panels with an area of 20.52 m^2 (GaAs solar cells). Battery: NiH_2, 60 Ah, 27 cells. The S/C design life is 12 years. The INSAT-2E communication payload carries 17 transponders, 12 operating in the normal C-band frequency and five in the lower extended C-band. Seven of the normal C-band transponders have wide beam coverage, the remaining ten have zonal coverages. [1113)]

Orbit: Geostationary, location at 83° eastern longitude.

S/C structure (Figure 194):

- The S/C main body is a cuboid of dimension: 1930 mm x 1770 mm x 2375 mm. The deployed S/C length is 25.78 m.
- The main cylinder of 1175 mm diameter and 2375 mm length accommodates fuel and oxidizer tanks
- North & south equipment panels are of size 2300 mm x 1930 mm. They are attached to the main cylinder using shear frames
- Cutouts are provided for the solar array drive assembly on the south deck and for the solar sail on the north deck

S/C mechanisms:

- Deployment mechanisms include single-sided solar panel, solar sail and shaped reflectors
- Two flaps offsetting seasonal variations in solar pressure; it is operated by a shaped memory alloy spring

Thermal system:

- The thermal design requires heat dissipation of about 1700 W
- Use of embedded heat pipes adopted to optimize the thermal system mass
- The thermal control system is complemented by the use of optical solar reflectors, multi-layer insulation blankets, thermal coatings, thermal tapes and heaters
- Appropriate thermal management for NiH_2 battery to maintain its temperature between $\pm 10°C$.

Power system:

- Single-sided deployable solar array (GaAs/Ge cells) on the south side of the S/C with four panels of size: 2.7 m x 1.9 m
- EOL power of 2143 W
- Two 60 Ah NiH_2 batteries provide the eclipse requirement of 1800 W for full communication payload operation
- Two independent power buses with cross-strap capability, each bus regulated to 42+0.5 V by sequential switching
- Bus selection scheme provided for platform subsystems

Telemetry, Tracking and Command (TT&C) system:

- C-band telemetry and telecommand links with the subsystems deriving their heritage from earlier INSAT-2 S/C
- Near omni radio visibility during transfer orbit phase
- Normal and dwell mode telemetry possible by ground command
- Simultaneous dwelling on any of the 16 channels
- The system is capable of providing 2048 on/off commands and 64 data commands
- Provision of variable execution pulse width command feature
- The ranging accuracy is 50 m during transfer orbit and 30 m during on-orbit phase

1113) ISRO brochure of INSAT-2E, provided by George Joseph

Attitude and Orbit Control System (AOCS):

- The AOCS subsystem is comprised of microprocessor-based electronics (Mil-Std-1750A), attitude sensors, inertial reference unit, momentum/reaction wheels, magnetic torquer, solar flap and RCS (Reaction Control System) thrusters
- The propulsion system uses bi-propellants with 440 N LAM (Liquid Apogee Motor) for orbit raising and a total of 16 RCS thrusters for orbit and attitude control (fuel for a mission life of 12 years). Anti Earth viewing thrusters provide a backup for LAM
- The attitude sensor configuration supports all requirements. The north-facing Earth sensor is used during transfer orbit to maintain Earth lock. Redundant Earth sensors serve the same purpose during the on-orbit phase. The digital sun sensor, used in the transfer orbit, has a FOV of 37°. Four CASS (Coarse Analog Sun Sensors) are used, four Pi sun sensors are backup for CASS.
- AOCS pointing accuracies: pitch and roll = 0.2°, yaw = 0.4°
- Some special features of AOCS are:
  - Fault detection and identification logic for change over to redundant system
  - Safe mode operation in case of Earth lock loss
  - Long pulse detection logic which prevents unduly long during thruster firing
  - Remote programming for unforeseen requirements
  - Inertial reference unit with three gyros in orthogonal axis provide accumulated angle and rate information
  - Two momentum wheels provide stiffness to the S/C along the pitch axis. Redundancy for one wheel failure is provided by the reaction wheel along the yaw axis.

Meteorological payload of INSAT-2E: [1114] [1115]

VHRR/2 (Very High Resolution Radiometer). This ISRO instrument is an enhanced version of the two-band VHRR flown on INSAT-2A/2B (see description above, F.6.2). A third band was added, the water vapor band [MWIR (Mid-Wave Infrared)], which is flown on the INSAT-2E spacecraft for the first time. The water vapor band is used for estimating water-vapor-tracer winds at the mid-troposphere, identifying subsidence/convection zones, and indirectly estimating the atmospheric correction to SST (Sea Surface Temperature) for the TIR channel. [1116]

Spectral band: VIS Spectral band: TIR Spectral band: MWIR (Water vapor)	0.55 - 0.75 µm; Integrated out-of-band response <3% Inter detector mismatch <5% 10.5 - 12.5 µm; Integrated out-of-band response <3% Out-of-band response peak <0.1% 5.7 - 7.1 µm
Spatial resolution VIS Spatial resolution TIR and MWIR	56 µrad (or 2 km x 2 km) 224 µrad (or 8 km x 8 km)
Radiometric performance: SNR Radiometric performance: NEDT	>6 for VIS at 2.5% albedo <0.25 K at 300 K for IR channel
Dynamic range of TIR/MWIR channels Dynamic range of VIS channel	4-340 K 0-100%
Misregistration between VIS and IR	<56 µrad
Modulation Transfer Function (MTF)	>21% for IR and TIR; >23% for VIS channel

Table 311: Specification of the VHRR/2 instrument

VHRR/2 consists of the following major subsystems: scan mechanism assembly, radiant cooler, Optics assembly, and camera electronics.

- Scan mechanism assembly: The scanner consists of a two-axis gimbal-mounted beryllium mirror. The gimbal is servo-driven in two orthogonal axes. The mirror swings me-

1114) V. S. Iyengar, C. M. Nagrani, et al., "Meteorological imaging instruments on-board INSAT-2E," Current Science, Vol. 76, No 11, June 10, 1999, pp. 1436-1443

1115) Special Section: INSAT-2E, Current Science, Vol. 76, No 11, June 10, 1999, pp. 1431-1450

1116) P. S. Desai, "Satellite Meteorology in India: Accomplishments and Challenges," Space Forum, Vol. 5, No. 1-3, 2000, pp. 203-216 chanically ±5°/s (E-W fast scan) to provide an optical sweep of 20° (N-S) to the detector FOV to generate one line/s. In the other axis (N-S), the mirror is stepped through 224 µrad for each scanning line. In addition to the 20° (N-S) stepping, there is a provision to rotate the mirror through 90° to view the blackbody. There is also an offset capability in the E-W direction to position the Earth disk at the center of the image for longitudinal S/C positions between 70-100° E.

- Radiant cooler. A passive system is used to cool the infrared detectors to an operating temperature of 105-115 K.
- Optics assembly: The system consists of a telescope, dichroic beam splitter, infrared collimating lens, infrared relay optics, and VIS band optical elements. The telescope is a Ritchey-Chretien type instrument with a 200 mm diameter concave hyperboloid primary mirror and a 56 mm diameter secondary mirror, the separation distance is 285 mm. The convergent radiation beam from the secondary mirror is intercepted by a dichroic mirror which reflects 85% of IR energy and transmits 75% of the visible energy without distortion of the wavefront. The transmitted VIS band energy is reflected by a fold mirror onto an array of CCD detectors, while the reflected IR energy is collimated prior to exiting the telescope. This permits independent performance optimization of the TIR and IR (water vapor) bands in the radiant cooler/detector system.
- Camera electronics: The subsystem provides the following functions: a) digitization of the detector outputs, b) control of the IR detector temperatures, c) generation of clock and other control pulses, d) formatting of data, and e) instrument monitoring. The source data is coregistered and digitized to 10 bits.

The 2-D image of the Earth is generated by sweeping the IFOV of the detectors by rotation of the scan-mirror gimbals in two orthogonal axes. For every sweep of the mirror, four contiguous lines of VIS, and one line of TIR and IR bands each, are generated in the E-W direction. At the end of a sweep, the mirror is stepped south through an angle equivalent to eight km in ground distance. Three operational modes are provided:

- Full frame mode. An image of 20° x 20° (TFOV) is generated within 33 minutes, covering the full disk of the Earth.
- Normal fram mode: Same coverage in E-W direction as in full frame mode; however, the N-S scan is 14° covering latitudes from 50° to 40° S in 23 minutes.
- Sector frame mode: This capability provides 4.5° coverage in N-S direction (about 2800 km with 351 scan lines) in 7.2 minutes. The E-W extent remains unchanged as in full and normal frame modes. The sector may be positioned anywhere within TFOV in steps of 0.5° (312 km or 39 lines) in N-S direction.

Instrument calibration: cold space views at the east/west ends are used for establishing reference radiance for all three bands. A full calibration of the TIR and IR bands is provided by swinging the mirror to view a blackbody cavity.

Status: The VHRR/2 instrument of INSAT-2E failed already in 1999. [1117)]

CCD Camera. The experimental instrument uses a linear CCD detector array with three spectral bands (see Table 312). Applications of the instrument data are in meteorology as well as in vegetation mapping. Of the three bands, the first two being similar to the NOAA AVHRR bands 1 and 2, provide "vegetation index" observation; the third band is used for snow-cover, snow-cloud discrimination and for aerosol measurements. A linear silicon array is used for the VIS and VNIR bands, while a InGaAs array is used for the SWIR band. The camera provides a nadir spatial resolution of <= 1 km in all three bands. The swath of the CCD Camera is 6300 km. The source data is coregistered and digitized to 10 bits. The data rate of the CCD Camera is 1.3 Mbit/s (QPSK modulation).

The CCD Camera is a first-time introduction on INSAT-2E. It permits greater accuracy in cyclone tracking and also affords monitoring of local severe storms. Plans for the third-gen-

1117) K. S. Jayaraman, "India Approves Nation's 1st Dedicated Weather Satellite," Space News, Dec. 11, 2000, pp. 3, 20 eration INSAT series call for a tandem operation, in which two members, INSAT-3A and -3D are earmarked for meteorology. While the INSAT-3A meteorological payload is still very similar to that flown on INSAT-2E (VHRR/2), the INSAT-3D payload will feature an enhanced VHRR instrument with six channels and a higher spatial resolution; two of the channels provide a split window for the TIR bands at 10.3-11.3 μm and 11.5-12.5 μm. This feature will be used to improve SST estimates.

Spectral bands	0.62 - 0.68 μm, VIS band 0.77 - 0.86 μm, VNIR (Visible Near Infrared) band 1.55 - 1.69 μm, SWIR (Shortwave Infrared) band
Spatial resolution	1 km x 1 km
MTF	<=0.23
Dynamic range	0-100% albedo
Noise performance	SNR <= 128 at 100% albedo
Coverage	6250 km x 300 km per scan line (max. number of lines = 31)
Image repetivity	One scan line per minute
Instrument mass, power	<55 kg, <50 W

Table 312: Specification of the CCD Camera

The CCD Camera consists of the following elements: scan mechanism assembly, optics and detector assembly, and camera electronics.

- Scan mechanism assembly: Features a gimballed scan mirror which sweeps FOV in two orthogonal axes. The fast sweep generates 300 video lines over a 6300 km east-west direction every minute (with a fast retrace capability). The 0.4° south stepping of the mirror after each east-west scan provides the generation of successive image strips. The TFOV (Total Field of View) of the mirror is ±13° in the E-W and ±10° in the N-S direction, while the actual image is ±5° in E-W and ±5° in the N-S direction. The actual image may be positioned anywhere within the mirror TFOV.
- Optics and detector assembly: The assembly consists of the scan mirror, telescope, dichroic beam splitters, fold mirror, lens doublet and bandpass filters. The front-end optics, namely scan mechanism and telescope, is identical to that of VHRR (except that the telescope performance has been upgraded to provide a FOV of ±0.25° at twice the spatial frequency of VHRR). The VIS and VNIR silicon CCD detectors have 2048 element linear arrays, as well as the InGaAs photodiode detector (cooled) for the SWIR band. The outputs of the three line arrays, providing a total of 900 pixels, are processed to construct 300 image pixels for each band.
- Camera electronics: The subsystem provides: a) clock and bias inputs to the detectors, b) processes and digitizes the detector outputs., c) formats the video signal along with auxiliary information, d) monitors the instrument; e) interfaces with other subsystems, and f) controls the SWIR detector temperature through a feedback loop.

Meteorological payload data transmission:

The VHRR data in PCM NRZ-S format is BPSK modulated, the data rate is 526.5 kbit/s. The data of the CCD Camera in PCM NRZ-L format is QPSK modulated, the data rate is 1.3 Mbit/s. Both meteorological payloads have a common up convertor and a redundant transmitter for a downlink through a 4.5 GHz carrier. A separate 0.9 m diameter Earth view antenna on-board the S/C is providing transmission services for the meteorological data.

Data from the INSAT series instruments are acquired, processed and disseminated by IMD (India Meteorological Department), the main data center and operator the INSAT series. A DCS (Data Collection System) is flown on all INSATs for collecting and relaying data from platforms in the ground segment.

Total number of channels	17
Normal C-band channels Uplink frequency Downlink frequency	12 5850-6410 MHz 3625-4185 MHz
Lower extended C-band channels Uplink frequency Downlink frequency	5 6450-6650 MHz 3425-3625 MHz
Downlink coverage: Uplink coverage:	Wide beam (channel 6, 8, 10, 12, 14, 15, and 17), zonal beam all other channels Wide beam
Transmit EIRP	36 dBW for all channels
RF output power	60 W (wide beam); 32 W (zonal beam)

Table 313: Characteristics of the INSAT-2E communication payload

F.6.3 INSAT-3 Satellite Series

The INSAT-3 series of ISRO is a multipurpose spacecraft with an all Ku-band transponder layout for direct-to-home television and roof-top data communication. The satellites incorporate state-of-the-art communication technology such as high power antennas, steerable beam and digital compression/decompression capabilities. The solar panel of the INSAT-3A S/C uses an advanced rigid deployable sun-oriented array of GaAs/Ge cells with an area of 26.5 m^2. The series is planned for launch from 2000 onwards.

Mission	Comment	Meteorological Payload	Power, Battery, Other
INSAT-3B	Launch March 21, 2000 on Ariane-5 from Kourou, co-located with INSAT-2E at 83° East	None	3.45 kW at equinox, NiH_2, 70 Ah, 27 cells
INSAT-3A	Launch planned for 2002	VHRR/2, DCS, SAS&R	
INSAT-3C	Launch planned for Sept. 2001		36 transponders, 24 in C-band, 12 in extended C-band
INSAT-3D	Launch planned	VHRR/3, DCS, SAS&R	

Table 314: Overview of INSAT-3 satellite series

F.6.3.1 INSAT-3B

INSAT-3B is the first of the five satellites that was launched in the INSAT-3 series (built by ISRO). The satellite structure resembles a cuboid of 1.93 m x 1.7 m x 1.65 m and, with the two solar panels deployed (total area of 23 m^2), it measures 14.7 m in length. The sun tracking solar panels generate 1.7 kW of power. A 24 Ah Ni-Cd battery supports the payload operations during eclipses. INSAT-3B is three-axis body-stabilized using momentum/reaction wheels, earth sensors, sun sensors, an inertial reference unit and magnetic torquers. It is equipped with unified bi-propellant thrusters. The satellite has two deployable antennas and three fixed antennas that carry out various transmit and receive functions. The antennas have a pointing accuracy of $\pm 0.2°$ in pitch and roll axes, and $\pm 0.4°$ in yaw axis. The satellite uses a passive thermal control system. The S/C mass is 2070 kg at launch, with 1100 kg of hydrazine propellant for orbit raising, station keeping and on-orbit attitude control. The S/C design life is 10 years. [1118)]

INSAT-3B was launched successfully on March 21, 2000 from Kourou on an Ariane-5 vehicle (along with a passenger payload called Asia Star of the US company World Space), putting INSAT-3B into GTO (Geostationary Transfer Orbit).

Orbit: Geostationary orbit at 83° East, co-located with INSAT-2E.

1118) http://www.isro.org/insat3b.htm

The INSAT-3B communication payload provides 12 extended C-band channels, each having a bandwidth of 36 MHz. The Ku-band payload provides three channels, having a bandwidth of 77/72 MHz. The Mobile Satellite Service transponders operate in C/S band frequencies. Compared to INSAT-2C/2D, the power of extended C-band transponders on board INSAT-3B has been increased from 10 W to 15 W and that of Ku-band from 20 W to 55 W. - The INSAT spacecraft series missions are operated by the MCF (Master Control Facility) at Hassan in Karnataka, India. MCF has a network of six Satellite Control Earth Stations (SCES) to acquire its data.

F.6.3.2 INSAT-3A

INSAT-3A is a multipurpose satellite like its predecessors supporting the functions of communications and meteorology. A launch of the spacecraft is planned for early 2002.

Meteorological payload:

The meteorological instruments are identical to those of the INSAT-2E instruments, namely VHRR/2 and the CCD Camera.

F.7 METEOSAT

Background: METEOSAT is the European meteorological program that was initiated by ESA in 1972 (actually the beginnings of European space meteorology goes back to 1968 under ESRO). Meteosat-1 was the first European meteorological geosynchronous satellite, launched on November 23, 1977. The next satellite in this preoperational series, Meteosat-2, was launched on June 19, 1981. The success of the Meteosat program led a large group of European nations to establish firm plans for sustained operations by an independent agency. On May 24, 1983, 16 countries signed the EUMETSAT convention, which came into force in June 1986. On January 1, 1987, responsibility for the operation of the Meteosat satellites was transferred from ESA to EUMETSAT. Since then ESA continues to build and operate the MOP (Meteosat Operational Program) series as a EUMETSAT contractor. In a 1993 agreement, EUMETSAT was given the full operational mandate. This means that EUMETSAT provides all operational functions and services of its satellites (with a dedicated EUMETSAT ground segment) established in 1995. [1119) 1120) 1121) 1122) 1123) 1124)]

The prime objective of EUMETSAT is to establish, maintain and exploit European systems of operational meteorological satellites. In 1994 EUMETSAT commenced two new programs: MSG (Meteosat Second Generation), a geostationary program with operational services starting in 2000; EPS (EUMETSAT Polar System) with the MetOp services starting in the year 2003. The space segment of both of these programs is being developed in cooperation with ESA.

Meteosat itself is part of a global network of geostationary meteorological satellites distributed around the equator. All activities regarding satellite and operations are coordinated by the CGMS (Coordination of Geostationary Meteorological Satellites) committee represented by WMO and the countries providing the satellites [Europe (EUMETSAT), USA (NOAA), Russia (NPO Planeta), Japan (JMA), India (ISRO), China]. - Meteosat satellites are normally positioned at 0° longitude (Meteosat-3 was reassigned and positioned in re-

1119) R. Tessier, "The Meteosat Programme," ESA Bulletin 58, May 1989, pp. 45-57
1120) ESA Information Note to the Press No. 4, Feb. 11, 1991, "MOP-2 Ready for Launch"
1121) "Current and Planned European Operational Meteorological Satellite Systems," John Morgan, Proceedings of the Twenty-Third International Symposium on Remote Sensing of The Environment, Bangkok, Thailand, April 18-25, 1990, ERIM, Ann Arbor, MI, Vol. I, pp. 107-116.
1122) 'The Meteosat Operational Programme - From Experiments to Exploitation,' Earth Observation Quarterly, No. 25, March 1989
1123) Introduction to the METEOSAT Operational System, ESA BR-32 ISSN 250-1589, Sept. 1987
1124) 'EUMETSAT Directory of Meteorological Satellite Application,' ISBN 92 91110 006 4, 1991, EUMETSAT

cent years at 75° western longitude in support of GOES coverage - Extended Atlantic Data Coverage Mission) . The coverage of Meteosat satellites is Europe, Africa and the Atlantic Ocean.

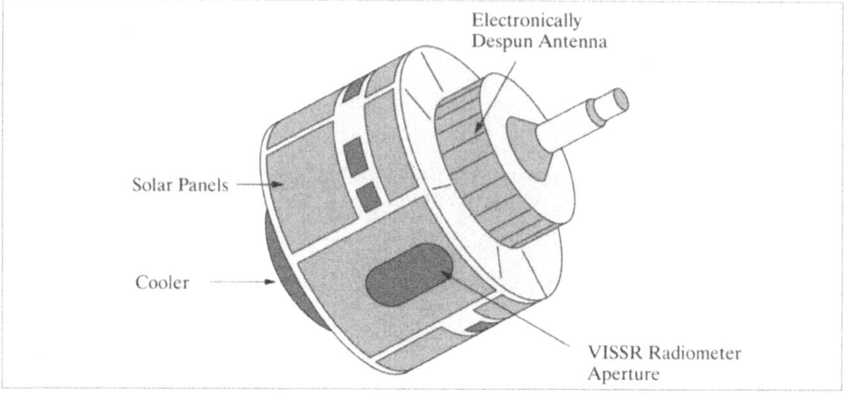

Figure 195: The Meteosat S/C model

Prototype satellites of the preoperational program:	
Meteosat-1	Launch: Nov. 23, 1977; service operation until Nov. 1979
Meteosat-2	Launch: June 19, 1981; service operation from Aug. 12, 1981 until Aug. 11, 1988 as prime satellite, re-orbiting in December 1991.
Meteosat-3	(initially designated as P2); Launch: June 1988 (2 years). Note: this satellite was also positioned over the West Atlantic as a temporary replacement for GOES.
EUMETSAT (MOP) Missions:	
Meteosat-4	MOP-1, Launch: March 6, 1989; first image of MOP-1 on April 19, 1989
Meteosat-5	MOP-2, Launch: March 2, 1991; 1998 repositioning to 65° East for INDOEX campaign support
Meteosat-6	MOP-3, Launch: Nov. 20, 1993
Meteosat-7	MOP-4, Launch: Sept. 3, 1997 (nominal life = 5 years), Meteosat-7 is part of the Meteosat transition program toward MSG (Meteosat Second Generation)

Table 315: Overview of Meteosat missions

Spacecraft: Meteosat is a spin-stabilized S/C (dual-spin configuration at 100 rpm). It features a cylindrical body of 2.1 m in diameter (toped by a drum-shaped section and further smaller cylinders), the overall length is 3.195 m. The main cylindrical body contains most of the payload. Its outer surface is covered with solar cells (more than 8000 cells mounted on five standard panels). The surface of the drum-shaped section is covered with an array of radiating dipole antenna elements. Electronics within the drum activate the individual elements in sequence, in reverse order to the S/C spin sense. The drum constitutes an electronically despun portion of the S/C, namely the antenna whose function is to ensure that S-band communications are always directed towards Earth. - Two Earth horizon sensors and two sun sensors provide attitude information. The S/C has four main thrusters (10 N) and two 2 N thrusters as actuators (hydrazine propellent) for orbit/attitude control. The S/C dry mass is 322 kg (plus 39 kg of hydrazine). [1125)]

Application: Weather forecast, meteorology, climatology (sea surface temperature monitoring, cloud cover statistics, wind and rainfall assessment). Collection of environmental data.

1125)"The METEOSAT System, EUM TD 05, March 1996, brochure provided by EUMETSAT

An image is provided in 1/2 hour periods in the visible range and in the IR and WV regions. The first three missions were part of the preoperational program. MOP (Meteosat Operational Program); the MOP phase extends to the end of 1995. MOP is considered the first Meteosat generation (nominal life=2 years for METEOSAT-4 to -6).

Sensor:

The observation payload consists of a high-resolution radiometer (image generation in the thermal infrared region (TIR), in the water vapor absorption bands (WV), and in the visible range (VIS).

VISSR (Visible and Infrared Spin Scan Radiometer, platform spin rate: 100 rpm), also referred to as **MSR** (Multispectral Radiometer) and **MVIR** (Meteosat Visible and Infrared Radiometer). Objectives/applications: Earth and atmospheric monitoring, operational meteorology, climatology. Basic climatological data sets and precipitation index are derived daily. Measurements: cloud coverage, cloud motion winds, cloud top heights, upper tropospheric humidity, precipitation and sea surface temperature.

The optical system of VISSR consists of a scanning Ritchey-Chretien telescope with a primary aperture of 400 mm diameter (140 mm secondary aperture) and focal lengths of 3650 mm for VIS and 535 mm for WV and IR ranges. A series of mirrors is used to collect the incoming radiation and to focus it onto the corresponding detectors (silicon photodiodes for VIS, HgCdTe detectors for IR and WV). The HgCdTe detectors are cooled to 90 K by a passive system. The S/C scans the full Earth disk within a 30 min period. Scanning from east to west is achieved by through the S/C spin. Scanning from south to north is achieved by small incremental steps in the pointing of the telescope.

Calibration is provided by two black bodies located on opposite sides of the main optical path, one is kept at 290 K, the other is kept at 340 K. In addition, cold space views are obtained during normal imaging operations.

Spectral ranges (3 bands)	0.5 - 0.9 μm (VIS), visible band 5.7 - 7.1 μm (WV), emission of water vapor 10.5 - 12.5 μm (TIR) thermal infrared
FOV	18° (0.314 rad)
IFOV	0.065 mrad (VIS), 0.14 mrad (WV, TIR)
Ground pixel size	2.5 x 2.5 km for VIS, (5 x 5 km for WV and TIR)
Image lines	5000 (VIS), 2500 (WV, TIR)
Pixel samples per line	5000 (VIS), 2500 (WV, TIR)
Line duration, line recurrence	30 ms, 600 ms
Number of pixels/image	25 Mega pixels (VIS), 6.25 Mega pixels (WV, TIR)
Image duration, image recurrence	25 min (scanning), 30 min (duty cycle = 100%)
Data transmission rate to ground	333kbit/s (normal mode), 2.7 Mbit/s (burst mode)

Table 316: VISSR parameter specifications

Meteosat Communications:

- S-band in the frequency range of 2098 - 2110 MHz.
  - Transmission of raw image data from the S/C to Fucino (primary ground station).
  - TT&C data
  - Transmission of DCP reports
  - Uplink of image dissemination data from the primary station to the S/C
  - Uplink of MDD (Meteorological Data Distribution) data from a ground station
- L-band in the frequency range of 1675 - 1696 MHz. The L-band is for user-related functions for the following services:
  - WEFAX analog image dissemination and DCP retransmissions (1691.0 MHz)

- HRI (High Resolution Image) digital image dissemination, with a few WEFAX transmissions (1694.5 MHz)
- Meteorological data distribution with up to four channels spaced at 30 kHz (1695.605 - 1695.935 MHz)

• UHF-band in the frequency range of 402.0 - 402.2 MHz. DCPs (Data Collection Platforms) mat be given access to one of the 66 uplink UHF channels at 3 kHz separation.

F.7.1 Meteosat Data Collection System (DCS)

The Meteosat DCS provides an operational collection service for a large number of data collection platforms within its field of view (about 80° great circle of arc of its subsatellite point). The overall system is composed of the following subsystems:[1126)]

1) The space segment: a DCS platform on Meteosat satellites consisting of: a UHF/S-band transponder, an electronically despun S-band antenna, a UHF receive antenna and, as a backup, an S-band toroidal pattern transmit antenna.
2) The Data Collection Platforms (DCPs) in the ground segment
3) The Meteosat ground station (ESA Odenwald station) and DCP processor at ESOC Darmstadt

The DCS on-board Meteosat provides a transponder with a pass band that is divided into 66 communication channels (3 kHz per channel). A message from a DCP in the ground segment is relayed via the DCS to the Meteosat ground station. Each DCP is allocated one of the 66 parallel channels. Since the DCP transmit either according to a schedule, or under 'alert' conditions, many platforms can use a single 3 kHz channel sequentially. The platforms (DCPs) transmit (uplink) their data in the 402 MHz UHF band to the Meteosat DCS, where it is converted to S-band (1675 MHz) and retransmitted (downlinked) to the Meteosat ground station. There, the signals received are checked for quality before being transmitted to ESOC for processing and distribution. At ESOC, all DCP messages are collected, processed, disseminated in near real time and distributed to the platform owners. The distribution of DCP data can also be achieved via the 'DCP Retransmission' service.

Data Collection Platforms.

Meteosat services about 1000 DCPs in the ground segment. The coordinated DCS design approach from Meteosat (Eumetsat), GOES (NOAA) and GMS (JMA) allows the reports from DCP carried on ships, aircraft and other mobile platforms, moving from one satellite coverage area to another, to be received and transmitted by one of the satellites in the network (and transmitted via terrestrial lines throughout the world).
Meteosat services three basic types of DCPs (Eumetsat requires certification and licensing of all DCPs):

• 'Self-timed' DCPs, which transmit their data at regular intervals based on an internal clock (this is the most common type of platform in use)
• 'Alert' DCPs, which transmit a small amount of data when a particular parameter has been exceeded. The transmission time of 'alert' DCP messages is limited to about 10 seconds in order to decrease the probability of overlapping messages on the same channel.
• 'Hybrid' DCPs, which are a combination of 'alert' and 'self-timed' DCPs.

DCP transmissions are limited to about 1 minute, each transmission is termed a DCP message. One DCP message may contain several sets of information or 'reports.' A message format consists of the following contiguous elements:

• unmodulated carrier for 5 seconds

1126) "Meteosat Data Collection System," March 1990, ESOC

- a 250-bit preamble (alternating '0' and '1' - to acquire the bit rate and lock onto it)
- a 15-bit synchronization word
- the DCP address (a 31-bit Bose-Chaudhuri-Hocquenghem (BCH) coded word)
- the data field (up to 5192 bits, - 'alert' DCPs are limited to 184 bits of data)
- the 31-bit End of Transmission (EOT) sequence.

The DCP RF uplink is in UHF band (between 402.001 and 402.199 MHz). The Meteosat DCS transponder permits up to 33 international [in common with GOES and GMS spacecraft (402.0025 - 402.0985 MHz)] and 33 regional channels (402.1015 - 402.1975 MHz) for DCP reports. The carrier is phase modulated by the signal bit stream, the modulation index being 60°.

'Self-timed' DCPs report according to a schedule agreed to by the DCP owner and the operations center. Depending on the application this may range from several times per hour to once every 24 hours. Each hour is divided into 40 time slots; each report must be completed within its given time slot. Since the maximum duration of a report is about one minute, at least 30 seconds are available as a guard time to accommodate any drift in the DCPs internal clock.

'Alert' DCPs transmit on channels which are free of self-timed DCPs. Although the duration of an alert message is limited to about 10 seconds, it is of course possible that two or more alert reports will overlap on the same frequency channel. The event of such an overlap entails the loss of both messages. In order to reduce this overlap probability, alert platforms repeat their messages several times at fixed time intervals.

DCS Access Method.

Data transmission from a DCP to the Meteosat DCS is generally self-triggered by the DCP. All DCS receivers are in listening mode trying to lock onto any message that is being assigned to them by the DCPs on a random access basis. The mean availability of the DCS is >95%, a user can expect that more than 95% of the messages transmitted by his DCP is being received, processed and distributed.

DCP Data Processing. The data handling system performs the following functions (starting in 1995):

- Completeness checks for constituting blocks and DCP address recognition
- Storing of DCP messages and processing information onto a database
- Monitoring of reporting DCPs with respect to a predefined schedule (check of transmission chain)
- A log of received and processed messages, including anomalies
- Processing of DCP data and bulletin preparation for distribution
- Archiving of DCP data. Retrieval for DCP users
- Generation/storage of a performance analysis.

All geostationary meteorological satellites series (METEOSAT, NOAA-GOES, GMS (JMA), etc. use DCS (or variations thereof. Note: the ARGOS-DCS concept on polar-orbiting NOAA-POES S/C with the corresponding PTTs (ground) differs in uplink frequency, message length and format).

F.7.2 Meteosat DCP Retransmission System

DCP users may also directly receive their (DCP and/or WEFAX) data if a Secondary Data User Station (SDUS) is operated on their home premises. This service is offered to users at any point within the Meteosat telecommunication range. DCP data received at the Meteosat ground station are preprocessed and routed to ESOC. Those DCP data intended for direct retransmission are selected and routed to a special processing unit in the ground station. The selection is made by using look-up tables of DCP addresses. All DCP data listed in these look-up tables are buffered and transmitted in a time-sharing mode over one of the Meteosat transponder channels used for image transmission (1691, 1694.5 MHz). DCP data is transmitted in time slots between image transmissions. The time slot is 27 seconds between two WEFAX transmissions. DCP data transmission may be continuous if there is no transmission of WEFAX images or ranging. The bit rate during transmission is 12.5 kbit/s; on average, there are 15 transmission slots of 27 seconds per hour.[1127)]

WEFAX (Weather Facsimile) transmissions are in a format compatible with the APT (Automatic Picture Transmission) service of the NOAA-GOES and POES series. The WEFAX transmission format consists of the following elements:[1128),1129)]

- Start signal (300 Hz), 3 seconds
- Phasing signal, 5 seconds
- Image (800 lines), 200 seconds
- Stop signal (450 Hz), 5 seconds

Consecutive WEFAX transmissions are scheduled in 4-minute slots. This leaves a break of 27 seconds between two WEFAX transmissions. In these 27 second intervals, the stored DCP reports are disseminated, the individual reports being packed into HDLC frames.

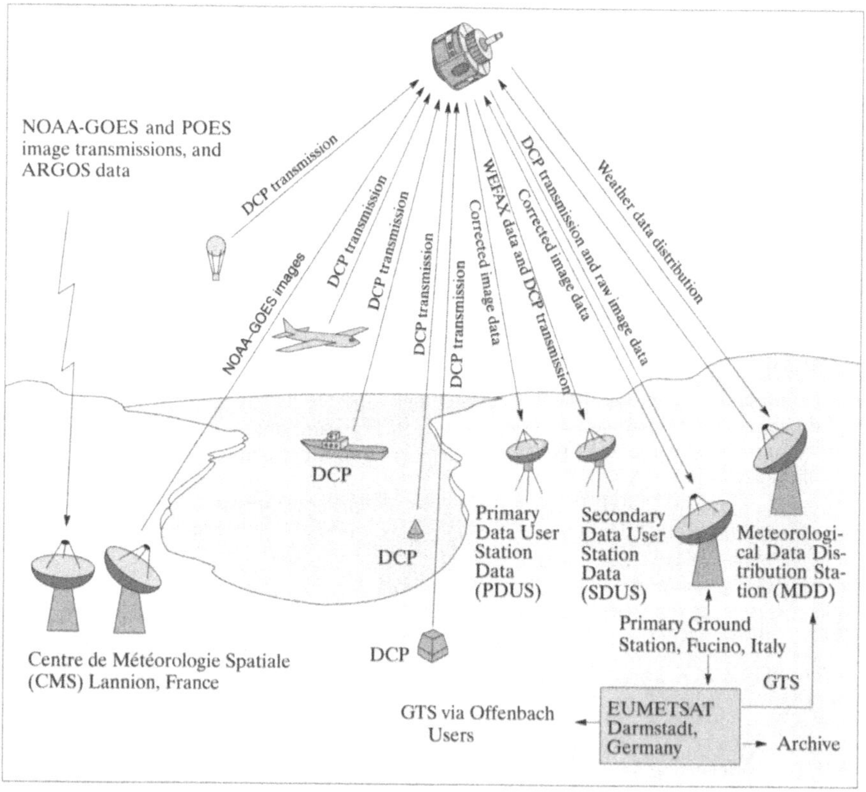

Figure 196: Basic configuration of the Meteosat Operations System (Dec. 1995)

1127) "Meteosat DCP Satellite Retransmission System," January 1990, ESOC
1128) "Meteosat WEFAX Transmissions," ESOC paper, March 1990
1129) "Meteosat High Resolution Image Dissemination," ESOC paper, Oct. 1989

F.7.3 MOSAIC

All weather services rely heavily on the timely distribution of meteorological observations and derived products. These requirements have stimulated the development of a meteorological data distribution service. The principal Meteosat missions can be combined by the user to provide an integrated data access system. This concept is known as 'Meteosat Operational Systems for data Acquisition and InterChange' (MOSAIC). Since 1991 users can readily combine satellite imagery, local reports from the data collection system, as well as conventional meteorological data and forecasts. Meteosat acts in this setup simply as a relay for most data distribution services (direct broadcast). The MDD (Meteorological Data Dissemination) system is one facet of MOSAIC; it provides a service with enormous distribution potential, particularly across Africa and the Middle East.[1130)]

MDD (operational since 1992) consists of three elements: the uplink sites, the Meteosat transponder (relay), and the user reception stations. The initial two operational uplink sites are at Bracknell (UK Meteorological Office) and at Rome (Italian Meteorological Service). A third uplink started operations in 1994 (Toulouse, France). The data sets disseminated via MDD channels consist of meteorological information in the form of alphanumeric characters, bit-oriented data streams, and graphical information - all of which use a common data transmission protocol.

MDD offers four downlink channels from Meteosat in S-band (1695 MHz), each with a transmission rate of 2400 bit/s. The channel spacing is 31.2 kHz with a bandwidth of ±10 kHz for each channel. The following center frequencies are either in use or reserved for future extension:

- Channel 1: 1695.6938 MHz
- Channel 2: 1695.7250 MHz (Bracknell, UK)
- Channel 3: 1695.7562 MHz (Toulouse, France)
- Channel 4: 1695.7874 MHz (Rome, Italy)

MDD transmission from the source to the receiving stations is as follows:

- The GTS (Global Telecommunication System of WMO) is used for the acquisition by the MDD uplink station (see Figure 196)
- The link protocol in the ground segment is X.25 (CCITT)
- The data blocks are received by the MDD uplink preprocessor (encryption and FEC)
- The data are uplinked via Meteosat relay function to the receiving stations
- Each MDD receiving station acquires the data blocks, preprocesses them and reassembles the data into WMO bulletins or meteorological charts.

Application: an extensive range of operational meteorological charts covering Africa, the Middle East, Europe, Asia, parts of the Americas, and all intervening oceans is broadcast through the MDD service from Bracknell.

F.8 MSG (METEOSAT Second Generation)

The first generation METEOSAT series of EUMETSAT is planned to be replaced by a second generation series (MSG) with a launch of the first satellite (MSG-1) in early 2002, to be followed by at least two more satellites, ensuring operational continuity for at least 12 years. The MSG programme definition started in 1993 with a phase C/D started in 1995. The programme is funded by the European Space Agency (ESA) and the European Organization for the Exploitation of Meteorological Satellites (EUMETSAT) according to the following share of responsibilities:

- ESA develops the MSG-1 prototype satellite according to EUMETSAT requirements and acts, on behalf of EUMETSAT, as procurement agent for MSG-2/-3 satellites

1130) "MOSAIC Meteorological Data Distribution," EUMETSAT, EUM UG 01

- EUMETSAT contributes one third of the MSG-1 funding, and funds procurement of MSG-2/-3, finalizes and maintains end user requirements, procures all launch services including launchers, develops the ground segment, ensures consistency between segments (space, ground, and launch service segments) and operates the MSG system.

The MSG system, as follow-on to the Meteosat first generation system, is designed to support nowcasting, very short and short range forecasting, numerical weather forecasting and climate applications over Europe and Africa, with the following mission objectives:

- Multispectral imaging of the cloud systems, the Earth surface and radiance emitted by the atmosphere, with improved radiometric, spectral, spatial and temporal resolution as compared to the first generation Meteosat
- Extraction of meteorological and geophysical fields from the satellite image data for the support of general meteorological, climatological and environmental activities
- Data collection of data from data collection platforms (DCP)
- Dissemination of the satellite image data and meteorological information upon processing to the user community in a timely manner for the support of nowcasting and very short range forecasting
- Support to secondary payloads of scientific or pre-operational nature (GERB) and Search & Rescue (called GEOSAR) in supporting operations which are not directly relevant to the MSG program
- Support to the primary missions (e.g. archiving of data generated by the system MSG, for successful operation of the system etc.).

Figure 197: Illustration of the MSG satellite

The MSG satellites are being built by a European industrial consortium led by Alcatel Space Industries (ASI) of France. [1131] [1132] [1133]

MSG satellites are spin-stabilized (cylindrical in shape, similar to the first generation S/C series) with a rotation speed of 100 rpm. The S/C body is a cylindrical-shaped solar drum, 3.2

1131) W. Veith, "The MSG Satellite and its Subsystems," Proceedings of the 2000 EUMETSAT Meteorological Satellite Data Users' Conference, Bologna, Italy, May 29-June 2, 2000, pp. 15-32

1132) S. Rota, "The METEOSAT Second Generation," Proceedings of the EUMETSAT Meteorological Satellite Data User's Conference, Copenhagen, Denmark, Sept. 6-10, 1999, pp. 25-32

1133) http://www.esa.int/msg/

m in diameter and 3.7 m in height (2.4 m body) with stepped cylinder. The major subsystems of the platform are: AOCS (Attitude and Orbit Control Subsystem), EPS (Electrical Power Subsystem), UPS (Unified Propulsion Subsystem), and DHSS (Data Handling Subsystem)

- AOCS: Attitude is measured by means of Earth sensors and sun sensors. AND (Active Nutation Damping) is used during GTO; there is no AND in the GEO phase. In addition, AOCS provides synchronization signals to keep the SEVIRI, GERB, and the electronically de-spun antenna Earth-pointing.
- EPS: Solar power of 600 W EOL is provided to the S/C bus 28 V power bus. Two NiCd batteries provide 1200 Wh for ecliptic operations support.
- UPS: A bi-propellant unified propulsion system of two 400 N apogee engines is used for a 3-burn insertion from GTO into GEO. Four tanks hold up to 976 kg of propellant. About 810 kg are used for the GEO insertion of MSG.
- DHSS: The system consists of three units: CDMU (Central Data Management Unit) and two RTUs (Remote Terminal Units).

The satellite itself is built in a modular way around three main sub-assemblies: The SEVIRI instrument is located in the central compartment; the communication payload, including antennas and transponders, are positioned in the upper compartment; while the platform support subsystems are located in the lower compartment. The S/C design life is seven years. Total S/C mass = 2040 kg, power = 600 W (EOL).

Orbit: Geostationary orbit at 0.0° longitude (held to within 1° by thrusters). Two satellites are simultaneously in orbit (one operating and one in cold redundancy) to assure availability. The MSG-2 launch is planned 18 months after MSG-1 to provide a two-satellite operational system.

Application: Operational meteorology and climate monitoring.

Sensor complement:

SEVIRI (Spinning Enhanced Visible and Infrared Imager), designed and built by Astrium SAS, France. [1134] SEVIRI is the principal on-board instrument, an imaging radiometer for imaging and sounding (12 channel instrument as defined in Table 317). Its operating principle is based on collecting radiation from a target area and focusing it on detectors sensitive to 12 different bands of the electromagnetic spectrum by means of a telescope. This is followed by the electronic processing of the signals provided by the detectors. Channels VIS 0.6 µm, VIS 0.8 µm, IR 1.6 µm and HRV are referred to as "warm", while channels IR 3.9 to IR 13.4 µm are referred to as "cold." The cylindrical instrument has a diameter of about 1 m and a height of 2.1 m along the spin axis of the satellite. Instrument mass = 260 kg, power = 153 W. The instrument functional architecture is based on four main assemblies:

- TSA (Telescope and Scan Assembly) including the calibration unit and the refocusing mechanism. TSA employes a three-mirror telescope of compact design. The primary mirror is concave a-spherical with a diameter of 510 mm. The secondary mirror is concave a-spherical of 200 mm diameter. The tertiary mirror is convex a-spherical of 60 mm diameter. The total length of the telescope structure is 1.3 m. All mirrors are lightweighted and manufactured from Zerodur. The rotating scan mirror assembly uses a linear spindle drive with a stepping motor, providing continuous bi-directional image scanning. North to south scan of the Earth: scan capability of 22° in N-S direction and 18° in E-W direction. At each satellite revolution, three image lines are acquired (9 lines for the High Resolution Visible channel) for a total of about 1250 lines in a repeat cycle of 15 minutes. Each nominal raw image consists of 3750 lines, each one containing about 3834 pixels except for HRV channel for which a nominal raw image consists of 3750 lines with 5751 pixels per line.

  The calibration unit permits calibration of the IR channels by inserting a CRS (Black-

1134) J. Schmid, "The SEVIRI Instrument," Proceedings of the 2000 EUMETSAT Meteorological Satellite Data Users' Conference, Bologna, Italy, May 29-June 2, 2000, pp. 23-32 body Calibration Reference) source into the optical beam at the focal point of the primary mirror. A flip-flop type mechanism is employed based on a DC voice coil motor. A refocusing mechanism is included in the SEVIRI telescope to correct, on an occasional basis, for potential defocus after launch and during lifetime. It consists of a stepper motor, a transmission gear box and a roller screw to provide the translation. [1135)]

- FPCA (Focal Plane and Cooler Assembly). FPCA is a passive two-stage cooler providing an 85 K environment for the IR channels.
- EUA (Electrical Unit Assembly), consists of FCU (Functional Control Unit), DE (Detection Electronics) including the MDU (Main Detection Unit), the PU (Preamplifier Unit) and the detectors. EUA controls SEVIRI and processes its data. The DE consists of the detectors and the front-end electronics.

Channel		Nominal spectral band (μm)	NEΔR or NEΔT	Max Dynamic range *
Name	Center of λ (μm)			
HRV	Broadband (silicon response)		1.07 W/(m² sr μm) at 1.3 W/(m² sr μm)	460 W/(m² sr μm)
VIS 0.6	0.635	0.56-0.71	0.53 W/(m² sr μm) at 5.3 W/(m² sr μm)	533 W/(m² sr μm)
VIS 0.8	0.81	0.74-0.88	0.49 W/(m² sr μm) at 3.6 W/(m² sr μm)	357 W/(m² sr μm)
IR 1.6	1.64	1.50-1.78	0.25 W/(m² sr μm) at 0.75 W/(m² sr μm)	75 W/(m² sr μm)
IR 3.9	3.90	3.48-4.36	0.35 K at 300 K	335 K
WV 6.2	6.25	5.35-7.15	0.75 K at 250 K	300 K
WV 7.3	7.35	6.85-7.85	0.75 K at 250 K	300 K
IR 8.7	8.70	8.30-9.10	0.28 K at 300 K	300 K
IR 9.7	9.66	9.38-9.94	1.5 K at 255 K	310 K
IR10.8	10.80	9.80-11.80	0.25 K at 300 K	355 K
IR 12.0	12.00	11.00-13.00	0.37 K at 300 K	335 K
IR 13.4	13.4	12.40-14.40	1.8 K at 270 K	300 K

Table 317: Channel definitions of the SEVIRI instrument

The overall SEVIRI design is based on a compact telescope and scan assembly, allowing the implementation of a large passive cooler which improves IR detector performances by lowering their operating temperature. The imaging SEVIRI radiometer is equipped with a patented three-mirror (3M) telescope of compact design (focal length of 5367 mm) which allows the insertion of a small black body for full-pupil calibration. The primary mirror (M1) is a concave circular ellipsoid (centered on the satellite axis). M1 is a 500 mm diameter mirror with a baffled central hole of 90 mm diameter. [1136)] It is followed by a magnifying two-mirror assembly including the secondary mirror (M2), which is a concave ellipsoid centered on the satellite axis, and the tertiary mirror (M3) which is a convex spherical mirror. The M2/M3 assembly is a compact "light-tight" configuration allowing easy alignment. - The 12 channel detectors are positioned at the focal plane using a total of 42 detectors. Each channel has an array of 3 detector elements, with the exception of the HRV channel which has 9 detectors. The eight IR channels have HgCdTe detectors, they are passively cooled. The VIS channels feature photo-voltaic silicon diodes while the NIR channels have InGaAs photovoltaic diodes. - Data quantization is done inside the MDU (Main Detection Unit) by a 12-bit ADC, for an effective 10-bit resolution at the electronic outputs, after digital dynamic offset and fine gain correction. Auxiliary data are added to the detector data for image processing on ground.

The Earth's radiation enters the instrument at every revolution through a 50 cm x 80 cm aperture. The nominal repeat cycle of 15 minutes was the driver in selecting the number of

1135) M. Huchler, M. Seibel, I. Köker, "The SCAN Assembly for the SEVIRI Instrument on MSG," DGLR-JT98-118, 1998, pp. 1027-1036

1136) D. M. A. Aminou., B. Jacquet, F. Pasternak, "Characteristics of the Meteosat Second Generation (MSG) Radiometer/Imager: SEVIRI", Proceeding of SPIE, EUROPTO series, Vol. 3221, pp 19-31, 22 September, 22, 1997 detectors per channel and the spin rate (100 rpm). Twelve minutes are allocated to the imaging phase, leaving three minutes for calibration, retrace and stabilization. The 1 km sampling at SSP of the High Resolution Visible (HRV) channel is achieved by using 9 broadband detection elements. The other channels are sampled at 3 km SSP by using 3 narrowband detection elements per channel.

Earth frame East-West	18.40° (321.5 mrad.), HRV: 9.2° (160.7 mrad)
Earth frame North-South	18.01° (314.4 mrad.)
Scan range North-South	20.0° (384 mrad., 1527 steps)
Scan line step	51.88 arcsec (251.5 µrad., 9 km at SSP)
Scan mechanism step	25.94 arcsec (125.8 µrad., 9 km at SSP)
Spin rate	100 rpm
Line cycle	0.6 s
Imaging time per line	30.672 ms (5%); HRV: 15.336 ms
Earth imaging time	12.5 minutes
Calibration, retrace and stabilization	2.5 minutes
Repeat cycle	15 minutes

Table 318: Some SEVIRI imaging parameters

SEVIRI employs the classical calibration approach using deep space as cold source and a known on-board source as a warm reference. The on-board blackbody temperature is used to determine the correction factor accounting for the different levels of background flux. The deep-space view is performed via the full optical path by commanding the acquisition of a sufficient number of samples during that part of the S/C revolution, when neither the Earth nor the sun (or moon) is in the FOV of SEVIRI. [1137)]

Channel groups	Scanning parameters	Data rate before stretching	Data rate after stretching
3 VNIR channels (2 VIS +1 IR)	3 detectors per channel 3834 pixels per line 10 bits per pixel	11.25 Mbit/s	0.5751 Mbit/s
8 IR channels	3 detectors per channel 3834 pixels per line 10 bits per pixel	30.00 Mbit/s	1.5336 Mbit/s
1 HRV (High Resolution Visible) channel	9 detectors per channel 5751 pixels per line 10 bits per pixel	33.75 Mbit/s	0.8627 Mbit/s
Total		75 Mbit/s	2.9714 Mbit/s

Table 319: Projected data rates of the SEVIRI instrument

GERB (Geostationary Earth Radiation Budget), [1138)] [1139)] [1140)] [1141)] PI: J. Harries, Imperial College, London. GERB is an AO (Announcement of Opportunity) instrument, and provided on a national funding basis by a consortium led by the UK (NERC), Belgium (OSTC) and Italy (ASI). RAL of UK provides overall instrument management, systems engineering and other services. The objective is to monitor the Earth's radiation budget (global climate change, food production and natural disaster prediction) measuring at the top of the atmosphere (continuous temporal sampling), in particular the shortwave and longwave regions

1137) P. Pili, "Calibration of SEVIRI," Proceedings of the 2000 EUMETSAT Meteorological Satellite Data Users' Conference, Bologna, Italy, May 29-June 2, 2000, pp. 33-39

1138) J. E. Harris, "The Geostationary Earth Radiation Budget experiment: Status and Science," Proceedings of the 2000 EUMETSAT Meteorological Satellite Data Users' Conference, Bologna, Italy, May 29-June 2, 2000, pp. 62-71

1139) J. E. Harries, D. Crommelynck, "The Geostationary Earth Radiation Budget experiment on MSG-1 and its potential applications," Advanced Space research, Vol. 24, 1999, pp. 915-919

1140) http://www.ssd.rl.ac.uk/gerb/

1141) P. Vogel, "The GERB Experiment," Proceedings of the EUMETSAT Meteorological Satellite Data User's Conference, Copenhagen, Denmark, Sept. 6-10, 1999, pp. 124-130 of the spectrum, essential for the understanding of the Earth's climate balance. The instrument is composed of two elements, the IOU (Instrument Optical Unit) and the IEU (Instrument Electronics Unit), featuring the following basic design:

- Three-mirror anastigmatic system + one rotating and one flat folding mirror
- Wide-band linear detector array
- Continuously rotating scan mechanism
- Channel separation via quartz filter
- Blackbody for thermal calibration
- Solar diffuser for shortwave calibration
- Passive thermal design
- Structure based on solid optical bench

At the core of the GERB instrument is a broadband, three mirror telescope housed in the IOU. This views the Earth with a black wideband detector array, providing measurements of the Earth's output radiation in a total band, and a shortwave band. Shortwave measurements are accomplished by using a quartz filter to block the wavelengths beyond 4 μm. The longwave band is obtained by subtraction. GERB removes the effect of S/C spin (100 rpm) by means of a rotating mirror, this increases the length of available exposure time per spin. The detector consists of a 256-element blackened thermoelectric array, mounted in the N-S direction. This arrangement provides an image column per S/C rotation; a complete image is obtained by successive measurements of columns. Every 15 minutes a complete dataset of both, solar and total spectral band, is obtained for the entire area visible from geostationary orbit. The great advantage of GERB is its ability to sample a large region of the globe with high time resolution. GERB instrument data provides close synergies with other instruments such as CERES, ScaRaB, and with SEVIRI on MSG (observations from both geostationary and polar-orbiting satellites). [1142)]

Spectral bands (2)	0.32 - 4.0 μm (Shortwave) Solar band	0.32 - 30 μm (Longwave) Total band
Radiometry: Absolute accuracy (each pixel) Noise (each pixel, 15 min av.) Dynamic range FSR (Full Scale Radiance)	 <2.4 Wm^{-2} sr^{-1} (i.e., $<1\%$) <0.8 Wm^{-2} sr^{-1} 0-380 Wm^{-2} sr^{-1} 240 Wm^{-2} sr^{-1}	 <0.4 Wm^{-2} sr^{-1} (i.e., $< 0.5\%$) <0.15 Wm^{-2} sr^{-1} 0-90 Wm^{-2} sr^{-1} 77 Wm^{-2} sr^{-1}
IFOV or pixel size (resolution)	44.6 km x 39.3 km (NS x EW) at nadir	
Coverage, cycle time	Full Earth disk, all channels in 5 minutes	
Co-registration	Spatial: 3 km with respect to SEVIRI at satellite sub-point Temporal: within 15 minutes of SEVIRI at each pixel	
Spectral and MTF	Performance specified by templates	
Sensor mass; power; data rate	25 kg; 35 W average; 50.6 kbit/s (L-band)	

Table 320: GERB instrument performance parameters

Instrument calibration: [1143)] [1144)] [1145)] Extensive pre-calibrations were provided on the ground. In-flight calibration is provided by a blackbody source and a solar transmission diffuser. For zero reference space views are used. An image is obtained by measuring the signal difference between views of the on-board blackbody and the Earth-view at every rotation of the satellite using the thermoelectric detector. Possible degradation of the shortwave spectral response can be corrected by means of occasional comparisons with an on-board solar-illuminated integrating sphere.

1142) J. Mueller, et al., "GERB: An Earth Radiation Budget Instrumentation on Second Generation Meteosat," Advanced Space Research, Vol. 24, No 7, 1999, pp. 921-924

1143) S. Dewitte, et al., "In-Flight Calibration of the GERB-1 Instrument," Proceedings of the EUMETSAT Meteorological Satellite Data User's Conference, Copenhagen, Denmark, Sept. 6-10, 1999, pp. 113-120

1144) J. Mueller, "Geostationary Earth Radiation Budget (GERB) Instrument Calibration Plans," Advances in Space Research, Vol. 19, No 9, 1997, pp. 1307-1316

1145) J. Mueller, et al., "Earth Radiation Budget Data from Geostationary Orbit," Proceedings IGARSS'99, Vol. 2, Hamburg, Germany, June 28 - July 2, 1999, pp.824-825

Parameter	Meteosat 1st generation (MOP)	Meteosat 2nd generation (MSG)
Visible channels	1	3 + HRV
Water vapor	1	2 channels
IR window (+absorption)	1 (+0)	6 (+2) channels
Sampling distance	VIS: 2.5 km, IR:5 km	VIS: 3 km, HRV:1 km, IR:3 km
Radiometric resolution	0.4 K	0.25 K
Image repeat cycle	30 minutes	15 minutes
Raw data rate	333 kbit/s	3.2 Mbit/s
Data collection system (DCS)	33 regional channels @ 0.1 kbit/s 33 international channels	210 regional channels @ .1 kbit/s 40 international channels
Primary dissemination	HRI: 166 kbit/s	HRIT: 1 Mbit/s
Secondary dissemination	WEFAX: analog	LRIT: 128 kbit/s
MDD (Met. Data Distribution)	MDD: up to 4x2 kbit/s	data in LRIT
DCP retransmission system	DRS: 12.5 kbit/s	data in LRIT

Table 321: Service comparison of 1st and 2nd generation Meteosat S/C [1146)]

S&R (Search and Rescue) payload within the S&RSAT/COSPAS system. Provision of transparent relay function for search and rescue operations (a 406 MHz transponder is carried by the MSG satellites).

DCS (Data Collection System) on-board collection system (see also F.7.1). The current data rate of 100 bit/s for each DCP (Data Collection Platform) in the ground segment will continue to be supported by MSG operations. MSG supports a significantly increased number of simultaneously relayed messaging channels.

Nr. of channels supported	Channel bandwidth	Frequency Range (MHz)	Usage
210	1.5 kHz	401.7025 - 402.0010	Neighboring satellites' regional DCP band (contingency relay only, no MSG processing)
33	3.0 kHz	402.0025 - 402.0985	International DCP band
224	1.5 kHz	402.1015 - 402.4360	MSG regional DCP band

Table 322: Some DCS performance characteristics of the MSG satellites

F.8.1 MSG Ground Segment (Stations)

The MSG ground segment is composed of:

- MCC (Mission Control Center) located at ESOC in Darmstadt, Germany
- PGS (Primary Ground Station) located in Usingen, Germany
- BRGS (Backup and Ranging Ground Station) located in Maspalomas, Gran Canary Island, Spain
- An application ground segment, which extracts meteorological and geophysical products from the calibrated and geolocated image data generated by MCC, and performs data management functions. The application ground segment is composed of MPEF (Meteorological Product Extraction Facility) and U-MARF (Unified Meteorological Archive and Retrieval Facility), both are located at EUMETSAT HQ in Darmstadt, and a distributed network of SAFs (Satellite Application Facilities). There are five SAFs using MSG data, they are:
  - SAF on Ocean and Sea Ice hosted by France
  - SAF on support to Nowcasting & VSRF hosted by Spain
  - SAF on Climate Monitoring hosted by Germany
  - SAF on Numerical Weather Prediction hosted by the United Kingdom

1146) R. Francis, V. Gärtner, "The Transition from MTP to MSG operations for the End-User Community," Proceedings of the EUMETSAT Meteorological Satellite Data User's Conference, Copenhagen, Denmark, Sept. 6-10, 1999, pp. 55-60

- SAF on Land Surface Analysis hosted by Portugal

- User Ground Segment. This comprises all MSG user stations. These are receive-only systems, operated by the users, which make use of either LRIT (Low Rate Information Transmission) or HRIT (High Rate Information Transmission) from the MSG satellites.

F.8.2 MSG Communication Services and Data Distribution

The MSG [1147] communication payload consists of three subsystems, namely the antenna subsystem, the transponder subsystem (including the S&R transponder), and the TT&C subsystem.

Parameter	Raw Data	HRIT	LRIT	DCP	S&R
Uplink frequency (MHz)	N/A	2015.65	2101.5	402.06	406.05
Downlink frequency (MHz)	1686.83	1695.15	1691.0	1675.281	1544.5
Useful signal bandwidth (MHz)	5.4	1.96	0.66	0.75	0.06
Bit rate	7.5 Mbit/s	2.3 Mbit/s	290 kbit/s	100 bit/s	400 bit/s
Modulation	QPSK	QPSK	BPSK	PM	PM

Table 323: Summary of MSG communication-link characteristics

Antenna	Type	Function	Reference Coverage	Polarization
L-band	EDA (Electrically Despun Antenna)	Raw data transmission HRIT & LRIT transmission DCP transmission S&R transmission	Elevation >5° Zone B Europe Europe	Linear horizontal
S/L-band	Slotted waveguide TPA (Toroidal Pattern Antenna)	HRIT & LRIT reception Raw data transmission (backup)	Europe	Linear horizontal
UHF-band	Crossed dipoles circular array EDA	DCP reception S&R reception	Full Earth coverage	Circular right-hand
S-band TT&C	Printed quadrifilar TPA	Telecommand reception Telemetry transmission Tracking	Azimuth 360° Elevation 120° from north axis	Circular

Table 324: Specification of some MSG antenna parameters

HRIT/LRIT Dissemination Service: A primary objective of the service is to deliver image data for nowcasting within a few minutes of the end of acquisition of each image, therefore the timeliness of data delivery is an issue of utmost importance.

Two dissemination channels are defined within the MSG system to broadcast data to end-users via (Low Rate and High Rate Information Transmission (LRIT/HRIT) schemes. Both channels multiplex image data from the SEVIRI and foreign satellites together with meteorological products and DCP data (link), within the limits of the defined packetized data rate. The limited channel capacity requires the use of data compression in order to maximize the amount of information to be transmitted. [1148]

The HRIT data stream has a capacity of 1 Mbit/s which allows the reception of the disseminated level 1.5 data in quasi real-time and full spatial and temporal resolution. It is segmented, encrypted and JPEG-compressed for the purpose of transmission. The compression is lossless for all channels and lossy for the HRV (High Resolution Visible) channel.

1147) K. D. McMullan, "The MSG Mission Communications Payload, including the Search and Rescue Transponder," Proceedings of the EUMETSAT Meteorological Satellite Data User's Conference, Copenhagen, Denmark, Sept. 6-10, 1999, pp. 106-113

1148) D. Just, Y. Buhler, "Data Compression for METEOSAT Second Generation," Proceedings of the EUMETSAT Meteorological Satellite Data User's Conference, Copenhagen, Denmark, Sept. 6-10, 1999, pp. 131-138

The LRIT data stream capacity is 128 kbit/s. The images are all compressed in a lossy manner (by about a factor of 8) in full spatial resolution with the image data rounded to 8 bits representation. The current baseline is that only every second image is disseminated.

Parameter	LRIT	HRIT
Length of coded VCDU	1020 octets	
Center frequency	1691.0 MHz	1695.15 MHz
Bandwidth	0.660 MHz	1.960 MHz
Polarization	linear horizontal	
Packetized data rate	128 kbit/s	1 Mbit/s
Total coded data rate	293.9 kbit/s	2.3 Mbit/s
Modulation	PCM/NRZ/BPSK	PCM/NRZ/QPSK
Pulse shaping	Raised cosine filter Roll-off factor 1.0	Raised cosine filter Roll-off factor <= to 0.7
Coding	Concatenated coding, Reed-Solomon (255,223)+convolutional (1/2 rate, k=7)	
Coding gain	9.4 dB	
Eb/No for required probability of frame loss	2.8 dB	
Achieved margins in case of nominal S/C and user station	Worst case 0 dB Nominal case 3 dB	

Table 325: Physical link parameters of the HRIT/LRIT service

Data Product	Description
Level 1.0	SEVIRI and GERB data as observed by the satellite (raw data)
Level 1.5	Geometrically corrected, navigated, Earth located and calibrated SEVIRI data
Level 2.0	Geophysical parameters extracted from level 1.5 SEVIRI data by MPEF and SAFs
Level 3.0	Further products derived from level 2.0 SEVIRI data by some SAFs

Table 326: MSG generic data products

F.9 MTSAT (Multifunction Transport Satellite)

MTSAT is a Japanese geostationary multi-purpose satellite program, procured by JMA (Japan Meteorological Agency) and JCAB (Japan Civil Aviation Bureau), and funded by the Japanese Ministry of Land, Infrastructure and Transport, with the following overall objectives: [1149] [1150] [1151]

1) **Aeronautical mission**. To augment the air-traffic control services through enhanced communication and position information to Japan's (and the Asia-Pacific region) air traffic capability. The objective is to provide a satellite-based service, referred to as CNS/ATM (Communication, Navigation, Surveillance / Air Traffic Management) by utilizing the GNSS and augmentation services for satellite navigation. The CNS/ATM concept makes use of the ADS (Automatic Dependent Surveillance) function which automatically reports the current aircraft position measured by GNSS to ATC (Air Traffic Control Center).
2) **Meteorological mission.** MTSAT is a successor program to the GMS (Geostationary Meteorological Satellite) series - representing the next-generation weather-watch satellite series. The advanced meteorological observation capabilities include data collection services for a large user community.

1149) "Meteorological Observation with the Multi-functional Transport Satellite (MTSAT)," brochure of JMA
1150) MTSAT, brochure of JCAB
1151) A press kit "Launch and Tracking & Control of the Multifunctional Transport Satellite (MTSAT) was provided by T. Moriyama of NASDA

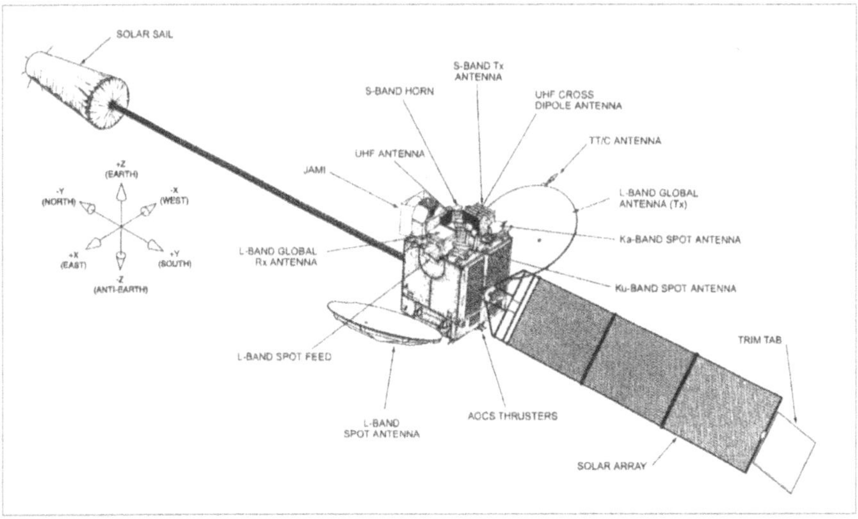

Figure 198: Illustration of the MTSAT-1R spacecraft

The MTSAT series uses the services of GPS satellites in order to provide (supplementary) enhanced aeronautical communication, air navigation, GPS augmentation and differential correction services. Several ground stations complement the space segment (There are two JCAB stations and a separate JMA ground station. The JMA station includes satellite monitoring and control functions, communications functions and image processing and dissemination functions.

The overall MTSAT space segment is composed of two spacecraft, MTSAT-1R and MTSAT-2. MTSAT-1R is scheduled for launch in 2003, MTSAT-2 in 2004.

Background: MTSAT-1 was launched on Nov. 15, 1999 with an H-2 vehicle from the Tanegashima Space Center, Japan. However, a launch failure of the H-2 vehicle (No 8) occurred due to an abnormal burning of the first stage resulting in a destruction of the entire vehicle. As of March 2000, Space Systems/Loral (SS/L) of Palo Alto, CA, received a follow-up contract from Japan's Ministry of Transport [1152] to build a replacement satellite, MTSAT-1R. In addition, Japan's Ministry of Transport ordered MTSAT-2 from MELCO (Mitsubishi Electric Company) of Tokyo, Japan.

F.9.1 MTSAT-1R

The MTSAT-1R spacecraft bus, built by SS/L, uses a three-axis, stabilized FS 1300 platform. The overall S/C configuration consists of a box-like main structure, a single-wing solar array mounted on the south face of the structure, a solar sail in the north direction, and two large L-band antennas mounted on the east and west faces. The overall length of the S/C is 33.05 m. The attitude control subsystem uses a bias momentum method to stabilize the S/C. Attitude is measured by an Earth sensor. A boom-mounted solar sail in the opposite direction of the solar array, serves as attitude actuator by providing a counterbalance to the solar pressure. A trim tab is mounted on the outer edge of the solar array offering a control mechanism. The solar array, consisting of three panels, uses GaAs cells and continuously tracks the sun in order to maximize power output. Power of 3087 W (EOL) is provided. The NiCd battery has 22 cells for eclipse phase power. The bi-propellant propulsion subsystem uses 12

1152) Note: As of Jan. 1, 2001, the Japanese Ministry of Transport changed its name to: Japanese Ministry of Land, Infrastructure and Transport (MLIT).

thrusters (22 N each) for attitude control. In addition a thruster of 490 N provides control for initial orbit raising. The planned S/C mass at launch is 3251 kg with 1883 kg of propellant mass. [1153]

Aeronautical Payload: Frequency bands and EIRP (Effective Isotropic Radiated Power)	L-band 6 spot beams: 43 dBW, (1.5/1.6 GHz) L-band 1 global beam: 40 dBW L-band GNSS: 31 dBW Ku-band 4 spot beams: 22-27 dBW, (12/14 GHz) Ka-band 3 spot beams: 22-27 dBW, (20/30 GHz)
Meteorological Payload: Frequency bands and EIRP	S-band global: 55 dBm UHF global: 44 dBm
TT&C (Telemetry Tracking & Command): Frequency bands and EIRP	USB (Unified S-band): 27.5 dBm (USB is used for orbit raising and contingency operations via the omni-directional telemetry and command antenna) S-band global: 35.5 dBm Ku-band spot: TBD, downlink 12 GHz, uplink 14 GHz
Spacecraft mass	About 1,400 kg of dry mass, about 3,300 kg of launch mass
Length of S/C, S/C power	33 m (total length), 3 kW
Spacecraft design life	10 years for aeronautical mission 5 years for meteorological mission

Table 327: Some parameters of the MTSAT-1R spacecraft

Orbit: Geostationary orbit between 130-140º eastern longitude.

F.9.1.1 Aeronautical Mission

The objective is to use a spaceborne augmentation system to improve the management of Japan's civil air traffic. Faced with the constant growth in air transport in South-East Asia and the special geographical factors in this region, the Japanese Ministry of Transport represented by the Japanese Civil Aviation Bureau (JCAB) has opted for a full-scale implementation of a satellite-based Communication, Navigation, Surveillance/Air Traffic Management (CNS/ATM) concept. This concept is based on the use of a geostationary satellite in order to overcome the limits imposed by traditional terrestrial navigation aids.

The MTSAT aeronautical mission provides three basic functions for its area of coverage: communications, navigation, and surveillance.

1) **Communications**

Voice/data interchange between aircraft and control tower. Air traffic communications use the L-band and the AMSS (Aeronautical Mobile Satellite Service) protocol which is specific to mobile telecommunications. Links between the Earth segment and satellite segment use the Ku-band and Ka-band.

MTSAT significantly improves the quality and capacity of voice and data communications to/from aircraft compared to conventional systems. In addition, it permits direct communication between any two aircraft in a region as well as to ground control centers.

2) **Navigation by satellite - referred to as SBAS (Satellite-Based Augmentation System)**

MTSAT serves as a geostationary wide-area augmentation system for all GPS signals in its large area of coverage by transmitting:

- GPS-like signals (ranging function)
- GPS health and integrity conditions obtained by ground monitoring stations (GMS)
- Ranging errors (differential correction function) - these are the conventional DGPS services

The GPS wide-area signal augmentation provided by MTSAT makes all regional navigation seamless, very reliable, available, more accurate, and continuous. This service is referred to

1153) Information provided by Ken Faller of Space Systems/Loral as **MSAS** (MTSAT Satellite-based Augmentation System). The MSAS architecture is composed of the following ground-based elements: MRS (Monitor and Ranging Stations), GMS (Ground Monitor Stations), and MCS (Master Control Stations). In addition, there are two GES (Ground Earth Station) in the system, namely the Kobe Aeronautical Satellite Center and the Hitachi-Ota Aeronautical Center), providing operational support.

MSAS information on the real-time condition of the GPS constellation is transmitted to each aircraft via the integrity function of MSAS, while the differential correction function provides ranging error data to each aircraft. MSAS uses advanced technologies such as satellite orbit ranging, and ionospheric and tropospheric delay estimation assumption to ensure the reliability of these functions.

MTSAT-1R navigation payload SBAS: The navigation payload is built by Alcatel Space Industries of France under subcontract to SS/L.

3) **Air traffic surveillance**

MTSAT-1R is the key element of the Japanese GNSS (Global Navigation Surveillance System). The S/C features an ADS (Automatic Dependent Surveillance) relay service with the objective to transmit accurate positioning data for aircraft obtained by GPS (or GLONASS) to achieve higher-quality control. The data is transmitted at regular intervals to ground control centers via MSAS. ADS enables positioning information of aircraft to be displayed on control consoles, permitting ground control personnel to verify the aircraft position anywhere along its flight path. Such data include aircraft identification, four-dimensional position, and supplementary data to the air traffic control facility. The links between the ground and the satellite use the Ku-band and Ka-band; links between the aircraft and the satellite use the L-band.

The overall design of the "aeronautical mission" with its three basic functions provides redundancies. However, these redundancies are deliberately designed into the space segment as well as into the ground segment to provide an operational system that satisfies every aspect of the ICAO (International Civil Aviation Organization) requirements in terms of reliability, availability, and accuracy.

F.9.1.2 Meteorological Mission

The Meteorological Mission of MTSAT succeeds the observational function of the current GMS (Geostationary Meteorological Satellite) series. GMS-5 (with VISSR), the latest in the series, was launched March 18, 1995 has exceeded its design life already in 2000. It may run out of fuel in less than a year and will very probably not be able to continue its service until the launch of MTSAT-1R.

JAMI (Japanese Advanced Meteorological Imager). The overall objective of the meteorological mission is to provide an improved functionality and continuous service of meteorological observables (timely, high-quality, full-disk multispectral imagery for operational weather needs) for Japan and for the countries in the area of MTSAT coverage (East Asia, and Australia). The JAMI instrument is being developed by Raytheon SBRS under contract to Space Systems/Loral. The JAMI performance enhancements include: [1154)]

- Coverage of the 0.55 μm to 12.5 μm spectral region using four infrared (IR) bands and one solar reflective band (VNIR)
- Large-format arrays enable faster full disk coverage rate with slower scan rate. The benefits of slower scan rate include longer dwell times, higher detector resolution, and a longer scanner life
- Full disk coverage with calibration scans in 24 minutes
- Onboard calibration system for all bands

1154) Information provided by Dominick M. DellaValle of SBRS, Goleta, CA

Figure 199: Illustration of the JAMI instrument

JAMI's design is based on advanced imager technologies that have already been space-qualified and flown in Raytheon-built hardware:

- Uses a two-axis gimbaled scan mirror with qualified optical and mechanical elements which have been derived from many Raytheon programs with demanding requirements for spaceflight operation including: MODIS, GOES/GMS (Imager/VISSR), Landsat (TM, ETM+), and TRMM/VIRS
- Photovoltaic HgCdTe focal plane arrays that incorporate qualified designs with direct flight experience in TRMM/ VIRS, Multispectral Thermal Imager (MTI), and MODIS. Active cooling of the focal plane detector arrays is provided.
- Processing hardware and software was derived from MODIS and incorporates processing algorithms which have already been proven on Landsat, Hubble Space Telescope, and many other spaceflight programs
- Equipped with an onboard blackbody calibrator that has already been qualified and spaceflight proven on TRMM/ VIRS and MODIS with four years of on-orbit operation

Spectral bands	VNIR: 0.55 - 0.90 µm (cloud cover) MWIR1: 3.5 - 4.0 µm (nighttime cloud cover) MWIR2: 6.5 - 7.0 µm (water vapor) TIR1: 10.3 - 11.3 µm (SST and water vapor) TIR2: 11.5 - 12.5 µm (SST and water vapor)
Scanner	Two-axis scan system
Detectors	Monolithic silicon (visible) Photovoltaic HgCdTe (all infrared regions)
Ground resolution	0.5 km in VNIR (Visible Near Infrared), 2 km in infrared regions
Data rate	2.8 Mbit/s
Instrument mass, power	163.7 kg (including 16.7 kg of interconnecting harness), 169 W

Table 328: JAMI instrument specifications

Communication of meteorological data: Imagery obtained is transmitted to a ground station, referred to as CDAS (Command and Data Acquisition Station) where some information, such as calibration tables, are added. In addition, HiRID, HRIT (High Rate Information Transmission), and LRIT (Low Rate Information Transmission) products, quite similar to the EUMETSAT products, are generated and distributed. The data are transmitted in S-band (downlink frequency: 2026-2035 MHz, uplink frequency: 1677-1695 MHz). An S-band slot array antenna is used for downlink transmissions while an S-band horn antenna is employed for uplink reception. The S-band downlink carries the raw imager data, the image products that are being retransmitted after generation by CDAS, the data received from the CDP (Data Collection Platforms), and spacecraft telemetry data.

MTSAT broadcasts processed data and imagery to users throughout the Asia-Pacific region, including airports, weather forecasting agencies and ships at sea.

F.9.1.3 DCS (Data Collection System)

The MTSAT series supports the collection of environmental data in its ground segment with a corresponding on-board DCS (Data Collection System), in the same manner as provided by the GMS series. The ground segment platforms are referred to as DCP (Data Collection Platform) which are automated remote weather stations.

The DCS is part of IDCS (International DCS) coordinated with other meteorological satellite operators. The DCP data are collected through the MTSAT spacecraft and rectified to the WMO codes at MSC and distributed to users through GTS.

- IDCP (International DCP). The IDCP, a self-timed system, supports mobile DCPs at various locations (on ships, aircraft and balloons) which may move from the coverage of one geostationary meteorological satellite to another.
- RDCP (Regional DCP). The MTSAT S/C supports regional DCPs located on moored buoys, on remote islands, on mountains, etc. - and mobile DCPs in its own coverage area. MTSAT provides 100 channels for the RDCP function. RDCPs are interrogation-type system.

There are two types of DCPs: 1) Self-timed DCPs which transmit observation data automatically on scheduled times. 2) The interrogation DCPs transmit observation data by an interrogation command from the MSC.

The interrogate command is uplinked to the satellite via CDAS in S-band, and then repeated by the satellite in UHF. The UHF signal is broadcast to DCPs in a global pattern using a dipole antenna. The same antenna receives the uplinked UHF signal from the DCPs. The DCP signal is retransmitted to CDAS via the same S-band downlink that carries the imager data, from where it is sent by terrestrial microwave link to MSC.

In addition, the DCS is used to transmit Tsunami Earthquake emergency information and seismic intensity data.

Frequency bands allocated	402.0 - 402.1 MHz (IDCP) 402.1 - 402.4 MHz (RDCP)
Data rate	100 bit/s
Power	43 - 46 dBm
Data modulation	PCM/PSK ($\pm 60°$)

Table 329: Specification of DCPs

Part G Meteorology - LEO (Low Earth Orbit) Missions

G.1 DMSP (Defense Meteorological Satellite Program)

Background:[1155),1156),1157)] DMSP is the meteorological long-term program of the US Department of Defense (DoD) which originated in the mid-1960s with the objective to collect and disseminate worldwide cloud cover data on a daily basis. DMSP was originally known as DSAP (Defense System Applications Program) and as DAPP (Defense Acquisition and Processing Program). [1158) 1159)]

From mid-1965 to early 1970, the series of DMSP satellites was known as the Block 4 series. S/C of this series were of the shape of a spin-stabilized octagon with a mass of about 150 kg. The payload consisted of two vidicon cameras for collecting high-resolution TV pictures (resolution: 2.4 km at image center to about 16 km at the edge of a picture). Some later S/C also had a low-resolution VIS (0.4 - 4 μm) and infrared (8.0 - 12 μm) sensor. From these, better sensors emerged with first attempts to combine reflected and emitted energy for the cloud analysis problem.

Beginning in early 1970, a new generation of satellites was inaugurated, the Block 5 series. These were three-axis stabilized spacecraft that initially employed a scanning radiometer as imaging sensor (two channels). The VIS channels had a spectral band from 0.4 - 1.1 μm and the lower resolution channel (approx. 3.6 km at nadir) also had a low-light level amplification system. These two design factors considerably enhanced the utility of the system.

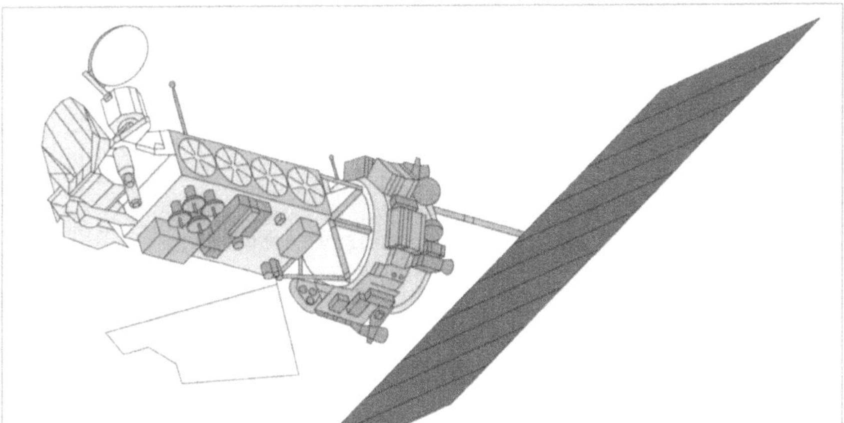

Figure 200: Schematic illustration of DMSP Block 5D-3 S/C model

1155) W.D. Meyer, "DMSP: Review of its Impact, in Monitoring Earth's Ocean, Land, and Atmosphere from Space," Volume 97, Progress in Astronautics and Aeronautics, AIAA, 1985, pp. 131- 147

1156) S. Ferry, "The Defense Meteorological Satellite System Sensors: An Historical Overview," May 1989

1157) R. B. Gomez, M. C. Colton, D. Boucher, F. P. Kelly, "The Defense Meteorological Satellite Program (DMSP)," ISSSR, Maui, HI, Nov. 16-20, 1992

1158) W. D. Meyer, "The Defense Meteorological Satellite Program: A Review of its Impact," Monitoring Earth's Ocean, Land, and Atmosphere from Space - Sensors, Systems, and Applications, Progress in Astronautics and Aeronautics, AiAA, Volume 97, 1985, pp. 129-149

1159) http://www.laafb.af.mil/SMC/SESS/

S/C Bus	S/C Series	Launch Date/ Mission End	Sensor Complement	S/C Mass (kg)
Block 4A	F-1	19.1.1965		150
	F-2	18.3.1965		130
	F-3	20.5.1965		130
	F-4	10.9.1965		130
	F-5	6.1.1966		130
	F-6	30.3.1966		130
	F-7	16.9.1966		125
	F-8	8.2.1967		130
	F-9	23.8.1967		195
	F-10	11.10.1967		195
Block 5A	F-1	23.5.1968		195
	F-2	23.10.1968		195
	F-3	23.7.1969		195
	F-4	11.2.1970		195
	F-5	3.9.1970		195
	F-6	17.2.1971		195
Block 5B	F-1	14.10.1971		195
	F-2	24.3.1972		195
	F-3	9.11.1972		195
	F-4	17.8.1973		195
	F-5	16.3.1974		195
	F-6	9.8.1974		195
Block 5C	F-1	24.5.1975		194
	F-2	19.2.1976		175
Block 5D-1	F-1	11.9.1976 / 17.9.1979	OLS, SSH, SSJ/3, SSB, Contamination Monitor	450
	F-2	4.6.1977 / 19.3.1978	OLS, SSH, SSJ/3, SSB, SSB/0, IFM, SSI/E, SSI/P	450
	F-3	30.4.1978 / Dec. 1979	OLS, SSH, SSJ/3, SSB, GFE-3R	513
	F-4	6.6.1979 / 29.8.1980	OLS, SSH, SSJ/3, SSI/E, SSM/T, SSC, SSD	513
	F-5	14.7.1980 (failed)	OLS, SSH-2, SSJ/3, SSI/E, SSB/O, SSR	513
Block 5D-2	F-6	20.12.1982 / 24.8.1987	OLS, SSH-2, SSI/E, SSJ/4, SSB/A	750
	F-7	18.12.1983/17.10.1987	OLS, SSM/T, SSI/E, SSJ/4, SSB, SSJ*, SSM	750
	F-8	18.6.1987 / 13.8.1991	OLS, SSM/I, SSM/T, SSI/ES, SSJ/4, SSB/X-M	750
	F-9	3.2.1988	OLS, SSM/T, SSI/ES, SSJ/4, SSK	750
	F-10	1.12.1990 /2. 1995	OLS, SSM/I, SSM/T, SSI/ES, SSJ/4, SSB/X-2	750
	F-11	28.11.1991 / 8. 2000	OLS,SSM/I, SSM/T,SSM/T-2,SSJ/4,SSI/ES-2, SSB/X-2	830
	F-12	29.8.1994	OLS, SSM/I, SSM/T, SSM/T-2, SSJ/4, SSI/ES-2, SSB/X-2, SSM	830
	F-13	24.3.1995	OLS, SSM/I, SSM/T, SSJ/4, SSI/ES-2, SSB/X-2, SSM, SSZ,	750
	F-14	4.4.1997	OLS, SSM/I , SSM/T, SSM/T-2, SSJ/4, SSI/ES-2, SSM,	750
Block 5D-3	F-15	12.12.1999	OLS, SSM/I, SSJ/4, SSI/ES-2, SSM-Boom, SSZ	1125
	S-20	projected for Nov. 2001 to become F-16	OLS, SSMIS , SSI/ES-3, SSJ5, SSM-Boom, SSULI, SSUSI, SSF,	1125
	S-17	projected for 2003 to become F-17	OLS, SSMIS, SSI/ES-3, SSJ5, SSM-Boom, SSULI, SSUSI, SSF,	1125
	S-18	projected for 2004 to become F-18	OLS, SSMIS, SSI/ES-3, SSJ5, SSM-Boom, SSULI, SSUSI, SSF,	1125
	S-19	projected for 2006 to become F-19	OLS, SSMIS, SSI/ES-3, SSJ5, SSM-Boom, SSULI, SSUSI, SSF,	1125
	S-16	projected for 2007 to become F-20	OLS, SSMIS, SSI/ES-3, SSJ5, SSM-Boom, SSULI, SSUSI, SSF,	1125

Table 330: Chronological overview of DMSP series satellites

The overall DMSP satellite program is composed of three elements: the space segment, the C3S (Command, Control and Communications Segment), and the user segment. DMSP is managed by SMC (Space & Missile Systems Center) at Los Angeles Air Force Base, El Segundo, CA.

DMSP employs the concept of total system integration from spacecraft operation through data reduction and interpretation. The space segment is controlled by the Air Force Space Command (AFSPC) through command sites centered in Colorado Springs, CO. By the end of 1992, 22 Air Force terminals, all equipped to receive Block 5 image data, were deployed worldwide.

Some DMSP satellite characteristics: The satellites are built by LMMS (Lockheed Martin Missile & Space) company. They are three-axis stabilized. Pointing of the satellites is maintained by three orthogonal reaction wheel assemblies. A precision mounting platform provides a pointing accuracy of 0.01°. The S/C structure has a diameter of about 1.2 m and a length of about 3.7 m. The S/C mass of Block 5D-2 is about 820 kg in orbit, including the sensor payload of 250 kg. The S/C mass of Block 5D-3 is about 1210 kg, including 310 kg of payload.

DMSP satellites are launched from Vandenberg AFB. Real-time commands and stored command files are relayed via commercial communication satellites to Command Readout Stations (CRSs) which then relay these commands to the DMSP satellites. Data may also be relayed via tracking stations [located in Hawaii, New Hampshire and Thule (Greenland)]. The meteorological data are processed by centralized users at the Air Force Global Weather Center, Fleet Numerical Meteorology and Oceanography Center (FNMOC), and deployed tactical assets.

The Block 5D-1 satellites were flown from 1976 to 1980. The first Block 5D-2 satellite was launched in 1982. Block 5D-2 satellites are the designated satellites until the late-1990s when the first of the Block 5D-3 satellite series was launched (F-15 launch on Dec. 12, 1999). The block 5D-3 satellites are projected to carry an upgraded complement of sensors to the 5D-2 configurations. The required pointing accuracy of 0.01° is achieved by star sensing and three orthogonal gyroscopes measuring short-term variations in the S/C attitude.

NPOESS (National Polar-orbiting Operational Environmental Satellite System): The DoD DMSP program and the NOAA-POES program convergence takes place in two phases.

- First phase: Starting in May 1998, all DMSP satellite operational command and control functions of AFSPC (Air Force Space Command) were transferred to a tri-agency Integrated Program Office (IPO) established within NOAA. NOAA was given the sole responsibility of operating both satellites programs, POES and DMSP (from NESDIS, Suitland, MD).
- During the second phase, the IPO will launch and operate the new NPOESS satellites that will satisfy both the DoD and Department of Commerce (DOC/NOAA) requirements.

Orbital Parameters of DMSP: [1160)]
Sun-synchronous orbits, altitude = 811 - 853 km (833 km nominal), inclination = 98.9°, period = 101.6 minutes, there are normally two satellites in operation at any one time (one in a morning and one in a late morning equatorial crossing time).

G.1.1 Description of Block 5D-2 and 5D-3 Sensors [1161)]

OLS = Operational Linescan System (primary sensor on each satellite, built by Northrop Grumman, Westinghouse Corporation). Objective: day and night cloud cover imagery. The OLS instrument consists of two telescopes and a photomultiplier tube (PMT). The detectors sweep back and forth in a "whiskbroom" fashion (a "flying spot design" is employed - a

1160) R. Massom, "Satellite Remote Sensing of Polar Regions," Applications, Limitations and Data Availability, Belhaven Press, London subset of whiskbroom scanners, O.3.2). The continuous analog signal is sampled at a constant rate so the Earth-located centers of each pixel are roughly equidistant, i.e., 0.5 km apart, 7,325 pixels are digitized in the cross-track direction.

Swath width = 3000 km from a nominal 833 km orbit altitude. OLS provides global coverage in both visible (L data) and IR (T data) modes. Fine resolution data with a nominal linear resolution of 0.56 km are collected as needed day and night by the IR detector, and as needed during daytime, by a segmented silicon diode detector (LF data). A high resolution photometer tube is used for nighttime visible imagery (used for fire detection).

- Band 1: VIS wavelength = 0.4 - 1.1 μm (0.58 - 0.91 μm FWHM)
- Band 2: TIR wavelength = 10.0 - 13.4 μm (8 - 13 μm, old prior to 1979), resolution = 0.56 km for fine resolution data (stored data is smoothed to 2.7 km resolution), continuous data collection, polar stereographic image products have ground resolution of 5.4 km. Swath = 3000 km. The IR system counts are automatically calibrated to vary between 190 and 310 K of effective blackbody or brightness temperatures.
- The PMT is sensitive to radiation from 0.47 - 0.95 μm (0.51 - 0.86 μm FWHM).

The scanning telescope of OLS is a f/5.8 Cassegrain design with a 20.3 cm clear aperture and an effective collecting area of about 185 cm^2. The telescope has an effective focal length of 122 cm. Two telescope calibration mirrors intercept the normal FOV at the edge of scan with hot and cold loads of known temperatures. The light from the telescope is split into two channels by a beam splitter, and sent via relay optics to the visible and infrared focal planes, as well as to the photo multiplier tube that provides useful nighttime visible imagery (approx. 0.5 - 0.85 μm half-power response point bandpass) down to a lunar illumination level of about a quartermoon. The telescope images over a scan angle of ±56.25° which corresponds to a swath width on the ground of 2960 km (some overlap at the equator from orbit to orbit). Glare suppression is built into the system by a variety of sun shades, field stops, low scatter surfaces, and aperture stops. This minimizes the amount of data lost in the orbit due to sunlight saturating the visible detectors. OLS provides the ability to command gain adjustments to overcome slow degradation of the thermal transmission of the system. The system is designed to produce near-constant high resolution imagery for most DOD applications because the location of clouds, fog, ice floes, etc., is much more important than accurate radiometry. The near constant resolution across the swath is accomplished through a combination of the natural rotation of the detector footprint and detector segment switching. On-board data smoothing (averaging) can be done to reduce the data rate by a factor of 25, smoothing electronically the pixels in the cross-track direction and digitally averaging in the along-track direction. However, smoothing is only done in cases to cope with current recorder limitations. When this mode is used, the original high-resolution imagery cannot be recovered.

OLS utilizes three types of detectors:

- A silicon photoconductive diode is used for daytime VIS imagery. Three segments in the detector provide for a nearly constant resolution across the swath. Two smaller segments are used for scan angles >41°±±°, all three segments are summed in the middle portion of the scan about nadir.
- A two-segment HgCdTe photoconductive detector is used for the TIR channel. The detector is cooled to 108 K by a cone cooler. The two detector segments are used on the far right and left parts of the scan and are summed over the middle portion of the scan (within ±41°).
- A single PMT detector is used for nighttime visible data [a GaAs opaque photocathode and multiple dynode PMT).

1161) J. Goyette, L. Belsma, J. Bohlson, D. Glackin, "Defense Meteorological Satellite Program capabilities through the end of this century and requirements for the converged DMSP NOAA Systems," European Symposium on Satellite Remote Sensing II, Paris, France, Sept. 25-28, 1995, published in SPIE Vol. 2578

SSM/I = Special Sensor Microwave Imager. [1162)]

The SSM/I is a seven-channel, four-frequency, linearly-polarized, passive microwave radiometer which measures atmospheric, ocean, and terrain microwave brightness temperatures (similar to NIMBUS-7 SMMR) which are converted into environmental parameters such as: sea surface winds, rain rates, cloud water, precipitation, soil moisture, ice edge, and ice age. SMM/I data is used to obtain synoptic maps of critical atmospheric, oceanographic and selected land parameters on a global scale. The archive data consists of antenna temperatures recorded across a 1400 km swath (conical scan), satellite ephemeris, Earth surface positions for each pixel and instrument calibration. The electromagnetic radiation is polarized by the ambient electric field, scattered by the atmosphere and the Earth's surface, and scattered and absorbed by atmospheric water vapor, oxygen, liquid water and ice. [1163)]

Channel Nr.	Center Frequency (GHz)	Wavelength	Footprint (km) along x cross	Polarization	NEDT (K)	Environmental Response
1	19.35	1.55 cm	68.9 x 44.3	Vertical	0.45	Ocean surface wind, land surface moisture
2	19.35	1.55 cm	69.7 x 43.7	Horizontal	0.42	
3	22.235	1.35 cm	59.7 x 39.6	Vertical	0.73	Ocean surface wind, land surface moisture
4	37.0	0.81 cm	35.4 x 29.2	Vertical	0.37	Rain, cloud water content, ice cover
5	37.0	0.81 cm	37.2 x 28.7	Horizontal	0.38	
6	85.0	0.35 cm	15.7 x 13.9	Vertical	0.69	
7	85.0	0.35 cm	15.7 x 13.9	Horizontal	0.73	

Table 331: Specification of some SSM/I parameters

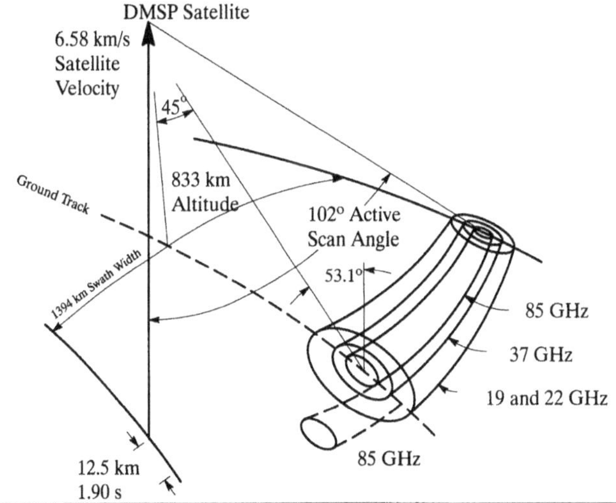

Figure 201: The scan geometry of the DMSP SSM/I sensor[1164)]

The SSM/I instrument was built by Hughes Space and Communications Co. It consists of an offset parabolic reflector (61 cm x 66 cm) that is fed by a seven-port horn antenna. The re-

1162) J. W. Sherman, "The Near-Term Suite of Satellite Sensors to Support Developing Countries' Climate and Global Change Photograms," Proceedings of the Twenty-Fourth International Symposium on Remote Sensing of the Environment, ERIM Ann Arbor MI., Volume I, 27-31 May 1991, pp. 27-28

1163) J. P. Hollinger, J. L. Pierce, G. A. Poe, "SSM/I Instrument Evaluation," IEEE Transactions on Geoscience and Remote Sensing, Vol. 28, No. 5, pp. 781-790, 1990

1164) The rotating antenna sweeps the surface in two alternating modes - one in which all four frequencies are recorded, and another in which only 85 GHz data are recorded. The use of a single antenna results in different ground resolutions for each frequency.

flector and feed are mounted on a drum which contains the radiometers, digital data subsystem, mechanical scanning subsystem, and power subsystem. The drum assembly rotates about the axis of the drum. A small mirror and a hot reference absorber are mounted on the assembly.

The SSM/I sensor executes a 45° conical scan of the Earth's surface from nadir. This gives a nominal incidence (zenith) angle of 53.1° to the Earth's surface from the nominal orbit. Only part of the possible 360° scan in azimuth is used to collect data. The active azimuth scan angle is 102.4° is ahead of the S/C for an afternoon ascending orbit and behind the S/C for a morning ascending orbit. The sensor electronics perform an integration and hold sequence on each channel, timed so that the odd 85 GHz reading is centered with the 37 GHz reading. The sensor conically scans the Earth and atmosphere at a scan rate of 31.9 scans/min (or 1.88 s/scan). The sampling scheme results in 128 samples (called scene stations) per scan for the 85 GHz channels, and 64 scene stations per two scans for the other channels. The footprint sizes are listed in Table 331. The footprints are sampled every 25 km for the 19, 22, and 37 GHz channels, and every 12.5 km for the 85 GHz channel.

Each scan consists of a cold reading, a warm load reading, and the scene stations. The cold reading utilizes a view to deep space (3 K black body), and the warm load temperature (variable over an orbit) is read by three precision thermistors. Each sensor has a different set of coefficients that enters into the retrieval algorithms.

SSM/T-1 (Special Sensor Microwave Temperature Sounder), in operation since 1979, first flown on F-4, built by the Aerojet Corporation. The SSM/T-1 is a seven channel passive microwave sounder consisting of a rotating 7 step antenna reflector (rotating once every 32 seconds), a Dicke-switched seven channel radiometer, a digitizer, and signal processor.. It measures the Earth's surface and atmospheric emission in the 50 to 60 GHz oxygen band. The SMM/T-1 is a cross-track nadir scanning radiometer having a FOV of 14.4°. At nominal altitude (833 km) the subtrack spatial resolution is an approximate circle of 174 km diameter at nadir elongating to an ellipse of 305 x 313 km at the extreme viewing angles toward the limb. There are seven total cross-track scan positions separated by 12° with a maximum cross-track scan angle of 36°. Swath width = 1500 km (data coverage gap between successive orbits). The SSM/T-1 is a step and stare type sensor which dwells on each of the seven scene stations, then observes a cold (3 K) reference, followed by a warm (300 K) reference. The daily data volume is 1.5 MByte per satellite.

Channel Nr.	Frequency (GHz)	Peaking altitude (km)	NEDT (K)
1	50.5	0	0.6 (window channel)
2	53.2	2	0.4
3	54.35	6	0.4
4	54.9	10	0.4
5	58.4	30	0.5
6	58.825	16	0.4
7	59.4	22	0.4

Table 332: Parameters of the SSM/T-1 instrument

SSM/T-2 (Special Sensor Microwave Water Vapor Profiler-2)
The SSM/T-2 is a modification of SSM/T-1 for water vapor sounding. It was built by the Aerojet Corporation. The first flight of the instrument took place on F-11 in 1991. The instrument is a cross-track scanning, five channel, passive total power microwave radiometer system which consists of a single, self-contained module with a step-scan motion in the cross-track direction of ± 40.5°. The SSM/T-2 scan mechanism is synchronized with the SSM/T-1 so that the beam cell patterns of the two sensors coincide. The observation rate is 7.5 scans/minute. There are 28 observations (beam positions) per scan for each of the five channels, with each observation having a spatial resolution of about 48 km. All five channels have coincident centers. The swath width is about 1500 km.

The instrument employs a single offset parabolic reflector with a 6.6 cm diameter projected aperture. The reflector is shrouded to eliminate the possibility of rays from the sun striking either of the calibration paths and causing unwanted thermal gradients. The feedhorn is a corrugated pyramidal horn with a flare designed to minimize phase center separation over the bandwidth (91 to 183.3 GHz), while providing a spherical wave illumination of the reflector. To achieve the cross-track scanning, the reflector alone rotates. The rotation of the reflector produces a rotation of the plane of polarization of the upwelling scene which is permitted provided that the polarization remains identical for the two window channels and 183.3 ± 7 GHz. These channels must have the same polarization characteristics because they measure contributions from both the atmosphere and the surface. During each scan period, and for all five channels at 28 discrete earth viewing positions, four discrete calibration measurements of a hot-load target (about 300 K), and cosmic background radiation (about 3 K) are monitored.

Channel Nr.	Frequency (GHz)	Peaking altitude (km)	NEDT (K)	Beam width
1	183.310 ±3.000	8	0.6	3.3°
2	183.310 ±1.000	6	0.8	
3	183.310 ±7.000	10	0.6	
4	91.655 ± 1.250	0	0.6 window channel	6°
5	150.0 ± 1.250	2	0.6	3.7°

Table 333: Parameters of the SSM/T2 instrument

The sensor has footprints at nadir (resolutions) ranging from 48 km (for highest frequency) to 120 km. The noise levels (NEDT) for the five channels are 0.6 K to 0.8 K. The peaking measurement heights from 0 to 10 km vary because they are dependent on the total atmospheric water content. This variation, together with the changing viewing geometry, complicates the retrieval of water vapor profiles. Like the SSM/T-1, the SSM/T-2 is a step stare system, but with 28 steps per scan versus 7. The scan period has been decreased to 8 seconds per scan from 32 seconds. Each scan consists of a warm load reading, a cold load reading, and 28 scene stations. - The data is used to determine water vapor mass in seven layers, and specific and relative humidity at six levels. When combined with the SSM/T-1 data, meteorological conditions can be analyzed, such as types of cloud formations, icing potential, and the location of weather systems, based on differences in air mass temperature and moisture.

SSMIS (Special Sensor Microwave Imager Sounder), built by the Aerojet Corporation, a subsidiary of GenCorp of Sacramento, CA. SSMIS = SSM/I + SSM/T-1 + SSM/T-2 (Block 5D-3 sensor). In the Block 5D-3 satellite era, the Block 5D-2 passive microwave sensor suite is combined into a single new sensor package - the SSMIS (see Ref. 1157). SSMIS measures microwave energy from the Earth's surface and atmosphere at 24 discrete frequencies from 19 to 183 GHz. The sensor generates a conical scan pattern at 45° from nadir. This scan pattern, combined with the 144° clear field of regard, results in a constant angle of incidence on the Earth of 53.1° and a swath width of 1,700 km. Improvements include 24 channels of data which are all coincident, increased resolution range, increased FOV, and enhanced ground processing software. The sensor adds one additional channel over the SSM/T to improve the measurement of the tropopause temperature and height. The frequencies chosen (channels 1-7 and 24) provide near uniform coverage in height to about 32 km. With the addition of five more channels (channels 19-23), the temperature retrievals are extended up to about 80 km. - The instrument has a mass of 96 kg, power = 135 W, the data rate is 14.2 kbit/s. The antenna has a diameter of 61 cm, the beam width is in the range: 0.4°-1.9°.

The scene spacing for the sounder channels has been improved from 120 km to 480 km of the earlier instruments to 50 km for the lower atmosphere and to 75 km altitude for the upper atmosphere measurements. SSMIS uses almost the same channels as SSM/I for the environmental parameter extraction. The frequency of 85 GHz was changed to 91 GHz to save an extra channel in the system. SSMIS augments the rain retrieval and cloud amounts with channel 8 at 150 GHz.

Channel Nr.	Center Frequency (GHz)	Footprint (km) along x cross	Polarization	NEDT (K)
1	50.3	25.8 x 17.5	Horizontal	0.4
2	52.8	25.8 x 17.5	Horizontal	0.4
3	53.596	25.8 x 17.5	Horizontal	0.4
4	54.4	25.8 x 17.5	Horizontal	0.4
5	55.5	25.8 x 17.5	Horizontal	0.4
6	57.29	25.8 x 17.5		0.4
7	59.4	25.8 x 17.5		0.5
8	150.0	14.4 x 13.1	Horizontal	0.7
9	183.31+7	14.4 x 13.1	Horizontal	1.2
10	183.31+3	14.4 x 13.1	Horizontal	1.0
11	183.31+1	14.4 x 13.1	Horizontal	1.0
12	19.35	70.1 x 42.4	Horizontal	0.7
13	19.35	70.1 x 42.4	Vertical	0.7
14	22.235	70.1 x 42.4	Vertical	0.7
15	37.0	44.2 x 27.5	Horizontal	0.5
16	37.0	44.2 x 27.5	Vertical	0.5
17	91.65	14.4 x 13.1	Vertical	0.3
18	91.65	14.4 x 13.1	Horizontal	0.3
19	63.283+0.235	25.8 x 17.5	H+V	1.9
20	63.793+0.358	25.8 x 17.5	H+V	1.9
21	63.793+0.358+0.002	25.8 x 17.5	H+V	1.4
22	63.793+0.358+0.006	25.8 x 17.5	H+V	1.0
23	63.793+0.358+0.016	25.8 x 17.5	H+V	0.6
24	63.793+0.358+0.050	25.8 x 17.5	H+V	0.7

Table 334: SSMIS frequency and resolution characteristics

Parameter	Accuracy goals (rms)
Lower atmosphere	4 K at 1000 mb (millibar)
Temperature profiles	3 K at 850 mb
Determine 15 temperatures from 1000–10 millibars; determine 14 thickness between levels)	1 K at 700-10 mb
Upper atmosphere	5 K at 10 - 0.2 mb
Temperature profiles	7 K at 0.2 - 0.3 mb
Tropopause temperature, pressure	3 K, 20 mb
Humidity profiles (specific and relative humidity at six levels and total water vapor mass and water between specific pressure levels	Specific humidity 1.5 grams/kg or ± 20 percent over clear sky or ocean
Ocean wind speed	<2 m/s
Rain rate, rain flag	5 mm/h, flag
Cloud water over ocean	0.10 kg m^{-2}
Soil moisture, ice concentration	± 10%, ± 10%
Ice edge and snow edge	Flag
Water vapor over ocean	Tropics ± 3 kg m^{-2}, mid-latitude ± 2 kg m^{-2}, polar ± 1 kg m^{-2}
Snow water content, land surface temperature	± 3 cm, ± 2.5 K

Table 335: Some parameter accuracy requirements of SSMIS

SSMIS employs a two point (hot and cold load) calibration taken once during each scan of 1.88 seconds duration in all 24 channels. The scan width has been increased from 104° to 144° (FOV) by positioning the sensor farther out board on the S/C. The six separate feed horns for each frequency require the flight software in the sensor to align the data prior to transmission to the ground (via OLS). SSMIS operates asynchronously with OLS, the data to OLS is transmitted in two blocks/s.

The sounding data is only partially smoothed on-board. The 183 GHz channels are downlinked at 12.5 resolution, they have to be averaged to the 50 km spacing for the sounding retrievals. The 12.5 km data is used for the display of water and vapor data.

SSMIS features redundant Gunn diode oscillators and phase lock loop oscillators. The sensor conically scans the Earth and atmosphere at a scan rate of 31.9 scans/min (or 1.88 s/scan).

G.1.2 Space Environment Sensors

The sensors of the space environmental suite provides data on the geophysical environment of the upper atmosphere and ionosphere.

SSI/ES-2 = Special Sensor Ionospheric Plasma Drift/Scintillation Monitor (PL sensor built by the University of Texas). Measurement of the ambient electron density and temperatures, the ambient ion density, and the average ion temperature and molecular weight at the DMSP orbital altitude. - The instrument consists of an electron sensor (Langmuir probe) and an ion sensor mounted on a 2.5 meter boom. The ion sensor is a planar aperture, planar collector sensor oriented to face the spacecraft velocity vector at all times. In addition to the Langmuir probe and planar collector which make up the SSI/E, the SSI/ES has a plasma drift meter and a scintillation meter. The data volume of SSI/ES-2 is 12 MByte/satellite-day.

SSJ/4 = Precipitation Electron/Proton Spectrometer (PL sensor built by Ampek Inc.). The SSJ/4 is a next generation sensor of the SSJ/3. SSJ/4 has been flown starting with F6 (launch in Dec. 1982) until F15 (launch Dec. 12, 1999). Objective: Measurement of transfer energy, mass, and momentum of charged particles through the magnetosphere-ionosphere in the Earth's magnetic field. The instrument looks toward the satellite zenith. - The SSJ/4 sensor consists of four electrostatic analyzers that record the flux of precipitating ions or electrons at 20 fixed energy channels between 30 eV and 30 keV. The curved plate detectors allow precipitating electrons and ions to enter through an aperture of about 20 x 10 (FWHM). Electrons and ions of the selected energy are deflected toward the target by an imposed electric field applied across the two plates. The two low energy detectors consist of 10 channels centered at 34, 49, 71, 101, 150, 218, 320, 460, 670, and 960 eV. The high energy detector measures particles in 10 channels centered at 1.0, 1.4, 2.1, 3.0, 4.4, 6.5, 9.5, 14.0, 20.5 and 29.5 KeV. Each detector integrates each channel for 0.09 s from high energy channel to low. A complete cycle is sampled each second. The primary sources of the particles precipitating into the upper atmosphere are the northern and southern auroral zones. The daily data volume is approximately 1 Mbyte per satellite. The sensor data also supports missions which require knowledge of the polar and high-latitude ionosphere, such as communications, surveillance, and detection systems that propagate energy off or through the ionosphere.

SSJ/5 = Precipitation Electron/Proton Spectrometer (first flight on F16)
Upgrade of SSJ/4. Detects and analyzes electrons and ions that precipitate in the ionosphere to produce an aurora display. The sensor data also supports missions which require knowledge of the polar and high-latitude ionosphere, such as communications, surveillance, and detection systems that propagate energy off or through the ionosphere.

SSM = Triaxial Fluxgate Magnetometer (PL sensor built by NASA/GSFC).
The SSM measures geomagnetic fluctuations associated with solar geophysical phenomena (i.e., ionospheric currents flowing at high latitudes). In combination with the SSI/ES (or SSI/ES-2) and the SSJ/4, the SSM provides heating and electron density profiles in the high-latitude ionosphere. SSM takes and reports 12 readings/s for the Y and Z axes. Only 10 readings of the 12 readings per second are reported for the X axis due to telemetry limita-

tions. The SSM's axes are aligned with the spacecraft's axes where X is downward and aligned to local vertical within 0.01 degree, Y is parallel to the velocity vector for spacecraft with ascending node in the afternoon/evening sector, and Z is away from the solar panel and anti-parallel to the orbit normal vector.) The measurement range is ±65535 nT for each axis, with a one-bit resolution of 2 nT. The first SSM flight started with F12 (launch 29.8, 1994). [1165]
Note: The magnetic field has three sources: 1) the magnetic field from the solid Earth, 2) the magnetic field from electrical currents flowing in the ionosphere and magnetosphere, and 3) the magnetic field from the spacecraft. Measurement of source 2 is the principal objective of the SSM, the measurement of source 1 is a secondary objective, and measurement of source 3 is a nuisance which is eliminated from the data as much as possible during data processing.

SSB/X-2 = Gamma Ray Particle Detector (built by Sandia National Labs).
The SSB/X-2 is an array-based system which detects the location, intensity, and spectrum of X-rays emitted from the Earth's atmosphere. The array consists of four identical and independent directional detectors.

SSZ = Laser Threat Detector (built by Aerospace)
SSZ is a static Earth-viewing sensor, monitoring electromagnetic radiation.

SSI/ES-3 = Enhanced Ionospheric Plasma Drift/Scintillation Monitor
An upgrade of the SSI/ES-2 sensor. The data supports a variety of HF and UHF communications missions, and is used for atmospheric drag calculations for low Earth orbit satellites.

SSM-Boom = Triaxial Flux Gate Magnetometer
Upgrade to SSM. Measures geomagnetic fluctuations associated with solar geophysical phenomena (i.e. ionospheric currents flowing at high latitudes). Combinations of data from SSI/ES (or SSI/ES-2), SSJ/4, and SSM will provide more complete specification of heating and dynamics of the high-latitude ionosphere and neutral atmosphere.

SSULI = Special Sensor Ultraviolet Limb Imager (built by JHU/APL and sponsored by NRL). The optical instrument is a spectrograph with the objective to measure extreme and far ultraviolet radiation (vertical profiles) from the Earth's limb. The primary observations, ranging from 80 - 170 nm, with 1.5 nm resolution, are of radiation from atomic oxygen and nitrogen, and molecular nitrogen, resulting in direct measurements of the electron density vertical profile as well as ion and neutral densities. The vertical profiles in the upper atmosphere and ionosphere are obtained by viewing the Earth's limb at a tangent altitude of approximately 50 km to 750 km. The LORAAS (Low Resolution Airglow/Aurora Spectrograph) instrument on ARGOS (Launch Feb. 23, 1999) is a SSULI prototype instrument. LORAAS data of ARGOS is being used to validate SSULI algorithms that convert raw measurements into useful environmental parameters that characterize the upper atmosphere. The first SSULI instrument will be flown on the F-16 S/C in Jan. 2001.

SSULI is mounted on the +x panel of the spacecraft. The sensor consists of a folded telescope, a scan mirror assembly with a mounting interface to the UV Wadsworth spectrograph, collimator, grating (focuses the radiation onto the detector and disperses the light into a spectrum from 80 - 170 nm, detector (imaging microplate detector with a wedge and strip anode that characterizes photon events). A scanning mirror sweeps the FOV across the Earth limb to obtain vertical profiles. The scan is performed in the orbital plane, looking either forward or aft. The primary scan mode extends from 10° to 27° below the satellite horizon (with the capability to go from 10° to 40° in an alternate scan mode), forming vertical images of the Earth's atmosphere from 750 km to 50 km every 90 seconds. The scan rate is faster at high tangent altitudes and slows down at lower altitudes where there is more

1165) http://www–vsbp.plh.af.mil/projects/dmsp/ssm12ins.html

structure in the airglow. The intensity at these wavelengths is so low, both sensors count individual photons on the focal plane using wedge and strip microchannel plate detector technology. 1166) 1167)

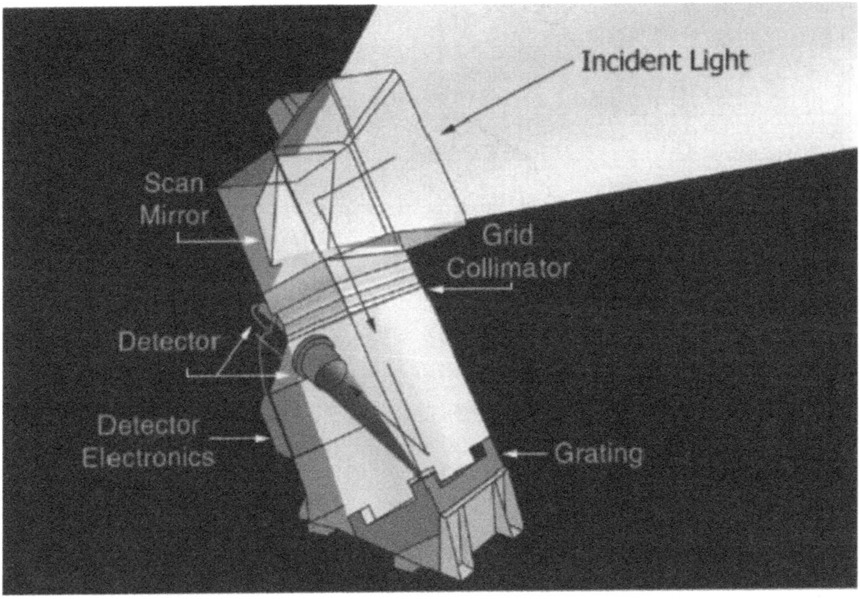

Figure 202: The SSULI optical path

Radiation into SSULI is reflected off the scan mirror to a grid collimator. The radiation is dispersed off a grating into its spectral elements, focused onto a detector, and converted into an electrical signal. The SSULI electronics use this charge to determine the wavelength and amplitude of the photons striking the detector.

Spectral range	80 to 170 nm with a resolution of 1.5 nm or less
IFOV (Instantaneous Field of View)	0.1° vertical and 2.4° horizontal
FOR (Field of Regard)	30° x 2.4°
Scanning range	10° - 40° below of spacecraft horizon (y-direction)
Scanning rate	up to 6° per second
Sensitivity	0.5 counts per second per Rayleigh at 83.4 nm

Table 336: SSULI parameter specification

Parameter	Species	Wavelength
Dayside ionosphere	O^+	834 Å
Dayside ionosphere	O	911 Å, 1304 Å, 1356 Å
Dayside neutral density	O	989 Å, 1304 Å, 1356 Å, 1641 Å
Dayside neutral density	N_2	1085 Å (N+), N_2 LBH bands
Dayside neutral density	O_2	Schumann-Runge, Absorption of N_2 LBH
Nightside neutral density	O	1356 Å
Temperatures		Emission scale heights

Table 337: SSULI observables

1166) Information of all Block 5D-3 sensors and update of Table 330 provided by Major J. Sorlin-Davis, Dept. of USAF, The Pentagon

1167) http://tipweb.nrl.navy.mil/Projects/ssuli/ssuli.html

Parameter	Full-scan Imaging	Reduced Imaging	Spectrograph
IFOV cross-track IFOV cross-track (wide slit) IFOV along-track	0.3° 0.74° 11.84°	0.30° 0.74° 11.84°	0.18° - 11.84°
Pixel IFOV Limb cross-track, along-track Earth cross-track, along-track	 0.30°, 1.48° 0.30°, 0.74°	 0.30°, 1.48° 0.30°, 0.74°	 0.18° 1.48°
Scanned FOV Limb cross-track, step resolution Earth cross-track, step resolution	 9.6°, 0.4° 124.8°, 0.8°	 9.6°, 0.4° 105.6°, 0.8°	
Spatial resolution at nadir cross-track, along-track	 10 km x 10 km	 10 km x 10 km	 2.3 km x 20 km
Pixel step period Limb, Earth	 0.112 s, 0.112 s	 0.156 s, 0.112 s	
Spectral resolution (nm) Narrow, slit, normal slit, wide slit	 1.2, 1.9, 4.2	 1.2, 1.9, 4.2	 1.2, 1.2
Sensitivity (counts/s/Rayleigh) 121.6 nm 130.4 nm 135.6 nm 140-150 nm 165-180 nm	 0.016 0.120 0.160 0.160 0.020	 0.016 0.120 0.160 0.160 0.020	 0.019 0.144 0.192 0.192 0.024
Data frame Date rate, data frame period	 3816 bit/s, 22 s	 3816 bit/s, 22 s	 3816 bit/s, 3.0 s
Spatial pixels per frame Limb cross-track, along-track Earth cross-track, along-track	 24, 8 156, 16	 24, 8 132, 16	 1, 6
Output word size: Limb, Earth	6 bits, 6 bits	7 bits, 7 bits	–,10 bits

Table 338: SSUSI performance parameters of imaging spectrograph (SIS)

SSUSI = Special Sensor Ultraviolet Spectrographic Imager (built by JHU/APL) [1168) 1169)] SSUSI is a nadir-pointing instrument that measures UV radiation from the Earth's atmosphere and ionosphere, it also measures visible radiation (airglow and terrestrial albedo). The instrument consists of three subassemblies:

- SIS (Scanning Imaging Spectrometer)
- NPS (Nadir-looking Photometer System)
- Support module

SIS in turn consists of a cross-track scanning mirror at the input to the telescope (folded design) and spectrograph optics. There are redundant 2-D photon-counting detectors at the focal plane (detector size: 16 pixels in along-track and 160 pixels in the cross-track direction). The detectors employ a position sensitive anode to determine the photon event location. The scan mirror sweeps the 16 pixel footprint from horizon to horizon, producing one frame in 22 seconds. The imaging mode performs simultaneous measurements in five wavelength bands from 115 - 180 nm. The imaging mode scan cycle consists of a limb-viewing section followed by an Earth viewing (nadir) section. Limb-viewing imagery is collected from -72.8° from nadir to -63.2° from nadir. The limb-viewing section has a cross-track resolution of 0.4° per pixel, it consists of 24 cross-track pixels and 8 along-track pixels (at five bands). The Earth-viewing section has a cross-track resolution of 0.8°.

NPS consists of three nadir-looking photometers. It operates in the visible portion of the electromagnetic spectrum, monitoring airglow at 427.8nm and 630nm and the terrestrial albedo near 630nm. NPS operates only on the nightside of the orbit. Its data determine the auroral oval location and provide information to help determine electron densities in the F-layer, energy deposition in the auroral region (day and night), photoelectrons, neutral composition, and equatorial electrojet. Each photometer unit includes an integrated detec-

1168) L. J. Paxton, et al., "Special Sensor Ultraviolet Spectrographic Imager (SSUSI): An Instrument Description," JHU/APL paper

1169) http://sd–www.jhuapl.edu/SSUSI/

tor package consisting of a photomultiplier tube, high voltage power supply, and pulse amplitude discriminator electronics.

Parameter	Unit Nr. 1	Unit Nr. 2	Unit Nr. 3
Pixel FOV:full angle (circular)	2°	2°	2°
Spatial resolution at nadir	25 km	25 km	25 km
Center wavelength	427.8 nm	630 nm	629.4 nm
Spectral bandwidth	5.0 nm	0.3 nm	0.3 nm
Optic diameter	1.27 cm	5.5 cm	5.5 cm
Pixel integration time	1.0 s	1.0 s	1.0 s
Sensitivity (counts/s/Rayleigh)	5	30	30
Maximum count per pixel	500,000	100,000	100,000

Table 339: NPS performance parameters

Parameter	Unit Nr. 1	Unit Nr. 2	Unit Nr. 3
Photocathode	Bi-alkali	Tri-alkali	Tri-alkali
Input window	glass	glass	glass
Cathode diameter	7 mm	7 mm	7 mm
Wavelength	427.8 nm	630 nm	629.4 nm
Output signal	pulse	pulse	pulse
Power	0.5 W	0.5 W	0.5 W
Operating temperature	-30° - -20° C	-30° - -20° C	-30° - -20° C

Table 340: NPS detector parameters

G.1.3 Early Sensors of the DMSP Program

These were the instruments of the early phase of the DMSP program. They are no longer operational, but kept for reference.

SSL = Lightning Detector (old series).
The SSL was a "one-of-a-kind" experiment. Night operation to detect lightning flashes in the 0.4 - 1.1 μm range. FOV = 1200 x 1200 nmi.

SSB = Gamma Tracker. (old series)
Used to track fallout and nuclear debris entrained in the atmosphere from 10 to 15 km. The sensor detected the fission gammas emitted by the debris.

SSE = Temperature Sounder. (old series)
Eight channel scanning filter radiometer. Six of the channels measure in the 15 μm carbon dioxide band, one in the 12 μm window, and the last in the rotational water vapor band near 20 μm. Radiance measurements of the Earth's atmosphere are processed to obtain vertical temperature profiles.

SSJ = Auroral Electron and Ion Spectrometer
The SSJ counted ambient electrons with energies ranging from 50 eV - 20 keV. It determined the number of electrons having energies within certain sub-ranges of the 50 eV to 20 keV spectrum by utilizing a time-sequenced variable electrostatic field to deflect the particles toward the channeltron detector.

SSJ/3 = Auroral Electron and Ion Spectrometers.
The SSJ/3 is a next generation sensor of the SSJ. The SSJ/3 was flown on all Block 5D-1 spacecraft with the exception of F-1. Objective: Measurement of transfer energy, mass, and momentum through the magnetosphere-ionosphere in the Earth's magnetic field.

SSI/E = Topside Ionospheric Plasma Monitor.
The SSI/E measures ambient electron density and temperatures, ambient ion density, and

average ion temperature and molecular weight at the DMSP orbital altitude. The instrument consists of an electron sensor (Langmuir probe) and an ion sensor mounted on a 2.5 m boom. The ion sensor is a planar aperture, planar collector sensor oriented to face into the spacecraft velocity vector at all times.

SSJ* = Space Radiation Dosimeter. The SSJ* measures the accumulated radiation dose produced by electrons in the 1 - 10 MeV energy range, protons of greater than 20 MeV, and the effects of the occasional nuclear interactions produced by energetic protons. Accumulated dose was measured over a period of one year (minimum).

SSD = Atmospheric Density Sensor.
The SSD provides a measure of major atmospheric constituents (nitrogen, oxygen, and ozone) in the Earth's thermosphere by making Earth limb observations of ultraviolet radiation from the thermosphere. The sensor measured the radiation emitted from excitation of molecular nitrogen by impinging solar radiation.

SSC = Snow Cloud Discriminator.
The SSC sensor is a pushbroom scan radiometer. Wavelength 1.51 - 1.63 μm (IR). A binary system, it was used to determine the presence of snow versus clouds. It is a proposed concept sensor intended to help determine if machine processing could make a snow/cloud distinction. The along-track scan is provided by the forward motion of the S/C, while the 40.2° cross-track scan is provided by a linear array of 48 detector elements at the image plane of a wide-angle lens.

SSB = Gamma X-Ray Detector.

SSB/A = X-Ray Spectrometer.
The SSB/A detects X-rays and gamma rays from bomb debris or those X-rays produced by the Bremsstrahlung process when electrons precipitate from the Earth's radiation belts. By sensing these X-rays, the SSB/A provides the location of the aurora as it orbits the Earth.

SSB/O = Omnidirectional Gamma Detector.
The SSB/O was an experiment to determine if more accurate atmospheric measurements could be obtained by measuring the co-orbiting particles and the upward flux and subtracting it from the subsatellite scene. The experiment was extremely successful. The SSB/O was sensitive to X-rays in the energy range of approximately 1.5 keV.

SSB/S = Scanning X-Ray Detector.
The SSB/S detects the location, intensity and spectrum of X-rays emitted from the Earth's atmosphere. The detector consists of three sensors.

SSB/X, SSB/X-M= Gamma Ray Detectors.
The SSB/X is an array-based system which detects the location, intensity, and spectrum of X-rays emitted from the Earth's atmosphere. The array consists of four identical and independent directional detectors. The SSB/X-M and SSB/X-2 follow from the SSB/X and are also gamma-ray and particle detectors. The SSB/X-M and SSB/X-2 consist of two identical and independent gamma-ray detectors and three particle detectors (capable of detecting gamma-ray bursts).

SSH = Infrared Spectrometer. The IR spectrometer is a multispectral sounder for humidity, temperature, and ozone measurements. The SSH provides soundings of temperature and humidity and a single measurement of ozone for vertical and slant paths lying under and to the side of the subsatellite track. The SSH made a set of radiance measurements in narrow spectral channels lying in the absorption bands of carbon dioxide, water vapor, and ozone. The radiance measurements were mathematically inverted to yield vertical temperature profiles, water vapor, and total ozone content.

SSH-2 = Infrared Temperature and Moisture Sounder.
The SSH-2 provides soundings of temperature and humidity for vertical and slant paths ly-

ing under and to the side of the subsatellite track. SSH-2 is physically identical to SSH, but with different (and tighter) filter bands. 16 frequency bands in the range of 3.7 - 30 μm. Resolution = 60 km. Swath width = 2240 km.

G.1.4 DMSP Data Availability - Visible and Infrared Imagery

NOAA/NESDIS-NSIDC[1170)] and -NGDC[1171)] (Boulder, CO) have established a collection of digital satellite imagery acquired from the DMSP series of the US Air Force under NOAA-NESDIS contract. Hence, DMSP imagery data are available for the general user community (among them the SSM/I, SSM/T and SSM/T-2 sensor data). These data are prepared from a global, digital intensity file used operationally by the Air Force in forecasting and are subsequently archived at NGDC [DMSP imagery are archived after operational use (usually 45 to 60 days)]. A digital archive of DMSP data has been operational at NGDC since April 1992 (NGDC receives two 5 GByte tapes in compressed format every day). Archival services are continually upgraded.

Parameter	Spatial Resolution (km)	Range of Values	Quantization Levels	Absolute Accuracy
Ocean surface wind speed	25	3 - 25	1	± 2 m/s
Ice: area covered Ice: age Ice: edge location	25 50 25	0 - 100 1. year, multiyear N/A	5 1 yr, >2 yr N/A	± 12% None ± 12.5 km
Precipitation over land ares	25	0 - 25	0,5,10,15,20, ≥25	± 5 mm/hr
Cloud water	25	0 - 1	0.05	± 0.1 kg/m^2
Integrated water vapor	25	0 - 80	0.10	± 2.0 kg/m^2
Precipitation over water	25	0 - 25	0,5,10,15,20, ≥25	± 5 mm/hr
Soil moisture	50	0 - 60%	1	None
Land surface temperature	25	180 - 340 K	1	None
Snow water content	25	0 - 50 cm	1	± 3 cm
Surface type	25	12 types	N/A	N/A
Cloud amount	25	0 - 100%	1	± 20%

Table 341: Environmental products of the SSM/I sensor[1172)]

The imagery collection consists of three positive transparency products produced by USAF.

1) Limited coverage (at 0.6 km resolution) is available for selected areas
2) Global coverage is available on single-orbit strips (at 2.7 km resolution)
3) Mosaics compiled from several orbits, with latitude and longitude grids added, are available globally (at a resolution of 5.4 km).

Mosaics are the only pre-gridded product available. Single-orbit strips can be custom gridded for an additional fee. Each of these products has been produced since 1973, except the mosaics, which were available from December 1975 on.

Parameter	AMSU-A (flown on NOAA-15)	SSM/I (flown on DMSP series)
Window channels (GHz)	23.8, 31.4, 50.3, 89.0	19.4, 22.2, 37.0, 85.5
Polarization	Mixed	V & H (V only for 22.2)
Scan geometry	Cross-track ±48.33°	Conical 45°
Field of View (FOV)	Varies with angle about 48 km at nadir about 150 km at limb	Varies with frequency about 15 km at 85.5 GHz about 60 km at 19.4 GHz
Swath width	about 2000 km	about 1400 km

Table 342: Comparison of AMSU-A and SSM/I channels

1170) "Defense Meteorological Satellite Program, Visible and Infrared Imagery Collection," NOAA-NSIDC, Feb. '84
1171) "Data Management Plan for the Archive of DMSP Digital Data at NGDC," April 28, 1992, Draft; Courtesy of W. Kroehl, NGDC
1172) J. Holloinger, "DMSP Special Sensor Microwave/Imager Calibration/Validation," Final Report, Vol. I and II, NRL, 1989

The AMSU-A (Advanced Microwave Sounding Unit) instrument of the NOAA POES series was first flown on NOAA-15 (launch May 13, 1998).[1173] It's primary objective is to provide significantly improved temperature soundings compared to its predecessor (MSU). However, since AMSU-A features similar window channels as the SSM/I sensor, it permits also the determination of surface and hydrological products similar to those derived from SSM/I.

NOAA/AVHRR-2 Sensor	System Parameter/ (Applications)	DMSP/OLS Sensor
1.1 km LAC 4.0 km GAC (degraded at edges) good good	spatial resolution (sea ice leads) (meteorology)	0.55 km 'fine' 2.7 km 'smooth' (const. across swath) better better
5 narrow channels Ch1: .55-0.68 μm Ch2: .725-1.0 μm Ch3: 3.55-3.93 μm Ch4: 10.30-11.30 μm Ch5: 11.50-12.50 μm good good	spectral resolution (sea surface temp.) (vegetation index)	2 broadband channels VIS: 0.4-1.1 μm IR: 10.5-12.5 μm marginal N/A
IR yes (10 bit) VIS no (pre-launch, drifts)	absolute calibration	IR yes (8 bit) VIS no (continued gain adjustment)
no N/A developmental N/A	visible-band dynamic range/ nighttime operation (auroral characteristics) (biomass burning,) NESDIS, NASA (moonlit clouds/snow)	yes yes added potential with unique visible band good
no N/A N/A	coincident passive micro-wave (snow/ice, rain rate) (surface wind/soil moisture)	yes good good
no no	coincident space environ-ment measurements (auroral image feature + electron flux)	yes yes

Table 343: Comparison of two sensors - NOAA AVHRR/2 and DMSP/OLS [1174]

G.2 EPS (EUMETSAT Polar System)

The EUMETSAT Polar System (EPS) consists of the ESA-developed MetOp (Meteorological OPerational) series of spacecraft and an associated ground segment for meteorological and climate monitoring from polar, low Earth orbit, providing "morning" service for operational meteorology. Within the framework of international agreements, the NOAA POES series will continue to provide the "afternoon" service. The MetOp series, although an independent development, is complementary to: a) the NOAA POES system, b) the EUMETSAT/ESA MSG (Meteosat Second Generation) system, and c) the ESA ENVISAT system, where MetOp completes the mission objectives of the original POEM-1 mission.

Background: European needs for meteorological observations in polar ("morning" and "afternoon") orbits have been generously provided by NOAA S/C and payloads (including some instruments developed in Europe) over the last quarter century. The EPS (EUMETSAT Polar System) is the European contribution to the joint European/US operational po-

1173) D. Moore, L. Zhao, et al., "The new NOAA Microwave Surface and Precipitation Product System," Proceedings of the EUMETSAT Meteorological Satellite Data User's Conference, Copenhagen, Denmark, Sept. 6-10, 1999, pp. 325-329

1174) Courtesy G. Scharfen, NOAA/NESDIS/NSIDC

lar system, through **IJPS (Initial Joint Polar System)**. A cooperation agreement between NOAA and EUMETSAT was signed in November 1998. The EPS and POES systems form together the IJPS to provide global meteorological data from the series. With EPS, EUMETSAT is committed to take over the morning orbit service from NOAA. On the other hand, NOAA will continue to provide the POES series afternoon service. Both services are coordinated and integrated, on the basis of exchange of data, instruments and operational services (see also G.15.3). 1175)

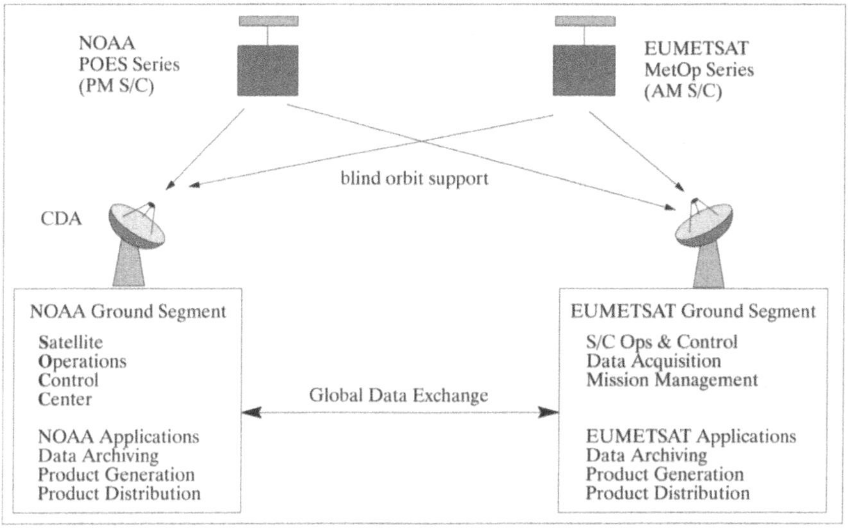

Figure 203: Integrated concept of NOAA/EUMETSAT meteorological polar satellites

The prime objectives of the EPS MetOp mission series are as follows:

- To ensure continuity and availability for operational purposes of polar meteorological observations from the "morning" orbit to the global user community
- To provide enhanced monitoring capabilities (complimentary to ENVISAT) to fulfil the requirements to study the Earth climate system as expressed in a number of international cooperative programs such as: GCOS (Global Climate Observing System), IGBP (International Geosphere and Biosphere Program), and WCRP (World Climate Research Program). The aim is to provide continuous, long-term data sets.

EPS is an end-to-end system composed of a space segment and a full ground segment. The third satellite in this series, MetOp-3, is not formally part of IJPS, because its planned launch date falls already into the era of the US NPOESS (National Polar-orbiting Operational Environmental Satellite System) era, representing in itself the merged US POES (civil) and DMSP (military) series.

Within the European framework, the ESA program developing MetOp-1, and the EUMETSAT procurement of MetOp-2/3, is under the responsibility of a joint ESA/EUMETSAT team. EUMETSAT is also directly responsible for the delivery of the following payloads: MHS, IASI, ARGOS/ADCS, SEM, AMSU-A HIRS/4, and AVHRR/3. The last four payloads are contributed by NOAA under the IJPS agreement: CNES develops the IASI instrument with joint funding from EUMETSAT, and provides the Argos/ADCS, and in

1175) A. Ratier, "The EUMETSAT Polar System," Proceedings of the 1999 EUMETSAT Meteorological Satellite User's Conference, Copenhagen, Denmark, Sept. 6-10, 1999, pp.3-10 addition, part of the S&RSAT (Search & Rescue) auxiliary payload. EUMETSAT is responsible for the definition of the overall EPS system, the development and operations of the ground segment, and for the operation of the space segment.

G.2.1 MetOp-1 Satellite

MetOp-1 is Europe's first polar-orbiting satellite dedicated to operational meteorology. The MetOp program was originally planned as a much larger satellite concept, called POEM, a successor mission series to ERS-1/2 on the Columbus Polar Platform (PPF design). However, this idea was abandoned at the ESA Ministerial Council in Granada, Spain, in 1992. Instead, Envisat and MetOp were born. [1176)]

The overall architecture of the MetOp-1 S/C comprises two largely independent modules, namely SVM (Service Module) and PLM (Payload Module).

- SVM provides the main satellite support/service functions. The SVM provides all the standard service functions of a S/C, like: attitude and orbit control, propulsion, power generation, and the on-board data handling and distribution systems. The SVM design is based on the SPOT MK3 bus (used on the SPOT series, ERS-1/2 and Helios-1A and 1B S/C). SVM is a box-shaped structure interfacing with the launch vehicle and the PLM.
- PLM provides accommodation and supporting subsystems (data handling, power, communications) to the payload complement. The instruments and antennas are mounted on the external panels, while most of the electronics systems are accommodated inside the PLM.

The S/C overall size is: 6.3 m x 3.4 m x 3.45 m (launch configuration) and 17.6 m x 6.7 m x 5.4 m (on-orbit configuration). The S/C is three-axis stabilized using reaction wheels. The pointing knowledge is: 0.07° (x-axis), 0.10° (y-axis), 0.17° (z-axis). The overall S/C mass at launch is about 4300 kg, including 316 kg of hydrazine. The PLM mass is 2232 kg, the SVM mass is 1486 kg. Single sided solar arrays provide a power of 3.890 kW (EOL), the average power over one orbit is 2.280 kW (EOL). Five 40 Ah batteries provide power during eclipse periods. The mission design life is 5 years.

A launch of MetOp-1 on Soyuz-ST from Baikonur is planned for 2005 (MetOp-2 launch in 2008, MetOp-3 launch beyond 2010). Full approval of the EPS (EUMETSAT Polar System) program was granted in September 1998. The prime S/C contractor is Astrium SAS (France), co-contractors are Astrium GmbH, Germany, Alenia, Italy, and Astrium Ltd., UK.

Communication: An omnidirectional S-band coverage provides TT&C support (uplink 2 kbit/s, downlink 4 kbit/s). Instrument data is downlinked in X-band at a rate of 70 Mbit/s. On-board storage capacity of 24 Gbit (solid state recorder with a data rate of 70 Mbit/s) is provided. In addition to on-board recording and X-band downlink capabilities, MetOp supports the real-time broadcast of instrument data to local users by the following means:

- LRPT (Low-Rate Picture Transmission) links with 72 kbit/s in VHF-band for selected instrument data.
- HRPT (High-Rate Picture Transmission) links with 3.5 Mbit/s in L-band. The HRPT broadcast service is very similar to the existing NOAA POES service and enables regional users to receive all data relevant to their area in real time.

The provided VHF low-rate digital direct broadcast service replaces the old analog APT (Automatic Picture Transmission) service of NOAA, employing data compression (modified JPEG compression scheme) to ensure high-quality images. The digital LRPT service retains the VHF frequency and bandwidth of the APT service, but provides three channels of AVHRR data at the full instrument spatial and radiometric resolution.

1176) P. G. Edwards, D. Pawlak, "MetOp: The Space Segment for EUMETSAT's Polar System," ESA Bulletin, No 102, May 2000, pp. 7-18

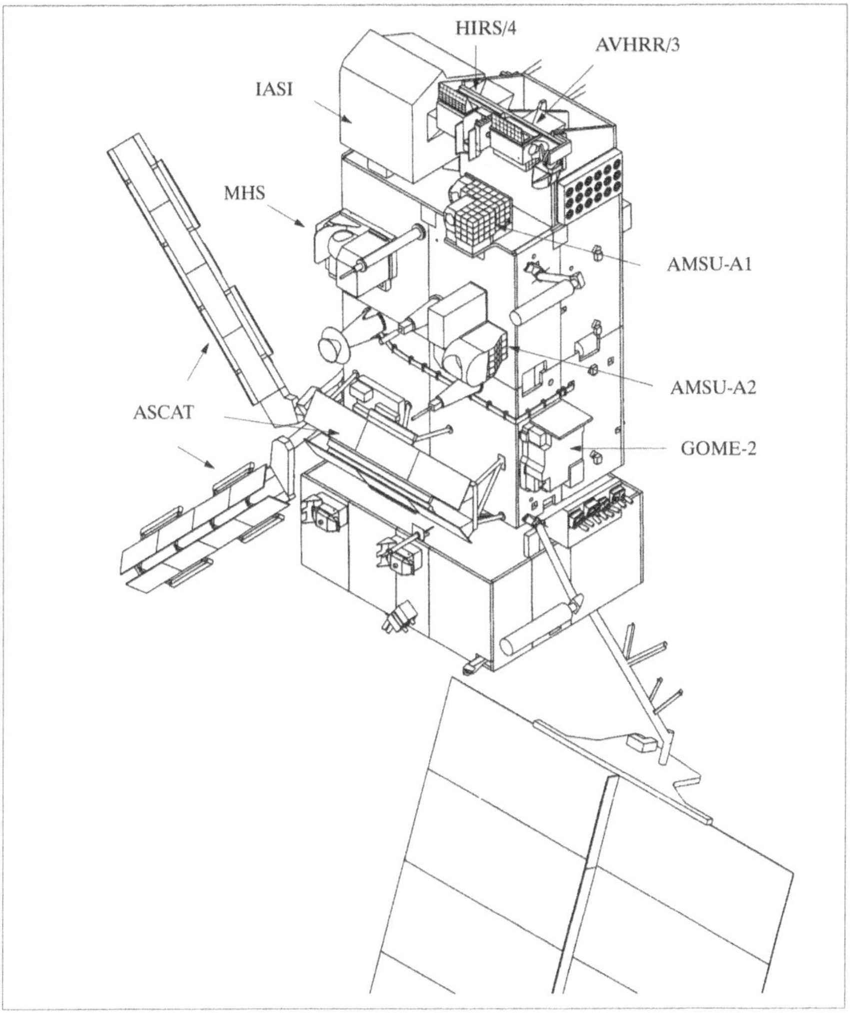

Figure 204: Illustration of the MetOp spacecraft

Data Type	Frequency Domain	Modulation Scheme	Data Rate
TT&C uplink	S-band, 2053.4 MHz	NRZ/PSK/PM	2 kbit/s
TT&C downlink	S-band, 2230 MHz	SP-L/PSK/PM	4 kbit/s
Global data dump	X-band, 7.750-7.900 GHz	QPSK	70 Mbit/s
LRPT downlink	VHF-band, 137.1 MHz	QPSK	72 kbit/s
HRPT downlink	L-band, 1701.3 MHz	QPSK	3.5 Mbit/s

Table 344: Summary of MetOp communication links with the ground segment

On-board data handling employs the CCSDS protocols. A selective encryption capability is used to ensure the commercial and data-denial needs of EUMETSAT and NOAA, respectively. Spacecraft operations are performed by EUMETSAT with the Kiruna ground station serving as prime.

Service provision: The MetOp program, as successor to the NOAA POES morning series, is required to provide a continuous broadcast of its meteorological data to the worldwide user community, so that any ground station in any part of the world can receive local data when the satellite passes over that receiving station. This implies continued long-term provision of LRPT and VHF downlink services. [1177)]

Orbit: Near-circular sun-synchronous polar morning orbit (9:30 AM on descending node), mean altitude = 825 km, inclination = 98.704°, repeat cycle = 5 days (71 orbits).

S/C attitude	Three-axis stabilized through reaction wheels Orbit maneuvers through a hydrazine propulsion system Pointing knowledge: 0.07° (x-axis), 0.10° (y-axis), 0.17° (z-axis)
Data handling	Instruments science data acquired as CCSDS packets Science data formatting and multiplexing, encryption for selected instruments Instruments and housekeeping data storage in a solid-state recorder (24 Gbit)
Communications	Omnidirectional S-band coverage (uplink 2 kbit/s, downlink: 4.096 kbit/s Instrument global data stream downlinked via X-band (70 Mbit/s) Real-time broadcasting of instrument data in HRPT: 3.5 Mbit/s via L-band for all instruments, and LRPT: 72 kbit/s via VHF for selected instruments
On-board power	2210 W from solar panel, average power over one orbit (EOL) Five 40 Ah batteries Unregulated power (22 37.5 V) and 50V regulated power lines for SVM/PLM units Unregulated power lines (22 37 V) for European instruments 28 V regulated power lines for NOAA instruments
Design life	5 years
S/C mass	4179 kg
Propellant mass	315.7 kg of hydrazine, stored in 4 tanks (including residual)
S/C operations	S/C controlled by EUMETSAT (Kiruna ground station Instruments X-band data downlinked nominally over two ground stations Recorded data downlinked not later than one orbit after recording S/C autonomy required for 36 hours without ground contact

Table 345: Summary of MetOp main features and performances

The MetOp operational meteorological mission objectives consist of:

- Global sounding: To provide information about 3-D temperature and humidity fields in support of operational numerical forecasting systems
- Global imagery: To provide cloud imagery for forecasting applications, sea surface temperatures (SST), radiation budget temperatures. To support the global sounding mission through the identification of cloud-free areas
- Data collection and location: To support WWW objectives by the reception and dissemination of in-situ observations from ocean buoys and similar data collection platforms
- Preoperational data: To provide access to data from instruments which have not yet been declared fully operational
- Global data access: To primarily support global-scale weather forecasting by providing global data to the meteorological services within 2 1/4 hours of observation
- Local data access (HRPT and LRPT): To support regional weather forecasting by providing broadcasted data to local receiving stations when the satellite is in visibility.

The MetOp climate monitoring mission contributions (for GCOS) consists of:

- Imagery and sounding
- Ocean measurements (including surfaces stress and winds)
- Clouds and Earth radiation budget: Radiation is the primary energy source of the climate system and principle heat input source to the oceans
- Sea ice information: The extent of sea ice is an important variable in connection with both ocean heat budget and radiation balance

1177) D. Pawlak, C. Bousquet, "MetOp: the space segment of the EUMETSAT Polar System," Proceedings of the 2000 EUMETSAT Satellite Data Users' Conference, Bologna, Italy, May 29-June 2, 2000, pp. 209-213

- Atmospheric minor constituents: Concern over the depletion of stratospheric ozone suggests the importance of maintaining a continuous data set of global total column ozone and vertical profiles
- Precipitation estimates.

The MetOp Earth sciences research mission objectives include data provision to the European science community to advance investigations in fields such as:

- Atmospheric physics: chemistry, radiation and energy balance, clouds
- Oceanography: general ocean circulation and fluxes of heat, momentum and gases; modeling
- Hydrology: water cycle, continental snow and mountain glaciers, land cover, soil moisture, vegetation
- Cryosphere: sea ice, continental ice, modeling

The MetOp surveillance mission contributions to the regular monitoring of application-oriented parameters:

- Environment: pollution control, natural disasters, renewable resources
- Marine: offshore activities, ship routing, fishing, sea ice routing

G.2.2 MetOp-1 Sensor Complement

For a description of AVHRR/3, HIRS/4 and AMSU-A, see chapter G.15.1).

Payload Sensor	Mission Objectives	Sensor Provider
AVHRR/3 (Advanced Very High Resolution Radiometer)	Global imagery, global sounding, ocean measurements (SST), clouds and Earth radiation budget, land measurements	NOAA
HIRS/4 (High Resolution Infrared Sounder)	Global sounding, atmospheric minor constituents, (ozone)	NOAA
AMSU-A (Advanced Microwave Sounding Unit-A - A1 & A2)	Global sounding, sea ice	NOAA
MHS (Microwave Humidity Sounder)	Global sounding, clouds and Earth radiation budget, sea ice	EUMETSAT
IASI (Improved Atmospheric Sounder Interferometer)	Global sounding, ocean measurements (SST), clouds and Earth radiation budget, some atmospheric trace constituents, land measurements	CNES/ EUMETSAT
ASCAT (Advanced Scatterometer)	Ocean measurements, surface stress and surface wind	ESA
GOME-2 (Global Ozone Monitoring Experiment-2)	Atmospheric trace gases (ozone content and profile)	ESA/ EUMETSAT
GRAS (GNSS Receiver for Atmospheric Sounding)	Atmospheric refractive index measurement in limb-sounding mode	ESA/ EUMETSAT
ARGOS (Remote Data Collection System),	Data collection and location; ADCS on-board, PTTs and DCPs in the ground segment	CNES
S&R = Search and Rescue System	Cooperative satellite-based radiolocation system for search and rescue operations. Relay of emergency radio signals to ground stations from aviators, mariners and land travellers in distress.	CNES/NOAA
SEM-2 (Space Environment Monitor-2)	Monitoring of the S/C environment (solar-terrestrial)	NOAA

Table 346: Overview of MetOp-1 payload complement

ASCAT = Advanced Wind Scatterometer [ESA instrument, developed by Astrium GmbH under contract to Astrium SAS, France]. Objective: Determining wind vector fields at sea surface by measuring the backscattering coefficient [normalized radar cross section sigma-zero, also referred to as NRCS] on a global basis. The requirement calls for the measurement of wind speeds in the range of 4-24 m/s with an accuracy of 2 m/s and a direction accuracy of ±20°. [1178) 1179) 1180) 1181)]

ASCAT is a real-aperture C-band (5.255 GHz) dual-swath and three-look radar instrument with high radiometric resolution and stability. The design of the system exploits the LFM (Linear Frequency Modulation) measurement principle on the basis of long transmit pulses (about 10 ms) with linear frequency modulation (chirps). This permits the application of low-peak transmission power (about 120 W) and frequency domain processing. ASCAT sequentially illuminates two strips (swaths) of the sea surface (two 550 km wide swaths, one on either side of the satellite ground track). This feature results in fast global coverage capability (twice as fast as ERS-1). Heritage of ERS-1 AMI-SCAT sensor. The major differences between ASCAT and AMI-SCAT are:

- Enhanced coverage due to double swath operation
- Spatial resolution: increased spatial resolution is provided on an experimental basis
- Use of solid-state technology
- Improved radiometric performance (accuracy and inter-beam stability)
- Reduced downlink data rate due to on-board data processing (from 1.4 Mbit/s on AMI-SCAT to 55 kbit/s on ASCAT)
- In addition to the processing of echo signals, the instrument also performs an internal calibration process within each pulse repetition interval
- Reduced power needed for the transmission of continuous-wave pulses (about 120 W of peak power are needed for ASCAT compared to 4.8 kW for AMI-SCAT on ERS-1).

Parameter	Nominal mode	High-resolution mode
Center frequency	5.255 GHz	
Swath width (full performance) Swath width (reduced performance)	500 km 550 km	
Swath length	continuous	
Incidence angle mid near at H min	25°	
Localization accuracy	4.4 km	
Polarization	vertical	
Cross polarization	>20 dB	
Spatial resolution	50 km	25 - 37 km
Spectral resolution	0.0195/km	
Sampling interval	25 km	12.5 km
Radiometric resolution at low wind (minimum backscattering)	2.5 - 7.1%	6.0 - 17.6%
Radiometric resolution at high wind (maximum backscattering)	2.0 - 2.7%	5.0 - 9.1%
Radiometric accuracy	0.47 - 0.55 dB pp	0.48 - 0.56 dB pp
Interbeam radiometric stability	0.33 - 0.41 dB pp	0.33 - 0.41 dB pp
Ambiguity under worst case scenario	0.34 - 3.3%	0.34 - 3.3%
Dynamic range (backscatter coefficients on the ground at near and far swath)	-8.6 - 4.3 (near) dB -28.6 - -8.8 (far) dB	-8.6 - 4.3 (near) dB -28.6 - -8.8 (far) dB
Aliasing error	0.08%	0.08%

Table 347: Performance parameters for ASCAT in nominal and high-resolution modes

The ASCAT instrument consists of the following elements: Antennas (6), SFE (Scatterometer Front End), RFU (Radio Frequency Unit), HPA (High Power Amplifier), DPU (Digital Processing Unit, PDU (Power Distribution Unit), and ICU (Instrument Control Unit).

1178) R. V. Gelsthorpe, E. Schied, J. J. W. Wilson, "ASCAT - MetOp's Advanced Scatterometer," ESA Bulletin, No 102, May 2000, pp. 19-27
1179) "ASCAT Advanced Scatterometer," ESA brochure
1180) F. Rostan, M. Kuntz, S. Schütz, "The Advanced Scatterometer (ASCAT) Ground Processing Prototype," Proceedings of IGARSS'99, Vol. I, Hamburg, June 28 - July 2, 1999, pp. 224-226
1181) H. Ebner, H. R. Schulte, H. Hölzl, D. Miller, P. Hans, "ASCAT - Advanced Wind Scatterometer," IGARSS'92, Volume I, pp. 435-439

ASCAT uses three antennas for each of the two swaths to obtain wind speed and wind direction measurements with ambiguity present. All six antennas are slotted waveguide aluminum arrays. One antenna in each set is looking sideward at 90° (cross-track direction), one forward at 45°, and one aftward at 135° (see Figure 205). All antennas form a fan beam with narrow azimuth pattern and a relative wide elevation pattern, resulting in a swath of about 550 km. The local incidence angles are between 25-65° (mid near at H min = 25°). In each swath a regular grid of points (nodes) is defined where the σ° values are determined. The internode distance is 25 km for nominal resolution, and 12.5 km for high-resolution observations. The wind data at each node are extracted from a wind model. [1182) 1183)]

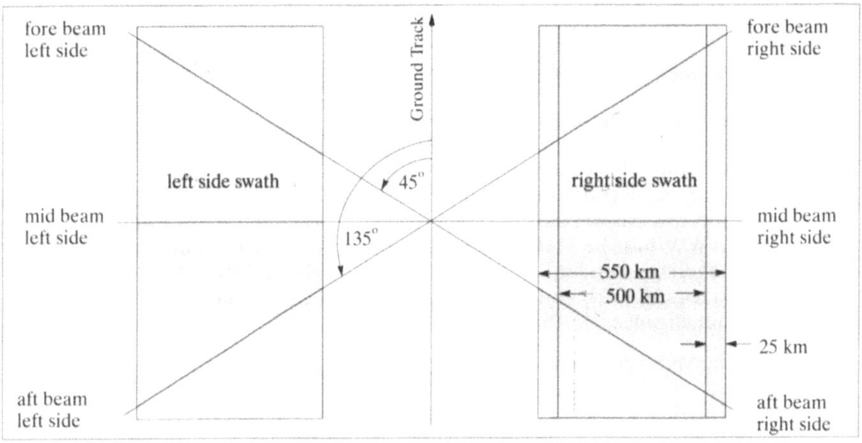

Figure 205: Top view of ASCAT observation geometries

The received echo signal is de-chirped with an image of the transmitted pulse, filtered and down-converted to baseband. The baseband signal is sampled and processed by the onboard digital signal processor. This processing consists of a power spectrum estimation (transformation to the frequency domain and square modulus detection) followed by (spatial) low-pass filtering. This results in echo power spectra, where each frequency corresponds to a specific slant range distance.

ASCAT is calibrated by using an instrument-internal calibration unit which monitors potential variations in transmitted power and receiver gain. External calibration is achieved by using ground transponders and the Amazonian rain forest. [1184)]

MHS (Microwave Humidity Sounder), a EUMETSAT instrument, built by Astrium UK. [1185)] MHS is a five-channel self-calibrating instrument, providing humidity profiling capability in the frequency range of 89 - 190 GHz. The channels 2-5 provide a humidity sounding capability (water vapor absorption line), while channel 1 measures the Earth's surface temperature and emissivity, in conjunction with the AMSU-A window channels, and detects cloud and precipitation contaminated pixels. The MHS instrument is a total power radiometer which measures the total noise power from the scene. MHS scans in the cross-track direction at a rate of 2.66 seconds (the scan is synchronized with AMSU-A). The Earth view

1182) F. Rostan, M. Kuntz, S. Schütz, "The Advanced Scatterometer (ASCAT) Ground Processing Prototype," IGARSS'99 Proceedings, Hamburg, June 28-July 2, 1999, Vol. 1, pp. 224-226

1183) A. Stoffelen, "Error Modeling and Calibration; towards the true surface wind speed," Journal of Geophysical Research, Vol. 103, 1998, pp. 7755-7766

1184) F. Rostan, "The Calibration of the MetOp/Advanced Scatterometer (ASCAT)," IEEE/IGARSS 2000, Honolulu, HI, July 24-28, 2000

1185) C. J. Bushell, et al., "Humidity Sounders for the 21st Century," Proceedings of the EUMETSAT Meteorological Satellite User's Conference, Copenhagen, Denmark, Sept. 6-10, 1999, pp.77-84 and two calibration views are scanned at a constant velocity, rapid acceleration/deceleration is during the intervening periods. The circular IFOV has a diameter at nadir of 30 km. FOV=±50°, resulting in a swath width of 2000 km.

Parameter	Channel 1	Channel 2	Channel 3	Channel 4	Channel 5
Center frequency (GHz)	89.0	157.0	183.311±1	183.311±3	190.311
Center frequency stability (MHz)	-0, +7	-40, +34	-27, +18	-27,+18	-5, +36
Polarization	V	V	H	H	V
Temperature sensitivity (k)	0.22	0.52	0.24	0.21	0.57
Calibration accuracy (abs K)	0.42	0.24	0.16	0.42	0.29
Beam efficiency (%)	94.6	99.0	95.4	95.4	94.9
Beamwidth (°)	1.12	1.17	1.02	1.02	1.05
Beam pointing (°)	0.01	0.01	0.01	0.01	-0.02
Cross polarization (%)	1.2	0.6	3.5	3.5	1.9

Table 348: MHS instrument performance parameters

MHS is a compact quasi-optical heterodyne radiometer, consisting of the following elements: reflector, receiver (quasi-optical front end, amplifiers, filters, video detectors), scan mechanism, on-board calibration target, and an electronics unit. The instrument mass is 63 kg, power = 66 W. While the MetOp-specific instruments use the European OBDH (On-board Data Handling) standard interface, MHS is using the MIL-STD-1553 interface. Both of these standards are high-level command and control interfaces, permitting the use of instrument intelligence. [1186)]

Note: Under the MOU between EUMETSAT and NOAA concerning cooperation for polar satellite systems, the MHS instrument is also being flown on the NOAA POES series (EUMETSAT-funded instrument) with a first flight planned on NOAA-N. MHS is used in conjunction with AMSU-A and replaces the AMSU-B in the POES series instrument baseline.

IASI = Improved Atmospheric Sounder Interferometer (CNES/ASI) [1187)] [1188)] [1189)] IASI is a Fourier Transform nadir-viewing imaging interferometer operating in the thermal infrared spectrum from 3.62 (2760 cm^{-1}) to 15.5 μm (645 cm^{-1}) at moderate spectral resolution. The instrument consists of a Fourier Transform Spectrometer (FTS), based on a Michelson interferometer (the flat mirrors are replaced with retroreflecting cube-corners), coupled to an integrated imaging system, permitting the characterization of cloudiness inside the FOV of FTS. Objectives: sounding tropospheric moisture and temperature, measuring column-integrated O_3, CO_2, CH_4, and NO_2, and measuring some trace gases which drive the budget of tropospheric chemistry and contribute to the greenhouse effect. IASI scans in the cross-track direction with a swath width of about 2052 km (FOV =±48.33°, provision of motion compensation). During the scan the mirror moves to 30 positions, one every 216 ms. IFOV=3.33° x 3.33° (about 48 km x 48 km at nadir). Within each mirror position four sub-FOV's are imaged in a 2 x 2 matrix form onto four detectors. Each sub-FOV corresponds to a circular field-of-view of 0.84° (diameter about 12 km at nadir). The total scan period of 8 seconds is coordinated with the AMSU-A scan cycle. A calibration (cold-space viewing and updating of calibration coefficients) is performed after each scan cycle. The calibration accuracy is ≤0.5 K @ 280 K; the inter-channel accuracy ≤0.2 K; NEΔT < 0.228 at 280 K for all bands. IASI is considered a preoperational instrument on MetOp-1. Band-1 and -2 detectors are of HgCdTe type (photovoltaic), while the band-3 detectors are of InSb type (photoconductive). Data quantization = 12 bit. On-board processing activities

1186) G. Shaw, C. J. Bushell, "The Radiometric Performance of the Microwave Humidity Sounder," Proceedings of the EUMETSAT Meteorological Satellite User's Conference, Copenhagen, Denmark, Sept. 6-10, 1999, pp.135-139

1187) D. Simeoni, C. Singer, G. Chalon, "Infrared Atmospheric Sounding Interferometer," IAF-96-B3.P212, Acta Astronautica, Vol. 40, No. 2-8, pp. 113-118, 1997

1188) "Improved Atmospheric Sounding Infrared," ASI/CNES brochure, April 1991

1189) http://www−projet.cst.cnes.fr:8060/IASI/index.html

performs the following functions: spike elimination, non-linearity corrections, filtering and resampling, Fourier Transform, phase correction, scan mirror reflectivity correction, radiometric calibration, and variable coding.

Band	Bandwidth, FWHM resolution	Comment
Band 1	645-1210 cm^{-1} with 0.35 cm^{-1} (unapodized resolution) 15.5 - 8.26 µm	Band 1 provides retrievals of temperature profiles and ozone 650-770 cm^{-1} - Temperature sounding (CO_2 band) 770-980 cm^{-1} - Surface and cloud properties 1000-1070 cm^{-1} - Ozone sounding 1080-1150 cm^{-1} - Surface and cloud properties
Band 2	1210-2000 cm^{-1} with 0.5 cm^{-1} (apodized resolution) 8.26 - 5 µm	Band 2 provides retrievals of humidity and some trace gases 1210-1650 cm^{-1} - Water vapor sounding, N_2O, CH_4, and SO_2
Band 3	2000-2760 cm^{-1} with 1.5 cm^{-1} (apodized resolution) 5 - 3.62 µm	Band 3 provides retrievals of temperature & some trace gases 2100-2150 cm^{-1} - CO column amount 2150-2250 cm^{-1} - Temperature sounding, N_2O column amount 2350-2420 cm^{-1} - Temperature sounding 2420-2700 cm^{-1} - Surface and cloud properties 2700-2760 cm^{-1} - CH_4 column amount

Table 349: Spectral parameters of IASI

Parameter	IASI (IASI pixel)	AMSU-A	MHS	AVHRR/3	HIRS/3 (4)
Mission (s) of instrument flown	MetOp-1, etc.	MetOp-1, NOAA-15,etc.	MetOp-1, NOAA-N	NOAA-15,etc. MetOp-1, etc.	NOAA-15,etc. MetOp-1, etc.
Scan type	Step and dwell	Step and dwell	Continuous	Continuous	Step and dwell
Scan rate	8 s (8 s)	8 s	2.667 s	0.167 s	6.4 s
Sampling interval	216 ms (216 ms)	200 ms	19 ms	0.025 ms	100 ms
Scan separation	52.69 km (23.81)	52.69 km	17.56 km	1.1 km	42.15 km
Pixels/scan	-, (120)	30	90	2048	56
IFOV	3.33° (0.84° circular)	3.3° circular	1.1° circular	0.0745° square	0.69° circular
IFOV at nadir	47.63 km, (12 km)	47.63 km	15.88 km	1.1 km	10.0 km
IFOV edge cross-track (km) along-track (km)	146.89, (39.14) 78.79, (20.31)	146.89 78.79	52.83 27.10	6.15 2.27	33.27 17.03
FOV	±48.33° (49.16°)	±48.33°	±49.44°	±55.37°	±49.5°
Swath (km)	2052, (1228)	2052	2134	2900	2160

Table 350: FOV and scan parameter comparison of major sounding instruments

GOME-2 (Global Ozone Monitoring Experiment-2), of GOME-1 heritage flown on ERS-2. The ESA/EUMETSAT jointly-funded instrument was designed and built by Officine Galileo/Alenia Difesa, Italy. The overall objectives of GOME-2 are: measurement of total column amounts and stratospheric and tropospheric profiles of ozone. In addition: measurement of column amounts of H_2O and other gases involved in ozone photochemistry (like NO_2, OClO, BrO, and possibly ClO in polar spring conditions, and pollutants like SO_2 and HCHO). GOME can also be used to investigate the distribution of atmospheric aerosols and clouds-plus-surface spectral reflectance. [1190)]

The GOME-2 instrument consists of the following modules or units:

- Spectrometer: almost identical to the GOME-1 instrument. The major change for the GOME-2 spectrometer resulted from the accommodation of the more complex PMU (Polarization Monitoring Unit) on GOME-2.
- FPA (Focal-Plane Assembly). There are a total of six FPAs, four are assigned to the spectrometer channels and two to the new polarization channels. Each FPA contains a

1190) J. Callies, E. Corpaccioli, M. Eisinger, A. Hahne, A. Lefebvre, "GOME-2 - MetOp's Second-Generation Sensor for Operational Ozone Monitoring," ESA Bulletin, No 102, May 2000, pp. 28-36 random-access linear silicon photodiode array, consisting of 1024 elements, which is reverse-biased and operates in charge accumulation mode. The detectors are cooled to -38° C for the spectrometer channels and to 0° C for the PMUs.

- PMU (Polarization Monitoring Unit): The new PMU provides polarization information (with sufficient accuracy to correct for the polarization dependence of the instrument) by monitoring the spectral range using 200 pixels with a spectral resolution of between 2.8 to about 40 nm, the integration time is 23 ms. The s- and p-polarized parts of the radiation are measured simultaneously. The s- and p-channels have an identical FOV with the main channel. Note: other major ozone measuring instruments like TOMS, SBUV and OMI disregard the measurement of polarization.

Spectrometer type	Double monochromator with pre-disperser prism and 4 holographic gratings
Spectral range	240 - 790 nm
FOV (Filed of View)	0.286° cross-track x 2.75° along-track, or 4 km x 40 km respectively
Entrance slit dimension	0.2 mm cross-track x 9.6 mm along-track
Channels and resolution	1: 240 - 315 nm 0.24 - 0.29 nm (resolution) 2: 311 - 403 nm 0.26 - 0.28 nm 3: 401 - 600 nm 0.44 - 0.53 nm 4: 590 - 790 nm 0.44 - 0.53 nm
Polarization	200 detector pixels
Monitoring unit	312 - 790 nm in 12 programmable bands Spectral resolution: 2.8 nm at 312 nm to 40 nm at 790 nm
Viewing modes Nadir cross-track Solar Lunar	 1920 km, 960 km, 480 km, 360 km, 240 km, 120 km Fixed angle once per day Fixed/varying angle, about 6 times per year
Spectral calibration White light source Dark signal	Fixed angle (once per day to once per month) Fixed angle (once per day to once per month) Fixed angle (night side of the orbit)
Spatial resolution	40 km x 40 km (with 960 km swath and integration time of 0.1875 s) 40 km x 5 km (for polarization monitoring)
Instrument data rate	400 kbit/s
Instrument mass, power	73 kg, 58 W
Instrument size	656 mm (zenith, nadir) x 848 mm (cross-track) x 468 mm (along-track)

Table 351: Major parameters of the GOME-2 instrument

- Calibration unit: The unit contains two light sources (spectral light and white). The white light source is used to monitor the etalon, present on the cooled detectors. The spectral light source is a hollow cathode lamp filled with a mixture of neon and argon. The calibration unit has in addition a diffuser in support of solar calibrations.

- SU (Scan Unit): SU features a rotating mirror, positioning in front of the spectrometer. The mechanical assembly is almost identical to that of GOME-1. The electronics assembly of GOME-2 allows five constant scan profiles and five new scans, compensating for the Earth's curvature and providing a constant linear scan speed on the ground. A new wide-amplitude scan corresponding to 1920 km on the ground, is also implemented. The wider swath permits global coverage in 1.5 days. All scans are reprogrammable. The basic scan timing is 4.5 s in the forward and 1.5 s in the flyback scan.

- CDHU (Command and Data Handling Unit): A processor is responsible for all ICU (Instrument Control Unit) functions such as command handling, maintenance of history file, monitoring of instrument parameters and format handling. CDHU provides 12 predefined timelines, each dedicated to a specific orbit sequence.

GRAS (GNSS Receiver for Atmospheric Sounding), manufactured by Saab Ericsson Space of Sweden and co-funded by ESA and EUMETSAT. The objectives are to provide atmospheric temperature and humidity profiles. The data is used in NWP (Numerical Weather

Prediction) models. Radio occultation of the atmosphere is a well-proven technique whose performances have been demonstrated for Earth atmospheres by GPS/MET and its successor TRSR, both instruments were developed at JPL. [1191] The measurement principle is based on the fact that microwave signals passing through the atmosphere up to a LEO satellite at the limb are bent and delayed due to the variations of the refractive index of the various layers of the atmosphere and the ionosphere. The refractive index depends upon the pressure, temperature and humidity of the atmosphere. [1192] [1193] [1194] [1195]

GRAS is a dual-frequency GPS high-performance instrument with semi-codeless operating capabilities. The instrument provides carrier, code phase and signal amplitude measurements. The second GPS frequency (L2) is used for ionospheric correction of the signals. GRAS features twelve dual-frequency channels, four for rising and setting occultations and eight for navigation. The navigation channels provide the position of the Metop satellite in real-time. Carrier and code phase data are also provided for the processing of the precise orbit determination of Metop (position and velocity) necessary for the ground processing of the occultation data and in particular the effects on the Doppler caused by the GPS signal passing through the atmosphere and the troposphere. A ground network of GPS receiver tracking stations is used for determining the actual orbits of the GPS satellites. This ground network can operate in differential mode in order to cancel clock error variations. - The GRAS instrument consists of the following elements:

- Antennas. The three GRAS antennas are pointed in the satellite's velocity, anti-velocity, and zenith directions. The velocity and anti-velocity antennas are phased arrays, each containing 18 dual-band patches with a shaped antenna pattern optimized for the occultation of the Earth's limb and its atmosphere.
- RFCU (RF Conditioning Unit). Each of the three RFCUs consists of two extremely steep bandpass filters, a low-noise amplifier and a single down-converter stage.
- GEU (GRAS Electronics Unit). The GEU is built around the ESA-developed AGGA (Advanced GPS/GLONASS ASIC). The AGGA simultaneously performs the final down-conversion, de-spreading, and correlation of four dual-frequency GNSS channels. A number of AGGA observables are delivered to the TSC2102OE DSP, an ESA-funded Digital Signal Processor which performs the functions of: measurement, signal tracking, navigation solution, time management, and measurement data processing.

The GRAS instrument has a mass of 20 kg (together with a 7 kg deployment mechanism for the anti–velocity occultation antenna), and a power consumption of 37 W. The average data rate is 27 kbit/s with a maximum rate of 60 kbit/s.

The GRAS instrument provides more than 500 occultations/day, with quasi-uniform geographical distribution, a vertical temperature sounding of 1K rms, with a vertical resolution of 150 m in the troposphere and 1.5 km in the stratosphere over an altitude range of 5 to 30 km. In addition, GRAS provides an opportunity to establish the height of the tropopause with a vertical accuracy of better than 1 km. The horizontal resolution of the radio-occultation technique, limited by the limb-sounding approach itself, is in the order of 100 km in the troposphere and 300 km in the stratosphere.

1191) W. G. Melbourne, et al., "The application of spaceborne GPS to limb sounding and global monitoring," JPL publication 94-18, 1994

1192) M. Loiselet, N. Stricker, Y. Menard, J.-P. Luntama, "GRAS - MetOp's GPS-based Atmospheric Sounder," ESA Bulletin, No 102, May 2000, pp. 38-44

1193) M. Loiselet, N. Stricker, Y. Menard, J.-P. Luntama, "GNSS Radio Occultation Receiver for Atmospheric Sounding," Proceedings of the 2000 EUMETSAT Meteorological Satellite Data Users' Conference, Bologna, Italy, May 29, June 2, 2000, pp. 192-200

1194) G. Bergeton Larsen, "Atmospheric Products from the GRAS Meteorology SAF (Satellite Application Facility)," Proceedings of the 2000 EUMETSAT Meteorological Satellite Data Users' Conference, Bologna, Italy, May 29, June 2, 2000, pp. 243-249

1195) Note: At the start of the GRAS project, the initial goal was to use both the GPS and the GLONASS constellations. However, during the course of the Metop phase B, GLONASS capabilities have been descoped resulting in an implementation which considers only the GPS tracking capabilities.

Level-1 products	Level-2 products
Residual phase data	Refractivity profiles
Sounding occultation table	Temperature profiles
Navigation data	Moisture profiles
Total bending angle as a function of the impact parameter at L1 and L2	Pressure profiles
Ionospheric corrected total bending angle	Integrated precipitable water vapor
TEC	Temperature and humidity retrieval parameters
Auxiliary data	Observation error characteristics
Earth orientation parameters	TEC
Observation error characteristics	

Table 352: Overview of GRAS level-1 and level-2 products

An enhanced version of the GRAS instrument, called GPSOS (GPS Occultation Sensor), will fly on the NPOESS series of NOAA. Other GRAS versions are also under consideration for other spacecraft such as the GCOM (Global Change Observation Mission) series of NASDA, the ACE (Atmosphere Climate Experiment) microsatellite constellation of ESA's Earth Explorer Program, and the WATS (Water-vapor in Atmospheric Troposphere and Atmosphere Sounding) microsatellite constellation of ESA.

ARGOS/ADCS (Advanced Data Collection System) an advanced ARGOS system (also flown on NOAA-N') provided by CNES (see C.1). The objectives are data collection and location from PTTs (Platform Transmitter Terminal) and DCPs (Data Collection Platform) in the ground segment PTTs and DCPs are both part of the ARGOS ground segment for environmental data measurement.

S&RSAT (Search&Rescue Satellite Aided Tracking System). The S&RSAT payloads are part of an international cooperative satellite-based radiolocation system to support search and rescue operations for aviators, mariners, and land travellers in distress (see I.6). The S&RSAT payload consists of S&RR (S&RSAT Repeater), provided by CRC/Canada; and S&RP (S&RSAT Processor), provided by CNES/France). [1196)]

The communication link specification for the ARGOS and S&RSAT services are:

- ADCS data reception at 401.65 MHz
- ADCS data transmission at 466 MHz
- S&RR beacon signal repetition at 121.5 MHz, 243 MHz, and 406.05 MHz
- S&RP-3 data reception at 406.05 MHz (common with S&RR)
- S&RR data transmission at 1544.5 GHz.

SEM-2 (Space Environment Monitor-2), a NOAA-provided instrument package (see G.15.2).

G.3 Feng-Yun-1 (Polar Orbiting Satellite Series)

Feng-Yun (Feng-Yun = wind and cloud) is a meteorological satellite series of China, organized by CMA (China Meteorological Administration). [1197) 1198)] Within the meteorological program, the odd-numbered satellites (Feng-Yun-1 or simply FY-1) refer to the polar-orbiting LEO series, while the even-numbered S/C (FY-2, FY-4, etc.) refer to the GEO se-

1196) **Note: Although the original acronym for 'Search and Rescue' is 'SAR' in the context of NOAA (or MetOp) missions, it was changed in this book consistently to 'S&R' in order to distinguish it from the other widely-used meaning of SAR, namely 'Synthetic Aperture Radar,' a sensor type. A consequence is the use of 'S&RSAT' (instead of SARSAT), to use S&RP instead of SARP, as well as to use S&RP-3 instead of SARP-3**

1197) W. Zhang, "Meteorological satellite program of China," Proceedings of SPIE, Vol. 3501, Optical Remote Sensing of the Atmosphere and Clouds, Sept. 15-17, 1998, Beijing, China, pp. 12-22

1198) W. Zhang, "Meteorological satellite program of China," Proceedings of the Asian Conference on Remote Sensing, HongKong, China, Nov. 22-25, 1999, pp. 299-305 ries. The first generation LEO satellite (FY-1) series consists of a total of four S/C. The funding body in China is the Ministry of Aerospace, while CMA (China Meteorological Administration) is the agency in charge of all functions, pertaining to the business of meteorological satellites (management of space and ground segments, etc.).

G.3.1 FY-1A, -1B

The FY-1 series satellite program was initiated in 1977. The S/C (FY-1A and FY-1B) were designed and built by the Shanghai Institute of Satellite Engineering, part of CAST (China Academy of Space Technology); their payloads were developed by the Shanghai Institute of Technical Physics (SITP), part of CAS (China Academy of Sciences).

Both S/C are similar in design. The satellite structure is a hexahedron of size: 1.4 m x 1.4 m x 1.2 m and a mass of 750 kg for FY-1A. The FY-1A bus was increased for the FY-1B satellite. Likewise, total spacecraft mass increased from 750 kg to about 880 kg. Both satellites are three-axis stabilized (attitude control is maintained by a combination of nitrogen cold gas thrusters and reaction wheels, attitude is sensed by gyros and horizon sensors) and powered by two solar arrays (about 3.5 m long each) with a combined rating of 800 W. Nickel-cadmium batteries are used for electrical power storage. The launch of both S/C was performed by Long March 4 (CZ-4) boosters from Taiyuan, China. Launch of FY-1A: Sept. 7, 1988; Launch of FY-1B: Sept. 3, 1990.

Orbit of FY-1A: Sun-synchronous polar orbit with an altitude of 900 km; inclination = 98.9°, equatorial crossing at 3:30 PM on a descending node, period = 102.86 min.

Orbit of FY-1B: Sun-synchronous polar orbit with an altitude of 900 km; orbital period = 102.86 minutes; inclination = 99.1°, eccentricity < 0.005; equatorial crossing at 7:50 AM.

Figure 206: The Feng-Yun-1 S/C model

Sensor complement:

MVISR (Multichannel Visible and IR Scanning Radiometer) developed and built by SITP (Shanghai Institute of Technical Physics). [1199] [1200] [1201] Two identical instruments were used on-board each S/C (which operate alternately). The objectives are:

- To acquire global visible and infrared cloud images for weather forecasting
- To take images of the land surfaces for the monitoring of crops, vegetation, snow coverage, floods, etc.

1199) H. Gong, Q. Zheng, W. Wang, "The Improvement of the Detecting Property and the Performance of the Very High Resolution Scanning Radiometer on FY-1B Meteorological Satellite," AAS 91-658, pp. 497-503

1200) SITP internal paper written by D. Kuang,, H. Gong, and Q. Zheng

1201) Q. B. Zheng, X. R. Xue, "Optical Design of the Remote Sensing Instrument for FY-1 Meteorological Satellite," Chinese Journal of Infrared & Millimeter Waves, Volume 9, Number 2, 1989

- To measure the sea surface temperature (SST)
- To perform an ocean color experiment for the retrieval of chlorophyll distributions (bands 3 and 4).

MVISR is a five-channel optomechanical scanning instrument operated at 360 rpm. The optical system consists of co-axial and co-focal paraboloid primary mirrors with a 20 cm diameter. The beam splitters of the VIS/IR channels divide the incident radiation into TIR, NIR and VIS beams which are projected onto the respective detectors (HgCdTe, single element Si detector, and three-element Si detector). The optical scanner consists of an elliptical Beryllium-based plane mirror and a driving motor. The mirror rotating axis is 45°. IFOV of 1.2 mrad (1.1 km spatial resolution) in HRPT mode and 4 km resolution in APT mode; FOV=55.4°; swath width = 2,860 km. Instrument calibration of the infrared channels is provided by a reference blackbody and by cold-space views in each scan period. The calibration parameters are transmitted as part of every data record. The TIR detector (HgCdTe) is operated at 105 K (cooled by a radiation cooler).

Channel	Wavelength (μm)	Objectives/Primary Use
1	0.58 - 0.68	Daytime cloud and surface imaging
2	0.725 - 1.10	Daytime cloud image and water, ice, snow, and vegetation observation
3	0.48 - 0.53	Ocean color imaging
4	0.53 - 0.58	Ocean color imaging
5	10.5 - 12.5 (TIR)	Diurnal cloud image surface observation and sea surface temperature

Figure 207: MVISR characteristics

Communications in real-time. Capability of on-board data storage. The satellites provided data in three modes: HRPT (High Resolution Picture Transmission) at 665.4 kbit/s, APT (Automatic Picture Transmission), and DPT (Delayed Picture Transmission). The data formats of HRPT and APT are very similar to those of the NOAA series satellites (satellite attitude is included).[1202)]

The FY-1 series ground segment consists of three ground stations located in Beijing, Guangzhou and Urumqi; a data processing center is at NSMC (National Satellite Meteorological Center). NSMC broadcasts the FY-1 orbital prediction via GTS so that users throughout the world may receive HRPT and APT data in real time. [1203)]

Carrier Frequency	Data Rate Subcarrier Frequency	Modulation Mode	Power
HRPT: 1695.5, 1704.5 MHz	0.6654 Mbit/s	PCM/PSK	5 W
APT: 137.035, 137.795 MHz	2.4 kHz	AM/FM	8 W

Table 353: HRPT/APT transmission channel characteristics of FY-1

Status of missions:

FY-1A (launch Sept. 7, 1988). The S/C went into its predicted orbit and started to make observations on the same day. The imagery was all nominal with an observed SNR higher than the design value. After 39 days of operations, the FY-1A satellite suffered serious malfunctions in its attitude control system and was lost thereafter as a consequence. Another problem was observed on the instrument infrared detectors in form of water vapor contamination, causing the loss of the IR signal.

FY-1B (launch Sept. 3, 1990). The S/C reached its predicted orbit and provided a considerably improved operations environment on the accuracy and stability of S/C attitude as well as for the observing instruments. However, after operations for half a year, the same atti-

1202) 'The Data Format and the calibration Parameters of FY-1 Meteorological Satellite,' Satellite Meteorology Center, SMA

1203) Ch. Weng, et al, "Remote Sensor on the FY-1 Satellite", Paper IAF-92-099, 43rd Congress of the International Astronautical Federation, August-September 1992 tude problems occurred as with the FY-1A S/C. This time around ground operations were able to rescue the S/C. Observations were continued until late 1992 when again attitude problems brought the mission to an end.

G.3.2 Feng-Yun-1C and -1D

The FY-1C/D S/C are upgraded versions of the FY-1B satellite. A major effort has been made to improve the S/C reliability. The satellite structure is a hexahedron with body dimensions of 1.42 m x 1.42 m x 1.2 m. The deployed length of the S/C is 10.5 m. There are two solar panels, each with a total collecting area of 9.58 m^2 providing power of 250 W. The S/C is three-axis stabilized with a pointing accuracy of 0.4°. Two MVISR instruments (one redundant) are mounted on the nadir-facing side of the S/C. The S/C mass is 954 kg, the S/C design life is two years. [1204] [1205] [1206]

FY-1C was launched May 10, 1999 [along with the science exploratory satellite named Shi Jian-5 (SJ-5)] on a CZ-4B vehicle from the Taiyuan launch site, China.

Orbit: Sun-synchronous polar orbit, altitude = 863 km, inclination = 98.8°, period = 102.3 minutes, eccentricity < 0.005. [1207] [1208] [1209]

A launch of FY-1D, planned for the fall 2001, was postponed to 2002.

Sensor complement:

MVISR (Multichannel Visible and IR Scanning Radiometer), developed and built by SITP (Shanghai Institute of Technical Physics). The upgraded MVISR is now a 10-channel instrument (the total number of bands was doubled for the -1C and -1D instruments form the previous -1A and -1B instruments). The optomechanical instrument has a scan mirror with a rotation rate of 360 rpm. The aperture of the optical system is 200 mm in diameter. The IFOV = 1.2 μrad, the nadir resolution of all channels is 1.1 km. The instrument FOV is ±55.4° (2800 km swath). The scan rate of MVISR is 6 lines/s, the total number of pixels of every scan line is 2048.

Band No	Spectral Range (μm)	Primary use of data
1	0.58 - 0.68 (VIS)	Daytime clouds, ice and snow, vegetation
2	0.84 - 0.89 (VNIR)	Daytime clouds, vegetation, water
3	3.55 - 3.93 (MWIR)	Heat source, night cloud
4	10.3 - 11.3 (TIR)	SST, day/night clouds
5	11.5 - 12.5 (TIR)	SST, day/night clouds
6	1.58 - 1.64 (SWIR)	Soil humidity, ice/snow distinguishing
7	0.43 - 0.48 (VIS)	Ocean color
8	0.48 - 0.53 (VIS)	Ocean color
9	0.53 - 0.58 (VIS)	Ocean color
10	0.90 - 0.965 (VNIR)	Water vapor

Table 354: Spectral parameters of MVISR

1204) X. Jianping, W. Caiying, "The Chinese Meteorological Satellite Programs," Proceedings of the 2000 EUMETSAT Meteorological Satellite Data Users' Conference, Bologna, Italy, May 29 - June 2, 2000, pp. 168-173

1205) Gong Huixing, Zheng Qinbo, Weng, Chuijin, "The FY-1C Meteorological Satellite and its Remote Sensor," Proceedings of the Asian Conference on Remote Sensing, HongKong, Nov. 22-25, 1999, pp. 1253-1257

1206) Y. Liu, W. Zhang, Y. Zongdong, "FY-1C Polar Orbiting Meteorological Satellite of China: Satellite Ground System and Preliminary Applications," Proceedings of the Asian Conference on Remote Sensing, HongKong,, Nov. 22-25, 1999, pp. 1261-1267

1207) Information provided by Wang Xinmin of the Shanghai Institute of Technical Physics, Shanghai

1208) http://www.cma.gov.cn/fy2/chnsmc.htm

1209) T. Pirard, "Earth Observation Technology", Spaceflight, January 1995, pp. 20-21

Optical aperture of telescope	200 mm
IFOV, FOV (swath width)	1.26 μrad (1.1 km spatial resolution), ±55.4° (3100 km swath)
Detector material	Si detector for bands 1,2, 7, 8, 9, 10 HgCdTe for bands 3, 4, 5 and 6
Cooler	Radiant cooler operating at 105 K
Radiometric sensitivity	NEDR <3 x 10^{-4}, bands 1, 2, 7, 8, 10 NEDR <1 x 10^{-3}, band 6 (Noise Equivalent Delta Radiance) NEDT < 0.27 K, band 3 NEDT <0.25 K, band 4, 5 (Noise Equivalent Delta Temperature)
Radiation calibration accuracy	>10% for VIS and IR bands, 1.0 K for TIR bands
Scan rate	6 lines/s
Pixels per scan line	2048 for CHRPT and DLPT data 1018 for DGPT data
Data quantization	10 bit
Instrument mass, power	55 kg, 45 W

Table 355: Specification of the MVISR instrument

Communication: The High Resolution Picture Transmission mode of FY-1C and FY-1D is named CHRPT (China HRPT or simply CHRPT), implemented in S-band. The data rate of CHRPT is 1.3308 Mbit/s, the modulation is BPSK/Bi-phase,m the polarization is right hand circular, the EIRP = 39.4 dbm. The transmission frequency is 1700.5 MHz. The data format is very similar to that of HRPT of the NOAA satellites. The on-board data storage capacity is increased to 300 minutes of data take (compared to 60 minutes on FY-1A/1B). This offers the provision of a parallel (to CHRPT) low-rate data stream, once per day, of four selected channels with reduced resolution (3.1 km) data, the low-rate transmission rate is 88.72 kbit/s. The low-rate data is referred to as DGPT (Delayed Global Picture Transmission), while the ten-channel data with 1.1 km resolution from anywhere in the world is referred to as DLPT (Delayed Local Picture Transmission). There is no support provided for APT mode communications.

G.3.3 Feng-Yun-3 (FY-3) Satellites

The FY-3 series represents the second generation of Chinese polar orbiting meteorological satellites. The first launch of this series is planned for 2004. The overall objectives of the FY-3 series are:

- To provide global measurements of 3-D thermal and moisture soundings of the atmosphere, and to measure cloud and precipitation parameters in support of NWP (Numerical Weather Prediction).
- To provide global imagery of large-scale meteorological and/or hydrological events and biosphere environment anomalies
- To provide geophysical parameters in support of global change and climate monitoring.

G.4 METEOR-1 Series

"Meteor" is the Soviet/Russian generic name for its Low-Earth-Orbiting (LEO) meteorological satellite family which started operation in 1969 (built by VNIIEM, Moscow). Sponsoring agency: ROSHYDROMET, the Russian Federal Service for Hydrometeorology and Environmental Monitoring. Prior to this series there was an experimental Cosmos series with the first orbiting meteorological satellite, "Cosmos-44," which was launched on August 25, 1964, and was followed by nine analogous Cosmos satellites until 1969, when the series was officially named 'Meteor-1.' The designations of the series Meteor-1, -2, and -3 define different payload configurations and different orbits; the Meteor-3 series is presently the most advanced payload generation and the operational series. A total of 25 Meteor-1 satellites were launched from 1969 to 1977 (three to four launches per year, see Table 361). Note:

Meteor-1-18 is also referred to as Meteor-Priroda-1; Meteor-1-25 is also named Meteor-Priroda-2, and Meteor-1-28 is the Meteor-2-2 satellite.[1210),1211)]

Background:[1212)] ROSHYDROMET is the national central agency providing relevant environmental and climate information to the public, various industrial organizations and decision-making bodies. Hydrometeorological services together with the provision of information on environmental pollution are based on the performance of the National Observing System for the State of the Environment and Climate (GSNP). GSNP operates data acquisition facilities (ground network of synoptic, hydrological and upper-air stations, weather radars, aircraft, rocket sounding stations, satellites and ocean ships), telecommunication and processing facilities, analysis and forecasts based on advanced atmospheric and hydrological models.

Application: Meteorological sciences, operational meteorology

Orbit: Non-sun-synchronous polar orbit, average altitude of 650 km for the first nine satellites and 900 km for the rest of the series, inclination = 81-82°.

Sensors:

- **TV** = TV optical instrument (MR-600A). Framing (vidicon technique). Spectral range: 0.4 - 0.8 μm; swath width = 1000 km; resolution = 1.25-3 km;
- **IR** = TV IR instrument Lastocha. Spectral range: 8 - 12 μm; swath width = 1000 km; resolution = 15 km;
- **AC** = Radiation Budget Sensor. AC is a narrow-sector device which measures the sun's radiation fluxes, the thermal radiation of the Earth's surface, cloud cover and the atmosphere. Spectral range: 0.3 - 30 μm; swath width = 2500 km; resolution = 40 km x 50 km.

The Meteor series, somewhat similar to the NOAA-TIROS series, provides a daily weather review for more than two-thirds of the globe, observation of clouds, ice cover and atmospheric radiation; these satellites also permit the study of weather fronts and jet stream currents and monitor of clear-air turbulence (for warning intercontinental airliners, etc.).

G.5 METEOR-2 Series

Soviet/Russian Earth observation satellite series in the field of meteorology. Sponsoring agency: ROSHYDROMET. The Meteor-2 series (developed and built by VNIIEM) came into use in 1975. Until 1990, 21 satellites of this type have been launched. The system comprises two to three satellites continuously operating in near polar orbit with an average altitude of 900 km.

Orbit: Non-sun-synchronous polar orbit, altitude between 850 and 950 km, inclination = 81-82°.

Sensor complement:
The on-board instrument package includes three television-type (frame technique) VIS and IR scanners, a five-channel scanning radiometer and a device (RMK) for measuring radiation flux densities in the near-Earth space (see sensor definition of Meteor-3 series).

The Meteor-2-22 (launch Aug. 31, 1993) satellite carried in addition to its regular payload a unique **Fizeau RRA (RetroReflector Array)** for SLR (Satellite Laser Ranging) applications. SLR tracking of this satellite was used for precise orbit determination until Oct. 1998. The

1210) "The Cambridge Encyclopedia of Space," Cambridge University Press 1990, p. 235
1211) The original text was reviewed by Y. V. Trifonow of VNIIEM, Moscow
1212) COSPAR-90-Paper by A. Karpov, USSR State Committee for Hydrometeorology, Moscow. Title of paper: "Hydrometeorological, Oceanographic and Earth-Resources Satellite Systems operated by the USSR."

retroreflector array consists of three corner cubes in a linear array with the two outer corner cubes pointing at 45° angles relative to the central cube. The central cube is made of fused silica and has a two-lobe FFDP (Far Field Diffraction Pattern) providing nearly equal intensities for compensated and uncompensated velocity aberration. Both outer reflectors have aluminum coating on the reflecting surfaces and near-diffraction-limited FFDPs. One of the end reflectors is made of fused silica with an index of refraction of 1.46; it provides partial compensation of the velocity aberration. The other end reflector is made of fused glass with an index of refraction of 1.62; it provides a perfect compensation of the velocity aberration. [1213]

G.6 METEOR-Priroda Series

USSR experimental mission series considered predecessor of the Resurs-O1 series (see D.32). Sponsoring agency: ROSHYDROMET. First Earth resources satellite series for continuous imaging from space. First launch: July 9, 1974. In 1983 Meteor-Priroda data were used to determine snow depths for all Soviet agricultural regions, to assist in planning the spring work season and to forecast the harvest. The Meteor-1 series universal platform was also used for the Meteor-Priroda series up to No. 4. Starting with Meteor-Priroda-5, a Meteor-2 platform was employed.

Orbit: Sun-synchronous polar orbit, altitude = 630 km, inclination = 98°, period = 97-98 minutes, repetition cycle = 16 days

Sensor complement:

- **MSS Fragment** (Multispectral Scanning System Fragment). [1214] [1215] Fragment was designed and developed at IKI (Space Research Institute) in Moscow. The instrument is an optomechanical scanning (whiskbroom) system of high resolution [analogous in function to MSS (Multispectral Scanner System) on Landsat]. The optical subsystem (built by VEB Kombinat Carl Zeiss, Jena) employs a Cassegrain telescope with an aperture of 24 cm diameter and a focal length of 1 m. The scan frequency is 13.02 Hz (cross-track direction). The detector system uses an array of 35 photodiodes. There are eight spectral bands: a) 5 bands are in the spectral range of 0.4-1.0 μm, all with a spatial resolution of 80 m; b) 2 bands in the SWIR range of 1.2 - 1.8 μm with a resolution of 240 m, and c) 1 SWIR band in the range 2.0 - 2.4 μm with a resolution of 480 m. The swath width is 85 km; the instrument data rate is 5.6 Mbit/s. Instrument mass = 280 kg; size: 1.66 m x 1.44 m x 0.73 m; power = 220 W. Extensive pre-launch calibrations were conducted in facilities especially built for Fragment. The first and only launch of MSS Fragment took place on Meteor-Priroda-5 on June 18, 1980. Fragment operated onboard the spacecraft for 4 years.
- **MSU-S** = Multispectral Scanner of moderate resolution (two-channels). Spectral range: 0.58 - 0.7 μm, and 0.7 - 1.0 μm, resolution = 140 - 240 m, swath width = 1380 km.
- **MSU-SK** = Multispectral Scanner of moderate resolution with conical scanning. Spectral ranges: 0.5 - 0.6, 0.6 - 0.7, 0.7 - 0.8, 0.8 - 0.9 μm. Spatial resolution = 175-243 m (in VIS) and 500 m (in IR); swath width = 600 km
- **MSU-E** = High-Resolution Multispectral Scanner with Electronic Scanning. Three spectral ranges: 0.5 - 0.7 μm, 0.7 - 0.8 μm, and 0.8 - 1.0 μm. Spatial resolution = 28 x 28 m; swath width = 28 km. Spaceborne CCD detector technology was introduced with

1213) V. P. Vassiliev, L .I. Gusev, V. D. Shadgorodsky, J. J. Degnan, "Experimental Verification of the Fizeau Influence on the Reflected Beam Direction in Satellite Laser Ranging," Ninth International Proceedings on Laser Ranging Instrumentation, 1994.

1214) "Atlas zur Interpretation von kosmischen Scanneraufnahmen, Multispektralsystem Fragment, Methodik und Ergebnisse," Akademie-Verlag, Berlin, Nauka Verlag, Moscow, 1989

1215) Information provided by Ian Ziman of IKI, Moscow

MSU-E, first flown on the Meteor-Priroda-5 (launch June 18, 1980) spacecraft of the former Soviet Union. 1216) MSU-E was built at ISDE (Russian Institute of Space Device Engineering) in Moscow; it featured a CCD line array of 1024 pixels, three parallel line arrays, each of 1024 elements, and provided pushbroom imagery in three spectral bands (visible range). MSU-E provided a nadir view with a FOV of 2.5°. A radiation cooler provided a detector temperature in the range of -30 to -50°C. Instrument mass of 17 kg. No in-flight calibration was performed.

- **MSU-M** = Multispectral Scanner of low resolution; swath width = 2000 km; resolution = 1 km x 1.7 km; four spectral ranges: 0.5 - 0.6, 0.6 - 0.7, 0.7 - 0.8, 0.8 - 1.0 μm.
- **SHF** = Passive MW Radiometer measuring the emission radiation of the atmosphere/ocean system at the following microwavelengths/frequencies: 0.8 cm (37.5 GHz), 1.35 cm (22.22 GHz), 1.55 cm (19.4 GHz), and 4 cm (7.5 GHz). The swath width = 800 - 900 km depending on orbit height. Resolution = 1.2 - 3.0°.
- **SI-GDR** = Spectrometer/Interferometer (German Democratic Republic). An infrared Fourier spectrometer (original designation: PM = Profile Module) with the objective of temperature profile measurements in the atmosphere. Spectral range: 6.25 - 25 μm; spectral resolution = 5 cm^{-1}; nadir pointing; resolution = 2.2 x 2.4°.
- **R10M** = Radiation Measurement Complex, analogous to RMK-2

Name of Satellite	Launch Date	Period (min)	Perig. (km)	Apog. (km)	I (°)	Remarks / Sensor Complement
Meteor-P.-1	9.7.1974	102.6	877	905	81.2	MSU-M, MSU-S, TV, IR,, SHF
Meteor-P.-2	15.5.1976	102.4	866	908	81.2	MSU-M, MSU-S, IR, SI-GDR, R10-M, SHF
Meteor-P.-3	29.6.1977	97.5	602	685	97.9	MSU-M, MSU-S, SI-GDR, R10-M, SHF
Meteor-P.-4	25.1.1979	97.5	628	656	98.0	MSU-M, MSU-S, SI-GDR, R10-M,
Meteor-P.-5	18.6.1980	97.3	589	678	98.0	MSU-M, MSU-S, Fragment, R10-M, MSU-SK, MSU-E
Meteor-P.-6	10.7.1981	97.6	611	688	97.9	MSU-M, MSU-S, R10-M, SHF

Table 356: Overview of the Meteor-Priroda missions

G.7 METEOR-3 Series

Russian Earth observation satellite series. Sponsoring agency: ROSHYDROMET. The first Meteor-3 launch was on Oct. 24, 1985 from Plesetsk using the F2 launch vehicle. The Meteor-3 series is the current generation of Russian meteorological satellites. Nominal lifetime = 2 years.

The spacecraft series incorporates three-axis stabilization (0.5° accuracy) and twin 10-m span solar panels. The orbit is adjusted by ion thrusters. Meteorological data is transmitted to four primary sites in Russia (USSR) in conjunction with about 80 other smaller sites. Internationally compatible Automatic Picture Transmission (APT) was made available on 137 - 138 MHz channels to ground workstations.

Both Meteor-2 and Meteor-3 satellite types provide main and regional centers of Russia and CIS with global data on the distribution of clouds, snow and ice in VIS and IR bands, radiation flux data at least twice daily, atmospheric temperature-humidity sounding data, data on the temperature of the ground surface, cloud-top heights and sea surface temperature.

1216) A. S. Selivanov, Y. M. Tuchin, M. K. Naraeva, B. I. Nosov, "Experimental Satellite System for Earth Monitoring," Issledovanie Zemli iz Kosmosa, W 5, 1981

Russian plans envisage the development of an integrated hydrometeorological satellite system in the 90s, comprising geostationary and low-orbit satellites. Instruments on-board the geostationary satellites will provide continuous observations of the Earth's disk (footprint) within 60° with respect to a stationary point over the equator.

Orbit: Non-sun-synchronous near-polar prograde orbit, drifting slowly with local time; altitude = 1230 km; inclination = 82.5°; period = 109 minutes; the higher altitude of Meteor-3 enables an extension of the instrument swath width, thereby providing complete coverage of the Earth's surface.

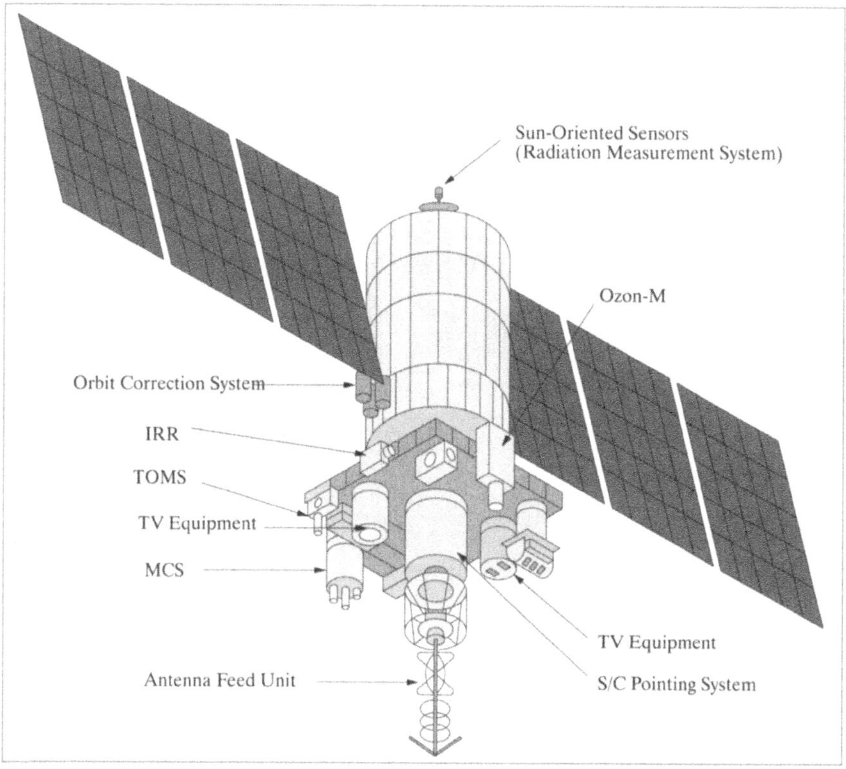

Figure 208: The Meteor-3 series S/C model

Sensor complement:

1) Television Systems

TV camera systems **MR-2000M** and **MR-900B**. Objective: Observation of daytime Earth cloud cover in the visible spectrum (0.5 - 0.7 μm) at a local solar angle not less than 5°. Spatial resolution of 0.7-1.4 km for MR-2000M and 1-2 km for MR-900B. The swath width = 3100 km (MR-2000) and 2600 km (MR-900). The MR-2000M camera provides storage and direct transmission operation. The MR-900B camera has no storage operation mode (only direct transmission). The data acquisition range is stable within a radius of 3000 km. Output products: individual images, photomosaics of images from 2-3 passes over receiving station within 300 km in radius. Global photomosaics of images of various regions of the globe (twice daily), cloud-free photomosaics of arctic and antarctic oceans once in five days.

Parameter	Real-time Mode	Store-and-forward Mode
Swath width (km)	2600	3100
Number of pixels/scan line	900	2000
Number of reproduced brightness levels	50	50

Table 357: TV system parameters

2) Optical Scanning Systems

Klimat= Infrared Radiometer (operational since 1988)[1217]. An electromechanical device with a scan angle of ± 48°, a swath width of 1300 km, a spatial resolution of 0.45 km x 0.9 km; IFOV=0.7 x 1.4 mrad; surface temperature range = 223-313 K; temperature difference at 300 K, background = 0.2 K. Measurement spectrum: 10.5 to 12.5 μm. Instrument mass = 75 kg. Detectors: CdHgTe cooled to 80 K. Output products: Global photomosaics of Northern and Southern Hemispheres, tropical zone, individual images; digital SST and top-of-cloud height charts, tropical cyclone coordinates, cloud amount data on regular grid over the globe.

SM = Multichannel Spectrometer, also known as **"Device 174 K"** (optical instrument for atmospheric observation). Electromechanical device, scan angle =± 24°. Scanning 10-channel IR radiometer for atmospheric thermal sounding. Spectral range: 9.65 - 18.7 μm (9.65, 10.60, 11.10, 13.33, 13.70, 14.25, 14.43, 14.75, 15.015, 18.70). Resolution = 42 km; swath width = 1000 km; output products: SATEM messages with atmospheric thermal sounding data (total ozone content)

3) Radiation Measurement System

RMK-2 = Radiation Measurement Complex. Objectives: Registration of flux densities of protons in the 5-90 MeV and electrons in the 0.15-3.0 MeV energy regions.

- Electron and proton flux density. Ranges: 2.5 -1 x 10^5 particles/cm^{2}s, and 0.111 - 4.4 x 10^3 particles/cm^{2}s
- Radiation dose exposition in the range: 1 x 10^{-7} - 5 x 10^{-4} particles/s
- Number of channels: 12

Meteor Data: There is a prime network of ground stations in Moscow, Novosibirsk, Chabarowsk, and Tashkent (operator/coordinator: NPO Planeta). There are also 80 APT (Automatic Picture Transmission, compatible with NOAA APT) stations in Russia/CIS receiving Meteor data (on 137-138 MHz). The APT stations provide data reception for Meteor and for NOAA series satellites.

A new feature of the Meteor program is the incorporation of sensors and instruments from other space agencies through international cooperation. The following foreign sensors/ payloads are flown or planned on Russian missions:

4) Sensors/systems from other agencies

TOMS (NASA sensor)
A Meteor-3 series S/C (Meteor-3-6) was successfully launched on August 15, 1991 with a NASA sensor on-board. [1218] [1219] [1220] A refurbished TOMS sensor (of the original engineering model for Nimbus-7) was part of this payload. The instrument transmits its data to ground stations in the US and Russia on a daily basis (US archive at GSFC, Russian archive in Dolgoprudny outside of Moscow). The orbital lifetime was projected at two years. Objective: Mapping of vertical ozone profiles. TOMS has a swath width of 3100 km. Six spectral bands at: 0.3125, 0.3175, 0.3313, 0.3398, 0.360, and 0.380 μm. The ground resolution is 47 x

1217) Y. V. Trifonov, "Meteor-3 space system for hydrometeorological observation," VNIIEM, Moscow, 1991
1218) 'Soviets to Launch U.S. Ozone Mapper,' Space News Aug. 5-18, 1991, p. 14
1219) 'TOMS Arrives Successfully in Space,' Space News Aug. 19-25, 1991, p. 2
1220) "TOMS Mission Declared Over by NASA Officials," Space News, February 20-26, 1995, p. 11

47 km at nadir and 62 x 62 km overall. Note: The TOMS instrument failed to provide operational service after December 27, 1994.

ScaRaB (Scanner for Radiation Budget), of CNES [France (CNES, LMD), Russia (Planeta, RKA), Germany (GKSS) are program partners]. ScaRaB is a joint development of a cross-track scanning radiometer. Its objective is the collection of data on shortwave and longwave radiation (reflected solar and emitted thermal radiation) to estimate the Earth's radiation budget at the top of the atmosphere on global and regional scales. The instrument features four channels. Channels 2 and 3 are considered the main channels, while channels 1 and 4 are auxiliary channels. The optical subsystem features four parallel telescopes, one telescope per channel, they are identical except for their filters. [1221) 1222) 1223)]

Nr.	Spectral band (channel)	Band description	Filter
1	0.5 - 0.7 µm	Visible channel: scene identification	Interference
2	0.2 - 4 µm	Solar channel: derivation of Earth radiation budget parameters	Fused silica
3	0.2 - 50 µm	Total radiation channel	None
4	10.5 - 12.5 µm	Atmospheric channel scene identification	Interference

Table 358: Spectral bands of ScaRaB

ScaRaB uses BARNES pyroelectric detectors for all bands (placed at the focus of a spherical aluminium mirror), which are sensitive only to the AC component of the signal (i.e. the modulated energy). Hence, chopping is needed for each pixel. This reduces the influence of the self radiation of the telescope and filters. Two mechanical choppers are used (one for two channels), providing a 10 Hz chopping frequency. The four channels, the two choppers, and a filter wheel dedicated to channel 2 and 3, are mounted on a scanning optical bench (rotor). The telescopes are swiveled by the optical bench so that no extra mirror for the scanning is needed. This reduces the likelihood of offsets dependent on the scanning angle.

The spatial resolution of ScaRaB data is 48 x 48 mrad, scan angle = 100°, swath width = 3200 km. ScaRaB points to nadir and scans the full field of view (FOV) within six seconds. In this cross-track mode data are generated continuously.

Parameter	Value	Parameter	Value
IFOV (spatial resolution)	48 mrad x 48 mrad (60 km x 60 km at nadir)	Sampling interval Sampling period	34 mrad 62.5 ms
FOV (swath)	100° (3200 km)	Scan period	6 s
Pixels per scan	51	Useful scan time	3.18 s
Dynamic range (solar)	up to 425 W m^{-2} sr^{-1}	Instrument mass, power	40 kg, 42 W (average)
Dynamic range (total)	up to 500 W m^{-2} sr^{-1}	Instrument size (mm)	614 x 512 x 320

Table 359: ScaRaB instrument parameters

Calibration subsystem: Gray lamps and blackbodies are used for on-board gain calibration; deep space is used for offset calibration. That subsystem comprises a set of two reference blackbodies for channels 3 and 4, and a set of gray calibration lamps for channels 1, 2 and 3. There is continuous thermal control of the blackbodies. The gray lamps are turned on during the calibration session (typically once per day). In addition, there are short wave references, consisting of two lamps for the calibration of channels 2 and 3 (typical use is once per month). On the ScaRaB/Meteor-3-7 mission, however, the lamp system was damaged so that actual calibration was performed by using the instrument temperature and a pre-

1221) http://scarab.cnes.fr:8020/

1222) J. L. Monge, R. Kandel, L. A. Pakhomov, B. Bauche, "ScaRaB Earth radiation budget scanning radiometer," SPIE, Vol. 1490, 'Future European and Japanese Remote Sensing Programs,' 1991

1223) J. Mueller, et al., "Ground Characterization of the Scanner for Radiation Budget (ScaRaB) Flight Model 1," Journal of Atmospheric and Oceanic Technology, Vol. 14, No 4, pp.802-813, 1997.

launch established gain-temperature law. The remaining lamps were then used to verify this calibration. During one year of operation, no significant sensor degradation was observed.

ScaRaB has a duty cycle of 100%, data rate=3 kbit/s, data volume=18 Mbit/orbit. An instrument mass memory provides data storage for up to 12 hours. The mass of the instrument is 40 kg, the maximum power use is 70 W.

The data processing system is based on algorithms for transforming the instantaneous measurements of radiances, filtered by the optics and detectors, into estimates of the monthly mean values of the radiant excitations in the solar and thermal domains, at the top of the atmosphere. This requires corrections for non-flat spectral response, anisotropic, and diurnal variations. The estimates are provided on a spatial grid of 250 km.

Parameter	Meteor-3 Series	Meteor-3M Series
Payload mass (kg)	700	900
Orbital altitude (km)	1200 - 1250	900 - 950
Orbit inclination (°)	82.5	82.5
Satellite pointing accuracy for all 3 axis (arcmin)	20 -30	10
Stabilization accuracy (°/s)	0.05	0.001
Power, average/day, (W)	600	1000
Payload voltage range (V)	24 - 34	27 ± 1
Nominal design life (year)	2	3-5
Geometric correction accuracy (km)	15	5
Transmitter frequency (MHz) real-time mode store-and-forward mode	 137-138 (analog) 466.5 (analog)	 137-138 (analog) 1690-1710 (digital)
Orbit correction engine	available	available

Table 360: Meteor-3 series satellite/observation characteristics

PRARE (German Microwave Tracking System, see chapter H.7.2). PRARE is flown on Meteor-3-7 as a passenger instrument.

Name of Satellite	Launch Date	Period (min)	Perig. (km)	Apog (km)	Incl. (°)	Remarks / Sensor Complement
Meteor-1-1	23.3.1969	97.9	644	713	81.2	TV, IR, AC
Meteor-1-2	6.10.1969	97.7	630	690	81.2	TV, IR, AC
Meteor-1-3	17.3.1970	96.4	655	643	81.2	TV, IR, AC
Meteor-1-4	28.4.1970	98.1	637	736	81.2	TV, IR, AC
Meteor-1-5	23.6.1970	102	863	906	81.2	TV, IR, AC
Meteor-1-6	15.10.1970	97.5	633	674	81.2	TV, IR, AC
Meteor-1-7	20.1.1971	97.6	630	679	81.2	TV, IR, AC
Meteor-1-8	17.4.1971	97.2	620	646	81.2	TV, IR, AC
Meteor-1-9	6.7.1971	97.3	618	650	81.2	TV, IR, AC
Meteor-1-10	29.12.1971	102.7	880	905	81.2	TV, IR, AC
Meteor-1-11	30.3.1972	102.6	878	903	81.2	TV, IR, AC
Meteor-1-12	30.6.1972	103	897	929	81.2	TV, IR, AC
Meteor-1-13	27.10.1972	102.6	893	904	81.2	TV, IR, AC
Meteor-1-14	20.3.1973	102.6	882	903	81.2	TV, IR, AC
Meteor-1-15	29.5.1973	102.5	867	909	81.2	TV, IR, AC
Meteor-1-16	5.3.1974	102.2	853	906	81.2	TV, IR, AC
Meteor-1-17	24.4.1974	102.6	877	907	81.2	TV, IR, AC
Meteor-1-19	28.10.1974	102.5	855	917	81.2	TV, IR, AC
Meteor-1-20	17.12.1974	102.4	861	910	81.2	TV, IR, AC
Meteor-1-21	1.4.1975	102.6	877	906	81.2	TV, IR, AC
Meteor-2-1	11.7.1975	102.5	872	906	81.3	TV, IR, SM, RMK-2 (experimental)
Meteor-1-22	18.9.1975	102.3	867	918	81.2	TV, IR, AC
Meteor-1-23	25.12.1975	102.4	857	913	81.3	TV, IR, AC
Meteor-1-24	7.4.1976	102.3	863	906	81.2	TV, IR, AC
Meteor-1-26	5.10.1976	102.5	871	904	81.3	TV, IR, AC
Meteor-2-2	6.1.1977	103	893	932	81.3	TV, IR, SM, RMK-2
Meteor-1-28	5.4.1977	102.5	869	909	81.2	TV, IR, AC
Meteor-2-3	14.12.1977	102.5	872	906	81.2	TV, IR, SM, RMK-2
Meteor-2-4	1.3.1979	102.3	857	908	81.2	TV, IR, SM, RMK-2
Meteor-2-5	31.10.1979	102.6	877	904	81.2	TV, IR, SM, RMK-2
Meteor-2-6	9.9.1980	102.4	868	906	81.2	TV, IR, SM, RMK-2
Meteor-2-7	15.5.1981	102.5	868	904	81.3	TV, IR, SM, RMK-2
Meteor-2-8	25.3.1982	104.2	954	976	82.5	TV, IR, SM, RMK-2
Meteor-2-9	15.12.1982	102	836	904	81.3	TV, IR, SM, RMK-2
Meteor-2-10	28.10.1983	101	780	901	81.2	TV, IR, SM, RMK-2
Meteor-2-11	5.7.1984	104	954	974	82.5	TV, IR, SM, RMK-2
Meteor-2-12	7.2.1985	104	950	975	82.5	TV, IR, SM, RMK-2
Meteor-3-1	24.10.19852	110.3	1235	1263	82.5	MR-2000M, MR-900B, IR, SM, RMK-2
Meteor-2-13	6.12.1985	104	952	975	82.5	TV, IR, SM, RMK-2
Meteor-2-14	27.5.1986	104.1	953	974	82.5	TV, IR, SM, RMK-2
Meteor-2-15	5.1.1987	104	950	973	82.5	TV, IR, SM, RMK-2
Meteor-2-16	18.8.1987	104.1	954	974	82.5	TV, IR, SM, RMK-2
Meteor-2-17	30.12.1987	104	952	975	82.5	TV, IR, SM, RMK-2
Meteor-2-18	30.1.1988	104.1	947	973	82.5	TV, IR, SM, RMK-2
Meteor-3-3	26.7.1988	109.4	1198	1221	82.5	MR-2000M,-900B,Klimat,SM, RMK-2,
Meteor-2-19	28.2.1989	104.1	951	974	82.5	TV, IR, SM, RMK-2
Meteor-3-4	25.10.1989	109.5	1191	1228	82.6	MR-2000M,-900B,Klimat,SM,RMK-2, IR
Meteor-2-20	28.6.1990	104	950	974	82.5	TV, IR, SM, RMK-2
Meteor-2-21	28.9.1990	104.1	953	975	82.5	TV, IR, SM, RMK-2
Meteor-3-5	24.4.1991	109.5	1195	1230	82.5	MR-2000M,-900B,Klimat,SM,RMK-2,
Meteor-3-6	15.8.1991	109.5	1195	1230	82.5	MR-2000M,MR-900B,Klimat,SM, RMK-2, IR, TOMS
Meteor-2-22	31.8.1993	104.0	944	979	82.5	TV, SM, RMK-2, RRA (+TEMISAT)
Meteor-3-7	25.1.1994	109.4	1186	1207	82.5	MR-2000M,MR-900B,Klimat, RMK-2, IR, SM,ScaRaB, PRARE, + TUBSAT-B
Cosmos-1066	23.12.1978	102.2	848	908	81.2	Exp. payload on Meteor-1 satellite technology,special geophysical experiment.
Intercosmos-Bulgaria 1300	7.8.1981	101.9	825	906	81.2	Bulgarian sensors on Meteor-2 satellite technology (ionospheric plasma and high-energetic fluxes of charged particles, electric and magnetic fields, etc.)

Table 361: Russian environmental/meteorological satellites (chronological order) [1224] [1225]

1224) Courtesy of B. S. Zhukov (IKI RAN), Y. V. Trifonov and Y. V. Dubrovinsky (VNIIEM), Moscow

1225) Meteor 2-22 was launched in honor of A. G. Iosiphyan, the founder and first director of VNIIEM and the designer of the Meteor-1, Meteor-2, and Meteor-Priroda satellite series

G.8 Meteor-3M Series

The Meteor-3M series represents a modernization of Russia's national meteorological satellite system. [1226] The US instrument SAGE-III is flown on the Meteor-3M-1 mission. A corresponding cooperative agreement was approved by the Gore-Chernomyrdin Commission on Dec. 16. 16, 1994. Plans for Meteor-3M are to combine the meteorological observations of the Meteor-3 series with the Earth-surface observations of the Resurs series, starting with Meteor-3M-2.

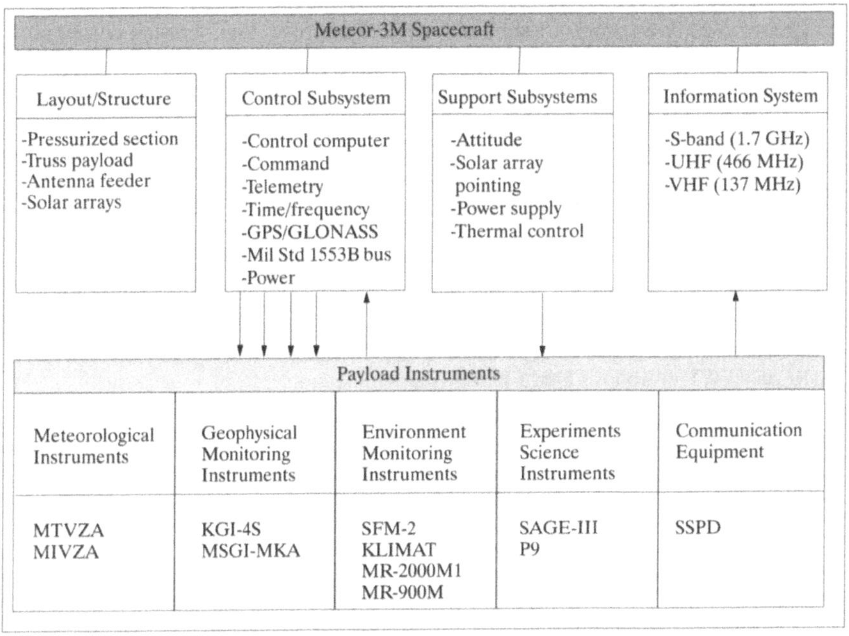

Figure 209: Functional layout of the Meteor-3M satellite

G.8.1 Meteor-3M-1

The Meteor-3M-1 spacecraft structure consists of: a) hermetic container with instrument rack , instrument platform, and c) solar arrays. Most electronics and some service system modules are contained in the hermetic container. Most payload instruments and some service modules are mounted to the external thermally stabilized platform. The spacecraft is three-axis stabilized. The S/C pointing accuracy 0.1°, the angular drift rate is 0.0005°/s. A navigation subsystem (GPS/GLONASS receiver) provides orbit determination and timing services. Solar power of 2 kW (BOL) is provided by two deployed panels which are continuously sun pointed for optimum power generation. The S/C mass is up to 800 kg (up to 370 kg of payload mass), the design life is seven years. The Meteor-3M-1 spacecraft was designed and built by NIIEM (Scientific Research Institute for Electromechanics) in Istra, Russia.

A launch of Meteor-3M-1 is planned for Nov. 2001 on a Zenit-2 launch vehicle from the Baikonur Cosmodrome, Kazakhstan. Secondary payloads on the flight are the Badr-2 satel-

1226) A. I. Bedritsky, V. V. Asmus, A. B. Uspensky, "Current and Future Russian Meteorological Satellite Systems and their Applications," Proceedings of the EUMETSAT Meteorological Satellite Data User's Conference, Copenhagen, Denmark, Sept. 6-10, 1999, pp. 17-23 lite of Pakistan, Maroc-Tubsat of Morocco, the Compass spacecraft of IZMIRAN, Moscow, and REFLECTOR (Retroreflector Ensemble For Laser Experiments Calibration Testing & Optical Research), a US/Russian nanosatellite, funded by AFRL and developed by ISDE (Institute of Space Device Engineering), Moscow. The objective of the passive REFLECTOR spacecraft is to support SLR (Satellite Laser Ranging) experiments. [1227) 1228)]

Orbit: Sun-synchronous circular polar orbit, mean altitude = 830 km, inclination = 98.85°, period = 102 minutes; local time of ascending node is 9:15 AM

Data: On-board information system (OBIS) for recording of instrument data streams up to 150 kbit/s from a single instrument. The solid-state recorder has a capacity of 100 Gbit. OBIS provides time compression of collected streams with the following output options (RF data downlinks). OBIS has a mass of 20 kg and a power consumption of 32 W.

- S-band (N1 digital data streams, 1.67-1.71 GHz broadcast to HRPT-compatible stations at a data rate of 665.4 kbit/s, NOAA-POES series compatible)
- S-band (N2 digital data stream, 1.67-1.71 GHz broadcast to a ground network of AIRS (Autonomous Information Reception Station) at a data rate of 66.54 kbit/s)
- VHF-band (analog data stream, 137.3-137.85 MHz broadcast compatible with APT standard)
- X-band (8.320, 8.064 or 8.192 GHz, digital recorder dump rate of 61.44 Mbit/s and 15.36 Mbit/s to ground stations).

The S/C is being operated by the RSA Mission Control Center in Korolev (Moscow Region), Russia. The SAGE-III data capture and processing is being performed at NASA/LaRC via a WFF (Wallops Flight Facility) ground station.

Sensor complement: The Meteor-3M-1 sensor complement consists of the following instruments: MTVZA, MSU-MR (2 instruments), MSU-SR (2 instruments), IKFS-2, MSGI-MKA, KGS-4S, a DCS (Data Collection System), and SAGE-III of NASA/LaRC.

MTVZA (Microwave Radiometer for Temperature Sounding of the Atmosphere). MTVZA is a passive 20-channel microwave radiometer (similar to NOAA's AMSU-A and -B radiometers). The objective is to provide atmospheric temperature and humidity soundings (water vapor) for cloudy and clear-sky conditions.

Observation region	Parameter	Range of measurements	Error
Ocean	SST		±1.0 K
	Surface wind velocity	4-30 m/s	±2 m/s
	Ice: age (gradations)		3-4
	concentration		20%
	boundary		10 km
Atmosphere	Temperature profile	up to 45 km	±1.5 K
	Column humidity	0-5 g/cm^2	±0.25 g/cm^2
	Clouds:		
	total amount, balls		
	water content	0-5 kg/m^2	±0.2 kg/cm^2
	dynamics		
	effec. temperature		±3.0 K
	Rain rate	0-25 mm/h	±5.0 mm/h
Land	Thickness of dry snow	up to 1.5 m	±0.2 m
	Boundary of snow		±10 km

Table 362: Geophysical parameters derived from MTVZA and IKFS-2

1227) L. Mauldin, R. Salikhov, S. Habib, A. Vladimirov, et al., "Meteor-3M/Stratospheric Aerosol and Gas Experiment (SAGE-III) Jointly Sponsored by the National Aeronautics and Space Administration and the Russian Space agency," SPIE International Asia-Pacific Symposium, Sept. 14-17, 1998, Beijing, China

1228) Information provided by Vladimir Kharitonov of VNIIEM, Moscow

Channel No	Center Frequency (GHz)	No of Passbands	Bandwidth (MHz)	Approx. Peak Sensitivity Altitude (km)
1	18.7	1	200	-
2	23.8	1	400	-
3	31.5	1	400	-
4	36.7	1	400	-
5	42.0	1	2000	-
6	48.0	1	2000	-
7	52.28	1	400	2
8	52.85	1	300	4
9	53.33	1	300	6
10	54.40	1	400	10
11	55.45	1	400	14
12	56.9682±0.1	2	50	20
13	56.9682±0.05	2	20	25
14	56.9682±0.025	2	10	29
15	56.9682±0.01	2	5	35
16	56.9682±0.005	2	3	40
17	89	2	4000	Surface
18	183.31±7.0	2	1500	1.5
19	1.83.31±3.0	2	1000	2.9
20	183.31±1.0	2	500	5.3

Table 363: Channel characteristics of MTVZA

Frequency (GHz)	18.7	23.8	31.5	36.7	42.0	48.0	52-57	89	183
Number of channels	2	1	2	2	2	2	5	2	3
Polarization	V, H	V	V, H	V, H	V, H	V, H	V	V, H	V
Spatial resolution (km)	75	68	45	41	36	32	30	18	12
Conical scanning period	2.52 s								
Viewing angle	51.3°								
Incident angle	65°								
Swath width	2800 km								
Instrument mass, power	80 kg, 80 W								

Table 364: Performance characteristics of MTVZA

MSU-MR (Low Resolution Multispectral Scanner). The instrument features six spectral bands in VNIR, SWIR, and TIR (0.5-0.7 μm, 0.8-1.1 μm, 1.6-1.8 μm, 3.5-4.1 μm, 10.5-11.5 μm, and 11.5-12.5 μm). The spatial resolution is <=1 km on a swath of 3000 km. The brightness temperature range is between 213-3313 K. The instrument has a mass of 30 kg and a power consumption of 100 W (max).

MSU-SR (Medium Resolution Multispectral Scanner). The instrument provides three spectral bands in VNIR (0.535-0.575 μm, 0.63-0.68 μm, 0.76-0.9 μm). The spatial resolution is 100 m on a swath of 1200-1600 km provided with two side-by-side instruments. The mass of one instrument is 5 kg with a power consumption of 15 W (max).

IKFS-2 (Infrared Fourier Spectrometer-2). The objective is atmospheric temperature and water vapor sounding. The instrument performs measurements in the spectral range of 5-15 μm; the spectral resolution is 0.5 cm^{-1}. Sampling steps of 150 km are provided on a swath of 2500 km. The pixel resolution at nadir is 35 km. A micro-cryogenic system is used to cool the infrared detectors. The instrument has a mass of 50 kg and a power consumption of 125 W.

Geophysical Monitoring System - consisting of two instruments: 1) **MSGI-MKA** (Spectrometer for Geoactive Measurements), and 2) **KGI-4S** (Radiation Monitoring System).

The **MSGI-MKA** instrument features four channels for the measurement of the following parameters:

- Electron fluxes in the energy range of 0.1-15 keV (high-sensitivity channel)
- Ion (proton) fluxes in the energy range of 0.1-15 keV (high-sensitivity channel)
- Electron fluxes in the energy range of 0.1-15 keV (low-sensitivity channel)
- Monitoring of integral electron fluxes with a threshold energy of 40 keV

The FOV (Field of View) is 10° x 10° for each channel (3) and 20° x 20° for the integral electron flux. The instrument has a mass of 5 kg and a power consumption of 6.8 W.

KGI-4S (Radiation Monitoring System). The objective is to monitor flux densities within the following threshold energy ranges:

- Total proton flux threshold energy of: 5, 15, 25, 30, and 40 MeV
- Total electron flux threshold energy of: 0.17, 0.7, 1.7, 2.0 and 3.2 MeV
- Proton fluxes with threshold energies of: 25 and 90 MeV

The KGI-4S instrument has a mass of 12 kg and a power consumption of 6.8 W (max).

SAGE III (Stratospheric Aerosol and Gas Experiment III), PI: M. P. McCormick, Hampton University, Hampton, VA (formerly NASA/LaRC). [1229] [1230] [1231] [1232] [1233] The instrument was built at Ball Aerospace of Boulder, CO, as prime contractor. SAGE-III was selected to fly on Meteor-3M and on ISS (International Space Station) in 2004. SAGE III is an Earth limb-scanning grating spectrometer to measure vertical profiles of aerosol, ozone and other constituents in the atmosphere. The SAGE-III instrument is part of the NASA EOS mission in characterizing the Earth system.

Background and heritage: SAM [SAM (Stratospheric Aerosol Measurement) was flown on ASTP (Apollo-Soyuz Test Project), July 15-24, 1975, to perform the first successful solar occultation measurement of stratospheric aerosol], SAM II (launch Oct. 24, 1978 on Nimbus-7, see M.17.7), SAGE I, SAGE II). Both SAM and SAM-II were single spectral instruments measuring the aerosol extinction near the 1000 nm wavelength region. Multiple spectral measurements began with SAGE-I, with a launch on the AEM-2 (Application Explorer Mission-2) satellite, Feb. 18, 1979 (see A.5). SAGE-II is an advanced version of SAGE-I with 7 channels at 385, 448, 453, 525, 600, 940, and at 1020 nm. SAGE-II was flown on ERBS with a launch on Oct. 5, 1984 (see A.13). The measurements of SAM-II, SAGE-I and SAGE-II have provided long-term observations of aerosol and ozone for over 20 years.

The SAGE-III instrument has the following science objectives:

- Retrieve global profiles (with 1 to 2 km vertical resolution) of atmospheric aerosols, ozone, water vapor, NO_2, NO_3, OClO, temperature and pressure in the mesosphere, stratosphere and troposphere
- Investigate the spatial and temporal variability of the measured species in order to determine their role in climatological processes, biogeochemical cycles, the hydrologic cycle, and atmospheric chemistry
- Characterize tropospheric and stratospheric aerosols and upper tropospheric and stratospheric clouds, and investigate their effects on the Earth's environment, including radiative, microphysical, and chemical interactions
- Extend the SAM II, SAGE I and SAGE II self-calibrating solar occultation data sets (begun in 1978), enabling the detection of long-term trends

1229) A. M. Larar, "Optical Spectroscopic Techniques and Instrumentation for Atmospheric and Space Research III," Proceedings of SPIE, Vol. 3756, July 19-21, 1999, Denver, CO, pp. 102-179

1230) W. P. Chu, R. Veiga, "Overview of the SAGE-III Experiment," Proceedings of SPIE, Vol. 3756, July 19-21, 1999, Denver, CO, pp. 102-109

1231) W. P. Chu, R. Veiga, "SAGE-III / EOS," Proceedings of SPIE, Vol. 3501, Sept. 15-17, 1998, Beijing, pp. 52-60

1232) R. E. Veiga, W. P. Chu, A. J. Ray, "The SAGE-III Instrument and Level-1 Data Corrections," ESAMS'99 European Symposium on Atmospheric Measurements from Space, ESTEC, Noordwijk, The Netherlands, Jan. 18-22, 1999,

1233) http://arbs8.larc.nasa.gov/sage3/

- Provide atmospheric data essential for the calibration and interpretation/correction of other satellite sensors, including EOS- and ground-based sensors.

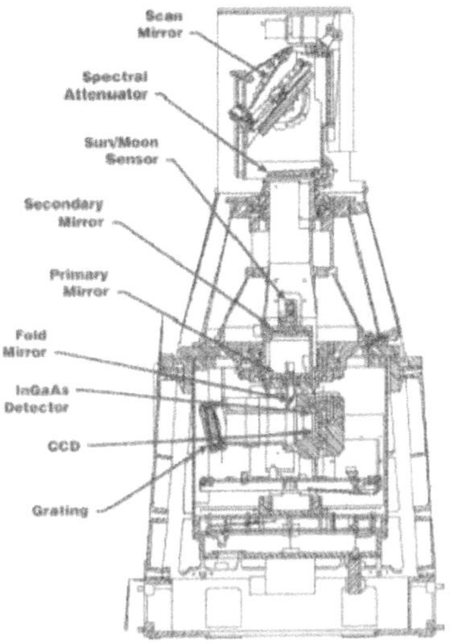

Figure 210: Illustration of the SAGE-III instrument

Measurement approach: Self-calibrating solar and lunar occultation, with nine spectral channels, from 280 to 1550 nm, to study aerosols, ozone, OClO, NO_2, NO_3, water vapor, temperature, and pressure. The instrument looks at the sun (or the moon) through the Earth's limb, utilizing a two-axis passive suntracker with a scan mirror,that scans the FOV across the solar disk (obtaining multiple samples at each altitude).

Spectral channels: 9 solar	290-1550 nm
FOV (Field of View)	± 185° in azimuth, -24.81 to -31.02° in elevation from local horizontal
IFOV	1-2 km vertical resolution; ±0.5 km vertical at 20 km tangent height
Thermal control	Passive heaters and thermal electric cooler
Thermal operating range	10-30° C
Data quantization, data rate	16 bit, 100 kbit/s for 8 minutes (three times per orbit)
Duty cycle	Measurements during solar and lunar occultations
Instrument mass, power, size	76 kg (total, 35 kg sensor), 80 W, 73 cm x 45 cm x 93 cm

Table 365: Some SAGE-III instrument parameters

SAGE-III consists of three major subsystems:

1) Pointing subsystem: It consists of a scan mirror which acquires the radiant target and performs vertical scanning across the target. The scan mirror is mounted on an azimuth drive which rotates over 360° for pointing in azimuth direction.

2) Imaging subsystem. The objective is to produce a focused image of the target at a focal plane where the instrument's field of view is situated. - A Dall-Kirkham telescope design is used with an f/4 ratio. A slit (1/2 x 5 arcmin), located in the focal plane, serves as aperture and as entrance slit to the grating spectrometer. The entire telescope assembly, including scan mirror, can rotate in azimuth to eliminate the problem of image rotation during azimuth rotation. The target sensor assembly is mounted on the back of

the telescope secondary and on the side of the telescope housing. The target sensor consists of a bi-level photodiode, capable to perform both solar and lunar acquisition with a change in dynamic range of six orders of magnitude.

3) Spectrometer subsystem. The objective is to measure solar radiation from 280 to 1040 nm and 1 to 2 nm spectral resolution. An additional photodetector measures radiation at 1550 nm. The spectrometer is a new design, utilizing a holographic, aberration-reduced grating to provide stigmatic imaging at 440 and 868 nm with 1 nm resolution below 450 nm and 2 nm resolution between 740 and 960 nm. The detector array consists of two elements, a Tektronix 800x10 pixel backside-illuminated CCD array (283-1030 nm range) and an infrared photodiode (InGaAs type for measurements at 1550 ±15 nm) that are spatially co-registered.

Cha. No.	Spectral center or range (nm)	Channel accuracy (nm)	No. of subchannels	Bandwidth at FWHM (nm)	Cut-off range (nm)	Integration time (ms)	SNR
Solar Channels							
1	290	±1	1	<5.0	<7	2.23	3725
2	385	±1	1	<14.5	<17	0.26	6926
3	430-450	±0.5	20	<1.1	<3	0.11	2084
4	525	±1	1	<10.5	<13	0.09	7844
5	600	±1	1	<10.5	<13	0.10	7865
6	740-780	±1	20	<2.2	<5	0.17	3083
7	920-960	±1	20	<2.2	<5	0.58	3033
8	1020	±1	1	<20.5	<23	1.43	9678
9	1550	±5	1	<32.5	<23		
Lunar Channels							
1	380-680	±0.5	300	<1.5	<17	62	160-300
2	740-780	±1	20	<2.2	<17	62	300
3	920-960	±1	20	<2.2	<17	62	150

Table 366: SAGE III channel specifications

		Solar Occultation		Lunar Occultation	
Wavelength (nm)	Constituent	Altitude (km)	Error (%)	Altitude (km)	Error (%)
280	O_3	50-85	10	-	-
385	Aerosol	15-35	10	-	-
380-420	OClO	-	-	15-25	25
430-450	NO_2, Aerosol	15-45, 10-35	10	20-50	10
470-490	O_3	-	-	15-35	10
525	Aerosol	6-35	10	-	-
600	O_3	6-60	5	-	-
640-680	NO_3	-	-	20-55	10
740-780	O_2, Aerosol, density	6-70, 6-35	2	-	-
920-960	H_2O, Aerosol	3-50, 3-35	10	-	-
1020	Aerosol	0-35	5	-	-
1550	Aerosol	0-35	5	-	-

Table 367: SAGE III measurement capability (single profile)

The spectrometer with the CCD array assembly provides continuous wavelength coverage between 290 and 1040 nm with a spectral resolution between 1.2 to 2.5 nm, permitting the measurement of multiple absorption features of each gaseous species and multi-wavelength measurement of broadband extinction by aerosols. Nine channels are routinely utilized in solar occultation measurements and three channels are used in lunar measurements. - Spectral calibration is continuous, combined with the self-calibrating nature of the occultation technique. In addition, the CCD array permits in-orbit wavelength and intensity calibration from observations of the exo-atmospheric solar Fraunhofer spectrum.

The SAGE-III retrieval algorithm is a procedure converting the instrument's response to solar or lunar flux (about 70-80 spectral bands) into vertical profiles of molecular density of gaseous species, aerosol extinction at eight wavelengths, temperature, and pressure.

Product name	Accuracy: Systematic/random	Vertical coverage
Aerosol extinction at 8 wavelengths (solar)	5% / 5%	0-40 km
H_2O concentration	10% / 15%	0-50 km
NO_2 concentration & slant path column amount	10% / 15%	10-50 km, 10-50 km
NO_3 concentration (lunar)	10% / 10%	20-55 km
O_3 concentration & slant path column amount	6% / 5%	6-85 km, 50-85 km
OClO concentration (lunar)	25% / 20%	15-25 km
Pressure	2% / 2%	0-85 km
Temperature profile (solar)	2 K / 2 K	0-85 km
Cloud presence	N/A	6-30 km

Table 368: Overview of some SAGE-III products

G.9 NPP (NPOESS Preparatory Project)

NPP is a joint NASA/IPO (Integrated Program Office) mission with a planned launch in late 2005 and a mission duration of five years. The two objectives are to demonstrate the performance of three advanced sensors and their associated Environmental Data Records (EDR), such as sea surface temperature, and to provide data continuity for key data series initiated by NASA's EOS Terra and Aqua missions. Because of this second role, it is sometimes known as a bridging mission. Two of the mission instruments are **VIIRS** (Visible/Infrared Imager and Radiometer Suite) and **CrIS** (Cross-Track Infrared Sounder). Both are under development by the IPO. NASA/GSFC develops a third sensor, namely **ATMS** (Advanced Technology Microwave Sounder). This suite of sensors is able to provide cloud, land and ocean imagery, covering the spectral range from the visible to the thermal infrared, as well as temperature and humidity profiles of the atmosphere. In addition, NASA is developing the NPP S/C and providing the launch vehicle (Delta-2 class). IPO is providing satellite operations and data processing for the operational community; NASA is supplying additional ground processing to support the needs of the Earth science community. [1234) 1235)]

Satellite system overview:

The ACS (Attitude Control Subsystem) provides 3-axis stabilization using 4 reaction wheels for fine attitude control, 3 torquer bars for momentum unloading, thrusters for coarse attitude control (such as during large-angle slews for orbital maintenance), 2 star trackers for fine attitude determination, 3 gyros for attitude and attitude rate determination between star tracker updates, 2 Earth sensors for safe-mode attitude control, and coarse sun sensors for initial attitude acquisition, all monitored and controlled by the spacecraft controls computer. ACS provides real-time attitude knowledge of 10 arcsec (1 sigma) at the S/C navigation reference base, real-time spacecraft position knowledge of 25 m (1 sigma), and attitude control of 36 arcsec (1 sigma). The EPS (Electrical Power Subsystem) uses GaAs solar cells to generate power of about 1.5 kW (EOL). The solar array rotates once

1234) Information provided by Raynor L. Taylor of NASA/GSFC
1235) http://jointmission.gsfc.nasa.gov/

per orbit to maintain a nominally normal orientation to the sun. In addition, a single-wing solar array is mounted on the anti-solar side of the S/C; its function is to preclude thermal input into the sensitive cryo radiators of the VIIRS and CrIS instruments. A regulated 28 ±6 VDC power bus distributes energy to all S/C subsystems and instruments. A NiH (Nickel Hydrogen) battery system provides power for eclipse phase operations. The C&DHS (Command & Data Handling Subsystem) collects instrument data (12 Mbit/s max total) via an IEEE 1394a-2000 "firewire" interface, and stores the data on-board. Upon ground command or autonomously, the C&DHS transmits stored instrument data to the communication system for transmission to the ground. Also, the C&DHS generates a real-time 15 Mbit/s data stream consisting of instrument science and telemetry data for direct broadcast via X-band to in-situ ground stations.

The spacecraft is designed to be highly autonomous. For satellite safety, the S/C controls computer monitors spacecraft subsystem and instrument health. It can take action to protect itself (for example, in the event of an anomaly that threatens the thermal or optical safety and health of the S/C, then it can enter into a safe or survival mode and stay in the mode indefinitely until ground analysis and resolution of the anomaly). In addition, the satellite is designed to require infrequent uploads of commands (the instruments operate mainly in a mapping mode and therefore require few commands even for periodic calibration activities, and a sufficiently large command buffer is available for storage of approximately 16 days of commands).

The spacecraft has an on-orbit design lifetime of 5 years (available consumables for 7 years). The S/C mass is about 1400 kg. NPP is designed to support controlled reentry at the end of its mission life (via propulsive maneuvers to lower the orbit perigee to approximately 50 km and target any surviving debris for open ocean entry). NPP is expected to have sufficient debris that survives reentry so as to require controlled reentry to place the debris in a pre-determined location in the ocean.

The launch of NPP is planned for late 2005 on a Delta-2 vehicle from VAFB.

Background: The satellite is in a competitive down-select phase, with preliminary designs presented by two spacecraft developers (Ball Aerospace of Boulder, CO, and SpectrumAstro of Gilbert, AZ). Two preliminary design contracts were awarded in March 2000, and preliminary design reviews took place in late 2000. Award of the spacecraft development contract is planned for 2002, and instrument deliveries to the spacecraft developer are planned for 2004.

Communication of data: The satellite collects instrument data, stores the data onto a solid-state recorder of about 200 Gbit capacity. All data are transmited either via Ku-band through the NASA Tracking and Data Relay Satellite System (TDRSS) network, or via X-band at 300 Mbit/s to polar ground stations. In addition, the satellite directly broadcasts all instrument data at 15 Mbit/s via an Earth-shaped antenna to in-situ direct broadcast receive terminals (similar to the service provided by the EOS Terra and Aqua satellites). The TT&C function uses S-band communications with uplink data rates of up to 16 kbit/s and downlink rates of up to 128 kbit/s. The NOAA network of polar ground stations will be used for mission operations (back-up TT&C services via TDRSS through S-band omni antennas on the satellite). The performance goal calls for EDR delivery within 3 hours of acquisition. NPP also focuses on ground segment risk reduction by providing and testing a subset of an NPOESS-like ground segment. Developed algorithms can be thoroughly tested and evaluated. This applies also to the methods of instrument verification, calibration, and validation.

Orbit: Sun-synchronous near-circular polar orbit, altitude = 824 km, inclination =98.7°, equatorial crossing time at 10:30 AM ± 10 minutes on descending node. The repeat cycle is 16 days (quasi 8-day).

Sensor complement: [1236)]

ATMS (Advanced Technology Microwave Sounder). A NASA-provided instrument (built by Aerojet of Azusa, CA) with the objective to combine the passive-microwave observation capabilities of three heritage instruments, namely AMSU-A1/A2 and AMSU-B/MHS, into a single instrument with a correspondingly reduced mass and power consumption and with advanced microwave-receiver electronics technologies. ATMS is a passive total power microwave sounder whose observations (measurement of microwave energy emitted and scattered by the atmosphere), when combined with observations from an infrared sounder (CrIS), provide daily global atmospheric temperature, moisture, and pressure profiles.

Can-nel	Center frequency (GHz)	Max. band-width (GHz)	Center frequency stability (MHz)	Temp. sensitiv-ity (K) NEΔT	Calibra-tion ac-curacy (K)	Static beam-width (°)	Quasi polariza-tion	Characterization at nadir (reference only)
1	23.8	0.27	10	0.9	2.0	5.2	QV	Window-water Vapor 100 mm
2	31.4	0.18	10	0.9	2.0	5.2	QV	Window-water Vapor 500 mm
3	50.3	0.18	10	1.20	1.5	2.2	QH	Window-surface Emissivity
4	51.76	0.40	5	0.75	1.5	2.2	QH	Window-surface Emissivity
5	52.8	0.40	5	0.75	1.5	2.2	QH	Surface air
6	53.596 ±0.115	0.17	5	0.75	1.5	2.2	QH	4 km ~ 700 mb
7	54.40	0.40	5	0.75	1.5	2.2	QH	9 km ~ 400 mb
8	54.94	0.40	10	0.75	1.5	2.2	QH	11 km ~ 250 mb
9	55.50	0.33	10	0.75	1.5	2.2	QH	13 km ~ 180 mb
10	57.290344	0.33	0.5	0.75	1.5	2.2	QH	17 km ~ 90 mb
11	57.290344 ±0.217	0.078	0.5	1.20	1.5	2.2	QH	19 km ~ 50 mb
12	57.290344 ±0.3222 ±0.048	0.036	1.2	1.20	1.5	2.2	QH	25 km ~ 25 mb
13	57.290344 ±0.3222 ±0.022	0.016	1.6	1.50	1.5	2.2	QH	29 km ~ 10 mb
14	57.290344 ±0.3222 ±0.010	0.008	0.5	2.40	1.5	2.2	QH	32 km ~ 6 mb
15	57.290344 ±0.3222 ±0.0045	0.003	0.5	3.60	1.5	2.2	QH	37 km ~ 3 mb
16	87-91	2.0	200	0.5	2.0	2.2	QV	Window H_2O 150 mm
17	166.31	2.0	200	0.6	2.0	1.1	QH	H_2O 18 mm
18	183.31±7	2.0	100	0.8	2.0	1.1	QH	H_2O 8 mm
19	183.31±4.5	2.0	100	0.8	2.0	1.1	QH	H_2O 4.5 mm
20	183.31±3	1.0	50	0.8	2.0	1.1	QH	H_2O 2.5 mm
21	183.31±1.8	1.0	50	0.8	2.0	1.1	QH	H_2O 1.2 mm
22	183.31±1	0.5	30	0.9	2.0	1.1	QH	H_2O 0.5 mm

Table 369: Channel characteristics of ATMS

ATMS will replace instruments currently flying on POES satellites. The new instrument is about one-third the size and mass of the existing microwave sounding instruments (on

1236) R. E. Murphy, R. Taylor, et al., "The NPOESS Preparatory Project: Mission Concept and Status," IGARSS 2001, Sydney, Australia, July 9-13, 2001

POES and on Aqua). This miniaturization of ATMS is enabled by the application of new technologies, principally in the area of microwave electronics. Also, this miniaturization enables the use of smaller spacecraft to fly ATMS and the other required instruments, thereby reducing the cost of future weather and climate research satellites.

ATMS is a cross-track scanning total power microwave radiometer, with a swath width of 2300 km and a spot size of approximately 1.5 km [the native observation resolution is finer than 1.5 km (in fact about 0.5 km), but ground processing performs a spatial averaging computation to increase the SNR]. Thus, the spatial resolution of the ATMS data products is 1.5 km. Two continuously rotating reflectors direct the microwave energy emitted by the atmosphere into four feedhorns, via tuned reflectors. Microwave detectors and associated electronics filter the microwave signal to measure 22 separate channels from 23 to 183 GHz, and convert the channels into electrical signals that are then digitized. The instrument includes on-board calibration sources viewed by the reflectors during each scan cycle. In addition, cold space is viewed during each scan cycle. Both calibrations provide for the highly accurate microwave sounding measurements required by the operational and science applications of ATMS data. The instrument data are transmitted to the spacecraft via a MIL-STD-1553B bus interface. ATMS has a mass of about 66 kg and consumes about 85 W of orbital average power.

On the ground, ATMS raw data are converted into brightness temperature measurements by channel, are radiometrically corrected using calibration data, and are ortho-rectified. ATMS brightness temperatures by channel are then used in conjunction with the corresponding data from the infrared sounder (CrIS) to retrieve atmospheric temperature and humidity profiles for use in data assimilation algorithms for operational or climate research use.

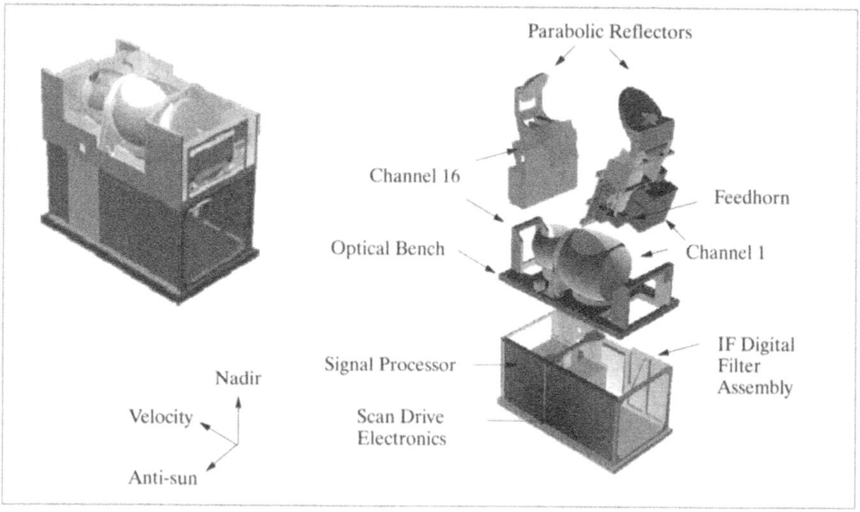

Figure 211: Schematic illustration of ATMS

The instruments **CrIS** (Cross-Track Infrared Sounder) and **VIIRS** (Visible/Infrared Imager and Radiometer Suite) are defined under NPOESS (R.147.3).

Technology demonstrations:

- Use of advanced low noise amplifier technology for atmospheric sounding (ATMS)
- S/C on-board processing using reconfigurable computing and RAM-based field-programmable gate arrays for generation of information products (option).

NASA invited ESA to propose payloads for a possible fight opportunity on the NPP. ESA plans to propose the following candidate instruments: GRAS (GNSS Receiver for Atmospheric Sounding), COALA (Calibration for Ozone through Atmospheric Limb Acquisitions), and SWIFT (Stratospheric Wind Interferometer For Transport studies).

G.10 NPOESS (National Polar-orbiting Operational Environmental Satellite System)

NPOESS is the planned next generation operational US polar-orbiting environmental satellite system, replacing the current generation POES and DMSP satellite constellations. The objective of the NPOESS program is to provide a single, national polar-orbiting remote-sensing capability for meteorological, oceanographic, climatic and space environmental data that will satisfy the operational demands of both the US civilian and military communities.

Background: In the fall of 1993, the US National Performance Review (NPR) and the subsequent Presidential Decision Directive/NSTC-2 (May 5, 1994) called on DOC, DoD and NASA, to "converge" the US civil and military operational meteorological satellite programs (POES of NOAA and DMSP of DoD), in order to reduce duplication of effort and to generate cost savings. In October 1994, an Integrated Program Office (IPO), consisting of a team made up of NOAA, NASA and DoD representatives, was established organizationally under NOAA (with the IPO Headquarters located in Silver Spring, MD) with the explicit objective to integrate their separate meteorological programs into a single program that includes: planning, development, management, acquisition, and operations. A tri-agency MOA (Memorandum of Agreement) was signed in May 1995. The merged program received the name of **NPOESS (National Polar-orbiting Operational Environmental Satellite System)**. The IPO is a tri-agency office reporting through NOAA to an Executive Committee of representatives from DOC, DoD and NASA.

The IPO plan for NPOESS is to have the first satellite ready for launch in the 2008 time frame. As a consequence, the previously planned POES O, P, and Q series satellites have been cancelled, as well as the previously planned follow-on system to DMSP, known as the Block-6 series. In May 1998, the US Air Force Space Command transferred its day-to-day operations of the DMSP spacecraft to NOAA. With this action, NOAA assumed full responsibility for the operation of both the POES and DMSP satellite constellations. NOAA maintains and conducts joint satellite control operations at a collocated operations center where it also conducts operations of NOAA's GOES (Geostationary Operational Environmental Satellite) series (from Suitland, MD).

G.10.1 NPOESS Transition Period Overview

The merging/evolution of the DMSP- and POES series has the following consequences:

- For the present and continuing into the latter part of the 1st decade of the 21st century, US polar-orbiting operational environmental/meteorological services will continue to be met by "flying-out" the remaining current generation satellites of each series (POES ATN and DMSP-5D3). This "fly-out" will be accomplished in a manner consistent with the current operation of these legacy programs (i.e. orbits, capabilities, etc. will remain unchanged).
- By 2005/6, the already planned merger of the civil US/European polar orbiting systems will consist of a total of two operational satellites (a US "afternoon" spacecraft with a European "mid-morning" spacecraft), a coordinated system, referred to as the IJPS (Initial Joint Polar System), see G.15.3. As stated previously, one orbital plane (the traditional civil "afternoon orbit") will be provided/operated by NOAA (remaining

POES series up to the NOAA-N' mission), the other orbital plane (the traditional civil "mid-morning" orbit, now maintained by NOAA spacecraft) will be provided by EUMETSAT with the start of the MetOp series (MetOp-1 launch in 2005). The Metop series satellites will be launched into a mid-morning orbit (with the following NOAA-provided instruments: AVHRR/3, HIRS/4, AMSU-A and SEM-2), while the POES series satellites will support the afternoon orbit.

- On the other hand, European-provided instruments are flown on NOAA-POES satellites since 1978, starting with the Argos DCS (Data Collection System) of CNES, and SSU (Stratospheric Sounding Unit), a UK sensor provided by BMO - on TIROS-N (launch Oct. 13, 1978) and subsequent NOAA satellites. The S&RSAT (Search & Rescue Satellite) payload, first flown on NOAA-8 (launch March 28, 1983), is provided by France and Canada. Three AMSU-B (Advanced Microwave Sounding Unit - B) instruments of UKMO are being flown on NOAA-K (NOAA-15, launch May 13, 1998), NOAA-L (NOAA-16, launch Sept. 21, 2000), and NOAA-M satellites. The MHS (Microwave Humidity Sounder), a EUMETSAT instrument, is planned to be flown on on the NOAA-N and N' satellites.
- For the NPOESS transition period, there are also operational polar orbiting satellites of the continuing DoD DMSP series (up to F-20), operated by NOAA that will continue to fly in their traditional orbital planes (early morning and mid-morning).
- The next-generation NPOESS series is planned to begin operations with an initial launch planned in the timeframe of 2008, and full operational capability being achieved around 2013. The overall polar orbiting environmental satellite constellation (beyond 2010) will consist of a total of three US satellites, namely with three NPOESS S/C provided and operated by the IPO, with continuing critical sounding data contributions to be provided by the mid-morning MetOp series S/C provided and operated by EUMETSAT. ARGOS Data Collection (DCS) and Search and Rescue (S&RSAT) systems will also continue to be flown as well.

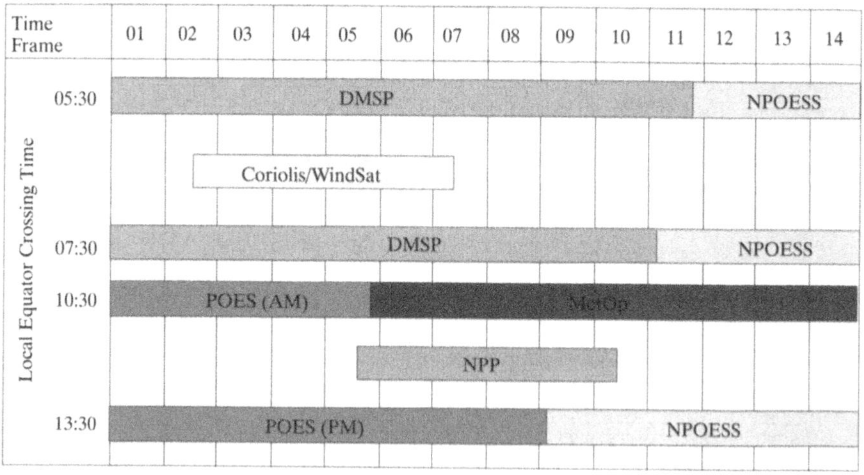

Figure 212: Overview of satellite series transition period

- The NPOESS program is preceded by the launch of a single satellite in 2005, referred to as NPP (NPOESS Preparatory Project) to test the performance of several new NPOESS instruments and to provide climatic data continuity from the Earth Observing System (EOS) satellites (EOS/Terra and EOS/Aqua), operated by NASA.

- The DoD Coriolis mission with the WindSat instrument, implemented and managed by NRL, is a passive microwave technology demonstration mission and considered a further risk reduction project for NPOESS (validation of the CMIS instrument). A launch of Coriolis is planned for early 2002.

The NPOESS system is required to provide an operational remote-sensing capability to acquire and receive in real-time at field terminals global and regional environmental imagery and meteorological, climatic, terrestrial, oceanographic and solar-geophysical and other data in support of mission requirements. The NPOESS data sets will contain a number of variables that are currently not included in operational measurements (such as: radiation budget, total ozone, wind speed and direction, ocean topography, and ocean color) and will offer improved quality for some variables now being measured (such as: atmospheric moisture and temperature profiles, all-weather SST, and vegetation indices).

G.10.2 The NPOESS Satellite

At the current time (2001) two competing contractor teams are accomplishing risk reduction and preliminary designs for the NPOESS spacecraft and ground systems. An engineering, manufacturing, and development contract is planned to be awarded in the 2002 timeframe. The S/C size and payload capacity will be similar to the current POES series. The S/C design requirements call for a 15 year design life (5 years in storage, 3 years of intermittent pre-launch operation, and 7 years of on-orbit design life performance). Requirements call also for up to 21 days of autonomous operation capability with a goal of 60 days. The requirements on RF data transmission call for a nominal data rate of about 12 Mbit/s, LRD (Low Rate Data) of 3 Mbit/s, with an overall system design capacity of up to 20 Mbit/s.

Orbit: Sun-synchronous orbit, nominal altitude = 833 km, inclination =98.7°. The orbit will be maintained with regard to altitude (±17 km), inclination (±0.05°) and nodal crossing times (±10 min).

G.10.3 NPOESS Sensor Complement

A key aspect of the convergence effort is the definition, design, development, and implementation of five critical sensor suites that will fly aboard NPOESS (competitive development of the sensor instruments). BATC (Ball Aerospace & Technologies Corp.) is the prime contractor for OMPS and is competing for the CMIS contract. Boeing (formerly HSAC) is also competing for the CMIS contract. ITT Aerospace/CD is the prime contractor for CrIS. Saab Ericsson Space is the prime for GPSOS and Raytheon SBRS is the prime contractor for the VIIRS. The CMIS and VIIRS instruments are planned to fly on the early morning, mid-morning and the afternoon satellites, while the CrIS and OMPS sensors will fly on the afternoon satellites.

VIIRS (Visible/Infrared Imager and Radiometer Suite). An Engineering, Manufacturing, and Development (EMD) contract has been awarded to Raytheon Santa Barbara Remote Sensing (SBRS). VIIRS is an advanced, modular, multi-channel imager and radiometer (of OLS, AVHRR/3, MODIS, and SeaWiFS heritage) with the objective to provide global observations (moderate spatial resolution) of land, ocean, and atmosphere parameters at high temporal resolution (daily). Typical data products of VIIRS include cloud/weather imagery, sea-surface temperature measurements, ocean color characterization, and land-surface vegetation indices (VIIRS is the primary data provider to a total of 24 environmental data records). A swath width of 3000 km is provided (corresponding to FOV=±55.84°) with a spatial resolution for imagery related products of no worse than 0.4 km to 0.8 km (nadir to edge-of-scan). The radiometric bands provide a resolution about twice in size to the imagery bands. Note: Most derived data products will be produced at somewhat coarser resolutions by aggregation of on-board data.

The VIIRS instrument employs an all-reflective optics assembly taking advantage of recent optics advances: a) single 4 mirror imager, b) 2 dichroics and 1 fold, c) aluminum DPT-bolt together technology (DPT = Diamond Point Turning). A rotating off-axis and afocal TMA (Three Mirror Anastigmatic) telescope assembly is employed [Note: The telescope rotates 360°, thus scanning the Earth scene, and then internal calibration targets.]. The aperture of the imaging optics is 19.1 cm in diameter, the focal length is 114 cm (f/5.97). The VIIRS optical train consists of the fore optics (TMA), the aft optics [an all-reflective FMA (Four Mirror Anastigmatic) imager], and the back-end optics, which include microlenses for the cooled focal planes. A total of 22 spectral bands have been selected as defined in Table 370. VIIRS features band-to-band registration for all bands (optical alignment of all FPAs). A total of three focal planes and four FPAs (Focal Plane Arrays) cover the spectral range of the instrument, one FPA for DNB (Day-Night Band), one for VNIR, SWIR/MWIR, and TIR. The DNB spectral range of 0.5-0.9 µm CCD detector features four light-sensitive areas (3 with TDI, one without) and near-objective sample spacing.

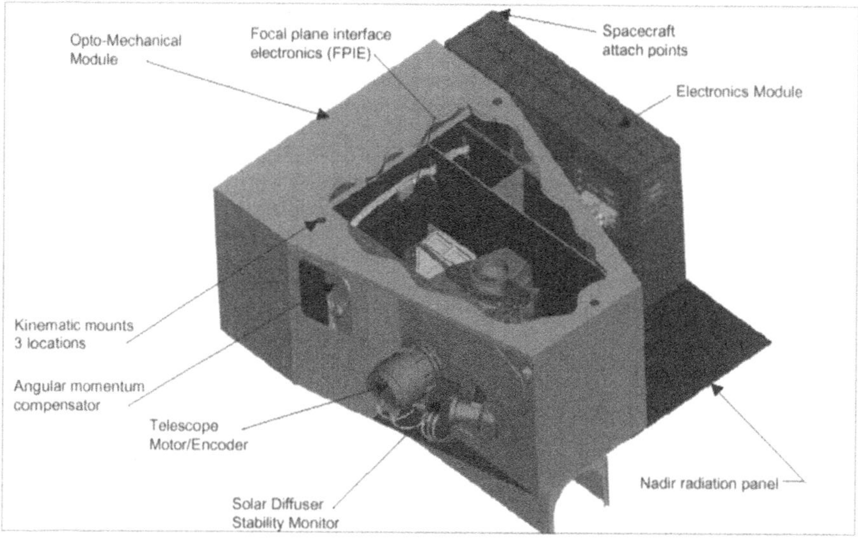

Figure 213: Illustration of the VIIRS instrument

The VNIR FPA employs a PIN (Positive Insulator Negative) diode array/ROIC (Readout Integrated Circuit) design collocated with the DNB monolithic CCD. All detectors in the SWIR/MWIR/TIR regions employ photovoltaic (PV) detectors with an element spacing of 12 µm. A readout IC (Integrated Circuit) at each FPA provides improved noise levels and built-in offset correction. A cryogenic module (three-stage radiative cooler) provides FPA cooling. Calibration is performed with three on-board calibrators: a) a solar diffuser provides full aperture solar calibration, a solar diffuser stability monitor, and c) a blackbody. A single-board instrument computer provides a processing capability including data aggregation, data compression [lossless (2:1 Rice compression) and lossy (JPEG) algorithms are used], and CCSDS data formatting. The VIIRS instrument has a mass of about 160 kg, power of about 134 W (operational average), peak power of 177 W, and a volume of <1.2 m^3. The data rate is 6.7-10.5 Mbit/s (high rate mode) and 220 kbit/s (low rate mode with 10:1 JPEG compression). The VIIRS instrument features a SBC (Single Board Computer) for all instrument operations and control; it communicates with the S/C via an IEEE 1394a cable interface.

VIIRS is an opto-mechanical instrument (like OLS, ETM, and MODIS), it is a whiskbroom scanning radiometer to cover a wide swath. The detector line arrays [16 detectors in each array for the SWIR/MWIR and TIR bands, 32 detectors in the array for the VNIR and DNB (Pan) bands] of the whiskbroom scanner are oriented in the along-track direction. This arrangement provides a parallel coverage of 11.87 km along-track in one scan sweep (cross-track direction). The wide along-track coverage permits sufficient integration time for all cells in each scan sweep. One cross-track scan period is 1.786 s in length. The data quantization is 12 bits (14 bit A/C converters for lower noise). [1237] [1238]

Band	Center wave (μm)	Bandwidth (μm)	Comment (driving EDR observation requirements)
VNIR (Visible Near-Infrared) spectral region, use of Si detectors in FPA			
DNB	0.70	0.40	Day Night Band, broad bandwidth maximizes signal (essential night-time reflected band)
M1	0.412	0.02	Ocean color, suspended matter, net heat flux, mass loading
M2	0.445	0.018	Ocean color, suspended matter, net heat flux, mass loading
M3	0.488	0.02	Ocean color EVI, surface type, aerosols suspended matter, net heat flux, mass loading
M4	0.555	0.02	Ocean color, surface type, suspended matter, net heat flux, mass loading
I1	0.645	0.05	Imagery, NDVI, cloud mask/cover, cloud optical properties, surface type, albedo, snow/ice, soil moisture
M5	0.672	0.02	Ocean color, aerosols, suspended matter, net heat flux, littoral transport, mass loading
M6	0.751	0.015	Ocean color, mass loading
I2	0.865	0.039	Imagery NDVI (NDVI heritage band), snow/ice, surface type, albedo
M7	0.865	0.039	Ocean color, cloud mask/cover, aerosols, soil moisture, net heat flux, mass loading
SWIR (Short-Wave Infrared) spectral region, use of PV (Photovoltaic) HgCdTe detectors			
M8	1.24	0.02	Cloud optical properties (essential over snow/ice), active fires
M9	1.378	0.015	Cloud mask/cover (thin cirrus detection), aerosols, net heat flux
M10	1.61	0.06	Aerosols, cloud optical properties, cloud mask/cover (cloud/snow detection), active fires, soil moisture, net heat flux
I3	1.61	0.06	Imagery snow/ice (cloud/snow differentiation), surface type, albedo
M11	2.25	0.05	Aerosols (optimal aerosol optical thickness over land), cloud optical properties, surface type, active fires, net heat flux
MWIR (Mid-Wave Infrared) spectral region, use of PV (Photovoltaic) HgCdTe detectors			
I4	3.74	0.38	Imagery (identification of low and dark stratus), active fires
M12	3.70	0.18	SST (Sea Surface Temperature), cloud mask/cover, cloud EDRs, surface type, land/ice surface temperature, aerosols
M13	4.05	0.155	SST (essential for skin SST in tropics and during daytime), land surface temperature, active fires, precipitable water
TIR (Thermal Infrared) spectral region, use of PV (Photovoltaic) HgCdTe detectors			
M14	8.55	0.3	Cloud mask/cover (pivotal for cloud phase detection at night, cloud optical properties
M15	10.762	1.00	SST, cloud EDRs and SDRs (Science Data Records), land/ice surface temperature, surface type
I5	11.45	1.9	Imagery (nighttime imagery band)
M16	12.013	0.95	SST, cloud mask/cover, land/ice surface temperature, surface type

Table 370: Definition of VIIRS spectral bands

Some operational features of VIIRS:

- All functions are individually commandable
- Macro commands (stored sequences, all macros are reprogrammable) simplify the commanding and reduce the uplink data

1237) http://npoesslib.ipo.noaa.gov/viirs_released_papers.htm

1238) C. P. Welsch, H. Swenson, S. A. Cota, F. DeLuccia, J. M. Haas, C. Schueler, R. M. Durham, J. E. Clement, P. E. Ardanuy, "VIIRS (Visible Infrared Imager Radiometer Suite): A Next-Generation Operational Environmental Sensor for NPOESS, IGARSS 2001, Sydney, Australia, July 9-13, 2001

- Time-tagged commands allow delayed execution (provides for 30 days autonomous operations)
- The swath widths and locations are individually programmable by band (improved resolution views of selected target near nadir)
- Diagnostic mode features improved versatility

DNB FPA	VNIR FPA	SWIR/MWIR FPA	TIR FPA
One broadband	9 bands	8 bands	4 bands, 1 with TDI
CCD detector	Si PIN detector	PV HgCdTe detector	PV HgCdTe detector
FPIE (Focal Plane Interface Electronics)	ROIC (Readout Integrated Circuit)	ROIC	ROIC
Filter/Bezel	Filter/Bezel	Si micro-lens array	Ge micro-lens array
T_{ops} = 253 K	T_{ops} = ambient	Filter/Bezel	Filter/Bezel
		T_{ops} = 80 K	T_{ops} = 80 K

Table 371: Overview of the FPA design of VIIRS

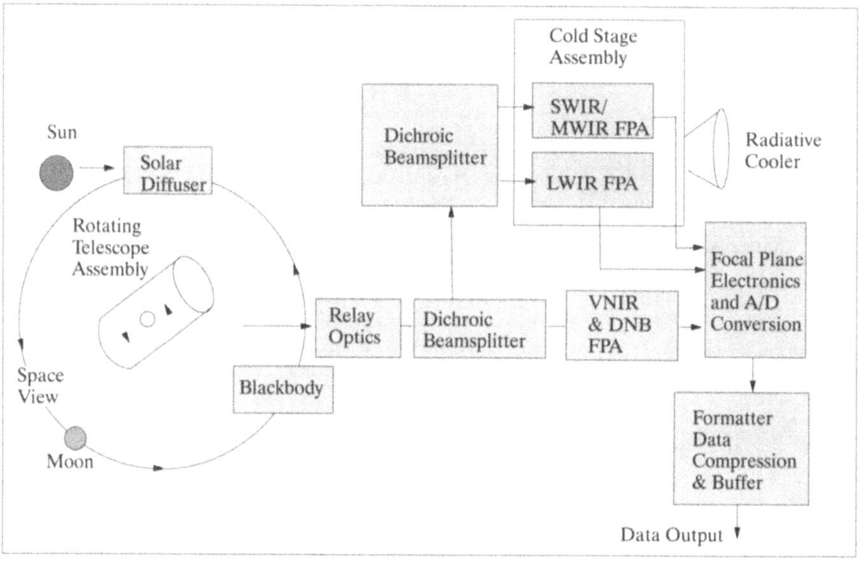

Figure 214: Major subsystems/components of VIIRS

CMIS (Conical-scanning Microwave Imager/Sounder). [1239] The CMIS instrument, of SSM/I (Special Sensor Microwave Imager), SSMIS (Special Sensor Microwave Imager Sounder) and TMI (TRMM Microwave Imager) heritage, is a dual- and single look passive polarimetric microwave radiometer. It is a sensor with a potential aperture just slightly over 2 m (notional) which simultaneously operates over the potential range from 1 to 250 GHz to provide an "all-weather" (i.e., not impacted by clouds) microwave imaging and sounding (profiling) capability. Many microwave radiation frequency ranges exhibit transparency to clouds and other atmospheric obstacles (fog, rain), thus providing forecasters and scientists with an "all-weather" measurement capability. In addition, microwave radiation doesn't need sun illumination for instrument observation. Hence, the objective of CMIS data is to collect microwave radiometry and sounding data to support a variety of environmental applications and to derive such products as:

1239) http://www.ball.com/aerospace/npoess_cmis.html

- Ocean wind speed/direction
- Atmospheric temperature and moisture profiles
- Ocean/water imaging and characterization (including SST)
- Cloud imaging and characterization
- Land imaging and characterization

CMIS data provides resolutions from 15 - 50 km at nadir depending on the frequency and the parameter measured. In the summer of 2001 a CMIS instrument contract was awarded to Boeing Satellite Systems of El Segundo, CA.

CrIS (Cross-Track Infrared Sounder). The instrument is being developed by ITT Aerospace/Communications Division of Ft. Wayne, IN, as the prime contractor. CrIS, of HIRS/4 heritage, is a high-spectral and high-spatial resolution infrared sounder for atmospheric profiling functions. The overall objective is to perform daily measurements of Earth's radiation to determine the vertical atmospheric distribution (surface to the top of the atmosphere) of temperature (profiles to better than 0.9 K accuracy in the lower troposphere and lesser accuracy at higher altitudes), moisture (profiles to better than 20-35% accuracy depending on altitude) and pressure (profiles to better than 1.0% accuracy) with an associated 1.0 km vertical layer resolution. The Michelson interferometer sounder has over 1300 spectral channels, it covers a spectral range of of 650-2550 cm^{-1} (or 3.5 to 16 μm), with a spectral resolution of 0.6525 cm^{-1} (LWIR), and a ground spatial resolution (IFOV) of 14.0 km. The IFOVs are arranged in a 3 x 3 array. The swath width is 2300 km.

Spectral bands	
SWIR	(2155 - 2550 cm^{-1}) or 4.64 - 3.92 μm
MWIR	(1210 - 1750 cm^{-1}) or 8.62 - 5.7 μm
TIR (also known as LWIR)	(650 - 1095 cm^{-1}) or 15.3 - 9.1 μm
Spectral resolution:	
SWIR	(<2.5 cm^{-1}) or 5.4 nm (at 4.64 μm) to 38.4 nm (at 3.92 μm)
MWIR	(<1.25 cm^{-1}) or 92.8 nm (at 8.62 μm) to 40.6 nm (at 5.7 μm)
TIR	(<0.625 cm^{-1}) or 146 nm (at 15.3 μm) to 51.7 nm (at 9.1 μm)
Number of IFOVs	3 x 3
IFOV diameter	14 km
IFOV motion (jitter)	< 50 μrad/axis
Mapping accuracy	< 1.45 km
Absolute radiometric uncertainty	<0.8% (SWIR), <0.6% (MWIR), <0.45% (TIR)
Radiometric stability	<0.65% (SWIR, <0.5% (MWIR), <0.4% (TIR)
Instrument size	61 cm x 40 cm x 40 cm
Instrument mass, power, data rate	76 kg, 86 W, 1.48 Mbit/s

Table 372: Key performance characteristics of CrIS

The baseline CrIS instrument design consists of nine independent single-function modules: [telescope, optical bench, aft-optics, interferometer, ICT (Independent Calibration Target), SSM (Scene Selection Module), detectors, cooler, processing and control electronics, and instrument structure].

- 8 cm clear aperture
- 4-stage split-patch passive cooler (81 K for LWIR patch temperature, 98 K for MWIR/SWIR patch)
- High-performance PV (photovoltaic) detectors
- 3 x 3 arrays (14 km IFOVs)
- Three spectral bands
- All-reflective telescope
- Proven Bomem plane-mirror Michelson interferometer with dynamic alignment
- Deep-cavity internal calibration target based on MOPITT design
- Two-axis scene selection module with image motion compensation
- A modular design (allowing for future addition of an active cooler and >3 x 3 arrays

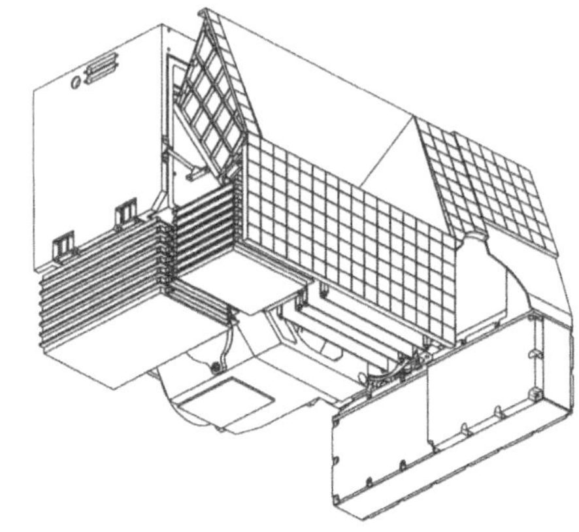

Figure 215: Schematic illustration of the CrIS instrument

OMPS (Ozone Mapping and Profiler Suite). The objective is to measure the total amount of ozone in the atmosphere and the ozone concentration variation with altitude. OMPS is of SBUV/2, TOMS and GOME heritage. Also, the OMPS limb-sounding concept/technology was already tested with ISIR flown on STS-85 (Aug. 7-19, 1997, see J.7) and with SOLSE/LORE flown on STS-87 (Nov 19 - Dec. 5, 1997). The system requirements call for an instrument mass of <=45 kg, an average orbital power of <=45 W, a peak data rate of <=40 kbit/s, and ozone vertical resolution of <=5 km. The vertical resolution requirement demands an instrument design to include a limb-viewing sensor in addition to a heritage-based nadir-viewing sensor. [1240] [1241]

- The nadir sensor wide-field telescope feeds two separate spectrometers, a) for total column observations (mapper) and b) for nadir profiling observations. The total column spectrometer (300-380 nm) has a 2800 km cross-track swath divided into 35 IFOVs of nearly equal angular extent. Measurements from this spectrometer are used to generate total column ozone data with a resolution of about 50 km x 50 km at nadir. - The nadir profile spectrometer (250-310 nm) has a 250 km cross-track swath corresponding to a single cell. Co-registration with the total column spectrometer provides the total ozone, surface and cloud cover information needed for nadir profile retrievals.
- The limb sensor uses a single prism to disperse three vertical slits directed along-track, each separated by 250 km at the limb tangent point. These slits are 150 km high in object space and are sampled at 1 km intervals. To accommodate the very high scene dynamic range, these slit images pass through a beam splitter to divide the scene brightness into three brightness ranges. As a result there are nine limb images of the dispersed slits on the CCD. The measured limb radiances in the ultraviolet, visible, and near-infrared provide data on ozone, aerosols, Rayleigh scattering, surface and clouds that are used to retrieve ozone profiles from the tropopause to 60 km.

OMPS calibration: Solar illuminated diffusers are used for radiometric and spectral calibrations (two diffusers for each sensor). The working diffuser is used weekly and the reference diffuser is used twice annually to monitor the on-orbit degradation of the working diffuser.

The OMPS program will create five ozone data products:

1240) P. H. Graf, I. Becker, et al., "The Preliminary Design of the Ozone Mapping and Profiler Suite (OMPS)," July 2000 as provided on: http://npoesslib.ipo.noaa.gov/omps_released_papers.htm

1241) http://npoesslib.ipo.noaa.gov/Released_papers/OMPS_Flyer2.pdf

- High performance total column environmental data record (EDR) product
- Heritage TOMS V7 total column data records
- High performance ozone profile product
- Heritage SBUV/2 nadir profile data records
- Infrared total column data records from CrIS (Cross-track Infrared Sounder) radiances.

Parameter	Nadir Total Column (Nadir Mapper)	Nadir Profile (Nadir Profiler)	Limb Soundings
Spectral range	300-380 nm	250-310 nm	290-1000 nm
Spectral radiance range [photons/(s cm^2 sr nm)]	9 el 3 (380 nm) 8 el 1 (308 nm)	2 el 3 (310 nm) 1.5 el 8 (252 nm)	9 el 3 (600 nm) 5 el 0 (300 nm)
Minimum SNR	1000	35 (252 nm) 400 (310 nm)	320 (290 nm at 60 km) 1200 (600 nm at 15 km)
Integration time	7.6 s	38 s	38 s
Spectral resolution	1 nm FWHM 2.4 samples/FWHM	1 nm FWHM 2.4 samples/FWHM	1.5-40 nm FWHM 1 sample/FWHM
FOV	110° x 0.3°	16.7° x 0.3°	8.5° x 2.7° (3 sets)
Cell size	49 km x 50 km (nadir)	250 km x 250 km	1 km vertical
Revisit time	Daily		4 days (average)
Swath	2800 km	250 km	3 vertical slits along-track and ±250 km

Table 373: Performance parameters of the OMPS spectrometers

GPSOS (GPS Occultation Sensor). The objective is to provide refraction measurement and monitoring of the GPS constellation to characterize the ionosphere. GPSOS is of GPS/MET and GRAS [GNSS (Global Navigation Satellite System) Receiver for Atmospheric Sounding)] heritage. GPSOS is a high-precision GPS instrument consisting of: [1242]

- One zenith antenna for navigation, together with two high-gain phased array occultation antennas (one for the velocity and one for the anti-velocity direction)
- Three RFCUs (Radio Frequency Conditioning Units) for removal of RFI (Radio Frequency Interference), low noise amplification and down-conversion
- A cold-redundant GEU (GPSOS Electronic Unit), including an ultra-stable oscillator, analog-to-digital conversion, digital down-conversion and GPS signal processing hardware, a DSP (Digital Signal Processor) and a MIL-STD-1553B interface.

SESS (Space Environment Sensor Suite). SESS is the complement of sensors and algorithms used to measure the characteristics of: auroral boundary, auroral energy deposition, auroral imagery, electric field, electron density profile, geomagnetic field, in-situ plasma fluctuations, in-situ plasma temperatures, ionospheric scintillation, neutral density profile, medium energy charged particles, energetic ions, and supra-thermal to auroral energy particles. The data provide information about the space environment necessary to ensure reliable operations of current and future space-based and ground-based systems, to facilitate the analysis of system anomalies that may be the result of space environmental effects, and to aid in the design and efficient operations of future systems that may be affected by the space environment.

Other Payloads: In addition to these payloads, a host of other sensors (to include DCS and S&RSAT) are envisioned to fly on-board NPOESS. Exact definition of these payloads is currently the subject of trade studies being conducted by the IPO's competing system contractors and their exact characteristics must wait until the NPOESS EMD contractor is selected (i.e. their proposed supplementary payload suite is selected). However, it is envisioned that these additional payloads will include an Earth Radiation Budget Suite (ERBS) similar to NASA's CERES instrument, as well as, a Radar Altimeter (Jason-1 heritage), and a TSIS (Total Solar Irradiance Sensor) instrument of TIM (Total Irradiance Monitor) heritage, flown on SORCE (Solar Radiation and Climate Experiment).

1242) Information provided by Peter Sinander of Saab Ericsson Space AB, Göteborg, Sweden

In addition, NASA is developing for flight on both NPP and NPOESS, the ATMS (Advanced Technology Microwave Sounder). ATMS is an advanced technology (for size/cost reduction) version of the current generation AMSU-A/AMSU-B microwave suite. Performance is planned to be similar to AMSU and as is currently done on POES, ATMS will be hosted with the CrIS to form a combined cross-track IR/MW sounding suite (heritage HIRS/AMSU; AIRS/AMSU).

G.11 TIROS Meteorological Satellite Series (with the POES Program)

The US meteorological satellite program started in 1960 with the launch of TIROS-1 (Television Infrared Observation Satellite), the first true weather satellite. The program proved to be immensely successful. The science community began quickly to realize the riches of information to be obtained from large-scale observation of the Earth. A new era was opened from space, for weather services in particular, as well as for Earth observation in general.

TIROS was a NASA-procured and operated program [other participants: US Army Signal Research and Development Lab, US Weather Bureau,US Naval Photographic Interpretation Center, RCA (S/C manufacturer and integrator of the series)]. The main observing instruments of the TIROS series were TV cameras which provided daily cloud cover pictures for the weather services. Initial downlink transmissions could only be performed offline (data was first stored on-board and later transmitted). Starting with TIROS-8, real-time observational data could be transmitted in APT mode to ground stations. Eventually, APT pictures could be received on fairly simple ground stations anywhere in the world. This broadcasting policy eventually entailed a global user community. Early spacecraft and instrument technologies were rudimentary (coarse resolution data) compared to those of the nineties. But with each succeeding generation of satellites, the remote sensing instruments, as well as the entire infrastructure, evolved and became more sophisticated. New orbital strategies of 'polar orbit' and 'sun-synchronous orbit' were introduced, first with Nimbus-1 (in 1964), then with TIROS-9 and TIROS-10 in 1965. The infrared cameras of the TIROS series permitted also the study of the Earth's heat distribution and provided an idea of the amount of heat reflection from the Earth.

With ESSA (Environmental Science and Services Administration), the second generation of spacecraft and instruments, there was a definite need for **'operational services.'** This applied in particular for continuous and frequent coverage and the need for high-resolution global data. The broadcast services of two satellites in orbit were used by an increasing global community.

The primary objective of the 1st-generation TIROS program (M. Tepper, program scientist) was to demonstrate the feasibility and capability of observing the Earth's cloud cover and weather patterns from space. This was done by testing experimental television techniques that led eventually to a world-wide meteorological satellite information system.

G.11.1 TIROS-1 (TIROS-A)

A NASA spacecraft launched on April 1, 1960 on a standard Thor-Able vehicle from Cape Canaveral, FL. The S/C was of cylindrical shape (18-sided polygon), about 1.1 m in diameter and 0.55 m in height, with a mass of 122 kg. The structure was made of aluminum alloy and stainless steel and was covered on the outside (cylinder and top) by 9200 silicon solar cells. The solar cells served to charge the 21 Ni:Cd batteries. TIROS-1 was spin-stabilized and space-oriented. Hence, observations could only be made when the cameras were pointing toward Earth and when that portion of the Earth was illuminated by the sun. Communication: A single monopole antenna for reception of ground commands extended out from the top of the cover assembly. A pair of crossed-dipole telemetry antennas (235 MHz) projected down and diagonally out from the baseplate. Mounted around the edge of the baseplate were five diametrically opposed pairs of small, solid-fuel thrusters that maintained the satellite spin rate between 8 and 12 rpm.

Orbit: near-circular orbit, apogee=750 km, perigee=693 km, inclination=48.4°, period=99.2 minutes.

Sensor complement:

Two vidicon television cameras, TV-WA and TV-NA (PI: H. I. Butler), were mounted into the baseplate of the S/C, with their optical axis parallel to the rotation axis of the satellite, providing low-resolution and high-resolution data respectively. The two sensor units were capable of either concurrent or independent operation. The TV cameras used 500 scan-line, 1.27 cm diameter vidicons. A magnetic tape recorder for each camera provided storage of up to 32 frames (images) while the satellite was out of range of the ground stations. Each camera could take 16 pictures per orbit at 128 second intervals. Transmission of the 32-frame sequence was accomplished in 100 s by a 2 W FM transmitter operating at a nominal frequency of 235 MHz. Images of daytime cloudcover were produced for a latitudinal region of ± 55°.

TV-WA (Television-Wide Angle) with a FOV of 104°. At nominal attitude and altitude (approximately 700 km), a picture taken by TV-WA covered a 1200 km x 1200 km at nadir with a spatial resolution of 2.5 to 3.0 km.

TV-NA (Television-Narrow Angle) with a FOV of 12°. The TV-NA instrument had a footprint of 120 km x 120 km with a resolution of 0.3 km to 0.8 km.

TIROS-1 was operational for 78 days (until June 17, 1960), when an electrical power failure prevented further useful TV transmission, producing over 22,000 cloud-cover photographs.

G.11.2 TIROS-2 (TIROS-B)

Launch on Nov. 23, 1960 on a three-stage Delta vehicle from Cape Canaveral, FL. Same S/C structure as for TIROS-1, S/C mass=127 kg. - A new attitude control system was tested which utilized the Earth's magnetic field for S/C orientation. In addition, an infrared horizon sensor was used for attitude sensing. Two infrared radiation experiments, a five-channel medium-resolution scanning radiometer and a two-channel nonscanning low-resolution radiometer for measuring radiation from the Earth and its atmosphere were included. The satellite spin rate was maintained between 8 and 12 rpm by the use of five diametrically opposed pairs of small, solid-fuel thrusters. The satellite spin axis could be oriented to within 1 to 2° accuracy by the use of a magnetic attitude control device consisting of 250 cores of wire wound around the outer surface of the spacecraft. The interaction between the induced magnetic field in the spacecraft and the Earth's magnetic field provided the necessary torque for attitude control. [1243)]

Orbit: near-circular orbit, apogee=742 km, perigee=609 km, inclination=48.57°, period=98.3 minutes.

Sensor complement:

The two television cameras, **TV-WA** (Television-Wide Angle) and **TV-NA** (Television-Narrow Angle) of TIROS-1 were flown with some minor changes. The cameras were automatically triggered into action only when they came in view of the Earth. The TV cameras used 500 scan-line, 1.27 cm vidicons. Deposits on the lens of TV-WA caused all its pictures to be of very poor quality. The remaining TV-NA camera operated nominally until February 1, 1961, and sporadically thereafter until September 27, 1961. Tape recorder technique same as for TIROS-1. Over 25,000 usable pictures were transmitted.

1243) A description of TIROS-2 is presented in the Journal of the British Interplanetary Society, Vol. 19, Pages 386–409, 1963-64.

Widefield Radiometer (PI: R. A. Hanel). The low-resolution, nonscanning, two-channel radiometer measured the thermal and reflected solar radiation from the Earth-atmosphere system. The radiometer consisted of two detectors - one black and one white thermistor bolometer. Each of the detectors was mounted in the apex of a highly reflective mylar cone. The black detector responded equally to reflected solar radiation and longwave terrestrial radiation (0.2 - 50 μm). The white detector reflected solar and visible radiation and measured only longwave thermal radiation (5 - 50 μm). The optical axis of each detector was parallel to the satellite spin axis. The field of view (FOV) of the detectors, when viewing the Earth directly below the satellite, was a circle of 832 km diameter (50° FOV). The area was within the field observed by TV-WA, and thus a direct measure of the heat balance of the Earth-atmosphere system viewed in any of the pictures was provided. The experiment performed normally, but the quality of the data was very poor because the sensitivity of the detectors was lower than expected.

Scanning Radiometer (PI: J. D. Barksdale). The sensor measured the emitted and reflected radiation of the Earth and its atmosphere. The five-channel radiometer scanned the Earth and space as the S/C spun about its axis. The radiometer's bi-directional optical axes were inclined to the satellite spin axis at angles of 45° and 135°. The sensor used bolometer detectors and filters to limit the spectral responses and to provide comprehensive data by measuring radiation intensities in selected portions of the spectrum. The spectral bandwidth of each channel and its associated parameter were: channel 1, 6.0 - 6.5 μm (water vapor absorption), channel 2, 8.0 - 12.0 μm (atmospheric window), channel 3, 0.25 - 6.0 μm (reflected solar radiation), channel 4, 8.0 - 30 μm (terrestrial radiation), and channel 5, 0.55 - 0.75 μm (reflected solar radiation). Initially, all channels performed normally. However, channels 1 and 4 gradually deteriorated and became useless by January 1961. The signal to noise ratio of channels 3 and 5 was extremely low, and the output was highly questionable. Five months of terrestrial radiation measurements were obtained for the latitudinal region of ± 55° before the radiometer chopper motor failed on April 22, 1961.

G.11.3 TIROS-3 (TIROS-C)

Launch on July 12, 1961 on a three-stage Thor-Delta vehicle from Cape Canaveral, FL. Same S/C structure and attitude control system as for TIROS-2, S/C mass=129.5 kg. The flight control system also optimized the performance of the solar cells and TV cameras and protected the five–channel infrared radiometer from prolonged exposure to direct sunlight.

Orbit: near-circular orbit, apogee=812 km, perigee=742 km, inclination=47.9°, period =100.4 minutes.

Sensor complement:

Low-Resolution Omnidirectional Radiometer (PI: V. E Suomi). Objective: Measurement of the amount of solar energy absorbed, reflected, and emitted by the Earth and its atmosphere. The sensor consisted of two sets of bolometers in the form of hollow aluminum hemispheres, mounted on opposite sides of the spacecraft, whose optical axes were parallel to the spin axis. The bolometers were thermally isolated from but in close proximity to reflecting mirrors so that the hemispheres behaved very much like isolated spheres in space. One bolometer in each set was painted black, and one was painted white. The black bolometer absorbed most of the incident radiation while the white bolometer was sensitive mainly to radiation with wavelengths > 4 μm. Reflected and emitted radiation could thus be separated. The sensor temperatures were measured by thermistors fastened to the inside of the hollow hemispheres. The sensor temperatures, taken every 29 s, were an average of the two temperatures from the matched thermistors. The experiment was a success, and usable data were received from July 12, 1961, to October 20, 1961.

Widefield Radiometer (PI: R. A. Hanel). Identical sensor as flown on TIROS-2.

Scanning Radiometer (PI: R. M. Rados). Identical sensor as for TIROS-2. Sensor mass = 2 kg, power = 1 W.

Television Camera System (same as for TIROS-2). The experiment consisted of two redundant pairs of TV cameras (2 **TV-WA** and 2 **TV-NA**), magnetic tape recorders, and TV transmitters. One of the TV-WA cameras failed 13 days after launch. The remaining camera produced useful data until January 23, 1962. The experiment was credited with observing all six major hurricanes of the 1961 season. During the operational lifetime of the experiment, over 24,000 usable pictures were obtained.

G.11.4 TIROS-4 (TIROS-D)

Launch on Feb. 8, 1962 on a three-stage Thor-Delta vehicle from Cape Canaveral, FL. Same S/C structure and attitude control system as for TIROS-2 and -3, S/C mass=129.5 kg. With the exception of the degraded response of the five-channel scanning radiometer, the S/C performed normally until May 3, 1962, when one camera failed. On June 10, 1962, the other camera's tape recorder failed. The scanning radiometer provided usable data until June 30, 1962.

Orbit: near-circular orbit, apogee=840 km, perigee=712 km, inclination=48.3°, period =100.0 minutes.

Sensor complement:

Low-Resolution Omnidirectional Radiometer (PI: V. E Suomi). Identical sensor as on TIROS-3. Usable data were received from February 8, 1962, to June 28, 1962.

Widefield Radiometer (PI: R. A. Hanel). Identical sensor as flown on TIROS-2.

Scanning Radiometer (PI: J. D. Barksdale). Identical sensor as for TIROS-2. Sensor mass = 2 kg, power = 1 W.

Television Camera System (same as for TIROS-2). The experiment consisted of two independent TV cameras (**TV-WA** and **TV-MA**), magnetic tape recorders, and TV transmitters. The TV-MA (Television-Medium Angle) featured a FOV of 80°, resulting in a footprint of 725 km x 725 km and a resolution of 2 km (note: the TV-WA FOV remained at 104°). The experiment performed normal until May 3, 1962, when TV-WA failed. The tape recorder on the remaining camera failed on June 10, 1962. However, limited real-time TV pictures were available up to June 18, 1962, at which time the system was deactivated.

The US Weather Bureau initiated an international facsimile transmission network to share the cloud pictures with weather services around the world.

G.11.5 TIROS-5 (TIROS-E)

Launch on June 19, 1962 on a three-stage Thor-Delta vehicle from Cape Canaveral, FL. Same S/C structure and attitude control system as for TIROS-2, S/C mass=129.5 kg. The larger orbital inclination of 58° extended the effective TV coverage to latitudes of ± 65°.

Orbit: elliptical orbit, apogee=972 km, perigee=586 km, inclination=58.1°, period =100.0 minutes.

Sensor complement:

Television Camera System (same as for TIROS-4). The experiment consisted of two independent TV cameras (**TV-WA** and **TV-MA**), magnetic tape recorders, and TV transmitters. The TV-MA (Television-Medium Angle) featured a FOV of 80°, resulting in a footprint of 725 km x 725 km and a resolution of 2 km. The TV-MA camera failed 17 days after launch.

The remaining TV-WA camera performed normal until May 14, 1963, when the shutter electronics failed. The TV experiment produced over 48,000 meteorologically useful pictures.

G.11.6 TIROS-6 (TIROS-F)

Launch on Sept. 18, 1962 on a three-stage Thor-Delta vehicle from Cape Canaveral, FL. Same S/C structure and attitude control system as for TIROS-2, S/C mass=127.5 kg.

Orbit: near-circular orbit, apogee=713 km, perigee=686 km, inclination=58.32°, period =98.7 minutes.

Sensor complement:

Television Camera System (same as for TIROS-4). The experiment consisted of two independent TV cameras (**TV-WA** and **TV-MA**), magnetic tape recorders, and TV transmitters. The TV-MA (Television-Medium Angle) featured a FOV of 80°, resulting in a footprint of 725 km x 725 km and a resolution of 2 km. The camera system performed normal after launch until November 29, 1962, when the TV-MA camera vidicon failed. The remaining TV-WA camera system failed on October 21, 1963. The experiment transmitted approximately 60,000 meteorologically useful pictures and furnished information leading to many storm advisories in both the U.S. and abroad. The detection of snow cover was tested for the first time.

G.11.7 TIROS-7 (TIROS-G)

Launch on June 19, 1963 on a three-stage Thor-Delta vehicle from Cape Canaveral, FL. Same S/C structure and attitude control system as for TIROS-2, S/C mass=134.7 kg. The S/C performed normal until December 31, 1965, and sporadically until February 3, 1967. The S/C was operated for an additional 1.5 years to collect engineering data. It was deactivated on June 3, 1968.

Orbit: near-circular orbit, apogee=749 km, perigee=621 km, inclination=58.23°, period =97.4 minutes.

Sensor complement:

Television Camera System (same as for TIROS-4). The experiment consisted of two redundant pairs of TV cameras (2 **TV-WA,** each with a FOV of 104°), magnetic tape recorders, and TV transmitters. The experiment yielded numerous meteorologically useful pictures and provided almost continuous hurricane coverage during its approximately 2.5 years of operational lifetime. Beginning early in 1965 and continuing until February 3, 1967, the S/C was programmed to take only the pictures required to supplement the photographic coverage acquired by other satellites in the series.

Low-Resolution Omnidirectional Radiometer (PI: V. E Suomi). Identical sensor as on TIROS-3. Usable data were received from June 19, 1963, to September 13, 1963.

Scanning Radiometer (PI: J. D. Barksdale). Identical sensor as for TIROS-2 -3, and -4. Sensor mass = 2 kg, power = 1 W.

Langmuir Probe (PI: L. H. Brace). Objective: Measurement of electron density and temperature. The cylindrical probe consisted of two concentric electrodes. The inner electrode, of 0.056 cm in diameter and 23 cm in length, was used as a collector. The outer electrode served as a guard electrode and was 0.168 cm in diameter and 10 cm long. The probe was swept through the voltage range 0 to 1.5 v in 2 s. The current at the collector was measured as the voltage was varied, and the signal was stored on a tape recorder and played back upon interrogation by a ground station. The experiment operated normal from launch until July 14, 1963, when an electrical failure prevented the tape recorder from operating normally.

G.11.8 TIROS-8 (TIROS-H)

Launch on Dec. 21, 1963 on a three-stage Thor-Delta vehicle from Cape Canaveral, FL. Same S/C structure and attitude control system as for TIROS-2, S/C mass=265 kg.

TIROS 8 was the first satellite to be equipped with **Automatic Picture Transmission (APT)** capabilities. The APT experiment provided real-time Earth-cloud pictures taken by the satellite to any properly equipped ground receiving station. In addition to the APT camera system, the satellite carried one TV-WA (104° FOV) camera. Pictures taken by TV-WA were transmitted directly or were stored in a tape recorder on board for subsequent playback, depending on whether the spacecraft was within or beyond communication range of either of two ground receiving stations. The S/C performed nominal after launch. Over 50 ground stations participated in the APT experiment, which was terminated by the end of April 1964 due to degradation of the APT camera. The TV-WA camera transmitted useful data until February 12, 1966. The satellite was deactivated on July 1, 1967, after being left on for an additional time period for engineering purposes.

Orbit: near-circular orbit, apogee=765 km, perigee=691 km, inclination=58.5°, period =99.3 minutes.

Sensor complement:

One **TV-WA** (Television-Wide Angle) with 104° FOV (same as for TIROS-4). The system included a two-track magnetic tape recorder, and a transmitter. The experiment performed nominally after launch, and good data were obtained until Aug. 31, 1965. Beginning early in 1965 and until Feb. 12, 1966, the satellite was programmed to take only the pictures required to supplement the photographic coverage acquired by other satellites in the series. After that time, the experiment was activated only for engineering checks until July 1, 1967.

APT (Automatic Picture Transmission - PI: C. H. Hunter). The system consisted of a camera and transmitter combination designed to test the feasibility of transmitting local daytime pictures of cloudcover conditions to properly equipped ground receiving stations on a real-time basis. The camera system consisted of a single camera with a 2.54 cm diameter vidicon. The camera used a 108° FOV and a f/1.8 objective lens, with a focal length of 5.7 mm, and was mounted on the satellite baseplate, with its optical axis parallel to the spacecraft spin axis. The actual picture taking required 8 s, the transmission of the image took 200 s. Earth-cloud images retained on the photosensitive surface of the vidicon were read out at four lines/s to produce an 800-line picture. A 5 W TV transmitter at 136.95 MHz relayed the pictures to local APT stations within communication range. The faceplate of the vidicon had reticle marks that appeared on the picture format to aid in relating the picture to its geographical position on the Earth's surface. At nominal satellite altitude of 700 km, a picture had a footprint of 1200 km x 1200 km with a horizontal resolution of 7.5 km at nadir. The experiment performed nominally, and good quality pictures were obtained until the experiment was terminated owing to degradation of the APT camera. The APT experiment demonstrated the feasibility of using weather satellites to provide meteorologists with local cloudcover data on a near real-time basis requiring only the use of a photo-facsimile machine and a relatively inexpensive antenna and receiver.

G.11.9 TIROS-9 (TIROS-I)

Launch on Jan. 22, 1965 on a three-stage Thor-Delta from Point Arguello. Same S/C structure and attitude control system as for TIROS-2, S/C mass=138.3 kg.

TIROS 9 was the first of the so-called '**Cartwheel**' meteorological TV satellites. That is, the S/C spin axis was maintained normal to the orbital plane. For the first time, the two TV-WA cameras were mounted into the sidewalls of the cylinder, 180° opposite to each other. This concept, also referred to as the **cartwheel configuration**, allowed each camera to observe the same ground target during each cycle of rotation. - The satellite was still equipped with small solid-fuel thrusters as in the case of previous TIROS S/C. However, the system was used only as a backup. The satellite spin rate and attitude were primarily determined by **QOMAC** (Quarter-Orbit Magnetic Attitude Control) system. QOMAC used the torque developed by interaction of the Earth's magnetic field with a current-carrying loop mounted in the satellite.

A failure in the S/C guidance system placed the satellite in an unplanned elliptical (700 to 2500 km) orbit. The TV system operated normally until April 1, 1965, when one of the TV-WA cameras failed. The other camera operated normally until July 26, 1965, and sporadically until Feb. 15, 1967. **TIROS-9 was the first S/C in the TIROS series to be placed in a near-polar orbit, thereby increasing TV coverage to the entire daylight portion of the globe.**

Orbit: sun-synchronous elliptical polar orbit, apogee=2582 km, perigee=705 km, inclination=96.4°, eccentricity = 0.11693, period =119.23 minutes.

Sensor complement:

The TV camera system was identical to that flown on all previous TIROS missions, two TV-WA (104° FOV) cameras equipped with 1.27 cm diameter vidicons. Unlike previous TIROS TV cameras, however, the cameras on TIROS-9 were mounted 180° apart on the side of the spacecraft and canted 64° from the S/C spin axis. The cameras were automatically triggered into action only when they came in view of the Earth. The TV system could operate in either real-time or tape recorder mode, depending on whether the S/C was within or beyond communication range of either of two ground receiving stations. The recorder could store up to 48 frames of pictures. Transmission of the 48-frame sequence was accomplished in 120 s using a 5 W FM transmitter operating at a nominal frequency of 235 MHz. At an altitude of 700 km, a picture had a footprint of 1200 km x 1200 km with a spatial resolution of 2.5 to 3.0 km at nadir. From a near-polar orbit, the camera system could provide complete TV coverage of the entire daylight portion of the globe. In spite of an unplanned elliptical orbit, the experiment produced over 70,000 meteorologically useful pictures.

G.11.10 TIROS-10

Launch on July 2, 1965 on a three-stage Delta vehicle from Cape Canaveral, FL. Same S/C structure and attitude control system as for TIROS-2, S/C mass=127 kg.

Orbit: near-circular sun-synchronous orbit, apogee=837 km, perigee=751 km, inclination=98.65°, period =100.76 minutes. Note: the sun-synchronous retrograde orbit drifted westward about 1° per day (the same rate and direction as the Earth moves around the sun), this provided maximum lighting conditions for observations as well as for battery charging.

Sensor complement: Two TV-WA cameras were mounted on the base plate (as with TIROS-1 to -8) providing more than 400 images per day (each with an area of 1300 km x 1300 km). The TV system operated nominally until Sept. 30, 1965, and sporadically through July 31, 1966, when the spacecraft was deactivated.

G.12 TOS/ESSA Satellite Series (2nd Generation)

TIROS-9 and -10 were test satellites providing improved configurations/technology for TOS (TIROS Operational Satellite) system. Operational use started in 1966. Once in orbit, the TOS satellites were given the name of ESSA (Environmental Science and Services Administration). S/C manufacturer was RCA. Note, ESSA was a predecessor organization of NOAA.

The prime objective of the TOS/ESSA program was to provide ESSA with the capability of observing the Earth's cloud cover and weather patterns from space. The operational system consists of two polar orbiting sun-synchronous satellites crossing the equator during local morning and afternoon. The even-numbered satellites of the ESSA series provided direct APT transmissions to the ground. The data was used by the meteorological community to prepare operational services for weather analysis and forecasts.

G.12.1 ESSA-1 (TOS-1)

Launch on Feb. 3, 1966 on a three-stage Delta vehicle from Cape Canaveral, FL. The S/C was the same as the one used for TIROS-9. S/C mass of 138 kg.

ESSA 1 was a spin-stabilized operational meteorological S/C. The satellite had essentially the same configuration as that of the TIROS series, i.e., an 18-sided right prism, 107 cm across opposite corners and 56 cm high, with a reinforced baseplate carrying most of the subsystems and a cover assembly (hat). Electrical power was provided by approximately 10,000 solar cells (1 cm x 2 cm) that were mounted on the cover assembly and by 21 Ni:CD batteries. Two redundant wide-angle cameras were mounted on opposite sides of the spacecraft and canted 75° from the S/C spin axis. A pair of crossed-dipole command and receiving antennas projected out and down from the baseplate. A monopole telemetry and tracking antenna extended up from the top of the cover assembly. The satellite was placed in a cartwheel orbital mode, with its spin axis maintained normal to the orbital plane. The satellite spin rate and attitude were determined primarily by MASC (Magnetic Attitude Spin Coil). MASC was a current-carrying coil mounted in the cover assembly. The magnetic field induced by the current interacted with the Earth's magnetic field to provide the necessary torque to maintain a desired spin rate of 9.225 rpm. Five small solid-fuel thrusters mounted around the baseplate provided a second means of controlling the S/C spin rate. The satellite performed normally after launch until October 6, 1966, when the camera system failed. The spacecraft was deactivated on May 8, 1967, after being left on for an additional time period for engineering purposes.

Orbit: near-circular sun-synchronous polar orbit, apogee=845 km, perigee=702 km, inclination=97.9°, period=100.35 minutes.

Sensor complement:

AVCS (Advanced Vidicon Camera System). A combination of camera, tape recorder, and transmitter that could record and store a series of remote daytime cloudcover pictures for subsequent playback to a ground station. The system was identical to those flown on previous TIROS missions, consisting of two redundant 500 scan-line TV cameras with 1.27 cm vidicons. However, on ESSA 1 the cameras were mounted 180° apart on the side of the S/C and were canted 75° from the spacecraft spin axis (cartwheel configuration for the entire ESSA series). The cameras were triggered into action only when they came into view of the Earth. Each tape recorder had two separate channels, one for storing video signals and one for sun-angle data, which served as a time reference. Up to 32 pictures consisting of five levels of gray could be stored for subsequent playback. At an orbital altitude of 1450 km the cameras covered a footprint of 1200 km x 1200 km with a spatial resolution of about 3.0 km at nadir. The sensor provided over 100,000 usable pictures.

G.12.2 ESSA-2 (TOS-2)

Launch on Feb. 28, 1966 on a three-stage Delta vehicle from Cape Canaveral, FL. The S/C structure was the same as for ESSA-1 with a mass of 286 kg.

Orbit: near-circular sun-synchronous polar orbit, apogee=1418 km, perigee=1356 km, inclination=101.0°, period=113.57 minutes.

Sensor complement:

APT (Automatic Picture Transmission). The camera system consisted of two redundant APT cameras with 2.54 cm diameter vidicons. Each camera had a 108° wide-angle f/1.8 objective lens with a focal length of 5.7 mm. The two cameras were mounted in cartwheel fashion. The cameras were programmed to take four or eight APT pictures per orbit. The actual photography required 8 s and the transmission 200 s. Earth-cloud images retained on the photosensitive surface of the vidicon were read out at four lines per second to produce an 800-line picture. Two 5 W TV transmitters (137.5 MHz) relayed the pictures to local APT stations within communication range.

The S/C remained operational for 1692 days until deactivated by NASA on Oct. 16, 1970.

G.12.3 ESSA-3 (TOS-3)

Launch on Oct. 2, 1966 on a three-stage Delta vehicle from VAFB (Vandenberg Airforce Base), CA. The S/C was of the same structure as ESSA-1 with a mass of 145 kg. Identical experiments were flown on ESSA-5, ESSA-7, and ESSA-9.

Orbit: near-circular sun-synchronous polar orbit, apogee=1493 km, perigee=1383 km, inclination=101.0°, period=114.6 minutes.

Sensor complement: No APT system. AVCS was essentially the same as flown on ESSA-5, -7, and -9.

Two **AVCS** (Advanced Vidicon Camera System) were flown (redundant camera and tape recorder combination) in cartwheel configuration. Spectral band = 0.45-0.65 µm. Both AVCS featured direct readout and delayed readout capabilities. The Earth-pointing camera used a 108° wide-angle lens (5.7 mm focal length) with an f/1.8 aperture and a 2.54 cm diameter vidicon with 833 scan lines. A video frame consisted of a 0.25 s period of blanked video, followed by 6.25 s of vidicon scan video (833 lines), and a final 0.25 s period of blanked video. Eleven pictures were taken at 260 s intervals to cover the sunlit portion of the Earth (sun elevation greater than 15°). The tape recorder could read out between photographic cycles without losing a picture or interrupting a sequence. At nominal satellite altitude of 1450 km, the AVCS image covered a footprint of 3000 km x 3000 km with a ground resolution of about 3 km at nadir. Over 90,000 AVCS images were transmitted.

FPR (Flat Plate Radiometer). FPR mass = 3kg, power = 4 W. FPR was designed to provide a measurement of the global distribution of reflected solar and longwave radiation leaving the Earth. The FPR system consisted of four detectors, an analog-to-digital converter, and a tape recorder. The detectors had a hemispheric FOV of 2 pi steradians and were mounted on the S/C baseplate facing earthward. The detectors used coated aluminum disks as a sensing element. Two of the disks were white and responded only to infrared energy (7 - 30 µm) radiated from the Earth and its atmosphere. The other two disks were painted black and had a broader band sensitivity (0.3-30 µm). Two disks (one of each type) had a thermistor bolometer mounted on the back surface to measure the disc temperature. The other two disks used thermopiles. An identical experiment was flown on ESSA-5, -7, -9 and NOAA-1. FPR failed on Jan. 20, 1967.

The S/C remained operational for 738 days, it was deactivated by NASA on Dec. 2, 1968.

G.12.4 ESSA-4 (TOS-4)

Launch on Jan. 26, 1967 on a three-stage Delta vehicle from VAFB, CA. The S/C was of the same structure as ESSA-2 with a mass of 285 kg.

Orbit: near-circular sun-synchronous polar orbit, apogee=1443 km, perigee=1328 km, inclination=102.0°, period=113.48 minutes.

Sensor complement: Same APT system in cartwheel configuration as flown on ESSA-2, -6, and 8. One camera failed almost immediately after launch. The other camera obtained data until Dec. 6, 1967.

ESSA-4 remained operational for 465 days (Dec. 6, 1967), it was deactivated by NASA on May 5, 1968.

G.12.5 ESSA-5 (TOS-5)

Launch on April 20, 1967 on a three-stage Delta Vehicle from VAFB, CA. The S/C was of the same structure as ESSA-1 with a mass of 145 kg.

Orbit: near-circular sun-synchronous polar orbit, apogee=1423 km, perigee=1361 km, inclination=101.97°, period=113.63 minutes.

Sensor complement: No APT system. Two **AVCS** (Advanced Vidicon Camera System) in cartwheel configuration (same system as flown on ESSA-3). AVCS functioned until October 8, 1969, when the satellite was placed in a standby mode.

FPR (Flat Plate Radiometer). Identical instruments were flown on the ESSA-3, -7, and -9 spacecrafts. FPR on ESSA-5 operated nominally until it failed on Sept. 22, 1967.

ESSA-5 remained operational for 738 days (Oct. 8, 1969), it was deactivated by NASA on Feb. 20 1970.

G.12.6 ESSA-6 (TOS-6)

Launch on Nov. 10, 1967 on a three-stage Delta vehicle from VAFB,CA. The S/C was of the same structure as ESSA-2 with a mass of 299 kg.

Orbit: near-circular sun-synchronous polar orbit, apogee=1488 km, perigee=1410 km, inclination=102.12°, period=114.8 minutes.

Sensor complement: Same APT system in cartwheel configuration as flown on ESSA-2, -4, and 8. The APT system operated nominally until Nov. 4, 1969.

ESSA-6 was able to transmit up to eight images daily to individual ground stations. The S/C remained operational for 465 days, it was deactivated by NASA on Dec. 3, 1969.

G.12.7 ESSA-7 (TOS-7)

Launch on Aug. 16, 1968 on a two-stage Delta vehicle from VAFB, CA. The S/C was of the same structure as ESSA-1 with a mass of 145 kg.

Orbit: near-circular sun-synchronous polar orbit, apogee=1476 km, perigee=1432 km, inclination=101.7°, period=114.98 minutes.

Sensor complement: No APT system. Two **AVCS** (Advanced Vidicon Camera System) in cartwheel configuration (same system as flown on ESSA-3). One AVCS camera failed almost immediately after launch. The remaining camera system failed on July 19, 1969.

Two arrays of radiometers (**FPR**=Flat Plate Radiometer) were also mounted in opposite directions (180° apart) with the objective to measure the global distribution of solar radiation reflected by the Earth's surface and atmosphere. FPR failed on June 23, 1969.

ESSA-7 remained operational for 571 days, it was deactivated by NASA on March 10, 1970.

G.12.8 ESSA-8 (TOS-8)

Launch on Dec. 15, 1968 on a two-stage Delta vehicle from VAFB, CA. The S/C was of the same structure as ESSA-2 with a mass of 297 kg.

Orbit: near-circular sun-synchronous polar orbit, apogee=1473 km, perigee=1410 km, inclination=101.9°, period=114.7 minutes.

Sensor complement: Same APT system in cartwheel configuration as flown on ESSA-2, -4, and -6. ESSA-8 was able to transmit eight to ten images per day to nearly 400 APT ground stations around the world.

The S/C remained operational for 2644 days, it was deactivated by NASA on Mar. 12, 1976.

G.12.9 ESSA-9 (TOS-9)

Launch on Feb. 26, 1969 on a three-stage Delta vehicle from Cape Canaveral, FL. The S/C was of the same structure as ESSA-1 with a mass of 145 kg. ESSA-9 remained operational for 1726 days, it was deactivated by NASA on Nov. 15, 1972.

Orbit: near-circular sun-synchronous polar orbit, apogee=1508 km, perigee=1427 km, inclination=101.79°, period=115.3 minutes.

Sensor complement: No APT system. Two **AVCS** (Advanced Vidicon Camera System) in cartwheel configuration (same system as flown on ESSA-3).

Two arrays of radiometers (**FPR**=Flat Plate Radiometer) were also mounted in opposite directions (180° apart) with the objective to measure the global distribution of solar radiation reflected by the Earth's surface and atmosphere. FPR operation was terminated in May 1970.

G.13 ITOS (Improved TIROS Operational System)

In 1970, a new agency, **NOAA** (National Oceanic and Atmospheric Administration), was created in the Department of Commerce (DOC) to operate the satellites and to improve services. Procurement of satellites and instruments, as well as the launch services, remained the responsibility of NASA/GSFC.

Spacecraft operations of this third-generation series was provided by NOAA, the newly-founded agency of DOC (Department of Commerce) in 1970. The ITOS series more than doubled the TOS series capability - the cameras were now able to observe during nighttime as well as during daytime.

G.13.1 NOAA-1 (ITOS-A, also known as ITOS-1 and TIROS-M)

A NASA spacecraft launched on Dec. 11, 1970 on a two-stage Delta-N vehicle from Point Arguello (VAFB), CA. The primary objective was to provide improved operational infrared and visual observations of the Earth's cloud cover for use in weather analysis and forecasting (double the daily coverage of the ESSA satellites, to provide infrared imagery, cloud top temperature and surface temperature twice a day to APT users). A secondary objective included providing both solar proton and global heat balance data on a regular daily basis. To accomplish these tasks, the S/C carried four cameras, two television cameras for Automatic Picture Transmission (APT) and two Advanced Vidicon Camera System (AVCS) cameras.

Spacecraft: a near-cubical shape (1 m x 1 m x 1.2 m) with three identical solar panels attached generating 250 W. Each panel was 4.2 m in length. The dynamics and attitude control system maintained the desired spacecraft orientation. Three-axis stabilization (two horizon scanners, sun sensors and Freon gas jets provided attitude control). Earth orientation of the satellite body was maintained by taking advantage of the precession induced by the momentum flywheel so that the satellite body precession rate of one revolution per orbit provided the required Earth direction. S/C mass = 352 kg, two APT (Automatic Picture Transmission) links in S-band.

The S/C and experiments performed nominally until the incremental tape recorder failed on May 29, 1971.

Orbit: Near-circular sun-synchronous polar orbit, apogee = 1472 km, perigee=1422 km, inclination = 101.9°, period =114.8 minutes.

Sensor complement:

AVCS (Advanced Vidicon Camera System). AVCS was identical to the systems flown on ESSA-3, -5, -7 and -9. AVCS functioned nominally until it was placed in a standby mode on June 22, 1971, when overheating developed in the S/C control system.

SR (Scanning Radiometer), built by SBRC. Instrument mass = 7.9 kg, power = 4.8 W. Two instruments were flown. SR1 resolution = 4 km, spectral band = 0.5-0.72 μm. SR2 resolution = 8 km, spectral band = 10.5-12.5 μm. Swath width of 3000 km. SR consisted of two scanning radiometers, a dual SR processor, and two SR recorders. The sensor permitted the determination of surface temperatures of the ground, the sea, or cloud tops viewed by the radiometer. The SR sensors measured reflected radiation from the Earth-atmosphere system in the 0.52 - 0.72 μm band (SR1) during the day and emitted radiation from the Earth and its atmosphere in the 10.5-12.5 μm band (SR2) during the day and night.

The SR mirror scanned the Earth's surface perpendicular to the satellite's orbital path at a rate of 48 rpm. As the S/C progressed along its orbital path, each rotation of the mirror provided one scan line of picture. Radiation collected by the mirror was passed through a beam splitter and spectral filter to produce the desired spectral separation. Up to two full orbits of data (145 min) could be stored on magnetic tape for subsequent transmission (1697.5 MHz) to an acquisition station. The data could also be transmitted in real-time to local APT stations. Once the signal was received by the ground station, a continuous picture was formed by using a facsimile recorder whose scan was in phase with the satellite's forward motion. At nominal spacecraft altitude (approximately 1450 km), the SR had a ground resolution of better than 4 km at nadir. The radiometer was capable of yielding radiance temperatures from 185 to 330 K to an accuracy of 4 to 1 K, respectively. The SR instrument functioned nominally until Jan. 5, 1971, when the number 1 scanning radiometer became inoperative owing to failure of the scanning mirror motor. The remaining radiometer was placed in a partial record mode at the end of June 1971 as a result of overheating in the spacecraft attitude control system.

FPR (Flat Plate Radiometer). See description under G.12.3. FPR data were obtained until May 29, 1971, when the incremental on-board tape recorder failed.

SPM (Solar Proton Monitor). Three solid-state detectors monitored the omnidirectional fluxes of solar protons with energies above 10, 30, and 60 MeV, respectively. Two telescopes consisting of solid-state detectors each measured directional fluxes of protons in three energy intervals between 0.27 MeV and 3.2 MeV, protons between 3.2 and 60 MeV, protons above 60 MeV, and alpha particles between 12.5 and 32 MeV. In the polar cap region which was of the greatest interest, the telescopes viewed parallel to, and perpendicular to, the local magnetic field direction. An additional solid-state detector measured directional fluxes of electrons of energies greater than 140 keV. This detector looked in a direction perpendicular to the orbit plane. The detectors worked well until tape recorder failure in May 1971.

APT (Automatic Picture Transmission). Essentially the same system as employed for ESSA-2, -4, -6, and -8. At the nominal satellite altitude of 1450 km, each picture covered approximately 3140 km across-track and 2400 km along-track with a ground resolution of about 3 km at nadir. The APT system functioned nominally until June 20, 1971.

Note: **ITOS-B** (launch on Oct. 21, 1971 on a Delta vehicle from VAFB) failed to achieve a successful Earth orbit. A malfunction in the second stage launch vehicle caused the S/C to reenter the Earth's atmosphere about 1 hour after lift-off. **ITOS-C** had been scheduled for launch in February 1972. However, because of long delays in equipping the launch vehicle with an onboard inertial guidance computer, the launch was cancelled and the satellite was placed in storage. It was replaced in the launch sequence by the more sophisticated NOAA-2 (ITOS–D) S/C. **ITOS-E** (launch on July 16, 1973 from VAFB, S/C mass =747 kg) experienced a launch failure.

G.13.2 NOAA-2 (ITOS-D)

The NASA S/C was launched on Oct. 15, 1972 on a Delta 300 vehicle from VAFB, CA. S/C mass = 306 kg. The satellite design was identical to that of NOAA-1. NOAA 2 was not equipped with conventional TV cameras (like AVCS). It was the first operational weather satellite to rely solely upon radiometric imaging to obtain cloudcover data. The VHRR, VTPR, and SR sensors were mounted on the S/C baseplate with their optical axes directed vertically earthward.

Orbit: Near-circular sun-synchronous polar orbit, apogee = 1458 km, perigee=1448 km, inclination = 101.8°, period =114.9 minutes.

Sensor complement: No APT system.

VTPR (Vertical Temperature Profile Radiometer). Instrument mass = 23 kg, power = 33 W. Two instruments were flown, resolution = 59 km, swath width = 1364 km. Four spectral bands for VTPR1: 14.92-14.99 µm, 14.65-14.87 µm, 14.29-14.49 µm, and 14.03-14.22 µm. Four spectral bands for VTPR2: 13.70-13.88 µm, 13.30-13.48 µm, 18.38-19.01 µm, and 11.93-12.08 µm. The VTPR instrument consisted of an optical system, a detector and associated electronics, and a scanning mirror. The scanning mirror looked at the Earth's surface perpendicular to the satellite's orbital path. As each area was scanned, the optical system collected, filtered, and detected the radiation from the Earth into the eight spectral intervals. VTPR operated nominally until March 18, 1974.

VHRR (Very High Resolution Radiometer). Instrument mass = 16 kg, power = 25 W. Measurement of SST (Sea Surface Temperature) surface temperatures of the Earth, and cloud tops in daylight as well as at night. Resolution = 0.87 km, swath width = 2580 km, two spectral bands = 0.6-0.7 µm, 10.5-12.5 µm. Revolving mirrors (at 400 rpm) scanned the Earth's surface 180° out of phase (one mirror at a time) and perpendicular to the orbit path. The experiment included two scanning radiometers, a magnetic tape recorder, and associated electronics. The two-channel VHRR operated similarly to the Scanning Radiometer (SR) but with much greater resolution (0.9 km compared to 4 km for the SR at nadir). One VHRR channel measured reflected visual radiation from cloud tops in the limited spectral range of 0.6 - 0.7 µm. This provided more contrast between the Earth and clouds than the SR by reducing the effect of haze. The second channel measured infrared radiation emitted from the Earth, sea, and cloud tops in the 10.5-12.5 µm region. This spectral region permitted both daytime and nighttime radiance measurements.

SR (Scanning Radiometer). Two instruments were flown. SR1 resolution = 3.7 km, spectral band = 0.52-0.73 µm. SR2 resolution = 7.4 km, spectral band = 10.5-12.5 µm. Swath width of 3000 km. Identical instruments were flown on NOAA-1, NOAA-3, and NOAA-4.

SPM (Solar Proton Monitor). An identical instrument was flown on NOAA-1.

All imagery was obtained with scanning radiometers. The last data available of NOAA-2 is from March 19, 1974, the S/C was deactivated on Jan. 30, 1975.

G.13.2.1 NOAA-3 (ITOS-F)

NOAA-3 was launched on Nov. 6, 1973 on a Delta 300 vehicle from VAFB, CA. The objective was to provide VIS and IR images of cloud cover, snow and ice, the sea surface, and to collect information on the vertical temperature and moisture profiles of the atmosphere.

The satellite design was identical to NOAA-1. NOAA-3 (mass of 746 kg) was the first satellite to permit direct broadcast of VTPR data.

Orbit: Near-circular sun-synchronous polar orbit, apogee = 1509 km, perigee=1500 km, inclination = 102.1°, period =116.1 minutes.

Sensor complement: Identical sensors as for NOAA-2, namely: VTPR, VHRR, SR and SPM. All imagery was obtained with scanning radiometers.

NOAA-3 remained operational for 1029 days until deactivated by NOAA on Aug. 31, 1976, the last data available is of Dec. 17, 1974.

G.13.3 NOAA-4 (ITOS-G)

Launch of S/C on Nov. 15, 1974 on a two-stage Delta 2310 vehicle from VAFB, CA. The S/C design and sensor complement was identical to that of NOAA-2. S/C mass of 340 kg.

Orbit: Near-circular sun-synchronous polar orbit, apogee=1457 km, perigee=1443 km, inclination=101.7°, period=114.9 minutes.

NOAA-4 remained operational for 1463 days until deactivated by NOAA on Nov. 18, 1978. Last data available is of Sept. 15, 1976.

G.13.4 NOAA-5 (ITOS-H)

Launch of S/C on July 29, 1976 on a two-stage Delta 2310 vehicle from VAFB, CA. The S/C design and sensor complement was identical to that of NOAA-2. One SR failed on Feb. 24, 1978, the other SR failed on March 16, 1978.

Orbit: Near-circular sun-synchronous polar orbit, apogee=1520 km, perigee=1502 km, inclination=102.1°, period=116.2 minutes.

NOAA-5 remained operational for 1067 days until deactivated by NOAA on July 16, 1979. Last data on March 1, 1979.

G.14 TIROS-N (4th Generation) Satellite Series

TIROS-N through NOAA-D (NOAA-12) are called the TIROS-N series (Being a new development, funded by NASA rather than NOAA, the first satellite in this series retained the NASA name "TIROS", with the designation "N" being the next letter in that series. NOAA-M was the first satellite in the 4th generation, ITOS, series). NOAA-D was launched out of sequence and later became NOAA-12. An overview of all US meteorological satellite generations is given in Table 374 on page 737. A communications overview of TIROS-N is given in Figure 216 on page 736.

The program has evolved over several generations of satellites (TIROS, TOS/ESSA, ITOS to the TIROS-N series), starting in 1960 with TIROS-1 to the last in this series being the NOAA-14 operational satellite (see Table 374). Along with the general program evolution, there has also been an enormous sophistication in sensor technology, hence in observation capability. A large number of daily atmospheric profiles or "soundings" find their way into global forecasting models.

Since 1978, the operational system consists of two polar orbiting sun-synchronous satellites (ATN series) crossing the equator during local morning (07:30) and afternoon (13:30). Continuous S-band broadcast is provided for an international user community. **Any user with a receiving station has access to the data.**

NOAA-8 through NOAA-14 are called the **Advanced TIROS-N (ATN)** series of the 4th generation meteorological satellites. The improvements of this series consisted of adding instrument capacity to the satellite (stretched platform structure) for growth potential. The S/C dimensions are: 4.18 m x 1.9 m. The size of the solar panel is 2.37 m x 4.91 m. The S/C mass at launch is 1718 kg at launch and 1034 kg in orbit. All S/C in this series are three-axis stabilized.

The TIROS-N and ATN series were a cooperative effort between NASA, NOAA, the UK, and France. for providing day and night global environmental and associated data for regular daily operations.

Background: The program was referred to as **POES** (Polar-orbiting Operational Environmental Satellites) only sometime during the 4th generation series in the 1980s, to distinguish it from the GOES (Geostationary Operational Environmental Satellite) program. The early TIROS series did not have polar orbits, they were not used in the early days of Earth observation.

The NOAA-POES series (Polar-orbiting Operational Environmental Satellites) and the NOAA-GOES series are regarded the backbone of the US meteorological program. Both types of satellites carry data collection systems for the readout of in situ environmental data. In addition to providing standard meteorological readings in remote locations, the data collection systems allow the monitoring of sea conditions through measurements on ocean buoys, readout of data from aircraft and balloons, and the relay of hydrological data on river levels, snow thickness, etc.

G.14.1 TIROS-N Satellite

The TIROS-N series S/C (design heritage of DMSP) features a box-like structure (3.7 m long, 1.9 m wide) of aluminum and titanium with one large solar panel (see Figure 217). The S/C is three-axis stabilized, nadir pointing with control to better than ± 0.1° using reaction wheels (with a motion rate of less than 0.035 °/s). A hydrazine propulsion system is used for precision orbital insertion, but is deactivated immediately there-after and is not used for station-keeping or orbit maintenance. The solar panel (2.37 m x 4.9 m) generates power over 1 kW. NiCd batteries provide power during eclipse. The total S/C mass is 1418 kg at launch and 735 kg in orbit. Data transmission with APT and HRPT (High Rate Picture Transmission) capability.

Launch of the experimental S/C on Oct. 13, 1978 on an Atlas E/F vehicle from VAFB, CA.

Orbit: near-circular sun-synchronous polar orbit, apogee=862 km, perigee=846 km, inclination=98.9°, period=102.3 minutes.

Sensor complement:

AVHRR (Advanced Very High Resolution Radiometer). Objective: global daytime and nighttime sea-surface temperatures (SST) and information about ice, snow, and clouds. This first AVHRR sensor is an across-track scanning instrument (built by ITT Aerospace of Fort Wayne, IN). The instrument provides for outputting data in five spectral ranges, however, the initial versions of the instrument supported only four (0.55-0.90, 0.725-1.10, 3.550-3.93, and 10.50-11.50 µm and the fourth channel is repeated in channel five (see description in chapter G.14.11). All four channels had a spatial resolution of 1.1 km, the two IR-window channels had a thermal resolution of 0.12 K at 300 K. The AVHRR and the associated S/C data system is capable of providing data in real-time and recorded modes simultaneously. Real-time or direct readout data were transmitted to ground stations both at low (4 km) resolution via APT and at high (1 km) resolution via high-resolution picture transmission (HRPT).

TOVS (TIROS Operational Vertical Sounder). This designation includes three instruments, namely HIRS/2, SSU, and MSU. The objective is to measure vertical profiles of temperature and water vapor from the Earth's surface to the top of the atmosphere.

- **HIRS/2** (High-Resolution Infrared Sounder). Objective: measurement of atmospheric temperature and humidity. Soundings are taken in a nadir scan from the ground up to altitudes of 40 km. Measurements are made in 19 IR bands and one VIS band with the following center wavelengths in μm: 1) 14.95, 2) 14.71, 3) 14.49, 4) 14.22, 5) 13.97, 6) 13.64, 7) 13.35, 8) 11.11, 9) 9.71, 10) 8.16, 11) 7.33, 12) 6.72, 13) 4.57, 14) 4.52, 15) 4.46, 16) 4.40, 17, 4.24, 18) 4.00, 19, 3.76, and 20) 0.69 μm. Spatial resolution = 30 km (FOV), swath width = 2240 km.
- **SSU** (Stratospheric Sounding Unit, a UK sensor provided by BMO). Temperature measurements in the upper atmosphere are derived from radiance measurements made in three channels using a pressure-modulated gas (CO_2) to accomplish selective bandpass filtration of the sampled radiances. Measurements of temperature profiles, top of atmosphere radiation (from 25 km to 50 km in altitude). Spectral ranges (cm^{-1}): 669.99, 669.63, and 669.36. Spatial resolution: 147.3 km at nadir. Swath width = 1474 km. Data output: 12-bit binary sampled at 0.48 kbit/s.
- **MSU** (Microwave Sounding Unit, a JPL sensor). See MSU description in chapter G.14.11.

SEM (Space Environment Monitor). The SEM package includes three sensors:**HEPAD** (High Energy Proton and Alpha Particle Detector), **MEPED** (Medium Energy Proton and Electron Detector), and**TED** (Total Energy Detector). See description of the SEM package in chapter G.14.12.

DCS (Data Collection System) Argos. See Argos description in chapters G.15.4 and C.1.

TIROS-N remained operational for 868 days until deactivated by NOAA on Feb. 27, 1981.

G.14.1.1 NOAA-6 (NOAA-A)

The S/C was launched on June 27, 1979 on an Atlas E/F vehicle from VAFB,CA. This was the first operational satellite of the TIROS-N series. Repeat cycle: 9 days for nadir viewing. The S/C prior to launch became new names in this series, starting with NOAA-A (not ITOS) to designate the new S/C design. The NOAA-6 S/C is identical to the TIROS-N S/C with a total mass of 723 kg.

Orbit: near-circular sun-synchronous polar orbit, apogee=843 km, perigee=833 km, inclination=98.7°, period=101.5 minutes.

Sensor complement: Same sensor complement as for TIROS-N with a different spectral channel 1 for AVHRR.

AVHRR (Advanced Very High Resolution Radiometer). This AVHRR sensor differs slightly in the first spectral channel from the TIROS-N version (Channel 1 = 0.550-0.68 μm), otherwise the instrument is identical to that of TIROS-N.

NOAA-6 was used as standby after the launch of NOAA-8 (1983). TOVS failed on Oct. 22, 1983. NOAA-6 remained operational until Nov. 16, 1986 and was deactivated by NOAA on March 31, 1987.

G.14.2 NOAA-B

Launch of S/C on May 29, 1980 on an Atlas E/F vehicle from VAFB, CA. The S/C (mass of 1405 kg) failed to achieve proper orbit.

G.14.3 NOAA-7 (NOAA-C)

Launch on June 23, 1981 on an Atlas E/F vehicle form VAFB, CA. Same S/C as NOAA-6 and the same sensor complement except for AVHRR (see AVHRR/2 configuration in Table 375. The S/C mass is 1405 kg.

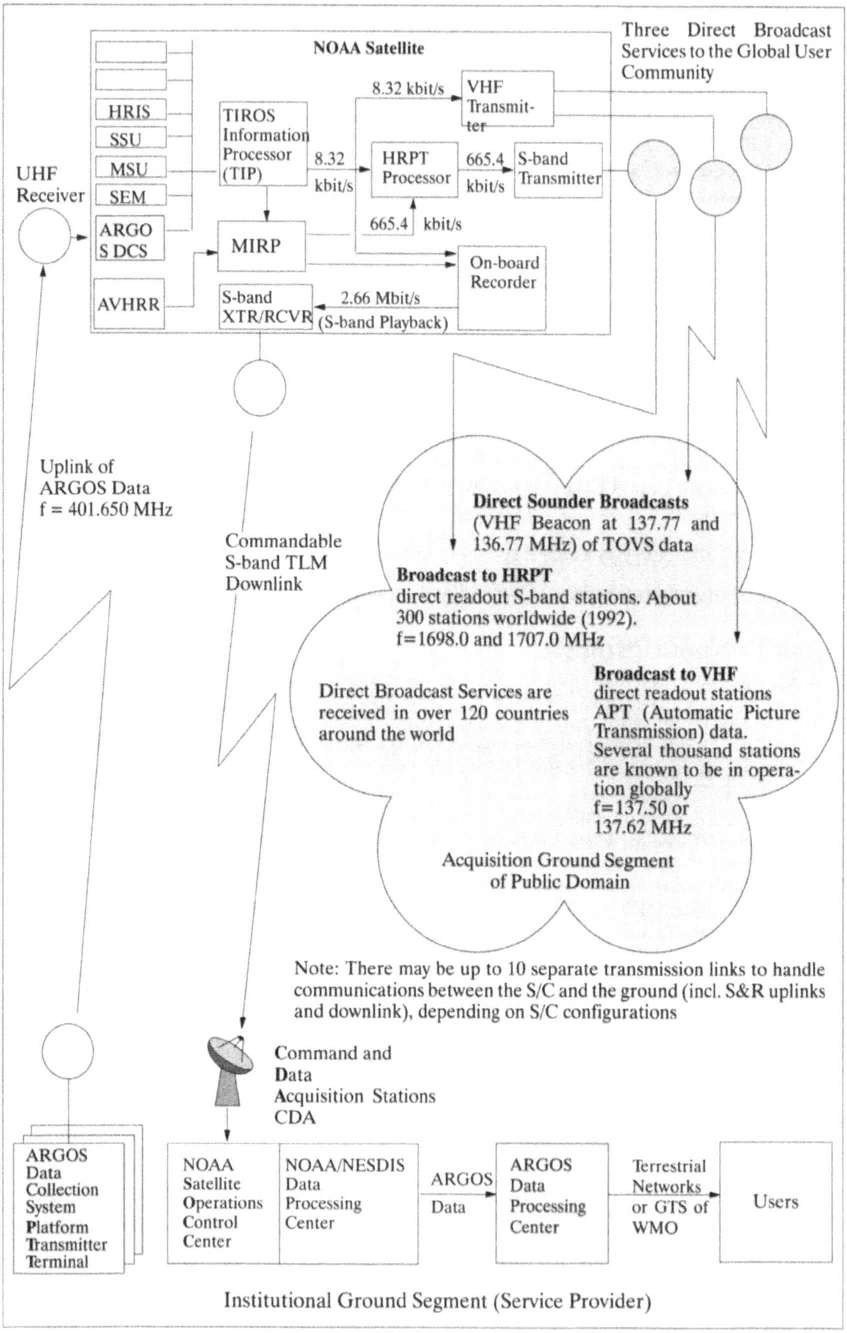

Figure 216: The ARGOS system concept within the NOAA POES/TIROS family

Satellite Name	Launch Date	Period (min)	Perigee (km)	Apogee (km)	Inclin. (°)	Remarks/Sensor Complement
TIROS-1	01 Apr 60	99.2	796	867	48.3	1 TV-WA and 1 TV-NA
TIROS-2	23 Nov 60	98.3	717	837	48.5	1 TV-WA, 1 TV-NA, passive & active IR scan
TIROS-3	12 Jul 61	100.4	854	937	47.8	2 TV-WA, HB, IR, IRP
TIROS-4	08 Feb 62	100.4	817	972	48.3	1 TV-WA, IR, IRP, HB
TIROS-5	19 Jun 62	100.5	680	1119	58.1	1 TV-WA, 1 TV-MA
TIROS-6	18 Sep 62	98.7	783	822	58.2	1 TV-WA, 1 TV-MA
TIROS-7	19 Jun 63	97.4	713	743	58.2	2 TV-WA, IR, ion probe, HB
TIROS-8	21 Dec 63	99.3	796	878	58.5	1st APT TV direct readout & 1 TV-WA
Nimbus-1	28 Aug 64	98.3	487	1106	98.6	3 AVCS, 1 APT, HRIR, three-axis stabilization
TIROS-9	22 Jan 65	119.2	806	2967	96.4	First "wheel," 2 TV-WA, global coverage
TIROS-10	02 Jul 65	100.6	848	957	98.6	Sun synchronous, 2 TV-WA
ESSA-1	03 Feb 66	100.2	800	965	97.9	1st operational system, 2 TV-WA, FPR
ESSA-2	28 Feb 66	113.3	1561	1639	101.0	2 APT, global operational APT
Nimbus-2	15 May 66	108.1	1248	1354	100.3	3 AVCS, HRIR, MRIR, APT
ESSA-3	02 Oct 66	114.5	1593	1709	101.0	2 AVCS, FPR
ATS-1	06 Dec 66	24 hr	41,257	42,447	0.2	SSCC (Spin Scan Cloudcover Camera)
ESSA-4	26 Jan 67	113.4	1522	1656	102.0	2 APT
ESSA-5	20 Apr 67	113.5	1556	1635	101.9	2 AVCS, FPR
ATS-3	05 Nov 67	24 hr	41,166	41,222	0.4	MSSCC (Multicolor Spin scan Cloud Camera)
ESSA-6	10 Nov 67	114.8	1622	1713	102.1	2 APT TV
ESSA-7	16 Aug 68	114.9	1646	1691	101.7	2 AVCS, FPR, S-band
ESSA-8	15 Dec 68	114.7	1622	1682	101.8	2 APT TV
ESSA-9	26 Feb 69	115.3	1637	1730	101.9	2 AVCS, FPR, S-band
Nimbus-3	14 Apr 69	107.3	1232	1302	101.1	HRIR, IRIS, MRIR, MUSE,SIRS, IDICS, IRLS, 2 APT, 2 AVCS, 2 SR, FPR, three-axis stabilization
ITOS-I	23 Jan 70	115.1	1648	1700	102.0	SIRS B, IRIS, SCR, THIR, , FWS, IDCS,
Nimbus-4	15 Apr 70	107.1	1200	1280	99.9	IRLS, MUSE,BUV
NOAA-1	11 Dec 70	114.8	1422	1472	102.0	2 APT, 2 AVCS, 2 SR, FPR
NOAA-2	15 Oct 72	114.9	1451	1458	98.6	2 VHRR, 2 VTPR, 2 SR, SPM
Nimbus-5	11 Dec 72	107.1	1093	1105	99.9	SCMR, ITPR, NEMS, ESMR, THIR
NOAA-3	06 Nov 73	116.1	1502	1512	101.9	2 VHRR, 2 VTPR, 2 SR, SPM
SMS-1	17 May 74	1436.4	35,605	35,975	0.6	VISSR, DCS, WEFAX, SEM (EPS,MAG,XRS)
NOAA-4	15 Nov 74	101.6	1447	1461	114.9	2 VHRR, 2 VTPR, 2 SR, SPM,
SMS-2	06 Feb 75	1436.5	35,482	36,103	0.4	VISSR, DCS, WEFAX, SEM
Nimbus-6	12 Jun 75	107.4	1101	1115	99.9	ERB, ESMR, HIRS LRIR, T+DRE, SCAMS,THIR TWERLE, PMR
GOES-1	16 Oct 75	1436.2	35,728	35,847	0.8	VISSR, DCS, WEFAX, SEM (EPS,MAG,XRS)
NOAA-5	29 Jul 76	116.2	1504	1518	102.1	2 VHRR, 2 VTPR, 2 SR, SPM
GOES-2	16 Jun 77	1436.1	35,600	36,200	0.5	VISSR, DCS, WEFAX, SEM (EPS,MAG,XRS)
GOES-3	15 Jun 78	1436.1	35,600	36,200	0.5	VISSR, DCS, WEFAX, SEM (EPS,MAG,XRS)
TIROS-N	13 Oct 78	98.9	849	864	102.3	AVHRR,HIRS/2,SSU,MSU,HEPAD, MEPED, TED
Nimbus-7	24 Oct 78	99.3	943	955	104.1	LIMS, SAMS, SAM-II, SBUV/TOMS, „ERB SMMR, THIR, CZCS, SMMR
NOAA-6	27 Jun 79	101.3	807.5	823	98.7	AVHRR,HIRS/2,SSU,MSU,HEPAD,MEPED,TED
GOES-4	09 Sep 80	1436.1	35,600	35,600	0.5	VAS, DCS, SEM (EPS,MAG,XRS,HEPAD),WEFAX
GOES-5	22 May 81	1436.1	35,600	35,600	0.5	VAS, DCS, SEM (EPS,MAG,XRS,HEPAD),WEFAX
NOAA-7	23 Jun 81	101.9	852	869	98.9	AVHRR,HIRS/2,SSU,MSU,HEPAD,MEPED,TED
NOAA-8	28 Mar 83	101.2	801	826	98.2	AVHRR,HIRS/2,SSU,MSU,MEPED, TED,S&R
GOES-6	28 Apr 83	1436.1	35,803	35,771	0.1	VAS, DCS, WEFAX, SEM (EPS,MAG,XRS)
ERBS	05 Oct 84	96.8	393	608	57	ERBE, SAGE-II
NOAA-9	12 Dec 84	102.0	842	862	98.9	AVHRR,(HIRS/2,SSU,MSU)=TOVS , S&R, SBUV/2, ERBE,
NOAA-10	17 Sep 86	101.3	803	824	98.7	AVHRR,(HIRS/2,SSU,MSU)=TOVS, MEPED, TED, S&R, SBUV, ERBE,
GOES-7	26 Feb 87	1436.1	35,759	35,826	0.1	VAS, DCS, WEFAX, SEM (EPS,MAG,XRS), S&R
NOAA-11	24 Sep 88	102.1	845	863	98.9	AVHRR,(HIRS/2,SSU,MSU)=TOVS, S&R, SBUV/2,
NOAA-12	14 May 91	101.3	806	825	98.7	AVHRR, HIRS/2, MSU,SEM (MEPED,TED)
NOAA-13	09 Aug 93	a S/C failure (power loss) occurred on Aug. 21, 1993				AVHRR, TOVS, SBUV, SEM, ARGOS, S&R, EHIC, MAXIE
GOES-8	13 Apr 94		Geosynchronous orbit			Imager, Sounder, SEM, DCS, Wefax, S&R
NOAA-14	30 Dec 94	102.1	847.7	860.8	98.9	AVHRR, SBUV/2, HIRS/2I, MSU, SSU, SEM, ARGOS (DCS), S&R
GOES-9	23 May 95		Geosynchronous orbit			Imager, Sounder, SEM, DCS, Wefax, S&R
NOAA-15	13 May, 98	101.3	803.7	817.8	98.70	AVHRR/3, HIRS/3, AMSU-A (A1+A2), AMSU-B, SEM-2, ARGOS (DCS-2), S&R
NOAA-16	21 Sept, 00	102.1	845.4	860.4	98.74	AVHRR/3, HIRS/3, AMSU-A (A1+A2), AMSU-B, SBUV/2, SEM-2, ARGOS (DCS-2), S&R

Table 374: Summary of U.S. civilian env./met. satellites (chronological order)

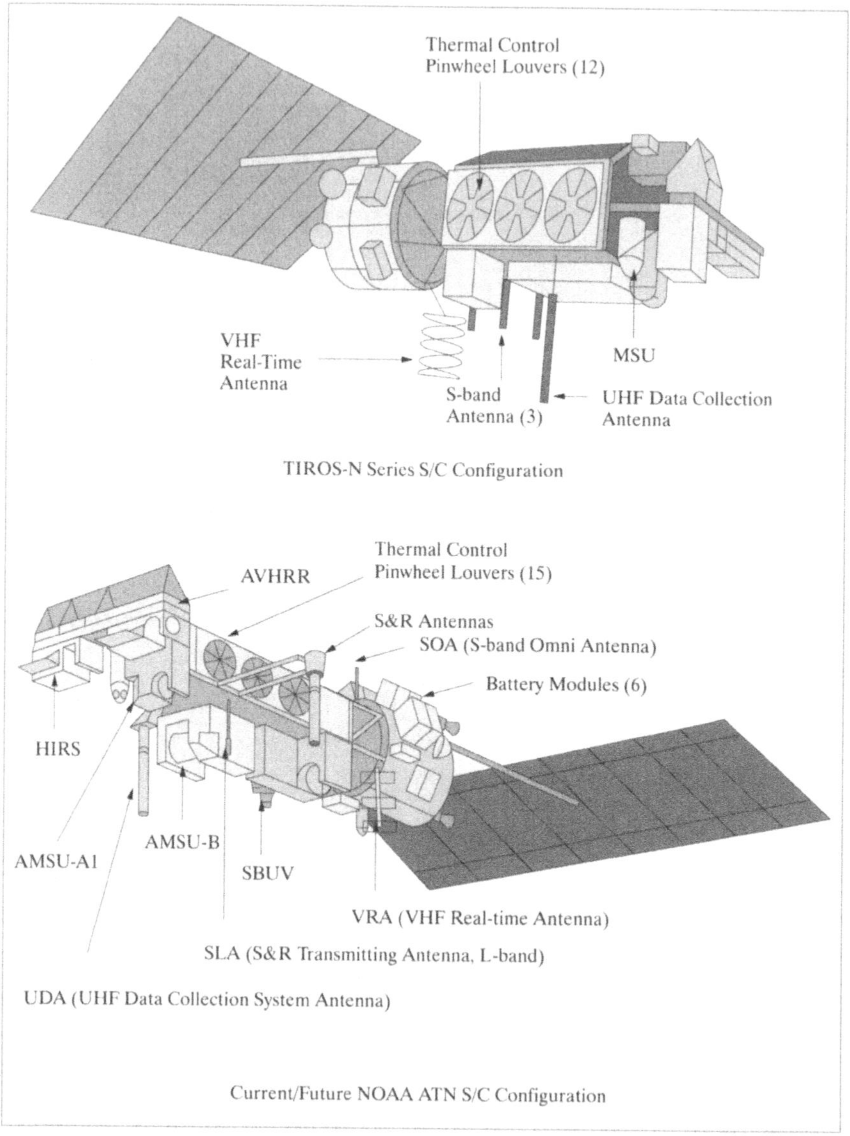

Figure 217: Typical NOAA POES series S/C models

Orbit: near-circular sun-synchronous polar orbit, apogee=869 km, perigee=852 km, inclination=98.9°, period=101.9 minutes.

TOVS failed on Feb. 7, 1985. The MEPED and HEPAD space environment instruments were turned off April 1, 1982 because of electrical interference with the command and control receiver. NOAA-7 remained operational for 2625 days and was deactivated by NOAA on June 7, 1986 due to a power failure.

G.14.4 NOAA-8 (NOAA-E)

Launch on March 28, 1983 on an Atlas vehicle from VAFB, CA. First of ATN satellite series. The system **S&RSAT** (Search and Rescue Satellite) is flown for the first time (see chapter I.6). Note: Although the original acronym for 'Search and Rescue' is 'SAR' in the context of NOAA missions, it was changed in this book consistently to 'S&R' in order to distinguish it from the other widely-used meaning of SAR, namely 'Synthetic Aperture Radar,' an active sensor type. A consequence is the use of 'S&RSAT' (instead of SARSAT), to use S&RR (Search and Rescue Repeater) instead of SARR, as well as to use S&RP (Search and Rescue Processor) instead of SARP.

Orbit: near-circular sun-synchronous polar orbit, apogee=829 km, perigee=806 km, inclination=98.9°, period=101.2 minutes.

Sensor complement: AVHRR, TOVS (HIRS/2,SSU,MSU), MEPED, TED. Same as for NOAA-6 with the addition of S&RSAT. - AVHRR was capable of operating in both real-time or recorded modes. Real-time or direct readout data were transmitted to ground stations both at low (4 km) resolution via APT, and at high (1 km) resolution via high-resolution picture transmission (HRPT).

NOAA-8 remained operational for 1016 days until deactivated by NOAA on Jan. 9, 1986, following a thermal runaway which destroyed the battery.

G.14.5 NOAA-9 (NOAA-F)

Launch on Dec. 12. 1984 on an Atlas E/F vehicle from VAFB, CA. First test flights for **ERBE** (Earth Radiation Budget Experiment, see description in chapter A.13) and **SBUV/2** (Solar Backscatter Ultraviolet) instruments (G.14.11). S/C mass = 1712 kg.

Orbit: near-circular sun-synchronous polar orbit, apogee=862 km, perigee=842 km, inclination=99.17°, period=102 minutes.

Sensor complement: AVHRR/2, TOVS (HIRS/2, MSU, SSU), ERBE, SBUV/2, S&RSAT, Argos/DCS.

Status 1992 (S/C in standby operation, the power system is considered to be very marginal). Digital tape recorder failed two months after launch. ERBE stopped operation in Jan. 1987. AVHRR has at times exhibited anomalous behavior in its synchronization in MIRP. MSU channels 2 and 3 have failed, and the power system is degraded. SBUV/2 is operating satisfactorily. The satellite is collecting, processing, and distributing SBUV/2 and ERBE nonscanner data. It is also providing S&R data.

NOAA-9 was deactivated on August 1, 1993. It was reactivated on Aug. 23. 1993 after the failure of NOAA-13.

Status 3/2001: S/C deactivated Feb. 20, 1998 in anticipation of the NOAA-15 launch.

G.14.6 NOAA-10 (NOAA-G)

Launch on Sept. 17, 1986 on an Atlas E/F vehicle from VAFB, CA. Same S/C structure as for NOAA-8. S/C mass = 1700 kg.

Orbit: near-circular sun-synchronous polar orbit, apogee=824 km, perigee=803 km, inclination=98.7°, period=101.3 minutes.

Sensor complement: AVHRR, TOVS (HIRS/2, MSU, SSU), ERBE, SBUV/2, SEM (MEPED, TED), S&RSAT, Argos/DCS. The TED on NOAA-10 displayed serious anomalies beginning in late 1987 and data acquired after February 29, 1988 are of marginal quality.

The solid state detectors in the electron portion of the MEPED became noisy after 01 January 1, 1991 and those data were much degraded for the remainder of the mission.

Status 1992 (S/C in standby operation utilizing STIP record mode): The ERBE non-scanner mode is performing well, while the ERBE scanner mode failed in 1989. It has exhibited a scan sticking anomaly that is apparently generic to the instrument. The S&R processor (406 MHz) has failed.
Status 1/1994: in standby operation. Operational instruments: AVHRR, MSU, DCS, HIRS/2, S&R. Status 1/1996: The NOAA-10 S/C was deactivated in March 1995.

G.14.7 NOAA-11 (NOAA-H)

Launch on Sept. 24, 1988 on an Atlas E/F vehicle from VAFB, CA. Same S/C structure as for NOAA-8. S/C mass = 1712 kg.

Orbit: near-circular sun-synchronous polar orbit, apogee=863 km, perigee=845 km, inclination=98.9°, period=102.1 minutes. Ascending orbit at 1:49 PM.

Sensor complement: AVHRR/2, TOVS (HIRS/2, MSU, SSU), SBUV/2, S&RSAT, Argos/DCS.

Status: The AVHRR scan motor failed in Sept. 1994, leaving the instrument inoperative. In Oct. 1994 the SBUV/2 diffuser failed. However, SBUV/2 continues to collect global ozone data. The S/C was placed in standby mode in March 1995.

G.14.8 NOAA-12 (NOAA-D)

Launch on May 14, 1991 on an Atlas E/F vehicle from VAFB, CA. Same S/C structure as for NOAA-N. S/C mass = 1416 kg.

Orbit: near-circular sun-synchronous polar orbit, apogee=825 km, perigee=806 km, inclination=98.7°, period=101.3 minutes. AM orbit.

Sensor complement: AVHRR, TOVS (HIRS/2, MSU, SSU), SEM (MEPED, TED). S&RSAT was not flown.

Status 3/2001: HIRS suffered a significant degradation on May 30, 1997. Otherwise, NOAA-12 supports all other services.

G.14.9 NOAA-13 (NOAA-I)

Launch on Aug. 9, 1993 on an Atlas E vehicle from VAFB, CA. Same S/C structure as for NOAA-8. S/C mass = 1712 kg.

Orbit: near-circular sun-synchronous polar orbit, apogee=825 km, perigee=806 km, inclination=98.7°, period=101.3 minutes.

Sensor complement: AVHRR/2, TOVS (HIRS/2, MSU, SSU), EHIC, MAXIE, SBUV/2, MEPED, TED, S&RSAT, Argos/DCS.

The S/C and its systems operated successfully for 12 days until a circuit failure resulted in a power loss. Hence, the S/C and payload could not be operated and must be regarded a total loss.

The regular payload of NOAA-13 was complemented with two special instruments, MAXIE and EHIC. Both were one-of-a-kind sensors, hence they were not available for NOAA-14 or follow-on missions.

MAXIE = Magnetospheric Atmospheric X-Ray Imaging Experiment (Lockheed and University of Bergen, Norway). MAXIE maps the intensities and energy spectra of X-rays pro-

duced by electrons that precipitate the atmosphere. With mechanical scanning, MAXIE will obtain new high-resolution X-ray imaging data on auroral and substorm processes at a temporal resolution and repetition rate so far not available.
Combined with other data, a more complete analysis of the radiation environment is obtained. Derived products include vertical profiles of ionization and electrical conductivity, as well as the imaging of the aurora in sunlight. These products are used to determine atmospheric scale height and other synoptic information.

EHIC = Energetic Heavy Ion Composition (U. of Chicago and NRC HIA of Canada) EHIC measures the chemical and isotopic composition of energetic particles between hydrogen and nickel over the energy range of 0.5 to 200 MeV/nucleon. The experiment will measure energetic solar flare particles in the polar regions where the Earth's magnetic field connects to the interplanetary field carried in the solar wind. It will also measure trapped energetic particles in the magnetosphere.
The primary objective of EHIC is to obtain elemental and isotopic composition data, which can be used to test models for solar flare ion acceleration, ion transport in interplanetary space, ion entry into the magnetosphere, and nucleosynthetic processes leading to the elemental and isotopic mix found at the sun.

G.14.10 NOAA-14 (NOAA-J)

Launch on Dec. 30, 1994 on an Atlas E vehicle from VAFB, CA. Same S/C structure as for NOAA-8. S/C mass = 1745 kg.

Orbit: near-circular sun-synchronous polar orbit, apogee=861 km, perigee=847 km, inclination=98.1°, period=102.1 minutes. PM orbit.

Sensor complement: AVHRR/2, SBUV/2, TOVS (HIRS/2I, MSU, SSU), SEM, Argos/DCS, S&RSAT

Status 3/2001: All instruments are functional except the S&R (Search & Rescue) repeater which failed immediately after launch. Also the SBUV/2 data is highly degraded due to failures in a grating drive and also due to orbital drift.

G.14.11 Sensor Descriptions of 4th Generation Series

AVHRR (Advanced Very High Resolution Radiometer) is a broadband, four or five channel (depending on model) cross-track scanning whiskbroom instrument (built by ITT Aerospace of Fort Wayne, IN). Data output = 10 bit, spatial resolution = 1.1 km at nadir and 6 km at maximum scan (at the swath edges). IFOV = 1.4 mrad (average), scanning rate = 360 per minute, sampling rate = 2048 per scan, sample step = 0.95 mrad (1.36 samples per IFOV), scan angle (max) = ±55.4° from nadir, swath width = 2400-3000 km (depending on S/C orbit). Note: the five-channel configuration is also referred to as 'AVHRR/2.'

AVHRR is comprised of five modules: Scanner module, electronics module, radiant cooler, optical system, and baseplate. - The optical system consists of an afocal reflective Cassegrainian telescope with an aperture diameter of 20.3 cm, combined with secondary optics which separate the radiant energy into discrete spectral bands; these are focused onto the respective field stops. The VIS and VNIR channels (channel 1 and 2) use silicon detectors to measure incident radiation. All IR detectors are cooled to 105 K. Channel No. 3 uses InSb detectors, while channels 4 and 5 use HgCdTe detectors. Radiant cooler: two-stage system.

All five channels of AVHRR are registered so that they all measure energy from the same target location at the same time. All five channels are also calibrated so that the signal amplitude in each channel is a measure of the scene radiance as received at the sensor aperture. The solar channels are pre-launch calibrated, the thermal channels are calibrated against a laboratory blackbody source. A passive blackbody mounted on the baseplate assists inflight calibration of the thermal bands. No inflight calibration of the VIS and NIR channels is available. Although the calibration varies from instrument to instrument, the design goals for the infrared channels are: NEDT of 0.12 K (@ 300 K) and an SNR of 3:1 @ 0.5% albedo.

Objectives: Measurement of radiance data for investigation of clouds, land-water boundaries, snow and ice extent, sea surface temperature, day and night cloud distribution, vegetation index. The benefit of AVHRR data lies in its high temporal frequency of coverage, global coverage at least once per day. Applications: Operational meteorology, oceanography, climatology, vegetation monitoring, land and sea ice observation.[1244)]

Channel No.	TIROS-N	NOAA-6, -8, -10	NOAA-7, -9, -11, -12, -14	IFOV (mrad)	Principal use of channel
1	0.550 - 0.90	0.550 - 0.68	0.550 - 0.68	1.39	Day cloud and surface mapping
2	0.725 - 1.10	0.725 - 1.10	0.725 - 1.10	1.41	Surface water delineation, vegetation mapping
3	3.550 - 3.93	3.550 - 3.93	3.550 - 3.93	1.51	SST and fire detection
4	10.50 - 11.50	10.50 - 11.50	10.30 - 11.30	1.41	SST and night time cloud mapping
5	repeat of Channel 4	repeat of Channel 4	11.50 - 12.50	1.30	Surface temperature and day/night cloud mapping
	AVHRR/1		AVHRR/2		Instrument configuration/designation

Table 375: Spectral channels of AVHRR

S/C	Launch Date	Ascending Node	Descending Node	S/C Service Period	
TIROS-N	Oct. 13, 1978	15:00	03:00	Oct. 19, 1978	Jan. 30, 1980
NOAA-6	Jun. 27, 1979	19:30	07:30	Jun. 27, 1979	Nov. 16, 1986
NOAA-7	Jun. 23, 1981	14:30	02:30	Aug. 24, 1981	Jun. 7, 1986
NOAA-8	Mar. 28, 1983	19:30	07:30	May 3, 1983	Oct. 31, 1985
NOAA-9	Dec. 12, 1984	14:20	02:20	Feb. 25, 1985	Aug. 3, 1995
NOAA-10	Sep. 17, 1986	19:30	07:30	Nov. 17, 1986	March 1995
NOAA-11	Sep. 24, 1988	13:30	01:40	Nov. 8, 1988	Sep. 9, 1994
NOAA-12	May 14, 1991	19:30	07:30	May 14, 1991	present
NOAA-13	Aug. 9, 1993	Power failure 12 days after launch, NOAA lost contact with the S/C			
NOAA-14	Dec. 30, 1994	13:40	01:40	Dec. 30, 1994	present
NOAA-15	May 13, 1998	19:26	07:26	Dec. 15, 1998	present
NOAA-16	Sep. 21, 2000	14:00	02:00	Sep. 21, 2000	present

Table 376: Temporal AVHRR coverage of NOAA POES series satellites

HIRS/2I (High Resolution Infrared Radiation Sounder, built by ITT)
Measurement of atmospheric temperature and humidity. This instrument detects and measures energy emitted by the atmosphere to construct a vertical temperature profile from the Earth's surface to an altitude of about 40 km. Measurements are made in 20 spectral channels in the infrared band (one frequency lies at the high frequency end of the visible range). Nadir scan, swath = 2240 km. Spatial resolution = 20.4 km.

Cha. No	Center frequency (µm)	Cha. No	Center frequency (µm)	Cha. No	Center frequency (µm)	Cha. No	Center frequency (µm)
1	14.95	6	13.64	11	7.33	16	4.40
2	14.71	7	13.35	12	6.72	17	4.24
3	14.49	8	11.11	13	4.57	18	4.00
4	14.22	9	9.71	14	4.52	18	3.76
5	13.97	10	8.16	15	4.46	20	0.69

Table 377: Center frequencies of HIRS/2 channels

1244) A. Schwalb, "The TIROS-N/NOAA A-G Satellite Series," NOAA Technical Memorandum NESS 95, March 1978

MSU = Microwave Sounding Unit (JPL sensor)
MSU detects and measures the energy from the troposphere to construct a vertical temperature profile to an altitude of about 20 km. Measurements are made by radiometric detection of microwave energy divided into four frequency channels. Each measurement is made by comparing the incoming signal from the troposphere with the ambient temperature reference load. Because its data are not seriously affected by clouds, the MSU is used along with HIRS/2 and HIRS/2I to remove measurement ambiguity when clouds are present. Data output: 12-bit binary at 0.32 kb/s. MSU has been operational since 1978.

Spectral ranges	50.3, 53.74, 54.96, and 57.95 GHz (4 channels)
FOV	95°
Swath width	2348 km
Spatial resolution	105 km at nadir

Table 378: MSU specification parameters

SSU = Stratospheric Sounding Unit, (UK sensor built by Matra Marconi)
Temperature measurements in the upper atmosphere are derived from radiance measurements made in three channels using a pressure-modulated gas (CO_2) to accomplish selective bandpass filtration of the sampled radiances. Measurements of temperature profiles, top of atmosphere radiation (from 25 km to 50 km in altitude). Spectral ranges (cm^{-1}): 669.99, 669.63, 669.36. Resolution: 147.3 km at nadir. IFOV: 10° .Data output: 12-bit binary sampled at 0.48 kbit/s.

ERBE = Earth Radiation Budget Experiment (on NOAA-9 and -10 only). Measurement of Earth radiation gains and losses. Duty cycle: 100%. See sensor description in chapter A.13.

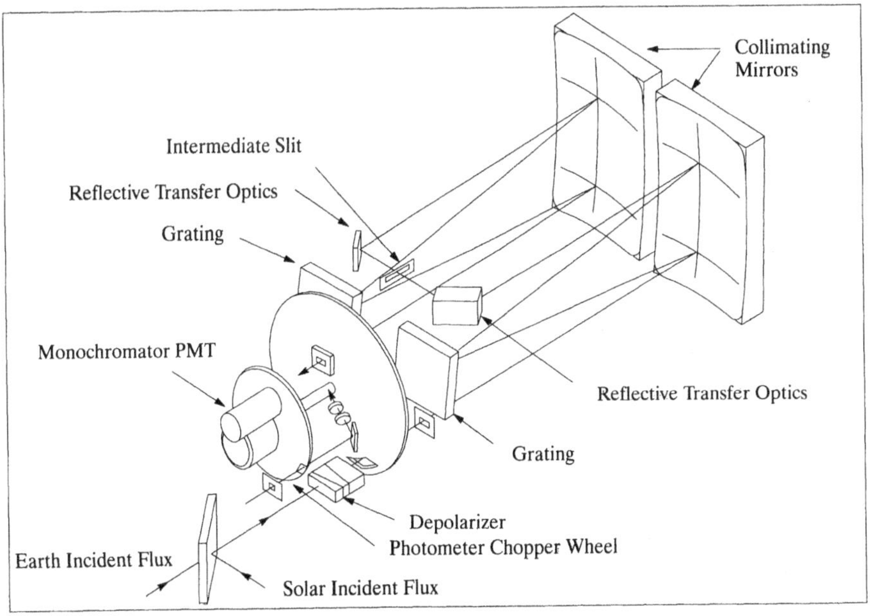

Figure 218: Schematic optical diagram of the SBUV instrument

SBUV/2 = Solar Backscatter Ultraviolet (on PM satellites only)
The SBUV/2 sensor is a spectrally scanning UV radiometer (nadir-viewing sensor). Measurement of solar irradiance and scene radiance (backscattered solar energy) over a spectral range of 0.16 to 0.40 μm (sweep mode). Resolution = 170 km (FOV=11.3°). The instru-

ment design is based upon the technology developed for the SBUV/TOMS flown on Nimbus 7. Application: atmospheric chemistry (measurement of trace gases). Sensor objectives:

- to make measurements from which total ozone concentration in the atmosphere can be determined to an absolute accuracy of 1 percent.
- to make measurements from which the vertical distribution of atmospheric ozone can be determined to an absolute accuracy of 5%.
- to measure the solar spectral irradiance from 0.16 to 0.40 μm.

Discrete mode	Measures from 252.0 to 339.8 nm in 12 discrete 1 nm bands
Sweep mode	Continuous scan from 160 to 400 nm
Cloud Cover Radiometer (CCR)	Measures brightness temperature of the Earth's surface at 380 nm

Table 379: Some parameters of SBUV/2

G.14.12 SEM (Space Environment Monitor)

This instrument package is different from the normal Earth-observing payload of the NOAA-POES satellite series. The objective is to provide "space weather" to the user community by measuring the solar wind particle flux and its variations. The package is provided by NOAA/SEC of Boulder, CO. SEM-1 instruments were initially introduced on satellites in geostationary orbit on 1974. SMS-1, a predecessor of the NOAA-GOES series, was the first satellite to carry SEM. NOAA introduced SEM on the POES series as well starting with TIROS-N in 1978.

SEM = Space Environment Monitor (POES series instruments)
SEM is a multichannel, charged-particle spectrometer that measures the population of the Earth's radiation belts and the particle phenomena resulting from solar activity (both of which contribute to solar/terrestrial energy interchange). SEM consists of two or three separate sensor units and a common DPU (data processing unit). The sensor units are TED, MEPED, and HEPAD (occasionally). The lower-energy sensors (TED, plus the proton and electron telescopes of MEPED) have pairs of sensors with different orientations because the direction of the particle fluxes is important for characterizing the energy interchanges taking place. Objectives:

- to determine the energy deposited by solar particles in the upper atmosphere
- to provide a solar storm warning system

TED = Total Energy Detector. Measurement of protons and electrons in 11 bands in the energy range:: 0.3 to 20 keV.

MEPED = Medium Energy Proton and Electron Detector. Performance: Proton: 30 to 2500 keV in five bands. Electron: >30 to > 300 keV in three bands. Ions: > 6 MeV. Omni-proton: > 16 MeV, > 36 MeV, > 80 MeV.

HEPAD = High Energy Proton and Alpha Particle Detector (see F.4.2).

G.15 5th Generation Satellites of NOAA-POES Series

The 5th generation series of polar orbiting NOAA satellites started with NOAA-K (NOAA-15) which was launched on a Titan II vehicle on May 13, 1998 from VAFB, CA into a morning orbit to replace the aging NOAA-12 satellite.

The KLM S/C have significant changes to the previous 4th generation satellites with regard S/C design and payload. The S/C structure comprises four assemblies: IMP (Instrument Mounting Platform), EBM (Equipment Support Module), RCE (Reaction Control Equipment), RCS (Reaction Support Structure), and SA (Solar Array) assembly. To meet the increased power requirements of new instruments, two additional solar panels have been added, the solar array has about 45% more output. The batteries, propulsion tank capacity, the size of the reaction wheels and magnetic coils used for momentum unloading and attitude control, have also increased in capacity. The spacecraft structure has been stiffened primarily to support the heavier AMSU instruments and improve launch vehicle load margins. Several antennas have been relocated and/or built with new materials and processes to improve performance. Flight computer memory has been doubled and the flight software modified to meet new requirements. [1245)]

1st generation 1960 - 1965	TIROS series [TIROS-1 (1960) to TIROS-10 (1965)]
2nd generation 1966 - 1969	TOS (TIROS Operational System) series as pre-launch designation. The in-orbit satellite designation was ESSA [ESSA-1 (1966) to ESSA-9 (1969)], after the S/C operating agency.
3rd generation 1970 - 1976	ITOS (Improved TIROS Operational System) series as pre-launch designation. The in-orbit satellite designation was NOAA [NOAA-1 (1970) to NOAA-5 (1976)]
4th generation 1978 - 1994	ATN [Advanced TIROS-N) series. After TIROS-N (1978) the pre-launch designation changed to NOAA-A (the corresponding inflight name was NOAA-6). The pre-launch letter designation was kept throughout. NOAA-14 was the last S/C
5th generation 1998 - 2008	NOAA-15 and -16 (launch May 13, 1998 and Sept 21, 2000, respectively), former NOAA-K and -L), NOAA-M, -N, -N'

Table 380: Overview of the US polar meteorological programs

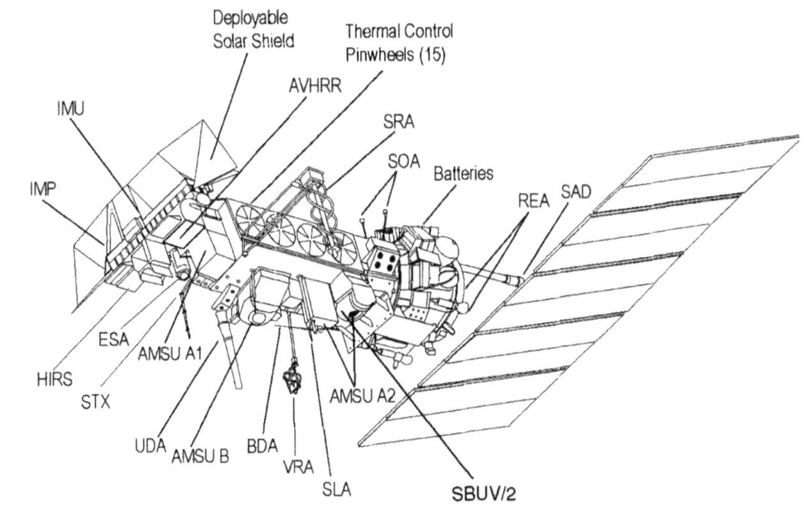

Legend:

BDA	Beacon Transmitting Antenna	SLA	S&R Transmitting Antenna (L-band)
ESA	Earth Sensor Assembly	SRA	S&R Receiving Antenna
IMP	Instrument Mounting Platform	STX	S-band Transmitting Antenna
REA	Reaction Engine Assembly	SOA	S-band Omni Antenna (2 of 6 shown)
SAD	Solar Array Drive	UDA	UHF Data Collection System Antenna
		VRA	VHF Realtime Antenna

Figure 219: Illustration of the 5th generation satellite configuration

The ADACS (Attitude Determination and Control Subsystem) provides in conjunction with RCS (Reaction Control Subsystem) and the CCS (Command and Control Subsystem)

1245) M. Mignogno, "NOAA Polar Program Plans for Continuous Satellite Coverage," Proceedings of the 1999 EUMETSAT Meteorological Satellite Data User's Conference, Copenhagen, Denmark, Sept. 6-10, 1999, pp. 11-15 the functions of the on-orbit attitude control and ascent guidance. It is a zero-momentum system consisting of reaction wheels and Earth, sun, and inertial reference sensors. In the subsystem's attitude–control mode, the Earth Sensor Assembly (ESA) and the sun sensor, together with rates derived from the IMU (roll, pitch and yaw)), furnish the primary attitude reference. Actuation is provided by reaction wheels and cold gas thrusters. Attitude accuracy is 0.2°, knowledge is 0.1° in post-processing. Attitude determination with 3-sigma variation >0.14° is provided onboard the satellite.

Satellite orbit	Sun-synchronous circular, altitude = 833±19 km or 870±19 km, inclination = 98.7° (retrograde)
S/C dimensions (main body)	4.18 m length, 1.88 m diameter, overall length = 7.4 m
S/C mass	1479 kg on orbit, 2232 kg at launch (756 kg propellant)
S/C power (solar array)	833 W min, array size: 2.73 m x 6.14 m (16.76 m^2)
On-board data recording	5 DTR (Digital Tape Recorder), each with a capacity of 0.9 Gbit. Solid-State Recorders (SSRs) will fly on NOAA-M (1 SSR, 4 DTR), and NOAA-N and N' with 3 SSRs each with a capacity of 2.7 Gbit
AKM (Apogee Kick Motor)	Star 37XFP (Thiokol Corporation), used to circularize the orbit after S/C separation. A thrust of 42.38 kN is provided (51 s burn)

Table 381: Some 5th generation NOAA spacecraft characteristics

Satellite communication subsystem: The on-board communication subsystem comprises 14 antennas, 9 transmitters, and redundant receivers, together with associated filters and other RF feed components. The subsystem provides the following functions: (refer to Figure 216)

- Reception and demodulation of S-band TT&C data
- Continuous transmission (broadcast) of TIP data via the VHF Beacon
- Continuous transmission (broadcast) of APT data at VHF-band
- Continuous transmission (broadcast) of HRPT data at S-band
- Transmission of stored LAC, GAC, TIP, and AIP (AMSU Information Processor) data upon command at S-band
- Reception and filtering of DCS signals
- Reception, processing and retransmission of S&R signals.

Type of Service	NOAA-9 through NOAA-N'
GAC (Global Area Coverage) Playback	Reduced resolution AVHRR data (4 km) plus stored TIP data at full resolution soundings. GAC data is derived on-board the NOAA satellite by subsampling and averaging the nominal 1 km AVHRR imagery. It provides daily global coverage recorded on-board and transmitted to a NOAA ground station. About 115 minutes of GAC data can be stored on-board.
LAC (Local Area Coverage)	LAC is nominally 1 km resolution AVHRR imagery (which is normal or 'high' resolution data) recorded with the on-board tape recorder for subsequent transmission to a NOAA ground station. About 11 minutes of LAC data can be accommodated on a recorder. LAC imagery can only be obtained from NOAA/NESDIS.
HRPT (High Resolution Picture Transmission)	Full resolution AVHRR & TIP data, Frequency: 1698 or 1707 MHz, data rate = 665.4 kbit/s, split phase PSK. The imagery is available in a format defined by the ground receiving station.
APT (Automatic Picture Transmission) (LRPT in OPQ era)	Reduced resolution geometrically corrected analog video from two channel of AVHRR, selected by user command. Frequency: 137.50 or 137.62 MHz
Beacon	HIRS, SEM, DCS, SBUV, & Eph

Table 382: Real-time and global data services current/future NOAA POES missions [1246)]

Sensor complement of NOAA-15: AVHRR/3, HIRS/3, AMSU-A (A1+A2), AMSU-B, SEM-2, ARGOS (DCS-2), and S&RSAT.

1246) In NOAA terminology, TIP (TIROS Information Processor) data refers to low-rate instrument data multiplexed with satellite housekeeping data. It contains all environmental instrument information except that from the AVHRR and the AMSU.

NOAA-15 orbit: Circular sun-synchronous near-polar orbit; altitude = 833 km; inclination =98.7°; orbital period =102.12 min; equator crossing at about 7:30 AM.

Spacecraft	Orbit	Launch or projected launch
NOAA-15 (K prior to launch)	AM orbit	May 13, 1998
NOAA-16 (L prior to launch)	PM orbit	Sept. 21, 2000
NOAA-M	AM orbit	2001 with Titan-2 vehicle
NOAA-N	PM orbit	2004 with Delta-2 vehicle (projected)
MetOp-1	AM orbit	2005 with Soyuz vehicle
NOAA-N'	PM orbit	2008 with Delta-2 vehicle
MetOp-2	AM orbit	2010 with Soyuz vehicle

Table 383: Projected launch dates of POES series and MetOp series S/C

G.15.1 Sensors for the POES K, L, M, N, N' Series

AVHRR/3, HIRS/3 (KLM, HIRS/4 starting with N), AMSU-A, AMSU-B, SEM-2, SBUV/2 (on all PM missions from K through N'), ARGOS (DCS-2), S&RSAT. [1247)]

HIRS/3 = High Resolution Infrared Sounder. [1248)] Heritage of HIRS/2I. Objective: measurement of temperature profiles, moisture content, cloud height and surface albedo. HIRS/3 scans the Earth's surface in twenty spectral bands in the range 0.69 - 14.95 µm. The sensor scans across-track in a 'stop-and stare' mode (discrete stepping) at a scan rate of 6.4 s (56 Earth views per scan). FOV = ±49.5° about nadir (swath = ±1080 km). Collected energy is separated by a beamsplitter into LWIR (> 6.5 µm) and SWIR (3.7 to 4.6 µm), passed through field stops and through a rotating filter wheel to cooled detectors. In the SWIR path, a second beamsplitter separates the visible channel to a silicon detector. Three detectors are used to sense the radiation. A silicon photodiode at the instrument temperature (nominally 15° C) detects the visible energy. An InSb detector and a MCT detector (mounted on a passive radiator and operating at 100 K) sense the SWIR and TIR energy. The NESR (Noise Equivalent Spectral Radiance) is given in $W/(cm^2 \; sr \; cm^{-1})$. [1249)]

Optical FOV (Field of View)	1.4° for VIS and SWIR; 1.3° for TIR
Telescope aperture	15 cm diameter
Channel to channel registration	TIR: <1.5% of 1.8° step size; SWIR: <1% of cha. 19 FWHM
Earth scan angle, scan steps	±49° from nadir, 56 increments of 1.8°
Step and dwell time	100 ms total
Total scan plus retrace time	6.4 s
Swath width	2160 km
Spatial resolution	20.3 km (1.4° FOV) at nadir; 18.9 km (1.3° FOV) at nadir
Data quantization	13 bit
Data rate	

Table 384: Performance parameters of HIRS/3

IR Calibration of the HIRS/3 is provided by programmed views of two radiometric targets: a warm target (blackbody at 290 K) mounted to the instrument base and a view of space. Data from these views provides sensitivity calibrations for each channel every 40 lines (256 seconds), if commanded. Internally generated electronic signals provide calibration and stability monitoring of the amplifier and readout electronics. Note: The calibration sequencing of HIRS/3 has changed (versus HIRS/2) to remove viewing of the cold internal reference. This allows an additional scan line of data each calibration period. Secondly,

1247) CEOS Summary Report, WGD-10 Meeting, Annapolis MD, April 16-18, 1991

1248) http://www2.ncdc.noaa.gov/docs/klm/html/c3/sec3-2.htm

1249) P. M. Taylor, B. A. Banks, "An overview of the NOAA/NESDIS data processing systems and derived products for NOAA-KLM," Earth System Monitor, Vol. 8, No. 4, June 1998, pp. 7-11 while the HIRS/3 is used primarily for temperature sounding, channel 20 has been upgraded to enhance generation of radiation budget products. The HIRS/3 instrument objective is to achieve greater overall detector performance and lower noise levels.

Channel No.	Center Wavenumber (cm^{-1})	Center Wavelength (µm)	Sensitivity NESR $W/(cm^2\ sr\ cm^{-1})$	FWHM Bandwidth (cm^{-1})
1	669	14.95	0.300	3
2	680	14.71	0.067	10
3	690	14.49	0.050	12
4	703	14.22	0.031	16
5	716	13.97	0.021	16
6	733	13.64	0.024	16
7	749	13.35	0.020	16
8	900	11.11	0.010	35
9	1,030	9.71	0.015	25
10	802	12.47	0.015	16
11	1,365	7.33	0.020	40
12	1,533	6.52	0.020	55
13	2,188	4.57	0.0006	23
14	2,210	4.52	0.0003	23
15	2,235	4.47	0.0004	23
16	2,245	4.45	0.0004	23
17	2,420	4.13	0.0002	28
18	2,515	4.00	0.0002	35
19	2,660	3.76	0.0001	100
20	14,500	0.690	0.10% albedo	1,000

Table 385: Spectral parameters of HIRS/3

HIRS/4 (High Resolution Infrared Sounder). This instrument will fly on NOAA-N, -N', and MetOp 1 and 2, but not on MetOp-3. It has essentially the same features as the HIRS/3, with one significant change. The sampling distance along a scan line remains at nominally 18 km, but the IFOV has been decreased to 10 km from 18 km. This was done to improve the utility of the data near cloud boundaries.

AVHRR/3 = Advanced Very High Resolution Radiometer (3rd generation of AVHRR, heritage of AVHRR/2, built by ITT Aerospace of Fort Wayne, IN). Objectives: day and night imaging in six spectral bands of land, water and top of cloud surfaces, sea surface temperature, ice snow and vegetation cover. The instrument scans in the cross-track direction with a FOV of ±55.37° about nadir (swath of 2900 km), IFOV of 1.1 km at nadir (1.3 mrad x 1.3 mrad for all channels). The detectors are passively cooled to < 105 K by a two-stage passive radiant cooler. AVHRR/3 uses a continuously internal rotational scanning mirror which also views deep space and an internal calibration source. Calibration of the IR channels is performed with four internal blackbodies every scan line.[1250) 1251)]

Telescope	20.3 cm diameter afocal reflective Cassegrainian type
Scan motor	360 rpm hysteresis - synchronous
Scan mirror	21 cm x 29.5 cm elliptical ribbed beryllium
Cooler	Two stage radiant cooler controlled @ 105 K
Data quantization	10 bit
Video sample rate	40 kHz simultaneous sample of all channels
No of samples per scan	2048
Output data rate	200 k word/s max (2 Mbit/s in burst mode, 621.3 Kbit/s average)
Instrument size	80 cm x 36.5 cm x 29 cm
Instrument mass, power	33 kg, 27 W (orbital average) at 28 V

Table 386: AVHRR/3 performance characteristics

1250) Information provided by G. A. Mandt of NOAA POES Program, "AVHRR/3 Instrument Technical Overview," March 15, 1995 of ITT A/CD

1251) http://www2.ncdc.noaa.gov/docs/klm/html/c3/sec3-1.htm

AVHRR/3 adds a sixth channel in the IR range at 1.61 μm (referred to as channel 3a, operating during the daylight part of the orbit). The objective of channel 3a is to aid in improved snow and ice discrimination and aerosol detection. Channel 3b corresponds to the previous channel 3 on AVHRR/2 and operates during the night portion of the orbit. Splitting channel 3 in this way maintains the HRPT format. The data from the six channels is simultaneously sampled at 40 kHz and converted to 10-bit binary values. The data samples are output in a non-continuous burst of 10 space samples, 2048 Earth samples, and 10 internal calibration target samples at 6 Hz, synchronized with the scan mirror.

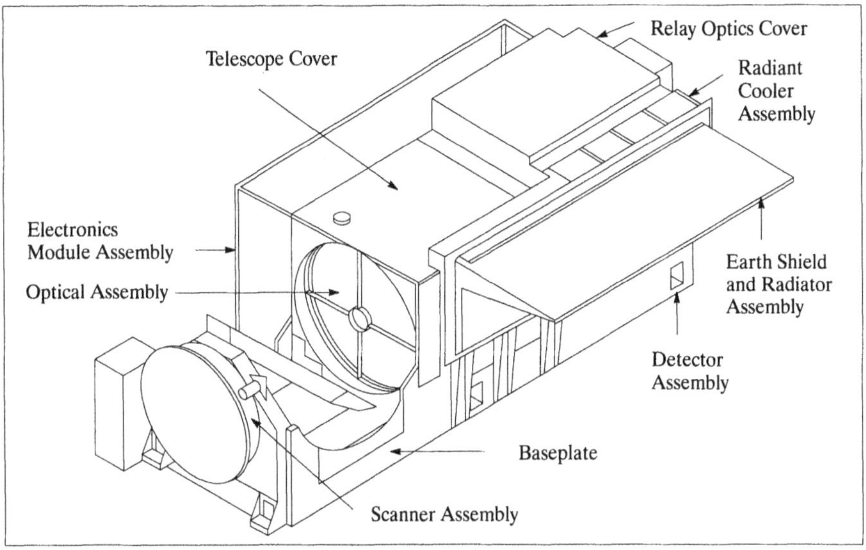

Figure 220: Illustration of the AVHRR instrument

Channel	Center wavelength (μm)	Spectral Range FWHM (μm)	Channel Noise	Detector Type
1	0.630 (VIS)	0.58 - 0.68	SNR ≥ 9:1 @ 0.5% albedo	Silicon
2	0.862 (VNIR)	0.725 - 1.00	SNR ≥ 9:1 @ 0.5% albedo	Silicon
3a	1.61 (NIR)	1.58 - 1.64	SNR ≥ 20:1 @ 0.5% albedo	InGaAs
3b	3.74 (IR-window)	3.55 - 3.93	NEΔT ≤ 0.12 K @ 300 K	InSb
4	10.80 (IR-window)	10.30 - 11.30	NEΔT ≤ 0.12 K @ 300 K	HgCdTe
5	12.00 (IR window)	11.50 - 12.50	NEΔT ≤ 0.12 K @ 300 K	HgCdTe

Table 387: Spectral parameters of AVHRR/3

AVHRR/3 (like AVHRR/2) is an optomechanical system comprised of five modules: scanner module, electronics module, radiant cooler, optical system, and baseplate unit. - The optical system consists of collecting telescope (an afocal reflective Cassegrainian type telescope with an aperture diameter of 20.32 cm), combined with secondary (relay) optics which separate the radiant energy into six discrete spectral bands; these are focused onto their respective field stops. Polarization effects have been minimized in channels 1, 2 and 3A by passing the optical beam, transmitted by the first VIS/IR beamsplitter, through a second beamsplitter oriented in such a way as to compensate for polarization introduced by the first beamsplitter (the split gain technique increases the sensitivity for snow and ice coverage, aerosol distribution and vegetation index calculation). - The scan motor design is based on an 80 pole hysteresis synchronous motor. It rotates the mirror at a rate of 360 rpm to produce a contiguous scan on the Earth's surface.

The **AMSU** (Advanced Microwave Sounding Unit) is a 20-channel microwave radiometer. Its primary mission objective is to obtain global temperature and humidity profiles. AMSU is comprised of three separate units: AMSU-A1 (channels 3-15), AMSU-A2 (channels 1 and 2), and AMSU-B (channels 16 - 20). Channels 3 - 14 use the 50 to 60 GHz oxygen band to provide data for vertical temperature profiles up to 50 km. The "window" channels (1, 2, 15, and 16) provide data to enhance temperature sounding by correcting for surface emissivity, atmospheric liquid water, and total precipitable water. Channels 18 - 20 use the 183.3 GHz water vapor absorption line to provide data for the humidity profile. The first flight of AMSU is on the NOAA-15 satellite.

AMSU-A = Advanced Microwave Sounding Unit - A (always consisting of AMSU-A1 and AMSU-A2), was built by Aerojet, a subsidiary of GenCorp of Sacramento, CA for NASA/GSFC. AMSU-A replaces the MSU and SSU instruments flown on previous missions. Objectives: atmospheric temperature profile measurements from the surface up to 50 km in 15 channels. Temperature resolution: 0.25 - 1.2 K. AMSU-A is configured in the following major subsystems: Antenna/Drive/Calibration Subsystem; Receiver Subsystem; Signal Processor Subsystem and Structural/Thermal Subsystem. The Antenna/Drive/Calibration subsystem consists of a conical corrugated horn-fed shrouded reflector, multiplexer, closed-loop antenna scan drive assembly and closed path calibration assembly. The shrouded reflector is rotated once every scan line (8 s) for:

- Each of 30 Earth viewing scene observations
- A view of the cosmic background (about 2.73 K)
- A view of a warm calibration load (about 300K)

During the rotation cycle, the shroud prevents solar reflections from interacting with the warm load and also ensures maximum coupling of the source radiation to the antenna feed. A complete end-to-end in-flight calibration is achieved in a through-the-antenna method, which provides maximum in-flight calibration accuracy. The accuracy of the warm calibration load brightness temperature is > ± 0.2 K. The closed loop antenna scan drive provides beam pointing accuracy within ± 0.2°. A resolver in the antenna drive assembly provides antenna beam position information.

Sensor Unit	Channel No.	Center Frequency (GHz)	Bandwidth (MHz)	Sensitivity NEΔT (K)	Polarization at Nadir
AMSU-A2	1	23.8	280	0.3	V
(2 channels)	2	31.4	180	0.3	V
AMSU-A1	3	50.300	180	0.4	V
(13 channels)	4	52.800	400	0.25	V
	5	53.596± 0.115 (2 bands)	170	0.25	H
	6	54.400	400	0.25	H
	7	54.940	400	0.25	V
	8	55.500	330	0.25	H
	9	57.290344 = Flo	330	0.25	H
	10	Flo ± 0.217 (2 bands)	78	0.4	H
	11	Flo ± 0.3222, (±0.048)	36	0.4	H
	12	Flo ± 0.3222, (±0.022)	16	0.6	H
	13	Flo ±0.3222, (± 0.010)	8	0.8	H
	14	Flo ±0.3222, (± 0.0045)	3	1.2	H
	15	89.000	6000	0.5	V
AMSU-B	16	89.0 (2 bands)	1000	0.37	90-θ
(5 channels)	17	150.0 ±0..9 (2 bands)	1000	0.84	90-θ
	18	183.31±1.00 (2 bands)	500	1.06	90-θ
	19	183.31±3.00 (2 bands)	1000	0.70	90-θ
	20	183.31±7.00 (2 bands)	2000	0.60	90-θ

Table 388: Spectral parameters of the AMSU-A and AMSU-B instruments [1252)]

The AMSU-A1 instrument has two 8 cm diameter antennas (reflectors without momentum compensation), each with a 3.3° nominal IFOV at the half power points (FWHM). Each

1252) Note: The polarization angle is defined as the angle from horizontal polarization (i.e., electric field vector parallel to satellite track) where is the scan angle from nadir. In this table, the polarization angle is horizontal when the angle indicated is θ and vertical when 90-θ antenna provides a cross-track scan of ±48.33° from nadir with a total of 30 contiguous Earth views (stepped scan positions) per scan line. The total scan period is 8 seconds. The footprint (resolution) at nadir is 50 km. The swath width is approximately 2000 km. Channels 11 through 14 contain 4 pass bands each. Data rate of 3.2 kbit/s. Instrument mass = 100 kg, power = 125 W.

AMSU-A2 has a single 17 cm diameter antenna (reflector with momentum compensation) with a 3.3° nominal IFOV. All other instrument/observation parameters are the same as for AMSU-A1.

AMSU-A utilizes an 8-second scan period with a step and settle scan across the Earth scene while AMSU-B utilizes an 8/3-second scan period with a constant speed scan across the Earth scene.

AMSU-B = Advanced Microwave Sounding Unit - B (three flight instruments are being built and provided to NOAA-K, -L and -M satellites by UKMO, Farnborough, PI: R. Saunders, AMSU-B is built by MMS Bristol, UK). The objective is to obtain humidity profiles in five channels spanning the height range from the surface to about 42 km. AMSU-B covers channels 16 through 20. The highest channels: 18, 19 and 20, span the strongly opaque water vapor absorption line at 183 GHz and provide data on the atmosphere's humidity level.

The AMSU-B consists of a scanning parabolic reflector antenna (30 cm diameter) which is rotated once every 8/3 seconds and focuses incoming radiation into a quasi-optic system which then separates the frequencies of interest into three separate feed horns of the receiver assembly. The receiver subsystem provides further demultiplexing of the 183 GHz signal in order to selectively acquire three defined double sided bands around the 183 GHz signal. The antenna provides a cross-track scan of ±48.95° (FOV) from nadir with a total of 90 Earth views per scan line. The instrument's IFOV is 1.1° and the separation between the center of one Earth view to the next is 1.1°. The total scan period is 8/3 seconds. Internal calibration is performed with an internal warm load and cold space (space calibration = 80° from nadir). All channels have a spatial resolution of about 16.3 km at nadir (1.1° IFOV). The swath width is approximately 2200 km. Data rate = 4 kbit/s. The instrument mass is 60 kg, power = 90 W. AMSU-B ensures a twice daily full global coverage. AMSU-B employs the Schottky planar whiskerless diode technology in the mixer of the receiver. - AMSU-B is the first moisture sounder flown in the NOAA polar series. In addition to providing high-resolution atmospheric moisture profiles, AMSU-B is used to produce precipitation and surface products from two window channels. [1253)]

MHS =Microwave Humidity Sounder. MHS is an EUMETSAT instrument provided by the UKMO (UK Meteorological Office). MHS is a five channel self-calibrating instrument providing humidity profiling capability in the frequency range of 89 - 190 GHz. The measured signals are also sensitive to a) liquid water in clouds (cloud liquid water content) and b) graupel and large water droplets in precipitating clouds (qualitative estimate of precipitation rate).

Channel	Center Frequency (GHz)	Max. Bandwidth (MHz)	Sensitivity (K)
1	89.0	2800	1.0
2	157.0	2800	1.0
3	183.311 ± 1.0	1000	1.0
4	183.311 ± 3.0	2000	1.0
5	190.311	2200	1.0

Table 389: Spectral Parameters of MHS

MHS scans in the cross-track direction at a rate of 8/3 seconds (2.67 s) in continuous mode. The instrument's IFOV is 1.1° and the separation between the center of one Earth view to

1253) http://www2.ncdc.noaa.gov/docs/klm/html/c3/sec3–4.htm

the center of the next is exactly 10/9° (1.111°). The circular IFOV has a diameter at nadir of about 16.3 km. The observed swath width is ±1078 km. MHS technology used: advanced quasi-optics, planar millimetric frontends, precision calibration target, scan mechanism and reflector are combined with low power and mass electronics for minimum mass and power. Thermal control is passive and isolated from the spacecraft. Instrument mass = 63 kg, power = 83 W. The MHS instrument replaces AMSU-B, it is planned to be flown on the NOAA-N and N' satellites, it will also be flown on the MetOp series of EUMETSAT.

SBUV/2 = Solar Backscatter Ultraviolet Radiometer (further development of Nimbus SBUV). This instrument is pointed toward nadir; there are 12 channels between 255 nm and 340 nm. Measurement of ozone profiles and of backscatter radiation. This sensor is used on PM and late AM missions (NOAA-M is planned for a 10 AM descending sun-synchronous orbit. NOAA-16, -N and -N' are planned for 2:00 PM ascending sun-synchronous orbits.)

G.15.2 SEM-2 (Space Environment Monitor-2)

Starting with NOAA-K (NOAA-15, launch May, 13, 1998), NOAA/SEC added an improved SEM package, referred to as SEM-2. SEM-2 has added in-flight calibration capability and improved charged particle energy coverage. The TED measurement range is now from 0.05 keV to 20.0 keV in 16 energy bands. The energy coverage of the proton detector telescope portion of the MEPED is expanded to 6 energy bands, 30 keV to >6900 keV, and a fourth omnidirectional detector has been added to the MEPED to extend the energetic proton energy coverage to >140 MeV

SEM-2 is a multichannel, charged-particle spectrometer that measures the population of the Earth's radiation belts and the particle phenomena resulting from solar activity (both of which contribute to solar/terrestrial energy interchange). SEM-2 consists of separate sensor units and a common DPU (data processing unit). The sensor units are TED, MEPED, and HEPAD (occasionally). The lower-energy sensors (TED, plus the proton and electron telescopes of MEPED) have pairs of sensors with different orientations because the direction of the particle fluxes is important for characterizing the energy interchanges taking place. Objectives:

- To determine the energy deposited by solar particles in the upper atmosphere
- To provide a solar storm warning system. Large, abrupt changes occur in the solar wind. These changes produce changes in the magnetosphere and the ionosphere. This can result in hazards to people such as astronauts, there is also the potential of increased radiation exposure to people in high flying aircraft. Other functions/facilities on Earth effected by solar wind changes are: a) disruption of navigation, b) absorption, even blackout, of radio waves so that radio communication is disrupted; c) induced voltages and currents in electric power circuits leading to circuit breaker trip, damage to equipment, and failure of transformers; d) induced currents in buried pipe lines causing accelerated corrosion; e) damage by electrons and protons to satellite circuits and solar panels.

TED = Total Energy Detector. TED measures proton and electron fluxes in 16 bands in the 0.05 to 20 keV energy range. Two independent measurements of the particle energy flux are made at zero and 30 degrees from the local vertical. The total energy measurement is divided into two ranges: 0.05 to 1 keV and 1 to 20 keV and each measurement is made independently for electrons and protons. TED also measures the maximum differential energy flux density and the energy at which it occurs for each direction and particle type (electron and proton). TED consists of eight Electrostatic Analyzers (ESA), pulse height discriminators (PHD), an In-Flight Calibrator (IFC), two high voltage (HV) supplies, a sweep voltage supply and housekeeping circuits.

MEPED = Medium Energy Proton and Electron Detector. MEPED provides both directional and omni-directional measurements. The directional sensors utilize telescopes which make independent measurements of the particle types. Directional measurements are made near the local vertical and near 90° to the local vertical. Protons are measured in 6 bands in the energy range: 30 to >6900 keV. Electrons are measured in thee bands: >30 to > 300 keV. The omni-directional sensors measure proton energy in the following ranges: >16 MeV, >35 MeV, >70 MeV and >140 MeV. Each sensor consists of a dome of moderating material which absorbs energy from the particle (and so sets the detection energy threshold), a silicon solid state detector, a preamplifier, and a level comparator which responds to particles with enough energy to go through the moderator and produce a pulse from the detector large enough to exceed the level in the comparator.

The MEPED instrument consists of two proton telescopes, (each containing two solid state detectors (SSDs)), two electron telescopes (each containing a single solid state detector and four omni-directional sensors (each containing a single solid state detector), charge-sensitive preamplifiers, analog signal processors, proton and electron coincidence logic, in-flight calibrator, low-voltage regulators, SSD bias supply and analog housekeeping.

G.15.3 IJPS (Initial Joint Polar System)

Since the early 1990s, NOAA and EUMETSAT have been discussing/planning future polar cooperation with increased European responsibility for the "morning orbit" to ensure continuity of the POES (Polar-orbiting Operational Environmental Satellites) services. The basic intent is to join the space segment of the emerging MetOp program of EUMETSAT with the existing POES program of NOAA into a fully coordinated service, thus sharing the costs of a program for synergetic reasons. The plans came to a common baseline and agreement, referred to as IJPS, in 1998. [1254) 1255) 1256) 1257)]

Background: The NOAA 4th-generation polar program (ATN), which started in 1978 with the launch of TIROS-N, is based on the services of two operational satellites flying in complementary sun-synchronous orbits, one in a "morning or AM" orbit, and the second in an "afternoon or PM" orbit. It so turns out that the data from the PM mission is primary in the USA, with the AM mission providing supplementary and back-up coverage. In Europe the converse is true, with the AM mission providing the most timely coverage.

IJPS comprises two series of independent, but fully coordinated polar satellite systems, namely POES and MetOp, to provide for the continuous and timely collection and exchange of environmental data from space. EUMETSAT plans to include its satellites MetOp-1, -2 and -3 for the morning orbit, while NOAA is starting with its NOAA-N and N' spacecraft for the afternoon orbit of the coordinated system. Plans call for MetOp-1 and NOAA-N to be launched in 2005 and 2004, respectively.

Instruments: Each IJPS satellite series carries a common set of instruments and additional instruments specific for each orbit. The common set of instruments (morning and afternoon orbits) include: AVHRR/3, HIRS/4, AMSU-A (A1+A2), MHS, ARGOS/ADCS, S&RSAT with instruments including S&RR (CRC, Canada)+S&PR-3 (CNES), and SEM-2. The MHS instrument is provided by EUMETSAT for the NOAA and MetOp series. It replaces the AMSU-B in the POES series instrument baseline. As part of the acquisition activities for NOAA satellites, NOAA will also provide all common instrument sets for MetOp satellites (MHS and ARGOS/ADCS excepted).

The MetOp series specific instruments (in addition to the common set) are:

1254) M. Mignogno, M, Langevin, "Cooperation in Polar Orbiting Environmental Satellites: NOAA and EUMETSAT Joint Plans for the next Decade," Proceedings of Information for Sustainability, 27th International Symposium on Remote Sensing of Environment, Tromsoe, Norway, June 8–12, 1998, pp. 555-559

1255) A. F. Durham, "Future Polar Satellite Program Plan for Global Environmental Observations," IAF 92-0083, 43rd Congress of the International Astronautical Federation, Aug. 28-Sept. 5, 1992 Washington, D. C.

1256) Bruce H. Needham, "Instrumentation and Services for the NOAA Polar-Orbiting Operational Environmental Satellites (POES) in the 21st Century," NOAA/NESDIS, Office of System Development, Washington D.C., '90

1257) "Pre-Phase-A Study of NOAA O, P, Q Spacecraft and Ground Segment LRPT and HRPT Data Handling and Transmission Subsystems" Draft Final Report, Oct. 16, 1990, Atlantic Research Corp. prepared for NASA/GSFC

- IASI (Infrared Atmospheric Sounding Interferometer)
- ASCAT (Advanced Scatterometer)
- GRAS (GNNS Receiver for Atmospheric Sounding)
- GOME-2 (Global Ozone Monitoring Experiment-2)

The NOAA series specific instrument (in addition to the common set) is:

- SBUV/3 (Solar Backscatter Ultraviolet Radiometer)

EUMETSAT data transmission services:
MetOp satellites provide the service of direct broadcasts using LRPT (Low Resolution Picture Transmission) and HRPT (High Resolution Picture Transmission) with selective virtual channel encryption capability for LRPT and HRPT data. EUMETSAT has adopted CCSDS (Consultative Committee on Space Data Systems) packet telemetry standards for LRPT and HRPT transmissions. The MetOp series will also use an X-band downlink for instrument data dumps from the on-board solid-state recorder.

NOAA data transmission services:
NOAA satellites will use APT (Automatic Picture Transmission) and HRPT direct broadcast services as implemented in the POES system.

IJPS satellite operations:
IJPS satellite operations are implemented separately by EUMETSAT and NOAA with the provision of backup/cross-support capabilities for each other in the ground segment. The global data set is shared. Each ground segment provides for S/C control, command and data acquisition, data processing, distribution, and archiving capabilities. Each agency is in full control of its S/C and ground segment operations.

G.15.4 ARGOS on NOAA-POES Satellites

ARGOS Data Collection System (DCS) on NOAA S/C (provided by CNES/France, installed on all ATN family satellites since 1978; the first satellite equipped with ARGOS was TIROS-N.

The ARGOS/DCS supports NOAA in its overall environmental mission objectives, collecting (ground and space) truthing data. The concept uses many ground segment platforms (fixed and moving), i.e. buoys, free-floating balloons, and remote weather stations, and equips them with a **P**latform **T**ransmitter **T**erminal (PTT) package. These PTTs collect and process relevant environmental data and transmit them to the NOAA-POES satellites. The on-board ARGOS DCS receives the incoming signal and measures both the frequency and relative time of occurrence of each transmission. The S/C retransmits these data via the CDA (Command and Data Acquisition) stations (one at Wallops Island VA, the other at Fairbanks, AK; there is in addition a downlink station (CMS) at Lannion, France, (see Figure 196) to a central processing facility. The DCS information is decommutated and sent to the ARGOS processing center where it is processed, distributed to the user community, and stored on magnetic tape for archival purposes. (see also Figure 216).

Each ARGOS PTT transmits encoded messages at regular intervals on a 401.650 MHz uplink. Messages transmitted by the various platforms within satellite visibility are received and selected for processing on a random access basis. The satellite DCS computes the Doppler shift on the receive frequency and generates the ARGOS telemetry message which includes PTT identification, sensor data, measured frequency, and time and date of measurement. A small portion of the S/C downlink is reserved for ARGOS data. Each time a satellite is within visibility of one of the three receiving stations, it downlinks the recorded data.

Some ARGOS system characteristics: (see also chapter C.1)

- Minimum platform/satellite elevation angle of visibility: 5°

- Percentage of platforms with four Doppler measurements per day: >85%
- Measured location accuracy: 350 m
- Message capacity for sensor data: 32 to 256 bits
- Messages are of duration <1 s and are transmitted at regular intervals by any PTT
- Uplink operational frequency: 401.650 MHz
- Typical power of PTT uplink: 200 mW at intervals of 90-15 seconds for location (drifting) PTTs and 200-300 seconds for data collection-only (fixed) platforms.

In 1988 an 'Argos World Service' was introduced providing five times daily location reports on vehicles and freight carrying a standard transmitter. The NOAA Argos packages receive all messages within a 5000 km diameter visibility circle at any instant (footprint); four PTTs can be processed simultaneously by NOAA-1 to 12 (and up to J) S/C.

From NOAA-K onwards (satellites and its successors, there is a modified Argos data collection system referred to as **DCS-2** (the predecessor of ADCS). The data transmission rate for DCS-2 changed from 1200 bit/s to 2560 bit/s. The PTT capacity was increased from four to eight, this means eight DRU (Data Recovery Unit) on-board. The data are formatted and stored, then dumped each time the satellites moves within visibility of one of the three ground stations (Wallops, Gilmore Creek, or CMS). VHF and S-band transmitters also perform real-time relay (broadcast) for any user station within visibility.

An advanced ARGOS system (ADCS) having increased receiver sensitivity, higher platform data rates, and "forward" (S/C-to-platform) command transmission capability is planned for NOAA-N' and for the METOP series.

GPSR Compliance

The European Union's (EU) General Product Safety Regulation (GPSR) is a set of rules that requires consumer products to be safe and our obligations to ensure this.

If you have any concerns about our products, you can contact us on ProductSafety@springernature.com

In case Publisher is established outside the EU, the EU authorized representative is:

Springer Nature Customer Service Center GmbH
Europaplatz 3
69115 Heidelberg, Germany

Batch number: 10761593

Printed by Printforce, the Netherlands

Observation of the Earth and Its Environment

Springer-Verlag Berlin Heidelberg GmbH

Herbert J. Kramer

Observation of the Earth and Its Environment

Survey of Missions and Sensors

With 522 figures and 857 tables

Springer

Dr. Herbert J. Kramer

Gautinger Strasse 7
82205 Gilching
Germany

e-mail: herb.kramer@gmx.net

Additional material to this book can be downloaded from http://extras.springer.com

ISBN 978-3-642-62688-3 ISBN 978-3-642-56294-5 (eBook)
DOI 10.1007/978-3-642-56294-5

Cataloging-in-Publication Data applied for

Kramer, Herbert J.: Observation of the earth and its environment : survey of missions and sensors ; with 857 tables / Herbert J. Kramer. - 4. ed.. - Berlin ; Heidelberg ; New York ; Barcelona ; Hong Kong ; London ; Milano ; Paris ; Tokyo : Springer, 2002
(Engineering online library)

Originally published by Springer-Verlag Berlin Heidelberg New York in 2002.
Softcover reprint of the hardcover 4th edition 2002
http://www.springer.de

Coverdesign: MEDIO AG, Berlin
Typesetting: Digital data supplied by author

SPIN: 10847323 62/3020 UW Printed on acid-free paper - 5 4 3 2 1 0 -

Deutsches Zentrum
für Luft- und Raumfahrt e.V.
(German Aerospace Center)

DLR

This documentation, in particular all previous editions, were made possible with the resources and support of the German Remote Sensing Data Center (DFD). My special thanks go to DLR/DFD for the opportunity to complete the 4th edition.

This book is again dedicated to my wife Mechtild
and to our daughters Monika, Ursula and Mechtild.

My family provided unequivocal support in my long-term book-writing efforts.
It is the reason why this edition could be completed.

Table of Contents

The sections P and Q are only part of the CD-ROM that comes along with the book. All information up to and including Appendix C are part of the 4th edition (a single volume) as well as part of the CD-ROM.

Part H Satellite Radionavigation Systems

H.1 Galileo

Since 1994 the European Union (EU) has been pursuing a strategy aimed at enabling Europe to the development of a Global Navigation Satellite System (GNSS). In early 1999 the EU proposed a strategy with the goal to design, implement and operate its own constellation of navigation satellites within a program by the name of Galileo (in honor of Galileo Galilei, Feb. 15 1564 - Jan. 8. 1642, Italian astronomer and physicist, founder of experimental physics and astronomy). On March 29, 1999, the EU Transport Ministers endorsed the proposed Galileo program at a meeting in Brussels. Major reasons for Europe's decision to build its own navigation system are:

- Current dependence on navigation systems of GPS and GLONASS that are run by military organizations without any means of international control.
- Europe wants its own civil-controlled navigation system for political and security reasons. To be in a position to compete for a fair share in a large global navigation and communication market. The commitment to build and operate the Galileo navigation system represents a strategic, economic, and technological venture for Europe.
- The requirement for safety-critical application services. Galileo should be able to provide a service with a certifiable performance level (which neither independent satellite navigation system can presently do) to support multimodal traffic, sufficient in particular in civil aviation , marine navigation and road transport.

Responsibility and funding for the definition phase of the Galileo System was given to the EC (European Commission) and to ESA. In this arrangement, the EC is in charge of the overall architecture, including the definition of end user requirements. ESA is responsible for the definition of the space segment and related ground segment required for the navigation satellites and their operation. The ESA program (space segment and related control ground segment) is referred to as **GalileoSat**.

The strategic goals of the overall program include:

- The implementation of a navigation satellite constellation and associated ground infrastructure, providing world-wide coverage, modernized-GPS compatible
- International cooperation, with US to assure system compatibility, and with the Russian Federation to benefit from the Russian industry know-how acquired in the framework of GLONASS
- An approach in phases (GNSS-1, GNSS-2) [1258) 1259)], including a definition phase until the end of 2000, a development/validation phase mainly financed through public funding (ESA, EC) to be completed by 2006, and an operational phase involving private funding, leading to full Galileo operational capability by 2008.

The main drivers for the Galileo baseline architecture definition are:

- Independence from other navigation systems
- Inter-operability with GPS and GLONASS
- Navigation (position, velocity) and timing services within the "global Earth coverage and space coverage"
- European integrity service for Galileo
- Independence of the integrity system from the ground control system

1258) GNSS-1 (Global Navigation Satellite System-1). The first generation GNSS-1 comprises the following elements: GPS, GLONASS and their augmentation systems [WAAS (Wide Area Augmentation System) of the US, EGNOS (European Geostationary Navigation Overlay System) of Europe, and MSAS (Multi-Transport Satellite Augmentation System) of Japan]. The three segments of GNSS-1 are expected to be operational by 2003.

1259) GNSS-2 (Global Navigation Satellite System-2). The second generation GNSS comprises all elements of GNSS-1 plus Galileo. GNSS-2 is planned to be operational by 2008.

- World-wide dissemination services for:
  - Search and Rescue (S&R) return link data
  - Two-way navigation-related data communication capability (relating to: weather data, alerts to traffic information and accident warnings, etc.)
- GPS+GLONASS integrity provided by the EGNOS MCCs (Mission Control Center)
- World-wide dissemination of additional Galileo/GPS+GLONASS integrity regional messages provided by NEIDS (Non-European Integrity Determination Systems).

On Dec. 21, 2000 the Transport Council made the decision for the Galileo development phase. The overall Galileo concept is conceived to be a cooperative Public-Private Partnership (PPP) program between industry and governments represented by the respective European institutions.

H.1.1 GalileoSat

According to current plans, in the operational phase the Galileo System will consist of at least 27 spacecraft in Medium Earth Orbit (MEO). First demonstration service is planned to start in 2006, and the system is expected to be fully operational by 2008. The overall architecture of Galileo is shown in Figure 221 (without the monitoring and control data flows).

The main design drivers (and some high-level system requirements) for the space segment are: [1260)]

SAS (Safety-of-life Access Service) with regard to:

- The satellite constellation design
- The constellation maintenance strategy
- Requirements for availability of any individual satellite

The satellite design is driven by:

- Use of two clock technologies: S-RAFS (Space-Rubidium Atomic Frequency Standard) and S-PHM (Space-Passive H Maser)
- Specified navigation signal frequency plan
- Provision of 6 CDMA (Code Division Multiple Access) integrity uplink channels per satellite
- HPA (High Power Amplifier) with 60 W RF output power (requirement)
- Time critical functions for dissemination of integrity data
- Capacity (data rate) for the dissemination of: a) navigation data, b) navigation-related data, c) S&R return link data, d) integrity data

Navigation Service	Navigation System Error (95%)		Mean Availability
	Vertical	Horizontal	
SAS (Safety-of-life Access Service) Dual frequency	7.5 m	4.5 m	99.6%
GAS (Governmental Access Service) Dual frequency	11.4 m	6.5 m	99.6%
OAS (Open Access Service) Single frequency	28.5 m	16.2 m	99.6%
OAS/CAS (Controlled Access Service) Dual frequency	7.2 m	4.2 m	99.6%

Table 390: Navigation system performance (reference receiver with a 10° masking angle)

Service classes provided:

1) A basic service, free of charge, for applications intended for the public at large. This service corresponds to **OAS** (Open Access Service) providing position, velocity and

1260) Information provided by Hans L. Trautenberg of Astrium GmbH, Munich (viewgraph package), representing the ESA baseline concept as of Feb. 2001 time. The OAS performance competes with GPS services. The S&R (Search & Rescue) service is part of OAS (in fact, S&R messages will be transmitted in all localization services).

2) A subscription service, with restricted access, for commercial and professional applications needing superior performance levels and a guarantee of service. This service corresponds to **CAS** (Controlled Access Service), this is also referred to as Commercial Access Service (it requires payment by the service user). CAS provides an improved service consisting of: a) accuracy and integrity service to professional markets, b) ranging and timing service to knowledgeable professionals (surveying, meteorological forecasting, time calibration, etc.).

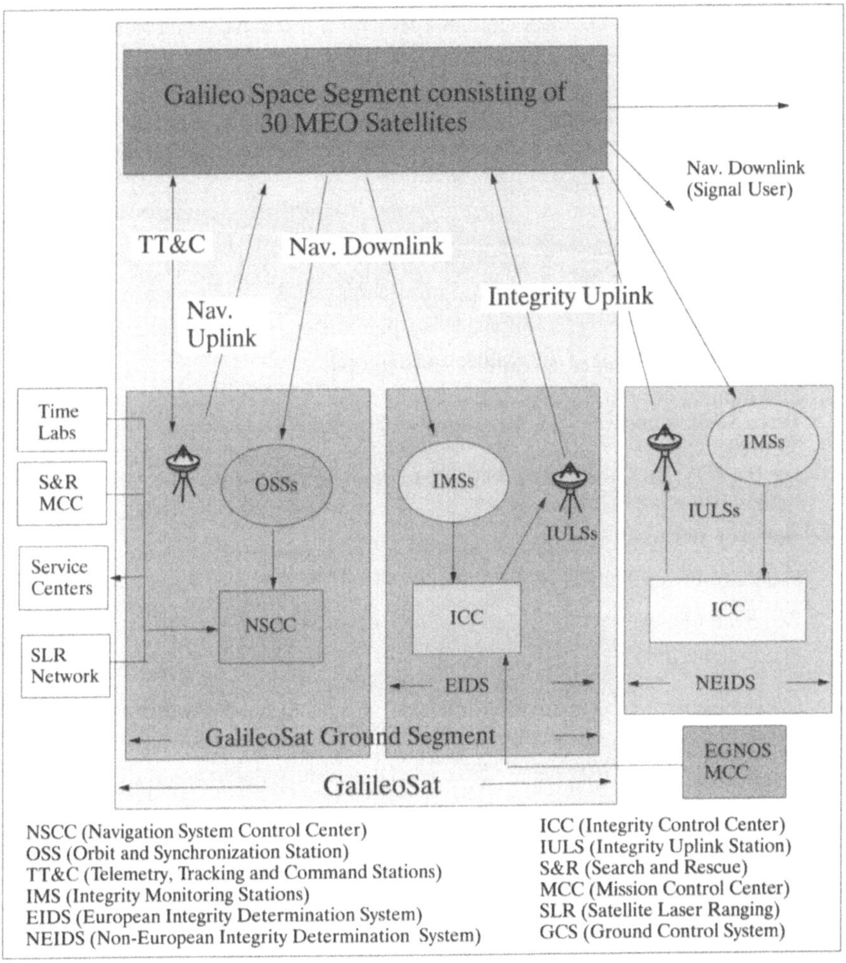

Figure 221: Overview of Galileo high-level architecture

3) A restricted, high-level subscription service for applications which must not suffer any interruption or disturbance for reasons of security. This corresponds to **SAS** (Safety-of-

life Access Service) and **GAS** (Governmental Access Service) providing a high integrity service for strategic applications. SAS will be used for all types of safety-relevant applications like: air-traffic assistance, railway and water traffic, network synchronization, power switching, etc. Parts of the SAS message may be encrypted to protect against spoofing.

Integrity SAS system performance

- Protection levels: horizontal = 12.1 m, vertical = 20.6 m
- Mean availability: 99.3%
- Continuity risk: 8×10^{-6} / 15 s [6×10^{-6} / 15 s for SIS (Signal-In-Space)]
- Integrity risk: 2×10^{-7} / 150 s (1.6×10^{-7} /150 s for SIS)
- Time-to-alert: 6 s (5.2 s for SIS)
- RAIM (Receiver Autonomous Integrity Monitoring)is used to cover local effects
- SISA (Signal-In-Space Accuracy)/SISA-IF is used to cover system contributions

Constellation baseline of 30 satellites in MEO

- 27/3/1 in Walker configuration (nominal state).
- Three S/C are reserved for in-orbit redundancy (quick recovery in case of failure)
- The average orbital altitude of the constellation is 23,616 km, inclination = 56°
- The orbital period is 5/3 revolutions/day.

Note: A Walker constellation [noted T/P/F (where T= number of satellites, P = number of orbital planes, and F characterizes the interplane phasing)] is a symmetrical constellation of T satellites; T/P satellites are evenly spaced on each orbital plane; and P orbital planes are evenly spaced through 360° of ascending node. In this case 27 satellites are in a symmetrical Walker configuration (in three orbital planes), plus one active in-orbit spare per plane.

Cost-effective deployment of the Galileo constellation

- Low number of orbital planes (3)
- Three Soyuz launches (2 S/C per launch)+ 3 Ariane-5 launches (8 S/C per launch) in 2006-2007
- For the IOV (In-Orbit Validation) phase, the two first Soyuz launches may be performed two years before.

Maintenance and spare strategy

- Maximize the probability of the nominal constellation state (27/3/1)
- Two main phases:
- Maintenance phase
  - In case of satellite failure: re-positioning of the in-orbit spare
  - In case of in-orbit spare use (or failure): launch of two S/C with Soyuz
- Renewal phase:
  - Satellite lifetime of 15 years in service; duration of 20 years
  - Anticipation of EOL failure by constellation renewal after about 10 years
- Evolution of nominal constellation state
  - Probability always >0.945

Constellation performance [1261)]

- Galileo stand-alone (elevation >10°)
- Nominal state:
  - Vertical accuracy better than 7.5 m (100% availability)
  - Horizontal accuracy better than 4.5 m (100% availability)

1261) G. Salgado, S. Abbondanza, R. Blondel, S. Lannelongue, "A New Model - Constellation Availability," Galileo's World, Spring 2001, pp. 30-35

- Degraded states:
  - In case of failures the availability still remains high >96-98%
- Galileo+GPS
  - Global improvement of performances and higher robustness to failures
  - Vertical accuracy <4 m with availability >99% over the major part of the world
  - Especially of interest for "urban canyon" environments (i.e.> 25°)
- Evolution of performances versus time
  - Orbital perturbations lead to drifts on S/C positions
  - Worst accuracy degradation kept below 5% with only 1 in-plane maneuver each 5 years

ODTS (Orbit Determination and Time Synchronization) requirements

- Galileo will be usable as a stand-alone system, independently from any other satellite navigation system
- The ODTS contribution to UERE (User Equivalent Range Error) is 65 cm (1 sigma) to meet the system performance requirements
- GalileoSat will establish its own reference system time, referred to as GST (Galileo System Time), and maintain is synchronized with TAI (Time Atomic International)
- Galileo will be synchronized with TAI within 50 ns (offset) on a long-term basis with an uncertainty of less than 33 ns.

ODTS basic principles

- ODTS is one common process to determine the orbit parameters of all satellites and all system clock comparisons (on-board and on-ground) with respect to GST
- Input is collected by a global network of stations performing one-way measurements (pseudo-ranges and carrier-phase observables)
- Processing is centralized and runs in the OSPF (Orbit and Synchronization Processing Facility)
- The ODTS algorithm is based on a dynamic least-squares batch processing offering good robustness and flexibility

ODTS operational scheme

- One-way measurements are collected every 10 s and transmitted with a maximum allowable network delay of 60 seconds
- The centralized processing facility, considering also the PTS's (Precise Time Station - a collection of cesium and hydrogen maser clocks) clock comparisons, computes:
  - The satellite Rubidium clock corrections of 100 minutes validity based on 160 minutes of measurement data
  - The S-PHM (Satellite-Passive H Maser) corrections of 8.3 hours validity based on 14 hours of measurement data
  - Satellite ephemeris of 12 hours validity based on 3 days of measurement data
  - GST using composite clocks
- The process runs periodically to maintain the UERE (User Equivalent Range Error) clock contribution within 1.5 ns (45 cm)
- Then the GST synchronization to TAI (Time Atomic International) is performed (on a daily period)
- All data are archived for offline trend analysis and maintenance

Space segment design features

- Constellation of 30 spacecraft including in-orbit spares
- MEO (Medium Earth Orbit) of 23,616 km with direct orbit injection as baseline
- The spacecraft transmit continuous ranging codes and navigation data
- Ground contact every 100 minutes for upload of timing parameters and navigation data and to perform S/C commanding and monitoring
- Integrity data are uplinked every second by dedicated stations; the data are passed via a separate on-board data path
- The spacecraft offer an autonomous capability in the event of loss of ground contact or other failures
- The spacecraft lifetime is 15 years, the system lifetime is 20 years.

Spacecraft design summary

- The baseline system is designed for the use of MEO spacecraft only (however, optional GEO spacecraft configurations have been studied)
- A common design basis for the MEO and GEO spacecraft has been adopted
- The design draws on experience of pervious programs (Globalstar) but makes use of the ESA technology programs for critical developments

MEO S/C parameters		GEO S/C parameters	
Mass	700 kg	Mass	1100 kg
Power	1500 W	Power	1700 W
Dimensions	2.7 m x 1.2 m x 1.1 m	Dimensions	2.8 m x 1.5 m x 1.2 m

Table 391: Satellite budget figures (as of 2001)

Platform structure	- Closed box with aluminum sandwich panels and two internal shear webs - All panels are primary structure providing I/F with appendages, equipment, etc. - Classical attachment of equipment/units with bolts screws and specific brackets - I/F between S/C and Ariane-5, Proton, and Soyuz launchers through a dispenser
Thermal control	- Mainly passive design supported by a heater system to maintain all equipment - Design based on use of standard technologies: heat pipes, radiators, MLI (Multi Layer Insulation), etc. - Monitoring of temperatures via ICDU (Integrated Control and Data Unit) - Specific accurate thermal control for payload equipment (e.g. clocks)
Platform propulsion	- Provides means to perform orbit corrections to: a) reach final orbit position, b) maintain pointing and position on station, c) achieve the graveyard orbit at EOL
Avionics and TT&C	- Avionics: a) centralized architecture, b) data transfer: standard serial bus and point-to-point serial link, c) AOCS: Earth pointing and yaw steering capability - TT&C: a) 2 transponders (TM/TC): 2 receivers in hot redundancy and 2 transmitters in cold redundancy, b) 2 integrity receivers in hot redundancy, c) integrity signal directly routed to the payload, d) 2 omnidirectional antennas
Platform power	- Standard silicon technology, 1560 W (EOL) of solar array - Lithium-Ion battery with 80 Ah capacity, 12 blocks in series (3400 Wh)
Navigation payload features	- Payload provides precise navigation services in L-band - Reception & storage of uplinked navigation data - Decoding & formatting of integrity data from the integrity link - Generation of ranging codes - Provision of messages with error correcting code - Assembly of navigation message in the agreed format - Broadcast of navigation, timing and integrity signals - Navigation payload mass = 80 kg, power = about 900 W
S&R payload	- Location of distress beacons (406 MHz legacy and new beacons) with acknowledgement to distress users through the Galileo navigation signal - S&R payload mass = 15 kg, power = 40-60 W

Table 392: Summary of MEO platform design

Spacecraft design overview

- Autonomy (data handling system + clocks)
- Centralized functions such as command/control, power distribution
- Standardized interfaces for ease of testing
- Separate payload for ease of AIT (Assembling, Integration and Test)
- Navigation service (messages, integrity, etc.) driven from ground to simplify S/C design and reduce outage risk
- Reliance on existing designs where possible but taking advantage of emerging technologies such as batteries, clocks, etc.

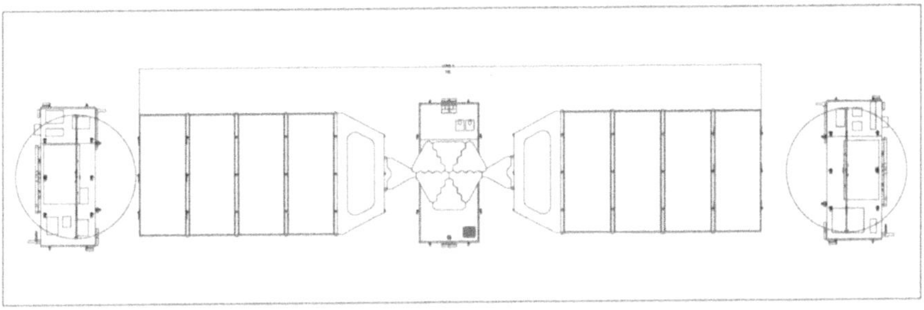

Figure 222: Schematic configuration of a Galileo spacecraft

H.1.2 System Integrity Concept

The concept is based on a split of integrity between the system and the user

- System provides timely warning if any errors are caused by either a) the satellite, b) the clock, c) the signal, and d) the navigation message - are larger than predicted via a combination of SISA (Signal-In-Space Accuracy) and the IF (Integrity Flag).
- The user has to assess all the hazards on the: a) signal propagation path, and b) all local effects - by using redundant signals and signal quality measures employing RAIM (Receiver Autonomous Integrity Monitoring).
- SISA (Signal-In-Space Accuracy) is a quantitative quality estimation of the orbit and clock prediction by the OSPF (Orbit and Synchronization Processing Facility) updated with every clock update.
- If an error occurs in the satellites, clocks, signal, navigation message or in the OSPF process, it has to be detected by the IPF (Integrity Processing Facility) and a warning (IF) has to be sent to the user.
- The check in the IPF has to be performed nearly instantaneously, with a sufficient number of IMS (Integrity Monitoring Station) reports for a significant statistical test; the option must also exist to exclude an IMS which exhibits a local signal disturbance.

User Integrity Algorithm

- Compute the protection level based on signals from satellites with non-set integrity flags
- If the protection level <alert limit - perform fault detection
- If no fault detected go to first step
- If fault detected, try fault identification

- If no fault isolation possible raise alert - otherwise after fault isolation go to first step
- This approach ensures maximum availability fo a navigation solution with integrity for users with a wide range of alert limits, if sufficient satellites are available for RAIM.

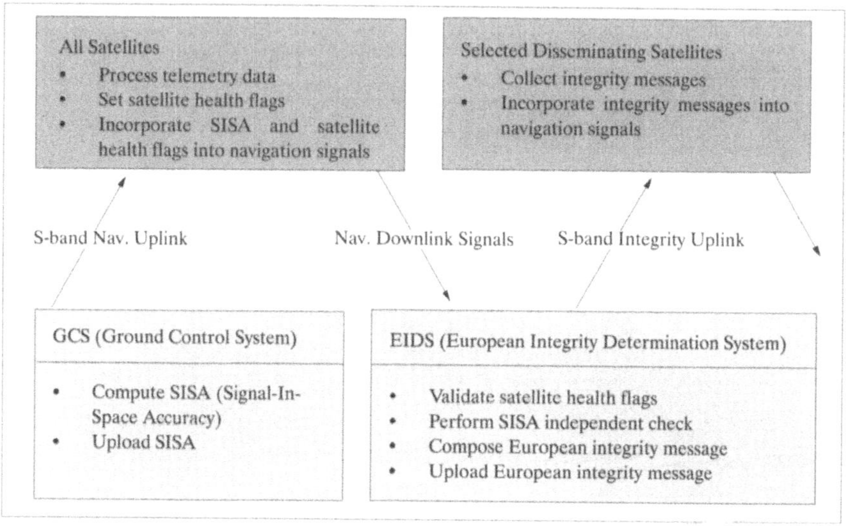

Figure 223: Schematic illustration of the system integrity concept

H.1.3 Signal Baseline

- European stand-alone concept
  - No overlay or bandwidth sharing with GPS and GLONASS
  - Low risk for frequency allocation process
  - Two scenarios considered: L-band only, and combined L- and C-bands
- L-band scenario: without use of C-band

Band	Center Freq. (MHz)	Band-width (MHz)	Service	Modula-tion Type	Subcarr-ier	Chip Rate (Mchip/s)	Power ratio	Data Rate (bit/s)	Symbol rate (S/s)
E5	1,202.025	24	OAS, CAS, SAS	QPN	I Q	15.345 15.345	50% 50%	~ 1000 (pilot)	2000 -
E6	1,278.750	40	GAS, CAS	BOC (10,5)	I Q	5.115 2.046	50% 50%	~ 1000 ~ 1000	2000 2000
E2	1,561.098	4.092	GAS	QPN	I Q	3.069 3.069	50% 50%	~ 1000 (pilot)	2000 -
E1	1,589.742	4.092	OAS, CAS, SAS	QPN	I Q	3.069 3.069	50% 50%	~ 1000 (pilot)	2000 -

Table 393: Summary of signal configuration

European WRC2000 (World Radiocommunication Conference 2000) scenario. Open, commercial and public services are allocated within E1, E2, E5, and E6 frequency bands.

H.2 GNSS-1 Augmentation Systems

The first generation Global Navigation Satellite System, GNSS-1, is an on-going project for satellite-based navigation, based on the US GPS and the Russian GLONASS satellite constellations. In addition to the primary constellations, there are regional augmentation systems (all in the development phase starting in 1998), such as WAAS (USA), EGNOS (Europe), and MSAS (Japan), providing enhanced performance services of integrity, availability, and continuity monitoring, to achieve the level of performance needed for civil aviation applications. The space segments of the three augmentation systems are also referred to by their generic name as: SBAS (Satellite-Based Augmentation System). - In addition to the three augmentation systems, Australia is considering the implementation of GRAS (Ground-based Regional Augmentation System) in support of its civil aviation needs. [1262)]

The augmentation systems are needed because the current constellations of GPS and GLONASS fail to meet the functional requirements critical to safety of flight. These are:

- **Accuracy.** It refers to the difference between the measured position at any given time to the actual or true position
- **Availability**. It refers to the ability of a system to be used for navigation whenever it is needed by the users, and its ability to provide that service throughout a flight operation
- **Integrity**. It refers to the ability of a system to provide timely warnings to users or to shut itself down when it should not be used for navigation.
- **Continuity of service.** It refers to the ability of a system to provide a service throughout any flight operation.

Obviously, there are also other areas of GPS signal applications outside of the "aviation safety regime" that need (or will need) the infrastructure and service spectrum provided by the augmentation systems.

H.2.1 WAAS (Wide Area Augmentation System)

WAAS is the US FAA (Federal Aviation Agency) program - to improve the accuracy, integrity, and availability of GPS signals for civil aviation to levels that support flight operations in the National Airspace System (NAS) from en route navigation through Category I precision approaches. IOC (Initial Operational Capability) of WAAS is planned for 1998. [1263)]

The overall WAAS architecture uses a network of ground-based DGPS stations [24 wide-area reference stations (WRS), two wide-area master stations (WMS), and two satellite uplinks] and several geosynchronous communications satellites to provide nationwide coverage. Differential corrections and integrity data derived from the ground-based network, as well as additional ranging data, are being broadcast to users from the geostationary satellites [Inmarsat-3 constellation, launch of Inmarsat-3 (F1, F2, F3, F4) in 1996 and 1997, Inmarsat-3-F5 launch on Feb. 5, 1998] using an L1-like signal. [1264)] [1265)]

Each WRS station in the network collects pseudorange and atmospheric measurements, and pools its data with a WMS (Wide-area Master Station), where estimates for individual corrections are computed for each of the major systematic error types (ionospheric delay, satellite clock and satellite position) for each GPS satellite in view. WAAS corrections are then relayed and broadcast from transponders of geostationary communications satellites to reach all users in a wide area.

The correction message update distinguishes between slow (ephemeris and long-term clock errors) and fast-changing corrections (SA clock errors). The slow correction update is no more often than every 5 minutes, while the faster corrections require a faster update rate. Any GPS receiver in the area of coverage may determine precise orbits from these

1262) G. K. Crosby, W. S. Ely, K. W. McPherson, et al., "A Ground-based Regional Augmentation System (GRAS) - The Australian Proposal," ION-2000, Salt Lake City, UT, Sept. 19-22, 2000, pp. 713-721

1263) R. Loh, V. Wullschleger, B. Elrod, M. Lage, F. Haas, "The US Wide-Area Augmentation System (WAAS)," Navigation ION, Vol. 42, No. 3, Fall 1995, pp, 435-465

1264) G. V. Kinal, O. Razumovsky, "Performance of the Inmarsat-3 Navigation Augmentation Payloads," Proceedings of ION GPS-97, Sept. 16-19, 1997, Kansas City, MO, pp. 1285-1294

1265) T. Walter, A. Hansen, J. Blanch, et al., "Robust Detection of Ionospheric Irregularities," ION-2000, Salt Lake City, UT, Sept. 19-22, 2000, pp. 209-218 corrections.[1266) 1267)] WAAS will improve the basic GPS accuracy to approximately 7 meters vertically and horizontally.

LAAS (Local Area Augmentation System)

The second augmentation to the GPS signal is LAAS (also developed by FAA), intended to complement the WAAS and function together to supply users of the US NAS (National Airspace System) with seamless satellite-based navigation for all phases of flight. LAAS applies to locations where the WAAS is unable to meet existing navigation and landing requirements (such as availability).

A primary design goal of WAAS and LAAS is to insure that signal-in-space failures are detected by ground facilities and affected measurements are excluded before differential corrections are broadcast to users. One such failure may be unintentional interference or intentional jamming in the GPS frequency band. To protect integrity, WAAS and LASS ground facilities must be able to quickly detect the presence of interference that fall within the restricted zone defined by WAAS and LAAS system requirements and thus may be hazardous to users. A system referred to as GIDL (Generalized Interference Detection and Localization System) provides these services. [1268) 1269)]

H.2.2 EGNOS (European Geostationary Navigation Overlay System)

EGNOS is the European regional augmentation system to GPS and GLONASS services. A consortium of ESA, EC (European Commission) of the EU, and Eurocontrol (European Organization for the Safety of Air Navigation, Brussels) are jointly developing EGNOS, Europe's regional augmentation system for satellite navigation. The objective is to provide GPS and/or GLONASS users with improved accuracy, integrity and availability, in particular to satisfy civil aviation requirements during all flight phases. Initial operational capability of EGNOS service is planned to start in early 2004 and, ultimately, expanding that system into Galileo, Europe's contribution to the next-generation GNSS-2 (Global Navigation Satellite System-2). [1270) 1271) 1272) 1273)]

EGNOS is considered an integral element of GNSS-1 (Global Navigation Satellite System-1) whose main service elements are the GPS and GLONASS constellations. As such, EGNOS must also be interoperable with WAAS-3 (Wide Area Augmentation System) of the FAA in USA, as well as with the Japanese MSAS-4 (Multi-Transport Satellite Augmentation System). MTSAT-1R (Multifunctional Transport Satellite) is the space segment of MSAS.

The EGNOS concept is designed to serve the needs of all modes of transport in the European Region ECAC (European Civil Aviation Conference). A total of 38 states belong to the ECAC service area. The built-in expansion capabilities of EGNOS allow the propagation of the EGNOS technology and service to other regions of the world (Africa, South America, and portions of Asia). EGNOS uses a ground infrastructure and geostationary

---

1266) J. Ceva et al., "Incorporation of Orbital Dynamics to Improve Wide-Area Differential GPS," Navigation ION, Vol. 44, No. 2, Summer 1997, pp. 171-213

1267) http://gps.faa.gov/Programs/WAAS/waas.htm

1268) K. Gromov, D. Akos, S. Pullen, P. Enge, B. Parkinson, "GIDL: Generalized Interference Detection and Localization System," ION 2000, Sept. 19-22, 2000, Salt Lake City, UT, pp. 447-457

1269) A. Manz, K. Shallberg, Peter Shloss, "Improving WAAS Receiver Radio Frequency Interference Rejection," ION 2000, Sept. 19-22, 2000, Salt Lake City, UT, pp. 471-479

1270) E. Copros, J. Spiller, T. Underwood, Ch. Vialet, "An Improved Space Segment for the End-State WAAS and EGNOS Final Operational Capability," Proceedings of ION GPS-96, Sept. 17-20, 1996, Kansas City, MO, pp.1119-1125

1271) S. Loddo, D. Flament, J. Benedicto, P. Michel, "EGNOS, the European Regional Augmentation to GPS and GLONASS," Proceedings of ION GPS-96, Sept. 17-20, 1996, Kansas City, MO, pp. 1143-1150

1272) J. Beale, P. Campagne, "European Commission Actions to Consolidate The European Contribution to a GNSS," Proceedings of ION GPS-96, Sept. 17-20,1996, Kansas City, MO, pp. 1467-1471

1273) http://www.esa.int/EGNOS/

(GEO) satellites to provide regional coverage with the essential functional augmentations (of accuracy, availability, integrity, and continuity of service) to the GPS and GLONASS systems as outlined under H.2.

The overall functions of EGNOS are: a) to collect raw data from GPS, GLONASS and GEO satellites, b) to compute the augmentation message, and c) to broadcast these messages (differential corrections, etc.) to users via the GEO satellites. In addition, EGNOS also disseminates accurate time to all users in the geostationary broadcast area, synchronized to UTC. EGNOS operates on the GPS L1 frequency; hence, it will be receivable with standard GPS receivers. [1274)]

The EGNOS system architecture is composed of the following elements:

1) The EGNOS space segment consists of navigation transponders, leased on-board two Inmarsat-3 satellites (AOR-E and IOR), which broadcast EGNOS signals to the user community. These satellites are in orbit and positioned at longitudes 65.5° East (IOR) and at 15.5° West (AOR-E). In addition, an EGNOS navigation payload is flown on ARTEMIS (Advanced Relay and Technology Mission Satellite) of ESA (15° W). EGNOS itself relies on the signals of the GPS an GLONASS constellations.

2) The EGNOS ground segment is composed of the following elements:
It consists of a number of RIMS (Ranging and Integrity Monitoring Station) which are connected to to a set of redundant control and processing facilities called MCCs (Mission Control Center). The system will deploy 34 RIMS and four MCCs, located at Torrejon (Spain), Gatwick (UK), Langen (Germany) and Ciampino (Italy). The MCC determines the integrity, pseudo-range differential corrections for each monitored satellite, and ionospheric delays and generates the GEO satellite ephemeris. The functions of RIMS are to perform one-way ranging of the GEO satellites (3). The RIMS data are transmitted to the MCCs for processing. The calculated positions are then transmitted in the form of messages to the EGNOS NLES (Navigation Land Earth Stations). The NLES uplink these GPS-like signals to the EGNOS transponders for broadcast.

- MCC (Mission Control Center), consisting of a CPF (Central Processing Facility) and a CCF (Central Control Facility)
- RIMS (Ranging and Integrity Monitoring Station). There are several RIMS in the system.
- NLES (Navigation Land Earth Stations). The system will deploy two NILES (one primary and one backup) per GEO navigation transponder as well as a NILES facility for test and validation purposes. The NILES are located in: Torrejon (Spain), Fucino (Italy), Aussaguel (France), Raisting (Germany), Goonhilly (UK), and Sintra (Portugal).
- EWAN (EGNOS Wide Area Network)
- EGNOS support facilities [PACF (Performance Assessment and System Checkout Facility), ASQF (Application Specific Qualification Facility), and DVP (Development Verification Platform)].

The functions of RIMS are to perform a) the acquisition of GPS and GLONASS raw data, b) acquisition of weather parameters. The RIMS data are then transmitted to the MCCs for processing (computation, distribution, validation and transmission of GEO ranging, Integrity (GIC) and Wide Area Differential (WAD) data). The calculated positions are then transmitted in the form of messages to the NLES. The NLES uplink these GPS-like signals to the EGNOS transponders (on GEO satellites) for broadcast. NLES also keeps accurate synchronization of GEO message transmission relative to GPS time.

3) The user segment consists of EGNOS standard receivers addressing applications in the fields of aeronautical, maritime, and land-mobile navigation.

1274) L Gauthier, P. Michel, J. Ventura-Traveset, J. Benedicto, "EGNOS: The First Step in Europe's Contribution to the Global Navigation Satellite System," ESA Bulletin, No. 105, Feb. 2001, pp. 35-42

The EGNOS space segment is being built by an industrial team led by ASI (Alcatel Space Industries) of France (participation of European and Canadian industry). The Earth stations of France Telecom in Aussaguel and of Deutsche Telecom at Raisting are being used as primary NLES.

The geostationary ranging service requires accurate positioning of the GEO satellites. The integrity monitoring service broadcasts range error estimates for each GPS, GLONASS or EGNOS navigation signal. The wide area differential position service broadcasts correction signals to improve the precision of satellite navigation.

EGNOS System Test Bed (ESTB):

In 1996 an ETS (Early Test System) study/campaign on behalf of ESA was started to investigate the conceptual prototype EGNOS performance at the user level in terms of accuracy and integrity as a functional end-to-end system.[1275)] This includes the following elements:

- Software emulation of RIMS with real data of GPS/GLONASS receivers
- Software simulation of different wide area differential algorithms at MCCs, to compute ephemeris and clock corrections and to generate the EGNOS navigation message.
- Analysis of latency effects
- Analysis of the effect of message error rates

ESTB became operational in January/February 2000. [1276)] A MOPS (Minimum Operational Performance Standard), compliant with SIS (Signal-In-Space), is available for testing through the Inmarsat AOR-E providing first continuous GPS augmentation services within Europe. It enables users to compute their positions to an accuracy of a few meters (3 m horizontal and 5 m vertical within 95% of the time).

To optimize the overall ESTB effort, existing assets have been taken into account in building the ESTB. These include the SAREF system of NMA (Norwegian Mapping Authority) and the EURIDIS ranging system of CNES. In early 2001, ESTB was connected with the Italian MTB (Mediterranean Test Bed), provided by ENAV (Italian Civil Aviation Authority).

The ESTB architecture is composed of: a) the space segment (comprising two Inmarsat transponders), and b) a ground segment (comprising a number of reference stations across Europe, a processing center and the Inmarsat uplink stations). The ESTB has already supported a number of application demonstrations during 2000. These included the landing of aircraft at several airports, the guiding of ships into harbors, and also the navigation of cars. ESTB will be operational on a 24-hour basis starting in early 2001. The service will embrace applications for service expansion (outside of Europe) and interoperability analysis with WAAS and MSAS. The EGNOS AOC (Advanced Operational Capability) Operational Readiness Review (ORR) is scheduled for December 2003, enabling operations to start in 2004.

H.2.3 MSAS (Multi-Transport Satellite Augmentation System)

The Japanese MSAS is implemented by JCAB (Japan Civil Aviation Bureau) and funded by the Japanese Ministry of Transport. The MSAS space segment is based on MTSAT (Multi-function Transport Satellite) with an aeronautical and a meteorological mission (see F.9 and a description of MSAS under F.9.1.1). MTSAT-1R is scheduled for launch in 2003.

The MSAS ground-based architecture is composed of the following elements: MRS (Monitor and Ranging Stations), GMS (Ground Monitor Stations), and MCS (Master Control

1275) J. Nieto, M. A. Molina, M. L. de Mateo, R. Roman, L. Andrada, "Assessment of EGNOS System and Performance: Early Test System," Proceedings of ION GPS-97, Sept. 16-19, 1997, Kansas City, MO, pp. 1345-1354

1276) A. Cruz, J. Cosmen, J. M. Legido, J. Caro, H. Secretan, N. Suard, "EGNOS System Test Bed: Achievements and Ongoing Upgrades," ION-2000, Salt Lake City, UT, Sept. 19-21, 2000, pp. 199-208

Stations). In addition, there are two GES (Ground Earth Station) in the system, namely the Kobe Aeronautical Satellite Center and the Hitachi-Ota Aeronautical Center), providing operational support.

The MSAS user segment consists of receivers addressing the applications in the fields of aeronautical, maritime, and land-mobile navigation. [1277)]

H.3 GLONASS

GLONASS [1278)] (Global Orbiting and Navigation Satellite System - Global'naya Navigatsionnaya Sputnikovaya Sistema). GLONASS is a Russian operational satellite navigation system in an experimental phase (designer: NPO-PM, Krasnoyarsk, Siberia; builder: Polet Production Association, Omsk). The system consists of a constellation of satellites capable of providing high-accuracy position information on a global scale for users on sea, land, or in space. Since September 1993 the GLONASS program is operated by VKS (Voenno Kosmicheski Sily), the Russian Military Space Forces, which has responsibility for deployment and on-orbit maintenance and certification of GLONASS user equipment.

The first GLONASS satellite series was launched on Oct. 12, 1982. Further launches were in 1983 (2), 1984 (2), 1985 (2), 1986 (1), 1987 (2), 1988 (1), 1994 (on Aug. 11). Each successful launch has always resulted in positioning of three satellites into orbit. As of November 20, 1995 there were 16 operational GLONASS satellites (full complement of 8 satellites in orbital plane 1)[1279)] in orbit. The GLONASS constellation will eventually be completed with a total of 24 satellites, and three additional spares. Full operational capability of the GLONASS system was expected by 1996, but so far this goal has not been achieved. In Aug. 2001 the Russian government approved funding to upgrade and complete its GLONASS program. [1280)] [1281)] [1282)]

Glonass No.	Kosmos No.	Slot/Nr of Frequency Channels	Launch Date	Service Introduction	Status ,or end of operation	Outage date
249	2111	05/19 to 23	08.12.1990	28.12.1990	withdrawn	15.08.96
750	2139	22/11	04.04.1991	28.04.1991	29.09.1994	14.11.1994
753	2140	21/20	04.04.1991	28.04.1991	06.01.1992	04.06.1993
754	2141	24/14	04.04.1991	04.05.1991	26.02.1992	16.06.1992
768	2177	03/22	30.01.1992	24.02.1992	09.01.1993	29.06.1993
769	2178	08/02	30.01.1992	22.02.1992	23.05.1997	24.06.1997
771	2179	01/17-23	30.01.1992	18.02.1992	25.10.1996	21.12.1996
756	2204	18-21/24	30.07.1992	19.08.1992	27.06.1997	05.08.1997
772	2205	21/08	30.07.1992	29.08.1992	29.06.1994	27.08.1994
774	2206	24/01	30.07.1992	25.08.1992	18.05.1996	26.08.1996
773	2234	02/05	17.02.1993	14.03.1993	09.03.1994	17.08.1994
759	2235	06/23	17.02.1992	25.08.1993	30.06.1997	05.08.1997
757	2236	03/12	17.02.1993	14.03.1993	27.07.1997	23.08.1997
758	2275	3/18	11.04.1994	04.09.1994	operational	
760	2276	3/17	11.04.1994	18.05.1994	30.07.1999	09.09.1999
761	2277	3/23	11.04.1994	16.05.1994	24.07.1997	29.08.1997
767	2287	12/22	11.08.1994	07.09.1994	05.11.1998	03.02.1999
770	2288	14/09	11.08.1994	04.09.1994	operating	
775	2289	16/22	11.08.1994	07.09.1994	operating	
762	2294	04/12	20.11.1994	11.12.1994	04.091999	19.11.1999
763	2295	03/21	20.11.1994	15.12.1994	27.07.1999	05.10.1999
764	2296	06/13	20.11.1994	16.12.1994	27.10.1999	30.11.1999

1277) M. Kawai, H. Nakao, K. Wakasa, "GPS/SBAS Receiver Flight Test in Japan," ION GPS 2000, Sept. 19-22, 2000, Slat Lake City, UT, pp. 266-276

1278) "Understanding Signals from GLONASS Navigation Satellites," International Journal of Satellite Communications', Vol. 7, 11-12, 1989, pp. 11-22

1279) "Russians Launch Trio of GLONASS Satellites," GPS World, January 1995, p. 15

1280) N. Yefimova, "Russia's GLONASS System Awaits Upgrade," Space News, Aug. 13, 2001, p. 24

1281) N. L. Johnson, "GLONASS Spacecraft," GPS World, Nov. 1994, pp. 51-58

1282) http://mx.iki.rssi.ru/SFCSIC/english.htm

Glonass No.	Kosmos No.	Slot/Nr of Frequency Channels	Launch Date	Service Introduction	Status ,or end of operation	Outage date
765	2307	20/01	07.03.1995	30.03.1995	10.09.1999	19.11.1999
766	2308	10/09	07.03.1995	05.04.1995	operating	
777	2309	19/03	07.03.1995	06.04.1995	17,07.1997	26.12.1997
780	2316	15/04	24.07.1995	26.08.1995	03.12.1998	06.04.1999
781	2317	10/09	24.07.1995	22.08.1995	operating	
785	2318	11/04	24.07.1995	22.08.1995	operating	
776	2323	09/06	14.12.1995	07.01.1996	operating	
778	2324	15/11	14.12.1995	26.04.1999	spare	29.01.01
782	2325	13/06	14.12.1995	18.01.1996	operating	
779	2364	01/02	30.12.1998	18.02.1999	operating	
784	2363	08/08	30.12.1998	29.01.1999	operating	
786	2362	07/07	30.12.1998	29.01.1999	operating	
783	2374	3/18	13.10.2000	05.01.2000	operating	
787	2375	3/17	13.10.2000	04.11.200021	operating	
788	2376	3/24	13.10.2000	04.11.2000	operating	

Table 394: GLONASS satellite constellations (May 2001)

A GLONASS S/C has a total mass of 1400 kg. The S/C consists primarily of a cylindrical pressure vessel equipped with a small aft payload platform and two large solar arrays (span of more than 7 m). The electrical power generating capacity is 1.6 kW. Each S/C is three-axis stabilized and equipped with a propulsion system for initial orbit acquisition and positioning.

Parameter	GLONASS	NAVSTAR/GPS
Orbital Parameters		
Period (minutes)	675.73 (~11.25 hours)	717.94 (~12 hours)
Inclination (degree)	64.8	55.0
Semimajor axis (km)	25,510	26,560
Orbital plane separation (degree)	120 (3 orbital planes)	60 (6 orbital planes)
Ground track repeat (orbits)	17	2
Ground track repeat (days)	8 sidereal days	1 sidereal day
Transmission Signal Parameters		
Signal separation technique	FDMA	CDMA
L-band carrier frequencies (MHz)	L1= 1602.5625-1615.5 L2= 1246.4375-1256.5	L1 = 1575.42 L2 = 1227.60
PRS clock rate (MHz) C/A code	0.511	1.023
(PRS=Pseudorandom Sequence) P code	5.11	10.23
PRS length (chips) C/A code	511	1.023
P code	5.11×10^6	6.187104×10^{12}
Navigation Message		
Superframe duration (minutes)	2.5	12.5
Superframe capacity (bits)	7,500	37,500
Word duration (seconds)	2.0	0.6
Word capacity (bits)	100	30
Number of words per frame	15	50
Satellite ephemeris specification	Geocentric Cartesian coordinates & their derivatives	Kepler elements and perturbation factors
Time reference	UTC (SU)	UTC (USNO)
Position reference	SGS 85	WGS 84

Table 395: Selected NAVSTAR/GLONASS parameters [1283]

Measurement principle: The user makes range and Doppler measurements to a number of navigation satellites within visibility. Position and velocity are part of the message content from the navigation satellite. Orbit: Each GLONASS spacecraft is in a near-circular orbit at an inclination of 64.8° and an altitude of about 19,100 km.

When fully deployed the GLONASS constellation is composed of 24 satellites in three orbital planes whose ascending nodes are 120° apart. Eight satellites are equally spaced in

1283) Y. Gouzhva, I. Koudryavtsev, V. Korniyenko, I. Pushkina, "GLONASS Receivers: An Outline," GPS World, January 1994, pp. 30-36 each plane with a 45° argument of latitude displacement. The planes themselves have a 15° argument of latitude displacement.

GLONASS[1284] and NAVSTAR-GPS have a many similarities with respect to: orbits, frequencies, and message formats. GLONASS employs an Earth-centered, Earth-fixed Cartesian coordinate frame of reference different from that used by GPS: GLONASS uses SGS 85 (Soviet Geodetic System); GPS uses WGS84 (World Geodetic System).

Note: GLONASS employed SGS 85 until 1993 when it switched to PZ-90 (also referred to as PE-90). Readers interested in the PE-90 coordinate system and its transformation to WGS84 should consult the MIT/LL GLONASS Group publication. [1285]

An improved version called GLONASS-M has been under development since 1990, the first satellite of this class is expected to be launched by 1996. This S/C offers improved timing and frequency accuracies as well as longer lifetimes (5 to 7 years).

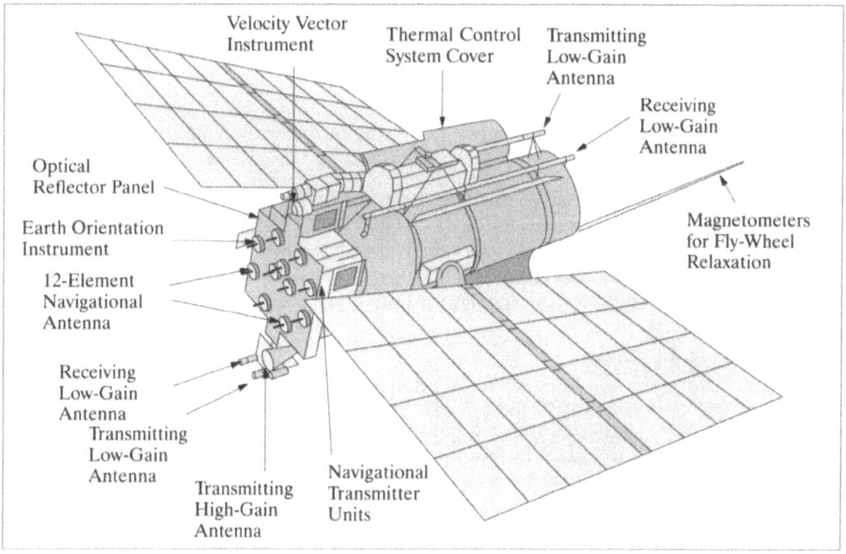

Figure 224: The GLONASS S/C model [1286]

GLONASS - GPS Differences:

- While GPS satellites operate with one L1 frequency for the entire constellation, GLONASS satellites each have their own discrete frequency. The frequency allocation scheme for GLONASS, which employs the Frequency Division Multiple Access (FDMA) technique, is **1602 + K(0.5625) MHz**, where K = the frequency channel number. Each GLONASS satellite is identified by a unique orbital slot number.
- The Russian government does not implement a policy of signal degradation on the L1 frequency. Hence, the stand alone accuracy of a GLONASS receiver is measurably better than that of a CA Code GPS receiver. L1 GLONASS rms accuracy is about 16 meters; L1 GPS is 100 meters rms with SA on.

1284) P. N. Misra, E. T. Bayliss, R. R. LaFrey, M. M. Pratt, R. A. Hogaboom, R. Muchnik, "GLONASS Performance in 1992: A Review," GPS World, May 1993, pp. 28-38

1285) P. N. Misra, et al., "Integrated use of GPS and GLONASS: Transformation between WGS 84 and PZ-90," Proceedings of ION GPS-96, The Institute of Navigation, pp. 307-314, 1996. (http://satnav.atc.ll.mit.edu/papers/PZ90-WGS84/PZ90-WGS84.html)

1286) Courtesy of A. Selivanov, ISDE and B. Zhukov, IKI, Moscow

H.4 GPS (NAVSTAR-GPS)

NAVSTAR [1287) 1288) 1289) 1290)] = NAVigation System with Time And Ranging. GPS = Global Positioning System. The NAVSTAR/GPS program of DoD was formally chartered in 1973 with the designation of the U.S. Air Force as Executive Agent and the establishment of an acquisition agency, the GPS Joint Program Office (JPO), [1291)] with representation from all the U.S. military services.

GPS is a space-based radio positioning/navigation system that provides three-dimensional position, velocity, and time information to suitably equipped users anywhere on or "near" the surface of the Earth. The total system consists of 21 operational satellites in near-circular orbit in six orbital planes, so that observers anywhere on the surface of the Earth have at least four satellites in view. In addition, there are three backup satellites in orbit. Initially the orbital inclination was defined to be 63°, but this figure has been changed to 55° to allow Shuttle launches. The orbital altitude is 20,200 km.

H.4.1 GPS Space Segment

The space segment consists of the GPS satellites with an operational constellation of 24 satellites: 21 operational satellites plus three active spares. The orbital parameters of GPS are given in Table 395.

As of 1998 there have been three distinct groups of GPS satellites so far, with one subgroup. The groups are designated as 'blocks.' The block-I satellites were used for system testing. Functional system capability was achieved with the block II series, featuring cesium clocks for timing, implementation of selective availability (SA), detection of error conditions, and radiation-hardened electronics for longer life times.

H.4.1.1 Block I Satellites

The block I prototype satellites were intended to validate the GPS concept; they were launched in the time frame 1978 to 1985 on Atlas E/F vehicles from VAFB, CA, with the corresponding designation of SVN (Satellite Vehicle Number) 1 through 11. GPS block I satellites were built by Rockwell International, Seal Beach, CA. They were able to sustain on-board operations for up to 3.5 days between navigation message uploads from the ground. Each S/C was capable of storing up to 14 days of navigation messages from the ground. However, the lack of an on-board momentum management required frequent updates from the ground to prevent the satellites from losing attitude determination capability. Some block I satellites operated for more than double their design life. [1292)]

The S/C is spin-stabilized (18 hydrazine thrusters in two modules); power:410W from two solar panels (total of 5 m^2) and 3 x 15 Ah NiCd batteries; mass=760 kg at launcher separation, 455 kg on-orbit; design life=5 years (propellant for 7 years). Atomic clocks: three rubidium, one caesium. Signal strength: +23.8 dBW for P signal, +26.8 dBW for C/A signal. The last remaining block I type satellite (SVN 10) was deactivated on Nov. 18, 1995. [1293)]

1287) B. W. Parkinson, J. J. Spilker Jr., P. Axelrad, P. Enge, "Global Positioning System: Theory and Applications, Vol. I and II, " AIAA, 1996

1288) "The NAVSTAR GPS System," AGARD Lecture Series No. 161, ISBN 92-835-04771, Sept. 1988

1289) "Understanding Signals from GLONASS Navigation Satellites," International Journal of Satellite Communications, Vol. 7 11-12, 1989, pp.11-22

1290) "Navstar," Jane's Spaceflight Directory 1988-89, 4th Edition, pp. 404-405

1291) B. W. Parkinson was the first director of JPO, located at SAMSO of the USAF in El Segundo, CA

1292) M. Shaw, P. Levin, J. Martel, "The DoD: Stewards of a Global Information Resource, the Navstar Global Positioning System," Proceedings of the IEEE, Vol. 87, No. 1, Jan. 1999, pp. 16-23

1293) Note: The block-I satellites were actually preceded by the NTS (Navigation Technology Satellite) experimental series. NTS-1 was launched on July 14, 1974 (the first satellite to fly atomic clocks: two rubidium oscillators) - NTS-2 was launched June 23, 1977 (first cesium clock in space).

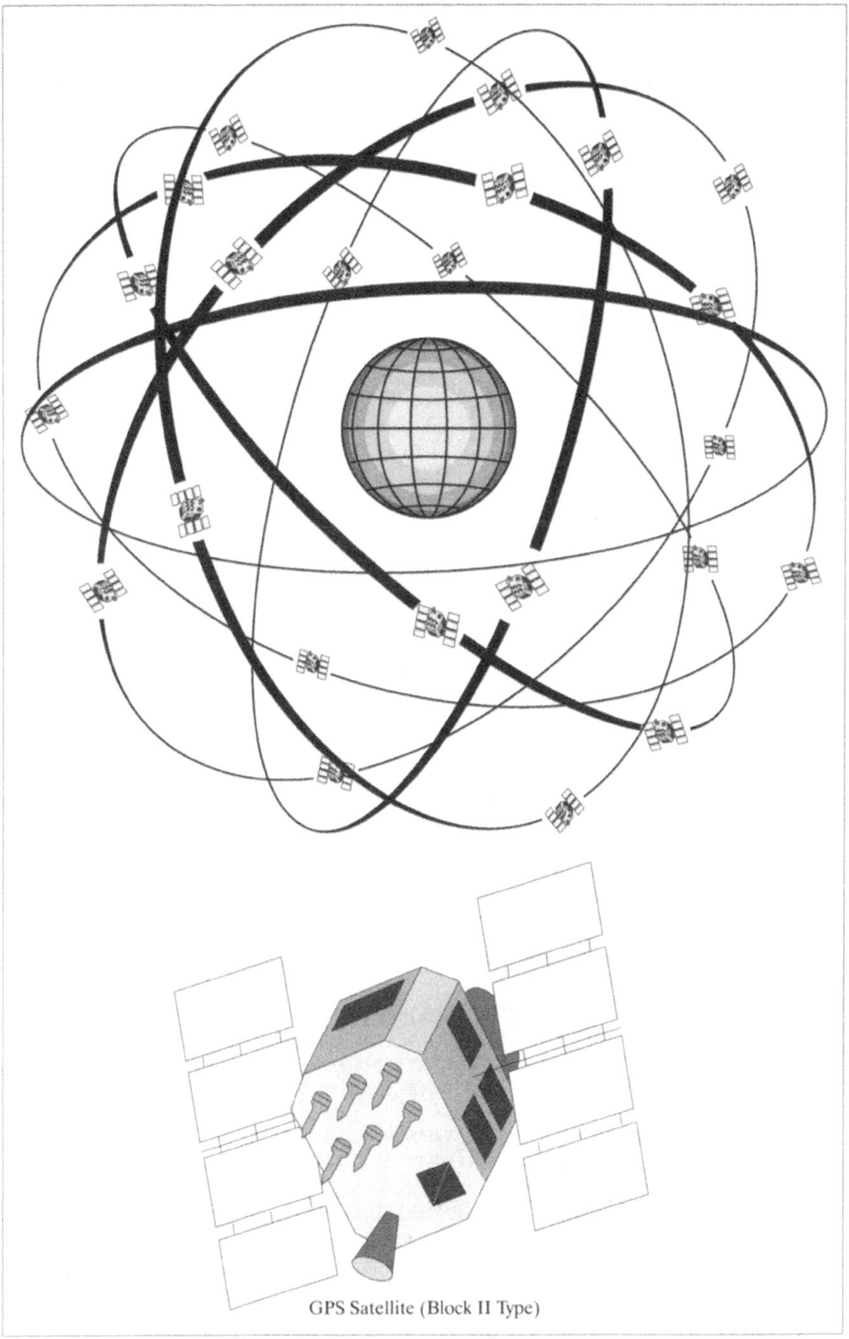

Figure 225: Global Positioning System (GPS)

SVN (Satellite Vehicle Number)	PRN (Pseudo Random Noise)	Launch Date	Date of Service Availability	Orbital Plane	Comment Block (NAVSTAR No.)	Deactivated or lost or failure
1	4	22. Feb. 78	29 Mar. 78		I	25 Jan. 80
2	7	13 May 78	14 July 78		I	30 July 80
3	6	6 Oct. 78	9 Nov. 78		I	25 Apr. 92
4	8	11 Dec. 78	8 Jan. 79		I	27 Oct. 86
5	5	9 Feb. 80	27 Feb. 80		I	28 Nov. 83
6	9	26 Apr. 80	16 May 80		I	10 Dec. 90
7		18 Dec. 80		launch failure		
8	11	14 Jul. 83		C3	I	04 May 93
9	13	13 Jun. 84		C1	I	28 Feb. 94
10	12	08 Sep. 84		A1	I	18 Nov. 95
11	3	09 Oct. 85		C4	I	27 Feb. 94
14	14	14 Feb. 89	15 Apr. 89	E1	II-1	
13	2	10 Jun. 89	11 Jul. 89	B3	II-2	
15	15	01 Oct 90	15 Oct. 90	D2	II-9	
16	16	18 Aug. 89	14 Oct. 90	E2	II-3	
17	17	11 Dec. 89	06 Jan. 90	D3	II-5	
18	18	24 Jan. 90	16 Feb. 90	F3	II-6	
19	19	21 Oct. 89	26 Nov. 89	A4	II-4	
20	20	24 Mar. 90	18 Apr. 90	B2	II-7	10 May 96
21	21	02 Aug. 90	22. Aug. 90	E2	II-8	
23	23	26 Nov. 90	10 Dec. 90	E4	IIA-10	
24	24	04 Jul. 91	30 Aug. 91	D1	IIA-11	
25	25	23 Feb. 92	24 Mar. 92	A2	IIA-12	
28	28	10 Apr. 92	25 Apr. 92	C2	IIA-13	
26	26	07 Jul. 92	23 Jul. 92	F2	IIA-14	
27	27	09 Sep. 92	30 Sep. 92	A3	IIA-15	
32	32	22 Nov. 92	11 Dec. 92	F1	IIA-16	
29	29	18 Dec. 92	05 Jan. 93	F4	IIA-17	
22	22	03 Feb. 93	04 Apr. 93	B1	IIA-18	
31	31	04 Apr. 93	13 Apr. 93	C3	IIA-19	
37	07	13 May 93	12 Jun. 93	C4	IIA-20	
39	09	26 Jun. 93	20 Jul. 93	A1	IIA-21	
35	05	30 Aug. 93	28 Sep. 93	B4	IIA-22	
34	04	26 Oct. 93	29 Nov. 93	D4	IIA-23	
36	06	10 Mar. 94	28 Mar. 94	C1	IIA-24	
33	03	28 Mar. 96	09 Apr. 96	C2	IIA-25	
40	10	16 Jul. 96	15 Aug. 96	E3	IIA-26	
30	30	12 Sep. 96	01 Oct. 96	B2	IIA-27	
42	12	17 Jan. 97	launch failure	F5	IIR-1	Booster
43	13	23 Jul. 97	31 Jan. 98	F5	IIR-2	
38	08	6 Nov. 97	18. Dec. 97	A3	IIA-28	
46	11	7 Oct. 99	3 Jan. 00	D2	IIR-3	
51	20	11 May 00	1 Jun. 00	E1	IIR-4	
44	28	16. Jul. 00	17. Aug. 00	B5	IIR-5	
41	14	10. Nov. 00	10. Dec. 00	F1	IIR-6	
54		30. Jan. 01	15. Feb. 01	E4	IIR-7	

Table 396: GPS launch dates and satellite constellations

Navigation payload: (on each block series)

The navigation payload is that part of the satellite which is responsible for the reception of data from the OCS (Operational Control System), intersatellite ranging (only on the block

IIR and block IIF versions) and the transmission of ranging codes and navigation data to the user community. The major elements of the navigation payload are shown in Figure 226. The frequency standards subsystem contains the atomic clocks. The code generator subsystem generates the C/A code and P(Y) codes for modulo-2 addition with the navigation message (sent to the L-band subsystem for transmission to the users). The processor memory also interfaces with the transmitter/receiver for intersatellite ranging (only on block IIR and IIF versions). [1294)]

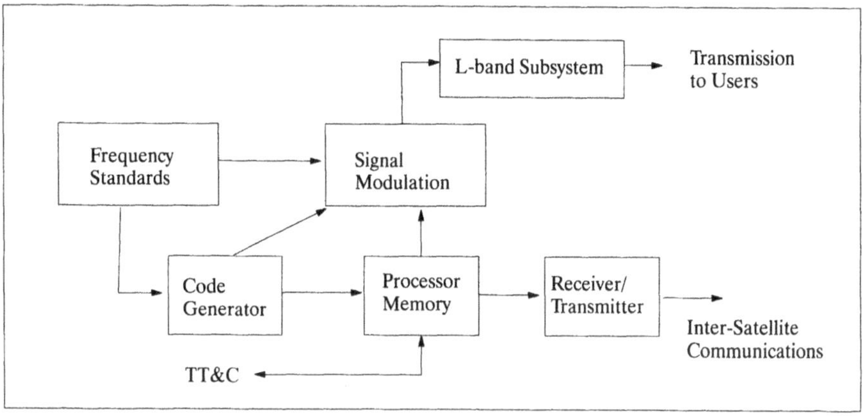

Figure 226: Schematic illustration of the navigation payload

The L-band subsystem contains the L1/L2 synthesizers and the antenna. The L-band antenna array is composed of twelve elements of hollow cylinders arranged in two concentric rings. The inner and outer rings are fed 180° out of phase with each other, with 90% of the power supplied by the outer array. This antenna configuration results in a "dimpled" pattern. The pattern is designed to supply even power across the face of the Earth.

All satellite signal radiation of the GPS constellation (at 20,000 km altitude) is directed toward Earth, providing coverage in particular for "Earth-bound" users, this includes also S/C in the envelope of LEO and MEO orbits. Naturally, coverage will decrease with altitudes approaching those of the GPS constellation and beyond (GEO, etc.).

H.4.1.2 Block II Satellites (NAVSTAR II-1 to II-8)

A total of nine satellites of this series were launched in the time frame April 1989 to Oct. 1990 on Delta 2 expandable launch vehicles from Cape Canaveral, FL. The satellites were built by Rockwell International. The S/C operations layout allows for minimal interaction from the ground without interruption of the signal broadcast. The incorporation of radiation-hardening capabilities is a significant improvement of the block-II design (protection against the harsh radiating environment). In addition, selective availability (SA) and anti-spoofing features are incorporated to enhance security. With no autonomous on-board momentum control, frequent ground contact is needed for for the spacecraft's momentum management.

The S/C structure is made of aluminum honeycomb panels, the total S/C span is 5.3 m with panels deployed. The S/C is spin-stabilized, with 60 kg hydrazine supply, sufficient for 10 years of normal operations. Power=700 W from two solar panels (total area of 7.26 m^2), and 3 x 35 Ah NiCD batteries; S/C mass=1667 kg at launch, 843 kg on-orbit; design life=7.5 years. Atomic clocks: two rubidium, two caesium. Signal strength: -160 dBW for C/A and -163 dBW for P on L_1; - 166 dBW for C/A or P on L_2, -165 dBW for C/A on L_3.

1294) L. F. Wiederholt, E. D. Kaplan, "Understanding GPS, Principles and Applications," Artech House Inc., Boston, 1996, Chapter 3

Sensor complement: (provided by Sandia National Laboratory of DOE)
Note: In addition to the GPS navigation payload, other instruments/experiments have been added to the GPS satellites over the course of time. Examples of the non-GPS-related instruments follow:

The block I satellites (starting with SVN 8) and the follow-up block II satellites incorporate devices by the name of **IONDS** (Integrated Operational Nuclear Detection) that allow monitoring of nuclear explosions either in the atmosphere or in space in support of the Nuclear Non-proliferation Treaty and the Atomic Test Ban (joint program of USAF and DOE). The information is routed via the ring-shaped UHF antennas from satellite to satellite within the GPS constellation. The instrument package includes X-ray sensors, background X-ray sensors, bhangmeters, and radio-wave detectors.

Nuclear explosions emit a very short X-ray flash, in the order of $< 1\ \mu s$. This flash can be seen by the X-ray detectors on several satellites. By measuring the time delay of arrival, the location of the explosion can be determined. The background X-ray sensors monitor the global X-ray environment. The bhangmeters are optical sensors (radiometers) trained of looking for the characteristic double flash of a nuclear explosion by an event trigger mechanism. - Nuclear explosions also emit radio waves which are referred to as EMP (Electromagnetic Pulse). The radio-wave detectors are able to record these EMPs.

H.4.1.3 Block IIA Satellites (NAVSTAR IIA-10 to IIA-27)

A total of 18 satellites of this series were launched in the time frame Dec. 1990 to Sept. 1996 on Delta 2 expandable launcher from Cape Canaveral, FL. The series was designed and built by Rockwell International. An enhanced survivability feature was introduced (storage capability of navigation messages and reference data for up to 180 days). In the event that ground stations are unable to upload navigation information, the satellites will continue to transmit the same navigation message for up to 180 days (graceful degradation). Normal operations were performed on a daily basis to account for orbital perturbations and to provide accurate ephemeris data. - The OCS is limited in its ability to forecast the satellite ephemeris and clock data because of the unpredictable nature of the orbital perturbations. Thus, the accuracy of the navigation message data degrades gracefully over time such that the URE (User Range Error) is bound by 10 km after 180 days. Typically, the URE is 5.5 m (1 sigma) based on fresh uploads of navigation message data on a daily basis. - An autonomous on-board momentum management capability was added to the series, reducing the ground contact requirements.

The S/C structure is made of aluminum honeycomb panels, the total S/C span is 5.3 m with panels deployed. The S/C is spin-stabilized, with 60 kg hydrazine supply, sufficient for 10 years of normal operations. Power=700W from two solar panels (total of 7.26 m^2), and 3 x 35 Ah NiCD batteries; S/C mass=1880 kg at launch, 930 kg on-orbit; design life=7.5 years. Atomic clocks: two rubidium, two caesium. Signal strength: -160 dBW for C/A and -163 dBW for P on L_1; - 166 dBW for C/A or P on L_2, -165 dBW for C/A on L_3.

H.4.1.4 Block IIR (Replacement Operational Satellites)

The satellite series (with 21 planned launches) is designed and built by Lockheed Martin (formerly GE Astro Space) of Valley Forge, PA; ITT Aerospace/Communications is providing the navigation payload. The contract was awarded in Dec. 1989. The first block IIR satellite experienced a launch failure (booster) on Jan. 17. 1997. The first successful launch of the series occurred on July 22, 1997 (the S/C was declared operational on Jan. 31, 1998 after successful on-orbit testing). The block IIR satellite series features enhanced autonomous satellite operations (AutoNav) for a period of up to six months without ground control corrections (reprogrammable processors that enable problem fixes, upgrades in flight, and increased satellite autonomy and radiation hardness). Orbital position is measured by UHF cross-link ranging between the individual block IIR satellites in the constellation. Orbit estimates (ephemeris and clock data) and updates are performed automatically. This AutoNav feature allows the system to maintain system accuracy for much longer periods between ground contacts. The atomic clocks, normally updated daily, offer five times the stability of their predecessors. [1295)]

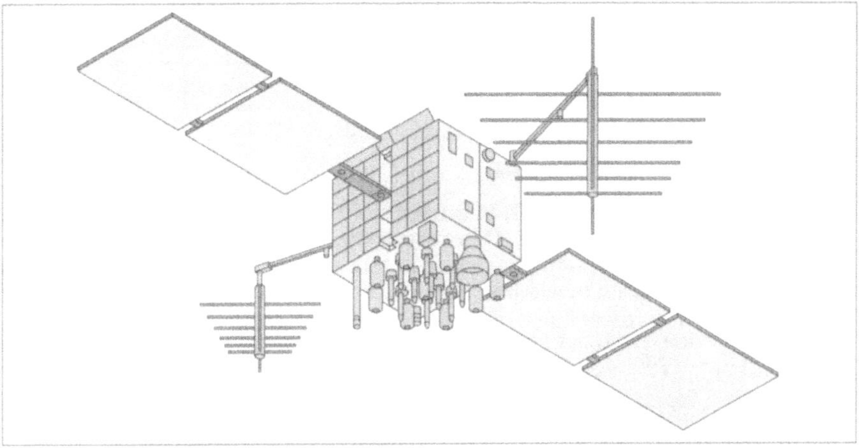

Figure 227: GPS Block IIR spacecraft

Each S/C is three-axis stabilized, using momentum wheel, magnetorquing, and 16 hydrazine thrusters (95 kg of total hydrazine); the payload is mounted on six honeycomb structural panels mounted on a central aluminum core; the thermal system: passive and heaters. Power=1136 W (EOL) from two solar panels with a total area of 13.4 m^2, NiCd batteries provide power during eclipses; S/C mass=2037 kg at launch, 1075 kg on-orbit; design life=10 years.

The new series satellites not only ensures a seamless continuance of service, but also introduces several new features when put into operation. One such feature is the GPS integrity channel (GIC). It is used to broadcast GPS and GLONASS S/C health information to aircraft users. GIC monitors the satellite signals and broadcasts "use/don't use" signals as well as coarse differential corrections to users. GIC uses geostationary satellites to warn aircraft worldwide in realtime (<10 seconds).

Upgrades to Block IIR spacecraft (for 12 out of 21 in the series): [1296)]

Modernized Block IIR (IIR-M) spacecraft feature enhancements in the following areas:

- Software enhancements. The five on-board processors provide a reprogramming capability. The SPU (Spacecraft Processing Unit) supports all bus-related activities such as TT&C attitude control and power management. The MDU (Mission Data Unit) performs functions related to navigation and sensing.
- Hardware enhancements. The upgrades are listed in Table 397. The new L-band HPA (High Power Amplifier) employ the GaAs and FET technologies. The new WGMIC (Waveform Generator/Modulator/IPA/Converter) is programmable using FPGA techniques. The WG supports all generation and routing of the baseband signals (C/A, P(Y), and M-codes). The WG employs a unique ITT proprietary implementation of Interplex modulation which enables the co-existence of three signal components (C/A, P9Y), and M) on a single constant envelope modulated carrier.

1295) S/C drawing courtesy of J. Keating, Lockheed Martin Astro Space, Valley Forge, PA

1296) T. Hartman, L. R. Boyd, D. Koster, J. A. Rajan, J. Harvey, "Modernizing the GPS Block IIR Spacecraft," ION GPS 2000, Sept. 19-22, 2000, Salt Lake City, UT, pp. 2115-2121

Component	Magnitude of change	Description of change
Antenna panel	Moderate redesign	Replace L-band elements with broadband proprietary elements
L-band subsystem	Technology upgrade	Replace five separate components with three (L1 and L2 HPAC and WGMIC) multifunction assemblies
L-band structural panel	New design	Higher power dissipation requires integral heat pies in the panel honeycomb structure
PCE (Payload Control Electronics)	Minor modification	Add power switching and fusing to accommodate additional power
PRU (POwer Regulation Unit)	Minor modification	Spare pins connected to additional power and return lines for higher power

Table 397: Block IIR-M hardware upgrades

- Signal enhancements. Block IIR-M spacecraft will support additional military-unique (M-code) signals on L1 and L2. Other capabilities include enhanced navigation software, C/A (Coarse/Acquisition) code on L2 and increased power on the present C/A and P(Y) code signals. In addition, the signal power level output of the Block IIR-M spacecraft can be redistributed on-orbit within C/A, P(Y), and M-code signal components as needed. The Block IIR-M navigation payload design provides the flexibility to support future GPS signal enhancements on-orbit.

Implementation time frame	Activity
May 2000	SA (Selective Availability) removed
2003 - 2006	GPS Block IIR-M enhancements - A civil C/A code is added on L2 - M-code (Military code) on L1 and L2
2005 - 2010	GPS Block IIF enhancements - C/A code on L2 - M-code on L1 and L2 - L5

Table 398: Overview of GPS signal modernization schedule

With the elimination of SA (May 2, 2000), the next largest contributor to the GPS positioning error is the signal delay caused by the Earth's atmosphere. Since the military has full access to two signals and frequencies through the PPS (Precise Positioning Service), military users can correct the ionospheric error by forming a linear combination of L1 and L2 pseudorange measurements to mathematically estimate and remove almost all the ionospheric bias from the L1 measurements. [1297)]
To compensate for the ionospheric error in limited civilian applications, some receiver manufacturers have developed innovative techniques for using components of the encrypted Y-code signal to calculate the ionospheric effects. However, to function effectively, these so-called "semi-codeless" receivers require an SNR for the L2 signal that is considerably higher than the SNR required by a military dual-frequency PPS receiver.

H.4.1.5 Block IIF Satellites

The 4th generation GPS satellites are referred to as block IIF. [1298)] [1299)] A USAF contract for 33 satellites was awarded to Rockwell International as prime contractor in April 1996 (as of Dec. 1996 part of Boeing North American Inc. of Seal Beach, CA, a subsidiary of The Boeing Company). Block IIF satellites will provide a second civil GPS frequency. First launches are planned to start in 2003 to replace the older block IIA and block IIR satellites.

1297) K. Sandhoo, D. Turner, M. Shaw, "Modernization of the Global Positioning System," ION-2000, Sept. 19-22, 2000, Salt Lake City, UT, pp. 2175-2183

1298) S. C. Fisher, K. Ghassemi, "GPS IIF - The Next Generation," Proceedings of the IEEE, Vol. 87, No. 1, Jan. 1999, pp. 24-47

1299) K. Ghassemi, S. C. Fisher, "Performance Projections of GPS IIF," Proceedings of ION GPS-97, Sept. 16-19, 1997, Kansas City, MO, pp. 407-415

The Block IIF system allows affordable technology insertion and block upgrades, while emphasizing compatibility and interoperability with the current space vehicles, ground control system, and user equipment.

Performance feature	Block II & IIA	Block IIR	Block IIF (requirements)
Nr. of navigation signals	3 signals	3 signals	up to 5 signals
Signal power to user (combined L1 and L2)	-158.2 dBW L1 -166 dBW L2	-158.2 dBW L1 -166 dBW L2	-156.4 L1 C/A, - 155.7 L1 P (dBW) -152.6 dBW L2
Space & ground URE	2-3 m	1-2 m (AutoNav)	0.75 - 1 m cross-link navigation
Civil performance (2 dRMS)	100 m with SA 25 m w/o SA	100 m with SA 25 m w/o SA	<10 m w/o SA (Selective Availability) and with ionospheric correction
Design life	7.5 years	10 years	15 years
Design flexibility	N/A	Auxiliary payload 45 kg, 90W	Auxiliary payload capacity up to 180 kg and 275 - 1000 W of power

Table 399: Performance parameter comparison of GPS satellite generations

The design features an independent WAAS data uplink and an independent cross-link to update WAAS data between all GPS satellites. Furthermore, the S/C series features advanced atomic clocks to greatly improve performance and give the control segment greater visibility into the health of the units. With four frequency standards composed of both cesium and rubidium technologies, these extremely accurate GPS atomic clocks must keep time to within 8 nanoseconds a day. The improvement involves converting the GPS cesium clocks from analog to digital - the first ever use of the commercially available digital cesium standard technology on spacecraft. The processor-controlled cesium clocks continuously optimize their performance by adjusting internal parameters and compensating for environmental effects. They also perform self-checkout diagnostics.

The block IIF satellite design is of block IIA heritage, consisting of a bus, subsystems, the navigation payload, and other auxiliary payloads. It features a modular design built around a composite "thrust cylinder." The orbital insertion subsystem, a solid apogee kick motor, is inserted into the thrust cylinder. "Intercostal" panels support the external structure, which in turn supports the electronics packages installed on two modular panels. Forward and aft bulkheads are supported by the trust cylinder and intercostals. The forward bulkhead supports the main L-band antenna array, the UHF crosslink array, the forward telemetry and commanding conical spiral and bicone antennas, and an array of four L-band antennas for auxiliary payload data transmission. The forward bulkhead also contains a combined (spinning and three-axis) Earth sensor and other auxiliary payload sensors.

S/C design life	15 years
L1 link L2 link L5 link (options for L5 civil signal)	1575.42 MHz, 29 dBW EIRP 1227.6 MHz, 29 dBW EIRP 1176.45 MHz, for ARNS (Aeronautical Radionavigation Services)
S/C operations	Autonomous operations for up to 60 days
S/C mass	2090 kg with growth potential for new payloads
S/C power	1560 W EOL with growth potential to 2.5 kW
S/C body dimensions	2.4 m x 1.75 m x 1.53 m
S/C data bus	1553B data bus for easy expendability
Uplink frequency, data rate (TT&C)	1783.74 MHz, 1 or 2 kbit/s
Downlink frequency, data rate (TT&C)	2227.5 MHz, up to 32 kbit/s

Table 400: Some S/C parameters of the block IIF series

Payload of Block IIF Satellites:

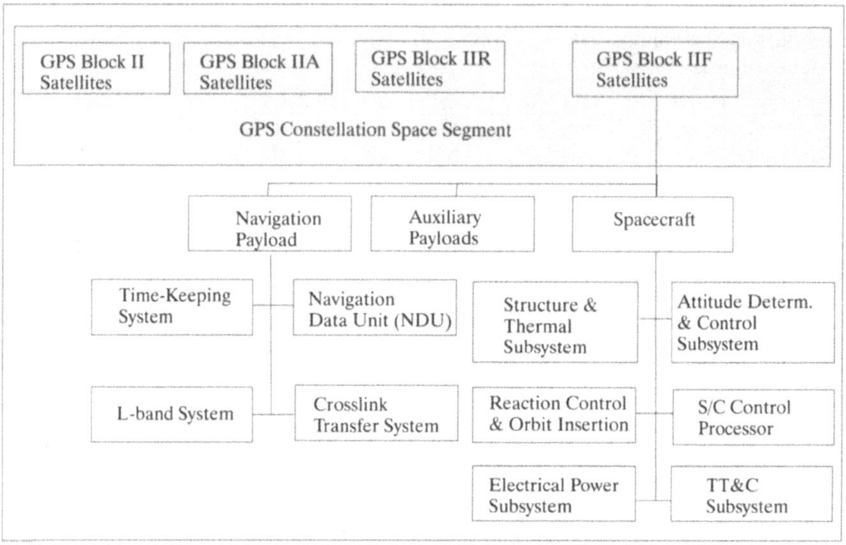

Figure 228: Configuration of the GPS block IIF S/C into the overall GPS constellation

H.4.2 GPS Control Segment

The control segment, called the Operational Control System (OCS), consists of five monitor stations located around the world [Colorado Springs, Hawaii and Kwajalein (Pacific), Ascension Island (Atlantic), and Diego Garcia (Indian Ocean)], and a master control station (MCS), located at Falcon AFB, near Colorado Springs (Note: in 1998, Falcon AFB was renamed to Shriever AFB). Operations are performed by the USAF/SC (Space Command). The monitor stations use a GPS receiver to passively track all satellites in view and thus accumulate ranging data from the satellite signals. The information from the monitor stations is processed at MCS to estimate satellite orbits (ephemeris) and to update the navigation message to each satellite. Updated information (including clock corrections) is transmitted to the satellites via ground antennas, which are also used for transmitting and receiving satellite control information. TT&C functions between MCS and the satellites are transmitted in S-band. [1300)]

H.4.3 GPS User Segment

The GPS user segment consists of the global user community with their GPS receivers. A receiver converts the satellite signals into position, velocity, and time estimates. Four satellites are required to compute position and time.

GPS/NAVSTAR is based on direct one-way range measurements with synchronized time references. The satellites share a common time system known as 'GPS time' and continuously transmit (broadcast) a precise time reference as a spread spectrum signal at two frequencies in L-band: L1 = 1575.42 MHz, L2 = 1227.6 MHz. Two spread spectrum codes are used: a civil coarse acquisition (C/A) code (1.023 MHz), and a precise (military) protected code (10.23 MHz), referred to as P-code (P). Both carriers, L1 and L2, are phase-modulated with the P-code. The L1 band is, in addition, phase-modulated by the C/A-code. Both signals, L1 and L2, are modulated with a PRN (Pseudo Random Noise) code - a repeated digi-

1300) Ch. Shank, J. W. Lavrakas, "Inside GPS: The Master Control Station," GPS World, September 1994, pp. 46-54 tal sequence of random bits, that permits a navigation user device to detect a start (i.e., "phase") of the repeated sequence. The PRN technique is also used for message security. - The GPS signals contain navigation data in the form of time of transmission and satellite position.

The accuracy of both codes is different. The receiver of the civil code cannot decode the military P code when the security status 'Selective Availability' in GPS satellites is turned on. [Note: With selective availability (SA) turned on, military users determine their location within 17.8 m, while civilian users determine their position within an accuracy of 100 m; hence, selective availability degrades the navigation information to all civil users]. In March 1990 SA was turned on by DoD, significantly degrading the accuracy of GPS for civilian users of the system.

In the meantime some civil users get centimeter-type accuracy through phase-tracking of the GPS carrier (even with the degraded GPS signals); this is also referred to as "realtime kinematic carrier-phase tracking (RTK)" or CDGPS (Carrier-phase Differential GPS).

Predictable Accuracy	Repeatable Accuracy	Relative Accuracy
PPS = Precise Positioning Service (reserved for military and selected user applications)		
Horizontal = 17.8 m Vertical = 27.8 m Time = 90 ns	Horizontal = 17.8 m Vertical = 27.7 m	Horizontal = 7.6 m Vertical = 11.7 m
SPS = Standard Positioning Service (available to general user community)		
Horizontal = 100 m Vertical = 156 m Time = 175 ns	Horizontal = 100 m Vertical = 156 m	Horizontal = 28.4 m Vertical = 44.5 m

Table 401: GPS accuracy characteristics

Measurement principle: The user makes a range measurement and a range delay measurement, resulting in a three-dimensional position (with four satellites this can be longitude, latitude, and elevation). The visibility of up to three satellites allows two-dimensional position determination (for instance longitude and latitude, which is sufficient for most Earth surface movement). GPS employs the following reference system: DoD World Geodetic System (WGS-84).

Initial Operating Capability (IOC) of the GPS system was declared in December 1993. On February 17, 1995, the FAA announces that GPS is now operational and is an integral part of US air traffic control system. The US Air Force Space Command declared the GPS system operational as of July 17, 1995 for the international user community. An anti-spoofing feature was enacted with the start of the operational period. This means that the P-code is henceforth encrypted (for the Block II configuration satellites) and referred to as Y-code. Any receivers not properly equipped cannot have access to the Y-code. Conventional receivers still have access to the coarse C/A code pseudo-range and L1 carrier phase, but not to the L2 pseudo-range, which is necessary for ionospheric corrections.

H.4.3.1 Fundamental GPS Observables

The full range of signals transmitted by the GPS satellites have expanded the capabilities for positioning moving vehicles, such as aircraft, to the sub-meter level. These applications have extended beyond the initial design of the system which was based principally on the decryption of the two codes. The three fundamental GPS observables used to precisely position a vehicle are: the pseudorange measurements, the carrier phase and the Doppler shift.

- The **pseudorange measurements** are the measured difference between the arrival time of a GPS signal (as measured on the receiver clock) and its transmission time (as determined by the satellite clock). By simultaneously observing four or more satellites, it is possible to determine the position of the receiver and correct for differences in time between the receiver's clock and the GPS satellite time system. Selective Availability (SA) and the inability for civilian users to access the P-code pseudorange measurements reduce the accuracy of GPS pseudorange positions. The ephemeris error and clock dithering introduced by SA reduces the pseudorange position accuracies to 60 m.

- **Carrier-phase measurements** (short for carrier-frequency phase measurements) are made by reconstructing the carrier signal. All carrier-phase tracking is differential, requiring both a reference and remote receiver tracking carrier phases at the same time (this use of GPS requires specially equipped carrier-tracking receivers). The carrier-phase process includes removing the bi-phase encoding and measuring the phase difference between the reconstructed carrier and a local oscillator within the receiver for both frequencies. As this carrier-beat phase rotates through cycles, the number of cycles is accumulated. The phase measurement is the accumulated phase change from the time the satellite is acquired by a receiver (locked on) until the receiver loses the signal (loss of lock). The carrier phase measurement has a noise level of a few millimeters. The critical parameters for carrier phase measurements are the resolution of the ambiguity and the necessity of maintaining continuous lock on the satellite signal. - Carrier-phase tracking of GPS signals has resulted in a revolution in land surveying and many other applications.

  In satellite formation flying applications, the CDGPS (Carrier-phase Differential GPS) technique of relative position measurements is used to provide a high-bandwidth, low-noise measurement of multiple vehicle states (relative position and attitude). [1301] Given GPS measurements at two nearby antennas, relative position between these antennas can be estimated to a high degree of accuracy based on tracking the relative phase. GPS transceivers mounted on the vehicles can allow relative positioning in such cases by augmenting the GPS satellite constellation, and can potentially act as a self-constellation when no GPS satellites are available. The same CDGPS technique can also be used for station keeping.

- The **Doppler shift** is most often determined from the time derivative of the carrier phase. The Doppler observable, which is rarely used for geodetic positioning, can be useful for connecting phase measurements across small gaps or cycle slips (<5 s) when the signal from a satellite is lost.

H.4.3.2 Availability of GPS/GLONASS Systems

The GPS system was offered for general civil use, worldwide - with no user fees attached for 10 years - by the FAA administrator James B. Busey at the Air Navigation Conference (ANC) in Montreal in Sept. 1991. This makes the GPS services available to anyone who can live with the 100 m accuracies of the SPS mode. The 10 year period starts at the beginning of 1995 when GPS is fully deployed (21 operational satellites plus three spares).[1302] GPS can provide a nearly 100% service availability (three-dimensional) with a 21 satellite constellation.

At the same conference, the Soviet Union (now Russia) offered its GLONASS service to the general worldwide civil user community for a period of 15 years after full deployment at no charge. The commitment to the SPS service accuracies of 100 m latitude/longitude and 150 m altitude were matched by Russia. Full GLONASS orbital capability of 24 satellites is expected to be achieved in 1996.

In the meantime the US Government has agreed to give 10 years notice should the US not be able to continue to meet the commitment. The PDD (Presidential Decision Directive) of

1301) F. H. Bauer, K. Hartman, J P. How, et al., "Enabling Spacecraft Formation Flying through Spaceborne GPS and Enhanced Automation Technologies," Proceedings of the ION-GPS Conference, Nashville TN, Sept. 15, 1999

1302) "GPS - the Next Generation," GPS World, Nov. Dec. 1991, pp. 12-16

March 29, 1996 required a joint military/civil task force, by DoD and DoT respectively, overseeing GPS policies and operations. The PDD also stated also that SA (Selective Availability) would be evaluated annually, starting in 2000. There was an underlying promise that SA would be turned off within a decade (i.e. in 2006). - The Interagency GPS Executive Board (IGEB) was set up as a consequence of PDD, jointly chaired by DoD and DoT, providing some formal civil agency participation in the GPS program.

On May 2, 2000 the SA (Selective Availability) feature was removed for the GPS constellation. The Presidential Directive (of US President Bill Clinton) permits civil users worldwide general access to the highest-possible accuracy of GPS signals. Elimination of SA represents a first step in GPS service modernization to the global user community.

The international community also realizes the need for an integrated long-range Global Navigation Satellite System (GNSS). GNSS has been proposed by the ICAO (International Civil Aviation Organization). GPS and GLONASS will certainly be key elements in this constellation.

In January 1999, US government officials, the Interagency GPS Executive Board, approved the frequencies for two additional civilian GPS signals. The second civilian signal is located at 1227.6 MHz, along with the existing coded military signal L2. It is going to be available starting with the launch of the Block-2F series in 2003. A third civilian signal, exclusively for civil aviation, is at 1176.45 MHz (L5) along with C/A code on L2 and M-code on L1 and L2.

H.4.3.3 GPS Applications

Both GPS and GLONASS provide basic services to large military and civil user communities by supplying instantaneous position, velocity, and time. These parameters are fundamental for navigation in general; they are also implicit to virtually all applications in the field of Earth observation and surveying. Current interest in GPS/GLONASS applications lies in the potential of the system's capabilities; the number of uses will be limited only by the imagination. Only civil applications are considered in this context.

Research Applications

GPS[1303] lends itself to many different scales of application, from the global level to the most discrete and site-specific. In between are regional and national projects. At the global level GPS applications are planned to help define the dimensions of the planet itself: creating geoidal models, establishing the terrestrial reference frame and Earth rotation parameters, mapping the dynamic topography of the oceans and monitoring plate tectonics. The list of ongoing and possible research applications of GPS services is long. Some examples are:

- Measurement of crustal motion in Central and South America (CASA).
- Tide gauge positioning along the Baltic Sea (unified height system to study the relative land uplift and sea level variations). Poland, Finland, Sweden, etc.
- Development of regional differential GPS (DGPS) systems to aid maritime vessel traffic services along coastlines and in harbors (US Coast Guard and several Scandinavian agencies).[1304]
- CNRS of France started a project using GPS positioning to help georeference early photos of glaciers in the French Alps. Growth and recession of ice sheets.
- NASDA is planning to integrate GPS into the navigation system of its H-II orbiting plane.
- Russian scientists use GPS for oceanic work, models for earthquake predictions, etc.
- A new earthquake forecasting research program in the USA is for instance based on GPS technology (1991). The proposed concept uses an array of GPS monitoring sta-

1303) Glen Gibbons, "What in the World!?!" GPS WORLD, April 1991, p. 21-24

1304) B. Tryggö, R. Bäckström, "Threading the Needle: Differential GPS on the Baltic Sea," in GPS World Sept. 1991, pp. 22-26 tions positioned along a fault; all of these stations are tied into a computer center which calculates Earth movements in the order of 1 cm.[1305)]

- SCIGN (Southern California Integrated GPS Network) is a collaborative project (NASA/JPL, USGS, UCSD, NSF, etc.) to install 250 permanent, continuously recording GPS receivers in and around the Los Angeles basin to provide accurate deformation information to improve estimates on earthquakes.[1306)]
- NASA/JPL with its extensive experience in interplanetary navigation and corresponding planet atmosphere interpretation is investigating its occultation techniques to be applied directly to GPS receiver measurements. This allows to determine the Earth's major atmospheric constituents and other parameters. See also GGI sensor definition under EOS sensors, and GPS/MET of Microlab-1.
- There are numerous projects and campaigns in solid Earth research. GIG '91 (GPS Experiment for the International Earth Rotation Service and Geodynamics).
- The Space Shuttle has flown GPS receivers many times.
- In the future GPS will be used for attitude determination on commercial satellites.
- GPS is being used as a precision timing reference in many applications, such as: synchronization of digital telecommunication networks, geodetic positioning, etc.
- There are also numerous questions that involve standards and procedures.
- etc.

Commercial Applications

The potential of GPS is finding its way into the commercial world, which it is just getting to know the technology and gearing up for new navigation and surveying products. The following applications could be supported:[1307)]

- Civilian air traffic control. GPS will serve as a guidance system for instrument landings. This can be very advantageous to all, especially to Third World Nations who have practically no guidance systems installed at all. The completion of the GPS and GLONASS constellations will be of utmost importance to the civil aviation community.
- Field demonstrations have been conducted of runway and approach guidance of aircraft with carrier-phase DGPS, and precision landing using a DGPS system combined with an inertial navigation system.
- Navigation on the oceans. Marine navigators are asking for a DGPS (see H.4.4 for DGPS) network around coast lines.
- Navigation systems for automobiles such as IVHS (Intelligent Vehicle/Highway Systems).
- Navigation systems for outdoors (tracking, boating, expeditions, etc.)
- Navigation systems for blind people!
- Commercial surveying. We are just beginning to explore the huge potential of the GPS surveying techniques in terms of accuracy, efficiency, and productivity. Example: an integrated GPS tachymeter system is being developed with an on-the-fly datalink and direct output to a Geographic Information System (GIS).
- There is seemingly not a single application in surveying and navigation that cannot be improved using GPS/GLONASS.
- etc.

The market offers a wide variety of GPS receivers/products to serve an increasingly diverse range of GPS applications. The list of manufacturers would be too long to be considered here. In general GPS applications technology is now both operationally capable and also becoming more affordable to the consumer market.

1305) "GPS is Newest Aid in Earthquake Forecasting," Space News, March 18-24 1991, pp. 22

1306) http://milhouse.jpl.nasa.gov/

1307) "Smart Policy: Make Best GPS Data Available to All," Space News, April 1-7 1991, pp. 15

H.4.3.4 Some GPS Orbit and Attitude Instruments

Basic concept of real-time 'interferometric' attitude measurements: An array of GPS antennas (e.g. four) are placed on the rigid structure of a spacecraft (fixed baseline configuration - a minimum of only two GPS satellites are required for an attitude fix). Software controlled multiplexing allows for signals being received at all antennas to reach a single receiver, or signals from different antennas are processed by different channels in the receiver. The carrier-phase differences allow the receiver to recover the angle between the antenna baseline and the incident signal direction. Additional on-board code computes attitude from successive carrier phase measurements for real-time determinations. - The accuracy of GPS receiver attitude determination approach is strongly dependent on antenna placement (baseline), the accuracy of the carrier phase measurements, and on-board data-processing techniques. For a 1 m antenna separation, the point solution accuracy is about 0.3°. A Kalman filter can be used to incorporate data about the vehicle dynamics to improve the attitude estimates. [1308) 1309) 1310)]

As of 1999/2000 the phase determination precision is typically 1 mm, so that with an antenna spacing of 1 m an angular accuracy of better than 0.1° can be obtained.

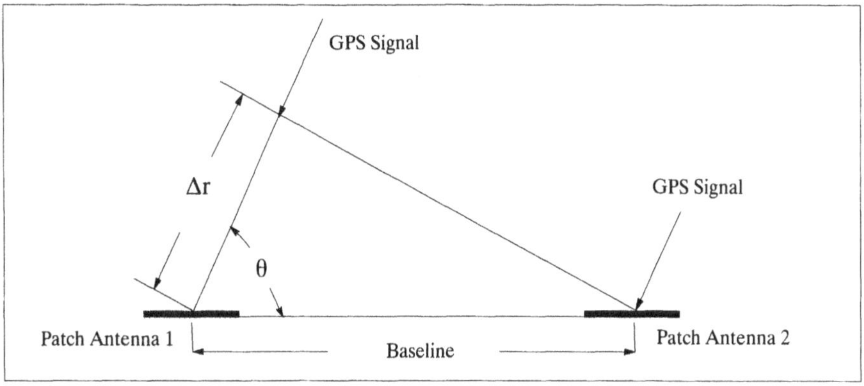

Figure 229: Schematic configuration for GPS interferometry

A great advantage of the GPS-based attitude determination method is its driftless source of attitude, it does not require any calibration. A divers set of applications is possible.

- **TANS** Vector [1311)] (Trimble Advanced Navigation Sensor Vector - of Trimble Navigation, Sunnyvale, CA). A C/A code, L1, SPS (Standard Positioning Service) receiver with real-time (10 Hz) attitude determination (azimuth, pitch and roll) capability; in addition it provides position, velocity, and time. The TANS Vector system comprises two major sub-assemblies: the RPU (Receiver Processor Unit) with bi-directional interface RS-422, and an array of four low-profile antennas. The nominal measured attitude accuracy is dependent on antenna separation and installation configuration. For a 1m baseline the attitude accuracy is 0.3° (rms), for a 3m baseline the accuracy is 0.10° (rms). The first spaceborne TANS Vector system was flown on the APEX mission of USAF (launch Aug. 3, 1994). The next mission was CRISTA-SPAS-1, which was deployed from the Shuttle Atlantis (SST 66, November 3-14, 1994); the use of the

1308) http://gauss.gge.unb.ca/grads/sunil/sgps.htm

1309) E. G. Lightsey, "Spacecraft Attitude Control Using GPS Carrier Phase," Chapter 16 of 'Global Positioning System: Theory and Applications, Vol. 2,' AIAA Volume 164

1310) C. E. Cohen, "Attitude Determination," Chapter 19 of 'Global Positioning System: Theory and Applications, Vol. 2,' AIAA Volume 164

1311) J. K. Brock, R. Fuller, et al., "GPS Attitude Determination and Navigation Flight Experiment," Proceedings of ION GPS-95, Sept. 12-15, 1995, Palm Springs, CA, Sept. 1995, pp. 545-554

TANS Vector receiver for a 24 hour period produced rms attitude variations around mean motion of 0.19° in roll, 0.15° in pitch and 0.26° in azimuth.

- **GPS Tensor** (a common product of Space Systems/Loral, Palo Alto, CA, and LABEN, Milan, Italy). A sophisticated receiver technology invented at Stanford University, Palo Alto, CA (C. E. Cohen), with licensed technology from Trimble Navigation and Stanford University. The Tensor receiver provides in real-time 13 output states of orbit/attitude data, enough for orbit and attitude determination of any space vehicle. The major instrument parameters provided are: roll, pitch, and yaw angles (and their rates), position, velocity, time **and filtered orbit determination** using a continuous-tracking, L1-, 9-channel C/A code receiver.[1312)] [1313)]

  The GPS Tensor instrument adds some advancements to Trimble's TANS Vector receiver. Like TANS Vector, it consists of two modules: the RPU (Receiver Processor Unit) with a dual, full duplex RS-422 interface, and an array of up to four low-profile and lightweight (0.2 kg) antennas. A special purpose ASIC was jointly designed by SS/L and Laben (a subsidiary of Alenia Spazio of Milan, Italy) to process signals of up to nine channels of GPS satellite data, including carrier phase. In addition, the ASIC measures the Doppler frequency of each GPS satellite in view and determines the code phase to form the pseudo-range measurements. An adapted Kalman filter smooths the position readings. A RAIM (Receiver Autonomous Integrity Monitoring) algorithm provides assurance for accuracy and reliability of the navigation system (autonomous detection of large range errors). - GPS Tensor was successfully test flown for the first time on the free-flyer payload ORFEUS-SPAS-2 (see Table 48).

- **GADACS** (GPS Attitude Determination and Control System). A NASA prototype instrument flown on the SPARTAN free-flyer payload of Shuttle flight STS-72 (Jan. 11-20, 1996). The prime objective is to demonstrate GPS attitude measurement technology in LEO. GADACS used two TANS Vector receivers (of Trimble Navigation, Sunnyvale, CA), along with an array of four L-band patch antennas mounted onto a special plate inside of SPARTAN, and DRIRU-II (Dry Rotor Inertial Reference Unit), a high precision gyroscope developed by Teledyne (for proof of GPS measured attitude). GPS data were recorded along with gyro measurements for verification and calibration of the GPS attitude measurements. [1314)] Note: GADACS was not able to perform attitude determination because it did not lock onto the four GPS signals necessary to obtain a position fix (a required first step to obtain attitude solutions. [1315)]

- **GANE** (GPS Attitude Navigation Experiment). A NASA prototype instrument (developed by: JSC, GSFC, and C. S. Draper Laboratory, Cambridge, MA) flown on Shuttle flight STS-77 (May 19-29, 1996) as a hitchhiker carrier payload on TEAMS (Technology Experiments Advancing Missions in Space). The prime objective was to mitigate the technology risk associated with the use of GPS to determine attitude, position, and velocity for ISS (International Space Station) and to evaluate the accuracy of GPS multipath models. GANE data were used to assess the performance of GPS-based navigation and attitude using an antenna configuration and orientation similar to that planned for ISS. The GANE hardware is comprised of a six-channel, four-antenna (micro pulse choke-ring antennas) system, mounted onto the platform (1.5 m x 3 m) in the cargo bay, and two Trimble TANS Vector receivers. Two instruments were available to calibrate the GPS attitude data: HAINS (High Accuracy Inertial Navigation System) of

1312) W. Johnson, "Attitude Adjustment, GPS Innovation keeps Satellites Oriented," Satellite Communications, June 1995, pp. 19-21

1313) R. Fuller, D. Hong, S. Hur-Diaz, J. Rodden, M. Tse, "GPS Tensor An Attitude and Orbit Determination System for Space," Proceedings of ION GPS-97, Sept. 16-19, 1997, Kansas City, MO, pp. 299-311

1314) F. Bauer, E. Lightsey, et al., "Pre-Flight Testing of the SPARTAN GADACS Experiment," Proceedings of ION GPS-94, Salt Lake City, pp. 1233-1241

1315) F. H. Bauer, J. R. O'Donnell, "Space-Based GPS 1996 Mission Overview," Proceedings of ION GPS-96, Sept. 17-20, 1996, Kansas City MO, pp. 1293-1302 the Shuttle which consists of a platform-type inertial measurement units, and a Honeywell ring-laser gyro used as the GANE IRU. - A total of four on-orbit operational test sequences were performed. The GANE requirements called for an attitude accuracy of 0.3° (3 sigma) and attitude rates to within 0.01°/s (3 sigma) per axis at 0.5 Hz. The post flight processing results provided accuracies of 0.1° in yaw and pitch and 0.2° in roll. [1316]

- **ARP** (ATV Rendezvous Pre-development). [338] ARP refers to ESA GPS relative navigation experiments flown on Shuttle flights STS-84 (May 15-24, 1997), STS-86 (Sept. 25 - Oct. 6, 1997) and on the MIR station, in preparation for the rendezvous maneuvers of ATV (Automated Transfer Vehicle) for ISS (International Space Station).
- **GADFLY** (GPS Attitude Determination Flyer). A NASA attitude experiment on the Lewis S/C (launch Aug. 23, 1997, the Lewis S/C could not be operated after launch and reentered the atmosphere on Sept. 28, 1997), consisting of the following hardware: four GPS antennas and pre-amplifiers, cross-strapped to two GPS Tensor (SS/L) receivers. The prime objective was attitude determination, a secondary objective was to demonstrate precise time distribution and to provide real-time navigation solutions to the S/C subsystems and other experiments. To improve the real-time navigation accuracy delivered by the GPS SPS service, the GPS Enhanced Orbit Determination Experiment (GEODE), a navigation filter, was developed for GADFLY. The GEODE algorithm executes a background Kalman filter and a foreground real-time state propagator, outputting position and velocity every second. [1317] [1318] GADFLY was also flown on the NASA SSTI-Clark spacecraft (1998).
- **SIGI** (Space Integrated GPS/INS). SIGI is a NASA instrument (Honeywell Space Systems as prime contractor) with the objective to provide attitude, position and time for ISS. The assembly consists of a GPS receiver and INS (Inertial Navigation System). The INS is a laser ring gyro package needed for sensing accelerations and blending data with GPS solutions to improve accuracy. The SIGI GPS receiver is a 12-channel, C/A code, avionics-grade instrument. Six channels are dedicated to C/A code tracking for navigation. The other six channels are multiplexed across four antenna inputs to provide differential carrier-phase measurements (attitude determination). SIGI combines the GPS attitude measurements in a dynamic filter with measurements from its RGA (Ring Gyro Assembly) to produce an estimate of ISS's attitude. - Russia has its own attitude determination system on its the Russian Service Module. It serves also as backup for SIGI. [1319] [1320] [1321]

H.4.3.5 IGS (International GPS Service for Geodynamics)

IGS is an international organization established in 1993 by IAG (International Association of Geodesy) with a central bureau at NASA/JPL for overall management and organization. The primary objective of IGS is to provide a service to support, through GPS products, geodetic and geophysical research activities (broad spectrum of operational activities performed by governmental organizations; services to develop standards and specifications, [and to provide Earth rotation parameters of high resolution for the International Terres-

1316) M. E. Lisano, J. R. Carpenter, S. Gomez, "Navigation, Attitude Determination, and Multipath Analysis Results from the STS-77 GPS Attitude and Navigation Experiment (GANE)," Navigation, Vol. 46, No. 3, Fall 1999, pp. 175-192

1317) R. C. Hart, K. R. Hartman, A. C. Long, T. Lee, D. H. Oza, "GPS Enhanced Orbit Determination Experiment (GEODE) on the SSTI Lewis Spacecraft," Proceedings of ION GPS-96, Sept. 17-20, 96, pp. 1303-1312

1318) J. R. O'Donnell, J. D. McCullough, E. G. Lightsey, R. G. Schnurr, L. Jackson, "Testing of GPS-Based Attitude Control Systems," Proceedings of ION GPS-96, Sept. 17-20, 1996, pp. 1063-1072

1319) E. G. Lightsey, G. C. Blackburn, J. E. Simpson, "Going Up: A GPS Receiver Adapts to Space," GPS World, Sept. 2000, pp. 30-34

1320) S. F. Gomez, "Attitude Determination and Attitude Dilution of Precision (ADOP) Results for International Space Station Global Positioning System (GPS) Receiver," Proceedings of ION, Sept. 19-22, 2000, pp. 1995-2002

1321) J. Simpson, C. Campbell, E. G. Lightsey, L. Jackson, "Testing Results of the X-38 Crew Return Vehicle GPS Receiver," Proceedings of ION, Sept. 19-22, 2000, Salt Lake City, UT, pp. 2038-2046 trial Reference Frame (ITRF) maintained by the International Earth Rotation Service (IERS), and to monitor the global deformations of the Earth's crust]. IGS collects, archives and distributes GPS observation data sets of sufficient accuracy to satisfy the objectives of a wide range of applications and experimentation. These data sets are used by the IGS to generate the following data products:[1322), 1323) 1324)]

- High accuracy GPS satellite ephemerides
- Earth rotation parameters
- Coordinates and velocities of the IGS tracking stations
- GPS satellite and tracking station clock information
- Ionospheric information

IGS accomplishes its mission through the following components:

- Networks of tracking stations. The networks consists of 30 - 40 core stations and 150 - 200 fiducial stations.
- Data centers. There are operational, regional, and global data centers. The operational data centers are in direct contact with the tracking sites. Their task is to reformat, archive, maintain and download their data to regional or global data centers. The regional data centers collect reformatted tracking data from several operational centers and transmit their data to global data centers. The global data centers are the main interfaces to the analysis centers and the user community.
- Analysis and associate analysis centers. Analysis centers generate IGS data products which are delivered to the global data centers

A number of campaigns (since 1992) by an IAG subgroup (G. Beutler, I. Mueller, etc.) have been conducted to test the feasibility of the program. Initial tracking service products (pilot program) are provided by the following data centers: JPL (central bureau), NASA's Crustal Dynamics Data Information System (CDDIS), Institut Geographique in France, and the Scripps Institution of Oceanography.

H.4.3.6 CIGNET

CIGNET = **C**ooperative **I**nternational **GPS** **Net**work of IAG (International Association of Geodesy). CIGNET continuously tracks the GPS satellites, providing a fiduciary data set that may be used for orbit computation and research. The data from all stations are sent weekly to the National Geodetic Survey (Rockville, MD.) where they are archived. The following SLR (Satellite Laser Ranging) ground stations are participating in the initial CIGNET program.[1325)]

- Kokee-Orbit, Hawaii, USA
- Mojave, California, USA
- Onsala, Sweden
- Richmond, Florida, USA
- Tromsö, Norway
- Tsukuba/Kashima, Japan
- Westford, Massachusetts, USA
- Wettzell, Germany
- Yellowknife, Canada

While the current CIGNET provides improved ephemeris data, the more expansive geodetic system would also provide positioning data. As for costs, operating a couple of com-

1322) "International GPS Services for Geodynamics," 1994 Annual Report, September 1, 1995, IGS Central Bureau, edited by J. F. Zumberge, R. Liu, and R. E. Neilan

1323) G. Beutler, E. Brockmann, "Proceedings of the International GPS Service for Geodynamics (IGS) Workshop," March 25-26, 1993, Astronomical Institute, University of Bern

1324) R. E. Neilan, J. F. Zumberge, G. Beutler, J. Kouba, "The International GPS Service: A Global Resource for GPS Applications and Research," Proceedings of ION GPS-97, Sept. 16-19, 1997, pp. 883-889

1325) CIGNET Report, CSTG Bulletin No. 11, Title: New Satellite Missions for Solid Earth Studies, June 1989, pp. 235-256 mercially available GPS/GLONASS receivers at a site that is part of a global CIGNET is orders of magnitude less expensive than building a new laser or VLBI facility.

Note: CIGNET was an important early activity coordinated by USGS for IAG's commission VIII. It become the core of the initial IGS network.

H.4.4 DGPS (Differential GPS)

A DGPS [1326] system employs a local reference station, which has a high-quality GPS receiver at a known, surveyed location. The reference station estimates the slowly varying error components of each GPS satellite range measurement and forms a (scalar) correction for each GPS satellite in view. This correction is broadcast to all DGPS users on a convenient communication link. With this concept higher position accuracies are achieved for users in the vicinity (typically up to 150 km) of a DGPS reference station. The differential correction greatly improves the accuracy for all users, regardless of whether selective availability (SA) is activated or not. This improvement arises because the largest GPS errors vary slowly with time, however, they are strongly correlated with distance.

A surveying application: By using GPS equipment on an aircraft – typically dual-frequency, 24 channel receivers with the antenna as close as possible to the camera lens itself – and a DGPS base station on the ground, the position of the plane itself can be recorded to an accuracy level of about two meters (this obviates the need to send in survey crews in advance to set visible ground markers).

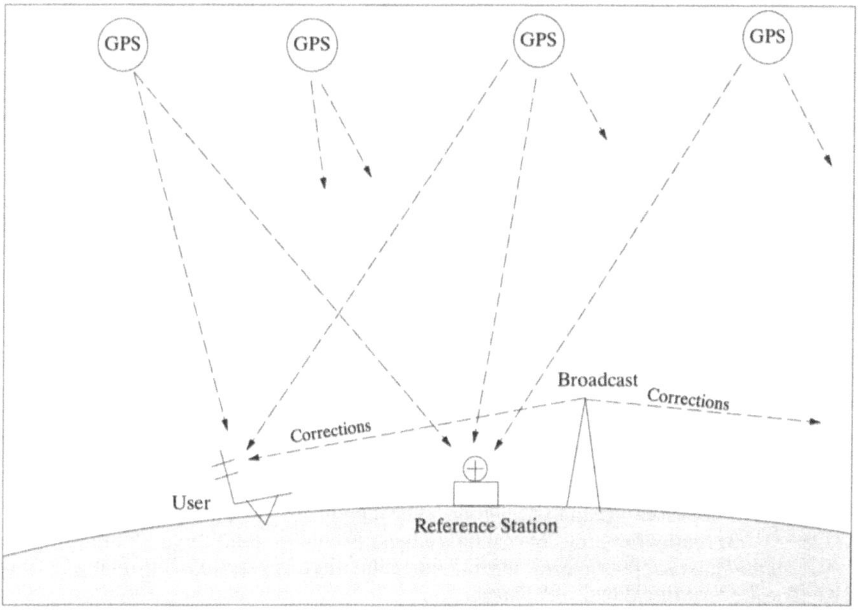

Figure 230: Differential operation of GPS with a reference station

In the USA there are plans to modify more than 200 radio beacons to enable broadcasts of DGPS signals along coastal regions. A full DGPS network is planned to be operational by 1997. The network will broadcast corrections to ranging data received from GPS satellites to improve vessel navigation up to a range of a few hundred km (from the reference station).

1326) P.K. Enge, R.M. Kalafus, M.F. Ruane, "Differential Operation of the Global Positioning System," IEEE Communications Magazine, July 1988, Vol. 26, No.7, pp. 48-59

Many radio beacon networks of other countries are also being upgraded for the broadcast of differential GPS corrections to marine users. The corrections will improve the accuracy of a mariner's GPS fix to <10 m, which allows harbor and harbor entrance navigation.

A DGPS positioning service was developed by Magnavox and CUE Network in the USA. This positioning service provides real-time, 1-5 m accuracy DGPS corrections, with broadcast over FM subcarriers from selected area radio stations.[1327)] The broadcast standard is referred to as RBDS (Radio Broadcast Data System); RBDS is compatible with the pager-based DGPS service, called ACC-Q-POINT (formerly known as: Pinpoint).
Another commercial service provider company 'Differential Corrections Inc. of Cupertino, CA. has plans to rapidly establish a worldwide network of FM RBDS stations.[1328)]

The concept of wide-area DGPS (WADGPS) networks of reference stations is used to form a vector correction for each satellite. These vector corrections are valid over much greater geographical areas, such as continents.

H.5 MTSAT (Multifunction Transport Satellite)

See the MTSAT description under F.9.

H.6 Transit - Navy Navigation Satellite System (NNSS)

A US Navy first generation satellite navigation system (the world's first), designed and built by JHU/APL - the entire development program (S/C hardware, ground support network, user equipment, etc.) was conducted within APL at Laurel, MD, RCA (later named: GE Astro Space, now Lockheed Martin) was the satellite integrator starting with Oscar-18 (Transit-O-18). The program was initiated in the spring of 1958, based on JHU/APL's Doppler tracking discoveries (the Sputnik orbit could be determined from RF Doppler data) and the realization of the "navigation problem" (it states: if the position of the satellite were accurately known, then Doppler data could tell an observer on the ground his unknown position). The Transit program was sponsored by ARPA. [1329)] [272)] [1330)]

Orbits: Transit satellite series orbits were generally at altitudes (apogees) of about 1100 km. The initial satellite launches (up to Transit-4B) were from Cape Canaveral, FL. All further launches took place from VAFB, CA, with polar orbit inclinations (Table 402).

The Space Segment: The first satellites (up to and including Transit-3B) were spherical in shape. Transit-1A was completed in seven months after initial program funding. Transit-4A and -4B were drumshaped (almost a cylindrical body with four solar panels) to provide more space for solar cells. In addition, operational frequencies of 150 and 400 MHz were used for the first time. The dual frequency corrected for ionospheric errors. On-board time keeping was done with cesium oscillators. Throughout the Transit system service period (1959 - 1996) four generations of cesium frequency standards and three generations of hydrogen masers were experienced, starting with a stability of 5 parts in 10^{10} to about 2 parts in 10^{15} for the best tuned hydrogen maser.

Transit-1A to -4B were regarded experimental satellites. The mass of each experimental satellite was about 136 kg. The TRAAC (Transit Research And Attitude Control) satellite, also built by APL and launched together with Transit-4B on an Ablestar vehicle (launch

1327) B. McGarigle, "'Top 40 Hydrography: Surveying with FM-based DGPS," GPS World April 1993 pp. 37-40
1328) "California-Based Firms Offer Highly Accurate GPS Services," Space News, Nov. 29-December 5, 1993, p. 7
1329) R. J. Danchik, "An Overview of Transit Development," Johns Hopkins APL Technical Digest, Vol. 19, No. 1, 1998, pp. 18-26
1330) W. H. Guier, G. C. Weiffenbach, "Genesis of Satellite Navigation," Johns Hopkins APL Technical Digest, Vol. 18, No. 2, 1997, pp. 178-181

Nov. 15, 1961), flew the first gravity-gradient boom in history. However, stabilization was not demonstrated since only about 1 m of the 18 m boom (total length) deployed. The TRAAC S/C lost signal on Aug. 12, 1963 due to the degradation of solar cells. [1331)]

The next series of Transit satellites (starting with -5A-1) were regarded as early operational prototype spacecraft. They were launched on a solid-fuel Scout vehicle from VAFB, CA, into polar orbits. The spacecraft mass of the first satellite, Transit-5A-1, was 55 kg. Three-5B series S/C were launched by Thor Ablestar vehicles, each with a piggyback 5E series S/C. The 5E S/C series had a solar power supply, while the 5B series S/C had nuclear power supplies. The Transit-5A series evolved into the Transit-5C-1 satellite with solar power supply.

The Transit Oscar series (Note: the Transit Oscar series should not be confused with the AMSAT Oscar series). The operational series of Transit satellites, more commonly referred to as "Oscar" series [or SOOS (Stacked Oscars on Scout)], closely followed the design of Transit-5C-1 with an important change: Hysteresis rods were installed on the solar panels to dampen the residual motion after the S/C despin maneuver in early orbit. The method of magnetic core storage (30 kbit) was used. Oscar-11 (Transit-O-11) was not used, it was later modified and launched as Transat on Oct. 27, 1977 from VAFB (Scout-D1 vehicle). The design life of the Oscar series was 3 years. Early Oscar series satellites had only short lifetimes. However, beginning with Transit-O-12 (Oscar-12), the satellites demonstrated an average orbital lifetime of more than 14 years. In fact, two S/C, Transit-O-13 and Transit-O-20, operated more than 21 years.

Transit Improvement Program (TIP). TIP was initiated in 1969 with the objective to provide a radiation-hardened satellite. Each TIP satellite was equipped with a minicomputer providing a memory of 64 kbyte. In addition the 150 and 400 MHz transmitters were redesigned, providing an output power of 3 W (150 MHz) and 5 W (400 MHz). A hydrazine thruster system was used to correct for orbital precession. - Each TIP satellite had the requirement to broadcast ephemerides for five days autonomously without input from the ground. This in turn required the development of a drag-free satellite series, referred to as "Triad" (Transit-improved DISCOS). [1332)] The DISCOS (Disturbance Compensating System) technology was spearheaded by Stanford University (the "ball-in-the-box" concept). Triad uses a three-body system [the actual satellite at bottom facing nadir, DISCOS at the center position, and a nuclear power supply (radioisotope thermal generator) at the top] connected by deployable booms. A major improvement in the TIP-satellite design was the introduction of the quadrifilar helix antenna. All major electronics subsystems were made redundant, radiation-hardened integrated circuits were interconnected on a ceramic substrate to obtain high-density packaging.

Triad-1 (launched Sept. 2, 1972) was the first satellite to fly a completely gravitational orbit, free from all surface forces such as drag and radiation pressure. The orbit could in fact be predicted for up to 60 days.

Triad-2 (TIP-2) and Triad-3 (TIP-3) satellites were each equipped with a redundant pulsed-plasma electric propulsion system (Isp = 225 kg s) used for drag compensation. The 1 kg Teflon, used for both thrusters, provided a fuel supply for 10 years.

Nova satellites. Three more spacecraft, nearly identical to the TIP series design, were built by RCA on Navy request. Improvements included: the addition of magnetic damping to the DISCOS and a stiffening to the boom assembly, reference clock, computer capacity, 6 m gravity boom, S/C mass was 136 kg. The NOVA satellites used the electric propulsion system (PPT system, Isp = 543 seconds, total impulse = 2450 Ns) of the TIP series satellites to

1331) Note: The very first Transit satellites transmitted signals at four frequencies: 54, 162, 216, and 324 MHz. The signals provided experimental data to evaluate ionospheric effects as a function of frequency. The final design is based on a two-frequency method for correcting ionospheric error.

1332) J. Dassoulas, "The TRIAD Spacecraft," Johns Hopkins APL Technical Digest, Vol. 12, No. 2, pp. 2-13, June 1973 compensate for drag with thruster firings. The NOVA satellites had operational lifetimes of 8 to 9 years. 1333) 1334)

The Control Segment:

Transit system operations and control are conducted at NAVSOC (Naval Satellite Operations Center, since 1962), Point Mugu, CA - working through tracking stations at Laguna Peak (near Point Mugu, CA), Prospect Harbor, ME, Rosemont, MN, Wahiawa, HI, Finegayan, Guam (since 1993) and at Falcon AFB, Colorado Springs, CO (since 1988). The tracking stations collected Doppler data and transmitted in the uplink the predicted ephemeris to each satellite. This data was stored on-board for continuous rebroadcasts to the user community. 1335) 1336)

When Transit navigation services were terminated/decommissioned on Dec. 31, 1996, NAVSOC retained other mission assignments, among them operations of GFO-1 and NIMS (Navy Ionospheric Monitoring System). On Jan. 1, 1997, the remaining Transit system constellation (of six Oscars in three orbital planes) became NIMS with a new application, namely to utilize the Transit system resources for computerized ionospheric tomography (CIT). In this setup, the NIMS satellites are being used as dual-frequency beacons by ground collection sites to determine the free electron profile of the ionosphere.

Satellite	Launch	Comment
Transit-1A	Sept. 17, 1959, Cape Canaveral (CC), Thor Able-2 launcher	Failed to achieve orbit (final stage failure)
Transit-1B	April 13, 1960, CC, Thor-Able-star, I= 66.7°, perigee = 626 km, apogee=1078 km	Operation for 89 days, first transmit on 2 frequency pairs to test ionospheric correction technique
Transit-2A Solrad-1	June 22, 1960, Thor Ablestar, I= 66.7°, pe.=626 km, ap.=1078	First dual-payload launch, Transit-2A was operational until Nov. 26, 1962
Transit-3A Solrad-2	Nov. 30, 1960, CC, Thor Ablestar	Failed to achieve orbit
Transit-3B Lofti-1	Feb. 22, 1961, CC, Thor Ablestar	First digital memory in space (384 bits of magnetic core shift register) to store the navigation message
Transit-4A Injun-1 Solrad-3	June 29, 1961, CC, Thor Ablestar	First triple-payload launch. First nuclear power supply tested (RTG). Memory storage capacity (magnetoresistive delay line) is increased to 2048 bits.
Transit-4B TRAAC	Nov. 15, 1961, CC Thor Ablestar I=32.4°, ap.=1100 km, pe.=951	TRAAC flew the first gravity-gradient boom, but only about 1 m of the 18 m long boom deployed
All further launches of Transit series satellites were conducted for VAFB, CA - requiring polar orbits		
Transit-5A-1	Dec. 18, 1962, VAFB, Scout	New technique to deploy solar panels
Transit-5A-2	April-5, 1963, VAFB, Scout	Failed to achieve orbit
Transit-5A-3	June 16, 1963, VAFB, Scout	A malfunction of memory occurred, hence it could not be used for navigation. Transit-5A-3 is the first satellite to achieve gravity-gradient stabilization.
Transit-5BN-1 Transit-5E-1	Sept. 28, 1963, VAFB, Thor Ablestar	Transit-5BN-1 was the first operational navigation satellite used by the Navy (service until Nov. 1964). Transit-5E-1 was operational for 11 years.
Transit-5BN-2 Transit-5E-2	Dec. 5, 1963, VAFB, Thor Ablestar	
Transit-5BN-3 Transit-5E-3	April 12, 1964, VAFB, Thor Ablestar	Failed to achieve orbit
Transit-5C-1	June 4, 1964, VAFB, Scout	Solar power supply

The Oscar (=Operational) satellite series.

1333) W. L. Ebert, S. J. Kowal, R. F. Sloan, "Operational NOVA Spacecraft Teflon Pulsed Plasma Thruster System," AIAA-89-2497, AIAA/ASME/SAE/ASEE 25th Joint Propulsion Conference, Monterey, CA, July 10-12, 1989

1334) Y. Brill, et al., "The Flight Application of a Pulsed Plasma Microthruster: the NOVA Satellite," AIAA-82-1956, 16th International Electric Propulsion Conference, Nov. 1982

1335) G. C. Kennedy, M. J. Crawford, "Innovations Derived from the Transit Program," Johns Hopkins APL Technical Digest, Volume 19, No. 1, 1998, pp. 27-35

1336) A. J. Tucker, "Computerized Ionospheric Tomography," John Hopkins APL Technical Digest, Vol. 19, No. 1, 1998, pp. 66-71

Satellite	Launch	Comment
Transit-O-1	Oct. 6, 1964, Thor Ablestar	Secondary payloads: Dragsphere-1 and -2
Transit-O-2	Dec. 13, 1964, Thor Ablestar	Secondary payload: Transit-5E-5
Transit-O-3	March 11, 1965, Thor Ablestar	Secondary payload: Secor-2
Transit-O-4	June 24, 1965, Thor Ablestar	
Transit-O-5	Aug. 13, 1965, Thor Ablestar	Sec. payload: Dodecapole-2, Temsat-1, Surcal-5, Long Rod-1, Calsphere-2
Transit-O-6	Dec. 22, 1965, Scout-A	
Transit-O-7	Jan. 28, 1966, Scout-A	
Transit-O-8	March 26, 1966	
Transit-O-9	May, 19, 1966, Scout-A	
Transit-O-10	Aug. 18, 1966, Scout-A	
Transit-O-12	Apr. 14, 1967, Scout-A	
Transit-O-13	Sept. 25, 1967, Scout-A	Transit-O-13 was declared non-operational Jan. 89 after 21 years and 8 months of continuous service.
Transit-O-14	May 18, 1967, Scout-A	
Transit-O-18	March 2, 1968, Scout-A	
Transit-O-19	Aug. 27, 1970, Scout-A	
Transit-O-20	Oct. 30, 1973, Scout-A1	
Transit-O-24 Transit-O-30	Aug. 2, 1985, Scout-G1	Transit-O-30 as secondary payload
Transit-O-27 Transit-O-29	Sept. 16, 1987, Scout-G1	Transit-O-29 as secondary payload
Transit-O-23 Transit-O-32	April 26, 1988, Scout-G1	Transit-O-32 as secondary payload
Transit-O-25 Transit-O-31	Aug. 25, 1988, Scout-G1	Transit-O-31 as secondary payload
Triad-1 (TIP)	Sept. 2, 1972, Scout-B1	Drag-free satellite technology was demonstrated
Triad-2 (TIP-2)	Oct. 11, 1975, Scout-D1	
Triad-3 (TIP-3)	Sept. 1, 1976, Scout-D1	
NOVA-1	May 15, 1981, Scout-G1	Orbit: 1170 km x 1187 km, inclination = 109°
NOVA-3	Oct. 12, 1984, Scout-G1	Orbit: 1150 km x 1198 km, inclination = 90.1°
NOVA-2	June 16, 1988, Scout-G1	Orbit: 1150 km x 1199 km, inclination = 90°

Table 402: Overview of Transit program satellites

The User Segment

The Transit navigation operates on the principle that a receiver's position can be determined on a single satellite pass by measuring the Doppler shift for a period of 10 to 15 minutes. A user's receiver on the ground measured the time history of the refraction-corrected Doppler data and recorded the orbital ephemeris as the satellite passed overhead. The user was able to calculate position initially within a few hundred meters, later within 15 to 25 meters. Some limitations: The user had to know his altitude and the satellite ephemeris; an error was introduced for an unknown vessel speed. A position fix with the Transit constellation was only available every 35 to 100 minutes. Navigation information could only be obtained by instruments of slow-moving users (like ships). For stationary users, such as surveying and oil platform location, integration of measurements from several passes yielded rms accuracies in the order of 5 m. Toward the end of Transit system operations in 1995, commercial receiver prices were as low as $1000. The 2-D system did not permit velocity determination.

In 1967 Transit navigation services were made available to the general public (commercial and private users). Commercial companies were permitted to manufacture and to sell low-cost receiving equipment. [1337] This action resulted in the use of more than 80,000 privately-owned Transit receivers (in particular in the field of commercial shipping). Also, oil-drilling platforms were among the first to use Transit to determine the boundaries of oil deposits.

During its 32 years of operation, the Transit Navigations System (or NNSS) provided accurate and reliable global navigation for the US Navy and for the civilian community. The system was continuously improved, in addition it contributed to many technology advances (see Table 42).

1337) L. J. Rueger, "Development of Receivers to Characterize Transit Time and Frequency Signals, John Hopkins APL Technical Digest, Vol. 19, No. 1, 1998, pp. 53-59

H.7 Summary of Microwave Tracking Systems

There are four spaceborne radiometric tracking systems in use addressing a variety of navigation requirements, among them precision orbit determination. These are: GPS, GLONASS, DORIS and PRARE.[1338]
While these systems differ in the source of the radio signals, they share common characteristics: they operate by determining the signal's "carrier" phase, its time derivative (i.e. Doppler shift) and/or its derivative with frequency (i.e. group delay and range).

GPS and GLONASS (one-way broadcast systems)

GPS and GLONASS are very similar general purpose positioning systems consisting of constellations of 21-24 satellites in roughly 12-hour orbits, continuously broadcasting dual frequency navigation signals. The carriers are modulated with pseudonoise ranging codes to permit both Doppler and one-way range measurements (also known as "pseudorange" measurements). The signals from each satellite blanket the Earth, extending roughly 3000 km beyond the Earth limb. At an altitude below 3000 km, any user of GPS and/or GLONASS will have four or more satellites (transmitters) in view. The combination of dual data types (phase and pseudorange) and continuous three-dimensional coverage makes it possible to employ a quasi-geometric positioning technique.

PRARE and DORIS

PRARE and DORIS are tracking systems requiring a host satellite (for the space segment instrument) and global ground-based tracking networks for precise orbit determination. Both systems (PRARE and DORIS) require an orbit determination process using conventional dynamic techniques (with physical models) whose accuracies depend on the quality of the models.

DORIS is a one-way Doppler system which broadcasts continuously at two frequencies: 401 and 2036 MHz (a two-way upgrade is planned). The DORIS instrument on the host satellite observes individual ground beacons in sequence and measures the Doppler frequency of the received signals.

PRARE is a two-way tracking system, broadcasting dual-frequency signals (2200 and 8500 MHz) from a transmitter on-board the host satellite to receiver-transponders on the ground. The signals are modulated by pseudonoise ranging codes to permit both range and range-rate measurements. The 8500 MHz signal is coherently transponded back to the host satellite (at 7200 MHz) for on-board range and range rate extraction. The 2200 MHz signal is received and tracked on the ground to provide an ionospheric correction.

H.7.1 DORIS Tracking System

See description of DORIS under E.21.1

H.7.2 PRARE Tracking System

PRARE (Precise Range And Range-Rate Equipment) is a microwave tracking system operating at centimeter accuracy levels for the measurement of satellite to ground range and range rate. The concept is based on an autonomous spaceborne two-way, dual frequency microwave tracking system with its own telemetry, telecommand, data storage, timing and data transmission capabilities. - PRARE is a DARA-funded project, defined/designed by

1338) See "Orbital Analysis" (Chapter 6.4, pp. 205-212) in 'The Interdisciplinary Role of Space Geodesy,' Springer Verlag, 1989,

Ph. Hartl (University of Stuttgart) and Ch. Reigber (GFZ Potsdam), and built by Kayser Threde of Munich, Time-Tech of Stuttgart, and Dornier of Friedrichshafen. [1339) 1340)]

PRARE flies on Meteor-3-7 (launch: Jan. 25, 1994) as a passenger payload; as such it is regarded a proof-of-concept mission for all PRARE subsystems. PRARE is also on ERS-2. Applications: PRARE precision orbit data sets can be used in the fields of geodesy, geophysics, and atmospheric sciences, and to supplement orbit-sensitive remote sensing systems for altimetry and SAR interferometry applications.

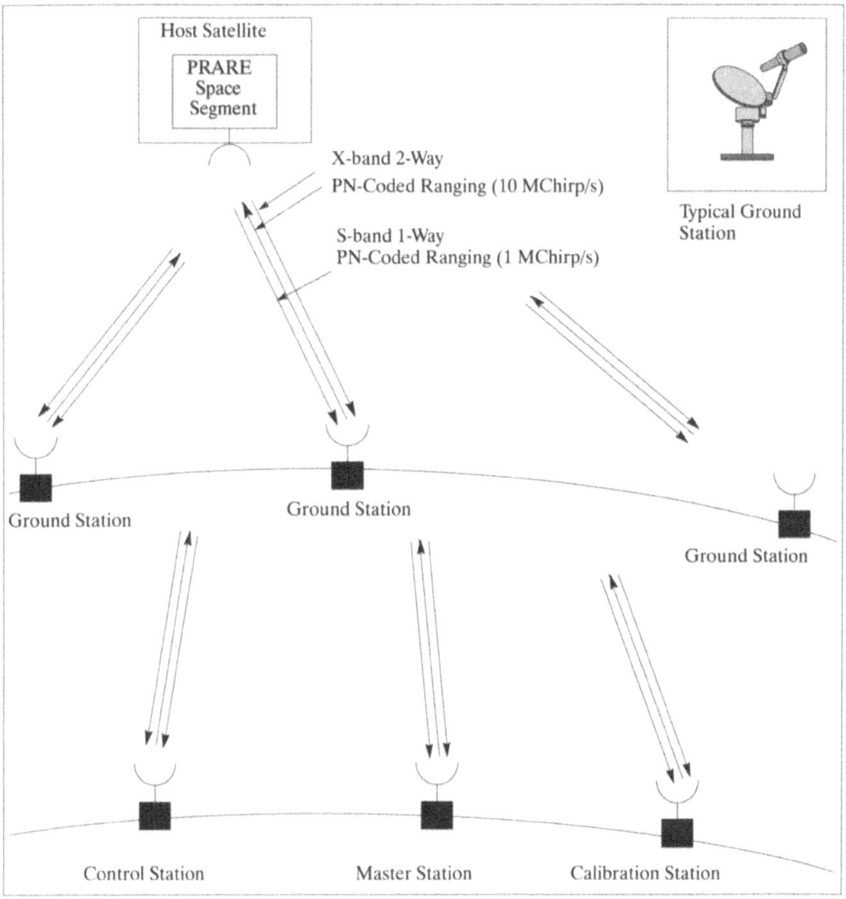

Figure 231: PRARE measurement principle with space and ground segments

PRARE consists of three major components:

1) Space Segment
This is a small self-contained hardware unit with a mass of 19 kg, power consumption of

1339) "The Precise Range and Range Rate Equipment PRARE: Status Report on System Development, Preparations for ERS-1 and Future Plans," Submitted by F. Flechtner, K. Kaniuth, Ch. Reigber, H. Wilmes of DGFI, Second International Symposium on Precise Positioning with the Global Positioning System (GPS '90), Sept. '90, Ottawa

1340) P. Hartl, C. Reigber "Das PRARE-System der ERS-1 Mission," Die Geowissenschaften, 9. Jahrgang, Heft 4-5, April-Mai 1991, pp. 156-162.

39 VA (operational), or 8 VA in standby mode, and communication links independent of the host satellite.

2) Ground Station Network
The network consists of small automated and transportable ground stations (whose positions are known or can be determined from PRARE data to high precision). At X-band a ground station works as a regenerative and coherent transponder, at S-band a station acts as a receiver for the transmitted signals and measures the difference of the time of arrival of both signals.

3) Control Segment (in the case of Meteor-3-7 it consists of:)
- a command station
- a master station (with a time control unit and a preprocessing computer in connection with a user center, eg., DFD). The master station may also command the space segment.
- a station for calibration in collocation with a third-generation laser tracking system.

PRARE Measurement Principle

Two continuous signals are emitted from the space segment to the ground, one of which is in S-band (2.2 GHz), the other in X-band. Both signals are modulated with a PN-code (pseudo random noise, 1 MChirp/s for the S-band and 10 MChirp/s for the X-band) used for the distance measurement and containing data signals (broadcast information) for ground station operation (prediction of orbit visibility, etc.). The time delay in the reception of the two simultaneously emitted signals is measured at the ground stations with an accuracy of better than 1 ns; the result is retransmitted and stored in the on-board memory of PRARE for later ionospheric correction of the data.

In the ground station the received X-band signal is transposed to 7.2 GHz, coherently modulated with the regenerated PN code (or with one of three orthogonal copies for code multiplexing) and including the two-frequency delay data as well as housekeeping and meteorological ground data is retransmitted to the PRARE space segment.

PRARE Control Segment

The PRARE control segment consists of a command station in Stuttgart, a master station at DLR/DFD in Oberpfaffenhofen, and a calibration station in Wettzell.

- Command Station. Controls eventualities and contingency situations. The functions are:
  - Monitoring and commanding the PRARE space segment
  - Simulation capabilities
  - Establishing spacecraft contact in contingency situations during visibility periods.
- Master Station. Functions as a central receiving and preprocessing station for all PRARE tracking data from the global network.
  - Planning and scheduling all tracking activities
  - Updating orbit prediction elements (uplink)
  - Referencing of PRARE space clock time to UTC time standard.
  - Processing of PRARE tracking data, and quality control.
- Calibration Station. Provides the capability to calibrate PRARE in flight with a precise ranging standard (laser station).

PRARE Ground Stations

A PRARE ground station is mobile and consists of the following components:
  - an antenna unit with an offset antenna of 60 cm diameter, fronted electronics and tracking system

- an electronics unit including RF-modules, station processor and power supply
- a monitor and test computer (PC) as user interface.

Organizations/Institutions owning/operating a PRARE ground station:

- ASI, Agenzia Spaziale Italiana, Italy (2)
- AWI, Alfred-Wegener-Institut, Germany (1)
- CSR, Center for Space Research, USA (3)
- DGFI, Deutsches Geodätisches Forschungsinstitut, Germany (1)
- DLR, Deutsche Forschungsanstalt für Luft- und Raumfahrt, Germany (1)
- DASA-Dornier, Germany (1)
- DUT, Delft University of Technology, The Netherlands (1)
- GFZ, GeoForschungsZentrum Potsdam, Germany (9)
- IfAG, Institut für Angewandte Geodäsie, Germany (2)
- INS, Institut für Navigation, University of Stuttgart, Germany (1)
- LAPAN, Lembaga Penerbangan dan Antariksa Nasional - Indonesian National Institute of Aeronautics and Space, Jakarta (1)
- NIPR, Nippon Institute for Polar Research, Japan (1)
- Statens Kartverk, Norway (3)
- University of Copenhagen, Denmark, (1)
- POL, Prowdman Oceanographic Laboratories, UK (1)

X-band uplink	7225.296 MHz
Ground transponder power	60 cm parabolic dish, 5 watts transmit
X-band downlink	8489 MHz; 1 Mbit/s BPSK (bandwidth = 1MHz); 1 W transmit power
Satellite antennas	Crossed dipoles at X-band and S-band
Noise values	±1.5 cm rms for X-band ranging (1 measurement/s) ±0.05 mm/s rms for X-band Doppler (30 s integration time, 90° elevation)
Bias values	< 1 cm for X-band; < 3 cm for S-band (after post-processing)
Range-error estimation	Tropospheric error: 2-7 cm; ionospheric error < 1 cm; Thermal noise and calibration error: 2-3 cm; Antenna phase center uncertainty < 1 cm
Total ranging accuracy	3-7 cm rms (1 s integration time without postprocessing of filtering)
Total range-rate accuracy	0.05 mm/s rms (30 s integration time and elevation of 90°)

Table 403: PRARE system characteristics and measurement precision parameters

Part I Satellite Emergency Services & Environmental Monitoring

Satellite services for emergency and environmental monitoring support are treated as a new service category. Such services are actually not new, in the past they were provided by a number of operational satellites with imaging instruments (SPOT series, etc.) capable of taking occasional snapshots of a disaster-stricken region.

But there is a general need for better services from space for rapid and frequent disaster monitoring capabilities. On a worldwide scale, natural and human-induced disasters (earthquakes, forest fires, floods, volcanic eruptions, etc.) claim around 100,000 human lives every year, the damages alone are in the order of several hundred billion dollars (or euros). In such extreme situations, emergency managers in the field must have access to information for effective crisis response. The information must be available rapidly and frequently. - The world population more than doubled during the past 40 years of the space-age, reaching 6 billion in 1998, the prediction for 2050 is about 9 billion. It implies that the toll in human lives as well as in total damages will increase in future catastrophes.

Dedicated spaceborne systems (constellations) could be the answer to such services, offering a portion of a global infrastructure for coordinated action. So far, a few space projects have been defined. It will take another decade to reach a certain maturity of services.

I.1 BIRD (Bi-Spectral Infrared Detection)

BIRD is a DLR microsatellite demonstrator mission with the objective to observe fires/hot spots on Earth (caused by lightning, volcanism, oil wells, smoke/water-cloud discrimination, and by man) to permit environmental impact studies. Secondary objectives involve technology demonstrations, namely the operation of a new type of infrared sensor system on a microsatellite and on-board preprocessing techniques. [1341) 1342) 1343) 1344)]

The S/C structure resembles a compact cube (side length of 62 cm, 1.6 m deployed length), consisting of the following segments: service, electronics, and payload, with one fixed and two deployable solar panels. The bus design is modular using low-mass aluminum and honeycomb structures. BIRD is three-axis stabilized, pointing toward nadir in observation mode with an accuracy requirement of ±5 arcmin, a pointing knowledge of ±0.2 arcmin per axis, and a slew rate of 1°/s. The S/C attitude is measured by two star sensors, pointing in directions that differ by 30°. The star sensor assembly is referred to as ASTRO-5 (Autonomous Star Sensor), it includes a processor for autonomous attitude determination. In addition, there is a 3-axis laser gyroscope providing a resolution of 2.7 arcsec with a drift rate of 1°/h. The actuators of ACS (Attitude Control System) are: magnetorquers (3 pairs) and 4 reaction wheels. The S/C has a command-controlled body-pointing capability ($\pm 30^{\circ}$) in any direction. A GPS receiver, GEM-S (GPS Embedded Module) of Rockwell-Collins, is used for orbit determination. It is a five-channel C/A- and P-code for L1 signal tracking. The mass of GEM-S is 0.4 kg, power = 6.5 W. An on-board navigation system (ONS) software package uses the GEM-S inputs to provide a) position information (<90 m) for WAOSS imagery, b) orbit information at 2 Hz sampling rates. ONS provides the instantaneous nadir and

1341) I. Walter, K Brieß, et al., "BIRD - Microsatellite for Hot Spot Detection," Proceedings of the 13th AIAA/USU Conference on Small Satellites, Aug. 23-26, 1999, Logan UT, SSC99-IX-4

1342) K. Brieß, H. Hahn, H. P. Röser, "BIRD - A DLR Small Satellite Mission for the Investigation of Hot Spots, Vegetation and Clouds," Proceedings of ISPRS International Symposium on Earth Observation System for Sustainable Development, Feb. 25-27, 1998, Bangalore, India, pp. 216-222

1343) K. Brieß, W. Bärwald, T. Gerlach, H. Jahn, F. Laura, H. Studemund, "The DLR Small Satellite Mission BIRD," Acta Astronautica, Vol. 46, No. 2-6, , 2000, pp. 111-120

1344) O. Montenbruck, E. Gill, H. Kayal, "The BIRD Satellite Mission as a Milestone Towards GPS-based Autonomous Navigation," ION GPS-2000, Sept. 19-22, 2000, Salt Lake City, UT, pp. 1968-1975 along-track direction for camera pointing as well as precise position values for real-time geocoding of the imagery. Total S/C mass is 88 kg, average power is 40 W (200 W max), 8 NiH_2 battery cells (12 Ah) provide power during orbit eclipse and peak power phases; the S/C design life is 1 year. The on-board recorder has a capacity of 2 x 1 Gbit.

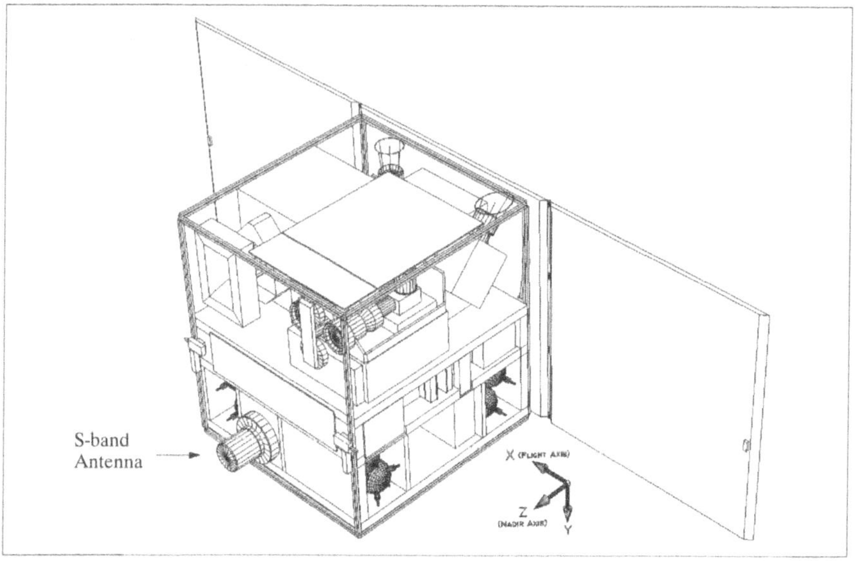

Figure 232: The BIRD S/C model

A launch of BIRD as a secondary payload is planned on an ISRO PSLV-C3 launcher for Sept. 2001, along with TES (Technology Experiment Satellite) of ISRO as the primary payload and PROBA of ESA as another secondary payload. [1345)]

S/C communications are in S-band. The downlink transmitter frequency is 2201,7 MHz, the downlink data rate is 2.2 Mbit/s for recorder dumps during a station pass, and the modulation is: PCM/BI-F-L/BPSK. The uplink frequency is 2032,5 MHz, the data rate is 19.2 kbit/s, and the modulation is PCM/GMSK/PM. - Spacecraft operations are provided by GSOC using one tracking station. A CCSDS protocol suite is used for communications with the ground segment.

S/C operation modes:

- Sun-pointing mode. Most of the time the S/C is operated in sun-pointing mode (solar panels are normal to solar radiation) for energy accumulation.
- Earth-pointing mode G (coarse, downlink of data during station pass)
- Earth-pointing mode F (fine, remote sensing)
- On-board processing (and experimental classification) of sensor data
- Remote sensing or downlink at night-time
- In-flight calibration mode (pointing into deep space or to the moon)

The S/C instrument observations are limited to about 10 minutes per orbit (duty cycle of 0.1). This is estimated to be enough for disaster monitoring coverage. Outside of activity periods (observation and/or communication with the ground segment) the S/C is operated in sun-pointing mode.

1345) K. Brieß, W. Bärwald, F. Lura, et al., "The BIRD Mission is completed for launch with PSLV-C3 in 2001," Proceedings of the 3rd International Symposium of IAA, Berlin, April 2-6, 2001, pp. 323-326

Orbit: Sun-synchronous circular orbit, altitude = 575 km, inclination = 97.7°, local equatorial crossing time at 10:30 AM.

Sensor complement:

HSRS (Hot Spot Recognition Sensor). 1346) HSRS is a two-channel (MWIR, TIR) infrared imaging instrument, of IRCAM heritage (an airborne prototype imager), with the objective to detect ground surface hot spots. A pushbroom design is employed. The two spectral bands of 3.4-4.2 μm and 8.5-9.3 μm of the system use HgCdTe detector line arrays, each of 2 x 512 pixels. The detector are arrays of HgCdTe (GEC Marconi, UK) loophole diodes hybridized on a silicon CMOS multiplexer chip, both ranges are with an individual layout. The detectors are cooled to 80 K by an integrated cooler-detector unit of Stirling design. The two camera heads are integrated into an optomechanical structure for reasons of co-alignment.

Parameter	HSRS	WAOSS-B
Spectral bands	3.4 - 4.2 μm (MWIR band) 8.5 - 9.3 μm (TIR band)	600 - 670 nm forward 840 - 900 nm, nadir backward
F-number	2.0	2.8
Focal length	46.39 mm	21.65 mm
Pixel element size	30 μm x 30 μm	7 μm x 7 μm
Nr of pixels	2 x 512 staggered	3 x 5184 (2884 illuminated)
IFOV (resolution)	2.22 arcmin	1.11 arcmin
FOV (cross-track)	19°	50°
FOV (along-track)	NA	+25°, 0°, -25°
Ground pixel size	372 m (orbit altitude of 575 km)	185 m
Swath width	190 km (orbit altitude of 575 km)	533 km
Data quantization	14 bit	11 bit
Data rate	957 kbit/s	500 kbit/s
Power consumption	90 W electronics unit inclusive	18 W
Instrument mass	7.3 kg camera head + 6.5 kg electronics unit	8.4 kg

Table 404: Specification of the HSRS and WAOSS-B instrument parameters

WAOSS-B (Wide-Angle Optoelectronic Stereo Scanner, BIRD version). The two-channel VNIR imager is a modified spare model of WAOSS flown on the Russian Mars-96 mission (launch on Nov, 17, 1996 from Baikonur; however, the S/C failed to obtain a Mars orbit). The WAOSS instrument is a compact monoblade device consisting of camera head, electronic modules, and interface boards in one unit. The along-track stereo imaging system is based on 3 CCD-lines (single optics and focal plate), operating in pushbroom mode, and taking images simultaneously in the forward-, nadir- and backward-pointing direction of the orbital ground track. The convergence angle is 25° (angle between forward- and nadir-pointing directions, and angle between nadir and backward-pointing directions). New lenses with integrated filter and wide FOV were introduced. Other modifications pertain to software updates. The main parameters of the instrument are defined in Table 404.

I.2 DEMETER (Detection of Electromagnetic Emissions transmitted from Earthquake Regions)

See the DEMETER description under M.15.1 (Myriade Microsatellite Program of CNES).

I.3 DMC (Disaster Monitoring Constellation)

The Disaster Monitoring Constellation (DMC) is an international project proposed and led by SSTL (Surrey Satellite Technology Ltd), Surrey, UK, to construct a network of five

1346) W. Skrbek, E. Lorenz, "HSRS - An Infrared Sensor for Hot-Spot-Detection," Proceedings of SPIE, Vol. 3437, 1998, San Diego, CA, pp. 167-176 affordable LEO microsatellites. The objective is to provide a daily global imaging capability at medium resolution (32-36 m) for rapid-response disaster monitoring and mitigation.

Since a microsatellite constellation represents the most affordable and practicable solution to this problem, SSTL proposed in 1999 a five microsatellite constellation to its customers - with a stipulation/agreement by each potential "subscriber" or buyer of a satellite that they work as a constellation in the event of natural or man-made disasters. This innovative proposal provides opportunities, especially to developing countries, to not only own their own satellite, but also to have access to a disaster monitoring and mitigating satellite system. The key to all of this is Surrey's philosophy of low-cost, rapid access to space.

Parameter	Description
Constellation	
Network	5 spacecraft in a single orbital plane, 5 member ground stations
Constellation orbit	Sun-synchronous circular orbit, altitude = 686 km, inclination =98°, ascending node at 10:30
Constellation revisit time	Daily revisit (imaging capability) to any point on the equator + more often to higher latitudes
Lifetime	5 years target in orbit operational life
Satellite Payload	
Payload	Two three-band linear array CCD cameras
Payload performance	Swath width of 600 km, Spatial resolution of 32 m at nadir, 34-35 m resolution at the swath edge
Sensor spectral bands	0.52-0.62 μm (green), 0.63-0.69 μm (red), 0.76-0.9 μm (NIR)
Data quantization	8 bit
Optics	Focal length = 150 mm, aperture diameter = 60 mm
On-board data storage	2 x 512 MByte SSDR (Solid-State Data Recorder), providing 1 Gbyte
Satellite Platform	
Platform size	About 60 cm x 60 cm x 60 cm, plus antennas
S/C launch mass, power	<100 kg, 4 GaAs body-mounted solar panels
Battery	NiCd cells, 8 Ah capacity
Orbit determination	GPS receiver
Orbit control	Pressurized cold gas or liquified gas system for constellation phasing maneuvers, 10 m/s delta-v, 0.05 N thrusters
Attitude determination	2 x three axis fluxgate magnetometer; 4 x two-axis sun sensors
Attitude control	3 x orthogonal dual-wound magnetorquers; Momentum wheel (yaw and pitch axis control) Gravity-gradient boom
On-board computer	2 x OBC (providing redundancy)
Communications	
TT&C link	Uplink: omni-directional patch antenna, downlink: monopole antenna S-band, CP-FSK modulation, data rate of 9.6 kbit/s
Payload downlink	Quadrifilar helix antenna, RF power of 4 W S-band link with B/QPSK modulation, data rate of 8 Mbit/s
Ground Segment	
S-band GSN (Ground Station Network)	3.5 m dish antenna, TT&C and payload imagery downlink support (each spacecraft provider is expected to have his own ground station)
Data requests	Via service providers over Internet to CCMPS (Centralized Constellation Mission Planning System)
Data distribution	Between GSNs and service providers via Internet
Data product	Rapid response, intelligible, north-oriented multispectral imagery
Service providers	The Reuter Foundation (to disaster relief agencies); CHEST (to higher education community) ESYS (for commercial evaluation and for data dissemination in schools)
Maximum System Image Capability	
Capacity/day per member	Capacity/day per member >80 images of size: 100 km x 100 km (non-cooperative mode)

Table 405: Technical specification of the DMC microsatellite constellation

As of June 2001, the customers for the first constellation are:

- CNTS (Centre National des Techniques Spatiales) of Algeria. Algerian technicians are currently building their own microsatellite, referred to as AlSat-1, at SSTL.
- BNSC (British National Space Centre) is funding its own microsatellite in the constellation.
- The government of Nigeria has signed a contract for a DMC microsatellite. Nigerian engineers are expected to arrive at SSTL and to commence their know-how training program.
- Contract negotiations with two more partners are at the negotiation stage.

The DMC is implemented through international cooperation focused on the Surrey Space Club. It is anticipated that the cooperative use of the constellation will benefit all the members. A launch of DMC is planned for the end of 2002/ start of 2003. [1347)]

The constellation has an imaging capability of about 500 images per day. Imagery requests are planned to be coordinated through a centralized mission planning system at SSTL, UK. It is estimated that up to 5% of the imagery will be made freely available to disaster relief organizations. A rapid on-line news and communications service is proposed to make the imagery available to the relief agencies.

Figure 233: Illustration of a DMC spacecraft

I.4 Fuego/FOC (Fire Observation Constellation)

Fuego/FOC is a European program initiative, defined and developed with EC support (within the IV Framework Program of the EC), coordinated by INSA (Ingeniería y Servicios Aeroespaciales) of Madrid, Spain, drawing on support/inputs of many participating institutions across Europe (government and academic). The overall objective is to provide a global capability of early fire detection and high-resolution fire-line monitoring by means of a minisatellite constellation - to be of service to the fire fighter community. Secondary applications include volcano monitoring - for civil protection measures, and general surveillance monitoring.

In Europe, forest fires account for the destruction of 200,000 to 600,000 hectares of forest every year; about 80% of this destruction occurs in the Mediterranean region where fires

1347) Information provided by Audrey Nice and by John Cooksley of SSTL become increasingly more damaging in dry and hot summers due to the abandonment of traditional agricultural activities. Between 1991 and 2000, 4 million hectares of land were destroyed in France, Greece, Italy, Portugal and Spain, causing enormous damage to the ecosystem, substantial economic losses, and human casualties. A space-based observation system is considered the proper approach to improve the current fire-fighting strategies.

Several studies on Fuego, sponsored by the EC, ESA, and by Spanish national funding have been carried out [The Fuego-1 project (feasibility and definition study) started in late 1996 and was completed in 1998. The Fuego-2 project (design of payloads, platform, ground system) lasted from late 1998 to the beginning of 2001.]. ESA is considering Fuego/FOC within its Earth Watch initiative. A submittal of the program to its member states, covering the development, construction, launch and operations of a demonstration satellite, is expected in 2001. The implementation proposal (2001) of the Fuego concept within ESA's Earth Watch program is taking a two-sliced approach: [1348) 1349) 1350) 1351) 1352)]

1) FuegoSat. This refers to the Fuego system demonstration and validation phase prior to the deployment of the full constellation (precursor mission). FuegoSat builds on the previous studies of the EC and industry. It comprises the following elements:
   - One demonstration satellite, identical to the full system S/C, to be launched in 2004
   - A ground system (minimum system) to permit satellite operations
   - Ten transportable user stations to permit simultaneous demonstrations of the Fuego service provision to multiple users in Europe and elsewhere
   - Seven years of operation
2) The implementation of the full constellation, co-financed by ESA and the industrial team exploiting Fuego, is on a 50-50 basis. The full Fuego system is based on a constellation of 12 minisatellites, with a design life of seven years, and with multiple user stations allowing direct reception of data located at the user control center at a regional level.

Users (fire fighting authorities) have been actively involved in the conception of Fuego since its very start of the program. The identification of user requirements (for fire detection and fire monitoring) has been carried out in successive iterations. This resulted in a corresponding service and data product provision concept proposal. - The user recommendations came up with the following primary objectives of the Fuego Earth Watch Mission:

- Automatic provision of an early fire detection product which identifies the fire outbreaks and their most relevant characteristics like position and intensities.
- Quick delivery of a reliable fire detection product to the user community
- Continuous visualization of the developed fires on request
- The products must be cost effective
- The service of fire detection and monitoring must be provided on a continuous basis, around the clock and throughout mission life.

I.4.1 Fuego System Concept

Each spacecraft of the Fuego series is three-axis stabilized. The total mass of a S/C is limited to 240 kg, including 62 kg of payload mass. The maximum volume limit of a S/C is defined as: 300 cm x 75 cm x 75 cm (the actual dimensions of the bus are: 180 cm x 75 cm x 75 cm without solar panels). The S/C is composed of a lower platform (propulsion and platform interface),

1348) C. Martin-Rico, J. Gonzalo, A. Mariani, W. Leibrandt, "The Fuego System Concept," Acta Astronautica, Vol. 48, No. 1, 2001, pp. 45-56

1349) Information provided by Ignacio Tourne and Jesus Gonzalo of INSA, Spain

1350) http://www.insa.es/fuego/

1351) "Proposal for an Earth Watch FuegoSat Element, submitted by CDTI," ESA Earth Observation Program Board, ESA/PB-EO (2001) 30, March 27, 2001. And: "FuegoSat Program Proposal, The Precursor Mission to the Fuego Earth Watch Program," ESA/PB-EO (2001) 30, Annex

1352) D. Escorial, I. F. Tourne, F. J. Reina, J. Gonzalo, "Fuego: A Dedicated Constellation of Small Satellites to Detect and Monitor Forest Fires," Proceedings of the 3rd International Symposium of IAA, Berlin, April 2-6, 2001, pp. 163-166 a middle platform, an upper platform (payload interface), the internal structure, the lateral panels, and the solar panel wings. The power requirements call for 500 W (EOL). A propulsion module ensures the proper constellation deployment and maintenance during the mission. The attitude system requirements for data acquisition call for a pointing accuracy of 0.15° in cross-track and 0.1° in the along-track direction with a stability of 6 arcsec [(1 sigma) from 0.5-2 Hz, and 3 arcsec (1 sigma) from 2-150 Hz]. In addition, each S/C is required to perform pitch maneuvers of 45° in 30 s, to permit extended observation periods onto an area of interest from its orbital path. The on-board data handling requirements call for the MIL-STD-1553B bus standard. The S/C design life is seven years.

A multi-satellite-launch strategy is selected with one launch per orbital plane. The Rockot launch vehicle of Eurockot is the selected baseline (Baikonur launch site). The four S/C will be installed on a single floor of the rocket fairing.

Orbit: Circular (non-sun-synchronous) orbit, altitude = 700 ±15 km, inclination = 47.5° ± 0.05°, period = 98.8 min. The configuration of the constellation consists of three orbital planes with four satellites per plane (Walker configuration of 12/3/2). This arrangement provides a global FOR (Field of Regard) observation capability with a revisit time between 23.8 to 25.8 minutes and a service provision around the clock.

In the general acquisition geometry, the fire detection limit is ±55° about nadir, corresponding to a total FOR (Field of Regard) of 2500 km. However, the fire monitoring limit is within ±39° about nadir, corresponding to a total FOR of 1250 km. The actual observation swath within each FOR is always ±7.2°, or 176 km (total) at nadir.

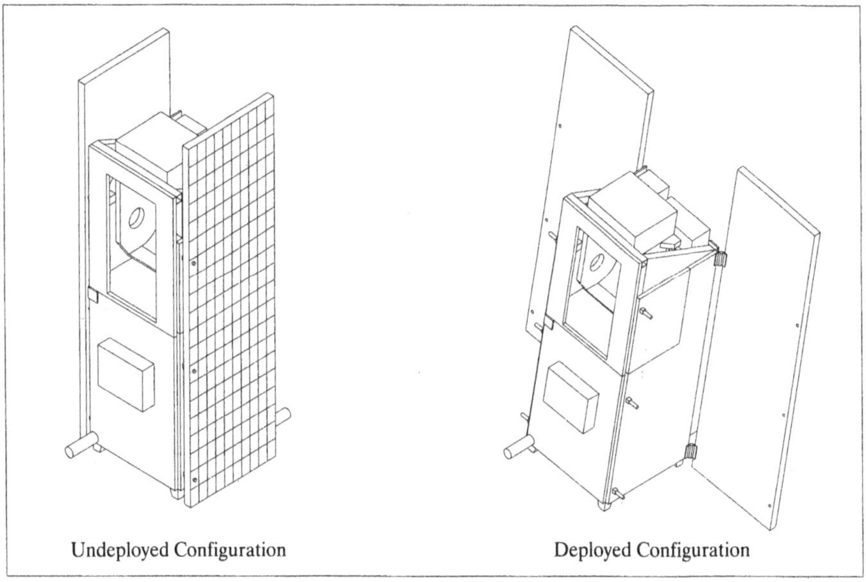

Figure 234: Illustration of a Fuego spacecraft

RF communications: TT&C communications are provided in S-band. The requirements call for an on-board data storage capability and date rates of 16 kbit/s (NRZ-M/PSK/PM modulation) in downlink and 2-4 kbit/s in uplink. The payload data transmission is in L-band at 4 Mbit/s. All communication protocols are in CCSDS standard.

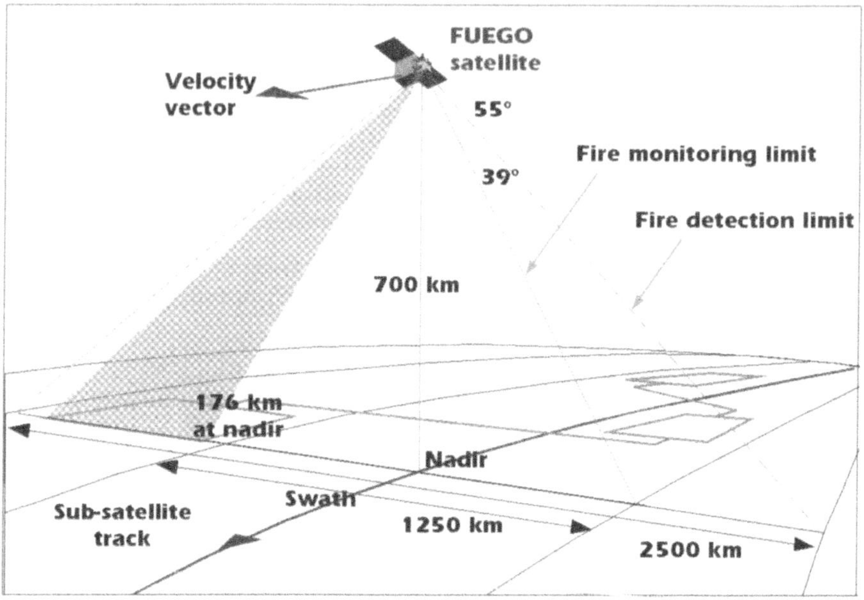

Figure 235: Sketch of the acquisition geometry

Ground segment requirements: The ground segment consists of two main elements:

- The system ground segment consists of a Mission Control Center (MCC) and two ground stations (primary and backup) for the acquisition/transmission of the constellation telemetry and command data (TT&C function). In addition, MCC receives the payload data.
- User ground segment: A distributed network of Primary User Stations (PUS) located at the user premises and operated by the fire-fighting authority of the region. A PUS consists of a PC, a 1.5 m antenna, a front end, and a connection to public networks (also contact with MCC). Each PUS is capable to acquire the real-time payload data.

Sensor complement:

FIS (Fuego Imager Suite). The Fuego instrument package consists of four pushbroom cameras in the four spectral regions: VIS (Visible), NIR (Near Infrared), TIR (Thermal Infrared) and MIR (Medium Infrared), also referred to commonly as MWIR (Mid-Wave Infrared). The cross-track pointing feature is implemented with a scan mirror [a steerable mirror is used to point to the areas of interest within FOR (Field of Regard)] common to all cameras. The WMA (Wide Mirror Assembly) is composed of a mirror, a motor and the driver electronics which deflects the optical axis of the four cameras into the desired direction. The payload power consumption is defined as: 180 W (peak), 140 W (average), and <75 W in standby.

Cross-track pointing capability	±54.1° (for fire detection), ±38.4° (for fire monitoring)
Swath width capability	2500 km (for detection), 1250 km (for fire monitoring)
Pointing knowledge	±20 arcsec (2 pixels)
Pointing stability	0.015°/s
Mirror moment of inertia	0.01 kg m^2
Allowed time for step movement	3 s

Table 406: Requirements of cross-track mirror

The MWIR camera is most appropriate to detect hot spots (mix of solar reflected and thermally emitted radiation), while the TIR camera is used to provide more information on fire presence (background temperature estimation, etc.) and to filter false alarms. The main objectives of the VIS and NIR cameras are to provide rejection of false alarms.

Parameter/Spectral Region	MWIR	TIR	VIS	NIR
Observation geometry				
Swath width FOR (Field of Regard)	2500 km	2500 km	2500 km	2500 km
Monitoring swath width (FOV)	176 km	176 km	176 km	176 km
Detector elements per row Total number of detector elements	1100 x 2 2200	240 at 45° 480	8800 8800	8800 8800
GSD (Ground Sampling Distance)	80 m	367 m	20 m x 4 m	20 m x 4 m
Optics				
FOV (Field of View)	14.41°	14.41°	14.41°	14.41°
IFOV (Instantaneous Field of View)	0.114 mrad	0.524 mrad	0.0286 mrad	0.0286 mrad
Aperture diameter	54 mm	60 mm	35 mm	35 mm
Focal length	87.5 mm	60.7 mm	227.5 mm	227.5 mm
F-number	f/1.6	f/1.0	f/6.5	f/6.5
MTF (Modulation Transfer Function)	0.28	0.28	0.28	0.28
Radiometry				
Spectral range	3.55 - 4.15 µm	8.0 - 12.0 µm	0.58 - 0.68 µm	0.76 - 0.90 µm
Spectral band center	3.80 µm	10.0 µm	0.63 µm	0.83 µm
Band width	0.70 µm	4.0 µm	0.10 µm	0.14 µm
Sampling time	12.49 ms	57.23 ms	3.122 ms	3.122 ms
NEΔT at 300 K	1.7 K	0.5 K	-	-
Data quantization	12 bit	12 bit	9 bit	8 bit
Radiometric accuracy (% reflectance)	-	-	5	5

Table 407: Specification of major FIS instrument parameters

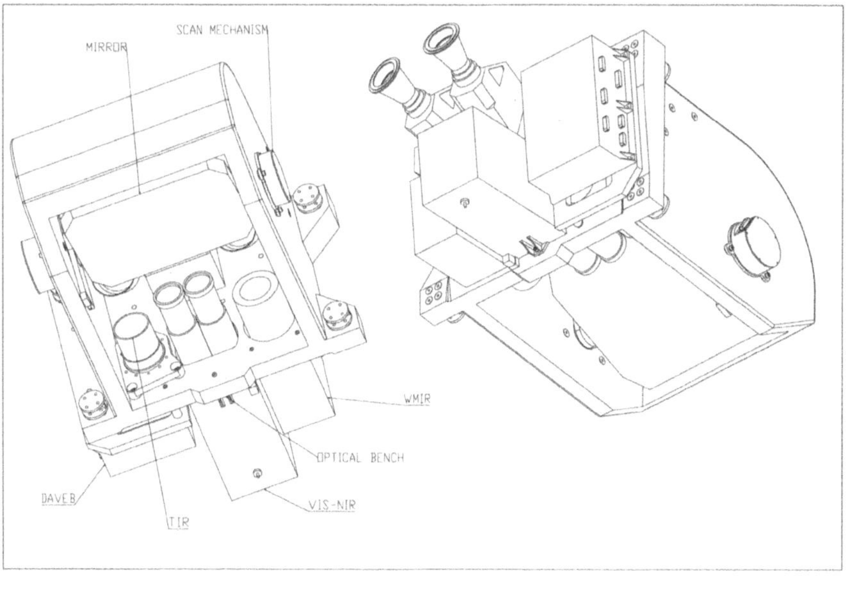

Figure 236: Schematic arrangement of the FIS instrument (two views)

Fire Monitoring Products:

A service subscriber receives a monitoring product with information related to the on-going fire in the specified region., and providing information on the fireline characteristics (geometry, position and intensity).

- Detection product: Provided with a nominal FOR (Field of Regard) of 2500 km. The imagery at the swath edge exhibits pixel distortion; however, this pixel distortion does not prevent from good fire detection.
- Monitoring product: Provided with a nominal FOR of 1250 km. The imagery of half the detection swath width offers a good fireline representation (with only small pixel distortions at the swath edges).

I.4.2 FuegoSat

FuegoSat represents the precursor mission to Fuego/FOC. The primary objective is to provide total confidence in the system prior to embarking on the full program. Its purpose is to demonstrate/validate the critical functionality of the mission. An evaluation of all mission (and economical) aspects resulted finally in a FuegoSat concept consisting of is a single spacecraft.

The FuegoSat spacecraft is identical in design, functionality, and payload configuration to the Fuego/FOC series. The system ground segment consists of MCC and a single ground station. The user ground segment is limited to 10 transportable user stations to permit simultaneous demonstrations of the system.

The implementation and operation of FuegoSat and Fuego/FOC is being managed by an entity referred to as FueCorp (Fuego Corporation). Current plans calls for FueCorp to be established/constituted in February of 2002.

A launch of FuegoSat is planned for 2004 on a Rockot launch vehicle.

I.5 GMES (Global Monitoring for Environment and Security)

GMES is a European initiative put forward by the EC (European Commission) in 1998 with the objective to determine Europe's global monitoring role in the field of the environment and security. The goal of GMES is to develop operational information services on a global scale, using both space- and ground-based monitoring systems, in support of environment and security policy needs. Key issues in this field include the monitoring of: [1353)]

- International environmental conventions (environmental issues of a global nature). The EU has signed more than 40 international treaties to bring forward a more global dimension to environment and security (e.g. biodiversity, global change, desertification). The "Kyoto Protocol" is an example of "treaty monitoring" requirements. NATO and its Member States are increasingly concerned with non-traditional threats to security, including the consequences of environmental change.
- Environmental stress (environmental issues of a regional nature). Environmental stress poses a potential threat to security at all geographic levels. Taking preventative action on environmental stress is the most appropriate approach to preventing environmental conflicts.
- Risks and natural disasters/hazards (including humanitarian help). The major topics of natural hazards are: earthquakes, landslides and avalanches, volcanoes, forest fires, and floods. The service of forest fire detection and monitoring (including risk evaluation) is an important activity in this category.

1353) "A European Approach to Global Monitoring for Environment and Security (GMES): Towards Meeting User Needs," ESA/PB-EO (2001) 56, Rev. 1, Annex, June 13, 2001, Version 2.01

The JRC/SAI (Joint Research Center/Space Applications Institute) of the EC in Ispra (Italy) is the main technical support office in charge of the GMES project management and chairs the so-called "GMES Partnership", a working group bringing together the partners of the project: government agencies, institutional bodies (ESA, EUMETSAT, national space agencies), and the space industry. [1354)]

The GMES Partnership was formed in late 2000. It involves the member states, the major space agencies, industry representatives and the EC. The mandate is to help formulate the GMES Proposal in response to the council request. GMES seeks to capture and focus existing industrial and political efforts, and thus to provide a common framework to bring activities together and to develop a common ownership and strategy.

The upcoming ESA Earth-Watch missions are an important element in the GMES monitoring strategy, providing long-term monitoring services in various aspects of Earth observation; this includes also long-term data sets. [1355)] [1356)]

The GMES approach is based on scientific understanding of the whole Earth system. It will benefit from research that links Earth system science to the social sciences in the search for means of measuring and modelling mankind's impact on the natural environment, and vice versa

Specific implementation proposals for GMES are due in the second half of 2001. The topic of GMES will be addressed at EU Ministerial level in November 2001. By the end of 2001 the EC will have produced a detailed Action Plan for implementing the initial period.

I.6 Search & Rescue (S&R) Satellite Systems

I.6.1 COSPAS-S&RSAT Constellation

COSPAS (Space System for the Search of Distressed Vessels) and **S&RSAT** (Search & Rescue Satellite Aided Tracking System)[1357)] payloads are part of an international cooperative satellite-based radiolocation system to support search and rescue operations for aviators, mariners, and land travellers in distress. The COSPAS-S&RSAT system provides distress alert and location data to RCCs (Rescue Coordination Centers) for 121.5 MHz beacons within the coverage area of COSPAS-S&RSAT ground stations, referred to as LUTs (Local User Terminals), and for 406 MHz beacons activated anywhere in the world. The overall system concept is illustrated in Figure 237. The nominal system configuration comprises two COSPAS and two S&RSAT payloads on polar-orbiting satellites. Russia provides two COSPAS satellites in near-polar orbits at an average altitude of 1000 km (83° inclination). The USA supplies two NOAA POES satellites with on-board S&RSAT payloads provided by Canada and France. The orbital period for each satellites is about 100 minutes, the footprint of the COSPAS-S&RSAT visibility is in the order of 4000 km in diameter. The first COSPAS payload was launched on June 29, 1982, while the first S&RSAT equipment was flown on NOAA-8, launched in March 1983.

Background: The satellite system was initially established under a MOU among agencies of the former USSR, USA, Canada and France, signed in 1979. The system was declared operational in 1985. In 1988 the four space segment providers signed an agreement which en-

1354) http://gmes.jrc.it

1355) "Earth Watch Initiative: The GMES Element, Initial Program Proposal," ESA/PB-EO (2001) 28, Rev.1, May 11, 2001

1356) "GMES (Global Monitoring for Environment and Security)," Draft by EC and ESA, ESA/PB-EO (2001) 56, April 2001, Version 1.2

1357) **Note: Although the original acronym for 'Search and Rescue' is 'SAR' in the context of NOAA missions, it was changed in this book consistently to 'S&R' in order to distinguish it from the other widely-used meaning of SAR, namely 'Synthetic Aperture Radar,' a sensor type. A consequence is the use of 'S&RSAT' (instead of SARSAT)** sures service continuity and availability of the system to all States on a non-discriminatory basis. As of 1997, the participants of this program include the four parties to the COSPAS-S&RSAT International Program Agreement [Canada (CRC), France (CNES), Russia (ROSHYDROMET), and the USA (NOAA)], and ground segment providers. About 30 countries have joined the COSPAS-S&RSAT program.

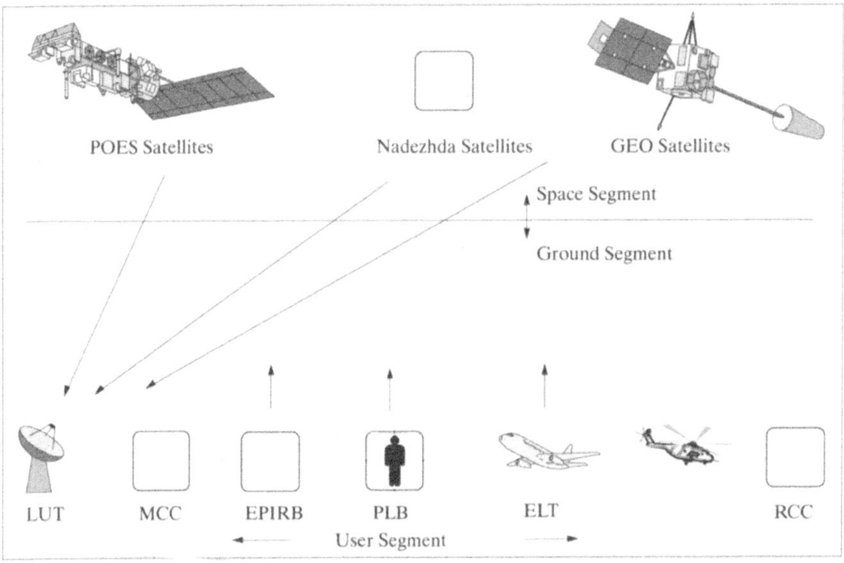

Figure 237: System elements of Search & Rescue Satellites

I.6.1.1 Alert Signal Devices (User Segment)

There are three types of radiobeacon devices (heritage of older rescue services) in use, namely maritime EPIRBs, aviation ELTs, and personal locator beacons (PLBs).[1358)]

EPIRB (Emergency Position Indicating Radio Beacon). A small battery-powered transmitting device which is carried on vessels (several nations require EPIRB devices on all vessels). It can be activated in times of trouble to send out help signals. EPIRB devices operate on a frequency of 121.5/243 MHz (the 243 MHz frequency is used only in some older beacons), they may also include the 406.025 MHz alert signal for global detection.

ELT (Emergency Locator Transmitter). The 121.5 MHz band service emergency beacons are required on many aircraft, referred to as ELT 121.5, with a smaller number carried on maritime vessels. The 121.5 MHz frequency band is used by an older type of beacons which do not transmit any encoded information. It is also the frequency used for low-power homing transmitters included in most beacons. The more recent ELT devices on aircraft are provided with dual frequency alert systems, 121.5 and 406.025 MHz.
Beacons with the 406.025 MHz signal transmit digitally encoded information which may include beacon identification (which allow S&RSAT services to access registration data bases providing additional information on the unit in distress), and beacon location data (determined by satellite navigation devices such as GPS or GLONASS).

PLB (Personal Locator Beacon). These are handheld or pocket size dual frequency (121.5 and 406 MHz) or single frequency (either 121.5 or 406.025 MHz) devices which transmit distress signals to search and rescue authorities via the COSPAS-S&RSAT satellite system.

1358) Cospas-Sarsat homepage: http://www.worldserver.pipex.com/cospas-sarsat

LEO Satellites in polar or near-polar orbits				
Payload	Spacecraft	Launch Date	Status/Comment (Dec. 1996)	Orbit (km)
COSPAS-1	Cosmos 1381	June 29, 1982	Decommissioned in March 1988	989 x1082, 83°
COSPAS-2	Cosmos 1447	Mar. 24, 1983	Decommissioned in Dec. 1989	959 x 1013, 83°
S&RSAT-1	NOAA-8	Mar. 28, 1983		801 x 826, 98.2°
COSPAS-3	Cosmos 1574	June 21, 1984		965 x 1005, 83°
S&RSAT-2	NOAA-9	Dec. 12, 1984	S&RP not operational for 406 MHz	842 x 862, 98.9°
S&RSAT-3	NOAA-10	Sept. 17, 1986	Operational	803 x 824, 98.7°
S&RSAT-4	NOAA-11	Sept. 24, 1988	Operational	845 x 863, 98.9°
COSPAS-4	Nadezhda-1	July 4, 1989	Operational	960 x 1014, 83°
COSPAS-5	Nadezhda-2	Feb. 27, 1990		956 x 1021, 83°
COSPAS-6	Nadezhda-3	Mar. 21, 1991	Operational	958 x 1018, 82.9°
S&RSAT-5	NOAA-13	Aug. 9, 1993	S/C failure 12 days after launch	
COSPAS-7	Nadezhda-4	July 14, 1994	Operational	954 x 1005, 82.9°
S&RSAT-6	NOAA-14	Dec. 30. 1994	Operational	848 x 861, 98.9°
S&RSAT-7	NOAA-15	May 13, 1998	Operational	833 km, 98.86°
COSPAS-8	Nadezhda-5	As required	Ready for launch	870 km, 98.74°
S&RSAT-8	NOAA-16	Sept. 21,2001	Operational	
GEO Satellites furnished with experimental Search & Rescue Equipment				
GEOS&R	GOES-7	Feb. 26, 1987		
SAS&R	INSAT-2	July 9, 1992		
SAS&R	INSAT-2B	July 22, 1993		
GEOS&R	GOES-8	Apr. 13, 1994		
COSPAS	GOMS-1	Oct. 31, 1994		
S&RSAT	GMS-5	March 18, 1995		
GEOS&R	GOES-9	May 23, 1995		
SAS&R	INSAT-2C	Dec. 6, 1995		
SAS&R	INSAT-2D	June 3, 1997		

Table 408: Overview of S&R payload launches

I.6.1.2 Satellite Payloads (Space Segment)

S&RSAT (Search and Rescue Satellite Payload)

The S&RSAT payload consists of S&RR (S&RSAT Repeater), provided by CRC/Canada; and S&RP (S&RSAT Processor), provided by CNES/France). The S&R problem requires two basic functions for effective S&R operations to take place, namely: 1) **alerting** and 2) **locating**. The alerting function only requires a low-capacity one-way communications system. The locating function, however, places far more demands on the system. The overall system uses low-powered battery-operated distress transmitters that are received by polar orbiting satellites with an S&R subsystem on-board. S&R exploits the Doppler shift resulting from relative motion between the distress transmitter and the polar orbiting satellite. A successful alert requires at least one satellite pass over the distress area to detect a signal and locate the position of the emergency transmitter. In some cases a second pass may be required to resolve ambiguity.

The S&RSAT payload consist of a 2-band (121.5 MHz and 406.05 MHz) repeater S&RR and a 406.025 MHz processor S&RP. The S&RR downlink is at 1544.5 MHz and, besides the two repeated bands, also includes the 2400 bit/s bit stream S&RP output. The 121.5 and 406 MHz bands are also serviced by two Russian COSPAS satellites which, together with the NOAA satellites, provide timeliness of response.

COSPAS (Space System for Search of Vessels in Distress)

The USSR began deploying the space segment with the launch of Cosmos 1383 on June 30, 1982 from Plesetsk into a 989 km x 1028 km, 83° inclination orbit (Tsikada navigational satellite series). Designated COSPAS-1, the 121.5 MHz band remained operational until December 1987, with 406 MHz utilized primarily for interference monitoring. Cosmos 1447 (launched March 24, 1983) and 1574 (June 21, 1984) adopted the roles of COSPAS 2 and 3, with a third vehicle available as replacement. The COSPAS payload normally shares the Nadezda (Hope) satellite platform with a Doppler navigation payload of the Tsikada system.

I.6.1.3 COSPAS-S&RSAT Ground Segment

The emergency signals are detected by the space segment (such as COSPAS-S&RSAT) and relayed to LUTs which process the signals to determine beacon location. Alerts are then relayed, together with location data, via MCC (Mission Control Center) to the appropriate search and rescue point of contact or to RCC (Rescue Coordination Center).

Doppler location is the means used for signal location. The 406 MHz devices include an identification code in the alert message. Most 406 MHz devices also include a 121.5 MHz homing transmitter to support search and rescue operations. [1359) 1360) 1361)]

The 406 MHz emergency beacon signals are immediately processed and stored on-board the satellite and are transmitted to the ground from a continuous memory dump, providing complete worldwide coverage. Around the world, ground station LUTs (Local User Terminal) acquire the processed data and unique beacon identification and send these located and identified alerts to MCCs (Mission Control Centers), which forward the alerts to appropriate Rescue Coordination Centers for action. The 406 MHz beacons are designed to work well with the satellite; the system nominally provides better than 4 km accuracy, 90% ambiguity resolution on first pass, and better than 90% location probability on one pass. Note: the US S&RSAT operational ground system facilities consist of S&RSAT, SOCC at Suitland, MD as the MCC, and three LUTs. In addition to the US facilities, many other cooperating nations operate their own LUTs and MCCs.

The 121.5 MHz emergency beacons, whose use predates the satellite system, have not been specified to work with the satellite; consequently the results are variable, depending on the quality of the beacon. Nominally, location accuracy is about 20 km. All the processing is accomplished within the LUT, and because the satellite does not store these data, only beacons with mutual view of the satellite and LUT will be detected. No identification is included with the 125.5 MHz transmissions. Consequently, many nonbeacon sources are also detected as beacons, increasing the difficulty of using these alerts. Even with these problems, the large number of beacons in the field have provided an impressive performance history. More than 1700 people have been saved by the S&R/COSPAS forces using satellite-derived locations, and for more than 300 of them, the satellite provided the only means of alert.

LUT locations: Goose Bay, Edmonton and Fort Churchill in Canada; Toulouse, France; Tromsö, Norway; Lasham, UK; Kodiak (AK), San Francisco (CA), Hawaii, Texas, Guam, and Puerto Rico, USA; Arkangelsk, Novosibirsk, Nakhodka, and Moscow, Russia; Albany and Bundaberg, Australia; Bangalore, and Lucknow, India; Hong Kong and Beijing, China; Brasilia and Recife, Brazil; Santiago and Punta Arenas, Chile; Lahore, Pakistan; Yokohama, Japan; Wellington, New Zealand; Bari, Italy; Singapore, Singapore; Callao, Peru; Ambon and Jakarta, Indonesia; Maspalomas (Canary Islands), Spain; Quargla, Algeria; Keelung, Taiwan.

1359) Advanced TIROS-N (ATN) NOAA-I, NASA /NOAA Bulletin 1991

1360) "Proceedings of the Twenty-Third International Symposium of Remote Sensing Environment," Vol. I, Bangkok, Thailand, April 18-25, 1990,, Erim, P.O. 8618 Ann Arbor Mich. p. 94

1361) Y. G. Zurabov, "The COSPAS-S&RSAT System: Results and Prospects," Space Bulletin, Vol. 1, No. 1 1993, pp. 11-13

I.6.2 GEOS&R (Geostationary Search & Rescue)

With the launch of GOES-H (GOES-7) on Feb. 26 1987, NOAA has started to introduce the S&RSAT payload also on its geostationary satellites. The use of geostationary satellites means that the alert signals can be received almost instantly. However, in order to automatically determine the coordinates of the emergency signal, it is necessary to wait for the system's LEO satellite (position determination can only be provided from a system that moves relative to the Earth). The system thus works in two stages. In the first stage only the emergency signal is received via the geostationary satellite (it is planned to equip GMS and GOMS satellites for this service as well). The received alert signal is transmitted to the search and rescue service to prepare for the operation. In the second stage, the site of the signal origin is determined by the LEO satellite. See also chapter F.4.3 on GEOS&R.

Part J Shuttle - Selected Missions and Payloads

J.1 ASTRO-SPAS (Astronomy Platform - Shuttle Pallet Satellite)

ASTRO-SPAS is the generic name of a reusable platform, designed and built by DASA (formerly MBB, Munich, Germany) under DLR contract, which is used as a self-contained and autonomous free-flyer service structure for special Shuttle payloads with free-flyer requirements for short-duration missions (up to the length of a Shuttle mission). The SPAS structure consists of low-weight, high-stiffness carbon fiber tubes with titanium nodes. Standardized mounting panels are provided for subsystem and payload equipment. The platform is deployed/retrieved by the Shuttle's robot arm RMS (Remote Manipulator System) for a free-flyer mission which may entail separations from the Shuttle up to 100 km. As a service structure, SPAS is particularly suited as a test bed for new science instrumentation and technology demonstrations in space. [1362)]

The platform overall size is 4.5 m x 1.75 m, its empty mass is about 1240 kg (including service subsystems), it can accommodate a payload up to a total satellite mass of 3600 kg. The platform offers the following service subsystems:

- Electrical power: Modular Li-SO_2 battery packs (up to 16), 110 kJh, with 40 kJh of energy available to the payload instruments.
- Thermal control: Passive thermal control via radiation and conduction through the platform surface and multilayer insulation blankets.
- Data management: An on-board computer provides all data management functions such as: telecommanding, storage of source data onto a recorder, telemetry data handling, attitude control, etc.
- Platform stabilization: A three-axis stabilization is provided. A precision star tracker serves as reference for pointing accuracies <3 arcseconds to astronomical targets. A GPS Tensor receiver system provides in addition orbit and attitude. Attitude control (actuator) is provided with a 12-nozzle cold gas thruster system (100 mN thrust).
- Operational modes: Two modes are provided, 'inertial pointing' and 'orbit motion.'
  - The inertial pointing mode serves mainly for astronomical observations. The star tracker (CCD camera) measures the position of three guide stars in its FOV of 4.5^o x 6°, the gyro package senses rotations.
  - The orbit motion mode is used for atmospheric research to point into a specific direction. One axis points into a constant, commandable altitude layer (stabilized to ± 2 km). The GPS Tensor instrument and the star tracker provide attitude, position and velocity of the platform.
- TT&C: A scheduled communications link via Shuttle is provided by an S-band transponder with uplink data rates of up to 2 kbit/s and downlink rates of up to 16 kbit/s.

The SPAS free-flyer platform concept was first demonstrated for the MOMS-01 imaging payload on two Shuttle missions: STS 7, June 18-24, 1983, and STS-41B, Feb. 3-11, 1984. The current ASTRO-SPAS platform is of SPAS heritage with four flights negotiated between NASA and DLR (DARA).

J.1.1 ORFEUS-SPAS-1

ORFEUS (Orbiting Retrievable Far and Extreme Ultraviolet Spectrograph) was a joint DARA/NASA science mission, flown on the ASTRO-SPAS free-flyer platform of Shuttle flight STS-51 (Sept. 12-22, 1993).

1362) R. Wattenbach, K. Moritz, "Astronomical Shuttle Pallet Satellite (ASTRO-SPAS)," Acta Astronautica, Vol. 40, No. 10, pp. 723-732, 1997

The ORFEUS instrument payload is an observatory containing three spectrographs to probe the cosmos. Two spectrographs share a single main telescope [1 m mirror, focal length = 2426 mm, mirror micro-roughness < 0.5 nm rms, mirror surface is coated with a layer of iridium]. The telescope tube consists of a carbon fiber compound, the mirror is a light-weight design (70kg). The entire telescope plus spectrograph assembly has a length of 3960 mm, a diameter of 1120 mm, and a total mass of 970 kg.

- **FUV** (Far Ultraviolet) Echelle spectrograph (U. of Tübingen and Landessternwarte Heidelberg, built by Kayser-Threde, Munich). Spectral range = 90 - 125 nm. Two reflection gratings and microchannel plate detectors offer a spectral resolution of λ= 10,000. [1363)]
- **EUV** (Extreme Ultraviolet) spectrograph of UCB. The EUV spectrograph is located at the prime focus of the f/2.4 normal incidence primary mirror with a spectral range of 39 - 120 nm (4-6 cm^2 effective area). Objective: provision of high-resolution (λ/5000) spectroscopy of point sources.

IMAPS (Interstellar Medium Profile Spectrometer) Princeton University. IMAPS is an objective grating, Echelle spectrograph in the spectral range of 95 - 115 nm, with the objective to record very high-resolution UV spectra of stars (resolving power of 100,000, detailed observations of the chemical and physical properties of interstellar gas detected around bright, nearby stars). IMAPS does not utilize the main telescope on ORFEUS, but is attached to the side of the platform.

SESAM (Surface Effects Sample Monitor). The objective is to explore the influence of atmospheric atomic oxygen on optical surfaces (used in UV astronomy, tests of "whiteness" samples, etc.). The instrument can accommodate, expose, and test up to 40 samples.

J.1.2 CRISTA-SPAS-1

CRISTA (Cryogenic Infrared Spectrometer and Telescopes for the Atmosphere) is the name of the main payload [STS-66 (Nov. 3-14, 1994)], a mission scientifically combined (co-manifested) with the US ATLAS-3 program (J.2). The orbital altitude was about 300 km at an inclination of 57° of the SPAS free-flyer platform, released and retrieved in an observational orbit between 40 to 70 km behind the Shuttle. As such, it operates independently, except during communication periods with the orbiter. During these periods, the instrument relays status information via the Shuttle to the ground. Dimensions of CRISTA-SPAS: 2 m in length x 4.6 m in height, mass = 3400 kg.

CRISTA is the prime payload of SPAS (PI: D. Offermann, Univ. of Wuppertal, Germany, CRISTA is sponsored by DARA). [1364)] It is an infrared instrument with three independent telescopes pointed at the limb (in three directions/dimensions simultaneously). Two telescopes are laterally directed (pointing angles = ±18° from center). Objective: analysis of dynamic processes (winds, wave interaction and turbulence) in the middle atmosphere with the detection of trace gases. The four spectrometers provide 26 spectral bands, 21 bands from 4.6 - 14.1 µm, and 5 bands from 15.2 - 71 µm. Each telescope has a short-wavelength spectrometer attached; the center telescope measures the longer wavelengths. The spectrometers take one spectrum per second and measure up to 15 trace gases in 26 channels during this time. The telescopes obtain complete altitude observations of these gases within about one minute as the lines-of-sight are scanned through the atmosphere. The high measurement speed and the sensitivity of the instrument are achieved by cryogenic cooling of the CRISTA optics and detectors with liquid helium. Spectral resolution power ($\lambda/\Delta\lambda$) = 500, spatial resolution: 500 x 650 km in the horizontal direction and 2-3 km in the vertical direction. The CRISTA instrument is contained in a vacuum container cooled.

1363) I. Appenzeller, et al., "Medium-Resolution Far-Ultraviolet Spectroscopy of PKS 2155-304," The Astrophysical Journal, 439: L33-L37, Feb. 1, 1995

1364) P. Barthol, K. U. Grossmann, D. Offermann, "Telescope design of the CRISTA/SPAS experiment aboard the Space Shuttle," SPIE, Vol 1331, Stray Radiation in Optical Systems, 1990, pp. 54-63

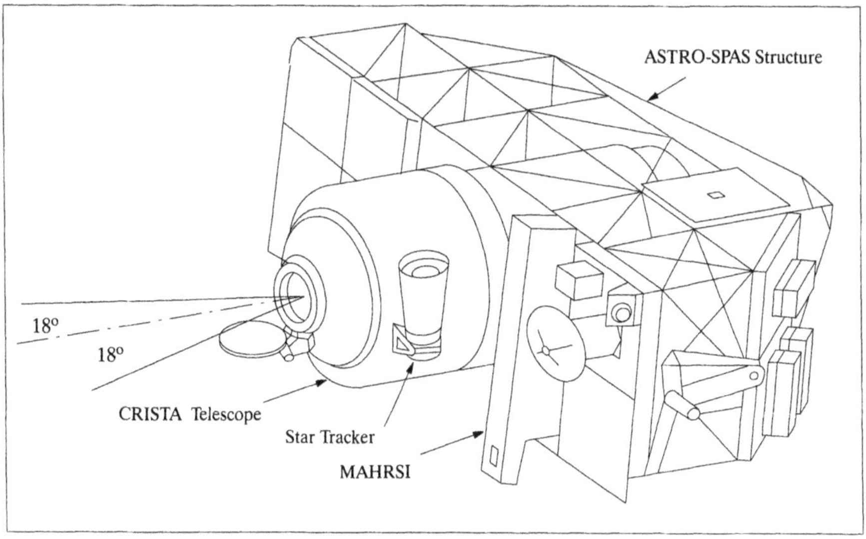

Figure 238: Configuration of ASTRO-SPAS platform with its sensors

The SPAS platform was three-axis stabilized to point its telescope near the Earth's horizon using mass expulsion thrusters and aiming at a point 62.9 km above the WGS-84 ellipsoid. Attitude determination was achieved by a star tracker-gyro inertial reference unit (IRU), position data was provided by an Alcatel/SEL GPS receiver. The IRU data determined attitude within 0.05° and served as "truth" reference for a comparison with GPS data. [1365]

Telescope type (total of 3)	Herschel, 120 mm aperture diameter, limb scan
FOV vertical, horizontal	3 arcmin (1.5 km at tangent point) FWHM, 29 arcmin FWHM
Scan range	±2.25° (lateral telescopes), +2.4° to - 1.6° (center telescope, short wave)
Vertical resolution	2-3 km
View directions	162°, 180°, 198° with reference to flight vector
Horizontal resolution	200 km x 650 km (maximum at equator)
Pointing accuracy (CRISTA-SPAS included)	less or equal of ±300 m at tangent point (vertical direction, absolute) ±80 m at tangent point (vertical direction, relative) ±1.5 km at tangent point (horizontal direction, absolute)
Spectrometer type (4)	Ebert-Fastie
Spectral range	4-71 μm (26 spectral channels)
Spectral resolution ($\lambda/\Delta\lambda$)	300-600
Integration time	1.2 s (for one spectrum)
Altitude scan	<1 minute
Mass of optics	230 kg
Cryostat, tank 1 Cryostat, tank 2	Volume of 725 liter of supercritical helium at 5-13 K Volume of 55 liter of subcooled helium at 2.5 - 4.2 K
Electronics mass, power, data	180 kg, 76 W (average power consumption), 123 kbit/s data rate
Instrument mass, size	1350 kg, 2980 mm in length and 1380 mm in diameter

Table 409: Specification of the CRISTA instrument

Definitions for Table 410:
1) The range limit is determined by filter transmission for values marked with a *. 2) The range limit is extended by the Gille and House method for all slashed wavelengths (example: 8.6/9.7). 3) Detector types: SiGa (Silicon Gallium), GeGa (Germanium Gallium), BIB

1365) L. Ward, P. Axelrad, A Combined Filter for GPS-Based Attitude and Baseline Determination," Proceedings of ION GPS-96, Sept. 17-20, 1996, Kansas City, MO, pp. 1047-1061

(Blocked Impurity Band - BIB was first used with CRISTA-2). In the spectrometer detector column (1), SL, SCS, and SR stand for left, center, and right telescopes and short wavelengths, respectively. SCL is for longer wavelengths in the center telescope only. [1366)]

Spectrometer Detector	Trace gases or parameter observed	Wavelength range (μm)	Detector Type
SL-1	CH_4, N_2O, N_2O_5	7.5 - 8.6/9.7	SiGa
SL-3	O_3	8.91 - 10.22/11.32	SiGa
SL-4	HNO_3, F12	10.43 - 11.78/12.88	SiGa
SL-5	T, O_3, F11, HNO_3, $ClONO_2$ (CCl_4)	11.55 - 12.88/13.98	SiGa
SL-6	H_2O, NO_2	6.07 - 6.73/7.28	SiGa
SL-8	T, P	12.79 - 14.1/15.2	SiGa
SCS-1	CH_4, N_2O, N_2O_5	7.5 - 8.6/9.7	BIB
SCS-2	CO_2, CO	4.18 - 4.81/5.36	BIB
SCS-3	O_3,	8.91 - 10.22/11.32	BIB
SCS-4	NO, H_2O	4.92 - 5.58/6.13	BIB
SCS-5	HNO_3, F12	10.43 - 11.78/12.88	BIB
SCS-6	T, O_3, F11, HNO_3, $ClONO_2$ (CCl_4)	11.55 - 12.98/13.98	SiGa
SCS-6R	T, O_3, F11, HNO_3, $ClONO_2$ (CCl_4)	11.55 - 12.98/13.98	SiGa
SCS--6L	O_3, F11, HNO_3, $ClONO_2$ (CCl_4)	11.55 - 12.98/13.98	SiGa
SCS-7	H_2O, NO_2	6.07 - 6.73/7.28	BIB
SCS-8	T, P	12.79 - 14.1/15.2	SiGa
SR-1	CH_4, N_2O, N_2O_5	7.5 - 8.6/9.7	SiGa
SR-3	O_3	8.91 - 10.22/11.32	SiGa
SR-4	NO, H_2O	4.92 - 5.58/6.13	SiGa
SR-5	HNO_3, F12	10.43 - 11.78/12.88	SiGa
SR-6	T, O_3, F11, HNO_3, $ClONO_2$ (CCl_4)	11.55 - 12.88/13.98	SiGa
SR-7	H_2O, NO_2	6.07 - 6.73/7.28	SiGa
SR-8	T, P	12.79 - 14.1/15.2	SiGa
SCL-1	O_3	9.29 - 10.30	BIB
SCL-2	T. P	14.43* - 15.58	BIB
SCL-3	O, HF, H_2O	59.0 - 65.0	GeGa
SCL-4	O, HF, H_2O	60.1 - 66.1	GeGa
SCL-5	N_2O, CO_2	16.21* - 17.42	BIB
SCL-6	H_2O, HCl	65.0 - 70.91	GeGa

Table 410: CRISTA detector specification

GPS-based attitude determination of SPAS. A TANS Vector (SS/L) receiver, modified by Stanford University software, provided independently attitude and position data. Four GPS patch antennas were mounted on the zenith face of SPAS. The carrier-phase differential GPS (CDGPS) measurements of the TANS Vector receiver served as input for a recursive Kalman filter attitude estimation algorithm. [1311)]

MAHRSI (Middle Atmospheric High Resolution Spectrograph Investigation), Naval Research Lab, PI: R. R. Conway. Measurement of dayglow in the 190 - 320 nm region with a resolution of 0.02 nm. The prime objective is to measure limb intensity profiles of the resonance fluorescent scattering of sunlight, OH and NO, in the mesosphere and thermosphere (40 - 160 km region in total). From these intensity profiles, vertical density profiles of OH and NO are inferred with a vertical resolution of 2 km. By measuring Rayleigh scattering intensity profiles, the experiment provides precise knowledge of the neutral density and temperature in the mesosphere. The MAHRSI instrument consists of four subsystems: the telescope/spectrograph assembly, MECA (MAHRSI Electronics Controller Assembly), MDEA (MAHRSI Detector Electronics Assembly), and HVPS (High Voltage Power Sup-

1366) D. Offermann, et al., "Cryogenic Infrared Spectrometers and Telescopes for the Atmosphere (CRISTA) experiment and middle atmosphere variability," Journal of Geophysical Research, Vol. 104, No D13, July 20, 1999, pp. 16,311-16,325 ply). The telescope/spectrograph assembly employs an f/7.5 Czerny-Turner spectrograph with a 75 cm focal length behind a 50 cm focal length telescope (57 cm^2 aperture) mounted on CRISTA-SPAS. The wavelength range of sensitivity is from 195-320 nm, the spectral resolution is 0.018 nm at 310 nm and 0.026 nm at 215 nm. A spectral bandwidth of about 4 nm is imaged on the focal plane at a given grating position. The grating can be scanned to cover the entire spectral range of the instrument. A highly polished plane scan mirror at the baffled aperture of the telescope controls the vertical motion of the FOV (0.01° x 1.13°). MDEA provides power distribution, operates the intensified CCD (ICCD) detector, and transfers data from the CCD to the microprocessor after each integration. MAHRSI is operated and controlled by MECA (management of the command data interface, control of the dust door, scanning mirror, rotating polarizing filter, and scanned grating). During observation periods, CRISTA-SPAS holds MAHRSI's FOV parallel to that of CRISTA's center telescope. [1367)]

SESAM (Surface Effects Sample Monitor). The objective is to explore the influence of atmospheric atomic oxygen on optical surfaces (used in UV astronomy, tests of "whiteness" samples, development stable of surface layers, etc.). The instrument can accommodate, expose, and test up to 40 samples.

J.1.3 ORFEUS-SPAS-2

ORFEUS (Orbiting Retrievable Far and Extreme Ultraviolet Spectrograph) is a joint DARA/NASA science mission, flown on the ASTRO-SPAS free-flyer platform of Shuttle flight STS-80 (Nov. 19 - Dec. 7, 1996). The ORFEUS-SPAS-2 mission is designed to study the spectra that reveal the physical conditions needed to understand the universe.

The ORFEUS-SPAS-2 payload is identical to that of ORFEUS-SPAS-1 with the following addition:

ARP (ATV Rendezvous Predevelopment). Deployment and retrieval maneuvers of SPAS release/approach with Shuttle. [1368) 1369)] The ORFEUS-SPAS-2 free-flyer on Shuttle flight STS-80 carried, in addition to its prime payload, an ARP secondary payload of ESA/ESTEC, consisting of a GPS Tensor receiver and R-GPS (Relative-GPS) navigation algorithms (referred to as ARP-Kernel, developed by MMS, France). The prime ARP contractor is MMS. ATV is the future Automated Transfer Vehicle, the ESA resupply vehicle for ISS (International Space Station), ARP is considered a prototype system.

[The overall ARP activities are split into two main phases: a) the "long-range" rendezvous leg which demonstrate relative GPS techniques (using Tensor), and b) the "short-range" leg, based on measurements of the Rendezvous Sensor (using a laser sensor of DASA Jena Optronics, Germany). The Rendezvous Sensor (RVS) employs the following principle:

- A laser beam (905 nm wavelength) is emitted from RVS (installed on the forward section of the ATV)
- The beam is reflected by a dedicated target pattern (composed of six retro-reflectors) installed in proximity of the docking port of the ISS Service Module.
- The reflected beam is detected by RVS on ATV, then processed to provide the required data to the on-board navigation function.

RVS is physically composed of two separate units, the sensor head, and the electronic unit.]

The STS-80 ASTRO-SPAS (ORFEUS-SPAS-2) mission is regarded as Flight Demonstration 1 (FD1) in the context of ARP. FD-1 (a joint ESA-NASA experiment) was considering

1367) R. R. Conway, M. H. Stevens, et al., "Middle Atmosphere High Resolution Spectrograph Investigation," Journal of Geophysical Research, Vol. 104, No D13, July 20, 1999, pp. 16,327-16348

1368) M. Cislaghi, U. Thomas, M. Lellouch, J. M. Pairot, "Development and Verification of Automated Rendezvous for ATV," Proceedings IAF-96-T.2.08, Oct. 7-11, 1996, Beijing

1369) M. Cislaghi, U. Thomas, M. Lellouch, G. Limouzin, "ATV - Pre-development Program - Flight Demonstrations," IAF-97.T.2.03

only the GPS aspects of the ARP-Kernel (ARPK). - The Shuttle orbiter carried also a GPS receiver (TANS Quadrex), implemented on the fixed part of WSF (Wake Shield Facility) within the cargo bay, and an optical sensor TCS (Trajectory Control Sensor). TCS (of NASA, a laser based sensor) verified the R-GPS measurements by tracking a retro-reflector mounted on the x-face of ORFEUS-SPAS.

J.1.4 CRISTA-SPAS-2

A reflight of the CRISTA-SPAS-1 mission and payload took place on STS-85 (Aug. 7-19, 1997), labelled as CRISTA-SPAS-2.

The scientific emphasis was on the study of small-scale tracer "filaments" (long, thin regions of differing composition, including temperatures) in the stratosphere. Data analysis of the two CRISTA-SPAS missions may indicate, how these filaments contribute to the transport of ozone. The STS-85 flight offered an opportunity for increased latitudinal coverage of the atmosphere beyond that of the STS-66 mission (data of summer conditions at northern latitudes and the polar night at high southern latitudes). A CRISTA/MAHRSI campaign (Appendix Q), employing balloons, research aircraft, etc. was conducted in Europe in parallel to the STS-85 mission, in support of CRISTA-SPAS-2.

MAHRSI = Middle Atmospheric High Resolution Spectrograph Investigation (Naval Research Lab, PI: R. Conway). Same instrument as in the STS-66 mission (CRISTA-SPAS-1). During STS-85 flight, MAHRSI gathered new vertical profile data on the distribution of OH in the mesosphere and upper stratosphere under very different conditions (both seasonal and diurnal) from its previous flight on STS-66. Also, more measurements of nitric oxide were conducted. Researchers hope to gain a better understanding of how hydroxyl (OH) behaves during different seasons around the globe and how it interacts with the behavior of ozone and other trace gases.

IPEX-II (Interferometry Program Experiment 2). A NASA/JPL experiment, the prime objective was to monitor on-orbit the microdynamic behavior of space-like booms (to serve as input for future interferometry missions). The IPEX-II boom is a 9-bay expandable ADAM mast (2.35 m x 0. 3 m x 0.3 m), developed by AEC-Able, with steel-bracing cables pre-loaded at about 112 kg. The boom was cantilevered to the side of the ASTRO-SPAS platform. The experiment consisted of three phases:

- Investigation of thermal snapping in joint-dominated structures due to temperature gradients. A suite of 24 high sensitivity micro-g accelerometers with collocated temperatures sensors, mounted on the boom, monitored normal operations aboard CRISTA-SPAS in free flight. Data was obtained for five sun-to-shade orbital periods.
- Investigation of structural boom behavior due to induced loads. Two shakers, mounted orthogonally at boom tip, performed random and step-sine tests between 9-300 Hz, with loads from 0.048 to 0.00055 kg to detect any possible nonlinear behavior in the boom.
- Investigation to isolate and to quantify on-board disturbance sources of the CRISTA-SPAS platform and to measure their effect on the boom. This involved gyro and thruster operations and in-between times.

J.2 ATLAS (Atmospheric Laboratory for Application and Science)

The ATLAS program is part of NASA's 'Mission to Planet Earth' (MTPE). The ATLAS missions (on the Shuttle) investigate specifically how Earth's atmosphere and climate are affected by the sun and by the products of industrial complexes and agricultural activities. Experiments flown on the three ATLAS missions gathered data throughout the sun's

11-year activity cycle.[1370),1371)] There is a "core" of instruments which was the same on all ATLAS missions, and additional instruments unique to several of the Shuttle flights. An important goal of the ATLAS program is to provide measurements that relate to and coincide with other instruments that are flying on other satellites. There is a coordinated underflight program for correlative measurements such as: UARS, NOAA-POES, ERBS, EURECA, TOMS (on Nimbus-7 and Meteor-3-6).

ATLAS-1 consists of 12 international experiments and supporting hardware mounted on a two-Spacelab-pallet train in the orbiter payload bay, and one experiment contained in two canisters mounted on one adapter beam assembly (SSBUV). These experiments study the chemical makeup of the atmosphere between approximately 15 and 600 km above the Earth's surface, measure the total energy contained in sunlight and how that energy varies, investigate how Earth's electric and magnetic fields and atmosphere influence one another, and examine sources of ultraviolet light in the universe. Many of the experiments are also scheduled for later ATLAS missions, so the data gathered during ATLAS-1 was the first in a series of long-range studies that measure changes in the atmosphere and the sun.

ATLAS-1 had periods of solar pointing, Earth-limb pointing, and additional special attitudes throughout the mission. The orbiter is the primary experiment pointing system; however, some of the experiments have limited fine-pointing capability for enhanced pointing accuracy.

Orbit: ATLAS-1 was launched March 24, 1992 by STS-45 (Atlantis) to a 296 km orbit, inclination = 57°, 8 day mission. For ATLAS-2 and -3 orbits see O.17 (Space Shuttle Mission Chronology).

Sensors (Core Instruments):

MAS = Millimeter-Wave Atmospheric Sounder (PI: G. Hartmann, MPAe, Lindau). Measurement of millimeter-wave emission from the atmosphere. As a limb emission instrument, it gets nearly global coverage. Determination of the following parameters:

- Upper troposphere: H_2O, O_3, and CO profiles
- Stratosphere: O_3 for trend analysis, ClO (globally), H_2O
- Mesosphere: O_3, H_2O, and CO
- Temperature and pressure of stratosphere and mesosphere

MAS is the predecessor of AMAS, a Heterodyne Limb Sounder (like MLS), observing in the channels: 61 GHz, 62 GHz, 63 GHz, 183 GHz, 184 GHz and 204 GHz. MAS determines pressure, temperature, ozone, water and ClO profiles (10-100 km altitude). MAS uses a dish-shaped antenna to scan the Earth's limb to collect spectral information at distinct altitudes. The ClO receiver on ATLAS-3 has been upgraded for better sensitivities. The instrument has an antenna aperture of 1 m x 1.3 m; it features 240 channels (12 bit), mass = 200 kg, power = 400 W, data rate = 86.4 kbit/s.

The MAS experiment is a joint project of: University of Bremen (K. Künzi, scientific investigation), University of Bern (hardware), MPAe Lindau (data processing and scientific investigation) and the Naval Research Lab (Washington).

ATMOS = Atmospheric Trace Molecule Spectrometer (JPL Sensor, PI: M. R. Gunson). This is an infrared absorption instrument which works during occultations (sunrise, sunset) and measures a wide variety of species with good vertical resolution. The ATMOS experiment was already very successful on Spacelab-2 (flown in 1985). It employs a high-resolution FT (Fourier Transform) interferometer and limb viewing in the bands of 2.2-16 μm. During ATLAS-3 the ATMOS instrument also had a video camera to record pictures of the sun during sunrises and sunsets (pointing confirmation).

1370) Jack Kaye, "Summary of ATLAS Shuttle Missions," Paper presented at the EOS-B Atmospheric Payload Panel Meeting Washington, D. C., Feb. 26-27, 1991

1371) Information provided by the Earth Science Application Division (ESAD Office) at NASA HQ, Washington

Configuration	compensated (cat's-eye) type: continuous scan
Spectral coverage	600 - 4800 cm^{-1} (2.2-16 μm)
Maximum resolution	0.013 cm^{-1} (unapodized)
Maximum OPD (Optical Path Difference)	±48 cm
Scan time	2.2 s
Detector type	HgCdTe at 77 K
Sensitivity	SNR >100:1 at 2500 cm^{-1} (source-noise limited)
Operating temperature	-5 to + 45 °C
Beamsplitter and compensator	potassium bromide
Beam diameter (internal)	2.5 cm
Telescope diameter	7.5 cm
IFOV	selectable: 1, 2, or 4 mrad
System étendue	≥ 1.39 x 10^{-4} cm^2 sr
Path difference control	stabilized He/Ne laser
Instrument mass, power, data rate	250 kg, 360 W, 16 Mbit/s

Table 411: Parameters of the ATMOS FTS

SUSIM =Solar Ultraviolet Spectral Irradiance Monitor (Naval Research Laboratory, US, PI: G. Brueckner).[1372)] Measures solar UV flux as a function of wavelength from 110 to 410 nm (compatible with SUSIM on UARS (see A.32), also for calibration of UARS instrument). SUSIM can also observe ozone and molecular oxygen profiles by occultation during solar observing orbits. Heritage: SUSIM flew on Spacelab-2 (STS-19), July 29-August 6, 1985.

SOLCON = Solar Constant Sensor (IRMB, Brussels, Belgium, PI: R. Crommelynck). Measurement of the solar constant. The SOLCON instrument is a cooperative effort of IRMB, Space Science Dept. of ESA, and LaRC. Measurement of the absolute value of the total solar irradiance (and long-term variations). The technique used is to compare the heating of a cavity exposed to sunlight with the temperature of another cavity that is not exposed to sunlight. Instrument accuracy to within 0.1% of the solar constant, and about 0.01% precision. SOLCON is a high-resolution, self-calibrating radiometer with a digital processing/converter unit.

ACRIM = Active Cavity Radiometer Irradiance Monitor (JPL, PI: R. C. Willson). See description under ACRIMSAT (A.2) and SMM (K.22). ACRIM is a pyrheliometer measuring the total solar irradiance (solar constant) with a technique similar to SOLCON's.

SOLSPEC = SOLar SPECtrum Measurement (CNRS, France, PI: G. O. Thuillier). Measures the solar irradiance from 180-3200 nm using three double spectrometers and an onboard calibration device. SOLSPEC can also observe abundances of ozone by measuring the backscatter of specific UV and VIS wavelengths during nadir operations. SOLSPEC heritage: flew on Spacelab-1 in November 1983 (see J.22 on page 853).

SSBUV[1373)] = Shuttle Solar Backscatter Ultraviolet Spectrometer (GSFC, PI: E. Hilsenrath). Uses UV backscatter in nadir to measure vertical profiles of ozone in the stratosphere and in the lower mesosphere from 180 to 450 nm. Note: SSBUV is not a core instrument; it is a separate ATLAS Shuttle payload (co-manifested with ATLAS and integrated into the ATLAS science plan).

SSBUV is a sensor of SBUV/2 heritage (from Nimbus-7 and NOAA-9 satellites onward). Objective: Calibration of long-term satellite ozone data sets with complementary Shuttle flights.

SSBUV is scheduled to fly eight times from 1989 to 1996. The first flight with SSBUV instrumentation occurred on October 19, 1989 on the Shuttle Atlantis. Throughout the

1372) SUSIM brochure of Naval Research Lab, available at NASA HQ's Document Resource Facility
1373) "Calibration of Long Term Satellite Ozone Data Sets Using the Space Shuttle," E. Hilsenrath, in Optical Remote Sensing of the Atmosphere, 1990 Technical Digest Series of the Optical Society of America, Vol. 4, pp. 409-412

Shuttle flight period coincident observations were taken with the SBUV on Nimbus-7 and the SBUV/2 on NOAA-9 and NOAA-11 satellites. The SSBUV spectrometer is located in a GAS canister attached to the side of the Shuttle's cargo bay. A motorized door assembly opens up to allow the SSBUV to view the Earth and the sun.

The SSBUV instrument is the SBUV/2 engineering model now flying on NOAA satellites. See also J.26 on page 861 for a description of SSBUV and G.13 (NOAA-POES) for a description of SBUV/2.

The ATLAS Program on Shuttle (about 1 week each time) has the following time schedule:

- ATLAS-1 Mission: March 24 - April 2, 1992 with Atlantis (correlative measurements to UARS). Flown mission on STS-45
- ATLAS-2 Mission: April 8 - 17, 1993 (correlative measurements to UARS). ATLAS-2 payloads gathered data on the relationship between the sun's energy output and the Earth's middle-atmosphere chemical makeup. Study of how these factors affect the Earth's ozone level. Flown mission on STS-56.
- ATLAS-3 Launch: Nov. 3-14, 1994 (with the CRISTA and MAHRSI payload on the ASTRO-SPAS free-flyer, see description under J.1). Flown mission on STS 66.

Additional ATLAS sensors are:

ALAE = Atmospheric Lyman-Alpha Emissions (CNRS, France). Uses on-board hydrogen and deuterium cells to measure thermospheric/exospheric H and D concentrations, as well as Lyman alpha amounts in the interplanetary medium. ALAE will fly on ATLAS-1.

ISO = Imaging Spectrometric Observatory (MSFC). The spectrometer measures 'low light observations' in daylight and on the night-side of the Earth. There are five spectral bands from 30-1300 nm. ISO will fly on ATLAS-1.

ENAP (Energetic Neutral Atom Precipitation), of the University of Texas, Dallas. ENAP is a study which will use data from ISO.

AEPI = Atmospheric Emissions Photometric Imaging (Lockheed, Palo Alto Research Lab). Optical emissions from the upper atmosphere/ionosphere and from the Space Shuttle environment. Images of natural and induced aurora and airglow. AEPI flew on ATLAS-1.

SEPAC = Space Experiments with Particle Accelerators (Southwest Research Institute, US). Uses electron beam accelerator and other instruments to carry out active and interactive experiments on and in the Earth's ionosphere. SEPAC was flown on ATLAS-1.

FAUST = Far Ultraviolet Space Telescope (University of California, Berkeley). Far-ultraviolet images of large-scale phenomena. FAUST flew on ATLAS-1.

J.3 Bitsy-SX (Bitsy-Spacecraft in Future-X)

Bitsy-SX is a Shuttle demonstration payload in NASA's Future-X program, referred to as SPASE (Small Payload Access to Space Experiment), designed and built around AeroAstro's Bitsy-SX spacecraft kernel.

Background: In 1998 NASA announced a new program, Future-X, managed by MSFC, with the aim to support a series of flight demonstrations to validate new technologies. The idea is to prove technologies that improve performance and lower development, production and operating costs of future earth-to-orbit and in-space transportation systems.
Bitsy™ is a trade name, representing a line of spacecraft "kernel" modules, designed and developed by AeroAstro of Herndon, VA. The "Bitsy™ concept" as such is an enabling technology, meaning that one does not fly a Bitsy (though among its design requirements is

that it can behave as a fully functional spacecraft); rather one uses Bitsy as the starting point in the development of a full spacecraft. The kernel approach applies COTS (Commercial-of-the-Shelf) and standardization concepts to the fullest extent to obtain significant cost and turnaround advantages over conventional bus development. 1374)

The first Bitsy-SX implementation (the total mass of the Bitsy-SX with a NASA microgravity payload is about 35 kg) will fly on the SPASE vehicle in Nov. 2001 (Shuttle flight STS-108). SPASE uses Bitsy-SX, perched upon an octagonal experiment structure, that fits inside a GAS (Get Away Special) canister on Shuttle. The outside of the experiment housing is covered with solar panels. Bitsy-SX contains lithium-ion batteries, providing 7 Wh of energy (average orbital power is 5 W) during orbital eclipse phases. The ACS (Attitude Control System) of SPASE is passive, using hysteresis rods and permanent magnets to keep the S/C body rates below 10^{-5} g for the crystal growth experiment. The payload interaction consists primarily of keeping the crystal growth cavity within a specified temperature range, commanding a digital camera to take images periodically, and downloading those images when a ground station contact occurs.

Deployment of the SPASE free-flyer vehicle from the GAS canister is attended by the Shuttle crew (Hitchhiker launch). It occurs at an altitude high enough above the atmosphere to remain under the 10^{-5} g requirement for the science experiment. Communications (in UHF at 9.6 kbaud) with SPASE as well as S/C operations are conducted from a ground station at the University of Alabama in Huntsville. SPASE uses internally RS-232 interfaces to communicate with the payload. Simple timed commanding is supported as well as threshold monitoring of all S/C and payload functions. The science mission is designed to last 6 months prior to vehicle reentry into the atmosphere.

J.4 CIRRIS (Cryogenic Infrared Radiance Instrumentation for Shuttle)

CIRRIS is a DoD/USAF instrument, a Michelson FTS (Fourier Transform Spectrometer) interferometer, in a program (AFP-675) of upper atmosphere and space research. The objectives are to develop sensors with the capability to detect and track such targets as aircraft, missiles and satellites against an atmospheric background. The intent is the development of predictive models of atmospheric chemistry and physics under conditions of quiescent and disturbed upper atmospheres by aurora, airglow, rocket fuel injection, and targets of opportunity.

CIRRIS-1A, designed and built by the Center for Space Engineering at USU (Utah State University) for AGFL (Air Force Geophysics Laboratory) of Hanscom AFB, MA, and flown on Shuttle flight STS-39 (Discovery, Apr. 28-May 6, 1991), collected these types of data on the Earth's limb. The orbit of STS-39 provided an altitude of about 260 km with an inclination of 57°. Specific interests were nocturnal, diurnal and latitudinal geomagnetic effects on infrared background radiation. Note: CIRRIS is a predecessor instrument of SPIRIT-III on MSX (see M.14). 1375) 1376) 1377) 1378)

The primary CIRRIS-1A experiment assembly consists of an FTS interferometer and a highly sensitive spatial radiometer (length of CIRRIS-1A assembly is 2.8 m, mass = 2000

1374) S. A. McDermott, D. J. Goldstein, "The Bitsy™ Spacecraft Kernel: Reducing Nanosatellite Mission Cost in the MSFC Future-X Program Through Miniaturized technologies," Proceedings of the 13th AIAA/USU Conference on Small Satellites, Aug. 23-26, 1999, Logan UT, SSC99-IX-8

1375) J. C. Kemp, et al., "Cryogenic Michelson Interferometer Spectrometer for Space Shuttle Application," Proceedings of SPIE, Vol. 686, 1986, pp. 151-159. Application: Infrared Detectors, Sensors, and Focal Plane Arrays

1376) C. L. Wyatt, "CIRRIS-1A interferometer: radiometric analysis," Applied Optics, Vol. 28, No. 23, Dec. 1, 1989, pp. 5069-5072

1377) G. E. Bingham, et al., "Cryogenic Infrared Radiance Instrumentation for Shuttle (CIRRIS-1A) Earth limb spectral measurements, calibration, and atmospheric O_3, HNO_3, CFC-12, and CFC-11 profile retrieval," Journal of Geophysical Research, Vol. 102, D3, Feb. 20, 1997, pp. 3547-3558

1378) D. K. Zhou, et al., Stratospheric CH_4, N_2O, H_2O, NO_2, N_2O_5, and $ClONO_2$ profiles retrieved from Cryogenic Infrared Radiance Instrumentation for Shuttle (CIRRIS-1A)/STS-39 measurements," Journal of Geophysical Research, Vol. 102, D3, Feb. 20, 1997, pp. 3559-3573 kg). These multi-detector instruments share the common collecting optics of a single high off-axis rejection telescope. Mounted on the outside of the telescope heat exchanger are two photometers, two TV cameras, a 16-mm film camera celestial aspect sensor (co-aligned with the primary sensor), and a horizon sensor - to provide position data on the sensor's field of view (FOV).

The interferometer is of standard Michelson design with flat mirrors and reference system path through the center of the infrared beam. Major FTS elements include the optical filter wheel (eight position with seven filters to reduce photon noise), the optical modulator (operating at 20 K contains the optical elements, the translating mirror system, the mirror position reference system, and the remote alignment system), the detectors, internal calibration sources, the FTS control electronics, and the FTS signal conditioning electronics. The spectral range of the system is from 2.5 - 25 µm. The five SiAs detectors provide a resolution of the interferogram ranging between 1-8 cm^{-1} over an eight order dynamic range. The five detectors, with a TFOV of 1.22° x 1.5° (or 21.3 mrad x 26.2 mrad), include one large detector for maximum sensitivity, one detector designed to provide high sensitivity at 1 cm^{-1} resolution, and three small detectors for spatial definition. The interferometer, radiometer, and telescope are cooled by supercritical helium at 20 K to minimize background noise.

CIRRIS-1A calibration sources include wideband sources, a quasi-blackbody source, an alignment source, an off-axis source, photon bias, and detector stimulator sources and passive cold sources.

The sensors are suspended in a two-axis gimbal, mounted in an equipment support system (ESS), to provide the required scanning capability.

Spectral range	2.5 - 25 µm (400 cm^{-1} - 4000 cm^{-1})		
Detectors	SiAs, five element array		
Sensitivity NESR (Wcm^{-2} sr^{-1} cm^{-1} @ 20 µm)	Detector 1: 1.3 10^{-14} Detector 3: 2.2 10^{-13} Detector 4: 5.0 10^{-13} Detector 5: 5.0 10^{-13} Detector 2: 1.1 10^{-13}	[8 cm^{-1} scan in 1.53 s] [8 cm^{-1} scan in 1.53 s] [8 cm^{-1} scan in 1.53 s] [8 cm^{-1} scan in 1.53 s] [1 cm^{-1} scan in 9.75 s]	
Dynamic range	8 orders considering the entire array		
Signal bandwidth	DC to 900 Hz (max signal frequency = 450 Hz)		
Focal plane optics	Ritchey-Chretien f/1.6		
Interface optical aperture	4.76 cm		
FOV	Detector 1: 9.6 mrad x 26.2 mrad Detector 2: 10.0 mrad x 10.0 mrad Detector 3: 2.1 mrad x 7.0 mrad Detector 4: 1.6 mrad x 5.0 mrad Detector 5: 1.6 mrad x 5.0 mrad Overall: 21.3 mrad x 26.2 mrad		2.50 10^{-4} sr 1.00 10^{-4} sr 1.48 10^{-5} sr 7.77 10^{-6} sr 1.00 10^{-6} sr

Table 412: Parameters of the CIRRIS-1A instrument

J.5 EURECA (European Retrievable Carrier)

An ESA spacecraft/platform[1379)] designed and developed to be recovered from orbit after completion of a particular mission, returned to Earth, refitted with a new sensor complement, and relaunched for another extended/new mission. Typical mission duration is of the order of six months to one year. The concept allows complete payload/sample recovery.

J.5.1 EURECA-1 Mission

Launch of EURECA-1 by Shuttle (Atlantis, STS-46) from KSC on July 31, 1992. Retrieval of EURECA-1 by Shuttle (Endeavour) during the STS-57 mission from June 21 - July 1,

1379) ESA Press Release, ESA/ESTEC, 1991

1993. Mission operations of EURECA-1 was provided by ESOC in Darmstadt; some payload operations were provided by MUSC (Microgravity User Support Center) of DLR in Cologne, Germany. The spacecraft was developed and integrated at DASA/ERNO, Bremen, Germany.[1380] [1381]

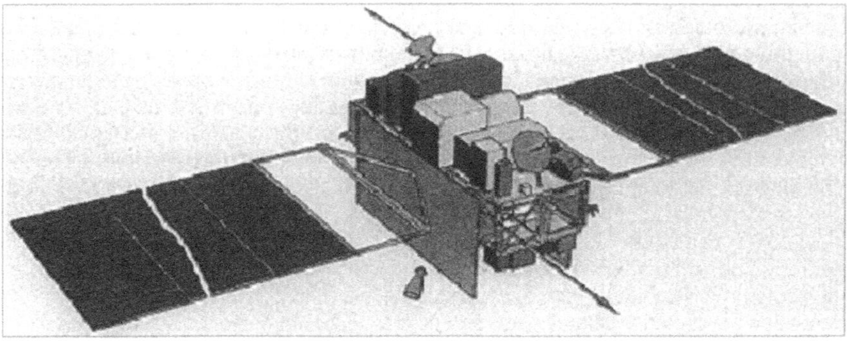

Figure 239: Illustration of the EURECA platform

The EURECA S/C structure had dimensions of 4.6 m x 2.6 m. The bus structure consisted of carbon fiber struts connected by titanium nodal joints. A grapple fixture permitted deployment/retrieval by the Shuttle Remote Manipulator System. EURECA was three-axis stabilized by magnetorquers, supported by a reaction control assembly of 6 x 21 mN nitrogen thrusters. Orbit transfers between altitudes of 400-500 km were performed by the orbit transfer assembly of redundant 4 x 21 N hydrazine thrusters (the propellant volume was sized for two transfers and 9 months on-orbit stay). The attitude rate was measured by an accelerometer package, gyros and IR Earth and sun sensors. Power was provided by twin deployable/retractable panels of silicon solar cells, generating 5 kW at 28 VDC (providing 1 kW average for payload operations). In addition, 4 NiCd batteries provided energy of 40 Ah. The launch mass of EURECA was 4490 kg (payload capacity of up to 1000 kg). Data was transmitted via S-band downlink at rates up to 256 kbit/s.

Orbit: Low Earth Orbit (LEO) circular; altitude = 502 km, inclination = 28°, retrieval altitude of 476 km.

Sensor Complement

Only those sensors are described and indexed that fit into the scope of this book, namely: "Observation of the Earth and its Environment." The other sensors (in the life and material sciences) are simply listed with a short commentary.

ERA = Exobiology and Radiation Assembly. A multiuser life science facility for experiments on the biological effects of space radiation. PI: H. Bücker, DLR, Germany.

AMF = Automatic Mirror Furnace. Optical radiation furnace for growing single, uniform crystals from the liquid or vapor phases, using the travelling heater or Bridgman methods. PI: K. W. Benz, Universität Freiburg, Germany.

SGF = Solution Growth Facility. A multiuser facility for growing monocrystals from solution. PI: J. C. Legros, Université Libre de Bruxelles, Belgium.

PCF = Protein Crystallisation Facility. A multiuser solution growth facility for protein crystallisation in space. PI: W. Littke, Universität Freiburg, Germany.

1380) P. Ferri, H. Hübner, S. Kellock, W. Wimmer, "The Joint ESA-NASA Operations for Eureca's Deployment and Retrieval, "ESA Bulletin, Number 76, November 1993, pp. 81-90

1381) F. Dreger, J. Fertig, D. Gawthrope, S. Martin, et. al., "Eureca: The Flight Dynamics of the Retrieval," ESA Bulletin, Number 76, November 1993, pp. 92-99

MFA = Multi-Furnace Assembly. A multiuser facility dedicated to material science experiments. PI: A. Passerone, National Research Council, Genoa, Italy

HPT = High Precision Thermostat. An instrument for long-term experiments requiring microgravity conditions and high precision temperature measurement and control (typical experiments are: 'caloric', 'critical point', and 'phase transitions'). PI: G. Findenegg, Ruhr Universität, Bochum, Germany.

SFA = Surface Forces Adhesion. Study of the dependence of surface forces and interface energies on physical and chemical-physical parameters such as surface topography, surface cleanliness, temperature, and the deformation properties of the contacting bodies. PI: G. Poletti, Università di Milano, Italy.

RITA = Radio-Frequency Ion Thruster Assembly. Study of the use of electric propulsion in space. PI: H. Bassner, DASA-MBB, Munich, Germany.

IOC = Inter-Orbit Communication Instrument. A technological experiment to provide a pre-operational in-flight test and demonstration of all required instrument functions and services. PI: R. Tribes, CNES, France.

ASGA = Advanced Solar Gallium Arsenide Array. A technological experiment for the performance testing of future solar arrays. PI: C. Flores, CISE SPA, Segrate, Italy.

WATCH = Wide Angle Telescope. Objective: detection of celestial gamma-ray burst sources and X-ray sources with photon energies in the range of 5 - 200 keV, measurement of source direction. PI: N. Lund, Danish Space Research Institute, Lyngby, Denmark.

TICCE = Timeband Capture Cell Experiment. Study of microparticle population distributions in near-Earth space, typically Earth debris, meteoroids, and cosmic dust. TICCE captures micron-dimensional particles with velocities in excess of 3 km/s and stores the debris for retrieval and post-mission analysis. PI: J. A. M. McDonnell, University of Kent, UK.

Particles detected by the instrument pass through a front foil and into a debris collection substrate positioned 100 nm behind the foil. Each perforation in the foil has a corresponding debris site on the substrate. The foil is moved during the mission in 50 discrete steps. The phase shift between the debris site and the perforation enables a determination of the impact time. Instrument mass = 8 kg, dimensions: 690 x 690 x 80 mm.

ORA = Occultation Radiometer Instrument. Objective: measurement of aerosols and trace gas densities in the Earth's mesosphere and stratosphere (vertical profiles between 20 -100 km of: ozone, nitrogen dioxide, water vapor, carbon dioxide, and background and volcanic aerosols) . PI: E. Arijs, BIRA (Belgisch Instituut voor Ruimte Aeronomie), Brussels, Belgium.

ORA measures the intensity of solar radiation during the sunrise and sunset phases of each orbit. It uses the SUN pointing capabilities of EURECA and makes measurements in ten narrow wavelength bands (UV/VIS, NIR) in the spectral region 250 - 1100 nm. ORA consists of a UV/VIS unit, an NIR unit and a control/electronics unit. The UV/VIS unit has 8 similar modules which measure the relative solar irradiance in narrow bands. Each module contains a quartz window, an interference window to select the appropriate wavelength, optics to limit the detector viewing angle, and a photodiode.

SOVA = Solar Constant and Variability Instrument (copy of SOLCON). Objective: measurement of the solar constant, its variability and its spectral distribution. PIs: D. Crommelynck, IRMB, Brussels, Belgium; and C. Fröhlich, PMOD/WRC Davos, Switzerland.

The instrument is composed of three boxes, two of which (SOVA1 and SOVA2) house the various devices dedicated to functions such as sun-pointing, sun photometers, and two different types of absolute radiometers. The absolute radiometers are very accurate instruments for the measurement of total irradiance. They are essentially cavities, in which the solar radiation is trapped and absorbed. Heat flux transducers convert the radiative power into electrical power and return a calibrated measurement of the sun's radiation. The high precision radiometer (RELOS) is used to measure solar oscillations in total irradiance. The sun-pointing resolution of SOVA2 is 2 arc seconds. The sun photometers use interference filters to select the wavelength band and silicon diode devices to measure the radiation. The complete set of photometers covers the spectrum from 303 nm to 865 nm.

SOSP = Solar Spectrum Instrument. Objective: study of solar physics and solar-terrestrial relationships in aeronomy and climatology. Measurement of absolute solar irradiance and its variations in the spectral range 170 - 3200 nm (accuracy of 1% in VIS/IR and 5% in UV). PI: G. Thullier, CNRS, Verriers le Buisson, France.

SOSP has three spectrometers, one for each spectral domain (UV/VIS/IR). The spectrometers select the spectral range using a double monochromator; irradiance measurement with a photomultiplier in UV and VIS, in IR with a cooled lead sulfide cell. Spectral resolution: 1 nm in UV/VIS, 20 nm in IR. In-flight calibration is provided.

J.6 IPS (Instrument Pointing System)

IPS is an ESA-sponsored instrument, built by Dornier, and flown on three Spacelab missions with the objective to provide instrument pointing with very high accuracy and stability at stars, the sun, the Earth or other targets of observation. IPS was first flown on Shuttle flight STS-51-F as Spacelab-2 (July 29 - Aug. 6, 1985), then on STS-35 as Astro-1 (Dec. 2-10, 1990), and again on STS-67 as Astro-2 (March 2-18, 1995). [1382)]

The IPS provides precision pointing for a wide range of payloads, including large single instruments (telescopes) or a cluster of instruments or a single small-rocket-class instrument. The pointing mechanism can accommodate instruments of diverse sizes and weights (up to 7,000 kg) and can point them to within 2 arcseconds and hold them on target to within 1.2 arcseconds. The IPS accuracy for a 2000 kg payload is 0.4 arcseconds lateral and 11.2 arcseconds for roll under star tracker control (0.5 arcseconds lateral and 41 arcseconds roll in sun mode). The overall mass of IPS is 1,180 kg, power demand of up to 1.25 kW, and a data rate of 16 Mbit/s.

IPS consists of a three-axis gimbal system mounted on a gimbal support structure connected to the pallet at one end and to the aft end of a payload at the other, a payload clamping system to support the mounted experiment elements during launch and landing, and a control system based on the inertial reference of a three-axis gyro package and operated by a gimbal-mounted minicomputer.

The basic structural hardware is the gimbal system, which includes three bearing/drive units, a payload/gimbal separation mechanism, a replaceable extension column, an emergency jettisoning device, a support structure and rails, and a thermal control system. The gimbal structure itself is minimal, consisting only of a yoke, an inner gimbal and an outer gimbal to which the payload is attached by the payload-mounted integration ring.

The three identical drive units are so arranged that their axes intersect at one point. From pallet to payload, the order of the axes is elevation, cross-elevation and azimuth. Each drive assembly includes three wet-lubricated ball bearings, two brushless dc-torquers and two single-speed/multispeed resolvers.

IPS is controlled through the Spacelab subsystem computer and a data display unit and keyboard. It can be operated either automatically or by the Spacelab crew from the pressurized module and also from the payload station on the orbiter aft flight deck. IPS has two operat-

1382) http://www.ksc.nasa.gov/shuttle/technology/sts–newsref/spacelab.html

ing modes, which depend on whether the gimbal resolver or gyro is used for feedback control of attitude. An optical sensor package - consisting of one boresighted fixed-head star tracker and two skewed fixed-head star trackers - is used for attitude correction and also for configuring the IPS for solar, stellar, or Earth viewing.

During a search phase, each star tracker searches a 2° x 2° FOV for one or two star images (stellar or solar missions) appearing in the optics. When a star is detected, the position of the star image on the FOV is determined by a scanning procedure. The tracker delivers this position to the Spacelab Subsystem Computer as electrical coordinates.

Acquisition modes: IPS can obtain optical hold on a specific target in two ways, automatic and manual. The automatic mode uses a procedure known as "Operational Identification" which uses a known set of "guidestars" to identify the star field of each tracker. The manual mode consists of a procedure known as "Manual Target Acquisition" to identify the boresight tracker region.

J.7 ISIR (Infrared Spectral Imaging Radiometer)

ISIR is a high precision, multispectral imaging radiometer of NASA/GSFC (built by Space Instruments Inc., Encinitas, CA) flown on STS-85 (Aug. 7-19, 1997) as a hitchhiker payload (ISIR was one of twelve Hitchhiker experiments that were carried in the Shuttle bay on two cross-bay Hitchhiker bridge assemblies). [1383] [1384] The objective of ISIR was to demonstrate/validate a new infrared imaging technology with an uncooled detector array for the measurement of the infrared brightness temperature of cloud tops as well as the Earth's surface. The new technology is based on microbolometer focal plane arrays, allowing infrared imaging without external cooling. The ISIR detector is a warm focal plane 327 x 240 pixel array of silicon micromachined microbolometers. ISIR demonstrated the potential of this new technology for future much more compact, lower cost and more reliable spaceborne infrared imagers. On STS-85, ISIR was operated in conjunction with the Shuttle Laser Altimeter (SLA-2), for a new type of cloud science observation through combined active and passive remote sensing.

ISIR is a compact spectral imager providing radiometrically calibrated infrared imagery in three narrow spectral band centers at 8.55, 10.8 and 11.8m, and in a broad-band window from 7 to 13 µm. ISIR's main scientific mission is to image clouds from Shuttle orbit and provide information on cloud-top temperature and classification. A science goal is discrimination of cirrus cloud particle type by split-window analysis. The split-window analysis involves the differential radiance between the 10.8 and 11.8 µm atmospheric window channels. The required NEDT (Noise Equivalent Difference Temperature) system sensitivity performance was 0.1 K. The target spatial resolution was 250 m with a total image swathwidth of 85 km from the shuttle altitude.

ISIR is as a pushbroom imager with an MBA (Microbolometer Array) detector. A scene is imaged in a fixed staring mode with the cross-track pixels providing the image swath. Imaging is done with a f/0.73 germanium lens of 50.8 mm focal length and 70 mm optical diameter. The cross track full FOV is approximately 15°. A filter wheel between the lens and detector array is driven for spectral selection. In order to obtain the necessary instrument calibration during operation, an on-board blackbody calibration target is included. The blackbody is viewed by swinging in a mirror in front of the lens. A curved focussing mirror is used in order to minimize the size of the blackbody target. The overall instrument includes electronics and tape recorders. Two standard Exabyte-8505c 8mm tape recorders were used for data storage.

1383) http://isir.gsfc.nasa.gov/main.html

1384) J. D. Spinhirne, et al., "Preliminary Spaceflight Results from the Uncooled Infrared Spectral Imaging Radiometer (ISIR) on Shuttle Mission STS-85," SPIE, 1998

Spectral bands (4)	8.2 - 9.0 μm; 10.3 - 11.3 μm; 11.5 - 12.5 μm; and 7.0 - 13 μm
Dynamic range	Scene temperatures from 0 K to 400 K
NEDT @ 300 K	0.01 K to 0.06 K
Spatial resolution	1 mrad (250 m on ground from Shuttle @ 250 km)
Field of View (FOV)	19.25°. (81 km swath on ground from shuttle @ 250 km)
Optics	f/0.73, 50 mm focal length, 68 mm diameter
Inflight calibration	Ambient & cold calibration capability
Instrument mass; power	13 kg; 20 W operating, 120 W during cold calibration
Instrument size	45 cm x 35 cm x 30 cm
Data rate	368 kbit/s (with 12 bit encoding)

Table 413: ISIR instrument performance parameters

The primary advantage of the MBA detector (built by Loral with DARPA funding) used in ISIR is that it is a large-format (327 x 240) array providing good sensitivity in the thermal IR at room temperature. A TDI (Time Delay Integration) operating mode was implemented providing a (x 40) cumulative imaging capability. This required a special flight operation mode for high resolution imaging with less than 1/8 pixel smearing. An instrument roll stability of 0.01°/s and a yaw accuracy to the ground velocity vector of ±0.1° were provided by the Shuttle.

During the STS-85 mission, three special surface tracking maneuvers were executed by the Shuttle for the ISIR experiment. The goal was to image the thermal emission of cloud tops in high resolution over a large range in angles. The results permit test of angular cloud radiation models that are a key element for cloud retrieval applications of current GOES and LEO satellite data. - There were three aspects of the ISIR experiment that were firsts from space for cloud science.

- The combination with direct laser measurements of cloud height
- A global data set of thermal infrared imagery at near ¼ km resolution was acquired
- The additional new observation from space was the use of the 8.5 μm wavelength, which has applications for cloud particle discrimination.

J.8 LDEF (Long Duration Exposure Facility)

LDEF is a NASA/LaRC[1385)] freeflying spacecraft which accommodates technology, science, and applications experiments on long-term exposure to the space environment, with the objective to collect small meteoroids and space debris in Low Earth Orbit (LEO) for post-mission impact analysis of surfaces and materials that have been exposed to the space environment. Directional resolution of the flux of meteoroids and space debris particles.

LDEF was deployed on April 7, 1984 on Shuttle flight STS-41 C. Retrieval was planned after 10 months. Due to delays caused by the Challenger accident (STS-25), it was finally recovered 69 months after launch on January 12, 1990 (STS-32). As a consequence of the delay, much more data had been gathered than planned. Post mission deintegration in SAEF-II (Spacecraft Assembly and Encapsulation Facility) at NASA/KSC.

The LDEF spacecraft is an open-grid, 12-sided, cylindrical structure. Length of cylinder = 9.1 m, diameter = 4.3 m. Gravity gradient stabilized spacecraft, with the longitudinal axis pointing toward the center of the Earth. Surface elements are fixed relative to LDEF's velocity vector. Magnetic actuators control the rotation around the longitudinal axis. The LDEF structure is configured with 72 equal-size rectangular openings on the sides and 14 openings on the ends (six on the Earth-facing end, and eight on the space-facing end) for mounting experiment trays. Total mass = 9700 kg.

1385) A. S. Levine (editor), "LDEF - 69 Months in Space, First Post-Retrieval Symposium," NASA Conference Publication 3134 (Part 1 and Part 2), Proceedings of a symposium sponsored by NASA at Kissimmee, Florida, June 2-8, 1991

Orbit: almost circular orbit, altitude = 477 km, inclination = 28.5°. At the time of retrieval the orbital altitude had decreased to 335 km.

Experiments: All LDEF experiments are self-contained in trays that are clamped to the facility structure. The LDEF has 72 peripheral and 14 end experiment trays. The 12 sides of the LDEF structure are numbered rows 1 through 12 in a clockwise direction when facing the end with the support beam (the Earth-facing end in orbit). The six longitudinal locations are identified alphabetically as Bay A through Bay F, starting at the end with the support beam. A tray location is designated by Bay and Row: A-1, B-5, F-8, etc. The Earth-facing end is designated by a G identifier; the locations have even-number clock-position identifications (G-12, G-2, G-4, G-6, G-8, and G-10). The space-facing end is designated by an H identifier, the locations also following a clock-position convention (H-12, H-1, H-3, H-5, H-6, H-7, H-9, and H-11).

The 57 LDEF experiments are extensively described in the references, only a few experiments follow with short descriptions. [1386]

Exp. No.	Experiment Title (Sponsoring Institution)	Tray Numbers (Exp. Location)
A0015	Free-Flyer Biostack Experiment, DLR (Inst. für Flugmedizin)	C2, G2
A0019	Influence of Extended Exposure in Space on Mechanical Properties of High-Toughness Graphite-Epoxy Composite Material (U. of Michigan)	D12
A0023	Multiple Foil Micro-Abrasion Package, MAP (U. of Kent, UK)	C3, C9, D12, E6, H11
A0034	Atomic Oxygen Stimulated Outgassing (Southern University/MSFC)	C3, C9
A0038	Interstellar Gas Experiment (NASA-JSC / University of Bern)	E12, F6, H6, H9
A0044	Holographic Data Storage Crystals for LDEF (Georgia Institute of Technology)	E5
A0054	Space Plasma High Voltage Drainage (TRW Space and Technology Group)	B4, D10
A0056	Exposure to Space Radiation of High-Performance Infrared Multilayer Filters and Materials Technology Experiments (University of Reading/British Aerospace)	B8, G12
A0076	Cascade Variable Conductance Heat Pipe (McDonnell Douglas Astronautics Co.)	F9
A0114	Interaction of Atomic Oxygen with Solid Surfaces at Orbital Altitudes (U. of Alabama in Huntsville/NASA/MSFC)	C3, C9
A0133	Effect of Space Environment on Space Based Radar Phased Array Antenna (Grumman Aerospace Corporation)	H7
A0134	Space Exposure of Composite Materials for Large Space Structures (NASA/LaRC)	B9
A0135	Effect of Space Exposure on Pyroelectric Infrared Detectors (NASA/LaRC)	E5
A0138-1	Study of Meteoroid Impact Craters on Various Materials (CERT/ONERA)	B3 These experiments are also referred to as: 'Frecopa'
A0138-2	Attempt at Dust Debris Collection with Stacked Detectors (CERT/ONERA)	
A0138-3	Thin Metal Film and Multilayers Experiment (CNRS/LPSP)	
A0138-4	Vacuum Deposited Optical Coatings Experiment (Matra S. A., Optical Division)	
A0138-5	Ruled and Holographic Gratings Experiment (Inst. SA/JOB INYVON Division)	
A0138-6	Thermal Control Coatings Experiment (CERT/ONERA, CNES/CST)	
A0138-7	Optical Fibers and Components Experiment (CERT/ONERA)	
A0138-8	Effect of Space Exposure of Some Epoxy Matrix Composites on Their Thermal Expansion and Mechanical Properties (Matra S. A., Space Division)	
A0138-9	The Effect of the Space Environment on Composite Experiments (Aerospatiale)	
A0138-10	Microwelding of Various Metallic Materials Under Ultravacuum (Aerospatiale)	

1386) W. Flury, "Europe's Contribution to the Long Duration Exposure Facility (LDEF) Meteoroid and Debris Impact Analysis," ESA Bulletin, Number 76, November 1993, pp. 112-118

Exp. No.	Experiment Title (Sponsoring Institution)	Tray Numbers (Exp. Location)
A0139A	Growth of Crystals from Solutions in Low Gravity (Rockwell Int. Science Center/Technical University of Denmark)	G6
A0147	Passive Exposure of Earth Radiation Budget Experiment Components (The Eppley Laboratory, Inc.)	B8, G12
A0171	Solar Array Materials Passive LDEF Experiment (NASA/MSFC, NASA/LeRC, NASA/GSFC, NASA/JPL)	A8
A0172	Effects of Solar Radiation on Glasses (NASA/MSFC, Vanderbilt University)	D2, G12
A0175	Evaluation of Long-Duration Exposure to the Natural Space Environment on Graphite-Epoxy Mechanical Properties (Rockwell Int. Corp., Tulsa Facility)	A1, A7
A0178	A High Resolution Study of Ultra-Heavy Cosmic Ray Nuclei, 'UHCRE' (Dublin Institute for Advanced Studies, Ireland, ESA/ESTEC)	A2, A4, A10, B5, B7, C5, C6, C8, C11, D1, D5, D7, D11, E2, E10, F4
A0180	The Effect of Space Environment Exposure on the Properties of Polymer Matrix Composite Materials (University of Toronto)	D12
A0187-1	Chemistry of Micrometeoroids (NASA-JSC, University of Washington, Rockwell Int. Science Center	A3, A11
A0187-2	Chemical and Isotopic Measurements of Micrometeoroids by Secondary Ion Mass Spectrometry 'SIMS' (McDonnell Center for the Space Sciences, MPI für Nuclear Physics Heidelberg, Munich Technical University, Dornier Co.)	C2, E3, E8
A0189	Study of Factors Determining the Radiation Sensitivity of Quartz Crystal Oscillators (Martin Marietta Laboratories)	D2
A0201	Interplanetary Dust Experiment, IDE (Inst. for Space Science and Technology, NASA/LaRC, North Carolina University)	B12, C3, C9, D6, G10, H11
M0001	Heavy Ions in Space (Naval Research Laboratory, Washington)	H3, H12
M0002-1	Trapped Proton Energy Spectrum Determination (AF Geophysics Laboratory)	D3, D9, G12
M0002-2	Measurement of Heavy Cosmic-Ray Nuclei on LDEF (U. of Kiel, Germany)	E6
M0003	Space Environment Effects on Spacecraft Materials (The Aerospace Corporation)	D3, D4, D8, D9
M0004	Space Environment Effects on Fiber Optics Systems (AF Weapons Laboratory)	F8
M0006	Space Environment Effects (AF Technical Applications Center)	C2
P0003	LDEF Thermal Measurement System (NASA/LaRC)	Center ring
P0004-1	Seeds in Space Experiment (George W. Park Seed Company, Inc.)	F2
P0004-2	Space-Exposed Experiment Developed for Students, SEEDS, (NASA/HQ)	F2
P0005	Space Ageing of Solid Rocket Materials (Morton Thiokol, Inc.)	Center ring
P0006	Linear Energy Transfer Spectrum Measurement (U. of San Francisco/MSFC)	F2
S0001	Space Debris Impact Experiment (NASA/LaRC)	A5, A6, A12, B1, B2, B6, B8, B11, C4, C7, D2, D6, E1, E4, E7, E11, F1, F3, F5, F7, F10. F11, G4, G8, H5
S0010	Exposure of Spacecraft Coatings (NASA/LaRC)	B9
S0014	Advanced Photovoltaic Experiment (NASA/LeRC)	E9
S0050	Investigation of the Effects of Long Duration Exposure of Active Optical System Components (Eng. Exp. Station, Georgia Institute of Technology)	E5
S0050-1	Investigation of the Effects of Long Duration Exposure on Active Optical Materials and UV Detectors (NASA/LaRC)	E5
S0069	Thermal Control Surfaces Experiment (NASA/MSFC)	A9
S0109	Fiber Optic Data Transmission Experiment (JPL)	C12
S1001	Low Temperature Heat Pipe (NASA/GSFC, NASA/ARC)	F12, H1
S1002	Investigation of Critical Surface Degradation Effects on Coatings and Solar Cells Developed in Germany (MBB)	E3
S1003	Ion Beam Textured and Coated Surfaces Experiment (NASA/LeRC)	E6

Exp. No.	Experiment Title (Sponsoring Institution)	Tray Numbers (Exp. Location)
S1005	Transverse Flat Plate Heat Pipe Experiment (NASA/MSFC, Grumman Aerospace Corporation)	B10
S1006	Balloon Materials Degradation (Texas A&M University)	E6

Table 414: Summary of LDEF experiment complement

Frecopa = French Cooperative Payload (PI: J. C. Mandeville, CERT/ONERA and others). Objective: determination of the number of impacts, and the size and chemical composition of the impacting cosmic dust and debris. A collection area of about 2000 cm^2 is exposed to the space environment (multilayer thin foil detectors). In addition a large variety of materials placed on the same tray (8500 cm^2) is exposed.

SIMS = Chemical and Isotopic Measurements of Micrometeoroids by Secondary Ion Mass Spectrometry (PI's: E. Zinner Washington University, St. Louis, E. K. Jessberger, MPI Heidelberg, and others). Objective: chemical and isotropic measurements of micrometeoroids by secondary ion mass spectrometry. The experiment consists of 237 capture cells, each measuring 8.6 x 9.4 cm, located on three different rows. The target material (germanium wafers) is of very high purity, essential for the determination of the composition of the deposit. The SIMS technique is sufficiently sensitive to allow analysis of deposits that are 0.5 to 5 monolayers thick.

MAP = Multiple-Foil Micro-Abrasion Package (J. A. M. McDonnell, U. of Kent, UK). Objective: measurements of impactor velocity, density, angle of incidence, and chemical composition. A capture cell experiment. Each detector consists of two foils and a pure, polished stop plate to catch fragments of impacting particles. Deployment of MAP detectors on leading and trailing surfaces, and the surfaces normal to them, of LDEF.

UHCRE = A High Resolution Study of Ultra-Heavy Cosmic Ray Nuclei (PI: D. O'Sullivan, Dublin Institute for Advanced Studies, Ireland, and others). The experiment consists of 18 m^2 of thermal blankets used to collect a large number of of meteoroid and debris impact records of various sizes.

J.9 LFC (LARGE FORMAT CAMERA)

Shuttle Missions (STS-2 and STS-41G) Nov. 12, 1981 and Oct. 5-13, 1984 (part of OSTA-3 payload)[1387)]

Orbit (STS-41G): The Shuttle mission was flown at an inclination of 57° at three different altitudes. Orbits 001 - 022 were flown at an altitude of 352 km; orbits 023 - 036 were flown at an altitude of 272 km; orbits 037 - 128 were flown at an altitude of 225 km.

Application: High-resolution mapping camera. Cartographic mapping to achieve imagery at a scale of 1:50,000. LFC (408 kg) was used at China's request to try to obtain data on geological faults in Asia, among other tasks.

Sensor:
LFC = Large Format Camera (NASA Experiment, built by Itek Optical Systems (Div. of Litton Systems Inc.), Lexington, MA).

1387) B.B. Schardt, B.H. Mollberg, "The Orbiter Camera Payload System's Large-Format Camera and Attitude Reference System," in Monitoring the Earth's Ocean, Land, and Atmosphere from Space, Volume 97 , AIAA, 1985, pp. 684 - 709

Spectral range	400 - 900 nm
Focal length	305 mm
Image format	23 x 46 cm
Image scale	1 : 738 000
FOV	73.7° across track; 41.1° along track
Swath width	170 x 340 km (at 225 km altitude)
Kodak B/W (Neg.)	Panatomic-X Aerocon 3412
Kodak B/W (Neg.)	High definition aerial 3412
Kodak Color (Pos.)	High definition aerial 3412
Kodak Color (Pos.)	Aerial Color SO-242
Data rate	225 frames/film

Table 415: Specification of the LFC instrument

J.10 LITE (Lidar In-Space Technology Experiment)

LITE[1388] = Lidar In-Space Technology Experiment (NASA/LaRC, PI: M. P. McCormick). The LITE-1 payload was flown on a Shuttle mission (STS-64) from September 9-20, 1994.

Orbit: Altitude = 258 km; inclination = 57°; the orbit allows lidar measurements over both land and water. - Shuttle underflights by aircraft with laser sensors, including the Wallops Electra and P-3 aircraft, are undertaken around the world for LITE data correlation and verification.

Objective: Detection of stratospheric and tropospheric aerosols; measurement of the planetary boundary layer, cloud top heights, atmospheric temperature and density in the range from 10 to 40 km; regional and diurnal studies.

Laser transmitter module:		Telescope	
Output wavelengths	1064, 532 and 355 nm	Primary mirror	0.956 m diameter
Output energy	486, 460 and 196 mjoule	Secondary mirror	0.311 m diameter
Beam divergence	0.9, 0.6 and 0.6 mrad	Focal length	4.825 m
Beam quality	5.0, 4.5, 5.5 X-diff. limit.	Obscuration ratio	f/5.1
Pulse rate frequency	10/s (all wavelengths)		
Pulse width	27, 27, 31 ns		
Aft Optics (Nighttime measurements)		Aft Optics (Daytime measurements)	
Optical throughput	58, 49, 49 %	Interference filter transmission	50, 40, 12 %
Color filter BW	675, 265, 60 nm	Overall optical throughput	29, 20, 6 %
Detector quantum eff.	33, 14, 21 %	Interference filer BW	1, 0.3, 1 nm
		Detector quantum eff.	33, 14, 21 %
		FOV selectable all wavelengths	1.7 , 3.4 , 5.6 mrad and blocked
Signal Processing Electronics (all channels)			
Baseband amplifier bandwidth		2 MHz	
Filter characteristics		7 pole Bessel lowpass	
Digitizer resolution		12 bits	
Digitizing rate		10 MHz during sampling period	
Data sample period		660 µs	

Table 416: Instrument parameters of LITE

Sensor: The instrument consists of a nominal 1 m diameter telescope receiver, a three-color neodymium YAG laser transmitter, a boresight unit, and the system electronics. LITE makes extensive use of the Shuttle resources for electrical power, thermal control, and command and data handling. The laser transmits energy at three harmonically related wave-

1388) "Lidar In-Space Technology Experiment (LITE): NASA's first In-Space Lidar System for Atmospheric Research," Optical Engineering, Jan. 1991, Vol. 30 No. 1 pp. 88-95 lengths into the atmosphere. The receiver collects the energy backscattered from the atmosphere and brings it into focus on three detectors.

Some experiments:

- Instrument measurements 'troposphere': aerosol backscatter cross-section; aerosol scattering ratio; Planetary Boundary Layer (PBL) height; PBL optical depth.
- Instrument measurements 'clouds': height; fractional cloud cover; reflectance/albedo; optical depth.
- Instrument measurements 'Stratosphere': density and temperature profiles to 40 km; aerosol backscatter cross-section; aerosol scattering ratio. These measurements were made at 532 nm and 355 nm wavelength only.
- Study of a wide variety of phenomena on regional and global scales.

J.11 MAPS (Measurement of Air Pollution from Satellites)

MAPS is a NASA/LaRC Shuttle payload with the objective to measure (nearly) global tropospheric carbon monoxide (CO) distributions in the middle troposphere in order to evaluate CO sources and chemistry, and to determine the seasonal variations of this key atmospheric trace gas.

The MAPS instrument is a gas filter correlation radiometer consisting of an electro-optical sensor, an electronics module, a digital tape data recorder, and an aerial camera. MAPS is mounted onto MPESS (Multi-Purpose Experiment Support Structure) in the forward end of the Shuttle cargo bay. The instrument makes nadir-viewing measurements from Shuttle orbits of approximately 220 km altitude.[1389)]

The electro-optical sensor incorporates two gas cells, one containing 350 hPa of CO and another containing 149 hPa of N_2O; their corresponding detectors; a direct radiation detector; an external balance and gain check system, and an internal balance system.

MAPS sensor	Gas filter correlation radiometer
Spectral band	2080 to 2220 cm^{-1} (4.5 - 4.8 μm)
Resolution (sensitivity)	1.7 10^{-7} W cm^{-2} sr^{-1} noise-equivalent radiance in the radiometer channel 5.7 10^{-9} W cm^{-2} sr^{-1} noise-equivalent radiance in the difference channel
Swath width	Footprint is 20 km at ground surface (corresponds to spatial resolution)
Data rate	43 bit/s (on-board recorded and downlinked during flight)

Table 417: Specification of some MAPS parameters

MAPS measures the absorption lines of CO and N_2O of incoming radiation. The spectrum is split into three beams; one beam passes through a cell containing CO and falls onto a detector - the CO gas cell acts as a filter. A second beam falls directly onto a detector. The observed signal differences are used for CO determination in the altitude range of 0-18 km. The third beam of the incident radiation passes through a cell containing N_2O and is measured by another detector. Again, the N_2O gas cell acts as a filter for the effects of N_2O present in the atmosphere. Since the global distribution of N_2O is well known, the measured values of N_2O may be used as an indicator for the presence of clouds in the field of view and to correct the simultaneous CO measurement for systematic errors.

The aerial camera mounted next to the MAPS electro-optical head provides information on cloud cover and terrain over which the sensor data are gathered.

MAPS was first flown on STS-2 (November 12.-14, 1981) as part of the OSTS-1 payload, then on STS-13 (October 5 -13, 1984) as part of the OSTA-3 payload. These flights revealed longitudinal variations in CO distributions and identified biomass burning in the Southern

1389) Information provided by V. Connors and D. O. Neil of NASA/LaRC

Hemisphere as a major global CO source (equal in magnitude to the amount of CO generated in the Northern Hemisphere by fossil fuel combustion). MAPS flew on STS-59 (April 1994) and STS-68 (October 1994, see Table in chapter O.17).

J.12 MOMS-01 (Modular Optoelectronic Multispectral Scanner)

MOMS is a German imaging payload/experiment of DLR, University of Munich, and others, built by MBB of Munich). MOMS-01 was a Shuttle payload (mounted on SPAS = Shuttle Pallet Satellite) on two missions (STS 7: June 18, 1983, a 6-day mission; and STS-41B): Feb. 3, 1984, an 8-day mission).

The MOMS-01 instrument consists of five major elements, mounted onto the carbon-fiber structure of the SPAS-01 free-flyer.

1) The optical module (scanner head) with four objectives, eight arrays, and associated shutters. Each module, representing one spectral band, consists of filters, dual-lens optics, four CCD detector arrays, and preamplifier electronics.
2) A power box for overall power conditioning and thermal and shutter circuitry
3) A logic box for all sensor function control (including real-time correction, and formatting of the source data stream from the optical module)
4) A HDDT (High Density Digital Tape) recorder (model: Bell&Howell MARS-1428 LT-3B) for digital data storage (72 Gbit capacity/tape)
5) A pressurized container for the recording system.

MOMS-01 is a two-channel system which uses the spectral bands of 575 - 625 nm for land observations, and 825 - 975 nm for vegetation detection. A double optical system (focal length = 237.2 mm, relative lens aperture = 1:3.5) is used per band for illuminating the CCD array consisting of four groups with a total length of 6912 pixels (16 μm pixel size). Swath width = 140 km (FOV=26.2°), spatial resolution (pixel size on ground) = 20 x 20 m (IFOV=67.2 μrad). [1390) 1391) 1392) 1393)]

Application: Observation of the Earth with configurable sensors. Imaging of different ground targets with low-to-high contrast and albedo (arid regions, regions with dense and sparse vegetation coverage, coastal zones, mountainous terrain, open ocean islands) to demonstrate the instrument capabilities for thematic mapping.

Orbit: Shuttle orbit, inclination = 28.5°; STS 7 = 292 km altitude; STS-41B = 289 - 300 km altitude; swath width = 140 km.

MOMS-01 is regarded an early imaging charge-coupled device (CCD) instrument operated in a civilian spaceborne application. CCD pushbroom type: Reticon CCPD 1728 (EG&G Reticon, Sunnyvale CA). Each CCD line array has 1728 detectors. Scan line extension beyond the one CCD array is achieved with a dual-lens configuration, linking all four CCDs to a total scan line of 6912 pixels; data quantization = 7 bits, data rate = 40 Mbit/s; the data were recorded on-board. The MOMS-01 missions yielded 450 individual scenes for thematic evaluation. The total experiment time was 26.5 minutes on STS-7, and 30 minutes on STS-41-B.

Note: Both optical systems observed in reality the entire swath width in parallel (not the split arrangement as shown in Figure 240). This double exposure concept was simply need-

1390) F. Ackermann, J. Bodechtel, F. Lanzl, D. Meissner, P. Seige, H. Winkenbach;"MOMS-02 - Ein multispektrales Stereo-Bildaufnahmesystem für die zweite deutsche Spacelab-Mission D2," Geo-Informations-Systeme, Zeitschrift für interdisziplinären Austausch innerhalb der Geowissenschaften, Wichmann Verlag, Jahrgang 2, Heft 3/1989, S. 5 - 11

1391) J. Bodechtel, D. Meißner, P. Seige, H. Winkenbach, J. Zilger, "The MOMS Experiment on STS-7 and STS-11 - First Results and Further Development of the Modular Optoelectronic Multispectral Scanner," Proceedings of the Eighteenth International Symposium on Remote Sensing of the Environment, Volume I, 1984, pp. 77-85

1392) "MOMS-01: First Results of STS-7 Mission," IGARSS'83

1393) J. Bodechtel, R. Haydn, J. Zilger, "MOMS-01: Missions and Results," Monitoring Earth's Ocean, Land, and Atmosphere form Space - Sensors, Systems, and Applications, edited by A. Schnapf, Progress in Astronautics and Aeronautics, AIAA, Vol. 97 1985, pp. 524-535 ed to provide a seamless linkage of all four CCD detector arrays for each spectral band across the swath.

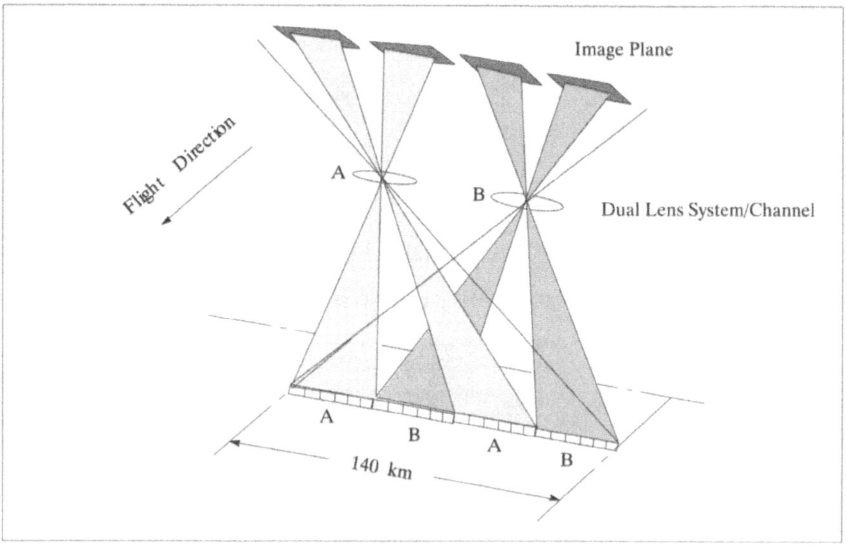

Figure 240: MOMS-01 schematic imaging configuration of CCD arrays

J.13 MOMS-02 (Modular Optoelectronic Multispectral Scanner)

MOMS-02 on the Spacelab D-2 Mission (STS-55, 10 day flight). Launch: April 26, 1993. MOMS-2 is an advanced version of MOMS-01 (PIs: F. Lanzl, F. Ackermann, J. Bodechtel). The D-2 orbit had an inclination of 28.5° and an altitude of 296 km (this means: observation of equatorial regions).

The data are used in the following applications: Land cover (vegetated areas: land use, biomass estimation; unvegetated areas: lithology, mineral prospecting, tectonic investigations); geomorphology, ecology, basic research in the spectral signatures of rocks, soil, vegetation, etc. [1394] [1395] [1396]

Objectives:

- Stereoscopic visual observation (high degree of interpretability)
- Provision of high-quality topographic regional maps (scale 1:50000) and digital terrain models (<5 m pixel size on ground)
- Test of a digital photogrammetric observation technique and processing system (prototype)
- Correlation of high-resolution panchromatic data with multispectral data.

Experiment: a) triple-stereoscopy, b) along-track-stereoscopy (forward and backward tilt), c) high-resolution imagery, d) multispectral observation (refined modelling of MS classification), e) combination of stereo and multispectral imagery.

1394) P. Seige, "MOMS-02 - Eine hochauflösende stereoskopische und multispektrale Kamera auf der zweiten deutschen Spacelab Mission D-2," DLR-Nachrichten, Heft 77, Februar 1995

1395) J. Bodechtel, S. Lutz, "Neue Wege der Erderkundung," aus Einsichten, Forschung an der LMU, pp. 38-43, 1992

1396) MBB Endbericht, "MOMS-02 auf D-2," die Entwicklung von EOS über MOMS-EM, MOMS-01 bis MOMS-02, Doc. No. MOMS-02.RP.0100.0, Dec. 20, 1993

The MOMS-02 payload has a total of five optical systems: three are used for stereoscopic imagery, two are employed for multispectral imagery, one is used for high-resolution data. Focal length (f): 220 mm (ch. 1-4); 237.2 mm (ch. 6,7), 660 mm (ch. 5). Quantization: 8 bit with seven gain steps (uncompressed for MS and stereo, 6 bit compressed for stereo). Seven variable observation modes of different band combinations are defined.

Channel	Mode	Orientation	Bandwidth	Ground Pixel	IFOV	Swath Width
1	MS	nadir	449-511 nm	13.5 m x 13.5 m	45.45 µrad	78/43 km
2	MS	nadir	532-576 nm	13.5 m x 13.5 m	45.45 µrad	78/43 km
3	MS	nadir	645-677 nm	13.5 m x 13.5 m	45.45 µrad	78/43 km
4	MS	nadir	772-815 nm	13.5 m x 13.5 m	45.45 µrad	78/43 km
5 (PAN)	HR	nadir	512-765 nm	4.5 m x 4.5 m	15.15 µrad	37/27 km
6 (PAN)	Stereo	+21.4°	524-763 nm	13.5 m x 13.5 m	42.16 µrad	78/43 km
7 (PAN)	Stereo	-21.4°	524-763 nm	13.5 m x 13.5 m	42.16 µrad	78/43 km

Table 418: Performance parameters of MOMS-02 (300 km orbit)

Channel	1	2	3	4	5	6	7
Mode1/No. of pixels					8304	2976	2976
Mode2/No. of pixels	5800	5800	5800	5800			
Mode3/No. of pixels,			5800	5800		5800	5800
Mode4/No. of pixels	5800		5800	5800		5800	
Mode5/No. of pixels	5800		5800	5800			5800
Mode6/No. of pixels		3220	3220	3220	6000		
Mode7/No. of pixels	3220		3220	3220	6000		

Table 419: Operational modes of MOMS-02

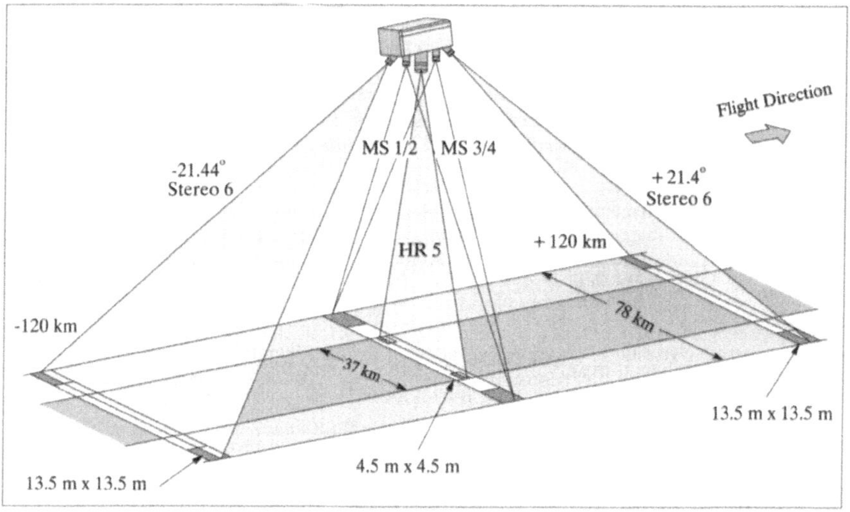

Figure 241: MOMS-02 imaging geometries [1397)]

Data: On-board recording onto HDT (uncompressed for stereo), maximum recording time is 5.5 hours; this amounts to about 2.5 x 10^{12} bit of data. The maximum recording data rate = 100 Mbit/s.

1397) Courtesy of P. Seige, DLR

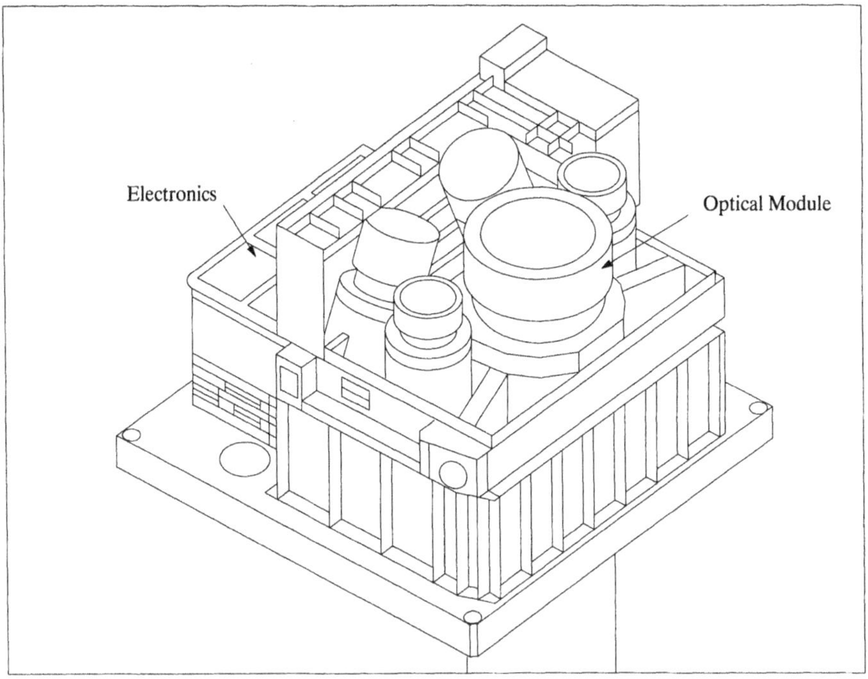

Figure 242: The MOMS-02 instrument

J.14 SAC (Satélite de Aplicaciones Científicas)

The SAC satellite program of CONAE (Comisión Nacional de Actividades Espaciales), the Space Agency of Argentina, started in 1994 with the approval of the National Space Program by the government of Argentina. CONAE devised a plan, taking into account the available resources, and came up with a policy of international cooperative missions for an optimum in return of the investment. SAC is a key element in this program. [1398)]

Satellite	Launch Date	Prime Mission Objective	Partners in Mission
SAC-B	Nov. 4, 1996 (launch failure)	Science	USA, Brazil, Italy
SAC-A	Dec. 4, 1998 (STS-88 launch)	Technology	USA
SAC-C	Nov. 21, 2000 from VAFB (secondary payload to EO-1)	Earth observation	USA, Denmark, Brazil, Italy, France

Table 420: Chronological overview of CONAE's SAC program

J.15 SAC -A (Satélite de Aplicaciones Científicas-A)

The SAC satellite program of CONAE (Comisión Nacional de Actividades Espaciales), the Space Agency of Argentina, started in 1994 with the approval of the National Space Program by the government of Argentina. CONAE devised a plan, taking into account the available resources, and came up with a policy of international cooperative missions for an optimum in return of the investment. SAC is a key element in this program.

1398) D. Caruso, "CONAE's Satellite Missions," IAA 2nd International Symposium on Small Satellites for Earth Observation, Berlin, April 12-16, 1999, pp. 375-381

SAC-A is a CONAE/NASA demonstration mission, an Argentine-built satellite, with the objective to test and characterize the performance of new equipment and technologies which may be used in future operational or scientific missions of CONAE. NASA provides the launch and the Hitchhiker equipment for SAC-A. S/C operations are performed by CONAE from Argentina. [1399)]

Satellite	Launch Date	Prime Mission Objective	Partners in Mission
SAC-B	Nov. 4, 1996 (launch failure)	Science	USA, Brazil, Italy
SAC-A	Dec. 4, 1998 (STS-88 launch)	Technology	USA
SAC-C	Nov. 21, 2000 from VAFB (secondary payload to EO-1)	Earth observation	USA, Denmark, Brazil, Italy, France

Table 421: Chronological overview of CONAE's SAC program

The SAC-A payload consists of the SAC-A satellite installed in a canister equipped with a Hitchhiker Pallet Ejection System (PES) and Hitchhiker Motorized Door Assembly (HMDA). For the STS-88 mission, SAC-A shared a common Hitchhiker Avionics Unit with MightySat I of AFRL (Kirtland AFB, NM). The SAC-A design life is six months. S/C mass = 64 kg, power = 36 W EOL. SAC-A is three-axis stabilized using momentum wheel, magnetic coils and a three-axis magnetometer. Attitude is measured by a DGPS receiver and an assembly of Si cells as a coarse sun sensor. A launch of STS-88 took place on Dec. 4, 1998. SAC-A was deployed on Dec. 14.

Communications: TT&C via S-band with backup systems for telecommand (TC) in VHF and telemetry (TM) in UHF. Note: The redundant TC/TM system is also being used in a whale tracker experiment. [1400)]

Orbit: near-circular orbit, altitude = 390 km, inclination = 51.6°.

Sensor/experiment complement (technology demonstration):

DGPS (Differential GPS Receiver), PI: R. Alonso. The objective is to provide real-time autonomous attitude measurements for the satellite, ultimately simplifying the amount of ground processing required to control an orbiting satellite. Special algorithms were developed for non-nadir pointing directions of the S/C.

CPC (Commercial Panchromatic Camera), PI: J. Yelos. The objective is to test the performance of a CCD camera for Earth imaging.

Solar cell experiment (PI: J. Duran). Use of an Si cell assembly for power generation and parallel use as a coarse sun sensor. In addition, a CONAE-developed deployment mechanism was tested for the solar panels.

J.16 SHIMMER (Spatial Heterodyne Imager for Mesospheric Radicals)

SHIMMER is a high-resolution UV interferometer, a cooperative development of NRL (Naval Research Laboratory) in Washington DC, UWM (University of Wisconsin at Madison), and St. Cloud State University in St. Cloud, Minnesota. The primary science objective is to acquire global profiles of hydroxyl (OH) in the upper stratosphere and mesosphere by measuring its resonance fluorescence in the UV. The flight of SHIMMER as a Shuttle mid-deck payload will serve as a proof-of-concept of this new and powerful technique, facilitating future high-resolution imaging instruments that are small, rugged, light-weight and

1399) D. Caruso, "CONAE's Satellite Missions," IAA 2nd International Symposium on Small Satellites for Earth Observation, Berlin, April 12-16, 1999, pp. 375-381

1400) http://sspp.gsfc.nasa.gov/current.html

have no moving parts. SHIMMER is planned to be flown on Shuttle in the time frame of 2002.[1401] [1402] [1403] [1404]

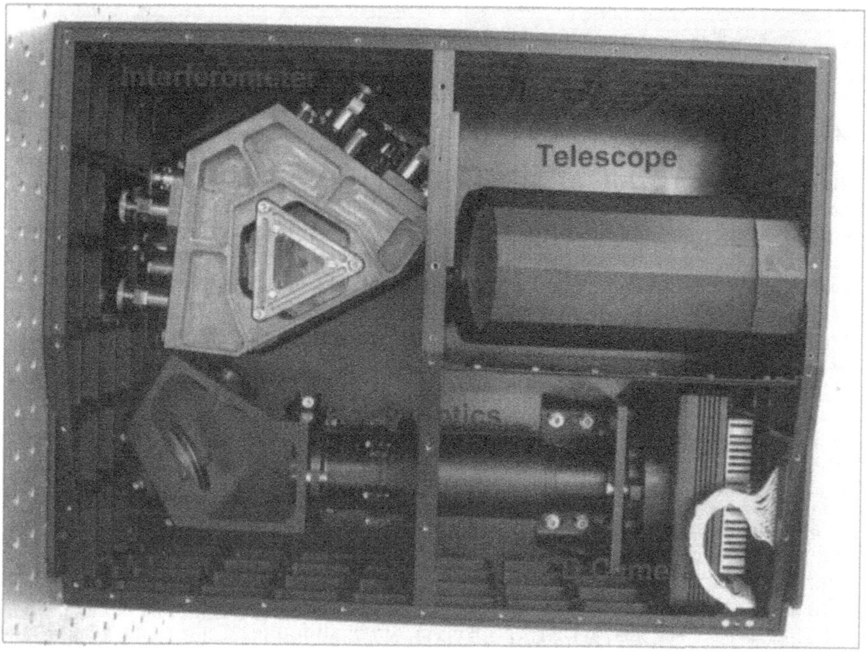

Figure 243: Photo of the SHIMMER instrument assembly

Parameter	Description
Type of measurement	Ultraviolet solar resonance fluorescence
Geophysical parameters observed	OH, Rayleigh scattering, Mie scattering
Viewing geometry	Earth Limb
Time required for a vertical image	2 seconds
Spectral coverage	307-310 nm
Interference filter	2.3 nm FWHM centered at 308.9 nm
Entrance optics	500 mm focal length telescope
Exit optics	7 element relay system
Detector (frame transfer device)	CCD array of 1024 x 1024 pixels at 24 µm
Gratings: Clear aperture Groove density Littrow angle Littrow wavelength	20 mm x 20 mm 1200 lines/mm 10.7° 307 nm
Field widening prisms: Clear aperture Wedge angle Incident angle Exit angle	 22 mm x 22 mm 13.02° 8.73° 10.7°
Beamsplitter: Clear aperture	28 mm x 28 mm

1401) Note: The technology of spatial heterodyne spectroscopy was developed by Fred Roesler and John Harlander of UWM in 1990. SHIMMER observations represent the first spaceborne demonstration of this technology.

1402) Information provided by Robert R. Conway and Christoph R. Englert of NRL

1403) http://uap-www.nrl.navy.mil/shimmer/shimmer.htm

1404) J. Harlander, H. T. Tran, F. L. Roesler, K. P. Jaehnig, et al., "Field-Widened Spatial Heterodyne Spectroscopy: correcting for Optical Defects and New Vacuum Ultraviolet Performance Tests," EUV, X-Ray and Gamma-Ray Instrumentation for Astronomy V, SPIE Proceedings 1994, Vol. 2280, p. 310-319

Parameter	Description
FOV (Field of View)	10° at gratings, 2.3° x 2.3° on sky
Spectral resolution	58 MB, or 0.0058 nm (resolving power of 53,000)
Non-aliased spectral range	2.95 nm
Achieved spectral range	Filter limited
Instrument mass, power, size	22 kg, 27 W (with cold finger), 52 cm x 42 cm x 23 cm
Image rate	150-750 images per orbit

Table 422: Summary of basic instrument specifications

The major instrument elements are the telescope, the relay optics, the interferometer, and the CCD camera. The telescope images the sunlit limb scene onto the gratings. The relay optics create an image of the fringe localization plane (gratings) at the CCD. The optical elements of the interferometer are held by a Vascomax steel structure which allows precise positioning of each element. Every pixel row of the CCD, parallel to the horizon, contains the spectral information (interferogram) for the corresponding tangent altitude of the FOV (Field of View). The rows (32 rows) are averaged to yield a vertical resolution of about 2.2 km. The SHIMMER middeck instrument is controlled by a laptop (see also O.11).

J.17 SLA (Shuttle Laser Altimeter)

SLA is a NASA/GSFC designed and developed instrument (within the MTPE program), a pathfinder for future operational space-based laser remote sensing devices and lidar over land surfaces, which is flown as a 'hitchhiker payload' on Shuttle flights. A schedule of 4 flights has been planned for a continuously improving the SLA sensor package. The first flight (SLA-1) took place on STS-72 (January 11-20, 1996). The second flight, SLA-2, was on STS-85 (Aug. 7-19, 1997). There are plans to fly a small-footprint, ultra high pulse repetition rate version of SLA-2 (namely SLA-3) in late December 1998 as part of the STS-96 mission (or on SRTM).

SLA has three major objectives:

- To provide an in-space engineering testbed for future spaceborne laser altimeters
- To acquire samples of land topography and vegetation data
- To measure cloud and aerosol layers in the atmosphere

J.17.1 SLA-1

The SLA instrument package consists of the LAC (Laser Altimeter Canister) and ASC (Altimeter Support Cannister) which are interconnected. The LAC is equipped with an opening motorized door assembly with an optical window. LAC contains a laser transmitter and a concave mirror/detector which collects the reflected laser radiation. The ASC contains control, telemetry, telecommand and power conditioning electronics. The SLA instrument consists of a 38 cm telescope, a diode-pumped Nd:YAG laser transmitter (Q-switched), an altimeter receiver electronics package, and a 250 MHz waveform digitizer coupled to the altimeter receiver for capturing the backscattered echoes from the interaction of short duration laser pulses (about 8 ns duration) with the ground surface.

The SLA altimeter sends out a series of short laser pulses and measures the time-of-flight and the spreading of the laser pulse by the Earth's surface. The pulse time interval measurement is to a precision of about five ns. Recorded is also the temporal shape of the laser echo from the Earth's surface for interpretation of surface height distribution, referred to as TVR (Total Vertical Roughness) within the 100 m diameter footprint. For example, tree heights can be determined by measuring the characteristic double-pulse signature that results from a separation in time of laser backscatter from tree canopies and the underlying ground.

SLA-1 demonstrated the viability of surface lidar sensing from Earth orbit, obtained a high-quality (meter-level accuracy) data set on ocean and land topography, and was a major factor in the selection of VCL (Vegetation Canopy Lidar) for the first ESSP (Earth Systems Science Pathfinder) misssion. Tree height data sets obtained on SLA-1 were the first measurements from space that can be tied directly to remote sensing of above ground biomass and forest canopy architecture. Only those echoes that were unsaturated were used to derive TVR values, which represent a total canopy assemblage height in forested areas where the local topography is of low slope (<3°).

J.17.2 SLA-2

Based on the SL-1 experience (saturation of the link margin over unvegetated areas due to the enormous dynamic nange changes of backscattered signal strengths), the SLA-2 instrument was modified to include an innovative VGA (Variable Gain state Amplifier) system in the altimeter receiver chain. This VGA permits the SLA ground operator to control the signal intensity from the silicon avalanche diode detector to the waveform digitizer to preclude saturation.

SLA-2 provided direct measurements of the height of clouds, and acquired profiles of land and surface vegetation canopies over continental and island targets with a 1-10 m rms surface elevation accuracy and sub-kilometer spatial resolution. SLA-2 also measured the complex shape of laser pulse echoes from land and vegetation with sub-meter vertical resolution and obtained grids of altimetry data for major topography study regions, including the continental United States. Other primary targets for SLA-2 surface research included: the Amazon Basin in Brazil; the Canadian Boreal Forest; the Kamchatka Peninsula; the Patagonian ice fields; West Africa Tropical Forests; high latitude steppe deserts in China; large inland seas including the Caspian and Aral Seas; large oceanic islands, including Tasmania, the Falkland islands, and Madagascar; the Northern Sahara Sand Seas (Grand Ergs) and the Mississippi Delta region of the U.S.

J.18 SPARTAN (Shuttle Pointed Autonomous Research Tool for Astronomy)

A continuous NASA/GSFC program which started in the early 1980s. SPARTAN is a Shuttle-deployed retrievable free-flyer platform (an autonomous sub-satellite, three-axis stabilized) providing short-term LEO observation opportunities for instrumentation in various disciplines and fields of applications, such as astronomy, remote sensing, and technology demonstrations. [1405] [1406]

The Spartan spacecraft is a small, rectangular, free-flying vehicle, measuring roughly 1 m x 1.25 m x 1.5 m. It is released from the shuttle and picked up after several days of conducting its experiments. Several SPARTAN platform configurations have been built so far.

Mission duration	2-5 days, depending on mission requirements
Deployed mass	up to 1400 kg
Platform dimensions	Mission dependent (up to a length of 2.3 m) a rectangular pallet
Power	Self-contained subsystem, 28±5 VDC, 3 A max, 7 kWh available
Instruments	Mission-dependent
Attitude control	Three-axis stabilized with cold gas thrusters, max slew rate : $1°/s^2$
Thermal control	Active and passive, SPARTAN electronics limited to 0° - 50°C
Data rates	Mission dependent
Data storage	Onboard recorder

Table 423: Some characteristics of the SPARTAN platform series 200

1405) SPARTAN Capabilities Statement, SP515, 1993, NASA/GSFC

1406) http://spartans.gsfc.nasa.gov/

Shuttle Flight	Payload	Comments
STS-51G, June 17 - 24, 1985	2 X-ray proportional counters	SP-A (mass of 1100 kg)
STS-51L, Jan. 28, 1986	**Challenger accident**	SP-Halley, 203, (UV-spectrometer)
STS-56, April 8 -17, 1993	UVCS, WLC, solar physics	SP-201-1, (mass of 1360 kg)
STS-64, Sept. 9 - 20, 1994	UVCS, WLC, solar physics	SP-201-2, reflight of SP-201-1 timed to coincide with Ulysses south polar pass
STS-63, Feb. 3 - 11, 1995	Far UV imaging spectrograph (FUVIS) of NRL	SP-204, galactic dust clouds
STS-69, Sept. 7-18, 1995	UVCS, WLC, solar physics	SP-201-3, mass = 1136 kg, reflight of SP-201-1 timed to coincide with Ulysses north polar pass
STS-72, Jan. 11 - 20, 1996	REFLEX, GADACS, SELODE, SPRE	SP-206 OAST flyer, technology experiments
STS-77, May 19 - 29, 1996	Inflatable Antenna Experiment	SP-207/IAE
STS-87, Nov 19 - Dec. 5, 1997	UVCS, WLC, TEXAS, VGS, solar physics	SP-201-4, reflight of SP-201-1
STS-95, Oct. 29 - Nov. 7, 1998	UVCS, WLC, TEXAS, VGS, solar physics	SP-201-5

Table 424: Overview of SPARTAN flights on Shuttle

J.18.1 SPARTAN-1

Spartan-1 (also referred to as Spartan-101), as the first of the series (launch June 17, 1985 on STS-51G), was intended to demonstrate the low-cost approach and its ability to achieve the desired technical performance in the STS environment. Pointing and stabilization were achieved by an attitude control system capable of three-axis stabilized pointing to any target within ± 3 arcmin. The scientific objective of the spacecraft was to investigate the photon emission processes in clusters of galaxies and to explore the center of our galaxy. The X-ray detector had an energy range of 1 to 12 keV, scanning its targets (various cosmic sources) with narrowly collimated (5 arcmin x 3°) gas scintillation proportional counters (two identical counters, each with an effective area of 660 cm^2). Counts were accumulated for 0.812 s into 128 energy channels. The energy resolution was 16% at 6 keV. - Total mass = 1,100 kg.

J.18.2 SPARTAN-Halley

This spacecraft failed when Space Shuttle Challenger exploded on launch (Jan. 28, 1986, STS-51L). Same platform and capabilities as SPARTAN-1. The main objective of this spacecraft was to obtain UV spectra of the coma and tail of Comet Halley in January 1986 shortly before its perihelion.

J.18.3 SPARTAN-201

The SPARTAN-201 series (five missions) is dedicated to NASA's solar physics program. Scientific data are collected during each mission using a tape recorder and, in many cases, film cameras. There is no command and control capability after deployment. Power during the deployed phase of the mission is provided by on-board batteries, and attitude control is accomplished with pneumatic gas jets. Retrieval of the SPARTAN platform at the end of each mission by the Shuttle.

- SPARTAN-201-1 (STS-56, April 8 -17, 1993, mass of 1360 kg). Objective: monitoring of solar-wind acceleration by observing the hydrogen, proton and electron temperatures and densities, and the solar-wind velocities in a variety of coronal structures at locations from 1.5 to 3.5 solar radii from the sun.
  UVCS (Ultraviolet Coronal Spectrometer) of SAO (Smithsonian Astrophysical Laboratory, Cambridge, MA). Measurement of line profiles/intensities of Lyman-alpha

(1215 Ä) and intensities of the Oxygen VI lines (1031.9 and 1037.6 Ä). The measurements are used to determine velocities, temperatures and densities of the coronal plasma in the regions observed.
WLC (White-light Coronagraph) of GSFC, built by NCAR. Measurement of polarization/ intensity of white-light corona. The measurements allow determination of electron densities in the coronal features observed. The WLC instrument is the most recent version of the spaceborne externally occulted Lyot coronagraph. WLC uses a rotating half-wave plate, a linear polarizer, a serrated occulting disk for occulting the disk at 1.25 solar radius, and a linear, wide dynamic range detector CCD camera to measure the intensity and polarization of broad band visible coronal radiation. The distance between pixel centers of the CCD camera is 22.5 arcseconds. Observations out to 6 solar radii from sun-center were obtained within a sector of 60° in width centered on the UVCS radial scan capability.

- SPARTAN-201-2 (STS-64, Sept. 9-20, 1994). Reflight of UVCS and WLC to study the solar wind. The SPARTAN platform flew about 80 km behind the Shuttle orbit for its measurement mission..

- SPARTAN-201-3 (STS-69, Sept. 7-18, 1995, mass = 1136 kg). Reflight of UVCS, WLC and an X-Ray instrument to study corona and galactic clusters during its two-day mission.

- SPARTAN-201-4 (STS-87, Nov 19 - Dec. 5, 1997) S/C mass = 1350 kg. The 201-4 mission was coordinated with the SOHO mission. Instruments: UVCS, WLC, TEXAS, VGS. All S/C operations were preprogrammed. No RF links until addition of TEXAS on SP201-04. Power supplied by Silver-Zinc batteries, 18 kWh/ 28 VDC.
  TEXAS (Technology Experiment Augmenting Spartan). TEXAS is an S-band communication system baselined for future Advanced Spartans. The system is used to allow real-time downlink of solar images and uplink of pointing corrections. The WLC instrument was modified for solar image downlink and pointing command uplink via interface to TEXAS.
  VGS (Video Guidance Sensor), an MSFC experiment with the objective to test the ranging and attitude measurement system being developed for the RLV automated docking system.

- SPARTAN-201-5 (STS-95, Oct. 29 - Nov. 7, 1998). Reflight of UVCS.

J.18.4 SPARTAN-204

SPARTAN-204 (STS-63, Feb. 3 - 11, 1995). Objective: observations of galactic dust clouds using **FUVIS** (Far UV Imaging Spectrograph), of NRL in the DoD Space Test Program (STP). FUVIS obtained far ultraviolet spectroscopy of diffuse sources, both natural and man-made [Shuttle surface glow and plume emissions from the RCS (Reaction Control System) thrusters]. - In addition to the FUVIS operations, Spartan 204 supported two space station Detailed Test Objectives (DTO). For the first DTO, Spartan 204 was fitted with six laser retro-reflectors to aid in testing the tracking control system.

J.18.5 SPARTAN-206

SPARTAN-206 (STS-72, Jan. 11 - 20, 1996). OAST (Office of Aeronautics and Space Technology) free-flyer with REFLEX, GADACS, SELODE and SPRE.

REFLEX (Return Flux Experiment). A technology experiment designed to verify computer generated contamination models used during design, fabrication, and operation of payloads. The primary objective of REFLEX is to investigate one important mechanism, referred to as molecular backscattering or “return flux,” associated with on-orbit molecular

contaminant transport (study of mixing contaminants with residual molecules of the atmosphere that are reflect back onto the S/C surface).
REFLEX consists of the following equipment: a mass spectrometer; three Temperature-Controlled, Quartz Crystal Microbalances (TQCMs); two electronics boxes; a pneumatics system; and a support structure. The spectrometer, built by the Physics Department at the University of Minnesota, is a Mattauch-Herzog type, and is equipped with an energy analyzer. It will measure the species, velocities, and densities of molecules in the Earth's residual atmosphere as well as return flux molecules from the onboard inert gas sources of Argon and Krypton. The shuttle's Low Earth Orbit (LEO) ensures that the atmospheric density is ample to provide sufficient backscattering of the molecules. The three TQCMs are placed on the exterior of the experiment to study contamination buildup and atomic oxygen erosion; one of the TQCMs is uncoated and the other two are coated, one with a film of carbon and one with a material called Kapton.

GADACS (GPS Attitude Determination and Control System). [1407] A NASA/GSFC prototype instrument. The prime objective is to demonstrate GPS attitude measurement technology in LEO (derivation of position, velocity, time, and attitude). GADACS used two TANS Vector GPS receivers (of Trimble Navigation, Sunnyvale, CA), along with an array of four L-band patch antennas (and pre-amplifiers) mounted onto a special plate inside of SPARTAN, and DRIRU-II (Dry Rotor Inertial Reference Unit), a high precision gyroscope developed by Teledyne (for proof of GPS measured attitude). GPS data were recorded along with gyro measurements for verification and calibration of the GPS attitude measurements. [1408] Note: GADACS was not able to perform attitude determination because it did not lock onto the four GPS signals necessary to obtain a position fix (a required first step to obtain attitude solutions.

SELODE (Solar Exposure to Laser Ordnance Device). The major objectives of SELODE are to verify if natural, low-orbit, solar energy can inadvertently fire laser pyrotechnics, and characterize how the LEO environments effect the reliability and/or performance of the pyrotechnics after flight. SELODE is the first experiment to test these devices on a shuttle mission. SELODE consists of four different Laser Initiators (LI), one kind of laser detonator (LD), and two Laser Diode Firing Units (LDFU). The initiators and detonators are exposed to sunlight three different ways: magnified solar exposure, direct solar exposure, and no solar exposure at all.

SPRE (SPARTAN Packet Radio Experiment), developed by the University of Maryland. Objectives: Test of GPS and packet radio based tracking system using a LEO satellite. Provision of an educational tool. SPRE is an amateur radio communications experiment. SPRE relays ground station positions and transmits telemetry containing the GPS location of the spacecraft and housekeeping data. The GPS data is generated by another Spartan experiment and sent to SPRE via a serial connection. Special software called APRtrak is used at SPRE ground stations to provide a mechanism to plot the positions of stations and objects world–wide using SPRE transmissions.
The SPRE experiment is constructed by integrating several specific subsystems. Some of these subsystems were designed and built by students, while other subsystems were purchased as a unit (such as the radios). The SPRE hardware consists of an electronics box and two antenna arrays. The electronics box contains the SPRE computer, packet terminal node controller (TNC - radio modem), and two radios. The computer receives GPS data from GADACS and communicates with the REFLEX mass spectrometer. The two antennas arrays are mounted on opposite sides of the spacecraft above the thermal louvers.

1407) F. H. Bauer, J. R. O'Donnell, "Space-Based GPS 1996 Mission Overview," Proceedings of ION GPS-96, Sept. 17-20, 1996, Kansas City MO, pp. 1293-1302

1408) F. Bauer, E. Lightsey, et al., "Pre-Flight Testing of the SPARTAN GADACS Experiment," Proceedings of ION GPS-94, Salt Lake City, pp. 1233-1241

J.18.6 SPARTAN-207

SPARTAN-207/IAE (Inflatable Antenna Experiment) on STS-77, May 19 - 29, 1996. The objective was to validate and characterize the mechanical function and performance of a 14 m diameter inflatable deployable antenna reflector structure in orbit. The payload consisted of two main pieces of hardware: the SP207/IAE (Spartan free flyer and experiment) and the Spartan Flight Support Structure (SFSS), which held the Spartan S/C in the Space Shuttle Cargo Bay. The struts are attached to the Spartan spacecraft, and the entire assembly was deployed and then recovered by the Space Shuttle. Once in low-Earth orbit, the Spartan became a platform for the antenna that, when inflated in space, was roughly the size of a tennis court. After inflation, an optical system surveyed the antenna and measured the accuracy of the surface at a variety of internal pressures and thermal conditions.

The antenna experiment was conducted in order to gather more definite information on other galaxies and on Earth's environment, there is a need for larger and more reliable space antennas.

J.18.7 SPARTAN-250 Carrier System

The Spartan 250 platform is an upgrade to the existing Spartan 200 series carrier. It is (Spartan 250) a medium sized, low cost, reusable, 3-axis stabilized, freeflying carrier that provides extended mission flight opportunities for a variety of scientific studies in low Earth orbit. Spartan 250 is designed to be a multipurpose carrier system which evolves in response to each mission. Spartan 250 is taken into orbit by the Shuttle, deployed, operated via ground commands as a sub-satellite, retrieved on the same mission, and returned to the ground for subsequent reuse. Because of the absolute necessity to recover the hardware, redundancy and back-up systems have been incorporated to assure this recovery.

Experiment mass	Up to 450 kg
Experiment power Experiment power voltage	Up to 175 W (orbital average), 245 W (peak) Unregulated +7/-6 VDC
Guidance, navigation, control	Three-axis stabilized (solar, stellar, nadir, zenith, and limb)
Experiment data storage	Up to 3 GByte
Data interfaces (experiment-S/C)	MIL-Std-1553, Up to 200 kbit/s for total experiment complement
Data RF communications	Up to 2.25 Mbit/s downlink in S-band, 30 Mbit/s in X-band

Table 425: Some characteristics of the SPARTAN platform series 250

J.18.8 SPARTAN-251

The XSS-10 (Experimental Spacecraft System) mission is a joint effort between AFRL and NASA (GSFC, JSC, KSC) in pursuit of space operations.

The SPARTAN 251 carrier for the XSS-10 mission consists of the carrier previously flown on the SPARTAN-204 and -207/IAE missions with significant enhancements. The SPARTAN carrier is upgraded from the 200-series carrier to the SPARTAN-250 series configuration with the addition of an RF command and telemetry link and an on-board computer.

J.18.9 SPARTAN-401

The SPARTAN-400 is design concept for a standardized, reusable bus that may be deployed from the Shuttle cargo bay and retrieved on a later Shuttle flight. The carrier itself is focused on providing experimenters with standard interfaces, capabilities, services and physical accommodations satisfying a wide range of requirements. The spacecraft provides a three-axis stabilized platform that supports solar or stellar inertial fine pointing, as well as nadir, offset nadir, and ground track.

As an option, the Spartan 400 spacecraft has the capability of incorporating a propulsion system. This system widens the range of mission orbit altitudes available from the Shuttle launch from 300 km to over 600 km. Higher apogee altitudes are possible with eccentric orbits.

J.19 SIR-A (Shuttle Imaging Radar)

SIR-A = Shuttle Imaging Radar with Payload A. A JPL payload on the Shuttle. Launch: Nov. 12, 1981 (duration: 3 days, second orbital test flight of the Shuttle Columbia, STS-2). The science payload aboard this Shuttle flight was referred to as OSTA-1 (Office of Space and Terrestrial Applications).

The success of the Seasat SAR prompted the first flight of a SAR on the Shuttle. The Shuttle was seen as an opportunity to progressively develop and fly increasingly more complex radar systems for short missions, allowing the hardware to be returned to Earth for reuse on follow-on missions.

SAR Application: Land use, geology, hydrology, forestry, and mapping.

Orbit: Shuttle orbit, 38° inclination, 257 km altitude, SAR swath width: 50 km.

Sensor complement:

SAR = Synthetic Aperture Radar in HH Polarization; L-band (freq. = 1.28 GHz); fixed look angle (incidence angle) = 47 degrees; pixel size = 40 x 40 m; optical data only.

Note: The system characteristics for SIR-A, were defined primarily by the spare parts available from the Seasat SAR. SIR-A, therefore, had essentially the same design as Seasat: L-band wavelength and HH polarization. The antenna look angle was fixed at 470 in order to optimize geologic mapping in topographically rough terrain. From the shuttle altitude, thc 2.16 m x 9.4 m antenna (spare Seasat SAR antenna panels were used) provided a 50 km swath. The antenna was mounted in a fixed open position on the starboard side of the payload bay, with the long axis parallel to the velocity vector when the shuttle was flying nose-forward. This allowed imaging only on the northern side of the shuttle nadir track. Because simplicity was a requirement for SIR-A, the data were recorded optically in the payload bay and were not accessed until the film was off-loaded after landing. The optical recorder on which the data for SIR-A were recorded was a spare from the Apollo 17 Lunar Sounder Experiment and was 14 years old when flown on SIR-A. [1409)]

Data: Image size: 50 km x length of active orbit. Optical image correlation, transparencies and photos available. Transmission: no transmission, optical recording on-board.

The SIR-A mission provided much improved image data for geologic analyses, since they are relatively free of the layover distortion that accompanied Seasat images of high-relief regions. SIR-A SAR data also led to the discovery of buried and previously uncharted dry river beds beneath the Sahara Desert in Sudan and Egypt. This demonstrated the ability of a L-band radar to penetrate up to several meters in hyper-arid sand sheets.

Other remote sensing instruments of the OSTA-1 payload were: (see also LFC in J.9)

SMIRR[1410)] = Shuttle Multispectral Infrared Reflectance Radiometer (NASA/JPL experimental instrument). SMIRR was also part of the science payload (OSTA-1)

Objectives of SMIRR: a) obtain radiometric measurements from orbit of land surfaces in 10 wavelength channels (rock and mineral identification). b) determine the spectral re-

1409) J. Way, "Spaceborne Imaging Radar – From Remote Sensing Science to Earth Science Questions," Launchspace Magazine, Volume 3.04, Aug/Sep 1998

1410) Manual of Remote Sensing, Second Edition, American Society of Photogrammetry, 1983, pp. 1707-1710 sponse of known rock types under different climatic conditions worldwide; c) establish the utility of orbital narrow-band radiometry in the 2.0 to 2.5 μm region; d) establish the requirements for future narrow-band imaging systems.

OCE = Ocean Color Experiment.[1411)]
The experiment consists of a multispectral image scanner of high sensitivity, dedicated to the measurements of water color and its interpretation in terms of major water constituents (chlorophyll, sediments, pollutants, etc.). The instrument is an 8-channel optomechanical image scanner. Total wavelength range from 464 - 773 nm. FOV = ±45°. IFOV = 3.5 milliradian.

Prior to and during the Shuttle mission with the OSTA-1 payload, various OCE-related experiments took place. The main objective of these activities was to perform comparative measurements from aircraft, ship and ground for correlation and verification of the Shuttle OCE data. A duplicate instrument, referred to as **OCS** (=Ocean Color Scanner) was shipped from GSFC to DLR and installed in aircraft for pre-Shuttle test flights in various regions of Europe.

The following organizations participated in OSTA-1/OCE related activities: GSFC (USA), DLR (Germany), University of Lisbon (Portugal), Hydrologic Institute of the Navy (Portugal), University of Lille (France), GKSS (Germany), INTA Madrid (Spain), IIP Barcelona (Spain).

J.20 SIR-B

SIR-B = Shuttle Imaging Radar with Payload B. A JPL payload on the Shuttle as part of the OSTA-3 payload. Launch: Oct. 5, 1984 (duration: 1 week). Application: Land use, geology, hydrology, forestry, mapping.

Orbit: Shuttle orbit, 57° inclination, 225 km altitude, swath width: 30 - 60 km

Sensor: **SAR** = Synthetic Aperture Radar in HH Polarization; L-band (freq. = 1.28 GHz); look angle = variable from 15 - 55°. SIR-B provided for the first time the ability to mechanically tilt the antenna over a range (15 - 55°) so that radar imagery from multiple angles of incidence could be obtained. Pixel size = 25m azimuth; radiometric resolution = 3-6 bit raw data. - The SIR-B radar was significantly upgraded. It had the additional flexibility of imaging geometry, particularly the effect of incidence angle on radar backscatter. One additional panel was added to the SIR-B antenna, bringing its total length to 10.7 m; again the antenna was mounted on the starboard side of the payload bay. The southern side of the track was also imaged during the SIR-B flight by rolling the shuttle 75° or yawing it 180°. A final improvement to the antenna was the ability to fold for launch in order to accommodate an additional deployable payload.

Data: Image size: 30 - 60 km x length of active radar. A digital data handling assembly (DDHA) converted the analog signal to a digital signal on-board the Shuttle. Data transmission: Shuttle downlink via TDRSS in Ku-band; data rate: 40 Mbit/s.

J.21 SIR-C/X-SAR

Shuttle Imaging Radar with Payload C (SIR-C).[1412)] [1413)] [1414)] [1415)] A NASA/JPL, DARA/DLR, and ASI payload on the Shuttle. This payload is also known under the name SRL (Space Radar Laboratory). Total payload mass = 11,000 kg, power consumption of payload sensors = 3 - 8.5 kW. Two Shuttle missions were conducted, each of 10 days duration.

1411) H. v.d. Piepen, V. Amann, H. Helbig, HH. Kim, W. Hart, et al. "The Promise of Remote Sensing," IEEE paper presented at IGARSS '82, June 1-4, Munich

The SIR-C/X-SAR is the first spaceborne radar system capable of obtaining simultaneous multifrequency and multipolarization radar imagery. - A multipolarization, distributed C-band system was added along with a more powerful multipolarization distributed L-band system. Instead of using a single, high-power transmitter, the distributed SIR-C radars consisted of numerous low-power solid-state transmitters distributed across the antenna aperture. Large power losses were avoided and as much as an eight-fold improvement in efficiency was possible. The distributed C- and L-band SIR-C radars allowed electronic beam steering in the range direction (123°) from a fixed antenna position of 38° (look angle), thus making it possible to acquire multi-incidence angle data without tilting the entire antenna.

1) SRL-1 (STS-59): April 9 - 20, 1994

2) SRL-2 (STS-68): September 30 - October 11, 1994

Application: Land use, geology, hydrology, oceanography, snow and ice, vegetation, calibration, and technological experiments.

Objectives:

- Conduct geoscience investigations that require the observational capabilities of orbiting radar sensors, alone or in conjunction with other sensors, that will lead to a better understanding of the surface conditions and processes on the Earth.
- Explore regions of the Earth's surface that are not well characterized because of vegetation, cloud, or sediment cover in order to better understand land and ocean surface conditions and processes on a global scale.
- Incorporate this new knowledge into global models of surface and subsurface processes.

Orbit: Shuttle orbit, 57° inclination, 225 km altitude.

Sensor complement: All three SAR instruments were flown on each mission. Microwave sensors are independent of the day/night cycle and mostly independent of the weather. - The SRL-2 orbit was nearly identical to that of SRL-1 permitting repeat cross-track interferometric SAR processing for measuring elevation as well as to detect change in the radar direction (a significant amount of data was collected at all three frequencies). - The total instrument/antenna mass of SIR-C/X-SAR is 11,000 kg, power consumption = 3 - 7.5 kW.

The dedicated payload filled the entire Shuttle cargo bay. The antenna was mounted in a tilted position and the shuttle rolled to a nominal 38° look angle. In order to acquire data at look angles other than 38° with the conventional X-band system, the X-band antenna, which was mounted along the upper portion of the array, was mechanically tilted. [1416)]

1) **L-band SAR** (1.250 GHz, wavelength = 23.5 cm), JPL. The L-band antenna is a planar array (12 m x 2.95 m in size) composed of a uniform grid of dual-polarized microstrip antenna radiators, active phased arrays. Further details are given in Table 426.

2) **C-band SAR** (5.3 GHz, wavelength = 5.8 cm), JPL.
The SIR-C payload comprised the L-band and C-band SAR antenna plus instrumentation. The SIR-C antenna boresights were steered electronically to provide coverage at varying distances from the Shuttle ground track. The SIR-C phased array also provided for broadening of the beam in the elevation direction from its minimum value of 5° to

1412) "X-band Synthetic Aperture Radar (X-SAR) and its Shuttle-Borne Application for Experiments," paper by Herwig Öttl and Francesco Valdoni

1413) R.L. Jordan, B. L. Huneycutt, M. Werner, "The SIR-C/X-SAR Synthetic Aperture Radar System," Proceedings of the IEEE, Vol. 33, No. 4, July 1995, pp. 829-839

1414) Special Issue on SIR-C/X-SAR, IEEE Transactions on Geoscience and Remote Sensing, Vol. 33, No. 4, July 1995

1415) R.L. Jordan, B. L. Huneycutt, M. Werner, "The SIR-C/X-SAR Synthetic Aperture Radar System," Proceedings of the IEEE, Vol. 79, No. 6, June 1991, pp. 827-838

1416) F. V. Stuhr, R. L. Jordan, M. U. Werner, "SIR-C/X-SAR A Multifaceted Radar," IEEE Aerospace and Electronic Systems Magazine, Vol. 10, No. 10, Oct. 1995, pp. 15-25

16° (selection of seven values). The SIR-C phased array enabled the operational modes of **ScanSAR** and **Spotlight**. In ScanSAR, the antenna pattern coverage on the ground was stepped in the cross-track direction during the synthetic aperture period to allow coverage over a wider swath, however, at the expense of azimuth resolution. The swath width ranged from 15 to 65 km for calibrated images and 40 to 90 km for mapping mode (ScanSAR) images. For the Spotlight mode, the boresight was positioned in azimuth to dwell on a particular area as the Shuttle flew by. This permitted an increase in azimuth resolution to a value of 7 m for the selected area, at the expense of the along-track swath.

3) **X-SAR** (SAR for X-band Measurement (9.6 GHz, wavelength = 3.1 cm), provided by DARA/DLR and ASI, built by Dornier and Alenia). X-SAR uses only vertical polarization (VV). The X-SAR instrument uses a passive slotted-waveguide antenna (12 m x 0.4 m) which is tilted mechanically to align the X-band beam with the SIR-C C-band and L-band beams. The X-SAR antenna has a fixed beamwidth of 5.5° in elevation and 0.14° in azimuth as opposed to the phased array, multi-polarization antenna of SIR-C. The instantaneous area illuminated by the antenna on the ground (footprint) is an ellipse of size 60 km x 0.8 km from an orbital altitude of 225 km. The electronics part of X-SAR is mounted on a cold plate structure and positioned underneath the antenna structure. A TWT amplifier is transmitting up to 1736 pulses/s at a peak transmit power of 3.35 kW. The pulses are frequency-modulated with a pulse length of 40 μs and a programmable bandwidth of 10 or 20 MHz.

Parameter	L-band Antenna	C-band Antenna	X-band Antenna
Frequency	1.25 GHz	5.3 GHz	9.6 GHz
Antenna aperture length Antenna aperture width	12.0 m 2.95 m	12.0 m 0.7 m	12.0 m 0.4 m
Architecture	Active Phased Array		Slotted waveguide
Phase control	4 bits	4 bits	N/A
Polarization	HH and VV	HH and VV	VV
Polarization isolation	>25 dB	>25 dB	39 dB
Antenna gain	36.4 dB	42.8 dB	43.5 dB
Mechanical steering range	N/A	N/A	±23°
Electronic steering range	±20°	±20°	N/A
Elevation beamwidth	5-16°	5-16°	5.5°
Azimuth beamwidth	1.0°	0.25°	0.14°
Transmit pulse length	33.8 μs	8.5 μs	40 μs
Radiometric resolution	1.5 dB	1.5 dB	2.5 dB
Peak radiated power	3200 W	1700 W	1500 W
System noise temperature	450 K	550 K	551 K
Mass of antenna structure	3300 kg		49 kg
Look angle (adjustable off-nadir angle)	15° - 55°	15° - 55°	15° - 55°
Swath width	10 km - 70 km	10 km - 70 km	15 km - 45 km
Azimuth resolution (4 look)	30 m	30 m	25 m
Range resolution with 10/20 MHz bandwidth	25 m/13 m	25 m/13 m	20 m/10 m
Data rate per channel	45 Mbit/s Total of 90 Mbit/s	45 Mbit/s Total of 90 Mbit/s	45 Mbit/s

Table 426: SIR-C/X-SAR instrument parameters

Data:

- The source data are digitally coded and recorded on-board.
- Transmission: only one SAR data stream at a time is possible. Ku-band via TDRSS to White Sands, NM, etc.
- Image size: 100 km (flight direction) x 50 km

- A SIR-C/X-SAR POCC was operated at NASA/JSC (Houston, TX).

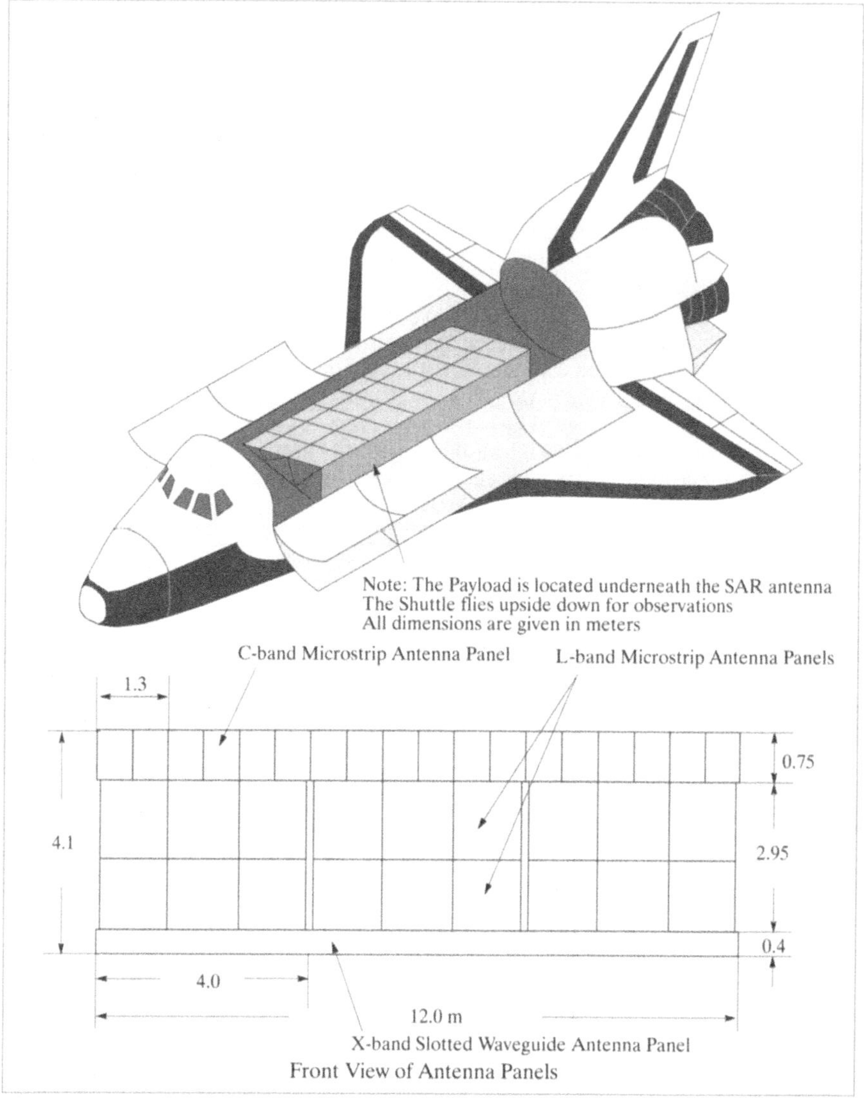

Figure 244: Configuration of the SIR-C/X-SAR antenna payload in the Shuttle bay

Nominally, 50 hours of SIR-C data (on each of the four channels) and 50 hours of X-SAR data were recorded by on-board tape recorders. In addition, a limited amount of SIR-C/X-SAR data was transmitted to the ground (via TDRSS Ku-band 50 Mbit/s data link). for real-time digital processing during the mission. Data "takes" were largely over experiment sites selected prior to launch, with some in-flight "targets of opportunity." The orbital altitude was trimmed for the last days of the second flight (SRL-2) to provide a repeat-track interferometric observation geometry. See also the campaign description of SRL-1 and -2 in Appendix Q.

The intent was to provide data calibrated in such a way as to allow comparisons with other spaceborne SAR data (eg., ERS-1, JERS-1, Radarsat, etc.) so that a time-series view of key geophysical parameters may be realized.

MAPS (Measurement of Air Pollution from Satellites). See description under J.11.

Also, on-board the SRL flights, was an ocean wave spectra processor, designed and built by Johns Hopkins Applied Physics Laboratory, which collected data on ocean surface wave length, direction, and height.

J.22 Spacelab-1

Shuttle Mission (STS-9, Columbia) Nov. 28, 1983 - Dec. 8, 1983 (10 day flight). Orbit: 240 - 257 km, inclination = 57°. NASA-ESA mission.

Sensor complement:

Metric Camera (German/ESA experiment; the camera was a slightly modified aerial survey camera of the Zeiss RMK A 30/23 type). Objective: to test the mapping capabilities of high-resolution space photography on a large film format.
Application: Topographic and thematic mapping. The metric camera provided high-resolution photographs and experimental results on planimetric and topographic mapping from outer space. [1417]

Spectral range	535 - 900 nm
Focal length	305 mm
Image size, Image scale	23 cm x 23 cm, 1:820,000
Image overlapping	60-80%
FOV (Field of View), Swath width	41.2°, 189 km
B/W film	Kodak Double-X Aerographic film 2045; pixel size = 12 x 12 m
Color film	Kodak Aerochrome infrared film 2443; pixel size = 12 x 12 m
Data rate	550 frames per film

Table 427: Some instrument parameters of the Metric Camera

For metric camera operations, the Shuttle flew with the open cargo bay oriented towards the Earth. In this flight attitude, the camera's optical axis looked vertically down to the Earth's surface. A total of 21 camera operations (exposure series) were taken which varied in duration from 2 to 18 minutes.

MRSE = Microwave Remote Sensing Experiment. MRSE operates in two modes, the scatterometer mode or the thermal radiometer mode. The two frequency scatterometer measured backscatter from the ocean surface at two adjacent frequencies. The thermal radiometer measured surface temperature.

GRILLE = Infrared spectrometer (French/Belgian sensor). GRILLE has a high resolution in the infrared region (2 μm); it measures vertical profiles of: CO, CO_2, NO, H_2O, CH_4, H_2O and HCl.

SOLSPEC = SOLar SPECtrum Measurement (CNES). Measures the solar irradiance from 180 - 3200 nm using three double spectrometers and an on-board calibration device.

SOLCON = Solar Constant Sensor (IRMB, Brussels, Belgium, PI: R. Crommelynck). Measurement of the solar constant. The SOLCON instrument is a cooperative effort of IRMB, Space Science Dept. of ESA, and LaRC. Measurement of the absolute value of the total solar irradiance (and long-term variations).

1417) "Spacelab-1 Metric Camera, User Handbook and Data Catalogue," compiled by M. Schroeder, E Suckfüll, G. Todd, and P. Lohmann of DLR, Oberpfaffenhofen, Dec. 1986

J.23 Spacelab-3

Shuttle (Challenger, STS-17) mission between April 29 and May 6, 1985.[1418)]
A total of 15 experiments on-board, most of them on microgravity (crystal growth).
Orbit: 352 km altitude, 57° inclination

Sensor:

ATMOS = Atmospheric Trace Molecule Spectrometer (JPL sensor). Same instrument as on ATLAS. Objective: Investigation of the distribution of neutral constituents in the Earth's atmosphere. The ATMOS sensor is a Michelson interferometer, with a response to radiation in the NIR to IR regions, designed to make observations from on board the Shuttle in the solar occultation mode.

ATMOS gathered data during 20 occultation events; 13 sunset occultations located between 26° - 34° N latitudes and 7 sunsets around 48° S.
ATMOS employs high-resolution FT (Fourier Transform) interferometry and limb viewing in the of 2.2 µm to 16 µm bands.

J.24 Shuttle EO Imaging Cameras

A special camera system was developed for Shuttle flights with the objective to obtain high-resolution images of special areas of interest from piloted space systems.

Applications: spot target observations.

ESC - Electronic Still Camera[1419)]
The system consists of the following components:

- a hand-held battery-operated high-resolution camera (fully digital and programmable)
- a lap-top computer-based playback/downlink unit for on-board image processing
- a ground station capable of receiving Shuttle data, processing images, producing hard copies, and distributing the data and hard copies to end users.

The camera is designed and built with astronomical-grade CCDs, which resulted in near-film-quality imaging. The downlink capability provides the means of investigator ground inspection and interaction during Shuttle flights. This interaction capability optimizes the yield of experiments and observations. The Orbiter's Ku-band downlink scheme permits image transmission within 15 seconds; however, ESC images may also be downlinked through any channel, regardless of bandwidth and bit rate. As of 1992 the Naval Research Laboratory is developing a latitude-longitude locating system, which when used with the ESC will identify the location of the Earth target imaged within 1 nautical mile. The camera flown on STS-48 was set up for a 1000 x 1000 CCD array.

ESC specifications:

- ESC spectral range: 0.4 - 1.1 µm; the spectral response allows for multispectral imaging. Human interaction on-board permits quick decisions: for instance on the best way to photograph a scene, (lenses may be exchanged, filters may be selected for multispectral imaging, a new data storage cartridge may be loaded, etc.).
- Ground resolution of about 40 m (at 470 km orbit with 180 mm lens); or of = 15 m (at 290 km orbit with 300 mm lens, theoretical limit)
- ESC is fully programmable to support a variety of sensors: 1000 x 1000, 2000 x 2000, and 1000 x 1000 color (note: a 2000 x 2000 array is planned)
- Dynamic range = 60 dB; sensor dynamic range = 80 dB

1418) "Overview of ATMOS Results from Spacelab-3," Optical Remote Sensing of the Atmosphere, 1990 Technical Digest Series of the Optical Society of America, Volume 4, pp. 64-66
1419) S. D. Holland, "The NASA Electronic Still Camera System," IEEE IGARSS '92 Volume I, pp. 149-151

- Removable image storage media for unlimited storage capability
- Adaptable to all Nikon lenses or any instrument that uses a Nikon mount
- Retains the functions of a Nikon F4, such as autofocus, autoexposure, and manual override.

The ESC was first used on mission STS-48 in September 1991. Since then, ESC has flown on STS-42 and is planned to fly on STS-45, STS-49, and STS-53.

Playback/downlink capabilities:

- Employs virtual imaging, which allows the user to display ESC images on a variety of monitors with a variety of solutions
- Provides standard and nonstandard image processing
- Allows the user to select and deselect desired images for downlinking
- Allows notes to be included with each downlinked image.

Ground station capabilities:

- Compatible with Shuttle digital downlink
- Primary image processing performed with Sun 3/470 and PIXAR imaging computers
- Capability of processing 4000 x 4000 true color images
- Data can be networked anywhere in the world
- Hardcopy capability for ground crew

J.24.1 Shuttle Film Camera Systems

Earth photographs are collected using 35 mm, 70 mm, and 5 inch (12.7 cm) format still cameras, equipped with a variety of lenses and film types.

- Modified Hasselblad single-lens reflex camera, model 500 ELX (prior to Jan. 1990 ELM).[1420),1421)] A 70 mm- format hand-held camera for photography by astronauts. An important accessory is the data recording module for the film magazine, which records the exact time when the photograph was taken (for post-flight identification of the scene). Data Recording Modules (DRMs) were installed on Hasselblad cameras in April 1985 with STS-51-B. Before DRMs were installed, determining the location of ocean features required identifying geographic references in the photograph, eg., coastlines or islands.

- Rolleiflex single-lens reflex camera, model 6008 (was flown several times during 1990-92, first flight on STS-32 in Jan. 1990). A 60-frame, 70 mm film magazine was devised. The camera features automatic exposure control.

- Linhof Aero Technika 5-inch (12.7 cm) format camera system. The camera is flown on a space-available basis. First flight on STS-41G in Oct. 1984. Lens focal lengths: 90 mm or 250 mm. Time and date recording of Linhof photographs began in Dec. 1988 with STS-27.

- Nikon camera systems F3 and F4. These cameras are being used primarily to document the Shuttle interior, they have also been used for Earth observation photography. Telephoto lenses of 180 mm and 300 mm focal lengths. Cataloging of these 35 mm Earth images has been instituted since STS-37 (April 1991).

Postflight Support:
A catalog is published after each Shuttle flight supplying for each frame the geographic location, geographical name, flight ephemeris data, and some descriptive feature information. A typical Shuttle flight produces 3000-6000 pictures.

1420) D. L. Amsbury, J. M. Bremer, "Recent Developments in Space Shuttle Remote Sensing, using hand-held Film Cameras," IGARSS '92, Volume I, pp. 152-154

1421) S. G. Ackleson, D. E. Pitts, "Global Distribution of hand-held Photographs of Ocean and Coastal Regions Taken during Space Shuttle Missions, 1981-1991," IEEE IGARSS '92 Volume II, pp. 1550-1552

Electronic Database:
The Space Shuttle Earth Observations Project (SSEOP) Photography Database,[1422] containing geographical references to more than 90,000 space photographs (over 26,000 from Shuttle flights up to 1991) is operational at NASA/JSC, Houston, TX. Access to this database is free-of-charge and available 24 hours a day.

J.24.2 IMAX Space Cameras

IMAX Corporation of Mississauga (Toronto), Canada, is the pioneer of large-format film and giant-screen projection technology. IMAX was created after Montreal's Expo 1967, when Canadian film makers Graeme Ferguson, Robert Kerr and Roman Kroitor impressed audiences with their multi-screen films. Aside from the entertainment business activities, there is also an engagement in the documentation of significant space activities, to promote NASA's educational goals using the IMAX film medium. [1423]

Background: The IMAX space camera project is a collaboration between IMAX Corporation, NASA, the Smithsonian Institution's National Air and Space Museum, and Lockheed Martin Corporation. The system, developed by IMAX Corporation of Toronto, Canada, uses specially designed 65 mm cameras and projectors to record and display very high definition color motion pictures which, accompanied by six-channel high-fidelity sound, are displayed on screens in IMAX and OMNIMAX (dome projection) theaters that are up to ten times larger than a conventional screen, producing a feeling of "being there."

IMAX space cameras have been flown on Shuttle missions STS-41-C (launch April 1984), 41-D, 41-G, -29, -34, -32, -31, -42, -46, -51, -61, -63, -71, -74, -88, etc. - to document crew operations in the payload bay and the orbiter's middeck and flight deck as well as to film spectacular views of space and Earth. Film from those missions was used as the basis for the IMAX productions "The Dream Is Alive," "The Blue Planet," and "Destiny in Space."

The STS-88 mission was the first space station assembly flight of the Shuttle to ISS (International Space Station) which took place from Dec. 4 to 15, 1998. The primary objectives of ICBC (IMAX Cargo Bay Camera) on STS-88 were to film the Node 1 installation onto the orbiter docking system (ODS), the functional cargo block (FGB) rendezvous, FGB docking, extravehicular activity (EVA) tasks, separation burn, and fly-around. ICBC is a space-qualified, 65 mm color motion picture camera system consisting of a camera, lens assembly, and a film supply magazine containing approximately 1100 m (3,500 feet) of film and an empty take-up magazine. [1424]

The 65 mm film from STS-88 was transferred to 70 mm motion picture film for use in a large-format feature film. An audio tape recorder with microphones in the crew compartment recorded middeck sounds and crew comments during camera operations. The audio was transferred to tapes and compact disks to accompany the motion picture.

J.24.3 IMAX-3D Space Cameras

In 1997, IMAX Corporation in association with NASA and Lockheed Martin, started a program to develop new IMAX-3D space cameras.

J.24.4 ICBC (IMAX Cargo Bay Camera)

The primary objectives of ICBC (IMAX Cargo Bay Camera) on STS-88 are to film the Node 1 installation onto the orbiter docking system (ODS), the functional cargo block

1422) R. M. Nelson, K. J. Willis, W. J. Daley, F. R. Brumbaugh, J. M. Bremer, "Cataloging and Indexing - The Development of the Space Shuttle Mission Data Base and Catalogs from Earth Observations hand-held Photography," IEEE IGARSS '92 Volume I, pp. 155-157

1423) http://www.imax.com/

1424) http://www.shuttlepresskit.com/STS–88/payload19.htm

(FGB) rendezvous, FGB docking, extravehicular activity (EVA) tasks, separation burn, and fly-around.

The ICBC is a space-qualified, 65 mm color motion picture camera system consisting of a camera, lens assembly, and a film supply magazine containing approximately 3,500 feet of film and an empty take-up magazine. The camera is housed in an insulated, pressurized enclosure with a movable lens window cover. The optical center line of the 30 mm camera lens is fixed and points directly out of the payload bay along the orbiter Z-axis with a 23-degree rotation toward the orbiter nose. Heaters and thermal blankets provide proper thermal conditioning for the camera electronics, camera window, and film magazines.

For STS-88, the delivery reel is loaded with 3,500 feet of film (nominally), enough for approximately 10.5 minutes of filming at normal camera speed (24 frames per second, fps). On this flight, the camera speed can be changed to 6 fps for photographing slower moving objects. The ICBC can also be loaded with a 2,200-foot film magazine. A single 30 mm wide-angle lens is mounted on the camera; lenses and film cannot be changed during the flight. ICBC operations are terminated when all film is exposed.

The ICBC is controlled from the aft flight deck with the enhanced GAS autonomous payload controller (GAPC) and uses orbiter dc power. A crew member can command the ICBC to turn main power on, go to a standby mode, adjust f-stop and focus, and film a scene. A spotmeter will be used by the crew to aid in setting the IMAX camera f-stops. By using the GAPC, the crew member can also determine the status of the camera, such as the current f-stop and the amount of film exposed. A light level measurement unit is used to set the lens aperture. A fixed focus zone and seven aperture settings are available for this flight. A tape recorder is also provided for crew documentation. All the GAS hardware, such as the GAS control decoders, status responder units, GAPCs, and the GAS signal and control cable, are owned, serviced, and certified by NASA's Goddard Space Flight Center.

The basic operational profile of the ICBC is as follows: enable the heaters within seven hours of launch or approximately 30 minutes before a planned payload activity to be filmed, maintain thermal conditioning of the camera and film magazine, perform a typical filming sequence, and return to thermal conditioning.

A typical filming sequence begins with powering the camera in standby mode. This consists of powering up the internal camera electronics, feed magazine and drive, take-up magazine and drive, IMAX interface electronics, and the lens drive to a standby mode. The f-stop, focus, and frame rate are adjusted to the desired settings. Actual filming occurs when the door motor and camera drive motor are operated. The camera then returns to standby until the end of the filming sequence.

Background: The IMAX space camera project is a collaboration between IMAX Corporation, NASA, the Smithsonian Institution's National Air and Space Museum, and Lockheed Martin Corporation. The system, developed by IMAX Corporation of Toronto, Canada, uses specially designed 65 mm cameras and projectors to record and display very high definition color motion pictures which, accompanied by six-channel high-fidelity sound, are displayed on screens in IMAX and OMNIMAX theaters that are up to ten times larger than a conventional screen, producing a feeling of "being there."

The 65 mm film from STS-88 will be transferred to 70 mm motion picture film for use in a future large-format feature film. An audio tape recorder with microphones in the crew compartment will record middeck sounds and crew comments during camera operations. The audio will then be transferred to tapes or compact disks to accompany the motion picture.

IMAX cameras have been flown on Shuttle missions STS-41-C (launch April 1984), 41-D, 41-G, -29, -34, -32, -31, -42, -46, -51, -61, -63, -71, -74, -88, etc. - to document crew operations in the payload bay and the orbiter's middeck and flight deck as well as to film spectacular views of space and Earth. Film from those missions was used as the basis for the IMAX productions "The Dream Is Alive," "The Blue Planet," and "Destiny in Space." The STS-88 mission was the first space station assembly flight of the Shuttle to ISS (International Space Station) which took place from Dec. 4 to 15, 1998.

The IMAX project is designed to document significant space activities and promote NASA's educational goals using the IMAX film medium.

J.25 SRTM (Shuttle Radar Topography Mission)

SRTM is a joint Shuttle mission on STS-99 (Atlantis) between NASA, DoD/NIMA (National Imagery and Mapping Agency), DLR (Deutsches Zentrum für Luft- und Raumfahrt, Germany), and ASI (Agenzia Spaziale Italiana). Funding for SRTM is from NIMA, while NASA is supplying the SIR-C hardware, Shuttle launch, ground systems, mission operations and data processing support. DLR and ASI are supplying the X-SAR hardware, ground systems, mission operations and data processing support. - SRTM represents a reflight of the basic SIR-C/X-SAR payload instruments on the SRL-1 (STS-59, April 9 - 20, 1994) and SRL-2 (STS-68, Sept, 30 to Oct. 11, 1994) missions (J.21). However, the SRTM configuration considers only two independent SAR systems, one in C-band (JPL instrument), the other in X-band (DLR/ASI). Both SAR systems operate with the main antenna of each instrument located in the open cargo bay of the Shuttle, and a second receive antenna mounted on a deployable outboard mast, respectively. The configuration (a bistatic system) for both radar systems represents a single-pass interferometer, also referred to as IF-SAR, with a baseline of 60 m (separation of the two antenna systems). The mission is managed by JPL. SRTM represents the first fixed baseline single-pass spaceborne IFSAR technology with wide-swath scanning SAR and dual-frequency (C-band and X-band) coverage. The 11-day STS-99 Shuttle mission took place Feb. 11-22, 2000. [1425] [1426] [1427] [1428] [1429]

The mission objective is to obtain single-pass interferometric SAR imagery to be used for DEM (Digital Elevation Model, also referred to as DTM = Digital Terrain Model) generation, i.e. topographic maps. Coverage of the Earth's land surfaces is provided between latitudes of -54° and +60°, representing nearly 80% of the land masses. The digital topographic map production objective (C-RADAR) calls for requirements which meet ITHD-2 (Interferometric Terrain Height Data -2) specifications, i.e., a spatial pixel (30 m x 30 m) posting with a 16 m absolute vertical linear accuracy and 20 m absolute horizontal radial accuracy at 90%. On-board data storage uses high-rate data recorders (300 cartridges each of 40 GByte capacity). - Applications: the topographic data are being used for such studies as: flooding, erosion, landslide hazards, earthquakes, weather forecasts, climate zones, etc.

Orbit: circular exact repeat orbit, altitude = 233 km, inclination = 57°. The Shuttle flies upside down and tail forward for simultaneous radar observations.
The look angle for the C-RADAR antenna system is 45°, for the X-RADAR antenna system = 52°. A nominal data collection period of 80 hours is assigned for both radars.

The on-board AODA (Attitude and Orbit Determination Avionics) consists of a suite of sensors (target tracker, a new GPS receiver system configuration, called BlackJack for orbit

1425) Note: While the L-band radar/antenna is flown as part of the overall structure, there is no interferometric L-band capability due to the large antenna dimensions needed at the outboard location. The L-band radar is not being operated, however, some of its electronic capabilities are used for the C-RADAR.

1426) M. U. Werner, "X-SAR/SRTM a Spaceborne Single Pass Interferometric SAR," Joint workshop of ISPRS WG I/1, I/3, and I/4: Sensors and Mapping from Space, University of Hannover, Germany, Sept 29 - Oct. 2, 1997

1427) SRTM Information Sheet of JPL

1428) R. L. Jordan, E. R. Caro, Y. Kim, Y. Shen, F. V. Stuhr, M. U. Werner., "Shuttle Radar Topography Mapper," Proceedings of the EUROPTO Conference: Symposium on Remote Sensing, Conference on Microwave Instrumentation for Remote Sensing of the Earth II, Taormina, Italy, Sept. 24-26, 1996

1429) P. Chien, "Around the World in 11 Days," Launchspace Magazine, Vol. 3.06, Dec. 1998 and attitude determination, distance meters, inertial reference units, and star tracker) to collect attitude and position information. The AODA data rate is 8.192 kbit/s. The baseline knowledge of the two inboard and outboard antenna pairs is known within 3mm, the angular alignment knowledge is better than 9 arcsec. The stiff and rigid mast (or boom) can experience length variations of up to 10 mm (thermal distortion) and bending/twist up to 0.1° during radar data acquisition. The mast itself consists of 87 cubical sections (referred to as bays) with a total mass of 360 kg. The total mast plus outboard antenna system has a mass of 1700 kg (AEC-Able Engineering is the manufacturer of the extendible boom). Since the long boom tends to drift into a gravity-gradient attitude (where one end of the S/C points toward Earth), this has to be counteracted by a thruster system mounted on the boom structure.

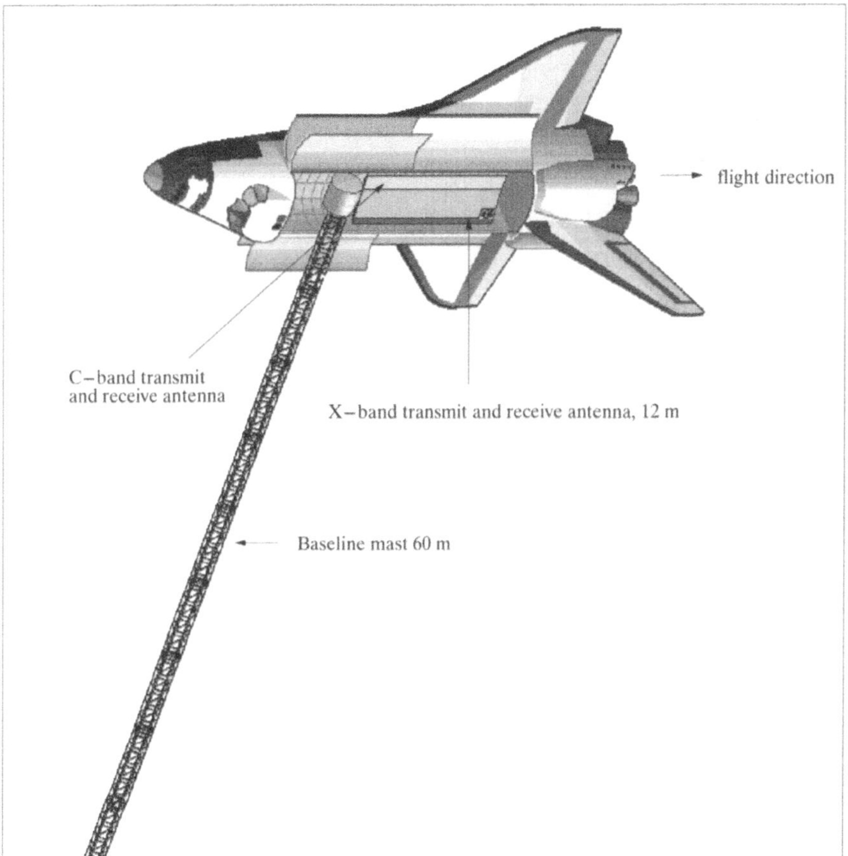

Figure 245: Configuration of the SRTM/X-SAR payload

Sensor complement: It consists of modified SIR-C/X-SAR instruments (plus added on structures) flown on SRL-1 and SRL-2 in April and October 1994, respectively.
The two single-pass interferometric SAR systems, namely **C-RADAR** and **X-RADAR**, are

configured in such a way that each radar is operating a transmit/receive antenna system inside the Shuttle cargo bay (primary or inboard system) and a receive-only antenna system at the boom tip (secondary or outboard system). The boom structure is oriented in the cross-track direction, perpendicular to the velocity vector of the Shuttle, with tip-mounted antennas, forming a parallel antenna system with the inboard system. Single-pass interferometry observation requires a beam overlap geometry of the primary and secondary antenna systems. This implies alignment measures for each antenna system to compensate for possible misalignments during the mission. 1430)

- The X-RADAR system provides a two-axis alignment capability. The inboard antenna can be steered mechanically (tilt of ±1.5°) in the cross-range direction to fine-tune the footprint of the outboard antenna. In addition, the outboard antenna can be steered electronically (±1°) in the azimuth direction to match the footprint of the inboard antenna.

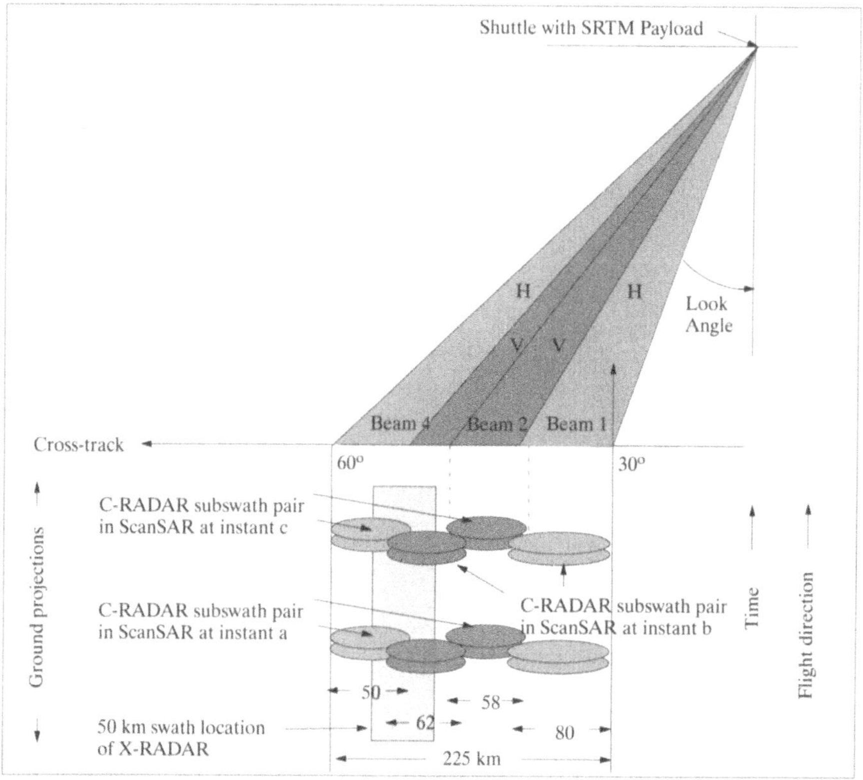

Figure 246: Illustration of SRTM observation geometries

- The phased array antenna system of C-RADAR provides 2-D electronic steering of the inboard and outboard systems for beam alignment. Four C-band subswaths of the inboard system are generated sequentially in ScanSAR mode, where two beams are forming an instantaneous ScanSAR pair, each pair producing a subswath with V and one with H polarization. Two parallel transmit/receiver channels are employed for this

1430) M. U. Werner, J. Heinstadt, "A Spaceborne X-band Single Pass Interferometric SAR Antenna System," ESTEC Workshop on Large Antennas for Radio Astronomy, Noordwijk, NL, Feb. 28-29, 1996 purpose, one is of the C-band system, the other channel is "borrowed" from the otherwise dormant L-band system.

C-RADAR (C-band Radar Instrument). C-RADAR (5.3 GHz, wavelength = 5.6 cm, bandwidth = 10 MHz) is provided by JPL, it offers a dual-polarization capability. The peak radiation power is 1.2 kW/polarization. In ScanSAR mode, C-RADAR employs two simultaneous beam pairs, as shown in Figure 246, which switch back and forth within the full swath width, creating a contiguous and overlapping SAR imagery pattern. The total swath width is 225 km providing a pixel size (resolution) of about 30 m (data point spacing of one arcsec). Instrument source data are collected at 180 Mbit/s (in four channels) and stored on the onboard recorders. Instrument telemetry data at 15.12 kbit/s is downlinked continuously.

X-RADAR (X-band Radar Instrument). [1431] X-RADAR (9.6 GHz, wavelength = 3.1 cm) is provided by DLR and ASI, it offers fixed beam (52° look angle) single polarization (VV inboard and V outboard antenna). The peak radiation power is 1.7 kW at a PRF of 1674 pulses/s. X-RADAR operates in a wide-swath mode of 50 km, the horizontal pixel resolution is 30 m x 30 m with a vertical height resolution of 6m relative and 10 m absolute. The X-SAR swath is positioned in the middle of the two outer subswaths of C-RADAR. The instrument science data rate is 90 Mbit/s and stored on the on-board recorder. Instrument telemetry data at 9.6 kbit/s is downlinked continuously.

Parameter	L-band Antenna	C-band Antenna	X-band Antenna
Frequency	1.25 GHz	5.3 GHz	9.6 GHz
Bandwidth		10 MHz	9.5 MHz
Inboard antenna: Aperture length x width	12.0 m x 2.9 m	transmit/receive 12.0 m x 0.75 m	transmit/receive 12.0 m x 0.4 m
Outboard antenna: Aperture length x width		receive only 8.0 m x 0.75 m	receive only 6.0 m x 0.4 m
Architecture	Active Phased Array Patch-type planar antenna with 2-D steering		Slotted waveguide array
Phase control	4 bits	4 bits	N/A
Polarization	4 subswaths transmit/receive H,V,V,H, inboard 4 subswaths receive only H,V,V, H, outboard		VV inboard antenna, (V outboard antenna)
Polarization isolation	25 dB	25 dB	39 dB
Antenna gain	36.4 dB	42.7 dB	44.5 dB
Mechanical steering range	N/A	N/A	±23°, fixed at 7° SRTM
Electronic steering range	±20°	±20°	±0.9°
Elevation beamwidth	5-16°	5-16°	5.5°
Azimuth beamwidth	1.0°	0.25°	0.14° (0.28°)
Transmit pulse width		34 μs	40 μs
Look angle (adjustable off-nadir angle)	15° - 55°	15° - 55°	52°, 17° - 60° possible
Peak radiated power	4400 W	1.2 kW/polarization	1.7 kW
System noise temperature	450 K	550 K	551 K
PRF		1344-1550 Hz	1440-1674 Hz
Swath width		225 km	50 km
Ground resolution		30 m x 30 m	30 m x 30 m
Height resolution relative		10 m	6 m
Quantization		8 bit (equivalent)	6 bits I and 6 bits Q
Data rate		180 Mbit/s(4x45 Mbit/s)	90 Mbit/s (2x45 Mbit/s)

Table 428: SRTM/X-SAR instrument parameters

J.26 SSBUV (Shuttle Solar Backscatter Ultraviolet Spectrometer)

SSBUV is a NASA/GSFC instrument (PI: E. Hilsenrath) using UV backscatter in nadir to measure vertical profiles of ozone in the stratosphere and in the lower mesosphere in the

1431) K. B. Klein, M. U. Werner, "System Performance Monitoring for X-SAR/SRTM," EUSAR'98, VDE-Verlag, May 25-27, 1998, Friedrichshafen, Germany, pp. 383-386 spectral range from 200 to 405 nm. The objective is to fly the SSBUV payload on numerous Shuttle missions to provide complementary calibration data for long-term satellite ozone data sets. The first flight with SSBUV instrumentation occurred on October 19, 1989 on the Shuttle Atlantis (STS-34). Throughout this Shuttle flight coincident observations were taken with the SBUV on Nimbus-7 and the SBUV/2 on NOAA-9 and NOAA-11 satellites.

The SSBUV instrument and its flight support electronics, power, data and command systems are mounted in the Shuttle's payload bay in two flight canisters (total mass of 410 kg). The instrument canister holds the SSBUV instrument, its aspect sensors and inflight calibration system. Once in orbit, a motorized door assembly opens the canister, allowing the SSBUV to view the sun and the Earth. The canister closes, providing contamination protection, while SSBUV performs inflight calibrations. [1432)]

The SSBUV instrument is the SBUV/2 engineering model now flying on NOAA satellites - see G.13 (NOAA-POES). The SBUV/2 sensor is a spectrally scanning UV radiometer (nadir-viewing sensor) measuring solar irradiance and scene radiance (backscattered solar energy) over a spectral range from 200 to 405 nm (note: SSBUV actually measures from 160-405 nm; however, there is no calibration below 200 nm because a vacuum is needed for this region). Resolution = 50 km.

The SSBUV program has pursued a vigorous laboratory calibration effort using NIST-traceable standards. Of particular importance are the direct comparisons conducted with the instruments for the TOMS-EP (TOMS), ERS-2 (GOME), and NOAA-14 (SBUV/2) missions. These instruments were calibrated using SSBUV standards. The SSBUV flight on STS-72 provided a first opportunity to compare space observation from instruments which have been intercalibrated on the ground.

Monochromator	0.25 cm diameter double Ebert Fastie design, F5
Detector	Biakali photomultiplier tube (PMT)
Grating	Holographic, 2400 lines/mm
Wavelength	160-405 nm continuous for solar radiance, 12 steps programmable for ozone measurements
Bandwidth	1.1 nm
Dynamic range	10^6
Linearity	<1%
FOV	11.3°

Table 429: Specification of the SSBUV instrument [1433)]

Shuttle Flight	Date	SSBUV coincident ozone observations with instruments on the following satellites
STS-34, Atlantis	Oct. 18-23, 1989	Nimbus-7 (SBUV/TOMS), NOAA-9, -11 (SBUV/2)
STS-41, Discovery	Oct. 6-10, 1990	NOAA-9 and -11 (SBUV/2)
STS-43, Atlantis	August 2-11, 1991	Meteor-3-6/TOMS, NOAA-9 and -11 (SBUV/2)
STS-45, Atlantis (ATLAS-1),	Mar. 24, to April 2, 1992	NOAA-11, Meteor-3-6/TOMS, UARS (CLAES, ISAMS, HALOE)
STS-56, Discovery (ATLAS-2)	April 8-17, 1993	NOAA-11, Meteor-3-6/TOMS, UARS (CLAES, ISAMS, HALOE)
STS-62, Columbia	March 4-18, 1994	NOAA-11, UARS (ISAMS, HALOE)
STS-66, Atlantis (ATLAS-3)	Nov. 3-14, 1994	Meteor-3-6/TOMS, NOAA-11, UARS (ISAMS, HALOE),
STS-72, Endeavour	Jan. 11-20, 1996	NOAA-11, -14, UARS (ISAMS, HALOE), ERS-2 (GOME),

Table 430: Survey of Shuttle flights with the SSBUV payload

1432) "Calibration of Long Term Satellite Ozone Data Sets Using the Space Shuttle," E. Hilsenrath, in Optical Remote Sensing of the Atmosphere, 1990 Technical Digest Series of the Optical Society of America, Vol. 4, pp. 409-412

1433) Information provided by E. Hilsenrath of NASA/GSFC, Greenbelt, MD

Part K Space Science/Solar-Terrestrial Missions

K.1 ACE (Advanced Composition Explorer)

A NASA/GSFC solar-terrestrial mission in the Explorer Program (Explorer-71) with the objectives to determine: the elemental and isotopic composition of matter, the origin of the elements, the formation of the solar corona and acceleration of the solar wind. S/C designer and builder: JHU/APL. The S/C structure has two octagonal decks, 1.6 m across and 1 m high; the S/C is spin stabilized with the spin axis Earth/sun pointing (star and sun sensors). Mission launch occurred on August 25, 1997 with a Delta II launch vehicle from Cape Canaveral, Florida. S/C mass = 785 kg (includes 189 kg of hydrazine fuel for orbit insertion and maintenance), power=443 W (EOL) from four fixed solar arrays, nominal life of the mission is 2 years with a five-year goal.[1434)]

Orbit: Halo orbit about the Lagrangian (or libration) point L_1 (250 Earth radii toward the sun, or about 1.5 million km from Earth).

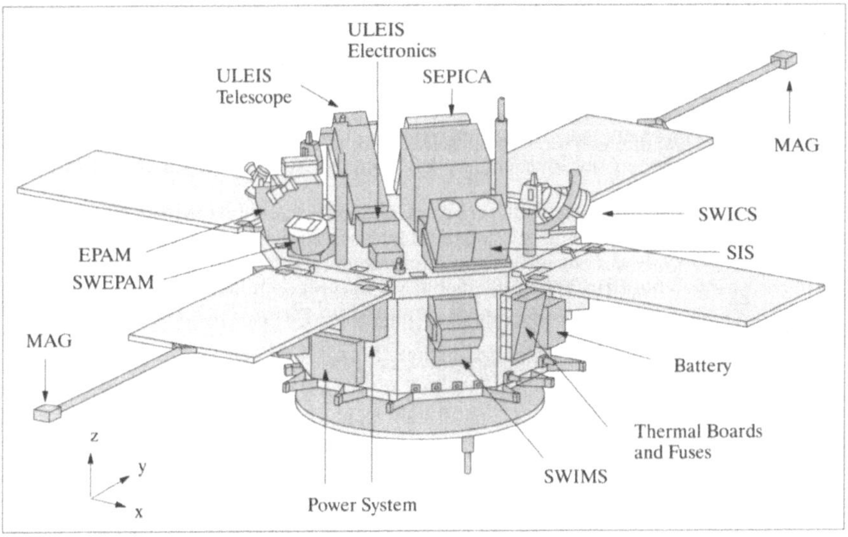

Figure 247: The ACE S/C model

Sensor complement: (PI for mission: E. C. Stone, Caltech/JPL).

- **SWIMS** = Solar Wind Ion Mass Spectrometer (PI: G. Gloeckler, U. of Maryland). Objectives: Measurement of solar wind composition data over a wide range of solar wind bulk speeds and for all solar wind conditions. Abundances of most of the elements and several isotopes in the mass range from 4 - 60 amu (atomic mass unit) every few minutes. SWIMS uses a time-of-flight (TOF) measurement technique to determine the mass of a solar wind ion with high accuracy. SWIMS consists of the Wide-Angle, Variable Energy/charge (WAVE) three chamber parallel-plate electrostatic analyzer, the time-of flight High-Mass Resolution Spectrometer (HMRS), high-voltage supplies, and analog and digital electronics.

1434) Information provided by D. L. Margolies of NASA/GSFC

- **SWICS** = Solar Wind Ion Composition Spectrometer (PI: G. Gloeckler, U. of Maryland). Objective: measurement of the elemental and ionic-charge composition and the temperature and mean speeds of all major solar wind ions from H through Fe at solar wind speeds ranging from 145 km/s (for protons) to 1532 km/s (for Fe^{+8}). The instrument, which covers an energy per charge range from 16 - 60 keV/Q, combines an electrostatic analyzer with post-acceleration, followed by a time-of-flight (TOF) and energy measurements.

- **ULEIS** = Ultra-low Energy Isotope Spectrometer (PI: G. Mason, U. of Maryland, R. Gold, APL). Objective: measurement of ion fluxes over the charge range from He through Ni from about 20 keV/n to 10 MeV/n (superthermal and energetic particle ranges). ULEIS is a time-of-flight (TOF) mass spectrometer which identifies incident ion mass and energy by simultaneously measuring the time-of-flight, τ, and residual kinetic energy, E, of particles which enter the telescope cone and stop in one of the six detectors in the telescope.

- **SEPICA** = Solar Energetic Particle Ionic Charge Analyzer (PI: E. Möbius, U. of New Hampshire and MPE Garching; D. Hovestadt, MPE Garching). Objective: measurement of the ionic charge state, Q, the energy, E, and the nuclear charge, Z, above 0.2 MeV/n. - Energetic particles entering the multi-slit collimator will be electrostatically deflected between the six sets of electrode plates which are supplied with variable high voltages up to 30 kV. The deflection, which is inversely proportional to energy per charge, E/Q, is determined in the back portion of the instrument (dE/dX device and a position-sensitive silicon solid-state detector). The residual energy of the particle, E_{res}, and the amount of electrostatic deflection is directly determined in the detector, thus yielding the energy per ionic charge, E/Q, of the incoming particle, and its energy, E.

- **SIS** = Solar Isotope Spectrometer (PI: A. Cummings, California Institute of Technology). Objective: measurement of elemental and isotopic composition of solar energetic particles, anomalous cosmic rays, and interplanetary particles from He to Zn over the energy range from 10 - 100 MeV/nucleon. Measurements by a technique considering a particle's energy loss ΔE in a detector (multiple ΔE versus residual energy E).

- **CRIS** = Cosmic Ray Isotope Spectrometer (PI: A. Cummings, California Institute of Technology; T. von Rosenvinge, GSFC; R. Binns, Washington U.; M. Wiedenbeck, JPL). Objective: measurements of all stable and long-lived isotopes of galactic cosmic ray nuclei from He to Zn over the energy range from ~ 100 to 600 MeV/nucleon. CRIS also provides limited measurements of low energy H isotopes and data for exploratory studies of the isotopes of "ultra-heavy" (UH) nuclei. Measurements by a technique considering a particle's energy loss ΔE in a detector (multiple ΔE versus residual energy E). CRIS is of CRRES, ISEE-3 and SAMPEX heritage.

- **EPAM** = Electron, Proton, and Alpha-particle Monitor (PI: R. Gold, JHU/APL). Objective: measurement of solar and interplanetary particle fluxes with a wide dynamic range and a directional coverage of nearly a full unit sphere. EPAM consists of five apertures in two telescope assemblies. It measures ions ($E_i \gtrsim 50$ keV) and electrons (E_e >30 keV) with essentially complete pitch angle coverage from the spinning ACE spacecraft. It also has an ion elemental abundance aperture using a ΔE versus E technique in a three-element telescope.

- **SWEPAM** = Solar Wind Electron, Proton, and Alpha Monitor (PI: D. McComas, LANL). Objective: high quality measurements of electron and ion fluxes in the low energy solar wind range (electrons: 1 - 1240 eV; ions: 0.26 - 35 keV). SWEPAM is of Ulysses mission heritage. SWEPAM makes simultaneous and independent electron and ion measurements with two separate sensors. Both sensors make use of curved-plate electrostatic analyzers which are spherical sections cut off in the form of a sector.

- **MAG** = Magnetic Field Monitor (triaxial fluxgate; PI: N. Ness, U. of Delaware). Objective: measurement of the three components of the magnetic field. MAG is boom-mounted.

Instrument	Mass (kg)	Power (W)	Data rate (bit/s)	Measurement Technique	Type. Energy (MeV/nucleon)
CRIS	29.3	22	462	dE/dX x E	~300
SIS	20.6	21.5	2000	dE/dX x E	~50
ULEIS	18	30.5	1000	TOF x E	~5
SEPICA	28.6	16	600	ΔE x E x E/Q	~1
SWICS	6.3	9	500	TOF x E x E/Q	~0.001
SWIMS	8.3	19.4	505	TOF thru special E-field	~0.001
SWEPAM	6.7	6.4	1000	Electrostatic Analyzer	~0.001
EPAM	6.4	6.5	160	dE/dX x E	~0.3
MAG	4.3	4.4	300	Triaxial Fluxgate	

Table 431: ACE instrument summary

Communications with ACE are at S-band. Science and engineering data are collected during one 3-4 hour pass per day. Real-time data is transmitted at a data rate of 6.9 kbit/s or 434 bit/s. Data may be recorded onto a 2 Gbit solid-state recorder (two recorders available) and played back at a rate of 78 kbit/s.

K.2 ACTIVE (AKTIVNY-IK)

Russian (IKI) solar-terrestrial mission within the Intercosmos program (International Space Plasma-Waves Laboratory, Intercosmos 24). Objectives: comprehensive study of VLF-wave (Very Low Frequency) propagation phenomena in the Earth's magnetosphere and wave interaction with the energetic particles of the radiation belts.[1435),1436)]

Mother-daughter pair of spacecraft. The mother craft is also designated as Intercosmos 24, the daughter spacecraft is provided by the Czech Republic and is designated 'C2-AK' or 'Magion-2'. Mother craft: three-axis spin-stabilized; S/C diameter = 2 m; S/C height = 3 m; mass = 270 kg; power = 270 W;

Launch on Sept. 28, 1989 with a Cyclone launch vehicle from Plesetsk, releasing the Magion-2 subsatellite. Magion-2 orbits around the mother in a controlled mode. Magion-2 separation from mother satellite was on October 3, 1989. Magion-2 began its science program in January 1991. ACTIVE is an operational mission as of 1993. Scientific program leader (since 1993): G. L. Gdalevich, (IKI, Moscow).

Orbit: apogee = 2500 km, perigee = 500 km, inclination = 82.5°, period = 116 minutes.

Sensors of mother S/C:
The mother craft carried a plasma generator for a modulated plasma environment around the antenna. A neutral xenon injector injected xenon around the S/C for monitoring the ionization of neutral gas flow and its propagation effects.

VLF = Very-Low Frequency Generator [PI: O. A. Molchanov, IFZ RAN (Institute of Earth Physics) and Yu. N. Agafonov, IFZ RAN]. Objectives: generation of VLF waves in the range 9-11 kHz via a 20 m diameter ORA-20 loop antenna (current about 10A). Note: ORA = O-formed loop antenna.

1435) "The ACTIVE International Space Plasma-Wave Laboratory," The Solar-Terrestrial Science Project of the Inter-Agency Consultative Group for Space Science, ESA SP-1107, November 1990, pp. 45-49
1436) Aktivny-IK, Interavia Space Directory, 1992-93, pp. 149

ORA-20 (transforms oscillations from the generator to electromagnetic emission). The tube from soft ductile aluminum alloy, with a wall thickness of about 1 mm, is used for the build-up of the antenna loop. The tube is tightly rolled up at launch and slowly unfolds in orbit (by its spring-load) on command into a loop antenna of 20 m diameter.

PVP = Generator of Electric Oscillations (PI: Yu. M. Mikhailov, IZMIRAN). Objectives: generation of VLF waves in the range 1.5 - 20.5 kHz. The PVP sensor consists of a self-opening band-type cylindrical dipole antenna of 15 m length and 25 mm diameter.

- Output voltage: 50 - 300 V
- Time of impulse: 0.25 s
- Time of pause: 0.75 s

NVK-ONCH = VLF Analyzer (PI: Yu. M. Mikhailov, IZMIRAN). Objective: measurement of magnetic and electric field components in the range from 8 Hz - 20 kHz, three magnetic and two electric components.

Sensitivity of electric components	10^{-7} V m^{-1} $Hz^{-1/2}$
Sensitivity of magnetic components	10^{-5} nT $Hz^{-1/2}$
Dynamic range	80 dB
Sensors for magnetic components	coil-type magnetic antenna with ferrite
Sensors for electric components	double spherical sondes with a diameter of 100 mm and with distances of 2.5 m between the sondes

Table 432: NVK-ONCH parameters

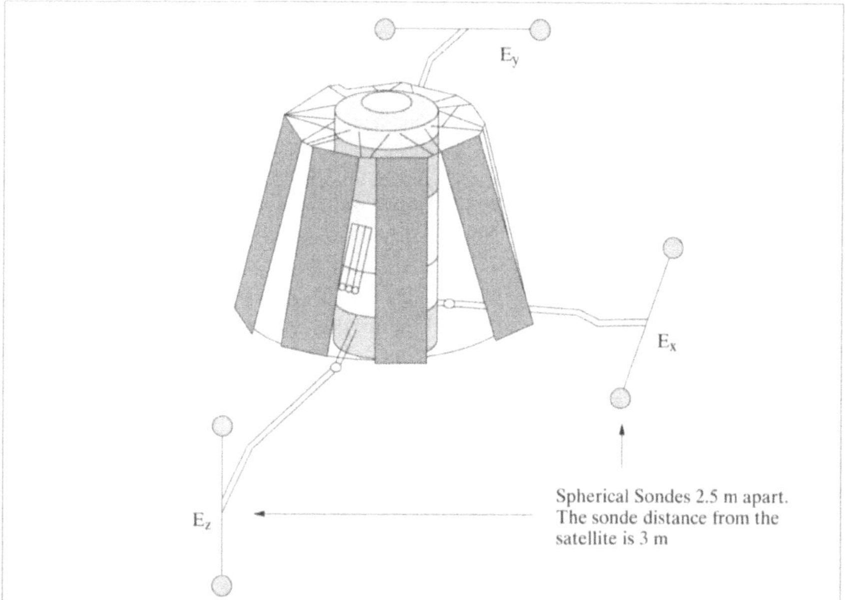

Figure 248: The ACTIVE S/C model

SHASH = VLF Spectroanalyzer (PI: J. Tarchai, Eötvös U. of Budapest). Objective: digital analysis of VLF signals and noise levels from the NVK-ONCH instruments in the range of 20 Hz - 20 kHz. Independent transmission of the analyzed information is provided by a transmitter (frequency = 460 MHz, transmission rate = 100 kbit/s).

VLF-2 = Very-Low Frequency Generator 2. (PI: Z. Klos, Center of Cosmic Research, Polish Academy of Sciences, Warsaw). Objective: provision of a parallel analysis of VLF spectra (one electric component with the use of 12 filters); measurement range = 20 Hz - 20 kHz. The VLF-2 sensor uses the same equipment as NVK-ONCH.

PRS = Plasma Wave Spectrometer (PI: Z. Klos, Center of Cosmic Research, Polish Academy of Sciences, Warsaw); Objective: measurements of HF-noise spectra in the range from 0.1 - 10 Hz; data rate = 10 kbit/s; mass = 5.1 kg. The sensor consists of a self-opening band-type antenna of 15 m length.

ZL-A = Langmuir Probe Experiment (PI: J. Rustenbach, MPE, Berlin). Objective: measurement of plasma electron/ion temperature and densities. Electron temperature range = 10^3 - 10^4 K, electron density range = 10^2 - 10^8 cm^{-3}. The sensor is a cylindrical sonde.

KM-6 = Cold Plasma Analyzer (PI: J. Shmilauer, Geophysical Institute, Prague). Objective: measurement of plasma density and temperature, electron distribution and drift velocity; data rate = 23 kbit/s; mass = 5.7 kg; electron temperature range = 10^3 - 10^4 K; ion density 10^2 - 5 x 10^6 cm^{-3}. The sensor is a plane sonde that is made up of 4 electrodes.

NAM-5 = Radiofrequency Mass-Spectrometer (PI: J. Shmilauer, Geophysical Institute, Prague). Objective: measurement of ions and neutral plasma composition. Range of mass = 1-60 amu; data rate = 29 kbit/s; mass = 6.1 kg

DME = Soft Electron Detector (PI: A. Melentiev, IKI). Objective: flux measurement of soft electrons in the energy spectrum of 0.01 - 10 keV; data rate = 6 kbit/s; mass = 6.1 kg.

ANAPURNA = Energy and Pitch Distribution Experiment (PI: A. Melentiev, IKI). Objective: measurement of energy and angular distributions of electrons in the range: 0.2 - 40 keV; data rate = 4 kbit/s; mass = 11 kg

SPE-1 = High Energy Particles Experiment (PI: K. Kudela, Institute of Exp. Physics, Kosice, CSFR. Objective: measurement of spectra and anisotropy of electrons and protons; E_e = 30-600 keV; E_p = 15-600 keV; data rate = 14 kbit/s; mass = 4.5 kg.

K.2.1 Subsatellite Magion-2 (C2-AK)

Magion = **Mag**netospheric and **Ion**ospheric research satellite (built by the Academy of Sciences of the Czech Republic, Geophysical Institute, Prague) with spring separation on command from mother craft. Objectives: monitoring of propagation VLF waves from the mother spacecraft. Magion 2 introduced the current Magion base configuration with a mass of about 52 kg and a diameter of 0.6 m. The octagonal bus is equipped with four small solar arrays as well as body-mounted solar panels.

Full deployment of booms and four solar panels was confirmed on Dec. 24. 1989. The science program began on January 21, 1990. Magion-2 was operational until November 20, 1990. Magion-2 mass = 51.7 kg (11.4 kg of scientific instruments); 26 face polyhydron; 560 mm diagonal (see Figure 249 for Magion S/C). Data: Downlink transmission in VHF-band (frequencies of 137 and 400 MHz). Selectable digital data rates of 5, 10, 20, and 41 kbit/s.

Orbit: perigee = 507 km, apogee = 2491 km, inclination = 82.59°

Sensors for the measurement of the electric and magnetic fields; VLF waves, plasma and energetic particles.

SGR-7 = 3-axis Fluxgate Magnetometer (PI: M. Ciobanu, IKI, Romania). Measurement range of ±50048 nT or 6256 nT, resolution = 16 nT or 2 nT, frequency range = 0 - 20 Hz. Measures magnetic field vector and serves as an attitude sensor.

- SGR6 = 1-component Fluxgate Variometer; dynamic range = ±156 nT; resolution = 50 nT; frequency range = 0.1 - 20 Hz.

KEM-1 = Magnetic and Electric Field Analyzer (PI: P. Triska, GFU, Czech Republic);

- ULF = Electric Field Experiment (three components of the quasi-static electric field); dynamic range = 0.005 - 8000 mV, frequency range = 0.1 - 20 Hz.
- VLF Wave Experiment (2-axis electric and 1-component magnet ELF-VLF field measurements, broadband waveforms, spectrum analyzer, filter bank;
  $E_{x,y,z}$: 0.1 Hz - 120 kHz, sensitivity of 10^{-7}V/mHz$^{1/2}$, 120 dB dynamic range.
  B_x : 10Hz - 40 kHz, sensitivity of 5 x 10^{-6} nT/Hz$^{1/2}$ at 2 kHz and 10^{-4} nT/Hz$^{1/2}$ at 100 Hz, dynamic range 120 dB.
- Data transmission modes selectable: broadband analog data 10 Hz - 60 kHz one channel, 10 Hz - 20 kHz broadband three channels, subcarriers of 1 kHz bandwidth four channels.
- Filter bank; 17 Hz - 15 kHz, eight filters, four independent sets
- Frequency analyzer: range of 1 - 220 kHz, 32 frequency steps, full spectrum/2s, selection of any frequency.

KM-12 = Cold Plasma Analyzer (PI: J. Shmilauer, GFU, Czech Republic); HF probe measures two components of the electron temperature in the range: Te = 10^3 - 10^5 K, spacecraft potential from -2 to +2V; Spherical ion trap measures ion density (N_i) in the range: 10^8 - 10^{13} /m^3, $\Delta N_i/N_i$ fluctuations f_{max} = 50 Hz.

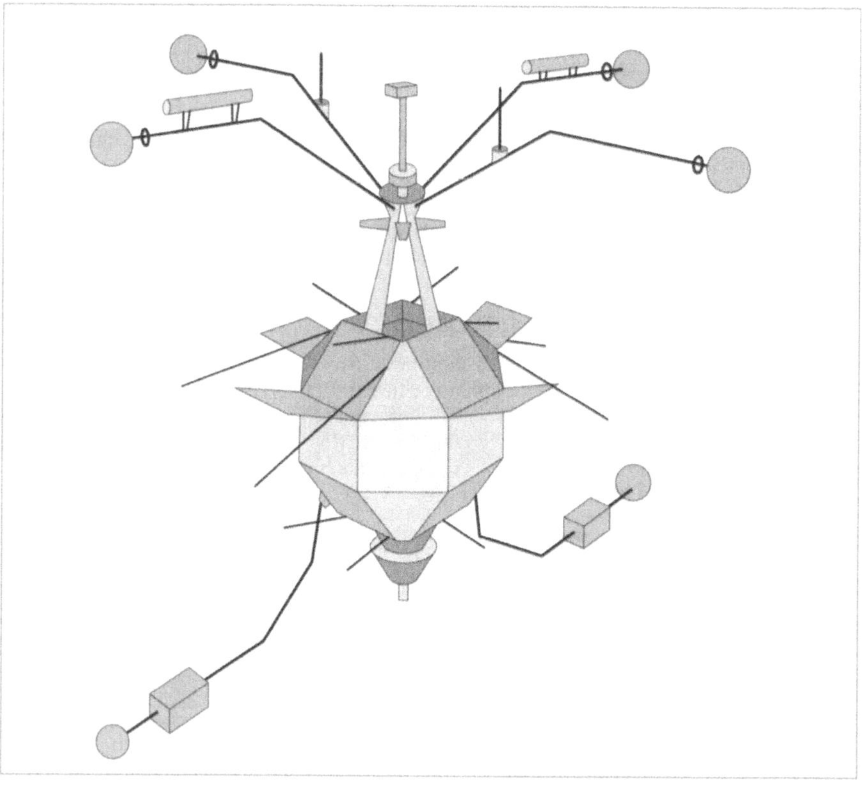

Figure 249: The Magion-2 S/C model

ZL-A-S = Langmuir Probe (PI: K. Sauer, MPE, Germany); measurement of electron and ion density in the range: 500 - 10^8 cm^{-3}; electron temperature: 0.05 - 3 eV; current range:

10^{-10} - 10^{-3} A; in the current-mode density fluctuations f= <200 Hz, $\Delta N_e/N_e$ resolution 10^{-3}.

PRS-2-C = Radiowave Spectrometer (PI: Z. Kloss, CBK PAN, Poland). Measurement of HF wave spectra. Frequency range = 0.1 - 10 MHz, dynamic range = 1 μV - 10 mV, 0.2 s/ spectrum using Δf=50 kHz or 2s/spectrum using Δf= 15 kHz; field fluctuations at fixed frequency (selectable), Δt= 1ms.

DOK-A-S =Silicon Detector Spectrometer (PI: K. Kudela, UEF SAV, Slovakia); two sensors (parallel and perpendicular to the magnetic field vector), measurement of electrons and ions (20 keV - 1 MeV, eight energy levels, geometric factor 10^{-2} cm^2 sr.

MPS SEA= Energetic Particle Spectrometer (PI: Z. Nemecek, Prague University, Czech Republic); electrostatic analyzers, measurement of electrons and positive ions: 0.2 - 20 keV in 16 energy levels, pitch angle resolution 30°, geometric factor is ~10^{-3} cm^2 sr.

FDS = Photometer (PI: N. Petkov, IKI, Bulgaria); measurement of optical ionospheric plasma emissions (630 and 577.7 nm).

K.3 ALEXIS (Array of Low-Energy X-Ray Imaging Sensors)

ALEXIS [1437)] is a sophisticated 'minisatellite technology demonstration mission' of Los Alamos National Laboratory, (LANL) Los Alamos, New Mexico, funded by DOE and built by AeroAstro Inc., Herndon, VA. Sandia National Laboratory, UCB/SSL (UC Berkeley/ Space Sciences Laboratory) are partners of LANL in the mission. The spacecraft was launched by a Pegasus air-launched booster (of OSC) on April 25, 1993. During powered flight a solar paddle was damaged; initial attempts to contact ALEXIS were unsuccessful. The satellite responded in June 1993 and soon was brought under control. By late July 1993, full satellite operations had been restored through implementation of new procedures for attitude control. S/C operations are performed from LANL. [1438)] [1439)] [1440)]

The S/C bus is a complex, robust, and flexible structure comprised of communications, power, and control subsystems. The bus processor comprises redundant 80C86-based CPUs and six on-board mass memory boards with 96 MByte (SRAM) total storage capacity. This bus assembly occupies less than 25% of the S/C volume with a mass of 45 kg. The S/C is spin-stabilized and uses solar pointing for orientation (fine and coarse sun sensors) and magnetic torque coils for attitude reconstitution. The bus points the payload in the anti-sun direction and rotates once per about 50 seconds about that sun-line. Sun sensors and a limb sensor provide attitude knowledge of 0.1°. Attitude is controlled using magnetic torque coils. Total S/C mass = 113 kg, power = 60 W. Power is produced by four deployable solar panels and a fifth fixed disk which also serves as a sun shield. Power is distributed to four 1.2 Ah NiCd 28V batteries.

The ALEXIS payload consists of an ultrasoft X-ray telescope array and a high-speed VHF receiver/digitizer with the name of Blackbeard.

S-band data transmission (uplink/downlink). Downlink data rate = 750 kbit/s, uplink data rate = 9.6 kbit/s.

Orbit: Apogee = 844 km, perigee = 749 km, inclination = 70°

1437) W. Priedhorsky, B. W. Smith, J. J. Bloch, D. H. Holden, D. C. A. Roussel-Dupré, R. Dingler, R. Warner, G. Huffman, R. Miller, B. Dill, R. Fleeter, "The ALEXIS Small Satellite Project: Initial Flight Results," AIAA Space Programs and Technologies Conference, Sept. 21-23, 1993/ Huntsville, Al.; and Proc. SPIE Vol. 2006, 1993, pp. 114-126

1438) D. Roussel-Dupre, J. J. Bloch, et al., "ALEXIS - the six Year Telescope Flight Experience," SPIE, Vol. 3764-34, 1999

1439) D. Roussel-Dupre, et al., "ALEXIS, the Little Satellite That Could - 4 Years Later," Proceedings of the 11th AIAA/USU Conference on Small Satellites, Sept. 15-18, 1997, Logan, UT, SSC97-IV-3

1440) http://nis–www.lanl.gov/nis–projects/alexis/

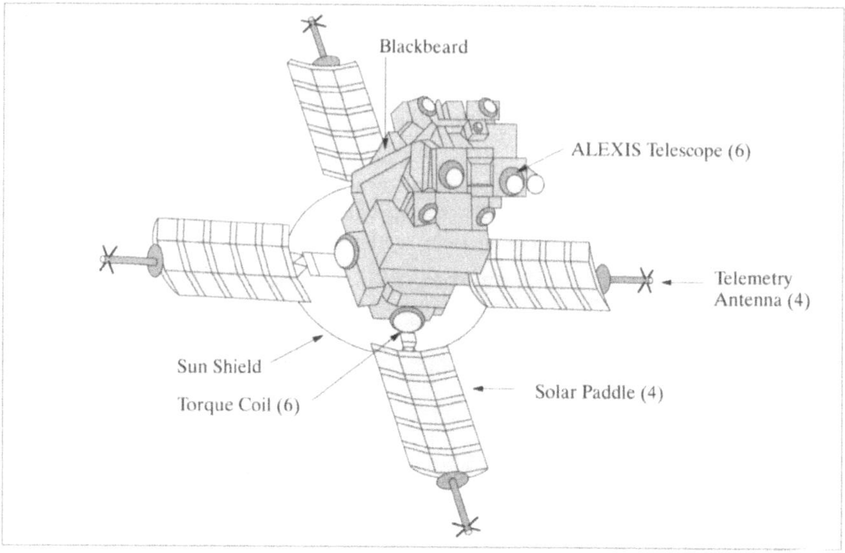

Figure 250: The ALEXIS S/C model

Status of S/C mission: Up to 1999 the ALEXIS mission was operating nominally, it has exceeded all expectations in spite of the serious launch incident with the solar panel array. A completely new software system (on the ground) is in place to calculate S/C attitude, critical to the on-board operation of the X-ray telescopes. Over 80 GByte of of mission data have been received thus far. The design life of one year has been exceeded by far.[1441)]

Dual-ported RAM failures in the fall of 1999 have resulted in data only available from telescope pairs #1 & #2 and from the Blackbeard experiment. Data quality in the telescopes has been degrading as solar maximum approached due to increased particle backgrounds.

Sensor complement:

ALEXIS = Array of Low Energy X-Ray Imaging Sensors (instrument has the same name as the S/C). Alexis is an outward-looking instrument for the detection of astrophysical signals. The instrument is an ultrasoft X-ray monitor, consisting of six compact normal-incidence telescopes tuned to narrow bands centered on 66, 71 and 93 eV. The 66 and 71 eV bandpasses are centered on a cluster of emission lines from Fe IX-XII. The 93 eV band, although designed as a continuum channel, includes Fe XXIII line characteristics of 10^7 K plasma.

The six ALEXIS EUV telescopes are arranged in pairs covering three overlapping 33° FOVs. During each 45-second spin of the S/C, ALEXIS monitors the entire antisolar hemisphere. Each telescope consists of a spherical mirror with a Mo-Si layered synthetic microstructure (LSM) or multilayer coating, a curved profile microchannel plate detector located at the telescope's prime focus, a UV background-rejecting filter, electron rejecting magnets at the telescope aperture, and image processing readout electronics. The geometric collecting area of each telescope is about 25 cm2, with spherical aberration-limiting resolution to about 0.25°s. The resolution of each telescope is limited to about 0.5° diameter. The ALEXIS detectors are double plate, curved front faced, microchannel plates (MCP) paired with wedge and strip-resistive anodes. The two MCPs, of 46 mm in diameter, are single photon imaging detectors, each with 12.5 µm diameter channels.

1441) Information provided by J. Bloch of LANL, Los Alamos, NM

The mass of the ALEXIS instrument is 45 kg, power = 45 W; the average data rate is 10 kbit/s of event data. Position and time of arrival are recorded for each event (detected photon, etc.). Measurement objectives: mapping the diffuse background in three bands, performing a narrow-band survey of point sources, searching for transient phenomena, and monitoring variable ultrasoft X-ray sources.[1442), 1443)]

Blackbeard. The instrument looks at signals emitted near the Earth. Blackbeard is a radio frequency (RF) experiment with the objective to study distortion and interference effects on transient transionospheric VHF signals, such as lighting and artificial pulses. The instrument senses perturbations to the ionosphere; it can make a distinction between multipath distortions resulting from large-scale coherent perturbations and from small-scale random perturbations to the ionosphere. The specific experiments of Blackbeard include:

- Broadband VHF measurements of transient signals originating from a controlled pulsed ground beacon, to characterize broadband ionospheric distortion.
- Narrowband VHF measurements of CW-signals from a multichord interferometry ground beacon array, to characterize the ionospheric structure contributing to transmission distortion.
- Surveying power envelopes of lightning and man-made interference in selectable VHF bands, for background rejection purposes.

Blackbeard operation consists of on-board 150 MHz digitization for 0.1 s in a broadband reception mode; or 50 kHz digitization for 320 s in a narrow-band reception mode; or 120 kHz effective digitization for 130 s in a power-envelope survey mode. - The broadband mode has selectable bandwidths up to 65 MHz within the ranges 25-100 MHz and 100-175 MHz, with a maximum of 30 dB SNR. The narrow-band mode has eight selectable 4 kHz bands between 32 and 36 MHz, with a maximum 40 dB SNR and 0.1 Hz Doppler resolution. Mixed-mode operation is available in which broadband and narrowband data are collected to allow correlated RF distortion and ionospheric structure analysis.

The Blackbeard instrument observed strange radio bursts called TIPPs (Trans-Ionospheric Pulse Pairs). These strange signals are the most intense radio sources from Earth which can be much stronger than typical lightning. Then in 1996, LANL researches reported the first simultaneous Blackbeard observations and multiple ground station measurements of TIPPs. The new evidence suggests that TIPPs come from thunderstorms and probably comprise an atmospheric event and its reflections off Earth. [1444) 1445)]

K.4 AMPTE (Active Magnetosphere Tracer Explorers)

AMPTE[1446),1447),1448)] is a three satellite cooperative mission of the United States, Germany and the United Kingdom. Objectives: studies of solar-terrestrial interactions, in particular the interaction processes between two cosmic plasmas. Further aims of this mission are a systematic exploration of the highly variable mass and charge composition of the natural plasma population and a detailed investigation of the magnetospheric boundaries.

1442) W. C. Priedhorsky, J. J. Bloch, S. P. Wallin, W. T. Armstrong, O. H. W. Siegmund, J. Griffee, R. Fleeter, "The ALEXIS Small Satellite Project: Better, Faster, Cheaper Faces Reality," IEEE Transactions on Nuclear Science, Vol. 40, No. 4, August 4, 1993, pp. 863-873

1443) J. J. Bloch, et al., "Design, Performance and Calibration of the ALEXIS Ultrasoft X-Ray Telescopes," SPIE, Vol. 1344, 1990 , pp. 154-165

1444) Dateline: Los Alamos, Current Missions, "ALEXIS: The Little Satellite That Could," Jan./Feb./Mar. issue 1998, pp. 12-14

1445) D. Roussel-Dupre, J. J. Bloch, E. M. Johnson, J. Theiler, "ALEXIS-the Six Year Telescope Flight Experience," Proceedings of SPIE, Vol. 3765, 1999 pp. 329-340, EUV, X-Ray, and Gamma-Ray Instrumentation for Astronomy X, Oswald H. Siegmund; Kathryn A. Flanagan; Eds.

1446) Ampte brochure of MPE Garching

1447) Special Issue on the Active Magnetosphere Particle Tracer Explorer (AMPTE), in IEEE Trans. on Geoscience and Remote Sensing, May 1985, Volume GE-23, No. 3, pp. 175-314

1448) A. Valenzuela, G. Haerendel, H. Föppl, F. Melzner, H. Neuss, E. Riegler, J. Stöcker, O. Bauer, H. Höfner, J. Loidl, "The AMPTE artificial comet experiments," reprinted from Nature Vol. 320, No. 6064, pp. 700-723, April 24, 1986

AMPTE uses artificial injection of rare ionic species for long-range tracing of mass transport into and through the magnetosphere system. Injection of lithium, barium and europium ions into the solar wind and magnetotail.

The AMPTE program consists of three spacecraft - the Ion Release Module (IRM) provided by Germany, the United Kingdom Subsatellite (UKS), and the US/NASA Charge Composition Explorer (CCE). All S/C were launched in a stack on a single Delta vehicle on Aug. 16, 1984 from Cape Canaveral.

The IRM S/C has a mass of 705 kg. It carries scientific instruments and 16 canisters, 8 of which are filled with 5.8 kg of a copper-oxide-lithium mixture, the remaining eight are filled with a copper-oxide-barium mixture. The canisters are released in pairs by ground command, and the copper-oxide thermite reactions, which vaporize the tracer elements, are initiated by internal timers triggered during the release. A total of seven releases were planned over a period of 8 months which could be monitored by the maneuverable UKS following 3 minutes behind IRM and by CCE in a much lower orbit.
The IRM S/C was developed and built by MPE Garching and operated by GSOC (of DFVLR, now DLR) at Oberpfaffenhofen. IRM failed in eclipse on Aug. 12, 1986.

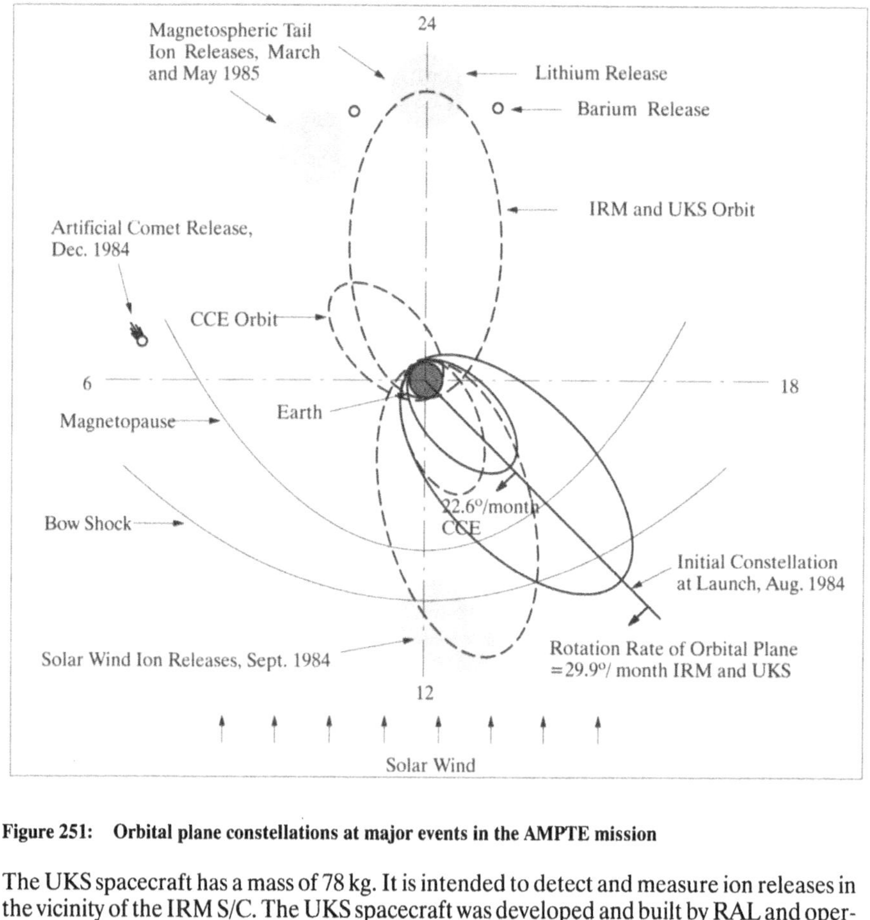

Figure 251: Orbital plane constellations at major events in the AMPTE mission

The UKS spacecraft has a mass of 78 kg. It is intended to detect and measure ion releases in the vicinity of the IRM S/C. The UKS spacecraft was developed and built by RAL and operated by SERC. UKS failed unexpectedly on Jan. 16, 1985.

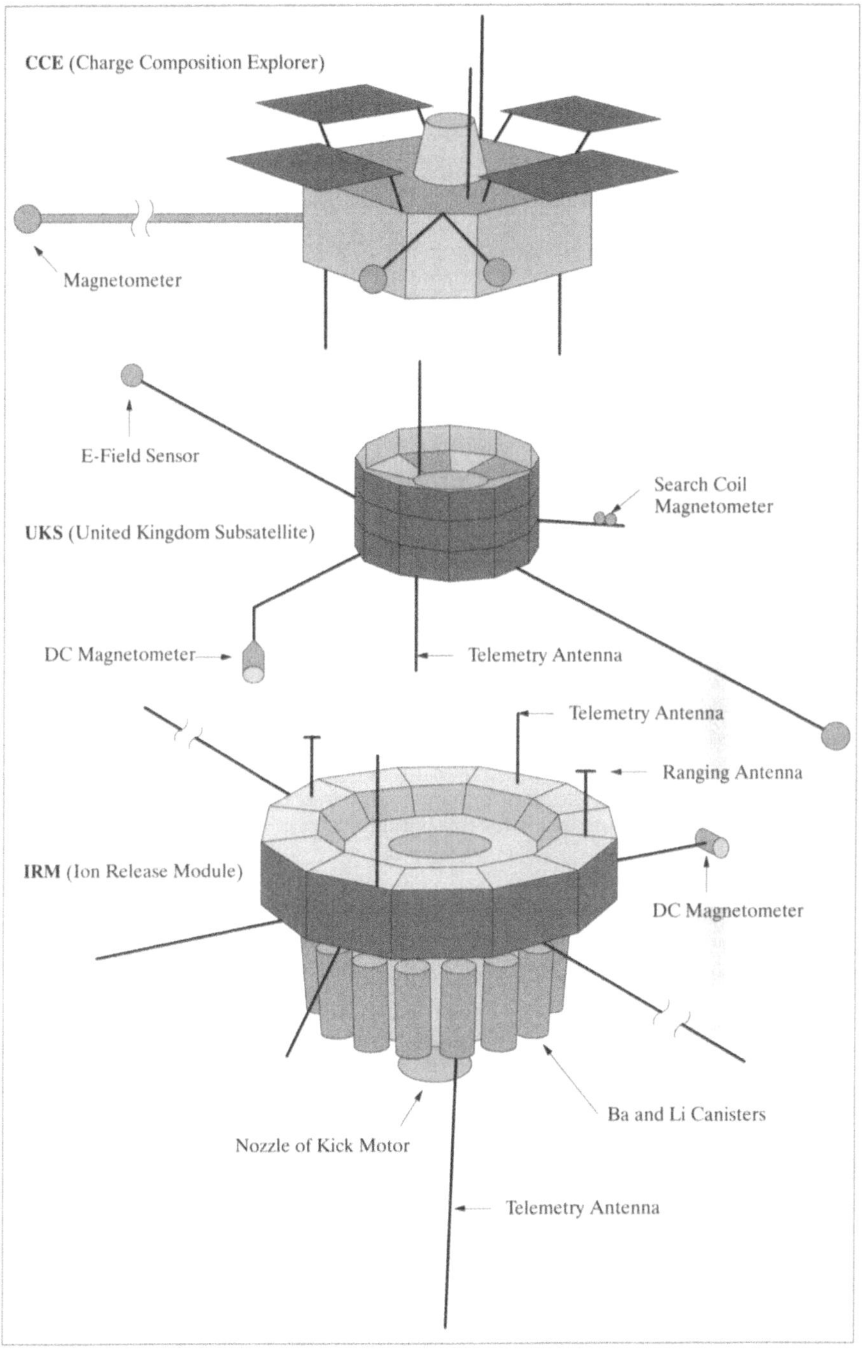

Figure 252: The AMPTE mission S/C models (launch stack configuration)

Final Orbits	CCE	IRM	UKS
Apogee (geocentric)	8.8 R_E	18.7 R_E	18.7 R_E
Perigee (height)	1113 km	552 km	552 km
Period	15.6 hrs	44.3 hrs	44.3 hrs
Inclination	4.82°	28.68°	28.68°
Releases (active injections)		7	
1st releases (Sept. 1984) of lithium tracer ions into the solar wind near the 'nose' of the magnetosphere and close to the Earth/sun line.			
2nd release (Dec. 27, 1984) called 'artificial comet' with IRM located in the dawn magnetosheath			
3rd and 4th releases (March and May 1985) with IRM located in the Earth's magnetotail region			

Table 433: Orbit parameters and major release periods of the AMPTE spacecraft

The CCE spacecraft weighs 240 kg. Its function is to detect and measure trace quantities of injected ion releases at far distances. APL of Johns Hopkins developed and built the CCE spacecraft under NASA/GSFC contract, JPL provided tracking and S/C operations support. CCE operated fully until Jan. 10, 1989 and then intermittently until July 17, 1989 when the mission was terminated.

In September 1984, lithium was released into the solar wind by IRM; preliminary results indicated that less than 1% of the solar wind gained access to the Earth's magnetosphere. On December 27, 1984, IRM released two barium canisters; they were observed to explode 10 minutes later by UKS, CCE, two high-flying aircraft and ground stations in the Pacific. The expanding cloud of the barium mixture (in a position about 103,000 km above the Pacific ocean) ionized within seconds by solar UV radiation, fluorescing in the dawn sky as seen from the western US. After about 12 minutes an 'artificial comet' of about 225 km diameter had formed, developing a tail longer than 9000 km.[1449)]

K.4.1 IRM Instrumentation (Sensors)

IPIP = IRM Plasma Instrument Package (3-D Plasma Analyzer). The instrument package consists of three sensors. Two of them measure the 3-D velocity distribution of ions and electrons. The third sensor is a retarding potential analyzer (RPA) for low-energy electron measurements; symmetrical quadrispheres. Energy ranges: ~0 eV - 25 eV, and 15 eV - 30 keV.

MSIS = Mass-Separating Ion Spectrometer. Objective: good mass-imaging characteristics and a wide fan-shaped angle of acceptance such that all ion species of interest can be measured simultaneously. Measurement technique: quadrispherical E/q analysis, magnetic analysis. Coverage: 0.01 - 12 keV/q.

SULEICA = Suprathermal Energy Ionic Charge Analyzer. Measurement of the ionic charge state and mass composition of all major ions from H through Fe for energies of the suprathermal plasma (~5 - 270 keV/q) by the use of electrostatic deflection, time-of-flight (TOF) measurement, and energy analysis in solid-state detectors. Measurement technique: electrostatic analyzer, time-of-flight and total E. Coverage: 10 - 300 keV/q. The data contain counting rate information for evaluating absolute particle fluxes, live pulse-height events, and matrix rates for selected ions.

Magnetometer. Measurement technique: vector fluxgate (triaxial). Objective: measurement of the magnetic variations during artificial plasma-cloud injections and study of magnetospheric boundary layers and the ring-current region. Coverage: DC - 50 Hz. Dynamic range = 0.1 - 60000 nT; resolution = 16 bit analog to digital conversion.

PWI = Plasma Wave Instrumentation (spectrometer). Measurement of plasma wave activity in the magnetosphere/solar wind environment and inside artificial plasma clouds. Mea-

1449) "AMPTE," Interavia Space Directory 1992-93, p. 149

surement technique: 42 m tip-to-tip dipole antenna, boom-mounted search coils. Coverage: E-field: DC - 5 Hz; B-field: 30 Hz - 1 MHz.

Lithium/Barium Experiments. Measurement technique: Copper-oxide thermite reaction. The injection of lithium and barium plasmas into the solar wind and geomagnetic tail constitute the active element of the AMPTE mission. The objective of the releases is ion tracing and the study of the interaction of two vastly different plasmas in space.

- 8 lithium release canisters (52 kg)
- 8 barium release canisters (108 kg)

K.4.2 UKS Instrumentation (Sensors)

3D-Ion Analyzer. Objective: measurement of the 3-D distribution function of positive ions. Measurement technique: electrostatic analyzer. Coverage: 10 eV/q - 20 keV/q.

3D-Electron Analyzer. Objective: measurement of electrons (count rates). Measurement technique: electrostatic analyzer. Coverage: 6 eV - 25 keV.

SPACE = S/C Particle Correlator Experiment. Objective: Measurement of particle modulations (ions and electrons) resulting from local wave-particle interactions. Measurement technique: electron and ion signal processing. Coverage: ~1 Hz - ~1 MHz.

Magnetometer (of ISEE1/2 heritage). Measurement technique: vector fluxgate (triaxial), boom-mounted about 1 m from the S/C body. Coverage: DC - 10 Hz.

Plasma Wave Spectrometer. Objective: Measurement of the electric and magnetic field components. Measurement technique: 7 m tip-to-tip probe. Coverage: E-field: 30 Hz - 132 kHz, 4 spot frequencies to 2 MHz; B-Field: 30 Hz - 50 kHz (boom-mounted search coil).

K.4.3 CCE Instrumentation (Sensors)

HPCE = Hot Plasma Composition Spectrometer (energetic ion-mass spectrometer and an electron background-environment monitor). Heritage of GEOS-1/2, ISEE-1 and DE-1 missions. Measurement technique: retarding potential electrostatic analyzer, ExB analyzer. Coverage of ion composition: 0 eV/q - 17 keV/q. Measurement technique of electrons: magnetic analyzers. Coverage: 50 eV - 25 keV.

CHEM = Charge-Energy-Mass Spectrometer. Measurement of energy spectra , pitch angle distributions, and ionization states. Measurement technique: electrostatic analyzer, time-of-flight and total E. Coverage: ion composition in the range from ~1 keV/q - 300 keV/q.

MEPA = Medium Energy Particle Analyzer. Measurement technique: time-of-flight and total E. Coverage: ion composition in the range of 10 keV/nucleon - > 1.0 MeV/nucleon.

Magnetometer. Measurement technique: vector fluxgate (triaxial). Coverage: DC - 50 Hz.

Plasma Wave Spectrometer. Objective: provision of first-order correlative information for studies of strong wave-particle interactions that develop close to the magnetic equator or have maximum effectiveness there. Measurement technique: electric dipole. Coverage: AC E-fields. 5 Hz - 178 kHz.

K.5 APEX (Active Plasma Experiment)

APEX [1450] [1451] is a cooperative solar-terrestrial mission within the Intercosmos program (Russia/CSFR/Hungary/Germany/Poland/Bulgaria/Romania). Objectives: study of terrestrial magnetospheric and auroral ionospheric relationships by the injection and monitoring of electron and ion beams by a parent/subsatellite pair. The injection of electron and ion beams into the magnetosphere is accomplished by the parent craft.

- Simulation and initiation of aurora and radio frequency radiation in an auroral region.
- Study of the dynamics of modulated beams and plasmoids in the near-Earth plasma.
- Study of the nature of electrodynamic relationships of electromagnetic waves in the magnetosphere and ionosphere.
- Determination of radio emission characteristics of modulated beams of charged particles and plasmoids.
- Search for nonlinear wave structures of the electromagnetic soliton type in a disturbed environment.

Launch: December 18, 1991 by SL-14 from Plesetsk. APEX mass = 1350 kg, payload mass = 340 kg (parent satellite), Magion-3 (subsatellite)= 52 kg. Science program leader: V. N. Oraevski (IZMIRAN). APEX is an operational mission as of 1993.

Orbit: apogee = 3071 km, perigee = 438 km, inclination = 82.56°.

Sensor complement:

UEM-2 = Electron Accelerator (PI: V. Dokukin, IZMIRAN). Objective: Injection of a modulated electron beam with an energy of 10 keV. The beam current is in the range from 10^{-2} A to 15 A. The frequency modulation is in the range from 10 Hz - 2.5×10^5 Hz.

UPM = Neutral Plasma Accelerator (PI: V. Dokukin, IZMIRAN). Objective: injection of xenon ions. The beam has an energy of 300 eV, the beam current is 2A; the frequency modulation range is from 60 - 1000 Hz.

PEAS = Electron and Ion Analyzer (PI: N. M. Shutte, IKI). Objective: measurement of the energetic and pitch-angle distributions of electrons and ions with energies in the range from 30 eV to 30 keV. The sensor is a toroidal electrostatic analysator.

DEP-2E = Electric Field Instrument (V. Chmirev, IZMIRAN, G. Stanev, IKI, Bulgaria). Objective: measurement of quasi-static electric fields. Sensor: three-component double-spheric sonde.

DEP-2R = AC Field Analyzer (V. Chmirev, IZMIRAN, G. Stanev, IKI, Bulgaria). Objective: measurement of the quasi-static electric field. The signals of DEP-2R are analyzed by DEP-2E.

DANI = Potential and Soft Particle Analyzer (PI: Tc. Dachev, IKI, Bulgaria; V. Temnyi, IZMIRAN). Objective: measurements of energetic and pitch-angle distributions of electron and ions in the range from 0.1 - 30 keV as well as the potential of the satellite and the electron density. Sensors: toroidal analysator, parabolic analysator (Rogoski belt).

KM-10 = Cold Plasma Measurements (PI: V. Afonin, IKI; J. Shmilauer, GFI CSAN, former CSFR now CR). Objective: measurement of the temperature distribution of electrons, the anisotropy of the temperature, the energetic distribution in a range from 0.1 - 10 eV, the density of ions, and the potential of the satellite. The sensor is a plane sonde with four electrodes.

NVK-ONCH = VLF Analyzer (PI: Ya. Sobolev, IZMIRAN). Same instrument as in ACTIVE mission.

UF-3K = Photometer (PI: Yu. Ruzhin, IZMIRAN). Objective: observation of optical effects at the injection time of the electron and ion beams. Working wavelengths: 3914 Å, 5577 Å, and 6300 Å. The dynamic range is 80 dB and the look angle is 8° (from nadir).

1450) "The Active Plasma Experiments in the Earth's Magnetometers," The Solar-Terrestrial Science Project of the Inter-Agency Consultative Group, ESA SP-1107, November 1990, pp. 55-60

1451) Information provided by Yu. M. Mikhailov, IZMIRAN

FS = Photometer (PI: Yu. Ruzhin, IZMIRAN). Objective: measurement of the illumination intensity (at the injection of the electron and ion beams) in the range from 3914-6563 Å.

SGR-5 = Fluxgate Magnetometer (PI: L. Zhusgov, IZMIRAN). Objective: measurement of the three components of the magnetic field in the range of ±64,000 nT with an accuracy of 1 nT.

MNCH = Search-Coil Magnetometer (PI: V. Chmirev, IZMIRAN). Objective: measurement of the magnetic field variations in the frequency range from 0.1 - 10 Hz and an amplitude range of 40 - 400 nT. The information transfer is provided on the DEP-2E sensor.

NAM-5 = Radiofrequency Mass-Spectrometer (PI: V. Istomin, IKI, J. Shmilauer, GFI CSAN, CSFR). Same instrument and same objectives as on ACTIVE mission.

AVCH-2T = HF-Field Analyzer (PI: Z. Klos, Center of Cosmic Research, Polish Academy of Sciences, Warsaw; S. A. Pulinets, IZMIRAN). Objective: Spectrum analysis of high-frequency electromagnetic radiation in the range of 0.1 to 15.1 MHz. The sensor is a self-opening band-type antenna of 15 m length.

K.5.1 APEX Subsatellite (Magion-3) Scientific Payload

Objective: monitoring the propagation of electron beams and plasma injected by the parent spacecraft as well as monitoring plasma of natural origin. Magion-3 was separated from its parent on December 28, 1991; the experiments are almost identical to those of Magion-2 (ACTIVE mission).

Orbit: perigee = 500 km, apogee = 3200 km, inclination = 83°

SGR-7 = 3-axis Fluxgate Magnetometer (PI: M. Ciobanu, IKI, Romania). Measurement range of ±50048 nT or 6256 nT, resolution = 16 nT or 2 nT, frequency range = 0 - 20 Hz. Measurement of magnetic field vector and serving as attitude sensor.

- SGR6 = one-component Fluxgate Variometer; dynamic range = ±156 nT; resolution = 50 nT; frequency range = 0.1 - 20 Hz.

KEM-1 = Magnetic and Electric Field Analyzer (PI: P. Triska, GFU, Czech Republic);

- ULF = Electric Field Experiment (three components of the quasi-static electric field); dynamic range = 0.005 - 8000 mV, frequency range = 0.1 - 20 Hz.
- VLF Wave Experiment (three-axis electric and one-component magnet ELF-VLF field measurements, broadband waveforms, spectrum analyzer, filter bank; $E_{x,y,z}$: 0.1 Hz - 120 kHz, sensitivity of 10^{-7}V/mHz$^{1/2}$, 120 dB dynamic range. B_x : 10Hz - 40 kHz, sensitivity of 5 x 10^{-6} nT/Hz$^{1/2}$ at 2 kHz and 10^{-4} nT/Hz$^{1/2}$ at 100 Hz, dynamic range 120 dB.
- Data transmission modes selectable: broadband analog data 10 Hz - 60 kHz one channel, 10 Hz - 20 kHz broadband three channels, subcarriers of 1 kHz bandwidth four channels.
- Filter bank; 17 Hz - 15 kHz, eight filters, four independent sets
- Frequency analyzer: range of 1 - 220 kHz, 32 frequency steps, full spectrum/2s, selection of any frequency.

KM-12 = Cold Plasma Analyzer (PI: J. Shmilauer, GFU, Czech Republic); HF probe measures two components of electron temperature in the range: Te = 10^3 - 10^5 K, spacecraft potential from -2 to +2V; spherical ion-trap measures the ion density (N_i) in the range: 10^8 - 10^{13} /m^3. $\Delta N_i/N_i$ fluctuations f_{max} = 50 Hz.

ZL-A-S = Langmuir Probe (PI: K. Sauer, MPE, Germany); measurement of electron and ion density, range: 500 - 10^8 cm^{-3}; electron temperature: 0.05 - 3 eV; current range: 10^{-10} - 10^{-3} A; in the current-mode density fluctuations f= <200 Hz, $\Delta N_e/N_e$ resolution 10^{-3}.

PRS-2-C = Radiowave Spectrometer (PI: Z. Kloss, CBK PAN, Poland). Measurement of the HF wave spectra. Frequency range = 0.1-10 MHz, dynamic range = 1 μV-10 mV, 0.2 s/spectrum using Δf=50 kHz or 2 s/spectrum using Δf= 15 kHz; field fluctuations at fixed frequency (selectable), Δt= 1 ms.

DOK-A-S =Silicon Detector Spectrometer (PI: K. Kudela, UEF SAV, Slovakia); two sensors (parallel and perpendicular to the magnetic field vector), measurement of electrons and ions (20 keV - 1 MeV, 8 energy levels, geometric factor 10^{-2} cm^2 sr.

MPS SEA= Energetic Particle Spectrometer (PI: Z. Nemecek, Prague University, Czech Republic); electrostatic analyzers, measurement of electrons and positive ions: 0.2 - 20 keV in 16 energy levels, pitch angle resolution 30°, geometric factor is $\sim 10^{-3}$ cm^2 sr.

K.6 ASTRID

ASTRID is a spin-stabilized microsatellite series designed and built by SSC (Swedish Space Corporation) of Solna, Sweden; the payload design is by the Swedish Institute of Space Physics in Kiruna (IRF-K).

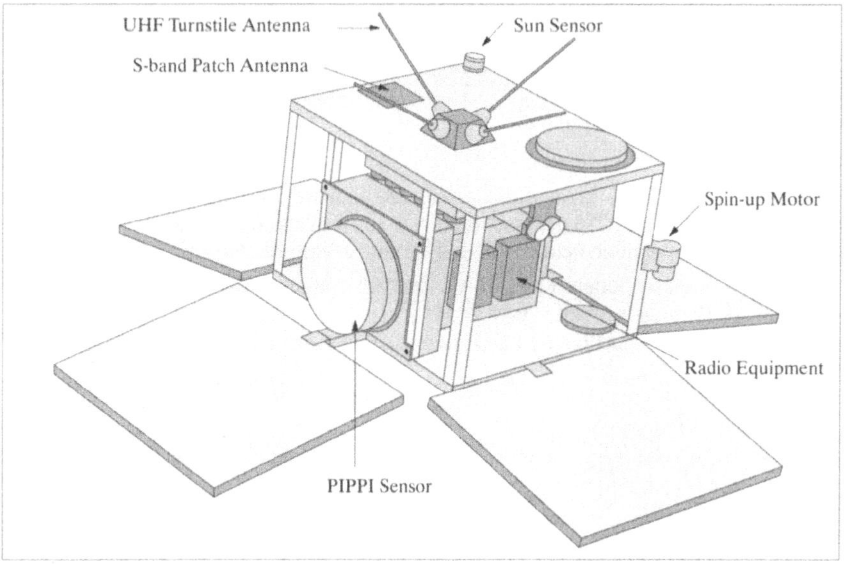

Figure 253: ASTRID-1 S/C Model

K.6.1 ASTRID-1

ASTRID-1 is a research mission in the field of solar-terrestrial energy transport physics and has the objective to investigate the Earth's near-space plasma with emphasis on neutral particle phenomena with high-resolution observations in the upper ionosphere and the lower magnetosphere.[1452)]

The S/C design marks the first use of the Freja-C compact satellite platform featuring a generic bus. The S/C structure consists of a perforated, corrosion-resistant Al-honeycomb

1452) Information provided by S. Grahn of SSC

core; it is a box-like structure (420 x 420 x 290 mm) with four solar panels. ASTRID-1 is spin-stabilized; attitude determination and control of the S/C is provided by a sun sensor, by two 2-axis flux-gate magnetometers, and two magnetotorquers. Antennas: S-band transmit and UHF transmit & command receive. RF transmitters: S-band frequency = 2200-2300 MHz, RF output power = 2 W, phase modulation; UHF: frequency = 400.15-401 MHz, RF output power = 2W, frequency modulation. Command receivers: frequency = 449.95 MHz, FM modulation. Power: solar array output = 42 W at 29.8 V plus battery. S/C mass = 27 kg (payload mass=4.5 kg).

ASTRID-1 was launched piggyback, together with FAISAT-1, on a Russian launch vehicle (Kosmos-3M rocket) from Plesetsk on January 24, 1995. Design life of S/C = 6 months. Mission control performed by the Swedish Institute of Space Physics in Kiruna via Esrange. Data rates: 131 kbit/s (S-band) and 8 kbit/s (UHF). Status: The ASTRID-1 payload was operational until March 1, 1995. The S/C platform worked until September 1995.

Orbit: Circular orbit, altitude = 1000 km (apogee=1026 km, perigee=966 km, inclination = 83°, nodal period = 105.07 minutes.

Sensors: all instruments are built by IRF (Swedish Institute of Space Physics, Kiruna.[1453])

PIPPI = Prelude in Planetary Particle Imaging (PI: O. Norberg, IRF, Kiruna)
A neutral particle imager (main sensor of ASTRID-1) with the objective to measure energetic neutral particles in the Earth's magnetosphere. The instrument consists of two cameras: an SSD (Solid State Detector) camera which resolves the energy of detected particles; and the MCP (Microchannel Plate) camera which uses a technique whereby incoming neutrals cause charged secondary particles to be emitted from a graphite target; the secondaries are then detected by MCP. Both cameras have deflection systems which can reject charged particles up to an energy of 140 keV. The sensor aperture plane is perpendicular to the spin plane; all directions are thus covered in half a spin period or ~1.5 s. PIPPI also serves as the data processing unit for EMIL and MIO.

Parameter	SSD Camera	MCP Camera
Energy range	13-140 keV	0.1-70 keV
Energy resolution	8 levels	-
Sampling time	31.25 ms	31.25 ms
Number of apertures	14	31
Angular resolution	23° x 5°	11° x 9°
Geometric factor	0.035 cm^2 steradian	0.08 cm^2 steradian
Bit rate	60.5 kbit/s	16 kbit/s
Mass, Power	3.1 kg, 4.0 W	

Table 434: PIPPI instrument characteristics

EMIL =Electron Measurements, In-situ and Lightweight
The electron spectrometer consists of a swept-energy toroidal electrostatic analyzer and a microchannel plate (MCP) detector with six angular channels in the satellite spin plane. The sensor measures the electron distribution in the magnetosphere at 62.5 ms or 125 ms resolution.

Energy range	50 eV - 40 keV, resolution ($\Delta E/E$) = 0.10
Energy steps	32 or 64
Sampling time per step	2 ms
Total geometric factor	4 x 10^{-4} cm^2 steradian keV/keV
Bit rate	24 kbit/s
Mass, Power	0.9 kg, 0.9 W

Table 435: EMIL instrument characteristics

1453) Information provided by O. Norberg of IRF, Kiruna

MIO =Miniature Imaging Optics
The instrument consists of two UV-photometers mounted in the satellite spin plane. MIO-1 measures the Lyman-α emission from the Earth's geo-corona; MIO-2 observes auroral emissions. Each photometer consists of optics mounted in a stainless steel tube with a ceramic channel electron multiplier at the opposite end.

Spectral passbands	MIO-1 (Lyman-α): 121 nm (MgF2 + oxygen gas filter + KBr MIO-2 (oxygen): 125-160 nm (CaF2 + KBr)
Focal width, FOV, sampling time	255 mm, 1°, 2 ms
Geometric factor	2 x 10^{-4} cm^2 steradian
Bit rate, Mass, Power	8 kbit/s, 0.3 kg, 0.16 W

Table 436: MIO instrument characteristics

K.6.2 ASTRID-2

ASTRID-2 is also a research mission in the field of solar-terrestrial energy transport physics with the objective to perform a) high-resolution E-field and B-field measurements in the auroral region, b) electron density measurements, c) high-resolution measurements of electron and ion distributions functions, and d) UV auroral imaging and atmospheric UV-absorption measurements. The S/C structural design and stabilization is identical to that of ASTRID-1 (spin-stabilized sun-pointing platform) Attitude is sensed by ASC (Advanced Stellar Compass). The total S/C mass is 30 kg with a payload mass of about 10kg. The downlink is in S-band, data rate of 128 kbit/s, BPSK modulation (convolutional encoded). The uplink is in S-band at 10 kbit/s, FSK modulation. A launch took place on a Russian vehicle (KOSMOS-3M, ASTRID-2 was a secondary payload to a Russian navigation satellite, named Nadezhda-5) from Plesetsk on Dec. 10, 1998.

Orbit: Circular orbit, altitude = 1000 km, inclination = 83°.

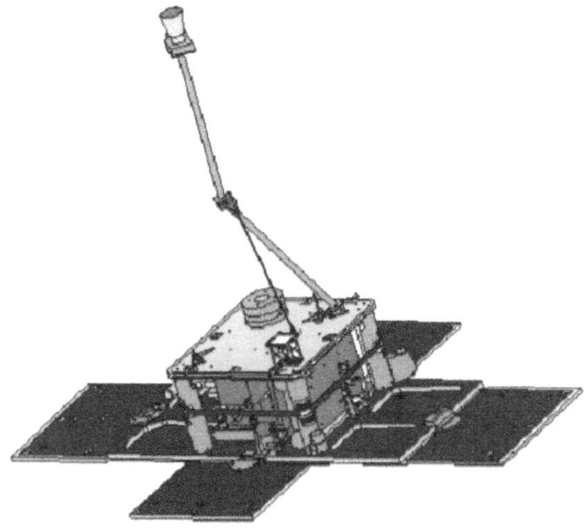

Figure 254: The ASTRID-2 spacecraft

Sensor complement:

EMMA (Electrical and Magnetic field Monitoring of the Aurora). The objective is to measure both the electrical and magnetic fields.

LINDA (Langmuir Interferometer and Density experiment for Astrid-2). The objective is to measure the fine structure of the plasma density irregularities down to 1 m scales and, by using two probes, distinguish between temporal and spatial effects. LINDA consists of two 10 mm diameter spherical probes mounted on two light-weight booms. The booms are 61 cm long and mounted on the outer tips of the solar panels giving a probe-to-probe separation distance of 2.9 m. The sampling rate is 32 k-samples/s by using the snapshot technique. The measured quantities are the plasma density (derived from the probe current) and the relative density variations, derived from the variations in probe current.

MEDUSA (Miniaturized Electrostatic Dual-tophat Spherical Analyzer), a joint instrument provided by Southwest Research Institute, San Antonio, TX, and IRF-K. The objective is to measure simultaneously electrons and ions with energies up to 18 keV/q. The resolution is 16 energy sweeps/s for electrons, and 8 seconds for ions. Particles are measured in 16 sectors in the plane of acceptance, which is almost parallel to the satellite spin plane. MEDUSA includes a CPU for instrument control and data compression; instrument mass=1.5 kg, power = 5.3 W, data rate = 32 kbit/s (includes PIA photometers).

PIA (Photometers for Imaging the Aurora), provided by MPAe, Katlenburg-Lindau, Germany, and by IRF-K. PIA consists of two spin-scanning photometers (PIA-1/2) for auroral imaging and one sunward-looking photometer (PIA-3) for atmospheric absorption measurements. PIA-3 is located under the sunward-facing platform and views the sun in Lyman-alpha (121 nm) via a diffuse reflector. PIA-1 and PIA-2 have four pixels each and a focal width of 250 mm. All photometers are sampled at 256/s. Instrument mass = 0.6 kg.

K.7 CLuster (Four S/C Mission in Concert with SOHO)

Cluster is a collaborative ESA/NASA mission within ESA's 'Solar Terrestrial Science Program' (STSP), part of ISTP (International Solar Terrestrial Physics Programme). Study of key interaction processes between two cosmic plasmas [study of small-scale structures (from a few to a few tens of ion Larmor radii) in the Earth's plasma environment].[1454),1455)] The goal is to study the physical processes involved in the interaction between the solar wind and the magnetosphere by visiting key regions like the polar cusps and the magnetotail (mapping in three dimensions the plasma structures contained in these regions). Other regions of measurement are: a) solar wind and bow shock, b) magnetopause, and c) auroral zone. The simultaneous four-point measurements (with four S/C) also allow the derivation of differential plasma quantities for the first time.

K.7.1 Cluster-I

Note: The launch failure of Ariane-5 on June 4, 1996 with the Cluster satellites onboard and the ESA decision of a recovery program - are the reasons for a division into Cluster-I and Cluster-II programs.

The four identical Cluster S/C are spin-stabilized (15 rpm, stringent requirements on electromagnetic cleanliness). Conductive surfaces and an extremely low S/C-generated electromagnetic background noise are mandatory for accurate electric field and cold plasma measurements.

1454) ISTP Global GEOSPACE Science - Energy Transfer in Geospace, ESA/NASA/ISAS brochure, 1992

1455) J. Credland, G. Mecke, J. Ellwood, F. Drigani, P. Ferri, et al., Special Section of the Cluster mission, spacecraft, payload, data, and mission operations, ESA Bulletin, No. 84, Nov. 1995, pp. 113-150

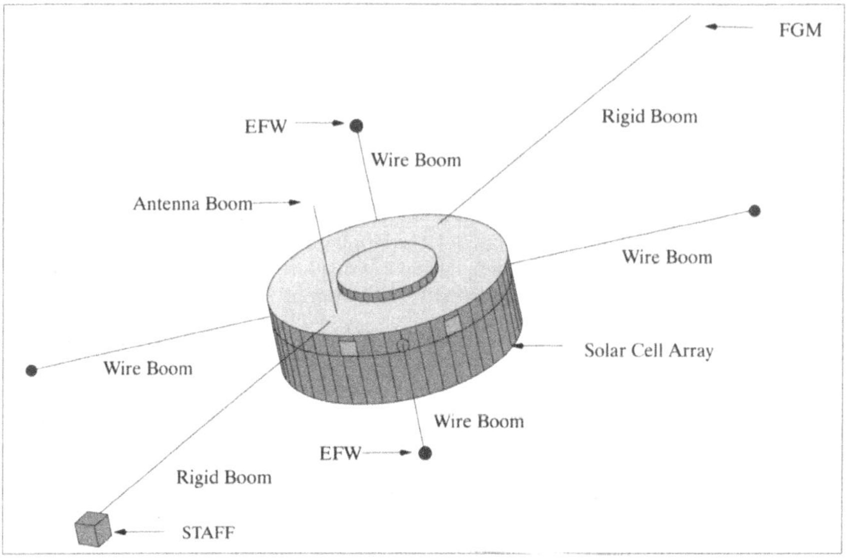

Figure 255: The Cluster S/C model

Each S/C is cylindrical in shape with a diameter of 2.9 m and a height of 1.3 m with conductive surfaces. The S/C structure is based on a central CFRP cylinder supporting the main equipment platform (an aluminum-skinned honeycomb panel reinforced by an outer aluminum ring). The solar array power is 224 W (47 W are allocated to the payload), the five AgCd batteries have a capacity of 80 Ah. The solar arrays consist of BSR (Back-Surface-Reflection) cells, arranged in self-compensating formations to minimize the generation of DC magnetic fields. The conductive coating on the cell cover glass minimizes the build-up of differential charge potentials. - S/C attitude and spin data are provided by an internally redundant star mapper and an internally redundant X-beam sun sensor. On-board attitude determination is better than 0.25° by the star mapper and the sun sensor. The reconstitution of attitude data, such as inertial attitude, spin rate and spin phase, is performed on the ground. Orbit and attitude maintenance are performed by using control thrusters, both semi-radial and axial, together with the main engine, which are used to perform the large orbital change maneuvers required to reach the polar mission orbit from geostationary transfer orbit. Launch masses: No 1 = 1183 kg; No 2 = 1169 kg; No 3 = 1171 kg; No 4 = 1184 kg. Each S/C has 650 kg of propellant and 72 kg of payload mass.

The platform accommodates on one side the instruments. Each satellite carries two high capacity redundant tape recorders (data return of 50% per orbit). Two rigid booms, each 5 m, carry the magnetometers. Two pairs of wire booms, each with a tip-to-tip length of 100 m, permit electric field measurements (background magnetic field of about 0.25 nT is aim). - Cluster operations are performed by ESA/ESOC with support provided by NASA's DSN (Deep Space Network). The telemetry downlink of each S/C supports the following data rates: a) 2 kbit/s of TT&C data, b) 22 kbit/s for nominal science data, c) 131 kbit/s for burst science data, and d) 262 kbit/s for the WBD (Wide Band Data) experiment.

Orbit: Near-polar elliptical orbits of the spacecraft with an apogee of 20 R_E (127,000 km) into the solar wind, and a perigee of 4 R_E (25,500 km), inclination of 90°. Six months later the apogee swings through the geotail. The inter-spacecraft distances are smaller ($<R_E$) on the dayside orbital phase, eg., when the apogee is turned into the solar wind, and up to 4 R_E in the nightside orbit phase, eg., when the apogee swings through the geotail.

The Cluster launch on June 4, 1996 with the first Ariane-5 flight from Kourou turned out to be a failure. Ariane-5 rose flawlessly to an altitude of 3.5 km, at which point a sudden swivelling of both solid-booster nozzles caused the vehicle to tilt sharply. The resulting intense aerodynamic structural loads caused the rocket vehicle to break up, prompting the onboard safety systems to initiate self-destruction. The entire four-spacecraft mission was lost.

Sensor complement: (each payload consists of 11 instruments) [1456] [1457] [1458]
A Wave Experiment Consortium (WEC) was formed to get maximum scientific return from the available spacecraft instruments. WEC comprises five coordinated experiments designed for measuring electric and magnetic fluctuations, and small-scale structures within critical layers in the Earth's magnetosphere. These WEC experiments are: STAFF, EFW, WHISPER, DWP, and WBD.

FGM = Fluxgate Magnetometer. (PI: A. Balogh, Imperial College, London, UK)
Study of small-scale structures and processes in the Earth's environment. Objective: Provision of intercalibrated measurements of the magnetic field vector **B** at the four Cluster S/C. Identical instrumentation on all S/C (two triaxial fluxgate sensors and a data processing unit). Measurement of B wave from DC to 10 Hz, resolution ≥6 pT. Four magnetometer ranges:

- ±256 nT resolution = ±0.015 nT
- ±1024 nT resolution = ±0.061 nT
- ±8192 nT resolution = ±0.5 nT
- ±65536 nT resolution = ±4 nT

STAFF = Spatio-Temporal Analysis of Field Fluctuations (N. Cornilleau-Wehrlin, CRPE/CNET, France)
Objectives: Study of wave-particle interaction in the region where the solar wind meets the Earth's magnetosphere. Measurement of B wave from up to 10 Hz, compressed data up to 4 kHz, cross-correlator for <E, B>. A STAFF instrument is provided on all four S/C; it consists of a three-axis search coil magnetometer (measurement up to 4 kHz), and a spectrum analyzer to perform the auto and cross correlations between electric and magnetic components on board each satellite. STAFF is one of the five complementary experiments which form the Wave Experiment Consortium (WEC).

EFW = Electric Fields and Waves [G. Gustavson, IRF-U (Swedish Institute of Space Physics), Uppsala, Sweden]

Objectives: measurement of electric field and density fluctuations, determination of electron density and temperature, study of nonlinear processes that result in acceleration of the plasma, study of large-scale phenomena of data from all four S/C. The instrument consists of double probes, mounted on two pairs of wire booms, each 100 m tip-to-tip. Measurement of E wave from 10 Hz, compressed data up to 100 kHz, sensitivity < 50 nV/m $(Hz)^{1/2}$. EFW measurement modes:

- Instantaneous spin plane components of the electric field vector, over a dynamic range of 0.1 - 700 mV/m, and with variable time resolution down to 0.1 ms.
- Low energy plasma density, over a dynamic range at least 1 to 100 cm^{-3}
- Electric field and density fluctuations in double layers of small amplitude, over dynamic ranges of 0.1 to 50 mV/m for the fields, and 1 - 50% for the relative density fluctuations, time resolution of 0.1 ms on some occasions
- Waves, ranging from electrostatic ion cyclotron emissions having amplitudes as large as 60 mV/m at frequencies as low as 50 mHz, to lower hybrid emissions at several hundred Hz and with amplitudes as small as a few μV/m.

1456) C. P. Escoubet, C. T. Russell, R. Schmidt [editors], "The Cluster and Phoenix Missions," published in a special Cluster issue of Space Science Reviews, Vol. 79, No. 1-2, 1997, Kluver

1457) http://sci.esa.int/missions/index.cfm?TypeID=8&page=about

1458) "The Cluster Mission - Scientific and Technical Aspects of the Instruments," ESA SP-1103, ISSN 0379-6566, Oct. 1988

- Time delays between signals from up to four different antenna elements on the same S/C, with a time resolution of 25 μs on some occasions.
- S/C potential

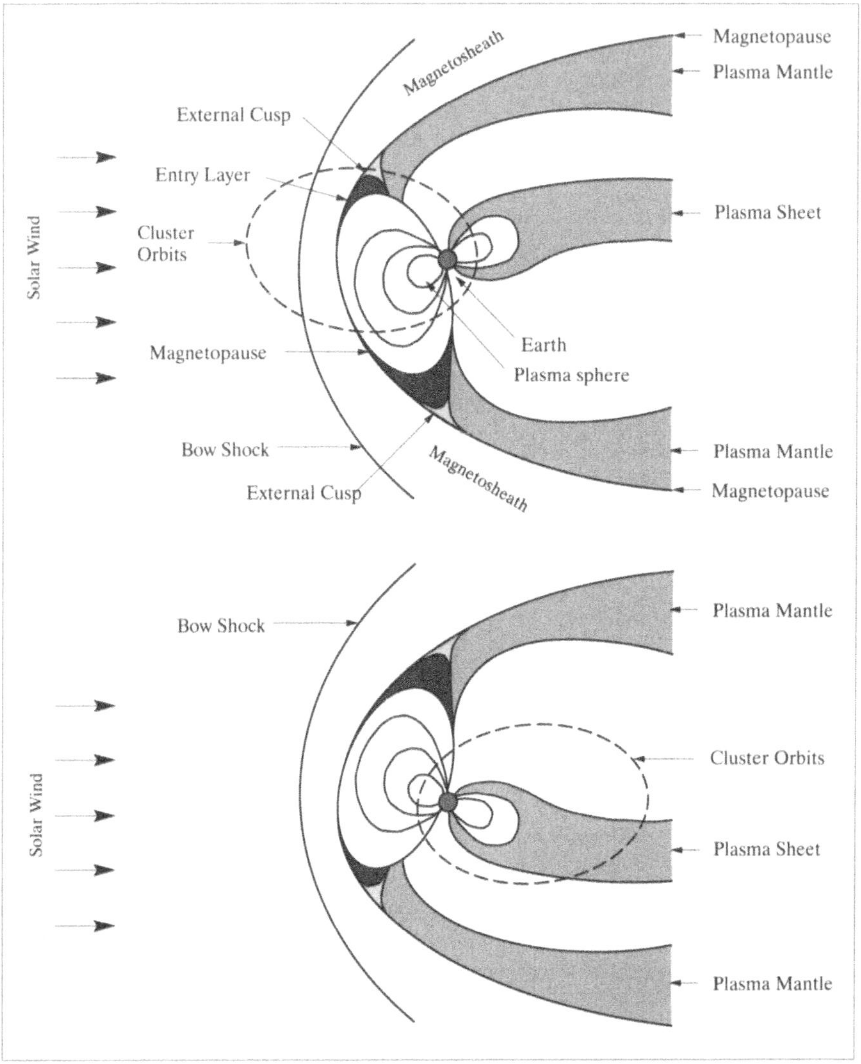

Figure 256: Cluster orbits in relation to the magnetosphere at six month intervals

WHISPER = Waves of High Frequency and Sounder for Probing of Density by Relaxation [P. M. E. Décréau, LPCE (Laboratoire de Physique et Chimie de l'Environnement), Orléans, France]. Objectives: accurate measurement of total plasma density within the range from 0.2 - 80 cm^{-3} (prime objective), continuous survey of one electric component of plasma waves in the frequency range from 4 - 80 kHz with an accuracy of about 160 Hz (secondary objective).

WHISPER employs the method of relaxation sounding (using parts of the EFW wire booms). The analysis of density variations is performed via active sounding of plasma resonances. Expected measurement results:

- identification of regions in space and mass transport
- spatial extension and drift speed
- density fluctuations
- wave mode identification
- coldest component of the electron density

Passive measurements of natural plasma waves up to 400 kHz.

WBD = Wide Band Data (D. A. Gurnett, University of Iowa, USA)
Objectives: Study of E-field wave form up to ~100 kHz in the Earth's magnetosphere. A wideband receiver system measures electric and magnetic fields over a frequency range from 25 Hz to 577 kHz. The Cluster wideband receiver is similar to the instruments flown on ISEE-1 and DE-1. WBD makes use of the EFW sensors (two electric dipole antennas, two search coil magnetometers); conversion frequencies: 0, 125 kHz, 250 kHz, 500 kHz; bandpass filter ranges: 1 kHz-100 kHz, 50 Hz-25 kHz, 10 Hz-10 kHz; frequency resolution: limited by FFT (75 Hz typical); time resolution: 10-20 ms (per FFT spectrum); The WBD measurements complement those of the other WEC instruments and also provide a unique new capability to perform very-long baseline interferometry (VLBI) measurements. - The wideband technique involves transmitting band-limited waveform data to a ground station using a high-rate data link (262 kbit/s R/T or 131 kbit/s in burst mode).

DWP = Digital Wave Processor (PI: H. Alleyne, University of Sheffield, UK)
Objectives: Correlation of wave/particle phenomena and wave/particle interactions. DWP is the on-board control system (multiprocessor unit based on the use of transputers with parallel processing and re-allocatable tasks to provide a high-reliability system.) of all WEC instruments performing data compaction and compression, event selection, particle/wave correlation, and control of WHISPER.

EDI = Electron Drift Instrument (Pi: G. Paschmann, MPE, Garching, Germany)
Objectives: accurate and highly sensitive measurements of the electric field. EDI measures the drift of a weak beam of test electrons which for certain emission directions return to the S/C after one gyration (the drift is related to the electric field and to the gradient in the magnetic field). EDI consists of two emitter/detector assemblies, each with a 2π FOV. E = (0.1 - 10 mV/m, <100 Hz), $\nabla \underline{B}$, $|\underline{B}|$ = 5-1000 nT, emission and tracking of two electron beams.

CIS = Cluster Ion Spectrometry [Pi: H. Rème, CESR (Centre d'Etudes Spatial des Rayonnement), Toulouse, France]. Objectives: Study of the dynamics of magnetized plasma structures in the vicinity of the Earth's magnetosphere (physics of the bow shock, the magnetopause boundary, the polar cusp, the geomagnetic tail, and the plasma sheet). CIS is a comprehensive ionic plasma spectrometry package on-board the four Cluster spacecraft capable of obtaining full 3-D ion distributions with good time resolution (one spacecraft spin) with mass per charge composition determination.
CIS consists of two instruments, CODIF and HIA to measure both the cold and hot ions from the solar wind, the magnetosheath, and the magnetosphere (including the ionosphere) with sufficient angular, energy, and mass resolution.
CODIF = Composition and Distribution Functions analyzer (energy range = 0.02 - 40 keV/q) to measure the 3-D distribution of major ion species (H^+, He^{2+}, He^+, and O^+). The instrument is a symmetrical hemispherical analyzer with RPA (Retarding Potential Analyzer) and TOF (Time-of-Flight) electronics, FOV = $2\pi \times 8^o$; split geometric factor.
HIA = Hot Ion Analyzer for high time resolution of the solar wind (energy range = 3 eV/q - 40 keV/q). The instrument is a symmetric quadri-spherical analyzer; FOV = $2\pi \times 8^o$ with high resolution. ($\geq 2.8^o$).

PEACE = Plasma Electron and Current Analyzer [Pi: A Fazakerley, MSSL (Mullard Space Science Laboratory), University College, London, UK]. Objectives: Study of the 3-D velocity distribution of electrons. PEACE consists of two electron sensor devices. LEFA = Low Energy Electron Analyzer (energy range = 0 - 100 eV). The instrument is a spherical electrostatic analyzer with a FOV of π x 3.8°. HEFA = High Energy Electron Analyzer (energy range = 0.1 - 30 keV). The instrument is a troidal electrostatic analyzer with a FOV = 2π x 4.6°.

RAPID = Research with Adaptive Particle Imaging Detectors [PI: B. Wilken, MPAe (MPI für Aeronomie), Lindau/Harz, Germany]. Objectives: Study of suprathermal plasma distributions in the energy range from 20 - 400 keV and 2 keV/n - 1500 keV for electrons and ions, respectively.
RAPID provides complete coverage of the unit sphere in phase space. It is made up of two instruments, the IIMS and IES. IIMS = Imaging Ion Mass Spectrometer for ion distribution. IIMS is a position-sensitive instrument with solid state detectors and TOF (Time-of-Flight) electronics. IES = Imaging Electron Spectrometer for measuring energetic electrons. IES has a position-sensitive solid state detector.

ASPOC = Active S/C Potential Control [PI: W. Riedler, IWF (Institut für Weltraumforschung), Graz, Austria]. Objective: control of the S/C charge potential to insure effective, complete measurement of the ambient plasma distribution functions. The basic approach involves the emission of charges from the S/C sufficient to balance the excess of charge accumulating on the satellite by the environment (photo emission of electrons drives the S/C potential positive relative to the plasma potential). - The method employed is field ionization with a Liquid Metal Ion Source (LMIS) of the "solid needle" type. LMIS uses Indium as charged material. The instrument consists of an electronics box and two cylindrical emitter modules with five needles each.

K.7.2 Cluster-II

In April 1997 ESA decided to go ahead with a recovery program, a new four-spacecraft mission, called Cluster-II, retaining all of the critical mission parameters. The Cluster-II mission comprises the Phoenix [1459)] spacecraft (built with spares from the four original Cluster satellites) and three identical new satellites (construction and test of four S/C in 18 months, the industrial consortium is led by Astrium GmbH of Friedrichshafen, Germany). The overall objective is to investigate the physical interaction between the sun and the Earth, with four spacecraft flying in tetrahedral formation. [1460)] [1461)] [1462)] [1463)] [1464)] [1465)]

Like their predecessor of Cluster-I, each Cluster-II S/C is cylindrical in shape with a diameter of 2.9 m and a height of 1.3 m (launch mass = 1200 kg, payload = 72 kg, S/C (dry) = 550 kg, propellant = 650 kg), solar array power = 224 W (payload power 47 W), battery capacity = 80 Ah.. Each Cluster-II S/C has a single main thruster (400 N) and eight smaller (10 N) thrusters. All thrusters use mono-ethyl hydrazine and mixed oxides of nitrogen for propellant. The large propellant mass is needed for an extensive series of maneuvers to reach the operational orbit and, during the course of the mission, to change the relative spacing of the

1459) Note: The new S/C was named after the mythical bird "Phoenix." In ancient Egypt and in classical antiquity, phoenix is a fabulous long-lived bird associated with the worship of the sun. As its end approached, the phoenix fashioned a nest of aromatic boughs and spices, set it on fire, and was consumed in the flames. From the pyre miraculously sprang a new phoenix, which, after embalming its father's ashes in an egg of myrrh, flew with the ashes to Heliopolis ("City of the Sun") in Egypt, where it deposited them on the altar in the temple of the Egyptian god of the sun, Re.

1460) "The Cluster-II Mission - Rising from the Ashes," The Cluster-II Project Team, ESA Bulletin, 102, May 2000, pp.47-53

1461) C. Ph. Escoubet, "Cluster-II: Scientific Objectives and Data Dissemination," ESA Bulletin, 102, May 2000, pp. 54-60

1462) M. Warhaut, S. Matussi, P. Ferri, "Cluster-II: Evolution of the Operations Concept," ESA Bulletin, 102, May, pp. 61-67

1463) J. Credland, R. Schmidt, "The Resurrection of the Cluster Scientific Mission," ESA Bulletin, No. 91, August 1997, pp. 5-10

1464) http://sci.esa.int/cluster/

1465) The Cluster mission is covered in considerable detail in Space Science Reviews, Vol. 79. pages 11-658. (Jan. 1997)

S/C. The four spacecraft are spin-stabilized at 15 rpm. The in-orbit configuration is characterized by two 5 m experiment radial booms, four 50 m experiment wire booms and two axial telecommunications antenna booms. The S/C attitudes ensure a solar aspect angle of about 90°.

More detailed objectives call for the mapping of plasma structures and current densities at:

- The solar wind bow shock: Study of the propagation of electric waves through the bow shock and magnetosheath
- The magnetopause, characterizing the motion and local geometry of the magnetopause, and identify the mechanism whereby plasma infiltrates the magnetopause
- The polar cusps, studying the behavior of postulated plasma vortices
- The magnetotail: observation of ion beams, and calculation of the magnitude of field aligned currents, in the plasma sheet boundary layer. Studies of the disruption of cross-tail currents during substorms, and the consequences for the plasma sheet
- The auroral zones. Determination of the sources of magnetospheric plasma, such as the polar wind, the cleft ion fountain, and nightside auroral zone.

The Cluster S/C are described under K.7.1.

The launch of the Cluster-II mission uses two Soyuz launch vehicles (with two S/C at a time) from Baikonur. The launch of the first mission took place on July 16, 2000. The second pair of S/C was launched on Aug. 9, 2000. The launch sequence called first for the injection of the S/C into an intermediate transfer orbit in pairs of two. Then, a series of propulsive maneuvers brought the four S/C from their initial transfer orbits into their mission orbits.

On launch day (Aug.9, 2000), the four Cluster spacecraft were named by ESA as: Rumba, Salsa, Samba, and Tango. The names of the dances were suggested by Raymond Cotton of Bristol, UK (the winner of a naming competition); they are supposed to reflect the way in which the four satellites are dancing in formation around the heavens during their mission. The two S/C of the first launch carry the names Salsa and Samba; the spacecraft of second launch have the names Rumba and Tango.

Orbit: Polar elliptical orbits, nominal apogee = 19.6 R_E (125,000 km), perigee = 4 R_E (25,000 km). The orbit for each S/C is selected so that each is located at a vertex of a predetermined tetrahedron when crossing the regions of interest within the magnetosphere. The size of this tetrahedron is varied from 200 km up to about 19,000 km during the course of the mission. The S/C, after release, use their own on-board propulsion systems to reach the final operational orbit. Since the Cluster-II orbit is fixed in the inertial system, the rotation of the Earth around the sun causes the S/C to cross the various near-Earth plasma regions, such as the Earth magnetotail soon after launch and the polar cusp and solar wind six months later.

Sensor complement:

The instrument complement remains identical to the original version of Cluster-I. Each of the four Cluster-II spacecraft carries an identical set of 11 scientific instruments as defined in K.7.1.

Ground Segment:

Operations of four S/C are provided by two centers: ESA/ESOC in Darmstadt, Germany for mission operations, and JSOC (Joint Science Operations Center, located at the Rutherford Appleton Laboratory, Didcot, UK. The function of JSOC is the coordination of science operations. Its main task is to merge the input from the PI teams into a command schedule. In addition, JSOC monitors the health of the payload instruments and disseminates information of the mission.

Communications: Science data are downlinked at a rate of 16.9 kbit/s real-time (105 kbit/s burst mode) or stored on one 2.25 Gbit solid-state recorder for replay (Note: the original two solid-state recorders were replaced by a single recorder of a new design). The S-band system operates at 2025-2110/2200-2290 MHz (up/down) at 10 W. Data from the four S/C is synchronized via a highly stable on-board clock. - The Villafranca ground station (Spain) of ESA is the prime station for all TT&C functions. Further support is provided by NASA's Deep Space Network (as well as other ground stations) during critical initial mission phases. About 1 GByte of science data per day is downlinked.

In the ground segment, the Cluster science data are available to the scientific community through CSDS (Cluster Science Data System). The CSDS consists of eight nationally funded and operated data centers. These are:

- Austrian Cluster Data Center, IWF, Graz (ASPOC)
- Chinese Cluster Data Center, Beijing (ASPOC)
- French Cluster Center, CNES, Toulouse (CIS, STAFF, WHISPER)
- German Cluster Data Center, MPE, Garching (EDI, RAPID)
- Scandinavian Data Center, KTH (Royal Institute of Technology), Stockholm (EFW)
- Hungarian Data Center, KFKI, Budapest
- UK Cluster Data Center, RAL, Didcot and QMW, London (DWP, FGM, PEACE)
- US Cluster Data Center, GSFC, Greenbelt, MD (WBD)

K.8 CORONAS-I

Coronas [Complex of Orbital (Near-Earth) Observations of the Activity of the Sun][1466),1467)] is a Russian Intercosmos project with the cooperation of several international partners [Ukraine (Dnepropetrovsk) for satellite construction, Poland and CSFR for sensor development, Germany for science data reception and data interpretation (DLR/DFD, and Astrophysikalisches Institut Potsdam (AIP)].

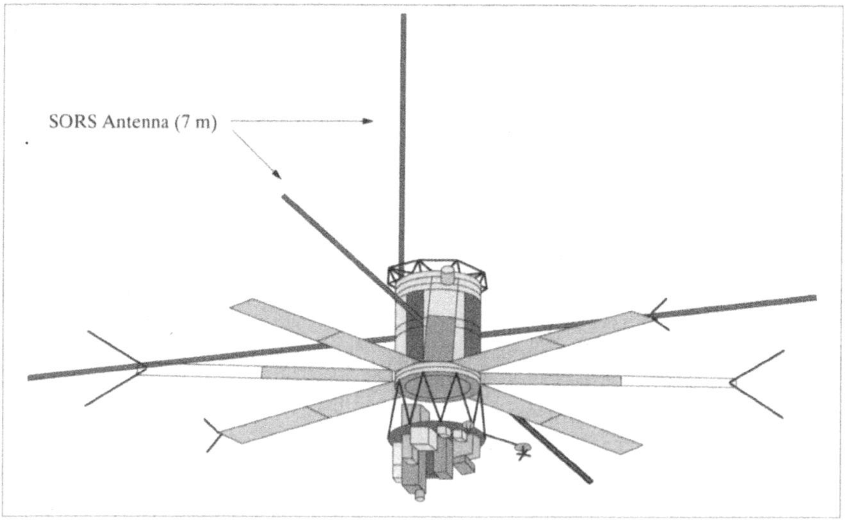

Figure 257: The CORONAS-I spacecraft model

1466) V. N. Oraevsky, Yu. D. Zhugzhda, "Project CORONAS-I - Orbital Observations of the Solar Activity and Oscillations, 1991," Coronas Information Series, paper by AIP, DLR/DFD and IZMIRAN, K Pflug (editor)

1467) I. Sobelman, I Zhitnik et al., " XUV and Optical Observations of the Sun by Means of the TEREK-C Telescope/Coronograph and the RES-C Spectroheliometer aboard the CORONAS-I satellite," 1992, Coronas Information Series, paper by AIP, DLR/DFD and IZMIRAN, K Pflug (editor)

Main objectives: study of the internal structure of the sun and the different processes of solar flares. Coronas-I (with IZMIRAN as project coordinator) is the first of three spacecraft (CORONAS-I launch: March 2, 1994 from Plesetsk); satellite mass= 2200 kg, diameter= 2 m, height= 6 m. There are a total of 10 scientific instruments on-board. The lifetime of a mission is expected to be 1.5 years. The CORONAS project is viewed as an important successor mission to NASA's most successful Solar Maximum Mission (SMM). The CORONAS-I mission provides observations during minimum solar cycle, while the -F mission will mostly be for maximum solar cycle observations.

Note: In July 1994 the sun sensor of the attitude system experienced a malfunction which resulted in permanent orientation problems of the S/C (all of the gas for the AOCS was consumed within the first six months of operation). This in turn made the data of a number of instruments useless that depended on proper sun orientation.
Status: The CORONAS-I mission was completed in late 2000. Sensor measurements were still being collected in the un-oriented mode of CORONAS-I to study the background radiation.

Data from the solar flare observations is considered for the study of:

- energy transfer from the solar interior to the surface and its subsequent release in solar flare events,
- explosive flare processes on the sun,
- the characteristics of solar cosmic rays, exit conditions, their propagation in the IMF and subsequent influence on the magnetosphere of the Earth,
- development of a theory of flares and techniques for forecasting their geophysical effects.

Orbit: Circular polar orbit, altitude of 500 km, inclination = 82.5°; quasi-synchronous orbit to assure 20 day observation periods in which the satellite will be outside the Earth's shadow.

Sensor complement:

No.	Instrument/Subsystem	Acronym	Mass (kg)	Power (W)	Principle Institut
1	Solar XUV Telescope/ Coronograph	TEREK-C	45	40	FIRAS
2	Solar X-ray Spectral Polarimeter	RES-C	35	30	FIRAS
3	Diagnostic of Energy Sources and Sinks in Flares	DIOGENESS	15.4	24	AI CSAV, CBK
4	Solar X-ray and γ-ray Scintillation Spectrometer	HELICON	35	10	FTI
5	Solar Burst Spectrometer	IRIS	30		FTI
6	Solar UV Radiometer	SUFR-Sp-C	3.6	12	IPG
7	Vacuum UV Solar Spectrometer	VUSS	18.5	20	IPG
8	Solar Flux Optical Photometer	DIFOS	6	18	IZMIRAN, GAO, FMI
9	Solar Radiospectrometer	SORS	18.5	21.5	IZMIRAN, CBK
10	Solar Cosmic Ray Spectrometer Complex	SKL	63	25	NIIYaF, IEP,SAV
11	Processing unit for amplitude and time analysis of input signals	AVS	12	22	MEPhI
12	System of compilation of scientific information	SSNI	60	80	OKB MEI

Table 437: Overview of CORONAS-I scientific payload

TEREK-C = Solar X-Ray Telescope/Coronograph. PIs: I. A. Zhitnik, I. Sobelman. Instrument is of TEREK heritage flown on PHOBOS-1 in 1988.
Objectives: studies of large and fine structures in the solar atmosphere, determining the characteristics of hot solar plasma in active regions, studies of coronal holes, etc. TEREK records images of the solar corona in the spectral ranges as defined in Table 438.

The detectors in all channels are CCD frame transfer image sensors with a dynamic range of $\sim 10^5$. Spatial resolution of ~ 7 seconds.

Spectral range	Description
5 - 25 Å, X-ray channel	Lines of ions: O VII-VIII, Mg XI, Fe XVII, Si XIII (Wolter-1 type grazing incidence objective with filters mounted on a 4-position wheel for subchannel selection)
170 - 180 Å	Lines of ions: Fe IX-XI
290 - 320 Å	Lines of ions: He II and Si XII
Near 52 Å	Lines of: Si XII
Near 130 Å	Lines of: Fe XXIII-XXIV
4000 - 6000 Å, channel K	Optical coronograph

Table 438: Spectral ranges of TEREK-C

RES-C = Solar X-Ray Spectral Polarimeter. Objective: Study of hot solar plasma structures in the range of 1-3 x 10^7 K. RES-C provides monochromatic images of the sun's X-ray radiation and linear polarization simultaneously in three spectral ranges:

- 180 - 205 Å (XUV channel) consisting of two subchannels with perpendicular dispersion planes. Each channel contains a grazing incidence plane grating (2400 lines) with multilayer coating, spherical mirrors working at normal incidence and intensified cooled CCD frame transfer image sensors. The spectral resolution is 0.03 Å, the angular resolution 6" x 90".
- 8.41 - 8.43Å (Mg XII channel). The Mg XII channel comprises the bent quartz (d=4.25 Å; 5 cm) and crystal spectrometer with CCD image sensor. The spectral resolution is about 7 x 10^{-4}Å; the angular resolution is 6" x 6".
- 1.85 - 1.87Å (Fe XXV channel). This channel contains the bent spherical quartz (d=1.18Å, R=30 cm) crystal with CCD image sensor. The spectral resolution is 2 x 10^{-4} Å; the angular resolution is 2'.

DIOGENESS = Diagnostic of Energy Sources and Sinks in Flares. Spectrometer and Photometer within the soft X-ray region. Objectives: provision of energy source and sink measurements that enable a study of the balance of the solar flare energy contained in its main thermal reservoir.

The DIOGENESS instrument consists of three independent units: a Bragg high-resolution spectrometer (BS), an X-ray spectrophotometer (BF), and a microcomputer for the control of BS and BF.

BS spectral range: 2.835 - 3.356Å. The most important lines are CaXVIII, CaXIX and CaXX. The BS time resolution is 0.1 - 10 s.

BF spectral range: 2 - 160 keV. The energy channels are: 2-4, 4-8, 10-20, 20-40, 4-80, and 80-160 keV. The time resolution is 1 second in the range of 2-8 keV and 0.1 second in the range of 10-160 keV.

HELICON = Solar X-ray and γ-ray Scintillation Spectrometer. Helicon has the following characteristics:

- number of detectors: 2 (solar and antisolar directions)
- energy range: 10 keV - 8 MeV
- energy windows for the investigation of time histories: 10-50 keV, 50-250 keV, 250-800 keV
- sensitive area of each detector: 314 cm^2 which corresponds to a sensitivity for flares with fluencies > 10^{-7} erg cm^{-2} s^{-1}
- statistical criterion for identification of flare: G_σ above the background
- number of PHA (Pulse Height Analyzer) channels: 64 for each detector
- number of spectra registered during one flare: up to 64 (depending on flare power)
- accumulation time for single spectra: 2 ms to 8 s (depending on the magnitude of the flare)
- number of time analysis channels: 4096
- time resolution: from 2 ms to 0.25 s
- background spectrum control: 2 minutes each

- duration of time history recording: up to 4 minutes

IRIS = Solar Burst Spectrometer. Objective: provision of measurements for the study of integral intensities and X-ray spectra with the following characteristics:

- energy channels: 2-4, 4-8, 8-12, 12-20, 20-30, 30-40, 40-60, 60-90, 90-120, 120-150, 150-200, and 200-300 keV
- studies of X-ray precursors of solar flares in the range of 2-20 keV with a sensitivity of 10 erg cm^{-2} s^{-1}
- studies of the dynamics of hard X-ray spectra during flare impulse phases in the range of 30-120 keV with a time resolution of about 0.01 s

SUFR-Sp-C = Solar UV Radiometer. Objectives: measurement of the solar radiation intensity in the UV range of $\lambda < 130$ nm. SUFR records the solar EUV radiation with the use of thermoluminescent material (24 specimens arranged on a rotating wheel). The thermoluminescent material $CaSO_4$ (Mn) is practically insensitive to radiation in the range of $\lambda >$ 130 nm. Dynamic range of the SUFR-Sp-C measurements: 0.1-30 erg cm^{-2} s^{-1}
The instrument contains 24 screens: one without filters measuring the total EUV radiation flux for $\lambda < 130$ nm, others with different filters (MgF, Al foil, Mylarfilm). The filters isolate the wave bands of soft X-rays (0.3-12 nm) and the Lyman alpha hydrogen line (121.6 nm). The absolute error does not exceed 15%. The exposure time for one screen and measuring time of its thermoluminescent signal is 18.5±0.5 s. The total measurement cycle last 8 minutes.

VUSS = Vacuum UV Solar Spectrometer. Objective: long-term measurements of the solar EUV radiation. The instrument performs absolute measurements of solar line intensities in the wavelength region of 20-58 nm. Its operation is based on the concept of collision photoionization spectroscopy. The maximal spectral resolution is $\lambda/\Delta\lambda \approx 400$. The threshold sensitivity is 5 x 10^6 quants x cm^{-2} s^{-1}.

DIFOS = Solar Flux Optical Photometer (PI: Yu. D. Zhugzhda). Objective: observation of global solar oscillations in the spectral range of 0.4-1.0 μm. A three-channel instrument at 0.550 μm (.1 μm band), 0.750 μm (0.10 μm band), and 0.4-1.1 μm range. Intensity measurements with a relative accuracy of about 10^{-5}. The time resolution is 16 seconds.

SORS = Solar Radiospectrometer. Objectives: investigation of solar radio bursts, diagnosis of surrounding ionospheric plasma near the satellite.
SORS works with two sweeping receivers in the frequency ranges of 50 kHz - 30 MHz, and 25 MHz - 300 MHz. A wide-band channel (Δf=4 MHz) is used to study the detailed structures of radio bursts in time scales of ms.
The SORS measurements of solar radio-emission in the low frequency range <25 MHz will complement the ground-based radioastronomical measurements with high time resolution. Direct measurements of UV emission and of local plasma density on the same satellite will improve ionospheric modelling and forecasting.

SKL = Solar Cosmic Ray Spectrometer Complex (PI: S. N. Kuznetsov). Objectives: measurement of charged and neutral components of energetic particle fluxes (energy spectra and elemental, isotopic and charge composition). SKL consists of three devices: SKE-3, MKL, and SONG.

- SKE-3 measures the fluxes of nuclei with charges ranging from z=1 to 10, and with energies from 1.5-19 MeV / nucl for ^{4}He and 3.7-4.6 MeV / nucl for ^{20}Ne.
- MKL is a cosmic ray monitor for measuring the spectra of electrons (0.5 - 12.0 MeV), protons (1 - 200 MeV), and alpha-particle fluxes (120 - 240 MeV), which includes five detectors.
- SONG measures the solar neutrons, gamma rays (continuum and lines) and detects charged particles, mainly electrons and protons. A Cs (Te) detector (200 mm diameter, h = 100 mm) surrounded by a 4π anticoincidence shielding of a plastic scintillator is used. The SONG instrument permits the detection of:

- neutrons with E ≥ 30 MeV (5 channels)
- gamma rays in the range of 0.1-100 MeV (9 chan.) and 0.3 - 16 MeV (236 chan.)
- high-energy protons in the ranges 200 - 500 MeV and 11 - 108 MeV (6 chan.).

SONG operates in two modes ("flare" and "monitoring") with a time resolution of 0.04 s, 2.5 s, or 120 s.

Data Transmission Parameters:

- Transmission power +9 dBW
- Antenna 2 V-dipole
- Antenna gain +0.5 dB
- Polarization right circular
- EIRP +5.5 dBW
- Carrier frequency 136.35 MHz / 137.45 MHz (VHF)
- Bandwidth 120 Hz
- Modulation BPSK, bi-phase
- Data rate 32 kbit/s / 64 kbit/s
- Word length 8 bit
- Frame length 128 byte

Coronas-I has an on-board data recording capability. The ground segment consists of Russian S/C control (TT&C function) and science data reception by the DLR/DFD ground station Neustrelitz (only science data reception station in the project). Neustrelitz is responsible for all science data reception, for intermediate archiving of all data, and for science data distribution to all parties involved. DLR/DFD Neustrelitz also processes/evaluates data from the SUFR and VUSS sensors. The AIP (Astrophysikalisches Institut Potsdam) receives data of the following sensors for evaluation: DIFOS, HELICON, SORS, and SKL.

IZMIRAN	Institute of Terrestrial Magnetism, Ionosphere and Propagation of Radio Wave Propagation -Russian Academy of Sciences, Troitsk (Moscow region)
FTI	Ioffe Physical Technical Institute (St. Petersburg)
GAO	Main Astronomical Observatory of the Ukrainian Academy of Sciences, Kiev
IKI - RAN	Space Research Institute - Russian Academy of Sciences, Moscow
OKB MEI	Special Research Bureau of Moscow, Power Engineering Institute
MEPhI	Moscow Engineering Physics Institute
MIFI	Moscow Institute of Physical Engineers (note: MIFI and MEPhI are the same things)
NIIYaF	Institute of Nuclear Research of Moscow State University (MGU)
IPG	Federov Institute of Applied Geophysics, Moscow
FIRAS	Lebedev Physical Institute of the Russian Academy of Sciences
FMI	Physical Mechanical Institute, Lvov (Ukraine)
	Scientific Production Facility for Space Devices (Dnepropetrovsk, Ukraine) - Coronas satellite development and integration
AI CSAV	Astronomical Institute of the Academy of Sciences of the Czech Republic, Prague
AI SAV	Astronomical Institute of the Slovak Academy of Sciences
IEP Kosice	Institute of Experimental Physics of the Slovak Academy of Sciences
CBK	Space Research Center of the Polish Academy of Sciences, Poland (also referred to as SRC)
DLR/DFD	Germany, Neustrelitz Station - reception of all science data
AIP	Astrophysikalisches Institut, Potsdam, Germany - processing of some sensor data

Table 439: Principal institutes involved in the cooperative project CORONAS

K.8.1 CORONAS-F

CORONAS-F is the second satellite of the international series with IZMIRAN as the prime lead and coordinator of the mission (V. N. Oraevsky). The Southern Design Office, Dnepropetrosvk, Ukraine [also referred to as YUZHNOYE (Research and Production Association)], as the spacecraft builder and integrator. The overall objective of CORONAS-F is the study of active sun and its neighborhood (small and large scale processes) by observations in the spectral range from gamma rays to radio waves. Specific science objectives are: 1468)

- To study the energy transport from the solar interior to the surface, its build-up in the upper atmosphere and subsequent release in non-stationary solar events (includes measurements of electrons, protons, and neutrons)
- The study of major dynamic phenomena of the active sun (sunspots, flares, plasma ejection's) in order to develop their theoretical interpretation and prediction methods
- The study of cosmic rays, accelerated in solar flares, as well as of other active phenomena, their escape, interplanetary propagation, and geophysical effect
- Seismological studies of the solar interior based on observed global oscillations

The CORONAS-F spacecraft is spin-stabilized and sun-pointed. The pointing accuracy of the longitudinal S/C axis to the sun is 10 arcsec, the pointing knowledge is 5 arcsec. The projection of the angular speed relative to the S/C coordinate system is known to 0.005°/s. The S/C has a total mass of 2260 kg, the payload mass is 395 kg, the design life is >1 year.

A launch of CORONAS-F took place on July 31, 2001 (launch site: Plesetsk, Russia).

Orbit: Circular orbit, altitude = 500 km, inclination = 82.5°. The orbit ensures continuous observations of the sun in recurrent intervals of about 20 days.

Instrument	Objectives
DIFOS (Solar Photometer), IZMIRAN	Investigation of the global oscillations of the solar irradiance
SORS (Solar Radiospectrometer), IZMIRAN	Registration of the radio emission of the solar active regions and flares
HELICON (Solar X-ray and γ-ray Scintillation Spectrometer), FTI	Study of the solar flare processes in ranges of soft X-ray to γ-ray region
IRIS (Solar Burst Spectrometer), FTI	Study of the solar flare processes in the X-ray range
TEREK-C (Solar X-Ray Telescope/Coronagraph), FIAN	Provision of solar imagery (and solar corona) in UV and soft X-ray ranges
RES-C (Solar X-Ray Spectral Polarimeter), FIAN	Measurement of monochromatic solar imagery in UV and soft X-ray ranges
SPR-N (Solar X-Ray Polarimeter), FIAN	Summary polarization measurement of the solar X-ray radiation
SCR (Cosmic Ray Spectrometer), NIIYaF MGU	Measurement of energetic particle fluxes
SUFR-Sp-C (Solar UV Radiometer), IPG	Measurement of the solar radiation intensity in the vacuum UV range
VUSS (Vacuum UV Solar Spectrometer), IPG	
DIOGENESS (X-Ray Photometer Spectrometer), SRC, Poland	Diagnostic of the energetic characteristics of the solar flares
RESIK (X-Ray Spectrometer), SRC	Spectral observations of the solar X-ray radiation
AVS (Amplitude-Time Spectrum Analyzer), MIFI	Investigation of the solar flare spectrum in the γ-ray region
RPS-1 (X-Ray Semiconductor Spectrometer), IKI	Registration and spectrometry of the X-ray solar flare radiation and its forerunners
IMAP-5 (Three-Axis Magnetometer), IZMIRAN	Registration of the Earth's magnetic field components
SSNI (Data Collection System), IZMIRAN	High-speed data collection from instruments in the ground segment

Table 440: Sensor complement of CORONAS-F

K.8.2 PHOTON

PHOTON is a Russian solar mission (third satellite in the CORONAS program) with the following objectives:

1468) Information provided by Vladimir Kuznetsov of IZMIRAN, Troitsk

- Study of the temporal variations in the energy spectra of hard radiation in the energy band from 15 keV to 2000 MeV
- Fast timing of the hard X-radiation in flares
- Measurement of the linear polarization of hard X-radiation fluxes
- Observation of neutrons with energies E > 5 MeV and measurement of their spectrum in the region of 5 - 300 MeV
- Study of energy spectra (charge and mass composition of ions accelerated in flares)
- Observation of energetic electrons, electrical fields and magnetic fields. [1469)]

The PHOTON mission is a cooperative project of FIRAS (P. N. Lebedev Physical Institute, Moscow), MEPhI (Moscow Engineering Physics Institute), YUZHNOE (Ukraine), TATA (Bombay), FTI (Ioffe Physical Technical Institute, St. Petersburg), AIP (Astrophysikalisches Institut Potsdam), FhG/IPM (Fraunhofer Gesellschaft/Institut für Physikalische Meßtechnik, Freiburg), and DLR/DFD (Neustrelitz). It uses the same S/C structure as for the CORONAS-I and -F missions. The PHOTON S/C is three-axis stabilized; the pointing accuracy to the sun is within 10 arcmin; payload mass = 650 kg; S/C mass = 2500 kg; payload power = 400 W (average); on-board data recorder, science data in S-band.

A launch of PHOTON is TBD (a new launch date of the mission will occur in the summer 2001).

Orbit: Circular high-inclination orbit, altitude = 500 km, inclination = 82.5°.

Sensor complement: [Sensors NATALYA-2M, RT-2, PENGUIN and FXM have two modes of operation, a routine mode and a flare mode (the flare mode exhibits higher activity levels, higher readout frequencies, and a higher number of energy channels)]

NATALYA-2M (Gamma-ray Spectrometer, FIRAS and MEPhI). The objective is the measurement of gamma rays in the energy range of 0.3 - 2000 MeV and of neutrons in the range of 10 - 300 MeV. The instrument consists of two spectrometers (upper and lower). The upper spectrometer is surrounded by a scintillation dome. Both spectrometers employ identical detectors, eight CsI (Tl) blocks and PMTs. Measurement of gamma rays in four energy bands (0.3 - 2 MeV, 1 - 10 MeV, 7 - 200 MeV, 0.1 - 2 GeV), and neutron-type events. The energy resolution for gamma rays is 14% (E = 0.660 MeV), 6.0% (E = 10 MeV), and 12% (E = 100 MeV). An α-source is used for absolute instrument calibration. Instrument mass = 530 kg.

RT-2 (X-ray Telescope), TATA Institute of Fundamental Research, Bombay, India). Objective: Measurement of hard X-rays in the energy range of 15 - 150 keV, hard X-rays and low-energy fluxes in the range of 150 - 10,000 keV, and gamma rays. RT-2 employs three identical detector units with three different FOVs (2°, 3°, and 5° respectively). NaI (Tl)/CsI (Na) detectors are used for the registration of X-rays. Both scintillators and collimators are surrounded by shielding. Instrument mass = 6 kg.

PENGUIN (Hard X-ray Polarimeter and Neutron Detector, FTI and MEPhI). Objective: Measurement of hard X-ray fluxes in the energy range of 20 - 150 keV (fast timing, energy spectra and linear polarization), of hard X-ray and low energy particles in the range of 20 - 1000 keV, and of neutrons in the range of 5 - 50 MeV (energy of recoil protons). Two types of scintillators are used for the registration of neutrons and photons, namely high-Z CsI (Tl) and low-Z monocrystal p-terfinil (p-TF, an organic scintillator). The assembly is covered by two plastic counters for rejection of incoming charged particles. Linear polarization measurements are based on analysis of the Compton scattering asymmetry. Instrument mass = 17.5 kg.

FXM (Fast X-ray Monitor, MEPhI). Objective: Measurement of X-ray radiation in six energy levels covering the region between 20 - 600 keV (20 - 30, 30 - 40, 40 - 50, 50 - 70, 70 - 130,

1469) Yu. D. Kotov, S. I. Nikolsky, V. I. Dranovsky, "Satellite Project PHOTON for the study of solar hard radiation," paper provided by Yu. Kotov

130 - 600 keV). FXM is a scintillation counter using a monocrystal of $YAlO_3$ (Ce) as scintillator with a fast readout capability. The minimum time (24 ns) interval for averaging the counting rate depends on the flare intensity (3 ms for large flares). The time resolution may be changed via telecommand. Instrument mass = 6 kg.

ELECTRON-5 (Charged Particle Spectrometer), MEPhI. Objective: measurement of charged particles in the following energy ranges: [electrons: 0.04 - 2 MeV, protons: 0.5 - 60 MeV, α-particles: 3 - 120 MeV, nuclei Z <8: 1.2 - 20 MeV/nucleon]. The instrument includes a dE/dx semiconductor telescope and two Geiger counters. The geometrical factor of the telescope is ~0.2 cm^2 sr. FOV = 20° for the spectrometer, FOV = 46° for the Geiger counters. Instrument mass = 15 kg.

REIS (Complex for the Measurement of Electromagnetic Fields, IZMIRAN). The complex consists of three instruments: FM-2, DMP-1, and DEP-1. The magnetometer FM-2 measures the magnetic field in the range ±65.5 μT with an accuracy of 15 nT. Complex mass = 30 kg.

Sensor	Type of Field	Components	Frequency Band	No. of Sensors
Magnetometer FM-2	magnetic	H_x, H_y, H_z	0 - 4 Hz	one
DMP-1	magnetic	H_x, H_y, H_z	5 - 1000 Hz	one
DEP-1	electric	E_x, E_y	0.1 - 1000 Hz	four

Table 441: Sensors in the REIS complex

DMP-1 Sensor		DEP-1 Sensor	
Frequency (Hz)	Sensitivity (A/m $Hz^{0.5}$)	Frequency (Hz)	Sensitivity (A/m $Hz^{0.5}$)
20	3 x 10^{-7}	0.1 - 10	10^{-4}
120	4 x 10^{-8}	20	10^{-6}
700	6 x 10^{-9}	200	10^{-7}
1000	5 x 10^{-9}	1000	10^{-8}

Table 442: Sensitivities of DMP-1 and DEP-1 sensors

EUV-PHOKA (EUV Photocathode Experiment), MEPhI, AIP, FhG/IPM. Objectives: Measurement of the solar EUV and X-ray spectrum (whole solar disk intensity variation in seven spectral bands). The instrument consists of the following elements:[1470)]

- Detector module: photodiodes (XUV silicon photodiode detectors), collimators, filters; preamplifiers, electrometers and logarithmic amplifiers in each channel; thermosensors)
- Electronic module: (input amplifiers, 12 bit A/D converter, processor unit, etc.)

Modes of operation: 1) Normal mode with a time resolution of 10 s allows an estimation of long-term variations (days, solar rotations and longer) and flare processes, 2) Flare mode with time resolution of 1 second, 3) Occultation mode with a time resolution of 0.2 s (scan of the Earth's atmosphere during the orbital exit from its shadow). Instrument data volume is 1024 - 1408 kbit/day depending on operational mode. Instrument mass = 2.5 kg

Channel No.	Wavelength (nm)	Emission Line
1	<10	
2	30.4	He II
3	58.4	He I
4	80	O II - O IV
5	102.6	Lyman-β
6	121.6	Lyman-α
7	VIS	

Table 443: Spectral parameter of EUV-PHOKA

1470) Yu. Kotov, K. Pflug, G. Schmidtke, "EUV-PHOKA,"paper provided by K. Pflug

K.9 Equator-S

Equator-S ('small') is a solar-terrestrial energy transport mission of MPE-Garching (DARA sponsored) for plasma and field measurements in several regions not covered by any of the ISTP missions, namely the low-latitude dayside magnetopause and its boundary layer, the magnetospheric boundary regions, the auroral source region, the equatorial ring current region, and the near-Earth equatorial plasma sheet. Equator - S is considered a complement to the 'Global Geospace Science (GGS) Program.' EQUATOR-S was launched from Kourou on an Ariane 44P vehicle on December 2, 1997, together with JCSAT-5. The S/C and payload are provided by Germany (PI for mission: G. Haerendel, MPE).[1471]

Within the GGS/ISTP context, simultaneous measurements in the solar wind (WIND and SOHO) and in the magnetospheric boundary regions (CLUSTER and EQUATOR-S) will allow a more quantitative understanding between input parameters and transfer efficiencies. Simultaneous measurements at equatorial (EQUATOR-S) and polar latitudes (POLAR and CLUSTER) at distances of 10 R_E will provide clues about the origin and low-latitude connection of entry layer and mantle, on the dayside, and between plasma injections from the tail and large-scale morphology and dynamics of auroral sheet currents, Alfven waves, particle beams and plasma flows, on the night side. The energy flow from the deeper tail as well as the sequence of events constituting a magnetospheric substorm can only be assessed with simultaneous measurements in the near-Earth plasma sheet (EQUATOR-S) and tail (GEOTAIL and INTERBALL).[1472) 1473)]

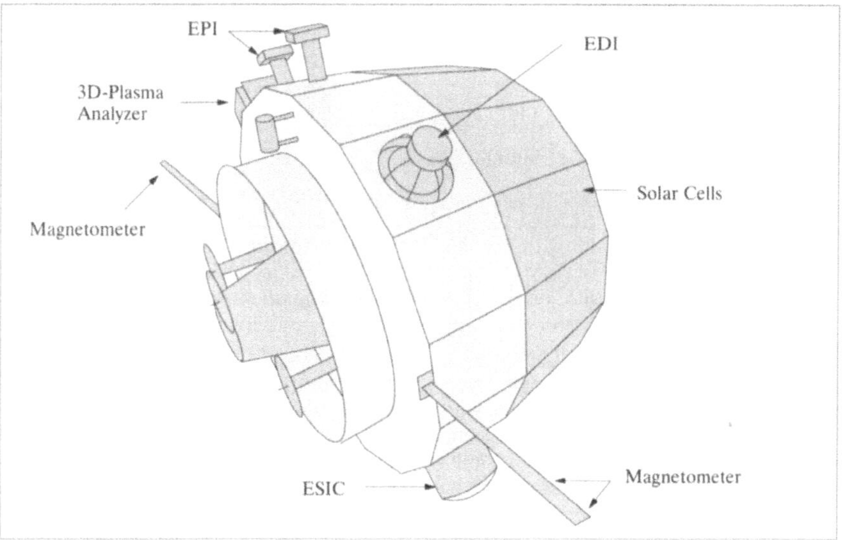

Figure 258: The Equator-S S/C model

The Equator-S spacecraft is spin-stabilized (attitude control with a magnetorquer, S/C attitude is determined from magnetometer and sun sensor data), with its spin axis oriented at

1471) "A Small Equatorial Satellite to Complement the Global Geospace Science Program - Equator - S," MPE Equator-S proposal, Sept. 30, 1991

1472) "EQUATOR-S - A Contribution to the Interagency Solar-Terrestrial Physics Programme," MPE paper provided by H. Höfner

1473) G. Haerendel, "Equator-S: The Mission and first Coordinated Measurements with GEOTAIL," Advances in Space Research, Vol. 25, No 7-8, 2000, pp. 1277-1286 right angles to the Earth-sun line, and with a spin rate of 60 rpm to allow for high temporal resolutions for the measurement of the particle distributions functions. Total S/C mass=235 kg, S/C power=70 W (16 planar solar cells mounted around the circumference); four NiCd batteries, 1 Ah each (power during eclipses); S/C design life = 2 years; payload mass = 48 kg, power = 46 W; downlink data transmission rates vary between 32 and 262 kbit/s, uplink data rate at 500 bit/s, a CRC code is used for error detection, 1.5 GByte of on-board memory storage for flexible data acquisition. RF systems: two redundant RF transmitters, S-band (2284.5 MHz); RF power=1W; two redundant RF receivers, S-band (2103.6 MHz); five switchable antennas with 4π steradians coverage, gain ≥ 3 dB.

Orbit: A near-equatorial orbit with a 10 R_E distance apogee and low perigee at 500 km altitude, inclination = 7°. Equator-S was launched into a geostationary transfer orbit (perigee=200 km, apogee=35,975 km, inclination=7°). Insertion into the final orbit with a one-impulse maneuver (perigee=500 km, apogee=65,600 km, inclination=7°). Equator-S has its own solid-propellant kick-motor (Thiokol Star 13A).

The S/C was designed, built and tested by MPE. Some assistance and in-kind contributions are received from ESA and NASA. Mission operations are provided by DLR/GSOC.

A spacecraft failure occurred on May 1, 1998, terminating all operations. The evidence indicates that the May 1 failure of the redundant processor system was caused by a latch-up in one of the memory chips of the processor system, as was the failure of the prime processor system on December 17 1997.

Sensor complement:

- **MAM** (Magnetic Field Instrument), Lead investigator: W. Baumjohann, MPE. The instrument consists of three-axes fluxgate magnetometers, one mounted at the end of a 1.8 m boom, the other 50 cm further inward, to reduce (and determine) the amount of interference from the S/C.
  The sampling rate is 128 vectors/s when only one of the magnetometers is used, and 64 vectors/s when both are utilized. The resolution is 16 bits, and ranges can be selected in steps of 4, between 250 and 64,000 nT.

- **EDI** (Electron Drift Instrument), Lead investigator: G. Paschmann, MPE. This instrument is identical in concept and design to the EDI sensor on CLUSTER. EDI measures the drift of a weak beam (<1 μA) of test electrons which is related to the ambient electric field and the gradient in the magnetic field. EDI is capable of measuring the electric field, local gradients in the magnetic field, and the magnetic field itself. EDI consists of two emitter/detector assemblies, each with a 2π FOV. E = (0.1 - 10 mV/m, <100 Hz), $|\underline{B}|$ = 5-1000 nT, emission and tracking of two electron beams.
  - Sample rate: 32 vectors/s
  - Mass: 10.8 kg
  - Power consumption: 10.3 W
  - Data rate: 4 (16) kbit/s

- **3D-Plasma Analyzer** (also referred to as 3D-A. The instrument is of WIND 3-D Plasma instrument heritage; lead investigator: G. K. Parks, U. of Washington, Seattle). The instrument consists of an electron electrostatic analyzer (EESA) and an ion electrostatic analyzer (PESA).
  Measurement of ion and electron distribution functions in 192 angle- and 56 energy-channels, in the energy-per-charge range from 10 eV to 25 keV; $\Delta E/E = 0.2$; geometric factor = 10^{-3} cm^2 sr; FOV = 180° x 12°; angular resolution = 5.6° x 5.6°. Instrument mass = 8.6 kg, power = 7.1 W, data rate = 8 (36) kbit/s.

- **EPI**=Energetic Particle Instrument (EPI is physically part of 3D-Plasma Analyzer). Lead investigator: T. Sanderson, ESA/ESTEC. The instrument consists of a pair of double-ended semiconductor detector telescopes providing 3-D measurements of electrons and ions with high sensitivity and high energy resolution. Measurement ranges: electrons from 20 keV to 226 keV; ions from 20 keV to 400 keV; instrument mass = 5.1 kg, power = 4.5 W, data rate = 2 (9) kbit/s.

- **ESIC** (Ion Composition Instrument). Lead investigator: L. Kistler, UNH, Durham, USA. This instrument is essentially a copy of CODIF (Composition and Distribution Functions Analyzer) on the CLUSTER mission (heritage of SULEICA and the 3-D Plasma Instrument flown on AMPTE). ESIC consists of a retarding-potential analyzer (RPA), followed by a toroidal energy-per-charge (E/q) analyzer with disk-shaped field of view, followed by a 20 keV post-acceleration into a time-of-flight analysis section. ESIC measures the 3-D distribution function of the major ion species ($H,^+$ $He,^{2+}$ $He,^+$ $O,^+$ and a combination of $O_2{}^+$ and NO^+). CODIF covers the energy range from about 15 eV to 40 keV/q. Instrument mass = 8.2 kg; power = 7.6 W, data rate = 6 (53) kbit/s.
- **PCD** (Potential Control Device). Lead investigator: K. Torkar, Space Research Institute, Graz. The instrument reduces the S/C potential (caused by the ambient plasma charging of the outer surfaces) to acceptable levels. The method employs beams of 6 keV indium ions to neutralize the S/C charging (see also 'ASPOC' of the CLUSTER mission). Instrument mass = 2.6 kg, power = 2.8 W, data rate = 128 bit/s.
- **SFD** (Scintillating Fiber Detector). Lead investigator: L. Adams, ESA/ESTEC. Measurement of the omnidirectional energy flux of protons and electrons. SFD makes use of the light emission property of some materials when hit by ionizing radiation. Optical fibers guide the emitted light to a photodiode operated in current mode. A logarithmic amplifier converts this detector current to an analog output voltage. Energy discrimination is achieved by using three differently shielded channels. This way electrons above 0.26, 0.4, and 1.9 MeV are measured, and photons above 6.3, 9.5, and 35 MeV, with a time resolution of 64 s. Mass = 0.4 kg, power=0.4 W, data rate= 128 bit/s.
- **GPS receiver** (Motorola Viceroy GPS receiver capable of L_1 band C/A code reception). Lead investigator: B. Eisfeller, Bundeswehr Universität, Neubiberg. A prime objective is to investigate the conditions of GPS reception above the GPS satellite constellation. Determination of S/C position. Mass=2.6 kg, power=8.3 W, data rate=512 bit/s.
- **Solar Cell Diagnostic** (ESTEC, NASA). Measurements of temperature, current/voltage characteristics (degradation by radiation). Mass=0.4 kg, data rate=128 bit/s.

K.10 EXOS (Exospheric Observations)

Japanese (ISAS) solar-terrestrial interaction program with a series of four satellites for the study of the magnetosphere and auroral phenomena.

K.10.1 EXOS-A (Kyokko)

ISAS mission with the objective to study the magnetosphere (auroral activity), plasma environment and VLF (Very Low Frequency) emissions. S/C mass = 126 kg. Launch: Feb. 4, 1978 by M-3H-2 vehicle from Kagoshima. End of EXOS-A mission in November 1979 due to the degradation of solar cells by radiation belt particle bombardment.

Orbit: apogee = 3970 km, perigee = 630 km, inclination = 65°.

Sensors:

UV-TV Camera. This wide-angle TV camera took an instantaneous auroral picture that covered almost the entire polar region, every 128 seconds during its operation over the northern polar region. About 10 to 20 pictures were obtained during one pass. More than 20,000 pictures of aurorae in UV were obtained in all.

UV Glow Spectrophotometer. Spectrophotometry of ultraviolet glows from the thermosphere, magnetosphere, and interplanetary space in the spectral range of 300 - 1300 Å.

Electron Spectrometer. Measurement of upward and downward electron fluxes along the magnetic field line of force in the energy range from several eV up to around 10 keV. Measured data show precipitation patterns with longitudinal as well as latitudinal variations.

Ion Mass Spectrometer. Objective: study of the light ion trough, atmospheric composition of ionized species of H, He, and O atoms in the thermosphere.

Ionospheric Plasma Probes. Measurements of density and temperature of ambient electrons by plasma probes, and plasma waves especially with reference to the emission associated with auroral phenomena.

ISAS also cooperated with Kyokko in an international experiment with a Scandinavian group during rocket and balloon campaigns at Kiruna, Sweden.

K.10.2 EXOS-B (Jikiken)

ISAS mission with the objective to explore geomagnetic field, plasma density, and wave particle interaction as part of the IMS (International Magnetosphere Study) program.[1474)] Launch of EXOS-B on Sept. 16, 1978. The mission lasted over 3 years.

Orbit: quasi-equatorial orbit, apogee = 30,100 km, perigee = 220 km, inclination = 31°, period = 524 minutes. Within three years the apogee gradually changed from the initial value of 30,100 km to 28,000 km, while the perigee oscillated in the range from 220 to 290 km.

Sensor complement:

SPW = Stimulated Plasma Wave Experiment (H. Oya, Tohoku University) in the plasmasphere and magnetosphere across the plasmapause.
Objective: study of plasma wave resonance mechanisms (upper hybrid resonance, plasma resonance, harmonics of electron cyclotron resonance, electrostatic resonance, sequence of diffuse resonance). Plasma waves are exited by RF pulses transmitted from the S/C to study nonlinear wave-plasma interactions.

SPW instrumentation consists of four systems: the signal generation system (high power RF transmitter for the stimulation of plasma waves in the frequency range from 10 kHz to 3 MHz), the power amplifier, the antenna system, and the receiver. The signal generator can produce two types of frequency variations that are swept frequency mode: Tx sweep, and the fixed frequency mode: Tx-Fix-n (Tx-Fix-1 = 21 kHz, Tx-Fix-2 = 90 kHz, Tx-Fix-3 = 390 kHz, and Tx-Fix-4 = 1.7 MHz).

NPW-A = Natural Plasma Wave Astronomy Mode (H. Oya, Tohoku University). Objective: observation of planetary (terrestrial kilometric radiation) and solar radio waves in the frequency range from 10 kHz - 3 MHz.

Instrumentation: the satellite is equipped with two sets of long dipole antennas arranged orthogonally to each other in the perpendicular plane of the S/C spin axis (length = 102 m tip-to-tip and 72 m tip-to-tip). The 102 m antenna is mainly used for plasma and radio wave reception, while the 72 m antenna is used for the plasma relaxation sounder and the impedance probe.

VLF = Very Low Frequency Wave Detectors (I. Kimura, Kyoto University). Objective: observation of wave particle interactions in the magnetosphere (electromagnetic and electro-

1474) Journal of Geomagnetism and Geoelectricity including Space Physics, Volume 33, No. 1, 1981, featuring EXOS-B, pp. 1-160 static waves in a frequency range between 150 Hz and 9.5 kHz, the measured range of electron is from 5 eV to 11 keV). The VLF receiver is normally connected to the 102 m dipole antenna.

DPL = Doppler Shift Measurement. Objective: detection of ionization ducts in the magnetosphere. DPL observes the Doppler shift (intensity, phase shift, and antenna capacitance at the same frequency) of an NWC (22.3 kHz) signal transmitted from North West Cape in Australia.

IEF = Impedance and Electric Field Measurement. Objectives: measurement of antenna impedance, DC electric field and AC electric field (low frequency) using a long cylindrical antenna system. Study of electron density and wave phenomena in the frequency range from 1 Hz to 450 Hz.

IEF has a cross-dipole antenna system with an originally designed length of 120 m; these antenna elements were not fully deployed due to an unstable satellite attitude. The actual length of each antenna element is as follows: A1 = 33.4 m, A2= 36.2 m, B1 = 51.7 m, B2 = 51.3 m. IEF consists of four instruments:

- IEF-1 is a frequency swept impedance probe using the A1 antenna and a frequency range from 10 kHz to 3 MHz with a swept period of 2 seconds or 8 seconds. An equivalent capacitance range measured is from 0.5 pF to 5000 pF.
- IEF-S and D are single and double Langmuir probes probe using A1 and A2 antennas, respectively.
- IEF-C is an electric field detector using A1 and A2 or B1 and B2 antennas with 1 Hz to 450 Hz electric field measurements.

MGF = Triaxial Fluxgate Magnetometer. Objective: measurement of the geomagnetic vector field. Study of relation between field-aligned currents and dynamics of the magnetosphere, and the generation and propagation mechanisms of hydromagnetic waves near the plasmapause. Resolution from 2 nT to 380 nT in four ranges. Sampling rate of either 1 second or 4 seconds. The dynamic range of MGF covers 247 nT to 46,840 nT.

ESP = Charged Particle Detectors. Objectives:

- investigate the dynamics of electrons and ions in the inner magnetosphere
- study wave-particle interaction in the inner magnetosphere together with wave measurements
- observe the response of magnetospheric plasma when an active experiment, that is SPW (high power radio wave emission) and/or CBE, is performed.

ESP is composed of an electron sensor, an ion sensor and electronic circuits. Measurements of electron and ion fluxes in the energy ranges from a few eV up to 10keV and of ion fluxes from 10 eV up to 30 keV, respectively, in the magnetosphere of L-shells from $L \cong 2.5$ to $L \cong 8$. Energy resolution $\Delta E/E = 3\%$ for electrons and 5% for ions.

CBE = Controlled Beam Experiment. Objectives: control of satellite potential by electron beam emission. Study of wave excitation (linear and nonlinear wave phenomena caused by beam-plasma interaction).

K.10.3 EXOS-C (Ohzora = Sky)

ISAS mission with the objective of upper-atmosphere exploration. Launch: Feb. 14, 1984 by a M-3S-4 vehicle from Kagoshima, Japan. S/C mass = 208 kg. The S/C is stabilized with the aid of a momentum wheel, whose spin axis is automatically controlled so that one side of the S/C faces the solar disk. Power = 160 W. The mission ended in 1987.

Orbit: apogee = 865 km, perigee = 354 km, inclination = 75°, period = 97 min.

Sensor complement:

- LAS=Limb Scanning Spectrometer (passive device) for the 1.27 μm infrared atmospheric band airglow to deduce ozone density in the altitude range of 70 - 90 km.
- UV Spectrometer for nadir observation of backscattered UV radiation (2500 - 3500 Å) to obtain ozone profiles in the altitude range of 25 - 60 km.
- Solar Image Radiometer. Measurements in several visible and near-infrared bands to detect limb absorption by stratospheric aerosols and ozone.
- IR Solar Spectrometer. Monitors stratospheric water vapor, methane, carbon dioxide, and ozone by the limb absorption technique.
- Topside Ionospheric Plasma Sounder (including a receiver). Measurement of plasma waves caused by plasma turbulence associated with precipitating particles.
- Receiver for global monitoring of electromagnetic waves radiated from power line networks and high-power electric equipment on the ground.
- Electron Temperature Probe[1475)] (for electron density and temperature measurements), consisting of four sensor electrodes (S1, S3, S5, S7) mounted at the ends of the solar cell paddles. The electrode plane is parallel to the satellite spin axis. Dayside cusp observation. Characteristics: frequency of the oscillator = 30 kHz; amplitude of the signal = 500 mV, 250 mV, 0 mV; input impedance of DC amplifiers =110 MΩ
- Energy spectrum analyzer for precipitating particles.

K.10.4 EXOS-D (Akebono)

ISAS mission for 'exospheric observations' with the objective to study the generation mechanism of auroral charged particle precipitation in the polar cusp region in comparison with the simultaneously observed electric field. The S/C is spin-stabilized at 7.5 rpm; the spin axis points toward the sun; attitude control is by magnetic torquing. Payload mass = 295.4 kg, science instruments = 97 kg (including antennas and booms). Surface conductivity: 98.9% of the S/C surface, including the solar cells, is made conductive so as to minimize electrostatic disturbances around the satellite.[1476), 1477)]
Launch: Feb. 22, 1989 with M-3SII-4 vehicle from Kagoshima, Japan. EXOS-D is operational as of the end of 1995 and is expected to continue operations for a few more years.

Orbit: elliptical polar orbit with an initial apogee into the southern hemisphere, apogee = 10,500 km, perigee = 275 km, inclination = 75°, period = 212 minutes.

Sensor complement:

EFD = Electric Field Detector (PI: H. Hayakawa, ISAS)
Objective: study of the quasi-static vector electric field along the spacecraft orbit, in particular: a) the electric field structure in the auroral ionosphere and its relationship to ion dynamics and auroral ion acceleration, b) global dynamics of the polar ionosphere, c) plasma wave instability, d) electric field in the "plasma bubble."

EFD measures the electric field by two techniques: namely the double probe and the ion beam technique. Double probe measurement from DC to 10 Hz; the ion beam measures the electric field twice per spin period.

1475) K. I. Oyama et al, "Electron Temperature Probe on Board Japan's 9th Scientific Satellite Ohzora," J. Geomagnetism and Geoelectricity, Volume 37, 1985, pp. 413-430

1476) EXOS-D (Akebono) - Japan's 12th Scientific Satellite - A Study of auroral particle acceleration processes, ISAS brochure

1477) Selected papers on EXOS-D (Akebono) Observations in Geophysical Research Letters, Volume 18, No. 2, Feb. 1991, pp. 293-352

Ion beam technique: the instrument consists of four lithium ion guns, four lithium ion detectors, a data control unit, an ion beam direction control unit, and a peripheral interface module.
Double probe technique: measurement of the electric field in the plane perpendicular to the satellite spin axis using two pairs of wire antennas (30 m). The sampling rates are 32, 8, and 2 samples per second.

MGF = Magnetic Field Detectors (PI: H. Fukunishi, Tohoku University) Objectives:

- study the relationship between fine structures of field-aligned currents and auroral forms by comparing magnetometer data with ATV data and ground-based auroral TV data.
- determine the current carriers using simultaneous magnetic field and particle data
- study the driving mechanism of field-aligned currents from simultaneous measurements of magnetic fields and electric fields
- identify plasma wave modes in the frequency range < 100 Hz

MGF is a triaxial fluxgate magnetometer (mounted at the tip of a 5 m mast) and a triaxial search coil magnetometer (mounted at the tip of a 3 m mast). The fluxgate magnetometer measures within four ranges: ±65,536 nT, ±16,384 nT, ±4,096 nT, and ±1024 nT with resolutions of: 2 nT, 0.5 nT, 0.125 nT, and 0.031 nT, respectively. Sensor drive frequency = 15 kHz; frequency response = DC to 50 Hz. Search coil magnetometer: triaxial search coil with 10^5 turns and permalloy cores, frequency response = 1-1000 Hz; sensitivity = 1 pT at 10 Hz.

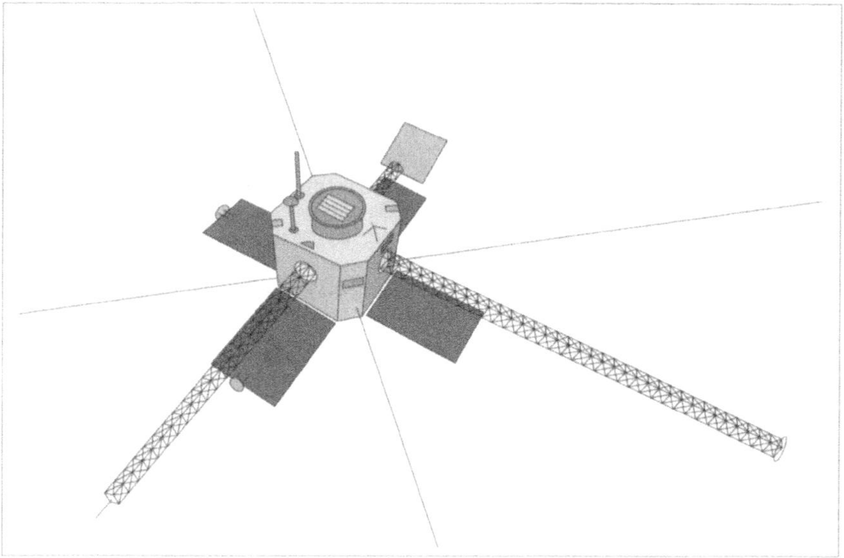

Figure 259: The Exos-D (Akebono) S/C Model

VLF = Very Low Frequency Wave Detectors (PI: I. Kimura, Kyoto University)
Objectives: study of the behavior of plasma waves associated with auroral particles, wave particles interaction mechanisms, and propagation characteristics of whistler mode and electrostatic mode waves in the magnetospheric plasma.

The instrument, consisting of the following subsystems, uses a pair of dipole wire antennas, 60 m tip-to-tip.

- WBA (wide band receiver for observation of VLF spectra), 50 Hz - 5 kHz E or B component
- MCA (multichannel analyzers), 16 channels for E, 16 channels for B
- PFX (step frequency receiver for the measurement of wave normal direction and pointing flux), band width = 50 Hz
- ELF (ELF band receiver, 4 channels), 100 Hz, 50 Hz (upper cut-off frequency)
- VIP (vector impedance measurement of the wire antennas)
- DUP (data processing unit)

Joint experiments with VLF and the HIPAS (High Power Active Stimulation) facility located in Fairbanks, Alaska, were undertaken in 1989. The ground VLF observations were successful in the sense that the VLF signals associated with the amplitude modulated HF HIPAS transmissions were clearly detected.

PWS = Plasma Wave Detectors and Sounder (Pi: H. Oya, Tohoku University)
Objectives: measure high frequency plasma waves (observe naturally occurring plasma and electromagnetic waves), conduct stimulated plasma wave experiments.

The instrument consists of:

- a dynamic spectrum analyzer (frequency range from 10 kHz to 5 MHz, frequency resolution = 1 kHz, dynamic range is about 100 dB)
- a polarization measurement system (frequency range from 20 kHz to 5 MHz)
- a pointing vector measurement system (measurement of wave amplitudes and phases of two electric field components and three magnetic field components, the analyzed wave frequency is stepped in frequency from 60 kHz to 1.3 MHz)
- a high power transmitter/sounder (coverage of a wide frequency range at high power from 20 kHz to 12 MHz, also operational modes to detect plasma density with high spectral resolution)
- an impedance measuring system (swept frequency impedance of the antenna in the plasma is measured in the same frequency range as the sounder)

The PWS sensors employ two sets of 60 m tip-to-tip antennas and a 70 cm diameter three dimensional loop antenna.

LEP = Low Energy Particle Detectors (PI: T. Mukai, ISAS). Objectives: comprehensive observations of energy and pitch angle distributions of electrons, ions, and mass per unit charge of the ions in the auroral magnetosphere. In particular:

- study of the characteristics of charged particles associated with various types of auroras
- study of particle acceleration due to electrostatic potential drops along magnetic field lines
- identification of charge carriers of the field-aligned currents
- study of wave-particle interactions
- study of particle modulations (<16 Hz) and determination of source locations
- direct detection of flux modulations of electrons and ions in the VLF and HF ranges
- study of the plasma transport between the polar ionosphere and the distant magnetosphere

LEP-S1 instrument (3-D energy/charge analyzer). Energy range = 10 eV - 16 keV for electron measurement and 13 eV/Q -20 keV/Q for ion measurement; scanning rate = 32 steps/second; energy resolution $\Delta E/E$ = 12%; FOV = 8° x 10° centered at 180°, 150°, 120°, 90°, 60° with respect to the solar direction.

LEP-S2 (3-D energy/charge analyzer). Identical to the LEP-S1. The view directions of LEP-S1 and LEP-S2 are symmetrical with respect to the spin axis of the satellite.

LEP-M (3-D energetic ion-mass spectrometer). Energy range = 1 - 25 keV/Q. Type of measurement = 135° spherical electrostatic analyzer, 40° magnetic analyzer and MCP; simultaneous measurements of mass and pitch angle distributions, differential energy spectra (32 steps) for ion species H^+, He^{++}, He^+, O^{++}, and O. Scanning rate = 16 steps/second for energy scanning; FOV = 3° x 34° with the longer dimension parallel to the spin axis of the satellite.

SMS = Suprathermal Ion Mass Spectrometer (PI: B. Wahlen, NCR, Canada)
Objectives: measurement of the distribution functions of the major as well as the minor ion constituents of the magnetospheric ion population.

SMS is a radio-frequency-type ion mass spectrometer measuring mass, energy and angular distributions of positively charged ions species. The instrument uses a time-of-flight technique to measure ion velocities and electrostatic deflection to define the ion energy per unit charge, thereby determining the ion mass per unit charge. Instrument characteristics:

- energy/unit charge (E/Q) range: - $0 < E/Q < 4$ kV
- mass/unit charge (m/Q) range: - $0.8 < m/Q < 60$ AMU/Q
- energy resolution (ΔE): - $0.05 < \Delta E < 0.2$ kV (programmable)
- mass resolution ($\Delta m/m$): - $0.05 < \Delta m/m < 0.2$ (programmable)
- angular resolution in plane to S/C spin axis ($\Delta\theta$): - $3° < \Delta\theta < 90°$ (FWHM, programmable and dependent on energy and geometric factor)
- angular resolution in plane to the scan ($\Delta\theta$): - $3° < \Delta\theta < 90°$ (FWHM, programmable and only available for low energy $E/Q < 0.1$ kV ions)

TED = Thermal Electron Detectors (PI: K. Oyama, ISAS). Objective: to measure the behavior of background plasma electrons associated with auroral phenomena. In particular:

- Detection of the field-aligned current carriers responsible for the downward current in the auroral region.
- Observation of the behavior of thermal electrons in the trough region and their relationship to double layers or to V-shaped electric fields.
- Observation of the velocity distribution function anisotropies of thermal electrons in the polar region.
- Study of the heating mechanism of thermal electrons due to wave-particle interaction.
- Observation of the behavior of thermal electrons at low altitudes in the South Atlantic geomagnetic anomaly.
- Observation of the equatorial "plasma bubble" and study of the nonthermal properties associated with this phenomenon.

TED-1 instrument characteristics (electron temperature T_e and floating point potential V_f): energy range: $T_e = 0 - 1.0$ eV, $-5V < V_f < 5V$; type of measurement: two planar probes with a shape of two semicircular disks, each mounted on the tip of a solar paddle. The surfaces of the two probes are arranged to be perpendicular to each other. Superposition of AC signal of 1 kHz.

TED-2 instrument characteristics (velocity distribution and electron density): energy range: 0 - 2.5 eV, 0 - 5 eV; density range: $10^2 - 10^6$ cm^{-3}; type of measurement: two planar probes of circular shape with their surfaces mutually perpendicular.

ATV = Visible and UV Auroral Television (PI: T. Oguti, University of Tokyo)
Objective: imaging the global aurora through two spectral windows in the VIS and UV ranges with highly sensitive devices which are essential to the observation of faint polar cusp auroras.

ATV produces auroral images in the VIS (557.7 nm) and UV (115-160 nm) spectral ranges using a despun mirror system. ATV uses a CCD imaging system of 488 (vertical) x 376 (horizontal) pixels with a FOV of 30° (vertical) x 40° (horizontal) for VIS, and 36° (vertical) x 36° (horizontal) for UV, respectively. The integration time of the CCDs may be set to 100 ms, 200 ms, 400 ms, or 600 ms.

Data:
On-board storage capability of 67 Mbit (bubble recorder). Data transmission in S-band

PCM at 65, 16, and 4 kbit/s; and UHF analog data transmission at 10 kHz and at 5 kHz. The EXOS-D (Akebono) science data will be placed in the public domain some years after acquisition.

K.11 FREJA

Swedish/German small-satellite mission with international cooperation in the instrumentation (follow-up to Viking, see K.29). The objective is: magnetospheric research, to make high-resolution measurements in the upper ionosphere and lower magnetosphere (auroral phenomena). The satellite features a magnesium structure, it was built by SSC (project management: Swedish Board of Space Activities) and weighs 256 kg (spinning disk with a diameter of 2.2 m and with a spin axis solar-oriented, spin rate = 10 rpm). An attitude knowledge of 0.5° is provided. Nominal lifetime = 2 years. Project scientists: R. Lundin of Swedish Institute of Space Physics, and G. Haerendel, MPE Garching.
Piggyback launch of Freja on Oct. 6, 1992 with a Long March (CZ-2C) vehicle from Jiuquan Spacecraft Launch Center, China. Operational mission until June 1995. FREJA was put into a standby mode as of June 30, 1995 (to be used for future special campaigns).

Orbit: apogee = 1790 km, perigee = 650 km, inclination = 65°

Sensor complement: The science payload includes six radial wire booms (1 < 15 m) and two stiff radial booms (1m and 2m).

F1: Electric Field Experiment (PI: Göran Marklund, Royal Institute of Technology, Stockholm). The objective is to further explore (and estimate the relative roles of) various suggested mechanisms for particle acceleration either by quasi-DC parallel electric fields (eg., double layers, anomalous resistivity and magnetic mirror effects) or by wave-resonance. Measurement of the electric field as potential difference between opposite probes. Two components can be measured since the probes are all in the spin axis of the satellite (the third component is deduced). The measurement parameters are as follows: maximum field = 1V/m; minimum field = 0.03 mV/m; accuracy ≈ 0.5 mV/m; sampling rate (normal mode) = 768 s^{-1}; sampling rate (burst mode) = 6144 s^{-1}. F1 measures in addition the satellite potential. The electric field is measured with the double probe technique. Six spherical probes are used, extended on three wire boom pairs in the spin plane. [1478] [1479]

F2: Magnetic Field Experiment (PI: L. Zanetti, APL, Johns Hopkins University, MD, USA). Objectives: static DC structure measurements of large and small scale current with full orbit data and 50 m resolution; vector wave measurements up to Nyquist frequencies of 64 Hz and spin axis measurements up to 256 Hz in the normal 256 kbit/s telemetry mode. Instrument characteristics:

- Triaxial ring core fluxgate sensor, 2m boom mounted
- Low noise (10 μV) analog, 16 bit A/D, S/C mounted
- Internal 1.3 Mbit RAM, event trigger, FFT S/W, oversampling
- Real-time data output: 14.3 kbit/s for 256 kbit/s TM rate, and 28.6 kbit/s for 512 kbit/s TM rate
- DC - B: 128 vector samples/s (vs/s), range ±65,000 nT, ±1 nT
- AC - B: 128 vs/s, bandpassed 1.5-128 Hz, range ±500 nT
- Spin axis: 1.5 - 256 Hz FFT, 2 Hz resolution
- Instrument mass: 3.5 kg (excluding boom and mount)
- Instrument power: 3.7 W (including DC/DC)

1478) M. André (editor) and the Freja Science Team, "The Freja Scientific Payload," Swedish Institute of Space Physics, Kiruna, May 1991

1479) "The Freja Scientific Satellite," brochure of Swedish Space Corporation

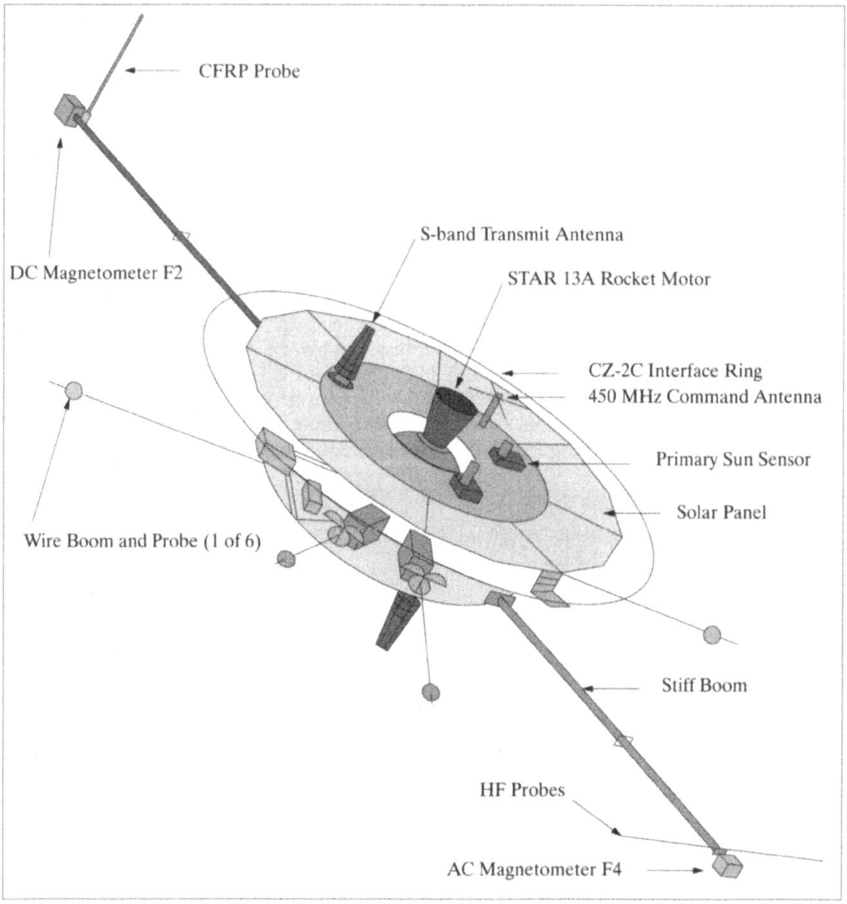

Figure 260: The Freja S/C model

F3C: Particles Experiment - Cold Plasma Analyzer (PI: B. A. Wahlen, NRC, Canada). Objectives: quantitative measurements of cold ionospheric plasma distribution; small scale auroral plasma structure investigation; identification of the plasma wave mode; investigation of low-altitude energization mechanisms; direct observation of the cloud plasma drift velocity; coordinated ground-based observations.
Instrument characteristics: boom-mounted sensor; technique: electrostatic analyzer;

- Species detected: negatively and positively charged particles (electrons and ions)
- Energy range: $0 \leq E/Q \leq 300$ eV
- Dynamic range: $10^2 \leq n \leq 10^5$ cm^{-3}
- Energy resolution: $0.05 \leq \Delta E/E \leq 0.20$ (programmable)
- Angular resolution: azimuth, $10^\circ \leq \Delta\phi \leq 90^\circ$ (programmable)
- Angular resolution: elevation, $5^\circ \leq \Delta\alpha \leq 90^\circ$ (programmable)
- Temporal resolution: ~10 ms (typical), maximum ~0.1 ms (on one selected parameter)
- Instrument mass: 5.5 kg (including boom)
- Instrument power: 8.4 W (normal)

- Telemetry rates: 16.384 kbit/s (nominal); 32.768 kbit/s (maximum)

F3H: Particles Experiment - Hot Plasma Analyzer (PI: L. Eliasson, Swedish Institute of Space Physics, Kiruna). Objectives: study of ion distributions of ionospheric origin (mainly H^+ and O^+) in the energy range from a few eV up to several keV transverse to the geomagnetic field, of electron distributions observed on auroral field lines, and study of auroral acceleration mechanisms. The instrument consists of three units: MATE (an electron spectrometer), TICS (a 3-D ion composition spectrometer, and DPU (data processing unit).

- **MATE** characteristics: Measurement of the angular and energy distributions of electrons with high temporal and spatial resolution; energy range: 0.1 keV - 120 keV; angular segments: 32; energy levels: 16; FOV/sensor head: 2° x 10°; energy resolution: 15-30% FWHM; minimum sampling time: 10 ms/energy-angle matrix; maximum data rate: 51.2 kByte/s (no data compression); normal data rate: ≈20 kbit/s; mass = 2.7 kg; power = 3.8 W.
- **TICS** characteristics: Measurement of 'hot' magnetospheric and 'cold' ionospheric ion distributions, study of the heating/acceleration of ions perpendicular to the magnetic field lines. TICS consists of a 90° spherical 'top hat' electrostatic analyzer with 360° FOV followed by a cylindrical sector magnet momentum analyzer. Energy range: 0.5-15000 eV/q; mass range: 1-40 AMu/q; angular segments: 32; energy levels: 32; energy resolution: 10% FWHM, FOV/sector head: 5° x 10°; sampling time: 10 ms/mass-angle matrix (32 x 32); FOV: 360°; geometric factor:1 - 10^4 (cm^2 s sr keV/keV) per 11° opening; time resolution: 0.5 spin period; maximum data rate: 102.4 kByte/s (no data compression); normal data rate: ≈ 20 kbit/s; mass: ≈3.3 kg; power: 4.5 W.

F4: Waves Experiment (PI: B. Holback, Swedish Institute of Space Physics, Uppsala). Objectives: study of wave modes and energy; measurement of electric, magnetic, density and electron temperature wave fields and turbulence; background plasma density and electron temperature.
The F4 instrument utilizes a number of sensors mounted on booms in the satellite spin plane. Three pairs of spherical probes (P1-P4 and P8-P9) are mounted on three wire boom pairs (switchable for electric field and plasma density measurements). A Cylindrical Langmuir Probe (CYLP) is mounted at the outer end of a stiff boom. The High Frequency (HF) experiment uses the P1 and P2 signals or the dedicated HF probes (HFa, HFb). The Search Coil Magnetometer (SCM) assembly consists of three search coils mounted at the end of a stiff boom. Measured quantities of F4 instrument:

- Electric wave fields (ΔE)
- Magnetic wave fields (ΔB)
- Plasma density ($\Delta n/n$) and temperature ($\Delta T_e/T_e$) variations
- Langmuir (DC) current to provide the cold plasma characteristics (N_e and T_e)

F5: Auroral UV Imager (PI: J. S. Murphree, University of Calgary, Canada). Objectives:

- Determination of the growth rate of distortions in discrete arcs
- Determination of the local time extent of optical substorm onsets
- Characteristics of discrete arcs from Maxwellian and non-Maxwellian source distributions (in conjunction with the particle observations)
- Determine whether episodic expansion or continuous propagation is the dominant form of substorm spiral motion
- Characterize substorm onset precursor activity.

F5 is of Viking mission heritage (see V5 in chapter K.29). F5 is an inverted Cassegrain Burch type camera [mass = 10.8 kg; power = 8.4 W (high); TM rate = 88.064 kbit/s or 44.032 kbit/s; image storage capability: 952 kByte]. F5 measurement characteristics:

- Camera optics: inverse Cassegrain, speed F/1; focal length = 22.4 mm; FOV = 25° x 20°; optical axes are 90° from spin axis.
- Spectral passband: camera 0: 134 nm - 180 nm (BaF_2 + CsI); camera 1: 125 nm - 160 nm(CaF_2 + KBr);

- Detector: intensified CCD; image size (max) 385 x 4598 pixels; pixel size = 22 μm x 22 μm; full well potential: 3 x 10^5 e^-
- Resolution: angular= 0.0783° x 0.0783°; spatial (from apogee with 2 x 2 pixels) ~ 5 km
- Exposure: 0.37 s (for 6 second spin period)

Branch	Sensor Signal	Sampling rate k-samples/s	Word (bits)	Telemetry (Word/s)	Duty Cycle (%)
HF 1 channel	HF Probes $e_1 - e_2$	8,000,000	8	1024	0.01
MF 1 channel	$e_{i,k}$ $b_{x,y,z}$	32.768	16	512	1.6
LF 4 channels	$e_{i,k}$ $b_{x,y,z}$; $\Delta n/n_{1,2,3,4}$	4.096	16	768	18.8
n up to 4 probes	$n_{1,2,3,4}$	0.128	16	128	100

Table 444: Summary of key parameters for the F4 wave analyzer

F6: Electron Beam Experiment (PI: G. Paschmann, MPE, Garching). Objectives: measurement of electric fields associated with the auroral acceleration region (method based on sensing the drift of weak electron beams). F6 is of GEOS heritage, see also EDI instrument on Cluster missions (K.7). F6 has three electron guns (three components of E-field, up to 700 samples/s) consisting of the following components:

- an electron gun with a magnetic deflection system
- a position-sensing detector
- the analog detection electronics
- the gun and deflector supplies
- interfaces with the electronics box

F7: Particle Correlator Experiment (PI: M. H. Boehm, MPE, Garching). Objectives: measurement of electron distributions, electron-electric field correlations, particle-particle correlations.

The F7 TESP/correlator instrument consists of a swept-energy electrostatic analyzer to measure the full electron distribution function at 64 ms time resolution, and wave/particle and particle/particle correlations (correlation with F4) measuring variations in the distribution function at frequencies up to several MHz. TESP measurement parameters:

- Angular channels: 32, evenly spaced
- Energy range: ~10 eV - ~30 keV
- Typical number of energy steps: 32 or 64
- Typical sampling time per step: 1 ms
- Sweep retrace time: < 1 ms
- Maximum entrance aperture deflection angle: ~30°, partially blocked by the satellite
- Total geometric factor (numerically calculated): 0.025 cm^2 ster keV/keV
- Approximate energy width (retarding grid off): 15% FWHM
- Approximate energy width (retarding grid on): ~7% FWHM at ~0.006 cm^2 ster keV/keV.

Freja Data:
On-board data storage capability. S-band downlink with data stream phase-modulated onto the carrier, data rates (direct transmission) = 262 kbit/s or 524 kbit/s, >2 Mbit/s (burst memory mode). Esrange ground station at Kiruna for S/C control and science data reception, Prince Albert station in Saskatchewan also for data reception.

Temporal resolution of Freja data:

- Fields: ≤ 0.1 ms
- Particles: ≈ 10 ms
- Images: ≈ 6 s

Spatial resolution of Freja data:

- Fields: ≤ 1 m
- Particles: ≈ 100 m
- Images: ≈ 2 km

K.12 Genesis (Solar-Wind Sample Return Mission)

A NASA mission into a halo orbit around the Lagrangian point L1 (outside of the Earth's magnetosphere) with the objective to monitor the solar wind and to collect particles of the solar wind by ion implantation into the collector materials for a duration of two years, and to return these samples for isotopic and chemical analysis in terrestrial laboratories. - The outer portion of the sun is believed to reflect the bulk composition of the original solar nebula. One way to measure the composition of the outer portion of the sun is to analyze the solar wind. The science objectives are:

- To obtain precise measures of solar isotopic abundances. Measurement of isotopic compositions of oxygen, nitrogen, and noble gases. These data enable scientists to better understand the isotopic variations in meteorites, comets, lunar samples, and planetary atmospheres.
- To obtain greatly improved measures of solar elemental abundances
- To provide a reservoir of solar matter for 21st century science research, eliminating the need for future solar wind sample return missions.

Overall mission scenario: At L1, Genesis unfurls its collector arrays and begins collecting particles of the solar wind that imbed themselves in specially designed high purity wafers. After two years, the sample collectors will be re-stowed and returned to Earth for a mid-air recovery of the sample return capsule. The samples will be stored and cataloged under ultra-pure clean room conditions and made available to the world scientific community for study. [1480)]

The Genesis mission is managed by JPL. The S/C and the SRC (Sample Return Capsule) were developed and built at LMA (Lockheed Martin Astronautics), Denver, CO. Other project partners are: LANL (Los Alamos National Laboratory - development of the sample concentrator and two solar wind monitors); NASA/MSFC [with AAIF (Advanced Analytical Instrument Facility)]; JSC (Johnson Space Center) - providing the class 10 cleanroom for assembly and post-flight disassembly of the solar wind collectors, as well as curation of the returned samples (Note: "class 10" refers to the fact that only 10 particles of contaminant per cubic meter are allowed). The head of the project is Principal Investigator Donald Burnett of the California Institute of Technology. Approximately 18 science co-investigators are from various institutions and countries, including Japan, Switzerland, and the U.K.

The Genesis S/C is sun-oriented and spin-stabilized with two fixed solar arrays. The spin rate on station is one revolution in 37.5 seconds. The S/C has three major components: the platform, the SRC (Sample Return Capsule), and the payload. The platform structure employs a truss of graphite-polycyanate composite tubes with titanium end-fittings. The truss supports a composite honeycomb-reinforced deck that forms a platform for equipment mounting. The SRC is attached to the platform via three bipod struts. The size of the platform, 1.7 m diameter x 0.5 m height, is dictated by the SRC diameter of 1.4 m, selected in turn to maximize the payload collector area. Attitude control is provided by thrusters; the attitude is sensed by redundant star sensors. The deployed length of the S/C is 7.9 m. Power of 202 W (EOL) is provided by two non-articulated solar arrays (each of 1 m^2); in addition, two batteries provide energy during trajectory correction maneuvers. The C&DH (Command and Data Handling) subsystem uses a RAD 6000 processor (redundant processors, uplink/downlink, I/O and payload/attitude control functions. A propulsion subsystem with

1480) Note: In the initial project phase (1995) the solar-wind sample return mission was called "Suess-Urey." It was renamed to "Genesis" in 1996 after the completion of Phase A.

monopropellant hydrazine thrusters (78 kg of propellant) is used for the cruise phase from Earth to the halo orbit. The S/C mass is 494 kg, the launch mass (including propellant) is 636 kg.

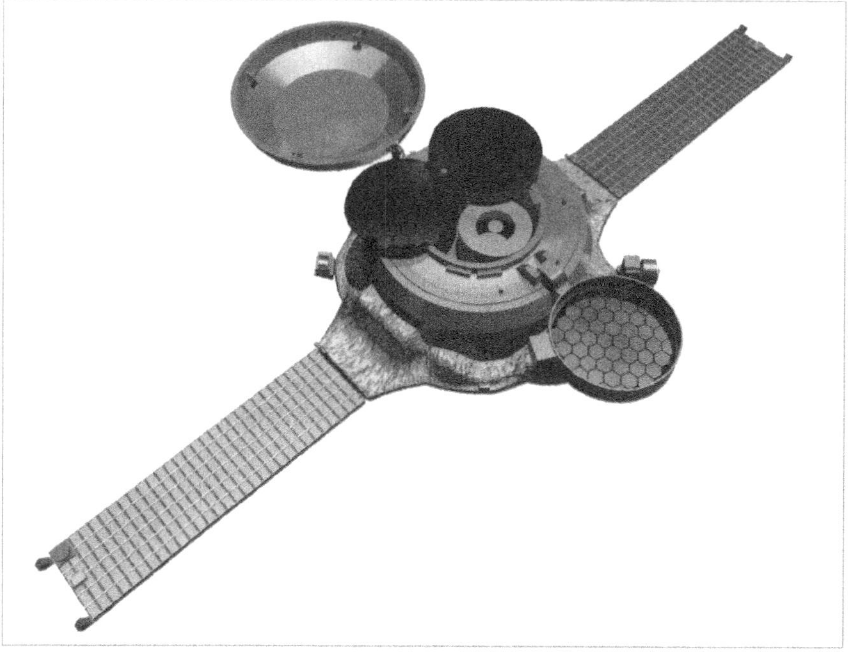

Figure 261: Illustration of the Genesis spacecraft

Communications: An on-board recorder of 1 Gbit storage capacity is provided. The uplink and downlink is in S-band at 15 kbit/s 120 bit/s during cruise phase). The low data rate of the solar wind monitors at <1 kbit/s allows infrequent (weekly) contacts with the S/C via DSN.

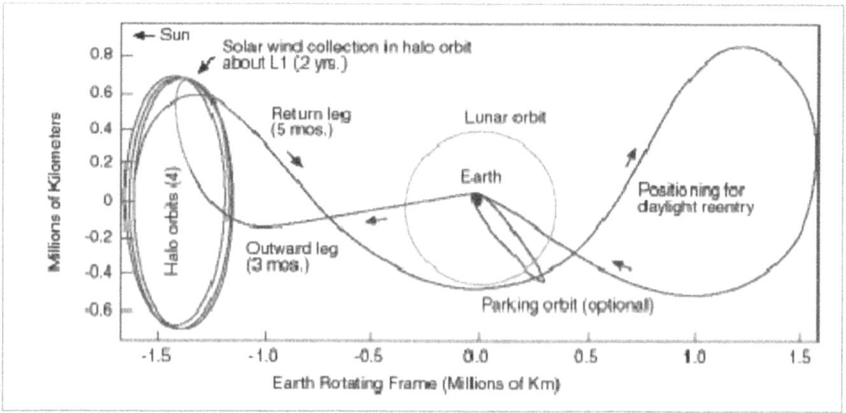

Figure 262: The Genesis trajectory

A launch of Genesis took place Aug. 8, 2001 on a Delta-7326 vehicle from Cape Canaveral Air Station, FLA.

Orbit: A halo orbit into the Lagrangian point L1, balanced gravitationally between the sun and the Earth. L1 is located about 1.5 million km from Earth, amounting to 1/100 the distance from Earth to the sun. The S/C takes approximately 3 months to arrive at the halo orbit insertion, after which it makes 4 orbits about the L1 point over a period of about 2 years. The spacecraft then moves out of the halo orbit on a trajectory taking it through the L2 point on the anti-sunward side of the Earth in order to position it for a daytime reentry. Reentry is currently scheduled for September, 2004 over the Utah Test and Training Range (UTTR), where the S/C can be tracked by the High Altitude Multiple Object Tracking System (HAMOTS). Two helicopters will be vectored to the spacecraft location and will catch the capsule as it descends on a parachute, thus avoiding a hard ground-landing. An optional three-day parking orbit allows alternate reentry dates if conditions prove to be poor for the initial reentry date. [1481] [1482] [1483]

Instrument/Payload Complement:

The Genesis mission has two primary instruments, a) collector arrays which passively collect solar wind samples and b) the concentrator. Both the collector arrays and the concentrator contain suites of ultra-high purity target materials, each of which is tailored to enable the analysis of a different element family. In addition, there are two solar wind monitors (electron and ion spectrometers) to measure and classify the solar wind. [1484]

Collector Arrays: A set of four panels (comprised of individual hexagon collectors), each capable of separate deployment to sample the different kinds of solar wind.

Each collector has the form of a flat pan with a diameter of 73 cm. In L1 halo orbit, after the canister lid opens, the stack of four arrays is rotated away from the canister, exposing the concentrator inside the canister. The topmost two of the four collector arrays, along with the array and the canister lid, are always exposed to the solar wind flux. The two lower collectors in the stack are exposed one at a time, depending on which of the solar wind regimes is being detected by the solar wind monitors. The collector retraction mechanism is inertial to the canister, consisting of a set of motors driving concentric cylindrical actuators which rotate the arrays individually about a common axis. - Each of the four collector arrays is composed of 55 hexagonal collector wafers, made of different materials and of ultra-high purity. The majority of the wafers are made of silicon; they are about 0.5 mm thick and 10 cm (point-to-point) hexagonal shape. Other wafer materials are aluminum and gold foils. The temperature of all collector arrays is kept at < 200 K.

Solar Wind Concentrator: The assembly consists of an electrostatic mirror which enhances the density of C, N and O ions from the solar wind to provide better signal-to-background ratios for these elements. An ion concentration by a factor of >20 is achieved, focusing them onto a target of 6 cm in diameter. The basic design is illustrated in Figure 263. Solar wind ions enter from the top through a series of three high transmission grids. The topmost grid is maintained at ground potential so that potentials inside the concentrator do not extend outside of it or affect the trajectories of particles prior to their entry. The next grid is biased at a variable positive high voltage. This voltage is set such that the bulk of the solar wind protons are rejected while essentially all of the heavier ions are transmitted. The third grid is mounted to a metal structure which completely encapsulates the inner region of the concentrator and is maintained at a constant negative high voltage of -8 kV. The target is a

1481) http://genesismission.jpl.nasa.gov/

1482) http://www.gps.caltech.edu/genesis/genesis3.html

1483) D. Rapp, F. Naderi, M. Neugebauer, D. Sevilla, D. Sweetnam, D. C. Wiens, D. Burnett, et al., "The Suess-Urey Mission (Return of Solar Matter to Earth)," Acta Astronautica 39, 1996, pp. 229-238

1484) A. J. Jurewicz, D. S. Burnett, R. C. Wiens, et al., "Genesis Solar -Wind Sample Return Mission: The Materials," Proceedings of the 31st Lunar and Planetary Science Conference, March 13-17, 2000, Houston, TX six centimeter disk mounted onto a structural spider that is in the plane of, and supports, the third (-8 kV) grid.[1485]

Heavy ions enter through the grounded grid, pass through the proton suppression grid, and are accelerated into the interior region by -8 kV on the accelerator grid. They are then reflected between the mirror grid and the solid electrode, and embedded deep into the target. The inset shows micro-machined steps in the electrode which serve to reflect light and heat back out into space rather than concentrating it also onto the target. Energetic ions from the sun penetrate into the collector materials and remain implanted.

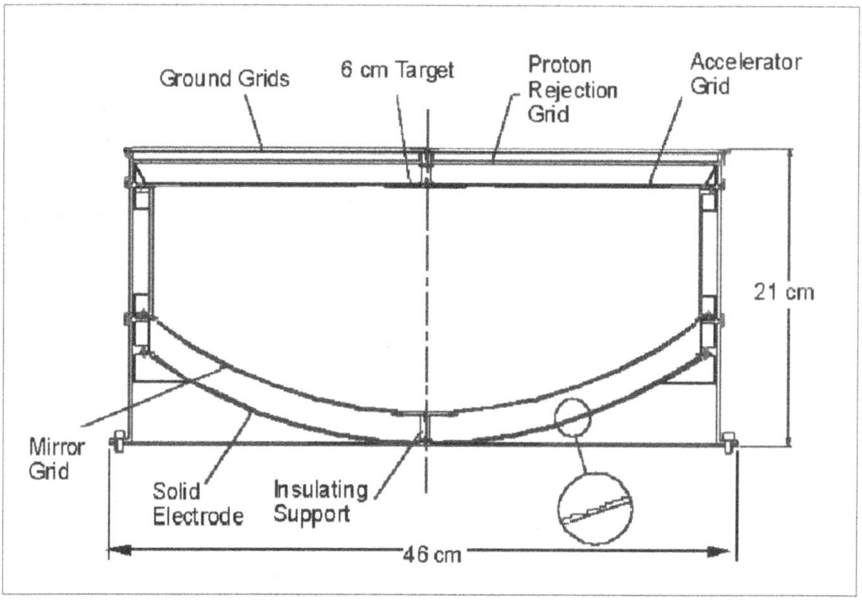

Figure 263: Schematic illustration of the solar wind concentrator

The solar wind collection system is built around collector arrays which are deployed from the SRC (Sample Return Capsule) which faces the sun.

Solar-Wind Monitors: A pair of solar wind monitor instruments (electron spectrometer and ion spectrometer) are used to detect the solar wind regime. These monitors determine the type of prevalent solar wind being measured [types: a) fast solar wind from coronal holes, b) slow interstream solar wind, and c) coronal mass ejections (CMEs)], to document the solar wind properties, and to determine the solar-wind speed. The information is needed to control the deployment of collectors and voltages on the solar wind concentrator. These monitors measure properties such as: 1) the H/He ratio, 2) the bulk wind velocity, 3) ion and electron temperatures, and 4) the angular distribution of super-thermal electrons. These data are fed into an on-board computer algorithm which determines in real-time the occurrence of a solar wind change.

- The **electron spectrometer** is a spherical-section electrostatic analyzer which uses an array of 7 channel electron multipliers, plus the spacecraft spin, to map out nearly the entire unit sphere in velocity space. The instrument has a FOV of 10°-170° with respect to the spin axis by 30° in azimuth. There are 24 available energy steps (analyzer plate

1485) D. J. McComas, B. L. Barraclough, R. C. Wiens, et al., "Solar Wind Concentrator," Measurement Techniques in Space Plasmas: Particles, Geophysical Monograph 102, AGU, 1998 voltages), logarithmically spaced from 1.5 eV to 1000 eV. Four energies are sampled per spin.

- The **ion spectrometer** is a spherical-section electrostatic analyzer which uses a Z-stack MCP (Micro Channel Plate) detector, plus the spacecraft spin, to ensure capture of the solar wind beam over a wide range of solar wind conditions. The FOV is 20° starting along the spin axis by <10° in azimuth. There are 40 energy steps per spectrum, chosen from an array of 180 available steps spanning 200 eV/q to 25 keV/q.

Sample Canister Assembly: The collector arrays and the solar wind concentrator are mounted within the sample canister. The overall configuration of the instruments has been designed to achieve this ultra-clean environment within the canister by the exclusion of practically all sources of outgassing. While the solar wind collectors reside within the canister, the mechanisms and electronics mount on the outside of the canister. The only exception to this is the collector array deployment mechanism, which by necessity must be attached to the deployable arrays within the canister. - The canister structure consists of a base and cover, each machined out of a billet of aluminum alloy. The canister measures 78 cm in outside diameter and 38 cm in height. This structure supports the collectors, mechanisms and electronic boxes for the launch and recovery loads, and transfer them to the SRC through six interface struts.

K.13 GEOTAIL

GEOTAIL is Japanese/USA (ISAS)/NASA) collaborative mission within ISTP (S/C built by ISAS, launch by NASA/GSFC). GEOTAIL inaugurates the Collaborative Solar-Terrestrial Research Program (COSTR). COSTR defines the NASA contributions to the GEOTAIL, SOHO, and Cluster missions. GEOTAIL objectives: study of the structure and dynamics of the geomagnetic tail. In particular:

- Determine the overall plasma, electric and magnetic field characteristics of the distant and geomagnetic tail.
- Determine the role of the distant and near-Earth tail in substorm phenomena and in the overall magnetospheric energy balance and relate these phenomena to external triggering mechanisms.
- Study the processes that initiate magnetic field reconnection in the near-Earth tail and observe the microscopic nature of the energy conversion mechanism in the reconnection region.
- Study plasma entry, energization, and transport processes in interaction regions such as the inner edge of the plasma sheet, the magnetopause and the bow shock, and investigate associated boundary layer regions.[1486)]

GEOTAIL launch on July 24, 1992 from Cape Canaveral with a Delta 2 vehicle.[1487)] S/C mass = 970 kg (330 kg propellant, 105 kg science payload). There is a pair of 100 m tip-to-tip antennas and two 6 m long masts. The S/C is a 20 rpm spin-stabilized cylinder of 2.2 m diameter and 1.6 m in height. Data are received at a rate of 64 kbit/s in the real time mode, and at the same time they are recorded on board at a rate of 16 kbit/s. GEOTAIL is operational as of the end of 1999. [1488)]

Orbit: GEOTAIL objectives require spacecraft measurement in two orbits: a nightside double lunar swingby GeoTail orbit to distances of 220 R_E, and a low inclination orbit at geocentric distances of about 8 to 30 R_E. GEOTAIL uses the gravity of the moon to assist its orbit on the night side of the Earth, where the magnetotail is stretched out as a result of the impact of the solar wind encountering the Earth. In this phase, GEOTAIL's orbit extends from 220 R_E (1,401,620 km) at its farthest point to 8 R_E (50,960 km) at its nearest point.

1486) "The GEOTAIL Mission," in NASAFacts, GSFC, June, 1992
1487) "Delta Launches GEOTAIL," Space News, July 27-Aug. 9, 1992, p. 12
1488) http://www.isas.ac.jp/e/enterp/missions/index.html

- Distant tail orbit: 1.75 years in distant tail configuration (double lunar swingby to an 8 x 220 R_E orbit). This orbit also allows us to study the boundary region of the magnetosphere as it skims the magnetopause at perigees. In the first two years the double lunar swing-by technique was used to keep apogees in the distant magnetotail.
- Near tail orbit: 1.45 years in near tail configuration (reduced to an 8 x 30 R_E orbit, 7.5° inclination).

The apogee was lowered down to 50 R_E in mid November 1994 and then to 30 R_E in February 1995 in order to study substorm processes in the near-Earth tail region. The present orbit is 9 Re x 30 Re with an inclination of -7° to the ecliptic plane.

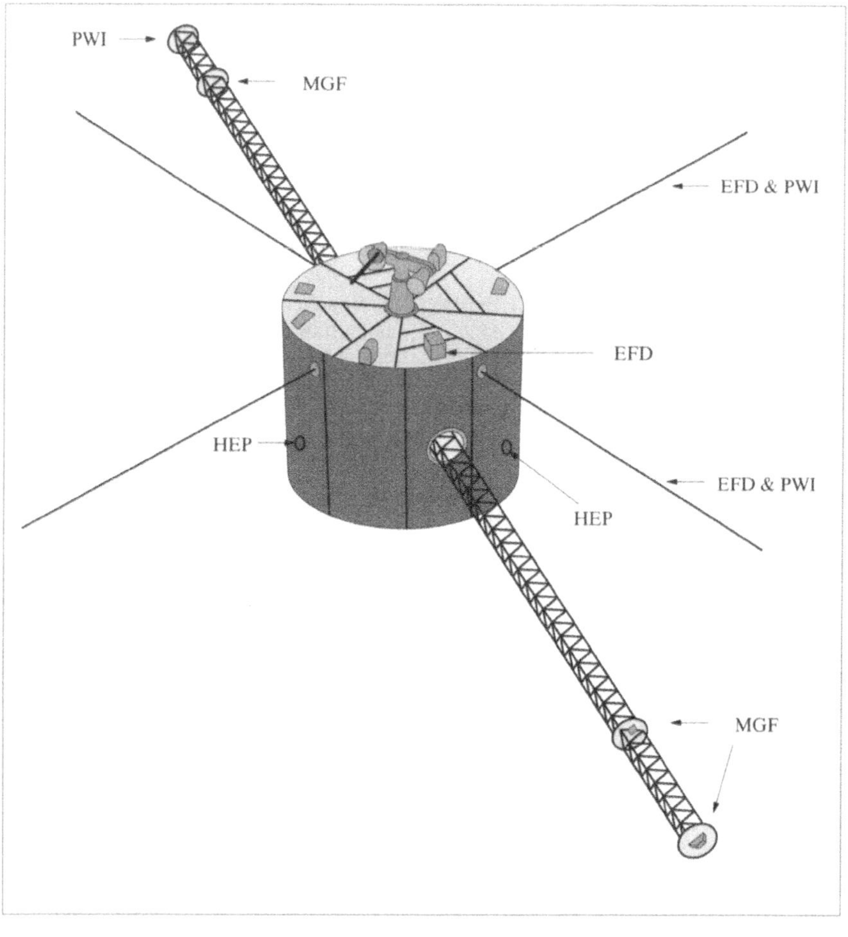

Figure 264: The GEOTAIL S/C model

Science background: The solar wind, emanating from the sun, injects plasma into the magnetosphere and transfers energy to it. Several times a day the magnetosphere undergoes a disturbance called a substorm. As the substorm grows, most of the solar energy is dissipated within the magnetosphere, ionosphere and upper atmosphere. This disturbance ultimately causes auroral displays, the acceleration of charged particles to high energies, the emission of intense plasma waves, and the generation of strong ionospheric currents that produce significant changes in the upper atmosphere. These waves and currents often result in severe problems on Earth with regard to communications, power supplies, and spacecraft electronics.

Sensor complement: [1489)]

- **EFD** = Electric Field Detector (PI: K. Tsuruda, ISAS). Objectives: study of the coupling of the E-Field in the near-Earth magnetosphere and in the ionosphere (in particular during substorms). EFD uses electric-field antenna sampling at 64 samples per second, and an electron beam technique at 2 samples per spin.

- **MGF** = Magnetic Fields Measurement (PI: S. Kokubun, U. of Tokyo, R. Lepping, GSFC, instrument sponsored by ISAS). Objectives: study of the transport dynamics of mass, momentum, and energy between the magnetospheric and ionospheric plasma. Study of merging in the magnetotail. Instrument: MGF also contains the GEOTAIL Inboard Magnetometer provided by the US.

- **HEP** = High Energy Particles Experiment (T. Doke, Waseda University, Tokyo, instrument sponsored by ISAS). Objectives: measurement of high energy particles up to 25 MeV for electrons, 35 MeV for protons, and 210 MeV/charge for ions. Measurements may indicate the plasma boundary surfaces and reflect whether magnetic field lines are open or closed.

- **LEP** = Low Energy Particles Experiment (PI: T. Mukai, ISAS). Objectives: study of the dynamics of the magnetotail plasmas, plasma circulation and its variability in response to fluctuations in the solar wind and in the interplanetary magnetic field. Measurement of electrons from 6 eV to 36 keV, and ions from 7 eV to 42 keV/ charge.

- **PWI** = Plasma Waves Investigation (PI: H. Matsumoto, Kyoto University, instrument sponsored by ISAS). Objectives: study of the wave phenomena related to plasma dynamics in the different regions on various scales (phenomena include magnetic-field-line merging, moving plasmoids, and particle acceleration). Measurement of plasma waves in the frequency range of 5 Hz - 800 kHz. PWI contains also the Multi-Channel Analyzer provided by the US.

- **EPIC** = Energetic Particle and Ion Composition Experiment (PI: D. J. Williams, APL, John Hopkins University, instrument sponsored by NASA). Objectives: measurement of the charge, mass, and energy of ions. Study of the relative importance of ion sources and mechanisms for acceleration, transport and loss of particles, the formation and dynamics of magnetospheric boundary layers. - The EPIC instrument is actually composed of two separate sensor and processing assemblies:
  - STICS (Supra-Thermal Ion Composition Spectrometer). Objective: Measurement of ions. STICS uses a quadrispherical electrostatic analyzer followed by a foil/solid state detector time-of-flight (TOF) telescope to measure charge state, mass and energy of ions with energies of 30 - 230 keV/charge. It uses an electrostatic analyzer with a geometry factor of 0.05 cm^2 sr, time of flight and energy analysis.
  - ICS (Ion Composition Subsystem). The objective is to measure mass and energy properties of energetic ions with energies of less than 50 keV to 3 MeV. ICS uses a pair of collimators with sweeping magnets to reject electrons, followed by TOF and energy analysis, with a geometry factor of 0.2 cm^2 sr. A thin foil/solid state detector electron telescope measures electrons higher than 30 keV.

- **CPI** = Comprehensive Plasma Investigation (PI: L. A. Frank, U. of Iowa, instrument sponsored by NASA). Objectives: measurement of the 3-D plasma in the Earth's mag-

1489) "GEOTAIL Instruments and Initial Results," Foreword by A. Nishida, Journal of Geomagnetism and Geoelectricity, ISSN 0022-1392, Vol. 46, 1994, netotail. The plasma data will be correlated with the magnetic field, plasma waves, electric particles, and auroral imaging data to determine magnetotail plasma dynamics. Instrument: measurement range of 1 eV - 50 keV for the Hot Plasma and Ion Composition Analyzer, and 150 eV - 7 keV energy /unit charge for the Solar Wind Analyzer. Plasma parameters, including heat flux and field-aligned current density, are measured.

Data: Two on-board recorders at 450 Mbit each. Real-time/playback transmission rates at 16.384 kbit/s, 65.536 kbit/s, or 131.072 kbit/s. Ground data reception at Usuda and Kagoshima stations (Japan) and NASA DSN.

K.14 HESSI (High Energy Solar Spectroscopic Imager)

A NASA SMEX mission, selected in Oct. 1997, and managed for NASA/GSFC by the Space Science Laboratory at UCB (University of California, Berkeley). The overall objective is to explore the basic physics of particle acceleration and energy release in solar flares. The prime observations performed are simultaneous, high resolution imaging and spectroscopy of solar flares from 3 keV X-rays to 20 MeV gamma rays with high time resolution. HESSI is a collaboration between the following institutions: GSFC, UCB (PI: Robert Lin), PSI (Paul Scherrer Institute, Villigen, Switzerland), and ETH Zürich (HESSI Experimental Data Center - HEDC). The S/C bus is designed and built by Spectrum Astro of Gilbert, AZ. [1490)]

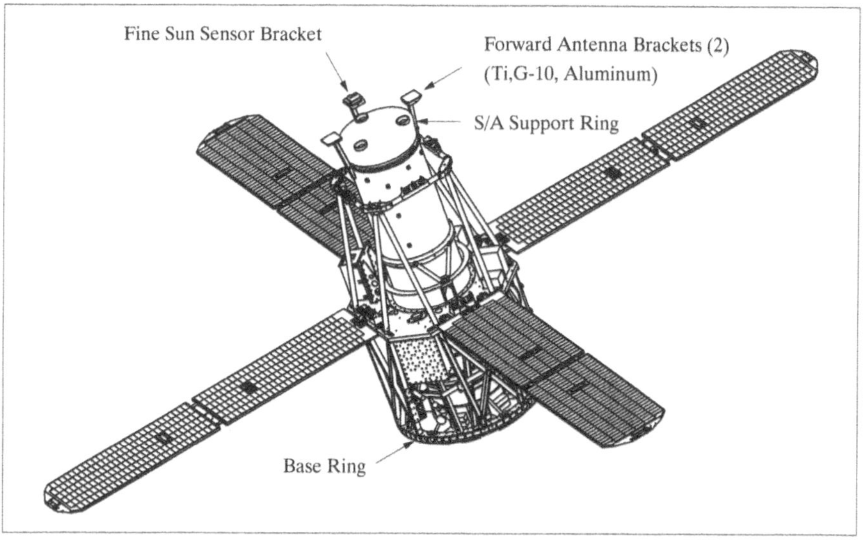

Figure 265: The HESSI S/C model

HESSI is a sun-pointing and spin-stabilized S/C spinning at 12-20 rpm. The S/C structure is 1.1 m in diameter (at base) and 2.1 m in length. Its attitude and control subsystem employs sun sensors (fine and coarse) and a magnetometer for attitude sensing and magnetic torque rods as actuators. The S/C is capable of performing autonomous sun acquisition and spin-up from any orientation. Sun pointing (precession) control is $<0.2^o$ provided by SAS (Sun Aspect System). The on-orbit mass properties adjustment direct the sun pointing error measurement to about 0.05^o. The S/C mass is 280 kg, power = 400 W. The power is provided

1490) http://hessi.ssl.berkeley.edu

by four deployable solar wings; in addition there is an NiH_2 battery energy of 15 Ah. The S/C design life is two years with a goal of three years. [1491] [1492]

S/C communication is in S-band. The downlink frequency is at 2215 MHz with selectable data rates of 4 Mbit/s, 1 Mbit/s or 125 kbit/s, with NRZ-M and BPSK data modulation. The uplink frequency is 2039.6458 MHz, the data rate is 2 kbit/s. Continuous S/C operations are supported through a UCB ground station and Mission/Science Operations Center. The data are distributed to SDAC (Solar Data Analysis Center) at GSFC and to HEDC at Zürich. There is also a complementary ground-based program supported by observatories throughout the world.

A launch on a Pegasus XL vehicle is planned for 2001 from Cape Canaveral, FL.

Orbit: circular orbit, altitude = 600 km, inclination = 38°

Sensor complement: (the instrument name is identical to the S/C name)

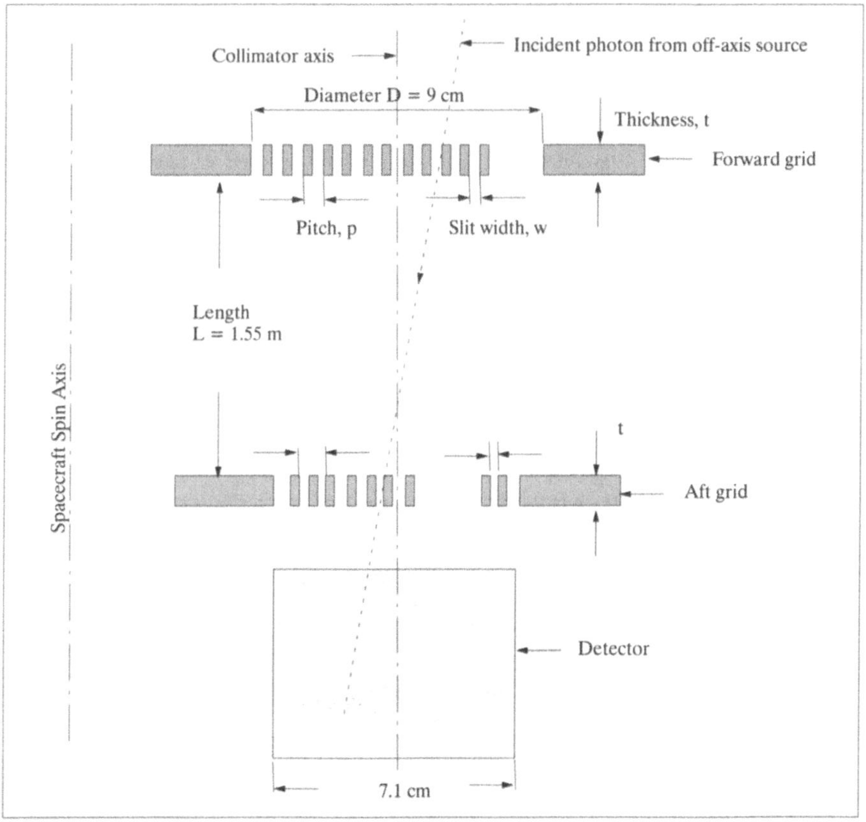

Figure 266: Schematic illustration of the HESSI grid pair parameters

HESSI (High Energy Solar Spectroscopic Imager). The objective is to obtain high fidelity color movies of solar flares in X-rays and gamma rays [imaging of solar flares in energetic

1491) http://hesperia.gsfc.nasa.gov/hessi/
1492) Information provided by Brian R. Dennis of NASA/GSFC

photons from soft X-rays (about 3 keV) to gamma-rays (up to about 20 MeV) and to provide high resolution spectroscopy up to gamma-ray energies of about 20 MeV]. The instrument employs two new complementary technologies: fine grids (molybdenum and tungsten grids with slits as fine as 20 μm wide) to modulate the solar radiation, and germanium detectors to measure the energy of each photon very precisely (about 1 keV FWHM). HESSI is an FTS (Fourier Transform Spectrometer) device using a set of nine RMCs (Rotational Modulation Collimators). Each RMC consists of two widely-spaced, fine-scale linear grids, which temporally modulate the photon signal from sources in the field of view as the S/C rotates about an axis parallel to the long axis of the RMC. The modulation can be measured with a detector having no spatial resolution placed behind the RMC. The modulation pattern over half a rotation for a single RMC provides the amplitude and phase of many spatial Fourier components over a full range of angular orientations but for a small range of spatial source dimensions. Multiple RMCs, each with different slit widths, can provide coverage over a full range of flare source sizes. An image is reconstructed from the set of measured Fourier components in exact mathematical analogy to multi-baseline radio interferometry.

Energy range	3 keV to 20 MeV
Energy resolution	<2 keV below 1 MeV to 5 keV at 20 MeV
Angular resolution	2.3 arcseconds from 3 to 100 keV, 7 arcseconds to 400 keV, 36 arcseconds above 1 MeV
Instrument mass, power	130 kg, 148 W
Instrument size	Grid support structure: 45 cm diameter, 1.7 m long Detector/cooler enclosure: 1 m diameter x 30 cm deep
Data storage capability	16 Gbit in 10 minutes (solid-state on-board recorder)

Table 445: HESSI instrument specification

The detectors are the largest currently available hyperpure (n-type) germanium detectors of size: 7.1 cm in diameter and 8.5 cm in length. They are cooled to 77 K by a single stage electro-mechanical cryocooler (an integral counterbalanced Stirling cycle cooler which provides up to 4 W of cooling at 77 K, with an input of 100 W). The detectors cover the entire X-ray to gamma-ray energy range from 3 keV to 20 MeV. The keV spectral resolution of germanium detectors is necessary to resolve all of the solar gamma-ray lines (with the exception of the neutron deuterium line, which has an expected FWHM of only 0.1 keV).

K.15 IMAGE (Imager for Magnetopause-to-Aurora Global Exploration)

The first MIDEX (Medium-class Explorer) mission of NASA/GSFC with the overall objective to study the global response of the Earth's magnetosphere to changes in the solar wind. Observations of high spatial and temporal resolution 3-D imagery of magnetospheric plasma motions. IMAGE uses ENA (Energetic Neutral Atom), ultraviolet, and radio imaging techniques to: [1493)]

- Identify the dominant mechanisms for injecting plasma into the magnetosphere on substorm and magnetic storm time scales
- Determine the directly driven response of the magnetosphere to solar wind changes
- Discover how and where magnetospheric plasmas are energized, transported, and subsequently lost during substorms and magnetic storms.

SwRI (Southwest Research Institute, PI: J. L. Burch,) of San Antonio Texas is the prime partner of NASA in this project. The S/C, built by LMMS (Lockheed Martin Missiles & Space), employs a spin-stabilized platform. It has the form of a regular octagon and measures 2.25 m in diameter and 1.52 m in height. Surface-mounted solar cells (high-efficiency, dual-junction $GaInP_2/GaAs/Ge$ cells) provide an efficiency of 20-21.5% with average pow-

1493) W. C. Gibson, et al., "IMAGE, the First of the New MIDEX Missions," Proceedings of the 13th AIAA/USU Conference on Small Satellites, Logan, UT, Aug. 23-26, 1999, SSC99-VII-2 er of 250 W (21 Ah NiCd batteries for eclipse operation). S/C mass = 496 kg. - IMAGE has a nominal spin period of 2 minutes (or a spin rate of 0.5 ± 0.01 rpm); its spin axis is perpendicular to the orbital plane. The AD&C (Attitude Determination & Control) subsystem consists of the following actuators/sensors: a magnetic torque rod, a passive nutation damper, a three-axis magnetometer for magnetic aspect information, an enhanced sun sensor, and AST (Autonomous Star Tracker) developed by LMMS. AST is mounted with its boresight 10° from the spin axis. The two-axis sun sensor provides spin rate and sun aspect angle information. MCS (Magnetic Control System) controls both spin axis orientation and spin rate. Attitude knowledge is accurate to within 0.1°. The C&DH (Command & Data Handling) subsystem design employs the MIL-STD-1553B bus which communicates with the instrument controllers over RS-422 interfaces. The CCSDS protocol is used for all internal and external communication. The 1553/CCSDS protocol design provides substantial advantages in terms of spacecraft/payload decoupling (the S/C serves mainly as a "bent pipe" for science data produced by the payload). All time synchronization between S/C and payload is accomplished exclusively over the 1553 bus. [1494) 1495)]

Mechanical subsystem	Aluminum honeycomb side panels (8), forward and aft panels, payload deck, and interior shear walls; two central load-bearing aluminum cylinders (forward and aft)
Mass	Spacecraft: 340 kg, Instrument Payload: 196 kg, Total: 536 kg
Thermal subsystem	Heat pipes in payload deck connected to radiators on S/Ct side panels; MLI blankets; thermostat- and CIDP/PDU-controlled electrical resistance heaters for payload and spacecraft operations and survival
Command & data handling subsystem	System Control Unit (SCU) consisting of RAD6000 flight computer plus modules for I/O, mass memory, command & telemetry, communications & memory, and power supply; VME backplane; 4.096 Gbit DRAM in mass memory module; Mil-Std-1553B interfaces to other observatory systems
Communications subsystem	S-band transponder; 1 medium-gain helix antenna; 2 low-gain omni-directional antennas; uplink data rate: 2 kbit/s; downlink data rate: 38 kbit/s (real-time mode); 2.28 Mbit/s (stored data playback mode)
Attitude determination & control subsystem	Spin-stabilized; closed-loop spin-rate control Sensors: Adcole 44690 sun sensor; Lockheed Martin ATC AST-201R autonomous star tracker; MEDA TAM-2A 3-axis magnetometer Actuator: 1 Ithaco magnetic torque rod
Solar arrays	Body-mounted dual-junction gallium indium phosphide/gallium-arsenide/germanium solar cell arrays
Battery	21 Ah super NiCd battery; operating range: 22-34 VDC

Table 446: Overview of S/C design features

Communication: three antennas for S-band communication with the ground. A medium-gain helix antenna and two low-gain omni-directional antennas. One of the omni antennas is mounted on the aft end panel of the spacecraft; the other is mounted together with the helix antenna on the forward panel. The helix antenna is used to transmit data from the spacecraft to the ground; the co-mounted omni antenna is used to receive uplinked commands and data. Uplink data rate = 2 kbit/s. Downlink of stored science, engineering, and housekeeping data at a rate of 2.28 Mbit/s. In addition to the playback of stored data, the S/C also continuously transmits real-time data at a nominal rate of 38 kbit/s. The real-time data is mainly for CRL of Tokyo, Japan and for NOAA (space environment weather forecasts).

A launch on a Delta II 7623-9.5 ELV vehicle took place from VAFB on March 25, 2000.

Orbit: HEO (Highly-elliptical Earth Orbit), a polar orbit, perigee = 1000 km, apogee = 7.2 R_E (45,922 km). The location of the apogee changes during the course of the two-year mission, both in latitude and, because of the Earth's revolution about the sun, in local time. At the beginning of the mission, apogee is at approximately 40° north geographic latitude and

1494) http://image.gsfc.nasa.gov/
1495) http://pluto.space.swri.edu/IMAGE/

at dusk local time. As the Earth moves around the sun, the plane of the orbit shifts relative to the Earth-sun line (by 30° of longitude each month). During IMAGE's two-year nominal mission, the line of apsides will precess over the pole and return to 40° north geographic latitude. This type of orbit permits the IMAGE instruments to image the inner magnetosphere on nearly a continuous basis. [1496)]

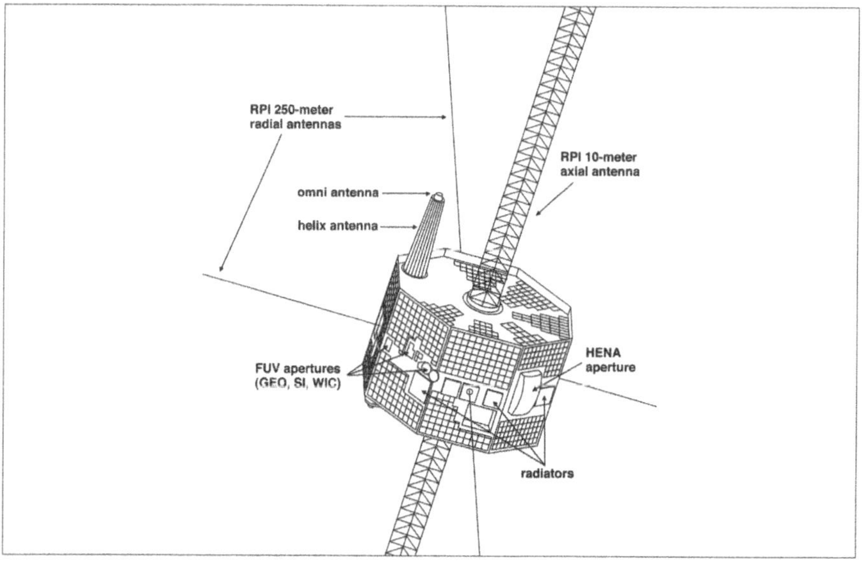

Figure 267: The IMAGE S/C model

Sensor complement:

The IMAGE S/C carries three ENA (Energetic Neutral Atom) imagers whose combined energy coverage permits the detection of ENAs with energies ranging from 1 eV to 500 keV per atomic mass unit (amu). Each neutral atom instrument generates images showing the intensity and spatial distribution of ENA emissions produced in the inner magnetosphere through charge-exchange reactions between geocoronal neutral hydrogen and various magnetospheric ion populations. Neutral atom imaging of the ionosphere and magnetosphere is possible because the Earth's geocorona acts like an imaging screen for magnetospheric and ionospheric ions. The interpretation and quantification of observed ENA signals depend upon the knowledge of the energy dependence and magnitude of the appropriate charge-exchange cross sections.

LENA (Low-Energy Neutral-Atom Imager), PI: T. E. Moore of GSFC (collaboration of GSFC, University of Maryland, University of New Hampshire, University of Denver, University of Bern, Lockheed Martin Co.). The objective is to detect ENAs in the energy range from 10-500 eV. LENA's primary role is to image the outflow of low-energy ions from the polar ionosphere. The specific objectives are to: [1497)]

- Measure neutrals without interference from electrons, ions, or UV
- Distinguish neutral protons from oxygen
- Determine ion outflow on five minute time scales over broad range of local times
- Measure energies as low as 10 eV with high counting statistics

1496) J. L. Burch, J. L. Green, S. A. Fuselier, "Mission Allows Magnetospheric Physicists to "See" the Invisible," EOS Transactions of AGU, Vol. 82, No 22, May 29, 2001, pp. 241 and 245

1497) T. E. Moore, et al., "The Low Energy Neutral Atom Imager for IMAGE," Kluver Academic Publishers, 1999

The LENA instrument consists of a collimator, a conversion unit, an extraction lens, a dispersive energy analyzer, and time-of-flight mass analyzer with position-sensitive particle detection. LENA is specifically designed of looking at and in the direction of the sun. Principle of operation: Neutral particles enter the instrument through a collimator which filters charged particles. LENA converts neutrals to negative ions through a near specular glancing reflection from a tungsten surface. Negative ions from the surface are then collected by the extraction lens which focuses all negative ions with the same energy to a fixed location. In the extraction lens, the ions are accelerated by 20 kV prior to entering the electrostatic analyzer (they are detected as they pass through a thin foil and strike an image plane detector). Finally, the ions pass into a time-of-flight/position sensing section where ion mass, energy, and angle are determined.

LENA uses electrostatic optics techniques for energy (per charge) discrimination and carbon foil time-of-flight techniques for mass discrimination. The instrument has a 90° x 8° FOV in 12 pixels, each nominally 8° x 8°. The S/C spin provides a TFOV of 90° x 360°, comprised of 12 x 45 pixels.

Energy range (incident neutral atom)	15-1250 eV
Energy resolution	E/delta E - 1 at FWHM
Mass range	1-20 amu (H^+ and O^+)
Mass resolution	M/delta M = 4 at FWHM
Angular coverage	Sampling: 8° x 8° x 12 sectors per spin: 360° x 90° in 45 x 12 samples
Angular resolution/response	8° x 8° = 12 steradians (sr)
TFOV (Total Field of View)	2.8 π sr
Pixels per image	3.2 k = 3 energy x 45 Az x 12 polar x 2 mass
Pixel physical aperture	1.0 cm^2 (A_{eff} <1 x 10^{-3} cm^{-2})
Pixel solid angle	0.02 sr per pixel x 12 pixels
Time resolution 1D polar x energy Time resolution 2D (Az x polar) x energy Time resolution 3D (Az x polar x energy)	2.7 s 120 s 1 spin period (120 s)
Dynamic range	10^4
Instrument mass, power, data rate	20.75 kg, 13.1 W (orbit averaged), 0.5 kbit/s

Table 447: Overview of LENA instrument parameters

MENA (Medium-Energy Neutral-Atom Imager), PI: C. J. Pollock of Southwest Research Institute (collaboration of SwRI, LANL, RAL, USC, UCB, University of West Virginia). Objective: Detection of ENAs in the energy range 1-30 keV. Provision of images of the ring current, near-Earth plasma sheet, and the nightside injection boundary. In addition, MENA images the ion populations of the cusp. The instrument determines the time of flight and incidence angle of the incoming ENAs; from these raw data it calculates their trajectory and velocity and generates images of the magnetospheric regions from which they are emitted. The MENA instrument (a slit-type imager) is composed of three identical sensors, mounted to a common DPU (Data Processing Unit) assembly. Each sensor is supported by two dedicated FFE (Front End Electronics) cards, time of flight (FEETOF), a pulse height analyzer (FEEPHA), and a triple-function (+10 kV, +4 kV, -1kV) high-voltage power supply. Each of the three sensors provides 1-D imaging of incident ENAs in the polar angle direction based on a spherical coordinate system. The second imaging dimension (azimuth) is obtained using collimation and S/C spin. The sensors are mounted to provide look directions that view a common azimuth (spin) angle, but are offset from one another in their polar angle field of view. The center of the sensor 2 look direction is perpendicular to the spin axis, at a polar angle of 90°. - ENAs, charged particles and photons incident from within the sensor's FOV enter through the collimator, where charged particles with energies up to 13 times the adjustable applied voltage are removed by electrostatic deflection. The remaining particles and photons must pass through a free-standing UV blocking grating, where the UV photons are removed around a very wide stop band by the optical properties of the grating. The grating structure eliminates the solar hydrogen Lyman-α (121.6 nm) light reflected from the geocorona. The detector system uses MCP detector arrays (The detector assembly consists of a Hamamatsu MCP and an anode that employs a novel "capacitive charge division" technique to determine the position of the ENA impacts on the MCP).[1498)]

The TOF for each ENA detected is determined by the FEE from the start and stop pulses triggered in the MCP. This value, together with the positions of the start and stop pulses on the MCP, is processed by the look-up tables to compute the incidence angle of an incoming ENA, the length of its path through the detector, and its velocity. The amplitude of the start and stop pulses i.e. their "pulse height," is also measured. Knowledge of both the spin phase and the incidence angle is needed to determine the position in the sky from which the detected ENAs are emitted and to produce the image of the ENA emission region.

HENA (High-Energy Neutral-Atom Imager), PI: D. G. Mitchell of JHU/APL [collaboration of APL, University of Maryland, University of Arizona, MPAe (Katlenburg-Lindau, Germany)]. Detection of ENAs in the 10-500 keV energy range. HENA imaging is focused principally on the ring current, inner plasma sheet, and substorm injection boundary. HENA is a modified version of the Cassini INCA instrument, it provides global images of ENA emissions from Saturn's magnetospheric ion populations. The two main HENA components are the sensor and the main electronics unit (MEU) The HENA sensor consists of alternately charged deflection plates mounted in a fan configuration in front of the entrance slit; three microchannel plate (MCP) detectors; a solid-state detector (SSD); two carbon-silicon-polyimide foils, one at the entrance slit, the other placed just in front of the back MCP; and a series of wires and electrodes to steer secondary electrons ejected from the foils (or the SSD) to the MCPs.[1499)]

HENA determines the velocity of the ENAs that it detects by measuring their time of flight (TOF) and trajectory through the sensor, i.e., from the entrance slit either to the back foil and 2-D imaging MCP detector or to the SSD.

Wavelength, bandpass	30.4 nm, 5 nm
Sensitivity	0.2 counts/s per Rayleigh per pixel
Field of View (FOV)	30° by 84° (instantaneous), 84° by 360° (total)
Spatial resolution	640 km x 640 km at apogee (8 Earth radii)
Imaging frequency	1 image accumulated every 10 min (= 5 S/C rolls)
Instrument mass, power	15.6 kg, 15.5 W
Thermal range	-25 to +40° C (operating), -50° to +60° C (non-operating)

Table 448: EUV instrument parameters

EUV (Extreme Ultraviolet Imager), PI: B. R. Sandel, University of Arizona. The objective is to detect solar EUV photons with a wavelength of 30.4 nm that are resonantly scattered by singly ionized helium in the plasmasphere, the torus of cold dense plasma surrounding the Earth in the inner magnetosphere. A sophisticated computer deconvolution technique is used to translate the EUV photon counts registered by the instrument into images of the plasmasphere. detect solar photons. The imager consists of three identical sensor heads mounted one above the other in a common bracket; a common electronics module. Each sensor head has a FOV of 30°, the three sensors are tilted so that their FOVs overlap, giving the imager a fan-shaped IFOV of 30° by 84°. With each rotation of the S/C, the imager completes a 360° sweep of the sky, resulting in a TFOV of 84° by 360°. Spatial resolution is about 0.6° at apogee (EUV can distinguish plasma-spheric features with scale sizes down to about 640 km). This resolving power enables the study of fine-scale density structures.

1498) C. J. Pollock, et al., "Medium Energy Neutral Atom (MENA) Imager for the IMAGE Mission," Space Science Reviews (Kluver), Vol 91, 2000, pp. 113-154

1499) http://pluto.space.swri.edu/IMAGE/HENA_description.html

FUV (Far Ultraviolet Imager), designed and built by UCB. The FUV instrument employs three detectors: [1500)]

- The Wideband Imaging Camera (WIC) to image the aurora in broad band for maximum spatial resolution day and night
- The Spectrographic Imager (SI) to measure different types of aurora and separating them by wavelength and to measure proton induced aurora by removing the bright geocorona emissions
- Geocorona photometers (GEO) to observe the distribution of the geocorona emissions to derive the magnetospheric hydrogen content responsible for neutral atom generation in the magnetosphere.

The **WIC** instrument is designed to image the whole Earth and the auroral oval from satellite distances greater than 4 Earth radii to the center of the Earth. A curved image intensifier is optically coupled to a CCD and the optics provides a FOV of 17° x 17°. Spectral range: 140-160 nm; resolution elements of < 0.1°; temporal resolution of 120 s; the size of the final images is 256 x 256 pixel elements, corresponding to spatial resolutions of < 100 km at apogee distances.

SI is designed to image the whole Earth proton aurora from satellite distances greater than 4 Earth radii to the center of the Earth. It uses a reverse Wadsworth design to select the Doppler shifted Lyman H-alpha line at 121.82 nm in the ultraviolet part of the optical spectrum and to reject the non-Doppler shifted Lyman H-alpha from the geocorona at 121.567 nm. The FOV is 15° x 15°. The temporal resolution between two images is 120 s and the size of the final images is 128 x 128 pixel elements, corresponding to spatial resolutions of < 100 km at apogee distances.

GEO is designed to measure the Hydrogen Lyman-alpha emission of the neutral atmosphere. The three photometers have a field of view of 1°x 1° and look into three different directions perpendicular to the spin axis of the satellite and tilted by ±28°. They are ultraviolet photon counters with filters giving the desired bandpass. An additional O_2 gas cell provides an excellent throughput of the Hydrogen radiation and simultaneously rejects the photons at 130.4 nm from excited neutral Oxygen. The temporal resolution between two complete measurements around 360° is 120 s.

RPI (Radio Plasma Imaging, PI: B. W. Reinisch, University of Massachusetts at Lowell). The objective is to characterize plasma in the Earth's magnetosphere utilizing imaging in the radio frequency range. The RPI instrument is a low-power radar which operates in the radio frequency bands which contain the plasma resonance frequencies characteristic of the Earth's magnetosphere (3 kHz to 3 MHz). RPI can locate regions of various plasma densities by observing radar echoes from the plasma that are reflected where the radio frequency is equal to the plasma frequency. By stepping through various frequencies for the transmitted signal, features of various plasma densities can be observed and, by fitting contours and/or magnetospheric models to the features, a 3-D specification of the shape of the magnetosphere can be created.
The RPI instrument consists of an electronics enclosure, four 250 m wire antennae with deployers (including switches and couplers), and a z-axis boom canister containing two 10 m lattice boom antennae and two preamplifiers. Instrument mass = 49.8 kg, power = 133.98 W (peak) and 30.8 W (average). [1501)]

K.16 IMP-8 (International Monitoring Platform)

This NASA/GSFC S/C is also known by the names of Explorer 50 and IMP-J. IMP-8 is the last in a series of ten IMP missions with the primary objective to perform detailed and near-

1500) http://sprg.ssl.berkeley.edu/image/sidescript.html
1501) http://image.gsfc.nasa.gov/rpi/

continuous observations of the sun/Earth environment (solar wind monitoring, measuring the plasma/field environment of the magnetosheath and the magnetotail). Launch on Oct. 26, 1973 from the Eastern Test Range by a Delta vehicle. - The objectives of the extended mission are similar to the original goals with emphasis on providing solar wind parameters as input for magnetospheric studies, as 1-AU baseline for deep-space studies, and to continue solar cycle variations studies with a single set of well-calibrated and understood instruments.

IMP-8 is a drum-shaped spacecraft of 135.6 cm diameter and 157.4 cm length. It is spin-stabilized (23 rpm). The telemetry data rate is 1.6 kbit/s (VHF-band). The S/C is being tracked by the following ground stations: Wallops Island, VA; Redu, Belgium; Tasmania, Australia, Santiago de Chile; and Hawaii.

Orbit: Geocentric elliptical, apogee = 247,267 km, perigee = 189,024 km, inclination varies between 0-55° with a periodicity of several years, period = 12.6 days, spin vector direction = normal to ecliptic. IMP-8 spends 60% or more of each 12-day orbit in the solar wind, and the rest of its time in the magnetosheath and magnetosphere.

The IMP-8 satellite is operational as of the end of 2000 (the longest NASA mission so far). The CPME and EPE instruments have been operating for 27 years. They generated a wealth of high-quality data that have led to new discoveries and have resulted in hundreds of publications. Both the CPME and EPE instruments continue to perform without problems.

Sensors:[1502)]

1) **Field Investigations**

- **GNF** = Magnetic Fields Experiment (PI: N. F. Ness, GSFC). Boom-mounted triaxial fluxgate magnetometer (mass=3.2 kg) in the range of ± 36 nT per sensor. The resolution per sensor is ± 0.3 nT. Vector measurements are performed every 40 ms.
- **GAF** = DC Electric Fields Investigation (PI: T. L. Aggson, GSFC). Instrument (mass=11.5 kg) utilizes two 60 m wire antennas to measure DC electric fields in the solar wind and magnetosheath.
- **IOF** = AC Electric and Magnetic Fields Experiment (PI: D. A. Gurnett, University of Iowa). The plasma wave instrumentation (mass=12 kg) consists of a 121.8 m tip-to-tip dipole antenna for electric field measurements (in the S/C spin plane) and a triaxial search coil for magnetic field measurements. Magnetic measurements are made in seven channels in the frequency range 40 Hz to 1.78 kHz (cycle time of 10.24 s for one set of measurements). The electric field is measured by a 15 channel spectrum analyzer operating between 40 Hz and 178 kHz, and a wide band receiver tunable to 2000, 500, 125, and 31.1 kHz.

2) **Plasma Investigations**

- **LAP** = Los Alamos Plasma Experiment (PI: S. J. Bane, Los Alamos National Laboratory). Solar Plasma Electrostatic Analyzer Experiment. The instrument (mass=6.3 kg) consists of a hemispherical plate electrostatic analyzer for the measurement of ion and electron distributions. Resolution time < 2 minutes.
- **IOE** = Low Energy Particles Investigation (PI: L. A. Frank, University of Iowa). The instrument (mass=2.6 kg) consists of a Low-Energy Proton and Electron Differential Energy Analyzer (LEPEDA) and a Geiger tube. LEPEDA measures fluxes of ions and electrons, separately and simultaneously, in each of 16 energy per charge channels (between 50 eV and 45 keV) in each of 16 azimuthal directions around the S/C spin vector. The Geiger tube measures electrons in the range > 45 keV. A full measurement cycle = 82 seconds.

1502) J. H. King, "Availability of IMP-7 and IMP-8 Data for the IMS Period," The IMS Source Book, GSFC, pp. 10-20,

- **MAP** = Solar Plasma Faraday Cup Experiment (PI: H. S. Bridge, MIT). The instrument (mass=6.5 kg) consists of a split collector Faraday cup measuring ions in the energy range of 50 eV/q to 7 keV/q in 24 steps, and electrons in the range from 22 eV/q to 2 keV/q in 21 steps. An angular distribution is taken at each energy step. The measurement sequence is controlled by the measured flux. Typical measurements are 30 seconds in the solar wind and magnetosheath.

3) **Energetic Particles Investigations**

- **EPE** = Energetic Particle Experiment (PI: D. J. Williams, NOAA). The instrument (mass=3.3 kg) is a solid-state telescope which measures fluxes of ions in four energy channels ranging from 0.05 - 0.20 MeV to 2.1 - 4.5 MeV, and electrons in the ranges 30-90 and 100-200 keV.

  The EPE particle detector assembly consists of a main magnet-detector assembly and two auxiliary detector heads. All detectors are fully depleted, surface barrier, solid state detectors, and are operated with bias voltages 1.25-1.5 times the values required for full depletion. To minimize radiation damage effects, all detectors directly exposed to ion fluxes are mounted with the aluminum contact exposed to the incoming beam. [1503)]

  The main detector assembly consists of (i) a three-element telescope (detectors A, B, and C), (ii) a sweeping magnet that keeps low energy electrons (Ee<=200-300 keV) away from the telescope, and (iii) two detectors (D and E) to detect the swept electrons. The telescope covers the proton energy range 50 keV<=Ep<=25 MeV and the alpha particle range 2.2 MeV<=Ea<=35 MeV. Detectors D and E cover the electron energy range 30 keV<=Ee<=200 keV.

- **MAE** = Ion and Electron Investigation (PI:G. Gloeckler, University of Maryland). The experiment consists of two detector systems (mass=7 kg). An electrostatic deflection spectrometer (EDS) measures the energy per charge of incident ions in several ranges between 37 and 1200 keV/q.

- **CPME** = Charged Particle Measurement Experiment (PI: S. M. Krimigis and R. B. Decker, JHU/APL). The instrument is a solid-state telescope (mass=3.9 kg) measuring fluxes of protons in 11 energy channels between 0.29 and 140 MeV, and alpha particles in 6 channels between 0.64 and 52 MeV/n. Time resolution for the measurement cycle is 10.24 s.

- **GME** = Solar and Cosmic Ray Particles Investigation (PI: F. B. Mc Donald, GSFC). The experiment (mass=11 kg) consists of three telescopes intended to measure the energy spectra and composition of solar and galactic electrons, protons, and heavier nuclei. The lowest energy, solid-state telescope measures electrons (>0.15, 0.35, 0.75 MeV), protons (between 0.05 and 25 MeV) and particles $2 \leq Z \leq 28$ (1.6-12 MeV/n). The second telescope measures protons (0.8-4 MeV) and 4-20 MeV/n nuclei with $1 \leq Z \leq 26$. The third telescope uses CsI scintillator elements to measure electrons (2-12 MeV) and Z=1 to 30 nuclei (20-500 MeV/n).

- **CAI** = Electron Isotopes Investigation (PI: E. C. Stone, JPL). The instrument (mass=8 kg) consists of an 11 element solid-state telescope for composition and spectra measurements of galactic and solar cosmic rays in the energy range of 1-40 MeV/n.

- **CHE** = Cosmic Ray & Solar flare Isotopes Investigation (PI: J. A. Simpson, U. of Chicago). The instrument (mass=7.4 kg) consists of a pair of solid-state telescopes. The main telescope measures nuclei in the energy range of 10 to 100's of MeV/n, and electrons in the range of about 2 to about 25 MeV. The second telescope measures protons and alpha particles in the 0.5-1.8 MeV/n range.

1503) http://www.jhuapl.edu/

Data: Telemetry primary transmitter (PCM) at frequency of 137.980 MHz, secondary transmitter at frequency 136.800 MHz. Science Data is archived at World Data Center A at GSFC.

The combined CPME-EPE data sets have been used to address a wide variety of scientific problems. These include studies of energetic particle activity in and around Earth's magnetosphere, solar energetic particle events, solar X-rays, shock accelerated ions and electrons in the interplanetary medium and near Earth's bow shock, iron group ions in high speed solar wind streams, the intensity gradients of galactic cosmic rays, and dynamical chaos in the magnetosphere. - The CPME measurements have been used to generate hourly averages of particle intensities published for over more than a decade in Solar Geophysical Data. These data have been used by the at-large science community for many correlative studies in a variety of scientific problems.

K.17 INTERBALL

Russian multi-spacecraft mission (IKI-RAN, Project Scientist: L. Zelenyi) with international cooperation (Austria, Bulgaria, Canada, Czech Republic, ESA, Finland, France, Germany, Hungary, Italy, Kirgizia, Poland, Romania, Slovakia, Sweden, UK, and Ukraine) on payload instruments within ISTP. Objectives: Study of solar-terrestrial transport phenomena (solar wind energy and interaction with the Earth's magnetosphere, its storage there and subsequent dissipation into the tail and auroral regions of the magnetosphere, ionosphere, and atmosphere during magnetospheric substorms). INTERBALL is coordinated with the following projects: GEOTAIL, WIND, POLAR, SOHO, FAST, and RELICT-2. [1504)] [1505)] [1506)] [1507)]

The program consists of four satellites (Prognoz series), in either a "Tail Probe" or an "Auroral Probe"configuration, each configuration is provided with a subsatellite; namely: "Tail C2-X" and "Auroral C2-A" (provided by Czech Republic). Nominal mission duration: 1 year. "Tail Probe" was launched August 3, 1995 with a Molniya-M vehicle from Plesetsk; the "Auroral Probe" launch occurred on August 29, 1996.

INTERBALL S/C data: Spin stabilized spacecraft with nominal spin rate of 0.5 rpm. Mass of Auroral Probe payload = 212 kg; mass of Tail Probe payload = 150 kg. Each satellite has three telemetry systems ; RTK (16 kbit/s max.), SSNI (32 kbit/s max.), and STO (60 kbit/s max.).
RTK = Radio Telemetrii Komplex
SSNI = System Sbora Nutschnoi Informasii
STO = System Telemetrii

Earth Orbits:

- Auroral Probe (coplanar with subsatellite C2-A, also referred to as Magion-5): Apogee = 19,196 km; perigee = 791 km; inclination = 62.8°; distance between Auroral Probe and C2-A = 10 - 1000 km.
  Status: The Auroral Probe operations stopped in Feb. 1999. The Magion-5 spacecraft is operational as of June 2001.
- Tail Probe (coplanar with subsatellite C2-X, also referred to as Magion-4): Apogee = 191,907 km; perigee = 793 km; inclination = 62.8°; distance between Tail Probe and C2-X = 10 - 10000 km.

1504) "INTERBALL - Study of Magnetospheric Plasma and Solar-Terrestrial Relations," Academy of Sciences of the USSR Space Research Institute, 1987

1505) J. Büchner, L. M. Seljenyi, "Interbol erforscht die Magnetosphäre," Astronomie und Raumfahrt, GDR, 25. Jahrgang, 1987, Heft 3, pp. 77-80

1506) "Interball Project - Magnetospheric System of 4 Spacecraft," The Solar-Terrestrial Science Project of the Inter-Agency Consultative Group for Space Science, ESA SP-1107, November 1990, pp. 61-73

1507) "Interball - Mission and Payload," document of 410 pages, by RKA, IKI, and CNES, May 1995

Status: The Tail Probe reentered the atmosphere on Oct. 16, 2000 in an operational state. The Magion-4 operations lasted until Oct. 1997.

The study objective of the subsatellites is aimed to distinguish the space and time dependence of the measured plasma parameters for moving satellite systems in moving plasmas (transport phenomena). Measurement of the correlation vector of the magnetic field, plasma 3-D flux and ion 3-D flux is from two points separated between 1 and 1000 km.

The simultaneous measurements of the same plasma parameters on the mother satellite and the subsatellites permits the determination of the space and time variations. The plasma measurements are done at very high time resolution (0.2-1 s), providing study of the fine structure of the magnetospheric boundaries. The overlapping plasma sensors further allow a study of 3-D distribution functions of ions (separately for several species) and electrons in a wide range of energy with good resolution and experiment reliability.

K.17.1 "Auroral Probe" Sensors

Project manager: R. A. Kovrazhkin, IKI, Moscow

1) **Auroral Plasma Experiments:**

- **SKA-3** = Electron/Proton Distribution Experiment (PIs: A. K. Kuzmin, F. K. Shuiskaya, IKI); objective: measurement of electron and proton distribution. Energy range = 0.03 - 15 keV; electron and ion anisotropy (M=1, 4, 16), E=30-500 keV/Q; time resolution = 1.5 and 0.2 s/sample; data rate = 3 kbit/s; mass = 32 kg

- **ION** =Ion Spectra Experiment (PIs: J. A. Sauvaud, CESR, France, R. A. Kovrazhkin, IKI, Moscow); objective: measurement of ion spectra and anisotropy (M=1, 4, 16), E=0.005-20 keV/Q; time resolution = 3.2 s/sample; data rate = 3 kbit/s; mass = 17 kg

- **PROMICS-3** = 3-D Spectrometer - Ion Composition Experiment (I. Sandahl, IRF, Kiruna, Sweden; N. Pissarenko, IKI, Moscow); ion energy range: 4 eV to 70 keV with mass separation; electron energy range: 12 eV to 35 keV; time resolution = 1.2 and 2.6 s/sample; data rate = 2 kbit/s; mass= 12.9 kg

2) **Magnetic, Electric Fields and Wave Experiments**

- **FM-3I** = Investigation of Magnetic Fields (PIs: A. Bochev, Bulgaria, V. Stiazhkin, Russia); Objective: measurements of three components in frequency range from DC to 10 Hz with two dynamic ranges: 200 nT with 0.1 nT resolution, and 1000 nT with 1 nT resolution; time resolution = 0.12 s/sample; data rate = 0.5 kbit/s; mass = 4.9 kg

- **IESP-2** = Electric Field and Intensity Experiment (PIs: M. Mogilevsky, Russia; G. Stanev, CLKI Bulgaria; S. Pezzant, France); measurement range = 0 - 50 Hz; time resolution = 0.03 s/sample; data rate = 7.2 kbit/s; mass = 6.7 kg

- **NVK-ONCH** = VLF Electromagnetic Waves Experiment (PIs: M. Mogilevsky (Russia); A. Goljavin, IZMIRAN); measurement range = 20 Hz - 20 kHz; data rate = 2 x 12 kHz; mass = 17 kg

- **MEMO** = Analyzer of Magnetic Waves (PIs: F. Lefeuvre, LPCE, France; M. Mogilevsky, IKI); measurement range < 2 MHz; time resolution = 0.25 and 120 s/sample; data rate = 20 kbit/s; mass = 14.5 kg

- **POLRAD** = Polish Radiometer - Auroral Kilometric Radio-radiation Experiment (PIs: I. Hanasz, Poland, M. Mogilevsky, IKI, Moscow); measurement range = 20 kHz - 2 MHz; data rate = 2.3 kbit/s; mass = 22.5 kg

3) **Thermal Plasma Experiments**

- **HYPERBOLOID** = Ion Mass-Analyzer (PIs: J. J. Berthelier, CRPE, France; T. Muliazchik, IKI); E = 0.0 - 100 eV; v = 0.1 - 20 km/s; species: H^+, He^+, O^+, O^{++}, N^+, N_2^+, NO^+, O_2^+; time resolution = 1 s/sample; data rate = 8 kbit/s; mass = 15 kg
- **KM-7** = Cold Plasma Experiment (PIs: J. Smilauer, Prague, Czech Republic; V. V. Afonin, IKI); measurement of of plasma electrons; temperature range < 10 eV; time resolution = 0.1 s/sample; data rate = 0.3 kbit/s; mass = 2.7 kg
- **ALPHA-3** = Trapped Ion Experiment (PI: V. Bezrukikh, IKI); N > 1 cm^{-3}; energy < 25 eV; time resolution = 1 s/sample; data rate = 0.16 kbit/s; mass = 3.5 kg
- **RON** = Ion Emitter Experiment (PIs: W. Riedler, Space Research Institute, Graz, Austria; R. Schmidt, ESA/ESTEC; Yu. Galperin, IKI); N_2^+, In^+; current = 1 - 10 μA; data rate = 0.2 kbit/s; mass = 7.5 kg

4) **Energetic Particles Experiment**

- **DOK-2A** = Electron and Proton Experiment (PIs: K. Kudela, Institute of Experimental Physics, Kosice, Czech Republic; V. M. Lutsenko, IKI); E_e = 10-400 keV; E_p = 15-1000 keV; time resolution = 1 s/sample; data rate = 0.8 kbit/s; mass = 5.5 kg

5) **Auroral Oval Image Experiments**

- **UFSIPS** = Radiation Emission Experiment (PIs:A. K. Kuzmin, IKI; K. Palazov, IKI-BAN Bulgaria;); measurement in the lines: 1304Å, 1356Å, and 1493Å; time resolution = 1 line per rotation; data rate = 0.2 kbit/s; mass = 25 kg;
- **UVAI** = UV Auroral Imager (PIs: L. L. Cogger, U. of Calgary, Canada; Yu. I. Galperin, IKI); range = 1400 - 1600Å; time resolution = 1 picture per rotation; data rate = 3 kbit/s; mass = 20 kg

6) **Subsatellite C2-A**

- C2-A (Magion-5) = Measurement of electric and magnetic fields, VLF waves, plasma and energetic particles (PIs: P. Triska, Ya. Voita, Geophysics Institute, Prague); data rate = 40 kbit/s; mass = 68.5 kg.

K.17.2 "Tail Probe" Sensors

Project manager: M. Nozdrachev, IKI, Moscow

1) **Plasma Experiments**

- **SKA-1** = 3-D Ion Distribution Measurement (O. Vaisberg, L. Avanov, IKI); energy range = 0.05 - 5 keV/Q; time resolution = 3.7 s/sample; data rate = 2 kbit/s; mass = 26.1 kg
- **ELECTRON** = 3-D Electron Distribution Function (PIs: J. A. Sauvaud, CESR, France; O. Vaisberg, N. Borodkova, IKI); range = 0.01 - 30 keV; time resolution = 120 or 3.7 s/sample; data rate = 1 kbit/s; mass = 6.5 kg
- **PROMICS-3** = 3-D Ion Composition Spectrometer (PIs: I. Sandahl, IRF, Sweden; N. Pissarenko, IKI; H. Koskinen, FMI, Finland); M = 1, 32; E = < 100 keV/Q; time resolution = 1.2 or 2.6 s/sample; data rate = 2 kbit/s; mass = 12.9 kg
- **VDP** = 3-D Ion Faraday Cups (PIs: G. Zastenker (IKI); Z. Nemechek, Charles University Prague, Czech Republic); E_i > 0 eV; time resolution = 0.1 s/sample; data rate = 1 kbit/s; mass = 4.9 kg. Omnidirectional fast measurements of the integral ion or electron flux and its direction.

- **AMEI-2** = Energy-Mass Analyzer (PIs: R. Koleva, STIL, Bulgaria; V. Smirnov, IKI); (M = 1-16); E = 0.1 - 10 keV/Q; time resolution = 120 s/sample; data rate = 0.2 kbit/s; mass = 9 kg. Measurement of energy spectra of H^+, He^+, He^{++}, O^+ and heavier ions in all directions.
- **MONITOR-3** = Solar Wind Analyzer (PIs: A. Feodrov, IKI; Ya. Shafrankoya, Charles University, Prague, CR); E = 0.4 - 15 keV/Q; time resolution = 1 s/sample; data rate = 8 kbit/s; mass = 7.8 kg. High-resolution measurements of the solar wind ion fluxes and energy-angular directions of H^+ and He^{++} in sun direction.
- **CORALL** = Wide-Range 3-D Ion Spectrometer (PIs: R. Himenez, Cuba; Yu. Yermolaev, IKI); range = 0.1 - 30 keV/Q; time resolution = 120 s/sample; data rate = 1 kbit/s; mass = 5 kg. Angular-energy ion distribution in all directions.

2) **Thermal Plasma Experiment**

- **ALPHA-3** = Ion Trap Experiment (PI: V. Bezrukikh, IKI); N > 1 cm^{-3}; E < 25 eV/Q; time resolution = 16 s/sample; data rate = 1 kbit/s; mass = 3.5 kg

3) **Energetic Particles and X-Ray Experiments**

- **SKA-2** = Spectrometric Device Complex (PIs: E. Morozova, IKI, S. Fisher, CR); low and energetic charged particle composition and anisotropy; E_e = 40 - 200 keV; E_i = 50 keV - 150 MeV; time resolution = 120 s/sample; data rate = 0.25 kbit/s; mass = 22.4 kg
- **DOK-2X** = Electron and Proton Experiment (PIs: K. Kudela, Institute of Experimental Physics, Kosice, Slovakia; V. Lutsenko, IKI); measurement of fluxes and anisotropy; E_e = 10 - 400 keV; E_i = 15 - 1000 keV; time resolution = 120 or 1 s/sample; data rate = 2 kbit/s; mass = 5.5 kg
- **RF-15** = Solar X-Ray Experiment (PIs: O. Likin, IKI; F. Farnik, Astr. Institute, CR; J. Silvester, Poland); range = 2 - 200 keV; time resolution = 0.1 s/sample; data rate = 0.01 kbit/s; mass = 10 kg
- **SOSNA-2** = Dosimeter (PIs: V. Bengin, V. Petrov, IMBP, Moscow; S. Kuznetsov, Moscow University).

4) **Fields and Wave Experiments**

Complex ASPI = Analysis of Plasma Spectra Instabilities (Project manager: S. Romanov, Science manager: S. Klimov, IKI). ASPI consists of the following instruments: OPERA, MIF-M, IFPE, ADS, PRAM; the subsatellite C2-X wave experiment is also included.

- **OPERA** = Onde di Plasma Et Radiazioni Aurorali (PIs: E. Amata, CNR/IFSI, Italy; S. Savin, IKI); measurement in frequency range of 0 - 150 kHz; time resolution = 2 or 64 sample/s; data rate = 3 kbit/s; mass = 3.7 kg. FGM-1 magnetic field data acquisition and processing.
- **MIF-M** = Multicomponent Investigations of Fluctuations of the Magnetic Field DC/AC (PIs: S. Romanov, M. Nozdrachev, IKI); frequency range = 0 - 40 kHz; time resolution = 1 or 64 samples/s; data rate = 2.5 kbit/s; mass = 8.5 kg
- **IFPE** = Investigations of Fluctuations of Protons and Electrons (PIs: J. Büchner, MPE, Berlin; H. Lehmann, DLR Berlin; S. Romanov, IKI); frequency range = 0.1 - 1000 Hz; time resolution = 1 or 64 sample/s; data rate = 2.5 kbit/s; mass = 4.9 kg.
- **ADS** = Analyzer of Dynamic Spectra (PIs: J. Juchnievicz, Space Research Institute, Poland; S. Klimov, IKI); frequency range = 0.1 Hz - 40 kHz; time resolution = 8 or 1 s; data rate = 1 kbit/s; mass = 5 kg. Nine-channel spectra-analyzer and cross-correlator, two-component measurements of current fluctuations.

- **FGM-1** = 3-D Magnetic Field (J. Rustenbach, MPE Berlin; S. Savin, IKI). Measurement range 0-25 Hz in the range of 0.25-256 nT.
- **PRAM** = Adaptive Processing of Wave Information (PIs: S. Romanov IKI); data rate = 4 kbit/s; mass = 2.4 kg, digital processing of MIF-M, IFPE and RF-15 information.
- **FM-3I**= 3-D Magnetometer (PIs: M. Nozdrachev, IKI; V. Styazhkin, IZMIRAN); measurement range = 0.2-200 nT; resolution = ±0.05 nT; time resolution = 0.12 s/sample; data rate = 0.5 kbit/s; mass = 4.9 kg.
- **AKR-2** = Analyzer of Kilometric Radiation (PIs: L. Fisher, Komensky, University of Bratislava, Slovakia; V. Grigorjeva, Astronomical Institute of Moscow University); frequency range = 100 kHz - 1.5 MHz; time resolution = 0.06 s/sample; data rate = 1.8 kbit/s; mass = 2.1 kg

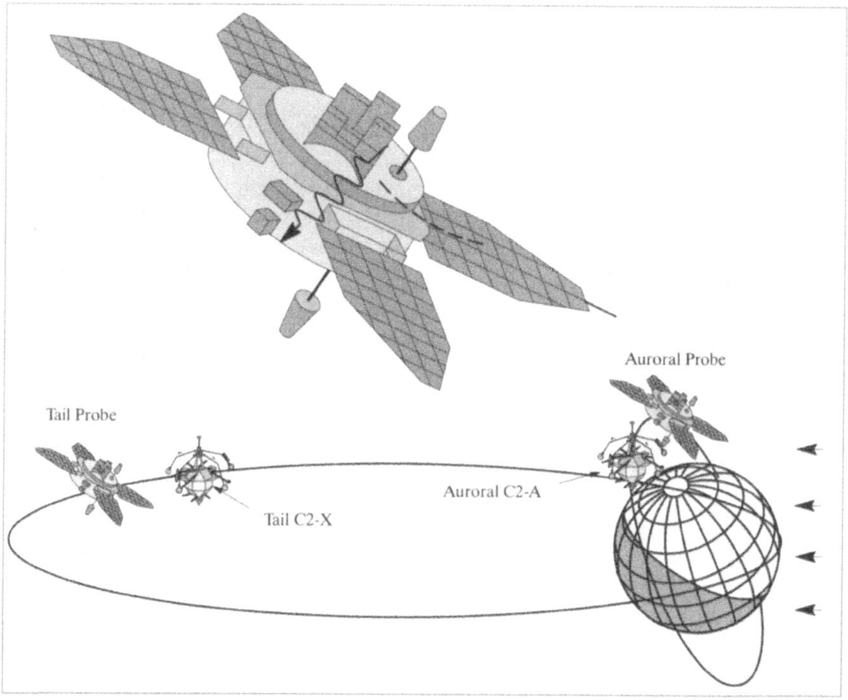

Figure 268: The INTERBALL S/C models and trajectories

5) **Subsatellite C2-X**
The subsatellite is referred to as Magion-4 (PI: P. Triska, see K.5.1). Its objective is to monitor the propagation of electron beams and plasma injected by the parent spacecraft as well as to monitor plasma of natural origin. Measurement of electric and magnetic fields, waves, plasma and energetic particles (PIs: P. Triska, Ya. Voita, Geophysics Institute Prague, Czech Republic); data rate = 20 kbit/s; mass = 58.7 kg; wave experiment (magnetic field: 0.1 Hz - 20 kHz; electric field: 0.1 Hz-250 kHz; plasma current: 0.1 Hz - 256 kHz). Note, C2-X is incorporated in the ASPI Complex.

Magion-4 (as well as all other Magion S/C) was operated and controlled via the Panská Ves Station, about 60 km north of Prague, Czech Republic. [1508]

K.18 ISEE (International Sun-Earth Explorer)

A NASA/ESA cooperative program involving three satellites. The ISEE-1 and ISEE-3 S/C were the principal contributions of NASA, while ISEE-2 was built and managed by ESA.

K.18.1 ISEE-1 and -2 Mission

Objectives: Observation of the near-Earth magnetosphere and its boundaries, better understanding of many phenomena, such as the Earth's bow shock, the magnetosheath and magnetopause, interactions between the tail and aurorae, and particle populations and flows in the tail.
Joint launch of ISEE-1 and -2 S/C on Oct. 22, 1977 (by NASA). Highly eccentric orbit around the Earth with an apogee of 23 R_E. Both S/C in the same orbit plane with a controllable separation distance, inclination = 28.9°. Both S/C reentered during Sept. 1987.

Sensors:[1509] Only a very brief overview is given; the reader is referred to the references for more information. Note: the ISEE sensor acronyms are 3-letter words: The first two letters refer to the PI followed by one letter designating the S/C: ISEE-1 = M (Mother), ISEE-2 = D (Daughter), ISEE-3 = H (Heliocentric).

ANM/AND = Electrons & Protons Instrument (ISEE-1 and -2, PI: K. A. Andersen, UCB). Objectives: Study of the varies energetic particle phenomena found in the Earth's magnetosphere, magnetopause, magnetosheath, bow shock, and upstream medium. Measurement over a wide range of energies, from ~1.5 to 300 keV for both electrons and protons.

BAM/PAD = Fast Plasma Experiment (on ISEE-1 and -2, PIs: S. J. Bame, Los Alamos Scientific Lab, G. Paschmann, MPI Garching). Three electrostatic analyzers (with 90° spherical section) provide electron and proton measurements. Each instrument uses a divided secondary emitter system to intercept the analyzed particles. ISEE-1 carries also a solar wind experiment (SWE) to measure solar wind ions with high resolution.

Protons	5 eV - 40 keV	ISEE-1
Electrons	5 eV - 20 keV	ISEE-1
Ions	50 eV - 40 keV	ISEE-2
Electrons	5 eV - 20 keV	ISEE-2

Table 449: Measurement ranges of BAM/PAD

FRM/FRD = Low Energy Protons & Electrons (ISEE-1 and -2, PI: PI: L. A. Frank, U. of Iowa) Objective: study of directional intensities of positive ions and electrons over a large solid angle. Energy range: 1 eV ≤ E/Q ≤ 50 keV in 63 bands with 17% resolution.

GUM/GUD = Plasma Wave Investigation (ISEE-1 and -2, PI: D. A. Gurnett, U. of Iowa). Objective: Study of wave/particle interaction in the Earth's magnetosphere and in the solar wind. The instrument on ISEE-1 uses three electric dipole antennas with lengths of 215 m, 73.5 m, and 0.6 m for the electric field measurements, and a triaxial search coil antenna for magnetic field measurements. The ISEE-2 instrument uses two electric dipoles with lengths of 30 m and 0.6 m, and a single-axis search coil antenna for magnetic field measurements.

1508) P. Triska, et al., "Ground-Based Control and Telemetry Station Panská Ves: 20 Years of Operation with Scientific Microsatellites," IAA 2nd International Symposium on Small Satellites for Earth Observation, Berlin, April 12-16, 1999, pp. 147-150
1509) Special issue on 'Instrumentation for the International Sun-Earth Explorer Spacecraft' in IEEE Transactions on Geoscience Electronics, Volume 16, No.3, July 1978

Magnetic field levels	10 - 100 kHz (3 axis, 16 channels)
Electric field levels	10 Hz - 10 kHz (3 axis, 12 channels)
Sweep frequency spectrum	10 kHz - 200 kHz (128 steps), analysis of the electric field signals

Table 450: GUM/GUD parameters

HAM = Plasma Density Experiment (ISEE-1, PI: C. C. Harvey, ESTEC)
Objective: Study of total electron density by means of radio techniques.
The propagation experiment measures the phase velocity of a radio wave of frequency 683 kHz and 272.5 MHz (transmitted from the Mother S/C ISEE-1 and received on the daughter S/C, ISEE-2).

RUM/RUD = Fluxgate Magnetometer Experiment (ISEE-1 and -2, PI: C. T. Russell, UCLA). Objective: Study of the dynamic plasma (magnetospheric phenomena) and field environment of the Earth.
Magnetometer is mounted on a 3 m boom. The instrument has two commandable ranges of $\pm 256\ \gamma$ and $\pm 8192\ \gamma$ with an accuracy of 0.025%.

EGD = Solar Wind Ion Experiment (ISEE-2, PIs: E. Egidi, G. Moreno, CNR Frascati). Objective: Study of the transient phenomena in the solar wind to obtain a spatial gradients of the interplanetary plasma.
The instrument is based on two identical hemispherical electrostatic energy selectors for the measurement of positive ions in two different energy windows.

- Ions: 50eV/q - 25 keV/q
- Electrons: 35 eV - 7 keV

HEM = VLF Wave Propagation Experiment (ISEE-1, PI: R. A. Helliwell, Stanford) Objective: Study of VLF-wave-particle interactions in the magnetosphere (note: VLF = Very Low Frequency in the 10 - 30 kHz range). A second goal is the determination of the effects upon energetic particles in the magnetosphere of electrical power transmission line radiation. The instrument setup consists of three separate elements:

- a broadband VLF receiver on ISEE-1
- a broadband VLF transmitter located at Siple station in the Antarctic
- ground stations in the Antarctic and Canada

HPM = DC Electric Field Experiment (ISEE-1, PI: J. P. Heppner, GSFC)
Objective: Study of the transfer mechanisms (mass, momentum, and energy at the magnetopause), in particular the spatial extent and variability of the zone of strong electric fields, or fast convection in adjacent magnetospheric regions.
Instrument: 8 channel spectrum analyzer. Measurement ranges: 0.1 Hz - 3200 Hz in 9 steps.

HOM = Low Energy Cosmic Ray Experiment (ISEE-1, PI: D. Hovestadt, MPI Garching). Objective: Measurement of elemental abundances, charge state composition, energy spectra, and angular distributions of energetic ions in the energy range of 2 keV/charge to 80 MeV/nucleon, and of electrons between 75 - 1300 keV. The instrument consists of three sensor systems:

- ULECA is an electrostatic deflection analyzer, its energy range from about 3 to 560 keV/charge
- ULEWAT is a double dE/dX versus E thin-window flow-through proportional counter/ solid-state detector telescope covering the energy range from 0.2 to 80 MeV/nucleon (Fe).
- the ULEZEQ sensor consists of a combination of an electrostatic deflection analyzer and a thin-window proportional counter. The energy range is 0.4 MeV/nucleon to 6 MeV/nucleon. Objective: collection of composition data in the trapped radiation zone.

MOM = Quasi-Static Electric Field Experiment (ISEE-1, PI: F. S. Mozer, UCB). Objectives:

- study of the quasi-static electric field over a dynamic range of 0.1 - 200mV/m
- study of wave electric fields at frequencies <1000 Hz with a sensitivity <1μV/m $(Hz)^{1/2}$ at all frequencies
- study of plasma density and temperature

Measurements are made of the potential difference between a pair of 8 cm diameter viterous carbon spheres which are mounted on the ends of wire booms and are separated by 73.5 m in the spin plane of the satellite.

OGM = Fast Electron Spectrometer Experiment (ISEE-1, PI: K. W. Ogilvie, GSFC) Objective: Study of three-dimensional plasma distribution in the solar wind, magnetosheath, outer magnetosphere, and near tail regions.
Instrument with three energy ranges: 7.5-512 eV, 11-2062 eV, and 109-7285 eV.

SHM = Ion Composition Experiment (ISEE-1, PI: R. D. Sharp, Lockheed, Palo Alto) Objective: Study of the composition of the hot magnetospheric plasma. Ion composition of the ring current, the plasma sheet, the plasmasphere, the magnetosheath, and the solar wind in order to establish the origin of the plasmas in the various regimes of the magnetosphere and to identify mass and charge dependent acceleration, transport, and loss processes.
The instrument consists of two ion mass spectrometers which can be operated independently. The spectrometers point 5° above and 5° below the ISEE-1 spin plane. Measurement ranges: 1 AMU to > 150 AMU in 64 channels at each of 32 energy channels covering the energy per charge range from 0 to ~17 keV/e.

WIM/KED = Medium Energy Particles Experiment (ISEE-1, -2, PI: D. J. Williams, NOAA, Boulder CO). Objective: Study and identify the physical mechanisms of medium energy particles associated with acceleration, source and loss processes, and boundary and interface phenomena throughout the orbits of ISEE-1 and -2.
The experiment consists of the WIM instrument (Wide Angle Particle Spectrometer and a Heavy Ion Telescope) on ISEE-1 and the KED instrument (five sensor systems mounted at various angular positions with respect to the S/C spin axis) on ISEE-2.

- Protons: 20 keV - 2 MeV in 8 channels, in 16 channels	ISEE-1
- Electrons: 20 keV - 1.2 MeV in 8 channels, in 16 channels	ISEE-1
- Protons: 20 keV - 2 MeV in 12 channels	ISEE-2
- Electrons: 20 keV - 300 keV (to 1.2 MeV for 90° unit)	ISEE-2

K.18.2 ISEE-3 Mission

Objective: Measurement of the ingredients of the interplanetary medium: the properties of the solar-wind plasma, magnetic and electric fields, solar and heliospheric charged particles and galactic cosmic rays. Exploration of the distant geomagnetic tail.[1510)]
ISEE-3 S/C: Launch Aug. 12, 1978 (NASA) by Delta vehicle from Cape Canaveral.

Orbit: ISEE-3 was first placed 1.5 million km (~240 Earth radii, R_e) sunward from the Earth in a halo orbit around the so-called Lagrangian Point L_1 (also referred to as liberation point). At L_1 the S/C co-rotated with the Earth around the sun during the course of each year.
ISEE-3 was redirected in mid-1982 from its L_1 position (which it had occupied for more than 4 years) to study the distant Earth's magnetic tail throughout most of 1983. After that ISEE-3 was redirected into a heliocentric orbit to fly past the Giacobini Zinner comet on Sept. 11, 1985, and upstream of Halley's comet on March 28, 1986. The S/C was renamed ICE (International Cometary Explorer) in 1984.

Sensor complement:
The ISEE-3 payload consisted of 13 instruments provided by both US and European groups.

1510) K. P. Wenzel, "Earth's Distant Geomagnetic Tail Explored by ISEE-3 Spacecraft," ESA Bulletin 37, 1984 pp. 46-50

- **ANH = X-Rays and Electrons Instrument** (PI: K. A. Anderson, UCB)
  - Measurement of solar flare X-ray bursts and transient cosmic gamma-ray bursts. A proportional counter and scintillation detector cover the energy range from 5 - 228 keV.
  - Measurement of electrons from ~2 keV to ~1MeV with high energy and angular resolution. (Study of interplanetary and solar electrons in the energy range between the solar wind and galactic cosmic rays).
- **BAH = Solar Wind Plasma Experiment** (PI: S. J. Bame, Los Alamos Scientific Lab) Two electrostatic analyzers (with 135° spherical section) provide electron and ion measurements. Each instrument uses a divided secondary emitter system to intercept the analyzed particles.

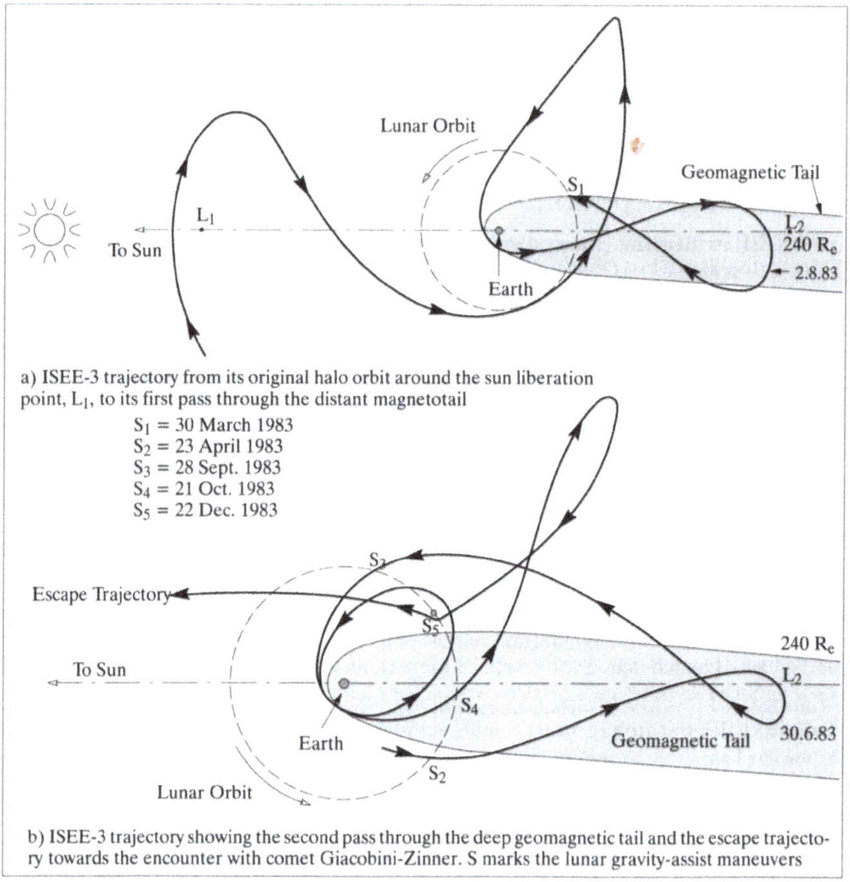

a) ISEE-3 trajectory from its original halo orbit around the sun liberation point, L_1, to its first pass through the distant magnetotail

b) ISEE-3 trajectory showing the second pass through the deep geomagnetic tail and the escape trajectory towards the encounter with comet Giacobini-Zinner. S marks the lunar gravity-assist maneuvers

Figure 269: ISEE-3 spacecraft trajectory overview

- **HKH = High Energy Cosmic Ray Experiment** (PI: H. H. Heckman, UCB) Multidetector cosmic ray experiment to identify the charge and mass of incident cosmic ray nuclei from H through F_e species (over energy ranges from 20 to 500 MeV/nucleon).

- **HOH = Low Energy Cosmic Ray Experiment** (PI: D. Hovestadt, MPI Garching)
Objective: Study of nuclear and ionic composition of solar, interplanetary, and magnetospheric accelerated and trapped particles.
Measurement of elemental abundances, charge state composition, energy spectra, and angular distributions of energetic ions in the energy range of 2 keV/charge to 80 MeV/nucleon, and of electrons between 75 - 1300 keV.
- **DFH = Low Energy Proton Experiment** (PI: R. J. Hynds, Imperial College, London).[1511] Objective: Study of low energy protons from a solar flare to relate particle fluxes measured near the Earth to fluxes in the upper corona (investigation of the gross scale of coronal control).
DFH experiment to measure low energy protons in the energy range from 35-1600 keV.
- **MEH = Cosmic Ray Electrons and Nuclei** (PI: P. Meyer, U. of Chicago). Objective: Study of the long and short-term variability of cosmic ray electrons and nuclei.
Measurement of the energy spectrum of cosmic electrons in the range of 5-400 MeV. In addition, determination of the energy spectra and relative abundances of nuclei from protons in the iron group (energies from 30 MeV/n to 15 GeV/n).
- **OGH = Plasma Composition Experiment** (PI: K. W. Ogilvie, GSFC)
Objective: Study of the dynamics and energetics of the solar wind acceleration region.
Ion mass spectrometer for the measurement of ionic composition of the solar wind.
- **SCH = Plasma Wave Instrument** (PI: F. L. Scarf, TRW, Los Angeles)
Objective: Study of interplanetary wave-particle interactions in the spectral range from 1 Hz to 100 kHz.

  Measurements of magnetic field and electric field components on long booms (90 m tip to tip). Magnetic field levels: 8 channels, 60 dB range, 20 Hz - 1 kHz. Electric field levels: 16 channels, 80 dB range, 20 Hz - 100 kHz.
- **SBH = Radio Mapping Experiment** (PI: J. L. Steinberg, Meudon Observatory, Paris)
Objectives: a) monitoring the solar wind flow and perturbations of the magnetic field in conjunction with simultaneous measurements on ISEE-1 and -2 (bow shock, magnetopause, neutral sheet), and b) propagation studies of particle fluxes and shock waves in the solar wind (large scale structure of the magnetic field).
Measurement of the interplanetary scintillation of natural radio sources using two dipole antennas, one in the spin plane (90 m tip to tip) and one along the spin axis (15 m tip-to-tip). Each of these antennas drives two radiometers (10 kHz bandwidth and 3 kHz bandwidth).
- **SMH = Helium Vector Magnetometer** (PI: E. J. Smith, JPL)
Objective: Continuous observation of the interplanetary magnetic field near 1 AU (structure, direction, polarity north-south component, magnitude, dynamic phenomena).
Boom-mounted magnetometer sensor (3 m) with the following characteristics:
  - 8 dynamic ranges of: ±4, ±14, ±42, ±144, ±640, ±4000, ±22000, ±140000 γ
  - frequency response: 0 - 3 Hz within three bands (0.1 - 1, 1 - 3, and 3 - 10 Hz) to measure fluctuations parallel to the S/C spin axis.
- **STH = Heavy Isotope Spectrometer Telescope, HIST** (PI: E. C. Stone, CIT)
Objective: measurement of the isotopic composition and energy of solar, galactic, and interplanetary cosmic ray nuclei for the elements Li through Ni in the energy range from ~5 to 250 MeV/nucleon.
Charge, isotope, and energy range: Z 3 - 28 (Li to Ni); A 6 - 64 (^{6}Li to ^{64}Ni)
Mass resolution: Li 0.065 - 0.83 proton masses; Fe 0.18 - -0.22 proton masses.

1511) A. Balogh, R. J. Hynds, J. J. van Rooijen, G. A. Stevens, T. R. Sanderson, K. P. Wenzel, "Energetic Particles in the Heliosphere - Results from the ISEE-3 Spacecraft," ESA Bulletin 27, 1981, pp. 4-12

- **TYH = Medium Energy Cosmic Ray Experiment** (PI: T. Y. von Rosenvinge, GSFC) Objective: measurement of the charge composition of nuclear energetic particles over the energy ranges from ~1 - 500 MeV/ nucleon, and charges from Z=1 to Z=28. The experiment consists of two telescopes. The combined charge, mass, and energy intervals covered by these two telescopes are as follows:
  - Nuclei charge of energy spectra: Z = 1-30, energy range 1-500 MeV/nucleon
  - Isotopes: Z=1, ΔM=1, from 4-70 MeV/n
Z=2, ΔM=1 from 1-70 MeV/n
Z=3-7, ΔM=1 from 30-140 MeV/n
  - Electrons: ~2-10 MeV
  - Anisotropies: Z=1-26 (1-150 MeV/n for Z=1,2); Electrons: 2-10 MeV

K.19 POLAR

NASA/GSFC [1512) 1513)] solar-terrestrial mission within GGS and ISTP with the objective to study the ionospheric role in substorm phenomena and in the overall magnetospheric energy balance (measurements of plasma entry and transport into the northern dayside cusp regions at high altitudes and over the southern polar cap at low altitudes, global imaging of the northern auroral zone).

The Polar launch occurred on February 24, 1996 with a Delta II vehicle from Vandenberg Air Force Base. Spacecraft mass = 1297 kg (269 kg propellant, 264 kg science payload). Nominal lifetime = 3 years. S/C is spin-stabilized at 10 rpm (a smaller platform of the S/C is despun and can be pointed to maintain the viewing field of certain instruments), 2.4 m diameter, 1.8 m height, cylinder with conductive surfaces. Solar arrays provide 440 W, including 186 W for payload. The S/C was built by Lockheed Martin Astro Space.

Orbit: Polar elliptical orbit with an apogee of 9 R_E (56,500 km) and a perigee of 2 R_E (11,500 km); inclination = 86°; period = 18 hours.

Sensor complement: [1514)]

1) Observation of local electromagnetic fields in the low frequency range

MFE = Magnetic Field Experiment (PI: C. Russell, UCLA). Objectives: Study of the coupling of the solar wind and the magnetosphere through currents driven in the polar cusp (energy and momentum exchange with the magnetosphere at the cusp-magnetosheath interface).

Instrument: two fluxgate magnetometers; sampling of the magnetic field within the frequency range from 0 - 50 Hz. Magnetic field strengths between 10^{-6} and 0.6 Gauss are detectable.

EFI = Electric Fields Instrument (PI: F. Mozer, U. of California, Berkeley). Objectives: Study of the electric field structure of the high-latitude magnetosphere, the cusp, and the plasma mantle.

EFI is a dual-probe instrument. Sampling of the electric field between 0 and 20 kHz. Electric field strengths between 0.1 and > 1000 mV per meter at a rate of 40 samples per second in the normal mode, more than 1000 samples per second in the burst mode. The burst will be coordinated with HYDRA and TIDE instruments.

PWI = Plasma Wave Instrument (PI: D. Gurnett, U. of Iowa). Objectives: Study of wave/ particle processes mediated by electromagnetic turbulence (momentum transfer in the

1512) "The Solar-Terrestrial Science Project of the Inter-Agency Consultative Group for Space Science," ESA SP-1107, November 1990, pp. 11-15

1513) "ISTP Global GEOSPACE Science - Energy Transport in Geospace," ESA/NASA/ISAS brochure, 1992 of GSFC

1514) http://www-spof.gsfc.nasa.gov/istp/polar/polar_inst.html

geospace system, particularly in the boundaries).
The PWI instrument samples electric field noise above the highest EFI frequencies well into the radio band. Magnetic-loop and search coils are used to sample the magnetic fluctuations above the highest frequencies detectable by MFE (identification of the characteristic modes of plasma behavior).

The PWI employs seven distinct sensors for detecting the electric and magnetic fields of plasma waves. These sensors consist of: a pair of orthogonal two-sphere electric antennas in the spin plane of the S/C with sphere-to-sphere separations of 100 m and 130 m respectively, shared with the EFI; a short two-sphere electric antenna aligned along the spacecraft spin axis with a sphere-to-sphere separation of 14 m, also shared with EFI; a triaxial magnetic search coil mounted on the end of a rigid stacer boom, shared with EFI; and a magnetic loop antenna mounted on the same boom and oriented parallel to the 100 m electric antenna in the spin plane. The spheres on the electric antennas are 9 cm in diameter and each contains a high-impedance preamplifier that provides signals to the boom deployment mechanisms. Amplifiers in the deployment mechanisms buffer signals to EFI and PWI independently. Each of the three magnetic search coils consists of two bobbins mounted on high permeability micro-metal cores 40 cm long. The loop sensitivity constant is 385 micro V/nT Hz and the resonance frequency is 45 kHz.

HYDRA = Hot Plasma Analyzer Experiment (PI: J. Scudder, GSFC). Objectives: Study of low-energy electrons in the global magnetic topology. Study of electron and ion signatures that accompany geomagnetic substorms, auroral arcs, field-aligned currents, and particle precipitation.

The HYDRA ensemble of electrostatic analyzers is one part of the HYDRA instrument. It measures electrons and ions in energy per unit charge between 1 eV and 30 keV and observes them in 12 directions simultaneously. In addition, a device with a position-sensitive array, called 'parallel plate analyzer,' images electrons within and near the magnetic field with a 1.5° angular resolution, energy ranges between 10 eV and 10 keV. Sampling rates: 2/second.

TIDE/PSI = Thermal Ion Dynamics Experiment / Plasma Source Investigation (PI: T. Moore, MSFC). Objectives: Study of low-energy ions (transport mechanisms) to evaluate the ionosphere as a source of plasma for the magnetosphere.
TIDE samples ions extracted from the ionosphere (whose mass is determined by a time-of-flight scheme). Energy range: 0.1 eV per charge to 100 eV per charge.

2) Observation of particle populations associated with electromagnetic fields

TIMAS = Toroidal Imaging Mass-Angle Spectrograph (PI: E. Shelley, Lockheed Palo Alto Research Lab.). Objectives: Study of the properties, location, and morphology of the polar cusp, which is the principal source region for entry of solar wind plasma and hot ionospheric plasma into the magnetosphere. The TIMAS instrument samples ions of resolved mass that are either energized ions of ionospheric origin or stored particles of solar wind origin. Energy range: 50 eV - 30 keV. Sampling rate is 10 times per minute (one per satellite spin). TIMAS data is used in combination with data from TIDE and from SWICS on the WIND satellite.

Parameter	Value	Parameter	Value
Energy range	0.015-32 keV/e	Time resolution	0.375 (s, 2-D); 3 (s, 3-D)
Energy resolution	0.08 (ΔE/E)	Instrument mass, power	15.7 kg, 14 W
Mass range	1-32 (AMU/e) 64 chan.	Data rate	4.1 kbit/s
IFOV	2° x 157° azimuth 10° elevation	Solid angle coverage Angular resolution	4 pi x 0.98 sr 11.25° x 11.25°

Table 451: Key TIMAS instrument parameters

CAMMICE = Charge and Mass Magnetospheric Ion Composition Experiment (PI: T. Fritz, Los Alamos National Lab.). Objectives: Study the mechanisms that control the energization, storage, and precipitation of particles in the high-latitude magnetosphere. CAMMICE permits determination of the composition of major ion constituents in the near-Earth plasma sheet and in the ring current. Energy ranges between 6 keV per charge and 60 MeV per ion. For full angular distribution, CAMMICE measures at a rate of 10 samples per minute, or once per spin of the S/C. The angular resolution approaches 0.2°.

CEPPAD (Comprehensive Energetic-Particle Pitch Angle Distribution), PI: B. Blake, The Aerospace Corporation, El Segundo, CA). Objectives: Investigation of quantitative information on the sources, energization, transport, and losses of energetic particles in the magnetosphere. Measurement of the rate of particle precipitation into the Earth's upper atmosphere. The CEPPAD investigation uses a variety of techniques to provide detailed energy spectra and angular distributions of energetic particles. The instrument consists of three packages. Two are spacecraft body mounted and the third is located on the despun platform. The first body-mounted package consists of the Imaging Proton Sensor (IPS) and the Digital Processing Unit (DPU). The second consists of the Imaging Electron Sensor (IES) and the High Sensitivity Telescope (HIST). The single despun platform package is the SEPS (Source/Lose-Cone Energetic Particle Spectrometer). The IPS, IES and HIST all use the body-mounted DPU. The SEPS sensor is independent of the body mounted sensor and contains a separate digital processing unit. This approach was taken because of the limitations inherent in communicating between the spacecraft body and despun platform.

- **IPS** (Imaging Proton Sensor). The IPS measures protons over the energy range from about 14 - 1500 keV in 16 energy bands over nine separate look directions. It uses a spectrometer which incorporates MSSD (Microstrip Solid State Detector) having a planar configuration with six individual elements. The nine detectors are arranged to be at the polar angles: 10°, 30°, 50°, 70°, 90°, 110°, 130°, 150°, and 170° with respect to the S/C spin axis. Each detector has an IFOV of 20° (polar direction) x 11.25° (azimuthal direction). The total IFOV is 180° x 11.25°. A full coverage of the unit sphere is obtained once every 6 second spin. [1515] Although designed to measure ions in-situ, the IPS detects also ENAs with high efficiency.
- **IES** (Imaging Electron Sensor). The IES measures electrons over the energy range from about 25 - 400 keV using a spectrometer which incorporates a MSSD with a 0.5 cm x 2.1 cm planar configuration and five individual elements. The MSSD forms the image plane for a sensor segment with a "pin-hole" aperture. The MSSD has a thick "dead layer" and thus does not respond to protons with energies below about 250 keV. The "pin-hole" aperture accepts electrons over a 60° angular segment in a plane containing the satellite spin axis. Each of the five detector elements in the MSSD detects electrons in a 12° angular subinterval of the 60° FOV. The complete IES system has a nominal geometric factor of 6×10^{-3} cm^2. The nominal angular resolution of a detector element is 12° x 12° and its elemental geometric factor is approximately 3.8×10^{-4}.
- **SEPS** (Source/Lose-Cone Energetic Particle Spectrometer). [1516] SEPS consists of two independent telescopes which measure both the energetic electron and ion fluxes in the vicinity of the magnetic field-aligned loss, and source cone regions with high sensitivity and with fine angular and time resolution. The electron telescope has twice the sensor area of the ion telescope and uses aperture wheels to vary its dynamic range. Particle angular imaging is obtained using pinhole camera apertures in front of the electron XY position-sensitive detectors. The ion telescope is similar to the electron telescope except for the reduced sensor area and the fact that the aperture wheels are replaced by magnets which sweep out electrons.

1515) M. G. Henderson, G. D. Reeves, A. M. Jorgensen, et al., "POLAR CEPPAD/IPS Energetic Neutral Atom (ENA) Images of a Substorm Injection," Advances in Space Research, Vol. 25, No. 12, 2000, pp. 2407-2416

1516) http://www.css.tayloru.edu/~ physics/seps.html

Parameter	Electron telescope	Ion telescope
Energy range (keV)	100-300	100 to > 1000
Energy resolution (FWHM keV)	30	30
No of energy channels (log/linear)	16	16
Field of View (FOV), °	48 x 24	20 x 20
XY image matrix (pixels)	256	128
Detector area (cm^2)	12.5	6
Proton window (keV)	100	about 10
Geometric factor (cm^2 sr)	0.02 - 0.0002	0.001
Total number of pixels	512	256

Table 452: Parameter definition of SEPS

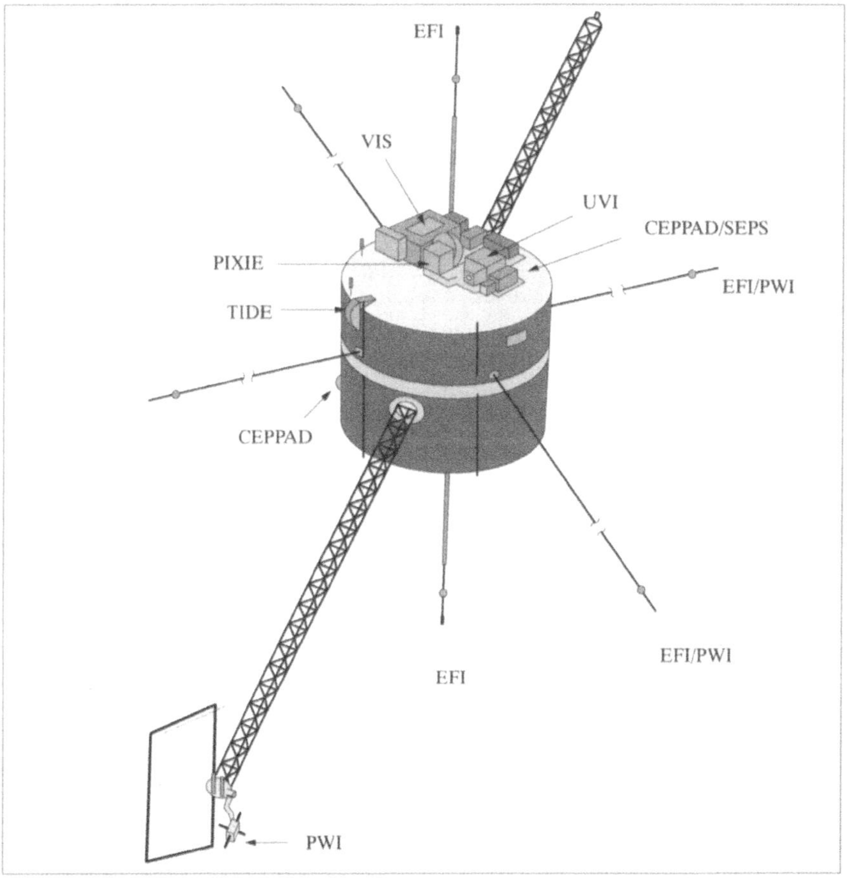

Figure 270: The POLAR S/C model

3) Imaging instruments for spatial observations

UVI = Ultraviolet Imager (PI: M. Torr, MSFC). Objectives: Study of the spatial and temporal descriptions of the aurora and images of total particle flux, characteristic energy, thermospheric neutral composition, and ionospheric conductances. UVI images the dayside and nightside auroras in the vacuum UV range using five specially designed filters. The detector is an intensified CCD used in conjunction with a fast reflective optical system to image an 8° FOV at a nominal rate of two frames per minute. The UVI is able to detect and provide images of very dim emissions with a wavelength resolution never achievable before. The resulting images permit to quantify the overall effects of solar energy input to the Earth's polar regions. The key wavelengths imaged are: 130.4 nm (OI 1304 filter), 135.6 nm (OI 1256), 140-160 nm (LBH short), 160-175 nm (LBH long), and 175-190 nm (solar). LBH stands for Lyman-Birge-Hopefield (spectral bands in the 140-180 nm range).

VIS = Visible Imaging System (PI: L. Frank, U. of Iowa). Objectives: Quantitative assessment of the dissipation of magnetospheric energy into the auroral ionosphere. Development of an energy flow model within the magnetosphere using VIS data in three ways: to illustrate the topology of the magnetosphere, to delineate the response of the magnetosphere and the magnetotail to substorms and solar-wind conditions, and to identify the locations of and mechanisms for suprathermal charged-particle acceleration.
The VIS instrument consists of a set of three low-light-level cameras. Two of these cameras share primary and secondary optics and are designed to provide images of the night-time auroral oval at visible wavelengths (FOV= 20°x 20°). A third camera is used to monitor the FOV directions of the sensitive auroral cameras with respect to the sunlit Earth. The two auroral cameras have spatial resolution of 10 km (0.011° x 0.013°) and 20 km (0.22° x 0.25°) respectively from an orbital altitude of 8 R_E. A CCD image has a size of 256 x 256 pixels and takes 12 s to acquire. VIS uses an image intensifier readout through 12 visible narrow-band filters producing five separate auroral images per minute.[1517)]

PIXIE = Polar Ionospheric X-ray Imaging Experiment (PI: W. Imhof, Lockheed Palo Alto Research Lab.). Objectives: Study of the morphology and spectra of energetic electron precipitation and its effect on the atmosphere. Derivation of the total electron energy deposition rate, the energy distribution of the precipitating electrons, and the altitude profile of ionization and electrical conductivity.
PIXIE uses a pinhole camera concept to measure the spatial distribution and temporal variation of X-ray emissions from the Earth's atmosphere. The time resolution of images is 60 seconds or better (the optimal radial distance to obtain the best images is 6 R_E).

Investigation	Experiment Acronym	Mass (kg)	Data Rate (Bit/s)
Magnetic Field Experiment	MFE	5.0	500
Electric Field Instrument	EFI	31.9	2500
Plasma Wave Instrument	PWI	18.4	2520
Hot Plasma Analyzer	HYDRA	14.4	4400
Thermal Ion Dynamics Experiment	TIDE	33.3	2520
Toroidal Ion Mass Spectrograph	TIMAS	16.5	3600
Charge and Mass Magnetospheric Ion Composition Experiment	CAMMICE	12.9	1280
Comprehensive Energetic Particle Pitch Angle Distribution	CEPPAD	14.4	4380
Ultraviolet Imager	UVI	18.0	12000
Visible Imaging System	VIS	24.0	11000
Polar Ionospheric X-ray Imaging Experiment	PIXIE	24.5	3500

Table 453: Instrument summary of the Polar S/C payload

Data: On-board recording capability of science data (1.3 Gbit digital tape recorder). Science data transmission of 56 kbit/s (RT) and 512 kbit/s (PB) to DSN (four 60 minute nominal contact periods per day). The Polar S/C provides on-board interconnection of instrumentation for data communication. Data sharing among the instruments can be triggered by pattern recognition schemes in on-board computers.

1517) L. A. Frank, et. al., "The Visible Imaging System (VIS) for the Polar Spacecraft," Space Science Reviews, Vol. 71, 1995, pp. 297-328

K.19.1 SAC-B (Satélite de Applicaciones Científicas-B)

SAC-B is a cooperative solar mission between CONAE (Comisión Nacional de Actividades Espaciales) of Buenos Aires (CONAE is the civilian space agency of Argentina) and NASA for the study of solar physics and astrophysics (measurement of solar flares, gamma-ray burst sources and the diffuse soft X-ray cosmic background). 1518) CONAE is responsible for the design and construction of the SAC-B satellite (the S/C is built by INVAP S.E. of San Carlos de Bariloche, Argentina), NASA provides two instruments and launch services on a Pegasus XL vehicle.

The launch of SAC-B (along with HETE) took place on Nov. 4, 1996 from NASA/WFF, VA. The third stage of the Pegasus XL failed to separate properly - both S/C are still attached to the third stage in low Earth orbit. About 12 hours of checkout could be obtained after launch on battery power. The solar panels of SAC-B remained in solar shadow by the attached 3rd stage of the launcher and its avionic system. Hence, no operational status could be achieved.

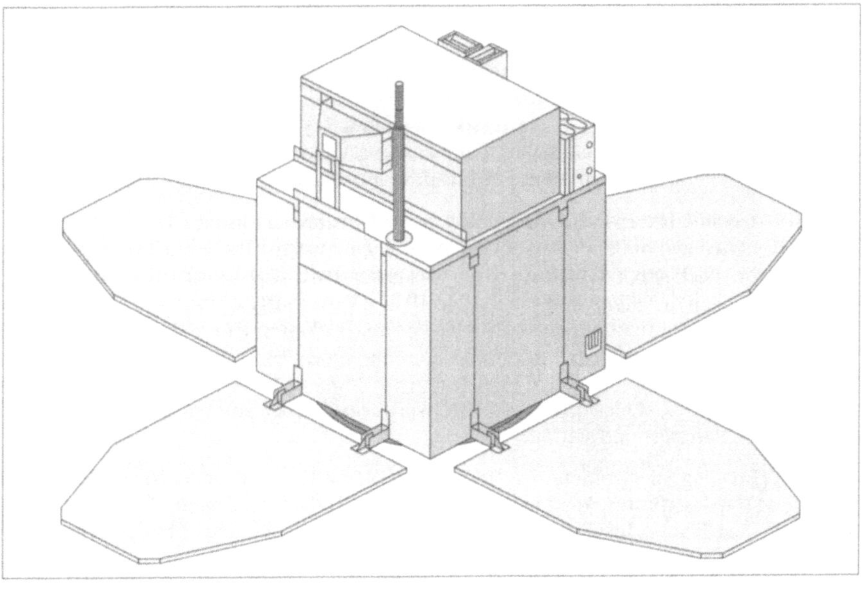

Figure 271: The SAC-B S/C model

The design life is three years or more. CONAE provides a single ground station (Buenos Aires) for mission operations and science data collection. NASA plans to provide DSN and Wallops Island station support for initial orbit monitoring and emergency backup throughout the mission life. Mission program managers: D. Caruso of CONAE and David Gilman of NASA/HQ.

The S/C mass is 191 kg; three-axis stabilized (two momentum wheels in a "V" configuration for roll and yaw control). Pitch axis control and momentum unloading is accomplished using magnetic torquers; attitude control by a coarse sun sensor, a two-axis Fine Sun Sensor (FSS) and a three-axis magnetometer (TAM); power storage with two 10 Ah NiCd batteries; all sensors are thermally controlled. The S/C body is basically a rectangular parallelepiped 62 cm x 62 cm wide by 80 cm high with four extended solar panels (62 cm wide by 76 cm long), the solar power is 256 W. The HXRS and SOXS instruments are mounted on the sun-

1518) Information provided by D. Gilman and J. L. LaBrecque of NASA/HQ facing end of the S/C along with the FSS with FOV toward the sun. The GRaBS sensors mount to the solar platform (bottom of S/C) with a FOV on the opposite sides of the S/C normal to the sun line. The CUBIC and Italian ISENA instruments are mounted on the top (anti-sun facing side) of the S/C. CUBIC's FOV is normal to the sun line. SAC-B will be sun-facing with various roll angles to accommodate the CUBIC experiment requirements. - Data: on-board storage of HK data in central memory and science data storage in instrument memories; data transmission in S-band at 100 kbit/s.

Orbit: Circular orbit, altitude = 550 km, inclination = 38°

Sensor complement:

HXRS (Hard X-Ray Spectrometer) provided by IAFE (Argentina Institute of Astronomy and Space Physics). Objective: observation of the solar flare hard X-ray spectrum between 20 and 320 keV of rapidly varying events on time scales as short as tens of milliseconds. In addition to the study of the temporal evolution of X-ray emissions during solar flares the instrument will provide the temporal evolution of nonsolar γ-ray bursts.

GXRE (Goddard X-Ray Experiment) provided by NASA/GSFC. The instrument has two sets of detectors (in separate modules).
The **SOXS** (Soft X-Ray Spectrometer) performs coordinated observations with HXRS by observing soft X-ray emissions from solar flares.
The other module is referred to as **GRaBS** (Gamma Ray Burst Spectrometer). It provides time profiles of the X-ray emission of nonsolar γ-ray bursts in the energy range 20 keV to >300 keV. There are two modules of GRaBS, one on each side of the SOXS instrument.

CUBIC (Cosmic Unresolved X-Ray Background Instrument using CCDs, PI: G. P. Garmire) designed and built by Penn State University, University Park, PA. The objective is to observe the spectrum of the diffuse X-ray background with high sensitivity and spectral resolution between the energy ranges of 0.2 to 10.0 keV over large portions of the sky in order to study the multiple components of this background radiation. The instrument consists of a pair of X-ray CCDs operated in photon-counting mode and exposed directly to the sky through an aperture with a 5° x 5° FOV for energy levels <1 keV, and 10° x 10° FOV for all other energy levels. - Once in orbit, CUBIC will be operated by students from Penn State via the Buenos Aires ground station.

ISENA (Imaging Spectrometer for Energetic Neutral Atoms) provided by CNR/IFSI (Institute of Physics of Interplanetary Science, Frascati, Italy, PI: S. Orsini). [1519] Partner institutions: Max-Planck Institut für Aeronomie, Katlenburg-Lindau; Physics Department, University of Arizona, Tucson, AZ. The objective of ISENA is detection of energetic neutral atoms emitted from the inner magnetosphere through charge-exchange between the magnetospheric energetic ions and the geo-coronal neutral gas. The instrument covers a 60° by 60° FOV(Field of View) using two sensors with an angular resolution of 15° by 3°. Mass and energy resolution (in the range between 5/8 keV to 150 keV) are made possible by the simultaneous detection of the flow velocity (using the time-of-flight technique) and of the energy (using a solid-state detector) of the incident energetic neutral atoms. Instrument mass = 6 kg; power = 6 W.

K.20 SME (Solar Mesosphere Explorer)

Atmospheric mission in NASA's Explorer Program (Explorer 64) operated by the University of Colorado at Boulder (JPL management of SME project, S/C was built by Ball Aerospace, Boulder, CO). NASA launch: Oct. 6, 1981 with a Thor Delta vehicle from VAFB, CA (UoSAT-1 was a passenger payload on this flight). The satellite was operational until April

1519) Information provided by R. Ibba of ASI, Rome, Italy

13, 1989. Objective: Study of atmospheric (mesosphere) ozone and the processes that form and destroy it. [1520) 1521) 1522) 1523) 1524) 1525)]

SME was a spin-stabilized S/C at 5 rpm; the spin axis was maintained nearly perpendicular to the orbit velocity vector by means of a magnetic torquing attitude control system. The spacecraft body was a cylinder approximately 1.7 m x 1.25 m (S/C mass = 417 kg) and consisted of two major modules: the observatory module that housed the scientific instruments, and the spacecraft bus. The command system was capable of executing commands in real time or from stored program control. Power was supplied by a solar cell array. The telemetry system was used either in a real-time or in a tape-recorder mode. The payload (instrument package) was developed at the Laboratory for Atmospheric and Space Physics (LASP) of the University of Colorado.

Orbit: Sun-synchronous near-circular orbit; altitude = 534 km, inclination = 97.5°, period = 95.5 minutes.

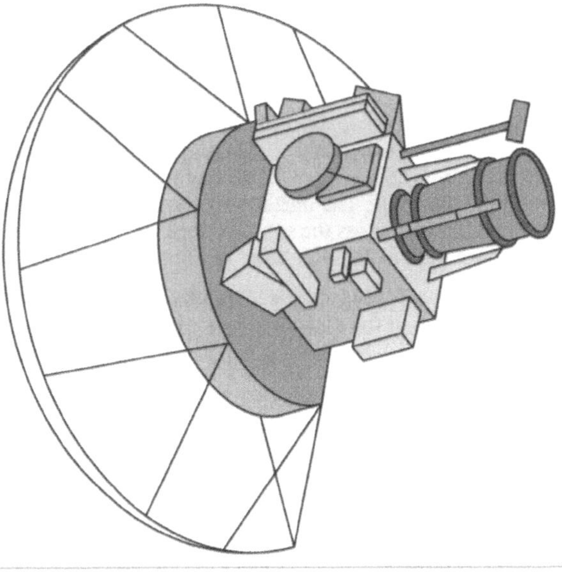

Figure 272: The SME S/C model

Sensor complement: (total of five instruments with four limb-scanning optical instruments)

UV Ozone Experiment. Objective: observation of ozone concentrations (ozone absorption of Rayleigh scattered sunlight at the limb) in the altitude region of 50 - 70 km. The instrument is a dual-channel Ebert-Fastie spectrometer operating in the spectral range of 1800 - 3100 Å (channel A) and 2230 - 3404 Å (channel B). There were 208 or 11 grating steps per scan, respectively. The Cassegrain telescope focal length = 250 mm. FOV = 0.1° x 1.5°.

1520) "Solar Mesosphere Explorer - Scientific Data & Publications," Final Report, LASP, December 1989

1521) J. R. Cowley, G. M. Lawrence, "Earth Limb Altitude Determination for the Solar Mesosphere Explorer," AIAA-83-0429

1522) Ch. Barth, "Solar Mesosphere Explorer to Study Ozone," Nature, Volume 293, Sept. 24, 1981

1523) J. R. Stuart, K. A. Gause, "Solar Mesosphere Explorer Mission," AIAA paper, 17th Aerospace Sciences Meeting, Jan. 15-17, 1979,

1524) "Solar Mesosphere Explorer - Experiment Description," LASP paper, University of Colorado at Boulder

1525) Ch. Barth, et al., "Solar Mesosphere Explorer: scientific objectives and results," Geophysical Research. Letters, Vol. 10, No. 4, p. 237, 1983

Infrared Radiometer (passive 4-channel device). Objective: radiance measurements in the infrared region. The radiometer/telescope had the following spectral ranges: 17.2 - 13.2 μm (carbon dioxide emission), 15.7 - 14.7 μm, 10.6 - 8.6 μm (ozone emissions), and 7.2 - 6.1 μm (water emissions). The full widths at half-maximum were 4.0, 1.0, 2.0, and 1.1 μm, respectively. All four channels utilized HgCdTe detectors. Wavelength separation was accomplished with multilayer bandpass filters. The instrument line of sight was normal to the spin axis. IFOV sweeps through a 4° angle that is equivalent to a limb scan from 20 - 150 km in altitude. Each channel is sampled 20 times per S/C revolution.

Airglow Instrument. Objective: measurement of the ozone altitude profile by limb scanning at 1.27 μm. The instrument consisted of a grating spectrometer (Ebert Fastie) and an uncooled detector system with a narrow FOV.

Visible Nitrogen Dioxide Experiment. Objective: measurement of nitrogen dioxide concentrations in the altitude region from 20 - 40 km (NO_2 absorption of Rayleigh scattered sunlight at the limb). The instrument is an Ebert-Fastie spectrometer (of same design as before). Solar radiation was measured between 4390 to 4420 Å and between 3200 to 6400 Å. There were 512 and 438 grating steps per scan, respectively. The instrument line of sight was normal to the spin axis.

Solar UV Monitor. Objective: monitoring of solar flux scattered from a diffusing screen. The instrument was an Ebert-Fastie spectrometer (same design as for UV Ozone Experiment). Operation of instrument in two modes: 1. scanning of the spectral range of 1600 Å to 3100 Å by rotation of the grating; 2. setting of the grating to a required wavelength with the instrument monitoring the solar flux. The instrument was mounted to the side of the S/C scanning once per S/C revolution across the sun. The look direction was 45° to the spacecraft axis of rotation. There were 512 grating steps per scan.

Data: The SME data are archived in the NSSDC (National Space Science Data Center) at GSFC and in the SME database at the University of Colorado.

K.21 SMEX (Small Explorer Program)

A NASA/GSFC satellite program (since 1989) with the objective to introduce innovative technologies into low-cost spacecraft design.

The design philosophy employs and end-to-end approach. Emphasis is placed on design simplicity (also a reduction of redundancies), functional independence at subsystem level, and flexibility to accommodate growth. The modular architecture applies to hardware, software, and to services of the system. Communication between subsystems flows via the MIL-STD-1553 data bus. The instruments and subsystems communicate with the computer using this bus. For instruments that produce high data rates, there are four EIA RS422 Direct Memory Access (DMA) interfaces to the computer memory.

A SMEX spacecraft consists of a one-piece cast aluminum structure (a thermally conductive composite S/C structure). The following on-board technologies are employed: the computer is a single-board 80386/80387 RISC processor (Loral RAD-6000 32-bit) featuring dynamic memory management; fiber-optic data bus; GaAs solar arrays (modular); high-density solid-state recorder; high torque reaction wheels, and NiCd batteries. Attitude sensors: coarse sun and fine sun sensors, star tracker, magnetometer (3-axis fluxgate), gyroscopes. Actuators: magnetic torquers and reaction wheels. S/C communications: RF transmitters with 4 Mbit/s X-band downlink, QPSK modulation and 3.5 W RF power; RF receivers with 4 kbit/s S-band uplink, BPSK modulated directly on the S-band carrier with Bi-phase L capability. All communications (format, procedure, etc) are in CCSDS standard. S/C baseline configuration: minimum operational lifetime of one year, S/C mass of 75 kg plus up to 225 kg of payload mass, power = 150-400 W for payload use. [1526)]

The overall SMEX missions include the following S/C: SAMPEX, FAST and SWAS (Submillimeter Wave Astronomy Satellite), TRACE and WIRE (Wide-Field Infrared Explorer).

K.21.1 SAMPEX (Solar Anomalous and Magnetospheric Particle Explorer)

SAMPEX (Explorer 68) is a NASA/GSFC SMEX mission with the objectives to measure energetic electrons as well as ion composition of particle populations from about 0.4 MeV/nucleon to hundreds of MeV/nucleon from a zenith-oriented satellite in near polar orbit; to study the energy, composition, and charge states of particles from supernova explosions, from the heart of solar flares, and from the depths of nearby interstellar space; also to monitor the magnetospheric particle populations which occasionally plunge into the middle atmosphere of the Earth. Mission design life = 1 year, with a goal of 3 or more years of operation. [1527)]

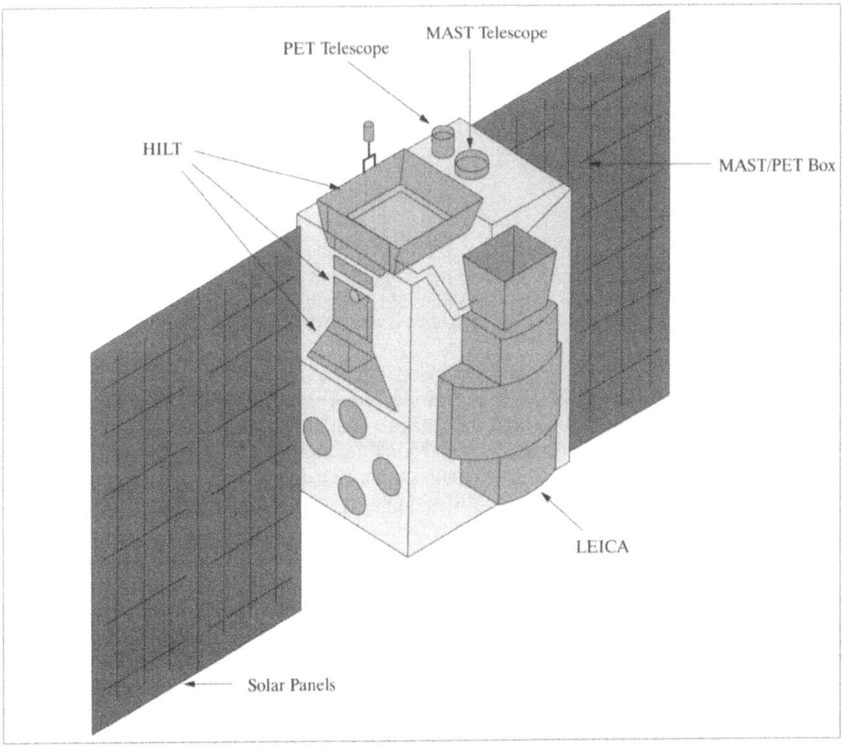

Figure 273: The SAMPEX S/C model

SAMPEX is a momentum-biased, sun-pointed S/C that maintains the experiment view-axis in a zenith direction. It points its solar arrays at the sun by aiming the momentum vector

1526) J. G. Watzin, "SMEX-LITE: NASA's Next Generation Small Explorer," Proceedings of the 10th Annual AIAA/USU Conference on Small Satellites, Logan, UT, Sept. 16-19, 1996

1527) D. N. Baker, G. M. Mason, O. Figueroa, G. Colon, J. G. Watzin, R. M. Aleman, "The Solar, Anomalous, and Magnetospheric Particle Explorer (SAMPEX) Mission," Preprint 93-128, U. of Maryland - see also (by the same authors): IEEE Transactions on Geoscience and Remote Sensing, Vol. 31, No. 3, May 1993, pp. 531-541 toward the sun and rotating the S/C one revolution per orbit about the sun/spacecraft axis. The ACS (Attitude Control System) uses a combination of three orthogonal torque rods to act against the Earth's magnet field, and one momentum-biased reaction wheel to provide the bias momentum. A tow-axis digital sun sensor, a three-axis magnetometer and a set of five coarse sun sensors are used for attitude determination (ACS estimation of sun location during orbit eclipse to continue three-axis attitude solution). Two solar arrays of 1.7 m^2 area provide an average orbit power of 100 W. One 9 Ah NiCd battery is used. The payload mass is 40 kg, the total S/C mass is 157 kg (aluminum structure, passive thermal control). The solid-state recorder has a capacity of 30 MByte. On-board fiber optic data bus (Mil-Std 1773). The S/C features two omni antennas and a near-Earth 5 W S-band transponder.

SAMPEX was launched on July 3, 1992 from NASA's Western Test Range/VAFB, Ca. from a Scout launch vehicle. As of the end of 1997 SAMPEX is fully operational.

Orbit: Polar orbit, perigee = 512 km, apogee = 687 km, inclination = 82°, period = 96.7 minutes.

Sensor complement:

LICA (Low-Energy Ion Composition Analyzer). Sensor mass = 7.4 kg, power = 4.9 W, bit rate = 1.3 kbit/s. PI: Mason. Objective: to measure 0.5 - 5 MeV/nucleon solar and magnetospheric ions (He through Ni) arriving from the zenith in twelve energy bands. LICA is a time-of-flight spectrometer that monitors incident ion mass and energy by simultaneously measuring the time-of-flight and the kinetic residual energy of particles that enter the telescope and stop in an array of four solid-state detectors. The time-of-flight is determined by start and stop pulses from chevron microchannel plate (MCP) assemblies that detect secondary electrons emitted from the entrance foil and a foil in front of the solid state detector, respectively, when the ion passes through them. These secondary electrons are accelerated to approximately 1 kV and deflected onto the MCPs by electrostatic mirrors. The measured energy and velocity are combined to yield the mass of the ion and the energy per nucleon.

HILT = Heavy Ion Large Telescope (MPE Garching/NASA). Sensor mass = 22.8 kg, power =5.7 W, bit rate = 0.9 kbit/s. Objectives: to measure the charge, energy, and mass of cosmic rays in the energy range of about 8.0 - 310 MeV/nucleon. Specifically, the energy ranges were: He: 3.9 - 90MeV/nucleon; C: 7.2 - 160 MeV/nucleon; O: 8.3 - 310 MeV/nucleon; Ne: 9.1 - 250 MeV/nucleon; and Fe :11 - 90 Mev/nucleon. The instrument consists of a three-element ion drift chamber with two thin multilayer entrance windows followed by an array of 16 solid-state detectors and a scintillation counter with photodiodes. HILT uses a flow-through isobutane system for the drift chamber.

MAST = Mass Spectrometer Telescope (Caltech/GSFC). Instrument mass = 7.5 kg, power = 3.3 W, bit rate = 1.4 kbit/s. MAST monitors the isotopic composition of elements from Li (Z=3) to Ni (Z=28) in the rang of about 10 MeV/nucleon - several hundred MeV/nucleon. MAST consists of a combination of surface barrier and lithium-drifted solid-state detectors (11 total). Combined matrix detector positions determine the particle trajectories, allowing accurate corrections to be made for the path length variation with angle and detector response for nonuniformities. MAST also performs measurements of stopping He isotopes from about 7 - 20 MeV/nucleon.

PET = Proton/Electron Telescope (Caltech/GSFC). Instrument mass = 7.5 kg, power = 1.2 W, bit rate = 0.5 kbit/s. PET complements MAST by monitoring the energy spectra and relative composition of protons (18 - 250 MeV) and helium nuclei (18 - 350 MeV/nuclei) of solar, interplanetary, and galactic origins, and the energy spectra of solar flare and precipitating electrons from about 0.4 to about 30 MeV. The instrument measures both trapped and precipitating energetic particles in different parts of the SAMPEX orbit.

Energy Range for:	LICA	HILT	MAST	PET
Electrons	-	-	-	0.4-30 MeV
H	0.76 - 6.1	-	-	18 - 250 MeV
He	0.45 - 6.1	4.3 - 38	7 - 20	18 - 350 MeV/n
C	0.44 - 11.4	7.2 - 160	14 - 210	34 - 120 MeV/n
Si	0.33 - 5.5	9.6 - 177	21 - 330	54 - 195 MeV/n
Fe	0.21 - 3.1	11 - 90	27 - 450	70 - 270 MeV/n
Charge range for elements	1 - 26	2 - 28	2 - 28	1 - 2 (1 -28*)
Charge range for isotopes	2 - 16	2	2 - 28	1 - 2 (1 - 10*)
Geometry factor (cm^2 sr)	0.8	60	7 - 14	0.3 - 1.6
FOV (°, full angle)	24 x 20	68 x 68	101	58
Mass (kg)	7.4	22.8	8.8	(incl. with MAST)
Power (W)	4.9	5.6	5.3	(incl. with MAST)
Telemetry (kbit/s)	1.3	0.9	1.4	0.5
* commandable low-gain mode				

Table 454: Performance overview of SAMPEX instruments

Data: On-board data storage is provided by solid-state memory (26.5 Mbyte of data storage). Data transmitted to the ground is formatted as a "Packet Telemetry" stream in accordance with CCSDS standards. S/C operation by GSFC on a continuous basis. Data are transmitted to the ground twice per day via the Wallops Flight Facility.

K.21.2 FAST (Fast Auroral Snapshot Explorer)

NASA/GSFC [1528) 1529)] mission within SMEX. The objectives are to measure and to study the rapidly varying electric and magnetic fields and the flow of electrons and ions in the aurora regions of the Earth, to study the physical causes of complex auroral displays, and to investigate how electrical and magnetic fields accelerate electrons, protons, and other ions in the auroral regions. The S/C instrument data will be correlated with observations by other S/C as well as by geomagnetic ground stations from Earth. The S/C is spin-stabilized with its spin-axis oriented parallel to the orbit axis. It spins in a cartwheel mode with a period of about 5 seconds (12 rpm). Spin rate and spin-axis orientation are maintained by two magnetic torquer coils, one spinning sun sensor, one horizon crossing indicator, and a magnetometer. The ACS provides a closed-loop spin-rate control. The body-mounted solar array contains 5.6 m^2 of solar cells 52 W of average orbit power). Total S/C mass of 190 kg, S/C length = 0.86 m, diameter = 1.17 m, design life = 1 year, payload of four instruments (payload mass of 51 kg). The communication system uses an S-band transponder, 2.25 Mbit/s downlink and 2 kbit/s uplink data rate. [1530)]

The launch date was August 21, 1996 (placing FAST into the ISTP program) with an enhanced Pegasus vehicle (XL) from Vandenberg, CA. PI for mission: C. W. Carlson of UCB (University of California at Berkeley).

Orbit: Non-sun-synchronous elliptical orbit, perigee = 351 km, apogee = 4175 km (apogee maximum at northern latitude), inclination = 83°, period = 133 minutes. The satellite crosses the auroral zones (which form ovals at about 65°-70° magnetic latitude North and South) four times an orbit. The orbit was designed to have a Northern apogee during January and February of 1997 for coordinated ground-based and optical observations.

Sensor complement:

ESA = Quadrispherical Electrostatic Electron Analyzer (UCB). The objective is to measure electrons and ions with high time resolution. ESA consists of an ion spectrometer and an electron spectrometer to make detailed distribution function measurements; in addition

1528) D. Baker, G. Chin, R. Pfaff, "NASA's Small Explorer Program," Physics Today, Dec. 1991, pp. 44-51
1529) C. W. Carlson, "The Fast Auroral Snapshot Explorer," EOS, Vol. 73, No. 23, 1992, pp. 249, 253, 254
1530) http://sunland.gsfc.nasa.gov/smex/fast/

there are electron-stepped ESAs to make very high time resolution electron measurements with lower energy and angle resolution. The electron-stepped ESAs provide a 64-step energy sweep, covering a range of 3 eV to 30 keV. Sixteen ESAs are configured in four stacks placed around the S/C (360 FOV). [1531]

TEAMS = Time-of-Flight Energy Angle Mass Spectrograph (U. of New Hampshire and Lockheed, Palo Alto Research Lab). TEAMS is a high sensitivity mass-resolving spectrometer that measures the full 3-D distribution functions of the major ion species with one spin of the S/C. TEAMS typically looks for H^+, He^{++}, He^+, O^+, and a combination of O^{++} and NO^+ over an energy range of a few eV to 5 keV. TEAMS combines the selection of ion by energy/charge by electrostatic deflection and subsequent time-of-flight analysis.

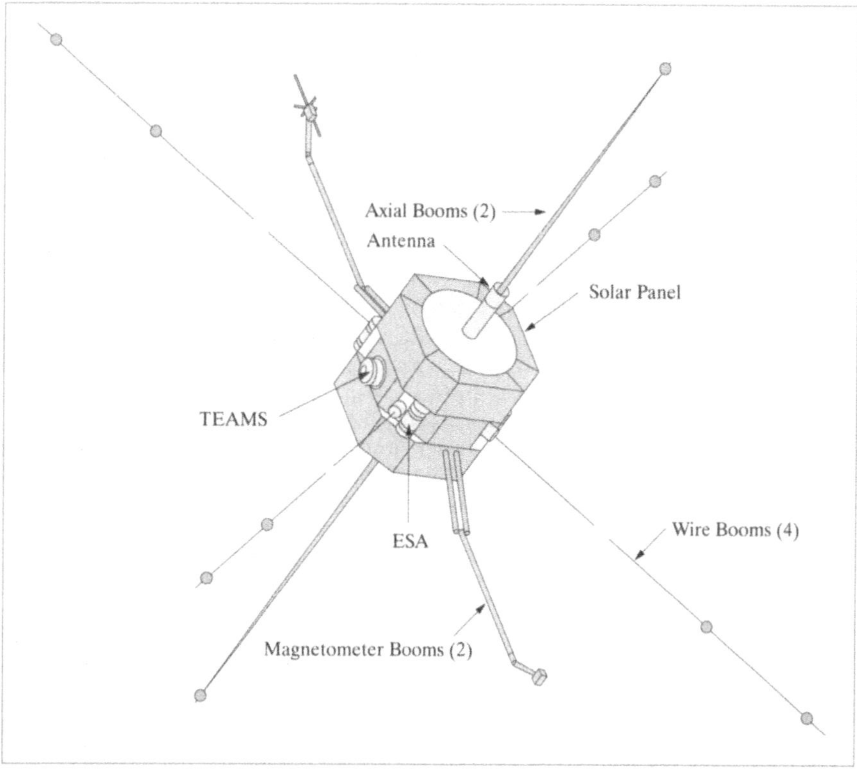

Figure 274: The FAST S/C model

EFLPI = Electric Field / Langmuir Probe Instrument (UCB). [1532] The objective is to measure the vector electric field from DC to 2 MHz, thermal plasma density and temperature, density fluctuations, propagation velocity of structures, and wavelength measurements using three orthogonal boom pairs, two of which have spherical probes at two locations. The electric field measurement has eight spherical sensors located on four wire booms (two each) in the spin plane and two that are on rigid axial booms. All of the spherical sensors can operate in "voltage mode" in which they measure the local plasma potential with respect to the payload. The electric field signals are measured by sensor pairs which form 56 m, 7.7 m,

1531) http://sprg.ssl.berkeley.edu/fast/

1532) R. E. Ergun, C. W. Carlson, F. S. Mozer, et al., "The FAST Satellite Fields Instrument," Space Science Review, 2000 or 5 m dipoles. Six of the sensors can operate in "current mode" as Langmuir probes which measure plasma current. In current mode, the probe is biased at a fixed potential. EFLPI includes a wave/particle correlator. The total mass of EFLPI is 28.45 kg, power = 15.9 W.

MFI = Magnetic Fields Instrument (UCLA). The objective is to measure vector DC and AC magnetic fields using a three-axis fluxgate magnetometer and a three-axis search-coil magnetometer (up to 20 kHz in 2 axis and 1.2 kHz in the 3rd axis). The two magnetometers are boom-mounted 180° apart.

Data: On-board storage capability of 1 Gbit solid-state memory. The instruments acquire data at a maximum rate of 8 Mbit/s (snapshot data collection, with a maximum period of about 100 s of data can be acquired per auroral pass). On-board data preprocessing capability. Continuous synoptic data are collected at a slower rate to put the "snapshots" into proper context. Real-time data are acquired at stations in the auroral zone, memory dumps can be acquired at other NASA stations.

Note: in the 'sample array' column of Table 456, the designation '4Mx48Ex64Ω' refers to: 4M (mass) x 48E (energy) x 64 Ω (solid state array).

Measurement	Components	Frequency Range	Sampling Resolution	Range	Resolution	Data Rate (max.) kbit/s
DC E-Field	3-axis	0-300 Hz	30 μs	±1.6 V/m	0.05 mV/m 16 bit	5000
Wave E-Field	3-axis	0.3-20 kHz	30 μs	±200 mV/m	0.05 mV/m 16 bit	5000
Swept Freq. Spectrum Analyzer	3-axis	0-2 MHz	32 μs	70 dB	8 bit	246
E-Field Rectifier Filters	3-axis	0.1-2 MHz	30 μs	0.1-1 mV/m	8 bit	960
AC E-Field Burst Memory	2 - 4	0 - 1000 kHz	0.5 μs	selectable impulse	10 bit	2000
Density	3-axis	0.2 - 1 MHz	0.5 ms	1-10^5 cm^{-3}	8 bit	8
Temperature	-	0 - 2 kHz	1 s	0.1 eV-1 keV	8 bit	0.08
MFI Fluxgate Magnetometer	3-axis	0-50 Hz	2 ms	10^{-5} - 0.6 G	16 bit	48
MFI Search Coil Magnetometer	3-axis	10 Hz - 2.5 kHz	0.1 ms	-	16 bit	240

Table 455: FAST field instruments [1533)]

Measurement (ESA)	Coverage	Energy Range (keV)	ΔE/E	Sampling Resolution	FOV (°)	Angular Resolution	Sample Array
Ion Mass Spectrometer	3-D	0-12	0.13	2.5 s (3D) 78 or 156 ms (2D)	360° x 11°	10x22.5° 3D 5.6 or 11.2°	4Mx48Ex64Ω (3D) 4Mx48Ex16α (2D)
Ion Spectrometer	2-D	0-24	0.20	78 ms	360° x 12°	11.2x12°	48Ex32α
Electron Spectrometer	2-D	0-30	0.15	78 ms	360° x 10°	11.2x10°	48Ex32α
Electron Stepped ESA	2-D	0-30	0.15	1.6 ms	360° x 10°	22.5x10°	6Ex16α

Table 456: FAST particle detectors

K.21.3 TRACE (Transition Region and Coronal Explorer)

A NASA/GSFC mission in the SMEX (Small Explorer) program with the objective to study the three-dimensional magnetic structures which emerge through the visible surface of the

1533) Information provided by C. Cattell of UCB

sun - the photosphere - and to define both the geometry and dynamics of the upper solar atmosphere - the transition region and corona. TRACE is the name of the prime instrument (as well as of the mission), which is mounted onto a SMEX spacecraft designed and built by GSFC. [1534), 1535)]

The spacecraft is three-axis-stabilized using the TRACE telescope as a fine sun sensor. A S/C pointing accuracy of 20 arcseconds is required. Mission design life = 1 year. S/C mass = 250 kg. The S/C was air-launched April 2, 1998 on a Pegasus XL vehicle, released off the central California coastline.

- The ACS (Attitude Control System) uses three magnetic-torquer coils, one digital sun sensor, six coarse sun sensors, four reaction wheels, one three-axis magnetometer, and three two-axis inertial gyros. The ACS use the S/C computer to perform closed-loop attitude determination and control.
- The command and data handling system uses a 32-bit on-board processor (80386/80387) with 300 MByte solid-state memory for science data.. A MIL-STD-1553 data bus is used to connect all subsystems and instruments. Science data is passed over the RS-422 interface at rates up to 900 kbit/s. A 5 W S-band transponder is used to downlink 2.25 Mbit/s of data up to six times a day. Commands are uplinked at 2 kbit/s.
- Power of 222 W is provided by GaAs solar cells, deployed in four panels with a total area of 2 m^2. A 28 V unregulated bus distributes the power, a 9 Ah super NiCd battery provides energy storage.
- S/C operations from GSFC; ground stations at Wallops, VA, and Fairbanks, Alaska.

Orbit: Sun-synchronous polar orbit (permitting continuous observation of uninterrupted image sequences of the sun), altitude=700 km, inclination=98.2°, period=96 minutes.

Sensor:

TRACE (Transition Region and Coronal Explorer, built at Lockheed of Palo Alto, CA; PI: A. M. Title, the science team is from diverse institutions). The objective is to collect comprehensive multispectral images of solar plasmas at temperatures from 10^4-10^7 K with 1 arcsecond spatial resolutions and excellent temporal resolution and continuity. An overall science TRACE goal is to explore the relation between diffusion of the surface magnetic fields and changes in heating and structure throughout the transition region and corona. The TRACE instrument is a high-resolution multispectral spectrometer (in the EUV and UV region) featuring a 30 cm diameter Cassegrain telescope (160 cm in length, 8.66 m focal length) and a filter system which feeds a CCD detector array (1024 x 1024 Lumigen coated, front illuminated, three-phase CCD). The detectors are passively cooled to -65°C.

Each quadrant of the primary mirror is coated for sensitivity to a different wavelength range. Light entering the instrument passes first through the entrance filter assembly which transmits only far and extreme UV. Visible and near UV radiation (hence, most of the solar energy) are reflected back into space. Radiation transmitted through the entrance filters passes to a shutter that blocks three quadrants of the aperture, so that only one quadrant of the telescope is illuminated at a time. Photons passing the shutter's open quadrant proceed to the primary mirror, encountering a multilayer coating for a narrow-band EUV quadrant, or a broadband coating for the UV quadrant. The segmented coatings on solid mirrors form identically sized and perfectly coaligned images.

The reflected beam from the primary mirror proceeds to the secondary mirror which reflects it towards the focal plane. The secondary mirror is active to correct for pointing jitter and has coatings matching those on the four quadrants of the primary mirror. - Further stations in the radiation path are the filter wheels and the focal plane shutter. The final element

1534) T. D. Tarbell, M. Bruner, B. Jurcevich, J. Lemen, K. Strong, A. Title, J. Wolfson, L. Gloub, R. Fisher, "The Transition Region and Coronal Explorer," Proc. of the Third SOHO Workshop, Estes Park, CO, September 26-29, 1994, pp. 375-384

1535) TRACE www page of GSFC and of Lockheed

in the optical train is the CCD camera. The on-board computer permits very flexible use of the CCD array, including adaptive target selection, data compression, and fast operation for a limited FOV. TRACE is of SXT heritage of Solar-A (Yohkoh).

Parameter	Value	Parameter	Value
Primary mirror diameter	30 cm	Image stabilization	±0.1 arcsec
Effective focal length	8.66 m	Offset pointing	Disk center ±1°
FOV	8.5 x 8.5 arc min	CCD read-out time	2.1 s
CCD size	1024 x 1024 pixels	Average data rate	3.6 Gbit/day
Pixel size	21 x 21 μm (.5x.5 arcsec)	Telescope dimensions	37 x 37 x 188 cm
Quantization	12 bit	Instrument mass	51 kg

Table 457: TRACE instrument parameters

The time frame of the TRACE mission has the advantage of operating simultaneously with the SOHO (ESA/NASA, November launch 1995) mission. This implies coordination of observations from both S/C and the merging of datasets collected from TRACE and SOHO.

TRACE applications: study of magnetic field structure and evolution; coronal heating and magnetic fields; on-set of coronal mass ejections; flaring X-ray bright points; etc.

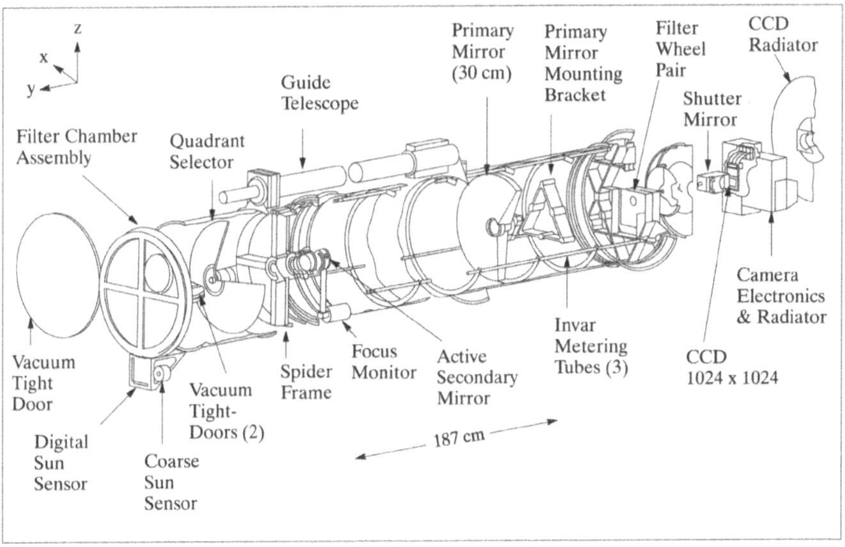

Figure 275: TRACE telescope layout

Band (Å)	Bandwidth (Å)	Ion	Region of Atmosphere	Solar Temp. Range Log (T)
2500	Broad	continuum (C)	Photosphere	3.6 - 3.8
1700	Broad	continuum	Tmin & Chromosphere	3.6 - 4.0
1570	30	C I, Fe II, continuum	Tmin & Chromosphere	3.6 - 4.0
1550	30	C IV	Transition Region	4.8 - 5.4
1216	84	H I, Lyman α	Chromosphere	4.0 - 4.5
171	11	Fe IX	Corona	5.2 - 6.3
195	14	Fe XII	Corona	5.7 - 6.4
284	14	Fe XV	Corona	6.1 - 6.6

Table 458: TRACE spectral bands

K.22 SMM (Solar Maximum Mission)

NASA/GSFC mission. Objectives: Observation of solar flares (trigger mechanisms) over a wide band of wavelengths in the UV, X-ray and gamma-ray regions of the spectrum; coverage of the maximum period of a sunspot cycle; better understanding of the violent nature of the sun and its effects on the Earth; measurement of radiation fluxes. Launch Feb. 14, 1980 by Delta vehicle from Cape Canaveral. Satellite mass = 2315 kg. End of mission: 1989. Launched during a period of maximum solar activity, SMM observed more than 12,000 flares and over 1,200 coronal mass ejections during its lifetime. [1536), 1537)]

SMM performed flawlessly until an attitude-control system failure (actually a failure of a fuse) occurred on Nov. 23, 1980, which kept the S/C from pointing precisely at the sun. In April 1984 SMM became the first satellite to be retrieved, repaired, and redeployed in orbit by a Shuttle crew (STS–41C). The opportunity was also used to replace in orbit a dish antenna, enabling SMM to transmit its data via TDRS-1. SMM collected data until November 24, 1989, and re-entered on December 2, 1989.

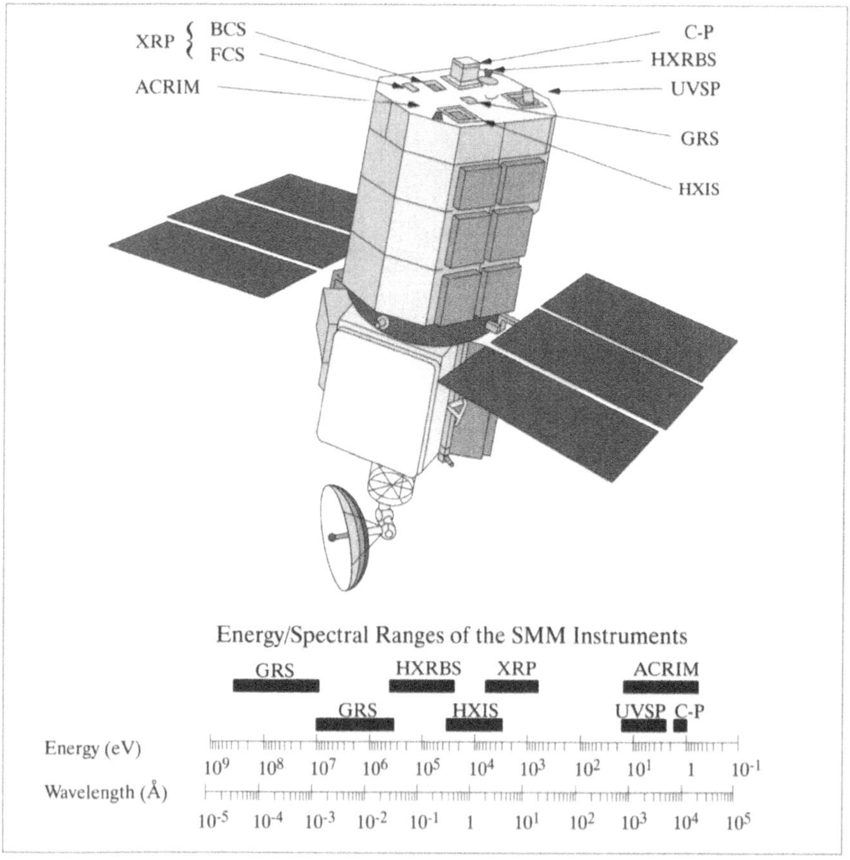

Figure 276: The SMM S/C model

1536) S. P. Maran, B. E. Woodgate, "A Second Chance for Solar Max," Sky & Telescope, June 1984, pp. 498-500
1537) A. Chaikin, "Solar Max: Back from the Edge," Sky & Telescope, June 1984, pp. 494-497

SMM is of modular design (built by Fairchild) measuring approximately 4 m in length, fitting into a circular envelope 2.3 m in diameter. The instrument module occupies the top 2.3 m and contains the solar payload instruments together with the fine-pointing sun-sensor system. Below the instrument module is the Multimission Modular Spacecraft (MMS) containing the systems for attitude control, power, communication, and data handling. Between the instrument module and the MMS is the transition adaptor, supporting two fixed solar paddles that supply between 1500 and 3000 W of power. [1538), 1539)]

Orbit: circular Earth orbit, apogee = 569 km, perigee = 566 km, inclination = 28.5°, period = 96 min (the S/C spends about 60 min of each orbit in full view of the sun).

Sensor complement:

ACRIM = Active Cavity Radiometer Irradiance Monitor (PI: R. C. Willson, JPL).[1540)]
Objective: regular observations of total solar irradiance. The goals of the experiment are:

- i) to begin a long-term climatological database on solar irradiance variability
- ii) to provide a shorter-term database for solar physics investigations.

ACRIM has three active cavity radiometer (ACR) type IV sensors. The sensors view the sun through a 5° (full angle) field of view. ACRIM uses black-body cavity to measure the total irradiance of the visible hemisphere of the sun.
ACRIM is a cavity pyrheliometer, with solar input during half of each 131.072 s measurement interval and servo-adjusted heat input during the rest of each cycle, when the shutter blocks the solar input. The solar irradiance at the Earth is therefore proportional to the difference between the heating rates with shutter open and closed. The accuracy of each irradiance measurement obtained in this way is about 0.1%.

C-P = Coronagraph - Polarimeter (PI: R. M. MacQueen, NCAR Boulder, CO). Objective: corona change monitoring over time, observation of the corona's large-scale magnetic structures and their evolution. The C-P instrument is an externally occulted Lyot coronagraph with a vidicon detector. Images of the corona with 10" spatial resolution are built up in four quadrants by means of sector mirrors (spectral range: 4465 - 6385 Å in 7 bands). In addition, with seven filters and three polaroids with principal axes 60° apart, the C-P is able to obtain measurements of coronal electron densities and distinguish ejected material from coronal features at chromospheric temperatures.

GRS = Gamma-Ray Spectrometer (PI: E. L. Chupp, U. of New Hampshire). The instrument includes seven integral-line calibrated NaI(Tl) detectors for nuclear lines between 0.3 and 0.9 MeV. The energy resolution (full width at half maximum) of these detectors is > 7% at 0.662 MeV. In addition, a thick CsI(Na) crystal, along with the NaI(Tl) spectrometers, is sensitive to γ-rays in the range 10-140 MeV and to neutrons with energies above 20 MeV. An auxiliary system of two NaI detectors yields hard X-ray fluxes from 10 to 140 keV. Time resolutions are typically about 2 s and 16 s, but can be as fine as 64 ms for the 300-350 keV energy band.
Among other things, GRS made important contributions to the international study of Supernova 87A, which in Feb. 1987 provided astronomers with their first local opportunity since 1604 to study such an explosion.

HXRBS = Hard X-Ray Burst Spectrometer (PI: B. R. Dennis, GSFC). Objectives:

- to determine the nature of the mechanisms which accelerate electrons to 20-200 keV in the first stage of the solar flare, and to >1 MeV in the second stage of many flares
- to characterize the spatial and temporal relation between electron acceleration, storage and energy loss throughout the solar flare.

1538) "NASA's Solar Maximum Mission: A Look at the New Sun," June 1987, NASA brochure, edited by J. B. Gurman
1539) "The Solar Maximum Mission Experiments," Solar Physics, Volume 65, pp. 5-109
1540) R. C. Willson, S. Gulkis, M. Janssen, S. H. Hudson, G. A. Chapman, "Observations of Solar Irradiance Variability," Science, Volume 211, Feb. 1981, pp. 700-702

The instrument is a collimated X-ray spectrometer. The detector consists of two primary scintillation crystal components. Scintillation events in its actively collimated CsI(Na) detector are read out every 128 ms in fifteen energy channels between ~25 to ~500 keV. A circulating memory is able to accumulate relatively brief periods of data during the more intense flares with time resolution down to 1 ms. The full width at half maximum field of view is ~40°.

HXIS = Hard X-Ray Imaging Spectrometer (PI: C. de Jager, Astronomical Institute, Utrecht, NL, and U. of Birmingham, UK). Objective: study of the energy problem associated with hard X-ray generating electrons. Study of flare models and mechanisms. The instrument consists of an imaging collimator of ten grid plates, each divided into 576 sections, and a position-sensitive detector system consisting of 900 mini-proportional counters. The grids form a coarse FOV 6.35 arcmin in extent and a fine field of view 2.67 arcmin across (with 8 arcsec square pixels). Each pixel is sampled in six energy bands ranging from 3.5 to 30 keV, with time resolution down to 1.25 s.
HXIS served as "flare alarm" to alert the other instruments electronically. Note: HXIS malfunctioned in June 1981 and could not be repaired in the SMM rescue mission in April 1984. The 1980 HXIS data set, however, forms a most impressive record of the spatial and energetic distribution of X-rays in solar flares, with a spatial resolution never before achieved at such high energies.

UVSP = Ultraviolet Spectrometer and Polarimeter (PI: E. Tandberg-Hanssen, MSFC). Objective: Study of UV radiation from the solar atmosphere, in particular from the active regions, flares, prominences, and the active corona. UVSP consists of an aplanatic Gregorian telescope of about 2" resolution, an Ebert-Fastie spectrometer, and five photomultiplier detectors. The telescope secondary is mounted on gimbals and may be rastered to build up an image of an area on the sun of up to 256 x 256 arcsec, in steps of as little as 1 arcsec. The entrance apertures range in size from 1 x 1 arcsec to 15 arcsec x 286 arcsec, and the exit slits range from 0.1 to 3.0 Å in second order.

A polarimeter may be inserted behind the exit aperture; it consists of two systems: a three-quarter wave retarder for those wavelengths where the spectrograph grating is an effective linear polarizer, and a quarter-wave retarder and linear polarizer for use at wavelengths at which the grating polarization is very low. The instrument is sensitive to wavelengths of 1170 - 1800 Å in second order and up to 3600 Å in first order.

In April 1985, the UVSP grating drive mechanism became stuck; the grating appeared to be fixed at a wavelength of about 1380 Å (second order; about 2760 Å in first order). During periods of solar activity, UVSP is used in second order to provide pointing and timing information for other SMM instruments; at other times, first order measurements of ozone concentration in the Earth's atmosphere are carried out at S/C sunrise and sunset (occultation measurements). These unique observations make possible a mapping of ozone concentrations at latitudes of ±50° at altitudes of 50 - 75 km. Approximately 20,000 separate ozone altitude profiles were obtained from late 1985 through March 1989.[1541)]

XRP = (Soft) X-Ray Polychromator (PIs: J. L. Culhane, U. College, London; K. J. H. Phillips, RAL, UK; K. T. Strong, Lockheed, Palo Alto, CA). Objective: investigation of those aspects of solar activity that lead to the production of plasma temperatures in the 1.5 to 50 million degree range.
The XRP consists of two distinct instruments, the **BCS** = Bent-Crystal Spectrometer (PI: J. L. Culhane) and **FCS** = Flat-Crystal Spectrometer. The FCS is able to rotate its crystals to provide its seven detectors with access to the spectral range 1.40-22.43 Å. The BCS, with a collimator field of view of 6', is able to obtain spectra in the range 2-3 Å, at a resolving power $\lambda/\Delta\lambda$ of $\approx 10^4$ with eight position-sensitive proportional counters.

1541) A. C. Aikin, W. Henze, D. J. Kendig, R. Nakatsuka, H. J. P. Smith, "Variations of Mesospheric Equatorial Ozone as observed by the Solar Maximum Mission," Geophysical Research Letters, Vol. 17, No. 3, March 1990, pp. 299-300

Data: The downlink for science data was considerably improved with the availability of TDRS in April 1983. Data from ground-based sun observations (collation of telescopic data and magnetograms with Solar Max's maps) served as forecasts event-predictions on the sun for instrument precision pointing setup of the flare watch on the actual flare site.

K.23 SOHO (Solar and Heliospheric Observatory)

ESA/NASA collaborative mission within ESA's 'Solar Terrestrial Science Programme' (STSP), part of ISTP (International Solar Terrestrial Physics Programme). So far, SOHO represents the most comprehensive space mission devoted to the study of the sun and of the heliosphere. The following overall mission objectives are pursued:

- Solar spectroscopy at soft X-ray and EUV wavelengths (study of the composition of the solar corona, of the structure and dynamics of the magnetic structures making up the corona, and of coronal holes, etc.)
- Study of the structure and dynamics of the solar interior through the observation of minute oscillations on the sun's surface (helio-seismology).
- Study of the solar wind and solar energetic particles, interaction with the Earth, plasma processes in both the solar and magnetospheric context.

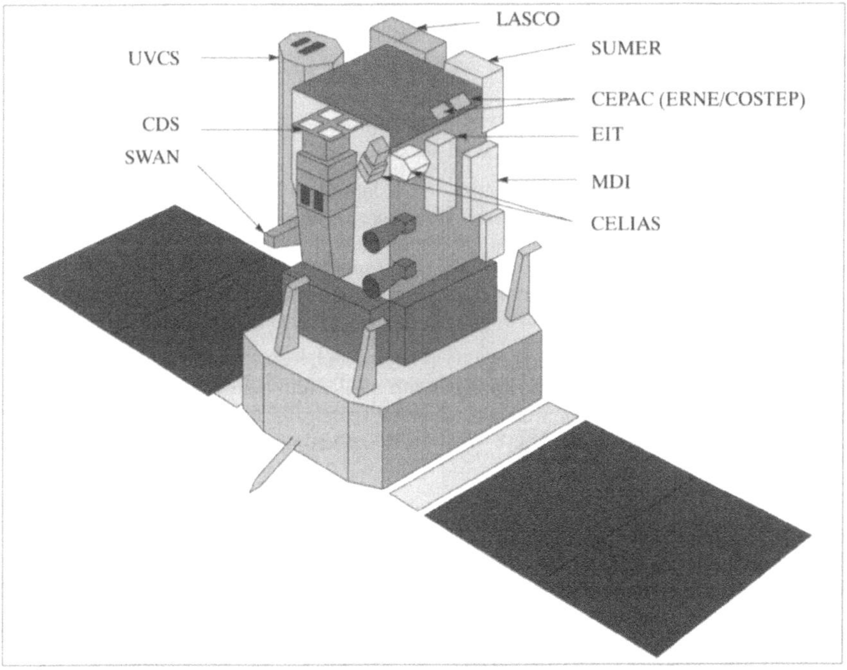

Figure 277: The SOHO S/C model

The spacecraft, built by Matra Marconi Space (MMS), has dimensions of 2.5 m x 2.9 m x 3.9 m (height, and 9.5 m span of the solar array). It is three-axis stabilized within 10 arcsec, with a sun-pointing stability of 1 arcsec over a period of 1.5 minutes. Actuators are 4 reaction wheels and two sets of eight thrusters. Attitude is measured by two fine-pointing sun sensors, two star trackers, and three gyros. Power = 1500 W (solar cells) and 950 W from two 20

Ah NiCd batteries. S/C launch mass = 1875 kg, scientific payload mass = 650 kg, nominal lifetime = 3 years. Within the collaborative mission, the functions are divided in the following way: ESA is responsible for S/C procurement, integration and testing; NASA provides the launch and mission operations from GSFC. A launch from Cape Canaveral occurred on December 2, 1995 aboard an ATLAS-II AS vehicle. [1542) 1543) 1544)]

Note: Operational control of the SOHO S/C was lost on June 25, 1998 (when SOHO spun out of control and communication was lost). The first successful attempts to regain control were conducted on Sept. 16, 1998. The S/C was directed to turn its face and solar power panels fully towards the sun. Instrument re-commissioning started on Oct. 5, 1998 with SUMER and ended with CELIAS on Oct. 24, 1998. Final recovery from ESR (Emergency Sun Re-acquisition mode) and full S/C operation of SOHO was regained Feb. 1, 1999. A gyroless mode of operation was devised and installed with a new software patch for a modification of the AOCS (Attitude and Orbit Control Subsystem), making SOHO the first three-axis-stabilized S/C of ESA to be operated without a gyro.

Orbit: Halo orbit around the sun/Earth Lagrangian Point (L1), about 1.5 million km from Earth (see Figure 394 and Table 547). The pointing accuracy of the S/C in the direction of the sun is ≤ 1 arcsec. SOHO was inserted into its halo orbit around L1 on Feb. 14, 1996 (six weeks ahead of schedule - it retained practically all of its hydrazine supply).

Sensor complement: (eight from Europe, three from USA) [1545) 1546)]

1) Solar Atmosphere Remote Sensing Instruments

SUMER = Solar Ultraviolet Emitted Radiation (PI: K. Wilhelm, MPAe, Katlenburg-Lindau, Germany)
Study of plasma flow characteristics (turbulent motions, waves. temperatures, and densities of the plasma in the upper atmosphere of the sun).

The instrument takes images of the sun in EUV light with high resolution in space, wavelength and time. The spatial resolution is 1.2 arcsec; the spectral resolving power is characterized by a $\lambda/\delta\lambda$ of up to 4.0×10^4 ($\delta\lambda$ corresponds to the pixel size). Spectral shifts can be determined with sub-pixel accuracy. Wavelength range = 500 - 1600 Å. Integration can be as short as 1 second. Data: on-line profiles, shifts and broadenings, ratios of temperature- and density-sensitive EUV emission lines in the range between 10^4 and 2×10^6 K.

CDS = Coronal Diagnostic Spectrometer (B. E. R. Harrison, RAL, UK)
Study of the transition region and corona temperature and density (intensity ratios of selected EVU line pairs, with spatial and temporal scales appropriate to the fine-scale features of the solar atmosphere). There will be close collaboration and correlation with SUMER. [1547)]

The instrument features a Wolter-Schwarzschild type II telescope [designed by MPE Garching, outer f-number = 9.48, focal length = 257.83 cm, FOV = 4 arcmin, on-axis resolution = 2 arcsec (FWHM)], a normal incidence spectrometer (NIS) with a wavelength = 2 bands in 240 - 800 Å range, and a grazing incidence spectrometer (GIS) with a wavelength = four bands in the 150 - 800 Å range.

1542) P. Lo Galbo, M. Bouffard, "SOHO - A Cooperative Scientific Mission to the Sun," ESA Bulletin, Aug. 1992, pp. 21-25

1543) "The Solar-Terrestrial Science Project of the Inter-Agency Consultative Group for Space Science," ESA SP-1107, November 1990, pp. 21-24

1544) J. Credland, F. Felici, M. Grensemann, J. A. Steinz, "Three Missions, Three Launches, Six Spacecraft for Science in 1995," ESA-Bulletin, No. 82, May 1995, pp. 36-47

1545) The SOHO Mission - Scientific and Technical Aspects of the Instruments, ESA SP-1104, ISSN 0379-6566, Nov. 1988

1546) F. Felici, F. C. Vandenbussche, C. Berner, R. Thomas, W. Worrall, et al., Special Section of the SOHO project, spacecraft, payload, and operations in ESA Bulletin No. 84, Nov. 1995, pp. 81-112

1547) K. A. Lidiard, P. F. Gray, "Optical design of the coronal diagnostic spectrometer (an instrument on the Solar and Heliospheric Observatory)," Optical Engineering, Vol. 36, No. 8, Aug. 1997, pp. 2311-2319

EIT = Extreme-Ultraviolet Imaging Telescope (J. P. Delaboudinière, IAS/CNRS, France). Study of the evolution of chromospheric and coronal structures (identification and interpretation of the spatial and temperature fine structures of the solar atmosphere). EIT provides full disk images in emission lines formed at temperatures that map the solar structures. Images in four narrow band passes are obtained using normal incidence multilayered optics deposited on the quadrants of a Ritchey-Chrétien telescope. Temperature range of solar structures to be measured: 6×10^4 - 3×10^6 K. Resolution = 1 arcsec (uniform over FOV = 50 x 50 arcmin). EIT is able to measure the solar transition region and inner corona in four, selected bandpasses in the extreme ultraviolet (EUV):

- Fe IX/X, 171 Å
- Fe XII, 195 Å
- Fe XV, 284 Å
- He II, 304 Å

Using either full-disk or subfield images, the EIT can image active regions, filaments and prominences, coronal holes, coronal "bright points," polar plumes, and a variety of other solar features. EIT was designed to be used in conjunction with other instruments, in particular LASCO, SUMER and CDS.

UVCS = Ultra-Violet Coronograph Spectrometer (J. L. Kohl, SAO, USA)
Study of electron and ion temperatures, densities and velocities in the corona (for interpretation of coronal heating, solar wind acceleration, and solar wind composition). Spectroscopic measurements of the solar corona out to 10 solar radii from the sun's center.

The UVCS instrument consists of an occulted telescope and a high resolution spectrometer assembly. Three off-axis paraboloidal mirrors focus co-registered images of the extended corona onto the three entrance slits of the spectrometer assembly. The spectrometer has three sections. One section is optimized for line profile measurements of H I Lyman-α; its wavelength range is 1148 - 1248 Å. Another section is optimized for line intensity measurements of O VI λ 1032; it is also used to observe Si XII λ 499 and 521 in second order. The first order range is 932 - 1068 Å. The Mg X doublet at λ 610 and λ 625 may be observed in first order. The third section is used to measure polarized radiance of the visible corona.
Telescope: off-axis parabolic mirror, focal length = 750 mm, image scale = 0.218 mm/arcmin, FOV = 42' (tangential) x 141'.
Spectrometer: gratings = 105 (disp) x 70 mm; nominal radius = 750 mm; radius ratio = 1.0259 (1216 Å), = 1.0215 (1032 Å, 625Å), = 1.0040 (visible); ruling frequency = 2400 mm^{-1}; reciprocal dispersion = 5.63 Å/mm (1st order), = 2.82 Å/mm (2nd order).

Spectral Line	Observed Quantity	Spectral Resolution FWHM (Å)	Spatial Resolution FWHM
H I 1216	Profile	0.2	10" x 10"
H I 1216	e^- Profile	2.0	1.6' x 1'
Fe XII 1242	Intensity	1.23	1' x 1'
Fe XII 1242	Profile	0.14	10" x 5'
O VI 1032/37	Intensity	1.23	1' x 1'
O VI 1032	Profile	0.14	10" x 5'
Si XII 499/521, 2nd order	Intensity	0.61	1' x 1'
Si XII 499, 2nd order	Profile	0.07	7" x 5'
Mg X 610/625, 2nd order	Intensity	0.61	1' x 2.5'
MG X 610, 2nd order	Profile	0.07	7" x 5'

Table 459: Observational parameter specifications for the UVCS instrument

LASCO = Large-Angle and Spectrometric Coronograph (PI: G. E. Brueckner, NRL, USA)
Study of structure evolution, mass, momentum and energy transport in the corona. Measurement of electron column densities from just above the limb, at 1.1 R_{sun}, out into deep

heliospheric space, at 30 R_{sun}. Spectral analysis of the inner corona with a high resolution scanning. imaging interferometer. Measurement of spectral profiles of three emission lines and one Fraunhofer line for each pixel (providing temperatures, velocities, turbulent motions and volume densities). Direction of coronal magnetic fields via polarization analysis. The LASCO instrument consists of the coronograph optical system (with three optical systems and CCD cameras) and the instrument control system.

SWAN = Solar Wind Anisotropies (J. L. Bertaux, CNRS, France)
Study of solar wind mass flux anisotropies and temporal variations. The SWAN instrument is a Lyman photometer which maps the sky's interplanetary hydrogen emission almost entirely every other day. From these sky maps, the latitude distribution of the solar wind mass flux from the equator to the pole can be measured, as well as the variation of this distribution (time scale ≈ one month).
The instrument consists of two identical sensor units placed on opposite sides of the S/C and driven by a common electronic box. Each sensor consists of a two-mirror scanning system, a light baffle (to keep out stray light), a hydrogen cell in which a tungsten filament is alternately heated to produce Lyman-α absorbing hydrogen atoms and cooled to produce the reference signal. Resolution of Lyman-α light distribution is 1°.

2) Solar Wind "In Situ" Instruments

CELIAS = Charge, Element and Isotope Analysis (D. Hovestadt, MPI, Garching, Germany). Study of ionic energy distribution and composition. Measurement of the mass, ionic charge and energy of the low and high speed solar wind, of suprathermal ions, and of low energy flare particles.

Sensor	Area, Geometric Factor	ΔE/E	Efficiency of the TOF unit
CTOF	0.08 cm^2	0.04	0.25 - 0.65
MTOF	0.013 cm^2	0.03-5	0.03 - 0.1
PM	7×10^{-5} cm^2	0.05	-
STOF	0.1 - 0.2 cm^2 sr	0.1	0.2 - 0.8

Table 460: Measurement capabilities of the CELIAS instrument

CELIAS includes three mass- and charge-discriminating sensors based on the time-of-flight technique: CTOF (Charge Time-of-Flight) for the elemental, charge and velocity distribution of the solar wind; MTOF (Mass Time-of-Flight) for the elemental and isotopic composition of the solar wind, MTOF also includes a proton monitor (PM); and STOF (Suprathermal Time-of-Flight) for the mass, charge and energy distribution of suprathermal ions. Mass resolution: $M/\Delta M > 100$ MTOF; charge resolution: $\Delta Q \approx 0.3 - 1$, $4 < M < 60$ (typical for CTOF and STOF).

CEPAC = COSTEP/ERNE Particle Analyzer. Collaboration (PI: J. Torsti, ERNE Finland, H. Kunow, U. of Kiel, Germany). Note: COSTEP = Comprehensive Suprathermal and Energetic Particle Analyzer. Three sensors are furnished by the COSTEP consortium: LION (Low Energy Ions and Electrons), MEICA (Medium Energy Ion Composition Analyzer), and EPHIN (Electron Proton Helium Instrument).
Study of energy distribution of particles, energy spectrum and composition. Measurement of particle emissions from the sun over a wide range of species (electrons through iron) and energies (60 keV/particle to 500 MeV/nucleon).

LION instrument consists of two sensor heads each containing a double telescope to measure energy spectra of ions and electrons in the range of 60 keV - 5 MeV for protons, and 60 keV - 300 keV for electrons.

EPHIN is a multielement array of solid state detectors with a plastic scintillator guard counter to measure energy spectra of electrons in the range of 150 keV to > 5 MeV, and hydrogen and helium isotopes in the range of 4 MeV/n to > 53 MeV/n.

3) Helio-seismology Instruments

GOLF = Global Oscillations at Low Frequencies (PI: A. Gabriel, IAS/CNRS, France)
Study of the internal structure of the sun by measuring the spectrum of free global oscillations [global velocity and magnetic field oscillations (low degree modes)]. GOLF measures both p and g mode oscillations, with the emphasis on the low order long period waves which penetrate the solar core (frequencies between 10^{-7} and 6 x 10^{-3} Hz, with a sensitivity of 1 mm/s). The method involves an extension to space of the ground-based technique for measuring the mean line-of-sight velocity of the viewed solar surface. A sodium vapor resonance scattering filter is used in a longitudinal magnetic field to sample the two wings of the solar absorption line.

VIRGO = Variability of Solar Irradiance and Gravity Oscillations [PI: C. Froehlich, WRC (World Radiation Center), Davos, Switzerland] [1548)]
Study of irradiance oscillations (low degree modes) and the solar constant. The total irradiance is measured with active cavity radiometers PMO6-V and DIARAD (Differential Absolute Radiometer), the spectral irradiance by three-channel Sunphotometers (SPM), and the radiance with 12 resolution elements on the solar disk using LOI (Luminosity Oscillations Imager). A scientific objective is probing the solar interior by helioseismology with p- and g-mode solar oscillations determined from spectral irradiance (SPM) and radiance (LOI) variations to the mission time on time scales of minutes.
VIRGO consists of a number of subunits. The optical system of the VIRGO assembly consists of a Ritchey-Chrétien type telescope with a 1300 mm focal length, diameter of 55 mm, and central obstruction 25 mm. Detector: Deep diffused silicon photodiode with 12 scientific and 4 guiding pixels.

- **PMO6-V** (Physikalisch-Meteorologisches Observatorium, 6th radiometer - VIRGO), built by WRC of Davos, is an active cavity radiometer (absolute instrument) measuring continuously the degradation of the active instrument. Some parameter specifications are: life variation: 30 ppm/year, absolute accuracy: 0.17%, sampling rate: 1 solar total irradiance/2 minutes, duty cycle: 17%, resolution: 50 ppm.
- **DIARAD** (Differential Absolute Radiometer), built by IRMB, Brussels. An active cavity radiometer. It is composed of two cylindrical cavities coated inside with diffuse black and mounted next to each other on the same heat sink. Some parameter specifications are: lifetime variation: 30 ppm/year, absolute calibration: 0.15%, sampling rate: 1 solar total irradiance/3 minutes, duty cycle: 11%.
- **SPM** (Sunphotometer), built by WRC. There are two instruments to measure to measure the spectral irradiance at 335, 500, and 865 nm with a bandwidth of 5 nm.
- **LOI** (Luminosity Oscillations Imager), built by ESA. Measurement of the sun diameter variations with a precision of about 0.3 arcseconds for a 1.5 hour integration time. The LOI wavelength and bandpass is chosen to be the same as the 500 nm channel of SPM. This allows the instruments to be calibrated to each other.

MDI = Michelson Doppler Imager (P. H. Scherer, Stanford University, CA, USA)
Study of velocity oscillations (high degree modes). Measurement of line-of-sight velocity by Doppler shift, transverse velocity by local correlation tracking, line and continuum intensity, and line-of-sight magnetic fields with both 4 and 1.4 arcsec resolution (2 and 0.7 arcsec pixels, respectively).
MDI provides velocity maps by sampling the solar Ni I 6768 Å line profile at four points and finding the Doppler shift by determining the phase of the first Fourier coefficient of the line shape with respect to a reference wavelength.

SOHO Data: On-board recording capability of 1 Gbit of data (requirement of 48 h autonomous operation without ground contact)on a tape recorder and 2 Gbit in solid-state memory. Science data transmission via S-band (2.245 GHz). Downlink date rate at 40 kbit/s (160 kbit/s when MDI is in high rate mode).

1548) http://virgo.so.estec.esa.nl/

Some SOHO Highlights

Since SOHO began its mission in 1995, the observatory has provided an unparalleled wealth of information about the sun, from its interior, through the hot and dynamic atmosphere, out to the solar wind. Only a few items are listed here. [1549)]

- Global structure and dynamics of the solar interior: The unprecedented accuracy of helioseismic data from SOHO's MDI, GOLF and VIRGO instruments has enabled substantial improvement in models of the solar interior, and has even shown the importance of considering mixing effects, who in turn solve existing riddles in the isotopic composition of the sun.
- Transition-region dynamics: Several types of transient events have been detected in the quiet sun. Explosive events have been studied using SUMER instrument data.
- Corona: LASCO and EIT are principal instruments. These instruments observed coronal mass ejections (CMEs) that generated subsequent disturbances. These disturbances were observed by other S/C to establish the cause-and-effect relationship for a solar system event that extended from the sun to the solar wind to Earth's magnetosphere and ionosphere. EIT has discovered large-scale transient waves (Coronal Moreton Waves) in the corona that propagate outward from the active regions below CMEs.
- Solar wind: The first tracing of the slow-speed solar wind near the equatorial current sheet
- Solar wind composition: The first detection of elements and isotopes in the solar wind (composition of calcium and nitrogen) observed from data of the CELIAS/MTOF instruments
- Comet hunter SOHO: On Feb. 4, 2000, SOHO discovered its 100th comet. On Aug. 25, 2000, astronomers announced that SOHO had recorded its 200th sun-grazing comet.
- The first image of the convection zone of a star

The scientific achievements of the continuing SOHO mission are the result of a concerted, multi-disciplinary effort by a large international community of solar scientists.

K.24 SOLAR-A (Yohkoh)

Japanese (ISAS) solar-terrestrial mission. The primary objective is the study of high energy phenomena on the sun through X-ray and gamma-ray observations.[1550)]

The S/C has dimensions of 1 m x 1 m x 2 m (height), mass = 420 kg (100 kg science payload), power supply = 450 W (solar cells). Launch on Aug. 30, 1991 by an ISAS M-3SII-6 vehicle; nominal life = 2 years; S/C is three-axis stabilized pointing toward the sun. Status of Yohkoh: As of 4/2001 nominal S/C operations are conducted.

Orbit: apogee = 792 km, perigee = 525 km, inclination = 31.3°, period = 97 minutes

Sensor complement:

- **HXT** = Hard X-ray Telescope. Multipitch bi-grid modulation collimators with NaI scintillators. HXT is a Fourier telescope with 64 collimators covering an energy range from 10 - 100 keV at 7 arcsec resolution for the whole solar disk. HXT mass = 48 kg.

- **SXT** = Soft X-ray Telescope[1551)] (US-Japan collaboration between Lockheed Palo Alto Lab and Tokyo Astronomical Observatory). A grazing incidence soft X-ray mirror with an X-ray CCD detector. SXT uses a 23 cm diameter Wolter-Nariai optics of 155 cm focal length; it covers the energy range from 0.25 - 3 keV at 2 arcsec resolution. Imaging

1549) B. Fleck, P. Brekke, et al., "Four Years of SOHO Discoveries - Some Highlights," ESA Bulletin, 102, May 2000, pp. 69-86

1550) "The SOLAR-A Mission," The Solar-Terrestrial Science Project of the Inter-Agency Consultative Group for Space Science, ESA SP-1107, November 1990, pp. 74-76 is provided by a 1024 x 1024 pixel CCD behind two six-position filter wheels. Mass = 27 kg. - SXT has produced more than 1 million images of the sun since the S/C was launched.

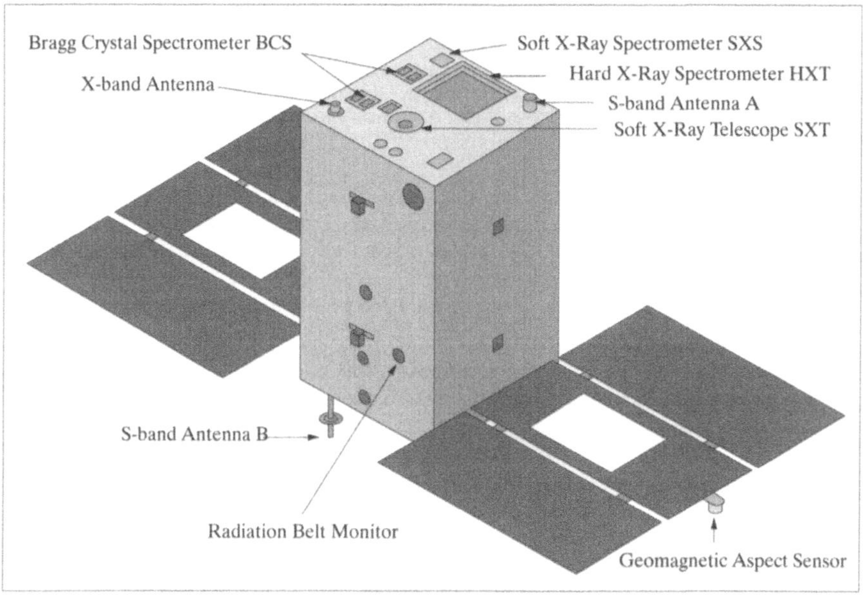

Figure 278: The Solar-A S/C model

- **BCS** = Bragg Crystal Spectrometer (UK-Japan collaboration, Mullard Space Science Lab and RAL, and Tokyo Astronomical Observatory). Bent crystals with position sensitive proportional counters. Mass = 13 kg
- **WBS** = Wide-band Spectrometer. A set of proportional counters, an NaI scintillator and two BGO scintillators. Measurement ranges: 2 - 30 keV, 20 - 400 keV, and 0.2 - 100 MeV. Mass = 16 kg

Data: On-board recorder (80 Mbit capacity, magnetic bubble). Real-time data transmission of 32 kbit/s, or 262 kbit/s playback data, both to Kagoshima. Downlink communication in S-band and X-band.

K.25 Solar-B

Solar-B is an approved ISAS solar physics mission, a successor to the successful Yohkoh/Solar-A mission of Japan/US/UK. The SOLAR-B mission will be conducted as a multi-institutional and international collaboration including Japan with ISAS (Institute of Space and Astronautical Science), the United States with NASA (GSFC and MSFC), and the United Kingdom with PPARC (Particle Physics and Astronomy Research Council). As the lead agency, ISAS is responsible for the spacecraft, spacecraft integration and the optical telescope. The Solar-B payload consists of a set of high-resolution solar telescopes and spectrometers. The overall goal is to study and understand the dynamic sun and its effect on terrestrial climate (global change) and space weather. Further major science objectives are:

1551) "Yohkoh's Prodigious Output helps Scientists Study Sun," Space News, June 7-13, 1993, p. 12

- Comprehensive study of the sun's magnetic fields and a new view into the magnetic dynamics of the plasma universe (first space-based observations of the sun's vector magnetic fields). Understanding basic plasma astrophysics such as the dynamo and magnetic reconnection processes.
- Study of the physics of the sun's luminosity and its variation
- Study of the processes which produce the UV and X-ray emission in the chromosphere and corona
- Study of solar wind energetics. Eruption and expansion of the sun's atmosphere.

The Solar-B spacecraft is three-axis stabilized pointed toward the sun. It is being developed/integrated by Mitsubishi Electric Corporation (MELCO) of Tokyo under ISAS management. Attitude is sensed by IRU (Inertial Reference Unit) gyros with a combination of two sun sensors and a star tracker providing absolute calibration; attitude actuation is provided by four reaction wheels. The pointing stability is in the order of 0.1 arcsec over 10 seconds while finer image stabilization of 0.02 arcsec is achieved for the optical telescope using its own correlation tracker/tip-tilt mirror system. Power of 500 W is provided by two solar arrays (total power of 1 kW). The S/C mass is 900 kg, the mission life is three years. [1552] [1553]

The SOLAR-B spacecraft is scheduled for launch in the summer of 2005 on an ISAS M-V-7 launch vehicle from the Kagoshima Space Center (KSC). This will keep the instruments in continuous sunlight, with no day/night cycling for nine months each year.

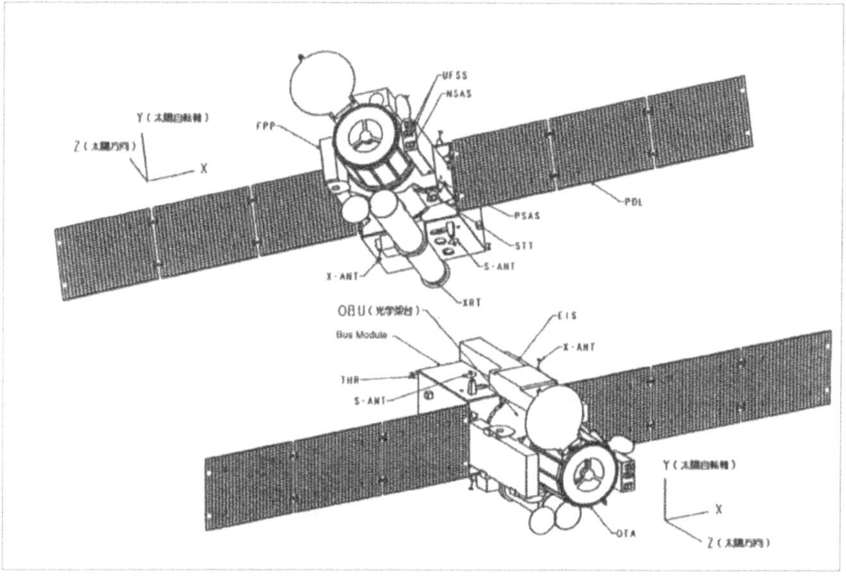

Figure 279: Illustration of the Solar-B spacecraft [1554]

Orbit: Sun-synchronous circular orbit, altitude = 600 km, inclination = 97.9°

An on-board solid-state recorder has a capacity of 5 Gbit. An on-board data compression technique (a special purpose, 12-bit JPEG/DPCM chip) provides either lossy or lossless image compression, dependent on the type of imagery processed. RF communications are provided in X-band and S-band. Downlink data rates are 4 Mbit/s (X-band).

1552) http://www.isas.ac.jp/e/enterp/missions/index.html
1553) http://science.msfc.nasa.gov/ssl/pad/solar/solar−b.htm
1554) Courtesy of Takeo Kosugi of ISAS

Sensor complement:

SOT (Solar Optical Telescope). SOT represents a joint development between Japan and the US, in which Japan (ISAS and NAO) provides the telescope optics and the US (NASA/GSFC) the FPP (Focal Plane Package). The objective is to obtain continuous high-resolution imagery of the sun. SOT is designed to achieve a consistent 0.2 arcsec spatial resolution.

SOT is a diffraction-limited, aplanatic Gregorian telescope with an aperture of 0.5 m. It provides an angular resolution of about 0.2 arcsec over the field of view of about 328 x 164 arcsec. The focal plane package of the optical telescope consists of a broad- and narrow-band filtergraphs (filter vector magnetographs) and a spectro-polarimeter. This combination allows to obtain, for the first time, a continuous series of high-precision vector magnetograms, Dopplergrams, and filtergrams with sub-arcsec resolution.

XRT (X-Ray Telescope). A cooperative instrument between NASA and ISAS. NASA provides the grazing-incidence mirror optics while ISAS provides the CCD camera. XRT images the high temperature (0.5 to 10 MK) corona with improved angular resolution, say approximately 1 arcsec, a few times better than Yohkoh's soft X-ray telescope.

EIS (EUV Imaging Spectrometer). PPARC provides the EIS instrument with NASA and ISAS participating with some hardware and software. EIS is primarily aimed at observing the dynamics of the solar corona. The objective of EIS is to determine velocity fields and other plasma parameters in the corona and the transition region, thus helping to relate coronal dynamic behavior, observed with the XRT, to the underlying photospheric magnetic field, observed with the optical telescope. The EIS instrument is also expected to unveil inflow and outflow of magnetic reconnection, and hence together with the two other telescopes, to finally solve long-standing controversies on coronal heating and dynamics.

K.26 STEREO (Solar-Terrestrial Relations Observatory)

The STEREO mission is a strategic element of NASA's Sun-Earth Connection program, a multi-institutional and international collaboration involving participants from France, Germany, the Netherlands, Belgium, Switzerland, the United States, and the United Kingdom. The overall objective is to increase the understanding of the origin and consequences of CMEs (Coronal Mass Ejections). The mission consists of two nearly identical spacecraft in heliocentric elliptical orbit in the ecliptic plane at approximately 1 AU from the sun: one drifting ahead of the Earth and one behind. Simultaneous image pairs are obtained by the satellites instruments at gradually increasing separations over the course of the two-year mission. A launch is scheduled for 2004. [1555) 1556) 1557) 1558) 1559)]

The STEREO S/C bus consists of a box-like structure of size 2.61 m x 1.85 m x 1.73 m. The S/C is built and integrated by JHU/APL with NASA/GSFC procuring the instruments and launch vehicle. The satellite is three-axis stabilized. Attitude sensing is provided by a star camera, two DSAD (Digital Solar Attitude Detector) [1560)] systems, an IMU (Inertial Management Unit), and a pointing error signal generated by the instrument guide telescope. Actuation is provided by four reaction wheels and 12 thrusters. A S/C pointing accuracy of 20 arcsec is provided. Two deployed solar arrays generate a power of 544 W (EOL). A Super

1555) G. L. Baer, J. E. Eichstedt, D. A. Ossing, "Solar Terrestrial Relations Observatory (STEREO) Mission and Concept of Operations," Proceedings of the 13th AIAA/USU Conference on Small Satellites, Aug. 23-26, 1999, Logan UT, SSC99-IIb-1

1556) "The Sun and Heliosphere in Three Dimensions," Report of the NASA Science Definition Team for the STEREO Mission

1557) J. Galloway, "STEREO Mission Design," NASA/GSFC paper, Sept. 1, 1999

1558) http://sd-www.jhuapl.edu/STEREO/index.html

1559) http://stp.gsfc.nasa.gov/missions/stereo/stereo.htm

1560) K. Strohbehn, M. N. Martin, S. E. Jaskulek, "Micro Digital Solar Attitude Detector," JHU/APL Technical Digest, Vol. 22, No 2, 2001, pp. 104-105

NiCd battery of 21 Ah is used for eclipse phase operations. The bus is designed around the IEM (Integrated Electronics Module) and the PSE (Power System Electronics) unit. The IEM contains the RF uplink and downlink cards, power converter cards, and an AIE (Attitude Interface Electronics) card. The cards within the IEM communicate over a PCI (Parallel Communication Interface). The Power System Electronics unit contains relays and control circuitry for the S/C power system. A MIL-STD-1553 bus architecture is used for on-board command and telemetry transmission between the IEM, the instruments, attitude sensors, and the PSE. The spacecraft has a mass of 335 kg and a design life of two years.

An on-board solid-state recorder has a capacity of 7.5 Gbit. RF communications are provided in X-band. The downlink data rate is variable up to about 357 kbit/s. The downlink data modulation is bi-phase shift keyed (BPSK). The uplink modulation is phase modulated (PM). Mission operations are conducted by JHU/APL in conjunction with each POC (Payload Operations Center) for instrument operations.

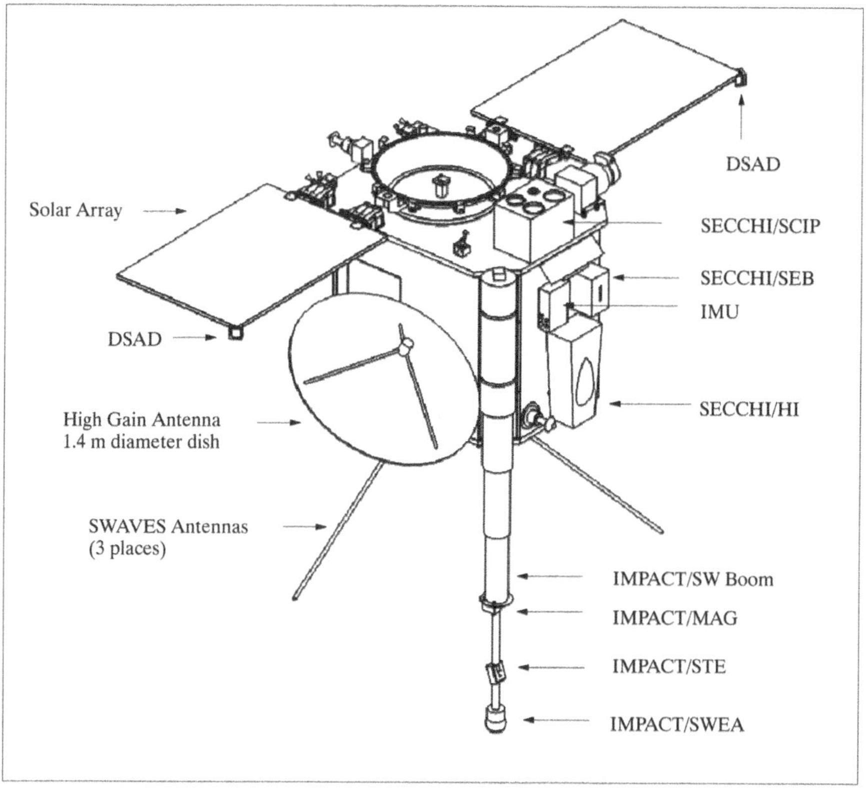

Figure 280: Illustration of a STEREO satellite [1561)]

S/C operations scenario. The design enables nearly autonomous operations with occasional uplink support and to recover science data on a daily basis. The instrument/satellite operations are decoupled, operating independently from each other. In this setup, the instrument science teams are responsible for the following tasks: a) planning, scheduling and generating of commands; b) instrument health; c) instrument calibration; and d) synchronization of instrument operations with the S/C.

1561) Image courtesy of Leslie M. Cusick, NASA/GSFC

The current standard for sun observations is the SOHO (Solar and Heliospheric Observatory) mission with its instrument package. The science objectives of the STEREO project are obviously set to improve on the SOHO observations and to find answers to such questions as:

- Are CMEs driven primarily by magnetic or non-magnetic forces?
- What configuration of the corona leads to CMEs?
- What mechanism initiates a CME?
- What accelerates CMEs?
- How does a CME interact with the heliosphere?
- How do CMEs cause space weather disturbances?

The STEREO mission makes use of simultaneous interplanetary spacecraft operation and stereographic image reconstruction to develop a comprehensive 3-D description of the sun and the heliosphere.

Orbit: Heliocentric elliptical orbit in the ecliptic plane drifting at about 22° from the Sun-Earth line (at nearly 1 AU). The mission profile calls for solar imaging and heliospheric environment sampling by two identical spacecraft (with identical sensor complements) at gradually increasing angular separations from Earth (one drifting ahead of Earth and one behind).

- STEREO Nr. 1 leads Earth by 22° after 1 year
- STEREO Nr. 2 lags Earth by 22° after 1 year

Sensor complement: [1562)]

All STEREO instruments involve participation and collaboration of teams from a umber of institutions. This applies to instrument design and development as well as to instrument operations, data sharing and analysis.

Phenomenon	Feature size resolved (and/or time step)	Physical properties
CMEs near the sun	40, 000 km = $2x10^{-4}$ AU (6 min)	Density, velocity, internal structure, extent
Flares	2,000 km	Position, density, structure
Moreton waves	5,000 km	Wave front shape, velocity, underlying magnetic field
Coronal loops	2,000 km	Temperature, density, structure, deflection by waves
Coronal streamers	40,000 km	Distortion by CMEs, extent
Coronal holes	2,000 km	Footprint, spreading
SEPs (Solar Energetic Particles)	2 minutes	3-D distribution function
CMEs near Earth	0.01 AU (images),1 minute (plasma)	Magnetic field, density, velocity, shape, extent, temperature
Interplanetary shocks	0.02 AU (5 seconds)	Extent, velocity, strength

Table 461: STEREO measurement objectives

The overall measurement strategy calls for:

- Imaging of the solar atmosphere and heliosphere from two perspectives simultaneously
- Tracking of disturbances in 3-D from their onset at the sun to beyond Earth's orbit
- Measurement of energetic particles generated by the disturbances
- Sampling of fields and particles in the disturbances as they pass Earth's orbit.

SECCHI (Sun-Earth Connection Coronal and Heliospheric Investigation), PI: Russell A. Howard of NRL. The instrument is named after one of the first solar physicists, namely A.

1562) http://sd–www.jhuapl.edu/STEREO/Inst/payload.html

Pietro Secchi (1818-1878), who used the new medium of photography to record solar eclipses. The overall objective is to study the 3-D evolution of CMEs from birth at the sun's surface through the corona and interplanetary medium to its eventual impact at Earth. SECCHI consists of a suite of remote sensing instruments: an EUVI (Extreme Ultraviolet Imager), two white-light coronagraphs (COR1 and COR2), and an HI (Heliospheric Imager) package. The EUVI+COR1+COR2 devices are collectively referred to as SCIP (Sun Centered Imaging Package). The SEB (SECCHI Electronics Box) supports all of the SECCHI instruments. The total mass of the SECCHI instrument is 48.1 kg (SCIP=30.2 kg, HI=6.0 kg, SEB=11.9 kg), total power = 44 W. [1563)]

Parameter	COR1	COR2	HI	EUVI
Instrument type	Internally occulted Lyot coronagraph	Externally occulted Lyot coronagraph	Externally occulted coronagraph	EUV narrow-band-pass Ritchey Chretien telescope
Observable	K-corona and CMEs	K-corona, F-corona and CMEs	K-corona, F-corona and CMEs	Emission line corona and upper chromosphere
FOV (Field of View)	1.25 - 4 R_{sun}	2-15 R_{sun}	12 to > 215 R_{sun} HI-1:12-84 R_{sun} HI-2: 66-318 R_{sun}	0 to 1.5 R_{sun}
Spatial scale	7.5 arcsec pixels	14 arcsec pixel	HI-1: 35.2 arcsec HI-2: 120 arcsec	1.6 arcsec pixels
FPA (Focal Plane Array)	1024 x 1024 (2kx2k array summed 2x2)	2048 x 2048	1024 x 1024 (2kx2k array summed 2x2)	2048 x 2048
Bandpass	650-750 nm	650-750 nm	450-750 nm	He II: 30.4 nm Fe IX: 17.1 nm Fe XII: 19.5 nm Fe XIV: 21.1 nm
Exposure times	1 s, 3 required for pB	3 s, 3 required for pB	HI-1: 12 s, HI-2: 60 s Exposures >8 required	Fe IX: 3 s Fe XII: 5 s Fe XIV: 7 s He II: 10 s
Maximum cadence	15 s	22 s	HI-1: 1 min HI-2: 8 min	11 s at full resolu. in Fe IX (4.75 s at half resolution)
Synoptic cadence	8 min	20 min	HI-1: 1 hour HI-2: 2 hours	20 min full resolu. 2.5 min half resolu.
Absolute pointing required	7 arcsec occulter positioning	30 arcsec occulter positioning	30 arcmin	3 arcmin FOV overlap
Pointing stability required	1.5 arcsec over pB sequence (7 s)	1.5 arcsec over pB sequence (17 s)	0.5 arcsec for HI-2 sequence (1 hr)	1.7 arcsec over 1 exposure
Long-term pointing required	7 arcsec over a month to obtain background subtraction (F-corona+ stray light) for total B determination		5.0 arcmin to obtain background model	N/A
Aperture diameter	36 mm	30.5 mm	HI-1: 16 mm HI-2: 21 mm	98 mm
EFL (Effective Focal Length)	f/20	f/6	HI-1: f/5 HI-2: f/2 - f/4	f/18
Stray light/disk light rejection	10^{-6} B_{sun}	10^{-11} B_{sun}	HI-1: 10^{-13} B_{sun} HI-2: 10^{-14} B_{sun}	10^{-12} ratio of visible/EUV
CCD detectors	EEV 42-40, 2kx2K, 13.5 μm pixels, backside illuminated, AR coated (except EUVI), >100 k e^- full well			
Camera	Common electronics, 1 MHz readout rate, 14 bit/pixel quantization			

Table 462: SECCHI instrument specification

The **SCIP** (Solar Centered Imaging Package) contains three compact telescopes to view the solar disk and solar corona. The SCIP structure provides the required alignments, stiffness, and modularity to the three instruments. SCIP also contains the GT (Guide Telescope) that provides fine-pointing information to the S/C attitude control system and to ISS (Instru-

1563) R. A. Howard, J. D. Moses, D. G. Socker, "Sun Earth Connection Coronal and Heliospheric Investigation (SECCHI)," Proceedings of SPIE, Vol. 4139-26 ment Stabilization System) of the EUVI. The pointing strategy is based on pointing the S/C to minimize the scattered light in COR1 and relying upon the SCIP structure to assure that COR2 and EUVI are within acceptable pointing tolerances. The SCIP and the HI assemblies have their own camera electronics inside their enclosures. These units are known as CEB (Camera Electronics Box). The SCIP camera electronics are composed of three circuit cards with a common backplane interface. The CCD (Charge Coupled Device) camera electronics are designed for independent control of all instrument (COR1, COR2, HI, and EUVI) CCDs, a camera readout rate of 1 Mpixel/s through either of two CCD output ports, and 14-bit data quantization.

- **EUVI** (Extreme Ultraviolet Imager). EUVI observes the chromosphere and innermost corona underlying the same portions of the corona and the heliosphere observed by COR1, COR2, and HI. Four EUV emission lines between 17.1 and 30.4 nm are observed. The EUVI design uses a small Ritchey-Chretien telescope of EIT (EUV Imaging Telescope) heritage on SOHO and of TRACE (Transition Region and Coronal Explorer) mission heritage. The improved mirror coatings, detector array, and telemetry allocation provide a higher sensitivity and image cadence of the EUVI instrument than previously possible. The optical system of EUVI, consisting of entrance filter, shutter, filter wheel, primary mirror and active secondary mirror, and a CCD array, provides pixel-limited imaging throughout the FOV in a compact envelope. The EUVI optics are fully baffled. The CCD detector assembly uses a backside-illuminated, backside-thinned and pacified array of size 2k x 2k with an EUV quantum efficiency >70%. The detector is passively cooled by a radiator. EUVI has its own ISS to satisfy its jitter requirements. The ISS includes the EUVI secondary mirror as the actuator and GT as the high-accuracy angular motion sensor. The EUVI secondary mirror is actuated to compensate for the S/C jitter. The GT is used to provide fine-pointing angular measurements to orient the boresights of the SECCHI telescopes to point toward the geometric center of the sun.

- **COR1** (Coronagraph1 Imager). The objective is to explore the inner corona in white light and pB down to 1.25 R_{sun}. COR1 observes the inner (1.25 - 4 R_{sun}) corona with a high frequency and polarization precision. The COR1 optical design is an all-refractive 7-element system. The instrument FOV is is 1.25-4 R_{sun} for a full field width of 2.13° and an aperture diameter of 36 mm. The spectral range is 650-750 nm. Images are collected at three different linear polarization angles. The occulter is attached to the second optical element in COR1, a doublet field lens. The final stage consists of two doublets, along with the filter and polarization optics. The stage takes the chromatically aberrated image of the corona, near the occulter, and relays it onto the CCD detector. The 2k x 2k detector data is summed on-board to 1k x 1k images to produce 7.5 arcsec pixels.

- **COR2** (Coronagraph2 Imager). The objective is to explore the outer corona (2-15 R_{sun}) in white light and pB with high spatial and temporal resolutions. The COR2 design is of LASCO heritage on SOHO. The optical design of the instrument uses a classical externally occulted Lyot design with an entrance aperture diameter of 30.5 mm and an overall length of 1348 mm. The spectral range and polarization technique is the same as COR1 to provide a seamless transition between the two regions.

- **HI** (Heliospheric Imager). The objective is to extend the concept of traditional externally occulted coronagraphs to a new regime, by observing the heliosphere from the sun to the Earth (12-300 R_{sun}). The HI assembly consists of two instruments, HI-1 and HI-2, the exterior baffle systems, the hinged baffle cover, the focal plane packages, the HI camera electronics, and the radiators for the CCDs. - HI-1 looks at the inner heliosphere to within 3.28° of the sun with an opening angle of 20°. The HI-1 camera uses a five-vane forward linear baffle system and a matching linear internal occulter, a 20° full field angle lens, and a 2048 x 2048 pixel format CCD.

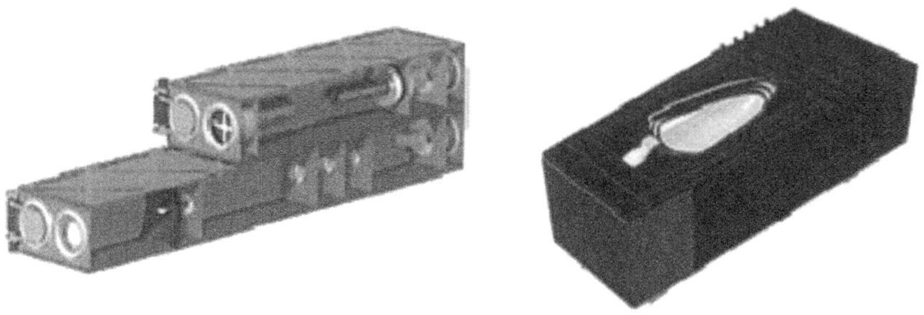

SCIP (Sun Centered Imaging Package) HI (Heliospheric Imager)

Figure 281: SECCHI instrument illustration

The HI-2 camera looks further out, from an elongation of 18.36°, with an opening angle of 70°. The HI-2 objective is set deep within the forward baffle system shadow at a diffraction angle of 16.5°. The HI-2 camera consists of a wide-angle fisheye lens and a CCD detector array. [1564]

The SEB (SECCHI Electronics Box) provides all command and data handling functions and is the only electronic interface with the S/C. SEB consists of a control computer, interfaces for the camera and CCD, fine-pointing and jitter control electronics, a housekeeping and data acquisition system, and power conversion. A RISC-based single board computer RAD6000 (32 MHz) of Lockheed Martin is used. The image processing and compression software employs such compression routines as Rice (lossless) and a lossy wavelet compression (H-compress) technique.

IMPACT (In-situ Measurements of Particles and CME Transients), PI: Janet G. Luhmann of the Space Sciences Laboratory of UCB (University of California at Berkeley). [1565] The IMPACT investigation science focuses on the magnetic connections to the sun and topology of interplanetary CME-related interplanetary disturbances, and the energetic particles that precede and accompany these disturbances as they move toward Earth. IMPACT addresses these with a combination of solar wind electron and solar energetic particle instrumentation that, with the PLASTIC solar wind ion measurements provide the comprehensive, identical in-situ measurements at the two STEREO spacecraft locations necessary to relate what is seen in the SECCHI images to the space weather environment at 1 AU. IMPACT's solar wind electron instruments and magnetometer are mounted on the STEREO boom, while the SEP (Solar Energetic Particles) instruments are on the spacecraft body. All of these instruments use the IDPU (IMPACT Instrument Data

Processing Unit), which serves as a single interface to the spacecraft. The PLASTIC solar wind ion instrument also uses the IMPACT IDPU. IMPACT's boom instruments consist of the electron instruments SWEA (Solar Wind Electron Analyzer) and STE (Suprathermal Electron Telescope) and the magnetometer. The SEP package is also a multi-instrument (4) system, but it is more physically integrated, and so included under one heading here.

SWEA (Solar Wind Electron Analyzer). The objective is designed to measure the distribution function of the solar wind core and halo electrons from about 1 eV to several keV, with high spectral and angular resolution over practically the full spherical range. This capability allows the distinction between these components in detail during both undisturbed periods

1564) http;//www.pxi.com/SECCHI

1565) http://sprg.ssl.berkeley.edu/impact/

and the passage of CME generated disturbances, when the interplanetary field rotates far out of the ecliptic plane.
SWEA consists of a hemispherical top hat ESA (ElectroStatic Analyzer) that provides a 360° FOV in a plane, combined with electrostatic deflectors to provide nearly 4π coverage when SWEA is mounted at the end of the STEREO boom. The inner plate radius is 3.75 cm and the plate separation is 0.28 cm. The resulting energy resolution dE/E is 18%, and the geometric factor is 0.01 cm^2 sr E (eV). SWEA compensates for the effects of spacecraft potential on the lowest energy particles by having an outer hemisphere that can be biased according to the plasma density measured by the PLASTIC solar wind ion instrument.

STE (Suprathermal Electron Telescope). STE is a new instrument that covers electrons in the energy range of about 2-20 keV with approximately 50 times the sensitivity of previous instruments. These higher energy electrons are frequently accelerated in flares or impulsive flare-like bursts, and they produce solar type III radio bursts as they escape the sun. They provide a unique tracer of the footpoints of CME and normal interplanetary magnetic fields, and allow the measurement of the length of those field lines. In addition, an electron superhalo is continuously present in the interplanetary medium at those energies. STE utilizes passively cooled SSDs (Silicon Semiconductor Devices) that measure all energies simultaneously. The STE consists of two detector arrays of four SSDs in a row, each about 0.1 cm^2 in area and about 500 µm thick. Each array looks through a rectangular opening that provides a FOV of about 20° x 80° for each SSD with the 80° direction perpendicular to the ecliptic. Adjacent FOVs are offset for a total FOV of about 80° x 80°. The two arrays are mounted back-to-back, looking in opposite directions, centered about 25° from the average Parker Spiral field direction. STE is located just inboard of SWEA on the STEREO boom to clear its FOV and remain in shadow.

MAG (Magnetometer). The MAG system is a simplified version of the magnetometers flown on Mars Global Surveyor and Lunar Prospector. It is a triaxial fluxgate design, mounted on the STEREO boom (about 4m) just inboard of the SWEA and STE instruments. The fluxgate sensors use a ring core geometry, with magnetic cores consisting of permalloy. The units are compact, low power, and ultra-stable. To optimize sensitivity at the low field values to be found in interplanetary space, the magnetometer dynamic range is divided into 8 ranges that are automatically switched whenever the field being measured exceeds or falls below predetermined levels. The maximum range is sufficiently large to allow IMPACT MAG magnetic field measurements in the Earth's magnetic field during the commissioning phase orbits.

SEP (Solar Energetic Particles) package. SEP consists of 4 separate instruments that cover the full solar energetic particle energy, flux and composition ranges needed to meet the STEREO science goals. The SEP instruments include the SEPT (Solar Electron Proton Telescope), SIT (Suprathermal Ion Telescope), LET (Low Energy Telescope), and the HET (High Energy Telescope). Together, the SEP instruments measure electrons, protons and heavier ions from several tens of keV to approximately 100 MeV (or MeV/nucleon).

- **SEPT** (Solar Electron Proton Telescope). SEPT consists of two dual, double-ended magnet/foil solid state detector particle telescopes that cleanly separate and measure electrons in the energy range 20-400 keV and protons from 20-7000 keV, while providing anisotropy information through the use of several FOVs. Each SSD detector in SEPT is 300 µm thick and 0.53 cm^2 in area. A rare-earth permanent magnet is used to sweep away electrons for ion detection, while a parylene foil transmits electrons but stops protons. SEPT is divided into two pieces for FOV reasons. The SEPT-E telescope is housed with the rest of the SEP package, located on the body of the S/C. It looks in the ecliptic plane along the Parker Spiral magnetic field direction, both forward and backward. SEPT-N/S is housed separately at a different spacecraft location and looks out of the ecliptic plane perpendicular to the nominal magnetic field, both north and south. The viewing cones for the SEPT telescopes are each about 60°.

- **SIT** (Suprathermal Ion Telescope). SIT is a time-of-flight ion mass spectrometer that measures elemental composition of He-Fe ions over the energy range of about 30 keV/nucleon to 2 MeV/nucleon. The FOV angles are 17° x 44°, with the 44° angle in the ecliptic plane, centered approximately 60° from the spacecraft-sun line to avoid sunlight while still intercepting significant numbers of Parker Spiral field controlled energetic ion fluxes. The telescope analyzes ions that enter through thin entrance foils and stop in a solid state detector. A time-of-flight approach for determining the composition utilizes start and stop times obtained from secondary electrons entering a MCP (MicroChannel Plate) detector system. The MCP and SSD areas are each 6.0 cm^2. The SIT geometric factor allows study of even small SEP events.

- **LET** (Low Energy Telescope). LET is a special double-fan arrangement of 14 solid-state detectors designed to measure protons and helium ions from about 1.5 to 13 MeV/nucleon, and heavier ions from about 2 to 30 MeV/nucleon. LET uses a standard dE/dx vs. E-technique, identifying particles that stop at depths of about 20-70 µm and about 70-2000 µm corresponding to two general energy ranges. The large FOV spans from 20° above to 20° below the ecliptic plane and extends 65° to either side of the forward and backward Parker Spiral field directions in the ecliptic plane. Like SIT, LET's large geometric factor also ensures the detection of even small SEP events.

- **HET** (High Energy Telescope). HET also uses the solid-state detector, dE/dx vs. E-approach, but in a six-detector, more traditional linear arrangement designed to measure protons and helium ions to 100 MeV/nucleon and energetic electrons to 5 MeV. HET identifies particles that stop at depths of 1 to 8 µm in the detectors. Some information will also be obtained on heavier nuclei up through Fe using the dE/dx vs. E signatures and ranges together, and penetrating particles will be analyzed. The FOV covers a 47.5° cone around the Parker Spiral field direction. SIT, LET and HET are all packaged together with the SEPT-E subsystem and a SEP CPU (Central Processing Unit) in the main SEP package on the spacecraft body, but they use the IMPACT IDPU to interface with the spacecraft together with the rest of IMPACT.

PLASTIC (PLAsma and SupraThermal Ion and Composition), PI: Antoinette Galvin of UNH (University of New Hampshire), cooperation of UNH, University of Bern, Switzerland, MPE Garching, Germany and NASA/GSFC. The objective of PLASTIC is to study coronal SW (Solar Wind) and solar wind-heliospheric processes. The science objective is to measure ions in the energy-per-charge range of 0.2 to 100 keV/e. The instrument performs three functions in one package: [1566)]

- Measurement of the distribution functions of solar wind protons and alpha particles (providing density, velocity, kinetic temperature and its anisotropy), with a time resolution of about one minute.
- The SW sector also provides, on at least five minute resolution, the elemental composition, charge state distribution, kinetic temperature, and velocity of the more abundant solar wind heavy ions (e.g., C, O, Ne, Mg, Si, and Fe).
- The PLASTIC Wide Angle Partition measures the distribution functions of suprathermal ions H through Fe, with a comparatively large geometrical factor that allows the study of suprathermal particles, including shock-accelerated particles and pick up ions. Together with IMPACT measurements, PLASTIC completes the required STEREO mission in-situ observations.

SWAVES (STEREO/WAVES), an interplanetary radio burst tracker. PI: Jean-Louis H. Bougeret of CNRS (Centre National de la Recherche Scientifique) Observatory of Paris. The goal of the SWAVES radio and plasma waves investigation is to obtain unique and critical observations for all primary science objectives of the STEREO mission, the generation of CMEs, their evolution and their interaction with Earth's magnetosphere. SWAVES can

1566) http://stereo.sr.unh.edu/

probe a CME from lift-off to Earth by detecting the coronal and interplanetary (IP) shock of the most powerful CMEs, providing a radial profile through spectral imaging, determining the radial velocity from about 2 R_{sun} (from center of sun) to Earth, measuring the density of the volume of the heliosphere between the sun and Earth, and measuring important in situ properties of the IP shock, magnetic cloud, and density compression in the fast solar wind stream that follows. SWAVES achieves these goals by measuring IP (Interplanetary) type II and type III radio bursts, both remotely and in-situ. [1567)]

- SWAVES antenna system. SWAVES uses three mutually orthogonal monopole stacer antenna elements as its prime sensors, each 6 meters in length. The three monopoles are deployed away from the sun so that they remain out of the fields of view of sunward looking instruments. The antenna design optimizes the radio burst tracking in the 16MHz - 30kHz range and maintains a high SNR for expected solar type II, type III and other solar and IP radio emissions.
- Preamplifiers. A high input impedance preamplifier is connected to each of the three electric monopoles.
- Radio receivers. There are five radio receivers in the SWAVES instrument to cover the following frequency ranges: 1) LFR Lo: 10-40 kHz; 2) LFR Hi: 40-160 kHz; 3) HFR: 0.125-16 MHz; 4) FFR1: 50 MHz fixed frequency; 5) TDS: 250,000 samples/s time series snapshots.
- Housekeeping and low-rate science. This system provides digitized and selectable housekeeping data to monitor the status and health of the various parts of the instrument.
- Calibration. The instrument contains its own internal calibration circuitry. An internal calibration sequence will be invoked about once per day.

The focal point of SWAVES instrument commanding and operations will be performed at the UMN (University of Minnesota).

Space weather data. "Real-time" in situ space weather data products from IMPACT and PLASTIC, consisting of approximately 1 minute cadence information on solar wind electrons and ions, solar energetic particles and magnetic fields are packaged in the data stream from the IMPACT IDPU with the regular coded science and housekeeping telemetry from the instruments.

K.27 TWINS (Two Wide-angle Imaging Neutral-atom Spectrometers)

TWINS is a collaborative Sun-Earth connection two-satellite mission of NASA, LANL (Los Alamos National Laboratory), JHU/APL, SwRI (Southwest Research Institute), USC (University of Southern California), WVU (West Virginia University), and The Aerospace Corporation. The overall mission objective is to demonstrate a new capability of stereoscopic imaging of the magnetosphere. By imaging the charge exchange neutral atoms over a broad energy range (about 1-100 keV) using two identical instruments on two widely spaced high-altitude and high-inclination spacecraft, TWINS will enable the 3-three-dimensional visualization and the resolution of large scale structures and dynamics within the magnetosphere for the first time. Areas to be studied by TWINS include: [1568)] [1569)]

- Neutral atoms and charged particles
- The Earth's magnetic field
- Geomagnetic storms and substorms
- Auroras
- The 11-year solar cycle
- Space weather - the effects of the sun on human life on earth and in space.

1567) http://www-lep.gsfc.nasa.gov/swaves/swaves.html

1568) D. J. McComas, "Two Wide-Angle Imaging Neutral-Atom Spectrometers,"

1569) http://nis-www.lanl.gov/nis-projects/twins/

The feasibility of magnetospheric imaging using ENAs (Energetic Neutral Atoms), which arise from the charge exchange process between cold geocoronal neutral hydrogen and local energetic ion populations, was first demonstrated in the early 1990s (ENAP on ATLAS-1 (STS-45 flight, launch March 24, 1992). The POLAR mission (launch Feb 24, 1996) flies the CEPPAD/SEPS (Comprehensive Energetic-Particle Pitch Angle Distribution / Source Loss Cone Energetic Particle Spectrometer) instrument package capable of detecting ENAs. The SAC-B mission (launch Nov. 4, 1996) had ISENA (Imaging Spectrometer for Energetic Neutral Atoms) of CNR/IFSI on-board. Unfortunately, the third stage of the SAC-B launch vehicle failed to separate from the S/C, thus preventing any operations. The IMAGE mission of NASA (launch March 25, 2000) employs three ENA imagers (LENA, MENA, HENA). It is expected that the imagery obtained by the TWINS constellation (two widely-spaced high-altitude spacecraft) may revolutionize the study of the magnetosphere.

Each S/C is three-axis stabilized and approximately nadir pointing.

The first TWINS S/C is planned to be launched by early 2002, and the second one by early 2004, providing a full two year stereo mission beginning in 2004.

Orbit: An EEO (Elliptical Earth Orbit) or a Molniya-type orbit, perigee = 300 km, apogee = 7.2 R_E or 46,000 km, inclination = 63.4°, period about 12 hours. The majority of the orbital period is spent near apogee.

Sensor complement: The instrument package consists of TEI and a Lyman imager to monitor the geocorona. Both instruments are mounted on a rotating actuator platform to allow 360° azimuthal viewing.

TEI (TWINS ENA Imager). TEI is of MENA instrument heritage flown on NASA's IMAGE mission. TEI measurements of neutral H, He, and O cover an energy range of about 1-100 keV with 4° x 4° angular resolution and a 1 minute time resolution. The sensor heads and the signal processing electronics of TEI, including signal amplification, trajectory calculation, time-of-flight (TOF) determination, and pulse height analysis, are identical to those of MENA on the IMAGE mission.

The primary differences between MENA and the TEI are the DPU and the way the viewing plane is moved across the sky, both of which are spacecraft-driven. For IMAGE, the S/C spin sweeps the MENA field of view (FOV) across the sky. Because the TWINS S/C are three-axis stabilized, the sensor heads are mounted on a rotatable actuator which sweeps the FOV over all pertinent look directions. The fundamental measurements made by TEI are:

- Trajectory measurement of an ENA using the detected positions of the ENA and secondary electrons emitted as the ENA transits an ultrathin foil
- TOF measurement between an ENA and its correlated secondary electrons
- Pulse height analysis of the detected signals.

Imaging is derived from the trajectory measurement. Speed is derived from the trajectory and TOF measurements, and the ENA mass is inferred from the speed and pulse height analysis.

The collimator plates have a 2.7 cm inner radius and a 9.8 cm outer radius. The gap between adjacent plates is 0.5 cm. This geometry yields a 4° FWHM azimuthal FOV and, with the plates alternately grounded and biased to +10 keV, rejects ions and electrons up to an energy of 130 keV/q, where q is the ion charge.

ENAs and associated secondary electrons strike a standard 10 cm x 10 cm MCP (Microchannel Plate) detector in a chevron (2-plate) configuration. An electric field, which is applied at the entrance surface of the MCP detector, using a grounded suppression grid and a +100 V bias on the front of the MCP detector, increases both the detection efficiency and the spatial resolution. TWINS employs MCPs with a 60:1 length-to-diameter channel ratio.

Parameter	Value	Comment
Geometric factor	0.038 cm^2 sr	Two heads
Polar FOV	±60° FWHM	Imaged
Polar resolution (mass and energy dependent)	4° FWHM	Imaged, best case
Azimuthal FOV	±2° FWHM	Collimated
Azimuthal resolution	4° FWHM	Collimated
Energy range	1-100 keV	TOF window for H
Mass identification	H, O	TOF+PHD
Energy resolution (ΔE/E)	0.4	TOF+mass
Time resolution	60 s	Actuator rotation period
UV rejection	4.7×10^{-9}	Worst case

Table 463: Performance parameters of TEI

Lyman-α detector. The objective is the measurement of the geocoronal H density. The instrument consists of the following elements: Baffle, BP filter, and Channeltron detector. Data is collected with 2° azimuthal angle spacing.

Aperture area	3.2 cm^2
FOV (Field of View)	4° FWHM, 2 detectors with lines-of-sight offset by ±40° with respect to rotation axis
Wavelength, wavelength resolution	121.584 nm, ±5 nm
Sensitivity	0.1-0.5 counts/s/R
Sampling	Every 2° azimuthal motion (0.667 s)

Table 464: Parameters of the Lyman-α detector instrument

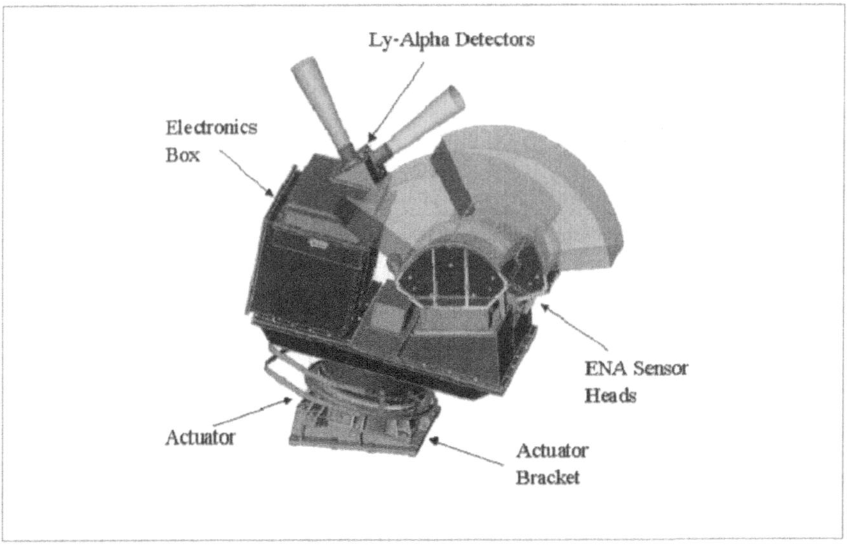

Figure 282: The TWINS instrument package

K.28 Ulysses

A joint ESA/NASA project which started in 1977 under the name of 'Out-of-ecliptic' mission; it was later renamed ISPM (International Solar-Polar Mission), and was given its final name 'Ulysses' in 1984. The overall objective of Ulysses is to explore the sun's polar regions (first-ever study of the particles and fields in the inner heliosphere at all solar latitudes, including the polar regions). Particular emphasis is on the study of the solar wind, the interplanetary magnetic field, and the complex wave and particle interaction phenomena that exist in the interplanetary medium. Naturally there is international collaboration and coordination of the Ulysses mission with other ISTP projects like: INTERBALL (K.17), SOHO (K.23), WIND (K.30), and SOLAR-A (Yohkoh, K.24). [1570] [1571] [1572] [1573] [1574]

The structure of the Ulysses S/C is a box-shaped aluminum bus. The S/C is spin-stabilized (5 rpm); a high-gain communication antenna (HGA) with a 1.65 m diameter dish is mounted onto the spin axis; the dish is pointing toward the Earth continuously. The AOCS (Attitude Control System) is comprised of the redundant sun-sensor system, the AOCE (Attitude and Orbit Control Electronics), the AME (Attitude Measurement Electronics) and the RCE (Reaction Control Equipment). Power is provided by RTG (Radio-isotope Thermoelectric Generator), with 285 W at start and 244 W at the end of the first solar cycle. Two sets of 4 x 2 N hydrazine thrusters (33 kg of propellant) provide spin control and trajectory corrections. The S/C is thermally controlled by passive means (radiators, thermal paints, and thermal blankets) and by heaters. S/C total mass of 370 kg, mass of science payload = 55 kg. Booms: radial dipole antenna of 72.5 m length, axial monopole antenna of 7.5 m length, and a radial magnetometer boom of 5.6 m length. Major contributions: ESA [funding of S/C, S/C operations, 50% of experiments by ESA member states]; NASA [funding of S/C launch, provision of RTG, use of DSN (Deep Space Network) and JPL control center, 50% of experiments]. The S/C was built by an industrial consortium with Dornier System as the prime contractor.

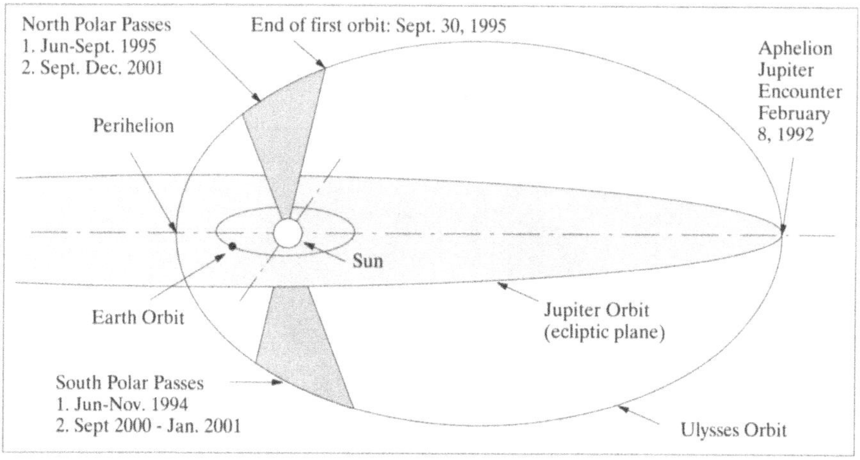

Figure 283: The Ulysses flight path as viewed from 15° above the ecliptic plane

Data communications: X-band downlink (20 W, 8.4 GHz frequency) and S-band uplink/downlink (5 W, 2112 MHz uplink and 2293 MHz downlink). Telemetry data at 1.024 kbit/s (real-time) and at 512 bit/s of storage data. Two redundant tape recorders have a capacity of

1570) ESA Bulletin No. 63, August 1990, Special Issue on Ulysses

1571) N. Angold, et al., "Ulysses Operations at Jupiter - Planning the Unknown," ESA Bulletin No. 72, November 1992, pp. 44-51

1572) R. G. Marsden, K. P. Wenzel, "The Ulysses Jupiter Flyby - The Scientific Results," ESA Bulletin No. 72, November 1992, pp. 52-59

1573) R. G. Marsden, "Ulysses Explores the South Pole of the Sun," ESA Bulletin No. 82, May 1995, pp. 48-55

1574) K. P. Wenzel, et al., "The Ulysses Mission," Astronomy and Astrophysics, Supplement Series, Vol. 92, January 1992, pp. 207-219

45.8 Mbit each. Ulysses is tracked 8 hours/day by the 34 m antennas of NASA's Deep Space Network. Mission operations are supported by a joint ESA/NASA team at NASA/JPL.

The Ulysses S/C along with the Inertial Upper Stage (IUS) was launched on the Shuttle mission STS 41 (Discovery), October 6, 1990, from Cape Canaveral, Florida.

Orbit: The Ulysses probe is the first S/C having left the ecliptic plane to observe the poles of the sun. [1575] [1576] [1577] The orbital change into a normal plane to the ecliptic was achieved with a sling-shot flyby, a gravity-assist maneuver, of Jupiter (Feb. 8, 1992). The S/C operates at heliocentric ranges which vary between 1.34 and 5.4 AU (where the solar flux varies between 1400 Wm^{-2} and 45 Wm^{-2}) inclined 79.1° to the ecliptic. The portions of the orbit when Ulysses is above 70° solar latitude have been designated as "polar passes." The first such polar pass occurred in 1994 (south) and 1995 (north). The S/C takes 6.2 years to complete one orbit of the sun. The second set of polar passes takes place from Sept. 2000 to Jan. 2001 (south) and Sept. to Dec. 2001 (north).

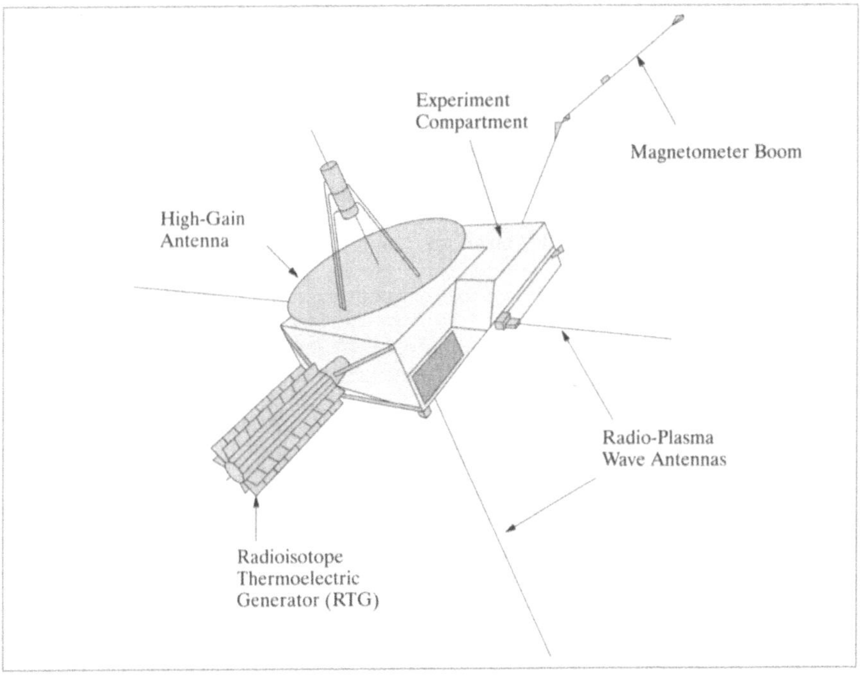

Figure 284: The Ulysses S/C model

Sensor complement:

COSPIN = Cosmic-ray and Solar Particle Investigation (PI: J. A. Simpson, U. of Chicago). Objective: Measurements of solar and interplanetary particles to characterize their intensities, energy spectra, anisotropy and composition as a function of solar latitude. The instrument comprises five solid-state detector telescopes and a double Cerenkov/semiconductor

1575) R. G. Marsden, " Ulysses at Solar Maximum and Beyond," ESA Bulletin, No 103, Aug. 2000, pp. 41-47

1576) A. Mc Garry, N. Angold, "Ulysses 7 Years On - Operational Challenges and Lessons Learned," ESA Bulletin No. 92, November 1997, pp. 69-74

1577) R. G. Marsden, K. P. Wenzel, "The Heliosphere in Perspective - Key Results from the Ulysses Mission at Solar Minimum," ESA Bulletin No. 92, November 1997, pp. 75-81 telescope; energy range = 0.3-600 MeV/n for nuclei and 1-300 MeV for electrons. In addition, one of the telescopes provides measurements of the isotropic composition of galactic cosmic-ray nuclei up to iron (transport of particles). Mass=14.6 kg, power=14.7 W, data rate=160 bit/s.

DUST = Dust Experiment (PI: E. Grün, MPIK, Heidelberg). Objective: Measurement of the properties (mass, speed, direction, and electric charge) and distribution of cosmic dust particles in the solar system. The instrument consists of a large hemispherical sensor and a data processing unit. The sensor is made up of a grid system for the measurement of particle charge, an electrically grounded and gold-plated target (hemisphere), and a negatively biased ion collector. Instrument mass=3.8 kg, power=2.2 W, data rate=8 bit/s.

EPAC/GAS = Energetic Particle Composition Experiment/Interstellar Neutral Gas Experiment (PI: E. Keppler, MPAe, Lindau). Objective: Measurement of the intensities, anisotropies, composition and energy spectra of energetic interplanetary ions in the energy range 0.4-15 MeV/n. Study of the acceleration and transport of low-energy ions of both solar flare and interplanetary origin, with emphasis on their characteristics at high solar latitudes. EPAC consists of a four-telescope detector system. Each telescope contains one epitaxial detector and two surface-barrier detectors. The telescopes are arranged in such a way that between them they cover 80% of a full sphere per S/C spin.
GAS Objective: Direct observations of interstellar neutral helium atoms that have penetrated the heliosphere. Measurement of the parameters (density, bulk velocity relative to the solar system, and temperature) of interstellar gas. The instrument consists of a sensor employing CEM (Channel Electron Multiplier) detector elements. The instrument is mounted on a rotating platform with a scan angle of 180° and is mechanically attached to an electronics box. Total mass=4.3 kg, power=4.1 W, data rate=16 bit/s.

Event		Date
Launch		October 6, 1990
Jupiter flyby		February 8, 1992
1st polar pass of sun (a polar pass is considered the period above 70° solar latitude)	Start Maximum latitude (80.2°S) End	June 26, 1994 September 13, 1994 November 5, 1994
Perihelion		March 12, 1995
2nd polar pass of sun	Start Maximum latitude (80.2⁰ N) End	June 19, 1995 July 31, 1995 September 29, 1995
Start of 2nd solar orbit		October 1, 1995
3rd polar pass of sun	Start Maximum latitude (80.2° S) End	September 8, 2000 November 27, 2000 January 16, 2001
Perihelion		May 26, 2001
4th polar pass of sun	start maximum latitude (80.2⁰ N) end	September 3, 2001 October 13, 2001 December 12, 2001
End of mission		December 31, 2001 (or later)

Table 465: Key dates during the Ulysses mission

FGM/VHM = Fluxgate Magnetometer/Vector Helium Magnetometer (PI: A. Balogh, Imperial College, London). Objective: In-situ observations of the helio-latitude dependence of the interplanetary magnetic field (structure, characteristics, topology, dynamic properties, etc.). Both magnetometers are boom-mounted to eliminate the magnetic effects of the S/C (resolution of 4 pT in the most sensitive range). The FGM instrument is based on three ring-core fluxgate sensors arranged in an orthogonal triad. The VHM sensor measures magnetic fields by their effect on the efficiency with which metastable helium can be optically pumped. Changes in optical pumping efficiency are dependent on the magnetic-field intensity and its angle with respect to the optical axis of the sensor. Mass=4.7 kg, power=5.4 W, data rate=80 bit/s.

GRB = Gamma-Ray Burst Experiment (PI: K. C. Hurley, UCB). Objective: Measurement of the intensity and the spectral characteristics of X-ray and gamma-ray bursts of solar and cosmic origin. Localization of the sources of bursts of high-energy photons. The instrument consists of a gamma-ray sensor and a soft X-ray sensor mounted on the S/C radial boom, and a data processing unit inside the S/C. The gamma-ray sensor consists of two hemispherical CsI scintillation crystals coupled to PMT detectors; FOV $=4\pi$ (nearly unobstructed view); energy range = 15-150 keV. The soft X-ray sensor measures X-rays in the energy range 5-15 keV; it consists of two Si surface-barrier detectors and front-end electronics passively cooled to -50°C. Mass=2.0 kg, power=2.6 W, data rate=40 bit/s.

HI-SCALE = Heliosphere Instrument for Spectra Composition and Anisotropy at Low Energies (PI: L. J. Lanzerotti, AT&T Bell Labs). Objective: Measurement of interplanetary ions and electrons at the lowest energies covered by the suite of energetic-particle experiments. Ions at energies > 50 keV and electrons above 30 keV are detected by five separate solid-state detector telescopes, oriented to give essentially complete pitch-angle coverage of the spinning S/C. Two of the telescopes contain magnets that prevent low-energy electrons from reaching the detectors, permitting clean ion measurements. Thin aperture foils that stop low-energy ions but allow electrons to pass freely are included in the 3rd and 4th telescopes. The 5th telescope performs ion-composition measurements (three detector elements). Mass=5.8 kg, power=4.0 W, data rate=160 bit/s.

SWICS= Solar Wind Ion Composition Spectrometer (PI: G. Gloeckler, U. of Maryland; J. Geiss, U. of Bern). Objective: Measurement of the chemical and ionic-charge composition of solar-wind ions, as well as their temperatures and mean speeds. Detection of all major solar-wind ions from H through Fe, at solar-wing speeds ranging from 145 km/s (protons) to 1352 km/s (Fe^{+8}). The instrument consists of an electrostatic analyzer (fan-shaped collimator enclosing the electrostatic deflection system), a time-of-flight telescope (contained in a drum-shaped high-voltage bubble), and solid-state proton/alpha detectors. Measurement of ion energy per charge, mass, and total energy, determination of the atomic mass (by combining the speed and total energy signals), and the ionization state of the particle (by combining the energy per charge and total energy values). Mass=5.6 kg, power=4.0 W, data rate=88 bit/s.

SWOOPS = Solar Wind Observations Over the Poles of the Sun (PI: J. L. Phillips, LANL). Objective: Observation of the solar-wind flow at middle-to-high solar latitudes for the construction of a representative model of the heliosphere. Measurement of the bulk flow parameters (density, speed and direction) of the solar wind in three dimensions at all heliocentric distances and heliographic latitudes reached by the S/C. SWOOPS consists of two independent instruments, each making use of a curved-plate electrostatic analyzer and CEM (Channel Electron Multiplier) detectors. The CEMs are arranged so that particle velocity distributions can be resolved for all orientations of the S/C spin axis. Electrons with energies between 1-900 eV are detected; the measurement range for ions is 35 to 257 eV/Q. The ion instrument incorporates a stepper-motor-driven aperture wheel with seven apertures. Mass=6.7 kg, power=5.5 W, data rate=160 bit/s.

URAP = Unified Radio and Plasma-Wave Experiment (PI: R. J. MacDowall, NASA/GSFC). Objective: Detection of distant radio wave emissions (via remote sensing) and detection of locally-generated plasma waves; tracking of solar radio noise bursts as they travel away from the sun for the study of the heliospheric magnetic field and solar wind flows. The antennas used by this experiment consist of a pair of wire booms (72 m tip-to-tip) lying in the S/C spin plane forming a dipole, and an axial boom (7.5 m long) which forms a monopole antenna along the S/C spin axis. Mass=7.4 kg (excluding antennas), power=10 W, data rate=232 bit/s.

GWE = Gravitational Wave Experiment (PI: B. Bertotti, U. of Padua). Objective: Search for low-frequency wide-band gravitational waves thought to be emitted from various sources. If Ulysses is exposed to a burst of gravitational waves, then the S/C will experience a minute perturbation in its orbital velocity, which in turn produces a characteristic noise component in the Doppler-shifted radio signal received on the ground.

SCE = Solar Corona Experiment (PI: M. K. Bird, U. of Bonn). Two radio-science experiments are being carried out with the Ulysses S/C by making use of the S/C and ground communication equipment (SCE and GWE). SCE conducts coronal sounding using dual-frequency ranging and Doppler measurements during times of superior conjunction in order to determine the electron flux density and solar-wind velocity in the solar corona (also mapping of the electron content in the high-latitude heliosphere).

Two radio science investigations, using the spacecraft and the ground communication equipment, were conducted during specific missions periods.

K.29 Viking

The Viking project is a Swedish satellite mission, funded by the Swedish Board of Space Activities, and managed by SSC (Swedish Space Corporation) with international cooperation on sensors. Objectives: Study of the physical processes associated with the interaction of the solar wind with the Earth's magnetosphere (establish the region for acceleration of auroral particles and the mechanisms involved, behavior of the resulting Aurora Borealis) in the high latitude medium height (1-2 R_e) regions. [1578) 1579) 1580)]

Launch Feb. 22, 1986 by Ariane V16 from Kourou (piggyback payload with SPOT-1). The satellite was attitude controlled (spin stabilized at 3 rpm, cartwheel mode, magnetic control), size approximately 1.9 x 0.5 m, satellite mass in orbit = 286 kg, payload mass = 44.7 kg (68 kg including booms and antennas). S/C size: 1.9 m diameter, 50 cm height. S-band downlink at 49.6 kbit/s data rate. Power from solar cells on spacecraft body totalling 2.2 m^2, providing 80 W average power. No onboard data storage.
The design life of Viking was 8 months, the satellite provided data for 15 months (end of mission May 12, 1987).

Orbit: Polar orbit, apogee = 13530 km, perigee = 817 km, inclination = 98.8°, orbital period = 262 min, drift of rising node = 0.17°/day, rotation of line of apsides = - 0.50°/day. It took 8 months for the apogee to move from a starting point with a latitude of about 40° N up over the polar region and down to the equator. After these 8 months the apogee was in the southern hemisphere. Note: there was no on-board storage capability and there was no receiving station in the Antarctic region; hence, measurements could only be made at low altitudes from the Esrange/Kiruna ground station.

Sensor complement:

V1: Electric Field Experiment (PI: Lars Block of RIT, also involved: ESTEC and UC Berkeley). Measurement of the DC electric field by three orthogonal pairs of spherical probes.

- Four wire booms, each 40 m long
- Two rigid booms, 4 m each
- 53 or 106 samples per second
- data rate = 4.3 kbit/s
- mass = 4.7 kg (without booms)
- power = 7.3 W

1578) B. Hultqvist, "The Swedish Satellite Project Viking," Journal of Geophysical Research, Vol. 95, No. A5,, May 1, 1990, pp. 5749-5752
1579) B. Hultqvist, "The Viking Project," Geophysical Research Letters, Volume 14, No. 4, April 1987, pp. 379-382
1580) "The Viking Program," EOS Transactions, American Geophysical Union, Volume 67, No. 42, Oct. 21, 1986, pp. 793-795

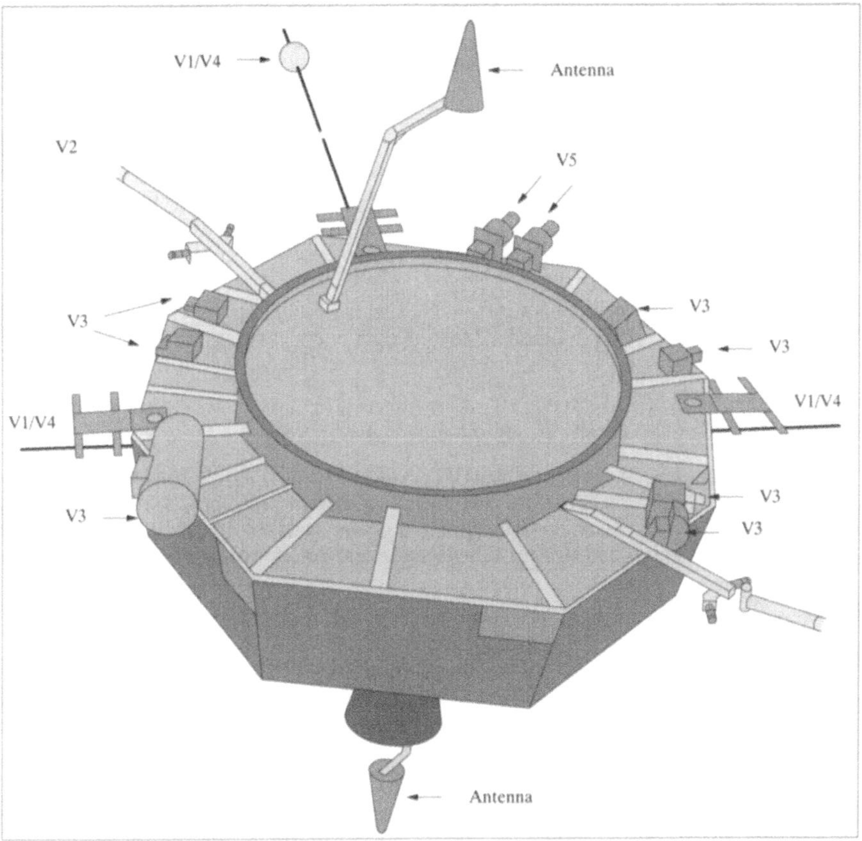

Figure 285: The Viking S/C model

V2: Magnetic Field Experiment (PI Tom Potemra of APL, also involved: IRF-U. of Sweden, GSFC/NASA, and U. of Kyoto). The V1 and V2 experiments have the same sampling rates to determine whether the variations in the electric and magnetic fields are due to fine-scale currents or waves.

- three-axial fluxgate magnetometer system on a 2 m long radial boom
- 53 samples per second
- four ranges of measurement: from ±65536 (±8) nT to ±1024 (±o.125) nT
- data rate = 2.2 kbit/s
- mass = 2.1 kg
- power = 1.1 W

V3: Hot Plasma/Energetic Particle Experiment (PI R. Lundin of IRF, also involved: U. of Bergen, MPI Lindau, Los Alamos National Laboratory, Aerospace, APL, RAL). Measurement objectives:

- electrons with spectral resolution $\Delta E/E \sim 0.05$, from 10eV to 40 ke
- electrons with angular resolution $\Delta\alpha \leq 2^{1}$ from 0.1 keV to 300 keV
- energy and pitch angle distributions of ions $\Delta E/E \sim 0.08$, $\Delta\alpha \leq 6$ from 40 eV to 40 keV
- three-dimensional distribution functions of ions from the satellite potential to 10 keV
- composition, energy, and pitch angle distribution of ions up to 10 MeV/nucleon.

All these measurements can be made simultaneously. The instrument consists of seven units covering different kinds of particles and different energy ranges.

- data rate = 12.4 kbit/s
- mass = 18.2 kg
- power = 13.7 W

V4L: Low-Frequency Wave Experiment (PI: G. Gustavsson of IRF-U, also involved IRF of Sweden, DSRI of Denmark, Cornell University, NY). The wave experiment is divided into two parts, depending on the frequency. V4L measures waves up to about 10 kHz. Two wave signals can be processed simultaneously that may come from any of the electric field antennas, from a magnetic loop antenna measuring fluctuations in the direction of the spin axis, or from the V1 experiment.
The experiment uses two wire booms from the Electric Field Experiment and in addition a magnetic coil antenna (mounted on a 2 m long rigid boom) to measure one component of the magnetic wave field.

- frequency range: 0 - 15 kHz
- measurement also of $\Delta N_e/N_e$ and the wave phase velocity

V4H: High-Frequency Wave Experiment (PI: A. Bahnsen of DSRI, also involved: CRPE of France, IRF-U of Sweden). V4H has two sets of eight filters each, with center frequencies from 4 kHz to 512 kHz, and a pair of stepped frequency analyzers from 10 kHz to 500 kHz. Analog signals are provided by one of the electric field antennas or from the magnetic loop antenna.

- the experiment uses the same sensors as the low-frequency experiment
- frequency range: 10 kHz - 500 kHz
- the experiment also contains an active resonance experiment which among other things gives the electron density from the plasma frequency.
- data rate = 25.6 kbit/s (for low and high frequency experiments together)
- mass = 11.4 kg (for low and high frequency experiments together)
- power = 21.2 W (for low and high frequency experiments together)

V5: Auroral Imaging Experiment (PI: C. Anger/S. Murphree of U. of Calgary, also involved: NRC of Canada, U. of York, Canada, U. of Saskatchewan, Canada, U. of Alberta, Canada, IRF of Sweden, MISU of Sweden). Study of the dynamic behavior of auroral forms. The instrument determines the auroral luminosity distribution over the polar cap and the auroral regions.

- field of view: 29° x 24°
- wavelength ranges: 1240-1500 Å and 1330-1900 Å (two different cameras)
- at apogee one pixel covers 20 x 20 km^2.
- an image can be produced every spin (in 20 seconds)
- the aurora can be seen with good contrast also on the sunlit side of the Earth
- data rate = 5.1 kbit/s (a higher bit rate of 30.7 kbit/s can be used at the expense of the wave experiments)
- mass = 8.5 kg
- power = 5.3 W

K.30 WIND

NASA/GSFC solar-terrestrial mission within the US GGS (Global Geospace Science) ISTP initiatives/programs with the objective to study sources, acceleration mechanisms and propagation processes of energetic particles and solar wind. Investigation of solar wind mass momentum and energy) with input first from the day-side double lunar swingby orbit, and later from a small halo orbit at L_1.

Launch: Nov. 1, 1994 with a Delta II vehicle from Cape Canaveral. S/C mass = 1250 kg (300 kg propellant, 195 kg science payload). Nominal lifetime = 3 years. Spin-stabilized S/C at 20 rpm around an axis within 1° of normal to the ecliptic. S/C size: 2.4 m diameter and 1.8 m height. Solar arrays provide 370 W, including 144 W for payload instruments. The Wind S/C was built by Martin Marietta Astro Space of Princeton, NJ.

Orbit: lunar swingby to a 250 R_E apogee and a perigee of at least 5 R_E during first two years; thereafter L_1 halo orbit (figure-eight orbit around the Earth and the moon on the sun side of the Earth).

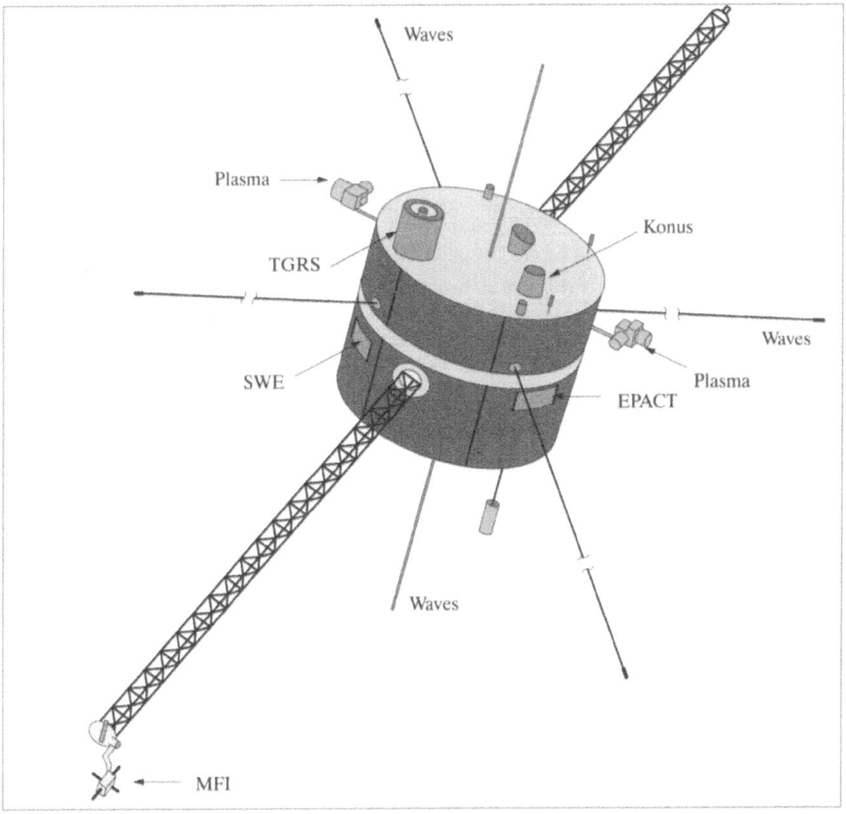

Figure 286: The WIND S/C model

Sensor complement:

MFI = Magnetic Field Investigation (PI: R. Lepping, GSFC). Objectives: investigation of the structure and fluctuations of the interplanetary magnetic field (transport of energy and acceleration of particles in the solar wind).
Instrument: Magnetometer measures the intensity and direction of magnetic field vector. Measurement rate of 44 vectors per second.

WAVES = Radio and Plasma Wave Experiment (PI: J. Bouqeret, Observatoire de Meudon, France). Objectives: measurement of the plasma over a very wide frequency range.

SWE = Solar Wind Experiment (PI: K. Ogilvie, GSFC). Objectives: measurement ions and electrons in the solar wind and the foreshock regions. Rates of once per minute for ions and 20 times per minute for electrons. Deduction of solar wind velocity, density, temperature, and heat flux.

EPACT = Energetic Particles Acceleration, Composition, Transport (PI: T. von Rosenvinge, GSFC). Objectives: investigation of the elemental and isotopic abundances of the minor ions making up the solar wind with energies in excess of 20 keV. Measurements at a rate of once per minute for ions and 20 times per minute for electrons. Deduction of solar-wind velocity, density, temperature, and heat flux.

TGRS = Transient Gamma Ray Spectrometer (PI: B. Teegarden, GSFC). Objectives: observation of transient gamma-ray events, spectroscopic survey of cosmic gamma-ray transients, measurements of gamma-ray lines in solar flares. Measurement ranges: 15 keV - 8.2 MeV.

SMS experiment, consisting of: **SWICS** (Solar Wind Ion Composition Study) and **STICS** (Mass Sensor, and Suprathermal Ion Composition Study); (PI: G. Gloeckler, U. of Maryland). Objectives: determine the abundance, velocity, spectra, temperature, and thermal speeds of solar-wind ions (plasma investigations in conjunction with EPACT).

PLASMA = 3-D Plasma and Energetic Particles Experiment (PI: R. Lin, U. of California, Berkeley). Objectives: measurement of ions and electrons with energies above that of the solar wind and into the energy particle range. Energy range: 0.03 - 30 keV, sampling rate: 20 times per minute; wide angular coverage, good directional sensitivity. Study of particles in the bow shock and in the foreshock regions.

KONUS = Gamma Ray Burst Investigation (PIs: E. Mazets/T. Cline, Ioffe Physical Technical Institute, St. Petersburg, GSFC, instrument is sponsored by Russia). Objectives: gamma-ray burst studies similar to the TGRS studies. KONUS has a lower resolution than TGRS but a broader coverage to complement TGRS. KONUS also performs event detection and measures time history.

Data: On-board recording capability (1.3 Gbit digital tape recorder). Transmission via DSN (Deep Space Network) for 2 hr nominal daily contact. Science data rates: 5.6 kbit/s realtime and 128 kbit/s playback data. The WIND S/C provides on-board interconnection of instrumentation for data communication. Data sharing among the instruments can be triggered by pattern recognition schemes in on-board computers.

Part L Space Stations

L.1 ISS (International Space Station)

The International Space Station is the largest space program that has ever been undertaken on the basis of international cooperation. It is being built and operated by five International Partners: the USA, Russia, Japan, Canada, and Europe. In Europe, ten member states of ESA are participating in the program: Belgium, Denmark, France, Germany, Italy, the Netherlands, Norway, Spain, Sweden, and Switzerland. In addition, Brazil is participating.

The built-up phase of ISS began with the launch of FGB (Functional Cargo Block) on Nov. 20, 1998, also referred to as Zarya (meaning 'sunrise'), on a Proton vehicle from the Baikonur Cosmodrome in Kazakhstan. "Permanent" habitation of ISS started with the launch of the first crew on Oct. 31, 2000 (see chapter 1.4 for the various phases of ISS assembly).

The ISS will consist of various modules linked to a truss segment holding the solar panels. The US is the overall integrator of the ISS, it has the responsibility for developing and operating major elements and systems. US-provided systems are: FGB (Functional Cargo Block) or Zarya, built for NASA by KhSC of Moscow, the US Laboratory, known as node 1 (Unity), the ESA/Italian built nodes 2 and 3, the truss, and solar arrays. A robotic arm, provided by Canada, will run along the truss. Russia provides, among other contributions, the Service Module Zvezda (meaning 'star'), including accommodation facilities, research modules and additional solar arrays. Three Japanese research elements [JEM-PM (Japanese Experiment Module-Pressurized Module), JEM-EF (JEM-External Facility), and ELM-PS (Experiment Logistics Module-Pressurized Section)], the European Columbus Laboratory, and three MPLMs (Multi-Purpose Logistics Module), provided to the ISS program by ASI of Italy, will be linked to the front of the US modules. There is also ELM-ES (ELM-Exposed Section), an unpressurized section of Japan which serves as a storage and carrier module for spare external hardware and user payloads. When completed the total mass of ISS will exceed 400,000 kg. The ISS power resources call for a minimum of 26 kW continuous support and 30 kW average power to payloads during standard and microgravity modes of operation.

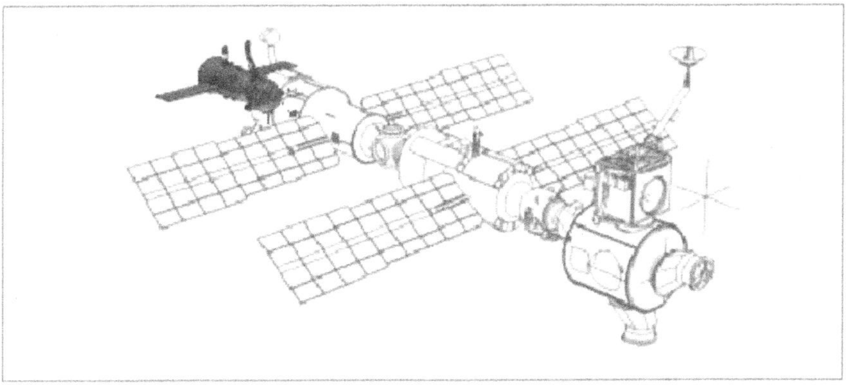

Figure 287: ISS configuration at start of permanent habitation in 2000

RF communications for payload data via TDRS: ISS provides an on-board command and data distribution network associated with a forward (uplink) and return (downlink) antenna communication. A 72 kbit/s forward S-band link is used to send commands to payloads. A 150 Mbit/s Ku-band system is used for the payload downlink. On-board data are distributed from the payloads APS (Automated Payload Switch). APS distributes the data to a high-rate modem for distribution to the Ku-band system. Three types of connections are available:

- MIL-STD 1553B payload bus (through a payload multiplexer/demultiplexer)
- Ethernet 802.3
- Fiber-optic HRDL (High-Rate Data Link)

A 216 Gbit communications outage recorder is provided to record research data during loss of signal with the communication system.

The MCC (Mission Control Center) at NASA/JSC and the POIC (Payload Operations Integration Center) at NASA/MSFC provide integrated station mission and payload control. The ISS partners will operate POCCs (Payload Operation Control Center) in their own countries.; they are networked into the integrated research operations of ISS. [1581)]

Orbit of ISS: circular orbit (non-sun-synchronous), target altitude of 410-460 km (about 350 km during the assembly phase), inclination = 51.6°, period of about 90 minutes.

Operational mode	Description
Standard	• Represents core operations when tended or preparing to support human presence • Provides a "shirt sleeve" environment • Internal and external operations supported, monitored and controlled
Reboost	• Used to obtain additional altitude while maintaining a habitable environment and supporting internal and external user payload operations • Attitude controlled propulsively
Microgravity	• Consists of capabilities required for microgravity research by user payloads in a habitable environment • Does not include effects of crew activity, but does include effects of crew equipment (e.g. exercise devices)
Survival	• Initiated upon command or when a warning of imminent thread (e.g., loss of attitude control, loss of thermal conditioning, available power out-of-range) is not acknowledged by the on-orbit crew, the Orbiter crew, or the ground • Autonomously attempts to correct the threatening condition and provides keep-alive utilities to ISS crew/core systems • Precludes support or commanding of external or internal operations
Proximity operations	• Provides capabilities related to supporting safe operations with other vehicles while maintaining a habitable environment and supporting internal and external user payload operations • Vehicle is actively determining and controlling attitude non-propulsively
Assured safe crew return	• Provides mitigation capability for life threatening illness, unrecoverable loss of ISS habitability, or extended problem requiring resupply/servicing, which is prevented from occurring due to launch problems • Consists of actions, operations and functions necessary to safely populate the CRV (Crew Return Vehicle), separate the CRV, return the CRV to Earth, and egress the CRV upon recovery on the ground
External operations	• Utilizes functionality related to supporting station-based external external operations while maintaining a habitable environment and supporting internal and external payload operations • Vehicle actively determining and controlling its attitude non-propulsively

Table 466: ISS modes of operations

L.2 ISS Utilization - Selected Payloads and Instruments

The ISS offers a great utilization potential for research including a wide range of applications in such fields as astronomy, microgravity, material science, biology, medicine, and

1581) J. D. F. Bartoe, L. Fortenberry, "One Year Old and Growing: A Status Report of the International Space Station and its Partners," Acta Astronautica, Vol. 47, No. 2-9, 2000, pp. 589-597 technology demonstrations in virtually all disciplines. ISS will also serve as a platform for the wide field of Earth observation. The following sub-chapters represent short descriptions of the first candidate ISS external payloads and instruments that are being considered by ESA, NASA, NASDA, etc., **the emphases of selected topics is on Earth observation.**

For station-internal research payloads, ISS adopted standard accommodations with ISPR (International Standard Payload Rack). The 37 ISPR slots for science payloads on ISS provide a common set of interfaces regardless of location. Non-standard services are also provided at selected locations to support specific payload requirements. Each ISPR provides 1.6 m^3 of volume. The rack mass is 104 kg and can accommodate a payload mass of up to 700 kg. In addition, the ISS offers a choice of several external sites for mounting payloads to be exposed to the surrounding space environment. These are:

- Four locations on the S3 segment of the US-provided Station Truss of 107 m in length
- Ten locations on the Japan-provided JEM (Japanese Experiment Module)
- Four locations for ExPAs (Express Pallet Adapter) on the Columbus module of ESA
- Mounting locations on the Russian segment.

Three options exist for US Attached Payload Resource Accommodations (Station Truss):

1) A standardized US-provided ExPA (Express Pallet Adapter), launched and retrieved by the Shuttle, allows payloads to be mounted at external sites on the ISS Truss structure. The pallets can be zenith- or nadir-pointing, allowing them to carry instruments requiring solar or celestial viewing (zenith) or Earth viewing (nadir). An ExPA facilitates the housing of six adapters, each capable of carrying a payload of up to 225 kg on a mounting surface of 1 m^2. An ExPA occupies an entire truss site using the truss attach system.

2) A US Express Pallet (allocation per pallet). The mounting platform dimensions are: 394 cm x 229 cm. The total payload mass may be up to 1360 kg.

3) A US Truss Site (payload allocation per site). Payload masses of up to 4990 kg can be accommodated with a maximum payload height of 3.1 m.

L.2.1 ACCESS (Advanced Cosmic-Ray Composition Experiment for Space Station)

ACCESS is a NASA space science instrument on ISS (program scientist: W. V. Jones) in collaboration with many universities and foreign institutions. [1582] The overall objective is to determine the highest-energy cosmic rays by measuring the characteristics of individual elements over a wide range of mass and energy (the cosmic ray energies of supernonae are estimated to be in the order of 10^{15} eV (the charge range Z for high energy is: $1 <= Z <= 28$). Cosmic rays are an important part of the dynamics of our Galaxy; their impact is largely hidden from us due to the Earth's magnetic field. Individual elemental spectra can be used to directly determine the source spectra of the cosmic-ray nuclei. ACCESS can study these spectra with unprecedented accuracy over more than 4 decades in energy in a single instrument. To collect enough of the very high-energy comic rays ACCESS must have a large effective collecting area of several m^2. In addition, ACCESS must also contain a calorimeter to provide an energy estimate for lighter cosmic rays. ACCESS is designed for a zenith-pointed site on ISS, consisting of the combination of a large-area transition radiation detector, referred to as TRD (Transition Radiation Detector) with a hadronic calorimeter (to analyze hydrogen, helium, and a number of heavier cosmic rays). The instrument mass is estimated to be 5900 kg (instrument size: about 2 m x 2 m x 3 m; power of 850 W; mission lifetime of 4 years). A Shuttle launch of ACCESS to ISS is planned for 2007.

1582) http://lheawww.gsfc.nasa.gov/ACCESS/

L.2.2 ACES (Atomic Clock Ensemble in Space)

ACES is an ESA-selected nadir-oriented instrument. The core of the ACES project, developed by CNES, is a laser-cooled cesium atomic clock, called PHARAO (Project d'Horloge Atomique à Refroidissement d'Atomes en Orbite), that is exploiting the microgravity conditions onboard ISS to reach unprecedented precision not achievable on Earth. In fact, a cold atomic clock works more accurately under weightlessness than under Earth's gravity. This is related to the principle of "atomic fountain" which is used in the clock. Under microgravity, the cold atoms cloud can traverse the microwave cavity with a slower speed than on Earth. This permits to increase the interaction time between the cesium atoms and the microwave field to 5-10 s which in turn reduces the atomic resonance linewidth. - The result would be an improvement of the frequency stability and accuracy of the atomic clock by a factor of 100 over conventional devices on Earth. [1583)]

The operation of the laser-cooled cesium atomic clock PHARAO opens opportunities in various fields of fundamental research and application fields. The ultra precise measurement of time will also allow relativistic measurements and tests, applications in atmospheric physics and geodesy, navigation and advanced telecommunications.

L.2.3 AMS (Alpha Magnetic Spectrometer)

AMS is a NASA-selected payload, the first sensitive spaceborne magnetic spectrometer, with international cooperation of China, Finland, Germany, Italy, Russia, Switzerland, and Taiwan (the team is led by Samuel C. Ting of MIT). The key science objective is to search for antimatter and of the missing matter in the universe (broadly stated, the AMS goal is to provide first detailed observations of charged particles outside the Earth's atmosphere - in the cosmic void). The search is basically for anti-helium and anti-carbon with a detector sensitivity 10^4 to 10^5 times better than current limits. A predecessor AMS experiment was initially flown on Shuttle flight STS-91 (June 2-12, 1998). The early experiment was particularly helpful in calibrating the instrument. In 2003 (Shuttle flight UF-4), AMS will be positioned on an external payload attach site of ISS and remain there for 3-5 years. [1584)] [1585)]

L.2.4 ARISS (Amateur Radio on the International Space Station)

ARISS is a US program that offers an opportunity for students to experience the excitement of Amateur Radio by talking directly with crew members of the ISS. ARISS is sponsored by ARRL (American Radio Relay League), AMSAT (Amateur Satellite Corporation), NASA, and FCC (Federal Communications Commission).
The ARISS communication package (initial radio station) was flown to ISS onboard STS-106 flight of Atlantis (Sept 8-20, 2000) to be available for the "Expedition One Crew." A Russian Soyuz rocket (launch Oct. 31, 2000) lifted the first expedition crew to ISS from Baikonur, referred to as ISS-2R (start of permanent habitation of ISS). The ARISS downlink VHF frequency is 145.80 MHz (worldwide for voice and packet); the voice uplink frequency is 144.49 MHz for regions 2 and 3, and 145.20 MHz for region 1. The packet uplink frequency is at 145.99 MHz (worldwide). [1586)] [1587)]

L.2.5 CRESPO (Coral Reef Ecosystem Spectro-Photometric Observatory)

CRESPO is a UnESS (University Earth System Science Project) instrument (PI: M. Atkinson, University of Hawaii). CRESPO is a four-channel narrow spectral bandwidth and high spectral resolution imager. The overall objectives are:

1583) R. D. Andresen, G. Peters, "European Payloads Selected for Early Utilization on the International Space Station," ESA Bulletin No 98, June 1999, pp. 76-86
1584) http://hpl3tri1.cern.ch/AMS/
1585) http://ams.cern.ch/AMS/ams_homepage.html
1586) http://spaceflight.nasa.gov/station/reference/radio/
1587) http://www.rac.ca/ariss.htm

- To develop the first comprehensive and systematic global survey of coral reef "health" and status
- To produce a systematic synoptic spectral observations of reef benthic communities
- To provide direct and significant participation by students in all aspects of the CRESPO mission.

The instrument will obtain key multispectral data for over 9000 mapped reefs and other reefs identified on the basis of an autonomous data acquisition algorithm. These data will be the only global uniform data set for coral reefs, and will also improve the georeferencing of reefs by a factor of 10 in remote areas in the Pacific. - CRESPO will be installed on ISS using the Window Observational Research Facility (WORF, see L.2.18).

L.2.6 EUTEF (European Technology Exposure Facility)

EUTEF is a multi-user support facility of ESA. The objective is to provide a modular accommodation for a variety of technology payloads requiring space exposure. It consists of a material-property laboratory allowing periodic on-board measurements of surface degradation, and a comprehensive environment-monitoring package to characterize the ISS space environment, including high-energy cosmic radiation, the natural and ISS-induced plasma environment, atomic-oxygen concentration, etc. The servicing of the payloads is provided by a robotic arm (ASI-provided). EUTEF is planned for launch in 2005 (UF5).

L.2.7 FOCUS (Fire Detection and Analysis Sensor System)

FOCUS is a DLR-proposed ESA prototype instrument for HTE (High Temperature Event) detection/monitoring (PI: H. Röser of DLR, Berlin, in cooperation of various institutes from Spain, Italy, Germany, France, Greece, and Russia). FOCUS is a demonstration package (to be used in future operational missions) with a set of infrared instruments and related electronics and software for autonomous on-board detection and monitoring/classification of high-temperature events on Earth such as vegetation fires (biomass burning) and volcanic activity/eruptions. The FOCUS design concept features a new combination of spatial and spectral co-registered infrared data specifically tailored for high-temperature events. The following geophysical variables and data products are the main interest for: [1588)]

1) Wildfire recognition

- Physical characterization of hot spots: fire area and temperature, precise geolocation
- Maps pf active fires and burnt areas
- Plume temperature, column content (density) and profile of the plume trace gases: CO_2, CO, NO, CH_4, H_2O, SO_2, CO/CO_2 ratio, CH_4/CO_2 ratio
- Optical depth of smoke, aerosol particle size distribution

2) Volcanic activity observation

- Surface temperature images of volcanic regions
- Maps of near-vent surface temperature
- Maps of geothermal anomalies
- Volcanic plume radiance maps and spectral signatures
- Volcanic gas species column content (in particular SO_2 flux estimates), estimation of non-SO_2 gas species flux/content, using the ratio with SO_2 flux/content.

3) Coal seam fire recognition

- Physical characterization of hot spots: fire area and temperature, precise geo-location
- Column content (density) and profile of CO_2, CO, including the CO/CO_2 ratio to characterize the combustion efficiency.

1588) FOCUS Phase-A Extension Executive Summary, prepared by OHB System, DLR, and Zeiss, Sept. 2000

4) Pre-operational requirements for HTE (High-Temperature Events)
- Near real-time on-board detection and messaging of the main parameters of HTEs
- Comparison of the wildfire detection capability of a detected spaceborne sensor system with ground-based fire alert systems under various illumination, seasonal and local conditions
- Feasibility and test of the plash plume tracking techniques in context with aviation security.

5) Scientific requirements to MTP (Moderate Temperature Phenomena)
- Observation of urban heat islands is a study example in the category.

The FOCUS instrument, as considered in the extended feasibility study, consists of a wide-angle FFS (Fore-Field Sensor) for autonomous hot-spot detection, MS (Main Sensor) for detailed observation and analysis of high-temperature phenomena, and an on-board data handling and control system. The FFS is a forward-looking wide-field imager with three channels in VIS, MWIR, and TIR. MS is a combination of spatially highly-resolving imagers in the VIS to TIR spectral region and a profiling high-spectral resolution IR spectrometer.

The MVIR and TIR channels of the instrument feature a very large dynamic range of 20 bit and 14 bit, respectively (in comparison to comparable existing sensors), permitting to retrieve the high-temperature event and area unrestricted by sensor saturation.

Spectral bands: FFS Spectral bands: MS-Imager Spectral range: MS-Spectrometer	VIS (0.63-0.70 μm), MVIR (3.2-4.3 μm), TIR (8-10 μm) VIS (0.63-0.70 μm), NIR (0.77-0.80), MVIR, TIR 3-15 μm
FFS MS-Imager MS-Spectrometer	Swath: 355 km Swath: 70 km Footprint diameter: about 7 km / 60 km in cross-track
Ground resolution	FFS = 340 m, MS-Imager = 70 m
Attitude knowledge (ISS); with star camera	0.53° or about 4 km; 0.06° or about 500 m
Thermal system	5 radiators on 3 sides
Source data rate	56 Mbit/s
Instrument mass; power;	221 kg; 111 W standby, 385 W operation;
Instrument size	115 cm x 85 cm x 101 cm

Table 467: FOCUS instrument parameters

The line detectors for the MVIR and TIR channels of FFS and MS-Imager consists of staggered arrays of 512 x 2 elements, both with an identical layout. Coolers provide detector temperatures of 100 K for MVIR and 80 K for TIR. - The forward-looking FFS functions as a scene survey instrument in front of the nadir-looking MS-Imager. This forward-looking observation, about 30 seconds ahead of the MS-Imager nadir observation, is required for the near real-time ADM (Autonomous Detection Mode) processing sequence, consisting of the following functions:
- Recording of the co-registered VIS, MVIR, and TIR channel data of FFS
- Autonomous hot spot detection
- Transmission of low-rate data with information of the hot spot clusters
- Selection of hot spot clusters for detailed observations by the MS-Imager
- Pointing of the MS-Spectrometer FOV toward the selected hot spot cluster (use of cross-track tilt mechanism).

A cross-track tilt mirror is the optical element providing a parallel cross-track positioning capability of ±30° (FOR) for the MS-Imager and the MS-Spectrometer. When funded FOCUS is expected to be installed in the timeframe of 2004.

L.2.8 GTS (Global Transmission Services)

GTS is a relatively small payload of ESA. The objective is to transmit highly accurate time and coded data signals for dedicated receivers on the ground. The GTS experiment, mounted on the Russian Service Module, uses a transmitter accommodated on-board the ISS for signal distribution via an externally mounted antenna with a transmission cone half-angle of about 70°. The signals transmitted, at two dedicated frequencies in the 400 MHz (UHF) and 1.4 GHz (S-band) ranges, can be received for ISS overflight periods of 5-12 minutes several times per day by ground receivers with sufficient sensitivity. Some expected applications are:

- Accurate time receipt and automated local time conversion for mobile uses on the ground (wrist watches)
- Car theft protection (electronic car keys)
- Coding and re-coding of electronic cards (chip cards, smart cards, credit cards)

L.2.9 LCDE (Laser Communications Demonstration Equipment)

LCDE is a CRL (Communication Research Laboratory) experiment. The objective is to demonstrate large capacity (multi-gigabit/s) two-way communications (inter-satellite and ISS to ground). Experiments in uplink and downlink will be performed at data rates of 2.5 Gbit/s (uplink and downlink rates) using 1.5 µm wavelength technology (1.552 µm for transmit and 1.562 µm for receive) in EDFA (Er-doped Fiber Amplifiers) and lasers. LCDE employs a coaxial Cassegrain telescope with an antenna diameter of 15 cm. The tracking system has two modes: 1) coarse mode: with two-axis gimbals and a sensitive CCD detector array, 2) fine mode: with a quadrant photo detector and a fine-steering mirror. The tracking error requirement calls for <1 µrad (micro radian). The experiment has two windows, one on the Earth-facing panel and the other on the tilted top panel. LCDE is planned to be launch in 2005. [1589] [1590]

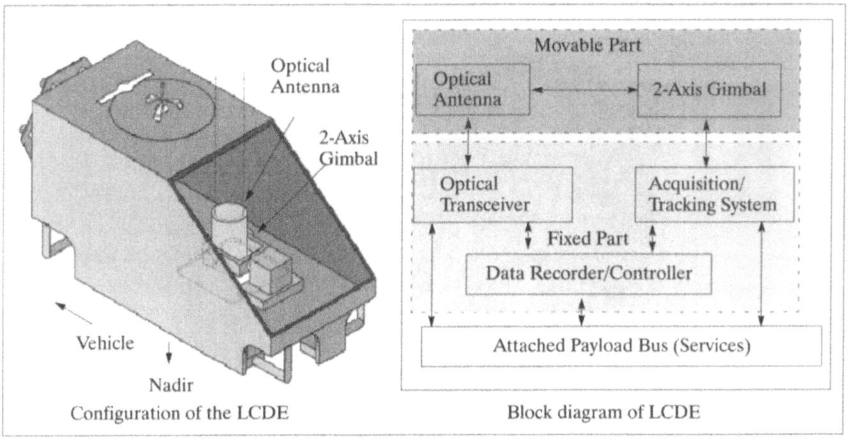

Figure 288: Illustration of LCDE

Antenna type	Coaxial Cassegrain type telescope with Coude optics
Antenna diameter	15 cm, f/D ratio 1.0 (TBD)
Tracking system - coarse	Two-axis gimbals and sensitive CCD detector, FOV more than 0.6°, Gimbal steering angle: azimuth: -30° to 210°, elevation: -60° to 90°
Tracking system - fine	Quadrant photodetector and fine-steering mirror, FOV > 0.02°
Tracking error	<1 µrad (rms, communication); <5 µrad (rms, debris detection, tracking of geodesic satellites

1589) Information provided by Araki.Kenichi of the NASDA Tsukuba Space center

1590) http://jem.tksc.nasda.go.jp/kibo/kibomefc/lcde_e.html

Required optical power	-95 to -65 dBm (TBD, on the telescope aperture)
Optical communication rate	2.48832 Gbit/s (on transmit and on receive)
Modulation scheme	Intensity modulation using Return to Zero pulse
Demodulation scheme	Direct detection with polarization maintaining EDFA
Communication wavelength	1.552 μm (transmit) and 1.562 μm (receive)
Output power	400 mW (high power EDFA output)
Debris detection wavelength Debris detection sensitivity	1.552 μm (same as communication transmitter) SNR > 10 for a 1 cm sized debris at 2 km distance
Optical filter bandwidth	0.68 μm ± 0.002 μm (receiving beacon wavelength in communication mode)
Power dissipation	<115 W for the LCDE mission equipment
LCDE mass	<90 kg

Table 468: Performance characteristics of LCDE

L.2.10 PARCS (Primary Atomic Reference Clock in Space)

PARCS (PI: Donald Sullivan, NIST) is a US ISS experiment, an advanced cesium atomic clock, planned for the time frame 2004/5 with a mission duration of a year. The objective of PARCS is to test a variety of predictions of the Theory of Relativity. One of these predictions, is that clocks tick slower in strong gravity than they do in weak gravity. Such tiny shifts have already been made in previous experiment. The aim of PARCS is to measure them about two orders of magnitude more accurate than ever before. This requires extremely accurate clocks, on ISS as well as on ground. The timing function of the GPS constellation is used to compare the frequencies of both clocks. The knowledge of the frequency differences between each clock and the GPS system permits the calculation of the frequency difference between the ground clock and the space clock. [1591)]

Figure 289: Overall view of the PET facility (Express Pallet Mounting)

L.2.11 PET (Photovoltatic Engineering Testbed)

PET [PI: Geoffrey Landis of OAI (Ohio Aerospace Institute)] is a facility planned to fly on the ISS (in 2004) to test advanced solar cell types in a space environment. This will reduce

1591) http://funphysics.jpl.nasa.gov/technical/lcap/parcs.html

the cost of validating new technologies and bring them to spaceflight readiness by measuring them in the in-space environment. PET is expected to provide routine access to a space-based calibration and measurement facility. The PET calibration facility is configured to fit the space-station as a self-contained payload with a standardized interface to the station. The PET facility is not limited to testing of solar cells, but will also be valuable for testing the space degradation of solar-cell encapsulants and coverglass materials, solar-cell interconnect materials, and solar cell substrates. [1592)]

L.2.12 RACE (Rubidium Atomic Clock Experiment)

RACE (PI: Kurt Gibble, Yale University) is a follow-up experiment of PARCS with a planned launch in 2006/7 with a mission duration of a year. RACE will be used to perform a series of experiments to test Einstein's Theory of Relativity (with much greater accuracy than PARCS). RACE builds on the technology of the PARCS experiment using advanced shutters and laser cooling techniques to produce ultra-cold atoms with a temperature of 1 x 10^{-6} K above absolute zero. The RACE design employs a "two clocks in one" configuration, featuring two-back-to-back sets of clock cavities, with a source of ultra-cold atoms between them. - RACE will be compared with SUMO. [1593)]

L.2.13 SAGE-III (Stratospheric Aerosol and Gas Experiment III)

A launch of the SAGE-III instrument (NASA/LaRC) to ISS is planned for 2004. See SAGE-III description under G.8.

L.2.14 SEDA-AP (Space Environment Data Acquisition equipment-Attached Payload)

SEDA-AP is a NASDA instrument package of the Exposed Facility (EF) of Japan's Experiment Module, called Kibo [Kibo consists of four major elements as shown in Figure 290: PM (Pressurized Module), EF (Exposed Facility), the logistics modules attached to each of them, and RMS (Remote Manipulator System)]. The objective is to measure quantitatively the space environment (neutrons, light high-energy particles, heavy ions, cosmic dust, atomic oxygen, plasma, etc.), to study the environmental effects on materials and electronic devices, and to investigate the interaction with and from the space environment at Kibo's Exposed Facility. The data acquired by this mission will be useful for space equipment design, space-related scientific research, ISS operation, and space weather forecasting. [1594)]

The SEDA-AP package is composed of SEDA and APBUS (Attached Payload BUS). The APBUS structure size is: 80 cm x 100 cm x 185 cm, mass = 216 kg, power = 132 W. SEDA is a generic term for a number of space environment sensors, such as the NEM (Neutron Monitor), HIT/PLAM (Heavy Ion Telescope/Plasma Monitor), SDOM (Standard Dose Monitor), AOM (Atomic Oxygen Monitor), EDEE (Electronic Device Evaluation Equipment), and MPAC &SEED (Micro-Particles Capture & Space Environment Exposure Device). The NEM sensor extends about 1m from the structural body and measures neutrons. [1595)]

NEM measures thermal neutrons with energies of about 100 MeV neutron (using a Bonner Ball detector and scintillation fiber detector). PLAM measures the density and electron temperature of space plasma. SDOM measures the energy distribution of high energy light particles (such as electrons, protons, alpha particles etc.). AOM measures the amount of atomic oxygen on the ISS orbit. The objective of EDEE is to determine the single event phenomena and radiation damage of electronic parts. MPAC captures on-orbit micro-particles. SEED exposes materials for space use to real space environment.

1592) http://oai.org/PET/index.html
1593) http://funphysics.jpl.nasa.gov/technical/lcap/race.html
1594) http://jem.tksc.nasda.go.jp/kibo/kibomefc/seda_ap_e.html
1595) http://jem.tksc.nasda.go.jp/kibo/kibomefc/index_e.html

A portion of he instrument package, MPAC&SEED, and a HDTV (High Definition Television) camera will be flown to ISS in July 2001 (on a Russian Progress resupply vehicle) and will be attached to the outside of SM (Service Module) of ISS. The planned launch date of SEDA-AP is 2005.

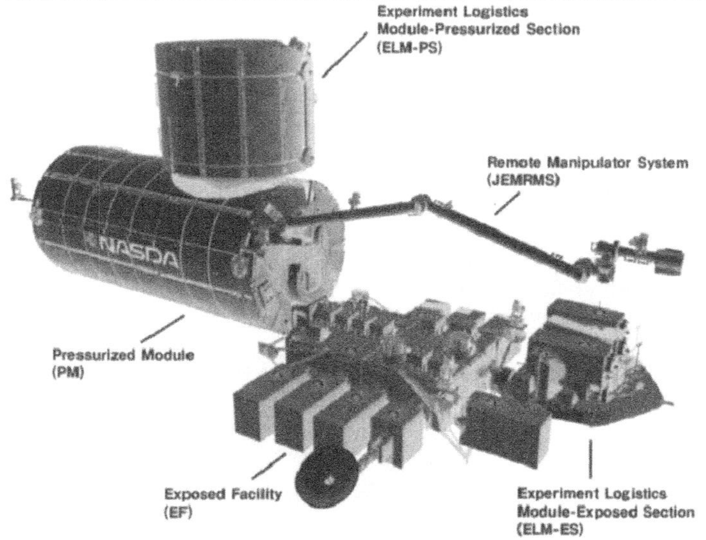

Figure 290: Illustration of the Japanese Experiment Module Kibo

L.2.15 SMILES (Superconducting Submillimeter-wave Limb-Emission Sounder)

SMILES is a cooperatively developed Japanese instrument of NASDA, CRL(Communications Research Laboratory), and NAO (Nobeyama Radio Observatory), projected to fly on ISS/JEM (Japanese Experiment Module). The overall objective of SMILES is to observe submillimeter-wave (thermal) radiation or limb-emission spectra of the atmosphere and to demonstrate an ultra-sensitive submillimeter-wave sounder technology. The monitoring of trace gas distributions is performed in an altitude range of 10 to 60 km at height resolutions of about 3.5 km. The species to be measured are: O_3, ClO, HCl, $HO^{37}Cl$, BrO, NO, HNO_3, N_2O, HO_2, H_2O_2, $O^{18}O$, and SO_2..

Species	LSB1-channel 624.32 - 625.52 GHz	LSB2-channel 625.12 - 626.32 GHz	USB-channel 649.12 - 650.32 GHz
O_3	625.37	625.37	-
ClO HCl HOCl	- 624.98 (^{37}Cl) 625.07	- 625.92 (^{35}Cl) -	649.45 - -
^{81}BrO	624.77	-	650.18
HO_2 H_2O_2	- 625.04	- -	649.70 -
HNO_3	624.48, 624.78	-	650.28
SO_2	624.89	625.84, 626.17	648.69, 649.05, 649.24

Table 469: Overview of trace gases and their observation frequencies measured by SMILES

Observations: SMILES on ISS/JEM, orbiting at an inclination of 51.6° at altitudes between 350 and 460 km, provides mappings of these molecules on a daily basis in a latitudinal range between 65° N and 38° S. Analysis of the spectra provides an altitude profile of molecular abundance for each species. The antenna beam width is 0.083° in vertical and 0.055° in horizontal direction. This leads to a best altitude resolution of about 3.5 km. Observations are conducted between tangent altitudes of 10-60 km. [1596) 1597) 1598) 1599)]

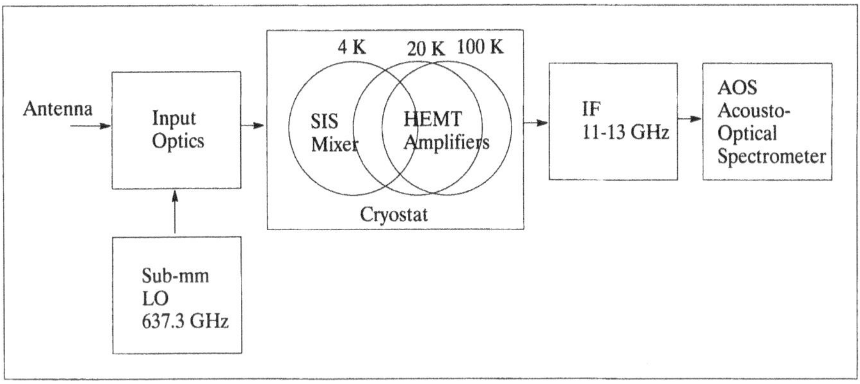

Figure 291: Schematic SMILES configuration

To achieve wideband heterodyne spectrometry, both of the LSB (Lower Side-Band) and USB (Upper Side-Band) are detected separately under SSB (Single Side-Band) conditions. The local frequency (LO) selected is 637.32 GHz. Three measurement channels are selected due to limited spectroscopic bandwidths and electromagnetic interference on ISS/JEM. Either the LSB1- or the LSB-2channel can be combined with the USB-channel in the measurement.

Observation method	Limb sounding
Observation altitude, height resolution	10 - 60 km, Δh=3.5 km
LO (Local Oscillator) frequency	637.3 GHz with 11-13 GHz IF (Intermediate Frequency)
Frequency bands (3) in GHz	624.32 - 625.52, 625.12 - 626.32, 649.12 - 650.32
Dynamic range, spectral resolution	3 - 300 K, 1.8 MHz
Total bandwidth	2,400 MHz
Antenna size, main beam size	400 mm (El) x 200 mm (Az); 0.083° (El), 0.17° (Az)
Antenna scan range	+5° to -40° in elevation
Sub-mm Mixer IF Amplifier Cryostat	4 K cooled SIS, mixer (Nb:AlOx:Nb) 20 K and 100 K cooled HEMT amplifiers 2 Stirling cycle + Joule Thomson cooler
Spectrometer; bandwidth resolution	2; 2.6 GHz, 1.8 MHz (3dB)
System noise System sensitivity Integration time	<500 K (SSB) <1 K rms 0.47 s
Data quantization, data rate	16 bit, about 200 kbit/s
Instrument mass; power, size	<500 kg; <820 W, 0.8 m (W) x 1 m (H) x 1.85 m (L)
Mission life	1 year

Table 470: Some instrument performance characteristics of SMILES

1596) H. Masuko, T. Manabe, et al., "Superconducting Submillimeter-wave Limb Emission Sounder (SMILES) Onboard Japanese Experimental Module (JEM) of International Space Station (ISS)," Proceedings of the IEEE/IGARSS 2000 Conference, Honolulu, HI, July 24-28, 2000

1597) http://smiles.tksc.nasda.go.jp/index1e.html

1598) J. Inatani. H. Ozeki, et al., Submillimeter Limb-emission Sounder JEM/SMILES aboard the Space Station," SPIE's 2nd International Asia-Pacific Symposium on Remote Sensing of the Atmosphere, Environment and Space, Sendai, Japan, Oct. 9-12, 2000

1599) H. Ozeki, Y. Kasai, S. Ochiai, S. Tsujimaru, et al., "Submillimeter-wave spectroscopic performance of JEM/SMILES," SPIE's 2nd International Asia-Pacific Symposium on Remote Sensing of the Atmosphere, Environment and Space, Sendai, Japan, Oct. 9-12, 2000

The SMILES instrument consists of the following basic elements: an antenna, receiver optics, a SIS (Superconductor-Insulator-Superconductor) receiver with a 637.32 GHz phase-locked oscillator and a low-noise InGaAS HEMT amplifiers, a cryogenic system, an IF subsystem with a bandwidth of 11-13 GHz, two acousto-optical spectrometers with a frequency resolution of 1.8 MHz, and a signal processing and control unit. In particular: [1600)]

- An offset Cassegrain-type antenna for decreasing sidelobes in the scanning direction. The main dish of the antenna has an elliptical feature with a long axis of 0.4 m to obtain a height resolution of 3.5 km. The antenna beam is tilted by 45° to the left-hand side, providing an observation coverage of up to 65° N latitude. The antenna is continuously scanned in the altitude range of -35 km to >80 km for tangent-height observations from 10-60 km.
- The local signal is produced by multiplying six times the signal of a phase-locked Gunn oscillator operated at 106.22 GHz. An new type device, referred to as MPI (Martin Puplett Interferometer) and a wire-grid polarizer combination, are used for the separate detection of the SSB, USB and LSB signals.
- A heterodyne-type SIS (Superconductor-Insulator-Superconductor) receiver with a wide-operation bandwidth of 30 GHz is used to achieve ultra-low noise temperature and conversion loss in the 637.3 GHz band..
- The SIS receiver (radiometer) is composed of SIS mixers operating at 4 K and HEMT (High Electron Mobility Transistor) amplifiers operating at 20 K and at 100 K. A Joule-Thomson mechanical cooler (developed by Sumitomo Heavy Industries, Ltd.), combined with a two-stage Stirling cycle, is utilized for cooling. The total mass and power of the cooler is about 60 kg and 250 W, respectively. The main body of the cooler (cryostat) has a size of 350 mm in diameter and 500 mm in length.
- Two AOS (Acousto-Optical Spectrometer) devices with a bandwidth of about 1.2 GHz and a spectral resolution of 1.8 MHz are used for channel combination.

JEM/SMILES is considered a "proof-of-concept" or demonstration mission before a similar instrument will be installed on ATMOS-C1 (Atmospheric Chemistry and Dynamics Mission). The SMILES instrument will be placed on the Exposed Facility of JEM (JEM-EF). A launch of JEM/SMILES is planned for 2005 by an H-IIA vehicle from the Tanegashima launch site in Japan.

L.2.16 Solar-A (Solar Monitoring Observatory)

SOLAR-A is an ESA experiment package consisting of three instruments, namely SOVIM (Solar Variability and Irradiance Monitor), SOLSPEC (Solar Spectral Irradiance Measurements), and SOL-ACES (Solar Auto-Calibrating EUV/UV Spectrophotometers). The overall objective is to measure the solar spectral irradiance with unprecedented accuracy. Knowledge of the "solar constant" and its variations is of great importance for atmospheric modelling, atmospheric chemistry, and climatology applications. The three instruments cover the combined wavelength range from 17-3000 nm in which 99% of the solar energy is radiated. The Solar-A package will be mounted on the ESA-developed CPD (Course Pointing Device), located on EPA of ISS. Integration of Solar-A on ISS is planned for 2005.

L.2.16.1 SOVIM (Solar Variability and Irradiance Monitor)

SOVIM is an ESA instrument, designed and developed at PMOD/WRC Physikalisch-Meteorologisches Observatorium Davos, World Radiation Center, Switzerland (PI: C. Fröhlich), in cooperation with investigators from Belgium, France, ESTEC, Switzerland, and USA. The overall objective is to measure TSI (Total Solar Irradiation) and SSI (Spectral Solar Irradiance). SOVIM is of SOVA heritage flown on EURECA-1 (launched July 31, 1992 and retrieved July 1, 1993).

1600) Y. Fujii, K. Kikuchi, et al., "Spaceborne 640 GHz SIS Receiver Based on 4-K Mechanical Cooler," UV, Optical and IR Space Telescope Instruments, Proceedings of SPIE, Vol. 4013, pp. 90-99, 2000

The objective of SOVIM is to monitor the sun's irradiance with high precision, stability and accuracy; the pointing accuracy is of the order of 1° to compensate for ISS motions. SOVIM observations are conducted during the sun-pointed periods of 10-15 minutes per orbit throughout the mission. TSI measurements are performed with active cavity radiometers (PMO6 and DIARAD) which are traceable to the SI scale of irradiance through direct comparison with cryogenic radiometers on the ground and in vacuum for the PMO6, and thorough characterization in the laboratory for the DIARAD. SSI measurements are carried out by SPM (Sunphotometers) of the type now operating successfully in VIRGO/SOHO. A detector based calibration method, directly traceable to the scale established by cryogenic radiometers, is used to calibrate the SPM with high accuracy. Specific TSI and SSI objectives are: [1601)]

- To obtain quasi-continuous high quality measurements of the solar irradiance variation
- To determine with high accuracy the amount of spectral redistribution of the solar output
- To search for the long periodicities or quasi-periodicities found in other solar parameters
- To study the influence of active regions and other large scale solar structures on the solar irradiance
- To investigate the energy storage in the convection zone in connection with the energy blocking of active regions
- To investigate the mechanisms of solar radiative forcing of climate change on seasonal to decadal time scales
- To continue the historical TSI monitoring record by linking the present SOLSTICE (UARS), VIRGO (SOHO), CERES (Terra), ACRIM-III (ACRIMSAT), and future CERES (Aqua), TIM, SOLSTICE (SORCE), TIM (NPP, NPOESS), and SOVAP (Picard) measurements.

The SOVIM instrument consists of three packages: SOVIM1, SOVIM2 and SOVIM3.

- SOVIM1, developed at IRMB, contains one absolute radiometer DIARAD (Dual Irradiance Absolute Radiometer). DIRAD is composed of two cylindrical cavities coated inside with diffuse black and mounted next to each other on the same heat sink. The flat bottom of the cavities are in fact heat flux transducers on which heating elements have been mounted. Both cavities see the same thermal environment through accurately know circular apertures. The operation is based on electrical substitution. The difference of the electrical power fed to the active channel when its shutter is open (exposed to the sun) and when it is closed is proportional to the incident solar irradiance. The SOVIM1 mass is 7.3 kg, power = 11.5 W.
- SOVIM2, developed at PMOD/WRC (Switzerland), contains two absolute radiometers, an IR Radiometer, four SPM with three independent channels each (type1: 402, 500 and 865 nm, type2: 310, 610 and 719 nm, all redundant) and a two axis pointing monitor (TASS). The SOVIM2 mass is 13.3 kg, power = 18.5W.
- SOVIM3, developed at PMOD, contains the microcomputer (80C86) based experiment controller, part of the data acquisition system and the power and data interfaces to ISS. The SOVIM3 mass is 3.4 kg, power = 3.5 W.

Data acquisition is performed by parallel VFCs (Voltage Frequency Converters) with a frequency of about 500 kHz. The basic sampling period is 10 s. Electrical calibrations are performed at zero, half scale and full scale. The pointing provided by CDP is $\pm 1^{\circ}$ (2 sigma) with a stability of 0.3° over 10 s. ISS position and attitude data are available for post-factum data analysis. SOVIM data are processed at SROC (Space Remote Operation Center) located at ESA.

1601) http://estirm2.oma.be/main_projects.html

L.2.16.2 SOLSPEC (Solar Spectral Irradiance Measurements)

SOLSPEC is an ESA instrument (PI: G. Thuillier of CNRS/SA, Verrieres le Buisson, France, with cooperation of Belgium, France, Germany, Switzerland, UK, and USA). SOLSPEC is of EURECA (1992-1993) and Spacelab-1 (Nov. 28- Dec. 8, 1983) heritage. The objective is the measurement of the solar irradiance in the spectral range from 180-3200 nm using three double spectrometers and an on-board calibration device.

L.2.16.3 SOL-ACES (Solar Auto-Calibrating EUV/UV Spectrophotometers)

SOL-ACES is an ESA instrument [1602)] (PI: G. Schmidtke of the Fraunhofer Institut für Physikalische Messtechnik, Freiburg, Germany, with co-investigators from Germany and USA). SOL-ACES is a new instrument covering the solar irradiance in the EUV/UV spectral range of 17-220 nm. Major elements of the instrument are: filter wheel, stepper motor, spectrometers, ionization chamber, gas supply, electronics unit (control, data acquisition, communication). Solar irradiance measurements (full disk) are obtained with four grating spectrometers. To obtain high radiometric accuracy of better than 10%, a double ionization chamber is assigned to each of the spectrometers as primary detection standard. Optical bandpass filters are mounted on a filter wheel and placed at the entrance apertures of the spectrometers and ionization chambers, they establish the radiometric link between the devices (instrument calibration uses the optical bandpass filters in conjunction with the double ionization chambers as radiometric standards). The grating spectrometers are designed as scanning monochromators operating at fixed incidence angles. The deflected radiation is monitored by rotating an assembly containing a parabolic mirror, an exit slit and a CEM (Channel Electron Multiplier). The dispersed radiation is focused through the spectrometer exit slit by a parabolic mirror and collected by CEM operated in a pulse counting mode. The optical length of the ionization chamber of 0.5 m is divided into two identical electrode sections. In addition, the transmitted radiation is measured by a silicon diode detector located at the end of the absorption path. The data permits the absolute quantification of the solar flux in the prescribed spectral range. - SOL-ACES has a mass of 23 kg and a size of 24 cm x 28 cm x 60 cm. The instrument will be installed on CPD (Coarse Pointing Device) of ISS for a mission length of about 1.5 years. The planned launch date is 2005.

L.2.17 SUMO (Superconducting Microwave Oscillator)

SUMO (PI: J. Lipa, Stanford University) is an ISS clock experiment consisting of a superconducting cavity-stabilized oscillator (SCSO) system. This is a distinctive type of clock that can be used for experiments in relativity. - A SCSO clock is built out of a microwave cavity constructed of a superconducting metal that operates at cryogenic temperatures. Since the timekeeping mechanism of SCSO is different from that of an atomic clock, a comparison between the two can give a fundamental understanding about changes in the universe. In addition, a comparison of SUMO against a more standard atomic clock may help to to answer question of whether they vary for different places and velocities. This experiment will also help to determine how gravity and other factors have an effect on the speed of time.

SUMO will be installed in LTMPF (Low Temperature Microgravity Physics Facility), part of JEM-EF (Japanese Experiment Module-Exposed Facility). The planned launch date for SUMO is 2006.

L.2.18 WORF (Window Observational Research Facility)

WORF is an experiment rack located in the US Laboratory module Destiny on ISS (flown on Shuttle flight STS-98, Feb.7-20, 2001). The objective of WORF is to support cameras

1602) F. G. Wienhold, J. Anders, B. Galuska, U. Klocke, M. Knothe, W. J. Riedel, G. Schmidtke, et al., "The Solar Package on ISS: SOL-ACES," Physical Chemistry of the Earth (C), Vol. 25, No 5-6, 2000, pp. 473-476 and other instruments that utilize the Earth-pointing window in the Lab (rack-mounted devices with a view through a porthole facing Earth). WORF is effect an active payload isolation system. The control architecture and hardware design for this system are based on proven technologies such as the Mast Mounted Sight (MMS) equipment, employed in Bell helicopters, and the STABLE platform, flown on STS-73. The control system concept consists of an inner fine-pointing loop for high-frequency isolation about the x and z rotational axes and an outer position loop for all six degrees of freedom (no contact between the WORF rack and the isolated platform). The performance analysis verified the close-loop requirements of pointing accuracies. The line-of-sight rotation requirements call for a pointing stability of 7.5 and 25 μrad for large (<150 kg) and small payloads (<15 kg), respectively. [1603)]

EarthKAM (Earth Knowledge Acquired by Middle school students). NASA's KidSat program of a camera on three Shuttle flights (STS-76, STS-81, STS-86) was renamed to EarthKAM. In Feb. 2001, EarthKAM was flown to ISS on Shuttle flight STS-98 and installed on WORF as a permanent resident. EarthKAM is one of several users on WORF permitting for more continuous photography of Earth on student request. The EarthKAM program is based at the University of California, San Diego (UCSD). NASA,UCSD and TERC are EarthKAM partners, in addition there are many sponsors of the program. [1604)]

L.3 MIR-1 Orbital Station

MIR (=Peace) is Russia's (operator: RKK Energia) modular orbital station (launch of the MIR core module: Feb. 20, 1986) of considerable dimensions and payload capacities. The station has been occupied continuously by cosmonaut crews from Sept. 1989 to Aug. 1999. [1605)]

On April 4, 2000, the 14-year-old MIR was reoccupied by a cosmonaut crew. The mission of the two cosmonauts is to bring the aging craft back to life after it circled the Earth unoccupied for eight months. MirCorp of Amsterdam, Holland, leased the commercial rights of the station from Rosaviakosmos, the Russian Aviation and Space Agency.

MIR is comprised of the following separate modules and a Soyuz (Sojus) TM spacecraft functioning as a service vehicle:

1) the **MIR Core module**, which contains the cosmonaut crew quarters (mass = 20,000 kg, 4.2 m diameter), existing

2) the **Kvant 1 module** for astrophysics (mass = 11,000 kg), existing

3) the **Kristall module**, which is used to manufacture new materials in microgravity (mass = 20,000 kg), existing

4) the **Kvant 2 module**, which houses the airlock used by cosmonauts to exit MIR for space walks (mass = 20,000 kg), existing

5) the **Spektr (Spectrum) module**, was launched May 20, 1995 (Proton launch vehicle)

6) the **Priroda module**, which is used for Earth observation (this module was launched on April 23, 1996 (see chapter D.28 for Priroda).

One of MIR's objectives is Earth observation (resources, environment). The presence of an on-board crew is a great advantage for the maintenance and repair of all on-board equip-

1603) S. A. Green, R. L. Boucher, J. T. Harduvel, "Active Precision Pointing for the Window Observational Research Facility on the International Space Station," Proceedings of the AIAA Space Conference and Exposition, Long Beach CA, Sept. 19,21 2000

1604) http://www.earthkam.ucsd.edu/public/iss/

1605) "Soviets to Set Record Pace for MIR Repairs," Space News June 10-16, 1991, p. 12

ment and for the collection of meaningful correlative information. The crew makes observations possible that require very intricate and complicated operational sequences or methods not easily susceptible to algorithmic definition. Between 1990 and 1993 a considerable amount of Earth observation (resources) activity was introduced and also new and more advanced sensor equipment.

Orbit: Altitude = 350 - 410 km; inclination = 51.6°.

The MIR station was deorbited in March 2001 with a final reentry into the Pacific Ocean on March 23, 2001. Like its Salyut predecessors, MIR had an edge over Skylab: It could be resupplied by unmanned Progress freighters carrying fuel, oxygen, and other essentials. And with the help of mission controllers, cosmonauts kept MIR going three times as long as the five years its designers had envisioned. Some 104 cosmonauts and astronauts from 12 countries were living for a total of 4,591 days - more than 12 years - aboard the station. They performed 23,000 separate scientific experiments and made 78 spacewalks totaling 352 hours. Again and again, MIR was the setting for new space endurance records, culminating in Valeriy Polyakov's marathon stay of 438 days.

Sensors: (Earth Observation equipment (sensors, etc.) is located in the MIR core module)

- Module Kvant 2 - Camera KAP-350 (350 mm focal length). Provision of high-quality images (format 180 x 180 mm, scale 1 : 1000 000, resolution = 30 m)
- Module Kristall (was introduced in 1990)
  Contains two cameras of the KFA-1000 type, which are mounted fan-like for extra-wide swath width (see D.31 for a KFA-1000 description). The two KFA-1000 cameras are also known under the designation: 'Priroda-5'.
- Module "I" is the "PRIRODA" complex, scheduled to be introduced at MIR in March 1996 (see D.28 for PRIRODA definition).
- Hasselblad camera 500-EL. Hand-held (portable) cameras for the crew to document their visual observations.
- Spectrometer - System "Gemma" (hand-held). Prototype testing on MIR as of 1989. Measurement of 456 narrow channels in the spectral range from 0.4 - 1.1 μm.
- MKS-M = Multichannel Spectrometer of DLR, consisting of the following systems:
  - MKS-M-AS = Atmospheric spectrometer (atm. corrections in the VIS range)
  - MKS-M-BS = Biospectrometer (determination of ocean phytoplankton).

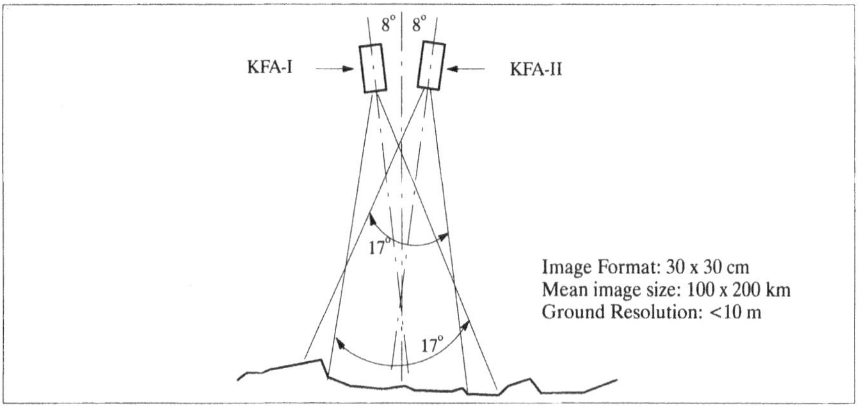

Figure 292: Optical geometries of the KFA-1000 Camera (Priroda-5) on MIR

Data:[1606] MIR data are being offered commercially to the user community by 'Energia USA' of Henderson, VA, a US subsidiary of RKK Energia, Moscow.

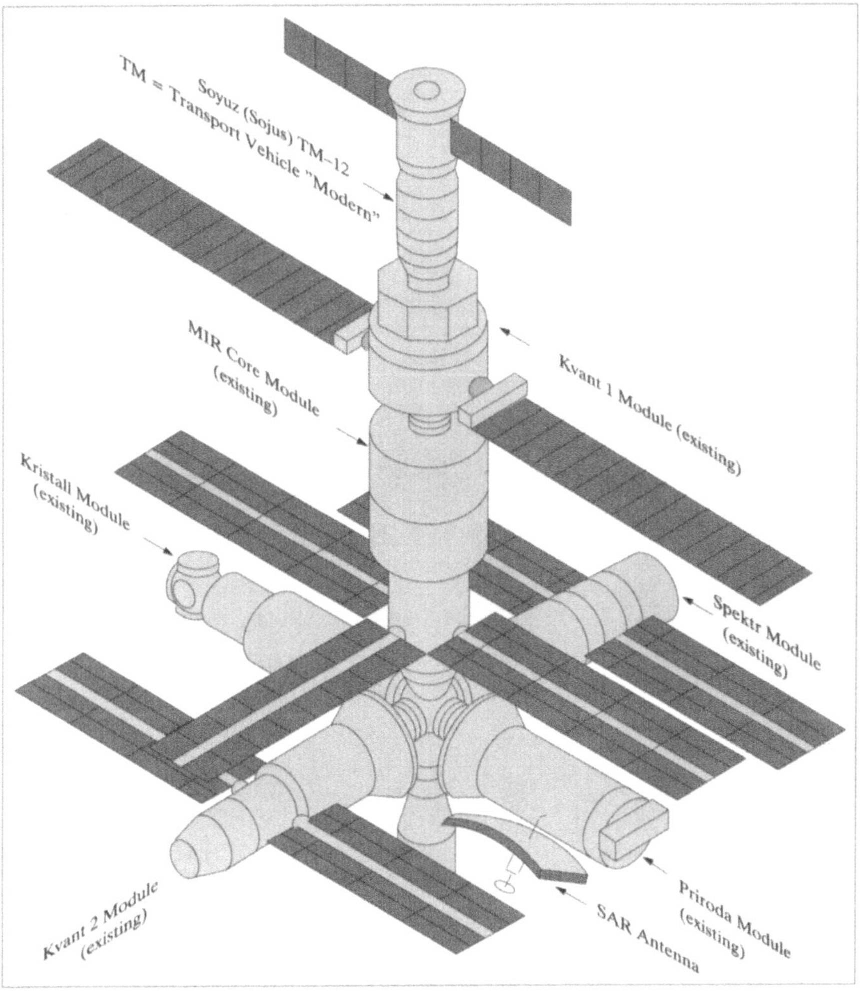

Figure 293: Overall configuration of the MIR orbital station

Instrument	Wavelength (cm)	Swath (km)	Pixel Size (km)	View Angle	MIR Module	Resolution
Spektr 256-Z	450-830 nm	8°	65x120 km^2 /1.5 nm			
Ozon-M	0.25-0.29μm 0.37-0.39μm 0.60-0.64 μm 0.99-1.03 μm	2' x 25'	15 km / 2-7Å 1 km height	Priroda	Occultation	
IR Radiometer Jausa-100-3 Neva-3 Neva-5	 1.8-3.0μ 1.8-3.0μ 3.0-5.0μ	 50' x 50' 10' x 10' 10' x 20'	 -/- -/- -/-	 Spektr Spektr Spektr		

1606) "Earth Imagery from MIR offered to Commercial Buyers," Space News, April 25-May3, 1992, p. 27

Instrument	Wavelength (cm)	Swath (km)	Pixel Size (km)	View Angle	MIR Module	Resolution
Matrix Spectrometer Zwet	1.5-3.3 μ	20' x 20'	-/0.02 μm	Spektr	Spectrum - 0.5 s	
Optical Complex - Scanner						
Name of Instrument	Spectral Range (nm)	Channels	Swath Width/FOV	Spatial/Spectral Resolution	MIR Module	Remarks
MOS-Obsor-A	757-768	4	80 km	1.4nm /2.7km	Priroda	Imag. Spectrometer
MOS-Obsor-B	457-1030	13	80 km	10nm /0.6km	Priroda	Imag. Spectrometer
MSU-SK	0.5-12.5 μm	5	350 km	100nm /120x300m	Priroda	Conical Scanner (39°)
MSU-E	0.5-0.9 μm	3	2 x 27 km	100nm/25m	Priroda	Pushbroom
Optical Complex - Lidar						
Balkan-1	532 nm		90"	3m vertical	Spektr	Imp. = 0.15 J, Freq.=1 x in 5.5 s
Alissa (France)	532 nm		3'	150m vertical	Priroda	Imp. = 40 mJ, Freq.= 8 Hz
Microwave (MW) Complex - Radiometer						
KR-05 (λ scan)	0.46-0.55 (Δf =1GHz)		1.5		Spektr	
Ikar. Complex	0.3-6	60-750	5-75		Priroda	
IKAR-N,R-30	0.3					
(Nadir) R-80	0.8					
R-135	1.35					
R-225	2.25	60	60	Nadir		0.15 K
R-600	6.0					
IKAR-P RP-225	2.25 (3 polar)	750	75	40°	Priroda	0.15 K
RP-600 (Panorama)	6.0 (3 polar)	750	75	40°	Priroda	0.15 K
IKAR-Delta (scanning)					Priroda	
RD-30	0.3	400	5	40°		1.5 K
RD-80	0.8	400	8	40°		0.5 K
RD-135	1.35	400	15	40°		0.4 K
RD-400	4.0	400	50	40°		0.15 K
Active Microwave (MW) Instruments						
SAR Travers	9.3 23	50 50	0.15	35°	Priroda	
Altimeter Greben	2.25	2.5	2.5 x 9.75	Nadir	Priroda	min. ±10 cm
Photographic Complex						
Name of Instrument	Spectral Range (nm)	Observation Surface	Spatial Resolution	MIR Module	Image Format/No. of Images per film	
Priroda-5	400-800	32° x 16° 100 x 200 km	5-10 m	Kristall	=30 x 30 cm²/1500	
KAP-350	400-800	40°	30-40 m	Kvant-2	= 18 x 18 cm²/600	
MKF-6MA	480-840	32° x 24°	30 m	Kvant-2	= 56 x 81 mm²/2400	
TV Complex						
KL-103 W	400-800	16° x 20°	250 m	Kvant-2	Color Camera	
Atlas	400-800	20° x 41°	200-10 m	Kvant-2		
Optical Complex - Spectrometer						
Instrument	Spectral Range	Field of View (FOV)	Spatial/ Spectral Resolution	MIR Module	Remarks	

Instrument	Wavelength (cm)	Swath (km)	Pixel Size (km)	View Angle	MIR Module	Resolution
Telespectrometer Phase	445-2200 nm	10'	2-/15, 100, 200, 250 m	Kvant-2	Impulse Frequency = 10 GHz	
IR Spectrometer Istok-1	3.6-16 µm 64 chan.	12' x 48'	0.8x6 km^2/ 125,250 nm	Priroda	θv =0-90°, Nadir and Occultation	
Telespectrometer ITS-7D	4000-8000 nm	0.2 x 0.2° 0.2 x 0.8° 60 x 240 km^2	-/150 and 300 nm	Kvant-2	Spectrum - 1 s	
Phönix	2.61-2.63 µm	1°24' x 25'	-/0.4 cm^{-1}	Spektr	Spectrum - 1 s	
Volchov-1	5-22 µm	10°	-/10 cm^{-1}	Spektr	Spectrum - 2 s	
Volchov-2	5-22 µm	20'	-/16 cm^{-1}	Spektr	Spectrum - 2 s	
Skif	400-1200 nm	0.87 x 0.17°	-/14 a. 3nm	Core		
MKS-M-AS	757-770 nm	1.15 x 0.09°	6.3 x 0.5 km^2/1.5nm	Core		
MKS-M2-BS	415-830 nm	0.46 x 0.46°	2.4x2.4 km^2 /10 nm	Core		
MKS-M2-AS	757-770 nm	1.15 x 0.09°	6.3 x 0.5 km^2/1.5nm	Kvant-2	Spectrum - 20 ms on pointable platform	
MKS-M2-BS	415-1030 nm	0.46 x 0.46°	2.4x2.4 km^2 /10 nm	Kvant-2	Spectrum - 20 ms on pointable platform	

Table 471: Earth observation instruments on the Russian space station MIR [1607) 1608) 1609)]

L.4 Salyut Space Station

USSR space station program. Salyut development began in the 1960s. The first launch occurred April 19, 1971. The different Salyut space stations were equipped with orbital maneuvering engines that permitted a controllable reentry capability (S/C debris could fall harmlessly into the oceans). The different space stations were raised periodically into higher orbits by visiting Soyuz spacecraft. [1610)]

Earth observation sensors (civil program) on Salyut-6 and -7 consisting of a floor-mounted photographic complex.
Note: there was also an astronomical and a material science instrument package on-board.

Name	Launch Date	Apogee (km)	Perigee (km)	Inclination (°)	Remarks
Salyut-1	April 19, 1971	222	200	51.6	reentry Oct. 11, 1971
Salyut-2	April 3, 1973	260 296	215 261	51.6	reentry May 28, 1973
Salyut-3	June 25, 1974	270 275	219 253	51.6	reentry Jan. 24, 1975
Salyut-4	Dec. 26, 1974	270 355 369	219 343 349	51.6	reentry Feb. 3, 1977
Salyut-5	June 22, 1976	260 275	219 256	51.6	reentry Aug. 8, 1977
Salyut-6	Sept. 29, 1977	275 377 521	219 335 470	51.6 83	crew occupation for a total of 676 days reentry July 28, 1982
Salyut-7	April 19, 1982	260	212	51.6	crew occupation for a total of 812 days, reentry 1989

Table 472: Overview of Salyut Space Stations

MKF-6 = Multispectral Camera (Carl Zeiss, Jena), first operation of this IKF sensor flown on Soyus-22 in 1976, afterwards in operation on Salyut-6 and -7. MKF-6 operated in six

1607) Overview paper provided by G. Zimmermann of DLR (IKF) Berlin, Aug. 1991
1608) Note: The sensors of existing modules are operational (Priroda and Spektr modules are planned)
1609) MIR Earth Images are sold by 'Energia Deutschland GmbH', a joint venture of NPO Energia, Moscow and Kayser-Threde of Munich, Germany - see Space News, Aug. 17-23, 1992, p. 13
1610) Interavia Space Directory 1990-91 (previously Jane's Space Flight Directory), pp. 122-124

spectral bands and could cover 10 million km^2 of the Earth's surface with each film cassette. Spectral bands: 0.46-0.50 μm, 0.52-0.56 μm, 0.58-0.62 μm, 0.64-0.68 μm, 0.7-0.74 μm, 0.79-0.89 μm. Number of lenses = 6; focal length = 125 mm; mass of camera = 73 kg; power = 100 W; FOV=40.5°.

KATE-140 = Camera System. Spatial resolution of 50 - 70 m.

L.5 SKYLAB Space Station

Skylab was a manned, orbiting NASA space station composed of five parts, the Apollo telescope mount (ATM), the multiple docking adapter (MDA), the airlock module (AM), the instrument unit (IU), and the orbital workshop (OWS). The Skylab was in the form of a cylinder, with the ATM being positioned 90 deg from the longitudinal axis after insertion into orbit. The ATM was a solar observatory, and it provided attitude control and experiment pointing for the rest of the cluster. It was attached to the MDA and AM at one end of the OWS. The retrieval and installation of film used in the ATM was accomplished by astronauts during extravehicular activity (EVA). The MDA served as a dock for the command and service modules, which served as personnel taxis to the Skylab. The AM provided an airlock between the MDA and the OWS, and contained controls and instrumentation. The IU, which was used only during launch and the initial phases of operation, provided guidance and sequencing functions for the initial deployment of the ATM, solar arrays, etc. The OWS was a modified Saturn 4B stage suitable for long duration manned habitation in orbit. It contained provisions and crew quarters necessary to support three-person crews for periods of up to 84 days each. All parts were also capable of unmanned, in-orbit storage, reactivation, and reuse. [1611)]

Skylab size: 26 m length and about 6.8 m in diameter, the total mass was about 74,700 kg. Skylab was launched on May 14, 1973 (Saturn 5) and subsequently visited by three Apollo astronaut crews, who lived and worked in it for periods of 28, 59, and 84 days respectively (1st crew launch: May 25, 1973 to Feb. 8, 1974 (last splashdown of a crew). After hosting three teams of astronauts, the unmanned Skylab reentered the Earth's atmosphere on July 11, 1979.

Skylab-1	Launch: May, 14, 73 (unmanned)	34,981 orbits	Re-entry: July 11, 1973
Skylab-2	Launch: May 25, 73 (manned)	404 orbits	Re-entry: June 22, 1973
Skylab-3	Launch: July 28, 73 (manned)	858 orbits	Re-entry: Sept. 25, 1973
Skylab-4	Launch Nov. 16, 73 (manned)	1,214 orbits	Re-entry: Feb. 3, 1974

Table 473: Overview of Skylab missions

Orbit: Circular orbit, altitude = 435 km, inclination = 50°, period = 93.4 minutes.

Sensor complement: (only the Earth Resources Experimental Package is considered. There were in addition many other sensors).

S-190A = Multispectral Photographic Camera System (used six identical cameras with different film/filter combinations in order to view the same ground area simultaneously in the visible region). The S190A experiment consisted of six high-precision 70 mm cameras. The matched distortion and focal length camera array contained forward motion compensation to correct for S/C motion. The f/2.8 lenses, with a focal length of 15 cm, had a FOV of 21.1° providing surface coverage of about 163 km x 163 km. The system was designed for the following wavelength/film combinations: 1) 0.5-0.6 μm, Panatomic-X B+W; 2) 0.6-0.7 μm, Panatomic-X B+W; 3) 0.7-0.8 μm, IR B+W; 4) 0.8-0.9 μm, IR B+W; 5) 0.5-0.88 μm, IR color; and 6) 0.4-0.7 μm, high-resolution color. The spectral regions designated were se-

1611) "Skylab," Jane's Spaceflight Directory 1988-89, 4th Edition, pp. 117-122 lected to separate the visible and photographic infrared spectrum into bands that were expected to be most useful for multispectral analysis of earth surface features. Further spectral refinements were made by using different filter combinations. The camera system provided photos with a ground resolution of 30 to 46 m in the visible wavelengths and 73 to 79 m in the infrared wavelengths.

S-190B = Earth Terrain Camera. The S190B camera utilized a single 45 cm focal length lens with 12 cm film. Its field of view of 14.2° provided a surface coverage of about 109 km x 109 km. This camera was designed to use high-resolution color film and was operated from the OWS window, producing photos with a ground resolution of 17 to 30 m. The camera compensated for S/C forward motion through programmed camera rotation. Shutter speeds were selectable at 5, 7, and 10 msec with a curtain velocity of 2.8 m/s.

S190A Camera System		S190B Camera System	
Spectral range:	400 - 900 nm	Spectral range	400 - 880 nm
Focal length:	152 mm	Focal length	457 mm
Image format:	5.7 x 5.7 cm	Image format	11.4 x 11.4 cm
Image scale:	1 : 2,850,000	Image scale	1 : 950,000
Image overlapping:	60%	Image overlapping:	60%
FOV	21.2°	FOV	14.24°
Image size	163 x 163 km	Image size	109 x 109 km
Ground resolution	100 - 260 m	Ground resolution	55 - 100 m

Table 474: Specification of S190A and S190B camera systems

S-191 = Infrared Spectrometer (camera system).[1612)] Objective: to make an evaluation of the applicability and usefulness of sensing earth resources from orbital altitudes in the visible through near-infrared and in the far infrared spectral regions. The S191 experiment was basically a visible and infrared spectroradiometer (hand-held device). Measurement of radiation flux in the bands from 0.4 - 2.4 µm, and from 6.2 - 15.5 µm. FOV = 1 mrad (0.435 km diameter circular foot print), with a spectral resolution of 1 to 5%.. Note: The S191 and S192 experiments were operated from May 1973 to Feb. 1974. Both systems provided useful data. The data return was limited by the amount of magnetic tape that could be transported in resupply flights.

S-192 = Multispectral Optomechanical Scanner. Objectives: to assess the feasibility of multispectral techniques, developed in the aircraft program, for remote sensing of Earth resources from space. Specifically, attempts were made at spectral signature identification and mapping of ground truth targets in agriculture, forestry, geology, hydrology, and oceanography. The S192 optical mechanical scanner utilized a 30 cm reflecting telescope with a rotating mirror. The telescope and mirror were mounted outside the multiple docking adapter. Spectral range: 0.4-12.5 µm; number of spectral bands = 13; use of silicon detectors for bands 1-12 (0.41 - 2.43 µm), while band 13 (10.2 - 12.5 µm) used photoconductive HgCdTe detectors; radius of circular scan = 42 km; swath width = 72.4 km; IFOV = 79 m (0.182 mrad.); instrument mass = 57 kg; power = 266 W (peak). The conical scan pattern was formed by giving the scanner a 9° forward tilt from nadir. Scanning was accomplished by a small flat fold mirror rotating on an arm to scan a circular zone of the image in the first focal plane of the telescope (primary spherical mirror of 51 cm diameter). The scan mirror scanned a zone of this known constant spherical aberration and subsequently corrected by the refocusing optics.

S-193 = Passive Microwave Radiometer/Active Scatterometer and Radar Altimeter (sometimes also referred to as RADSCAT). Objectives: simultaneous measurements of radar backscatter and radiometric brightness temperature in a number of scanning modes, primarily for the purpose of studying surface winds and precipitation over the oceans, and to provide engineering data for use in designing space radar altimeters.

1612) P. Slater, "Remote Sensing," Optics and Electronics Systems, Addison-Wesley Publishing Co., 1980, pp. 456-462

This design represents a first implementation of a combined passive/active microwave sensor. All three instrument components operated at the same frequency of 13.9 GHz and shared a common gimballed antenna (mounted on the outside of the multiple docking adapter) and scan system. Spatial resolution = 16 km (circular footprint at nadir); mechanically scanning parabolic antenna (1.15 m diameter reflector) with dual polarization and a 2° IFOV (swath width = 180 km). The antenna was gimballed, permitting along-track and cross-track scanning. - Using a pulse width of 0.1 μs this system was able to get a resolution of 15 m. It operated over short orbital segments only but it was able to demonstrate the measurement of coarse features of the marine geoid such as major ocean trenches. [1613)]

The scatterometer measured the backscattering coefficient of ocean and terrain as a function of incidence angle ranging from 0 to 48°. The radiometer was a passive sensor which measured the brightness temperature, from a cell on the Earth's surface, as a function of incidence angle from the surface. The altimeter was a compressed-pulse radar system to measure average ocean-surface elevation variations with a resolution of about 0.9 m. The S-193 ground coverage was 48° forward and 48° to either side of the spacecraft ground track. All data were recorded on magnetic tape on one digitized channel. The radiometer/scatterometer data were recorded at 5.33 kbit/s, the altimeter data at 10 kbit/s.

S-194 = Passive Microwave Radiometer. Objectives: to measure the brightness temperature of the Earth's surface along the S/C track, which would provide ocean surface features, varying winds over ocean areas, and Earth surface features information (soil moisture). S-194 was a modified Dicke-type radiometer with tuned RF receiver and gain modulation. Frequency = 1.4 GHz (21 cm wavelength, L-band); fixed nadir viewing phased-array antenna. The system utilized a built-in calibration scheme that sampled known sources. The spatial characteristics were: an antenna half-power beam width of 15°, first null beam width of 37° (97% of power) and a circular footprint of about 124 km diameter (half-power). All data were recorded on magnetic tapes. The data output was at 200 bit/s.

1613) E. G. Njoku, "Passive Microwave Remote Sensing of the Earth from Space," Proceedings of the IEEE, Vol. 70, No. 7, July 1982, pp. 728-750

Part M Technology Missions

M.1 ARGOS (Advanced Research and Global Observation Satellite)

ARGOS (also referred to as P91-1) is a large DoD research and development satellite mission, managed by the Tri-Service Space Division at Kirtland AFB (SMC/TE), Albuquerque, NM. It is part of the DoD Space Test Program (STP) with the objective to demonstrate several new space technologies and to fly payloads for global Earth sensing and celestial observations. [1614] [1615] [1616]

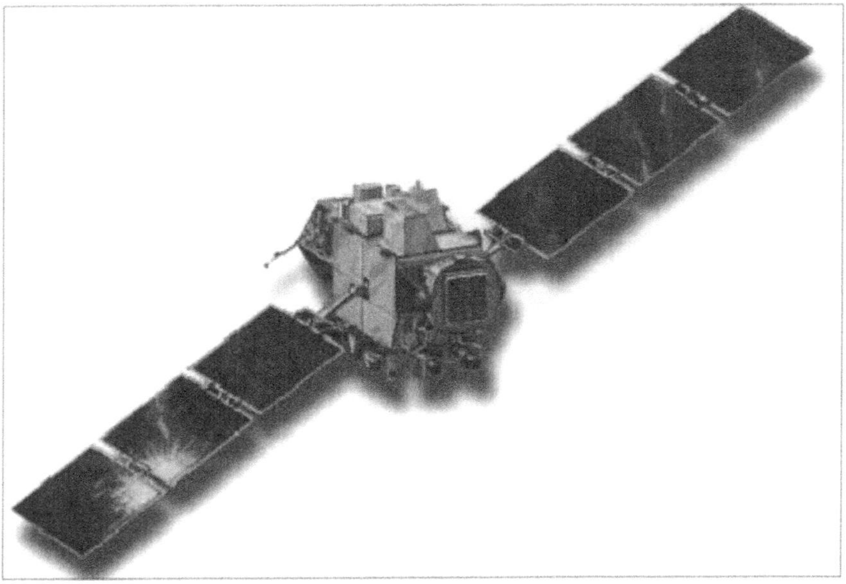

Figure 294: Illustration of the ARGOS satellite

The S/C was designed/built at Boeing North American, Anaheim and Seal Beach, CA. The ARGOS structure of modular panel design, it is three-axis stabilized. Navigation and attitude determination are supported by horizon sensors (to calibrate gyro drifts), three-axis gyros (ring laser gyro package), and an embedded GPS receiver. The attitude pointing accuracy requirement for the S/C is 0.05° (1 sigma). At the core of space vehicle flight computer processing is the Integrated Electronics Unit (IEU), which acts as the "brains" for space vehicle operations. An on-board automated mission planning system, developed by the University of Colorado, is used to optimize on-board data handling and power. The on-board solid-state recorder provides 2.6 Gbit of data storage. S/C mass = 2718 kg, power = 2.2 kW (average = 1 kW), design life = 3 years.

A launch of ARGOS on a Delta-II vehicle from VAFB (along with SUNSAT (University of Stellenbosch, South Africa) and Ørsted (Denmark) as secondary payloads) took place on Feb. 23, 1999.

1614) http://www.laafb.af.mil/SMC/PA/Fact_Sheets/Argos.htm

1615) http://www.te.plk.af.mil/stp/argos/argos.html

1616) http://www.pxi.com/brochures/argos/index.html

Communication: The downlink data rate is 4 kbit/s or 128 kbit/s on carrier I, and 1, 4 or 5 Mbit/s on carrier II. The uplink data rate is 2 kbit/s.

Orbit: Sun-synchronous circular orbit, altitude = 835 km, inclination = 98.7°, equatorial crossings at 14:00 (daytime orbit) and 2:00 hours (night side orbit).

Sensor complement:

HTSSE-II (High Temperature Superconducting Space Experiment II), sponsored by NRO (National Reconnaissance Office), Navy, DARPA, BMDO, NASA, Canadian Government, and German Government. HTSSE-II is an NRL instrument with the objective to demonstrate and validate the performance of eight HTS (High Temperature Superconductor) materials/components in a microwave system setting (semiconductors and RF devices) with an accuracy sufficient to detect long-term changes. Measurement of power consumption, speed, space radiation effects, etc. at temperatures of 77 K (Note: HTS is an important emerging technology for the 21st century with the potential of major breakthroughs in commercial and military applications). Power reduction factors of two to three orders of magnitude are expected. The HTSSE-II instrument was designed and built in a partnership program between NRL industry and academia. The goal is the demonstration of an advanced mechanical cryocooler that may be used for focal plane arrays and cooled semiconductors as well as for HTS devices. 1617) 1618) 1619)

Device category	Major design features	Supplier
Channelizers/filters	4 channel input multiplexer @ 4 GHz 4 channel filter @ 4GHz 5 channel input multiplexer @ 8 GHz	ComDev Westinghouse Space Systems/Loral
Receivers	60 GHz communication receiver Wideband cueing receiver, >2 GHz chirp bandwidth Hybrid 9 GHz chan. receiver with MMIC mixer Low-noise HTS/GaAs down-converter: 7 to 1 GHz	TRW MIT/LL NRL NASA/JPL
A/D converter	Digital logic using Josephson junctions	Conductus
Digital IFM	5 bit, 16 MHz resolution, f_c=4GHz, bandwidth of 500 MHz	Conductus
Digital multiplexer	Logic using HTS SQUID	TRW
Delay line	40 ns delay line	Westinghouse
Antenna array	Adaptive nulling design, 4 elements, f_c=5 GHz	Univ. of Wuppertal
HTS Material Environment Effects Monitor	Measure of space radiation effects on T_c, J_c, R_s and λ	NRL

Table 475: Summary of electronic devices selected for HTSSE-II demonstration

The BAe (British Aerospace - now part of Astrium Ltd) Stirling-cycle cryocooler (Oxford design with demonstrated reliability; nominal cooling capacity of 780 mW at 75 K) was selected for the HTSSE-II cryogenic bus cooling source. The HTSSE-II payload is divided into five temperature-controlled zones: 1) the electronics deck, 2) the BAe cryocooler, 3) the cryogenic cold bus, 4) CBSS (Cold Bus Support Structure), and 5) the TRW experiment package. Operating temperatures for the electronic deck and the BAe and TRW cryocoolers are separately controlled. The HTSSE-II cryogenic subsystem design is unique to space-deployable cryogenically cooled electronics in general, and microwave applications in particular. The central cold bus is the heart of the cryogenic subsystem where seven of the eight HTS experiments are mounted and cooled by the BAe cryocooler. The 8th device is a stand-alone subsystem where the HTS device is integrated with its own cryogenic refrigerator.

HTSSE-II is instrumented to monitor and diagnose the in-flight cryocooler and cryogenic payload status. On-orbit measurements are performed in combination with a ground-

1617) Special issue of IEEE Transactions on Microwave Theory and Techniques, Vol. 44, No 7, July 1996, pp. 1193-1392

1618) R. A. McKnight, M. F. Bahrain, et al., "On-Orbit Status of the High Temperature Superconductivity Space Experiment (HTSSE-II), AIAA-99-4486, 1999

1619) http://ssdd.nrl.navy.mil/www/htsse/htsse.htmlx

based, calibrated signal source. This is referred to as "bent-pipe" mode of operation. In this approach, test signals are transmitted from an NRL ground facility; they are received by the payload, passed through the device under test, detected, digitized, and stored for a later downlink to the receiving site. Experiments are characterized using on-board parameter measurement equipment. In characterizing an experiment (e.g., multiplexer, channelizer, etc.) a received pulse-modulated signal is amplified, filtered and applied to the HTS experiment. Simultaneously, the same signal is applied to the reference channel (C-band or X-band). Both the experiment output and the reference signal are detected and digitized providing comparative amplitude outputs. Since the two signals are from the same antenna and RF front-end hardware, the pulse amplitude difference in the two paths is a measure of the response of the experiment.

Device or component	Major design features	Supplier
RF input/output cables	Low thermal loss stainless steel, coaxial cables of 0.53 mm diameter	Gore
MLI (Multi-Layer Insulation blanket for cryogenic cold bus	Aluminized mylar layers separated by silk net	Lockheed
Mechanical cryocooler using Stirling cycle	Maintain cryogenic cold bus at 77 K with 500 mW load Stand-alone cryogenic system at 65 K with 250 mW load CDE (Cryocooler Drive Electronics) with closed loop	BAe TRW Lockheed
Digital instrumentation	Digitization of data from from RF processing system	Aeronics
RF and video instrumentation	Detection/processing of RF test signals	ITT Government Systems
HRT (HTSSE Remote Terminal)	Satisfy all command and data telemetry needs	Gulton
Receive antenna	Receive C- and X-band signals transmitted from ground	ACA (Antenna Corp. of America)

Table 476: Device summary of HTS experiments of the common cryogenic cold bus

The HTSSE-II RF payload consists of a dual-band antenna, associated payload receiver modules, a VPM (Video Processor Module), and IFMM (Instantaneous Frequency Measurement Module). The receiver module and the VPM have RF and video interfaces with the cold bus containing the HTSSE-II experiments. The VPM performs parameter measurements on the detected RF signals from the receiver. Additional interfaces are VPM to digitizer and to the HRT (HTSSE Remote Terminal). All receiver control functions, such as redundancy switching, attenuator controls, and channel selection, are received through the VPM from the HRT. The HRT receives this information from the ARGOS S/C command and control and data bus, which uses MIL-STD-1553 bus architecture.

The receiver consists of a superconducting X-band four-channel demultiplexer with 100 MHz-wide channels, four commercial monolithically integrated GaAs mixers, and four custom-designed hybrid-circuit detectors containing heterostructure ramp diodes. The superconducting frequency demultiplexer is fabricated from thin films of YBCO (Yttrium-Barium-Copper-Oxide) which were pulse laser deposited onto an MgO substrate. The composite receiver unit has been integrated into the payload of HTSSE-II. - The HTSSE-II instrument has a mass of 132 kg, the power is 98 W.

EUVIP (Extreme Ultraviolet Imaging Photometer, sponsored by the US Army and built by UCB). The objective is to investigate the upper atmosphere and plasmasphere for safe military communication system's design, prediction of magnetic storms, and characterization of the aurora. EUVIP also observes the Earth's horizon and stellar environment by measuring background radiation for future sensor design. The EUVIP experiment is mounted to the S/C body and oriented to view tangent altitudes from 200 to 550 km. EUVIP is also oriented to view the much of the same region of the sky measured by the HIRAAS limb scan.

The EUVIP experiment is essentially a telescope with a digital focal plane. The telescope design is based on a grazing incidence Wolter-Schwarschild Type I optics. The entrance aperture is a concentric ring with 40.3 cm outer diameter and 37.5 cm inner diameter, and 156 cm^2 geometric area and 80 cm^2 effective area in the EUV. The experiment detector uses a stack of microchannel plates and a wedge-and-strip anode for two dimensional imaging with 256 x 256 pixels. The overall instrument FOV is roughly 5¼ in diameter and aligned with the optical axis of the telescope. The instrument resolution is about 1.5 arcmin corresponding to 0.2 mm resolution at the detector focal plane. EUVIP dimensions are 89 cm x 58 cm x 53 cm, the instrument mass is 70.5 kg.

The detector is covered with three filters to isolate separate spectral regions. The three filter passbands measure the following constituents: 1) Lexan/B (7 - 20 nm); 2) Al/C (16-30 nm) and 3) Al/Sn (80 - 90 nm). The first and second filters measure astronomical sources and the Earth's magnetospheric emission from the He^+ line at 30.4 nm. The third filter passband contains the dayside ionospheric emission from O^+ at 83.4 nm, also measured by the HIRAAS and GIMI experiments.

USA (Unconventional Stellar Aspect), also referred to as NRL-801 experiment.[1620)] USA is an NRL payload (226 kg) with the objective to provide a research platform (testbed) for X-ray timing, time-resolved spectroscopy, and also to explore applications of X-ray sensor technology. A further objective of USA is to use **X-ray detectors to provide autonomous timekeeping and navigation capabilities** (use of X-ray binaries and pulsars as clocks and observing horizon crossings of bright X-ray sources). Observations of atmospheric occultations of X-ray sources, as in GIMI (Global Imaging Monitor of the Ionosphere) with stellar occultation observations in the far ultraviolet, can provide unique information about the composition and structure of Earth's upper atmosphere. Unlike in the UV, the atmospheric attenuation of X-rays is relatively insensitive to the chemical composition of the gas (i.e. whether atomic or molecular, nitrogen or oxygen), and can be used in both day and night conditions. Therefore, atmospheric occultation measurements with the USA experiment complement the upper atmospheric measurements made with GIMI and HIRAAS (High Resolution Airglow / Aurora Spectrograph).

USA conducts feasibility tests of X-ray satellite navigation and new computational approaches to autonomous parameter estimation that includes GPS inputs and a variety of redundant truth measures. X-ray attitude determination consists of directing an X-ray sensor at a field in the sky and comparing the detected sources with expectations based on the known maps of that field. The offset between the map and the observed pattern gives an error signal. Attitude is determined in inertial space. USA employs the method of two gimballed sensors to obtain the full attitude solution with X-ray sources. -Satellite position determination is made in X-rays by observing transitional events such as horizon crossings or occultations of stars with known positions. [1621)]

The USA instrument design is based on the use of large-area proportional-counter detectors in the energy range of 1-15 keV. The detector assembly is mounted on a two-axis pointing system, which allows the detector to be pointed at celestial targets of interest during the space flight mission. The USA instrument is a reflight of the X-ray detectors on Spartan-1 (Shuttle flight STS-51G, June 17-24, 1985). Key characteristics of the USA instrument are:

- Long observing times on bright X-ray objects
- Large-area detectors with high time-resolution capability (area: 2000 cm^2, telemetry: 40 kbit/s with 128 kbit/s available for short periods; 1 μs time resolution (time is provided by the GPS receiver); USA also receives the output of the ring laser gyro package.
- Low-energy response (down to 1 keV)
- High flexibility in data handling.

1620) K. S. Wood, et al., "The USA Experiment on the ARGOS Satellite: A Low Cost Instrument for timing X-ray Binaries," SPIE Proceedings, Vol. 2280, 1994 p. 19

1621) J. H. Beall, T. Crandall P. S. Ray, "Innovative Satellite Navigation Exercises Utilizing the USA Experiment and the ARGOS Satellite," http://www.pxi.com/brochures/argos/ARGOS_exp.pdf

There are reasons why X-ray navigation might prove to be attractive in the future. The advantages are associated mainly with drawbacks of optical methods or with potential advantages of X-ray characteristics that have no exact analogs in the optical wavelengths. These include:

- The ability to operate in wavelengths in which the ground cannot be seen, thereby removing any hazard of blinding by ground-based lasers
- Freedom from difficulties of dealing with scattered sunlight or the bright Earth
- Capability of performing multiple tasks (position, attitude, and timekeeping) with a single sensor
- Horizon sensing in wavelengths where only the upper atmosphere is traversed by a ray grazing the effective horizon, removing weather problems and making possible simple (smart) horizon sensors
- Feasibility of high-accuracy attitude sensing without optics
- Exploitation of a simpler sky in which many bright sources are available without distraction from nearby sources that are only slightly fainter
- Exploitation of spectral features of bright non-optical sources, such as the periodic intrinsic variability of many bright X-ray sources
- Straightforward exploitation of lunar occultations of celestial sources.

In addition, three radiation-hardened 32-bit space computers [RH32 (of Rome Laboratory, Rome, NY) and RH3000 (Harris)] are part of the USA platform. They are being space-qualified by performing real-time fault-tolerant applications. Both boards have full access of the full downlink telemetry data stream. A software package provides fault-tolerant features for the detection of in-flight transient errors.

ESEX (Electric Propulsion Space Experiment). The objective is to address the DoD spacelift and maneuvering requirements for global surveillance and communications orbits and the application of advanced technology in order to reduce life cycle costs. ESEX (built by TRW and managed by Phillips Laboratory of USAF) demonstrates high-power arcjet propulsion technology (reliable arcjet thruster operation) for orbit transfer, maneuvering capability, and attitude adjustment. With a power input of 26 kW and a mass of 450 kg, ESEX is so far the largest electric propulsion system to orbit. ESEX operations are being tracked with ground radars and optical sites to allow characterization of the arcjet's electromagnetic and visual signature.

SPADUS (Space Dust Experiment). A small ONR (Office of Naval Research) instrument with the objective to measure and to characterize orbital debris and dust in the LEO environment of highly populated sun-synchronous orbits. Generation of 3-D survey maps of the present dust distribution. SPADUS also includes ancillary diagnostics to measure the local radiation environment. The experiment results are expected to help spacecraft shielding and electronics design, extending space vehicle lifetimes. Instrument mass of 23 kg.

CIV (Critical Ionization Velocity). A PL (Phillips Laboratory) experiment with the objective to release Xenon and Carbon Dioxide to study ionization caused by plasma and collision processes in the upper atmosphere. CIV makes also use of the gas releases by the reactive control system of ARGOS to measure/identify the plumes and atmospheric wakes.

HIRAAS (High-Resolution Airglow/Aurora Spectroscopy). [1622] Objective: test of a new technique for ionospheric and thermospheric remote sensing. An NRL instrument package with the objective to measure the naturally occurring thermospheric and ionospheric airglow (environmental model improvement for predicting effects of the ionosphere on communications). HIRAAS is a multi-instrument experiment package containing three UV spectrographs (HITS, LORAAS, and ISAAC) to measure naturally occurring atmospheric emissions. The three instruments look into the same direction and operate in parallel. The

1622) http://ltpwww.gsfc.nasa.gov/ISSSR–95/hyperspe.htm entitled: "Hyperspectral Imaging of the Global Ionosphere from the ARGOS Satellite," science objectives of HIRAAS experiment are to measure, by remote sensing, the densities and temperatures of the various upper-atmospheric atoms, molecules, and ions, and how they vary with altitude, position over the Earth, and solar activity. The gases to be measured include O_2, O, O+, N_2, N, NO, and H.

A limb scan is performed in the aft orbit direction from -10° to -37° relative to the S/C horizon corresponding to tangent altitudes from 750 km to 50 km (90 s per downward scan, 15 s flyback). The imagery of the ionosphere obtained is "hyperspectral" at wavelengths from 50 to 310 nm. An essential element of these investigations involves comparisons with ground-based ionosondes and radars to provide ground truth for remote sensing observations.

- **HITS** (High Resolution Ionospheric/Thermospheric Spectrograph). The goal is to measure FUV (Far UV) radiation in the spectral range is from 50 - 170 nm at a resolution of 0.03 nm. HITS features a 1 m focal length Rowland Circle spectrograph fed by a telescope consisting of a parabolic mirror coated with silicon carbide for maximum reflectivity in the EUV. The spectrograph uses a 4800 l/mm grating. A grating drive mechanism rotates and translates the grating to select 10 nm segments within the 50-170 nm total passband and maintains focus and resolution at each position. The HITS spectrograph uses a MCP (Microchannel Plate) detector with a wedge-and-strip readout anode and a cesium iodide photocathode. The detector format provides a one-dimensional spectrum with 1024 pixels in the dispersion direction once per second. The HITS spectrograph resolution of 0.03 nm is sufficient to resolve the fine structure of atomic and ionic multiplets (a multiplet refers to a spectrum line of several components) and the rotational structure of molecular bands. For example the O^+ 83.4 nm multiplet is a triplet emission (83.267, 83.333, 83.447 nm) and with the HITS resolution all three components are resolved. Separating the three lines provides additional information for ionospheric remote sensing since the multiple scattering opacity varies with each line. Each line provides a solution to the ion density in a different opacity regime.

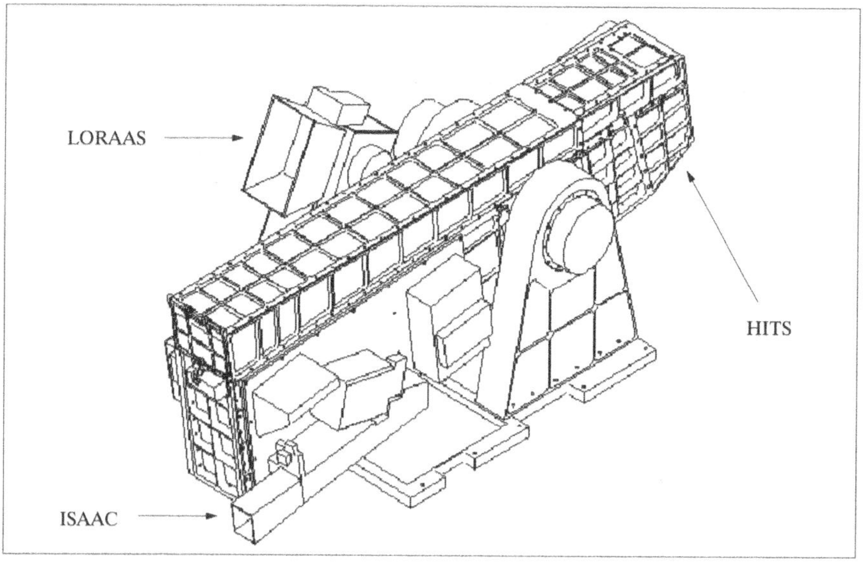

Figure 295: Illustration of the HIRAAS instrument

- **LORAAS** (Low Resolution Airglow/Aurora Spectrograph). Measurement of the FUV (Far UV) in the 80-170 nm range. The LORAAS instrument is a 0.25 m focal length

Wadsworth spectrograph with a stepper motor controlled scan mirror, mechanical grid collimator, grating and wedge-and-strip detector. Both the scan mirror and grating are coated with silicon carbide for maximum efficiency in the EUV. The scan mirror sweeps the FOV across the limb and is synchronized with the HIRAAS scan mechanism. The detector is coated with CsI and produces a linear array spectrum with 256 pixels in the dispersion direction once per second. LORAAS provides the entire 80-170 nm spectrum at moderate resolution (1.5 nm) while HITS measures subsets at high resolution. The LORAAS instrument is an exact copy of the **SSULI** (Special Sensor Ultraviolet Limb Imager) spectrograph that flies operationally on the DMSP Block 5D3 weather satellite series (G.1). Since the ARGOS satellite has an orbit similar to the DMSP satellites, LORAAS and HITS provide a validation and engineering test of the SSULI concept.

- **ISAAC** (Ionospheric Spectroscopy & Atmospheric Chemistry spectrograph). ISAAC measures in the mid-UV band from 180-310 nm. The ISAAC instrument is a modified 1/8 m Ebert-Fastie spectrograph fed by an off-axis telescope. The ISAAC detector is an intensified diode array which uses a microchannel plate stack, Cs2Te photocathode, phosphor and fiber optic reducer. The spectrograph uses a stepper motor to select a 25 nm portion of the 180-310 nm passband. The ISAAC spectral resolution of 0.25 nm is sufficient to resolve the rotational structure of a number of molecular bands from which the temperature of the lower thermosphere and ionosphere can be determined. Additionally, the major molecular ion in the F1- and E- region of the ionosphere is NO^+, and ISAAC can measure several bands of NO from which models will be used to infer the lower ionospheric distribution.

The HIRAAS measurements test a new technique for ionospheric and thermospheric remote sensing. Extreme UV imaging of the ionosphere is required to improve a number of DoD systems that depend on radio and microwave propagation through the upper atmosphere and ionosphere. Neutral density measurements support standing NORAD operational requirements to improve satellite drag forecasting and the ability to predict orbital life and re-entry impact locations. The end product is an improved environmental model for predicting effects of the ionosphere on communications and eventually a "weather" prediction of the upper atmosphere for operational use.

GIMI (Global Imaging Monitor of the Ionosphere).[1623) 1624)] An NRL sensor with the primary objective to map and monitor the ionospheric O^+ and electron density on a global basis, day and night, by wide-field imaging of ionospheric far-ultraviolet emissions. A secondary objective is to map and monitor neutral upper atmospheric constituents, specifically N_2, O_2, and NO. The first two are measured by observing the occultation of UV-bright stars. NO is measured by observing the UV night airglow emissions. Astronomical objectives of GIMI include obtaining an all-sky survey of both point and diffuse sources in the spectral range of 131-200 nm, and of point sources in the range of 92-110 nm. GIMI is mounted on the Earth-facing (nadir) panel of the ARGOS satellite to view the Earth limb and other directions in the zenith angle range 70.25 to 135.25°. The instrument consists of two wide-field UV imaging cameras, using electron-bombarded CCDs (EBCCDs), mounted on a two-axis gimbal system for nadir observations. The two cameras are co-aligned to permit simultaneous UV observations.

- GIMI camera 1 observes in the 75-110 nm range (EUV) and has a 10° square FOV. It is used primarily for observing O^+ emissions at 83.4 nm in the day airglow, and stellar occultations in the 92-110 nm range for measurements of upper atmospheric N_2. It is also used for observing stars and other astronomical objects in the 92-110 nm range.
- The GIMI camera 2 observes in the 131-160 nm and 131-200 nm spectral ranges (FUV). The camera photocathode is split into two regions (using different materials).

1623) G. R. Carruthers, T. D. Seeley, "Global Imaging Monitor of the Ionosphere (GIMI): a Far Ultraviolet Imaging Experiment on ARGOS," SPIE Proceedings, Vol. 2831, 1996, p. 65

1624) http://spacescience.nrl.navy.mil/Branches/gimiwebpage.htm

The 131-200 nm range provides a FOV of 3.3° x 10° when viewing Earth, whereas the 131-160 nm range portion uses a FOV of 6.7° x 10°. Both spectral ranges are used to measure O_2 by observing stellar occultations. The longer range near 190 nm is used to observe NO nightglow emissions, the shorter range at 135.5 nm observes ionospheric O^+ recombination emissions.

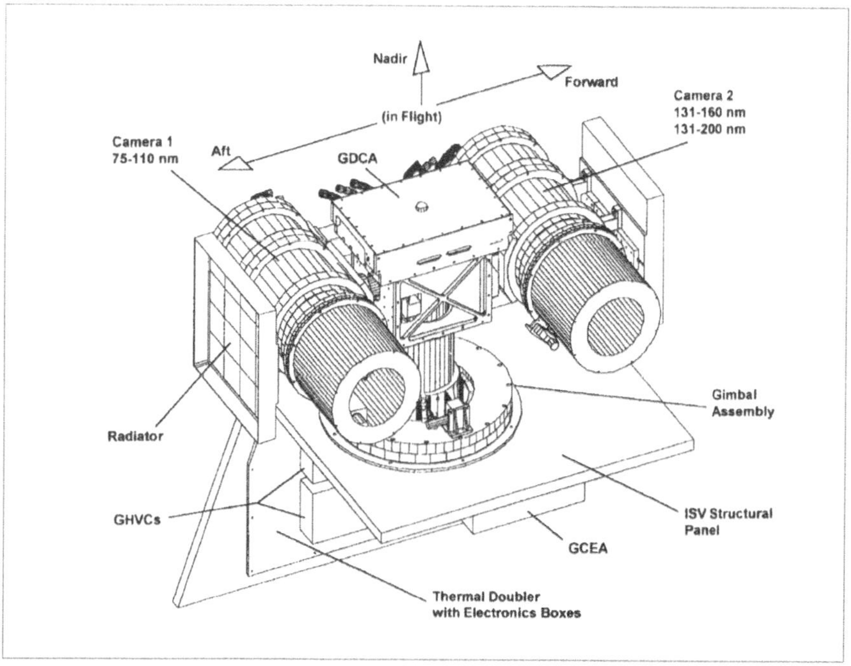

Figure 296: Illustration of the GIMI instrument

Astronomical sky surveys are made using both ranges of camera 1. Camera 2 is used only at night (in Earth shadow), while camera 1 can be used day and night. The GIMI gimbal system allows the cameras to point in a zenith angle range from 75° to 135° in the orbit plane (forward or aft). The cameras may also be gimbaled in the yaw direction (cross track) to observe auroras and other phenomena of interest. Observations of the moon and known stars are also used for in-flight calibration of GIMI.

For observations of the upper atmosphere and ionosphere, the cameras are pointed at a zenith angle of 112° (22° below horizontal), in which case the line of sight has a 300 km tangent altitude (altitude of closest approach of the line of sight to Earth). Repeating sequences of exposures are used to cover a continuous strip of up to 120°.

The HIRAAS, GIMI and EUVIP instruments on the ARGOS S/C provide a powerful remote sensing exploration of the global ionosphere. This unusual combination of hyperspectral imagers and imagers provides a unique database of global structure and variability of ionospheric structure.

Students (high school and college) have been involved in nearly all aspects of the design, development, and testing of GIMI prior to launch, and will likely continue to be involved after launch (in mission operations and data analysis). During the actual mission, it is planned that students will be able to access GIMI data by way of the Internet, in a manner similar to that of the Moonlink education and public outreach program associated with the Lunar Prospector mission, which is implemented by Space Explorers, Inc.

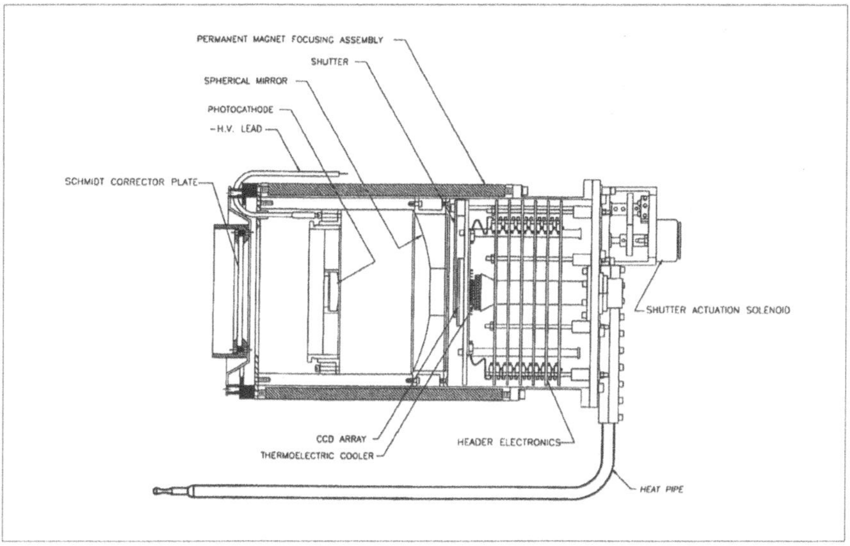

Figure 297: Diagram of the EBCCC cameras

CERTO (Coherent Radio Topography Experiment). An NRL sensor with the objective to test and develop tomographic algorithms for reconstruction of ionosphere densities and irregularities. The experiment consists of a three-frequency radio beacon and radiating antenna mounted on the S/C. Receivers on the ground use differential phase techniques to derive the integrated electron density. CERTO data analysis permits an impact assessment on navigational accuracy, communication systems, and remote sensing by radar.

M.2 ARTEMIS (Advanced Relay and Technology Mission Satellite)

ARTEMIS is ESA's first GEO data relay communication satellite with the objective to demonstrate new communication technologies, principally for data relay and mobile services. The technology demonstrations include an optical intersatellite link, first European operational use of an electric ion propulsion system, and a transponder for the support of EGNOS (European Geostationary Navigation Overlay Service) for signal enhancement of the GPS/GLONASS navigation satellite constellations. [1625)] [1626)] [1627)]

The S/C structure consists of a box-shaped three-axis bus of Italsat heritage (Alenia Spazio bus family, Alenia is also the prime contractor to ESA). The primary structure consists of the central cylinder (aluminum honeycomb skinned with carbon fiber), the main platform, the propulsion platform, and four shear panels. The secondary structure is made up of the N/S radiators, the E/W panels, and the Earth-facing panel. The central propulsion module houses the propellant tanks, LAE (Liquid Apogee Engine), the East panel with the L-band

1625) A. Dickinson, G. Oppenhäuser, et al., "The Artemis Program," ESA Bulletin, No. 91, August 1997, pp. 32-39

1626) A. Wilson, "ARTEMIS," ESA publication BR-142 with the title: More than Thirty Years of Pioneering Space Activities, 1999, pp. 156-161

1627) A. Dickinson, S. Greco, I. La Rosa, M. Protto, "The ARTEMIS Program: Near Term Advanced Communications Technology," Proceedings of 47th AIAA Congress, Beijing, China, Oct. 7-11, 1996 antenna feed, the West panel with the IOL (Inter Orbit Link) antenna. The two antenna reflectors (2.85 m diameter) for IOL support are dominant features of the S/C structure.

Figure 298: Illustration of the ARTEMIS S/C

S/C attitude is measured by Earth/sun sensors and gyros. Reaction wheels serve as actuators. Thrusters of an RCS (Reaction Control System) are used for wheel off-loading. The UPS (Unified Propulsion System) employs a conventional bi-propellant system of a single 400 N LAE for insertion into GEO. The propellants are stored in two Cassini-type 700 liter tanks. E/W positioning is maintained by RCS. N/S positioning is maintained by electric ion thrusters. The IPS (Ion Propulsion System) comprises two thruster assemblies, RIT (RF Ion Thruster) and EIT (Electro-bombardment Ion Thruster). Each is powered and monitored separately, but a common propellant supply is used (40 kg of xenon), 600 W of input power is needed for operation.. S/C electric power of 2.8 kW (at equinox after ten years to a 42.5 VDC bus) is provided by two solar wings. BSR (Back Surface Reflecting) solar cells are mounted on each of the two solar wings. Two NiH_2 batteries provide energy of 60 Ah for eclipse protection. A S/C design life of 10 years is provided. The S/C launch mass is 3100 kg (550 kg payload, 1538 kg propellant).

A launch of ARTEMIS, along with BSAT-2B of BSS (Broadcasting Satellite System Corp. of Tokyo, Japan as co-passenger, took place on July 12, 2001 on an Ariane-5 vehicle from Kourou. An upper stage malfunction of Ariane-5 resulted in a lower orbit than GEO (the two satellites were left in an orbit with a perigee of 592 km and an apogee of 17,528 km instead of the intended GTO of 858 km x 35,853 km). ESA is planning recovery actions to get ARTEMIS into GEO using the kick motor as well as two experimental electric propulsion systems.

Orbit: Geostationary orbit at 21.4° eastern longitude.

Payload/experiment complement:

The S/C data relay payload provides feeder links between Artemis and the ground as well as IOLs (Inter Orbit Links) between ARTEMIS and the S/C in LEO (SPOT-4). The feeder links operate at 20/30 GHz, while the IOLs can operate in S-band (2 GHz), Ka-band (23/26 GHz), and optical frequencies.

The feeder link, S-band and Ka-band elements jointly comprise the SKDR (S/Ka-band Data Relay) payload, while the optical IOL payload element is called SILEX (Semiconductor Intersatellite Link Experiment). ARTEMIS data relay service support (via RF links) is planned to be provided to ENVISAT of ESA.

SKDR (S/Ka-band Data Relay). The objective of the IOL antennas (2.85 m diameter) is to track a LEO user satellite via either loaded table values and/or error signals - and to receive up to 450 Mbit/s of data in the Ka-band, or up to 3 Mbit/s in S-band for relay via the feeder link to Earth (return link operation). Up to 10 Mbit/s in Ka-band and 300 kbit/s in S-band may be transmitted by ARTEMIS to the LEO satellite (forward link operation). [1628] In addition, ARTEMIS broadcasts a 23.540 GHz beacon to help the LEO satellite to track it.

- A single Ka-band transponder (plus one backup) provides return/forward frequencies of 25.25 - 27.5/23.2 - 23.5 GHz links in Rx/Tx, adjustable EIRP (Effective Isotropic Radiated Power) of 45-61 dBW, G/T of 22.3 dB/K, up to 150 Mbit/s each of the three channels LEO to ARTEMIS (return link), and up to 10 Mbit/s from ARTEMIS to LEO (forward link). RH/LHCP on command.
- One S-band transponder (plus one backup) provides return/forward frequencies of 2.200-2.290/2.025-2.110 GHz links in Rx/Tx, adjustable EIRP 25-45 dBW, G/T of 6.8 dB/K. The bandwidth is 15 MHz. Up to 3 Mbit/s of data can be transmitted in a single channel from LEO to ARTEMIS (return link), and up to 300 kbit/s can be transmitted from ARTEMIS to LEO (forward link). RH/LHCP on command.

Feeder link of SILEX and SKDR: Three transponders (plus one backup) act as ground-ARTEMIS links for SILEX and SKDR. The feeder Ka-band frequencies are: 27.5-30/18.1-20.2 GHz for Rx/Tx. The EIRP is 43 dBW, G/T of 0 dB/K, use of 234 MHz bandwidth, linear vertical polarization.

Technology Demonstrations:

SILEX (Semiconductor Intersatellite Link Experiment), an ESA laser experiment built by MMS, France. SILEX consists of two optical terminals, namely OPALE (Optical Payload for Intersatellite Link Experiment) located on ARTEMIS, and PASTEL (PAssager SPOT de Técommunication Laser) on-board SPOT-4. The objective is to beam data at rates of 50 Mbit/s (bit error rate of $<10^{-6}$) from the transmitter terminal on SPOT-4 in LEO - to the receiver (OPALE) on ARTEMIS for subsequent relay via feeder link to the SPOT ground segment in Toulouse. The SILEX terminal on-board ARTEMIS is also being used to support a second LEO experiment, namely an IOL between ARTEMIS and OICETS (Optical Inter-orbit Communications Engineering Test Satellite) of NASDA.

- **PASTEL** (PAssager SPOT de Técommunication Laser). A joint ESA/CNES passenger demonstration experiment. PASTEL is a prototype high data-rate intersatellite transmission system based on laser technology. The objective is to transmit imaging data from SPOT-4 to ARTEMIS. The aim of the experiment is to validate the PASTEL concept design in an operational environment. PASTEL is a gimbal-mounted assembly consisting of a telescope, an optical bench with a fine pointing system, communication detectors with avalanche photodiodes, a thermal control system for precision temperature control, a two-axis gimbal mechanism, and the launch locking mechanisms needed during the launch phase. The telescope mirrors and main structural elements are made of Zerodur. The acquisition and tracking sensors use CCD detectors. The laser diodes are of the GaAlAs type. The SPOT-4 - ARTEMIS optical links operate at wavelengths of 830 nm. The peak output power is 160 mW (60 mW continuous operation), the beamwidth is 0.0004°. Data to be transmitted include: HRVIR image data, pseudo-noise (PN) code, PASTEL telemetry.

1628) Note: In very elaborate communication systems with intermediate geostationary transmission satellites, the term 'uplink' is usually replaced by 'forward link' to avoid confusion. Similarly, the term 'downlink' is usually replaced by 'return link.'

- **OPALE** (Optical Payload for Intersatellite Link Experiment) terminal, mounted on the geostationary satellite ARTEMIS (a GEO terminal). The receiver employs Si-APD (Silicon Avalanche Photodiode) detectors and a low-noise trans-impedance amplifier of 1.5 nW useful receiver power.

Each SILEX terminal features a telescope of 25 cm diameter (which is mounted on a coarse pointing mechanism), and provides an 'antenna' gain of well above 100 dB. The disadvantage of these extreme antenna gains is the very narrow width of the transmitted beam, requiring very accurate pointing. The divergence tolerance of the optical communication beam for the SILEX configuration is 8 µrad (or about 0.00046°). PASTEL and OPALE use a dedicated acquisition sequence. Initially, both terminals (OPALE and PASTEL) coarsely point to each other. This is done when OPALE scans a wide-angle (750 µrad) beacon beam in the direction of PASTEL. On illumination of PASTEL by the beacon beam, it rapidly corrects its line of sight and directs in turn a narrow communication beam towards OPALE. Similarly, OPALE detects the incoming PASTEL signal, aligns its line of sight, and transmits its narrow communication beam towards PASTEL. The two terminals then remain locked on each other in closed-loop tracking, permitting subsequent communication.

LLM (L-band Land Mobile) payload. The objective of the communications payload is to permit two-way communications, via satellite, between fixed Earth stations and land mobiles, such as trucks, trains or cars, anywhere in Europe and North Africa. The LLM package is fully compatible with the EMS (European Mobile System) payload developed by ESA and flown on Italsat-2. Hence, full redundant support is provided. - The LLM receives the signals transmitted by the fixed users in Ku-band (14.2 GHz) and transmits them at L-band (1550 MHz) to the mobile users (forward link). The return link establishes the connection from the mobile user at L-band (1650 MHz) to the S/C, and at Ku-band (12.75 GHz) from the S/C to the fixed user in the ground segment. About 400 bi-directional user links can be established simultaneously.

ARTEMIS carries two antennas of 2.85 m diameter and a multiple element feed for pan-European coverage and three European spot beams. Three 1 MHz plus three 4 MHz SSPA (Solid-State Power Amplifier) channels, provide 400 2-way circuits with an EIRP > 19 dBW. The on-board L-band transmits to terminals (users) in the ground segment at 1550 MHz and receives data at 1650 MHz. A Ku-band feeder link at 14.2/12.75 GHz Rx/Tx transmits the data to the home stations. All channels are fully tunable and most commandable for LH/RHCP support.

IPP (Ion Propulsion Package). The electric propulsion system of ESA on-board ARTEMIS is being used for inclination control throughout the lifetime of the S/C (i.e., N/S station keeping maneuvers). IPP consists of a redundant pair of thruster assemblies, one mounted on each of the north and south faces. Each assembly comprises ITAM (Ion Thruster Alignment Mechanism) upon which two redundant thrusters from different sources are mounted. The entire IPP (2 RITA+ 2 EITA) assembly has a mass of 84 kg (without propellants). [1629)]

- **RITA** (Radio-frequency Ion Thruster Assembly) of DASA, Germany. The RIT-10 thruster for ARTEMIS has a beam diameter of 9.8 cm and delivers 15 mN thrust with an Isp of 3500 s and a system power of 560 W. The system provides 15,000 hours of operation at 15 mN (see O.14.2). A single RITA unit has a mass of 13.9 kg.
- **EITA** (Electron-bombardment Ion Thruster Assembly). [1630)] [1631)] [1632)] EITA is an ion engine system mainly developed by DERA, UK. The UK-10 has an exit diameter of 10 cm diameter, it provides a maximum thrust of 23 mN at an Isp of 3400 s. The input pow-

1629) Information provided by R. Killinger of DASA (now Astrium GmbH)

1630) D. G. Fearn, "Low Cost Missions Using Ion Propulsion," Proceeding of the British Interplanetary Society Symposium on 'The search for life on Mars,', London, Nov. 11, 1998

1631) Information provided by C. Edwards of DERA

er is 700 W. The EITA version for ARTEMIS is provided by MMS, UK. It delivers 18 mN of thrust. A single EITA unit has a mass of 15.2 kg.

EGNOS (European Geostationary Navigation Overlay Service) payload. [1633] The objective is to provide enhanced navigation performance in terms of accuracy and integrity (with the required levels of availability and continuity) over the ECAC (European Civil Aviation Conference) [1634] region. The service may later be extended to neighboring regions.

The EGNOS payload on ARTEMIS uses the Ku-band in the uplink and downlink (for S/C - fixed user communication in the ground segment). The uplink frequency is allocated at 13.875 GHz, and separate from the LLM feeder link frequency, while the downlink of the navigation payload is shared with the LLM channels (12.748 GHz). The transmitted EGNOS wide-area service signal is the GPS L1 frequency at 1575.42 MHz (L-band).

The total mass of the navigation payload, including structure, thermal control hardware and the DC harness, is 25 kg. Its total power consumption is about 110 W.

Receive frequency	13.875 GHz (Ku-band)
Transmit frequencies	12.748 GHz (Ku-band), 1575.42 MHz (L-band)
Useful bandwidth	4 MHz
G/T [(receiver) Gain / (noise) Temperature]	>-2.3 dB/K
EIRP (Effective Isotropic Radiated Power)	>17 dBW for Ku-band; >27 dBW for L-band
Polarization	LP for Ku-band, RHCP for L-band
Frequency stability	2x 10^{-12} (10s); 10^{-9} (24 h); 2 x 10^{-7} (life)

Table 477: Some performance characteristics of the ARTEMIS navigation payload

The EGNOS payload on ARTEMIS serves as a geostationary wide-area augmentation system for all GPS signals in its large area of coverage by transmitting:

- GPS-like signals (ranging function)
- GPS health and integrity conditions obtained by ground monitoring stations. This is the RAIM (Receiver Autonomous Integrity Monitoring) function.
- Ranging errors (differential correction function) - these are the conventional DGPS services.

M.3 Bitsy-SX (Bitsy-Spacecraft in Future-X)

Bitsy-SX is a Shuttle demonstration payload in NASA's Future-X program, referred to as SPASE (Small Payload Access to Space Experiment), designed and built around AeroAstro's Bitsy-SX spacecraft kernel.

Background: In 1998 NASA announced a new program, Future-X, managed by MSFC, with the aim to support a series of flight demonstrations to validate new technologies. The idea is to prove technologies that improve performance and lower development, production and operating costs of future earth-to-orbit and in-space transportation systems.
Bitsy™ is a trade name, representing a line of spacecraft "kernel" modules, designed and developed by AeroAstro of Herndon, VA. [1635] The "Bitsy™ concept" as such is an enabling technology, meaning that one does not fly a Bitsy (though among its design requirements is that it can behave as a fully functional spacecraft); rather one uses Bitsy as the starting point in the development of a full spacecraft. The kernel approach applies COTS (Commercial-

1632) H. L. Gray, "Development of Ion Propulsion Systems," GEC Review, Vol. 12, No 3, 1997, pp. 154-168

1633) S. Badessi, C. F. Garriga, J. Ventura-Traveset, J. M. Pieplu, "The European ARTEMIS Satellite Navigation Payload: Enhancing EGNOS AOC Performance," ION GPS 1998, Nashville, TN (USA), Sept. 15-18, 1998.

1634) The ECAC coverage area is from 30° W to 45° E and from 25° N to 75° N

1635) S. A. McDermott, D. J. Goldstein, "The Bitsy™ Spacecraft Kernel: Reducing Nanosatellite Mission Cost in the MSFC Future-X Program Through Miniaturized technologies," Proceedings of the 13th AIAA/USU Conference on Small Satellites, Aug. 23-26, 1999, Logan UT, SSC99-IX-8 of-the-Shelf) and standardization concepts to the fullest extent to obtain significant cost and turnaround advantages over conventional bus development.

The first Bitsy-SX implementation (the total mass of the Bitsy-SX with a NASA microgravity payload is about 35 kg) will fly on the SPASE vehicle in Nov. 2001 (Shuttle flight STS-108). SPASE uses Bitsy-SX, perched upon an octagonal experiment structure, that fits inside a GAS (Get Away Special) canister on Shuttle. The outside of the experiment housing is covered with solar panels. Bitsy-SX contains lithium-ion batteries, providing 7 Wh of energy (average orbital power is 5 W) during orbital eclipse phases. The ACS (Attitude Control System) of SPASE is passive, using hysteresis rods and permanent magnets to keep the S/C body rates below 10^{-5} g for the crystal growth experiment. The payload interaction consists primarily of keeping the crystal growth cavity within a specified temperature range, commanding a digital camera to take images periodically, and downloading those images when a ground station contact occurs.

Deployment of the SPASE free-flyer vehicle from the GAS canister is attended by the Shuttle crew (Hitchhiker launch). It occurs at an altitude high enough above the atmosphere to remain under the 10^{-5} g requirement for the science experiment. Communications (in UHF at 9.6 kbaud) with SPASE as well as S/C operations are conducted from a ground station at the University of Alabama in Huntsville. SPASE uses internally RS-232 interfaces to communicate with the payload. Simple timed commanding is supported as well as threshold monitoring of all S/C and payload functions. The science mission is designed to last 6 months prior to vehicle reentry into the atmosphere.

M.4 DODGE (Department of Defense Gravity Experiment)

A DoD engineering technology satellite, built and operated by JHU/APL, with the objective to study a number of advanced biaxial and triaxial gravity-gradient stabilization techniques at near-synchronous altitudes. Secondary objectives included obtaining measurements of the Earth's magnetic field at near-synchronous altitudes and black-and-white and color TV photography of the entire Earth's disk. [1636)]

The S/C structure was in the form of an octagonal aluminum shell with a truncated pyramid at the top and a 25.4 cm diameter cylindrical mast extending 1.57 m from the satellite base. The satellite body was 2.41 m long and 1.22 m in diameter (S/C on-orbit mass of 102 kg). A total of 10 knobbed booms were carried on board. Upon radio command, these booms could be independently extended or retracted along three axes to various limits out to 45.75 m. The cylindrical mast housed a 4.6 m boom that extended through the end of the mast, two 15.25 m long damper booms that extended in the x-y plane, and triaxial vector magnetometer sensors. The remaining seven booms were contained in the satellite body along with a two-camera (one color and one black-and-white) vidicon camera system. The command system consisted of a dual command receiver, dual command logic, and power switching circuitry. The telemetry system included two directional antennas mounted on the mast, two 38-channel commutators for housekeeping data, and a dual transmitter system that transmitted analog data at a frequency of 240 MHz and TV data at 136.8 MHz. The satellite was successfully stabilized 12 days after launch by means of the gravity-gradient booms and libration dampening systems. It was oriented with its base and mast directed toward the center of the Earth's disk.

DODGE was launched on July 1, 1967 on a Titan III-C vehicle (multiple S/C launch along with DATS 1, LES 5, and IDCSP 16, 17, and 18.) from Cape Canaveral, FL. The satellite was operational for over 3 years.

Orbit: apogee = 33,659 km, perigee = 33,270 km, inclination = 6.2°, period = 1318.9 minutes.

1636) http://nssdc.gsfc.nasa.gov/nmc/sc–query.html

Sensor complement:

Dual Vidicon Cameras. Objectives were to determine the S/C alignment with respect to the Earth and to measure the amount of solar and gravitationally induced bending of the downward-pointing stabilization boom. In addition, the cameras were capable of providing information on the structure and dynamics of global cloud systems, cloud heights, airglow, and auroras. The cameras were mounted in the base of the satellite and viewed down the 1.57 m long cylindrical mast toward the Earth. - The camera system consisted of two vidicon cameras, one with a 22° FOV, the other with a 60° FOV, and associated electronics and power converters. Both cameras were equipped with a 2.54 cm vidicon tube (512 lines/scan) and a special slow scan (200 sec/scan) video pickup. The 60° FOV camera took black-and-white pictures only and used a simple blade-type shutter. The 22° FOV camera, however, took both color and black-and-white pictures. It used a rotating eight-channel color wheel placed in front of the camera, which provided shuttering action. Three of the channels were equipped with blue, green, and red filters, one channel was left blank, and the remaining four had various shortwave cutoff (haze or Rayleigh) filters. High-purity quartz cover plates were placed over the optical filters for protection against radiation damage. On July 25, 1967, the 22° FOV camera took the first color picture of the Earth ever made from a near-synchronous altitude. In all, over 25, 000 pictures were obtained.

Three-Axis Fluxgate Magnetometer. Objectives: study of the Earth's magnetic field and geomagnetic micro-pulsations and other temporal variations. The detector consisted of a triaxial fluxgate magnetometer mounted at the end of a 158 cm cylindrical mast which protruded from the bottom of the S/C. The magnetometers each had a range of ± 250 gamma, and a sensitivity of .2 gamma. Inflight calibration was performed once each hour in a cycle occurring at the start of each hour for which the magnetometers were commanded to record continuously. A 20 Hz low pass filter with zero attenuation between 0 and 0.5 Hz and a 3 db attenuation between 0.5 and 10 Hz limited the frequency response of the satellite. After nine months in orbit, the operations schedule permitted the acquisition of vector magnetic data on a regular basis. The observations were limited to the region between 0 - 150° west and ± 6.3° latitude.

M.5 DS1 (Deep Space 1)

DS1 is the first satellite mission in NASA's New Millennium Program (NMP). A major objective of NMP is to identify and to test new technologies in spacecraft and instrument design as well as in spacecraft operations, to validate new and high-risk concepts in the spaceborne environment, thereby advancing the horizons of future missions. DS1 is in particular a low-cost minisatellite technology demonstration mission of NASA/JPL with the following objectives: [1637] [1638] [1639]

1) Demonstrate the in-space flight operations and quantify the performance of the following five advanced technologies:
   - Solar electric propulsion (SEP)
   - Solar concentrator arrays
   - Autonomous navigation
   - Miniature camera and imaging spectrometer
   - Small deep-space transponder

In addition, the objectives called for the testing of any three of the following six advance technologies:

1637) M. D. Rayman, Ph. Varghese, D. H. Lehman, L. L. Livesay, "Results from the Deep Space 1 Technology Validation Mission," 50th International Astronautical Congress, Amsterdam, The Netherlands, Oct. 4-8, 1999, IAA-99-IAA.11.2.01, the paper is also published in Acta Astronautica, Vol. 47, No 2-9, pp. 475-487, Sept. 2000

1638) http://nmp.jpl.nasa.gov/ds1/

1639) R. M. Nelson, "Deep Space One: Preparing for Space Exploration in the 21st Century," EOS, Vol. 79, No. 41, Oct. 13, 1998, pp. 493-496

- Ka-band solid-state power amplifier
- Beacon monitor operations
- Autonomous remote agent
- Low-power electronics
- Power actuation and switching module
- Multifunctional structure

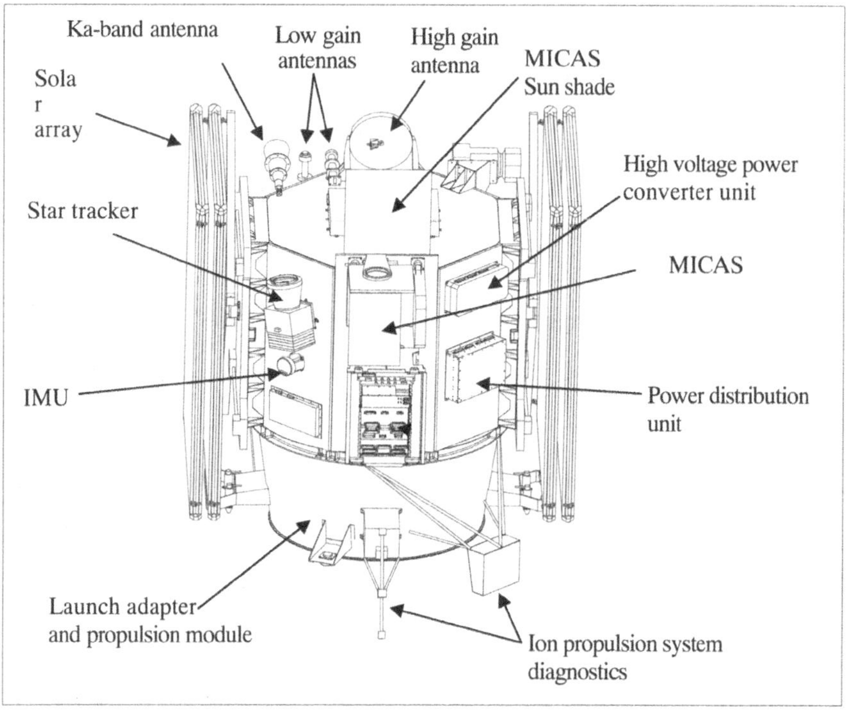

Figure 299: The DS1 S/C in stowed configuration

2) Acquire the data necessary to quantify the performance of these advanced technologies by the end of the primary mission (Sept. 1999).

3) Utilize the on-board IPS to propel the DS1 spacecraft on a trajectory to encounter an asteroid in 1999.

4) Assess the interaction of the IPS operations with the S/C and its potential impact on charged particle, radio waves and plasma, and other science investigations on future SEP-propelled missions. - A twelfth technology, a miniature integrated ion and electron spectrometer, PEPE, was not included in the mission success criteria, but it was on board and received a thorough evaluation.

The S/C structure consists of an octagonal aluminum space frame (1.1 m x 1.1 m x 1.5 high), the overall stowed S/C dimensions are: 2.5 m high, 2.1 m deep, and 1.7 m wide, based on the MSTI (Miniature Seeker Technology Integration) spacecraft, built by Spectrum Astro, Inc. of Gilbert, AZ (SA-200HP bus). With most of the components mounted on the exterior of the bus, their accessibility simplifies replacement during integration and test. The S/C is powered by batteries and two solar panels attached to the sides of the frame which span of

11.75 m when deployed. The solar panels, designated SCARLET II (Solar Concentrator Arrays with Refractive Linear Element Technology) constitute one of the technology tests on the spacecraft (see SCARLET below). The pair of solar arrays produced 2.5 kW of power at 1AU. Each array is comprised of four panels and a single-axis gimbal controls to point the panels in the more sensitive longitudinal axis toward the sun. - The total mass of the S/C is about 486.3 kg, composed of 373.7 kg dry spacecraft, 31.1 kg of hydrazine, and 81.5 kg of xenon for IPS.

Communications are via a high gain antenna, three low gain antennas, and a Ka-band antenna, mounted on top of the spacecraft. One low-gain antenna is mounted on the service boom and points toward Earth. The CCSDS protocol suite is used for all data communications.

The launch of DS1 took place on October 24, 1998 on a Delta launcher (7326-9.5 with three strap-on solid propellant rockets) from Cape Canaveral Air Station, FL (secondary payload: SEDSAT-1). Its first destination was the near-Earth asteroid "1992 KD" with an estimated diameter of 2-5 km. The primary mission of DS1 ended on September 18, 1999, completing its mission of demonstrating the new technologies.

Orbit: The SEDSAT-1 microsatellite (University of Alabama, Huntsville) was ejected from the second stage. The Delta vehicle entered a 185 km parking orbit, then fired again to enter an orbit: 174 km x 2744 km, inclination of 28.5°. The third stage separated and accelerated to a solar orbit with the DS1 satellite, while the second stage burned again with SEDSAT-1 for an orbit of: 556 km x 1042 km, inclination of 31.5°.
DS1 was injected into a solar orbit with the third stage burn. Orbital period: 453.00 d, inclination = 0.4°, periapsis = 0.99 AU, apoapsis = 1.32 AU, eccentricity = 0.14300, epoch start date/time = 1998.297:12:08:00 (24 Oct.).

M.5.1 Advanced technology payload complement

The DS1 payload consists of 12 technologies, two of them happen to be science instruments.

IPS (Ion Propulsion System). Objective: validation of system performance. IPS, provided by NSTAR [(NASA SEP Technology Application Readiness), where SEP is the acronym for Solar Electric Propulsion], uses a hollow cathode to produce electrons to ionize xenon by collision. The Xe^+ is electrostatically accelerated through a potential of up to 1280 V and emitted from the 30 cm diameter thruster through a pair of molybdenum grids. A separate electron beam is emitted to produce a neutral plasma beam. The power processing unit (PPU) of the IPS can accept as much as 2.5 kW, corresponding to a peak thruster operating power of 2.3 kW and a thrust of 92 mN. Throttling is achieved by balancing thruster and Xe feed system parameters at lower power levels; and at the lowest PPU thruster power, 525 W, the thrust is 19 mN. The specific impulse decreases from 31,400 m/s at high power to 19,000 m/s at the minimum throttle level. - A comprehensive diagnostic system is also on the spacecraft to validate system performance. The diagnostic instrument suite includes a retarding potential analyzer, two Langmuir probes, search-coil and fluxgate magnetometers, a plasma wave sensor, and two pairs of quartz-crystal microbalances and calorimeters.
At the end of the primary mission in Sept. 1999, IPS had operated for about 3000 hours. The IPS operated over a broad range of its 112 throttle levels, from input power levels of 580 W to 2140 W, with corresponding specific impulses of 19,375 m/s and 31,200 m/s, respectively (determined by radio navigation). Up to June 30, 1999, there were 1799.4 hours of thrusting by IPS, the total Xe consumption was 11.4 kg, providing 699.6 m/s. By August 31, 2000 IPS had accumulated > 5200 hours of operations, a longer time period than any propulsion system on any S/C so far. There were a total of 34 IPS starts in the mission from Nov. 24 1998 to June 30 1999. All other S/C systems operated normally during IPS thrusting.

AutoNav (Autonomous navigation). [1640] Objective: evaluation of autonomous on-board orbit determination (determination of S/C location in the solar system and in its flight path)

and control capability with the intent to free ground-segment resources. The AutoNav system (a software package including a baseline trajectory, a star catalog, the ephemerides of the asteroids, operations procedures, sequences, etc.) photographs reference asteroids against the background of fixed stars. With the knowledge of time, AutoNav computes the position of the asteroids. By measuring where the asteroids appear relative to the stars, it computes where the S/C must be. It then can project its path to its destination and use its propulsion system to make any course changes that are required.
AutoNav began functioning immediately upon activation of the S/C. The ACS (Attitude Control Subsystem) used a commercial star tracker to determine its attitude. Then the real-time part of AutoNav correctly provided ACS with the position of the sun as reference so that ACS could turn the S/C to the attitude required to illuminate the solar arrays. About once per week throughout the mission, AutoNav was invoked by the operating sequence to acquire optical navigation images. The system then issues commands to ACS and the integrated camera and imaging spectrometer (MICAS) to acquire visible-channel images, each with one beacon asteroid and known background stars. A heliocentric orbit is computed with a sequence of these position determinations combined with estimated solar pressure, and on-board knowledge of the thrust history of IPS. The trajectory then is propagated to the next encounter target, course changes are generated by the maneuver design element. - Typical AutoNav heliocentric orbit determinations differed from radiometric solutions (references to test AutoNav) by <1000 km in position and <0.4 m/s in the velocity vector. Later refinements improved these values substantially.

MICAS (Miniature Integrated Camera Spectrometer). MICAS (heritage of Pluto Integrated Camera Spectrometer) was designed and built by a team from USGS, SSG Inc., the University of Arizona, Boston University, Rockwell Science Center, and JPL. The MICAS package, combines the functionality of a framing camera (staring mode) with that of an imaging spectrometer. It features two visible-range imaging channels (APS and CCD - labeled after the detector type used), an ultraviolet imaging spectrometer, and an infrared imaging spectrometer, plus thermal and electronic control. The CCD array of size 1024 x 1024 pixels is the prime detector, operated in staring mode, for obtaining frame images. The CMOS active pixel sensor (APS) includes the timing and control electronics on-chip with the detector. The UV spectrometer has a spectral range of 80 to 185 nm with 2.1 nm spectral resolution. The infrared spectrometer covers the range from 1200 to 2400 nm with spectral resolution of 12 nm. The imaging spectrometers operate in pushbroom mode. - All four detector systems share a common optical system, namely a telescope with an aperture of 10 cm diameter. With a structure of a highly stable SiC, no moving parts are required. Spacecraft pointing directs individual detectors at the desired targets.

Parameter	UV Spectrometer	VIS (APS) Spectrometer	VIS (CCD) Spectrometer	IR Spectrometer
Spectral range	80 - 185 nm	500 - 1000 nm	500 - 1000 nm	1200 - 2400 nm
Focal length	17.1 cm	67.7 cm	67.7 cm	75.2 cm
Detector elements	35 x 164	256 x 256	1024 x 1024	256 x 256
Pixel size	54 μm	12 μm	9 μm	40 μm
FOV	0.63° x 0.03°	0.26° x 0.26°	0.69° x 0.78°	0.7° x 0.003°

Table 478: Some parameters of the MICAS instrument [1641)]

MICAS serves three functions on Deep Space 1. First, as with all the advanced technologies, tests of its performance to establish its applicability to future space science missions (demonstration of lightweight imaging technology). Second, the visible CCD channel is used to gather images for AutoNav's use. Third, it collects scientific data during this mission at the asteroid flyby and possibly the two comets. The MICAS instrument mass is 12 kg.

1640) D. Normile, "NASA Craft to Take the Controls in Flight," Science, Vol. 282, Oct. 23, 1998, pp. 604-605

1641) J. Oberst, B. Brinkmann, B. Giese, "Geometric Calibration of the MICAS CCD Sensor on the DS1 Spacecraft: Laboratory versus In-flight Data Analysis," Proceedings of ISPRS, Amsterdam, The Netherlands, July 16-23, 2000, Vol. XXXIII B1, pp. 221-230

PEPE (Plasma Experiment for Planetary Exploration). The objective was to assess the performance of this highly integrated, low-mass and low-power instrument. The instrument provided measurements of the three-dimensional plasma distribution over its field of view (2.8 π steradians). PEPE, built by Southwest Research Institute (SwRI) and Los Alamos National Laboratory (LANL), combines multiple instruments into one compact 5.6 kg package. The instrument includes a very low-power-consumption, low-mass microcalorimeter, provided by Stanford University, to help understand the plasma/surface interactions. - During the cruise phase, PEPE measures the solar wind energy spectrum of electrons and ions from 8 eV to 33 keV per unit charge with at least 5% resolution. Instead of using moving parts, it electrostatically sweeps its field of view, achieving a resolution of 45° in azimuth and 5° in elevation. PEPE also measures ion mass in the range of 1 to 500 amu per unit charge at a mass resolution of 5%.
PEPE validates the design for a suite of plasma physics instruments in one package; it assists in determining the effects of the ion propulsion system on spacecraft surfaces and instruments and on the space environment, including interactions with the solar wind; and it may make scientifically interesting measurements during the cruise phase and the encounter with the asteroid and possibly the two comets. During the asteroid flyby, the instrument was used to search for material outgassed from, or produced by photo-ionization of the asteroids surface.
PEPE made measurements of the solar wind with the IPS on and off. The main result is that SEP (Solar Electric Propulsion) can be used on future missions without interfering with the science payload.

SDST (Small Deep Space Transponder), built by Motorola and a **Ka-band solid-state power amplifier**, developed by Lockheed Martin. The SDST package combines the devices of: receiver, command detector, telemetry modulator, exciters, beacon tone generator, as well as the control functions into one unit with a mass of 3 kg (less than half the mass than would be required without this new technology). SDST allows X-band uplink and downlink as well as a Ka-band downlink. The Ka-band signal is amplified by a solid-state power amplifier to 2.3 W with an overall efficiency of 13%.
In addition to characterizing the operation of the Ka-band solid-state power amplifier, DS1 provided Ka-band signals for DSN use in verifying systems for acquiring, demodulating, decoding, and processing telemetry as well as in producing 2-way Doppler and ranging data. As the Earth-S/C range increased , certain tests were repeated to assure that the transition through threshold in a selected Ka-band region was observed.

SDST has also the ability to generate beacon signals in **Beacon Monitor Operations** (Beacon Monitor includes a software that diagnoses the spacecraft's condition). The system then transmits one of four tones to indicate to the operations team the urgency of the spacecraft's need for DSN (Deep Space Network) coverage. The four tones correspond to 1) the S/C is healthy and doesn't need any assistance, 2) reporting the occurrence of an unusual but not threatening event permitting network scheduling procedures, 3) alerting the ground that intervention is needed to prevent the loss of important data or to assist in resolving problem, and 4) requiring immediate assistance because the S/C has encountered a mission-threatening emergency. In each case, when tracking is initiated, the data summarization system provides a synopsis of the pertinent S/C data.

ARAEX (Autonomous Remote Agent Experiment). This is a software package consisting of the following three modules: PS (Planner/Scheduler), EXEC (Smart Executive), and MIR (Mode Identification and Recovery). The objective of this ambitious experiment involves turning over the responsibility (i.e. control) for the S/C to an autonomous agent. The concept involves a new architectural approach, taken by a team from JPL, AMES and the Carnegie Mellon University (Pittsburgh, PA), which uses an agent of the ground team onboard the spacecraft. This remote agent is tested in a restricted case on DS1, in preparation for more ambitious experiments on subsequent flights. The Remote Agent includes an onboard mission manager that carries the mission plan, expressed as high-level goals. A plan-

ning and scheduling engine uses the goals, comprehensive knowledge of the spacecraft's state, and constraints on spacecraft operations, to generate a set of time-based or event-based activities, known as tokens, that are delivered to the executive. The executive expands the tokens to a sequence of commands that are issued directly to the appropriate destination on the spacecraft. The executive monitors the response to these commands (through the mode identification and reconfiguration module) and reissues or modifies them if the response is not what was anticipated. Several faults are being simulated during the remote agent experiment on DS1.
On May 17, 1999 the primary S/C command was given over to **Remote Agent** for three days of S/C operations. In this period, Remote Agent successfully planned DS1 activities on-board and then carried out the plan without ground intervention. [1642)] The software detected, diagnosed and fixed simulated problems, showing that it can make decisions to keep the mission on track. This Remote Agent capability, a precursor to self-aware and self-controlled robots, will reduce the cost of future S/C operations as computers become "decision-making partners" along with humans.

Low Power Electronics Experiment (MIT/LL). The objective is to validate the performance of electronic devices throughout the life of the mission, with particular interest in the effects of radiation. A low-power electronics experiment contains a ring oscillator, multiplier, and discrete transistors to test 0.9-volt logic and 0.25 nm gate lengths (achieved with 248 nm lithography).

PASM (Power Actuation and Switching Module), a joint development of JPL, Lockheed and Boeing. The objective is to determine the working performance of PASM (internal test load of up to 40 V and 3 A). PASM contains two sets of four power switches, each set is controlled by its own mixed signal ASIC, providing voltage and current sensing and current limiting. High-density packaging technology quadruples the packing density over the current state of the art.

Multifunctional Structure Experiment, provided by USAF/PL and Lockheed Martin Astronautics. Objective: performance test of a multifunctional structure (attached to the S/C bus) consisting of electronic connection systems for embedded devices and the thermal control of a test panel. This new packaging technology combines load-bearing elements with electronic housings and thermal control, thus greatly reducing the mass of S/C cabling and traditional chassis.

SCARLET-II (Solar Concentrator Array with Refractive Linear Element Technology), built by AEC-Able Engineering Co., of Goleta, CA. SCARLET is the high-power solar array of DS1; it uses cylindrical silicone Fresnel lenses to concentrate sunlight onto a strip of photovoltaic cells and acts to protect the cells. Each array is composed of four panels of size 160 cm x 113 cm. The multijunction $GaInP_2$/GaAs/Ge photovoltaic cell modules are interconnected in series to produce about 2500 W (at 100 V) at the beginning of the mission. - The dual-junction cells achieved an average efficiency of 22.5% established during in-flight analysis.

M.5.2 Major events and status of extended mission in 2000

On July 29, 1999, DS1 successfully performed a close flyby of asteroid 9969 Braille using the AutoNav system. [1643)] At about 27 km separation, it was by far the closest flyby of an asteroid ever attempted.

Two months after the end of its extremely successful primary mission, the Star Tracker of DS1, responsible for the spacecraft's orientation, ceased operating. NASA decided it could

1642) It turned out that remote agent did have a problem (which, of course, is the reason for testing it!) that prevented it from continuing for the entire three days. The experiment was successful in that the bug was found. However, the bug did not present a risk to the spacecraft, so another experiment was designed and allowed the remote agent to complete all of its test objectives.

afford to go ahead with a risky extended mission for a comet encounter. Thereafter, JPL engineers devised a way (a long-distance rescue mission) to restore the spacecraft's sense of direction by writing new computer programs to use the camera (MICAS) as a substitute Star Tracker. The rescue also involved developing new operational procedures (new methods of flying the spacecraft) that went along with the new software. The challenging task was completed in June 2000 to resume thrusting in time to give DS1 a chance to encounter a comet Borrelly in September 2001.

There were formidable obstacles prior to achieving this feat. For one, the camera's field of view is 100 times smaller than the Star Tracker's. Also, while the Star Tracker could estimate the probe's orientation in space four times every second, MICAS produces a computer file that takes more than 20 seconds to transfer to the computer for analysis. Updates from the camera would be at least 80 times slower than the original guidance system.

After all preparations were completed and tested successfully, the ion engine was turned on June 21, 2000 (after a hiatus of 7 months) for tests of operating it with the new control system. On June 28 thrusting began in the direction required to reach comet Borrelly. - At launch, the plan for a possible extended mission was to go to Borrelly. After the spacecraft had been flying for about half a year, JPL decided that the mission was going so well and the advanced technologies were working so well that comet Wilson-Harrington could be added to the extended mission proposal. When the star tracker failed, project management reverted to the earlier plan.

M.6 EO-1 (Earth Orbserving-1)

EO-1 is a mission in NASA's NMP (New Millennium Program) Earth science program (a Landsat-7 follow-up and a Landsat-8 predecessor) with the overall objectives to perform Landsat-like measurements and to explore new remote sensing technologies (instruments, spacecraft, ground segment) that advance and enhance capabilities [evaluation/validation of technologies and performance, intersatellite calibration (comparison of data), evaluation of lunar calibration, autonomous navigation/instrument operation, use of proven new technologies for other missions]. A number of baseline validation scenarios are defined (generation of test scenes, ground proofs in conjunction with major field campaigns, airborne underflights, tandem flights with Landsat-7, etc.) in support of the overall objectives. S/C design life = 1 year. The paradigm of "faster, better and cheaper" applies to all aspects of the mission. A launch from VAFB on a Delta 7320-10 launch vehicle took place on Nov. 21, 2000 (along with SAC-C (CONAE, Argentina), and Munin (Sweden) as secondary payloads). [1644] [1645]

The EO-1 spacecraft, managed and operated by GSFC, was built and integrated by Swales Aerospace of Beltsville, MD as prime contractor. The S/C bus (of MIDEX heritage) features an Al structure, hexagonal in shape with 1.25 m diameter (across flats) and 0.73 m high (bus mass of 370 kg, payload mass of up to 110 kg). The S/C is three-axis stabilized for inertial and nadir pointing, an autonomous star tracker provides three-axis attitude knowledge. Four 1 N thrusters of a hydrazine propulsion system provide a pointing accuracy of 0.03°, the jitter is < 5 arcseconds. The PPT, a secondary propulsion system, represents a new technology which may eventually replace the reaction wheels. S/C orbit and attitude adjustment is based on GPS receivers and ACS (Attitude Control System) sensors (demonstration of autonomous maneuver capabilities initiated by GPS). A versatile ACS software permits S/C slewing to celestial bodies; this provides also a cross-track pointing capability to observe adjacent tracks. A hydrazine propulsion system with 22.3 kg propellant mass is used

1643) Note: The asteroid, discovered in 1992, was only recently (1999) named in honor of Louis Braille (1809-1852), the Frenchman, who invented the alphabet for the blind.

1644) S. G. Ungar, "Technologies for Future Landsat Missions," PE&RS, Vol. LXII, No. 7, July 1997, pp. 901-905

1645) http://eo1.gsfc.nasa.gov/Technology/Documents/InstrumentOverview.html

for orbit maintenance and formation flying. An on-board solid-state recorder provides science data storage of up to 40 Gbit. S/C total mass =529 kg, power= 600 W EOL (300 W orbit average) and 50 Ah super NiCd battery (28±7 V DC power). The OBC uses a Mongoose V processor with 12 MHz and 1.8 Gbit of TT&C data storage. A lightweight fiber optic data bus with ATM (Asynchronous Transfer Mode) protocol is used to transfer on-board data.

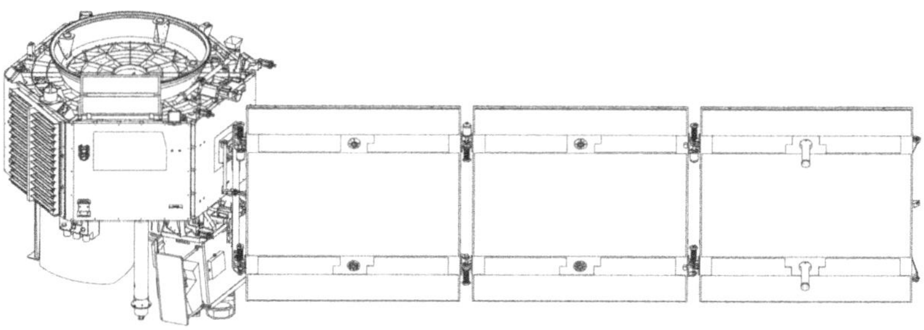

Figure 300: The EO-1 S/C model

Communications: The EO-1 antenna complement is comprised of two transmit/receive S-band omni antennas and a transmit-only electronically steerable XPAA (X-band Phased Array Antenna). The 64-element XPAA is nadir pointing. The XPAA is LHCP polarized with a 3 dB beam width which varies from 18°-30° depending on scan angle. It scans a 360° azimuth angle within a 65° half cone angle from boresight. Each of the 64 XPAA transmit elements contains its own solid-state power amplifier. One of the omni antennas is zenith facing, the other is nadir facing. Use of CCSDS protocols in S/C communications. The science data downlink is in X-band: frequency = 8225 MHz, data rate=105 Mbit/s, minimum EIRP = 22.0 dBW, encoding scheme= modulo-4 gray code differential encoding, modulation=QPSK. The TT&C links are in S-band, either direct or via TDRSS. The downlink data rate is selectable from 2 kbit/s to 2 Mbit/s (2, 32, 1000, or 2000 kbit/s), the uplink data rate is 2 kbit/s. The ground network consists of the following stations: SGS (Spitzbergen Ground Station) at Svalbard, Norway; AGS (Alaska Ground Station) at Fairbanks, AK; WOTS (Wallops Orbital Tracking Station) at Wallops Island, VA; and MSG (McMurdo Ground Station) at Antarctica. SGS is the primary station. In addition, there is the TDRSS White Sands Station for links via TDRSS. EO-1 mission operations are provided at NASA/GSFC.

Orbit: Sun-synchronous circular polar orbit, altitude = 705 km, inclination = 98.7°, period = 99 minutes, descending nodal crossing time of 10:15 AM. A coordinated tandem orbit within a minute of Landsat-7 is prescribed for reasons of data calibration and synergetic use of data.

M.6.1 Sensor Complement

ALI (Advanced Land Imager), designed/built by MIT/LL [focal plane system: Raytheon/Santa Barbara Remote Sensing (SBRS), optical system: Sensor Systems Group, Inc. (SSG)]. ALI is a hybrid multisensor system, a technology verification instrument, for the measurement of Earth surface reflectance. A prime objective is to provide continuity with data from earlier Landsat missions at reduced sensor/operating costs (introduction of a new measurement approach).

The ALI design features a WFT (Wide Field Telescope) and a highly integrated multispectral and panchromatic instrument. The telescope employs a reflective Cooke triplet design with a 12.5 cm unobscured aperture diameter and a field of view (FOV) of 15° (across-track) x 1.26° (along-track) and is providing a ground swath of 185 km nominally centered along the nadir ground-track. The design uses four mirrors; the primary is an off-axis asphere, the secondary is an ellipsoid, and the tertiary is a concave sphere; the fourth mirror is a flat folding mirror. This technology enables the use of large arrays of detectors at the focal plane for covering an entire swath of 185 km. The optical design features a flat focal plane and telecentric performance, which greatly simplifies the placement of the filter and detector array assemblies. The design uses silicon carbide mirrors and an Invar structure with appropriate mounting and attachment fittings. The detector array assembly has 10 spectral bands in VNIR and SWIR. [1646]

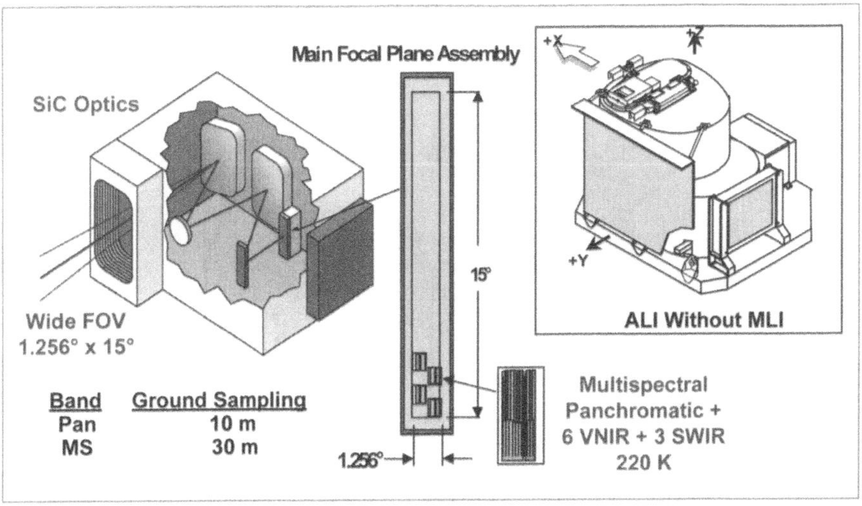

Figure 301: ALI FPA and optics schematic illustration (left) and instrument (right)

ALI (EO-1)			ETM+ (Landsat-7)		
Band No.	Wavelength (µm)	GSD (m)	Band No.	Wavelength (µm)	GSD (m)
PAN	0.480 - 0.690 VNIR	10	PAN	0.50 - 0.90 VNIR	13 x 15
MS-1'	0.433 - 0.453 VNIR	30			
MS-1	0.450 - 0.515 VNIR	30	MS-1	0.450 - 0.515 VNIR	30
MS-2	0.525 - 0.605 VNIR	30	MS-2	0.525 - 0.605 VNIR	30
MS-3	0.630 - 0.690 VNIR	30	MS-3	0.630 - 0.690 VNIR	30
MS-4	0.775 - 0.805 VNIR	30	MS-4	0.760 - 0.890 VNIR	30
MS-4'	0.845 - 0.890 VNIR	30			
MS-5'	1.200 - 1.300 SWIR	30			
MS-5	1.550 - 1.750 SWIR	30	MS-5	1.550 - 1.750 SWIR	30
MS-7	2.080 - 2.350 SWIR	30	MS-7	2.080 - 2.350 SWIR	30
			MS-6	10.40 - 12.50 TIR	60

Table 479: Spectral parameter comparison of ALI and ETM+ instruments

The focal plane assembly (FPA) for ALI is populated with four sensor chip assemblies (SCA) providing a continuous FOV of 3° x 1.625° for a fully populated focal plane, each SCA providing a cross-track coverage of 37 km. [There are four SCAs per module. Each module covers a ground swath swath of about 37 km. A fully populated focal plane would have five populated modules, i.e., 20 SCAs in all. ALI has only one module as a low-cost

1646) http://eo1.gsfc.nasa.gov/Technology/ALImultispectral.htm

technology demonstration.] The system operates in pushbroom fashion providing panchromatic and multispectral imagery. A single hybridized SCA is employed for PAN/MS (panchromatic multispectral) detection in the VNIR/SWIR spectral range (0.4 - 2.4 µm). This includes a PAN band, six VNIR bands (identical with Landsat ETM+), plus two additional VNIR bands (primed in Table 479), and three SWIR bands. The focal plane is only partially populated with SCAs resulting in only partial coverage by the various SCAs.

- Four MS SCAs are used side by side. Each SCA contains nine rows consisting of 320 MS detector cells in the cross-track direction. The dimension of each MS cell is 39.6 µm x 39.6 µm, providing a GSD (Ground Sample Distance) of 30 m
- The PAN detector cells are integrated directly into each MS SCA as a 960-element row, providing overlapping coverage to the MS bands, but at a 10 m GSD.

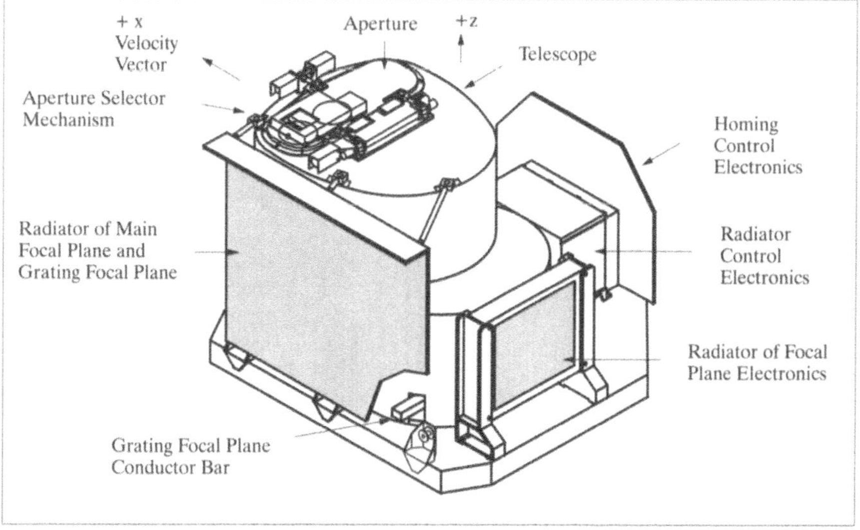

Figure 302: Illustration of the ALI instrument

The HgCdTe SWIR detectors provide high performance over the 0.9 to 2.5 µm wavelength region at temperatures which can be reached by passive or thermoelectric cooling. The nominal focal plane temperature is 220 K and is maintained by the use of a radiator. Application of detectors of different materials to a single readout integrated circuit (ROIC) enables a large number of arrays covering a broad spectral range to be placed close by together. This technology is extremely effective when combined with the WFOV optical design being used on ALI. - The VNIR bands feature a new SiC (Silicon Carbide) detector technology. SiC offers the advantage of a very high stiffness-to-density ratio and a very high conductivity to heat capacity ratio. These characteristics are superior to currently used materials for reflective optical systems. The high stiffness-to-density ratio of silicon carbide (SiC) allows mirrors of very low weight to be designed and still maintain the necessary surface figure to provide the performance required for high-resolution optical imaging. SiC optics and structures have the potential of lowering optical system weights an order of magnitude over current conventional systems.

ALI is the prime instrument of EO-1. It is mounted to and aligned to an interface plate, bolted to the S/C. The instrument-to-S/C alignment is within 20 arcseconds. A comparison of ALI/ETM+instruments (see also Tables 479, 480, and 482) reveals the following: ALI can gather higher-quality imagery than ETM+ but is smaller and less expensive (one-forth); its mass is one-forth, with one-half its size it requires only one-fifth the power of ETM+.

Parameter	ALI (EO-1)	ETM+ (Landsat)
Instrument mass	106 kg	425 kg
Instrument power	100 W	545 W
Instrument size	0.2 m^3	1.4 m^3
Nr. of VNIR/SWIR bands	10	7
Detectors per band	6200	16
Thermal bands	None	1
Data rate	300 Mbit/s	150 Mbit/s
Pan resolution	10 m	15 m
Relative SNR	4 x	1 x

Table 480: EO-1/Landsat instrument parameter comparison

Hyperion. The Hyperion instrument is of HSI (Hyperspectral Imager) heritage on the Lewis S/C (launch Aug. 23, 1997, control of the S/C was lost Aug. 26, the S/C was lost and reentered the atmosphere Sept. 28, 1997) and was built by TRW of Redondo Beach, CA. The instrument is a grating imaging spectrometer, the design includes a telescope, two grating spectrometers with the supporting focal plane electronics, and the cooling system.

The fore-optics design is based on the EOC (Electro-Optical Camera) instrument of the KOMPSAT mission (KARI of Korea). The telescope provides for two separate grating image spectrometers to improve SNR. A focal plane array provides separate VNIR and SWIR detectors (HSI heritage). The SWIR detectors are cooled by a cryocooler. A dichroic filter in the system reflects the band from 400 to 1,000 nm to one spectrometer and transmits the band from 900 to 2,500 nm to the other spectrometer. The SWIR overlap with the VNIR from 900 to 1000 nm permits cross calibration between the two spectrometers. Both spectrometers use a JPL convex grating design in a 3 reflector Offner configuration. The VNIR spectrometer uses a 60 (spectral) x 250 (spatial) pixel detector array, which provides a 10 nm spectral bandwidth over a range of 400-1000 nm. The SWIR spectrometer has HgCdTe detectors in an array of 160 (spectral) x 250 (spatial) channels, cooled to 120 K. The SWIR spectral bandwidth is 10 nm. The telescope is a three-mirror astigmatic design, the high resolution hyperspectral imager is capable of resolving 220 spectral bands (from 0.4 to 2.5 μm) with a 30 meter resolution. All of the mirrors in the system are constructed from coated aluminum; the structure holding the optical elements is also constructed from aluminum so that the mirrors and housing all expand and contract at the same rates.

A common calibration system is provided for both the VNIR and SWIR spectrometers. Dual calibration lamps produce reference signals to monitor detector performance following image acquisition. Calibrations are also performed with the sun, ground target sites, and the moon. The solar calibration utilizes a diffuse reflector on the backside of the optical cover to provide uniform illumination across the focal plane arrays. Direct viewing lunar calibration is accomplished by scanning the instrument across the lunar surface. Solar and in-flight calibration data is used as the primary source for monitoring radiometric stability, with ground site and lunar calibration as secondary alternatives.

Spectral coverage	400 - 1000 nm (VNIR); 900 - 2500 nm (SWIR)
FOV	0.624° (i.e., 7.5 km swath width from a 705 km orbital altitude)
IFOV (spatial resolution)	42.55 mrad (30 m)
Detectors	60 x 250 silicon CCD array in VNIR; 160 x 250 HgCdTe array
Spectral resolution	10 nm in VNIR and SWIR (220 bands)

Table 481: Hyperion instrument parameters

The instrument typically images surface scenes of 7.5 km (swath) by about 100 km in length, providing detailed spectral mapping across all 220 channels with high radiometric accuracy.

The data is typically processed into cubes (19.8 km long by 7.5km wide) to facilitate data handling in current desktop computers. Each image cube consists of 75 MByte of data. A typical acquisition consists of multiple image cubes.

LAC (LEISA Atmospheric Corrector). The objective is to provide atmospheric correction data (water vapor) for ALI and ETM+ on Landsat-7. LAC is a high spectral, moderate spatial resolution hyperspectral imager based on wedge filter technology (heritage of WIS and LEISA). The spectral coverage is from 0.89 - 1.6 μm; in addition there is a separate channel at 1.380 μm to detect cirrus clouds. Spatial resolution = 250 m. LAC is coaligned with the ALI WFT assembly to cover the full swath of 185 km.

The instrument uses three 256 x 256 pixel InGaAs infrared detector focal plane assemblies in a single module. Each array is placed behind a lens covering a FOV of 5° to obtain a swath width of 185 km (15°). A wedged dielectric film etalon filter (a linear variable etalon) is placed in very close proximity to a 2-D detector array. This produces a 2-D spatial image that varies in wavelength along one dimension. The filter size is 1.024 cm x 1.024 cm and covers the 890 to 1600 nm spectral region at a resolution of 30 to 40 cm^{-1}, with a linear dependence of wavenumber on position. Reflective ¼-wave stacked layers placed on both sides of one, or more, ½-wave etalon cavity(s) provide the spectral resolution. Order-sorting of the etalon is accomplished with lower resolution filter layers. In operation, the 2-D spatial image is formed by a small, wide field of view lens. Unlike the grating spectrometer that captures the spectra at a point "instantaneously," the spectrum for the LAC is obtained as the orbital motion of the spacecraft scans the image across the focal plane in wavelength, thereby creating a 3-D spectral map. The spatial resolution of a pixel is 360 mrad x 360 mrad, or about 250 m x 250 m at nadir.

Parameter	Multispectral Instruments		Hyperspectral Instruments	
	ETM+(Landsat-7)	ALI (EO-1)	Hyperion (EO-1)	LAC (EO-1)
Spectral range	0.4 - 2.4 μm (excluding TIR)	0.4 - 2.4 μm	0.4 - 2.5 μm	0.89 - 1.6 μm
Spectral resolution	30 m	30 m	30 m	250 m
Swath width	185 km	37 km	7.5 km	185 km
Spectral resolution	Variable	Variable	10 nm	2-6 nm
Spectral coverage	Discrete	Discrete	Continuous	Continuous
PAN band resol.	13 m x 15 m	10 m	N/A	N/A
Nr of bands	7	10	220	256

Table 482: Comparison of some instrument parameters on EO-1 and Landsat-7

M.6.2 Demonstration of seven new technologies on EO-1

XPAA (X-band Phased Array Antenna), built by Boeing Co. XPAA is a communication experiment with the objective to demonstrate link-pointing capability with the use of a body-fixed low-mass and low-cost phased array antenna. The antenna is mounted on the Earth-facing side of EO-1 to allow communications with ground stations (downlink of high-rate data from the EO-1 solid-state recorder). XPAA is composed of a flat grid of 64 radiating elements whose transmitted signals are combined spatially to produce the desired antenna directivity. The phases of each of the radiating elements are varied by computer to point the beam in the desired direction. The antenna's mass is 5.5 kg. It has an Effective Isotropic Radiated Power (EIRP) of approximately 160 W and transmits data at 105 Mbit/s. An inherent advantage of the body-fixed design is to permit simultaneous capture and transmission of data, avoiding perturbations to instrument measurements.

CCR (Carbon Carbon Radiator). A material science experiment with the objective to demonstrate various uses of carbon-carbon. The composite material is used in the primary structure of EO-1, serving as both an advanced thermal radiator and a load bearing structure. The advantages of carbon-carbon are: high thermal conductivity (including through thickness) and good strength and weight characteristics. The use of CCR technology may simplify thermal radiator design in some applications (e.g., no need for actively cooled radiators) and support increased science payload mass ratios for future missions.

LFSA (Lightweight Flexible Solar Array). The objective of this photovoltaic system is to demonstrate significant improvements in the power to weight ratios of solar arrays. LFSA features copper indium diselinide ($CuInSe_2$ or CIS) solar cells and shape memory alloys for the hinge and deployment systems. The shockless deployment technique could improve the S/C dynamics during deployment. It is also much safer than conventional solar array systems using conventional pyrotechnics for deployment. The goal of LFSA is to achieve >100 W/kg power efficiency ratios (conventional solar arrays of the late 1990s provide <40 W/kg).

WARP (Wideband Advanced Recorder Processor). WARP) is a high-rate solid-state recorder on EO-1. The objective is to demonstrate a number of high density electronic board advanced packaging techniques. WARP utilizes advanced integrated circuit packaging (3-D stacked memory devices) and "chip on board" bonding techniques to obtain very high density memory storage per board (24 Gbit/memory card). It also includes a Mongoose V processor which can perform on-orbit data collection, compression and processing of land image scenes. WARP consists of a high-rate (up to 840Mbit/s capability), high-density (48 Gbit storage), low mass (< 25 kg) solid-state recorder/processor with X-band modulation capability. It is in fact the highest rate solid-state recorder NASA has ever flown.

PPT (Pulsed Plasma Thruster), built by General Dynamics (formerly Primex Aerospace) for NASA/GRC). The system is also referred to as PRS-101 (Plasma Rocket System-101), its model number. The objective is to demonstrate on-orbit electromagnetic propulsion technology and to provide a S/C precision-pointing capability. [The EO-1 primary propulsion is provided by the hydrazine system for orbit maintenance and formation flying. The PPT is being used during limited periods of the mission to demonstrate the capability of a PPT system by performing the function of a momentum wheel. When in operation, the pitch axis momentum wheel is being idled and the PPT takes over control of the pitch axis.] The PPT consists of a coiled spring to feed the Teflon propellant, an igniter plug to initiate a small trigger discharge and an energy storage capacitor and electrodes (also: on-board electronics to convert the low voltage from the S/C to appropriate charging voltage for the capacitor, to provide energy to fire spark plugs, and to isolate PPT discharge from the S/C power bus). Plasma is created by the ablation of the Teflon propellant from discharge of the storage capacitor across the electrodes. The plasma is accelerated by the Lorentz force in the induced magnetic field to generate thrust. Impulse levels of 10-1000 μNs are achieved; the specific impulse is 880-1170 m/s; the average power consumption of PPT is between 1-100 W. PPT is used to maintain S/C attitude (fine pitch attitude control pointing requirements while meeting stringent electromagnetic and contamination constraints for the mission). A series of fine pitch pointing maneuvers are planned after the end of the primary imaging mission. The total mass of the PPT is 4.87 kg. [1647)]

FODB (Fiber Optic Data Bus). The objective is to demonstrate high aggregate bandwidth data transfer capabilities on-board EO-1 - with low mass and low power requirements. FODB was developed as a standard high-speed (Gbit/s) data interface for future spacecraft. It complies with the IEEE P1393 (Spaceborne Fiber Optic Data Bus) standard. The bus concept is based on a ring topology of 2 to 128 nodes which provides flexibility to meet differing payload requirements. The master node is a Controller Fiber Bus Interface Unit (CFBIU) and up to 127 slaves are implemented with a Fiber Bus Interface Unit (FBIU). The interface between nodes is implemented in fiber optics, it reduces weight and provides for an EMI/EMC problem-free system. The FODB on EO-1 is implemented with four nodes. In this configuration, the WARP data system contains the master controller node

1647) Information provided by Andrew Hoskins of General Dynamics

(CFBIU) and one slave node (FBIU). An external instrument terminal box, mounted on the spacecraft, contains two slave nodes, glue logic, power, and ALI instrument interface connectors. The FODB bus employs an ATM (Asynchronous Transfer Mode) with minimum overhead which simplifies the node interface design. Transfer rates between 200 Mbit/s and 1 Gbit/s are supported by FODB.

EFF (Enhanced Formation Flying), developed at a.i. solutions Inc. The objective is to demonstrate an on-board software package capable of autonomously planning, executing, and calibrating routine spacecraft maneuvers to maintain satellites in their respective constellations and formations. The long-term goal of this technology is to enable many small and inexpensive spacecraft to fly in formation and gather concurrent science data in a "virtual satellite". This "virtual satellite or virtual platform" concept lowers total mission risk, increases science data collection and adds considerable flexibility to future Earth and space science missions.

The EFF features include an innovative use of fuzzy logic decision making capabilities and natural language to resolve multiple conflicting constraints; scripting environment enables algorithm updates without software changes; flight wrapper interfaces directly with the command and data handling subsystem for input and output; multiple operating modes allow execution control; generic closed-loop formation flying control algorithms applicable to many missions; modular architecture design is flexible enough to control execution of multiple and varying algorithms from several partners.

The EO-1 spacecraft is using a Tensor GPS receiver of Space Systems/Loral (SS/L) for active on-board navigation and attitude control.

M.7 EO-3 (Earth Observing-3, GIFTS-IOMI Mission)

EO-3 is the second Earth-observing technology mission in NASA's New Millennium Program (NMP). The joint NASA, Navy, and NOAA mission consists of a single instrument, namely GIFTS (Geosynchronous Imaging Fourier Transform Spectrometer) to be flown on a geosynchronous satellite. The goal of GIFTS is to serve as a testbed for the following planned next-generation instruments: [1648] [1649]

- Demonstration/validation of advanced Earth observation capabilities from GEO using the newly developed instrument GIFTS
- Support of the Navy's IOMI (Indian Ocean METOC Imager) project
- Demonstration of the required ABS (Advanced Baseline Sounder) capabilities for NOAA's GOES program (technology infusion, demonstration of forecast utility).

The S/C is three-axis stabilized and employs the T310 bus of TRW (prime S/C integrator). The bus is a cube of 170 cm side length; the GIFTS instrument is mounted to the nadir face of the spacecraft. The overall deployed configuration has a length of 8.5 m (2 solar panels). The boresight pointing is $<0.017^{\circ}$ (3 sigma), the attitude stability is <10 µrad rms per axis. A star tracker assembly is the main attitude sensor. The bus has the capability to accommodate up to 379 kg of payload mass and to provide 585 W of payload power. The design life of the spacecraft is 9 years.

The mission is planned to be launched in late 2004 on a Delta-4 vehicle from Cape Canaveral, FL., as part of the US Air Force's Space Test Program.

Orbit: Geosynchronous orbit at three successive locations: Eastern Pacific and Western Atlantic during 2005 and 2006. The Indian Ocean location is planned for the time frame 2007-2012 in support of the Navy's IOMI project.

1648) W. L. Smith, et al., "Geostationary Fourier Transform Spectrometer (GIFTS)–The New Millennium Earth Observing-3 Mission," IRS '00: Current Problems in Atmospheric Radiation, edited by W. L. Smith and Y. U. Timofeev, A. Deepak Publishing, Hampton Virginia, 2001.

1649) http://nmp.jpl.nasa.gov/eo3/index.html

RF communications: TT&C data employ the S-band, the uplink data rate is 1 kbit/s (encrypted), the downlink has a variable rate of 8-512 kbit/s (encrypted). The payload data are downlinked in X-band at a rate of 80 Mbit/s.

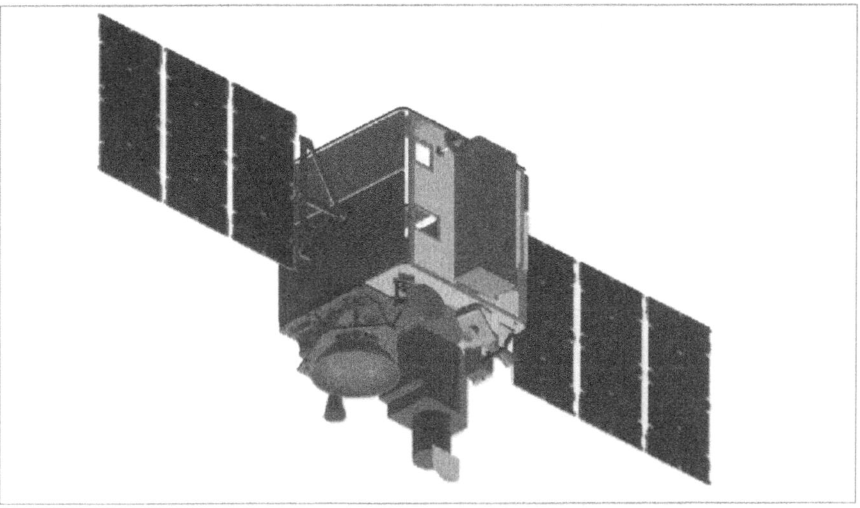

Figure 303: Illustration of the EO-3 spacecraft

Sensor complement:

GIFTS (Geosynchronous Imaging Fourier Transform Spectrometer) of NAST and HIS heritage. [1650) 1651) 1652)] The GIFTS instrument design includes advanced technologies in imaging, active cooling, fast data processing, pointing and control, radiation protection, and lightweight materials. The overall objective is to space-qualify the new technologies and to introduce the new concepts into next-generation instruments of operational weather observing systems for enhanced prediction analysis and service capabilities. The instrument is being designed/developed in partnership by NASA/LaRC (prime), SSEC (Space Science and Engineering Center) of the University of Wisconsin, Madison, and the Space Dynamics Laboratory (SDL) of Utah State University. GIFTS is an advanced geosynchronous sounding instrument with the following measurement capabilities:

- Regional to full-disk visible and multispectral infrared imaging with 1 to 5 minute temporal frequency
- Full-disk temperature, moisture, and tracer wind sounding with 1 to 3 hour temporal frequency for global numerical weather prediction
- Regional half-hourly high-resolution sounding for mesoscale intense weather observation and forecasting
- Soundings of chemical composition for monitoring pollutant and greenhouse gas episode evolution and transport.

The science objectives are: a) to measure temperature, water vapor, tracer wind, and chemical composition distribution with high spatial and temporal resolution for considerable improvements in observational weather observation and air quality monitoring; b) to provide dynamic observations of cloud spectral radiance and associated atmospheric properties in support of NASA's ESE missions; c) to observe the transport of radiantly active pollutants and greenhouse gases in support of the Aura atmospheric chemistry mission.

1650) http://oea.larc.nasa.gov/PAIS/GIFTS.html
1651) http://its.ssec.wisc.edu/~ bormin/GIFTS/
1652) http://danspc.larc.nasa.gov/GIFTS/

The technology objectives are to validate advanced technologies in support of improved remote sensing capabilities of future missions from geosynchronous orbit.

- FTS (Fourier Transform Spectrometer). An LFPA of 128 x 128 elements is employed. The FTS feature enables the simultaneous measurement of infrared radiation spectra by each detector element with a spectral resolution (up to 0.6 cm^{-1}) sufficient for resolving the structure of the atmosphere with high vertical resolution.
- LFPA (Large area format Focal Plane detector Array). The LFPA component provides the nearly continuous observation of large geographical areas (512 km x 512 km - anywhere on the visible disk) with high horizontal resolution (4 km x 4 km in the infrared, and 1 km x 1 km in the visible region) within time intervals of 10 s.
- Data readout and digital signal processing electronics. Demonstration of a signal and data processing architecture capable of handling the massively parallel data stream. This involves a new generation of data processing and compression hardware and algorithms to reduce the data rates transmitted.
- Radiation shielding
- Lightweight structures and optics
- Redundant miniature cryogenic coolers. Validation of low-power, high-efficiency mechanical cryocoolers and thermal switches.

The GIFTS optics subsystem uses a 24 cm diameter aperture yielding diffraction limited spatial resolutions. The footprint size is 4 km but the resolution ranges from 4 km at the shortest wavelength to about 8 km at the longest wavelength of the GIFTS spectrum. Nyquist spatial sampling is obtained by using four times as many detector elements (twice as many in each areal dimension) than required for a 4 km spacing of the measurements.

GIFTS uses two detector arrays to cover the spectral bands 685 to 1130 cm^{-1} (14.6 - 8.85 µm) and 1650 to 2250 cm^{-1} (6.06 - 4.45 µm). The FTS approach (of the Michelson interferometer) for GEO applications allows spectral resolution to be easily traded for greater area coverage or higher temporal resolution. Full disk sounding coverage can be achieved in about 7 minutes. High vertical resolution soundings and atmospheric chemistry measurements of GIFTS require a spectral resolution of 0.6 cm^{-1} and a longer stare time, thereby reducing the area coverage and/or frequency of observation relative to the imagery mode of operation. Nevertheless, GIFTS can cover a major portion of the visible disk with high vertical resolution soundings in less than 0.5 hour. This feature is important for obtaining wind profiles from GEO temperature and moisture sounding data. A relatively long dwell time and a more limited area coverage "self-validation mode" of operation enables spectral resolutions of 0.6 cm^{-1} with high radiometric precision. - Periodic views of on-board references and cold space are used to achieve high calibration accuracy.

A low visible light sensitive camera, using 512 x 512 active pixel detector arrays, provides quasi-continuous imaging of clouds at 1 km spatial resolution. Extended Earth coverage is achieved by step-scanning the instrument FOV in a contiguous fashion across any desired portion of the Earth. The radiance spectra observed at each time step are transformed to high vertical resolution (1-2 km) temperature and water vapor mixing ratio profiles.

Mode of operation	Spectral resolution	Spatial coverage	Observation time (scan speed of 0.17 m/s and 1 s telescope step time)
Regional imaging	36 cm^{-1}	6,000 km x 6,000 km	3 minutes
Global sounding	18 cm^{-1}	10,000 km x 10,000 km	7 minutes
Regional sounding and chemistry	0.6 cm^{-1}	6,000 km x 6,000 km	25 minutes
Self validation (radiometric precision>0.1 K)	0.6 cm^{-1}	1,000 km x 1,000 km	60 minutes

Table 483: GIFTS operation modes in GEO orbit

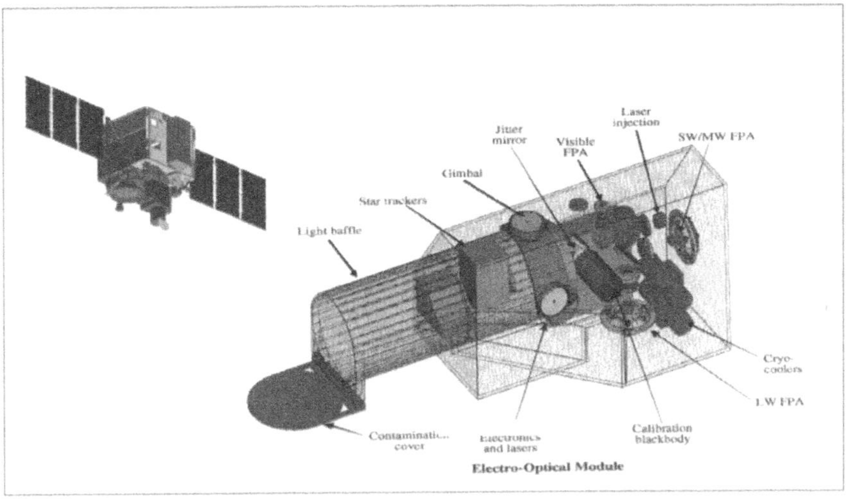

Figure 304: Schematic illustration of the GIFTS instrument

Parameter	GOES (I-M) Sounder	GIFTS
No of channels	19	1724
Detector elements	4	16,384
Spatial footprint/spacing	8/10	4/4
Coverage rate per 3000 km x 3000 km area	about 42 minutes	about 7 minutes
Spectral coverage	0.70 - 14.71 µm	4.45 - 6.06 µm 8.85 - 14.6 µm

Table 484: Comparison of GIFTS performance parameters with current sounders in GEO

The profiles are obtained on a 4 km grid and converted to relative humidity profiles. Images of the horizontal distribution of the relative humidity for the various layers, vertically separated by about 2 km, are constructed for each spatial scan. The sampling period ranges from minutes to an hour, depending on spectral resolution and area coverage. Successive images of clouds and the relative humidity for each layer are then animated to reveal the motion of small-scale thermodynamic features of the atmosphere. Automated auto-correlation feature tracking programs are then used to compute the speed and direction of the features, thereby providing a measure of the wind velocity distribution. The net result is a dense grid of temperature, moisture and wind profiles which can also be used for atmospheric analysis and operational weather prediction. Feature tracking can also be used for mixing ratio profiles of O_3 and CO, derived from their spectral radiance features observed by the instrument, providing a direct measure of the transport of these pollutant and greenhouse gases. It is the imaging interferometer technique (i.e. the combination of the FTS and the LFPA) in GEO that enables the tracer wind profile and trace-gas transport measurements.

GIFTS has a mass of 130 kg and a power consumption of 320 W (max). The instrument data rate is 60 Mbit/s (max).

M.8 ETS (Engineering Test Satellite)

NASDA's ETS satellite series is aimed at developing and demonstrating common base technologies. The program started with the launch of ETS-I (Kiku-1), NASDA's first satellite, on Sept. 9, 1975, from Tanegashima Space Center using an N-I launch vehicle.

S/C	Launch/ Service life	Objective/Comment
ETS-I	Sept. 9, 1975/ April 28, 1982	Circular orbit of 1000 km, 47° inclination, S/C mass = 82.5 kg
ETS-II	Feb. 23, 1977/ Dec. 10, 1990	Japan's first GEO S/C at position 130° E, S/C mass =130 kg, S/C diameter = 1.4 m
ETS-III	Sept. 3, 1982/ March 8, 1985	Data for designing earth observation satellites with high power requirements; testing of three-axis attitude control; S/C mass = 385 kg; Circular orbit of 1000 km altitude, 45° inclination.
ETS-IV	Feb. 11, 1981/ Dec. 24, 1984	GTO orbit with 225 km perigee and 36,000 km apogee, inclination=28.5°; S/C mass = 640 kg
ETS-V	Aug. 27, 1987/ March 31, 1989	Test of GEO three-axis bus system technology and experiments for mobile communications satellite. GEO orbit at 150° E. S/C mass of 550 kg
ETS-VI	Aug. 28, 1994/ July 9, 1996	Confirm the technology required for a GEO three-axis satellite bus system for high performance S/C and test of advanced communication equipment. S/C mass of 2000 kg (660 kg payload), 10 year design life for S/C bus. Also feature of a high precision attitude control system Injection into GEO failed due to an apogee engine malfunction
ETS-VII	Nov. 28, 1997	Rendezvous-docking and space robotic experiments
ETS-VIII	2003	Demonstrate support of mobile communications with S/C in GEO

Table 485: Overview of early ETS series satellites

M.8.1 ETS-VII (Engineering Test Satellite VII)

ETS-VII (Kiku-7) is a NASDA technology demonstration satellite with the objective to conduct rendezvous-docking and space robotic experiments. In the rendezvous-docking experiment, the chaser satellite conducts rendezvous-docking with the target satellite by both automatic and remotely piloted controls, and in the space robotic experiments, unmanned space work is carried out by teleoperation. In addition, the space robotic experiments by MITI/ETL, CRL and NAL (National Aerospace Laboratory) are carried out.

The ETS-VII system consists of two spacecraft: a chaser satellite (Hikoboshi) with a total mass of 2,540 kg, and a target satellite (Orihime) of mass 410 kg. Both satellites are three-axis stabilized. The mission design life is 1.5 years.

A launch of ETS-VII took place on Nov. 28, 1997 from Tanegashima Space Center on the H-2 launch vehicle. ETS-VII experienced an attitude stability problem on Nov. 30, 1997 that could be corrected.

Orbit: Circular orbit, altitude = 550 km, inclination = 35°, period = 96 minutes.

Rendezvous-docking experiment:

The overall objective was to acquire/retrieve materials to/from a satellite in orbit. The following experiments were carried out:

- Demonstration of rendezvous-docking equipment in orbit
- Evaluation of both automatic and remotely piloted rendezvous-docking in orbit
- Demonstration of simultaneous operation of two spacecrafts.

To enable the unmanned autonomous rendezvous docking experiment, the Rendezvous docking equipments: GCC (Guidance Control Computer), DM (Docking Mechanism), PXS (Proximity Sensor), RVR (Rendezvous Laser Radar), and GPSR (GPS Receiver) are developed and used in the Rendezvous-docking demonstrations.

For each experiment, the chaser satellite detaches the target satellite and drifts away to distances of up to 9 km); subsequently, a rendezvous-docking maneuver brings the chaser satellite back close to the target resulting in a final docking maneuver.

The rendezvous-docking operations were conducted from the NASDA ground stations using NASA's TDRS (Tracking and Data Relay Satellite) system. [1653] [1654] [1655]

- On July 7, 1998, the first experiment was successfully performed. In this flight, autonomous docking from 2 m was demonstrated. The Chaser detached the Target based on "separation command" from the ground. The Chaser automatically conducted formation flight for 15 minutes at the range of 2 m from the Target and docked again. The Chaser conducted six degree-of-freedom (6DOF) control using the PXS navigation.
- The second rendezvous docking experiment was initiated on Aug. 7, 1998. In this flight attitude anomalies occurred, and the Chaser made a flight to 12 km in range. As a result, the Chaser successfully docked with the Target on Aug. 27 using not only the PXS navigation but also the relative GPS navigation and the RVR navigation.
- The third experiment flight was performed in Oct. 26-27, 1999 with the following test sequence: 1) Flight remotely controlled from the ground; 2) Collision Avoidance Maneuver (CAM) flight; 3) R-bar approach flight (approaching from the side of the Earth).

Through these experiment flights, the autonomous rendezvous docking technologies from the relative approach to the docking (from 10 km to docking) were verified on orbit.

Figure 305: Illustration of the ETS-VII spacecraft

Space robotic experiments:

The objective was to acquire the basic capability to teleoperate a space robot and to analyze the engineering data for the development of future advanced space robots.

- After the successful launch and initial check out of the satellite, various space robot experiments were conducted between Mar. 1988 to May 1999. All planned robotic experiments were successfully conducted. Further experiments were additionally conducted until Nov. 1999.

1653) I. Kawano, M. Mokuno et. al., "Result of Autonomous Rendezvous Docking Experiment of Engineering Test Satellite VII, " Journal of Spacecraft and Rockets, Vol.38, No.1, Jan.-Feb, 2001. p.105

1654) I. Kawano, M. Mokuno, T. Miyano, T. Suzuki, "Analysis and Evaluation of GPS Relative Navigation Using Carrier Phase for RVD Experiment Satellite of ETS-VII," ION GPS-2000, Sept. 19-22, 2000, Salt Lake City, UT, pp. 1655-1660

1655) M. Mokuno et. al., "Experimental Result of Autonomous Rendezvous Docking on Japanese ETS-VII satellite," Proceedings of the Annual AAS Guidance and Control Conference, AAS99-022

- On March 16, 1999 NASDA conducted successfully the telemanipulation experiment of the robot arm on ETS-VII. Objective: to evaluate (compare) the difference of operation of the Shuttle remote manipulator and the ETS-VII onboard robot arm.
- Joint NASDA ESA experiments: In the period April 6-9, 1999, ESA demonstrated autonomous robot manipulations and robot user interactions with VIABLE (Vision Interactive Autonomy Bi-Lateral Experiment). [1656] The ESA experiments concerned advanced schemes for planning, commanding, controlling and monitoring the activities of a space robot arm system. One set of experiments tests an operational mode called "interactive autonomy", whereby the robot motions are split into typical "tasks" of medium complexity. Ground operators can interact with the tasks (parameterizing, commanding, rescheduling, monitoring, interrupting them as needed), relying on the fact that each task is autonomously executed using appropriate sensor-based control loops. The second group of experiments dealt with vision-based robot control. Using the Japanese-provided on-board vision system (including a hand camera and a scene-overview camera). It has been demonstrated that reliable automatic object localization and grasping can be performed even without the artificial markers which are typically used to guide telemanipulation.
- Joint NASDA DLR and IRF experiments: Teleoperation experiments of the onboard robot arm on ETS-VII were conducted April 19-21, 1999 using NASDA's robot operation facility at Tsukuba Space Center and GETEX (German Teleoperation Experiment) of DLR/IRM and of IRF (Institute for Robotics Research), University of Dortmund).

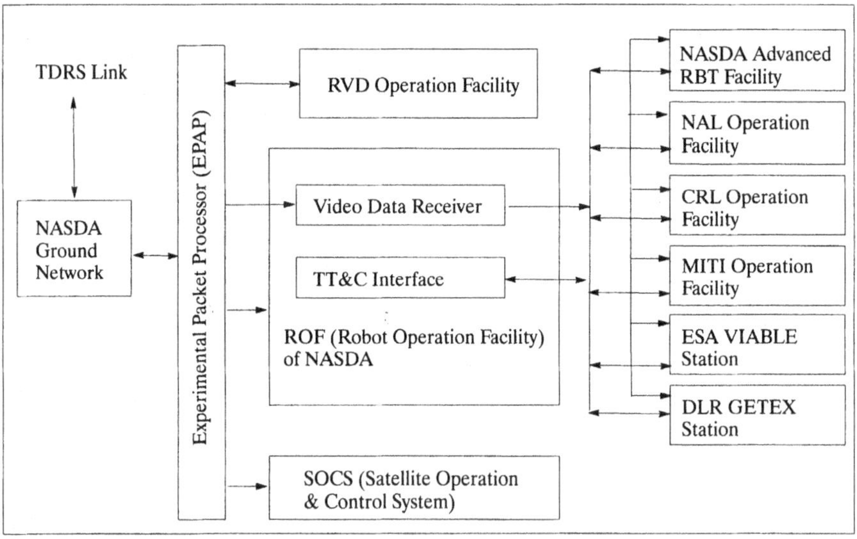

Figure 306: ETS-VII ground segment configuration with guest experiment stations

- On Sept. 1, 1999 the first successful Target Capture experiment of ETS-VII took place. In the experiment, the satellites' docking mechanism was first loosened, allowing Orihime to float freely within certain limits; specifically, Orihime was made to float about 200 mm from the robot arm aboard Hikoboshi. Starting from this state, the robot arm then automatically approached Orihime without assistance from the ground, and successfully grasped the handle on Orihime using the finger attached to its tip. The suc-

1656) G. Visentin, F. Didot, "Testing Space Robotics on the Japanese ETS-VII Satellite," ESA Bulletin No. 99, Sept. 1999, pp. 61-65 cessful release, tracking, and capture of the satellite without help from the ground was a first in space history. Previous automatic docking activities, such as those conducted with Russian Soyuz and Progress spacecraft, have used either operator assistance or ground control computers. The United States has not conducted a robotic docking in space.

M.8.2 ETS-VIII (Engineering Test Satellite VIII)

ETS-VIII is an advanced NASDA satellite with the aim to develop and verify the world's largest geostationary satellite-bus technology, considered to be necessary for space missions at the beginning of the 21st century. The mission objectives are to conduct orbital experiments on the following structures/systems: (the first three are required to realize mobile satellite communications with hand-held terminals, similar to popular cellular phones).

- LDR (Large-scale Deployable Reflector) developed by NASDA, a system needed for large-scale space structures. The LDR structure features a 2.4 mm rms surface precision and antenna diameter expendability. It consists of 14 hexagon-shaped modules. ETS-VIII is equipped with two LDR modules, one of data transmission and the other for reception.
- HPT (High-Power Transponder), developed by ASC (Advanced Space Communications Research Laboratory)
- OBP (On-Board Processor), provided by ASC and CRL (Communications Research Laboratory)
- Moreover, the ETS-VIII S/C carries a high precise clock system for satellite positioning experiments, developed by NASDA and CRL.

The S/C shape is a rectangular box with deployable solar paddles. The satellite design employs a number of features such as:

- A light structure to improve the payload to bus system ratio to 40% from the conventional 30%
- The bus system consists of several modules to reduce development time through concurrent work to integration
- The bus power supply voltage was changed from the conventional 50V to 100V for greater electrical power
- The MIL-STD-1553B data bus standard is used for commonality
- In addition, the CCSDS packet protocol suite was chosen for all data communication with the ground segment
- The thermal subsystem uses heat pipes to connect the north and south panels of the satellite thereby expanding the effective radiation surface
- Use of an attitude control system with fault-tolerant functions and in-orbit reprogramming capability.

The satellite consists of three modules: bus, payload, and propulsion module. [1657) 1658)] It is three-axis stabilized, the accuracy of the AOCS (Attitude Orbit Control System) is: roll/pitch axes = ±0.05° max., yaw axis = ±0.15° max. The AOCS employs an Earth sensor assembly, a sun sensor and an integrated rate gyro assembly for attitude sensing. Actuation is provided by four reaction wheels, each of 50 Nms, in a skewed arrangement. In addition, 13 accelerometers are installed to monitor the dynamic behavior of the deployed structure. The S/C mass is 2900 kg at the beginning of mission life with a payload mass is 1100 kg. S/C electrical power is 7.5 kW with a regulated 100 V of bus voltage. The NiH_2 batteries provide a capacity of 100 Ah for eclipse operations. The overall length of the spacecraft is 40 m measured along the axis of the solar panels, and 37 m measured along the axis of the two LDRs.

1657) M. Homma, S. Yoshimoto, N. Natori, Y. Tsutsumi, "Engineering Test Satellite-8 for Mobile Communications and Navigation Experiment," 51st IAF Congress, Rio de Janeiro, Brazil, Oct. 2-6, 2000, IAF-00-M.3.01

1658) A. Meguro, A. Tsujihata, N. Hamamoto, M. Homma, "Technology status of the 13 m aperture deployment antenna reflectors for Engineering Test Satellite VIII, Acta Astronautica, Vol. 47, No 4-7, 2000, pp. 147-152

Two propulsion systems are used: a) an apogee (bi-propellant) engine in the 500 N class with 22 N thrusters (used for attitude control and E-W station-keeping), and b) a Xenon ion engine providing a thrust of 25 mN, two ion engines are used for N-S station-keeping. The S/C bus has a design life of 10 years, the mission design life is three years.

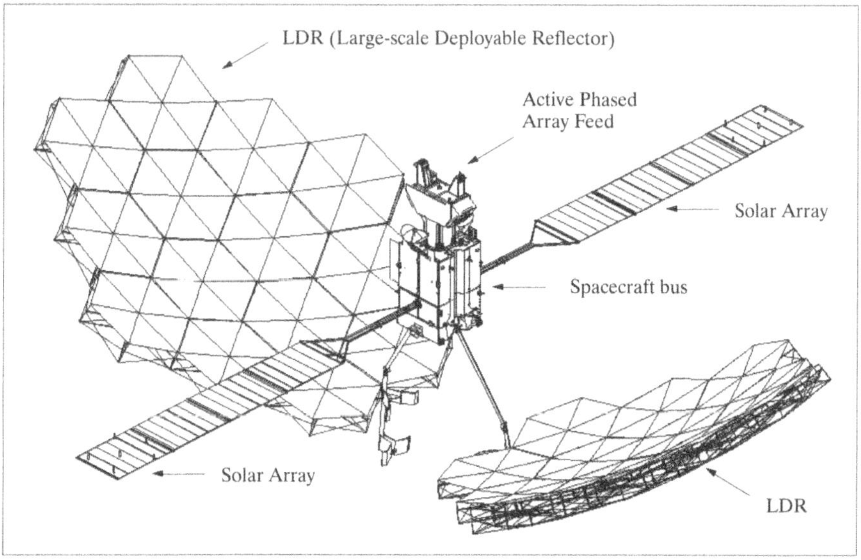

Figure 307: The ETS-VIII deployed S/C configuration

The launch of ETS-VIII with an H-IIA launch vehicle from Tanegashima Space Center is planned for 2003.

Orbit: Geostationary orbit of ETS-VIII at 146° E longitude.

TT&C communications are provided in S-band. All data transmission is done with CCSDS protocols. The MIL-STD-1553B data bus is used for all on-board data handling. The OBP design supports autonomous FDIR (Fault Detection, Isolation and Reconfiguration) functions.

Parameter	OBP (Onboard Processor)	OPS (Onboard Packet Switch)
Modulation	35 k symbol/s, $\pi/4$ QPSK	512 k symbol/s, $\pi/4$ QPSK
Access method	MC-TDMA (5 channels/carrier)	Slotted ALOHA
Information rate	5.6 kbit/s (voice)	4 - 32 kbit/packet
Number of channels	500	128 packets/carrier (max)
Satellite EIRP/Carrier and G/T	49 dBW, 13 dB/K	52 dBW, 13 dB/K
Ground terminal EIRP and G/T	9 dBW, -24 dB/K	18 dBW, -21 dB/K

Table 486: Specification of the onboard processors

Parameter	Feeder Link	Mobile Link
Frequency band	Ka-band	S-band
Antenna aperture	0.8 m diameter	13 m diameter
EIRP (Effective Isotropic Radiated Power)	44 dBW @ NPR=20 dB	65 dBW @ NPR=16 dB
G/T [(receiver) Gain/(noise) Temperature]	15 dB/K	15 dB/K
Beam scanning range	N/A	±2°

Table 487: RF subsystem performance

Mobile satellite communications and broadcasting experiments:

S-band communications and broadcasting experiments are being conducted to demonstrate a mobile satellite communications and broadcasting system technologies with small-scale ground terminals such as hand-held terminals. These experiments employ the following devices:

- Large-scale Deployable Reflector: The entire spacecraft structure is dominated by the two large deployable antenna reflector structures, each of size 19.2 m x 16.7 m with a 13 m aperture. Each mesh deployment antenna consists of 14 modules. Each module in turn consists of a gold-plated Molybdenum mesh surface, a spatially determined cable network, and a deployable truss as supporting structure. The entire antenna system fits in a stow-volume of 1 m diameter x 4 m in height.

Antenna layout	Offset parabola
Aperture size, F/D	13 m, 0.8
Offset angle, surface error	52.5°, less than 2.4 mm rms (S-band: $\lambda/50$)
Beam pointing error	Less than 0.83° (static)/0.0045° (dynamic)
Antenna mass	<400 kg (2 reflectors, include supporting booms, adjustment mechanisms and deployment drive electronics)
Antenna stiffness	>0.1 Hz (in orbit) >40 Hz (during launch in longitudinal direction) >20 Hz (during launch in lateral direction)

Table 488: Design specifications of the antenna reflectors

- Active-Phased-Array Feeder: The phased array feeder systems are combined with the LDRs to create multiple steerable MSS (Mobile Satellite communication Systems) and to support mobile BSS (Broadcasting Satellite System) experiments (3). The transmission side feeder consists of 31 solid state power amplifiers (300W in gross output), and the receiving side feeder of 31 low noise amplifiers. The on-board beam-forming network controls the amplitude and phase of the signals to/from the feeder elements to create multiple steerable beams and to form the desired beam patterns.
- On-board signal switching: Two sets of onboard baseband switching equipment are installed to realize onboard signal routing. One is an on-board processor (OBP), which is designed to perform voice channel switching among mobile users or between mobile users and ground network users and has 500 channel capacity. The other is an onboard packet switch (OPS) for high-speed packet data communications.

Satellite positioning experiments:

Positioning experiments (in S- and L-band) are conducted to study basic satellite positioning system characteristics and to demonstrate the feasibility of new satellite positioning system concepts (consisting of MEO and GEO constellation elements). A high-accuracy time signal, which is similar to the GPS data, created by an atomic cesium frequency standard on the satellite is transmitted. Furthermore, the TCE (Time Comparing Equipment) is used to conduct precise time comparison between the on-board atomic clock and the ground reference clocks. High-accuracy orbit determination is supported by SLR (Satellite Laser Ranging) stations.

Frequency	L-band (1598.88±7.5 MHz), S-band (2491.005±7.5 MHz)
EIRP	38 dBW
Transmit signal	Spread spectrum signal with navigation message (50 bit/s)
PN code	1.023 Mchirp/s, 1.705 Mchirp/s, 3.41 Mchirp/s, 5.115 Mchirp/s
Objective accuracy	<100 m, orbit determination < 10 ns, clock synchronization without TCE (Time Comparing Equipment) <10 ps, clock synchronization with TCE

Table 489: Specification of the positioning system

In the experiment, time signals from two sources, namely the GPS constellation and from the on-board atomic clock are used to determine the position of the satellite. The intend is to reduce the number of spacecraft in the constellation of a future navigation system (if sufficient accuracy can be achieved by this method of orbit determination).

M.9 FedSat-1 (Federation Satellite One)

FedSat-1 is an Australian microsatellite to commemorate the centenary celebration of Australian Federation in 2001. [1659] [1660] [1661] [1662] The program is managed by CSIRO (Commonwealth Science and Industrial Research Organization - Canberra, Australia) with participation from universities and industry (Auspace Ltd, Canberra; and Vipac Engineers and Scientists Ltd, Canberra are the prime contractors). The Australian Government created a new entity in 1997, CRCSS (Cooperative Research Center for Satellite Systems). The specific mission of CRCSS is to promote Australian space research and to create a favorable and sustainable environment for Australian industry, government agencies and universities, involved in services, applications, and research. As of 2000, there are more than sixty CRCs (Cooperative Research Centers) in Australia active in engineering, science, medicine, and technology. The main objectives of the FedSat mission are in the fields of: communications, space science, remote sensing, engineering research, and education/training.

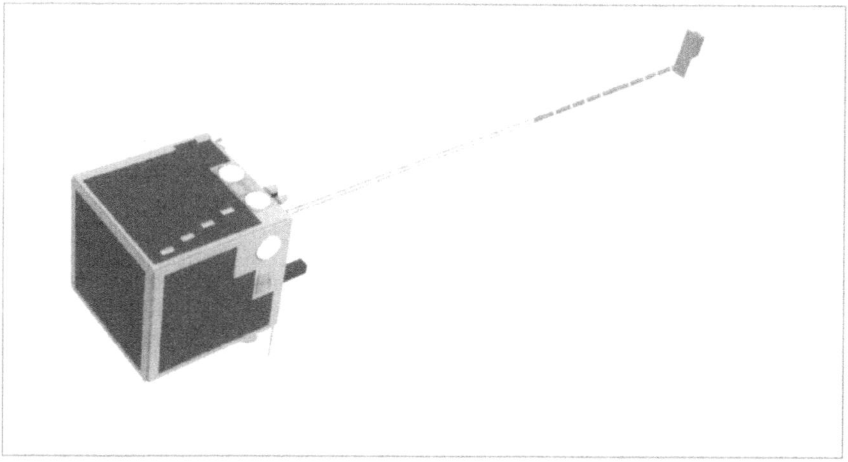

Figure 308: Illustration of the FedSat spacecraft

The S/C structure resembles a cube of 50 cm side length and is based on a variant of the standard MicroSIL bus, built by SIL (Space Innovations Limited) of Newburry, Berks, UK, a commercial platform provider (SSTL of Surrey purchased SIL as of 2000). The design is based around six honeycomb outer panels, and an interior double shelf, dividing the platform subsystems on the lower shelf from the payload instruments on the top shelf. FedSat-1 is three-axis stabilized. The S/C attitude is measured by three digital sun sensors (linear array of 256 photodiodes) and a three-axis fluxgate magnetometer. Attitude is controlled by

1659) S. Russell, M. Vesely, C. Graham, M. Petkovic, "Progress Towards FedSat 2001 A'stralian Space Odyssey," Proceedings of the 13th AIAA/USU Conference on Small Satellites, Aug. 23-26, 1999, Logan UT, SSC99-IX-6

1660) http://www.crcss.csiro.au/

1661) A. J. Barrington-Brown, A. N. Wicks, et al., "FedSat - an advanced microsatellite based on a MicroSIL bus," Proceedings of the 12th AIAA/USU Conference on Small Satellites, Logan, UT, Aug./Sept., 1998

1662) E. C. Graham, "FedSat: An Australian research microsatellite mission," IAF Congress, Melbourne, Australia, 1998, in Acta Astronautica, 1999 three precision reaction wheels, three magnetorquers, and three solid-state rate gyros as actuators. The attitude knowledge accuracy is ±1°. The ACS (Attitude Control System) is provided by CSA (Canadian Space Agency) and built by Dynacon Enterprises Ltd. of Mississauga, Ontario. [1663] Power (60 W max) is provided by surface-mounted solar cells (19 strings of GaAs cells on Ge substrates). In addition, there are 16 NiCd battery cells, each with a capacity of 6 Ah for eclipse phase operations. On-board data storage of 2.56 Gbit is provided. The S/C mass is 58 kg. The S/C design life is three years with a goal of five years of operations.

Communications: The S/C features two S-band patch antennas, plus a UHF quarter-wave whip, and Ka-band isoflux antenna. The S/C TT&C communications are in S-band. The S-band transmitter has an output power of 2 W. Uplink frequency = 2205 MHz, downlink frequency = 2030.4375 MHz. Downlink data rates of up to 1 Mbit/s are supported. Suppressed carrier BPSK modulation is employed. The on-board receiver accepts telecommand data, BPSK modulated, on to a subcarrier, with uplink rates between 8 bit/s to 4 kbit/s. Redundancy is provided through the experimental communications payload, with a UHF uplink at 4 kbit/s, and a Ka-band downlink with 125 kbit/s or 250 kbit/s capability. The communications protocol is known as the Packet Utilization Standard. It is basically an ESA augmentation of CCSDS.

Spacecraft operations are provided by CRCSS (Cooperative Research Center for Satellite Systems) through its participating organization ITR (Institute for Telecommunications Research) at the University of South Australia in Adelaide. All communications with the S/C are supported by one ground station, located at Adelaide, offering 2-3 passes per day. A piggyback launch of FedSat-1 is planned on the H2A vehicle of NASDA (second flight) along with ADEOS-II in early 2002.

Orbit: Sun-synchronous circular orbit, altitude = 802.9 km, inclination =98.67°, equator crossing time on ascending node at 10:30 AM.

Sensor complement:

CPE (Communications Payload/Experiment), developed at ITR of UniSA (University of South Australia) and at UTS (University of Technology, Sydney) in cooperation with CRCSS. The objective is the study of new Ka- L-, and UHF-band techniques for communications and data delivery (modulation, coding and multiple access techniques, on-board data reduction, regeneration and processing). The following subsystems are part of the communications payload/experiment: [1664]

- **ADAM** (Advanced Data Acquisition and Messaging System) of UniSA. A portion of ADAM's functionality include the "old-fashioned" DCS (Data Collection System) services. ADAM consists of a BBP (Baseband Processor) unit and a UHF transceiver unit. It provides the functionality of two-way packet communications for remote environmental monitoring and forward messaging.
- Ka-band payload module (CSIRO)
- Ka-band Earth terminals (UTS)
- Ground control station for FedSat TT&C services (UniSA).

The Ka-band unit is used for bent-pipe and regenerative communications, and as a testbed to space-qualify an GaAs MMIC (Microwave Monolithic Integrated Circuit) device of CRCSS. BBP provides on-board processing of the Ka- and UHF-band units; it also acts as a low-power single modem with flexible operations support (study of communication protocols). Key features of the BBP include turbo coding on the uplink (QPSK modem with a bit error rate of 10^{-10}, it operates on a signal SNR as low as 2 dB, and a data rate of 2 Mbit/s), highly flexible operation including reconfigurable field programmable gate arrays

1663) S. C. O. Grocott, "Modular Attitude Control System for Microsatellites with Stringent Pointing Requirements," Proceedings of the 14th AIAA/USU Conference on Small Satellites, Logan, UT, Aug. 21-24, 2000, SSC00-VIII-6

1664) http://www.crcss.csiro.au/reports/rept2000/Part05.htm

(FPGAs), low power consumption (<5 W), and burst demodulation. A copy of ADAM will be flown on KAISTSAT-4 (launch in 2002) of KAIST/SaTReC of Korea (see D.19.3) as its DCS (Data Collection System). ADAM is also planned to be flown on a satellite by ATSB of Malaysia.

Potential applications of the UHF and BBP units include ARGO (Array for Real-time Geostrophic Oceanography) and mobile communications. [1665] The ARGO program involves data collection services, i.e. uploading of data from a large number of floating buoys in the Southern Ocean and forwarding this information to a ground station. A portion of the ARGO array consists of submergible buoys called PALACE floats (these are Lagrangian buoys which may float at any altitude in the Ocean, measuring temperature and salinity). Two-way link capabilities (with multiple access scheme and error control techniques) permit a reconfiguration of the floats; turbo-coding techniques permit much larger messages per packet. Mobile applications include communications with land-mobile, laptop and hand-held terminals in remote areas for data and messaging experiments. [1666]

NewMag (Magnetometer Experiment). A space science experiment in solar-terrestrial physics, developed at the University of Newcastle, New South Wales, Australia, with the objective to measure electrical currents and perturbations in the Earth's magnetic field in the frequency range of 0.1 Hz to 1 kHz. The aim is to observe naturally occurring electromagnetic wave signatures (oscillating wave fields, including ULF hydrodynamic waves and lower ELF-band ion cyclotron waves) and field-aligned currents in the ionospheric F-region and the exosphere (plasma-wave model improvement of the ionosphere and exosphere). The NewMag and BlackJack ionospheric measurements contribute to this objective. NewMag is a three-axis fluxgate magnetometer. It is designed and built at UCLA (C. T. Russell) and is of MFI (Magnetic Fields Instrument) heritage, flown on the FAST mission (launch of FAST on Aug. 21, 1996, K.21.2). Other missions of MFI heritage are ISEE-1/2 (launch Oct. 2, 1977), AMPTE-UKS (launch Aug. 16, 1984), POLAR (launch Feb. 24, 1996), and Galileo (launch 1989). NewMag has a dynamic range of ±65536 nT with 16 bit resolution (±0.2 nT), and a bandwidth of 10Hz (vector sample rates: internal = 1000/s, standard = 10/s, burst mode = 100/s). The instrument is mounted at the end of a 2.55 m boom (to avoid magnetic interference from the satellite itself), it has a mass of 3 kg and a power consumption of 3 W. [1667]

BlackJack (GPS Flight Receiver), built by SpectrumAstro of Gilbert, AZ, under the designation **AstroNav**. BlackJack is provided by NASA/JPL (cooperative research agreement with CRCSS) in exchange for BlackJack data availability to US researchers. [1668] On the Australian side, the project is conducted in cooperation with QUT (Queensland University of Technology) and the Department of Physics of La Trobe University. There are two major measurement objectives: 1) determination of precise S/C timing, position and attitude, and 2) to obtain dual-frequency ionospheric soundings, permitting the derivation of total ionospheric electron density profiles, by the method of refractive occultation monitoring. The TEC (Total Electron Content) occultation observations of BlackJack provide slant range measurements which can be converted into vertical profiles. The instrument is a single-antenna navigation version of JPL's BlackJack receiver. A small patch antenna, facing backward in the anti-velocity direction, is mounted onto FedSat. The instrument is configured to track up to 16 GPS satellites simultaneously (3 channels per satellite). The tracking modes supported are: Code Phase L1 (1575.42 MHz) C/A code, L1 P Code/Codeless, L2 (1227.6 MHz) P Code/Codeless & Carrier Phase (Integrated Doppler) on both L1 & L2. The BlackJack assembly has a mass of 2.4 kg and a power consumption of about 11 W.

1665) Note: ARGO is a component of the international Global Ocean Data Assimilation Experiment (GODAE).

1666) C. Graham, M. Petkovik, S. Russell, E. S. Seumahu, M. Vesely, "The FedSat Microsatellite," Proc. of ICICS'99 (International Conference on Information Communications and Signal Processing), Singapore, Dec. 7-10, 1999

1667) B. J. Fraser, C. T. Russell, J. D. Means, F. W. Menk, C. L. Waters, " FedSat - An Australian Research Microsatellite," Advances in Space Research, Vol 25, Issue 7-8, pp.1325-1336, 2000.

1668) Information provided by Thomas P. Yunck of JPL and Bill Falkenberg of SpectrumAstro Inc.

The receiver operation has a duty cycle imposed (10 minutes on, 40 minutes off, repeated indefinitely) due to FedSat power limitations. Hence, it is planned to take data only at regular navigation measurements (1 sample every 30 s), this rate may be exceeded for short periods only.

HPCE (High Performance Computer Experiment). HPCE is a cooperative experiment between QUT and JHU/APL. The objective is to study reconfigurable computing in space. The HPCE design employs a modular and reconfigurable FPGA (Field Programmable Gate Array) logic. The hardware configuration of wire connections are determined by downloading from memory a stream of configuration bits. These reconfigurations can also be uplinked during the mission. Each FPGA is capable to reconfigure itself in a very short time. The reconfigurable technology permits a change of physical FPGA circuits via software control.

M.10 MDS (Mission Demonstration Satellite)

MDS is a NASDA technology program, initiated in 1997, with the objective of short-term and economical satellite development, implementation of advanced missions and space demonstration of mission equipment. The strategy is to use a conventional bus and existing support equipment in this mission scenario.

M.10.1 MDS-1

The objectives of MDS-1 are to verify the use of commercial parts in orbit, to verify minimization technology for components, and to collect space environment data (e.g., radiation on equipment, etc.). The MDS-1 design is based on the concept of integrating proven technologies. The bus, an aluminum honeycomb structure, consists of box of size: 1.2 m x 1.2 m x 1.5 m. The spin axis of the S/C is perpendicular to the solar panels and to the boom. A magnetometer is mounted onto the deployable boom (3 m in length) to measure the Earth's magnetic field. The nominal spin rate is 5 rpm with a general boom orientation toward the sun (the angle between the sun direction and spin axis is 10°). A S/C power of >900 W (EOL) is provided by two deployable solar panels (3.3 m^2 total area and GaAs cells). The overall length of the satellite is 7.8 m from tip to tip. Two NiMH battery sets provide energy of 20 Ah for eclipse operations. Attitude sensing is provided by SAS (Sun Aspect Sensor), a horizon sensor, and a magnetometer. Occasional actuation is provided by thrusters to keep sun orientation (but MDS-1 is basically spin-stabilized). A reaction control subsystem utilizes hydrogen for initial spin-up. The S/C mass is 480 kg. The S/C design life is one year.

MDS-1 will be launched into the geostationary transfer orbit by an H-IIA vehicle from the Tanegashima Space Center in 2001/2. The co-passenger on this flight is VEP#3 (H-IIA Vehicle Evaluation Payload). [1669] [1670] [1671] [1672] [1673]

Orbit: Geostationary transfer orbit (GTO), perigee = 500 km, apogee = 36,000 km, inclination = 28.5°.

RF communications: The on-board communication subsystem provides a USB (Unified S-band) command reception, telemetry transmission, and ranging (turn-around transmission of the ranging tone from/to the ground station) function. The downlink frequency is 2.212 GHz with data rates ranging from 128 bit/s (housekeeping data) to 16 kbit/s (mission data);

1669) M. Usui, M. Takei, K. Arai, R. Kuramasu, "MDS Project: New Challenge of Japanese Satellite Development," Proceedings of the 51st IAF Congress, Rio de Janeiro, Oct. 2-6,. 2000, IAF-00-U.1..06

1670) http://yyy.tksc.nasda.go.jp/Home/Projects/MDS/index_e.html

1671) http://oss1.tksc.nasda.go.jp/mds/what-e.html

1672) Information provided by Akio Yamamoto of NASDA

1673) T. Nagai, M. Mokuno, "Small Satellite Development and Future Vision of NASDA," Proceedings of the 6th ISU Symposium on Smaller Satellites: Bigger Business?, Strasbourg, France, May 21-23, 2001 the uplink frequency is 2.036 GHz with a data rate of 500 bit/s. The on-board recorder has a capacity of 1 Gbit.

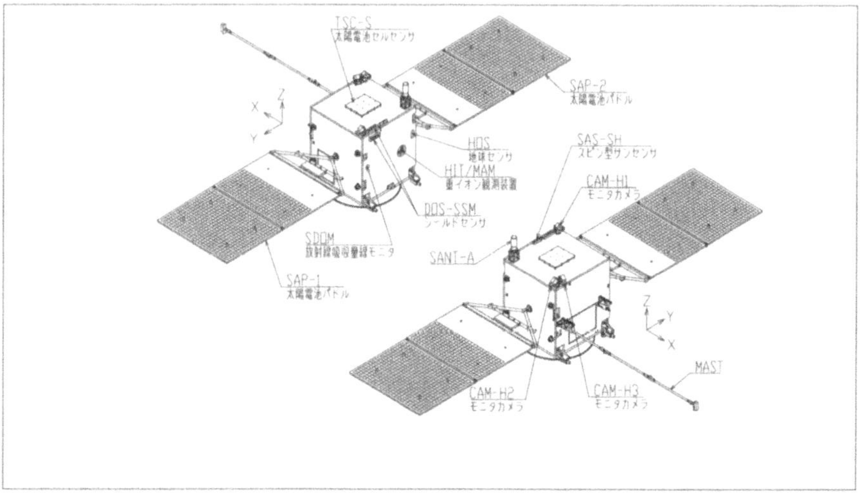

Figure 309: Illustration of the MDS-1 spacecraft (two views)

Sensor complement / demonstration of commercial devices or components

SEDA (Space Environment Data Acquisition). Objective: measurement of the space environment for performance evaluation of the demonstration equipment. The data obtained is also used to improve environmental models. SEDA is an instrument package consisting of the following instruments:

- SDOM (Standard Dose Monitor). SDOM measures the energies of electrons, protons and alpha particles.
- DOS (Dosimeter). The objective is to measure the radiation absorption.
- HIT (Heavy Ion Telescope). HIT measures heavy ions incident on the sensor with regard to type, energy and direction.
- MAM (Magnetic Monitor). MAM measures the magnetic flux in the MDS-1 orbit.

CSD (Commercial Semiconductor Device). This package represents a collection of commercial semiconductor devices, fabricated with different processes and materials, exposed to the radiation environment. The objective is to obtain data on the "total ionizing dose effect" and the "single event effect." The results in space are also compared with ground test data. The aim is to establish evaluation methods for ground tests by comparing ground and space tests. CSD contains the following test devices: 4 Mbit SRAM, 64 Mbit DRAM, 1 Mbit EEPROM, 4 Mbit FLASH memory, FPGA, and 1 M Gate Array.

TSC (Terrestrial Solar Cells). The experiment consists of a collection of ground-used solar cells exposed to the radiation environment. The objective is to utilize the data obtained for the design of future solar cells with such features as: higher efficiencies, superior radiation tolerance, low mass and low cost. The following solar cell types are part of the experiment: Poly Si cell, n-type base Si cell, InGaP/GaAs tandem cell, CIS (Copper Indium di-Selenide) cell, NRS/BSF (Non-Reflective Surface/Back Surface Field) Si cell (space use), GaAs/Ge cell (space use).

Demonstration and space qualification of miniature lightweight components of high functionality and reliability intended for installation in future spacecraft or instruments.

CPV (Common Pressure Vessel type battery). The objective is to test a new battery type in space that might lead to substantial reductions in size and mass of the electrical power subsystem of a S/C. Demonstration of CPV type NiH_2 batteries with regard to battery charge/discharge control using high-frequency switching devices. The CPV battery consists of two cells (there are 16 mini cells to each cell); the capacity of each CPV cell is 5 Ah. The energy/unit mass of a CPV system is in the order of 35 Wh/kg. - Note: The Clementine spacecraft (launch Jan. 25, 1994, orbit to the moon, the S/C experienced a malfunction on May 5, 1994 after its lunar mission was completed) of DoD and NASA was probably the first mission to demonstrate the CPV battery technology in space.

SSR (Solid State data Recorder). The objective is to evaluate the SSR performance by using commercial semiconductors. The instrument consists of a control unit (64 Mbit DRAM) and stack memory modules in 3D high-density layout as memory unit, the storage capacity is 32 Gbit.

PCS (Parallel Computer System). Objective: demonstration and evaluation of parallel processing in combination with the "fault tolerant technique" with the use of MPU (Multi Processing Unit) to accomplish higher performances and reliabilities.

M.11 MightySat

MightySat is a long-term, multi-mission, small satellite program (started in 1994) of the Space Experiments Directorate of Phillips Laboratory (USAF/PL) at Kirtland Air Force Base, Albuquerque, NM (as of 1998 designated as AFRL (Air Force Research Laboratory). The overall program objectives are to provide an environment for frequent, inexpensive, on-orbit demonstrations of emerging space system technologies and to accelerate their transition into operational use. The MightySat spacecraft are modular and functionally standardized platforms and buses capable of supporting a wide range of experimental payloads. Launches are considered from STS (Shuttle) or with multi-service launch systems. The emphasis is on low-cost projects, with fast building periods from contract to launch, with high-risk and high-payoff technologies. [1674] [1675] [1676]

M.11.1 MightySat I

MightySAT I is a single mission project which uses an all-composite bus (similar in configuration to its aluminum XSAT predecessor of NASA) with significant mass reduction. [1677] The S/C was built by OSC of McLean, VA (formerly CTA Space Systems). The S/C structure is characterized by three decks supported by six structural frames which make up a hexagonal prism body. The S/C mass is 63 kg, the payload mass is 17 kg (average power of 12 W), orbit average power = 14-27 W. The S/C is spin-stabilized (3 rpm) with the spin axis oriented normal to the orbit plane. Attitude sensing and actuation is provided by a three-axis magnetometer, two coarse sun sensors, and three torque coils. A coarse attitude knowledge of $\pm 5^{\circ}$ is required. S/C design life of one year.

The S/C has four subsystems: C&DH (Command & Data Handling) made up of nine electronic boards, EPS (Electrical Power Subsystem) consisting of seven solar panels and a

1674) J. Freeman, C. Rudder, P. Thomas, "MightySat II: On-orbit Lab Bench for Air Force Research Laboratory," Proceedings of the 14th Annual AIAA/USU Conference on Small Satellites, SSC00-I-2, Aug. 21-24, 2000

1675) R. J. Davis, J. F. Monahan, T. J. Itchkawich, "MightySAT I: Technology in Space for about a Nickel," Proceedings of the 10th Annual AIAA/Utah State University Conference on Small Satellites, Sept. 16-19, 1996

1676) B. Braun, R. Davis, T. Itchkawich, T. Goforth, "MightySat-I: In Space," Proceedings of 13th Annual AIAA/USU Conference on Small Satellites, Logan Utah, Aug. 23-26, 1999, SSC-99-I-3

1677) http://www.vs.afrl.af.mil/factsheets/msat.html

single 21-cell NiCd battery (4 Ah), RF (Radio Frequency - communications) consisting of a receiver, transmitter, transmit/receive switch, and a four-blade antenna, and ADACS (Attitude Determination and Control Subsystem). Data transmission is in UHF-band (306.775 MHz) at 9600 bit/s (downlink, BPSK modulated) and at 2400 bit/s (uplink, FSK modulated). The link margin is about 20 dB for uplink and downlink above a 5° elevation angle.

MightySat I was launched by ejection from the Space Shuttle using HES (Hitchhiker Ejection System) on Dec. 14, 1998. The entire satellite, HES, and avionics were stowed away into a standard Hitchhiker canister of 0.14 m^3 in volume prior to launch. Deployment was initiated by the Shuttle crew on Dec. 14, 1998. The entire MightySat I deployment sequence took about 10 minutes. The launch of Shuttle flight STS-88 took place on Dec. 4, 1998 (landing of STS-88 on Dec. 15 at KSC). [1678)]

MightySat mission operations are conducted by SMC/TEO at Kirtland AFB. All communications are via AFSCN (Air Force Satellite Control Network). The operations concept is largely based upon experience from similar satellite programs, most notably RADCAL.

Orbit: altitude = 385 km (initial altitude), inclination = 51.6°. The MightySat I S/C reentered the Earth's atmosphere on Nov. 16, 1999 due to its relatively low orbital altitude. All of the mission objectives were accomplished.

Payload complement: MightySat I has five AFRL-developed advanced technology demonstration experiments. Two of the demonstrations are tests of experimental bus components.

ACS (Advanced Composite Structure). Objective: technology test. The S/C structure (a hexagon with a diameter of 50 cm and a height of 53 cm with a mass of about 8 kg) consists of a composite frame, three decks, and seven solar panel substrates. The composite material is graphite fiber. However, ACS has no data interfaces to the S/C (the structure was ground-tested). The S/C frames were fabricated by using the so-called "SnapSat" approach (developed by Composite Optics Inc.), in which the elements are cut from cured flatstock layups and fitted together using a mortise&tenon technique.

ASCE (Advanced Solar Cell Experiment). Objective: advance of in-space power generation technology. Use of dual-junction solar cells (average efficiency of 21%) which provide a 15% performance gain over conventional GaAs cells. The cells are covered with a GaInP (gallium indium phosphate) layer which captures and converts the short wavelength spectrum. ASCE consists of 13 strings of 40 GaAs cells (2cm x 4 cm) and six strings of 18 GaInP cells (2 cm x 2 cm) for comparison of cell performance. The cells are bonded directly to the solar panel substrates. An objective is also to determine aging effects in the space environment.

MAPLE-1 (Microsystem and Packaging for Low Power Electronics). The objective is to provide an on-orbit demonstration of advances in low-power electronics including performance (emerging electronics and packaging technologies in space). The MAPLE-1 experiment suite is a collection of five sub-experiments, brokered by a central controller that supplies power and communications through interfaces to the MightySat I host satellite. MAPLE sub-experiments explore flight-worthiness, issues associated with low-power, commercial and radiation tolerant microelectronics, advanced microelectronics packaging, MEMS (Micro-Electromechanical Systems) devices, circuits, associated subsystems, and components. Particular experiments are:

- In-situ operation of MEMS commercial cantilever beam accelerometers. Two accelerometers are used, a local 5 gram full-scale unit and a remote 2 gram full-scale unit.
- Monitoring of advanced packaging structures through reliability monitoring integrated circuit die (monitoring of moisture, dust, and material property changes in electronic components), developed by SNL. The SSRB (Solid State Recorder Board) was

1678) B. Braun, R. Davis, T. Itchkawich, T. Goforth, "MightySat-I: Transitioning Space Technology to the Warfighter," AIAA-99-4484, 19999 designed to evaluate the performance of HCSM (High Capacity Spaceborne Memory) SRAM module in the space environment. The EMB (Environment Monitor Board) was designed to measure the temperature inside the Maple-1 enclosure.

- Total ionizing dose dosimetry
- Advanced thin film high density interconnect multi-chip modules
- Comparison of military grade commercial and radiation hardened bulk silicon anti-fuse-based field-programmable gate arrays.

SMARD (Shape-Memory Actuated Release Device). Objective: demonstration of a new class of low-shock release devices. SMARD devices are based upon a shape-memory alloy (Nitinol) which is used as the driving force to actuate the release of a fastener. - The SMARD payload consists of four release devices mounted on a common, instrumented deck (a conventional pyrotechnic device, a linkwire device, and two shape-memory actuated devices are used). The SMARD experiment involved measuring the response of a three-axis MEMS accelerometer to the shock wave generated when a test "separation bolt" was fired.

MPID (Micro-Particle Impact Detector) of NASA/LaRC. [1679] The objective is to collect information on orbital debris. Measurement of direction and time of impact of spaceborne micro-particles with time of impact resolution of 0.1 s. The primary element in this experiment consists of two MOS (Metal Oxide Semiconductor) discharge capacitor detectors that discharge upon hypervelocity particle impact. MPID consists of two small plates (4 cm x 8 cm, each the size of a credit card), mounted on the bottom outside surface of the satellite, providing indications of micro-particle impacts (recording of time of impact). Each particle impact causes an impact event record that is stored in the S/C control unit for later downlink. Each impact event record stores time of impact and output from two coarse sun sensors. Data from the coarse sun sensors is used to help determine attitude of the spacecraft.

M.11.2 MightySat II.1 (Sindri)

The MightySat II program, initiated in March 1996, represents a series of up to five small satellite missions over a decade. The S/C structure is of modular design (built by Spectrum Astro Inc. of Gilbert, AZ). The size of the payload envelope is: 61 cm x 61 cm x 46 cm. The S/C is three-axis stabilized (inertial pointing); attitude knowledge/control = 0.15°/0.18°. The S/C has deployable solar arrays (Si), and a total mass of 121 kg (57 kg of payload mass); power = 330 W (BOL, 100 W average) with 28 V unregulated bus. In addition, there are three NiCd batteries, each with 4 Ah, for eclipse operations. A RAD6000 CPU is used. The S/C design life is 1 year. [1680] [1681] [1682]

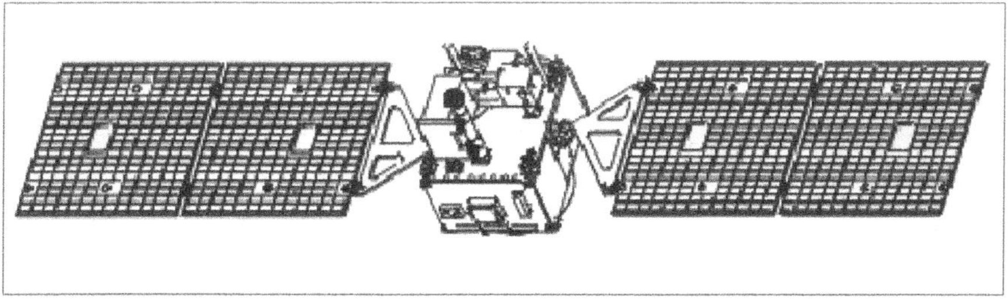

Figure 310: The MightySat-II.1 spacecraft illustration

Attitude determination is provided with a star tracker and an interferometric fiber optic gyro, attitude control (actuation) uses three orthogonal reaction wheels. Secondary control

1679) http://setas-www.larc.nasa.gov/mpid/mpid.html
1680) http://www.vs.afrl.af.mil/vsd/mightysatII/index.html
1681) http://www.spectrumastro.com/
1682) http://www.vs.afrl.af.mil/factsheets/msat2.html

may be achieved with three torque rods, the torque rods serve also to dissipate reaction wheel momentum. The ACS (Attitude Control Subsystem) also autonomously controls solar array articulation. The command and data handling subsystem provides 380 MByte of solid state storage at data rates of 20 MByte/s. Communications are in S-band, data rates of 2 kbit/s uplink, 1 Mbit/s downlink for payload data (encrypted uplink/downlink). S/C operations are conducted by SMC/TEO from Kirtland AFB. A launch of MightySat II.1, dubbed "Sindri," on an OSC Minotaur vehicle (a converted Minuteman-2 missile motor combined with the upper stages of the Pegasus-XL booster), took place on July 19, 2000 from VAFB, CA.

Orbit: Sun-synchronous circular orbit, altitude = 550 km, inclination = 97.6°, local crossing at 11:15 on a descending node.

Sensor/experiment complement:

MightySat II.1 carries a total of ten new technologies, including both experimental bus components and stand-alone experiments.

FTHSI (Fourier Transform HyperSpectral Imager) designed and built by Kestrel Corporation of Albuquerque, NM, and the Florida Institute of Technology, Melbourne, FL, heritage of airborne version of FTVHSI. The objective is to demonstrate spaceborne hyperspectral imaging technologies (FTHSI is the first functioning spaceborne hyperspectral imager). The FT-approach is considered to be promising for spaceborne hyperspectral concepts (FTHSI demonstrates the advantage of Fourier systems over dispersive hyperspectral imagers, in that it can record the full spectra without any time delay and can decouple the spatial and spectral signatures). The FTHSI instrument contains three major optical subsystems: a monolithic Sagnac interferometer which produces the spatially modulated interferogram; a Fourier transform lens, which frees the spectral properties of dependence on aperture geometry and allows the wide FOV; and a cylindrical lens, which re-images one axis of the input aperture onto the detector array providing the one dimension of imaging. The 1-D image is passed through the interferometer where the rays are split, slightly separated, and recombined to create an interference pattern (interferogram) in one dimension. From the interferometer, a Fourier lens collimates the light and a cylindrical lens images the energy onto the detector array, preserving the one by n spatial dimension. [1683)]

Spectral range	475 - 1050 nm
Spectral resolution	87 or 44 cm^{-1} (corresponding to: 15-95 nm or 10-48 nm)
Number of usable spectral bands	145 or 290
FOV (Field of View)	3.0 °
IFOV (Instantaneous Field of View)	0.0058° or 0.0029°
Spatial resolution (best cross/along track)	28 m x 30 m
Swath width	7-29 km
Scene length range of imagery	10 km to 473 km
Instrument pointing (control/knowledge)	0.15°/0.15°
Instrument mass, power	20.45 kg, 66/60 W (peak/average)
Source data rate	20 MByte/s

Table 490: Some performance characteristics of FTHSI

The optical system of FTHSI employs a Ritchey-Chretien telescope with a clear aperture of 165 mm. The camera is an adaptation of a commercial camera made by Silicon Mountain Devices. The system has an f/3.4 number in the spatial and an f/5.3 number in the spectral dimension. A solid-block design is used to maintain alignment of the Fourier optics and cylinder lens. The detector assembly consists of a large-format CCD array (1024 x 1024) with a

1683) L. J. Otten III, A. D. Meigs, et al., "The engineering model for the MightySat II.1 hyperspectral imager," Proceedings of the Sensors, Systems and Next Generation Satellites, SPIE Vol. 3221-54, Sept. 1, 1997, London, UK, pp. 412-420

12 bit full-frame readout capability at up to 120 frames/s. Four operating modes are available to vary on command the spectral and spatial imagery resolutions, respectively: 512 x 512, 512 x 1024, and 1024 x 1024.

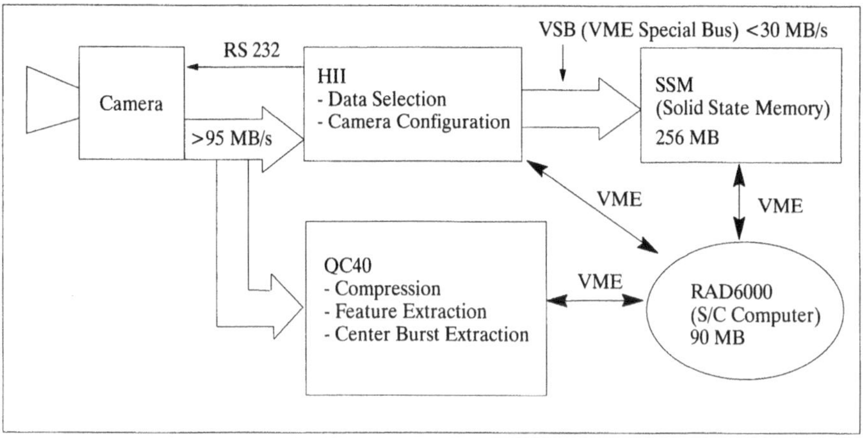

Figure 311: Block diagram of data-handling system of FTHSI

The instrument is operated with a maximum of 137 usable bands out of the 256 available, required by the limiting Nyquist sampling rate (to improve SNR). Note: an FT-type sensor creates data from zero wavenumbers to the Nyquist cutoff. Since the sensor cannot see wavenumbers lower than about 9523 cm^{-1} (1050 nm), there is information lost due to the sensor detectivity. The higher spectral resolution mode that operates the detector in the completely unbinned mode is not being used.

Data management: [1684] The FTHSI instrument takes advantage of the VME satellite bus services by storing source data directly in an allocated RAM of 134 MByte. The data transfers are provided by the custom-built HII (Hyperspectral Imager Interface) card, which handles packing and delivery to the S/C memory, it serves also as the RS 232 controller interface to the camera. A typical data collect (observation) lasts from 6-30 s. Total scene sizes are limited by the resolution selected and the storage capacity of 134 MByte RAM.

Function	Selections available	Default mode
Camera frame rate (frames/s)	15, 30, 60, 110	60
Camera gating time (s^{-1})	1, 1/125, 1/250, 1/500, 1/1000	1
Data bit depth	12 or 8 bit	12 bit
Nr. of imaging tiles per record	1, 2 or 4	4
Camera binning modes	1 x 1, 1 x 2, 2 x 1, 2 x 2	2 x 2
Data block length	0 to 134 MByte	134 MByte
Camera gain	1 or 4	1

Table 491: FTHSI camera operating modes and performances

SAC (Solar Array Concentrator). SAC focuses more light on each solar cell, increasing the solar energy available and reducing the number of cells needed to produce the same amount of power. The objective is to test the new solar panel array design. The steerable SAC is capable to focus more light on each solar cell, increasing the solar energy available. SAC covers one-third of one of the four solar panels. The cells are made of GaAs material (22-24% efficiency). The SAC provides a concentration ratio of 3:1 and has a 20° pointing tolerance toward the sun. SAC generates 37 W (BOL, 27.5 W EOL), its mass is 0.74 kg.

1684) Courtesy of Leonard John Otten III of Kestrel Corporation, Albuquerque, NM

MFCBS (Multi-Functional Composite Bus Structure). The spacecraft design incorporates new materials to be flight-tested as a lighter, more flexible alternative to the traditional aluminum bus structure. The MFCBS design includes an integrated thermal management system, robust structural integrity, high-attitude control accuracy, and precision three-axis stabilization.

NSX (NRL SGLS Transponder), with SGLS referring to "Space-to-Ground Link System." NSX is a miniaturized satellite communications unit, about 70% smaller and lighter than the current industry product, the objective is to test its functionality. NSX has a mass of 1.5 kg and a size of 14.5 cm x 14.5 cm x 7 cm. It consists of a COMSEC (Communications Security) transmitter and receiver segment and is connected to the VME card. An encryptor/decryptor feature is provided in COMSEC. NSX requires 6 W in receive mode and 23.2 W in transmit/receive mode.

QC40 (Quad-TMS320C40 processor). Objective: test of the space radiation susceptibility of critical new forms of microelectronic components through an advanced, and high-speed processor. In particular, QC40 performs on-board processing and compression of the FTHSI raw data (also interferogram conversion), increasing the number of images that may be downlinked to a ground station. Feature extraction involves comparing and matching collected interferograms with on-board interferograms of known materials. Real-time center burst extraction allows ground operators to obtain a quicklook of the collected image to assess its further handling. - QC40 consists of two VME cards. One board is comprised of four COTS TMS320C40 microprocessors (120 Megaflop), while the other board provides the interface between the processor board and the S/C electronics. QC40 has a mass of 1.4 kg, the power is 17 W.

SMATTE (Shape Memory Alloy Thermal Tailoring Experiment). Objective: an experiment involving actuation of a bi-modal composite sheet with respect to performing vibration isolation, structural control, deployment, and separation functions. SMATTE can change its physical properties, such as stiffness, damping, and shape, as a function of tailored thermal input signals. However, when SMATTE is heated above its transition temperature, it returns to its memorized shape. SMATTE consists of a layer of polymer matrix composite with a thin strip of shape memory alloy on each side; it is interfaced with the VME card and uses 5.3 W of power. Stress induced from thermal warping of the composite is automatically relieved by opposing stress in the shape memory alloy film. Optical fiber strain gauges monitor the performance of SMATTE.

SAFI (Solar Array Flexible Interconnect). SAFI incorporates copper leads embedded in a flexible, composite film reducing the weight and complexity of traditional hard wiring (SAFI connects the solar cells and routes the current off the panel). The objective is to test a technology which might lead toward multi-functional bus technology. SAFI is bonded to the solar panel, it has a thickness of <0.25 mm.

PICOSAT1.1. MightySat II.1 flies the Aerospace Corporation's designed and built launcher assembly and two tethered sub-satellites. They are referred to as PICO20 and PICO22, according to their nodal positions in a prescribed array of picosats. Both PICOSATs are of the same hardware design as the PICOSAT1.0 assembly, flown on OPAL (launch of OPAL on Jan. 26, 2000, see chapter N.11.3). The software of PICOSAT1.1 is updated to reflect the lessons learned by the ground operations from the OPAL flight. The overall objectives are:

- To extend the PICOSAT1.1 mission life (to about a week) by accurate ejection knowledge and improved coordination with the USAF SSN (Space Surveillance Network)
- Test of improved flight software, designed to reduce power usage - to extend mission life
- To demonstrate the system functionality after long-term storage on a host S/C. The PICOSAT1.1 system is planned to be ejected after one year of the MightySat II.1 launch, in July 2001.

- To demonstrate repeatability and robustness of a low-cost platform for advancing MEMS in space systems applications
- To demonstrate radically new operating concepts for future systems. The PICOSAT1.1 experience will serve as a low-cost risk mitigator for the release and flight of small bodies from a primary S/C, in support AFRL and DARPA small-satellite programs like TechSat 21, XSS (Experimental Spacecraft System), and MEPSI (MEMS-based PicoSat Inspector), the launch of the latter one is planned for 2003.

Each PICOSAT, a box of size: 25 mm x 75 mm x 100 mm and a mass of 0.275 kg, uses a small, battery operated, very low power radio with a ground-link and a cross-link (PICOSAT to PICOSAT) capability. The tether, of 32 m length, contains a gold wire, serving as radar 'target' to facilitate ground based radar tracking (with SSN). A sensor board in each PICOSAT has one chip containing 4 MEMS RF switches (a 100 V charge pump on the sensor board is needed by the switches) that are operated in space after deployment. [1685)]

SOR (Starfire Optical Reflectors). The Starfire Optical Reflectors are utilized by the Starfire Optical Range, an optical research facility at Kirtland Air Force Base, to develop optical sensing, imaging, and propagation technologies to support Air Force aerospace missions. Two COTS optical reflector mirrors are affixed to the coarse sensor boom on the S/C top deck (the reflectors are passive with a mass of 1.76 kg). The Starfire Optical Range at Kirtland performs active ranging of satellite position via ground-based laser.

M.12 MINISAT

MINISAT is a national space program of Spain, funded by the Inter-Ministerial Committee of Space Science and Technology (CICYT) and by INTA, supported by the Center for the Development of Industrial Technology (CDTI), and managed by INTA (National Institute for Aerospace Research). The program started in 1990 with first feasibility studies, the design phase was initiated in 1993. The major objectives of the program are to develop a national (INTA, industry, academia, and private sector) capability and to create an environment for state-of-the-art space age technology and research - to mobilize human resources. The emphasis is an affordable technology in the minisatellite and microsatellite segments, along with all the associated activities for all phases of successful space missions. [1686)] [1687)]

The first stage of the MINISAT program concluded with the development and qualification of the MINISAT-0 platform. Phase A of MINISAT-01 started in 1995. - The MINISAT-1 modular satellite family of the program is dedicated mainly to LEO EO mission applications. The third stage of the MINISAT program, the MINISAT-2 family, is an adaptation of the MINISAT-1 family for GEO mission applications. [1688)]

M.12.1 MINISAT-01

MINISAT-01 is an INTA satellite (platform built by CASA, payload built by INTA), comprised of a low-cost multipurpose bus with a payload consisting of three experiments and a technology demonstrator. The science objective of the mission is the study of background radiation in the extreme ultraviolet spectrum, low energy gamma radiation, and the behavior of liquid bridges in microgravity. The S/C consists of a service module (or bus) and a payload module (payloads up to 300 kg can be accommodated in the architecture). The satellite measures 1145 x 1005 x 1170 mm and has four deployable solar panels (each 60 W, size of

1685) B. Iannotta, "SWARM," Smithsonian Air & Space, August/September, 2000, pp. 44-49

1686) Paper provided by M. A. Garcia Primo of INTA

1687) M. A. Garcia Primo, "Spanish MINISAT Program - Objectives and Operational Results," Proceedings of the 4th International Symposium on Small Satellites Systems and Services, Sept. 14-18, 1998, Antibes Juan les Pins, France

1688) F. Cerezo Martinez, " MINISAT-01 (One Year After)," Proceedings of the 4th International Symposium on Small Satellites Systems and Services, Sept. 14-18, 1998, Antibes Juan les Pins, France

550 mm x 800 mm). The spacecraft average power requirement is 60W. Uplink and downlink communication is via S-band, 5W transmitter/receiver at 1 Mbit/s, the on-board data storage capability is 32 MByte. MINISAT-01 is a sun-pointing, momentum-bias (spin) stabilized satellite with a capability to convert to a 3-axis stabilized one by employing torque rod control (three orthogonal torque rods and one momentum/reaction wheel as actuators for attitude control). The pointing accuracy is ±3°. The attitude is sensed by two two-axis magnetometers and two coarse sun sensors with overlapping FOVs. The S/C bus has a mass of 105 kg, payload mass is 90 kg (total mass of 195 kg). The design life is 4-5 years.

The S/C was launched on a Pegasus-XL vehicle from the Canary Islands (Spain) on April 21, 1997. MINISAT-01 is operated by INTA and tracked via the Maspalomas ground station (15° 37' 45" W, 27° 45' 49" N).

Orbit: Circular orbit, altitude = 576 km, inclination =151° (29° retrograde), period =96 minutes.

Sensor complement:

EURD (Espectrógrafo Ultravioleta extremo para la observación de la Radiación Difusa - Extreme UV Spectrograph for the Study of Diffuse Radiation), developed by INTA and UCB (University of California at Berkeley). The objectives are to conduct spectrographic observations of the diffuse EUV region of radiation, to study the nature of the interstellar medium, the airglow in the upper atmosphere (atomic oxygen and oxygen lines), and to search for dark matter in the form of massive (about 10 eV), long lived (> 1024 s) neutrinos. The instrument employs two spectrometers. An important feature of these spectrometers is their ability to identify or eliminate systematic effects that have compromised previous efforts to detect diffuse UV radiation. Each spectrometer is equipped with a filter wheel at its entrance slit, which is stepped through four operating positions to evaluate internal background, glare, starlight, and true interstellar plasma emission. These positions include an open position transmitting all wavelength, a metal shield that blocks all radiation and provides a measurement of the internal background, a magnesium fluoride filter giving a measurement of the Lyman alpha radiation, and an aluminum filter that transmits most of the EUV radiation and strongly blocks any scattered Lyman alpha radiation. The EURD instrument is observing continuously during eclipse periods with the S/C pointing into the anti-sun direction.

Spectral band	300-1050 Å
FOV	26° x 8°
Grating	8 cm diameter, 18 cm focal length, holographically ruled 2460 lines/mm
Grating overcoat	long wavelength: silicon carbide; short wavelength: boron carbide
Detector	Multichannel Plate (MCP) with wedge and strip encoding
Detector photocathode	long wavelength: chemically treated; short wavelength: magnesium fluoride
Size (each spectrograph)	40 cm x 40 cm x 13 cm
Mass (each spectrograph)	11 kg

Table 492: Key instrument parameters of EURD

CPLM (Column of Liquid Bridge in Microgravity), developed at the Polytechnic University of Madrid. Objective: to study the behavior of axis-symmetric liquid bridges in microgravity conditions. Accelerations are induced by satellite maneuvers. The CPLM instrument consists of the following components mounted inside a cylindrical container: a test cell with a liquid bridge cell, optical detectors to measure the liquid bridge deformation, motor control to allow the liquid bridge formation and the modification of the liquid bridge configuration trough its length, an accelerometer unit to measure the accelerations supported by the liquid bridge, a power supply unit to deliver the appropriate voltages to the different subsystems, a temperature sensor to monitor the CPLM temperature, and pressure sensors to monitor the pressure in the test cell and outside of the test cell. The liquid bridge is oriented perpendicular to the z axis (sun to satellite direction) and also to the z spin rate direction. It means when the satellite is spinning from -0.375 to +0.375 rpm some axis-symmetric accelerations are induced in the liquid bridge. CPLM is operating about once per week (5 minute cycle).

LEGRI (Low Energy Gamma Ray Imager), developed by a Spanish/British team from the University of Valencia, INTA, CIEMAT, Birmingham University, and RAL (Rutherford Appleton Laboratory). The objective is to demonstrate the technological feasibility of a new generation gamma-ray telescope detector design optimized for low-energy radiation (in the energy range of 10 - 200 keV), using solid-state crystals as detector elements. The actual measuring objective is to detect the gamma-ray radiation emitted by various galactic sources such as neutron stars, binary systems, black hole candidates, and active galactic nuclei.
The initial LEGRI imaging array design employed 100 HgI_2 detectors, a new and emerging detector material, providing an excellent response in the 10-200 keV energy range, with the best presently achieved efficiency/volume ratio. In addition, the material works efficiently at room temperature and exhibits a very high resistance to radiation damage. CIEMAT (Centro de Investigaciones Energéticas y Medioambientales - Environmental and Energetic Research Center) has been working on this material, a production line for such detectors is under development.
The final imaging array design consists of 80 HgI_2 by 20 CdZnTe detectors. The intent is to compare the performances of both detector types, using the same FEE and working under the same background fluxes. - LEGRI is observing in the day part of the orbit mainly and in whatever direction perpendicular to the sun direction.

The LEGRI instrument consists of the following elements:

- A detector unit, comprising a detector plane formed by an array of 80 HgI2 and 20 CdZnTe elements, the associated front end electronics and a mechanical collimator, all within a mechanical assembly acting as a passive shield.
- A mask unit, located at 540 mm from the detector plane and parallel to it, consisting of a coded mask made by tungsten elements on a honeycomb plate and a support structure.
- A power supply and digital processing units, which act as direct interfaces to PLM for power and data transmission.
- A high voltage supply unit, which provides the high voltages needed to operate the detectors.
- A start sensor, which is used to determine precisely and continuously the attitude of the satellite, allowing for the reconstruction on ground of the gamma-ray images without spatial blurring caused by platform drift or jitter.

ETRV (Experiencia Tecnológica de un Regulador de Velocidad - Speed Regulator Technology Demonstrator), developed by CASA. ETRV is a mechanical speed regulator and deployment mechanism intended to deploy in the future such items as antennas, panels, and booms. The instrument consists of the following elements: a motor and torsion spring; gears to reduce the speed; a momentum wheel to simulate the deployable structure; a pyrotechnic nut to maintain the configuration from launch to the time to fire the pyro and to deploy the dummy mass (momentum wheel); a magnetic Reed switch to measure the angular movement of the momentum wheel and it rates; and associated electronics to manage all the signals. - For the technology demonstration, the speed regulator ETRV must deploy the momentum wheel at a constant rate and with a total deployment angle of 180° in 180 seconds.

M.13 MITA (Minisatellite Italiano di Technologia Avanzata)

The Minisatellite for Advanced Technology (MITA), developed by a team of small Italian companies (prime contractor: Carlo Gavazzi Space S.p.A. of Milan - with Contraves Italiana, FIAR Spazio, Laben, etc. as sub-contractors), is funded by the Italian Space Agency (ASI). The objective of the MITA satellite project is to design and implement technology (demonstrate the capabilities of the Italian aerospace industry) of a low-cost bus for a wide range of Earth mission applications.

The bus design is modular (composed of tubular elements and closing sandwich panels), based on a cubic-shaped structure in the 100 kg (plus) mass range of a minisatellite. The S/C is three-axis stabilized, it is intended in particular for LEO missions with a lifetime of 5 years (MITA-1 has an expected lifetime of three years due to its low orbital altitude). The bus employs dual-redundant data handling and telemetry/telecommunication subsystem chains (the other subsystems are partially redundant). The attitude control system measures attitude with the following sensors: two monoaxial horizon sensors, a triaxial magnetometer (redundant) and five coarse sun sensors. A momentum wheel and three magnetic coils are used as actuators. Electrical power is provided by two solar panels featuring GaAs solar cells. Each panel is able to produce 200 W (EOL). NiH_2 batteries serve as a secondary energy source during eclipse phases of the orbit with a capacity of 150 Wh. Power conditioning and distribution is performed by a power electronics unit. The thermal control system is passive.[1689) 1690) 1691)]

S/C dimensions	1800 mm x 1400 mm x 700 mm
S/C mass	170 kg
Average power consumption, peak power	80 W, 125 W
Attitude control	Three-axis stabilized, Earth pointing
Pointing accuracy	$\pm 1^{\circ}$ in each axis (knowledge)
S/C communications	S-band
Data rates: telemetry, telecommand; protocol	512 kbit/s, 4 kbit/s; CCSDS
Mass memory	64 MByte

Table 493: Specification of the major elements of the MITA-1 bus

Communication: The on-board data handling is based on transputer with DSP processor embedded. On-board storage capacity of 64 Gbit. The TM/TC protocol is CCSDS compatible. The downlink is in S-band with a frequency of 2.2 GHz, data rate: 1 Mbit/s, transmitter power: 1 W. The uplink is in S-band with a frequency of 2.1 GHz, data rate: 4 kbit/s.

MITA-1 was launched July 15, 2000 (along with CHAMP) aboard a Russian Cosmos-3M vehicle from Plesetsk, Russia.

Orbit: Circular orbit, altitude = 450 km, inclination = 87.275°.

Sensor complement:

NINA-2 (New Instrument for Nuclear Analysis). NINA-2 is of NINA heritage flown on RESURS-O1-4 (launch July 10, 1998).[1692)] NINA-2 is a compact telescope, developed by INFN (Italian National Institute of Nuclear Physics) of Rome, Italy and MEPhI (Moscow Engineering and Physics Institute). The objective is to measure fluxes of charged particles, in particular to detect cosmic ray nuclei of galactic, solar, or other origin from hydrogen to iron, between 10 and 200 MeV/n (contained particles) and 1 GeV/n (outside containment). The silicon detector telescope is composed of 16 X-Y planes, giving information on the energy of the crossing particle and its incident angle. Each of the 32 sensitive elements consists

1689) P. Sabatini, R. Aceti, et al., "MITA: In-Orbit Results of the Italian Small Platform and the first Earth Observation Mission, HYPSEO," Proceedings of the 3rd International Symposium of IAA, Berlin, Apr. 2-6, 2001, pp.71-74

1690) P. Sabatini, T. Lupi, "The MITA satellite: an Italian bus for small missions," IAA 2nd International Symposium on Small Satellites for Earth Observation, Berlin, April 12-16, 1999, pp. 35-37

1691) M. Casolino, et al., " Continuation of the mission NINA: Nina-2 experiment on MITA satellite," Proceedings of 26th ICRC, Salt Lake City, 1999, OG 4.2.17

1692) R. Sparvoli, et al., "Launch in orbit of the telescope NINA for cosmic ray observations: preliminary results," Proceedings of The Sixth Topical Seminar on 'Neutrino and Astro–Particle Physics,' Centro Studi 'I Cappuccini' in San Miniato al Todesco, Italy, May 17-21, 1999 of two n-type silicon detectors, 60 mm x 60 mm, divided in 16 strips and connected to a supporting ceramic frame under lateral strips (1 and 16). Each couple of detector is glued orthogonal in order to provide X and Y independent view information. The thickness of the detector is 150±15 μm for the first plane, and 380±15 μm for the remaining 15 planes. The geometric factor of the instrument ranges from 8.6 cm 2 sr for low energy particles to 1 cm 2 sr for particles crossing the detector. The instrument mass is 40 kg, power = 40 W.

MTS-AMOS (Micro Tech Sensor-Attitude and Orbit Measurement System). An ESA technology sensor for combined attitude and orbit measurements, consisting of three elements: an APS (Active Pixel Sensor) Camera with beamsplitter-optic, a Magnetic Field Sensor, and an Angular Rate Sensor.

M.14 MSX (Midcourse Space Experiment)

The MSX satellite is an observatory-class observation platform and global surveillance system of the US DoD; it is funded and managed by BMDO (Ballistic Missile Defense Organization) with APL (Applied Physics Laboratory of Johns Hopkins University, Baltimore, MD) as the S/C integrator. The primary objectives of MSX are to detect, acquire and track targets and to discriminate lethal from nonlethal objects (detailed characterization and modeling of target objects and their associated phenomenology of the terrestrial, Earth-limb, and celestial backgrounds).

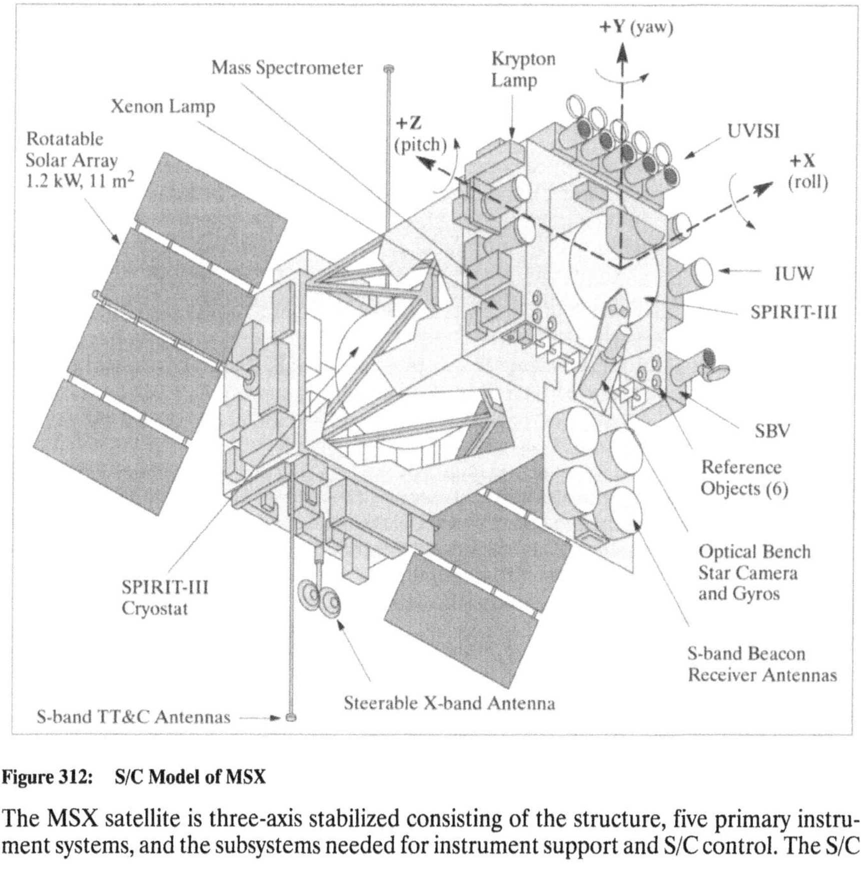

Figure 312: S/C Model of MSX

The MSX satellite is three-axis stabilized consisting of the structure, five primary instrument systems, and the subsystems needed for instrument support and S/C control. The S/C has a design life of five years, its body measures 150 x 150 x 510 cm, and its mass is about 2700 kg. Power: 1.2 kW from solar array, 28 VDC, 50 Ah NiH_2 battery. - The attitude control system consists of four reaction wheels and three magnetic torque rods. Any three of the four wheels can provide three-axis control of the satellite. Attitude sensors include two three-axis ring laser gyro systems, a star camera, two horizon sensors, five digital sun-angle detectors, and a three-axis magnetometer. The system achieves a real-time pointing accuracy of better than 0.1° and post-processing knowledge of 9 µrad over instrument integration durations of about 1s. - The S/C is operated by APL; its S-band is also monitored by AFSCN (Air Force Satellite Control Network). APL also receives the prime science data and provides processing and archiving for all data. MSX was launched April 24, 1996 on a Delta II vehicle from VAFB, CA.

The MSX communicates on three bands: L-band (1827.8 MHz), S-band (2282.5 MHz), and X-band (8475 MHz). The L-band is used exclusively for commanding and memory loads (data encryption). The S-band has two transponders, diplexers, and antenna pairs for the transmission of 16 kbit/s housekeeping data and either ranging or 1 Mbit/s transmission of compressed or sampled science data. The X-band (two transmitters) transmits 25 Mbit/s prime science data from the S/C data recorders (54 Gbit capacity each). The S-band beacon receiver is a passive radar tracker that can acquire a target with an initial pointing uncertainty of ±5° at a maximum range of 8000 km. It provides calibrated angle tracking with a residual error of 0.1°.[1693)]

Some of the sensor data (mostly concerning targets and satellites) will be classified, but the majority are unclassified. These latter data will eventually be released to the general science community after the normal period (~2 years) reserved for exclusive exploitation by the principal investigators.

Orbit: Nearly sun-synchronous orbit, altitude = 898 km, inclination = 99.16°.

Sensor complement:[1694)]
The suite of state-of-the-art sensor systems include: SPIRIT-III (a cryogenic scanning radiometer and FTIR spectrometer), the UVISI system, SBV, and a suite of instruments to monitor contamination on and around the S/C. MSX is a fully steerable S/C providing a range of pointing capabilities (0.1° pointing accuracy): limb scans, nadir scans, or limb point and stare, to cross-track scans. Of importance to the science community are stellar occultation and daytime limb observations for the retrieval of trace gases: ozone, aerosol, pressure, temperature and NO_2 in the stratosphere.

SPIRIT-III (Spatial Infrared Imaging Telescope), built by SDL (Space Dynamics Laboratory) of Utah State University. SPIRIT-III is the primary sensor system on MSX with the objective to perform midcourse surveillance functions, collect target and background phenomenology data, and to demonstrate advanced cryogenic sensor technologies. The instrument consists of an off-axis re-imaging telescope with a 35 cm diameter unobscured aperture, a six-channel Fourier-transform spectrometer, a five-band scanning radiometer, and a cryogenic dewar/heat exchanger (the dewar cools the telescope, radiometer and spectrometer, the cryogen is expected to last 18 months). Instrument mass = 967 kg; size = 107 cm diameter x 360 cm in length. Note: SPIRIT-I and -II were instruments (FTIR spectrometers only) flown on sounding rockets in the late eighties and in 1991.

The telescope has three sections: the afocal foreoptics, the radiometer re-imaging optics, and the spectrometer collimating optics. The foreoptics use an off-axis all-reflective design. An auto-collimator measures telescope alignment with respect to the S/C attitude system optical bench to an accuracy of 5 µrad. The Fourier-transform spectrometer has six Si:As

1693) John D. Mill, et al., "Midcourse Space Experiment: Introduction to the Spacecraft, Instruments, and Scientific Objectives," Journal of Spacecraft and Rockets, Vol. 31, No. 5, September-October 1994, pp. 900-907

1694) J. F. Carbary, E. H. Darlington, K. Heffernan, T. J. Harris, C. I. Meng, M. J. Mayr, P. J. McEvaddy, K. Peacock, "Aerial Surveillance Sensing Including Obscured and Underground Object Detection," Proceedings of SPIE, April 4-6, 1994, Orlando Florida, Volume 2217 detectors operating at 10.5 to 11.0 K. It collects double-sided interferograms in six spectral bands with programmable resolution of 2, 3.9, or 20 cm^{-1} over sample times of 4.2 s, 2.2 s, and 0.55 s. - The radiometer has five Si:As focal plane detector arrays of 8 x 192 pixels each, operating between 11 and 12 K. It collects data in six passbands with a spatial resolution of 90 μrad. The scan mirror can remain fixed or can operate at a constant scan rate of 0.46°/s with programmable scan fields of regard of 1° x 0.75°, 1° x 1.5°, or 1° x 3°. The radiometer focal plane assembly employs a combination of dichroic and bandpass filters to allow simultaneous measurements in bands A, D and E and or in bands B and C. Table 1363 lists the four operational modes and provides actual values for the radiometer field of regard (FOR). - Earthlimb observations of SPIRIT-III include an experiment to measure chlorofluorocarbons and nitric acid. Channel 5 (10.6-13 μm) of the FT spectrometer collects data in the altitude range from 10 to 40 km. Dominant emissions in this channel include nitric acid, CFC compounds, ozone, and the thermal signature of stratospheric aerosols.

FT Spectrometer Passbands		Radiometer Passbands			
Channel	Passband (μm)	Band	Passband FWHM (μm)	Active Columns (array size)	Sensitivity (NEFD) 10^{-18} W/cm^2
1	17.2 - 28.0	A	6.0 - 10.9	8	1.1
2	2.6 - 4.9	B_1	4.22 - 4.36	2	10
3	5.8 - 8.9	B_2	4.24 - 4.45		
4	4.0 - 28.0	C	11.1 - 13.2	4	0.8
5	10.6 - 13.0	D	13.5 - 16.0	4	0.7
6	2.5 - 24.0	E	18.1 - 26.0	2	1.7

Table 494: SPIRIT-III instrument parameters

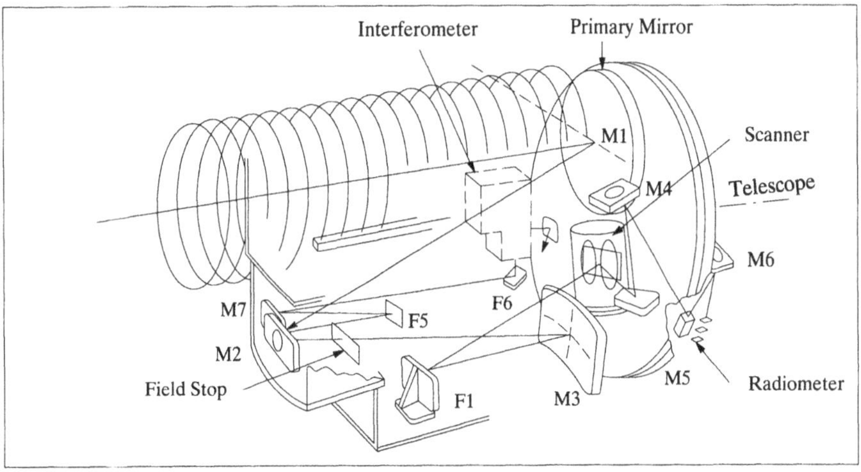

Figure 313: Schematic diagram of the SPIRIT-III instrument

Parameter/Function	Mirror-hold	Mirror-scan 1 (0.75°)	Mirror-scan 2 (1.5°)	Mirror-scan 3 (3.0°)
Scan rate	0.05868°/s (orbit drift rate)	0.46°/s	0.46°/s	0.46°/s
Scan period (one full, double scan cycle)	N/A	3.66 s	7.32 s	14.65 s
5-color scan FOR	drift position limited	0.426° x 0.99°	1.268° x 0.99°	2.953° x 0.99°
Color-set scan FOR	drift position limited	0.842° x 0.99°	1.685° x 0.99°	3.369° x 0.99°
Sampling rate	72/360	72/360	72/360	72/360

Table 495: Operational scan modes of the SPIRIT-III radiometer

UVISI (Ultraviolet/Visible Imaging and Spectrographic Imaging, a suite of nine instruments, designed by APL). The primary objective of UVISI is to collect data on celestial and atmospheric backgrounds; secondary objectives include target characterization and contamination observations in conjunction with the contamination instruments. The UVISI sensor system consists of five SPIM (Spectrographic Imager) instruments and four imagers. The imagers include two WFOV (Wide-Field-of-View) and two NFOV (Narrow-Field-of-View) sensors in the UV and VIS spectrum respectively. Together the SPIMs cover a spectral range from 110 nm (UV) to 900 nm (NIR). - The nine instruments share a common boresight with each other, with the SPIRIT-III IR radiometer, and with the SBV instrument.

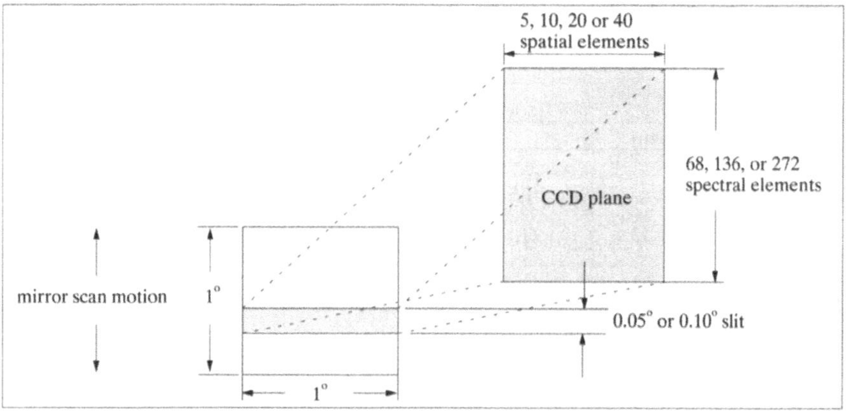

Figure 314: Schematic of spectrographic imaging of UVISI SPIMs

Instrument6	Passband Range (nm)	Resolution Δλ (0.10° slit)	Resolution Δλ (0.05° slit)	No. of Bins
SPIM1	113-173	0.8 nm	0.5 nm	68, 136, or 272
SPIM2	162-252	1.2 nm	0.9 nm	68, 136, or 272
SPIM3	251-388	1.8 nm	1.5 nm	68, 136, or 272
SPIM4	377-582	2.8 nm	2.1 nm	68, 136, or 272
SPIM5	589-902	4.3 nm	2.9 nm	68, 136, or 272

Table 496: UVISI spectrographic-imager (SPIM) characteristics

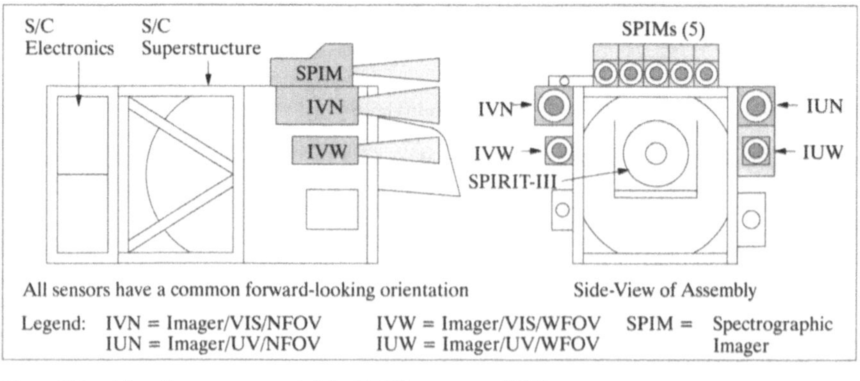

Figure 315: Mounting arrangement of the UVISI sensors on MSX

UVISI SPIMs. All five SPIMs feature an all-reflective off-axis parabolic design (external sunshade, scanning mirror, slit/filter, collimating mirror, dispersive grating, and an intensified CCD focal plane) in which selectable slits provide spectral resolutions between 0.5 nm to 4.3 nm. The SPIM image planes have programmable spectral dimensions with 68, 136, or 272 pixels and programmable spatial dimensions with 5, 10, 20, or 40 pixels.[1695] A scan mirror sweeps the slit through a second spatial dimension and generates a spectrographic image once every 5, 10, or 20 seconds. The SPIM CCD detector plane collects an image of the slit length in one direction and spectral information in the other (Figure 134). One spatial dimension is resolved along the slit and the other spatial dimension information is obtained by moving the FOV either 0.05° or 0.1° before recording another observation. It takes about 10 s to complete a 1° x 1° image in either 10 or 20 steps. The five-position slit/filter mechanism provides two slit sizes (1.0° x 0.10° and 1.0° x 0.05°) and various blocking filters that eliminate extraneous spectral orders and long-wavelength contaminants.

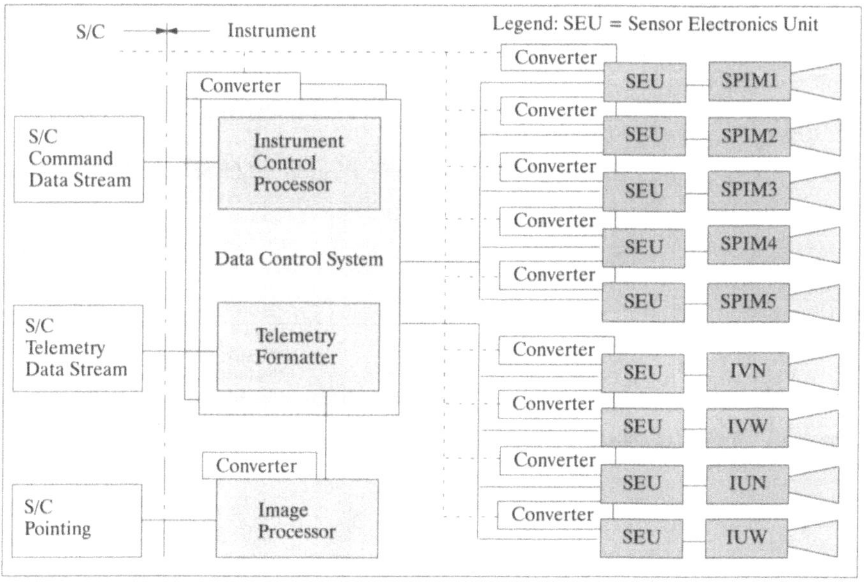

Figure 316: UVISI instruments and major subsystems/interconnections

SPIM electronics perform pixel-summing operations (programmable) on the 40 x 272 element SPIM focal plane array, yielding 40, 20, 10, or 5 bins in the spatial direction, and 272, 136, or 68 bins in the wavelength direction.[1696] The pixel resolution is 14 bit. - With a dynamic range of 10^{11}, UVISI has the capability to obtain data ranging from noontime surface-reflected visible radiation to observations of a diffuse UV celestial background. Objective and applications: generation of spatial/spectral maps of various objects and scenes; prime SPIM observations include the atmospheric dayglow and nightglow, aurora, stars,

1695) Note: The spatial resolution of the SPIMs is driven by the point-spread function in one direction (along the slit) and by the point-spread function <u>and</u> the mirror step size in the other direction. For the 0.05° mirror steps one can assume that it is driven by the point-spread function in both directions, and is about 0.85 mrad. The spatial resolution is diminished by using the 0.1° steps or by reducing the number of bins in the readout, by co-adding 2, 4, or 8 adjacent pixels. This is to reduce the bandwidth requirement by trading spatial resolution, spectral resolution and frame rate. The nadir resolution is 0.85 mrad x 900 km ≈ 770 m. Nadir FOV is 17 mrad (1°) x 900 km ≈ 15 km x 15 km.

1696) Note: The bins are formed in the SPIM electronics by co-adding 1, 2, or 4 adjacent pixels; this is done to reduce the data bandwidth requirement in cases where UVISI is not the principal instrument, or higher frame rates are needed which can be traded off against resolution. For the case of 136 and 272 bins, the bins overlap; for the case of 68 bins, the bins are noncontiguous.

zodiacal light, and plume contrails. When suitably inverted, SPIM radiance measurements can reveal atmospheric properties such as species concentrations, temperatures, and altitude profiles. SPIM investigations of auroral radiances can provide estimates of fluxes and energies of precipitating particles that cause auroral emissions.

UVISI Imagers. The four imager suite consists of: external sunshade, imaging optics, filter wheel and drive motor, and intensified CCD focal planes. Three of the four imagers employ all-reflective optics. Both NFOV imagers use a Cassegrain design; the UV WFOV imager uses an off-axis three-mirror design. - The commandable filter wheel in each imager houses three bandpass filters on a neutral-density (ND) filter. In case of the visible WFOV imager, a lens is substituted for one of the filters. This lens is used to focus the near-field backscattered emissions from the xenon flashlamp as part of the MSX contamination experiment. The filter bands were chosen for specific objectives. For example, the UV-NFOV 200-230 nm filter is used to observe the NO_y bands in airglow measurements. The UV-WFOV 117-127 nm filter is used for viewing Lyman-α emission from hydrogen in the geo-corona, from auroral hydrogen precipitation, or from outgassing of the SPIRIT-III cryogen. Each imager has a focal plane CCD array of 256 x 244 pixels with 12 bit resolution. The imager nadir resolutions are:

- NFOV: 0.09 mrad x 900 km $\approx$ 80 m
- WFOV: 0.82 mrad x 900 km $\approx$ 750 m

The limb resolutions can be calculated from geometry, given a tangent height.

Instrument	UV-NFOV (IUN)	UV-WFOV (IUW)	VIS-NFOV (IVN)	VIS-WFOV (IVW)
FOV	$1.28° \times 1.59°$	$10.5° \times 13.1°$	$1.28° \times 1.59°$	$10.5° \times 13.1°$
Resolution (IFOV)	90 µrad	820 µrad	90 µrad	820 µrad
Passband (nm)	180-300	110-180	300-900	380-900
ND filter WB1 filter WB2 filter WB3 filter	10^{-3} 200-230 230-260 260-300	10^{-3} 117-127 128-138 145-180	10^{-4} 305-315 350-440 470-640	10^{-4} 426-429 529-631 (lens)

Table 497: UVISI imager characteristics

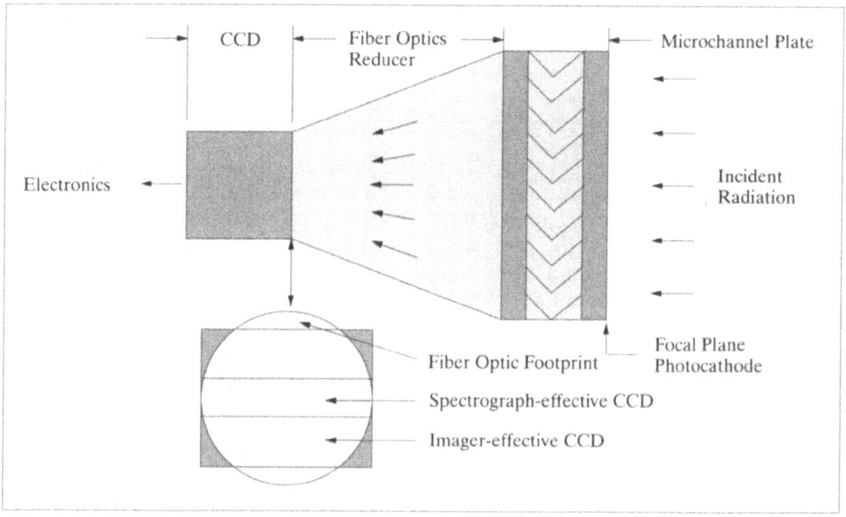

Figure 317: Schematic diagram of the UVISI focal plane unit

The UVISI image processing system can acquire and isolate likely targets in an image FOV and communicate their positions to the MSX flight processor, which in turn points the satellite in a "closed-loop" fashion. The term "target" may refer to either a point source such as a star, to another satellite, or to an extended source such as an auroral or cloud-top feature. The processor software performs initialization and tracking functions that include filtering, smoothing, thresholding, and centroiding. The image processor seeks targets based on an a priori target description file containing weights for numerous target features such as size, brightness, and location. The image processor then transforms target locations from UVISI pixel coordinates to S/C coordinates and passes them on the MSX flight processor, which performs Kalman filtering of other targeting inputs to select "true" targets for observation.

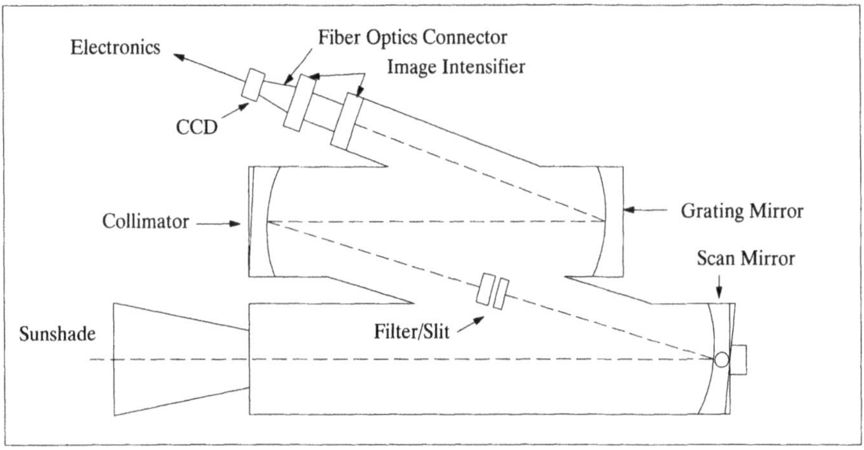

Figure 318: Optical diagram of the UVISI spectral imagers (SPIMs)

Spectral range	0.3 - 0.9 μm
Spatial resolution (IFOV)	60 μrad
FOV	1.4° x 6.6°
Aperture, focal ratio	15 cm, f/3
Focal Plane Array (FPA, four CCDs)	420 x 1680 pixels
Frame times	0.4 s, 0.5 s, 0.625 s, 1.0 s, 1.6 s, 3.125 s

Table 498: SBV camera parameters

SBV (Space Based Visible, designed by MIT Lincoln Lab) camera. The objective is to collect data on celestial target signatures (stars) in the VIS range and to perform surveillance demonstrations (cataloging of resident space objects). The instrument incorporates a 15 cm aperture off-axis, re-imaging, all-reflective telescope, a thermo-electrically cooled (front-illuminated) bare CCD focal plane, and electronic subsystems. The focal plane consists of four CCDs with 420 x 420 (27 μm) pixels each. SBV instrument mass = 78 kg, power = 68 W.

MSX carries in addition a suite of contamination instruments with the objective to characterize the molecular and particulate environment around the S/C (all APL sensors). This suite includes the following instruments:

- **NMS** (Neutral Mass Spectrometer). A quadrupole RF instrument with two independently programmable filaments which can be operated at either high or low emission currents. It measures species ranging in mass from 1-150 amu (atomic mass units).
- **IMS** (Ion Mass Spectrometer). A Bennett RF positive ion mass analyzer with sampling port oriented towards the ram velocity vector (+X axis) during the park mode of the MSX S/C. It measures ions ranging in mass from 1-60 amu. Gas densities as low as 10^6 molecules/cm^3 and 10^3 ions/cm^3 can be measured with NMS and IMS, respectively.

- **CQCM** (Cryogenically-cooled Quartz Crystal Microbalance). Five sensors for measuring depositions on the SPIRIT-III primary mirror.
- **KF/KR** (Krypton Flashlamp/Radiometer). KF/KR works in conjunction with the UVISI SPIM3 to measure water vapor concentrations.
- **TQCM** (Temperature-controlled Quartz Crystal Microbalance). Four sensors are distributed about the MSX instrument section with the objective to view the mass flux arriving from a particular direction of interest.
- **TPS** (Total Pressure Sensor). Measurement of ambient pressures ranging from 1×10^{-10} to 1×10^{-5} Torr.

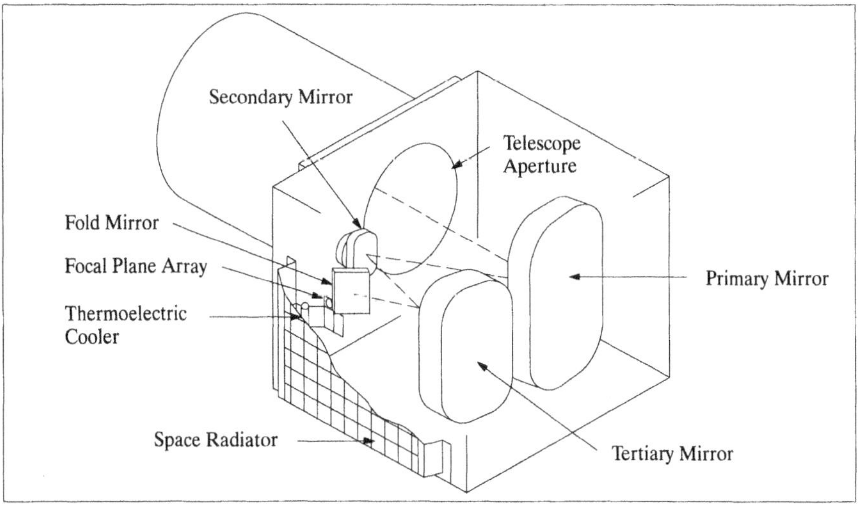

Figure 319: SBV telescope schematic

- **XF** (Xenon Flashlamp). Produces an intense VIS photon beam for detection of backscattered radiation by the UVISI wide field visible imager (IVW). The photon beam is pulsed at 1/2 s, the beam intersects the imager line-of-sight at 200 cm. The imager collects a series of frames of the backscattered emissions, from which the size and velocity of particles as small as 0.5 µm can be detected.

M.15 Myriade (CNES Microsatellite Program)

At the start of the 21 century the Myriade program represents the new microsatellite concept of CNES (France), taking advantage of low-cost satellites for technology demonstrations and to serve specific needs of the scientific community. The goal is to provide missions of about 10 MEuro (launch included) at a rate of two missions/year. In general, there are simply more frequent launch opportunities available for microsatellites as secondary payloads on a number of launch vehicles. The objectives are to take advantage of these new enabling capabilities (innovation and introduction of technologies, operations, management methods, building partnerships, etc.). Myriade offers the opportunity for a rearrangement of agency/industry functions such as to relieve CNES of the recurrent provisioning and integration tasks. It is also considered a structure for international cooperation (training for prime contractorship). [1697) 1698)]

1697) Information provided by Bernard Tatry of CNES

1698) J. P. Aguttes, "High Resolution (metric) SAR Microsatellite Based on the CNES Myriade bus," Proceedings of IGARSS-2001, July 9-13, 2001, Sydney Australia

M.15.1 DEMETER

DEMETER (Detection of Electromagnetic Emissions transmitted from Earthquake Regions) is an approved French microsatellite mission (first mission of the CNES microsatellite program referred to as MYRIADE) with the objective to observe geophysical parameters of the terrestrial environment. [1699] This involves the study of ionospheric perturbations (measurement of electromagnetic waves and their effects), caused by natural phenomena, such as Earth tremors and volcano eruptions. Of particular interest is the time behavior (occurrence) of these electromagnetic perturbations in relation to Earthquakes. The hypothesis is that they might offer an early warning function prior to eruptions. The participants in the DEMETER project are: CNES, LPCE (Laboratoire de Physique et de Chimie de l'Environnement d'Orléans), CETP (Centre d'etude des Environnements Terrestre et Planetaire) in Velizy/ Saint-Maur, IPG-Paris (Institut de Physique du Globe de Paris) and CESR (Centre d'Etude Spatiale des Rayonnements) in Toulouse. LPCE is responsible for the science payload.

Figure 320: Illustration of Picard (left) and DEMETER (right) satellites

The satellite is of a box-like structure with dimensions of 60 cm x 60 cm x 80 cm. A single solar panel is used for power generation. The S/C is three-axis stabilized with a pointing accuracy of a few degrees. Attitude sensing is provided by a star sensor, attitude actuation is provided by reaction wheels and magnetic actuators. The S/C mass of DEMETER is about 120 kg, the design life is two years.

A launch of DEMETER is planned for mid 2003 as a secondary payload on a Dnepr launcher.

The payload recorder has a capacity of 8 Gbit. RF communications are provided in X-band (frequency = 8.2 GHz) with a downlink data rate of up to 17 Mbit/s. The data modulation scheme is NRZ-L/MDP8-2/3. The CCSDS communication protocol is used.

Orbit: Sun-synchronous orbit, altitude = 700 km, inclination = 98°.

Sensor complement:

IMSC (Instrument Magnetometre Search Coil), or Search Coil Magnetometer of CNRS/ LPCE. The objective is to measure the magnetic field components in the frequency range of 10 Hz to 18 kHz. The magnetic vector is measured by the current induced by a coil that is wound around a micro metal bar and by an orthogonal three-antenna system, mounted on a boom at 1 m distance.

IPD (Instrument Detecteur de Plasma), or Plasma Detector Instrument of CNRS/CESR. The objective is to measure the energy spectrum of electrons, at right angles to the magnetic

1699) DEMETER brochure of CNES, provided by Bernard Tatry

field, in the energy range of 30 keV to 2 MeV. The information is needed to estimate the disturbances in the radiation belts induced by earthquakes.

ISL (Instrument Sonde de Langmuir), or Langmuir Probe Instrument of ESA/ESTEC. The instrument is used to measure the total plasma density in the range of 100 to 5×10^6 particles/cm^3, the electron temperature in the 500-3000 K range, and the satellite potential in the range of ± 5 V. The instrument consists of two Langmuir probes, one is cylindrical the other is spherical with a 6 cm diameter segmented probe. The plasma density and temperature are determined from the Langmuir probe current-voltage curve.

ICE (Instrument Champ Electrique), or Electric Field Instrument of CNRS/CETP. The objective is the measurement of the vector components of the Earth's electric field in the frequency range from DC to 3 MHz. EFI consists of an antenna system featuring four spherical electrodes placed at the end of 4 m booms.

IAP (Instrument Analyseur de Plasma), or Plasma Analyzer Instrument of CNRS/CETP. The objective is the measurement of ionospheric plasma parameters (density, temperature, and energy distribution). IAP is a two-analyzer instrument.

M.15.2 Microscope

Microscope is an approved French gravity-research microsatellite mission, put forward by ONERA and CERGA (Centre d'Etudes et des Recherches en Geodynamique et Astrometrie) in Grasse, France. The objective is to conduct a fundamental physics experiment, namely to test the first law of thermodynamics (underlying the general theory of relativity). This concept is also known as the EP (Equivalence Principle) test. The best accuracy value of the EP achieved by experiments on Earth is 10^{-13}, the Microscope mission wants to improve this value to 10^{-15}.. Leading-edge technologies are implemented, these include: ultra-sensitive accelerometers, ionic thrusters, and simultaneous spin control and drag-free systems. [1700)]

The CNES microsatellite bus consists of a box-like honeycomb-and-plate structure of approximate size: 60 cm x 60 cm x 80 cm. The solar panels (AsGa cells) are body-mounted on three sides of the S/C, providing a power of 100 W (equally shared between the payload and the electric propulsion system). Particular attention is paid to the thermo-elastic behavior of the S/C as well as to possible magnetic disturbances to the test masses. The satellite is spin-stabilized with the spin axis directed normal to the orbit plane. Attitude sensing is provided by a sun sensor and a star tracker. Attitude actuation and drag compensation is provided by DAOCS as described below. In addition, actuation may be provided by magnetorquers and reaction wheels (when needed like in the orbit injection phase). The S/C mass is < 120 kg, the mission design life is one year.

A launch of Microscope is planned for late 2004 on a Dnepr launch vehicle as a secondary payload.

The on-board solid-state recorder has a capacity of 1 Gbit (the total source data rate, including housekeeping data, is about 1 kbit/s). RF communications are provided in S-band with a downlink data rate of 400 kbit/s (max). Satellite operations are conducted at CNES. The science mission center is at ONERA.

Orbit: Sun-synchronous quasi-circular orbit, altitude = 700 km, inclination = 98°.

Sensor/payload complement:

ONERA is the designer/builder of the two similar differential payload accelerometers. These accelerometers are of STAR and SuperSTAR (Super Space Three-axis Accelerome-

1700) P. Touboul, B. Foulon, L. Lafargue, G. Metris, "The Microscope Mission," Proceedings of the IAF Congress, Rio de Janeiro, Brazil, Oct. 2-6, 2000, IAF-00-J1.06

ter for Research mission) heritage flown on the CHAMP and the GRACE missions, respectively. They comprise two concentric electrostatic differential accelerometers. The test mass of each accelerometer is maintained along the three orthogonal axes at the center of the fused silica instrument cage by electrostatic forces. Electrodes, engraved in the cage wall, are used for the capacitive sensing of the mass position and attitude. Both test masses are controlled with respect to the cage frame: the sum of the forces is maintained to a null value by the satellite drag compensation system; the thrusters of the electric propulsion system act to move the instrument silica frame following the masses. The difference of the electrostatic forces is then observed in the orbital path for the search of the EP-violating signal. The electrode configuration allows six-degree-of-freedom measurements and control of each test mass.

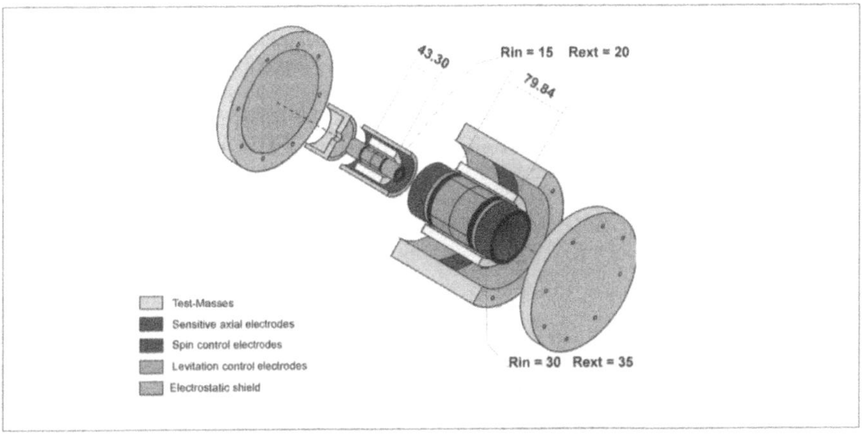

Figure 321: Schematic illustration of the electrostatic differential accelerometer

Both differential accelerometer cores are integrated in tight vacuum housings with thermal insulation and magnetic shielding provided. The instrument temperature variation is $< 1^\circ$ C per orbit. The entire instrument is positioned at the center of mass of the spacecraft to reduce any torque demand onto the electric propulsion system. The sensitive axes of the accelerometer are oriented in the orbital plane of the S/C x-axis, with the center of the test masses on the rotating axis of the S/C (normal to the orbital plane). The total instrument mass is 25 kg, the power is 25 W.

Operations of the accelerometers: After all initial tests and calibrations, the EP experiment is realized with the first differential accelerometer in inertial and spinning attitudes and with two angular phases along the orbit. The test is concluded with a calibration. The same procedure is repeated with the second differential accelerometer.

Rotational axis of S/C	About y and z or along y, z		About x or along x	
	Max value at DC	Stability at f_{ep}	Max value at DC	Stability at f_{ep}
Angular velocity	10^{-5} rad/s	10^{-5} rad/s $Hz^{-1/2}$	10^{-5} rad/s or $2x10^{-3}$ rad/s (spin)	10^{-5} rad/s $Hz^{-1/2}$
Angular acceleration	10^{-5} rad s^{-2}	$3x10^{-8}$ rad s^{-2} $Hz^{-1/2}$	10^{-5} rad s^{-2}	10^{-7} rad s^{-2} $Hz^{-1/2}$
Linear acceleration	$3x10^{-9}$ ms^{-2}	10^{-9} ms^{-2} $Hz^{-1/2}$	$3x10^{-9}$ ms^{-2}	10^{-9} ms^{-2} $Hz^{-1/2}$

Table 499: GG velocity and acceleration specifications

DAOCS (Drag, Attitude and Orbit Control System). A FEEP (Field Emission Electric Propulsion)-type system with 8 or 12 thrusters. Four pods of two or three thrusters are installed on the S/C corners of two opposite faces. One of the pods contains the star tracker. Each thruster provides a thrust range of 15-100 μN controllable in steps of 1 μN. The power consumption for each one of the four pods is about 8 W (mean value on the orbit corresponding to a mean thrust of 15 μN on the orbit). The mass of each pod is around 3 kg (thruster, electronics and propellant).

M.15.3 PARASOL

PARASOL (Polarization and Anisotropy of Reflectances for Atmospheric Science coupled with Observations from a Lidar) is a French science mission, put forward by CNRS/LOA [Laboratoire d'Optique Atmosphérique (University of Lille, France)], with the general objective to study the Earth's atmosphere. The PARASOL mission flies the POLDER instrument as its main payload with the objective to improve the microphysical and radiative property characterization of clouds and aerosols for model improvement. Clouds and aerosols play an important role in direct and indirect radiative forcing of the Earth's energy budget. The study of the dynamic processes and interactions between liquid and solid particles in the atmosphere is therefore a major goal. PARASOL (POLDER) complements PICASSO-CENA (Lidar) and NASA's Aqua and Aura (MODIS, CERES) missions, it also makes use of the CloudSat (CPR) radar data. The instruments form a so-called "train of satellites," all in the same orbits and in a PM sequence (a loosely-coupled constellation), provide a unique opportunity of near-simultaneous observations, by collecting data in different spectral domains (optical and microwave regions) and at various FOVs. [1701] [1702]

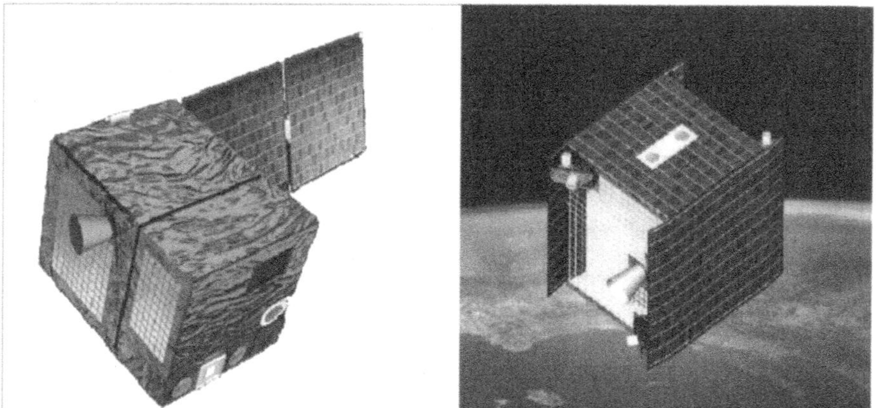

Figure 322: Illustration of the PARASOL (left) and Microscope (right) satellites

The PARASOL microsatellite of CNES consists of a box-like structure (size: 60 cm x 60 cm x 80 cm) with deployable solar panels. The S/C is three-axis stabilized, providing a pointing accuracy (knowledge) of 0.02°. Attitude sensing is provided by a star sensor, actuation is provided by reaction wheels. The solar power is about 80 W (average orbit value). The S/C mass is 120 kg, the mission design life is one year with two years as goal. A launch of PARASOL is planned as a secondary payload on Ariane-5 in 2004.

A payload mass memory provides a capacity of 16 Gbit. RF communications are provided in S-band (frequency = 2.2 GHz) at a downlink data rate of 400 kbit/s and an uplink rate of 20 kbit/s. In addition, there is an X-band downlink for payload data. The modulation scheme is

1701) http://www–projet.cst.cnes.fr:8060/PARASOL/index.html

1702) Information provided by Bernard Tatry of CNES

QPSK. The communication protocol is in CCSDS standard. Mission operations are performed at CNES (MYRIADE Control Center).

Orbit: Sun-synchronous circular orbit, altitude = 700 km, inclination = 98.21°, the local equator crossing time is at 13:30.

Sensor complement:

POLDER (Polarization and Directionality of the Earth's Reflectances), a CNES instrument of ADEOS and ADEOS-II heritage. The objective is the measurement of bidirectional and polarized solar radiation reflected by the atmosphere, and aerosols. POLDER consists basically of a digital staring camera composed of a 284 x 364 pixels CCD detection array, wide-field-of-view telecentric optics (± 51° cross-track and ± 43° along-track) and of a rotating wheel carrying spectral and polarized filters.

POLDER has nine spectral channels, three of which are implemented with polarized filters (total of 15 channels, three channels are needed for each polarized band). The spectral region is from 443 to 1020 nm. For the polarized wavelengths, 3 filters measure the linear polarization of the incoming radiation in three directions separated by 120°. The observed data permit the derivation of the Stokes parameters and the total irradiance. The acquisition of an image sequence takes 20 seconds, enabling the observation of one natural target from different directions.

A scene (image) has a size of about 1440 km (along-track) x 2200 km (cross-track) with a ground spatial resolution of 7 km x 6 km at nadir. POLDER has a mass of about 30 kg, the power consumption = 77 W in imaging mode (mean power of 29 W). The data rate is 880 kbit/s at 12 bit quantization.

M.15.4 Picard

Picard is a CNES solar-terrestrial microsatellite mission with French multi-institutional and international cooperation. [1703] The overall objective is to monitor the solar diameter, the differential rotation and the solar constant and to study the long-term nature of their relations. The mission is named after Jean Picard (1620-1682), a French astronomer, who measured the solar diameter, observed sunspots, and determined the rotational velocity of the sun. Picard was also the first who accurately measured the length of a degree of a meridian (longitude line) and from that estimated the size of the Earth.

The Picard microsatellite consists of a box-like structure (60 cm x 60 cm x 80 cm) with a single solar panel. The platform is three-axis stabilized with a pointing accuracy of 0.01° and a pointing stability of 0.01°. Attitude sensing is provided by a star sensor and/or by the payload telescope, actuation is provided by a reaction wheels. Solar power of 80 W (min) and batteries of 9 Ah provide power all along the orbit (including the eclipse phases). The platform mass is 65 kg, the S/C mass is about 110 kg. The mission design life is two years with a goal of six years. A launch of Picard is planned for 2006 on a Russian launcher.

The on-board recorder has a capacity of 1 Gbit. RF communications are provided in S-band. The downlink data rate is 400 kbit/s. Mission operations are preformed at CNES.

Orbit: Sun-synchronous orbit, altitude = 700 km, inclination = 98°.

Sensor complement:

SODISM (Solar Diameter Imager and Surface Mapper), a CNRS/SA (Verrières-le-Buisson, France) instrument in collaboration with ESTEC. The objective is to observe the full image of the solar disk (limb). The instrument consists of a telescope with an aperture diam-

1703) L. Dame, M. Meissonnier, B. Tatry, "Picard Microsatellite Program," 5th International Symposium on Small Satellites Systems and Services, La Baule, France, June 19-23, 2000 eter of 110 mm. SODISM measures the solar radiation in four spectral bands and two calibration bands (flat field and star field), using two cascading filter wheels (each with five positions). Nominal mode observations are done in the UV band (230 nm), corresponding to a mostly flat UV continuum formed in the photosphere. The visible band (548 nm) is used for the solar diameter measurement. The 160 nm band and the Lyman alpha filters are used for identification of active regions and prominences. Diffusion plates are used to monitor the CCD response and sensitivity (flat field). The detector array employs thinned CCD arrays in a back-illuminated mode of operation to increase the short-wavelength quantum efficiency. The "star field" band is used to image stars. The measurements provide a reference used to calibrate the structural changes in the focus or CCD dimensions which could affect the quality of the sun diameter measure. The instrument design takes great care of component mechanical and thermal stability (materials, structure, alignment, etc). The instrument mass is 17.5 kg.

Measurement parameters	Solar diameter, differential rotation of the full sun
Spectral bands	6 bands: 230, 548, 160 nm, Lyman alpha, flat field & star field
Bandwidth	8 nm for 230 band, 100 nm for 548 band,
Telescope focal length, solar image	2650 mm, 25 mm diameter
Telescope optics	Primary diameter of 120 mm (110 mm are used) Secondary diameter of 34 mm (25 mm are used)
Detector array (EEV-4280 back-thinned CCD)	2048 x 4096 square pixels of 13.5 µm (frame transfer mode, 2048 x 2048 pixels of the array are used)
Guider acquisition range, Guider nominal pointing range Guider servo bandwidths	1.2° ±30 arcsec 0 to a few Hz (platform), <50 Hz (prime mirror fine guiding)
Quad-cell image displacement	$<10^{-2\pi}$ sensitivity
Piezo displacement range	±6 µm (±1 arcmin)
Absolute solar shape precision	Better than 10 marcsec (milli-arcseconds)
Relative semi-diameter precision	Better than 1 marcsec

Table 500: SODISM instrument parameters

SOVAP (Solar Constant Variability, Picard) of IRMB (Institut Royal de Météorologie Belgique) or Royal Meteorological Institute of Belgium in Brussels. SOVAP is of SOVA and SOLCON heritage, flown on eight Shuttle flights (EURECA, 1992-93) in the time frame 1983 to 1998 (see Table 38). SOVAP is a differential absolute solar radiometer with the objective to measure the solar constant. The SOVAP radiometric core is formed by two blackened cavities mounted side-by-side on a common heat sink. A heat flux transducer is mounted between each cavity and the heat sink. The difference between the ouputs of the two transducers is the differential heat, in which the common part of the thermal surrounding, seen by the two cavities, is eliminated. The instrument mass is 5.8 kg.

Measured parameter, instrument noise	Total irradiance ($W\ m^{-2}$), $<0.1\ Wm^{-2}$
Number of channels, number of reference voltages	2, 6
Cavity type	Cylindrical, diffuse black
Diameter of precision aperture, slope angle	1 cm, 2.5°
Solar sampling period, duty cycle	3 minutes, 50%

Table 501: SOVAP instrument parameters

Measured parameter (center wavelengths)	Spectral irradiance ($W\ m^{-2}$) at 230, 311, 402 and 548 nm
Bandwidth (FWHM)	8 nm (the 402 nm band has a bandwidth of 5 nm)
Number of photometers	12 (3 sets of 4)
Cavity type	Cylindrical, diffuse black
FOV (Field of View), slope angle	2.5°, 0.7°
Diameter of precision aperture	3 mm
Accuracy of aperture area, cross talk	$<10^{-3}$, $<10^{-5}$

Table 502: PREMOS instrument parameters

PREMOS (Precision Monitoring of Solar variability) of WRC (World Radiation Center) of Davos, Switzerland. The instrument package consists of four "filter radiometers," the absolute cavity measurement technique is based on the principle of substitution of electrical power for radiative power, but with the preselection of a known and reduced spectral bandwidth. The sun radiation measurements are in four bands in the UV and visible range; two bands (230, and 548 nm) are identical with those of SODISM. Three sets of four photometers are used, one set serves as reference to determine the aging effects. The 230 nm band has a dual function: 1) it monitors the UV flux and 2) it is used to estimate possible degradations of the SODISM CCD detectors. The instrument mass is 4.1 kg.

M.16 NEMO (Navy EarthMap Observer)

The NEMO program (started in 1997) is a cooperative US government/industry satellite program, sponsored by ONR (Office of Naval Research) and DARPA as well as commercial investments, between the Navy's NRL (Naval Research Laboratory) on the government side as the program manager in partnership with commercial developers, a consortium led by STDC (Space Technology Development Corporation) of Alexandria, VA. In 1999, STDC was acquired by Earth Search Sciences of McCall, Idaho. [1704] [1705] [1706] [1707]

The overall objective is the development/demonstration of hyperspectral technology and the collection/processing of moderate-resolution hyperspectral imagery for military and commercial use. The emphasis is on broad-area, synoptic, and unclassified hyperspectral imagery. The prime military interest is to obtain imagery of the coastal regions on a global scale (littoral modeling, shallow water bathymetry, topography, bottom type composition, detection of underwater hazards, water clarity and visibility, etc.), while the civil needs call for imagery supporting such applications as land use management, agriculture, environmental studies, and mineral exploration. Contract agreements call for STDC and its industry partners to provide the commercial satellite bus, selected flight avionics components, launch services, and long-term flight operations. NRL provides the design and integration of the NEMO sensor imaging payload with the commercial satellite bus, bus modifications, the on-board processor, ORASIS (Optical Real-time Adaptive Signature Identification System), and systems engineering. NRL has responsibility for calibration software and calibrated radiances; development of at-launch algorithms including those for atmospheric correction; on-board data compression software and related software for the ground data system; some data processing; and data archiving. The planned products of at-launch algorithms are water clarity (probably diffuse attenuation coefficient at 490 nm), concentration of chlorophyll, absorption coefficient of colored dissolved organic matter, concentration of suspended sediments, bathymetry, and bottom characteristics.

The satellite consists of eight hardware subsystems and a software subsystem that collectively accommodate the payload and meet the mission and science requirements (Table 220). The satellite platform selected is a commercial bus (LS-400) from SS/L. [1708] [1709] The satellite is three-axis stabilized and consists of a trapezoidal main body and two deployable solar arrays. STDC has responsibility for ground data processing system architecture, some data processing and merging, data archiving, and a web-based database interface. The S/C design life is five years (minimum of three year on-orbit mission life).

1704) T. Wilson, C. Davis, "Naval EarthMap Observer (NEMO) Satellite," Proceedings of SPIE, Vol. 3753, Denver, CO, July 19-21, 1999, pp. 2-11

1705) Note: In Dec. 1999, STDC was acquired by Earth Search Sciences Inc. of Alexandria, VA

1706) http://nemo.nrl.navy.mil/public/index.html

1707) C. O. Davis, K. Carder, "Requirements Driven Design of an Imaging Spectrometer System for Characterization of the Coastal Environment," Proceedings of SPIE, Vol. 3118, San Diego, CA, 1997

1708) C. O. Davis, "The Hyperspectral Remote Sensing Technology (HRST) Program," NRL White Paper, 1997

1709) C. O. Davis, K. Carder, "Requirements Driven Design of an Imaging Spectrometer System for Characterization of the Coastal Environment," Proceedings of SPIE, Vol. 3118, San Diego, CA, 1997

Subsystem	Short Description
ADCS (Attitude Determination and Control Subsystem)	3-axis stabilized for all modes; 0.07° control Geolocation: 30 m with a Circular Error of Probability (CEP) of 0.9 2 star trackers and 1 inertial reference unit for attitude determination 4 reaction wheels and 2 electromagnetic torquers for attitude control
EPS (Electrical Power Subsystem)	2 gimbaled (single axis) solar array panels 4 for 3 redundant power control box 64 Ah nickel hydrogen battery Approximately 1500 W power
RCS (Reaction Control Subsystem)	Monopropellant hydrazine blowdown system 1 propellant tank (76.5 kg capability) 5 x 1N thrusters for orbit correction/orbit maintenance
TCS (Thermal Control Subsystem)	Passive thermal control with heater augmentation Battery and payload panels thermally isolated from bus
Structural Subsystem	Primary structure is a combination of a rigid aluminum tubular frame and aluminum honeycomb shear panels Kinematically decoupled optical bench S/C mass: 496 kg dry; 547 kg wet; payload mass: 292 kg
Mechanisms Subsystem	Optical bench launch restraints Contamination covers for sensors Solar array drive mechanisms for gimballing arrays Launch vehicle separation Two redundant main electrical umbilical connectors
CT&DH (Command, Telemetry, and Data Handling Subsystem)	On-Board Processing Electronics (OBPE) primary controller: MIL-STD-1750A radiation-hardened processor Payload controller: R3000 processor (12 MIPs) Imagery On-Board Processor (IOBP): multiple Super Harvard Architecture RISC Computing (SHARC) processors in a parallel array Solid state data recorder with a capacity of 56 Gbit
Communications Subsystem	Data downlink in X-band, data rate of 131 Mbit/s; 20 watt; 8025-8400 MHz, with embedded housekeeping telemetry Command uplink in S-band, data rate of 2 kbit/s, 2074.1 MHz Low-rate downlink in S-band, data rate at 1.024 Mbit/s, 2252.5MHz, supports tactical imagery demonstrations
Software Subsystem	On-board task scheduling capability, FDIR (Embedded fault Detection, Isolation, and Recovery)

Table 503: Spacecraft subsystem characteristics

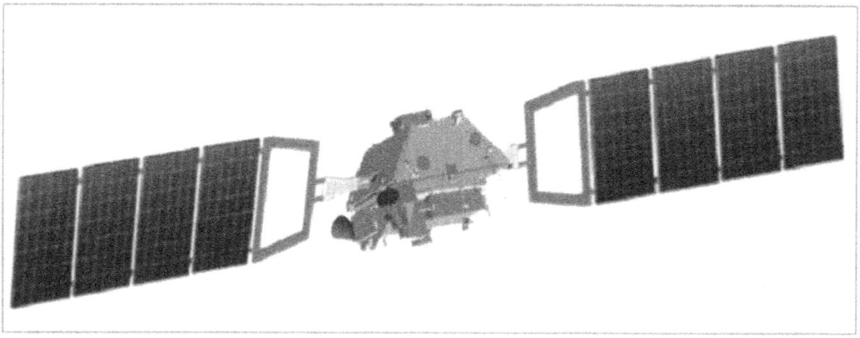

Figure 323: Illustration of the NEMO S/C

A launch of NEMO is planned for 2003. Launch vehicle is TBD.

Orbit: Sun-synchronous circular orbit, altitude = 605 km, inclination = 97.81°, nodal equator crossing at 10:30 AM (ascending node), repeat cycle = 7 days with a 2.5 day global average re-access capability when S/C pointing (±30°) in the cross-track direction is used.

Sensor complement:

COIS (Coastal Ocean Imaging Spectrometer), a hyperspectral (ocean color - land surface) imager provided by NRL (built by SAIC). The instrument employs a TMA (Three-Mirror-off-Axis) anastigmatic design, a 15 cm diameter aperture telescope, a dichroic beam splitter, and two spectrometers (VNIR and SWIR). The overall spectral range of COIS is from 400 to 2500 nm with 10 nm spectral resolution (210 bands). A FOV of 2.86° provides a swath width of 30 km. COIS uses WFOV (Wide Field of View) grating spectrometers. The two spectrometers are:

- A Visible Near Infrared (VNIR) spectrometer that disperses the 400 to 1000 nm radiation into 60 spectral bands (10 nm wide) and onto a Focal Plane Array (FPA). This provides a resolution of 1.0 - 1.5% of the band and 60 spectral bins.
- A Short-Wave Infrared (SWIR) spectrometer that disperses the 1000 to 2500 nm radiation into 150 spectral bands (10 nm wide) and onto an FPA. This provides a resolution of about 0.6% of the band and 150 spectral bins. A cryocooler provides active cooling at 110 K.

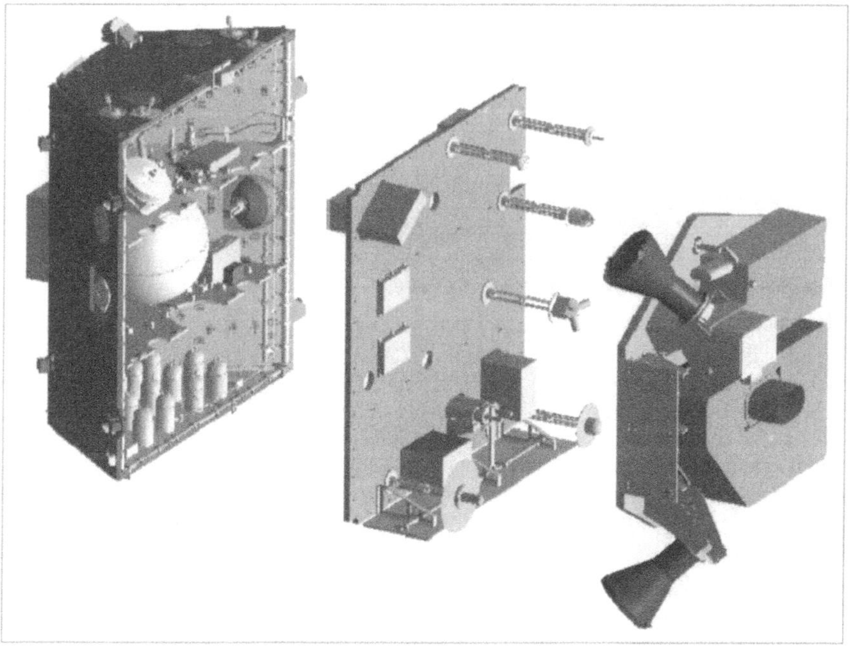

Figure 324: Exploded bus view with COIS instrument

Spectral coverage	400 - 2500 nm with a 10 nm spectral resolution (total of 210 spectral bands)
Spatial resolution	30 m or 60 m (for high SNR data products)
SNR (Signal-to-Noise)	>200 for VNIR data (5% reflectance target which is typical of the ocean) >100 for SWIR data at 30% reflectance (typical)
Pointing capability	±30° in the cross-track and along-track directions (body pointing) The pointing is used to increase integration time when sampling in the 30 m mode
Detector arrays	FPA of 1024 x 1024 pixels for VNIR (silicon CCD) and SWIR (MCT detector)
Pixel summing	Pixels are summed by 6 in VNIR and SWIR to achieve the desired spectral resolution
Data volume	Estimated average daily imagery is about 227 Gbit

Table 504: Some performance parameters of COIS

The COIS instrument provides very high SNR environmental products for imaging the low-reflectivity ocean surface. The NEMO platform can implement Ground Motion Compensation (GMC) sufficient to reduce the apparent ground speed by a factor of 5. This provides the required dwell time to give a high SNR at a GSD of 30 m. COIS is also capable of producing 60 meter GSD, high SNR data products without GMC by using spatial binning of the hyperspectral FPAs. - COIS calibration is maintained by on orbit by monthly imaging of the full moon and weekly imaging of uniform ground reference targets. 1710)

A typical COIS scene is 30 km wide by 200 km in length, with 30 m by 30 m pixel resolution, yielding nine image strips per orbit (number of strips limited by on-board data storage capacity). An alternate pixel size is 60 m x 60 m, yielding a 30 km wide swath as long as 8000 km per orbit.

Parameter	COIS - VNIR	COIS - SWIR	PIC
GSD (Ground Sample Distance)	30 m with 5:1 GMC 60 m with no GMC	30 m with 5:1 GMC 60 m with no GMC	5 m (with or without GMC)
Spectral range	0.4 - 1.0 µm	1.0 - 2.5 µm	0.45 - 0.69 µm
Spectral bands	60	150	1
FOV, swath width	2.86°, 30 km	2.86°, 30 km	2.86°, 30 km
Swath length	200 km at 30 m GSD Continuous at 60 m GSD	200 km at 30 m GSD Continuous at 60 m GSD	Continuous
Aperture diameter	15 cm	15 cm	16.4 cm
Focal length, f/Nr.	36 cm, f/2.4	36 cm, f/2.4	120 cm, f/7.32
Pixel size	18 µm	18 µm	10 µm
Nr. of pixels/band	6	6	N/A
Data quantization	12 bits	12 bits	12 bits
Detector material	Silicon	MCT	Silicon

Table 505: Characteristic parameters of COIS and PIC

PIC (Panchromatic Imaging Camera). The instrument uses an off-axis telescope to simultaneously image the same 30 km swath as the COIS instrument onto a 6000 pixel CCD linear array (PIC data is co-registered with COIS data). The spectral range is from 0.45 - 0.69 µm. The panchromatic imagery has a resolution of 5 m GSD. The moderate COIS GSD matches the spatial scale of natural objects in the littoral zone, while the high resolution PIC provides simultaneous context and sharpening and supports civil land imaging requirements. The 30 km ground swath of COIS and PIC spans the near-shore littoral region of land, beach, and ocean. COIS and PIC use a single optical bench to minimize optical boresight drift. 1711)

Mission operations:

NEMO uses a combination of commercial/industry operations and an existing DoD infrastructure ground segment to provide the user communities (military and civil) with data services. The ground segment consists of the following elements:

- ASOC (Advanced S/C Operations Center), located in Washington, DC area. ASOC is the central hub for all mission planning, S/C operations, and data processing
- TAGS (Transportable Autonomous Ground Stations), located at Fairbanks, AK and elsewhere. Downlinked data is maintained on mass storage devices at the TAGS until final receipt of the IPC is verified.
- IPC (Image Processing Center). The IPCs perform quick-look data processing to level 1B and data archive functions.
- NCC (NRL Control Center). NCC is the science hub for the NEMO program. Also data distribution function for the military side.

Adaptive Spectral Signature Recognition.

The NEMO S/C employs **ORASIS** (Optical Real-time Adaptive Signature Identification System) developed by NRL, an automated and patented end-to-end HSI (Hyperspectral

1710) M. Corson, "Calibration of the NEMO sensor imaging payload," SPIE Proceedings, Vol. 3437, 1998
1711) A. Myers, "NEMO satellite sensor imaging payload," SPIE Proceedings, Vol. 3437, 1998

Imaging) algorithm and advanced processing system that performs on-board data compression, thereby significantly reducing the capacities needed for on-board storage and for downlink transmission.

ORASIS identifies the spectral signatures corresponding to physical objects in the scene without supervision or a priori knowledge. The approach is to analyze each spectrum in the scene sequentially, discarding duplicate spectra, and working only with the unique spectra. Using convex set methods and orthogonal projection techniques, each observed spectrum is then analyzed in terms of the set of spectra that represent the physically meaningful patterns or end-members that have combined to make the observed spectrum. Matched filters (Filter Vectors) are created, they can be used to de-mix the image. [1712)]

The ORASIS processor, hosted on IOBP (Imagery On-Board Processor), is based on the SHARC (Super Harvard Architecture Computer), it has 32 processors for data handling and a processing capability of up to 2.5 GFLOPS. The on-board processing minimizes subsequent ground processing and data exploitation, and enables the on-board production of data products, such as maps of identified features. An important benefit of ORASIS processing is a **greater than tenfold data compression**, relieving the bottlenecks of on-board data storage, transmission to the ground, and beyond.

M.17 Nimbus

A NASA/GSFC meteorological research-and-development satellite program (parallel to the operational TIROS program) which started in the early 1960s with the prime objective to test new instrument concepts, a secondary objective was to provide atmospheric data for improved weather forecasts. The series grew more into a major Earth sciences program (study of oceans, land surfaces and atmosphere) with successive S/C and with the availability of better sensing instrumentation. The ERTS (Earth Resources Technology Satellite, later renamed to Landsat) program, developed in the early 1970s, is a direct descendent of the Nimbus program.

The basic Nimbus series S/C structure is butterfly-shaped (some call it 'ocean-buoy shaped') when deployed in orbit, designed to serve as a stabilized, Earth-oriented platform. It consisted of three major elements: 1) a sensory ring, 2) solar paddles, and 3) the control system housing. The solar paddles and the control system housing were connected to the sensory ring by a truss structure. The Nimbus-1 S/C was nearly 3.7 m tall, 1.5 m in diameter at the base, and about 3 m across with solar paddles extended. The sensory ring at the base of the satellite housed the electronics equipment and battery modules. The lower surface of the sensory ring provided mounting space for sensors and telemetry antennas. An H-frame structure mounted within the center of the torus provided support for the larger sensors and tape recorders. Mounted on the control system housing were sun sensors, horizon scanners, gas nozzles for attitude control, and a command antenna. Pitch and roll stability was sensed by horizon scanners, while a gyroscope controlled stability in the yaw axis. Use of a stabilization and control system allowed the spacecraft's orientation to be controlled to within $\pm 1^\circ$ for all three axes. [1713)] [1714)]

All Nimbus series S/C were built and integrated by GE Astro Space. The series with their orbits and sensor complements are also listed in Table 374. The Nimbus program, consisting of eight satellite launches, seven of which were successful in achieving orbit, carried a total of 48 instruments and experiments.

1712) J. Bowles, et al., "Hyperspectral Data Compression and Science Algorithms for the NEMO Satellite," Proceedings of 1st EARSeL Workshop on Imaging Spectroscopy, University of Zürich, Switzerland, Oct. 6-8. 1998, pp. 183-190

1713) http://nssdc.gsfc.nasa.gov/nmc/sc–query.html

1714) I. S. Haas, R. Shapiro, "The Nimbus Satellite System: Remote Sensing R&D Platform of the 1970s," Monitoring Earth's Ocean, Land, and Atmosphere from Space - Sensors, Systems, and Applications, Progress in Astronautics and Aeronautics, AiAA, Volume 97, 1985, pp. 71-95

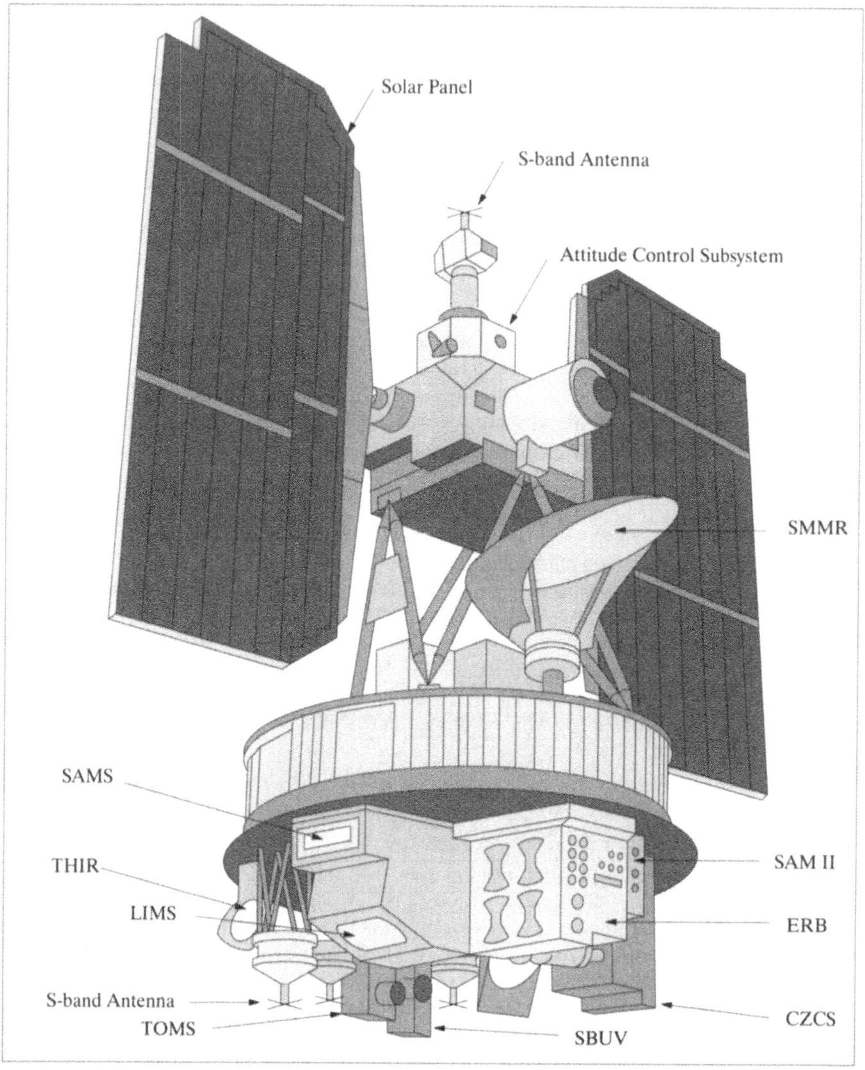

Figure 325: The Nimbus-7 spacecraft model

M.17.1 Nimbus-1

The S/C was launched Aug. 28, 1964 on a Thor-Agena-B vehicle from VAFB, CA. Nimbus-1 S/C mass = 375 kg. The satellite was fully operational for 3 1/2 weeks when a failure of the solar-array power system stopped further experimentation.

Orbit: Polar elliptical orbit, perigee = 429 km, apogee = 937 km, inclination = 98.66°, eccentricity = 0.03595, period = 98.42 minutes. Note: A short burn of the Agena rocket resulted in an elliptical orbit instead the intended circular orbit of 1100 km.

Sensor complement:

HRIR (High-Resolution Infrared Radiometer). HRIR mass = 5.1 kg, power = 4 W. Objective: to map the earth's nighttime cloudcover (to complement the daytime AVCS coverage) and to measure the radiative temperatures of cloud tops and surface terrain. The HRIR instrument consisted of the following basic components: an optical subsystem, an infrared detector (lead selenide photoconductive material), an electronics subsystem, a magnetic tape recorder, and a filter to minimize attenuation effects of water vapor and carbon dioxide. Measurement of thermal radiation in the spectral region of 3.5 - 4.1 μm. IFOV = 1.5° (corresponding to 8 km ground resolution). The radiometer was capable of measuring radiance temperatures from 210 to 330 K. The HRIR system was operational for 3 1/2 weeks.

AVCS (Advanced Vidicon Camera System). The AVCS consisted of three cameras, a tape recorder, and an S-band transmitter. The cameras were mounted on the satellite sensory ring, facing earthward and deployed in a fan-like array to produce a three-segment composite picture. Each camera covered a FoV of 37° with the center camera pointing straight down. The optical axes of the other two cameras were directed 35° to either side. Each of the cameras employed an f/4 lens with a focal length of 16.5 mm. A potentiometer attached to the solar array controlled the lens opening from f/16 when the S/C was over the equator to f/4 when it was near the poles. The 800 scan-line, 2.54 cm diameter vidicon pickup tubes yielded a linear resolution of better than 1 km at nadir from an altitude of 800 km. At this altitude, the camera array produced a composite picture covering an area of 830 by 2700 km. Up to 192 pictures (two full orbits of data) or 64 pictures per camera could be stored on recorder for subsequent playback to an acquisition station. AVCS provided near-global, high-resolution cloudcover pictures ever assembled. Subsequently, the same basic camera system was also used for the TOS/ESSA series program (G.12).

APT (Automatic Picture Transmission). Heritage of APT on TIROS-8 with a launch on Dec. 21, 1963. A vidicon camera and transmitter combination designed to transmit local daytime, slow-scan television pictures of cloudcover conditions to properly equipped ground receiving stations on a real-time basis. The camera used a 108° wide-angle f/1.8 objective lens with a focal length of 5.7 mm. Images retained on the photo-sensitive surface of the 2.54 cm diameter vidicon system were read out at four lines/s to produce an 800 line picture. A 5 W TV transmitter (136.95 MHz) relayed the pictures to local APT stations within communication range. The faceplate of the vidicon had reticle marks that appeared on the picture format to aid in relating the picture to its geographical position on the Earth's surface. At nominal satellite altitude, a picture covered a footprint of about 1660 km x 1660 km with a horizontal resolution of around 3 km at nadir. Over 1600 pictures were obtained in the 3 1/2 weeks of mission lifetime.

M.17.2 Nimbus-2

Nimbus-2 was launched May 15, 1966 on a Trust Augmented Thor-Agena-D vehicle from VAFB, CA. S/C mass = 414 kg. The Nimbus-2 S/C and its sensors performed nominally after launch until July 26, 1966, when the S/C tape recorder failed. Its function was taken over by the HRIR tape recorder until November 15, 1966, when it also failed. Some real-time data were collected until January 17, 1969, when the mission was terminated.

Orbit: Sun-synchronous polar near-circular orbit, perigee = 1103 km, apogee = 1179 km, inclination = 100.35°, period = 108. 15 minutes.

Sensor complement:

HRIR (High-Resolution Infrared Radiometer). Identical sensor as flown on Nimbus-1.

AVCS (Advanced Vidicon Camera System). Identical sensor as flown on Nimbus-1. At the orbital altitude of 1100 km, the camera array could produce a composite picture covering an area of 720 by 3400 km.

APT (Automatic Picture Transmission). Identical instrument as flown on Nimbus-1. A picture covered a footprint of 1200 km x 1200 km with a horizontal resolution of > 3 km at nadir. Over 300 APT stations in 43 countries were able to this data.

MRIR (Medium-Resolution Infrared Radiometer). Sensor mass = 9.5 kg, power = 7.5 W. Objective: measurement of the intensity and distribution of electromagnetic radiation emitted by and reflected from the Earth and its atmosphere in five selected wavelength ranges from 0.2 to 30 μm. Data for heat balance of the Earth-atmosphere system were obtained, as well as measurements of water vapor distribution, surface or near-surface temperatures, and seasonal changes of stratospheric temperature distribution. MRIR featured a flat scanning mirror inclined at 45° to the optical axis. The mirror rotated at 8 rpm and scanned in a plane perpendicular to the direction of motion of the satellite. Each of the five channels contained a 4.33 cm diameter folded telescope with a 2.8° IFOV (55 km ground resolution) and a thermistor-bolometer. The collected energy was modulated by a mechanical chopper to produce an ac signal. The signal was then amplified and recorded on magnetic tape for subsequent playback to a ground acquisition station.

Band	Spectral Range	Description
1	6.4 - 6.9 μm	which covered the 6.7 μm water vapor absorption band
2	10 - 11 μm	atmospheric window
3	14 - 16 μm	which covered the 15 μm carbon dioxide absorption band
4	5 - 30 μm	which provided the emitted long-wavelength infrared energy for heat budget estimations
5	0.2 - 4 μm	which yielded information on the intensity of reflected solar energy (albedo)

Table 506: Spectral parameters of MRIR

M.17.3 Nimbus-3

Nimbus-3 was launched April 14, 1969 on a Thor-Agena vehicle from VAFB, CA. S/C mass = 576 kg. Nimbus-3 performed nominally until July 22, 1969, when the IRIS experiment failed. The HRIR and SIRS experiments were terminated on January 25, 1970, and June 21, 1970, respectively. The remaining experiments continued operation until September 25, 1970, when the rear horizon scanner failed. All S/C operations were terminated on January 22, 1972.

Orbit: Polar near-circular orbit, perigee= 1075 km, apogee = 1135 km, inclination = 99.9°, period = 107.4 minutes.

Sensor complement:

MRIR (Medium-Resolution Infrared Radiometer). Identical sensor as flown on Nimbus-2, but with slightly changed spectral ranges.

Band	Spectral Range	Description
1	6.5 - 7 μm	which covered the 6.7 μm water vapor absorption band
2	10 - 11 μm	atmospheric window
3	14.5 - 15.5 μm	which covered the 15 μm carbon dioxide absorption band
4	20 - 23 μm	which covered the spectral region containing the broad rotational absorption bands of water vapor
5	0.2 - 4.0 μm	which yielded information on the intensity of reflected solar energy (albedo)

Table 507: Spectral parameters of MRIR

HRIR (High-Resolution Infrared Radiometer). A modified version HRIR as flown on Nimbus-1 and -2. It used a dual band-pass filter which transmitted reflected solar radiation in the 0.7 - 1.3 μm band as well as emitted thermal radiation in the 3.4.- 4.2 μm band. By detecting reflected solar radiation in the 0.7 - 1.3 μm band, the radiometer could also map the Earth's cloud cover during the daytime. Radiant energy from the earth was collected by a flat scanning mirror inclined at 45° to the optical axis. The mirror rotated at 48 rpm and scanned in a plane normal to the S/C velocity. The radiation reflected from the scan mirror was chopped at the focus of a 10.2 cm f/1 modified Cassegrain telescope. The modulated energy was then refocused on a lead selenide detector cell that transformed the received radiation into an electrical output. The output was amplified and recorded on magnetic tape for subsequent playback to a ground acquisition station. Using the direct readout infrared radiometer (DRIR) system, nighttime and daytime data could be transmitted by the real-time transmission system (RTTS) to ground APT stations. A ground resolution of 8.5 km could be obtained at nadir. The HRIR measured radiance temperatures between 210 and 330 K to a general accuracy of 1 K.

IRIS (Infrared Interferometer Spectrometer). Sensor mass = 13 kg, power = 16 W, average bit rate = 3.75 kbit/s. Objective: measurement of the vertical structure of the atmosphere and the emissive properties of the earth's surface by measuring the surface and atmospheric radiation in the 5.0 - 20 μm band using a modified Michelson interferometer. Incoming radiation was reflected into the instrument from a plane mirror. The radiation was split into two beams that recombined and interfered after reflection on a fixed mirror and a moving Michelson mirror. The recombined beam was then focused on a bolometer detector. Interference effects resulted from the optical path difference between the two beams as the mirror moved. The moving mirror traveled about 2 mm in 11 s to give an interferogram, which was recorded on magnetic tape. The interferograms were transmitted to an acquisition station, where a Fourier transform was performed to produce a thermal emission spectrum of the Earth. From these spectra, vertical profiles of temperature, water vapor, and ozone, as well as other parameters of meteorological interest, could be derived. The IFOV provided a resolution circle on the ground of 144 km in diameter.

SIRS (Satellite Infrared Spectrometer), built by SBRC. Sensor mass = 41 kg, power = 24 W, average bit rate = 0.1 kbit/s. Objective: indirect determination of the vertical temperature profiles of the atmosphere by measuring the infrared radiation emitted from the Earth and its atmosphere in seven spectral intervals in the carbon dioxide band (11 - 15 μm) and one interval in the atmospheric window centered at 11.1 μm. - A Fastie-Ebert fixed-grating spectrometer was used consisting of the following components: 1) a plane, light-collecting mirror to provide a single Earth-viewing beam fixed in the vertical; 2) a rotating chopper mirror; 3) a spherical mirror; 4) a 12.7 cm diffraction grating with 1250 lines per inch (or 492 lines per cm); 5) a set of eight exit slits with a single interference filter; 6) eight wedge-immersed thermistor bolometers; 7) a blackbody radiation source for calibration, and 8) eight preamplifiers and eight operational amplifiers. FOV = 11.5° x 11.5° centered at nadir (a footprint of about 220 km x 220 km on the ground). Data from the 11.1 μm channel yielded surface and/or cloud-top temperatures. Data from the carbon dioxide band could be used to generate temperature-pressure profiles by a mathematical inversion technique.

IDICS (Image Dissector Camera System). Sensor mass = 11.2 kg, power = 18 W. Objectives: provision of daytime cloudcover photographs. IDICS was a shutterless electronic scan and step tube mounted behind a wide-angle (108°), 5.7mm focal length lens. Scanning and stepping functions occurred continuously while the satellite progressed along its orbital path. The FOV of the optics was 73.6° in the direction of flight and 98.2° in the cross-track direction. The image was focused by the optics on a photosensitive surface of the image dissector tube. A line-scanning beam scanned the photosensitive surface at 4 Hz with a frame period of 200 s. A picture was about 1400 km x 1400 km with a resolution of 3 km at nadir.

IRLS (Interrogation, Recording, and Location System). The objective was to collect and retransmit meteorological, geophysical, and other experimental data from remote data collection stations (platforms) deployed on a global scale. The IRLS could also determine the location and track the movement of properly equipped objects such as balloons, ocean buoys, and ships to within an accuracy of 2 km.

IRLS consisted of the following components: 1) a 466 MHz receiver, 2) a 401.5 MHz transmitter, 3) decoding and coding circuits, 4) a range detector, and 5) a 20 kByte satellite data memory capable of storing data measurements during each orbit for up to 20 different interrogations. When within range of an acquisition and command station, the satellite command memory was programmed to communicate with selected platforms during the coming orbit. The satellite stored both the address (number) of each platform and the desired time that each should be contacted. At the appropriate time in orbit, the satellite interrogated each platform, measured the satellite to platform distance by determining the round trip propagation time of the RF signal, received the analog data from the platform, converted it to digital form, and stored it. Upon return to the locale of the ground station, the station commanded the satellite to transmit the stored data and to accept new commands for the next orbit. IRLS functioned nominally until Sept. 1970, when the operational capacity was severely restricted by spacecraft yaw problems. All data acquisition ceased on January 22, 1972, when spacecraft operations were terminated.

MUSE (Monitor of UV Solar Energy). The objective was to measure the solar flux, look for temporal variations in the solar UV flux, and to measure the atmospheric attenuation by viewing into the setting sun after the S/C. The instrument had five broad bands in the range of 1150 to 3000 Ä with maximum response at 1216 Ä, 1600 Ä, 1800 Ä, 2000 Ä, and 2600 Ä. MUSE consisted of the following components: five vacuum photodiodes, housed in an electronics package and a sensor package, mounted in the rear of the Nimbus spacecraft. All sensors except the 1216 Ä sensor had semi-transparent photocathodes that were deposited on the windows. The 1800, 2000, and 2600 Ä sensors had aluminum oxide windows, while the 1216 Ä and 1600 Ä sensors had MgF2 and CaF2 windows, respectively. The five spectral regions were determined by the transmittance of the filter or window materials on the short wavelength side, while the long wavelength cutoffs were produced by the varying degrees of opacity of the different photocathode materials. The solar aspect system measured the angle of incidence of the solar rays. One sample/s was measured. Sensor FOV=90°. Solar acquisition began at 45° prior to the Earth day/night terminator and ceased completely at the satellite day/night transition. The instrument had only an inflight electrical calibration sequence because there were no known suitable UV sources that could provide an inflight optical calibration.

SNAP-19 (Radioisotope Thermoelectric Generator). The objective was to assess the operational capability of radioisotope power for space applications.

Nuclear Auxiliary Power.

M.17.4 Nimbus-4

Nimbus-4 was launched on April 8, 1970 on a Thor-Agena-D vehicle from VAFB, CA. S/C mass = 675 kg. The spacecraft performed nominally until April 8, 1971, when attitude problems started. On April 14, 1971, the spacecraft turned around and flew backwards in orbit. On May 12, 1971, the spacecraft was successfully rotated 180°. The experiments operated on a limited time basis after that time until September 30, 1980.

Orbit: Sun-synchronous near-circular polar orbit, perigee = 1092 km, apogee = 1108 km, inclination = 80.1°, period = 107.2 minutes.

Sensor complement:

IDICS (Image Dissector Camera System). Similar system as flown on Nimbus-3. IDICS data could be transmitted via APT or stored and transmitted to ground acquisition stations. The experiment performed nominally until April 8. 1971. After May 12, 1971 limited data were obtained until February 5, 1972.

THIR (Temperature-Humidity Infrared Radiometer). Sensor mass = 7.6 kg, power = 8.2 W, bit rate = 2 kbit/s. THIR was built by Hughes SBRC. Objective: detection of emitted thermal radiation in two spectral ranges, measurement of daytime and nighttime surface and cloud-top temperatures as well as the water vapor content of the upper atmosphere. The instrument consisted of a 12.7 cm Cassegrain system, a scanning mirror common to both channels, a beam splitter, filters, and two germanium-immersed thermistor bolometers. Incoming radiant energy was collected by a flat scanning mirror inclined at 45° to the optical axis. The mirror rotated through 360° at 48 rpm and scanned in a plane normal to the spacecraft velocity vector. The energy was then focused into a dichromatic beam splitter, which divided the energy spectrally and spatially into two channels (10.5 - 12.5 μm IR window, and 6.5 - 7.0 μm for water vapor). Direct readout infrared radiometer (DRIR) data could be transmitted to APT ground stations for both day and night portions of the orbit. The IR window channel had a ground resolution of about 7 km and the water vapor channel about 22 km at nadir. THIR operated nominally until Jan. 11, 1971. A similar experiment was flown on Nimbus 5, 6 and 7.

IRIS (Infrared Interferometer Spectrometer). Similar system as flown on Nimbus-3. Sensor mass = 17 kg, power = 12 W, bit rate = 3.75 kbit/s. Spectral range of modified Michelson interferometer = 6.25 to 25 μm. The instrument had a FOV of 5° and a spectral resolution of less than 0.4 μm (nominally 1.4 reciprocal cm). The IRIS instrument was turned off on January 25, 1972 to conserve spacecraft power.

SIRS (Satellite Infrared Spectrometer). Similar system as flown on Nimbus-3. Sensor mass = 31.5 kg, power = 30 W. The instrument measured the infrared radiation (11 to 36 μm micrometers) emitted from the Earth and its atmosphere in 13 selected spectral intervals in the carbon dioxide and water vapor bands plus one channel in the 11 μm atmospheric window. The main components of SIRS consisted of: 1) a plane, light-collecting mirror to provide one fixed and two variable Earth-viewing angles; 2) a rotating chopping mirror that served alternately to collect space radiation and Earth radiation; 3) a 6.3 cm diffraction grating with 1250 lines per inch (or 492 lines per cm); 4) 14 slits with associated interference filters; 5) 14 thermistor bolometers; and 6) a blackbody source for calibration purposes. The a scan mirror could observe the ground surface 12.5° to either side of the subsatellite track. FOV = 215 km. The SIRS experiment performed nominally for several months after launch but began to deteriorate in early 1971. The experiment operated on a limited time basis until March 6, 1973.

MUSE (Monitor of UV Solar Energy). Identical system as flown on Nimbus-3.

BUV (Backscatter Ultraviolet). Sensor mass = 14.5 kg, power = 7 W. Objective: to monitor the vertical distribution and total amount of atmospheric ozone on a global scale by measuring the intensity of UV radiation backscattered by the Earth/atmosphere system during day and night in the spectral band of 2500 - 3400 Ä. The instrument consisted of a double monochromator containing all reflective optics and a photomultiplier detector. The double monochromator was composed of two Ebert-Fastie-type monochromators in tandem. Each monochromator had a 52 mm x 52 mm grating with 2400 lines per mm. Light from a 0.05 sr solid angle (subtending approximately a 222 km^2 area on the ground) entered the nadir-pointing instrument through a depolarizing filter. A motor-driven cam step rotated the gratings to monitor the intensity of 12 ozone absorption wavelengths. The detector was a photomultiplier tube. For background readings, a filter photometer measured the reflected UV radiation in an ozone-free absorption area near 3800 Ä. Signals from both units were read by separate range-switching electrometers with seven ranges. Cycle time of 6144 s. Each cycle, in turn, was divided into 192 BUV frames of 32 s duration. Calibration by on-board light sources was performed in 26 of the 192 frames. During each of these data frames, the monochromator measured the intensity of the UV radiation in each of the 12 wavelength bands, while the photometer measured the UV intensity in a single wavelength band. The dwell time at each wavelength was 1.8 s, and, during this interval, four analog UV intensity measurements were taken at 400 ms intervals in addition to an integrated pulse count measurement of the UV intensity and energetic particle flux. Once each orbit, the FOV was changed to monitor the sun or moon directly.

FWS (Filter Wedge Spectrometer). Objective: to determine the radiance from the Earth-atmosphere system as a function of wavelength by measuring the emitted and reflected infrared radiation in the 1.2 - 2.4 μm and 3.2 - 6.4 μm bands. FWS consisted of the following components: 1) a telescope; 2) a rotating disk chopper; 3) a rotating (3.75 rpm) circular interference filter wheel; and 4) a lead selenide detector. The filter wheel was a two 180°-segment (one per passband) 100 layer interference filter with the layer thickness linearly increasing as a function of angular position, causing the bandpass to shift toward longer wavelengths. Incoming radiation was reflected off a surface mirror and was collected by a telescope oriented normal to the earth's surface. The telescope had a 3° FOV toward nadir, and a swath about 57 km. The telescope focused the collected radiation onto the edge of the multi-toothed chopper wheel that chopped the energy at 333 Hz. The energy was refocused onto the edge of the circular variable filter at an aperture that acted as both spectrometer slit and a system field stop. The energy was then reimaged on a lead selenide detector radiatively cooled to 175 K. The incident radiation was sampled 20 times/s, resulting in a spectral intensity plot of 158 points for each passband per revolution. Onboard calibration was accomplished by alternate viewing of the earth and calibration standards by the detector. On June 8, 1970, the FWS suffered a mechanical failure.

SCR (Selective Chopper Radiometer). Sensor mass = 15.3 kg, power = 5 W. Objective: to observe the emitted infrared radiation in the 15 μm absorption band of carbon dioxide. The SCR had six channels, which were arranged in three units of two. The four lower channels were called single-cell channels. The optics of each channel consisted of a cantilever-mounted blade shutter that oscillated at 10 Hz and successively chopped the FOV between Earth and space. The chopped radiation was then passed through a 10 cm path length of carbon dioxide, the pressure being set for each channel to define the viewing depth of the atmosphere. Behind the carbon dioxide path was a narrow-band filter, the centers of which were different for each channel, and a light pipe which focused the radiation on a thermistor-bolometer detector. To obtain adequate height resolution in the upper layers of the atmosphere, the upper two channels operated on a slightly different principle and were known as double-cell channels. The technique consisted of switching the radiation between two half-cells, which were semicircular in shape and of 1 cm path length, and which contained different pressures of carbon dioxide. A movable 45° mirror replaced the oscillating shutter used in the lower four channels. - Inflight calibration was carried out by viewing of an internal reference blackbody of known temperature prior to the view of space. The output of each channel was sampled once every second. The upper two channels had a circular FOV of about 160 km in diameter, and the lower four had a rectangular FOV about 112 km square.

IRLS (Interrogation, Recording, and Location System). Instrument mass = 7 kg, power = 12 W. Identical instrument as flown on Nimbus-3.

M.17.5 Nimbus-5

Nimbus-5 was launched on Dec. 11, 1972 by a Delta vehicle from VAFB. S/C mass = 770 kg.

Orbit: Sun-synchronous polar orbit, perigee = 1089 km, apogee = 1101 km, inclination = 99.9°, period = 107.2 minutes.

Sensor complement:

ITPR (Infrared Temperature Profile Radiometer). Sensor mass = 15.9 kg, power = 16 W, bit rate = 0.24 kbit/s. Objective: to measure the three-dimensional temperature field in the Earth's atmosphere with a spatial resolution of 32 km. The radiometer sensed four intervals in the 15 μm CO_2 band, one interval in the water vapor rotational band near 20 μm and two spectral intervals in the atmospheric window regions near 3.7 and 11 μm. The ITPR viewed the Earth successively at various angles distributed symmetrically about nadir in a plane normal to the orbital track. Forty-two geographically independent scan spots were taken along a single strip. As the satellite progressed along its orbital path, the radiometer observed 10 such 42-spot strips to form a matrix of independent scan spots. Each matrix was produced in 222 s with the whole scanning sequence repeated every 240 s. Matrix measurements taken in the CO_2 and water vapor absorption bands were used to calculate temperature profiles and total water vapor content in the troposphere and lower stratosphere.

SCR (Selective Chopper Radiometer). Sensor mass = 11.3 kg, power = 15 W. Similar instrument as flown on Nimbus-4. Objectives: to observe the global temperature structure of the atmosphere up to 50 km in altitude, to make supporting observations of water vapor distribution, and to determine the density of ice particles in cirrus clouds. SCR measured emitted radiation in 16 spectral intervals separated into the following four groups: 1) four CO_2 channels between 13.8 and 14.8 μm; 2) four channels at 15.0 μm; 3) an IR window channel at 11.1 μm, a water vapor channel at 18.6 μm, two channels at 49.5 and 133.3 μm, and 4) four channels at 2.08, 2.59, 2.65,and 3.5 μm. SCR viewed a 48 km circle on the Earth's surface with a ground resolution of about 25 km.

NEMS (Nimbus-E Microwave Spectrometer). Sensor mass = 22.6 kg, power = 35 W, bit rate = 0.06 kbit/s. Objectives: to demonstrate the capabilities and limitations of microwave sensors for measuring tropospheric temperature profiles, water vapor abundances, cloud liquid water content, and Earth-surface temperatures. NEMS frequencies: 22.235, 31.4, 53.65, 54.9 and 58.8 GHz. The three channels near the 5 μm oxygen absorption band were used primarily to determine the atmospheric temperature profiles. NEMS provided measurements even in cloudcover conditions that normally restrict the usefulness of conventional IR data in such situations. The two water vapor channels near 10 μm permitted the water vapor and cloud liquid water content over oceans to be estimated and also to yield an estimated temperature once the surface emissivity had been calibrated by comparison with direct measurements. The three oxygen channels shared a common signal and reference antenna. Both water vapor channels had their own signal and reference antennas. NEMS viewed a 180 km diameter circle on the earth's surface.

ESMR (Electrically Scanning Microwave Radiometer). Sensor mass = 28 kg, power = 40 W, bit rate = 0.40 kbit/s. Objectives: 1) to derive the liquid water content of clouds from brightness temperatures over oceans, 2) to observe differences between sea ice and the open sea over the polar caps, and 3) to test the feasibility of inferring surface composition and soil moisture. ESMR frequency at 19.36 GHz (1.55 cm wavelength). A phased-array antenna system (83.3 cm x 85.5 cm), deployed after launch, scanned the Earth successively at various angles in cross-track direction, producing a brightness-temperature map of the surface of the Earth and its atmosphere (half-power beam width = 1.4° x 1.4°, beam scan angle = ±50°, polarization = horizontal). The scanning process was controlled by a computer, it consisted of 78 symmetrically distributed independent scan spots extending 50° to either side of nadir. Angular separation of the scan spots allowed for an 8.5% overlap between view positions. ESMR had an accuracy of about ± 1° C with a spatial resolution of about 25 km at nadir.

SCMR (Surface Composition Mapping Radiometer). Sensor mass = 25 kg, power = 7 W, bit rate = 0.25 kbit/s. Objective: to measure 1) terrestrial radiation in the 8.3 - 9.3 μm and 10.2 - 11.2 μm intervals, and 2) reflected solar radiation in the 0.8 - 1.1 μm range. Surface composition and sea surface temperatures could be obtained from these measurements. The SCMR had an IFOV of 0.6 mrad, equivalent to a ground resolution of 660 m at nadir. The scan mirror rotated at 10 rps to provide scan lines 800 km wide across the spacecraft track. SCMR operation until Jan. 4, 1973. A modified instrument, heat capacity mapping radiometer, was flown on the Heat Capacity Mapping Mission (HCMM).

THIR (Temperature-Humidity Infrared Radiometer). Similar instrument as flown on Nimbus-4.

M.17.6 Nimbus-6

Nimbus-6 was launched on June 12, 1975 by a Thor-Delta vehicle from VAFB, CA. S/C mass = 586 kg. The complement of advanced sensors was capable of: 1) mapping tropospheric temperature, water vapor abundance, and cloud water content; 2) providing vertical profiles of temperature, ozone, and water vapor; 3) transmitting real-time data to a geostationary spacecraft (ATS-6); and -4) yielding data on the earth's radiation budget.

Orbit: Sun-synchronous circular polar orbit, perigee = 1093 km, apogee = 1101 km, inclination = 100°, period = 107.3 minutes.

Sensor complement:

ERB = Earth Radiation Budget. Sensor mass = 18 kg, power = 30 W, bit rate = 0.50 kbit/s. Objective: measurement of reflected and emitted terrestrial radiation fluxes in conjunction with solar radiation. The results were used: 1) to determine the Earth radiation budget, 2) to determine the angular distribution of terrestrial radiation for various meteorological and geographic regimes, and 3) to correlate measurements made using identical but independent channels calibrated to the same standard.

The multichannel radiometer employed a bi-axial scanning mechanism which enabled measurements to be obtained from the forward horizon to the aft horizon in a 64 s interval. Each axis of the scanning mechanism contained four shortwave channels (0.2 - 4.0 μm) and four longwave channels (4.0 - 50 μm) with a FOV of 0.25° x 5.14°. The channels were oriented in a directional fan to cover 20° to each side of nadir. The 64 s scan period allowed an area to be measured from up to 17 different angles as the spacecraft passed overhead.

Incoming solar radiation from 0.2 - 50 μm was normally monitored in 10 spectral intervals as the satellite orbited over the Antarctic, just before it started its northward trip on the daylight side of the earth. Terrestrial radiation measurements were taken continuously in the 0.2 - 4 μm, 0.7 - 3 μm, and 4 - 50 μm intervals. The measurements were taken in two ways. Four channels, using fixed wide-angle optics (FOV = 133.3°), measured the total outgoing radiation integrated over the entire disk of the Earth. The second set of measurements was obtained from eight high-resolution narrow-angle scanning channels that measured the terrestrial radiation emanating from a relatively small area over a range of zenith and azimuth angles. A similar instrument was flown on Nimbus 7. The solar and wide-angle channels operated successfully and provided good quality data. The scanning channels developed mechanical scan problems in August 1975 and operated only in the nadir position after March 1976.

ESMR (Electrically Scanning Microwave Radiometer). Similar instrument as flown on Nimbus-5. Sensor mass = 36.7 kg, power = 70 W, bit rate = 2.40 kbit/s. Objectives: to measure the Earth's microwave emission to provide the liquid water content of clouds, the distribution and variation of sea ice cover, and gross characteristics of land surfaces (vegetation, soil moisture, and snow cover).
The two-channel scanning radiometer operated in a 250 MHz band centered at 37 GHz. One channel was used to measure the vertical polarization and the other measured the horizontal polarization. The antenna beam array, a 90 cm x 20 cm x 12 cm box-like structure, was mounted on top of the S/C sensory ring and was pointed in the direction of the spacecraft's forward motion and tilted down 45° from the satellite antenna axis. The antenna beam scanned the earth in 71 discrete steps for various angles extending up to 35° on either side of the orbital plane. The deduced brightness temperatures were expected to be accurate to within 3 - 5 K. Spatial resolution was 20 km in the cross-track direction and 45 km in the direction parallel to the subpoint track.

HIRS (High-Resolution Infrared Radiation Sounder). Sensor mass = 13.5 kg, power = 10 W, bit rate = 10.8 kbit/s. Objectives: to provide vertical temperature profiles twice daily on a global basis, to extend up to about 40 km, and to provide information on the water vapor distribution in the troposphere. HIRS measured radiances primarily in five spectral regions: 1) seven channels near the 15 μm CO_2 absorption band, 2) two channels (11.1 and 3.7 μm) in the IR window, 3) two channels (8.2 and 6.7 μm) in the water vapor absorption band, 4) five channels in the 4.3 μm band, and 5) one channel in the visible 0.69 μm region. The sounder consisted of a Cassegrain telescope, scanning mirror, dichromatic beam splitter, filter wheel, chopper, and associated electronics. HIRS scanned the Earth's surface in a plane normal to the spacecraft's orbital path with a maximum scan angle of 30° to either side of nadir to provide data with a spatial resolution of 25 km.

LRIR (Limb Radiance Inversion Radiometer). Sensor mass = 18 kg, power = 36 W, bit rate = 15.68 kbit/s. Objective: to provide calibrated radiance versus altitude profiles by intercepting radiation emanating from an atmospheric path which is tangential to a particular geocentric height. LRIR sensed radiation in four spectral intervals: 1) the 14.6 - 15.9 μm CO_2 band, 2) the 14.2 - 17.3 μm CO_2 band, 3) the 8.8 - 10.1 μm ozone band, and 4) the 20 - 25 μm water vapor rotational band. - Measurements taken in the two CO_2 channels and the water vapor channel were used to calculate global temperature and water vapor profiles in the stratosphere and lower mesosphere. In addition, values of the geostrophic wind up to 1 mb (approximately 48 km) were derived analytically from the deduced temperature profiles. The radiometer included an optical system, a scanning mirror, choppers, and associated electronics and employed an ammonia-methane cooler system for three of the four detector channels. While the deduced temperature profiles had an rms accuracy of 3° at heights above 15 km, the values for ozone were accurate to within 20% at 1 mb. Water vapor values at the same height were within 50%.

PMR (Pressure Modulated Radiometer). Sensor mass = 2.3 kg, power = 3 W, bit rate = 0.01 kbit/s. Objective: to obtain radiometric measurements in the 15 μm CO_2 band at altitudes between 45 and 70 km on a global scale. The PMR comprised two similar radiometer channels, each consisting of a plane scanning mirror, reference blackbody, pressure-modulator cell, and detector assembly. The plane mirror was gold coated and mounted at 45° on a 90° stepping motor so that the FOV of the channel could be directed to space or to the internal reference blackbody for inflight range and zero calibration. The motor was mounted on a pair of flexible pivots so that the mirror could be rotated through ± 7.5° from its rest position to give the required Doppler scan. Major components in the pressure-modulator cell were a movable piston, a diaphragm, and a magnetic drive coil. The detector assembly consisted of a field lens, a condensing light pipe, and a pyroelectric flake bolometer. Each radiometer had a field of view that was 20° whole-angle across the spacecraft's line of flight and 40° whole-angle parallel to the line of flight. The derived temperature values were within 2 K at 65 km and about 0.2 K near 50 km with a vertical resolution of 10 km.

SCAMS (Scanning Microwave Spectrometer). Sensor mass = 23.4 kg, power = 46 W, bit rate = 0.12 kbit/s. Objective: to map tropospheric temperature profiles, water vapor abundance, and cloud water content to be used for weather prediction even in the presence of clouds, which block conventional satellite infrared sensors. SCAMS is of NEMS heritage of Nimbus-5. SCAMS continuously monitored emitted microwave radiation at frequencies of 22.235, 31.65, 52.85, 53.85 and 55.45 GHz. The three channels near the 5.0 μm oxygen absorption band were used primarily to deduce atmospheric temperature profiles. The two channels near 10 μm permitted water vapor and cloud water content over calm oceans to be estimated separately. The instrument, a Dicke-superheterodyne type, scanned ±45° normal to the orbital plane with a 10° FOV. The three oxygen channels shared common signal and reference antennas. Both water vapor channels had their own signals and reference antennas. The absolute rms accuracy of the oxygen channels was better than 2 K, that of the water vapor channels better than 1 K. The dynamic range for all channels was 0–400 K. The ground resolution was approximately 145 km near nadir and 330 km at the scan limit.

THIR (Temperature-Humidity Infrared Radiometer). Sensor mass = 7.6 kg, power = 8.2 W. Similar instrument as flown on Nimbus-4 and -5.

TWERLE (Tropical Wind Energy conversion and Reference Level Experiment). Sensor mass = 12 kg, power = 17 W. TWERLE is a GARP experiment. Objectives: 1) to measure the upper atmospheric winds in the tropics, 2) to study the relative air motion along isobaric surfaces to determine the rate of conversion of atmospheric potential energy into kinetic energy, and 3) to provide direct measurements of various meteorological parameters that served as reference points in adjusting indirect temperature soundings made from satellites. - The experiment consisted of two basic components: 1) approximately 300 constant-level meteorological balloons to yield measurements of winds, temperature, and pressure in the tropics and at southern hemisphere midlatitudes at 150 mb (about 13.6 km altitude), and 2) the Nimbus 6 random access measurements system (RAMS) to provide data collection and location determinations from the balloons. The balloons were equipped with a transmitter-oscillator, solar power supply, digitizer/modulator, and sensors. RAMS merely detected each balloon signal (401.2 MHz) and extracted the carrier frequency, balloon identification, and sensor data. This information, along with time references, was stored in digital form for subsequent relay to a ground acquisition station. The balloon's position and velocity were derived from the relative motion between the platform and the satellite by measuring Doppler shifts in the carrier signal received from the balloon. TWERLE was capable of a location accuracy of 5 km and a platform velocity accuracy of 1 m/s.

T+DRE (Tracking and Data Relay Experiment). This experiment provided the Nimbus portion of a communication link from Nimbus to ATS to a ground station. The purpose of the experiment was to gain information on the use of such a link for range and rate communications (for satellite geodetic purposes) and for data communication from a LEO S/C through a GEO S/C to a ground telemetry station. The instrumentation included an S-band transponder, a command detector/decoder, an antenna programmer, a digital evaluation module, an S-band antenna, and an antenna gimbal assembly. Initial experiment operation was nominal.

M.17.7 Nimbus-7

NASA/GSFC Earth observation spacecraft. NIMBUS-7 was launched Oct. 24, 1978 on a Delta vehicle from VAFB, CA. The S/C was three-axis stabilized and had a mass of 965 kg (832 kg on-orbit dry mass). The attitude control subsystem maintains S/C alignment with the local orbital reference axis to within 0.7° of the pitch axis and 1° of the roll and yaw axis. [1715] [1716]

Application: Oceanography, pollution of the atmosphere, meteorology; measurement of trace gases and particles in the atmosphere; distribution phenomena of air pollution; observation of the ocean surface color and of the temperature; ice observation, in particular in the coastal regions.

Orbit: Sun-synchronous polar orbit, apogee = 954 km, perigee = 941 km, inclination = 99.15°, period = 104.16 minutes, ascending node at 12 noon equator crossing time.

Sensor complement:

CZCS = Coastal Zone Color Scanner (NASA, built by Ball Brothers Research Corp., Broomfield, CO - the predecessor of Ball Aerospace and Technologies Corp.). Instrument mass = 27 kg, power = 11.4 W. Objectives: a 'proof-of-concept' experiment for the measurement of ocean color (chlorophyll concentration, sediment distribution, gelbstoff concentrations as a salinity indicator), and the temperature of coastal waters and open ocean (all parameters were optimized for ocean color sensing).

1715) "The NIMBUS-7 User's Guide," NASA/GSFC, Prepared by The Landsat/Nimbus Project, Aug. 1978
1716) "NIMBUS-7, Observing the Atmosphere and Oceans," NASA pamphlet Dec. 1983

CZCS is a multispectral radiometer (6 bands) utilizing a rotation plane mirror at a 45° angle to the optical axis of a Cassegrain telescope. Band wavelengths: 0.433-0.453 µm (chlorophyll absorption), 0.51-0.53 µm (chlorophyll concentration), 0.54-0.56 µm (Gelbstoff concentration), 0.66-0.68 µm (aerosol absorption), 0.70-0.80 µm (land and cloud detection), and 10.5-12.5 µm (TIR, surface temperature). The mirror scanned 360° but only the 80° of data centered on nadir were collected for ocean color measurements. The instrument viewed deep space and calibration sources during the remainder of the scan. The incoming radiation was collected by the telescope and divided into two streams by a dichroic beam splitter. One stream was transmitted to a field stop that was also the entrance aperture of a small polychromator. The radiance that entered the polychromator was separated and reimaged in five wavelengths on five silicon detectors in the focal plane of the polychromator. The other stream was directed to a cooled mercury cadmium telluride detector in the thermal region (10.5-12.5 µm). A radiative cooler was used to cool the thermal detector. To avoid sun glint, the scanner mirror was tilted about the sensor pitch axis on command so that the line of sight of the sensor was moved in 2° increments up to 20° with respect to the nadir.

CZCS instrument characteristics: multispectral cross-track scanning radiometer, spatial resolution: 825 m x 825 m (each band) at nadir or IFOV = 0.865 mrad, swath width = 1600 km. Data quantization = 8 bits, data rate = 800 kbit/s. Data products: global maps of chlorophyll concentration, sediment distribution, gelbstoff concentrations as a salinity indicator, and temperature of coastal waters and the open ocean. Prelaunch calibration used an integrating sphere and a blackbody source for the TIR channel. Inflight calibration used an incandescent light source for the first five bands. The TIR channel was calibrated by viewing the blackened housing of the instrument.- The CZCS instrument operated on an intermittent schedule due to overall power demands. The TIR channel failed within the first year after launch. CZCS was operational for eight years before it was turned off in Dec. 1986.

ERB = Earth Radiation Budget. Sensor mass = 32.7 kg, power = 36.2 W, bit rate = 4.5 kbit/s. Similar to the instrument flown on Nimbus-6. Objective: to determine the Earth radiation budget on both synoptic and planetary scales by simultaneous measurements of incoming solar radiation and outgoing Earth-reflected (shortwave) and emitted (longwave) radiation. - The ERB instrument consisted of a 22-channel radiometer containing separate subassemblies to perform the required solar, Earth-flux (wide angle), and scanned Earth radiance (narrow angle) measurements. The systems used optical filters for spectral discriminations, and for uncooled thermal detectors, thermopile detectors in the solar and fixed-Earth-flux channels, and pyroelectric detectors in the scanning channels. The 10 solar channels observed the sun in front of the observatory in the X-Y plane. The solar channels obtained usable solar data only during a period of about 3 minutes in each orbit when the spacecraft was over the Antarctic region. Their full response FOV was 0.18 rad. The solar channel subassembly was pivoted ±0.35 rad in the X-Y plane to compensate for sun-angle deviation when required. The channel 10c (solar channel) was a model H-F (Hickey-Frieden) self-calibrating cavity thermopile used for monitoring the total solar irradiance in the spectral range 0.2 - 50 µm. The technology for long-term solar radiation monitoring utilized is referred to as ESCC (Electrically Self Calibrating Cavity). The four fixed Earth-flux channels (numbered 11-14) were mounted so that they could continuously view the total Earth disk, and record data at 0.25 s intervals. The eight narrow FOV scanning channels were mounted in the scanning head. The NFOV channels consisted of four shortwave (0.2 - 4.8 µm) and four longwave (4.5 - 50+µm) channels.
The scanning head was gimbal-mounted in the radiometer unit main frame. The FOVs of the telescopes were asymmetric (4.4 x 89.4 mrad) and those of the shortwave and longwave channels were coincident. The 89.4 mrad FOVs of the four pairs of channels were not contiguous, but covered only alternate 89.4 mrad angular intervals along the horizon. ERB operation was stopped on Jan. 4, 1994.

LIMS = Limb Infrared Monitor of the Stratosphere. LIMS is of LRIR heritage flown on Nimbus-6. Sensor mass = 68.4 kg, power = 27 W, bit rate = 4 kbit/s. Measurement of vertical gas concentrations and temperature profiles in the stratosphere. Observables: O_3, H_2O, NO_2, HNO_3 and temperature. Channel center wavelengths: 6.25, 6.75, 9.65, 11.35, 15.25 μm and one broad channel from 13.3 - 17.2 μm. The LIMS instrument was a six-channel radiometer consisting of two electronic boxes and the radiometer unit (Hg:Cd:Te detectors were used, cooled by a two-stage solid cryogen cooler). Radiance from the Earth's limb entered the OMP (Optical Mechanical Package) aperture, reflected off the scan mirror to the 18 cm diameter off-axis parabolic primary mirror where the radiation was focused and chopped at 945 Hz. The radiation was re-collimated by the secondary mirror and directed through a Lyot stop to a folding mirror and into the detector capsule assembly (DCA). The radiation was then focused through a cadmium telluride lens and through interference filters, which defined the FOVs, and onto an array of discrete mercury cadmium telluride detectors. The detectors were maintained at about 63 K temperature by the cryogen. The LIMS began a scan near 153 km altitude, taking about 12 s to move near 38 km below the solid limb, then retraced its motion upward. After every second scan pair, the scan mirror was placed in a position to observe radiation from a small cavity blackbody for inflight warm calibration after which the instrument viewed space to obtain a cold calibration point.

SAM-II = Stratospheric Aerosol Measurement II (NASA/LaRC). Sensor mass = 17 kg, power = 0.8 W, bit rate = 4 kbit/s. The instrument is a sun photometer that views a small portion of the sun through the Earth's atmosphere during S/C sunrise and sunset. SAM-II measures aerosol extinction and extinction ratio profiles, and stratospheric optical depth as a function of altitude, latitude and longitude. Channel wavelength: 0.98-1.02 μm. Spatial resolution is 1 km (vertical); swath width is 5 - 40 km vertical. Measurement calibration technique: the Langley technique is used to calculate zero air mass solar intensity. By measuring the intensity of the sun over a several-hour period on a clear, optically stable day and using the calendar day and latitude/longitude of the measurement location, one can plot intensity versus air mass.

The SAM-II instrument package consisted of optics and electronics subassemblies. The optical assembly consisted of gimbals, a flat entrance window (which filters out UV radiation), Cassegrain optics, a flat scanning mirror, sun acquisition sensors, and a sun-photometer detector package. Solar radiation was reflected from the scan mirror into the Cassegrain telescope forming a solar image at the slit plate, which contained two solar edge sensors for monitoring solar limb crossings on either side of the detector aperture. Solar radiation passed through the aperture, was collected by a field lens, passed through an interference filter for wavelength discrimination, and finally measured by a silicon photodiode detector. The optics assembly was gimbaled in azimuth. After acquisition in azimuth, the mirror servo scanned in elevation until the Sun was acquired. The Sun was then scanned back and forth. The photometer viewed a portion of the solar disk with a 0.145 mrad IFOV and a sampling rate of 50 samples/s. SAM-II operation was stopped on January 4, 1994.

Data products: global maps of the concentration and optical properties of stratospheric aerosols as a function of altitude, latitude, and longitude, which prove valuable for studies on radiative transfer and climatic effects; aerosol transport sources and sinks in the stratosphere; seasonal variations and sudden warming phenomena; and volcanic injection phenomena. When no clouds were present in the instrument's IFOV, then tropospheric aerosols were also mapped.

SAMS = Stratospheric and Mesospheric Sounder. Sensor mass = 23.6 kg, power = 20 W, bit rate = 25 kbit/s. Measurement of vertical gas concentrations (H_2O, CH_4, CO and NO) and temperature profiles in the stratosphere and mesosphere. Observation of 'resonant scattering of solar radiation.' Channel wavelengths: nine channels defined by gas cell modulation 4.1 to 15 μm and 25 - 100 μm. Radiation from the limb of the atmosphere was incident on a scan mirror in front of a 15 cm aperture telescope. The scan mirror scanned the limb, viewed space for calibration, and viewed the atmosphere obliquely to obtain vertical profiles. There were three adjacent FOVs, each 28 mrad x 2.8 mrad (corresponding to 100 km by 10 km at the limb). The FOVs were focused onto a field-splitting mirror by the telescope which directed radiation to six detectors. Separation into channels was accomplished through dichroic beam splitters. There were seven pressure modulator cells (PMC), two containing CO_2, the remainder N_2O, NO, CH4, CO, H2O. Pressure in the cells could be varied on command by changing the temperature of a small container of molecular sieve material attached to each PMC. The spectral parameters for the H_2O channel were 2.7 μm and 25 - 100 μm. All other channels lay within the range 4.1 to 15 μm. A chopper operating at 250 Hz within the telescope, allowed the measurement of two separate signals from all detectors, one at 250 Hz and one at the PMC frequency. SAMS was operational until April 9, 1985.

SBUV/TOMS = Solar Backscatter Ultraviolet/Total Ozone Mapping Spectrometer. Instrument mass = 20 kg, power = 20 W, bit rate = 4 kbit/s. Measurement of vertical O_3 profiles, total column amounts of atmospheric O_3, solar irradiance, and terrestrial radiances.

SBUV was a double Ebert-Fastie spectrometer and filter photometer, similar to the BUV on Nimbus-4. SBUV used three detectors: a photomultiplier tube (PMT) and a photodiode for the monochromator, and one photodiode for the photometer. Both the monochromator and the photometer have chopper wheels operating at 25 Hz. The SBUV used a depolarizer to eliminate the sensitivity of the grating monochromator to polarization of the backscattered radiation. The instrument's field of view (FOV) at nadir was 0.20 rad. A roughened aluminum diffuser plate viewed the sun for solar-spectral irradiance measurements and for calibration by viewing a mercury-argon lamp. In one mode, SBUV serially monitored 12 selected narrow wavelength bands in the spectral region from 0.250 - 0.340 μm. - The SBUV spectrometer had a second mode of operation that allowed a continuous solar-spectral scan from 0.16 - 0.4 μm for detailed examination of the extraterrestrial solar spectrum and its temporal variations. A parallel photometer channel at 0.343 μm measured the reflectivity of the atmosphere's lower boundary in the same 0.21 rad FOV. Spatial resolution of SBUV = 11.3° in each band. Swath width = 200 km, separated by the 26° longitude interval between successive orbits. SBUV was operational until 1990.

TOMS was a single Ebert-Fastie spectrometer with a fixed grating and an array of exit slits. TOMS step-scanned across the orbital track 51° from the nadir in 3° steps with a FOV of about 0.052 rad. At each scan position, the Earth radiance was monitored at six wavelengths between 0.31 and 0.38 μm to infer the total ozone amount. The TOMS completed a cross scan in eight seconds, with one second for retrace, to record 35 scenes per scan. At each scene, a chopper sequentially sampled all six wavelengths four times. TOMS used the same type of PMT as SBUV, and had a separate mercury-argon lamp for wavelength calibration and a separate depolarizer. TOMS shared the diffuser plate with SBUV. Both SBUV and TOMS had five scanner modes and a shared electronics module. TOMS measured backscattered radiation sampled at six wavelengths from 0.312 - 0.340 μm. Spatial resolution is 3° (scanned through the satellite subpoint and perpendicular to the orbital plane). Swath width = 2700 km. The TOMS instrument failed in May 1993.

Data products: vertical distribution of ozone, global maps of total ozone and 200-mb height fields; incident solar ultraviolet irradiance and ultraviolet radiation backscattered from the Earth. SBUV helps to determine the total amount of atmospheric ozone in a vertical column above the subsatellite point; vertical profile of ozone above the ozone maximum; measurements of ultraviolet solar spectral irradiance and its temporal variability over the 160 - 400 nm range (with a spectral resolution of 1 nm). TOMS exploits the polar orbit to yield global, contiguous maps of total ozone concentrations.

SMMR = Scanning Multichannel Microwave Radiometer. Sensor mass = 52.3 kg, power = 60 W, bit rate = 25 kbit/s. Observation of sea-ice parameters, ocean surface conditions, atmospheric conditions, land parameters, glacial features. Microwave brightness temperatures were observed with a 10-channel (five-frequency dual polarized) scanning radiometer operating at frequencies of 37 (0.81 cm), 21 (1.42 cm), 18 (1.66 cm), 10.69 (2.8 cm), and 6.6

(4.54 cm) GHz. Six Dicke-type radiometers were utilized. Those operating at the four longest wavelengths measured alternate polarizations during successive scans of the antenna; the others operated continuously for each polarization. The antenna was a parabolic reflector offset from the nadir by 42 deg. Motion of the antenna reflector provided observations from within a conical volume along the ground track of the spacecraft. An identical instrument was flown on SEASAT (launch June 28, 1978). SMMR operations on Nimbus-7 ended on July 6, 1988.

THIR = Temperature Humidity Infrared Radiometer (built by Hughes SBRC). Similar instrument as flown on Nimbus-4, -5, and 6. The Nimbus-7 THIR operation ended on Nov. 30, 1987.

M.18 OICETS (Optical Inter-orbit Communications Engineering Test Satellite)

OICETS is a NASDA technology satellite with the objective to conduct, with ESA cooperation, on-orbit demonstrations of high-accuracy pointing, acquisition and tracking technologies (key elements of optical inter-orbit communications), important for future space activities such as global data reception from Earth observation satellites. The OICETS mission goal is to conduct inter-orbit laser communications experiments with ARTEMIS of ESA. The advantages of optical communications are seen in the support of much higher data rates and transmission capacities than in conventional radio links, lower risk of link interference with other optical communication systems (there can be no interference between optical and radio transmissions), as well as in more compact and lighter designs of communication equipment.

The configuration of the overall system for optical inter-orbit communication experiments consists of the following space segment elements: OICETS in LEO, DRTS (Data Relay Test Satellite) of NASDA in GEO, ARTEMIS of ESA in GEO. The ground segment consists of NASDA's TACC (Tracking and Control Center) and domestic ground stations as well as of an ESA ground station (experiments between OICETS and ARTEMIS). OICETS is controlled by TACC via an S-band inter-orbit link, or by an S-band direct link with TACC. The technical challenge of the technology of the optical experiment demonstration between LEO and GEO satellites involves alignment and stabilizing issues, it requires pointing errors of a few µrad by each S/C.

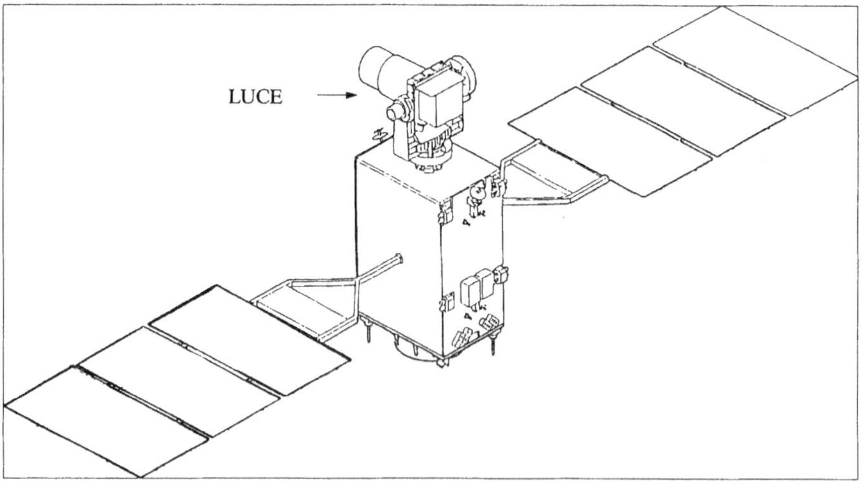

Figure 326: Illustration of the OICETS spacecraft

The OICETS satellite is three-axis stabilized with a box-shaped bus (of dimensions 0.78 m x 1.1 m x 1.5 m) and two solar panels (the S/C has a span of 9.4 m with both panels deployed). The S/C is built by NEC of Tokyo. The spacecraft power is 1220 W (EOL), NiMH batteries of 13 Ah provide power during the eclipse phase of the orbit. S/C mass = 570 kg, the mission life time of about one year. The attitude control subsystem (AOCS) consists of reaction wheels and magneto-rods for actuation, gyros, fine and coarse accuracy sun sensors, and Earth horizon sensors provide the attitude sensing. The attitude determination accuracy is specified within 0.2° (3-sigma). AOCS maintains the attitude of OICETS within specified limits, caused by LUCE movements/perturbations, by exchanging some data with LUCE. The pointing for inter-orbit communication is done at the LUCE level.

A launch of OICETS is planned on a J-1 rocket from the Tanegashima Space Center, Japan, for Feb. 2002.

Orbit: altitude = 610 km (BOL) to 550 km (EOL), inclination = 35°, period = 95 minutes.

Sensor complement:

LUCE (Laser Utilizing Communications Equipment), a NASDA-sponsored instrument built by NEC Corporation. The main objectives of LUCE in-orbit experiments are: [1717) 1718) 1719) 1720)]

- To evaluate the performance of optical devices in the space environment
- To evaluate the ATP (Acquisition Tracking and Pointing) capabilities of the system by using planets as targets
- To conduct inter-orbit communication experiments with ARTEMIS (including evaluation of atmospheric effects and micro-vibration effects caused by on-board equipment)
- To conduct optical communication experiments between OICETS and the CRL (Communication Research Laboratory) ground station.

Parameter	Forward link (ARTEMIS-OICETS)	Return link (OICETS-ARTEMIS)
Wavelength	819 nm 801 nm beacon	847 nm
Polarization type	LHCP	LHCP
Data rate	2.048 Mbit/s	49.3724 Mbit/s
Signal format	2 PPM (Pulse Position Modulation)	NRZ
Required bit error rate	10^{-6}	10^{-6}

Table 508: Laser link communication parameters

The optical part of LUCE (LUCE-O) is installed on the anti-Earth panel, the electronics unit of LUCE (LUCE-E) is located inside the S/C. The LUCE power consumption is about 230 W during communications support services.

- LUCE-O consists of three modules: an optical antenna, the optical system, and of two-axis gimbals. The optical antenna is a center-feed Cassegrain mirror telescope. The optical system, located behind the antenna and mounted on a thermally-controlled optical bench, consists of a laser transmitter, a receiver, ATP sensors and mechanisms, relay optics, and some electronics. The optical antenna and the optical system are mounted on two-axis gimbals, referred to as the "coarse pointing mechanism." The mass of LUCE-O is about 110 kg.
- LUCE-E includes the ATP control electronics and the communications electronics. The mass of LUCE is about 40 kg.

1717) T. Yamawaki, T. Jono, M. Toyoshima, K. Nakagawa, A. Yamamoto, K. Shiratama, Y. Koyama, " Development of LUCE for OICETS," 51st IAF Congress, Rio de Janeiro, Brazil, Oct. 2-6, 2000, IAF-00-M.2.05

1718) OICETS brochure of NASDA

1719) K. Nakagawa, A. Yamamoto, M. Toyoda, "Performance test result of LUCE engineering model," Photonics West, Free Space Laser Communications Technologies XII of SPIE, San Jose, CA, Jan. 22-26, 2000

1720) http://yyy.tksc.nasda.go.jp/Home/Satellites/e/oisets_e.html

The ATP system performs the followings tasks:

- In the acquisition phase, ATP compensates for the initial LOS (Line of Sight) error, caused mainly by the host S/C (OICETS) attitude error and ephemeris error.
- In the tracking phase, ATP compensates dynamic disturbances caused by the host S/C and the terminal itself within a few micro-radians.
- In the pointing phase, ATP points the beam in the direction of the partner S/C and compensates for relative motion of the S/C during the finite travelling time of the communication beam.

ATP subsystem	Tracking accuracy Pointing accuracy	<±2 µrad (3 sigma) for acquisition <±1 µrad (3 sigma) for tracking <±3.6 µrad (3 sigma) for acquisition <±2.6 µrad (3 sigma) for tracking
	Coarse acquisition and tracking components	Detector: CCD (672 x 488 pixels) Drive mechanism: 2-axis gimbals Actuator: direct drive motor
	Fine acquisition and tracking components	Detector: QD (Quadrant Detector) Drive mechanism: 1-axis mirror drive Actuator: multi-layered piezo-electric
	Receiving beam wavelength	Beacon: 797 to 808 nm Communication: 815 to 825 nm
Transmission subsystem	Laser diode Date rate Modulation	Wavelength: 843 to 853 nm about 50 Mbit/s NRZ (Non Return to Zero)
Receiver subsystem	Detector Wavelength Data rate Modulation	APD (Avalanche Photo Diode) 815 to 825 nm about 2 Mbit/s 2 PPM (Pulse Position Modulation)
Optical antenna	Diameter Wave front error FOV	26 cm (effective) <λ/20 (rms) < 1 mrad
MVE (Micro-Vibration Measurement Equipment)	Frequency range Range Resolution	0.5 to 1000 Hz ±100 mgal <50µgal
Items of pointing subsystems	Coarse-pointing mechanism	Fine-pointing mechanism
Structure	2-axis gimbals	Pair of independent 1-axis mirror drives
Angular coverage	Azimuth: 10-370° Elevation: 0-120°	±500 µrad
Tracking accuracy	±0.01°	±1 µrad
FOV (Field of View)	±0.2°	±200 µrad
Bandwidth	2 Hz	200 Hz

Table 509: Specification of the LUCE instrument

The ATP system supports the above functions by a coarse-pointing system, a fine-pointing system, and a point-ahead system. The coarse- and the fine-pointing systems perform the acquisition and tracking tracking of the incoming beam - while the point-ahead system maintains the pointing direction. The coarse-pointing mechanism consists of the two-axis gimbals, two optical encoders, and two direct-drive motors. A CCD detector array (672 x 488) is used as the coarse-pointing sensor for initial acquisition and tracking of the incoming beam. The fine-pointing mechanism consists of two assemblies leading the incoming beam for x- and y-axis respectively. Angular errors of the incoming beam misalignments of ±1 µrad are detected in the tracking phase. The point-ahead mechanism uses the same design as the fine-pointing mechanism. Quadrant silicon photo diodes (QD) are used to sense misalignments.

In addition to LUCE, OICETS is carrying MVE (Micro-Vibration Measurement Equipment) to sense the vibration environment of the spacecraft.

M.19 PICOSat (STP)

PICOSat is the name of an off-the-shelf microsatellite for the USAF Space Test Program (STP), in support of DoD space experiments. PICOSat represents a new procurement approach around a proven and low-cost microsatellite bus. In May 1997, a DoD contract was awarded to SSTL of Surrey, UK, calling for the acquisition of a microsatellite with all needed integration services, launch interfaces, on-orbit operations and payload data retrieval and distributions services by SSTL. All satellite experiments are furnished by DoD (managed by USAF/SMC). The mission is officially designated as STP mission P97-1. The PICOSat name was derived from the first letter of the initial four experiments. [1721)]

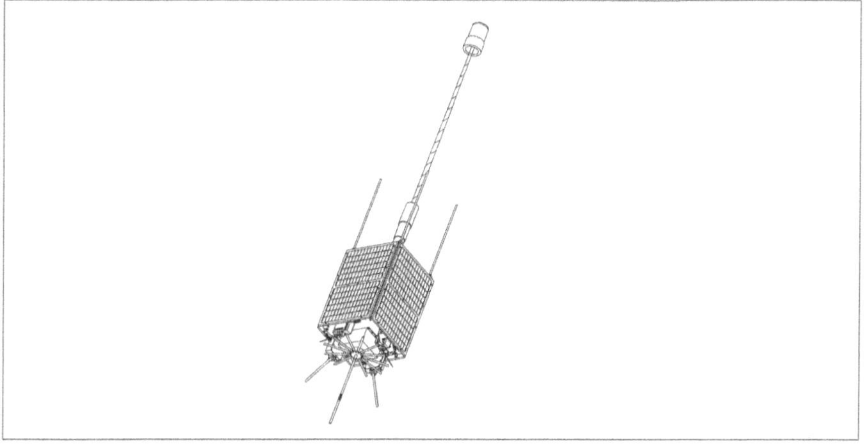

Figure 327: Illustration of the PICOSat spacecraft

The S/C platform is based on the latest version of the 'Micro-Bus' of SSTL (the platform, a modular tray design, developed in 1988 and first flown on UoSAT-3 and -4 in 1990). The Micro-Bus platform is a boxlike three-axis stabilized (Earth-pointing) structure with dimensions of 69 cm x 36 cm x 36 cm, comprising the following modules: battery (10 NiCd, 6 Ah), solar power (50 W max, 22 W average), uplink/downlink, telemetry/telecommand, onboard computer (OBC), and attitude determination and control (ADCS). The four outside surfaces are covered with solar panels (GaAs). ADCS employs a combination of passive gravity-gradient stabilization and active closed-loop damping (three-axis magnetorquer which acts against the magnetic field) and boom to maintain pointing within $\pm 1^{\circ}$ (nadir pointing of $\pm 5^{\circ}$ in pitch/roll). Attitude is determined by a combination of sun and Earth sensors and a magnetometer. OBC consists of a 80C186 CPU, 16 MByte RAM primary processor and a 80386EX CPU, 32 MByte RAM co-processor. Uplink communication is in VHF-band at 9.6 kbit/s, downlink in UHF-band at 9.6 kbit/s and 384 kbit/s. Both modulations are in CPFSK. The total S/C mass is 67 kg. Design life of one year with a three year goal.

A launch of PICOSat is planned for Sept. 2001 on an Athena-1 vehicle from Kodiak Island, AK. Other payloads on the same launch vehicle are: STARSHINE-3 (NRL), PCSat (Prototype Communications Satellite) of USNA, and SAPPHIRE of Stanford University. [1722)]

On-orbit mission control is conducted by the SSTL ground site in Guildford, UK. The US Air Force Academy at Colorado Springs, CO operates a backup ground station for PICOSat to increase the amount of experimental data.

1721) M. Tobin, et al., "Off-the-shelf Microsatellites for Science and Technology Missions," Proceedings of the 11th AIAA/USU Conference on Small Satellites, Sept. 15-18, 1997, Logan, UT

1722) "PICOSat Launch Set for August 2001," Space News, Dec. 18, 2000, p. 2

Orbit: Circular orbit, altitude = 800 km, inclination = 67°.

Sensor complement: a technology demonstration payload

PBEX (Polymer Battery Experiment), designed and developed at JHU/APL and sponsored by USAF/RL (Rome Laboratory). Objective: first demonstration of the charging and discharging characteristics of polymer batteries in the space environment. PBEX validates the use of lightweight, flexible battery technology to decrease cost and weight for future military and commercial space systems. In PBEX, both electrodes and the electrolyte are made of polymers. The all-plastic battery, composed of nontoxic materials, has the advantage that it can be recharged hundreds of times and can be operated in the temperature range of +50° C to -45° C. PBEX features a sandwich design offering more flexibility than conventional batteries. The anode and cathode are made of a foil-like sheet. The mass of PBEX is 0.4 kg, the size is 14 cm x 14 cm x 2.8 cm. PBEX can be commanded to alternate between charge and discharge cycles. Two charge and two discharge levels are available which can be selected from the ground. During an operational life of 18 months, PBEX will undergo more than 5000 charge/discharge cycles over a temperature range from -20° to +50° C. [1723)]

Parameter	Lead-Acid	Nickel-Cadmium NiCd	Lithium-Ion Li-Ion	Polymer Battery
Specific energy (Wh/kg)	10-35	20-37	100-130	20-90
Vol. specific energy Wh/l	50-80	40-90	300	25-40
Cycle life	200-150	250-2000	100-400	1000
Cell voltage	2.0	1.2	1.4-2.4	1.5-3.1
Discharge profile	Flat	Flat	Sloping	Sloping
Self discharge (%/month)	4-8	10-20	<1	2-5

Table 510: Parameter comparison of rechargeable battery types

IOX (Ionospheric Occultation Experiment) of USAF/SMC with instrument integration by The Aerospace Corporation. [1724)] Objective: to demonstrate ionospheric electron profile monitoring (content and densities) using GPS satellite occultation measurements. IOX is of TRSR (TurboRogue Space Receiver) heritage as flown on SUNSAT and Ørsted with improved codeless performance on 2 of the 8 channels. IOX permits the tracking of 8 GPS channels in parallel in the modes C/A, P1, P2 (if anti-spoof is off) or in C/A, P2 (effectively if anti-spoof is on). The IOX nominal sampling time is 0.1 Hz for navigation data and 1 Hz for ionospheric occultation data (there is also a 50 Hz sampling mode to measure ionospheric scintillation as well as observables in the troposphere). The receiver performs phase measurements of both GPS frequencies L1 and L2, and code pseudoranges. The on-board orbit determination function is not supported by IOX (although some orbit data are provided to the CERTO instrument).

The instrument consists of four elements: a dual-frequency receiver (Allen Osborne), a S/C interface electronics module (built by The Aerospace Corporation), an outboard low noise amplifier/filter (by Delta Microwave), and a dual-frequency GPS patch antenna (BATC). The patch antenna is mounted onto the outer face of one of two dummy sun-sensor blocks that are attached to the nadir-facing instrument platform of PICOSat. The IOX antenna points into the aft direction. Ionospheric observations by IOX are made in concert with ground-based measurements of the CERTO beacon whenever possible. Although ionospheric science is the primary focus of IOX, tropospheric occultation measurements will also be made if telemetry rates permit. The instrument mass is about 3 kg, power is about 10 W, the data rate is 6-50 Mbyte/day if operated continuously.

CERTO (Coherent Electromagnetic Radio Tomography), developed at NRL. The objective is to determine ionospheric electron density by using a three-frequency beacon; coop-

1723) Information provided by Joseph. J. Suter of JHU/APL, Laurel, MD

1724) Information provided by Paul R. Straus of The Aerospace Corporation, El Segundo, CA erative ionospheric observations with fixed ground receivers. The instrument is of CERTO heritage flown on ARGOS of DoD (launch Feb. 23, 1999). CERTO provides a global ionospheric map to aid the prediction of radio-wave scattering. This knowledge will improve navigation accuracy and communications capacity for military and commercial systems.

OPPEX (Optical Precision Platform Experiment), developed at AFRL. Objective: to demonstrate 20 dB vibration isolation from S/C to payload using active controls. OPPEX, also referred to as UQP (Ultra-Quiet Platform), aims to provide a 10:1 reduction in vibration isolation over a 100 Hz bandwidth between the spacecraft bus and a science payload. The intend is to reduce the launch cost and to improve the performance of spaceborne sensors.

M.20 PROBA (Project for On-Board Autonomy)

PROBA is microsatellite technology demonstration mission in ESA's General Study Program with the objective to address issues of on-board operational autonomy of a generic platform. The following functional capabilities and/or techniques are to be demonstrated: [1725] [1726]

- Command management coordination of on-board resources and house-keeping functions
- Scheduling, preparation, and execution of instrument observations (coordination of sensor settings, pointing angles, etc.)
- Source data handling functions (data collection, processing, storage, distribution)
- Source data communications management
- Performance evaluation and estimation of drifts, trends
- Failure detection and failure handling procedures

Further (secondary) objectives of PROBA are space environment investigation and Earth observation. The scientific interest relates to the use of the imaging spectrometer CHRIS. [1727]

PROBA was designed/developed by a consortium led by Verhaert Design and Development N. V. of Kruibeke, Belgium and sponsored by OSTC (Federal Office for Scientific, Technical, and Cultural Affairs) of Belgium. The S/C structure resembles a box (60 cm x 60 cm x 80 cm) of conventional honeycomb design with body-mounted solar panels (GaAs) on five sides. The satellite is three-axis stabilized. The ACNS (Attitude Control and Navigation Subsystem) performs autonomous on-board navigation. Attitude measurements are provided by an autonomous star tracker [referred to as ASC (Advanced Stellar Compass), a GPS-based attitude sensor (SRG-20), gyroscopes (one resonating gyro, built by Sagem of France, flies for the first time) and four three-axis magnetometers; attitude control is provided with four reaction wheels and four magnetorquers. All ACNS sensors and actuators are controlled by the ACNS software package (developed at the Université de Sherbrouk) running on the central ERC-32 RISC processor. There is no propulsion capability for orbit correction. S/C power = 90 W (peak) provided by body-mounted GaAs solar cells with integrated diode, a 36 Li-ion cell battery of 9 Ah capacity is used for energy storage. A centrally switched 28 V regulated bus distributes power to all sensors and subsystems. PROBA has a Memory Management Unit (MMU) with a 1.2 Gbit capacity for data storage. S/C mass = 110 kg. The S/C provides nadir and inertial pointing capabilities; along-track body pointing is up to 38°, while cross-track pointing is up to 30°. Yaw steering of up to 24° is only used for calibration purposes. The S/C design life is 2 years.

1725) D. Bernaerts, F. Teston, J. Bermyn, "PROBA (Project for On-Board Autonomy)," Proceedings of the 6th ISU Symposium on Smaller Satellites: Bigger Business?, Strasbourg, France, May 21-23, 2001

1726) F. Teston, R. Creasey, J. Bermyn, D. Bernaerts, K. Mellab, "PROBA: ESA's Autonomy and Technology Demonstration Mission," Proceedings of the 13th AIAA/USU Conference on Small Satellites, Logan UT, Sept. 23-26,1999, SSC99-V-8

1727) F. Teston, R. Creasey, J. van der Ha, "PROBA: ESA's Autonomy and Technology Demonstration Mission," IAA-97-1.3.05, 48th International Astronautical Congress, Oct. 6-10, 1997, Turin, Italy

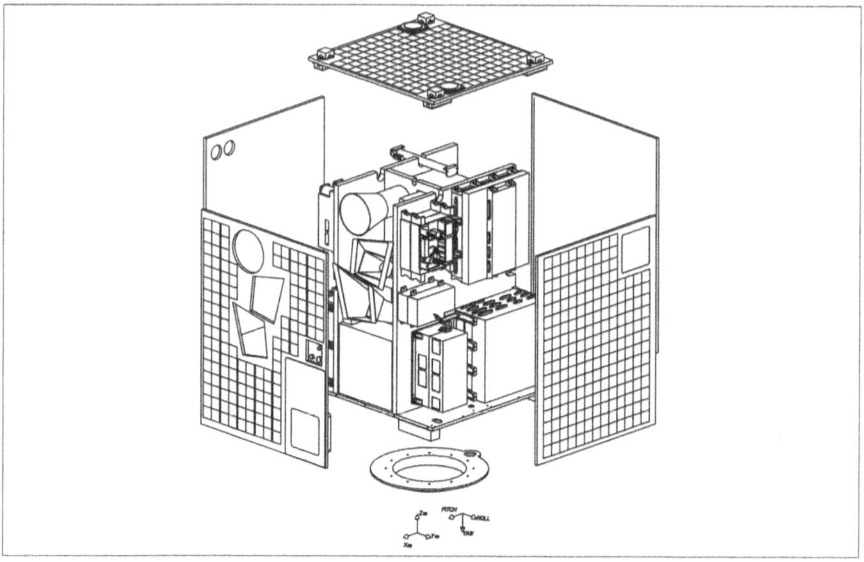

Figure 328: Illustration of the PROBA S/C

A launch of PROBA is planned on the PSLV launcher of ISRO (secondary payload to TES of ISRO and BIRD of DLR) for Sept. 2001.

Communications: CCSDS-compatible uplink (2-4 kbit/s, PSK/PM modulation) and downlink (up to 1 Mbit/s, transmitter power of 2 W, BPSK modulation) packetized communications in S-band.

Orbit: Sun-synchronous elliptical polar orbit, perigee = 575 km, apogee = 725 km, inclination = 97.7º, period = 101.3 min, repeat cycle = 16 days (approximately), equator crossing time at 10:30 AM on descending node. The pointing capability of PROBA permits a CHRIS imagery repeat cycle of approximately 2 days at mid latitudes.

Sensor complement:

CHRIS (Compact High Resolution Imaging Spectrometer), an AO instrument funded by BNSC of UK and built by Sira Electro-Optics Ltd of Chislehurst, Kent, UK. Objective: collection of BRDF (Bidirectional Reflectance Distribution Function) data for a better understanding of spectral reflectances. CHRIS is the prime instrument of the PROBA mission. The technology objective is to explore the capabilities of imaging spectrometers on agile small satellite platforms. CHRIS provides 19 spectral bands in the VNIR range (410 - 1050 nm) at a spatial resolution of 25 m and a swath width of 19 km. It can be reconfigured to provide 62 spectral bands at a spatial resolution of 50 m. There is a very broad range of application areas, including vegetation mapping, agricultural crop forecasting, forestry, water quality, air quality and pollution monitoring.[1728) 1729) 1730)]

The CHRIS instrument design comprises a catadioptric telescope, an imaging spectrometer, and an area detector array at the focal plane of the spectrometer. The focal length of the telescope is 746 mm, the aperture diameter is 120 mm (f/6). All refracting elements in the

1728) M. A. Cutter, D. R. Lobb, T. L. Williams, R. E. Renton, "Integration & Testing of the Compact High-Resolution Imaging Spectrometer (CHRIS)," Proceedings of SPIE, Vol. 3753, Denver, CO, July 19-21, 1999, pp. 180-191

1729) M. A. Cutter, D. R. Lobb, R. A. Cockshott, "Compact High Resolution Imaging Spectrometer (CHRIS)," IAA 2nd International Symposium on Small Satellites for Earth Observation, Berlin, April 12-16, 1999, pp. 205-208

1730) "Exploitation of CHRIS data from the PROBA Mission," Experimenters Handbook, Issue 4, Oct. 18, 1999 design are made of fused quartz. The telescope is axially symmetrical and has only spherical surfaces. The spectrometer uses "prisms" with curved surfaces integrated into a modified Offner relay. The dispersion of the spectrometer varies from about 2 nm to 10 nm across the spectral range, with the highest dispersion at 415 nm and the lowest at the high end (1050 nm) of the spectral range. A pixel registration of better than 5% is provided in the spectral and spatial directions, with resolution limited essentially by the detector pixel size. An area-array CCD detector at the focal plane provides pushbroom imagery. The detector is a thinned, back-illuminated, frame-transfer CCD (1152 rows and 780 columns) - the rows are assigned to separate wavelengths, while the CCD columns are used to separate resolved points in the image. The detector array operates in a frame transfer mode, with image and masked storage zones. The spectrometer fills <200 of the CCD rows, part of the nominally-unexposed area is used to calibrate for stray light and CCD smear effects.

Spectral range	410 nm to 1050 nm
Spectral resolution	1.25 nm to 11 nm
Spectral bands	19 band readout at 25 m spatial resolution, nominal mode for land studies 62 band readout at 50 m spatial resolution, nominal mode for aerosol studies
Spatial resolution	25 m at nadir, integration to 50 m, 75 m, etc.
Image size or FOV	18.6 km x 18.6 km at nadir (748 x 748 pixels)
Radiance range	albedo 1
Radiometric resolution	0.5% at 20% albedo
Body pointing capability	±30° in cross-track and ±50° in along-track direction
CCD detector area array	748 nominal resolved elements per swath width (cross-track direction) 576 lines in the along-track direction (about 200 lines are used for spectral resolution - the others are used for smear/stray-light correction) The total frame time for 25 m ground sampling is 12.7 ms
Electronics features	Programmed line integration and dumping on chip for spectr. band selection Pixel integration on chip for spatial resolution control Correlated double sampling (noise reduction circuit) Dynamic gain switch for optimum usage of the ADC resolution 12 bit data quantization (ADC)
Instrument size	about 200 mm x 260 mm x 790 mm
Instrument mass, power	14 kg, about 10 W

Table 511: Some specification parameters of the CHRIS instrument

The CHRIS imagery is processed on-board, i.e. analyzed and compressed, by a TEMIC DSP (Data Signal Processor), a space-qualified TSC 2102OE device, [1731] and stored in MMU (Memory Management Unit), both part of PPU (Payload Processing Unit).

The platform/instrument can be commanded to perform the following functions:

- Target location - requiring roll maneuvers to point cross-track
- Viewing directions for each target in one orbit - requiring pitch maneuvers to point along-track
- Spectral bands and spectral sampling interval in each band
- Spatial sampling interval

Atmospheric science objectives of CHRIS focus on aerosols, which as well as being important for weather and climate, are also a consideration for accurate atmospheric correction of satellite data. Operational plans call for a total of 30 test sites: 15 for aerosol/atmosphere studies, 10 for land surface studies and 5 for coastal studies. Aerosol studies include a number of different continental, marine, urban and desert test sites. Land surface sites include temperate agricultural areas, boreal forests and semi-arid areas.

1731) TEMIC was a daughter of Daimler-Benz until 1997 when it was acquired by Vishay and in 1998 sold to Amtel (with plants in Heilbronn, Germany and Nantes, France)

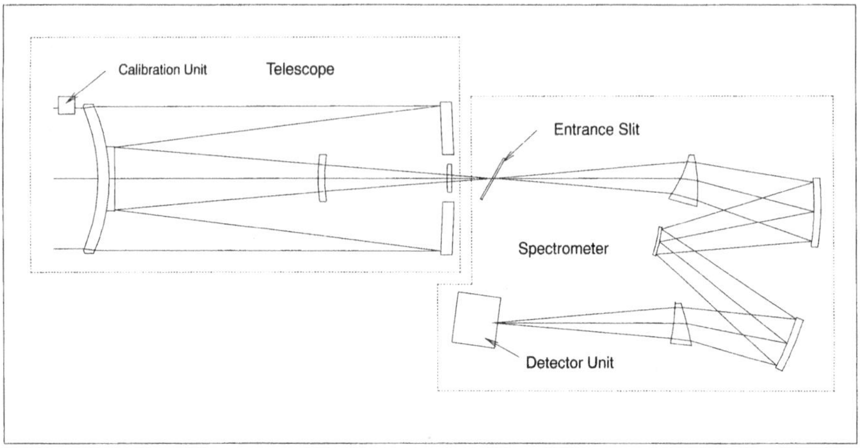

Figure 329: Instrument optical design

Figure 330: Illustration of the CHRIS instrument

Performance: a) 3 particle detectors for directional electron and proton spectroscopy (measurement error < 1%) b) Detection and counting of cosmic ray events, c) Internal and external Radfets for total radiation dose measurement d) Processor based autonomous operation for several days
Design parameters: a) Instrument mass: 2.5 kg, including external Radfets, b) Instrument size: 95 mm x 122 mm x 217 mm, c) Use of standard OBDH/RTU interfaces, d) Instrument power: 20 - 50 V primary power bus input voltage

Table 512: SREM instrument performance/design parameters

SREM (Standard Radiation Environment Monitor), an ESA/ESTEC instrument of REM heritage flown on STRV-1b and the MIR space station (Oerlikon-Contraves of Switzerland is the prime contractor of the instrument). The objective of SREM is to measure energetic electrons (0.3-6 MeV), protons (8-300 MeV), and heavy ions, as well as the total accumulated dose, encountered during the mission.

DEBIE (DEBris In-orbit Evaluator), built by Metorex International and Patria Finavitec, both of Oy, Finland. The objective is to monitor sub-millimeter sized particles which impact the detector surfaces - measurement of mass ($>10^{-14}$ g), impact speed and penetration power. The instrument consists of a central processing unit and up to 4 separate sensor units (each 10 cm x 10 cm) which can be placed on different spacecraft surfaces (two are on the ram side and two on the deep space face). DEBIE uses a combination of impact ionization, momentum and foil penetration detection. The instrument is designed as a standard detector to be flown on different spacecraft and missions with little or no modifications (planned to fly on ISS). DEBIE on PROBA uses two impact detectors located on the panel; one detector is looking in the flight direction, the other detector is looking in the cross-track direction. The DEBIE data is used in risk assessment models and for the design of protective shielding.

Demonstration of Autonomy Technologies:

The attitude control and avionics subsystems accommodate the core technologies for S/C autonomy. The particular subsystems are:

- **SGR-20** (Space GPS Receiver 20), built by SSTL, UK (heritage of UoSat-12 and TMSat). Objective: demonstration of autonomous operations for orbit and attitude determination (alternatively to ASC). The instrument consists of a 24 channel GPS L1 and C/A receiver, and four antennas for position and medium-accuracy attitude determination. The SRG–20 instrument is designed to be tolerant to radiation effects.

- **ASC** (Advanced Stellar Compass), developed and built by DTU (Technical University of Denmark). ASC is already flown on Ørsted, a Danish geomagnetic research microsatellite mission that was launched Feb. 23, 1999 (see also ASC description under Ørsted E.18 and CHAMP E.1). ASC is an autonomous star tracker, the main attitude sensor of PROBA, with the objective to provide full-sky coverage and to achieve a high pointing accuracy (precision of a few arcseconds) for Earth observation. ASC is capable to reconstruct autonomously the S/C inertial attitude starting from the condition "lost in space." The camera of the star imager is a 752 x 588 pixel CCD device. ASC consists of a camera head unit (CHU) connected to a DPU (Data Processing Unit), i.e. a microcomputer fitted to a frame-grabber. The CHU acquires star images within its FOV, while the DPU provides the processing power to perform image analysis, pattern recognition, data reduction, and communication.

- **PASS** (Payload Autonomous Star Sensor), developed and built in collaboration by Sira Electro-Optics Ltd and MMS (UK) for a demonstration flight on PROBA. The instrument is an autonomous star tracker with a FOV of 19.3° x 14.4°, capable of immediate recovery from a "lost-in-space" condition. The PASS elements are: a) an optical head (50 mm focal length, f/2) with a CCD detector array (770 x 576) and an ADC unit (12 bit quantization); b) a DSP-based processor (radiation-tolerant STAR DSP board), which reads relevant pixels from the head buffer, calculates centroids of star events, and hosts the star identification software; c) star identification software, which identifies stars in each camera frame and estimates the pointing direction from this data. PASS provides pointing accuracies of 3 arcseconds in pitch and yaw and 30 arcseconds in roll at update rates of 5 Hz (up to 20 Hz with reduced accuracy). The instrument has a mass of about 2.4 kg, power consumption of 9 W, and standard RS 422 / IEEE 1355 interfaces. [1732)]
There are several names for the instrument: a) WASS (Wide Angle Star Sensor), the original name used for the first customer, DERA; b) PASS (Payload Autonomous Star Sensor) is

1732) R. Cockshott, D. Purll, N. Fillery, V. Lewis, "The UK Wide Angle Star Sensor (WASS)," Presented at the poster session of the 4th ESA International Conference on Spacecraft Guidance, Navigation and Control Systems, Oct. 18-21, 1999, Noordwijk.

the name used in the PROBA technology demonstration; and c) AST20 (Autonomous Star Tracker 20) is the name chosen for commercial applications.

- **ERC-32 high-performance RISC processor**, funded by ESA, is provided by TEMIC Semiconductors (the DHS uses two of these processors). ERC-32 is a space version of a standard commercial processor, a high-performance radiation tolerant (>80 krad) SPARC V7 processor, providing 10 MIPS and 2 MFLOPS with a floating point unit. A memory controller includes all peripheral functions. The ERC32 computer is able to support a number of on-board processing functions including science data processing. This capability enhances the S/C autonomy with respect to data distribution through the selective use of the downlink and the on-board mass memory.

- **VMC** (Visual Monitoring Camera) is a stand-alone wide-angle digital camera system, based on CMOS APS imaging technology, and designed and built by DSS/OIP and IMEC of Belgium. The camera is modular, offering a choice of various support options with regard to interfaces, detectors, color or grey-scale, and exchangeable optics. VMC accepts image capture and exposure control commands, performs image acquisition, and is capable of storing one image in an on-board buffer to facilitate readout at a lower data rate. In addition, VMC accepts commands from, and transmits imagery to, standard serial digital interfaces such as TTC-B-01 synchronous and RS-422-like asynchronous.

Spectral range	400 - 650 nm, providing color imagery or B&W
Focal length, F-number	12.2 mm, f/5
FOV (Field of View)	40° x 31° - using IRIS-1
Distortion	<0.5%
Resolution	10 cycles/mm in detector plane
Detector size	640 x 480 pixels; 14 μm pitch (IRIS-1)
Data quantization	7 bits
Image capture speed	200 ms
Image download speed	1 image/s (max)
Operational modes	autonomous or command-interactive operation
Interfaces	Synchronous serial TTC-B-01-like up to 1 Mbit/s, or Asynchronous serial RS-422 up to 3.125 Mbit/s
Power supply	either 28 VDC or 10 VDC
Power consumption	3.0 W (28 V version); 2.0 W (10 V version)
Instrument size (mm)	65 (width) x 60 (height) x 103/108 (depth)
Instrument mass	0.430 kg (excluding mounting bracket)

Table 513: VMC instrument parameters

VMC employs IMEC's IRIS-1 (Integrated Radiation-tolerant Imaging System), a single-chip integrating CMOS imager. IRIS-1 was developed for general purpose imaging applications in space, including visual telemetry, robot vision, planetary imaging, and land/rover imaging. The following functions are integrated on one chip: a focal plane detector array, double sampling readout structures, ADC (Analog Digital Converter), and control/interfaces. VMC flown on PROBA implies a technology demonstration.

- **HRC** (High Resolution Camera), an ESA/ESTEC instrument built by DSS/OIP of Belgium. HRC is of VTS (Visual Telemetry System) heritage, test-flown on TEAMSAT and XMM (X-Ray Multi-Mirror Mission) of ESA. The objective is to demonstrate a high-resolution imager. HRC is a miniaturized black and white imager with 10 m ground resolution. The telescope is of Cassegrain type with an aperture size of 115 mm and a focal length of 2296 mm. The CCD detector array uses 3D packaging technology. It contains 1024 x 1024 pixels of 14 μm size. The field of view (along the diagonal of the detector) is 0.504°. Images are digitized to 10 bits before transmission to the spacecraft.

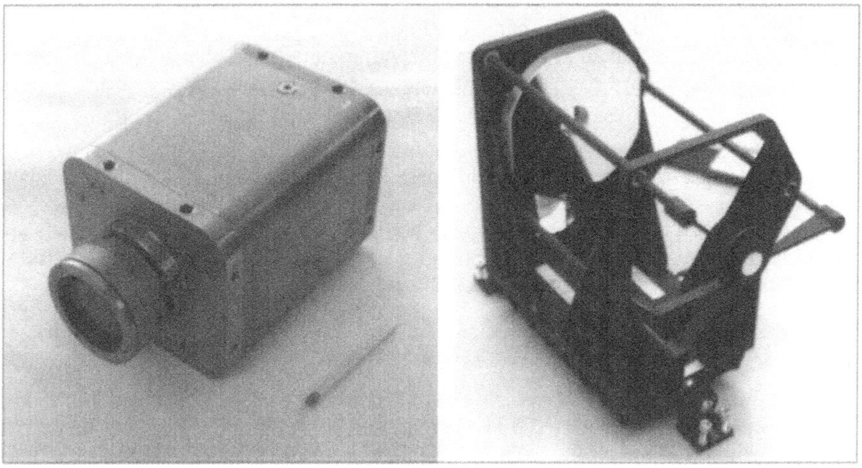

Figure 331: Illustration of the VMC (left) and HRC (right) instruments

- **Mission operations concept.** PROBA software provides considerable flexibility in the allocation of on-board resources and in scheduling of operations when compared to the relatively rigid concepts used in conventional missions. In particular, the following operations functions are being planned for implementation on PROBA:

1) On-board housekeeping: PROBA monitors autonomously all routine housekeeping and resource management tasks. This includes also the decision-making process in the case of (foreseeable only!) anomalies, i.e. failure detection, failure identification and first-level recovery actions. A summary of the information available on-board is downlinked to the control centre at regular intervals.

2) On-board data management: PROBA takes care autonomously of all management tasks related to the on-board data handling, storage, and downlinks. The S/C uses a 1 Gbit mass memory for data recording and a tuneable 2 Kbit/s to 1 Mbit/s downlink. As a worst case, two passes of about 10 min every 12 hours are available for downlink of telemetry data (as well as for command uplinks).

3) On-board resources usage: PROBA takes care autonomously of all management tasks related to the on-board power usage. Any excess power and energy (above the basic spacecraft control requirements during daylight and eclipse phases) is allocated to the instruments and to the spacecraft subsystems supporting the specific operations of the instruments, for instance to the wheels for attitude maneuvers. The allocation is performed on a dynamic basis, resolving task constraints and priorities. Constraints include for each activity the power and data storage area needed, the pointing requested, etc.

4) Instrument commanding: All preparatory, commanding, and data processing activities related to the instrument operations are performed on-board (after an appropriate initialization period). This includes the planning, scheduling, resource management, navigation, and instrument pointing as well as the downlinks of the processed data. In the case of CHRIS, the main EO instrument, a specific attitude (pointing and rates) of the S/C is required to perform its Earth observation function. The calculations of the relevant slew characteristics are based on a request file (containing the coordinates of the target area and the observation duration) which is uplinked from ground.

5) Science data distribution: The collected science data are normally downlinked to the (nominal) ground station from where they may be routed automatically or on request

to a user's site using Internet links. Furthermore, it is planned to demonstrate an automatic direct data distribution capability to different user ground antennae upon their requests without human involvement and with minimum possible delay. The optimal downlink times may be uplinked in the request file or calculated on-board.

The ground segment of PROBA is based on a portable ground station, located at Redu, Belgium, and additional support from small stations to validate the concept of distributed users. Users around the world can ask the satellite to take pictures by sending a request through internet. A web server at the ground station handles the request and uplinks it to the satellite, which schedules autonomously and takes the imagery. The picture data are stored in the on-board mass memory unit, they are downlinked during a ground station pass, after which it becomes available to the user.

M.21 RADCAL (Radar Calibration Satellite)

A microsatellite built by CTA Space Systems of McLean, VA for the USAF under the Small Test and Small Launch Vehicle (ST&SLV) program. The satellite is hexagonal in shape with 762 mm diameter, it has a total mass of 87kg. It is passively stabilized to a nadir-pointing attitude by a gravity gradient boom and magnetic nutation dampers (magnetorquers and magnetometer). A prime objective is to provide space-based radar cross-sectional area calibration (via two C-band transponders) for more than 70 military and civilian radars operating in C-band (the S/C position is known to a fair degree of accuracy, thus allowing ground-based users to calibrate their systems by acquiring the RADCAL satellite and comparing their position with the "known" position). A further objective is to demonstrate spaceborne GPS-based attitude determination. RADCAL also carries a peak power tracker and a UHF Store and Forward payload. The design mission life is three years.

The S/C was launched on June 25, 1993 on a Scout launch vehicle from VAFB, CA. Mission design life is 1 year with a 3-5 year goal. The satellite transmits/receives at 306.775 MHz. [1733) 1734)]

Initial S/C operations were performed from SMC (USAF/Space & Missile Systems Center at Los Angeles AFB). Since November 1996 RADCAL is controlled from Kirtland AFB, NM (SMC/TEO).

Orbit: Near-circular polar orbit, altitude = 815 km x 765 km, inclination = 89.5°.

Sensor complement:

TANS Quadrex GPS receiver (Trimble Navigation). Objective: to demonstrate post processing attitude determination Two receivers (channels: 6+6, L1 C/A) were modified by Stanford University to measure the differential phase of GPS signals arriving at four microstrip patch antennas mounted on the zenith face of the vehicle and canted outward 17.5°. The phase differences between each of three slaves and a master antenna are downloaded for post-processing.

SSPSR (Small Satellite Power System Regulator), built by PL. The objective is to test improved methods for battery charging.

M.22 SJ (Shi Jian Program)

Shi Jian (SJ, meaning "practice" or "experimental") is a scientific as well as a technology demonstration minisatellite series of CAST (Chinese Academy of Space Technology)

1733) L. M. Ward, P. Axelrad, "A Combined Filter for GPS-Based Attitude and Baseline Estimation," Navigation: Journal of The Institute of Navigation, Vol. 44, No. 2, Summer 1997, pp. 195-213

1734) L. M. Ward, P. Axelrad, "Spacecraft attitude estimation using GPS: Methodology and results for RADCAL,." Navigating the 90s: Technology, Applications, and Policy, Proceedings of The Institute of Navigation, National Technical Meeting, Anaheim, Calif., 18-20 January, The Institute of Navigation, Alexandria, Va., pp. 813–825.

which started in the early 1970s. It includes spin-stabilized spacecraft and an a three-axis stabilized spacecraft. An overview of the missions is given in Table 514. For SJ-2, all three spacecraft were launched by a single launch vehicle. The SJ-4 launch was from the Xichang launch site with a Long March-IIa vehicle. [1735)]

SJ-4: The objective was to monitor the space environment and its effects. The S/C, built by CAST, has a mass of 397 kg with dimensions: 1.6m diameter and 2.1m high. The main instruments are: Semiconductor high energetic electron detector, Semiconductor high energetic protons, heavy ions detector, Electrostatic analyzer, Electric potential meter, Static Single Events Upset monitor, and Dynamic Single Events Upset monitor.

Spacecraft	Launch Date	S/C Mass	S/C Orbit	Payload
SJ-1 (Shi Jian)	Mar. 3, 1971	221 kg	266 km x 1826 km, I= 69.9°	Particle detectors
SJ-2 SJ-2A SJ-2B	Sept. 20, 1981	250 kg 480 kg 257 kg	232 km x 1598 km, I= 50.9°/5° 232 km x 1615 km, I=69.9° 232 km x 1608 km, I=69.9°	Ionospheric studies and other experiments
SJ-4	Feb. 8, 1994	397 kg	210 km x 36125 km, I=28.6°	Particle detectors
SJ-5	May 10, 1999	298 kg	870 km, sun-synchronous	Particle detectors, etc.

Table 514: Overview of SJ series satellite launches

M.22.1 SJ-5 (Shi Jian - 5)

SJ-5 is a Chinese scientific and technology demonstration satellite. It is the first modern small satellite of China. The satellite platform was developed by CAST, CSSAR (Center for Space Science & Applied Research) of CAS was responsible for the development of the payload and the ground application system. The overall objectives of the mission include:

- Detection of SEU (Single Event Upset) and investigation of the corresponding countermeasures. In addition, dose measurment of the high-energy particles in the space environment
- Demonstration of a space science fluid experiment
- Demonstration of advanced space technology experiments including the operation of multiple-mode attitude control, integrated onboard electronic system, a unified S-band TT&C system, S-band data transmitter, on-board data management system and large-capacity solid state storage.

The SJ-5 spacecraft structure resembles a box with overall dimensions of 1.2 m x 1.1 m x 1.04 m. The total deployed length of SJ-5 is about 5 m. The spacecraft features two modes of attitude stabilization, namely three-axis as well as spin-stabilized attitude control. The spin rate of S/C is between 10-15 rpm during the spin control mode (a comparison of the two control modes is an experiment objective). Operations in spin control mode are conducted about the axis of least inertia. The attitude measurement information is provided by a sun sensor and an infrared Earth sensor. Actuation is provided by a momentum wheel, magnetorquers and a cold-gas propulsion system. The total S/C mass is 298 kg; the payload mass is 58 kg; an average power of 165W is provided by the solar array of 3.4m^2 in size. NiCd batteries of 17Ah capacity are used for night phase power supply. The S/C design life is three months.

The SJ-5 satellite was launched on May 10, 1999 from the Taiyuan launch site in China as secondary payload together with the primary payload of FY-1/C (meteorological satellite of China) on the LM-4B launch vehicle. User requirements called for the following mission scenario: a) the first three months of the mission were dedicated to the science goals of the mission (these were achieved successfully), and b) the subsequent mission life was dedicated to technology demonstrations. As of spring time 2001 CAST is still able to receive data from SJ-5.

1735) Information provided by Lihua Zhang of CAST, Beijing, China

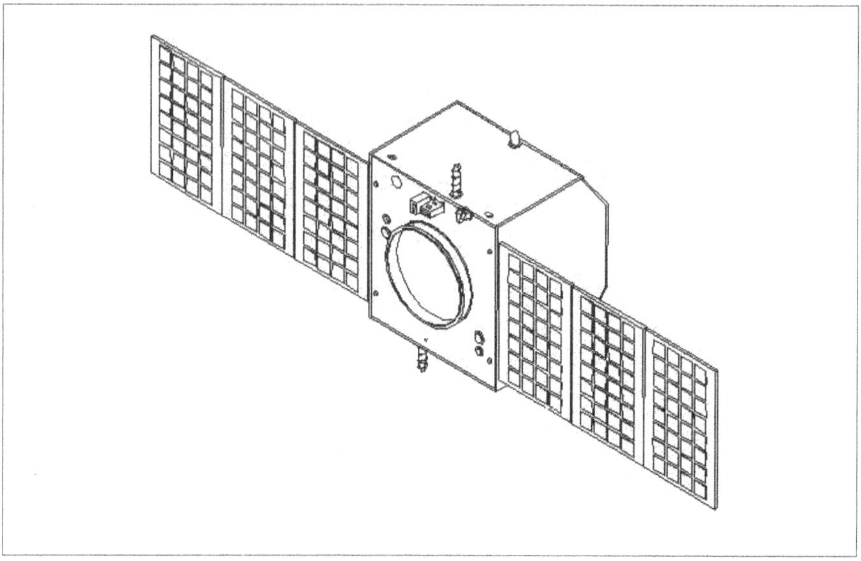

Figure 332: Illustration of the SJ-5 spacecraft

Orbit: Sun-synchronous circular orbit, altitude = 870 km, inclination = 98.8°, nodal period = 102.35 minutes.

Sensor/experiment complement: There are 11 science experiments onboard, developed by CSSAR and by the Institute of Mechanics (joint program). The experiments or instruments are:

- SEP (Single Event Phenomena) Monitor
- SEP Shielding Experiment Device
- SEU (Single Event Upset) Detector
- SEP synthetical detector
- SEL (Single Event Latch-up) Experiment
- High-energy Proton / Heavy Ion Detector
- High-energy Electron Detector
- PMOS Dosimeter (a radiation dose detector)

The eight instruments are used to measure the SEU effect and the high-energy particle environment. The observation of SEU, the investigation of the countermeasures and the measurement of the radiation in the space charged particle environment are more comprehensive compared with those in the SJ-4. With eight instruments, the investigation of SEU countermeasures includes almost all the methods known today. The measurement range of the energy spectra of the high-energy charged particles is the most completed so far, and the measurement of the radiation total dose is the first ever taken in china .The charges particles distributions in the abnormal area of the south Atlantic and in the subsiding areas of the two poles in the orbit have been obtained for the first time with the instrument developed by CSSAR. The collected data and the research results are significant for improving the design of space electronic instruments, raising the reliability and extending satellite's service life.

The experiments "Solid-state Transmitter" and "Solid-State Memory" are technology demonstrations.

- Space Fluid Experiment Device. A micro-gravity experiment in the field of fluid mechanics. The objective is to investigate the coupling between heat capillary convection and buoyancy convection, and other phenomena under the condition of micro-gravity. The study of these phenomena and the laws governing them will advance the development of fluids. There are close relations among micro-gravity mechanics of fluids, space material science (crystal growth), space life science (cell breading) and other disciplines. Multi-layer heat capillary convection is one of frontier subjects in the micro-gravity fluid science research. The technology is very important not only for developing the methods of space crystal material growth, improving the quality of crystals, but also for a better understanding of the characteristics and the mechanisms of the convection movement of the fluid with a free surface or a boundary surface on the ground, which is a fundamental problem in the physics of fluids. This experiment in microgravity fluid mechanics is the first ever carried out by Chinese scientists, and also the most complex one so far.

M.23 SMART-1 (Small Mission for Advanced Research in Technology)

SMART-1 is an approved ESA minisatellite technology mission to the moon (part of the ESA Horizons 2000 Science plan) with the objective to demonstrate innovative and key technologies for scientific deep-space missions. The key objective is to qualify solar electric propulsion (ion drive, but also the peculiar flight dynamics and orbit control concepts) as the primary means of propulsion in deep-space missions, tested in a transfer and gravity-assisted orbital trajectory to the moon.. A mission goal is to spend six months in orbit around the moon, using a suite of instruments to study its composition. Science objectives are to address key questions about the moon: the origin of the Earth-moon system and the role of accretionary processes, long-term volcanic and tectonic activity on Earth's natural satellite, the thermal and dynamic processes responsible for lunar evolution and the external processes on the moon's surface such as erosion, and ice deposition. SMART-1 demonstrates also a new approach in terms of implementation strategy (low-cost) and procurement for ESA. [1736) 1737) 1738) 1739) 1740) 1741)]

The prime contractor for the SMART-1 platform is the Science Systems Division of SSC (Swedish Space Corporation) of Solna, Sweden. The ASC (Advanced Stellar Compass) of the attitude system is manufactured by DTU, Denmark (heritage of Astrid-2, TEAMSAT, CHAMP, SAC-C, ADEOS-II, GRACE, and PROBA). A launch of the satellite is planned for 2002 as an auxiliary payload on an Ariane-5 vehicle.

S/C is three-axis stabilized	Two autonomous star trackers referred to as ASC (Advanced Stellar Compass), four reaction wheels, four angular rate sensors, eight 1 N attitude control hydrazine thrusters One two-axis gimballed platform for the PPS-1350 system
Solar power	1875 W BOL with multi-junction cascade cells
Batteries	Five Li-ion battery cells, each with 130 Ah
Xenon propellant mass	<82 kg (for PPS-1350)
Communication	Two S-band TT&C transponders, 65 Kbit/s data rate, the mass of the TT&C system is <10kg
S/C launch mass, payload mass	<350 kg, <15 kg

Table 515: Characteristics of the SMART-1 satellite

Orbit: The SMART-1 S/C is intended to reach lunar orbit starting from GTO (Geostationary Transfer Orbit) by using electric propulsion. The overall mission can therefore be divided into the following phases:

1736) G. D. Racca, A. Elfving, A. Marini, et al., "SMART-1 mission description and development status", Submitted to Planetary and Space Science, MS-No: PSS 79, October 30, 2000

1737) G. D. Racca, G. P. Whitcomb, B. H. Foing, "The SMART-1 Mission," ESA Bulletin 95, Aug. 1998, pp. 72-81

1738) http://www.estec.esa.nl/spdwww/smart1/html/overview.html

1739) http://sci.esa.int/smart/

1740) B. H. Foing, G. Racca, A. Marini, et al., "Status of SMART-1 ESA Mission to the Moon," 31st Lunar and Planetary Science Conference, March 13-17, 2000, Houston, TX

1741) http://www.ssc.se/ssd/smart1.html

- Launch and early GTO phase. The standard GTO of Ariane has an apogee of 35781 km and a perigee of 622 km.
- The cruise phase to the moon will last for about 12 months
- Moon capture phase and descent to the operational orbit
- Lunar observation phase. The present baseline plan considers a lunar orbit with an apocenter of 10,000 km and a pericenter of 300 km to 1800 km. A six month observation period in lunar orbit is followed.

Figure 333: Illustration of the SMART-1 S/C

Communication: The downlink data rate is 65 kbit/s or 2 kbit/s. The modulation scheme is PCM/PSK/PM for the low data rate and PCM/SP-L/PM for the high data rate. The ranging signal in the ranging channel of the transponder directly phase-modulates (PM) the downlink carrier. The uplink frequency is 2058.2 MHz, the downlink frequency is 2235.1 MHz.

Sensor/Technology Demonstration Complement:

PPS-1350 (Propulseur Plasmique Stationaire - or Stationary Plasma Thruster-1350), developed by SNECMA of France (thruster) and ETCA of Belgium (power processing unit). The 1350 label designates the nominal input power (in Watt). The PPS-1350 is an evolution of the SPT-100 (Stationary Plasma Thruster), built by the Fakel Design Bureau in Kaliningrad, Russia. The demonstration of the electric propulsion and attitude control actuator functions are primary objectives for the SMART-1 mission (the S/C achieves lunar orbit by means of the electric propulsion). The duration to reach the moon after injection into a standard Ariane-V GTO (apogee=35781 km, perigee=622 km), is of the order 12 months. Together with the electric propulsion, a series of gravity assists by the moon are also exploited to reduce the amount of fuel consumed.

Principle of operation of the ion engine which uses Xenon as propellant: Electrons from an external cathode enter a ceramic discharge chamber (10 cm in diameter), attracted by an anode. On their way to the anode, the electrons encounter a radial magnetic field created between inner and outer coils, causing cyclotron motion around the magnetic field lines. Collisions between drifting electrons and xenon propellant create the plasma. The ions created are then accelerated by the negative potential existing near the exit of the chamber due to the Hall effect. The external cathode acts also as a neutralizer, injecting electrons into the thrust beam to maintain zero-charge equilibrium both in the beam and on the spacecraft. This type of thruster provides a nominal thrust of 70 mN at 16,000 m/s specific impulse (Isp) and 1350 W of nominal input power from the solar array.

EPDP (Electric Propulsion Diagnostic Package), provided by Laben/Proel of Florence, Italy, is a technology payload to study the plasma environment around the SMART-1 spacecraft and the contamination on its external surfaces, when the electric propulsion system is both on and off. EPDP consists of a suite of sensors, comprising a proximity Langmuir probe, a Retarding Potential Analyzer and two Deposition sensors (a Solar cell and a Quartz Crystal Micro-balance). EPDP performs key measurements to characterize the electric propulsion system and its influence on the spacecraft environment. The total mass of EPDP is about 2 kg.

SPEDE (Spacecraft Potential, Electron and Dust Experiment), jointly developed by FMI (Finnish Meteorological Institute), Helsinki, ESA/ESTEC, IRFU (Uppsala Sweden) and KTH (Stockholm, Sweden). SPEDE consists of two electric sensors mounted on the ends of 60 cm booms. These sensors can be driven either as a Langmuir probe or as an Electric Field probe. The instrument complements the EPDP diagnostic package in monitoring the solar electric propulsion, by measuring potential and induced currents by means of a Langmuir probe accommodated on the tips of the booms. It also performs cruise science and lunar orbit science by monitoring the space/time variations of the plasma and electron environment, especially coupled to solar wind effects. The mass of SPEDE is 0.8 kg.

RSIS (Radio Science Investigation Series), a set of radio-science investigations conceived by the University of Rome, Italy. RSIS aims at validating the technique of measuring the rotational state of a celestial body from the orbit (in particular the physical librations in latitude) by combining orbit and attitude determination with accurate imaging. RSIS will test this method on the well-known moon librations making use of the capabilities KaTE Deep-space transponder for precise Doppler and ranging measurements, preparing also for future high precision geodesy and relativity experiment. In addition, RSIS will monitor the dynamic performance of the electric propulsion system, by precision tracking during thrusting periods.

KaTE (X/Ka-band TT&C Experiment), built by Astrium GmbH and managed by ESA/ESTEC. KaTE is an experimental deep-space telecommunication subsystem operating in the X-band (8 GHz) and Ka-band (32/34 GHz) frequencies. KaTE will demonstrate telecommunication capabilities with superior data rate and high accuracy, will characterize the Ka-channel and will perform experiments of all the key TT&C functions (uplinking commands in X-band, downlinking data in X- and Ka-band, ranging, tracking etc.). A new encoding/decoding technique (Turbo code) is also being tested, and possibly also the operation of VLBI (Very-Large Baseline Interferometry). In addition, KaTE supports the RSIS experiment, thereby permitting the characterization of the electric propulsion. With its high-data rate downlinking capability, KaTE will contribute to the transmission of science data from the SMART-1 spacecraft.

D-CIXS (Demonstration of a Compact Imaging X-ray Spectrometer), the design is led by RAL (Rutherford Appleton Laboratory) of UK (the D-CIXS team includes members from the University of Helsinki; University of Sheffield; MPAe of Katlenburg-Lindau, Germany; IRF of Kiruna, Sweden; and NOAA of Boulder, CO). The objective is to demonstrate a radically new approach to greatly reduce the mass and volume of the instrument. D-CIXS represents a new generation of X-ray imagers for planetary observation, based on the use of an advanced dual microstructure collimator and X-ray detector technologies. D-CIXS includes also **XSM** (X-ray Solar Monitor), provided by the Observatory of the University of

Helsinki, for absolute calibration and solar X-ray flux monitoring. The D-CIXS instrument images high-throughput fluorescence X-rays from the SMART-1 target object, as a means to carry out spatially resolved material analysis of the moon's surface. Mapping of the composition of the moon, in particular key elements like iron, silicon, magnesium, sodium, carbon, and oxygen with a spatial resolution of 30 km in the energy range from 0.5 to 10 keV. This permits to test formation theories of the Earth-moon system. D-CIXS provides the first high-resolution global map of moon-surface chemistry and the relative abundances of the major lunar rock types. The instrument mass is 3 kg including its electronics. [1742] [1743] [1744] [1745]

Telescope FOV	12° x 24°
Detector array, SCD pixel size	24 SCDs (Swept Charge Detectors), 10 mm x 10 mm
Quantum efficiency	40% @ 0.5 keV to 80% @ 5 keV
Operating temperature	270 K
Energy band, energy resolution	0.5 keV to 10 KeV, 140 eV
Sampling rate	50 Hz
Source data rate	2.74 kbit/s

Table 772: Parameters of the D-CIXS instrument

The spectrometer design is based on "swept charge detectors" (SCD), developed by EEV, Chelmsford, UK, and "micro-collimator X-ray optics." Each of the 24 detectors has a 10 mm x 10 mm sensitive area with an energy resolution > 200 eV, they operate at near room temperature and have a higher tolerance to radiation doses than conventional CCDs. The other innovative feature of the spectrometer is its "micro-collimator X-ray optics" which ensures that only X-rays from a single well defined direction reach the detectors. The collimator with its associated filters also rejects solar wind particles and radiated heat from the lunar surface.

The XSM (X-ray Solar Monitor), covering the spectral range 1-20 keV, is an auxiliary instrument of D-CIXS with the objective to monitor throughout the whole mission duration the X-ray activity of the sun, and to calibrate D-CIXS measurements when it is pointed to a celestial object, and during the final phase of the mission, to the moon. The electronics of the XSM consists of a) pre-amplifier stages, which are identical in the two sensor units, and b) an electronics board in the main D-CIXS instrument box, which includes further stages of the signal processing electronics.

SIR (SMART-1 Infrared Spectrometer) of MPAe (Max Planck Institut für Aeronomie), Katlenburg-Lindau, Germany. SIR is a technology demonstration instrument with the primary objective to space-qualify and to miniaturize an existing, commercially available, small and very low mass near-infrared (NIR) monolithic grating spectrometer (Carl Zeiss, Jena). SIR is a highly compact monolithic grating point spectrometer, consisting of an optical entrance or 'slit' for selecting its view of a target source or area, a collimator which focuses the infrared rays into a parallel beam, a grating which disperses the light and a linear array InGaAs detector which sequentially records the incident light, creating a spectrum. The instrument covers the spectral range of 0.9 to 2.4 micron, the spectral resolution is 60 nm, the field-of-view is about 1 mrad providing 300 m spatial resolution at peri-selene. SIR provides the capability of mineral mapping, including olivine, pyroxene, and feldspar. The total instrument mass is about 2 kg. [1746] The SIR and AMIE instruments have the same alignment allowing simultaneous and complementary observations.

AMIE (Asteroid Moon micro-Imager Experiment). AMIE is developed in the framework of ESA's Technology Research Program in conjunction with CSEM (Swiss Center for Elec-

1742) http://sci.esa.int/content/doc/10/19216_.htm

1743) http://sspg1.bnsc.rl.ac.uk/Share/d–cixs.htm

1744) M. Grande, et al., "Lunar Elemental Composition and Investigations with D-CIXS X-Ray Mapping Spectrometer on SMART-1," 31st Lunar and Planetary Science Conference, March 13-17, 2000, Houston, TX

1745) S. K. Dunkin, M. Grande, et al., "The D-CIXS X-Ray Spectrometer on ESA's SMART-1 Mission to the Moon: Science Objectives," 31st Lunar and Planetary Science Conference, March 13-17, 2000, Houston, TX tronics and Microtechnology) of Neuchatel, Switzerland (there are AMIE partners in France, Italy, Finland, and Switzerland). The overall objectives are to study the moon's morphology, topography and surface texture. The specific objectives of the tiny camera are to: a) provide imagery of the lunar south pole (Aitken basin), b) to obtain imagery for the study of the permanently shadowed regions of the moon (ice deposits), c) to investigate eternal light (crater rims), and d) to obtain imagery of high latitudes regions (south) mainly at far side of the moon.

The AMIE micro-imager employs a very compact design, a package of dedicated optics, electronics and mechanical interfaces that weighs but 450 grams (all included). This feat has been achieved by using new microtechnology in terms of micro-mechanics, micro-optics and 3D-stack packaging. AMIE features a FOV of 5.3° x 5.3°. At closest orbital distance of 300 km (pericenter), a footprint of 27 km x 27 km is provided. The pixel size for the 1024 x 1024 CCD array is therefore 27 m x 27 m at pericenter. The main system functions are:

- To take images with color information
- To keep image data in a memory buffer
- To perform image processing (data compression, image subtraction)
- To allow the downloading of the images by the OBC (On Board Computer)

The AMIE electronics are composed of:

- A microprocessor board (CSEM) for the instrument management
- A micro DPU (Digital Processing Unit) for data compression
- A S/C interface board.

AMIE features also three filters: in the visible (750 nm) and infrared (900, 950 nm), thus providing multispectral data. AMIE will work closely in conjunction with the SIR and D-CIXS instruments. Its infrared filter overlaps and complements the SIR spectrometer (900 nm - 2400 nm) permitting studies of the different lunar rock types. Information obtained by AMIE and D-CIXS will help establish abundance maps of the elements present on the lunar surface.

In addition, AMIE is supporting three other SMART-1 experiments/technology demonstrations:

- LaserLink experiment: involving the demonstration of acquisition of a deep-space laser link from the ESA OGS (Optical Ground Station), located at the Observatorio del Teide at Tenerife, and validation of advanced atmospheric communication technique (sub-aperturing). The OGS transmitter uses a Ti:Sapphire laser pumped by a 28 W Argon laser. The LaserLink experiment will be performed by pointing the spacecraft's AMIE camera towards the Earth. Apart from a laser line filter which covers a section of the CCD imaging array of the camera, there is no specific LaserLink equipment aboard SMART-1.
- Measurement of lunar libration in coordination with the radio-science experiment RSIS (Radio Science Investigation Series).
- OBAN (On-Board Autonomous Navigation). OBAN validates navigation algorithms by planetary body tracking. It makes use of the S/C star trackers and of AMIE images. The overall objective is to test navigation algorithms with respect to their workability in a real-mission environment. The OBAN experiment is designed to function in what is termed an 'open loop', obtaining all the data an autonomous navigation system would require, but instead of being processed onboard, this information is downlinked to be processed on Earth. The experiment involves the spacecraft looking at certain celestial objects or at 'beacons', taking images of them with the AMIE camera. These time-tagged pictures, together with information from the attitude and control system of SMART-1, allows the OBAN ground system to determine the lines of sight to the beacons, and to calculate the precise trajectory. Subsequently all the parameters that

1746) http://disr01.mpae.gwdg.de/wuttke/SIR/SIR.htm

would be required to carry out a correction in the spacecraft's course can be obtained. The picture analysis algorithms and computations are rather complex, but the principles of the experiment are those of elemental triangulation geometry.

M.24 SPORT (Small Payload Orbit Transfer)

AeroAstro of Herndon, VA, has developed the SPORT ™ space access capability, enabling small payloads to use low-cost secondary launch opportunities and still reach their desired final orbits. The concept allows small payloads to effectively use a wider variety of launch opportunities, including numerous under-utilized geosynchronous transfer orbit (GTO) slots. Considering the available launch opportunities, it turns out that about 90% of today's the small satellites plan for LEO observations/operations. However, about 90% of the large launch vehicle capacity is destined for GEO. The SPORT concept tries to provide a solution to this asymmetric launch situation with its orbit transfer capability.

The SPORT concept is a flexible orbit transfer vehicle designed to go from intermediate orbits, such as GTO, to final LEO ones (normally from higher to lower orbits) by utilizing a combination of chemical propulsion and aerobraking technology. With aerobraking, SPORT can achieve orbit changes that are many times greater than that achievable with direct propulsion. After using the propulsion system to lower the orbit perigee, the aerobrake gradually slows SPORT via atmospheric drag. After the orbit apogee is reduced to the target level, an apogee burn raises the perigee and ends the aerobraking phase. At the conclusion of the orbit transfer maneuver, either the aerobrake or SPORT can be shed, as desired by the payload. A variety of low-cost core electronics (Bitsy, see M.3), propulsion and power options provide the mission developer with the ability to tailor SPORT to individual needs. 1747) 1748) 1749)

The SPORT system is equipped with a large deployable aerobrake, consisting of an inflatable umbrella (several radially-oriented tubular spars, which harden after inflation, stretch the aerobrake cloth). After about 20-90 days, depending on upon atmospheric conditions and target altitude, the aerobraking phase concludes with a thruster firing at apogee, which raises the perigee above aerobraking altitude. Additional thruster firings are made to trim the orbit to the desired parameters. Upon completion of the orbit maneuver, SPORT can separate from the payload and use residual propellant to lower the perigee so that additional atmospheric drag will deorbit the spent stage. The SPORT/payload system is spin-stabilized during the transfer phase.

Technically, all low LEO spacecraft experience multi-pass aerobraking orbit profiles as they gradually decelerate from atmospheric drag. However, in general, these spacecraft take many years for substantial orbit changes to occur (say, from a 700 km circular orbit). To be useful in an orbit transfer vehicle, this deceleration must be enhanced so that substantial orbit changes occur within time scales on the order of weeks to months, at most. This enhancement can be achieved by increasing the spacecraft profile area by at least one to two orders of magnitude. When deployed, the SPORT aerobrake increases the SPORT profile area by a factor of fifty. - For eccentric orbits such as GTO, only the perigee portion of the orbit experiences significant atmospheric drag. This asymmetrical drag reduces the orbit apogee, while leaving the perigee altitude relatively unchanged. Over time, the cumulative energy loss at perigee produces large altitude changes in the orbit apogee. SPORT is designed to operate with a perigee altitude in the 150 to 200 km range to achieve GTO-LEO aerobraking.

1747) P. W. Gloyer, D. J. Goldstein, "Small Payload Orbit Transfer (SPORT) System: Lowering Launch Cost Without Increased Risk," IEEE Aerospace Conference, Big Sky, Montana, March 10-17, 2001, 0-7803-6599-2/01

1748) P. W. Gloyer, D. J. Goldstein, " Small Payload Orbit Transfer (SPORT) System: An Innovative Approach to Lowering Missions Costs without Increased Risk," Proceedings of the 14th Annual AIAA/USU Conference on Small Satellites, LOgan, UT, Aug. 21-24, 2000, SSC00-IV-6

1749) Information provided by Aaron Jacobovits of AeroAstro

A first demonstration flight of the SPORT concept is scheduled with a Malaysian microsatellite of ATSB (Astronautic Technology S B) of Kuala Lumpur (an agreed partner project as of Nov. 2000). [1750] A GTO launch on Ariane-5 ASAP (Ariane Structure for Auxiliary Payloads) is planned for 2003 with a subsequent GTO-LEO aerobraking maneuver. Like many countries near the equator, Malaysia's most desired orbit is a low-inclination low-Earth orbit, from which their geographically advantaged location provides revisits on every orbit.

SPORT can go more places than just LEO. With the advantage of the low orbital velocity at apogee, plane changes of 30° or more can be provided, thus extending the region of coverage considerably from GTO injections. Piggybacked to Molniya, SPORT will be capable to insert into sun-synchronous and other polar inclinations. [1751]

M.25 STRV (Space Technology Research Vehicle)

STRV is a multi-microsatellite cooperative technology program, initiated by DERA (Defence Evaluation and Research Agency, UK) in 1992. The prime objective of the program is to provide opportunities for the evaluation of several new technologies. The first two S/C of the mission were launched in June 1994. Two more microsatellites are planned to be launched in 2000, with further missions anticipated beyond these. [1752] [1753] [1754] [1755]

M.25.1 STRV-1a and -1b

STRV-1a and -1b are two microsatellites of DERA (funding by MoD) at Farnborough, UK (with contributions from: ESA, BMDO/JPL, USAF/PL), which were launched on ASAP, Ariane-4 flight V64, on June 17, 1994. The objective was to evaluate new technologies in the high radiation dose environment of GTO (GEO Transfer Orbit) - Test of advanced structural materials, radiation hardened computers, sensors, solar cells, and microelectronics. Both S/C are cubical in shape, spin-stabilized at 5 rpm (magnetorquer attitude control and cold gas thruster system, attitude sensing with Earth and sun sensors), with a structure consisting of carbon/peek thermoplastic skinned aluminum honeycomb panels. The GaAs solar cells are body-mounted. In addition, there are 16 NiCd cells providing 46 Whr of power storage. The S/C design life is one year. The RF system conforms to ESA standards using dual redundant transmitters/receivers at S-band. The data transmission (uplink rate: 125 bit/s, 2.025-2.12 GHz, and downlink rate: 1 kbit/s, 2.2-2.3 GHz) uses the CCSDS packet TM/TC concept implementation of ESA with custom chip-sets.

Orbit: Elliptical GTO orbit, perigee = 300 km, apogee = 36000 km, inclination = 7°, period = 10 hours and 35 minutes. The orbit crosses the Van Allen Belts each orbit with a maximum of ionization radiation seen in near-Earth orbit, suffers electrostatic charges close to apogee, and atomic oxygen erosion at perigee.

Mission operations were performed from the DERA/Lasham ground station and the NASA Deep Space Network from June 1994 to June 1996, when routine orbital maintenance of the S/C was passed to LASP (Laboratory of Atmospheric and Space Physics) at the University of Colorado, CO, USA. Note: During the initial orbital period, up to about the

1750) http://www.aeroastro.com/pressroom.html

1751) R. Fleeter, "New Propulsive Module for Nanosatellites," Proceedings of the 6th ISU Symposium on Smaller Satellites: Bigger Business?, Strasbourg, France, May 21-23, 2001

1752) N. Wells, J. Eves, P. Mace, "Space Technology Research Vehicles STRV-1A and -1B, Final Report (DRA, ESA, BMDO) Vol. 1, August 1995

1753) N. Wells, "The Space Technology Research Vehicles STRV-1A and -1B: Mission Update," paper provided by the author

1754) R. Blott, N. Wells, " The Space Technology Research Vehicles: STRV-1A,B,C&D," Proceedings of the AIAA/USU Conference on Small Satellites, Sept. 16-19, 1996, Logan, UT

1755) http://www.dra.hmg.gb/html/case/strv/menu.htm

end of August 1994, STRV-1a experienced in-flight anomalies (low power due to failure of at least one battery cell) resulting in a delay for some experiment operation until September 1994. [1756]

Despite the very high radiation dose received by the platform systems during the mission, the majority have performed exceptionally well. With one or two small exceptions, the spacecraft electronics were built entirely from military specification components. Although only one receiver on each spacecraft remains usable, the redundancy philosophy adopted has prevented the termination of the mission.

Parameter	STRV-1a	STRV-1b
Total mass	52 kg	54 kg
S/C dimensions	440 mm x 440 mm x 420 mm	
S/C structure	Carbon/PEEK thermoplastic skinned Al honeycomb	
S/C thermal control	Passive	
Power generation and storage	Four body-mounted GaAs solar panels, 46 Whr NiCd battery	
S/C power (at start of mission)	31 W (at 28 V)	33 W (at 28 V)
Power available to payload	19 W (average), 120 W (peak)	
Stabilization	Spin at 5 rpm (provided by the launcher at separation)	
Attitude sensing subsystem	V-slit sun/Earth sensors and analog sun sensors	
Attitude control subsystem (ACS)	Magnetorquer coils & cold gas thruster	Magnetotorquer coils
Onboard computer (OBC)	Dual-redundant GEC/Plessey Mil Std SOS 1750a microprocessor	
Memory	Dual-redundant 128 kBytes SOS RAM, 64 kBytes ROM	
Communications	ESA packet TM/TC CCSDS standard, 12 m antenna at DERA Lasham ground station (26 m DSn antennas)	

Table 516: Overview of STRV-1a and -1b design parameters

M.25.1.1 STRV-1a Sensor/Experiment Complement

AOE (Atomic Oxygen Experiment), provided by DERA and the University of Southampton. Objective: Measurement of atomic oxygen (AO) by silver loss determination in different material specimens - consideration as a future monitoring method of AO environments. AOE is comprised of twelve silver resistance sensors. As silver is converted to its non-conducting oxide the resistance increases. This increase can be measured and converted into silver loss. The erosion resistance of test materials is measured by coating the thin silver films with overlays of the test material. When this coating is breached the silver starts to oxidize, a resistance increase is observed. Knowledge of the fluence experienced and the thickness of the coating enables calculation of the erosion rate. Silver sensors coated with polyethylene, PTFE, carbon, and silica were flown.

BRE (Battery Recharge Experiment), provided by ESA/ESTEC. Objective: Demonstration of control techniques using conventional constant current battery charging, followed by current tapering, instead of simple voltage level triggered end-of-charge cut-off for LEO S/C. Selection of an end-of-charge voltage level to avoid both undercharge and overcharge throughout the lifetime of the S/C.
The basic problem was selecting an end-of-charge voltage level which would avoid both battery undercharge and overcharge throughout the S/C lifetime. BRE was activated in September 1994. The results of the test have implications for S/C power subsystem designers.

CAE (Charge Alleviation Experiment) provided by DERA. Objectives: 1) Demonstration of an active S/C surface charge alleviation system, and 2) Test of several new technologies in orbit, in particular the effects of operation of other S/C subsystems such as that of an ion

1756) N. Wells, "Space Technology Research Vehicles (STRV-1a and -1b): Lessons Learned After four Years in GTO," Proceedings of the 4th International Symposium on Small Satellites Systems and Services, Sept. 14-18, 1998, Antibes Juan les Pins, France propulsion system (UK-10 hollow cathode) and that of a cold gas ACS (Attitude Control Subsystem) on the bulk and differential charging of surface materials. [1757]

The CAE system is comprised of the following components: (total mass = 3.25 kg, power = 30W during steady-state operation, power = 80 W for initial heater operation of about 30 s).

- Fill and drain valve
- Xenon storage tank
- Five standard series 107 solenoid valves
- Miniature tank pressure transducer
- Plenum tank (100 cm^3)
- Miniature plenum tank pressure transducer
- Plenum tank orifice
- Spin-up thrusters and associated pipe work
- UK-10 ion thruster hollow cathode
- Two DC-to-DC power supplies and associated electronics
- Interconnecting pipe work and weld fittings.

The CAE system relies on the ability to produce a diffuse cold plasma of inert xenon gas around the S/C thereby neutralizing the high differential surface potentials. In the CAE design, the diffuse plasma envelope is provided (produced) by the operation of three subexperiments with objectives of their own.

- First flight test of the UK-10 ion thruster hollow cathode assembly.
- First flight test of the xenon gas flow control system, developed for the UK-10 IPS (Ion Propulsion System), with associated solenoid valves, orifices, and valve actuating electronics.
- The provision of a cold gas attitude control subsystem (ACS), a backup for the magnetorquers for control of S/C precession.

SCDE (Surface Charge Detector Experiment) provided by DERA. Objective: Non-contact measurement of surface potentials. The Pockels effect technique (an electro-optical method) was chosen which relies on special crystal properties. The experiment collects charge on a piece of kapton (an insulator) which is exposed to the space environment. Sensing crystals are placed under the kapton, one of the crystals is shielded from electric fields by a thin piece of foil (the monitor). A plane-polarized laser beam is passed through each crystal and analyzed (the strength of the electric field is deduced from the polarization state). The monitor crystal does not respond to the electric field, thus providing a reference. SCDE had a mass of 0.6 kg, power of 2 W, and dimensions of about 100 mm x 150 mm x 30 mm.

CID (Cold Ion Detector). A DERA/MSSL instrument with the objective to measure the energy spectrum of incident, low-energy ions in the interplanetary plasma. A further objective of CID is to act as a diagnostic tool - to directly sense the cold ions produced by the CAE neutralizer. CID is an electrostatic deflection differential energy analyzer consisting of the following elements:

- A collimator accepting a beam of conical shape with a half angle of 5.7°
- An electrostatic energy analyzer with a passband of DE/E = 0.14
- A microchannel plate detector which records the impact of the particles.

The CID instrument points perpendicular to the spin axis of the S/C, protruding some 70 mm from the surface.

LPE (Langmuir Probe Experiment), provided by DERA. Objective: Xenon plasma diagnostic instrument for CAE. A conducting surface is biased with a sweeping voltage, attract-

1757) Note: Differential charging of surface material occurs when these materials are electrically isolated, either from the S./C structure or from neighboring surfaces, or when the surfaces are insulators themselves. In these situations charge build-up cannot leak away, resulting in large potential differences between neighboring surfaces.

ing either positive ions or electrons, which are measured as a current in the experiments electronics. The LPE consists of a conducting planar probe of 20 mm diameter with a surrounding guard ring of 100 mm diameter. In the presence of plasma a sheath forms over the conducting surface.

CREDO-II (Cosmic Radiation Environment and Dosimetry Experiment). A DERA/ UKAEA experiment with the objective to characterize the radiation dose received by the spacecraft [a) total dose, b) cosmic ray and solar heavy ions, c) protons trapped in the Van Allen belts]. The instrument is the latest of an evolving series which have flown on aircraft [the Concorde (BA) and Boeing-767 (SAS)], on the Shuttle (STS-48, -44, -53, -56, -68, and -63), in sun-synchronous orbit (UoSat-3, since Jan. 23, 1990), and on the USAF APEX satellite (orbit: 359 km x 2544 km, 70° inclination).

CREDO employs a telescope technique to detect coincidences between parallel planes of p-i-n (p-intrinsic-n) diodes to provide directional information and to define the pathlength of particles through the diodes to within 29%. The square diodes (each with a path length of 200 μm, area of 10 mm x 10 mm) are configured in two orthogonal arrays with their axes aligned with two S/C axes. Pairs of diodes are placed in a telescope arrangement with a separation of 1.75 mm, each array consists of four such pairs. - The upper and lower arrays of each telescope system are connected via separate charge amplifier systems into hardware discriminators with thresholds close to 0.33 pC (7.4 MeV) of charge deposited per event. Events are coincident when both discriminator thresholds are exceeded within a 10 μs gate time. Upon sensing a coincident event, the CREDO-II micro-controller performs pulse-height analysis upon the output of one of the planes of each telescope. Fifteen channels of pulse-height analysis are provided. The telescope system does not respond to particles arriving with oblique angles of incidence. Only ions are detected by this arrangement. - Total dose is measured at three locations within the CREDO box using radFET dosimeters. These are MOSFET devices are designed to trap positive charges and shift the threshold voltage of the radFET. Each radFET is co-located with a temperature sensor to permit temperature correction of voltage shift. - Non-coincident events detected by the telescope system are processed similarly, but separately from the coincident events. The detector arrangement is sensitive to protons which can deposit sufficient energy via ionization along oblique paths through the diodes.

RDRS (Radiation Dose Rate Sensor), provided by DERA. Objective: Monitoring of gamma ray bursts.

M.25.1.2 STRV-1b Sensor/Experiment Complement

CVSE (Cryocooler Vibration Suppression Experiment), provided by BMDO and managed by JPL. Objective: To demonstrate and to qualify a new vibration suppressing design by employing piezo materials. (Cryocoolers are used to cool IR or CCD detectors to lower the thermal noise threshold. Since Stirling coolers have a mechanical compressor, there is a significant vibration problem introduced by the cooling of optical instruments). The intent is to use these low-power flight piezo drivers and control systems in zero-g environments.[1758)]

To meet stringent power, weight, and space constraints, the experiment makes use of the tiny 0.2 watt 80 K tactical Stirling coolers (of Texas Instruments). Two different vibration-cancellation actuator techniques are being demonstrated: 1) applique ceramic piezoelectric actuators that are bonded to the coldfinger and stretch the coldfinger to cancel tip motion, and 2) commercial low voltage piezoelectric translators that similarly cancel tip motion by moving the entire cryocooler in three axes. Motion of the coldfinger tip is measured in all three axes to 10 nm accuracy using eddy-current transducers. Two types of control systems are also being demonstrated: 1) an analog control system that uses a bandpass filter to track the drive signal and suppress it, and 2) a unique narrow-band adaptive feed-forward

1758) http://www.jpl.nasa.gov/adv_tech/coolers/Integ.htm

system that continually updates a steady-state command signal to each actuator to cancel the tip vibration. Either control system can be used with either actuator.

Using the cryocooler as a vibration source, the response of the satellite was determined from accelerometer data. The translators with digital control reduced the vibration by a factor of about 75.

SEE (Space Environmental Effects) experiments. The overall objective is to measure and evaluate the effects of the high radiation environment on electronic devices. The device sets tested are:

- Neural network integrated circuits
- HIP (Heterojunction Internal Photoemission) infrared sensors
- CMOS (Complementary Metal-Oxide Semiconductor) radiation and SEU (Single Event Upset) monitors

Two identical sets of devices are used, one with minimal shielding to radiation in the spin axis of the S/C, and the other set with significant shielding to provide a basis for comparison. Unfortunately, SEE data were only obtained up to early October 1994 when a system failure occurred.

- **NNE** (Neural Networks Experiment). Objectives: To demonstrate fault-tolerant and graceful degradation characteristics of analog neural network VLSI chips (JPL) in the space environment. For NNE, a special feed-forward "learning" architecture was chosen (signal flow from one layer of neurons to the next through a connecting mesh of synaptic weights) with a suitable pattern recognition problem as a bench-mark for learning.

- **HIP** (Heterojunction Internal Photoemission) Infrared Sensors. The objective of this experiment is to determine if exposure of the HIP sensor to the complex spectrum of incident radiation might result in increased sensitivity. - The experiment measures the dark current to obtain an indication of the radiation damage to the specimen.

- **SEU** (Single Event Upset)/**RADMON** (Radiation Monitor) Experiment. Objective: To develop a small, low-cost, low-power SEU and TD (Total Dose) RADMON using standard CMOS processes. RADMON consists of 16 SRAMs (4-kbit each) designed to detect protons, alpha particles, and heavy ions (SEU-SRAMs). In addition, RADMON contains four p-FETs for detecting total dose (TD-FETs). The SEU-SRAMs and TD-FETs are shielded behind primary and control shields.

REM (Radiation Environment Monitor). Objectives: To study the GTO radiation environment, its spatial, temporal and directional variability for the development of improved models. The results are intended to be used in the design of electronic components for future S/C.
REM consists of two independent shielded silicon detectors (and charge-sensitive amplifiers) with different types of shielding. Energetic particles impacting on the detectors generate pulses which are counted. Each REM detector (300 Tm silicon diode detectors) has 16 channels for pulse magnitude discrimination. The 'electron detector' is 25 mm^2 in surface area with a 3 mm Al shield; the 'photon detector' has a surface area of 150 mm^2, it is shielded by 3 mm Al and an additional 0.75 mm of tantalum. The instrument accumulates energy deposit spectra over varying integration periods, adapted to the speed of the spacecraft and the available on-board memory. These energy deposits can be converted fairly easily into total absorbed dose.

SCTE (Solar Cell Technology Experiment). The objective is to measure the current/voltage characteristics of many new or improved types of solar cells. A total of 47 cell types are exposed to the radiation environment and operated whenever the sun direction is within 5° of the normal to the panel.

M.25.2 STRV-1c and -1d

These are DERA follow-up microsatellite technology missions based on the STRV-1a & -b design (they incorporate a sample 'multi-function structure' designed to achieve significant savings in future spacecraft mass and cost). [1759) 1760)] The overall objectives are to demonstrate emerging technologies on the premise of 'faster, better, cheaper' in a collaborative environment of participating institutions. A particular objective is to enhance the capabilities of future communications, navigation, and surveillance systems.

The S/C of STRV-1c/-1d are slightly larger than those of STRV-1a/-1b to accommodate more experiments (otherwise same architecture). The only significant enhancement is the introduction of full dual-redundancy throughout all the platform subsystems. [1761) 1762)]

S/C total mass, power	100 kg, 65 W (BOL), GaAs solar cells
S/C size	Cube with 700 mm length of each side
Payload mass	25 kg
Battery	NiCd cells providing 59 Whr
Power bus	Regulated 28V ± 2%
Computers	Military specification 31750a (dual–redundant silicon-on-sapphire)
Platform memory	2 MByte per computer (dual-redundant Military Specification RAM)
Communications Uplink Downlink	ESA CCSDS Packet TM & TC (with optional encryption on STRV 1d) S-band at 1 kbit/s S-band at 10 kbit/s
Attitude control	Spin stabilized control to ± 3°. Magnetorquer & cold gas precession thruster maneuvers
Nominal attitude	Sun aspect angle = 90° ± 10° & orbit normal ± 10°

Table 517: S/C parameters of STRV-1c/-1d

A simultaneous ASAP (Ariane Structure for Auxiliary Payloads) launch of both STRV spacecraft took place on Nov. 16 2000 on an Ariane-5 launch vehicle from Kourou. The STRV satellites and the AMSAT Phase 3-D satellite were secondary payloads to the PAS-1R (PanAmSat-1R) communications satellite. [1763)]

Orbit: Elliptical GTO (Geosynchronous Transfer Orbit), perigee = 620 km, apogee = 36,000 km, inclination = 7° (the orbital period exceeds 10 hours).

M.25.2.1 STRV-1c Sensor/Experiment Complement

AOE-2 (Atomic Oxygen Experiment-2), provided by the University of Southampton and ESA. Heritage of STRV-1a. Objective: To characterize and to flight test a novel zinc-oxide sensitive renewable atomic oxygen flux sensor based on semiconducting thin films. These films have the ability to be regenerated after atomic oxygen exposure which offers a significant advantage over the 'use-once' silver films used in the AOE-1 flown aboard STRV-1a.

The AOE-2 instrument is designed to measure the flux of atomic oxygen (AO) as the spacecraft enters the Earth's atmosphere near perigee, typically at an altitude of around 600km. The main aim is to monitor the AO flux at various stages during a perigee pass. This is to be achieved by monitoring the resistance of several metallic and semiconductor thin films upon which the AO is impinging. The metallic (silver) resistance elements were flown on the previous mission (STRV-1a); the secondary aim is to test the new semiconductor resistance elements in the orbital environment and to compare the results obtained with those

1759) A. Cant, H. Simpson, "STRV-1c & -d Satellite Architecture Design Document," March 1998, provided by DERA

1760) N. Wells, "STRV-1c & -d Mission Definition Specification," Feb. 1998, provided by DERA

1761) N. Wells, R. Blott, "STRV-1c&d Program Update," Proceedings of the 11th AIAA/USU Conference on Small Satellites, Sept. 15-18, 1997, Logan, UT

1762) http://www.dera.gov.uk/html/case/strv/high.htm

1763) N. Wells, Countdown to launch of the first microsatellites qualified for flight on Ariane-5 ASAP," Proceedings of the 14th Annual AIAA/USU Conference on Small Satellites, Logan, UT, Aug. 21-24, 2000, SSC00-I-7 from the silver elements. In particular, the sensitivity of the semiconductor elements to AO (which will otherwise decrease with time/AO fluence) is to be recovered by heating to approximately 80° C; this takes place periodically during a "refresh" phase.

CDMS (Cosmic Debris and Micrometeoroid Sensor), provided by the University of Kent at Canterbury. Objective: To characterize the debris and micrometeoroid environment of GTO. In particular:

- Measurement of separate components of space debris and micrometeoroids (both as interplanetary a and b meteoroids) in terms of flux distribution
- Definition of the anisotropy of space debris, including altitude distributions and approximate orbital characteristics
- Identification of space debris sources, especially temporal factors
- Exploration of GTO orbit space as a source of micro debris from satellite transfer operations to GEO.

The CDMS consists of two sensor head units mounted externally on the satellite, with one sensor on the top panel (+Z face), and on a spinning face (+Y face), and a control unit located within the satellite. The sensor units combine impact momentum detection on particles (using piezo-electric crystals mounted on the carrier for a thin aluminium foil), with penetration measurement (of the thin foil) and impact plasma detection using plasma sensing wires mounted above and below the foil. The instrument is designed to provide a continuous, in situ measurement of individual impact events. The control unit adds both a time tag and a spin phase tag to the data packet for each impact event. The spin phase is derived using an analogue sun sensor (solar cell) mounted on the +Y face sensor. The pointing direction for each impact event is reconstructed on ground from the event time, spin phase and the spacecraft orbital and attitude data files.

CEASE (Compact Environmental Anomaly Sensor Experiment), provided by AFRL, Hanscom Air Force Base, Bedford, MA. Objective: To demonstrate on-orbit autonomous monitoring of the S/C environment and to provide "alerts" when satellite charging, SEUs or other radiation events are likely. See description under M.29.2.

CREDO-II (Cosmic Radiation Environment and Dosimetry Experiment), provided by DERA, MoD, heritage of STRV-1b. Objective: To monitor those aspects of space radiation which cause single event effects in microelectronic components. The measurements comprise energetic proton fluxes from the inner radiation belt and solar flares, and the linear energy transfer spectra of heavy ions in cosmic rays and solar particle events. The instrument is as a compact, low power monitor housed on a single board. Two telescopes are employed, each comprising two pin diodes of area 3cm2 in coincidence, in order to define the arrival directions of the particles and their pathlength through the detector. In one telescope the charge depositions are amplified and pulse-height analyzed to give the linear energy transfer spectra, while in the other the counts above a low threshold are used to monitor the much higher proton fluxes. Counts are accumulated into programmable time bins to reflect the finer resolution required at perigee compared with apogee and to make optimum use of the limited data storage and telemetry.

DDM (Deep Dose Monitor), provided by DRE (Defense Research Establishment, Ottawa, Canada). Objective: To record the space radiation environment absorbed dose behind various thin shields and to estimate the dose due to low energy electrons. The DDM instrument is a small, low power, real-time dosimetry system consisting of four radiation sensitive dual MOSFETs (Metal Oxide Semiconductor Field Effect Transistors) which measure total dose during the mission. Each MOSFET has a unique shield covering the sensitive device to monitor the wide range of doses. The MOSFETs are capable of measuring high energy electrons and protons with dose levels between rad (radiation dose of energy) and Mrad.

GAGE (GPS At GEO Experiment), provided by MoD, DERA, DoD, JPL, and ESTEC. Objective: To characterize the GPS signal availability at selected altitudes up to GEO with particular emphasis on investigating the feasibility of GPS utilization on GEO satellites. The primary objective is to measure the GPS signal strengths at the L1 and L2 frequencies, to enable the definition of an operational GPS package for future spacecraft in high orbits up to GEO. GAGE consists of a patch antenna and GPS sampling device, known as 'Bit-grabber', both provided by JPL. The device captures the GPS RF environment for short (up to 400 ms), predetermined periods of time and downlinks this information for subsequent processing. The ground segment of GAGE consists of observable extraction and orbit determination software to process the data obtained from the spacecraft and provide the mission results.

GFSE (Gas Flow Sensor Experiment), provided by DERA. Objective: To demonstrate in space a gas flow control and sensing principle (advanced xenon gas flow control and monitoring system) to test the range of UK ion thrusters developed by DERA. The flow sensor uses a thermal technique to measure the flow rate of gas. The gas flows through a small bore stainless steel tube around which is wound two identical coils of wire whose resistance is highly dependant on temperature. A small current is then passed through the wire which heats the coil. - The new system uses a novel thermal flow controller and an accurate, and highly sensitive, flow sensor. These units allow the system to be significantly smaller, cheaper, more flexible and, potentially, more reliable.

LCE (Linear Concentrator Experiment), provided by BMDO, NASA, USAF/PL. Objective: To measure the IV curves and to determine the pointing sensitivity of a new linear concentrator solar cell assembly.

MTB (Microwave Test Bed), provided by MMS, DERA, MoD. Objectives: To evaluate microwave SHF (Super High Frequency) technologies for future military space communication systems. MTB is a single channel transparent SHF (X-band) transponder, built using microwave and electronic components developed under the Matra Marconi Space (MMS) EXPERT satellite project.

PM (Proton Monitor), provided by Thomson & Nielsen Electronics Ltd, Canada, and sponsored by DRE, Ottawa. Objective: To flight-prove a monitor which measures the upset rate due to high energy protons and to compare the results with other on-board monitors and ground-based tests. Secondary objectives include:

- Correlation of the PM proton data with data from other radiation experiments and the AOE-2 experiment
- Correlation of the PM upset monitor data with single event effects/upsets data from other memory devices of STRV-1c
- To distinguish between solar protons, the trapped radiation environment and incident high energy cosmic rays.

The PM instrument consists of a dynamic RAM which is sensitive to proton-induced single event upsets. The experiment is designed to measure both the proton environment and the solar activity.

SCTE (Solar Cell Technology Experiment), provided by Mod, DERA, ESTEC, NRL, USAF/PL, EEV(English Electric Valve Co., UK), heritage of STRV-1b. Objective: To measure the IV curves of up to 47 new solar cell and coverglass technologies. In particular to:

- Demonstrate the use of different types of advanced solar cells in the space environment
- Determine the true absolute electrical performance of the cells
- Compare their relative resistance to the space radiation environment
- Compare actual flight degradations with those predicted from ground testing
- Provide flight data for advanced cells entering production to give satellite program managers confidence of their reliability and suitability for use on large projects.

There are two parts to the experiment: 1) The flight demonstration of advanced cells on the power generating panels of STRV-1c/d. Panel performance data is provided from the solar array current monitors in the spacecraft power system. 2) The in-flight measurement of the current-voltage characteristics of 38 individual solar cells. The temperature coefficients of some cells are measured throughout the mission.

SMX-1 (Sparc Microprocessor Experiment), provided by MoD, ESA, MMS, DERA and SEI (Space Electronics Inc.) San Diego, CA. The overall objective is to evaluate the ESA ERC32 chip set and advanced processing technologies and shielding experiments in the harsh space environment. SMX-1 consists of the ESA ERC32 chip set and components (some components are shielded). Three RADFETS to monitor the radiation dose on the SMX-1 board. Note: SMX-2, a similar ERC32 board, is flown on STRV-1d in a more shielded location.

SREM (Standard Radiation Environment Monitor), provided by ESTEC, SREM is of REM heritage flown on STRV-1b and the MIR space station.. SREM is also being flown on PROBA. Objective: To map the radiation environment of GTO with high temporal and spatial resolution at post-solar minimum conditions. SREM detects and counts electrons, protons, and cosmic rays with a coarse spectral resolution. The measurement is based on examining the sizes of charge pulses generated by charged particles in silicon diodes polarized in reverse by a high voltage. Two of the three diodes are in-line and operate in 'coincidence', where a charged particle passing though both is detected and identified. SREM also measures the total radiation dose encountered by SREM itself and at different remote locations onboard the satellite, by using RADFETS.

M.25.2.2 STRV-1d Sensor/Experiment Complement

CERTO/PLUS (Coherent EM Radio Tomography & Profiling the Limb with UV Sensors), provided by NRL. Objective: To map the ionospheric electron densities in the satellite's orbital plane and to study the evolution of 1-100 km sized low latitude irregularities using radio/optical tomography techniques. The CERTO/PLUS experiment consists of a three-frequency radio-beacon, antenna and an UV photometer. With these flight instruments and ground radio/telemetry receivers, integrated ionospheric densities are determined near the satellite orbit plane. This provides a two dimensional mapping of the ionosphere which assists in the understanding of the distortion and attenuation of radio frequency (RF) signals caused by naturally occurring ionospheric anomalies.

ECSE (Encrypted CCSDS Space Experiment), provided by MoD, DERA, MMS. Objective: To demonstrate a prototype secure CCSDS packet TM/TC system in an operational environment. a) Demonstration of ESA Packet Telecommand encrypt/decrypt, authentication, validation and anti-replay attack functionality; b) ESA Packet Telemetry encrypt/decrypt functionality, c) Extraction of security management functions onboard the spacecraft and simplified processing of these security management functions. The ECSE ground segment consists of telecommand and telemetry security layer software integrated into the primary STRV TM/TC processor workstation.

ECSE is composed of two segments: a space segment and a ground segment. The ECSE space segment consists of a number of switches, several electronic devices, and interfaces to the Sparc processor with the relevant software loaded.

QWIP (Quantum Well Infrared Photodetector), an infrared camera system provided by JPL and sponsored by BMDO. Objective: To characterize the radiation detection performance of QWIP technology in the spectral range of 11-17 μm by imaging of the moon. The QWIP detector is mounted directly to the 55 K coldfinger of a 1 W tactical cooler (Texas Instruments) using an IDA (Integral Dewar Assembly). The test objectives include determining cooler/battery compatibility and measuring power train efficiency, inrush current levels, and current ripple.

SAS (Satellite Attitude Sensor), provided by CREST (Canada). Objective: To demonstrate the capabilities of a new type of infrared Earth sensor over a range of altitudes from LEO to GEO in a harsh radiation environment on a slow-spinning satellite.

SCL (Spacecraft Command Language), provided by NASA, ICS (Interface Control Systems, US). Objectives: To monitor spacecraft systems.

SCPS (Space Communications Protocol Standards), provided by USAF, NASA, MoD, DERA. Objectives: To demonstrate and test a full space implementation of the Space Communications Protocol Standards.

SMX-2 (Sparc Microprocessor Experiment-2), provided by MMS, DERA, ESTEC. Objective: To demonstrate the ERC-32 Sparc chip-set in a harsh radiation environment and to demonstrate its ability to meet the platform OBC function and to support the experiments: ECSE, SCPS, SCL, and OBPE. The SMX-2 experiment consists of two Single electronic boards with the ESA Sparc ERC32 chip set incorporated into them. SMX-2 has 4 MByte of RAM and 2 MByte of EEPROM. SMX-2 is housed in the same box as ECSE.

Demonstration technologies:

ETB (Electronics TestBed), provided by AFRL, Kirtland AFB, NM and sponsored by BMDO. ETB is a Data Handling System (DHS), providing a general purpose interface between multiple space flight experiments and the spacecraft. The DHS design simplifies some of the complexities by providing a simplified interface for operating the space experiment, recording the results, and communicating the results to the spacecraft. It also reduces the risk to the spacecraft mission by providing a degree of separation between the spacecraft and less-proven new technologies, and it provides a similar degree of separation between experiments.

BEPT (Bradford Engineering Pressure Transducer), provided by Bradford Engineering and sponsored by NIVR. The principal objective is to flight demonstrate BEPT. The instrument is used to measure the pressure in the operational cold gas attitude control system on STRV-1d.

InP/Si Solar Panel, provided by NRL.

GaAs/Ge Solar Panels, provided by ESTEC.

Lithium Ion Battery, provided by BNSC, DERA, AEAT (Atomic Energy Authority Technology, UK).

M.26 TEAMSAT

TEAMSAT (Technology, science and Education experiments Added to Maqsat) is an ESA/ESTEC low-cost satellite demonstrator mission, the first S/C delivered to orbit by the Ariane-5 launcher on its second qualification flight (launch on Oct. 30, 1997 from Kourou). The TEAMSAT spacecraft turned out to be an embedded payload (350 kg) of the Ariane-5 upper instrumented test platform, referred to as Maqsat-H, carrying five experiments provided by various European universities. Very tight experiment development and S/C schedules of only one year (from project proposal to launch) resulted due to the late decision to grab a free-flight opportunity on Ariane 502 (along with the acceptance of all flight constraints) for 'hands-on' experimentation. In the end, over 40 young trainees and students from ten European countries along with experienced ESA engineers were actively involved in the TEAMSAT project. [1764) 1765) 1766)]

The TEAMSAT S/C structure consisted of a closed aluminum box of octagonal form (0.755 m high and a diameter of 0.944 m) containing the electrical system, the experiments and the

1764) M. Bandecchi, W. J. Ockels, "The TEAMSAT Experience," ESA Bulletin 95, Aug. 1998, pp. 132-143

1765) A. Bradford, F. Müller-Stute, B. Sarti, "Engineering TEAMSAT - From Concept to Delivery," ESA Bulletin 95, Aug. 1998, pp. 144-147

1766) S. Habinc, D. Hardy, P. Sinander, C. Smith, "TEAMSAT's Data-Handling Systems," ESA Bulletin 95, Aug. 1998, pp. 148-151 ejection mechanism for YES. The TEAMSAT mass of 350 kg was part of the Maqsat-H structure. [Total mass of the Maqsat-H platform was 2300 kg; another platform on this flight was Maqsat-B]. TEAMSAT electrical power was provided by three NiCd batteries. Mission operations was provided through ESA/ESOC. From a technology point of view, TEAMSAT's achievements included: [1767) 1768) 1769)]

- TEAMSAT is the first ESA S/C flown with telemetry and telecommand systems both fully compatible with ESA/CCSDS standards and the first S/C to exploit the adaptive asynchronous telemetry capabilities they support. Use of new chips for telemetry transfer frame generation. Dynamic bandwidth allocation without the use of an on-board computer. First ESA S/C to use the standard packet telecommand protocols.
- Use of a quadrifilar helix antenna for 100 MHz to 3 GHz links. The same system is scheduled to be flown on Metop-1.
- Demonstration of GPS reception from above the GPS satellite constellation.

The TEAMSAT mission ended on Nov. 2. 1997. After three days of intensive operations the available battery energy had been consumed as planned.

Orbit: GTO (Geostationary Transfer Orbit) achieved: Perigee = 525 km, apogee = 27,000 km, inclination = 7.76°, period = 7 hours, 40 minutes.

Sensor/experiment complement:

AVS (Autonomous Vision System), provided by the Danish Technical University (DTU), Lyngby, Denmark. AVS is of ASC (Advanced Stellar Compass) heritage flown on Ørsted. AVS is an advanced star compass which automatically recognizes the star field and tracks non-stellar objects (like satellites, comets or asteroids). AVS consists of two units: a lightweight CCD camera (0.18 kg, 0.4 W) and a DPU (Data Processing Unit) based on 486 technology (0.75 kg, 5 W). The microcomputer of the DPU is radiation tolerant. More than 11,000 stars are stored in memory. AVS is capable of:

- Identifying guide stars in spin-stabilized mode
- Autonomously determine the attitude in a three-axis stabilized or slow spin mode
- Identify non-star objects
- Track or image a detached probe
- Compress image/data for ground analysis

AVS was used in particular during the initial phases of the mission to determine the attitude and orbit of the S/C and to study the autonomous operations during severe radiation doserates in GTO. A thin shielding of the AVS was used to achieve a full lifetime dose during the battery-limited lifetime of TEAMSAT. [1770) 1771)]

FIPEX (Flux Probe Experiment), provided by the University of Stuttgart (FIPEX is of TEXUS sounding rocket campaign heritage). The objective is the measurement of atomic oxygen concentration at different orbital altitudes (> 200 km up to 1000 km) to study erosion and degradation of optical surfaces in orbit; the data are also valuable for the verification of numerical models used in re-entry aero-thermodynamics. Oxygen is measured on the principle of electrolysis using a solid electrolyte (potentiometric oxygen sensor). The FIPEX instrument consists of:

- An array of five identical oxygen sensors, used to provide data redundancy and eliminate aerodynamics effects

1767) M. Jones, B. Melton, M. Bandecchi, "TEAMSAT's Low-Cost EGSE and Mission Control Systems," ESA Bulletin 95, Aug. 1998, pp. 152-157

1768) http://www.estec.esa.nl/teamsat/

1769) C. Smith, "Low-cost, ASIC-based Telemetry and Telecommand Systems - The TEAMSAT Experience," Proceedings of the 4th International Symposium on Small Satellites Systems and Services, Sept. 14-18, 1998, Antibes Juan les Pins, France

1770) J. L. Joergensen, et al., "Radiation Impacts on Star-Tracker Performance and Vision Systems in Space," IAA 2nd International Symposium on Small Satellites for Earth Observation, Berlin, April 12-16, 1999, pp. 393-396

1771) M. Betto, et al., "The Determination of the Attitude and Attitude Dynamics of TEAMSAT," IAA 2nd International Symposium on Small Satellites for Earth Observation, Berlin, April 12-16, 1999, pp. 397-400

- Each sensor head contains an oxygen sensor, a sealed reference gas volume and both a reference pressure and temperature sensor.

The sensors are pre-heated to around 600°C to create sufficient ion conductivity within the electrolyte. The heating is feedback controlled so as to provide temperature stabilization for vacuum operation.

VTS (Visual Telemetry System), an ESA/ESTEC instrument built by MMS (UK), IMEC (Inter-university MicroElectronics Center), Leuven, Belgium and DSS/OIP (Delft Sensor Systems/Optronic Instruments & Products) in Oudenaarde, Belgium. VTS provides a new on-board visualization technology for monitoring S/C activities (solar array deployment, etc.). VTS consists of three cameras and a master unit (up to eight off camera modules may be supported), providing JPEG-compressed (and buffered) image sequences. The master unit is a flight-spec TSC-21020 based DSP computer. Each camera module has a maximum imaging capability of 512 x 512 pixels [using APS (Active Pixel Sensor) technology with associated electronics for camera imaging control], which are digitized to 8 bits. Full-size images can be acquired at up to 1 image/s. APS provides 6 decades of dynamic range. A standard 60° FOV is provided.

A flexible system of controlling imaging sequences is provided in the master unit which includes sub-imaging and interleaving to give higher imaging rates; typically up to 16 images/s. The master unit also stores images (up to 64 full size images, 128 MByte buffer capacity) prior to acquisition by the spacecraft Data Handling System. Image compression is provided to minimize the impact on the spacecraft telemetry downlink. In normal operation two image grades are provided (100:1 and 15:1). Custom compression ratios can be provided based on trade-off of image quality and telemetry bandwidth/downlink periods. The mass of VTS is 1.7 kg.

ODD (Orbiting Debris Device), provided by ESTEC. ODD tested the capabilities of optical and radar instruments on the ground to detect and track objects in space. For this purpose the upper Ariane 502 satellite (Maqsat-H) was covered with contrasting black and white paints of well known properties (75 % white and 25 % black). The other payloads have been left unchanged for comparison and also form part of the experiment.

YES (Young Engineers' Satellite), provided by the Technical University of Delft (Netherlands). YES is a free-flying subsatellite, intended to deploy a 35 km tether attached to an inert counter-balance mass (referred to as TORI), to study the dynamics of tethered satellites. YES is physically the lid of the TEAMSAT box. -The YES subsatellite contained additional experiments:

- A GPS receiver was installed to evaluate GPS reception outside the GPS constellation and to provide information on the tether dynamics
- The SF (Scintillating Fiber) detector. The SF nuclear detector module consists of three radiation measuring channels for fiber-optic-coupled SF probes. Detected radiation are fast protons and electrons. A new radiation-hardened experimental SF probe and a doped spinel crystal were flown as the third detector.
- RADFET (Radiation-sensitive Field Effect Transistor) experimentation. Three RADFETs were used to measure the total ionizing dose behind thin shielding (so-called "skin-dose").
- Sun senors. Seven SAS (Solar Aspect Sensors) measure the S/C attitude. Each of the seven sun sensors consists of two light-sensitive transistors mounted in an aluminium housing. On top of each transistor two baffles are mounted, forming narrow slits through which the transistor is looking out. This makes the FOV wide in one direction (about 90°) and narrow in the other (1 to 2°). As the slits on each sun sensor are perpendicular to each other, the sensors are always looking for the sun in two perpendicular planes. Together, the seven sun sensors have a 360° FOV in all three perpendicular planes through the S/C.

- Accelerometers. Test of new technologies [MNT (Micro & Nano-Technologies)] which are being considered as a possible future way to miniaturize systems and subsystems in space (measurement of vibrations).

Two on-board computers were installed, JORIS (student-designed and programmed) for control of the tether experiment, and a backup PC104 (a small version of a PC providing also the interface to a commercial camera (QuickCam) for imagery acquisition during YES ejection.

In June 1997, the international Steering Committee for Space Debris gave its no-go decision on the deployment of the YES mission's 35 km long tether (posing a high collision and debris risk). As a consequence the tether was not deployed. Nevertheless, a YES separation procedure (without tether) was conducted, the counter-balance mass remained mated to TEAMSAT. The ejection of YES was performed during the 8th orbit on Nov. 2 (ejection speed of 1.7 m/s) in preparation of future flights. Many useful results were obtained despite the restrictions imposed.

M.27 Tether Missions/Experiments

A space tether is a long cable used to couple spacecraft together as they orbit the central body (i.e. Earth). [1772] Tethers are usually made of thin strands of high-strength fibers such as Spectra or Kevlar. Any space tethered system is intimately connected to the gravitational force field. Conducting tethers offer the additional capability to interact with the magnetic and electrical force fields. [1773] [1774] [1775] [1776]

The tethered system demonstrates gravity gradient attitude control. This is a very low cost attitude control system and if suitable for the mission, typically orients the system to within ± 10 degrees of the vertical, both in and out the orbit plane.

Because the space tether makes it possible to transfer energy and momentum from one object to another, it can legitimately be called a form of space propulsion. There are two general categories of tethers.

Momentum-exchange tethers (nonconductive tethers representing passive propulsion). They allow momentum to be transferred between objects in space, such as two spacecraft (tethers may redistribute momentum of a system from one body to another, but overall momentum is always conserved). The principle is based on the gravity gradient force.

Two objects, separated by a distance but tied together by a tether, are "pulled" apart by the gravity gradient force [this causes vertical (radial) alignment between the two objects]. Due to irregularities in the central body's gravitational field, the nearly radially aligned tether system actually librates, or oscillates, in a pendulum-like motion, about the system's center of mass. This swinging motion may be used to raise or lower the orbit of a tandem system without using any propellant.

- A tethered system release of an end-body from the remaining end-body and tether causes a momentum gain for the released end-body, resulting in a higher orbit for the released end-body orbit and a lower orbit for the remaining end-body and tether.
- A bolo tether rotates end-over-end in the orbit plane. This system could propel a payload attached to one end into a different orbit. The bolo could conceivably catch a payload.

1772) D. A. Arnold, "The Behavior of Long Tethers in Space," The Journal of the Astronautical Sciences, Vol. 35, No. 1, January-March 1987, pp. 3-18

1773) I gratefully acknowledge the major review, revision and addition of this chapter provided by Michael Zedd of NRL, Washington, D. C.

1774) R. L. Forward, J. Davis, "Doing the Do-Si-Do," Launchspace Magazine, April/May 1998

1775) M. L. Cosmo, E. C. Lorenzini (editors), "Tethers In Space Handbook," Third Edition, Dec. 1997, NASA/MSFC, download from http://cfa-www.harvard.edu/spgroup/handbook.html

1776) L. Johnson, R. D. Estes, E. Lorenzini, et al., "Electrodynamic Tethers for Space Propulsion," Proceedings of AIAA, Reno, NV, Jan. 12-15, 1998, AIAA 98-0983

- A stationary tether refers to two end-bodies connected by a tether of constant length. The system may be used to drag the lower end-body payload through the higher atmosphere (for sampling) and simultaneously lowering the system's orbit.

Electrodynamic tethers (conductive tethers representing active propulsion and power generation). An electrodynamic tether is essentially a long conducting wire extended from the main end-body (spacecraft). There may be a second end-body to help deploy the bare tether. The gravity gradient field of the "S/C and string system" tends to orient the tether in a vertical position. A tether in Earth orbit interacts with the Earth's magnetosphere due to the relative orbital velocities (in LEO systems the orbital velocity of the tether is greater than the rotational velocity of the geomagnetic field, this force is drag on the tether); the motion of the conductor across the magnetic field induces a voltage along the length of the tether. This voltage can be up to several hundred volts per km.

The interaction of the tether system with the magnetosphere can be used in the design of the system to act either as an "electrodynamic power or thrust system" to boost the orbit of the S/C; or it may be used to act as an "electrodynamic drag system" to lower the orbit of the S/C. The direction of the current flow in or out of an electrodynamic tether system determines if the interaction contributes to drag or to propulsion.

- **Electrodynamic power or tether drag system.** In this case the tether converts the orbital energy of the host S/C into electrical power. The drag force is used to decrease the orbit of the tether and its host S/C. In this design scheme current is produced by the interaction of the tether ends with the external plasma (a current loop consists of the tether, the external plasma, and the ionosphere). The collection of electrons from the plasma creates a net positive charge. This means that when electric power is generated by the tether for on-board use or simply for dissipation, it is generated at the expense of orbital energy.
- **Electrodynamic thrust tether system.** In this case, an EMF (Electromotive Force) acts to induce an electric potential along the tether system; the tether is used to boost the orbit of the host S/C. This concept requires power to be injected into the tether system from an on-board power supply, causing the system to gain altitude. The tether interaction with the Earth's magnetic field provides a "push" resulting in an orbit boost of the S/C.

A side effect of tether current flow is the production of electromagnetic waves (ULF, ELF, VLF) into the ionosphere, resulting from the electron transfer from the tether back into the plasma. This feature may give rise to use the tether also as an antenna.

Mission (Agency)	Launch (Timeframe)	Orbit	Tether Length	Comment
Gemini-11(NASA)	Sep. 11-15,'66	LEO	30 m	Spin-stabilized, 0.15 rpm
Gemini-12(NASA)	Nov. 11-15,'66	LEO	30 m	Local vertical, stable swing
H-9M-69	1980	Suborbital	500 m	Partial deployment
S-520-2	1981	Suborbital	500 m	Partial deployment
Charge-1	1983	Suborbital	500 m	Full deployment
Charge-2	1984	Suborbital	500 m	Full deployment
ECHO-7	1988	Suborbital		Magnetic field aligned
Oedipus-A (NRC/ NASA,CRC, CSA)	Jan. 30, 1989	Suborbital	958 m	Spin-stabilized, 0.7 rpm (Canada/USA)
Charge-2B	1992	Suborbital	500 m	Full deployment
TSS-1 (STS-46) (ASI/NASA)	Jul. 31-Aug. 8, 1992,	LEO	< 270 m	Electrodynamic tether, partial upward deployment and retrieval of tether
SEDS-1 (NASA)	Mar. 29,1993	LEO	20 km	Nonconducting tether, downward deployment, swing and cut
PMG (DoD) Plasma Motor Generator	Jun. 26, 1993	LEO	500 m	Electrodynamic tether, upward deployment, tether current in both directions. Demonstration of hollow cathode plasma contactors for current collection and emission

Mission (Agency)	Launch (Timeframe)	Orbit	Tether Length	Comment
SEDS-2 (NASA)	Mar. 9, 1994	LEO	20 km	Local vertical stable, downward deployment
Oedipus-C (NRC/ NASA,CRC, CSA)	Nov. 6, 1995	Suborbital	1 km	Spin-stabilized, 0.7 rpm (Canada/USA)
TSS-1R (STS-75) (ASI/NASA)	Feb. 22-Mar. 9, 1996	LEO	19.6 km	Electrodynamic tether, severed prior to full deployment
TiPS (NRO/NRL)	May 12, 1996 Jun. 20, 1996 deployment	LEO	4 km	Long life tether (tracking of one year)
YES (Technical Univ. of Delft)	Oct. 30, 1997	GTO	35 km	YES (of TEAMSAT) was not deployed to avoid a collision with other S/C (M.26)
STEX/ATEx (NRO/NRL)	Oct. 3, 1998 Jan. 16 1999 deployment	LEO	6 km	Nonconductive tether, planned demonstration of tether system stability and control. However, a deployment malfunction caused an experiment failure
PICOSAT1.0 of the Aerospace Corporation	Jan. 27, 2000 on OPAL of Stanford	LEO	30 m	PICOSAT1.0 was ejected OPAL on Feb 6, 2000. The system of two S/C operated for 3 days when battery power decayed.
PICOSAT1.1 of the Aerospace Corporation	July 19, 2000 on MightySat II.1 of AFRL	LEO	30 m	PICOSAT1.1 will be ejected from Mightysat II.1 one year after launch (in Sept. 2001) to test storage effects on the system.
METS (MIR Electrodynamic Tether System)	2001	LEO	5 km	First practical application of non-rocket propulsion in space
ProSEDS (NASA)	2001	LEO	25 km	Upward deployment (producing drag and electrical power)
TSE (ESA)	2002	LEO	35 km	Demonstration of ISS sample return

Table 518: Chronology of tether missions

The drag feature of tethers may be used to deorbit a body (a malfunctioning or out-of-service satellite). A multitude of low-cost applications can be designed around tether systems besides propulsion (such as communications, atmospheric studies, gravity experiments, power generation, etc). [1777)]

History: The first two tethers (each 30 m in length) flew in 1966, during the Gemini-11 and -12 manned orbital missions. Orbital tether experiments began again in 1980 and have continued through the present. Tether missions in the 1990s have included several important milestones, such as: (see also Table 518)

- First retrieval of a tether in space (TSS-1, 1992)
- Successful deployment of tethers 20 km long (SEDS-1, 1993)
- First operation of an electrodynamic tether (PMG, 1993) with current driven in both power mode and thrust mode.

M.27.1 ASTOR (Advanced Safety Tether Operation and Reliability)

ASTOR is a proposed Shuttle hitchhiker payload (HH-S) of NASA/MSFC and the Michigan Technic Corporation, Holland, MI. The objective is to demonstrate the performance of ETD (Emergency Tether Deployment). ETD is designed to overcome the safety hazard caused by snags during tether deployment. ASTOR itself consists of three elements: the satellite upper unit, the satellite lower unit, and the tether. Each unit is hexagonal in shape with a length of 66 cm and 46 cm in diameter. The mission involves ejection of HH-S and deployment of ASTOR. ETD provides for accurate control and deployment of the tether. [1778)]

M.27.2 BOLAS (Bistatic Observations with Low Altitude Satellites)

BOLAS is a proposed Canadian/USA satellite technology mission (tethers, microsatellites, the use of GPS, etc.) with CSA and NASA/MSFC as contributing partners (the BOLAS mis-

1777) http://www.tethers.com

1778) http://www.airseds.com/astor.html

sion team includes institutes, Canadian and US universities, and industry). As of 2000, the BOLAS mission is postponed indefinitely due to funding problems.

Sensor complement:

HF Receiver, provided by CRC (Communications Research Center).

SII (Suprathermal Ion Instrument), provided by the University of Calgary.

TECHS (Thermal Electron Capped Hemisphere Spectrometer), provided by SwRI (Southwest Research Institute), San Antonio, TX.

GPS Receiver, provided by the University of New Brunswick.

M.27.3 METS (MIR Electrodynamic Tether System)

Another electrodynamic tether, nicknamed Firefly, is expected to be attached to the MIR space station to reboost the altitude decaying MIR, thereby adding more years to its lifetime. The METS initiative/study was sponsored by FINDS (Foundation of the International Non-Governmental Development of Space) and carried out by Energia Ltd. and Tether Applications Co. of San Diego, CA. If successful in 2001, this will be the first practical application of non-rocket propulsion in space. METS is mounted on the EVA ladder at the end of the Kvant-2 module. The tether consists of a 5 km anodized aluminum wire deployed downward from MIR. The tether is kept nearly vertical by gravity gradient effects acting on the wire and a 200 kg ballast mass at the end of the wire. The 1 km section of tether furthest from MIR is bare metal and this strip will begin collecting electrons that are moving within the Earth's magnetic field. As the electrons move along the wire toward the space station, a hollow cathode emitter mounted on MIR will spray them back into space, completing the circuit and letting a current flow. This will result in a net force increasing the station's orbit speed. [1779] [1780]

M.27.4 OEDIPUS

Various Canadian organizations (NRC, CRC, CSA) have successfully deployed two tethered systems from sounding rockets. These suborbital missions for "Observations of Electric Field Distribution in the Ionospheric Plasma - a Unique Strategy" were accomplished as OEDIPUS-A in January 1989 and OEDIPUS-C in November 1995.

OEDIPUS-C has two spinning payloads connected by a 1,174 m tether. The scientific objectives addressed natural and artificial waves in the ionospheric plasma. The mission provided perspectives on plane and sheath waves and their interaction with space plasma. The science experiments were also designed to help to understand how charged particles associated with aurora affect satellite transmissions. [1781]

M.27.5 PMG (Plasma Motor Generator)

PMG is a tether payload of a DoD satellite that was launched as a secondary payload on a Delta-2 launch vehicle on June 26, 1993. The objectives of PMG were to test: 1) the performance of the HCA (Hollow Cathode Assembly) to provide a low impedance bipolar electrical current between a S/C and the ionosphere, and 2) demonstrate the functions of space propulsion and electrical current generation (and current reversal in tether using a bias voltage). The PMG consisted of four basic elements: FEP (Far-End Package), NEP (Near-

1779) J. Oberg, "Saving MIR With A Rope Trick," IEEE Spectrum, July 2000, Vol. 37, No. 7

1780) http://www.finds-space.org/METS.html

1781) A. Jablonski, F. Vigneron, G. Tyc, and H. G. James, "OEDIPUS-C Mission Tether Dynamics Results," paper presented at the Tether Technology Interchange Meeting, MSFC, Sept. 9-10, 1997, NASA/CP-1998-206900

End Package), an electronics box called SEDS (Small Expendable Deployer System), and PDP (Plasma Diagnostic Package) of NASA/LeRC. The tether was 500 m in length and consisted of a number 18 AWG (American Wire Gauge) Teflon-insulated copper wire. After third-stage separation, the PMG was left in an elliptical orbit (193 km x 869 km, inclination = 25.7°). The FEP was ejected upward with an initial velocity of 2-3 m/s. Each HCA was equipped with a 1 liter gas bottle, on/off solenoid, gas metering block power supplies to produce a weekly ionized xenon cloud. Both end platforms carried a 28 V silver cell battery for a nominal lifetime of 3-6 hours. [1782)]

The PMG current showed to be fully reversible, operating either as a generator system with electron current flow down the tether, or as a motor with electron current driven up the tether. The induced EMF was measured with total voltage biased from +150 to -90 V, and also with the bias turned off by using only the induced EMF.

M.27.6 ProSEDS (Propulsive Small Expendable Deployer System)

A NASA/MSFC tether demonstration experiment, planned to fly in June 2001. ProSEDS is based on the SEDS deployment scheme, with the objective to deploy an electrodynamic tether from a Delta-II rocket's third stage (launching a GPS satellite) to achieve about 1 N drag thrust for stage deorbiting. The tether is composed of 10 km of nonconducting Spectra tether for ballast and to begin the deployment and 5 km of aluminum wire (1.2 mm diameter) covered with thermal coatings next to ProSEDS. Normally, these stages slowly spiral back to Earth over the next half year as atmospheric drag nibbles away at their speed. By generating an electrical current, ProSEDS will turn itself into an electromagnetic brake. An orbital decay of at least 5-15 km in altitude per day is expected with this deorbiting technique. It's not quick compared to a retrorocket, but it is much faster than natural decay. And it's being done without the use of any propellant. [1783)] [1784)] [1785)]

Orbit: circular orbit, altitude = 400 km, inclination = 35°

M.27.7 SEDS (Small Expendable Deployer System)

SEDS is a NASA/MSFC tether experiment demonstration program, developed by MSFC and by Tether Applications Co. of San Diego. Two flight experiments (nonconducting momentum-exchange tethers) have been flown as Delta secondary payloads on the second stage of GPS satellite launches. Each SEDS sub-satellite was deployed by means of a deployer system. [1786)]

SEDS-1, launched March 29, 1993, was a 20 km downward deployment with the objective to demonstrate the use of a tether to place a payload in a deorbit trajectory (study of reentry after the tether was cut). The Delta second stage was in an elliptical orbit (195 km x 705 km, inclination of 34°), the tether deployment was initiated at apogee. Telemetry was obtained from the deployer through the Delta second stage telemetry system and from the tether end-mass via an S-band transmitter. The deployment took about one orbit and the tether brake was applied. The tether with its end-body payload swung to local vertical at which time the tether was cut at the deployer end and the end-body, trailing the tether, reentered the atmosphere and burned to destruction off the coast of Mexico. The reentry was video taped by a team of observers.

SEDS-2, launched on March 9, 1994, was also a 20 km downward deployment, but the mission objective was to demonstrate the use of a modified SEDS to deploy and stabilize a teth-

1782) J. McCoy, et al., "Plasma Motor-Generator (PMG) Flight Results," *Proceedings of the Fourth International Conference On Tethers In Space*, Science and Technology Corp., Hampton, VA, Apr. 1995, pp. 57-82

1783) L. Johnson, J. Ballance, "Propulsive Small Expendable Deployer System (ProSEDS) Space Demonstration," paper presented at the Tether Technology Interchange Meeting, MSFC, Sept. 9-10, 1997, NASA/CP-1998-206900

1784) http://stp.msfc.nasa.gov/astp/proseds.html

1785) L. Johnson, "The Tether Solution," IEEE Spectrum, July 2000, Vol. 37, No. 7

1786) http://std.msfc.nasa.gov/astp/tethers_electetherprop3.html

er system along the local vertical. The orbit of the second stage was circular at an altitude of 350 km. The tether was not intentionally cut and instrumentation on both the deployer end and end-body measured performance for several hours until battery depletion. System stability was within 4°. The tether was cut by a micrometeoroid or space debris about 5 days after launch. Approximately seven km of tether remained attached to the second stage until reentry into the atmosphere two months later. [1787] [1788]

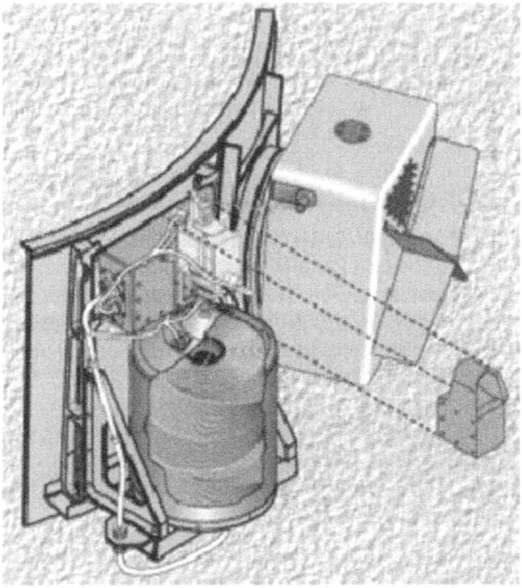

Figure 334: Illustration of SEDS on the Delta second stage

For both payloads, the end-body instrument package was designed and built by LaRC. It consisted of a three-axis tensiometer, a three-axis magnetometer, and a three-axis accelerometer. The end-body measured tether system performance and end-body attitude.

M.27.8 STEP-AIRSEDS

STEP-AIRSEDS (Space Transfer Using Electrodynamic Propulsion - Atmospheric Ionospheric Research Small Expendable Deployer System) is a proposed electrodynamic tether mission of NASA, designed and built by the Michigan Technic Corporation, Holland, MI. The objective is to demonstrate new technologies (mitigation of orbital debris) including an electrodynamic tape tether and a deployer that can retract the tether. This 1,000 kg satellite is required to conduct boost, de-boost, plane-change and station-keeping operations during the mission. The satellite operates in an altitude range between 350 km and 1,100 km for a minimum of one year. It is targeted for a launch in 2004. Currently, The Michigan Technic Corporation is conducting the Phase B development of the satellite.

M.27.9 STEPS (Station Tethered Express Payload System)

STEPS is a NASA/MSFC project in the SBIR Phase II program to increase sample return opportunities between regular service flights to ISS (of Shuttle, Soyuz, or Raduga). The

1787) H. F. Smith, "The First and Second Flights of the Small Expendable Deployer System (SEDS)," *Proceedings of the Fourth International Conference On Tethers In Space*, Science and Technology Corp., Hampton, VA, Apr. 1995, pp. 43-56.

1788) http://ixeab8.larc.nasa.gov/seds/

STEPS design uses a return capsule of 78 cm diameter with a payload capacity of 100 liter. STEPS is both deorbited and oriented for reentry by a 2.5 kg expendable tether. The tethered deorbit and kite-tail-orienting effect before reentry were both demonstrated on the SEDS-1 (Small Expendable Deployer System) mission in March 1993. [1789)]

The STEPS deployer is a small easily reloadable version of the SEDS deployer. It is part of a reusable ejector/deployer assembly that remains on the Station (ISS). The empty capsule plus its tether weigh 10 kg and take only 30 liter of storage space. The ejector/deployer stow has a volume of 20 cm x 40 cm x 100 cm.

A STEPS precursor flight is planned as a Delta secondary payload. It is intended to test the heat shield, modified SEDS deployer, tether control law, and a soft mid-air recovery technique.

M.27.10 TiPS (Tether Physics and Survivability)

TiPS is a low-budget quick reaction experiment (a secondary payload) sponsored by NRO [National Reconnaissance Office (of DoD, USA)]. It was launched May 12, 1996 on a DoD satellite from VAFB, CA (Titan-4 vehicle), and deployed on June 20, 1996 [built and operated by NCST (Naval Center for Space Technology) of NRL]. The experiment objective is to increase knowledge about gravity-gradient tether dynamics and the long-term survivability of tethers in space. A deployment sequence jettisoned the TiPS satellite from a host vehicle and then separated its two end-bodies from each other (initial separation velocity of 5.1 m/s). [1790) 1791) 1792)]

The TiPS payload consists of two end-bodies (of box-like shape, each covered with 18 laser retroreflectors), with a total mass of 54 kg [37.7 kg for the larger box (dubbed 'Ralph') [1793)] and 10.8 kg for the smaller upper box (dubbed 'Norton'), tether mass of 5.5 kg], were separated by four km of a nonconducting tether. The cylindrical shape of the tether is due to strands of Spectra woven around yarn. Ralph contains a SEDS (Small Expendable Tether Deployer System) box from NASA which in turn contains all the electronic components (telemetry system, turn-count recorder, and temperature sensors). The telemetry system is powered by a non-rechargeable battery, which operated for the first eight hours of the satellite's life after deployment. Telemetry generated during the separation of the two end-bodies confirmed full tether deployment. NRL, NASA and an international network of SLR (Satellite Laser Ranging) stations (in addition ground-based radar data and optical data from several telescopes) provided global tracking of the system for the duration of one year (history of the librational and rotation motion dynamics).

Retroreflectors are mounted on the exterior surfaces of both Ralph and Norton, for long-term passive monitoring of the tethered system. End-body discrimination is accomplished by coating the retroreflectors of Ralph to reflect only one of the two transmitted laser wavelengths. The uncoated retroreflectors on Norton reflect both transmitted wavelengths.

Orbit: Near-circular orbit, mean altitude = 1022 km, inclination = 63.4°, period = 105 min.

The TiPS orientation is controlled by gravity gradient forces. The passive TiPS satellite orbits the earth with a nominal vertical orientation, with Ralph closest to the earth. About this nominal orientation, the tethered system undergoes libration. The various tracking information was used to determine the orbit and attitude of the tethered system. The most signif-

---

1789) http://std.msfc.nasa.gov/astp/tethers_reentrycaps.html

1790) W. Barnds, S. Coffey, M. Davis, et. al., "TiPS: Results of a Tethered Satellite Experiment," paper presented at the AAS/AIAA Astrodynamics Conference in August 4-7, 1997, Sun Valley, Idaho, AAS 97-600.

1791) http://hyperspace.nrl.navy.mil/TiPS/home.html

1792) K. T. Alfriend, W. J. Barnds, et al., "Attitude and Orbit Determination of a Tethered Satellite System," AAS/AIAA Astrodynamics Specialist Conference, Halifax, Nova Scotia, Aug. 14-17, 1995, AAS 95-351

1793) Note: Ralph and Norton are two characters of the Honeymooner's television fame (mainly during the 1950-60s), representing adequately the satellite mass ratio icant finding from the experiment was that the attitude and rotational motions of the tether system have damped significantly since initial deployment (from about 40° in-plane and 33° cross-plane to about 7.5° and 5° - 7°, respectively).

Tether length, nominal diameter	4023 m, 2 mm (approximately)
Tether construction	The outer layer is Spectra 1000 braid for strength, the core is acrylic yarn which will puff the other braid out to the 2mm diameter and give it a larger cross section to improve its resistance to debris and small micrometeoroids.
Tether breaking strength	90 kg (approximately)
TiPS total mass	54 kg
Mass of Ralph (lower end body)	37.7 kg
Mass of Norton (upper end body)	10.8 kg
Mass of tether	5.5 kg
TiPS initial spin rate	4 rpm

Table 519: TiPS specification parameters

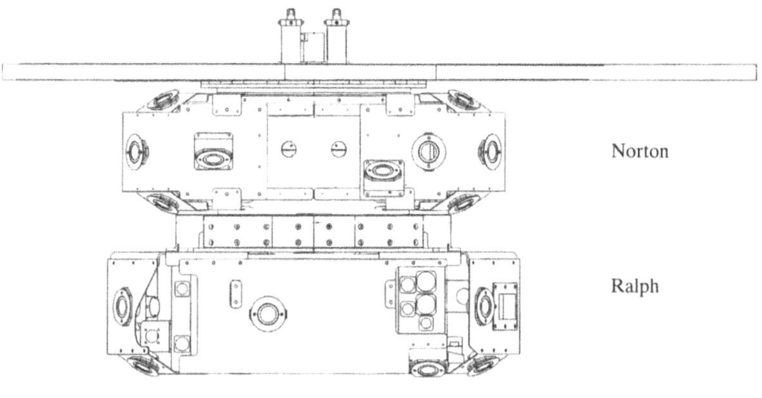

Figure 335: Illustration of the stowed TiPS S/C on the host vehicle

M.27.11 TSE (Tether System Experiment)

TSE is a proposed ESA tether project, initiated in 1997. The study is carried out by an international project team, led by Kayser-Threde GmbH of Munich, Germany. The launch of the TSE demonstration mission is planned for 2002. The objectives of the TSE mission are to: [1794)]

- Demonstrate the operation of a tether-deployment mechanism suitable for ISS sample return and the ability to operate properly in the ISS (International Space Station) environment
- Test a robust tether-deployment control strategy for re-entry initiation, and demonstration of an accurate landing scheme of future sample-return capsules
- Demonstrate the critical technologies associated with the re-entry capsule of an ISS tethered sample return system
- Collect experimental data on system dynamics to support the validation of models used for tether-deployment simulations.

The TSE design is conceived as a low-cost mission using existing hardware whenever possible. The mission scenario calls for the TSE subsatellite to be installed on a Russian Prog-

1794) J. G. Izquierdo, H. Rozemeijer, S. Müncheberg, "The Tether System Experiment - Preparing for ESA's First Tether Mission,," ESA Bulletin No 102, May 2000, pp. 139-143 ress cargo vehicle, conducting a service mission to ISS. The ejection of the subsatellite from Progress and tether deployment is planned to occur after completion of the service mission in a safe distance away from ISS.

The tether deployment strategy consists of two phases: In the first phase, the tether is deployed to 3 km vertically downward of Progress to stabilize the capsule and to reduce sensitivity to ejection errors. After synchronization with the target landing site, the second phase is initiated with the full deployment of the tether to 35 km length. At the end of deployment, the tether motion is stopped by a brake, causing a swing-back into the direction of the local vertical. The tether is cut by the deployer system at a predefined instant (close to the local vertical), resulting in a momentum exchange (sling shot) that releases the subsatellite onto its re-entry trajectory, with the tether in tow.

Thirty minutes later, the subsatellite reaches the Earth's upper atmosphere and decays above the Pacific Ocean. From activation to burn-up, the subsatellite collects experiment data such as position and velocity (GPS receiver), angular rates and accelerations (IMU), and temperatures. The information is transmitted to the tether deployer system on-board Progress for storage and subsequent downlink.

System mass Tether deployer and electrical subsystems Subsatellite Experiment container	143 kg (total) 60 kg 42 kg 41 kg
TSE dimensions (without container)	1265 mm x 588 mm x 588 mm
Power consumption	88 W during tether deployment
Tether type	Dyneema 4 x 400 denier (momentum exchange tether)
Tether diameter, length	0.5 mm, 35 km

Table 520: Technical data of TSE

The tether deployer system provides the following functional support:

- Tether storage during launch and pre-experiment phases
- Ejection/acceleration of the subsatellite at initiation of the TSE release
- Tether guidance throughout the experiment to prevent tether damage
- Monitoring of tether system behavior
- Control of tether release (tether tension and velocity)
- Tether cutting in nominal and non-nominal modes
- Reporting of status information to the data handling system

The actual TSE deployment was preceded by a test campaign (in 2000) with a breadboard system to cover all aspects of functional system testing.

M.27.12 TSS (Tethered Satellite System)

TSS is a cooperative program of ASI and NASA for a reusable multi-disciplinary facility to conduct space experiments in Earth orbit. The system consists of a satellite, a conductive tether (approx. 20 km long) and a deployer in the Shuttle's cargo bay.

Background: The TSS program traces its birth to 1974 when Guiseppe Colombo of Padua University (Italy) came up with the idea of using a long tether to support a satellite from an orbiting platform. Later in the 1970s, G. Colombo and Mario Grossi of the Smithsonian Astrophysical Laboratory (SAO), Cambridge MA, proposed a mission to study the possible scientific applications for long tethers in space. A NASA-ASI memorandum of understanding was signed in 1984, in which NASA agreed to develop a deployer system and tether and ASI agreed to develop a special satellite for deployment.

- **TSS-1**. The first TSS mission was conducted on STS-46, July 31 - Aug. 8, 1992. The objective of the TSS-1 mission was to demonstrate the feasibility of deploying and controlling long tethers in space, and to evaluate some of the unique applications of the TSS as a tool for research by conducting exploratory experiments in space plasma physics. An electrically conductive satellite, a sphere of 1.6 m diameter with a mass of 518 kg, was deployed; it was tethered to the Orbiter by a conductive tether (2.54 mm diameter using Kevlar and Nomex with 10 strands of 34 AWG copper wire and a Teflon sheath). A maximum upward separation of 267 m between the Orbiter and satellite was achieved. Several unsuccessful attempts were made to extend the tether to its full planned extension of 20km. Deployed operations lasted nearly one full 24 hour period - then the tether was retrieved. Even though the primary mission goals were not achieved many interesting results were obtained.

The **SETS** (Shuttle Electrodynamic Tether System) was one of several instrument packages included in TSS-1. It includes five instruments, a flight computer, and a power interface box. **TCVM** (Tether Current Voltage Monitor) measured the tether current and voltage. **FPEG** (Fast Pulsed Electron Gun) provided removal of negative charge from the Orbiter electrical ground. The **CCP** (Charge Current Probe) used a dual-sensor combination to measure the differential changes in the vehicle potential and return current from the ionosphere to the Orbiter structure. The **SRPA/LP** (Spherical Retarding Potential Analyzer and Langmuir Probe) instrument determined the thermal plasma characteristics relative to the ionosphere. **AMAG** (Aspect Magnetometer) measured the magnetic field in the Orbiter's payload bay.

- **TSS-1R**. A reflight mission, TSS-1R, flew on STS-75 Feb. 22 - March 9, 1996. The TSS-1R mission suffered a broken tether (the cable suddenly snapped near the top of the deployment boom, it was burned through by an electrical arc prior to maximum deployment of 20 km). Tether dynamics were verified up to a length of 19.6 km providing a total of five hours of deployed electrodynamic measurements (largest man-made electrodynamic structure placed in orbit so far). Additional TSS-1R satellite payload instruments were: a triaxial accelerometer and a current meter, and the 2.4 m extendable instrument booms.

Both TSS missions confirmed concept validation of long gravity-gradient tethers; moreover the data gathered during the second mission confirmed the possibility to use the system for the conversion of orbital energy to electrical energy. [1795) 1796)]

M.28 TOPSAT

TOPSAT is a UK collaborative small-cost satellite mission of DERA (QinetiQ), SSTL, RAL and InfoTerra, funded by the UK Ministry of Defense (MOD) and by the BNSC (British National Space Centre) small satellite initiative (Mosaic). The overall objectives are to demonstrate the capabilities of small spacecraft on a high-resolution imaging mission, including an end-to-end ground segment with services for a commercial user community (direct user access, innovative solutions). In this setup, the mission is designed and managed by DERA (QinetiQ), SSTL (Surrey Satellite Technology Limited) is building and integrating the microsatellite bus, RAL (Rutherford Appleton Laboratory) is developing the imaging instrument, while InfoTerra is responsible for data distribution (rapid delivery) and the marketing services. The data is also being used by institutional users (government agencies, military, etc.). The demonstration mission is planned for one year. There will be an evaluation in terms of all aspects of performance.

The spacecraft structure of SSTL employs a truncated pyramid design which offers a high volume efficiency in multiple launches and more solar power. TOPSAT is three-axis stabi-

1795) N. Stone, et al., "A Review of Scientific and Technological Results from the TSS-1R Mission," paper presented at the Tether Technology Interchange Meeting, MSFC, Sept. 9-10, 1997, NASA/CP-1998-206900

1796) B. Strim, M. Pasta, and E. Allais, "TSS-1 vs. TSS-1R," *Proceedings of the Fourth International Conference On Tethers In Space*, Science and Technology Corp., Hampton, VA, Apr. 1995, pp. 27-42 lized. Attitude is maintained by reaction wheels, the momentum is being off-loaded through magnetorquers. The reaction wheel/magnetorquer combination provides also a S/C pointing capability of ±30°. Attitude is sensed by Earth and sun sensors. A high-precision three-axis gyro permits the offtrack and pitch compensation maneuvers required of the mission and image rectification if needed. An attitude stability of 0.002° is provided for image acquisition (1.5 s), the control accuracy is ±0.2° with a knowledge of ±0.1° in all axes. A GPS receiver provides orbit determination and on-board clock updates. Solar power of 55 W (orbital average) is provided by three body-mounted solar panels (GaAs cells). NiCd batteries provide 168 Wh of power storage capacity. The command and data handling subsystem employs the 80386 processor architecture providing a semi-autonomous control of the satellite. The spacecraft has a mass of about 125 kg. The design life is one year, although there are no life limiting factors in the design (e.g. fuel). A launch of TOPSAT is planned as a secondary payload in 2003. [1797) 1798) 1799)]

Orbit: Sun-synchronous circular orbit, altitude = 600 km, inclination = 97°, period = 100 min, 10:30 local equator crossing time. A four day target revisit capability is provided.

RF communications: An on-board storage capability in Flash memory of up to four full and uncompressed scenes (= 2 Gbit) is provided. Standard VHF and UHF subsystems are provided. The VHF uplink is used for spacecraft commanding including file transfer. The UHF downlink is used for telemetry. In addition, there is an X-band downlink for the high-rate image data dumps with a data rate of 25 Mbit/s developed by DERA (QinetiQ). Mission operations are provided by QinetiQ. Two mobile ground stations will be used, one at QinetiQ and one at BURS (Bradford University Remote Sensing), UK. An S-band tracking beam is provided to enable spacecraft tracking with the mobile ground stations.

Sensor complement:

HIROC (High Resolution Optical Camera), developed by RAL. The objective is to provide panchromatic (2.5 m resolution) and multispectral imagery (5 m resolution) over a wide swath. The optical system employs a folded focal length of 1680 mm and a collecting aperture of 200 mm diameter. A three-mirror off-axis, aspheric design is used to provide flat image projections free of first order aberrations. Two of the mirrors are off-axis sections resulting in an obstruction-free aperture system. The design features a nearly diffraction-limited wide FOV (Field of View) of 1.4° in the cross-track direction and a sufficiently wide FOV in along-track to accommodate a number of detector array configurations (pushbroom operation). The telescope, the optics, all sensors and their associated front-end electronics are mounted on a CFRP (Carbon Reinforced Polymer) optical bench providing the needed stiffness, low thermal expansion, and alignment qualities. The FPA (Focal Plane Assembly) uses two separate CCDs, one for each channel (i.e., PAN and MS image detection). TDI (Time Delay and Integration) is provided through pitch compensation (by rotating the spacecraft counter to the satellite track velocity vector), increasing the integration time of the imagery and enhancing the SNR (TDI by motion compensation). The objectives call for up to 8 times TDI with an SNR of 50 in the prime band. The system is designed for a 90% probability of achieving 15% MTF at the Nyquist frequency after manufacture, assembly, integration and launch and including factors for the optical system, detector sampling, temporal sampling, jitter and smear.

The camera is mounted externally to the spacecraft platform, it uses heaters and MLI blankets to maintain temperature within limits. It is mounted to the platform using blade flexures that accommodate some difference in thermal expansion between the platform and payload that may otherwise warp the camera. Three internal, non-structural baffles protect

1797) P. Brooks, "TOPSAT – High Resolution Imaging From a Small Satellite," Proceedings of the 3rd International Symposium of the IAA, Berlin, April 2-6, 2001, pp. 319-322

1798) Camera image courtesy of RAL, spacecraft image courtesy of DERA

1799) K. M. Wallace, I. Parker, "The Topsat Satellite," IGARSS 2001, Sydney, Australia, July 9-13, 2001 the system from stray light. A further baffle covers the top of the monocoque and provides an attachment point for an aperture cover.

Spectral range of panchromatic band Spectral range of multispectral bands	0.5-0.7 μm 0.4-0.5 μm, 0.5-0.6 μm, and 0.6-0.7 μm
Pan detectors Multispectral detectors	Kodak 6000 element array (Si) at 7 μm pitch Kodak 3x 2000 element array (Si) at 14 μm pitch
Spatial resolution	Pan = 2.5 m, MS = 5 m
Swath width	Pan = 15 km, MS = 10 km
Maximum scene length	About 30 km
FOR (Field of Regard)	Body pointing capability of ± 30° in two axes
On-board data compression	The DERA data handling unit (DHU) provides a DSP which can be programmed with different image handling techniques
Instrument mass	<30 kg including FPA, electronics, thermal control, aperture cover and power conditioning
Instrument power	30 W operational, 10W standby (thermal control)
Instrument size	70 cm x 55 cm x 35 cm

Table 521: Specification of the HIROC instrument

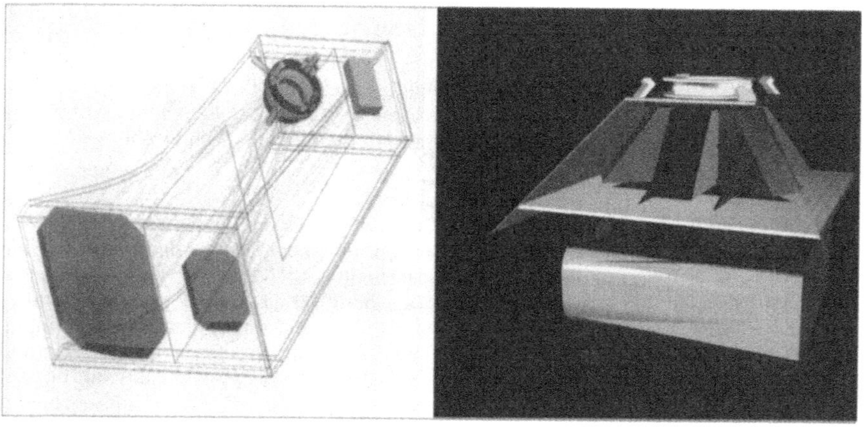

Figure 336: Illustration of HIROC optics (left) and the TOPSAT S/C (right)

M.29 TSX-5 (Tri-Service Experiments Mission 5)

A US DoD mission of AFRL (Kirtland AFB) within STP (Space Test Program) with the objective to operate two experimental payloads - **STRV-2** (Space Technology Research Vehicle-2) and **CEASE** (Compact Environmental Anomaly Sensor Experiment) - for a period of six months with a one year goal. TSX-5 was launched June 7, 2000 on a Pegasus XL vehicle over the Pacific, some 80 km west of VAFB (OSC's L-1011 jumbo jet - dubbed Stargazer - dropped the launch vehicle and payload from its belly).

The STRV-2 spacecraft structure, built and integrated by OSC of Dulles, VA, uses a bus of the STEP (Space Test Experiment Platform) program heritage (a nominal 95 cm point-to-point 12-sided structure). The all-aluminum structure consists of a honeycomb core plate with sheet metal panels mounted to milled longerons or stringers. The 0.82 m long STRV-2 payload is mounted to the top of the spacecraft structure. The S/C subsystems include: EPS (Electrical Power Subsystem, C&DH (Command and Data Handling), communications, and an attitude determination and control subsystem. The electronic components of the subsystems and the CEASE experiment are mounted internal to the S/C on the -x or wake oriented core plate. The S/C mass is 249.5 kg.

Orbit: Elliptical orbit of 410 km x 1750 km, inclination =69°.

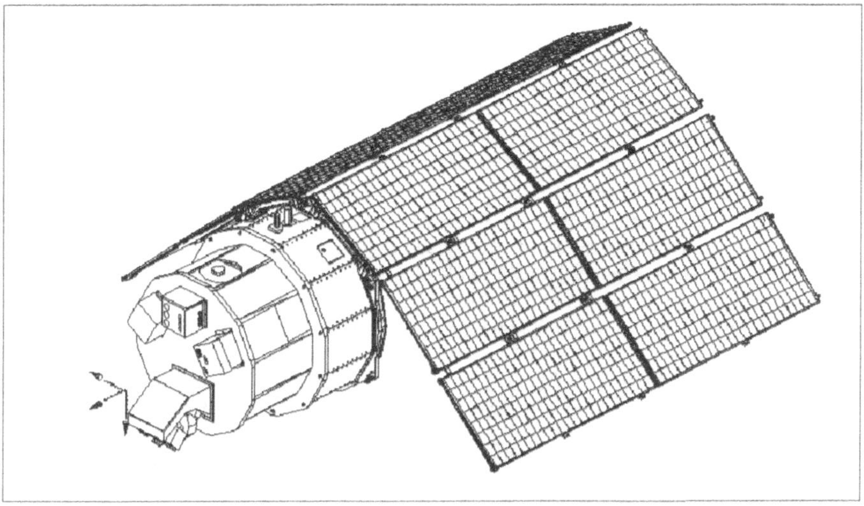

Figure 337: The TSX-5 S/C model

M.29.1 STRV-2 (Space Technology Research Vehicle-2)

STRV-2, sponsored by BMDO and DERA (UK), is a multi-national, highly integrated suite of experiments designed to push the envelope of space-based imaging technology, satellite vibration suppression, and material science. The prime objectives include demonstration of operation of optical instruments (MWIR and Lasercom) on a low-cost, non-precision platform, downlinking of data via Lasercom, obtaining infrared background data in selected wavelength bands at low and high altitudes, and assessing performance of candidate components in the space environment. Additional objectives of STRV-2 are the testing of the following experiments: ACCESS, SAMMES, VISS, and ETB; they include characterization of the space environment (radiation, micrometeoroid and debris), vibration isolation and suppression, assessment of contamination transport and effects aboard a low-cost satellite, and the demonstration of a modular architecture for autonomous command, control and data handling. The minimum success criteria for STRV-2 include collection of infrared background data for phenomenal analysis, demonstration of a laser communication downlink, demonstration of vibration isolation and suppression technology, and collecting six months data on the space environment and its effects on candidate components. [1800)]

All the data obtained from the BMDO STRV-2 experiments, applicable to the design and operation of civilian and commercial spacecraft, are made available to scientific investigators for analysis and interpretation.

Sensor/experiment complement:

MWIR (Medium Wave Infrared Imager), a DERA instrument with a mass of 23 kg and power consumption of <60 W. The instrument main features are:

- The optics of the Dall-Kirkham reflective telescope comprise: a 200 mm aperture primary mirror, a secondary mirror and a refractive relay lens system to provide a fully

1800) http://www.te.plk.af.mil/stp/tsx5/strv2/strv2.html

cold shielded thermal imagery. A filter wheel, containing 6 filters, is used to select the required waveband. The contains two relative calibration sources. They are imaged at the pre- and post-image instant to give a "flat field" calibration across the focal plane.
- An FPA (Focal Plane Assembly) of two staggered CMT arrays, each of size 2 x 512 elements, the detectors are separated in the along-track direction. The instrument operates in a dual pushbroom fashion, obtaining two near-simultaneous images in parallel.
- A Stirling cycle cryocooler, developed by Texas Instruments Inc., provides 1 W of cooling to the FPA; the operating temperature is 80 K. This is a COTS design, the STRV-2 mission represents its first use on a spaceborne platform.
- Carbon fiber composite telescope structure of DERA.
- Control electronics, pre-amplifier and memory of DERA.

The imager filter wheel carries filters for spectral ranges from 4.3 to 5.15 μm and 5.4 to 6.0 μm and a reference source system. The imager is designed for nadir Earth viewing; however, the 5.4 to 6.0 μm band is in the water absorption region, hence unable to see to the ground. The detector arrays produce a ground resolution of 35 m. - DERA acquires MWIR data to evaluate the instrument's capability to detect aircraft movement (non-afterburning aircraft from space using passive detection). At the resolution of 35 m, most aircraft will still be subpixel in size. This means that the thermal contrast may not be visible in the natural variation of the background radiance or clutter. The two near-simultaneous images are co-registered and subtracted from each other to reduce this clutter. The resultant "frame-differenced" image removes the background features.
Measurements are made to determine the detectable limits for a variety of backgrounds (ocean, urban, rural). The data provide missile designers information necessary to develop effective detection/tracking algorithms. A secondary objective of the MWIR experiment is to collect earth background data as a function of seasonal variation. [1801)]

Lasercom (Laser Communication Experiment), developed by Astro-Terra, Inc.. The prime objectives are to demonstrate high data rate laser communications between a satellite and other platforms, and to establish the practicality and benefits of free-space laser communications. Secondary objectives are to downlink stored data from the MWIR instrument at data rates between 50 and 100 Mbit/s and to demonstrate data rates in a repeater mode between 270 - 1000 Mbit/s.

ACCESS (All Composite Experiment Spacecraft Structure). The objective is to demonstrate the performance of composite spacecraft, and show the advantages relative to conventional designs.

SAMMES (Space Active Modular Materials Experiment System), developed by Physical Sciences Inc. The objectives are to: a) demonstrate the utility of on-board monitoring of spacecraft subsystem performance, including the effects of contamination and the ambient environment; b) to obtain environmental degradation data for selected spacecraft materials and components in lower MEO orbit; c) to improve the quantitative knowledge of the energetic radiation environment in lower MEO orbit. SAMMES is designed to provide long-term, time-variant engineering performance data on materials exposed on the exterior of spacecraft and to quantify the phenomena of contamination deposition and effects.

VISS (Vibration, Isolation, Suppression and Steering System). VISS is an AFRL instrument, a self-contained precision vibration control device designed to provide an ultra-quiet environment for sensitive optical sensors, laser transmitters, and other detection and measurement devices. VISS is designed to isolate a precision payload from spaceborne disturbances using passive isolation in combination with voice coil actuators. VISS is a platform which is mounted to a satellite bus via six actuators arranged in a hexapod (Stuart platform) configuration. In the low-frequency range, the passive design is supplemented by a voice-coil-based system mounted in parallel to the D-strut. This portion is actively controlled to

1801) S. J. Cawley, S. Murphy, A. Willig, P. S. Godfree, "The Space Technology Research Vehicle 2 Medium Wave Infrared Imager," Proceedings of the 3rd International Symposium of IAA, Berlin, April 2-6, 2001 provide vibration isolation (in all six degrees of freedom) from the satellite to the payload (optics, etc.), suppression of vibration generated within the payload (cryocooler, etc.), and steering about any or all of the three rotational axes. The objective is to achieve the following goals: >20 dB vibration isolation, >20 dB vibration suppression, ± 0.3° steering of the payload at a rate of 2 Hz. The payload for VISS in the STRV-2 configuration is the DERA-built MWIR telescope. Accelerometer data is recorded to determine the system performance. [1802)]

ETB (Engineering Test Bed). This includes the MDIM (Meteoroid and Debris Impact Monitor) and the Meteoroid Impact Sensor of BMDO. ETB enables demonstrating the durability of advanced micro-electronic components and COTS (Commercial Off-The Shelf) devices in the space environment. ETB provides system designers with an opportunity to test device designs proposed for incorporation in future space systems.

MDIM (Meteoroid and Debris Impact Monitor) of NASA/LaRC (an element of the SEE program). The objective is to monitor populations of small-mass particles as a function of altitude. The MDIM experiment is of OMDC (Orbiting Meteoroid & Debris Counting) heritage flown on the Clementine S/C. MDIM utilizes 48 MOS detectors which are mounted on the exterior of the STRV2 spacecraft. When space debris or meteoroids impact the detector, the sensor location and time of impact is stored and down linked for data analysis. From this information, space debris flux and directionality can be assessed. [1803)]

M.29.2 CEASE (Compact Environmental Anomaly Sensor Experiment)

CEASE, sponsored by AFRL, Hanscom Air Force Base, Bedford, MA, is an environmental scanner, providing the S/C with essential knowledge about the surrounding space. CEASE uses this flight to prove its near spacecraft environmental assessment capabilities. The overall objectives are to develop and demonstrate on-orbit, autonomous, compact, light weight, low-power instrumentation to monitor the environment around a spacecraft, and to provide alerts whenever the environment is likely to cause any of the following anomalies: surface charging, deep dielectric charging, single event upsets (SEU's), or radiation dose effects.

The instrument also provides on request detailed data on particle fluxes incident on the spacecraft over the 72 hours prior to the request. This feature allows the S/C operator, once an anomaly has occurred, to have sufficient data to analyze and understand the cause of the anomaly. CEASE's output can be used to distinguish between natural effects and those caused by other (possibly hostile) actions. By on-board analysis of its measured space-environment data using decision making algorithms in its microprocessor, CEASE provides alerts/warnings in terms of a series of ascending order flags, indicating the likelihood and severity of the forthcoming anomalies. [1804)]

The instrument has a size of 10 cm x 10 cm x 8.2 cm, a mass of 1 kg, and power consumption of <1.5 W. An RS-422 or 1553B interface is used. The instrument is flown on the following missions: TSX-5, STRV-1c, and DSP.

1802) K. K. Denoyer, R. S. Erwin, R. R. Ninneman, "Advanced SMART Structures Flight Experiments for Precision Spacecraft," Acta Astronautica, Vol. 47, No 2-9, 2000, pp. 389-397

1803) http://setas-www.larc.nasa.gov/strv2/strv2.html

1804) http://www.vsbs.plh.af.mil/projects/cease/cease.html

Part N University/Student-Developed Satellites & Payloads

During the 1990s, satellite and payload development projects have become the program of choice for challenging (multi-year) training courses in quite a few engineering departments at universities throughout the world. The intent is always to enrich the student training program, to stimulate interest in a problem-solving multi-disciplinary technical environment, to be imaginative and resourceful, and to take some risks – with ample and essential help from mentors and partners (industry, institutional, or otherwise). Cooperation on many levels and active participation/publication within the international space science community are important ingredients in the overall objectives of research and development. In some instances, project-sharing among engineering departments of several universities is being practiced in order to handle the demanding and complex project goals in a certain time frame. In general, a good amount of enthusiasm and lots of volunteer work by all parties involved are needed to bring such low-cost program activities to maturity – an invaluable amount of professionalism is gained for all students in such programs. Some of the student-involved projects are presented here.

One of the biggest problems of small satellite building is to get a launch opportunity as a secondary payload with some primary payload. Many things have to be considered for such a piggyback flight, including the orbit and the price tag.

Note: The UoSAT program of SSTL [Surrey Satellite Technology Ltd (University of Surrey, UK)] with its many microsatellite missions is included in Part D under D.40.
The documentation of **UNP** (University Nanosatellite Program), sponsored by DoD (AFRL, DARPA), NASA and US industry, and consisting of ten university teams, could not be completed due to a lack of information. I managed a draft description of about 15 pages. But in the constellation of military-sponsored programs, my review and information requests were generally simply ignored by AFRL contacts as well as by most university contacts. - The UNP initiative is only one aspect of AFRL's TechSat 21 (Technology Satellite of the 21st Century) program which was started in 1998. The overall idea is to explore distributed mission concepts of clustered micro- and nanosatellite constellations, able to cooperate and to share resources and functions (processing, communications, payload, and/or observation and mission functions) with each other, in order to perform the functions of a larger single satellite. In the time frame Dec. 1998 to spring 1999, UNP (University Nanosatellite Program) selected 10 universities to participate in the development of university-class nanosatellites. In this concept, AFRL is providing the launches (Shuttle) for the nanosatellite constellations. NASA/GSFC is providing advanced crosslink communications and navigation hardware along with flight algorithms to demonstrate formation flying. Numerous industry partners are also supporting the universities with hardware, design expertise, and test facilities.

UNP includes the following projects (launches are projected for 2002):

1) ION-F (Ionospheric Observation Nanosatellite-Formation). A three-satellite mission consisting of: USUSat of Utah State University (USU), Logan, UT; DawgStar of the University of Washington (UW), Seattle WA; and HokieSat of Virginia Polytechnic Institute and State University (VT), Blacksburg, VA.
2) 3CSat (Three Corner Satellite). A three-satellite mission of Arizona State University (ASU), Tempe, AZ; University of Colorado (CU), Boulder, CO; and New Mexico State University (NMSU), Las Cruces, NM.
3) Emerald (Electromagnetic Radiation and Lightning Detection). A two-satellite constellation of Stanford University, Stanford, CA, and Santa Clara University, Santa Clara, CA.
4) CPM (Constellation Pathfinder Mission). Boston University (BU), Boston, MA.
5) Solar Blade Heliogyro Nanosatellite. Carnegie Mellon University (CMU), Pittsburgh, PA

N.1 ASUSat-1 (Arizona State University Satellite 1)

ASUSat-1 is a student-designed, built, and operated nanosatellite mission of ASU at Tempe, AZ, sponsored by industry and government. The objectives call for an Earth imaging and technology demonstration mission, as well as an audio transponder service for the amateur radio community. The ASUSat project started in 1993, went through several different launch opportunity versions and configurations, and a long waiting waiting period before the satellite lifted from the ground. A launch took place on Jan. 27, 2000 (UTC, as a secondary payload to JAWSAT, OPAL, FalconSat and OSCE) on a OSC Minotaur vehicle from VAFB, CA. ASUSat-1 was the first S/C deployed from JAWSAT.

The S/C design employs a composite 14-sided structure (almost a cylinder with 25 cm in height and 32 cm in diameter), capped by two composite bulkheads (one fixed, the other removable). The solar cells (GaAs) are body-mounted to the drum-shaped outer surface and to the top bulkhead (70%). ASUSat-1 is an Earth-pointing, three-axis stabilized S/C, using a deployable gravity-gradient boom (2 m) with a small tip mass (135 grams). Damping is provided by a new **fluid-damper** design, referred to as the "ball-within-a-ball concept." The inner ball has three substantially different inertias, causing it to align itself always with the gravity and velocity vectors. The movement of this ball within the fluid dissipates sufficient energy to damp the S/C oscillations within a reasonable amount of time. A small (bakkup) torque coil is also integrated into the S/C, capable to reposition the satellite in the event of an uncontrollable flip-over. Attitude information is provided by clusters of sun- and Earth sensors. For this purpose, sixty tiny sensors are mounted into the exterior S/C surface. Each sensor cluster houses three VIS sensors for sun radiation detection and one IR sensor for Earth radiation detection. Periodic readings from these sensor clusters and their known positions provide an attitude determination with an accuracy of about $\pm 10^{\circ}$ for each axis. A GPS receiver (Trimble Navigation) provides position and velocity measurements. S/C mass = 5.9 kg, solar power = 8-14 W [a 6-cell Sanyo NiCd battery pack (5 Ah) is used for eclipse operations], passive thermal system. The OBC is built around the Intel 80C188EC embedded processor. [1805] [1806] [1807] [1808]

Communications: 5 W transmitters/receivers. Transmission in digital mode for TT&C services, in analog mode for all AMSAT services. Use of HDLC frame protocol (packet) at 9.6 kbaud with GMSK (Gaussian Minimum Shift Keying) modulation for all services.

- TT&C services: Bi-directional RF links between S/C and ground station at ASU. The on board transmitter uses direct frequency modulation of a 436.5 MHz carrier using the modem's output to produce narrow band frequency modulation (NBFM).
- AMSAT services: The uplink frequency for voice communication is 145.990 MHz. The downlink frequency is 436.7 MHz.

Orbit: Circular sun-synchronous (terminator) orbit, altitude = 750 km, inclination = 100°, period = 99 minutes.

Status: After a successful launch, ASUSat-1 telemetry was sporadically received at various stations around the world, in particular by amateur radio ground stations. The first contact confirmed that the S/C had been successfully deployed and was functioning in orbit. During initial passes, the ASUSat students successfully commanded and controlled the satellite. Initial systems checkout confirmed that the communications, commands, power regulation, data acquisition, and structures/deployment subsystems were performing as expected. The last report received from ASUSat1 was by the SUNSAT team (Stellenbosch University,

1805) A. Friedman, B. Underhill, et al., "ASUSat-1: Low-Cost Student-Designed Nanosatellite," Proceedings of the 14th AIAA/USU Conference on Small Satellites, Logan, UT, Aug. 21-24, 2000, SSC00-V-2

1806) J.D. Rademacher, H. L. Reed, J. Puig-Suari, "ASUSat 1: An Example of Low-Cost Nanosatellite Development," Acta Astronautica, Volume 39, Number 1–4, pp. 189 -196, 1996

1807) S. Ferring, D. Waller, J. D. Rademacher, A. Friedman, H. L. Reed, "ASUSat-1: The Development of a Low-Cost Nano-Satellite," Proceedings of the 11th AIAA/USU Conference on Small Satellites, Sept. 15-18, 1997

1808) http://nasa.asu.edu/asusat/

South Africa) at fourteen hours into the mission. This reception included a telemetry frame that confirmed that the satellite did indeed have a critical problem in the power system. Unfortunately, this problem prevented the solar arrays from supplying power. Predicted lifetime of the satellite on battery power alone was estimated to be fifteen hours.

With only two telemetry frames available for analysis it was not possible to draw any firm conclusions. Even with such a short mission, the ASUSat team has gathered an immense amount of experience and insight into the building, launching and operation of a satellite.

Sensor complement:

ASUSat Imager. The instrument package consists of two Dycam modular cameras. Each camera has its own electronics board (processor and 1 MByte of memory), taking snapshots on command. They are also able to provide compression and storage services. The cameras feature a 2-D CCD detector array (496 x 365), providing a resolution of about 0.5 km from a 550 km orbital altitude (FOV=18°). Camera-1 uses a red pass filter and operates in the spectral range of 600 - 800 nm (VNIR), providing color imagery. Camera-2 uses a short pass filter in the spectral range of 420 - 550 nm (VIS), providing gray-scale imagery. The two cameras are daisy-chained inside a single anodized alumina housing (about 5 cm x 6.5 cm x 16 cm) with a total mass of about 0.5 kg. Imagery applications are in vegetation indexing, coastal mapping, water-body penetration, etc.

ASUSat technology demonstrations include: the mostly composite satellite structure, the student-designed array of attitude determination sensors, the Trimble GPS unit, the Maxon transmitter, the gravity gradient deployment mechanism, gravity gradient fluid damper, satellite deployer, and the student-designed CPU and power boards.

N.2 BREM-SAT 1

BREM-SAT 1 is a microsatellite built by the German University of Bremen's Center of Applied Space Technology and Microgravity (ZARM) in cooperation of OHB System of Bremen under the sponsorship of the German Space Agency (DARA). The satellite was deployed from a GAS (Get Away Specials) container on Shuttle flight STS-60 (Feb. 3-11, 1994). Deployment occurred on Feb. 9, 1994 with a standard door assembly and a modified GAS Carrier Ejection System. The S/C had a 12-sided cylindrical structure with a diameter of 48 cm and a height of 52 cm, total mass of 63 kg. The S/C was spin-stabilized with an average spin of 1 rpm (use of magnetic torquers and a momentum wheel for stabilization), attitude measurement with a star sensor, sun sensor, and magnetometer. The exterior was covered with solar cells, average power = 25-40 W. Two battery sets provided energy during the nighttime portion of the orbit. OBC: T-800 processor with 1 MByte RAM and 512 kByte EEPROM.

The scientific mission objectives were to measure heat conductivity, obtain residual acceleration forces, estimate the in-orbit on-board microgravity quality, investigate the density distribution and dynamics of micrometeorites and dust particles in low–Earth orbit, map atomic oxygen, measure the exchange of momentum and energy between the molecular flow and the rotating satellite, and measure pressure and temperature during satellite reentry. [1809] [1810] [1811] [1812]

S/C communication: VHF communication in uplink (137 MHz) and downlink (145 MHz), with data rates of 1.2 kbit/s in the uplink and 9.6 kbit/s in the downlink. Modulation: bi-

1809) H. J. Königsmann, H. Oelze, H. J. Rath, "BREMSAT - First flight Results," Proceedings of the 8th AIAA/USU conference on small satellites, USU, Logan UT, 1994

1810) Abschlußbericht BREM-SAT 1, University of Bremen/ ZARM, April 29, 1996

1811) M. Wiegand, H. J. Königsmann, "A Small Re-entry Capsule - Bremsat-2," Proceedings of the 10th AIAA/USU conference on small satellites, USU, Logan UT, 1996

1812) H. J. Königsmann, J. R. Wertz, S. D. Dawson, "BREM-SAT: A Reducing Space Mission Cost Case Study," Microcosm Directory of Space Technology Data Sources, 1997, Section L, pp. L-1 to L-21

phase modulation in the downlink and FM/FSK in the uplink. The satellite's four antennas are omnidirectional. Downlink power of 2 W, uplink power of 5 W.
BREM-SAT 1 was operated by the ZARM tracking station (4-6 passes per day - total of 1114 contact periods) for over a year. It reentered the atmosphere on Feb. 12, 1995. In addition, there was a backup station at Iowa State University and several mobile stations for reentry data reception.

Orbit: The initial orbit was that of STS-60, altitude = 360 km, inclination = 57°.

Sensor complement:

The BREM-SAT 1 mission consisted of three phases. A total of six experiments were conducted during these phases.

- Microgravity phase. Defines the orbital phase on-board the Shuttle prior to BREM-SAT 1 deployment.
- Orbital phase. Defines the free-flying phase of BREM-SAT 1.
- Reentry phase.

Experiment-1: Measurement of heat conduction of light-viscous fluid mixtures.

Experiment-2: Acceleration measurements (absolute and rest acceleration) during reentry. A set of commercially available one-axis analog micro-g accelerometers of type Q-Flex QA-1400 (Sundstrand) were tetrahedron-mounted and arranged normal to each other in a way as to measure accelerations in three axis (the major S/C axis and accelerometer axis were intentionally displaced for reasons of redundancy).

Experiment-3: Measurement of cosmic dust/micro-meteoroid distributions. The instrument has the name of **MDC** (Munich Dust Counter) and is a cooperation of the Institute of Space Technology at the Technical University of Munich and ZARM of the University of Bremen. Objective: Measurement of impact charges and deduction of particle mass, velocity, and approximate flight direction. MDC consists of an impact box with a target area and two side-mounted charge collectors, biased by positive and negative high voltages, generating an electric field. This electric field separates the impact plasma into positively and negatively charged components. The negatively charged components of the plasma (elections, negative ions) are accelerated toward the positively biased collector, while the positive ions are accelerated toward the negatively biased collector. Charge-sensitive amplifiers convert the input charge into an output voltage.
MDC was mounted on the outer cylinder of the S/C, offering a FOV of ±76° for particle collection, with the spin axis oriented toward the north pole of the ecliptic. This configuration provided measurements in the ecliptic plane. [1813)]

Experiment-4: Aerodynamic reentry measurements - pressure densities. The measurement of normal and tangential pressures was obtained with strain gauges.

Experiment-5: Aerodynamic reentry measurements - Patterson pressure, heat flow, pressure and temperature of the satellite surface. Use of 14 Ni-Cr-Ni thermo-elements.

Experiment-6: Measurement of atomic oxygen. [1814)] **OXFLUX** (Oxygen Flux, an ESA-provided instrument). The OXFLUX detector consists of two coated QCM (Quartz Crystal Micro-balances) with the associated electronics and interfaces. [1815)] OXFLUX is located underneath MDC in the plane of the velocity vector. The instrument is connected to the

1813) R. Münzenmayer, H. Iglseder, H. Swedhem, "The Munich Dust Counter - An Experiment for the Measurement of Micro-meteoroids and Space debris," Proceedings of the First European Conference on Space Debris at ESA/ESOC Darmstadt, Germany, April 5-7, 1993

1814) H. P. Willemsen, M. v. Eesbeek, "OXFLUX, Atomic Oxygen Sensor on BREM-SAT 1," Executive summary of the Final Report of Contract ESA No 9037/90/NL/JG

1815) Note: QCM measurements are based on the principle that the resonance frequency of a crystal controlled oscillator varies inversely proportional to the mass change of the crystal. Hence, a QCM may be used to measure the erosion of a pre-deposited coating under exposure to an atomic oxygen environment.

OBC by a 5 V RS-232 serial interface (data rate = 9.6 kbaud). The two crystal oscillators transmit their frequency signals of approx. 10 MHz to the two 16-bit counters. The frequency is measured every second. The temperatures at each crystal are measured continuously with a resolution of 0.5° C. The atomic oxygen flux can enter the detector cavity when the angle between the ram-direction and cavity opening is between ±16.7° (view factor). OXFLUX measurements were obtained throughout the free-flying period of BREM-SAT 1.

Parameter	Value	Parameter	Value
Crystal frequency	approx. 10 MHz	Total fluence measured	2.5 10^{20} O-atoms
Measuring range	approx. 65 kHz	Crystal temperature	±50° C
Resolution	1 Hz	Resolution	0.5° C
Instrument mass	0.557 kg	Instrument size (mm)	110 x 110 x 60

Table 522: Specification of the OXFLUX detector

N.3 CHIPSat (CHIPS Satellite)

CHIPSat is a mission of UCB (University of California at Berkeley, PI: M. Hurwitz), supported by NASA's UNEX (University-class Explorer) program, with the objective to obtain spectral sky maps of the scientifically critical EUV (Extreme Ultraviolet) band between 90-260 Å. The CHIPS full-sky survey helps to determine the electron temperature, ionization conditions, and cooling mechanisms of the so-called "local interstellar bubble," a cloud of hot gas surrounding our solar system that extends about 300 light-years from the sun. [1816)]

Background: The CHIPS (Cosmic Hot Interstellar Plasma Spectrometer) project was selected by UNEX in 1998. The CHIPS mission (i.e. the instrument) was initially proposed as a secondary payload aboard a FAISat communications S/C. This approach was dropped in favor of a small spacecraft with a single instrument payload. Thus the CHIPS project mutated to CHIPSat.

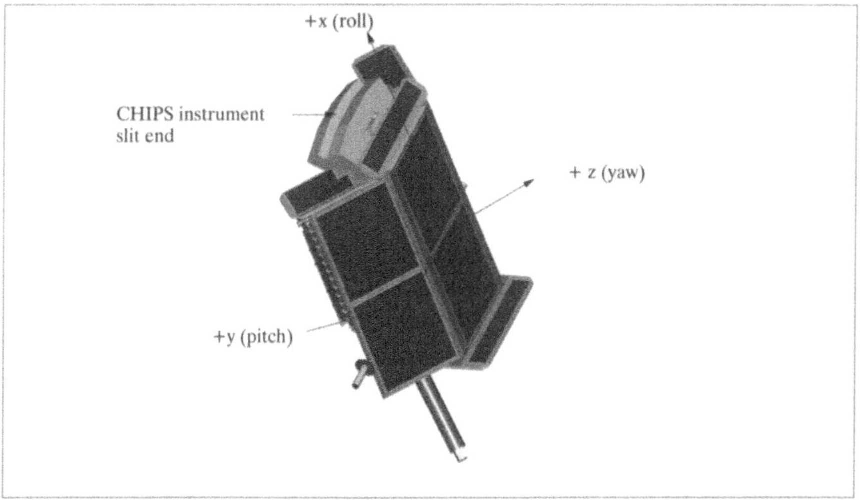

Figure 338: Illustration of the CHIPSat spacecraft

1816) W. Marchant, E. Riddle Taylor, "Status of CHIPS: A NASA University Explorer Astronomy Mission," Proceedings of the 14th AIAA/USU Conference on Small Satellites, Logan, UT, Aug. 21-24, 2000, SSC00-V-6

CHIPSat [1817] is a dedicated microsatellite built by SpaceDev, Inc. for UCB/SSL (Space Sciences Laboratory). CHIPSat is a three-axis stabilized S/C using 4 momentum wheels, three torque coils, two sun sensors, magnetometer, a moon sensor, and rate sensors to provide an attitude pointing accuracy of about ±2°. The S/C is nominally sun-pointing with complete freedom to yaw about the solar array normal vector allowing the CHIPS instrument to obtain a full-sky survey within six months, while avoiding pointing the instrument at the sun, Earth, moon and the orbital Ram direction. The design permits access to all points on the celestial sphere within the one year mission lifetime. The S/C structure employs a milled aluminum transition adapter and aluminum honeycomb panels with facesheets for structural integrity. Power (106 W EOL) is provided by body-mounted solar arrays using dual-junction GaAs/InP/Ge solar cells, NiCd batteries are used during solar eclipses. In addition, small keep-alive arrays are positioned on the other five sides of the S/C providing enough power to run critical subsystems regardless of the S/C attitude. A passive thermal system is used for CHIPSat. The C&DH (Communications and Data Handling) subsystem employs a single-board computer (Motorola Power PC 750 CPU, memory, and I/O for distributed processors). The S/C mass is 85 kg, the design life is 18 months (one year mission).

The RF communications are in S-band. The data rates are: 4 kbit/s in the uplink, the downlink supports data rates of up to 1 Mbit/s. Modulation: the uplink is PM carrier, suppressed carrier BPSK on a subcarrier; the downlink uses suppressed carrier BPSK. The TCP/IP and UDP/IP (User Datagram Protocol/Internet Protocol) protocol suite is used to communicate all data between the S/C and the ground user directly. Data is received, archived, and monitored at MCC (Mission Control Center) at SpaceDev and then sent to SOC (Science Operation Center) at UCB/SSL. [1818]

The UDP/IP is protocol is selected for real-time monitoring and real-time commanding (it de-couples both directions) and presents much less overhead. The setup permits the reception of engineering and status packets (telemetry) in case the uplink isn't working. Conversely, the setup permits also to command "into the blind" by uplinking UDP packets in case the telemetry isn't working.
Note: The UDP service of the TCP/IP protocol permits to send discrete packets of information called "datagrams" that aren't guaranteed to get there and may arrive out of order depending upon their routing through the IP system. A two-way communication isn't needed in this setup because the data are broadcast. So, if a guarantee is needed that at least some packets get through, even if one direction of the communication link fails, then UDP may be used. - TCP deals with making sure that all the packets arrive and are in the correct order. TCP implies a two-way connection and a higher level of communications overhead to assure that all the packets arrive and are in the correct order.

A launch of CHIPSat is planned as a secondary payload to a NAVSTAR GPS satellite on a Delta-2 rocket in April 2002.

Orbit: circular orbit, altitude of 190 km x 1200 km (initial) and 600 km (final), inclination = 38°. - Note: The 190 km x 1200 km orbit is the initial orbit in which the Delta-2 2nd stage, with CHIPSat still attached, is left when the 3rd stage separates. But then, the 2nd stage is fired twice again, first to lower the apogee, then about 45 minutes later, to raise the perigee. Only then does CHIPSat separate from the 2nd stage.

Sensor complement: [1819]

CHIPS (Cosmic Hot Interstellar Plasma Spectrometer) is a low-cost instrument with high sensitivity and spectral resolution in the spectral band near 170 Å, built at UCB/SSL. The science objectives are to carry out EUV spectroscopy to determine how the million degree

1817) http://chips.ssl.berkeley.edu/chips.html
1818) Courtesy of Will Marchant of UCB/SSL
1819) http://chips.ssl.berkeley.edu/instrument.html

gas-cloud surrounding our solar system, cools. - The CHIPS instrument uses an array of grazing-incidence optics to achieve a peak resolving power of f/150 for diffuse emission in a field of view (FOV) of 5° × 26.7° (the gratings are aligned in one dimension on the sky). Light enters the spectrograph through the array of six entrance slits (channels). The entrance apertures are narrow slits, each covered by a rotating mechanism with closed, narrow, and wide settings to protect the interior from contamination during launch. The gratings disperse and focus the diffuse extreme ultraviolet radiation onto a single detector through a filter assembly. The detector is a planar, photon-counting MCP (Micro Channel Plate) with a crossed delay line anode. In-band photon locations are determined from the anode, which converts the light into analog electronic pulses.

Note: The use of variable line-spaced gratings in instruments is of EUVE and ORFEUS (Orbiting Retrievable Far and Extreme Ultraviolet Spectrograph) instrument heritage. The ORFEUS payload flew twice with the ASTRO-SPAS missions on STS-51 (Sept. 12-22, 1993)) and on STS-80 (Nov. 19 - Dec. 7, 1996). The NASA satellite EUVE (Extreme Ultraviolet Explorer) was launched June 7, 1992.

Parameter	Value	Parameter	Value
Entrance slits	0.027 cm x 6 cm	Nr. of spectral channels	6
Grating graze angle	14°	FOV (Field of View)	5° x 26.7°
Groove density	1625-1900 mm^{-1}	Bandpass	90 - 260 Å
Grating curvature	148 cm	Spectral resolution	45-155 ($\lambda/\Delta\lambda$)
Grating dimensions	8 cm x 6 cm x 1.2 cm	A_{eff} x Ω	(1.3-8.8) x 10^{-4} sr
Optical coatings	Rhodium	Payload mass	<25 kg (spectrograph) 10 kg (electronics box)
Detector area	5.0 cm x 6.0 cm	Payload power	21.2 W (average) 30 W (peak)
Photocathode	NaBr	Attitude control	±2°
EUV filters	720 Å Polyimide; 500 Å Boron, 1000 Å Al; 300 Å C	Data rate	>16 MByte/24 hours

Table 523: CHIPS instrument performance parameters

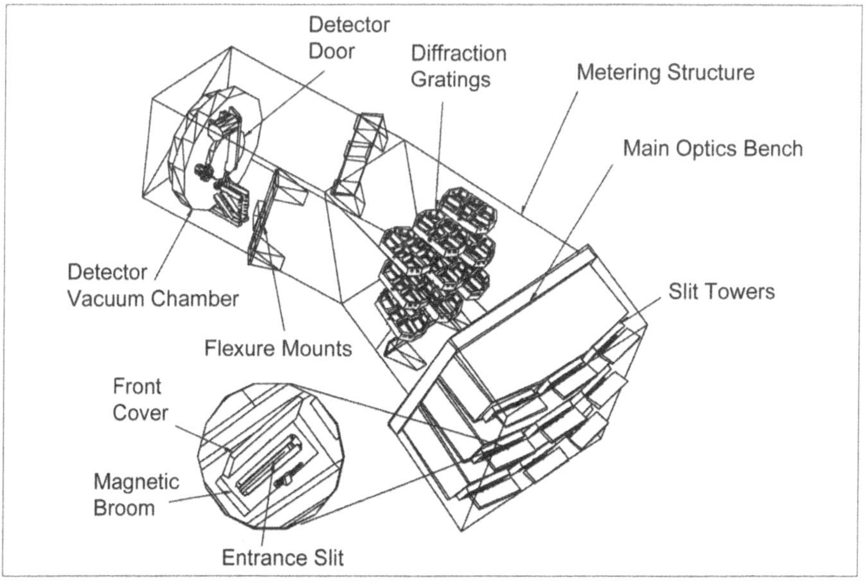

Figure 339: Schematic concept of CHIPS

N.4 CX-I (Citizen Explorer-I)

CX-I is a microsatellite mission of the Colorado Space Grant Consortium (CSGC), designed and built by students at the University of Colorado in Boulder, CO (project partners/sponsors are NASA, and industry such as Ball Aerospace, Boeing, Lockheed Martin, Analytical Spectral Devices Inc., CIRES, etc.). [1820] [1821] The spacecraft objectives are to advance radio communications and to determine atmospheric ozone levels by measuring direct, reflected and transmitted sunlight at various wavelengths. The following overall mission goals are pursued:

- The educational objectives are to: a) directly downlink science and S/C data to participating K-12 schools worldwide, b) enable receipt of data at schools with low-cost ground stations (<$800), c) provide the means for students to share data over the internet, d) provide a means for K-12 students to take an active part in data collecting and analysis with student-built ground instruments.
- The technology objectives are to: a) demonstrate S/C autonomy and the ability to evolve and migrate autonomy, b) demonstrate the "beacon mode" communications concept.
- The science objectives are to: a) provide global total ozone observations, b) provide a network of UV monitoring sites, c) provide observations of aerosol optical depth to assist in modeling surface level UV flux.

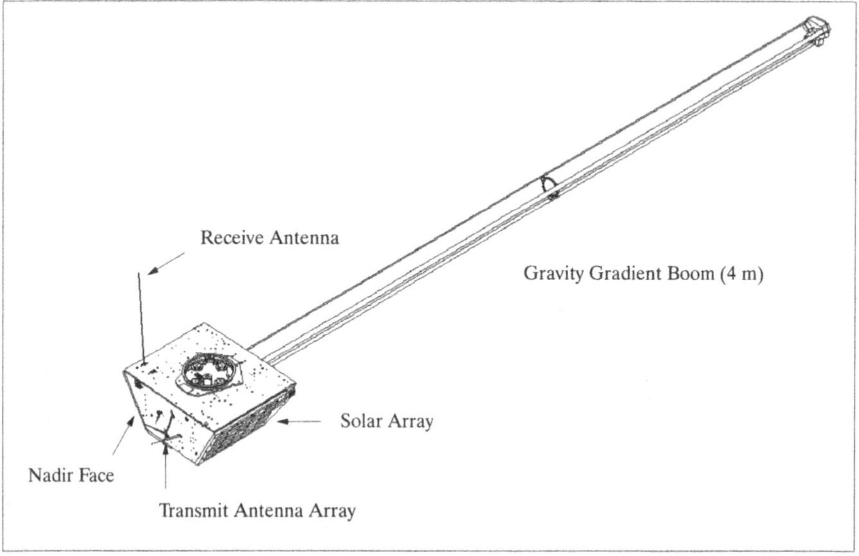

Figure 340: Illustration of the CX-I spacecraft

The S/C employs an aluminum plate construction as a basic platform for all instruments and components. CX-I uses a gravity-gradient-boom design (passive stabilization with a 4 m long deployable boom and a tip mass of 2 kg, pointing in the zenith direction) to keep the satellite facing the Earth in the same direction at all times. Additional attitude control is provided by three orthogonal magnetic coils. The attitude is sensed by two 2-axis magnetometers and a fine sun sensor. In addition, a set of six coarse sun sensors is mounted on all sides of the satellite providing full spatial coverage. S/C power is provided by four sur-

1820) E. Hansen, "Advancing Radio Communications technology with the Citizen Explorer Mission," Presented at the AMSAT Conference, New Orleans, Oct. 1998

1821) http://citizen-explorer.colorado.edu/index.html

face-mounted solar panels; the panels are comprised of triple-junction GaAs cells with an efficiency of 23%. A peak-power tracking system is used to maximize the solar panel efficiency. A NiCd battery stack (1.5 Ah capacity) is employed for energy during orbital eclipse phases. The C&DH (Command and Data Handling) subsystem is based on a standard PC/104 architecture. C&DH has also the capability to store on-board data. The microsatellite has a mass of 45 kg, power = 40 W at 30 V (average power of 27 W), design life = 1 year.

A launch of CX-I is planned for early 2003.

Communications: CX-I is equipped with two transmitters, a receiver, and a modulator/demodulator. AMSAT communication standards are used. Data are transmitted in UHF-band at a frequency of 436.759 MHz in downlink at a data rate of 9.6 kbit/s. Mission operations are conducted by students at MOCC (Mission Operations Control Center) of the University of Colorado, Boulder.

Orbit (requested/desired): Sun-synchronous circular orbit, altitude = 705 km, inclination = 98.7° period = 99 minutes, descending nodal crossing time of 10:15 AM.

N.4.1 On-board Sensor Complement

The science sensor complement consists of two flight instruments and two hand-held ground instruments. The flight instruments include a spectrophotometer (informally termed "speck") that measures UV radiation, and a photometer that measures visible reflectance. The ground instruments include a UV detector and an aerosol meter. Data from the four instruments complement each other to provide information about ozone levels in the stratosphere.

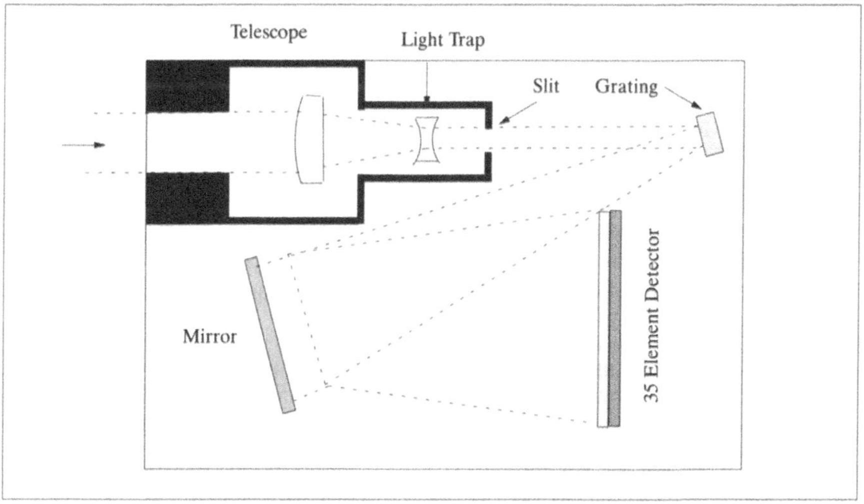

Figure 341: Schematic illustration of radiation path optics in the Speck instrument

SPECK (Spectrophotometer). The objective is to measure atmospheric reflected radiation in the UV spectral range of 280-350 nm. The instrument consists of an aperture with a FOV of 6.9°, a broadband visible light filter, two lenses, a light trap, a grating, a mirror, and a 35-element (pixel) detector. Radiation entering the aperture is passed through a broadband filter which is opaque for visible light. The beam is then collimated by a lens system and dispersed by a grating into 16 bands (or channels), reflected off a UV-coated flat mirror, and detected by a 35-pixel linear diode array. Speck provides observations in 5 of the 6 channels used by TOMS. The channels between 280-340 nm are used for total column ozone monitoring; the 340-350 nm channels measure the albedo. The on-board data is quantized to 16 bits. Instrument mass = 1 kg, power = 2 W, data rate = 2.56 kbit/s.

Photometer. The objective is to measure visible atmospheric reflectance around 550 nm. The instrument consists of an aperture, lens, filter, and a photodiode. The light passes through a bandpass filter which allows only the visible light through and is detected with a photodiode. The photometer measures visible (yellow-green) light reflected off the Earth's atmosphere or surface. The data of the photometer are used to identify cloud tops and ice/snow versus dry land, ocean, and vegetation reflectance.

N.4.2 Ground Instruments

Ground truth data are collected by a network of participating student observers. There are two hand-held instruments used by the CX-I project. The first is an aerosol detector that can either be purchased from CSGC (Colorado Space Grant College) or built by a website instruction set. The second instrument is an UV detector. The UV detector must be purchased directly from Sunsor, Inc. The information produced by the two hand-held instruments is used, in conjunction with the satellite data, to create global maps of ozone coverage.

Sunsor UV-Detector. This is a bandpass photometer using a UV-sensitive photodiode detector. A band-pass filter is integrated onto the surface of the detector so that only UV light (in the 290-314 nm range) contributes to the signal.

Aerosol Detector. The objective is to measure the aerosol content in the atmosphere by measuring solar intensity (visible light range). The aerosol detector measures the intensity of green light, which is used to determine the number of aerosol particles in the atmosphere.

N.4.3 Technology Demonstrations

Beacon mode operations. A new spacecraft operations/communications concept, called the "beacon mode," is being demonstrated. The technique provides a way to operate a spacecraft and monitor its health without extensive day-to-day attention by ground controllers. Instead, the S/C uses an onboard software package, capable of monitoring onboard health functions and making autonomous decisions and/or alerting ground control. The objectives of CX-I beacon mode are to demonstrate the following concepts:

- A summarization of S/C status with a limited number of messages to indicate its need for contact with the ground.
- The ability and interest of amateurs and K-12 students to participate in space missions and to extend the traditional MOCC functions. The functions include:
    - Monitoring and evaluation of S/C health and status by amateurs and K-12 students
    - The receipt of these messages by participating amateurs and K-12 students and relay of these messages to the MOCC.
    - The relay of telemetry data received at a participating school to the MOCC.
- A significant reduction in time that a MOCC needs to be staffed
- The applicability of this concept to other missions (of NASA and the commercial world).

The beacon mode experiment was designed by a team at the University of Colorado, with help from experts at JPL.

On-board autonomy and autonomy migration. These technologies consist of:

- An intelligent onboard system capable of monitoring health and status and alerting other onboard software.

- Onboard scripts to determine if an alarm is valid based on context checks
- Actions generated by onboard scripts including the selection of the appropriate beacon message for downlink, reconfiguring of the S/C, and/or safing the S/C.
- Data summaries to enable ground users to diagnose the situation.

These key technologies are part of the CX-I design. Several mission simulations are scheduled where a select set of schools and amateur sites are participating in mission simulations to train users and to verify all functions/components in an end-to-end environment.

On-board autonomy is accomplished through the use of several software tools/techniques: a) SCL (Systems Control Language) performs onboard control and monitoring; b) Scripts, written in SCL by students, accomplish the data summaries used in the Beacon Mode; c) Rules and constraints, written in SCL by students, provide on-board autonomy - or the ability to detect situations and to automatically respond to these situations.

PPP (Point-to-Point Protocol). The objective is the use of off-the-shelf computer networking protocols to communicate with the S/C over the RF link. The TCP/IP protocol running of PPP is being used for CX-I, i.e., CX-I piggybacks the TCP/IP PPP packets onto standard amateur radio packets to change the asynchronous nature of PPP packets to the synchronous nature of packet radio.

N.4.4 Data Distribution Scheme and User Involvement

CX-I data and data products are communicated/distributed via the internet with all participating parties (schools, citizens, and other users). MOCC maintains a website of satellite data, processed data, and composite maps of the atmosphere built up from user contributions, ground measurements, etc. Users at participating schools are able to share data with other users and to generate regional and global maps of the atmosphere and of UV radiation.

N.5 FS-1 (FalconSat-1)

FalconSat-1 is the first of a series of small satellites, designed, built and operated by cadets at the US Air Force Academy, Colorado Springs, CO. The overall science objective of the mission is to determine the effects of S/C charging in the LEO environment. Other objectives are to demonstrate/validate the S/C design and to provide a hands-on experience for all students involved. [1822] [1823]

The S/C structure is nearly cubical in shape with dimensions: 46 cm x 46 cm x 42.5 cm. The solar arrays are body-mounted providing 24 W of power, in addition there is a battery with 10 NiCd cells (4.4 Ah). The bus design features a stack assembly into which the subsystems are mounted. The S/C is spin-stabilized at about 10 rpm pointing toward nadir (a minimal pointing of $\pm 25^{o}$ is required in the direction of S/C velocity vector). Attitude is measured by a three-axis magnetometer; a torque rod functions as an actuator, damping is provided by hysteresis rods. The launch vehicle imparts an initial spin on S/C separation. In addition, nylon solar radiation paddles are used to maintain the initial spin rate. The mass of the microsatellite is 50 kg.

A launch of FalconSat-1 took place on Jan. 27, 2000 (UTC) with a converted Minuteman-II missile (Minotaur) from VAFB, CA.. The Minotaur of OSC is a four-stage vehicle with the first and second stages being Minuteman-II stages; the two upper stages come from OSC's Pegasus launcher. FalconSat-1 was part of the JAWSAT multi-payload adapter and

1822) R. Chari, "Pre-Flight Characteristics of the US Air Force Academy's FalconSat-1," Proceedings of the 13th AIAA/USU Conference on Small Satellites, Aug. 23-26, 1999, Logan UT, SSC99-VII-8

1823) http://www.usafa.af.mil/dfas/Research/FalconSat1/falconsat1.htm

launched with the following satellites: ASUSat-1, OPAL (PicoSat, Artemis, StenSat), OSCE, MASat, and JAWSAT (see N.7).

S/C communications are provided in UHF (downlink) at 400.475 MHz and in VHF (uplink) at 148.030 MHz. The data rate for transmit and receive is 9.6 kbit/s. The signal modulation scheme employs GMSK. - Status: The FalconSat-1 spacecraft failed after about a month in orbit, apparently due to a power failure.

Orbit: Sun-synchronous orbit, altitude = 750 km, inclination = 100°, period = 99 minutes.

Sensor Complement:

CHAWS-LD (Charging Hazards and Wake Studies - Long Duration experiment). The objective is to measure electric charge characteristics of the S/C. Voltage and current sensors, made of sheets of stainless steel (and an electric circuit board stacked together with aluminum and Teflon spacers), are installed on the four sides of the S/C. As FalconSat moves through the space plasma, it creates a wake region behind it, in which primarily electrons accumulate. Hence, the electrically isolated sections of the voltage sensors on the wake side are negatively charged. The current sensors reject electrons and collect ions from the space plasma, providing a current that is correlated to the ambient plasma density. The relative amount of current collected by each sensor provides in addition information about the S/C attitude relative to the plasma flow. Data measurements can be made at 5 or at 10 Hz. A series of measurements is taken over the course of one or more orbits. - One voltage and one current sensor are center-mounted at side of the S/C. The surface of the voltage sensors, exposed to the plasma, is the metalized surface of the circuit board. The exposed surface of the current sensors is an electro-formed stainless steel mesh. The data collected from each sensor are stored in the memory of the flight computer. They can be transmitted to the ground during station passes.

N.6 NanoSat

NanoSat is a low-cost technology demonstration nanosatellite of INTA (Instituto Nacional de Tecnica Aerospacial), the Spanish Space Agency. INTA partners in the project are the universities and students of: UPM (Universidad Politecnica de Madrid) and CNM (Centro Nacional de Microelectronica' de Barcelona). The overall objective is to get involved and familiarized with all aspects in the development of micro- and nano technology: a) on the spacecraft and instrument level, b) to demonstrate new types of magnetic and solar sensors, c) to space-qualify new ASIC designs and components, d) to allow store-and-forward communications, and e) to obtain hands-on experience with NanoSat operations and to conduct experimentation in the space environment. [1824)]

The satellite is spin-stabilized. The S/C structure consists of two hemispheres (top and bottom), each one has one hexagonal and six trapezoidal shaped sides. The overall structure approximates roughly a sphere. The satellite has the following subsystems: PDU (Power Distribution Unit), OBDH (On-Board Data Handling), Communications and Experiments. Power (17 W average) is provided by surface-mounted solar cells. NiCd batteries provide 4.8 Ah of energy for eclipse operations. The OBDH provides all S/C control, processing functions and experiment interfacing (based on Motorola's microprocessor MC68332, 4 MB of storage capacity, 8 kB PROM, 512 kB EEPROM, 768 kB of protected RAM). The S/C uses a magnetometer for attitude sensing. The communications subsystem employs omnidirectional antennas. Two digital modems are implemented for experimentation purposes; one uses a single DSP chip, the other is based on an ASIC design. The UHF band (387.1 MHz downlink, 400 MHz uplink, GMSK modulation and Viterbi encoding) is

1824) A. Martinez, I. Arruego, M. T. Alvarez, J. Barbero, et al., "Nanosatellites Technology Demonstration," Proceedings of the 14th Annual AIAA/USU Conference on Small Satellites, Logan, UT, Aug. 21-24, 2000, SSC00-II-2 used for all data communications with the ground. The downlink data rate is 24 kbit/s. The satellite mass is about 13 kg. All ground station access is based on the TDMA protocol using a Slotted Aloha access version.

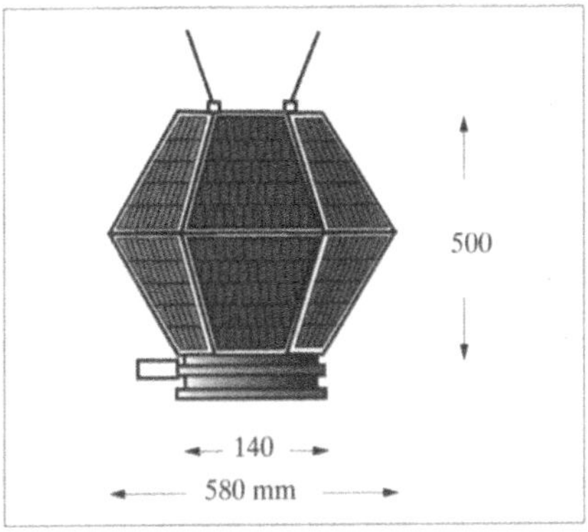

Figure 342: Illustration of the NanoSat spacecraft

A launch of NanoSat is planned for 2003 on an Ariane-5 vehicle (as an ASAP launch) from Kourou.

Orbit: Sun-synchronous orbit, altitude = 645 km, inclination = 98°, period = 97 minutes.

Sensor complement/experiments:

Solar sensors: Porous silicon-based (PS) photodiodes are used with the objective to: a) test the performance of solar cells based on PS-layer technology, and b) provide a means of sun aspect angle sensing for spin control. Since the output signal of each solar detector is proportional to the cosine angle of incidence of the solar radiation, the S/C spin axis determination is made by using the output from three out of six PS-based solar sensors positioned at known locations of the outer S/C surface. This method permits a nutation angle determination to within 5°.

Magnetometer. The objective is the measurement of the Earth's magnetic field (magnitude and direction). The magnetic field measurements have a resolution of 1 mGauss scalar in the range of ±0.1 Gauss. The instrument has a total mass of 220 grams (200 g control electronics) and a power consumption of <2W.

N.7 JAWSAT (Joint Airforce Academy / Weber State University Satellite)

The JAWSAT mission is a collaboration between the United States Air Force Academy (USAFA) and the Center for Aerospace Technology (CAST) at Weber State University in Ogden, Utah. The JAWSAT project was initially designed (1994) with a pulsed-plasma thruster to train Air Force Academy cadets. In its new dedication (April 1998), JAWSAT is mainly functioning as a multi-payload adapter for the following satellites: [1825] [1826] [1827]

- ASUSat-1 (Arizona Sate University Satellite-1)

1825) J. L. Smith, D. Richards, M. Wood, G. Sharp, W. Clapp., "The JAWSAT Mission: Final Report and Lessons Learned," Proceedings of the 14th AIAA/USU Conference on Small Satellites, Logan, UT, Aug. 21-24, 2000, SSC00-V-4

1826) http://cast.weber.edu/jawsat/jawsat.html

1827) http://www.osss.com/

- FalconSat of the USAF Academy, Colorado Springs
- OPAL (Orbiting Picosat Automatic Launcher) of Stanford University with the following picosats:
  - PicoSat of the USAF (DARPA), Aerospace Corporation (2)
  - Artemis of Santa Clara University (3)
  - StenSat (1)
- OCSE (Optical Calibration Sphere Experiment) of AFRL
- MASat (Miniature Amateur Radio Satellite)
- JAWSAT of Weber State University

JAWSAT includes two other payloads which remain attached to the S/C structure. These are: PEST (Plasma Experiment Satellite Test), provided by NASA/MSFC, and ACP (Attitude Control Platform) of Weber State. This includes the efforts of several universities, local aerospace companies, the Air Force Academy, the Air Force Research Laboratory, and NASA.

The JAWSAT **MPA** (Multi-Payload Adapter) technology, in a sense "a mass-transport" capability to space, was developed at CAST of Weber State University and built by OSSS (One Stop Satellite Solutions) and of Ogden, UT. The objective is the deployment of a number of small satellites. The design elements/features of MPA are:

- Complete three-axis attitude determination and control
- Lightweight isogrid space frame
- Flight computer with EDAC memory module
- Onboard multi-channel imaging system
- Onboard power system with battery charge regulation

Figure 343: Illustration of the JAWSAT MPA spacecraft model

MPA for JAWSAT consists of six waterjet-cut aluminum isogrid pieces assembled in a "windmill" design, a frame of 71 cm x 71 cm x 76 cm (the isogrid space frame is composed for four vertical panels, a top and a bottom panel that can be arranged in a number of configurations). MPA can be quickly and easily proportioned for multiple applications and payloads. JAWSAT is three-axis stabilized. Since attitude determination and control are fundamental to this mission, JAWSAT is equipped with several systems for attitude determination and a 3-axis stabilization system. The overall S/C structure has dimensions of: 90 cm x 90 cm x 107 cm, its mass is 230 kg, 80 W of average power are provided from solar panels.

Note: The small satellites of normally cubic shape are also being referred to "cubesats" (a generic name). In the same context the MPA, a platform to integrate multiple payloads, is also referred to as a CubeSat launcher.

The ACP (Attitude Control Platform) of CAST uses tiny reaction wheel canisters (4), and a magnetometer; ACP is used to orient JAWSAT during the deployment of its payloads. ACP may also be used for a free-flying satellite with all the attributes needed to operate a mission. On this testbed spacecraft, a number of attitude sensors are available for comparison and possible integration.

JAWSAT was launched on Jan. 27, 2000 (UTC) with an OSC Minotaur vehicle (a converted surplus Minuteman-2 motor combined with the upper stages of the Pegasus-XL booster) from the commercial California Spaceport at Vandenberg, operated by SSI (Spaceport Systems International). [1828] All satellites were deployed successfully from the JAWSAT mothership.

Orbit: Circular sun-synchronous (terminator) orbit, altitude = 750 km, inclination = 100°, period = 99 minutes.

JAWSAT payload complement:

DPA (Deflection Plate Analyzer). DPA is flown as part of **PEST** (Plasma Experiment Satellite Test), provided by NASA/MSFC, on JAWSAT. The objective of PEST is to study plasma (electrically charged gases), at orbital altitudes. DPA uses an new analysis technique that can measure multiple ion streams and determine the intensity, flow direction, and energy and mass distributions for each stream. The space test requires comparing the DPA with instruments that have been successfully used in space many times.

OCSE (Optical Calibration Sphere Experiment) of AFRL. [1829] OCS is a 3.5 m diameter kapton/aluminum balloon that is designed to orbit the Earth for up to two years. The highly-reflective silver-colored balloon was built by L'Garde, Inc. of Tustin, CA. The overall objective is to further refine the ability to track high-flying satellites. The Air Force uses its SLR (Satellite Laser Ranging) network of ground stations to periodically track OSCE. - The balloon was deployed from a tiny canister shortly after orbit injection and inflated.

N.8 NavGold

NavGold [1830] is a project of the University of Colorado at Colorado Springs (UCCS) involving the design, development, launch, and operation of two nanosatellites. The objectives of NavGold include the establishment of an on-orbit test bed for demonstrating and validating critical nanosatellite technologies that enable formation flying. Some specific objectives are: [1831] [1832]

- To demonstrate autonomous relative-navigation accuracies of < 100 m, with a goal of 10 m accuracy or better

1828) S. Schoneman, S. J. Buckley, G. Stoller, L. M. Marina, C. B. Morris, "Demonstration of a New Smallsat Launch Vehicle: The Orbital/Suborbital Program (OSP) Space Launch Vehicle Inaugural Mission Results," Proceedings of the 14th AIAA/USU Conference on Small Satellites, Logan, UT, Aug. 21-24, 2000, SSC00-I-1

1829) http://www.lgarde.com/programs/ocse.html

1830) The name NavGold is derived from Navigation, emphasizing the relative navigation focus of the project, and Gold, one of the colors of the University of Colorado at Colorado Springs.

1831) D. Sipple,.J. Torley, F. Chavez, et al., "NavGold – An On-Orbit Test Bed for Experiments in Formation Flight," Proceedings of the 13th AIAA/USU Conference on Small Satellites, Aug. 23–26, 1999, Logan UT, SSC99–II–5

1832) http://mae.uccs.edu/fdcl/navgold/home.html

- To demonstrate autonomous relative-position maneuvering and stationkeeping capability with relative position accuracy equal or better than 100 m
- To demonstrate autonomous relative-attitude maneuvering and stationkeeping capability with a relative attitude accuracy of < 2°.
- To provide an educational, multidisciplinary team-building experience.

The S/C structure consists of a hexagonal prism with a diameter of 45 cm and a height of 31 cm. Each of the satellites is three-axis stabilized by an attitude control system. Attitude is sensed by digital sun sensors, a three-axis magnetometer and a horizon sensor (all developed at CU). The accuracy of each sensor is about 0.5°. Four magnetic torque rods and cold gas thrusters are used as actuators. Power of 18 W is provided by surface-mounted GaAs solar arrays. There are NiCd batteries (15 cells) for the eclipse phase of the orbit. Regulated power distribution is provided. A mission life of 3-5 months is expected prior to rapid orbital decay. The S/C mass is about 14 kg.

A Shuttle launch in 2003 is planned using the HES (Hitchhiker Ejection System).

A cold gas thruster system is used as on-board propulsion system, consisting of a propellant tank and six thrusters. The mass of the system is about 1.5 kg. The truster system is used for attitude control as well as for orbit maneuvers.

Orbit: Circular orbit, altitude = 400 km, inclination = 51.6°, period = 92 min.

Navigation, Guidance, and Control (NGC):

NGC is the primary payload package consisting of the attitude sensor suite and the actuator/propulsion elements.

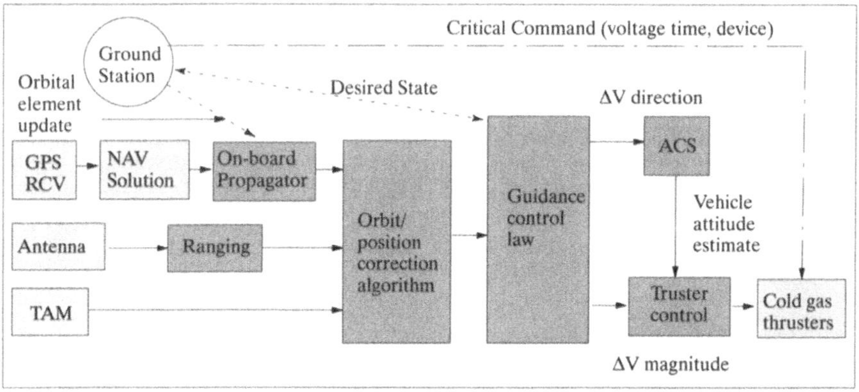

Figure 344: Block diagram of navigation and guidance function

Relative navigation in formation flight: [1833)]

Formation flight involves the use of an active control scheme to maintain the relative states of the formation flying satellites. The leader-follower concept is chosen for the two nanosatellites of NavGold (other possible types of formation flight are: a) same ground track, and b) side by side), representing the simplest configuration of in-orbit-plane formation flight. In this formation, the orbital elements of each S/C are identical except for the true anomaly.

All S/C maneuver experiments are to be performed autonomously. This involves the following classes of maneuvers:

1833) F. Chavez, D. K. Schmidt, "Formation Flying and Relation Navigation - A Nanosatellite Research Mission," Proceedings of the 23rd Annual AAS Guidance and Control Conference, Feb. 2-6, 2000, Breckenridge, CO, AAS 00-061

- Relative re-positioning. - a) relative phasing maneuver (increase/decrease in relative separation or phase angle), and b) relative attitude maneuver (increase/decrease in relative attitude separation)
- Relative position stationkeeping
- Relative re-orientation
- Relative orientation stationkeeping.

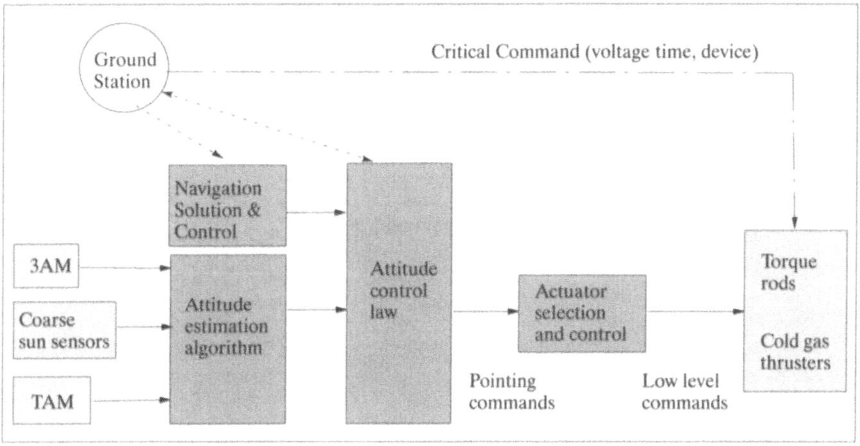

Figure 345: Block diagram of attitude determination and control function

Communications architecture:

1) S/C to ground station (TT&C) communication. The link utilizes the UHF-band according to US radio amateur assigned frequencies. The downlink data rate is 9.6 kbit/s.

2) Inter-satellite communication. The link utilizes the UHF band. Operations of this link are non-concurrent to ground transmissions of the leader S/C to avoid link interference. The link performs two functions: a) to relay status of the "follower" S/C to the "leader" S/C, b) to provide relative S/C ranging measurements. The data of the inter-satellite link includes GPS information, payload status, bus status and range based beacon data.

Control architecture of the constellation

The selection of an appropriate augmented GPS receiver system is under consideration as of 2001. Augmentation refers to a parallel measurement technique to GPS, such as tonal ranging in which a pseudo-random code is transmitted on an RF crosslink between the two spacecraft.

N.9 Munin

Munin is a joint nanosatellite pilot project between IRF (Swedish Institute of Space Physics), students at Umeå University (RYP), students at Luth (Luleå University of Technology), the former institutions are located in Sweden, and SwRI (The Southwest Research Institute) in San Antonio, Texas, USA. The scientific objective of Munin is to collect data on the auroral activity on both the northern and southern hemispheres (and distribution of electrons, ions and neutral particles), such that a global picture of the current state of space weather activity can be made available on-line.

The name of the satellite comes from Nordic mythology: "The ravens Munin and Hugin flew out and brought back news from every corner of the world. Sitting on the God Odin's shoulders, they whispered all the news in his ears. Munin represented the memory and Hugin the intelligence... and they were his embodied soul..."

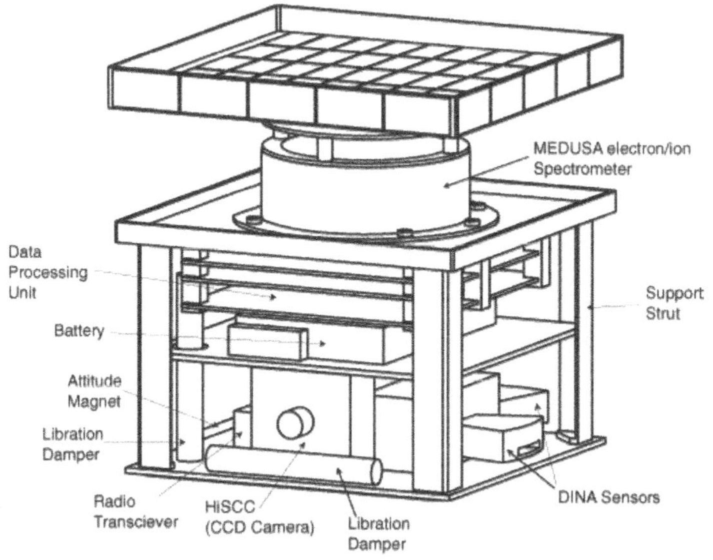

Figure 346: The Munin S/C model (only one solar array is shown)

The satellite is of size 21 cm x 21 cm x 22 cm, it uses a passive magnetic stabilization system, a permanent magnet holds the satellite aligned along the geomagnetic field lines (like a compass needle). Since the S/C does not require high precision pointing, the attitude problem of restoring torque is solved by the use of a permanent magnet. Average pointing accuracies of $\pm 5^{\circ}$ are achieved. [1834] There are no actuators and traditional orientation sensors for attitude control. There are also no moving parts, such as deployable booms, etc. The S/C attitude is determined by the Kiruna ground station from the following inputs: a) current measurements of the six solar arrays which cover each side of the S/C body, b) the data of two single-axis on-board magnetometers which measure the **H** vector, and c) images of the Earth's limb and of stars by a CCD camera. The passive magnetic attitude stabilization system of Munin, developed by IRF in cooperation with the Keldish Institute of Applied Mathematics, Moscow, employs a magnetic hysteresis damper. [1835] [1836]

The satellite is covered with solar cells (Si) on all surfaces, which provide enough power when the satellite is in sunlight and recharge a battery (lithium-ion) for power during eclipse periods. S/C mass = 6 kg. On-board data storage of 2 MByte of RAM. The data is downlinked with a radio modem [a TEKK KS-1000 UHF transceiver at 405.55 MHz downlink and 449.95 MHz uplink, data rate between 4.8 - 19.2 kbit/s] during passes over the Kiruna

1834) M. Ovchinnikov, V. Pen'kov, O. Norberg, S. Barabash, "Attitude Control System for the First Swedish Nanosatellite "Munin," Acta Astronautica, Vol. 46, No 2-6, 2000, pp. 319-326

1835) http://munin.irf.se/frames/technology_index.html

1836) O. Norberg, W. Puccio, J. Olsen, et al., "Munin: A Student Nanosatellite for Space Weather Information," Proceedings of the COSPAR Colloquium on Scientific Microsatellites, "Microsatellites as Research Tools," Tainan, Taiwan, 1997 ground station (68° N, 20° E). It is made available to all on Internet immediately after reception for space weather forecasts (auroral oval, geomagnetic storms, etc.). Munin was launched as a secondary payload on a Delta 7320-10 launcher from VAFB on Nov. 21, 2000 (primary payloads were EO-1 and SAC-C).

Orbit: Elliptical polar orbit of 705 km x 1700 km, inclination =98.2° (for Munin only, the other two S/C are separated and put into a sun-synchronous orbit at 707 km altitude).

Sensor complement:

MEDUSA (Miniaturized Electrostatic Dual-tophat Spherical Analyzer), a joint instrument (of Astrid-2 heritage) provided by SwRI and IRF-K. MEDUSA is a combined electron and ion spectrometer. Objective: simultaneous measurement of electrons and ions with energies up to 18 keV/q with a resolution of 16 energy sweeps/s for electrons, and 8 sweeps/s for ions. Particles are measured in 16 sectors in the plane of acceptance, which is aligned with the Earth's magnetic field. Particles enter the spectrometer aperture at any angle in the plane of incidence, electrons and ions are then deflected into their respective spectrometer unit by a spherical electrostatic analyzer. The particles hit a microchannel plate after being filtered in energy in the electrostatic analyzer, the hits are counted by preamplifiers connected to registers, and the hits per sample interval are then further processed by the data processing unit (DPU). The mass of MEDUSA is 0.6 kg, the power consumption is 1000 mW.

DINA (Detector of Ions and Neutral Atoms), a solid-state neutral particle detector of IRF (heritage of PIPPI on Astrid-1). The objective of DINA is to extend the measurements of the precipitating ions to the higher energy range 20 - 500 keV complementing the MEDUSA experiment (also measurement of ENA). The instrument consists of two sensors with an aperture of 5° x 30° each (angular resolution: 2.5° x 25°). The sensor DINA-0 is looking at 180° pitch angle, DINA-90 is looking at 90° pitch angle. The mass of DINA is 0.340 kg, the power consumption is 500 mW. DINA performs alternate measurements of ions and energetic atoms. Mass identification is performed by (Delta E) / E detectors). For the energy range 20 - 100 keV, the front detector provides integral flux of all masses.
Over the northern hemisphere DINA-0 provides measurements of the ion precipitating flux during the auroral oval crossing. While the S/C moves through the polar cusp, DINA-90 makes measurements of ENA (Energetic Neutral Atoms) flux from the exosphere at one local time sector. Over the southern hemisphere DINA-0 pointing down in the southern hemisphere is aimed to detect outflowing ENA in the precipitation region. The information about input ion flux can be obtained from the DINA-90 measurements because the precipitating ion distribution can be considered approximately isotropic.

N.10 NUSAT (Northern Utah Satellite)

NUSAT is a collaborative satellite project designed and built by students at Weber State University (Ogden, Utah) with outside participation from academia (Utah State University, New Mexico Sate University), industry, and government agencies. The objective and primary mission was to solve a timely problem of FAA (Federal Aviation Agency), namely to calibrate antennas from orbit used for the purpose of air traffic control radar antenna pattern mapping. [1837] [1838]

The NUSAT S/C structure consists of a 26-sided multi-hedron with a size of 49 cm in diameter and a mass of 68 kg. Power is provided by solar panels and a 10.5 V battery. NUSAT was deployed from a modified GAS (Get Away Specials) canister on Shuttle flight STS-51B

1837) J. L. Smith, et al., "Low-Cost Attitude Determination and Control for Small Satellites," Proceedings of the 10th Annual AIAA/USU Conference on Small Satellites, pp. 1-20, Sept. 16-19, 1996

1838) http://cast.weber.edu/nusat/index.html

(Spacelab-3 mission, April 29 - May 6, 1985) along with GLOMR (Global Low Orbiting Message Relay), a DARPA payload developed by DSI (Defense System Inc.). However, GLOMR failed to eject from the GAS canister on this flight. The mishap hindered NUSAT to support its very mission objectives. The NUSAT microsatellite orbited Earth for 20 months (reentry in Dec. 1986) and demonstrated that satellites could be built small, simple, and at low cost for special applications.

Orbit: Altitude = 411 km, inclination = 57°, period = 90 minutes.

NUSAT Experiment:

Conventional antenna calibration methods involved antenna signal tracking of the setting sun for long periods of time. The NUSAT project was to demonstrate a new antenna calibration technique of FAA antennas based on the following principle: Since a satellite moves across the sky much faster than the sun, and since it can respond to signals on the ground, antennas anywhere on the ground can be calibrated in only a few minutes as opposed to the several hours that it would take to use the sun. FAA antennas would then be free to resume monitoring aircraft.

N.11 OPAL (Orbiting Picosat Automatic Launcher)

OPAL is a student-designed microsatellite of SSDL (Space Systems Development Laboratory) at Stanford University, Stanford, CA. The objective is to test the feasibility of launching several picosatellites (daughterships) from a mothership microsatellite. The picosats and the mothership, OPAL, are instrumented with magnetometers to measure their attitude relative to the magnetic field. The intent is to simultaneously sample a volume of space for denser magnetic field measurements. OPAL uses a mechanical launching system to eject six hockey-puck sized picosats. OPAL's secondary payload consists of a suite of miniature accelerometers and a fluxgate magnetometer. The OPAL design was started in 1995 and was completed in May 1999. [1839) 1840) 1841)]

The OPAL S/C structure resembles a hexagonal prism, consisting of 4 hexagonal horizontal trays connected by 4 vertical rods. The space between the trays is used by each subsystem to place its components. The components are mounted in aluminum boxes, the boxes are fastened to the trays. The trays are surrounded by 6 side panels, which double as structural members and solar array surfaces (GaAs and Ge). Solar cells are also mounted on the outside of the top and bottom trays. All panels and trays are made out of 6 mm aluminum honeycomb, except for the bottom tray, which is 12 mm thick to support the carrier attachment and to be able to handle the heavier launch loads. Some S/C parameters: body height = 23.5 cm, radius = 21 cm, total mass = 25 kg. The on-board computer consists of a Motorola 68332 microcontroller (16 MHz, 1 MB RAM). Power storage is provided by NiCd batteries (1.2 V, 5 Ah). Power is regulated to 5V and 8V, the average power consumption is 2.4 W (12 W max). Secondary power is provided by NiCd batteries. Note: An attitude control system was not implemented since the mission objective did not require any attitude control. However, the attitude can be estimated from the magnetometer readings as well as from the solar panel currents.

Communications: OPAL has four 1/4 wavelength monopole antennas (each about 17.5 cm) which are placed on the top and bottom panels (right-hand polarized). OPAL uses packet radio transmissions over amateur radio frequencies (uplink/downlink at 420-450 MHz,

1839) J. Cutler, G. Hutchins, R. Twiggs, "OPAL: Smaller, Simpler, and Just Plain Luckier," Proceedings of the 14th AIAA/USU Conference on Small Satellites, Logan UT, Aug. 21-24, 2000, SSC-VII-4

1840) D. S. Clarke, M. T. Hicks, et al., "Picosat Free Flying Magnetometer Experiment," Proceedings of the 10th Annual AIAA/USU Conference on Small Satellites, Sept. 16-19, 1996

1841) http://ssdl.stanford.edu/opal/index.html

transmitter of 2.2 W, AX25 protocol, data rate = 9.6 kbit/s). The IP (Internet Protocol) is used between users and the OPAL ground system. [1842)]

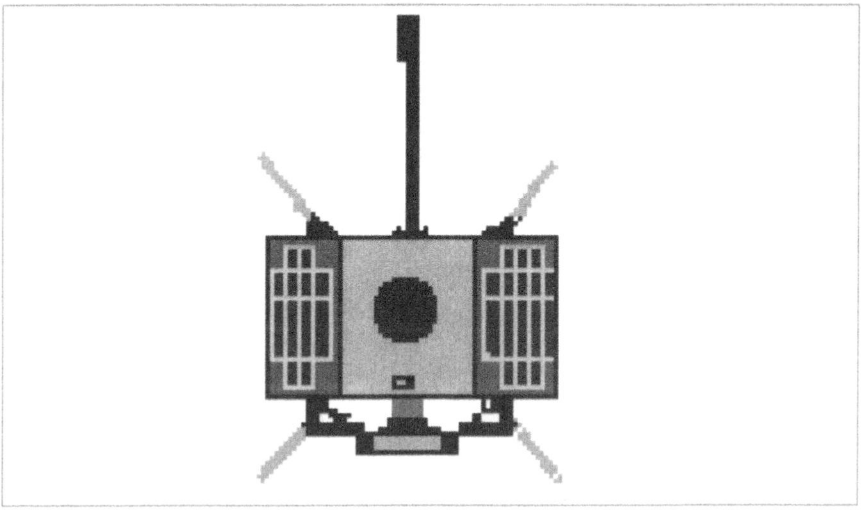

Figure 347: Illustration of the OPAL microsatellite

A launch as a secondary payload to JAWSAT, ASUSat-1, OPAL (StenSat, PICOSAT1.0, Artemis), OCSE, and FalconSat, on an OSC launch vehicle (Minotaur) took place on Jan. 27, 2000 from VAFB, CA. OPAL was attached to the multi-payload adapter built by OSSS of Ogden, UT. In addition to OPAL, three other free-flying satellites (ASUSat-1, OSCE, and FalconSat) were attached to the adapter and deployed within minutes of reaching orbital altitude.

Status: OPAL achieved its primary mission goal, all six picosats were successfully deployed (ejection was confirmed by telemetry), the component testbeds are fully operational. Extended mission operations focus on long-term characterization of the bus and component (magnetometer) testbeds. After launch and initial operation, OPAL has been designated as: OPAL OSCAR-38 (OO-38) by AMSAT, the Radio Amateur Satellite Corporation.

Orbit: Circular sun-synchronous orbit, altitude = 750 km, inclination = 97.5°, nodal crossing at 6 AM/6 PM.

N.11.1 Sensor/payload complement

OPAL has three payloads: a) the Picosat Launcher and Payload as an end-to-end demonstration of a mothership/daughtership mission architecture and technology (the objective is to explore distributed sensing in space), b) the Accelerometer Testbed, and c) the Magnetometer Testbed. - OPAL mission operations are conducted from the SSDL ground station.

Background: The original OPAL mothership system was designed as an FFM (Free-Flying Magnetometer) mission, sponsored by JPL. In this version, the daughterships carried precision magnetometers and were similar in shape to a hockey puck. A lack of funding terminated this approach. - New sponsors were found for the picosatellite mission. The launcher was redesigned to suit the new requirements.

1842) J. Cutler, G. Hutchins, C. Kitts, R. Twiggs, "Infrastructure for Internet-Based Operations," Proceedings of the 14th AIAA/USU Conference on Small Satellites, Logan, UT, Aug. 21-24, 2000, SSC00-IX-4

Picosat Launcher and Payload. [1843] OPAL carried and ejected a total of six picosatellites: StenSat (1), PICOSAT1.0 (2), and Artemis (3). The picosat launcher consists of a total of four picosat launch tubes. Each is capable of holding two short (7.5 cm x 10 cm x 2.5 cm) or one long (7.5 cm x 20 cm x 2.5 cm) picosat. Six picosat flew on OPAL, four short and two long.

A latched door with a dual-release mechanism was used to eject the picosats (single shot release and replacement of actuators). The firing was ground commandable.

Accelerometer Testbed. Objective: Test of COTS instruments in space. Three accelerometer types, working on different measurement principles, are being tested, consisting of: a micromachined capacitor, piezoelectric crystal, and magnetic inductor.

- Capacitive sensor: The ADXL05 from Analog Devices measures changes in capacitance in a circuit etched on a silicon die.
- Piezoresistive sensor: The PCB 336M27 from PCB Piezotronics uses the piezoelectric properties of a quartz crystal to provide a measurement of acceleration (sensitivity of 1 mgal).
- Inductive sensor: The Geophones (GS-11D & GS-30CT) from GeoSpace Corporation use coil's motion with respect to a magnet to generate a signal.

Two accelerometers of each type were flown in the testbed. A stimulation source was provided to exercise the accelerometers. Characterization of the accelerometers is defined as determining short-term sensor degradation due to launch and initial exposure to the space environment.

Magnetometer Testbed. Objective: to measure OPAL's attitude (S/C alignment within 2°). A miniature 3-axis fluxgate magnetometer designed by Applied Physics Systems (model APS533). The magnetometer is enclosed in a fiberglass package and is very compact (19 mm diameter x 4 cm long cylinder, 18 grams of mass). It provides direction information to better than 0.1°, while consuming only 200 mW. It generates three voltages proportional to the magnetic field in three perpendicular directions, with an accuracy of ± 0.1%. Resolution of the magnetometer is limited by OPAL's 12 bit A/D conversion. The magnetometer can be read with a frequency of up to 1 kHz.

Characterization of the magnetometer is defined as determining short-term magnetometer degradation due to launch and initial exposure to the space environment. Also, determination of the long-term magnetometer performance degradation due to extended exposure. The magnetometer performance is measured by comparing the vector magnitude of the Earth's magnetic field as measured by the sensor to the predicted vector magnitude of standard geomagnetic modeling software.

N.11.2 StenSat

StenSat is a picosatellite (260 cm^3 in volume, mass = 0.235 kg), developed by a group of amateur enthusiasts in the Washington DC area. The satellite structural components consist of an aluminum chassis (of size 10 cm x 7.5 cm x 2.5 cm), a magnet, top and bottom printed circuit board solar cell panels, two internal printed circuit boards (receiver/transmitter and controller), and black and white paint. One NdFeB disk magnet provides alignment and attitude control. The painting of different sides of one of the dipole antennas with black and white paint introduces a slow radiometric spin. Power is supplied by the top and bottom panel of six GaAs solar cells. Further subsystems are controller, sensors and antenna.

1843) Note: The three picosats are the main payload of OPAL, a technology demonstration project funded for JPL. The ultimate goal of this project is to be able to launch hundreds of picosatellites from a mothercraft in low Earth orbit - to obtain simultaneous measurements of the magnetic field over a large volume. Each picosat could measure the magnetic field and then transmit the data back to the mothercraft.

The picosatellite is used by amateur radio operators world wide. It operates as a single channel mode "J" FM voice repeater. The uplink frequency is 145.84 MHz and the downlink is 436.625 MHz. StenSat periodically transmits 1200 baud AX.25 for broadcasting telemetry. Additionally, amateur radio operators are able to "PING" the satellite by transmitting a six digit DTMF command to the receiver uplink. [1844)]

Status: StenSat was released from the OPAL launcher on February 10th. Unfortunately, no confirmed signals have been received.

N.11.3 PICOSAT1.0

The PICOSAT1.0 mission of the USAF (sponsored by DARPA) was developed and built by the Aerospace Corporation, El Segundo, CA. PICOSAT1.0 is a free-flyer technology demonstration mission consisting of two tethered PICOSATs with the following objectives:

- To demonstrate the basic functional elements of a low-power LEO "swarm" or formation PICOSAT array
- To communicate from space using node-type radios and to report the results of MEMS-switched tests.

The two PICOSATs, tiny boxes of size: 25 mm x 75 mm x 100 mm, each with a mass of 0.275 kg, were ejected from OPAL's spring-loaded launcher on Feb. 6, 2000, in sight of the SRI (Menlo Park) ground station. Each PICOSAT uses a small, battery operated, very low power radio with a cross-link capability. The PICOSATs are referred to by their nodal names as PICO21 and PICO23. The communication system of each PICOSAT employes a receiver/transmitter system featuring: a) a spread-spectrum DCT (Digital Cordless Telephone) chip set, a COTS product of Conexant, b) a CPU/RF board design. The system is operated at 64 mW output transmit power. The store and forward communication protocol is a random-access packet system operating at a frequency of 915 MHz spread-spectrum (20 channels) with a data rate of 80 kbit/s. The satellite computer is a 6502 system (Rockwell Science Center, Thousand Oaks, CA) with a simple operating system for data collection , system management and the network protocol. Each PICOSAT is powered by three tiny Lithium Thionyl Chloride batteries with a capacity of 1 Ah each for a total of 3 Ah and a voltage of 3.65 V. The simplicity of the PICOSAT design does away with such standard spacecraft components as attitude sensors and a propulsion system.

The tether system of PICOSAT1.0, of 32 m in total length, uses a polyethylene 3-braid thread which contains 26 gold wire dipoles along its length, tuned to 430 MHz and 1 GHz. The tether was released from a tether spool right after the PICOSAT1.0 ejection from OPAL. The tether objectives are: a) to enhance system detection (the tether system serves as a dipole, enlarging the radar cross section of the assembly for a more distinctive signature and thus permitting ground detection), b) to keep the PICOSATs within range of each other (32 m max) for communication reasons (with each other and to the ground). A sensor board on each PICOSAT has a chip containing 4 MEMS RF switches (a 100 V charge pump on the sensor board is needed by the switches) that are operated in space after deployment. An A/D converter measures the resistance across each switch every time a switch occurred and records the results.

Status: [1845)] In retrospect, the tether release mechanism worked nominally, communication was established between each PICOSAT and the ground. The ground station was able to call PICO21 or PICO23, upload instructions and download housekeeping data (temperature and MEMS switch test results). The crosslink capability was also tested, communicating from the ground indirectly with PICO23 via PICO21. Detection of the PICOSAT1.0 system from a ground-based radar system, referred to as SSN (Space Surveillance Network),

1844) http://users.erols.com/hheidt/

1845) Information provided by Ernest Y. Robinson and by David A. Hinkley of the Aerospace Corporation.

was verified. The radar tracking provided also ephemeris data for orbit determination. The PICOSAT1.0 mission lasted until Feb. 9 when battery power decayed.

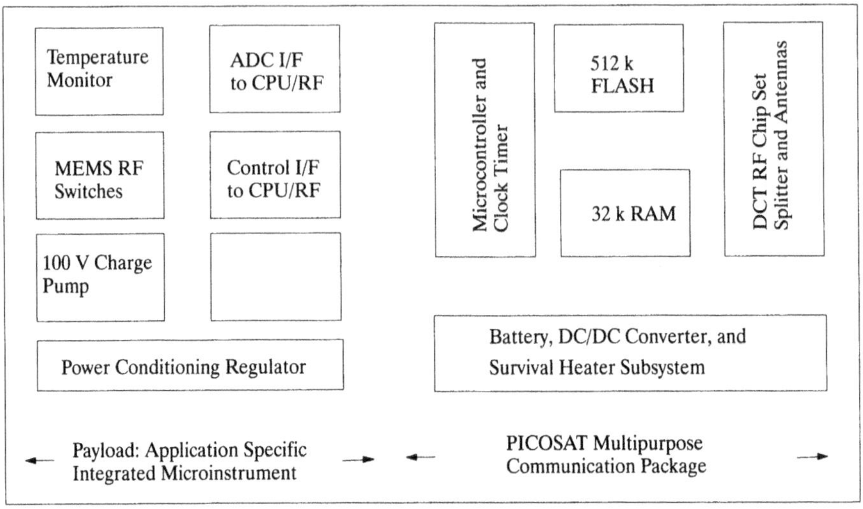

Figure 348: PICOSAT1.0 system block diagram

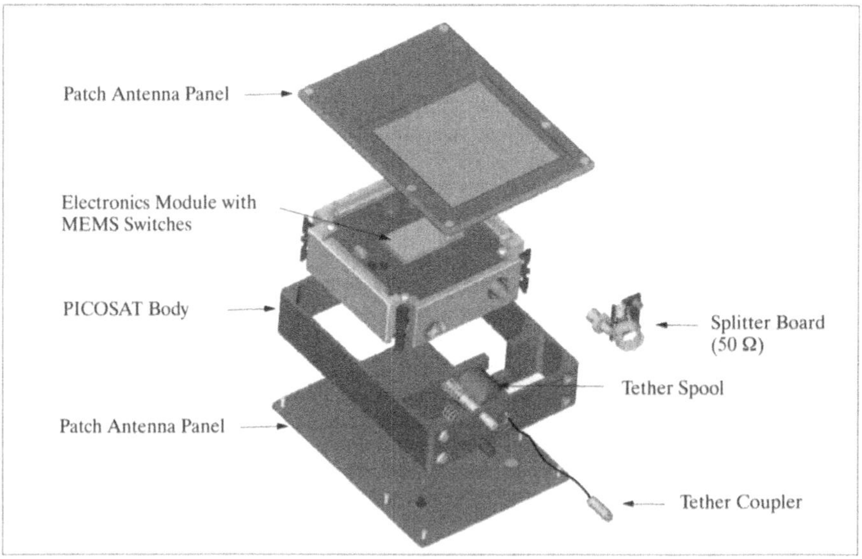

Figure 349: Illustration of major PICOSAT1.0 components for a single spacecraft

N.11.4 Artemis

Artemis is a set of three picosatellites provided by undergraduate engineering students of Santa Clara University, Santa Clara, CA. The mass of Artemis is < 1kg by utilizing a commercial microprocessor, transmitting on amateur radio frequencies from the picosatellite to ground, and using commercial off-the-shelf batteries to power the picosatellite. The primary mission of the OPAL picosatellite payload is to provide an end-to-end mission demonstration of mothership and daughtership technologies.[1846)] [1847)] [1848)]

The structure of two Artemis picosats is of size: 20 cm x 2.5 cm x 7.5 cm with a mass of 0.611 kg. The two picosats consist of the following subsystems:

- VLF (Very Low Frequency) Receiver. The VLF receiver board is the main payload aboard Artemis. The objective is to research the effects of lightning on the outer ionosphere. Measurements are taken by using the VLF radio signals traveling between the ground and the S/C. Lightning discharges and the effects on the Earth's near space environment can be recorded with the VLF receiver. The VLF receive frequency range is 0.1-12 kHz, with 5 kHz considered typical of a lightning strike. The two picosatellites have different VLF thresholds, namely 100 mV and 10 mV respectively. Measurements are taken simultaneously by both picosatellites. Comparison of the data permits the extraction of the occurrence and amplitudes of horizontal and vertical lightning.
- Power is provided by GaAs solar cells and two NiCd batteries, rated at 1.25 V and 800 mAh.
- A microcontroller provides the necessary functionality to check for adequate battery supply, read sensors, read and log pulses from the VLF receiver, transmit using FSK modulation, and receive commands sent from ground.
- Attitude is determined by four VNIR phototransistors, mounted perpendicular to each other. Rubber ferrite magnets are used as actuators to control the spin axis, aligning it relative to the Earth's magnetic field.
- The communications subsystems transmits and receives over amateur radio frequencies. The data rate is 1.2 kbit/s.

The third picosat, named JAK, was a simple beacon transmitter designed to test the ground reception ability of spaceborne picosat broadcasts.

Mission operations: The picosatellites deploy their communication and VLF antennas after ejection from the mothership OPAL. Both picosatellites broadcast their call-signs once per minute for tracking purposes. Upon a command reception, the CPU initiates the VLF experiment by enabling the VLF receiver. The experiment is carried out in one minute during which the CPU counts the number of pulses the receiver detects above the threshold voltage. The measured data are downlinked in the next (and follow-up) broadcast until the next command is received.

Status: JAK was ejected with StenSat on Feb. 10 (they were stored in the same launch tube). No reception of JAK's beacon signals was confirmed. The limited battery life of JAK ended soon all communication attempts. The two VLF picosats were released on Feb. 12, 2000. However, no signals could be confirmed.

N.12 PANSAT (Petite Amateur Navy SATellite)

PANSAT (Petite Amateur Navy SATellite) was designed and built at the Naval Postgraduate School (NPS) in Monterey, CA. as an educational project for officer students. It is a low-cost microsatellite for messaging or relay services. The overall objective is to demonstrate digital communications using direct-sequence spread-spectrum and store-and-forward packet radio technology. The S/C design is kept very simple. There is neither an attitude control system nor an on-orbit propulsion system. This makes PANSAT a tumbling S/C. The

1846) M. F. Breiling, C. Y. Hu, et al., "The ARTEMIS Project: Picosatellite-Based Missions to Study VLF Phenomenon," Proceedings of the 13th AIAA/USU Conference on Small Satellites, Aug. 23-26, 1999, Logan UT, SSC99-VIII-3

1847) http://screem.engr.scu.edu/artemis/

1848) A. Valdez, C. Hu, C. Kitts, et al., "The Artemis Project: Picosatellites and the Feasibility of Smaller, Faster, Cheaper Approach," Proceedings of the IEEE Aerospace Conference, Snowmass, Co, March 6-13, 1999 microsatellite has a total mass of 57 kg. It was launched as a Hitchhiker (HH) payload on Shuttle flight STS-95 (Oct. 29, 1998; use of HH Pallet Ejection System). [1849)]

Orbit: LEO orbit, altitude = 555 km, inclination = 28.4°

Communications payload:

A UHF operating center frequency of 436.5 MHz is used, a bit rate of 9600 bits per second and 4 MByte of message storage. Amateur radio ground stations are able to utilize PANSAT for store-and-forward communication and to test spread-spectrum communications. In addition, PANSAT provides many potential applications for low-cost communications. The low probability-of-intercept is an important feature for the military in downed-pilot-rescues. The pilot could obtain his/her location through a GPS system and uplink the information to the orbiting satellite at low risk. Examples of civilian uses include emergency rescue and communication to remote areas. A modified amateur satellite ground station is needed to communicate with PANSAT. The NPS ground station utilizes off-the-shelf software, is microcomputer controlled, and is equipped with a spread-spectrum modem. The NPS ground station is similar to a typical amateur radio user station, except it has spacecraft command capability. The NPS ground station is also utilized as a classroom instructional laboratory.

N.13 SAPPHIRE (Stanford AudioPhonic Photographic IR Experiment)

SAPPHIRE is a graduate-student project within the framework of SQUIRT (Satellite Quick Research Testbed) in SSDL (Space Systems Development Laboratory) at Stanford University. The overall objective is to design, construct and operate a microsatellite. Sapphire's primary mission is to space-qualify a micromachined infrared sensor (PI: T. Kenny).

The S/C structure consists of four stacked aluminum honeycomb trays, one to a subsystem (from bottom: Power, Communications, CPU, and Sensors), with eight external panels (six sides, top and bottom) upon which the solar cells are mounted. The entire hexagonal cylinder, fully loaded and on the launch interface, has a mass of about 19.6 kg. Solar cells (GaAs) provide 16 W of peak power (8 W average). Eclipse power is provided by ten NiCd "D" cells in series.

Figure 350: Illustration of the SAPPHIRE satellite

The ADC (Attitude Determination and Control) subsystem follows in the footsteps of the early Oscar satellites (AMSAT) to provide a "controlled tumble" with a spin of about 1/3 to 1/2 rpm. Four permanent (ALNICO-V bar) magnets are aligned with the vertical S/C z-axis

1849) http://www.sp.nps.navy.mil/pansat/pansat.html

(providing the primary means of attitude control), forcing the S/C to line up with the local North direction of the Earth's magnetic field. This in turn generates near-nadir pointing in the northern hemisphere (the top face of the S/C is pointing to nadir), a prerequisite for the camera system. In the southern hemisphere, the top face of the S/C looks to the horizon, while the bottom face of the S/C is pointing toward nadir, giving the IR sensors full view of the Earth in this region. [As the latitude of the S/C increases, the S/C pitch angle increases causing SAPPHIRE to "flip" over at each pole crossing]. - A minute spin moment (torque) is induced by the four turnstile transmit antennas on the outside of the S/C, which are painted black on one side, white on the other, to create a spin effect about the Z axis. [1850] Hysteresis rods are placed perpendicular to the Z-axis to damp nutation about that axis as well as to create a maximum spin rate. S/C attitude is sensed by Earth sensors.

Communication: The subsystem uses modified Hamtronics kits. The transmitter broadcasts with 2 W of power at 437.100 MHz (UHF). The receiver is at 145.945 MHz (uplink in VHF). Data is transmitted at 1200 baud and uses AFSK encoding, the AX.25 packet protocol, and standard Mode J Amateur Satellite frequencies. The voice downlink utilizes frequency modulation and operates at variable speeds. [1851] [1852]

The SAPPHIRE project started in 1994 (it is in effect the first spacecraft designed and fabricated by SSDL). It was flight-readied in 1998, and is planned to be launched as a secondary payload in Aug. 2001 (PICOSat, STARSHINE-3, PCSat) on Athena-1 from the Kodiak Launch Complex on Kodiak Island, AK.

Orbit: circular orbit, altitude = 500 km, inclination = 67°.

Sensor complement:

THD (Tunneling Horizon Detector). Objective: to test a new infrared sensing technique for Earth horizon detection on microsatellites. The THD experiment uses a micromachined IR detection system of JPL on-a-chip, employing the electron-tunneling principle, and operating at room temperature. The sensor is constructed with roughly 1 micron of separation between the membrane and tip. To pull the membrane close enough to enable tunneling currents, an electrostatic force is created by applying a large voltage (100s of Volts) between the membrane and the deflection electrodes. An external circuit controls this "deflection voltage" to maintain a constant tunneling current.

THD is a fixed narrow-FOV horizon detector. Its primary output is an analog signal that triggers on the Earth/space and space/Earth horizon crossings. The THD assembly contains a TIS (Tunneling Infrared Sensor) with 2 independent sensing elements and corresponding electronic circuitry. THD is sensitive to the entire infrared spectrum with an average sensitivity of 1,500 V/W. The assembly is fastened to the payload tray and looks out a slit aperture from the side of the spinning S/C, perpendicular to the spin axis ("scanning" is provided by the spinning S/C). The slit aperture limits the FOV to 6° in the transverse spin plane and to 22° in the vertical plane. The tunneling element outputs are proportional to the level of variation in incident infrared radiation on TIS and consist of a 3.0 V signal + 1/400 of the tunneling high voltage + the tunneling signal disturbance.

Digital Camera. A commercially available B&W CCD camera (model "Fotoman Plus" of Logitech). The objective is to obtain imagery in the northern hemisphere of about 1 km spatial resolution. The capabilities of the digital camera include JPEG image compression, storage in DRAM of up to 32 (496x360) images with 8-bit gray scale in compressed form. A compressed image has a volume of about 23 kbytes. The Digital Camera has a mass of 0.450 kg and an average power consumption of 2 W.

1850) Note: The solar pressure creates a minute, however constant torque, resulting from black and white painted antennas

1851) http://aa.Stanford.EDU:80/~ ssdl/

1852) R. Twiggs, M. Swartwout, "SAPPHIRE - Stanford's First Amateur Satellite," AMSAT-NA 16th Space Symposium, Vicksburg, MS, Oct. 16, 1998

Voice Synthesizer. The objective of the "Digitalker" is to test and illustrate a new communication approach which might be of interest to the Amateur Radio community. There are certainly also some applications seen in the field of education. The Digitalker can be received by radios (hand-held or whatever) when the satellite is overhead, capable to "talk" to an audience, giving for instance a lesson on a particular subject. The on-board voice synthesizer is based on the commercial V8600 Speech Synthesizer of RC Systems with some modifications by the student team.

Telemetry Experiment. A pseudo-payload, designed by students, with the objective to assess attitude determination by using the solar panels as a differential sun sensor.

N.14 SEDSAT-1 (Students for the Exploration & Development of Space)

SEDSAT-1 is a microsatellite of the University of Alabama in Huntsville, AL (UAH). The objectives are to 1) provide hands-on experience for students in the full lifecycle development of space systems, 2) advance the state of the art in low-cost attitude stabilization and control for microsatellites, 3) collect space performance data on advanced electronics technologies, 4) collect and distribute imaging data, and 5) provide packet S&F communications service for the amateur radio community. The project is supported by NASA, DARPA, and industry. A launch of SEDSAT-1 took place October 24, 1998 as a secondary payload on JPL's DS-1 mission.

The box-shaped S/C has dimensions of 30 cm on each side. The structure consists of honeycomb-machined aluminum panels forming a nearly solid cube. The solar cells are in integral part of the external panels (five sides are covered with solar cells. The S/C is stabilized by an active control system using three mutually perpendicular magnetorquers as control actuators (test of new control algorithms). The OBC is of type 80186 with a 1 MByte ECC SRAM and 32 MByte DRAM solid state disk. In addition, there is a transputer-based image processor (Inmos T805) with integral frame grabber and 128 MByte of solid state memory. S/C mass = 36 kg, power = 30 W (depending on mode). SEDSAT-1 operation relies on 16 NiMH (Nickel Metal Hydride) recharge batteries (each rated at 1.5 V) during the eclipse periods of the orbit. A total of six antennas are installed on SEDSAT-1; three are fixed and three are deployable antennas. The S/C design life is three years.

Orbit: Elliptical orbit, perigee = 556 km, apogee = 1042 km, inclination = 31.5°. Note: The orbit is not optimal, but there was no better choice to obtain as a secondary payload.

Sensor/experiment complement:

SEASIS (SEDSAT Earth Atmospheric and Space Imaging System). SEASIS consists of two camera systems, PAL and Telephoto, and an image processor (Inmos T805) with corresponding subsystems (I/O devices, control logic, digitizers, buffers, mass memory, etc.). The overall objective is to use one camera for an attitude determination experiment, to collect imagery of the Earth's surface with the second camera, and to make the data available to a large user community via the Internet. The two cameras are mounted perpendicular to each other, observing in the S/C x-axis and y-axis (the z-axis is nadir pointing). Both digital camera systems are Sony products, employing a 2-D CCD array detector assembly (Si detectors), and featuring a 12-filter interferometric filter wheel between the optical train and the camera imaging chip (i.e., CCD array). Five spectral bands at 10 nm bandwidth are defined at the following center wavelengths: 440, 540, 610, 670, and 777 nm. The source data of each camera system is output in NTSC format (B&W) with a resolution of 768 x 484 pixels (data rate of 20 Mbit/s from each data stream) and stored into mass memory. Both cameras can be operated in parallel (snapshot imagery). Control is provided for camera operation including filter wheel selection. The fixed mounting of the cameras into the S/C structure permits observation (of the Earth's surface) by the mechanism of body-pointing. The S/C is turned into the desired direction.

Camera operation. Only a single spectral band at a time can be used for observation. The normal operating mode takes a snapshot in each band while moving the filter wheel (sequential collection of "multispectral" data). Snapshots at up to two hertz can be taken.

- Camera-1 is a digital video camera featuring a Panoramic Annular Lens (PAL) with a FOV of 360° and ±40° on optical equator. Objective: demonstration of wide-angle attitude determination and wide-area Earth observation. The PAL system is always pointed at the Earth's horizon, imaging a panoramic "doughnut" (a ring-sized shape) projected onto the camera. The PAL images are then used with corresponding algorithms for attitude determination.
- Telephoto camera. The instrument features a 50 mm focal length lens. With Telephoto pointed at nadir, the Earth appears centered on the equator of the PAL camera, providing attitude information of the S/C relative to Earth as well as relative to the Earth/sun system. The spatial resolution on the ground is between 100 - 300 m for the orbital altitudes of 500 - 1000 km. A corresponding image scene has the size of 100 km x 140 km. Data quantization is 12 bit.

  A further objective of the Telephoto camera is to detect lightning activity by counting the lightning strikes through a timed series of images. The filter is centered at a wavelength of 777 nm for this observation mode.

Ground truth. SEASIS observation imagery is paralleled by several **PGAMS** (Portable Ground-based Atmospheric Monitoring System) stations in the continental USA, permitting synchronous measurements of atmospheric parameters in up to 512 wavelength bands. The SEASIS filters were selected to match the PGAMS bands for some of the most interesting atmospheric absorption bands.

Mode-A Transponder. Objective: to provide analog relay capabilities to the amateur radio community. Data is received at 147 MHz and reflected at 29 MHz. In addition, an FSK transmitter is provided on this link, permitting send-only communication at 300 baud. It is intended for transmission of S/C housekeeping data.

Mode-L Transponder. Objective: to provide digital packet store-and-forward capabilities to the amateur radio community. Data transmission rates in uplink and downlink at 9.6 and at 57.6 kbit/s. The uplink frequency is in L-band at 1.2 GHz, the downlink frequency is in UHF at 430 MHz.

Technology experiments include testing of NiMH batteries and specialized electronic parts.

N.15 Sputnik-II

Sputnik-II is a nanosatellite, a commemorative functioning replica of the original Sputnik (40th anniversary), built by French high-school students from the l'Aeroclub of France (radio transmitter) and staff from the Russian Aeronautical Federation (structure). The project was funded by various sponsors in the space industry. Sputnik-II was deployed from the MIR space station during an EVA by Russian cosmonauts Anatoly Solovyov and Pavel Vinogradov on November 3, 1997. Sputnik-II is 1/3 scale of the original with a mass of 3 kg. 1853) 1854)

The initial orbit of Sputnik-II was identical to the MIR orbit at deployment, namely a 383 km x 391 km orbit inclined at 51.6°. The satellite carries a low-power (100 to 200 mW) 2 m VHF-FM beacon transmitter operating on 145.820 MHz that emits a pulsating tone with duty cycle of 5, with 1.3 kHz frequency as a function of the internal temperature of the

1853) http://www.ee.surrey.ac.uk/SSC/SSHP/nano/nano1997.html

1854) http://www.oceanes.fr/~ fr5fc/angspoutnik.html

spacecraft, similar to the original Sputnik. The antenna polarization is circular, however no satellite attitude control exists to control the polarity sense. Its lifetime in orbit is expected to be between one and two months as it employs non-rechargeable batteries. Orbital lifetime is expected to be about 18 months. The SSHP author still observed the satellite near the 20th November. The spacecraft was reported to be spinning rapidly (8-13 rpm), and finally stopped transmitting on December 29, 1997 (55 days of radio transmission). Sputnik-II decayed on May 21, 1998.

N.16 STARSHINE (Student-Tracked Atmospheric Research Satellite for Heuristic International Networking Equipment)

STARSHINE is a US cooperative program of small optically reflective spherical student satellites, designed and built by NRL/NCST (Naval Center for Space Technology). They are being deployed by NASA from Hitchhiker canisters in Space Shuttle cargo bays, as well as from an Athena unmanned launch vehicle, into highly inclined low earth orbits at a rate of once every year or so. [1855) 1856) 1857)] The principal objectives of project STARSHINE are educational and motivational. If students help "build" the spacecraft (by polishing its mirrors), they should be more excited about tracking it and using it to measure upper atmospheric density and the response of that region of the atmosphere to solar storms. - Self-contained mirror polishing kits, containing two mirror blanks each, were prepared by project volunteers and mailed to 1050 schools around the world. Nearly 1800 mirrors were polished by 25,030 students in some 700 schools in 18 countries.

N.16.1 STARSHINE-1

STARSHINE-1 is an optically reflective small passive satellite with no moving parts or electrical components. It is spherical with an outer diameter of 47.5 cm, the S/C mass is 39 kg. The shell is made of aluminum, the exterior is covered with 878 aluminum mirrors that are 25 mm in diameter. The Starshine structure consists of the following basic components: two hemispherical domes (two top dome has 499 mirrors, the bottom dome 399 mirrors), and an equatorial disk (to join the two hemispheres).

STARSHINE-1 was launched by Shuttle flight STS-96 (launch May 27, 1999 - deployed June 5, 1999), with the objective to provide educational observations for students around the world. More than 1,000 schools across America and the world have helped construct the satellite and participated in the project. Although only slightly larger than a basketball, STARSHINE-1 is covered by almost 900 highly polished mirrors that make it visible from the ground. It was released from Discovery on June 5, 1999, near the end of the STS-96 mission after the Shuttle had left ISS (International Space Station).

Orbit: The initial orbit of the STARSHINE after release was 387 km in altitude and 51.6° in inclination (decaying circular orbit), orbital period of 90 minutes.

Satellite operation: The twinkling satellite was visible to the naked eye against the star background, during certain recurring morning and evening twilight periods, to observers around the world between the latitudes of ± 60°. Student observers around the world recorded the position, of the tumbling (sunlight reflecting) satellite and provide this information to the project's website. They measured right ascension and declination at precise times by reference to known stars, and they were also recording the precise time of their observations by the use of stopwatches synchronized with international time signals. Some observers used

1855) B. Braun, C. Butkiewicz, J. Vasquez, G. Moore, "The Starshine Satellite From Concept to Delivery in Four Months," Proceedings of the 13th Annual AIAA/USU Conference on Small Satellites, Aug. 23-26, 1999, Logan UT, SSC99-I-7

1856) http://www.azinet.com/starshine/index.html

1857) Information provided by Gil Moore, Director of the STARSHINE project

GPS receivers to measure the latitude, longitude and altitude of their observing sites. They posted their observations and station locations on the STARSHINE web site to permit computation of the classical elements of the satellite's orbit by the angles-only-method of Laplace. The changes in the decaying orbit were used to calculate the density of the upper atmosphere. Student investigations were also made to the effect as to how the density of the upper atmosphere varies with solar activity.

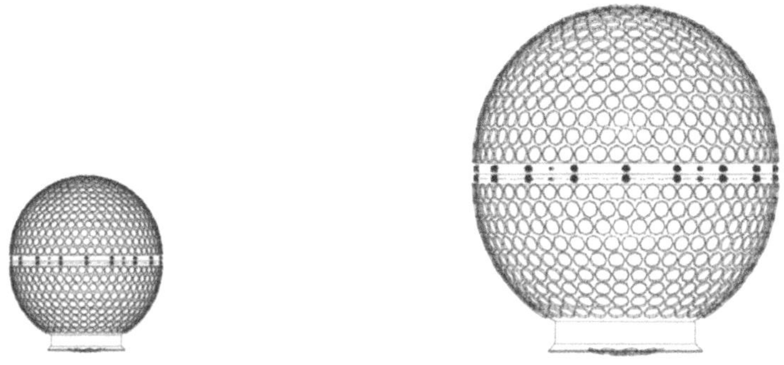

Figure 351: Illustration of STARSHINE spacecraft

It turned out that Starshine did initially not tumble as expected after deployment from HES (Hitchhiker Ejection System). The rate of visible flashes produced by the mirrors varied from once per 15 seconds to once or twice per pass. The situation improved in the latter phases of the mission as the satellite descended into the atmosphere (the denser air causes a torque on the S/C). Re-entry of STARSHINE-1 occurred on Feb. 8, 2000 when it was consumed by aerodynamic heating at an altitude of about 80 km above the Atlantic Ocean off the coast of Brazil. The project STARSHINE observers webpage is at NASA/HQ under: http://spacekids.hq.nasa.gov/starshine/

N.16.2 STARSHINE-2

As of early 2001, STARSHINE-2 is under construction to be deployed from STS-108, an Endeavour mission to ISS in October 2001. The 858 mirrors that will cover the outside surface of this satellite have already been polished by students in 26 countries. These mirrors have been coated with a scratch-resistant, anti-oxidizing layer of Silicon Dioxide by optical technicians at the Hill Air Force Base. At NRL, the mirrors are installed onto the spacecraft. The satellite has the same size and mass as STARSHINE-1. This time around, it also contains a special cold-gas spin system to rotate the satellite at 5°/s to enhance the rate at which sunlight will flash from its mirrors. In addition, the satellite carries twenty laser retroreflectors, distributed evenly across its surface, to permit tracking by the International Satellite Laser Ranging Network.

N.16.3 STARSHINE-3

The structure of STARSHINE-3 is a sphere of 94 cm in diameter and a mass of 91 kg (microsatellite class). It is covered with 1000 student-polished mirrors; in addition, twenty laser retro-reflectors are mounted on its surface. The spacecraft is built and integrated by NRL with assistance from Calhoun Community College in Decatur, Alabama, and the C. F. P. Paul Rousseau school in Drummondville, Canada. STARSHINE-3 is of higher technical sophistication featuring a nano-g drag accelerometer of SwRI (Southwest Research Institute), and an experimental integrated power supply, consisting of combined solar cells and thin film batteries provided by NASA/GRC. The power is needed for two-way RF communications (amateur standard) between the ground and the spacecraft, consisting of a telemetry transmitter, a command receiver, a rechargeable battery, a secondary solar array, signal-conditioning circuitry, and an antenna. The data is transmitted to ground receiving stations at the University of Alaska, Fairbanks, the US Naval Academy, Annapolis, MD, Santa Clara University in Santa Clara, CA, and to numerous amateur radio stations around the world. The downlink is in UHF-band.

A launch of STARSHINE-3 is targeted for Aug. 31, 2001 on an Athena-1 vehicle from the Kodiak Launch Complex on Kodiak Island, Alaska. The entire launch payload consists of four small satellites: STARSHINE-3, PICOSat, a technology demonstration microsatellite of DoD/AFRL, PCSat (Prototype Communications Satellite) a microsatellite designed and built by Midshipmen of the United States Naval Academy (USNA), Annapolis, MD, and SAPPHIRE (Stanford AudioPhonic Photographic IR Experiment) of Stanford University. The launch of all satellites represents also the inauguration of launch services from this site.

Orbit: circular orbit, altitude = 500 km, inclination = 67°. The higher orbit implies also a longer lifetime of the satellite in the order of several years.

N.17 STEDI (Student Explorer Demonstration Initiative)

The STEDI program is an USRA-managed (Universities Space Research Association) cooperative initiative (a competitive pilot program) among US universities to define, build and operate small and low-cost spacecraft and sensors as part of the civil space program. The overall STEDI objective is to give US universities an opportunity to conduct research in space - through cooperative agreements - which is related to scientific disciplines, eg., astrophysics, Earth sciences, life and biomedical sciences and applications, microgravity sciences and applications, solar system exploration, and space physics. STEDI is funded by NASA. There are three phases to the STEDI program:

1) Phase 1: Mission definition
2) Phase 2: Mission implementation
3) Phase 3: Mission operations and data analysis

In February 1995 two university teams were selected from many applicants by USRA to implement their particular proposals (phase 2); in addition a third team was selected to act as backup. The University of Colorado at Boulder is one team in this scenario with the SNOE (Student Nitric Oxide Explorer) mission; the other team is headed by Boston University, with a satellite project named TERRIERS (Tomographic Experiment using Radioactive Recombinative Ionospheric EUV). The backup team is headed by the University of New Hampshire at Durham with the CATSAT mission (Cooperative Astrophysical and Technology Satellite).

N.17.1 SNOE (Student Nitric Oxide Explorer)

The SNOE spacecraft and its instrument complement are being designed, built, and operated at the University of Colorado at Boulder [Laboratory for Atmospheric and Space Physics (LASP); PI: Ch. Barth; student involvement in all aspects of the project]. There is also collaboration with Ball Aerospace, NCAR, JPL, and Arapahoe High School of Littleton, CO. The science objectives of SNOE are to measure nitric oxide density in the terrestrial lower thermosphere (90-200 km) and to analyze the energy inputs to that region from the sun and magnetosphere that create this compound and cause its abundance to vary dramatically. The S/C is a compact hexagonal modular structure, 0.95 m high and 1 m across its widest dimension. S/C mass = 115 kg; spin rate = 5 rpm normal to the orbit plane; attitude knowledge provided by horizon crossing indicators, magnetometers and GPS; attitude control with torque rods and nutation damper; passive thermal control; power = 35 W; data rate = 6 MByte/day. The science data are made available to the science community. The mission design life is one year. The SNOE S/C was launched on a Pegasus XL vehicle (along with Teledesic LLC's T1 satellite - the world's first commercial Ka-band LEO satellite) on Feb. 26, 1998. The launch originated from VAFB, CA. [1858] [1859] [1860] [1861] [1862]

RF communications are provided in S-band. The S-band system consists of a transmitter, receiver/demodulator, hemispherical antennas, coupler, RF switch, and filter. The downlink data rate is 128 kbit/s, the modulation is PCM/PSK/PM. The uplink data rate is 2 kbit/s NRZ-M using PCM/PSK/PM modulation. The communications protocol employs the CCSDS standard. Mission operations for SNOE are being performed at LASP of the University of Colorado at Boulder with tracking and communication services provided by NASA.

Orbit: Sun-synchronous polar circular orbit, altitude = 556 km, inclination = 97.5°, equator crossing at 10:30 AM.

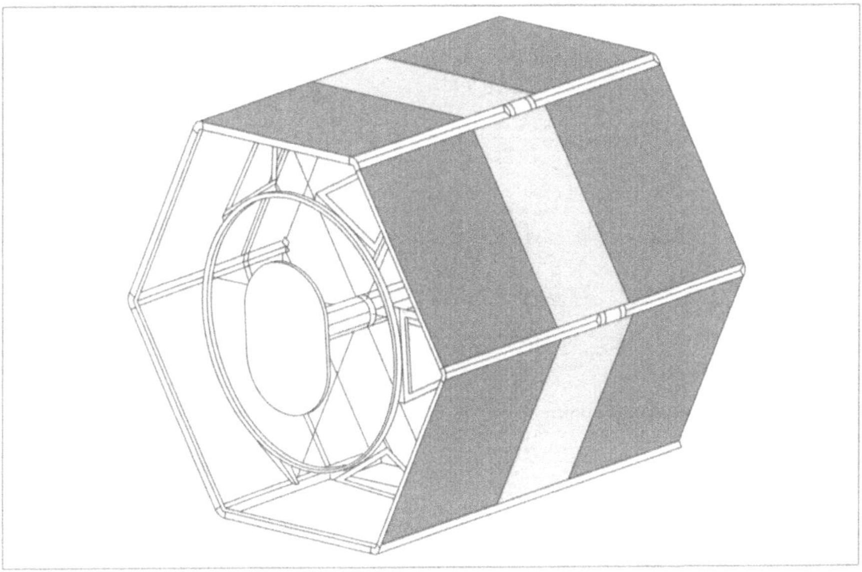

Figure 352: The SNOE S/C model

Sensor complement:

UVS (Ultraviolet Spectrometer). A limb-viewing instrument with the objective to measure the densities of nitric oxide in the altitude range of 50-200 km by observing the (1,0) and (0,1) gamma bands. UVS is of SME (Solar Mesosphere Explorer, see K.20) heritage; it con-

1858) S. C. Solomon, S. M. Bailey, Ch. A. Barth, et al., "The SNOE Spacecraft: Integration, Test, Launch, Operation, and On-orbit Performance," Proceedings of the 12th AIAA/USU Conference on Small Satellites, Logan, UT, 1998

1859) Information provided by S. C. Solomon, University of Colorado at Boulder

1860) S. C. Solomon, et al., "The Student Nitric Oxide Explorer," Proceedings of the 9th Annual AIAA/USU Conference on Small Satellites, Utah State University, Logan, Utah, 1995

1861) S. M. Bailey, et al., "Science Instrumentation for the Student Nitric Oxide Explorer," Proceedings of the 9th Annual AIAA/USU Conference on Small Satellites, Utah State University, Logan, Utah, 1995

1862) http://lasp.colorado.edu/snoe/overview.html

sists of an Ebert-Fastie spectrometer, an off-axis telescope, and two Hamamatsu PMT detectors. The spectrometer has a focal length of 125 mm and uses 3600 lines/mm mechanically ruled plane grating, which produces a dispersion of 2.15 nm/mm at the detectors. The grating in the spectrometer is set to place the (1,0) gamma band (215 nm wavelength) on one detector and the (0,1) gamma band (237 nm wavelength) on the other detector. Both channels have a sensitivity of ~ 1.1 counts/second/Rayleigh. - UVS is mounted with its optical axis perpendicular to the S/C spin axis. Its telescope images the entrance slit of the spectrometer on the limb. The image of the slit is 7 km high (resolution). The integration time is 28 ms for 2x spatial oversampling.

AP (Auroral Photometer). [1863] AP is a two-channel broadband nadir-viewing instrument with the objective to observe UV emissions from the aurora and to determine the energy deposited in the upper atmosphere by energetic auroral particles. The channels (A and B) consist of two Hamamatsu PMT detectors, a US window/filter for each channel, and a FOV limiter. FOV = 11° (full cone) for each channel. Spectral range observed = 125-180 nm for channel A and 135-180 nm for channel B. The sensitivity of channel A at 130.4 nm is ~70 counts/second/Rayleigh; the sensitivity of channel B at 135.6 nm is ~27 counts/second/Rayleigh. - The AP instrument is mounted with its optical axis perpendicular to the S/C spin axis. The integration time for each channel is 187 ms, providing 32 samples per channel per spin.

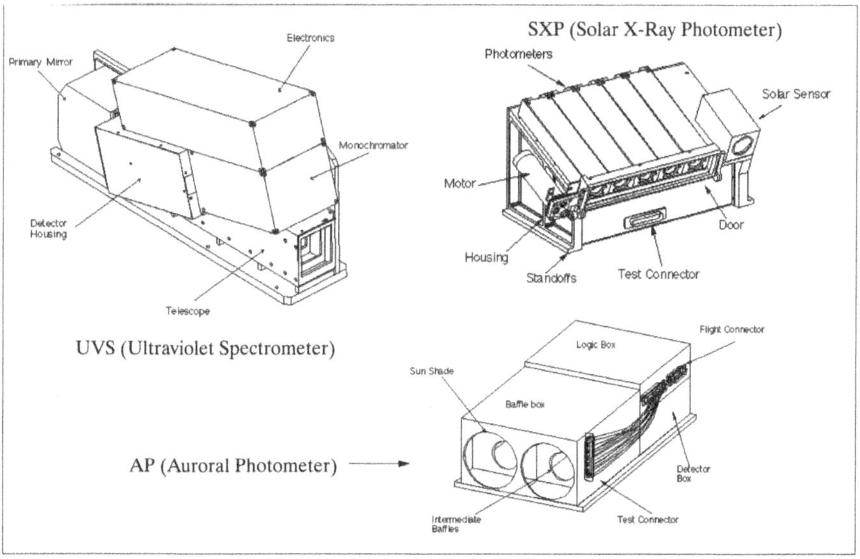

Figure 353: Illustration of the SNOE instruments

SXP (Solar X-ray Photometer). [1864] A sun-viewing instrument with the objective to measure the solar soft X-ray flux within a FOV of ±35° of the local zenith direction in the spectral range of 2-31 nm. Each of the five photometer channels contains a silicon photodiode detector; the wavelength selection is accomplished by thin metallic films deposited directly onto the diode surface. The integration time is 62.5 ns. A small two-axis sun sensor is coaligned with SXP to measure the solar incidence angle for the instrument.

1863) S. C. Solomon, Ch. Barth, S. M. Bailey, "Auroral production of nitric oxide measured by the SNOE satellite," Geophysical Research Letters, Vol. 26, No 9, May 1, 1999, pp. 1259-1262

1864) S. M. Bailey, T. N. Woods, Ch. A. Barth, S. C. Solomon, "Measurements of the solar soft x-ray irradiance from the Student Nitric Oxide Explorer," Geophysical Research Letters, Vol. 26, No 9, May 1, 1999, pp. 1255-1258

BGSR (microGPS - Bit Grabber Space Receiver). The receiver is provided by JPL for orbit determination of SNOE (cooperation with the University of Colorado at Boulder). Instrument size: 6.3 cm x 11.5 cm x 5 cm, mass =0.68 kg, power = 2.1 W (while operating, but 0.02 W on orbit average). The GPS orbit is determined about three times per orbit for short time intervals, otherwise the receiver is in hibernation. The information is used for post-factum ground orbit determination.

N.17.2 TERRIERS

The TERRIERS (Tomographic Experiment using Radioactive Recombinative Ionospheric EUV) satellite project is a collaboration between Boston University (Center for Space Physics), Aero Astro of Herndon, VA, University of Illinois at Urbana-Champain (UIUC), NRL, MIT's Haystack Observatory, USAF Phillips Laboratory (Hanscom Field, Bedford, MA), and Cleveland Heights High School. The primary goal of the mission is to obtain individual 2-D (altitude-latitude) profiles of ionospheric electron density using EUV (Extreme Ultraviolet), visible and radio instruments. A further step in the analysis is to use these individual cuts (similar to CAT scan projections) to construct eventually a global 3-D image of the ionosphere. The overall approach permits detailed studies of solar and geomagnetic storms for a better understanding of the ionospheric structure and its relationship to external forces. D. Cotton of Boston University is the overall PI of the TERRIERS program.

TERRIERS is a spin-stabilized S/C with a spin rate of 10 rpm who's design is based on HETE. Satellite mass = 122 kg, total payload mass = 49 kg, S/C power = 21 W (average), design life = 1 year. The S/C was launched May 18, 1999 by a Pegasus XL launch vehicle off the coast of California (the aircraft started from VAFB, CA). Another satellite on the same launcher was MUBLCOM (Multiple Path Beyond Line of Sight Communications), a demonstration satellite for DARPA of DoD. [The objective of MUBLCOM is to demonstrate a unique new capability to provide space-based digital voice and data communications to combat forces or commercial users that were previously considered out of range of standard radio communications systems.] [1865)]

Orbit: Sun-synchronous circular polar orbit, altitude = 550 km, inclination = 97.5°, ascending node at 9 AM, period of about 96 minutes.

Mission and payload operations for TERRIERS are being performed by students at the Payload Operations Center of Boston University. The S/C is serviced by one ground station; all communications are in S-band.

Status: After orbit injection the TERRIERS S/C failed to orientate itself to allow the solar panels to work fully and its battery lost power after launch. A review board found that a wiring error caused the spacecraft's attitude control to malfunction and its solar panels to point away from the sun.

Sensor complement:

TESS (Tomographic EUV Spectrograph System). TESS is the prime instrument package of the TERRIERS payload. [1866)] The objective is to image the strength of radiative EUV emission profiles in the ionosphere with good spatial (50 km) and spectral resolutions (20 Å). The overall system consists of five spectrographs, four identical nightglow instruments (for redundancy and added sensitivity), and one spectrograph for daytime operation, featuring a smaller aperture slit to reduce sensitivity and to increase spectral resolution (10 Å). Each spectrograph is referred to as a SEIS (Single Element Imaging Spectrograph). The instrument features a diffractive optical design combining the spectral qualities of a Rowland-mount spectrograph with the imaging capabilities of a Wadsworth-mount spectro-

1865) Information provided by S. Chakrabarti of Boston University, Boston, MA

1866) D. M. Cotton, et al., "A single-element imaging spectrograph," Applied Optics, Vol. 33, 1994, p. 1958 graph with a unique toroidal diffraction grating. The detectors employed are MCP (Micro Channel Plate).

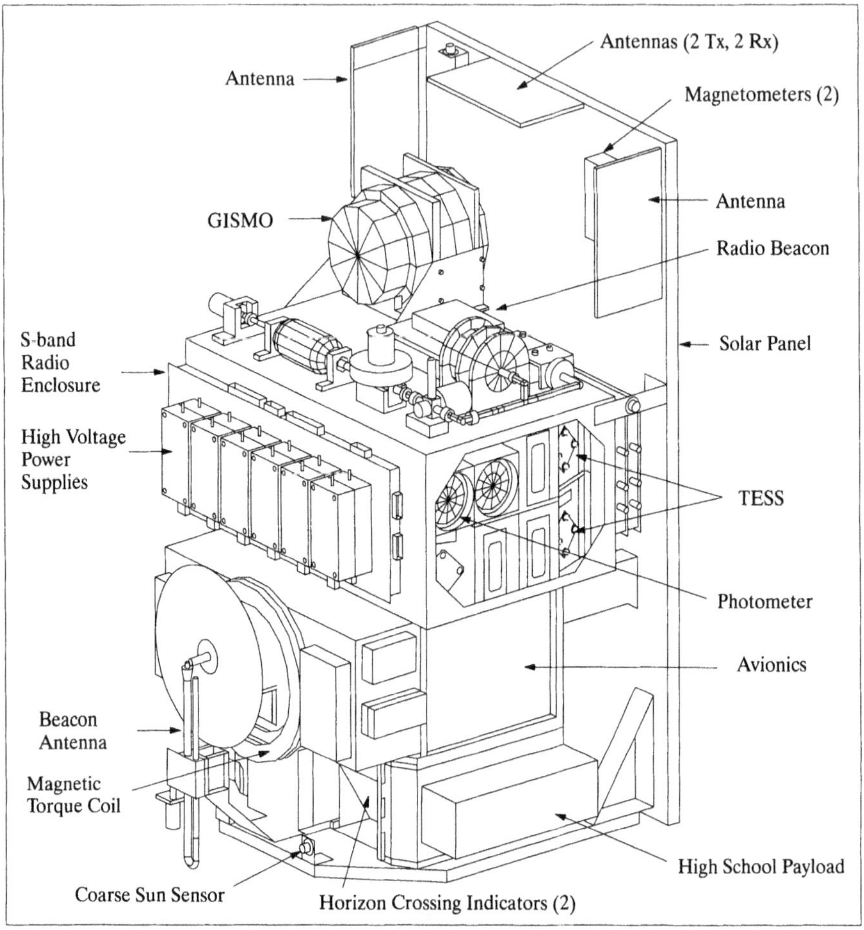

Figure 354: The TERRIERS S/C model

Parameter	Night (day) value
Number of spectrographs	4 (1) SEIS
Spectral range, spectral resolution	800 - 1400 Å, 10 - 20 Å
Slit size	0.5 mm (0.1 mm) x 40 mm (10 mm)
Grating surface figure line density blank size ruled area radii blaze wavelength reflective coating	Toroidal 3200/mm 30 mm x 75 mm 25 mm x 65 mm 150 mm x 290.4 mm 900 Å Si C
Detector type photocathode format	MCP with wedge strip KBr 0.20 mm x 0.20 mm (256 x 256 pixels)
FOV	10° x 10° with 0.5° spatial resolution across the limb

Data volume	~340 Mbit/day
On-board data storage	20 MByte in memory storage with error correction (2 systems)

Table 524: TESS parameter specification

GISSMO (Gas Ionization Solar Spectral Monitor).[1867] The instrument is an optics-free solar EUV spectrometer with the objective to measure solar EUV emissions in the spectral range from 7 to 40 nm. The instrument design features a gas cell in which solar EUV photons ionize neon, producing an ion and a photoelectron of energy equal to the ionizing photon's energy minus the ionizing potential of neon. The subsequent measurement of the photoelectron's energy, coupled with a knowledge of the neon's cross section for photon absorption, yields information about the original photon's energy and flux. - The spectral resolving power ($\lambda/\Delta\lambda$) of the GISSMO spectrometer varies from 25 to 50, depending on photon energy. The instrument sensitivity remains constant over large (solar cycle) time scales - normally a problem for conventional instruments.

BEACON. BEACON is an on-board dual-frequency transmitter (150/400 MHz frequency), developed by NRL, with the objective to measure electron density profiles through the upper atmosphere in conjunction with a ground-based network of radio beacon receivers. This ground network is provided and operated by NRL, UIUC and by Phillips Laboratory with the intent to observe electron densities from line-of-sight TEC (Total Electron Content) measurements.

Photometers. Two lightweight photometers, each with a FOV of 0.4° in diameter, will be flown aboard TERRIERS. One photometer will measure atomic oxygen emissions at 630 nm (to obtain photoemission profiles); the second instrument is intended to measure the off-line signal to obtain an estimate of background radiation. Both instruments use interference filters and integrated photon detectors consisting of channeltron detectors and associated electronics.

N.17.3 CATSAT (Cooperative Astrophysical and Technology Satellite)

CATSAT is a small satellite mission designed and built by the University of New Hampshire (UNH, Durham, NH) with support from the University of Leicester (Leicester, UK), sponsored by USRA (Universities Space Research Association and NASA in the STEDI (Student Explorer Demonstration Initiative) program. Its scientific mission is to study the nature and distribution of cosmic gamma-ray bursts by observations in the soft X-ray band down to energies of 500 eV. Its technology mission will be to demonstrate that an all university team can construct, launch, and operate a mission of this extent. [1868] [1869] [1870]

The S/C is three-axis stabilized with an orbital zenith and solar reference position accuracy of ±5°. CATSAT is a minisatellite with a boxlike structure of dimensions: 100 cm x 100 cm x 103 cm, S/C mass = 168 kg, power = 90 W. The attitude system employs a biased momentum system; a redundant set of reaction wheels with magnetic torque coils for momentum dumping and backup (the torque coils are used to offload the reaction wheel de-spin to prevent saturation, hence: 'momentum dumping'). Attitude measurement is provided by two three-axis magnetometers, two coarse and two fine sun sensors and two horizon sensors, providing ±5° control and ±1° of attitude knowledge. TT&C communication uses S-band in the uplink and downlink using CCSDS protocols. The uplink is PM modulated at 4 kbit/s, the downlink uses BPSK modulation at 16 kbit/s or 1 Mbit/s. The science data volume is about 50 MByte/day and is downlinked about twice per day. Planned design life is one year

1867) J. S. Vickers, et al., "Gas ionization solar spectral monitor (GISSMO)," Optical Engineering, Vol. 32, 1993, p. 3126

1868) Information provided by D. Forrest of the University of New Hampshire at Durham

1869) C. Wood, D. Forrest B. McKinnon, D. Nelson, "CATSAT Structural Design," Proceedings of the AIAA/USU Conference on Small Satellites, Sept. 16-19, 1996, Logon, UT

1870) http://www.catsat.sr.unh.edu/mission/index.html

with a goal of two years of operation. A launch is planned for Dec. 2001 on a Delta-II 7310 from VAFB in a dual payload configuration with NASA's ICESat mission.

Orbit: Sun-synchronous dawn/dusk orbit, altitude = 590 km, inclination = 94°, equator crossing at 6 AM on the ascending node.

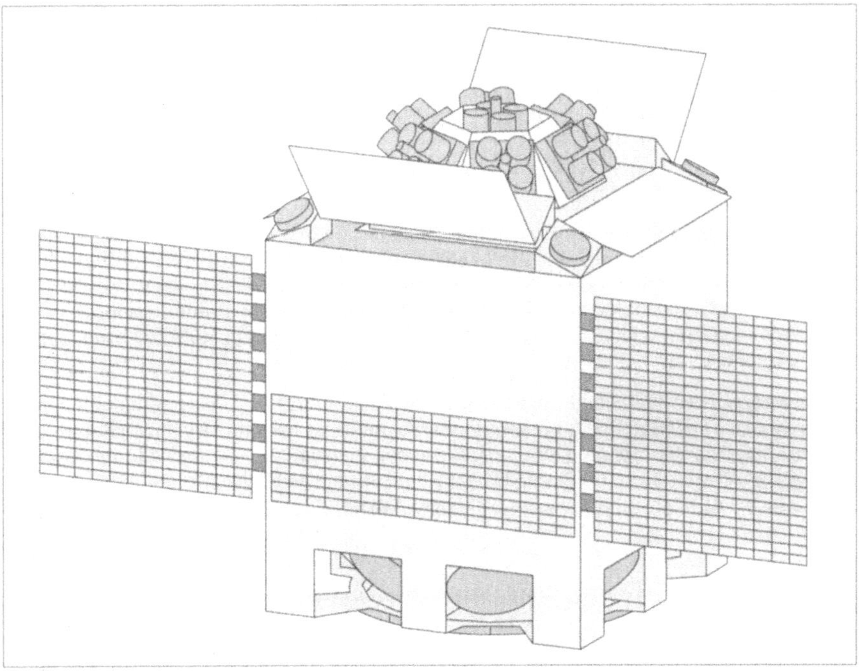

Figure 355: The CATSAT S/C model

Sensor Complement:

SXR = Soft X-Ray spectrometer. The objective is to measure burst spectral information in the energy range from 500 eV to about 15 keV. The instrument uses APD (Avalanche Photodiode) detectors, material = Si, size = 1.7 cm^2, with a 300 eV FWHM energy resolution near 1 keV. SXR consists of seven modules each containing 16 APDs and collimated to a 1 steradian FOV (the detectors are arranged in four groups of four on each module which are jointly collimated to yield a FOV of about 1 steradian. The pointing directions for the modules are offset so that together they view nearly 2pi steradians). The instrument detectors are radiatively cooled to -40°C; the energy loss information is recorded in seven separate spectra at a resolution of one second.

The use of APDs for direct soft x-ray detection is a new application for the devices, which are usually used to detect optical emission, and represents a key enabling technology for low-cost missions. APDs are constructed from high purity n-type silicon wafers that are doped to form a p-n junction. A very high reverse bias (about 1.5 kV) is applied across the diode so that charge carriers liberated by x-ray interactions in the silicon migrate into the depletion region, and are rapidly accelerated, generating a cascade of electron-hole pairs as they collide with silicon atoms.

Parameter	Value	Parameter	Value
Detector type	Silicon (ADP)	Energy range	0.5 to 20 keV
Operating temperature	-30° C	Energy resolution	E/D E > 1 at 500 eV
Quantum efficiency	> 80% at 2 keV	Active area	> 190 cm^2
Maximum count rate	10^4 cts s^{-1}	Limiting burst fluence	about 10^{-6} ergs cm^{-2} s^{-1}
Angular resolution	23.6°	Total FOV	about 2 pi steradians

Table 525: Parameters of the SXR instrument

Parameter	Value	Parameter	Value
Detector type	4 cm f x 1 cm CaF2(Eu) + 3" f PMT	Operating temperature	0° C
Energy range	15 to 380 keV	Energy resolution	E/D E > 3 at 30 keV
Quantum efficiency	> 80% at 2 keV	Active area	80 cm^2
Maximum count rate	10^5 cts s^{-1}	Limiting burst fluence	about TBD ergs cm^{-2} s^{-1}

Table 526: Parameters of the HXR instrument

HXR = Hard X-Ray spectrometer, built by UNH. The objective is to measure burst spectral distribution in the energy range from 15 to 380 keV. The instrument uses four independent CaF2 scintillation detectors of size: 7.6 cm in diameter x 1 cm, viewed by PMTs (of 7.6 cm diameter) with 20% FWHM resolution at 60 keV. The four detectors are located on the corners of the SXR assembly and are canted at 45° so that altogether they observe 2pi steradians of sky centered on the zenith direction. The energy loss information is recorded in four separate spectra at a one second time resolution.

DGS = Directional Gamma-ray Spectrometer. The objective is to measure the burst spectral information in the 250 keV to 5 MeV energy range. The sensors are NaI scintillators 7.6 cm x 7.6 cm in size, viewed by PMT detectors with 7% FWHM energy resolution at 662 keV. The instrument consists of three closely packed detectors located near the mass center of the spacecraft. Burst directional information is obtained by self shielding (i.e., the detector response function makes it possible to calculate the direction of the incident radiation with the help of a shadow technique applied to the particular geometric arrangement of the three detectors). The energy loss information is recorded in four separate (three singles and one multiple) spectra at sampling rates of one second.

There are two ground stations for CATSAT operations, one at UNH and one at the University of Leicester, UK.

N.18 SUNSAT (Stellenbosch University Satellite)

SUNSAT-1 is an advanced microsatellite program (start in 1992), developed by graduate students of the Department of Electrical and Electronic Engineering at Stellenbosch University, Stellenbosch, South Africa (Stellenbosch University is located about 50 km east of Cape Town).[1871) 1872)] The overall objective is to demonstrate proven and new technologies in the fields of remote sensing, spacecraft control, the solid Earth sciences, and to provide services (mail box, speech and data relay experiments) to the AMSAT community. The S/C structure features a tray-based design of proven flexibility. The trays are sandwiched (stacked) between base and top plates of the S/C structure, containing all subsystems and payload instruments. The dimensions of the box-like S/C are: 45 cm x 45 cm x 62 cm; S/C mass = 64 kg; power = 30 W from body-mounted solar panels and rechargeable NiCd batteries; design life = 5 years; two on-board computers and an attitude processor.

1871) G. W. Milne, A. Schoonwinkel, et al., "SUNSAT - Launch and first Six Month's Orbital Performance," Proceedings of the 13th Annual AIAA/USU Conference on Small Satellites, Aug. 23-26, 1999, Logan Utah, SSC99-1-4

1872) A. Schoonwinkel, G. W. Milne, et al., "Pre-Flight Performance of SUNSAT, South Africa's First Remote Sensing and Packet Communications Microsatellite," Proceedings of the 10th Annual AIAA/USU Conference on Small Satellites, Sept. 16-19, 1996

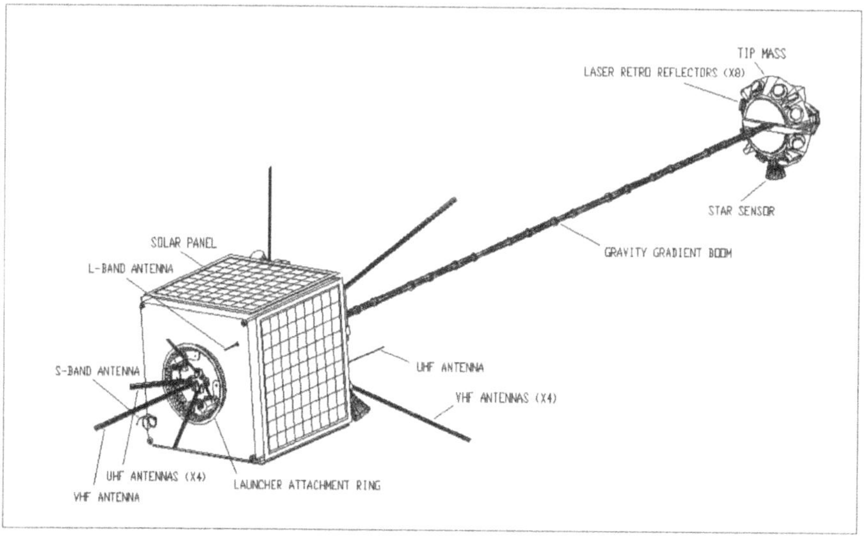

Figure 356: SUNSAT-1 S/C model in operational configuration

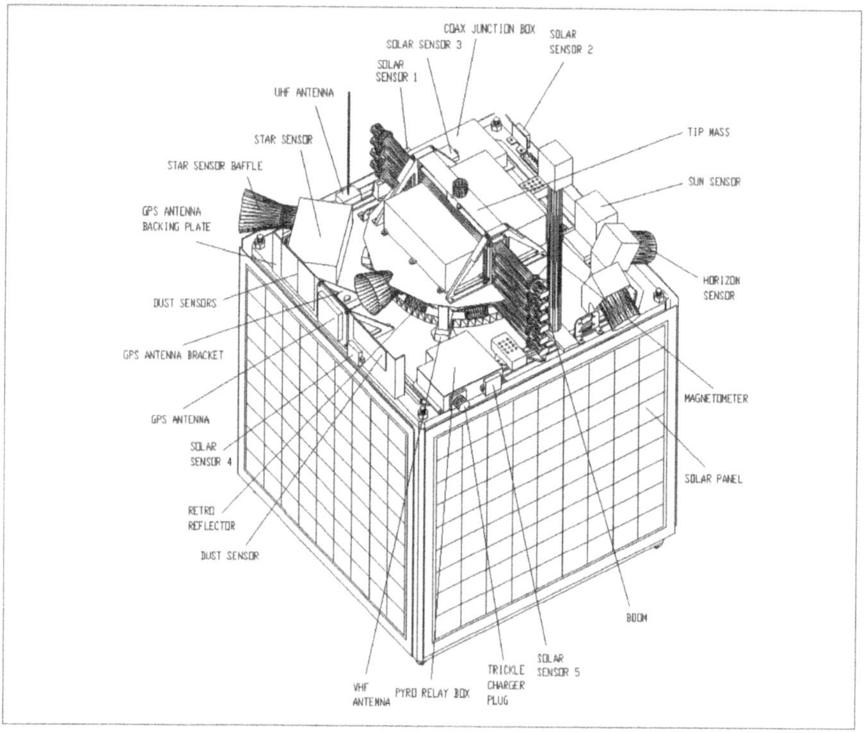

Figure 357: SUNSAT-1 instrumentation on zenith-pointing face

SUNSAT is stabilized with a gravity-gradient boom (2.2 m boom with a 4.2 kg tip mass consisting of 8 laser retro-reflectors, a star sensor and a magnetometer). The S/C is pointing toward nadir. Attitude actuators are: 3-axis reaction wheels (to provide fine-pointing during imaging), 3-axis magnetic torquer coils (continuous control capability). Attitude sensors are: 3-axis fluxgate magnetometer, 2-axis horizon sensor (CCD array,to obtain pitch and roll to an accuracy of 0.02°), a fine slit sun sensor (CCD array, to obtain yaw information to an accuracy of <0.1°), a second star sensor to measure a star positions with accuracies of <12 arcseconds. A small S/C spin is provided during normal operations; however, not during imaging periods.

SUNSAT's communications employ the store-and-forward concept and the PACSAT protocol suite to provide the services for amateur radio communications (VHF, UHF, L- and S-bands), data downlinking (VHF, UHF, S-band), data collection (VHF, UHF), and spacecraft control. Different frequencies are at times required for different services. The power amplifiers have 5 W output at VHF and S-band and 10 W output at UHF. The S-band downlink permits QPSK modulation.

SUNSAT-1 was launched on a Delta-II vehicle from VAFB, CA on Feb. 23, 1999 as a secondary payload along with the Ørsted satellite of Denmark. The prime satellite was ARGOS of DoD. SUNSAT has been allocated the Amateur designation of SO-35 (SUNSAT Oscar 35) and is supporting Amateur radio FM repeater operations in South Africa and the USA.

Orbit: Sun-synchronous orbit, altitude = 857 km x 655 km, inclination = 96.4°.

Sensor complement:

HRI (High Resolution Imager). A three-color CCD imager, developed jointly by Stellenbosch University, CSIR (Council for Scientific Research, South Africa), and KAIST (Korean Advanced Institute of Science and Technology). The imager consists of a single 10 cm diameter optical tube assembly, comprising a lens system, penta-prism with dichroic color splitter, three vertically mounted linear CCD detectors (TC104 of Texas Instruments with 3456 pixels/line) and their clock drivers and output buffers. The optical tube is mounted diagonally across the bottom of the S/C on bearings, and can be rotated by a stepper motor. Stereo images are taken with the optical tube horizontal and normal to the velocity vector (forward or backwards pitching of up to 24° to obtain various stereo base/height ratios). Images to the left or right of the ground track can be taken by placing the optical tube parallel to the velocity vector. A 64 MByte RAM enables an image to be stored for subsequent downlinking. The spectral band selection is identical to that of TM of the Landsat missions (LS-2, -3, and -4). A ground pointing accuracy of the imagery within 1 km is the goal.

Imagery from a PAL TV camera is transmitted in real-time via S-band to a ground station at Stellenbosch. The HRI downlink is via the VHF and UHF transmitters, because of a malfunction of the high-speed modem, HRI transmission could not occur in S-band, as planned originally.
The reaction wheels of the attitude control system provide 3-axis stabilization during imaging periods. Fast target pointing (body pointing of the S/C) is accomplished with an axis-rotation algorithm.

Parameter	Value	Parameter	Value
Spectral band 1	0.52 - 0.61 μm	Lens focal length	570 mm
Spectral band 2	0.61 - 0.70 μm	Aperture diameter	100 mm
Spectral band 3	0.76 - 0.87 μm	Spatial resolution	15 m (800 km orbit)
Linear CCD detectors	3 x TC104 (one/band)	Swath width	51 km (800 km orbit)
Nr. of pixels/detector	3456 (active), 3490 total	FOV	3.715°
Sensor pixel size	10.7 μm	Line scan rate	400 lines/s
Total data rate	70.5 Mbit/s	Quantization	8 bit

Table 527: Summary of Imager parameters

TRSR (TurboRogue Space Receiver). A NASA/JPL-provided advanced GPS receiver (TurboRogue, built by Allen Osborne Associates) to perform experiments in atmospheric, ionospheric, and gravity mapping. The receiver is designed to overcome many of the anti-spoofing difficulties experienced by a GPS receiver on TOPEX/Poseidon. The receiver performs phase measurements of both GPS frequencies L_1 and L_2, and derives pseudoranges from these measurements. The dual-frequency occultation technique employed measures retarded signals (atmospheric propagation delays) which permit the derivation of atmospheric profiles of density, pressure, and temperature (or moisture) in an altitude range from 85 km to the ground. The co-location of the GPS receiver and the satellite laser retroreflector permit a check of the two reference frame systems for GPS and SLR geodesy.

- Objectives for atmospheric mapping: to improve occultation measurements of the troposphere to measure temperature (resolution of 1°C) and water vapor. Continuous GPS operations in occultation mode provides over 300 globally distributed measurements per day.
- Ionospheric objectives: continuous monitoring of the ionospheric structure. TurboRogue measures the arrival times of the GPS signals at two frequencies. Hence, the electron content of the ionosphere is measured in terms of the differential arrival times of these two signals.
- Geodesy and gravity objectives: the GPS and SLR systems of SUNSAT are being used for precise orbit determination. There are also plans to monitor the long wavelength gravity field and its variation in time.

MIS (Meteoroid Impact Sensor) of NASA/LaRC within the SEE program. The objective is to monitor the meteoroid and small mass man-made debris population. The experiment, which utilizes MOS detectors, is similar in concept and design to OMDC (Orbiting Meteoroid & Debris Counting) experiment flown on the Clementine spacecraft. When space debris or meteoroids particles impact the sensor, then the sensor location and time of impact is stored and down linked for data analysis. From this information, space-debris flux and directionality can be assessed.

Magnetometer. The magnetometer (provided by Hermanus Observatory, South Africa), collocated with the star camera in the tip mass of the gravity boom, provides a complementary measurement of the magnetic field to the Ørsted geomagnetic satellite.

Laser retroreflector. The passive laser retroreflectors (boom and S/C-mounted, provided by NASA/GSFC) offer SLR (Satellite Laser Ranging) measurements from many ground stations. The design features eight encapsulated retroreflectors inserted in a ring configuration azimuthally about the tip mass.

N.19 SURFSAT (Summer Undergraduate Research Fellowship Satellite)

SURFSAT-1 is a student summer project at the California Institute of Technology and 14 other universities in cooperation with JPL (NASA funding). The project began in 1987 with six students, each summer a new group of students took over, through 1994 a total of 61 students participated. The objective was to design, build, launch and operate a low-cost, low-power microsatellite for the purpose of testing the performance of space communication in Ka-band at frequencies of approximately 32 GHz. A further objective is to test NASA's VLBI ground stations (the stations communicate with two VLBI spacecraft in orbit: Japan's VSOP and Russia's RadioAstron). Surfsat-1 was launched November 4, 1995 on a Delta II vehicle as a secondary payload on the Radarsat satellite from Vandenberg AFB, CA.[1873)]

The microsatellite consists of two aluminum devices, each the size of a box of 30 cm edge length, which remain attached to the 2nd stage of the launch vehicle as it orbits the Earth.

1873) http://sgra.jpl.nasa.gov/html_surfsat/SURFSATHomePage.html

The Surfsat-1/Delta 2 system is gravity gradient stabilized. Power is provided by body-mounted solar panels providing about 15 W orbit average power. The boxes are electrically connected by an interconnecting wire harness. Surfsat-1 contains low power radio beacons and transponders, radiating over hemispherical antennas in three microwave bands, for testing new NASA ground-tracking facilities. Three omni antennas (two on the primary box, one on the secondary box) are used for uplink and downlink communication. Three independent command detector units are used to control the operational state of the S/C. The satellite uses passive thermal control. The S/C has no propulsion system, batteries or attitude control/determination.

Orbit: Near circular polar orbit, perigee = 1000 km, apogee = 1400 km, inclination = 98.6°

Payload complement: Solar-powered mW-level microwave beacons and transponders with hemispherical coverage. Each "box" carries a transponder. The primary transponder operates at X-band and Ka-band. The secondary transponder operates at Ku-band. Both transponders radiate milliwatt (mW) level signals through the omni antennas.

- Coherent beacons for deep space communication tests: 10 mW at 8.4 GHz (X-band), 1 mW at 32 GHz (Ka-band)
- 10 mW output coherent microwave transponders (7.2 GHz up and 8.4 GHz down) for space VLBI (Very Long Baseline Interferometry) for RadioAstron (S/C of Russia)
- 20 mW output coherent microwave transponders (15.3 GHz up and 14.1 GHz down) for space VLBI tests for VSOP
- Three independent X-band command detector units for determining S/C states: On/Off, Acquisition/Normal, Ka-band/Ku-band
- Three (X-band, Ku-band, and Ka-band) short corner-mounted stub antennas with hemispherical coverage.

N.20 TechSat/Gurwin-II

TechSat/Gurwin-II [1874] [1875] [1876] is a microsatellite, built by the Haifa-based Technion (Israel Institute of Technology) with industrial and government support (the S/C is also referred to as TechSat-1B). The Earth-pointing satellite was launched on July 10, 1998. A Russian Zenit-2 vehicle carried the Resurs-O1-4 satellite and five piggyback payloads (IRIS-1, TMSAT, TechSat/Gurwin-II, FASat-Bravo, and SAFIR-2) from the Baikonur Cosmodrome into orbit. [1877]

Status: The spacecraft is operational and functioning as of June 2001. All its subsystems were tested under various operational conditions. Of the payload complement, SOREQ and SUPEX offered non-stop operations; two instruments (ERIP and XDEX) were periodically turned off; OM-2 failed after 10 months of operations; while SLRRE, the laser retro-reflector, was only used in the initial phase of the mission.

Background: TechSat is a microsatellite family of Asher Space Research Institute (ASRI) of Technion. As a university, Technion is involved in the development of space-qualified systems based on advanced and innovative technologies. TechSat is an academic program at Technion, it is also supported by Israel industry. The TechSat development started in 1993. TechSat-1a, a technology demonstration microsatellite (50 kg), was launched March 23, 1995 on a Russian Start launcher (a newly converted Russian intercontinental ballistic missile) and experienced a launch failure.

1874) Note: The microsatellite is named in honor of Joseph and Rosalind Gurwin whose long-term support for space research at Technion enabled the TechSat mission.

1875) Information provided by R. Waller of Technion.

1876) http://www.technion.ac.il/shell/Research/Space–Institute.html

1877) M. Guelman, F. Ortenberg, A. Shiryaev, R. Waller, "Microsatellites for Science and Technology: Gurwin-TechSat in-flight Experiments Results," Proceedings of the 3rd International Symposium of IAA, Berlin, April 2-6, 2001, pp. 67-70

The TechSat/Gurwin-II satellite is of cube shape with a size of 445 mm x 445 mm x 445 mm. It is three-axis stabilized, it has a momentum wheel and three magnetorquers as actuators, a three-axis magnetometer as attitude sensors; all attitude instruments have a total power consumption of about 3 W. The power consumption for all housekeeping functions is less than 10 W (including transmitters, receivers, on-board computer, and power conditioning. Initial attitude stabilization (after injection) used “Coarse Cruise” algorithms that utilize magnetometer data only with pointing accuracies within 5°. This was followed by the ”Fine Cruise” phase. The solar cells are mounted on four sides of the six outer aluminum panels (a NiCd battery is provided for eclipse operations). The fifth panel, pointing toward Earth, includes antennas, the retroreflector, the UV spectro radiometer (OM-2) and a camera. The sixth panel is opposite the Earth-pointing panel 5; it is not exposed to the sun and therefore does not include solar cells. The structure plays a major role in the thermal design. The heat flows from the solar illuminated panels to all parts of the structure that are used as radiators. The S/C mass is 48 kg, the total payload mass is 6.6 kg, power = 20 W. The S/C design life is one year.

Satellite operations are conducted from a ground station at Technion. The S/C features a digital store and forward multi-user system, compatible with existing store and forward facilities already in use on microsatellites (use by the international amateur radio electronic community). Communication is realized via receive and transmit antennas. Three uplinks in the 145 MHz VHF band (2 m), three uplinks in the 1270 MHz L-band (23 cm), and one downlink in the 435 MHz UHF-band (70cm). Data is transmitted at two available rates: 1200 bit/s and 9600 bit/s. At 1200 bit/s the carrier modulation is BPSK (downlink) and FM (uplink). At 9600 bit/s the carrier modulation is FM (downlink & uplink). [1878] [1879]

Orbit: Sun-synchronous circular orbit, altitude = 820 km, inclination = 98.8°, period = 101 min, local time of equator crossing is at 10 AM in descending node.

Sensor complement:

OM-2 (Ozone Meter-2). Objective: Measurement of the ozone concentration in the Earth's atmosphere (vertical distribution of ozone and the total ozone amount in the nadir direction). Study of latitudinal, seasonal and planetary-scale ozone variability. OM-2 is a UV spectroradiometer with a total mass of 1.80 kg (optical head of 1.55 kg and the microcontroller of 0.25 kg), “a tiny SBUV instrument,” measuring in the spectral range of 252 - 340 nm. The instrument uses a filter-wheel photometer that measures the SBUV (Solar Backscattered UV) radiance.

Spectral region	252 - 340 nm
Number of spectral bands	7 fixed wavelengths with 1.0 nm bandwidth located at: 252.0, 273.5, 283.0, 292.2, 301.9, 320.0, 340.0 nm
Measurement height of atmosphere	0 - 55 km
Swath width	170 km
Spatial resolution (IFOV), vertical resolution	170 km x 70 km, 5 km
Precision of ozone density profile determination	10-15%
Instrument mass, power	1.80 kg, 3 W

Table 528: OM-2 parameter definition

The optical sensor head consists of the following subsystems: a single lens objective, a filter wheel, a set of apertures, a baffle, and a photomultiplier detector (Rb_2Te, 26 mm in diameter). The mounted objective has an aperture of 10 cm in diameter and an effective focal length of 80 mm. OM-2 conducts sequential measurements of SBUV radiation in seven wavelengths (each of 1 nm width), the sampling time is 30 ms. A total measurement se-

1878) http://techsat.internet–zahav.net/

1879) http://www.technion.ac.il/pub/projects/techsat/asher/asri.html

quence lasts 5 s. The total footprint size is 70 km (along-track) x 170 km (cross-track), the corresponding FOV is 3° x 12°. The data volume of one day of contiguous measurements is about 50 kByte after data compression. [1880] [1881]

Spectral range	0.5 - 0.8 µm
Spatial resolution	52 m along-track x 60 m cross-track
Image size	25 km (along-track) x 31 km (cross-track)
Size of CCD detector array	12 mm
SNR	50 dB
Light sensitivity	0.5 lux
Exposure	1/60 to 1/1000 seconds
Instrument mass, power, data rate	1.0 kg, 4.5 W

Table 529: Instrument parameters of ERIP

ERIP (Earth Remote-Sensing Imaging Package). Objective: Collection of snapshot panchromatic imagery in the vicinity of ground stations. The instrument consists of a CCD video camera unit (VCU) and an Image Processing and Control card (IPC). ERIP uses a Nikkon objective, f= 135 mm, and a PUL NiX CCD TM-720. On command, a video image is captured, digitized, compressed and stored in an image buffer for later transmission. The compressed image data is transferred to the OBC (On Board Computer) and transmitted to the ground station, where the captured image is decompressed and displayed. Each image contains about 250 kByte of data, or about 60 kByte after compression.

SOREQ (Single Event Monitor for Detecting Protons and Heavy Ions in Space). Objective: measurement of the solar charged particle environment (protons and heavy particles) for a better understanding of the changing radiation environment (interest in the hazardous nature of the environment to long-term effects on S/C electronic systems). SOREQ measures SEU (Single Event Upset) and SEL (Single Event Latch-up) occurrences in six HM65162 (2k x 8 SRAM) devices. [Note: single-event latch-ups manifest themselves in a sudden increase of power consumption.] The devices are arranged in two groups, allowing for the examination of chips from different date codes having different sensitivities. The TechSat computer initializes, twice per minute, a read-write circle (about 1 ms in duration) and serially reads the shift register. This enables mapping of events with a resolution of about 2°. The instrument mass is 0.23 kg, its power consumption is < 30 micro W. [1882]

SUPEX (Superconductivity Experiment). [1883] [1884] Objective: Conduction of a series of operational tests of the instrument in flight - consideration for later use of power generation. The superconducting device is based on thin film technology (developed at the Physics Department of Technion) made of $Y_1Ba_2Cu_3O_7$. It uses superconducting filters to separate the channels. The experimental assembly on-board TechSat comprises the HTS (High Temperature Superconductor) device, a cryocooler and electronic instrumentation. An automatic electronic technique is used to measure the transition temperature and critical current in the superconducting state. Cyclic measurements (about 15 minutes) are conducted once a week. The device is mounted into an insulating housing, and is thermally attached to a K-508 miniature cryocooler of Ricor, Ltd., Kibbutz Ein-Harod Ihud, Israel. The overall

1880) A. Devir, F. Ortenberg, “”Space-based small ultraviolet photometer for the measurement of the ozone concentration in the Earth's atmosphere,” Proceedings of SPIE, Vol. 3110, 1997, pp. 161-170

1881) M. Guelman, F. Ortenberg, B. Wolfson, “Flight Tests of the novel TechSat Satellite Ozone Meter: Algorithms and Measurement Processing Results,” Proceedings of the 40th Israel Annual Conference of Aerospace Sciences, 2000, pp. 299-310

1882) J. Barak, E. Adler, M. Murat, et al., “ The SOREQ Radiation Monitor for Detecting Protons and Heavy Ions in Space and its Preliminary Flights Data on Gurwin II TechSat,” Proceedings of the 14th AMSAT-UK Colloquium Space-Communication-99, University of Surrey, July 23-25, 1999, pp. 2-9

1883) E. Polturak, G. Koren, et al., “Design and Performance of a Space Based High Temperature Superconductivity Experiment,” Proceedings of the 14th AMSAT-UK Colloquium Space-Communication-99, University of Surrey, July 23-25, 1999, pp. 10-14

1884) E. Polturak, G. Koren, M. Ayalon, “Space Based High Temperature Superconductivity Experiment,” Proceedings of the 40th Israel Annual Conference of Aerospace Sciences, 2000 instrument mass is 0.63 kg, the power consumption is 12 W. The nominal cooling power is 0.5 W at 77 K.

XDEX (X-Ray Detector Experiment). Objective: 1) test of high performance, sensitive X-ray detectors (CdZnTe) with high energy resolution; 2) test of a focal plane array where photon counting and signal processing can be performed and stored; and 3) test of degradation of the detectors/focal plane array assembly in a high-energy particle environment. The instrument consists of solid-state detectors, a sensitive preamplifier, a microcontroller and memory. The experiment is a step towards development of an X-ray telescope based on CdZnTe detectors. The instrument mass is 1.6 kg.

SLRRE (Satellite Laser Ranging Retroreflector Experiment). Objective: High-precision laser ranging measurements from the ground for orbit determination (5-10 cm range). SLRRE is a passive on-board experiment consisting of an array of laser retroreflectors, corner-cube mounted on the Earth-viewing panel (panel 5) of the microsatellite. The SLRRE mass is 0.65 kg.

N.21 TUBSAT (Technical University of Berlin Satellite)

TUBSAT is a low-cost and fast-turnaround microsatellite program series defined, designed and built by TUB (Technical University of Berlin, with DLR funding). The objective is to explore technical capabilities in microsatellite design (in particular in the field of attitude determination) and space-related applications. The following fields of study are of particular interest:[1885) 1886)]

- S&F (Store & Forward) communication techniques in conjunction with DCS (Data Collection Systems)
- Test of attitude control systems
- Test of disaster monitoring technologies

Satellite	Payload Class	S/C Mass	Launcher	Launch Date
TUBSAT-A	S&F communications	35 kg	Ariane-4	July 17, 1991
TUBSAT-B	Earth observation	45 kg	Tsyklon	Jan. 25, 1994
TUBSAT-N, N1	S&F communications	8 kg, 3 kg	SHTIL	July 7, 1998
DLR-TUBSAT	Earth observation	45 kg	PSLV	May 26, 1999
MAROC-TUBSAT	Earth observation	40 kg	Zenit	2001

Table 530: Overview of the TUBSAT program

N.21.1 TUBSAT-A

The microsatellite has a cube structure with dimensions of 38 cm x 38 cm x 38 cm and a mass of 35 kg. It is three-axis stabilized, control is provided by three magnetorquers and a magnetometer. Attitude is sensed by star and sun sensors. The objectives are to study S&F techniques and to test an attitude control system. TUBSAT-A was launched piggyback July 17, 1991 from Kourou by Ariane V44/ASAP together with ERS-1 and three other microsatellites [UoSAT-5 (SSTL), SARA (French University), DATA-X (OSC)]. As of 1/1996 TUBSAT is operating nominally.

Orbit: Sun-synchronous polar orbit, altitude = 780 km, inclination = 98.42°, equator crossing time at 10:30 AM.

S&F communication is provided in VHF-band (143.075 MHz) at a data rate of 1200 baud. The satellite uses an ALOHA-type access system for ground communications; it transmits a

1885) Information provided by U. Renner of TUB

1886) U. Renner, "Small Satellites at the Technical University of Berlin," IAA 2nd International Symposium on Small Satellites for Earth Observation, Berlin, April 12–16, 1999, pp. 253–256 set of telemetry data plus one kByte of text data every 4.5 minutes. On-board storage capacity is 32 kByte. Typical experiments are with lightweight (hand-held) ground stations such as manned expeditions in arctic and antarctic regions, tracking of drifting buoys and location experiments with animals (deer). TUBSAT is operated by a single ground station at TUB; it comprises a steerable cross-polarized yagi-antenna and standard radio amateur receive and transmit equipment.

Status: TUBSAT-A is fully operational as of June 2001 (10 years in orbit as of July 17, 2001).

N.21.2 TUBSAT-B

A microsatellite with dimensions of 38 cm x 38 cm x 50 cm and of 45 kg mass with the objective of Earth imaging. The S/C is three-axis stabilized with three reaction wheels. Attitude is sensed by star and sun sensors. TUBSAT-B was launched on January 25, 1994 as a secondary payload(together with Meteor-3-7) by a Russian Tsyklon launcher from Plesetsk.

Orbit: Polar orbit, altitude = 1250 km, inclination = 82.3°.

On March 5, 1994 contact with TUBSAT-B was lost after 39 days of successful operation. Transmission frequency in VHF-band (143.075 MHz) at data rates of 1200 and 2400 baud.

Sensor:
A video camera system in the VNIR spectral range capable of taking snapshots of the ground. System parameters: Zeiss Jena telescope with 1 m focal length (folded optics), aperture = f/5.6, CCD array: Thomson-CSF, 288 x 384 pixels. FOV = 0.35° x 0.5°; spatial resolution = 23 m x 30 m.

N.21.3 TUBSAT-N (Technical University of Berlin Satellite-Nano)

TUBSAT-N and -N1 are two communication (data collection) nanosatellites of the Technical University of Berlin (Institute of Aeronautics and Astronautics) which were tandem-launched July 7, 1998 on a Russian SS-N-23 military launch vehicle (also referred to as SHTIL). [1887)] The submarine launch, from the Delta-IV class nuclear vessel 'Novomoskovsk' of the Northern Fleet, took place in the western Barents Sea (in the vicinity of Murmansk). Both satellites were separated from the launch vehicle's payload capsule after orbit injection. The launch represents the world's first underwater/space launch of a satellite into Earth orbit on the basis of a commercial service. Both satellites are used for mobile communications with the following objectives and service provision (a consistent extension of the TUBSAT-A package): [1888)]

- Demonstration of low-cost access to space without performance reduction
- Bidirectional data transfer between autonomous environmental stations and the satellite (e.g., drifting buoys in the ocean, arctic and antarctic meteorological stations, etc.)
- Tracking of medium-sized and large mammals
- Worldwide positioning and deactivation services of stolen cars and support of very mobile communications.

TUBSAT-N has the shape of a small box with dimensions of 320 mm x 320 mm x 104 mm; its mass is 8.5 kg. A carbon fiber structure was chosen for both satellites for stiffness and lightweight design (passive thermal stabilization). Implementation of a single-board design, functional integration of modules/subsystems (e.g., TT&C with communication module), there is no physical separation between platform and payload. The ACS (Attitude Control

1887) Note: SS-N-23 is the NATO designation for the Russian RCM 54 missile, built by the Makeyew State Rocket Center of Miass (a town in the Ural Mountains).

1888) R. Schulte, "TUBSAT-N, A Global Communication Satellite System, Based on Nanosatellites," Proceedings of the 4th International Symposium on "Small Satellites Systems and Services," Sept. 14-18, 1998, Antibes Juan les Pins, France

Subsystem) consists of a CCD star sensor, a 3-D magnetic compass, two magnetic coils and a reaction wheel. Note: ACS on TUBSAT-N is simply flown as an experiment to space qualify the star sensor and the reaction wheel (mobile communication with omnidirectional antennas does not require such attitude control). Electrical power is provided by 9 NiCd battery cells of 5 Ah. They are charged by two strings of solar cells (34 cells per string) with cell dimensions of 6 cm x 4 cm. The solar cells are mounted on the top and bottom surface of the box. TUBSAT-N has an on-board storage capacity of over 6000 messages at 134 bytes in length. The messages are transferred several times per day to the TUB ground station.

TUBSAT-N1 is of same design but somewhat smaller (dimension: 320 mm x 320 mm x 34 mm; mass = 3 kg; 9 NiCd battery cells of 2.8 Ah). The design life for each S/C is seven years. The solar generator consists of two parallel strings with 36 cells on each side with cell dimensions of 4 cm x 2 cm. Nine NiCD cells of 2.8 Ah provide battery power.

The tandem launch of the two satellites represents a demonstration of the launch and separation technology of the SS-N-23 launch system. In addition, the communication coverage of the overall system is improved and redundancy provided by the two-satellite configuration.

Orbits: elliptical orbits, both satellites are in the same orbital plane, apogee = 780 km, perigee = 404 km, inclination = 78.9°.

Status of mission: Duc to rclatively low orbital perigees the satellites ended their service lives; TUBSAT-N1 after about one year of operation, and N after two years of operation. The lower mass of N1 (3.0 kg) was more affected by the drag forces than the N spacecraft (8.5 kg) with more than double the mass of N1.

Communication configuration for data collection:

TUBSAT-N uses four independent communication channels (two VHF, two UHF), all with FFSK modulation (optional GMSK). TUBSAT-N1 uses two UHF communication channels, all with FFSK modulation (optional GMSK). Data collection function: Access to the satellites is via MP-TDMA (Modified Preassigned - Time Division Multiple Access). The satellite antenna radiation pattern is sufficiently omnidirectional to guarantee a link budget in any direction. The S/C UHF and VHF transmitter power is 10 W.

A mobile communications terminal in the ground segment consists of the following components: 16-bit CPU, RAM and FLASH memory, real-time clock, DC/AC converter for GPS receiver, FFSK or GMSK ASIC and a UHF/VHF transceiver, based on YAESU (manufacturer) chip sets . Several I/O pins, a serial communications interface and eight analog channels provide connection to external devices (e.g., LCDs, keyboards, measurement devices, etc.). With the knowledge of position, time and orbital parameters of the satellite, provided at each contact, the mobile terminal is able to calculate the upcoming contact times with the space segment for elevations above 5°. Each terminal has predefined time slots for communicating with the satellites (including update orbital parameters). The data collection rates between S/C and terminal are either 1200 or 2400 baud. An additional downlink transmitter with 9.6 kbaud and GMSK modulation is used to transmit the collected messages to the central ground station.

The two nanosatellites collect also data from a number of Lagrangian buoys that are automatically submerged to depths of 2500 m to measure salinity and other parameters, they only rise to the ocean surface after predefined measurement periods for the purpose of data transmission.

N.21.4 DLR-TUBSAT

A joint venture between TUB and DLR, Berlin Adlershof. The prime objective of the mission is to test a newly developed attitude control system (S/C attitude recovery from hibernation); a secondary objective is to test a TV camera system to demonstrate disaster monitoring. The microsatellite is three-axis stabilized, it has dimensions of 32 cm x 32 cm x 32 cm and a mass of 45 kg. The bus consists of four aluminium shells, one for each subsystem. The first shell contains ACS (Attitude Control Subsystem), the NiH_2 batteries and the TT&C subsystem are part of shell number two. The payload is part of the third shell, and OBDH (On-Board Data Handling), along with an S-band transmitter, is contained in the fourth shell. [1889] [1890] [1891]

Attitude is sensed by the camera assembly; the actuators consist of three reaction wheels (TUB design) and three fiber optic laser gyros providing a pointing knowledge accuracy of ±1 arcsecond. The agile S/C is pointable into any direction and able to stabilize itself on short notice within a few seconds. Tracking modes for inertial pointing (S/C attitude) and/or target pointing (Earth observation) are supported. In its normal standby (hibernation) mode, the S/C tumbles in its orbit at rates of about 0.1 rpm. An acquisition sequence starts with a rate reduction command, followed by a coarse sun acquisition maneuver. This may be followed by a TV transmission.

A S/C passenger launch took place on an ISRO PSLV launcher from the SHAR lunch site, India, on May 26, 1999 (along with IRS-P4 as the primary payload and KITSAT-3 as another secondary payload).

Orbit: Sun-synchronous polar orbit, altitude = 720 km, inclination = 98.28°, period = 99.3 min.

Status: DLR-TUBSAT is fully operational as of June 2001.

Figure 358: The DLR/TUBSAT S/C model

1889) M. Steckling, U. Renner, H. P. Röser, "DLR-TUBSAT, a Multipurpose Microsatellite for Varying Earth Observation Applications," IAA 2nd International Symposium on Small Satellites for Earth Observation, Berlin, April 12-16, 1999, pp. 347-350

1890) U. Renner, Earth Observation with TUBSAT-C," 4th International Symposium: Small Satellites Systems and Services, Antibes, France, Sept. 14-18, 1998

1891) S. Roemer, U. Renner, "Flight Experiences with DLR-TUBSAT," Proceedings of the 3rd International Symposium of IAA, Berlin, April 2-6, 2001, pp. 75-78

Sensor complement:

Camera Assembly. The system (a Sony CCD array system with Nikon optics) consists of three commercially available TV cameras: a wide-angle camera with 16 mm focal length with a black-and-white chip, a standard-angle optic with 50 mm focal length and color CCD chip, and a narrow-angle camera (a tele lens) with 1000 mm focal lens and black-and-white chip. The objective is to provide a snapshot imaging capability of medium or high-resolution ground scenes of locations stricken by disaster (hazardous or unexpected environmental events such as wild fires, floods, and earthquakes). Each camera head of the camera assembly can be selected as the active one. All cameras are aimed in the same direction (nadir direction by default), but offer different fields of view (image sizes). One of the three cameras may provide imagery at any time (16 mm camera by default). The camera system automatically adjusts its gain to prevailing illumination conditions. A body-pointing capability is provided by ACS of the S/C, allowing the camera assembly to look at targets anywhere up to the horizon. The camera assembly has a video output as well as a digital output. Both of them can be selected to feed the S-band transmitter. The analog signal is coded as a standard FM PAL signal which may be received by a 3-m antenna dish and standard equipment. The digital data is bi-phase level encoded with a transfer data rate of 125 kbit/s. A 1.2 m antenna dish is sufficient for reception.

Spacecraft operation: The S/C is normally operated from a single ground station at TU Berlin (TT&C operations in UHF-band, the UHF transceiver receives and transmits data at a rate of 1200 baud, FFSK modulation). Most of the time (certainly over 95%) the satellite is orbiting in standby mode, slowly tumbling (a few rotations per orbit); all on-board systems are in hibernation (including ACS) to save power; only the UHF receiver is listening for potential commands. In case of an event, the S/C attitude may be restituted within a few minutes. However, this wake-up capability requires the S/C to be in view of a commanding UHF station. TUB has portable UHF ground stations from its TUBSAT-A experience to handle such events; it is also relying on the use of existing infrastructures of the remote sensing community.

Parameter	16 mm camera	50 mm camera	1000 mm camera
CCD array size	750 x 580 pixels	750 x 580 pixels	750 x 580 pixels
Pixel size	8 μm x 8 μm	8 μm x 8 μm	8 μm x 8 μm
IFOV		9.2×10^{-3} degrees	0.46×10^{-3} degrees
FOV		$6.9^{\circ} \times 5.336^{\circ}$	$0.345^{\circ} \times 0.267^{\circ}$
Nadir image size (1000 km altitude)	110 km x	120 km x 92.8 km	6 km x 4.6 km
Nadir ground resolution (IFOV)	370 m x 370 m	120 m x 120 m	6 m x 6 m
Focal length, f/number	16 mm	50 mm, f/1.8	1000 mm, f/11
Spectral range	Panchromatic	Panchromatic	Panchromatic

Table 531: Definition of Camera Assembly

There is no on-board storage capability for video data from the Camera Assembly. Real-time transmission of the video data is in S-band (frequency = 2.206 GHz, bandwidth = 8 MHz, data rate at 128 kbaud using analog video, the beamwidth of the S-band antenna is 70°) to a ground station in the vicinity of the event. Video transmission beyond the TU Berlin ground station has to be negotiated on a case-by-case basis. A possible alternative for video data reception coverage may also be found in the widespread use of portable commercial S-band transmitter systems and their corresponding receiver systems. The S-band transmitter system on DLR-TUBSAT happens to be of the same type as those regularly used in TV coverage of sports events, such as in the helmet of a downhill skier.

N.21.5 Maroc-TUBSAT

Maroc-TUBSAT is a cooperative microsatellite project between CRTS (Centre Royal Teledetection Spatiales) of Rabat, Morocco, and the Institute of Aeronautics and Astronautics of the Technical University of Berlin (TUB). In this joint venture, the Moroccan side is responsible for the development of the payload and launch of the satellite while the German side provides the satellite bus. The overall objectives of the mission are in Earth remote sensing (in particular with regard to vegetation detection at medium-scale resolutions), and in the field of store-and-forward communications for mobile localization. A further goal is to develop attitude control strategies for high-resolution Earth observations. [1892)]

The project selected an adaptation of the existing TUBSAT-C bus, a three-axis stabilized box structure of size: 320 mm x 340 mm x 362 mm. The main elements of the bus are: the attitude control module, the power module, and the payload module. The modular design provides functional freedom and flexibility during all phases of development. S/C power is provided by four surface-mounted solar cells (60 W max) and by 4 NiH_2 batteries with a capacity of 12 Ah. The attitude control subsystem (ACS) employs a star sensor (inertial attitude determination), a three-axis magnetic field sensor, and three fiberoptics gyros for attitude sensing. In addition, single solar cells on the surface-mounted solar panels are used for coarse sun direction determination. Attitude actuation is provided by three reaction wheels and magnetic coils. The S/C mass is 47 kg.

A launch of Maroc-TUBSAT is planned for Nov. 2001 on a Zenit-2 launch vehicle from the Baikonur Cosmodrome, Kazakhstan. The prime payload is Meteor-3M-1. Secondary payloads on the flight are the Badr-2 satellite of Pakistan, Maroc-TUBSAT of CRTS, the Compass spacecraft of IZMIRAN, Moscow, and REFLECTOR (US/Russia).

Orbit: Sun-synchronous circular polar orbit, mean altitude = 830 km, inclination = 98.85°, period = 102 minutes; local time of ascending node is 9:15 AM.

RF communications: Store-and-forward communications of TT&C data in UHF and VHF (AMSAT standard) are provided at data rates of 1.2 kbit/s (uplink and downlink, FFSK modulation). The RF output power is 3.5 W for UHF and 5.0 W for VHF. The payload data is transmitted in S-band (BPSK modulation, FEC, 2 W RF output) at data rates of up to 250 kbit/s. Spacecraft operations are conducted at (TUB, CRTS), ground stations are at TUB and at CRTS.

The mission scenario of spacecraft activity/operations is event-driven as practiced on DLR-TUBSAT. This implies that Maroc-TUBSAT spends most of its time in hibernation mode. Event recognition (time) or a command for a ground station cause the spacecraft to return to observation mode. - The following general start-up and acquisition sequence describes the steps involved in S/C recovery/initiation from the hibernation mode (ACS is off) into an operations mode for recording specific areas of the Earth by the payload camera:

- At a preprogrammed event time turns the ACS on. This causes a reduction in tumbling motions by a rate damping process, provided by the closed-loop system of gyros (3) and reaction wheels (3). The residual motion amounts to <0.5°/s.
- In the event that the star sensor is blinded by Earth or the sun, a slew maneuver is enacted to align the sensor to a star (information is provided by the solar cells and by orbit model inputs). The star sensor is now capable to determine the inertial attitude of the system.
- Meanwhile, the ground station software compares the current S/C position with the intended acquisition time. It seeks a minimum solution for the three different slew maneuvers (controlled by the gyros).
- The predefined attitude of the S/C is reached after slew maneuver completion. The pointing accuracy in this mode is better than ±0.05° (limited only by the star sensor performance).
- The hibernation mode is turned on again by the operator after observation and data transmission of the pass are completed. The batteries are recharged in hibernation mode.

1892) "MAROC-TUBSAT, A Microsatellite for Earth Observation," TU-Berlin, Ref. 2 EB-MT-R01

Sensor complement:

EIC (Earth Imaging Camera), developed at RAL, UK (same camera as on BADR-2). The instrument optics have a focal length of 72 mm with f/6 number. Observations are performed in the visible/near-infrared range. A filter is used for the near-infrared range. The FOV is ±8.5° x ±8°, providing an image size of about 190 km x 144 km at a resolution of 250 m. A CCD matrix array of 770 x 576 pixels is used (of EEV, Chelmsford, UK, pixel size of 32.5 μm) is used with frame transfer of imagery. The nominal integration time for a snapshot is 28 ms. A 12 bit data quantization is used.

Star Sensor (of ACS). The instrument consists of the following elements: A CCD array camera, optics with 16 mm focal length and f/0.95, analog electronics and a microcontroller for star tracking, a microcontroller for star recognition (relative and inertial attitude determination), a RS422 serial interface to the host computer, a mechanical interface for mounting and heat transfer, and a power supply. The CCD array is of size: 288 x 384 pixels with frame-transfer capability (8 bit data quantization). The microcontroller uses a star catalog of 6000 stars and a pattern recognition database. The FOV of the sensor is 21° x 31°, an inertial pointing accuracy of ±0.02° is provided. The instrument has a mass of 0.68 kg, power = 5 W, data rate = 38.4 kbit/s.

N.22 UniSat (University Satellite)

UniSat is an experimental microsatellite project of the University of Rome (Universita di Roma "La Sapienza", Scuola di Ingegneria Aerospaziale). Education and hands-on experience of students and faculty in the field of aerospace engineering is the overall theme of the program (involving a number of Italian universities), funded by the Italian Government MURST (Ministry of University and Technological Research) and by ASI (Italian Space Agency). In the framework of the UniSat program, a laboratory for satellite design and construction and an amateur ground station called SPIV (San Pietro in Vincoli), were established at the University of Rome. [1893)] [1894)]

The UniSat-1 development phase started in 1997 at GAUSS (Gruppo di Astrodinamica dell' Universita degli Studi "la Sapienza"), sponsored by ASI in cooperation with other Government Research Centers and with industry. The S/C body structure consists of an octagonal prism, 40 cm in diameter and a height of 25 cm. The modular tray design employs five Al/Al honeycomb plates, kept together by four stainless steel bars. Eight lateral panels with surface-mounted solar panels make up the external surface of the spacecraft. The satellite is spin-stabilized at 5 rpm with the spin-axis normal to the orbital plane. Attitude is sensed with two three-axis magnetometers and the use solar panels as a sun-sensing system. Actuation is provided by an active magnetic control system for spin-axis orientation and for spin rate control. One equatorial coil is used for spin-axis orientation and two transverse magnetic coils are used for spin-up/spin-down maneuvers. An average power of 10 W is provided by the solar panels. In addition, NiMH batteries of 2.4 Ah capacity are used for eclipse operations. The OBC (On-Board Computer) consists of a distributed network of microcontrollers (master/slave architecture) providing flexibility and reliability. UniSat-1 has a mass of 12 kg, qualifying almost as a nanosatellite. The S/C design life is one year.

The RF communications are implemented according to AMSAT standards with VHF (145.848 MHz) in uplink and UHF (436.450 MHz) in the downlink using the AX.25 packet protocol at data rates of 9.6 kbit/s. The UHF transmitting antenna consists of four dipoles providing circular polarization. The VHF receiving antenna is a dipole.

1893) F. Graziani, M. Ferrante, G. B. Palmerini, F. Santoni, P. Tortora, "UniSat program: a University Tool for Space Education," Proceedings of the 51st IAF Congress, Rio de Janeiro, Brazil, Oct. 2-6, 2000, IAF-00-P.2.07

1894) http://pcgauss5.ing.uniroma1.it/entra/attivita/attivita.htm

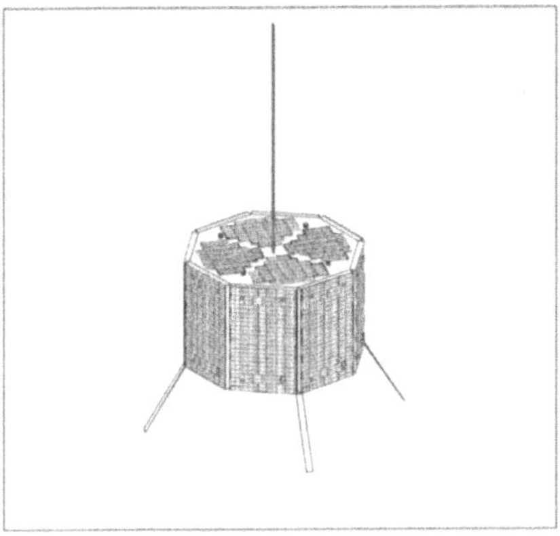

Figure 359: Illustration of the UniSat spacecraft

A launch of UniSat-1 took place on Sept. 26, 2000 from the Baikonur Cosmodrome on a Dnepr launch vehicle. UniSat-1, along with MegSat of Italy and TiungSat 1 of Malaysia, and SaudiSat-1A and -1B of SISR (Saudi Institute for Space Research), was launched in a multiple launch configuration (no primary payload). UniSat-1 is being operated by students.

Orbit: circular orbit, altitude = 650 km, inclination = 65°, period = 98 minutes.

Technology experiments:

The power subsystem represents two experiments; one involves a solar panel environment test, the second experiment involves a functional NiMH (Nickel Metal Hydride) battery test. The new batteries offer a much better capacity/mass ratio than conventional NiCd batteries. - The OBC distributed network architecture is a proof-of-concept test.

N.23 WeberSat

WeberSat is a joint project of CAST (Center for Aerospace Technology) at Weber State University in Ogden, Utah with AMSAT-NA (Radio Amateur Satellite Corp. - North America), a scientific, non-profit organization of amateur radio operators that have been active in building small satellites for a number of years. [1895] In the late 1980s, AMSAT-NA asked CAST to participate in the fabrication of four microsatellites. For their contribution, CAST received ownership one of the four microsatellites which was named WeberSat. The four microsats were launched Jan. 21, 1990 on the Ariane ASAP from Kourou, French Guiana on Ariane flight V35 (Spot-2 was the primary payload). All microsats were spin-stabilized and offered store-and-forward amateur packet communications. WeberSat operated nominally until 1998. The AMSAT name of WeberSat is WeberSat-Oscar 18.

WeberSat is a small microsatellite, equipped with five radio receivers, two transmitters, a color video camera, spectrometer, and a micrometeorite impact detector. The S/C structure is a cube of size: 22.2 cm x 22.2 cm x 31.75 cm. The solar panels are surface mounted, providing 6 W of average power in combination with NiCd batteries. The S/C mass is 12.25 kg. WeberSat is spin-stabilized, attitude is controlled by a gravity-gradient boom.

Communications: L-band AM receiver for Amateur radio, video S&F (Store & Forward).

Orbit: Sun-synchronous polar orbit, altitude = 800.0 km, inclination = 98.58°, equator crossings at noon and midnight.

1895) http://cast.weber.edu/webersat/index.html

Sensor complement:

Color Video Camera. The CCD camera has a resolution of 700 pixels by 400 lines (a snapshot image provides a coverage of about 200 km x 270 km), and can be viewed with Weberware software running on a personal computer having adequate graphics display capability. Digitized NTSC video from the camera is assembled into packets that are sent as unnumbered information UI frames. Ground stations must receive this data over several passes to capture a complete image. Each image contains about 200 kilobytes of data. An on-board 1265 MHz ATV uplink receiver allows ground stations to load WeberSat's memory using a groundbased video camera for later transmission by the spacecraft. WO-18 can also transmit CCD images on 70 cm using an analog "turbo" mode that is much faster than using the AX.25 packet downlink. However, no interference protection is afforded by the analog transmission mode since information received in this form does contain a checksum.

Visual Light Spectrometer. A light spectrometer employing an NEC linear CCD measures the spectrum of sunlight reflected from the atmosphere.

Micrometeorite Impact Detector. A piezoelectric Particle impact detector measures the micrometeorite environment.

Part O Reference Data and Definitions

The following table contains the "EO" missions (in alphabetical order according to mission acronym) that were launched after the 3rd edition was published in early 1996. Excepted are Shuttle launches and GPS/GLONASS launches, each of these series has its own launch table in this volume. There is no claim for completeness.

Mission	Launch Date	Comment
ACE	Aug. 25, 1997	Advanced Composition Explorer, NASA/GSFC
ACRIMSAT	Dec. 21, 1999	NASA minisatellite to monitor the amount of total solar energy
ADEOS	Aug. 17, 1996	Advanced Earth Observing Satellite, NASDA,
ARGOS	Feb. 23, 1999	DoD technology satellite, launch from VAFB
ARTEMIS	July 12, 2001	ESA communication/technology spacecraft, Ariane-5 launch
ASTRID-2	Dec. 10, 1998	Swedish solar-terrestrial energy transport mission (IRF-K)
CBERS	Oct. 14, 1999	Chinese/Brazil imaging mission
CHAMP	July 15, 2000	Challenging Minisatellite Payload (GFZ/DLR)
Cluster-1	June 4, 1996	ESA, launch failure (4 spacecraft)
Cluster-2	July 16, 2000 Aug. 9, 2000	ESA, two S/C were launched at a time ESA, two S/C were launched at a time
Coronas-F	July 31, 2001	Russian and NKAU (Ukraine) solar physics mission
CRISTA-SPAS-2	Aug. 7-19, 1997	STS-85
Deep Space-1	Oct. 24, 1998	First mission with ion propulsion, NASA/JPL
DMSP, 5D-2 F-14	Apr. 4, 1997	USAF launch from VAFB
DMSP, 5D-3 F-15	Dec. 12, 1999	launch aboard a Titan 2 rocket from VAFB,
EarlyBird	Dec. 24, 1997	EarthWatch controllers lost contact with the S/C on Dec. 28, '97
EO-1	Nov. 21, 2000	Earth Observing-1 of NASA
Equator-S	Dec. 2, 1997	MPI Garching, DLR
EROS-A	Jan. 22, 1998	West Indian Space Ltd., Cayman Islands, Shavit launcher failure from Palmachim, Israel
EROS-A1	Dec. 5, 2000	Commercial imagery of ImageSat International
ETS-VII	Nov. 27, 1997	Technology mission of NASDA
FAISAT-2V	Sept. 24, 1997	Data collection, communication
FASat-Bravo	July 10, 1998	Microsatellite of Chile built by SSTL (secondary payload to RESURS-O-4)
FAST	Aug. 21, 1996	NASA/GSFC mission within SMEX
FORTE	Aug. 29, 1997	Fast On-Orbit Recording of Transient Events, LANL, SNL
FY-1C	May 10, 1999	China (along with SJ-5)
FY-2B	June 10, 1997	GEO weather satellite of China (CASC)
FY-2 (Fen Yun-2)	June 26, 2000	GEO weather satellite of China (CASC)
Genesis	Aug. 8, 2001	NASA, Solar wind sample return mission
GFO-1	Feb. 10, 1998	GEOSAT follow-on Satellite, US Navy
GOES-10	Apr. 25, 1997	NASA launch on Atlas-1 vehicle from Cape Canaveral
GOES-11 (-L)	May 3, 2000	NASA launch of NOAA weather satellite
GOES-12 (M)	July 23, 2001	NASA launch of NOAA weather satellite on Atlas-II vehicle from VAFB
Helios-1B	Dec. 3, 1999	Military reconnaissance S/C of France, Italy and Spain, Ariane-4 launch, Clementine (built by SSTL) was a secondary S/C
IKONOS-1	Apr. 27, 1999	Space Imaging, launch failure
IKONOS-2	Sept. 24, 1999	Space Imaging
IMAGE	March 25, 1999	DOE S/C launch from VAFB
INSAT-2D	June 3, 1997	GEO meteorological/communication satellite of ISRO
INSAT-2E	April 2, 1999	GEO meteorological/communication satellite of ISRO
Interball-2	Aug. 29, 1996	Auroral Probe of Interball (Molniya-M launch from Plesetsk) Magion-5 as subsatellite
IRS-1D	Sept. 29, 1997	ISRO
IRS-P3	Mar. 21, 1996	ISRO
IRS-P4	May 26, 1999	ISRO, referred to as OceanSat-1, along with KITSAT-3 and DLR-TUBSAT

Mission	Launch Date	Comment
JAWSAT	Jan. 26, 2000	Weber University, secondary payloads: FalconSat, ASUSat-1, OPAL, StenSat, PICOSAT-1, -2, Artemis
KITSAT-3	May 26, 1999	KAIST (Korea) secondary payload to IRS-P4 of ISRO
KOMPSAT-1	Dec. 21, 1999	Cooperative minisatellite technology-training program of KARI with TRW, EOC (Electro-Optical Camera) and OSMI (Ocean Scanning Multispectral Imager) instruments
Landsat-7	April. 15, 1999	NASA S/C with ETM+imaging instrument
Lewis	Aug. 23, 1997	NASA, The Lewis S/C reentered the atmosphere Sept. 28, 1997. The Clark mission was cancelled in Feb 1998 due to cost overruns and launch delays
MegSat-0	April 28, 1999	Italian data collection satellite
Meteosat-7	Sept. 3, 1997	ESA/EUMETSAT
MightySAT-1	Dec. 4, 1998	AFRL technology S/C, launched as HH payload on STS-88, Dec. 14, 1998
MightySat II.1	July 19, 2000	AFRL technology S/C with ten technologies and FTHSI
MINISAT-01	April 21, 1997	INTA of Spain,
MITA	July 15, 2000	MITA (Minisatellite Italiano di Technologia Avanzata)
MSX	April 24, 1996	DoD satellite launch from VAFB
MTI	Mar. 12. 2000	DOE satellite launch from VAFB
MTSAT	Nov. 15, 1999	NASDA, launch failure
Nadezhda	June 28, 2000	Russian search and rescue satellite in COSPAS-S&RSAT series
NEAR	Feb. 17, 1996	NASA S/C launch on a Delta vehicle from Cape Canaveral. Flyby of asteroid 433 EROS on Dec. 23, 1998. Next rendezvous with EROS in mid-Feb. 2000
NOAA-15 (-K0	May 13, 1998	NOAA weather satellite (POES series)
NOAA-16 (-L)	Sept. 21, 2000	NOAA weather satellite
Munin	Nov. 21, 2000	IRF (Sweden) nanosatellite, secondary payload to EO-1
Odin	Feb. 20, 2001	Swedish satellite for astronomic and atmospheric research
Oersted	Feb. 23, 1999	DTU of Denmark S/C, measurement of the Earth's magnetic filed
OKEAN-O-1	July 17, 1999	Ukrainian/Russian satellite
Orbview-2 (SeaStar)	Aug. 1, 1997	Orbimage of OSC, Dulles, VA; SeaWiFS sensor for ocean color data
ORFEUS-SPAS-2	Nov. 19 - Dec. 7, 1996	STS-80
POLAR	Feb. 24, 1996	NASA/GSFC solar-terrestrial mission
Priroda	April 23, 1996	Russia, module of MIR station
QuickBird-1	Nov. 20, 2000	Earthwatch, commercial imaging, launch failure
QuikSCAT	June 19, 1999	NASA/JPL, wind measurements
RESURS-F	Nov. 18, 1997	launch from Plesetsk
RESURS-F-1M	Sept. 28, 1999	RKA
RESURS-O1-4	July 10, 1998	RS satellite launched from Baikonur on a Zenit-2 vehicle Secondary payloads: LLMS/IRIS, TMSat (Thailand), TechSat/Gurwin-II (Israel), FASat-Bravo (Chile), SAFIR-2 (OHB), WESTPAC (Australia)
REX-2	March 9, 1996	DoD S/C launch on a Pegasus vehicle from VAFB
ROCSAT-1	Jan. 27, 1999	First S/C of Taiwan with an ocean color sensor
SAC-A	Dec. 4, 1998	CONAE, Argentina, launched as HH payload on STS-88
SAC-B	Nov. 4, 1996	CONAE, Argentina, Pegasus XL launch failure
SAC-C	Nov. 21, 2000	CONAE, NASA, launched as secondary payload to EO-1
SACI-1	Oct. 14, 1999	INPE, secondary payload to ZY-1 of CASC
SCD-2	Oct. 23, 1998	Data collection satellite of INPE
SCD-2A	Nov. 2, 1997	INPE, VLS launch vehicle failure from Alcantara
SEDSAT-1	Oct. 24, 1998	Secondary payload to DS-1, S/C of Univ. of Alabama
Shenzhou	Nov. 20, 1999	Chinese test launch for manned space flight
SMEX/FAST	Aug. 21, 1996	NASA/GSFC, (Fast Auroral Snapshot Explorer)
SMEX/TRACE	Apr. 2, 1998	NASA/GSFC, (Transition Region and Coronal Explorer)
SNAP-1	June 28, 2000	SSTL technology S/C
SNOE	Feb. 25, 1998	University of Colorado at Boulder

Mission	Launch Date	Comment
SPOT-4	March 23, 1998	CNES, a new instrument "VEGETATION" is flown
SRTM	Feb. 11-22, 2000	NASA/DLR/ASI mission (Shuttle Radar Topography Mission)
STEDI/SNOE	Feb. 26, 1998	(Student Nitric Oxide Explorer), NASA, Univ. of Colorado on a Pegasus XL vehicle from VAFB, CA
STEX/ATEX	Oct. 3. 1998	DoD payload launched from VAFB with tether experiment
STARSHINE	May 27, 1999	Student satellite on STS-96
STRV-1c/d	Nov. 16, 2000	DERA satellites
SUNSAT	Feb. 23, 1999	Student-built satellite of Stellenbosch University, SA
TDRS-H	June 30, 2000	NASA Data Relay satellite
TEAMSAT, YES	Oct. 30, 1997	ESA technology demonstrator payload on Ariane-5 test flight
TechSat/Gurwin-II	July 10, 1998	Technion Israel Institute of Technology, Haifa, Israel
Terra (EOS/AM1)	Dec. 19, 1999	NASA S/C with MODIS, ASTER, etc.
TERRIERS	May 18, 1999	Boston University/NASA
TiungSat-1	Sept. 26, 2000	Microsatellite of ATSB, Kuala Lumpur, Malaysia (built by SSTL)
TMSat	July 10, 1998	Microsatellite of Thailand built by SSTL
TOMS-EP	July 2, 1996	NASA S/C, Ozone measurements
TRACE	April 2, 1998	NASA/GSFC mission in the SMEX (Small Explorer) program
TRMM	Nov. 27, 1997	NASDA/NASA (Tropical Rainfall Measuring Mission)
Tsinghua-1	June 28, 2000	Tsinghua University, Beijing - Imaging Demonstrator Mission Three S/C in Cosmos launch from Plesetsk (SNAP-1, Nadezhda (Hope) and Tsinghua-1
TSX-5	June 6, 2000	DoD with STRV-2 payload package and CEASE
TUBSAT-N+N-1	July 7, 1998	Data collection satellites of TU Berlin
DLR-TUBSAT	May 26, 1999	Earth observation, TU Berlin
UniSat	Sept. 26, 2000	Dnepr launch from Baikonur with SaudiSat-1A/1B (SISR), Uni-Sat, MegSat, and TiungSat 1 of Malaysia
UoSat-12	Apr. 21, 1999	First minisatellite of SSTL
WESTPAC	July 10, 1998	Australian research S/C of Electro Optic Systems Pty Limited (EOS)
ZY-1 (Zi Yuan-1)	Oct. 14, 1999	ZY-1 (Resource) was launched on a Long March 4B vehicle from the Taiyuan Satellite Launch Center.
ZY-2	Sept. 1, 2000	Chinese CZ-4B launch vehicle with the Zi Yuan 2 remote sensing satellite from the Taiyuan launch facility.

Table 532: EO satellite launches since submission of 3rd edition to the publisher (Feb. 1996)

Mission	Launch Date	Comment
Aqua (EOS/PM-1)	Dec. 2001	NASA satellite on Delta-2 launcher from VAFB
CBERS-2	Dec. 2001	CAST (China) and INPE (Brazil) satellite from Taiyuan
DMSP-16	Nov. 2001	US DoD weather satellite, Titan-2 launch
ENVISAT	Nov. 2001	ESA, Ariane-5 launch from Kourou
GRACE	Nov. 2001	US-German dual-minisatellite mission
HESSI	Oct. 2001	NASA solar physics mission, Pegasus XL launch
ICESat/CATSAT	Dec. 2001	NASA launch on Delta-2 from VAFB
Jason/TIMED	Dec. 2001	NASA launch (Delta-2-7925) from VAFB
METSat	Oct. 2001	ISRO (India) from SHAR (replacement of INSAT-2E)
Meteor-3M-1	Nov. 2001	Russian weather S/C along with BADR-2, Compass, etc.
OrbView-4/ QuikTOMS	Sept. 2001	OrbImage and NASA S/C on Taurus launch vehicle from VAFB
PICOSat/Star-shine-3	Sept. 2001	Launch on Athena-1 vehicle from Kodiak Island, AK - including Sapphire of Stanford, PCSat of USNA, and Bitsy-SX
QuickBird-2	Oct. 2001	EarthWatch imaging satellite, Delta-2 launch from VAFB
TDRS-I	Oct. 2001	NASA Tracking and Data Relay Satellite
TES, PROBA, BIRD	Oct. 2001	ISRO prime payload TES with secondary payloads: PROBA of ESA and BIRD of DLR

Table 533: EO satellite launches projected for 2001

O.1 Definitions, Concepts, Summaries

Remote Sensing

Remote sensing, in the broadest sense, is the measurement or acquisition of information of some property of an object or phenomenon that is not in physical contact with the object or phenomenon under study. Remote sensing instruments, such as optical imagers or radar sensors, acquire information about an object of interest by detecting and measuring the changes that the object imposes on the surrounding electromagnetic field. For instance, in the case of visible and near-infrared (VNIR) sensors the reflected sunlight is used to acquire information about the chemical composition and physical structure of the object being observed. In the case of thermal infrared and passive microwave sensors, the emitted field is used to acquire information about the thermal properties as well as the composition of the object. In the active microwave region, information about the object's physical structure and electrical properties is acquired by analyzing the reflected field when the sensor illuminates the object with a well-defined generated field of electromagnetic waves.

Electromagnetic energy in its broad spectrum is the supreme messenger, the basis for all remote sensing. It conveys information about its source (structure, position, speed, temperature), about the medium it traverses, and about the medium it intercepts (by absorption and/or reflection). This information is coded by pattern, color, and intensity.

There are four elements of a radiation-based remote sensing system:

1) The radiation source (i.e., the sun, radar). In Earth observation the sun is the radiation source most commonly exploited. A radar system uses its own illumination source in the microwave region of the electromagnetic spectrum.

2) The transmission path (i.e., atmosphere, vegetation canopy). Atmospheric attenuation may cause some or all of the radiation to be absorbed or scattered. The transmission path length is a function of the geometric location of the source, target and sensor.

3) The target (i.e., surface area such as soil or water, a resolution cell such as a volume of the lower atmosphere). The target is the instantaneous object of observation of a sensor.

4) The sensor (i.e., multispectral scanner, film, radiometer, etc.). The instrument (usually flown on a platform) detecting and measuring the radiation of the target.

O.1.1 Remote Sensing across the Electromagnetic Spectrum

Electromagnetic waves interact with matter by a variety of mechanisms which are related to the composition and structure of the object. These mechanisms modify the wave characteristics (spectral, content, polarization intensity, direction, etc.) in such a way that the object properties can be identified. The emission or absorption spectral lines of a gas in the infrared spectrum will uniquely identify the nature and concentration of its constituents as well as its temperature. The reflectance spectral lines of a solid in the visible and near-infrared may allow the identification of its chemical composition, and, to some extent, its crystalline structure.

The depolarization and spectral reflectivity of the radar echo from a surface provides information about the roughness structure, geometric structure, morphology, and dielectric constant of the surface and immediate subsurface. The time of arrival and spectral contents of the radar echo allow a location of position of the reflecting object and its speed to extremely high accuracy (few cm in location and few cm/s in speed). The intensity of the blackbody radiation field in the thermal and microwave regions allows the derivation of the thermal properties of the object.

O.1.2 Types and Classes of Remote Sensors and Sensing Data

The survey of spaceborne and airborne instruments represents a great variety of sensor types that need an ordering scheme for a better understanding and overview by the reader. There are many ways of sensor classifications (some of them are trivial or simply problem specific), for instance by:

- the type of application (field of research or operation such as: meteorology, atmospheric research, solid Earth, climatology, biosphere, etc.)
- the type of primary observation targets (oceans, land, coastal regions, ice sheets, or atmosphere.
- the observed frequency range (UV, optical, infrared, microwave, etc.),
- the type of sensing mechanism (passive or active source),
- the type of instrument (imager, altimeter, sounder, spectrometer, radiometer, etc.)
- the type of measurement precision (resolutions)
- etc.

Another classification scheme can be derived from the type of information being sought (Ch. Elachi, see Ref. 1896)). If we consider for instance the three fundamental information parameters as being: spatial, spectral, and intensity information, each located diametrically at the extremities of a triangle, as illustrated in Figure 360, then a number of basic relationships and commonalities are self-evident or may be directly extracted.

The primary sensor types are allocated in the triangle to the primary information parameters. The measurement of two prime parameters results in an instrument type (like an imaging spectrometer) which is located somewhere between the two extremities and is also designated accordingly. There are already attempts to combine all three fundamental information parameters into a single instrument, such as MISR (Multi-Angle Imaging Spectro-Radiometer) in EOS. Not everything can be characterized and explained by this method, but the scheme illustrates the fundamental principles involved, even beyond the fancy naming conventions of some sensors. [1896)]

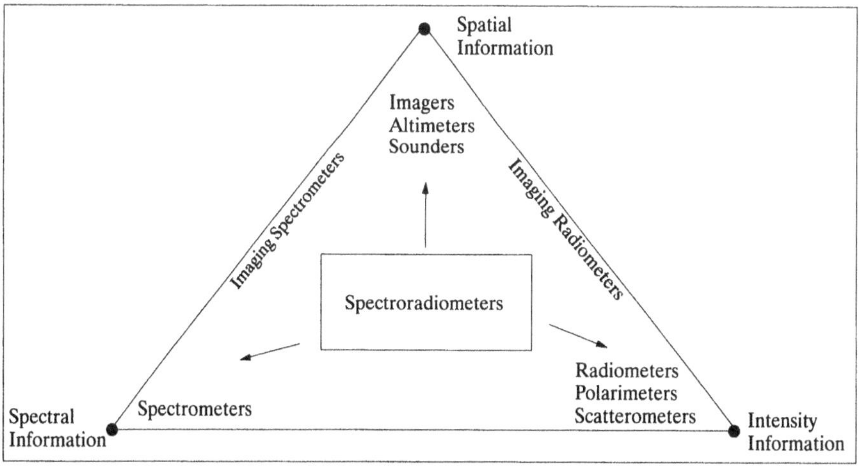

Figure 360: Overview of sensor class relationships

1896) Charles Elachi, "Introduction to the Physics and Techniques of Remote Sensing," John Wiley & Sons, 1987, Chapters 1-1 and 8-8.

1) Spatial Information

Spatial information throughout the spectrum (2-D or 3-D) is acquired with an imager (this may be a camera, a mapper, a scanner, etc. or combinations thereof). The major application is in topographic mapping.

a. Optical imaging techniques[1897] (optical systems are limited by cloud cover, etc.)
   - Stereophotogrammetry - photographic methods have the advantage of very good geometric resolutions; the disadvantages are in the handling of the film material, the data require geometric corrections, detection of reference points and conversion of brightness into digital information
   - Electromechanical line scanner - 2-D images are acquired line by line by scanning a mirror across track. Where the swaths of two adjacent subsatellite tracks overlap, stereo pairs are obtained. For polar orbiting satellites, overlappings are normally small near the equatorial regions and large over the polar regions. Examples: MSS and TM (of Landsat)
   - Electronic line scanner - in this method an array of CCDs (Charge-Coupled Device) is used instead of a single detector and scanning mirror. The advantages are: absence of moving elements, longer integration times per pixel, and better geometric performance. The disadvantages: the number of detectors is limited, which limits the resolution. Examples: HRV (Spot), MSU-E, etc.
   - Stereo line scanner - Three lines of semiconductors are arranged in the focal plane of the pushbroom scanner perpendicular to the direction of flight. The three lines scan simultaneously in forward-, down-, and aft-direction. Three images of the same scene are acquired from the same swath. The stereo images are taken simultaneously. Example: MOMS-02.

b. Sounding techniques - Sounders are used for field measurements to map subsurface structures, or to map atmospheric profiles of temperature, composition and pressure as a function of altitude. Nadir sounders scan across the satellite suborbital track to provide good horizontal resolution (in the suborbital plane), but they usually have poor vertical resolution. Limb sounders look at the horizon (the limb) and scan vertically, producing good vertical resolution but poor horizontal resolution (see Figure 363).

   The effect of the electromagnetic wave interactions on the Earth's atmosphere are relatively complex due to the multiplicity of mechanisms: scattering, absorption, emission and refraction. The atmospheric sounding technique is based on the fact that carbon dioxide (CO_2) is the same percentage of the total composition of the atmosphere at all altitudes - this is called the constant mixing ratio. The atmosphere has an emissivity that varies with wavelength from about 0% at 10 μm to about 100% at 16 μm. An infrared sounder with about five channels in the region from 11-16 μm probes the radiation of the atmosphere to different depths. Since the emissivity is high at 16 μm, only the top of the atmosphere is sensed by the 16 μm channel. The 11 μm channel, however, senses the radiation from all altitudes. The other channels are somewhere in between.

   All sounders examine either the emission or backscattered radiation spectrum within a narrow field of view.

   In the case of emission, the radiation source is the atmosphere itself. The sensor measures the spectral characteristics and intensity of the emitted radiation. If atmospheric parameters are measured, a down-looking geometry (nadir) is used for the spacecraft sounder.

   In the scattering technique, the approach is to measure the characteristics of the scattered wave field in the direction (or directions) away from the incident wave direction. The source can be the sun or a star.

1897) See chapter 5 in 'The Interdisciplinary Role of Space Geodesy,' Springer Verlag, 1987, pp. 164-165

Some sounding techniques/applications are:

- Temperature sounding - measurement of radiance variation as a function of frequency near a spectral line. For instance, the temperature is measured by examining a small part of the CO_2 spectral band at 15 μm or microwave emission from O_2.
- Composition sounding - identification of atmospheric constituents (by spectral techniques)
- Pressure sounding - technique: columnar absorption measurements of a homogeneously mixed gas constituent
- Density sounding - technique: refraction profile measurements
- Sounding of atmospheric motion (wind fields) - technique: measurement of the spectral line shift by the Doppler effect.

In the occultation techniques, the approach is to measure the changes that the atmosphere impinges on a signal of known characteristics as the signal propagates through the atmosphere. The signal source can be the sun, the moon, a star, or a man-made source. Usually a limb-looking geometry is used to sound the upper atmosphere by occultation or by emission.

Another sounding technique is called Lidar (see chapter O.8.5). This is an active technique, similar to radar. A laser pulse is generated and transmitted outward. The light is scattered back from the atmosphere. A time-gated detector senses the return. Since the sensor is not active immediately after the generation of the pulse, the detector is not 'blinded' by the backscatter from the mirror. In fact, since it is only active for a very brief period of time (the pulse length), it senses the return from only a small portion of the atmosphere at a specific range. Light travels about 0.33 m in 1 ns; thus, segments of the atmosphere in the order of 1 m in depth can be probed this way. The use of several wavelengths permits a limited amount of constituent identification.[1898)]

c. Altimetry

In a number of applications, the information required is strongly related to accurate distance measurements or to the 3-D spatial characteristics of a surface of an object. This is the wide area of topographic measurements with the use of altimeters.

- Radar altimeters (nonimaging radar sensors) use the ranging capability of radar sensors to measure the surface topographic profile (parallel to the subsatellite track). There are beam-limited altimeter types as well as pulse-limited altimeters.
- Scanning Laser Altimetry (see lidar ranging techniques under O.8.5). In the across-track scanning technique 3-D topography can be acquired by a series of neighboring profiles. Beam scanning can be achieved electronically by using a phased array antenna or mechanically by scanning the antenna or by spinning the spacecraft.
- All spaceborne radar altimeters to date have been wide-beam systems, limited in accuracy by their pulse duration.[1899)] Such altimeters are useful for smooth surfaces (oceans), but are ineffective over relatively high relief continental terrain. - A fundamental problem in narrow-beam spaceborne radar altimetry is the physical constraint of antenna size. Large antennas are required for small radar footprints.

d. SAR (Synthetic Aperture Radar) Mapping

SARs generate microwave images of a surface. By using the overlapping images from adjacent tracks, one can produce stereo images. However, observation of the two overlapping tracks must occur from the same side. SARs have the disadvantages of high data reduction cost and poor vertical resolutions.

1898) Encyclopedia of Physical Science and Technology, Academic Press, 1987
1899) NASA 'Topographic Science Report,' 1988, p. 46

e. Radar Interferometry
The radar interferometer applies to two radar receiver units and a baseline across track. The phase differences as a function of location are used for direct measurement of the elevation of the backscattering surface.

2) Spectral Information

Spectral information is acquired with a **spectrometer** (examples: Fourier Transform Spectrometers, grating instruments, Fabry-Perot instruments, or gas correlation spectrometers). Spectrometers are used to detect, measure and analyze the spectral content of the incident electromagnetic field. This information plays a key role in identifying the chemical composition of the object being sensed (for instance atmospheric constituents). In the case of atmospheric studies, the spatial aspect is less critical than the spectral aspect due to low spectral variation in the chemical composition. In the case of surface studies, both spatial and spectral information are essential, leading to the need imaging spectrometers (see chapter O.6).

The selection of the number of spectral bands, the bandwidth of each band, the imaging spatial resolution, and the instantaneous field of view leads to trade-offs based on the object being sensed, the sensor data-handling capability, and the detector technological capabilities.

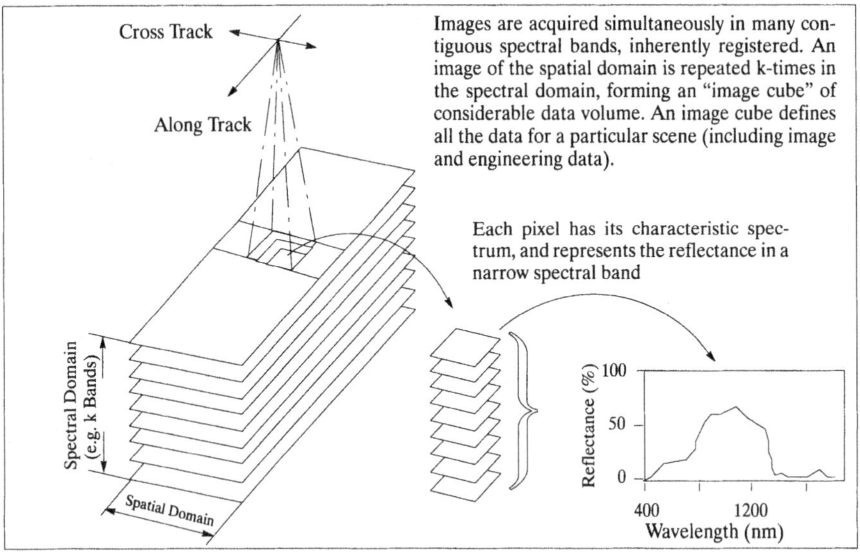

Figure 361: The imaging spectrometry concept

3) Intensity Information

In a number of applications the information needed is mainly contained in the accurate measurement of the intensity of the spectrum over a wide region.

- Radiometers - measurement of radiative intensities (flux, polarization, thermal properties, temperature profiles of the atmosphere and of sea surfaces, etc.). Imaging radiometers are used to spatially map the variation of these parameters.

- Scatterometers - are used in active remote sensing to accurately measure the backscattered field when the surface is illuminated by a signal with a narrow spectral bandwidth. Measurement of surface roughness.
- Polarimeters - these are a special type of radiometer. They are used for applications in which the key radiative information is embedded in the polarization state of the transmitted, reflected, or scattered wave. The polarization characteristic of reflected or scattered sunlight provides information about the physical properties of planetary atmospheres and other target signatures.

Remote Sensor Types	Part of Spectrum utilized	Remarks/Use Note: 1 Å = 10^{-10} m; 1 μm = 10^{-6} m
Scintillation Counters Gamma-ray spectrometers Geiger counters	<0.03 - 0.100 Å	Measurement of emitted natural radiation by gamma-ray detectors or Geiger counters
Scanners with photomultipliers Image orthicons and cameras with filtered IR film >2900 Å	100 Å - 0.4 μm	Records incident natural radiation. Imaged ultraviolet spectroscopy available.
Cameras		
Using conventional B&W and color film	4,000 - 7,000 Å (0.4 - 0.7 μm)	B&W film for high spatial detail
Using infrared film (B&W and IR color)	6,000 - 9,000 Å (0.6 - 0.9 μm)	Improved spatial detail through contrast. Greater reflectance gradients useful for vegetation surveys
Multispectral units	3,000 - 9,000 Å (0.3 - 0.9 μm)	Individual narrow band scenes available with multi-camera systems
Lidar Laser radar	4,000 - 11,000 Å (0.4 - 1.1 μm)	Monochromatic active systems for measuring backscattered radiation from the atmosphere (particles)
Radiometers	Usually IR and microwave bands	Generally measures total radiation in a wide band in the infrared or microwave regions. Imagery obtained by scanning techniques.
Photometers	4,000 - 7,000 Å (0.4 - 0.7 μm)	Measures luminous flux in various bands of the optical region for distribution, color, etc.
Spectrometers	In any spectral region	Narrow-band data available sequentially - Electromagnetic radiation amplitude vs frequency
Solid State Detectors Single detectors, Line arrays Matrices	1 μm - 1 mm	Single detecting element used in scanners, radiometers. 1-D and 2-D arrays for sequential data gathering.
Radars	1 mm - 0.8 m	Narrow band active systems. Both analog and imagery available.
Radiometers (microwave)	1 mm - 0.8 m	Passive systems. Both analog and imagery available.

Table 534: Survey of remote sensor types and their applications[1900)]

O.2 Some Aspects of Radiometric Instrument Calibration

In remote sensing, radiometry or spectroscopy is the measurement of radiation that is incident upon a sensor's receiver/detector after having passed from a target source through an intervening medium to the sensor. The objective of such a measurement is to characterize the source, in terms of its size, shape, location, composition, temperature, and maybe other parameters of interest. Target attributes can be inferred from the instrument response based upon adequate calibration.[1901)]

For the limited case of noncoherent and noninterfering radiation, the target can be characterized in terms of three nearly independent measurement domains:

1900) Manual of Remote Sensing, Second Edition, American Society of Photogrammetry, 1983, p. 41
1901) C. L. Wyatt, "Radiometry," in Encyclopedia of Physical Science and Technology, Vol. 11, 1987, pp. 738-749

- spatial (geometric extent of measurement)
- spectral (distribution of energy flux as a function of wavelength or frequency)
- polarization (orientation of the electromagnetic vector)

The sensor calibration must reflect the sensor response in these measurement domains. The calibration objective is to determine a functional relationship between the target source flux and the sensor output signal. This relationship is also referred to as sensor 'responsivity.' Inherent to sensor responsivity is the complete characterization of the instrument (such as IFOV, spectral bandpass, time constant, polarization, sensor subsystems, the signal conditioning, throughput of the instrument, etc.).

The quality of the measurement is the most difficult problem of calibration. Optically pure measurements are only approximated with practical instruments, which are always non-ideal. Consequently, the interpretation of field data is subject to some uncertainty and is often dependent upon assumptions that must be made about the target. Some of the measurement problems/objectives in the measurement domains are as follows:

- The sensor is bombarded by unwanted flux that originates from outside the instrument field of view (such as radiation contributions from the Earth's atmosphere, or from the sun, or from other background sources). Objective: The sensor output for a 'spatially pure measurement' is a function of the radiant flux originating from the target within the sensor IFOV and is completely independent of any flux arriving from outside this region (see Figure 362 for a general scenario).
- The sensor is also bombarded with unwanted flux that is out-of-band that is, outside the spectral region of interest. Objective: The sensor output for a 'spectrally pure measurement' is a function of the radiant flux originating from within the sensor spectral bandpass and is completely independent of any flux arriving at the sensor from any spectral region outside the bandpass.
- The flux originating from the target may be polarized. The sensor may be sensitive to the polarization of incident flux. Objective: The polarization characteristics may be considered as target attributes for discrimination against the background, or they may be related to physical properties of the target.

A major (ideal) objective of the calibration of a sensor type (radiometer, etc.) is to make the measurements independent of the instrument. This means that when a particular physical entity is to be measured at different times and places or with different instruments, the results should always be the same. It also means that attributes inferred by the measurements are target attributes and not instrument attributes.

Unfortunately, the complexity of the physical conditions of the instrument and source parameters existing at the time of measurement make this ideal of calibration somewhat difficult, if not impossible, to obtain. A basic rule of good performance is that the calibration should be conducted under conditions that reproduce, as completely as possible, the conditions under which the measurements are to be made.

A common practice for spectral-response comparison is the use of a calibration source. Obviously, a sensor in orbit cannot be calibrated as extensively as the same sensor in the laboratory, but high-quality sensors have generally built-in standards (along with well-defined calibration procedures), which permit a periodic comparison at least for the primary measured values. Depending on the quantity to be calibrated, such a 'standard' may turn out to be an external source of known reference (star, moon, sun, etc.) or an internal source, a blackbody or a calibration lamp (e.g., a high-temperature gray-body source).

Often temperature variations around the orbit may affect the spectral stability of the instrument. In such cases, spectral calibration needs to be repeated from time to time by either looking onto a built-in source with well known spectral features or by evaluating well known features in the spectrum itself. Calibration sources can be small semiconductor laser-diodes or atomic line-lamps such as low-pressure Hg or PtCrNe lamps. Optical instruments working in the visible or ultraviolet range of the spectrum often use the well known Fraunhofer-structures in the solar spectrum for spectral calibration.

If absolute radiometric accuracy is important, then spaceborne measurements may also be checked against underflying airborne measurements that are made with the same sensor type. [1902]

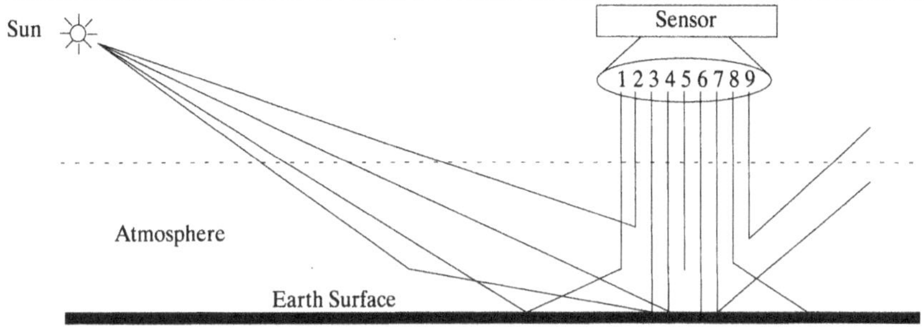

All first-order (direct) and second-order (single-scatter) radiance components are identified.

1) Sunlight scattered by atmosphere into sensor FOV
2) Adjacent ground radiance scattered into sensor FOV
3) Atmospheric emission reflected into sensor FOV
4) Sunlight reflected into sensor FOV
5) Direct atmospheric emission
6) Direct ground emission
7) Atmospheric emission reflected by ground
8) Ground emission scattered into sensor FOV
9) Atmospheric emission scattered into sensor FOV

Figure 362: Generalized scenario of radiative contributions in remote sensing

O.2.1 GNSS Radio Occultation Sounding

Note: The term GNSS (Global Navigation Satellite System) is used here because the method applies equally well to signals/receivers of the GPS and GLONASS constellations.[1903] A new approach to the field of radio occultation (limb sounding) techniques was introduced with the use of the GPS receivers as a science instrument in Earth observation - to perform atmospheric sounding measurements, gravity measurements, and ionospheric sounding measurements. [1904] - This new receiver occultation technique was introduced with GPS/MET on Microlab-1 (launch April 3, 1995) - tracking on C/A code, P code, and carrier phase.

High resolution atmospheric soundings can be retrieved, when the radio path between a LEO GPS receiver and a GPS satellite traverses the Earth's atmosphere. The GPS signals experience refractive bending and retardation in the atmosphere. Accurate receiver measurements of carrier phase changes permit to determine the atmospheric refractive index as a function of altitude. Pressure and temperature profiles can then be derived using the gas law.

1902) Courtesy of J. M. Palmer, University of Arizona, Tucson; published in: J. M. Palmer, "Calibration of satellite sensors in the thermal infrared," 108 / SPIE Vol. 1762 Infrared Technology XVIII, 1992

1903) S. Riley, N. Howard, E. Aardoom, P. Daly, P. Silvestrin, "A Combined GPS/GLONASS High Precision Receiver for Space Application," Proceedings of ION GPS-95, Sept. 12-15, 1995, Palm Springs, CA, pp. 835-844

1904) Note: The radio occultation technique was initially invented in the early 1960s by NASA/JPL in its planetary exploration programs to Venus, Mars and later to the outer planets.

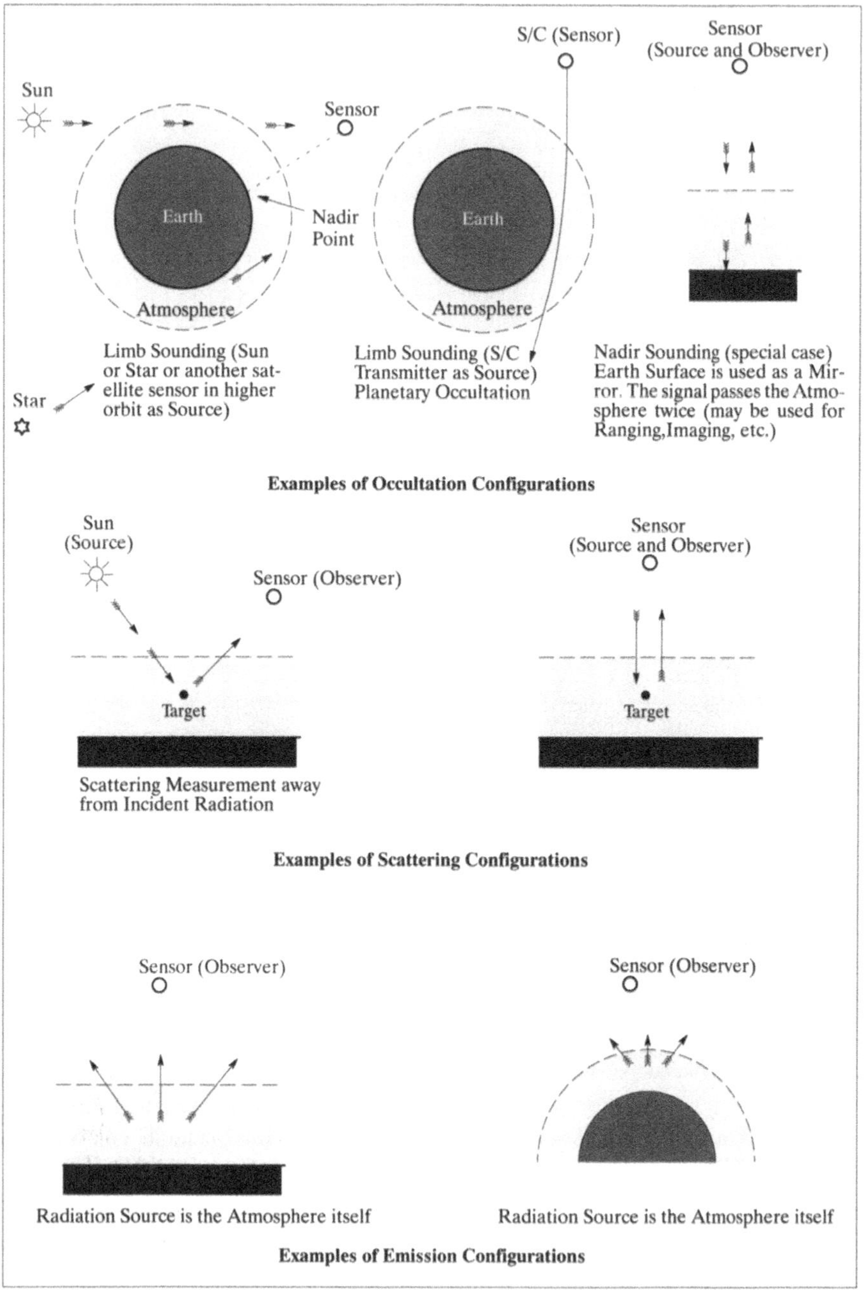

Figure 363: Different configurations for atmospheric sounding [1905)]

1905) Courtesy of Ch. Elachi, JPL

When the path of the GPS signal begins to transect the mesopause at about 85 km altitude, it is sufficiently retarded so that a detectable delay in the order of 1 mm in the dual frequency carrier phase observations is obtained by the LEO GPS receiver. As the signal path descends through successively denser layers of the atmosphere, the delay increases to approximately 1 km at the Earth's surface. Thus the atmosphere creates a unique signal with over six orders of magnitude in dynamic range. - GPS/MET temperature retrieval accuracies of better than 1K have been demonstrated in the altitude range of 10-30 km during the proof-of-concept phase of the mission with vertical resolutions of better than 1 km.

The possibility to extend these measurements into the troposphere (below 10 km) and into the upper stratosphere (up to about 45 km) is the subject of worldwide research. A single orbiting GNSS receiver can observe hundreds of such occultations per day. Hence, the all-weather radio occultation technique has the potential of providing an abundance of data for a wide range of applications, in particular for the fields of operational meteorology and climatology.

The GNSS radio occultation technique relies on precise time delay measurements. This method is generally not as involved with calibration problems usually associated with the other radiative sounding techniques. Hence, more independence of calibration gives radio occultation sounding a further advantage for the collection of long-term data sets.

O.2.2 Correction/Calibration Methods for Sensor Data

Once the functional relationships between the target source flux and the sensor output signals are known, then this knowledge can be applied for data correction in the ground processing algorithms.

- **Atmospheric correction.** Applied to sensor data in order to separate and eliminate the atmospheric path radiance contribution from the ground-reflected signal (i.e., surface contribution, also referred to as ground radiance). Path radiance is energy that may originate from the surface outside the 'field of view'; it is scattered into the beam by the atmosphere. In the visible region and at shorter wavelengths, Rayleigh scattering caused by density fluctuations in atmospheric gases is the dominant contributor to path radiance. Light scattered by particles or aerosols creates both path radiance as well as extinction.

  The algorithms employed are based on model assumptions about the atmospheric state and on supplementary measurements. So far only approximate methods are available. Much effort is given to the development of strict algorithms. The application of atmospheric correction (the atmospheric path radiance contribution of the signal is eliminated by 'masking') to sensor data permits the experienced user to compare different data sets of the same sensor (and/or of different sensors), measured at different times and different locations (mosaicking).

- **Geometric correction.** The removal of sensor, platform, or scene-induced geometric measurement errors in such a way that the data conform to a required projection. This process involves the creation of a new digital image by resampling the input digital data. - Geometric correction is performed by giving each pixel of the image the correct place on Earth. This process is known as **geocoding**, i.e. matching pixel coordinates with the proper Earth coordinates in the required projection.

- **Radiometric correction.** Processing techniques on sensor data to calibrate and correct the particular radiometric measurement errors of the sensor detectors. A typical error that may affect the actual radiation measurement of a sensor is the brightness intensity variation across the image plane. For sensor data in the optical range, the far-side data is usually brighter than near-side data, thereby reducing the contrast of the image; for SAR data, however, the opposite is true, e. g. the near-side data is brighter than the far-side data. Another error source may result from the fact that an output signal of a particular detector cell is not exactly proportional to the brightness intensity of the input signal. - All radiometric corrections are performed on the digital gray values of the pixels of an image, resulting in 'calibrated gray values.'

O.2.3 Electron-scanned Imaging Devices

Electron-beam imager system (vidicon, return-beam vidicon).[1906] A TV imaging technique, employed in the very early period of the space age (a framing system), consisting of the following elements: foreoptics, shutter, photoconductive surface (detector), electron gun, and beam-deflection system. The optical beam of the instrument focuses the image on a photoconductive surface in a fashion similar to that in a photographic system. However, the back side of photoconductive surface is negatively charged by the electron-scan beam prior to shutter opening. After shuttering, the image is retained as a charge pattern on the photoconductive surface. An electron beam is then used to produce an electrical signal by scanning the back side of the photoconductive surface. - The resulting signal of scan beam location is synchronized with the measurement value (signal coming off the transparent conductor) permitting reconstruction of the image.[1907] [1908]

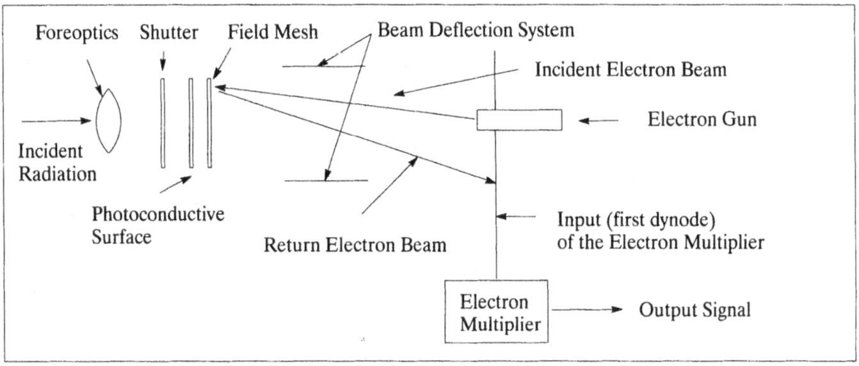

Figure 364: Schematic illustration of an RBV system

The RBV (Return-Beam Vidicon) also uses a photoconductive surface as a detector. However, in this arrangement, the electron beam is reflected from the back side of the photoconductive surface (weaker signal). The intensity of the electron beam, discharged from the electron gun, varies with the intensity of the light. The values of the intensity are then converted into digital information.

The RBV system instantaneously imaged an entire scene, (had greater inherent cartographic fidelity than imagery acquired by the early Landsat MSS sensor), and contained a reseau grid in the image to facilitate geometric correction of the imagery. This resulted in an array of tick marks that were precisely placed in each image.

O.3 Scanners

Many imaging instruments employ a scanning mechanism in order to cover a wide field of view in object space while utilizing narrow FOV (Field- Of View) telescope optics. The size

1906) Note: The name "vidicon" was coined in the early 1950s at RCA Laboratories to distinguish the electron-beam tubes from the photoemissive tubes.
1907) P. H. Swain, S. M. Davis, "Remote Sensing: The Quantitative Approach," McGraw-Hill, 1978, pp. 106-107
1908) P. K. Weimer, et al., "The Vidicon Photoconductive Camera Tube," Electronics, May 1950, pp. 70-73

and mass of instrument scanning mechanisms are determined geometrically by the telescope's aperture size and the desired object space field of view. Other factors, such as the need to view calibration reference sources, can also have a significant effect on the size of scanning subsystems.

A scanning system for the collection of imagery is a complex arrangement of optical, mechanical, electrical and electronic subsystems. Since the field of scanning applications is very broad, only those scanner systems are considered, that pertain to the classes of Earth observation instruments like imagers, imaging spectrometers, and imaging radiometers. The scanners listed in the following chapters are of solid-state technology.

The method of body pointing is yet another scanning alternative and is in particular suitable for single-sensor small satellites. Body pointing implies that the entire S/C (along with its imaging instrument) is pointed into the desired direction. The technique of body pointing may be employed to extend the regular FOV into a particular side-viewing direction to cover an occasional event of interest. With an agile S/C, body pointing may also be used in the along-track direction (forward and backward pointing on a cyclic basis) to collect imagery for stereo applications. It should be pointed out, however, that the body-pointing technique puts additional demands on the attitude control system of the satellite as well as on its power system.

O.3.1 Line Scanners

In remote sensing terminology a **line scanner** is a (nonphotographic) device that forms instantaneous exposure geometries between a target and a detector, with the sensor optics in between, thus forming an image by the successive addition of picture elements. The feature of 'scanning' adds picture elements in one direction (the cross-track direction), thereby increasing considerably the width of the observation path [this is different from 'sounding,' which considers only a line or surface (plane) observation of the immediate orbital path]. The received radiation is then dispersed into its spectral components (bands) using prisms, gratings, dichroics, or filters. An array of detectors senses the dispersed radiation. The signal from each detector is then amplified, processed and stored (on-board recorder).

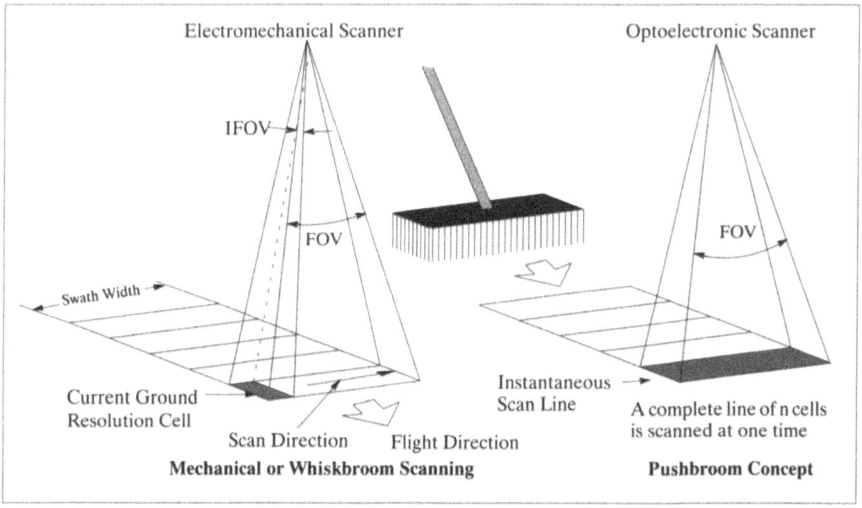

Figure 365: Schematic scanning/imaging geometries on a ground surface

The forward motion of the satellite or aircraft in the along-track direction provides the second image dimension. [1909)]

Basic scan patterns are oscillatory scan motion or rotating scans (continuous rotary scan motion). Key scan characteristics are: high precision, linearity, and high-speed.

By its very nature a scanner directs a moving beam of radiative energy. The direction of the beam is determined either by reflection with a mirror or by refraction with a prism. Movement of the beam is affected by either:

- An electromechanical or piezoelectric device that moves the mirror or prism in the line of path (cross-track direction)
- An electro-optical (optoelectronic) or acousto-optical devicc by applying electric waves or sound waves to change the refractive index of a photoelastic material.

A rotary scanner (polygon or disk scanner) is a high-inertia device in which the optical element is a refractive prism, a reflective polygon, or a disk of lenses or mirrors that rotates continuously in one direction. In such a system, a prism or a number of mirrors or lenses are arranged concentrically around an axis. The rotating device moves each facet, mirror, or lens through the energy path, creating a regular, repetitive scanning pattern.

Polygon scanners nearly always rotate continuously in one direction at a constant speed. Thus, high scan rates are easily attained, relatively large optical elements can be used. Typical design types are: inverted (inward-facing) polygons, pyramidal polygon scanners, and like mirror polygons (refractive or prism scanner). A typical scan pattern of a conical scanner is shown in Figure 381.

An oscillatory scanner is a low-inertia device in which the optical element is either an oscillating mirror (electromechanical - the scanner moves a mirror in a limited angular direction, rather than rotating it continuously), or an electro-optic (optoelectronic) device, or an acousto-optic deflector.

The effect of a scanning motion may also be provided by a spinning spacecraft. The world's first weather satellite, TIROS-1 (launch April 1, 1960) and the follow-up TIROS and ESSA series (up to ESSA-9, launch Feb. 26, 1969) were spin-stabilized. The first generation of Meteosat spacecraft employed also the spinning concept to collect imagery. From a historical point of view the electromechanical (or optomechanical) line scanner (whiskbroom) is one of the first solid-state designs in use of remote sensing. Since the early 1980s the conventional whiskbroom design has been succeeded by the introduction of optoelectronic (pushbroom) devices.

The viewing geometries remain constant for circular orbiting satellites which feature a nadir-viewing imaging instrument. This is in general not the case for nadir-viewing imagers on aircraft where the scanning geometries vary. An important relationship of an aircraft imager design is for instance the velocity-to-height ratio (v/H) because it determines the scan rates, the number of detectors scanned in parallel, and the electrical signal bandwidth. The v/H ratio describes the angular rate of image motion in the focal plane (rad/s). The maximum value of v/H is used to start the design trade-off process.

O.3.2 Electromechanical Line Scanner

The scanner records within its IFOV (instantaneous field of view) a very small area (ground cell), and builds up a line of recorded signals of these small areas referred to as pixels (picture elements). In this serial arrangement, each pixel requires a certain dwell or integration time for the radiation to be detected; thus, the time is additive. The next scan line depends

1909) The interested reader is referred to chapter 3 of the book: "The Infrared & Electro-Optical Systems Handbook," W. D. Rogatto, editor, Vol. 3, Electro-Optical Components, copublished by ERIM and SPIE, 1993 on the forward motion of the satellite. The electromechanical line scanning concept is also referred to as **whiskbroom**, creating an observation swath on the ground surface in obvious analogy to the successive sweep motion of a whiskbroom (in the cross-track direction) to clear a path (Figure 365). Note: The early whiskbroom designs with a single scan line improved considerably with the introduction of parallel scan lines in the along-track direction; thus, providing a wider along-track coverage in a single cross-track scan sweep.

A disadvantage of the electromechanical scanner design is the large sampling frequency per line needed for an unbroken coverage of a sufficiently wide swath. A typical Earth observation platform moves at about 6.5 km/s, hence, a 100 m resolution requires 65 lines/s. This in turn implies a sampling rate of 65,000/s for a swath of a 1000 pixels.

A whiskbroom scanner may provide panchromatic or multispectral imagery in the optical region of the spectrum with medium to high spatial resolutions and medium spectral resolutions. There are single-line scanners in use as well as multi-line scanner arrangements in the along-track direction. **A very important advantage of the single-camera whiskbroom concept: the whiskbroom scan arrangement permits large FOVs.** It means that rather wide swaths can be obtained without severe optical distortions at the swath edges. Disadvantage: the available time to dwell on each detector element (or ground cell) is very brief resulting in less spatial/spectral resolution and/or radiometric resolution. Scanning methods: scanners may use on-axis optics or telescopes with a flat scan mirror located in front of the collecting aperture (the scanning mechanism may be implemented with an oscillating mirror, rotating mirror, fiber scanner, etc.). Examples of line scanners with whiskbroom electromechanical technology are: TM and MSS on Landsat (also the ETM+ instrument on Landsat-7), MSU-SK on Resurs and Meteor series, AVHRR on NOAA series, OCTS on ADEOS, SeaWiFS on Orbview-2 (SeaStar), and CZCS on Nimbus-7. Airborne whiskbroom instruments are AVIRIS and MERES.

Flying Spot. The flying spot scanner is a subset of whiskbroom scanners, in which the focal plane consists of only a single detector. It scans with the same geometry as a line-array whiskbroom (e.g., unidirectional barrel roll or bidirectional sinusoidal scan), but it must scan relatively quickly in order to cover the swath in the along-track direction in a contiguous manner with only a single detector. One advantage to the flying spot is the lack of a potential for image striping due to detector-to-detector sensitivity variations that are not perfectly accounted for by radiometric calibration corrections. An example implementation of the flying spot concept is realized on OLS (Operational Linescan System) flown on the DMSP satellite series since 1976. [1910)]

O.3.3 Optoelectronic Scanners

Electro-optic (or optoelectronic) devices are parallel line-scanning systems with no moving parts (electronic steering of beam motion). The beam deflectors are either analog or digital. In analog deflectors, an applied voltage varies the phase delay across the incident beam; the phase front is tilted when the beam emerges, and the beam direction is changed by the same angle. In digital deflectors the beam direction is controlled by modulation of the polarization direction of the incident radiation.

The radiation of the total field of view (TFOV of n cells in a swath) in the cross-track direction is being imaged simultaneously by a corresponding detector line array (with n cells) positioned in the focal plane of a sensor. This parallel sensing arrangement allows naturally for more integration time of each individual pixel [in the serial whiskbroom concept the integration time for a long line of pixels (hundreds or even thousands) had to be very short]. Successive lines are imaged and sampled by the multiplexer as the platform moves along the orbital path. The time that elapses between the imaging of two successive lines can be as

1910) Flying Spot information provided by David L. Glackin of The Aerospace Corporation, El Segundo, CA long as the platform takes to move the distance in the scene. The general sweep motion in this scanning concept is now in the along-track direction and is referred to as **pushbroom** (see Figure 365). Commonly, the system is designed so that the instantaneous field of view (IFOV) moves through its own width during the frame period, the distance moved in the detector integration period is called GSD (Ground Sampling Distance). [1911] Examples of optoelectronic scanners are: HRV on Spot series, MSU-E on Almaz-1B, on Priroda, and on Resurs, MOS on IRS-P3 and Priroda, MOMS on Shuttle, LISS-III on IRS-1C/D, etc.

Advantages: Longer dwell times lead to low noise in the received signals. There are no moving parts on a pushbroom sensor; hence, scan speeds of less then 1 µs access time are possible. The design remains attractive for high-resolution imaging systems with a FOV <15°. Disadvantages: a very large number of detectors are needed for high resolution images (in the meantime the microelectronics industry is quite capable of producing large detector arrays); in addition, the pushbroom scheme requires a wide field-of view optics system to obtain the same same swath as for a corresponding whiskbroom scanner.

The detector arrays associated with such a line-scanning pushbroom system are usually of the type CCD (Charge-Coupled Device), but they may of course also be of type CID (Charge-Injection Device), CMOS/APS, etc. However, the long association of 'pushbroom' and 'CCD' has resulted in treating the two terms practically interchangeably as synonyms. In any case, a CCD detector is never associated with a whiskbroom (serial) sensing concept.

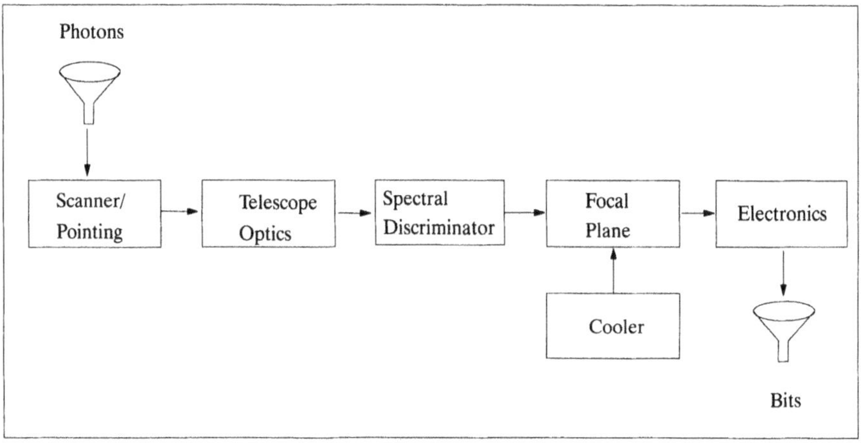

Figure 366: Illustration of the generic optoelectronic sensor

O.3.4 Observation Schemes

The most common detector formats and architectures in imaging sensor technology are: single elements in circular, rectangular, cruciform, and other geometries for rectile systems, the **line detector array** and the **area detector array** as illustrated in Figure 367. [1912]

Line detector arrays are in principle one-dimensional whiskbroom and pushbroom systems oriented in the cross-track direction. In both systems an image is formed on a line-by-line basis in the scan direction (identical to the along-track direction).

1911) Note: In a pushbroom line array, the "frame" is represented by a single line of CCD detectors covering the swath width. In a multispectral or hyperspectral design, the "frame" is represented by an area array, where each line of the CCD detectors covers the swath width of a particular spectral band.

1912) "The Future of Remote Sensing from Space: Civilian Satellite Systems and Applications," Office of Technology Assessment, US Congress, OTA-ISC-558, ISBN 0-16-041884-4, July 1993, pp. 142-143, original source: SBRC

Area arrays may be operated in either **scanning** or **staring** mode.

- Scanning mode operation: The 2-D array (say 2048 x 16 - is usually much longer in the cross-track direction than in the along-track direction) is scanned sequentially, on a line-by-line basis if it is a parallel pushbroom system (one-dimensional scan in the along-track direction). Usually, the detector line arrays (in this case 16) are utilized for different tasks, by dividing them up between panchromatic and multispectral bands, or using them for TDI imaging of a single band, or a combination thereof. Naturally, a large square size area array can also be operated in scanning mode. In fact, most 2-D arrays are operated in scanning mode.
- Staring mode operation: No scanning is needed for the 2-D array, although periodic shuttering may be used. The entire array (frame) is clocked and sampled periodically (see O.3.5).

	Detector Format		
Imaging System	Single Element	Linear Array	Area Array (2-D)
Scanning 2 axes	x	TDI	
Scanning 1 axis		x	TDI
Staring			x

Table 535: Correlation of imaging system and detector formats

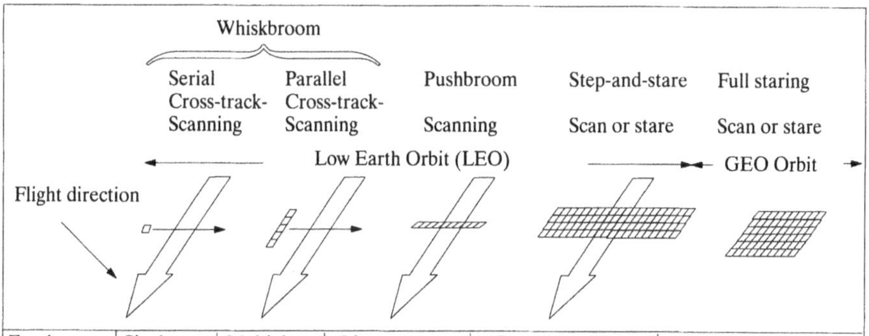

Focal Plane	Single detector per band	Multiple detectors per band	Linear array for each band	Two-dimensional array for each band	Two-dimensional array for each band
Example	AVHRR SeaWiFS	TM MSS	HRV (Spot)		SDI trackers Missile seekers
Cross-track scan speed	high	medium	zero	medium	zero
Scan motion	continuous	oscillatory	none	discontinuous	none

Figure 367: Some optical remote sensing observation schemes

O.3.4.1 Line (or linear) Detector Array

The conventional CCD line array consists of a single row of small, photosensitive cells (or pixels, usually square in size) onto which the incident radiation is focused. The line array observes a target with its length side positioned in the cross-track direction (normal to the forward motion of the satellite). An image is formed on a line-by-line basis in the along-track direction. Such a 'line' has in reality area - with the corresponding field of view (FOV), the array length translates to the swath width, the narrow stripe is one pixel wide. An image is made up of a number of contiguous stripes. In analogy to the older serial whiskbroom scan (where each imaging sweep is in the cross-track direction), this alternate parallel scan arrangement is referred to as **pushbroom** (Figure 365), with the sweep in the along-track direction. All systems with a pushbroom configuration are classified as electronic (or opto-electronic) scanning systems, while a whiskbroom system refers to the older mechanical scan technology (serial cross-track scan).

The parallel pixel observation scheme of the line array provides for relatively long dwell times which in turn leads to low noise in the received signals. There are no moving parts on a pushbroom sensor. Modern CCD line arrays with panchromatic and multispectral capabilities feature a separate detector line for each spectral band.

The current market (1997) offers line arrays of up to 12,000 pixels per line. This configuration provides a fairly good swath width and multispectral measurements with high radiometric resolution in parallel. Line arrays are used in camera systems, including stereoscopic applications.

Parameter	CCD Line Array	CCD Area Array
Number of pixels (max)	14,000	7,000 x 9,000
Pixel size FW (Full Well) QE (Quantum Efficiency)	6.5 μm x 6.5 μm 100,000 e 0.3 at 600 nm	12 μm x 12 μm 500,000 e 0.7 at 600 nm
TDI (Time Delay Integration) sensor		possible but operated in pushbroom mode
Geometric accuracy	high	very high
Stereo capability	two or three-line implementation	overlapping stereo
Spectral performance	sequential only (e.g. filter wheel)	one filter per line
Shutter	not required	required
Operation mode	pushbroom	pushbroom or staring

Table 536: Some configuration/performance capabilities of CCD line and area arrays

O.3.4.2 Area Arrays

The microelectronics integration technology lead eventually to the development of area arrays (CCD, CID, etc.) with the ability to observe in two dimensions simultaneously (an image per integration time). The resulting speed-up of the imaging process requires an alternate observation scheme such as step-and-stare. In this scheme, observations are made intermittently, i.e., after a certain step in the along-track direction occurs due to platform motion. An overlapping step-and-stare method is normally used for sensors in LEO orbit.

The step-and-stare imaging scheme in LEO requires motion compensation to allow the array to stare for a certain period at the same ground target as the platform is moving forward. The array is then stepped to a new location and held again until the cross-track imaging process is completed. The advantages of this scheme are an increased dwell time, only moderate field-of-view optics are needed. Area arrays are particularly suitable for TDI (Time Delay Integration) operations to increase the dwell times. In this case the integration times can be electronically adjusted to the conditions of illumination simply by commanding the number of TDI steps. The disadvantage is that a larger and more complex focal plane array (FPA) than the linear pushbroom scheme is needed.

O.3.5 Staring Array Systems

The staring concept was first formulated in the early 1970s with the advent of large mosaic focal plane arrays. The earliest staring systems were placed into geostationary orbit. The same area of the Earth's surface (footprint) is continuously in view, providing a fixed-position relation between the satellite and its target area. However, this concept has its limitations, since small motions induced by drift or jitter of the satellite cause clutter-induced noise. The field of space surveillance (the detection and tracking of remote point-source objects or targets located within a background scene) is a typical application of a staring system. The prime objective is to extract very weak target signals from a relatively bright and heavily cluttered background.

Staring FPAs (Focal Plane Arrays) are available in the UV, VNIR, SWIR, MWIR and TIR regions. The photon detection scheme is identical to that of non-staring systems (charge-carrier pattern). This pattern is usually integrated in parallel on a storage surface. Before integration, the photocurrent can first be amplified in an intensifier to provide extremely low-light-level detection. Readout of almost all staring sensors is done serially.

The photon flux levels from natural backgrounds are generally much lower at lower wavelengths. The use of staring sensors is desirable because photons imaged by the optics are collected with essentially 100% efficiency due to complete population of the image plane with detector elements.

The step-and-stare concept was later introduced for LEO satellites. In this concept, a sensor with a small FOV can be stepped periodically to new (overlapping or non-overlapping) staring positions as the satellite progresses in its flight path.

O.3.6 Time Delay Integration (TDI)

The TDI technique refers to a cumulative exposure concept of each ground image line by a CCD detector array (or any equivalent array). The main objective is to improve the SNR (signal-to-noise ratio) value which is one of the most important issues for high resolution imaging. The simplest TDI case is the superpositioning and co-addition of two independently acquired images, assuming precise co-registration. TDI is a focal plane technology which sums signals collected by a selectable long column of individual detector elements in the along-track direction, permitting controllable increased sensitivity of the imaging system appropriate to ground scene and illumination conditions. Multiple imaging of a scene with defined sub-pixel shifts will lead to improved spatial resolution, if the displacement between successive optical images can be kept fixed in the sub-pixel range. [1913) 1914) 1915)]

For a 2-D CCD array (say 14,000 x 32) moving with the satellite in the along-track direction, multiple one-dimensional images can be taken of each cross-track stripe as illustrated in Figure 368. Motion compensation is not needed if the detector row shifts are synchronized with the forward motion of the satellite. - In general, the TDI function may be performed off focal plane (normally after digitization) or in the readout circuitry. On-chip TDI requires temporary frame storage for each TDI stage and a multiplexer to transport the signal for the summing process. TDI may be done with a large array of storage capacitors; however, it is most commonly implemented on a CCD line array.

The phrase "time-delay" is used because the scanning process which naturally projects one long single line of detectors (of a total of 32 single lines, in our case) upon a given very narrow ground area (FOV line), projects the next single line upon the same area only after the time interval of one "integration time" has passed. Therefore the sensor electronics is set up to "delay" the input of data produced from photons encountering each single line until the next single line's data is ready, and so on. This permits summation (integration) of the signals produced for the same area element (ground pixel) by the total contributions of the detectors receiving that area element's photons, even though each individual signal was originally received by a detector on a separate single line.

1913) G. Schwarz, M. Datcu, "High Resolution Imaging Methods: A Comparison of Optical and SAR Techniques," The European Symposium on Remote Sensing, Barcelona, Spain, 21-25 Sept. 1998, SPIE, Vol. 3500

1914) R Nagura, "SN Improvement Ratio by Time Delay and Integration for High-Resolution Earth Observation System," Electronics and Communications in Japan, Part 1, Vol. 78, No. 3, 1995

1915) Information provided by S. Kilston of Space Imaging Eosat of Thornton, CO

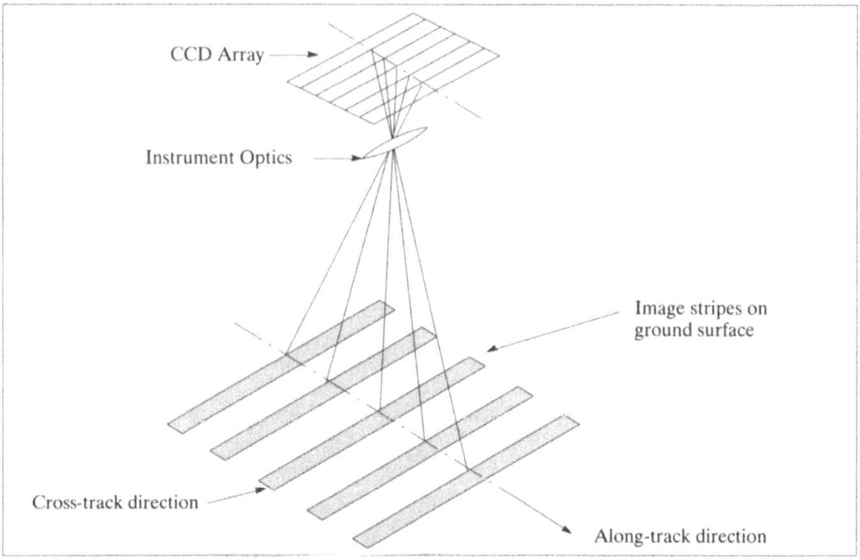

Figure 368: Illustration of the TDI concept

The "total integration time" is then the sum of individual integration times, for a longer effective exposure, and therefore a larger effective signal. The signal builds up in direct proportion to the number of TDI stages, whereas the random noise builds up only in proportion to the square root of that number, so the signal-to-noise ratio is greatly enhanced by this process. The technique is far superior to the use of one single line (say, 14,000 x 1 in size) of detectors exposed 32 times longer, because such detectors would offer far less spatial resolution in the along-track direction.

The TDI method is of course applicable for the entire optical spectrum. It can accommodate different scene and lighting conditions (image integration requires a large dynamic range of 10 to 11 bits). TDI is particularly suitable for weak infrared detection due to its signal enhancement capability.

The TDI concept was developed along with the early CCD technology at several places in the US industry and at NRL (DoD) in the later 1970s as well as at RSRE (Royal Signal and Radar Establishment, Malvern, UK) and first introduced into an infrared detector system by the name of SPRITE (Signal Processing In The Element). The first SPRITE detector had the following specification: eight intrinsic photoconductive detector elements each with a nominal sensitive area of 62.5 μm x 62.5 μm, HgCdTe detector material in the range 8-14 μm, cooling to 77 K, FOV = f 2.5. [1916) 1917)] - The SeaWiFS instrument of SeaStar (Orbview-2, launched Aug. 1, 1997) and the OSA instrument of the Ikonos-1 (CRSS) satellite are examples of more recent (1998) TDI implementations in 2-D CCD line-array technology in the visible range of the spectrum.

1916) C. T. Elliott, D. Day, D. J. Wilson, "An Integrating Detector for Serial Scan Thermal Imaging," Infrared Physics, Vol. 22, 1982, pp. 31-42

1917) A. Blackburn, et al., "The Practical Realization and Performance of SPRITE Detectors," Infrared Physics, Vol. 22, 1982, pp. 57-64

O.4 Sensor Detector Systems

The quality of all radiation sensing depends directly on the quality/capability of the observing instrument, in particular its detector technology. This realization has generated a lot of research and development in the field of detector technology.

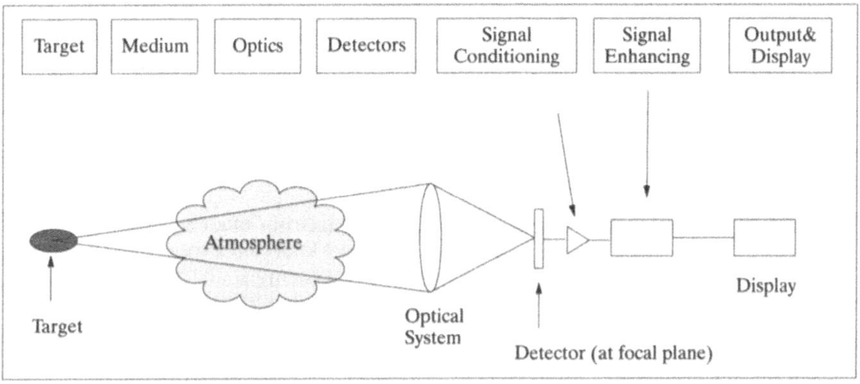

Figure 369: Schematic illustration of an optical system

The principal detection methods used in optical spectroscopy are photographic (e.g., film), photoemissive (photomultipliers), and photoconductive (semiconductor). Prior to about 1940, most spectra were recorded with photographic plates or film, in which the film is placed at the image point of a grating or prism spectrometer. An advantage of this technique is that the entire spectrum of interest can be obtained simultaneously, and low-intensity spectra can be easily taken with sensitive film.

O.4.1 Definitions

Detector. A detector is a device that detects and linearly transduces radiative power into an electrical signal. It refers to the unit cell comprising the smallest resolution element and the first stage of signal processing electronics with that pixel.

A detector typically takes an average value of the signal over a length of time (from about 1 ns to about 1 ms), namely the integration time. Thus, it is not possible to measure any signal variations at the rate of either the optical or the infrared frequency spectrum below the integration time limit. The electrical signal is usually proportional to the average power or photon rate or the integral over the integration time.

Photon detector. Refers to the class of detectors (also called photodetector or quantum detector), where the responsive element is sensitive to changes in the number of free electron carriers, namely electron-hole pairs, that are brought about by the absorption of incident photons. Incident radiation excites electrical charge carriers, causing them to move from one energy level to another energy level within the crystal lattice of the detector. Charge carriers which are at lower energy levels are said to exist in the valence band of the crystalline detector. If a charge carrier is excited into an energy level such that it can move freely throughout the crystalline structure of the detector, it is said to exist in the conduction band.

Photon detectors are usually semiconductor devices on silicon basis or compound materials, where the incident photon flux interacts directly with electrons in bound states, exciting them to a free (conducting) state. In item a) of the energy-level diagram (Figure 370), there is a band of many filled levels (valence band), separated by a gap of a band of many vacant levels (conduction band). In item b) there is only a very small bandgap between the two bands.

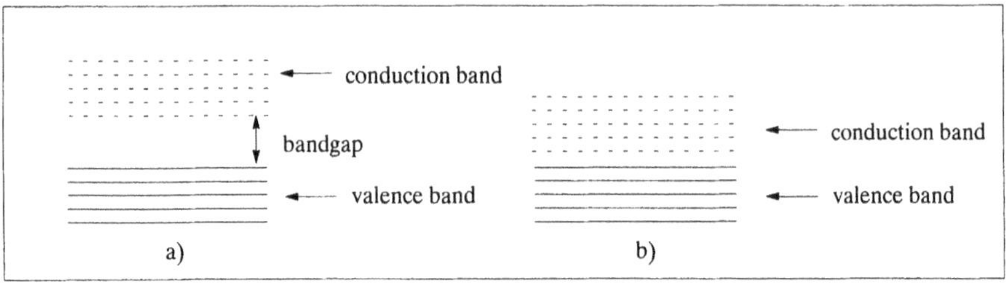

Figure 370: Energy level diagram

If a material is to conduct electricity, it must contain electrons that can have their energy increased by a small amount, thus producing a current. Materials with an energy level as in item b) are conductors. There are vacant levels available near the filled band - the electrons are thus able to accept a small amount of energy. Materials with energy levels as in item a) are insulators. It takes a certain amount of energy to cross the bandgap. Thus, a small increment in energy will not cause a current to flow.

The division of 'internal' and 'external' photon detectors in table 537 is according to whether the interactions are confined to the unit cell (pixel) or are in part external to the detector cell.

Intrinsic detector. Refers to narrow bandgap energy transfer (a current may flow due to a small energy change; however, the electron flow is very small). Detector materials of InSb and HgCdTe are in this class. Intrinsic arrays are those in which the detection and readout are accomplished using an intrinsic semiconductor substrate material.

Photon Detectors		Thermal Detectors	Electromagnetic Detectors
External	Internal	Bolometers	Heterodyne detection
Photocathode: Conventional Neg. electron affinity Gain multipliers: Photomultiplier tube Microchannel plate Gas avalanche	Photoconductive: Intrinsic Extrinsic Photovoltaic: P-N junction Avalanche diode P-I-N diode Schottky diode Heterojunction Graded junction	Thermistors Metal Superconductor Superinductor Cryogenic semiconductor	Metal-metal oxide-metal photodiode
	Photoelectromagnetic Dember effect	Pyroelectric	Josephson junction
	Phototransistors	Thermoelectric	
	Photon drag	Golay cell	
	Hot electron bolometer		

Table 537: Classification of detectors

Extrinsic detector. Refers to wide bandgap energy transfer. Detectors of the type silicon (Si), silicon compounds (SiGa, SiAs, SiPt, SiIn, etc.), doped silicon, photodiodes, and Schottky diodes are in this class. The extrinsic silicon detector type relies on photoexcitation of impurity levels within the bandgap of silicon. The spectral response depends on the energy level of the particular impurity and density (the long-wavelength spectral cutoff is a function of the doping density) state as a function of energy in the band to which the bound charge carrier is excited. Extrinsic silicon detectors operate either as high-impedance photoconductors or as IBC devices. An example of an extrinsic silicon IBC detector is SiAs.

	I Narrow (intrinsic) bandgap	II Wide (extrinsic) bandgap
Photoconductor	PbS HgCdTe	SiIn SiGa SiAs
Photodiode	InSb HgCdTe	SiX (impurity band) SiPt (Schottky diode)

Table 538: Detector types and representative examples

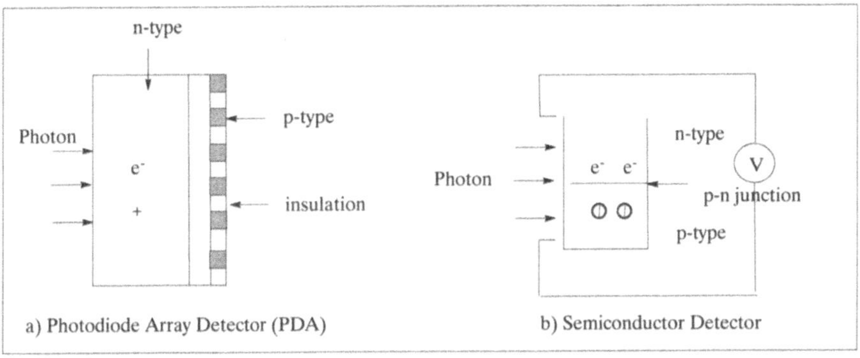

Figure 371: Schematic illustration of a PDA and a semiconductor detector

Quantum efficiency (QE). A measure of the efficiency with which incident photons are detected (such as a photodiode). Some incident photons may not be absorbed due to reflection or may be absorbed where the electrons cannot be collected. The QE is the ratio of the number of detected electrons divided by the product of the number of incident photons times the number of electrons each photon can be expected to generate. Visible wavelength photons generate one electron-hole pair. More energetic photons generate one electron-hole pair per each 3.65 eV of energy.

Thermal detector. Refers to the class of detectors, where the responsive element is sensitive to changes in temperature, brought about by changes in incident radiation. Thermal detectors include: thermocouples, bolometers, Golay cells, and pyroelectric devices. [1918)]

- Thermocouple. Refers to a junction of dissimilar metals. A change of temperature causes a current to flow across the junction. When a scene is optically focused onto a thermocouple, its temperature changes as the incident radiation flux changes. The change in flux can be detected by monitoring the voltage generated by the thermocouple. - A thermopile is a series of thermocouples connected together to provide increased responsivity. Thermopiles are used in applications where DC or staring performance is required (typical of Earth sensors that "ride" the horizon and respond to changes in the FOV due to changes in the "dip" angle).
- Bolometer. A bolometer is a thermal detector that makes use of the change in electrical resistance of certain materials when their temperature is changed. The change in resistance can be detected as a voltage signal by passing a small current through the bolometer (they operate as thermal equilibrium devices, whereas quantum detectors operate as nonequilibrium devices). Bolometer materials are selected for their small thermal capacity. Usually bolometers are cooled with liquid helium (to 4 K) to provide the needed efficiency. Bolometers are suitable detectors for the infrared and microwave regions. A thermistor is an example of a bolometer. Thermistor bolometers have been used for many years as the primary sensor in horizon crossing indicators, scanning Earth sensors, and payload instruments such as ERBE, CERES, HALOE, etc.

1918) M. Schlessinger, "Infrared Technology Fundamentals," 2nd edition, 1995, Marcel Dekker Inc., New York, Basel, Hong Kong

- Golay cell. A gas thermometer with an infrared window and membrane acting as an indicator. Absorption of radiation results in an increase of temperature and pressure, causing the membrane to move.
- Pyroelectric device. Pyroelectric sensors respond to the rate change of temperature. As such they are first derivative sensors and usually require that the scene be "chopped" or scanned. Incident infrared radiation increases the temperature of the crystalline responsive element. This temperature change alters the dipole moment producing an observable, external, electric field. Pyroelectric detectors consist of polarized material which, when subjected to temperature changes, change polarization. These detectors operate in a chopped system. The fluctuation in the exposure to the scene generates a corresponding fluctuation in polarization and thus an alternating current that can be measured.

At the start of the 21 century, the MEMS (Micro Electronic Machined Structures) technology makes it possible to be introduced into high-performance thermal detectors. MEMS has the advantage to allow a high-volume production of closely-packed thermal isolation structures required of high-performance thermal detectors. Prior to the use of MEMS technology , array of thermal detectors were difficult and expensive to produce and required larger geometry, pixel spacing and fill factors. MEMS produces higher performance devices, reduced crosstalk between pixels and better thermal management (Table 539). 1919)

Thermistor Bolometers		
Parameter	Conventional Packaging	MEMS Packaging
Min. element size and spacing	0.1 mm x 0.1 mm	0.025 mm x 0.25 mm
Linear array	0.1 -0.05 mm	0.025-0.004 mm
Matrix array	Not practical	0.025 mm x 0.025 mm
Integrated electronics	Nor practical	Yes
Thermopiles		
Min. element size and spacing	0.1 mm x 0.1 mm	0.025 mm x 0.25 mm
Linear array	0.1-0.05 mm	0.025-0.004 mm
Matrix array	Not practical	0.025 mm x 0.025 mm
Integrated electronics	Not practical	Yes
Pyroelectrics		
Min. element size and spacing	0.05 mm x 0.05 mm	0.025 mm x 0.25 mm
Linear array	0.1-0.05 mm	0.025-0.004 mm
Matrix array	Not practical	0.025 mm x 0.025 mm
Integrated electronics	Not practical	Yes

Table 539: Comparison of conventional/MEMS manufacturing techniques of IR sensors

Photoconductive (PC) effect (discovered by W. Smith in 1873). The electrical conductivity of a material depends on the number of free charge carriers (i.e. free electron-hole pairs) per unit volume and on the rate at which carriers move under the influence of an electric field. The photoconductive effect occurs when a bias voltage is applied across a uniform piece of detector material. The photocurrent is then proportional to the density of electrons excited into the conduction band by the incoming photons.

Photoemissive effect. Incident photons with sufficient energy striking a detector surface may cause an ejection of electrons from the surface thereby causing an electric current flow. Energetic radiation in the spectral ranges from gamma-ray to VNIR (up to wavelengths of about 1.5 μm) is responsible for the photoemissive effect. - Electrons are also emitted from a detector surface when it is bombarded with high-velocity electrons. This process of secondary emission with high internal amplification is utilized in photomultipliers (PMT).

1919) A. Doctor, "MEMS Technology Based Sensors for Payload Instruments and Attitude Control for Small Satellites," Proceedings of the 14th AIAA/USU Conference on Small Satellites, Logan, UT, Aug. 21-24, 2000, SSC00-III-4

Photoemissive detectors (photodiode, photomultipliers, image tubes, microchannel plate).

- A photoemissive diode (photodiode, see Figure 371) consists of a surface (cathode) permitting the ejection of electrons and a separate electrode (anode) on which the electrons are collected (both are often sealed within an evacuated glass envelope).
- A photomultiplier tube (PMT) is a sensitive radiation detector for low intensity applications (spectral range from EUV to NIR or from 0.1 to 1.1 μm). A PMT consists of a photoemissive cathode, a series of electrodes (dynodes), and an anode sealed within a common evacuated envelope. Photons that strike the cathode emit electrons due to the photoelectric effect. Voltages applied to the cathodes, dynodes, and anode cause electrons that are ejected from the photoemissive cathode to make collisions with the dynodes in succession, knocking out more electrons at each collision. In this arrangement the dynodes provide signal amplification up to factors of 10^6. The amplified signal is taken off the anode.

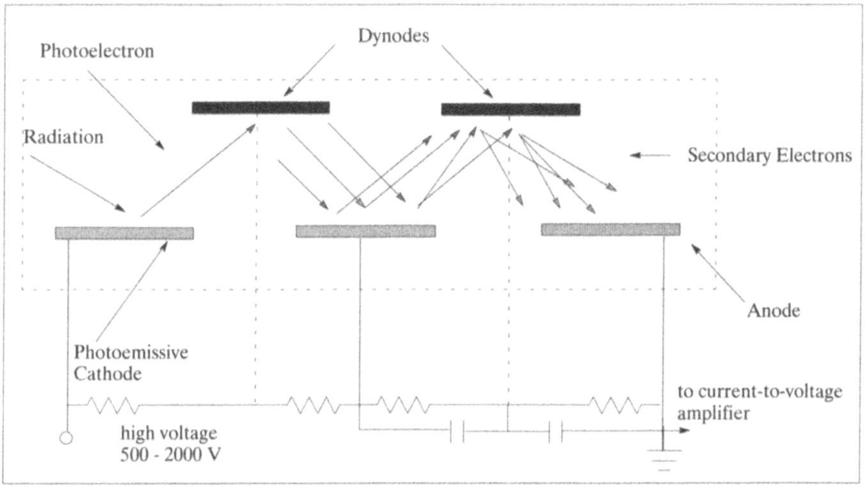

Figure 372: Schematic illustration of a PMT

- Microchannel plate (MCP). A detector system, covering the spectral range from UV to x-ray and to gamma-rays and particles, employing large-area electron multipliers that provide spatial resolution as well as amplification. Depending on requirements MCPs may be coupled with a scintillator, a film, or a CCD array to provide imaging in the spectral range. MCPs consist of a matrix of hollow glass tubes. The hole diameter and spacing is typically 5-50 μm. The surface of a plate is usually coated with a photocathode material, while the interior of the holes is coated with a material of a high secondary emission coefficient. MCPs are used for plasma particle, ion, photon, and/or electron counting applications (measurement of ionospheric fluxes in the solar wind, etc.). They exhibit a high count-rate capability; a stack of two or more MCPs allow ray-tracing for angular measurements (position-sensitive). Among photomultiplier detectors of single photons, MCPs have the best intrinsic detector resolution, in the order of 20 ps. The detection efficiency of a microchannel plate for photons is a function of the angle of incidence and photon energy.

Photovoltaic (PV) and Photoconductive (PC) detectors. These are often referred to as photon detectors since the effect is to convert the incident photon energy directly into the excitation of electrons in the conduction band. The two types differ in the method of sensing the photo-generated electrons electrically.

- The photovoltaic (PV) effect is achieved when a photon-produced electron hole pair is separated by a space charge field [examples: a p-n junction, Schottky barrier devices, IBC (Impurity Band Conduction) devices], thus producing a photocurrent. [Photons with sufficient energy create electron-hole pairs, they are separated by the internal field existing at the p-n junction. The valence-to-conductance bandgap in InSb is 0.23 eV at liquid nitrogen temperature. This accounts for the detector's sensitivity cutoff at a wavelength of 5.5 μm]. PV devices operate in the diode's reverse bias region; this minimizes the current flow through the device which in turn minimizes power dissipation. In addition, PV detectors are low noise because the reverse bias diode junction is depleted of minority carriers.
- The photoconductive (PC) effect occurs when a bias voltage is applied across a detector. The photocurrent is then proportional to the density of electrons excited into the conduction band by the incident photons. [The presence of electrons in the conduction band lowers the resistance. Intrinsic electron-hole pairs are created by raising an electron from a valence band to the conduction band. Extrinsic excitation refers to to electrons raised to levels within the forbidden band of the semiconductor]. PC detectors require readout amplifiers which bias the detectors with a constant current; this causes higher on-chip power dissipation. PC detectors can respond to fluctuation in radiation at operating temperatures greater than PV detectors with equivalent semiconductor bandgaps.

A variety of semiconductors can be used to exploit either or both effects. Intrinsic semiconductors of the type HgCdTe use both effects. InSb and InAs are commonly used as photovoltaic detectors. Typically, an InSb detector, operated at 77 K (liquid nitrogen), has a spectral response from 1 - 5.5 μm, peaking in responsivity at 5.3 μm.

IBC (Impurity Band Conduction) devices use a thin, heavily doped detection layer grown epitaxially on a lightly doped silicon substrate. Photons enter through the substrate and are absorbed in the heavily doped region. A bias applied to the surface detector contact depletes the detection layer of ionized donors. The impurity band holes are swept from the region that becomes the active region of the detector. Photodetection occurs when a photon ionizes a neutral donor site in the detection region, thus producing a free electron that is collected by the surface detector contact.

Schottky barrier detector (named after Walter H. Schottky). A photon detector, which utilizes internal photoemission of 'hot' carriers (holes) from metal or metal silicide electrodes into silicon for the detection of radiation. A back-illuminated Schottky barrier detector operates by photoemission of holes over the so-called Schottky barrier from a silicide into the valence band of silicon. Photons with energies between the bandgap and the barrier height are transmitted through the silicon substrate and absorbed by the metal electrode of the junction.

A Schottky diode has a metal-semiconductor contact (e.g., an Al layer in intimate contact with an n-type silicon substrate). It is electrically similar to a p-n junction, though the current flow in the diode is due primarily to carriers having an inherently fast response. It is used for high-frequency, low-noise mixer and switching circuits. Schottky implementations are limited to the SWIR range, they require low operating temperatures and display low quantum efficiencies. Their ease of fabrication and excellent uniformity may offset the disadvantages.

From a historic point of view, the first Schottky sensors were vidicons with a spectral range between 1-2 μm. These devices used P_2Si electrodes on n-type silicon. The first solid-state Schottky sensors had Pd_2Si electrodes on p-type silicon together with integral CCD multiplexers (spectral range: 1-3.5 μm). Since 1980 Schottky sensors use PtSi electrodes providing a spectral response from 3-5 μm at 80 K cooling and a range of 8-12 μm with 45 K cooling.

QWIP (Quantum Well Infrared Photodetector). Infrared photons excite ground state electrons out of a stack of quantum wells in a QWIP, producing a current in an applied electric field. QWIPs are extrinsic devices (for the 6-14 μm range, TIR) in which dopant concentrations are limited by the epitaxial growth processes. The technology of GaAs and AlGaAs multiple quantum well detectors is currently (later 1990s) the most mature among the different types of QWIPs. The potential advantages of QWIPs (GaAs and AlGaAs) include the use of standard manufacturing methods based on GaAS growth and processing techniques [monolithic integration of these detectors with GaAs field effect transistors (FET), CCDs and high-speed signal processing electronic is possible], highly uniform and well-controlled molecular beam epitaxy (MBE) growth on large GaAs wafers, high yield and thus low cost, thermal stability, and extrinsic radiation hardness. The QWIP limitations are: low quantum efficiency and high dark current. [1920) 1921)]

SLS (Strained Layer Superlattice). SLS (InAs, InGaSb) photodetectors employ transitions across a fundamental gap. Thus carrier lifetime is limited by radiative and Auger recombination processes rather than the much faster hot-electron processes in QWIPs. SLS technology is based on the type II interface where the conduction band of one material (e.g. InAs) is lower than the valence band of the second material (e.g. InGaSb). Because of this unusual band lineup, the superlattice can have a bandgap smaller than of either constituent material. Absorption of infrared radiation occurs between the valence band of the first material and the conduction band of the second one. The resultant effective bandgap is dependent upon the composition, well width, and strain within the superlattice layer. SLSs detectors are intrinsic devices (absorption occurs via interband transitions) that are utilized in the TIR spectral range.

Photographic detectors (plates and films). A photographic system is a framing system, it means that all of the data of an image are acquired simultaneously. The film serves as the radiation detector (photochemical surface). Photographic films have the disadvantage of a very limited spectral range (VNIR spectrum only), however, they feature very good spatial resolutions.

Josephson junction detectors. Make use of the nonlinear interaction between the incident photon's electromagnetic field and the high-frequency oscillating currents and voltages in a biased Josephson junction to perform millimeter-wave mixing and direct detection. The detectors are particularly useful for millimeter- and submillimeter-wave detection.

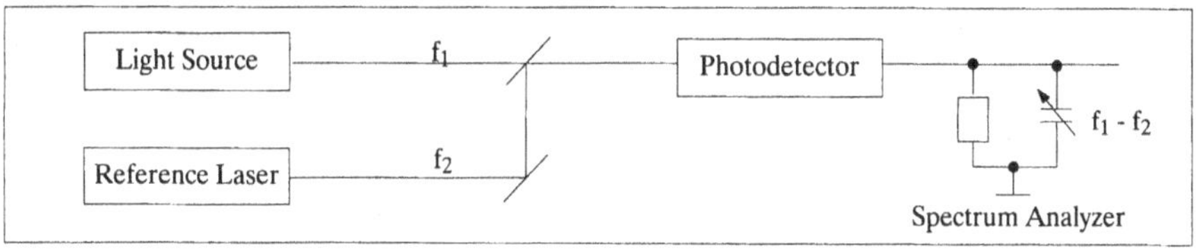

Figure 373: Optical heterodyning principle - measurement of difference frequencies

Heterodyning (optical heterodyning). A process in which the incoming frequency f_1 (to be analyzed) in an optical receiver system is superimposed over a second stable reference frequency f_2 prior to measurement of the total intensity with a photodetector. The measured intensity contains a term which is modulated by the frequency difference (f_1-f_2) of the two light waves. The heterodyning technology is particularly suitable for the detection of very weak infrared and microwave radiation sources; heterodyning instruments (such as microwave limb sounders) play a crucial role for the monitoring performance of the Earth's atmosphere (chemical composition of the atmosphere, monitoring/deduction of the fine structure of the spectra) as well as in the study of astronomical objects. The heterodyning tech-

1920) A Rogalski, "Comparison of the performance of quantum well and conventional bulk infrared photodetectors," Infrared Physics & Technology, Vol. 38, 1997, pp. 295-310

1921) J. R. Jensen, et al., "The Application of Quantum-Well Modulators in Satellite Instrument Design," Johns Hopkins APL Technical Digest, Vol. 15, No. 1, 1994, pp. 7-17 nique is applicable in the spectral range from about UV (0.3 μm) to the microwave region of about 10 mm wavelength.

- **Coherent detection** (optical). In coherent detection, a complete frequency band at the observed wavelength is converted with a local oscillator (LO) frequency to a much lower center frequency, where it can be amplified and spectroscopically analyzed without further degradation in SNR, and where the spectral resolution is being limited only by the spectral purity of the LO. The coherent technique combines the weak (observed) optical signal at nominal frequency f_1 with a strong optical reference beam, the frequency-stable LO at frequency f_2, on a wideband square-law detector, thereby producing radio frequency beats at the frequency difference $f_1 - f_2$. The sources must be narrowband, hence single-mode (both longitudinal and transverse) lasers are preferred. The power in the beat signal is proportional to both the signal power and the LO power.[1922)] Quantum-limited detection performance can be obtained, even on thermally noisy detectors, if the LO "shot" noise can be made to dominate the thermal noise. Frequency resolutions of better than 10^{10} can be obtained, enabling full spectral analysis of the optical signal. The first three spectral moments of signals backscattered from either solid or distributed targets in the lidar's scattering/receiving volume measure reflectivity, velocity and velocity spread, respectively.
- **Homodyne detection.** A form of coherent detection where the LO is derived from the transmitting laser; hence, nominally $f_1 = f_2$. This form of detection is often used on CW (Continuous Wave) systems where f_1 is Doppler-shifted slightly from f_2, and radio frequency beats occur at both $+(f_1-f_2)$ and $-(f_1-f_2)$.
- **Heterodyne detection.** A form of coherent detection where the LO is derived from an independent source that is locked to the transmitting source at some offset frequency; hence, $f_1 \neq f_2$. Beat frequencies occur in a band around (f_1-f_2), known as the intermediate frequency (IF). The frequency offset permits one to discriminate approaching from receding velocities, and is often used in conjunction with pulsed transmitters.
- **Incoherent detection.** Refers to direct detection of an optical signal on an optical detector, with no LO present. In the direct detection method, the optical signal field is analyzed and dispersed in an interferometric filter (or in diffraction grating) prior to detection. The measurement accuracy of a direct-detection interferometric system depends only on the total scattered signal, it is not dependent on the energy of individual pulses, but on the total laser energy. Incoherent detection is normally the method of choice at shorter wavelengths, where the thermal background emission gets rapidly less important due to the rapid drop of the radiation intensity on the Wien side of the blackbody curve, and at low to moderate spectral resolution.

In this context: **Superheterodyning** is primarily used to detect signals entirely in the radio-frequency domain, beating the beat signal again with a third stable source, the second LO, at another frequency, f_3.

The principal configuration of a heterodyne receiver consist of the following elements: quasi optics [a low power signal is combined with a closely spaced strong fixed frequency LO (Local Oscillator) signal], mixer [non-linear detector element; due to the non-linearity of the mixer the frequency difference of the signal and LO radiation which is called IF (Intermediate Frequency), can be obtained at the mixer output port], low-noise amplifier, band-pass filter, spectrometer, data acquisition and instrument control.[1923)]

Photoelectromagnetic (PEM) effect. Refers to incident radiation (photons) being absorbed at or near the surface of a semiconductor, thereby producing free electron-hole pairs. These pairs diffuse from the surface into the crystal. The presence of a magnetic field B directed

1922) M. J. Post, "Development of Coherent Laser Radar," Proceedings of IGARSS'94, August 8-12, 1994, Pasadena, CA, Volume II, pp. 923-925

1923) G. W. Schwab, "Heterodyne spectrometers," Infrared Physics & Technology, Vol 40, 1999, pp. 207-218 along the y-axis causes the electrons to separate from the holes as they diffuse away from the surface. The separation of charges produces a voltage across the terminals. If the external circuit is shorted, a current flows as long as incident photons arrive at the surface. If the exterior circuit is left open, a voltage appears across the open terminals and remains as long as the surface is irradiated.

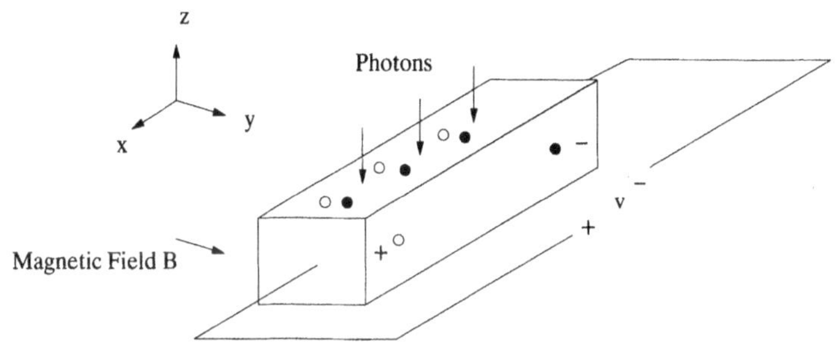

Figure 374: Illustration of the PEM effect

Array or detector type	Useful spectral range
Scintillation counters	0.03 - 0.100 Å
MCP (Microchannel Plate)	γ-rays to UV (..nm - 0.3μm)
Silicon photodiodes and photoemissive devices such as photomultiplier tube (PMT)	0.3-1.3 μm
InSb (indium antimonide)	1.3 - 5.5 μm
InGaAs (Indium Gallium Arsenide)	
HgCdTe (mercury cadmium telluride, also referred to as MCT) or photoconductors	1.3 - 35 μm
Semiconductor bolometer	>30 μm
Avalanche photodiode	
CdZnTe (Cadmium Zinc Telluride also referred to as CZT)	
PbS (Lead Sulfide)	
PbSi (Lead Silicon)	
PtSi (Platinum Silicon)	
QWIP (Quantum-Well Infrared Photodetector)	(6 - 25 μm)
Sb (antimonide)	
Si:As (Arsenic-doped silicon - Impurity Band Conduction (IBC) detector	infrared
AlGaAs/GaAS (Aluminum Gallium Arsenide)	
InSb/InAsSb long-wavelength strained-layer superlattice devices	
HIP (Heterojunction Internal Photoemission)	IR detector arrays
SiGe/SiHIP detectors	
Si:As IBC arrays	4-28 μm
Infrared tunneling sensor array	
Electron-tunneling Golay cell detector	
Superconducting micro bolometers	100 - 1000 μm
Resonant Infrared Photoconductor	

Table 540: Typical detector types utilized for spectral ranges

O.4.2 Charge-Transfer Devices

The collection/readout of photon-generated charge and the conversion of this charge into proper imagery is the purpose of charge-transfer devises in remote sensing systems. The three dominant charge-transfer devices are CCD (Charge-Coupled Device), CID (Charge Injection Device), and CMOS/APS (Complementary Metal-Oxide Semiconductor/Active Pixel Sensor). Charge-transfer devices provide storage and manipulation capabilities for signal sequences. All three device types depend on silicon detector material.

For Earth observation applications, CCD is by far the most widely utilized and most mature technology. High-speed CCDs are available with extremely high pixel counts, ultra-low noise, almost perfect linearity, and with quantum efficiencies of 90% or more. - CIDs with x-y addressability and row/column readout amplifiers offer advantages in applications, where radiation hardness is required (outward-looking sensors), but they are not widely used due to their high readout noise. - CMOS/APS with readout amplifiers resident in each photosite are an emerging technology with the availability of sub-micron semiconductor technology. The transistor sizes are now small enough to cover about 75% of the pixel area, allowing for the incoming radiation to get through a portion of the pixel. The drawbacks of CMOS/APS devices are: inherently higher readout noise, non-uniformity, and low quantum efficiency.

Parameter	CMOS/APS		CID	CCD
Manufacturer	Photobit	JPL	CID Technologies	Pixel Vision Inc.
Product	Pb256G			Kino
Format	256 x 256	1024 x 1024	1024 x 1024	2049 x 1024
Pixel size	11.9 μm x 11.9 μm		27 μm x 27 μm	12 μm x 12 μm
Pixel type	Photogate APS	Photodiode APS	Photogate CID	Back-illuminated CCD
Output	Analog	Analog	Analog	Analog
Number of outputs	1	1	1	8
Peak QE	20%	45%	30%	90%
Fill factor	40%	25%	25%	100%
Read noise	17 e^- rms	5 e^-	350 e^-	2 e^- rms
Saturation	88,400 e^-	300,000	1,000,000	240,000
Dynamic range	5,200:1	60,000:1		120,000:1
Operating voltage	3.3 V	3.3 V	±15 V	±15 V
On-chip quantization	No	No	No	No
Power	2 mW/Mp/s	30 mW/Mp/s		500 mW/Mp/s
Max pixel rate	1 Mp/s	1 Mp/s	21 Mp/s	160 Mp/s
Max frame rate	15 Hz	1 Hz	21 Hz	160 Hz
Dark current	815 pA/cm^2	500 pA/cm^2	500 pA/cm^2	25 pA/cm^2
Timing&control	On-chip	Off-chip	Off-chip	Off-chip
Window readout	Yes	Yes	Yes	Yes
Electronic shutter	No	No	No	Yes
Variable integration time	Yes	Yes	Yes	Yes
Fastest shutter speed	1/15 s	1/15 s	1/500 s	1/1000 s

Table 541: Comparison of 1997 imaging technologies/products of leading sensor suppliers

O.4.2.1 Charge-Coupled Device (CCD)

A CCD is a photosensitive solid-state silicon-based imaging sensor (detector) implemented with large-scale integration technology based on MOS (Metal Oxide Semiconductor) capacitors that collect and transfer photon-generated charge. A MOS capacitor is a three-layer sandwich formed by positioning a metal electrode, insulated by a layer of silicon dioxide, onto a silicon substrate (buried channel). Typically, the device is built on a p-type substrate. An n-type region (about 1 μm in thickness) is formed on the surface. [1924)] The gate electrode consists of a thin silicon dioxide layer either on a metal surface or on a heavily doped polycrystalline silicon layer.

1924) Note: CCD fabrication is based on **NMOS technology** because the surface layer consists of an n-type region.

The buried channel capacitor exhibits the following features:[1925)]

- The ability to create a potential well in a local region beneath a single electrode
- The ability to modulate or control the potential under the gate
- The storage location (channel minimum) is positioned away from the silicon - silicon dioxide interface
- Low dark current makes it possible to store signal charge for long periods of time
- The charge that collects can be generated optically, injected electrically, or created by charged particles such as cosmic rays, protons, or high-energy photons
- The ability to move charge from a position beneath one electrode to a second neighboring electrode rapidly and with very low loss.

Image capture. Each CCD pixel absorbs a number of incident photons, converting their energy into an electrical charge which is trapped in the depletion region of the substrate. The isolated charge packets are transported by manipulating potential wells (place of minimum potential) within the substrate. CCD readout techniques employ clock-controlled circuits which transfer these charges to a matching grid of elements and shift all charges by one row at a time (shift register).

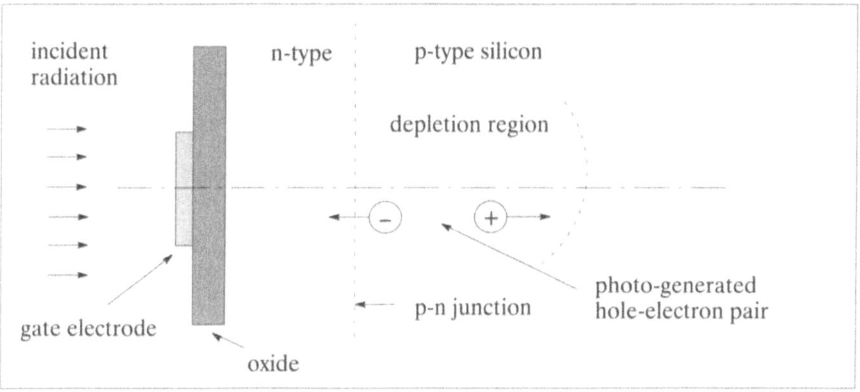

Figure 375: Schematic illustration of a buried MOS capacitor

Imaging arrays. Conceptually a CCD is a two-dimensional array of MOS capacitors (M columns by N rows) that collect and transfer photon-generated charge. An additional independent linear register (readout register) is placed next to the array with its charge transfer direction orthogonal to that in the array. This serial register is arranged so that there is a single pixel adjacent to each of the M columns, and is terminated in a charge-detection output amplifier. - Following an integration period, array readout involves simultaneous clocking of all rows of charge packets one pixel toward the serial register. The transfer process causes the bottom row to transfer into the serial register. The charge packets are then transferred along the serial register toward the output amplifier where they are detected. [The difference between a CCD and other charge-device technologies is that the charge from the CCD must be transferred down the parallel column and across the horizontal shift register to the amplifier, before it can be read out of the array.] The resulting data stream is a pixel-by-pixel, row-by-row representation of the image being detected on the CCD. - In a frame transfer configuration the detected image may first be transferred into an opaque temporary storage array by a proper clocking circuitry. From there it is shifted line by line into the serial readout register. [1926)]

1925) "An Introduction to Scientific Imaging Charge-Coupled Devices," of SITe, http://www.site–inc.com/tutorial.htm
1926) Information received from James Janesick of Pixel Vision Inc.

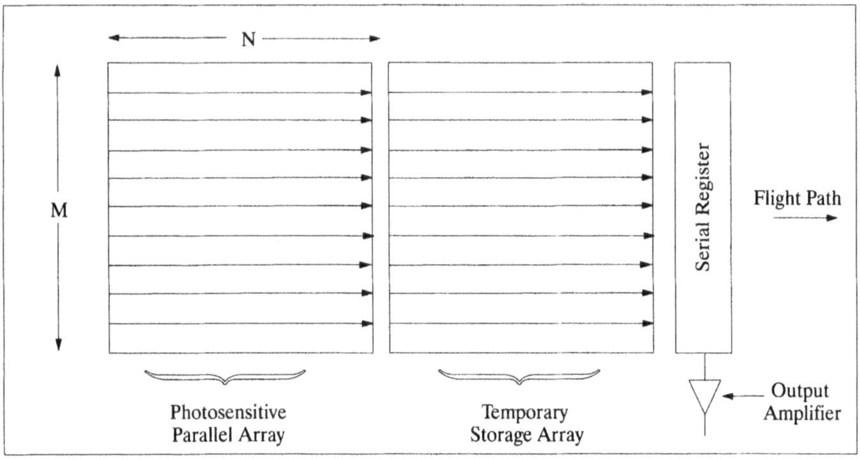

Figure 376: The CCD imaging array (back-illuminated frame transfer)

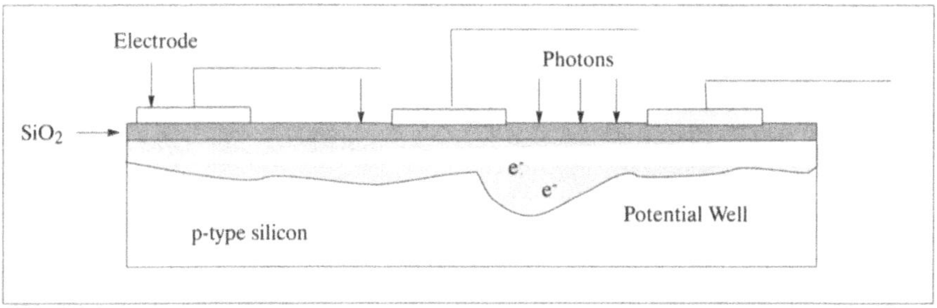

Figure 377: Schematic illustration of a CCD

An advantage of the CCD technology is that its readout scheme transfers the charge actually collected to the output amplifier. It means, there isn't any unwanted extraneous capacitance or resistance to degrade the signal integrity, the reason for its very low-noise behavior. Anti-blooming can be implemented by placing anti-blooming drains in the column stops adjacent to each pixel.

One of the limitations of conventional CCD structures are the polysilicon gate electrodes that are used to clock the charge down a column also obscure incident light - especially in the UV region of the spectrum. Back-illuminated operation overcomes this limitation by imaging through the thinned back surface of the CCD - away from the imaging electronics. Another limitation of the CCD scheme is that the charge must be transferred down a column before readout. If no shutter is used, there is a finite amount of time that each packet will integrate other parts of the scene while it is transferred from the active light sensitive region. This is often referred to as transfer smear. Frame transfer CCDs minimize this effect by using a storage region to protect the charge from incident radiation, they rely on a very fast transfer mode from the active to the storage region. Image smear decreases with the speed of the transfer mode.

In front-illuminated mode of operation, incident photons must pass through a passivation layer as well as the gate structure in order to generate signal electrons. Photons will be absorbed in these layers and not contribute to the signal. The QE (quantum efficiency) of front-illuminated CCDs is poor in the blue and UV regions due to the high absorption coefficient for short wavelength photons in silicon. - First generation linear CCD arrays are usually front-illuminated devices.

In back-illuminated mode of operation, the silicon wafer is thinned to increase short wavelength QE. Light is incident on the exposed, thinned surface. The incident photons do not pass through the front surface electrodes and passivation layers. Rather, an enhancement layer at the back surface creates an electric field that forces photo-generated electrons toward the potential wells.

CCDs are manufactured from wafers. Silicon wafer sizes in 1980 were of 100 mm diameter; in 1990, wafers of 150 mm (6 inches) diameter were introduced. Current (1998) maximum wafer sizes of 200 mm diameter are a limiting factor for large CCD arrays. The year 2000 may see the introduction of 300 mm wafers in the program of major wafer manufacturers.

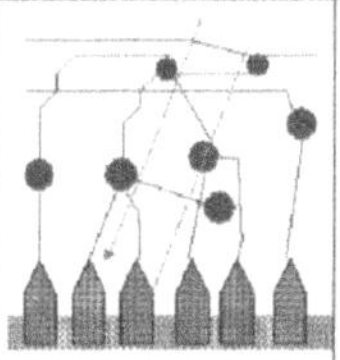
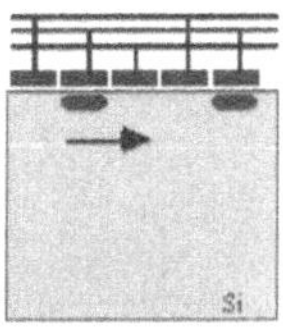

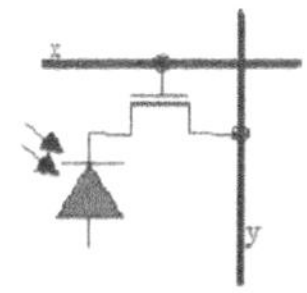

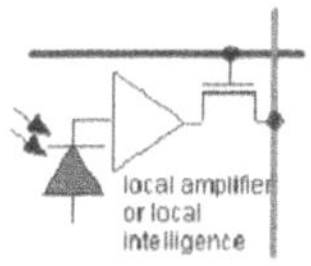

Eye	CCD	MOS diode array	MOS active pixel	Film
Photo-chemical detection, transfer by neurons	Multiplexing of charge packets by loss-less transfer	Multiplexing by switches on photodiode node	Multiplexing of the local amplifier output	Photo-chemical detection and development
Rod and cone signal levels are "processed" by ganglion cells	One charge sense amplifier at the end of the CCD register	One charge sense amplifier on the output bus	(Charge sense) amplifier in every pixel	Pure local change in optical absorption

Figure 378: Principles of the eye, CCD, and CMOS

O.4.2.2 Charge-Injection Device (CID)

A photo-sensitive image sensor (detector) implemented in large-scale integration technology. Charge packets are typically measured by injecting them into a substrate or by shifting charge packets under an electrode to induce a voltage on the capacitance formed by the electrode and the substrate. A CID can be randomly addressed. The pixel structure is contiguous with maximum surface to capture incident light which is useful for sub-pixel measurements.

CID imagers have pixels constructed of the same basic structure of CCD imagers - MOS capacitor integrating site. The difference is how the collected photon-generated charge is being read. In a CID each addressable x-y photosite contains two coupled MOS-charge storage areas, referred to as photogates. Simultaneous selection of one or more rows and columns defines the coordinates of the pixels to be read out. Driving one photogate transfers charge to a second neighboring photogate within the same pixel. The charge is then read out. In most implementations, a single amplifier is located per row or per column. Therefore the charge remains intact to allow further integration or adaptive exposure control. The photosite sampling occurs by a row being selected and making the connection to the readout amplifier. Then, a column is biased to transfer the collected charge from the column to the row at each pixel site. The collected charge at each pixel site remains at the pixel site during the readout operation. Only the pixel with its row connected to the output amplifier is read. The column can be rebiased to return the charge into the column of any other pixel in the row to be read or any other row if selected to the output amplifier.

The benefits of the CID structure are: random access (CIDs can perform fast windowing of selected regions of interest), front side sensitive, non-blooming, radiation hard (CIDs have more than 100 times greater tolerance to ionizing particle radiation than CCDs), non-destructive read, and adaptive exposure control. The random access and radiation hardness features make the CID technology quite suitable for outward-looking sensors (in particular for astronomy applications).

The disadvantage of CID detectors, compared to CCD detectors, is noise due to the readout method as described above. The charge collected at the pixel site is divided by the bus capacitance of the row, noise is introduced by the bus resistance. For this reason, CIDs have not been widely used for most applications.

Criterion	**Eye**	**CCD**	**CMOS APS**	**Film**
Spectral response	400-700 nm peaked at 555 nm	400-1000 nm smooth	400-1000 nm	300-700 nm
Peak quantum efficiency	<20%	>50%	>50%	
Dynamic range	10^6 logarithmic 10^2 linear	10^4 linear	$2x10^3$ linear, also 10^6 logarithmic	10^3 - 10^4 non-linear
Dark limit	0.001 lux 10^{-6} W/m^2	typ: 0.1 lux <0.0001 possible	typ: 1 lux 0.001 possible	virtually zero
Noise photons	10	10	100	100
Integration time (room temperature)	0.3 s	40 ms typical 5 min possible	40 ms typical 10 s possible	virtually unlimited
Maximum frame rate	ca. 15 Hz	10 kHz	>>10 kHz	1 snapshot only
AIC (Automatic Illumination Control) delay	1 minute	1 s typical	<1 s	
Resolution (packing density)				
Number of pixels	120 M	800 k typical 60 M record	800 k typical	>10^8 grains, with 100-1000 gr./pixel
Pixel pitch	2-3 μm	5-10 μm	5-10 μm	10-20 μm effect.
Focal plane size	3 cm	1 mm - 11 cm	1 mm - 2 cm	limited by film size
Operating conditions				
Radiation hardness	1 mrad	10 krad	10 krad (Mrad is assumed)	
Operating temperature	36°C	-200 to +200°C	0 K to +200°C	0 K to 100°C
Power dissipation	<1 mW	500 mW typically	50 mW typically	none
On-board image processing				
Cosmetic quality	perfect	very good	not so good	almost perfect
Color	ideal	poor (only RGB)	poor	poor or print quality
Absolute photometry	not possible	easy	easy	possible
Focal plane processing	extensive	none	typically none	none
Access method	data driven (focus of attention)	serial only	serial, random access	optical only
Data path	5 M nerves	8-10 bits	8 bits	none

Table 542: The human eye versus silicon - some performance parameters [1927)]

O.4.2.3 CMOS/APS Detectors

The availability of the sub-micron semiconductor technology in the 1990s was needed to develop the CMOS/APS technology (the transistor sizes are now small enough to cover

1927) Courtesy of Bart Dierickx of IMEC, Leuven, Belgium [presented at the IEEE Workshop on CCD&AIS (Advanced Imaging Sensors), June 4, 1997]. Note: Noise photons is defined as: noise photons = noise electrons / (quantum efficiency x fill factor)

only about 75% of the imager area - allowing for radiation to get through a portion of the pixels). The method utilizes readout amplifiers at each pixel site (photosite), thereby eliminating the bus capacitance and bus resistance problems of a CID.

Because the CMOS/APS technology can be manufactured on conventional CMOS production lines, they offer the potential for significantly lower cost, they also offer the capability to be integrated with other functions such as timing logic and A/D conversion. The radiation hardness of CMOS/APS sensors is exceptionally good at nearly 1 Mrad. Typical CCD detectors have a comparatively low radiation hardness of approximately 10 krad.

O.4.3 Infrared Detection

Infrared radiation can be detected by its interaction with matter either through the process of direct photon absorption at the detector surface, or through the photon-to-photon exchange heating effect. In the latter case, the radiation is absorbed as heat, resulting in a temperature rise. The detectors measuring the change in resistance, are referred to as thermal detectors, or energy detectors, or square-law detectors. In the former case (photon absorption), the valence band (or impurity level) electrons in a bandgap semiconductor are elevated to the conduction band by direct photon absorption. The increase in conduction band electrons causes a current flow that can be measured. The devices measuring the direct photon absorption (and thereby generation electron-hole pairs) are referred to as photon detectors, or as quantum detectors, due to the quantum-mechanical interaction between the photons and the detector material. Note, the “Definitions” of chapter O.4.1 apply also to the infrared region.

Photodetector materials	Spectral Range (μm)	Comments
Si (silicon)	0.2 - 1.1 (UV, VNIR)	1.1 μm is the cut-off wavelength for detection of light imposed by the fundamental band-gap of Si
Ge	0.4 - 1.8	
HgCdTe (also referred to as MCT - mercury cadmium telluride)	1 - 20	MCTs are the most important material for IR detectors and arrays. MCT operates in photovoltaic mode. Recent advancements in LPE p-on-n backside-illuminated detector technology have extended the cut-off wavelength to 17 μm
PtSi	3 - 5 (MWIR)	Schottky-barrier diode detector
InAs	1 - 3.8	
InSb (77K)	1 - 5.5 (SWIR, MWIR)	InSb operates in photovoltaic mode
HgZnTe		
QWIP	3 - 19	For low background, long wavelength applications

Table 543: Some detector materials for spectral ranges

Detector / Operating Temperature	300 K	190 K	80 K	1.5 - 60 K
PbS	3.0 μm	3.3 μm	3.6 μm	-
PbSe	4.4 μm	5.4 μm	6.5 μm	-
InSb	7.0 μm	6.1 μm	5.5 μm	5.0 μm
HgCdTe (PV mode)	1-3 μm	1-5 μm	3-12 μm	10-16 μm
HgCdTe (PC mode)	1-11 μm	3-11 μm	5-25 μm	12-25 μm
Extrinsic Si	-	-	-	8-32 μm
Extrinsic Ge	-	-	-	7-200 μm

Table 544: Detector spectral cutoff range for various detector materials

Table 544 summarizes the spectral response coverage of common IR detectors in four temperature ranges where 300 K refers to room temperature, 190 K is representative of a thermoelectric cooler, Freon-13, or dry ice; 80 K corresponds to the temperature of liquid nitrogen; 1.5 to 60 K refers to liquid Ne, H, or He. [1928) 1929) 1930) 1931)]

A major drawback of IR photodetectors is the need for cooling to suppress thermal generation of free carriers resulting in noise. Hence, conventional IR semiconductor imagers use cryogenic or thermo-electric coolers, complex IR optics, and expensive sensor materials. Advances in micromachining and materials science have also lead to the development of uncooled infrared detectors for some applications.

Thermal detectors (active or passive) require cooling to increase sensitivity (this principle was first demonstrated in Germany during World War II). Infrared radiation is 'thermal' by nature, hence the detector is affected by the medium that is measured. As a rule of thumb, the longer the IR wavelength that is to be measured, the colder the detector must be. In the VNIR region the detector element temperatures rarely need to be below 200 K. From 1.1 - 17 μm, temperatures are typically in the range 50-80 K. The longer wavelengths of the microwave region usually demand temperatures below 20 K. (Note: The detectivity of a cooled detector is much higher than one operating at room temperature The noise contribution from background radiation at 300 K is several orders of magnitude higher than that of the 4 K surroundings of the detector).

Photon detectors are fundamentally limited by statistical fluctuations in the background and signal photon flux. To reach this fundamental limit, thermal fluctuation noise in the detector element must be reduced by cooling the detector. The detection of the longer wavelength (lower energy) photons requires more cooling to avoid performance degradation due to thermal noise.

O.4.3.1 Detector Arrays and Focal Plane Assemblies (FPAs)

Detector technology can influence some basic sensor design trade-offs, resulting in size and weight reductions. Increasing the number of detector elements can permit reductions in optical aperture size, until aperture size becomes governed by spatial resolution - and not by the optical collecting area required to achieve the requisite radiometric sensitivity. In essence, the availability of integrated high-density detector arrays allows designers to reduce the size of the optics to dimensions that are close to those set by the theoretical diffraction limit.

Technology now permits the design and production of integrated detector arrays that include first–level preamplification and multiplexing as part of the focal plane; this in turn permits simplification of the sensor's other electronic subsystems.

Most conventional focal plane assemblies are built by using detector arrays (chips) with many detecting elements [in line array (pushbroom scanning) or area array (staring or scanning) operation mode]. These can be either monolithic (detector and signal processor in a single semiconductor crystal) or **hybrid** (detector and signal processor in separate elements).

The hybrid approach to imaging in the infrared region consists of coupling a silicon readout (CCD shift register or otherwise) to any one of the various types of IR photodetectors such as InSb, HgCdTe, etc. The silicon CCD acts as a signal processor (not as a detector). In direct-injection hybrid devices the photo-generated charge from a conventional infrared array is immediately transferred into a CCD shift register. Indirect injection devices employ a buffer stage between the photodetective elements and the CCD shift register, which may be a single MOSFET (Metal Oxide-Silicon Field Effect Transistor) or a number of off-chip

1928) W. D. Rogatto, "The Infrared &Electro-Optical Systems Handbook, Vol. 3, Electro-Optical Components" Co-published by ERIM and SPIE, 1993, p. 251

1929) Special Section: "Semiconductor Infrared Detectors," Optical Engineering, May 1994, pp. 1392-1510

1930) Special Section: "Infrared Technology - Part 2," Optical Engineering, March 1994, Vol. 33, No. 3

1931) Special Section: "Infrared Technology - Part 1," Optical Engineering, January 1994, Vol. 33, No. 1

amplifiers. The entire hybrid structure is often called a "focal plane array" (FPA). Hence, infrared detector arrays are "sandwiched" structures whose layers are separated by an insulator, a grid of electrical connections in the form of tiny raised sections, referred to as "bumps." At this point, it should be pointed out that infrared detectors are not based on the charge-coupling principle of the silicon CCD.

Two modes of operation are characteristic of monolithic infrared arrays: inversion mode and accumulation mode devices.

Inversion mode: These devices are constructed and operated like the basic silicon CCD, which is based on the generation of an inversion region at the insulator/semiconductor interface to which the photon-generated minority carriers migrate and are collected in potential wells (intrinsic mode operation).

Accumulation mode: This operation results from a proper polarity-gate voltage biasing that includes the same majority carriers at the semiconductor/insulator interface. The proper applied gate voltage is just the negative of that used for the more usual inversion mode devices. The accumulation mode enables extrinsic materials, whose photon-absorption cutoff wavelength is determined by the dopant impurity ionization levels, to be used for monolithic infrared CCDs.

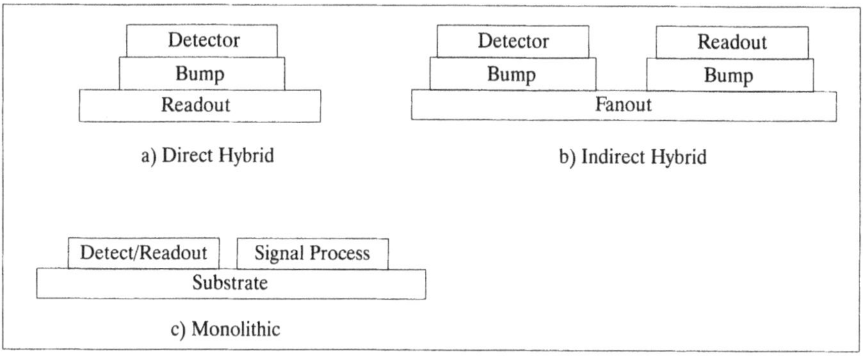

Figure 379: Some FPA detector readout architectures

The Schottky-barrier detection in monolithic infrared arrays is based on internal photoemission from metal/semiconductor Schottky barrier arrays onto a silicon substrate. Hot carriers are generated in the metal by absorption of photons with wavelengths greater than cutoff that penetrate through the semiconductor. The carriers with proper energy and momentum transit the contact barrier potential, they are emitted into the semiconductor, where they neutralize the fixed charges at the edge of the depletion region, induced by application of a reverse-bias voltage. Signal readout may be accomplished by a MOS transfer gate conversion of majority carriers to minority carriers required by the silicon CCD. Schottky barrier devices have a highly uniform responsivity across an array due to photodetection in a metal film. However, they exhibit a relatively low photoemissive yield in comparison to other CCDs.

In the PtSi (Schottky-barrier) detector FPA, radiation is transmitted through the p-type silicon and is absorbed in the metal PtSi (not in the semiconductor), producing hot holes that are then emitted over the potential barrier into the silicon substrate, leaving the silicon charged negatively. The negative charge of silicide is transferred to a CCD by the direct charge injection method.

O.4.4 Radiation Detection Limits

Within the EMS (Electromagnetic Spectrum) the spectral radiance (i.e. energy per unit area and solid angle) varies considerably and with it the ability to detect (measure) radiation. In general, photons are considerably more energetic at shorter wavelengths than at longer wavelengths.

The photon/frequency conversion relations in the EMS are:

1eV corresponds to 2.418 x 10^{14} Hz; or about 1.24 μm

$E[eV] = \frac{1.24}{\lambda[\mu m]}$ photon energy related to wavelength ranges is shown in Table 545

Spectral region	Range of spectral region	Photon energy level range
UV (Ultraviolet)	0.01 - 0.38 μm	124 - 3.4 eV
VIS (Visible)	0.4 - 0.7 μm	3.1 - 1.8 eV
NIR (Near infrared)	0.7 - 1.3 μm	1.8 - 1 eV
SWIR (Short-Wave infrared)	1.3 - 3 μm	1 - 0.4 eV (or 400 meV)
MWIR (Mid-Wave infrared)	3 - 5 (or 6) μm	400 - 250 meV
TIR (Thermal infrared)	6 - 14 μm	200 - 90 meV
FIR (Far infrared) also referred to as SMMW (Sub-millimeter Wave) region	about 10 - 1000 μm (1 mm) 100 μm (3 THz) 200 μm 300 μm 500 μm 800 μm	124 - 1 meV 12 meV 6 meV 4 meV 2.5 meV 1.4 meV
MW (Microwave region)	1 mm - 1 m or (<300 GHz frequencies < 300 MHz)	1 meV - 1 μeV

Table 545: Photon energy levels within some defined regions of the EMS

Sensitivity of the detector is a measure of its ability to discriminate between small signal differences. Single photons can easily be detected by photomultipliers (PMT) in the visible and near-infrared range (also referred to as VNIR). This is so because of the relatively high energy level of single photons in these particular regions - the transformation from a single photon, striking a detector surface, to many electrons within the detector, works well within VNIR.

The same detection method does not work anymore for photons of longer wavelengths since the signals are too weak for detection. For instance, in the SMMW (Submillimeter-Wave) region, also referred to as FIR, the photon energy range is somewhere between 1-100 meV. For comparison, in the VIS region the photon energy is about a thousand times higher.

The FIR region is of great interest to spectroscopic research, because it covers the rotational spectra of molecules and vibrational spectra of solids, liquids, and gases. There are many applications in atmospheric research, astronomy and biochemistry.

Up to about the year 2000, most observation studies in the FIR region were hampered by the absence of sensitive detectors. So far, measurements in FIR have been made with either bolometers, which measure the total absorbed power, or heterodyne detectors, which use reference frequencies to mix the absorbed radiation down to lower, more readily managed frequencies. Neither approach comes close to the single-photon sensitivity, possible with photomultipliers in VNIR. Even with superconducting bolometers, the attainable sensitivities in FIR are far below the level of single-photon detection.

First experimental results of single-photon detection in FIR are reported by S. Komiyama et. al. in the wavelength range of 175-210 μm. [1932)] [1933)] [1934)] This is done by using a SET (Single Electron Transistor), an element of nanotechnology, consisting of a semiconductor

1932) S. Komiyama, O. Astafiev, V. Antonov, T. Kutsuwa, H. Hirai, "A single-photon detector in the far-infrared range," Nature, Vol. 403, Jan. 27, 200, pp. 405-407

1933) L. Kouwenhoven, "One photon seen by one electron," Nature, Vol 403, Jan. 27, 2000, pp. 374-375

1934) R. Fitzgerald, "A Photon-Activated Switch Detects Single Far-Infrared Photons," Physics Today, March 2000, pp. 20-21 quantum dot in a high magnetic field. The detection mechanism measures cyclotron resonance between two Landau levels (the Landau levels are narrow bands of allowed electronic states). Photon absorption can occur only when the photon energy equals the energy difference between the Landau levels. The sensitivity is a consequence of the measurement technique, in which one absorbed photon leads to a current of 10^6 -10^{12} electrons through the quantum dot. By contrast, mechanisms of conventional detectors in single electron transistors produce only a few electrons per incident photons.

O.4.5 Acousto-Optic Devices

Acousto-optics is the science of the interaction of sound and light. Sound propagating through a solid or liquid creates compressions and rarefactions in the material that in turn pulls out color features from the light passing through it. This acousto-optic interaction has created a new optical filtering technology, referred to as AOTF (Acousto-Optic Tunable Filter), employed for spectral separation in imaging applications. One feature of AOTF is its electronically tunable center frequency. The center frequency is displaced by a change in sound frequency used to activate the filter. In this way, the AOTF sorts out radiation features emanating from an object. Since all objects have characteristic and distinctive optical signatures, an effective methodology exists for remote acquisition and identification applications.

Examples of acousto-optic devices include optical modulators, deflectors, scanners, Q-switches, isolators, spectrometers, and frequency shifters. These devices have many applications in high-speed laser printers, laser lithography, optical communications and computing, large-screen laser projectors, frequency shifting, particle inspection, optical spectrum analysis, signal processing, and radar, rangefinder, and target designation. [1935) 1936)]

Progress in acousto-optics has been stimulated by the development of growth methods for acousto-optic crystals and by new methods for fabricating piezoelectric transducers that can efficiently convert electrical energy into acoustic energy at frequencies ranging from a few tens of megahertz up to several Gigahertz. One promising development in acousto-optic devices is the recent commercial availability of acousto-optic tunable filters (AOTF). In these filters, the interaction between an ultrasonic wave and light in an acousto-optic crystal is used to spectrally filter the light. In operation, tunable filters resemble interference filters and can be used in applications that require a filter wheel, grating, or prism.

Principle of operation: The AOTF is based on acoustic diffractions of light in an anisotropic medium. The device consists of a piezoelectric transducer bonded to a birefringent crystal. When the transducer is excited by an applied RF signal, acoustic waves are generated in the medium. The propagating acoustic wave produces a periodic modulation of the index of refraction. This provides a moving phase grating that, under proper conditions, will diffract portions of an incident beam. For a fixed acoustic frequency, only a limited band of optical frequencies can satisfy the phase-matching condition and be cumulatively diffracted. As the RF frequency is changed, the center of optical passband is changed accordingly so that the phase-matching condition is maintained.

In a quartz collinear AOTF, the incident light, the diffracted filtered light and the acoustic wave all interact collinearly in a birefringent crystal. As a result of the acousto-optic interaction, part of the incident light beam within the filter spectral passband is coupled to the diffracted light beam. The polarization of the incident light beam is orthogonal to that of the diffracted light beam. Because of the zero-order beam and the diffracted beam are collinear, polarizers must be used to separate them.

1935) V. E. Pozhar, V. I. Pustovoit, "Acousto-optical spectrometers," Proceedings of the XIV International Conference on Geomagnetic Electronics and Electrodynamics, Section of Spin-Electronics, Moscow, Firsanovka, Nov. 13-16, 1998, Vol. 2, pp. 365-381

1936) http://www.brimrose.com/press_1.html

In a tellurium dioxide (TeO_2) non-collinear AOTF, the acoustic and optical waves propagate at quite different angles through the crystal. In this configuration, the zero-order and diffracted beam are physically separated, so that the filter can be operated without polarizers. Also, the two orthogonally polarized beams do not separate until they exit from the crystal, and the angle of diffracted beam is absent for the change in the first order with a change of wavelength. This implies that only a single fixed detector is necessary during a spectral scan.

Most AOTF devices are designed with two types of birefringent crystals depending upon operational wavelength. TeO_2 is preferred AOTF material because of its high acousto-optic figure of merit. The crystal, although useful in the visible and infrared region up to 4.5 μm, is not suitable for ultraviolet applications due to its short-wavelength transmission cutoff at 350 nm. For ultraviolet spectroscopy, crystalline quartz is used.

Spectral imaging: Because of its large field-of-view angle and high spatial resolution (>100 lines/mm), an AOTF can also be used for spectral imaging applications. It can be used for acquiring spatial, spectral, and polarization information from fluorescent probes in biological cells, space and terrestrial observation, and image sensing. Spectral imaging at orthogonal polarizations can give the spectral properties and size distribution of gases and aerosols, in addition to revealing the nature of solid surfaces from their reflectance spectra.

The spectral resolution of the AOTF depends on the design of the AOTF and the wavelength, which is proportional to the wavelength squared (λ^2). This is quite different from a traditional grating system in which the spectral resolution is proportional to the wavelength.

For the collinear diffraction geometry, a 100% diffraction efficiency can be obtained corresponding to the full conversion of the incident wave energy into the only diffraction mode, having an orthogonal polarization. In quartz crystals, used in instruments with AO interaction length of 10 cm, the required acoustic power is a few Watt, and in TeO_2 with a length of 3 cm, it is much less than a Watt. [1937)]

Since the cross-section sizes of collinear AO (Acousto-Optic) interaction are limited only by the sound beam diameter, the spatial aperture (pupil) of AOTF far exceeds the aperture of a slit on other spectral devices, e.g. gratings. A large AOTF aperture, both angular (up to dozens of degrees) and spatial (up to 1 cm^2) permits a large optical throughput. In a collinear diffraction, the interaction length is limited only by the dimensions of the crystal, this provides the high diffraction selectivity ($\Delta\lambda/\lambda$ is about 10^{-2} to 10^{-4}). The electronic control of the sound frequency makes fast AOTF spectrum tuning (10^{-5} to 10^{-6} s) possible.

Some AOS (Acousto-Optic Spectrometer) advantages. AOS offers some advantages when compared to the traditional optical instruments (such as the grating and Fourier-transform spectrometers). These are:

- High optical throughput due to the use of a bulk phase grating (having large cross-section sizes) produced in the crystal by the ultrasonic wave
- Rather high spectral resolution (in the order of 0.4 Å in the UV spectral region)
- High SNR because the measurement technique based on amplitude modulation and synchronous demodulation. High SNR means also that rather weak optical signals can be measured.
- High-speed performance of the instrument. AOS tuning time from any spectral point to another arbitrary spectral point is only limited by the ultrasound transit time (about 10^{-5} s). The random spectral access is a unique feature that enables principally new and most general data collecting algorithms
- The use of an adaptive measurement mode implemented via optimum fitting of the instrument function to the spectral object and changing this function during the measurement process. It is useful for detection of unknown objects (mixtures)

1937) S. E. Harris, R. W. Wallace, "Acousto-optic tunable filter," Journal of the Optical Society of America, Vol. 59, 1969, pp. 744-747

- High accuracy of the radiation intensity measurement (spectral brightness) due to exact analytical knowledge of the instrument apparatus function based on detailed mathematical analysis
- Absence of mechanical moving parts and insensitivity to vibrations, shocks and impacts
- Light mass and small size of the instrument, necessary for use of AOS in heavy exploitation conditions, e.g. on-board of aircraft and spacecraft
- One of the most important advantages of collinear AO systems is their ability to transmit color images of an object without any spatial distortions. This can be extremely useful in designing new imaging systems for the detection and identification of a variety of object monitoring. [1938)] [1939)] [1940)]
- Low cost of commercial production. The cost of AOS devices is several times cheaper in comparison with conventional grating-based instruments.

Some AOS instrument applications are:

- Environment monitoring of the Earth surface (land and ocean) from spacecraft, aircraft and ship; measurement of minor gases in the atmosphere; monitoring of vegetation characteristics (projective factor and vegetation index); detection of sea water bio-productivity and pollution.
- Utilization in industrial/technological processes: automatic control of technological processes in microelectronics, quality output control, monitoring of hazardous gases discharged to the atmosphere and impurities released into waste waters; detection of organic chemicals inside industrial facilities.
- Monitoring of natural disasters and accidents: earthquakes, forest fires, volcanic eruptions, etc. Detection of technogenic catastrophes, including military activity.
- Attempts of forecasting of natural disasters. Investigation of volcanic activity, detection of earthquake precursors for different tectonic regions of land- and sea floors.
- Chemical spectral analysis, in particular Raman spectroscopy, fluorescence, transient processes spectroscopy.

O.4.6 Resolution (for Visible and Infrared Imagery)

The amount of data/information contained within an image depends in large part on the image resolution. [1941)] Definitions of spatial, spectral and radiometric resolution are not unique. There are interdependencies between spatial, spectral, and radiometric resolutions for each remote sensor affecting the various compromises and trade-offs for particular applications.

Spatial (or Geometric) Resolution

Spatial resolution is the smallest unit of distance (usually a side length of a square area in an image) that can be discriminated by a sensor measurement of the target. Spatial resolution is a function of geometry (scale) between sensor and target for the instant of measurement. The image in a mechanical scanner setup is not formed at once over the swath width (as it would be the case in a camera setup), rather it is built up by a sequence of individual small images across the track (see scanners below). Each small image taken by the scanner represents an instant of measurement; hence it is the unit area of observation (actually a 'point') which is also referred to as 'the instantaneous field of view' (IFOV; note IFOV is a function of altitude, detector size, and focal length of the optical system). The output value of the sensor detector provides the value for one 'point' (smallest square area in an image), which is referred to as the picture element or simply pixel, in the overall image. Successive pixels in

1938) V. E. Pozhar, V. I. Pustovoit, "Main features of image transition through acousto-optical filter," Photonics and Optoelectronics, Vol. 2, 1997, pp. 67-77

1939) J. Yu, T. H. Chao, L. J. Cheng, "AOTF imaging spectrometer for NASA applications: System issues", Proceedings of SPIE, Vol. 1347, 1990, pp. 644-654

1940) D. R. Suhre, M. Gottlieb, L. H. Taylor, N. T. Melamed, "Spatial resolution of imaging noncollinear acousto-optic tunable filters", Optical Engineering, Vol. 31, 1992, pp. 2118-2121

1941) Manual of Remote Sensing, Second Edition, American Society of Photogrammetry, 1983, pp. 20-26 the scan line are generated by data from successive positions. The advancement to the next scan line is achieved by the orbital motion of the satellite (or by a mirror tilt).

IFOV, the unit area of measurement of a scanner, may be given as an angle (in mradians or in degrees) or directly as a length (edge length of the spatial resolution). The IFOV is chosen depending on the type of application. For instance: observation of small-scale variations with relatively small-area coverage (in this case regional patterns may be difficult to characterize, because generalization is required of the high frequency information in such images), or large-scale variations with frequent large-area coverage (regional patterns may be readily observed, however, details may be averaged within a pixel, implying a loss of information). Sensor arrangements on-board Earth observation satellites are possible that provide specifically high and low resolution images. The high resolution images are used for regional coverage and detail, while the low resolution images offer large-area coverage as well as high temporal coverage for quick synoptic views of an area. The price of spatial resolution is paid with correspondingly high data rates, having direct consequences on the dimension of all follow-up processes (communications in the space and ground segments, processing, storage, and user access costs).

Naturally, there is a tendency and requirement for ever improving spatial resolutions. The Landsat-5 MSS sensor has a spatial resolution of 80 m, adequate for many land-based applications. The Landsat-5 TM sensor has a spatial resolution of 30 m. The Spot HRV sensor has a spatial resolution of 20 m (10 m panchromatic). The trend goes toward a 5 m spatial resolution.

Spectral Resolution

The spectral resolution of a dispersive remote sensing instrument is determined by the bandwidth of the channels used. High spectral resolution is achieved by narrow bandwidths which, collectively, are likely to provide a more accurate spectral signature for discrete objects than are broad bandwidths. A disadvantage of broad-band instruments is the tendency of increased system noise with bandwidth (low signal-to-noise ratio), lowering the system's radiometric resolution. This problem may be somewhat reduced if relatively long look times are used during imaging.

Radiometric Resolution

Radiometric resolution refers to the resolving power of a system in wavelength and energy. The limiting factor for a radiometric measurement is the signal-to-noise ratio (SNR) of the instrument receiver (see chapter O.7). Considering the effects of varying illumination, the radiometric dynamic range of a sensor is determined by the maximum radiance value that the sensor can experience for a given band.

A sensor does not have the capability to obtain measurements of reflectivity as a continuous function of wavelength for all pixels in an image obtained with a multispectral scanner. All that is obtainable are integrated (i.e. averaged) reflectivities over the wavelength bands used in the receiver.

On the output side of a sensor, the measurement is converted into a number of discrete digital levels in a process also referred to as 'quantization'. This secondary measurement effect of quantization (the primary effect is SNR) is sometimes referred to as 'radiometric resolution.' In this terminology, 10 bit radiometric resolution signifies that a detector's converted digital output signal may range from 0 to 1023 (or 1024 quantization levels). - Some applications do not require very high radiometric resolution (for instance sea surface temperature measurements), while other observations (in particular for vegetation, chlorophyll, etc.) require as many spectral channels as possible for good interpretation of the data.

O.4.7 SQUID Sensors in Magnetometry

SQUID (Superconducting Quantum Interference Device). [1942] A new cryogenic particle detector type, based on SQUID technology, offers promising results in such applications as absolute magnetic field measurements. The SQUID technology provides the detection of magnetic signals with a sensitivity in the order of fT (femto Tesla or 10^{-15} T), making them the most sensitive detectors of magnetic signals so far. The overall SQUID measurement range stretches from about 500 nT to the fT level, thus providing a wide spectrum of application potential in such fields as geophysics, biomagnetism and nondestructive testing of materials. Note: the best sensitivity of fluxgate magnetometers is in the order of nT (nano Tesla). The HTS (High Temperature Superconductivity) technology (which started in the 1980s) is favoring the development of SQUID sensors.

A SQUID consists of a tiny superconducting ring (< 1mm in diameter), which is interrupted by one or two Josephson junctions, they act as a "weak" superconductor. Only a small current (usually several microamps) can flow through such a weak link without dissipation. Above a certain current, the so-called critical current, a voltage drop develops across the junction. A magnetic flux through the SQUID, induced by an external magnetic field, alters the value of the critical current. Specifically, it changes the phase of the Cooper pair wave function across the junction(s).

In case of the so called DC (Direct Current) SQUID (a superconducting ring with two junctions), the variation in critical current with varying magnetic flux can be detected by simply passing a DC current through the SQUID. The value of the current is chosen to be slightly larger than the sum of the critical currents of the two Josephson junctions (the critical current of the SQUID). Then a DC voltage drop develops across the SQUID, which varies periodically with varying magnetic flux. The periodicity of this so-called flux-to-voltage transfer function of the SQUID is the flux quantum (about 2 fWb [femto Weber]). If the SQUID ring contains only a single junction (the so-called RF SQUID), it is usually coupled to a resonant LC circuit, which is driven by an RF current. The amplitude of the resulting RF voltage across this circuit oscillates with varying flux through the SQUID, similar to the DC voltage in case of the DC SQUID.

The principle governing the operation of SQUID devices is flux quantization in superconducting loops and the Josephson effect. The SQUID detector functions as a transducer whose output voltage is a periodic function of the magnetic flux. This periodic response is linearized using flux-locked loop electronics based on 100 kHz flux modulation and phase sensitive detection.

Although a SQUID primarily measures magnetic flux, it can measure all physical quantities which can be transformed into magnetic flux, like electrical currents, voltages or displacements. SQUIDs therefore play an essential role in countless applications, both in science as well as in industrial measurements.

O.5 Cryocooling Techniques

Space cryocoolers are miniature refrigerators designed to cool sensitive spacecraft components to cryogenic temperatures. [1943] Cryogenic temperatures are necessary for the operation of many modern devices such as infrared detectors and focal planes, solid-state gamma-ray detectors, and a number of emerging superconducting technologies. Typical temperatures range from the boiling point of liquid Nitrogen at 77 K, down to near the boiling point of liquid Helium at 4 K.

1942) The SQUID was invented by James E. Zimmermann and Arnold Silver of the Scientific Laboratory of Ford Motor Co. in 1964

1943) http://www.jpl.nasa.gov/adv_tech/coolers/summary.htm

O.5.1 Stirling Cycle Cooler

A cycle named after Robert Stirling (1790 - 1878), a Scottish engineer. In this concept a confined volume of gas expands at high pressure and contracts at low pressure, thereby doing work on the surrounding environment. The expansion and contraction of the gas is driven by the absorption and rejection of heat at the engine's hot and cold heat exchanger. The Stirling cycle uses an air engine regenerator concept, having for its indicator diagram two isothermals and two lines of constant volume. The Stirling cycle can be applied as a heat engine cycle in which heat is accepted at a high temperature, rejected at a lower temperature, and net work or power is produced. Or the cycle can be applied as a refrigeration/heat pump cycle in which net work is done on the Stirling device and, heat is accepted at a low temperature and rejected at a higher temperature. If the useful heat is that which is accepted at the low temperature, then the device is being used as a refrigeration or cooling device. If the useful heat is that which is being rejected at the higher temperature, then the cycle is being used as a heating device. [1944] [1945]

Most mechanical cryocoolers operated so far in space are based on the Stirling cycle concept. The Oxford cryocooler (developed at Oxford Instruments Plc. in Eynsham, UK) and the Ball Aerospace multi-stage Stirling cycle coolers are examples of Stirling cycle coolers.

O.5.2 Pulse Tube Cooler

A pulse tube cooler is a mechanical device for cryogenic cooling. It is a closed system that uses an oscillating pressure at one end (typically produced by a compressor) to generate an oscillating gas flow in the rest of the system. This gas flow (usually helium gas) can carry heat away from a low temperature point (cold heat exchanger) if the conditions are right. An orifice controlling the flow at the other end of the cooler can provide the right condition for cooling to occur. A single pulse tube cooler can cool from room temperature to 70-80 K and multi-stage systems can cool much lower. The amount of heat they can remove is only limited by their size and the power used to drive them. Their efficiency is comparable to other systems such as Stirling coolers. [1946] [1947]

The principal benefits are greater reliability and lower cost compared to the Stirling cooler and an order of magnitude lower mass, lower cost, and longer life than the current state-of-the-art coolers. In 1994 TRW delivered its first pulse tube built under contract to the Air Force's Phillips Laboratory. The success of this cryocooler (1 W at 35 K with 200 W of input power) lead to TRW being selected to build a pulse tube cryocooler for the AIRS instrument of the Aqua spacecraft in the EOS program (planned launch in 2001). The TES (Tropospheric Emission Spectrometer) instrument of NASA/JPL (planned for launch on the Aura mission in late 2002) is another example of pulse-tube cryocooler technology. TES uses two coolers to cool two separate focal planes to 62 K. The two coolers are identical and are a variant of the TRW AIRS pulse-tube cooler, but configured with the pulse tube hard mounted to the compressor.

O.5.3 Hybrid Cryogenic System: CSE (Cryo System Experiment)

CSE was developed at Hughes Aircraft Co. and funded by NASA/JPL and successfully flown aboard the Shuttle Discovery (STS-63, February 3-11, 1995). The flight experiment validated and characterized the on-orbit performance of a hybrid cryogenic cooling system integrating two advanced cryogenic technologies: 1) a state-of-the-art Hughes ISSC (Improved Standard Spacecraft Cryocooler) 65 K long-life, low-vibration Stirling cooler, and 2) a Hughes experimental diode oxygen heat pipe (Note: a cryogenic diode heat pipe en-

1944) http://www.grc.nasa.gov/WWW/tmsb/stirling.html
1945) http://www.oxford-instruments.com/
1946) http://irtek.arc.nasa.gov/WhatisOPT.html
1947) http://ranier.oact.hq.nasa.gov/Sensors_page/Cryo/CryoPT/CryoPTHist.html

ables large physical separation between the cryocooler and its thermal load, and provides on-off switching to limit reverse heat flow when the cooler is turned off).

The CSE cooler provided 1.2 W of cooling at 65 K and is the first U.S.-built long-life Stirling cooler to operate in space. Its thermal performance over the course of the eight-day mission equaled its thermal performance during ground testing.

The second important cryogenic component successfully demonstrated in the CSE flight was an oxygen diode heat pipe. The cryogenic heat pipe is designed to provide a high thermal conductance path between the cryocooler and its intended cryogenic load; the diode nature of the heat pipe has the ability to limit heat flow in the reverse direction when the cooler is turned off. In space, the heat pipe performance met and exceeded ground test measurements.

O.5.4 Optical Cooling

Optical cooling by fluorescence is an alternate and new concept in refrigeration that represents possible applications in small cryocooler technology, in particular for infrared devices and spectrometers. The technique was first demonstrated in the timeframe 1994-1995 at LANL (Los Alamos National Laboratory) using laser-induced cooling of a solid by net anti-Stokes fluorescence. [1948] [1949]

The principle of fluorescent cooling makes use of anti-Stokes fluorescence, a phenomenon in which a substance that is excited by radiation at one wavelength fluoresces at a shorter wavelength. This results in more energy being radiated than is absorbed for each photon. Successful fluorescent cooling requires that a negligible fraction of the energy-level decays from the excited- to the ground-state groups occurs through non-radiative, heat-generating processes. Each fluorescent photon carries off, on average, thermal energy equal to the difference between the pump-photon and the mean fluorescent-photon energies.

A first-generation fluorescent cryocooler was designed at LANL utilizing a combination of diode lasers coupled with an ytterbium-fiber laser (ytterbium-doped ZBLANP) to efficiently produce pump radiation at the desired wavelength. The cooling element design employs an Yb^{3+} doped glass which has an exceedingly simple energy-level structure.

LASSOR (Los Alamos Solid-State Optical Refrigerator). [1950] [1951] The objective is the design of a cooler capable to produce 0.5 W of cooling at 77 K with an efficiency comparable to that of small commercial mechanical cryocoolers. The first-generation fluorescent cryocooler design is based on a cooling element of Yb^{3+}-doped ZBLANP (a heavy metal fluoride glass containing zirconium, barium, lanthanum, aluminum, sodium, and lead). High-power diode lasers are used to produce an effective pump radiation at the required wavelength. An optical fiber carrying the pump radiation passes through the wall of the cryocooler vacuum chamber to a cylindrical block cooling element. Dielectric mirrors for reflecting the pump radiation are deposited on both ends of the cooling element. The pump radiation enters the cooling element through a small hole in one of the mirrors and is largely beamed along the axis of the cylinder. This light is repeatedly reflected between the dielectric mirrors and from the sides of the glass cylinder until it is entirely absorbed by Yb^{3+} ions. The ions subsequently fluoresce isotropically in a broad spectral band. Most of the fluorescent radiation escapes from the cooling element and is absorbed by the warm walls of the cooling chamber. The object to be cooled (cold finger) is mounted on one of the mirrors of the cool-

1948) R. I. Epstein, M. I. Buchwald, B. C. Edwards, T. R. Gosnell, C. E. Mungan, "Observation of Laser-Induced Fluorescent Cooling of Solids," Nature, Oct. 12, 1995, p. 500

1949) T. R. Gosnell, C. E. Mungan, et al., "Laser Cooling of Solids," Proceedings of the joint Conference on Lasers & Electro-Optics (CLEO) and Quantum Electronics & Laser Science (QELS), May 25, 1995, Baltimore, MD

1950) B. C. Edwards, M. I. Buchwald, R. I. Epstein, "Development of the Los Alamos solid-state optical refrigerator," Review of Scientific Instruments, Vol. 69, No 5, May, 1998

1951) B. C. Edwards, M. I. Buchwald, et al., "Development of a Fluorescent Cryocooler," Proceedings of the 9th Annual AIAA/USU Conference on Small Satellites, Sept. 20, 1995, Logan, UT ing element and is thus shielded from the pump and fluorescent radiation. - A realistic mass estimate of such an instrument is in the order of 1 kg.

Some characteristics of the optical (solid-state) cooling technology are:

- No vibrations are produced by system operations (there are no moving parts)
- The system neither generates nor is it affected by electromagnetic interference
- Cooling temperatures of <=77 K (from room temperature) can be achieved
- A conversion efficiency of about 0.5% is achieved (DC power to cooling power)
- An instrument mass of <2 kg/W of cooling power is attainable
- Continuous operating life times of several years are possible

O.6 Imaging Spectrometers

The imaging spectrometer concept can be applied to the following spectral regions: VIS, NIR, SWIR and TIR (i.e. from about 0.4 µm to about 12 - 13 µm). However, different optical components and detectors have to be used for the different regions. The collecting optics and an imaging slit are used to image one line (one pixel wide) across the swath on an intermediate focal plane. A dispersive element (grating or prism) is then used to spectrally disperse the 'line' signal into a series of lines, each corresponding to a spectral band (see Figure 380). A 2-D detector array is then used to measure the dispersed light. Thus, each array column will give the spectrum corresponding to one pixel. The whole process is repeated as the platform moves by the equivalent of one spatial resolution element along the track.[1952)]

Imaging spectrometers allow the simultaneous acquisition of images in many contiguous, narrow spectral bands. Spaceborne sensors have relatively few and broad bands, while airborne sensors may have hundreds of narrow bands, the high spectral dimensionality of these airborne data enables the construction of laboratory-like spectra for each pixel in the scene (see Figure 361).

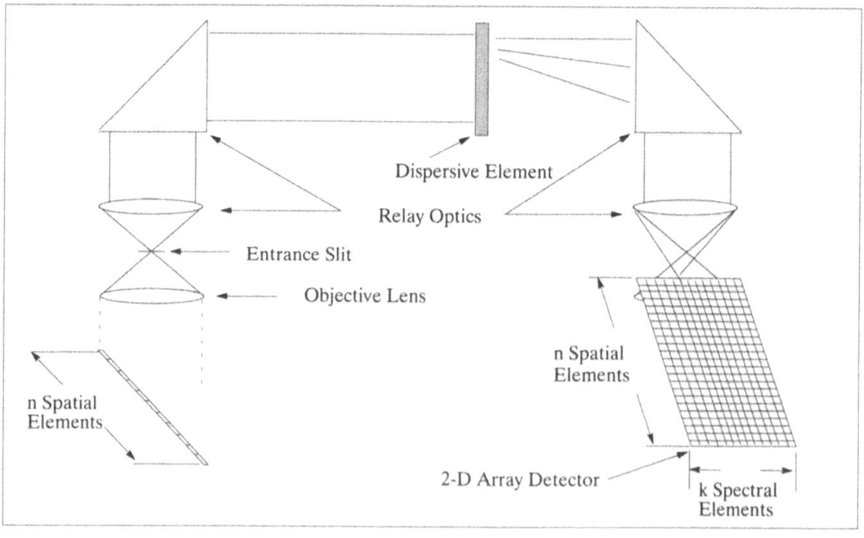

Figure 380: Collecting optics of an imaging spectrometer

1952) Ch. Elachi, "Earth Surface Sensing in the '90s," Progress in Imaging Sensors, Proc. ISPRS Symposium, Stuttgart, September 1-5, 1986, ESA SP-252, November 1986, pp. 1-9

Two approaches to the design of imaging spectrometers are in use. The wiskbroom approach includes an optomechanical fore-optics from which light is passed into a spectrometer, where it is dispersed and focused onto a linear detector array. This concept provides as many contiguous spectral bands as there are detector elements in the linear array. It is also possible to use 2-D (area) detectors instead of linear arrays in this configuration.

An alternate solution is the pushbroom approach using 2-D detector arrays at the focal plane of the spectrometer, thereby eliminating the need for the mechanical scanning device. In this arrangement, one dimension of the array corresponds to the cross-track pixel length while the spectral information is co-registered in the other direction.

The collecting optics of imaging spectrometers (2-D airborne devices), usually designed with many contiguous spectral and spatial bands, generates very high data volumes, which requires in turn very high-rate recording devices (sometimes in the order of 50-100 Mbit/s) for contiguous strip imagery acquisition. These recorders are usually very expensive, or not available at all for the intended application. In some cases a solution is provided by means of an interval acquisition technique employed by the sensor detector. The sensor abandons the concept of contiguous strip imagery in favor of discrete area or frame imagery. The data rate constraint at the recorder interface is circumvented in this way, because the next set of data frames is provided to the recorder at the instant when the previous set of frames was stored away.

O.7 Passive Radiometry (MW/MMW)

Radiometry is the measurement of microwave (MW) radiation generally considered to be in the wavelength region from about 1 mm to 1 m (the mm-wave region is a subregion of the microwave region and generally considered to be from 1mm to 10 mm wavelengths). Microwaves and mm-waves (MMW) have the known benefit of penetrating the Earth's atmosphere under conditions where visible and infrared waves display rather limited optical depth. The attenuation properties of the atmosphere have been a hindrance to the development of MMW scene imagers, while at the same time this attenuation has made it possible to sound the atmosphere. The MMW spectrum contains two principal gaseous absorbers: oxygen and water vapor. In addition to gaseous absorption, aerosols (solid and liquid) produce significant attenuation.

A passive system is restricted to measuring the incoming radiation of a wave spectrum in question. Hence, a passive system is restrained to that portion of radiation that is emitted with a reasonable intensity from the observed objects (Earth surface, or atmospheric volume, or whatever).

Microwave radiation can directly measure the dielectric properties of the Earth's surface. Any changes in these properties directly affects the reflectivity or emissivity measured by the microwave system. The dielectric property of the Earth's surface layer is in turn strongly dependent on the surface moisture content. Hence, an analysis of the microwave data can be related to the moisture content of the soil surface layers. Microwave radiation interacts also with atmospheric water in its various states (gaseous, liquid, and solid) in distinctly different ways. This property can be used to study the water cycle and the energy balance of the Earth. For instance, microwave techniques are used to detect humidity, rain, and hail in the atmosphere, as well as soil moisture, snow, and permafrost on land. Over the ocean, microwave sensors are capable of measuring sea surface wind speed and direction, sea ice, and sea surface temperature and salinity.

In general, passive microwave radiometry is limited by its poor spatial resolution, which depends on range, on the wavelength of the radiation used, on the aperture of the antenna system, and on the SNR (Signal-to-Noise Ratio). The SNR is in turn influenced by the strength of the signal produced by the target and by the temperature and sensitivity of the receiver.

The theoretical performance limit on radiometric sensitivity provides a useful framework for examining, the interdependencies among performance and design parameters. Radiometric sensitivity, i.e. SNR, depends on the optical aperture area, the detector solid angle (spatial resolution), signal integration time, and spectral bandwidth (spectral resolution), among other parameters (this applies to any imaging device that receives electromagnetic energy). Improvements in either spatial or spectral resolution thus come at the expense of radiometric sensitivity unless these are offset by a larger aperture, better optical transmission, or increased integration time. Improvements in technology can influence the relative attractiveness of these alternatives. For example, effective integration time (and hence radiometric sensitivity) can be increased by adding more detectors and slowing the effective scan rate or by using additional detectors to take multiple samples of the same point in object space and coherently summing the outputs by employing TDI (Time Delay Integration). These strategies offer improved performance at the expense of added complexity in the detector arrays and associated signal processing electronics; on the other hand, technology improvements in these areas have made it possible to exploit such design strategies with relatively low cost and risk. [1953)]

Passive sensors measuring radiation in the microwave spectrum have the advantage of collecting data during day and night cycles (measurement of emitted radiation rather than reflected solar radiation), the presence of cloud cover does not significantly degrade measurement accuracy. The spatial resolution of passive microwave sensors is fairly poor compared to VIS and IR sensors (longer wavelengths, by a factor of 100, and much lower intensity of microwave radiation results in poorer resolution); the advantage is global coverage capability. Passive microwave radiometry involves the detection of thermally generated microwave radiation. The characteristics of the received natural radiation, in terms of the variation in intensity, polarization, frequency and observation angle, is dependent on the nature of the surface being observed and on its emissivity. The spectrum of passive microwave radiometry is usually from 200 GHz - 1 GHz [or from 0.15 cm to 30 cm in wavelength; the newer planned sensors go to frequencies in the order of 600 GHz (0.5 mm wavelength)].

A passive microwave radiometer (or simply a 'radiometer') is different from traditional radar receivers (i. e. active microwave receivers) in two important respects. First, the input signal processed by a radar receiver may be coherent and nearly monochromatic, while secondly, natural radiation emitted by a surface (target or source) is phase-incoherent and extends over the entire electromagnetic spectrum. This means, it is noiselike in character and similar to the noise power generated by the receiver components.

The second difference relates to SNR at the receiver output. With the condition: SNR>>1, there is good confidence that the information can be extracted from the noise. However, the radiometric signal to be measured in a passive microwave receiver is usually much smaller than the receiver noise power. Radiometers are highly sensitive receivers that are configured to measure very small input levels with a high degree of precision. Several different configurations have been developed. Most of them use synchronous modulation at the input and demodulation at the output, and integrate the desired signal over as long a period of time as possible (integration time is set equal to the dwell time of the IFOV).

By inspection, optical wavelengths are about five orders of magnitude (10^5) shorter than microwaves; this implies that the resolution capability of an optical radiometer is inherently superior to the resolution of a microwave radiometer by about the same factor. For instance, the IFOV (spatial resolution) of an optical radiometer is of the order of tens of meters, compared to tens of kilometers for a microwave radiometer (for the same-size aperture and same orbital parameters). This means that the multiple-detector configurations

1953) http://www.nationalacademies.org/ssb/smallsatch3.htm

employed in optical scanners are not feasible at microwave frequencies unless very large antenna structures (apertures) are used.

Radiometric imaging of a scene of interest is accomplished by scanning the main beam of the antenna. For a moving platform, scanning in the cross-track dimension is sufficient to produce an image. Both mechanical and electronic (beam-steering) scanning techniques are used in microwave radiometry. In mechanical scanning, the direction of the antenna beam is changed by mechanical rotation or angular movement of the radiating aperture of the antenna system. Alternatively, phased array antennas can be used to steer the direction of the antenna beam electronically (no mechanical motion in the scanning process).

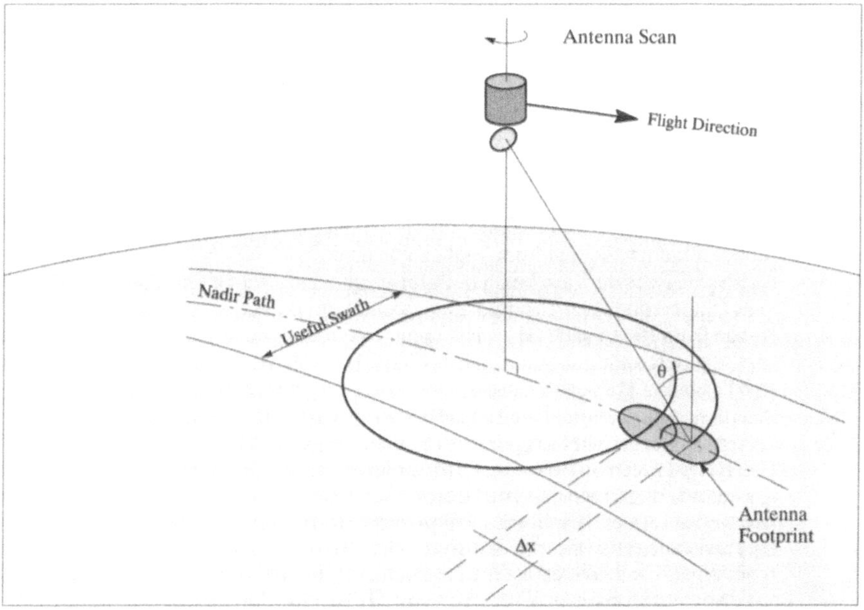

Figure 381: Typical geometries of a conical scanning passive MW radiometer

In the 1970s, detector technology and signal processing electronics were less well developed than today; intelligent designs of that era minimized the number of detectors via the use of scanning mechanisms and relatively large optical apertures. Now that higher density detector arrays and sophisticated signal processing electronics are readily available, aperture size can be (and has been) significantly reduced for many classes of instruments. Still, for imaging sensors of relatively high spatial resolution, aperture diameter is clearly the dominant sizing parameter because diffraction-limited performance is a fundamental physical limit that is not amenable to technological improvement. Technology advances have nonetheless made it feasible to design radiometers whose performance is close to theoretical limits.

O.7.1 Radiometer Instruments

Two generic types of radiometers have evolved, profiling instruments (or sounders, also referred to as broadband radiometers), and surface imaging instruments (multichannel radiometers).

1) **Atmospheric sounders** provide information about vertical profiles of temperature and molecular constituent concentrations in the atmosphere by making measurements

near the molecular resonance frequencies (resonance method with nadir pointing). Another approach to atmospheric constituent monitoring is the limb sounding technique. This method can provide substantially improved sensitivities and vertical resolutions (profiles) over those of the nadir-looking instruments.

Sounders typically have broad beams to minimize spatial noise caused by atmospheric structure, the largest possible pre-detection bandwidths to minimize detection thresholds (NEΔT); they tend to accommodate short post-detection integration time dictated by the platform motion. Typical sounding systems yield beam footprints at nadir of 10 to 150 km.

The superheterodyning detection scheme, used by microwave radiometers, provides excellent frequency resolution which permits to resolve pressure broadened emission lines. In this case the limb-sounding instrument scans the atmospheric limb with a sufficiently narrow antenna beam (typically a few tens of a degree).

At the end of the 1990s there are four types of receiver technologies applied to microwave radiometers. These are: [1954)]

- HEMT (High Electron Mobility Transistor). HEMT receivers are available to frequencies of 110 GHz. The sensitivity depends on the operational temperature of the HEMT amplifier. For 300 K operational temperature, a noise temperature of about 50 times the quantum limit can be achieved.
- Schottky. Receivers are being built for frequencies from about 200 to 2500 GHz.
- SIS (Superconductor Insulator Superconductor). Receivers cover the frequency range from 100 to 1200 GHz. The upper frequency limit is given by the energy gap of the superconducting material; it is 700 GHz for Nb (Niobium) and 1200 GHz for NbTi. Since a SIS device can carry currents up to those of the superconducting gap, there is no fundamental limit to the bandwidth of the IF (Intermediate Frequency). A typical operational temperature of a SIS device is 4 K.
- HEB (Hot Electron Bolometer). In a bolometric detector the resistance is changed by heating due to the absorbed energy. Superconducting materials are most sensitive to this effect. Applications of two signals with different frequency modulate the resistance at an intermediate frequency (IF). Useful IF bandwidths require fast cooling and low heat capacity of the detector. In HEB this is done by diffusion of the hot electrons out of the detector. IF bandwidths of several GHz can be achieved with a HEB mixer. Conventional HEB devices require cooling down to 4 K for optimum performance. The newer HTS (High Temperature Superconductivity) HEB mixer is capable to operate at 77 K.

2) **Surface imaging sensors**, on the other hand, operate primarily at window frequencies, where atmospheric absorption is low and surface features can be imaged or measured quantitatively. For most of these surface features high spatial resolution is desirable (see also Ref.1956).
The nadir-viewing technique is employed for surface imaging. Radiometric measurements are affected to some extent by water vapor, clouds and rainfall. Hence, most surface sensing radiometers include frequency channels sensitive to atmospheric water vapor and liquid water, to measure global distributions of these parameters and to correct for their effects on the measurement of the surface parameters.

A radiometer system for microwave remote sensing typically consists of the following subsystems:

- An antenna and scan subsystem, which receives incoming radiation from specified beam-pointing directions

1954) P. Hartough, Solar System Research with Microwaves," Proceedings of 32nd ESLAB Symposium on 'Remote Sensing Methodology for Earth Observation and Planetary Exploration,' ESA/ESTEC, Sept. 15-18, 1998 (SP-423 Dec. 1998), pp. 23-31

- A receiver subsystem (detectors plus the electronics subsystem) which frequency down-converts, amplifies, and detects the received signals in the assigned frequency bands
- A signal processor and housekeeping and control subsystem, which provides timing and sequencing signals for the antenna and radiometer subsystems, and performs digitizing and various other functions for the output of a digital data stream.

A radiometer responds to thermal radiation (noise-like signals) emitted and/or reflected from the atmosphere, the Earth's surface, or other bodies. It's output is expressed in terms of brightness temperature [T_B (f)], where T_B (f) is a temperature at frequency f of a black body that produces an equivalent power. The values of T_B typically range between about 90 K over moderately reflective oceans reflecting the atmosphere and cold space (3 K), and about 305 K over highly emissive tropical forests. The change of very small temperatures (ΔT), measured by a radiometer, are expressed as: [1955)]

$$\Delta T = C\,(T_A+T_R)\,/\,(B\,\tau)^{1/2}$$

where C is a constant (typically 1 - 2), T_A is the equivalent antenna noise-temperature, T_R is the receiver noise-temperature, B is the prediction bandwidth, and τ is the post detection integration time.

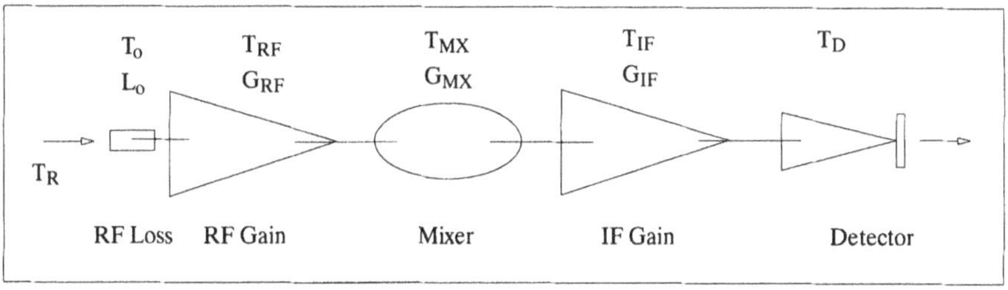

Figure 382: Generalized radiometer receiver block diagram

A radiometer receiver subsystem is illustrated in Figure 382. All receiver components are at temperature T_o = 300 k, with individual gain "G" (loss "L"), and individual noise-temperature "T". The radiometer's hardware design challenge rests primarily with minimizing the receiver's noise-temperature (T_R).

Many applications have become well established in the field of passive microwave radiometry, such as temperature sounding in operational meteorology, sea-ice mapping for navigation in the polar regions, measurements of ocean surface temperatures and surface wind speeds.

1955) W. T. Kreiss, I. Galin, "Millimeter wave sounders/imagers for spaceborne Earth observations and reconnaissance," Proceedings of SPIE, April 21-22, 1997, Orlando, FL, Vol. 3064 pp. 54-62

Launch Year	Spacecraft	Sensor	Frequencies (GHz)	Antenna Type	Swath Width (km)	Spatial Resolution (km)	Prime parameters measured or inferred (i.e. objectives)
1962	Mariner 2 (V. fly-by)		15.8, 22.2	Mech. scanned parabola		1300	Limb darkening of planetary emission, temperature
1968 1970	Cosmos 243 Cosmos 384		3.5, 8.8, 22.2, 37, same	Nadir-viewing parabola same		13	Atm: water vapor, liquid water content; Sfc: temp., sea ice concentration
1972	Nimbus-5	ESMR NEMS	19.3 22.2,31.4, 53.6,54.9,58.8	Electrically scanned array 5 lens-loaded horns, nadir	3000	25 200	Atm: rain rate; Sfc: sea temp., ice concentration Atm: temp. profile, water vapor; Sfc: ice classification, snow cover
1973	Skylab	S 193 S 194	13.9 1.4	Mech. scanned parabola nadir-viewing	180	16 115	Surface: winds, precipitation Surface: soil moisture
1974	Meteor		37	Dual-polariza. 35° from nadir			Atm: Liquid water content
1975	Nimbus-6	ESMR SCAMS	37 22.2,31.6,52.8 53.8, 55.4	Elec. scanned array, dual pol. 3 rot. hyperb. mirrors	1300 2700	20 x 43 150	Same as Nimbus-5 ESMR Same as Nimbus-5 NEMS
1978	DMSP	SSM/T	50.5,53.2,54.3 54.9,58.4,58.8 59.4	Single rotating mirror	1500	174	Atm: Temp. profile
1978	Tiros-N	MSU	50.3,53.7,55.0 57.9,	Dual rotating mirrors	2300	110	Atm: Temp. profile
1978 1978	Seasat, Nimbus-7	SMMR same	6.6,10.7,18, 21, 37	Offset-fed osc. parabola, dual polarization	600 800	18 x 23 22 x 35	Atm: water vapor, rain rate, Sfc: sea temp. wind speed, ice concentration, snow cover,
1979	Cosmos 1076, 1151	Device v Device π	3.53, 9.37, 22.2, 37.5 9.37	Nadir-viewing parabola (no scanning)	Swath=18 km at 0.8 cm (37.5) Swath = 80 km at 8.5 cm (3.53 GHz)		Sea surface: temp, wind speed, vapor in clouds, etc.
1979 1981	Bhaskara-I (ISRO) Bhask.-II (ISRO)	SAMIR SAMIR	19.1,19.6,22.2 19.35,22.235, 31.4	Fixed antenna spinning S/C same	150 & 1000 same	150 125	Atm: water vapor , liquid water content Atm: same, surface rain, wind
1987	DMSP	SMM/I	19.35,22.235, 37,85.5	Offset-fed rot. parabola, dual-polarization	1400	16 x 24	Atm: water vapor, liq. content; surface wind, sea ice, snow cover, soil moisture
1987	MOS-1	MSR	23.8, 31.4	Offset Cassegrain, mech. scan, dual-pol.	317	23	Atm: water vapor content, surface ice, snow,
1988	Okean-O1	RU-08	37.5	Scanning	550	15 x 20	Atm. water content
1991	ERS-1	ATSR-MWR	23.8, 36.5	Offset antenna		20	Atm. water content (vapor and liquid), MW emission
1991	UARS	MLS	63,183,205	3 mirror, scanning at 5.5 sec/ scan	N/A	3 km	Atm:limb sounding, concentration of ClO, H_2O, O_3, and atm. pressure
1992	ATLAS-1	MAS	61,62,63,183, 184, 204			10 (vert.) 4 for higher freq.	Atm: limb sounding, pressure, temp., concentration of ozone, water, etc.
1992	TOPEX/ Poseidon	TMR	18, 21, 37	Offset oscillating parabola, dual oscillator	N/A	30 km	Total water content, in nadir to correct the altimeter path for water vapor

Table 546: Survey of some early passive microwave radiometers [1956] [1957]

1956) The table of E. G. Njoku has been updated: "Passive Microwave Remote Sensing of the Earth from Space - A Review," in Proceedings of the IEEE, Vol. 70 No. 7, July 1882, pp. 728-750

1957) "The Multi-Frequency Imaging Microwave Radiometer," Instrument Panel Report, ESA SP-1138, Aug. 1990

O.7.2 Aperture Synthesis in Radiometry

Long integration times were a considerable disadvantage of good-quality radiometer imagery in the early phases of spaceflight. This lack of performance resulted from the inherently weak signals of passive radiometry to be measured (and the lack of adequate technology). As a consequence, a considerable amount of radiometer development concentrated on improving and obtaining near real-time radiometric imaging. It requires the addressing of two overriding problems: **sensitivity** and **resolution**. Sensitivity implies a large total collecting area, while resolution demands a large aperture of an antenna. - Eventually, integration times were considerably reduced with the introduction of focal plane arrays and increased aperture sizes. Aperture synthesis improves on this technology by offering a parallel readout scheme (simultaneous sampling), thereby eliminating the serial scanning requirement of the antenna beam. [1958)]

Aperture synthesis is a technique (pioneered in radio astronomy - with the luxury of long observation times) of generating high spatial resolution images by dividing the collection area of a telescope (or antenna) into smaller apertures spread out in a pattern covering several baselines. Specifically, the "sparse aperture" or the "partial aperture" technique can provide diffraction-limited performance that is comparable to that of a "filled aperture" - albeit at the expense of photon-collecting area and therefore, SNR. This may be a reasonable trade-off for systems in which resolution is more important than radiometric sensitivity. In microwave radiometry the concept employs an interferometric technique in which the product from antenna pairs is sampled as a function of pair spacing. Substantial reductions in the antenna aperture needed for a given spatial resolution can be achieved with this technique. However, the performance leap in resolution must be paid for with higher requirements for instrument precision sensing and stabilization. ESTAR (P.87) and MIRAS (P.132) are examples of airborne synthetic aperture microwave radiometers (both instruments operate in L-band).

The basic idea of aperture synthesis is to replace the conventional large aperture antenna of a radiometer with a thinned array of single small aperture antennas and to coherently correlate the input signals in pairs. The technique measures in parallel the normalized correlation function in the plane of the antenna array. For distant sources the spatial correlation function is proportional to the spatial Fourier transform of the source intensity distribution (Van-Cittert-Zernike theorem). The brightness temperature distribution is obtained by a Fourier inversion. - In such an aperture synthesis design, the resolution is determined by the transverse dimensions of the array, the sensitivity is that of a pencil-beam scanner with an area equal to the total area of the antennas.

O.8 Active Radiometry

An active instrument is restricted to wavelength ranges in which reasonable intensities of radiation must be generated by the remote sensing device in order to obtain a measurable return echo radiation (signal attenuation is minimized by choosing a suitable atmospheric "window," see Figure 414).

An active microwave system can improve the 'poor' spatial resolution associated with a passive microwave system. An active microwave system measures not only the intensity of radiation, but also:

- The time taken for the emitted pulse of radiation to travel from the satellite to the ground and back.
- The Doppler shift in the frequency of the radiation echo as a result of relative motion of the satellite on the ground

1958) A. R. Harvey, et al., "Optical up-conversion for passive millimeter-wave imaging," Proceedings of SPIE, April 21-22, 1997, Orlando FL, Vol. 3064, pp. 98-109

- The polarization of the radiation (note: polarization can also be measured by passive devices)

O.8.1 Types of Radar Sensors

There are three general categories of radar sensors (i.e. active microwave instruments) that are flown on satellites, depending on the type of geophysical measurement, namely:[1959)]

Altimeters

Altimeters are used for surface height measurement (distance) along the satellite track. They are used for land topographic mapping and for very high resolution (few cm) ocean topographic mapping (ocean circulation studies). This is achieved by very accurate measurement of the time delay for a radar pulse to propagate from the sensor to the surface and back. Scanning altimeters can provide topographic measurements across a wide swath, thus providing 'altimetric images.'

An accurate distance measurement (usually vertically down to the sea surface, i.e. nadir pointing) requires in turn very accurate knowledge of the satellite orbit. The size of the "footprint" of a radar altimeter pulse on the water surface is the area of the surface of the sea that contributes to the return pulse (echo) received by the altimeter.
Accurate measurements of the ocean surface mean level contribute to the detection and measurement of ocean currents, tides and storm surges, and to the accurate mapping of underwater features. Altimeters can also measure the surface wave height.

Scatterometers/Spectrometers

Scatterometers/spectrometers are used to measure very accurately surface reflectivity (back scatter) as a function of the frequency, polarization, and illumination direction of the sensing signal. They are used to characterize quantitatively the surface roughness. Spatial resolution is not of specific importance and is usually sacrificed at the expense of amplitude measurement accuracy. Scatterometry can be combined with high resolution imaging capability within the same sensor, leading to 'imaging scatterometry.'

Scatterometers use a complicated arrangement of radar beams (usually four which enables the determination of direction as well as of speed of the wind.

Imaging SARs (Synthetic Aperture Radars)

An imaging radar works very much like a flash camera in that it provides its own light to illuminate an area on the ground and take a snapshot picture, but at microwave wavelengths. A flash camera sends out a pulse of light (the flash) and records on film the light that is reflected back at it through the camera lens. Instead of a camera lens and film, an imaging radar uses an antenna and recorder (or tapes) to record its images. In a radar image, one can see only the radiation that was reflected back towards the radar antenna.[1960)]

A typical radar (Radio Detection and Ranging) system measures the strength and round-trip time of the microwave signals that are emitted by a radar antenna and reflected off a distant surface or object (target area). The radar antenna alternately transmits and receives pulses at particular microwave wavelengths (in the range of 1 cm to 1 m, which corresponds to a frequency range of about 300 MHz to 30 GHz) and polarizations (waves polarized in a single vertical or horizontal plane). For an imaging radar system, about 1500 high-power pulses per second are transmitted toward the target or imaging area, with each pulse having a pulse duration (pulse width) of typically 10-50 microseconds (μs). The pulse normally covers a small band of frequencies, centered on the frequency selected for the radar. Typical

1959) Ch. Elachi, "Spaceborne Radar Remote Sensing: Applications and Techniques," IEEE Press, 1988

1960) http://southport.jpl.nasa.gov/desc/imagingradarv3.html (courtesy of Anthony Freeman, JPL)

bandwidths for an imaging radar are in the range 10 to 200 MHz. At the Earth's surface, the energy in the radar pulse is scattered in all directions, with some reflected back toward the antenna.

This backscatter returns to the radar as a weaker radar echo and is received by the antenna in a specific polarization (horizontal or vertical, not necessarily the same as the transmitted pulse). These echoes are converted to digital data and passed to a data recorder for later processing and display as an image. Given that the radar pulse travels at the speed of light, it is relatively straightforward to use the measured time for the roundtrip of a particular pulse to calculate the distance or range to the reflecting object. The chosen pulse bandwidth determines the resolution in the range (cross-track) direction. Higher bandwidth means finer resolution in this dimension.

The length of the radar antenna determines the resolution in the azimuth (along-track) direction of the image: the longer the antenna, the finer the resolution in this dimension. Synthetic Aperture Radar (SAR) refers to a technique used to synthesize a very long antenna by combining signals (echoes) received by the radar as it moves along its flight track. Aperture means the opening used to collect the reflected energy that is used to form an image. In the case of a camera, this would be the shutter opening; for radar it is the antenna. A synthetic aperture is constructed by moving a real aperture or antenna through a series of positions along the flight track.

As the radar moves in its orbit, a pulse is transmitted at each position; the return echoes pass through the receiver and are recorded in an 'echo store.' Because the radar is moving relative to the ground, the returned echoes are Doppler-shifted (negatively as the radar approaches a target; positively as it moves away). Comparing the Doppler-shifted frequencies to a reference frequency allows many returned signals to be "focused" on a single point, effectively increasing the length of the antenna that is imaging that particular point. This focusing operation, commonly known as SAR processing, is now done digitally on fast computer systems. The objective in SAR processing is to correctly match the variation in Doppler frequency for each point in the image; this requires very precise knowledge of the relative motion between the platform and the imaged objects (which is the cause of the Doppler variation in the first place).

Because a SAR provides its own illumination (the radar pulses), it can image at any time of day or night, regardless of sun illumination. And because the radar wavelengths (microwaves) are much longer than those of visible or infrared light, SARs can also "see" through cloudy and dusty conditions that visible and infrared instruments cannot.

Imaging radars are used to acquire high resolution (from a few meters to cm range) large scale images of the surface. They are employed in the study of surface features such as geologic structures, ocean surface waves, polar ice cover, land use patterns. Conventional SAR systems are characterized by very high data rates, relatively large antennas, and large power requirements.

Hybrid radar sensors

Advanced imaging radar sensors simultaneously operate at a number of frequencies, in all polarization states. They are calibrated combining the imaging, scatterometry, and spectrometry functions. Stereo imagers and scanning radar altimeters are used to provide topographic maps over large swaths, thus combining the altimetry and imaging functions. Rain mapping radars use imaging, altimetry (sounding), and scatterometry functions to provide a calibrated three-dimensional volumetric image of rain regions, but at lower resolution than required with surface imagers or altimeters.

O.8.2 SAR Terminology and Definitions

An imaging radar system is generally arranged in a side-looking configuration.[1961] The radar antenna illuminates a surface strip (footprint) to one side of the nadir track. The side-looking configuration is necessary to eliminate right-left ambiguities from two symmetric equidistant points and to get better range resolution on the ground. As the platform moves in its orbit, a continuous strip of swath width is mapped in the along-track direction of the satellite. The look angle θ is the angle from the vertical to the center line of the radar beam direction (normal to the flight direction)

Actual imaging SARs use a series of pulses instead of a continuous signal. Each transmitted wave front hits the target surface at near range and sweeps across the swath to far range. The pulse bandwidth B determines the cross-track or range resolution. The echos are sensed coherently.

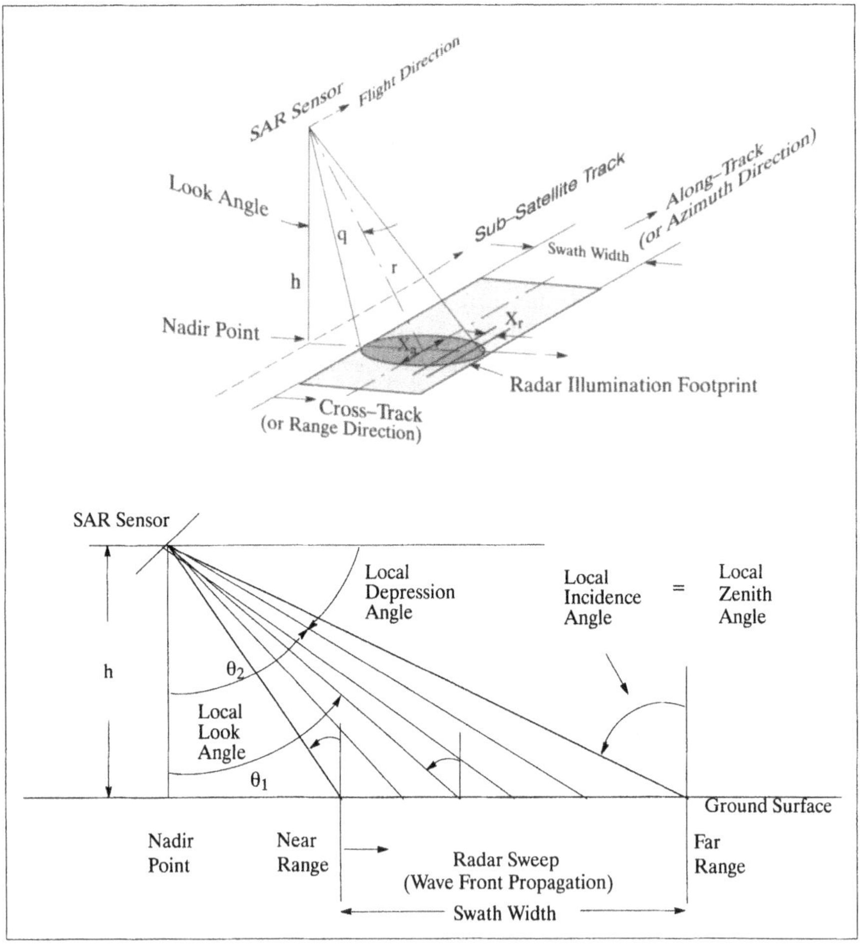

Figure 383: Characteristic geometry definitions for a side-looking imaging radar

1961) Ch. Elachi, "Spaceborne Radar Remote Sensing: Applications and Techniques," IEEE Press, 1988, pp. 63-64

SAR Antennas

In SAR observations the antenna beam casts an elliptical footprint on the ground surface with an effective rectangular aperture of typically 10 m in the along-track direction and 3 m in the cross-track direction (flat panel antenna). Multi-frequency observation requires the use of multi-frequency antennas. In this setup the along-track antenna dimension remains the same regardless of frequency while the cross-track dimension is approximately inversely proportional to frequency. Multi-polarization requires dual polarization antennas with good cross-polarization isolation. The use of electronic scanning (such as ScanSAR) provides flexibility in beam pointing and beam shape control for swath coverage.

SAR RF Electronics

The RF (Radio Frequency) portion of a SAR consists of signal generators, high power transmitters, low noise receivers, and the signal conditioning elements for amplification, filtering, and frequency conversion. Important RF features of a SAR instrument are: large dynamic range with good linearity and low noise floor, good amplitude and phase stability over time and temperature, and high power efficiency.

SAR Digital Electronics

The main functions provided by the digital electronics in SAR system are: radar configuration and timing control, radar signal digitization and processing, and radar system health telemetry generation. Advanced SARs perform in addition some digital processing to reduce the immense data rates.

SAR Images

Synthetic aperture radar images are acquired in an azimuth-range coordinate system; the sensor measures the Doppler history (time delay) of sequential wave fronts in the illuminated footprint. (Note: this is fundamentally different from VIS and IR imagery, which is acquired in an angle-angle coordinate system). The generation (reconstruction) of SAR imagery therefore involves an extensive amount of computation (with corrections applied for a number of effects) for the simple reason that the formation of each 'pixel' involves the combination of data from many thousands of echoes (i.e. synthetic array elements).

Most of the data handling in a SAR processor can be thought of as a correlation of the received signal with a two-dimensional reference function. The major processing step involves the accurate modeling of the SAR response to a point target and a continuous field of targets. Pixels as such exist only after the reconstruction of an image (representing the smallest distance unit that can be discriminated by computational methods) in their conventional raster pattern.

Range Resolution (across-track direction). The range resolution (X_r) corresponds to the minimum distance between two points on the surface which are separable. (B= bandwidth of signal (Hz); c = speed of light; θ = look angle)

$X_r = c/(2B \sin \theta)$ (Ground range resolution for SAR)

Azimuth Resolution (along-track direction).. The azimuth resolution (X_a) for a real aperture radar (non-SAR) corresponds to the two nearest separable points along the azimuth line, i.e. on a constant delay line. This is identical to the azimuth antenna footprint extension, because the echos returned from all the points along the line spanning that width are returned at the same time. (h = height above ground; λ = wavelength; L = antenna length)

$X_a = h\lambda/(L \cos \theta)$; (for a real aperture radar)

For a focussed SAR, however, X_a is independent of range.

$X_a = L/2$ (for a Synthetic Aperture Radar) one look highest azimuth resolution

Hence, the spatial azimuth resolution of the SAR is not dependent on the altitude of the sensor. This can be explained by the fact that the imaging mechanism uses the Doppler shifts in the echo and the differential time delays between surface points, neither of which is a function of the distance between the sensor and the surface. Note: the altitude still plays a factor in the power requirements to acquire a detectable echo.

O.8.3 SAR Imaging Modes

In airborne SAR operations there are generally three common imaging modes for data collection; they are referred to as: stripmap, spotlight, and ScanSAR. Note: the stripmap and ScanSAR modes are also used in spaceborne applications. [1962] [1963] [1964]

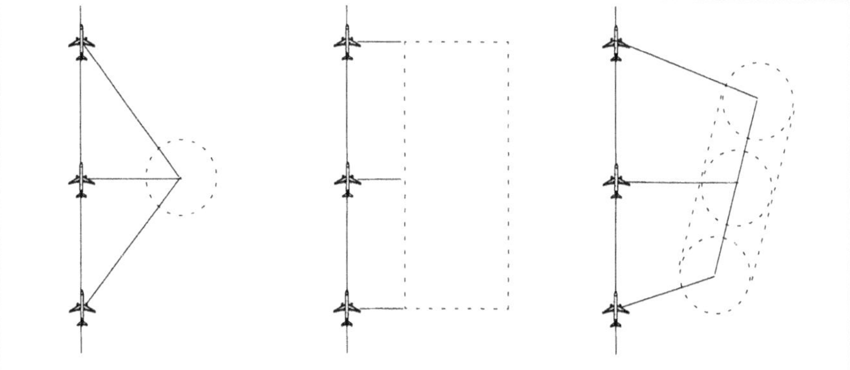

Figure 384: Illustration of common SAR imaging modes

- Stripmap: Antenna pointing is fixed relative to the flight line (usually normal to the flight line). The result is a moving antenna footprint that sweeps along a strip of terrain parallel to the path motion. The stripmap mode is normally used for the mapping of large areas (usually with coarse-resolution data).
- Spotlight: The sensor steers its antenna beam to continuously illuminate a specific (pre-determined) spot or terrain patch being imaged while the aircraft flies by in a straight line. The spotlight mode is a practical choice when the mission objective is to collect fine-resolution data from one or more localized areas. - The following attributes distinguish spotlight and stripmap modes. First, the spotlight mode offers finer azimuth resolution than that achievable in stripmap mode using the same physical antenna. Second, spotlight imagery provides the possibility of imaging a scene at multiple viewing angles during a single pass.
- ScanSAR: The sensor steers the antenna beam to illuminate a strip of terrain at any angle to the path of aircraft motion. The stripmap and spotlight modes are special cases of the ScanSAR mode.

O.8.4 Looks, Speckles and Radiometric Resolution of SAR Images

The number of neighboring pixels averaged in the processing of a SAR image is referred to as 'the number of looks N'. This technique is used to improve the noisy appearance of a ra-

1962) W. G. Carrara, R. S. Goodman, R. M. Majewski, "Spotlight Synthetic Aperture Radar," Artech House, Boston, 1995

1963) J. L. Walker, "Range-Doppler imaging of rotating objects," IEEE Transactions of Aerospace and Electronic Systems," Vol. 16, 1980, pp. 23-52

1964) D. A. Ausherman, A. Kozma, J. L. Walker, H. M. Jones, E. C. Poggio, "Developments in Radar Imaging," IEEE Transactions on Aerospace and Electronic Systems, Vol. 20, No. 4, July 1984, pp. 363-400 dar image. A noisy appearance results from the backscatter contributions of the many scattering points of the surface area. Two neighboring areas with the same backscatter cross section may be different in fine detail of the two surfaces, but otherwise homogeneous - the returned signals from the two areas will be different. There will be brightness variations in the resulting image from one resolution element to the next. This is referred to as 'speckle'.

The radiometric resolution in SAR images is mainly governed by the speckle phenomenon (coherent scattering) and not so much by the number of possible signal quantization steps. One way of describing radiometric resolution (Q) of a SAR image is by the signal-to-noise ratio. The radiometric resolution represents the logarithmic difference of two discernable backscatter coefficients which can be expressed in dB (neglecting thermal noise).

$$Q = 10 \log\left[1 + \frac{1}{\sqrt{N}}\right]$$

Small values of Q define good radiometric resolution. This is the case for higher values of N (looks). For N=1 → Q = 3 dB; N=16 → Q = 1 dB (approximately).

O.8.5 Lidars (Laser-Based Remote Sensing)

Lidar = **L**ight **D**etection **a**nd **R**anging. Lidar refers to laser-based remote sensing, i.e. radar principle applied in the optical (and IR) regions of the electromagnetic spectrum (0.3 - 10 μm wavelength range, or a frequency range of about 1000 - 30 THz). This wave spectrum (shorter wavelengths than the microwave spectrum) implies and promises to be the next level of observation technology and of information interpretation (in the number of phenomena as well as in detail and accuracy, the targets can be much smaller), due to the use of a much finer scale of measurement. Lidars are based on the principle of a laser-light pulse being sent into the atmosphere to probe the distance, physical state, or chemical composition of the backscattering medium (atmospheric layers, etc.). Lidar is also a generic term for a variety of sensors operating on different concepts, like: [1965) 1966)]

Incoherent and Coherent Lidars

In general the same principles apply to lidars in the optical spectrum as to radars in the microwave region. A lidar is said to be 'incoherent' if only a portion of the backscattered signal, the amplitude information, is recovered and processed. A lidar is 'coherent' when amplitude and phase information is recovered (permitting a spectral analysis of all echoes). Most current lidars are of the incoherent type (to be able to handle the volume of data).

O.8.5.1 Backscatter Lidar

In this technique information on a remote target is derived from the manner in which the transmitted energy is backscattered, reflected or reradiated by the target. The technique offers data on the scattering and extinction coefficients of the various atmospheric layers such as extent, height distribution, and optical thickness of aerosol and cloud layers. A typical backscatter lidar is the Neodymium doped Yttrium Aluminum Garnet laser (or simply Nd:YAG laser at 1.064 μm wavelength).

A variety of incoherent detection backscatter intensity measurement techniques have been utilized to measure the atmosphere. These include:[1967)]

- Direct time gated measurement of the molecular or aerosol laser backscatter intensity to measure particle concentrations

1965) H.Lutz, E. Armandillo, "Laser-Based Remote Sensing from Space," ESA Bulletin 66, May 1991, pp. 73

1966) R.T.H. Collis, P.B. Russell "Laser Applications in Remote Sensing," chapter 4 of 'Remote Sensing for Environmental Sciences,' Springer Verlag, 1976, pp. 110-146

1967) Special Issue on Laser Radar, Proceedings of the IEEE, Vol. 84, No. 2, Feb. 1996

- Differential attenuation measurement comparing the backscatter from the transmitter wavelength operating in an absorption band of the trace gas to that of a second transmitted wavelength not absorbed
- Raman shifted scattering where the radiated laser energy is transferred to a molecule which scatters the energy at a different wavelength than that of the original radiation
- Fluorescence where the atoms or molecules absorb the transmitted radiation and provide a detectable fluorescent emission. As an example, in time-resolved fluorescence spectroscopy, a UV laser beam excites a chemical being sensed; the emitted radiation at the chemical's characteristic wavelength gives a direct measure of the chemical's concentration.

O.8.5.2 Differential Absorption Lidar (DIAL)

In this path absorption technique information is derived relating to the path along which energy is transmitted or received by comparing the lidar echoes in a tuneable multiwavelength laser system (measurement of the differential ratios of the lidar returns with those of the transmitted ratios by tuning the laser wavelength to the specific absorption features of atmospheric trace constituents). The technique offers the capability to determine the densities of specific atmospheric constituents as well as water vapor and temperature profiles at better accuracies than obtainable with passive sounders.

O.8.5.3 Raman Lidar

The Raman Lidar technique involves detecting transmitted laser radiation which has been shifted in wavelength due to interaction with the scattering molecule. This wavelength shift, or Stokes shift, is equal in energy to a vibrational-rotational or rotational transition in the scattering molecule. The backscattered power of the wavelength shifted signal is proportional to the concentration of scattering molecules and inversely proportional to l-4. Thus, the primary advantage of Raman Lidar of DIAL is that it offers a direct measure of species concentration or mixing ratio by comparing the Raman signal of the scatterer to the Raman signal of N_2 or O_2. [1968)]

However, Raman scattering is a very weak process and the signal can be two to four orders of magnitude weaker than the elastic backscattered signal. Also, the weak scattering cross section typically limits Raman lidar to nighttime measurements at ranges of less than 10 km. To increase the Raman signal and make daytime measurements, high power lidar systems have been developed to operate at wavelengths from 248.5 to 268.5 nm. Unfortunately at these wavelengths absorption by molecular oxygen and ozone can attenuate the transmitted beam and solar irradiance can obscure the backscattered Raman signal.

O.8.5.4 Doppler Wind Lidar (DWL)

This technique is based on the principle of measuring the Doppler shift of the light backscattered from aerosol particles transported by the wind (it is assumed that the suspended aerosol particles are moving at the same speed as the wind). The technique offers the capability to measure global wind fields from a spaceborne platform. Range resolution is obtained by dividing the return signal into sequential time intervals, this technique is referred to as range gating. The returned signal is Doppler shifted due to the motion of the platform (aircraft or spacecraft) and the line-of-sight (LOS) component of the wind. Scan techniques (with the telescope) and processing algorithms must be employed to obtain the desired wind field. - Most wind lidars are based on the carbon-dioxide (CO_2) gas laser. Such an instrument generally needs a highly coherent and powerful laser beam , combined with a complex receiver and Doppler processor.[1969)] Coherent detection of the Doppler shift of the lidar backscatter is an extremely sensitive technique, but it must operate in the near in-

1968) http://daacdev1.stx.com/ELF/docs/PER/rall16.html

1969) W. E. Baker, et al., "Lidar-Measured Winds from Space: A Key Component for Weather and Climate Prediction," BAMS, Vol. 76, No. 6, June 1995, pp. 869-888 frared, and consequently is dependent on atmospheric aerosols for backscatter. The free troposphere, from about 2 to 20 km in altitude, is particularly important for long-range energy transport, it is characterized by very low aerosol content; hence, it is very difficult to measure for coherent Doppler wind lidar. [1970)]

Non-coherent, or "direct" detection of the wind speed Doppler shift, using optical interferometry, is an emerging and alternate approach to global measurements. Direct detection can employ the UV spectrum, where Rayleigh (molecular) backscatter is strong, thereby avoiding the dependence on atmospheric aerosols. In the "fringe imaging" technique, the lidar backscatter signal is passed through an interferometer (Fabry-Perot etalon), producing circular interference fringes. An imaging detector records the fringe images, and wind speed Doppler shifts are obtained from the measurement of small fringe displacements.

O.8.5.5 Ranging and Altimeter Lidar

Laser ranging and altimetry can provide accurate measurements of the distance from a reference height (i.e. the satellite orbital height) to precise locations on the Earth's surface. It can improve understanding and knowledge of many processes and phenomena in such solid Earth Sciences such as geodesy, geodynamics, ice dynamics, land topography, and Earth resources. These observational requirements are met by different types of instruments in spaceborne applications:

- A laser ranging system for accurate point positioning in geodynamics.
  This laser ranging concept from space could enhance the capabilities of the present ground-based satellite laser ranging systems (SLR's) to such high-altitude satellites as LAGEOS or Starlette. A single-wavelength nanosecond pulse laser is sufficient for centimeter-level accuracy. Subcentimeter accuracies require a dual-wavelength picosecond laser to compensate for atmospheric propagation effects (beam refraction).
- A laser altimeter for altimetry measurements over land, oceans, and ice.[1971)]
  Laser altimetry is the only way to obtain elevation data with centimeter-level vertical precision and horizontal resolution.
  The laser altimeter, primarily for land and ice applications, is an extension of the backscattering lidar instrument. The main characteristics of a laser altimeter are: the small diffraction-limited footprint (proportional to the laser wavelength), which permits topographic mapping of ice sheets, terrain, forestry or waters with high spatial resolution; insensitivity to speckles, since direct detection can be successfully employed in a visible-wavelength laser; high single pulse measurement accuracy, since in principle no averaging is required; and the insensitivity of beam propagation to atmospheric water vapor.

Current (end of 1990s) lidar sensor technology is in an experimental state.[1972)] The first space lidar LITE (Lidar In-space Technology Experiment) of NASA/LaRC was flown on Shuttle (STS-64, Sept 9-20, 1994). On the US side space lidar programs such as LAWS (Lidar Atmospheric Wind Sounder) have been cancelled due to severe budget cuts in the EOS program. In Europe the ESA sensors ATLID (Atmospheric Lidar) and ALADIN (Atmospheric Laser and Doppler Instrument) are considered prime candidates as backscatter lidar and as Doppler wind lidar, respectively, for later missions in the ENVISAT series. ALISSA (France) is planned to be flown on PRIRODA.

O.8.5.6 Lidar Principle

In the principal radar-type or "lidar" applications of laser energy, the following basic elements make up the system:

1970) J. A. McKay, D. Rees, "Design of a Direct Detection Doppler Wind Lidar for Spaceflight," European Symposium on Remote Sensing, Conference on Laser Radar Techniques (Ranging and Atmospheric Lidar) II, Barcelona, Spain, 21-24 Sept. 1998, SPIE, Vol. 3494, pp. 250-258

1971) "Topographic Science Working Group Report" to the Land Processes Branch, Earth Science and Applications Division, NASA Headquarters, 1988, p. 46

1972) "Lidar in Space," in Optical Remote Sensing of the Atmosphere, 1990 Technical Digest Series of the Optical Society of America, Volume 4, pp. 67-70

- Laser transmitter
- Collector (receiving telescope) The telescope acts as an optical antenna (analogous to radar) with a corresponding detector system (heterodyne, etc.) in the focal plane.
- Electronics and data processing resources

The laser transmitter, which is the essence of the active lidar technique, provides all the energy available for the determination of the remote target's characteristics (including distance). By far the most effective approach in all forms of active lidar systems is the use of laser energy in high intensity pulsed form, although range information can also be derived from continuous wave (CW) transmissions. At present two somewhat different laser ranging methods are used:[1973)]

1) the phase comparison technique using high-frequency modulation of one of the laser light's parameters (amplitude, polarization, wavelength, etc.)
2) the short pulse direct travel time measurement technique (delays are measured either by the 'single photon detection method,' or by the 'multiphotoelectron detection method').

The transmitted laser pulse from the lidar is propagated through the atmosphere and scattered, absorbed, or reflected by the various components of the atmosphere and surface (e.g. clouds, air molecules, aerosols, or the Earth's surface) at range R. Some portion of the scattered or reflected signal is directed back toward the lidar (receiving telescope) where it is detected. By measuring the intensity, polarization, and/or spectral properties of the return signal as a function of time (t), one can obtain information on the properties of the atmosphere.

The high degree of spatial coherence of laser energy makes it possible to achieve a good measure of directionality, and very narrow transmitted beams (typically less than a milliradian in angle) are used to achieve maximum concentration of energy and high resolution. Returned energy collected in the optical system of the receiver is converted to electrical signals by a photo detector. Due to the monochromaticity of laser energy, light of other wavelengths can readily be excluded from the detector by filters, thus reducing extraneous noise in the received signal.

Lidar Applications

- Lidar applications in atmospheric research:
  Meteorology, climatology, boundary layer physics, pollution, visibility, radiative budget, atmospheric dynamics and chemistry (composition)
- Lidar applications in solid-Earth research:
  Crustal movement, gravity field, Earth kinematics and terrain mapping; laser ranging will provide advances in precise orbit determination.

Lidar Observation Capabilities

Spaceborne lidars have the potential to observe directly or indirectly a number of atmospheric parameters, such as:

- Altitude (layer) measurements of features (such as cloud tops, planetary boundary layers, temperature inversions, subvisible cirrus clouds and aerosol particles, backscattering profiles and extinction coefficients,). This is the domain of backscatter lidars.
- Horizontal extent and density distribution of scattering layers (e.g. clouds and aerosols)
- Vertical humidity, temperature and pressure profiles throughout the depth of the atmosphere (depth-resolved measurements) and trace gas monitoring; the domain of the DIAL technique lidars.

1973) J. Gaignebet, G. Lund, "The Potential of two Color Laser Ranging," in 'Radars and Lidars in Earth and Planetary Sciences,' International Symposium Sept. 2-4, 1991, ESA SP-328, pp. 123-127

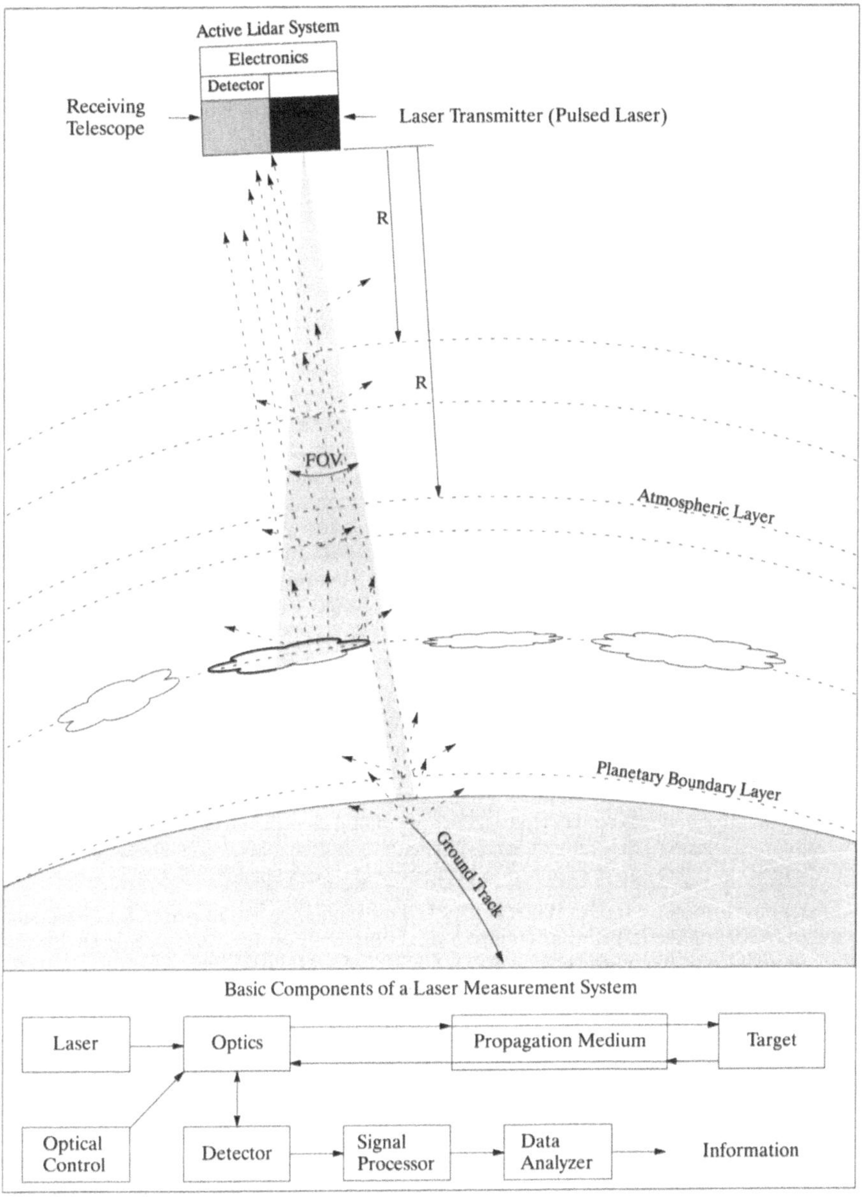

Figure 385: Spaceborne lidar and principle of backscattering lidar measurements

- Wind fields at many altitude layers, the domain of Doppler wind lidars (DWLs). A DWL is an active instrument which fires laser pulses towards the atmosphere and measures the Doppler shift of the collected return signal, backscattered at different levels (volume cells) in the atmosphere. The frequency shift results from the relative movement of the scatter elements along the line-of-sight of the instrument. This movement relates to the mean wind observed in the volume cell.

Conceptually lidar measurements provide good vertical resolutions for altitude/height assignments of atmospheric features (the data is considered to be superior to that obtained by employing sounding or imaging techniques). Lidars are active instruments; as such they can operate during the day and night cycle of an orbit. A disadvantage: lidars cannot penetrate optically thick layers (e.g. clouds).

Lidar concepts proved operational for extended periods throughout the past decade in ground-based and airborne equipment.[1974] [1975]

O.9 Fourier Transform Spectrometer (FTS)

An optoelectronic (or optomechanical) instrument, usually for the infrared region of the spectrum, providing high spectral resolution and sensitivity for remote-sensing applications (Earth surface imaging, atmospheric soundings of temperature and water vapor versus altitude, environmental monitoring of trace-gas concentrations and distributions, pollutants, atmospheric process studies, characterization of optical backgrounds, etc.). The technique employs the mathematical concept of Fourier series analysis as a means of converting a detector signal output - referred to as the interferogram - into a form useful for spectral analysis. There are several ways in which the detector signal can be created from the incident radiation. The approach taken by the majority of FTS instruments is to use an interferometer (Michelson, Sagnac, etc.) and corresponding foreoptics (lenses) to create the interference pattern in such a way that each optical frequency is coded as a unique electrical signal output of the detector. The amplitude of each frequency is proportional to the incident radiation (see also chapter O.10 and Figure 387).

The simultaneous observation of a wide spectral range is a way of increasing the signal bandwidth, an SNR (Signal-to-Noise Ratio) advantage occurs when compared to dispersive systems which observe one wavelength at a time. The result is a 'multiplex or Fellgett advantage' of FTS instruments (note: the term 'multiplexing' has its origin from the fact that each optical frequency is converted to a corresponding electrical frequency by 'frequency-division multiplexing'). The advantage is particularly noticeable for low-energy conditions or where scale expansion is required to bring out very weak signals.

In early interferometry times (1960s and 1970s) the challenge was in computer processing power converting the interferogram into a spectrum by a Fourier transform. In the meantime there have been considerable advances in FFT (Fast Fourier Transform) methods. In the mid-1990s FTS technology has become a viable concept for a variety of airborne and spaceborne applications because of the availability of increased computational power, better data processing algorithms, and transmission/recording capacities sufficient to handle large data volumes.

The functionality of an FTS (like a Michelson interferometer) is governed by the displacement of the movable mirror, which modulates the radiant intensity $I(\lambda)$ as a function of mirror distance (the pattern obtained is the sum of a collection of monochromatic interference patterns, each with its own amplitude)[1976]. The intensity distribution captured by the detector (array) is referred to as the 'interferogram' (representing the sum of all modulated waves). The interferogram is not directly usable; it requires a Fourier transformation to ob-

1974) H. Lutz, E. Armandillo, "Laser Sounding from Space," Report of the ESA Technology Working Group on Space Laser Sounding and Ranging, esa SP-1108, Jan. 1989

1975) P. Betout, D. Burridge, Ch. Werner, "Doppler Lidar Working Group Report," esa SP-1112, June 1989

1976) Note: For every position of the moving mirror, the radiation that reaches the detector is the sum of components of different amplitudes and different frequencies, and for each frequency there is a different phase difference causing the various states of interference. This is the interferogram (or interference pattern) - a sum of sine waves with different amplitudes.

tain the spectrum. An interferogram value (measurement) of the broadband radiation may be taken at each instant of maximum stroke distance of the movable mirror. - The spectral resolution of an FTS instrument is governed by the optical path length (the mirror position at zero distance is generally referred to as the 'zero path difference;' the spectral distribution becomes more detailed with growing mirror distance). All spectral elements are simultaneously present. A long stroke provides generally high resolutions. A key requirement of Michelson interferometry is to assure constant parallelism of the moving mirror surface during translation.[1977)]

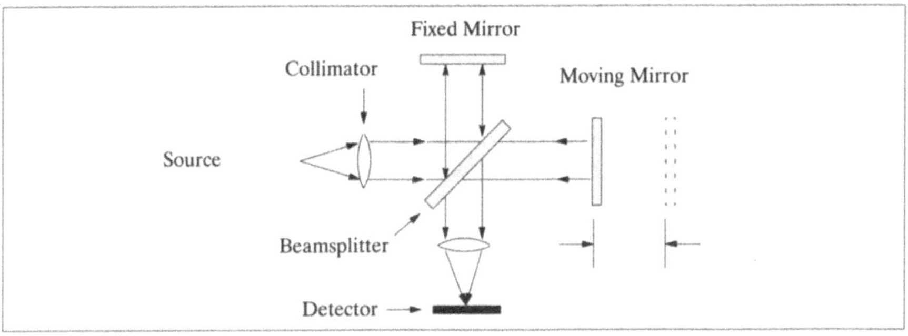

Figure 386: Conceptual diagram of a Michelson interferometer

The spectral range of an FTS system is mainly governed by the detector system design. If a wide spectrum is to be analyzed, then the detector system design must conform to the detector material limitations for the individual spectral regions. Low energy conditions may in addition require active cooling of the detectors for the infrared region.

A comparison of FTS instrument technology with that of a dispersive spectrometer brings out the following points:[1978)]

- The throughput of an FTS can be orders of magnitude greater than that of a dispersive spectrometer.
- The spectral resolution of an FTS is usually much better than that of a dispersive spectrometer; it is in fact constant across the spectrum.
- The frequency coverage of an FTS can be very large, 10:1 is routine. Dispersive spectrometers rarely cover more than 2:1.

Tables 17 and 18 give a chronological introduction of airborne and spaceborne FTS instruments, respectively.

O.10 Interferometry

Interferometry is the technology of combining radiation (visible light or microwave signals) from spatially distributed telescopes or antennas to measure distances with high precision. Radio astronomy turned first to interferometry to improve both spatial resolution and sensitivity. Astronomers combined the radiation received from many single telescopes into such systems as VLA (Very Large Array) and VLBA (Very Long Baseline Array). Interferometry allows us to use the virtual size of the array (of apertures) to achieve the desired degree of spatial resolution.

1) Spatially separated collecting apertures

1977) Courtesy of P. Haschberger, DLR

1978) R. Beer, "Remote Sensing by Fourier Transform Spectrometry," John Wiley & Sons, New York 1992, Chapter 4.1.5

An interferometer is an observational instrument that combines the output of a collection of spatially distributed "small" collecting apertures to synthesize the performance of a single larger aperture. The achievable resolution depends on the maximum distance separating any two apertures in the instrument. An interferometer using two small apertures (e.g. 10 cm diameter) separated by a baseline of 100 m offers a resolution ten times that of a 10 m diameter single aperture.

2) Single filled collecting aperture

The resolving power in a single filled aperture scales with the aperture diameter. The technique determines the relative phase of two (or more) wave fronts as a function of spatial location by observing interference fringes. Radiation is split into two or more beams which traverse different path lengths. The beams are reflected by mirrors and recombined for interference analysis. - Interferometers are used for the precise measurement of wavelengths, spectral fine structure, indices of refraction, and very small linear displacements (for instance to measure the angular width of sources, or to determine the angular position of sources, etc.).

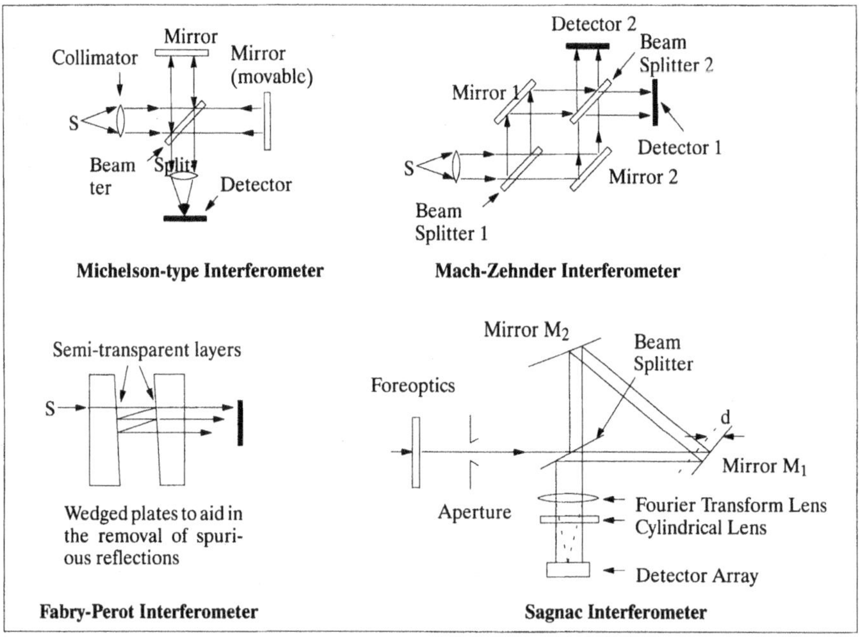

Figure 387: Illustration of some interferometer concepts

The design of interferometric instruments has a long history. The most important are listed.

- **Michelson interferometer** (Albert Abraham Michelson, US physicist, 1852-1931) [1979)] Incident radiation from source S is partly reflected and partly transmitted by a beam splitter to two mirrors. Both mirrors reflect the radiation back to the beam splitter, where interference occurs. In Fourier spectrometry one of the mirrors is movable.
- **Mach-Zehnder interferometer**. Variation of Michelson-type - incident radiation moves in only one direction with the use of two beam splitters. A test object is traversed only once.

- **Fabry-Perot interferometer** (Ch. Fabry, French physicist, 1867-1945). A multibeam interferometer consisting of two parallel glass plates that are coated with highly reflecting and partially transparent films on their inner surfaces. A radiation source S from a spectrometer generates multiple reflections between the two surfaces, producing concentric interference rings.
- **Sagnac interferometer**. (Sagnac, French physicist, 1869-1928). A variation of the Michelson interferometer. A triangle-path or common-path interferometer: the two beams emerging from the beamsplitter, one in reflectance and one in transmission, follow the same path in opposite directions. These two beams, observed beyond the interferometer, form two images of an input source.

O.10.1 Radar Interferometry

Interferometric measurements are not only performed in the optical and infrared regions, but also in the microwave region of the electromagnetic spectrum. For instance, conventional SAR image mapping of the Earth's surface suffers from poor height resolution to distinguish terrain features.

Background: In a conventional (single antenna) synthetic aperture radar, images are acquired in an azimuth-range (i.e., along-track - cross-track) coordinate system. Two principles are applied to achieve a high spatial resolution in the illuminated footprint. In the azimuth direction, a synthetic aperture is generated through the coherent addition of the radar echo. In the cross-track direction, however, the Doppler history (time delay) of sequential wave fronts is determined. The latter technique of time delay measurements leads to ambiguities, because two points with equal distance in the instantaneous footprint, but with different position and elevation, cannot be separated since their echoes are received simultaneously at the SAR antenna.
Single-pass SAR interferometry, on the other hand, employs two antennas, separated by the baseline. This 3-D observation geometry, similar in function to stereoscopic observation, is able to resolve the surface position ambiguities through slightly different delay times. In stereoscopic analysis, the parallax is determined through direct distance measurements. However, single-pass SAR interferometry employs signal coherence and phase difference measurements of both antennas to resolve the position ambiguities.

1) Cross-track SAR interferometry. The technique in either single-pass or two-pass (multi-pass) mode is the most widely used configuration.

The problem of terrain feature detection (elevations) is either overcome with a single-pass interferometric SAR using a dual-antenna system whose elements are positioned at a distance from each other (the so-called baseline), or by a regular (single antenna in monostatic mode) SAR instrument that covers the same ground swath a second time, possibly in a repetitive pass; the latter technique is also referred to as 'two-pass or repeat-pass interferometry' (the term "pursuit dual monostatic mode interferometry" is also used).

- Two-pass interferometry (use of a single antenna SAR system; however, the SAR platform must pass the same terrain in the same direction along a trajectory parallel to the first one). The interferometric image is obtained in post-processing by superposition of

1979) Albert A. Michelson is regarded as the father of interferometry. He is the first American who received the Nobel Prize in physics in 1907. In 1881, Albert A. Michelson did an experiment to try to detect a difference in the speed of light in two different directions: parallel to, and perpendicular to, the motion of the Earth around the sun. However, to his dismay, he found no difference. In 1887, Michelson repeated the measurement with Edward Morley. As they turned their apparatus (an interferometer), there was no measurable difference between the speed of light in the two directions. The Michelson-Morley experiment was an attempt to detect the velocity of the Earth with respect to a hypothetical luminiferous ether, a medium in space proposed to carry light waves. The experiment lead eventually to the deduction that the motion of the Earth through space has no effect on the velocity of light and that the absolute motion of the Earth is not measurable. This null result seriously discredited the ether theories and ultimately led to the proposal by Albert Einstein in 1905 that the speed of light is a universal constant.

both scenes (or multiple scenes) and by calculating the phase differences from slightly different viewing angles to obtain the surface elevation geometries in a rather tedious process. The image of two or more superimposed scenes is referred to as an 'interferogram.' - The concept of two-pass interferometry was first demonstrated with an airborne SAR (C/X-SAR) in 1991.

An extension of the two-pass interferometry technique is the so-called "differential SAR interferometry or **exact repeat-pass interferometry**." It is used to detect very small elevation changes (in the order of one cm or less) while preserving the same ground range resolution and swath width as that of a conventional SAR interferometer. The basic idea is to use one antenna flown twice over the same scene. If the second flight exactly duplicates the trace of the first, the time dependent phase changes can be measured. There would be no phase changes between the images at all unless there was a physical change in the scene, such as ground swelling and buckling in fault zones, residual displacements from seismic events, etc. If the two passes are made from flight tracks that are separated, it is no longer possible to distinguish surface changes from the parallax caused by topography. However, a third image made at some other baseline may be used to remove the topography and to leave only the surface changes. [1980)]

- Single pass interferometry. The interferometric SAR technique (referred to as **IFSAR** or **INSAR**), on the other hand, generates the interferogram in a single pass (dual antenna configuration) by relating the signals from both antennas in such a way that signal phase differences can be detected (the phase difference indicates the angle of arrival of the echo for each resolution cell in the SAR image). Across-track or along-track dual-antenna configurational directions or modes are possible in this context. - There are many key factors in producing accurate IFSAR maps, including accurate motion compensation, accurate baseline length and attitude measurement, phase-preserving SAR processing, 2-D phase unwrapping, and measurement of the absolute phase difference. - First airborne IFSAR (TOPSAR) was flown in 1994. The first spaceborne IFSARs, namely C-RADAR and X-RADAR on SRTM, are planned for launch in 1999.

2) Along-track SAR interferometry. It offers the capability to detect the motion of the illuminated targets. One application is the measurement of ocean surface currents. The SAR system must use two antennas, spaced along the radar velocity direction. The signals from each antenna are separately processed into two images, where the phase difference between the images is calculated. The phase difference of a resolution element is proportional to the radial distance moved by the resolution element in the time required for the rear antenna to move to the position formerly occupied by the forward antenna. - Early along-track SAR interferometry measurement demonstrations of ocean surface currents were conducted by JPL (in the time frame 1985-86) with an L-band SAR system aboard a NASA Convair-990 aircraft. In this configuration, one antenna was mounted near the rear cabin door, the other near the front, forming a baseline of 18.5 m (both antennas were pointing in the flight direction with separate, closely matched receiver channels). [1981)]
It should be pointed out that motion-sensing SAR systems (using two or more antenna systems aligned with the platform flight vector) provide the capability of detecting and/or measuring the radial component of velocity within the observed scene. Potential applications of motion sensing include such wide fields as oceanography, air traffic monitoring, land vehicle monitoring (detection of traffic jams on major highways), etc. [1982)]

1980) J. Moreira, "SAR Interferometry," http://daacdev1.stx.com/ELF/docs/PER/moreira3.html

1981) R. M. Goldstein, H. A. Zebker, "Interferometric radar measurement of ocean surface currents," Nature, Vol. 328, Aug. 20, 1987, pp. 707-709

1982) R. Saper, "Motion Sensing Synthetic Aperture Radar," http://daacdev1.stx.com/ELF/docs/PER/saper5.html

O.10.2 VLBI (Very Long Baseline Interferometry)

VLBI is a geometric (triangulation) technique, originally introduced in the field of astronomy (1960s), which measures the time difference between the arrival of radio signals, emitted from a distant source (mostly a quasar), at two or more widely-separated antennas (up to an Earth diameter and more). A precision synchronization of a ground-based VLBI-network of antennas permits the measurement of time differences to a few picoseconds. This in turn determines the relative positions of the antennas in the VLBI-network to a few millimeters, and the angular quasar position to fractions of a milliarcsecond. The high-resolution VLBI measurement technique has found many applications in Earth-based sciences, in particular geodesy.

VLBI is unique in its ability to define an inertial reference frame and to measure the Earth's orientation in this frame. Since the antennas are fixed to the Earth, their locations track the instantaneous orientation of the Earth in the inertial reference frame. Relative changes in the antenna locations from a series of measurements indicate tectonic plate motion, regional deformation, and local uplift or subsidence.[1983)]

Between 1979 and 1991 NASA's Crustal Dynamics Project (CDP) adopted the VLBI and SLR (Satellite Laser Ranging) techniques into a network of ground-based stations. This was followed by the Space Geodesy Program (SPG) with a substantial increase in sites through international cooperation. GPS techniques have also become part of SGP (the VLBI technique can be applied to GPS signals, in which case relative carrier phase measurements can be obtained without knowledge of either P-code or C/A-code). Of course, GPS receivers can also take advantage of the coded information for determining point or relative positions. - Sustained VLBI measurements (in one-day sessions) provide three-dimensional point positions; this data permits an estimation of velocities.

The VLBI technology offers also solutions for highly-accurate satellite position measurements using a ground-based network. The concept employs delta-differential one-way ranging (delta DOR) which fixes the spacecraft direction angle relative to that of a quasar. Accuracies of 30 nrad may be achieved. This relative positioning is useful only if the absolute position of the quasar is known to a similar level of accuracy.

Starting in 1997, NASA/JPL is defining the Space Interferometry Mission (SIM) to be launched in 2004. The SIM spacecraft will fly an optical interferometer (utilizing multiple telescopes) in a circular sun-synchronous orbit. The main objective is to improve angular measurement precision by two orders of magnitude - down to the level of a few microarcseconds. The technological challenges of such a mission are considerable.

O.11 Spatial Heterodyne Spectroscopy (SHS)

Interferometric spectroscopic instruments offer sensitivities typically 100 times those of conventional grating spectroscopic instruments of similar size in many applications. In 1990, Fred Roesler and John Harlander at the Department of Physics of the University of Wisconsin at Madison conceived and developed an unusual and novel interference spectroscopic technique called Spatial Heterodyne Spectroscopy (SHS).[1984)] [1985)] [1986)] It is an interferometric Fourier transform technique, but unlike conventional Fourier transform spectroscopy (FTS) it requires no moving parts for obtaining a spectrum. Moreover it can

1983) http://lupus.gsfc.nasa.gov/brochure/bintro.html

1984) J. M. Harlander, "Spatial Heterodyne Spectroscopy, Interferometric Performance at any Wavelength without Scanning," Ph.D. Thesis, University of Wisconsin-Madison, 1991

1985) J. Harlander, R. J. Reynolds, F. L. Roesler, "Spatial Heterodyne Spectroscopy for the Exploration of Diffuse Interstellar Emission Lines at Far-Ultraviolet Wavelengths," The Astrophysical Journal, Vol. 396, 1992, pp. 730-740

1986) Note: The invention was patented (US patent No 5059027, issued Oct. 22, 1991), the patent was assigned to the Wisconsin Alumni Research Foundation.

be field-widened without moving parts to provide additional gains of typically 100. As a result of these gains, the SHS instrument can be made small and still achieve a level of performance equal or superior to grating instruments of practical dimensions. At the same time it avoids many of the mechanical problems associated with conventional field-widened FTS techniques.

SHS provides the first practical approach to extend interference spectroscopy into the FUV (Far Ultraviolet, 1200-2000 Å) spectral range. The technique appears to be offering high-resolution spaceborne spectroscopy applications in astronomy (detection of faint interstellar emission lines, study of the dynamics of hot interstellar gases) as well as in Earth observation. Examples in this field include the measurement of vertical density profiles of the hydroxyl (OH) radical in the middle atmosphere (30-100 km), study of the distribution of aerosols in the mesosphere, and to investigate the role of OH in the photochemistry of water vapor and ozone in the presence of aerosols in the mesosphere.

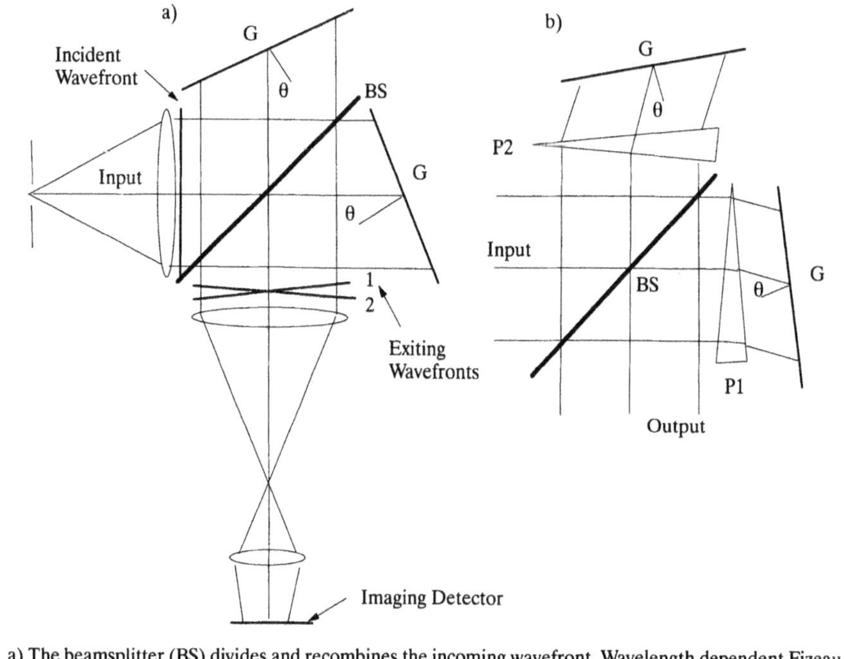

a) The beamsplitter (BS) divides and recombines the incoming wavefront. Wavelength dependent Fizeau fringes produced by diffraction gratings G are recorded by a position sensitive detector.
b) The same SHS configuration but with field widening prisms P1 and P2 added to each arm. The prism angles are chosen so that from a geometrical optics point of view the gratings appear coincident.

Figure 388: Schematic diagram of the SHS configuration

In the basic SHS design, Fizeau fringes of wavenumber-dependent spatial frequency are produced by a modified Michelson interferometer in which the return mirrors are replaced by conventional blazed diffraction gratings (see Figure 388 part a). The fringes are recorded on a position sensitive detector and Fourier-transformed to recover the spectral content of the source. Zero spatial frequency corresponds to the Littrow wavenumber of the gratings, which can be chosen by adjustment of the interferometer. Since zero spatial frequency corresponds to a finite wavenumber, SHS measures differences between the source and alignment wavelengths, and high resolution spectra over a limited spectral range can be recovered with modest requirements on the spatial resolution of the detector. In this process, no element is mechanically scanned.

The resolving power of an SHS design is equal to the theoretical resolving power of the dispersive (i.e. grating) system while the field of view of the system is characteristic of interferometric spectrometers (conventional Michelson and Fabry-Perot). The interferometric field of view gives SHS systems a 100-fold gain in sensitivity for diffuse source spectroscopy over diffraction grating spectrometers of the same size and resolving power. Furthermore, field widening techniques can be applied to SHS systems which enable SHS to view even larger fields of view. Gains associated with field widening are typically two orders of magnitude in solid angle over conventional interferometers (10^4 larger than diffraction grating spectrometers).

Field widening is accomplished by inserting prisms into the arms of the interferometer (see Figure 388 part b). The prism apex angles are chosen so that from a geometrical optics point of view the gratings appear coincident. The geometrical path difference in the system is then near zero for a wide range of input angles and the system behaves much like a conventional Michelson interferometer at zero path difference. Aberrations introduced by the prisms ultimately limit the field of view, but not before large gains can be achieved in many applications.

The SHS optical system is similar to the Twyman-Green interferometer that is often used for optical testing of surfaces. In the SHS system, diffraction gratings replace the test and reference surfaces of the Twyman-Green; hence grating figure errors, if present, distort the Fizeau fringe pattern. Such fringe distortions degrade the instrumental profile and reduce the signal-to-noise ratio in the recovered spectrum. However SHS fringe distortions for monochromatic light provide a measure of the grating figure errors that can be used in software to correct broadband interferograms without loss in signal-to-noise ratio.

Fringe distortions resulting from figure errors and index inhomogeneities in the beamsplitter and field widening prisms can be corrected if these elements are of good quality over the area sampled by one spatial resolution element on the detector. In the SHS system, these elements are nearly focused on the detector; hence the area that is sampled at each detector pixel is relatively small. This is to be contrasted with conventional interferometers where a single channel detector collects light over the full aperture of the critical optical components. Figure errors in conventional interferometers will result in a reduction in the contrast of the fringes and a reduced signal-to-noise ratio in the recovered spectrum.

Due to the heterodyne nature of SHS interferograms, figure errors distort the individual (carrier) fringes to a greater extent than the envelope (visibility) of the fringes. The envelope provides physical information on the spectral line shape; so therefore, a correction technique is needed which corrects the carrier fringes without distorting their envelope.

The optical concepts that lead to the predicted high performance for the SHS were also demonstrated and validated by Harlander and Roesler in various field tests.

NRL (Naval Research Laboratory) in Washington, DC has been cooperating with UWM (University of Wisconsin-Madison) for the last years to develop an instrument. A first implementation of the SHS concept is realized in SHIMMER (Spatial Heterodyne Imager for Mesospheric Radicals). The instrument is planned of be flown on Shuttle in 2002.

O.12 Orbital Concepts and Terminology in Remote Sensing

Ascending node. Point where a satellite orbit crosses the equator from south to north. The ‘descending node’ refers to an equator crossing from north to south.

Fictitious mean sun. An imaginary body in the equator plane with mean tropical motion of the sun. The concept is needed to define mean solar time [and in this respect the universal time (UT) and UTC (Universal Time Coordinated) respectively].

Local time. Defines the angle in the equator plane between the local meridian (of the place or the region of interest on the Earth's surface, for instance, a subsatellite point) and the fictitious mean sun. The local time is shown in Figure 389. The equator plane is subdivided into 24 hours (one period of rotation or one full day), with the morning period to the left of the local noon (fictitious mean sun position) and the afternoon period to the right of local noon.

Mean solar time. A measure of time based conceptually on the diurnal motion of the fictitious mean sun, under the assumption that the Earth's rate of rotation is constant.

Inclination (I) is defined as the angle between the Earth's equatorial plane and the satellite's orbital plane [which is for most inclinations slowly variable and always passes through the center of mass (the Earth)]. The different orbital periods of the Earth and the satellite provide longitudinal precession for successive orbits. An inclination angle of 30° means for instance that the observation area of the satellite is bounded by ± 30° latitude (tropical region). An inclination of 0° (or near 0°) refers to an equatorial orbit. **Polar orbits** have an inclination near 90°. For this case the Earth's rotation provides successive swaths of observation with every orbit. If we consider about 14.5 satellite orbits per day, then the instrument FOV must be about 25° for contiguous global coverage.

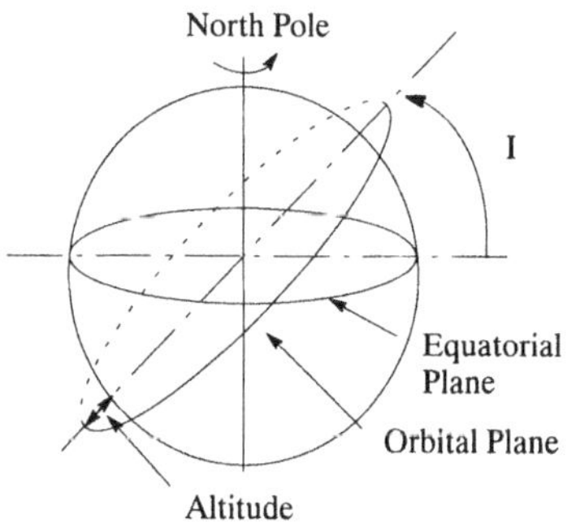

Orientation of the orbit in space

Mean motion. Refers to the average speed of a satellite in its orbital plane (units are usually given in revolutions per day). Satellites in circular orbits travel at constant speed. Satellites in non-circular (elliptical) orbits move faster when closer to Earth, and slower when they are farther away.

Nutation (Latin: nutare, "to nod"). Refers to the short-period oscillations in the motion of the pole of rotation of a freely rotating body which is subjected to a torque from external gravitational forces (the British astronomer James Bradley announced his discovery of nutation in 1748). In the case of the Earth, it is a small irregularity in the precession of the equinoxes. Nutation superimposes a small (wobbling) oscillation, with a period of 18.6 years upon the very slow motion of precession [nutation in longitude (of the ecliptic) with an amplitude of 17.2 arcsec, and a nutation in the obliquity of the ecliptic with an amplitude of 9.2 arcsec]. The cause of nutation lies mainly in the fact that the plane of the moon's orbit around the Earth is tilted by ~5° from the plane of the Earth's orbit around the sun (the ecliptic). The moon's orbital plane precesses around the Earth in 18.6 years, and the effect of the moon on the precession of the equinoxes varies with this same period.

Precession. A phenomenon associated with the action of a gyroscope or a spinning top, it consists of a comparatively slow rotation (gyration) of the axis of rotation of a spinning body about a line intersecting the spin axis and describing a cone, caused by the application of a torque (e.g. external gravitational forces), and tending to change the direction of the rotation axis. It is a motion continuously at right angles to the plane of the torque producing it. The smooth, slow circling of a spinning top is precession, the uneven wobbling is nutation.

Precession of the equinoxes. A slow westward motion of the equinoctial points along the ecliptic (the Earth's orbital plane) caused by the gravitational influence of the sun and the moon acting on the Earth's equatorial bulge in connection with its diurnal rotation (the planets exert an influence as well, but to a much lesser extent). Precession is the slow, top-like oscillation of the spinning Earth, with a period of about 26,000 years.

Prograde orbit (or direct orbit). A satellite orbit with an inclination of $<90^{\circ}$ (note: most satellite orbits are 'prograde'). For an Earth-orbiting satellite the tropical motion of the equator crossings (or nodal points) of successive orbits is from east to west, i.e., in the opposite direction of Earth rotation.

Retrograde orbit. A satellite orbit with an inclination $>90^{\circ}$. For satellite (circular or elliptical) orbits the tropical motion (of the nodal points) of successive orbits is from west to east, in the same direction of Earth rotation. All sun-synchronous orbits are retrograde orbits. For geosynchronous transfer orbits (which are 'prograde' from an inertial point of view) the apparent motion of the satellite is retrograde in the vicinity of the apogee; this is due to the smaller satellite velocity with respect to the Earth's rotation.

Right ascension of ascending node. Refers to the angle in the equatorial plane which is measured from the vernal equinox to the ascending node. Note, the inertial coordinate representation (of right ascension and declination) provides simply one angle (right ascension) for the orbital plane of the spacecraft which is inertial.

Precise orbit determination. Refers to the analysis of satellite tracking data (of various sources, such as: altimeter, GPS, SLR, Doppler, etc.) in order to model the position (or ephemeris) of the satellite at a given moment.

Tropical motion. Motion of a body (planet or satellite) related to the vernal equinox which is the intersection point of the Earth's equator with the ecliptic.

O.12.1 Sun-synchronous Orbit

An orbit is said to be sun-synchronous if the orbit precession of the satellite compensates for the Earth's revolution around the sun (about 0.9856° per day)[1987]. The daily node rotation of 0.9856° is eastward such that the angle between the orbital plane and the sun remains constant. For near-Earth satellite orbits (altitude < 1300 km) sun-synchronous orbits are polar orbits at an orbital inclination and altitude such that a satellite passes over a given site always at the same local time. The orbital nodes maintain a near constant solar time. The satellite always crosses the equatorial plane at the same local time [from an equator crossing to the next equator crossing (one orbital period), the Earth has rotated exactly as far as to make the local time of day the same]. This means that the sun position (lighting conditions) is always the same (within certain limits) for a repeat observation of any given site (consistent sensor-target-sun relationships). Sun-synchronous orbits may be placed so as to see points of the Earth at anywhere from local sunrise/sunset to local noon; due to this they are often referred to as 'morning' (AM) or 'afternoon' (PM) orbits. - There are advantages and disadvantages to the measurement at constant sun-angle depending on the application. Varying sun angles may be important for geologists to reveal subtle structural details not visible at higher sun angles.

There are two limiting cases for sun-synchronous orbits, depending on the choice of local time for the ascending nodal equator crossings:

1) A dawn/dusk orbit with ascending nodal equator crossings at 6 AM or 6 PM local time. For both cases the satellite orbital plane is perpendicular to the mean sunlight direction. This means the satellite orbit is normally sunlit, but always close and parallel to the terminator plane.

1987) Note: The daily rotation of the orbital satellite plane (with respect the the equatorial plane) is identical to the mean motion of the fictitious sun around the Earth - which in turn is identical to the mean motion of the Earth around the sun. The effect of the rotation of the orbital satellite plane is due to the oblateness of the Earth.

2) A full-sun / full-shadow orbit with ascending nodal equator crossings at local 12 AM (noontime or 12:00 hours) or at local midnight (0:00 hours). For both cases the mean sun direction is in the orbital plane. This means that about 60% of the orbit is in full sunlight and the other portion is in full shadow of the sun. The choice of the 12 AM ascending orbit provides first a sunlight phase followed by a shadow phase, while the midnight orbit reverses these phases.

All other choices of ascending nodal equator crossings at specific local times are positioned between these two limiting cases for sun-synchronous orbits.

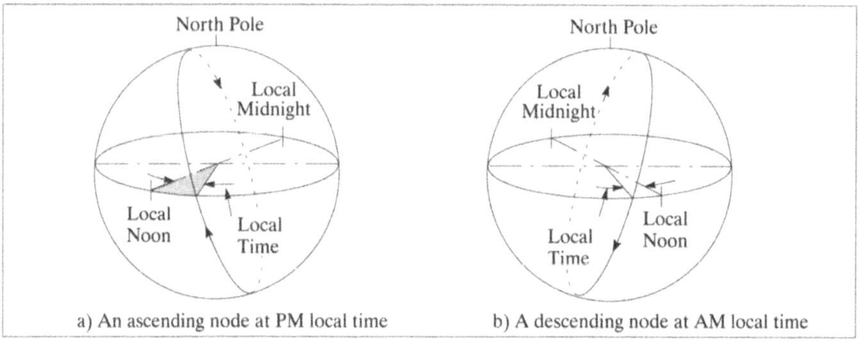

Figure 389: Examples of relative orientation of sun-synchronous orbits with local time

O.12.2 Geosynchronous Orbit

An orbit in which the satellite's orbital period is identical to the orbital period of the Earth. Thus, a geosynchronous satellite has the Kepler period of one sidereal day, which is equal to 23 hours, 56 minutes, and 4.09 seconds (or 1436.07 minutes). The semi-major axis of such a geosynchronous orbit is 42,164.182 km from the center of the Earth (or approximately 35,786 km above the Earth's surface). Two limiting cases of the geosynchronous orbits are of particular interest:[1988)]

1) Inclination = 0°. A geosynchronous orbit with zero inclination is conventionally referred to as **geostationary** or **GEO**. The ideal geostationary orbit is circular, its orbital plane coincides with the equator plane, and its orbital period is one sidereal day (23 hours, 56 minutes, 4 seconds). In practice, the orbits have a small eccentricity and inclination as a consequence of perturbations. Each satellite in GEO is usually assigned a longitude/latitude box, within which it has to stay, typically to within ±0.1° for direct-TV communication satellites. - The subsatellite point of a zero-inclination orbit remains ideally "stationary" over a particular meridian at the Earth's equator. Perturbations to the orbit of a satellite [due to a) anomalies in the Earth's gravity field, b) gravitational effects of the sun and moon, c) solar radiation pressure] are normally counteracted by the process of 'station-keeping'.

2) Inclination ≠ 0° [IGSO (Inclined Geosynchronous Orbit)]. A geosynchronous satellite with an inclination different from zero is not "stationary" when observed from the Earth. The subsatellite track described by such a constellation is referred to as 'lemniscate,' meaning 'the locus of the foot of the perpendicular from the center of a conic on its tangent'. The subsatellite-track projection onto the Earth's surface describes a path

1988) Courtesy of F. Jochim, DLR/GSOC

as outlined by the number 8 in the period of one day. An inclination of 60° causes a latitudinal displacement of ±60° during one orbital period (just as is the case for any non-geosynchronous orbit). However, the subsatellite-track projection (i.e. the Figure 8) of an inclined geosynchronous orbit has the property of remaining stationary over a particular meridian. Hence, there is no longitudinal precession for successive geosynchronous orbits. An inspection of the subsatellite track reveals zero longitudinal precession at the equator and increasing precession or regression with increasing latitude.

In general, a geosynchronous orbit is not necessarily a circular Earth orbit, but may also be defined as an elliptical HEO orbit. This can be done as long as the semi-major axis is 42,164 km and the orbital period is equal to a sidereal day. An example of such a geosynchronous HEO mission is ISO (Infrared Space Observatory) of ESA with a launch on Nov. 17, 1995. The orbit parameters of ISO are: perigee = 1,000 km, apogee =70,500 km, inclination = 5.25°, eccentricity = 0.825, period = sidereal day.

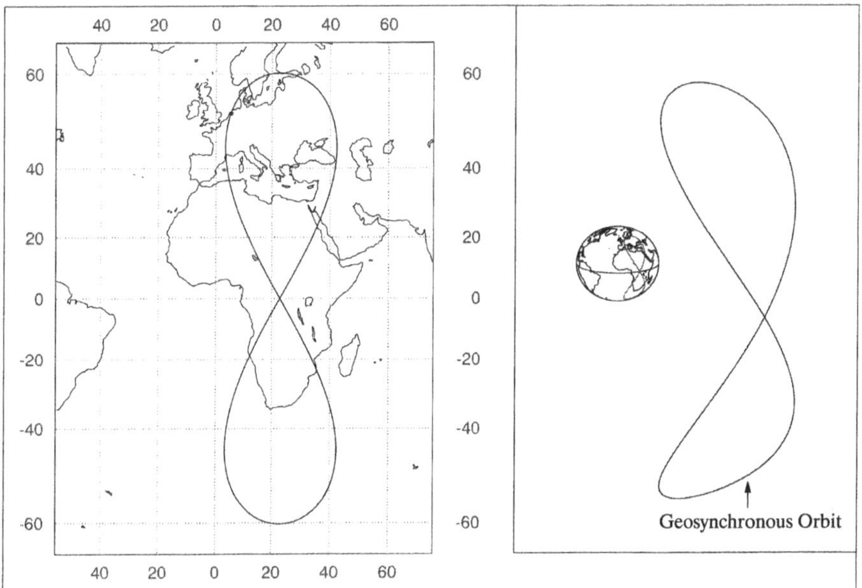

a) Projection of lemniscate at 24° eastern longitude

b) Geosynchronous orbit and projection of lemniscate onto the Earth at actual scales

Figure 390: A lemniscate of a geosynchronous orbit with an inclination of 60°

Item 1 above is simply a special case of item 2. There are numerous applications for geosynchronous orbits of spacecraft in such fields as communications, meteorology, tracking and data relay services to LEO satellites, etc. The first geostationary communications satellite in orbit was Early Bird (later renamed to Intelsat I) which was launched in April 1965. The first geostationary weather satellite was ATS-1, launched on December 7, 1966. Some advantages of a geostationary satellites are:

- The same area of the Earth's surface (footprint) is continuously in view, providing a fixed-position relation between the satellite and its ground segment (stations, etc.). This is favorable for signal recovery of low-power communications. Broadcasts are tied to a geographical region.

- In remote sensing the continuous view of the same portion of the Earth's surface provides an opportunity to observe such things as diurnal phenomena or the history of entire weather systems (study of lighting in thunderstorms, etc.).
- The footprint of a geostationary satellite is fairly large , due to its high altitude position. providing a coverage of about 1/3 the Earth's surface.

A disadvantage of a geostationary satellite in a voice communication system is the round-trip delay of approximately 250 ms. A further drawback of the geostationary orbit is its exclusivity over the equator at zero degree inclination. It is the most wanted orbit for many applications; hence, it is also the most crowded orbit, with problems of signal interference and other problems. The coverage is of course optimal for regions close to the equator; problems usually arise at high latitudes due to very low grazing angles. - On the other hand, the geosynchronous orbit with an inclination of, say, 60°, may be of interest in the future, in particular for meteorological applications, since it provides much better observation constellations for the high-latitude regions, including the polar caps. With nodal crossing times of, say, 6:00 AM and 6 PM, there is the opportunity to observe more intensely a portion of the illuminated northern half of a continent in combination with the dark southern half (or vice versa).

O.12.3 Repeat Coverage or Temporal Resolution

Time may be regarded as the fourth variable of resolution. A dynamic Earth environment changes with time; this applies in particular to meteorological observations and vegetation (crops) observations, etc. Repeat coverage (or temporal resolution) designates the period of time that elapses until the next observation of the same geographic area can take place. Total coverage of the globe can only be provided by polar-orbiting satellites. Their orbit and swath width are usually chosen in such a way that Earth rotation accounts for a continuous and contiguous observation swath (with some overlapping) of the globe. In general each sensor aboard a satellite has a different swath width.

There is a fairly simple trade-off between spatial resolution (or IFOV) and frequency of coverage. Constraints are imposed by the sensor design, the on-board electronics, and the amount of source data that can be handled. Thus the smaller the IFOV, the more data there is to be handled for any given area on the ground and the less frequently data will be available for a given area.

Examples of repeat coverage requirements for some remote sensing applications:

- Operational meteorology (weather forecast) requires a high repeat coverage of large areas at relatively low spatial resolutions (1-5 km); the emphasis is on the measurement of progressive daily cycles. About 20 - 30 min/image of the same area for meteorological geostationary satellites (such as METEOSAT, GMS, GOES and INSAT series). Geostationary satellites permit the tracking of continuous dynamic processes with the desired viewing frequency (for example: viewing of Earth surface targets at any sun illumination angle, and viewing each point on the ground at exactly the same zenith and azimuth angles over time).

- Monitoring of renewable resources (vegetation, agricultural crops) requires a medium repeat coverage (in the order of several days and up to 2 weeks) but relatively high spectral and spatial resolutions (30 m) on a seasonal basis. Polar orbiting satellites (Landsat, POES, SPOT, Resurs, etc.) realize these requirements.
  Note: There are also a number of transient phenomena that cannot be adequately seen from polar orbiting satellites, such as the diurnal variability and dynamic behavior of:
  - precipitation and evaporation
  - atmospheric water vapor and wind
  - vegetation color

- terrestrial ecosystems and land processes
- land-ocean-atmosphere energy fluxes
- tropospheric pollutant generation and transport
- ocean color (biomass productivity)
- soil moisture and net radiation balance
- etc.

- Cartographic applications (maps) of the Earth's surface require a relatively low repeat coverage but high resolutions (spatial and or spectral). Imagery for urban growth patterns is appropriate in time intervals of a year or so.

O.12.4 LEO (Low Earth Orbit)

The term LEO is used for Earth observation satellites as well as for communication satellite constellations. These are usually circular or near-circular orbits at altitudes of less than 2000 km above the Earth's surface. The orbital period in the altitude range of 300 - 2000 km varies between 90 - 120 minutes. A satellite pass (i.e. the visibility period of a satellite above the horizon) for an observer (or a ground station) may vary from about 5 - 20 minutes, depending on a horizon pass or a zenith pass of the satellite with respect to the observer.

Below 500 km the orbital lifetime is very much reduced by atmospheric drag; however, the observational resolution of Earth surface imagery increases. With increasing altitude the orbital life increases, along with the sensor footprint, while the resolution decreases. Increasing altitudes also provide a harsher radiation environment, thereby sharply reducing the lifetime of electronic components, this in turn may be counteracted by extra shielding or component hardening.

From a communications point of view the near circular LEO orbit is providing a nearly-constant signal strength between ground and satellite. Advantages of LEO transmissions are experienced in particular with much reduced signal power needs (and path losses) and much shorter signal time delays as compared to a GEO constellation. [1989] Another advantage is that LEO satellites avoid congestion (physical as well as the spectrum congestion) experienced in the spatially very limited GEO environment. Naturally, some LEO communication characteristics are disadvantageous, such as increased Doppler shifts, multipath propagation, variable signal path, and variable pass times; frequent hand-overs are needed.

The radius of a footprint of a communications satellite in LEO orbit varies from about 3000 to 4000 km. Hence, global real-time coverage can only be achieved with a satellite constellation. Several service concepts have emerged in the communications industry with regard to LEO systems. [1990]

- **Little LEO systems**. Refer to mobile satellite communication systems for non-voice messaging and data relay (i.e. store & forward) services. Satellite communication is usually in VHF and/or UHF bands. Examples: Faisat, GE Starsys, SAFIR, Orbcomm, Temisat, Leo One USA, Kitcomm, Gonets-D, etc.
- **Big LEO systems.** Refer to mobile satellite communication systems for global hand-held telephone service as well as mobile fax and data services (support of packet-oriented services). Satellite communication is usually in L and/or C bands (sometimes also Ka band). Examples: Iridium, Globalstar, Odyssey, etc.

1989) Note: The required signal power varies with the square of the distance. Hence, reducing the orbit from GEO to LEO translates into a multifold increase in available signal strength, allowing operation of a LEO constellation at lower power levels.

1990) Note: In a LEO communications system with handheld phone service (implying weak signals) the entire burden of link completion is placed on the satellite. Such a connection can physically be achieved only if the satellite employs the concept of very narrow-spaced spot beams (FOV of about 1°), each covering a "cell" on the Earth. Many of these beams must be employed to provide sufficient coverage of the intended service area (total footprint). The employed cell technique requires automatic handoff when a subscriber is passed from the coverage of one spot beam to that of another (the cells are moving while the subscriber is essentially stationary). In addition to the beam-to-beam handoff there is also the management of satellite-to-satellite handoff to be taken care of.

- **Broadband LEO systems.** Refer to mobile satellite communication systems for high-capacity data communications such as high-speed computer links and video telephony. Satellite communication is usually in Ku or Ka bands. Examples: Teledesic, Skybridge, M-Star.

O.12.5 MEO (Medium Earth Orbit)

Refer to near-circular orbits with altitudes in the range of about 5000 and 25000 km above surface. The orbit period may vary from about 3 to 12 hours. The GLONASS and GPS constellations are MEO examples with periods of 11.25 and 12 hours respectively. Other MEO satellite systems are LAGEOS with about 6000 km altitude, and the Inmarsat-P (communication satellite) series with altitudes of 10350 km. Maximum MEO contact times may vary from an hour to a few hours. Compared to a LEO system, hand-over is less frequent, while propagation delay is greater.

O.12.6 HEO (Highly-Elliptical Earth Orbit)

Refer to orbits with a typical perigee of 500 - 1000 km above the Earth's surface and apogees up to 20 R_E (or about 140,000 km) or more. HEO orbit periods may vary from about 10 hours to two days. Examples are the IRM and UKS satellites in the AMPTE project, each with apogees of 18.7 R_E and a period of 44.3 hours. GEOTAIL has an extreme HEO orbit with a perigee of 8 R_E and an apogee of 220 R_E. The satellites of the INTERBALL program are: 'Auroral Probe' with a perigee of 500 km and a apogee of 20,000 km, and 'Tail Probe' with a perigee of 500 km and an apogee of 30 R_E or about 190,000 km. - The high eccentricity of these orbits causes the satellite to spend about two thirds of the orbital period near apogee (referred to as apogee dwell), and during this time it appears to be almost stationary for an Earth observer. Free space loss and propagation delay in HEO are even greater than in GEO systems.

O.12.7 EEO (Elliptical Earth Orbit)

EEO refers to orbits chosen in such a fashion that maximum service provision (or viewing time) can be provided to a particular region of the Earth. This is accomplished with elliptical orbits where the apogee (by far the longest duration period of the orbit) occurs always in the intended region. Typical EEO examples are Molniya-type orbits (Russia) with apogees over the big land mass of Asia. They are characterized by an orbital period of 12 hours, apogee height of 39350 km over the northern hemisphere, perigee height of about 1000 km over the southern hemisphere, and an inclination of 63.4°. In this case the apogee is inertially fixed. The Molniya satellite spends about 11 hours (92%) of the 12 hour orbital period over the northern hemisphere. [1991]

An EEO could for instance also be selected for a communications system with apogees over the northern hemisphere, taking advantage of the fact that most of the world's population lives in the northern hemisphere for required service provision.

O.12.8 The Interferometric Cartwheel Orbit

The concept of the interferometric cartwheel orbit was initially proposed by D. Massonnet in 1997 (CNES internal memo); it is an R&D project of CNES since 1998. [1992] [1993] The study objective is in the framework of finding cost-effective solutions to the field of active microwave observations. Particular emphasis is on satellite constellations carrying an active

1991) Information provided by F. Jochim of DLR/GSOC

1992) D. Massonnet, "Capabilities and Limitations of the Interferometric Cartwheel," IEEE Transactions on Geoscience and Remote Sensing, Vol. 39, No 3, March 2001, pp. 506-520

1993) S. Ramongassie, L. Phalippou, E. Thouvenot, D. Massonnet," Preliminary Design of the Payload for the Interferometric Cartwheel," Proceedings of IEEE/IGARSS, July 24-28, 2000, Honolulu, HI, pp. 29-32 instrument (SAR) on one spacecraft, and passive receivers (radiometers) on other (passive) spacecraft in the formation in order to obtain interferometric imagery.

Orbital cartwheel geometry:

The overall orbital geometry consists of a spacecraft with a transmitter and a receiver function (SAR or Doppler radar instrument) and one or more spacecraft with passive receivers (radiometers) in formation on the same orbit, and close enough to be able to point into the footprint illuminated by the transmitter spacecraft (master slave relation). All spacecraft are co-planar, but the eccentricity of each passive spacecraft is slightly modified. This results in other orbits with the same periodicity which describes an ellipse relative to the original orbit. Three passive satellites with the same additional eccentricity but perigees distributed along the orbit lead to a configuration as shown in Figure 391. The perigees of the three spacecraft are separated by 120° (evenly distributed). The major axis of the ellipse is in the orbit plane, the minor axis is co-planar in the vertical direction. The minor semi-axis of the ellipse is the "vertical radius of the wheel." The major axis of the ellipse is twice the vertical radius. For a LEO constellation, a typical value of the vertical wheel radius may be in the range of 500-1000 m.

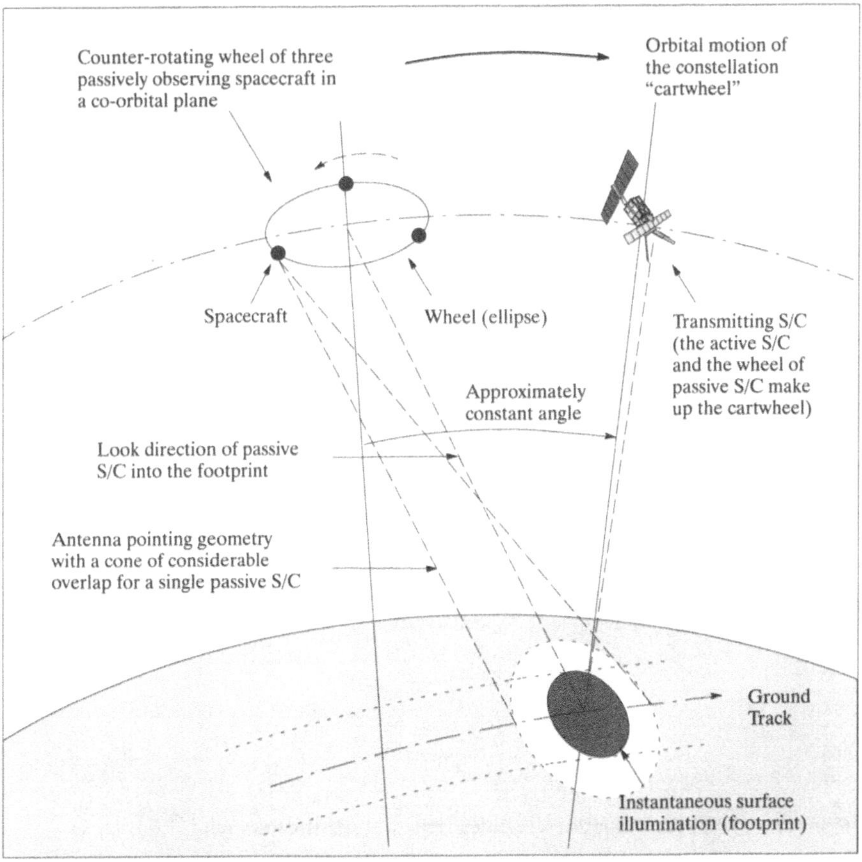

Figure 391: Orbital geometry schematics of a backward-moving cartwheel constellation

There are no severe pointing nor positioning requirements imposed by a cartwheel formation. Each passive spacecraft of the cartwheel can be provided with a small antenna whose footprint contains at least the footprint of the active SAR instrument. Thus, S/C pointing may be done with reaction wheels. For safety reasons only, on-orbit propulsion of the passive constellation is needed: a) to keep a proper distance to the transmitting S/C, and b) for wheel configuration keeping.[1994] [1995]

An advantage of the wheel configuration is the inherent stability of its vertical and horizontal baselines. Without going into orbital mechanics, it has been shown that a wheel of passively observing satellites, placed into the same orbit of a single actively observing spacecraft, remain in natural relative formation. In fact, the tandem linear motion of the counter-rotating wheel of passive spacecraft and that of the active spacecraft, resemble the typical motion of a "cartwheel," drawn backwards across the sky. The name cartwheel was therefore natural to the CNES inventors of this particular orbital constellation.

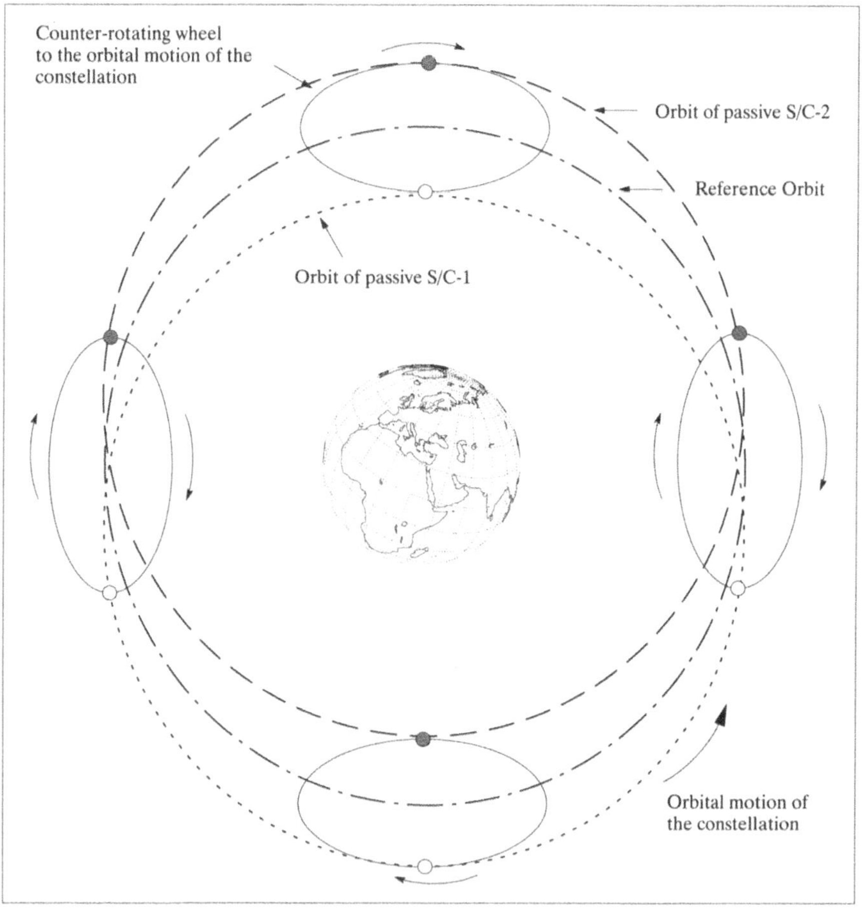

Figure 392: Schematic illustration of the wheel orbits of two passive spacecraft [1996]

1994) D. Massonnet, E. Thouvenot, S. Ramongassie, L. Phalippou, "A wheel of passive radar microsats for upgrading existing SAR projects," Proceedings of IEEE/IGARSS, July 24-28, 2000, Honolulu, HI

1995) D. Massonnet, French Patent No 236910D17306RS, "Roue interferometrique," April 30, 1998

1996) Courtesy of Friedrich Jochim of DLR/GSOC

Naturally, the wheel may also be positioned in a forward-moving cartwheel constellation. Figure 392 illustrates how the wheel of two spacecraft rotates along a complete orbit. The active spacecraft is not shown here for reasons of clarity.

The bistatic observation concept of a cartwheel constellation permits practically the simultaneous collection of SAR imagery from various vantage points. A footprint illuminated by only one SAR instrument on a single spacecraft can be viewed, and therefore the echo detected by the receivers (radiometers) of spacecraft in the vicinity of the transmitting spacecraft. The combination of the imagery from all members of the constellation is interferometric by its very nature. The spatial resolution of the imagery will improve considerably. - The cartwheel concept provides a great potential for cost-effective missions. A wide spectrum of applications can be expected with the introduction of this new technology.

In the cartwheel observation scheme, the slave radar instruments (receivers on the wheel S/C) are not synchronized with the receiver of the SAR instrument of the master spacecraft. The duty of each slave receiver is simply to point into the footprint illuminated by the transmitter, to collect continuously the data stream of echoes emanating from the illuminated footprint with sufficient bandwidth, and to store/transmit the results (in combination with data compression). Obviously, a ground-based re-synchronization process is needed to combine the data collected by all slaves.

Signal processing background: A pair of two images acquired simultaneously from two separate points in along-track is characterized by the critical horizontal baseline. It is such that the range difference between the receivers and a given ground target changes by one wavelength from one pulse to the next. For a coherent combination of the passive images made by each receiver, the along-track separation of the receivers must stay below this critical horizontal baseline. In the same way, a critical vertical baseline is defined as an adaptation of the conventional orthogonal baseline of active interferometric systems to the passive, vertically separated systems. It is such that the range difference seen by the second receiver, across the range pixel of the first receiver, differs from the range pixel size by one wavelength. Thus, for coherent combinations of the passive images made by each receiver, the vertical separation of the receivers must stay below this critical vertical baseline.
The combination of the critical horizontal (azimuth) and the critical vertical baselines creates a rectangle called the "interferometric area." The resolution of an interferometric product results from the coherent part common to the intersecting interferometric areas of the two receivers involved in the combination. The size of the critical baselines are inversely proportional to the associated resolutions in azimuth and range.

O.12.9 Some Orbit Selection Requirements

There are a number of factors which may influence the particular orbit selection:

- Global coverage requirement: implies a polar or near-polar orbit
- Constant illumination requirement: implies a sun-synchronous orbit
- Continuous observation requirement: implies a geosynchronous orbit
- Measurement of gravity anomalies: implies a low orbit
- Requirement to minimize the atmospheric drag: implies preference for a high orbit
- Requirement to minimize a radar sensor's power: implies preference for a low orbit

O.12.10 Walker Constellation

The Walker constellation is named after J. G. Walker of RAE [Royal Aircraft Establishment - later renamed to DERA (Defence Evaluation and Research Agency), as of July 2, 2001 DERA was renamed to QinetiQ]. Walker was a pioneer in studying Earth-coverage problems (in particular equal-area distributions) with various satellite constellations.[1997) 1998)]

A typical Walker constellation configuration is denoted by the symbols: T/P/F, where T= number of satellites, P = number of orbital planes, and F characterizes the interplane phasing (relative from plane to plane). In a symmetrical constellation of T satellites, T/P satellites are evenly spaced on each orbital plane, and P orbital planes are evenly spaced through 360° of ascending node. Example 1 of a Walker constellation 27/3/1: In this case 27 satellites are in a symmetrical Walker configuration (in three orbital planes). Example 2: 18/6/1. In this case 18 satellites are distributed in six planes (Δ=60°), the phasing ϕ=20°.[1999)]

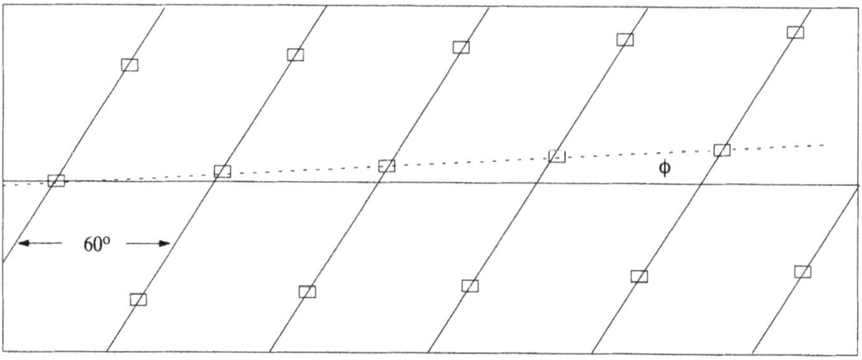

Figure 393: Equal distribution of a Walker constellation 18/6/1

Walker introduced a measure referred to as PU (Pattern Unit) = 360°/T to calculate the phase angle (phase angle = F x PU). Example: T/P/F = 54/9/3; then PU =6.667°; and the phase angle = 20°.

O.12.11 Libration Points/Lagrange Points

Libration points (also referred to as "Lagrangian points"[2000)]) in space are defined as points where the gravitational forces due to two or more bodies (the sun, or planets) balance, and thus where a spacecraft can be positioned in equilibrium at zero velocity.

Lagrangian points are particular solutions of the equations of motion applied to the "Problem of Three Bodies" in an orbital plane of two massive bodies in circular orbits around a common center of gravity and a third body of negligible mass. The equations of motions always yield five points at which the third body can remain at equilibrium. Three of the points are on the line passing through the centers of mass of the two massive bodies - L_3 beyond the most massive body, L_2 beyond the less massive body, and L_1 (the point through which the mass transfer occurs) between the two bodies. The other two points, L_4 and L_5, are located at the two points in the orbit of the less massive component which are equidistant (equilateral triangle) from the two main components, as illustrated in Figure 394.[2001) 2002)]

A satellite positioned at L_4 or L_5 is regarded as stable, i.e. it will remain at the point and move along in the specific reference system (here either in the sun-Earth system or in the

1997) J. G. Walker, "Continuous Whole-Earth Coverage by Circular-Orbit Satellites," IEE Conference publication No. 95, 1973, in Satellite Systems for Mobile Communications and Surveillance, pp. 35-38.

1998) J. G. Walker, "Continuous Whole-Earth Coverage by Circular-Orbit Satellite Patterns," Technical Report 77044, Royal Aircraft Establishment, Farnborough, UK, March 24, 1977 (full report of 80 pages)

1999) Information provided by Friedrich Jochim of DLR/GSOC.

2000) The first three libration points were discovered by Leonard Euler (a Swiss mathematician), the other libration points were discovered by Joseph Louis Lagrange (1736 - 1813), French mathematician

2001) Fred P. J. Rimrott, "Introductory Orbit Dynamics," Vieweg, Braunschweig/Wiesbaden, 1989, pp. 156-158

2002) Karl Stumpff, "Himmelsmechanik," Band II, VEB Deutscher Verlag der Wissenschaften, Berlin 1965

Earth-moon system), while a satellite located at L_1 or L_2 or L_3 is unstable, i.e. it will wander off if slightly disturbed. However, small orbit corrections by a thruster engine can provide compensation.

Within the sun-Jupiter system the L_4 and L_5 points are known to contain a number of asteroids referred to as the 'Trojans'. The first Trojan (Achilles) was discovered in 1906. By now, more than 50 Trojans are known.

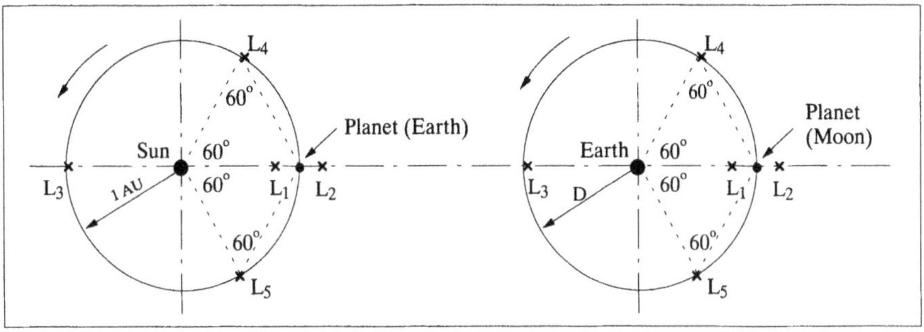

Figure 394: Lagrangian points of the sun-Earth and Earth-moon systems

Lagrangian Points	Sun - Earth System		Earth - Moon System	
	Distance from Sun	Distance from Earth	Distance from Earth	Distance from Moon
L_1	0.99 AU = 1.48 x 10^8 km	0.00997 AU = 1.49 x 10^6 km	0.85 D = 326740 km	0.15 D = 57660 km
L_2	1.01 AU = 1.51 x 10^8 km	0.010 AU = 1.50 x 10^6 km	1.17 D = 449748 km	0.17 D = 65348 km
L_3	0.999 AU = 1.496 x 10^8 km	1.999 AU = 2.992 x 10^8 km	0.99 D = 380556 km	1.99 D = 764956 km
L_4	1.00 AU	1.00 AU	1.00 D	1.00 D
L_5	1.00 AU	1.00 AU	1.00 D	1.00 D
	d_{Sun} = 1 392 000 km	d_{Earth} =12 740 km (average diameter)		d_{Moon} = 3476 km
Sun-Earth distance (center-center) = 1 AU (Astronomical Unit) = 1.496 x 10^8 km			Earth-Moon distance D = 384 400 km (center-center)	

Table 547: Values of the Lagrangian points in the sun-Earth and Earth-moon systems [2003)]

Mass	5.976 10^{24} kg
Equatorial radius	6,378.14 km
Mean density	5.515 g/cm^3
Mean distance to sun	149,600,000 km
Rotational period	0.99727 days (23.9345 h)
Orbital period	365.256 days
Mean orbital velocity	29.79 km
Orbital eccentricity	0.0167
Tilt of axis (obliquity of ecliptic)	23.45°
Equatorial escape velocity	11.18 km
Equatorial surface gravity	9.78 m/s^2
Visual geometric albedo	0.37
Mean surface temperature	15°C
Atmospheric pressure	1.013 bar
Atmospheric composition	
Nitrogen	77%
Oxygen	21%
Other (trace gases)	2%

Table 548: Some Earth parameters

2003) Table values: courtesy of F. Jochim of DLR/GSOC

L_1 in the sun-Earth system is located about 1.5 million km from the Earth towards the sun (this is 1/100 of the way from Earth to the sun). Both L_1 and L_2 are ideal positions from which to look toward the universe. L_1 in particular is a good vantage point to study the Earth as well as the sun (solar wind, etc.). However, a disadvantage at L_1 is the ferocious solar radiation which might overwhelm the S/C signal. At L_2, Earth's shadow blocks the solar radiation which is needed for power generation. A solution to this problem is the so-called "**halo orbit**" around the Lagrangian points. A S/C in a halo orbit around L_1 describes huge loops (the orbits resemble a so-called "halo" or disk about L_1) which are perpendicular to the Earth-sun axis, endlessly falling toward the balance point. Such halo orbits around L_1, although in the order of one million km in diameter, are considered as "small" compared with cosmological distances. The view angles from Earth to such a S/C in halo orbit remain fairly constant.

O.13 Observational Scales in Modeling

Consistent space and time scales are a necessity for modeling purposes in the atmospheric sciences or for other (large-scale) phenomena such as ocean circulation, in characterizing the dynamic behavior and structure of the system under investigation. The scale provides an ordering concept (and grid size or resolution) for the parameterization of observational data. In the literature pertinent to campaign descriptions there seem to be different meanings (or vague interpretations) to such definitions as 'macroscale' (sometimes also referred to as 'large-scale'), 'mesoscale,' and 'microscale.' Without going into details and into arguments of justification, I would like to refer to the article of I. Orlanski, who proposed his views on scales in 1975 (with a growing number of followers).[2004] While the definitions of Figure 395 were intended for meteorological modeling, they could also serve as an evolving structure for other fields of application. [2005]

Process	Variables	Time	
		Day	Year
Weather and Climate			
Damaging frost	Air temperature at 0 and 2 m levels	≤ 1	
Precipitation	Intensity, wind velocity		
Evaporation	mm/day or mm/year	≤ 1	
ENSO	Atmospheric pressure,		
	Water temperature	≤ 100	
Drought cycles			10
Ice Age	Glacial extent, air temperature		≥ 300
Hydrological Cycle			
Flood	Discharge, stage, snow cover	≤ 10	
Spring ice breakup	Stage	10	
Snowmelt	Discharge, stage, snow cover	≤ 100	
Glacier	Altitude		10
Biosphere/Ecosystem			
Burning	Biomass density	≤ 100	
Crop planting, harvesting			1
Recovery of temperate climate forest	Height, species composition		300
Recovery of damaged tundra or desert	Species composition		300

Table 549: Durations of events of significance to BAHC

A further extension of spatial and temporal scales (besides atmospheric sciences) for modeling hydrological and ecological phenomena (moisture and energy fluxes at the soil-vegetation-atmosphere interface) was proposed by A. Becker. His definition is being used for the modeling of BAHC (Biospheric Aspects of the Hydrological Cycle) data, an IGBP core

2004) I. Orlanski, "A Rational Subdivision of Scales for Atmospheric Processes," Bulletin American Meteorological Society, Vol. 56, No. 5, May 1975, pp. 527-530

2005) Sellers, et al., "Remote Sensing of the Land Surface for Studies of Global Change: Models-Algorithms-Experiments," Remote Sensing of the Environment, Vol. 51, 1995, pp. 3-26 project. The scale definition of I. Orlanski for the atmospheric sciences and meteorology is basically adopted in this presentation, as outlined in Figure 396 and Table 549, while the scales for hydrology and ecology are generally an order of magnitude smaller than those for the atmospheric sciences. [2006) 2007)]

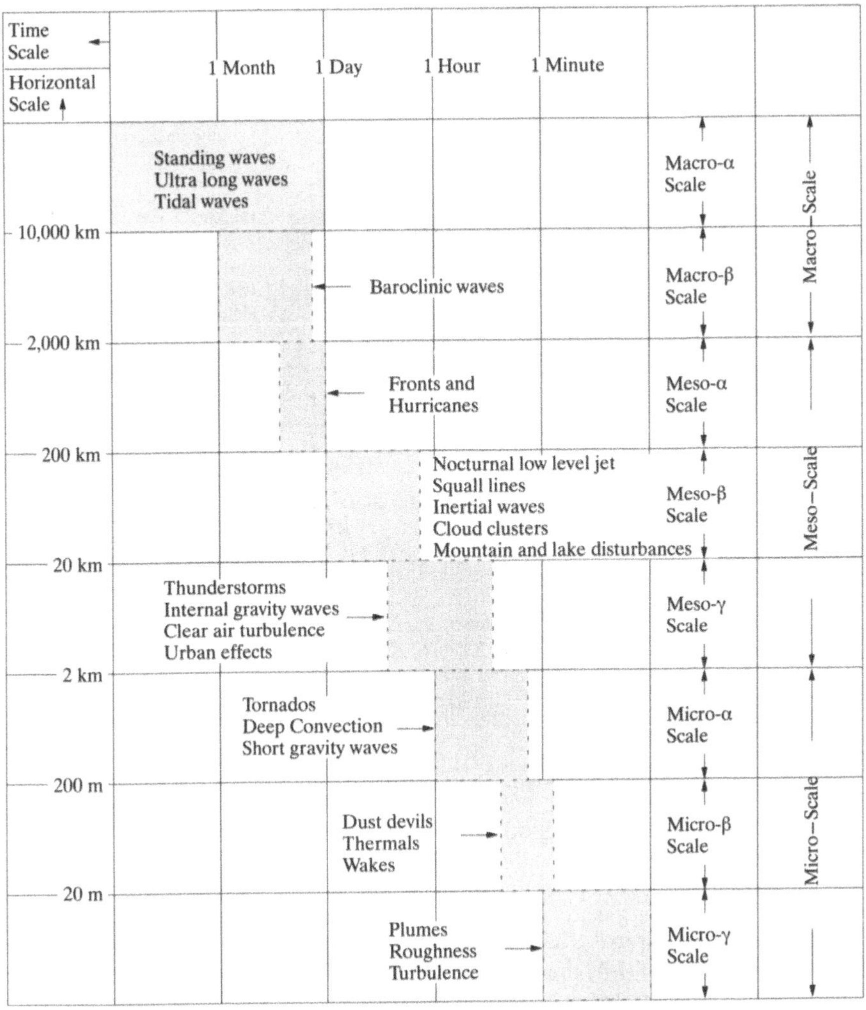

Figure 395: Scale definitions for processes with characteristic time/horizontal scales

2006) A. Becker, "Criteria for a hydrologically sound structuring of large scale land surface process models." In: Advances in Theoretical Hydrology: A Tribute to James Dooge; Ed. J. P. O'Kane, European Geophysical Society Series on Hydrological Sciences, 1992, pp. 97-111

2007) IGBP Global Change Report No. 27, Biospheric Aspects of the Hydrological Cycle (BAHC), The Operation Plan, 1993, pp. 26-28

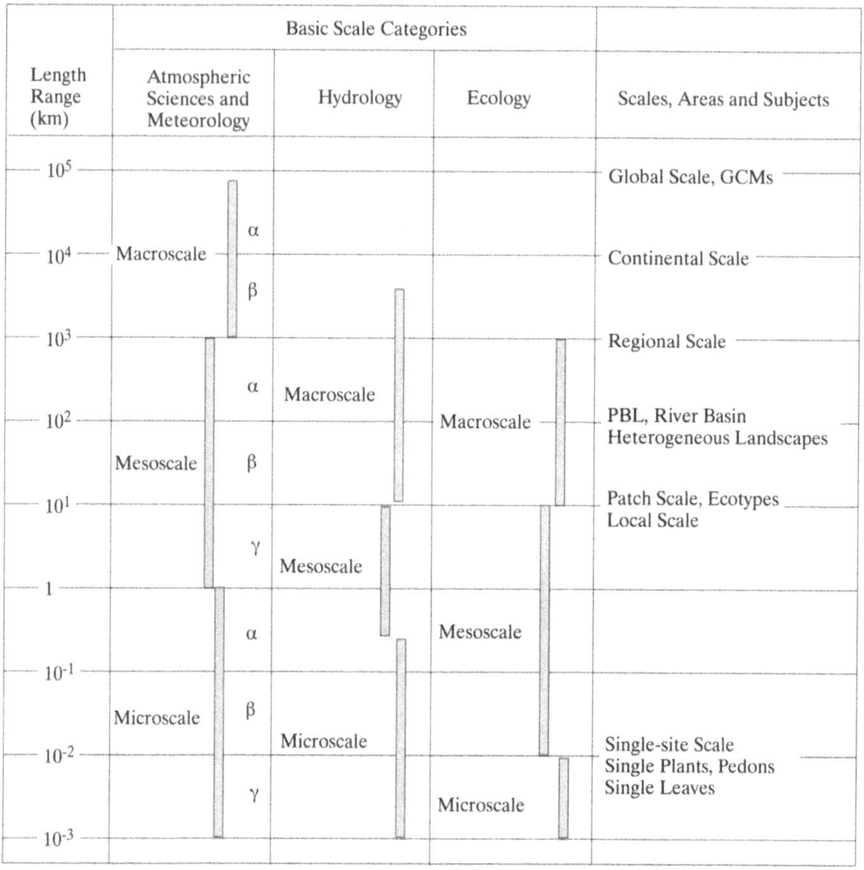

Figure 396: Spatial scales of significance to BAHC

O.14 On-Orbit Electric Propulsion

Since about the mid-1990s a new era of satellite propulsion technology is upon us with the introduction of on-orbit electric propulsion (EP) systems. Recent technology advances are mainly due to progress in the field of orbital energy supplies. The requirement of longer operational life times of orbital systems is also a design driver. - The EP technology provides higher specific impulse (Isp) than conventional chemical propulsion, however, at the expense of lower performance (i.e. much lower thrust levels). Typical EP systems provide spacecraft accelerations of 0.001 g or less. This implies that EP devices must thrust for much longer periods of time than a higher thrust (chemical) device to provide a given impulse. Some advantages/capabilities which make EP systems desirable for many tasks other than launching satellites.[2008)] [2009)] [2010)]

- Provision of platform attitude maintenance. A considerable improvement of spacecraft pointing accuracy can be achieved.

2008) D. F. Robertson, "Electric Propulsion: Here at Last," Space & Communications, May-June 1997, pp. 11-15
2009) S. Ashley, "Electric rockets get a boost," Mechanical Engineering, December 1995, pp. 61-65
2010) G. Saccoccia, J. Gonzalez del Amo, D. Estublier, "Electric Propulsion: A key Technology for Space Missions in the New Millennium," ESA Bulletin, No 101, Feb. 2000, pp. 62-71

- Raising of a satellite's orbit. This may either extend a mission's life considerably, or it may provide a capability of achieving a new orbit altogether (from LEO to GEO).
- Orbit repositioning and maneuvering (station-keeping, etc.). Flexibility for unanticipated maneuvering.
- Considerable reduction in propulsion system wet mass. Advanced EP systems offer a considerable reduction in mass.
- On-board safety is improved through the elimination of hazardous fuels (inert propellants can be used with EP systems).
- For microsatellites the benefits of simplicity and economy of a particular EP system can even outweigh the higher performance of another type EP system.
- Electric propulsion provides low thrust levels for extended periods - months, or even years - resulting in considerable final velocities. Motors may be developed to propell a microsatellite from a LEO orbit into an interplanetary trajectory.

O.14.1 Basic Thruster Concepts

All EP systems generate thrust by using electrical energy (from on-board solar arrays, a nuclear reactor, or batteries) to accelerate a propellant, thereby converting it to spacecraft kinetic energy. There are three basic classes of thruster technologies (differing mainly in the type of acceleration mechanism). But prior to venturing into these thruster classes, a definition of specific impulse is due.

O.14.1.1 Specific Impulse (Isp)

In the International System of Units [SI], the unit of **specific impulse (Isp) is newton-seconds per kilogram or Ns/kg.** Hence, Isp is the impulse (a force applied for a second) exerted with 1 kg of propellant. It follows that the higher the specific impulse, the greater the performance of the engine and the less propellant required. The numerical value of the specific impulse also corresponds to the effective exhaust velocity (**m/s**) of the gas exiting the thruster in a vacuum.

Classical chemical thrusters have a low specific impulse (in the order of 3,000 Ns/kg, with gases ejected at 3,000 m/s) which entails the need for considerable quantities of propellant. Electric thrusters eject their particles at very high speed (15,000 - 100,000 m/s). Hence, their Isp of 15,000 to 100,000 Ns/kg can be 5 to 30 times greater than a conventional thruster.

In the United States, the Isp is widely expressed in units of pound-thrust per pound-weight per second, with a remaining unit of "seconds for" Isp. When comparing these "Isp second" values with those of the SI system, then the "Isp second" values have to be multiplied by the factor 9.81 m/s^2 to account for the pound-force to pound-mass ratio and to obtain the unit m/s of the SI system. In this book all Isp values are given in the SI system.

A trust of <100 mN produced by a typical electric propulsion system is equivalent of holding an A4 paper in your hand. But functioning for long periods, such a system can ultimately provide a high velocity increment to a S/C. Electric propulsion thus has many advantages. For instance, half the mass of a typical geostationary telecommunications satellite is propellant. [2011)]

O.14.1.2 Electrothermal thrusters

The propellant is heated by either an electric arc or a resistance heater (electrical heating through a wall). The hot propellant is then exhausted through a nozzle to produce thrust.

- Arcjets. The system consists of three elements: the engine (two electrodes), the power source and processor, and the propellant subsystem. An arc discharge is sustained be-

2011) http//:sci.esa.int/content/doc/5b/9819_.htm

tween a cathode and anode that also serves as expansion nozzle. Typical propellants are hydrogen (Isp = 10,000 - 20,000 m/s), ammonia, or hydrazine (5000-7000 m/s). Hydrazine arcjet systems have been operating on orbit since 1994, primarily for North-South stationkeeping on geosynchronous communication satellites.

- Resistojets. A propellant gas (not plasma) is fed through a resistive heat exchanger prior to expansion through a nozzle. Many propellants can be used, such as: H_2, CO_2, CH_4, N_2H_4, NH_3, N_2, He, Ar, H_2, and H_2O. Typical thrusts are between several mN to several N (at efficiencies of up to 80%).
- RF-heated thruster. Use of radio-frequency waves to heat a plasma in a chamber.

ATOS (Arcjet Thruster on OSCAR Satellite). [2012] ATOS is an example of an arcjet implementation developed at IRS (Institute for Space Systems) of the University of Stuttgart, Germany. The system consists of the arcjet thruster, the fuel supply unit and the power supply electronics. The power supply transforms the 28 VDC from the satellite power bus into the 800 W maximum current regulated arcjet power. The fuel supply unit regulates the flow of gaseous ammonia from the two fuel tanks to the thruster. Its ammonia resistant valves and ammonia vaporizer were supplied by a Russian space company. ATOS is flown on the AMSAT Phase 3D spacecraft, launched Nov. 15, 2000 from Kourou (AMSAT-3D and the STRV-1c/d satellites were secondary payloads to the PanAmSat-1R communication satellite). ATOS, with an energy input of 750 W, is an ammonia arcjet with a specific impulse Isp of 4600 m/s and a thrust level of 115 mN.

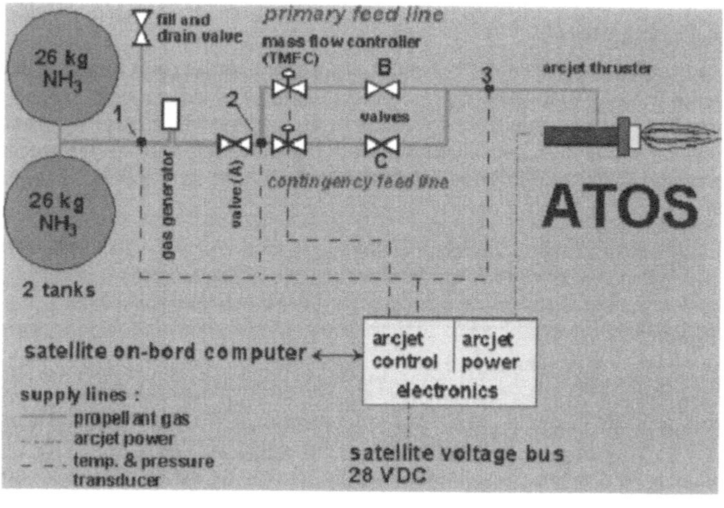

Figure 397: Schematic illustration of the ATOS system

O.14.1.3 Electrostatic thrusters

Use of electric fields to accelerate an ionized gas. This class has the following members: the ion engine, Hall-effect thrusters, and field-emission thrusters.

- Gridded ion engines (or the ion thruster). A gaseous (vaporized) propellant is ionized and accelerated to high exhaust velocity in an electric field. The electric field is created by applying high voltages between a series of electrically charged grids. The ions are then neutralized and leave the engine as high velocity neutral particles. Early ion engines used mercury or cesium for fuel, while newer designs use argon, krypton, or xe-

2012) http://www.irs.uni-stuttgart.de/RESEARCH/EL_PROP/ATOS/e_atos.html

non. The Isp of the xenon ion engine is about 30,000 m/s at 0.5 kW input power with 55% thrust efficiency.

- Hall-effect thrusters (free electron drift propulsion).[2013] The axial electric field developed between the discharge electrodes interacts with the radial magnetic field to generate a "Hall" current in azimuthal direction. This current, in turn, reacts against a magnetic field to generate a force on the propellant in the downstream direction. Isp from 10,000 - 20,000 m/s; lifetimes of 6000 hours have been achieved by Russian-built motors.
- Field-emission thrusters. Provision very low thrust levels (from 1 μN - 2 N) with very high accuracy and controllability (Isp from 60,000 - 100,000 m/s). In FEEP thrusters, ions are directly extracted from the liquid phase of the propellant. Thrust is obtained by exhausting a beam, mainly composed of singly-ionized atoms, produced by field emission.

O.14.1.4 Electromagnetic thrusters

In electromagnetic propulsion, thrust is generated by accelerating an ionized plasma or a conducting fluid (such as liquified Teflon) by electromagnetic fields.

- Pulsed Plasma Thruster (PPT). A PPT is an electromagnetic device which utilizes a high current discharge to ablate a solid propellant (such as Teflon) and accelerates the ionized material (plasma) through a nozzle (exhaust velocities of 10-20 km/s are obtained). A key advantage of PPTs is their extremely short pulse time (two orders less than chemical thrusters) which translates to more precise control for attitude and station-keeping maneuvers. Isp = 10,000-20,000 m/s.[2014]
- Stationary Plasma Thruster (SPT). The SPT design represents a synthesis of Hall ion thruster and MPD thrusters. Plasma is being generated by a discharge of electrons from a ring-like discharge chamber and a magnetic field. A voltage applied to the anode causes the ions to accelerate. Thrusts between 40 and 200 mN may be obtained (Isp is about 16,000 m/s at 1.4 kW input power with 50% thrust efficiency).

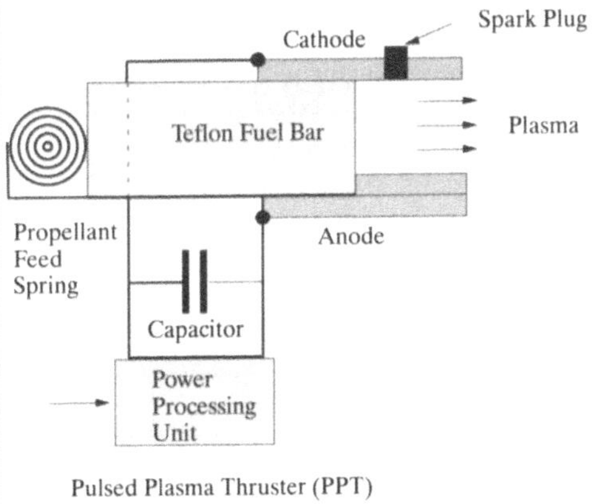

Figure 398: Schematic illustration of a PPT system

2013) The 'Hall effect' refers to the development of a transverse electric field in a solid material when it carries an electric current and is placed in a magnetic field that is perpendicular to the current. The electric field, or Hall field, is a result of the force that the magnetic field exerts on the moving positive and negative particles that constitute the electric current. This phenomenon was discovered in 1879 by E. H. Hall, a US physicist.

2014) R. J. Cassady, et al., "Pulsed PLasma Thruster Systems for Spacecraft Attitude Control," Proceedings of the 10th Annual AIAA/USU Conference on Small Satellites, Sept. 16-19, 1996

- Magneto Plasma Dynamic (MPD) thruster. The MPD device utilizes the Lorentz [2015] force arising from the interaction of discharge current with self-induced and/or an applied magnetic field.

Operation: Xenon gas flows at 3 mg/s into the ionization chamber through an annular plenum. The xenon atoms are ionized with energetic electrons emitted from a thermionic hollow cathode. The cathode filament emits electrons that are accelerated toward the chamber wall by a high positive-wall potential. In addition, a magnetic field is applied by concentric magnets, to increase the likelihood of electron-atom collisions. The magnetic field causes the electrons to spiral toward the walls, increasing travel time. The collisions produce a plasma equal in number of positive ions and negative electrons.
The positive ions in the plasma flow toward the extraction electrodes where they are accelerated by the negative grid potential. The grids create thousands of beamlets as the ions pass through the screens. The continuous ejection of positive ions requires neutralization to prevent gradual vehicle charging. The ion engine is controlled (throttled) by reducing the propellant mass flow.

	Resistojet	Arcjet	SPT	Ion Thruster
Propellant	N_2H_4	N_2H_4	Xenon	Xenon
Input electric power (kW)	0.2 - 0.8	0.5 - 2.0	0.15 - 1.5	0.4 - 2.0
Thrust (mN)	200 - 800	200 - 250	40 - 200	15 - 40
Specific Impulse, Isp (m/s)	2,700 - 3,000	4,500 - 6,000	16,000	28,000 - 35,000
Efficiency	0.9	0.33	0.48	0.75
Mission life (hours)	>390	>830	>4000	>8000

Table 550: Typical performance values of some electric propulsion systems

O.14.2 Some Developed Thruster Systems

IPS (Ion Propulsion System), developed by the NSTAR (NASA SEP Technology Application Readiness) team, where SEP is the acronym for Solar Electric Propulsion. NSTAR is a NASA technology demonstration program the started in 1992.; the prototype system was developed jointly at LeRC and JPL (IPS was built at Hughes Electronics Corp.). Note: The IPS system is also referred to as **NSTAR**. IPS is an ion engine (electrostatic thruster), consisting of three main components: the ion source, the ion accelerator, and the neutralizer.

Parameter	Description
Ion thruster exit diameter	30 cm
Input power range	0.5 - 2.5 kW
Thrust level	20 Nm at 0.5 kW; 92 Nm at 2.3 kW
Isp (specific impulse)	21,000 m/s at 0.5 kW; 30,000 m/s at 2.3 kW
Control	Throttleable in 16 discrete steps to accommodate the available power
Instrument mass	41.3 kg (without propellant)
Propellant mass	75 kg of xenon
Design operating life	8000 hours of full power

Table 551: IPS parameters for DS1

The system uses a hollow cathode to produce electrons to collisionally ionize xenon. The $Xe+$ is electrostatically accelerated through a potential of up to 1280 V and emitted from the 30 cm thruster through a molybdenum grid. A separate electron beam is emitted to produce a neutral plasma beam. The power processing unit (PPU) of the IPS can accept as much as 2.5 kW, corresponding to a peak thruster operating power of 2.3 kW and a thrust of 92 mN. Throttling is achieved by balancing thruster and Xe feed system parameters at lower

2015) H. A. Lorentz (1853-1928) Dutch physicist.

power levels, and at the lowest thruster power, 500 W, the thrust is 20 mN. The specific impulse decreases from 3100 s at high power to 1900 s at the minimum throttle level. The input power of 2.3 kW can be throttled to 0.5 kW. The thruster with an exit plane aperture of 30 cm diameter generates 90 mN of thrust at 1.8 kW input power.

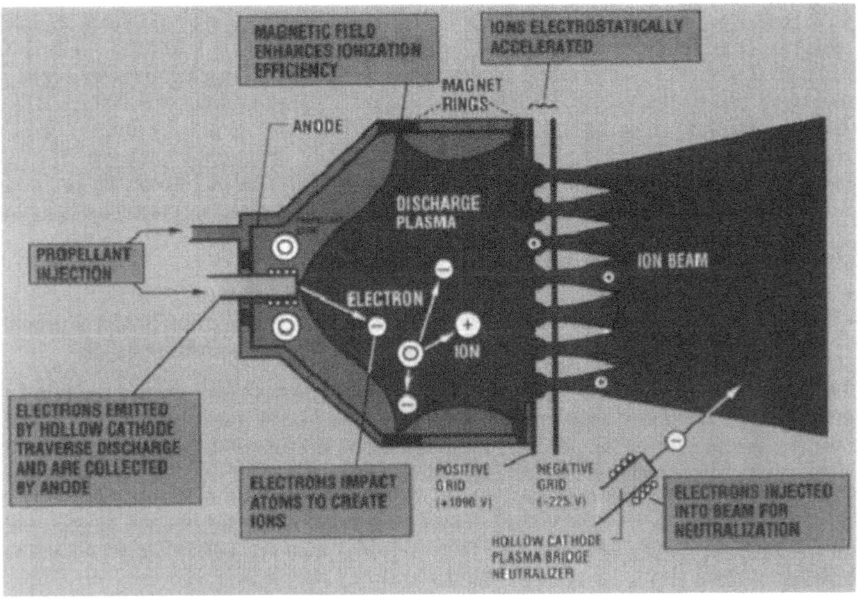

Figure 399: Schematic illustration of IPS

FEEP (Field Emission Electric Propulsion). FEEP is an electrostatic thruster technology capable of delivering very low thrust with very high accuracy and controllability. In FEEP systems a liquid propellant (cesium) is fed to the needle-like emitter; intense local electric fields cause charged liquid droplets to form spontaneously. The charged liquid droplets are extracted from the surface and accelerated by the electrostatic fields.

The thruster can accelerate a large number of different liquid metals or alloys; cesium proved to be the best choice, due to its high molecular weight and its low ionization potential. The power conversion efficiency reaches values as high as 98%. Specific impulse may be easily adjusted to meet the mission requirements. The thrust level is finely tunable, and the instantaneous switching capability permits pulsed mode operation. - FEEP features low system mass and small overall dimensions (absence of moving parts, there are no valves or pressurized gases). The FEEP propulsion concept is very suitable either for drag-free control or for high accuracy, noiseless pointing.

The FEEP concept is known since the late 1970s and was first proposed by ESTEC. Centrospazio of Pisa, Italy with ESA/ESTEC and ASI funding, developed some of the technology. Since FEEP is a generic term, an engine (engineering model) with the name of **RMT** (Radiofrequency with Magnetic field ion Thruster) was developed by Laben, an advanced electrostatic thruster, capable of providing very low thrust levels (from 1 μN - 2 N) with very high accuracy and controllability (Isp from 60,000 - 100,000 m/s). In FEEP thrusters, ions are directly extracted from the liquid phase of the propellant. Thrust is obtained by exhausting a beam, mainly composed of singly-ionized atoms, produced by field emission. - In the 1990s, the interest in low-thrust propulsion systems generated new research and development efforts in the FEEP technology. It is evident that the potential field of use of this thruster is not limited to the μN-level drag-free missions, but includes several applications like attitude and orbit control of small spacecraft and mutual positioning control in constellations.

The RMT propulsion system of Laben merged with **EMITS** (Electric Microthruster Test in Space) system of Centrospazio of Pisa, Italy. A demonstration is to be flown on the Shuttle GAS (Get-Away Special) system in 2001. [2016) 2017)] The EMITS flight experiment includes two thruster assemblies, the thruster power and control electronics (developed by LABEN, Milan, Italy), several probes and diagnostic devices, a battery, and a computerized experiment control/data storage unit (developed by Techno System Developments, Naples, Italy). The total duration of the experiment will be 2 days. During that period, the thrusters will be repeatedly switched on and off and cycled through their operational envelope. The top plate of the EMITS payload is mounted just behind the GAS MDA (Motorized Door Assembly). Once actuated by the Shuttle crew, the MDA exposes to space the following items:

- A 1 mN thruster (70 mm slit length, about 60 W)
- A 30 μN thruster (2 mm slit length, less than 1 W)
- A low power neutralizer, of the thermionic oxide cathode type
- A set of diagnostic devices, including quartz crystal microbalances (QCMs) as propellant deposition sensors, electrostatic ion beam probes, and a pressure gauge.

XIPS (Xenon Ion Propulsion System), built by HSC of Hughes Electronics Corp., Los Angeles, CA (production at Electron Dynamics Division). The instrument consists of the following elements: propellent (xenon) reservoir, a power processing unit, the ion thruster (ionization chamber, gridded electrodes), and a neutralizer. As of 1997 two XIPS versions exist to serve the HS-601 and HS-702 communication satellite platforms, designed for 12 to 15 years of operation. SATMEX-5, a communications S/C of Mexico (built by Hughes with an HS-601 bus), was launched Dec. 6, 1998 from Kourou equipped with XIPS for station keeping. SATMEX-5 is operated by Satellites Mexicanos S.A. de C. V.

Thruster	Nozzle diameter	Isp	Thrust	Power Input
HS-601 HP	13 cm	25,000 m/s	18 mN	0.5 kW
HS-702	25 cm	37,000 m/s	165 mN	4.5 kW

Table 552: XIPS thruster parameters

A typical satellite is furnished with four XIPS thrusters (two primary, two redundant) for stationkeeping and S/C momentum control, all connected to the same xenon reservoir. The autonomous XIPS operations provide stationkeeping within ±0.005° limits. The XIPS model HS-702 offers in addition the option of orbit raising. The XIPS model was first flown on two commercial GEO communication satellites (HS-601 platforms) for PanAmSat Corp. of Greenwich, CT: PanAmSat-5 (PAS-5), launched July 30, 1997, then on Galaxy VIII-i, launched on Dec. 8, 1997.

RIT-10 (Radio-frequency Ion Thruster), also referred to as **RITA** (Radio-frequency Ion Thruster Assembly). An electric propulsion system of Astrium GmbH (formerly DASA), Germany. RIT systems are in the category of gridded ion engines. The xenon propellant flows inside a ceramic discharge chamber through a gas distributor. The discharge chamber itself is surrounded by an induction coil connected to an RF (Radio Frequency) generator. Free electrons within the xenon gas collect energy from the RF-induced electric field and ionize the neutral propellant atoms via inelastic collisions. The discharge is ignited by the injection of electrons from the neutralizer. Thrust is generated by the acceleration of ions in the electrostatic field applied to an extraction system consisting of the extraction anode and grids. The negative potential of these grids grids accelerates the positive ions out of the stat-

2016) http://www.centrospazio.cpr.it/EMITS.html

2017) A. Genovese, S. Marcuccio, D. D. Pozzo, M. Andrenucci, "FEEP Thruster Performance at High Background Pressure," IEPC-97-186, Proceedings of the 25th Electric Propulsion Conference, Cleveland, OH, 1997 ic plasma. A neutralizer injects electrons into the beam to maintain its zero-charge equilibrium. 2018) 2019)

The RIT-10 system was successfully test-flown on-board the EURECA-1 platform (launch July 31, 1992 - retrieval July 1, 1993). Since then, four more RIT thruster models were designed/built at DASA or are in an advanced development phase:

- The RIT-10 thruster for ARTEMIS has a beam diameter of 9.8 cm and delivers 15 mN thrust with an Isp of 34,000 m/s and a system power of 560 W. The system provides 15,000 hours of operation at 15 mN.
- The further developed RIT-10 EVO (Evolution) thruster provides a maximum thrust of 35 mN (adjustable from 1 to 35 mN) at an Isp between 28,500 m/s and 44,000 m/s. A special beam controller was developed for this thruster, permitting thrust adjustment for air drag compensation purposes between 1 and 12 mN at a resolution of 25 μN at a response time of 0.05 s.

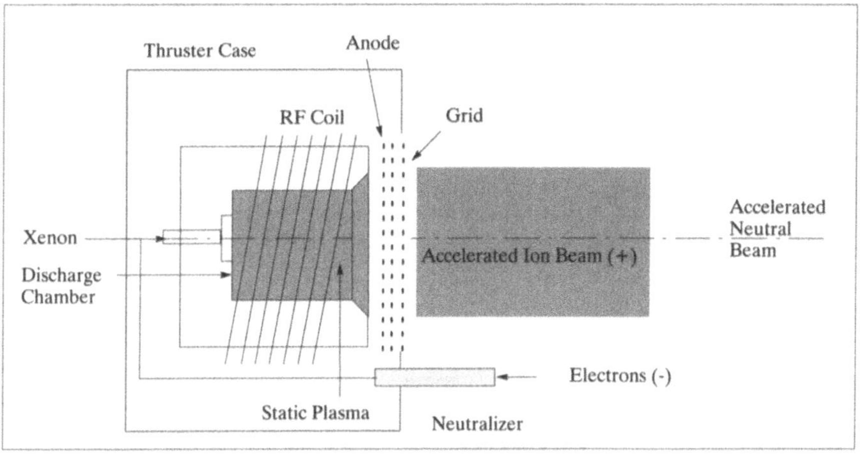

Figure 400: Schematic illustration of the RIT system

- The RIT-XT extended power thruster has a beam diameter of 206 mm and provides an adjustable thrust from 25 to 150 mN at an Isp between 28,500 m/s and 55,000 m/s. The input power for 100 mN is 2800 W for an Isp = 30,000 m/s. The objective is to provide sufficient power for orbit raising applications of small satellites in LEO constellations.
- The ESA-XX thruster (breadboard model as of 2000) is a radio frequency thruster built by DASA with an extraction grid built by AEA-Culham, UK. The ESA-XX has a beam diameter of 280 mm and was tested at thrust levels between 21 and 240 mN at an Isp of up to 53,000 m/s. The input power for 240 mN at maximum Isp is 8450 W. 2020)
  The ESA-XX thruster is a synergy of two European ion propulsion technologies, combing the discharge chamber of the RF ion thruster (RIT-35), with the extraction grid system of the EIT (Electron-bombardment Ion Thruster), the UK-25 system.

2018) R. Killinger, H. Bassner, J. Müller, "Development of an High Performance RF-Ion Thruster," Proceedings of 35th AIAA/ASME/SAE/ASEE Joint Propulsion Conference and Exhibit, June 20-24, 1999, Los Angeles, AIAA-99-2445

2019) G. D. Racca, G. P. Whitcomb, B. H. Foing, "The SMART-1 Mission," ESA Bulletin 95, Aug. 1998, pp. 72-81

2020) H. Bassner, R. Bond, V. Thompson, H-P Harmann, K. Groh, "Development and Performance Testing of the ESA-XX Ion Thruster," DASA paper

Parameter	EURECA	RIT-10 Artemis	RIT-10 EVO	RIT-XT
Thrust level	10 mN	15 mN	1 - 35 mN	25 - 150 mN
Thruster diameter	16 cm	16 cm	16 cm	28.9 cm
Discharge chamber	10 cm	10 cm	10 cm	22 cm
Beam diameter	85 mm	85 mm	85 mm	206 mm
No of holes	253	499	1483	8160
Beam voltage	1500 V	1500 V	1300 V	1200 V
Main bus power	440 W	560 W	100 - 1050 W	2600 W at 100 mN
Optimum Isp (m/s)	32,000	33,000	30,000 - 32,000	30,000 - 33,000
Thruster specific power	41 W/mN	35 W/mN	25-27 W/mN	22-24 W/mN

Table 553: Performance comparison of various RIT models

UK-10 (Electron Bombardment Ionization Thruster) IPS, also referred to as **EITA** (Electron-bombardment Ion Thruster Assembly). UK-10 is an ion engine system mainly developed by DERA, UK. The UK-10 provides a nominal thrust of 18 mN and a maximum thrust of 23 mN at an Isp of 33,000 m/s. The input power is 700 W. [2021)]

In the electron-bombardment-ionization scheme (Figure 401), the xenon propellant flows inside a metallic discharge chamber through a gas distributor. This gas is ionized in a DC discharge between the cathode and a concentric cylindrical anode. The efficiency of the process is enhanced by the application of an azimuthally symmetrical divergent magnetic field to the discharge chamber, generated by solenoids spaced around the periphery. Thrust is generated by the acceleration of ions in the electrostatic field applied to an extraction system (grids). The negative potential of the grids accelerates the positive ions of the static plasma. Ion beam velocities of 30-50 km/s are achieved. A neutralizer injects electrons into the beam to maintain zero-charge equilibrium in the beam.

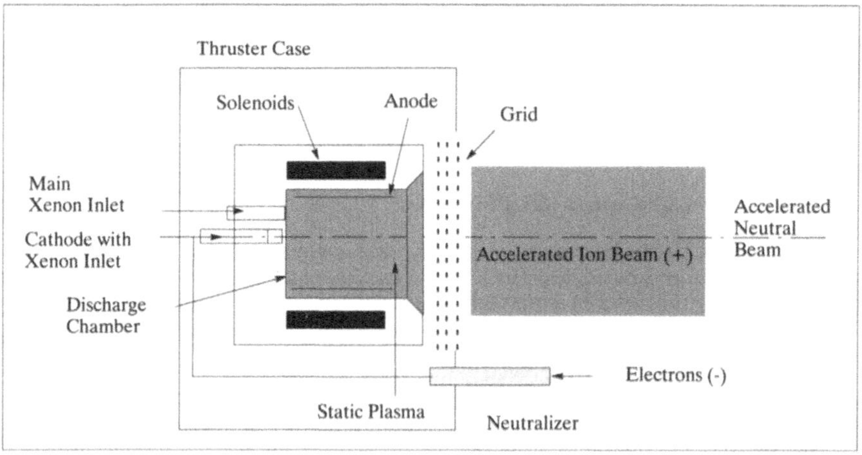

Figure 401: Schematic illustration of the UK-10 system

Parts of the UK-10 ion engine were successfully operated in space for the first time on STRV-1b (launch June 17, 1994) of DERA and used to conduct experiments associated with plasma discharging.

- The UK-10 ion propulsion system (IPS) is based on the 10 cm nominal diameter T5 thruster, coupled with PCCE (Power Conditioning and Control Equipment) and

2021) D. G. Fearn, P. Smith, "A Review of UK Ion Propulsion - A Maturing Technology," IAF-98-S.4.01, Sept/Oct. 1998, pp. 1-11

PSME (Propellant Supply Monitoring Equipment). The PCCE mass is about 12 kg, the PSME mass is 1.6 kg. The total input power at 18 mN thrust is 570 W. The initial application of the IPS is for ESA's ARTEMIS satellite.

- UK-25 IPS. The thruster is intended for such applications as orbit-raising and other transfer maneuvers. The design is very similar to the T5 configuration, featuring a 25 cm exit diameter. The thrust can be varied from 50 to 300 mN, requiring an input power of up to 10 kW at maximum thrust.

ESEX (Electric Propulsion Space Experiment). ESEX (built by TRW and supported/managed by Phillips Laboratory of USAF) demonstrates high-power ammonia arcjet propulsion technology (reliable arcjet thruster operation, power input = 26 kW) for orbit transfer, maneuvering capability, and attitude adjustment.

EPDM (Electric Propulsion Demonstrator Module). EPDM (sponsored by BMDO, managed and built by NRL in cooperation with NASA/GRC and TsNIIMASH) is flown on the STEX (Space Technology Experiment) mission of DoD (launch Oct. 3. 1998). EPDM is the first operational Hall thruster propulsion system of the USA. It's heritage is from the **RHETT** (Russian Hall Effect Thruster Technology) program of BMDO. The TAL-D55 (Thruster with Anode Layer) Hall effect thruster used in EPDM was developed by the Russian Central Research Institute of Machine Building (TsNIIMASH) in Korelev, Russia. The objective is to provide orbit-raising and stationkeeping for the spacecraft. EDPM features a 600 W, 15,000 m/s Isp xenon Hall thruster system, providing 40 mN of thrust. It consists of four subassemblies: [2022)]

- XFS (Xenon Flow System). XFS is a high-pressure storage system with regulated pressure and mass flow. The xenon is stored in a 20 cm spherical tank at a pressure of 64 kg/cm^2. Two solenoid valves upstream of the pressure regulator control the flow. A high-pressure transducer accurately determines the amount of xenon remaining. Thermostatically controlled heaters provide thermal conditioning. The mass of XFS is 5.7 kg (dry).
- PPU (Power Processing Unit) built by Primex Aerospace Co of St. Petersburg, FL. PPU consists of five separate but interrelated power supplies. The anode supply provides a regulated 300 V to the TAL-D55 thruster. The PPU has an electrical conversion efficiency of about 90% with an input voltage of 30 V. The 65 W of waste heat is dissipated through the baseplate of the unit, and the spacecraft rejects it to space. The mass of PPU is 10.3 kg.
- AIU (Auxiliary Interface Unit). Conditions discrete commands and telemetry (analog and discrete) and provides thermal conditioning of the XFS, the engine assembly, and XFS valve drivers. The AIU changes pulsed discrete commands into steady-state commands required by the PPU. An electronic thermostat accurately controls the temperature of the orifice heaters in the xenon flow system. The mass of AIU is 1.9 kg,it consumes an average of 6 W.
- EA (Engine Assembly). It consists of the TAL-D55 thruster which has a nominal discharge of 1.35 kW at 300 V. In flight, the engine is used at 600 W of thrust (limited by the spacecraft's power). The thruster is mounted on top of a titanium thermal isolator to minimize the conductive heat path back to the spacecraft. The engine and isolator are mounted on a small aluminum honeycomb panel along with the orifice block and the cathode assembly. The orifice block splits the flow of the low-pressure xenon between the TAL-D55 thruster and the cathode. Precision orifices and tight temperature control on the orifice block itself maintain the accurate single point mass flow setting. During system operation, a set of stored spacecraft commands is executed to initialize and thermally condition the system, and to fire the thruster.

SPT (Stationary Plasma Thruster), developed by the Fakel Design Bureau in Kaliningrad, Russia [since 1992, a joint venture with the name of IST (International Space Technology,

2022) http://home.att.net/~ spctch/Stex/epdm.html

Inc.) was formed with Atlantic Research Corp. of Gainesville, GA, USA]. Stationary plasma thrusters (Hall-effect thrusters) operate by ejecting xenon ions, accelerated and focused by electrical and magnetic fields, at very high speed.

Instrument Type	SPT-70	SPT-100	PPS-1350	SPT-140
Thrust (max)	40 mN	80 mN	70 mN	320 mN
Input power	750 W	1350 W	1350 W	4500 W
Applications	Over 60 systems in service on Russian S/C as of 1998	24 systems in service on Russian S/C	Under development	Under development

Table 554: Performance parameters of some SPT instrument types

The **PPS-1350** system is a Hall-effect thruster developed by SNECMA of Paris, France (thruster) and ETCA of Belgium (power processing unit). PPS stands for "Propulseur Plasmique Stationaire;" or Stationary PlasmaThruster, and 1350 is the nominal input power. The system uses xenon gas as propellant to provide a thrust of 70 nN. The first PPS-1350 system is scheduled to fly on the Astra-1K communication satellite of SES of Luxembourg (launch in 2001) for GEO station keeping. The Skybridge broadband mobile communication LEO satellite constellation of ASI (Alcatel Space Industries) uses also PPS-1350 thrusters.

Thruster Model	Category	Manufacturer	Status
PPS-1350	Electrostatic, Hall-effect	SEP, France	Qualified for GEO station keeping and LEO S/C
ROS-99	Electrostatic, Hall-effect	Matra Marconi Space (MMS), UK, France	Qualified for GEO S/C station keeping
HPHET	Electrostatic, Hall-effect	Matra Marconi Space (MMS), UK, France	Developed for orbit transfer of GEO and LEO S/C, qualification in 2001
RIT-10	Electrostatic, gridded ion engine	DASA, Germany	Under qualification for GEO S/C station keeping (2000)
T5	Electrostatic, gridded ion engine	MMS, UK	Under qualification for GEO S/C station keeping (2000)
T6	Electrostatic, gridded ion engine	DERA, UK	Engineering model for GEO station keeping
RMT	Electrostatic, gridded ion engine	Laben-Proel, Italy	Engineering model
RIT-15	Electrostatic, gridded ion engine	DASA, Germany	Advanced breadboard
ESA-XX	Electrostatic, gridded ion engine	DASA, Germany, AEA, UK	Breadboard
FEEP / EMITS	Electrostatic field emission	Centrospazio, Italy	System ready, flight qualification in 2000
Indium LMIS	Electrostatic field emission	ARC (Austrian Research Center), Austria	Breadboard
Ammonia arcjet	Electrothermal	IRS, Germany	Qualified for Amsat P3-D
Hydrazine arcjet	Electrothermal	FIAT-Avio	Engineering model
MPD thrusters	Electromagnetic	Centrospazio, Italy, IRS, Germany	Laboratory models

Table 555: European developments in electric propulsion (table from reference 2010)

HPHET (High-Power Hall-Effect Thruster), an ESA/industry co-funded development designed and built by MMS and its subcontractors (DERA, Centrospazio). The system is in the 3 kW class.

O.15 Summary of World Data Centers (WDCs)

The ICSU (International Council of Scientific Unions) Panel on World Data Centers (Geophysical, Solar & Environment) was established at it's 12th General Assembly in 1968. The

Panel oversees the operation of about 40 WDCs, which are maintained by their host countries and are responsible for collecting, archiving, exchanging, and distributing a wide range of data. These data provide baseline information for research in many ICSU discipline, especially for monitoring changes in the geosphere and biosphere.[2023]

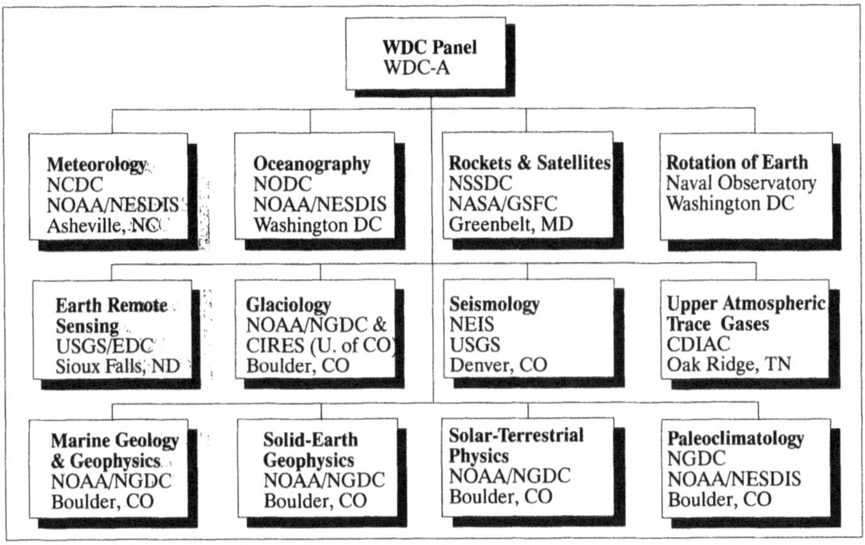

Figure 402: Structure of World Data Center A (USA)

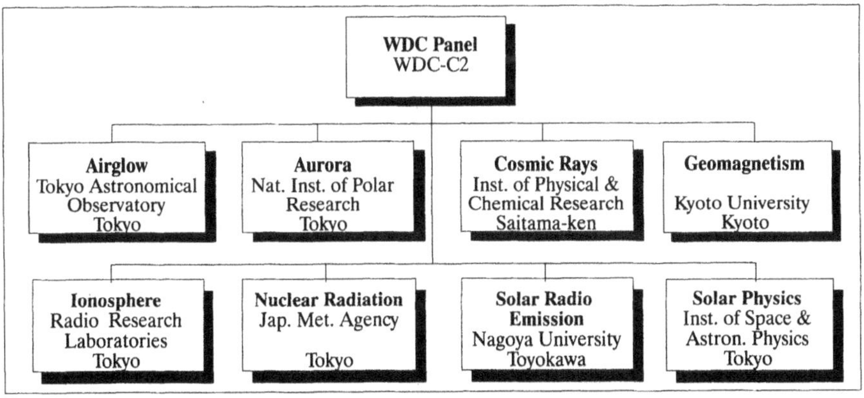

Figure 403: Structure of World Data Center C2 (Japan)

The WDC system is funded by the national host countries and grouped as follows:

WDC-A in the USA	WDC-B in Russia and CIS
WDC-C1 in Europe	WDC-C2 in Japan
WDC-D in China	

Table 556: Overview of WDC locations

2023) Brochure of WDC Publications Office, J. H. Allen, NOAA/NGDC, E/GCT2, Boulder, CO 80303, USA

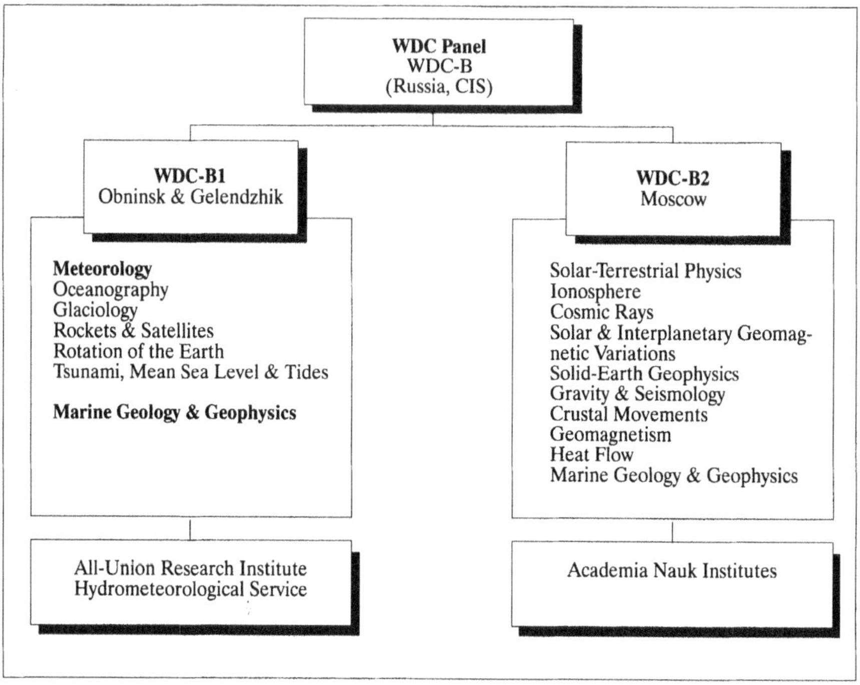

Figure 404: Structure of the World Data Center B (Russia, CIS)

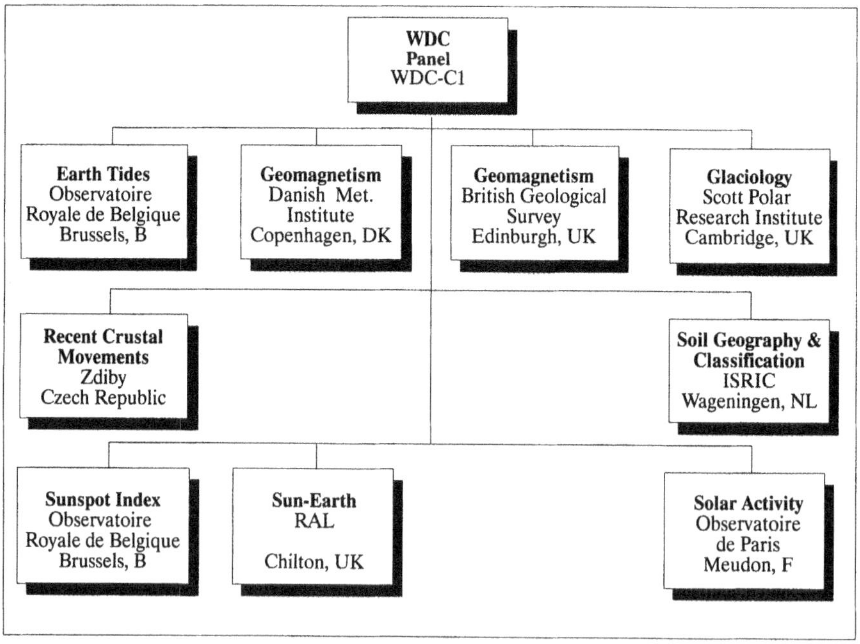

Figure 405: Structure of World Data Center C1 (Europe)

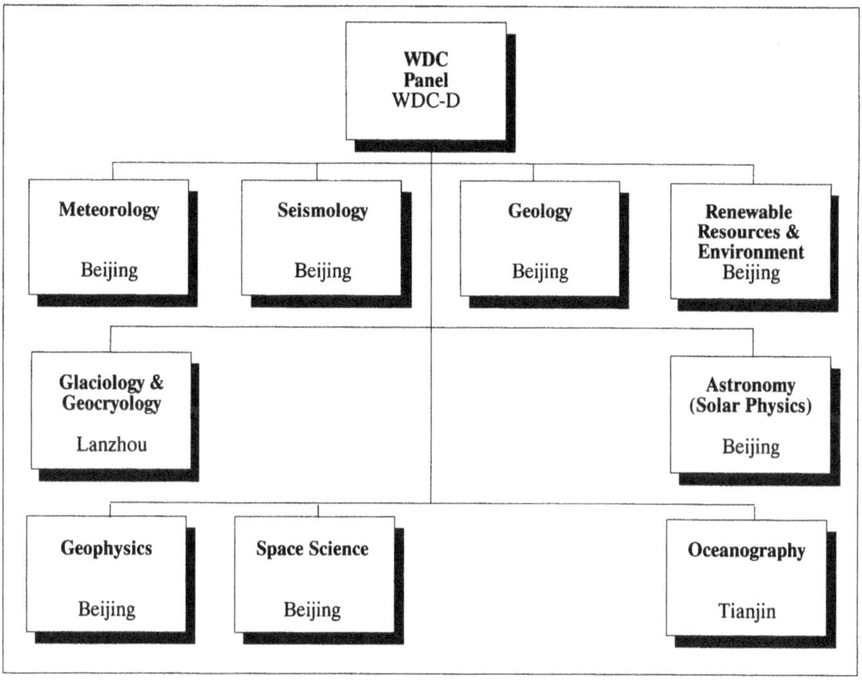

Figure 406: Structure of World Data Center D (China)

O.16 Committee on Earth Observation Satellites - CEOS

CEOS was created in 1984 as a result of the international Economic Summit of Industrialized Nations (G-7 Versailles Summit of 1982) and serves as the focal point for international coordination of space-related, Earth observation activities.[2024)]

Policy and technical issues of common interest related to the whole spectrum of Earth observation satellite missions and data received from such are addressed. CEOS encourages complementarity and compatibility among spaceborne Earth observing systems through coordination in mission planning, promotion of full and nondiscriminatory data access, setting of data product standards, and development of compatible products, services, and applications. The user community benefits directly from this international coordination.

Members are those national and international government agencies with funding and program responsibilities for a satellite Earth observation program currently operating or in the later stages of system development. CEOS members include (1995): ASI (Italy), BNSC (United Kingdom), CAST (China), CNES (France), CSA (Canada), CSIRO (Australia), ESA and EUMETSAT (Europe), DARA (Germany), INPE (Brazil), ISRO (India), NASA and NOAA (USA), NRSCC (China), NSAU (Ukraine), RKA and ROSHYDROMET (Russia), SNSB (Sweden), STA (Japan).

Governmental entities with a space-based Earth observation program in the early stages of development or with significant ground segment activity that supports CEOS member agency programs may be invited to participate as observers. Current (1995) observers are: BOST (Belgium), CCRS (Canada), CEC (Commission of the European Communities), CRI (New Zealand), and NSC (Norway).

2024) L. Moodie, "Committee on Earth Observation Satellites (CEOS)," CEOS Newsletter No. 1 , Summer 1993

Affiliate members of CEOS are: International Council of Scientific Unions (ICSU), France; International Geosphere-Biosphere Program (IGBP), Sweden; Intergovernmental Oceanic Commission (IOC), France; World Climate Research Program (WCRP), Switzerland; World Meteorological Organization (WMO), Switzerland; United Nations Environment Program (UNEP), Kenya; Global Climate Observing System (GCOS), Switzerland); Global Ocean Observing System (GOOS), France.

The CEOS plenary meets once a year. There are two working groups, each with four active subgroups as illustrated in Figure 407. The chairmanship of the working groups and subgroups is indicated.

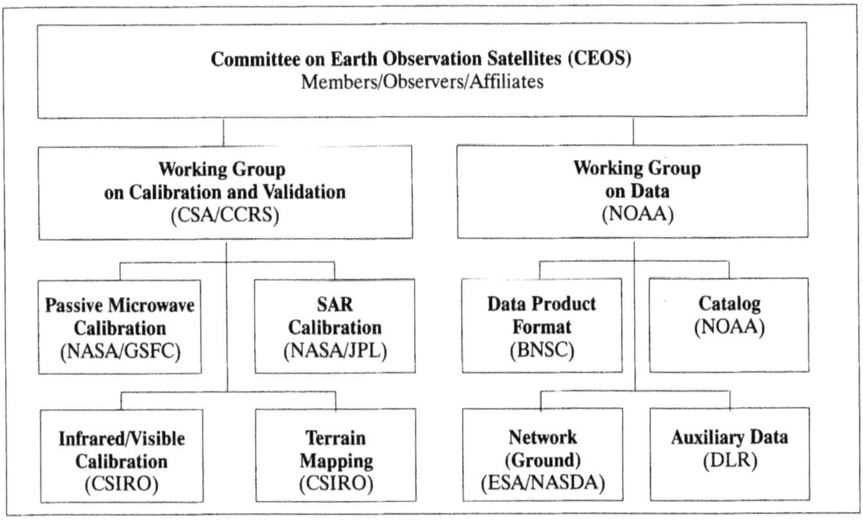

Figure 407: CEOS working group structure

Name of Launch Site	Resp. Organization (Country)	Location	Remarks
Baikonur Cosmodrome (Tyuratam Leninsk)	CIS (Kazakhstan)	45.6° N, 63.4° E	since 1957, launch of manned S/C, interplanetary S/C and geostationary S/C
Plesetsk (northern Cosmodrome)	Russia/CIS	62.8° N, 40.1° E	since 1966, launch of polar orbiting satellites
Kapustin Yar (Volgograd Cosmodrome)	Russia/CIS	48.4° N, 45.8° E	since 1962, launch of small satellites and interplanetary space probes
Cape Canaveral Air Force Station (CCAFS), FL	USAF/NASA (USA)	28.5° N, 80.0° W	since 1958, launch of manned S/C, interplanetary S/C, GEO satellites, etc.
Kennedy Space Center, FL	NASA (USA)	28.5° N, 80.0° W	since 1967, Complex 39 (launch of Saturn V vehicle, adapted to launch the Shuttle)
Vandenberg AFB, CA	USAF/NASA (USA)	34.4° N, 120.35°W	since 1959, launches of S/C into polar orbit
Vandenberg AFB, CA Commercial spaceport	Spaceport Systems International L.P.	34.4° N, 120.35°W	since 1998 (a joint venture of ITT Industries and CA Commercial Spaceport Inc.) launches of S/C into polar orbit
Wallops Island, VA	NASA/GSFC WFF (USA)	37.8° N, 75.5° W	since 1961, launch of sounding rockets and Scout launcher for small satellites, commercial launches since 1997
Edwards AFB, CA	USAF (USA)	35.0° N 118.0° W	Space Shuttle returns since 1981
Tanegashima Kagoshima Space Center	ISAS, U. of Tokyo, (Japan)	31.2° N, 131.1° E	since 1970, launch of scientific satellites
Tanegashima Space Center (TNSC)	NASDA (Japan)	30.4° N, 131.0° E	since 1975, launch site for scientific and commercial satellites
SHAR Sriharikota	ISRO (India)	13.9° N, 80.4° E	since 1980, launch of ISRO satellites
San Marco Platform (off the coast of Kenya)	CRA, U. of Rome (Italy)	2.9° S, 40.3° E	since 1967, launch of Scout rockets for NASA/ASI
Guiana Space Centre Kourou (Fr. Guiana)	CNES/ESA	5.2° N, 52.8° W	since 1970, launch of scientific and commercial satellites
Woomera Range	WRE (Australia)	31.1° S, 136.8° E	since 1967, launch of sounding rockets since 1999, launch of LEO S/C (Kistler)
Jiuquan SLC	China	40.6° N, 99.9° E	since 1970, launch of experimental satellites
Xichang SLC	China	28.25° N, 102.0° E	since 1984, launch of satellites into geost. and polar orbits
Taiyuan/Wuzhai	China	37.5° N, 112.6° E	since 1988, launches of CZ-4A and -4B into polar orbits for remote sensing
Esrange, Kiruna	SSC (Sweden)	67.9° N, 21.0° E	launch of sounding rockets and balloons
Alcántara Launch Center	CTA/INPE (Brazil)	2.3° S, 44.4° W	in preparation for satellite launches
Palmachim/Yavne	Israel	31.5° N 34.5° E	since 1989, launch of LEO satellites into westward direction (Ofeq series)
Svobodny Cosmodrome	Russia	51.4° N, 128.3° E	since 1996, Start-1 launcher (Zeya, etc. EarlyBird)
Sea Launch	Partnership of Boeing Co. USA, Yuzhnoye, Ukraine RKK Energiya, Russia, and Kvaerner Maritima A.S. Oslo, Norway	Pacific Ocean	since March 27,1999, inaugural launch of two-stage Zenit (Zenit-3SL and Zenit-2) rockets of the Yuzhnoye Design Bureau. The first commercial launch was on Oct. 9, 1999. The Sea Launch home port is Long Beach, CA for the command ship and launch platform
Poker Flat Research Range (PFRR), AK	University of Alaska (USA)	65.117° N 147.60 W	since 1968 rocket launch facility (auroral and meteorological research), located about 50 km north of Fairbanks
Kodiak Launch Complex on Kodiak Island, AK	Alaska Aerospace Development Corporation	57.445° N 152.34°W	The first launch is scheduled for Aug. 2001, PICOSat of AFRL, Athena 2 vehicle (along with Starshine).

Table 557: Coordinates of satellite launch sites around the world

O.17 Space Shuttle Mission Chronology

The following table contains brief information on all Shuttle flights with regard to orbits, launch/landing dates, orbiter type, and payload complement.[2025) 2026) 2027) 2028)] All payload abbreviations are defined in Appendix B (Acronyms and Abbreviations) starting with page 1393; however, no descriptions are given for these Shuttle payloads (which is beyond the scope of this book). The interested reader may consult the corresponding NASA home-pages on internet.

Note: Following STS-9 (Space Transport System) the flight numbering system was changed. Thus, the next flight, instead of being designated STS-10, became STS 41-B. The new numbering system was supposed to be more specific; the first numeral stood for the fiscal year of the launch, the "4" for 1984. The second numeral represented the launch site, 1 for KSC, and 2 for VAFB. The letter "B" designated the launch in the fiscal year. Following the Challenger accident (Jan. 28, 1986) NASA reestablished the original numbering system.

Mission Incl., Alt.(km)	Launch/ Landing Date	Orbiter	Mission Highlights / Payload Complement
Legend: I = Inclination (°)		A = Nominal Altitude (km)	
STS-1 I:40.3°, A:307	12. April 1981 14. April 1981	Columbia	Major systems tested, orbiter sustained tile damage
STS-2 I:38°, A:290	12. Nov. 1981 14. Nov. 1981	Columbia	OSTA-1 payload (including SIR-A, MAPS, OCE, SMIRR), see chapter J.19
STS-3 I:38°, A:272	22. Mar. 1982 30. Mar. 1982	Columbia	Get Away Special (GAS), Spacelab pallet mounted, OSS-1 payload, MLR, EEVT, HBT, SSIP
STS-4 I:28.5°, A:364	27. June 1982 4. July 1982	Columbia	DoD Payload, Get Away Special (GAS), CFES, MLR, IECM, SSIP
STS-5 I:28.5°, A:340	11. Nov. 1982 16. Nov. 1982	Columbia	ANIK C-3 and SBS C satellites, Get Away Special, SSIP
STS-6 I:28.5°, A:330	4. April 1983 9. April 1983	Challenger	1st EVA, TDRS-1, CFES, MLR, RME, NOSL, Get Away Specials
STS-7 I:28.5°, A:360 km	18. June 1983 24. June 1983	Challenger	ANIK-C2 and PALAPA-B1 satellites, Get Away Specials, 10 experiments mounted on SPAS-01 (MOMS-01), OSTA-2 payload, CFES, MLR, SSIP
STS-8 I:28.5°, A:353	30. Aug. 1983 5. Sept. 1983	Challenger	INSAT-1B satellite, CFES, SSIP, ICAT, RME, Get Away Specials
STS-9 I:57°, A:287	28. Nov. 1983 8. Dec. 1983	Columbia	Spacelab-1 mission (ESA), see J.22
STS-41-B (10) I:28.5°, A:300 km	3. Feb. 1984 11. Feb. 1984	Challenger	WESTAR-6 and PALAPA-B2 satellites, GAS, SPAS-2, MOMS-01, Cinema-360 camera, ACES, MLR, RME, IEF
STS-41-C (11) I:28.5°, A:580	6. April 1984 13. April 1984	Challenger	SMM satellite repair in orbit, LDEF deployment (57 experiments), IMAX camera, RME, Cinema-360, SSIP
STS-41-D (12) I:28.5°, A:340 km	30. Aug. 1984 5. Sept. 1984	Discovery	3 satellites (SBS-D, SYNCOM IV-2, TELSTAR), OAST-1, CFES III, RME, SSIP, IMAX camera, CLOUDS
STS-41-G (13) I:57°, A:352	5. Oct. 1984 13. Oct. 1984	Challenger	ERBS satellite,OSTA-3, LFC, IMAX camera, CANEX, APE, RME, TLD, SIR-B, Get Away Specials
STS-51-A (14) I:28.5°, A:342	8. Nov. 1984 16. Nov. 1984	Discovery	TELESAT-H (ANIK) and SYNCOM IV-1, retrieval of PALAPA-B-2 and WESTAR-VI, DMOS, RME
STS-51-C (15) I:28.5°, A:407	24. Jan. 1985 27. Jan. 1985	Discovery	DoD payload
STS-51-D (16) I:28.5°, A:527	12. April 1985 19. April 1985	Discovery	TELSAT-I (ANIK C-1) and SYNCOM IV-3 satellites, CFES III, SSIP, AFE, Get Away Specials, PPE
STS-51-B (17) I:57° A:411 km	29. April 1985 6. May 1985	Challenger	Spacelab-3 mission ESA, see J.23 [material sciences, life sciences, fluid mechanics, atmospheric physics (ATMOS), and astronomy], Get Away Specials (NUSAT)
STS-51-G (18) I:28.5°, A:405	17. June 1985 24. June 1985	Discovery	MORELOS-A, ARABSAT-A and TELSTAR-3D satellites, SPARTAN-1, Get Away Specials, HPTE, ADSF

2025) Information taken from NASA publication: "Space Shuttle Mission Chronology 1981-1992," and from Internet
2026) The acronyms of Shuttle payload complements are listed in Appendix C "Acronyms and Abbreviations"
2027) Shuttle orbits provided by J. Gass of NASA/GSFC
2028) "NASA Long-Term Plan For Shuttle Missions Details Station Timing," Space News March 7-13, p. 26, 1994

Mission Incl., Alt.(km)	Launch/ Landing Date	Orbiter	Mission Highlights / Payload Complement
STS-51-F (19) I:49.5° A:383 km	29. July 1985 6. Aug. 1985	Challenger	Spacelab-2 mission, [life sciences, plasma physics (PDP), astronomy (XRT, IRT), high-energy astro-physics, solar physics (SUSIM, HRTS, CHASE, SOUP), atmospheric physics, and technology research IPS]
STS-51-I (20) I:28.5°, A:514	27. Aug. 1985 3. Sept. 1985	Discovery	ASC-1, AUSSAT-1, and SYNCOM IV-4 satellites, PVTOS
STS-52-J (21) I:28.5°, A:590	3. Oct. 1985 7. Oct. 1985	Atlantis	DoD mission
STS-61-A (22) I:57°, A:383	30. Oct. 1985 6. Nov. 1985	Challenger	Spacelab D1 (German mission), GLOMR satellite of DARPA deployed from Get Away Special canister
STS-61-B (23) I:28.5°, A:416 km	26. Nov. 1985 3. Dec. 1985	Atlantis	MORELOS-B, AUSSAT-2, and SATCOM satellites, EASE, ACCESS, CFES, DMOS, MPSE, OEX, Get Away Special, IMAX cargo bay camera (ICBC)
STS-61-C (24) I:28.5°, A:392	12. Jan. 1986 18. Jan. 1986	Columbia	SATCOM KU-1 satellite, CHAMP, MSL-2, Hitchhiker-G1, IR-IE, IBSE, HPCG, SSIP, Get Away Specials
STS-51-L (25)	28. Jan. 1986	Challenger	Accident !!!! (catastrophic failure) TDRS-B, SPARTAN-Halley
STS-26 I:28.5°, A:376	29. Sept. 1988 3. Oct. 1988	Discovery	TDRS-3 satellite, PVTOS, PCG, IRCFE, ARC, IFE, MLE, PPE, ELRAD, ADSF, SSIP, OASIS-1
STS-27 I:57°, A:460	2. Dec. 1988 6. Dec. 1988	Atlantis	DoD payload, deployment of USA-34, this code name designates the first high-resolution imaging radar S/C of the US military (also known under the name Lacrosse-1). Orbit of Lacrosse-1 was raised to 668 km x 703 km.
STS-29 I:28.5°, A:340	13. Mar. 1989 18. Mar. 1989	Discovery	TDRS-4 satellite, OASIS-1, SHARE, PCG, CHROMEX, SSIP, AMOS, IMAX camera
STS-30 I:28.8°, A:340	4. May 1989 8. May 1989	Atlantis	Magellan/Venus radar mapper S/C, MLE, FEA, AMOS
STS-28 I:57°, A:	8. Aug. 1989 13. Aug. 1989	Columbia	DoD payload
STS-34 I:34.3°, A:342	18. Oct. 1989 23. Oct. 1989	Atlantis	Galileo S/C to Jupiter, SSBUV, GHCD, PM, STEX, MLE, IMAX camera, AMOS, SSIP
STS-33 I:28.5°, A:	22. Nov. 1989 27. Nov. 1989	Discovery	DoD payload
STS-32 I:28.5°, A:389	9. Jan. 1990 20. Jan. 1990	Columbia	SYNCOM IV-F5 satellite (DoD), retrieval of LDEF, CNCR, PCG, FEA, AFE, L3, MLE, IMAX, AMOS
STS-36,A:811 I:62-65°,	28. Feb. 1990 4. Mar. 1990	Atlantis	DoD payload (AFP-731)
STS-31 I:28.5°, A:703	24. April 1990 29. April 1990	Discovery	HST satellite, IMAX (ICBC), APM, PCG, RME III, IPMP, SSIP, AMOS
STS-41 I:28.5°, A:340	6. Oct. 1990 10. Oct. 1990	Discovery	Ulysses spacecraft (ESA/NASA), SSBUV, ISAC, CHROMEX, VCS, SSCE, IPMP, PSE, RME III, SSIP, AMOS
STS-38	15. Nov. 1990 20. Nov. 1990	Atlantis	DoD payload
STS-35 I:28.5°, A:520	2. Dec. 1990 10. Dec. 1990	Columbia	ASTRO-1 observatory (consisting of HUT, WUPPE, UIT, BBXRT, and IPS), SAREX-2, AMOS
STS-37 I:28.5°, A:518	5. April 1991 11. April 1991	Atlantis	GRO (Gamma Ray Observatory), CETA, APM, SAREX II, PCG, BIMDA, RME III, AMOS
STS-39 I:57°, A:298	28. April 1991 6. May 1991	Discovery	DoD classified mission including unclassified payload: AFP-675, IBSS, CIV, CRO, SPAS-II, STP-1, RME III, CLOUDS-1
STS-40 I:28.5°, A:340	5. June 1991 14. June 1991	Columbia	SLS-1 (Spacelab Life Sciences-1), Get Away Specials, MODE, OEX, GBA-2
STS-43 I:28.5° A:340 km	2. Aug. 1991 11. Aug. 1991	Atlantis	TDRS-5 satellite, SHARE II, SSBUV, TPCE, OCTW, APE-B, PCG III, BIMDA, IPMP, SAMS, SSCE, UVPI, AMOS
STS-48 I:57°, A:657	12. Sept. 1991 18. Sept. 1991	Discovery	UARS satellite, APM, MODE, SAM, CREAM, PARE, PCG II-2, IPMP, AMOS
STS-44 I:28.5° A:416 km	24. Nov. 1991 1. Dec. 1991	Atlantis	DoD mission, DSP (Defense Support Satellite), plus unclassified payloads: IOCM, AMOS, CREAM, SAM, RME III, VFT-1, UVPI
STS-42 I:57° A:348 km	22. Jan. 1992 30. Jan. 1992	Discovery	IML-1 (International Microgravity Laboratory-1), Get Away Specials, GOSAMR-1, IMAX, IPMP, RME III, SSIP, GBA-3

Mission Incl., Alt.(km)	Launch/ Landing Date	Orbiter	Mission Highlights / Payload Complement
STS-45 I:57°, A:340	24. Mar. 1992 2. April 1992	Atlantis	ATLAS-1 payload (ATMOS, MAS, SSBUV, ACRIM, SOLCON, SOLSPEC, SUSIM), Get Away Special, IPMP, STL-01, RME III, VFT-2, CLOUDS, SAREX II
STS-49 I:28.4°, A:390	7. May 1992 16. May 1992	Endeavour	EVA, Intelsat-VI capture and redeployment, ASEM, CPCG, UVPI, AMOS
STS-50 I:28.5°, A:350	25. June 1992 9. July 1992	Columbia	USML-1 (US Micrograviy Lab-1), IPMP, SAREX II, UVPI
STS-46 I:28.5° A:502 km	31. July 1992 8. Aug. 1992	Atlantis	EURECA-1 satellite, TSS-1 (Tethered Satellite System), EOIM-III/TEMP, CONCAP II and III, ICBC, LDCE, AMOS, PHCF, UVPI
STS-47 I:57°, A:346	12. Sept. 1992 20. Sept. 1992	Endeavour	Spacelab-J (Japan), Get Away Specials, ISAIAH, SSCE, SAREX II, AMOS, UVPI, GBA-4
STS-52 I:28.5°, A:302 km	22. Oct. 1992 1. Nov. 1992	Columbia	LAGEOS-II, USMP-1, CANEX-2, SVS, MELEO, QUELD, PARLIQ, OGLOW-2, SATO, CTA, CMIX, CPCG, CVTE, HPP, PSE, SPIE, UVPI, TPCE, ASP
STS-53 I:57°, A:370 km	2. Dec. 1992 9. Dec. 1992	Discovery	DoD payload GCP, Get Away Specials, CRYOHP, OD-ERACS, BLAST, CLOUDS, CREAM, FARE, HER-CULES, MIS-1, RME III, STL, VFT-2
STS-54 I:28.5°, A:296	13. Jan. 1993 19. Jan. 1993	Endeavour	TDRS-6 satellite, DXS, CGBA, CHROMEX, PARE, SAMS, SSCE
STS-56 I:57°, A:296	8. April 1993 17. April 1993	Discovery	ATLAS-2 payload (ATMOS, MAS, SSBUV, ACRIM, SOLCON, SOLSPEC, SUSIM), SPARTAN-201, SUVE, CMIX, PARE, STL-1, CREAM, HERCULES, RME III, AMOS
STS-55 I:28.5°, A:296	26. April 1993 6. May 1993	Columbia	Spacelab D2 (MOMS-02, MEDEA,WL, HOLOP, USS, MAUS, AOET, AR, BB, BA, ROTEX, MMA), SAREX
STS-57 I:28.5°, A:474 km	21. June 1993 1. July 1993	Endeavour	SPACEHAB-1 (22 experiments), EURECA-1 retrieval, Get Away Specials, CONCAP IV, SHOOT, FARE, SA-REX-II, AMOS
STS-51 I:28.5° A:296 km	12. Sept. 1993 22. Sept. 1993	Discovery	ACTS satellite, ASTRO-SPAS deployment of ORFEUS, IMAX, LDCE, CPCG, CHROMEX-04, HRSGS-A, APE-B, IPMP, RME III, AMOS
STS-58 I:39°, A:283	18. Oct. 1993 1. Nov. 1993	Columbia	SLS-2 (Spacelab Life Sciences-2), DEEFD, OARE, SA-REX-II, PILOT
STS-61 I:28.5°, A:594	2. Dec. 1993 13. Dec. 1993	Endeavour	Hubble Space Telescope (HST) repair mission (optical correction), IMAX
STS-60 I:57°, A:404	3. Feb. 1994 11. Feb. 1994	Discovery	WSF-01, Spacehab-2, COB/GBA, SAREX-2, APE-B, ODERACS, BREMSAT, CPL
STS-62, I:39°; A:260	4. Mar. 1994 18. Mar. 1994	Columbia	USMP-2, OAST-2, DEE, SSBUV/A, LDCE, APCG,PSE, CPCG, CGBA, BDS, MODE, AMOS, BSTC, EDO
STS-59, I:57°, A:222	9. Apr. 1994 20. Apr. 1994	Endeavour	SRL-01 (Space Radar Laboratory, also referred to as SIR-C/X-SAR, see J.21), MAPS (J.11), CONCAP-4, SA-REX-2, STL, TUFI, VFT-4, GAS (x3)
STS-65 I:28.5°, A:260	8. July 1994 23 July 1994	Columbia	IML-2, APCF, CPCG, AMOS, OARE, SAREX-2, EDO, MAST
STS-64 I:57°, A:258	9. Sept. 1994 20. Sept. 1994	Discovery	LITE-1 (see J.10), ROMPS, SPARTAN-201, TCS, SPI-FEX, GAS, SAFER, SSCE, BRIC, RME, MAST, SA-REX-II, AMOS, GBA-7
STS-68 I:57°, A:222	30. Sept. 1994 11. Oct. 1994	Endeavour	SRL-02 (SLR02 is comprised of: SIR-C/X-SAR, see J.21), MAPS (J.11)), CPCG, BRIC, Chromex, CREAM, MAST, GAS (5)
STS-66, I:57°, A:305	3. Nov. 1994 14. Nov. 1994	Atlantis	ATLAS-03 (ATMOS, MAS, SSBUV, ACRIM, SOLCON, SOLSPEC, SUSIM), CRISTA-SPAS (MAHRSI), ES-CAPE-II, PCG, SAMS, HPP
STS-63 I:51.6°, A:310	3. Feb. 1995 11. Feb. 1995	Discovery	Shuttle-MIR rendezvous and fly-around; IMAX, Pay-loads: Spacehab-3, SPARTAN-204, CGP/ODERACS, SSCE, AMOS, CONCAP-II, ICBC, CGP
STS-67, I:28.5°, A:400	2. Mar. 1995 18. Mar. 1995	Endeavour	Astro-2 observatory (consisting of HUT, UIT, WUPPE and IPS), GAS, CMIX, PCG, MACE
STS-71,I:51.6° A:395	27 June 1995 7 July 1995	Atlantis	MIR Docking, Spacelab-MIR, SAREX-II, IMAX
STS-70,I:28.4° A:295	13. July 1995 21 July 1995	Discovery	TDRS-7, MAST, RME-III, STL-B, WINDEX, VFT-4, BRIC, CPCG, HERCULES, MIS-B, NIH-R

Mission Incl., Alt.(km)	Launch/ Landing Date	Orbiter	Mission Highlights / Payload Complement
STS-69,I:28.5° A:370	7. Sept. 1995 18 Sept. 1995	Endeavour	WSF-2, deployment of SPARTAN-201-3, CAPL-2, TES-2, GAS, GBA, IEH-1, CONCAP-IV-3
STS-73, I:39° A:271	20.Oct. 1995 5. Nov. 1995	Columbia	Microgravity Laboratory 2 (USML-2), OARE-06, SAMS, 3DMA, STABLE
STS-74 I:51.6°, A:395	12. Nov. 1995 20. Nov. 1995	Atlantis	MIR flight 2, ICBC-05, IMAX, GLO-4, PASDE-1, MCSA, SAREX, GAS
STS-72 I:28.5°, A:455	11. Jan. 1996 20. Jan. 1996	Endeavour	Space Flyer Unit (SFU) retrieval, SSBUV-8, SLA-1, GAS, SPARTAN-206 (with REFLEX, GADACS, SELODE, SPRE), NIH-R3, NIH-C5, PCG,
STS-75 I:28.5°, A:298	22. Feb. 1996 9. Mar. 1996	Columbia	TSS-1R, USMP-3
STS-76 I:51.6°, A:292	22. Mar. 1996 31. Mar. 1996	Atlantis	MIR flight 3 (MEEP), Spacehab-SM, GAS, (EarthKAM)
STS-77 I:39°, A:284	19. May 1996 29. May 1996	Endeavour	Spacehab-4, Spartan-207 (IAE), TEAMS (GANE, PAMS, VRTE), GBA-9 (11 GAS)
STS-78 I:39°, A:278	20. June 1996 7. July 1996	Columbia	Spacelab/LMS, SAMS-D, OARE, BDPU (TMIBD,SIE), SAREX-II
STS-79 I:51.6°, A:360	16. Sept. 1996 26. Sept. 1996	Atlantis	MIR flight 4, IMAX, Spacehab (MACEK), SAREX-II
STS-80 I:28.5°, A:470	19. Nov. 1996 7. Dec. 1996	Columbia	ORFEUS-SPAS II, WSF-3, SEM-1
STS-81 I:51.6°, A:440	12. Jan. 1997 22. Jan. 1997	Atlantis	MIR flight 5, Spacehab, SAREX-II, KidSat (EarthKAM)
STS-82 I:28.45°,A:592	11 Feb. 1997 21 Feb. 1997	Discovery	Hubble Servicing Mission 2 (installation of two new instruments, STIS and NICMOS)
STS-83 I:28:45°,A:297	4 Apr. 1997 8. Apr. 1997	Columbia	MSL-1 (mission was terminated due to electrical power problems) MMA was part of MSL-1
STS-84 I:51.6°, A:	15. May 1997 24. May 1997	Atlantis	MIR flight 6, Spacehab-5,
STS-94 I:28.45°,A:295	1 Jul.1997 17 Jul. 1997	Columbia	MSL-1 reflight, CRYOFD, OARE, SAREX, MSX
STS-85 I:57°, A:290	7. Aug. 1997 19. Aug. 1997	Discovery	CRISTA-SPAS-2, MAHRSI, TAS-1 (SOLCON, ISIR, SLA-2, CVX, SEM-2, TPF, CFE, SAAMD/WBSAAMD), MFD (TPFLEX, ESEM), IEH (UVSTAR, SEH, DATA-CHASER, GLO), PCG-STES, MSX-8, GAS (2), BDS-03, SIMPLEX, SWUIS, BRIC-10, SSCE-7
STS-86 I:51.6°, A:296, A:395 dock	25. Sept. 1997 6. Oct. 1997	Atlantis	MIR flight 7, Spacehab, MEEP, SEEDS-II, RME's, KidSat, CPCG, CREAM, CCM-A, MSX, SIMPLEX
STS-87 I:28.45°,A:276	19. Nov. 1997 5. Dec. 1997	Columbia	USMP-4, Spartan-201-4, OARE, SOLSE/LORE, LHP, TGDF, NaSBE, GAS, MGBX, CUE, MSX, CUE, SIMPLEX, AERCam/Sprint
STS-89 I:51.6°, A:296 A:395 dock	22. Jan. 1998 31. Jan. 1988	Endeavour	MIR flight 8, Spacehab-DM (ADV XDT, ADV CGBA, EORF, MGM, RME, SAMS), MPNE, SIMPLEX, CEBAS TMIP, GPS DTO, OSVS, BIO3D, EarthKAM
STS:90 I:39°, A:280	17. Apr. 1998 3. May 1998	Columbia	Neurolab (Spacelab module), GAS (G-722), SVFE
STS-91 I:51.6°, A:385	2. June 1998 12. June 1998	Discovery	MIR flight 9, AMS, Spacehab-SM, GAS, SEM-3,
STS-95 I:28.4°, A:555	29. Oct. 1998 7. Nov. 1998	Discovery	Spacehab, HOST, IEH-03 (UVSTAR, SEH, SOLCON, STAR-LITE, PANSAT, CODAG), SPARTAN-201-5, SEM-4, GAS, IVHM-1,
STS-88 I:51.6°, A:390	4. Dec. 1998 15. Dec. 1998	Endeavour	ISS-1 (1st Space Station Assembly Flight) element called Unity, MightySat I, SAC-A, AMTEC, AWCS, ICBC, GAS (G-093), SEM-07
STS-96 I:51.66°,A:390	27. May 1999 6. June 1999	Discovery	ISS-2 flight, IVHM-2, Starshine (educational satellite), ICC, SVFE (GAS), SPACEHAB
STS-93 I:28.4°, A:290	23 July 1999 28 July 1999	Columbia	Chandra/AXAF-1 (deployment of X-Ray Observatory, a S/C with its own integral propulsion system), MSX, SIMPLEX, SWUIS, SAREX-II, GOSAMR, BRIC, MEMS, CGBA, MEMS, PGIM, LFSAH

Mission Incl., Alt.(km)	Launch/ Landing Date	Orbiter	Mission Highlights / Payload Complement
STS-103 I:28.4°, A:590	17 Dec. 1999 26 Dec. 1999	Discovery	Hubble Servicing Mission 3 (replacement of all six gyroscopes, a guidance sensor, and a S/C computer). Installation of a voltage/temperature kit for the S/C batteries. Installation of a new transmitter, solid state recorder (12 Gbit), and thermal insulation blankets.
STS-99 I:57°, A:233	11 Feb. 2000 22. Feb. 2000	Endeavour	SRTM (Shuttle Radar Topography Mission), EarthKAM
STS-101 I:51.6°, A:340	19. May 2000 29. May 2000	Atlantis	ISS-2A.2a assembly/maintenance flight Payloads: ICC (Integrated Cargo Carrier), MARS (Mission to America's Remarkable Schools), SEM-6 (Space Experiment Module-6), SPACEHAB/DM (Double Module), BioTube Precursor Experiment. - The current orbit of ISS is 368 km x 342 km. The average orbit decay of ISS is about 1.5 to 2 km/week. Atlantis raises the ISS orbit by about 30 km.
STS-106 I:51.6°, A:320	8 Sept. 2000 20. Sept. 2000	Atlantis	ISS-2A.2b assembly flight. Payloads: ICC, SPACEHAB/ DM,
STS-92 I:51.6°, A:320	11. Oct. 2000 24. Oct. 2000	Discovery	ISS-05-3A assembly flight. Payloads: Z-1 Truss/SLP, CMG (Control Moment Gyros), Ku/S-Band, PMA-3/SLP, DDCU (heat pipes), IMAX
STS-97 I:51.6°, A:310	30. Nov. 2000 11. Dec. 2000	Endeavour	ISS-04-4A flight. PV Module P6 (solar arrays and batteries)
STS-98 I:51.6°, A:310	07. Feb. 2001 20. Feb. 2001	Atlantis	ISS-07-5A assembly flight. Payloads: US Lab Destiny, ORU, PDGF
STS-102 I:51.6°, A:225	08. Mar. 2001 21. Mar. 2001	Discovery	ISS-07-5A1 assembly flight. Payload: Leonardo Multipurpose Logistics Module (MPLM)
STS-100 I:51.6°, A:310	19. Apr. 2001 01. May 2001	Endeavour	ISS-09-6A assembly flight. Payload: Raffaello MPLM-1, Racks/SLP, Canadarm2 (SSRMS,SPDM), LCA,RU, UHF Antenna
STS-104 I:51.6°, A:225	12. July 2001 24.July 2001	Atlantis	ISS-7A assembly flight. Payload: Airlock and support systems
STS-105	10. Aug. 2001 22. Aug. 2001	Discovery	ISS-7A.1 logistics flight. Payload: Donatello (MPLM-3)

O.18 Solar Wind and the Magnetosphere - An Introduction

Throughout history mankind has always known that life on Earth is dependent on light and heat from the sun, but beyond our senses, electromagnetic forces also link the sun and the Earth in a dynamic interplay that generates and characterizes our protective near-Earth environment and sparks the displays called the northern and southern lights. The space missions and space science of modern times have greatly improved knowledge of these electromagnetic forces and the overall energy transport in the sun-Earth system.[2029)]

The regions of space defined by the electromagnetic link include the sun and its sphere of influence - the heliosphere - and the Earth and its much smaller sphere of influence - geospace. Geospace includes the near-Earth space and reaches toward the sun, where the sun's heliosphere is disturbed by Earth's magnetic field.

Energy streams out from the sun toward the Earth in the form of a solar wind of electrified particles (1, see Figure 408). This hot, ionized gas, called plasma, streams toward Earth at about 1.5 million km per hour, carrying particles and magnetic fields from the sun outward past the planets (with the sun-Earth distance of 150 million km, the plasma needs about 100

2029) Taken from the introduction of the NASA/GSFC brochure: "ISTP Global GEOSPACE Science - Energy Transfer in Geospace (ESA/NASA/ISAS)," 1992, Courtesy of NASA/GSFC, Greenbelt, MD hours of travelling time). Earth is shielded from the full blast of these particles by its magnetosphere, the region around the Earth dominated by the Earth's magnetic field.

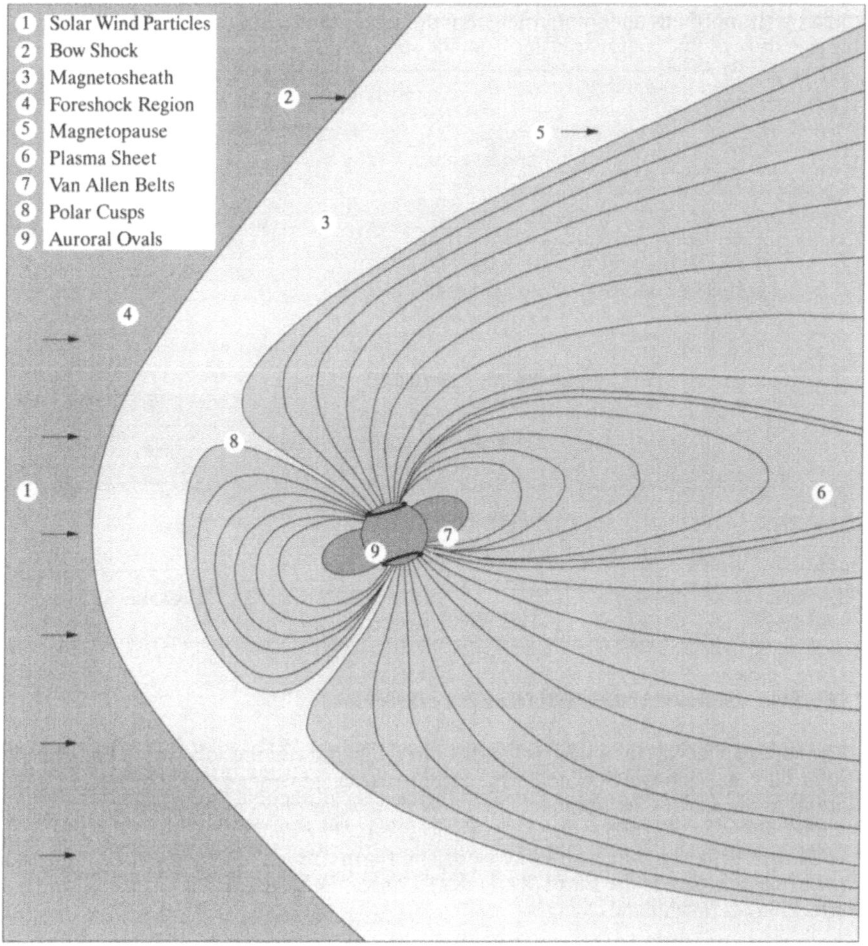

Figure 408: Characteristic model of the Earth's magnetosphere

As the solar wind approaches the Earth's magnetic field, a highly supersonic shock wave is created sunward of the Earth, similar in shape to the shock wave created when a jet plane breaks the sound barrier, but much stronger. This shock wave is called the bow shock (2). Most of the solar wind particles are heated and slowed down at the bow shock and detour around the Earth through a volume of space called the magnetosheath (3). Some particles are actually reflected back from the bow shock into the solar wind stream in a region of turbulence called the foreshock (4).

As the solar wind flows around the Earth, it stretches the Earth's magnetosphere out into a long tail, the magnetotail (not shown in Figure 408). Some of the particles being carried past the Earth leak through the barrier at the boundary of the Earth's magnetic field, called magnetopause (5), and are trapped inside the magnetosphere and stored in the plasma sheet (6) and Van Allen radiation belts (7). Some particles rush through funnel-like openings at the poles, called polar cusps (8). Some energetic particles come down along magnetic field lines and enter into the Earth's upper atmosphere. Particles accelerated in the magnetotail excite atoms and molecules in the Earth's atmosphere. These atoms and molecules then emit light known as the northern and southern lights (or auroras) in the auroral ovals (9), giving a visible signature of this energy transfer from the sun to the Earth.

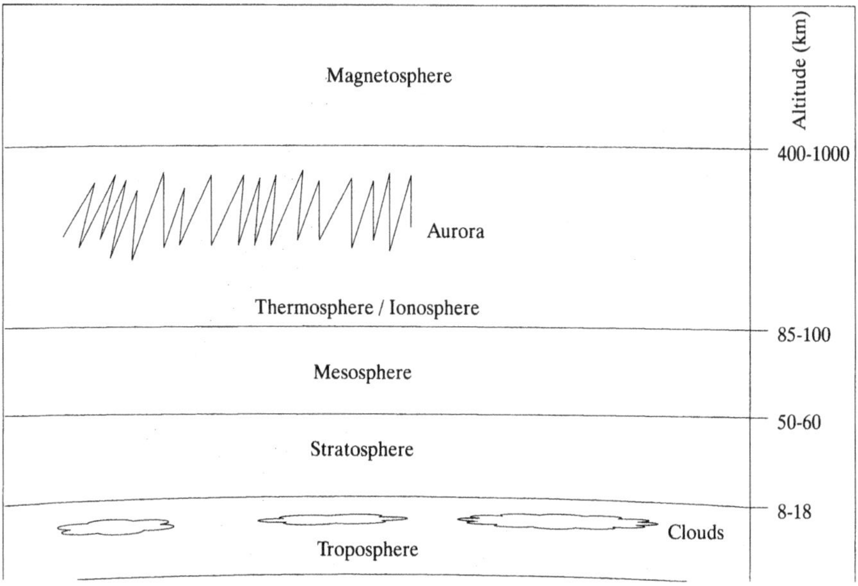

Figure 409: Schematic of the layers of the Earth's atmosphere

The sun is an active star whose variability affects the flow of the solar wind. For example, solar-flare explosions, associated with sunspots, can cause strong gusts of solar wind. Alterations in the Earth's environment caused by these solar phenomena happen on different time scales from less than a minute to over a century. The sun's variations (for example, solar X-ray bursts) can affect specific regions on Earth within the time required for light to travel from the sun to the Earth (8 minutes). Longer timescale, global solar variations may affect long-term climatic changes.

The best known terrestrial effects of solar activity are the geomagnetic storms and auroras that occur within a few days following major solar flares. In turn, the auroras contribute to the heating and ionizing of the upper atmosphere that generate the ionosphere, located at about 150-250 km above the Earth, where the neutral atmosphere gives way to ionized plasma.

Above our atmosphere, ions and charged particles bounce along and spiral around magnetic field lines, deflected from direct impact on the atmosphere and the people below. Thus, the geomagnetic field forms a mantle protecting us from harmful cosmic radiation.

Events on the sun can trigger changes in the electrical and chemical properties of the atmosphere, the ionosphere, the magnetosphere, the ozone layer, and high-altitude temperatures and wind patterns. These changes cause magnetic storms, communications static, power blackouts, and navigation problems for ships and airplanes with magnetic compasses. Also, satellites and spacecraft can be damaged or can reenter Earth's atmosphere prematurely because of solar storms.

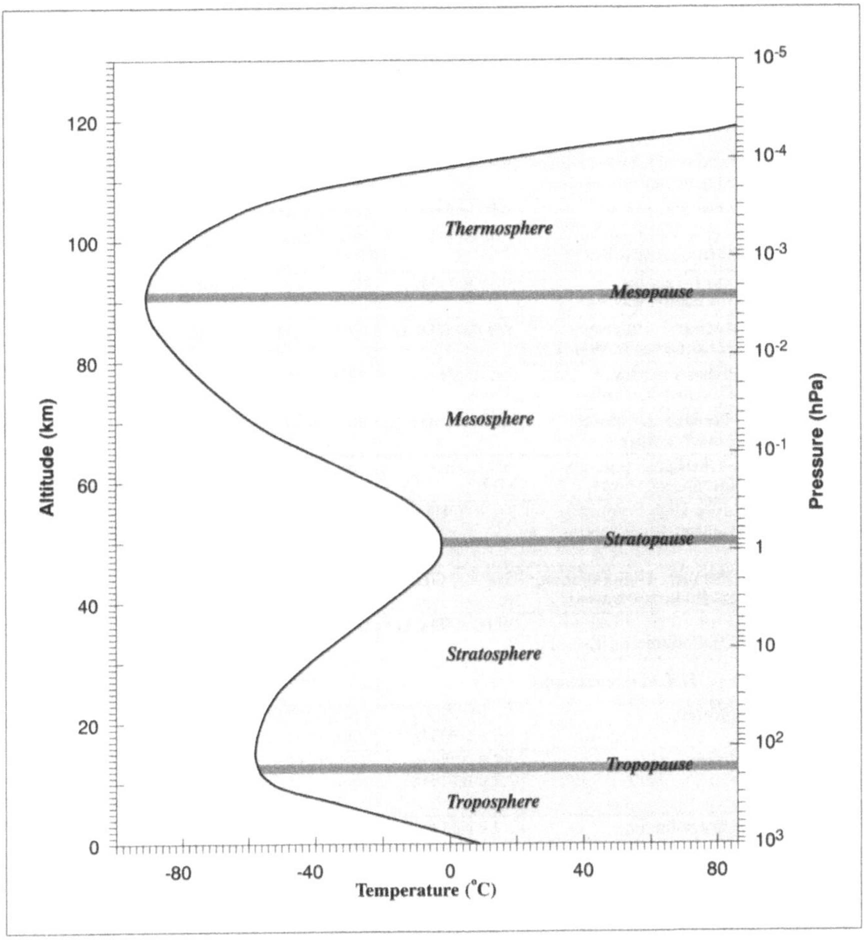

Figure 410: Temperature profile of the standard atmosphere

O.19 Frequency Designations

ITU provides general frequency band allocations (10 bands) for the electromagnetic spectrum as illustrated in Table 558 and in Figure 413 (ELF to EHF). However, these ranges are rather wide for practical RF applications (which are normally limited by waveguide designs). Radar engineers still use the old letter designations to denote the general frequency band at which a radar operates. The original letter designations (P, L, S, X and K) were introduced Before and during World War II. The nomenclature P-band is no longer in use since the 1960s. It has been replaced by UHF.

It turns out there is no uniform standard for the letter designations around. There seem to be quite a few different bandwidths definitions, originating from various sides of the communication industry and from research. New frequency band standards were introduced by such bodies as NATO. According to this scheme the spectrum between 0.1 - 100 GHz was given the letter designations from A to M in alphabetic and ascending frequency order.

The remote sensing community seems to adhere to the old frequency band designations in spite of these new developments. The sensor names and parameter definitions in this book are manifestations of the old standards.

ITU Designation	Name of Frequency Band	Frequency Band	Wavelength (λ)	Remarks/Use
ELF	Extremely Low Frequency (Megametric waves)	30 < 3000 Hz	10^4 - 100 km	Alternating current (50 - 60 Hz)
VF	Voice Frequency	300 < 3000 Hz	10^3 - 100 km	Voice
VLF	Very-low Frequency (Myriametric waves)	10 < 30 kHz	30 - 10 km	Radiotelegraphy
LF	Low Frequency (Kilometric waves)	30 < 300 kHz	10 - 1 km	Radiotelegraphy, radio
MF	Medium Frequency (Hectometric waves)	300 < 3000 kHz	1000 - 100 m	Radio, (AM radio)
HF	High Frequency (Decametric waves)	3 < 30 MHz	100 - 10 m	Radiotelephony, radio navigation, amateur radio
VHF	Very-High Frequency (Metric waves)	30 < 300 MHz	10 - 1 m	Radio, TV, radio navigation, radiobeacon, (FM radio)
UHF	Ultra-High Frequency (Decimetric waves)	300 < 3000 MHz	1 m - 0.1 m	TV, radiobeacon, satellite control, radiolocation, radio
SHF	Super-High Frequency (Centimetric waves)	3 < 30 GHz	10 - 1 cm	Radar, radiobeacon, satellite broadcast, Maser, MW heating
EHF	Extremely-High Frequency(Millimetric waves)	30 < 300 GHz	10 - 1 mm	
-	(Decimillimetric waves)	300 < 3000 GHz	1 - 0.1 mm	
← ITU Radio Spectrum			← Microwave Spectrum	
IR	Infrared	3×10^{11} Hz - 3.8×10^{14} Hz	1 mm - 0.78 µm	Heat sounding, laser communication
VIS	Visible Light	3.8×10^{14} Hz - 7.9×10^{14} Hz	0.78 - 0.38 µm	Light telephony, laser, electrooptical distance measurements
UV	Ultraviolet	7.9×10^{14} Hz - 3×10^{16} Hz	0.38 - 0.01 µm	Light telephony, laser, electro-optical distance measurements
X-Ray	X-ray (Soft band)	5×10^{15} Hz - 3×10^{19} Hz	60 - 0.1 nm	X-ray diagnostics, -medical therapy
X-Ray	X-ray (Medium band)	3×10^{19} Hz - 3×10^{20} Hz	10^{-2} - 10^{-3} nm	Materials testing
X-Ray	X-ray (Hard band)	3×10^{20} Hz - 2×10^{25} Hz	10^{-3} - 10^{-8} nm	Atomic particle transitions
γ-Ray	Gamma rays	8×10^{17} Hz - 4.7×10^{21} Hz	0.4 - 10^{-4} Hz	Radiology, material testing

Table 558: ITU frequency band allocation of the electromagnetic spectrum [2030)]

The Alphabet Soup of Frequency Bands

I am using a web-text by Jorn Christensen (http://satsys.etri.re.kr/tech/FreqBand.html), published in 1996 in "Satellite Industry Directory," Phillips Business Information, Inc. In Table 559, I have added to his IEEE definitions the ASPRS definitions as well. A further web-reference is: http://www.naval.com/radio–bands.htm

The frequency bands used for space radiocommunications are often denoted by various letters such as C-band, L-band, S-Band, etc. This practice of denoting frequency bands by letters has its origin in radar (radio detecting and ranging). Table 559 shows the standard letter

2030) "Reference Data for Radio Engineers," ITT (International Telephone and Telegraph Corp.), Sixth Edition 1982 designations for radar frequency bands of ASPRS (American Society for Photogrammetry and Remote Sensing) and of IEEE (Institute of Electrical and Electronic Engineers). There are many more definitions around, depending on author or institute.

Radar Band	Wavelength Range	Frequency Range			
P-band	136 - 77 cm	220 - 390 MHz			
UHF	100 - 30 cm	300 - 1 GHz	Radar Band	Wavelength Range	Frequency Range
L-band	30 - 15 cm	1 - 2 GHz	L-band	30 - 15 cm	1 - 2 GHz
S-band	15 - 7.5 cm	2 - 4 GHz	S-band	15 - 7.5 cm	2 - 4 GHz
C	7.5 - 3.75 cm	4 - 8 GHz	C	7.5 - 3.75 cm	4 - 8 GHz
X	3.75 - 2.4 cm	8 - 12.5 GHz	X	3.75 - 2.5 cm	8 - 12 GHz
Ku	2.4 - 1.67 cm	12.5 - 18 GHz	Ku	2.5 - 1.67 cm	12 - 18 GHz
K	1.67 - 1.18 cm	18 - 26.5 GHz	K	1.67 - 1.11 cm	18 - 27 GHz
Ka	1.18 - 0.75 cm	26.5 - 40 GHz	Ka	1.11 - 0.75 cm	27 - 40 GHz
V		40 - 75 GHz	V	0.75 - 0.40 cm	40 - 75 GHz
W		75 - 110 GHz	W	0.40 - 0.275 cm	75 - 110 GHz
			mm-band	2.75 - 1.0 mm	110 - 300 GHz
			Sub-mm-band	1.0 - 0.1 mm	300 - 3000 GHz
ASPRS Definitions			**IEEE Definitions**		

Table 559: Radar band letter designations of ASPRS and IEEE

The following comments pertain to Table 559:

- The official ITU designation for UHF extends from 300 to 3,000 MHz. In radar practice, however, the upper limit is usually taken as 1,000 MHz with the L- and S-band being used to describe the higher UHF region.
- The radar UHF band is sometimes called the P-band
- The designation “mm-band” is derived from “millimeter wave radar”, and is also used to refer to V- and W-bands, when general information relating to the region above 40 GHz is to be conveyed.

The ASPRS radar band designations can also be found in the following reference: “Principles & Applications of Imaging Radar,” 1998, edited by Floyd M. Henderson & Anthony J. Lewis, published by ASPRS in cooperation with John Wiley & Sons, Inc., p. 138, ISBN: 0-471-29406-3. The book is volume 2 of the 3rd editition (series): Manual of Remote Sensing.

Allocation of Frequency Bands

The ITU does not define a specific service such as “radar”, but defines the various services that use radar such as radiolocation, radionavigation, meteorological aids, earth exploration satellite and space research. Only small parts of the above bands are allocated to services that use radar since radar does not share frequencies well with other radiocommunication services.

Frequency bands are allocated to various types of services at international conferences held every two years, called World Radiocommunication Conferences, or WRCs (previously these conferences were called World Administrative Radiocommunication Conferences, or WARCs). There is always contention between the various service proponents to gain access to the maximum amount of spectrum, and therefore bands are often allocated to more than one service. When this happens, all of the services must share the frequencies under certain operating constraints, such as maximum power flux density limits. For the purpose of allocating frequencies to various radiocommunication services, the world has been divided into three regions.

When allocating frequencies the objective is to make the same allocation in all regions, i.e., worldwide. This lessens the potential for interference between regions and brings economies of scale to the manufacturing of radiocommunication equipment. However, it is not always possible to make a frequency allocation worldwide since the use of the spectrum has developed differently in the various regions. Therefore, allocations often vary from region to region. Within a region it is not always possible to get a consensus among administrations on what the allocations are to be for certain frequency bands. Even within an administration many interest groups are involved in setting the agenda for a WRC.

Since the ITU functions by consensus, a "consensus" is often archived by putting the exceptions into footnotes that apply to a frequency allocation. For example, at WARC 92 a world wide allocation was made for the broadcasting satellite service (BSS) for digital audio broadcasting but some countries chose alternative frequencies for this service or decided that this allocation should not be effective until the year 2007. Such conditions are reflected in footnotes to the appropriate frequency allocations.

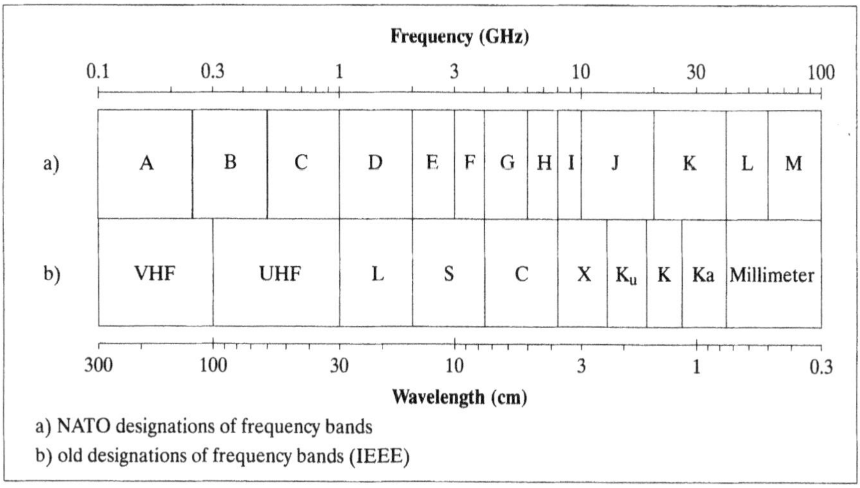

Figure 411: Illustration of frequency band allocations by different bodies

Radio Wave Propagation

Electromagnetic waves propagate in space at the speed of light (c), approximately 300,000 km/s. The wave amplitude varies periodically in time with frequency (f) measured in periods per second (Hz). The distance from crest to crest is the wavelength (λ), measured in meters. The relationship between these three parameters is $f = c/\lambda$.

Radio waves with frequencies of the order of 1 megahertz (MHz) can be reflected by the ionized atmosphere, or refracted when entering or coming out of these ionospheric layers, or even attenuated when crossing them. These effects, which influence the ray path, depend on the wavelength (or frequency), and the angle of incidence. Thus, radio messages can be sent to considerable distances, over the horizon and to the antipodes. Yet, some frequency limitations remain; the highest frequencies are not reflected at all. This last property makes space telecommunication possible. However, at the higher frequencies, radiation is affected by the atmospheric particles and molecules encountered, such as ice, rain drops, water vapor, and oxygen (see also Figure 412).

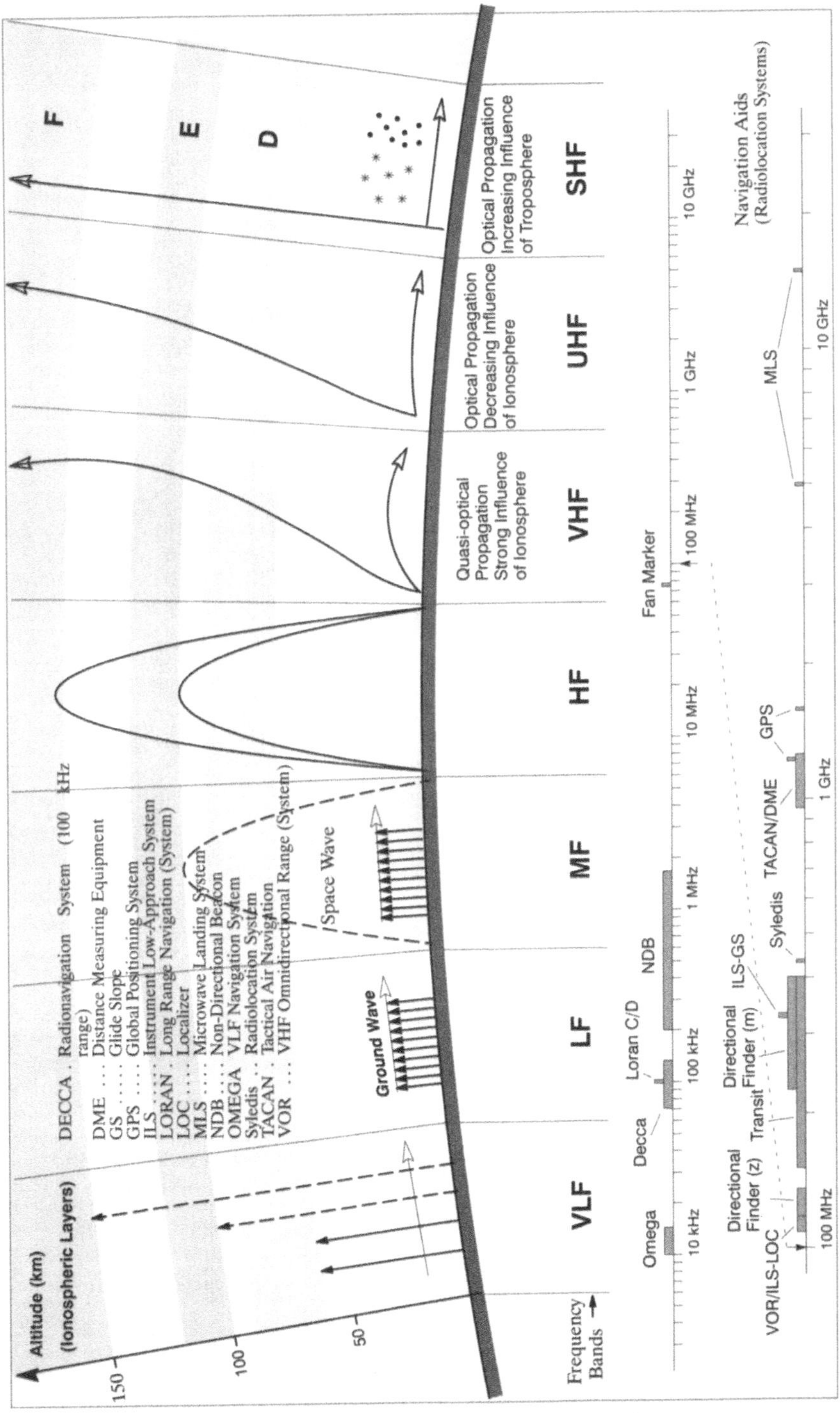

Figure 412: Radionavigation and radiolocation systems/radio wave propagation behavior in their bands

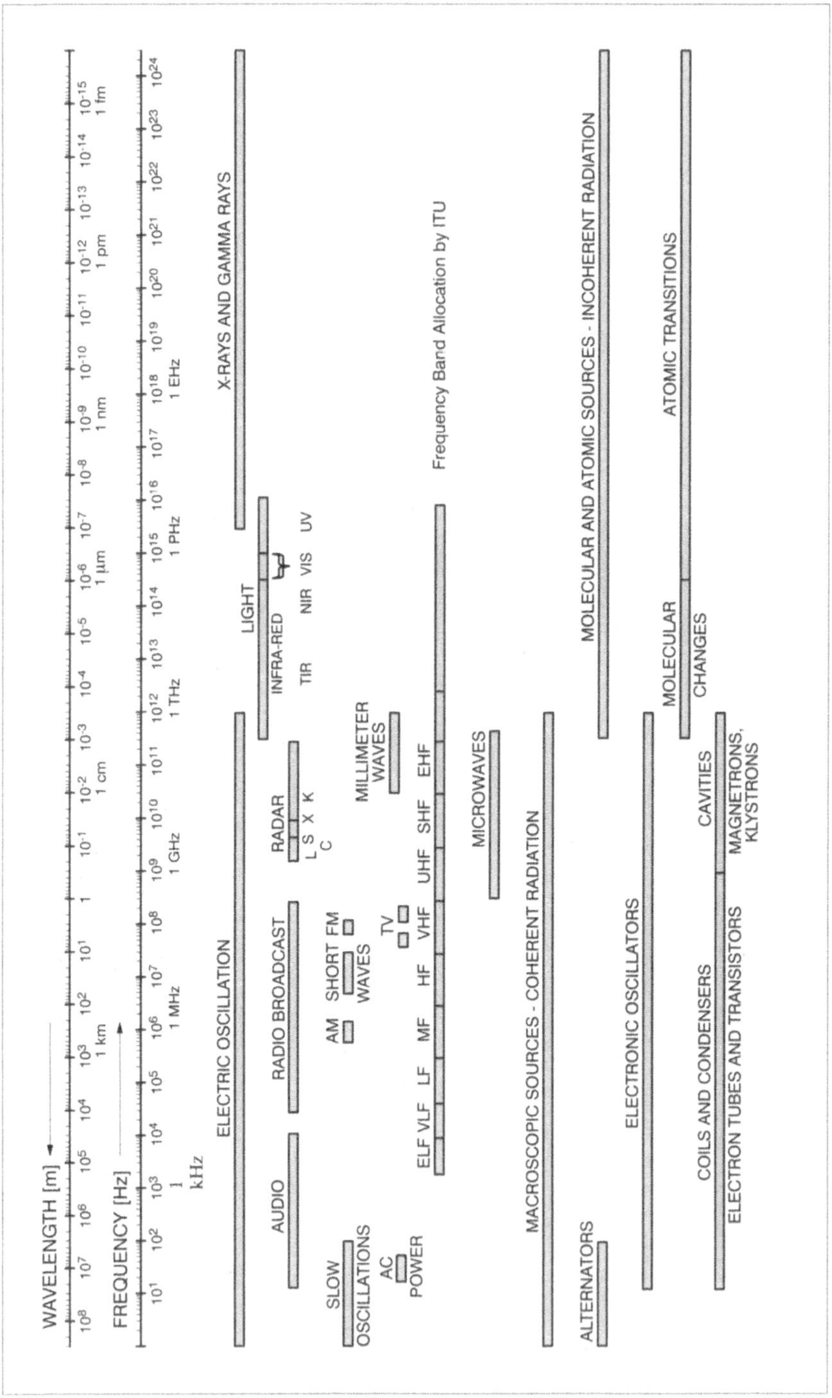

Figure 413: Electromagnetic spectrum with characteristic sources/frequency bands

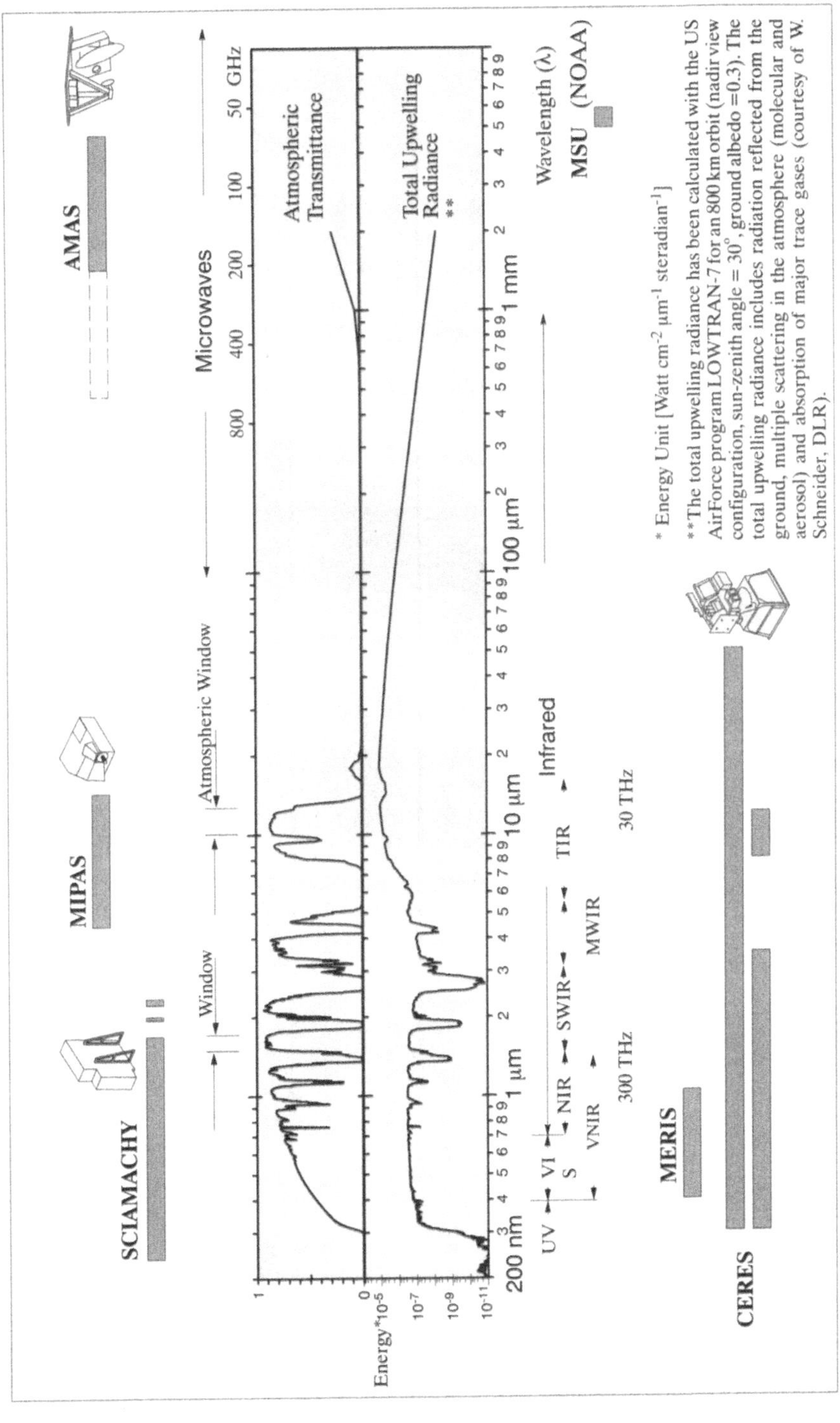

Figure 414: Atmospheric parameters and spectral ranges of some sensors

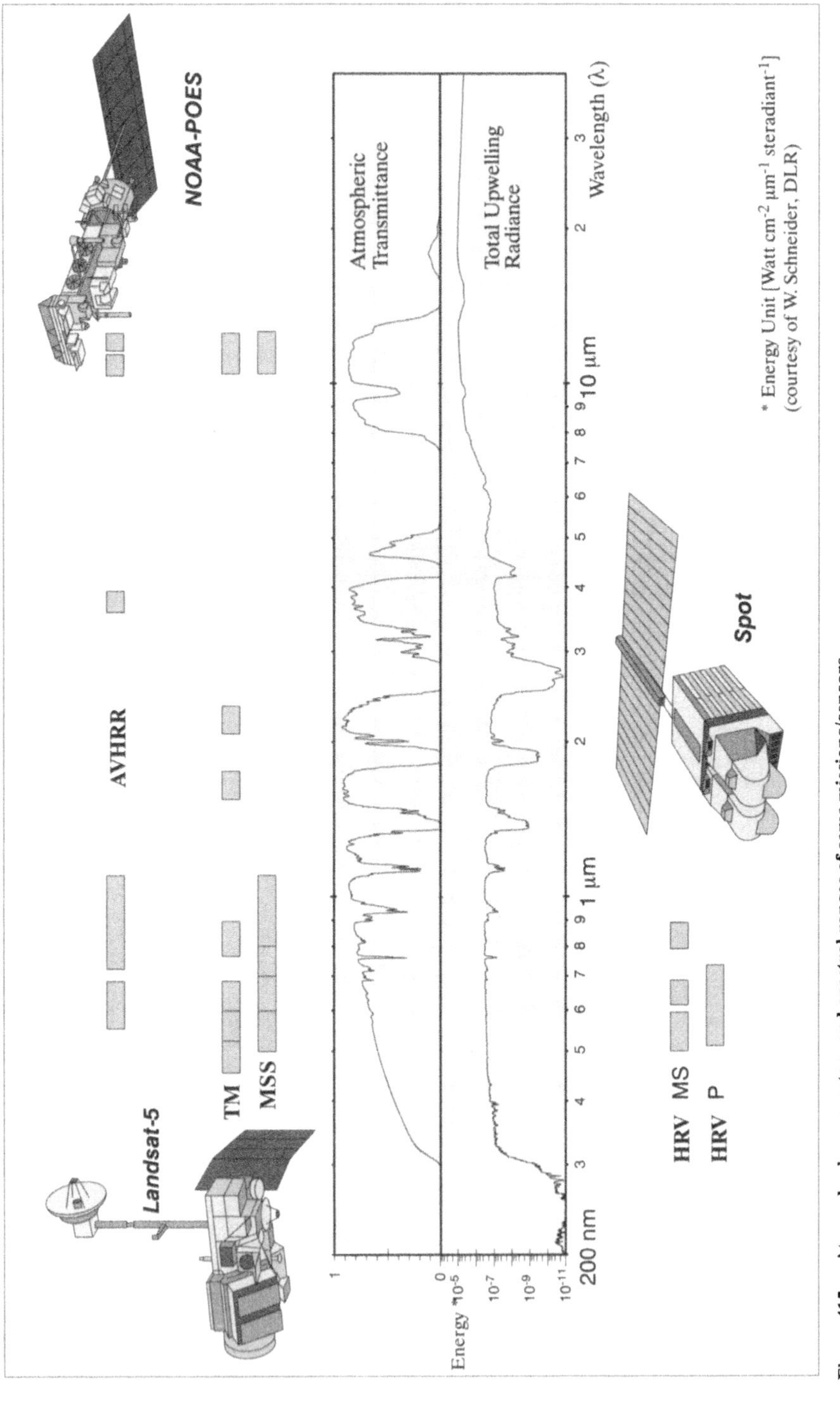

Figure 415: Atmospheric parameters and spectral ranges of some missions/sensors

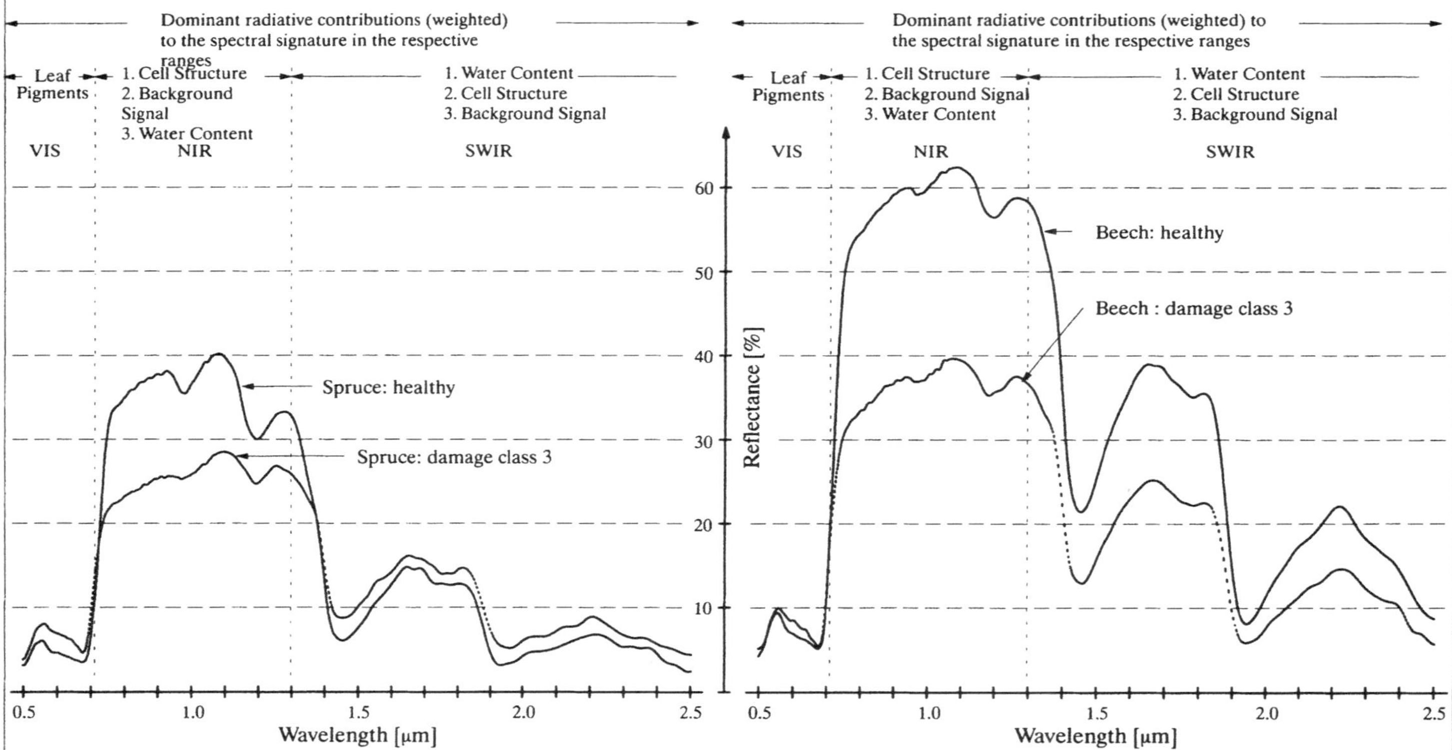

The curves represent measured spectral reflectances of healthy and heavily damaged beech and spruce branches in the laboratory. Such 'signatures' may be employed as references for image data interpretation. Lab sensor: IRIS Mark IV spectroradiometer. (courtesy of W. Kirchhof and H. Hoffmann, DLR)

Figure 416: Spectral signatures of vegetation in the electromagnetic spectrum

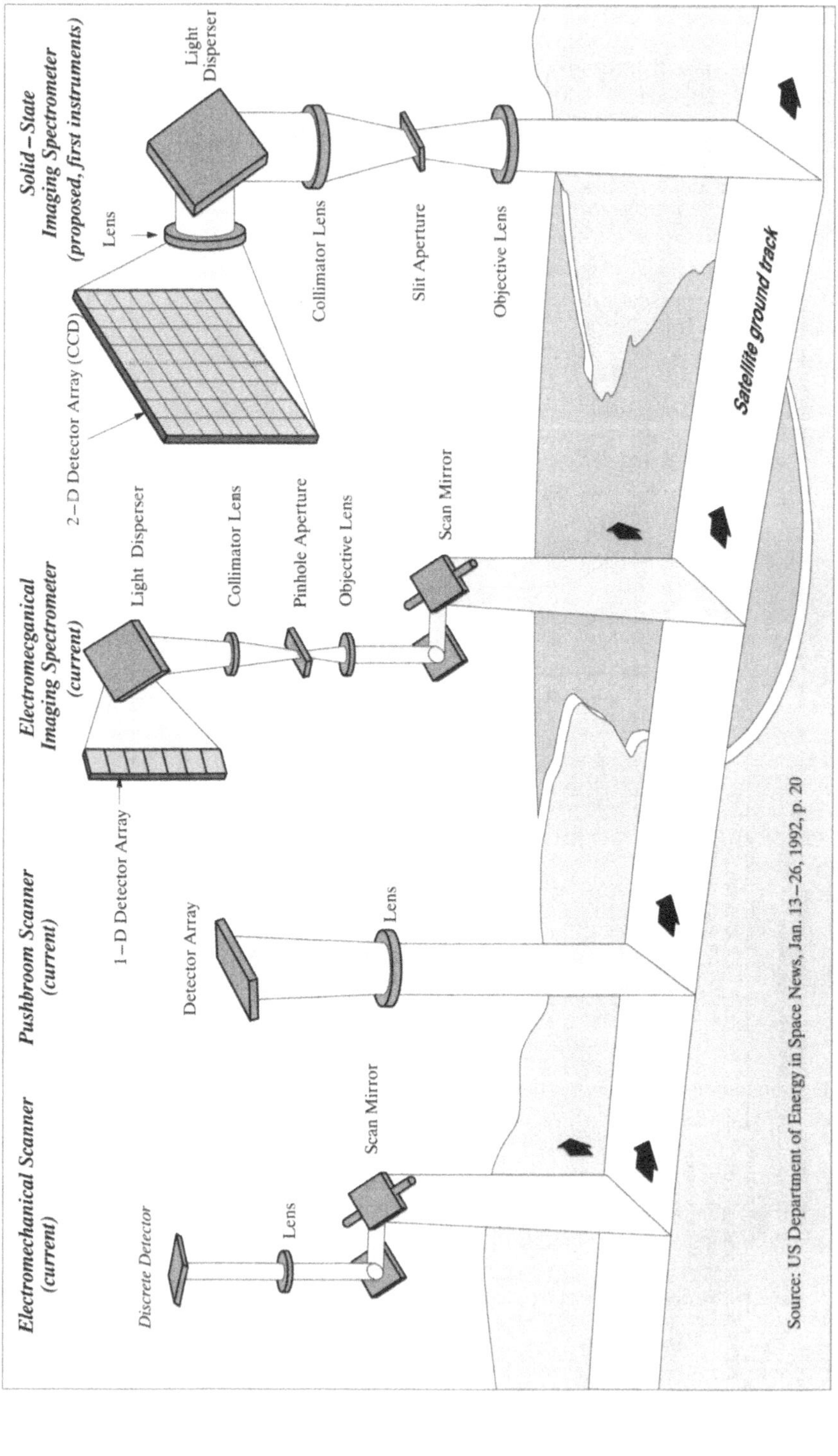

Source: US Department of Energy in Space News, Jan. 13–26, 1992, p. 20

Figure 417: Evolution of imaging scanner/spectrometer concepts

Appendix A Glossary

Only brief definitions are given to provide a quicklook reference for the reader with regard to terminology. Bold face type within a definition refers to a subject-related entry.[2031) 2032) 2033)]

Aberration. Geometrical errors in imagery whereby a perfect image is not formed. Typical aberrations include spherical aberration, astigmatism, coma, and chromatic aberrations. Lens bendings, locations, powers, materials, and increasing the number of lenses and aperture stop positions are all used to minimize aberrations.

Absorption band. A range of wavelengths, or frequencies, in the electromagnetic spectrum within which radiant energy is absorbed by a substance (gas, liquid or solid). In the gaseous phase, absorption lines are much narrower than in liquids or solids. In a polyatomic gas, an absorption band is actually composed of discrete absorption lines which appear to overlap. Each line is associated with a particular mode of vibration and rotation induced in a gas molecule by incident radiation. Examples:

- Ozone (O_3) has several absorption bands. They are: a) the Hartley bands (2000-3000 Å in UV with max absorption at 2550 Å); b) the Huggins bands (weak absorption between 3200-3600 Å); c) the Chappuis bands (weak and diffuse at 4500 Å and at 6500 Å in VIS); d) IR bands at 4.7, 9.6, and 14.1 μm.
- Molecular oxygen (O_2) also has several absorption bands. They are: a) Hopefield bands (between 670 and 1000 Å in UV); b) diffuse between 1019 and 1300 Å (UV); c) the Schumann-Runge continuum (between 1350-1760 Å); d) Schumann-Runge bands (between 1760 - 1926 Å); e) the Herzberg bands (between 2400-2600 Å); f) the atmospheric bands (between 5380-7710 Å, VIS); g) IR at about 1 μm ($=10^4$ Å).

Absorptivity. Ratio of the absorbed to the incident electromagnetic radiation on a surface.

Accuracy. Refers to an estimate of how well a certain parameter or measurement is known; it is a measure of the absolute truth of a measurement requiring absolute (traceable) standards. Accuracy is a quality that characterizes the ability of a measuring instrument to give indications equivalent to the 'true value' of the quantity measured. The quantitative expression of accuracy may also be given in terms of uncertainty. The actual or 'true value' of a quantity cannot be determined; it can only be said to exist within tolerance limits of a measured value. The measurement error is the algebraic difference between the measured (or indicated) value and the true value. Hence, accuracy is by its very nature only an estimation of the true value, taking into account all aspects of measurement.

Acid rain. Rain that is more acid than normal because the raindrops contain dissolved acid gases and/or dust particles (aerosols) from the atmosphere. The principal gases responsible for increased acidity are oxides of sulfur and nitrogen. Generally, rain with a pH below 4.5 is considered environmentally harmful.

Actinometer. A generic term for any instrument used to measure the intensity of radiant energy, in particular that of the sun.

Active sensor. A sensor having its own source of EMR (Electromagnetic Radiation); it transmits a series of signals to the target and detects the echo. A SAR instrument, a lidar, a radar altimeter, etc., are examples of active sensors.

Actuators. Refer to a class of on-board devices or techniques employed in particular for attitude control. Some examples are: reaction wheels, momentum wheel, magnetorquer coil/

2031) Portions of this glossary are taken from: "Glossary and list of Acronyms/Abbreviations," Earth Observation System (EOS), July 1992, Courtesy of EOS Project Science Office (V. V. Salomonson), GSFC, Greenbelt, MD.

2032) Jeanne Hopkins, "Glossary of Astronomy and Astrophysics," The University of Chicago Press, Second Edition, 1985

2033) R. J. Gurney, J. L. Foster, C. I. Parkinson (editors), "Atlas of satellite observations related to global change," Cambridge University Press, 1993 rod, permanent magnets, gravity-gradient boom, nutation damper, control moment gyros, cold gas thrusters, solid thrusters, ion thrusters, mono- or bi-propellant engine, etc.

Adaptive optics. A technique which tries to compensate for the atmospheric degradation of the incoming signal of optical imaging systems. However, the compensations achieved, regardless of method, are never "perfect." The following approaches are in use:

- An on-line adaptive optics system (hardware solution). In this version, the adaptive optics system is capable of compensating for the distortion of electromagnetic radiation as it passes through the turbulent atmosphere and the optical system. Elements of an adaptive optical system are a high-speed wavefront sensor (sensing the turbulence-induced aberrations), a flexible mirror system whose surface can be electronically controlled to correct for aberrations, and a computer controller that converts the wavefront measurements into deformable mirror commands. - More economic solutions are suggested by the use of LCPM (Liquid Crystal Phase Modulator) to deform the mirror, and by the use of the pupil masking technique. [2034) 2035)]
- A post facto approach (software solution), where the data are acquired by an imaging instrument and processed off-line with suitable iterative algorithms to increase the overall resolution obtained by the optics system. Speckle imaging is an example of such a technique. [2036)]
- A hybrid approach using an adaptive optics system in combination with post facto processing.

Advection. Refers to a change in property of a moving air parcel from one region to another (say, from a warm region to a cool region, thereby changing its temperature). Commonly, advection is divided into horizontal and vertical components; it differs from convection only in scale. Convection is transport by random thermally induced currents, whereas advection is transport by steady vertical currents.

Aerosol. Aerosols are a suspension of fine (solid or liquid) particles such as dust, smoke particles, water droplets, etc. in the atmosphere. The smallest aerosols are the atoms of the various atmospheric gases. The range of sizes varies from a few nanometers (molecules) to tens of micrometers (wind-driven sand). Some aerosols (sea salt and haze) occur naturally and some (smoke) are man-made (anthropogenic). The ocean is a significant source of natural tropospheric aerosols. Once in the atmosphere, aerosols may be transported away from their place of origin, sometimes over great distances. - Aerosols can directly and indirectly affect the radiation budget of the atmosphere. Their direct radiative effect is due to their scattering, absorption, and emission properties. Their indirect effect is a result of their ability to act as condensation nuclei in the formation of clouds. Both tropospheric and stratospheric aerosols play an important role in global climate change.

Airglow. A nighttime glow from the upper atmosphere, occurring over middle and low altitudes, due to the emission of light from various atoms, molecules, and ions.

Albedo. The fraction of the total solar radiation incident on a body (or a natural surface such as the ground, ice, snow, water, clouds, etc.) that is reflected by it (commonly expressed as a percentage, see also **planetary albedo**). Measured albedo information may be applied to cloud analysis, it may also serve in calculations of inherent contrast between targets and background. By definition, white surfaces (i.e. an all reflective Lambert surface) have albedos close to 1, black surfaces have albedos close to zero.

Aliasing. A term in data processing referring to two or more distinctly different signals having identical sample values.

2034) S. R. Restaino, D. M. Payne, "Adaptive Optics on a shoe string," SPIE Vol. 3494, 1998, pp. 152-160

2035) D. Dayton, S. Browne, J. Gonglewski, "Control Loop Analysis for a Nematic Liquid Crystal Spatial Light Modulator Used in an Adaptive Optics System," SPIE Vol. 3494, 1998, pp. 161-160-167

2036) J. C. Christou, et al., "Physically Constrained Iterative Deconvolution of Adaptive Optics Images," SPIE Vol. 3494, 1998, pp. 161-175-190, Proceedings of the SPIE EUROPTO Series, Sept. 23-24, 1998, Barcelona, Spain

Altimetry. Altimetry can be described as run-of-time distance measurements taken from a vertically tracking satellite to the geoid. A spaceborne radar altimeter maps the sea surface, which is complex, but which largely conforms to the equipotential surface known as the geoid. The ocean surface departs significantly (at the meter level) from the geoid - this departure is of great interest to the science community (oceanographers, meteorologists, geophysicists, etc.) for it reflects the global circulation patterns and also, in the form of tides and changes in sea level, fundamental climatic and solid Earth phenomena. Furthermore, altimetry provides approximate estimates of the geoid itself, in the form of a mean surface obtained by averaging measurements taken over many satellite passes over the same locations. More detailed explanations on altimetry are given below.

- A radar altimeter on-board a satellite permanently transmits signals at high frequency to the Earth's surface, and receives the echo from the sea surface. This is analyzed to derive a precise measurement of the round-trip time between the satellite and the sea surface. By averaging the estimates, say, over a second, this produces a very accurate measurement of the satellite-to-ocean range. However, as electromagnetic waves travel through the atmosphere, they can be decelerated (by water vapor) or accelerated (by ionization). Once these phenomena are corrected for, the final range R is estimated within 2 cm. [2037)]
- Satellite orbit: The ultimate aim is to measure sea level relative to a terrestrial reference frame. This requires independent measurements (accurate tracking) of the satellite orbital trajectory, i.e. exact latitude, longitude and altitude coordinates. The critical orbital parameters for satellite altimetry are altitude, inclination and period.
- Sea surface height (SSH): SSH is the range at a given instant from the sea surface to a reference ellipsoid. SSH is simply the difference between the satellite height (S) and the altimetric range (R): SSH = S-R. The SSH value takes account of such effects as:
  - The sea surface height which would exist without any disturbances (wind, currents, tides, etc.). This surface, called the **geoid**, is due to gravity variations around the world, which are in turn due to major mass and density differences on the sea-floor. For example, a denser rock zone on the sea-floor may deform the sea level in the order of tens of meters; this is visible as a hill on the geoid.
  - The ocean circulation, or dynamic topography. The ocean circulation, which comprises a permanent stationary component (permanent circulation linked to Earth's rotation, permanent winds, etc.) and a highly variable component (due to wind, tides, seasonal variations, etc.). The mean effect is on the order of one meter.

Amplitude modulation (AM). The baseband signal is caused to vary the amplitude of the carrier wave to create the wanted information content.

Analog data. Data represented in continuous form, as contrasted with digital data having discrete values.

Ångström (Å, after A. J. Ångström, a Swedish physicist). A unit of length used in the measurement of short wavelengths (X-rays, gamma rays, etc.) and in the measurement of molecular and atomic diameters. 1 Å = 10^{-10} m or 10^{-4} μm.

Antenna. A sender and receiver system of electromagnetic radiation (in remote sensing terminology the antenna may be part of the sensing instrument, or may be regarded as a coupling device between the target and the sensing instrument). The term antenna refers a) to that part of a transmitting system that converts electrical energy to electromagnetic waves; and conversely b) to that part of a receiving system that converts electromagnetic waves to electrical energy (current) in the receiver (a duplexer automatically switches the antenna from a transmitting function into a receiving function). Physically, an antenna consists of metal surfaces that provide conducting paths for oscillating electric currents and charges. The radiated power of an antenna depends on the shape and size of its geometric contour and on the amplitude and frequency of its oscillation. Some antenna designs:

2037) http://www–projet.cnes.fr:8170/HTML/information/frames/general/sitemap_uk.html

- **Dielectric rod antenna.** An antenna consisting of a dielectric cylinder that is partially inside a circular waveguide (pipe). It is possible to have an electric field applied to this device - no currents flow through it - although energy passes from one end to the other, thereby generating electromagnetic waves.
- **Dipole antenna.** A thin metal cylinder or wire excited by an alternating current generator at its center so that the ends are oppositely charged. - A dipole antenna is a form of open circuit in which the current oscillates between the ends of the conductor. A dipole antenna may also be a type of array consisting of a system of dipoles. A dipole antenna differs from a dish antenna in that it consists of many separate antennas that collect energy by feeding all their weak individual signals into one common receiving set.
- **Helix antenna.** A helical wire wound with a circumference of about one wavelength and a pitch of 1/4 wavelength over a ground plane with a 1 wavelength minimum diameter.
- **Lens antenna.** Several types are in use: a) dielectric lens - the aperture of the antenna is equal to the projection of the rim shape; b) artificial dielectrics; c) strip antenna - metal strips are used as waveguides to increase the phase velocity by acting as parallel-plate waveguides.
- **Loop antenna.** The current circulates around or oscillates within the closed loop. The most important application of the loop antenna is reception. The shielded loop antenna is useful as a probe for measuring the magnetic field.
- **Microstrip patch antenna.** A printed circuit antenna consisting of a radiating patch supported by a dielectric layer over a ground plane.
- **Monopole antenna.** A thin metal cylinder or wire erected vertically over a conducting plane and excited by an alternating current generator connected between the base of the cylinder and the conducting plane.
- **Pencil beam antenna.** An antenna whose radiation pattern consists of a single main lobe with narrow principal plane beamwidths and sidelobes having relatively low levels.
- **Phased-array antenna.** Phased arrays are inherently random-access devices consisting of multiple antenna elements (fixed dipoles) which are fed coherently and use variable phase- or time-delay control at each element to scan a beam to given angles. Arrays are sometimes used in place of fixed aperture antennas (e.g. reflectors or lenses) because the array arrangement allows more precise control of the radiation pattern (lower sidelobes). The primary reason for using arrays is to produce a directive beam that can be scanned (repositioned) electronically in two dimensions without any mechanical movement (see also **Phased-array Technology**).
- **Stick antenna.** Also referred to as a fan-beam antenna, it produces a major lobe whose transverse cross section has a large ratio of major to minor dimensions.
- **Slot antenna.** A slot in a metal sheet with dimensions $\lambda/2$ in length and width w ($w \ll \lambda$) provides a means for achieving efficient directional energy radiation and reception. The radiation leaving the slot antenna is polarized in the direction normal to the major slot dimension (if the slot is horizontal, then the polarization is vertical, and vice versa). Array arrangements of slots permit the radiation of higher energies and consequently the illumination of larger target areas.

Antenna reflectors. A reflector antenna is a large-aperture (gain) directional antenna. A parabolic reflector (mirror) has the property of transforming rays emerging radially from a point source at its focus into a bundle of parallel rays. The reflected parallel rays are all in phase in any plane perpendicular to the axis of the parabola. The laws of optics apply with respect to the radiation geometries and projections (focusing, bundling, redirection of energy, diffraction effects, etc.). Characteristic parameters of a reflector antenna are a function of the properties of the feed and the ratio of the reflector's focal length (F) to the diameter (d) of its circular aperture.

Reflector antennas are widely used in the microwave range of the electromagnetic spectrum in the fields of remote sensing, telecommunication, and radio astronomy. Reflectors are inherently broadband instruments; the bandwidth and polarization are determined by the feed antenna. Reflector examples: paraboloidal, parabolic cylinder, dual (Cassegrain, Gregorian), offset-fed, corner, dichroic.

- **Cassegrain dual-reflector antenna (telescope)**. N. Cassegrain, a French scientist, proposed the design in 1672. The basic properties of the Cassegrain dual-reflector are determined from the principles of ray optics. A small convex hyperboloidal subreflector is placed between the point source feed and the prime focus of the parabolic dish. Rays from the feed are transformed by the subreflector into rays that appear to be emerging from the paraboloid focus. The rays reflect from the parabolic reflector parallel to its axis.
- **Offset aperture reflectors**. The blocking of portions of the aperture of a reflector antenna by its feed, supporting structures, or by the subreflector generally degrades the radiation distribution. Hence, high-performance systems employ offset-fed antennas to eliminate the effects of aperture blocking.
- **Horn antenna**. A horn is an aperture antenna fed from a waveguide mode in an expanded waveguide. The bandwidth is determined by the feed waveguide.

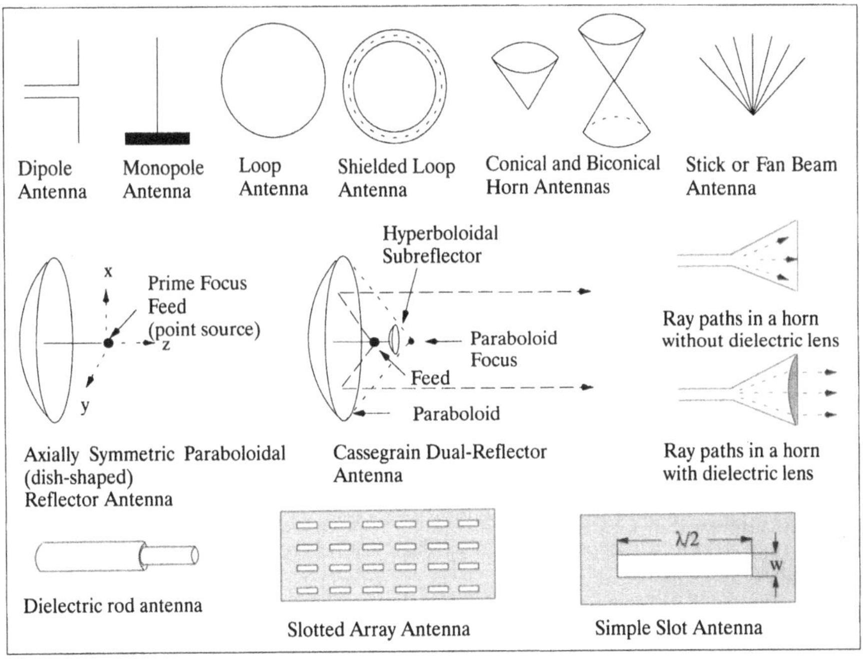

Figure 418: Some antenna geometries

- **Dichroic reflector**. The term dichroic (chros = color in Greek) implies selective absorption in crystals of electromagnetic radiation vibrating in different planes. - Refers to frequency-selective surfaces designed to exhibit different ratio-frequency properties. In its simplest form, a dichroic surface is virtually perfectly reflecting at one frequency and virtually transparent at another. The most common application of dichroic reflectors has been as subreflectors in Cassegrain antenna systems. Besides reflectors the dichroic principle is also applied to beamsplitters and filters.

- **Antenna retroreflector**. Retroreflection is defined as radiation that is returned in angular directions which are very close to those angular directions from which it came. This (incoming/outgoing) property is maintained over wide variations of the incident radiation. Retroreflector devices come in a variety of forms and have many uses. In remote sensing, a retroreflector uses total internal reflection from three mutually perpendicular surfaces. This kind of retroreflector is usually called a 'corner cube retroreflector'. A corner cube reflector is normally used for radar measurements having a known radar cross section.

Antenna aperture. Surface area (size) of a reflector or horn that is illuminated by the outgoing and/or incoming radiation.

Antenna bandwidth. Range of frequencies within which the performance of an antenna, with respect to some characteristic, conforms to a specified standard. There are many different types of bandwidths, including gain, VSWR (Voltage Standing Wave Ratio), and polarization. For example, a typical bandwidth specification might be: The antenna gain with isotropic must be $>$ 10 dB $\pm$10 MHz from a center frequency of 2090.0 MHz.

Antenna beamwidth. A planar cut through the radiation pattern containing the direction of the maximum of a lobe, the angle between the two directions in which the radiation intensity is one-half the maximum value and one tenth the maximum value, or in which zero is defined to be the half-power (3 dB), tenth-power (10 dB), or null beamwidth.

Antenna depression angle. The angle between the local horizontal and the center line of the antenna beam pointing at the target.

Antenna directivity. Directivity is a measure of the concentration of radiation in the direction of the maximum. It is the ratio of the radiation intensity in a given direction from the antenna to the radiation intensity averaged over all directions.

Antenna feed. The device in an antenna system that transmits or receives energy to or from the antenna aperture and the radio system.

Antenna footprint. Instantaneous projection of a directional antenna beam illumination on a surface.

Antenna gain. Ratio of the transmitted radiation intensity in a given direction to the radiation intensity that would be received if the power accepted by the antenna were radiated isotropically. In this context, 'peak antenna gain' refers to the maximum radiated intensity expressed as a ratio to the radiation power intensity of a hypothetical isotropic antenna fed with the same transmitting power.

Antenna - intermediate frequency (IF). In microwave systems, a frequency that is common to all channels at which amplification takes place, interconnections are made, and/or automatic gain is adjusted.

Antenna - isotropic. A theoretical antenna of infinitesimal size in which it is assumed that all of the energy is radiated (point source). This concept serves as a reference basis for other antennas of finite dimensions.

Antenna noise temperature. Refers to the increase of the receiver input noise temperature of an antenna system.

Antenna polarization. Spatial orientation of the electric/magnetic field radiated by an antenna. The vector electric/magnetic fields of free space traveling waves are perpendicular to the direction of travel. Polarization describes how fields behave in time and space. For example, a "circularly polarized" wave can be thought of as having the electric field rotating about the direction of travel. The direction of the filed rotates one turn per period of the wave. A "linearly polarized" wave radiated from an antenna into a specific direction has the electric field direction fixed with time - the only variations are in the instantaneous magnitudes of the electric/magnetic fields.

Antenna scanning techniques.[2038] For SAR observations the antenna beam casts an elliptical footprint on the ground with an effective rectangular aperture antenna of typical size 10 m (along-track) and 3 m (across track).

- **Electronic scanning.** Defines a method of positioning an electromagnetic beam in space or scanning across a target surface by electronic means. The antenna aperture remains fixed; no mechanical mechanism is involved in the scanning process.
  - Phase scanning (moved the beam by controlling the phase of the antenna illumination, using phase shifters or delay lines - see also **Phased-array technology**)
  - Frequency scanning (moves the beam by changing the carrier frequency of the transmitter and receiver)
  - Electronic feed switching
- **Mechanical scanning.** Defines a method of positioning an electromagnetic beam in space by mechanical rotation or angular positioning of the radiating aperture of the antenna system.
- **Electronic/mechanical scanning.** A hybrid method that employs electronic scanning in one dimension, say in elevation, and mechanical scanning in azimuth.

Antenna sidelobes. Undesired directions in which a directive antenna also receives or radiates power. Sidelobes are generally much weaker than the mean beam in the desired direction.

Antenna waveguide. Usually a hollow metal structure (pipe or other profile) intended to guide or to conduct along its path an electromagnetic wave in a given microwave range (a waveguide is usually attached to a horn; it may also directly serve as a feed for a reflector). The internal dimensions of a waveguide are related to the transmission efficiency of specific frequencies. The cutoff frequency refers to that frequency below which a particular waveguide cannot satisfactorily transmit the wave.

Antenna waveguide modes. Refer to the wave propagation distribution patterns that may exist within a waveguide. They depend on the shape and size of the waveguide with respect to the length of the wave traversing the guide. Each mode has a specific topology, velocity, and energy distribution along and across the guide cross section. The cross section of the waveguide may be square, rectangular, circular, or elliptical.

Anthropogenic gases.'Human-induced' gases emitted into the atmosphere and interacting with the environment. In a wider sense the term refers to all gases emitted as a result of human activities (e.g., chlorofluorocarbons from technical combustion processes, carbon dioxide and methane from livestock farms, rice paddies, biomass burning, etc.).

Aperture. Refers to the maximum diameter of a radiation beam that can pass through a system (either an optical lens or mirror system or an antenna system) on a telescope or satellite. The radiation-gathering power of such a system is proportional to the square of the diameter (or aperture) of the lens (mirror or antenna). Hence, sensors with wider apertures are able to capture more information. Apertures are used to restrict the field of view (FOV) of the responsive element (such as a detector). This is often done to reduce noise (cooled detectors are photon-noise limited) from extraneous sources.

2038) P. J. Kahrilas, "Electronic Scanning Radar Systems, Design and Architecture," in 'Practical Phased-Array Antenna Systems,' E. Brookner, Editor, Artech House, Boston, MA

Aperture stop. Location within a lens system where the principle ray passes through and crosses the optical axis. The presence of a mechanical limiting aperture (hole, slot, etc.) typically creates a limiting size.

Aperture synthesis. A technique (pioneered in radio astronomy) of generating high spatial resolution images by dividing the collection area of a telescope (or antenna) into smaller apertures spread out in a pattern covering several baselines. In microwave radiometry the concept employs an interferometric technique in which the product from antenna pairs is sampled as a function of pair spacing. Substantial reductions in the antenna aperture needed for a given spatial resolution can be achieved with this technique. However, the performance leap in resolution must be paid for with higher requirements for instrument precision sensing and stabilization. ESTAR (P.87) and MIRAS (P.132) are examples of airborne synthetic aperture microwave radiometers (both instruments operate in L-band).

Apodizing/unapodizing.[2039] The terms are used in the context of data processing in a FTS (Fourier Transform Spectrometer). The actual lineshape of a FTS interferogram is close to (sin x)/x, which is a function with intense side-lobes (also referred to as "feet"). This is the shape that the spectrum of an intrinsically sharp line (e.g. a laser line) would have if the spectrum were untreated, or "unapodized"; to apodize means literally (Greek) "to cut off the feet". Apodizing consists of treating the spectrum to reduce the sidelobes at the expense of degraded resolution (usually by a factor of about two). Apodizing can be done either by tapering the interferogram prior to transforming, or by algebraically filtering the spectrum after transforming.[2040]

Area Array Camera. Refers to a solid-state imaging device (CCD technology) with an array (rows and columns) of pixels producing a 2-D image. The Area Array Camera is also referred to as Matrix Array Camera.

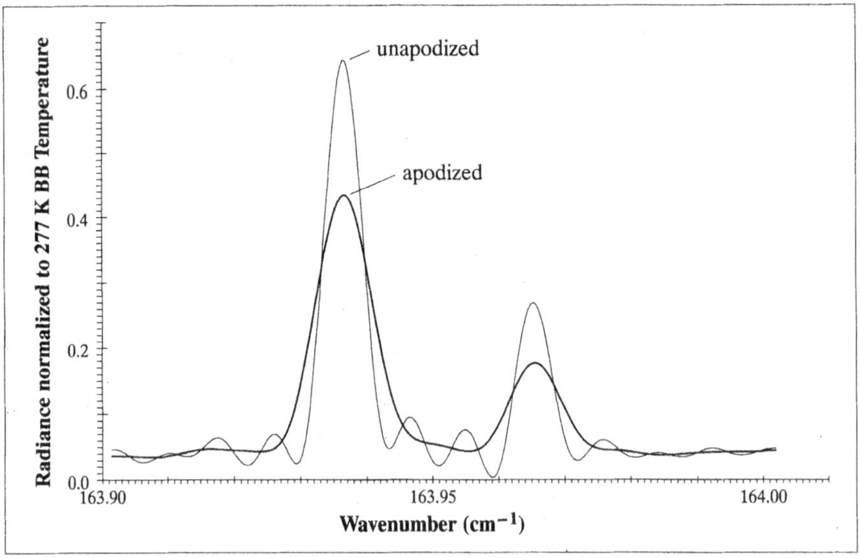

Figure 419: Sample illustration of an apodized and an unapodized radiation curve

2039) Typical sources observed by a spaceborne or airborne FTS are extended such that (due to FOV) the rays through the interferometer are not collimated, leading to side lobes.

2040) Courtesy of K. C. Chance of the Harvard-Smithsonian Astrophysical Observatory, Cambridge MA

Astigmatism. Refers to an aberration in which the light in one plane (for instance the plane of the paper) focuses at a different location from light in the orthogonal plane.

Atmosphere. The envelope of gases surrounding the Earth and bound to it by the Earth's gravitational attraction. Studies of the chemical and radiative properties, dynamic motions, and physical processes of the Earth-atmosphere system constitute the field of meteorology.

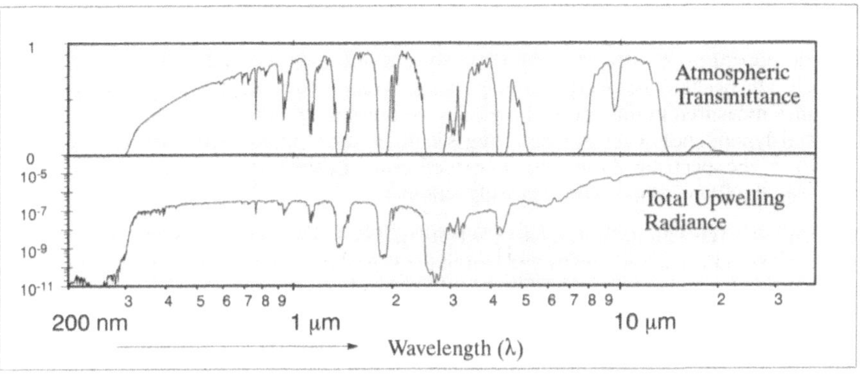

Figure 420: Atmospheric transmittance and radiance for UV to TIR regions

Atmospheric absorption. A process whereby some or all of the energy of electromagnetic radiation is transferred to the constituents of the atmosphere. Absorption by atmospheric gases is dominated by that of water vapor (H_2O), carbon dioxide (CO_2), and ozone (O_3) with smaller contributions from methane (CH_4), carbon monoxide (CO), and other trace gases. Water vapor is rather variable across the Earth's surface (location) as well as throughout the atmosphere (altitude). CO_2 and CH_4 are essentially uniformly mixed in the atmosphere, hence predictable in their effects (see also **Absorption bands**). - The sum effect of absorption results in atmospheric opaqueness in many spectral regions as illustrated in Figure 420. Hence, observational techniques must by tailored to utilize those atmospheric windows through which the surface can be viewed.[2041)]

Atmospheric attenuation. A process whereby some or all of the energy of electromagnetic radiation is absorbed and/or scattered when passing through the atmosphere. The amount of radiant energy that the atmosphere either removes or adds to that emitted or reflected from the Earth's surface depends on:

- The constituents of the atmosphere
- The path length of radiation (a function of geometry of the source, surface, and sensor)
- The reflectance of the surface surrounding scene or target area

Atmospheric boundary layer (also referred to as **Planetary boundary layer, PBL**). This boundary layer includes the bottom part of the atmosphere within which the energy exchange processes (between the Earth's surface and the atmosphere) occur mainly through vertical transport mechanisms of momentum and heat (production of wind shear turbulence plus convective heat turbulence). The whole PBL is heated by convection. Temperature gradients are strongest near the surface, because there convective 'eddies' are relatively small and inefficient in carrying heat upward. The thickness of the daytime PBL is typically in the order of a kilometer; however, it may be three times as high with strong heating. The PBL may also be limited by a top inversion. On a clear day the whole PBL is turbulent; it increases rapidly in the morning and only very slowly after the time of maximum heating, until it decreases to a minimum during the night (sensitivity to diurnal cycle).

2041) A. F. Goetz, J. B. Wellman, W. L. Barnes, "Optical Remote Sensing of the Earth," Proceedings of the IEEE, Vol. 73, No. 6, June 1985, pp. 950-969

Atmospheric correction. A problem with spaceborne surface observations is that a large portion of the received signal (about 80%) originates in the atmosphere. Much of this atmospheric signal is due to Rayleigh (or molecular) scattering, primarily from stratospheric ozone. Corrections attributed to Rayleigh scattering are normally estimated taking into account the geometry of a particular scene as well as the extraterrestrial solar radiation, ozone concentration, and atmospheric pressure. Aerosol scattering, primarily encountered in the marine boundary layer, represents another variable in the signal estimation. Since it is not possible to make direct measurements of these aerosols and their contribution to atmospheric optical properties, the remote-sensing community has relied on an indirect approach. The ocean is assumed as largely "black" in the VNIR portion of the spectrum; any radiance measured in this spectral region is assumed to originate in the atmosphere. The spectral dependence of aerosol scattering is in this manner propagated into the UV and VIS portion of the spectrum. Atmospheric correction algorithms try to account for of all the various signal influences in their processing schemes.

Atmospheric refraction. As a signal (electromagnetic radiation) traverses the atmosphere, it experiences propagation delay and bending due to the variable characteristics of the medium through which it is passing. The Earth's ionosphere introduces significant systematic perturbations on all microwave tracking data. At lower frequencies (150 MHz) daytime ionospheric biases can easily reach several kilometers in range, several meters per second in range rate, and up to two or three milliradians in position. Since most of the effects decrease as the inverse square of frequency, modern radiometric systems track at dual and well separated higher frequencies. As a signal traverses the troposphere it experiences a varying refractive index resulting primarily from spatial variations in atmospheric pressure, temperature and humidity. For radiometric technologies, variations in water vapor content is of chief concern. Optical signals have a much weaker dependence on water vapor. The varying refractive index influences the propagating signal in several ways. For optical signals, the most important effect is the varying group velocity where the pulse speeds up as it travels from the ground station to low pressure regions at higher altitudes. This change is a consequence of Snell's law of refraction, which predicts the bending and speed of a light ray as it moves through atmospheric layers with differing refractive indices.[2042)]

Atmospheric sciences. Study of the dynamics and structure of the Earth's atmosphere. There are three main areas:

- **Meteorology**. Primary concern is short-term weather variations in the lower regions of the atmosphere, in particular the troposphere.
- **Climatology**. Primary concern is long-term weather conditions on a global scale.
- **Aeronomy**. Involves research of the atmospheric regions above the lower stratosphere, dealing with such phenomena as ionospheric physics, photochemical processes of the upper atmosphere, aurorae, magnetospheric storms, etc.

Atmospheric window. Spectral bands for which atmospheric attenuation is relatively low (i.e., the bands for which the atmosphere presents minimal interference).

Aurora. Light radiated by ions in the Earth's upper atmosphere, mainly near the geomagnetic poles, stimulated by bombardment of energetically charged particles of the solar wind. Aurorae appear about two days after a solar flare and reach their peak about two years after a sunspot maximum. The northern aurora is also referred to as the 'aurora borealis' while the southern aurora is also called 'aurora australis.' [2043)]
For many centuries the aurora was referred to as the "northern lights" because it is a polar phenomenon and lies to the north when viewed from Europe. An aurora is a visible man-

2042) Ivan I. Mueller, S. Zerbini, "The Interdisciplinary Role of Space Geodesy," Lecture Notes in Earth Sciences, Springer Verlag, 1989, p. 187

2043) L. J. Paxton, C.-I. Meng, "Auroral Imaging and Space-Based Optical Remote Sensing," Johns Hopkins APL Technical Digest, Vol. 20, No 4, 1999, pp. 556-569 ifestation of "space weather" – the highly variable interaction between the sun and the Earth's magnetosphere, upper atmosphere, and ionosphere. The aurora is visible because of interaction of electrons and protons that are accelerated along the Earth's magnetic field lines from the magnetosphere (the cavity in the solar wind created by the Earth's magnetic field) into the Earth's atmosphere, where they undergo collisions with the background gas.

Auroral oval. Refers to the approximately circular band in the northern or southern hemisphere where aurora are most intense. The near-midnight portion of the oval, where some of the brightest emissions occur, is located about $\pm 65^{\circ}$ latitude. The mean diameter of the oval is about 4000 km.

Azimuth plane (direction). Observation by an instrument in the along-track direction, i. e. in the direction of the subsatellite track. In general the azimuth is the angle of horizontal deviation, measured clockwise, of a bearing from a standard direction.

Backscatter. Scattering of radiation (or particles) through angles greater than 90° with respect to the original direction of motion.

Band. A specification of a spectral range (say, from 0.4 - 0.5 μm) that is used for radiative measurements. The term 'channel' is also in common use with the same meaning as 'band'. In the ITU convention (see Table 558) for the electromagnetic spectrum, the term 'band' refers to a specific frequency range, designated as L-band, S-band, X-band, etc.

Band-to-band registration (also referred to as co-registration). Refers to multispectral image resolution, i.e. how well the same scene is recorded in different spectral bands. Co-registration of spectral bands is measured by the displacement of corresponding pixels in two different bands from their ideal relative location. Two pixels are "corresponding" if their footprints should ideally coincide or if the footprint of one should ideally lie within a specific region of the footprint of the other.

Bandpass. Defined as the frequency band(s) over which a microwave radiometer detects radiation. The equivalent for IR radiometers is the filter response.

Bandpass filter. A filtering device that allows transmission of only a narrow band of frequencies (the other frequencies are blocked out). The spectral width of this filter is characterized by its bandwidth.

Bandwidth. Range of frequencies over which an instrument (or communication link) can be used. It is usually specified in terms of 3 dB points that is, frequencies at which the response has fallen by 3 dB or 30% from the mid-frequency response. Bandwidth may also refer to the width of a spectral feature as measured by a spectroscopic instrument.

Baroclinic waves (disturbances). Any migratory cyclone more or less associated with strong baroclinity of the atmosphere, as evidenced on synoptic charts by temperature gradients in the constant-pressure surfaces, vertical wind shear, and concentration of solenoids in the frontal surface near the ground.

Baseband. Band of frequencies, usually the lowest frequencies in a microwave communications system, where basic information is assembled. This spectrum is generally that provided to the microwave system to be delivered to a distant point in the same format and information content.

Bathymetry. Measurement of water body depths, in particular ocean floor surveys. Generally, bathymetry surveys cannot be directly performed from a satellite. However, there are some areas of satellite applications: - bathymetry surveys in coastal regions (with a SAR instrument) or of shallow bodies of water; the other method is the interpretation and correlation of radar altimeter data (the technique relies on the assumption that the relationship between the gravity field and bathymetry is uniform over relatively small areas ($\approx$200 x 200 km).

Beamsplitter. A mirror, sometimes built into a prism, with the ability of reflecting part of a beam of radiation and transmitting the other part. The 'splitting' or diverting is performed on the energy level (frequency-selective surfaces), not on the spectral level (no spectral separation by dispersion). Such a partially diverted beam may be used in color separation cameras, or for the superposition of images in special cameras or in Fourier Transform Spectrometers (FTS), generating an interferogram in combination with an interferometer.

Bias. In electronics the term refers to the application of a voltage between two terminals (electrodes) resulting in current flow (also referred to as 'forward bias'). The term 'reverse bias' refers to the application of a voltage in such a way that no current can flow. The term 'unbiased' designates that no voltage is applied.

Beat wave. A composite wave formed by the superposition of two waves having different frequencies (f_1, f_2) and wavenumbers (k_1, k_2). Beat waves form at the sum and difference frequencies ($f_1 \pm f_2$) and wavenumbers ($k_1 \pm k_2$). See also **heterodyne detection**.

Biological productivity. The amount of organic matter, carbon, or energy that is accumulated during a given time period.

Bioluminescence. Refers to the production of light from chemiluminescent reaction in living organisms. Although bioluminescence is very dim, it features prominently in the ecology of the seas, occurring in all oceans; it is produced by a wide variety of marine plankton and nekton.

Biomass. The total dry organic matter or stored energy content of living organisms that is present at a specific time in a defined unit (community, ecosystem, crop, etc.) of the Earth's surface.

Biomass burning. A recognized major source of trace gases, including CO_2, NO_2, CO, CH_4, and of aerosol particles. It takes on many forms: burning of forested areas for land clearing, extensive burning of grasslands and savannas to sustain grazing lands, burning of harvest debris, use of biomass fuel for heating, forest fires induced by lightning or other hazards. The emissions of biomass burning represent a large perturbation to global atmospheric chemistry, especially in the tropics.

Biosphere. The portion of the Earth and its atmosphere that can support life. The part (reservoir) of the global carbon cycle that includes living organisms (plants, animals,) and life-derived organic matter (litter, detritus). The terrestrial biosphere includes the living biota (plants and animals), litter and soil organic matter on land; the marine biosphere includes the biota and detrius in the oceans.

Bistatic system. The bistatic remote-sensing concept refers to a measurement arrangement in which the transmitter and receiver locations are separated by a distance comparable to that of the target distance. In contrast, monostatic radars employ a transmitter and receiver at the same location (often using the same antenna) and measure the backscattered radiation. The great majority of all radars (SAR instruments, Doppler radars, etc.) in use today are monostatic.

Blackbody (BB). An idealized body that absorbs all the radiation incident upon it and re-radiates it according the Planck's law.

Blaze wavelength. The wavelength of the highest efficiency for a ruled diffraction grating, the "blaze" being the controlled shape of the rulings on the grating.

Blooming. Refers to the saturation effect in image detection devices, like CCDs.

Body-pointing. Refers to the pointing technique of an instrument within the field of regard. The instrument is pointed along with its platform (satellite) into the desired direction.

Bolometer. A detector type making use of the change in electrical resistance of certain materials (with small thermal capacity) when their temperature is changed. The resistance of most conductors varies with temperature, this change in resistance is measured by the bolometer. Bolometers are suitable detectors for the infrared and microwave regions.

Boresight. A technique for aligning sensors or detectors on a target.

Bragg scattering theory. According to this theory the normalized radar cross-section (NRCS) is proportional to the spectral energy density of the Bragg waves, i.e. of those surface waves with wave numbers k_B that satisfy the Bragg resonance condition:

$k_B = \frac{4\pi \sin \theta}{\lambda_o}$ where λ_o denotes the radar wavelength and θ the incidence angle.

BRDF (Bidirectional Reflectance Distribution Function). Specifies the behavior of surface scattering as a function of illumination and view angles at a particular wavelength. BRDF is defined as being the ratio of the reflected radiance to the incident flux per unit area.[2044)]

Brightness temperature. A concept referring to the equivalent blackbody temperature for a given frequency (range) according to Planck's law. The term brightness temperature is often employed for data of radio/microwave observations where the radiation is in the Rayleigh-Jeans tail of the thermal distribution. It means that the source emits (at the frequency of interest) the same amount of radiation as a blackbody at the brightness temperature. Intensities are measured in terms of brightness temperature (that is, the temperature a blackbody would have if it emitted an equal intensity of radiation at the same frequency). Examples of brightness temperature applications:

- Measurement and/or computation of the 'top of the atmosphere brightness temperature' or 'troposphere moisture content' from data of particular channels of such sensors as: AVHRR, TOVS, etc.
- The measurement of ocean surface roughness at microwave frequencies (with a radiometer) permits estimates of ocean-surface wind speeds. The method employs the concept of brightness temperature anisotropy which increases with ocean-surface roughness. The SSM/I sensor of the DMSP series uses this principle to map ocean surface wind speeds.
- The Earth's surface brightness temperature can be measured by channel 6 (TIR) of the Landsat TM sensor, or by the MWR instrument of ERS missions.

Cadastre. An official (governmental) land registry defining the ownership of a parcel along with ancillary information (description of parcel location, boundaries, shape and size, inventory of actual features and structures, value for taxation, etc.). Positional accuracies are an important issue of such registries. - Cadastral mapping takes many forms around the world, based on current and historic land registry, land reform policies, and available funding levels. In Europe the cadastre is linked to the legal land registration system within the context of a national geodetic reference system. In the US and Canada there is no central land registry (land registry information is maintained at multiple levels). In all parts of the world, however, land ownership information plays a key role in defining local and national economies and in managing natural resources and handling environmental issues.

Calibration. Characterization of a sensor (radiometer, spectrometer, etc., see also chapter O.2) in the spatial, spectral, temporal and polarization responsive domains. The term 'calibration' is being used so often by different people that it has several additional meanings, such as:
1. The activities involved in adjusting an instrument to be intrinsically accurate, either before or after launch (i.e. 'instrument calibration').
2. The process of collecting instrument characterization information (scale, offset, nonlinearity, operational and environmental effects), using either laboratory standards, field standards, or modeling, which is used to interpret instrument measurements (i.e. 'data calibration').

2044) See ERIM BRDF tutorial at URL: http://www.erim.org/on-line-docs/GUIDE/guide.frm.html

Candela (cd). A unit of luminous intensity equal to one sixtieth of the luminous intensity of one square centimeter (1 cm^2) of a blackbody surface at the solidification point of platinum.

Carbon cycle. A sequence of conversion processes from matter into energy. All reservoirs and fluxes of carbon; usually thought of as a series of the four main reservoirs of carbon interconnected by pathways of exchange. The four reservoirs - regions of the Earth in which carbon behaves in a systematic manner - are the atmosphere, terrestrial biosphere (usually includes freshwater systems), oceans, and sediments (includes fossil fuels). Each of these global reservoirs may be subdivided into smaller pools ranging in size from individual communities or ecosystems to the total of all living organisms (biota). Carbon is exchanged from reservoir to reservoir by various chemical, physical, geological, and biological processes.

Carrier. An electromagnetic wave in a communication path (basic center frequency of a signal) which does not carry information but is generally modulated by another wave (subcarrier) which contains the information.

Carrier phase. Refers to the fraction of a cycle, often expressed in degrees (360° to a cycle). Carrier phase can also mean 'the number of complete cycles plus a fractional cycle.' In GPS terminology, carrier phase refers to a receiver capable of locking onto a GPS signal and keeping track of the whole number of cycles of the carrier; this method creates a cumulative phase of the signal which is also known as 'integrated Doppler.' Much higher ranging accuracies can be obtained with carrier phase than without carrier phase tracking.

Catadioptric telescope (see also Schmidt telescope under Telescopes). Refers to a telescope design with a large FOV to eliminate image distortions (a catadioptric telescope design incorporates the best features of both the refractor and reflector, i.e., it has both reflective and refractive optics. The Schmidt telescope has a spherically shaped primary mirror. Since parallel light rays, that are reflected by the centre of a spherical mirror, are focused farther away than those reflected from the outer regions, Schmidt introduced a thin lens (called the correcting plate) at the radius of curvature of the primary mirror. Since this correcting plate is very thin, it introduces little chromatic aberration. The resulting focal plane has a field of view several degrees in diameter.

Charge-Coupled Device (CCD). A CCD is a photosensitive solid-state imaging sensor (detector) implemented with large-scale integration technology (normally based on MOS technology). A MOS capacitor is a three-layer sandwich formed by positioning a metal electrode, insulated by a layer of silicon dioxide, onto a silicon substrate. Incident radiation into the system is sampled by photodetectors, converted into an electronic charge and trapped in the depletion region of the substrate. The isolated charge packets are transported by manipulating potential wells (place of minimum potential) within the substrate. The ability to store a charge is fundamental to the operation of CCDs. It corresponds to a memory device storing analog quantities. - CCD readout techniques employ clock-controlled circuits which transfer these charges to a matching grid of elements and shift all charges by one row at a time. - The CCD technology was first demonstrated in 1969 at the Bell Laboratories. See also chapter O.4.2.1.

Charge Injection Device (CID). A photo-sensitive image sensor (detector) implemented in large-scale integration technology. Charge packets are typically measured by injecting them into a substrate or by shifting charge packets under an electrode to induce a voltage on the capacitance formed by the electrode and the substrate. A CID can be randomly addressed. The pixel structure is contiguous with maximum surface to capture incident light which is useful for sub-pixel measurement (O.4.2.2).

Chemiluminescence. Emission of light as a consequence of a chemical reaction, the result of thermal generation of electronic excited states. Chemiluminescence can be seen when occurring in the dark. - A number of chemical reactions generate products not in their lowest energy states, but rather in upper levels. That is, some of the exothermicity of the reaction is channeled internally into electronic, vibrational, or rotational energy of one or more of the products, rather than being released as heat. The excited product molecules may emit this energy as light, known as chemiluminescence because of the chemical source of energy. - The term of **surface chemiluminescence** belongs also into this context. In this scheme air is passed over a chemiluminescent plate causing the gas (eg., ozone) molecules to diffuse into the coating of the plate and in turn generating a chemiluminescent reaction. The reaction sequence causes the emittance of light (radiation) whose intensity is proportional to the ozone concentration.

Chirp principle. A microwave modulation technique in which the frequency of the transmitted microwave pulse is not constant but linearly changed in a positive sense (up-chirp) or in a negative sense (down-chirp). A frequency-modulated chirp is a signal with a (linear) increase in frequency or pitch.

Chlorofluorocarbons (CFCs). A family of inert, nontoxic, and easily liquefied chemicals used in refrigeration, air conditioning, packaging, and insulation, or as solvents or aerosol propellants. Because they are not destroyed in the lower atmosphere, they drift into the upper atmosphere, where - given suitable conditions - their chlorine components destroy ozone.

Chlorophyll. A green pigment essential for photosynthesis found in plants. It usually occurs in discrete bodies (chloroplasts) in plant cells, and is what makes green plants green. In remote sensing of aquatic ecosystems, the reflected radiance is related to the concentration of chlorophyll and other associated pigments. Since chlorophyll is green, the color of reflected light changes from blue to green as the concentration of chlorophyll increases. The concentration of chlorophyll is used to estimate the abundance of phytoplankton in ocean waters, and hence the abundance of ocean biota.[2045)]

Climate. The statistical collection and representation of the weather conditions for a specified area during a specified time interval, usually decades, together with a description of the state of the external system or boundary conditions. The properties that characterize the climate are thermal (temperatures of the surface air, water, land, and ice), kinetic (wind and ocean currents, together with associated vertical motions and the motions of air masses, humidity, cloudiness and cloud water content, groundwater, lake winds, and water content of snow on land and sea ice), and static (pressure and density of the atmosphere and ocean, composition of the dry air, salinity of the oceans, and the geometric boundaries and physical constants of the system). These properties are interconnected by various physical processes such as precipitation, evaporation, infrared radiation, convection, advection, and turbulence.

Climate change. The long-term fluctuations in temperature, precipitation, wind, and all other aspects of the Earth's climate. External processes, such as solar-irradiance variations, variations of the Earth's orbital parameters (eccentricity, precession, and inclination), lithosphere motions, and volcanic activity, are factors in climatic variation. Internal variations of the climate, e.g., changes in the abundance of greenhouse gases, may also produce fluctuations of sufficient magnitude and variability to explain observed climate change through the feedback processes interrelating the components of the climate system.

Correction/Calibration methods for sensor data (see chapters O.2 and O.2.2)

Cloud. A visible mass of condensed water vapor particles or ice suspended above the Earth's surface. Clouds may be classified by their visual appearance, height, or form.

Cloud albedo. Reflectivity that varies from less than 10 to more than 90 percent of the insolation and depends on drop sizes, liquid water content, water vapor content, thickness of the cloud, and the sun's zenith angle. The smaller the drops and the greater the liquid water content, the greater the cloud albedo, if all other factors are the same.

2045) Courtesy of W. Esaias of NASA/GSFC

Cloud feedback. The coupling between cloudiness and surface air temperature in which a change in surface temperature could lead to a change in clouds, which could then amplify or diminish the initial temperature perturbation. For example, an increase in surface temperature could increase evaporation; this in turn might increase the extent of cloud cover. Increased cloud cover would reduce the solar radiation reaching the Earth's surface, thereby lowering the surface temperature. This is an example of negative feedback and does not include the effects of longwave radiation or advection in the oceans and the atmosphere, which must also be considered in the overall relationships within the climate system.

Cloud microphysics. Study of cloud and precipitation particles (individual or populations) and their interactions with the environment. Of key importance are mass exchange processes such as nucleation, growth, and fallout leading to the broad characteristics of clouds and precipitation.

Code Division Multiple Access (CDMA). Refers to an access scheme which employs spread-spectrum modulations and orthogonal codes to share a communication link among its users. In the CDMA scheme, all users transmit simultaneously and at the same frequency, with each being assigned a unique pseudorandom noise code. Usually, the data is first phase-modulated by a carrier and then the carrier is bi-phase-modulated with a pseudorandom noise (PNR) code. This concept generates a wide bandwidth, low-energy spread spectrum signal.

Coherence. A fixed relationship between the phases of waves in a beam of radiation of a single frequency. Two beams of light are coherent when the phase difference between their waves is constant; they are **noncoherent** if there is a random phase relationship. In active measurement systems like radars, coherence refers to the availability of phase and amplitude measurements of the radar cross section of the recovered signals. Coherence provides the ability to maximize SNR and to measure other features like target radial velocity.

Coma. An off-axis aberration whereby the outer periphery of a lens system has a higher (or lower) magnification then the central portion of the lens. The image typically is comet shaped.

Contrast. The ratio of a certain quantity of radiation between the brightest and darkest part of an image or between two arbitrary places of an image, where the contrast is to be determined.

Convection (meteorology). Vertical wind motions and associated horizontal circulations associated with buoyancy.

Convolution. Mathematical process, appearing in linear or circular form, that models the input-output filtering process.

Convolution filter. A linear filter type as used in digital image processing of which the window operation has a mathematically linear character (weighted summation). Examples are low-pass filter, high-pass filter, gradient filters, Laplacian filters, etc.

Corona. Refers to the outer atmosphere of the sun whose structure is controlled by solar magnetic fields. The corona has temperatures between one and three million degrees. It merges into the solar wind at its upper boundary about 1-2 solar radii above the visible surface of the photosphere.

Crossover (difference). A crossover is defined as the intersection points of the satellite ground track with itself (due to Earth rotation). At this location, the two crossing passes (one ascending and one descending) provide independent subsatellite ground track measurements at the same location but at different times. In altimetry crossover differences contain information about uncertainties in the satellite ephemeris and therefore enable correction of radial orbit error.

Crossover point. Refers to radar measurements concerning the curves showing the dependency of the radar backscatter behavior on the incidence angle of the radar transmission signals onto surfaces of differing roughness. The crossover point is the incidence angle where the diffuse region changes into a specular region on the curves (or: the incidence angle where the effect of soil roughness vanishes and the radar backscatter value is determined by the presence of soil moisture).

Cryosphere (from the Greek word 'Kryos', icy cold). The Earth's cryosphere consists of four main elements: sea ice, seasonal snow on land, land ice (including glaciers, ice sheets, and ice shelves), and permafrost. The time scales on which these elements impact human activity range from daily to seasonal for sea ice and snow, while ice shelves respond in the range of 10-100 years, ice sheets (Antarctic and Greenland) have periods in the order of 1000-10,000 years. Land ice occupies about 11% of the continental surfaces. Sea ice and ice shelves spread around 7% of the total oceanic area.

Decibel (dB) - named in honor of Alexander Graham Bell. A measurement of signal strength, properly applied to a ratio of powers. For the signal power P compared by a ratio to a reference power P_{ref}, the definition is: $P_{db} = 10\log_{10} (P/P_{ref})$. As an example, the power ratio of 1/2 corresponds to "3 dB", derived from: $\log_{10} (0.5) = 0.3010$.

Deforestation. The removal of forest stands by cutting and burning to provide land for agricultural purposes, residential or industrial building sites, roads, etc., or by harvesting the trees for building materials or fuel. Oxidation of organic matter releases CO_2 to the atmosphere, with possible regional and global impacts.

Densiometer. A photometer designed for measuring the optical density of a material, generally a photographic image by visual or photoelectric effects.

Depolarization ratio. The ratio of intensities of light scattered perpendicular and parallel to the E-vector of the incident radiation.

Detector. A device that detects and linearly transduces radiative power into an electrical signal. Direct detectors may be categorized as photon detectors (an electrical signal is produced by free charges on the detector surface from the incident photons) or thermal detectors (an electrical signal is produced due to the temperature change). Detectors may also be classified according to their arrangement: single line detectors, array detectors. Thermal detectors usually require cooling (active or passive). Infrared radiation is 'thermal' by nature, hence the detector is affected by the medium that is measured. As a rule of thumb, the longer the IR wavelength that is to be measured, the colder the detector must be. In the VNIR region the detector element temperatures rarely need to be below 200 K. From 1-17 μm, temperatures are typically in the range 50-80 K. The longer wavelengths of the microwave region usually demand temperatures below 20 K. The detectivity of a cooled detector is much higher than one operating at room temperature The noise contribution from background radiation at 300 K is several orders of magnitude higher than that of the 4 K surroundings of the detector (see O.4).

Detector types for spectral ranges.[2046] From the UV to VNIR (0.3-1 μm), silicon photodiodes and photoemissive devices such as photomultiplier tube (PMT) are normally used.[2047] Between 1-12 μm, two technologies dominate: InSb (indium antimonide) from 1-5.5 μm, and HgCdTe (mercury cadmium telluride, also referred to as MCT). Both, InSb and MCT operate in a photovoltaic mode. For the spectral range of 12-35 μm, photoconductors or MCT are being used. Beyond 30 μm, the only available technology in use is the semiconductor bolometer.

Dewar [after Sir James Dewar (1854-1928) a Scottish chemist and physicist]. The term denotes a vessel to store hot or cold substances over long periods of time. It is a container with

2046) R. Beer, "Remote Sensing by Fourier Transform Spectrometry," John Wiley & Sons, Inc., New York, 1992, Chapter 4.1.2

2047) Note: Silicon is transparent in the spectral ranges of 1.4 - 7 μm and from 25 μm to well beyond 100 μm at least two walls and a space between the walls evacuated so as to prevent the transfer of heat. There are various techniques in use for minimizing the heat transfer in spaceflight for liquid helium, like: multilayer insulation, multiple reflective surfaces in vacuum, vapor-cooled shields, passive orbital disconnect struts, etc.

Dielectric. An insulating material or a very poor conductor of electric current. When dielectrics are placed into an electric field, practically no current flows in them because, unlike metals, they have no loosely bound, or free, electrons that may drift through the material. Instead, electric polarization occurs, reducing the electric field within the dielectric. A vacuum is the only perfect dielectric. - A dielectric gas is a nonconductor of electricity to high applied electrical stress; a gas with a high breakdown voltage.

Dielectric constant. A property of an insulating material (a dielectric) equal to the ratio of the capacitance of the capacitor filled with the given material to the capacitance of an identical capacitor in a vacuum without the dielectric material ($\varkappa = C/C_0$).

Diffraction. A process by which the direction of radiation is changed so that it spreads into the geometric shadow region of an opaque or refractive object that lies in a radiation field. Diffraction is an optical "edge effect," (differing only in degree from scattering) caused by particles with diameters of the same order of magnitude as, or larger than, the wavelength of radiation; scattering is caused by smaller objects. Diffraction causes a modification which light undergoes in passing by the edges of opaque bodies or through narrow slits or in being reflected from ruled surfaces, and in which the rays appear to be deflected producing fringes of parallel light and dark or colored bands.

Diffraction grating. A system of close equidistant and parallel lines or bars (also grooves) on a polished surface used for producing spectra by diffraction.

Diffraction-limited system. An optical system in which aberrations are negligible with respect to diffraction effects.

Diffuse radiation. Radiation propagating in many different directions through a given small volume of space (converse is 'specular radiation'). The ideal form of diffuse radiation is isotropic radiation (uniform radiation in all directions).

Digital count. Refers to the total number of pixels occurring in an image for each possible data value.

Digital Earth. A vision/initiative of Vice President Al Gore, presented in a speech at the California Science Center in Los Angeles, on Jan. 31, 1998. [2048] The proposed concept model of "Digital Earth"refers to a multi-resolution, 3-D representation of Earth, into which geo-referenced data can be embedded. A "Digital Earth" could, for instance, provide a mechanism for users to navigate and search for geospatial information, etc. - Obviously, such an objective is so vast, that no one organization in government, industry or academia could undertake such a project. A vast standards infrastructure is needed to make it happen! The benefits of such a seamless system are apparent to the entire Earth Observation community.

Digital filter. A digital device (or a mathematical procedure) capable of altering the magnitude, frequency or phase response of a digitally encoded input signal (it may also selectively transmit digital signals).

Digital Terrain Model (DTM). Refers to a land surface represented in digital form by an elevation grid or tables of three-dimensional coordinates to form surface contours. The DTM definition is identical to that of a **DEM** (Digital Elevation Model). A DTM or DEM forms the basic building block for combining other data for analysis. For instance, digitized

2048) G. W. Fuller, "A Vision for a Global Geospatial Information Network (GGIN) Creating, Maintaining and Using Globally Distributed Geographic Data, Information, Knowledge and Services," ASPRS, May 1999, pp. 524-538 spatial data (images) can be draped onto a DEM and analyzed using a GIS. The quality of such a DEM depends on the spatial resolution (in particular the topographic accuracy) of image data available. Interferometric SAR data and/or altimeter data are currently the best sources for DEM generation.

Diode. A semiconductor diode consists of a crystal (two terminals), part of which is n-type (negative charge) and part p-type (positive charge). The boundary between the two parts is called a p-n junction. There is a population of holes on the p-type side of the junction and a population of electrons on the n-type side. [2049] The p-n junction of the diode conducts current with one polarity of applied voltage but not with the other polarity. Rectification (current flow only in one direction) is a very important characteristic of the p-n junction. Another characteristic of the p-n junction is its direct conversion capability of radiant energy into electrical energy (optoelectronic effect). An incident photon, striking a p-n junction, has the same effect as a hole (positive charge); it is absorbed thereby creating electron-hole pairs. The resulting current can be detected (see **photodiode**).

Dipole. An electric system composed of two equal charges of opposite sign, separated by a finite distance; e.g. the nucleus and orbital electron of a hydrogen atom. An ordinary bar magnet is a magnetic dipole.

Dipole antenna. A type of array consisting of a system of dipoles. A dipole antenna differs from a dish antenna in that it consists of many separate antennas that collect energy by feeding all their weak individual signals into one common receiving set.

Discrete Fourier Transform. A mathematical method of transforming a time series into a set of harmonics in the frequency domain and vice versa.

Dispersion of spectra. The following methods are used to separate radiation (light) into its component spectra (colors):

- Refraction. Historically, prisms were first used to break up or disperse light into its component colors. The path of a light ray bends (refracts) when it passes through the prism, i.e. from one transparent medium to another (from air to glass).
- Diffraction. Diffraction gratings are composed of closely spaced transmitting slits on a flat surface or alternate reflecting and non-reflecting grooves. - In any grating spectrometer, if a slit aperture is moved along the surface where the spectral lines are focussed, the lines are transmitted successively through the aperture.
- Interference. An interferometer divides a wave front by semitransparent surfaces. This allows the beams to travel different paths. The beams are then recombined generating interference patterns.
- AOTF (Acousto-Optic Tunable Filter). The AOTF principle is based on acoustic diffractions of light in an anisotropic medium. An AOTF device consists of a piezoelectric transducer bonded to a birefringent crystal. When the transducer is excited by an applied RF signal, acoustic waves are generated in the medium. The propagating acoustic wave produces a periodic modulation of the index of refraction. This provides a moving phase grating that, under proper conditions, will diffract portions of an incident beam. - In operation, acousto-optic tunable filters resemble interference filters and can replace a filter wheel, grating, or prism in many applications (see O.4.5).

Diurnal cycle (Lat. diurnalis). A 24-hour (daily) cycle associated with solar heating during the day and radiative cooling during the night. A periodic cycle affecting nearly all meteorological variables.

Doppler effect (after Christian J. Doppler, Austrian physicist, 1803-1853). The alteration in frequency of a wave of radiation caused by relative motion between the observer and the

2049) Note: The term "hole" refers to a fictitious particle which carries a positive charge and moves, under the influence of an applied electric field (bias), in a direction opposite to that of an electron. The motion of electrons and holes in semiconductors is governed by the theory of quantum mechanics.

source of radiation.- The acoustic Doppler effect applies to the propagation of source waves; the optical Doppler effect depends on the relative velocity of the light source and the observer; the thermal Doppler effect causes a widening of the spectral lines.

Doppler radar. A radar system which differentiates between fixed and moving targets by detecting the change in frequency of the reflected wave caused by the Doppler effect. The system can also measure target velocity with high accuracy.

Doppler shift. Displacement of spectral lines (or difference in frequency) in the radiation received from a source due to its relative motion in the line of sight. Sources approaching (−) the observer are shifted toward the blue; those receding (+), toward the red. Used to determine radial distance.

Delay Doppler radar altimeter. An evolving technique which exploits signal processing algorithms borrowed from SAR processing schemes. When applied to an ocean-observing altimeter, real-time on-board processing achieves an integration level of the received signal that is about a magnitude higher (tenfold) than that achieved in conventional radar altimeters. This in turn translates into a tenfold reduction in required radiative power of the transmitter (or into a smaller radar antenna or a combination of both effects), improving instrument performance considerably. [2050)]

Downlink. Refers to the communication direction from a satellite (or aircraft) to a ground station. The prime information in this link is usually referred to as 'telemetry.' There may be different logical links in a downlink for instrument data and for the return (verification) of the telecommand data. In very elaborate communication systems with intermediate geostationary transmission satellites, the term 'downlink' is usually replaced by 'return link' to avoid confusion.

Dryline. A meteorological term referring to a boundary which separates moist and dry air masses. The dryline is an important factor in the frequency of severe weather in the Great Plains of the continental USA. It typically lies north-south across the central and southern high Plains states during the spring and early summer, separating moist air from the Gulf of Mexico and dry desert air from the southwestern states.

Dual spin. Refers to a spacecraft design whereby the main body of the satellite is spun to provide attitude stabilization. In this concept the antenna assembly is despun by means of a motor and bearing system in order to continually direct the antenna earthward. The dual-spin configuration thus serves to create a spin-stabilized satellite.

Duty cycle. Fraction of orbital period in which a sensor (or a sensor mode) is actually operational, determined by the overall power limitations of the payload. The concept of a duty cycle applies in particular to sensors with large power requirements such as active sensors, in particular SAR instruments.

Dwell time. The short period of time during which a detector collects radiation from a target area or volume. - A very short dwell time usually results in a low (i.e. poor) signal-to-noise (SNR) ratio with all its problems of proper signal recognition and discrimination. A small dwell time also implies 'fast' detectors and electronics.

Dynamic range. The dynamic range of a sensor system is determined by the ratio of the maximum observable energy (Q_{max}) and the minimum still-useful energy (noise level Q_{min}); it is defined in decibels (dB) as 10 log (Q_{max}/Q_{min}). All radiant energy $< Q_{min}$ vanishes into noise, while the energy above Q_{max} disappears into the saturation of the detector (see also **Signal-to-Noise-Ratio**).

Electromagnetic spectrum (EMS). The total range of wavelengths or frequencies of electromagnetic radiation, extending from the longest radio waves to the shortest known gam-

2050) G. H. Fountain, Robert E. Gold, et al., "A Technology Path to Distributed Remote Sensing," Small Satellites for Earth Observation, 2nd International Symposium of IAA, Berlin, April 12-16, 1999, pp.189-193 ma rays. - EMS energy for passive remote sensing, as derived from the sun, is either **reflected sunlight** or **re-emitted thermal radiation**. The transfer mode of reflected radiation is dominant in the window associated with VNIR and SWIR, while the preferred mode of long-wave transfer (thermal radiation) is in the TIR window (see also Figure 422 on page 1365).

- In the VNIR and SWIR (0.4 - 3 μm) wavelength regions, the predominant mode of energy detection is that of reflected sunlight
- In the MWIR region (3-6 μm) the detected energy is a mixture of solar reflected and thermally emitted radiation
- In the TIR window (6-13 μm), practically all energy received (detected) is attributed to thermal emission.

Note: Although the sun, with a brightness temperature of about 6000 K, is much hotter than the Earth's surface at about 290 K, the dominance of the Earth's thermal energy at longer wavelengths is a result of geometry (the sun subtends only about 0.5° at the Earth - the sun's disk angle is actually about 32'), and that the solar energy is subsequently scattered by the Earth's surface into 2π space.

Electron tube. The term is the generic name for a class of devices that includes: vacuum tubes, phototubes, gas-filled tubes, cathode-ray tubes, and photoelectric tubes. An electron tube typically consists of two or more electrodes enclosed in a glass or metal-ceramic envelope, which is wholly or partially evacuated. Its operation depends on the generation and transfer of electrons through the vacuum from one electrode to the other. Electron tubes have properties that cannot be surpassed by solid-state devices for particular applications. Their thermal ruggedness, operating efficiency, and high-power capabilities are features well beyond those of solid-state devices. As components of electronic systems, electron tubes are used as amplifiers, rectifiers, signal generators, and switches (in particular in the microwave region).

Electron volt (eV). A unit of energy used in atomic and nuclear physics; the kinetic energy acquired by one electron in passing through a potential difference of 1 volt in vacuum. 1 eV = 1.602×10^{-12} erg (or = 1.602×10^{-19} Joule)[2051]. An electron with an energy of 1 eV has a velocity of about 580 km/s. The wavelength associated with 1 eV is 12,398 Å. The eV is a convenient energy unit when dealing with the motions of electrons and ions in electric fields; the unit is also the one used to describe the energy of X-rays and gamma-rays. Nuclei in cosmic rays typically have energies ranging from a about 1 MeV (or less) to many GeV per nucleon.

Electrooptics. An imaging technique which uses optics, such as the collimation of light beams and the magnification of images by lenses and mirrors, rectification by prisms, and diffraction by gratings (usage: generally in the VNIR spectrum). - The newer devices of electrooptics make use of more electronic components such as the generation of light by lasers and solid state devices, featuring electronic scanning of the images and data presentation on electrically activated displays. To an increasing extent, computers are employing electrooptical techniques for their own operation and display. See also chapter O.10.2.

Emissivity (ε). The ratio of radiative energy (power) emitted by a body to that emitted by a blackbody at the same temperature. For all cases: $\varepsilon(\lambda) \leq 1$.

Energetic particles. These are electrons, ions, or atoms that have much higher energies than expected for the temperature of the gas in which they are transported (the solar wind is such a transport medium).

Energetic Neutral Atom (ENA). ENAs are created in the inner magnetosphere when charge exchange collisions occur between energetic ions and the cold neutral population of

2051) The energy unit Joule is named in honor of James Prescott Joule (1818-1889), a British physicist who established that the various forms of energy - mechanical, electrical, and heat - are basically the same and can be changed into one another. Thus he formed the basis of the law of conservation of energy, the first law of thermodynamics.

the Earth's extended atmosphere. ENAs travel in approximately straight-line trajectories away from the charge-exchange sites (because gravitational forces are negligible for typical energies of interest and they are unaffected by the Earth's electric and magnetic fields) and carry with them valuable information about the pitch angle and energy distributions of the ion population from which they were emitted. As of the 1990s the ENA emissions can actually be sensed remotely by appropriate imagers. Several NASA missions have such instrumentation: POLAR, IMAGE, and TWINS.

ENSO (El Niño Southern Oscillation). ENSO is regarded as a large-scale interannual climate variability (anomaly) especially with regard to precipitation regimes and sea surface temperature (SST) changes in the tropical Pacific Ocean. The warming effect of ENSO can dramatically alter precipitation patterns over much of the Pacific basin. Its recurrence every three to seven years provides a clear signal of climate variability on a global scale. Its land surface manifestations are illustrated by heavy torrential rains on the west coast of South America and droughts in Sahelian Africa, southern Africa, Australia, and eastern Brazil. The forecasting of ENSO events can greatly benefit the peoples and economies of the impacted areas.
Scientists believe **El Niño** conditions between Australia and South America are sparked when the steady westward trade winds weaken and even reverse direction. This wind shift moves a large mass of warm water, normally situated near Australia, eastward along the equator, pushing it toward the coast of South America. The transportation of such a large body of warm water affects evaporation, causing rain clouds to form that, in turn, alter typical atmospheric jet stream patterns around the world.

Background: In the 16 century, the El Niño climatic phenomenon was first observed by Peruvian fisherman as a warm current running along the coast of Peru. Since this recurrent event happened around the Christmas season, they christened the warm current as El Niño (meaning Christmas in Spanish). This virtually periodic phenomenon appears at two- to seven-year intervals, it has economic and environmental consequences that are sometimes catastrophic. The entire climate is thoroughly disturbed. Due to the warming of the Peruvian waters, normally the most productive in the world, the ocean is depleted of nutrients, causing rarefaction of the phytoplankton. A consequence of the disappearance of the aquatic life is a massive destruction of oceanic bird life.

Eötvös experiment. An experiment performed in 1909 by the Hungarian physicist Roland Eötvös (1848-1919) to establish that the gravitational acceleration of a body does not depend on its composition - i.e. that inertial mass and gravitational mass are exactly equal (later a major principle of Albert Einstein's general theory of relativity). In Einstein's version, the principle asserts that in free-fall the effect of gravity is totally abolished in all possible experiments and general relativity reduces to special relativity, as in the inertial state. Today, the linear gradient of gravity is defined in units of Eötvös, where 1 Eötvös = 10^{-9} s^{-2}; i.e. difference of 10^{-9} ms^{-2} acceleration per meter. The vertical gradient of gravity at the Earth's surface is about 3100 Eötvös.

Ephemeris. A tabular statement of the spatial coordinates of a celestial body or a spacecraft as a function of time.

Equatorial electrojet current. Current created around the Earth's equator by counter-rotating electrons and protons. This electrojet current causes ionization, and subsequently UV radiation, similar to auroral phenomenon usually associated with the Earth's F-Layer.

Equinox. Either of two points on the celestial sphere where the celestial equator intersects the ecliptic. At these two instances the sun is exactly above the equator and day and night are of equal length (see also **vernal equinox**).

Equivalence Principle (EP). A fundamental law of physics that states that gravitational and inertial forces are of a similar nature and often indistinguishable. EP in fact states that two fundamentally different quantities, inertia and passive gravitational mass, always be exactly proportional to one another. This is usually interpreted as implying that the two quantities are equivalent measures for a single physical property, the quantity of **mass** of an object; hence, the term Equivalence Principle. A direct consequence of this Equivalence Principle is the 'universality of free fall' such that all objects fall with exactly the same acceleration in the same gravity field. EP turned out to be a major principle of Albert Einstein's general theory of relativity. The Strong or Einstein Equivalence Principle states that all of the laws of physics (not just the laws of gravity) are the same in all small regions of space, regardless of their relative motion or acceleration.

In the Newtonian form, EP implies that, within a windowless laboratory freely falling in a uniform gravitational field, experimenters would be unaware that the laboratory is in a state of nonuniform motion. All dynamical experiments yield the same results as obtained in an inertial state of uniform motion unaffected by gravity. Experiments with ordinary pendulums test the principle of equivalence to no better than about one part in 10^5. The Hungarian physicist Roland Eötvös suggested (1909) that the attraction of the sun upon test masses could be compared with the inertial forces of the Earth's orbital motion about the sun. In the 1960s a series of careful observations (employing up-to-date methods of servo control and observation) were conducted by the American physicist Robert H. Dicke and his colleagues. They found that the weak equivalence principle held to about one part in 10^{11} for the attraction of the sun on gold and aluminum. A later experiment, with very different experimental arrangements, by the Russian researcher Vladimir Braginski, gave a limit of about one part in 10^{12} for platinum and aluminum. - A future NASA/ESA mission, called STEP (Satellite Test of the Equivalence Principle) or MiniSTEP, with a probable launch in 2005, has the objective to test EP to a precision of 1 part in 10^{18}.

Equivalent (or effective) Isotropic Radiated Power (EIRP). A measure of power radiated by an antenna in the direction of a receiver, expressed as the equivalent power that would have to be radiated uniformly in all directions.

Erlang. A measure of communication (telephone) traffic load expressed in units of hundred call seconds per hour (CCS). One Erlang is defined as the traffic load sufficient to keep one trunk busy on the average and is equivalent to 36 CCS. The measure is also used for DCS (Data Collection Satellite) access capabilities.

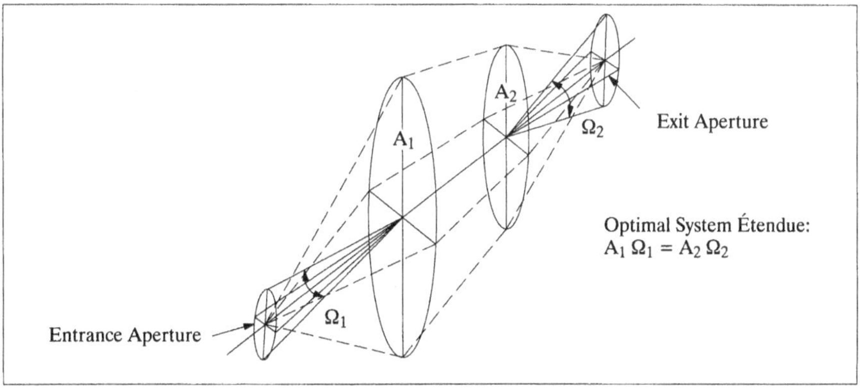

Figure 421: Definition of étendue

Étendue. The term (French for 'extent') describes a very fundamental property of an optical system, namely A Ω, the product of the area A of a light beam and the solid angle Ω contained within a beam. The importance of étendue lies in the fact that it is one of the major factors in determining the SNR (Signal-to-Noise Ratio) of a system; it is in fact a design parameter which can be maximized by the proper choice of configuration. Étendue has the units: cm^2 sr. It is a property that is at best conserved through an optical train, i.e., the system étendue will be that of lowest value in any part of the system. [2052)]

Excimer. In photochemistry a molecular aggregate formed by loose association of an excited state and a ground state of the same compound, where such association does not occur between two ground state molecules.

Extinction. The superimposed effect of two radiation effects that cancel each other, e.g. absorption and scattering.

F-Layer. One of the three regions (D, E, and F) of the Earth's ionosphere. The F-Layer constitutes the highest region, ranging from about 160-500 km. Within the F-Layer precipitating electrons from the magnetosphere cause ionization critical to long-wave radio communication. This ionization is typically caused by electrons with a kinetic electron energy < 1 keV. Ionization within the F-Layer may be characterized by sensors designed to measure UV radiation. F-Layer electron density is usually two to four orders of magnitude higher that that of the D- and E-Layers.

False color. A color imaging process which produces an image that does not correspond to the true color of the scene (as seen by the eye).

Fast Fourier Transform (FFT). An algorithm that is often used to implement Discrete Fourier Transforms.

Feedback mechanisms. A sequence of interactions in which the final interaction influences the original one. Negative feedback: An interaction that reduces or dampens the response of the system in which it is incorporated. Positive feedback: An interaction that increases or amplifies the response of the system in which it is incorporated.

Layer	Altitude	Major Components	Production Cause
D	70 - 90 km	NO^+, O_2^+	Lyman Alpha, X-rays
E	95 - 160 km	O_2^+, NO^+	Lyman Beta, Soft X-rays, UV continuum
F	160 - 500 km	O^+, N^+, NO^+	HE II, UV continuum

Table 560: Ionospheric layers and physical causes of ionization

Field of Regard (FOR). The pointing capability of a sensor with regard to the cross-track direction and/or along-track direction may provide additional coverage to the sensor in its orbit. This can be an advantage for monitoring (or imaging) events that are outside the swath width [or FOV(Field of View)] of a regularly nadir-pointing instrument. - An imaging sensor with a cross-track and an along-track pointing capability may have the potential of stereo imaging by taking the same image from different along-track positions in the same orbit.

Field of View (FOV). The total range of viewing of a sensor into the direction of the target. The cross-track component of FOV is equivalent to the swath width (see also IFOV).

Filter (in optical sensors). A device that - by interference absorption or reflection - selectively modifies the radiation transmitted through an optical system (see also 'convolution filter').

Fluorescence. Refers to the absorption of a photon of one wavelength and re-emission of one or more photons at longer wavelengths (known as Stoke's shift), especially the transformation of ultraviolet radiation into visible light. Plants re-emit a portion of the absorbed radiant energy in the visible region into the red and near-infrared region (0.65-0.75 μm). The distribution of wavelength-dependent emission intensity caused by a given wavelength

2052) R. Beer, "Remote Sensing by Fourier Transform Spectrometry," John Wiley & Sons, New York, 1992 excitation is known as the emission spectrum. The method of fluorescence has its advantages over other spectroscopic methods mainly due to its high sensitivity (applications in cell biology, photochemistry and the environmental sciences). - Note, since lasers are used as the excitation light source to induced fluorescence, this active remote-sensing method has become known as the laser-induced fluorescence (LIF) technique.

Focal length (f). Distance measured along the optical axis from the image to the plane, where the axial imaging cone of light intersects the input light bundle. The F-number or F-stop is defined as f/d (the focal length divided by the diameter of the lens opening). In antenna design the term focal length refers to the distance from the center feed to the center of the dish.

Footprint. Refers to the projection of the instantaneous area of coverage of a sensor (or antenna) onto the Earth's surface. A footprint may also be the instantaneous area of visibility of a data collection platform on a satellite.

Forward error correction (FEC). A transmission scheme which adds unique codes to the digital signal at the source so errors can be detected and corrected at the receiver.

Fourier Transform Spectrometer (FTS). An optoelectronic (or an optomechanical) instrument, usually for the infrared region of the spectrum, providing high spectroscopic resolution and sensitivity for remote-sensing applications (Earth surface imaging, atmospheric soundings, etc.). The technique employs the Fourier series concept as a means of converting a detector signal output - referred to as the interferogram - into a form useful for spectral analysis. There are several ways in which the detector signal can be created from the incident radiation. The approach taken by the majority of FTS instruments is to use an interferometer (Michelson, Sagnac, Fabry-Perot, etc.) and corresponding foreoptics (lenses) to create the interference pattern in such a way that each optical frequency is coded as a unique electrical signal output of the detector. The amplitude of each frequency is proportional to the incident radiation (see also chapter O.9 on page 1258).

Frequency Division Multiple Access (FDMA). A process that shares a spectrum of frequencies among many users by assigning to each a subset of frequencies in which to transmit signals. Each user is assigned a unique center frequency within the operational bandwidth.

Full Width Half Maximum (FWHM). In the microwave region the resolution is often stated in terms of 'half-power points' or 'half-power response width' of the measuring system. The half-power width of a response is easier to describe than true resolution because the concept does not involve the contrast of the target. - The true resolution for targets of any character can be derived from the actual response and estimated from the half-power width (resolution) of microwave instruments.

Gain. A measure of the amplification of the input signal in an amplifier. The gain of an amplifier is often measured in decibels (dB), which is ten times the common logarithm of the ratio of the output power of the amplifier to the input drive power [= 10 log (Q_2/Q_1)].

Galilei (Gal or gal). A unit of acceleration, 1 gal = 1 cm/s^2; 1 milligal = 10^{-3} cm/s^2. The term is commonly used in gradiometry.

Geocoding. Registration of images to the reference geometry of a map. In this process the imagery is corrected for all source-dependent errors and transformed to the desired map projection, and resampled to a standard pixel size. This is also called georeferencing.

Geodesy. The science of measuring the shape (and size) of the Earth, the geoid. A geodesic is the shortest line between two points on a mathematically derived surface (as a straight line on a plane or an arc of a great circle on a sphere). In general relativity the terms geodesic (noun) and geodetic (adjective) refer to paths followed by massive bodies ('geodesics') and light rays ('null geodesics') in curved four-dimensional space-time.

Geographical Information System (GIS). A combination of mutually referring data sets of various kinds of position-bound thematic data (database) and the necessary software to visualize this database, to manipulate it interactively and to analyze it in order to attain meaningful results.

Geoid. The Earth's conceptual gravitational equipotential surface (an ellipsoid) near the mean sea level, used as a datum for gravity surveys (a hypothetical ocean surface at rest). The geoid also serves as a reference surface for topographic heights, for example, as they are shown on maps. The geoid itself represents the surface of zero height.

Geomorphology (from Greek 'morph,' form). The explanatory science dealing with the form (shape) and surface configuration of the solid Earth (land and submarine relief features).

Geospatial information. Defined as information that identifies the geographic location and characteristics of natural or constructed features and boundaries on the Earth, including: statistical data; information derived from, among other things, remote sensing, mapping and surveying techniques; and mapping, charting and geodetic data, including geodetic products. [2053)]

Geosynchronous orbit. An orbit in which the satellite's orbital period is identical to the orbital period of the Earth. A geosynchronous orbit, unlike a geostationary orbit (with zero inclination, where the satellite motion relative to the Earth is at rest), does not impose any any restrictions on the orbit's eccentricity or inclination (see O.12.2 on page 1268).

Gradiometry. Study of the spatial gradient of the Earth's gravitational field.

Grating. See **diffraction grating**.

Grazing angle. Angle between the instantaneously transmitted signal of an active sensor (a SAR) and the local horizontal of the target. In other words, the grazing angle = $90 - \theta$, where θ is the incidence angle. Grazing incidences occur usually at very shallow angles. It's like skipping stones across a stream (the rock will skip only if it glances off the surface at a small angle).

Gravity-gradient boom. A deployable extension of a spacecraft (a rod fixed to the S/C with a small mass at its other end) intended to give the spacecraft elongated mass properties to contribute to gravity-gradient stability (the concept was first successfully demonstrated by JHU/APL on the Transit 5A-3 satellite with a launch on June 16, 1963). The principle: The attractive force F1 of mass 1 (satellite) about the common center of mass exceeds the attractive force F2 of mass 2. Hence, a torque arises to align the satellite to the vertical. - An elongated dumbbellshaped spacecraft is the most gravity gradient stable configuration with the long axis oriented vertically in orbit, i.e. (usually) the smaller mass is always pointing toward the center of the Earth. The gravity-gradient torque, small even for LEO S/C, decreases with the cube of the orbital radius. In GEO, gravity-gradient stabilization can barely be achieved.

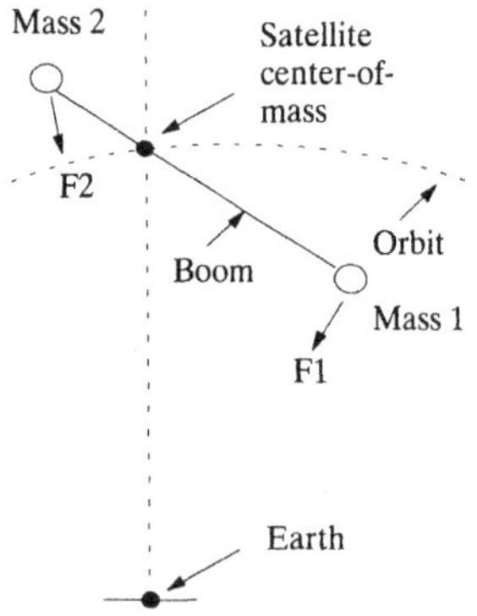

Many gravity-gradient stabilized LEO S/C use extendable booms to achieve a favorable moment-of-inertia distribution providing two-axis (pitch and roll) attitude control. The addition of an actuator, such as a momentum wheel, to a gravity-gradient stabilized S/C provides gyroscopic stiffness to passively stabilize the third (yaw) axis. The advantages of gravity-gradient stabilization are simplicity of control, long life and low power requirements.

2053) J. C. King, "Keynote Address to the ASPRS 2000 DC Annual Conference," PE&RS, Sept. 2000, pp. 1043-1053

Gravity wave (or gravitational wave). A wave disturbance in which buoyancy acts as the restoring force on parcels displaced from hydrostatic equilibrium.

Greenhouse effect. Refers to the trapping of heat from the sun by the atmosphere (mainly by its water vapor, which absorbs and reemits infrared radiation), in the same manner that the sun's heat is trapped by the glass walls of a greenhouse. The atmosphere, like the glass, is largely transparent to the sun's radiation, but it absorbs the longer wavelength radiation from the Earth's surface into which the sun's radiation is converted.

Greenhouse gases. Those atmospheric trace gases, such as water vapor, carbon dioxide, methane, and CFC's that are largely transparent to solar radiation but opaque to outgoing longwave radiation. Their action is similar to that of glass in a greenhouse. Some of the longwave (infrared) radiation is absorbed and reemitted by the greenhouse gases. The effect is to warm the surface and the lower atmosphere of the Earth.

Gridding. Use of a uniform system of rectilinear lines superimposed on imagery (such as photographs, mosaics, maps, charts, or other representations of the Earth's surface).

Ground pattern. Any specific identifying feature of the land surface which can be used for classification purposes.

Ground sampling distance (GSD). GSD is defined as the distance moved on the ground (in the along-track direction of the target area) during the integration period of the detector line array of an imaging instrument. Normally, the GSD is equated with the spatial resolution of a pixel or simply with IFOV (Instantaneous Field of View). However, this need not be the case. If the radiometric and electronic performance of a sensor allow, the GSD can be made smaller than IFOV to achieve better image quality because of the reduction of smear.

Ground track. Refers to the vertical projection of the actual flight path of a satellite (or aircraft) onto the surface of the Earth.

Ground truth. Reference data which is collected in the field (generally on or near the Earth's surface). The objective is to verify remotely-sensed primary sensor data against a typical 'reference.' Ground truth data is generally used in support of the analysis of remotely-sensed data. This ground truth may be gathered either by a single ground station, or by a network of ground stations (including buoys, remote terminals, etc.) whose data is collected by a 'data collection' satellite, or by airborne underflights of a spaceborne sensor. The ground reference that is being sought depends very much on the application. For instance, it may be a simple visual verification of the vegetation types in a particular scene, or it may be a particularly prepared ground patch with known radiative characteristics (reflectance, etc.) that is being used for the calibration of an airborne sensor. A very prominent ground truth station is for instance MOBY (see chapter P.139 on page 1731), which makes in situ measurements of ocean color near the Hawaiian island of Lanai.

Hall effect. Edwin H. Hall (1855 - 1938, American physicist) discovered what became known as the Hall effect in 1879 when investigating the nature of force acting on a conductor carrying a current in a magnetic field. Today, Hall effect measurements are used to characterize the electronic transport properties of semiconductors and metals. - In a model Hall effect measurement system, a uniform current density flows through a uniform slab of electrically conducting material in the presence of an applied perpendicular magnetic field. The Lorentz force then acts on the moving charge carriers, deflecting them to one side of the sample to generate an electric field perpendicular to both the current density and the applied magnetic field. The ratio of the perpendicular electric field to the product of current density and magnetic field is the Hall coefficient. The ratio of the parallel electric field to the current density is the resistivity.
The Hall effect is also employed in Hall plasma thrusters (electric propulsion).

Heterodyning (see chapter O.4)

High-pass filter (in optical sensors). An absorption filter pervious to electromagnetic radiation above a certain wavelength only.

Humidity. The water vapor content of air. The term is commonly used to mean "relative humidity," the dimensionless ratio of vapor that a given quantity of air can contain at a given temperature, expressed as a percentage. Perfectly dry air has a relative humidity of 0%; totally saturated air has a relative humidity of 100%.

Huygens's principle (Christian Huygens, Dutch physicist, 1629-1695). A very general principle applying to all forms of wave propagation which states that every point on the instantaneous position of an advancing wave/phase front may be regarded as a source of secondary spherical 'wavelets'. The position of the phase front an instant later is then determined as the envelope of all secondary wavelets. - This principle is extremely useful in understanding the effects due to refraction, reflection, diffraction, and scattering of all types of radiation. At his time, Huygens was the leading proponent of the wave theory of light.

Hydrological cycle. Virtually all the water of the hydrosphere is in constant circulation, moving through the hydrological cycle, a vast series of interchanges of geographic position as well as of physical state. Broadly speaking, the hydrologic cycle involves the transfer of water from the oceans through the atmosphere to the continents and back to the oceans. The processes involved are complex combinations of evaporation, transpiration, condensation, precipitation, runoff, infiltration, subterranean percolation, and others. The mean residence time of water in the atmosphere, on the land surface, and in the oceans is an important climate parameter.

Hydrometeors. These are cloud scatterers (when measured by a lidar or radar instrument) in different phases such as rain or frozen precipitation with a definite fall velocity.

Hydrosphere. All liquid and frozen surface waters (including the oceans, lakes and streams), groundwater held in soil and rock, and atmospheric water vapor. The Earth's total water budget is estimated at roughly 1.385×10^9 km^3. Over 97% of the Earth's total water supply is contained in the oceans and other saline bodies of water. Of the remaining 3%, a little over 2% is tied up in glaciers and ice caps and, along with atmospheric (0.001%) and soil moisture, is inaccessible. Hence, approximately 0.62% of the total amount of water may be regarded as 'freshwater' in rivers, lakes and groundwater supplies.

Hyperspectral. In remote sensing the term implies a spectral signature of narrow, continuous and contiguous spectral bands per pixel, i.e. a fine spectral resolution ($\Delta\lambda/\lambda$ = 1 - 5%), over an extended spectral range. A hyperspectral sensor has a minimum of 20 spectral bands (normally 30 - 200 bands). On the other hand, "multispectral" usually implies fewer, spectrally broader bands, which may be noncontiguously spaced color bands per pixel. Hyperspectral imaging implies a hyperspectral signature for each pixel of the image. The detailed spectral information captured in a hyperspectral image allows detailed examination of the observed scene.

Image. Remotely sensed imagery (collected by CCD instrument, camera, or radar - in a wide range of the electromagnetic spectrum) is measured data which is transformed by numerical algorithms into an image. This is referred to as "computed imaging." [2054] An image is a two-dimensional grid of data; each of its elements is a pixel (picture element) whose coordinates are known and whose light intensity has a DN (Digital Number) value. The coordinates of the pixels and their DN values describe the image in terms of rows, called 'lines', and columns, called 'samples'. An 8-bit pixel provides up to 256 brightness levels (level 0 is set to black, while level 255 is set to white); the brightness levels are also referred to as 'grayscale levels.' In 'false color image processing', those pixels which have the same DN value are assigned an arbitrary color. This enhancement technique is used, for example,

2054) Other forms of computed imaging (i.e. a restructured image of measured data) are in such fields as: tomography, x-ray crystallography, electron microscopy, seismic imaging, and radio astronomy.

to differentiate between various types of terrain or species of vegetation - to reveal changes which are otherwise not perceptible to the human eye.

Image correction. The adjustment of an image for errors: geometric, radiometric, etc.

Image correlation. The ability to locate or match a region of an image with a corresponding region of another image which can be taken with a different sensor at a different viewing angle.

Image degradation. Loss of resolution due to modulation transfer function defects including motion blur, nonlinear amplitude response, shading and vignetting and channel noise.

Image enhancement. The improvement of images to facilitate better interpretation (false color processing is an example), or further digital processing to develop a specific theme or to highlight certain features in an image series.

Image motion compensation. Algorithms (hardware and/or software) counteracting image motion during integration time, thereby reducing image blur. The blur effect increases when relative velocities of the sensor platform are noticeable with respect to the image integration time. The sensor platform attitude parameters are also important inputs for motion compensation. For low-flying aircraft motion compensation is normally in the forward direction. There are also algorithms in use for antenna motion compensation. For instance, the short-term scanning motion of the antenna is measured by IMU (Inertial Measurement Unit). This data is used as input to compensate the pulse-to-pulse phase of the radar to maintain coherence during synthetic aperture.

Image quality. Refers to the apparent central core size of the observed image, often expressed as an angular image diameter that contains a given percentage of the available energy. Sometimes it is taken to be the full width at half maximum (FWHM) value of the intensity versus the angular radius function. A complete definition of image quality would include measurements of all image distortions present, not just in size or projection. But this is frequently difficult to do, hence the approximations.

Image rectification. A process by which the geometry of an image area is made planimetric. Image rectification doesn't remove relief distortion or perspective distortion.

Image resampling. A technique for geometric correction in digital image processing. Through a process of interpolation, the output pixel values are derived as functions of the input values combined with the computed distortion. Nearest neighbor, bilinear interpolation and cubic convolution are commonly used resampling techniques.

Image space. The mathematical space describing the position (coordinates) of pixels in the image.

Imaging array. A solid-state imaging array consists of a 1-D or 2-D set of photodetectors onto which an image can be focused, together with an integral electronic readout scheme. The devices are classified according to the particular readout scheme and detector type. A linear array is an imaging array consisting of a single line of detectors. The introduction of detector arrays has resulted in a new type of spectrometer - the imaging spectrometer - capable of generating images in much narrower bands (hyperspectral imagers) than was possible with conventional spectrometers.

Imaging sensors. Instruments that produce a 2-D image of the target area. Imaging sensor systems may by subdivided into 'framing systems' (such as camera or vidicon) and 'scanning systems.'

Imaging spectrometry. The simultaneous acquisition of images in many contiguous spectral bands.

Incidence angle. Angle formed between the instantaneous line of measurement of a sensor to a target and the local vertical of the target (or object).

Inclination. In general the angle between two planes or their poles; usually the angle between an orbital plane and a reference plane (i.e. the equator plane). Inclination is one of the standard orbital elements specifying the orientation of the orbit plane (see also O.12 on page 1265).

Infrared radiation. Refers to electromagnetic radiation lying in the wavelength interval from 0.7 μm to 1000 μm (or roughly between 1 μm and 1 mm wavelength). Its lower limit is bound by visible radiation, and its upper limit by microwave radiation. Most of the energy emitted by the Earth and its atmosphere is at infrared wavelengths. Infrared radiation is generated almost entirely by large-scale intermolecular processes. The triatomic gases, such as water vapor, carbon dioxide, and ozone, absorb infrared radiation and play important roles in the propagation of infrared radiation in the atmosphere. - The phenomenon of infrared radiation was first discovered by the English astronomer Sir William Herschel (1738-1822) in 1800 (he dispersed the solar spectrum with a prism and found thermal effects of invisible radiation beyond the red).

In-situ soundings. An observation method (using sensors on such platforms as aircraft, balloons, ships, buoys, towers, spacecraft, on the ground, etc.) with the objective to measure parameters in the immediate environment. From a historical point of view 'in-situ' observation predated 'remote' observation by ages. Thermometers, barometers, thermocouples, hygrometers, air samplers, etc. are in-situ sensors, as are in fact most sensors in the fields of meteorology, atmospheric chemistry, hydrology, etc. By far the largest percentage of ground-truth observations are in-situ measurements (i.e., measurements of parameters at a particular location and at a particular time). Remote sensing data of a particular sensor and target area may be compared (calibrated) against in-situ data of that target area. Spaceborne data collection systems, like ARGOS on polar orbiting satellites, remotely collect data from many thousands of 'in-situ measurement systems' in the ground segment on a routine basis. Examples of spaceborne in-situ observations are: sensors measuring magnetic or electric field parameters, solar wind particles, etc.

By their very nature, in-situ measurements are local measurements; hence, they offer a very low observation efficiency with regard to coverage and timeliness (repeat periods). It would take a fleet of spacecraft to obtain in-situ data with sufficient spatial and timely resolution on a global scale.

Instantaneous Field of View (IFOV). A term denoting the (angular) aperture within which the sensor is sensitive to electromagnetic radiation The IFOV may be expressed either as a small solid angle (in mrad or μrad, in this case the value is independent of the orbital altitude of the sensor), or as a unit area (e. g. 6 m x 6 m), or simply as the pixel size. Hence, the IFOV actually represents the spatial resolution of a sensor measurement.

Integration time. Refers to the short time period allocated for the radiative measurement of the instantaneous area of observation by the detector of a sensor (see also **dwell time**). Depending on sensor type the integration time may be very short [as is the case with electromechanical scanning systems which measure each individual cell (IFOV) across the swath sequentially], while the entire swath width (FOV) is measured by a CCD detector array in a single measurement.

Interference. Signals that arise from sources extraneous to the measurement system and result in errors in the measured value.

Interference filter. A filter reflecting radiation selectively in a narrow spectral band.

Interferogram. An image of interference phenomena such as phase differences (as patterns of interference fringes) measured by an interferometer, FTIR spectrometers, or generated by SAR interferometry. Interference fringes form through the interaction of two beams. Hence, for a given wavelength, the signal on the detector is either strong or weak. Two cases are of interest:

- Constructive interference. This occurs when the optical path difference (OPD) between the two light paths from the collectors to the detector is zero or a multiple of the observing wavelength
- Destructive interference. It occurs when the OPD is a half-integer number of wavelengths. The destructive interference is also referred to as 'nulling mode'.

Interferometer. An instrument class for dispersing spectra. The technique determines the relative phase of two (or more) wave fronts as a function of spatial location by observing interference fringes. Radiation is split into two or more beams which traverse different path lengths. The beams are reflected by mirrors and recombined for interference analysis. (see chapter O.10).

Interlaced scanning. Refers to a subsampling readout technique from a detector array of a camera or from some other high data-rate instrument. The advantage of such a measure is to reduce the bandwidth for image transmission. However, the disadvantage is a delay in total image recovery on the receiving side, resulting in distortions for fast moving objects in successive frames. The interlaced scanning technique is used for NTSC (National Television System Committee, a US TV display standard) television camera readout, in which each frame is scanned in two successive fields, each consisting of all the odd or all the even horizontal lines. The technique is also employed in other camera systems.

Intermediate frequency (IF). A common microwave frequency in an instrument (say, 80 MHz) for all channels at which considerable amplification takes place, interconnections are made, automatic gain adjustment is provided, and channels may be disabled upon command (squelch). The IF concept of a relatively low frequency is used to simplify the design of all functions compared to performing them at much higher frequencies in the GHz region.

Interplanetary magnetic field. Designates the magnetic field carried out from the sun by the solar wind, which permeates the entire heliosphere.

Intertropical conversion zone. A low pressure trough and minimum east wind, lying between the trade regions of the two hemispheres, that are nearly continuous around the world on climatological charts.

Inversion (atmosphere). A positive temperature gradient or increase in temperature with elevation, resulting in adverse conditions for the dispersion of pollutants.

Ionosphere. Layer of the Earth's atmosphere, between approximately 60 and 1000 km in altitude (a highly variable and complex physical system), that is partially ionized by solar x-rays, ultraviolet radiation, and energetic particles from space. The process of ionization is controlled by chemical composition and transport by diffusion and neutral wind. The region between about 160 to 1000 km, known as the F-region of the ionosphere, contains the greatest concentration of electrons (the peak electron concentration is around 250-300 km). The presence of the ionized zone disturbs the propagation of electromagnetic waves, particularly at the lower wave frequencies. Below at certain frequency, the wave may be totally reflected by the ionosphere. - At times, the F-region of the ionosphere becomes disturbed and small-scale irregularities develop. When sufficiently intense, these irregularities scatter radio waves and generate rapid fluctuations (or scintillation) in the amplitude and phase of radio signals.
The ionosphere was discovered in the early 1900s when radio waves were found to propagate "over the horizon." If radio waves have frequencies near or below the plasma frequency, they cannot propagate throughout the plasma of the ionosphere and thus do not escape into space; they are instead either reflected or absorbed. At night the absorption is low since little plasma exists at the height of roughly 100 km where absorption is greatest. Thus, the ionosphere acts as an effective mirror, as does the Earth's surface, and waves can be reflected around the entire planet much as in a waveguide.

Ionospheric disturbances. Refer to transient changes in ionospheric densities or currents, or the appearance of electron density irregularities, usually in association with the arrival of x-rays or UV bursts from solar flares.

Irradiance. Received radiative power per unit area (or the incident radiative flux onto a surface expressed in Wm^{-2}).

Isotropic radiator (antenna). A theoretical radiator of infinitesimal size in which it is assumed that all energy is distributed evenly (point source). Such a concept serves as a reference for other antennas of finite dimensions.

Lambertian radiator. A target or an object having the radiative property of directional independence of its (emitted and reflected) energy. With respect to reflection the object or target may be regarded to be a diffusely reflecting surface.

Langmuir probe. An instrument employed to measure the current-voltage characteristics of a plasma (single and double probes are in use) in order to determine plasma density.

Laplacian filter. A linear window operation (digital filter) concerning second derivatives of the pixel values within a window, either unidirectional or bidirectional (orthogonal).

Laser (Light Amplification by Stimulated Emission of Radiation). A source of light that is highly coherent (spatially and temporarily) and emitted in one or more wavelengths. A typical laser consists of two essential elements: gain and feedback. A beam of light passing through the gain, or amplifying, medium stimulates it to release its stored energy in the form of additional light that adds to, or amplifies, the beam. Feedback is achieved by placing the gain medium within the resonator (a set of mirrors that reflects the beam back and forth through the gain medium). The light from such a laser is composed of a number of discrete wavelengths corresponding to different resonant frequencies, or modes, of the resonator. There are two groups of lasers which operate either in a pulsed mode or in a continuous mode. The spectral range of lasers extends from the UV to the TIR (90 nm to 12 μm). Types of lasers: [2055)]

- **Gas lasers**: Usually receive their energy input via collisions of gas atoms with high-energy electrons. This energy is provided by applying a high voltage between electrodes located within the gaseous medium. The most common types of gas lasers are:
  - **He-Ne laser**: Uses a gas discharge of helium and neon as the gain medium (this was the first laser to emit a continuous output beam).
  - **Argon and krypton ion laser**: Uses a gas discharge containing ions as a gas medium. First lasers to operate in the blue and green regions of the spectrum.
  - **CO_2 laser**: One of the most powerful lasers, operating mostly in the spectral region of about 10.5 μm. They range from small versions with a few mW of continuous power to large pulsed versions.

    TEA (Transverse Excitation-Atmospheric pressure): These are generally pulsed CO_2 lasers; both the gas flow (about 1 atmosphere) and the electric discharge are transverse to the optical axis. Such conditions yield tremendous population inversions for short times. Commercially available TEA lasers deliver 100 to 200 ns pulses of several Joules/pulse at a repetition rate of 50 Hz; they are used for welding and cutting.[2056)]
  - **Rare-Gas-Halide excimer laser**: Operate primarily in the UV region in mixtures of rare gases, such as argon, krypton or xenon, with halide molecules such as chlorine and fluorine.
  - **Chemical lasers**: In these lasers the molecules undergo a chemical reaction. The hydrogen-fluoride laser fits into this class.
- **Metal Vapor lasers**: These lasers are actually a type of gaseous laser, since the laser action occurs in the atomic or molecular vapor phase of the species. The two best-known types are the helium-cadmium ion laser and the pulsed copper vapor laser.

2055) Encyclopedia of Physical Science and Technology, Academic Press, 1987, Vol. 7 pp. 153-160
2056) Courtesy of L. Zink of NIST, Boulder, CO

Parameter	Laser type	Wavelength (nm)	Energy range
CW (Continuous Wave)	Argon Ion	488 and 514	1 μW to about 1 W
	He-Ne	633	1 μW to about 20 mW
	Diode	830	100 μW to about 20 mW
	Nd:YAG	1064 1319	100 μW to about 450 W 100 μW to about 10 mW
	HeNe	1523	100 μW to about 1 mW
	CO_2	10600 (or 10.6 μm)	1 μW to about 1 kW
Pulsed	KrF Excimer	248	10^{-3} to about 200 mJ/pulse 50 μW - 9W average power
	ArF Excimer	193	10^{-3} - 3 mJ/pulse 50 μW - 3 W average power
	Nd:YAG	1064	1 - 50 mJ/pulse 10 nW - 100 μW 10^{-3} - 10 nJ/pulse

Table 561: Overview of some laser characteristics [2057)]

- **Solid-State lasers**: These laser generally consist of transparent crystals or glasses as "hosts" within which ionic species of laser atoms are interspersed or 'doped.' Typical host materials include aluminum oxide (sapphires), garnets, and various forms of glasses, with the most common lasing species being neodymium ions and ruby ions. - The energy input in these lasers is provided by a light source that is focused into the crystal to excite the upper laser levels. The light source is typically a pulsed or continuously operating flash lamp.
  - **Nd:YAG laser**: A laser whose gain medium consists of a neodymium-doped yttrium aluminum garnet crystal. The laser emits in the NIR region at 1.06 μm.
  - **Ruby laser**: This laser is produced by implanting chromium ions into an aluminum oxide crystal host and then irradiating the crystal with a flash lamp to excite the laser levels.
  - **Color center laser**: This laser uses a different form of impurity species implanted in a host material (usually one part per ten thousand). Color lasers typically operate in the 0.8 - 4 μm region and are tunable by using different crystals having different emission wavelengths.
- **Semiconductor lasers**: Semiconductor or diode lasers are the smallest lasers yet devised (about the size of a grain of salt). They consist of a p-n junction formed in an elongated gain region, typically in a gallium-arsenide crystal, with parallel faces at the ends to serve as partially reflecting mirrors. The light output of semiconductor lasers can be directly modulated using bias current, they can be tuned in wavelength using both temperature and bias current. Semiconductor diode lasers range in wavelength from 0.7 to 1.8 μm with typically continuous output power of up to 10 mW. Two types of semiconductor diode lasers are in wide use: [2058)]
  - **EEL** (Edge-Emitting Laser). A horizontal-cavity laser with an optical output beam emitting from the edge of the laser chip.
  - **VCSEL** (Vertical-Cavity Surface-Emitting Laser). A VCSEL's cavity is perpendicular to the wafer plane (the beam is guided in the vertical direction). The VCSEL is used for wavelength engineering, in optical fiber communications, etc.
  - **Quantum cascade laser (QC laser).** [2059)] Refers to a laser-based semiconductor sensor that operates at room temperature and at high power to detect minute amounts of trace gases (ppb) or pollutants by scanning for their optical-absorption "fingerprints." The lasers' high peak power, of 50-60 mW at 300 K, allows the use of uncooled detectors and enables LIDAR applications. They are particularly well suited for portable, robust sen-

2057) M. Dowell, "Pulsed-Laser Metrology at NIST," Optics&Photonic News, Feb. 2001, pp. 30-33
2058) C. J. Chang-Hasnain, "VCSELs Advances and Future Prospects," Optics & Photonics News, May 1998, pp. 34-39
2059) http://www.bell-labs.com/news/1997/may/21/4.html

sors in applications such as the point detection of trace gases and remote sensing applications.
The QC laser technology was invented by Jerome Capasso, Jerome, Faist, Sivco, Carlo Sirtori, Hutchinson and Cho of Bell Labs (Murray Hill, NJ) in 1994, who demonstrated continuously tunable, single-mode, QC distributed-feedback lasers operating at mid-infrared wavelengths (5 and 8.5 μm) in pulsed mode. The single-mode tuning range is typically 50 nm in wavelength, and the peak powers are 60 mW.
QC lasers are made using the technique of MBE (Molecular Beam Epitaxy), featuring layered structures of only a few atoms thick. The QC laser's emission wavelength is determined initially by quantum-confinement effects: the fact that its layers are so thin that electrons are squeezed and change their quantum-mechanical properties, allowing a range of possible wavelengths. The distributed-feedback lasers incorporate a grating that makes it possible to further refine the laser's wavelength, making them continuously tunable. - The operation of a QC laser is unlike that of other laser types. They operate like an electronic waterfall: when an electric current flows through a QC laser, electrons cascade down an energy staircase; every time they hit a step, they emit an infrared photon. At each step, the electrons make a quantum jump between well-defined energy levels. The emitted photons are reflected back and forth between built-in mirrors, stimulating other quantum jumps and the emission of other photons. This amplification process enables high output power.
Bell Labs has built built QC lasers operating throughout the mid-infrared region from 4.5 to 11.5 μm in both pulsed mode at room temperature and in continuous wave (CW) mode at temperatures up about 110 °C.

- **Liquid (dye) lasers**: Dye lasers are similar to solid-state lasers in that they use a host material in which the laser (dye) molecules are dissolved. Different dyes have different emission spectra or colors, thus allowing dye lasers to cover a broad wavelength range (320 - 1500 nm). A unique property of dye lasers is the broad emission spectrum (typically 30-60 nm) over which gain occurs. The dye laser is tunable over a frequency range of 10^{13} Hz.
  - **Dye laser**: Laser in which the gain medium consists of an organic dye dissolved in a liquid solvent. Applications in areas where tunability of the laser frequency is required. Dye lasers are also used for producing ultra-short pulses, a technique which is referred to as 'mode-locking.'
- **Free-Electron lasers**: These lasers are significantly different from any other type of laser in that the laser output does not result from discrete transitions in atoms or molecules of gases, liquids, or solids. Instead, a high-energy beam (in the order of 1 MeV) of electrons is directed to pass through a spatially varying magnetic field that causes the electrons to oscillate in a direction transverse to their beam direction. This laser type can be used over a wide range of wavelengths from the UV to FIR.

Layman-alpha radiation. The radiation emitted by hydrogen at 1,216 Ä, first observed in the solar spectrum by rocket-borne spectrographs.

Leads. Leads are transient areas of open water and/or very new ice, created in response to convergence/divergence phenomena (deformation processes) in the polar ice pack (see also **nilas**).

Limb/Occultation sounding. A horizon-looking (or edge-looking i.e. outer edge of the apparent disk of a celestial body) observation technique that uses a distant object [(for occultation sounding) sun, star, or a sensor on another satellite in a different Earth orbit, (see Figure 363)] as a source to observe the signal on its path through the atmosphere that is essentially tangential to the Earth's surface. Two types of occultation techniques have been used in the past to determine the composition and structure of the atmosphere:

1) **Extinctive occultations:** These occur because atmospheric constituents absorb or scatter the incoming radiation. Since extinction cross sections are generally wavelength-de-

pendent, spectral measurements - as the star (sun) sets deeper into the atmosphere - are diagnostic of the atmospheric composition. Hence constituent profiles may be determined from the relative transmission (i.e., the ratio of occulted to un-occulted spectra0. As a result, extinctive occultation measurements are self-calibrating and ideal for long-term trend monitoring. - The technique provides measurements that are commonly referred to as trace gas monitoring. Examples of spaceborne instruments employing extinctive occultations are: SAGE (AEM-2, ERBS, Meteor-3M, ISS), POAM (SPOT-3,4,5), HALOE (ATLAS-1,2), ATMOS (ATLAS-1, Spacelab-3), ILAS (ADEOS, ADEOS-II), and GOMOS (ENVISAT).

2) **Refractive occultations:** They occur because density gradients in the atmosphere lead to refraction of the incoming radiation, causing it to follow curved paths through the atmosphere. Relative measurements of the degree to which the path of the incoming radiation is changed provide the bulk of atmospheric properties (density, pressure, temperature). Usually, this occultation technique employs dual-frequency carrier phase observations of retarded signals (atmospheric propagation delays) from GPS or GLONASS satellites which permit the derivation of atmospheric profiles of density, pressure, and temperature. The GPS/MET instrument of Microlab-1 and TRSR (TurboRogue Space Receiver) of CHAMP, SUNSAT, and other missions are examples of refractive radio occultation monitoring.

Very long atmospheric paths (up to 4000 km) with high sensitivities or dynamic ranges can be obtained in this 'limb-sounding' configuration. The refractive technique takes advantage of the precise knowledge of GPS satellite' positions and timing of GPS radio signals. Instruments like GPS/MET and TRSR measure the extra time it takes for a GPS signal to enter Earth's atmosphere obliquely, pass through, and re-emerge to strike the LEO S/C - compared to an otherwise un-refracted direct ray path. The time delays of the GPS signal due to such atmospheric passage during the course of the occultation are used to derive the corresponding bending angles of the ray path, which in turn are converted to the refractive index profile of the atmosphere.

Some key advantages of limb sounding are:

- It maximizes the reception of the signal emitted by the atmospheric layer at the viewed tangent height provided that the receiver antenna has a very narrow beam
- The background temperature (that of the deep space) is much colder than that of the atmosphere, which guarantees very low biasing of the measured atmospheric emissions
- Vertical profiles with a high resolution can be obtained by limb scanning with a very narrow antenna beam
- A global coverage is achieved with LEO polar orbiting satellites.

A major disadvantage of limb sounding is: Low horizontal resolution due to the measurement geometry (long path-length) and high speed of the spacecraft (signal integration along the moving tangent point).

Lithosphere. Earth's outer rigid crust composed of rocks rich in silicon, aluminum, calcium, sodium, potassium, and some other elements, and hence less dense than the underlying mantle. The lithosphere varies from less than 30 km thickness under the ocean to over 100 km under continents.

Local oscillator (LO). A receiver oscillator that produces a reference sinusoid for comparison with the noisy received sinusoid.

Look angle. The direction in which an antenna is pointing when transmitting and receiving from a particular cell (for an active instrument). Refers to the instrument pointing direction from nadir. - In the current literature the terms 'look angle,' 'illumination angle,' 'pointing angle,' 'off-nadir angle,' and 'viewing angle' all have the same meaning.

Looks (in SAR imagery) see chapter O.8.4 on page 1252.

LORAN (Long-Range Navigation). A hyperbolic radionavigation system (the term 'hyperbolic' refers to the reflected ionospheric transmissions) which uses the difference in the time of arrival of signals from individual transmitters to establish position. Historically, LORAN was developed at MIT (DoD funded) during World War II as a navigation aid, known as LORAN-A. The operation of LORAN-A was in the 1850-1950 kHz radio band. LORAN-A had a range of 1000 km. Later LORAN-B was developed to improve the accuracy of LORAN-A. In 1958 LORAN-C became operational; it was used commercially for marine navigation. LORAN-C was transferred to civil authority (DOT) in 1974. Later, FAA extended LORAN-C coverage to include the continental USA. - LORAN-C is a long-range (in excess of 1850 km), low-frequency (90 -110 kHz) hyperbolic navigation system. Loran-C transmissions consist of groups of eight or nine accurately timed and phase-coded pulses at a carrier frequency between 90 and 110 kHz. Each chain consists of a master and a number of slave transmitters. - In the mid 1960's LORAN-D (a low power transportable system with a range of 1100 km) was developed by the USAF. - Loran signals propagate as a ground wave, but sky waves reflected from the ionosphere are also received. - In the early 1990s, DoD declared that by the end of 1994, there would be no further military requirement for the system and authorized the transfer of LORAN-C assets to host nations for civil use. The USA plans to terminate Loran-C operations on Dec. 31, 2000.

A rebirth of LORAN-C in Europe was initiated in 1992 as result of an international agreement between the US and six European countries (Denmark, France, Germany, Ireland, the Netherlands, and Norway), known as NELS (Northwest European LORAN-C System). The NELS network, consisting of nine stations with a control center at Brest (France), started operations in 1999. The performance characteristics of the NELS network are: absolute location accuracy of 100-460 m; repeatable accuracy of 20-100 m; availability per station of 99.9%; and availability per chain of 99.7%. A policy recommendation of IALA (International Association of Marine Aids and Lighthouse Authorities) states that the future use of radionavigation be based on complementary satellite and terrestrial systems. [2060)]

Low Noise Amplifier (LNA). A preamplifier between antenna and receiver. It is usually attached directly to the antenna receive port. Its main function is to reduce the thermal noise of the received signal.

LOWTRAN. LOW-resolution TRANsmittance - a computer code (model of USAF Geophysics Laboratory, also referred to as AFGL, Hanscom AFB, MA) which predicts the atmospheric transmittance and thermal radiation emitted by the atmosphere and the Earth. LOWTRAN programs are applicable for wave numbers ranging from 350 cm^{-1} (28.5 μm) in the infrared to 40,000 cm^{-1} (0.25 μm) in the UV region. The LOWTRAN code calculates atmospheric transmittance and radiance, averaged over 20 cm^{-1} intervals in steps of 5 cm^{-1}. The succeeding models are more advanced, building on the capabilities and options of the previous LOWTRAN models. [2061)]

- LOWTRAN-2 (1972)
- LOWTRAN-3 (1975)
- LOWTRAN-3B (1976)
- LOWTRAN-4 (1978)
- LOWTRAN-5 (1980)
- LOWTRAN-6 (1983) includes solar/lunar scattering, new spherical refractive geometry subroutines, and an improved water vapor continuum model. Other modifications include a wind-dependent maritime aerosol model, a vertical structure aerosol model, a cirrus cloud model, and a rain model.

2060) U. Klinge, "A European Approach for an Integrated System - LORAN-C/Eurofix," Galileo's World, Vol. 1, No 1, Winter 2000

2061) F. X. Kneizys, et al., "Atmospheric transmittance/radiance: computer code LOWTRAN-6, Report AFGL-TR-83-0187, Bedford, MA, NTIS Report ADA137786

- LOWTRAN-7 (1988). This code has been extended to include the microwave spectral region.
- MODTRAN (1989)[2062] (MODerate-resolution LOWTRAN). MODTRAN is an extended version of LOWTRAN-7 (six additional subroutines) to increase its spectral resolution from 20 cm^{-1} to 2 cm^{-1} (FWHM). Further objectives were: a) to model molecular absorption of atmospheric molecules as a function of temperature and pressure; b) to calculate band model parameters for twelve LOWTRAN molecular species; c) to integrate LOWTRAN-7 capabilities into the new algorithms, maintaining compatibility with the multiple scattering option. For MODTRAN, molecular absorption is calculated in intervals of 1 cm^{-1} bins, the other parts of the calculation remain unchanged. The molecular species affected are: water vapor, carbon dioxide, ozone, nitrous oxide, carbon monoxide, methane, oxygen, nitric oxide, sulfur dioxide, nitrogen dioxide, ammonia, and nitric acid. The MODTRAN spectral region is from 0 - 17,900 cm^{-1} (with 2 cm^{-1} spectral resolution), calculations at larger wave numbers, i.e. the VIS and UV regions, are performed at the lower spectral resolution of 20 cm^{-1}.

Magnetometer. An instrument for measuring changes in the Earth's magnetic field.

Magnetopause. The surface defining an interface between the magnetic field of a star and matter in the disk; a surface where the average magnetic pressure of the magnetic field is in pressure balance with the plasma pressure. Earth's magnetopause is the region in the ionosphere where the magnetosphere meets the solar wind (see Figure 408).

Magnetosphere. The region of space surrounding a rotating, magnetized sphere. Specifically, the outer region of the Earth's ionosphere, starting at about 1000 km above Earth's surface and extending to about 60,000 km (or considerably farther, such as 100 R_E on the side away from the Sun).

Marginal ice zone (MIZ). The critical region in which polar air masses, ice, and water masses interact with the temperate ocean and climate systems (an important geophysical boundary zone involving energy exchanges). The transition zone is characterized by large horizontal gradients in the properties of the ice, ocean, and atmosphere.

Maser. An amplifier utilizing the principle of 'microwave amplification by stimulated emission of radiation.'

Measurement mode - duty cycle. The fraction of available time during which an instrument is actively performing Earth measurements and producing meaningful data, including incidental calibration and overhead (such as scan retrace). High data rate, high power consumption, and steerable instruments may have small duty cycles. Daylight-only instruments may have measurement mode duty cycles averaging 50 percent.

Medium Earth Orbit (MEO). Refers to all satellite orbits between LEO and GEO. MEO orbits have larger and longer footprints than LEO orbits. Navigation systems, like GPS and GLONASS, are examples of MEO orbits. MEO orbits are also attractive to a number of communication satellite networks due to their relatively small transmission delay times (in the order of 0.1 s). The shorter radial distance (compared to GEO) translates into improved signal strength at the ground which means better reception and ultimately smaller terminals.

Mesosphere. Region of the atmosphere between approximately 50 and 85 km in altitude.

Meteoroid. A small particle in space (sources are comets, detritus from asteroid collisions, and interstellar dust). A meteoroid that survives the passage through the Earth's atmosphere and reaches the Earth's surface is known as a meteorite.

2062) A. Berk, L. S. Bernstein, D. C. Robertson, "MODTRAN: A moderate resolution model for LOWTRAN-7," AFGL-TR-89-0122, Air Force Geophysics Laboratory, Hanscom AFB, MA

Microwave radiation. Electromagnetic radiation generally considered to be in the wavelength range from approximately 1 mm to 1 m (three orders of magnitude). In microwave radiometry polarization is often used as a discriminant parameter, because microwave antennas are easily built with a single polarization direction. - (On the other hand, most optical sensors are relatively independent of polarization; a special effort must be made to polarize the signal prior to detection). - The largest portion of the microwave region is also the 'radar' region, hence both terms are used interchangeably (this also applies to the microwave instruments), see also O.8.2 and O.8.1.

Microwave rainfall monitoring. Microwave radiation with wavelengths in the order of 1mm to 3 cm results in a strong interaction between the raindrops and the radiation (the drop size is comparable to the wavelength). Passive microwave data is helpful in locating the leading and trailing edges of rain areas (extent), however, the actual measurement of rainfall and rain rates provides unsatisfactory results. So far passive microwave systems function promisingly over sea surfaces but not satisfactorily over land surfaces. The general consensus is that a passive (multichannel and multipolarized radiometer)/active (radar) instrument complement can provide a better characterization of rain systems.
Radar measurement of rainfall is based on Rayleigh scattering caused by the interaction of rain and the radar signals. The PR (Precipitation Radar) instrument on TRMM with a transmission frequency of 14 GHz (2.15 cm wavelength) is expected to improve on the problem of rainfall measurement.

Microwave signal penetration. Most natural terrain materials, with the exception of water, are partially transparent to microwave frequencies. The energy of a (radar) wave incident upon a terrain surface is partially scattered back into the atmosphere; the remainder is transmitted across the boundary into the terrain medium. - The most important parameters governing the depth to which microwaves can penetrate natural materials such as soil, snow, or vegetation are the wavelength, the moisture content of the material (soil), and the shape and sizes of the scattering elements (such as the leaves in a vegetation canopy, or ice crystals on a snow surface). Radar observations in L-band (2-1 GHz) and P-band (1.0 - 0.3 GHz) frequencies are providing first results in penetration measurements, in particular with regard to soil moisture content and canopy penetration. - Note: The microwave permittivity of water is an order of magnitude higher than that of any natural dry material. [2063)]

Millimeter-Wave (MMW) region. Refers to the spectral region from 1 mm (300 GHz) to 10 mm wavelength (300 MHz). The MMW region is part of the microwave region of the spectrum which extents conventionally from 1mm to 1 m wavelength. The MMW region is in particular of interest to radiometry (passive sensing) applications. Millimeter waves are able to penetrate many types of inclement weather, as well as opaque solids, and offers a lot of contrast. The emissivity of objects in this region is about 10 times higher that that in the infrared region.
Submillimeter-wave (SMMW) region. Refers to the spectral region from 0.1 mm (3000 GHz or 3 THz) wavelength to 1 mm (300 GHz) wavelength. The SMMW is also being considered as part of the microwave region. SMMW observations are of particular interest to atmospheric science. Submillimeter-wave limb sounding has some advantages compared with limb sounding in other frequency ranges. In the UV and VIS region only daytime measurements are possible, while in the infrared region some important species like ClO and HCl are not detectable.

Modeling. An investigative technique that uses a mathematical or physical representation of a system or theory that accounts for all or some of its known properties. Models are often used to test the effects of changes of system components on the overall performance of the system.

2063) The penetration depth of microwaves decreases with increasing frequency. The atmosphere produces frequency-dependent distortions which set an upper frequency limit due to attenuation. This limit is about 90 GHz for airborne radars and about 15 GHz for spaceborne radars.

Modulation. A process of manipulating the characteristics (usually frequency or amplitude) of a carrier in relation to another wave or signal.

Modulation Transfer Function (MTF). [2064] A function measuring the reduction in contrast from object to image (actually from each pixel), that is, the ratio of image-to-object modulation for sinusoids of varying spatial frequencies. The reason: any optical system reduces the contrast in the image compared to the contrast of the objects imaged; this is expressed as MTF. Thus, the MTF provides the response of an optical sensor as a function of object scene contrast and spatial frequency. MTF is also a measure of how accurately the actual radiance from a pixel (IFOV) is measured (a lower MTF indicates contributions from other pixels to the pixel of observation). A radiometrically accurate IFOV is one for which MTF >0.95.

Note: EIFOV (Effective Instantaneous Field of View) is defined as the resolution corresponding to a spatial frequency (ground resolution) for which the system MTF is 50%.

Moire Interferometry. A method to determine 3-D profile information of an object or scene, using interference patterns. Two identical gratings of known pitch are used. The first create a shadow of parallel lines of light projected on the object. The second is placed in the imaging train, then superimposed on the shadow cast by the first grating, forming a moire fringe pattern. Varying the gap between the lines changes the sensitivity.

Mosaic. An assemblage of overlapping airborne or spaceborne photographs or images whose edges have been matched to form a continuous pictorial representation of a portion of the Earth's surface.

Multilook technique. A technique of averaging a number of independent samples per pixel. applied to radar in order to reduce speckle.

Multiple access techniques. There are three basic multiplexing schemes that allow a number of simultaneous transmissions over a single circuit. See also **CDMA**, **FDMA**, and **TDMA** in the glossary text.

- **Code Division Multiple Access (CDMA)**. Refers to an access scheme which employs spread-spectrum modulations and orthogonal codes to share a communication link among its users.
- **Frequency Division Multiple Access (FDMA)**. A process that shares a spectrum of frequencies among many users by assigning to each a subset of frequencies in which to transmit signals.
- **Time Division Multiple Access (TDMA)**. A process that shares the time domain of a single carrier among many users by assigning to each time intervals in which to transmit signal bursts.

Multipath (GPS multipath). [2065] The term multipath is derived from the fact that a signal transmitted from a GPS satellite can follow a 'multiple' number of propagation 'paths' to the receiving antenna. This is possible because the signal can be reflected back to the antenna off surrounding objects, including the Earth's surface (land and/or ocean). Some characteristics of the multipath signal are: a) a multipath signal arrives always at a later time than the direct-path signal, b) a multipath signal is normally weaker than the direct-path signal due to the reflection loss, c) if the delay of the multipath is less than two PRN code chip lengths, the internally generated receiver signal will partially correlate with it. If the delay is > 2 chips, the correlation power will be negligible.

Multiplet. A spectrum line having several components.

Multispectral. In remote sensing the term implies two or more broad spectral bands in which a sensor detects radiation (see also **Hyperspectral**). The multispectral concept im-

2064) G. Joseph, "How well do we understand Earth observation electro-optical sensor parameters?" ISPRS Journal of Photogrammetry & Remote Sensing, Vol. 55, 2000, pp. 9-12

2065) B. Townsend, J. Wiebe, A. Jakab, M. Clayton, T. Murfin, "Analysis of the Multipath Meter Performance in Environments With Multiple Interferers, " ION GPS 2000, Salt Lake City, UT, Sept. 19-22, 2000, pp. 480-488 plies image analysis based on spectral characteristics. Note: The MSS (Multispectral Scanner System) sensor on Landsat-1 was probably one of the first spaceborne multispectral instruments with four spectral bands in VNIR.

Nadir. Direction toward the center of the Earth. Opposite of zenith.

Nilas. Thin and elastic ice sheets of a rather dull surface and up to a thickness of about 10 cm. Nilas form in quiet sea water and bend easily in swells. Transverse pressure causes nilas to pile up. **Leads** may be frozen over by nilas; but nilas are not bound to be present in leads; they may also occur outside of leads.

Noise. Any unwanted or contaminating signal competing with the desired signal. In a SAR instrument, two common kinds of noise are additive, receiver noise and signal-dependent noise, usually additive or multiplicative. The relative amount of additive noise is described by the signal-to-noise ratio (SNR). Signal-dependent noises, such as azimuth ambiguities or quantization noise, arise from system imperfections, and are dependent on the strength of the signal itself.

Noise-Equivalent-Equivalent-Width (NEEW) (sometimes shortened to 'noise-equivalent-width'). A spectral resolution defining the smallest line area (in cm^{-1}) that can be measured, or the equivalent area of the average noise bump.

Noise-Equivalent Flux Density (NEFD), sometimes also referred to as 'noise-equivalent irradiance.' Defined as the in-band entrance aperture irradiance on one pixel, from a point source, that equals the sensor rms noise (units are W/cm^2). NEFD is related to a more familiar quantity, the Noise-Equivalent Radiance (NER), by the equation: NEFD = NER x pixel subtends (in steradians). NEFD can also be defined as the noise-equivalent power (NEP) per unit aperture area: NEFD=NEP/A.

Noise-Equivalent Radiance (NER). Defined as the in-band entrance aperture radiance ($W/cm^2/sr$) equal to the sensor rms noise. NER is the preferred figure of merit for an extended (pixel-filling) source. It is also defined as the noise-equivalent power (NEP) per unit aperture area, per pixel subtense ($NER=NEP/A/\Omega$).

Noise-Equivalent Spectral Radiance (NESR). Defined as the radiance ($W\ cm^{-2}\ sr^{-1}\ cm^{+1}$) density that corresponds to the rms value of the spectral noise of a calibrated spectrum (or: the radiance change corresponding to SNR=1). $NESR = NEP/(\mu \times \tau \times E \times (t)^{1/2} \times \Delta\sigma)$, with NEP = Noise Equivalent Power of the detector, μ = modulation efficiency, τ = optical efficiency, E = Etendue, t = scan time (1 instance), and $\Delta\sigma$ = spectral resolution.

Noise figure. Ratio of total output noise power of a system to that part of the output noise power due to the signal source.

Nonimaging sensors. Instruments which measure directly such quantities as radiant flux, irradiance, and radiance, which describe the intensity of a radiation field or the optical properties of a surface or a region of space. The sensors are nonimaging in the sense that they do not produce an image (a picture), but rather, integrate over time, space, and wavelength to produce a spectral curve, or a set of numbers that characterize the electromagnetic radiation. Typical measurement products are profiles, such as flux profiles, temperature profiles, moisture profiles, etc. Typical nonimaging instruments are: radar altimeters, sounders, scatterometers, spectroradiometers, radiometers, (note: there are also imaging radiometers which use scanning techniques, and imaging spectroradiometers using discrete filter-wheel systems), lidars, etc.

Nowcasting. The term is used in meteorology and refers to the development of atmospheric features on time scales between 0 and 3 hours over regional and local areas. Nowcasting is closely linked to very short-range-forecasting which covers developments over the time scale of 3 to 12 hours over regional areas.[2066] A typical example of an application of nowcasting is severe weather incidents, where small-scale features undergo rapid development. Nowcasting requires the (almost) continuous monitoring of the area(s) of interest. Furthermore, as many of the features to be observed are fairly small, quite high spatial resolutions are needed.

Nyquist sampling rate (Henry Nyquist, a US physicist and a pioneer in the field of communication theory, was borne in 1889 in Sweden).[2067] - Refers to a sampling rate above which a band-limited signal can be reconstructed from its sample value. If a signal s(t) contains no frequency components at or above f_N Hz [s(t) is then said to be band-limited to f_N Hz], then s(t) can be completely reconstructed from its sample values, provided the samples are taken at a rate equal to or in excess of $f_S = 2 f_N$samples/s. **This condition is known as the Nyquist or Shannon sampling theorem**; f_S is referred to as the Nyquist sample frequency, and f_N is sometimes called the Nyquist frequency. The Nyquist theorem essentially states that the sampling rate must be twice that of the highest frequency to be represented.

Observational reference frames and models:

- **Lagrangian experiments** (Joseph Louis Lagrange, 1736-1813, French mathematician and physicist). Refer to a physical system that changes as time goes on from one configuration to another as it is progressing along a particular evolutionary path (path with the smallest result). In this concept observer and observed object have zero velocity relative to each other (Lagrangian coordinates are also referred to as 'material coordinates;' they do not vary with time). The strategy is to observe a reference volume (a cell of air) of interest as it moves through space. All measuring devices move along with the reference object. The concept is generally very complex with respect to instrumentation. Drifting buoys in the ocean or constant-pressure balloons in the atmosphere are Lagrangian-type experiments of relatively small complexity.
- **Eulerian experiments** (Leonard Euler, 1707-1783). In Eulerian coordinates the properties of a fluid are assigned to points in space at a given time, without attempt to identify the individual fluid cells from one time to the next. The observer moves relative to the observed object. Example: observation of air masses. The vast majority of experiments, particularly in meteorology, are of the Eulerian type; a sequence of synoptic charts is a Eulerian data representation.
- **Transect experiments**. The observer takes snapshot measurements by transecting a large reference volume into different directions. Example: an aircraft or a ship observes many different small-scale air masses as it moves through them.

Occultation. Distortion or interruption of a direct observation path between the observer (sensor) and a target by an intervening medium (such as an atmosphere or a celestial body). The occultation technique may for instance be used to study the Earth's atmosphere (or planetary atmospheres) by remote sensing (in a limb-viewing configuration). The atmosphere causes signal propagation delay and bending [between a transmitter (a GPS satellite) and a receiver (a GPS receiver on a LEO satellite)] due to the variation in the index of refraction in different shells of the atmosphere. - In the conventional sense, occultation refers to light path obstruction by an astronomical body, such as a star, by another astronomical body, as seen from Earth. In this context, a solar eclipse is the occultation of the sun by the moon.

Ocean color. The color of the ocean reveals information on the presence and concentration of phytoplankton, sediments, and dissolved organic chemicals. By studying the color of the light scattered from the oceans, optical sensors can quantify the amount of chlorophyll and other constituents in the various regions of the ocean. - Ocean color gives a quantitative measure of the spectral radiance of light reflected from beneath the ocean surface. Since

2066) "Meteosat Second Generation Programme Proposal," ESA/PB-EO (92), 57, 9 November 1992, p. 14
2067) F. J. Taylor, "Digital Signal Processing," Encyclopedia of Physical Science and Technology, Academic Press, 1987, Vol. 12. p. 600

phytoplankton dominates the optical characteristics of most ocean waters, it permits the estimation of marine plant biomass, ocean optical properties, and marine photosynthesis (primary production), see also 'chlorophyll.'

Ocean mixing. Processes that involve rates of advection, upwelling/downwelling, and eddy diffusion and that determine, for example, how rapidly excess atmospheric carbon dioxide can be taken up by the oceans.

Oceans of the world and climate. The oceans play a decisive role in the evolution of the climate. Through irradiation of the sun and exchanges with the atmosphere, they receive considerable quantities of heat, in particular at the intertropical latitudes, which they store due to their high thermal capacity, and which the ocean currents redistribute from the equatorial regions to the polar regions. - Current estimates account for 30-50% of the meridian transport which makes the climate of the middle latitudes more hospitable. Fluctuations in the circulation and elevation of the average sea level, under the combined effect of thermal expansion and the melting of the icepack, are indicators of climatic anomalies. - Thus, the world's oceans represent a major regulating factor of the climatic system. Their circulation and evolution must be understood (via satellite altimetry or other means of observation) to account for the climate variability. Some figures of the oceans illustrate the importance of the oceans to our environment. [2068]

- About 70% of the Earth's surface are covered with oceans, this amounts to a total area of 360×10^6 (million) km^2, the total ocean volume is estimated to be 1.46×10^9 (billion) km^3; hence, the total mass is about 1.46×10^{18} metric tons.
- The average depth of the oceans is about 3800 m
- The mass of the oceans is about 300 times larger than the mass of the atmosphere; the heat storage capacity of the oceans is 1200 times the heat capacity of the atmosphere; the oceans provide 70 times the carbon storage capacity of the atmosphere. See also "altimetry" and "hydrosphere" in the glossary.

In addition to the normal ocean-atmosphere heat exchanges, the winds blowing on the sea surface, contribute to the productions of surface marine currents. These currents travel much more slowly than the winds; however, these ocean currents can store a large quantity of heat. This property enables the ocean to stabilize the Earth's temperature. An example of such an ocean current is the Gulf Stream which forms in the west Atlantic seaboard (mainly in the Gulf of Mexico) and travels in the direction of northern Europe.

Omega. A long-range, worldwide, all weather, day and night radionavigation system operating in the VLF (Very Low Frequency) band of the radio spectrum. The Omega network consists of eight atomic-clock-controlled transmitters transmitting sequentially on assigned frequencies between 10.2 and 13.6 kHz. The Omega network transmitters are located in Argentina, Australia, Japan, Liberia, Norway, Reunion, and USA (Hawaii and North Dakota). Users of Omega are commercial airlines, ships, land vehicles, meteorology (tracking of balloons), etc. Omega, like LORAN, uses phase differences of continuous-wave radio signals. The receiver of a user synchronizes to the transmitter frequency and measures the phase relationship of the receiver's location. Two or more 'line-of-position' measurements define the receiver location. Omega (of World War II vintage, developed by the USA) provides positioning within 2 to 4 nautical miles at a 95% confidence level with 95% availability. - The USA terminated permanently its Omega operations on Sept. 30, 1997.

Operational sensor. In Earth observation an instrument is said to be "operational" if the following services are provided: continuity of observations, timeliness of data delivery to the costumer, and several usable data products. A number of environmental/meteorological missions, like NOAA/POES and GOES, the METEOSAT series, etc., with their major instruments (AVHRR, etc.) and services, are considered "operational." Routine service

2068) http://www.cnes.fr/WEB_UK/activities/programmes/Topex/top_posd.html

provision and regular use of the data by a user community are key ingredients for operability.

Optical depth. Refers to the negative logarithm of the extinction [ln I/I_0], where I is the radiation intensity at the back plane of the absorbing medium, and I_0 the incident intensity. The optical depth is the product of the extinction coefficient, the density of the medium, and the length of the transmitted medium layer (Lambert Beer Law).

On-orbit electric propulsion systems. (see chapter O.14)

Optical spectrum. Refers to electromagnetic radiation of frequency (or wavelength) that can be focused, dispersed, and detected using optical components such as lenses, mirrors, and gratings. This includes more than the narrow visible region; in general terminology the optical spectrum extends from 0.01 to 1000 μm (UV to FIR inclusive).

Optoelectronics.[2069)] The term is a contraction of 'optical electronics' and refers to the photon effects (interaction/conversion, transmission) with a medium and vice versa. Optoelectronics is one of the foremost research fields affecting many areas of solid-state computer technology [microprocessors, data storage, communication (photon guidance between a source and a detector, optical fibers for signal transmission in optical communication, lasers), visual display methods using LEDs (Light Emitting Diode) and LCDs (Liquid Crystal Display)], energy detection and conversion [photon conversion into electrical energy (photodetection, photovoltaics), etc.]. Optoelectronics has also a wide filed of applications in imaging sensor design.
Note: Often the terms electro-optical, electrooptical, and electrooptics are used with the identical meaning of optoelectronics. From a detection standpoint, the term optoelectronics seems to be more logical, because a detection sequence goes from the optics to the electronics.

Optoelectronic devices. Refer to systems in which the photon, the basic unit of light, is affected. There are four basic groups of optoelectronic devices:

- Photodetectors and solar cells - that convert photons into an electrical current
- Light-emitting diodes (LEDs) and semiconductor lasers - that convert an applied voltage into emitted photons
- Optical +s - that guide light between a light source and a detector
- Liquid-crystal displays - that use an applied voltage to change the reflection of light.

Optoelectronic detection (pushbroom scanner). The scanner uses a line detector to scan the cross-track direction of a scene, the total field of view is detected (imaged) simultaneously. The number of pixels is equal to the number of ground cells for a given swath. The motion of the platform (airborne or spaceborne) provides coverage in the along-track direction. When a 2-D line detector is used (several lines of detectors in the cross-track direction), then one dimension (cross-track) represents usually the spatial dimension, while the other is used for different spectral bands (multispectral imaging). Examples of pushbroom scanners are: HRV on SPOT series of CNES, LISS on the IRS satellite series of ISRO, AVNIR on the ADEOS S/C of NASDA, MSU on the RESURS series of Russia, ALI of the EO-1 satellite of NASA.

Optomechanical detection (whiskbroom scanner). A form of radiation detection, employing an oscillating or rotating mirror to a line-scanning whiskbroom scanner. On-axis optics or telescopes with scan mirrors sweep from one edge of the swath to the other. The FOV of the scanner can be detected by a single detector or an along-track line-detector. Typical examples of optomechanical scanners are: TM and ETM+ of the Landsat series, AVHRR on the NOAA/POES series, SeaWiFS on Orbview-2 (formerly named SeaStar), ASTER and MODIS of the Terra satellite.

2069) Special issue on Optoelectronics Technology, Proceedings of the IEEE, Vol. 85, No. 11, November 1997

Laser scanners (i.e. active lidar systems) utilize also **optomechanical scanner assemblies** just as many multispectral whiskbroom scanners. However, as active sensing systems, they are using a laser beam as a sensing carrier. Two optical beams must be considered, namely the emitted laser beam and the received portion of that beam.

Orbit types and terminology. See chapter O.12 on page 1265.

- LEO (Low Earth Orbit)
- MEO (Medium Earth Orbit)
- HEO (Highly Elliptical Orbit)
- GEO (Geostationary Orbit)
- Sun-synchronous orbit

Orographic phenomena. Meteorological events (precipitation, special winds, clouds, fronts, etc.) associated with the disposition and character of hills and mountain ranges (distribution effects the linked to the form of the terrestrial relief). Orographic precipitation results from the lifting of moist air over an orographic barrier; however, it is not limited to the ascending ground, but may extend for some distance windward of the base of the barrier. Orographic lifting refers to the deflection of an air current up and over mountains. Examples of orographic winds are: Föhn, Mistral, Bora, Santa Ana.

Orthographic projection. A projection in which the projecting lines are perpendicular to the plane of projection (also referred to as orthogonal projection).

Orthophotography. A digital orthophoto is a raster image, which has been accurately scanned and rectified with the aid of geodetic surveying and photogrammetry.

Orthorectification. (orthorectified imagery)

Ozone. A molecule made up of three atoms of oxygen (O_3). Ozone strongly absorbs UV radiation in the wavelength range of 290 - 300 nm. In the stratosphere, it occurs naturally and provides a protective layer (ozone layer between ~ 12-30 km in which ozone is relatively concentrated $> 10^{12}$ molecules/cm^3) shielding the Earth from ultraviolet radiation and subsequent harmful health effects on humans and the environment. In the troposphere, it is a chemical oxidant and major component of photochemical smog. Ozone is an effective greenhouse gas especially in the middle and upper troposphere and lower stratosphere. - The **depletion of ozone** in polar latitudes is attributed to a sequence of chemical reactions involving chlorine and bromine compounds. These sources are simple organic compounds containing chlorine, e.g. chlorofluorocarbons (CFCs), and/or bromine (e.g. halogens). Nearly all of the chlorine and about half of the bromine in the stratosphere originates from human activities.

Paleoclimatology.[2070] Science which deals with past climate periods of the Earth in very large space-time scales. Long-term baselines of past climate changes are studied and reconstructed to understand climate processes and predict future climate change (climate models). Paleoclimate data are derived from ice cores, tree rings, marine and lake sediments, fossil pollen, plant macrofossils, paleovegetation, past sea surface and lake level data, terrestrial ice sheet height and extent, land surface properties, etc.

Panchromatic channel. A channel of a sensor-detector system which covers the entire width of a spectral range, in particular the visible range (VIS).

Parallax. Apparent change in the position of an object due to an actual change in the point of view of observation. Example application: In optical stereo observation the parallax between two images, viewed from different angles, is used to derive the third dimension of altitude. This topographic information is used for map-making.

2070) D. M. Anderson, R. S. Webb, J. T. Overpeck, B. A. Bauer, "The NOAA Paleoclimatology Program," NOAA Earth System Monitor, Vol. 3, No. 3, March 1993, pp. 6-8

Particle precipitation. Refers to the release of charged particles, stored in the Earth's magnetosphere, into the atmosphere. The particles follow magnetic field lines. They cause a glow (creating an aurora) when they strike the atoms of the upper atmosphere.

Passive sensor. A sensing system that detects or measures radiation emitted by the target. Such sensing systems do not emit any power to the target for purposes of measurement. Hence, passive sensors are sensitive to radiation of natural origin, usually reflected sunlight or energy emitted by an object. Examples of passive sensors are: cameras, multispectral scanners, and radiometers.

Peltier effect coolers. They work on the principle of the thermoelectric effect. Such coolers are thermodynamically reversible low impedance devices, operating at a high current from a DC power supply. A single stage cooler can typically achieve a temperature of -40°C, and lower temperatures can be achieved using several stages. A six stage device may achieve -100°C and give a cooling power of around 1 mW at -80°C. Peltier coolers are by their very nature vibration-free.

Perigee. The point in an orbit at which the spacecraft is nearest to the Earth.

Phase modulation (PM). Angle modulation in which the phase of a sine wave carrier is caused to depart from the carrier phase by an amount proportional to the instantaneous value of the modulating wave.

Phased-array technology. Phased arrays are random-access devices (antennas) employed for electronic beam-steering applications in the microwave and/or optical regions of the spectrum. The technology of electronic beam-steering overcomes many limitations of mechanical beam steering (in particular the motion of masses), offering such capabilities as very precise stabilization ($< \mu$rad), rapid random-access pointing over a wide field of regard (inertialess steering of beams), programmable multiple simultaneous beams, and other capabilities. In an active phased array system (such as a microwave antenna) individual transmit elements form and direct a beam into a particular direction (2-D steering). The field intensity across the aperture of an active microwave array is generally tapered at the edges to achieve low sidelobe levels. In some radar (SAR) applications the phased array concept is also referred to as **ScanSAR** with the capability to extend the regular swath width (see also **phased-array antenna** under **antenna**).

Photodetector. A semiconductor device that transforms radiation (light) into an electrical signal. There are two basic types of photodetectors: photodiodes and photoconductors.

Photodiode. Refers to a semiconductor diode which receives incident radiation thereby becoming a photodetector. Principle of operation: Photons (energy) incident on the photodiode (p-n junction) form electron-hole pairs in the detector material (silicon, for instance) when they are absorbed. Manipulation of the electron-hole pairs produce an output signal proportional to the amount of energy received. The time-varying signal represents the total amount of energy it receives. An important measure of how well the device converts photons to electrons is the **quantum efficiency** (QE). - Silicon is virtually transparent to radiation in the IR range. However, in the UV and VNIR range (0.2 - 1.1 μm), radiation has enough energy (i.e. photon absorption can take place) to create electron-hole pairs.

Photoelectric cell (or photocell). A detector (transducer) which converts electromagnetic radiation from the UV, VNIR regions of the spectrum into electrical quantities such as voltage, current, or resistance.

Photogrammetry. A remote sensing application using image surveys (initially photographs) from airborne sensors and producing (topographic) maps from these images (along with position data). The ASPRS definition is: "Photogrammetry is the art, science and technology of obtaining reliable information about physical objects and the environment through processes of recording, measuring, and interpreting photographic images and patterns of

electromagnetic radiant energy and other phenomena." [2071]
In the past, spaceborne imagery was not (or hardly) used for map making due to insufficient spatial resolution; this certainly changed with the availability of high-resolution imagery (<3m) starting from about 1998. Spatial imagery resolutions from satellites are classified as:

- Very low: for pixel sizes greater than or equal to 300 m
- Low: for pixel sizes between greater than or equal to 30 m and < 300 m
- Medium: for pixel sizes between greater than or equal to 3 m and <30 m
- High: for pixel sizes between greater than or equal to 0.5 m and <3 m
- Very high: for pixel sizes <0.5 m

Photomultiplier. A photoemissive detector in which amplification is obtained by secondary emission.

Photon. A particle description of electromagnetic radiation, which can exhibit the behavior of either waves or particles.

Photosphere. Refers to the layer surrounding the sun from which visible light is emitted into space.

Photosynthesis. The conversion of inorganic matter into organic matter by plants, using light as the energy source. Light energy absorbed by **chlorophyll** is used to manufacture carbohydrates (sugars) and oxygen from carbon dioxide (CO_2) and water. Photosynthesis is dependent upon favorable sunlight, temperatures, plant nutrients, and additionally, on soil moisture and carbon dioxide concentration for terrestrial plants. Increased atmospheric levels of carbon dioxide can increase photosynthesis in many land plants. Photosynthesis is responsible for the generation of all atmospheric oxygen. The photosynthetic generation of organic matter is also called 'primary production.'

Photovoltaic effect, photoconductive effect. Refers to the direct conversion of radiation into electrical energy. The photovoltaic effect is achieved when a photon-produced electron-hole pair is separated by a space charge field (p-n junction diode) thus producing a photocurrent. Solar cells are photovoltaic devices. The photovoltaic effect was first reported by the French scientist Edmund Becquerel in 1839. He observed a voltage between two electrodes in a beaker of electrolyte when the beaker was exposed to sunlight. - The photoconductive effect occurs when a bias voltage is applied across a uniform piece of detector material. The photocurrent is then proportional to the density of electrons excited into the conduction band by the incoming photons.

Phytoplankton. The assemblage of microscopic algae (diatoms, flagellates, etc.) in aquatic ecosystems that drift passively with currents (plankton = wandering). The "grass" of the sea, upon which virtually all marine life depends (see also **chlorophyll**).

Pixel (Picture Element). The smallest area unit of an image which is generated by a single digital measurement (see also **image**). A pixel has a unique position within an image raster. Its numerical value is taken artificially from complete or partial resolution cells, sometimes also referred to as 'image point.'

Pixel value. The digital radiation value of a pixel, expressed as a digital number (DN) or digital count (DC), radiance value, reflectance or other radiation value.

Planetary albedo. The fraction of incident solar radiation that is reflected by a planet and returned to space. The planetary albedo of the Earth-atmosphere system is approximately 30%, most of which is due to backscatter from clouds in the atmosphere.

Planetary Boundary Layer (PBL). Defined as the atmosphere between the Earth's surface and the free atmosphere. It is directly affected by the properties of the Earth's surface and

2071) E. M. Mikhail, "Is Photogrammetry still Relevant?," PE&RS, Vol. 65, No 7, July 1999, pp. 740-751 surface forcings, such as frictional drag, evapotranspiration, heat transfer, pollutant emission, and topography. See also **Atmospheric Boundary Layer**.

Plasma. A completely ionized gas (a low-density gas), the so-called fourth state of matter (besides solid, liquid, and gas) in which the temperature is too high for atoms as such to exist. A plasma is an electrically neutral gas consisting of charged particles such as ions, electrons, and neutrals. While the temperature of a plasma is very high, its density is very low.

Plasmoid. Bubble of plasma. Refers to the merging of magnetic lines in the magnetotail which is thought to produce a bubble of plasma, called a plasmoid that flows down the tail during active solar periods.

Polarimetry.[2072)] Polarimetry deals with the vector nature of polarized electromagnetic radiation throughout the frequency spectrum. The electromagnetic field is a traveling wave (at the velocity of light) with electric and magnetic vector fields perpendicular to each other and to the direction of wave travel. A change in the index of refraction (or permittivity, magnetic permeability, and conductivity) causes the polarization state of a single frequency wave to be transformed, i.e. to be repolarized. Hence, a reflected (scattered) polarized wave from an object such as a radar target must contain some innate information about the object. The interpretation of the behavior of these complex signatures (in particular the direction of the electric field vector of the reflected polarized radiation) is in effect a major objective in polarimetry. In remote sensing polarization measurements are mostly performed in the microwave region of the spectrum. The incorporation of coherent polarimetric phase and amplitude into radar signal and image processing is indeed very promising.

Polarization. Defines the spatial orientation or alignment of the electric (and magnetic) fields of an electromagnetic wave (radiated by an antenna). Horizontal (H) / vertical (V) polarization refers to the electric field (magnetic field) vector's being parallel / normal to the surface of the medium that the wave is incident upon.

- Like polarization: HH or VV (one component for the transmit and one for the receive signal, as is the case for **active sensors**)
- Cross polarization: HV or VH. Cross polarization requires multiple scattering by the target and therefore results in weaker backscatter than like polarization.
- Alternating dual polarization: alternate transmit and/or receive polarization so that two polarization combinations are measured - e.g. HH and HV or HH and VV.

Polarization is established by the antenna, which may be adjusted to be different on transmit and on receive. Reflectivity of microwaves from an object depends on the relationship between the polarization state and the geometric structure of the object. Possible states of polarization in addition to vertical and horizontal include all angular orientations of the E vector, and time varying orientations leading to elliptical and circular polarizations.

Polarization knowledge offers an additional capability in detecting object characteristics and in discriminating between them, especially in the microwave region of the electromagnetic spectrum (for passive and active sensors). See also **Radar polarimeter**.

Although VV, HH, VH, HV are common terms in "polarimetric radar," the generally accepted terms differ in **polarimetric radiometry**. Here, the four scalar brightness temperatures are used that make up the complete (modified) Stokes' vector: $T_v = |E_v|^2$, $T_h = |E_h|^2$, $T_U = 2Re\,(T_v T_h^*\}$, and $T_V = 2Im\,(T_v T_h^*\}$.[2073)] The modified Stokes vector is related to the (unmodified) Stokes vector as follows: $T_1 = (T_v + T_h)/2$, $T_2 = (T_v - T_h)/2$. Some authors use T_3 and T_4 in place of T_U and T_V to eliminate confusion between the vertical (v) and fourth Stokes parameter (V) indices.

2072) W-M. Boerner, H. Mott, E. Lüneburg, et al., "Polarimetry in Remote Sensing: Basic and Applied Concepts," Chapter 5 (94 p) in R. A. Reyerson, ed., The Manual of Remote Sensing, 3rd edition, ASPRS Publishing, Bethesda, MD, 1997

2073) Courtesy of Al Gasiewski of NOAA/ETL in Boulder, CO

Polar vortex. Refers to the swirling mass of air that appears each year over the Earth's poles.

Polynya. Polynyas (like leads) are openings in polar region sea ice, they range in size from a few hundred meters to hundreds of kilometers. Polynyas may be formed by two mechanisms: a) forming ice may be continually removed by winds or currents (an example is a shore polynya with offshore winds), b) oceanic heat may enter the region in sufficient quantity to prevent local ice formation.

Precision. The precision of a measurement is a measure of the reproducibility or consistency of measurements made with the same sensor. (The effect of random errors can be reduced by repeated measurement or by averaging, which increases the precision of a measurement).

Preprocessing. Commonly used to describe the correction and processing of sensor data prior to information extraction. For imaging data preprocessing includes geometric and radiometric correction, mosaicking, resampling, reformatting, etc.

Primary productivity. The rate of carbon fixation by marine photosynthetic organisms (phytoplankton). Primary productivity results in the reduction of dissolved inorganic carbon to form organic carbon, with concomitant release of oxygen.

Pseudolites. Refer to auxiliary ground-based transmitters that broadcast GPS-like signals to supplement those generated by the satellites. The transmitters were initially called "pseudo-satellites," which was abbreviated to "pseudolites." System developers used pseudolites as direct replacement for satellites that had not yet been launched, to facilitate system tests. A pseudolite transmits a signal with code-phase, carrier-phase, and data components with the same timing as the satellite signals. A GPS receiver acquires this signal and derives its measurements to be used in a navigation algorithm. Applications are particularly useful for equipment (receiver) tests for such functions as fault detection and isolation, for pseudorange correction (identical to DGPS), and many other practical solutions. [2074) 2075)]

Pseudorandom Noise (PRN). Refers to deterministic binary sequences which are used in spread spectrum communication systems and in ranging systems such as GPS and GLONASS. Two PRN codes are continuously broadcast by GPS satellites: the C/A code and the P-code, both codes modulate the navigation signals. The modulation appears to be random but is, in fact, predictable; hence the term 'pseudorandom'. The PRN technique allows the use of a single frequency by all GPS satellites and also permits a broadcast of a low-power signal. [Note: The PRN technology was first employed in the 1950s by radio astronomers. With them researchers were able to measure the time delay in the weak radar reflections from the surface of a distant planet - by finding the instant, when the received signals and the transmitted PRN sequences seemed to match most closely.] [2076)]

Pseudorange. Refers to the range between the antenna phase centers of a GPS satellite and a receiver, measured by the receiver's delay-lock loop using the C/A-code or P-code. The range is biased by the offset of the clock in the receiver from that in the satellite and by atmospheric propagation delays.

Pulse Code Modulation (PCM). Any modulation that involves a code made up of pulses to represent binary information. This is a generic term; additional specification is required for describing particular cases.

Quantization. The process of converting continuous values of information to a finite number of discrete values. A 10 bit quantization means that the measured signal can be represented by a total of 1024 digital values, say from 0 to 1023.

2074) S. Cobb, M. O'Connor, "Pseudolites: Enhancing GPS with Ground-based Transmitters," GPS World, March 1998, pp. 55-60

2075) J. M. Stone, et al., "GPS Pseudolite Transceivers and their Applications," Proceedings of ION National Technical Meeting, San Diego, CA, Jan. 25-27, 1999

2076) T. A. Herring, "The Global Positioning System," Scientific American, Feb. 1996, pp. 32-38

Quantum efficiency (QE). A measure of the efficiency with which incident photons are detected (such as a photodiode). Some incident photons may not be absorbed due to reflection or may be absorbed where the electrons cannot be collected. The QE is the ratio of the number of detected electrons divided by the product of the number of incident photons times the number of electrons each photon can be expected to generate. Visible wavelength photons generate one electron-hole pair. More energetic photons generate one electron-hole pair per each 3.65 eV of energy.

Radar (Radio Detection and Ranging). A method, a system, or a technique for using beam, reflected, and timed electromagnetic radiation to detect, locate, and track objects, to measure distance (range), and to acquire terrain imagery (see also chapter O.8.2). The term 'radar' in remote sensing terminology refers to active microwave systems (from about 1 GHz - 100 GHz; most current instruments operate below 10 GHz). The terms 'Doppler delay' (range), 'Doppler gradient' (range rate), and 'Doppler frequency analysis' are important parameters in the formulation of the range-Doppler radar imaging principles. The motion of the sensor-bearing vehicle provides the relative motion between sensor and target required to perform imaging.

Radar systems may be classified by the signal measurement technique employed - there is the pulsed radar class and the FM/CW (Frequency Modulation/Continuous-Wave) radar class. The pulsed radar is the most widely used type of radar systems. It is so called because the transmitter sends out pulses of microwave energy with relatively long intervals between pulses. The receiver picks up the echoes of the returned signals; the elapsed time (or run time) is a measure of the distance travelled.

FM/CW radar transmits continuous microwave energy - the resultant continuous echo cannot be associated with a specific part of the transmitted signal (hence, range information cannot be obtained). However, the system can determine the speed of a target by measuring the Doppler shift (change in frequency). A more sophisticated continuous-wave instrument, known as 'frequency-modulated radar,' is also able to measure range. This is done by tagging each part of the transmitted microwave signal (by continuously changing the frequency), rendering it recognizable upon reception. With the rate of frequency change known, the difference in frequency can be interpreted as a range measurement.

Radar instruments consist of the following major elements: RF electronics, antenna,digital electronics, and recorder. Radar instruments are built for a specific transmission frequency in the microwave spectrum, such as P-band, L-band, S-band, C-band, X-band, Ka-band, etc.; some very advanced instruments offer observation in multiple frequencies.

- **RAR** (Real-Aperture Radar). The term RAR is used because the along-track resolution of a surface image is determined by the actual length of the antenna aperture. In general, the larger the aperture of the antenna (in terms of wavelengths), the narrower will be the beam (along-track resolution is given by the width of the antenna sweep; across-track resolution is determined by the range-resolving capability of the instrument). RAR systems are usually much simpler than SAR systems in design and data processing. RAR-pulsed signals, based on the range-Doppler principle, are not required to be coherent (only the signal amplitude information is recovered and processed), representing and displaying backscatter characteristics from the surface sweep that are recorded on film or on magnetic tape. Microwave energy reflected from the surface terrain (target) is converted by the RAR instrument into electrical signals and recorded as a function of distance (along-track and across-track direction). The radar returns from the different positions in the sweep (and at the different ranges) are separated in time by the radar receiver (the across-track range measurement is a function of signal return time). After reception and recording of the previous pulse, a new pulse is transmitted for a new radar sweep.- The density of the image varies with the surface properties (roughness, moisture content, etc.). The image can be interpreted in terms of the topographical features of the terrain.

- **SAR** (Synthetic Aperture Radar), see also O.8.2. This radar type permits high-resolution imagery at long ranges (a SLAR device from a satellite orbit is of limited use due to the poor resolution obtained by the angular geometry constraints of the radar beam). A SAR instrument is also referred to as a 'coherent SLAR.' - SAR is a concept for complex signal processing techniques to recover an image by the coherent processing of all return signals of all targets (cells) in a single sweep. The cross-track Doppler range of all targets is determined by the signal return time. The pulse bandwidth determines the cross-track or range resolution. Coherence [2077] in this SAR context refers to the fact that the phase as well as amplitude information of the **radar cross section** is measured for all recovered signals. The SAR technique provides resolutions that would normally be associated with an antenna with an aperture far wider than that actually used. - A disadvantage of the SAR observation/processing technique is the generation of very high data rates (between 20 and 100 Mbit/s and more); this implies high communication rates and large storage volumes. On-board recorders are strained to their very limits to handle SAR data. First attempts are being made at real-time on-board preprocessing for the purpose of data reduction. The main elements of a SAR instrument are:
  - RF electronics. The RF portion of a SAR system consists of signal generators, high power transmitters (single or combined), low noise receivers, and the associated signal conditioning elements: amplification, filtering, and frequency conversion. Important RF characteristics to a SAR instrument are: large dynamic range with good linearity and low noise floor, good amplitude and phase stability over time and temperature, and high power efficiency.
  - SAR antenna. For an Earth observation SAR, the preferred antenna beam casts an elliptical footprint on the ground with an effective rectangular aperture of typical size 10 meters (along-track) and 3 meters (cross-track). Multi-polarization requires dual polarization antennas with good cross-pol isolation. Practically all SAR antenna designs feature a solid flat aperture (antenna and support structure are a dominant mass and volume factor of a SAR instrument).
  - Digital electronics. The main functions provided by the digital electronics in SAR are: radar configuration and timing control, radar signal digitization and formatting, radar housekeeping telemetry generation. Sometimes digital processing is performed to reduce the data, and the often coded radar illumination pulse is generated digitally.

  Imaging radars (SAR instruments) operate at a specific wavelength or frequency in the microwave region. [This is different from optical instruments which observe radiation in a spectral band (a region of frequencies) or in many spectral bands]. A radar system records the signal response from the ground target at a single specific wavelength (e.g. 15 cm).

  Background: The microwave region of the electromagnetic spectrum is generally considered from 1 mm to 1m wavelengths. In analogy to the optical band designations (red, green blue), the microwave spectrum was also given band designations in form of letter references. The military introduced these letter designations in the early days of radar research (mostly for reasons of security during World War II). The remote sensing community seems to adhere to these old "standards" as a means of "ball park reference" designation. Hence, the microwave region includes today such band designations as P, L, S, C, X, K, Ka, etc., for radar (SAR) instruments.

  Naturally, the rule of specific wavelength (for frequency) operation of a SAR instrument is not affected by this scheme of letter band designations. Newer SAR instruments operating at multiple frequencies are actually an agglomeration of single-frequency instruments.

2077) Note: As coherent pulses transmitted from the radar source reflect from the ground (target) to the advancing SAR instrument (on an aircraft or a spacecraft), the target acts as though in apparent (relative) motion. This motion results in changing frequencies which give rise to variations in phase and amplitude in the returned signals. - Offline processing of these data involves the analysis of the moderated pulses.

- **IFSAR** (Interferometric SAR) - see **Interferometric measurements**
- **SLAR** (Side-Looking Airborne Radar) - an active sensor with RAR technology.
- **Radar altimeter**. An active device observing the vertical distance between the instrument and the ground by measuring the elapsed time between the emitted and returned signals of electromagnetic pulses. Determination (mapping) of the height profile of the surface (topographic applications, in particular of ocean height surfaces).
- **Scatterometer**. A scatterometer is a nonimaging radar, distinguished from other radars by its ability to measure radiation amplitude. A radar scatterometer is an active device measuring the backscattering coefficient of the illuminated cell (area or volume under observation) at a specified configuration of incidence angles, wavelengths, and wave polarization orientations. The backscattering (or scattering) coefficient σ^0 describes the target backscattering characteristics (it is defined as the intensity of the power scattered by a 1 m^2 area of a target back toward the radar, relative to the incident power density) and varies as a function of surface roughness, moisture content, and dielectric properties. A rough ocean surface returns a weak pulse because sea surface waves scatter the energy of the microwave pulse in different directions. The scattering reduces the amount of energy which is received back at the satellite. On the other hand, a smooth ocean surface returns a strong pulse because there is very little wave effect. The surface roughness is related to the wind speed. High wind speeds disturb the smooth ocean surface and produce many waves of several cm in size while low wind speeds do not disturb the ocean surface as much and produce much smaller waves. In addition to windspeed, scatterometers (and SARs) measure the direction that waves are moving in relation to the satellite. The direction that the waves are oriented with respect to the radar pulse has an effect on the polarization of the returned signal. - Application: the surface backscattering coefficient may be used to derive the surface wind vector (in particular over oceans). See also **sigma naught**.
- **Lidar** (Light Detection and Ranging), see O.8.5. A lidar instrument is also referred to as an 'optical radar' [or a 'laser radar' - it also goes by the name of 'ladar' (laser detection and ranging)] since it utilizes the optical (and TIR) portion of the electromagnetic spectrum (0.3 - 10 μm wavelength range, or a frequency range of about 1000 - 30 THz). A very narrow beam (pulse) of laser light is emitted, the echo is analyzed. Lidar beam divergence is two to three orders of magnitude smaller compared to conventional 5 to 10 cm wavelength radars. This characteristic permits unambiguous velocity measurements near clouds and surface features.
- **Radar polarimeter**.[2078] This radar instrument type measures the complex (amplitude and phase) scattering matrix (i.e. the full polarization signature: VV, HH, VH and HV for transmit and receive signals) for every resolution element in an image. Radar polarimetry is therefore an extension of scatterometry, in which the received power of an echo is typically measured for one or more fixed polarization states and a single, or two orthogonal, transmit states. - Knowing the full scattering matrix permits calculation of the receive power for any possible combination of transmit and receive antennas; this process is called polarization synthesis. Hence, the information content derived from a polarimetric radar instrument is far superior to the information yielded by nonpolarimetric devices. Typical airborne polarimetric radar instruments are: ARMAR, CASSAR, C/X-SAR, DO-SAR, EMISAR, HUTSCAT, MMW-SAR, NUSCAT, P-3/SAR, IMARC, RAMSES, PHARUS, etc. (see Table 8); a typical spaceborne polarimetric radar instrument is the L/C- Band SAR (JPL) of the SIR-C payload.
  - Note: Conventional imaging radars operate with a single, fixed-polarization antenna for both transmission and reception of radio frequency signals. In this way a single scat-

2078) H. A. Zebker, J. J. van Zyl, "Imaging Radar Polarimetry: A Review," Proceedings of the IEEE, Vol. 79, Nr. 11, November 1991, pp. 1583-1606 tering coefficient is measured for a specific transmit and receive polarization combination for many thousands of points in a scene. A result is that only one component of the scattered wave, itself a vector quantity, is measured, resulting in a scalar characterization of the wave, and any additional information about the surface or volume contained in the polarization properties of the reflected signal is lost.

Radar albedo. Ratio of a target's radar cross-section in a specified polarization to its projected area; hence, a measure of the target's reflectivity.

Radar backscatter. Refers to the radar echo; a scattering process of microwave energy by an object/target in the direction of the radar antenna, after actively being irradiated by the radar source.

Radar cross section. A hypothetical area of an object of such an extent that if the power intercepted by this area were distributed isotropically over the space, it would render the same power density at the receiving antenna as the power density brought about in reality by the presence of the object or target. Usually, the radar cross-section concerning compound objects (distributed targets) is normalized: either as a radar cross section per unit area (differential scattering cross section or backscatter coefficient σ^o), or as a radar cross section per unit of area projected in the direction of transmission (gamma or scattering cross section). [2079)]

Radar meteorology. A discipline that uses backscattered electromagnetic radiation within the microwave band to gain information about the state of the atmosphere, especially with respect to clouds and precipitation. The return signal allows the interpretation of four fundamental properties of the spectrum: amplitude, phase, frequency, and polarization.

Radian. The size of angles in classical mechanics is expressed in radians. The concept of radians permits a simple mathematical relationship between the length L of the arc of a circle (i.e. a segment of the circumference) and the angle subtended (enclosed) at the axis by the arc. The arc length L is given by: $L=R\,\theta$. This means that when the length of the arc is equal to the radius ($L=R$), then θ is one radian (or the angle at the center of the circle subtended by an arc equal to the radius is one radian). In one revolution the arc length is equal to the circumference, so that $L=2\pi R$, or $\theta=2\pi$ radians, or 1 radian = $2\pi/360^\circ \approx 57.28^\circ$. Example: for R=1000 m and $\theta=0.2$ radians (FOV), then L=200 m (in case of a sensor, the swath width).

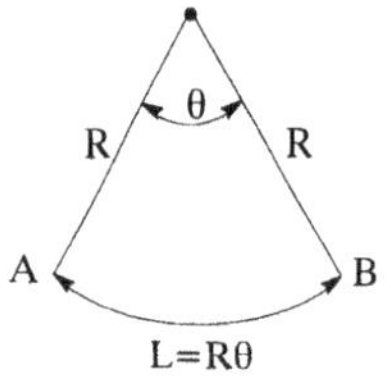

Angular size with no distance information is usually the only information available in observational astronomy. Some sample angular sizes are:

- The full moon subtends 30 arcminutes
- The Andromeda galaxy subtends about five degrees

Radiance. Energy per unit area and solid angle. Measure of energy radiated by an object. In general, radiance is a function of viewing angle and spectral wavelength.

Radiation laws.

- **Planck's law** (Max Planck, 1858-1947). $E_{b\lambda} = 2\,h\,c^2\,\lambda^{-5} / [e^{(hc/\lambda kT)} - 1]$, where $E_{b\lambda}$ = monochromatic emissive power (or spectral radiance, or brightness) per unit wavelength interval, λ = wavelength, T=absolute temperature, h = Planck's constant, k = Boltzmann's constant, and c = speed of light.
  In words, Planck's law states that the temperature of a blackbody is related to the

2079) H. J. Buiten, J. Clevers, "Land Observation By Remote Sensing," Gordon and Breach Science Publishers, ISBN 2-88124-936-6, 1993, p. 608

emitted radiance as a function of the wavelength (or of the frequency). Planck's equation is plotted in Figure 422 for several temperatures.

Note: The term 'spectral radiance' is commonly used for remote sensing instruments operating at optical wavelengths, while the term 'brightness' is mainly used for the microwave region of the spectrum. Then Planck's law is expressed in units of power density per frequency bandwidth (Hz^{-1}) rather than per unit wavelength interval (m^{-1}). Spectral brightness B_f is related to spectral radiance $E_{b\lambda}$ by: $B_f = E_{b\lambda}\,|d\lambda/df|$, which results in: $B_f = 2\,h\,f^3\,c^{-2} / [e^{(hf/kT)} - 1]$.

In the microwave region ($f < 300$ GHz), the term $hf/kT \ll 1$ for the range of physical temperatures commonly encountered in the Earth's surface and atmosphere. Consequently, the spectral brightness equation above reduces to a simpler form: $B_f = 2kT/\lambda^2$ which is known as the Rayleigh-Jeans law, a special case of Planck's blackbody radiation law (see also **brightness temperature**).

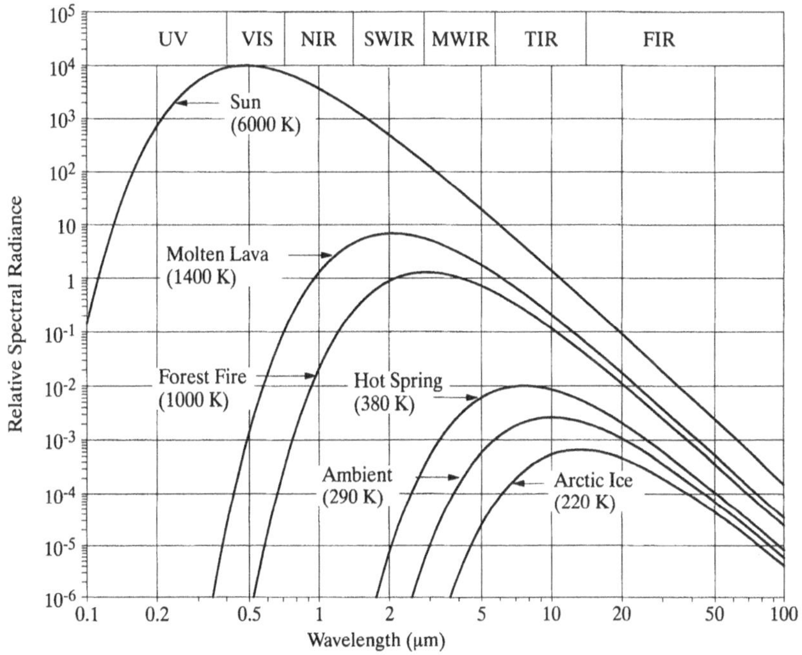

Figure 422: Hemispherical radiation emitted by objects at typical temperatures

- **Wien's displacement law** (Wilhelm Wien, 1864-1928). A law indicating that the wavelength at which the emitted amount of energy by a blackbody is maximal is inversely proportional to the absolute temperature of that body.

- **Kirchhoff's law** (Gustav Robert Kirchhoff, 1824-1887). A law stating that under conditions of thermal equilibrium, the absorption spectrum of an arbitrary body must be equal to its emission spectrum. Kirchhoff's identity: ε (emissivity)$=\alpha$ (absorptivity).

- **Lambert's law** (Johann Heinrich Lambert, 1728-1777). A law stating that the radiant intensity (flux per unit solid angle) emitted in any direction from a unit radiating surface varies as the cosine of the angle between the normal to the surface and the direction of radiation. The radiance of a radiating surface is therefore independent of direction. This law is also satisfied (by definition) by the distribution of radiation from a perfectly diffuse radiator.

Radiation hardening. Exposed satellite and instrument components, such as detectors, are constantly subjected to space irradiation effects. In particular, long-term exposure may cause radiation damage to electronic components by altering the properties of a material arising from exposure to ionizing radiation (penetrating radiation), such as X-rays, gamma rays. neutrons, or heavy-particle radiation. With proper hardening processes applied, the components may be turned into radiation-tolerant products.

Radio beacon. A type of radio transmitter with wide-angle coverage. It may emit signals continuously or, like the transponder, may respond to input energy before operating. Beacons are use primarily in navigation and radio-detection finding. In meteorology a beacon is used in rawinsonde observations.

Radiodetermination. Refers to the determination of position, velocity and/or other characteristics of an object, or to obtaining information relating to these parameters, by means of the propagation properties of radio waves. The radiodetermination service has two parts to it: the **radionavigation service** and the **radiolocation service** (see also Figure 412).

1) Radionavigation Systems. Radionavigation is used for the purpose of navigation (aeronautical, maritime, land, and space), including obstruction warning.
   - LORAN-C (Long Range Navigation) operates on 100 kHz; it is used in maritime and aeronautical applications (see **LORAN**)
   - Omega is a worldwide CW system; it is used for maritime and aeronautical navigation. System operation in the VLF band (9-14 kHz) on four discrete frequencies. (see **OMEGA**)
   - VOR/DME (VHF Omnidirectional Range/Distance Measuring Equipment). VOR operates in the 108-118 MHz band, providing azimuth readings to aircraft. DME is collocated with VOR providing distance; it operates in the 960-1215 MHz band.
   - TACAN (Tactical Air Navigation) is the US military version of DME. It operates in the 960-1215 MHz band.
   - ILS (Instrument Landing System) for precision navigation. ILS consists of a localizer operating in the 108-112 MHz band and a glidescope operating in the 328.6-335.4 MHz band.
   - MLS (Microwave Landing System), operating in the 5000-5150 MHz range with associated DME in the 960-1215 MHz range. MLS was initially considered a successor to ILS. It probably may be succeeded by GPS systems.
   - GPS (Global Positioning System), a US satellite-based system operating in the 1215-1240 MHz and 1559-1610 MHz bands. GPS was officially integrated into the US National Airspace System on February 17, 1994. In the future, GPS is expected to replace such systems as Omega, LORAN-C, and perhaps VOR/DME.
   - GLONASS (Global Orbiting and Navigation Satellite System), a Russian satellite-based system operating in the 1215-1260 and 1559-1626.5 MHz bands. By the year 2005 the second bandwidth of GLONASS is expected to be the same as that of GPS. There is a major trend towards increased use of the GPS and GLONASS satellite-based system for many navigation applications.
   - etc.

2) Radiolocation Systems: The service is used by pulsed and CW radar systems for a number of applications, such as determining precise location, search or surveillance, target tracking, weapons control, ground mapping and target identification, or combinations

of these applications. The military is by far the largest user of this service, but there are also a number civil users (NASA, NOAA, CNES, CRC, Russia, etc.[2080])

- S&RSAT (Search & Rescue Satellite, CNES/CRC) flown on NOAA-POES satellites. The S&R instruments consist of a 3-band (121.5, 243, and 406.05 MHz) repeater S&RR and a 406.025 MHz processor. The system may receive three types of radiobeacons, namely aviation ELTs (Emergency Locator Transmitter), maritime EPIRBs (Emergency Position Indicating Radio Beacon), and PLBs (Personal Locator Beacon). S&RSAT was declared operational in 1985. The COSPAS-SARSAT agreement was signed in 1988.[2081]
- COSPAS (Space System for Search of Vessels in Distress, Russian system). The system is flown on Cospas series satellites (named Nadezda) and administered by Russia, US, France, and Canada. Distress alert and location data to RCCs (Rescue Coordination Centers) for 121.5 MHz beacons within the area of COSPAS-S&RSAT ground stations (Local User Terminals - LUTs), and for 406 MHz beacons activated anywhere in the world.
- IVHS (Intelligent Vehicle Highway Systems) or ITS (Intelligent Transport Systems)
- etc.[2082]

Radiometer. An instrument for the quantitative measurement of the intensity of electromagnetic radiation in some band of wavelengths in the spectrum. Usually a radiometer is characterized with a prefix, such as IR-radiometer, or microwave-radiometer, to indicate the spectrum to be measured.

Radiometer (absolute radiometer). An instrument based on the measurement of a heat flux by an electrically calibrated transducer. Optical radiation absorbed in a black cavity is substituted by electrical heating during a shaded reference phase. For practical use of the instrument, an electronic circuit keeps the heat flux constant by controlling the power fed to a cavity heater (this is also referred to as substitution radiometry - directly relating the optical watt to the electrical watt). Absolute radiometers are used to measure the Solar Constant or TSI (Total Solar Irradiance).

Note: electrical substitution radiometry at cryogenic temperatures is also the basis of detector calibrations in which a detector's response to optical flux is measured as a function of wavelength.

Radiometric resolution. See **resolution**.

Radiosonde. A balloon-borne instrument which measures (by means of transducers) and transmits meteorological data (temperature, pressure, humidity). Various types of transmission schemes exist.

- **Rawinsonde** (Radio-Wind-Sonde). A balloon-borne instrument tracked by radar or radio direction-finder and operating on the same principle as a radiosonde, but with the additional capability to measure wind speed and direction.
- **Dropsonde**. A radiosonde or a rawinsonde dropped by parachute from an aircraft for the purpose of obtaining soundings of the atmosphere below. The radio signals of the dropsonde/rawinsonde are tracked for data evaluation.

Radio occultation principle. Fundamentally, the technique relies on the simple fact that a planet's atmosphere acts much like a spherical lens, bending and slowing propagation of microwave signals passing through it tangent to the surface. The lens effect results from decreasing atmospheric density with altitude. If the positions of transmitting and receiving satellites are precisely known, then the atmospheric delay can be measured precisely, the

2080) US Spectrum Requirements, URL: http://www.ntia.doc.gov/openness/sp_rqmnts/radar4.html
2081) NOAA S&RSAT homepage, http://psbsgil.nesdis.noaa.gov:8080/SARSAT/homepage.html
2082) COSPAS-S&RSAT homepage, http://www.worldserver.pipex.com/cospas-sarsat/

time derivative of which (Doppler) can be inverted to give atmospheric density versus altitude. See also **Limb/Occultation sounding** and **Occultation.** - Note: The radio occultation technique was first developed (in the 1970s) at Stanford University Center for Radar Astronomy and at NASA/JPL for the study of solar system planetary atmospheres (Venus, Mars, Jupiter, Saturn, Uranus and Neptune).

Raman spectroscopy. [2083)] A technique to investigate molecular properties using scattered light resulting from photon-molecule collisions. When a monochromatic light beam is incident on systems such as transparent gases, liquids, or solids, most of it is transmitted without change. However, a very small portion of the incident light is scattered. Although most of the scattered light has the same wavelength as the incident radiation, a small part of it has different wavelengths. The scattering of light at different wavelengths is called **Raman scattering** (Indian scientist Sir C. V. Raman, who, with K. S. Krishnan, first reported the phenomenon in 1928). The physical origin of Raman scattering lies in inelastic collisions between the molecules composing the system (e.g. the liquid) and photons, the particles composing the light beam. 'Inelastic collision' means that there is an exchange of energy between the photon molecule with a consequent change in energy, and hence wavelength, of the photon.

Range. The distance between two objects, usually between an observation point and a target (object under observation). Slant range: same as range - the line-of-sight distance between two objects.

Range direction. Observation of an instrument in the cross-track direction (normal to the subsatellite track). See also **azimuth direction**.

Range error. The (small) error in radar range measurement caused by the propagation of radio energy through a nonhomogeneous atmosphere. This error is due to the fact that the velocity of radio-wave propagation varies with the index of refraction, and that ray travel is not in straight lines through actual atmospheres (see also **Atmospheric refraction**).

Range resolution. Resolution characteristic of the range dimension, usually applied to the image domain, either in the slant range plane or in the ground range plane. Range resolution is fundamentally defined by the system bandwidth in the range channel. See also **SAR**.

Raster image. Refers to a matrix of row and column data points. Each data point is a pixel.

Rayleigh criterion. It states that the resolution of a system is directly proportional to the wavelength. Thus, with perfect optics, an imaging system working at 50 nm (EUV wavelengths) would have an order of magnitude better resolution than one working with visible light (about 500 nm wavelength).

Rayleigh scattering. Scattering by particles small in size compared with the wavelengths being scattered (e.g., the blue color of the sky and ocean is caused by Rayleigh scattering of the air and water molecules respectively). Rayleigh scattering is also caused by density fluctuations in atmospheric gases (it increases toward the shorter wavelengths proportional to λ^{-4} where λ is the wavelength). In a sensor energy balance, Rayleigh scattering adds to the radiation received by a sensor; this is most pronounced at shorter wavelengths.

Reaction/momentum wheels. These are actuators (fly wheels) which may be used for three-axis reaction control or momentum bias applications. By adding or removing energy from the flywheel, torque is applied to a single axis of the S/C, causing it to rotate (reaction). By maintaining flywheel rotation (momentum), a single axis of the spacecraft is stabilized. Such an assembly provides a reliable source of reaction torque and angular momentum storage for attitude control of medium to large spacecraft. Accelerating or decelerating a flywheel with an integral motor provides a means of controlled momentum exchange with the spacecraft platforms, which is advantageous for a variety of attitude control schemes.

2083) Encyclopedia of Physical Science and Technology, Academic Press, 1987

Real-Time Kinematic (RTK). Refers to a DGPS process where carrier-phase corrections are transmitted in real-time from a reference receiver at a known location to one or more remote "rover" receivers. RTK has become a preferred method for surveying applications since it provides real-time positions with high accuracy.

Reflectance. Refers to the fraction of the total radiant flux incident upon a surface that is reflected and varies according to the wavelength distribution of the incident radiation.

Reflection. The scattering of electromagnetic radiation by an object. Diffuse reflection causes the radiance of the reflected radiation to be equal in all directions (e.g. reflection from a rough surface). Specular reflection has a direction of preference (e.g. the reflection of a smooth surface). The use of the terms 'smooth' and 'rough' is independent of wavelength.

Reflectivity. A property of illuminated objects to reradiate a portion of the incident energy. For SARs, backscatter is the observable portion of the energy reflected. Backscatter, in general, is increased by greater surface roughness.

Refraction. A process by which the direction of energy propagation is changed due to a change in density within the propagating medium (smooth bending), or due to a discontinuity between two media (abrupt bending). - Atmospheric optical phenomena are produced by continuous and discontinuous refraction: scintillation, mirages, astronomical refraction, anomalous propagation of radio waves and the bending of sound waves are examples of refraction within a single medium.

Refractive index (in a medium). The inverse ratio of the wavelength (or velocity) of electromagnetic radiation in the medium to that in vacuum. A measure of the amount of refraction (a property of the dielectric constant). See also **Atmospheric refraction** and **Occultation**.

Registration. Geometric rearrangement of the pixels in an image for image matching by superposition - often to the reference geometry of a map (geocoding). Image registration is the process of matching (overlaying) two or more images so that corresponding coordinate points in the images correspond to the same physical region of the scene being imaged. The technique is used for a number of applications:[2084)]

- Integration of information taken from different sensors (sensor or image fusion)
- Analysis of changes in images taken at different times (temporal registration and change detection).

In a wider sense image registration tries to combine image data with different spatial, spectral and radiometric characteristics to improve the information extraction process from available imagery. Typical registration processing steps are: feature identification, feature matching, spatial transformation, and interpolation.

Relative aperture. For a photographic or telescopic lens system, the ratio of the equivalent focal length to the diameter of the entrance slit. It is expressed as f/45 or f/5.6, and is also called the 'f-number,' speed of lens, or the 'focal ratio.'

Renewable energy. Refers to energy technologies that generate electricity, fuels, and/or heat through the use of resources which are continually replenished, such as sunlight (photovoltaic), heat from the sun (solar thermal), wind, naturally occurring underground steam and heat (geothermal), plant and animal waste (biomass), and water (hydropower).

Repeat period (or cycle). Time interval between successive satellite observations of the same area of the Earth's surface.

Resampling. The rearrangement of the resolution cells of each scanned line of an image into geometrically equal terrain elements (geometric rearrangement) by creating artificial

2084) L. M. G. Fonseca, B. S. Manjunath, "Registration Techniques for Multisensor Remotely Sensed Imagery," PE&RS, Vol. LXII, No. 9, Sept. 1996, pp. 1049-1056 pixels whose spectral radiation data are computed from the original values proportional to the area coverage by the new pixels with respect to the resolution cells. In general a resampling process follows after a geometric rearrangement (registration) of the pixels because of the matching of two different images of the same region by means of a mathematical transformation. The resampling then involves the assignment of artificial pixel values to the newly formed pixels according to the selected sampling algorithm.

Resolution. A term defining the smallest discernable physical unit of an observed signal by a sensor.

- Spatial or geometric resolution defines the minimum (spatial) separation between two measurements in order for a sensor to be able to discriminate between them. Spatial resolution defines the size of an image resolution cell in the target area, or the size of pixels. The spatial resolving power is determined by the aperture dimensions of a lens or the antenna of a sensor. - Some spatial connotations are: GSD, IFOV, FOV, look angle of the sensor, shape and size of the object, position, site, distribution, texture.
- Spectral resolution refers to the resolving power of a system in terms of wavelength (or wavenumber) or frequency.
- Radiometric resolution refers to the resolving power of a system in terms of the signal energy [detection of energy differences (reflection and emission) in terms of temperature, intensity and power]. The radiometric resolution is the Noise Equivalent Delta Radiance (NEΔR), or the Noise Equivalent Delta Temperature (NEΔT), depending on the spectral measurement range. This can be defined as the minimum change in reflectance (or temperature) that can be detected by a sensor. The value depends on a number of parameters, such as SNR, the saturation radiance setting, and the number of quantization bits. The important parameter of an instrument is the SNR. The resolution capability of an instrument in terms of quantization does not necessarily give an idea of its precision or accuracy with which it can measure. Nevertheless, a higher number of bits increases the dynamic range of the instrument, permitting the measurement of very variable targets, without a gain change. See also **Full Width Half Maximum**.
- Temporal resolution concerns the time lapse between two successive images of the same area (by the same spaceborne sensor, at the next revisit time).

Rheology. A science dealing with the deformation and flow of matter.

S/A (Signal-to-Ambiguity ratio). In SAR instruments the ratio of the receiving power of the signal scattered and reflected from the observed (target) area to the power leaking into the observation area from the non-observed area.

SAR (Synthetic Aperture Radar), see SAR under **Radar**.

Satellite surface charging. All bodies which are placed in a plasma in thermal equilibrium acquire a negative electrostatic charge. The negative potential depends on the plasma temperature. At altitudes of 300 to 500 km, the average kinetic energy of the plasma is low (< 1 eV), hence, satellites become only weakly charged. At high altitudes (geostationary orbit and further out) the kinetic energy of the plasma is considerably larger (the plasma is referred to as 'hot'), hence, satellites acquire a high potential with respect to it (in the order of several keV). The electrostatic charge on satellite surfaces can pose a hazard, in particular when differential charging is leading to potential gradients. In some cases this potential build-up causes discharge arcing. Electrostatic charging by the natural space radiation environment is an accepted source of many anomalies of S/C electronics.

Satellite classes. Satellites may be categorized by a number of different criteria such as mass (large, small, mini, micro), or functions and services (EO, communication, space science, data collection, navigation, orbit, etc.), or by other criteria. Within the last years advances in digital microelectronics resulted in achieving sophisticated functions within ever smaller constraints of mass, volume, and power. This in turn brought about a miniaturization trend in platforms and instruments, and a demand for low-cost projects. The following classification has become widely accepted:

Satellite Class		Mass
Large satellite (observatory, etc.)		> 1000 kg
Medium size satellite		500 - 1000 kg
Minisatellite	Small Satellite Class (or LightSats)	100 - 500 kg
Microsatellite		10 - 100 kg
Nanosatellite		1 - 10 kg
Picosatellite		< 1 kg

Table 562: Satellite classification by mass criterion

Satellite Laser Ranging (SLR). Very precise range measurements from ground reference stations to geodynamic satellites (like Lageos, Starlette, Stella, Geo-IK, Etalon, EGS, etc.). The SLR technique employs short pulse lasers from the ground to retroreflectors on satellites. While the above listed geodynamic satellites are dense reflector-covered spheres (dedicated to laser ranging), there may also be configurations where a satellite flies a retroreflector arrangement as an experiment. The quantity of interest is time-of-light (round trip) corrected for ranging system internal delay (calibration), atmospheric refraction (delay), retroreflector offset to the S/C center-of-mass, and network epoch synchronization. The short wavelengths of visible light result in a single-shot precision of about 2 cm. SLR techniques are a strong contributor to advances in precision orbit determination.
The applications of SLR data from geodetic satellites includes detection and monitoring of tectonic plate motion, crustal deformation, earth rotation, and polar motion; modeling of the spatial and temporal variations of the earth's gravitation field; determination of basin-scale ocean tides; monitoring of millimeter-level variations in the location of the center of mass of the total earth system (solid earth-atmosphere-oceans); establishment and maintenance of the International Terrestrial Reference System (ITRS); detection and monitoring of post-glacial rebound and subsidence; monitoring the response of the atmosphere to seasonal variations in solar heating.

Satellite structure: Basic Elements. Satellite structures must survive launch, meet outgassing and other mission-specific requirements, provide stiffness, dimensional stability and thermal control, and allow equipment mounting and containment. Remote sensing satellites are comprised of a number of subsystems. The actual number of subsystems depends on the complexity of the mission and the overall design of the spacecraft. The trend is in the direction of standardized modular subsystems with high functional autonomy.

- Satellite structure. Refers to the basic platform or "bus" (design, body, shape, etc.) and subsystem accommodation.
- Thermal control subsystem (passive and/or active). Orbital temperatures may vary considerably due to varying solar irradiation. The subsystem provides the proper thermal environment for a number of subsystems (in particular electronic or optical equipment). Thermal balance may be maintained by using an exterior finish that absorbs or emits radiation; this is referred to as a 'passive system.' An active system may use louvers to achieve a required environment
- G&C (Guidance and Control) subsystem. G&C is responsible for all functionality associated with spacecraft attitude (sensing and control), a basis for proper S/C pointing. G&C is sometimes simply referred to as 'attitude control.' See also **Spacecraft stabilization**.
- Power subsystem. The subsystem is responsible for providing continuous power for all subsystems throughout the mission. The two most common power sources are solar cells and high performance batteries. The solar energy may vary depending on satellite orbit (due to sun eclipses or varying sun elevation angles). Batteries (such as NiCd or NiH_2) are used as a supplemental on-board energy source.
- Power distribution. Refers to the spacecraft cabling system to all subsystems. Sometimes this electrical distribution function is integrated into the spacecraft bus.
- Antenna subsystem. The subsystem is responsible for receiving and transmitting telecommunication signals between ground and spacecraft (maybe in several bands).

- C&DH (Command and Data Handling) subsystem. The subsystem is responsible for command processing, data management, health and status management, telecommunications management, and power management.
- Spacecraft bus. A shared communications medium for all subsystems (like serial busses or parallel backplane busses). This requires a common interface definition. Some S/C series of agencies or companies offer standardized systems capable of accommodating a variety of payloads and subsystems. Newer designs consider the S/C bus as the physical structure for distribution of all on-board services (data, electricity, etc.) to the payload along with the integration of all service subsystems (attitude and control, timing, thermal control, etc.).
- Spacecraft computer. Depending on spacecraft complexity there may be a S/C computer and/or subsystem computers.
- Data recorder. Responsible for recording data streams during non-contact periods of the S/C. This may be an independent device (high volume and high data rate) or solid state memory storage in a S/C computer.
- Payload instruments. A suite of sensors performing assigned observations. Such instruments may be imagers, sounders, radiometers, etc.
- Timing subsystem. Responsible for giving a uniform time stamp to all required interfaces.
- GPS receiver. Ever more satellites are carrying such a system for orbit determination.

Scales (macro-, meso-, and microscales). See observational scales in modeling chapter O.13 on page 1278.

Scanning. The sweep of a mirror, prism, antenna, or other element across a track (normal to the direction of flight); the footprint may be a straight line, a circle or any other shape. In general, the process of scanning is a programmed motion that can be used either for measuring angular location of a target, or it can be used to extend the angular range of an antenna beam. There are two basic ways of classifying scanning methods:

- From the viewpoint of the type of beam motion introduced to scan a volume, the methods are described as: raster scan, helical scan, etc.
- From the viewpoint of beam steering, the methods are described as mechanical, electromechanical, or as electronic.

Scanner. An instrument that scans and by this means produces an image. A two-dimensional image is generated by the forward motion of the satellite platform. The addition of single pixels in combination with cross-track scanning (whiskbroom) or of a cross-track line of pixels (CCD line array) are the basic elements of such an image. Common scanner types are: a) whiskbroom (cross-track multispectral imaging with discrete detectors), b) pushbroom (cross-track multispectral scanner with CCD line arrays), c) hyperspectral scanning with area arrays (see chapter O.3).

ScanSAR. A SAR imaging technique permitting acquisition of a larger observation swath than what would normally be possible due to range-Doppler ambiguity limitations, but at the expense of reduced resolution. The technique, based on phased array antenna technology with a rapid electronic steering capability of the elevation beam pattern, permits a high degree of flexibility in ground observation coverage. The principle of this mode of operation is to illuminate an area on the ground long enough to acquire imagery (synthetic aperture) for the desired resolution and then move the illuminated beam to a different area across the swath to increase coverage. Hence, the operational time of the SAR beam pattern is shared between two or more subswaths in such a way as to obtain full image coverage of each. However, a contiguous subswath coverage implies shorter integration times for each footprint, resulting in shorter integration times – consequently, the resolution of the resulting image is degraded.
The ScanSAR technique may also serve to cover an event of interest, positioned close-by

but still outside the normal coverage of the current orbit. The required beam pointing for such an event can be done on command.

Scattering. Light absorbed and subsequently re-emitted by particles suspended in a medium in all directions at about the same frequency. In scattering no energy transformation results, there is only a change in the spatial distribution of the radiation. - Scattering varies as a function of the ratio of the particle diameter to the wavelength of the radiation. When this ratio is less than about one-tenth, **Rayleigh scattering** occurs in which the scattering coefficient varies inversely as the fourth power of the wavelength. At larger values of the ratio of particle diameter to wavelength, scattering varies in a complex fashion described by the Mie theory (particle size is comparable with the wavelength dimension). At a ratio of the order of ten, the laws of geometric optics begin to apply and this serves to mark the somewhat diffuse upper boundary of the realm of scattering (where diffraction begins). Primary scattering of the Rayleigh type, largely by air molecules, is responsible for the blue sky and the polarization of the sky's light. On the other hand, Mie scattering occurs by the interaction of radiation (light) with aerosols or cloud particles.

Scattering matrix. An array of complex numbers that describes the transformation of the polarization of a wave incident upon a reflective medium to the polarization of the backscattered wave. See also **radar polarimeter** under **radar**.

Scatterometer types. There are two basic designs of scatterometers: the traditional fan-beam Doppler scatterometer (examples: NSCAT, AMI-SCAT), and the scanning pencil-beam scatterometer. The fan-beam Doppler scatterometer requires multiple antennas to achieve the target illumination pattern (sticklike antennas are used to broadcast long, narrow radar footprints). The FOV requirements of the antennas are very strict making fan-beam scatterometers very difficult to accommodate on S/C. - The design of the newer scanning pencil-beam instrument is more compact; they offer long dwell times which result in better SNRs. SeaWinds on ADEOS-II will be a scanning pencil-beam scatterometer.[2085)]

Schottky diode (named after Walter H. Schottky). A diode that has a metal-semiconductor contact (e.g., an Al layer in intimate contact with an n-type silicon substrate). The Schottky diode is electrically similar to a p-n junction, though the current flow in the diode is due primarily to carriers having an inherently fast response. It is used for high-frequency, low-noise mixer and switching circuits.

Scintillation. Variations in the brightness of starlight (i.e. 'twinkling') caused by turbulent strata very high in the Earth's atmosphere (ionosphere). Also, the emission of sparks or flashes. In general, scintillation refers to the fluctuation of amplitude and/or phase of a signal caused by the irregular structure of the propagating medium.

Scintillation counter. A device that uses a photomultiplier tube to detect or count charged particles (which produce scintillations of radiation when they impact upon phosphor) or γ-rays.

Sea Surface Salinity (SSS). SSS is an important variable in oceanography. In polar oceans, SSS intrusions with a low salinity influence the deep thermohaline circulation and the meridional heat transport. Variations in salinity also influence the oceans near surface dynamics in the tropics where rainfall modifies the buoyancy of the surface layer and the tropical ocean-atmosphere heat fluxes (warm surface pool dynamics). [2086)] - The physical basis for SSS remote sensing is the microwave brightness temperature (low frequency range in L-band around 1.4 GHz) which is directly linked to the dielectric constant of the target area (i.e., moisture or salinity); hence, proportional to moisture or salinity. SSS retrieval requires

2085) D. G. Long, M. W. Spencer, "Radar Backscatter Measurement Accuracy for a Spaceborne Pencil-Beam Wind Scatterometer with Transmit Modulation," IEEE Transaction on Geoscience and Remote Sensing, Vol. 35, No. 1, Jan. 1997, pp. 102-114

2086) G. Lagerloef, C. Swift, D. LeVine, "Sea Surface Salinity: The next remote sensing challenge," Oceanography, 8, 1995, pp. 44-50 knowledge of sea surface temperature and sea roughness and, additionally, it also requires a very high sensitivity from the sensor. [2087)]

Semiconductor junctions. Semiconductors whose principal charge carriers are electrons are called n-type (negative). If the charge carriers are mainly holes (a vacancy with positive charge), the material is p-type (positive).

Sensor. An instrument (generic term), usually consisting of optics, detectors, and electronics that collects radiation and converts it to some other form. The form may be a certain pattern (an image, a profile, etc.), a warning, a control signal, or some other signal. - The photographic camera is one of the best known examples of a 'remote sensor' which has been around since the first half of the nineteenth century.

Sensor characteristics. The ability of a sensor to detect and to resolve incoming radiation. For imaging sensors a very prominent characteristic is 'ground resolution,' its ability to distinguish objects on the Earth's surface. Other sensor characteristics are: scene size, spectral range, spectral resolution, radiometric resolution, pointing accuracy (location knowledge), and timeliness (in which images are returned to the user, the frequency at with which a given target can be revisited, the fraction of time that the sensor requires for taking an image).

Shielding. Refers to a technique of enclosing an object or a device within a container specifically designed to attenuate or otherwise exclude electromagnetic radiation.

Sidelobes. See **antenna sidelobes**.

Sigma (σ). The conventional measure of the strength of a radar signal reflected from a geometric object (the target area). Sigma designates the strength of reflection in terms of the geometric cross section of a conducting sphere that would give rise to the same level of reflectivity. See also **radar cross section**.

Sigma naught (σ^0). Scattering coefficient, the conventional measure of the strength of radar signals reflected by a distributed scatterer, usually expressed in dB. It is a normalized dimensionless number, comparing the strength observed to that expected from an area of one m^2. Sigma naught is defined with respect to the nominally horizontal plane, and in general has a significant variation with incidence angle, wavelength, polarization, as well as with the properties of the scattering surface itself.

Signal-To-Noise Ratio (SNR). The ratio of the level of information-bearing signal power to the level of noise power. The maximum SNR of a device is called the 'dynamic range.' In general, the higher the value of an instrument's SNR, the better the signal quality for recognition (detection) and interpretation.

Signature. The response of electromagnetic radiation to particular objects in the target area. Signatures may be used for pattern recognition which may in turn lead to target identification.

- The radar signature is the radar response (differential radar cross-section or the scattering cross section) of a particular material or object as a function of frequency, angle, polarization, or time.
- The **spectral signature** is the radiation response of an object as a function of wavelength.

Solar absorption technique. A method for measuring atmospheric constituents. As sunlight passes through the Earth's atmosphere, certain wavelengths are selectively absorbed by gaseous constituents. In the infrared region, nearly all gases have characteristic, discrete absorptions, whose positions and relative strengths are known from laboratory measurements of pure gas samples. This permits gaseous atmospheric constituents between the sun and an observer to be identified and quantified from high resolution solar spectra.

2087) http://www–sv.cict.fr/cesbio/smos/

Solar cell. An optoelectronic device (invented in 1954) that converts the radiant energy of sunlight directly into electrical power, based on photovoltaic principles. The solar cell is a large-area photodiode that detects the solar emission spectrum rather than a specific wavelength, as do photodiodes. The solar cell is unbiased, the load is connected directly across the two terminals of the p-n junction. Conversion efficiency, radiation hardness, and EOL (End Of Life) power are very important properties of solar cells. They are usually arranged in arrays or panels for spacecraft powering. During the 40 years of space technology, three generations of solar cells have been introduced:

- Silicon (Si) solar cells dominated the field until the early 1990s
- Gallium arsenide (GaAs) solar cells arrived in about 1990. They have better conversion efficiencies and radiation resistance in comparison with Si cells. GaAS cells can be manufactured on lightweight germanium substrates.
- The third generation of solar cells is the multifunction cell, or cascade cell. Current multijunction cells are based on GaInP (Gallium Indium Phosphide) material and GaAs on Ge substrate.

Solar cycle. The 9.5 - 11 year period between maxima (or minima) of solar activity (usually measured by the number of sunspots on the solar surface). About every 11 years the magnetic field of the sun reverses polarity; hence the more basic period may be 22 years. It is generally accepted that the solar cycle is maintained by a dynamo driven by the differential rotation of the sun's envelope.

Solar sail. A low-thrust propulsion technology (in the experimental/demonstration phase at the turn of the century) whose concept relies on the momentum transfer of photons (solar radiation pressure) on large, highly reflecting sails in space for passive propulsion such as orbit transfer functions. The concept involves the deployment and control (orientation) of a large sail in orbit on lightweight structures. The technology of such solarcraft is of interest for interplanetary missions.

Solar wind. A radial outflow of plasma from the solar corona, carrying mass and angular momentum away from the sun (see chapter O.18). The solar wind consists of a flux of particles, chiefly protons and electrons together with nuclei of heavier elements in smaller numbers, that are accelerated by the high temperatures of the solar corona, or outer region of the Sun, to velocities large enough to allow them to escape from the sun's gravitational field. At 1 au (astronomical unit) the solar wind contains approximately 1-10 protons/cm^3 moving outward from the sun at velocities of 350 to 700 km/s (or about 1.26 -2.52 million km/h); this creates a positive ion flux of 108 to 109 ions/(cm^2 s), each ion having an energy equal to at least 15 eV (electron volts). During solar flares, the proton velocity, flux, plasma temperature, and associated turbulence increase substantially.

Sounder. A remote sensing instrument that measures state parameters (like temperature, pressure, moisture, etc.) in a particular plane of observation for the derivation of profiles. A sounder may be a passive device by measuring the incoming radiation, it can also be an active device, transmitting signals (echo sounding) and receiving the echo information. Two basic configurations are in use:

- In the nadir-viewing configuration the observation plane is the orbit plane of the platform (series of footprints along the suborbital track). The scan technique provides good horizontal resolution of the measurements, but usually poor vertical resolutions.
- Limb sounders look at the horizon (the limb) and scan vertically, producing good vertical resolution but poor horizontal resolution.

Sounding. To 'sound' (to find bottom) originally referred to the measurement of water depths by sounding methods (sounding line, echo sounding, etc.) in shallow coastal waters and in rivers. The technique was much later extended to measure also the conditions of another medium, namely the atmosphere, at various heights. The first devices used were balloons with self-registering instruments (referred to as sondes) to record meteorological data. As new technologies became available, radiosondes, dropsondes from aircraft, rawinsondes, sounding rockets, ground-based, airborne and spaceborne instruments of a great variety appeared.

Spacecraft/platform attitude sensing and control devices.

1) A S/C attitude or pointing direction is determined by comparing information from various on-board sensors with the positions of known references. Attitude knowledge may be derived from the following orientation instruments:

- Magnetometers (measuring the known magnetic field components)
- Sun and/or star sensors or trackers (measuring of known celestial body directions)
- Earth horizon sensors (various types, mostly in the IR region; a horizon crossing indicator may determine the attitude of a spin-stabilized S/C with respect to the Earth; another horizon sensor may measure one component of the attitude of a three-axis stabilized S/C with respect to the Earth; there are scanning IR Earth horizon sensors, etc.).
- Gyroscopes (measuring inertial reference)
- GPS receiver (capable of measuring attitude). These GPS attitude instruments provide attitude and attitude rate data to actuators for real-time, autonomous attitude determination.
- Telescope (instrument guide telescope)
- etc.

2) The following instruments (or combinations thereof), referred to as actuators, provide attitude control:

- Momentum gyros
- Reaction/momentum wheels
- Thrusters (cold gas thrusters, solid thrusters, ion thrusters, mono- or bi-propellant engine, etc.)
- Magnetic torque coil/rods (magnetorquers)
- Permanent magnets
- Gravity gradient boom
- Nutation damper
- etc.

The simplest attitude control system is passive stabilization, either magnetically (a magnetometer as sensor in combination with a magnetic torque rod as actuator) or by gravity gradient methods. Passive stabilization can also be combined with active components, e.g. gravity gradient systems with magnetic torquers are quite common. Simple spinners, and momentum biased satellites represent the next advanced level of attitude control system, requiring at least for the momentum biased system some active stabilization about the angular momentum axis (typically the pitch axis). Zero-momentum systems with either reaction wheels or thrusters are the most complex systems, they require constant stabilization and become unstable if control is lost only for a short period of time. Simple and passively stable systems have a low pointing performance, and complex systems using reaction wheels are highly accurate pointing systems. [2088)]

Spacecraft/platform and instrument pointing. Good location knowledge of a target (of the ground surface, of a celestial body, etc.) by instrument pointing is an ever-present requirement of many missions (in particular for astronomy instrument pointing, also when imagery of the Earth's surface is used for cartographic applications). Precision pointing capability is the result of spacecraft stability through suitable attitude sensing and control mechanisms (some systems may include vibration control, elimination of alignment errors due to ther-

2088) H. J. Koenigsmann, G. Gurevich, "AttSim, Attitude Simulation with Control Software in the Loop," Proceedings of the AIAA/USU Conference on Small Satellites, Aug. 23-26, 1999, Logan UT, SSC-IIa-5 mal distortions, etc.). In general structural stiffness of the platform is an important prerequisite for a stable pointing environment. There are several classes of instruments with regard to pointing capability:

- Rigid-body instrument pointing or simply **body pointing**. This refers to a no-instrument-pointing capability relative to the platform. Most observation instruments on a spacecraft platform are fixed, they point into a constant direction (nadir, off-nadir, limb, zenith, etc.), their FOV (Field of View) provides a sufficient scan capability (for instance in the cross-track direction, and/or in the height direction) to measure all resolution cells in a swath.
- Some S/C with body-pointed (fixed) instruments are able to employ maneuvers to turn the entire S/C into a desired direction (example: IKONOS-1) thereby extending the field of regard considerably for observations outside of the normal swath width. S/C with relatively small masses (microsatellites) are most suited for this choice of pointing implementation.
- Instrument pointing relative to the platform. This class of sensors performs inertial pointing/tracking of a star or simply pointing/tracking of the sun or the moon. Some observation instruments need to be kept pointed for relatively long periods of time with extraordinary precision at faint celestial bodies. However, most instruments in this class are attitude sensors (such as: gyroscopes, magnetometers, horizon sensors, star or sun sensors, star trackers, accelerometers), their measurements serve as input for the on-board attitude control subsystem. The pointing knowledge bounds of a platform are always smaller than the actual pointing control bounds.
- Instrument pointing capability relative to the platform. These are observation instruments (imager, etc) performing fairly quick slew maneuvers, for instance in the along-track direction, to obtain stereo imaging.

Platform or Instrument	Pointing Knowledge	Pointing Accuracy (Control)
TIROS-N		0.1°
Spot-1 to -3		0.1°
ENVISAT	<0.03°	< 0.1° (3 sigma)
GP-B (relativity mission)		< 20 milliarcseconds (rms)
UoSAT		2-3° (rms) gravity-gradient boom system
DMSP (Block 5D-3)		0.01° (three orthogonal gyros)
SOHO		1 arcsecond (sun pointing over a period of 1.5 min)
TRACE		20 arcseconds (correction for pointing jitter)
Landsat-7	45 arcseconds	180 arcseconds
MSX		<0.1° (post-processing knowledge of 9 μrad)
GFO-1		0.25° (3 sigma)
Microlab-1		±2° gravity gradient boom system
IPS (Shuttle)		±1.2 arcseconds
CERES (on EOS)	180 arcseconds	
MISR (on EOS)	90 arcseconds	
SeaWinds	500 arcseconds	
OSA (on CRSS, also referred to as Ikonos-1)	Rms ground location accuracy: 2 m relative (with ground control points) 12 m absolute (without the use of control points)	

Table 563: Typical pointing parameters of a few satellites/instruments

SAR's capability to form good imagery relies significantly on the stability of the platform and, if the stability is not satisfactory, the precise knowledge of attitude information can be used to correct for orbital effects. On-board accurate and precise attitude/position determination is required, and in case of interferometry, most demanding. The baseline knowledge required is in the order of millimeter and the attitude of the baseline in the order of several arcseconds.

In an effort to achieve very precise aiming, ESA built a Spacelab system by the name of IPS (Instrument Pointing System - first flown on STS-51-F as Spacelab-2 in July/Aug. 1985).[2089] This three-axis gimbal pointing system provides precision pointing and tracking capabilities by establishing an inertially stable base from which stellar, solar, and Earth observations can be made (maintenance of pointing stability is within ±1.2 arcseconds). - In the same context, MACE (Middeck Active Control Experiment) is a NASA precision pointing system (built by MIT, LaRC, LMSC, et al.) flown on Shuttle flight STS-67 in March 1995, with the objective to explore high precision pointing and vibration control of future spacecraft and satellites. MACE extends conventional rigid-body instrument pointing to include flexible modes. Tests were conducted on the free-floating MACE platform to measure how disturbances caused by a payload impacts the performance of another nearby payload which is attached to the same supporting structure. MACE accomplishments: a) About 50 LaRC control systems were experimentally evaluated on-orbit, b) a reduction of at least 19 dB was achieved in the vibration levels, and c) MACE was able to synthesize and evaluate new control designs during the STS-67 flight. [2090] [2091] [2092]

Spacecraft/platform stabilization. Techniques that control the orientation (attitude) of the spacecraft in orbit with respect to certain known references. Several of the methods in use are:

- Single-spin stabilization. The whole spacecraft body rotates about the axis of the principal moment of inertia (acting like a gyroscope). These satellites cannot have oriented antennas, a severe drawback for certain applications.
- Dual-spin stabilization. A configuration in which the spacecraft consists of two parts: the platform, which is oriented toward the Earth, and the rotor, which rotates about the principal axis of the S/C thereby providing gyroscopic stiffness (example: Meteosat).
- Three-axis stabilization. A configuration in which the entire spacecraft is oriented toward a particular direction (usually toward the Earth in one dimension and aligned to the flight path in the other dimension). The control torques for attitude control are provided by a combination of reaction/momentum wheels, magnetotorquers, torque rods, gimbal system, and/or thrusters. In this concept, the rotating **reaction wheels** are able to absorb torque and momentum, while magnetic torquers or thrusters are used of allowing the wheels to slow their rotation rate. The same attitude control function may also be provided by an all-thruster system.
- Gravity gradient stabilization (passive stabilization method). A spacecraft consisting of two masses (main mass and small mass) that are connected by a rod or a boom. This two-mass arrangement produces a gravity gradient along the boom axis and an associated small torque which is employed for spacecraft orientation. This technique is normally used along with magnetic torquing (yet another passive stabilization method) for better attitude control of small satellites (mini, micro, or nanosatellites).

Space weather. This term refers to the conditions in space that affect the Earth and its space environment. Space weather is a consequence of sun behavior, the nature of Earth's magnetic field and atmosphere (in particular the ionosphere and magnetosphere), and Earth's location in the solar system. The solar wind, propagating against the Earth's magnetic field and interacting with it, shapes the near-Earth space environment. The response of the Earth's space environment to the solar wind is termed 'space weather.' Space weather can influence the performance and reliability of spaceborne and groundbased technological systems (power and communication systems) and can endanger human life and health

2089) http://www.msfc.nasa.gov/mol/description/ips/ips.html

2090) http://sun-valley.stanford.edu/users/howjo/mace.html

2091) K. K. Denoyer, R. S. Erwin, R. R. Ninneman, "Advanced SMART Structures Flight Experiments for Precision Spacecraft," Acta Astronautica, Vol. 47, No 2-9, 2000, pp. 389-397

2092) J. A. Woods-Vedeler, L. G. Horta, "On-Orbit Application of H-Infinity to the Middeck Active Controls Experiment: Overview of Results," AAS,1996-189 (spacewalks of astronauts). Other effects are: aurora and changes of climate. The space environment hazards that spacecraft and mission designers and operators need to be concerned with are: [2093)]

- Solar environment. The solar environment, directly or indirectly, effects all the other hazard environments. The solar activity levels, which follows an 11-year cycle, is a directly contributing factor which interacts to the radiation, thermal and plasma environments. The increased energy output from the Sun during its active periods heats the Earth's atmosphere and causes it to expand, which can effect the impact and neutral atmosphere environments, as well.
- Magnetic environment. The fields generated by the magnetic environment can directly interact with spacecraft. This is often taken advantage of in the attitude control subsystems, which can employ magnetometers and magnetic torque rods. The magnetic environment is also a major factor in determining the radiation and plasma environments around the Earth.
- Radiation environment. The radiation environment is principally composed of naturally occurring charged particles trapped in the Earth's magnetic field (also known as the Van Allen belts). Energetic solar particles and galactic cosmic rays also contribute to the natural radiation environment.
- Thermal environment. The thermal environment consists of thermal energy flux from the sun, the solar energy reflected back into space (and towards the spacecraft) from the Earth, referred to as albedo and the direct longwave thermal emission of the Earth due to its temperature, sometimes referred to as Earthshine.
- Impact environment. The impact environment consists of material from natural occurring micrometeoroids and from man-made debris flux. Due to the high relative velocities, even tiny particles can cause direct physical damage to the satellite structure and solar panels and can also induce damaging electrostatic discharges.
- Plasma environment. The plasma environment is mostly composed of charged particles (electrons) with energies too low to be a radiation hazard. However, these particles can
  strike and deposit themselves on external surfaces of the spacecraft or penetrate through the surface and deposit on internal components, causing electrostatic charge build-up. This charge can build up to high enough levels to create electrostatic discharge hazards that can damage spacecraft electronic components.
- Neutral atmosphere environment. The neutral atmospheric environment is the residual atmosphere remaining at spacecraft altitudes. The neutral atmosphere can contain atomic oxygen, which can damage the materials used on the spacecraft. Other residual atmospheric chemicals can also react with materials or be a source of contamination for optical systems.

Spatial frequency. Representation of an object or an image as a superposition of sinusoids (Fourier components).

Specific impulse. The specific impulse (Isp) of a thruster is the impulse (a force applied for a certain time) exerted with 1 kg of propellant. Therefore the units for specific impulse are Newton-seconds per kilogram (Ns/kg). By inserting the units of a Newton (1N = 1 kgm/s^2), the numerical value of the specific impulse also corresponds to the effective exhaust velocity (**m/s**) of the gas exiting the thruster in a vacuum.

Speckle. Refers to the phenomenon of a strong variation of echo signals from one resolution cell to another occurring in radar imaging (it is sort of a granular noise that affects the SAR images). Speckle occurs because the echo received consists of the sum of contributions of point targets in a each resolution cell, in continuously changing combinations (see also chapter O.8.4). Speckle is caused by the random interference of wavelets scattered by the microscopic fluctuations of the object surface within a resolution cell. The presence of

2093) M. Enoch, et al., "An Integrated Space Environment Analysis Tool (SEAT)," Proceedings of the 13th AIAA/USU Conference on Small Satellites, Aug. 23-26, 1999, Logan UT, SSC99-IIa-6 speckle in an image decreases the radiometric resolution, and thus reduces interpretability of the image (reduction in detail). Usually, filters are used to reduce the effects of speckle.

Spectra (of dispersion).Common methods are: (see also Table 812)

- Refraction: prisms are used to break up or disperse electromagnetic radiation into its component colors. The path of the radiation bends (refracts) when it passes from one medium into another.
- Diffraction: a lightwave breaks up into waves travelling in all directions as it strikes a surface. Diffraction gratings are composed of closely spaced transmitting slits on a flat surface (transmission gratings), or alternately reflecting and nonreflecting grooves on a surface (reflecting gratings).
- Interference: see **Interferometer**.
- Filter (electronically tunable filters)
- Filter (mechanical)
- Filter (mask)

UV	Ultraviolet: 0.01 - 0.38 μm
FUV	Far Ultraviolet: 90 - 125 nm
EUV	Extreme Ultraviolet: 40 - 90 nm (also abbreviated as XUV)
VIS	Visible:0.4 - 0.7 μm
NIR	Near infrared: 0.7 - 1.3 μm
VNIR	Visible/Near infrared: 0.4 - 1.3 μm, - the predominant mode of energy detection is that of reflected sunlight
SWIR	Short-Wave infrared: 1.3 - 3 μm - the predominant mode of energy detection is that of reflected sunlight
MWIR	Mid-Wave infrared: 3 - 6) μm - the detected energy is a mixture of solar reflected and thermally emitted radiation
TIR	Thermal infrared: 6 - 14 μm (also referred to as LWIR) - practically all energy received (detected) is attributed to thermal emission
VLWIR	Very Long-Wavelength Infrared (14 - 30 μm)
FIR	Far infrared: 10 - 1000 μm (note: 1000 μm = 1 mm). The TIR range is practically the lower portion of the FIR range.
MW	Microwave region: 1 mm - 1 m (<300 GHz frequencies < 300 MHz) - the detected energy is of microwave (thermal) emissions. Since microwave sensors do not depend on solar illumination the observations are virtually independent of aerosols and also much less affected by clouds (cirrus) than sensors operating in IR or VIS.

Table 564: Spectral regions of frequently-used acronyms

Spectral and spatial purity. An evaluation of the quality of radiometric measurements in the spectral and spatial domains.

Spectral band. An interval in the electromagnetic spectrum defined by two wavelengths, two frequencies, or two wavenumbers. Note: The so-called **optical spectrum** extends from 0.01 μm to 1000 μm, i.e., from the UV to the FIR region inclusively. This is followed by the microwave region.

Spectral resolving power. Ratio of $\lambda/\Delta\lambda$ (see also an example under **Wavenumber**).

Spectral signature. Quantitative measurement of the spectral properties of an object at one or several wavelength intervals.

Spectrometer. An instrument connected to a telescope that separates the light signals into different wavelengths or frequencies, producing a spectrum (thus permitting an analysis of the spectral content of the incident electromagnetic radiation). Usually, only a relatively small portion of the spectrum is measured by an instrument. Some spectrometer types are:

1) **Dispersive systems:** A class of spectrometers using the dispersive principle to separate radiation into its narrow-band components (spectral discrimination). A dispersive imaging spectrometer can only support one dimension of imaging (along the slit); the

other dimension is used for spectral dispersion. Images are built up by making successive exposures; hence, images are stacked side by side. This is done either by using the motion of an aircraft or spacecraft (pushbroom imaging) or by the use of a sideways-scanning mirror (whiskbroom imaging). See also O.3 and O.6.

- Prism spectrometer. From a historical point of view, glass prisms were first used to break up or disperse light into its component colors. The path of a light beam bends (refracts) as it passes from one transparent medium to another, e.g., from air to glass. A prism is used, along with collimating and re-imaging optical and mechanical components, to disperse light for spectral discrimination.
- Grating (diffraction) spectrometer. A grating is used (along with collimating and re-imaging optical and mechanical components) to disperse light by diffraction for spectral discrimination. The spectral dispersion is stated, for example 2-4 nm/mm at 300 nm, and the resolution is 0.5 nm.
- In a wedge spectrometer spectral discrimination occurs in a focused beam

2) **Filter Spectrometers (nondispersive systems).** Filters are used to control the spectral bandwidth of the radiation that is allowed to reach the detector system. Narrowband filters are in the order of 1-2 cm^{-1}.
   - Filter-wheel technique. Allows the selection of up to n discrete spectral bands.
   - Bandpass filter technique. Allows the transmission of only a narrow band of frequencies (the other frequencies are blocked out). The spectral width of this filter is characterized by its bandwidth. A typical bandpass filter instrument is TM on Landsat
   - Filter mask technique in which the spectral separation filters are mated to the detector array to achieve two-dimensional sampling of the combined spatial/spectral information passed by the filter. A typical instrument of this type is WIS (Wedge Imaging Spectrometer)
   - Dichroic systems. A filter method allowing selective absorption in crystals of electromagnetic radiation vibrating in different planes (usually filtering is based on wavelength). The dichroic principle is applied to beam splitters and filters.
   - Interference filter. A filter reflecting radiation selectively in a narrow spectral band

3) **Fourier Transform Spectrometers (FTS, nondispersive systems).** An FTS system provides a conventional spectrum, but with greater speed, resolution and sensitivity. This class of spectrometers separates the incoming broadband spectrum into narrow-band components with the use of an interferometer. An incoming wavefront into the interferometer is divided by a beam splitter (semitransparent surfaces). Beams produced in this way travel two different paths, then recombine (superposition principle), creating an interferogram. This interferogram (a function of signal intensity versus time) is normally digitized and converted to an absorption spectrum by means of a Fourier transform. Instruments with high resolving power often use interferometers in series with grating instruments. FTS can be designed to cover all spectral regions from the radio frequency to the UV.

4) **Correlation Spectrometers**, also referred to as NDIR (Non-Dispersive Infrared) spectrometers. A correlation spectrometer is a device for a gas-specific investigation that correlates the spectral signatures of the species to be analyzed with reference spectra.

5) **AOS (Acousto-Optical Spectrometer)**. The principle of an AOS is based on the diffraction of light at ultrasonic waves. A piezoelectric transducer, driven by the RF-signal (from the receiver), generates an acoustic wave in a crystal (the so called Bragg-cell). This acoustic wave modulates the refractive index and induces a phase grating. The Bragg-cell is illuminated by a collimated laser beam. The angular dispersion of the diffracted light represents a true image of the RF-spectrum according to the amplitude

and wavelengths of the acoustic waves in the crystal. The spectrum is detected by using a single linear diode array (CCD), which is placed in the focal plane of an imaging optics.

6) **Heterodyne Spectrometers** (nondispersive systems). See **Heterodyning**.

7) Lidar Spectrometers.

8) etc.

Spectrometry. In remote sensing the detection and measurement of radiation spectra of a target (area or volume). Each spectra has a characteristic pattern of absorption and emission bands. Comparison of these spectra against reference spectra provide information on the target's material composition. Imaging spectrometry refers to the simultaneous acquisition of images in many contiguous spectral bands.

Spectroradiometer. A combination of spectrometer and radiometer for measuring the energy distribution of emitted radiation.

Spectroscopy - differential absorption spectroscopy. A technique that uses two frequencies emitted by the same laser or by different lasers to perform measurements of the concentration of a gas along a given line of sight. The frequency of one laser signal is tuned to the frequency of the center line of the absorption feature; the frequency of the other laser signal is tuned aside from this feature. The difference in the amount of transmitted light at these two frequencies is the quantity that is being sought.

System Technology	Spectral Resolving Power $\lambda/\Delta\lambda$	Wavelength Range	Moving Parts	Simultaneous Acquisition of all spectral bands	Throughput
Grating (CCD detectors)	10^2 - 10^5	Narrow (optics-limited)	no	yes	low
Prism	10^2 - 10^3	Narrow (optics-limited)	no	yes	low
Fourier Transform Spectrometer (FTS)	10^6	Broad (detector-limited)	yes (no, depending on type)	yes	very high
Filter (electronically tunable)	10^2	Narrow (optics-limited)	no	no	very high
Filter (mechanical)	10^3	Broad (detector-limited)	yes	no	very high
Filter (mask)	10^2	Narrow (optics-limited)	no	yes	very high
Filter (mask) WIS	10^2	Broad (detector-limited)	no	no	very high

Table 565: Overview of some spectrometer technology characteristics

Spectroscopy - imaging. Imaging spectroscopy is the acquisition of images, where for each spatial resolution element (pixel) in the image a spectrum of the energy arriving at the sensor is measured. These spectra are used to derive information based on the signature of the interaction of matter and energy expressed in the spectrum.

Spectrum. Refers generally to the intensity distribution of electromagnetic radiation as a function of wavelength, wavenumber, or frequency (see Figure 413).

Spread-spectrum technology. A transmission technique that allows multiple senders and receivers to share the same portion of the spectrum (bandwidth) by having each sender encode its transmission in a unique way decipherable by only its intended receiver. Part of this technique is used in GPS (see chapter H.4) and GLONASS (see chapter H.3) communication. Spread spectrum technology is also used for PCS (Personal Communication Services) via satellite on such systems as 'Iridium' and 'Globalstar.' The spread-spectrum technology allows communication satellites to capture and transmit signals that normally would be lost because the original signals were too weak or had too much interference. The wide bandwidth of the technology (about three orders of magnitude higher than normal radio frequencies) make it difficult to intercept the signal by an unauthorized party. The feature of low interception probability is attractive for many communication applications.

Squint. The term is used to describe an oblique pointing geometry of a sensor. For instance, a typical SAR pointing geometry is in the cross-track direction, normal to the flight path. Squinting occurs when the antenna beam is pointed forward or backward from this orthogonal direction.

Standing wave. A wave that is stationary with respect to the medium in which it is embedded, e.g., two equal gravity waves moving in opposite directions.

Station keeping. Refers to the maintenance of a geostationary satellite in its assigned orbital slot with regard to position and orientation (attitude). Orbital drifts are due to small gravitational effects of the sun and the moon as well as to an inhomogeneous Earth. The physical mechanism for station keeping is the controlled ejection of hydrazine gas by command from a control center.

Steradian (sr). A unit of solid angle measure in the International System, defined as the solid angle of a sphere subtended by a portion of the surface, whose area is equal to the square of the sphere's radius. The total solid angle about a point is 4π steradians. The term steradian is derived from the Greek for 'solid' and 'radian' - a steradian is, in effect, a solid radian.

Stereoscopy. The spatial three-dimensional or 'stereo' observation of related 2-D images, showing the same object under different viewing angles. Stereo images are very appropriate for map-making and for many other applications (flight simulators, etc.). The image combination of a target area may either result from, say, three cameras of an instrument pointing into the forward, nadir and aft directions, respectively, of a subsatellite track, or from a single gimbaled camera, performing along-track imaging by pointing into the forward, nadir and aft directions successively. Stereo images offer better surface relief mapping capabilities than do regular 2-D images.

Store-and Forward (S&F). A non-real-time communication technique between a LEO satellite and its ground segment (often used for Data Collection Systems, e-mail systems, etc.). In this setup the originating ground station (or terminal) sends a digitized message to the LEO satellite; the satellite intermittently stores the message in an on-board storage system, and the destination ground station later receives the message when the satellite footprint is in its view. Multiple small satellites in polar LEO increase the message traffic capacity and reduce delivery delays.

Stratopause. Stratosphere-mesosphere boundary (at about 50-55 km in altitude) where a relative temperature maxima is found (see Figure 410).

Stratosphere. Region of the atmosphere between the troposphere and mesosphere, having a lower boundary of approximately 8 km at the poles and 18 km at the equator, and an upper boundary of approximately 50 km. Depending upon latitude and season, the temperature in the lower stratosphere can increase, be isothermal, or even decrease with altitude, but the temperature in the upper stratosphere generally increases with height due to absorption of solar radiation by ozone. - The importance of the stratosphere stems from the absorption of the bulk of the solar UV radiation, in particular in the wavelength regions of 290-320 nm. Penetration of this UV radiation to the Earth's surface may be harmful to life. The component in the stratosphere absorbing the bulk of the UV radiation is ozone (O_3).

Subcarrier. Refers to a second signal "piggybacked" onto the main signal (carrier) to carry an information channel.

Sunspot. A temporary disturbed area in the solar photosphere that appears dark because it is cooler than surrounding areas. Sunspots are concentrations of strong magnetic flux (2000 - 3000 gauss), with diameters less than about 50,000 km and lifetimes of a few weeks.

Sun-synchronous orbit. On orbit is said to be sun-synchronous when the precessing rate of the orbital plane of a satellite, caused mostly by Earth flattening at the poles, is the same as the apparent motion of the sun in the celestial sphere, namely 0.9856°/ day. Such an orbital configuration results in a (nearly) constant local time of ascending node (resulting in observations of a given area on the Earth's surface that are always made at the same local time of the day and the same solar incidence angle). See also chapter O.12.1.

Superconducting Tunnel Junctions (STJs). Initially under development as efficient detectors of x-rays, they are now being used as single photon detectors in the visible spectrum. STJ (developed at ESA/ESTEC) operates in the range 200 - 1000 nm with a spectral resolution of 45 nm. Unlike a silicon-based CCD, the niobium-based STJ generates a number of electrons (in the thousands) that depends on the incoming photon's energy. This property eliminates the need for filters or diffraction gratings that lower the overall efficiency.

Superconductivity is the ability of a material to carry electricity with no resistance. Superconductivity was discovered in 1911 by H. K. Onnes in Leiden, Netherlands, just three years after he had succeeded in liquifying helium. Onnes discovered the abrupt and complete disappearance of resistance in certain metals when they were cooled below the critical temperature Tc of 4.2 K using liquid helium. [Note: instrumentation at liquid helium temperatures is referred to as LTS (Low Temperature Superconductivity) devices]. - The value of Tc has changed ever since. The search for a higher Tc began in particular in the 1980s to save the enormous cooling costs at cryogenic temperatures leading eventually to **HTS** (High Temperature Superconductivity). [2094]

- Tc = 35 K (April 1986). Karl Alexander Müller and Johannes Georg Bednorz (IBM Research Laboratory, Switzerland) discovered superconductivity in $(La\text{-}Ba)_2$ CuO4. In 1987, the Nobel Prize in physics was awarded to both researchers. [2095]
- Tc = 77 K (end of 1986). P. C. W. Chu (University of Texas at Houston) discovered superconductivity in the liquid-nitrogen temperature range.
- Tc above 90 K (January 1987). M. K. Wu, Chu's former student, achieved stable and reproducible superconductivity above 90 K in $Y1Ba_2Cu3O7\text{-}d$ (Y123), with Tc close to 100 K.
- Tc = 110 K and 125 K (1988) for bismuth and thallium superconducting systems respectively
- Tc = 164 K (1993) for mercury-based compounds under pressure (University of Texas, Houston).
- etc.

The first SQUID (Superconducting Quantum Interference Device) instrumentation appeared in 1964 and was widely used in the field of cryogenics. In the late 1980's, the discovery of high-temperature superconductor materials opened the possibility of introducing the technology in superconducting instruments. Commercial applications of HTS technology in fields such as electric power, transportation, electronics and medicine are appearing in the 1990s. Current applications of HTS include thin-film technology, magnetic resonance imaging (MRI), wireless communication filters, and ultra-fast computer chips. Modern discoveries in superconductivity go far beyond piece-meal improvements in electric devices. They have opened the door on a totally new technology and stretch the imagination to the discovery of new applications.

Surface charge. A satellite immersed in an ambient plasma will come to equilibrium with that plasma by developing surface charges of the proper sign and magnitude to reduce the net current between the satellite and the ambient plasma to zero. The net current consists of a) currents from the environmental flux, b) secondary backscattered electrons and ions, and c) by photoelectrons from any illuminated areas on the spacecraft. As a result of these three processes contributing to the charged particle fluxes, a potential distribution exists about

2094) Special issue: "Superconductivity," Physics Today, March 1986

2095) http://www.physnet.uni-hamburg.de/home/vms/reimer/HTC/HTC.html

the spacecraft so that the net current to the satellite is zero. The potential distribution about a satellite may be rather asymmetric; this depends very much on the satellite geometry, it is also due to the anisotropic distribution of the particle fluxes. [2096)]

Surface roughness. Variation in surface height within an imaged resolution cell. A surface appears "rough" to microwave radiation when the height variations become larger than a fraction of the radar wavelength.

Synchronization (sync). Refers to the process of orienting the transmitter and receiver circuits in the proper manner in order that they can be synchronized. Usually a data format is preceded by a sync pattern which is recognized by the receiver.

Synoptic view. A large (inclusive) scene of the Earth's surface, or of an object/target under investigation, allowing a large-scale overview of features or phenomena or relations of a scene in a wider context.

Swath. Width of the imaged scene in the range direction.

Telemetry. A space-to-ground data stream of measured values (normally including instrument science data, instrument engineering data, and spacecraft engineering data) that does not include commands, tracking, computer memory transfer, audio, or video signals.

Telemetry, Tracking and Command (TT&C). Refers to the function of spacecraft operations (monitoring and control of all vital system parameters and tracking of the orbit) by a control center. These TT&C functions are normally completely separate from the spacecraft's user signal (communication satellite) or the measured source (or instrument) data (in case of an Earth observation satellite). Hence, they are also transmitted in a separate band.

Telescience. A technique referring to the control of scientific and/or engineering experiments/instruments from a remote location. Applications include various configurations such as Earth-Earth connections as well as Earth-spaceborne support.

Telescopes (types). Optical telescopes are of two basic types, refractors or reflectors that use lenses or mirrors, respectively, for their light collecting elements. The Galilean (1564-1642) and Keplerian (1571-1630) telescopes are of the refractive type. The Cassegrain and Gregory telescopes are of the reflective type. Reflectors are used in the UV, VIS and IR regions of the electromagnetic spectrum. The name of this type of instrument is derived from the fact that the primary mirror reflects the light back to a focus instead of refracting it.

- **Cassegrain telescope** (design was proposed in 1672 by N. Cassegrain, a French scientist). A reflective telescope in which a small hyperboloidal secondary mirror reflects the convergent beam from the paraboloidal primary mirror through a hole in the primary mirror to an eyepiece in back of the primary mirror (see also Cassegrain antenna). The Cassegrain telescope design is the most frequently used two-mirror system.
- **Dall-Kirkham telescope.** A Cassegrain-type instrument where the primary mirror geometry is an ellipse, while the secondary mirror is a sphere.
- **Ritchey-Chrétien telescope.** A Cassegrain-type instrument. The telescope design reduces the 'coma' (image aberrations) by modifying the primary and secondary surfaces of a Cassegrain telescope. The Ritchey-Chrétien telescope employs a hyperboloidal figure for both the primary and secondary mirror thereby providing excellent resolution over a large FOV.
- **Gregorian telescope.** James Gregory (a 17th century Scottish mathematician) devised an arrangement of two concave mirrors. Gregory placed a concave secondary mirror

2096) E. A. Bering, III, R. Kabadi, B. McIntyre, "High Voltage Spacecraft Charging: Theory and Measurement," Proceedings of the AIAA 2000 Space Conference and Exposition, Long Beach, CA, Sept. 19-21, 2000 outside the prime focus to reflect the light back through a hole in the primary mirror. The spacecraft of the SMM (Solar Maximum Mission), launched in 1980, flies a Gregorian telescope.

Type of Telescope	Primary Optic	Secondary Optic	Configuration 1 - Primary Optic 2 - Secondary Optic 3 - Eyepieces/Correctors 4 - Focus (usually also the image plane)
Newtonian Examples: ALF, UV-DIAL	Parabola	Diagonal Flat	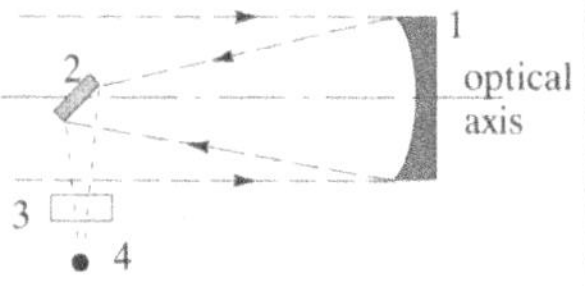
Gregorian Example: UVSP	Parabola	Ellipse	
Cassegrain Ex: ALISSA, GOMOS, ALEX	Parabola	Hyperbola	
Dall-Kirkham Ex: CAR, CAMS, LASE, MCR	Ellipse	Sphere	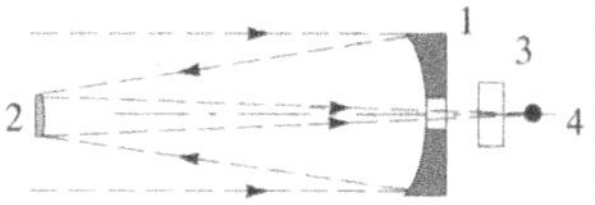
Ritchey-Chrétien Ex: SEVIRI, EIT, LEANDRE	Modified Parabola	Modified Hyperbola	
Schmidt Ex: LFS, OLS, HYDICE	Aspherical Refractor	Sphere	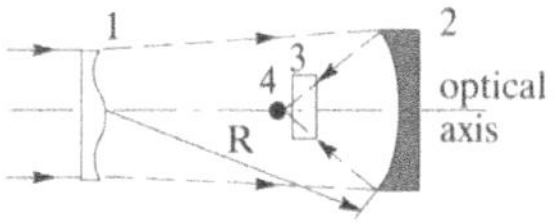

Figure 423: Basic optical configurations for common types of reflective telescopes

Type of Image Defect	Description
Spherical aberration	Light focuses at different places along the optical axis as a function of radial position
Coma	Image size (magnification) varies with radial position in the focal region.
Field curvature	Off-axis images are not focused on the ideal surface, usually a plane
Astigmatism	Light focuses at different places along the optical axis as a function of angular position in the aperture
Distortion	Focused off-axis image is closer or further from the optical axis then intended
Chromatic aberration	Shift in the focused image position as a function of wavelength

Table 566: Definition of some basic image aberrations occurring in telescopes [2097)]

- **Newtonian telescope**. The primary mirror is of parabolic geometry; the secondary mirror may be a flat plat or a refractive prism. The light beam is diverted to one side for observation.
- **Schmidt telescope**. Bernhard V. Schmidt (1879-1935). In 1930, Bernhard V. Schmidt of the Hamburg Observatory in Bergedorf, Germany, designed a catadioptric telescope

2097) Encyclopedia of Physical Science and Technology, Academic Press, 1987, Vol. 9 pp. 730-732 with a large FOV to eliminate image distortions (a catadioptric telescope design incorporates the best features of both the refractor and reflector, i.e., it has both reflective and refractive optics). This design compensates for most of the spherical aberration by means of an aspherical refractor at the center of curvature.

Theodolite. A surveying instrument used to measure horizontal and vertical angles.

Thermal control system. An on-board system which maintains all satellite components within allowable temperature limits for all operating modes of the satellite when exposed to the varying thermal environments throughout its lifetime. Typical thermal loads (forms of environmental heating) are: a) direct solar radiation, b) reflected radiation from the Earth's albedo, c) emission of long-wave IR radiation from the Earth, d) free molecular heating, and e) charged particle heating.

Thermal Infrared (TIR). Electromagnetic radiation in the spectral range of 6-20 μm. Many remote sensing applications utilize the 6-12 μm range. TIR is emitted energy, whereas NIR (Near Infrared) is reflected energy.

Thermal noise. The 'noisy' detector signal of an infrared sensor caused by the thermal heating of the detector itself. Thermal noise occurs when the system is not sufficiently cooled.

Thermistor. A semiconductor device (sensor) whose electrical resistance varies with temperature. Its temperature coefficient of resistance is high, nonlinear, and usually negative.

Thermoluminescence. A property of certain minerals which causes them to emit light when moderately heated, after electrons are excited into traps by ionizing radiation.

Thermosphere. Outermost layer of the atmosphere, above the mesosphere.

Thin-film technology. Thin films are an important ingredient in all surface technology applications. Surfaces play an important role in nature as well as in technology where small-scale exchange processes take place. Nanotechnology offers a new realm for surface technology in which to operate, providing also new analytical methods for much clearer windows onto nanoscale surface structures. In solar cell applications, thin-film technology refers to dielectric layers for optical anti-reflective coatings, electrical passivation and diffusion barriers. Applications of thin-film technology abound in such fields as: lithography, deposition, etching, epitaxy, diffusion, optics, sensor technology, etc.

Time Division Multiple Access (TDMA). A process that shares the time domain of a single carrier among many users by assigning to each time intervals in which to transmit signal bursts. In this scheme, all users transmit on the same frequency, each is assigned the total available bandwidth for a limited amount of time. TDMA systems segment time into frames, each frame is further partitioned into assignable time slots.

Tomography (optical tomography).[2098] A diagnostic technique permitting the mathematical reconstruction of 3-D images from a set of 2-D measurements. A typical application are the medical CAT (Computer Aided Tomography) scans which yield the structure of a human body from a set of X-rays. The tomography technique is also finding its way into high-speed applications (high temporal resolution), such as in aero-optical measurements of dynamic turbulent media (simultaneous measurements of the flow field and the optical field provide information of the flow structure in space and time). An application of this technique is the study of phenomena causing degradations in laser beam propagation through atmospheric boundary layer turbulence. In laser transmissions through turbulent media, adaptive optics systems are being used to correct for phase distortion. Adaptive optics systems rely on accurate measurements of the turbulent media and on its ability to distort the beam.

2098) Note: Tomographic methods were first formulated in the1970s as a means of remotely mapping inaccessible regions of the human body.

Naturally, the microwave region of the spectrum may also be used for tomographic studies. For instance, the basic concept in CIT (Computerized Ionospheric Tomography) research is to use LEO satellites as moving transmitters and an array of ground receivers to measure TEC (Total Electron Content) in the ionosphere.

Total Solar Irradiance (TSI). The total solar irradiance along with Earth's global average albedo determines Earth's global average equilibrium temperature. Because of selective absorption and scattering processes in the Earth's atmosphere, different regions of the solar spectrum affect Earth's climate in distinct ways. To place the 11-year sun cycle into perspective, the sun's TSI is about 1370 Wm^{-2} in space (i.e. in low Earth orbits of spacecraft). Since the intercepted radiation is distributed over the surface of the Earth, the average solar radiation at the top of the atmosphere is 1/4 of this, or about 340 Wm^{-2}; hence, a variation of 0.1% corresponds to 0.34 Wm^{-2}.[2099)] Planetary albedo scattering reduces this further to about 0.24 Wm^{-2} (approximately 20-25 % of the TSI is absorbed by atmospheric water vapor, clouds, and ozone, by processes that are strongly wavelength dependent. Ultraviolet radiation at wavelengths below 300 nm is completely absorbed by the Earth's atmosphere and contributes the dominant energy source in the stratosphere and thermosphere, establishing the upper atmosphere's temperature, structure, composition, and dynamics). Even small variations in the sun's radiation at these short wavelengths lead to corresponding changes in atmospheric chemistry. Radiation at the longer visible and infrared wavelengths penetrates into the lower atmosphere, where the portion not reflected is partitioned between the troposphere and the Earth's surface, and becomes a dominant term in the global energy balance and an essential determinant of atmospheric stability and convection. Thus it is important to accurately monitor both the TSI and its spectral dependence.[2100)]

Trace gas. A minor constituent of the atmosphere. The most important trace gases contributing to the greenhouse effect are water vapor, carbon dioxide, ozone, methane, nitrous oxide, and chlorofluorocarbons. Other trace gases include ammonia, nitric oxide, ethylene, sulfur dioxide, methyl chloride, carbon monoxide, and carbon tetrachloride.

Tracking system. Tracking is the process of following a moving object. Tracking system is a general name for an apparatus, such as a tracking radar, used to follow and record the position of objects (airborne or spaceborne). A theodolite and an observer form, for instance, an optical tracking system which is used in pilot balloon runs.

Transceiver. A term made up of the words 'transmitter' and 'receiver' of a signal transmission system. Since each side of a two-way system requires both functions, they are provided in one unit.

Transducer. A device chancing one form of signal energy into another, such as a microphone, a thermocouple, a photocell, etc.

Transmittance (transmissivity). The ratio of power transmitted through a layer of a medium to the power incident upon it.

Transmitter. An electronic device consisting of an oscillator, modulator and other circuits which produce a wave signal for radiation by an antenna.

Transponder. A combined receiver and transmitter system (usually part of a communications system of a satellite) whose function is to transmit signals automatically when triggered by an interrogating signal.

Traveling wave tube. A microwave power generating tube that accelerates electrons by varying a magnetic field between cathode and anode to set up waves of electron density.

2099) G. C. Reid, "Solar Variability and the Earth's Climate: Introduction and Overview," pp. 1-11 in Solar Variability and Climate, Editors: E. Friis-Christensen, C. Fröhlich, J. D. Haigh, M. Schüssler and R. von Steiger, Kluwer Academic Publishers, ISBN 0-7923-6741-3, 2000

2100) http://lasp.colorado.edu/sorce/

Tropical year. The interval of time between two successive vernal equinoxes. It is equal to 365.242 mean solar days.

Tropopause. Boundary between the upper troposphere and the lower stratosphere that varies in altitude between approximately 8 km at the poles and 18 km at the equator. The temperature gradient of the tropopause goes to zero (a relative temperature minima exists).

Troposphere. Lowest atmospheric layer, between the surface and the tropopause (lowest 8-15 km of the atmosphere, depending on latitude). The troposphere is characterized by decreasing temperature with height, large vertical motion, and large water vapor content. This is the region where most of the 'weather' occurs.

Uplink. Refers to the communication path direction from a ground station to a satellite. The information in this uplink is usually used for commanding of a subsystem (in general there are also uplinks in tracking systems, etc). In very elaborate communication systems with intermediate geostationary transmission satellites, the term 'uplink' is usually replaced by 'forward link' to avoid confusion.

Upwelling. The vertical motion of water in the ocean by which subsurface water of lower temperature and greater density moves toward the surface of the ocean. Upwelling occurs most commonly along the western coastlines of continents, but may occur anywhere in the ocean. Upwelling results when winds blowing nearly parallel to a continental coastline transport the light surface water away from the coast. Subsurface water of greater density and lower temperature replaces the surface water, and exerts a considerable influence on the weather of coastal regions. Carbon dioxide is transferred to the atmosphere in regions of upwelling. This is especially important in the Pacific equatorial regions, where 1 to 2 gigatons of carbon per year may be released to the atmosphere. Upwelling also results in increased ocean productivity by transporting nutrient-rich waters to the surface layer of the ocean. - The term 'upwelling' is also used in the context of 'upwelling radiation.' This refers to radiation (from the Earth's surface and from the atmosphere) observed from an airborne or spaceborne sensor.

UTM (Universal Transverse Mercator Projection). A widely used map projection which employes a series of identical projections around the world in the mid-latitude areas, each spanning six degrees of longitude and oriented to a meridian. The UTM projection preserves angular relationships and scale, it easily allows a rectangular grid to be superimposed on it.

Van Allen Belt. Regions or belts in the Earth's magnetosphere (at about 1.4-1.5 R_E and 4.5-6 R_E) where many energetically charged particles from the solar wind are trapped in the Earth's magnetic field.

Vegetation index. A mathematical algorithm of reflection values (reflectances, digital pixel values) in different spectral bands, used to estimate vegetation characteristics. Such an algorithm also serves to correct undesirable influences, such as differences of soil reflectance, atmospheric influences, etc. - In physical terms vegetation indices are radiometric measures of vegetation usually involving a ratio and/or linear combination of the red and NIR regions. Vegetation indices serve as indicators of relative growth and/or vigor of green vegetation, and are used as intermediaries in the assessment of various plant biophysical parameters, such as leaf area index, percent green cover, green biomass, and fractional absorbed photosynthetically active radiation (FPAR).
The spectral vegetation index concept and its universal generality has its roots in Landsat data interpretation.[2101)] As a consequence there are now a multitude of defined spectral vegetation and soil indices in existence. They are all derived, at least in part, by considering the contrast between visible (VIS) and near-infrared (NIR) spectral reflectance from land

2101) S. N. Goward, D. L. Williams, "Landsat and Earth Systems Science: Development of Terrestrial Monitoring," PE&RS, July 1997, pp. 887-900 surfaces. For typical broadband VIS/NIR measurements, green vegetation foliage produces stepped reflectance, with low VIS and high NIR reflectance, a result of pigment absorption in the VIS region and strong light scatter from cell walls in NIR.

Vernal equinox. The point of intersection between the ecliptic and the celestial equator, where the sun crosses from the south to the north (it is in fact the ascending node of the sun's orbit). The vernal equinox marks the beginning of spring for the northern hemisphere.

Very Long Baseline Interferometry (VLBI). In radio astronomy, the use of a system of two or more antennas placed several hundred or even several thousand kilometers apart, which are operated together as an interferometer. VLBI techniques are also employed in the field of Solid Earth Physics (geodynamics) for determining plate motions with accuracies of better than 1 cm/year. VLBI systems offer a superb tie to an inertial celestial reference frame based on extragalactic radio sources, which in turn is used to maintain the terrestrial reference frame.

Video. In general used to mean television, or a system used to communicate the television image. Specifically, pertains to the bandwidth and spectrum position of the signal which results from television scanning and which is used to reproduce a picture. Video images do not have the detailed resolution of film, but offer the advantage of immediate processing capability. This is particularly important in time-sensitive applications.

Vidicon. A generic name for a camera tube of normal light sensitivity. It outputs an analog voltage stream corresponding to the intensity of the incoming light.

Viewing angle. See **look angle**.

Water vapor. A very important constituent of the atmosphere. Its amount varies widely in space (vertical and horizontal) and time. The troposphere is the domain of water vapor, with about half of all the atmospheric water vapor in a layer below 2 km (only a minute fraction is above the tropopause). Water vapor is the major vehicle of atmospheric energy transport, a regulator of planetary temperatures and of rainfall. - The amount of water vapor in a given air sample may be determined in a number of different ways, involving such concepts as absolute humidity, mixing ratio, dew point, relative humidity, specific humidity, and vapor pressure.

Water-vapor absorption bands. Wavelength bands where water vapor - free or bound - absorbs radiation to a high degree. Absorption bands in the IR region are near 1.4 μm, 1.8 μm, 2.7 μm, 6.3 μm (strong), 11 μm, and 30 μm.

Wavefront. A three-dimensional surface in space for which the field radiated by the antenna has the same phase at all points. At a large distance R from the antenna, the wavefront is a spherical surface with radius R over the angular window established by the antenna pattern. For most geometries encountered in remote sensing, the wavefront may be approximated by a plane tangent to the spherical surface.

Wavenumber (ν). The number of waves per cm (the reciprocal of the wavelength). $\nu = 1/\lambda$. A wavenumber of 10,000 corresponds to a wavelength (λ) of 10^{-4} cm^{-1}, or to a wavelength (λ) of 1 μm. Conversion of a wavenumber resolution $\Delta\nu$ into a wavelength resolution $\Delta\lambda$:
$\nu = 1/\lambda \Rightarrow |\Delta\nu| = |\Delta\lambda| \; /\lambda^2 \Rightarrow |\Delta\lambda| = |\Delta\nu| \; \lambda^2$
Example: convert the wavenumber resolution $\Delta\nu = 20\ cm^{-1}$ into a wavelength resolution for the spectral range of 400-800 nm.
$|\Delta\lambda| = |\Delta\nu| \; \lambda^2 = 20\ cm^{-1} \times (400\ nm)^2 = 2 \times 10^{-6}\ nm^{-1} \times 160000\ nm^2 = \underline{0.32\ nm}$ for $\lambda = 400$ nm.
$|\Delta\lambda| = |\Delta\nu| \; \lambda^2 = 20\ cm^{-1} \times (800\ nm)^2 = 2 \times 10^{-6}\ nm^{-1} \times 640000\ nm^2 = \underline{1.28\ nm}$ for $\lambda = 800$ nm.

In this context belongs also the **spectral resolving power**, which is defined as the following ratio: $\lambda/\Delta\lambda$. Example: find the spectral bandwidth for a spectral range from 1.0 - 2.5 μm when the spectral resolving power ($\lambda/\Delta\lambda$) of the instrument is given as 250.

$\Delta\lambda = \lambda/250 = 1\ \mu m/250 = 0.004\ \mu m$ or 4 nm at the lower bound of the spectral range
$\Delta\lambda = \lambda/250 = 2.5\ \mu m/250 = 0.010\ \mu m$ or 10 nm at the upper bound of the spectral range.

WGS-72 (World Geodetic System 1972) an Earth-centered datum of DMA (Defense Mapping Agency of US DoD). WGS-72 was preceded by WGS-60 and WGS-66. A principal objective of WGS is to allow referencing of local geodetic systems to a single geocentric system. WGS-72 is the result of an extensive effort (extending over 3 years) to collect satellite, surface gravity, and astrogeodetic data available throughout 1972. These data were combined using a unified WGS solution (a large-scale least squares adjustment).

WGS-84 (World Geodetic System 1984).[2102] WGS-84 was also developed by DMA and is an improvement and a replacement for WGS-72. Radar altimeter data from GEOSAT was used to deduce geoid heights from oceanic regions of latitude ±70°. Geoid heights were also deduced from a large number of ground-based Doppler stations and ground-based laser satellite-tracking data, as well as surface gravity data.
Definition: WGS-84 is a set of parameters for determining geometric and physical geodetic relationships on a global scale. The system includes a geocentric reference ellipsoid, a coordinate system, and a gravity field model. The ellipsoid is essentially that of the International Union of Geodesy and Geophysics Geodetic Reference System 1980. The coordinate system is a realization of the conventional terrestrial system, as established by the International Earth Rotation Service. [2103]

Wind shear. The rate of change of the wind velocity components with distance.

Window (electromagnetic). Wavelength or frequency region in which the atmosphere is largely transparent to electromagnetic radiation (e.g. optical window, microwave window, etc.).

Window operation. Processing of the (radiation) values of pixels within a predefined window, mostly limited to one spectral band - also called a 'filter.' Examples are convolution filters, variance filters, etc. . The filter output is assigned to the central pixel of the window.

Whiskbroom scanner. A line-scanning optomechanical sensor system.

World Geodetic System 1984 (WGS-84). Refers to a set of parameters established by the US Defense Mapping Agency (DMA) for determining geometric and physical geodetic relationships on a global scale. The system includes a geocentric reference ellipsoid, a coordinate system, and a gravity field model. The ellipsoid is essentially that of the International Union of Geodesy and Geophysics (IUGG) 'Geodetic Reference System 1980.' The coordinate system is a realization of the conventional terrestrial system, as established by the International Earth Rotation Service.

Zenith angle. The angular distance of any celestial object from a given observer's zenith, measured along the great circle of the celestial sphere from zenith to object.

Zodiacal light. A faint glow that extends away from the sun in the ecliptic plane of the sky, visible to the naked eye in the western sky shortly after sunset or in the eastern sky shortly before sunrise. Its spectrum indicates it to be sunlight scattered by interplanetary dust. The zodiacal light contributes about a third of the total light in the sky on a moonless night.

2102) "World Geodetic System 1984," DoD DMA TR 8350.2, September 1987
2103) R. B. Langley, "A GPS Glossary," GPS World, October 1995, pp. 61-63

Appendix B Acronyms and Abbreviations

Units of Measure and some Physical Constants

A .... ampere - unit of electric current [named after André M. Ampère (1775-1836), French physicist]
Ah .... ampere hour
Å .... angstrom - unit of length (used in particular for the short wavelength spectrum); 1Å= 10^{-10} m [named after Anders Jonas Ängström (1814-1874), Swedish physicist and astronomer]
amu .... atomic mass unit (1.6605402 10^{-27} kg)
are .... unit of area (1 are = 100 m^2)
arcmin .... arcminute [1' = (1/60)°]
arcsec .... arcsecond [1" = (1/60)']
au .... astronomical unit - unit of length, namely the mean Earth/sun distance [=1.495978706 10^{13} cm, which is the semimajor axis of the Earth's orbit around the sun (or about 150 million km)]
bar .... pressure, (1 bar = 10^5 Nm^{-2})
c .... speed of light in vacuum (299,792,458 m/s)
cd .... candela (unit of luminous intensity). The candela is the luminous intensity, in a given direction, of a source that emits monochromatic radiation of frequency 540 × 1012 Hz and that has a radiant intensity in that direction of 1/683 watt per steradian.
cm .... centimeter (unit of length) 1 cm = 10^{-2} m
C.............. coulomb - unit of electrical charge; 1 C = 1 As [named after Charles-Augustin Coulomb (1736-1806), French physicist. The coulomb is the quantity of electricity transported in 1 second by a current of 1 ampere.
°C .... degree Celsius [named after Anders Celsius (1701-1744), Swedish astronomer]
dB............. decibel - a unit for expressing the signal strength [named after Alexander Graham Bell (1847-1922), Scottish-born American inventor]
dm .... decimeter (length) 1 dm = 10^{-1} m
E.............. Eötvös (1 E = 10^{-9} s^{-2}). The linear gradient of gravity is defined in units of Eötvös, named in honor of the Hungarian physicist Roland Eötvös (1848-1919)
eV .... electron volt (1.60217733 10^{-19} J). A unit of energy, equal to the energy an electron (or proton) would gain when accelerated by 1 volt.
F.............. farad - a unit of capacitance [named after Michael Faraday (1791-1867), English physicist and chemist]. The farad is the capacitance of a capacitor between the plates of which there appears a difference potential of 1 volt when it is charged by a quantity of electricity equal to 1 coulomb.
f .... focal length
f/d .... focal-length-to-diameter ratio
GHz .... Gigahertz (10^9 Hz)
gal............. unit of acceleration (used in particular in gravity measurements): 1 gal = 10^{-2} m s^{-2} = 1 cm s^{-2}; 1 mgal = 10^{-5} m s^{-2} [named after Galilei Galileo (1564-1642), Italian mathematician, astronomer and physicist]
gauss .... unit of magnetic induction [named after Carl Friedrich Gauss (1777-1855), German mathematician]
H .... henry - unit of magnetic inductance; 1 H = 1Wb/A or 1Vs/A [named after Joseph Henry, a nineteenth-century US physicist]
Hz .... hertz - a measure of frequency; 1Hz = 1/s [named after Heinrich Hertz (1857-1894), German physicist]
h .... hecto (10^2)
h (or hr) .... hour

h .... Planck's constant = 6.6260755 x 10^{-34} Js (joule second)
ha .... hectare (1 ha = 10^4 m^2)
hPa .... hectopascal (international standard of pressure, 1 hPa = 100 Pa)
Isp .... specific impulse with a unit Ns/kg. The numerical value of the specific impulse also corresponds to the effective exhaust velocity (**m/s**) of the gas exiting the thruster in a vacuum. See also Glossary.
J .... joule - unit of work or energy; 1 J = 1 Nm = 1 Ws [named after James Prescott Joule (1818-1889), British physicist]
K .... degree Kelvin [named after Sir William Thomson (Lord Kelvin, 1824-1907), Scottish engineer, physicist and mathematician]. The degree Kelvin is the unit of temperature determined by the Carnot cycle with the triple-point temperature of water defined as 273.15 K (corresponds exactly to 0 °C).
k .... kilo (10^3)
kbit/s .... kilobit per second (10^3 bit/s)
keV .... kiloelectron volt (10^3 eV)
kg .... kilogram (10^3 g)
kg/m^3 .... density
kHz .... kilohertz (10^3 Hz)
km .... kilometer (10^3 m)
krad .... kilorad (see rad below)
kW .... kilowatt (10^3 watt)
kWe .... kilowatt electric (used to distinguish electrical power from thermal power)
L .... liter (volume) 1l = 1 dm^3 [the symbol for liter is capitalized (when alone by itself) to avoid confusion with the number 1]
lm .... lumen (cd sr) luminous flux. The lumen is the luminous flux emitted in a solid cangle of 1 steradian by a uniform point source having an intensity of 1 candela.
lx .... lux (lm/m^2) illumination
M .... Mega (10^6)
Mbit/s .... Megabit per second (10^6 bit per second)
MeV .... Megaelectron volt (10^6 eV)
MHz .... Megahertz (10^6 hertz)
m .... meter
m .... milli (10^{-3})
m^2 .... area (square meter)
m^3 .... volume (cubic meter)
mb (mbar) ...... millibar
min .... minute
mg .... milligram (10^{-3} g)
mgal .... milligal (10^{-3} gal)
mJ .... millijoule (10^{-3} J)
ml .... milliliter (10^{-3} l)
mm .... millimeter (unit of length) 1 mm = 10^{-3} m
mN .... millinewton
mrad .... milliradian [2104)]
ms .... millisecond
m/s .... meter per second (velocity)
μ .... micro (10^{-6})
μm .... micrometer (10^{-6} m)
μrad .... microradian
μs .... microsecond (10^{-6} second)
N .... newton - unit of force; 1N = 1 kgm/s^2 [named after Sir Isaac Newton (1643-1727), English natural philosopher and mathematician]

2104) An example is given to better visualize the plane angle of a milliradian. The apparent sun disk angle as seen from Earth is 32' 26" (max, or about **30.7 mrad**), and 31' 31" (min).

Nm .... newton meter (work or energy)
n .... nano (10^{-9})
nm .... nanometer (10^{-9} m)
nm .... nautical miles [1 nm = 1852 m (international)]
nT .... nanotesla (10^{-9} tesla) unit of magnetic flux density
Ω .... ohm - unit of electrical resistance; 1 Ω = 1 V/A [named after Georg Simon Ohm (1789-1854), German physicist
Pa .... pascal - unit of pressure; 1 Pa = 1 N/m^2 [named after Blaise Pascal (1623-1662), French mathematician and physicist]
p .... pico (10^{-12})
pC .... picocoulomb (10^{-12} coulomb)
pT .... picotesla (10^{-12} tesla)
ppb .... parts per billion (10^{-9})
ppbv .... parts per billion, by volume
ppm .... parts per million (10^{-6})
ppmv .... parts per million, volume
pps .... pulses per second
ppt .... parts per trillion (10^{-12})
pptv .... parts per trillion (10^{-12}), by volume
R_E .... Earth radius = 6378.140 km (mean equatorial radius)
rad .... radian - a unit of plane angular measurement equal to the angle at the center of a circle subtended by an arc equal in length to the radius
rad/s .... radian per second (angular velocity)
rad .... In the context of radiation shielding, the term "rad" is also used for energy accumulated in matter (dosimetry for the energy absorbed per unit mass of material, usually by ionization processes). A rad is the amount of particle radiation that deposits 10^{-2} J/kg of target material. Besides the "rad" is the "Gray." 1 rad = 1/100 Gray. Note: A Gray is the radiation absorbed dose unit of SI (Systeme Internationale). 1 Gray = 1 J/kg (=100 rad). Or 10 Gray = 1000 rad = 1krad. [2105)]
rms .... root mean square
rpm .... revolutions per minute
rps .... revolutions per second
S .... siemens - unit of electrical conductance; 1 S = 1 A/V [named after Werner von Siemens (1816-1892), German electrical engineer]
s .... second
sr .... steradian - a unit of measure of solid angles expressed as the solid angle subtended at the center of a sphere by the portion of the surface whose area is equal to the square of the radius of the sphere
T .... Tera (10^{12})
TB .... TeraByte (10^{12} Byte)
THz .... Terahertz (10^{12} hertz)
tesla (T) .... unit of magnetic flux density. 1 T = 1 Wb/m^2 which corresponds to 10^4 gauss [named after Nikola Tesla (1856-1943), Croatian-born American inventor]
V .... volt - unit of electrical potential [named after Alessandro Volta (1745-1827), Italian physicist]
W .... watt - unit of power; 1 W = 1 J/s [named after James Watt (1736-1819), a Scottish mechanical engineer and inventor]
Wb .... weber - unit of magnetic flux [named after Ernst Weber (1901-), Austrian-born US engineer
Wh .... watt hour (work or energy)
Ws .... watt second (work or energy)

2105) Typical CMOS devices can tolerate 1-10 krad/year. Dose rates for a silicon target are usually stated in g/cm^2 or in thickness of aluminum shielding for a given orbit. For a sun-synchronous orbit, about 0.8 g/cm^2 (or 4 mm silicon thickness) is needed for a 1-year lifetime, and about 3 g/cm^2 (13 mm silicon) for a 10 year lifetime.

General conventions of unit representations:

The symbol "m" is used with various meanings depending on its position and occurrence in a unit. In single-digit instances, the symbol m stands simply for meter. This is also the case in double symbol instances, when m is in last position, like in Nm (newtonmeter), nm (nanometer), or mm (millimeter). When m is used in double-digit symbols in first place, like mm (millimeter), ml (milliliter), ms (millisecond), mN (millinewton), etc., then the first small "m" is always used in a diminutive sense referring to "milli" (10^{-3}).

The term small "k" stands for kilo (10^3) as in km (kilometer), kg (kilogram), kW (kilowatt), or kbit (kilobit). The capital letter "K," on the other hand, has the meaning of Kelvin, referring to a degree temperature on the absolute temperature scale.

The designations M (Mega), G (Giga), T (Tera), or µ (micro), n(nano), p (pico), etc., in combinations with other units, follow the same logic as outlined above and in Table 568.

Quantity	**Unit name**	**Unit symbol**
Length	meter	m
Mass	kilogram	kg
Time	second	s
Electric current	ampere	A
Thermodynamic temperature	kelvin	K
Luminous intensity	candela	cd
Amount of substance	mole	mol

Table 567: Symbols for the seven basic units in the SI system

The basic SI units come in all sizes. Since the SI system is built upon the base 10, the different sizes are base 10 multiples of the basic units as illustrated in Table 568.

Prefix	Symbol	Multiplication factor	Examples
Exa	E	10^{18} = 1,000,000,000,000,000,000	
Peta	P	10^{15} = 1,000,000,000,000,000	
Tera	T	10^{12} = 1,000,000,000,000	TByte
Giga	G	10^{9} = 1,000,000,000	GHz, GByte,
Mega	M	10^{6} = 1,000,000	MHz, Mbit/s,
kilo	k	10^{3} = 1,000	km (kilometer), kg (kilogram),
hecto	h	10^{2} = 100	hl (hectoliter), ha (hectare)
deca	da	10^{1} = 10	
		10^{0} = 1	
deci	d	10^{-1} = 0.1	dg (decigram), dl (deciliter)
centi	c	10^{-2} = 0.01	cm (centimeter), cl (centiliter)
milli	m	10^{-3} = 0.001	mm (millimeter), ml (milliliter)
micro	µ	10^{-6} = 0.000001	µm (micrometer), µg (microgram)
nano	n	10^{-9} = 0.000000001	nm (nanometer), ns (nanosecond)
pico	p	10^{-12} = 0.000000000001	ps (picosecond), pf (picofarad)
femto	f	10^{-15} = 0.000000000000001	fs (femtosecond)
atto	a	10^{-18} = 0.000000000000000001	

Table 568: Commonly used prefixes of SI multiples and submultiples

Quantity	**Unit name**	**Unit symbol (derivation)**
Force	newton	N ($kgms^{-2}$)
Energy	joule	J (Nm) or (Ws) or (kgm^2s^2)
Energy	kilowatt hour	kWh (3.6 10^6 J)
Energy	electron volt	eV (1.6 10^{-19} J)
Power	watt	W (Js^{-1}) or ($kgm^2\ s^{-3}$)

Quantity	Unit name	Unit symbol (derivation)
Frequency	hertz	Hz (s^{-1})
Electrical potential	volt	V (JC^{-1}) or (WA^{-1})
Electrical charge	coulomb	C (As)
Electrical resistance	ohm	Ω ($V A^{-1}$)
Electrical conductance	siemens	S ($A V^{-1}$)
Electrical capacitance	farad	F ($C V^{-1}$) or ($A s V^{-1}$)
Magnetic inductance	henry	H ($Wb A^{-1}$) or ($V s A^{-1}$)
Magnetic flux	weber	Wb (Vs)
Magnetic flux density	tesla	T ($Wb m^{-2}$)
Area	square meter	m^2
Volume	cubic meter	m^3
Volume	liter	L ($10^{-3} m^3$)
Velocity (speed)	meter per second	ms^{-1}
Temperature	degree Celsius	°C
Pressure	pascal	Pa (Nm^{-2}) or ($kg m^{-1} s^{-2}$)
Pressure	standard atmosphere	atm ($1.01325 10^5$ Pa)
Torque		Nm (newton meter)
Electric field strength		V m^{-1} (volt per meter)
Magnetic field strength		A m^{-1} (ampere per meter)
Plane angle (arc length)	degree arcmin (minute) arcsec (second)	1° = (π/180) rad 1' = (1/60)° 1" = (1/60)'

Table 569: Derived units commonly used in science and engineering [2106) 2107)]

Numbers

3DMA .... Three-Dimensional Microgravity Accelerometer (Shuttle payload)

A

AAAS .... American Association for the Advancement of Science (Washington DC)
AAE .... Austrian Aerospace GmbH, Vienna, Austria
AAOE .... Airborne Antarctic Ozone Experiment (1987)
AARGOS ...... A340 Atmospheric Research Global Observation System (MOZAIC)
AARI .... Arctic and Antarctic Research Institute (St. Petersburg, Russia)
AAS .... American Astronomical Society
AASE .... Airborne Arctic Stratospheric Expedition (see campaign survey)
ABLE .... Atmospheric Boundary Layer Experiment (campaign)
AC .... Alternating Current
ACC .... Anthropogenic Climate Change (CLIVAR subprogram)
ACCESS .... Assembly Concept for Construction of Erectable Space Structure (Shuttle)
ACE .... Advanced Composition Explorer (NASA, APL, etc., see K.1)
ACE-1, -2 ...... Aerosol Characterization Experiment (campaigns)
ACE .... Atmosphere Climate Experiment (an ESA mission, A.1)
ACEChem ...... Atmospheric Chemistry Explorer (a proposed ESA misson as of 2001)
ACES .... Acoustic Containerless Experiment System (Shuttle payload)
ACORN .... Airborne Composition Observations in the Region of the North-Atlantic-Corridor (P.40.2)
ACRES .... Australian Centre for Remote Sensing (Belconnen, Australia)
ACSYS .... Arctic Climate System Study (WCRP program)
ACT .... Applied Coherent Technology, Herndon VA (commercial provider of remote sensing products, operator of satellites, etc.)

2106) "Symbols and Abbreviations for Electrical and Electronic Engineering," IEE, 1980
2107) R. A. Nelson, "Guide for Metric Practice," Physics Today, Supplement to August 1997 issue, pp. 13--14

ACTS .... Advanced Communications Technology Satellite in GEO (NASA, Launch: Sept. 1993 by Shuttle Discovery). ACTS advanced considerably the use of Ka-band technology and paved the way for high-speed commercial Ka-band communications.
A/D .... Analog/Digital converter
ADCP .... Acoustic Doppler Current Profilers [(U. of Florida, Tokai University, Hiroshima University, Kyushu University, Japan, and CSIRO), subsurface upward-looking moorings]
ADEN .... ALOS Data European Node [an ESA initiative involving a number of distributed acquisition facilities capable of receiving ALOS data (SAR and optical) for European users: a) Toulouse (France) with upgraded X-band stations, b) DLR Neustrelitz (Germany) and Libreville (Gabon), c) TSS Tromsoe (Norway) and SSC Sturup (Sweden, d) ASI Mateira (Italy and Maspalomas (Spain)]
ADEOS .... Advanced Earth Observation Satellite (NASDA, D.1, D.2)
ADM .... Atmospheric Dynamics Mission (ESA Earth Explorer Core Mission)
ADPCM .... Adaptive Differential Pulse Code Modulation (a lossy data compression technique)
ADS-B .... Automatic Dependent Surveillance-Broadcast [an FAA system installed in aircraft (first prototypes as of 2000). When coupled with GPS, an aircraft's ADS-B unit can continuously broadcast its identification, position, altitude, direction, speed, rate of climb or descend, etc.] ADS-B is a key technology to free flight.
ADSF .... Automated Directional Solidification Furnace (Shuttle payload)
ADV CGBA .... Advanced Commercial Generic Bioprocessing Apparatus (Shuttle)
ADV XDT ..... Advanced X-Ray Detector (Shuttle payload)
AEAP .... Atmospheric Effects Aircraft Program (NASA)
AEB .... Agencia Espacial Brasileira - Brazilian Space Agency, Brasilia, Brazil (since 1994)
AEM-1 .... Applications Explorers Mission-1 (see HCMM A.18)
AEM-2 .... Applications Explorers Mission-2 (A.5)
AERCam/Sprint . Autonomous Extravehicular Activity Robotic Camera Sprint [Shuttle free-flying camera, first flown on STS-87 (Nov. 19 - Dec. 5, 1997)]
AEROCE ...... Atmospheric/Ocean Chemistry Experiment (campaign)
Aerospace Corp. . 'The Aerospace Corporation' (since 1960), a US private nonprofit research and development center with HQs in El Segundo, CA. Aerospace operates a Federally Funded Research and Development Center (FFRDC) for the Department of Defense (DoD). The primary customer is the Space and Missile Systems Center (SMC) of the US Air Force Materiel Command. The Aerospace Corporation provides engineering services and space technology expertise to DoD space programs and other US government agencies. Other company locations are in the Washington DC area, Colorado Springs, CO, Albuquerque, NM, Sunnyvale CA, VAFB, CA, and at KSC (Kennedy Space Center), FLA.
Aerospatiale .... A French aerospace conglomerate with 38,000 employees, HQ in Paris. Builder of the main stages of Ariane 4 and 5. Manufacturer of satellites and sensors. Three major divisions: Aircraft, Helicopters, and Space & Defense. Spacecraft platforms: Spacebus series.
AES .... Atmospheric Environment Service (of Environment Canada)
AESA .... Atmospheric Effects of Stratospheric Aircraft (NASA)
AF .... US Air Force
AFB .... Air Force Base (US Air Force)
AFC .... Affiliated Data Center (these are institutional facilities that are affiliated with EOSDIS, in particular NOAA facilities are AFCs)
AFE .... American Flight Echocardiograph (Shuttle payload)
AFGL .... Air Force Geophysics Laboratory (Hanscom AFB, Bedford, MA, USA) Note: in 1998 AGFL was renamed to "Hanscom Research Site"

AFP-675 .... Air Force Program 675 (Shuttle payload)
AFNOR .... Association francaise de normalization (French standards institute)
AFOSR .... Air Force Office of Scientific Research (an AFRL directorate)
AFRL .... Air Force Research Laboratory (USA). The nine AFRL sites are located at: Wright Laboratory, Wright-Patterson AFB, Ohio (AFRL HQs, directorates of: Air Vehicles, Propulsion, Directed Energy, and Materials & Manufacturing); Hanscom AFB, MA (Sensors directorate); Phillips Research Site, Kirtland AFB, Albuquerque, NM (Space Vehicles directorate); Rome Laboratory, Griffis AFB, Rome, NY; Edwards AFB, Edwards, CA; Brooks AFB, TX; Eglin AFB, FL; Tyndall AFB, FL; Bolling AFB (AFOSR directorate), Washington DC.
AFSCN .... Air Force Satellite Control Network (USA)
AFSK .... Amplitude Frequency Shift Keying (modulation technique)
AGARD .... Advisory Group for Aerospace Research and Development. AGARD is a NATO agency (with HQ in Neuilly-sur Seine, France), formed in 1954, with the objective to enhance the exchange of aerospace technology within NATO.
AGASP .... Arctic Gas and Aerosols Sampling Project (airborne campaign)
AGC .... Antenna Gain Control
AGGA .... Advanced GPS/GLONASS ASIC (ESA/ESTEC development). As of the end of 2000 the AGGA chip set is available to European industry, it is manufactured by ATMEL of Nantes, France. It is used in GRAS, in the LAGRANGE GNSS receiver of Laben SpA, Italy, and in the RIMS stations of the EGNOS program.
AGILE .... Astro-rivelatore Gamma ad Immagini LEggero (Gamma-ray Astronomical Low-Mass Detecor), an approved ASI mission with a planned launch in 2003
AGL .... Above Ground Level (usually the altitude of aircraft)
AGU .... American Geophysical Union (a society with over 35,000 members in over 115 countries. The objective is to advance progress in the Earth, atmospheric, oceanic, hydrologic, and space and planetary sciences.)
AIAA .... American Institute of Astronautics and Aeronautics (Reston, VA)
AIDJEX .... Arctic Ice Dynamics Joint Experiment (campaign)
Airbus Industrie . A consortium of European aerospace companies, founded in 1970. (partners are: Aerospatiale of France, DASA Airbus of Germany, British Aerospace, and Spain's CASA). Italy's Alenia, Fokker of the Netherlands, and Belairbus in Belgium are associate members who participate in selected programs. Some 32,000 people work directly for Airbus Industrie within the partner companies. Airbus Industrie is headquartered near Toulouse, France. Builder of civil aircraft (Airbus).
AIRS .... Autonomous Information Reception Station (see Meteor-3M series)
AIP .... American Institute of Physics
AIP .... Astrophysikalisches Institut Potsdam (Germany)
AKR .... Auroral Kilometric Radiation (ionospheric phenomenon)
ALACE .... Autonomous Lagrangian Circulation Explorer (free-floating ocean buoys designed to seek a pre-programmed depth; they drift with the ocean currents of that depth, and pop up periodically to report their position to a satellite), see also PALACE
ALE/GAGE .... Atmospheric Lifetime Experiment/Global Atmospheric Gas Experiment (campaign)
Alenia Spazio ... Alenia Aerospazio S.p.A. is a company of the Finmeccanica IRI group, an Italian consortium in aerospace, defense, energy, transportation and automation markets. Partner in many space programs (2500 employees), builder of COSMO-SkyMed. Subsidiaries: Laben S.p.A. (Laboratori Elettronici Nucleari) in Vimodrone (Milano, Italy) since 1958; SSI (Space Software Italia S.p.A. in Taranto, Italy; QSW (Quadrics Supercomputer World Ltd.) in Rome, Italy; HCSA (Hellenic Company

for Space Applications S.A.) in Paradisos Amarousiou, Italy; EuroSkyWay in Rome, Italy

ALEXIS .... Array of Low Energy X-Ray Imaging Sensors (LANL, K.3)
ALISSA .... l'Atmosphere par LIdar Sur SAliout (the French sensor was at first proposed by CNES for a Salyut flight)
ALMAZ .... ALMAZ = 'rough diamond' (Earth observation series, Russia), D.4
ALOHA .... One of several communication access methods
ALOHA .... Airborne Lidar and Observations of the Hawaiian Airglow (campaign)
ALOS .... Advanced Land Observing Satellite (D.3)
ALPEX .... Alpine Experiment (campaign)
AM .... Amplitude Modulation (modulation technique of the main carrier)
AM .... Ante Meridiem (US time notation designating morning hours, to distinguish from PM)
AMBIACE ..... Amazon Biogeochemistry and Atmospheric Chemistry Experiment (campaign)
AMEX .... Australian Monsoon Experiment (campaign)
AMM .... Antarctic Mapping Mission (Radarsat)
AMOS .... Air Force Maui Optical Station (Shuttle experiment). AMOS is located at the summit of Haleakala, on the island of Maui, Hawaii. The Air Force experiment is using the Shuttle orbiter as a calibration target for a ground-based experiment (research for electro-optical sensors)
AMPTE .... Active Magnetosphere Particle Tracer Explorers (cooperative mission of US/ NASA, Germany and UK, K.4)
AMS .... Alpha Magnetic Spectrometer (Shuttle payload) AMS was first flown on STS-91 (June 2 - 12, 1998). It is an anti-matter demonstration, an experiment with international cooperation from: USA, China, Finland, Germany Italy, and Switzerland
AMS .... American Meteorological Society
AMSAT .... The Radio Amateur Satellite Corporation (worldwide groups of Amateur Radio Operators (volunteers, normally organized by country), building, launching and communicating with each other through noncommercial amateur satellites, since 1969, also the name of satellites)
AMTEC .... Alkali Metal Thermal-to-Electric Converter (Shuttle payload)
ANL .... Argonne National Laboratory (Argonne, IL, USA, a DOE facility, operated by the University of Chicago)
ANSI .... American National Standards Institute
ANSTO .... Australian Nuclear Science and Technology Organization
Antrix Corp. .... Bangalore, India (the commercial marketing arm of ISRO, Antrix is the distributor of IRS data, etc.)
AO .... Announcement of Opportunity (usually for a sensor on a particular mission)
AOCS .... Attitude and Orbit Control System
AOET .... Atomic Oxygen Exposure Tray (Shuttle D2 mission)
AOS .... Acousto-Optical Spectrometer
AOTF .... Acousto-Optic Tunable Filter (an imaging dispersion technique)
APARE .... Asia/North Pacific Regional Study (campaign)
APCF .... Advanced Protein Crystallization Facility (Shuttle, see also PCF)
APCG .... Advanced Protein Crystal Growth (Shuttle, see also PCG)
APD .... Avalanche Photodiode (detector type)
APDA .... Arctic Precipitation Data Archive
APE .... Airborne Polar Experiment (campaign)
APE .... Auroral Photography Experiment (Shuttle payload)
APEX .... Active Plasma Experiment (Intercosmos, K.5)
APFO .... Aerial Photography Field Office (Salt Lake City, UT, USA)
APL .... Applied Physics Laboratory, since 1942, a facility of Johns Hopkins University (JHU), in Laurel, MD, USA
APM .... Ascent Particle Monitor (Shuttle experiment)

APRS .... Automatic Position Reporting System (a protocol used by the Amateur Radio community)
APS .... Active Pixel Sensor
APSC .... Asia Pacific Space Center
APT .... Automatic Picture Transmission (one type of NOAA downlink transmission; APT transmits data from two channels of the AVHRR at a reduced resolution of 4 km in the VHF frequency band (at 137.50 and 137.62 MHz)).
APV .... Autonomously Piloted Vehicle (Condor)
AR .... Anthrorack (Shuttle D2 mission)
ARAT .... Avion de Recherche Atmosphérique et de Télédétection (Atmospheric Research and Remote Sensing Aircraft), ARAT is jointly operated by INSU-CNRS, CNES, DMN (French National Weather Center), and IGN (Institut Géographique National). The aircraft is IGN property. ARAT is a Fokker 27 MK pressurized twin turboprop aircraft (service altitude = 5800 m, cruising speed = 350 km/h, flight endurance = 5 hr; on-board computer systems: HP1000 A900, recordings on high-capacity digital video cassette, two Exabyte 2.5 GByte recorders) .
ARC .... Ames Research Center (NASA facility at Moffett Field, CA, and at the Dryden Flight Research Facility in Edwards, CA, USA)
ARC .... Aggregation of Red Blood Cells (Shuttle experiment)
Archimedes I, II . Coordinated European airborne campaigns in the North Sea region (start in 1983, Archimedes IIa took place in April 1988)
ARCS .... Austrian Research Center Seibersdorf (since 1956, with sites at Seibersdorf, Leoben, Ranshofen, Vienna, Graz, Dornbirn, Wiener-Neustadt, and Budapest)
ARCSS .... Arctic Center of System Science (at NSIDC of U. of Colorado, Boulder, CO, USA)
ARESE .... ARM Enhanced Shortwave Experiment (campaign)
ARGO .... "Array for Geostrophic Oceanography," a global array of buoys [an international ocean program, part of GCOS/GOOS and CLIVAR - it consists of an array of 3000 free-drifting (Lagrangian) profiling floats that measure the temperature and salinity of the upper 2000 m of the ocean; start of deployment in 2000].
ARGOS .... Argos (CNES System) is a data collection and location system with a space segment and a ground segment. ARGOS is operational on NOAA polar-orbiting S/C. G.15.4, C.1
ARGOS .... Advanced Research and Global Observation Satellite (DoD, M.1)
ARIANESPACE A commercial launch service provider of Europe with HQ in France (since 1980, first commercial operator of launchers in the world). Twelve European countries participate in the Ariane program.
ARIES .... Australian Resource Information and Environment Satellite
ARISTOTELES . **A**pplications and **R**esearch **I**nvolving **S**pace **T**echniques **O**bserving **T**he **E**arth's Field from **L**ow **E**arth Orbiting **S**atellite (planned but cancelled ESA Mission)
ARM .... Atmospheric Radiation Measurement (campaign program of DOE)
ARMCAS ...... Arctic Radiation Measurements in Column Atmosphere-Surface System (campaign)
ARNS .... Aeronautical Radionavigation Service
ARTEMIS ...... Advanced Relay and Technology Mission Satellite (ESA)
ARPA .... Advanced Research Project Agency (US)
ARQ .... Automatic-Repeat Request
ASA .... Austrian Space Agency (Vienna, since 1972)
ASAP .... Adaptive Sensor Array Processing (MIT/LL)
ASAP .... Advanced Sensors Application Program (US Navy)
ASAP .... Airborne Science and Application Program (USGS, NASA)

ASAP .... Ariane Structure for Auxiliary Payloads (ASAP provides launch opportunities for microsatellites on a commercial basis, the ASAP-5 ring structure can accommodate up to 8 microsatellites with a volume restriction of 60 cm x 60 cm x 80 cm)
ASCOT .... Atmospheric Studies in Complex Terrain (campaign)
ASCS .... Agricultural Stabilization and Conservation Service (USA)
ASDAR .... Aircraft to Satellite Data Relay (wind observations are reported from commercial aircraft at cruising altitude via meteorological satellite communication links at 7 minute intervals)
ASE .... Automatic Air-Sampling Equipment, see P.40.4
ASEM .... Assembly of Station by EVA Methods (Shuttle demonstration)
ASES .... American Solar Energy Society
ASF .... Alaska SAR Facility in Fairbanks, Alaska (DAAC of NASA EOS Program. ASF is located at the Geophysical Institute of the University of Alaska at Fairbanks. Position: 65°N, 148°W. ASF is in effect a US-PAF for ERS-1/2 data as well as for JERS-1 and RADARSAT data.)
ASHOE .... Airborne Southern Hemisphere Ozone Experiment (campaign)
ASI .... Agenzia Spaziale Italiana (formerly PSN). ASI is the Italian Space Agency, Rome (since 1988)
ASI/CGS .... ASI/Centro di Geodesia "Guiseppe Colombo" in Matera, Italy, for Space Geodesy, Remote Sensing and Space Robotics. CGS hosts the I-PAF (Italian Processing and Archiving Facility), a multimission facility for archiving, processing and distributing remote sensing data.
ASI .... Alcatel Space Industries, France, since 1998 (ASI represents the merger of four space hardware development divisions from Alcatel, Dassault, Thomson and Aerospatiale)
ASIC .... Application Specific Integrated Circuit
ASIM .... Application Specific Microinstrument
ASP .... Attitude Sensor Package (Shuttle payload of ESA)
ASPRS .... American Society for Photogrammetry and Remote Sensing (Bethesda, MD, since 1934)
ASRI .... Asher Space Research Institute (of Technion Israel Institute of Technology, Haifa, since 1986)
ASRI .... Australian Space Research Institute, Elizabeth, SA
ASTEX .... Atlantic Stratocumulus Transition Experiment (airborne campaign at the Azores in 1992)
ASTP .... Apollo-Soyuz Test Project (1975)
ASTRE .... Accéléromètre Spatial Triaxial Electrostatique [an ESA accelerometer built by ONERA and part of ESA's MMA (Microgravity Measurement Assembly) flown on Shuttle flights STS-83 and STS-94]
Astrium .... Astrium is the name of a new European space company, a daugther of EADS, formally created in 2000. Astrium is a merger of Aerospatiale Matra of Paris, France, DASA of Munich Germany, and Marconi Electronic Systems of Stanmore, UK. German Astrium facilities are at Bremen, Ottobrunn, Friedrichshafen, Lampoldhausen, Rostock and Trauen. The French/British MMS (Matra Marconi Space) facilities are located at Portsmouth and Stevenhage, UK, and at Toulouse and Vélizy, France.
ASTRO-SPAS ... Astronomy Platform - Shuttle Pallet Satellite
ASU .... Arizona State University (Tempe, AZ)
ATEx .... Advanced Tether Experiment (NRL)
ATEX .... Atlantic Tropospheric Experiment (campaign)
ATI .... Along-Track Interferometry
ATLAS .... Atmospheric Laboratory for Application and Science (NASA program, payload series on Shuttle), J.2
ATLAS .... Autonomous Temperature Line Acquisition System (NOAA/PMEL mooring system measuring surface wind, air temperature, SST, ten sub-

surface temperatures and two subsurface pressures; all data are monitored by ARGOS)

ATLID .... Atmospheric Lidar (Sensor), an ESA backscatter lidar

ATN .... Advanced TIROS-N Series (NOAA, launched from 1983 on)

ATS .... Application Technology Satellite (NASA GEO satellite series prior to GOES)

ATSB .... Astronautic Technology SB, Kuala Lumpur, Malaysia

ATTAS .... Advanced Technology Testing Aircraft System (VFW-614 of DLR)

ATV .... Roton Atmospheric Test Vehicle (of Rotary Rocket Company, Redwood City, CA). Roton ATV is a fully reusable, single-stage-to-orbit, commercial launch vehicle. Roton is powered by a rotary engine burning liquid oxygen and jet fuel. ATV made its first successful flight on July 23, 1999.

ATV .... Automated Transfer Vehicle (ESA cargo resupply vehicle for ISS to be launched by Ariane-5, payloads of up to 7,500 kg can be delivered)

AU .... Astronomical Unit, Sun-Earth distance = 1.496 x 10^8 km (average)

AURA .... Association of Universities for Research in Astronomy [Washington DC, since 1957, AURA/STSI (Space Telescope Science Institute) is the operator of the Hubble Space Telescope for NASA]

AVHRR .... Advanced Very-High Resolution Radiometer (NOAA Sensor, AVHRR/3 on NOAA-K,L,M,N is to be renamed in VIRSR for NOAA-O,P,Q)

AVISO .... Archivage Validation and Interprétation des données des Satellites Océanographiques [Archiving, Validation and Interpretation of Satellites oceanographic data (CNES data center for GEOSAT, Topex/Poseidon, ERS-1/2, ENVISAT, Jason-1, etc.)]

AWCS .... Automated Wafer Cartridge System (Shuttle payload)

AWG .... American Wire Gauge (the higher the number the thinner the wire)

AWI .... Alfred Wegener Institut for Polar and Marine Research, Bremerhaven (since 1980) and Potsdam since 1992 (Germany)

AWS .... Automated Weather Station

AXAF .... Advanced X-ray Astrophysics Facility, a NASA satellite mission in a high elliptical Earth orbit, deployed by Shuttle STS-93; in the spring of 1999 AXAF has been renamed to "Chandra X-ray Observatory" in honor of the late India-American Nobel Laureate Subrahmanyan Chandrasekhar

AZBS .... Avionik Zentrum Braunschweig (Germany)

B

BA .... Baroreflex (Shuttle payload on D2 mission)

BAC .... Block Adaptive Quantization (a SAR raw data compression method)

BADC .... British Atmospheric Data Center (at RAL, Chilton, UK)

BAEX .... Baltic Aerosol Experiment (campaign)

BAHC .... Biospheric Aspects of the Hydrological Cycle (IGBP core project since 1994)

BALTEX .... Baltic Sea Experiment (campaign)

BAMS .... Bulletin of the American Meteorological Society (a periodical)

BAS .... British Antarctic Survey (Cambridge, UK)

BATC .... Ball Aerospace and Technologies Corporation (Aerospace Systems Division in Boulder, CO, and Telecommunication Products Division in Broomfield, CO) formerly: Ball Brothers Research Corporation, since 1956, [manufacturer of satellites such as: Seasat, SIR-C, COBE (Cosmic Background Explorer), CGRO (Compton Gamma Ray Observatory), ERBS, CRRES, GFO-1; and builder of instruments: CZCS, GHRS (Goddard High Resolution Spectrograph), STIS (Space Telescope Imaging Spectrograph), and NICMOS (Near-Infrared Camera and Multi-Object Spectrometer), all on HST, etc.]

BASE .... Beaufort and Arctic Storm Experiment (campaign)
BATERISTA ... Biosphere-Atmosphere Transfer and Ecological Research, In situ Studies in Amazonia (campaign)
BATGE .... Biosphere-Atmosphere Trace Gas Exchange in the Tropics (IGBP/IGAC campaign)
BATS .... Bermuda Atlantic Time-Series Study (campaign)
BB .... Biolabor (Shuttle D2 mission)
BBXRT .... Broad Band X-Ray Telescope (part of ASTRO-1 observatory, Shuttle)
BCP .... Ball (or BATC) Commercial Platform (BCP 2000 series bus, BCP 4000 series, BCP 5000, etc.)
BCRS .... Netherlands Remote Sensing Board (Delft, The Netherlands)
BCSC .... Boeing Commercial Space Co. (a subsidiary of the Boeing Co, chartered to commercialize space technologies)
BDPU .... Bubble, Drop and Particle Unit (Shuttle experiment)
BDS .... Bioreactor Demonstration System (Shuttle payload)
BEST .... Bilan Energétique du Système Tropical (Tropical System Energy Budget), a proposed CNES mission
BGR .... Bundesanstalt für Geowissenschaften und Rohstoffe (Hannover, Germany)
BIB .... Blocked Impurity Band (detector type)
BIBEX .... Biomass Burning Experiment (program of IGBP/IGAC)
BIMDA .... Bioserve/Instrumentation Technology Associates Materials Dispersion Apparatus (Shuttle payload)
BiCMOS .... Bipolar Complementary Metal-Oxide Semiconductor
BIL .... Band Interleaved by Line (image organization)
BILTEN .... TUBITAK-METU BILTEN - Information Technologies and Research Institute of the Scientific and Technical Council of Turkey [located on the campus of the Middle East Technical University (METU), Ankara, Turkey]
BIO3D .... Biochemistry of 3-D Tissue Engineering (Shuttle Payload)
BIP .... Band Interleaved by Pixel (image organization)
BIPM .... Bureau International des Poids et Mesures (Paris)
BIRA .... Belgisch Instituut voor Ruimte Aeronomie (Brussels, Belgian Institute of Space Aeronomy)
BLAST .... Battlefield Laser Acquisition Sensor Test (Shuttle experiment)
BLM .... Bureau of Land Management (USA)
BMBF .... Bundesministerium für Bildung, Wissenschaft, Forschung und Technologie (German Ministry of Education, Science, Research and Technology, the successor to BMFT, since 1994)
BMDO .... Ballistic Missile Defense Organization [US, Division within DoD, formerly SDIO (Strategic Defense Initiative Organization)]
BMFT .... Bundesministerium für Forschung und Technologie (German Ministry of Research and Technology, prior to 1994)
BMO .... British Meteorological Office (same as UKMO, HQs in Bracknell, Remote Sensing Instrumentation branch in Farnborough)
BMRC .... Bureau of Meteorology Research Centre (Melbourne, Australia)
BMV .... Bundesministerium für Verkehr (German Ministry of Transportation)
BMVg .... Bundesministerium für Verteidigung (German Ministry of Defense)
BNL .... Brookhaven National Laboratory (Upton, NY, USA)
BNSC .... British National Space Centre (London, UK) since 1985
BOC .... Binary Offset Coding (modulation technique)
Boeing Co. ..... Seattle, WA, USA. A conglomerate (over 200,000 employees) of Boeing + Rockwell International (purchase of Rockwell's aerospace and defense business in Dec. 1996) + McDonnell Douglas Corp. (merger with Boeing in Aug. 1997). Boeing is also a large manufacturer of telecommunication satellites. In October 2000, The Boeing Company acquired three units within Hughes Electronics Corporation: Hughes

Space and Communications Company, Hughes Electron Dynamics, and Spectrolab, Inc., in addition to Hughes Electronics' interest in HRL (Hughes Research Laboratory). The four are now part of Boeing's newest subsidiary, Boeing Satellite Systems, Inc.
BOREAS ...... Boreal Ecosystem-Atmosphere Study (campaign)
BOST .... Belgian Office of Science and Technology
BPDF .... Bidirectional Polarization Distribution Function
BPOT .... Bioluminescence Potential
BPSK .... Bi-Phase Shift Keying (modulation technique)
BRDF .... Bidirectional Reflectance and Distribution Function
BREMSAT ..... University of Bremen Satellite (Shuttle payload)
BRIC .... Biological Research in Canister (Shuttle experiment)
BrO .... Bromine monoxide
BSH .... Bundesamt für Seeschiffahrt und Hydrographie (Hamburg, Germany)
BSI .... British Standards Institution
BSPO .... Belgian Science Policy Office
BSRN .... Baseline Surface Radiation Network (WCRP/GEWEX)
BSTC .... Biotechnology Specimen Temperature Controller (Shuttle)

C

C/A .... Coarse Acquisition (a GPS and GLONASS code)
CAAC .... Civil Aviation Association of China
CAFE .... Central Australian Fronts Experiment (campaign)
CAM .... Centre d'Aviation Météorologique (France)
CAMAREX .... Carbon in the Amazon River Experiment (campaign)
CAMEX .... Convection and Atmospheric Moisture Experiment (airborne campaign conducted at NASA Wallops Flight Facility, Wallops Island,VA)
CAN .... Controller Area Network (used in embedded systems)
CANEX .... Canadian Experiments (Shuttle payload)
CAO .... Central Aerological Observatory (Moscow)
CAPE .... Convection and Precipitation Electrification Experiment (campaign)
CAPL .... Capillary Pumped Loop (Shuttle experiment of Hitchhiker payload, see also 'CPL')
CARIBIC ...... Civil Aircraft for Remote-Sensing and In-Situ-Measurements in Troposphere and Lower Stratosphere Based on the Instrumentation Container Concept (P.40.3)
CART .... Cloud and Radiation Testbed [field measurement component of the DOE ARM program; the three CART sites are: SGP (Southern Great Plains) near Billings in northern Oklahoma, TWP (Tropical Western Pacific on Manus Island, Papua, New Guinea), and NSA (North Slope of Alaska)]
CAPTEX .... Cross-Appalachian Tracer Experiment (campaign)
CAS .... Chinese Academy of Sciences (Beijing, China, since 1949)
CAS/CSSAR .... CAS/Center for Space Science and Applied Research, Beijing, China
CAS/IRSA ...... CAS/Institute for Remote Sensing Applications, Beijing, China
CAS/SITP ...... CAS/Shanghai Institute of Technical Physics, Shanghai, China
CASA .... Construcciones Aeronauticas S.A. (Madrid, Spain). In July 1999 CASA merged with DASA (DaimlerChrysler Aerospace AG)
CASC .... China Aerospace Science & Technology Corporation (Beijing, since 1993, also referred to as CAC). CASC, as a large state-owned enterprise, exerts primary control over the national space program on a day-to-day basis (handling of internal matters). CASC specializes in developing, building and suppling launch vehicles, satellites, various types of strategic and tactical missiles as well as satellite ground application systems and providing commercial launch services.
Over 130 Chinese organizations are subordinate to CASC, including five large academies [CALT (Chinese Academy of Launch-Vehicle

Technology), CAST (Chinese Academy of Space Technology), SAST (Shanghai Academy of Space-Flight Technology), CASET (Chinese Academy of Space Electronic Technology), and the Academy of Space Chemical Propulsion Technology]

CASI .... Canadian Aeronautics and Space Institute
CASP .... Canadian Atlantic Storms Program (campaign)
CAST .... Center for Aerospace Technology (Weber State University, Ogden, Utah)
CAST .... Chinese Academy of Space Technology (Beijing, China, since 1968)
CATSAT .... Cooperative Astrophysical and Technology Satellite (part of STEDI program, see N.17.3)
CBE .... Chemical Beam Epitaxy (a growth technique)
CBERS .... China/Brazil - Earth Resources Satellite, D.7. The satellite is also referred to as Ziyuan-1, meaning 'resource' in Chinese.
CCAFS .... Cape Canaveral Air Force Station (Cape Canaveral, FL, USA)
CCD .... Charge-Coupled Device (solid-state detector type)
CCE .... Charge Composition Explorer (S/C of AMPTE mission, K.4.3)
CCIR .... Comité Consultatif International des Radiocommunications (International Consultative Committee for Radio Communications , an organ of ITU). As of 1990 CCIR was renamed to ITU-R.
CCITT .... Comité Consultatif International Téléphonique et Télégraphique (one of three bodies for the definition of OSI. CCITT is a permanent organ of ITU). As of 1990 CCITT was renamed to ITU-T (ITU-Telecommunications)
CCM-A .... Cell Culture Module-A (Shuttle experiment)
CCN .... Cloud Condensation Nuclei
CCPD .... Charge-Coupled Photo Detector
CCRS .... Canada Center for Remote Sensing (Ottawa, Ontario; established in 1972, part of 'Department of Energy, Mines and Resources,' Canada)
CCSDS .... Consultative Committee for Space Data Systems
CD .... Compact Disk (introduction in 1982)
CDA .... Command and Data Acquisition (NOAA Antenna, downlink concept)
CDDIS .... Crustal Dynamics Data Information System (database at GSFC)
CDGPS .... Carrier-phase Differential GPS (a relative position measurement technique)
CDMA .... Code Division Multiple Access (a communication access scheme)
CDP .... Crustal Dynamics Program (NASA)
CdZnTe .... Cadmium Zinc Telluride (a detector material - also referred to as CZT)
CD-ROM ...... Compact Disk - Read Only Memory (storage capacity up to 650 MByte)
CD-R/W .... Compact Disk - Read/Write
CDTI .... Center for Technological and Industrial Development, Madrid, Spain
CDWL .... Coherent-detection Doppler Wind Lidar
CEA .... Commissariat à l'Energie Atomique
CEAREX ...... Coordinated Eastern Arctic Experiment (campaign)
CEBAS .... Closed Equilibrated Biological Aquatic System (Shuttle payload)
CEC .... Commission of the European Communities (Brussels, Belgium)
CEES .... Committee on Earth and Environmental Sciences (US interagency committee)
CEMAGREF ... Centre d'Etude du Machinisme Agricole du Genie Rural et des Eaux et Forests (France)
CEOS .... Committee on Earth Observation Satellites
CEPEX .... Central Equatorial Pacific Experiment (campaign)
CEPT .... European Conference of Postal and Telecommunications Administrations (Montreux, Switzerland, since 1959). CEPT comprises 43 European countries and is charged with representing Europe on such items as spectrum issues, etc.

CERFACS ...... Centre Européen de Recherche et de Formation Avancée en Calcul Scientifique (Toulouse, France, since 1987) European Center for Research and Advanced Training in Scientific Computation
CERGA .... Centre d'Etudes et des Recherches en Geodynamique et Astrometrie (in Grasse, France)
CERISE .... Caracterisation de l'Environnement Radioelectrique par un Instrument Spatial Embarque, (French S/C), D.40.11
CERN .... Centre Européen de Recherche Nucléaire (Geneva, Switzerland)
CES .... Committee on Earth Studies - a standing committee of the Space Studies Board within the National Research Council (NRC), USA
CESAR .... Cooperacion Española-Argentina (satellite of INTA and CONAE)
CESBIO .... Centre d'Etudes Spatiales de la Biosphère (Toulouse, France)
CESR .... Centre d'Etude Spatiale des Rayonnements (Toulouse, France, part of CNRS)
CETA .... Crew and Equipment Translation Aids (Shuttle experiment)
CETP .... Centre d'etude des Environnements Terrestre et Planetaire (Velizy/ Saint-Maur, France, CNRS Lab)
CEU .... Commission of the European Union (successor of previous CEC)
CEV .... Centre d'Essais en Vol (French Test Flight Center)
CfAO .... Center for Adaptive Optics, UCSC (University of California at Santa Cruz)
CFCs .... Chlorofluorocarbons
CFC-11 .... CCl_3F, trichlorofluoromethane, Freon-11
CFC-12 .... CCl_2F_2, dichlorodifluoromethane, Freon-12
CFES .... Continuous Flow Electrophoresis System (Shuttle payload)
CFRP .... Carbon Fiber Reinforced Plastic (also: CFRM for Material)
CGBA .... Commercial Generic Bioprocessing Apparatus (Shuttle experiment)
CGWIC .... China Great Wall Industry Corporation (Beijing, since 1980), provider of Long March launch services to the world market
CGMS .... Coordination Group for Meteorological Satellites [since 1972; active CGMS members are: EUMETSAT (Europe), JMA (Japan), China, Russia, NOAA (USA), WMO]. The global network of meteorological satellites constitutes a major portion of the space-based GOS (Global Observing System) of WWW (World Weather Watch).
CGP .... Shuttle payload consisting of: [CSE (Cryo System Experiment), GP (Glow Phenomenon)]
CGWIC .... China Great Wall Industry Corporation (launch service provider)
CH_3Cl .... Methyl chloride
CH_4 .... Methane
CHAMP .... Challenging Minisatellite Payload (E.1)
CHAMP .... Comet Halley Active Monitoring Program (Shuttle experiment)
CHASE .... Coronal Helium Abundance Spacelab Experiment (Spacelab-2)
CHEOPS .... CHEmistry of Ozone in the Polar Stratosphere (airborne campaign)
CHORUS ...... Chemistry of Ozone Reduction in the Lower Stratosphere (first Strato-2C mission)
CHROMEX .... Chromosomes and Plant Cell Division (Shuttle experiment)
CHRPT .... Chinese High Resolution Picture Transmission (downlink mode)
CID .... Charge-Injection Device (a charge-tranfer detection technology)
CID .... Collision-Induced Dissociation (a measurement technique in the atmospheric sciences for studies of ion-molecule reactions, etc.)
CIDESON ...... Centro de Investigacion y Desarrollo de los Recursos Naturales de Sonora (Hermosillo, Mexico)
CIEMAT .... Centro de Investigaciones Energéticas y Medioambientales (Environmental and Energetic Research Center), Spain
CIESIN .... Consortium for International Earth Science Information Network (a private nonprofit corporation in Ann Arbor, Michigan (University Center). CIESIN serves scientific, policy-making, educational, and

public access data and information needs. CIESIN developed and is operating SEDAC (Socio-Economic Data and Applications Center) as part of one of nine data centers of EOSDIS.
CIGNET .... Cooperative International GPS Network of IAG (International Association of Geodesy), H.4.3.6
CIGS .... Copper Indium Gallium Diselenide (solar arrays based on thin film technology)
CIMS .... Chemical Ionization Mass Spectrometry (a measurement technique frequently used for atmospheric measurements)
CIMSS .... Cooperative Institute for Meteorological Satellite Studies (University of Wisconsin, Madison)
CINDE .... Convection Initiation and Downburst Experiment (campaign)
CIR .... Color Infrared (video images)
CIRA .... Centro Italiano Ricerche Aerospaziali (Italian Aerospace Research Center) since 1984, Capua, Italy
CIRAC .... Canadian Institute for Research in Atmospheric Chemistry
CIRES .... Cooperative Institute for Research in Environmental Sciences (University of Boulder, and at NOAA, Boulder, CO, USA)
CIS .... Commonwealth of Independent States (part of former Soviet Union or USSR)
CIT .... California Institute of Technology (Pasadena, CA)
CIT .... Computerized Ionospheric Tomography
CITE .... Chemical Instrumentation Test and Evaluation (campaign)
CIV .... Critical Ionization Velocity (Shuttle experiment)
CIVEX .... Cloud Instruments Validation Experiment (campaign)
CLARA .... Cloud And Radiation (campaign)
CLASS .... Cross-chain LORAN Atmospheric Sounding System (NCAR ground-based sounding stations)
CLEOPATRA ... Cloud Experiment Oberpfaffenhofen and Transports (campaign)
CLIVAR .... Climate Variability and Predictability (WCRP campaign program)
CLIVAR-ACC .. CLIVAR - Anthropogenic Climate Change
CLIVAR-DecCen CLIVAR - Decadal-to-Centennial time-scales
CLIVAR-GOALS CLIVAR - Global Ocean-Atmosphere-Land System
CLOUDS ...... Cloud and Radiation Monitoring Satellite (a proposed ESA mission as of 2001, A.7)
CLOUDS ...... Cloud Logic to Optimize Use of Defense Systems (Shuttle payload)
CLUSTER ..... ESA/NASA Solar-Terrestrial Mission (K.7)
ClO .... Chlorine monoxide
$ClONO_2$ .... ($ClONO_3$) Chlorine nitrate
CLS .... Collecte, Localisation, Satellites (with HQ in Toulouse, France) CLS was set up in 1986 to market the Argos (data collection) system.
C-MAN .... Coastal-Marine Automated Network [NOAA/NWS/NDBC moored buoy network (over 100 buoys) with hourly reports via GOES DCS]
CMA .... China Meteorological Administration, Beijing (government agency)
CMA/NSMC .... CMA/National Satellite Meteorological Center, Beijing, China
CMC .... Canadian Meteorological Centre
CME .... Coronal Mass Ejection (of the sun)
CMESS-95 ..... Cooperative Multiscale Experiment Spring/Summer 1995 (campaign)
CMIX .... Commercial Materials Dispersion Apparatus Instrument Technology Associates Experiments (Shuttle experiment)
CMOS .... Complementary Metal-Oxide Semiconductor (solid-state microprocessor technology)
CMS .... Centre de Météorologie Spatiale (Lannion, France)
CMT .... CdHgTe (detector type)
CN .... Condensation Nuclei
CNCR .... Characterization of Neurospora Circadian Rhythms (Shuttle payload)

CNES .... Centre National D'Etudes Spatiales (Space Agency of France, Paris, Toulouse, Evry, and Kourou, since 1962). Employment (1999) of 2500 scientists and engineers; of these, about 1700 employees are at Toulouse. CNES/HQ is in Paris with about 250 employees.

CNES/AVISO ... CNES/Archiving, Validation and Interpretation of Satellites oceanographic data (CNES data center for GEOSAT, Topex/Poseidon, ERS-1/2, ENVISAT, Jason-1, etc.)

CNET .... Centre National d'Etudes des Télécommunications (France Télécom)

CNIE .... Comision Nacional de Investigaciones Espaciales (former Space Agency of Argentina)

CNR .... Consiglio Nazionale delle Ricerche (National Research Counsil of Italy, Rome). CNR is a government agency which promotes and coordinates institutional research in the interests of Italy. CNR was founded in 1923 and reorganized in 1945 and 1979. CNR funds/maintains 157 institutes, 117 study centers, and 16 research groups throughout Italy. Research is supported in the natural and human sciences. In 1980 PSN (National Space Program) was created within CNR. Some space projects supported by CNR are: Italsat, TSS (Tethered Satellite System), Iris (propulsion system for the transfer of useful loads from the Space Shuttle's "hold" to a higher orbit), Lageos-2, and Sax (X-ray astronomy). CNR maintains a number of cooperations with various space agencies. In 1988 ASI (Agenzia Spaziale Italiana) was founded which succeeded CNR in relations concerning matters of planning and administrative nature. Nevertheless, CNR continues to follow specific aspects of research within the context of its own bodies.

CNR/DCAS .... CNR / Direzione Centrale Attività Scientifiche (Rome, Italy)

CNR/FISBAT ... CNR / Istituto per lo Studio dei Fenomeni Fisici e Chimici della Bassa ed Alta Atmosfera (Institute of Physics and Chemistry of the Lower and Upper Atmosphere, Bologna, Italy)

CNR/IFA .... CNR / Istituto di Fisica dell' Asmosfera (Institute for the Physics of the Atmosphere, Frascati, Italy)

CNR/IFAM ..... CNR / Istituto di Fisica Atomica e Molecolare (Pisa, Italy)

CNR/IFCTR .... CNR / Istituto die Fisica Cosmica e Teccnologie Relative (Milano)

CNR/IFSI ...... CNR / Istituto de Fisica dello Spazio Interplanetario (Frascati, Italy)

CNR/IROE ..... CNR / Istituto di Richerca sulle Onde Elettromagnetiche (Florence, Italy)

CNR/IMGA .... CNR / Istituto per lo Studio delle Metodologie Geofisiche Ambientali (Bologna, Italy)

CNR/ITRE ..... CNR / Istituto di Technologie e Studie della Radiazioni Extraterrestri (Bologna, Italy)

CNR/LARA .... CNR / Laboratorio Aereo per Ricerche Ambientali (Laboratory for Airborne Environmental Studies, Rome, Italy)

CNR/PSN ...... Consiglio Nazionale delle Ricerche / Piano Spaziale Nationale (Italy)

CNRM .... Centre National des Recherches Meteorologiques (France)

CNRS .... Centre National de la Recherche Scientifique (National Research Center of France). CNRS is a government-funded basic-research organization which employs about 26,000 people, including more than 11,000 research scientists. The agency maintains facilities throughout France. There are over 1500 CNRS laboratories active in all fields of science. Most CNRS laboratories rely for their research on partnerships with French universities. There are also many CNRS cooperations and exchanges with other research organizations on a national and international level as well as with French industry. Only a few facilities (dealing mostly with the sciences of the universe, such as: oceanography, geophysics, climatology, hydrology, volcanology, seismology, astronomy, astrophysics, etc.) are listed below.

CNRS/CESR ... CNRS/Centre d'Etude Spatiale des Rayonnements (Toulouse, France)

CNRS/CERGA . CNRS/Centre d'Etudes et des Recherches en Geodynamique et Astrometrie (Grasse, France)
CNRS/CETP .... CNRS/Centre d'Etude des Environnenments Terrestre et Planétaires, (sites at: Vélizy, Issy-les-Moulineaux, and Saint-Maur des Fossés, France)
CNRS/IAS ...... CNRS/Institut d'Astrophysique Spatiale (Orsay, France)
CNRS/INSU .... CNRS/Institut National des Sciences de l'Univers (Paris, France)
CNRS/LAS ..... CNRS/Laboratoire d'Astronomie Spatiale (Marseille, France)
CNRS/LMD .... CNRS/Laboratoire de Météorologie Dynamique (Palaiseau, France)
CNRS/LOA ..... CNRS/Laboratoire d'Optique Atmosphérique (University of Lille, France)
CNRS/LPCA ... CNRS/Laboratoire de Physique et Chimie de l'Atmosphère (University of Strasbourg, France)
CNRS/LPCE .... CNRS/Laboratoire de Physique et de Chimie de l'Environnement (Orleans-la-Source, France)
CNRS/SA ...... CNRS/Service d'Aeronomie (Verrières-le-Buisson, France)
CNRSC .... China National Remote Sensing Center (since 1981)
CNSA .... China National Space Administration (Beijing, since 1993). The principal role of CNSA is to serve as China's policy organization and interface with other national space agencies.
CNS/ATM ...... Communication, Navigation and Surveillance/Air Traffic Management)
CNTS .... Centre National des Techniques Spatiales (Algiers, Algeria)
CO .... Carbon monoxide
CO_2 .... Carbon dioxide
COARE .... Coupled Ocean Atmosphere Response Experiment (campaign, see TOGA/COARE)
COAST .... Coastal Oxidant Assessment for Southeast Texas (campaign)
CODAG .... Cosmic Dust Aggregation (Shuttle payload)
CODAR .... Coastal Ocean Dynamic Application Radar (a ground-based, over-the-horizon radar which reflects off of the ionosphere to measure sea surface roughness and currents)
CODE .... Coastal Ocean Dynamics Experiment (campaign)
COF .... Columbus Orbital Facility (ESA module on ISS)
COHMEX ..... Cooperative Huntsville Meteorological Experiment (campaign)
COMETS ...... Communications and Broadcasting Engineering Test Satellite (prototype data relay satellite of Japan)
CONAE .... Comisión Nacional de Actividades Espaciales, Buenos Aires, Argentina (National Commission on Space Activities, since 1992) - Space Agency of Argentina
CONCAP ...... Consortium for Materials Development in Space Complex Autonomous Payload (Shuttle experiment)
CONUS .... Continental United States ('lower 48 states')
COPE .... Coastal Ocean Probing Experiment (campaign)
COPS-91 .... Cooperative Oklahoma Profiler Study-1991 (campaign)
CORONAS ..... Complex of Orbital Observations of the Activity of the Sun (Satellite of the Russian Space Agency, K.8)
COSMIC .... Constellation Observing System for Meteorology, Ionosphere and Climate (A.24)
COSMOS ...... The term 'Cosmos' or 'Kosmos' is used in Russia to designate any of a series of unmanned satellites that were launched starting in 1962 with Cosmos-1 (the counting in 1988 was up to 1800, in 1993 it is around 2200). The Cosmos satellite series has been used for a wide variety of purposes, including scientific research, Earth observation, experimental/technological payloads, preoperational meteorological satellites, navigation satellites, etc. There are also many satellites with military payloads under the Cosmos designation.

'Cosmos' is also the name of a Russian launch vehicle (first launch in 1964, from 1970-'87 there were 371 successful flights of the Cosmos launcher). The Cosmos launch vehicle remains in service (1996).

COSPAR .... Committee on Space Research (of ICSU, since 1958). COSPAR is an interdisciplinary scientific organization concerned with international progress in all areas of scientific research carried out with space vehicles, rockets, and balloons.

COSPAS .... Space System for the Search of Distressed Vessels (Russia's equipment flown on polar-orbiting S/C).

COSSA .... CSIRO Office of Space Science and Applications (since 1984, Canberra, Australia)

COTES .... Conventional Terrestrial Reference System (an IERS program for the specifications of positions on or near the Earth's surface)[2108]

COTS .... Commercial-Off-The-Shelf (products)

CPCG .... Commercial Protein Crystal Growth (Shuttle experiment)

CPFSK......... Continuous Phase Frequency Shift Keying (a modulation method)

CPL .... Capillary Pumped Loop Experiment (Shuttle)

CPMA .... Code Position Multiple Access (communication access concept)

CPR .... Cloud Profiling Radar (GEWEX)

CPRA .... Control of the Reception Pattern multi-element Antenna

CRA .... Centro Ricerche Aerospaziali (University of Rome, Italy)

C-RAM .... Chalcogenide - Random Access Memory

CRC .... Communication Research Center (an institute of Industry Canada, located at Shirleys Bay, west of Ottawa)

CRCSS......... Cooperative Research Center for Satellite Systems (Canberra, Australia, the new Australian space agency, as of Jan. 1, 1998 - it is also referred to as simply "CRC"). CRCSS, under the Cooperative Research Centres Program of the Commonwealth of Australia, is a union of 12 Australian organizations, including government, university and industry. Some of the participants are: CSIRO, University of South Australia, Queensland University of Technology, University of Newcastle, University of Technology, Sydney, Auspace Ltd. of Mitchell, ACT [Note: since 1990, Auspace has been a subsidiary of MMS (Matra Marconi Space) of France]

CREAM .... Cosmic Ray Effects and Activation Monitor (Shuttle payload)

CREST .... Center for Research in Earth and Space Technology (North York, Ontario, Canada). Formerly known as ISTS (Institute of Space and Terrestrial Science)

CRI .... Crown Research Institute (New Zealand)

CRISP .... Center for Remote Imaging, Sensing and Processing (since 1992, National University of Singapore, Singapore)

CRL .... Communications Research Laboratory, Tokyo, a division of the Ministry of Posts and Telecommunications (MPT) of Japan. Note: the former name of CRL (until 1987) was RRL (Radio Research Laboratories)

CRO .... Chemical Release Observation (Shuttle experiment)

CRP .... Cloud Radiation Program

CRPE .... Centre de Recherches en Physique de l'Environnement Terrestre et Planetaire, at the following sites: Vélizy, Issy-les Moulineaux, and Saint-Maur-des-Fossés, France (Lab was part of CNRS and of CNET, starting in January 1994 CRPE was reorganized and renamed CETP, there is no more dependence on CNET)

CRPSM .... Centro di Ricerca Progetto San Marco (San Marco ground receiving station and processing/archiving facilities located at Malindi, Kenya), CRPSM is owned and operated by the University of Rome, Italy

2108) See: "The International Earth Rotation Service," in 'The Interdisciplinary Role of Space Geodesy,' Springer Verlag, 1989, pp. 229-232

CRREL .... Cold Regions Research and Engineering Laboratory (US Army research facility in Hanover, NH, USA)
CRRES .... Combined Release and Radiation Effects Satellite (A.11)
CRSS .... Commercial Remote Sensing System (B.6); the S/C was renamed to IKONOS
CRSS .... Canadian Remote Sensing Society (since 1973); CRSS is part of CASI (Canadian Aeronautics and Space Institute)
CRT .... Cathode Ray Tube
CRTS .... Centre Royal Teledetection Spatiales, Rabat, Morrocco
CRV .... Crew Return Vehicle (or X-38 CRV of NASA, used for ISS evacuation in case of an emergency)
CRYOFD ...... Cryogenic Flexible Diode (Shuttle payload)
CRYOHP ...... Cryogenic Heat Pipe Experiment (Shuttle payload)
CRYOTSU ..... Cryogenic Thermal Storage Unit (Shuttle payload)
CRYSYS .... Use of the Cryospheric System to Monitor Global Change in Canada (campaign program)
CSA .... Canadian Space Agency (since 1989; CSA HQs and control center at Saint-Hubert, Québec)
CSCE .... Commercial Space Center for Engineering (established under contract with NASA/JSC, located on the Texas A&M University campus; CSCE supports industry development of palletized commercial payloads on external platforms on ISS)
CSE .... Consortium for Superconducting Electronics (USA) involving Bell Labs, IBM, MIT, MIT/LL, etc.
CSEM .... Centre Suisse d'Electronique et de Microtechnique (or: Swiss Center for Electronics and Microtechnology), Neuchatel, Switzerland
CSER .... Center for Satellite Engineering Research (University of Surrey, UK)
CSGC .... Colorado Space Grant Consortium - a NASA-funded institution which supports student-designed satellites
CSIC .... Consejo Superior de Investigaciones Cientificas (Spanish Research Council, Madrid)
CSIR .... Council for Scientific and Industrial Research, Pretoria, South Africa. CSIR is Africa's largest scientific and technological research, development and implementation organization.
CSIRO .... Commonwealth Science and Industrial Research Organization (Canberra, Australia)
CSMA/CD ..... Carrier Sense Multiple Access / Collision Detection (commercially known under Ethernet)
CSMT .... Center for Space Microelectronics Technology (NASA/JPL facility, since 1987)
CSOC .... Consolidated Space Operations Contract (NASA/Lockheed Martin contract for Shuttle operations, etc.). The objective is to achieve a low-risk, commercially-based space operations program for Shuttle.
CSR .... Centro de Sensores Remote (Italy)
CSSR .... Chinese Society of Space Research
CST .... CORE Software Technology, Pasadena, CA [developer of the world's first commercial on-line geo-spatial (image, cartographic, & demographic) indexing and distribution system]
CSTG .... Commission on International Coordination of Space Techniques for Geodesy and Geodynamics (since 1979), (Commission VIII of the International Association of Geodesy)
CSU .... Colorado State University, Fort Collins, CO
CTA .... Canadian Target Assembly (Shuttle payload)
CTA .... Centro Tecnico Aerospacial (Sao José dos Campos, S.P., Brazil)
CTA .... CTA Space Systems, McLean, VA, (since 1979) manufacturer of small satellite systems (Clark, EarlyBird, REX, etc.) and instruments; CTAST (CTA Space and Telecommunications) is the parent company

of CTA Space Systems. Note: CTA Space Systems was acquired by OSC of Dulles, VA, in Aug. 1997

CTD .... Conductivity-Temperature-Depth profilers (a buoy type used in a number of campaigns like NORSEX, TOGA/COARE, etc.)
CTIV .... Processing Center for VEGETATION Imagery (operated by Vito in Mol, Belgium, VEGETATION is a SPOT-4, 5 instrument
CTP .... Cloud Top Pressure
CUE .... Collaborative Ukrainian Experiment (Shuttle experiment)
CUZK .... Czech office for Surveying, Mapping and Cadastre
CVF .... Circular Variable Filter (filter technology)
CVTE .... Chemical Vapor Transport Experiment (Shuttle payload)
CVX .... Critical Viscosity of Xenon (Shuttle payload)
CW .... Continuous Wave
CZT .... Cadmium Zinc Telluride (a detector material - also referred to as CdZnTe)

D

DAAC .... Distributed Active Archive Center (NASA EOSDIS Program)
DAB .... Digital Audio Broadcasting
DARA .... Deutsche Agentur für Raumfahrtangelegenheiten, Bonn (German space agency (from 1989 to Sept. 30, 1997, DARA was integrated into DLR effective Oct. 1, 1997)
DARPA .... Defense Advanced Research Projects Agency (US DoD agency, since 1958, DARPA started as ARPA with an early focus on space research). Technological innovations such as the Transit navigation system, Internet (in 1969 ARPANET started which become later Internet), stealth technology, and many activities in the space program were sponsored by DARPA.
DASA .... DaimlerChrysler Aerospace AG, Munich (HQ), Germany (with 45,000 employees). Prior to Nov. 1998, DASA stood for 'Daimler-Benz Aerospace AG.' Prior to January 1995 the meaning of the acronym DASA was 'Deutsche Aerospace AG.' DASA/DSS (Dornier Satellitensysteme GmbH) is a DASA business unit responsible for all satellite-related activities with facilities in Friedrichshafen and Ottobrunn. DASA (founded in 1989) is a conglomerate of the previous companies: Dornier, MBB (Messerschmitt-Bölkow-Blohm), MTU (Motoren- und Turbinen-Union), and TST (Telefunken Systemtechnik). - In addition, DASA is a partner in many alliances such as: Airbus, Ariane, Eurocopter, etc. Today, the three independent business entities of DASA are: DASA/Airbus, DASA/DSS, and DASA/MTU. - As of 2000, DASA is called **Astrium GmbH** (see Astrium)
DAT .... Digital Audio Tape (a high-volume data recording technique, helical scan tape storage)
DATA-CHASER Distribution and Automation Technology Advancement - Colorado Hitchhiker and Student Experiment of Solar Radiation (Shuttle)
DBMS .... Database Management System
DBS .... Direct Broadcasting Satellite
DBSI .... Direct Broadcasting Satellite Industries Inc. of Mill Valley, CA
DC .... Direct Current
DCRS .... Danish Center for Remote Sensing (at EMI of TUD, Lyngby, Denmark)
DCRS .... Digital Cassette Recorder System
DCP .... Data Collection Platform (ground segment platform for environmental data measurement, Meteosat, GOES, GMS)
DCPI .... Data Collection Platform Interrogation (GOES)
DCS .... Data Collection System (NOAA- GOES series, Meteosat series, GMS series, geostationary satellites).

DCT .... Discrete Cosine Transformation (compression technique)
DCW .... Digital Chart of the World (a vector map database by DMA, Fairfax, VA, USA)
DEBITS .... Deposition of Biogeochemically Important Trace Species
DECAFE ...... Dynamics and Chemistry of the Atmosphere in Equatorial Forest (campaign)
DEE .... Dexterous End Effector (Shuttle)
DEM .... Digital Elevation Model (also referred to as DTM = Digital Terrain Model)
DEOS .... Delft Institute for Earth-Oriented Space Research [at Delft University of Technology (DUT), Delft, The Netherlands]
DERA .... Defence Evaluation and Research Agency [Farnborough, UK, an agency of MoD (Ministry of Defense)]. DERA was established in April 1995 from elements of the former RAE (Royal Aerospace Establishment).
DESPA .... Départment de Recherche Spatiale de L'Oservatoire de Paris/Meudon (France)
DFD .... Deutsches Fernerkundungsdatenzentrum (German Remote Sensing Data Center, DLR, Oberpfaffenhofen)
DFG .... Deutsche Forschungsgemeinschaft (German National Research Council)
DFN .... Deutsches Forschungsnetz
DFVLR .... Deutsche Forschungs- und Versuchsanstalt für Luft- und Raumfahrt (predecessor name of DLR from 1969 until 1989). History: In 1969 (April 1) a merger of the following German research facilities occurred, resulting in DFVLR with HQ in Köln-Porz: AVA (Aerodynamische Versuchsanstalt, founded 1907 in Göttingen), DFL (Deutsche Forschungsanstalt für Luftfahrt, founded 1936 in Braunschweig), DVL (Deutsche Versuchsanstalt für Luftfahrt, founded 1912 in Berlin-Adlershof, after WW-II in Mühlheim-Ruhr, since the 1960s in Köln-Porz). FFO (Flugfunkforschungsinstitut Oberpfaffenhofen), founded in 1937, was integrated into DVL (Köln-Porz) in 1965. FFM (Flugwissenschaftliche Forschungsanstalt München) joined DVL in 1963.
DGASP .... Dye 3 Gas and Aerosol Sampling Program (IGBP/IGAC program)
DGFI .... Deutsches Geodätisches Forschungsinstitut (Munich, Germany)
DGGTN .... Direction General de Geografica del Territorio Nacional (Mexico)
DGLR .... Deutsche Gesellschaft für Luft- und Raumfahrt - Lilienthal-Oberth e. V., Bonn
DGON .... Deutsche Gesellschaft für Ortung und Navigation (Düsseldorf, Germany - German Institute of Navigation)
DGPF .... Deutsche Gesellschaft für Photogrammetrie und Fernerkundung
DGPS .... Differential GPS, H.4.4
DHI .... Deutsches Hydrographisches Institut (Hamburg, Germany)
DIAL .... Differential Absorption Lidar (lidar technique)
DIN .... Deutsches Institut für Normung (German Institute for Standardization)
Discoverer II .... A US (military) technology demonstration program of DARPA, USAF and NRO, started in 1998, with the objective to develop a high-resolution interferometric SAR system (IFSAR) for surveillance and reconnaissance
DISCOS .... Database and Information System Characterizing Objects in Space (ESA/ESOC database for space debris and meteoroids, since 1990)
DLR .... Deutsches Zentrum für Luft- und Raumfahrt e.V. (German Aerospace Center, with HQ in Köln; DLR is also the German Space Agency). On Oct. 1, 1997 DARA was re-integrated into DLR. Prior to Oct.1.1997 the meaning of DLR was: Deutsche Forschungsanstalt für Luft- und Raumfahrt e.V.

DLR/DFD ...... DLR/Deutsches Fernerkundungsdatenzentrum (German Remote Sensing Data Center), Oberpfaffenhofen and Neustrelitz
DLR/FB .... DLR/Flugbereitschaft (aircraft operations; FB provides the services of flying sensors for other institutes of DLR)
DLR/GSOC .... DLR/German Space Operations Center, Oberpfaffenhofen
DLR/IOE ...... DLR/Institut für Optoelektronik (Institute of Optoelectronics), Oberpfaffenhofen
DLR/IPA .... DLR/Institut Physik der Atmosphäre (Institute of Atmospheric Physics), Oberpfaffenhofen
DLR/IRF ...... DLR/Institut für Hochfrequenztechnik (Institute of Radio Frequency Technology, Oberpfaffenhofen)
DLR/IRM ...... DLR/ Institut für Robotik und Mechatronik (Institute of Robotics and Mechatronics), Oberpfaffenhofen
DLR/ISST ...... DLR/Institut für Weltraumsensorik (Institute of Space Sensor Technology and Planetary Exploration, Berlin-Adlershof). There is also the abbreviation: DLR/IWS
DLR/MUSC .... DLR/Microgravity User Support Center (Cologne, Germany)
DMA .... Defense Mapping Agency (Fairfax, VA, USA, mapping, charting & geodetic products & services to the military, since 1972 - since 1996 DMA is an integral part of NIMA)
DMC .... Disaster Monitoring Constellation (a 5 S/C constellation under construction by SSTL, UK, planned launch in 2002)
DMI .... Danmarks Meteorologiske Institut (Danish Meteorological Institute, founded in 1872) Copenhagen, Denmark
DMN .... Direction de la Météorologie National (France)
DMOS .... Diffusive Mixing of Organic Solutions (Shuttle payload)
DMS .... Dimethylsulphide
DMSP .... Defense Meteorological Satellite Program (USA), G.1
Dnepr .... Russian/Ukrainian launch vehicle for satellites. As part of a nuclear disarmament agreement, former Soviet SS-18 ICBMs (Intercontinental Ballistic Missiles), were renamed to Dnepr. They are either being used for commercial launches, or destroyed by Dec. 31, 2007
DOAS .... Differential Optical Absorption Spectroscopy
DOC .... Department of Commerce (USA)
DoD .... Department of Defense (USA)
DOE .... Department of Energy (USA). Some major laboratories of DOE are:
ANL (Argonne National Laboratory), Argonne IL
BNL (Brookhaven National Laboratory), Upton, NY
FNAL (Fermi National Accelerator Laboratory), Batavia, IL
LANL (Los Alamos National Laboratory), Los Alamos, NM
LBL (Lawrence Berkeley Laboratory), Berkeley, CA
LLNL (Lawrence Livermore National Laboratory), Livermore, CA
ORNL (Oak Ridge National Laboratory), Oak Ridge, TN (since 1948)
PNL (Pacific Northwest Laboratory), Richland, WA
SLAC (Stanford Linear Accelerator Center), Stanford, CA
SNL (Sandia National Laboratory), Albuquerque, NM and Livermore, CA
DODGE .... Department of Defense Gravity Experiment (M.4)
DOM .... Dissolved Organic Matter
DORIS .... Determination Orbite Radiopositionnement Integres Satellite (CNES one-way tracking system for the measurement of precision orbits); another name convention is: Doppler Orbitography and Radiopositioning Integrated by Satellite, E.21.1
DoT .... Department of Transportation (USA)
DPCM .... Differential Pulse Code Mudulation (compression technique)
DRA .... Defence Research Agency [Malvern, Farnborough, etc., UK, with over 6000 employees; DRA was established in 1991, it is the successor orga-

nization of RAE (Royal Aerospace Establishment), ARE (Admiralty Research Establishment), RARDE (Royal Armament Research & Development Establishment), and RSRE (Royal Signal and Radar Establishment)]. As of April 1995 DRA was regrouped again and integrated as a division into DERA (Defense Evaluation and Research Agency). Another DERA reorganization in April 1997 dissolved DRA altogether.

DRAM .... Dynamic Random Access Method

Draper Lab ..... Charles Stark Draper Laboratory Inc. of Cambridge, MA. An MIT lab founded in the 1930s; an independent non-profit research lab since 1973. Focus on GN&C (Guidance, Navigation & Control) technologies.

DRB .... Defense Research Board, Canada

DREO .... Defense Research Establishment, Ottawa, Canada

DRI .... Desert Research Institute (of the University of Nevada)

DRS .... Data Relay Satellite (ESA system to relay information from the European space plane)

DRTS .... Data Relay Technology Satellite (Japan, Ka-band transmission)

DSB .... Double Sideband

DSN .... Deep Space Network (NASA/JPL)

DSP .... Defense Support Program (USA, DoD S/C series in GEO using infrared sensors to detect missile plumes against the Earth's background, to detect and report missile launches, space launches, and nuclear detonations) DSP S/C operate since the 1970s.

DSP .... Digital Signal Processing (computer, technology)

DSRI .... Danish Space Research Institute (Lyngby, Copenhagen, Denmark)

DSS .... Dornier Satellitensysteme GmbH (of DASA, Germany)

DSS .... Delft Sensor Systems (provider of optoelectronic instruments). DSS has been created by the integration of OIP (Optronic Instruments & Products), located in Oudenaarde, Belgium - and DIEO (Delft Instruments Electro-Optics, located in the Netherlands

DSSP .... Danish Small Satellite Program

DSSS .... Direct Sequence Spread Spectrum (communication technique)

DTE .... Digital Terrain Elevation

DTM .... Digital Terrain Model (also referred to as DEM = Digital Elevation Model)

DTU .... Technical University of Denmark, Lyngby

DUT .... Delft University of Technology (Delft, The Netherlands)

DVD .... Digital Versatile Disk [some standard DVD formats are: DVD-5 (4.7 GByte storage capacity, one layer per disk), DVD-9 (8.5 GByte, two layers per disk on one side, one layer is semi-permeable), DVD-10 (9.4 GByte, one layer per side and disk), DVD-18 (17 GByte, two layers per side and disk, one layer per side is semi-permeable)]

DWD .... Deutscher Wetterdienst [German Weather Service, with seven forecast centers in Offenbach (HQ), Hamburg, Potsdam, Leipzig, Essen, Stuttgart, and Munich]. DWD employs over 3000 people in over 150 localities throughout Germany.

DWL .... Doppler Wind Lidar (a active laser instrument based either on coherent heterodyne receiver technology or on incoherent direct receiver technology)

DYCOMS ...... Dynamics and Chemistry of Marine Stratocumulus Experiment (campaign)

E

EA .... Environment Agency (of Japan)

EADS .... European Aeronautic, Defense and Space Company (a holding company of DASA and a French pool group with Lagardere as the major partner). Merger announcement of DASA (Germany) and Aerospatiale Matra (France) in Oct. 1999 - the merger was realized July 10, 2000 with DASA, Aerospatiale Matra, and CASA of Spain. The following units are part of EADS: Airbus, Ariane, Astrium (75%), Dassault, Eurocopter, Eurofighter, and Rocket Systems.

EarlyBird .... Commercial imaging satellite (B.4.1)

EARSEC .... European Airborne Remote Sensing Capabilities [program since 1990 between CEC (JRC in Ispra, Italy) and ESA]

EARSeL .... European Association of Remote Sensing Laboratories (since 1976)

EarthCARE .... Earth Clouds Aerosol and Radiation Explorer (a proposed ESA core mission)

EarthKAM ..... Earth Knowledge Acquired by Middle school students (a NASA education program) The camera program started in 1996 as KidSat on Shuttle.

EARTHNET ... ESA Program since 1977. Earthnet refers to an ESA organization responsible for the ground segment of Earth Observation. Functions: acquisition, archiving and distribution of Earth science data.

EarthWatch Inc. . A US Earth observation company in Longmont, CO. EarthWatch was formed in January 1995 and is a joint venture of Ball Aerospace and WorldView Imaging Corporation (builder of EarlyBird and QuickBird)

Earth Watch .... ESA program [these are the operational (or pre-operational) service-oriented missions addressing specific application areas of Europe]. The Earth Watch Initiative started in 2001 with the goal to secure for Europe an independent sustainable capability in operational Earth observation

EASE .... Experimental Assembly of Structures in Extravehicular Activity (Shuttle)

EASOE .... European Arctic Stratospheric Ozone Experiment (campaign)

EBCCD .... Electron-bombarded CCD array

EC .... European Commission (since 1995: CEU (Commission of the European Union)

ECD .... Electron Capture Detector

ECLIPS .... Experimental Cloud Lidar Pilot Study (campaign)

ECMWF .... European Centre for Medium-Range Weather Forecasts (located in Reading, UK, founded in 1973). ECMWF is an international organization supported by the following European states: Austria, Belgium, Denmark, Finland, France, Germany, Greece, Ireland, Italy, Netherlands, Norway, Portugal, Spain, Sweden, Switzerland, Turkey, United Kingdom. ECMWF has working arrangements with WMO, EUMETSAT and ACMAD (African Centre for Meteorological Applications for Development).

ECS .... EOSDIS Core System (USA)

EDAC .... Earth Data Analysis Center (NASA contractor center at the University of New Mexico, Albuquerque, NM, since 1964)

EDAC .... Error Detection and Correction (information processing term)

EDC .... EROS Data Center of the US Geological Survey (Sioux Falls, SD, DAAC of NASA EOS Program for Land Processes)

EDI .... Electronic Data Interchange, (Format Specification according to ANSI Standard X.12; (an existing but non-ISO Protocol)

EDIFACT ...... Electronic Data Interchange for Administration, Commerce, and Transport

EDO .... Extended Duration Orbiter (Shuttle)

EEA .... European Environment Agency (since 1990, located in Copenhagen since 1993, Denmark)

EECF .... Earthnet ERS-1 Central Facility (ESA facility at ESRIN, Italy)

EELV .... Extended Envelope Launch Vehicle (US Air Force launcher)

EEP .... Earth Explorer Program (ESA)
EEV .... English Electric Valve, Chelmsford, UK (manufacturer of detectors)
EEVT .... Electrophoresis Equipment Verification Test (Shuttle)
EFEDA .... European Field Experiment in Desertification-threatened Areas (campaign)
e.g. .... abbreviation (Latin: exempli gratia) for example
EGNOS .... European Geostationary Navigation Overlay System (planned ESA complementary system to GPS and GLONASS to provide Europe with GPS/GLONASS service availability, continuity and signal integrity)
EGS .... European Geophysical Society
EGS .... Experimental Geodetic Satellite of NASDA, (Ajisai, E.3)
EHF .... Extremely High Frequency (30 - 300 GHz band)
EHIC .... Energetic Heavy Ion Composition Experiment
EIRP .... Effective Isotropic Radiated Power
EISAC .... European Imaging Spectrometry Aircraft Campaign (1989-90)
EISCAT .... European Incoherent Scatter Radar
EIT .... Electro-bombardment Ion Thruster (electric proplusion system of MMS, France)
EIT .... Electromagnetically Induced Transparency
ELAC .... European Lidar Airborne Campaign
ELDO .... European Launcher Development Organization (since 1962) ELDO is, along with ESRO, a predecessor organization of ESA
ELDP .... European Lake Drilling Project (campaign under PANASH)
ELF .... Extremely Low Frequency (30 - 3000 Hz)
ELINT .... Electronic Intelligence (used in the context of DoD missions)
ELITE .... European LITE (campaign) LITE = Lidar In-space Technology Experiment (Shuttle payload)
ELOISE .... European Land-Ocean Interaction Studies (campaign)
El-Op .... El-Op Electro-Optics Industries of Rehovot, Israel (as of 2000 El-Op is part of Elbit Systems Ltd. of Haifa, Israel)
ELRAD .... Earth-Limb Radiance Experiment (Shuttle payload)
ELT-121.5 ...... Emergency Locator Transmitter (see COSPAS-S&RSAT, I.6)
EMAC .... European Multi-Sensor Airborne Campaign (in the framework of ESA/JRC collaboration)
EMBRAER .... Empresa Brasileira de Astronautica SA (aircraft and space payload manufacturer, Saò José dos Campos, SP, Brazil)
EMEX .... Equatorial Mesoscale Experiment (campaign)
EMF .... Electromotive Force
EMP .... Electromagnetic Pulse
EMSL .... Environmental Monitoring Systems Laboratory (Las Vegas, NV, EPA facility)
ENA .... Energetic Neutral Atoms [neutral atoms or molecules created by charge exchange between energetic ions (such as the Earth's radiation belts) and a cold neutral gas (such as the Earth's exosphere)]
ENEA .... Ente per le Nuove tecnologie l'Energia e l'Ambiente (Rome, Italy)
ENSO .... El Niño Southern Oscillation
ENVISAT .... Environmental Satellite (ESA, see D.9)
EO .... Earth Observation
EO-1 .... Earth Observing-1 (NASA S/C)
EOCAP .... NASA's Earth Observations Commercial Applications Program, since 1987 (NASA's intent is to commercialize remote sensing technology originally developed to support scientific exploration)
EOL .... End of Life
EOPP .... Earth Observation Preparatory Programme (of ESA)
EORF .... Environment Measurements by the Real-Time Radiation Monitor (Shuttle payload)
EOS .... Earth Observing System (NASA), D.11

EOS .... European Optical Society
EOSAT .... Earth Observation Satellite Company (Commercial distributor of Landsat imaging data, located in Lanham, MD, since 1985, EOSAT is a joint venture of Lockheed Martin and Hughes Aircraft). Space Imaging Inc. (since 1994) of Thornton, CO of LM and E-Systems, acquired EOSAT in 1995. The new company was subsequently renamed into: Space Imaging EOSAT [distributor of IKONOS imagery, ERS-1/2, JERS and Radarsat data (USA), global distributor of IRS-1C/D imagery]. Since 1998 the company name is: Space Imaging. The owners of Space Imaging are: LM, E-Systems (of Raytheon Co, Lexington, MA), Mitsubishi, Vander Horst (Singapore), Halla Heavy Industries (Korea).
EOSDIS .... EOS Data and Information System
EP .... Electric Propulsion (of spacecraft)
EPA .... Environmental Protection Agency (USA, since 1970)
EPIRB .... Emergency Position Indicating Radio Beacon (on COSPAS and S&RSAT payloads)
EPOCS .... Equatorial Pacific Ocean Climate Studies (campaign)
EPOCS .... European Committee on Ocean and Polar Sciences
EPOP .... European Polar Platform (old name, now POEM)
EPOS .... European Proximity Operations Sensor (ESA test of GPS Tensor receivers and an optical rendezvous sensor for Shuttle-Mir docking maneuvers on STS-84 and STS-86)
Equator-S ...... Solar Terrestrial Mission (K.9)
ER-2 .... Extended Range U-2 (US research aircraft of NASA/ARC)
ERA .... European Robotic Arm (of ESA on ISS)
ERB .... Earth Radiation Budget
ERBS .... Earth Radiation Budget Satellite (NASA), A.13
ERICA .... Experiment on Rapidly Intensifying Cyclones over the Atlantic (campaign)
ERIM .... Environmental Research Institute of Michigan (HQ in Ann Arbor, MI). ERIM is a nonprofit contract research organization in the field of remote sensing. In May 1997, ERIM was transformed into a profit-seeking company and changed its name to "ERIM International." History: The Willow Run Laboratories were founded in 1947. In 1973 the Willow Run Laboratories team separated from the University of Michigan and became ERIM. - In 2000 ERIM International Inc. became part of Veridian Systems, the new company is called: Veridian ERIM International
EROS .... Earth Resources Observation Systems (Data Center of USGS in Sioux Falls, SD, archive for Landsat and other data)
EROS .... Earth Remote Observation System (B.5)
ERS-1,2 .... European Remote Sensing Satellite (ESA program), D.13 and D.14
ERS .... Earth Resource Satellite
ERSDAC .... Earth Remote Sensing Data Analysis Center (Tokyo, Japan)
ERTS-1 .... Earth Resources Technology Satellite (NASA satellite, in 1975 ERTS-1 was renamed to Landsat-1 and the entire ERTS program was renamed to Landsat)
ESA .... European Space Agency (since 1975), ESA-HQ in Paris (ESA member states are: Austria, Belgium, Denmark, Finland, France, Germany, Ireland, Italy, Netherlands, Norway, Spain, Sweden, Switzerland, and the United Kingdom; Canada is a cooperating country)
ESA/ESTEC .... ESA/European Space Research and Technology Centre (ESA facility in Noordwijk, Netherlands)
ESA/ESOC ..... ESA/European Space Operation Centre (ESA facility in Darmstadt, Germany)
ESA/ESRIN .... ESA/European Space Research Institute (ESA facility, Frascati, Italy)

ESA-IRS .... ESA - Information Retrieval Service (online database at ESRIN)
ESAC .... Earth Sciences Advisory Committee (ESA)
ESA/PB-EO .... ESA/Programme Board - Earth Observation
ESCAP .... (UN) Economic and Social Commission for Asia and the Pacific, Bangkok, Thailand
ESCAPE .... Experiment of the Sun for Complementing the ATLAS Payload and for Education (Shuttle Payload)
ESE .... Earth Science Enterprise [NASA program with the previous designation of MTPE (Mission to Planet Earth)]
ESEM .... Evaluation of Space Environment Effects on Materials (Shuttle payload of NASA/LaRC) ESEM experiments are focused on cosmic dust collection
ESF .... European Science Foundation (Strasbourg, France)
ESDIS .... Earth Science Data and Information System (NASA/GSFC)
ESIC .... Earth Science Information Center (USGS operates a network of ESICs to distrubute Earth science data and related products)
ESIS .... European Space Information System (ESA data system)
ESOC .... European Space Operation Centre [ESA facility in Darmstadt, Germany, since Sept. 1967; formerly ESDAC (European Space Data Center) under ESRO]
ESRIN .... European Space Research Institute (ESA facility in Frascati, Italy)
ESRO .... European Space Research Organization (founded in 1962 by ten European countries; predecessor organization of ESA)
ESSA .... Environmental Science and Services Administration (this was a predecessor organization of NOAA)
ESSP .... Earth System Science Pathfinder (small-scale, low-cost, and quick-turnaround NASA missions like QuikTOMS, VCL, GRACE, SORCE, ESSP-3 (formerly PICASSO-CENA), CLOUDSAT, VOLCAM, etc.)
ESTEC .... European Space Research and Technology Centre (ESA facility in Noordwijk, Netherlands)
ETALON ...... Russian passive satellite series for geodetic measurements, E.4
ETHZ .... Eidgenoessische Technische Hochschule, Zürich (Swiss Federal Institute of Technology, Zürich)
ETHZ/IGP ..... ETHZ/Institute of Geodesy and Photogrammetry
ETL .... Electrotechnical Laboratorium (of MITI, Japan)
ETS .... Engineering Test Satellite (NASDA technology series, Japan)
ETSI .... European Telecommunications Standards Institute (since 1988)
EU .... European Union (formerly EC = European Community)
EUCREX ...... European Cloud and Radiation Experiment (campaign)
EUMETSAT .... European Organization for the Exploitation of Meteorological Satellites (Darmstadt, Germany, since 1986 – operation of the Meteosat and the future MetOp systems). EUMETSAT member states are: Austria, Belgium, Denmark, Finland, France, Greece, Germany, Ireland, Italy, Netherlands, Norway, Portugal, Spain, Sweden, Switzerland, Turkey, and the United Kingdom.
EurasSpace ..... EurasSpace GmbH, Munich; Euro-Asian Space venture between DASA of Germany and CASC (Chinese Aerospace Corp. of Beijing); builders of telecommunication satellites (Sinosat-1, etc.)
EURIMAGE ... European Consortium for Satellite Image Dissemination (Rome, Italy - a commercial data distributor. The consortium is made up by the following companies: SSC, MATRA, NRSC and Dornier)
EURECA ...... European Retrievable Carrier (platform deployed and retrieved on Shuttle) J.5
EURISY .... European Association for ISY (one of two ISY organizers in Europe, see SAFISY)
Eurockot .... Eurockot Launch Services GmbH, Bremen, Germany. A joint venture company between Russia's Khrunichev and Germany's DASA. The

launcher is the Rockot vehicle, built by KhSC (Khrunichev State Research and Production Space Center), Moscow. Rockot is a modified version of Russia's SS-19 missile. The first launch demonstration of a Rockot vehicle occurred on May 16, 2000 from Plesetsk with Simsat-1 and -2, two dummy payloads.

EUROLAS ..... European Laser Stations (ground network of SLR stations)
EUROPTO ..... A joint venture between EOS (European Optical Society) and SPIE (Society of Photo-Optical Instrumentation Engineering)
Eurospace ...... The association of European space industry, Paris, since 1961
Eutelsat .... European Telecommunications Satellite Organization
EUV .... Extreme Ultra Violet (spectral range), see also: XUV
EXOS .... Exospheric Observations, ISAS program (K.10)
EXPRESSO .... Experiment for Regional Sources and Sinks of Oxidants (campaign)

F

FAA .... Federal Aviation Administration (regulatory agency for all civil aviation in the Department of Transportation, USA)
FACH .... Fuerza Aerara de Chile (Chilenian Air Force)
FAGS .... Federation of Astronomical and Geophysical Services
FAISAT .... Final Analysis Inc. Satellite (C.2)
FAO .... Food and Agriculture Organization (of the UN)
FARE .... Fluid Acquisition and Resupply Experiment (Shuttle)
FASat-Alfa ..... Fuerza Aerea Satellite - Alfa (D.40.12)
FASINEX ...... Frontal Air-Sea Interaction Experiment (campaign)
FAST .... Fast Auroral Snapshot Explorer (GSFC mission, K.21.2)
FAST .... Fore-Aft Scan Technique (radar)
FASTEX .... Fronts & Atlantic Storm Track Experiment (campaign)
FATE .... FIRST ATSR Tropical Experiment (campaign)
FBG .... Functional Cargo Block (first element of ISS also referred to as Zarya)
FCC .... Federal Communications Commission (Washington, DC, USA)
FDDI .... Fiber Distributed Data Interface
FDMA .... Frequency Division Multiple Access (access scheme)
FDP .... Fluorescent Dye Particles (a tracer technique in lidar observations)
FEA .... Fluid Experiment Apparatus (Shuttle)
FEC .... Forward Error Correction (transmission protocol technique)
FEEP .... Field Effect Electric Propulsion
Feng-Yun (FY) .. Chinese meteorological satellite series, G.3
FFT .... Fast Fourier Transform
FET .... Field-Effect Transistor (JFET = Junction Field-Effect Transistor)
FFSK .... Fast Frequency Shift Keying (modulation technique)
FGAN .... Forschungsgesellschaft für Angewandte Naturwissenschaften (German Defense Research Laboratory, Wachtberg, Germany)
FGGE .... First GARP Global Experiment (campaign)
FhG .... Fraunhofer Gesellschaft (in honor of Joseph von Fraunhofer, 1787 - 1826), a leading organization of applied research in Germany (HQ in Munich). FhG operates 47 research institutes in Germany with about 8500 employees. About 2/3 of FhG research is through contracts for industry and government. There are also FhG institutes in USA and Asia. Only a few institutes are listed here:
FhG/IAF .... Fraunhofer Gesellschaft/Institut für Angewandte Festkörperphysik (development of detectors), Freiburg, Germany
FhG/IFU .... Fraunhofer Gesellschaft/Institut für Atmospäheren und Umweltforschung (Institute of Atmospheric and Environmental Research), Garmisch-Partenkirchen, Germany
FhG/IOF .... Fraunhofer Gesellscahft/Institut für Angewandte Optik und Feinmechanik (Institute of Applied Optics and Precision Engeneering), Jena, Germany

FhG/IPM .... Fraunhofer Gesellschaft/Institut für Physikalische Meßtechnik (Institute of Physical Measurement Techniques), Freiburg, Germany
FHT .... Frequency Hopping Telemetry (a communication access method)
FIRAS .... P. N. Lebedev Physical Institute of the Russian Academy of Sciences (RAS), Moscow. FIRAS was established in 1967 as part of IKI. Since 1991 it is named AKTs FIRAS (radio astronomy)
FID .... Flame Ionization Detector
FIFE .... First ISLSCP Field Experiment (campaign)
FILE .... Feature Identification and Location Experiment (part of OSTA-1 payload on Shuttle STS-2 in Nov. 1981)
FIMR .... Finnish Institute of Marine Research (Helsinki, Finland)
FINDS .... Foundation of the International Non-Governmental Development of Space (USA, created in 1997)
FIR .... Far infrared: from about 10 - 1000 μm (note: 1000 μm = 1 mm)
FIRE .... First ISCCP Regional Experiment (campaign)
FIRESCAN .... Fire Research Campaign Asia-North (IGBP-IGAC-BIBEX campaign)
FIRESCHEME . Fire Information Systems Research in the Socio-Culture, History and Ecology, of the Mediterranean Environment (campaign)
FLINN .... Fiducial Laboratories for an International Network (a global network supporting Crustal Dynamics Test Sites)
FLIR .... Forward Looking Infrared (sensor)
FM .... Frequency Modulation (modulation technique of the main carrier)
FMC .... Forward Motion Compensation
FM/CW .... Frequency Modulation/Continuous Wave (a radar measurement technique to obtain range information - a sequence of FM/CW echoes contains both, range and Doppler information)
FMI .... Finnish Meteorological Institute (Helsinki, Finland)
FMS .... Flight Management System (avionics)
FOA .... Försvarets Forskningsanstalt (National Defense Research Establishment, Department of Information Technology, Linköpping, Sweden)
FOG .... Fiber-Optic Gyroscope (an angular rate gyro)
FÖMI .... Hungarian Remote Sensing Center, Budapest, Hungary
FOO .... Flight of Opportunity
FOR .... Field of Regard (total with of a ground imaging surface that is within the pointing potential of a sensor. Note: the FOV (or swath width) is always contained in the FOR)
FORTE .... Fast On-Orbit Recording of Transient Events (LANL, A.16)
FOV .... Field of View
FPA .... Focal Plane Array (also: Focal Plane Assembly - detector assembly of an imager instrument)
FPGA .... Field Programmable Gate Array
FRAM .... Ferroelectric Random Access Memory (a chip technology)
FREJA .... Swedish Solar-Terrestrial Mission (K.11)
FSK .... Frequency Shift Keying (modulation technique)
FTAM .... File Transfer Access and Management (OSI File Transfer Method)
FTFPV .... Flexible Thin-Film Photovoltaic (technology)
FTIR .... Fourier Transform Infrared (radiometer or spectrometer)
FTS .... Fourier Transform Spectrometer
FUV .... Far Ultraviolet (spectral region 90 - 125 nm)
FWG .... Forschungsanstalt der Bundeswehr für Wasserschall und Geophysik (Kiel, Germany)
FWHM .... Full-Width-Half-Maximum (of distribution curve)
FZJ .... Forschungszentrum Jülich (Germany, old name was KfA)
FZK .... Forschungszentrum Karlsruhe (Germany, old name was KfK)
FZK/IMK ...... FZK (Forschungszentrum Karlsruhe)/Institut für Meteorologie und Klimaforschung (Institute of Meteorology and Climate Research)

G

Ga .... Gallium (detector material)
GaAs .... Gallium Arsenide (a material used for solar panels, for detectors, and for fast computer chips)
GaIn .... Galileo Industries SA, located in Brussels, Belgium (a joint venture of Astrium, Alenia Spazio, and Alcatel Space, founded May 25, 2000)
$GaInP_2$ .... Gallium Indium Phosphide (solar cell type)
GaN .... Gallium Nitride (used in GaN photoconductive detectors)
GAC .... Global Area Coverage (the term is used for AVHRR data of NOAA)
GAF .... Gesellschaft für Angewandte Fernerkundung, Munich (since 1985, German commercial distributor of Earth observation data, such as Resurs data, Landsat data, IRS-1C/D data (via EOSAT), representative of EURIMAGE and SPOT-IMAGE in Germany, distributor for SOVINFORMSPUTNIK data, Radarsat data distributor for Germany, etc.
GABLE .... Global Atmospheric Backscatter Lidar Experiment (campaign)
GADACS ...... GPS Attitude Determination and Control Experiment (a GSFC GPS instrument package on Shuttle SPARTAN)
GADFLY .... GPS Attitude Determination Flyer (experiment on Lewis S/C)
GAIM .... Global Analysis, Interpretation and Modeling (IGBP project)
GALE .... Genesis of Atlantic Lows Experiment (airborne campaign in 1986)
Galileo Industries Galileo Industies SA is a European joint venture of the following companies(to define and build the Galileo System): Alenia Spazio of Rome, Alcatel Space of Paris, Astrium Ltd. of Stevenage, UK, and Astrium GmbH of Friedrichshafen, Germany.
GAME .... GEWEX-related Asian Monsoon Experiment (campaign)
GAMES .... Gravity and Magnetic Earth Surveyor (a NASA/GSFC mission)
GANDER ...... Global Altimeter Network Designed to Evaluate Risk (an SSTL, UK constellation planned to be launched in 2002)++++
GANE .... GPS Attitude Navigation Experiment (NASA Shuttle payload)
GARP .... Global Atmospheric Research Program (of WMO, since 1968)
GAS .... Get-Away Special (Shuttle canisters)
GATE .... GARP Atlantic Tropical Experiment (campaign)
GAUSS .... Galaktische Ultraweitwinkel Schmidt System, Shuttle payload (Galactic super wide angle Schmidt system)
GAW .... Global Atmosphere Watch (WMO)
GBA .... GAS Bridge Assembly (Shuttle payload)
GBRN .... Global Baseline Radiation Network (WCRP)
GC .... Gas Chromatograph
GCIP .... GEWEX Continental-Scale International Project
GCM .... General Circulation Model (atmospheric and climate research)
GCMD .... Global Change Master Directory (at NASA/GSFC since 1989)
GCOM .... Global Change Observation Mission (NASDA)
GCOS .... Global Climate Observing System (of WMO, IOC, UNEP, and ICSU)
GCOS/JSTC .... GCOS/Joint Scientific and Technical Committee (Geneva, Switzerland)
GCP .... Glow Cryoph Payload (DoD Shuttle payload)
GCTE .... Global Change and Terrestrial Ecosystem (IGBP core program)
GE .... General Electric Co., Fairchild, CT, USA
Ge .... Germanium (detector material)
GeGa .... Germanium Gallium (detector)
GEIA .... Global Emissions Inventory Activity (IGBP/IGAC focus 6 activity)
GEMINI .... NASA program of the 1960s
GEMINUS ..... Galileo European Multimodal Integrated Navigation User Service
GEMS .... Global Environment Monitoring System (of UNEP)
GENESIS ...... Galileo European Network of Experts to Support the European Commission

GEO .... Geostationary Earth Orbit (or geosynchronous orbit with zero inclination, the altitude is about 35,786 km)
GeoLITE .... Geosynchronous Lightweight Technology Experiment (NRO funded communications satellite)
GEO-IK .... Russian S/C for solid Earth research, E.5
GEOKhl RAN .. Vernadskiy Institute for Geochemistry and Analytical Chemistry of RAN, Moscow; since 1947, participation in programs: Luna, Venera, Salyut, MIR, Vega, Phobos, Voyager, Magellan, Mars Observer
GEOMAR ..... Research Center for Marine Geosciences (U. of Kiel, Germany)
GEOS .... Geostationary Satellite (ESA experimental program) E.6
GEOS .... Geodetic Earth Orbiting Satellite, E.7.1, E.7.2
GEOS-3 .... Geodynamics Experimental Ocean Satellite, E.7.3 (GEOS-3 is the first radar altimeter mission, end of mission in 1978)
GEOS&R ...... Geostationary Search and Rescue (system)
GEOSAT .... US Navy satellite (altimeter mission), E.8
GEOTAIL ...... Japanese (ISAS) mission to study the structure and dynamics of the geomagnetic tail (part of ISTP), K.13
GEOWARN .... Global Emergency Observation Warning and Relief Network (in planning phase by NASA/MSFC, etc.)
GER .... Geophysical & Environmental Research Corp. (Millbrook, NY, USA)
GEWEX .... Global Energy and Water Cycle Experiment (WMO program)
GFLOPS .... Billion Floating Point Operations per Second (10^9 - a measure of computer processing power)
GFO-1 .... Geosat Follow-On (Satellite), E.9
GFU .... Geophysical Institute of the Academy of Sciences of the Czech Republic, Prague
GFZ .... GeoForschungsZentrum (Potsdam, Germany, since 1992)
GGS .... Global Geospace Science (US program within ISTP with two spacecraft: Wind and Polar)
GGSE .... Gravity Gradient Stabilization Experiment. A technology satellite series (GGSE-1 to GGSE-5) launched by the US military (NRL of DoD) from Vandenberg AFB aboard Thor Agena-D rockets. GGSE-1 (39 kg mass): launch Jan. 11, 1964 into a 900 km altitude orbit with an inclination of 69.9°; GGSE-2 and GGSE-3 (each S/C of 4 kg mass): launch March 9, 1965; GGSE-4 and -5 (each S/C of 4 kg mass): launch May 31, 1967
GGTS-1 .... Gravity Gradient Test Satellite-1 (of the USAF was launched June 16, 1966 from Cape Canaveral)
GHCC .... Global Hydrology and Climate Center (at NASA/MSFC, Huntsville)
GHCD .... Growth Hormone Crystal Distribution (Shuttle experiment)
GIAC .... GPS Interagency Advisory Council
GIF .... Graphics Interchange Format of Compuserve (8-bit color format, used in HTML, etc.)
GIM .... Global Integration and Modeling (IGBP/IGAC focus 6 activity)
GIMEX .... Greenland Ice Margin Experiment (campaign)
GIPME .... Global Investigation of Pollution in the Marine Environment
GIS .... Geographic Information System (an archive in particular for forestry data)
GISP .... Greenland Ice Sheet Project
GISS .... Goddard Institute for Space Studies (New York, NY, since 1961 - a NASA/GSFC facility at Columbia University)
GKNPT Khrunichev Moscow; Leading company in the development, production, testing, and operation of launch vehicles and spacecraft, utilization of Proton. Participation in programs: Venera, Mars, Luna, Kosmos, Phobos, Vega, Gorizont, Salyut, MIR, Almaz, Energia-Buran, Zond, etc.
GKSS .... Gesellschaft für Kernergieverwertung in Schiffbau und Schiffahrt (Geesthacht, Germany)

GLAS .... Geoscience Laser Altimeter System (previously GLRS)
GLIS .... Global Land Information System (an online land data directory guide, a public information system operated by USGS at EROS Data Center)
Glavkosmos ..... Russian space organization agency with the objective to develop the commercial side of space activities (created in 1985)
GLO .... Glow Experiment (Shuttle payload)
GLOBE .... Global Backscatter Experiment (campaign)
GLOBSAT ..... Proposed Earth Observation Satellite by the French Earth Science Community.
GLOCARB ..... Global Tropospheric Carbon Dioxide Network (IGBP/IGAC program)
GLOCHEM .... Global Atmospheric Chemistry Survey (IGBP/IGAC program)
GLOMR .... Global Low Orting Message Relay (DARPA S/C flown on STS-61A)
GLONASS ..... Global Orbiting and Navigation Satellite System (USSR), H.3
GLONET ...... Global Tropospheric Ozone Network (IGBP/IGAC program)
GLOSS .... Global Sea Level Observing System (of IOC)
GLRS .... Geoscience Laser Ranging System (EOS Sensor), renamed in 1992 GLAS = Geoscience Laser Altimeter System
GMES .... Global Monitoring for Environment and Security (European initiative)
GMS .... Geostationary Meteorological Satellite, Operational Program of JMA (Japan Meteorological Agency), F.3
GMSK .... Gaussian Minimum Shift Keying (modulation technique)
GN&C .... Guidance Navigation and Control
GNSS .... Global Navigation Satellite System (a future civil satellite navigation system)
GOALS .... Global Ocean-Atmosphere-Land System (CLIVAR subprogram)
GOBEX .... Gotland Basin Experiment (campaign)
GOCE .... Gravity Field and Steady-State Ocean Circulation Experiment (core mission in ESA's Earth Explorer Program)
GOES .... Geostationary Operational Environmental Satellite (NOAA Series), F.4
GOFS .... Global Ocean Flux Study (program)
GOMS .... Geostationary Operational Environmental Satellite (Russian geostationary meteorological satellite series (at longitude 76 deg. East), F.5
GOOS .... Global Ocean Observing System [a joint program of the Intergovernmental Oceanographic Organization, WMO (World Meteorological Organization), UNEP (United Nations Environmental Program), and the International Council for Science]. GOOS integrates real-time in-situ and satellite observations with numerical model to form model-based information products for a variety of applications.
GORC .... Global Ocean Carbon Research Program
GORS .... General Organization of Remote Sensing (since 1986, Damascus, Syria), Space Agency of Syria
GOS .... Global Observing System (WWW)
GOSAMR-1 .... Gelatin of Sols: Applied Microgravity Research-1 (Shuttle experiment)
GOSIP .... Government Open System Interconnection Profile (US Government Standard, GOSIP is a subset of OSI)
GOSNIIAS ..... State Research Institute of Aviation Systems (Moscow, Russia)
GP-B .... Gravity Probe-B Relativity Mission (E.12)
GPCC .... Global Precipitation Climatology Center, (since 1988, located at the German Weather Service (DWD) in Offenbach, Germany, collection of raingauge-measured monthly precipitation data, worldwide)
GPCP .... Global Precipitation Climatology Project (by ICSU and WMO)
GPS .... Global Positioning System, H.4
GPS DTO ...... GPS Development Test Objective (Shuttle payload)
GRACE .... Gravity Recovery and Climate Experiment

GRAS .... Ground Regional Augmentation Ssytem (within the framework of EGNOS)
GRDC .... Global Runoff Data Center (Bundesanstalt für Gewässerkunde - Federal Institute of Hydrology, Koblenz, Germany)
GRGS .... Groupe de Recherches de Géodésie Spatiale (Grasse and Toulouse, France)
GRID .... (UNEP) Global Resources Information Database (at EDC) for the purpose of analyzing environmental data
GRIP .... Greenland Icecore Project
GRSS .... Geoscience and Remote Sensing Society
GSC .... Geological Survey of Canada
GSD .... Ground-Sampling Distance (spatial resolution).
GSFC .... Goddard Space Flight Center in Greenbelt, MD (DAAC of NASA EOS Program)
GSI .... Geological Survey Institute (Japan)
GSLV .... Geosynchronous Satellite Launch Vehicle (a three-stage ISRO launcher, since 1999, of PSLV heritage)
GSM .... Global System for Mobiles (digital cellular standard of ETSI)
GSOC .... German Space Operations Center (DLR facility in Oberpfaffenhofen)
GSTDN .... Ground-Station Tracking and Data Network (old NASA network)
G/T .... (receiver) Gain / (noise) Temperature
GTCP .... Global Tropospheric Chemistry Program (NSF program)
GTE .... Global Tropospheric Experiment (a NASA program)
GTE/CITE ..... Global Tropospheric Experiment/Chemical Instrumentation Test and Evaluation (campaigns)
GTO .... Geosynchronous Transfer Orbit
GTOS .... Global Terrestrial Observing System (WMO, UNESCO, IOC, FAO, ICSU)
GTS .... Global Telecommunications System (of the World Meteorological Organization (WMO))
GULFEX ...... Gulf Experiment (campaign)

H

H_2O .... Water
H_2O_2 .... Hydrogen peroxide
HALE .... High Altitude Long Endurance (campaign)
HAPEX .... Hydrologic and Atmospheric Pilot Experiment (campaign)
HaRP .... Hawaiian Rainbow Project (campaign)
HBr .... Hydrogen bromide
HBT .... Heflex Bioengineering Test (Shuttle)
HCMM .... Heat Capacity Mapping Mission (NASA sensor), A.18
HCHO .... (CH_2O) Formaldehyde
HCl .... Hydrogen chloride
HCT .... HgCdTe (detector type, see also MCT)
HDDT .... High Density Digital Tape
HDP .... Human Dimensions Programme (of ISSC)
HDT .... High Density Tape
HDLC .... High-Level Data Link Control (bit-oriented protocol)
HEB .... Hot Electron Bolometer (receiver type used in microwave spectrometers, etc.)
HELCOM ...... Helsinki Commission (since 1974, an intergovernmental organization of all countries surrounding the Baltic Sea to protect the Baltic Sea)
HELIOS-1 ...... A European military reconnaissance satellite program (Earth observation) sponsored by France (78.9%), Italy (14.1%) and Spain (7%). Helios-1A was launched July 7, 1995. Helios-1B was launched Dec. 3, 1999 on an Ariane 4 vehicle from Kourou. Both satellites were built by MMS

of Toulouse. Helios-1B, nearly an identical twin of Helios-1A, has a launch mass of 2544 kg (design life of 5 years, power = 2.2 kW). The Helios S/C bus is almost identical to the SPOT-4 platform. Attitude is measured by star sensors and two-axis gyros, actuators are reaction wheels and magnetic torquers. Both S/C are in a sun-synchronous orbit (altitude = 680 km, inclination =98°, period = 98 minutes), 180° apart to optimize coverage. The optical imaging system is referred to as EPV (Ensemble de Prise de Vues), built by Alcatel Space, it uses CCD line array detectors and provides a spatial resolution of about 1 m. Onboard storage is provided by two digital tape recorders for each S/C, each with a capacity of 120 Gbit. Helios-1B has in addition a solid state memory of 9 Gbit. All imagery is encrypted and downlinked in X-band at 50 Mbit/s (TT&C encrypted in S-band at 2 kbit/s). CNES provides S/C operations from Toulouse. The Helios ground segment comprises three user centers at Creil (Italy), Madrid (Spain), and CPFH (Main Helios Center France). Imagery is received at ground stations of the three partner countries [Maspalomas (Spain), Colmar (France), and Lecce (Italy)]. [2109)]

HELSTF .... High Energy Laser Systems Test Facility [a US DoD national test facility at WSMR (White Sands Missile Range), NM, supporting laser research, development, test and evaluation. HELSTF was established in 1985 as a tri-service test and evaluation facility for all high energy laser work. MIRACL (Mid-Infrared Advanced Chemical Laser) is located at WSMR]

HEMT .... High Electron Mobility Transistor (receiver type for microwave spectrometers)

HEO .... Highly Elliptical Orbit

HERCULES .... Hand-held, Earth-oriented, Real-time, Cooperative, User-friendly, Location-targeting and Environmental System (Shuttle experiment)

HES .... Hitchhiker Ejection System. HES provides a capability to eject a payload from a GAS (Get Away Special) canister on Shuttle.

HESS .... High-Latitude Ecosystems as Sources and Sinks of Trace Gases (IGBP/IGAC)

HETE .... High Energy Transient Experiment (MIT payload, built by AeroAstro of Herndon, VA)

HF .... High Frequency (3 - 30 MHz band)

HF .... Hydrogen fluoride

HgCdTe .... Mercury Cadmium Telluride (mercadtelluride, a detector material) also referred to as MCT and HCT

HgI_2 .... Mercury Iodine (a detector material)

HGF........... Hermann von Helmholtz-Gemeinschaft Deutscher Forschungszentren, Bonn (named after Hermann von Helmholtz, 1821 - 1894). Sixteen German research centers are members of HGF, an association with the objective to coordinate and foster interdisciplinary research, to share expensive technical equipment of their infrastructure, to cooperate on long-term system solutions, and to transfer new technology for industrial applications. All HGF centers are government-funded, they employ a total of about 23,000 persons with a budget of 3.6 billion DM in 1996. The following institutions are members of HGF:
AWI (Alfred-Wegener-Institut für Polar- und Meeresforschung, since 1980, Bremerhaven and Potsdam)
DESY (Deutsches Elektronen Synchrotron, Hamburg, since 1959)
DKFZ (Deutsches Krebsforschungszentrum, Heidelberg, since 1964)
DLR (Deutsche Forschungsanstalt für Luft- und Raumfahrt)
FZK (Forschungszentrum Karlsruhe)
GBF (Gesellschaft für Biotechnologische Forschung, Braunschweig)

2109) "Helios, Europe's eye in the sky," CNES Magazine, No 7, Nov. 1999

GFZ (GeoForschungsZentrum Potsdam, since 1992)
GKSS (Gesellschaft für Kernergieverwertung in Schiffbau und Schiffahrt, Geesthacht)
GMD (Gesellschaft für Mathematik und Datenverarbeitung, since 1968, German National Research Center of Information Technology, St. Augustin, and Darmstadt)
GSF (Forschungszentrum für Umwelt und Gesundheit, Neuherberg)
GSI (Gesellschaft für Schwerionenforschung, Darmstadt)
HMI (Hahn-Meitner-Institut, Berlin)
IPP (Max-Planck-Institut für Plasmaphysik, Garching)
KFA (Forschungszentrum Jülich)
MDC (Max-Delbrück-Zentrum für Molekulare Medizin, Berlin)
UFZ (Umweltforschungszentrum Leipzig-Halle)

HH .... Hichthiker (a Shuttle flight carrier system offered by NASA for small payloads, offering the provision of extended functional features) HH-S stands for 'sidewall mounting,' HH-C stands for 'cross bay mounting'

HH .... Horizontal transmit - Horizontal receive polarization

HIP .... Heterojunction Internal Photoemission (detector technology)

HNO_3 .... Nitric acid

HO_x (HOx) ..... Odd hydrogen (OH, HO_2, H_2O_2)

HOCl .... Hypochlorous acid

HOLOP .... Holographic Optics Laboratory (Shuttle D2 mission)

HOST .... HST Orbital Systems Test Platform (Shuttle payload)

HPCG .... Hand-held Protein Crystal Growth (Shuttle payload)

HPP .... Heat Pipe Performance (Shuttle experiment)

HPT .... Heterojunction Phototransistor (optoelectronic component which combines both optical detection and electrical gain in a single element)

HPTE .... High Precision Tracking Experiment (Shuttle payload)

HRPT .... High Resolution Picture Transmission (NOAA broadcast technique in S-band at frequencies of 1698.0 and 1707.0 MHz; data from all AVHRR channels (plus TOVS and SEM) is provided at full 1.1 km resolution)

HRSGS-A ...... High Resolution Shuttle Glow Spectroscopy (Shuttle payload)

HRTS .... High Resolution Telescope and Spectrograph (Shuttle, Spacelab-2 sr, a 30 cm, f/15 Gregorian telescope, spectrograph in UV range 1170-1700 Å, and a spectroheliograph observing at 1550 Å)

HSC .... Hughes Space & Communications Company (since 1961), an operating unit of Hughes Electronics Corporation, Los Angeles, CA. HSC is a manufacturer (world leader) of communication satellites (over 40% of market share). Provider of several standard platforms like HS 376 for spin-stabilized satellites, the HS 601 series is body-stabilized; in 1995 HSC introduced the body-stabilized HS 702 platform. Manufacturer of Syncom (first communications satellite, launch 1963), ATS-1 (first GEO weather satellite, launch in 1966), Pioneer (Venus Probe, 1978), Galileo (Jupiter Probe, launch 1989). Military satellite builder. - In January 2000, the HSC along with subsidiaries Hughes Electron Dynamics and Spectrolab were sold to the Boeing Company. They were reorganized into a business unit called "Boeing Satellite Systems."

HSCT .... High-Speed Civil Transport (USA)

HST .... Hubble Space Telescope (Shuttle launch)

HSRP .... High-Speed Research Program (NASA)

HTML .... HyperText Markup Language

HTS .... High-(Tc) Temperature Superconductivity, refers to material temperature Tc levels above those of liquid helium [the technology is employed in sensor design, thin-film applications, MRI (Magnetic Resonance Imaging), wireless communication filters, and ultra-fast computer chips]

HTSQUID ..... High-(Tc) Temperature SQUID (Superconducting Quantum Interference Device)
HTTP .... HyperText Transfer Protocol
Hughes .... Hughes Electronics Corporation, a worldwide operating company with HQ in Los Angeles, CA (a wholly owned subsidiary of General Motors Corporation founded in 1985). The conglomerate consists of: Hughes Aircraft Company, Hughes Telecommunications & Space (largest manufacturer in the world of telecommunication satellites), Hughes Network Systems, DIRECTV Inc., and Delco Electronics Corporation. HSC is part of Hughes Telecommunications & Space. In 2000, Hughes Electronics Corporation sold its satellite manufacturing business to Boeing Company.
Hughes (HAC) .. Hughes Aircraft Company, (since 1932, founded by Howard Hughes), part of Hughes Electronics Corporation, with HQ in Arlington, VA, a technology company with three major operating units: Information Systems (Reston, VA), Sensor & Communications Systems, and Weapons Systems. SBRC (as of 1996 SBRS, builder of Landsat instruments, MSS, TM, monolithic infrared focal plane arrays, etc.) is part of Sensor & Communications Systems
HUT .... Helsinki University of Technology (Helsinki, Finland)
HUT .... Hopkins Ultraviolet Telescope (part of Shuttle ASTRO observatory)
HV .... Horizontal transmit - Vertical receive polarization
HYDROMET .. Committee for Hydrometeorology (USSR/CIS agency in the field of Meteorology)
HypSEO .... HyperSpectral Earth Observer (an ASI mission in preparation, planned flight in 2003)
HWRP .... Hydrology and Water Resources Programme (WMO)

I

IAA .... International Academy of Astronautics (Paris, France)
IAE .... Inflatable Antenna Experiment (Shuttle)
IADC .... Inter-Agency Space Debris Coordination Committee
IAF .... International Astronautical Federation (Paris)
IAFE .... Institute of Astronomy and Space Physics (Argentina)
IAG .... International Association of Geodesy
IAHS .... International Association of Hydrological Sciences
IAI .... Israel Aircraft Industries (government-owned company, of Lod, Israel)
IAI/MBT .... IAI/Mifal Beth. MBT stands for the Hebrew translation of MIFAL BETH (or its abreviation of MABAT) which means 'the second plant,' since it was the second plant established by IAI in the 1960s. The Hebrew name of MABAT remained with the corresponding English acronym of MBT.
IAIN .... International Association of Institutes of Navigation (since 1975)
IAMAP .... International Association of Meteorology and Atmospheric Physics
IAMAS .... International Association of Meteorology and Atmospheric Sciences
IAP .... Institute of Atmospheric Physics, Moscow
IAPSO .... International Association for the Physical Sciences of the Oceans (one of seven associations of IUGG, which in turn is a union of ICSU)
IAS .... Institut d'Astrophysique Spatiale (Verrières-le-Buisson, France, lab is part of CNRS)
IASC .... International Arctic Science Committee (Arctic Centre, University of Lapland, Finland)
IASIS .... Interbranch Association Sovinformsputnik (commercial distributor of imagery from Russian defense satellites, Moscow)
IAU .... International Astronomical Union
IBAMA .... Instituto Brasileiro do Meio Ambientes Dos Recursos Naturais Renovaveis (Brazil)

IBC .... Impurity Band Conduction (detector technology)
IBFRA .... International Boreal Forest Research Association (since 1991)
IBFRA-SRF .... IBFRA - Stand Replacement Fire (working group)
IBSE .... Initial Blood Storage Experiment (Shuttle payload)
IBSFC .... International Baltic Sea Fishery Commission
IBSS .... Infrared Background Signature Survey (satellite of the USAF deployed on STS-39) IBSS was retrieved by the Shuttle on May 2, 1991.
ICA .... International Cartographic Association
ICAE .... International Conference on Atmospheric Electricity
ICAO .... International Civil Aviation Organization
ICAT .... Incubator-Cell Attachment Test (Shuttle)
ICBC .... IMAX Cargo Bay Camera (Shuttle), a 65 mm color motion picture camera
ICBM .... Intercontinental Ballistic Missile. Russia offers commercially four types of converted ICBMs for satellite launches. The types "Rockot" and "Strela" are based on the SS-19 Stiletto missile; "Dnepr" is based on the SS-18 Satan missile; "Start" is a converted SS-20 missile. The Rockot launch vehicle Rockot is a joint venture of Eurockot Launch Services GmbH, Bremen, Germany and of KhSC (Khrunichev State Research and Production Space Center), Moscow. ISC Kosmotras of Moscow markets the Dnepr vehicle. The Start (Start-1) vehicle is marketed by Puscovie Uslugi of Moscow.
ICC .... Instrument Control Center (EOSDIS Facility)
ICC .... Integrated Cargo Carrier (Shuttle payload, first flown on STS-96). ICC is an unpressurized flat bed pallet and keel yoke assembly. Constructed of aluminum, it is 2.5 m long, 4.5 m wide and 25 cm thick and has the capability to carry cargo (up to about 1350 kg) on both faces of the pallet, both atop and below. The ICC is used by astronauts throughout the construction of the Space Station as it transports hardware from locations on the station's exterior to work sites on the truss assemblies.
ICE .... International Cirrus Experiment (campaign)
ICE .... International Cometary Explorer (renamed ISEE-3 mission), K.18.2
ICES .... International Council for the Exploration of the Sea
ICET .... International Center for Earth Tides
ICIC .... Intercalibrations/Intercomparisons (IGBP/IGAC focus 7 activity)
ICSU .... International Council of Scientific Unions (HQs in Paris, France. ICSU is a non-governmental body created in 1931 to promote international science and its applications. It has a membership of international organizations (Scientific Unions), national science academies and research councils, and Scientific Associates. Some committees of ICSU are: IGBP, SCOPE, SCAR, COSPAR, etc.)
ICWG-EO ...... International Coordination Working Group for Earth Observation
IDA .... Institute of Defense Analysis (since 1957, a DoD nonprofit corporation)
IDEAL .... International Decade of of East African Lakes (campaign)
IDHT .... Instrument Data Handling and Transmission (ERS-1 S-band antenna)
IDN .... International Directory Network (CEOS-defined for databases, former designation 'PID')
ie .... abbreviation (Latin: id est) that is
IECM .... Induced Environment Contamination Monitor (Shuttle)
IEE .... Institution of Electrical Engineers (London, UK)
IEEE .... Institute of Electrical and Electronics Engineers (USA)
IEF .... Isoelectric Focusing (Shuttle payload)
IEH .... International EUV Hitchhiker (Shuttle payload)
IEICE .... Institute of Electronics, Information and Communication Engineers, Tokyo, Japan
IELV .... Intermediate Expendable Launch Vehicle (EOS program)

IEOS .... International Earth Observing System (Committee dealing with the policies, principles of data exchange, etc.; partner agencies are: CSA (Canada), ESA (Europe), NASA (USA), and STA (Japan). Delegations from agencies with operational environmental monitoring satellites: NASDA, MITI, JMA (Japan), EUMETSAT (Europe), NOAA (USA), AES (Canada). Typical IEOS missions are: ENVISAT (ESA), EOS/AM-1 (NASA), NOAA-N (NOAA), ADEOS (NASDA), and TRMM (NASA/NASDA).

IERS .... International Earth Rotation Service (Central Bureau in Paris, since 1988)

IERS .... International Earth Reference System

I/F .... Interface

IF .... Intermediate Frequency

IFAG .... Institut für Angewandte Geodäsie [Institute of Applied Geodesy - a federal agency under the jurisdiction of the German Ministry of the Interior (BMI) with research in the fields of geodesy, cartography and photogrammetry]. IFAG maintains a central office in Frankfurt/Main and branch offices in Leipzig, Potsdam, and Berlin. Note: In the late 1990s, IFAG was renamed to **BKG** (Bundesamt für Kartographie und Geodäsie.

IFARS .... Institute for Applied Remote Sensing (Wedel, Germany)

IFE .... Isoelectric Focusing Experiment (Shuttle payload)

IFEOS .... International Forum on Earth Observations Using Space Station Elements (since 1986)

IFOV .... Instantaneous Field of View

IFREMER ..... Institut Francais de Recherche pour L'Exploration de la Mer (French Ocean Agency in Brest, France). IFREMER/CERSAT is a processing and archiving facility for satellite data and is part of the "Département d'Océanographie Spatiale" at IFREMER.

IFSAR .... Interferometric SAR (measurement technique using two antennas, sometimes also referred to as 'InSAR')

IFTI .... Ioffe Physical Technical Institute (St. Petersburg)

IGAC .... International Global Atmosphere Chemistry (IGBP core program)

IGAP .... International Global Programme on Atmospheric Particles

IGARSS .... International Geoscience and Remote Sensing Symposium - since 1981, sponsored by GRSS (Geoscience and Remote Sensing Society)

IGBP .... International Geosphere-Biosphere Programme of ICSU (IGBP is closely linked, directly or through ICSU, to other international organizations involved in global change research, including: GCOS, IOC, IPCC, ISSC, SCOPE, UNEP, WCRP, WMO. Over 50 countries have national IGBP committees and supporting bodies. The IGBP Secretariat is in Stockholm, Sweden)

IGEB .... Interagency GPS Executive Board [IGEB (Presidential Decision Directive as of March 1996) offers some formal civil agency participation in the GPS program. It is jointly chaired by the DoD and DoT, with oversight and management of the dual use component of the GPS]

IGEX .... International GLONASS Experiment, a campaign under the auspices of IAG (International Association of Geodesy)

IGFOV .... Instantaneous Geometric Field of View

IGN .... Institut Géographique National (French National Geographic Institute, Paris)

IGOS .... Integrated Global Observing Strategy (for synergetic effects)

IGRF .... International Geomagnetic Reference Field

IGS .... International GPS Service for Geodynamics (since 1993)

IGSO .... Inclined Geosynchronous Orbit

IGU .... International Geographical Union

IGY .... International Geophysical Year [created in 1952 by the ICSU plenary meeting; the first IGY was planned for 1957/58 (a year of expected maximum solar activity), it coincided also with the start of the space age, the launch Sputnik-1 on Oct. 4, 1957]
IHP .... International Hydrology Programme (UNESCO)
IHO .... International Hydrographic Organization
IIP .... International Ice Patrol
IJSSE .... International Journal of Small Satellite Engineering (electronic journal on internet, edited at the University of Surrey, UK)
IKF............ Institut für Kosmosforschung, Berlin-Adlershof, in former East Germany. Note: as of Jan. 1992 the IKF was renamed 'Institute of Space Sensor Technology (ISST),' it is part of DLR)
IKI RAN .... Space Research Institute (of the Russian Academy of Sciences, RAN (or RAS, depending on the alphabet), Moscow, Russia; extraterrestrial physics and remote sensing, since 1965)
IKI-BAN .... Space Research Institute, Bulgarian Academy of Sciences (Sofia, Bulgaria)
ILS .... Instrument Landing System
ILS .... International Launch Services [a joint commercial venture between Lockheed Martin Corp. (USA), Khrunichev Space Center (KhSC) and RKK Energia (Russia), offering of Atlas and Proton launch systems. The first ILS launch occurred in Sept. 1996 (Inmarsat-3 from Baikonur); since April 15, 1993 all commercial contracts, involving the Proton launch vehicle, are handled by ILS.
IMAGES .... International Marine Global Change Study (IGBP project)
IMAU .... Institute for Marine and Atmospheric Research Utrecht (University of Utrecht, The Netherlands)
IMAX .... Image Maximum (a large screen motion picture camera/format used by the NASA/Smithsonian project to document significant space activities)
IMEC.......... Inter-university MicroElectronics Center, Leuven, Belgium. IMEC is a Flemish government initiative to bundle all microelectronics-related efforts of the three scientific universities into one independent nonprofit super-lab.
IMET.......... Improved Meteorological Instrumentation (WHOI buoy type)
IMEX .... Inner Magnetosphere Explorer, a mission of UMM (University of Minnesota at Minneapolis)
IMF .... Interplanetary Magnetic Field
IMK........... Institute für Meteorologie und Klimaforschung (Institute for Meteorology and Climate Research - a cooperative institute of the Nuclear Research Center Karlsruhe (KfK) and of the University of Karlsruhe, Germany)
IML .... International Microgravity Laboratory (Shuttle payload)
IMO........... International Maritime Organization
IMP .... International Monitoring Platform, K.16
IMS .... Information Management System at GSFC (The top-level function of EOS DAACs)
IMTA.......... Instituto Mexicano de Tecnologica del Agua (Cuernavaca, Mexico)
IMU........... Inertial Measurement Unit (navigation instrument on aircraft)
INCA.......... Indian National Cartographic Association
INDEX .... Indian Ocean Experiment (campaign)
INDEX .... Innovative Technology Demonstration Experiment (of ISAS, Japan)
INDOEX....... Indian Ocean Experiment (campaign)
INDREX....... Indonesian Radar Experiment (campaign)
INFN .... Istituto Nazionale Fisica Nucleare (Italian National Institute of Nuclear Physics), Rome, Italy
ING .... Istituto Nazionale di Geofisica (Rome Italy)

InAs .... Indium Arsenide (detector type for IR spectrum)
InGaAs .... Indium Gallium Arsenide (a detector type for IR spectrum)
InGaP/GaAs .... Indium Gallium Phosphorus/Gallium Arsenide (solar cell type)
INM .... Instituto Nacional de Meteorologica (Spanish Weather Service)
Inmarsat .... International Maritime Satellite Organization (London, UK)
InP .... Indium Phosphide (solar cell type)
INPE .... Instituto de Pesquisas Espaciais (National Institute of Space Research, Sao José dos Campos, S.P., Brazil, since 1971)
INQUA .... International Union for Quaternary Research (of ICSU)
INR .... Image Navigation and Registration (GOES Second Generation S/C)
INRA .... Institut National de la Recherche Agronomique (Grignon and Montfavet, France)
In-RIMT .... Indian Resources Information and Management Technologies Pvt. Ltd, Hyderabad, India
INS .... Inertial Navigation System (for aircraft navigation)
INS .... Institute of Nuclear Physics, (New Zealand)
INSA .... Ingenieria y Servicios Aeroespaciales, Madrid, Spain (Fuego mission coordinator)
INSAT .... Indian National Satellite (series, employed for meteorology and communication), F.6
IN-STEP .... In-Space Technology Experiments Program (NASA)
INSU .... Institut National des Sciences de l'Univers (Paris, part of CNRS)
InSb .... Indium antimonide (detector type material for infrared region)
INTA .... Instituto National de Técnica Aeroespacial (Space Agency of Spain)
Intelsat .... International Telecommunications Satellite Organization (Washington, DC)
INTERBALL ... IKI mission program (solar-terrestrial interaction) within ISTP, K.17
Intercosmos ..... USSR/CIS space program for collaborative science projects among its nine members and with other nations. Intercosmos was created in 1967 inviting the former Soviet-affiliated countries (like, East-Germany, Hungary, Bulgaria, Poland, etc.) to participate in the Soviet space program with their own national contributions (one area of participation was in remote sensing, building sensors for specific missions, dissemination and scientific interpretation of data, etc.). Activities in international manned space flight missions were also under the label of Intercosmos. Satellites in the Intercosmos program are named 'Intercosmos-n', like Intercosmos-19 (launched Feb. 27, 1979).
IOC .... Initial Operating Capability (GPS, GLONASS,)
IOC .... Intergovernmental Oceanographic Commission (of UNESCO)
IOCM .... Interim Operational Contamination Monitor (Shuttle payload)
ION .... Institute of Navigation (Washington, DC, since 1945)
IOP .... Intensive Observation Period (within a campaign)
IOS .... Institute for Ocean Sciences (Sydney, British Columbia, Canada)
IOW .... Institut für Ostseeforschung Warnemünde (Institute for Baltic Sea Research, Warnemünde, Germany)
IPCC .... Inter-Governmental Panel for Climate Change (set up by WMO and UNEP in 1988)
IPG .... Institute of Applied Geophysics (Moscow, Russia)
IPG-Paris .... Institut de Physique du Globe de Paris
IPMP .... Investigations into Polymer Membrane Processing (Shuttle experiment)
IPO .... Integrated Program Office (Silver Spring, MD), consisting of a team made up of NOAA, NASA and DoD representatives for the development of the NPOESS spacecraft series
IPOMS .... International Polar-Orbiting Meteorological Satellite
IPS .... Instrument Pointing System (Spacelab-2, built by ESA, structure for mounting telescopes)

IPS .... Ion Propulsion System
IRCFE .... Infrared Communications Flight Experiment (Shuttle)
IR&D .... Independent Research & Development (company internal funding)
IRD .... Institut de Recherche pour le Développement (Paris, France, successor organization to ORSTOM)
IRE RAN ...... Institute of Radioengineering and Electronics (of the Russian Academy of Sciences, RAN, in Moscow; founded in 1953, IRE is involved in remote sensing, etc., also providing general management services)
IRF .... Swedish Institute of Space Physics [(Institutet för rymdfysik), a governmental research institute with the following divisions: IRF-K (Kiruna), IRF-Um (Umea) with a Laboratory of Mechanical Waves and a Space Physics Group at Umea University, IRF-U (Upsalla), IRF-STL (Solar Terrestrial Physics) Lund Division]
IR-IE .... Infrared Imaging Experiment (Shuttle payload)
IRIS .... International Radio Interferometric Surveying (Subcommittee of the International Association of Geodesy)
IRIS .... Italian Research Interim Stage (upper stage used in conjunction with NASA's Shuttle to place payloads up to 900 kg into geo-transfer orbit)
IRLS .... Interrogation, Recording and Location Subsystem (French-US Eole experiment flown on Nimbus-3 in 1969)
IRM .... Ion Release Module (S/C of the AMPTE mission, K.4.1)
IRMB .... Institut Royal de Météorologie Belgique (Royal Meteorological Institute of Belgium, Brussels) also referred to as KMI/IRM
IROE - CNR .... Istituto Ricerca Onde Elettromagnetiche - Consiglio Nazionale delle Ricerche (Florence, Italy)
IRS .... Information Retrieval System (ESA data system)
IRS .... Indian Remote Sensing Satellites (ISRO), D.17 (IRS-1A, 1B, 1C, 1D, 1E, etc.)
IRS .... Inertial Reference System
IRS .... Institut für Raumflugsysteme (University of Stuttgart, Germany)
IRSA .... Institute for Remote Sensing Applications (of JRC, Ispra, Italy. In 1996 IRSA was renamed to SAI = Space Applications Institute)
IRSA .... Institute for Remote Sensing Applications, since 1980 (Beijing, Chinese Academy of Sciences)
IRSC .... Iranian Remote Secing Center, Tehran, Iran (funded by the Ministry of Posts and Telecommunications)
IRT .... Infrared Telescope (Spacelab-2 instrument, a 15 cm f/4 Herschelian telescope)
IRU .... Inertial Reference Unit
ISA .... Institute of Space Aeronomy (Brussels, Belgium)
ISA .... Israel Space Agency (since 1983 -within the framework of the Ministry of Science and Technology)
ISAC .... Intelsat Solar Array Coupon (Shuttle experiment)
ISAC .... ISRO Satellite Center (Bangalore, India)
ISAIAH .... Israeli Space Agency Investigation about Hornets (Shuttle experiment)
ISAL .... Investigation of STS Atmospheric Luminosities (Shuttle)
ISAS .... Institute for Space and Astronomical Science (University of Tokyo, Japan)
ISCCP .... International Satellite Cloud Climatology Project (by ICSU & WMO)
ISDE (RNII KP) Institute of Space Device Engineering, Moscow; a leading Russian company in the design and development of sensors; participation in programs: Venera, Vega, Phobos, Luna, Mars, Prognoz, Granat, Resurs, Okean, Glonass, etc.
ISDE .... International Symposium on Digital Earth
ISDN .... Integrated Services Digital Network
ISEE .... International Sun Earth Explorer (3 S/C mission), K.18

ISIR .... Infrared Spectral Imaging Radiometer (Shuttle payload)
ISIS .... Intelligent Satellite-Data Information System (a DLR/DFD archival system and service)
ISLR .... Integrated Side Lobe Ratio
ISLSCP .... International Satellite Land-Surface Climatology Project (by ICSU and WMO)
ISN .... Institute of Satellite Navigation at the University of Leeds, UK
ISO .... International Standards Organization (one of three bodies responsible for the definition of OSI)
ISOPS .... International Space Conference of Pacific-Basin Societies
ISRO .... Indian Space Research Organization (HQ at Bangalore, since 1969)
ISRO/IISU ..... ISRO Intertial Systems Unit
ISRO/ISAC ..... ISRO Satellite Center (Bangalore, India)
ISRO/ISTRAC .. ISRO Telemetry, Tracking and Command Network
ISRO/LPSC .... ISRO Liquid Propulsion Systems Center
ISRO/MCF ..... ISRO INSAT Master Control Facility
ISRO/SAC ...... ISRO Space Applications Center (Ahmedabad, India)
ISRO/SHAR .... ISRO Sriharikota Range (ISRO launch site, East Coast of India)
ISRO/VSSC .... ISRO Vikaram Sarabhai Space Center (launch vehicle development)
ISPRS .... International Society for Photogrammetry and Remote Sensing
ISPR .... International Standard Payload Rack (adopted by the ISS program), each ISPR provides 1.6 m^3 of space, the rack has a mass of 104 kg and can accommodate up to 700 kg of payload mass
ISS .... International Space Station
ISSC .... International Social Science Council (UN)
ISSI .... International Space Science Institute, Bern, Switzerland
IST .... Instrument Support Terminal (EOSDIS Facility)
ISTP .... International Solar-Terrestrial Physics Program [involves a total of 12 satellites provided by ESA (SOHO, CLUSTER), NASA [GGS (POLAR, WIND), IMP-8, FAST], IKI (Interball, ECOS-A), ISAS (Geotail)]
ISTRAC .... ISRO Telemetry and Command Center (Bangalore, India)
ISTS .... Institute for Space and Terrestrial Science (North York, Ontario, Canada) Note: A name change to CREST (Center for Research in Earth and Space Technology) took place on Sept. 24, 1997
ISTS .... Institute of Space and Astronautical Science (Tokyo, Japan)
ISY .... International Space Year (1992)
ITAR .... International Traffic in Arms Regulation (US regulations related to the export of satellite and rocket technology)
ITCZ .... Inter Tropical Convergion Zone
ITEX .... Island Thunderstorm Experiment (campaign)
ITIR[2110] .... Intermediate Thermal Infrared Radiation (EOS sensor); ITIR was renamed in 1990 ASTER = Advanced Spaceborne Thermal Emission and Reflection Radiometer
ITO .... Indium Tin Oxide (a light sensitive sensor type)
ITOS .... Improved TIROS Operational System (NOAA S/C)
ITRF .... International Terrestrial Reference Frame (established by IERS)
ITT A/CD ...... ITT (international Telephone and Telegraph Corporation) Aerospace/ Communications Division (Fort Wayne, IN), builder of remote sensing instruments (AVHRR, HIRS, GOES-series instruments, etc.). The parent company is ITT Industries Inc., headquartered in New York, NY.
ITU .... International Telecommunication Union (since 1865 founded as International Telegraphy Union, since 1934 as ITU, since 1947 ITU is a UN agency to cover standards for a wide range of telecommunication ser-

2110) "The Earth Observer," Vol. 2 No. 10, Dec. 1990, pp. 3

vices, including frequency allocations standards for fax, ISDN, JPEG, MPEG, ATM, etc., Geneva, Switzerland)

ITU-R .... ITU-Radiocommunication standardization sector (formerly known as CCIR - responsible for managing efficient use of the radio-frequency spectrum)

ITU-T .... ITU-Telecommunication standardization sector (formerly CCITT)

IUGG .... International Union of Geodesy and Geophysics (a union of ICSU)

IVHM .... Integrated Vehicle Health Monitoring (Shuttle payload, terchnology demonstration)

IVHS .... Intelligent Vehicle/Highway Systems

IWF .... Institut für Weltraumforschung, Graz, Austria

IZMIRAN ...... Institute of Terrestrial Magnetism, Ionosphere and Radiowave Propagation (of Russian Academy of Sciences, Troitsk, Moscow region)

J

JAMSTEC ...... Japan Marine Science and Technology Center (Tokyo)

JAROS .... Japan Resources Observation System Organization

JAFIC .... Japan Fisheries Information Center

JCAB .... Japanese Civil Aviation Bureau (JCAB is an agency/organization within the Japanese Ministry of Transport)

JEA .... Japan Environmental Agency

JEM .... Japanese Experiment Module (Japan's pressurized module directly attached to the Space Station Freedom)

JEM-EF .... JEM-External Facility

JEOS .... Japanese Earth Observation System

JERS .... Japanese Earth Resources Satellite, D.18

JFET .... Junction Field-Effect Transistor

JGOFS .... Joint Global Ocean Flux Study (IGBP program)

JGR .... Journal of Geophysical Research (a publication of AGU)

JGPSC .... Japan GPS Council (over 80 manufacturers, major users, research institutes, etc.)

JHU .... Johns Hopkins University (Baltimore, MD, USA)

JHU/APL ...... JHU/Applied Physics Laboratory, Laurel, MD, USA, since 1942; APL is a major space research institute (staff of 2700) and the designer and builder of satellites (Transit series, ACE, AMPTE/CCE, MSX, NEAR, TIMED, etc.), instruments, S/C engineering, technical innovations, etc.

JMA .... Japan Meteorological Agency (JMA is an agency/organization within the Japanese Ministry of Transport)

JODC .... Japan Oceanographic Data Center

JOWIP .... Joint Canada-US Ocean Wave Investigation Project (campaign)

JPEG .... Joint Photographic Experts Group (a compressed image format standard, 24-bit color; note: JPEG is a lossy compression technique based on DCT)

JPEG-LS .... JPEG lossless - use of a 2-D edge-detection predictor. JPEG-LS is the new (1998/9) lossless/near-lossless compression standard for continuous-tone images, ISO-14495-1/ITU-T.87. The standard is based on the LOCO-I algorithm (LOw COmplexity LOssless COmpression for Images) developed at Hewlett-Packard Laboratories.

JPL .... Jet Propulsion Laboratory, Pasadena, CA, since 1944 (DAAC of NASA EOS Program). JPL is the only NASA center that is managed by a university, namely the California Institute of Technology

JPO .... Joint Program Office (GPS)

JPOP .... Japanese Polar Platform

JRC .... Joint Research Centre (umbrella agency of CEU coordinating eight research institutes at five sites (Geel, Belgium; Karlsruhe, Germany; Petten, Netherlands; Ispra, Italy; Seville, Spain). IRMM (Institute for Reference Materials and Measurements) is located in Geel; ITU (Institute of Transuranium Elements) is in Karlsruhe; IAM (Institute of Advanced Materials) is in Petten; IPS (Institute for Prospective Technological Studies) in Seville. The following institutes are located in Ispra: ISIS (Institute for Systems, Informatics and Safety), EI (Environment Institute), SAI (Space Applications Institute), IHCP (Institute for Health and Consumer Protection). - The JRC Program Directorate is located in Brussels.

JSC .... Johnson Space Center (Houston, TX, USA)

JSC .... Joint Scientific Committee (of WCRP)

JST .... Japan Science and Technology Corporation (Tokyo. a Japanese government corporation promoting new technologies and basic research)

JUSREX .... Joint US/Russian Internal Wave Remote Sensing Experiment (campaign)

JWGA .... Joint Working Group ATMOS

K

KACST .... King Abdulaziz City for Science and Technology (Riyadh, Saudi Arabia, since 1977), home of SRISA (Space Research Institute of Saudi Arabia)

KAIST .... Korean Advanced Institute of Science and Technology (Seoul, Korea, since 1981)

KAIST/SaTReC . KAIST/ Satellite Technology Research Center (Taejon, Korea, since 1989, SaTReC is a university based research center)

KARI .... Korea Aerospace Research Institute (Taejon, Korea, since 1989)

KAO .... Kuiper Airborne Observatory (C-141 aircraft of NASA/ARC)

KAPEX .... Cape of Good Hope Experiments (campaign)

KARI .... Korea Aerospace Research Institute, Taejon, Korea

KACST .... King Abdulaziz City for Science & Technology, Riyadh, Saudi Arabia

KEOC .... Korean Earth Observation Center, Seoul, Korea

KfA .... Kernforschungsanlage Jülich (Nuclear Research Center, Jülich, Germany)

KfK .... Kernforschungszentrum Karlsruhe (Nuclear Research Center, Karlsruhe, Germany; KfK was renamed to FZK (Forschungszentrum Karlsruhe as of 1995)

KFKI .... Hungarian Research Institute for Particle and Nuclear Physics

KH .... Keyhole (a code name designating a DoD reconnaissance satellite series as well as the principal camera system of the S/C)

KhSC .... Khrunichev State Research and Production Space Center, Moscow

KidSat .... A NASA-sponsored program (start in 1995, the first Shuttle flight of Kidsat was on STS-76 in March 1996) to encourage the student and educator community in space technology involvement, to bring space exploration into the classrooms. Activities may encounter interpretation of remotely-sensed images, the development of imaging instruments as well as their on-orbit operation. Further Shuttle flights of KidSat on STS-81 (Jan. 12-22, 1997) and on STS-86 (Sept. 25 - Oct. 6, 1997). Access to the program is via Internet. KidSat observation missions are carried out on Space Shuttle flights and on the future Space Station.

KITSAT .... Korea Institute of Technology Satellite (D.19, D.40.6, D.40.10,)

KNMI .... Koninklijk Nederlands Meteorologisch Instituut (Royal Netherlands Meteorological Institute) De Bilt, Netherlands

KOMPSAT ..... Korea Multi-Purpose Satellite, D.20

KSC .... Kennedy Space Center (NASA facility at Cape Canaveral, FL, USA)

KTH .... Kungl Tekniska Högskolan (Royal Institute of Technology) Stockholm, Sweden

L

L3 .... Latitude/Longitude Locator (Shuttle experiment)
LAAS .... Local Area Augmentation System (GPS). LAAS is FAA's ground-based augmentation system for local area DGPS.
LABEN S.p.A. .. Laboratori Elettronici Nucleari, of Vimodrone (Milano, Italy), Lab of Alenia Spazio (a Finmeccanica company). LABEN was founded in 1958, it produces electronic systems, transducers, LAGRANGE (LABEN GNSS Receiver for Advanced Navigation), etc.
LAC .... Local Area Coverage (NOAA downlink mode)
Lacrosse/Vega ... A DoD/NRO radar imaging satellite reconnaissance program. Lacrosse-1 was launched Dec. 2, 1988 by Shuttle (STS-27) and went into a 57° orbit with an altitude of 680 km. Lacrosse-2 was launched from VAFB on March 8, 1991. Lacrosse-3 was launched from VAFB on Oct. 24, 1997. Lacrosse-4 was launched from VAFB on May 22, 1999.
LADAR .... Laser Detection and Ranging
LAEFF .... Laboratorio de Astrofisica Espacial Fisica Fundamental (Villafranca, Spain, Laboratory for Space Astrophysics and Theoretical Physics, since 1990)
LAGEOS-I,II ... Laser Geodynamics Satellite (NASA/ASI), E.15
LAMBADA .... Large-scale Atmospheric Moisture Balance of Amazonia using Data Assimilation (campaign)
LAN .... Local Area Network
LandSat .... Land (Remote Sensing) Satellite, US EO program, D.21
LANL .... Los Alamos National Laboratory (Los Alamos NM, DOE facility, operated by the University of California). Builder of satellites (ALEXIS, FORTE, MTI, etc) and instruments for space research (solar wind, lightning detection). Los Alamos played (and plays) a key role in monitoring treaty compliance with satellite sensors (detecting atmospheric nuclear tests).
LAPAN .... Lembaga Penerbangan dan Antariksa Nasional (Indonesian National Institute of Aeronautics and Space, Jakarta)
LAP-B .... Link Access Protocol (for B Channels)
LaRC .... Langley Research Center (Hampton VA, DAAC of NASA EOS Program)
LASP .... Laboratory for Atmospheric and Space Physics at the University of Colorado, Boulder, CO
LASER .... Light Amplification by Stimulated Emission of Radiation
LASSO .... Laser Synchronization from (Geo)Stationary Orbit (ESA, Meteosat)
LAT .... Laboratoire d'Astrophysique de Toulouse (France)
Lavochkin ...... Lavochkina Scientific Production Association, Khimky, Russia
LBH .... Lyman-Birge-Hopefield (spectral bands in the 140-180 nm range)
LCD .... Liquid Crystal Display (a device acting as a valve through which polarized light passes unless blocked by the application of a low voltage)
LDCE .... Limited Duration Space Environment Candidate Materials Exposure (Shuttle experiment)
LDCM .... Landsat Data Continuity Mission (of NASA, an LDCM launch is considered for the 2005/6 time frame)
LDEF .... Long Duration Exposure Facility, NASA S/C, J.8
LDEO .... Lamont-Doherty Earth Observatory (Columbia University, New York, NY, USA, since 1949)
LDR .... Linear Depolarization Ratio
LEADEX ...... Arctic Leads Experiment (campaign)
LED .... Light-Emitting Diode (a semiconductor device which becomes luminescent on application of a low voltage)
LEDA .... Landsat On-Line Earthnet Data Availability (ESA database file)
LEED .... Low-Energy Electron Diffraction

LEGOS .... Laboratoire d'Études en Géophysique et Océanographie Spatiale (Toulouse, France, affiliated with CNES, CNRS and the Université Paul Sabatier in Toulouse; research in geophysics, oceanography and glaciology)
LEO .... Low Earth Orbit (usually for all satellite orbits up to 1000 or 2000 km altitude; in contrast to geostationary (GEO) orbits at altitudes of about 36000 km)
LEOP .... Launch and Early Orbit Phase
LeRC .... NASA Lewis Research Center (Cleveland, OH, USA). Note: On March 1, 1999, LeRC was renamed to NASA's John H. Glenn Research Center (GRC) at Lewis Field, OH.
LERTS .... Laboratoire d'Etudes et de Recherches en Télédétection Spatiale (Toulouse, France, belongs to CNES/CNRS, renamed to CESBIO as of 1995)
LES .... Lincoln (Laboratory) Experimental Satellite. A DoD microsatellite series (up to LES-4) and minisatellite series (LES-5 to LES-9) designed and built at MIT/LL (test of communication technologies). Launch of LES-1 on Feb. 11, 1965; launch of LES-9 on March 15, 1976
LETI .... Laboratoire d'Electronique de Technologie et d'Instrumentation (at Grenoble, France)
LEWEX .... Labrador Extreme Wave Experiment (campaign)
LF .... Low Frequency (30 - 300 kHz band)
LFC .... Large Format Camera, J.9
LFSAH .... Light Weight Flexible Solar Array Hinge (Shuttle payload)
LHCP .... Left Hand Circural Polarization
LHP .... Loop Heat Pipe (Shuttle Experiment)
LH Systems ..... LH Systems LLC, with company HQ in San Diego, CA (airborne cameras). In 1997, Leica AG of Heerbrugg (photogrammetry and aerial camera systems), Switzerland, formed a joint venture with BAE SYSTEMS, Inc. of San Diego, CA, and with Helava Associates Inc. a subsidiary of GDE Systems. The new company is called "LH Systems LLC" in San Diego and LH Systems GmbH in Heerbrugg, Switzerland
Li-Ion (or LI) ... Lithium-Ion (battery type)
LIDAR .... Light Detection and Ranging
LIF .... Laser-Induced Fluorescence (active remote sensing method)
LIGA .... LIthographie, Galvanoformung und Abformung (lithography, electroplating and moulding)
LIGO .... Laser Interferometric Gravitational-wave Observatory
LIMEX .... Labrador Ice Margin Experiment (campaign)
LISA .... Laser Interferometer Space Antenna (a three S/C mission of NASA, ESA, etc., proposed for 2005 and beyond). The objective is to study low-frequency gravitational waves from galactic and extra-galactic binary systems. The three S/C are separated some 5,000,000 km apart, forming an equilateral triangle (a giant interferometer).
LISS .... Linear Imaging Self-Scanning Sensor (ISRO sensor series)
LITE .... Lidar In-space Technology Experiment, Shuttle mission, J.10
LLNL .... Lawrence Livermore National Laboratory (Livermore, CA, a DOE lab managed by the University of California)
LLV1 (or 2) ..... Lockheed Launch Vehicle 1 (or 2)
LM .... Lockheed Martin Corporation, HQ at Bethesda, MD. The world's largest space company resulted in 1995 as a merger of the former Lockheed Missiles and Space Co. with the former Martin Marietta Astronautics and Martin Marietta Astro Space (which itself is based on former GE Astro Space). The new LM structure has five sectors, each with operating units and subsidiaries. The sectors are: Aeronautics, Electronics, Energy, Information & Services, and Space & Strategic Missiles. LMMS (see below), LM Astronautics (Denver, CO), LM Telecommu-

nications (Sunnyvale, CA) are units of the Space & Strategic Missiles sector. Total LM employment is about 170,000.

LMD .... Laboratoire de Météorologie Dynamique, Palaiseau (Lab of CNRS)

LMI .... Lockheed Martin Intersputnik, a joint venture company (since 1997) of Lockheed Martin Corporation and the Intersputnik International Organization of Space Communications

LMLV .... Lockheed Martin Launch Vehicle [after its first successful flight, Aug. 23, 1997 (Lewis S/C), LMLV was renamed to Athena the Greek goddess of wisdom)]

LMMS .... Lockheed Martin Missile & Space Company (HQ at Sunnyvale, CA). LMMS is a major builder of satellites and sensors for civil (TIROS, AM-1, ISS, HST, Gravity Probe-B, Wind, Polar, Landsat-7, TRACE, etc.) and military (DMSP, GPS, etc.) US space programs as well as for commercial Earth observation programs (CRSS, etc.). LMMS has a workforce of about 19,000 employees and maintains facilities at the following locations: Huntsville, AL; Cape Canaveral, FL; Kings Bay, GA; East Windsor, NJ; Valley Forge, PA; Charleston, SC; Magna, UT; Bangor, WA; and Sunnyvale, Santa Cruz, Palo Alto and VAFB, all in CA. LMMS is also the manufacturer of the following standard platform series (communication satellite buses): S3000, S4000, S5000, S7000, and A2100; and the manufacturer of Motorola's Iridium system.

LMS .... Life and Microgravity Spacelab (Shuttle mission)

LNA .... Low Noise Amplifier

LNETI .... Laboratorio Nacional de Engenhario e Technologia Industrial (PoSAT consortium, Portugal)

LO .... Local Oscillator

LOA .... Laboratoire d'Optique Atmosphérique, (of CNRS, at the University of Sciences and Technology, Lille, France)

LOICZ .... Land-Ocean Interactions in the Coastal Zone (core program of IGBP)

LORAN .... Long Range Aid to Navigation (a radionavigation system as well as an instrument name). LORAN-C operates on 100 kHz and is a maritime and aeronautical radionavigation system.

LOS .... Loss of Signal

LOS .... Line of Sight

LOTREX ...... Landoberflächen-Traversen Experiment (campaign)

LOWS .... Lake Ontario Winter Storms (campaign)

LOWTRAN .... LOW-resolution TRANsmittance a computer code (model of USAF Geophysics Laboratory), see Glossary.

LPCE .... Laboratoire de Physique et de Chimie de l'Environment (CNRS), Orleans, France

LPCM .... Laboratoire de Physique et Chimie Marines (CNRS), Villefranche-sur-mer, France

LRPT .... Low Resolution Picture Transmission (NOAA downlink technique in S-band)

LS .... Landsat Satellite Series of NOAA

LSPIM .... Land Surface Processes and Interactions Mission (in ESA's Earth Explorer Program), see SPECTRA

LST .... Land Surface Temperature

LTAN .... Local Time Ascending Node (orbit parameter)

LTER .... Long-Term Ecological Research (NFS program that started in 1981, there are 19 major sites within LTER spread throughout the US)

LTS .... Low Temperature Superconductivity (refers to conductor material levels at liquid helium temperatures, Tc = 4 K)

LUCC .... Land-Use/Cover Change (IGBP program)

LUT .... Local User Terminal (NOAA concept for S&R reception)

LWIR .... Long-Wavelength Infrared (6-14 μm) same range as TIR

M

MAB .... Man and Biosphere Programme (UNESCO, since 1989)
MABL .... Marine Atmospheric Boundary Layer
MAC .... Multiphase Atmospheric Chemistry (IGBP/IGAC program)
MACE .... Middeck Active Control Experiment (of NASA and AFRL on Shuttle). MACE and MACE-II (AFRL) are designed to to investigate modeling and control issues (high precision pointing and vibration control)
MAC-Europe ... Multisensor Airborne Campaign - Europe
MACSI .... Microwave Airborne Campaign over Snow and Ice (campaign)
MAESA .... Measurement for Assessing the Effects of Stratospheric Aircraft (campaign)
MAESTRO ..... Multiple Airborne Experiments Towards Radar Observations (campaign)
MAGE .... Marine Aerosol and Gas Experiment (campaign)
Magnolia/MFE .. (MFE = Magnetic Field Experiment) A joint French/US program (proposal status) for long-term (>5 years) monitoring of the Earth's magnetic field and its temporal variations (objectives: main field model, secular variations, core motion determination, electrical conductivity of the mantle)
MAGS .... Mackenzie River GEWEX Study (campaign)
MAHLOVS .... Middle and High-Latitude Oceanic Variability Study
MAMA .... Multi-Anode Michrochannel Array (detector type)
MAP .... Mesoscale Alpine Programme (campaign)
MAP .... Microwave Anisotropy Probe (NASA S/C mission within the MIDEX program, measurement of the full sky cosmic microwave radiation)
MAPS .... Measurement of Air Pollution from Space Radiometer (Shuttle OSTA-1 experiment during STS-2 in Nov. 1981, and STS-59), J.11
MASER .... Microwave Amplification by Stimulated Emission of Radiation
MAST .... Military Application of Ship Tracks (Shuttle)
MAST .... Monterey Area Ship Tracks (campaign)
MAST .... Marine Science and Technology (campaign)
MASTEX ...... Mediterranean Aircraft-Ship Transmission Experiment (campaign)
MAUS .... Material Science Autonomous Payload (Shuttle D2 mission)
MBA .... Microbolometer Array (detector type)
MBARI .... Monterey Bay Aquarium Research Institute, Monterey, CA
MBB .... Messerschmitt Bölkow & Blohm (Munich, Germany, since 1989 MBB was integrated into the DASA conglamorate)
MBE .... Molecular Beam Epitaxy [a technique (developed by Bell Labs in 1968) to grow perfect crystals, atom by atom, over areas vast on an atomic scale. Applications: the production of photodiode arrays, quantum wells, heterojunction structures, etc.]
MCC .... Mission Control Center
MCHIP/s .... CHIP stands for Yes/No sequences in data transmissions. One MCHIP/s = 1 million information sequences/s
MCP .... Meteorological Communications Package (Meteosat). MCP permits direct data access to the operational meteorological instruments in full resolution during a pass. MCP allows in addition the transmission of global data sets for central ground stations.
MCP .... Microchannel Plate (detector)
MCSA .... MIR Cooperative Solar Array (installation on MIR by STS-74 crew)
MCT .... Mercury Cadmium Telluride (detector material, HgCdTe, also referred to as HCT detector)
MCTEX .... Marine-Continental Thunderstorm Experiment (campaign)
MDA .... MacDonald Dettwiler Associates Ltd, Richmond, BC, Canada (MDA is a developer of SAR processors, builder and operator of Radarsat-2)
MDL .... Multi-use Data Link (GOES Second Generation S/C)
MDT .... Mean Down Time

MEDALUS ..... Mediterranean Desertification and Land Use (campaign)
MEDEA .... Material Science Experiment Double Rack for Experiment Modules and Apparatus (Shuttle experiment)
MEDS .... Marine Environmental Data Service (Ottawa, Ontario, Canada)
MEEP .... MIR Environmental Effects Payload (Shuttle payload)
MEG .... Magneto-Encephalography (medical X-ray imagery)
MEI .... Moscow State Aviation Institute (Department of Spacecraft Electric Propulsion and Power Plants)
MELCO .... Mitsubishi Electric Company, Tokyo, Japan
MELEO .... Materials Exposure in Low Earth Orbit (Shuttle experiment)
MELV .... Medium Expendable Launch Vehicle (EOS program)
MEMS .... Micro-Electromechanical System (sensor technology), also Shuttle payload
MEO .... Medium Earth Orbit (altitude range of about 5000 - 25000 km)
MFD .... Manipulator Flight Demonstration (Shuttle payload, JEM flight demo)
Megha-Tropiques A CNES/ISRO minisatellite EO mission considered for launch in 2005. Note: Megha is the Hindi word for clouds.
MEPhI .... Moscow Engineering Physics Institute
MERIT .... Measure Earth Rotation and Intercompare the Techniques (an International Earth Rotation Service Program)
Meteo-France ... Meteorological agency of France (Toulouse, Brest, etc.)
METEOR ...... Russian meteorological satellite family, G.4 - G.8
METEOSAT .... European meteorological satellite series of EUMETSAT, F.7
METOC .... Meterology & Oceanography [a US Navy program considering everything from weather observation (instruments), operations of the system, GIS services, to oceanography applications and the combination of both functions]
MetOp .... EUMETSAT Meteorological Operational satellite series, G.2.1
MeV .... Mega-electron volt
MF .... Medium Frequency (300 - 3000 kHz band)
MFLOPS .... Million Floating Point Operations per Second (a measure of computer power)
MGBX .... Microgravity Glovebox Facility (Shuttle payload)
MGM .... Mechanics of Granular Materials (Shuttle payload)
MHS .... Message Handling System (MOTIS is the ISO definition of MHS)
MHT .... Matra Hautes Technologies, France, (MHT's parent company is the Ladardere Groupe; Matra Marconi Space (MMS) is a unit of MHT)
Microlab .... OSC satellite renamed to OrbView-1, B.8.1
MIDEX .... Medium-class Explorers (NASA program). A series of cost and schedule-capped programs, led by a PI and funded by NASA.
MIGITS .... Miniature Integrated GPS/INS Tactical System (a familiy of GPS-related receiver systems of Boeing Co.)
MILOX .... Mid-Latitude Ecosystems and Photochemical Oxidants (IGBP/IGAC)
MIL-STD-1553B A spacecraft communications bus standard. The structure of the bus consists of a single bus controller connected to remote terminals (up to 31 max can be used).
MILSTAR ...... Military Strategic and Tactical Relay (heritage of STP). MILSTAR is a series of advanced US military (DoD) communication satellites. The first two Block 1 spacecraft, launched in 1995, will eventually be replaced by the Block 2 Milstar 3 through 6, which are scheduled for launch beginning in 1999.
Minotaur .... An OSC (Dulles, VA) launch vehicle. The Minotaur is a four-stage vehicle with the first and second stages being Minuteman-II stages; the two upper stages come from OSC's Pegasus launcher. OSC's Minotaur is also known as the "Orbital/Suborbital Program Space Launch Vehicle." The US Air Force developed the Orbital/Suborbital Program as

	a way to cheaply launch small military payloads. OSC integrates the Minotaur launch vehicles and conducts launch operations under an Air Force contract.
MIR	Russian Space Station, L.3
MIRAS	MIR Infrared Spectrometer (note: this is a modified GRILLE sensor by ISA on the Shuttle ATLAS-1 mission)
MIRSL	Microwave Remote Sensing Laboratory (U. of Massachusetts at Amherst, MA)
MIRP	Manipulated Information Rate Processor (NOAA S/C subsystem)
MIS-1, -B	Microcapsules in Space-1 (Shuttle experiment)
MIST	Microbursts and Severe Thunderstorms (campaign)
MISU	Meteorological Institute of Stockholm University (Stockholm, Sweden)
MIT	Massachusetts Institute of Technology (Cambridge, MA)
MIT/ERL	MIT/Earth Resources Laboratory (Cambridge, MA, since 1982)
MIT/LL	MIT/Lincoln Laboratory (Lexington, MA, since 1951)
MITA	Microsatellite Italiano a Tecnologia Avanza (Italian Advanced Microsatellite platform), ASI standard platform
MITI	Ministry of International Trade and Industry (Japan)
MIZ	Marginal Ice Zone
MIZEX	Marginal Ice Zone Experiment (campaign)
MLE	Mesoscale Lightning Experiment (Shuttle payload)
MLML	Marine Light-Mixed Layers (campaign program and a moored site)
MLOPEX	Mauna Loa Observatory Photochemistry Experiment (campaign)
MLR	Monodisperse Latex Reactor (Shuttle experiment)
MLS	Microwave Landing System (cancelled by FAA in 1994)
MLTI	Mesosphere and Lower-Thermosphere/Ionosphere (altitude from about 60 to 180 km)
MMA	Microgravity Measurement Assembly (ESA payload on Shuttle)
MMIC	Microwave Monolithic Integrated Circuit (also: Millimeter-wave Monolithic Integrated Circuit)
MMS	Magnetospheric Multi-Scale (planned misson of NASA)
MMS	Matra Marconi Space [of France (HQ at Velizy, and major assembly plant at Toulouse) and UK (Bristol, Portsmouth, Stevenage)]. MMS was formed in 1990 by Matra Espace of France (Lagadere) and Marconi Space Systems (GEC) of UK (since 1994). MMS employs 5,000 people, 2300 in France and 2700 in the UK. MMS covers science (SOHO, Giotto, Hipparcos), Earth observation (Spot series, ERS, Polar Platform for Envisat, Metop), communications (builder of the Eurostar and Leostar platforms) launch vehicles, military reconnaissance S/C (Helios), etc. MMS is also an EO instrument builder (HRV on Spot series, ASAR, GOMOS, AASTR, SEVIRI, etc.) - As of 2000 MMS is called **Astrium SAS** in France and **Astrium Ltd**. in the UK (see Astrium)
MMW	Millimeter Wave (spectral range of 1mm to 10 mm)
MOBILHY	Modélisation du Bilan Hydrique (HAPEX campaign)
MOBLAS	Mobile Laser System (USA)
MOCE	Marine Optical Characterization Experiment (campaign)
MoD	Ministry of Defence (London, UK)
MODE	Middeck 0-Gravity Dynamics Experiment (Shuttle payload)
MO Disk	Magneto-Optical Disk
MODTRAN	Moderate-resolution LOWTRAN (see glossary under LOWTRAN)
MOEM	Micro Optoelectronic-Mechanical (device)
MOMS	Modular Optoelectronic Multispectral Scanner (Shuttle payload of 1983 and 84), J.12 and J.13
MONEX	Monsoon Experiment (campaign)

MOP .... Meteosat Operational Programme (European series of weather satellites from EUMETSAT)
MOS .... Marine Observation Satellite (NASDA Satellite, MOS-1 Launch: 1987, MOS-1b launch: Feb. 1989), D.23
MOS .... Metal-Oxide Semiconductor (solid-state technology); CMOS = Complementary MOS
MOSFET ...... Field-Effect Transistor (FET) using MOS technology
MOSES .... Molecules in Outer Space and Earth Stratosphere (Swedish Mission, renamed ODIN), A.20
MOVPE .... Metalorganic Vapor Phase Epitaxy (a growth technique)
MOZAIC ...... Measurement of Ozone by Airbus In-Service Aircraft, P.40.1
M2P2 .... Mini-Magnetospheric Plasma Propulsion (a new propulsion system/ technique, funded by NASA. The technology creates an electromagnetic bubble around a S/C and lets the solar wind push the S/C) [2111)]
MPAe .... Max-Planck-Institut für Aeronomie (Katlenburg-Lindau, Germany)
MPE .... Max-Planck-Institut für Extraterrestrik (Garching, Germany)
MPEG .... Motion Pictures Experts Group [compression/decompression standard for data, MPEG-1 is a video coding standard for small images on internet (since 1993), MPEG-2 is a standard for high-quality video images (since 1996)]
MPEI .... Moscow Power Engineering Institute, builder of EO instruments like radiometers [also known as SRB/MPEI (Special Research Bureau of MPEI)]
MPG .... Max-Planck-Gesellschaft zur Förderung der Wissenschaften e.V. (Germany). MPG is the single largest government-funded research organization in Germany. MPG is the successor of the Kaiser-Wilhelm-Gesellschaft founded in 1911. MPG maintains 68 research centers (and extensions), referred to as MPIs (Max Planck Institutes), throughout Germany. The organization employs about 11,000 people, including some 3000 scientists. In addition to its workforce MPG hosts a large number of (more than 5000 mostly on a yearly basis) research fellows, doctoral candidates, and guest scientists from other institutions. Basic research in the natural and human sciences is emphasized in all MPIs. Major fields of research are: physics, chemistry, biology, physical chemistry, astronomy, mathematics, computer science, and medicine.
MPI .... Max-Planck-Institut (generic)
MPIA .... Max Planck Institut für Astronomie (Heidelberg, Germany)
MPICh .... Max-Planck-Institut für Chemie (Mainz, Germany)
MPIfM .... Max-Planck-Institut für Meteorologie (Hamburg, Germany)
MPIK .... Max-Planck-Institut für Kernphysik (Heidelberg, Germany)
MPNE .... Microgravity Plant Nutrient Experiment (Shuttle payload)
MPSE .... Morelos Payload Specialist Experiments (Shuttle payload)
MPT .... Ministry of Posts and Telecommunication (Tokyo, Japan)
MRF .... Meteorological Research Flight (UK)
MRI .... Magnetic Resonance Imaging
MS .... Multispectral (data)
MSAS .... Multi-Transport Satellite Augmentation System (GNSS-1 element of Japan)
MSFC .... Marshall Space Flight Center (Huntsville, AL, DAAC of NASA EOS Program; Note: MSFC/DAAC closed as of March 31, 1997 due to reduced NASA budgets)
MSK .... Minimum Shift Keying
MSL .... Material Science Laboratory (Shuttle payload)
MSS .... Mobile Servicing System [a robotics system consisting of the elements: SSRMS (Space Station Remote Manipulator System), SPDM (Special Purpose Dextrous Manipulator), and MBS (Remote Mobile Server

2111) http://www.niac.usra.edu/Topics/Summary_Pages/Robert_Winglee.html

	Base System), all systems are built by Canada, that will be used to assemble and maintain the ISS (International Space Station)]
MSSL	Mullard Space Science Laboratory (University College London, UK)
MSTI	Miniature Sensor Technology Integration (a Phillips Laboratory technology demonstration program, Kirtland AFB, Albuquerque, NM)
MSU-E	Multispectral Scanner - Electronic Scanning
MSU-K	Multispectral Scanner - Circular Scanning
MSU-M	Multispectral Scanner - Low Resolution
MSU-S	Multispectral Scanner-Moderate Resolution
MSU-SK	Multispectral Scanner-Moderate Resolution,Conical Scanning
MSW	Medium and Short Wave (spectrum)
MSX	Midcourse Space Experiment (DoD mission, M.14, Note: MSX experiments are also performed from several Shuttle missions in conjunction with the MSX spacecraft)
MTBF	Mean Time Between Failure
MTF	Modulation Transfer Function
MTI	Moving Target Indication
MTPE	Mission To Planet Earth [US program, see D.12, Note: As of January 1998 MTPE was renamed by NASA to "Earth Science Enterprise" (ESE)]
MW	Microwave (spectral region with wavelengths from 1mm to 1 m)
MWIR	Mid-Wavelength Infrared (about 3 - 5 μm)
MWR	Microwave Radiometer

N

N_2O	Nitrous oxide
N_2O_5	Nitrogen pentoxide
N/A	Not Applicable (Not Available)
NABE	North Atlantic Bloom Experiment (campaign within JGOFS)
NAC	Narrow-Angle Camera
NACA	National Advisory Committee on Aeronautics (USA, 1915-1958, predecessor organization of NASA)
NADC	Naval Air Development Center (Warminster, PA, USA)
NAE	National Aeronautical Establishment of NRC (National Research Council, Canada)
NAL	National Aerospace Laboratory, Japan
NAPP	National Aerial Photography Program (of USGS). NAPP was initiated in 1987 with the objective to acquire and archive aerial photography (using either color or black-and-white film) on a five-year cycle at a scale of 1:40,000. NAPP is a program jointly funded by federal agencies and states that choose to participate. Data are available through the EROS Data Center in Sioux Falls, SD, USA
NARE	North Atlantic Regional Experiment (campaign)
NARSS	National Authority for Remote Sensing and Space Sciences, (Cairo, Egypt)
NAS	National Academy of Sciences (USA)
NAS	National Airspace System (FAA, USA)
NASA	National Aeronautics and Space Administration (USA, since 1958)
NASA/ARC	NASA/Ames Research Center (Moffett Field, CA, since 1939)
NASA/DFRC	NASA/Dryden Flight Research Center (Edwards AFB, CA, since 1946)
NASA/GSFC	NASA/Goddard Space Flight Center (Greenbelt, MD, since 1959)
NASA/HQ	NASA/Headquarters (Washington, DC)
NASA/JPL	NASA/Jet Propulsion Laboratory (Pasadena, CA, since Dec. 3, 1958)
NASA/JSC	NASA/Johnson Space Center (Houston, TX, since 1961)
NASA/KSC	NASA/Kennedy Space Center (Cap Canaveral, FL, since 1967)

NASA/LaRC .... NASA/Langley Research Center (Hampton, VA, since 1917)
NASA/LeRC .... NASA/Lewis Research Center (Cleveland, OH, since 1941). Note: LeRC was renamed to John H. Glenn Research Center (NASA/GRC) on March 1. 1999
NASA/GRC .... NASA/John H. Glenn Research Center
NASA/MSFC ... NASA/Marshall Space Flight Center (Huntsville, AL, since 1960)
NASA/SSC ..... NASA/Stennis Space Center (Pearl River, MS). Testing of rockets and engines (Shuttle); colocation of US Navy facilities, Naval Oceanographic Office, Naval Research Laboratory, National Data Buoy Center (NDBC, a NOAA/NWS facility), etc.
NASDA .... National Space Development Agency (of Japan, since 1969)
NASDA/EOC ... NASDA/Earth Observation Center (Tokyo, Japan, since 1978)
NASDA/EOPD . NASDA/Earth Observation Planning Department
NASDA/EORC . NASDA/Earth Observation Research Center (Tokyo)
NASDA/EOSD .. NASDA/Earth Observation Satellite Department
NATAC .... North Atlantic Chemistry Experiment (campaign)
NAVCEN ...... Navigation Center (US Coast Guard, Alexandria, VA - NAVCEN is responsible for gathering system status information on GPS, DGPS, Omega, and Loran-C)
NAVSOC ...... Naval Satellite Operations Center (US Navy, NAVSOC HQ is at Point Mugu, CA, since 1962. NAVSOC facilities strecht across the USA)
NAVSTAR-GPS . Navigation System with Time and Ranging - Global Positioning System (Precision real-time position determination system of the US AirForce, H.4)
NAWC .... Navy Air Warfare Center (Point Mugu, CA)
NaSBE .... Sodium Sulfur Battery Experiment (Shuttle payload)
NBIOME ...... Northern Biosphere Observation and Modelling Experiment (campaign)
NBS .... National Bureau of Standards (USA, since 1901, predecessor of NIST)
Nb:AlOx:Nb .... Niobium:Aluminum Oxyde:Niobium (tunnel junction material)
Nd:YAG .... A neodymium-doped yttrium aluminum garnet crystal (solid-state) laser
NCAR .... National Center for Atmospheric Research (Boulder CO, NCAR is managed and operated by the University Corporation for Atmospheric Research (UCAR) under the sponsorship of the National Science Foundation (NSF), NCAR has two laboratory sites in Boulder: Mesa Laboratory since 1966, Foothills Laboratory since 1992)
NCAR/ATD .... NCAR / Atmospheric Technology Division
NCAR/ACD .... NCAR / Atmospheric Chemistry Division
NCAR/RAF .... NCAR / Research Aviation Facility
NCAR/MMM ... NCAR / Mesoscale & Microscale Meteorology Division
NCAR/CGD .... NCAR / Climate and Global Dynamics Division
NCAR/HAO .... NCAR / High Altitude Observatory
NCC .... National Climatic Center (USA)
NCDC .... National Climatic Data Center (of NOAA/NESDIS, Asheville, NC)
NCDS .... NASA Climate Data Center (at GSFC, Science data archive for atmospheric chemistry and climate (ERBE, etc.)
NDBC .... National Data Buoy Center [a NOAA/NWS facility at Stennis Space Center (SSC), MS, since 1982; between 1970-1982 NDBO (NOAA Data Buoy Center) was the predecessor of NDBC at SSC]
NDIR .... Non-Dispersive Infrared (Spectrometer)
NDOC .... National Oceanographic Data Center (USA)
NDSC .... Network for the Detection of Stratospheric Change
NDTP .... North Dakota Thunderstorm Project (campaign)
NDVI .... Normalized Difference Vegetation Index
NEAT .... Near Earth Asteriod Tracking (NASA/JPL ground-based program to track NEO asteroids)

NEC .... Nippon Electric Company, Tokyo, Japan
NEDRES ...... National Environmental Data Referential Service (NOAA service)
NEIS .... National Earthquake Information Service (USGS, Denver, CO)
NEAR .... Near Earth Asteroid Rendezvous S/C (of NASA with a launch Feb. 17, 1996, the mission is managed and operated by JHU/APL). As of March 2000, NASA renamed the satellite to "NEAR Shoemaker" in honor of Eugene M. Shoemaker, a geologist.
NEΔR .... Noise Equivalent Delta (or Differential) Radiance (system sensitivity)
NEΔT .... Noise Equivalent Delta (or Differential) Temperature (system sensitivity), also referred to as NEDT
NEFD .... Noise-Equivalent Flux Density (see Glossary)
NEMO .... Navy EarthMap Observer (D.25)
NEMS .... Nano-Electromechanical System (sensor technology)
NEO .... Near Earth Object
NEP .... Noise-Equivalent Power
NER .... Noise Equivalent Radiance
NERC .... Natural Environment Research Council (Swindon, UK)
NERSC .... Nansen Environmental and Remote Sensing Centre (Bergen, Norway), formerly known as NRSC, a non-profit research institute affiliated with the University of Bergen.
NESR .... Noise-Equivalent Spectral Radiance (see Glossary)
NESDIS .... National Environmental Satellite Data and Information Service (NOAA centers at Suitland, MD, and Boulder, CO)
NEWS .... NOAA Earth Watch Service (information system)
NEXRAD ...... Next-Generation Weather Radar (a US ground-based system with the name of WSR-88D (Weather Surveillance Radar-1988 Doppler)
NFOW .... Narrow Field of View (sensor)
NGDC .... National Geophysical Data Center (NOAA facility at Boulder, CO, since 1965)
NGST .... Next Generation Space Telescope [NASA satellite (an infrared observatory positioned at L2) with a planned launch in 2009 to replace HST (Hubble Space Telescope)]
NH_3 .... Ammonia
NH_4 .... Ammonium (ammonia radical)
NiCd .... Nickel Cadmium (battery)
NiH_2 .... Nickel Hydrogen (battery)
NiMH .... Nickel Metal Hydride (battery)
NNMRS .... National Natural Resources Management System [an agency of DOS (Department of Space), India]
NO .... Nitric oxide
NO_2 .... Nitrogen dioxide
NO_3 .... Nitrate radical
NO_x (NOx) ..... Nitrogen oxides (NO, NO_2, NO_3)
NO_y (NOy) ..... Total active nitrogen
NOHRSC ...... National Operational Hydrologic Remote Sensing Center (of NOAA/NWS at Chanhassen, MN, USA)
NIAC .... NASA Institute for Advanced Concepts
NICMOS .... Near-Infrared Camera and Multi-Object Spectrometer (Hubble sensor, built by Ball Aerospace)
NIES .... National Institute of Environmental Studies, Tsukuba, Japan
NIH-R .... National Institute of Health (Shuttle experiment)
NIIEM .... Scientific and Research Institute of Electromechanics, Istra (Moscow Region), Russia; NIIEM was founded in 1960 by VNIIEM. In 1992 the institute NIIEM became an independent entity. Development of LEO meteorological satellites.
NIIR .... State Radio Scientific Research Institute, Moscow; developer/builder of communication equipment in the widest sense, participation in pro-

grams: Orbita, Ekran, Ekran-M, Moskva, Gorizont, Gals, Express, Interkosmos, Intersputnik, Apollo-Soyuz, Vega, Phobos, etc.

NILU .... Norwegian Institute for Air Research (Lillestrom, Norway)

NIMA .... National Imagery and Mapping Agency (Arlington, VA, a US government agency established in Oct. 1996). NIMA incorporates the Defence Mapping Agency (DMA), the Central Imagery Office, and the Defense Dissemination Office as well as CIA's Photographic Interpretation Center. NIMA is also the principal buyer of commercial imagery for all DoD organizations.

NIMBUS .... NASA EO missions series, M.17

NIMS .... Navy Ionospheric Monitoring System (H.6)

NIPR .... Nippon Institute for Polar Research, Japan

NIR .... Near Infrared (spectrum, from 0.75 to about 1.3 μm)

NIST .... National Institute of Standards and Technology (USA, an agency of DOC, formerly National Bureau of Standards, since 1901)

NIVR .... Nederlands Instituut voor Vliegtuigontwikkeling en Ruimtevaart (Netherlands Institute for Air and Space Development, Delft, The Netherlands, since 1946)

NKAU .... National Space Agency of Ukraine (since 1992), also referred to as NSAU

NLO .... Nonlinear Optics (NLO is widely used in solid-state laser technology)

NLR .... Nationaal Lucht- en Ruimtevaartlaboratorium (National Aerospace Laboratory, Amsterdam and Noordoostpolder, Netherlands) since 1961. NLR is of NLL (Nationaal Luchtvaart Laboratorium) heritage which was founded in 1937.

NMC .... National Meteoroligical Center (USA)

NMHC......... Non-methane hydrocarbons

NMOS .... N-channel MOS (Metal-Oxide Semiconductor)

NMP .... New Millennium Program (NASA/JPL)

NNSS .... Navy Naviation Satellite System (USA, also known as the 'Transit' system, was the world's first satellite navigation system, H.6)

NOAA .... National Oceanic and Atmospheric Administration (NOAA is an agency of the US Department of Commerce, established in 1970 (predecessor ESSA), it has the following major divisions: NOS (National Ocean Service), NWS (National Weather Service), NMFS (National Marine Fisheries Service), NESDIS (National Environmental Satellite, Data and Information Service), OOAR (Office of Oceanic and Atmospheric Research), and ONCO (Office of NOAA Corps Operations).

NOAA/AL....... NOAA/Aeronomy Laboratory (Boulder CO)

NOAA/AOML .. NOAA/Atlantic Oceanographic and Meteorological Laboratory, Miami, FL. The HRD (Hurricane Research Division) is part of AOML.

NOAA/ARL .... NOAA/Air Resources Laboratory, Silver Spring, MD. Note: ARL consists of the HQ-Division in Silver Spring, MD, the ATTD in Oak Ridge TN, the ASMD (Atmospheric Sciences Modeling Division) in Research Triangle Park, NC, the FRD (Field Research Division) in Idaho Falls, ID, and the SRRB (Solar Radiation Research Branch) in Boulder, CO.

NOAA/AOC .... NOAA/Aircraft Operations Center, MacDill AFB, Tampa, FL. Note: AOC was created in 1983 [initially known as OAO (Office of Aircraft Operations)] to manage NOAA aircraft, personnel, budget and facilities in support of NOAA aircraft programs. AOC is under ONCO.

NOAA/ATDD .. NOAA/Atmospheric Turbulence and Diffusion Division, Oak Ridge, TN

NOAA/CDC .... NOAA/Climate Diagnostics Center (Boulder, CO)

NOAA/CMDL .. NOAA/ Climate Monitoring and Diagnostics Laboratory, Boulder CO.

NOAA/ERL .... NOAA/Environmental Research Laboratories, headquartered in Silver Spring, MD. (under OOAR). All NOAA laboratories are run through OOAR/ERL, these are: AL, AOML, ARL, CDML, ETL, FSL, GFDL, GLERL, NSSL, PMEL, SEL, CDC, and the Joint Institutes.
NOAA/ETL .... NOAA/ Environmental Technology Laboratory, Boulder, CO, (formerly WPL = Wave Propagation Laboratory)
NOAA/FSL ..... NOAA/Forecast Systems Laboratory (Boulder, CO)
NOAA/GFDL ... NOAA/Geophysical Fluid Dynamics Laboratory, Princeton, NJ.
NOAA/GLERL . NOAA/Great Lakes Environmental Research Laboratory, Ann Arbor, MI.
NOAA/NSSL ... NOAA/National Severe Storms Laboratory, Norman, OK.
NOAA-NESDIS . NOAA/National Environmental Satellite Data and Information Service, Suitland, MD. - NESDIS functions are: Satellite Operations, Satellite Data Processing and Distribution, Research and Applications, Systems Development, National Climatic Data Center (NCDC), National Oceanic Data Center (NODC), National Geophysical Data Center (NGDC).
NOAA/NCDC .. NOAA-NESDIS/National Climatic Data Center, Asheville, NC.
NOAA/NDBC .. NOAA-National Data Buoy Center (a NOAA/NWS facility at Stennis Space Center, MS)
NOAA/NGDC .. NOAA-NESDIS/National Geophysical Data Center, Boulder, CO
NOAA/NODC .. NOAA-NESDIS/National Oceanographic Data Center (Silver Spring MD)
NOAA/NOS .... NOAA/National Ocean Service - NOS functions are: coast and geodetic survey, ocean resources conservation and assessment, ocean and coastal resources management, ocean and earth sciences.
NOAA/NSIDC .. NOAA/National Snow and Ice Data Center, Boulder, CO (NSIDC is located at the University of Colorado at Boulder)
NOAA/NWS .... NOAA/National Weather Service - NWS functions are: meteorology, hydrology, systems operations, systems development, national meteorological center, national data buoy center
NOAA/OAO .... NOAA/Office of Aircraft Operations, Miami, FL (old designation)
NOAA/OOAR .. NOAA/Office of Oceanic and Atmospheric Research - OOAR functions: oceanic research program, environmental research laboratories.
NOAA/PMEL ... NOAA/Pacific Marine Environmental Laboratory (Seattle, WA, since 1973)
NOAA/SEC .... NOAA/Space Environment Center (Boulder, CO)
NOAA/SEL ..... NOAA/Space Environment Laboratory (Boulder, CO), Note: NOAA/SEL changed its name to NOAA/SEC in 1997
NODS .... NASA Ocean Data System (located at JPL; Measurements in the archive are related to altimetry, scatterometry, and microwave radiometry. NODS archives and distributes data products for TOPEX/Poseidon)
NOPEX .... Northern-Hemisphere Climate Processes Land-surface Experiment (campaign)
NOPP .... National Oceanographic Partnership Program (USA, since 1997, NOPP has a mandate from Congress). The objective is to foster cooperation and partnerships among federal agencies, academia, industry and other members of the oceanographic scientific community.
NORCSEX ..... Norwegian Continental Shelf Experiment (campaign)
NORDA .... Northern Oceans Research and Development Activities (Canada)
NORSEX ...... Norwegian Remote Sensing Experiment (campaign)
NOSC .... Naval Ocean Systems Center (San Diego, CA)
NOSL .... Night/Day Optical Survey of Lightning (Shuttle experiment)
NOSS .... Naval Ocean Surveillance Satellite, also referred to as "Whitecloud," "White Cloud" or "Classic Wizzard" (a US Navy S/C series, sponsored

by NRO, and launched from VAFB, CA on Atlas vehicles). [2112] NOSS is a wide area ocean surveillance system used to determine the location of radio and radar transmissions, using triangulation. - Each NOSS launch placed a cluster of one primary satellite and three smaller sub-satellites (that trail along at distances of several hundred m apart in a triangle formation) into low polar orbit. This satellite array can determine the location of radio and radars transmitters, using triangulation, and the identity of naval units, by analysis of the operating frequencies and transmission patterns. NOSS used the ELINT technique called TDOA (time difference of arrival), rather than true interferometry. NOSS-1 launch April 30, 1976 (1100 km altitude, inclination = 63.5°), NOSS-2 launch Dec. 8, 1977, NOSS-3 launch March 3, 1980, NOSS-4 launch Feb .9, 1983, NOSS-8 launch May 15, 1987 (also referred to as USA-22), NOSS-9 launch Sept. 5, 1988 (also known as USA-32). - Second generation NOSS satellites were launched starting in 1990. There are three groups of the 2nd generation NOSS satellites each having three satellites in close proximity to one another. The first NOSS-2-1 triplet was launched on June 8, 1990 on a Titan-IV vehicle from Cape Canaveral; the second NOSS-2-2 triplet was launched on Nov. 8, 1991; and the third triplet of NOSS-2-3 was launched May 12, 1996 from VAFB.

NOWES .... Northern Wetlands Study (campaign)

NO_x .... Nitrogen oxides

NO_y .... Total reactive nitrogen

NPL .... National Physical Laboratory (Teddington, Middlesex, UK; NPL is an agency of the Department of Trade and Industry)

NPO .... Naulshno Proizwodstwennoje Objedijenie (Scientific/Research Production Association, Russia)

NPO AP .... NPO for Automation and Instrument Engineering, Moscow; since 1947; participation in the following programs: Venera, Mars, Luna, Soyuz, Proton, Zenit, Energia-Buran; builder of on-board guidance and navigation systems

NPO Geofizika .. Moscow; since 1908, a major enterprise for the development of automatic and visual opto-electronic instruments; participation in national programs: Vostok, Salyut, Soyuz, MIR, Energia-Buran, etc.

NPO Machinostroyenia Russian company, Reutov, Moscow Region, builder/integrator of S/C (ALMAZ series), participation in programs: Kosmos, Proton, Polyot, Salyut, etc.

NPO Planeta .... Scientific and Research Center on Space Hydrometeorology (Moscow, since 1974), operators of satellites (Meteor, Okean, Resurs, GOMS series) along with corresponding ground segments, providers of services to the user community in the areas of meteorology/climate, oceanography, Earth resources, and ecological monitoring. From an organizational point of view, NPO Planeta is an agency positioned under ROSHYDROMET, the 'Committee for Hydrometeorology and Environmental Monitoring'

NPO PM .... Research and Production Association of Applied Mechanics (Prikladnoi Mekaniki), Krasnoyarsk (a closed city until 1991) Siberia. NPO PM was founded in 1959, since 1977 it is builder/integrator of communication satellites (Gorizont, Express, Molniya-1, -2, -3, Raduga-1, Ekran, Ekran-M, Luch, Radio, etc.), navigation satellites (GLONASS, Tsikada), and geodetic satellites (GEO-IK, Etalon); advanced programs (Express-M, Gonets, Arkos, Mayak, Gals)

NPO Vega ...... Russian space/defense industry consortium, Moscow, designers and builders of SAR instruments, etc., operators of airborne instruments

2112) A. Andronov, "The US Navy's "White Cloud" Spaceborne ELINT System," in Zarubezhnoye Voyennoye Obrozreniye (Foreign Military Review), ISSN 0134–921X, No. 7, 1993, pp. 57–60, translated by Allen Thomson

NPO Yuzhnoye . . Design Office Yuzhnoye, in Dnepropetrovsk, Ukraine (builder of OKEAN S/C series)
NPOESS . . . . National Polar-orbiting Operational Environmental Satellite System (merged POES and DMSP series, with launches projected for 2008 and beyond)
NPOP . . . . NASA Polar Platform
NPP . . . . NPOESS Preparotary Project
NPS . . . . Naval Postgraduate School (Monterey, CA)
NRC . . . . National Research Council (Washington, DC, USA)
NRC (NRCAN) . Natural Resources Canada (Ottawa, Canada)
NRCS . . . . Normalized Radar Cross-Section (an aspect of ocean surface reflectivity, also referred to as σ^{o})
NRCT . . . . National Research Council of Thailand
NRL . . . . Naval Research Laboratory (Washington, DC). NRL is the US Navy's corporate research and development laboratory, created in 1923 with over 4000 personnel (among them 1500 scientists) in the 1990s. NRL maintains 15 research sites throughout the US. The three main NRL sites are at: Washington DC, NRL/SSC (Stennis Space Center in Bay St. Louis, MS), and NRL/MRY (Monterey, CA).
NRL/NCST . . . . . NRL/Naval Center for Space Technology
NRL/RSD . . . . . . NRL/Remote Sensing Division
NRLM . . . . National Research Laboratory of Meteorology (Japan)
NRO . . . . National Reconnaissance Office (agency of DoD, Chantilly, VA, USA). NRO sponsors and operates US reconnaissance S/C (Corona series, etc.). The primary user of the imagery is NIMA.
NROSS . . . . Navy Remote Ocean Sensing System (satellite)
NRSA . . . . National Remote Sensing Agency (since 1975, Balanagar, Hyderabad, India)
NRSC . . . . National Remote Sensing Centre (UK, this agency was privatized in 1989, commercial sale of remote sensing data, operator of UK-PAF for ESA)
NRSCC . . . . National Remote Sensing Center of China (Beijing)
NRZ . . . . Non-Return to Zero (communication signal parameter)
NSBF . . . . National Scientific Balloon Facility (NASA-owned facility in Fort Sumner, NM)
NSC . . . . Norwegian Space Centre (Oslo, Norway)
NSERC . . . . Natural Sciences and Engineering Research Council (Canada)
NSF . . . . National Science Foundation (Arlington, VA, USA; since 1950; NSF is an independent govenment agency responsible for promoting science and engineering). About 20,000 programs per year are supported by NSF.
NSI . . . . NASA Science Internet - an international dual protocol (TCP/IP and DECnet) network (successor to SPAN)
NSIDC . . . . National Snow and Ice Data Center (Boulder, CO, NOAA facility at University of Colorado, established in 1982). NSIDC is co-located with WDC-A (World Data Center A for Glaciology). NSIDC is also a DAAC site of the EOS Program. NSIDC has extensive holdings of cryospheric and polar ocean surface-flux data and routinely produces sea ice maps from SSM/I sensor.
NSMC . . . . National Satellite Meteorological Center [since 1971, NSMC is the research and operational facility of CMA (China Meteorological Administration)]. NSMC has ground stations in Beijing, Guangzhou, and Urumqi.
NSPO . . . . National Space Program Office (Hsin-Chu City, Taiwan)
NSSDC . . . . National Space Science Data Center (at NASA/GSFC)
NSSL . . . . National Severe Storms Laboratory (Norman, OK)
NSTAR . . . . NASA Solar Electric Power (SEP) Technology Application Readiness

NSW .... New South Wales (Australia)
NTIS .... National Technical Information Service (USA)
NTS .... Navigation Technology Satellite (DoD/NRL program of the 1970s also referred to as Timation which predated the GPS program)
NTSC .... National Television Standards Committee (US TV display standard which is also adopted by a number of other countries. This is a 525-line video signal with a 3.58 MHz chroma subcarrier at 60 Hz)
NTT .... Nippon Telegraph and Telephone Corporation (Japan)
NWP .... Numerical Weather Prediction
NWS .... National Weather Service (USA)

O

O_2 .... Molecular oxygen
O_3 .... Ozone
O_x (Ox) .... Odd oxygen ($O+O_3$)
OACT .... Office of Advanced Concepts and Technology (NASA, formerly OAST)
OAI .... Ohio Aerospace Institute, Cleveland, OH [consortium of nine Ohio universities, NASA/GRC (Lewis Field in Cleveland), AFRL (Dayton), and private industry]
OARE .... Orbital Acceleration Research Experiment (Shuttle payload)
OACES .... Ocean-Atmosphere Carbon Exchange Study (campaign)
OASIS-1 .... Orbiter Autonomous Supporting Instrumentation System (Shuttle payload)
OASIS .... On-Line Data Access and Service Information System (Catalog system at NOAA-NCDC)
OAST .... Office of Application and Space Technology (NASA, Shuttle payloads are also designated by this name - OAST-1, OAST-2, etc.)
OBC .... On-Board Computer
OBS .... Observatoire Paris-Mendon (France)
OCE .... Ocean Color Experiment (Shuttle payload)
OCEAN .... Ocean Color Environment Archive Network (ESA Program)
OClO .... (ClO_2) Chlorine dioxide
OCOS .... Ocean Climate Observing Study (campaign)
OCTW .... Optical Communications Through Windows (Shuttle experiment)
ODERACS ..... Orbital Debris Radar Calibration System (Shuttle payload)
ODIN .... Proposed Swedish astronomy and aeronomy mission (A.20, in Norse mythology Odin (also called Woden or Wotan) is one of the principal gods)
OECD .... Organization for Economic Cooperation and Development
OEDIPUS ...... Observations of Electric-field Distributions in the Ionosphere Plasma - a Unique Strategy (Canadian sounding rocket missions from Andoya, Norway and Poker Flats, Alaska)
OES .... Office of Earth Science (NASA/HQ, since 1998, formerly Office of Mission to Planet Earth (OMTPE))
OEX .... Orbiter Experiments (Shuttle)
OGLOW .... Sun Orbiter Glow (Shuttle experiment)
OH .... Hydroxyl radical
OHB-System .... Orbital- und Hydrotechnologie Bremen System GmbH (since 1958, originally known as: Otto Hydraulik Bremen). A mid-sized aerospace and telecommunication company, located in Bremen, Germany - with a number of company participants and subsidiaries in Germany and Italy. OHB-System is part of the Fuchs Gruppe (since 1981). Satellites built by the Fuchs Gruppe are: BremSat, SAFIR-1, -2, ABRIXAS, DIAMANT, MITA. Note: The company CARLO GAVAZZI SPACE S.p.A, Milan, Italy was taken over by the Fuchs Gruppe in 1996; OHB-Teledata was founded in 1996.

OICETS .... Optical Interorbit Communications and Engineering Test Satellite (of NASDA, Japan)
OIP .... Optronic Instruments & Products [OIP is trading under the trade name 'Delft Sensor Systems' (DSS)], located in Oudenaarde, Belgium
OKEAN .... Ukrainian/Russian satellite series, D.26
OMNI .... Operating Missions as Nodes on the Internet. OMNI is the first end-to-end demonstration of operating NASA missions as nodes on IP.
ONERA .... Office National d'Etudes et de Recherches Aérospatiales - The French Aeronautics and Space Research Center (Chatillon, Meudon, Palaiseau, Avrieux, Mauzac, Toulouse, Lille, France) ONERA reports to the French Ministry of Defense. CERT (Centre d'Études et de Recherches de Toulouse) is a center of ONERA. It carries out research for and with the aeronautics, space and defense industries.
ONR .... Office of Naval Research (HQ in Arlington, VA). ONR coordinates the science and technology programs of the US Navy and Marine Corps. NRL is a technical department of ONR.
OPO .... Optical Parametric Oscillator (laser type)
O-QPN .... Offset Quadriphase Pseudo-Noise
ORFEUS ...... Orbiting Retrievable Far and Extreme Ultraviolet Spectrograph (German/US Shuttle payload)
OREGIN .... Organization of European GNSS Equipment and Services Industry (an industry association to support development of Galileo equipment and services)
ORI .... Ocean Research Institute (University of Tokyo, Japan)
ORNL .... ORNL (Oak Ridge National Laboratory), Oak Ridge, TN (of DOE)
Ørsted .... Danish research satellite, E.18
ORSTOM ...... Office de la Recherche Scientifique et Technique Outre-Mer (Paris, Montpellier, Orleans, etc., France) also: L'Institut francais de recherche scientifíque pour le développement en coopération (French scientific research institute for development in cooperation). In 1998 OSTROM was renamed to IRD (Institut de Recherche pour le Développement)
OSA .... Optical Society of America
OSDPD .... Office of Satellite Data Processing and Distribution (of NOAA)
OSC .... Orbital Sciences Corporation (Dulles, VA, USA, since 1982, builder of small satellites and instruments, owner/operator of commercial launch services for small payloads, Pegasus vehicle, etc.). ORBCOMM, ORBIMAGE and Magellan (GPS receivers) are affiliates of OSC, so are CTA Space Systems (McLean, VA) and MacDonald Dettwiler Associates Ltd (MDA, Vancouver, BC).
OSI .... Open System Interconnect (a standard for open communication)
OSKAR .... Orbiting Satellite Carrying Amateur Radio (initially a satellite series of a USA-based group of amateur radio enthusiasts; OSKAR-1 was launched Dec. 12. 1961; in 1969 AMSAT was founded to give amateur radio satellites an international base)
OSS .... NASA's Office of Space Science (Shuttle payloads, etc.)
OSSS .... One Stop Satellite Solutions (Ogden, UT, since 1996, a spin-off commercial company of CAST at Weber State University). OSSS built MPA (Multi-Payload Adapter) for JAWSAT. OSSS is also the US contact/partner for the Dnepr launch vehicle of ISC Kosmotras of Moscow.
OSTC .... Federal Office for Scientific, Technical, and Cultural Affairs of Belgium [also referred to as SSTC (Services Fédéraux des Affaires Scientifiques, Techniques et Culturelles, Belgium)]
OSTA .... Office of Space and Terrestrial Applications, NASA (a designation that was also given to the early Shuttle payloads)
OSVS .... Orbiter Space Vision System (Shuttle payload)
OTTER .... Oregon Transect Ecosystem Research (campaign)

OWWS .... Operational Windshear Warning System (NCAR)

P

PACSAT .... A Protocol suite first developed by SSTL. PACSAT uses packet radio techniques in the microsatellite system to transmit its data over the satellite RF link. Several layers of protocol are implemented in the PACSAT suite, at the lower level HDLC (High-Level Data Link Control) and X.25 provide the functions of packet multiplexing, error detection and ARQ (Automatic-Repeat Request) error correction. PACSAT is a point-to-multipoint protocol (broadcast); small ground terminals in the satellite footprint receive/send the data. The PACSAT protocol suite is also supporting data communications within the radio amateur community.

PAF .... Processing and Archiving Facility (ESA facilities for the ERS-1 mission in Europe: D-PAF at DLR/DFD, Oberpfaffenhofen, Germany; F-PAF at CERSAT, Brest, France; I-PAF at ASI Matera, Italy; UK-PAF at RAE, Farnborough, UK)

PAGES .... Past Global Changes (IGBP core program)

PAL .... Phase Alternation Line (German TV display standard). PAL has 625 scan lines per frame at 50 Hz.

PALACE .... Profiling ALACE (Autonomous Lagrangian Circulation Explorer) of NOAA/AOML. PALACE is a later version of ALACE, first deployed in 1997. PALACE buoys have the added capability of data storage. They cary a sensor package providing measurements of various parameters such as conductivity and temperature. In the late 1990s, hundreds of PALACE floats in the Atlantic Ocean are reporting to data collection satellites on subsurface currents as well as profiles of salinity and temperature.

PALE .... Paleoclimates for Arctic Lakes and Estuaries (campaign)

PAM .... Portable Automated Mesonet (weather stations of NCAR)

PAMS .... Passive Aerodynamically-Stabilized Magnetically-Damped Satellite (Shuttle payload)

PAN .... Panchromatic (data)

PAN .... Peroxyacetylnitrate

PANASH .... Paleoclimates of the Northern and Southern Hemispheres (IGBP/PAGES program under focus 1)

PANSAT .... Petite Amateur Naval Satellite (S/C of Naval Postgraduate School, Monterey, CA, ejected from Shuttle)

PARE .... Physiological and Anatomical Rodent Experiment (Shuttle experiment)

PARLIQ .... Phase Partitioning in Liquids (Shuttle experiment)

PAS .... PanAmSat Corporation of Greenwich, CT (a daughter of Hughes Electronics Corporation of Los Angeles, CA. PanAmSat is the world leader of commercial satellite-based communications services, launch of first satellite (Galaxy-1) in 1983, launch of PAS-1 in 1988)

PASC .... Polar Atmospheric and Snow Chemistry (IGBP/IGAC project)

PASDE .... Photogrammetric Appendage Structural Dynamics Experiment (Shuttle)

PBL .... Planetary Boundary Layer

PbS .... Lead Sulfide (detector material)

PbSi .... Lead Silicon (detector material)

PC .... Photoconductive (detector)

PCG .... Protein Crystal Growth (Shuttle experiment)

PCG/STES ..... Protein Crystal Growth / Single-Locker Thermal Enclosure System (Shuttle experiment)

PCM .... Pulse Code Modulation

PD .... Photodiode (detector)

PDA .... Photodiode Array (detector)
PDP .... Plasma Diagnostics Package (Spacelab-2 sensor, studies of the interaction between the Earth's magnetic field and charged particles in the ionosphere)
PE&RS .... Photogrammetric Engineering & Remote Sensing (ASPRS journal)
PEACAMPOT .. Perturbation by East Asia Continental Air Mass to Pacific Oceanic Troposphere (campaign)
PEM-West ...... Pacific Exploratory Mission - West (campaign)
PEP .... Pole-Equator-Pole (transect of PANASH campaign)
PGIM .... Plant Growth Investigations in Microgravity (Shuttle experiment)
PHCF .... Pituitary-Growth Hormone Cell Function (Shuttle experiment)
PHOTON ...... Russian solar-terrestrial mission (K.8.1)
PL .... Phillips Laboratory of USAF (PL is headquartered at Kirtland Air Force Base, Albuquerque, NM, and has locations at Hanscom AFB, Bedford, MA, and Edwards AFB, CA)
PI .... Principal Investigator
PID .... Prototype International Directory (CEOS-defined Directory Interchange Format (DIF)); CEOS members operating an archive with PID capability are: CCRS, DLR/DFD, ESA/ESRIN, NASA, NASDA, NOAA, RAE, etc. . Hence, standardized archival access is possible (see: IDN).
PIDC .... Precision Instrument Development Center (of the National Science Council, Taiwan), Hsinchu, Taiwan ROC
PIK .... Potsdam Institut für Klimaforschung (Potsdam Institute for Climate Impaxct Research, Potsdam, Germany)
PILOT .... Portable Inflight Landing Operations Trainer (Shuttle experiment)
PILPS .... Project for Intercomparison of Landsurface Parameterization Schemes (WCRP/GEWEX project)
PIN .... Positive Insulator Negative (diode)
PIPOR .... Program for International Polar Ocean Research
PIXEL .... Picture Element
PLB .... Personal Locator Beacon (COSPAS and S&RSAT)
PLL .... Phase Locked Loop (communication technique to enable integration of voice and data)
PLO .... Phase Locked Oscillator
PM .... Phase Modulation (modulation technique of the main carrier)
PM .... Polymer Morphology (Shuttle experiment)
PM .... Post Meridiem (refers to the afternoon time designations in the US; a time of 5 PM is equivalent of 17:00 hours in international notation)
PMA .... Pressurized Mating Adapter (Shuttle)
PMAP .... Paleoenvironment Multiproxy Analysis and Mapping Project (see PANASH campaign)
PMG .... Plasma Motor Generator (Q.36.5)
PMOD/WRC ... Physikalisch-Meteorologisches Observatorium Davos, World Radiation Center (Switzerland)
PMS .... Particle Measuring Systems Inc. (of Boulder CO)
PMT .... Photomultiplier Tube (detector)
PMV&D .... (Plume Model Validation and Development (campaign)
PN .... Pseudo Noise (code)
PNEDC .... Programme National d'Etude de la Dynamique du Climat (France)
PNL .... Pacific Northwest Laboratory (Richland, WA, USA) of DOE, operated by Batelle Memorial Institute
PNR .... Pseudo Noise Number (a GPS series designation)
PNRA .... Italian National Programme for Antarctic Research
PRN .... Pseudo Random Noise
POCC .... Payload Operations and Control Center
POEM-1 .... Polar-Orbit Earth-Observation Mission (planned ESA Series) D.9

POES .... Polar-orbiting Operational Environmental Satellites (NOAA series of operational polar orbiting satellites), G.13
POGO .... Polar-Orbiting Geophysical Observatory
POL .... Prowdman Oceanographic Laboratories (UK)
POLAR .... NASA/GSFC Solar-Terrestrial Mission (K.19)
POLARIS ...... Photochemistry of Ozone Loss in the Arctic Region in Summer (campaign)
POLINAT ...... Pollution from Aircraft Emissions in the North Atlantic Flight Corridor (campaign)
PoSAT .... Portuguese Satellite (D.40.9)
PPARC .... Particle Physics and Astronomy Research Council, UK
PPE .... Phase Positioning Experiments (Shuttle payload)
PPF .... Polar Platform (ESA Columbus program, PPF is utilized for POEM payloads)
PPS .... Precise Positioning Service (GPS)
PRARE .... Precision Rate and Range-Rate Equipment, H.7.2
PRF .... Pulse Repetition Frequency
PRIMA .... Piattaforma Riconfigurabile Italiana Multi-Applicativa (Reconfigurable Italian Platform for Multiple Applications), ASI platform for a total S/C mass of 300-1000 kg
PRIRODA ..... Research module of the Space Station MIR (D.28)
PRN .... Pseudo Random Noise
PROBE .... Prototype Radiation Observation Experiment (campaign)
ProSEDS .... Propulsive Small Expandable Deployer System (tether experiment)
PROTEUS ..... Platforme Reconfigurable pour l'Observation, les Telecommunications et les Usages Scientifiques (French minisatellite bus for a S/C mass less than 500 kg)
PROTEUS ..... Profile Telemetry of Upper Ocean Currents [a NOAA/PMEL mooring system, a taut-wire surface mooring with a toroidal float similar to ATLAS]
PSI .... Paul Scherrer Institute, Villigen, Switzerland (database of space environmental data)
PSC .... Polar Stratospheric Clouds
PSE .... Physiological Systems Experiment (Shuttle)
PSE .... Polar Sunrise Experiment (campaign)
PSK .... Phase Shift Keying (a modulation technique)
PSLR .... Peak Side Lobe Ratio
PSLV .... Polar Satellite Launch Vehicle (ISRO launch vehicle)
PSN .... Piano Spaziale Nationale (previous name of Italy's Space agency , now ASI)
PSRC .... Polish Space Research Center, Warsaw, Poland
PtSi............ Platinum-silicide (detector material)
PTT .... Platform Transmitter Terminal (data collection platform for ARGOS system)
PTT .... Public (Postal) Telephone and Telegraph (utility company). Refers to operating agencies directly or indirectly controlled by governments in charge of telecommunication services in most countries of the world.
PTTI .... Precise Time and Time Interval
PV .... Photovoltaic (detector)
PVTOS .... Physical Vapor Transport of Organic Solids (Shuttle experiment)
PWV .... Precipitable Water Vapor (atmosphere)
PYREX .... Pyrenean Experiment (campaign)

Q

QinetiQ .... New name of DERA (Defence Evaluation and Research Agency), Farnborough, UK, pronounced as "kin-et-tik" (as of July 2, 2001)
QMW .... Queen Mary and Westfield College (London, UK)

QPN .... Quadra Pseudo Noise (modulation technique)
QPSK .... Quadra-Phase Shift Keying (4-PSK is a modulation technique and a data transmission standard). Soon 8-PSK and higher modulations for such applications as DBS (Digital Broadcast System) will be used.
QuickBird ...... Commercial imaging satellite (B.4.2)
QUT .... Queensland University of Technology, Australia
QWIP .... Quantum Well Infrared Photodetector (for applications in the range from 6 - 25 μm)

R

RADAR .... Radio Detection and Ranging
RADARSAT .... A Canadian (CSA/CCRS) EO mission with a SAR instrument (D.29)
RADCAL ...... Radar Calibration Satellite (A microsatellite of USAF, launch June 25, 1993 from VAFB. It provides space-based radar cross-sectional area calibration for more than 70 radars operating in the C-band, and carries two GPS receivers with the aim to demonstrate GPS based attitude determination.)
RADFET ...... Radiation-sensitive Field Effect Transistor
RAE .... Royal Aerospace Establishment [Farnborough, UK, (in the early 1990s RAE was renamed into 'DRA' - Defense Research Establishment)]
RAIM .... Receiver Autonomous Integrity Monitoring (a GPS and GLONASS technology - RAIM requires a minimum of five visible satellites for fault detection and six satellites for fault detection and exclusion)
Rosaviakosmos .. Russian Aviation and Space Agency (RASA), Moscow. The name of Rosaviakosmos was adopted by decree (No 1186) on Oct. 25, 1999. The previous name was RKA (Russian Space Agency) which in turn was created Feb. 25, 1992.
RAL .... Rutherford Appleton Laboratory (Chilton, Oxon, UK)
RAM .... Random Access Memory
RAN (RAS) .... Russian Academy of Sciences
RASS .... Radio-Acoustic Sounding System (a ground-based system of wind and temperature vertical profiles is used in meteorology and atmospheric research).
RBDS .... Radio Broadcast Data System
RCVR .... Receiver
R&D .... Research & Development
RDL .... Research and Development Laboratories, Culver City, CA
REBAL .... Radiation and Energy Balance for Imagery and Electromagnetic Propagation (campaign)
REFLEX .... Radiation and Eddy Flux Experiment (campaign)
REFLEX .... Return Flux Experiment (Shuttle SPARTAN payload)
REM .... Release/Engage Mechanism (Shuttle, used for Spartan flights)
RENE .... Rehearsal ERS-1 Validation Northern Europe (campaign)
RESTEC .... Remote Sensing Technology Center, Tokyo, Japan (since 1975)
Resource21 ..... Commercial imaging satellite (under development by Resourse21)
Resurs .... Russian satellite series for resource monitoring, D.31, D.32
RF (R/F) .... Radio Frequency (of active sensors, also data transmission link, etc.)
RFI .... Radio Frequency Interference
RGB .... Red, Green, Blue (color code)
RHCP .... Right Hand Circular Polarization
RICE .... Regional Interactions of Climate and Ecosystems (IGBP/IGAC program)
RIN .... Royal Institute of Navigation (UK)
RIRT (RIRV) ... Russian Institute of Radionavigation and Time, St. Petersburg, since 1957 (before 1993 the institute was called: Leningrad Scientific and Research Electrotechnical Institute); participation in programs: Tsikada, Glonass, Cospas-S&RSAT

RIS .... Resonance Ionization Spectroscopy (a laser technique)
RISDE .... Russian Institute of Space Device Engineering
RIT-10 .... Radio-frequency Ion Thruster (electric propulsion sytem of DASA)
RIT .... Royal Institute of Technology, Sweden
RITS .... Radiatively Important Trace Species (campaign)
RKA (RSA) .... Russian Space Agency, Moscow, since Feb. 25, 1992 (by decree issued by the President of the Russian Federation). RKA has centralized control of Russia's civilian space program, including all manned and unmanned nonmilitary space flights. - On Oct. 25, 1999, RKA changed its name officially to "**Rosaviakosmos**" (Russian Aviation and Space Agency). -- The prime contractor used by Rosaviakosmos is RKK Energiya, which owns and operates the Mission Control Center in Kaliningrad and operates the Mir space station.
RKK Energia ... Rocket Space Corporation, Kaliningrad, Moscow region (also referred to as RSC, since 1946); responsibility for all Russian manned space projects; builders of launch vehicles (Proton) and of S/C (i.e. MIR space station), payloads, sensors, etc.
RLG .... Ring Laser Gyroscope (an angular rate gyro)
RLSBO .... Radiolokazionnaja Sistema Bokowo Obzora (side view radar system)
RME .... Radiation Monitoring Experiment (Shuttle payload)
RMS .... Remote Manipulator System (robot arm of Shuttle)
RMS .... Royal Meteorological Service (UK)
RNII KP (ISDE) Russian Institute of Space Device Engineering, Moscow; a leading company in the design and development of sensors; participation in programs: Venera, Vega, Phobos, Luna, Mars, Prognoz, Granat, Resurs, Okean, Glonass, etc.
ROCSat .... Republic of China Satellite (Taiwan)
ROIC .... Readout Integrated Circuit (silicon device for readout of infrared detector photodiodes)
ROM .... Read Only Memory
ROMPS .... Robot Operated Materials Processing System (Shuttle payload)
ROSHYDROMET Committee for Hydrometeorology and Environmental Monitoring (Russian Government Agency, similar in functions and services to EUMETSAT and NOAA)
ROTEX .... Robotic Experiment (Shuttle experiment)
RSCC .... Russian Satellite Communication Company (Moscow)
RSI .... Radarsat International Ltd. (Richmond, BC, Canada, established in 1989 by a consortium of Canadian aerospace companies and Lockheed Martin of USA, RSI is the distributor of Radarsat data)
RSIF .... Rain-Sea Interaction Facility (at NASA/GSFC/WFF, established in 1993; RSIF provides a controlled environment for studies of a) microwave scattering from rain-generated features, and b) physical processes at the air-water interface and in the adjacent boundary layers)
RSRE .... Royal Signals and Radar Establishment (Great Malvern, Worcestershire, UK)
RTCA .... Radio Technical Commission for Aeronautics (Washington, DC)
RTCM SC-104 .. Radio Technical Commission for Maritime Services [the RTCM Special Committee 104 established the worldwide standard for meter-level differential GNSS (Global Navigation Satellite System) broadcasts]
RTG .... Radioisotope Thermoelectric Generator (a nuclear propulsion system first flown on Transit-4A, also on Ulysses K.28). Deep space missions in particular depend on RTG propulsion
RTK .... Real-Time Kinematic (a DGPS technique)
R/V (or RV) .... Research Vessel
RVSN .... Russian Strategic Missile Force (agency responsible for launching most of Russia's military satellites)
RWS .... Rijkswaterstaat (Rijswijk, Netherlands)

Rx/Tx .... Receiver/Transmitter

S

S/A .... Signal to Ambiguity ratio
SA .... Selective Availability (GPS)
SAA .... South Atlintic Anomaly
SAAMD/WBSAAMD Stand Alone Accelleration Measurement Device/Wide Band Stand Alone Acceleration Measurement Device (Shuttle payload)
SAAMEX ...... Surface & Atmospheric Airborne Microwave Experiment (campaign)
SABLE .... South Atlantic Backscatter Lidar Experiment (campaign)
SAC/CSIR ...... Satellite Application Center [of CSIR (Council for Scientific and Industrial Research), South Africa]. The SAC ground receiving station (Landsat, Spot, NOAA/POES series, ERS series, Radarsat, etc.) is located at Hartebeesthoek south-west of Pretoria, South Africa. Initial SAC tracking services started in 1961.
SAFARI .... Southern African Fire-Atmosphere Research Initiative (campaign)
SAFER .... Simplified Aid for EVA Rescue (Shuttle system)
SAFIR .... Satellite for Information Relay, C.5
SAFISY .... Space Agency Forum for the International Space Year in Europe (in 1992)
SAGA .... Soviet-American Gases and Aerosols Experiment (campaign)
SAGE .... Stratospheric Aerosol and Gas Experiment (NASA mission, G.8)
SAI .... Space Applications Institute (of JRC, Ispra, Italy)
SAI .... Spectrum Astro Inc. of Gibert, AZ
SAIC .... Science Applications International Corporation (HQs in San Diego, CA, since 1969, with over 35,000 employees worldwide)
SAIR .... Synthetic Aperture Interferometric Radiometer
SALRO .... Saudi Arabian Laser Ranging Observatory, located some 45 km northwest of Riyadh, Saudi Arabia (tracking of SLR systems)
SALSA .... Semi-Arid Land-Surface-Atmospheric Program (campaign). The SALSA program is a multi-agency, multi-national global-change research effort that seeks to evaluate the consequences of natural and human-induced changes in semi-arid environments.
SALT .... Savannas on the Long Term (IGBP program of France)
SALT .... Strategic Arms Limitation Treaty (cold war agreement)
SAM .... Shuttle Activation Monitor (Shuttle experiment)
SAMIR .... Satellite Microwave Radiometer (ISRO sensor on Bhaskara S/C)
SAMPEX ...... Solar Anomalous and Magnetospheric Explorer (GSFC mission, K.21.1)
SAMS .... Space Acceleration Measurement Systems (Shuttle experiment)
SAMSO .... Space and Missile System Organization (USAF in El Segundo, CA)
SAN MARCO .. Cooperative Italian/NASA mission (A.25)
SAO .... Smithsonian Astrophysical Laboratory (Cambridge, MA, USA)
SAR .... Synthetic Aperture Radar (a high-rate imaging technique)
SAREX-2 ...... Shuttle Amateur Radio Experiment (Shuttle payload)
SAREX-92 ..... South American Radar Experiment (ESA airborne campaign)
S&R .... Search and Rescue (Emergency System on NOAA S/C)
S&RSAT .... Search and Rescue Satellite Aided Tracking System (Canada/France/NOAA). I.6
SAS-1 .... Small Astronomy Satellite-1 (DoD S/C, launched Dec. 12, 1970)
SAS .... Synthetic Aperture Sonar
SAS&R .... Satellite Aided Search & Rescue (INSAT-2 system)
SASS .... Subsonic Assessment (program, NASA)
SaTReC .... Satellite Technology Research Center (Taejon, Korea, since 1992, SaTReC is a university based research center of KAIST) SaTReC performs KITSAT operations, etc.

SatReCi .... SaTReC Initiative (SaTReCi Co. Ltd was established in January, 2000 by former SaTReC engineers, Taejon, Korea)
SATO .... Space Adaptation Tests and Observations (Shuttle experiment)
SAXON-FPN ... Synthetic Aperture Radar and X-band Ocean Nonlinearities - Forschungsplatform Nordsee (campaign)
Sb .... Antimonide (detector type material)
SBAS .... Satellite Based Augmentation System (element of GNSS)
SBIR .... Small Business Innovation Research (a NASA-sponspred program)
SBIRS .... Space Based Infrared System (a US DoD 10-year development program that was approved in Oct. 1996 to include HEO/GEO and LEO satellite constellations along with a corresponding ground segment. The SBIRS mission is to develop, deploy, and to operate space-based surveillance systems for missile warning, missile defense, battlespace characterization, and technical intelligence). The SBIRS program office is at SMC, Los Angeles AFB, CA.
SBRC .... Santa Barbara Research Center (of Hughes Aircraft Company in Goleta, CA - The name (SBRC) was valid until 1996; the facility was renamed to SBRS)
SBRS .... Santa Barbara Remote Sensing (of Hughes Aircraft Company in Goleta, CA, since 1996). Note: in Dec. 1997 Raytheon merged with the defense operations of Hughes Electronics. The merger outcome was the "Raytheon Systems Company" with HQ in Washington DC, consisting of the following units: Raytheon Electronic Systems, Raytheon E-Systems, Raytheon TI Systems and Hughes Aircraft Company. SBRS instruments include: multispectral imagers (MSS and TM), radiometers, spectrometers, polarimeters, and sounders.
S/C .... Spacecraft
SCAPE .... Shenandoah Cloud and Photochemistry Experiment (campaign)
SCAR .... Smoke/Sulfates Clouds and Radiation (campaign)
SCAR .... Scientific Committee on Antarctic Research (of ICSU)
SCARLET ..... Solar Concentrator Array with Refractive Linear Element Technology (a patented solar cell technology of AEC-Able Engineering Co., Goleta, CA, sponsored by BMDO and NASA/LeRC)
SCATHA .... Spacecraft Charging at High Altitude (satellite of the USAF)
SCATT .... (Wind) Scatterometer (ESA)
SCCCAMP ..... South Central Coast Cooperative Aerometric Monitoring Program (campaign)
SCD .... Swept Charge Detector
SCD-1 .... Satélite de Coleta de Dados (Data Collection Satellite of Brazil), C.6
SCE .... Superconducting Electronics
SCISAT/ACE ... Science Satellite/Atmospheric Chemistry Experiment, A.26
SCIGN .... Southern California Integrated GPS Network
SCMS .... Small Cumulus Microphysics Study (campaign)
SCOPE .... San Clemente Ocean Probing Experiment (campaign)
SCOPE .... Scientific Committee on Problems of the Environment (ICSU)
SCPS .... Space Communications Protocol Standard (A standardization initiative by NASA, DoD, DERA and others with the objective to complement and expand the current CCSDS standards) Although the CCSDS packetised standards provide the underpinning for the automated, error-free exchange of data between space and ground stations, it is limited to basic data transfer. SCPS will provide the additional capability to aggregate both telecommand and telemetry data into recognisable files and transport them end-to-end through the data networks containing space links in a reliable and secure manner.
SCRS .... Saudi Center of Remote Sensing, Riyadh, Saudi Arabia
SCS .... Soil Conservation Service (USA)
SCSMEX .... South China Sea Monsoon Experiment (campaign)

SEACAT .... type of buoy (made by Sea-Bird Electronics), temperature and conductivity sensor
SEADEX ...... Shoreline Environment Atmospheric Dispersion Experiment (campaign)
SEAFIRE ...... South-East Asia Fire Experiment (campaign)
Sea Launch ..... A sea-going launch system, based at Long Beach, CA. Sea Launch is a joint venture of The Boeing Commercial Space Co., Seattle, USA, KB Yuzhnoye/PO of Dnepropetrosvk, Ukraine (provider of the Zenit rocket), RSC Energia of Korolev, Russia (builder of an upper stage of the rocket), and Kvaerner Maritime A/S, Lysaker, Norway and London, UK (builder of the self-propelled launch platform and the Sea Launch command and assembly ship). The Sea Launch venture was announced in June 1994. The first launch of a demonstration satellite with a Zenit-3SL rocket took place March 27, 1999 from the floating Sea Launch platform, positioned at the equator. Sea Launch has a capacity to put up to 5000 kg of launch mass into a geostationary transfer orbit (GTO).
Seasat .... NASA/JPL EO mission (D.35)
SEASOAR ..... Towed profiling CTD and ADCP system (TOGA/COARE campaign)
SeaStar .... An ORBIMAGE mission with the SeaWiFS sensor (B.8.2). In 1997 OSC renamed the SeaStar mission to Orbview-2)
SeaWiFS .... Sea Wide Field Sensor (this sensor is considered the CZCS successor)
SECAM .... Sequential Color and Memory [European (French) video standard]. SECAM has an image format of 4:3, operating with 625 lines per picture frame at 50 Hz and 6 MHz video bandwidth with a total of 8 MHz video channel width.
SECDED ...... Single Error Correction - Double Error Detection
SEDAC .... Socio-Economic Data and Applications Center (DAAC at CIESIN)
SEDIS .... SeaWiFS European Data Information System (ESA/ESRIN)
SEDS .... Students for the Exploration and Development of Space (since 1980, international student organization)
SEE .... Société des Electriciens et des Electroniciens
SEE .... Space Environments and Effects program since 1995 [NASA (US government, industry and university participants), also international participation]
SEEDS .... Seeds in Space Experiment (Shuttle payload)
SEH .... Solar Extreme Ultraviolet Hitchhiker (Shuttle payload)
SEI .... Space Electronics Inc., San Diego, CA
SEL .... Space Environment Laboratory (NOAA, Boulder CO, real-time processing of all SEM package data, space environment forecasts)
SEL .... Surface-Emitting Laser (a conventional diode laser with a horizontal cavity, beams are emitted in the direction parallel to the wafer plane)
SELODE .... Solar Exposure to Laser Ordnance Device (Shuttle experiment on SPARTAN)
SEM .... Space Environment Monitor (NOAA Sensor package on GOES and POES series; Note: the GOES series SEM package arrangement differs considerably from the POES series SEM package)
SEM .... Space Experiment Module (Shuttle structure for small experiments)
SEMAPHORE .. Structure des Echanges Mer–Atmosphere, Proprietes des Heterogeneites Oceaniques (French airborne campaign)
SERC .... Science and Engineering Research Council (UK, the Mullard Space Science Laboratory of SERC)
SERON .... South Eastern (US) Regional Oxidant Network (field program to study atmospheric chemistry, July-August 1991)
SES .... Saab-Ericsson Space, Sweden
SES .... Societe Europeenne des Satellites (Luxembourg, owner and operator of the ASTRA satellite series)

SESAME .... Second European Stratospheric Arctic and Midlatitude Experiment (campaign)
SESAME .... Severe Environmental Storms and Mesoscale Experiment (campaign)
SESO .... Société Européenne de Systèmes Optiques, (Aix en Provence, France)
SET .... Single Electron Transistor
SETAS .... Space Environments and Technology Archive System (NASA/LaRC)
SEU .... Single Event Upset
S&F .... Store-and-Forward (a non-real-time communication technique)
SFDU .... Standard Format Data Unit (a CCSDS format concept)
SGG .... Satellite Gravity Gradiometry
SGGM .... Superconducting Gravity Gradiometer Mission, NASA (SGGM was cancelled by NASA in the 1990s due to budget constraints)
SGLS .... Space-to-Ground Link System (satellite communications)
SGS 85 .... Soviet Geodetic System 1985
SGS .... Svalbard Ground Station, located (78° N, 20° E) on the Norwegian Svalbard archipelago (Spitzbergen is the largest island of the Svalbard archipelago) near the town of Longyearbyen, is owned by the Norwegian Space Center (Norsk Romsenter), Oslo, Norway, and operated by the Tromsø Satellite Station. The high latitude makes SGS (just 960 km from the North Pole) a very sought-after link for polar-orbiting satellites. In 1999, NASA built its own receiving station (two 11m antennas in X- and S-band) right next to SGS (part of Lockheed's Consolidated Space Operations Contract) to receive data from Earth observing satellites (Landsat-7, Terra, etc.). - SGS can provide S/C contact for all orbits of polar orbiting satellites having altitudes above 500 km.
SHAR .... Sriharikota Range (ISRO launch site, India)
SHARE .... Space-Station Heat Pipe Advanced Radiator Experiment (Shuttle)
SHEBA .... Surface Heat Budget in the Arctic (campaign)
SHELS .... Shuttle Hitchhiker Experiment Launch System
SHF .... Super High Frequency (3 - 30 GHz band)
SHOM .... Service Hydrographique et Océanographique de la Marine (French Naval Hydrographic and Oceanographic Service) since 1971, with HQ in Brest, France. SHOM is a public service and a defense support agency - providing science and technical services (data acquisition, bathymetry, cartography, geophysics, oceanography).
SHOOT .... Super Fluid Helium On Orbit Transfer (Shuttle experiment)
SHS .... Spatial Heterodyne Spectroscopy
Si .... Silicon (detector material)
SiAs .... Arsenic-doped silicon detectors
SiGa .... Silicon gallium (detector)
SI .... Système International d'Units (International System of Units)
SiC .... Silicon carbide (example: SiC-type ceramic mirrors and structures are components in opto-mechanical systems)
SICH .... Owl (in Ukrainian, see SICH-1 under OKEAN)
SIL .... Space Innovations Limited, Newburry, Berks, UK [founded in 1983, since 1998 a subsidiary of SpaceDev Inc., San Diego, CA; SSTL (Surrey) purched SIL in 2000]
SIM .... Space Interferometry Mission (NASA)
SIMMS .... Seasonal Sea Ice Monitoring and Modeling Site (campaign)
SIMPLEX ...... Shuttle Ionospheric Modification with Pulsed Local Exhaust (Shuttle payload)
SIO .... Scripps Institution of Oceanography (part of UC at San Diego, La Jolla, CA)
SIPT .... Société Internationale de Photogrammétrie et de Télédétection
SIR .... Shuttle Imaging Radar (SIR-A with Payload A; SIR-B with Payload B, etc.), see J.19 - J.21
SIRTF .... Space InfraRed Telescope Facility (NASA)

SIS .... Superconductor-Insulator-Superconductor (tunnel junctions, also a microwave spectrometer receiver type)
SITe .... Scientific Imaging Technologies Inc. (US company in Beaverton, OR, CCD imaging products)
SITP .... Shanghai Institute of Technical Physics (of the Academy of Sciences of China), founded in 1958. Development of optical and infrared sensors since 1964 as well as radiometers.
SIZEX .... Seasonal Ice Zone Experiment (campaign)
SJ .... Shi Jian (meaning "experiment" or "experimental"). A spin-stabilized scientific minisatellite series of CAST, China; launch of SJ-1 on March 3, 1971; SJ-2 (2A and 2B) launch Sept. 19, 1981, S/C mass = 257 kg for each S/C (note: three satellites were launched by a single launch vehicle); SJ-4 launch on Feb. 8, 1994 (orbit: 210 km x 36125 km, inclination = 28.6°), S/C mass = 396 kg; SJ-5 launch on May 10, 1999
SKYLAB .... Sky Laboratory, NASA Space Station of the 1970s (L.5)
SL .... Spacelab - a modular general purpose laboratory. An integral element of NASA's Space Shuttle Program provided by ESA (build by MBB/ERNO). Spacelab itself comprised several elements that could be mixed-and-matched to suit mission requirements. A typical launch mass of a Spacelab was in the order of about 10 tons. SL-1 totalled a PM (Pressurized Module) mass of 8,145 kg plus a Pallet mass of 3,386 kg (including 1392 kg of payload mass). Spacelab is the first European manned space project. A total of 22 missions were flown with Spacelab starting with STS-9 (Nov. 28, 1983) until STS-90 (April 17, 1998). The Spacelab program provided numerous investigators from many countries an opportunity, to fly their instruments. Experiments conducted were generally in the fields of Earth observation, astronomy, atmospheric physics, life sciences, and material sciences under microgravity conditions.
SLA .... Shuttle Laser Altimeter (Shuttle payload)
SLAR .... Side-Looking Airborne Radar (an active sensor with Real Aperture Radar technology)
SLC .... Satellite Launch Center (complex)
SloshSat-FLEVO A small satellite of the Netherlands to study fluid dynamics in low gravity with FLEVO (Facility for Liquid Experimentation and Verification in Orbit). Shuttle payload
SLR .... Satellite Laser Ranging (a network of ground stations providing services of laser range measurements)
SLS .... Space Life Sciences (Shuttle payload)
SLS .... Strained Layer Superlattice (infrared detector type)
SMART .... Small Missions for Advanced Research in Technology (ESA Horizons 2000 mission)
SMC .... Space and Missile Systems Center, part of Air Force Materiel Command, with HQs located at Los Angeles AFB, El Segundo, CA (since 1954). SMC has operating sites throughout the USA, including the operating location detachment at NASA's Johnson Space Center, Houston, Texas; Detachment 2 at Onizuka Air Station in Sunnyvale, CA; and Detachment 9 at Vandenberg Air Force Base, CA. SMC is also the parent center of the host unit at Kirtland Air Force Base, Albuquerque, NM. SMC's work force totals over 9,500 employees. Some major programs of SMC are GPS/NAVSTAR, DMSP, SBIRS, etc.
SMC/TE .... Space and Missile Systems Center / Test & Evaluation Directorate. A tri-service (Army, Navy, Air Force) S/C division with locations at Kirtland AFB, Albuquerque, NM; Falcon AFB, Colorado Springs, CO; VAFB, Vandenberg, CA; Los Angeles AFB, El Segundo, CA; and at NASA/JSC, Houston TX. SMC/TE was established in 1992.

SMC/TEL ...... Space and Missile Systems Center / Space and Missile Test Evaluation Directorate. The Air Force serves as the executive agent for the Space Test Program (STP).
SMC/TEO ...... SMC / Orbital Telemetry, Tracking and Commanding Operations Division
SME .... Solar Mesosphere Explorer (NASA, K.20)
SMEX .... Small Explorer Program (NASA/GSFC program since 1988 supporting disciplines in astrophysics, space physics and upper atmospheric science; SMEX missions are SAMPEX, FAST, SWAS, TOMS, etc.)
SMHI .... Sweriges Meteorologiska och Hydrologiska Institut (Swedish Meteorological and Hydrological Institute), Norrköping
SMM .... Solar Maximum Mission (NASA,K.22)
SMOS .... Soil Moisture and Ocean Salinity (ESA mission, D.36)
SMS .... Synchronous Meteorological Satellite (designation of the first US weather satellites (1974); this series was later renamed GOES (NOAA)
SMTP .... Simple Mail Transfer Protocol
SNAP .... Surrey Nanosatellite Applications Program (D.40.16)
SNCMP .... Service National des Champs Magnetique Pulses (Toulouse, France)
SNL .... Sandia National Laboratories (Albuquerque, NM - since 1945, and Livermore CA, USA; SNL is part of DOE and operated by AT&T since 1949). Since Oct. 1, 1993, SNL ismanaged by Martin Marietta Corp., now Lockheed Martin. Part of SNL is now part of LANL.
SNOE .... Student Nitric Oxide Explorer (N.17.1)
SNR .... Signal-to-Noise Ratio
SNSB .... Swedish National Space Board (RYDSTYRELSEN), Solna Sweden
SO_2 .... Sulphur dioxide
SO_4 .... Sulphur radical
SOCC .... Satellite Operations and Control Center (NOAA)
SOCEX .... Southern Ocean Cloud Experiment (campaign)
SODAR .... Sound Detection and Ranging (system)
SODERN ...... Societe Anonyme d'Etudes et Realisations Nucleaires (French instrument company)
SOFIA .... Stratospheric Observatory For Infrared Astronomy (P.188). A cooperative NASA and DLR astronomy observatory. A Boeing 747-SP aircraft, a modified airliner, is the platform of SOFIA. Flights start in late 2002, long-term observations for up to 20 years are planned. The telescope of SOFIA, provided by DLR, has an effective diameter of 2.5 m. The mass of the telescope is 18,000 kg.
SOFIA .... Surface of the Ocean, Fluxes and Interaction with the Atmosphere (campaign)
SOHO .... Solar and Heliospheric Observatory (see K.23)
SOI .... Silicon-On-Insulator (thin film technology). In SOI devices the electronic active layers are fabricated on the insulator layer, while in conventional bulk devices the active layers are fabricated on the silicon layer. SOI is the technology of choice for radiation-critical applications (immunity to single-event latch-up from high-energy particles).
SOLAR-A ...... ISAS Solar-Terrestrial Mission, K.24
SOLSE/LORE .. Shuttle Ozone Limb Sounding Experiment/Limb Ozone Retrieval Experiment (Shuttle payload)
SONEX .... SASS Ozone and NOx Experiment (NASA campaign in planning)
SOP .... Special Observation Period (in campaigns)
SOP .... Standard of Practice (refering to those technologies which are mainstream and in common use)
SORCE .... Solar Radiation and Climate Experiment, A.27
SOS .... Southern Oxidants Study (campaign)
SOUP .... Solar Optical Universal Polarimeter (Spacelab-2 sensor)

SpaceDev ...... SpaceDev Inc. is a commercial company with HQ in Poway (San Diego), CA, offering fixed-price (deep-space) missions.

SPACEHAB .... A concept for commercially sponsored and procured payloads and services on Shuttle. SPACEHAB Inc., of Vienna, VA, has a NASA contract leasing Shuttle space on a commercial basis in the so-called 'Commercial Middeck Augmentation Module' (CMAM), a pressurized research lab owned by SPACEHAB® (an extension of the Shuttle orbiter middeck in the Shuttle cargo bay). SPACEHAB in turn sells its services, providing the needed support for commercial development of space payloads as well as physical and operational integration, and all services (training, etc.) for these payloads. Once in flight, SPACEHAB payloads are crew-tended on request. The SPACEHAB contract was awarded in Nov. 1990, the first SPACEHAB flight took place on STS-57 in June 1993. - SPACEHAB-1, -2 identifies also a series of Shuttle payloads.

SPACELAB .... Space Laboratory on NASA Shuttle missions (J.22 - J.23)

Space Imaging .. Space Imaging Inc. (since 1994) of Thornton, CO, acquired EOSAT in 1995 [distributor of IKONOS imagery, ERS-1/2, JERS and Radarsat data (USA), global distributor of IRS-1C/D imagery]. The owners of Space Imaging are: LM, E-Systems (of Raytheon), Mitsubishi, Vander Horst (Singapore), Halla Heavy Industries (Korea).

SPADE .... Stratospheric Photochemistry, Aerosols and Dynamics Experiment (campaign)

SPDM .... Special Purpose Dextrous Manipulator

SPAN .... Space Physics Analysis Network (based on the DECnet protocol). [The US - SPAN (NASA) service was discontinued at the end of 1990; the E-SPAN (ESA) service will be continued]. SPAN permits user access to data archives. The successor of SPAN in the US is NSI (NASA Science Internet), a dual protocol (TCP/IP and DECnet) network.

SPARC .... Stratospheric Processes and their Role in Climate (WCRP project, successor to STIB)

SPARTAN ...... Shuttle Pointed Autonomous Research Tool for Astronomy (Shuttle). SPARTAN is a small free-flying vehicle (about 1 x 1.25 x 1.5 m) for a variety of experiments (managed by OAST)

SPAS .... Shuttle Pallet Satellite (a Shuttle retrievable free-flyer platform for payloads, SPAS was built by MBB), SPAS-1 on STS-7 in 1983, ASTRO-SPAS is a direct successor of SPAS, ASTRO-SPAS-1 on STS-51 in Sept. 1993

SPECTRA ...... Surface Processes and Ecosystems Changes through Response Analysis (a proposed ESA Core Mission), in 2001 SPECTRA is the new name and successor of PRISM (Processes Research by an Imaging Space Mission), an instrument, and LSPIM (Land Surface Processes and Interactions Mission)

SPECTRE ...... Spectral Radiance Experiment (campaign)

SPICE .... Sensors Performance in Cloud Experiment (campaign)

SPIE .... Society of Photo-Optical Instrumentation Engineering (international)

SPIE .... Shuttle Plume Impingement Experiment

SPIFEX .... Shuttle Plume Impingement Flight Experiment

SPIN-2 .... Space Information-2 Meter. SPIN-2 is a joint venture (company) of Interbranch Association SOVINFORMSPUTNIK (Moscow, Russia), Aerial Images, Inc. (Raleigh, NC), and Central Trading Systems, Inc., (Huntington Bay, NY). The objective is to market high-resolution panchromatic imagery data (2 m) of past Russian missions (Resurs-F series). See KFA-1000 camera system under RESURS-F (the camera is also known by the name KVR-1000).

SPORT .... Small Payload Orbit Transfer (an AeroAstro concept)

SPOT .... Système Pour l'Observation de la Terre (French Earth Observing Satellite), (D.37)

SPOT-IMAGE .. Spot data distributor (Toulouse, France, and Reston, VA, USA)

SPRE .... SPARTAN Packet Radio Experiment (an amateur radio experiment on Shuttle SPARTAN)
SPS .... Standard Positioning Service (GPS)
SPST .... Single Pole Single Throw (Switch)
SPT .... Stationary Plasma Thruster (method of electric on-orbit propulsion)
SQUID .... Superconducting Quantum Interference Device (detector type, most sensitive device for magnetic field detection in particular with superconducting technology)
SQPSK......... Staggered Quadrature Phase Shift Keying (modulation type)
SRAM .... Static Random Access Method
SRB .... Surface Radiation Budget (GEWEX project)
SRC .... Space Regatta Consortium (Konsorsium Kosmicheskaya regata) since 1990, the association is based on the premises of RSC Energia
SRI .... Stanford Research Institute (original designation, founded in 1946), now: 'SRI International' at Menlo Park, CA. The institute separated from the University for legal reasons, - SRI International is a nonprofit organization funded by contract research. About 2700 employees.
SRI .... Système de Référence Inertielle
SRL .... Space Radar Laboratory (Shuttle missions of SIR-C/X-SAR payloads)
SRON .... Space Research Organisation Netherlands (Stiching Ruimlteonderzoek Nederland, Utrecht, Groningen - the Netherlands), since 1983, builder of scientific instruments (HXIS, SCIAMACHY, HIFI, etc.)
SRSC .... Siberian Remote Sensing Center, Novosibirsk, Russia
SRTC .... Savannah River Technology Center (DOE facility in Aiken, SC, USA)
SRTM .... Shuttle Radar Topography Mission, J.25
SS/CPMA ...... Spread Spectrum/Code Position Multiple Access (communication concept)
SSB .... Single Sideband
SSBUV .... Shuttle Solar Backscatter Ultraviolet (Shuttle Experiment)
SSC .... Swedish Space Corporation (Solna, Sweden; a government-owned limited corporation under the Ministry of Industry, established in 1972)
SSC .... Stennis Space Center (a NASA center in Bay St. Louis, MS)
SSCE .... Solid Surface Combustion Experiment (Shuttle payload)
SSDL .... Space Systems Development Laboratory, since 1994 (at the Department of Aeronautics and Astronautics of Stanford University, Stanford, CA)
SSEOP......... Space Shuttle Earth Observation Project
SSH .... Sea Surface Height (measured by satellite altimetry)
SSI .... Spaceport Systems International, operators of the commercial California Spaceport at Vandenberg, CA
SSIP .... Shuttle Student Involvement Program
SS/L .... Space Systems/Loral, Palo Alto, CA (major US builder of communication satellites, consortium leader of Globalstar series, sensors, etc.)
SSMA .... Spread Spectrum Multiple Access (communication transmission technique) Spread-spectrum modulation is emerging as the technology of choice to provide secure, interference-tolerant transmission.
SSM/I .... Special Sensor Microwave/Image (US Department of Defense, US Air Force Sensor)
SSN .... Space Surveillance Network (of the US Space Command, Colorado Springs, CO)
SSP .... Sub-Satellite Point
SSPA .... Solid-State Power Amplifier
SSPM .... Solid-State Photomultiplier (detector type)
SSPP .... Shuttle Small Payloads Project
SSRMS .... Space Shuttle Remote Manipulator Arm (since 1981, also referred to as Canadarm), built by Spar Aerospace of Canada
SSS .... Sea Surface Salinity

SST .... Space Solar Telescope (planned Chinese satellite mission in LEO with a 1 m diameter telescope using a 2048 x 1024 CCD detector array)
SST .... Satellite-to-Satellite Tracking (a technique employed with two or more S/C in various orbits for determining the Earth's gravity field)
SST .... Sea Surface Temperature (a physical parameter derived from radiometer data)
SSTI .... Small Spacecraft Technology Initiative (a NASA program started in 94)
SSTL .... Surrey Satellite Technology Ltd (University of Surrey, UK). SSTL is a commercial company whose principal shareholder is the University of Surrey. SSTL was set up in 1985 to provide a commercial outlet for the University's S/C engineering research.
SSU .... Stratospheric Sounding Unit (UK sensor on NOAA S/C)
STA .... Science and Technology Agency (of Japan)
STA .... Space Transportation Association [Washington DC, In March 1998, a NASA study on space tourism was released ("General Space Travel and Tourism"). In response to the report's findings, STA has created a new "Space Travel und Tourism Division" (under DOC coordination) to promote public and private space travel]
STABLE .... Suppression of Transient Accelerations By Levitation Evaluation (Shuttle payload)
STADAN .... Space Tracking and Data Acquisitions Network (NASA/GSFC)
STALO .... Stable Local Oscillator
STARE .... Southern Tropical Atlantic Regional Experiment (campaign)
Starlette .... CNES 'Solid Earth' mission, a passive satellite for geodetic studies with SLR observations (E.19)
STARLink ...... Satellite Telemetry and Return Link (ER-2 telemetry link, see P.80)
STAR-LITE .... Spectrograph/Telescope for Astronomical Research (Shuttle payload)
START .... System for Analysis, Research and Training (WCRP, IGBP, HDP)
State Center Priroda Moscow; Scientific and production enterprise for Earth remote sensing, commercial distributor of imagery; participation in programs: Resurs-F1, -F2, Salyut, MIR
STC .... Sensitive Time Control (SAR antenna parameter)
STC .... Star Tracker Camera
STCUI-RAS .... Scientific Technological Center of Unique Instruments - Russian Academy of Sciences (Moscow)
STDN .... Standard Tracking and Data Network (NASA)
STEDI .... Student Explorer Demonstration Initiative (N.17)
Stella .... CNES experiment on-board Spot-3 for gravity field studies of the Earth (E.20)
STEP .... Satellite Test of the Equivalence Principle, an ESA/NASA program proposal (1989). A MiniSTEP mission resulted due to economic constraints.
STEP .... Science and Technology for Environmental Protection (CEC program)
STEP .... Solar-Terrestrial Energy Program (International Program)
STEP .... Space Test Experiment Platform (a minisatellite bus of TRW Inc. for the DoD STP program)
STEP .... Stratosphere Troposphere Exchange Project (campaign)
STERAO .... Stratosphere-Troposphere Experiments: Radiation, Aerosols, and Ozone (campaign)
STEREO .... Solar-Terrestrial Relations Observatory, K.26
STEX .... Sensor Technology Experiment (Shuttle)
STEX .... Space Technology Experiment (USA, NRO satellite launched Oct. 3, 1998)
STIB .... Stratosphere Troposphere Interactions and the Biosphere (Program)
STIS .... Space Telescope Imaging Spectrograph (new Hubble sensor since Feb. 1997)
STJ .... Superconducting Tunnel Junction

STL-1 .... Space Tissue Loss-1 (Shuttle experiment)
STOIC .... Stratospheric Ozone Intercomparison Campaign
STORM-FEST .. Stormscale Operational and Research Meteorology - Fronts Experiment Systems Test (campaign)
STP .... Space Test Program [of DoD, the USAF manages STP, since 1965; As of 2001, STP has flown more than 420 experiments on more than 130 missions (STEP, POAM-III on SPOT-4, FORTE, REX-II, ARGOS are some current missions of STP)]
STP .... Solar Terrestrial Probes (NASA program with such missions as TIMED, SOLAR-B, STEREO, MMS)
STP-1 .... Space Test Payload-1 (Shuttle)
STRAT .... Stratospheric Tracers of Atmospheric Transport (campaign)
STREAM ...... Stratosphere and Troposphere Experiments by Aircraft Measurements (campaign)
STS .... Space Transport System (Shuttle)
STSI .... Space Telescope Science Institute (Washington DC, see AURA)
STSP .... Solar Terrestrial Science Program (ESA). STSP comprises the SOHO and CLUSTER missions
SUCCESS ...... Subsonic aircraft: Contrail and Clouds Effects Special Study (campaign)
SUMMiT .... Sandia Ultra-planar Multi-level MEMS Technology (a MEMS fabrication process developed at Sandia National Laboratories, Albuquerque, NM)
SUNY .... State University of New York (Albany, Binghamton, Brockport, Buffalo, Stony Brook, etc.)
SUPARCO ..... Space and Upper Atmosphere Research Commission (Pakistan)
SURFSAT-1 .... Summer Undergraduate Research Fellowship Satellite (NASA/JPL)
SUVE .... Solar Ultraviolet Experiment (Shuttle experiment)
SVAT .... Soil-Vegetation-Atmosphere Transfer (models)
SVFE .... Shuttle Vibration Forces Experiment (Shuttle payload on STS-90 and STS-96)
SVGA .... Super Video Graphics Adapter
SVHS .... Super Video Home System (a tape recorder system)
SVI .... Spectral Vegetation Index
SVN .... Satellite Vehicle NAVSTAR (a GPS series numbering system)
SVS .... Space Vision System (Shuttle camera system for ISS assembly)
SWADE .... Surface Wave Dynamics Experiment (campaign)
SWAS .... Submillimeter Wave Astronomy Satellite (NASA/GSFC)
SWIR .... Short Wave Infrared (spectrum, from about 1.3 μm to 3 μm)
SwRI .... Southwest Research Institute (San Antonio, Texas, an independent, nonprofit, applied research and development organization with more than 2,700 employees)
SWUIS .... Southwest Ultraviolet Imaging System (Shuttle payload)

T

TACAN .... Tactical Air Communication and Navigation System (a navigation aid, primary Shuttle navigation device for landing, TACAN navigation is provided for Shuttle within 300 miles of the landing site)
TAMEX .... Taiwan Area Mesoscale Experiment (campaign)
TANS .... Trimble Advanced Navigation Sensor ('TANS Vector' is a solid state GPS attitude-determination and position-location system)
TAO .... Tropical Atmospheric Ocean (TOGA campaign)
TAS .... Technology Applications and Science (Shuttle payload)
TAS .... Thallium Arsenic Selenide (Tl_3AsSe_3)
TBD .... To be defined (or: To be determined)
TCIPO .... TOGA/COARE International Project Office (at UCAR, Boulder, CO)

TCP/IP .... Transmission Control Protocol/Interchange Protocol. TCP/IP represents a communication framwork for other protocols such as: email, FTP, HTTP, SSH (Secure Shell), voice over IP, other multimedia protocols, teleoperation of remote systems.
Note: the TCP/IP represents two layers of protocol: the TCP part and the lower level IP part. IP deals with how the data gets routed around the network. TCP deals with making sure that all the packets arrive and are in the correct order. TCP implies a two-way connection and a higher level of communications overhead to assure that all the packets arrive and are in the correct order.

TCS .... Thomson-CSF Semiconducteurs Spécifiques, Orsay, France

TCS .... Trajectory Control Sensor (Shuttle payload)

T&DR .... Tracking and Data Relay (NOAA)

TDI .... Time Delay Integration (a cumulative exposure concept for CCD imaging)

TDL .... Tunable Diode Laser (spectrometer; TDLs are suited for detection of trace gases by optical absorption)

TDLAS .... Tunable Diode Laser Absorption Spectrometer

TDMA .... Time Division Multiple Access

TDRSS .... Tracking and Data Relay Satellite System (NASA)

TEA .... Transverse Excitation Atmospheric (pressure) laser

TEAMS .... Technology Experiments Advancing Missions in Space (Shutte)

TEC .... Thermoelectric Cooler

TEC .... Total Electron Content (of ionosphere)

Technion .... Israel Institute of Technology, Haifa, Israel

TEMISAT ...... Telespazio Micro Satellite (see C.7)

TeO_2 .... Tellurium dioxide

TerraServer ..... A joint venture of Aerial Images Inc., Raleigh, NC; Microsoft Corp., Redmond, WA; Compaq Computer Corp., Houston, TX; and Eastman Kodak Co., Rochester, NY. TerraServer is a commercial service of spaceborne and airborne imagery provision via internet. The imagery offered comes from a variety of sources (commercial and institutional).

TERRIERS .... Tomographic Experiment using Radioactive Recombinative Ionospheric EUV (STEDI mission, N.17.2)

TERS .... Tropical Earth Resources Satellite [a joint program conceived by the Netherlands (NIVR) and Indonesia (LAPAN) in 1985, the program got stalled after phase A because of a lack of funds]

TERSS .... Tasmanian Earth Resources Satellite Station (Hobart, Australia)

TES .... Thermal Energy Storage (Shuttle payload)

TES .... Transition-Edge Sensor (calorimeter)

TFOV .... Total Field of View

TGDF .... Turbulent Gas-Jet Diffusion Flames (Shuttle Experiment)

TIFF .... Tagged Image File Format (a raster format in pixel representation used for scanned images)

TIMED .... Thermosphere, Ionosphere, Mesosphere Energetics and Dynamics (A.28)

TIP .... TIROS (or Telemetry) Information Processor (on-board POES S/C, also a downlink data stream of NOAA S/C)

TIPPs .... Trans-Ionospheric Pulse Pairs (These strange signals, observed on ALEXIS, are the most intense radio sources from Earth which can be much stronger than typical lightning)

TIR .... Thermal Infrared (spectrum, from 6 μm to about 14 μm)

TIROS .... Television and Infrared Observation Satellite (US Environmental/Meteorological Remote Sensing Program; TIROS 1-10 = 1st generation, ESSA 1-9 = 2nd generation, ITOS (TIROS-M) = 3rd generation,)

TIROS-N ...... TIROS-NOAA (4th generation TIROS satellite series, starting with NOAA-6, -7, -8, etc.)

TKSC .... Tsukuba Space Center, located Tsukuba Scince City, Japan (since 1972)

TLD .... Thermoluminescent Dosimeter (Shuttle payload)
TLM .... Telemetry
TMA .... Three Mirror Anastigmatic (telescope off-axis design method). Note: the term 'anastigmatic' refers to lenses that are able to form approximately point images of target (object) points.
TMIBD .... Thermocapillary Migration and Interaction of Bubbles and Droplets (Spacelab experiment)
TMIP .... TeleMedicine Instrumentation Pack (Shuttle payload)
TMSAT .... Thai MicroSatellite, was renamed to Thai-Phutt (D.40.15)
TNO/FEL ...... Netherlands Organization for Applied Scientific Research/Physics and Electronics Laboratory (The Hague and Delft, The Netherlands)
TNSC .... Tanegashima Space Center (NASDA's launch site at Tanegashima Island, Japan, located at 30.4° N, 131.0° E)
TOA .... Top-of-Atmosphere
TOF .... Time-of-Flight (measurement)
TOGA .... Tropical Oceans and Global Atmosphere Experiment (Program)
TOGA/COARE . Tropical Oceans and Global Atmosphere Experiment / Coupled Ocean Atmosphere Response Experiment
TOGA/TAO .... TOGA/Tropical Atmosphere-Ocean (array of wind and upper ocean thermistor chain moorings in the Tropical Pacific)
TOGA/WOCE .. TOGA/World Ocean Circulation Experiment
TOMS .... NASA missions (A.29)
TOPEX/Poseidon Topography Experiment for Ocean Circulation (NASA/CNES EO Mission), E.21
TOS .... TIROS Operational System (NOAA)
TOVS .... TIROS Operational Vertical Sounder (NOAA, a three instrument system consisting of : HIRS-2; SSU; and the MSU)
TPCE .... Tank Pressure Control Equipment (Shuttle payload)
TPF .... Two Phase Flow (Shuttle payload)
TPFLEX .... Two-Phase Fluid Loop Experiment (Shuttle payload)
TPFO .... TOPEX/POSEIDON Follow-On (mission, was renamed to Jason)
TRACE-A ...... Transport and Atmospheric Chemistry near the Equator - Atlantic (campaign)
TRAGEX ...... Trace Gas Exchange: Mid-Latitude Terrestrial Ecosystems and Atmosphere (IGBP/IGAC program)
TRANSHAB .... An inflatable system NASA is considering for use on the ISS starting in 2004
TREE .... Tropical Rain-Forest Ecology Experiment (campaign)
TREES .... Tropical Ecosystem Environment Observation by Satellites (Joint CEC, JRC and ESA program
TRIAD .... Transit-Improved DISCOS (US Navy S/C built by APL) H.6
TRMM .... Tropical Rainfall Measuring Mission (NASA-NASDA Mission), A.31
TRSC .... Thailand Remote Sensing Center, Bangkok
TRW .... Thompson, Ramo and Wooldridge [TRW Space & Electronics Group is located at Redondo Beach, CA]. Manufacturer of communication satellites (TDRS, Odyssey series), military spacecraft (STEP, AXAF, etc.), and remote sensing satellites (Lewis, EOS/PM-1, TOMS/EP, KOMPSAT-1, ROCSat-1, etc.)
TSI .... Total Solar Irradiance
TSIM .... Total Solar Irradiance Mission
TsNIIMASH .... Central Research and Scientific Institute of Machine Building, Korelev (Moscow Region), Russia (there is also the spelling of TzNIIMASH)
TsSKB-Progress . the Russian acronym for "Central Specialized Design Bureau Progress," Samara, builder of Resurs-F (and Resurs-RK) satellite series
TSS-1R .... Tethered Satellite System (ASI payload on Shuttle)

TT&C .... Telemetry, Tracking & Command (Data for S/C Operations)
TTFM .... Two-Tone Frequency Modulation (a measurement technique for trace gases)
TTL .... Transistor-Transistor Logic (semiconductor technology of the 1960s and 1970s - the microprocessor revolution began in 1973)
TUB .... Technical University of Berlin, Germany
TUBSAT .... Technical University of Berlin Satellite (N.21)
TUD .... Technical University of Denmark (Lyngby, Denmark)
TUFI .... Toughened Uni-Piece Fibrous Insulation (Shuttle payload)
TVA .... Tennessee Valley Authority (USA)
TWTA .... Traveling Wave Tube Amplifier (communication, amplification of a microwave frequency)

U

UAH .... University of Alabama in Huntsville, AL
UARP .... Upper Atmospheric Research Program (NASA)
UARS .... Upper Atmosphere Research Satellite (NASA satellite, launch: Sept. 1991) A.32
UAV .... Unmanned Aerial Vehicle (PERSEUS, CONDOR, etc.)
UC .... University of California [a nine campus university across the state, UCLA (Los Angeles), UCB (Berkeley), UCSD (San Diego), UCSB (Santa Barbara), UCI (Irvine), UCR (Riverside), UCSC (Santa Cruz), UCD (Davis), etc.]
UCAR .... University Corporation for Atmospheric Research (Boulder, CO, UCAR is sponsored by NSF - there are over 60 member institutions in UCAR)
UCB .... University of California, Berkeley
UCCS .... University of Colorado at Colorado Springs
UCLA .... University of California, Los Angeles
UDP/IP .... User Datagram Protocol/Internet Protocol. Note: UDP/IP does not need any handshaking to transfer data. TCP/IP requires bi-directional handshaking prior to data transfer.
UHB .... User Home Base
UHF .... Ultra High Frequency (300 - 3000 MHz band)
UIT .... Ultraviolet Imaging Telescope (part of ASTRO-1 payload on Shuttle)
UIT .... Union Internationale des Télécommunications
UKAEA .... United Kingdom Atomic Energy Authority
UKDoE .... United Kingdom Department of the Environment
UKMO .... United Kingdom Meteorological Office (same as BMO, HQs and Hadley Centre for Climate Prediction & Research are located in Bracknell, Remote Sensing Instrumentation branch in Farnborough)
UKS .... United Kingdom Subsatellite (S/C of the AMPTE mission, K.4.2)
ULIRGs .... Ultra-Luminous IR Galaxies
UMTS .... Universal Mobile Telecommunications System (standard)
UNAM .... Universidad Nacional Autónoma de México, Mexico City
UNAM-CE ..... Universidad Nacional Autónoma de México - Centro de Ecologica, Mexico
UNAM-IG ..... Universidad Nacional Autónoma de México - Institoto de Geologica
UNAVCO ...... University Navstar Consortium (a US Earth sciences community initiative to foster GPS applications in particular in the area of surveying)
UNCED .... United Nations Conference on Environment & Development
UNDP .... United Nations Development Programme
UNAVCO ...... University NAVSTAR Consortium (USA)
UNCOPUOS ... UN Committee on the Peaceful Uses of Outer Space
UNEP .... United Nations Environmental Programme (since 1972)
UNEP/GRID ... UNEP Global Resource Information Database

UNESCO ...... United Nations Educational Scientific and Cultural Organization (based in Paris, France)
UN-ESCAP ..... UN-Economic and Social Commission for Asia and the Pacific, Bangkok, Thailand
UnESS .... University Earth System Science (a NASA initiative with the objective to involve the student community in Earth science projects)
UNEX .... University-class Explorer [(mission) - A NASA program supporting university-designed/developed missions. The UNEX program is designed to provide frequent flight opportunities for highly focused and relatively inexpensive science missions whose total cost to NASA is limited to $13 million. The program is managed by NASA/GSFC.]
UNH .... University of New Hampshire, Durham, NH
UNISPACE ..... United Nations Conference on the Exploration of the Committee on the Peaceful Uses of Outer Space (UNISPACE-III took place in Vienna, Austria (July 19-30, 1999 - the first two UNISPACE conferences were held in 1968 and 1982)
UNOLS .... University National Oceanographic Laboratory System (USA)
UNS .... Universal Navigation System
UNOOSA ...... United Nations Office for Outer Space Affairs
UoSAT .... University of Surrey Satellite (UK, D.40)
UPC .... Universidad Politecnica de Barcelona (Spain)
UQPSK .... Unbalanced Quadrature Phase-Shift Keying (technique)
URE .... User Range Error (of GPS position service)
URL .... Uniform Resource Locator (WWW) for 'file:', 'http:', 'news:', and 'telnet:'
URSI .... Union Radio Scientifique Internationale (International Union of Radio Science)
USA .... United States of America
USA .... United Space Alliance LLC [of Houston, TX, a joint venture of Rockwell International (now The Boeing Company) and Lockheed Martin] - USA is the NASA prime contractor for all Space Shuttle operations/management at MSFC and at KSC, since Oct. 1996)
USACE .... US Army Corps of Engineers
USAF .... US Air Force
USAFA .... United States Air Force Academy (Colorado Springs, CO)
USAFB .... US Air Force Base
USAF/PL .... USAF/Phillips Laboratory, Kirtland AFB, Albuquerque, NM [part of AFRL (Air Force Research Laboratory), note: in 1998 the Phillips Laboratory was renamed: "Phillips Research Site"]
USAF/RL ...... USAF/Rome Laboratory, Griffis AFB, Rome, NY [part of AFRL]
USAF/SMC ..... USAF/Space & Missile Systems Center (see SMC/TE)
USCG .... US Coast Guard
USCON-CICTUS Universidad de Sonora - Centro de Investigaciones Cientificas y Tecnologicas de la Universidad de Sonora, Hermosillo, Mexico
USDA .... US Department of Agriculture
USDA/ARS .... USDA/Agricultural Research Service (Beltsville, MD and Tucson, AZ)
USEPA .... US Environmental Protection Agency
USES .... Universal Source Encoder for Space (a NASA developed chipset)
USFS .... US Forest Service
USFWS .... US Fish and Wildlife Service
USGCRP ...... US Global Change Research Program (since 1990). USGCRP sponsors global change research in a large number of institutions (over 300).
USGS .... United States Geological Survey (the science and technology agency of the Department of the Interior, DOI; USGS was established in 1879). The mission of USGS is to provide geologic, topographic, and hydrographic information to contribute to the management of the Nation's natural resources.

USML .... US Microgravity Laboratory (Shuttle payload)
USMP .... US Microgravity Payload (Shuttle payload)
USRA .... Universities Space Research Association, Columbia, MD [a nonprofit corporation organized in 1969 by NAS (National Academy of Sciences) at the request of NASA; as of 1995 there are 78 member universities]
USS .... Unique Support Structure (Shuttle)
USSR .... Union of Soviet Socialist Republics (former)
USU/SDL ...... Utah State University / Space Dynamics Laboratory (Logan, UT, Bedford, MA, and Albuquerque, NM). SDL is a non-profit organization owned by USU.
USWRP .... US Weather Research Program
UTA .... University of Texas at Austin
UTA/CSR ...... UTA/Center for Space Research (since 1981)
UTC .... Universal Time Coordinated
UTM .... Universal Transverse Mercator (coordinate reference system for large-scale maps)
UV .... Ultra Violet (spectral range from 0.01 - 0.38 μm)
UVCS .... Ultrviolet Coronal Spectrometer (a SAO instrument flown on the SPARTAN-201 series)
UVPI .... Ultraviolet Plume Instrument (Shuttle experiment)
UVSTAR .... Ultraviolet Spectrograph Telescope for Astronomical Research (Shuttle payload)
UWB .... Ultra Wideband (involves multi-octave frequency coverage of a sensor such as a radar system for the purpose of ground penetration)

V

VAFB .... Vandenberg Air Force Base, Vandenberg, CA
VCL .... Vegetation Canopy Lidar Mission, D.39
VCO .... Voltage Controlled Oscillator
VCR .... Video Cassette Recorder (also: Video Color Recorder)
VCS .... Voice Command System (Shuttle)
VCSEL .... Vertical Cavity Surface-Emitting Laser (type of semiconductor diode laser; the cavity is perpendicular to the wafer plane, thus the optical beam is guided in the vertical direction)
VDA .... VHF Collection System Antenna (NOAA)
VDC .... Volt Direct Current
VENTEX ...... Venting Experiment (campaign)
VFT-1 .... Visual Function Tester-1 (Shuttle experiment)
VH .... Vertical transmit - Horizontal receive polarization
VHF .... Very High Frequency (30 - 300 MHz band)
VHS .... Video Home System
VI .... Vegetation Index
Viking .... Swedish satellite mission for the study of the Earth's magnetosphere, K.29
VIR .... Visible Infrared (spectrum)
VIS .... Visible (spectrum 0.4 - 0.7 μm)
VITA .... Volunteers in Technical Assistance (a humanitarian organization in Arlington, VA, USA, providing communication services on a global scale)
VITO .... Vlaamse instelling voor technologisch onderzoek (Flemish institute for technological research), located in northern Belgium. One of its centers is the image processing/archiving center of the VEGETATION instrument on SPOT.
VLBI .... Very Long Baseline Interferometry
VLDS .... Very Large Data Store
VLF .... Very Low Frequency (frequency band of 10 - 30 kHz)
VLS .... Veiculo Lancador de Satellites (Brazil's launch vehicle)
VLSI .... Very Large Scale Integration (solid-state technology)

XPD .... X-ray Photoelectron Detection
XRT .... X-Ray Telescope (Spacelab-2 sensor, energy detection 2.5-25 keV)
XSS .... Experimental Spacecraft System (US AFRL minisatellite series)
XTI .... Cross-Track Interferometry
XTR .... Transmitter
XUV .... Extreme Ultra Violet (same as EUV, i.e. 1 - 130 nm spectral range)

Y

YAG .... Yttrium Aluminum Garnet (a type of solid-state crystal laser)
YBCO .... Yttrium-Barium-Copper-Oxide ($YBa_2Cu_3O_7$)
YBLCO .... Yttrium-Barium-Lanthanum-Copper-Oxide
YLF .... Yttrium Lithium Fluoride (a laser type)
YUZHNOYE ... Research and Production Association, Dnepropetrosvk, Ukraine (there is also the spelling of YUZHNOE - builders of two launch vehicles: Zenit and Cyclone; builders of OKEAN series satellites)

Z

ZAMG .... Zentralanstalt für Meteorologie und Geodynamik, with HQs at Vienna, Austria, since 1851 (Austrian Institute for Meteorology and Geodynamics)
ZARM .... Zentrum für angewandte Raumfahrttechnologie und Mikrogravitation (Center of Applied Space Technology and Microgravity - since 1985), an institute of the University of Bremen, Germany
ZUP .... Flight Control Center, Kaliningrad, Russia (TT&C function for MIR station along with RKK Energia)

Appendix C Index of Sensors

Numbers

A

B

C

D

E

F

G

H

I

J

K

L

M

N

O

P

Q

R

S

T

U

V

W

X

Y

Z

GPSR Compliance

The European Union's (EU) General Product Safety Regulation (GPSR) is a set of rules that requires consumer products to be safe and our obligations to ensure this.

If you have any concerns about our products, you can contact us on ProductSafety@springernature.com

In case Publisher is established outside the EU, the EU authorized representative is:

Springer Nature Customer Service Center GmbH
Europaplatz 3
69115 Heidelberg, Germany

Batch number: 10761593

Printed by Printforce, the Netherlands